a manual of

ACAROLOGY

Contributing Authors

Valerie Behan-Pelletier
Agriculture and Agri-Food Canada

David R. Cook
Paradise Valley, Arizona

Mark S. Harvey
Western Australian Museum

James E. Keirans
Georgia Southern University

Gerald W. Krantz
Oregon State University

Evert E. Lindquist
Agriculture and Agri-Food Canada

Roy A. Norton
State University of New York
College of Environmental Science and Forestry

Barry M. OConnor
University of Michigan

Ian M. Smith
Agriculture and Agri-Food Canada

David E. Walter
University of Alberta

Third Edition

a manual of ACAROLOGY

G. W. Krantz and D. E. Walter,
EDITORS

Texas Tech University Press

This book is typeset in Adobe Garamond Pro. The paper used in this book meets the minimum requirements of ANSI/NISO Z39.48-1992 (R1997).♾

Designed by Diane Luopa-Filimonov

Library of Congress Cataloging-in-Publication Data

A manual of acarology / G.W. Krantz . . . [et al.].—3rd ed.
p. cm.
Includes bibliographical references and index.
ISBN 978-0-89672-620-8 (litho case : alk. paper) 1. Mites—Classification. 2. Ticks—Classification. 3. Arachnida—Classification. I. Krantz, G. W.
QL458.K7 2009
595.4'2—dc22

2008035888

Printed in the United States of America
09 10 11 12 13 14 15 16 17 /
9 8 7 6 5 4 3 2 1

Texas Tech University Press
Box 41037
Lubbock, Texas 79409-1037 USA
800.832.4042
ttup@ttu.edu
www.ttup.ttu.edu

Contents

Preface to the Third Edition

The first edition of *A Manual of Acarology* was published in 1970 to meet a demand from the scientific community for basic, up-to-date information on the taxonomy and biology of mites. The demand was largely fueled by the discovery, during the twenty-five-year period that followed World War II, of thousands of previously unknown mite taxa taken from a bewildering variety of microhabitats worldwide and by the recognition that many of these taxa might be of biological or economic interest. An updated second edition of the *Manual* appeared in 1978, and it has served since that time as the basic laboratory text for acarological instruction throughout the world. Its usefulness notwithstanding, the many advances made since 1978 in acarine systematics and phylogeny, molecular biology, morphology, physiology, ecology, and behavior have totally eclipsed the content of the second edition. These advances have profoundly altered our understanding of the Acari and have inspired new and innovative approaches to resolving many basic and applied acarological problems. The number and significance of these developments, along with the discovery and recognition of a multitude of new taxa and the corollary need for major modifications to acarine higher classification, made it imperative that an updated revision of the *Manual* be provided for students and professional plant and animal scientists. Access to an abundance of new information, coupled with our decision to expand coverage in the third edition beyond the basic level found in earlier versions, has led to the creation of what we believe to be a landmark acarological text. Our primary goal in producing this book is to furnish a solid acarological foundation for those who wish to explore the complex and often astonishing world of mites and provide background and impetus for new research initiatives in acarology and related fields. We are hopeful that the appearance of the third edition in this first decade of the new millennium will signal a period of revitalization and renewal in the field of acarology.

Like the preceding editions of the *Manual*, the third edition has remained primarily taxonomic in its approach, but it also provides detailed, nontaxonomic information on subjects including phylogeny, biology, morphology, systematics, ecology, and behavior. Collection and rearing techniques are discussed in detail, as are specimen preparation and methods of preservation. Taxonomic diagnoses for the 124 presently recognized superfamilies of Acari are included in their appropriate systematic chapters; feeding habits, host range, and the distribution of member families and representative species are discussed under each superfamilial heading. Text information is complemented by numerous labeled figures, keys to families (the Ixodida are keyed to genus), a bibliography comprising more than four thousand entries, and a detailed index. Unlike previous versions, the new edition is the product of a team effort by ten authors whose contributions have been amalgamated into a (hopefully) seamless text.

Many individuals and agencies have contributed in significant ways toward making the third edition a reality. We are especially indebted to Marilyn Houck, Texas Tech University, who was instrumental in persuading the senior editor that the time for a revision of the *Manual* had indeed come, who served as liaison-in-residence to the director of Texas Tech University Press from 1999 to 2003, and who established the protocols that so effectively facilitated interactions between the authors and the editors. Marilyn also undertook the onerous task of scanning and sanitizing illustrations from the second edition, many of which appear in the pages to follow. We are pleased to acknowledge Michel Bertrand, Directeur Gérant, *Acarologia* (Montpellier, France), and Vikram Prasad, Editor-in-Chief, *International Journal of Acarology* (West Bloomfield, Michigan, USA), who generously granted us blanket permission to use needed illustrations from their respective journals. Editors and representatives of the following publications and institutions kindly provided permission to use specific illustrations: *Records of the South Australian Museum* (Adelaide); *Zoologisches Jahrbuch, Anatomie* (Jena); *The Canadian Entomologist* (Ottawa); Institut royal des Sciences naturelles de Belgique (Brussels); Museum and Institute of Zoology, Polish Academy of Sciences (Warsaw); Muséum national d'Histoire Naturelle (Paris); Royal Society of South Australia

(Adelaide); Elsevier (Amsterdam); University of Hawaii Press (Honolulu), Academic Press, Ltd. (London); CAB International Publishing (Wallingford, UK); Urban & Fischer Verlag Niederlassung (Jena); *Annual Reviews* (Palo Alto); *Bulletin of the National Science Museum* (Tokyo); Cambridge University Press (Cambridge, UK); Natural History Museum (London); Bishop Museum Press (Honolulu); *Zoologica* (Stuttgart), CSIRO Publishing (Collingwood, Australia); Wiley-Liss Inc. (New York); Ministry of Agriculture, Fisheries and Food (London); and Schweizerbart'sche Verlagsbuchhandlung (Stuttgart). Special thanks to Michelle MacKenzie and Sarah Seiter for their invaluable assistance in preparing the illustrations for Chapters 13 and 16, Joel Hallan for organizing and assembling our formidable bibliography, and Jerod Sapp for his technical wizardry in dealing with our many computer files. We are grateful to Heinrich Schatz and Gerd Weigmann, and to Hans Klompen and the 1998-2004 Acarology Summer Program classes at The Ohio State University, Columbus, for reviewing and testing some of the taxonomic keys included in this work.

Many colleagues have reviewed earlier drafts of the text and keys and have provided unpublished information, advice, specimens, and general assistance to the authors and coeditors during preparation of the book draft. With apologies to those whose names may have been inadvertently omitted, they are listed here in alphabetical order: F. Akashi, G. Alberti, J. W. Amrine, H. M. André, N. Ayyildiz, J.-I. Aoki, W. T. Atyeo, A. M. Avanzati, P. Balogh, A. Bochkov, Y. Coineau, J. C. Cokendolpher, M. Clayton, B. Eamer, E. Ebermann, C. M. Engelbrecht, A. Fain, R. B. Halliday, R. W. Husband, M. Ito, J. S. H. Klompen, W. H. Knee, W. Knülle, G. Krisper, S. Lindsay, W. L. Magowski, S. Mahunka, O. L. Makarova, L. Miko, M. Minor, G. J. de Moraes, M. L. Moraza, R. Ochoa, G. N. Oldfield, Z. Olszanowski, R. Peredes-León, A. D. Paschoal, C. Péréz-Iñigo, Jr., H. C. Proctor, P. J. A. Pugh, T. K. Qin, F. J. Radovsky, R. Schuster, O. D. Seeman, D. Sillman, S. Shimano, J. Travé, S. Woas, and Y. Yamamoto.

A project of this magnitude and complexity could not have been undertaken without the collaboration of our remarkably talented fellow authors: Valerie Behan-Pelletier, Dave Cook, Mark Harvey, Jim Keirans, Evert Lindquist, Roy Norton, Barry OConnor, and Ian Smith. Heartfelt thanks to them and to our collective spouses and families for their unflagging support and encouragement during the decade-long period that went into the preparation of this volume.

G. W. Krantz
Oregon State University

D. E. Walter
University of Alberta

CHAPTER ONE

G.W. KRANTZ

Introduction

If morphological and ecological diversity in an animal group can be considered a measure of its success, then the assemblage of invertebrates composing the arachnid subclass Acari would have to be among those accorded top honors. Unlike other arachnids, the mites have evolved far beyond saprophagy and predation. Some feed on plants, bacteria, or fungi, while others have developed obligate symbiotic relationships with vertebrate and invertebrate animals. Thanks to their remarkable evolutionary plasticity and relatively small size, mites have succeeded in colonizing a range of terrestrial, marine, and aquatic habitats that far exceeds those occupied by any other arthropod group, including insects. Mites may be found worldwide in virtually any locality capable of supporting life: from the windswept arctic tundra to the hot desert sands of the Sahara, from the icy depths of Pacific oceanic trenches to the hair follicles of our eyebrows. Mites abound in rivers, lakes, and streams and are a significant component of the arboreal fauna of tropical and temperate forests. They often occur in enormous numbers in the litter and humus layers that cover forest, grassland, and agricultural soils and may be found in the soil itself, often at depths of several meters. Because they are so small, many mites are easily dispersed on air currents and consequently are a common component of aerial plankton. Their diminutive size also may allow some mites to hitchhike from place to place on larger animals such as insects, birds, and mammals.

Many mite species are beneficial to humans in that they prey on invertebrate pests of agricultural and ornamental crops, often reducing or obviating chemical control measures. Others have been found to feed on noxious plants and have been used successfully in weed control programs. Nonpredaceous litter mites may be effective nutrient recyclers in forest floor ecosystems, comminuting organic litter to a size appropriate for use by other decomposers. While many mites are beneficial, a number of species are serious pests of crops and of humans and animals, damaging their hosts through their feeding activity and through transmission of disease organisms (see chapter 6).

The diversity of habitats in which mites are encountered is no more remarkable than their range of morphological and behavioral traits. Based on these characteristics, approximately 55,000 species of mites have thus far been recognized and described (Walter and Proctor 1999). Estimates of the actual number of extant mite species vary from 500,000 to 1,000,000, although recent hypotheses that greatly increase estimates of the extant insect fauna (Gaston 1991; Erwin 1991) suggest that the total number of mite species, many of which have established mono- or oligoxenous associations with insects, may be much greater than currently imagined. Estimates of a million or more mite species seem excessive until one considers that new mite species are routinely encountered even in previously well-collected substrates (Walter and Proctor 1999). A new and unimagined assemblage of mites may be revealed in a handful of litter taken from a tropical forest in Cameroon or a deciliter of interstitial water extracted from the tidal sands of the Irrawaddy Delta. Recent exploration of exotic realms such as the phylloplane of Australian rain forests (Walter 1996), the amber deposits of the Paleozoic and Mesozoic eras (Poinar 1992), and the nests of termites in the southwestern desert of Turkmenistan (Petrova-Nikitina 1987; Gordeeva, Niemi, and Petrova-Nikitina 1996) have revealed mite faunas whose existence had previously been unsuspected. In short, our current concepts of acarine systematics appear to be based on little more than a fragmentary knowledge of the fauna. Nevertheless, the great diversity of known acarine species has led to recognition of nearly 5,500 genera and 1,200 subgenera representing some 540 families (J. Hallan, pers. comm., 2006) in 124 superfamilies (see chapter 8).

The recognition of mites as discrete entities occurred well before the science of acarology emerged as a discipline. A reference to "tick fever" was found on an Egyptian papyrus scroll dated 1550 BC (Zaher and Hanna 1982), Homer mentioned the occurrence of ticks on Ulysses' dog in 850 BC, and Aristotle described a mite parasite of locusts (probably *Eutrombidium*) in *De Animalibus Historia Libri* some 500 years later. Other early references to Acari appear in the writings of Hippocrates, Plutarch, Aristophanes, and Pliny, among others (Oudemans 1926b; Emmanuel 1982; Gorirossi-Bourdeau 1995). Gorirossi-Bourdeau (1995) described a series of sculptures representing what appear to be ticks or trombidioid mites carved into the ceiling of an elegant Roman temple dedicated to Bacchus at Baalbek, Lebanon, at the beginning of the first millennium, probably the earliest known work of art depicting acarines.

Appellations for mites in the early literature took a number of forms. Emmanuel (1982) noted that Aristotle used the Greek terms *Kroton* or *Kynoraistis* for ticks, and *A-kari* (without head) to describe nonixodoid mites. Another possible derivation of the word *Acari* may have been from the Greek word *akares*, which means short or small. During medieval times, mites were commonly referred to as *lice*, *beesties*, or *little insects*. As in Aristotle's time (and until relatively recently), ticks were considered an entity separate from mites and were referred to as *ticia* (Old English) or *zecken* (Middle High German). The Old English word *mite*, meaning very small, may well have been derived from the *widow's mite*, an early Flemish coin of very small value (Walter and Proctor 1999). Rediscovery of the term *Akari*, or Acari, appears to have occurred about 1650, but it was not until the early twentieth century that the term *acarology* began to appear regularly in the literature (Krantz 1996).

Linnaeus used the generic name *Acarus* in the first (1738) edition of the *Systema Naturae*, for which he later named the type species, *A. siro*. The tenth edition of the *Systema* included fewer than 30 mite species, all of which were assigned to *Acarus*. During the 100 years that followed, several schemes of higher classification were introduced by workers such as DeGeer (1778), Latreille (1806–1809), Leach (1815), Dugès (1839), C. L. Koch (1842), and others to accommodate an ever-growing number of newly recognized mite taxa. Michael (1884a) summarizes these and other pioneer works.

The emergence of acarology as a modern science during the late nineteenth and early twentieth centuries was centered primarily in Europe and North America with the historic contributions of Michael (1884a), Kramer (1877), Mégnin (1876), Canestrini (1891), Banks (1904), Oudemans (1906), Reuter (1909), Jacot (1925), Trägårdh (1946), Sig Thor (1929), and Vitzthum (1940–43). Special mention should be made here of Antonio Berlese (1863–1927), an economic entomologist whose many contributions to acarine systematics and classification were instrumental in establishing acarology as a discrete discipline (Van der Hammen 1977a, 1979), and of François Grandjean (1882–1975), who produced a brilliant body of work on the systematics, morphology, phylogeny, and ontogeny of mites over a period of nearly 50 years (collected in Van der Hammen 1972–76). The work of these few pioneers provided the basis for virtually all acarological research until after the end of World War II.

The post–World War II awakening of acarology was triggered in part by the return to civilian life of many American, Japanese, Australian, and British medical corps personnel who wished to continue their wartime studies on mite-transmitted diseases such as scrub typhus and tick-borne hemorrhagic fever, as well as on the mites themselves. Among those who participated in the postwar acarological renaissance was George W. Wharton, a specialist on chiggers (TROMBICULIDAE) who not only produced an array of significant systematic papers on trombiculids in the years immediately following the war but also recognized the need for a broader approach to the field in the form of a general text that would update the classic and relatively unavailable earlier work of Vitzthum (1931a, 1940–43). Wharton's aspirations were shared by Dr. Edward W. Baker, acarologist with the Agricultural Research Service, USDA, and their *Introduction to Acarology* was published in 1952. Supplemented by the *Guide to the Families of Mites* by Baker and colleagues (1958), the *Introduction* served as the standard acarological text worldwide for more than a decade. Its importance as a major factor in the emergence of modern acarology cannot be overemphasized.

Several general reference and classificatory works in acarology have been published since the appearance of the *Introduction* and the *Guide* (Hughes 1959; Radford 1950; Evans, Sheals, and MacFarlane 1961; Sasa 1965; Hirschmann 1966; Krantz 1970b, 1978; Flechtmann 1976; Doreste 1984; Woolley 1988; Van der Hammen 1989; Evans 1992; Alberti and Coons 1999; Coons and Alberti 1999; Walter and Proctor 1999, 2001), as have many important research papers and books dealing with particular taxonomic groups or with specific topics in mite biology, physiology, ecology, and systematics. Many of these works will be cited in the chapters to follow.

The content and bibliography presented in this work are current through 2006. Selected 2007 references have been included where considered essential or appropriate.

CHAPTER TWO

G.W. KRANTZ

Origins and Phylogenetic Relationships

Small size and lack of a substantial exoskeleton have limited the availability of fossil evidence relating to the origins of most terrestrial Arthropoda. However, the information presently available suggests that arachnids first appeared on land in the late Silurian or early Devonian periods (ca. 425 mya), perhaps in the form of a now-extinct relative of a pulmonate trigonotarbid (Arachnida, Trigonotarbida). This momentous event took place more or less on the heels of the remarkable period of arthropod evolution and diversification that occurred during the Cambrian period, approximately 200 million years earlier (Gould 1989; Walter and Proctor 1999). The Arachnida has traditionally been considered a monophyletic assemblage, with the Scorpionida representing its most primitive member group (Kaestner 1980). However, scorpions possess an array of morphological characters not shared by other arachnid lineages, nor is there any evidence in the fossil record suggesting that an intermediate scorpion-like arachnid ever existed. In fact, it now appears that scorpions invaded the land as relatively large predators and found themselves in a well-established ecosystem already supporting a variety of smaller arachnids, including the primordial Trigonotarbida (Walter and Proctor 1999). Fossil evidence suggests that terrestrial arachnids acquired air-breathing respiratory organs at least three times during the course of their ascent from the primitive marine habitat onto land (scorpions, trigonotarbids, and the unusual tracheate eurypterid *Baltoeurypterus*), an observation that would seem to invalidate any argument in favor of the monophyletic origin of the Arachnida (Bergstrom 1979). Where, then, do the mites fit into arachnid phylogeny?

Based on their fossil record, it is known that the Acari had already achieved a moderate level of diversity by the early to mid-Devonian (Hirst 1923; Norton, Bonamo, et al. 1988; Kethley et al. 1989; Evans 1992; Subías and Arillo 2002), which means that ancestral mites may have invaded the terrestrial landscape as early as the late Silurian. All of the early Devonian fossil mites are members of the superorder Acariformes (chapters 8, 13–16), whose modern-day descendants feed primarily on fungi, algae, and organic detritus in soil and litter (Walter and Proctor 1999). Their small size and their present-day predilection for subsurface edaphic habitats suggest that the early derivative acariform mites may have invaded land via the sand interstices and soil pores of the littoral zone. In contrast to the Acariformes, the ancestors of the relatively large mites composing the superorder Parasitiformes (chapters 8, 9–12) probably made their terrestrial debut as predators in surface habitats of the littoral zone. Unlike the Acariformes, the known fossil record for the Parasitiformes does not commence until the late Cretaceous (Witalinski 2000; Poinar and Brown 2003). However, based on biological, biogeographical, and phylogenetic evidence, Kim (2004) estimated that the parasitiform Trigynaspida may have originated as early as the upper Triassic (ca. 220 mya). While the known fossil record for the Opilioacarida is limited to a recent find in Baltic amber (Dunlop, Wunderlich, and Poinar 2004), Dunlop (1995) earlier raised the possibility that the fossil Phalangiotarbida of the Carboniferous may in fact be a subgroup of the Opilioacarida.

Fossil evidence indicates that a major adaptive breakthrough may have occurred among the Acari during the late Mesozoic and early Cenozoic eras (60–100 mya), well after the major extinction events of the Permo-Triassic and after the appearance of most of the modern acarine lineages. Acarine radiation during the great speciation explosion of that time reflected the development of a level of morphological and ecological diversity that made possible their exploitation of many co-evolving plants and animals (*evolutionary synergism* of Lindquist 1975a). This was a period of great diversification of angiosperm plants and of the Insecta, a group with which the Acari have close present-day ecological ties. Phoretic mites have been identified on a variety of terrestrial and aquatic insects both in Baltic and Dominican amber from about 20–40 mya (Poinar 1992), and an association between a parasitic mite and its insect host is known to have occurred 40 million years earlier, during the late Cretaceous (Poinar, Pike, and Krantz 1993; Poinar et al. 1997). A reciprocal radiation similar to that observed between mites and insects may have been occurring between parasitic mites and their evolving vertebrate animal hosts in mid-Tertiary times (Poinar 1995; Klompen et al. 1996). Based on fossil evidence, early derivative members of the acarine suborder Oribatida (chapter 15) were already engaged in consuming vascular plant tissues as early as the late Carboniferous period (Labandeira, Phillips, and Norton 1997), some 200 million years before the Mesozoic-Cenozoic expansion.

The morphological and behavioral dichotomy between acariform and nonacariform mites offers reason enough to question whether there was a single evolutionary event in some ancient preacarine arachnid lineage that led to the appearance of the Acari as a monophyletic group, or whether two or more mitelike lineages evolved independently and came to resemble one another through convergence. Speculation regarding the ancestry of the Acari has been lively and prolonged, with proponents of monophyly and diphyly presenting a range of morphological, developmental, and behavioral arguments to support their respective positions (Zachvatkin 1952; Weygoldt and Paulus 1979a, b; Lindquist 1984; Van der Hammen 1989; Walter and Proctor 1999). The comparative morphological evidence for diphyly presented by Zachvatkin (1952) and Van der Hammen (1989) has proven controversial (Dunlop and Alberti 2007), which has led to broader support in recent years for the less ambiguous concept of monophyly in the Acari. In this connection, Lindquist (1984) established transformation series for 40 morphological, physiological, and developmental characteristics of 12 extant arachnid orders, including those composing the Acari (the Opilioacariformes, Parasitiformes (with the Holothyrida considered a separate order), and Acariformes), and concluded that all mites are derived from a common, albeit remote, ancestral stock. Based primarily on the joint possession of a movable gnathosoma and of a hexapod larva and three octopod nymphal instars (see chapters 3 and 5), Lindquist further concluded that the monophyletic Acari and the order Ricinulei (Kaestner 1980) are sister groups.

While a strong case can be made for monophyly in the Acari, spermatological evidence recently compiled by Alberti (2000) on the Chelicerata does not necessarily support a common origin for acariform and nonacariform mite lineages. The limited arachnid fossil record, coupled with continued conjecture regarding the utility of morphological and behavioral characters invoked in evolutionary studies of arachnid taxa, leaves room for continued speculation regarding the ancestral origins of the Acari.

Dunlop and Alberti (2007) have provided an in-depth review of the morphological, developmental, and molecular evidence that supports or refutes monophyly in the Acari. Their paper is recommended as a logical starting point for further exploration into acarine origins and evolutionary history.

CHAPTER THREE

G.W. KRANTZ

Form and Function

Mites are members of the Arthropoda, a vast assemblage of terrestrial and marine invertebrates that are considered to compose a monophyletic taxon (Weygoldt 1998) and that share the characteristics of jointed legs and a chitinous exoskeleton. Mites also possess the arthropodan features of an open circulatory system, ventral nerve cord, alimentary canal, striated muscles, and (in many cases) Malpighian tubules that collect and sequester excretory products for eventual elimination. Among the Arthropoda are two groups that, unlike other arthropods, lack antennae and mandibles. These are the Pantopoda (sea spiders) and the large and diverse group Chelicerata, which, based on recent molecular studies, is considered to be the sister taxon of a clade comprising the Crustacea and the Insecta (Telford and Thomas 1998a). The largest chelicerate group is the class Arachnida, a primarily terrestrial assemblage that includes such diverse forms as scorpions, spiders, vinegaroons, whipscorpions, and the mostly long-legged harvestmen (Opiliones) so common to both temperate and tropical ecosystems (Kaestner 1980). Of the 10 extant divisions of Arachnida recognized by Weygoldt and Paulus (1979a, b), only 3 are known to use foods in addition to, or other than, live prey. Some Opiliones and Araneae have been observed to scavenge on dead insects and on organic matter (Kaestner 1980; Foelix 1982), and some spiders are known to be kleptoparasites (Cangialosi 1997) and occasional feeders on nectar and pollen, but it is only in the Acari that parasitism, phytophagy, mycophagy, and saprophagy rival predation as major feeding strategies (see chapter 6).

Arachnids display a range of morphological and behavioral adaptations well suited to their lifestyles, but many of these adaptations reflect patterns that are essentially fixed at the ordinal or subordinal level. For example, the mouthparts of spiders always are adapted for piercing and sucking although they may use a variety of behavioral strategies for trapping their prey (Foelix 1982). Unlike that of other arachnids, the remarkable range of lifestyles seen in the Acari is more than matched by their diverse form and behavior.

Differentiating the Acari from Other Arachnids

While phylogenetic relationships within the Arachnida and between arachnid assemblages is not always clear, morphological differentiation of the two presently recognized mite superorders from other terrestrial arachnids is relatively straightforward. A key to the superorders and orders of Acari may be found in chapter 8 (Classification).

Key 3.1. Major Arachnid Orders

1a. Opisthosomatic spinnerets absent — **2**

1b. Somatic segments X and XI with 1–4 pairs of ventral spinnerets (Fig. 3.1K) — **Order Araneae** (Fig. 3.1J)

2a. With conspicuous primary somatic segmentation marked by sclerotized tergites — **3**

2b. Without conspicuous primary segmentation or segmentally arranged tergites — **Subclass Acari 5**

3a. With a terminal spine or a compact or whiplike flagellum, posterior 3–5 somatic segments distinctly narrowed — **Orders Scorpiones** (Fig. 3.1A), **Uropygi** (Fig. 3.1B), **Palpigradi** (Fig. 3.1C), **Schizomida** (Fig. 3.1D)

3b. Without a terminal spine or a flagellum, posterior somatic segments normally developed — **4**

4a. Palpi chelate — **Orders Ricinulei** (Fig. 3.1E), **Pseudoscorpionida** (Fig. 3.1F)

4b. Palpi simple or raptorial **Orders Amblypygi** (Fig. 3.1H), **Solifugae** (Fig. 3.1L), **Opiliones** (Figs. 3.1I, M)

5a. Coxae I–IV fused to podosomatic body wall so that the first completely free leg segment is the trochanter; tarsi of legs entire, without fissures (see key in chapter 8 for additional characters) **Superorder Acariformes** (Figs. 3.2B, D, E)

5b. Coxae of legs partially or completely articulated to body wall; tarsi II–IV each with a peripodomeric fissure separating the proximal basitarsus from a distal telotarsus (Fig. 3.2C) (see key in chapter 8 for additional characters) **Superorder Parasitiformes** (Figs. 3.2A, C, F)

External Morphology

The Exoskeletal Envelope

The integument of the typical acarine begins its development as undifferentiated tissue covered by a thin layer of **cuticulin** and separated from the underlying single-layered nucleated epidermis by a thin, poorly defined deposition zone called the **Schmidt layer** (Fig. 3.3A). As development proceeds, the cuticulin differentiates into a thick, chitinous **procuticle** and a thin, overlying **epicuticle**. The procuticle is composed of an underlying **endocuticle** and an outer **exocuticle**, both of which may become sclerotized to varying degrees in various regions of the body through orthoquinone tanning. New cuticle is laid down in the endocuticular layer and is preceded by secretion of granules, which coalesce into microfibers. These fibers then consolidate to form laminations as the cuticle matures (Brody 1970). The width of the laminations decreases toward the cuticular surface so that the lamina in the exocuticle and in the intervening "mesocuticle" become progressively compacted so as to resemble thin plates. **Pore canals** first appear in the endocuticular layers and move in a helical fashion toward the surface through the exocuticle, where they assume their typical linear, often branched, appearance (Norton et al. 1997). The canals terminate in a profusion of micropores in the cuticulin layer just beneath the epicuticle.

The epicuticle (Fig. 3.3B) is a multilayered envelope consisting of an inner epicuticle and an overlying secretion layer, or **cerotegument**, that is formed by secretions carried from the epidermis by the pore canals and that appears shortly after ecdysis (Vitzthum 1940–43). The inner epicuticle comprises an underlying homogenous layer and a covering cuticulin layer (Alberti, Storch, and Renner 1981; Norton et al. 1997). The cerotegument, which may or may not be conspicuous, consists of an underlying outer epicuticle covered by what appears to be a wax layer. A dense, often highly sculptured **cement layer** completes the epicuticular complex. The wax and cement layers on the body surface may provide a degree of protection against excessive water loss or absorption. Brody (1970) felt that the cement layer was of great importance in maintaining water balance in immatures of the acariform mite *Oppia coloradensis* Dolan. The underlying cuticle (or some portion of it) was reported by Wharton and Devine (1968) to be a major pathway for the sorption of water vapor in the parasitiform mite *Echinolaelaps echidnina* (Berl.), although oral or anal uptake of water has been identified as the means of maintaining critical equilibrium humidity in other acarines that have been studied (Knülle 1984).

In addition to the micropores that mark the termini of pore canals, a variety of **macropores** may be found in the cuticular surface of the body and appendages of mites. Like micropores, some of these apertures appear to have a secretory function; others may serve as sites for insertions of sensory structures, while still others are thought to be the external manifestations of **proprioceptors** that mediate physical or chemical conditions within the mite itself (also see Sensory Structures, below). Athias-Henriot (1969c, d) developed a classification of cuticular openings for mesostigmatic parasitiform mites in which three categories of apertures were recognized:

1. **Poroidotactic**, which includes the lyriform and rounded pores usually referred to as **lyrifissures** or **cupules** (Figs. 3.2F, 3.4D, E) common to both the body and appendages in many mite groups. They are considered to be proprioceptors.
2. **Setal**, which includes the insertions of tactile, olfactory, gustatory, and chemosensory hairs and pegs (Fig. 3.23). Certain specialized setal sensory structures of ixodid ticks may be secretory (Schulze 1942).
3. **Adenotactic**, which includes the openings of a variety of secretory structures basically composing a subcuticular gland cell with a distal sclerotized corolla or **calyx**, and a duct that connects the gland to the surface opening, or **solenostome**. The sperm induction pores found in many Mesostigmata (Fig. 3.22), the lateral dermal glands of uropodine Mesostigmata (Woodring and Galbraith 1976), and the dorsal idiosomatic glands exemplified by those described for the parasitiform families MACROCHELIDAE (Krantz and Redmond 1987) and ZERCONIDAE (Johnston and Moraza 1991) are typical of cuticular apertures in the Acari. Cuticular glands are referred to as **crobylophores** by Athias-Henriot (1975a). Additional information on acarine secretory systems may be found on pages 20–21.

Major Body Divisions

The absence of primary segmentation in most acarines, the inherent physical plasticity of the exoskeletal envelope, and the early ontogenetic development in many mite taxa of a posteroventral idiosomatic curvature or **caudal bend** (Fig. 3.4E), have complicated efforts to establish somatic derivations for most Acari (Evans 1992). However, setal/pore/gland patterns (see above) have proven useful in helping to pinpoint somatic origins and structural homologies for some acarine groups. Estimates of the number of primitive segments in the Acari have varied from 12 to 22, depending on the taxon being studied and on interpretation (Zachvatkin 1952; Van der Hammen 1970d; Sitnikova 1978; Jesionowska 1991). Coineau (1974a, b) presented a somatic map of a hypothetical precursor of the prostigmatic family CAECULIDAE based on an earlier construct of Grandjean (1954c), in which the archetypical acariform adult was considered to consist of 16 somites plus a cheliceral segment (Fig. 3.5A). The larval idiosoma was thought to comprise 11 somites, with a segment being added terminally at each postlarval molt (**anamorphosis**). Jesionowska (1991) discounted anamorphosis for larval Eupodoidea and Endeostigmata, suggesting instead that they already have a full complement of segments and that their development is based instead on growth. However, acariform mites typically undergo anamorphic addition of terminal hysterosomatic body segments during ontogeny, with the protonymph adding terminal segment AD to the six segments recognized in the larva (C, D, E, F, H, and PS). Segment AN is added in the deutonymphal stase, and PA in the tritonymph (Figs. 3.4B–E). Concomitant additions may be seen in the number of **genital papillae** (Fig. 3.14C), the number of cupule pairs that open on the hysterosoma (four in the larva, five in the protonymph, and six in succeeding stases), and the number of setae associated with the genital and anal fields.

Segmental amalgamation, or **tagmosis**, in the Acari has resulted in the formation of two major body divisions: an anterior **gnathosoma** (more commonly called the **capitulum** in ticks and water mites), which is derived from the first two primitive somatic segments and which carries the primary organs of food acquisition; and the posterior **idiosoma**, which is the site for virtually all other life functions including locomotion, postoral digestion, reproduction, respiration, and secretion (Fig. 3.2A). The two tagmata articulate by means of a **circumcapitular furrow** (Fig. 3.2D that may be obscured in some groups by secondary somatic development. A **dorsosejugal furrow** (sometimes referred to simply as a sejugal furrow) may be found at a level between legs II–III in opilioacariform and acariform mites (Figs. 3.2D, 15.2J), and a **postpedal furrow** may occur just behind legs IV. The positions of these sutures can provide convenient landmarks for defining portions of the acarine body, even in those actinotrichid groups where somatic plasticity often has resulted in migration and obfuscation of basic somatic boundaries, and in the appearance of novel body regions and demarcating fissures. The suborder Prostigmata (see chapter 13) is especially rich in examples of somatic plasticity, as exemplified by the family CAECULIDAE. Coineau (1974b) postulated that development of the caeculid habitus (Fig. 3.5B) entails movement of the mouthparts (somites 1–2) from a primitive hypognathous condition to a forward-directed (prognathous) condition and the extension of gnathosomatic elements dorsally and posteriorly to form a roof or **aspis** (also referred to as an aspidosoma) over the anterior portion of the idiosoma. The circumcapitular suture is consequently obscured by an overlying **abjugal furrow**, which terminates posterodorsally at the sejugal furrow. Based on the position of the anal aperture and the lateral cupules of the typical caeculid, displacement of opisthonotal elements proceeds both anteriorly and posteriorly: anteriorly with the disappearance of the postpedal furrow and concomitant appearance of a **disjugal furrow**, and posteriorly with the development of the caudal bend and movement of the anal aperture from a terminal to a ventral position. As may be seen from Fig. 3.5B, integumental plasticity in the CAECULIDAE has led to the expression of a podogastric "box" bordered by abjugal and disjugal furrows and carrying the insertions of coxae I–IV. Based on the locations of external structural signposts, similar migrations may be postulated for other acariform groups, including many brachypyline Oribatida (see chapter 15).

Special problems arise in determining primitive segmentation and tagmatic migration in many parasitiform mites, in which extensive development of dorsal and ventral shields often obscures even the most basic body divisions (Van der Hammen 1964). Here, the circumcapitular suture between the gnathosoma and idiosoma often is the only recognizable tagmatic landmark, and theories on the number of putative postpedal segments in parasitiform mites vary considerably (Van der Hammen 1989; Evans 1992). Clues as to primitive segmentation in the Mesostigmata may lie in the location and patterns of idiosomatic dorsal pores, setal insertions, and gland openings (Lindquist and Evans 1965; Athias-Henriot 1969d; Krantz and Redmond 1987; Johnston and Moraza 1991), although lack of embryological evidence makes verification of such connections highly speculative.

Gnathosoma

The gnathosoma of the Acari resembles the head of a generalized arthropod only in that the mouthparts are appended to it. The brain lies in the idiosoma behind the gnathosoma, and the **ocelli** (when present) are situated dorsally or dorsolaterally on the prosoma (Figs. 3.2F, 3.26C). While it might, therefore, seem to represent little more than a conduit through which food is relayed to the esophagus, the gnathosoma is a complex and highly specialized body region on which may be found a bewildering variety of adaptations for sensory reception, food gathering, and preoral digestion (Alberti and Coons 1999). The gnathosoma may play important roles in mating and sperm transfer in some taxa, and in silk production in others. In short, diversity in gnathosomatic form and function is profound when compared to homologous segments of other arachnids (Van der Hammen 1970b).

The gnathosoma is considered to be formed from the cheliceral segment and the derived paired biramous appendages of the second somite (Manton 1977; Telford and Thomas 1998a). The palpal coxae, derived from the primitive paired somatic protopods, have coalesced and expanded to contribute to the formation of the gnathosomatic floor and walls (i.e., the **subcapitulum** and the suboral **hypostome** (Fig. 3.6B)), providing a channel for the paired **chelicerae** that lie above it and an entrance to the buccal cavity that opens just below the chelicerae. Chelicerae have long been thought to have their derivations in the tritocerebrum with the loss of the deutocerebral segment, but it has recently been shown that the deutocerebrum has in fact not been lost in chelicerates and that the acarine cheliceral segment actually is homologous with the first antennal segment of insects (Telford and Thomas 1998a, b). Malar processes (**corniculi** (Fig. 3.6B) or **rutella** (Figs. 3.6C, 15.8A)) that are thought to be derived from terminal hypostomatic setae may flank the **lateral lips** of the hypostome. In some acarine groups, a rooflike epistome, or **gnathotectum** (Figs. 3.6A, E), extends anteriorly over a portion of the gnathosomatic dorsum to partially cover the cheliceral shafts. An apodemal **subcheliceral plate** provides a base over which the chelicerae can glide and serves as a site of muscle attachment for the underlying preoral **labrum** or labrum-epipharynx, an extension of the dorsal pharyngeal wall (Snodgrass 1948; Gorirossi 1950; Woodring and Galbraith 1976; Evans 1992; Alberti and Coons 1999). The labrum may function as a prepharyngeal valve in the Opilioacarida, the Mesostigmata, and the Ixodida, either closing off the buccal cavity when the pharyngeal muscles constrict to prevent food loss (Hughes 1959; Sonenshine and Gregson 1970) or serving as part of a sieve mechanism to prevent ingestion of solid food particles (Evans and Loots 1975). The highly variable **palpi** (see below) are derived from the paired postcoxal endopodal rami of somite 2, and elements of the median paired protopodal endites (*sensu* Manton 1977) of primitive arthropods may be retained in some taxa as secondary feeding structures. More specifically, the median **subcheliceral groove** of parasitiform mites is considered endital in origin and may function in concert with the **tritosternum** of somite 3 (Fig. 3.6B) as a fluid transport mechanism, directing overflow prey fluids to the prebuccal region (Wernz and Krantz 1976).

Internally, the buccal cavity of the gnathosoma opens into the **pharynx** (Figs 3.6E, 3.7A), which serves as a suction pump for ingesting food materials. The pharynx is innervated by several sets of constrictor and dilator muscles, which, along with those muscles that control the movement of retractile chelicerae and the palptrochanters, virtually fill the gnathosomatic capsule (Hughes 1959; Mitchell 1962b, c). Distinctive differences in basic pharyngeal shape have been observed at the ordinal level in the Acari. A cross section of the pharynx in the Mesostigmata reveals a triradiate form (Fig. 3.7A), while that of ticks (*Ixodida*) may be tetraradiate anteriorly and triradiate posteriorly. On the other hand, a basically crescentic or reniform shape has been found in pharyngeal cross sections of acariform mites (Hughes 1959; Nuzzaci 1976; Alberti and Coons 1999). Pharyngeal shape has been invoked as a family and generic character among several taxa of the prostigmatic subcohort Heterostigmata (Cross 1965; Lindquist 1973, 1986) (see also chapter 13).

Idiosomatic **salivary glands** may open into paired ducts located in the buccal cavity beneath the subcheliceral plate (Sonenshine and Gregson 1970; Evans 1992), or they may feed into paired grooved stylets that debouch preorally (Fig. 3.6A). Salivary glands produce the enzymes necessary for preoral digestion of food in a variety of acarines (Moss 1962; Roshdy 1972) and may also supply the cementlike substance used by ixodid ticks to affix themselves to their hosts while feeding (Chinery 1973). Other idiosomatic gland systems open into the buccal region of the gnathosoma of acariform mites, including some that produce silk and others that may be involved in osmoregulation (Alberti 1973; Brody, McGrath, and Wharton 1976). Glandular systems are discussed in the section on the idiosoma (p. 9).

Chelicerae

The chelicerae, along with the palpi, are the primary organs of food acquisition in the Acari and usually are adapted for chewing, piercing, tearing, or sucking. Chelicerae are considered to be derivative endopodal appendages of the first somite, with the primitive condition resembling the chelate form seen in some early Devonian acariform mites (e.g., Endeostigmata) and later in the free-living Parasitiformes. Grandjean (1947e) examined the chelicerae of *Pterochthonius angelus* (Berl.), a primitive arthronotine mite, and concluded that they evolved through a progressive series of segmental fusions and setal modifications from a leglike to a mandible-like appendage (Fig. 3.8). Grandjean suggested that the basal cheliceral segment was derived from the endopodal trochanter of somite 1, that the attached fixed digit or **digitus fixus** represents an amalgamation of the femur, genu, tibia, and tarsus and that the opposing terminal digit is pretarsal in origin.

The range of variation in cheliceral morphology both within and between superordinal acarine taxa is striking and reflects in large part the diversity of lifestyles and feeding habits within the subclass (de Lillo, Di Palma, and Nuzzaci 2001; Akimov, Badanin, and Wita 2002). While the primitive mandibulate form described above is retained in many parasitiform and acariform mites (Fig. 3.9B), cheliceral modifications in phytophagous and parasitic acarines often involve fusions or reductions of the fixed digits and modifications or extensions of the movable digits to form styliform, hooklike, or finely toothed organs for piercing plant or animal tissues or for scraping bacterial films (Figs. 3.9A, C–H). Karg (1993a) noted that there is a correlation in the free-living Mesostigmata between cheliceral structure and the type of food taken; that is, mites with short cheliceral digits armed with long teeth feed primarily on nematodes, while those with longer, slender cheliceral digits and retrorse teeth feed on other mites or on Collembola. However, Walter and

Ikonen (1989) showed that generalizations on feeding habits of mesostigmatic mites based on limited biological information can be misleading. For example, although the chelicerae of *Pergamasus* species (PARASITIDAE) clearly fit the profile of predators on mites and collembolans (Fig. 3.9D), they also have been observed to feed on nematodes. Similar misconceptions based both on cheliceral structure and on long-standing presumption of feeding habits have occurred in sarcoptiform groups. While the prostigmatic family STIGMAEIDAE (p. 296) had long been considered to be exclusively predaceous, Gerson (1972a) found that members of the genus *Eustigmaeus* feed on mosses. Similarly, Walter et al. (1986) observed that members of the cosmopolitan astigmatine genus *Tyrophagus* (ACARIDAE), long known to be fungivores and saprophages, will feed voraciously and develop successfully on a diet of nematodes. Their blunt, chelate-dentate chelicerae (Fig. 3.9B), therefore, function as well for omnivory as they do for fungivory.

Cheliceral retraction in acarines is mediated by well-developed retractor muscles that arise from the dorsal wall of the gnathosoma (Fig. 3.9G) or from the idionotum (Fig. 3.10B), while cheliceral extrusion is largely driven by hydrostatic pressure generated by contraction of the powerful dorsoventral idiosomatic muscles. Dorsoventral muscle insertions often may be identified by symmetrically placed sigillary scars located either in the soft integumental surface or in the sclerites that reinforce the integument (Fig. 3.10A).

Male Parasitiformes lack primary copulatory organs and usually use their chelicerae in sperm transfer (Walter and Proctor 1999), although direct venter-to-venter contact has been observed in some species. Ticks and some early derivative Mesostigmata (Uropodina, Sejida, Trigynaspida) extrude a flask-shaped or globular spermatophore from their genital openings. The male then flexes his gnathosoma ventrally and extends one of his chelicerae toward the neck of the spermatophore. Depending on the taxon, the palpi and/or the first pair of legs may be used to help remove the spermatophore from the genital opening in these groups.

Cheliceral structures that aid in the transfer of sperm have evolved in at least four lineages of Mesostigmata. Male Parasitiae have a median foramen in the movable cheliceral digit, the **spermatotreme** (Fig. 3.10C), which is formed through terminal fusion of the digit with a process that arises at its base. The spermatophore casing is extruded through the spermatotreme in its collapsed state and then filled with sperm and accessory gland products so that the spermatotreme ends up holding the balloonlike spermatophore by a narrow neck (Hyatt 1980). Three groups of Mesostigmata (the Dermanyssiae, the Heterozerconina, and the Celaenopsoidea (Antennophorina)) have chelicerae with semidetached or fused processes modified for sperm transfer and a male genital opening that has migrated anteriorly to the base of the tritosternum. Dermanyssine males have an often elaborate, species-specific outgrowth on the movable cheliceral digit called a **spermatodactyl** (Fig. 3.10D) that is used to transfer sperm material to the female. In some dermanyssines, males produce spermatophores with distinct casings (e.g., RHODACARIDAE; Lee 1974), while others extrude seminal droplets from their genital openings (e.g., MACROCHELIDAE; Krantz and Wernz 1979).

Ticks and many early derivative Mesostigmata deposit spermatophores directly on or just inside the female primary genital opening (**tocospermy** of Athias-Henriot 1969c), using either the chelicerae or direct venter-to-venter contact (see below). Male Dermanyssiae, however, transfer sperm to a porelike secondary genital opening, or **sperm induction pore**, usually near the base of the legs of the female. The term **podospermy** has been applied to this phenomenon by Athias-Henriot (1969c) (also referred to as porospermy). Rare variations are known, as in the semiaquatic uropodine mite *Caminella peraphora*, where sperm is transferred through a gelatinous secretion of the female (Compton and Krantz 1978) (see also Sperm Transfer, below).

Palpi

The palpi are derived from the leglike endopodal appendages of the second somatic segment and typically serve as platforms for an array of terminal chemosensory and thigmotactic sensory receptors, much like the antennae of insects and other Mandibulata. However, the palpi may be modified as raptorial structures in some predatory groups (Fig. 3.6D), as holdfast organs in certain parasitic taxa (Fig. 3.12D), or as filtering mechanisms in the microbivorous family HISTIOSTOMATIDAE (Fig. 3.9E). The simple palpi of predatory Mesostigmata may be used to manipulate the prey so as to bring it to the buccal area (Wernz and Krantz 1976). Subterminal palpal tarsal claws or **apoteles** (Fig. 3.10G) may also play a role in prey manipulation. Other less obvious modifications have led to segmental fusions and reductions so that the number of free palpal segments may vary from one or two for some Astigmatina and Prostigmata (Fig. 3.7B), to three or four for Ixodida (Fig. 3.10E), and to five for many Mesostigmata (Fig. 3.10G) and Oribatida (Figs. 3.6C, 3.10F). Incremental addition of palpal setae through ontogeny provides a useful means for identifying immature stages in the majority of Mesostigmata (Evans 1963d), and variations on the general ontogenetic pattern may occasionally be helpful in defining taxa (e.g., the genus *Blattisocius* of the family ASCIDAE) (Lindquist and Evans 1965).

Idiosoma

The acarine idiosoma comprises the entire body posterior to the gnathosoma and consequently assumes functions parallel to those of the abdomen, thorax, and parts of the head of other arthropods. The great diversity in idiosomatic form, chaetotaxy, and degree of sclerotization is evident in the illustrations that accompany the keys to families in subsequent chapters.

Typically, the acariform idiosoma includes an anterior **propodosoma** and a posterior **hysterosoma** (Fig. 3.2A), which may or may not be separated by a sejugal furrow. The anterior two pairs of legs are inserted ventrally or ventrolaterally on the propodosoma, while legs III (and legs IV in postlarval stages) are inserted in the adjacent portion of the hysterosoma. The latter region is referred to as the **metapodosoma**; it, along with the propodosoma, constitutes the leg-bearing portion of the idiosoma, or the **podosoma**. A postpedal furrow may separate the podosoma from the **opisthosoma**, which is that portion of the hysterosoma behind legs IV. The suffixes **notum** and **gaster** are useful in defining dorsal and ventral idiosomatic regions (viz. **podonotum**, **opisthogaster**, etc.). Other useful terms refer to the gnathosoma plus podosoma (**prosoma**), and to the gnathosoma plus the propodosoma (**proterosoma**).

The idiosoma characteristically carries one or more sclerotized shields or plates that may be either weakly tanned and difficult to discern as in some acariform taxa or strongly developed and heavily tanned as in many parasitiform groups. Retention of larval shield characteristics by sexually mature adults (**neoteny**) occurs commonly in the Astigmatina and some other Oribatida, but the degree of sclerotization in most Acari usually increases through ontogenetic development. Idiosomatic shields not only provide a degree of protection from predation and environmental perturbation but also serve as sites of muscle attachment. Because cheliceral extension in mite groups with retractile chelicerae is hydrostatically driven, and because forward gnathosomatic extension in all Acari is controlled to some extent by hydrostatic pressure, strong dorsoventral muscles are required to achieve necessary movement (Fig. 3.10B). Consequently, the anterodorsal and corresponding anteroventral portions of the idiosoma often develop shields or shield complexes to support attachment sites for these muscles. The dorsal shield component may cover only a small portion of the propodosoma (Fig. 3.2D), or it may extend laterally and posteriorly to cover all of it. Further extension to include the entire podonotum occurs commonly in the Mesostigmata, with the resulting **podonotal shield** abutting an **opisthonotal shield** (Fig. 3.11B). The opisthonotum may instead be armed with a reduced **mesonotal shield** or two or more **mesonotal scutellae** and a **pygidial shield** (Fig. 3.11D), from which cheliceral retractor muscles often arise in uropodine Mesostigmata. In some Oribatida, the proterosoma is covered by a **proterosomatic shield** that completely obscures the gnathosoma dorsally and abuts the anterodorsally extended opisthosoma posteriorly (Figs. 3.2B, 3.12A). The opisthosomatic shield not only forms a covering dorsally but often extends lateroventrally to produce a shell-like, circumgastric **notogaster** (Fig. 3.12C). Fusion of podonotal/hysteronotal shield elements into a **holonotal shield** occurs commonly in the Mesostigmata (Fig. 3.11A). Many of the smaller edaphic Acari have multiple dorsal idiosomatic platelets and reduced ventral sclerotization that allow for greater flexibility and ease of interstitial movement through closely packed soil or sand particles. Dorsal and ventral hysterosomatic shield complexes may be entirely absent in many Prostigmata, including representatives of the Eupodides and Raphignathina. In place of shields, the body wall of the tiny sand-inhabiting mite *Gordialychus tuzetae* Coineau et al. (NEMATALYCIDAE) is covered by an interlacing pattern of platelike "palettes" that work in concert with an underlying muscle layer to move the mite between the sand particles of its interstitial habitat (Haupt and Coineau 1999).

Relative size or structure notwithstanding, dorsal shields generally display patterns of ornamentation that often prove useful as diagnostic characters in taxonomic studies. A terminology for these patterns, derived from that applied to seed surface ornamentation by botanists, has been recommended by Evans (1992) for use in acarology. In addition to cuticular ornamentation, cerotegumental secretions carried from the epidermal layer to the surface by pore canals (Fig. 3.3) may produce distinctive and often bizarre shield ornamentation both in the Parasitiformes and Acariformes.

Ventrally, the idiosoma may be virtually unsclerotized, or it may carry a variety of shields. In the Acariformes, ribbonlike shield remnants, or **epimera**, often mark the attachment sites for the legs (Fig. 3.12B). These may be modified in various groups to form **epimeral plates**. Epimeric furrows separate the epimera and may extend internally to form **apodemes** (Evans 1992). In general, basal leg articulations and genital-anal apertures in acarines rarely are without some type of surrounding sclerite. Ventral shield expansion, reduction, and fusion into more or less fixed patterns are widespread, especially in the Holothyrida, Mesostigmata, and Oribatida, providing useful taxonomic indicators for identification of species and for establishing higher category relationships. Ventral shield reduction and concomitant development or retention of apodemes as major sites of muscle attachment are commonly observed in many acariform mite groups.

While there are notable exceptions (Norton 1994), the degree of idiosomatic sclerotization in adult Acari may reflect their position in the r/K selection spectrum. Briefly, a mite species at or near the K end of the spectrum is approaching the **carrying capacity** of its relatively stable habitat, which means that its success in that habitat depends on efficiency in utilization of limited food resources and production of relatively few progeny with high survival potential (Price 1997). Limited production obviates development of the physical plasticity necessary for simultaneous production of multiple progeny. Despite the physical limitations imposed by heavy sclerotization, K-selected coprophilous eviphidoid Mesostigmata (e.g., MACROCHELIDAE, PACHYLAELAPIDAE, EVIPHIDIDAE) may develop large local populations in a relatively short period of time by moving from deteriorating to nearby fresh dung pads on co-occurring insect carriers (Krantz 1983a). Phoresy provides these mites with habitat stability despite the ephemeral nature of individual dung substrates. The short generation time seen in many *Macrocheles* species, as little as 35 hours from egg to adult (Cicolani

1979), offsets the few progeny that they produce in a given generation, and at the same time assures that mated females will be available for dispersal to new loci when their current site can no longer support the population. Mite species at or near the r-end of the selection spectrum may develop evolutionary strategies that favor production of large numbers of progeny in their relatively unstable but resource-rich environments. Consequently, r-selected taxa tend to produce several generations per year (**multivoltinism**) at fixed sites with large numbers of progeny per generation. High fecundity and multivoltinism in these taxa may offset selective pressure to develop or retain strong idiosomatic sclerotization such as found in many K-selected Mesostigmata and Oribatida because sclerotization restricts the expansion necessary for production of multiple embryos. The relatively weak sclerotization in the graminivorous astigmatine genus *Acarus* (ACARIDAE) is typical of many r-selected mite taxa in which production of large populations (Solomon 1946a, b) can more than offset losses occasioned by vulnerability to predation or desiccation.

The primary external structures of the idiosoma are involved in **locomotion**, **respiration**, **sperm transfer**, **sensory reception**, **secretion**, and **osmoregulation**.

Locomotion

With some exceptions in the Prostigmata and Astigmatina, adult and nymphal Acari possess four pairs of jointed legs, while the larva has only three pairs. Typically, the legs comprise seven primary segments. Beginning with the most proximal, these are the **coxa**, **trochanter**, **femur**, **genu** (patella of some authors), **tibia**, **tarsus**, and **pretarsus**. Primitive secondary division of trochanters III–IV is retained in the Opilioacarida (Fig. 3.2F), while general retention of ancestral basitarsal and basifemoral podomeric sutures in parasitiform mites may give the impression of even further primary division (Fig. 3.2C). However, the resulting pseudosegments lack the innervation common to primary segments (Van der Hammen 1970c). Similarly, tarsus I in certain early derivative Mesostigmata (e.g., the Microgyniina, Sejida, Heterozerconina, and Cercomegistina) may have a separate distal podomeric **acrotarsus** (also referred to as apicotarsus) (Fig. 12.4) on which the pretarsus is inserted. Deutonymphs of some PARASITIDAE also have acrotarsi (Trägårdh 1911; Evans 1992; Moraza 2005c).

The tarsi of the prostigmatic family ANYSTIDAE may display multiple podomeric divisions (Otto 1999b). Subdivision of the femora is suppressed in the larval and protonymphal stases of early derivative acariform taxa, but divisions in the form of vestigial sutures may be seen between basi- and telofemora III–IV of the Heterostigmata (Lindquist 1984). Podomeric fusion of two (occasionally more) postcoxal primary segments may occur in the Acariformes (Hughes 1959; Evans 1992). Amalgamation of the femur and genu or of the tibia and tarsus occasionally results in formation of a femurogenu or a tibiotarsus.

Leg flexion in the Acari is made possible by posttrochanteral intrinsic and extrinsic intersegmental muscles (Fig. 3.13A), while leg extension—like that of the chelicerae in many taxa—relies heavily on hydrostatic pressure created by contraction of the dorsoventral idiosomatic muscles (Fig. 3.10B). The coxae in parasitiform mites have extrinsic levator and depressor muscles and also serve as sites of origin for trochanteral protractor and retractor muscles. Basifemoral levator and depressor muscles also may originate in the coxae. Movement of coxae in the Ixodida is limited by virtue of their close connections to the body wall to simple retraction and extension (Balashov 1972). Because of the absence of free coxae in the Acariformes, the origins of some of the above muscle groups are found instead in the trochanters, which have become the most proximal free leg segments (Van der Hammen 1985; Evans 1992; Alberti and Coons 1999).

Range of movement of individual leg segments is governed by intersegmental monocondylar or dicondylar articulations. A ventral monocondylar articulation usually is present between the coxa and trochanter in the Parasitiformes, while the articulation of the trochanter with the body wall in the Acariformes may either be mono- or dicondylar (Evans 1992). Lateral dicondylar articulations between trochanter and femur allows levation as well as depression at that joint, while the femur/genu, genu/tibia, and tibia/tarsus articulations are monocondylic and dorsal, providing pivot points for the downward flexion made possible by the associated flexor muscles (Fig. 3.13A).

When present, the terminal pretarsus typically comprises a **basilar sclerite** and a set of distal structures that may include a median **empodium**, a padlike or suckerlike **pulvillus**, and paired **claws** (Fig. 3.13C). The pretarsus generally is referred to as an **ambulacrum** when median elements are present. The basilar sclerite serves as a fulcrum that articulates with the tarsus via paired, ribbonlike **condylophores** contained in a terminal tarsal stalk. The sclerite is activated by extensor and depressor muscles that arise in the tibia or in the tarsus (Fig. 3.13A). The ambulacrum itself is a highly variable complex in which some elements may be greatly modified or lost. For example, among other options, the empodium may be clawlike or padlike (Figs. 3.13C, 3.25B) or armed with hairlike excrescences (Fig. 3.13D), or it may be absent (Fig. 3.24B). One or both of the paired claws may be absent (Fig. 3.12C), and the pulvillus often assumes the form of a sucker in various parasitic and commensal acariform taxa (Fig. 3.14A). Absence of ambulacra on tarsi I–II often is correlated with the presence of a well-developed complex of terminal sensory setae. Ambulacral morphology often provides useful guidelines in identification of species and higher taxa.

While the legs are primarily locomotory, some may be modified to serve other functions as well, including anchoring/grasping (Fig. 3.14A), jumping (Fig. 3.13E), and swimming (Fig. 3.14B) (Fain 1965e; Coineau 1974b; Krisper

1991; Fashing, OConnor, and Kitching 1996). Some or all of the segments of legs I and of other legs may be equipped with spinelike or spurlike apophyses to aid in capturing prey, or in the case of certain parasitic and phoretic families, they may be modified for clasping hairs or for adhering to the skin of their host animals. Paired hooks on femora I of the oribatid *Indotritia* sp. (ORIBOTRITIIDAE) provide part of a spring and catch mechanism that enables the mite to somersault out of harm's way. It is propelled by the release of the femoral hooks from the decurved aspis (see chapter 15) and simultaneous application of internal hydraulic pressure (Wauthy et al. 1998). Legs III and/or IV of male astigmatine mites (Oribatida) may be modified and enlarged for guarding or holding potential mates and for fighting or fending off male rivals (see chapter 16).

The tarsi and pretarsi of legs I–II are especially subject to modification in that initial contact with the food substrate generally is made there. Consequently, these segments may be equipped with various sensory and/or food-gathering structures (Fig. 3.13B). In some cases, legs I may become elongated and assume an antenniform appearance and function. Barker and Smith (1968) demonstrated that the mesostigmatic mite *Hypoaspis aculeifer* (Can.) (LAELAPIDAE) uses only legs II–IV in an alternating tripod configuration for locomotion, with legs I serving as surrogate antennae. Feider and Mironescu (1973) noted that certain nasal mites of the mesostigmatic family RHINONYSSIDAE (*Rhinonyssus, Mesonyssus*) use the claws of legs I as surrogate chelicerae, employing them to pierce and tear the nasal mucous membrane of their bird hosts. The chelicerae in these groups are weakly membranous and appear to function only as a part of the gnathosomatic pathway for food uptake. Legs II and/or IV of male Dermanyssiae (Mesostigmata) often have spurlike apophyses that aid in grasping the female during sperm transfer (Fig. 3.2C). Males of *Macrocheles muscaedomesticae* (Scopoli) (MACROCHELIDAE) have been observed to use legs II–IV as weapons against competing males (Yasui 1988) in much the same fashion as heteromorphic males of some ACARIDAE and TARSONEMIDAE, which employ their highly modified legs III or IV for the same purpose (Woodring 1969a, b; Garga, Proctor, and Belczewski 1997). Male *Glyptholaspis americana* (Berlese) (MACROCHELIDAE) that mount the dorsa of unreceptive females to initiate the premating ritual often are catapulted from the dorsum by the greatly enlarged legs IV of the female, which are brought forward and up under the encroaching male (Krantz and Wernz 1979).

The legs of Acari may be smooth or variously ornamented and possess a number of tactile and sensory setae that generally follow fixed patterns of ontogenetic appearance in a given taxon. In some orders (Mesostigmata, Sarcoptiformes), tactile setae often occur in recognizable whorls or **verticils** (Fig. 3.15) around each leg segment in numbers small enough to establish their identity among related taxa (Grandjean 1939a, 1941a; Evans 1969b, 1972). The setal distribution on any given segment in these groups may be reduced to a formula (Fig. 12.4) that can prove useful in establishing systematic and, in some cases, phylogenetic relationships (Evans 1963b; Moraza 2005c). The most widely used system of leg chaetotaxy was first developed for oribatid mites by Grandjean (1935c, 1940, 1941a, 1942a, 1946c, 1958f), reviewed by Norton (1977b), and outlined in Fig. 3.15. Grandjean's system of setal nomenclature has been adapted for use in other acariform taxa as well. Most of the leg segments in other orders (Opilioacarida, Holothyrida, and postlarval Ixodida) have so many tactile setae that recognition of individual verticils or setae may be difficult or impossible (Evans 1992). However, a leg chaetotactic scheme derived from the Grandjean verticil system has been developed for larval Ixodida, where the comparatively modest number of setae allows for its application (Edwards and Evans 1967; Edwards 1975).

Respiration

Exchange of carbon dioxide and oxygen in the Acari is accomplished in ways that are so diverse as to rule out any theory of single line evolution of respiratory systems. While respiration in some groups (e.g., Endeostigmata, and some Oribatida and Prostigmata) is thought to be cuticular, internal transport and exchange of oxygen and carbon dioxide in other acarine taxa usually are mediated by a branched tracheal system that opens externally through spiracular ports, or **stigmata** (Fig. 3.16). The presence or absence of stigmata and their relative positions provide a major diagnostic feature for identifying acarine orders and suborders. Thus, the presence of stigmata on the median portion of the idiosoma characterizes the postlarval stases of all four orders of the superorder Parasitiformes (Figs. 3.16A–D), while paired stigmata behind or between the cheliceral bases (Figs. 3.16F–H) or on or near the lateral humeral angles of the propodosoma (Fig. 3.16E) are typical of the acariform suborder Prostigmata. Members of the parasitiform order Opilioacarida (Fig. 3.16A) are unusual in having multiple pairs of dorsolateral stigmata (two to four pairs depending on the level of ontogenetic development) (Evans 1992; Alberti and Coons 1999).

Stigmatic openings in the Parasitiformes and Trombidiformes often are associated with discrete tubular **peritremes** (Figs. 3.16B, C, F, G) or with respiratory ducts that function in concert with a simple tracheal system (Figs. 3.16E, H). Peritremes in the Ixodida take the form of paired encircling plates around the stigmata (Fig. 3.16D), while those of Mesostigmata and Holothyrida generally extend anteriorly (or anteriorly and posteriorly) from the stigmatic opening as an external groove of varying length and complexity (Figs. 3.16B, C). The peritrematic groove or plate in these mites may be lined with cuticular **microvilli** reminiscent of the filtering devices found in the spiracular atria of insects (Chapman 1982). In addition to serving as a filter for preventing debris from clogging the groove (Radovsky 1969; Alberti and Coons 1999), these fingerlike hydrofuge structures may provide a means for conserving an air layer for plastron respiration in situations where submersion in ground or sea water may limit normal respiration (Hinton 1971; Krantz 1974; Pugh and King 1985a). Plastron respiration also occurs in the aquatic oribatid family HYDROZETIDAE, where a layer of air is trapped on narrow, rugose bands of integument on the dorsal and ventral margins of the opisthosoma (Fig. 3.18A). These bands connect anterolaterally and are contiguous with the internal respiratory system (Krantz and Baker 1982).

Tracheal ducts in the Oribatida (Fig. 3.16I) often open via obscure stigmata located in the acetabular invaginations between trochanters I–IV and the adjacent body wall, in the sejugal groove, on the appendages, or in the prodorsal bothridia (see section on Mechanoreceptors, and chapter 15), where they take the form of short **brachytracheae** (Norton et al. 1997). The tracheae themselves may be derived from internalized **areae porosae** (Figs. 3.17A, B), which develop from surface pores as a correlate of increased sclerotization and concomitant loss of gas permeability (Norton and Alberti 1997). Areae porosae also occur on the gnathosomatic dorsum of female ticks and may play a role in producing a waxy secretion that is deposited on their eggs (see Reproduction, below).

The paired stigmata of ixodid ticks are represented by slitlike **ostia**, each of which is surrounded by a heavily sclerotized peritrematic plate comprising a pigmented portion called the **macula**, and a sieve plate in which are contained a number of fine pores referred to as **aeropyles**. Woolley (1972) described these aeropyles as gates through which air passes into the atrial and subatrial chambers and thence into the tracheae. However, according to Roshdy and Hefnawy (1973), the pores do not connect with the atrial cavity and are not involved in respiration, although Evans (1992) points out that the peripheral aeropyles of *Ixodes ricinis* (L.) do communicate with the underlying atrial chamber and may be considered a form of peritreme. Aeropyles do not occur in soft ticks (family ARGASIDAE), which instead have a simple ostium set in a hinged flap that opens directly into the atrial cavity (Hinton 1967).

Stigmata generally connect to a well-developed tracheal system, but exceptions may be found in the prostigmatic family TROMBICULIDAE and in certain water mites (subcohort Hydrachnidiae, chapter 13) in which tracheae either do not occur or in which they form a subcuticular network with independent tracheal strands serving particular organs (Mitchell 1972). Ultrastructural studies of the cuticular pores of the marine mite *Copidognathus curtus* Hall (family HALACARIDAE) by Crowe and Camara (1973) have suggested a possible respiratory function for these openings.

Lack of discrete stigmata does not necessarily signify lack of a tracheal system. For example, the astigmatine mite *Gohieria fusca* (Oudemans) (GLYCYPHAGIDAE) has paired trachea-like structures that open into the genital atrium (Hughes 1959). Other glycyphagid mite species possess similar structures that often terminate in the genital or anal region (Evans 1992). It is possible, however, that these astigmatine "tracheae" may represent glandular rather than respiratory structures (see chapter 16). Certain early derivative Prostigmata and Oribatida have genital tracheae that have been considered as possible remnants of an earlier, better-developed respiratory system (Lindquist 1984).

Sperm Transfer

Sperm transfer in the Acari is highly diverse and may involve either indirect or direct strategies (Alberti and Coons 1999). In those mite groups where a ventral or dorsal male intromittent organ, the penis or **aedeagus,** is present (Figs. 3.19A–E), transfer of sperm may be made directly to the female genital opening, as in various Prostigmata (Alberti and Crooker 1985), or to a special extragenital **bursa copulatrix** (Fig. 3.2D) located terminally or posterodorsally on the female idiosoma, as in the Astigmatina and some Prostigmata. The bursa connects internally to a seminal receptacle from which paired efferent ducts lead into the ovaries (Griffiths 1970; Witalinski, Szlendak, and Boczek 1990; Witalinski 1991; Evans 1992). Here, the sole function of the genital aperture is that of an oviposition port, or **oviporus** (also referred to as an ovipore). An unusual method of sperm transfer has evolved in many families of astigmatine Oribatida, in which the female rather than the male possesses an intromittent **spermaduct** (Fig. 3.20H) in the form of an elongate pointed bursa copulatrix (Gaud, Atyeo, and Berla 1972; Gaud and Atyeo 1996a, b) (see also chapter 16). Sperm transfer occurs when the duct is inserted into the male genital opening. In those groups where an aedeagus is absent (e.g., the Parasitiformes and Opilioacariformes), direct sperm transfer to the female may be accomplished through venter-to-venter transfer of a sperm packet or through indirect means involving use of male and female extragenital structures (see below).

Noncopulatory sperm transfer (i.e., transfer without benefit of a male intromittent organ) takes many forms in the Acari, ranging from an exchange of sperm material during male/female pairing (Walter and Proctor 1999) to the production by the male of stalked sperm packets or **spermatophores** (Figs. 3.20A–G) that are deposited on the substrate in such a way as to make them available and attractive to passing females. Production of spermatophores is typical of a variety of arachnid and antennate arthropods but appears to have arisen independently in these diverse lineages (Witte and Döring 1999). Spermatophore morphology is highly distinctive within mite families and genera (Fernandez, Alberti, and Kümmel 1991) and often is characteristic of individual species as well. Spermatophores may vary from elongate, often intricately sculptured structures as seen in the prostigmatic family BDELLIDAE (Fig. 3.20A), to minuscule, leaflike structures as in the ERIOPHYIDAE (Fig. 3.20B). They also are produced by many water mites (Mitchell 1958; Efford 1966; Lanciani 1972; Proctor 1992a), although some Hydrachnidiae engage in direct sperm transfer. Branched spermatophore trees, each carrying two or more sperm capsules, have been described from laboratory cultures of members of the prostigmatic family ANYSTIDAE (Fig. 3.20F) (Otto 1999d, f). Here, the male attaches a new spermatophore stalk onto a stalk previously deposited by another male. Similar stacking of spermatophores has been described for other prostigmatic mite groups including the TYDEIDAE (Schuster and Schuster 1970), the NANORCHESTIDAE (Schuster and Schuster 1977), and the Hydrachnidiae, where stacking has been suggested as a possible strategy in sperm competition (Proctor 1992a). The widespread occurrence of spermatophores in the marine mite family HALACARIDAE (Kirchner 1967; Bartsch 1977a; Newell 1984) leads to speculation as to the chemical nature of an adhesive that would be strong enough to allow the spermatophore to maintain its position and integrity on exposed algal surfaces in a turbulent marine microenvironment.

Deposition of a spermatophore typically begins with the lowering of the male idiosoma to the substrate surface and the production of a droplet of heterogenous glandular material produced in the testes and supplemented by secretions from the accessory glands (Witte 1975a). When the male raises its idiosoma, the droplet is drawn out into a stalk that hardens upon exposure to air or water, and a sperm packet (with or without a surrounding sheath) is deposited at its tip (Pauly 1952; Woodring 1970; Alberti 1974; Theis and Schuster 1974; Alberti, Fernandez, and Kümmel 1991; Fernandez, Alberti, and Kümmel 1991; Witte 1991a; Oldfield and Michalska 1996). The packet either is picked up in the genitalia of the female (Lipovsky, Byers, and Kardos 1957) or placed there through manipulation by the male (see below), and the bare stalk is abandoned.

Spermatophore deposition may occur only when females are close to or actually paired with males (proximal), as in many of the terrestrial and aquatic parasitengonine Prostigmata such as the Trombidioidea, Eylaoidea, Arrenuroidea, and Anystoidea (Lanciani 1972; Witte 1984, 1991a; Proctor 1992a, 1994). Conversely, deposition may be dissociative and proceed in the absence of females. Dissociative spermatophore deposition occurs in certain Oribatida (Pauly 1952; Cancela da Fonseca 1969; Woodring 1970; Alberti, Fernandez, and Kümmel 1991) and in many terrestrial and aquatic prostigmatic families including the ERIOPHYIDAE and BDELLIDAE (Oldfield, Hobza, and Wilson 1970; Oldfield and Michalska 1996; Wallace and Mahon 1972), and the water mite families HYDRACHNIDAE and HYGROBATIDAE (Proctor 1992a). Spermatophores may be deposited singly, or they may be laid down in masses or rows of more than 100. Whether the transfer mode is dissociative or proximal, the male enhances the probability that the female will make contact with his spermatophores by means of physical, chemical, and/or behavioral signals. A series of elastic threads or a trail of secretions on which a sex attractant may be secreted is laid down by males of several prostigmatic taxa as a means of guiding females to their spermatophore deposition sites (Witte 1984). In *Linopodes* sp. (EUPODIDAE) the sperm droplet is affixed to a thread that then serves as the spermatophore support (Ehrnsberger 1988). On the other hand, Schuster (1972) observed no causal connection between the production of silk threads by males of *Lorryia schusteri* (Andre and Naudo) (TYDEIDAE) and the deposition of their spermatophores.

Attraction of female water mites of the families HYDRACHNIDAE and LIMNESIIDAE to spermatophores may rely on generation of a pheromone plume through the water from the site of male deposition (Evans 1992). In the Prostigmata (Witte 1984; Proctor 1992a, 1994) and occasionally in the Oribatida (Schuster 1962a, b; Alberti and Schuster 2005), a ritualistic "dance" consisting of a series of movements and caresses by the male often helps direct the female to the site of spermatophore deposition and may be a necessary prelude to spermatophore production (see also chapter 15). Even in the absence of females, the male of the prostigmatic mite *Anystis baccarum* (L.) (ANYSTIDAE), performs a "round dance" when ready to deposit stalked spermatophores on a substrate and then performs a similar maneuver following spermatophore deposition to attract females (Schuster and Schuster 1966).

The female of the water mite *Arrenurus manubriator* Marshall (family ARRENURIDAE) mounts the male prior to male deposition of a spermatophore, the contents of which are guided into the female genital opening by subsequent male movement (Proctor and Smith 1994). A similar sequence has been recorded for the terrestrial mite species *Saxidromus delamarei* Coineau (SAXIDROMIDAE), in which the male carries the female over the deposited spermatophore and lowers her over it to effect insemination (Coineau 1976a). A novel strategy of spermatophore transfer has been observed in a species of the water mite genus *Koenikea* (UNIONICOLIDAE). Here, the male mounts the dorsum of the female and produces 2–4 webbed spermatophores that are attached and stretched between the terminal segments of legs IV. The female then captures these segments and the spermatophores with her legs IV and draws them under her body to her genital opening, where sperm transfer is effected (Proctor 1992b). A similar strategy has recently been observed in other male unionicolids and may prove to be common in this group (Proctor and Wilkinson 2001). Finally, a spermatophore may be attached directly to the body of the female, either haphazardly as appears to be the case in some HALACARIDAE (Bartsch 1977a; Newell 1984), or in the region of the female genitalia as reported for a species of the oribatid genus *Pilogalumna* (GALUMNIDAE) (Estrada-Venegas, Norton, and Moldenke 1996). Although a complex premating ritual has often been observed between males and females of the oribatid mite *Collohmannia gigantea* Sellnick (COLLOHMANNIIDAE) (Schuster 1962b; Alberti and Schuster 2005), the mode of sperm transfer in this species remains a mystery in that spermatophores have not been observed in laboratory settings, nor has mating been verified. However, insemination may occur when the female accepts a nuptial food offering from the closely attending male (see chapter 15).

Direct venter-to-venter sperm transfer occurs in some parasitiform taxa where the male genital opening is well removed from the anterior border of the sternal shield and more or less at the same level as the female genital aperture (Fig. 3.21E). For example, in *Trichouropoda orbicularis* (C. L. Koch) (UROPODIDAE), a complex premating ritual is followed by a period of venter-to-venter contact during which a sperm packet is produced at the male genital aperture and affixed to the anterior margin of the female genital shield. Assimilation of the contents of the packet occurs soon afterward, leaving only an empty, shriveled envelope (Radinovsky 1965). A similar sequence has been observed in the acarology laboratory at Oregon State University in a species of the uropodine genus *Trachytes* (Fig. 3.21F). The male chelicerae may be used to manipulate the sperm packet during the transfer process in both the Uropodina and in the suborder Sejida (see p. 125), the males of which also have a medially placed genital opening and engage in venter-to-venter sperm transfer.

Male dermanyssine Mesostigmata have evolved an indirect method of sperm transfer that, as observed for some Sejida and Uropodina, involves use of the chelicerae. As noted in an earlier section of this chapter, a sperm packet is retrieved from the anteriorly placed genital aperture by one or both chelicerae and transferred directly to the female genital opening (tocospermy of Athias-Henriot 1969c), or sperm may be introduced via a cheliceral sperm transfer organ or spermatodactyl into paired extragenital solenostomes, or sperm induction pores, which arise at a variety of sites on the idiosoma (Fig. 3.22) and which are analogous to the bursa copulatrix of astigmatine females. The term podospermy has been applied to this phenomenon by Athias-Henriot (1969c). Not surprisingly, the male genital aperture in both tocospermous and podospermous Mesostigmata is located at the anterior margin of the sternal shield behind the tritosternum, near the cheliceral field (Fig. 3.21G).

Tocospermy in the Mesostigmata has been observed in the family PARASITIDAE (p. 141; Michael 1892; Micherdzinski 1969; Evans and Till 1979; Schwarz and Walzl 1996), in *Cercoleipus coelonotus* Kinn (CELAENOPSIDAE, p. 129; Kinn 1971), and in a variety of other mesostigmatic groups (tocospermy is presumed to occur also in the Opilioacarida and Holothyrida) (Alberti 1991). As noted earlier in this chapter, parasitids are unique in that the movable cheliceral digit of the male has a median spermatotreme (Fig. 3.10C), through which the neck of the sperm packet is secured before its transfer to the female genital opening (Hyatt 1980). The spermatotreme is formed through terminal fusion of the digit with the spermatodactyl that arises at its base.

A type of tocospermy also occurs in ticks (order Ixodida), in which some or all of the mouthparts of the male (hypostome, chelicerae, palpi) may be inserted into the female genital aperture to effect sperm transfer (Oliver, Al-Ahmadi, and Osburn 1974). A pretransfer insertion of the male mouthparts into the female genital aperture is followed by the production of a bulbous spermatophore at the male genital opening. The spermatophore is then transferred to the female genital aperture by the male mouthparts and attached there by the narrow spermatophore neck. The sac may contain one to several spermatid-bearing endospermatophores, which are introduced into the female reproductive tract by evagination of the ectospermatophoric sac and concomitant extrusion of the endospermatophores.

Based on the morphology of parasitid sperm (Fig. 3.21B), Alberti (1991) concluded that the tocospermous Parasitiae occupy an intermediate, derived position between typical podospermous forms that produce ribbonlike spermatozoa and the more primitive tocospermous taxa that have vacuolated spermatozoa (Fig. 3.21A), exemplified by the groups mentioned above. Thus, tocospermy *sensu* Athias-Henriot 1970 describes two basically different sperm transfer systems. The intermediate position of the Parasitiae was further underscored by Alberti's (1988) discovery of ribbonlike sperm cells in the hemocoel of a *Pergamasus* sp. female, which led him to suggest that fertilization in parasitids may occur via the hemocoel following penetration of the female tissues surrounding the oviduct. Alberti (1988, 2002a) and Alberti, Gegner, and Witalinski (2000) refer to the production of vacuolated spermatozoa seen in the early derivative groups Sejida, Uropodina, and the Epicriides as **architocospermy** (or archispermy), and production of the apotypic ribbonlike spermatozoa of the subcohorts Dermanyssiae and Parasitiae as **neospermy**. The latter two subcohorts are considered to be sister groups, with podospermy describing the reproductive system of Dermanyssiae, and **neotocospermy** describing that of the Parasitiae.

Sperm induction pores of podospermatic Mesostigmata are located at a variety of sites, including the soft cuticle between the bases of coxae III–IV, the posterodorsal faces of coxae III or IV, the trochanters of femora III, the endopodal or metasternal shields, and rarely on the sternal shield (Athias-Henriot 1970; Lee 1970). Alberti and Hänel (1986) and Alberti (1988) speculated about how a sperm access system as complex as that found in podospermous taxa may have evolved. The discovery, mentioned earlier, of sperm cells in the hemocoel of a *Pergamasus* female led them to suggest that sperm induction pores may have arisen as a female response to an earlier derived sperm transfer strategy involving male penetration of the weak cuticle adjacent to coxae III–IV.

The presence of female sperm induction pores is linked to the presence in the male of a distally free or totally fused, grooved or tubate spermatodactyl on the movable cheliceral digit. Sperm material passed to the chelicera(e) from the male genital opening moves into the spermatodactyl and is carried through its tip into the sperm induction pore of the female (Young 1968b; Krantz and Wernz 1979; Alberti, Di Palma, and Krantz 2004. Sperm transfer from the male genital opening to the spermatodactyl has not been observed frequently in tocospermous and podospermous Mesostigmata, but existing accounts of the process are diverse enough to suggest that sperm transport may vary considerably between taxa. In most cases, a premating ritualized series of movements and subsequent dorsal mounting by the male has been found to precede relocation of the male to a ventral position, where clasping of the female is followed by production of a sperm packet at the male genital opening. Lee (1974) observed that the male of *Euepicrius filamentosus* Womersley (OLOGAMASIDAE) apparently recovered its sperm packet in backwardly flexed spermatodactyli or associated movable cheliceral digits and then transferred the contents to the female sperm induction pore by alternate insertion of the spermatodactyl tips. However, the male of *Athiasella dentata* (Womersley) (OLOGAMASIDAE) was observed to pick up a sperm packet between its chelicerae following a backward flexion of the gnathosoma and to hold it by the digits of only one of its chelicerae prior to transfer. The sequence of events leading to sperm transfer in the family MACROCHELIDAE follows a different pattern (Krantz and Wernz 1979). A sperm droplet passes directly to the tip of the hypostome by means of a membranous, extrusible genital tube that extends over the tritosternum and retracted gnathosoma. The chelicerae are extruded so that the arthrodial brush of one of the chelicerae captures the droplet, the contents of which are then transferred to the open median groove of the opposing chela. Sperm passes from the groove into the contiguous closed canal of the spermatodactyl and thence to the female sperm induction pore. Only one spermatodactyl is used in a given transfer episode. Spermatodactyl ultrastructure was described for three species of Dermanyssiae (Alberti, Di Palma, and Krantz 2004) and was further elucidated in an in-depth functional morphological study of the mouthparts of a male *Veigaia* species by Di Palma et al. (2006).

Sperm transfer and mating in the Acari appear to be mediated in large part by pheromones, which are produced by receptive adult or preadult immature females as a means of ensuring successful capture of sperm material from available males. Sonenshine (1985) recognized three types of sex pheromones in Acari: (1) **arrestants**, which are produced by immature females to stimulate precopulatory guarding by adult males; (2) **attractants**, which are emitted by adult females to attract males; and (3) **contact pheromones**, which are secreted by adult females to stimulate copulatory behavior in sexually competent males. Sex pheromones in ticks are secreted by the foveal glands and by certain dermal glands, but accessory and coxal glands may also be involved in sex pheromone production (Sonenshine 1991). Arrestant pheromones are known for the predatory mesostigmatic families PHYTOSEIIDAE and MACROCHELIDAE (Hoy and Smilanick 1979; Yasui 1992) and have been reported to occur in the prostigmatic spider mite family TETRANYCHIDAE. Regev and Cone (1975) found that males of *Tetranychus urticae* Koch were strongly attracted to farnesol, a sesquiterpene alcohol that is secreted by female nymphs. In addition, the sex pheromones nerolidol and citranellol were found in secretions of pharate female *T. urticae* (Regev and Cone 1976, 1980).

Response to an attractant sex pheromone by males of the ixodid ticks *Amblyomma americanum* (L.), *A. maculatum* Koch, and *Dermacentor variabilis* (Say) was demonstrated by Berger, Dukes, and Chow (1971), and the pheromonal substance produced by *A. americanum* was later identified by Berger (1972) as 2-6 dichlorophenol. Pheromonal response has since been verified in *D. variabilis* and recorded in *D. andersoni* Stiles by Sonenshine et al. (1974) and by Sonenshine et al. (1976); in *Ixodes ricinis* (L.) by Graf (1976); and in *Hyalomma dromedarii* Koch by Khalil, Nada, and Sonenshine (1981). While attractant and contact pheromonal responses have been recorded only for the suborder Ixodida (Sonenshine 1985, 1991; Yasui 1992; Coons and Alberti 1999), behavioral and morphological evidence suggests that female deutonymphs and newly emerged adults of members of the mesostigmatic genera *Macrocheles* and *Glyptholaspis* (MACROCHELIDAE) also produce arrestant and/or attractant sex pheromones (Krantz and Wernz 1979; Krantz and Redmond 1987, 1988; also see Secretion, below). Contact pheromones appear to be involved in the premating ritual of the water mite *Neumannia papillator* Marshall (UNIONICOLIDAE). Upon contacting a female, or even after walking over a site where a female had previously rested, the male begins moving in a circular path and initiates a trembling response before depositing its spermatophores (Proctor 1992a). H. C. Proctor (pers. comm.) believes that the females of many water mites produce contact pheromones.

Sensory Structures

The idiosoma is equipped with a variety of sensory receptors, most of which are setal in nature. The vast majority of these setae are simple mechanoreceptors that respond to tactile stimuli, but some have been identified as chemoreceptors, thermoreceptors, hygroreceptors, and even as photoreceptors (Evans 1992). While tactile setae often are little more than hollow cuticular excrescences innervated by basal nerve cells, Grandjean (1935a) observed that most tactile and chemosensory setae of acariform mites contain an optically active iodophilic material that exhibits birefringence in polarized light. This material, **actinopilin** (or actinochitin), may occur as a solid core or as a layer of material surrounding a protoplasmic extension of the basal setal nerve cell (Figs. 3.23C, D). Members of the acarine orders that compose the superorder Acariformes (i.e., the Trombidiformes and Sarcoptiformes) (see chapter 8) possess optically active (anisotropic) setae and appendages derived from these setae and are commonly referred to as the Actinotrichida. Members of the acarine orders composing the superorder Parasitiformes (i.e., the Opilioacarida, Holothyrida, Mesostigmata, and Ixodida) have optically inactive or isotropic setae and also are known as the Anactinotrichida (Van der Hammen 1961; Lindquist 1984; Evans 1992).

Because of space limitations, the following account of sensory structures can be little more than introductory. More detailed discussions may be found in Evans (1992), Alberti and Coons (1999), and de Lillo and Nuzzaci (2005).

Mechanoreceptors

Most mechanoreceptors are simple tactile setae comprising an isotropic or anisotropic shaft, a membranous socket that allows a modicum of shaft movement through enervation of muscles of the peripheral skeletal system (Nuzzaci and Alberti 1996), and a complex of basal receptor cells that transmit signals from the seta to the central nervous system. Tactile setae (Fig. 3.23A) usually are simple and spinose, but they may be variously expanded or ornamented and often form distinctive patterns on the idiosoma that can prove useful in comparative taxonomic studies. In edaphic habitats, these setae may provide a measure of protection from potential predators, but the selective advantage of retaining a full setal complement (**holotrichy**) may be reduced in prey species that forage in more open situations such as plant leaf surfaces. Based on examples gleaned from relationships between predatory phytoseiid mites (PHYTOSEIIDAE) and their spider mite prey (TETRANYCHIDAE), Sabelis and Bakker (1992) have concluded that loss of setae in predatory phytoseiid plant foragers (**hypotrichy**) is a disadvantage in maneuvering through the dense webbing that protects some spider mite prey (Gerson 1985b).

Specialized tactile sensilla referred to as **trichobothria** (Fig. 3.23D) are found in small numbers on the body and appendages of some actinotrichid mites and on the legs of some members of the Opilioacarida (see chapter 9). Trichobothria may be hairlike and greatly elongate in some groups (e.g., the Bdelloidea), with broad, flat insertions that are unlike those of the typical tactile setae adjacent to them. Specialized prodorsal trichobothria may be inserted in complex, cuplike cavities referred to as **bothridia**, and their associated trichobothrial sensilla or modified sensilla are called **bothridial setae** or bothridial organs (Fig. 3.18C). One or two pairs of simple, pectinate, or clavate bothridial setae occur on the prodorsum of many prostigmatic, endeostigmatic, and oribatid mites (Kethley 1990; Evans 1992), and some members of the prostigmatic families EREYNETIDAE and EUPODIDAE have an additional pair posteriorly on the hysteronotal margin. A single pair of prodorsal bothridial setae is present in most Oribatida, and these are inserted in distinctive bothridia (Fig. 3.18C) (Alberti, Moreno, and Kratzmann 1994; Alberti 1998). Prostigmatic and oribatid bothridial setae usually appear in the larva and are retained throughout ontogeny. However, in families of the Heterostigmata, they arise from previously formed setiform setae only in the adult stase, and often only in the female (Lindquist 1986). Bothridial setae are highly variable in appearance, but they resemble typical acariform tactile setae in having a solid core of actinopilin. They are absent in the oribatid cohort Astigmatina.

While the function of the elongate leg trichobothria is thought to be anemotactic (Evans 1992), the trichobothrium and two associated modified setae found at the tip of tarsus I of the endeostigmatic mite *Grandjeanicus gabonensis* Coineau, Theron, and Valette (GRANDJEANICIDAE) appear to function in concert as an acoustic organ (Coineau, Theron, and Valette 1997). Pauly (1956) suggested that the prodorsal bothridia of the oribatid family DAMAEIDAE are vibro- and anemoreceptors, a conclusion shared by Alberti, Moreno, and Kratzmann (1994) in the case of *Acrogalumna longipluma* (Berl.) (GALUMNIDAE) and other Oribatida, and by de Lillo and Aldini (2002) for the heterostigmatic mite *Pediculaster mesembrinae* (Canestrini) (SITEROPTIDAE) (Fig. 13.86). The clavate bothridial organs of arboreal Oribatida are thought to function in geotaxis (Aoki 1973), or they may serve to reduce sensitivity in an environment where continuously moving air could overstimulate these receptors (Norton and Palacios-Vargas 1982). Wergin et al. (2000) observed regular and repeatable oscillations of the frozen bothridial setae of *Archegozetes longisetosus* Aoki (TRHYPOCHTHONIIDAE) and *Galumna jacoti* Marshall, Reeves, and Norton (GALUMNIDAE) and those of the prostigmatic mite *Pyemotes tritici* LaGrèze-Fossot and Montagné (PYEMOTIDAE) when imaged under low temperature SEM and suggested that these setae may serve as capacitors that are deflected by electromagnetic signals generated by their host or food substrate. Beklemishev (1969) and Evans (1992) mention the possibility that bothridial setae are involved in maintaining balance.

In addition to setate sensory structures, the cuticular surface of acarines is ornamented with a variety of distinctive porelike openings, some of which are considered to serve a sensory function. Among these are cuticular and intracuticular organs that apparently receive information on changes in physical tension within the body and then mediate the appropriate response of adjacent subsurface sensory neurons (Penman and Cone 1974; Alberti and Crooker 1985). These putative proprioceptors generally take the form of small, membrane-covered clefts and are referred to as lyrifissures, slit sense organs, or poroids. Cupules (Fig. 3.4E) are rounded, cup-shaped structures that lie in cuticular depressions and are considered by Alberti (1998) to be homologous with lyrifissures. Proprioceptors may be found on the legs, palpi, chelicerae (in Opilioacariformes and Parasitiformes), and idiosoma throughout the subclass (Grandjean 1935b; Sonenshine et al. 1984; Hess and Vlimant 1986; Evans 1992; Alberti 1998; Coons and Alberti 1999). Although proprioceptors are numerous on the body and/or appendages in other arachnid orders (Kaestner 1980), they tend to occur singly or in small numbers in the Acari. Opilioacarid mites are exceptional in this respect, having multiple rows of slit organs in their soft idiosomatic integument (Fig. 3.2F).

Chemoreceptors

Mites possess a variety of setiform chemosensory organs, most of which are located on the anterior appendages. Chemosensory setae of the Acariformes generally are anisotropic and are found on the terminal segments of legs I–II and on the palpi. While they share the internal characteristic of a nerve-innervated protoplasmic core surrounded by a sheath of optically active actinopilin, the external appearance of these setae varies considerably. **Eupathidia** (acanthoides and pseudacanthoides of Grandjean 1946a, identified by the symbol ζ) are spinose (Fig. 3.23C, rightmost), apically rounded or divided structures often found on the extremities of tarsi I–II (rarely on tarsus III) and on the tip of the palptarsus. Eupathidia apparently carry only a single terminal pore and may function as contact, or gustatory, chemoreceptors (Alberti 1998; Alberti and Coons 1999). Grandjean (1943a) noted that eupathidia are never replaced by normal setae in the course of ontogenetic development. Although their singular structure and presumed chemosensory function suggest that they are derived from some simpler ancestral organ, eupathidia are hypothesized by André (1979) to be phylogenetic precursors of optically active tactile setae.

Single or pseudosymmetrically paired eupathidia often are found on leg tarsi I–II (Fig. 3.25) and on the palptarsi of Prostigmata and some Oribatida, but they are absent in the Astigmatina (Grandjean 1939a; Norton 1977b; Lindquist 1985c). Evans (1992) feels that the location of eupathidia on the palpi and legs supports a gustatory-mechanosensory function for these structures. In this connection, it should be noted that the special hollow seta found at the tip of the palp of web-spinning tetranychine spider mites (TETRANYCHIDAE, p. 298) is generally considered to be a eupathidium, although it functions in delivery of silk received from an associated underlying gland or series of glands (Mills 1973; Alberti and Crooker 1985). Based on what is known about eupathidia in other acariform groups, their function in spider mites raises the possibility of another provenance. It is significant that neither palpal glands nor terminal hollow palpal hairs are present in tetranychid subfamilies that do not produce silk. The pair of eupathidia described at the terminus of the larval and protonymphal palptarsi of *Anystis* sp. (ANYSTIDAE) by Grandjean (1943a) and observed in nymphs and some larvae of the anystid genus *Erythracarus* by Otto (1999d) are likewise thought to mark the openings of silk ducts. Otto hypothesized that the ducts are associated with production of the silken cocoons in which the mites enclose themselves before molting. Lindquist (2000) has noted that a palpal silk gland system may have been ancestral in the Acariformes and a precursor to those systems in which silk is produced in podocephalic glands and delivered via the buccal cavity (see Gland Systems, below).

The **famulus** (Figs 3.23C, 3.24A, B (ε)) is similar to the eupathidium in having a terminal pore, but it tends to be smaller than surrounding setae and generally is found on tarsus I (sometimes on tarsus II) in the Acariformes (a famulus-like bifid seta occurs on acrotarsus II in the Opilioacarida, p. 104). Famuli may be represented by a globular head surrounded by bracts (Figs 3.23C, 3.24B), the ancestral condition according to Van der Hammen (1980), or by a falcate (Fig. 3.23C) or terminally branched (Fig. 3.24A) seta, but most are simple and peglike. They are often greatly reduced and may be invaginated into depressions or pits (Fig. 3.13B) (Alberti and Coons 1999; de Lillo and Nuzzaci 2005). The latter characteristic is atypical of eupathidia and other setal types and led Grandjean (1961b) to refer to the famulus as an "organe mystérieux." Its function remains unclear, although its structure suggests that the famulus may respond to gustatory stimuli.

In addition to the eupathidia and famuli, the legs and palpi of acariform mites often carry porose, thin-walled **solenidia** (Fig. 3.23B), terminally rounded or pointed setiform or peglike structures that lack the core or sheath of actinopilin characteristic of eupathidia and famuli. Solenidia are found on the palpal tarsus and may also occur on the tarsus and tibia, less frequently on the genu, and occasionally on the femur of legs I–IV. They are usually inserted in broad and immovable bases and often appear striated because of their internal structure (Alberti and Coons 1999). Unlike other setate acariform sense organs, solenidia contain no actinopilin and thus are not birefringent (de Lillo and Nuzzaci 2005). While they generally are erect, solenidia on legs I–II of eupodine Prostigmata may be recumbent or recessed into a pit on the segment on which they are inserted (Fig. 3.24E). Solenidia are considered by Evans (1992) to be olfactory in function and may be sexually dimorphic, for example, appearing on legs III–IV of only male Raphignathoidea.

A system of notation for mechanosensory and chemosensory setae and setate structures found on the appendages of actinotrichid mites was developed by Grandjean (1935c, 1939d, 1958f) to alleviate confusion and to pinpoint homologies where a given structure might assume a variety of forms in disparate taxa. Based on this system, the solenidia of femur, genu, tibia, and tarsus, respectively, carry the Greek symbols theta, sigma, phi, and omega (θ, σ, φ, ω) (Fig. 3.24C). As noted earlier, famuli and eupathidia are identified as ε and ζ, respectively. Analogous notation systems for the legs and palpi of the Mesostigmata have been developed by Evans (1963b, d). Other systems of notation used for identifying idiosomatic setae and sensory structures in various acarine groups are discussed in the chapters where these taxa are treated.

Unlike the Acariformes, all of the normal setae and setiform sensory structures of the Parasitiformes are optically inactive (**isotropic**) and cannot easily be homologized with sensory organs of acariform suborders. Setiform structures of several types form sensory fields on the appendages, especially on the palpi and on tarsi I. In his paper on the comparative morphology of tarsus I of various Mesostigmata, Haarløv (1943) described a group of setiform structures (the tarsal organ) that strongly resemble the solenidia of acariform mites. Jalil and Rodriguez (1970b) later described eight of these blunt, peglike structures at the terminus of tarsus I of *Macrocheles muscaedomesticae* (Scopoli) (MACROCHELIDAE) and experimentally established that some or all of them are involved in olfaction. Coons and Axtell (1973) referred to these porose "ursolenidia" as **sensilla basiconica** and corroborated the earlier findings of Jalil and Rodriguez on the basis of ultrastructural evidence. Meng et al. (1984) found similar tarsal sensilla in members of the mesostigmatic family LAELAPIDAE that likewise responded to olfactory stimuli. More recently, Pugh (1996b) observed that sessile setiform organs in the sensory field of tarsus I in the genus *Halarachne* (HALARACHNIDAE) (Fig. 12.60A), respiratory tract endoparasites of seals, may function either in thermo/hygroreception, olfaction, or gustatory chemoreception, while sensilla with movable bases may instead have a mechanoreceptive function.

Tarsus I of ticks (Ixodida) carries a distinctive dorsal sensory field commonly known as **Haller's organ** (Fig. 3.24D). The complex comprises an anterior pit (a) and a more posterior capsule (b), each of which contains a variety of sensilla that respond to different stimuli or sets of stimuli (Ivanov and Leonovich 1983; Hess and Vlimant 1986; Klompen and Oliver 1993a; Steullet and Guerrin 1994a, b; Coons and Alberti 1999). Porose, thin-walled olfactory receptors are located in the capsule portion, which is relatively deep and partially covered by a cuticular extension of the cavity rim. Response to aggregation-attachment pheromones (see Sperm Transfer, above) apparently is mediated by one or more of these receptor organs. A similar response has been identified as a function of one type of thick-walled sensilla located in the anterior pit of Haller's organ, along with heat and humidity reception (Lees 1948; Haggart and Davis 1980). A sensory field similar to Haller's organ in ticks has been observed in various Mesostigmata and in members of the holothyroid family ALLOTHYRIDAE (p. 107). Based on these similarities, Lehtinen (1991) suggested a close phylogenetic relationship between ticks and holothyrids and postulated a sister group relationship of these taxa with the Mesostigmata (however, see chapter 10).

Photoreceptors

Eyes, eyespots (aggregates of pigmented granules in typical eye loci), and other photoreceptive organs have been identified in many acarine groups. One or two pairs of simple ocelli or lateral photoreceptors may be present on the anteromedian or anterolateral aspect of the prodorsum of Opilioacarida, Holothyrida, Ixodida, Prostigmata and—exceptionally—of the Oribatida, though rarely on the hysteronotum in the latter suborder (Binnington 1972; Alberti 1975, 1998; Van der Hammen 1983a, b; Balogh and Mahunka 1983; Evans 1992; Coons and Alberti 1999; Alberti and Coons 1999). A typical ocellus consists of a biconvex or convex-concave cuticular lens with an underlying photoreceptor whose axons form the optic nerve (Evans 1992). The lateral prodorsal ocelli in some mites may be divided into two components, each of which may function in a different way. For example, in *Tetranychus urticae* Koch (TETRANYCHIDAE), the biconvex anterior eye is thought to act as a scanning-point detector for green and UV light, while the slightly convex posterior element receives only near-UV signals and apparently is nondirectional (McEnroe 1969; McEnroe and Dronka 1969; Alberti and Crooker 1985). Response of *T. urticae* to that portion of the far red spectrum near infrared may actually be mediated by other receptors (McEnroe 1971). Based on fine structure of the retina, Alberti, Kaiser, and Fernandez (1991) determined that the photoreceptors of *Neocarus texanus* (Chamberlain and Mulaik) (Opilioacarida), of *Cyta latirostris* (Hermann) (Prostigmata), and of several Oribatida represented three different organ types, with those of *N. texanus* being the most complex.

Grandjean (1958a) considered the primitive number of ocelli in the Actinotrichida to be six (two lateral and one median pair). While paired lateral ocelli are indeed common in many prostigmatic taxa (e.g., the water mites, the superfamilies Bdelloidea, Tetranychoidea, and Anystoidea, and the erythraeid subfamily Erythraeinae), paired anteromedian ocelli or eyespots are not often encountered. An unpaired anteromedian ocellus or pigmented light-sensitive zone may be found in various acariform mites (e.g., the HALACARIDAE, HYDRYPHANTIDAE, and ACARONYCHIDAE), but the median eye or pair of eyes is more commonly carried on the underside of a **naso** (Figs. 3.18D–E), an anteromedian prodorsal projection seen in a number of prostigmatic and early derivative oribatid families (Grandjean 1943c, 1958a; Van der Hammen 1969a; Coineau 1970; Alberti 1998). In addition to the inferior eye, Coineau (1970) identified a distinctive clear area on the upper surface of the naso in the prostigmatic family CAECULIDAE that may represent a second light-sensitive zone. Hygrobatoid water mites have light-sensitive lenses that are completely sunk beneath the cuticle, but they may be turned to face in different directions by means of muscles attached to spurs arising from the lenses (H. C. Proctor, pers. obs.).

In the Astigmatina, a single pair of anterolateral, hysterosomatic, lenslike organs occurs in deutonymphs of the genera *Bonomoia* and *Copronomoia* (HISTIOSTOMATIDAE) and in various genera of the WINTERSCHMIDTIIDAE (Fig. 16.11A), on the prodorsal apex of deutonymphs of *Hemisarcoptes* (HEMISARCOPTIDAE), and rarely on adults of *Carpoglyphus* (CARPOGLYPHIDAE) (Zakhvatkin 1959; Evans 1992). Anterolateral ocelli may also be found in many families of Prostigmata, in some Ixodida (Fig. 3.26C) and Holothyrida (ALLOTHYRIDAE), and in the Opilioacarida. Opilioacarids may have two or three pairs of prodorsal ocelli (Van der Hammen 1969b), but their homologies with actinotrichid ocelli (or with those of ticks) have not been established (Alberti 1998).

Many higher (brachypyline) Oribatida have one or more light-sensitive clear spots or elevated, lenslike lenticuli (singular: **lenticulus**) (Fig. 3.18A) located just behind the prodorsum on the opisthonotum. The cuticular cornea of the lenticulus is convex-concave in some of the oribatids that have been studied (*Hydrozetes lemnae* (Coggi) and *Scutovertex sculptus* Michael) but is biconvex in *Scapheremaeus argentinensis* Travé and Fernandez (Alberti and Fernandez 1990; Alberti, Kaiser, and Fernandez 1991). Light-sensitive clear spots such as those seen in the brachypyline families ORIBATELLIDAE, CHAMOBATIDAE, and EUZETIDAE differ from lenticuli in having their lamellated receptor poles directly attached to the underlying synganglion rather than being subcuticular and well removed from the synganglionic mass (Alberti 1998). Because of their unusual location and their appearance in derived taxa, both the lenticulus and the clear spot are regarded as having been secondarily evolved (Grandjean 1961b; Alberti and Fernandez 1990).

Light sensitivity may not necessarily be confined to those acarines that possess discrete photoreceptive organs. For example, some ixodid ticks have no external indications of eyes but nonetheless are capable of responding to changes in light intensity (Lees 1948). These responses are mediated by supracoxal sensory cells that have connections to terminal branches of the optic nerve (Binnington 1972). Evans (1992) mentions the likelihood that other eyeless ticks, and indeed other eyeless Acari, may eventually be shown to have an optical nervous system component that responds to light. An unusual form of light sensitivity in an eyeless mite was described by Camin (1953a), who demonstrated that the snake mite, *Ophionyssus natricis* (Gervais) (MACRONYSSIDAE), perceives light and shadow by means of a pair of photosensitive spots on the pulvillar membrane of tarsus I.

Secretion

It was noted earlier in this chapter (see Mechanoreceptors, above) that various porelike openings may be found in the acarine cuticle. Some of these openings have been found to mark the exits of ducts that carry the products of subcuticular gland cells (Lees 1947; Wharton, Parish, and Johnston 1968; Athias-Henriot 1969c, d, 1971, 1975a, b; Krantz and Redmond 1987, 1988; Johnston and Moraza 1991; Alberti and Seeman 2004). These products may vary from the cementlike or waxy substances such as those laid down by the dermal glands of ticks (Balashov 1972; Evans 1992; Coons and Alberti 1999), to arrestant or attractant sex pheromones (Hoy and Smilanick 1979), and to noxious defensive secretions for warding off would-be predators (Kerfoot 1982). An aggregation pheromone with probable cuticular origins may be largely responsible for the clustering behavior of recently fed chicken mites, *Dermanyssus gallinae* (DeGeer) (DERMANYSSIDAE)—a response that may help ensure concentration of mites in loci favorable for host encounter (Entrekin and Oliver 1984). Cuticular glands may also be responsible for the production of molting fluid in some ticks (Lees 1947) and of a substance in *Ceratozetes cisalpinus* (Berlese) (CERATOZETIDAE) and other oribatid mites that prevents debris from adhering to the cuticle (Woodring and Cook 1962). Alberti et al. (1996) report that the saclike bothridial setae of the oribatid *Eulohmannia ribagai* (Berlese) (EULOHMANNIIDAE) may have a secretory function and that the lateral cuticular areae porosae of other oribatids of the genera *Acrogalumna* and *Mixacarus* secrete a substance presumed to be a lipid. The humerosejugal porose areas common to many Oribatida may secrete substances used in the formation of waterproofing cerotegument, while some notogastric porose areas may produce sex pheromones (Norton and Alberti 1997). Species of the brachypyline oribatid genus *Scheloribates* produce alkaloids (pumiliotoxin and related compounds) that may be sequestered in the skin of the poison dart frogs (Dendrobatidae) that eat them (Takada et al. 2005). A lipid-like oily substance of unknown function also is secreted on the body and legs of some uropodine Mesostigmata (Woodring and Galbraith 1976; Evans 1992), and a similar material probably is produced by the dermal glands of other Mesostigmata as well (see Athias-Henriot 1975b; Krantz and Redmond 1987; and Johnston and Moraza 1991). The identity of these compounds is unclear, but Sakata et al. (1997) have found that the dorsal gland openings of *Uroactinia hirschmanni* Hiramatsu (UROACTINIIDAE) produce a range of volatiles including the naphthoquinones plumbagin and 7-methyljuglone, typically found only in plants and occasionally recovered from thrips. Both compounds are reported to have antibacterial and antifungal properties and may act as antimicrobials in *U. hirschmanni*. Similar or identical compounds are likely to be secreted by other anactinotrichid mites as well. Experimental evidence suggests that the paired cribral glands of *Macrocheles perglaber* are sites of sex pheromone production, with the cribrum itself serving as a dispersal platform for their products (Krantz and Redmond 1987).

Members of a number of derived oribatid taxa possess a pair of single-celled lateral **opisthonotal glands** (Fig. 3.2D) (also referred to as oil glands or expulsory vesicles), which may appear as large black, brown, yellow, or red spots where the integument is transparent enough to observe them (Evans 1992; Alberti and Coons 1999). The gland duct exits via a single pore in these "glandulate Oribatida" (Norton 1998) and is protected by a cuticular flap that may serve to control the delivery of gland products to the surface. These secretions contain a variety of volatile compounds (monoterpenes, hydrocarbons, esters, and aromatics) that may vary between species or supraspecific groups and that may serve as alarm pheromones, aggregation and/or sex pheromones, or as defense substances against fungal or predatory attacks (Kuwahara 1991; Kuwahara, Ishii, and Fukami 1975; Kuwahara et al. 1980; Kuwahara, Leal, and Suzuki 1989; My-Yen et al. 1980). Neryl formate has been identified as the key alarm pheromone produced in the opisthonotal glands of the several astigmatine Oribatida that have been tested (Kuwahara 1991), and it is also a component of the chemical bouquet of alarm and sex pheromones secreted by the crotonioid oribatid mites *Hydronothrus crispus* Aoki (MALACONOTHRIDAE) and *Platynothrus peltifer* (C. L. Koch) (CAMISIIDAE) (Sakata, Tagami, and Kuwahara 1995; Raspotnig et al. 2005). Mizoguchi et al. (2003) determined that neryl formate is produced in concert with the sex pheromone α-acaridial in *Rhizoglyphus robini* Claparède (ACARIDAE), where it is a major compound. *A*-acaridial is widely distributed in the Astigmatina. Presence or absence of certain chemical constituents in opisthonotal gland products may provide a chemical basis for establishing or verifying phylogenetic relationships within the higher Oribatida (Sakata and Norton 2001, 2003; Raspotnig, Krisper, and Schuster 2004; Raspotnig, Schuster, and Krisper 2003).

The lateral opisthonotal secretory glands, or **sensilla sagittiformia** (Fig. 3.10E), of ixodid ticks resemble astigmatine opisthonotal glands in position, but their major secretory product is squalene, a lipid defense pheromone (allomone) against ant predation. Yoder et al. (1996) have suggested that squalene in tick secretions is derived from the blood of their hosts, inasmuch as ticks cannot synthesize this substance.

Other acarine gland systems are discussed later in this chapter under Internal Morphology. A broader range of glandular/secretory systems is reviewed by Alberti and Coons (1999).

Osmoregulation

The larvae of many acariform mites possess a pair of **urstigmata**, or organs of Claparède, ventrally on coxae I or II or in the integument between them (Fig. 3.18B), although developmental studies by Thomas and Telford (1999) have shown conclusively that these variously shaped structures are derived from the bases of legs II. Urstigmata are considered homologous to the genital papillae of later ontogenetic stases (Grandjean 1946a; Knülle 1959; Van der Hammen 1969a; Alberti 1979). Genital papillae in most acariform taxa display a sequential development during postlarval ontogeny that often is referred to as **Oudemans' rule**. A full complement of papillae develops in the genital field of many Oribatida and in some Prostigmata, with one pair being present in the protonymph, two in the deutonymph, and three in the tritonymph and adult (Figs. 3.4F–I). There is often an associated incremental development in the number of genital setae, correlated with the addition of an opisthosomatic segment at each postlarval molt (anamorphosis; also see Major Body Divisions, above). Postprotonymphal stages in the astigmatine Oribatida and many terrestrial Prostigmata have a maximum of two papillar pairs, while some prostigmatic water mites (Hydrachnidiae) may have over 100 papillae tightly grouped in the genital area. Genital papillae are traditionally referred to as genital acetabula by water mite specialists (see chapter 13, Parasitengonina).

Based on examination of representative Prostigmata, Vercammen-Grandjean (1975a) concluded that the function of the urstigmata and genital papillae is respiratory. Subsequent studies by Alberti (1979) and Alberti and Bader (1990), however, suggested that these organs may instead be important sites for osmoregulation, at least in aquatic Actinotrichida. An osmoregulatory function also has been ascribed to the larval epimeral pores and the genital papillae of prostigmatic marine mites (HALACARIDAE, p. 239) and to the dorsal axillary organs of members of the aquatic/semiaquatic astigmatine family ALGOPHAGIDAE (p. 574) (Bartsch 1973, 1974a; Fashing and Marcuson 1996). Osmoregulation is an important function of the salivary glands of ixodid ticks and of the coxal glands in argasid ticks, effecting removal of excess water from the blood meal (Hagan, Oliver, and Pound 1984; Bowman et al. 1997; Coons and Alberti 1999) (also see Nonintegumental Gland Systems, below).

Internal Morphology

The thin idiosomatic hemocoel of acarines surrounds a surprisingly complex series of internal organ systems bathed in a relatively clear hemolymph that contains a variety of hemocytes, the functions of which may include clotting, macrophagy, and tissue dissolution during apolysis. Prohemocytes may serve as germinal cells in the development of certain types of hemocytes (Evans 1992; Coons and Alberti 1999; Alberti and Coons 1999). Generic substances that have been identified from acarine hemolymph include amino acids (Boctor 1972), lipids (Hajjar 1972; Woodring and Galbraith 1976), and glucose (Aboul-Nasr and Bassal 1971).

The hemolymph circulates freely throughout the hemocoel primarily as a result of body movement. Circulation to the legs and other extremities is facilitated by contraction of the dorsoventral idiosomatic muscles (Fig. 3.10B), an action that also brings about leg extension and (in many acarine taxa) cheliceral extrusion. Thus, although the postcoxal leg segments are equipped only with flexor muscles (Fig. 3.13A), hydrostatic pressure of the hemolymph induces the countermovement of extension. A dorsomedian heart provides additional means for hemolymph circulation in the Ixodida, Holothyrida, and some Mesostigmata (Winkler 1888; Thon 1906; Obenchain and Oliver 1976; Evans 1992).

The major internal organ complexes of the idiosoma are the **digestive/excretory**, **reproductive**, **glandular**, and **nervous systems**.

Digestion and Excretion

Although histological evidence remains largely conjectural, many mites are believed to use a preoral digestive process involving secretion of salivary enzymes onto or into a candidate food substrate and ingestion of the resultant liquefied product (Mitchell 1962c; Evans 1992; Alberti and Coons

1999; Coons and Alberti 1999). Preoral digestion of blood by ticks (Ixodida) has not been demonstrated (Akov 1982), but a number of salivary gland secretions that have a direct bearing on the feeding process have been identified in these acarines. Included here are anticoagulants (Balashov 1972) and various cytolytic and proteolytic enzymes that increase permeability of skin capillaries and cause host tissue destruction at the feeding site (Tatchell and Binnington 1973; Higgs et al. 1976). Salivary glands are discussed later in this chapter.

While the postoral digestive system assumes a variety of forms in the Acari, certain features are more or less constant (Alberti, Seniczak, and Seniczak 2003). A buccal cavity or mouth opens into a muscular ectodermal pharynx (see earlier section on the gnathosoma, p. 7), which leads posteriorly into a prosomatic **esophagus** (Fig. 3.6E). The narrow esophagus may pass through a portion of the anterior salivary or silk glands in some taxa (Alberti and Crooker 1985), continuing through the ganglionic mass or **synganglion** into a broader **ventriculus** (Figs. 3.6E, 3.26D). The ventriculus may be a simple, saclike structure, or it may have two or more gastric diverticula or **caeca** (Fig. 3.26B) arising from it (Blauvelt 1945; Young 1968a; Brody, McGrath, and Wharton 1972; Evans 1992). These caecal extensions have generally been thought to provide additional capacity for holding and processing ingested nutrients. However, Alberti, Seniczak, and Seniczak (2003) observed that the caecae of *Archegozetes longisetosus* Aoki (Oribatida, TRHYPOCHTHONIIDAE) never contained solid food and probably served instead as loci for digestive enzyme secretion and for resorption. The ventriculus passes directly into the postventricular midgut or **colon**, which in some groups may be divided by a narrowed, valvular **intercolon** into an anterior element and a **postcolon** (anterior hindgut of authors) (Bücking 2000). In the Anactinotrichida and in some families of Astigmatina, the posterior midgut (or possibly the postcolon or the intercolon, when present) may give rise to one or two pairs of **Malpighian tubules** (Jakeman 1961; Prasse 1967b; Young 1968a; Woodring and Galbraith 1976; Bücking 2002). These tubules collect and store insoluble nitrogenous waste, typically in the form of guanine or uric acid. Resorption of water from this waste may reduce it to the form of a pastelike substance by the time it is finally excreted (Hughes 1959). The colon or postcolon empties into an **anal atrium** (posterior hindgut of authors) that finally debouches through an anal opening.

In general, trombidiform mites have a considerably larger ventriculus than do parasitiform mites (Woodring and Galbraith 1976) and usually have a single pair of large caeca. The smaller ventriculus of most parasitiform mites may give rise to two or more large and often very long caecal pairs. Ticks are unusual in that they may have five or more primary caecal pairs in addition to several smaller diverticula (Balashov 1961; Sonenshine and Gregson 1970), a modification that allows for ingestion of large blood meals (Fig. 3.26A). Mesostigmatic mites may have two or three pairs of caeca, while oribatid mites typically have a single pair of short, thick caeca that contrast with the relatively narrow ventriculus (Hughes 1959). The single caecal pair found in the Opilioacarida (p. 104) is weakly developed and diminutive when compared to the ventriculus (Evans 1992).

In many cases, feeding habits in a given taxon appear to be correlated with the nature and size of the caeca in its representatives. For example, mites that ingest large volumes of liquid food often have gastric caeca that expand during feeding to extend the entire length of the idiosoma (e.g., the phytophagous spider mites and the hematophagous Ixodida) and may be long enough in some smaller mesostigmatic parasites to intrude into the leg cavities (Hughes 1959). Enlarged caeca not only increase the overall volume of the midgut for fluid reception and storage but provide a greatly increased gut surface for digestion and absorption of these fluids.

Digestion in the Acari is largely intracellular, at least in the anterior portion of the ventriculus. Here, the gut wall may be lined with vacuolated epithelial digestive cells that absorb soluble nutrients and secretory cells that release their products into the midgut. These cells, or portions of them, may be sloughed into the lumen and incorporated into the food bolus. In ticks, digestion appears to follow a two-phase process in which type 1 digestive cells take up fluid and particulate components of the blood, followed in later stages of engorgement by absorption of host hemoglobin by type 2 digestive cells. In acarines with large gut surface areas (such as found in ticks), most or all of the major internal organ systems may be in direct contact with at least one midgut or caecal lobe, so that nutrients accumulated in digestive cells may be moved directly from sites of absorption to sites of utilization. Two types of secretory cells have been found in ticks (Hughes 1954b; Agbede and Kemp 1985), one of which may produce a hemolysin in the early stages of attachment and engorgement, and the other a mucopolysaccharide. Digestion in other mite groups is largely mediated by the nature of the food ingested. Liquefied nutrients ingested by mesostigmatic mites appear to be subjected to postoral digestive processes similar to those of ticks, although digestive and secretory cells in the Mesostigmata are not always easily differentiated (Hughes 1952; Bowman 1984). Strong parallels may be seen between the digestive cell physiology of ticks and that of the hematophagous mesostigmatic mite *Spinturnix myoti* Kolenati (SPINTURNICIDAE, p. 156). Digestive cells in both taxa accumulate hematin and then are released from the ventricular wall to pass into the gut lumen (Evans 1968). Trombidiform mites also ingest food in a liquefied form, but the nature of digestion appears to vary considerably between the groups that have been studied (Evans 1992). Recognizable digestive and secretory cells are present in the anterior midgut of BDELLIDAE, but members of the TETRANYCHIDAE apparently have only one type of midgut cell. These cells absorb nutrients in the midgut but assume an excretory function at the juncture of the ventriculus and colon (Alberti 1973; Mothes and Seitz 1981b).

Members of the suborder Oribatida ingest largely solid food that cannot be assimilated in the same fashion as liquids, necessitating another strategy for digestion. Thus, while typical intracellular digestion and secretion appear to occur in the ventriculus and caeca of astigmatine mites (Hughes 1950; Prasse 1967a), Brody, McGrath, and Wharton (1972) found that *Dermatophagoides farinae* Hughes (PYROGLYPHIDAE), an astigmatine mite that feeds largely on particulate matter in house dust, engages in extracellular digestion as well. Solid food particles passing into the ventriculus form an amorphous, often jagged bolus that could easily tear the delicate epithelial layer of the gut wall as it passes posteriorly into the colon and hindgut. However, protection of the wall is provided by a **peritrophic membrane**, which envelops the food bolus as it moves into the postcolon. The peritrophic membrane remains in place around the bolus as digestion proceeds, with the resulting fecal pellet finally passing into the hindgut (Wharton and Brody 1972). Other astigmatine mites also produce peritrophic membranes and engage in extracellular digestion, and a similar membrane is found in nonastigmatine Oribatida (Evans 1992). Some mesostigmatic mites ingest solid food (ICHTHYOSTOMATOGASTERIDAE (Figs. 12.7A, B) and fungivorous ASCIDAE and AMEROSEIIDAE are examples), but they do so without benefit of a peritrophic system.

The fate of solid fecal by-products of the digestive process in adults of some Parasitengonina is complicated in that these forms have a discontinuous alimentary tract. Indigestible residues (primarily chromolipids) are moved into the gastric caeca and stored there until the pressure of accumulated fecal material on the adjacent opisthosomatic integument reaches a critical point. The integument then tears and, through a process called **schizeckenosy** (Mitchell and Nadchatram 1969), the caecal lobe or lobes containing feces are extruded and pinched off. The integumental split quickly heals, leaving only a faint scar to commemorate the event. Other Prostigmata also have lost the connection between the ventriculus and postventricular elements so that the alimentary canal can no longer operate in normal fashion (Reuter 1909). However, it does retain an excretory function. Where the gut is discontinuous, the proctodeal port may be referred to as a uropore rather than an anal opening.

Reproduction

Reproductive systems in male and female Acari vary considerably between higher taxa and may consist of paired, fused, or multiple elements (Fig. 3.27). In males, the basic components of the reproductive system are the paired or fused **testes** and **vasa deferentia**, one or more **accessory glands**, and an **ejaculatory duct** (Figs. 3.27A, B, E). A **seminal vesicle** may be found between the accessory glands and ejaculatory duct in the Parasitengonina, but it may arise directly from unpaired fused testes in members of the eupodidid superfamily Eriophyoidea and in the raphignathine family DEMODICIDAE. Conversely, the vesicle may arise directly from the unpaired testis (Nuzzaci 1976; Desch 1984). Accessory glands are absent in females of the Oribatida, dermanyssine Mesostigmata, and in most trombidiform and astigmatine mites, but they have been identified in Ixodida, the Parasitengonina, the mesostigmatic suborder Gamasina (Parasitiae) and the Uropodina (Hughes 1959; Evans 1992; Coons and Alberti 1999; Alberti and Coons 1999).

As noted in the earlier discussion on sperm transfer, an extrusible aedeagus is present in some groups of Acariformes. The basic female system (Figs. 3.27C, D, F, G) comprises a paired or fused *ovary* and **oviducts**, a median **uterus** (considered an unpaired portion of the oviducts), a **progenital chamber**, a **vagina**, and a terminal **genital aperture**. In the Astigmatina, sperm from the male may instead be transferred to the female via a terminal **bursa copulatrix**, which leads internally into a **seminal receptacle** and thence to the ovary (Hughes 1959; Witalinski, Szlendak, and Boczek 1990; Witalinski 1991). Here, the sole function of the genital aperture is that of an oviporus. A more complex form of extragenital insemination involving transfer of sperm to a remote induction pore occurs in many Mesostigmata (see earlier section Sperm Transfer, and below).

Male System

The testes are primitively paired in the Holothyrida and in some Mesostigmata (Fig. 3.27B), in the Astigmatina, and in some Trombidiformes. They are unpaired (or partially fused) in other acarine groups, including those prostigmatic taxa whose exceptionally small size limits development of complex internal systems (Nuzzaci 1976; Desch 1984; Woolley 1988). The elongate testes of ixodid ticks are unusual in that only their proximal portions are fused, while the testes of the prostigmatic family BDELLIDAE and the Parasitengonina often are multiple, multilobed, or acinous (Fig. 3.27E) (Hughes 1959). Accessory glands assume a variety of roles in mite taxa where they occur, including that of providing the seminal fluid matrix for transport of sperm material from the testes to the ejaculatory duct (Woodring and Galbraith 1976). Accessory gland secretions may also provide the biochemical means for spermatid capacitation in ticks and the prevention of spermatophore stalk desiccation during spermatophore deposition by higher Prostigmata (Oliver 1982; Witte 1975a; Evans 1992).

The male ejaculatory duct may exit into an aedeagus that passes sperm material directly to the female genital opening (Alberti and Storch 1976; Alberti and Crooker 1985), or it may be injected into a female extragenital pore (Athias-Henriot 1968). Noncopulatory deposition of spermatophores by male acariform mites may be mediated by an extrusible **spermatopositor** (Figs. 15.7E, F), a homologue of the female ovipositor described below (Alberti and Coons 1999). In cases of direct insemination, the sperm may be ejected by action of a muscular sperm pump located between the vasa deferentia and ejaculatory duct and passed through the aedeagus into the female seminal receptacle (Alberti and

Storch 1976; Alberti and Crooker 1985; Evans 1992). Where a discrete ejaculatory duct is absent, as in uropodine Mesostigmata, the ejaculatory function is taken over by a muscular genital atrium (Woodring and Galbraith 1976). Acarine sperm transfer in the Trombidiformes often involves the deposition of a packet of sperm or a spermatophore at a site that is well removed from the female (see Sperm Transfer, above).

Alberti (1980a, b, 1991, 2000) has found that acarine sperm ultrastructure appears to be highly derivative when compared to that of other arachnids, and that sperm types within the Acari tend to remain almost invariably taxon-specific at subcohortal levels and higher. This morphological predictability offers a potentially powerful tool for determining systematic interrelationships (Alberti 1984). Unlike less derived chelicerate groups, all known acarine spermatozoa are aflagellate, although the spermatozoa of the opilioacarid genus *Neocarus* do possess a median filamentous perforatorium. Acarine spermatozoa exhibit considerable diversity in form (see earlier section on sperm transfer), but it is clear that there is an overriding difference between the basic sperm morphology of the Anactinotrichida (large, complex, vacuolated, or ribbonlike) and that of the Actinotrichida (small, simple, nonvacuolated, diverse in form) (Fig. 3.21). Ribbonlike sperm are considered to be a derived state within the Anactinotrichida (Alberti 1991), and because ribbonlike sperm are characteristic of podospermous Mesostigmata, extragenital sperm transfer may likewise be considered derived.

Female System

The ovary is unpaired in the Parasitiformes, the Trombidiformes, and most acariform Oribatida, although it may be paired in the Astigmatina (Fig. 3.27F). Oocytes developing on the surface of the ovary of ticks and of the Parasitengonina often give the ovary (or portions of it) the appearance of a cluster of grapes (Brinton and Oliver 1971; Balashov 1972; Witte 1975b; Alberti and Coons 1999; Coons and Alberti 1999). As they develop and enlarge, oocytes finally create sufficient pressure on the basement membrane of the ovary to force themselves into the lumen. The adjoining oviducts are typically paired, but a single unpaired oviduct is found in some groups of Mesostigmata and Trombidiformes (Evans 1992). Included here are the diminutive species of the family ERIOPHYIDAE (Fig. 3.27G) and the tarsonemine TARSONEMIDAE, in which the unpaired condition probably has evolved as a consequence of small size. Secretions produced in the oviducts are thought to produce the eggshell that protects the developing embryo as it passes through the uterus and into the ectodermal vagina (Alberti 1974; Woodring and Galbraith 1976). The vagina often gives rise to an **ovipositor**, which may be a simple pouch- or tubelike eversion of a portion of the vaginal wall (Prasse 1970; Balashov 1972) or a complex extrusible muscular structure equipped with terminal lobes and sensory setae (Michael 1884a; Woodring and Cook 1962) that apparently is capable of placing individual eggs in precise niches (Figs. 3.27H–J). Ovipositors are rarely present in the Mesostigmata, although a reduced ovipositor may be found in other parasitiform taxa.

Oviposition in the Ixodida is preceded by application of a waterproof waxy coating to individual eggs by **Gene's organ**, an eversible dorsal gland located just posterior to the female gnathosoma (Booth, Beadle, and Hart 1984; Coons and Alberti 1999). The organ inflates over the adpressed gnathosoma to cover each egg as it passes from the female genital tract and then deflates to allow the extending gnathosoma to add the newly waxed egg to an ever-growing egg mass on the female prosoma for later deposition (Evans, Sheals, and MacFarlane 1961). Secretions from the adjacent dorsal gnathosomatic areae porosae in ixodid ticks may also play a role in the egg waxing process (Atkinson and Binnington 1973).

A more detailed discussion of oviposition in the Acari may be found in chapter 5.

In those dermanyssine mesostigmatic mites where extragenital insemination or podospermy occurs, each of the paired external sperm induction pores or solenostomes (Athias-Henriot 1968) opens into a small chamber referred to as the **sacculus vestibulus** (Fig. 3.28C), which leads in turn to a ringed spermathecal tube or **tubulus annulatus** (Krantz and Mellott 1968; Lee 1970). The organ complex that receives sperm from the tubulus assumes two major forms in podospermous Mesostigmata (Alberti and Di Palma 2002). Sperm deposited in a sacculus vestibulus of the laelapid or type A complex passes through the tubulus and connecting hornlike **ramus sacculi** into a median pouchlike or bilobed **sacculus foemineus** and then through a narrow **sperm duct** into an appended spermatheca or **receptaculum seminis** (Michael 1892; Young 1968a; Akimov and Yastrebstov 1984). Sperm material then moves from the receptaculum via a minute lumen (Pound and Oliver 1976) to the unpaired ovary. In the phytoseiid or type B complex (Fig. 3.28D), the solenostome opens into a duct that could be considered homologous with the tubulus annulatus in type A systems, but which is referred to as the **major duct** (Dosse 1958; Schicha 1987). The duct passes into an **atrium**, which in turn opens into a **calyx** and thence into an apparently blind vesicle that may be homologous with the type A sacculus foemineus. An evanescent **minor duct** arises from the embolus of the major duct, terminating in the region where a bipartite nutrimentary lyrate organ meets the ovary (Alberti and Di Palma 2002). Based on its morphology, the minor duct seems an unlikely route for passage of spermatozoa to the ovary (Di Palma and Alberti 2001). Consequently, the route of passage of sperm material from the type B organ system to the ovary remains unresolved. Alberti and Di Palma (2002) note the possibility that sperm cells may penetrate the epithelium of the vesicle and reach the ovary via the hemocoel, rather than via the minor duct.

Nonintegumental Gland Systems

Coxal Glands

Coxal glands (= nephridia, tubular glands, and supracoxal glands of authors) are common to most arachnids and to all acarine orders. They are represented by a pair of excretory structures presumably derived from primitive, segmentally arranged nephridial structures and consist of a coelomic sacculus and a coiled duct, or labyrinth (Beklemishev 1969; Kaestner 1980; Alberti, Kaiser, and Klauer 1996). Coxal glands may develop in different body regions in the various arachnid subgroups, but they typically debouch in the coxal regions of the legs. The supracoxal complex of structures associated with the bases of legs I of various astigmatine Oribatida (Figs. 3.28B, 16.9A, 16.16A) overlie a pair of glands long considered to be coxal glands, but there is no histological evidence that they have nephridial origins (Alberti and Coons 1999). The glands generally open via a small slit at the base of the supracoxal seta (*elc1*). While most acarine coxal glands arise in the prosoma in proximity to the legs, those of parasitine and dermanyssine Mesostigmata may be located in the opisthosoma, each connecting to a ventral pore on coxa I by means of a long duct that traverses most of the idiosoma (Alberti, Kaiser, and Klauer 1996).

Coxal glands are thought to be involved primarily in osmoregulation but may assume other functions in some taxa, including ion/water balance and excretion (Balashov 1972; Woodring 1973; Alberti, Kaiser, and Klauer 1996). In soft ticks (Ixodida, ARGASIDAE), the coxal organs contain elaborate canal systems that function in ion concentration regulation and in water balance (Araman and Said 1972; Obenchain and Oliver 1974; Hagan, Oliver, and Pound 1984). Coxal glands are absent in hard ticks (IXODIDAE), which instead void water via salivary secretions (Sonenshine 1991). Alberti, Kaiser, and Klauer (1996) suggest that dermal or accessory glands associated with the coxal glands or the coxae may be involved in osmoregulation and ion/water balance in the Astigmatina.

Podocephalic Gland Complex

The coxal glands in many acariform groups are the most posterior elements in a complex of propodosomatic glands that empty into paired lateral secretory ducts or **podocephalic canals** (Fig. 3.28A), which serve to gather gland products and discharge them into the gnathosomatic region (Grandjean 1938b, 1944c, 1968, 1970, 1971; Van der Hammen 1968a; Coineau 1974b; Evans 1992; Alberti and Coons 1999). Podocephalic canals often take the form of open or partially open external gutters, but they may be entirely internal in the Parasitengonina (Prostigmata). They are short and distinct in the Astigmatina (Fig. 3.28B) but are long, sinuous, and often obscure in the Oribatida, extending from coxae I to the pedipalpal bases.

The paired podocephalic ducts receive the secretions of the coxal glands and one to three additional gland pairs, and their products appear to serve diverse purposes. In the Oribatida, some of these glands may secrete an endocrine substance associated with molting, while others appear to function as salivary glands (Woodring 1973). Alberti (1973) and Alberti and Storch (1973) studied the podocephalic system in the BDELLIDAE and concluded that certain of the appended glands, in concert with the independent infracapitular glands (Fig. 3.28A), are involved in the production of silk. Mills (1973) considered the entire podocephalic complex to be involved in silk production in the raphignathine spider mite *Tetranychus urticae* Koch, with the tubular coxal glands of the female secreting a pheromone additive to attract nearby males to the female webbing. However, Alberti and Storch (1974) felt that the production of silk in tetranychids is entirely the function of large paired palpal unicellular glands that open into the tip of a hollow eupathidial hair at the terminus of each palp. Other acariform mite families produce silk from salivary gland secretions that are delivered via the buccal opening (Gerson 1985b). Included here are the TYDEIDAE (Schuster 1972), BDELLIDAE (Wallace and Mahon 1972), CUNAXIDAE (Walter and Kaplan 1991), CHEYLETIDAE (Furman and Sousa 1969; Mori, Saito, and Tho 1999), RHAGIDIIDAE (Ehrnsberger 1974), ALICORHAGIIDAE (Walter 1988b), and ANYSTIDAE (Otto 1999f; Lindquist 2001). Ehrnsberger (1979) demonstrated that only the first and third podocephalic glands are used in silk production in *Rhagidia*.

Moss (1962) identified additional idiosomatic glands in the prostigmatic genus *Allothrombium* (Family TROMBIDIIDAE), including a salivary gland that is independent of the podocephalic system, and two venom glands associated with the chelicerae and the hypostome.

Salivary Glands

Two basic salivary gland patterns have been identified in the Acari. The simpler of these is found in the Parasitiformes and consists of paired acinous glands typically located in the prosoma. The gland components feed into paired common ducts that empty in the buccal or prebuccal area. In Mesostigmata, the ducts may open externally into elongate or abbreviated preoral styli (**siphunculi** of Van der Hammen 1980) that often follow the dorsal adaxial angles of the corniculi and terminate well anterior of the hypostome (Fig. 3.6A). Their products are thought to assist only in preoral digestion, although identification of five different types of salivary cells in certain PARASITIDAE (Bowman 1984) points to the possibility of a more complex functional purpose for these organs. The siphunculi often are enlarged in hematophagous species (Evans and Loots 1975) and may be absent in the cohorts Cercomegistina and Antennophorina (Camin and Gorirossi 1955).

In ticks (Ixodida), the multicellular salivary glands feed into paired ducts that open internally into a **salivarium** located ventrad of the subcheliceral plate in the buccal cavity. Tick salivary gland cells or acini display a variety of different physiological characteristics and appear to be involved in a variety of functions, including feeding, excretion, and osmotic balance (Chinery 1965; Roshdy 1966; Sonenshine 1991; Coons and Alberti 1999; Coons and Roshdy 1973; Ribeiro 1995; Bowman et al. 1997). Precursors to a cementlike substance that serves to secure the tick to its host are produced by certain of these cells in *Boophilus microplus* (Can.) (IXODIDAE) (Binnington 1978) and other hard tick species. The cement plug may serve as a reservoir not only for viral tick-borne pathogens but also for bactericidal compounds (Alekseev et al. 1995) whose activity may be changed in the presence of these pathogens. An anticoagulant apparently is secreted by one of three type II cell masses found in the salivary glands of the soft tick *Argas persicus* (Oken) (ARGASIDAE) (Roshdy 1966, 1972). Salivary anticoagulants contain elements that enhance blood flow by thwarting platelet aggregation and vasoconstriction in the host (Bowman et al. 1997). Meredith and Kaufman (1973) suggest that type II acini in *Dermacentor andersoni* Stiles (IXODIDAE) may secrete a fluid derived from the hemolymph that passes to the salivary glands during feeding.

In the Acariformes, discrete salivary glands have been described for some bdelloid and parasitengonine Prostigmata (Michael 1896; Moss 1962; Mitchell 1964a) and for certain Oribatida (Woodring and Cook 1962). The eight to nine cells extending posterodorsally on each side of the cheliceral ganglion in *Sancassania berlesei* (Michael) (ACARIDAE) are considered by Prasse (1968) to be salivary gland cells. It seems likely that the salivary function in at least some actinotrichid groups is mediated through one or more glands composing the podocephalic complex. Woodring (1973) has suggested that podocephalic glands 1 (the rostral glands) might function as salivary glands in earlier derivative Oribatida.

Nervous System

Lying around the esophagus is a densely coalesced, compact central nervous system or **synganglion** consisting of a protocerebral **supraesophageal ganglion** and a tritocerebral **subesophageal ganglion** (Figs. 3.6E, 3.26B) and comprising an outer cortex and an inner neuropile (Blauvelt 1945; Jakeman 1961; Pound and Oliver 1982; Alberti and Crooker 1985; Nuzzaci and Alberti 1996). The cortex in ixodid ticks includes a perineurium, glial cells, and neuronal cell bodies (Coons, Roshdy, and Axtell 1974) that transmit nerve impulses and produce secretions that mediate these transmissions. A thin extracellular sheath rather than a complex neural lamella surrounds the synganglion of *Macrocheles muscaedomesticae* (Scopoli) (MACROCHELIDAE) (Coons and Axtell 1971). Unlike other arthropods where discrete dorsal and ventral ganglionic elements are joined by circumesophageal commissures and where a ventral chain of ganglia may extend well into the podosoma and beyond (Beklemishev 1969), the acarine brain displays no obvious dorsal or ventral discontinuities and is confined entirely to the prosoma. While absence of antennae presupposes secondary loss of the deutocerebral segment in the acarine synganglion (Woolley 1988; Obenchain 1974; Evans 1992), DNA studies of the oribatid mite *Archegozetes longisetosus* Aoki (TRHYPOCHTHONIIDAE) suggest that the deutocerebral segment in chelicerate arthropods is actually retained (Telford and Thomas 1998a).

The supraesophageal portion of the synganglion in the ixodid tick *Dermacentor variabilis* (Say) carries the optic and cheliceral lobes, along with the posterior stomodeal bridge and the pedipalpal ganglia (Ivanov and Leonovich 1983). Pedipalpal ganglia in astigmatic mites may innervate not only the palpi but the pharyngeal musculature as well (Hughes 1959). Although it has not been noted in ticks, an anterior unpaired rostral nerve that appears to serve the capitulum has been observed in the spider mite *Tetranychus urticae* (Blauvelt 1945; Alberti and Crooker 1985). Also, unlike the arrangement in ticks, the pedipalpal ganglia in *T. urticae* arise from the ventral subesophageal portion of the synganglion as in some mesostigmatic mites; for example, *Macrocheles muscaedomesticae* and *Varroa jacobsoni* (Coons and Axtell 1971; Akimov et al. 1988). The pedal ganglia and an unpaired posterior median or opisthosomatic ganglion also have their origins in the subesophageal synganglion, and an unpaired abdominal nerve may arise from the median ganglion to extend posteromedially and ventrally beyond the podosoma. The number of additional peripheral nerves described as arising from the synganglion varies greatly among acarine suborders, as does the terminology used to describe them. Included here are scutal, tritosternal, stomodaeal (one to several pairs), splanchnic, genital, and rectal nerves (Hughes 1959; Woodring and Galbraith 1976; Woolley 1988).

In addition to the neuronal cells that carry nerve impulses from the synganglion to the peripheral nervous system, the central nervous system contains neurosecretory cells that produce a range of compounds thought to be involved in mediating a variety of physiological processes, including molting, oviposition and ovipositional diapause, salivation, and nymphal cuticular apolysis (Hughes 1964; Ioffe 1965; Binnington and Obenchain 1982; Hänel 1986). A degree of sexual dimorphism may be seen in the level of neurosecretory development in parasitiform mites, with ephemeral males often having fewer cells than normally found in females. A further correlation between neurosecretory activity and gender was observed by Akimov et al. (1988), who found no neurosecretory activity in the brains of male *Varroa jacobsoni* that had completed copulation and concluded that the process of sperm transfer is under the control of neurosecretory processes. In *Dermacentor variabilis*, neural secretions apparently follow axon pathways to neuroendocrine complexes, the synganglionic lamella, and to various effector sites via the peripheral nervous system (Binnington and Obenchain 1982). The peripheral route of neurosecretions in other acarines has not been clearly defined.

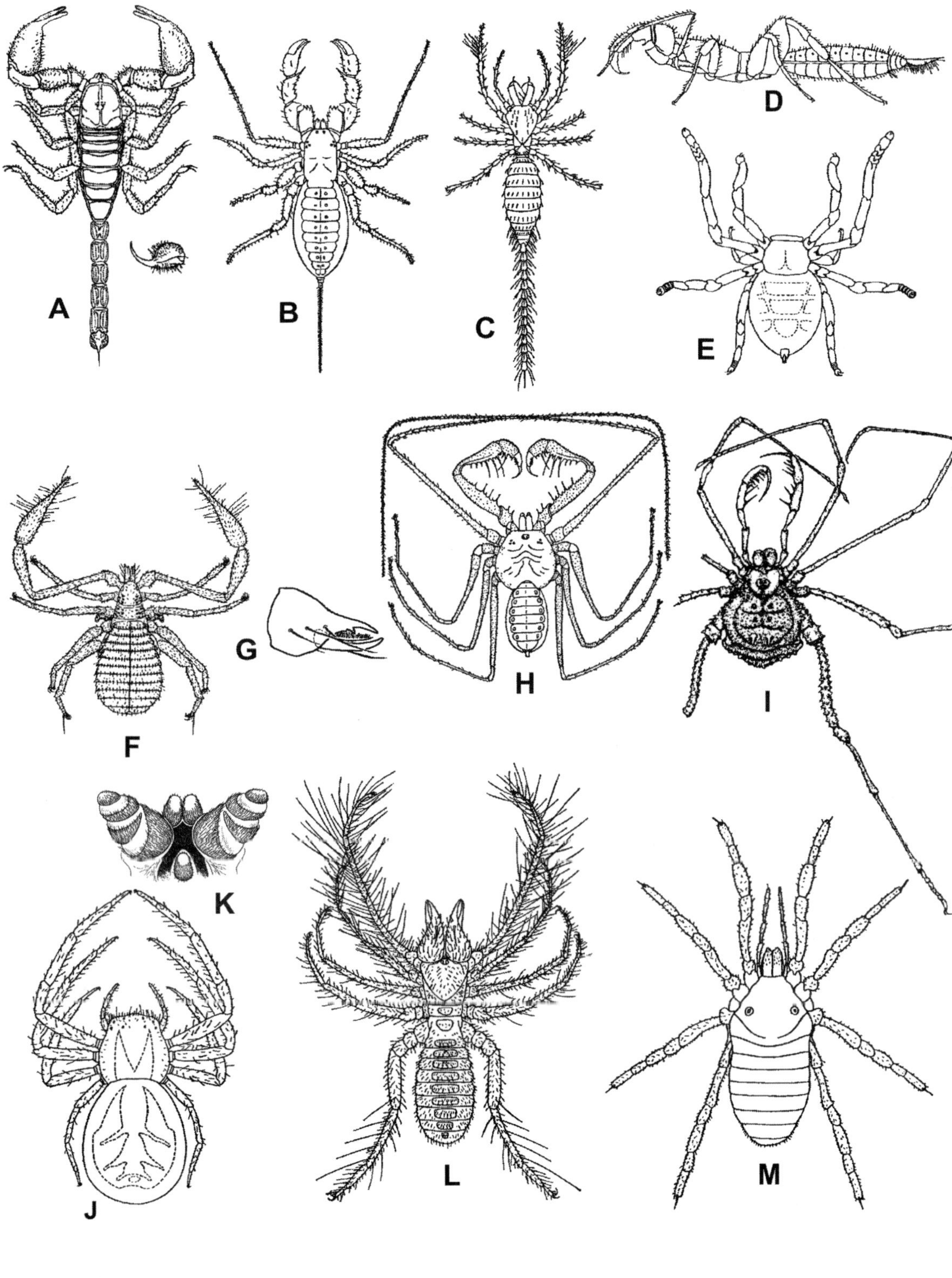

Fig. 3.1. Arachnid higher categories (orders) other than the Acari. **A**, Scorpiones with detail of terminal sting; **B**, Uropygi; **C**, Palpigradi; **D**, Schizomida; **E**, Ricinulei; **F**, Pseudoscorpionida; **G**, chelicera of pseudoscorpion; **H**, Amblypygi; **I**, Opiliones; **J**, Araneae; **K**, spinnerets of spider; **L**, Solifugae; **M**, Opiliones, Cyphopthalmi. (**A–F**, **H**, **J–M** from Savory 1964, with permission, Academic Press, London).

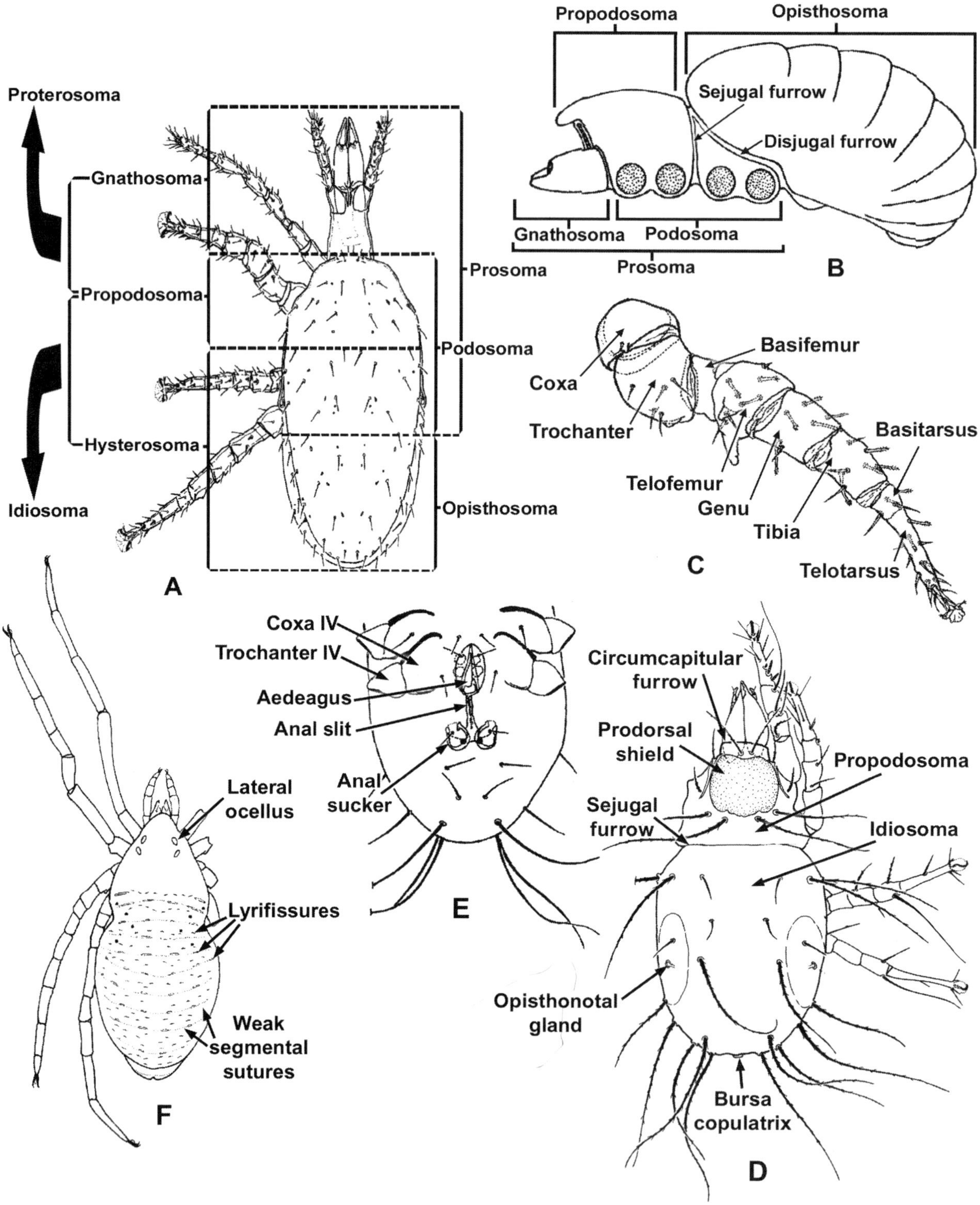

Fig. 3.2. External anatomy of Acari. **A**, *Macrocheles merdarius* Berlese (Mesostigmata, MACROCHELIDAE), dorsum showing major body divisions; **B**, lateral aspect of an acariform mite (diagrammatic) showing major body divisions (after Grandjean 1969b); **C**, *M. nemerdarius* Krantz and Whitaker (Mesostigmata, MACROCHELIDAE), leg IV of male (from Krantz and Whitaker 1988); **D**, *Tyrophagus* sp. (Oribatida, ACARIDAE), dorsum of adult female; **E**, *Sancassania* sp. (Oribatida, ACARIDAE), posteroventral region of an adult male; **F**, *Opilioacarus* sp. (Opilioacarida, OPILIOACARIDAE), dorsum of adult female.

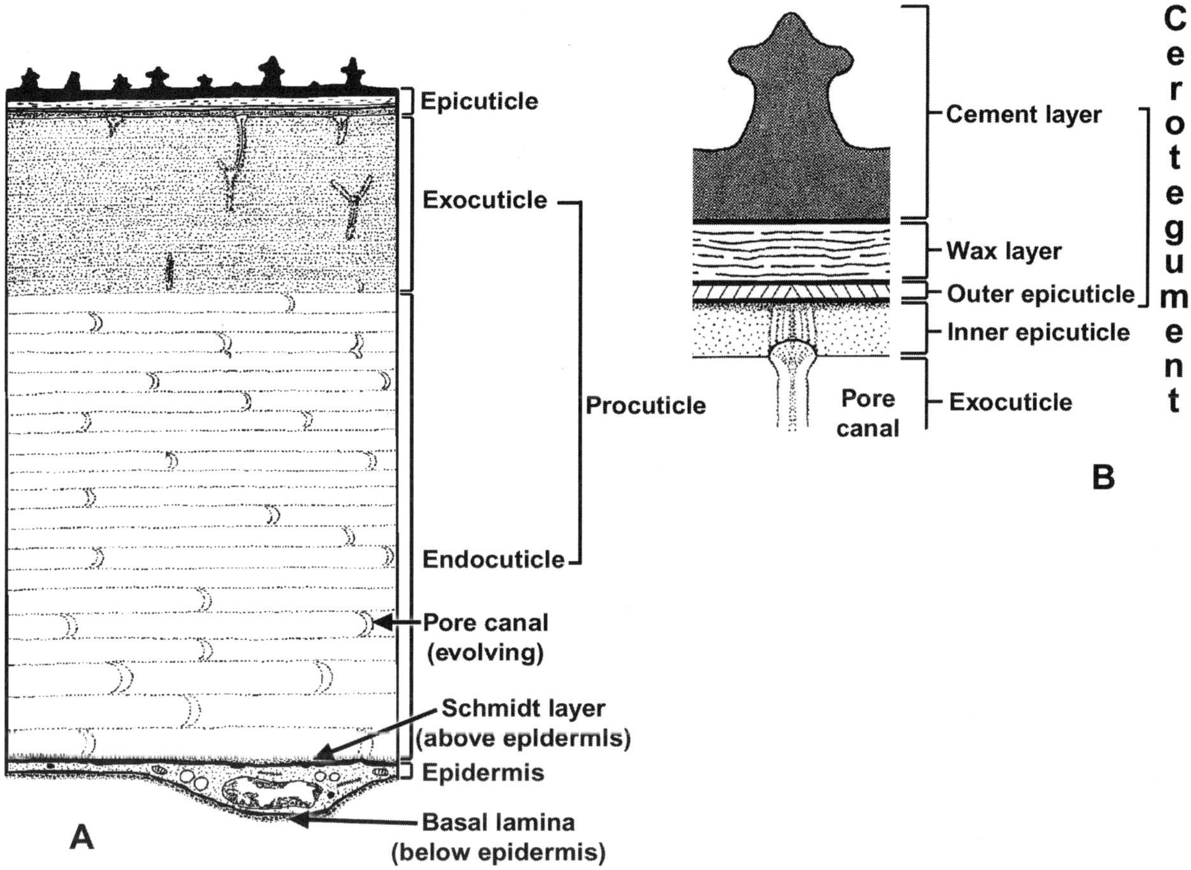

Fig. 3.3. The acarine cuticle. **A**, schematic cross section of adult oribatid integument (Acariformes, Oribatida); **B**, detail of epicuticular layers (from Norton et al. 1997, with permission, E. Schweizerbart'sche Verlagsbuchhandlung, Stuttgart).

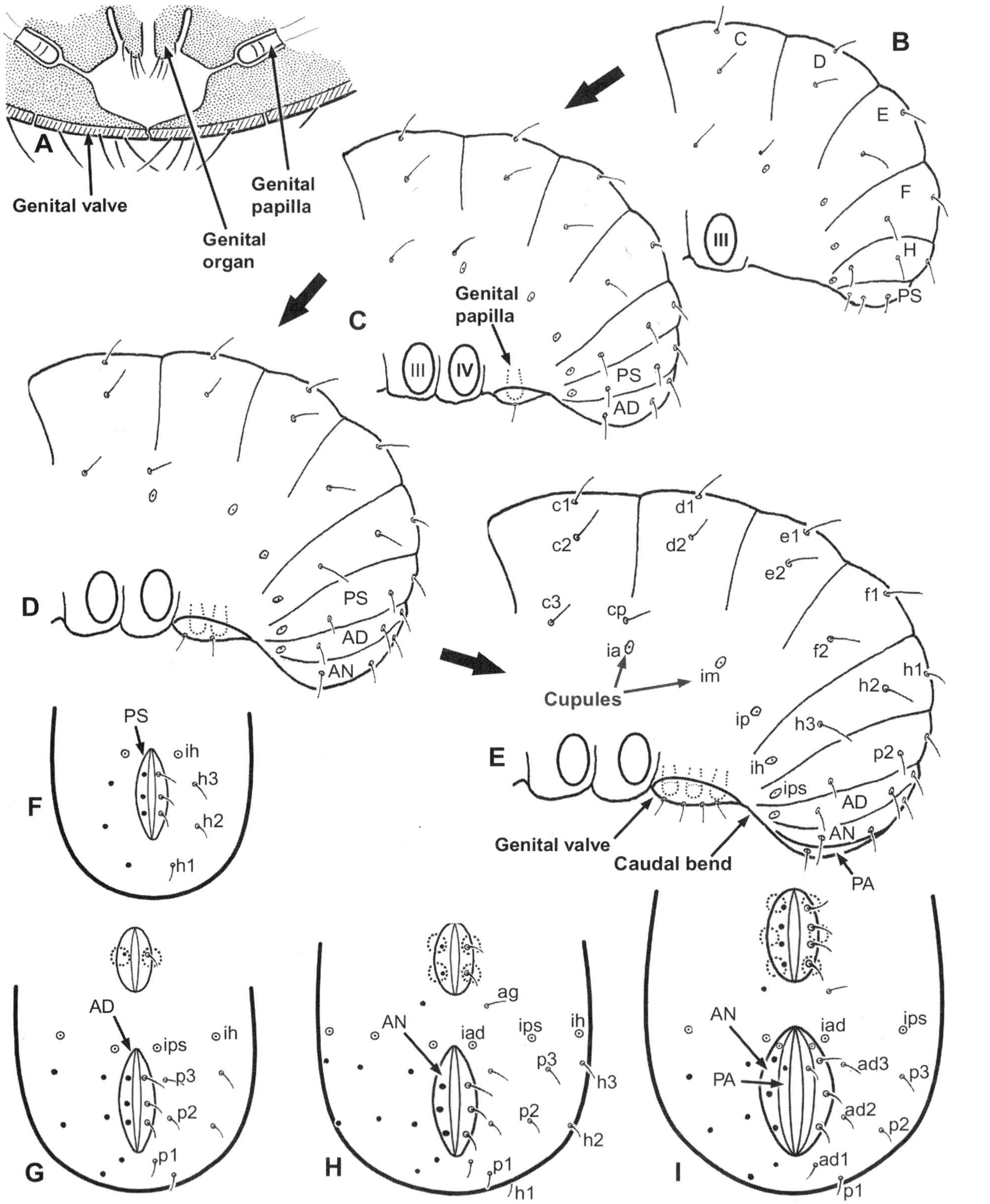

Fig. 3.4. Addition of segments (anamorphosis), setae, cupules, and genital papillae during ontogeny of acariform mites. **A**, schematic section of adult progenital chamber, showing ancestral arrangement of valves, genital papillae, and genital organ; **B**, schematic hysterosoma of larval acariform mite (left lateral view) showing ancestral segmentation, setae, and cupules; **C**, same, protonymph; **D**, same, deutonymph; **E**, same, tritonymph; **F**, schematic hysterosoma of larva in ventral view, without segmental boundaries; **G**, same, protonymph; **H**, same, deutonymph; **I**, same, tritonymph. (**A** after Grandjean 1969; **B–E** based on Grandjean 1939e; genital setal counts not meaningful). (Courtesy of R. A. Norton).

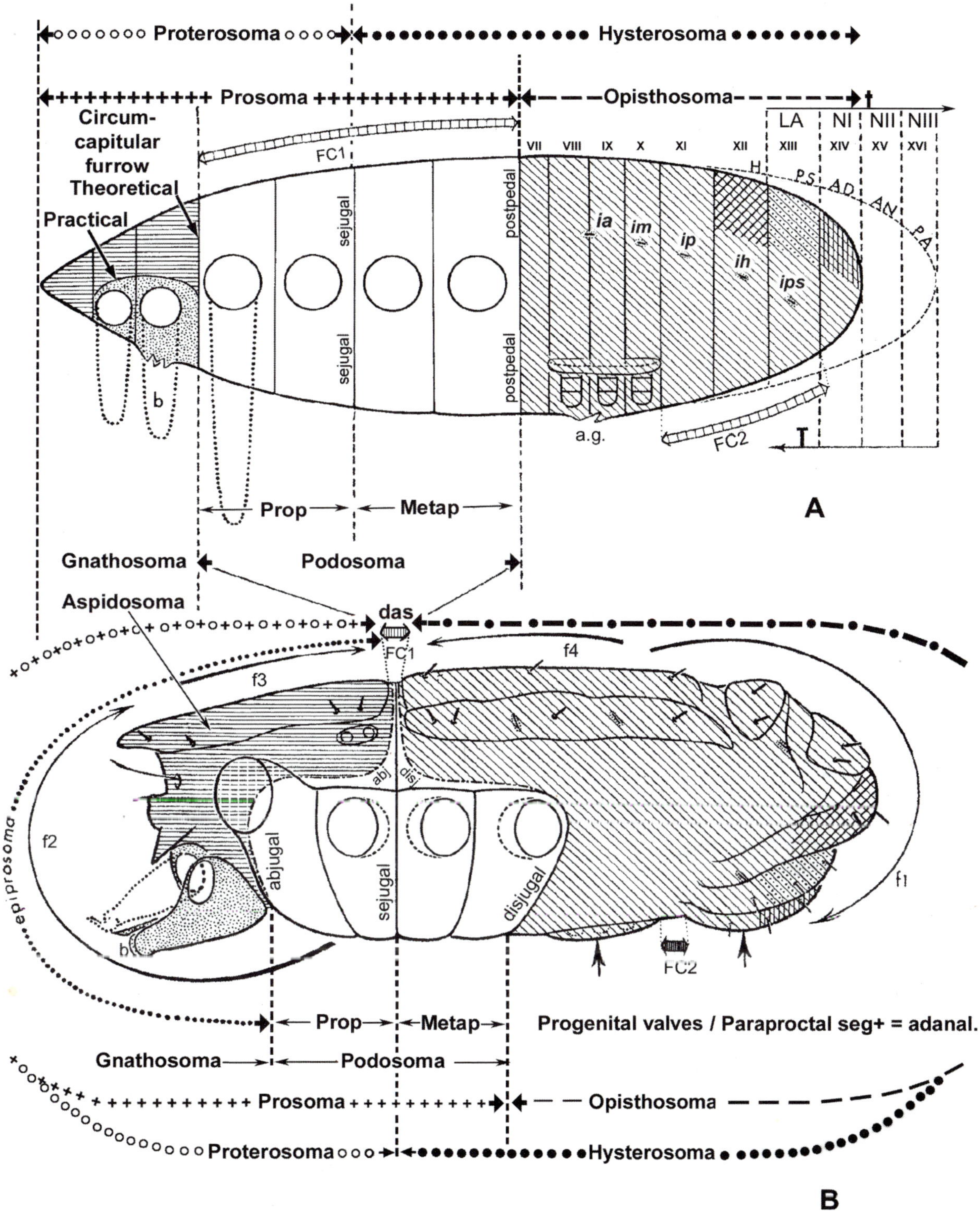

Fig. 3.5. Primitive segmentation and evolved modifications in the Acariformes (from Coineau 1974b). **A**, primitive acariform condition; **B**, segmental fusions and migrations in a caeculid mite (Prostigmata, CAECULIDAE).

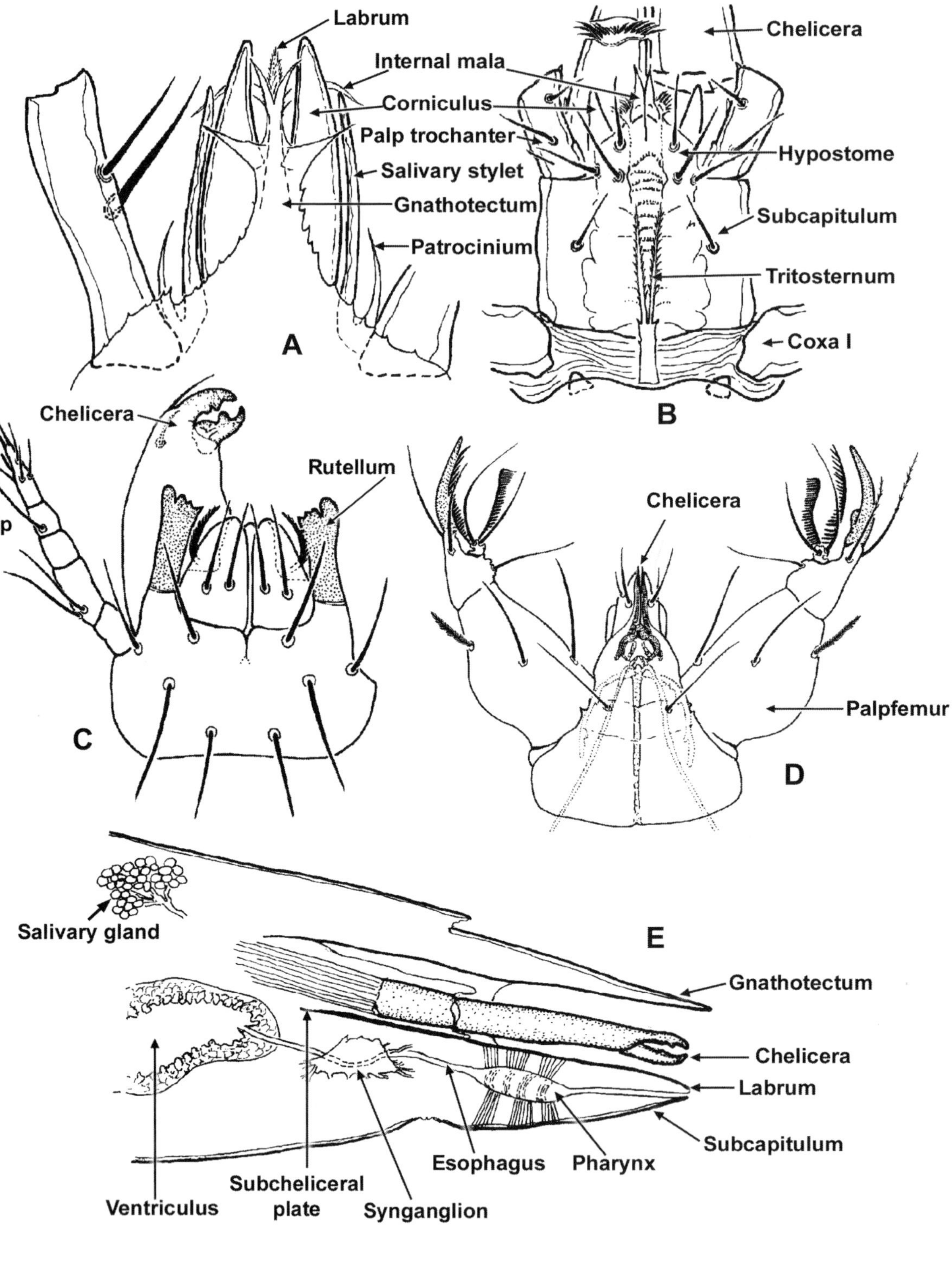

Fig. 3.6. Gnathosomatic morphology. **A**, *Macrocheles mycotrupetes* Krantz and Mellott (Mesostigmata, MACROCHELIDAE), dorsal aspect of female gnathosoma (after Krantz and Mellott 1968); **B**, *Parasitus coleoptratorum* (L.) (Mesostigmata, PARASITIDAE), ventral aspect of deutonymphal gnathosoma (after Wernz and Krantz 1976); **C**, *Aphelacarus* sp. (Oribatida, APHELACARIDAE), gnathosomatic venter; **D**, *Cheletomorpha lepidopterorum* (Shaw) (Prostigmata, CHEYLETIDAE), dorsum of gnathosoma and associated structures; **E**, *Caminella peraphora* Krantz and Ainscough (Mesostigmata, DITHINOZERCONIDAE), midsagittal section of gnathosoma and propodosoma (after Ainscough 1960).

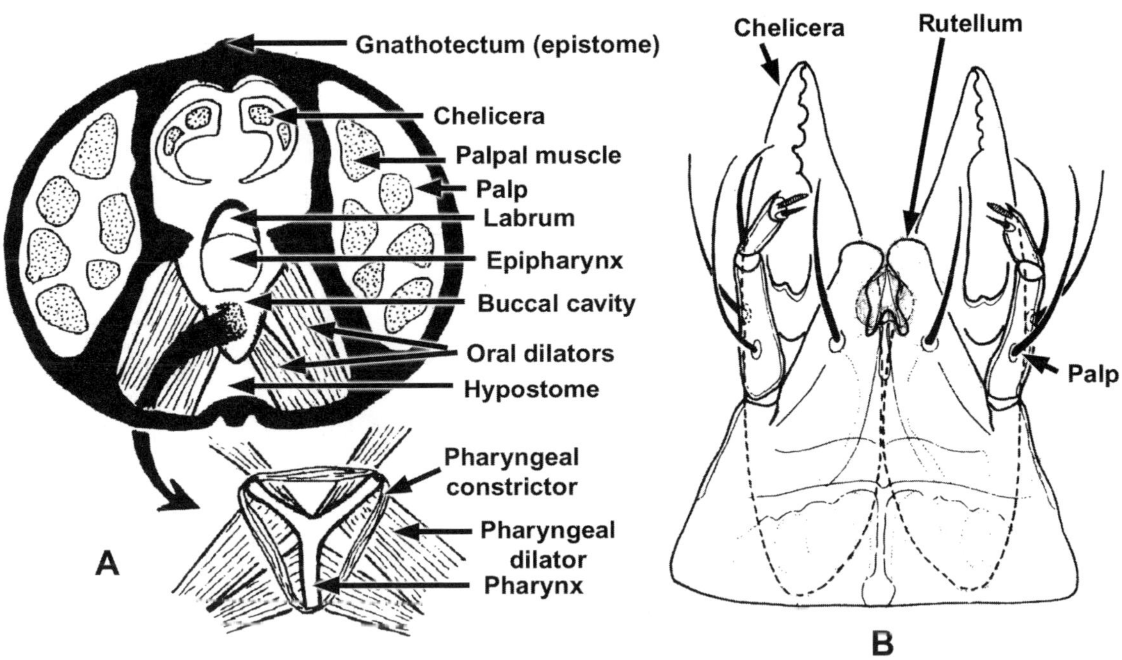

Fig. 3.7. Gnathosomatic morphology. **A**, *Ornithonyssus bacoti* (Hirst) (Mesostigmata, MACRONYSSIDAE), cross section of gnathosoma at level of hypostomatic setae 2–3, with detail of pharynx (after Hughes 1959); **B**, *Glycyphagus* sp. (Oribatida, GLYCYPHAGIDAE), venter of gnathosoma.

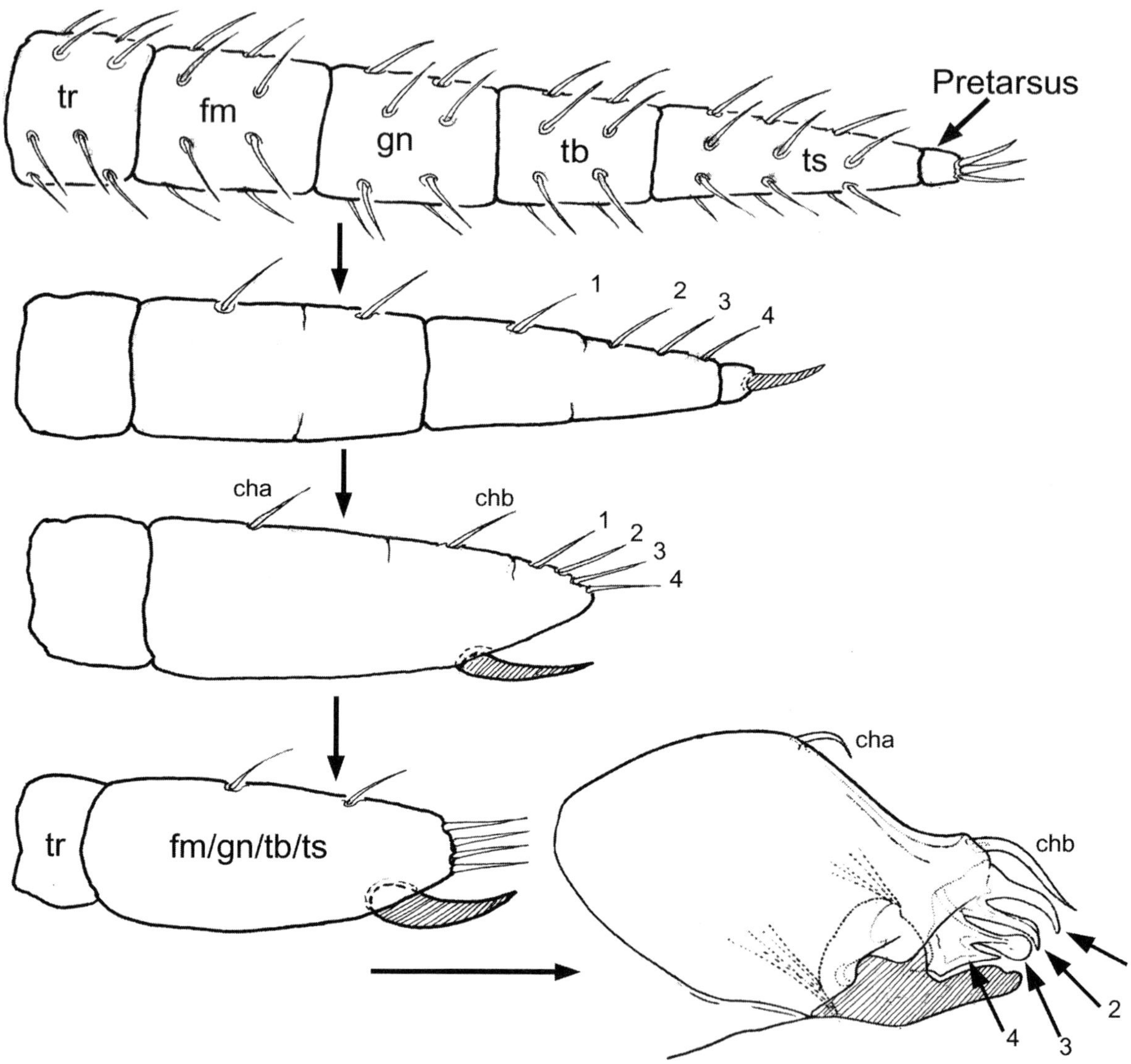

Fig. 3.8. *Pterochthonius angelus* (Berlese) (PTEROCHTHONIIDAE), an arthronotine mite, hypothetical derivation of the chelicera from a primitive leglike appendage (after Grandjean 1947d; Evans, Sheals, and MacFarlane 1961).

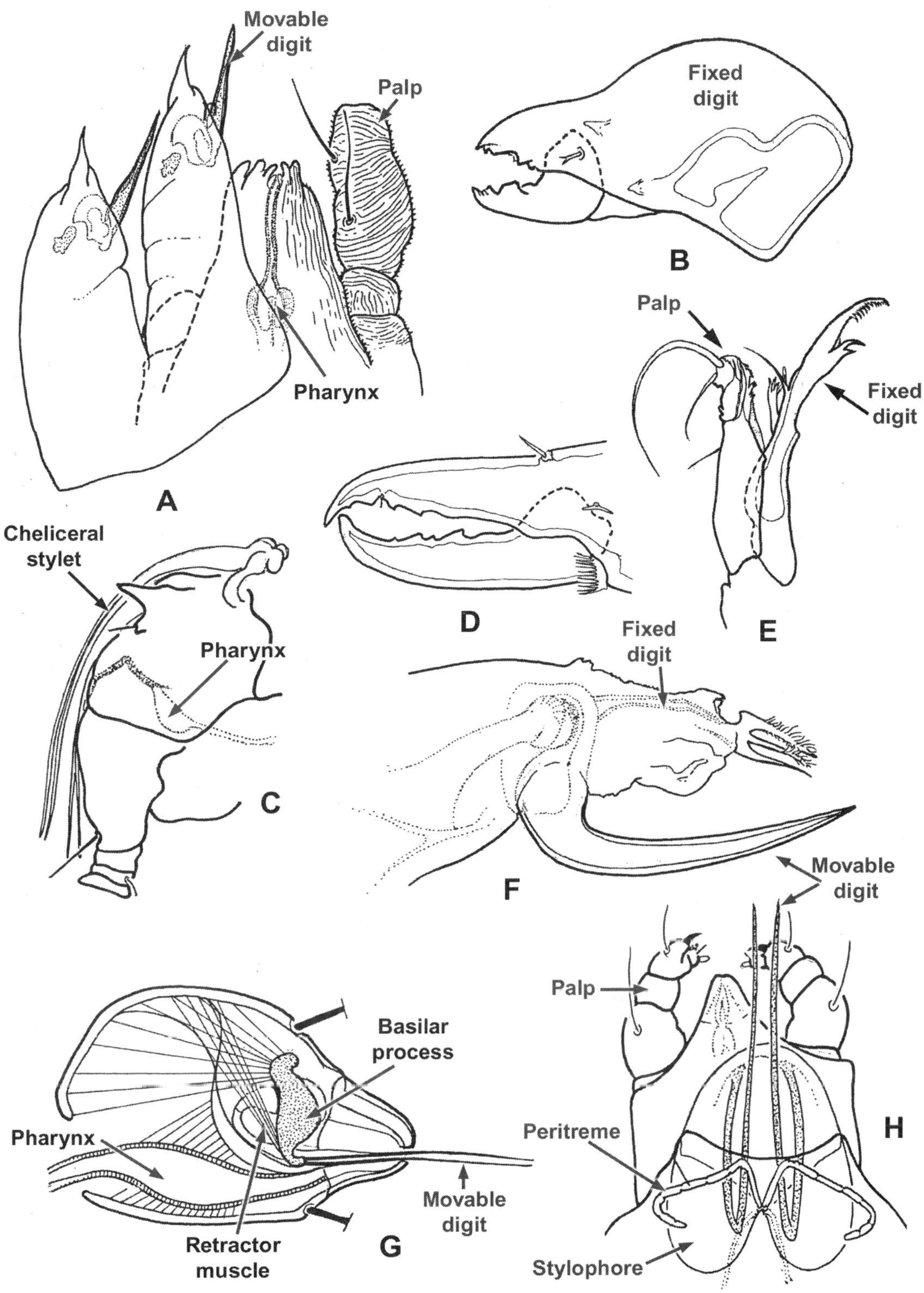

Fig. 3.9. A, *Paralorryia* sp. (Prostigmata, TYDEIDAE); **B**, *Tyrophagus putrescentiae* Schrank (Oribatida, ACARIDAE); **C**, *Diptacus* sp. (Prostigmata, DIPTILOMIOPIDAE); **D**, *Pergamasus brevicornis* Berl. (Mesostigmata, PARASITIDAE); **E**, *Histiostoma julorum* (Koch) (Oribatida, HISTIOSTOMATIDAE); **F**, *Penthaleus major* (Dugès) (Prostigmata, PENTHALEIDAE); **G**, midsagittal schematic of a tarsonemid-type gnathosoma (Prostigmata, TARSONEMIDAE); **H**, *Tetranychus pacificus* McGregor (Prostigmata, TETRANYCHIDAE). (**A**, **C**, **F**, **G**, and **H** from Krantz and Lindquist 1979, with permission, Annual Reviews, Palo Alto; **B** after Akimov 1974; **D** after Holzmann 1969; **E** after Hughes and Jackson 1958).

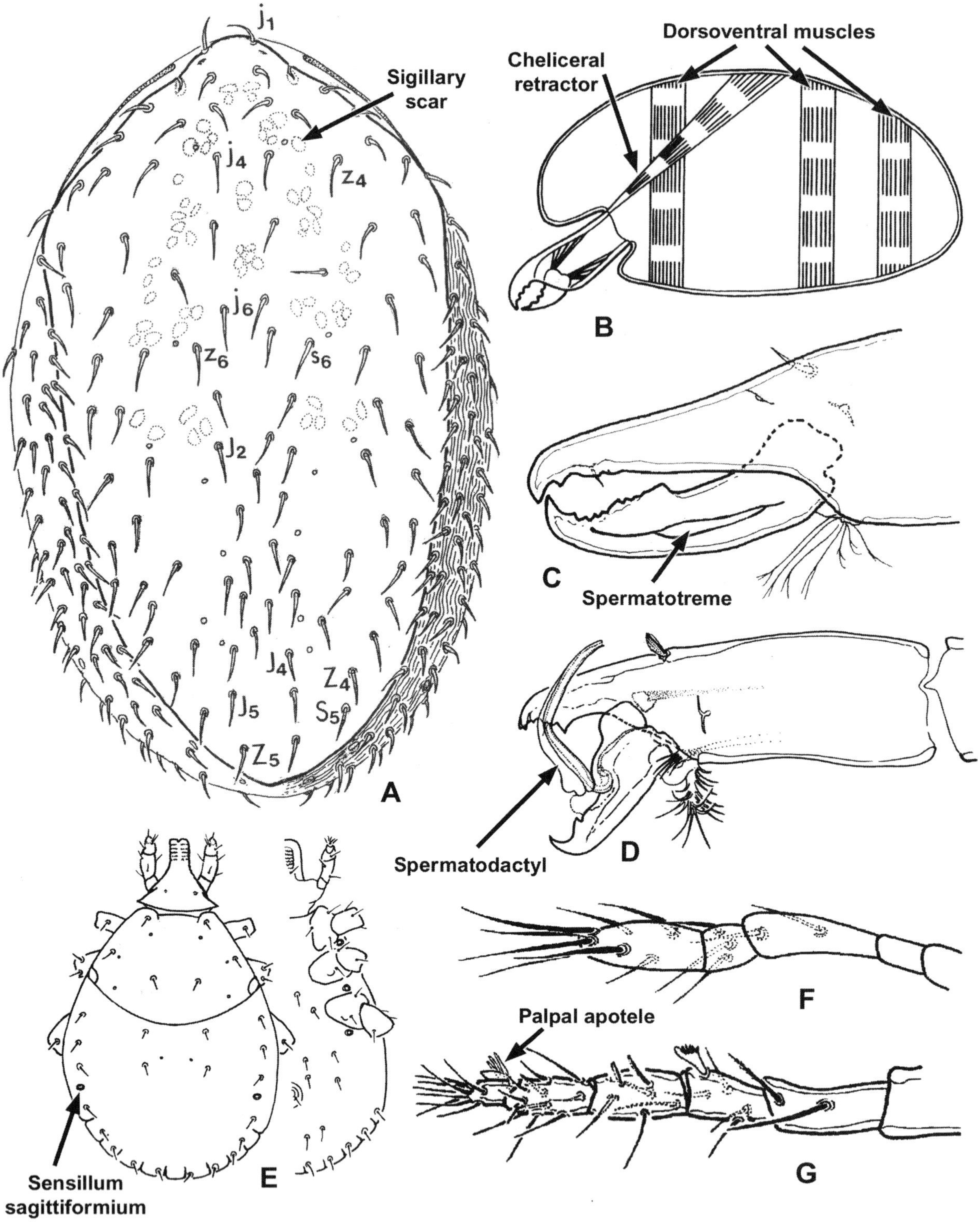

Fig. 3.10. A, *Mitonyssoides stercoralis* Yunker et al. (Mesostigmata, MACRONYSSIDAE), dorsum showing sigillary scar pattern on dorsal shield (after Radovsky and Krantz 1998); **B**, diagram of basic dorsoventral and cheliceral retractor musculature in a generalized mite (after a drawing by R. D. Mitchell); **C**, *Pergamasus crassipes* (L.) (Mesostigmata, PARASITIDAE), distal portion of chelicera of male; **D**, *Macrocheles peregrinus* Krantz (Mesostigmata, MACROCHELIDAE), male chelicera (after Krantz 1981); **E**, *Dermacentor variabilis* (Say) (Ixodida, IXODIDAE), dorsum and venter of larva showing opisthonotal position of sensilla sagittiformia; **F**, *Rhagidia* sp. (Prostigmata, RHAGIDIIDAE), palp; **G**, *Veigaia* sp. (Mesostigmata, VEIGAIIDAE), palp.

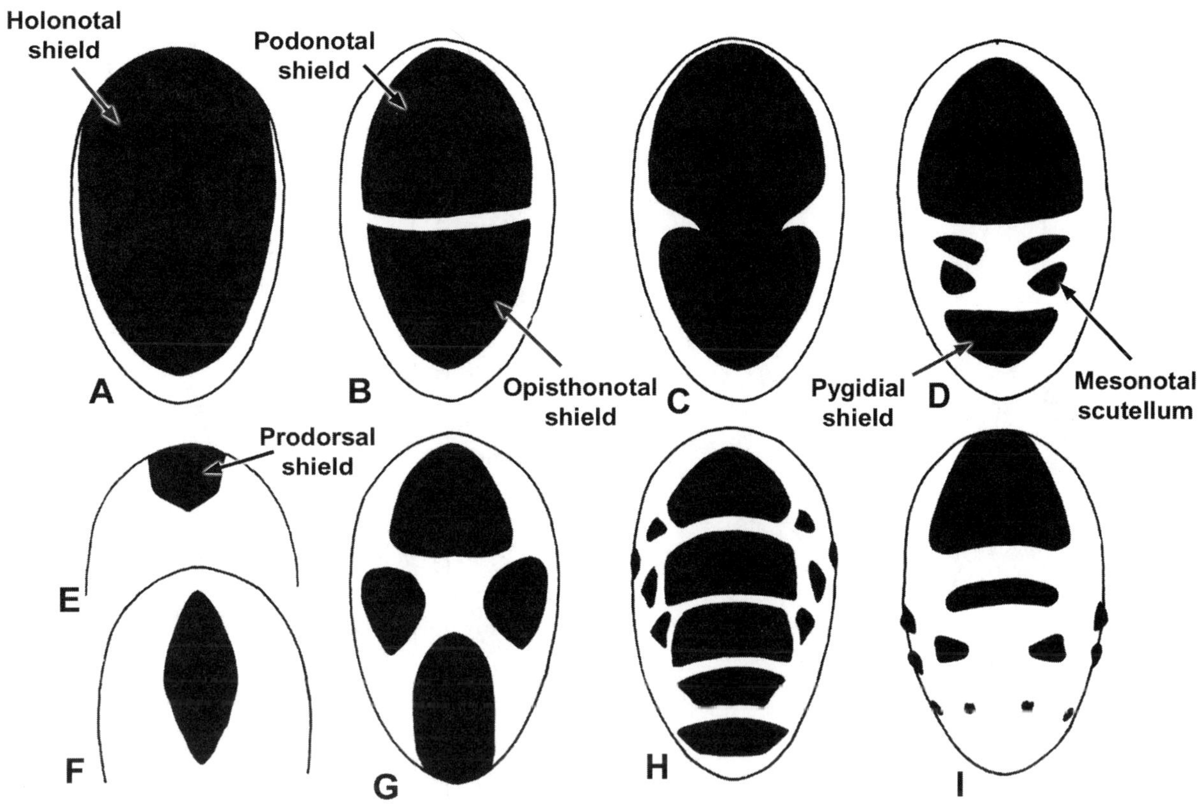

Fig. 3.11. Common dorsal shield configurations in the Acari. **A–D**, **F**, Mesostigmata; **B**, **G–H**, Prostigmata; **E**, Oribatida (Astigmatina); **I**, Oribatida (Palaeosomatides).

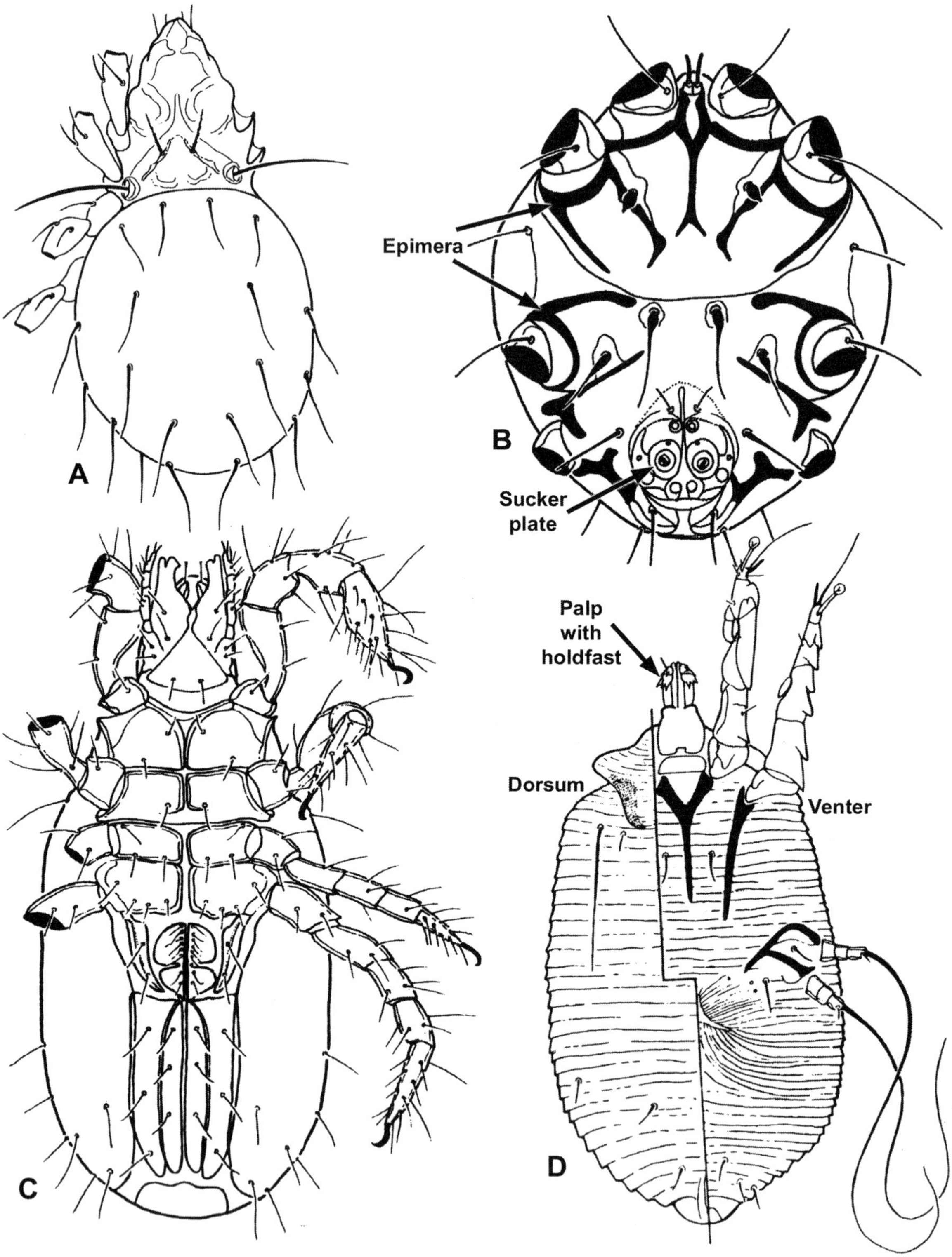

Fig. 3.12. Some oribatid body types. **A**, *Eremobelba flagellaris* Jacot (EREMOBELBIDAE), dorsum; **B**, *Chaetodactylus* sp. (CHAETODACTYLIDAE), venter of heteromorphic deutonymph; **C**, *Perlohmannia* sp. (PERLOHMANNIIDAE), venter; **D**, *Rhyncoptes anastosi* Fain (RHYNCOPTIDAE), dorsal/ventral composite.

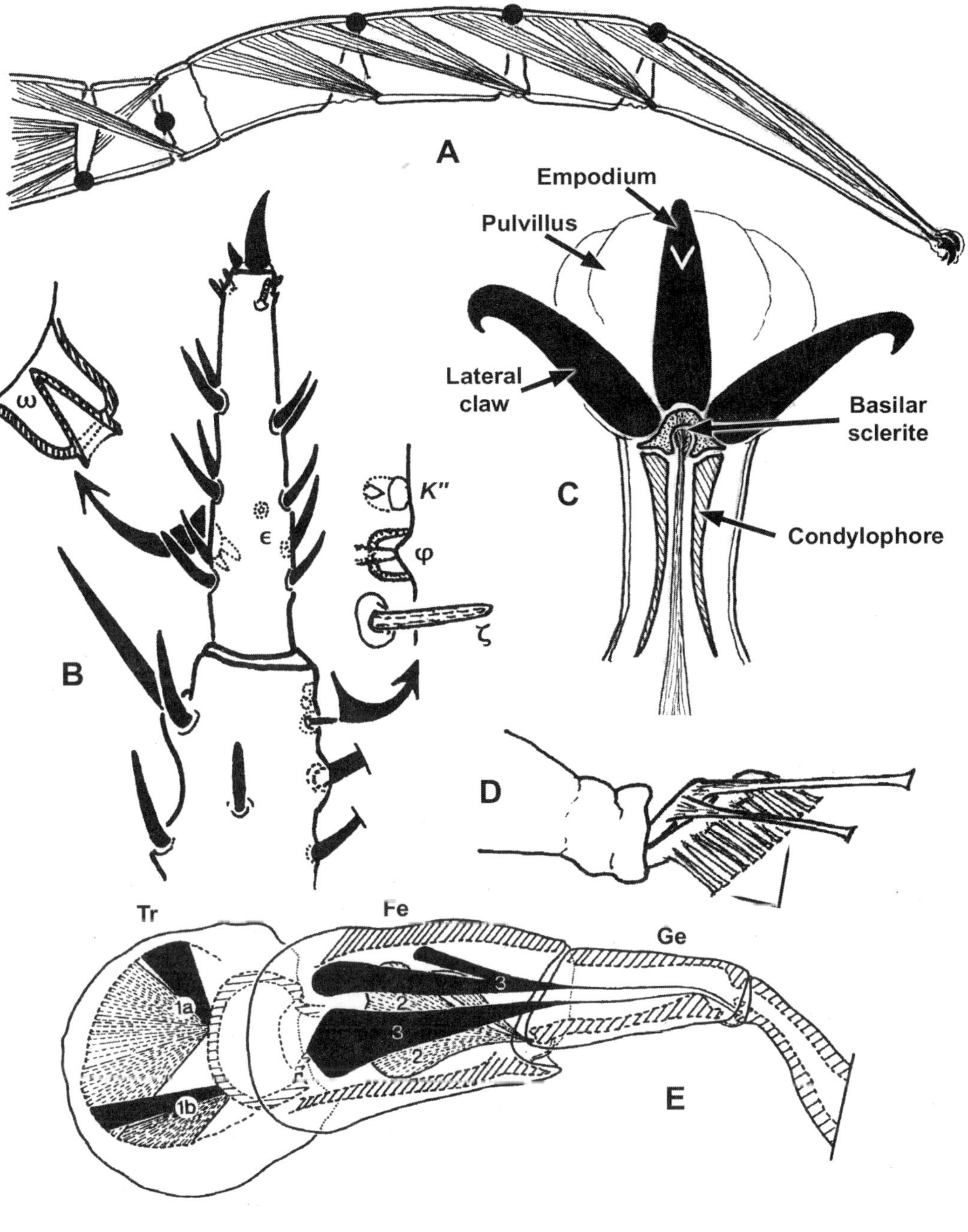

Fig. 3.13. **A**, leg musculature of a typical mesostigmatic mite, with major articulations indicated by solid black circles (after Evans 1992); **B**, *Caeculus liguricus*? Vitzthum (Prostigmata, CAECULIDAE), tibia and tarsus I with details of the recessed tarsal solenidion (ω left) and the tibial complex of recessed famulus (ε) and solenidion (φ) flanked by a eupathidium (ζ) and recessed tarsal seta *K''* (after Grandjean 1944a); **C**, diagram of a generalized acarine ambulacrum and associated tarsal elements; **D**, ambulacrum of a hystrichonychine spider mite (Prostigmata, TETRANYCHIDAE); **E**, *Zetorchestes falzonii* Coggi (Oribatida, ZETORCHESTIDAE), leg IV showing the greatly developed femoral levator (*1a*) and tibial flexor (*3*) muscles that play a major role in jumping (from Krisper 1990, with permission, G. Krisper and Urban and Fischer Verlag, Jena).

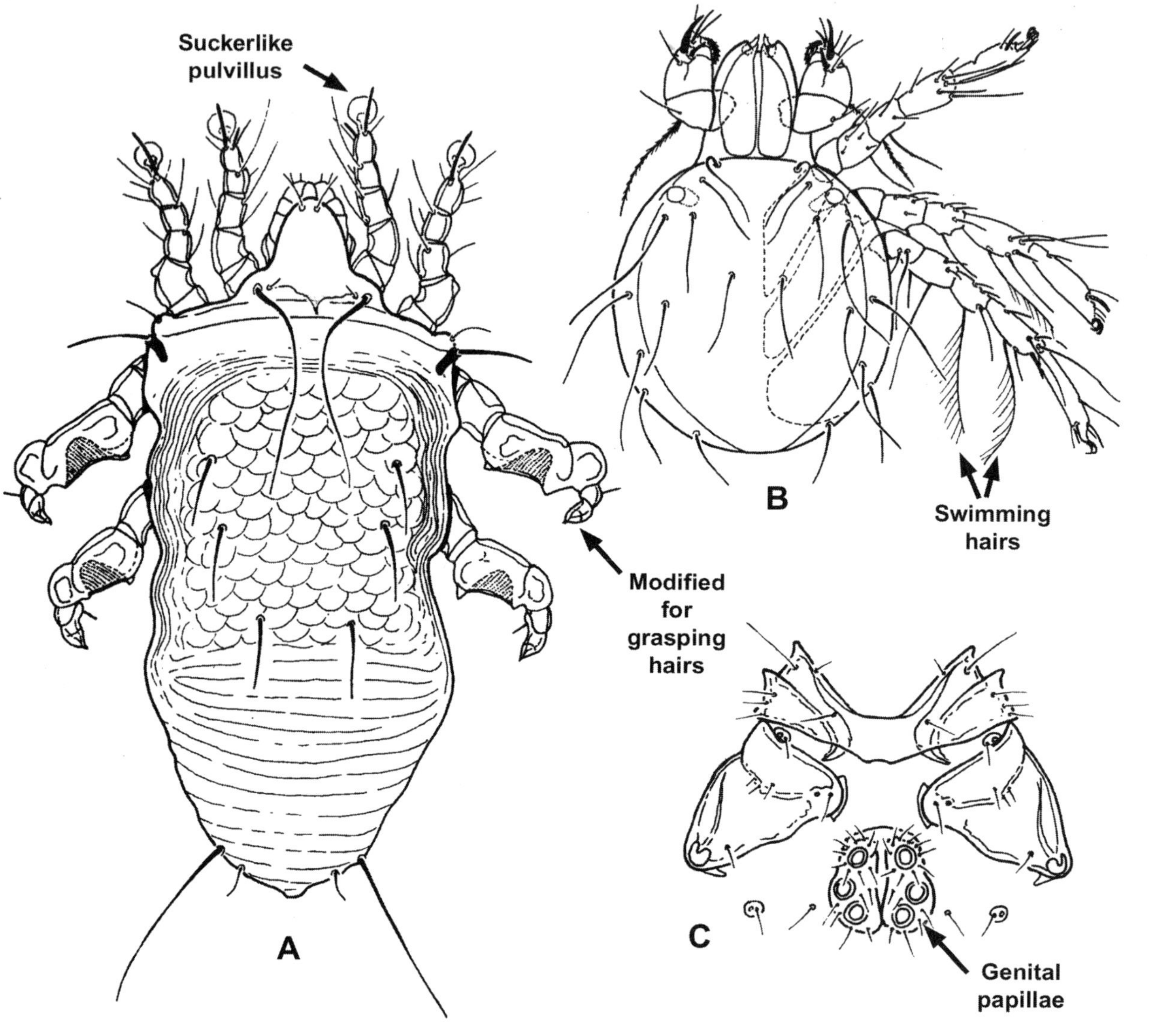

Fig. 3.14. A, *Myocoptes musculinus* (Koch) (Prostigmata, MYOCOPTIDAE), female dorsum showing suckerlike ambulacra I–II and grasping modifications of legs III–IV; **B**, *Arrenurus* sp. (Prostigmata, ARRENURIDAE), dorsum of larva showing swimming hairs on legs III–IV (after Cook 1974b); **C**, *Limnesia lambangensis* Piersig (Prostigmata, LIMNESIIDAE), coxal fields and genital opening with papillae (= genital acetabula) (after Cook 1974b).

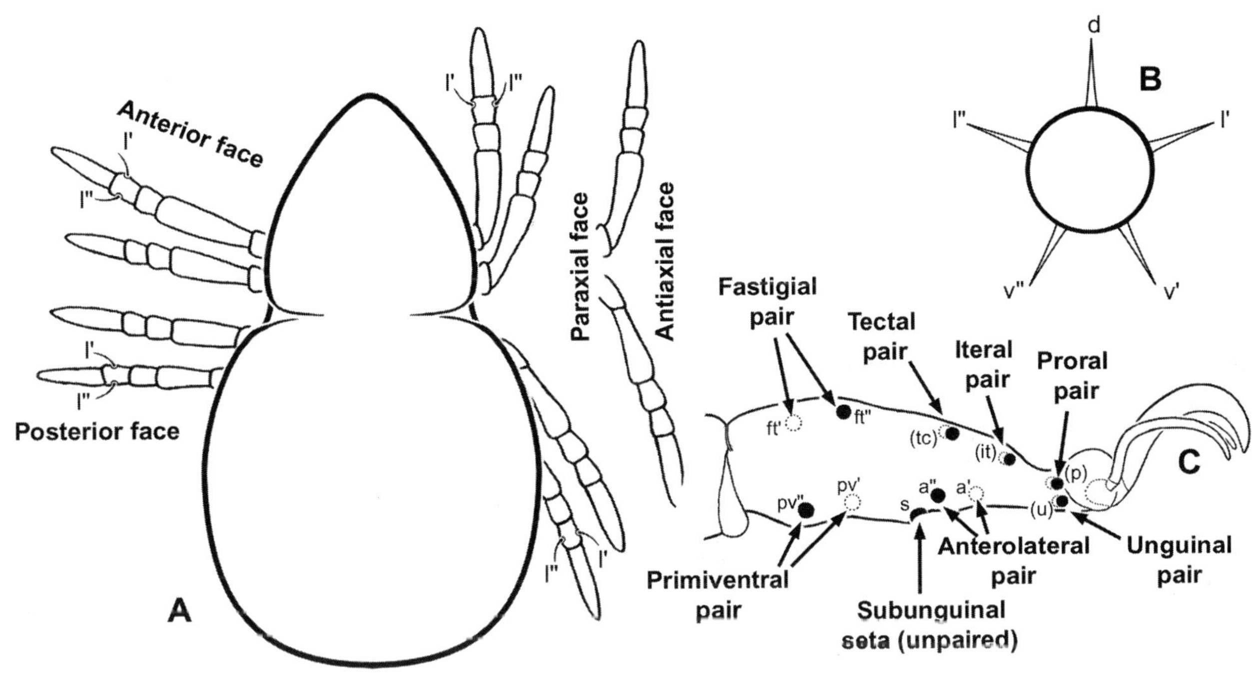

Fig. 3.15. Leg chaetotaxy in the Acariformes. **A**, schematic acariform mite, dorsal aspect, with legs arranged hypothetically on left and as naturally positioned on right, to show different positions relative to central body axis (paraxial, antiaxial) of metamerically homologous setae according to leg direction (prime indicates anterior face; double prime indicates posterior face); **B**, schematic verticil of 5 setae showing unpaired dorsal, paired lateral, and paired ventral setae; **C**, schematic depiction of unique nomenclature and notations of tarsal setae (setal pairs indicated by parentheses). (Courtesy of R. A. Norton)

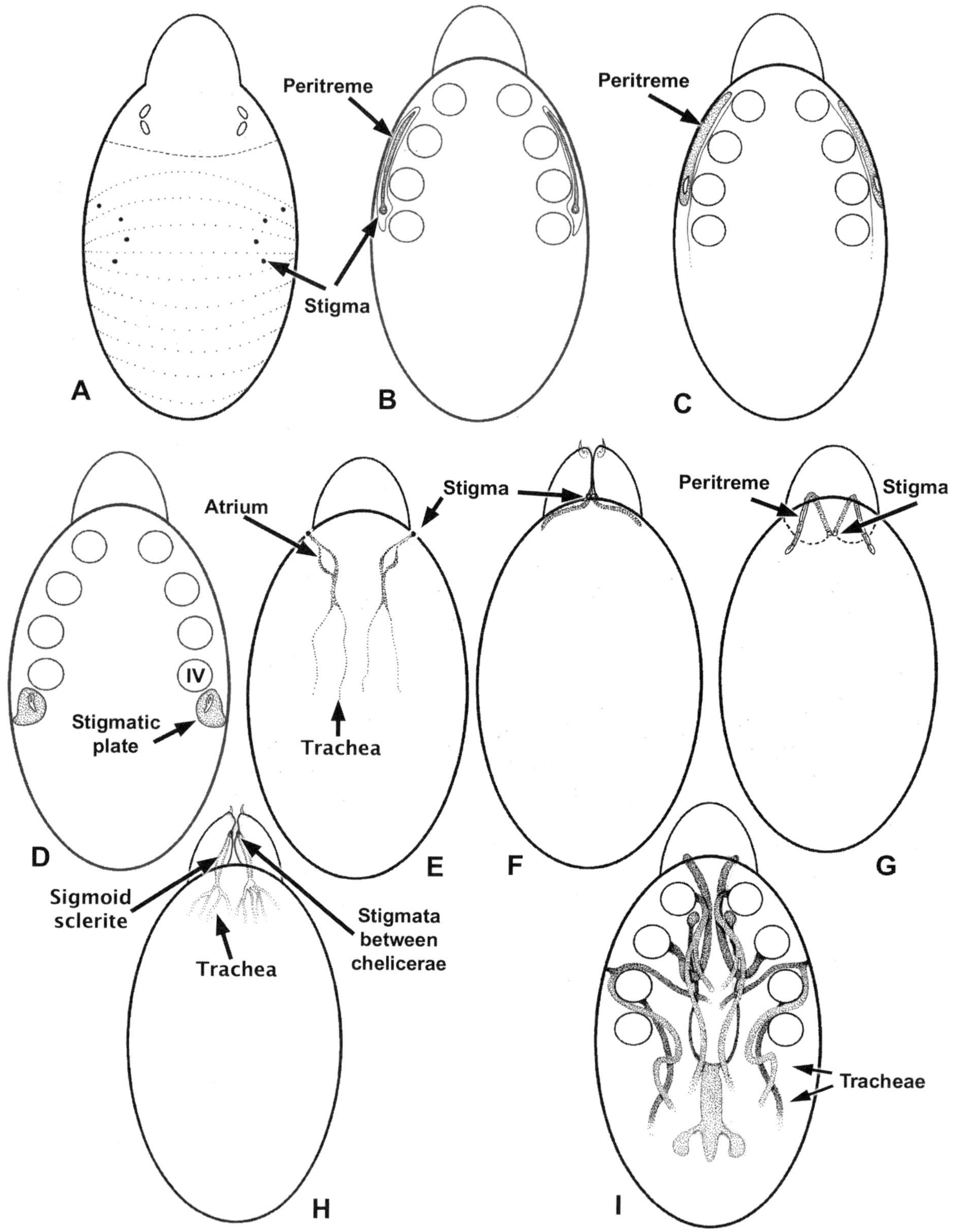

Fig. 3.16. Diagrammatic representations of acarine respiratory structures. **A**, Opilioacarida; **B**, Mesostigmata; **C**, Holothyrida; **D**, Ixodida (IXODIDAE); **E**, Prostigmata (Heterostigmatina); **F**, Prostigmata (Raphignathoidea); **G**, Prostigmata (Tetranychoidea); **H**, Prostigmata (Parasitengonina); **I**, Oribatida (composite idiosomatic tracheal system).

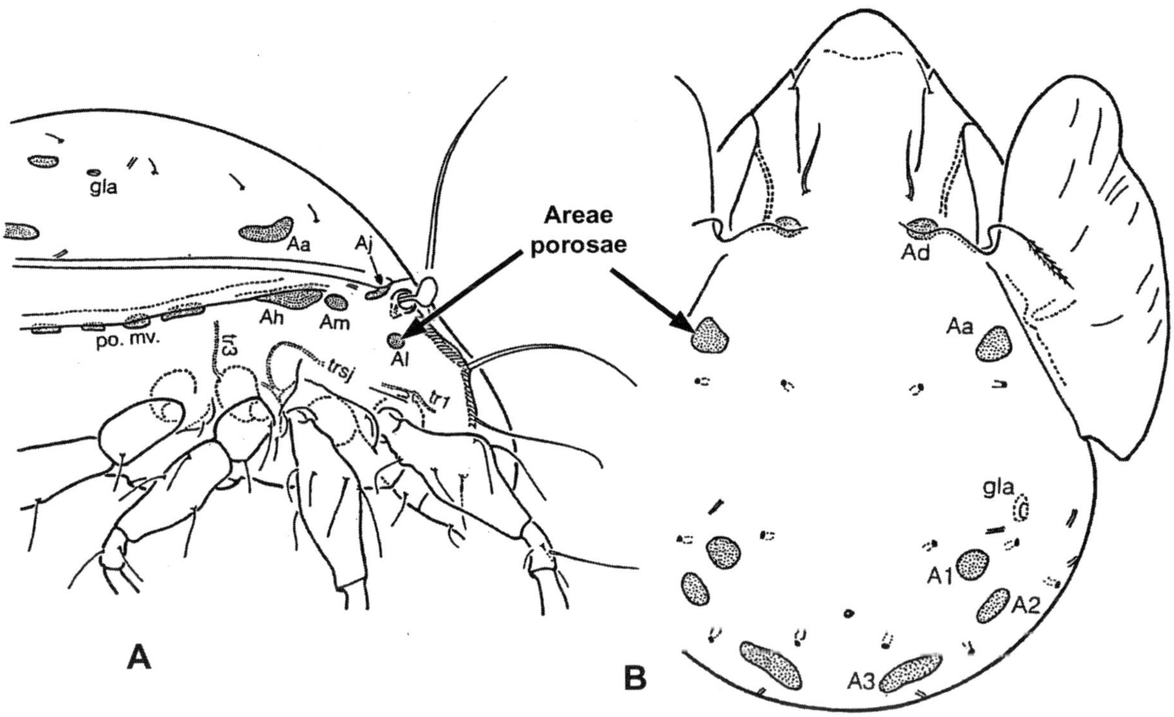

Fig. 3.17. Areae porosae and related structures in Oribatida (after Grandjean 1956, 1971; Norton et al. 1997). **A**, lateral area porosa (stippled) of a poronotic oribatid mite; **B**, dorsal area porosa (stippled) of a galumnid oribatid mite, including median acronotic porose organ (arrow) (GALUMNIDAE).

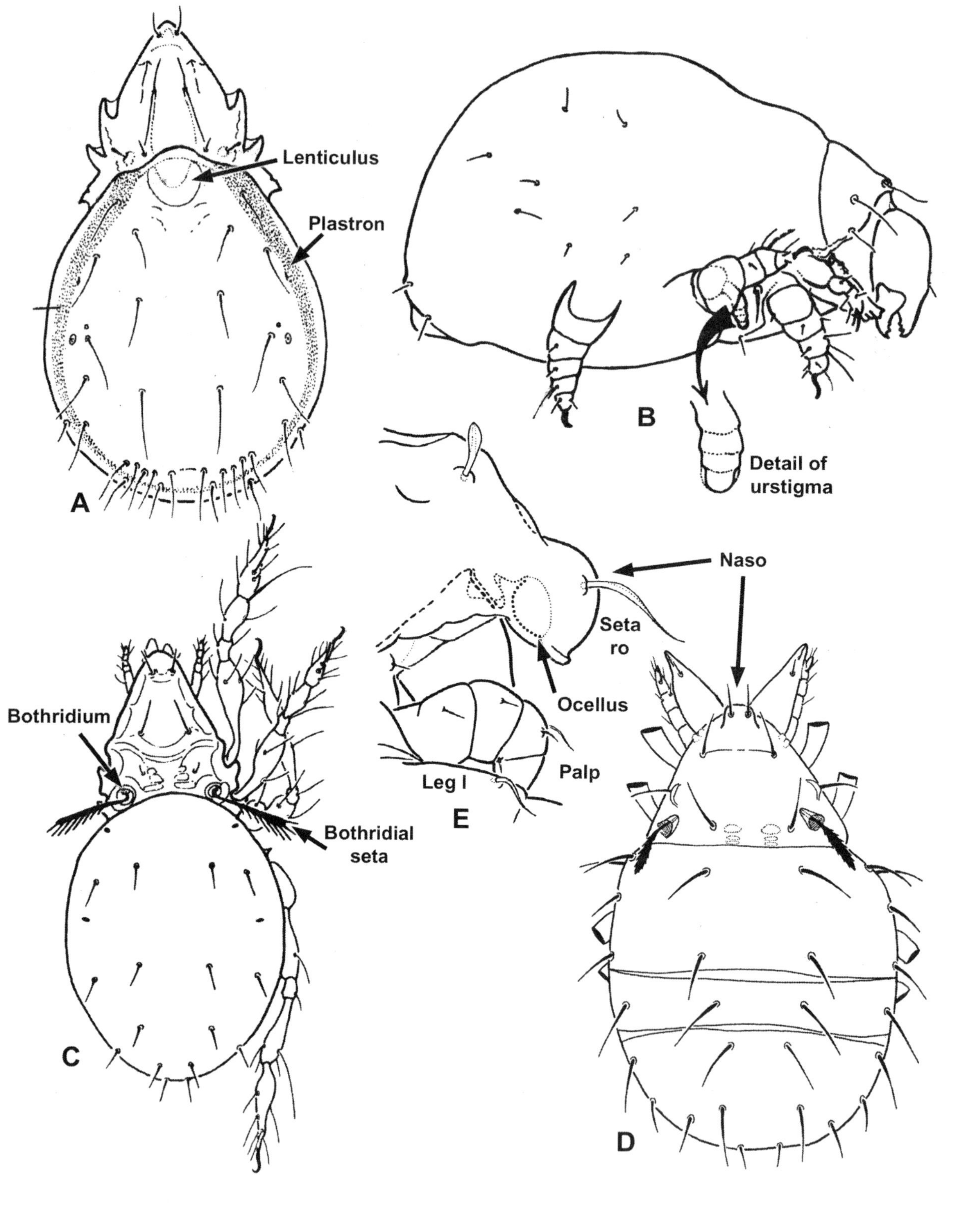

Fig. 3.18. Some oribatid facies and structures. **A**, *Hydrozetes parisiensis* Grandjean (HYDROZETIDAE), dorsum of adult (after Balogh 1972); **B**, phthiracaroid larva with detail of urstigma; **C**, *Multioppia* sp. (OPPIIDAE), dorsal aspect; **D**, **E**, *Brachychthonius* sp. (BRACHYCHTHONIIDAE); **D**, position of naso; **E**, position of inferior ocellus under naso (after Grandjean 1958a).

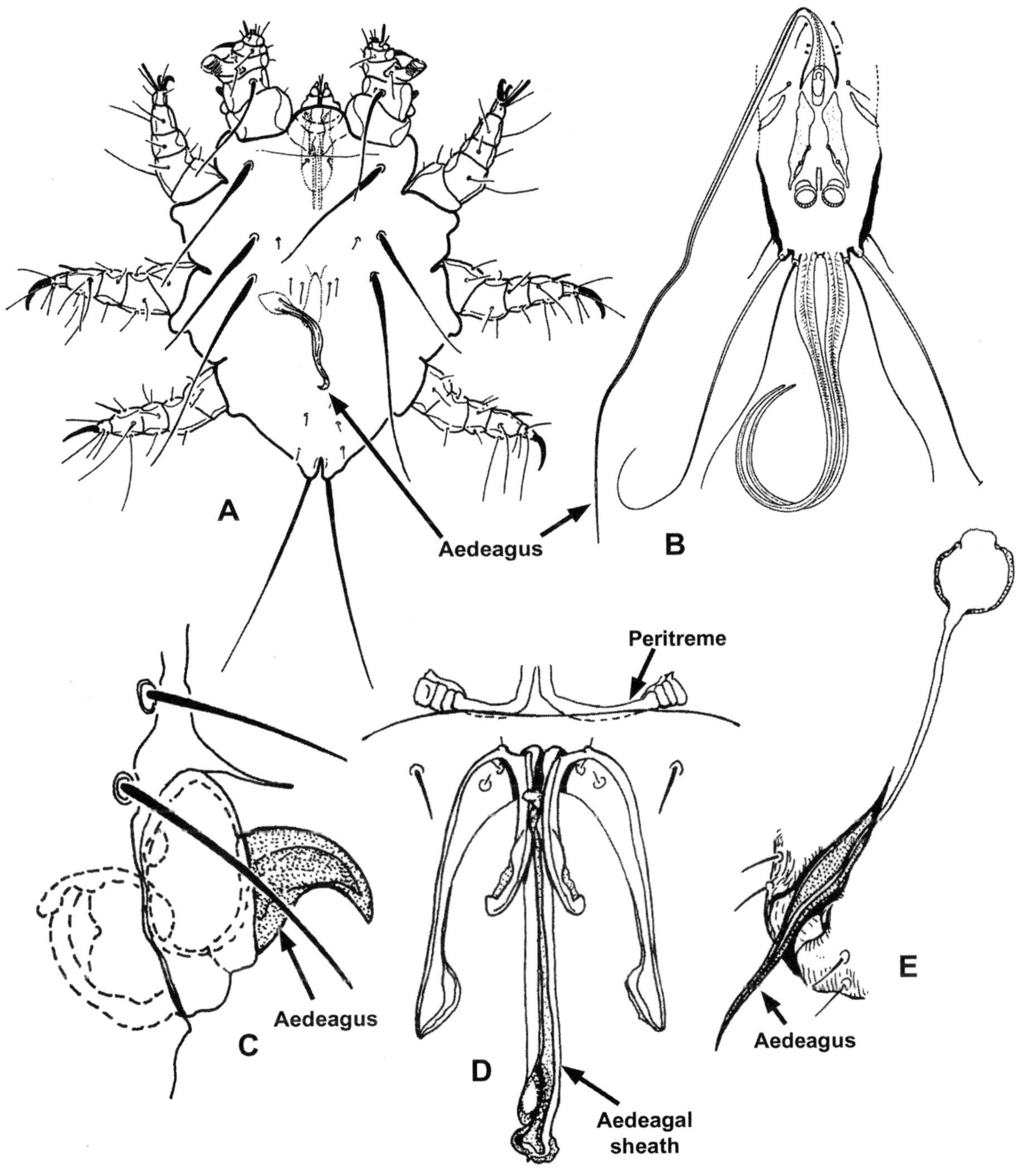

Fig. 3.19. Some male acariform facies and structures. **A**, *Protomyobia* sp. (Prostigmata, MYOBIIDAE), dorsum; **B**, *Proctophyllodes longiphyllus* Atyeo and Braasch (Oribatida, PROCTOPHYLLODIDAE), posteroventral aspect; **C**, *Glycyphagus destructor* (Schrank), aedeagus; **D**, *Harpyrhynchus* sp. (Prostigmata, HARPIRHYNCHIDAE), genital apparatus; **E**, *Eotetranychus carpini* (Oud.) (Prostigmata, TETRANYCHIDAE), aedeagus and associated structures.

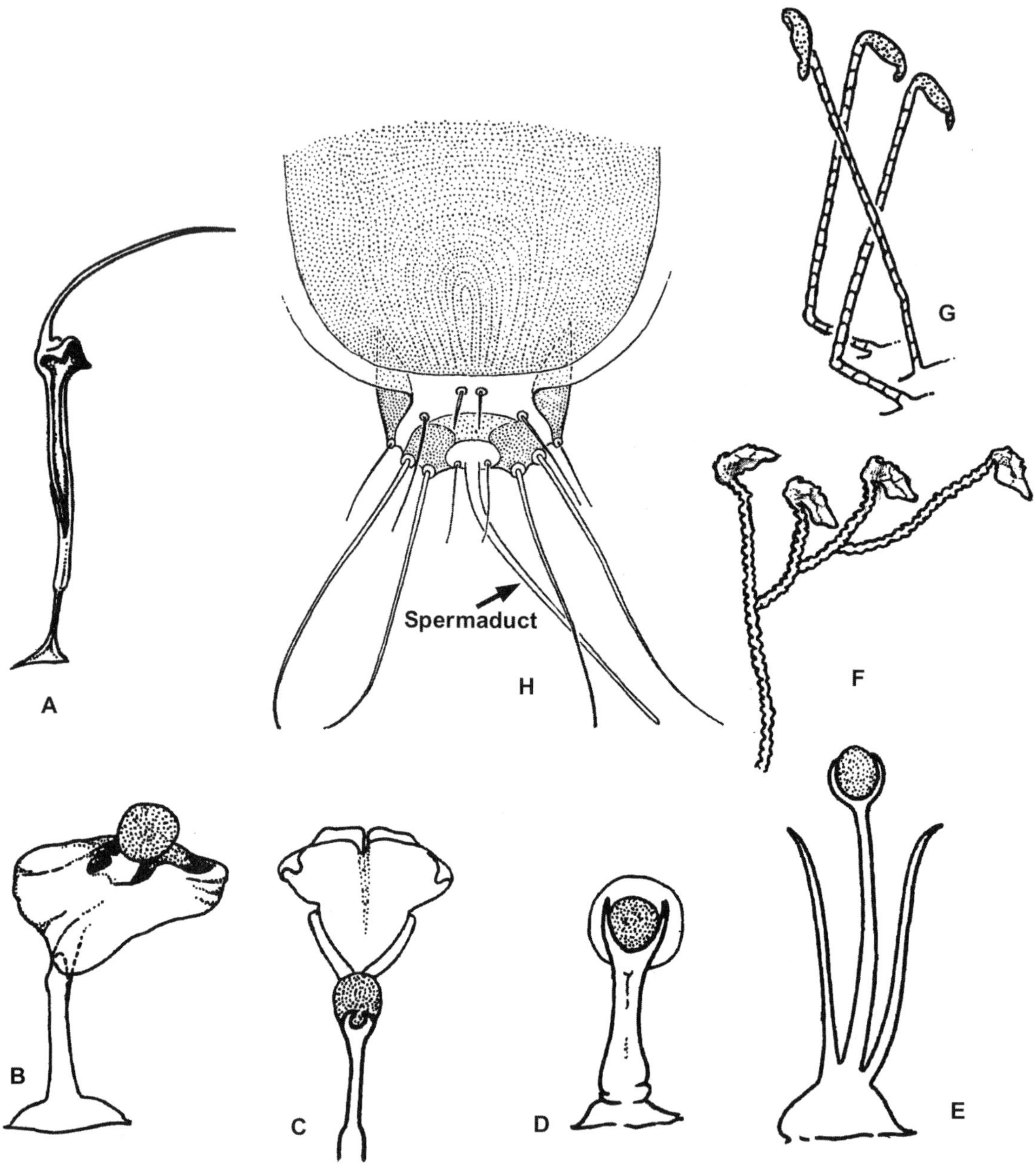

Fig. 3.20. Representative acariform spermatophores (sperm packets are stippled). **A**, *Neomologous littoralis* (L.) (Prostigmata, BDELLIDAE) (after Alberti 1974); **B**, *Phyllocoptruta oleivora* (Ashmead) (Prostigmata, ERIOPHYIDAE) (after Jeppson, Keifer, and Baker 1975b); **C**, *Abrolophus passerinii* (Berl.) (Prostigmata, ERYTHRAEIDAE) (after Witte 1975a); **D**, *Pergalumna emarginata* (Banks) (Oribatida, GALUMNIDAE); **E**, trombiculid type (Prostigmata, TROMBICULIDAE); **F**, *Erythraeus katherinae* Otto (Prostigmata, ANYSTIDAE) (after Otto 1999d); **G**, *Unionicola* sp. (Prostigmata, UNIONICOLIDAE) (after Proctor 1992b); **H**, *Mesosathes tetrasetosus* (Novaes) (Oribatida, CRYPTUROPTIDAE), posterodorsal aspect of female showing terminal sperm duct (from Gaud, Atyeo, and Berla 1972).

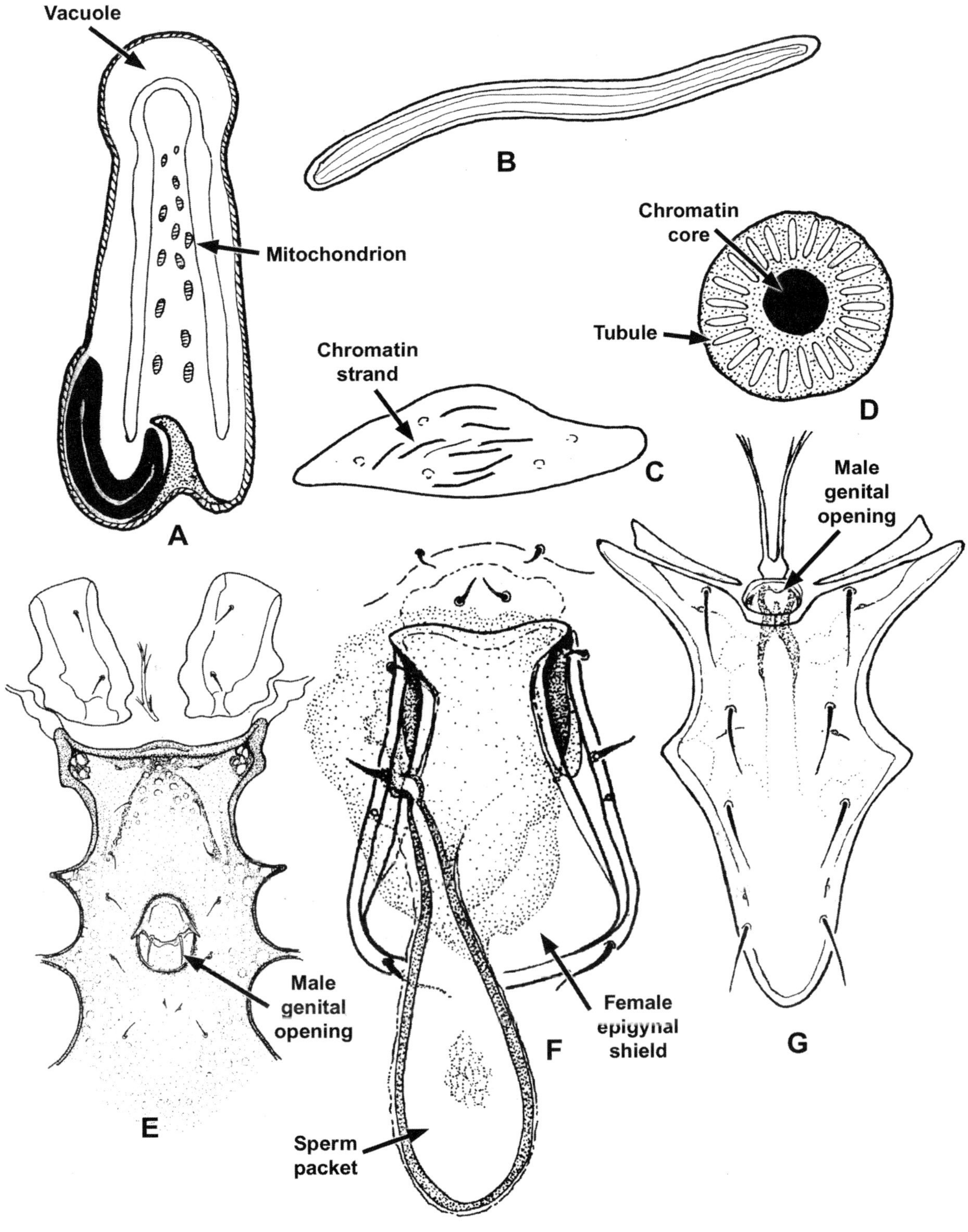

Fig. 3.21. Acarine spermatozoa. **A**, *Cilliba cassidea* (Hermann) (UROPODIDAE), vacuolated type, **B**, *Parasitus* sp. (PARASITIDAE), ribbon type; **C**, *Acarus* sp. (ACARIDAE), with strands of chromatin; **D**, *Tetranychus urticae* (Koch) (TETRANYCHIDAE), with chromatin core (after Alberti 1991).

Examples of sternitigenital structure in Mesostigmata. **E**, *Phaulodinychus mitis* (Leonardi) (UROPODIDAE), portion of male holoventral shield showing median position of genital aperture (after Krantz 1974); **F**, *Trachytes* sp. (TRACHYTIDAE), sperm packet attached to margin of female epigynal shield; **G**, *Veigaia* sp. (VEIGAIIDAE), male sternitigenital shield showing anterior genital aperture.

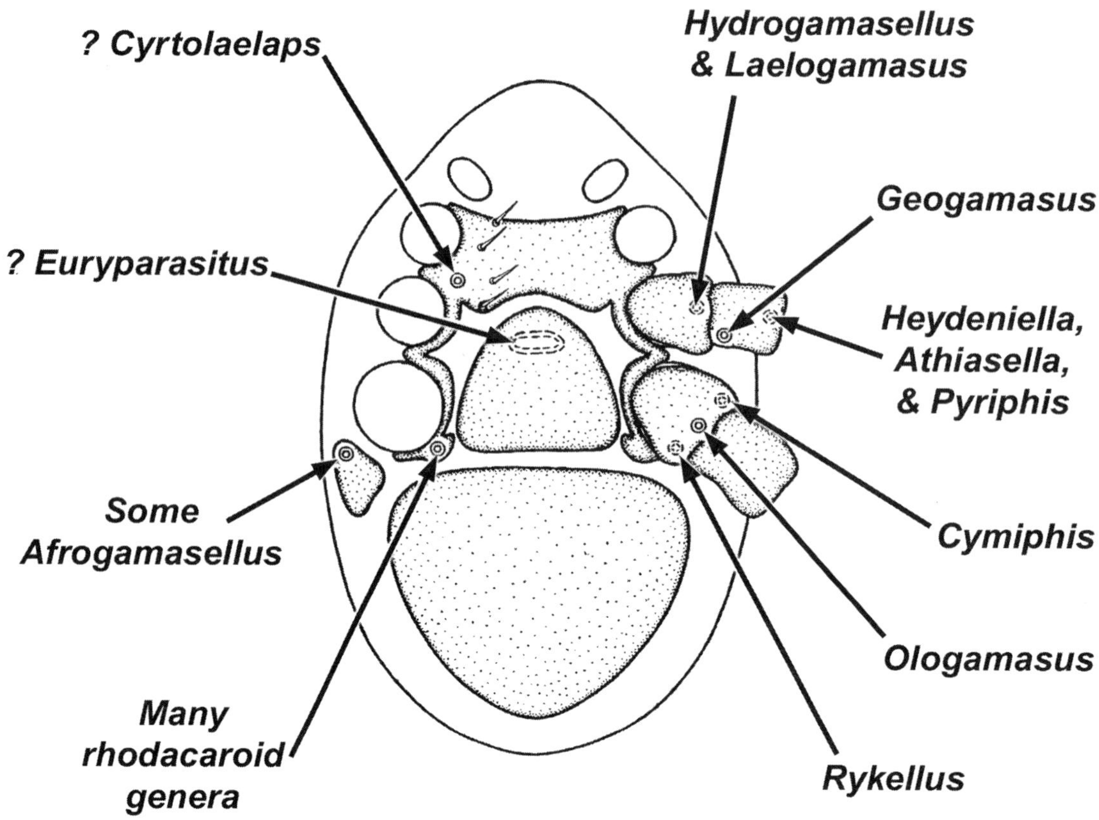

Fig. 3.22. Established and presumptive sites of sperm induction pores in female Rhodacaroidea (from Lee 1970, with permission, South Australian Museum, Adelaide).

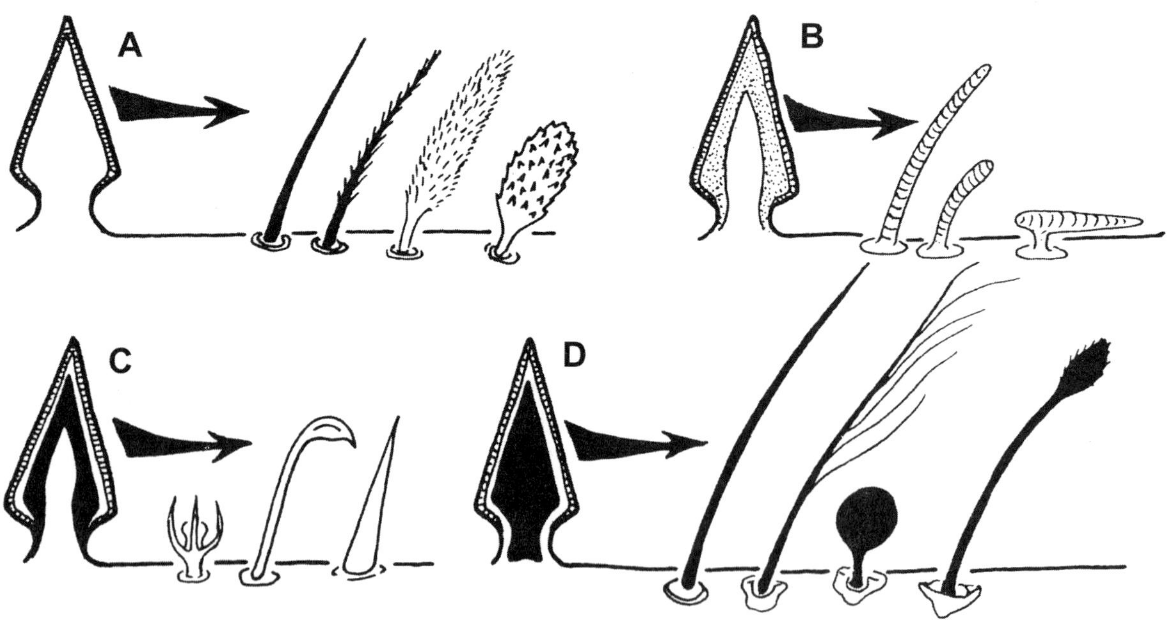

Fig. 3.23. Sectioned models and examples of acarine setal types. **A**, tactile setae (hollow); **B**, solenidia (protoplasmic core); **C**, famuli (two closest to section) and eupathidium (spinelike structure on right) with a sheath of actinopilin (black in section); **D**, trichobothria (actinopilin core).

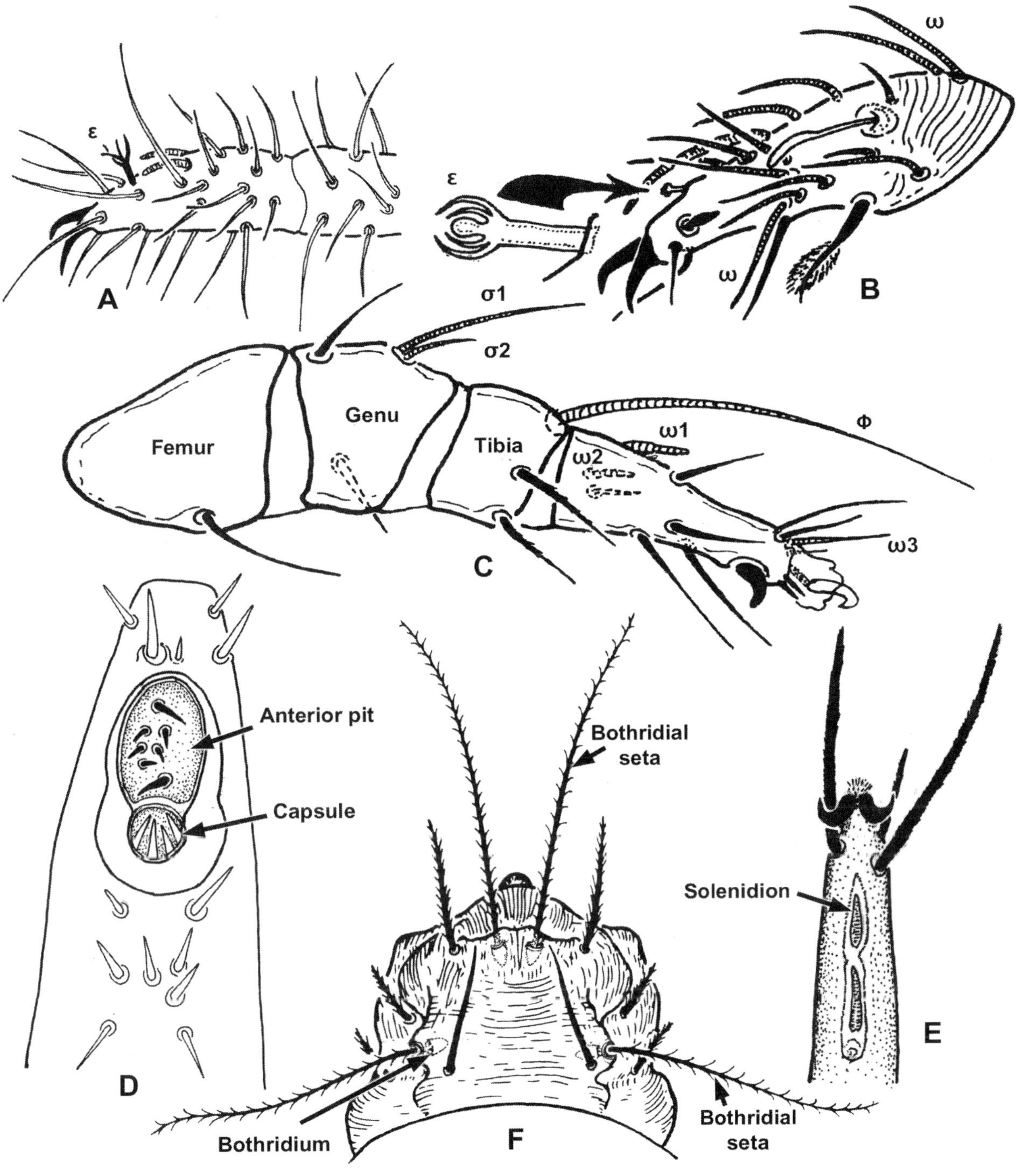

Fig. 3.24. A, *Labidostoma* sp. (Prostigmata, LABIDOSTOMATIDAE), tarsus I with branched famulus (ε); **B**, *Sphaerolichus barbarus* Grandjean (Sphaerolichida, SPHAEROLICHIDAE), with detail of primitive bracteate famulus (ε); **C**, *Acarus siro* L. (ACARIDAE), showing locations and signatures of solenidia (femoral solenidia absent); **D**, Haller's organ on dorsum of tarsus I of an adult ixodid tick (Ixodida, IXODIDAE); distal anterior pit sensilla are opaque, and those on the proximal capsule are shown in outline; **E**, *Eupodes longisetatus* Strandtmann (Prostigmata, EUPODIDAE); tarsus I with dorsal recumbent solenidion (after Shiba 1969); **F**, *Nanorchestes* sp. (Prostigmata, NANORCHESTIDAE), showing two pairs of bothridia and associated bothridial setae.

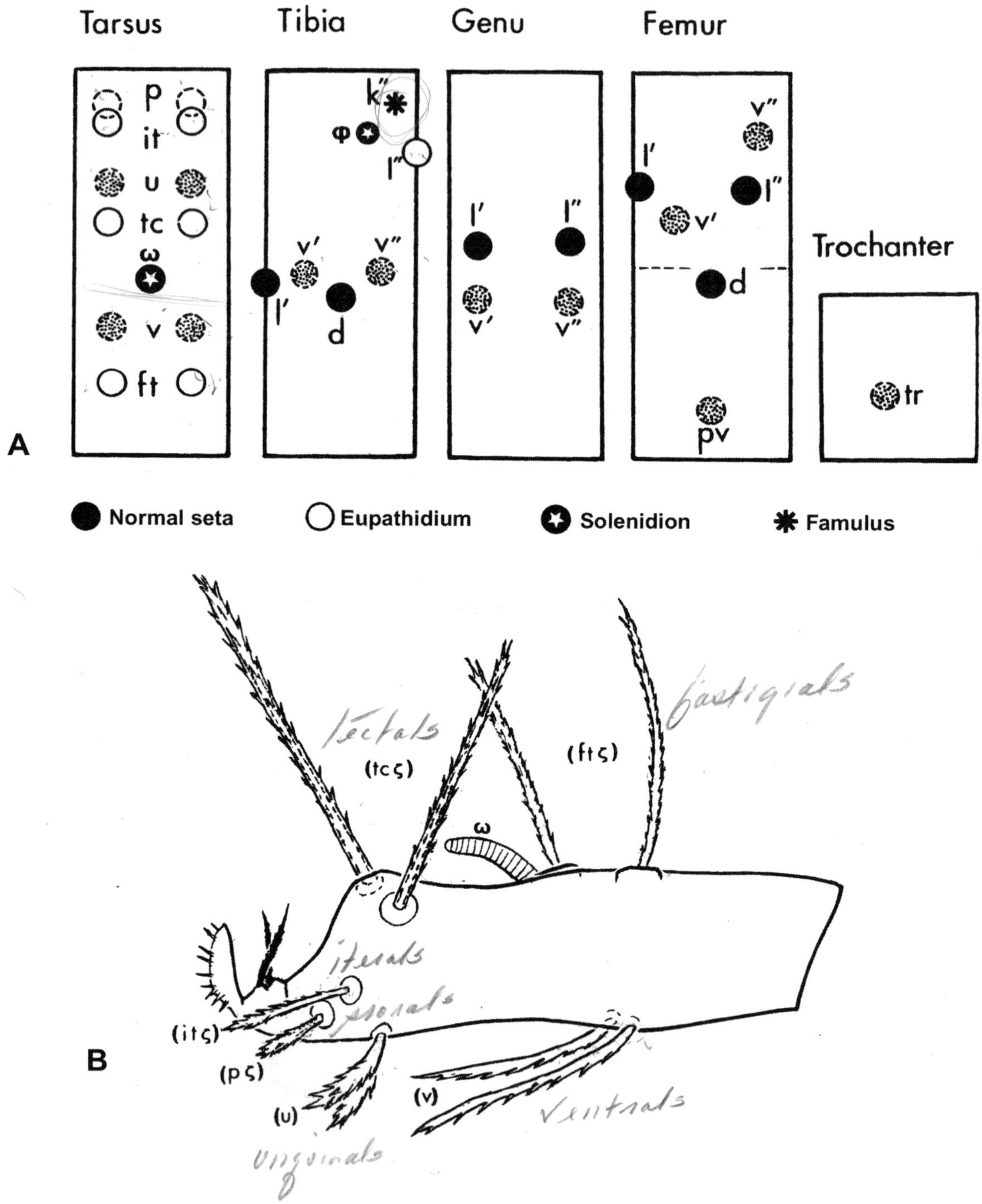

Fig. 3.25. Leg segment setal types in an archetypic tydeid mite (Prostigmata, TYDEIDAE) (after André 1981b). **A**, eupathidia include tarsal fastigials (*ft*), iterals (*it*), prorals (*p*), and tectals (*tc*) and a single lateral eupathidium (*l"*) on the tibia; **B**, *Meyerella marshalli* André, antiaxial aspect of tarsus I of the tritonymph.

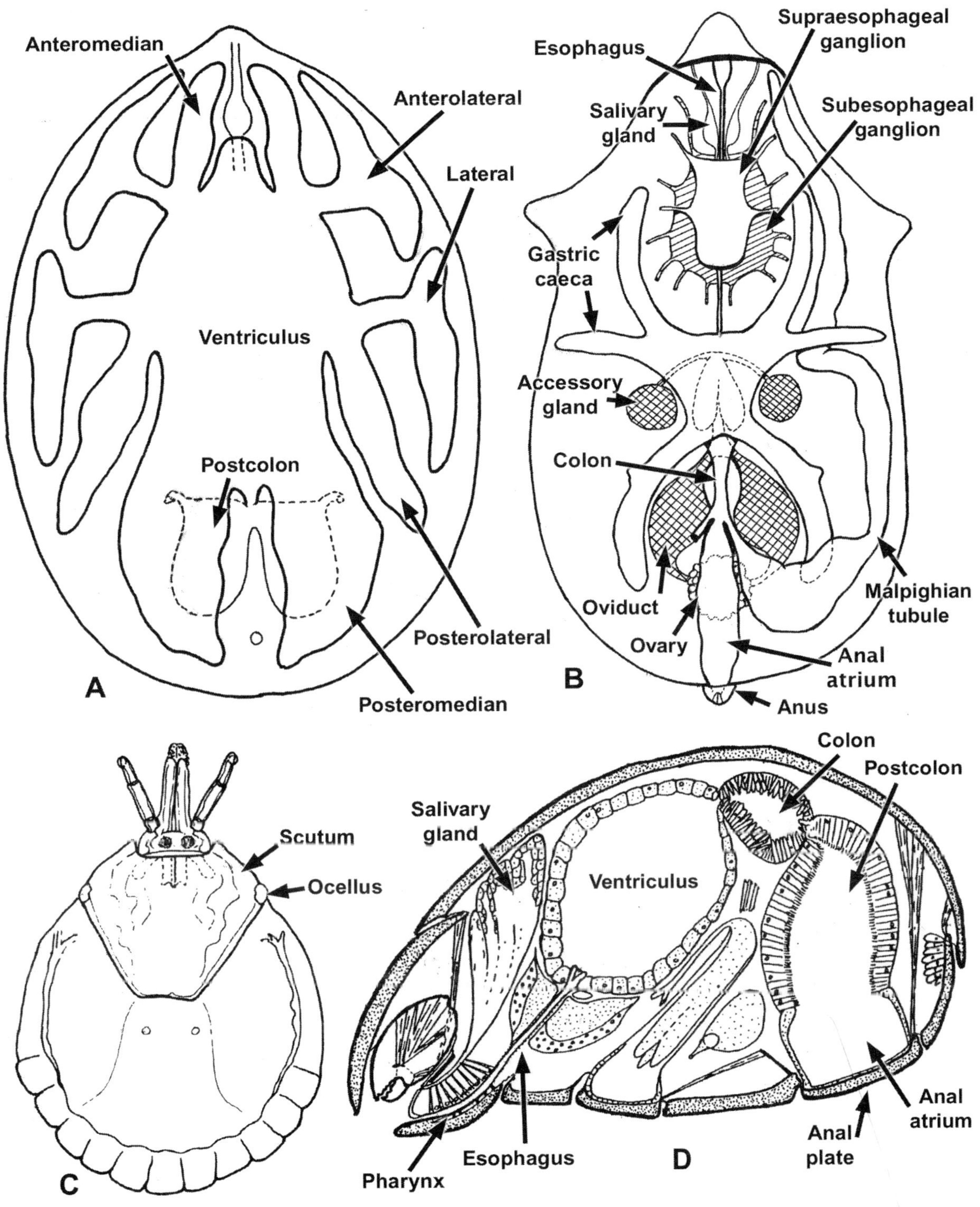

Fig. 3.26. **A**, *Ornithodorus kelleyi* Kohls (Ixodida, ARGASIDAE), dorsal internal aspect illustrating multiple (5) primary caecal pairs (after Sonenshine and Gregson 1970); **B**, *Caminella peraphora* Krantz and Ainscough (Mesostigmata, DITHINOZERCONIDAE), internal aspect of female; **C**, *Amblyomma* sp. (Ixodida, IXODIDAE), dorsum of female with lateral ocelli; **D**, *Euzetes* sp. (Oribatida, EUZETIDAE), schematic midsagittal section (from Evans 1992, with permission, CAB International, Wallingford, UK).

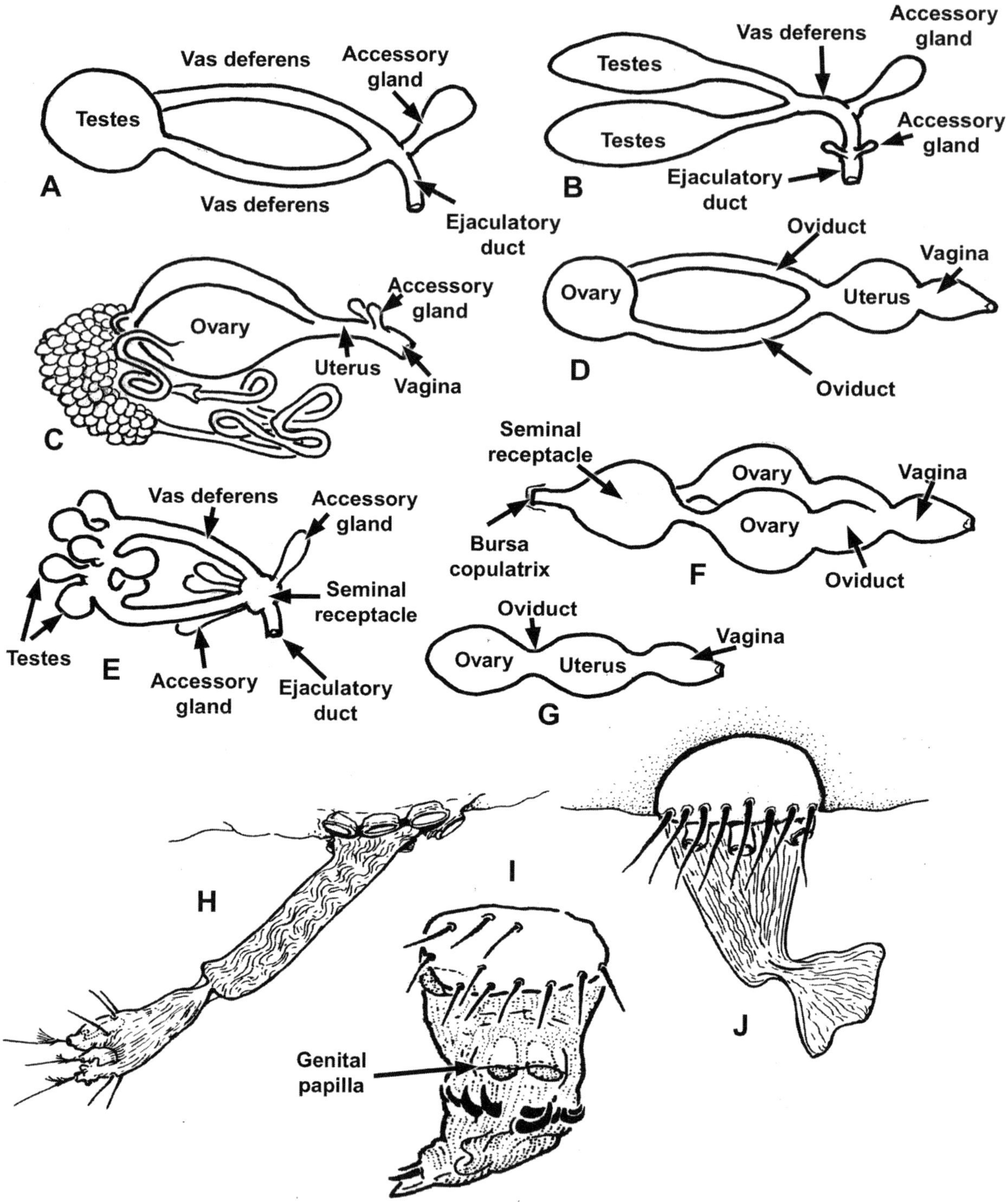

Fig. 3.27. Schematics of acarine reproductive systems (after Hughes 1959; Sonenshine 1970). **A**, PARASITIDAE (Mesostigmata), male; **B**, UROPODIDAE (Mesostigmata), male; **C**, ARGASIDAE (Ixodida), female; **D**, generalized female system; **E**, ERYTHRAEIDAE (Prostigmata), male; **F**, ACARIDAE (Oribatida), female; **G**, ERIOPHYIDAE (Prostigmata), female. Sarcoptiform ovipositor types. **H**, *Nanorchestes* sp. (NANORCHESTIDAE) (Prostigmata); **I**, *Acaronychus tragardhi* Grandjean (ACARONYCHIDAE) (Oribatida) (after Grandjean 1954c); **J**, *Nanhermannia* sp. (NANHERMANNIIDAE) (Oribatida).

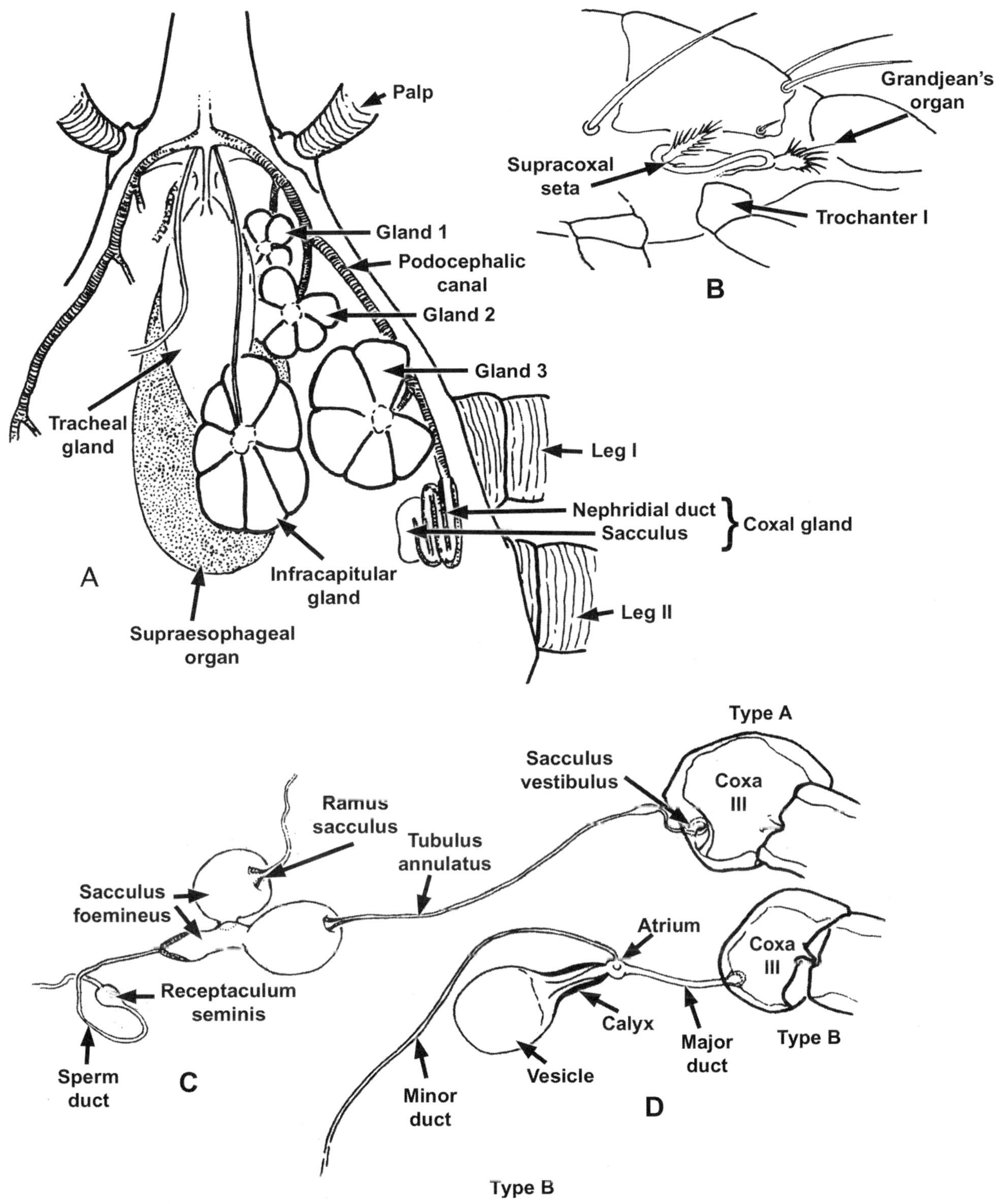

Fig. 3.28. A, idiosomatic gland complex of the bdellid type (BDELLIDAE) (after Grandjean 1938b; Alberti 1973; Alberti and Storch 1973); **B**, supracoxal complex (leg I) overlying a non-nephridial gland system in an astigmatine oribatid mite (after Hughes 1976); **C–D**, sperm access systems in the Mesostigmata (Dermanyssiae); **C**, type A complex (laelapid); **D**, type B complex (phytoseiid).

CHAPTER FOUR

D.E. WALTER

Reproduction and Embryogenesis

Sex Ratios and Genetic Systems

Mites often occur in approximately equal proportions of adult males and females in natural field populations. Such sexual equilibrium is referred to as the **apparent sex ratio**, one of the many types of population measurements that are used in interpreting life history. In terms of the evolution of mating and genetic systems, however, the most important is the **primary sex ratio,** or the relative proportion of male and female offspring in a brood at fertilization. Because of factors such as differential mortality of the two sexes in the egg or subsequent stages or a tendency for one sex to disperse more than the other, the apparent sex ratio may not always reflect the primary ratio. Because differentiating males from females at fertilization or at oviposition (**secondary sex ratio**) is rarely possible, the **progeny** or **tertiary sex ratio** at the adult molt is usually used as an estimate of the primary sex ratio.

Diplodiploidy, in which males and females each have two sets of chromosomes, with one pair contributed by each parent, appears to be the ancestral genetic system in mites (Norton, OConnor, and Johnston 1993; Wrensch, Kethley, and Norton 1994). The basal acariform lineages in both the Sarcoptiformes (e.g., Oribatida) and Trombidiformes (e.g., Parasitengonina) are diplodiploid, as are a number of parasitiform groups. For example, ticks (Ixodida) usually are diplodiploid, rarely triploid (Oliver 1977, 1983), and diplodiploidy appears to be basal in the Mesostigmata. However, supporting data are sparse in that only 4 of the 11 cohorts have species whose cytogenetics have been characterized (Norton, OConnor, and Johnston 1993; Cruickshank and Thomas 1999). Members of the parasitiform order Holothyrida have 1:1 field sex ratios, but nothing is known of their genetic system (Walter and Proctor 1998). Mite chromosomes are typically very small (< 0.003 mm), lack defined centromeres, and are considered by Wrensch, Kethley, and Norton (1994) to be holokinetic. Parasitiform mites usually have a haploid chromosomal complement (n) of 3–9, while that of Acariformes ranges from 2 to 12 (Norton, OConnor, and Johnston 1993; Weeks, Marec, and Breeuwer 2001).

While progeny sex ratios of approximately 1:1 would be expected to occur in diplodiploid mites, acarine sex ratios are often highly female-biased (e.g., 60%–95% females) among the species that have been studied (including some with equal field sex ratios). Examination of the genetic systems of these mites reveals that many have males with only half the number of chromosomes of females (**haplodiploidy**). The best-known haplodiploid genetic system is referred to as **arrhenotoky**, where males are produced parthenogenetically from unfertilized eggs (females arise from fertilized eggs). Arrhenotoky occurs in 20% of all animal species (Crozier and Pamilo 1996) and, among the Acari, appears to have evolved independently at least twice in the Prostigmata, once in the Astigmatina (HISTIOSTOMATIDAE), and at least four times in the Dermanyssiae (MACROCHELIDAE, Dermanyssoidea, ASCIDAE, and BLATTISOCIIDAE) (Norton, OConnor, and Johnston 1993; Cruickshank and Thomas 1999; Walter and Lindquist 2001). An undefined haplodiploid system has been found in one species of ant-associated Antennophorina (Franks, Healy, and Byrom 1991).

A second class of haplodiploid systems involves initial fertilization of eggs destined to produce males, followed by elimination or deactivation of the paternal chromosomes. Paternal genome loss, or PGL (Bull 1983), is exhibited by members of the mesostigmatic family PHYTOSEIIDAE in the form of **parahaploidy** or **pseudoarrhenotoky**. Early in embryogenesis, the paternal genome (or at least the major part of it) is discarded, leaving the zygote in a haploid state (Helle et al. 1978; Hoy 1979; Nelson-Rees, Hoy, and Roush 1980; Toyosima and Amano 1999). Male chromosomes may also be eliminated during spermatogenesis, resulting in the production of sperm with only the maternal genome (Proctor 1996; Cruickshank and Thomas 1999).

Obligate parthenogens make up less than 1% of the animal kingdom, but **thelytoky** (all-female parthenogenesis) is found in many groups of mites and is especially common in soil-inhabiting Mesostigmata, Endeostigmata, and Oribatida (Norton and Palmer 1991; Norton, OConnor, and Johnston 1993; Walter and Lindquist 1995; Walter and Proctor 1999). Hundreds of species representing several families of oribatid mites are obligate parthenogens. Clear examples of **amphitoky** (or deuterotoky), in which both sons and daughters are produced asexually, are rare in the Acari. However, there are exceptional cases where nonfunctional (spanandric) males are produced in thelytokous species, and functional males may occur in some thelytokous Mesostigmata and Astigmatina (Walter and Kaplan 1990b; Norton, OConnor, and Johnston 1993).

Among the Acariformes, the early derivative families collectively known as the Endeostigmata (p. 421) are more or less evenly divided between thelytokous and bisexual species. Thelytoky is known or suspected in many groups of Prostigmata, including some agricultural pests (e.g., *Penthaleus* species (PENTHALEIDAE) and species of the tetranychid subfamily Bryobiinae in the TETRANYCHIDAE). One especially bizarre example in the tetranychoid family TENUIPALPIDAE is the false spider mite, *Brevipalpus phoenicis* (Geijskes), which has been shown to consist entirely of haploid females (Weeks, Marec, and Breeuwer 2001). As mentioned previously, oribatids include a large number of thelytokous representatives (about 10% of known species), and one of these asexual lineages may have produced the large and primarily bisexual oribatid radiation comprising the cohort Astigmatina (Norton and Palmer 1991; Palmer and Norton 1992; Norton, OConnor, and Johnston 1993). Thelytoky is common among soil litter–inhabiting Mesostigmata, but it has a sporadic distribution across taxa (Walter and Lindquist 1995). Most thelytokes appear to have close bisexual relatives and only a few taxa (e.g., the genus *Protogamasellus* and some species of *Asca* (ASCIDAE)) appear to have radiated as asexual lineages.

All bisexual mites are fertilized internally, but the means employed by males to introduce sperm to the female are varied and often bizarre (Proctor 1998; Walter and Proctor 1999). Sperm transfer is discussed in detail in chapter 3.

In summary, both the Acariformes and Parasitiformes appear to have originated as diplodiploid lineages with relatively balanced sex ratios. However, in both superorders haplodiploid genetic systems and female-biased progeny sex ratios appear to have arisen independently many times. A tendency for the female bias to become absolute (thelytoky) also is apparent in many lineages of Acariformes, but the evolution of all-female parthenogenesis is more scattered in the Parasitiformes.

Embryogenesis

The eggs of mites have a chorion and a vitelline membrane and typically have a central nucleus surrounded by cytoplasm with an internal reticulum to which yolk granules are attached. Cleavage usually is superficial, with nuclear divisions occurring as synchronous waves during migration toward the periplasm. Each migrating nucleus is surrounded by a halo of cytoplasm not enclosed by a cell membrane. Nuclei move to the surface of the yolk mass, where they form a uniform blastoderm that later gives rise to the ectoderm. Cells are initially spindle shaped and well separated but eventually form a layer of cuboidal cells.

During the early stages of gastrulation, a ventral germ disc (interpreted as a telson) appears and develops as a funnel-shaped band in the yolk mass, eventually giving rise to the endo-mesoderm layers. The germ band elongates and gives rise to the cephalic lobes, after which the prosomatic segments with limb buds become distinct (leg segments I–III, followed by pedipalpal, leg IV, and cheliceral segments). Opisthosomatic metamerism is obscure but may represent five segments and the telson. The germ band then contracts longitudinally (**blastokinesis**), shortens perceptibly, and finally concentrates on the ventral surface of the yolk mass (Aeschlimann and Hess 1984).

In the Ixodida and Mesostigmata, the prosomatic limbs, labrum (an unpaired lobe anterior to the mouth), and mouth (stomodeum) develop as differentiation of the embryo proceeds, although external signs of segmentation may disappear (Aeschlimann and Hess 1984; Hutu 1991; Mautz, Hirschmann, and Kemnitzer 1986; Yastrebtsov 1991, 1992). The postoral chelicerae migrate to a position anterior to the mouth, and the pedipalpi move to either side of the stomodeum. Two poststomodeal lobes (lips) develop and sometimes fuse with the well-developed lobes on the pedipalpal coxae. The latter fuse ventrally and extend dorsally to produce the capitular ring or infracapitulum. The midgut and stomodeum join and displace the synganglion dorsally, and a hypoderm forms that will later give rise to the embryonic cuticle. Legs IV do not regress but remain as buds as the other limbs begin to grow. Although the basal musculature and nerves of legs IV develop normally, the larval cuticle covers the undeveloped limb buds. Larval ticks undergo a molt within the egg (Aeschlimann and Hess 1984), but it is unclear if this represents a prelarval stage.

Our understanding of embryogenesis in acariform mites has been greatly advanced by recent works using scanning electron microscopy on house dust mites (PYROGLYPHIDAE) (Walzl 1991), by differential interference contrast (DIC) and transmission electron microscopic (TEM) studies on *Sancassania berlesei* (Michael) (ACARIDAE) (Walzl, Gutweniger, and Wernsdorf 2005), and by scanning electron microscopy (SEM) and analyses of homeobox gene expression in the oribatid mite *Archegozetes longisetosus* Aoki (TRHYPOCHTHONIIDAE) (Telford and Thomas 1998a; Thomas and Telford 1999). Genetic evidence supports the interpretation that the chelicerae are homologous to antennae 1, the palpi to antennae 2, and legs I to the mandibles of crustaceans (Telford and Thomas 1998a, b). As in parasitiform mites, acariform chelicerae begin their development posterior to the mouth, but they migrate to a preoral position as the mouth moves between the palpi. The labral lobe grows to cover the mouth in its new position. Although the large palpal lobes form the subcapitulum as in the Ixodida and Mesostigmata, they do not extend dorsally to cover the cheliceral bases. Limb buds for legs IV are present in acariform mites but are less well expressed than in the Parasitiformes, and external evidence for their existence soon disappears. Finally, *A. longisetosus* clearly shows that the urstigmata (Claparède's organ) develop from the bases of legs II (Thomas and Telford 1999; see chapter 3).

In summary, little is known about acarine embryogenesis, and until recently, most of what was known came from the study of fairly large parasitiform mites such as ticks. Recent work on the molecular embryology of acariform mites, however, has illuminated many aspects of acarine development and has established homologies for prosomatic structures that are valid for the entire Chelicerata.

CHAPTER FIVE

D.E. WALTER
G.W. KRANTZ

Oviposition and Life Stages

Embryonic or postembryonic ovoviviparity has been observed in some mites (Strandtmann and Wharton 1958; Filipponi and Francaviglia 1963; Cross 1965; Mitchell 1968; Egan and Hunter 1975), but the majority of acarines that have been studied deposit one or more eggs. Oviposition is accomplished in several ways, but typically the oval, ovoid, elongate, or flattened eggs are passed through the genital valves and dropped either singly or in clusters. The eggs are highly elastic and resilient at the time of oviposition, which helps prevent egg damage during passage through the relatively small genital aperture of the female. Acarine eggs may be smooth or sculptured, and many are variously ornamented with striations, reticulations, or waxy extrusions. These characteristics, along with differences in shape, often are useful in determining their identity. For example, the eggs of some members of the prostigmatic family BDELLIDAE may be identified to species by their characteristic shape or sculpturing, and others by their color or the nature of the hairlike projections that cover them (Wallace and Mahon 1972, 1976). The elongate eggs of the feather mite genus *Fainalges* (XOLALGIDAE) often have ornamentation patterns that are species specific (Perez 1996).

Many prostigmatic and endeostigmatic mites produce silk to protect or anchor their eggs. These silken constructs vary from simple loose strands, as seen in many TETRANYCHIDAE and some species of CUNAXIDAE (e.g., *Coleoscirus simplex* (Ewing)), to the tightly woven webs of some EUPODIDAE, TETRANYCHIDAE (*Tetranychus* spp.) and ALICORHAGIIDAE (*Alicorhagia fragilis* Berlese) (Saito; Walter 1988b). Some species of the families BDELLIDAE (e.g., species of *Bdella* and *Spinibdella*) and CHEYLETIDAE (e.g., species of *Hemicheyletia* and *Oudemansicheyla*) spin a protective silken cocoon around each egg, which typically is smooth and featureless (Wallace and Mahon 1972; Alberti 1973; Walter and Proctor 1999). In contrast, eggs of the bdellid genera *Cyta* and *Bdellodes* often have a covering of fine hairlike projections or spinose capitate protrusions, presenting a pincushion appearance (Wallace and Mahon 1972, 1976). A waterproof waxy coating or thin lipid layer may be applied to each egg prior to deposition (Lees and Beament 1948; Beament 1951).

Mites that feed on plants or plant parts stored in bulk often follow an r-selected lifestyle model (see chapter 3) and oviposit freely on their plentiful food substrate, countering losses occasioned by heavy predation or exposure with high fecundity rates and/or multivoltinism (Solomon 1945; Huffaker, Van de Vrie, and McMurty 1969; Hussey, Read, and Hesling 1969). Similarly, high fecundity and the placement of eggs at sites where hatching larvae are most likely to encounter a host are common strategies employed by some parasitic acarines with transient hosts. For example, ticks (Ixodida) may deposit several thousand eggs at a time (Drummond, Casagrande, and Logan 1971; Sonenshine 1991), ensuring at least a minimal chance of survival to the adult stage (see chapter 3). Relatively few eggs may be produced by some females of the prostigmatic family TROMBICULIDAE (p. 276), but ovipositing individuals tend to congregate in the runways of their small mammal hosts so that the emerging parasitic larvae often occur in large concentrations known as "mite islands" (Audy 1968)—sites where their hosts are most numerous. Acarines that must find a limited food source in a broader habitat such as soil or water are often able to secrete their relatively few eggs so as to minimize losses to predators and desiccation. To this end, members of the acariform Endeostigmata, Oribatida, and many Prostigmata possess an extrusible ovipositor (Figs. 3.27H–J, p. 52) that holds the egg while simultaneously probing for a suitable niche in which to place it. Among the Parasitiformes, both the Opilioacarida and Holothyrida possess a short ovipositor, but these structures are absent in Ixodida and most Mesostigmata (some Uropodina have a tubular genital process that may represent an ovipositor). Even so, female mesostigmatic mites may expend considerable time and energy in placing their eggs in ways that enhance optimal survival of emerging progeny (Walter and Proctor 1999). For example, females of *Pergamasus quisquiliarum* Can. (PARASITIDAE) attach their eggs to the rootlets of plants, a protected site where hatching larvae may more quickly contact the symphylans on which they prey (Berry 1973). Parasitic Mesostigmata often are larviparous, but those that are oviparous typically place their eggs in protected situations where access of hatching larvae to the host will be virtually assured (Camin 1953a; Yunker 1973). *Dicrocheles phalaenodectes* (Treat) (LAELAPIDAE), a gamasine parasite of noctuid moths, prepares its oviposition site by kneading the tissue on which the egg is to be deposited (Treat 1958, 1975).

Members of the genus *Cheyletus* (Prostigmata, CHEYLETIDAE) often brood their clusters of eggs in concavities or caverns in their habitats (Summers and Witt 1972). Females of *C. eruditus* (Schrank), a grain-inhabiting predatory mite, have been observed to drive other cheyletids or larger arthropods from the nest area. Hatching larvae, however, are eaten by the mother if they tarry too long after eclosion (G. W. Krantz, pers. obs.). Many spider mites (TETRANYCHIDAE) deposit their eggs in silken webbing on their plant hosts, which provides a degree of protection from mite and insect predators (Saito 1985). Hazan, Gerson, and Tahori (1975) found that removal of eggs of *Tetranychus cinnabarinus* (Boisd.) from the surrounding webbing reduces their hatchability under conditions of extremely low or high humidity. The simultaneous development of large numbers of eggs is possible in some Prostigmata (Heterostigmatina) by means of **physogastry**, in which the idiosoma of feeding reproductive or prereproductive females may swell to many times its normal size in order to accommodate simultaneously developing embryos (Kaliszewski, Athias-Binche, and Lindquist 1995). Extreme physogastry occurs in some heterostigmatines, with internally sequestered embryos developing within the egg until the emergence of adults that then escape from the dying mother (Herfs 1926; Moser, Cross, and Roton 1971). Fully developed progeny of the mealworm egg parasitoid *Acarophenax mahunkai* Steinkraus and Cross (ACAROPHENACIDAE) were observed to move actively within the mother as little as 72 hours after her attachment to the host egg (Steinkraus and Cross 1993). Species of the pygmephorid genus *Trochometridium* oviposit their eggs in the normal fashion, but ontogenetic development is entirely within the egg, and hatching does not occur until the adult stage is reached (Cross 1965). Although typical oviparity is the general rule in the Sarcoptiformes, a peculiar form of viviparity known as **aparity** has been observed in the sarcoptiform cohorts Brachypylina and Astigmatina (Jacot 1933; Lipa and Chmielewski 1966). Eggs eclose within the dead body of the mother, and the developing progeny feed on her tissues until they have eaten their way out. Similarly, female red-legged earth mites (PENTHALEIDAE) die as spring approaches, and their bodies are filled with diapausing eggs. The eggs remain within the female's body, often being blown across pastures for long distances, until they hatch the following autumn (Wallace 1970; Ridsdill-Smith 1997).

Mites may be uni- or multivoltine, and some require more than a year to complete their development. Multiple generations are common in phytophagous, coprophagous, nidicolous, and parasitic groups for which a food source is available for extended periods of time (Hussey, Read, and Hesling 1969; Yunker 1973; Jeppson, Keifer, and Baker 1975b). Predatory acarines that feed in these habitats also tend to be multivoltine (Chant 1959; Axtell 1969; Walter and Proctor 1999). Univoltine acarines occur commonly in soil, litter, aquatic, and marine habitats (Hartenstein 1962b; Newell 1971a; Bartsch 1972; Alberti 1973; Pugh and King 1986; Green and MacQuitty 1987), although some marine mites (HALACARIDAE) are known to be multivoltine (Abé 1998). Univoltinism is suspected or documented among certain mite ectoparasites of insects (Anderson 1968; Treat 1975).

Development from egg to adult may require only two to three days in some Prostigmata and Mesostigmata, but more often acarine developmental time extends to one or more weeks or months and to over a year in many Oribatida (Walter and Proctor 1999). The common astigmatine mange mite, *Sarcoptes scabiei* (DeGeer) (SARCOPTIDAE), may complete an entire life cycle in 10 days (Yunker 1973), while species of the mesostigmatic family MACROCHELIDAE require as little as 34 hours to develop from egg to adult (Axtell 1969; Ruf 1996). Life cycles of 20 weeks to a year or more are commonplace for various Prostigmata (Vistorin-Theis 1975) and Oribatida (Haarløv 1960; Hartenstein 1962b; Shereef 1972; Norton 1994), and late-hatching tick larvae may require nearly two years to reach adulthood (Hunter and Bishopp 1911). Species occurring in inhospitable circumstances may require even longer periods to complete their development. For example, development of *Ixodes uriae* White, a widespread tick parasite of seabirds, may take four to five years in cooler latitudes where host availability is problematic (Eveleigh and Threlfall 1975). Similarly, *Alaskozetes antarcticus* (Michael), a brachypyline denizen of the maritime antarctic, often requires more than five years for development from egg to egg (Convey 1994; Block and Convey 1995). Seasonally abnormal temperatures and humidities in more equable climates can also produce prolonged developmental times in affected populations. Temperature, humidity, and food availability may also play a role in mite longevity, with more optimal conditions usually leading to shorter life spans. Astigmatine mites that inhabit grain storages where the food supply is virtually inexhaustible and where ambient conditions are ideal often have a life span of less than a month (Hughes 1976), while tick species that must locate and successfully parasitize more than one host animal to complete their life cycle (see chapter 11) are known to survive for several years (Balashov 1972). Predatory soil Mesostigmata may have an adult life span as short as a week when food is plentiful and oviposition continuous. However, when food is scarce, females become inactive and can fast for months with little mortality (Walter and Proctor 1999).

Acari Instars

Acari may pass through up to six (occasionally more) instars after eclosion. These are the **prelarva** (also referred to as prolarva or deutovum), **larva**, **protonymph** (nymphochrysalis), **deutonymph** (deuteronymph, hypopus, or wandernymph), **tritonymph** (teleiochrysalis), and **adult**.

Prelarva

Typically, the prelarva is a nonfeeding form that occurs not only in the Acari but in other arachnids as well. Grandjean (1938a) theorized that the ancestral prelarva was an active stage that routinely appeared prior to the larval molt, although most known contemporary acarine prelarvae are quiescent and sequestered within the egg chorion (Walter and Proctor 1999). In some cases, the prelarva appears to be little more than a featureless sac without legs or mouthparts (Fig. 5.1C). Extreme inhibition of this type is referred to as **calyptostasis** and is a feature of prelarvae of some Endeostigmata (e.g., species of *Terpnacarus* and *Alycus*), Oribatida (Grandjean 1962c; Lions 1967, 1973), and certain higher Prostigmata (Johnston and Wacker 1967; Walter and Proctor 1999). In other acariform groups, the inactive prelarva may possess three pairs of legs, recognizable mouthparts, and setae (Robaux 1971; Coineau 1974a, 1976b; Walter 1998b; Kethley 1992), although mouthpart development and leg segmentation may be inhibited. Grandjean (1957a) refers to such incomplete calyptostasis as **elattostasis** (Fig. 5.1E). Extreme examples of elattostasis are found in some NANORCHESTIDAE (Schuster and Pötsch 1989), ADAMYSTIDAE (Coineau 1979), and ANYSTIDAE (Otto 1997), in which the prelarva is not only highly developed but is also able to walk (but not feed). Whatever its state of development, the prelarva is considered by Coineau (1974a, b) to be an inhibited rather than a regressive form because its occurrence does not necessarily signify regressive tendencies in subsequent stages.

Acariform prelarvae share certain characteristics, including urstigmata (Figs. 5.1C–E, G) and a nonfunctioning stomodeum. Other features such as pretarsi, an anal opening, and distinct coxal glands may or may not be present in a given group. An extremely primitive prelarval stage with well-developed and apparently functional mouthparts and legs has been described for *Phalangiacarus brosseti* Coineau and Van der Hammen, an opilioacarid mite from Gabon (Coineau 1973a; Van der Hammen 1976; Figs. 5.1A, B).

The prelarva molts to the larval stage, sometimes with the aid of an anterior opening mechanism (Ehrnsberger 1974). Passage from the prelarval to the larval stage in prostigmatic mites may require as little as 36 hours (Coineau 1976b) or as much as 14–15 days (Robaux 1971).

Larva

The typical acarine larva is a hexapod form with little or no sclerotization and without indication of external genitalia (Fig. 3.18B, p. 49). Idiosomatic sclerotization, when present, often is confined to the region of the podosoma. Ventral shields are absent or obscure in larval Astigmatina and greatly restricted in other suborders. The urstigmata seen earlier in acariform prelarvae persist in the larval stage and are strongly produced in some groups (Fig. 3.18B). Some larvae tend to be weak, sluggish, nonfeeding forms (many free-living Mesostigmata, for example), while others may be voracious predators (as in the family CHEYLETIDAE) or aggressive parasites (as in the family TROMBICULIDAE). Unless they have been reared from known adults, lack of development or differentiation of many key structures in larval forms often makes their identification in collections difficult, except perhaps to higher categories. Exceptions may be seen in the Prostigmata and Ixodida, where key adult structures often are apparent throughout ontogeny. Members of parasitengonine families (Prostigmata) are routinely identified through larval rather than adult characteristics (Brennan and Jones 1959; Vercammen-Grandjean and Langston 1971a, b; Welbourn 1983; Southcott 1999).

The larva usually completes its development with little or no change in form other than that caused by engorgement. An exception is found in the larvae of sea snake parasites of the genus *Vatacarus* (TROMBICULIDAE), which display an unusual increase in overall size and the formation of new external structures. This phenomenon is known as **neosomy** and results (at least in part) from the secretion of new cuticle during the larval stasis (Nadchatram and Radovsky 1971; Audy, Radovsky, and Vercammen-Grandjean 1972). Suppression of a free larval instar is characteristic in some prostigmatic families such as the PODAPOLIPIDAE (p. 318), but it may occur in other groups only during periods of unfavorable conditions or when food quality is low. For example, when food quality is optimal, mating of sexually mature males and females of the fungivorous mite *Siteroptes cerealium* Kirchner (SITEROPTIDAE) results in the production of free, active larvae that molt directly into males and females. During the dry suboptimal summer months, however, the life cycle is abbreviated, and mated females produce no free larval forms. Females and males appear following physogastric development within the fertilized female (Rack 1972a).

Although acarine larvae typically have three pairs of legs, Coineau (1973a) noted distinct vestiges of legs IV in the larva of an African opilioacarid (Opilioacarida), a character also present in Australian Opilioacarida (Walter and Proctor 1999). Although vestiges of legs IV have been observed in embryological studies (Aeschlimann 1958), their appearance in the larval instar is unusual. However, the suppression of legs III and IV is characteristic of all stages of the

Eriophyoidea (gall and rust mites) and of many PODAPOLIPIDAE (Husband 1984a; Lindquist 1996b).

Nymph

With the exception of hard ticks (IXODIDAE) and certain higher Prostigmata, a single free nymphal stage is uncommon in the Acari. Two or three nymphal instars usually appear between the larval and adult stages, with as many as eight occurring in some ARGASIDAE (Ixodida) (Sonenshine 1991). Nymphs are generally octopod and undergo a progressive differentiation of shields with each molt until the adult stage. Setae of the idiosoma and appendages may be added with each nymphal molt. These interstadial changes often provide useful characters for taxonomic diagnoses. Primordial genital structures appear in the nymphal stadia of some taxa (Acariformes, Holothyrida) but not in others (Mesostigmata). Rarely, ontogenetic structures may regress in the final molt, as, for example, the peritremes of adult ZERCONIDAE (Mesostigmata), which are reduced compared to those of the preceding deutonymph.

Active protonymphal, deutonymphal, and tritonymphal stages occur in the Holothyrida, Opilioacarida, and Oribatida and in many Endeostigmata, Prostigmata, and Astigmatina. A maximum of two nymphal stadia occur in Mesostigmata (the protonymph and deutonymph), while in some Prostigmata (e.g., most Heterostigmatina), there may be no free nymphal stage. Only one active nymphal instar, the deutonymph, occurs in certain Prostigmata, with the proto- and tritonymphal stages passing as calyptostatic forms within the skin of the preceding stages (Johnston and Wacker 1967). Development of all the nymphal stadia within the larval skin is known to occur in certain TROMBICULIDAE (Audy, Radovsky, and Vercammen-Grandjean 1972).

Protonymph

The first nymph, or protonymph, usually is a free, active instar that may or may not feed but is typically adapted to a substrate that is similar or identical to that of later instars. Radovsky, Jones, and Phillips (1971) described an exceptional case of independent protonymphal adaptation in three species of *Radfordiella* (MACRONYSSIDAE), which feed in the mouth tissues of phyllostomid bats but only as protonymphs. The protonymphs (and tritonymphs) of the family PTERYGOSOMATIDAE, as well as those of most parasitengonine Prostigmata, are pharate forms that develop within the skin of the preceding instar (Johnston and Wacker 1967; Newell 1971b). Thus, while only one active nymphal instar occurs in these groups, there are two quiescent, sequestered calyptostases: one between the larva and active deutonymphal instar (nymphochrysalis) and the other between the deutonymph and adult (teleiochrysalis).

Deutonymph

The second nymphal stage, or deutonymph, often assumes the general nonsexual characteristics of the adult, differing from it only in size and sclerotization pattern. Some deutonymphs, however, are exceptional in being completely unlike the preceding and succeeding nymphal instars, both in morphology and behavior. The appearance of a facultative, heteromorphic deutonymph, often referred to as a **hypopus**, in the oribatid cohort Astigmatina (Fig. 5.2) in a given generation or population may be triggered by environmental, trophic, or genetic factors (Knülle 1987, 1991, 1995; Corente and Knülle 1996). The hypopus has been completely suppressed in some taxa (e.g., the acarid genus *Tyrophagus*, all Psoroptida). Hypopodes lack mouthparts, have a closed gut, are highly resistant to environmental stresses, and commonly have ventral suckers or claspers with which they secure attachment to passing animals (Michael 1901; Fain 1971b). The relationship generally is restricted to phoretic dispersal in that these nymphs lack functional mouthparts, but those of the family HEMISARCOPTIDAE are able to absorb nutrients derived from reflexed hemolymph secreted by their lady beetle hosts, possibly through the action of ventral sucker plate elements (Houck 1994). Heteromorphic deutonymphs of certain GLYCYPHAGIDAE sequester themselves in the subdermal tissues of their animal hosts (Fig. 5.2A) and apparently are capable of absorbing nutrients from host tissues (Fain 1965c, 1967b, 1969a; Fain et al. 1973; Lukoschus, Fain, and Driessen 1972). Inert deutonymphs lack suckers or claspers and largely rely on air currents for dispersal (see chapter 6).

Some parasitic Mesostigmata may feed during the deutonymphal stage (members of the LAELAPIDAE and DERMANYSSIDAE are examples), but deutonymphs are nonfeeding in more highly derived parasitic taxa such as the MACRONYSSIDAE and HALARACHNIDAE (Evans and Till 1965; Radovsky 1994). Two distinct deutonymphal morphotypes, one of which is phoretic, may be produced in some Uropodina and Sejida under suboptimal environmental conditions (Evans and Till 1979; Athias-Binche 1984, 1994; Athias-Binche and Evans 1981; Walter and Proctor 1999). Phoretic deutonymphs of the mesostigmatic cohort Uropodina usually do not feed while attached to their carriers, but there is at least one exception in the UROPODIDAE where all of the immature stages, including the deutonymph, are parasitic on ants (Krantz, Gomez, and Gonzalez 2007). Omission of the deutonymphal stage may occur in male parasitic mites, accelerating their developmental time and allowing them to appear before female ecdysis (Evans 1961; Treat 1975). Some authors consider the homology of the second nymphal stage across the Acari to be problematic and may use different terms (e.g., deutonymph vs. deuteronymph) for this stage in different acarine orders.

Tritonymph

Where it occurs, the tritonymph usually is an active instar (and sometimes a phoretic stage), but it is absent in the Mesostigmata and absent or represented by a pharate calyptostatic stage (teleiochrysalis) in many Prostigmata. In those

Astigmatina where hypopody occurs, the second homeomorphic nymph is considered by some authors to be the tritonymph, since it is the third postlarval preimaginal instar. When the tritonymphal stage is not expressed, the final molt to the adult occurs at the conclusion of the second nymphal stage.

Molting during the adult stage is rare but has been observed in some groups. Michener (1946), Imamura (1952), and Furumizo and Wharton (1975) have reported cases of postimaginal molting in species of TROMBIDIIDAE and ARRENURIDAE (Prostigmata) and in PYROGLYPHIDAE (Astigmatina). Adult molting is also known to occur in at least three genera of the parasitiform family OPILIOACARIDAE (Coineau and Legendre 1975; Klompen 2000a).

Paedomorphosis, Progenesis, and Neoteny in Acarine Ontogeny

With the exception of those mites with postimaginal molting, acarines have determinate growth and cease development when they reach the sexually mature adult stage. In most mites, the adult is easily recognized by the appearance of fully formed genital regions (often with distinctive shields and chaetomes) and sometimes by the appearance of an ovipositor, spermatophoric organs, or secondary sperm transfer systems. When the adult resembles an earlier ontogenetic stage, however, the concept of **paedomorphosis** often is invoked. Gould (1977) suggested that adults resembling immature stages of related organisms may arise as a consequence of either of two ecologically distinct selective regimes. In species exploiting ephemeral resources, accelerated development (**progenesis**) may result in the truncation of developmental sequences or the suppression of characters that are normally added ontogenetically. For example, development in *Siteroptes cerealium* is faster in offspring produced physogastrically, and they have fewer larval setae than those that develop from eggs laid outside the mother (Rack 1972a). In contrast, slowly developing species competing for stable resources (e.g., palaeostomatidid Oribatida) may carry immature characters into the adult stage (**neoteny**), as seen in the larviform axolotls of some salamanders. Other than the use of the term *larviform* in describing hexapod adults—for example, PODAPOLIPIDAE, CHEYLETIDAE (*Metacheyletia*), or nymphs (DEMODICIDAE)—the concept of paedomorphosis has been little noted in acarology. Chant (1993), however, demonstrated that the setae present in adult PHYTOSEIIDAE are primarily those that first appear in the larval stage, and setae added in later stages of related mites are those that tend to be absent. As a result of this "serial progenesis," the setation of adult phytoseiid mites resembles that of the larvae or nymphs of the closely related but generally more setose BLATTISOCIIDAE. More important, Chant's documentation of this pattern of setal suppression in the PHYTOSEIIDAE and related Gamasina has allowed the relatively unambiguous determination of setal homologies, even in highly deficient species, and may well apply to hypotrichous forms in all gamasine groups.

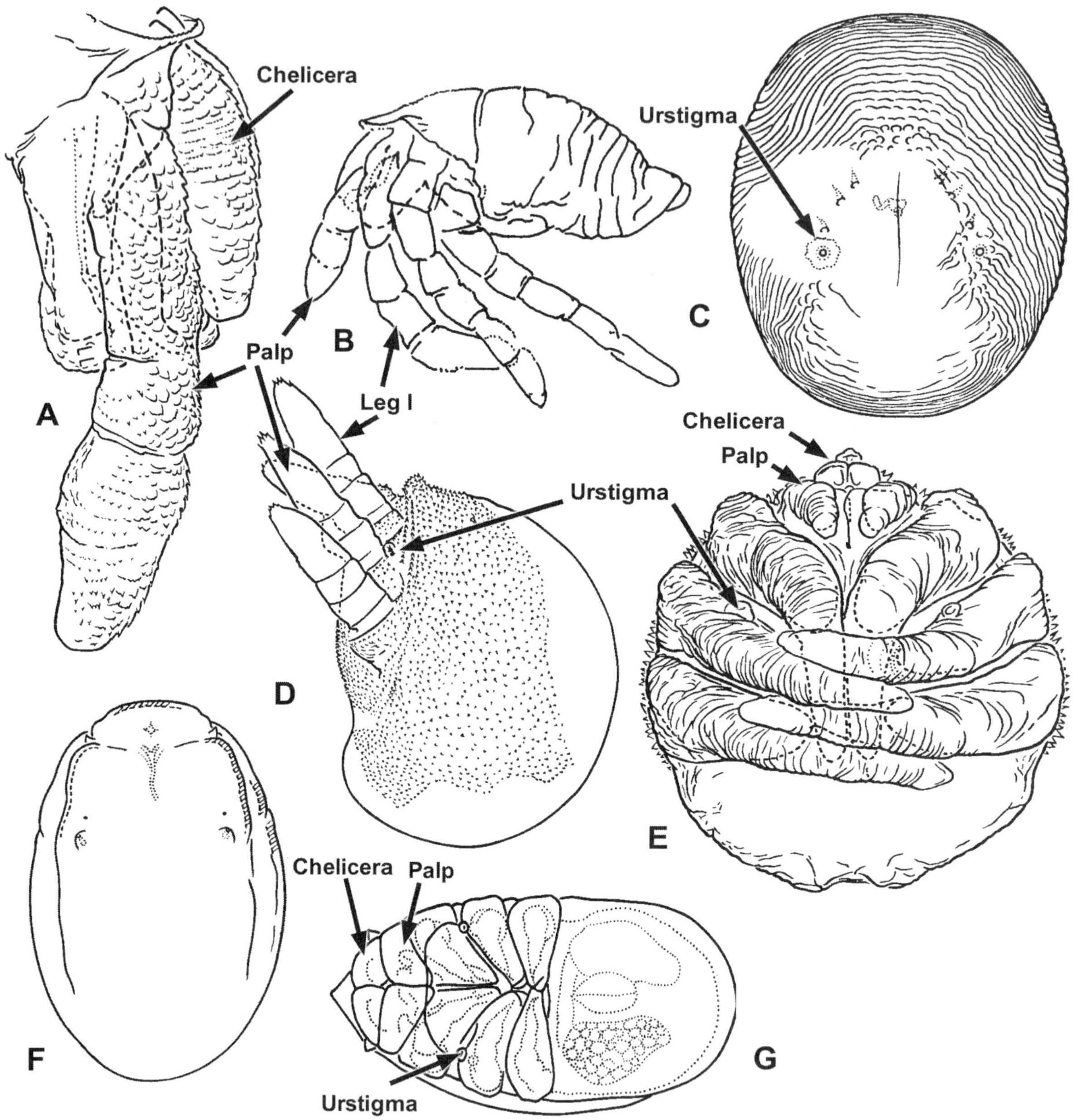

Fig. 5.1. Acarine prelarvae (from Coineau 1973a, 1974a, b; Van der Hammen 1976). **A**, *Phalangiacarus brosseti* Coineau and Van der Hammen (Opilioacarida, OPILIOACARIDAE), gnathosoma; **B**, *P. brosseti*, lateral aspect; **C**, *Balaustium florale* (Grandjean) (Prostigmata, ERYTHRAEIDAE), venter; **D**, *Anystis* sp. (Prostigmata, ANYSTIDAE), lateral aspect; **E**, *Microcaeculus steineri delamarei* Coineau (Prostigmata, CAECULIDAE), venter; **F**, *Pilogalumna allifera* (Oudemans) (Oribatida, GALUMNIDAE), venter; **G**, *Trhypochthonius tectorum* (Berlese) (Oribatida, TRHYPOCHTHONIIDAE), venter.

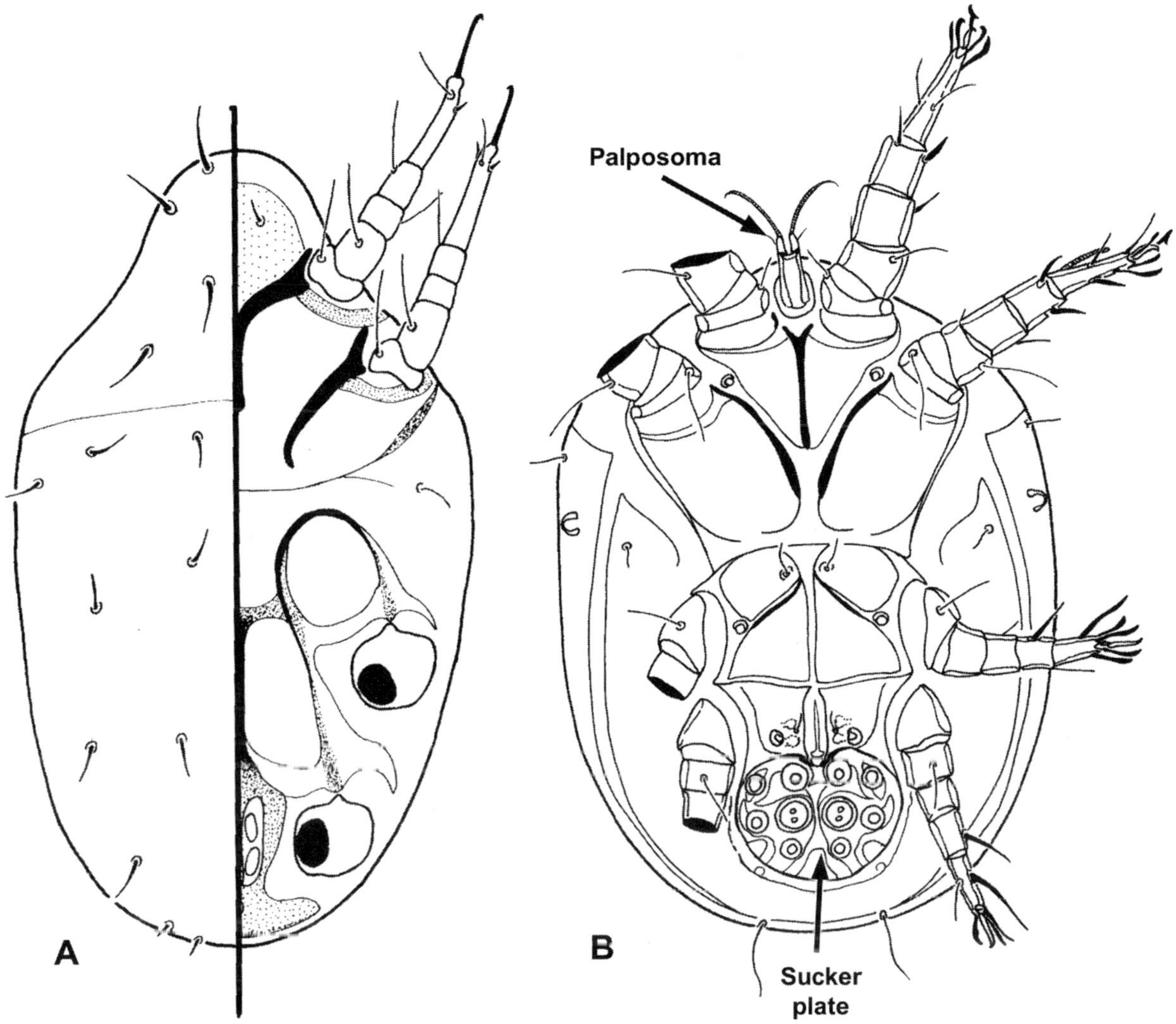

Fig. 5.2. Astigmatine deutonymphal types. **A**, *Rodentopus sciuri* Fain (GLYCYPHAGIDAE), endofollicular type (after Fain 1965c); **B**, *Sancassania* sp. (ACARIDAE), dispersive type.

CHAPTER SIX

G.W. KRANTZ

Habits and Habitats

The remarkable diversity in acarine morphology is reflected in the variety of ecological and behavioral patterns that mites have adopted. Because morphological specialization often is paralleled by specialization in choice of niche, an understanding of the ecology and behavior of mites is essential to understanding their classification.

Acarines may be free living or symbiotic, terrestrial or aquatic, arboreal or nidicolous. Most free-living mites occupy niches that are primarily or secondarily soil- and litter-related, a likely reflection of their primitive origins. Radiation of edaphic mites into terrestrial plant and animal systems and into aquatic and marine habitats throughout most of the world has left few microhabitats without an acarine fauna and has led to what may best be described as an "evolutionary revolution" in the subclass. The following discussion and cited examples illustrate the complexity of that revolution.

Free-Living Mites

The vast majority of known acarines are free-living forms that have adopted a wide spectrum of lifestyles. The group includes predaceous mites of infinite variety, herbivores that consume tissues or fluids of live plants, saprophages that use dead plant or animal tissue, and microbivores that feed on fungi, yeasts, or bacteria. Most mites may be categorized on the basis of trophic preference, but these categories are not mutually exclusive. Crossovers often occur, with predators feeding on live plants or plant products or dead animals, saprophages feeding on live microorganisms or attacking disabled or weakened animals, and fungivores or algivores accidentally or intentionally ingesting nematodes and bacteria (Walter 1987c; Luxton 1991; Walter and Proctor 1999). Furthermore, free-living mites may become parasitic or phoretic on other animals during the course of their development (Evans and Till 1966; Athias-Binche 1994) and therefore also could be considered symbionts. Of special interest in this regard are the terrestrial and aquatic members of the prostigmatic cohort Parasitengonina, which are typically parasitic on insects or vertebrates as larvae but revert to predation in their active postlarval stages (Welbourn 1983). As with trophic preferences, habitat types also may be obscured by crossovers—for example, when litter-inhabiting mites move onto plants or into human-made storage facilities that intersect their habitat (see below). With these limitations in mind, free-living mites may be categorized in terms of their proven or perceived primary habitats and trophic associations.

Soil Fauna

Microscopic examination of a handful of soil and overlying detritus from virtually any field, forest, or pasture is likely to reveal a surprising number and variety of mite species (Wallwork 1976; Walter and Proctor 1999). Greater concentrations of mites are found in the underlying **rhizosphere**, where a rich milieu of microflora and fauna living on and around plant roots and rhizomes provides a concentrated food resource for predators, fungivores, herbivores, and microbivores. Smaller, surface-litter forms may routinely migrate into underlying soil layers (Coineau 1974a; Krantz 1986), but the deeper strata (i.e., > 10–200 cm) often shelter unique cryptozoic mite faunas that never appear in the humus layer (Price and Benham 1976; Kethley 1990). While mite densities of 50,000–250,000 or more have been reported from a square meter of surface and subsurface soil (Peterson 1982), it is likely that improved extraction methods will show that the actual number of soil acarines far exceed present estimates. Wallwork (1959) employed an extraction method using heat separation (see chapter 7) and found mite populations in a square meter of relatively rich hemlock mor exceeding 400,000 during the winter months. Using a flotation sampling technique, André, Noti, and Lebrun (1994) estimated that the density of microarthropods in a nutritionally poor coastal sand-dune habitat exceeded 1,000,000 individuals per square meter, which suggests that wider use of flotation methods for soil sampling, such as the one described by André and Noti (1993), may necessitate a significant revision upward of earlier density estimates for soil acarines. A vast and generally distinctive acarine fauna may also be found in the suspended soil and humus layers that cover tree limbs in temperate and tropical forest canopies (Walter and Behan-Pelletier 1999).

Because of the inherent difficulties in definition occasioned by trophic crossovers, Walter and Proctor (1999) invoked the functional group concept of Cummins (1974) to define and describe clusters of edaphic, free-living species that include mites as members. A refinement of Root's (1967) guild community concept (also see Simberloff and Dayan 1991), the functional group is defined primarily in terms of similarity in size and nature of the resource used and only secondarily in terms of the taxonomic provenance of group members. A functional definition seems especially appropriate for nonpredatory soil and litter organisms, which often use a range of organic resources without regard for maintaining the trophic roles (e.g., saprophagy, phytophagy, and/or mycophagy) generally assigned to them. Some of the mites in these functional groups may even engage in facultative predation (Walter, Hunt, and Elliot 1988; Norton, Graham, and Alberti 1996), but truly predatory soil and litter species tend to retain their trophic identity even when alternate resources are available. They probably have minimal ecological impact on the larger soil arachnids and insects that share their predatory functional group because the latter tend to use prey resources of larger size.

Fungivores, Microbivores, and Detritivores

Several acarine groups use fungi either directly as a food source or indirectly as source habitats for mycetophilous prey (OConnor 1984a); however, ingesting and processing of particulate food require a combination of morphological and physiological attributes that is limiting for many of these groups. There is only one record of fungivory in the Opilioacarida (Walter and Proctor 1998), and relatively few are known in other parasitiform groups. These are mostly restricted to species of the early derivative sejoid genus *Asternolaelaps* (Athias-Henriot 1972; Walter and Proctor 1998) and to some of the higher Uropodina, including *Uropoda orbicularis* (Müller) and *Uroobovella marginata* (C. L. Koch). In addition, a species of the gamasine genus *Ameroseius* (AMEROSEIIDAE) was observed to feed on fungal mycelia (B. M. OConnor, pers. comm.), and *Proctolaelaps hypudaei* (Oudemans) (ASCIDAE) has been observed in fungal cultures and may feed on fungi in addition to preying on other mites (Mathys and Tencalla 1959; Muma 1975). However, the great preponderance of known grazers and browsers is found among the acariform Oribatida (Norton 1985; Walter and Proctor 1999), although several families of non-oribatid Sarcoptiformes (Endeostigmata) have been found to ingest solid fungal tissue (Walter 1988b). Like other Oribatida, many astigmatine taxa feed preferentially or incidentally on fungi (OConnor 1984a), but relatively few of them are represented in soil habitats. The acarids *Tyrophagus palmarum* Oudemans and *Schwiebea* spp. are examples of such species (Hughes 1976; Fain and Wauthy 1979).

Based on earlier work by Schuster (1956), Luxton (1972) recognized three major feeding categories among the Oribatida observed in soil and litter in a Danish beech forest. Macrophytophages were identified as feeding on decaying higher plant materials, while oribatids that consumed fungi, yeasts, bacteria, and algae were considered to be microphytophages. Mites that were thought to combine the feeding attributes of both macro- and microphytophages were classified as panphytophages. The utility of Luxton's categories led to their being incorporated in later works on oribatid ecology. However, subsequent observations on feeding by soil-inhabiting sarcoptiform mites have indicated that food choice is often more a matter of size than substance. Thus, oribatids that have been identified as primary macrophytophages may also be inadvertent predators, swallowing nematodes and protozoans along with plant material (Walter and Proctor 1999). Similarly, primary fungivores may ingest bacteria and other microbes that live in the fungal mat food resource. In short, most oribatid mites in soil and litter share the functional feeding characteristic of ingesting particulate matter, with little apparent regard for the provenance of that matter. Particulate feeders typically have massive chelate chelicerae that are well adapted for masticating bits of organic detritus (**detritivory**) or decaying plant or animal material (**saprophagy**) and for biting off fungal spores and hyphal strands (**fungivory** or **mycophagy**), pieces of algal filaments (**algivory**), or mosses (**cryptogamy**). Particulate feeding is considered ancestral in the Oribatida, an ancient taxon whose phylogenetic roots can be traced back to the late Carboniferous period (Norton 1985; Labandeira, Phillips, and Norton 1997) and beyond (also see chapters 2 and 15). It is likely, then, that the many morphological and behavioral adaptations to predation and to plant and animal parasitism seen in modern sarcoptiform Acari had their origins in the basic "bite and swallow" strategy of ancestral particle feeders.

Saprophagy and mycophagy occur widely in the Acari either as a way of life or as a facultative or sporadic phenomenon. Saprophagous mites may be found in highly diverse habitats, but most are found in soil and litter, where they feed on dead and dying plant or animal tissues. Saprophages may also graze on fungi and may incidentally or accidentally ingest bacteria, algae, or lichens during the feeding process. They are especially common in highly organic surface litters and underlying decomposition layers, but they may also be found grazing on decaying root and stem tissues in the deeper soil strata of the plant rhizosphere. Litter- and soil-inhabiting Oribatida are mostly known as particle-feeding saprophages and mycophages (Fig. 6.2). According to Norton (1985), saprophagy may have evolved in the Oribatida as a consequence of a taste for subsurface microbes, a predilection that led to the incidental ingestion of overlying detritus as well. Such indiscriminate particulate feeding also may lead to incidental ingestion of fungal spores and/or hyphae, although many oribatid and prostigmatic soil mites feed preferentially on fungi (Luxton 1972; Wallwork 1983;

Walter and Proctor 1999). In this connection, it should be noted that there is a positive correlation between the distribution of some oribatid species and the availability of fungal substrates (Klironomos and Kendrick 1995). The intimacy of these associations may play a key role in fungal distribution. For example, dispersal of ectomycorrhizal fungal spores may depend in large part on fungivorous Oribatida, which not only carry spores on their bodies as they move through the rhizosphere but also may pass viable fungal propagules in their feces (Behan-Pelletier and Hill 1978; Lilleskov and Bruns 2000).

The reduction of detritus by saprophagous oribatids has been demonstrated as a significant factor in comminuting forest litter for eventual breakdown by fungi and microorganisms, and it may play a limited role in direct (ingestion) and indirect nutrient recycling (Crossley 1977; McBrayer 1977). Examples of oribatid saprophages are legion, reflecting the fact that oribatids constitute the largest assemblage of edaphic acarines inhabiting forest soils and litter habitats (Hartenstein 1962a; Walter and Behan-Pelletier 1999). Decaying surface and rhizospheric plant and animal litter provide browsing niches for the great preponderance of soil Oribatida, including representatives of the superfamilies Crotonioidea, Phthiracaroidea, Oribatelloidea, Achipterioidea, Ceratozetoidea, and others (Luxton 1972; also see chapter 15). Decomposing outer root sheaths are a favored feeding site for *Rostrozetes flavus* Woodring (HAPLOZETIDAE) (Woodring 1965) and for many other oribatid species as well. *Perlohmannia dissimilis* Hewitt (PERLOHMANNIIDAE), a mixonomate oribatid, was reported to damage intact roots of potatoes, strawberries, and tulips (Evans, Sheals, and MacFarlane 1961).

While most oribatid saprophages graze on decaying substrate surfaces, others may excavate decaying plant tissues to reap the benefits of underlying food resources. For example, members of the superfamily Phthiracaroidea burrow into decomposing woody twigs and vascular leaf tissue, where they may consume the microbial or fungal organisms associated with decomposition, or they may feed on the vascular tissue itself. Nymphs of a *Hermannia* species (HERMANNIIDAE) were observed to excavate chambers in the bark lenticels of decaying yellow birch in Michigan, USA, producing "corky" fecal pellets that suggested the lenticel tissue itself was a major food component (Wallwork 1976). *Steganacarus magnus* (Nicolet) (PHTHIRACARIDAE) also burrows in decaying yellow birch, but it is found primarily in the heartwood, where it produces a complex network of burrows.

Unlike particulate feeders, mites that derive liquid nutrients from their soil and litter substrates must pierce the food surface to obtain nutrients. With few exceptions, nonpredatory liquid feeders in the edaphon (and other habitats) are members of the trombidiform suborder Prostigmata. They represent a diverse group of taxa that possess needlelike or attenuate styliform chelicerae better suited to piercing fungal spores and sucking up their contents than to ingesting the spore itself. Prostigmatic liquid feeders commonly encountered in the edaphon are typically mycophagous and include representatives of the eupodidid families EUPODIDAE and TYDEIDAE, and the heterostigmatine families TARSONEMIDAE, SCUTACARIDAE, and PYGMEPHORIDAE (Lindquist 1986; Kethley 1990; Walter and Proctor 1999). Many heterostigmatine fungivores display varying degrees of specificity in their choice of substrate, with some species of *Tarsonemus* apparently preferring *Alternaria* and *Penicillium* to other fungi (Karl 1965b; Gurney and Hussey 1967; Hussey and Gurney 1967). Members of the genus *Siteroptes* (SITEROPTIDAE) have been observed to feed only on particular fungi and to ignore others, suggesting a high level of feeding specificity in this genus (Smiley and Moser 1976; OConnor 1984a). Feeding specificity may be expressed in the siteroptid genera *Trochometridium* and *Siteroptes* to the point that the mites actually carry the preferred fungi in special **sporothecae** located between the bases of legs III–IV or behind coxisternals IV (Lindquist 1985b; Ebermann and Hall 2003). Sporothecae have also been described from the TARSONEMIDAE (Moser 1985; Lindquist 1985b) and more recently from two species of the scutacarid genus *Imparipes* (Ebermann and Hall 2003).

Algivory and Cryptogamy

Consumption of algae, lichens, and mosses may occur either by accident or design among various sarcoptiform assemblages that generally feed on microbes and detritus in soil. Although it cannot be classified as a true soil form, *Alaskozetes antarcticus* (Michael) (AMERONOTHRIDAE), a large and extremely common antarctic littoral oribatid species, grazes on algae and lichens in ice-free patches on the thin soil surface. Cryptogams are the dominant surface plant life on the otherwise largely barren landscape of Antarctica, and they serve to attract a substantial mite fauna in addition to *A. antarcticus* (Gressitt and Shoup 1967; Goddard 1982; Pugh 1993; Walter and Proctor 1999). Some of these species may be using cryptogamic patches primarily as a means of shelter, but others are reported to graze routinely on these plants. Included here are *Halozetes belgicae* (Michael) and other ameronothroid species, and the tiny (< 300 μm), globose, primitive endeostigmatic mite *Nanorchestes antarcticus* Strandtmann (NANORCHESTIDAE). Schuster and Schuster (1977) observed *N. amphibius* Topsent and Trouessart feeding on algae in the laboratory, confirming earlier suppositions that nanorchestids feed on microflora (Ryke 1966; Janetschek 1967) and lending support to the possibility that algivory may be the preferred feeding strategy for all nanorchestids. The bladelike labral process seen in *Nanorchestes* species, considered a piercing organ in a related genus (Trägårdh 1909), was thought by Krantz and Lindquist (1979) to function in piercing individual algal cells. Algae and mosses may also serve as food substrates for some members of the family PENTHALODIDAE, a relatively early derivative eupodoid assemblage characterized by somewhat distorted but clearly chelate chelicerae.

Lichens and mosses on soil or rock surfaces in other areas of the world, including northern temperate European seashores, also provide food and shelter for a variety of mites, including representatives of the oribatid genera *Ameronothrus*, *Maudheimia*, *Edwardzetes*, and *Trimalaconothrus* (Gerson 1973; Gjelstrup and Søchting 1984; Pugh 1996a). The prostigmatic mite *Tydeus tilbrooki* Strandtmann was reported to feed on lichen apothecia (along with algae and fungi) in the South Orkney Islands (Goddard 1982).

Predators

Predaceous acarines occur at all levels in the soil-litter column (Evans, Sheals, and MacFarlane 1961; Price 1973), where they feed primarily on nematodes and small arthropods or their eggs. Although most litter predators are confined to the surface and underlying decomposition layers, some may also be found foraging for prey on the stems and foliage of plants that overlie their primary habitats (Aoki 1973; Muma 1975); others may move into carrion or animal or plant manure deposits on the soil surface (Krantz 1998c), and still others may invade grain storages to feed on granary mites and small insects (Hughes 1976; Murphy and Sardar 1991). The vagility of many litter predators suggests that some of them will likely be found in aerial litter accumulations of the largely unexplored tropical and temperate forest canopy habitats throughout the world (Paoletti et al. 1991), although canopy niches that have been investigated appear to support largely unique faunas (Walter and Behan-Pelletier 1999).

Predatory soil mites generally have chelate dentate or highly modified, needlelike chelicerae well suited for macerating or piercing their prey, and some have raptorial palpi or enlarged spinelike setae on legs I that serve to capture and hold the prey during feeding. Surface-inhabiting mesostigmatic predators often are well sclerotized, long legged, and fast moving, while those that occur in subsurface niches tend to be better adapted to an interstitial existence. Coineau (1974a) observed that the relative size of soil mesostigmatic mites often diminishes with increases in the depth at which they occur. Thus, relatively large predators such as veigaiids (VEIGAIIDAE) tend to be common in the litter-humus layers, the smaller more weakly sclerotized DIGAMASELLIDAE at the humus-soil interface, and the diminutive RHODACARIDAE in the deeper interstitial layers of the soil itself. Commonly encountered mesostigmatic predators in temperate soil and litter habitats include representatives of the families PARASITIDAE, MACROCHELIDAE, DIGAMASELLIDAE, LAELAPIDAE, ASCIDAE, PARHOLASPIDIDAE, and ZERCONIDAE, while tropical soils are especially rich in OLOGAMASIDAE, UROPODIDAE, and ASCIDAE.

Some soil and litter Mesostigmata are considered beneficial because they feed on arthropods and other invertebrates that affect humans and animals. For example, some species of the family MACROCHELIDAE have been used successfully as control agents against synanthropic fly species that breed in dung (Axtell 1963; Rodriguez, Singh, and Taylor 1970; Krantz 1983a, 1998c), and others have shown promise as effective predators in laboratory observations (Halliday and Holm 1987). *Uroobovella marginata* (C. L. Koch) (UROPODIDAE) is a common predator of early instar house fly larvae (O'Donnell and Axtell 1965), as is *Parasitus coleoptratorum* (L.) (PARASITIDAE) (Wernz and Krantz 1976). *P. quisquiliarum* Canestrini is considered an important predator of the garden symphylan *Scutigerella immaculata* (Newport) in western Oregon, USA (Berry 1973). *Gaeolaelaps aculeifer* (Canestrini) (LAELAPIDAE), a common soil predator in lily bulb plantings, effectively suppressed populations of the bulb mite *Rhizoglyphus robini* Claparède (ACARIDAE), an important pest of lilies, and greatly reduced populations of the plant nematode *Tylenchorhynchus dubius* (Butchli) in laboratory tests (Sharma 1971; Lesna, Sabelis, and Conijn 1996). Ascid predators of plant nematodes in the rhizosphere (e.g., *Lasioseius scapulatus* Kennett) may have a significant impact on nematode populations, but field evaluations of their effectiveness are yet to be conducted (Imbriani and Mankau 1983; Gerson and Smiley 1990).

Most of the known soil-inhabiting prostigmatic mite families are exclusively predatory (Kethley 1990), but members of some of these families feed at other trophic levels. Among these are the TARSONEMIDAE, which are primarily fungivorous, phytophagous, or parasitic (Jeppson, Keifer, and Baker 1975c; Lindquist 1986), and the TYDEIDAE, which feed mostly on fungi or scale-produced honeydew (Brickhill 1958; Mendel and Gerson 1982). A species of the predaceous family CAECULIDAE and members of the endeostigmatic genus *Alicorhagia* (ALICORHAGIIDAE), known predators of nematodes, may also feed on fungi (Crossley and Merchant 1971; Walter 1988b). Commonly encountered prostigmatic predators in shallow soil and overlying litter layers include the families EUPODIDAE, RHAGIDIIDAE, BDELLIDAE, LABIDOSTOMATIDAE, STIGMAEIDAE, and CHEYLETIDAE. Deeper mineral soils (> 10 cm) often harbor a variety of predatory prostigmatic species (Price 1973, 1975; Kethley 1990), including representatives of the BDELLIDAE and RHAGIDIIDAE.

Predation is also known to occur among soil-inhabiting Astigmatina (Oribatida). Species of the genus *Tyrophagus* (ACARIDAE), an assemblage that earlier had been considered as basically fungivorous or graminivorous (Sinha and Mills 1968; OConnor 1982b), have since been observed to feed voraciously on nematodes, sometimes in preference to yeast or algal alternatives (Walter et al. 1986). *T. putrescentiae* (Schrank) also has been found to consume the eggs of the southern corn rootworm in corn and peanut fields in North Carolina, USA (Brust and House 1988). The bulb mite *Rhizoglyphus echinopus* (Fumouze and Robin) feeds on plant parasitic nematodes and may play a significant role in regulating their populations (Sturhan and Hampel 1977; Gerson and Smiley 1990). Live prey also serves as food for some species of Oribatida, especially members of the genus *Pergalumna* (GALUMNIDAE) (Rockett 1980). *P. omniphagous* Rockett and Woodring supplements its normally saprophytic diet with nematodes (Rockett and Woodring 1966), and *Aquanothrus* sp. (AMERONOTHRIDAE) feeds on live rotifers (Norton, Graham, and Alberti 1996). Other records of predation by Oribatida are cited by Luxton (1972).

Phytophages

Relatively few soil-inhabiting acariform mite groups feed on live plant tissues. Most of them feed on root tissue, corms, or bulbs in the rhizosphere, and some may be responsible for economic injury to ornamentals and vegetables. *Rhizoglyphus robini* Claparède, generally considered a detritivore, may contribute to bulb rot in Easter lilies grown in southwestern Oregon, USA, although evidence points to their being secondary invaders of mechanically damaged or pathogen-invaded bulbs (Ascerno, Pfleger, and Wilkins 1981). Other species of *Rhizoglyphus* feed on tulip, onion, and hyacinth bulbs and on *Gladiolus* corms and potatoes. At least one oribatid mite species, *Perlohmannia dissimilis* Hewitt (PERLOHMANNIIDAE), is known to feed on root tissue of a variety of vegetable and ornamental plants (Evans, Sheals, and MacFarlane 1961).

While acariform soil phytophages have chelate-dentate chelicerae suitable for grinding and macerating plant tissues, trombidiform plant feeders in soil habitats generally have short needlelike or long styliform chelicerae similar to those found in mycophagous trombidiform groups and use them to pierce individual underground or surface plant cells. Among these are some TARSONEMIDAE, such as *Steneotarsonemus laticeps* (Halbert), which feeds on amaryllis bulbs (Lindquist 1986), and certain members of the tetranychoid family TUCKERELLIDAE that are associated with grass or tree roots (Baker and Tuttle 1975b; Meyer and Ueckermann 1997; Walter and Proctor 1999). Tuckerellids typically are aboveground phytophages that feed on the aerial leaves or fruit of their plant hosts (Ochoa 1989a; Ochoa, Aguilar, and Vargas 1994a). Grass crowns and underlying rootlets appear to be the preferred habitat for the related tetranychoid families LINOTETRANIDAE and ALLOCHAETOPHORIDAE (Meyer and Ueckermann 1997).

Aerial Fauna

Predators

Predatory mites are commonly encountered on aerial portions of plants, where they feed on small insects or on phytophagous and mycophagous mites and their eggs (Wallace and Walters 1974; Jeppson, Keifer, and Baker 1975a; Santos and Laing 1985; Gerson 1985a; Gerson and Smiley 1990). Others may be found subcortically in the heartwood of trees attacked by bark and wood-boring beetles or in the fleshy or woody sporocarps that invade dead and dying trees (Pielou and Verma 1968; OConnor 1984a; Lindquist 1970a, 1995). As in soil-inhabiting mesostigmatic predators, the idiosomatic sclerotization of most phytophilous and mycetophilous Mesostigmata tends to be extensive, although the shields are often weakly tanned and poorly defined. Prostigmatic predators are numerous on most plants and often are brightly colored in hues of red, yellow, or green, often assuming the color of the prey species on which they feed. Most of these predators actively seek their prey on leaf and stem surfaces, but some phytophilous CHEYLETIDAE are ambush predators that lie in wait until the prey is close enough to capture (Muma 1975).

Because some of its member species are important in the control of spider mites (TETRANYCHIDAE) and eriophyoid mites in commercial orchard and vine crops, the mesostigmatic family PHYTOSEIIDAE is probably the most widely studied and best-known assemblage of phytophilous mite predators (McMurtry 1983; Croft 1990; McMurtry and Croft 1997). Many of these same species may feed and develop successfully on other foods, including insects and their eggs, and some may supplement their arthropod diet with plant juices, pollen, and/or honeydew (Kamburov 1971; Overmeer 1985; McMurtry and Croft 1997). Similarly, predatory members of the genus *Proctolaelaps* (ASCIDAE) may be facultatively pollenophagous or saprophagous on plants (Krantz and Lindquist 1979). The ascids *Asca citri* Hurlbutt and *A. tarsalis* De Leon, slow-moving species found in orchard citrus trees, are associated with scale insects (Muma 1975), but it is not clear that the mites actually prey on them. Other species of *Asca* collected on plants in the tropics are common in dead foliage on the plant or beneath it (Moutia 1958; De Leon 1967) and may be feeding primarily as saprophages rather than as predators.

Many trombidiform families have representative members that feed on plant-associated arthropods (Kethley 1990). Among these, the STIGMAEIDAE, TYDEIDAE, CHEYLETIDAE, ANYSTIDAE, and ERYTHRAEIDAE include species that have been shown to be effective predators of orchard pests (Laing and Knop 1983; Welbourn 1983). However, *Zetzellia mali* (Ewing), an important stigmaeid predator of spider mite and eriophyid eggs on apple trees (Santos 1976), also feeds on the eggs of predatory phytoseiid species when spider mite populations are low (Santos 1976; MacRae and Croft 1996). Members of the genus *Acaronemus* (TARSONEMIDAE) prey on the eggs of tenuipalpid and tetranychid mites (Smiley and Landwehr 1976), and species of the arboreal tarsonemid genus *Dendroptus* are suspected predators of eriophyids with which they associate (Beer 1963; Lindquist 1986). Eggs and postembryonic stages of larger insects may serve as food sources for some plant-inhabiting trombidiform predators. For example, adult female tarsonemid mites of the genus *Iponemus* feed as parasitoids on the eggs of ipine bark beetles in Europe and North America (Lindquist 1969b, 1983), and a species of *Balaustium* (ERY-

THRAEIDAE) contributes significantly to preadult mortality of the pierid butterfly *Colias alexandra* through predation on its eggs (Hayes 1985). Deutonymphs and adults of the erythraeid *Lasioerythraeus johnstoni* Welbourn and Young feed on nymphs of the tarnished plant bug (Young and Welbourn 1987), and other erythraeids are known to prey on aphids, scale insects, and the eggs of Lepidoptera.

Relatively few acariform mites are aerial predators. Among these are members of the astigmatine genera *Hemisarcoptes* (HEMISARCOPTIDAE) and *Winterschmidtia* (WINTERSCHMIDTIIDAE). Species of the latter genus often are associated with fungivorous beetles (OConnor 1984a) and may be predatory on the eggs of subcortical beetles (B. M. OConnor, pers. comm.). *Hemitarsonemus* species prey on diaspidid scale insects worldwide and are considered to be promising control agents for these pests (Laing and Knop 1983; Houck and OConnor 1990; Gerson and Smiley 1990).

Phytophages

Phytophagy, pollenophagy, and nectarivory occur sporadically among the Parasitiformes and the sarcoptiform Acariformes (Johnston and Bruce 1965; Del Fosse, Cromroy, and Habek 1975; Krantz and Lindquist 1979; Marshall, Reeves, and Norton 1987; Colwell 1995; Naskrecki and Colwell 1995; Seeman and Walter 1995; McMurtry and Croft 1997), but phytophagy is widespread enough among the trombidiform Acariformes to suggest that there was an early evolutionary commitment to plant feeding by several primitively predaceous and saprophagous trombidiform lineages. The genesis of this commitment may date from the Devonian period with the development of an association between early derivative nonrutellate Endeostigmata (OConnor 1984c) and an already burgeoning intertidal and riparian microflora (Hirst 1923; Kevan, Chaloner, and Savile 1975; Krantz and Lindquist 1979). The subsequent explosive rise and differentiation of angiosperm plants coincided with a period of radiation among trombidiform mite groups in the evolving plant habitat, resulting in a myriad of morphological, physiological, and behavioral adaptations for feeding and flourishing on these new resources. Today, over 4,000 trombidiform species representing over 300 genera are known to engage in obligate phytophagy (Amrine and Stasny 1994; Bolland, Gutierrez, and Flechtmann 1998), including some of the world's most important pests of commercial plant ecosystems.

Phytophagous trombidiform mites tend to be slow moving or sedentary, and most are weakly sclerotized. Many are conspicuously colored in red, yellow, or green hues, while others may appear translucent or opaque white. Like their edaphic plant-feeding counterparts, aerial trombidiform phytophages may feed directly on plant tissues (at least incidentally) or on subepidermal liquids derived from plant cells and associated tissues. Surface tissue injury may occur through feeding by *Penthaleus major* (Dugès) and *Halotydeus destructor* (Tucker) (PENTHALEIDAE), economically important eupodidid species that possess modified chelate chelicerae (Fig. 3.9F, p. 35) suitable for both piercing and rasping the leaf tissues of their many herbaceous hosts (Jeppson, Keifer, and Baker 1975a; Krantz and Lindquist 1979; Di Palma 1995). However, most trombidiform phytophages feed exclusively on the subepidermal liquid components of their host tissues, a strategy made possible by an earlier modification or loss of fixed cheliceral digits in their predaceous progenitors and development of the movable digits into piercing structures.

Phytophagous species in primarily nonphytophagous taxa such as the TYDEIDAE and STIGMAEIDAE tend to have relatively short, spearlike movable digits inserted in free or partially fused cheliceral bases (Fig. 3.9A, p. 35), similar to many predaceous or fungivorous species. The bryophagous members of the stigmaeid genus *Eustigmaeus* exemplify these forms. The extrusible styletlike movable digits of phytophagous TARSONEMIDAE, a diverse assemblage that also includes fungivores, algivores, predators, parasitoids, and parasites (Lindquist 1986), are contained in a gnathosomatic capsule composed of modified cheliceral bases and associated rostral elements (Nuzzaci et al. 2002). Phytophagous tarsonemids tend to be specific on particular vascular plant families or genera. For example, most *Steneotarsonemus* species are restricted to monocotyledonous plants, and several species apparently are specific to particular monocotyledonous hosts (Lindquist 1986). Instead of a gnathosomatic capsule, species in the exclusively phytophagous raphignathine lineage exemplified by the TETRANYCHIDAE and the TENUIPALPIDAE have a pair of elongate, extrusible cheliceral stylets inserted in a fused suprarostral pod or **stylophore** (Figs. 3.9G, H, p. 35) comprising the fused cheliceral bases. Cell contents or other subepidermal substances are sucked up via the channel created by close appression of these cheliceral elements (Krantz and Lindquist 1979, André and Remacle 1984; Lindquist 1985c). The cheliceral and auxiliary stylets of the highly host-specific eriophyoid families ERIOPHYIDAE, PHYTOPTIDAE, and DIPTILOMIOPIDAE are inserted in a knoblike **motivator** that lies over the rostrum (Figs. 3.9C, 13.18, pp. 35, 344) (Lindquist 1996c).

Visible symptoms of subepidermal feeding by phytophagous trombidiform mites include leaf stippling, spotting, bronzing, blistering, and curling and distortion of buds, flowers, and stems (Tomczyk and Kropczynska 1985; Lindquist 1986; Westphal and Manson 1996). Defoliation is common in plants heavily infested by tetranychids, while feeding by some eriophyoids results in the appearance of a variety of galls and other teratological abnormalities that usually are localized on the host but nevertheless may cause overall changes in growth patterns of the host plant (Keifer et al. 1982; Westphal and Manson 1996). Non-gall-forming eriophyoids include rust mites and russet mites that produce leaf discoloration and leaf vagrants that produce no obvious symptomatology (Amrine and Stasny 1994). It should be noted that eriophyoid mites transmit several important plant diseases, including wheat streak and ryegrass mosaic viruses, black currant reversion, and peach mosaic (Oldfield and Proeseler 1996). Additional information on tetranychoid and eriophyoid phytophages may be found in Helle and Sabelis (1985b); in Lindquist, Sabelis, and Bruin (1996); and in chapter 13 of this book.

Relationships between phytophagous mites and their predators often are influenced by chemical cues (**infochemicals**) that alert the potential prey to the presence of their enemies. These infochemicals may be produced directly by the predators in their exuviae, excreta, pheromones, or other products and are recognized by the prey species. They may also be generated indirectly as chemical signals released by alarmed, injured, or dead conspecifics in the prey population or as allelochemicals produced by other potential prey species in the immediate vicinity (Kats and Dill 1998; Grostal and Dicke 2000; Dicke and Grostal 2001).

Aerial Grazers

Along with acarine predators and phytophages, the aerial portions of plants are home to many mites that scavenge there for fungi and algae. Plant tissue that has been damaged by pathogenic fungi and subsequently colonized by other microorganisms is especially attractive to these mites, as are leaf surfaces on which homopteran-secreted honeydew produces dense fungal blooms (Walter and Proctor 1999). Decaying leaf and bud tissue also provides nutrients for an array of incidental nonpathogenic fungi with which mites are associated. For example, the cosmopolitan astigmatine mite *Czenspinskia transversostriata* (Oudemans) (WINTERSCHMIDTIIDAE) is common in the leaf axils and bud brackets of commercial hazelnut trees (*Corylus avellana*), where it apparently feeds on spores and hyphae of a *Cladosporium* fungus commonly associated with *Corylus* in western North America (G. W. Krantz, pers. obs.). In addition to their roles as phytophages and predators, some foliar and corticolous species of the family TYDEIDAE are thought to be fungivorous (Kazmierski 1998b), although evidence of fungivory in many tydeids is fragmentary and conflicting (André 1986). Species of several genera of TARSONEMIDAE are known fungivores in protected sites on leaves and fruit (Lindquist 1986). The fleshy, fruiting bodies of fungi that invade dead and dying plant tissue also serve as a resource for many acariform mites and as a niche for a variety of parasitiform predators and putative mycophages (OConnor 1984a; Lindquist 1995).

Although grazing on higher plants by parasitiform, trombidiform, and astigmatine mites is at most a sporadic phenomenon, it is a way of life for many species of nonastigmatine Oribatida, a group primarily associated with soil and litter habitats (André 1984a; Norton 1985; Walter and Behan-Pelletier 1999). Representatives of over 25 oribatid families are routinely found in moss, algae, and crustose and fruticose lichens on the bark of tree trunks and secondary stems. Arboreal oribatids often are characterized as having short clavate or globose prodorsal bothridial setae (see chapter 3), and many species have unusually heavy ceroteguments. Both of these modifications may be related to dealing with exposed habitats. Bothridial setae are considered anemotactic, and the heavy air currents on exposed leaf or bark surfaces would overstimulate elongate, ornamented bothridial sensilla of the type typically found in their edaphic counterparts (Norton and Palacios-Vargas 1982; Alberti, Moreno, and Kratzmann 1994). A heavy, waxy cerotegument probably aids in water conservation (Walter and Proctor 1999).

Acarodomatia

Predatory phytophilous species may patrol both leaf surfaces in their search for prey, but most foliar mites confine their daytime activities to the undersides of leaves, where they are better protected from the desiccating effects of sunlight and air currents. Plant mite distribution on leaf undersurfaces often is influenced by leaf architecture, with hairy leaves often supporting significantly greater numbers and kinds of foliar mites than the smooth leaves of co-occurring plant species (Walter 1992). Fine hairs along vein ribs and the tufts of hairs commonly found in the axils of intersecting leaf veins of woody trees and shrubs in temperate regions may provide additional shelter for mites. In place of tufts, many tropical and subtropical woody plants have leaf axillary pits or pouches protected by a triangle of tissue that covers the leaf angle. Lundström (1887) first observed the association of mites with the cottony axillary hair tufts in the undersurface vein axils of lime trees. He referred to these and to the leaf axillary pits as **acarodomatia**, comparing them to the larger, plant-produced myrmicodomatia that house certain tropical ants.

Predatory, fungivorous, and microbivorous mites compose over 90% of arthropods in plant domatia, and leaves with domatia have significantly more predators than those without domatia. These may include representatives of the families PHYTOSEIIDAE, CHEYLETIDAE, CUNAXIDAE, and STIGMAEIDAE, among others (Walter 1996; Walter and Proctor 1999). Where leaf domatia are present, phytoseiid mites use them rather than other portions of the leaf for oviposition and molting. Retention of wandering predatory mites on plant leaves with domatia may provide these leaves with a significant level of defense against attack by phytophagous mites (Walter and O'Dowd 1992; Grostal and O'Dowd 1994; Agwaral and Karban 1997). Similar results with tydeid mites feeding on leaf pathogenic fungi have been demonstrated in grapes (Norton et al. 2000).

Many soil and aerial acarines are subject to attack by a variety of bacterial, fungal, protozoan, and viral pathogens, some of which have been found to reduce population densities of economically important phytophagous species. Mite diseases, and the biotic and abiotic factors that may play a role in their activity, are reviewed by Van der Geest et al. (2000).

The Littoral, Marine, and Freshwater Fauna

Halbert (1920) determined that the intertidal zone of the western coast of Ireland comprises a series of distinct bands, each of which is dominated by particular algae or lichens. Evans, Sheals, and MacFarlane (1961) recognized these divisions and designated two additional zones—the tidal debris and estuarine—with each zone characterized by its own assemblage of mite predators and prey. Many other researchers, including Schuster (1957), Pugh (1993, 1995, 1996c), and Schatz (1998), have noted similar zonation at land-water interfaces in other regions of the world. According to Costa's (1974a) observations on the distribution of littoral-intertidal mesostigmatic fauna in Israel, the degree of habitat specificity in these zones may be substantial.

Invasion of the intertidal and other aquatic and semiaquatic zones by terrestrial arthropods is well documented among the Insecta (Cheng 1976; Resh and Solem 1996), but it is rare in most other terrestrial arthropod groups. Among the Arachnida, relatively few species of spiders (*Dolomedes*, some *Pardosa*, and the riparian orb web spiders of the genus *Tetragnatha*) are capable of walking and running on water surfaces, and only the diving bell spider, *Argyroneta aquatica*, can swim and walk underwater (Foelix 1982). However, in contrast to spiders, there have been at least 10 independent invasions of water habitats by different lineages of Acari (Walter and Proctor 1999). The infrequency of aquatic representatives in some of these lineages (e.g., the UROPODIDAE, ASCIDAE, and CAMISIIDAE) suggests that aquatic invasions by these taxa were relatively recent in geological time. Conversely, the prostigmatic subcohort Hydrachnidiae and superfamily Halacaroidea are highly diverse and include thousands of obligatorily aquatic and marine species that occupy a broad spectrum of freshwater and marine microhabitats. This diversity suggests early invasions of the aquatic milieu by their progenitors, followed by rapid radiation and speciation.

Members of several primarily terrestrial mite taxa occur in or near freshwater and marine habitats, and some of these have virtually abandoned their ancestral nonaquatic lifestyles. For example, the littoral-intertidal-estuarine milieu has been exploited by a variety of predatory Mesostigmata and Prostigmata that feed on arthropods and other invertebrates in accumulations of organic matter at the land-water interface. Members of the mesostigmatic families OLOGAMASIDAE, PARASITIDAE, HALOLAELAPIDAE, and LAELAPIDAE are common predators in these habitats (André 1934; Lee 1970; Costa 1974a; Hyatt 1980), as are representatives of the prostigmatic families BDELLIDAE, RHAGIDIIDAE, and ERYTHRAEIDAE (Evans, Sheals, and MacFarlane 1961; Womersley and Strandtmann 1963; Schuster 1979; Pugh and King 1985b; Ernst 1996).

Walter and Proctor (1999) listed 12 families of mites with close ties to the aquatic milieu, and these include representatives of all acarine orders except the Opilioacarida. Tree holes, leaf axils, and temporary pools are especially prone to invasion by subaquatic and aquatic Mesostigmata and Astigmatina, while more permanent impoundments are favored by other Oribatida and by Prostigmata. Mesostigmatic species include *Platyseius italicus* (Berlese) (ASCIDAE), an interesting subaquatic species that frequents sewage bed filters and uses a peritrematic type of plastron respiration to survive in this oxygen-poor habitat (Hinton 1971; also see chapter 3). Members of the astigmatine genera *Naiadacarus* (ACARIDAE), *Algophagus* (ALGOPHAGIDAE), and *Hormosianoetus* (HISTIOSTOMATIDAE) are tree hole and leaf axil specialists, feeding, respectively, on decaying leaves, the fungi associated with these leaves (although *A. pennsylvanicus* Fashing and Wiseman may also feed on particulate matter), and fine particulate matter filtered from the water column (Fashing 1994, 1998). A species of the mesostigmatic genus *Cheiroseius* (ASCIDAE) occurs as a transient in tree holes in southeast Queensland, Australia, but it feeds only occasionally on algophagids (*Lamingtonacarus* spp.). The water mite *Arrenurus kitchingi* Smith and Harvey (ARRENURIDAE), a generalist predator in the same habitat, likewise appears to have little interest in algophagids as a prey item. Unlike most algophagids, *Algophagopsis pneumatica* Fain and Johnston frequents algal mats on submerged rocks in stream riffles in California, USA (Fain and Johnston 1975), while an undescribed species of the same genus is found in the benthos of Crater Lake, Oregon, USA, in moss, algal mats (*Cladophora*), and epilithic crusts consisting of lichens, fungi, cyanobacteria (*Nostoc* sp.), and diatoms (Anderson and Krantz 1993). Fashing (1998) considers this mite to be a diatom feeder. Littoral species of the endeostigmatic genus *Nanorchestes* (NANORCHESTIDAE) use marine algae as food (Schuster and Schuster 1977; Schuster 1979), as do members of the marine intertidal astigmatine family HYADESIIDAE (OConnor 1982a).

Hydrophilous oribatid mites have adapted to a variety of freshwater and saltwater habitats, including ephemeral pools (Norton, Graham, and Alberti 1996), the benthos and seepages of streams and springs (Norton, Williams, et al. 1988; Norton, Behan-Pelletier, and Wang 1996; Schatz and Gerecke 1996), intertidal algae and lichens, and exposed littoral zones where heavy surf and storms may routinely cover the beaches and rocky surfaces on which many of these mites may normally be found (Schuster 1962a, 1979; Schulte, Schuster, and Schubart 1975; Pugh 1993; Schatz 1998). Littoral Oribatida often display behavioral rhythmicity, retreating from encroaching high tides and seeking shelter in rock crevices (Luxton 1992). Respiration during these periods of prolonged tidal submersion may be by means of cuticular plastron mechanisms (see chapter 3). Members of the brachypyline oribatid superfamily Hydrozetoidea (Grandjean 1948c) also use plastrons (Fig. 3.18A, p. 44) and commonly are found on submerged aquatic plants (Krantz and Baker 1982; Cromroy 1983). Immatures and adults of a species of *Hydrozetes* (HYDROZETIDAE) were collected in considerable numbers from within the aerated roots of *Cryptocoryne* sp., a popular aquarium plant (G. W. Krantz, pers. obs.).

A substantial and diverse continental and littoral invertebrate stygobiontic fauna inhabits subsurface soil and sand interstices at the subterranean groundwater interface (Delamare Deboutteville 1960; Botosaneanu 1986). Many crustaceans, amphipods, nematodes, isopods, and copepods are found in these cryptic habitats, along with a variety of mites that includes species of the mesostigmatic families OLOGAMASIDAE, RHODACARIDAE, and PACHYLAELAPIDAE (Willmann 1935; Krantz 1986; Lee 1986), representatives of several families of hydracarine water mites (Schwoerbel 1986a, b; Smith and Cook 1991), and freshwater interstitial members of a number of prostigmatic genera including *Hydrothrombium*, *Cerberothrombium*, and *Hygrothrombium* (TROMBIDIIDAE) (Schwoerbel 1986b). Not surprisingly, the marine mite family HALACARIDAE is a major component of the littoral interstitial fauna. Species representing at least a dozen halacarid genera in five subfamilies are found in stygobiontic habitats (Bartsch 1986a).

Aside from a few intertidal water mite groups that have adapted to a marine existence (the hygrobatoid genera *Pontarachna* and *Litarachna* are examples), only the HALACARIDAE have succeeded in colonizing the world's oceans and seas (Viets 1939, 1940b; Newell 1947b, 1984; Bartsch 1986a, 1989b; Green and MacQuitty 1987; Abé 1998). While freshwater mites have made relatively few inroads into the marine habitat, over 60 halacarid species representing nearly 20 genera and eight subfamilies occur commonly or sporadically in freshwater situations, including river and pond sediments, cave pools, hot springs, driven wells, and springs (Imamura 1968, 1970a; Petrova 1971; Mitchell 1974; Bartsch 1996). The two known species of the unusual monogeneric halacaroid family PEZIDAE (*Peza ops* Harvey and *P. daps* Harvey) are likewise confined to freshwater habitats (Harvey 1990a). Most halacarids are predatory or algivorous in a wide range of intertidal, subtidal, and abyssal habitats. They do not swim but rather crawl on their diverse substrates, perhaps reflecting an evolutionary response to the potential hazards faced by small swimming invertebrates in habitats constantly swept by tidal currents.

The Hydrachnidiae (Hydracarina, Hydrachnidia, or Hydrachnellae of authors) are typical of the broader lineages of the Parasitengonina in that the larvae are generally parasitic on arthropods and occasionally on vertebrates, the protonymphal and tritonymphal stases are calyptostatic (see chapter 3), and the deutonymphal and adult stases are mostly predaceous on the eggs and active instars of other mites and on small crustaceans, isopods, and insects (Schwoerbel 1987; Smith and Cook 1991; Walter and Proctor 1999). A common but not constant feature in the active stases of this highly diverse assemblage is the modification of some of the leg setae into swimming hairs (Fig. 3.14B, p. 40) that allow the mite to move easily in the water column. Many species are brightly colored in red, orange, green, or blue tones (e.g., see plates 41–46 in Soar and Williamson 1929), suggesting possible aposematic protection against predation. Among the many aquatic niches occupied by water mites, few are more unusual than hot springs. Members of the genera *Thermacarus* (THERMACARIDAE), *Wandesia* (*Partnuniella*) (HYDRYPHANTIDAE), and *Tyrrellia* (LIMNESIIDAE) have successfully adapted to thermal springs habitats, where water temperatures often exceed 40°C (Young 1969; Mitchell 1974; Smith 1991d).

Dispersal

Movement within or between habitats is an important component of acarine life histories (Walter and Proctor 1999) and may often be key to individual or population survival. Movement within a habitat generally depends on simple locomotion and is largely triggered by overcrowding, the need to locate food resources or appropriate hibernation and aestivation sites, or the need to find a more tolerable microclimate. Drastic changes in diurnal temperature or humidity levels, such as those occurring in desert or tropical habitats, may result in vertical movement from surface litter layers during the night to cooler or moister subsurface layers during the day (Tarras-Wahlberg 1961; Lees 1969; Wallwork 1976; Short and Norval 1981; Whitford et al. 1981). Seasonal environmental variation may also stimulate vertical movement within a habitat, as exemplified by the largely arboreal algivorous oribatid mite, *Humerobates rostrolamellatus* Grandjean (CERATOZETIDAE). *H. rostrolamellatus* was observed to move from the lower to the upper portions of apple trees prior to oviposition in spring and to move downward in late summer and fall (Murphy and Balla 1973). Seasonal vertical movement in *H. rostrolamellatus* and other mites may be environmentally triggered, but it may also ultimately lead to locating oviposition sites where developing food resources better insure success for emerging progeny.

Passive Dispersal

While within-habitat movement by mites in response to food or environmental cues generally involves little more than walking from one locus to another, travel between habitats can be complicated by the fact that mites lack wings and must rely on other means of dispersal. When a given habitat becomes suboptimal because of factors such as food depletion, habitat deterioration, or overcrowding, mites must seek and find a better situation or perish. Passive dispersal by means of air currents is a common strategy for aerial phytophagous forms such as tetranychine spider mites (TETRANYCHIDAE) and eriophyoids (Kennedy and Smitley 1985; Lindquist and Oldfield 1996) and for the co-occurring species of PHYTOSEIIDAE that prey on them (Johnson and Croft 1981; Hoy 1982; Sabelis and Dicke 1985). Thanks to their small size and natural buoyancy, aerial mites may be carried easily to adjacent plants or to far distant sites on passing breezes. Many of them display special predispersal behavioral patterns that further improve their chances for successful transport. For example, dispersing eriophyid mites may attach their caudal lobe to a leaf surface and raise their body into the air, actively "testing" passing air gusts with their legs. In laboratory observations, wind speeds of less than 10 mph were sufficient to dislodge virtually all questing protogynes of the filbert rust mite *Aculus comatus* (Nalepa) (ERIOPHYIDAE) from their host leaves (Krantz 1973b). Nonquesting protogynes and other stages were unaffected and remained attached. Air currents of less than 4 mph were found to dislodge 50% of questing adult female *Tetranychus urticae* Koch that had adopted the predispersal behavior of raising legs I and facing into the wind (Kennedy and Smitley 1985). Buoyancy of migrating spider mites may be further enhanced by a form of ballooning in which the migrating mite spins down from a leaf on a strand of silk that breaks off under sufficiently breezy conditions and acts as a sail for the attached mite. Unlike other members of the genus for which information is available, *T. urticae* raises legs I and the anterior portion of the body while producing a silken thread that is carried upward into the air column. When air current drag on the line exceeds gravitational pull, the attached mite becomes airborne (Smitley and Kennedy 1985). The ecological and evolutionary implications of ballooning strategies in spider mites and other arthropods are discussed in detail by Bell et al. (2005).

Phoresy

Passive dispersal on air currents may be an efficient means for mite transport, but only where target substrate destinations are numerous and where the dispersing population is large enough to offset the inevitable loss of individuals that fail to arrive at an appropriate site. For example, airborne eriophyoids may number in the thousands in a given orchard during periods of dispersal, but they often are carried beyond the orchard to substrates on which they cannot feed or survive. Unlike eriophyoids, K-selected mite species that have few progeny can ill afford to rely on passive dispersal strategies. Many have resolved this problem by using other animals for dispersal to habitats favored by both the mite and its carrier, thus ensuring a high rate of success in arriving at an appropriate substrate. Dispersal via other animals is referred to as **phoresy**, a phenomenon defined by Farish and Axtell (1971) as one in which "one animal actively seeks out and attaches to the outer surface of another animal for a limited period of time during which the attached animal (termed the phoretic) ceases both feeding and ontogenesis, such attachment presumably resulting in dispersal from areas unsuited for further development, either of the individual or of its progeny" (p. 17). Phoresy may be facultative, that is, induced by the density-dependent and/or environmental perturbations commonly encountered in transient habitats, or it may be obligatory, in which the mites follow an established seasonal migration cycle within a relatively stable habitat and use traditional coexisting carriers (Athias-Binche 1994). Phoresy typically is restricted to the deutonymphal or adult female stage.

There is a continuum of phoretic relationships in nature that range from those predicated on seemingly loose liaisons with any of a number of potential carriers available at a time of critical need, to specific associations based on recognition by the phoretic of a unique set of physical, chemical, and behavioral attributes that define its host (Costa 1969b; Binns 1982; Krantz 1983a, 1998c; Krantz et al. 1991; Athias-Binche 1994). An increasing commitment to phoresy by genera of the family MACROCHELIDAE—that is, from casual to highly specific—has been found to be correlated with an evolutionary progression from early derivative lineages to highly derived taxa (Krantz 1998c). Phoretic specificity often makes it possible to identify a given phoretic simply by identifying its carrier (Krantz 1967). However, a carrier may serve as host to more than one related phoretic species, and individual mite species may attach to multiple congeneric hosts (Ochoa and OConnor 2000). The broad mite *Polyphagotarsonemus latus* (Banks) (TARSONEMIDAE) offers an interesting example of phoresy largely confined to a single host family (Aleyrodidae), in which attachment preference differs strongly from one species to another. *P. latus* is phoretic on a variety of whitefly species, but it displays a strong preference for *Aleyrodes singularis* Danzig over three other tested candidate species (Palevsky et al. 2001).

Deutonymphs and adults of many families of Mesostigmata have established close phoretic relationships with other arthropods and, less commonly, with vertebrates. Included here are the PARASITIDAE, OLOGAMASIDAE, ASCIDAE, AMEROSEIIDAE, DIGAMASELLIDAE, LAELAPIDAE, EVIPHIDIDAE, PACHYLAELAPIDAE, MACROCHELIDAE, UROPODIDAE, SEJIDAE, and several families of the trigynaspid cohort Cercomegistina (Krantz 1967, 1998c; Kinn 1971; Treat 1975; Binns 1982; Lehtinen 1987; Athias-Binche 1994; Walter and Proctor 1999). Phoretic uropodine deutonymphs may attach to their carriers by means of an **anal pedicel** (Fig. 6.1), a liquid strand that is produced at the anal opening from a postventricular gland and hardens on contact with the air (Evans 1992). Some uropodines and microgyniines produce phoretic deutonymphs that differ from co-occurring normal or sedentary deutonymphs both morphologically and behaviorally (Athias-Binche 1978, 1984; Evans and Till 1979). These special dispersing forms are referred to as phoretomorphs. An exceptional case of adult **phoretomorphy** in the Mesostigmata has been found in *Antennoseius janus* Lindquist and Walter (ASCIDAE), which produces both nondispersing and morphologically distinctive phoretic adult female morphs (Lindquist and Walter 1989). While bee-associated LAELAPIDAE commonly disperse on the thoracic hairs of their hosts, members of the genus *Dinogamasus* are carried in the pouchlike propodeal acarinaria of female xylocopine carpenter bees (Skaife 1952; Gerling, Velthuis, and Hafetz 1989; OConnor 1993b; Krantz 1998a; Lundqvist 1998). Dorsolateral thoracic acarinaria also occur in some xylocopines and provide shelter for other mites (e.g., members of the cheyletid genus *Cheletophyes*) (OConnor 1993b), while thoracic atria of cerambycid sawyer beetles may accommodate phoretic species of *Mucroseius* (ASCIDAE) (Lindquist and Wu 1991).

Phoresy by mesostigmatic mites on vertebrates is relatively uncommon and is often precipitated by chance encounters, but there are some well-documented examples of predictable phoretic responses in this group. For example, nectarivorous and pollenophagous species representing four genera of ASCIDAE (*Rhinoseius, Tropicoseius, Proctolaelaps,* and *Lasioseius*) frequent flowers in the New World tropics and subtropics and commonly disperse on hummingbirds (Fain, Hyland, and Aiken 1977; Colwell and Naeem 1994; OConnor, Colwell, and Naeem 1997; Naskrecki and Colwell 1998). Many of these associations present clear evidence of coevolution between the hummingbirds, the mites, and the flowers they exploit (Colwell 1989). Other species of *Proctolaelaps* have been found on honey possums (*Tarsipes spencerae* Gray) and on the mouse *Pseudomys nanus* (Gould) in Australia, and five species of AMEROSEIIDAE representing the genera *Ameroseius* and *Hattena* have been collected from the heads of nectar-feeding birds in Australasia (Allred 1970; Domrow 1979a; Noske 1993; Seeman 1996).

Like the Mesostigmata, many free-living acariform mites have established close phoretic associations with arthropods or vertebrates in transient habitats. The heteromorphic deutonymph or hypopus of many astigmatine families attaches to its invertebrate carrier by means of posteroventral suckers contained in a distinctive plate (Fig. 5.2B, p. 63) (Griffiths et al. 1990). Hypopodes phoretic on hymenopteran hosts may secrete themselves instead in a propodeal cavity, or *acarinarium*, common to many female hymenopterans (Walter and Proctor 1999; OConnor and Klompen 1999). The wasp-associated winterschmidtiid *Kennethiella trisetosa* (Cooreman) is unique in that its deutonymphs secrete themselves in an acarinarium provided by the male of its wasp host, *Ancistrocerus antilope* (Cooper 1955; Cowan 1985). Vertebrate-associated deutonymphs of GLYCYPHAGIDAE (Fig. 5.2A, p. 63) have paired posteroventral claspers with which they grasp individual hairs of the host pelage (Fain 1969a). Other astigmatine hypopodes associated with vertebrates have neither developed suckers nor claspers, secreting themselves instead at the bases of hair follicles or in endofollicular spaces and engaging in activities that go well beyond simple phoresy (see below). Both genetic and environmental factors can affect hypopus induction (Knülle 1995; Corente and Knülle 2003; also see chapter 16).

Oribatid mites often engage in passive phoresy on nest-building vertebrates, especially rodents and birds. Oribatids of at least three families (MESOPLOPHORIDAE, OPPIIDAE, and ORIBATULIDAE) have established close phoretic relationships with insects, mostly passalid beetles (Norton 1980). Mesoplophorids capture individual host hairs in the angle created by depressing the prodorsal shield over its legs and epimera, creating a configuration employed by other ptychoid Oribatida, or box mites, for protection against desiccation and predators (see chapter 15). Phoresy also plays a significant ontogenetic role in a variety of trombidiform taxa, primarily in the Heterostigmatina. Phoretic adults (typically females) of several families, including the PYGMEPHORIDAE, SCUTACARIDAE, PYEMOTIDAE, and TARSONEMIDAE, occur on a broad spectrum of arthropod carriers, and some may be found on vertebrates (Krczal 1959; Cross and Moser 1971; Mahunka 1973c, 1975d; Binns 1982; Lindquist 1969b, 1986; Clift and Larsson 1987; Evans 1992; Athias-Binche 1994; Palevsky et al. 2001). Many phoretic heterostigmatines feed on the eggs or immature instars of their carriers, either as parasitoids or parasites (Kaliszewski, Athias-Binche, and Lindquist 1995). The deutonymphal phoretomorphy noted earlier for some mesostigmatic families is paralleled by adult female phoretomorphy in the superfamilies Pygmephoroidea and Scutacaroidea. Many astigmatine Oribatida also display deutonymphal polymorphism, both in phoretic and nonphoretic forms.

Phoretic morphs may possess thickened legs I, a more compact body than do nonphoretic females (or a longer body as in *Scutacarus baculitarsus agaricus* Norton (SCUTACARIDAE)), and greatly enlarged tarsal claws I with which they attach to their carriers (Moser and Cross 1975; Norton 1977b; Walter and Proctor 1999). Gravid females of the scutacaroid mite *Perperipes ornithocephala* Cross (MICRODISPIDAE) (Figs. 13.90E, F, p. 416), suspected parasitoids of their army ant carriers, bear a striking physical resemblance to the larvae of their ant host (Cross 1965), suggesting that female mites may be able to escape detection by ant workers through larval mimicry. A similar relationship has been documented between the wormlike adults of the dermanyssoid family LARVAMIMIDAE and their army ant hosts (Elzinga 1993; see chapter 12).

A number of atypical relationships between mites and their hosts offer exceptions to the Farish-Axtell definition of phoresy. To cite a few examples, phoretic deutonymphs of *Poecilochirus carabi* G. and R. Can. (PARASITIDAE) apparently feed on oral secretions of their silphid beetle carriers while in transit (Springett 1968), and a species of the genus *Proctolaelaps* (ASCIDAE) preys on phoretic nematodes while being transported on their mutual host, an oil palm weevil (Krantz and Poinar 2004). The associations of many species of Trigynaspida and monogynaspid Uropodina (Mesostigmata) with their hosts also appear to be more intimate than simple phoresy. *Antennophorus grandis* Berlese (ANTENNOPHORIDAE) attaches in the gular area of its ant host and strokes the ant's mouthparts with its legs to induce disgorgement of a droplet of fluid, which is then eaten by the mite

(Evans, Sheals, and MacFarlane 1961). The activity of the mite seems to inhibit the ant's social behavior, including its ability to tend ant larvae (Franks et al. 1991). A more benign commensal behavioral pattern is seen in the trigynaspid mites *Echinomegistus wheeleri* (Wasman) (PARAMEGISTIDAE) and *Micromegistus bakeri* Trägårdh (PARANTENNULIDAE) that scavenge food particles or feed on dermal secretions of their carabid beetle carriers (Nickel and Elzinga 1970a, b). Other antennophorines may have similar commensal relationships with diplopods, chilopods, and insects—especially with ants and beetles (Berlese 1904; Trägårdh 1950; Womersley 1958; Kethley 1977c). In these cases, phoretic transport to new sites may be subordinate to securing an appropriate feeding site on the host.

A unique form of phoresy or mutualism involving a specialized myrmecophilous oribatid mite *Aribates javensis* Aoki, Takaku, and Ito (ARIBATIDAE) and its host ant *Myrmecina* sp. has been described by Ito and Takaku (1994). *A. javensis* apparently is incapable of independent movement and relies on its ant host for protection and transport within the nest chamber and to new nest sites. In return, the mite secretes a substance on which the ants feed.

Although astigmatine deutonymphs are generally considered phoretic in terms of the Farish-Axtell definition, those of at least seven lineages of the superfamilies Glycyphagoidea and Hypoderatoidea experience pronounced growth while in the skin or hair follicles of their mammal and bird hosts, even though they lack a functional mouth (OConnor 1994). Deutonymphs of the genus *Hemisarcoptes* (HEMISARCOPTIDAE), whose adults are generalist predators of diaspidid scale insects, are thought to take up reflexed hemolymph exuded into the subelytral region from the pronotal area of *Chilocorus* spp., their coccinellid beetle hosts. The hemolymph is delivered via an atrium in the region of the genital and anal openings, possibly with the aid of the caudal ventral suckers (Houck 1994; Houck and Cohen 1995) (see also chapter 16). Deutonymphs of *Kennethiella trisetosa* (see earlier) may derive nutrients from the eggs of its solitary wasp host in much the same way (Cowan 1985). An interesting example of atypical phoretic behavior by a nonacarine arachnid involves the pseudoscorpion, *Cordylochernes scorpioides*, which not only feeds on phoretic mites while dispersing on the giant harlequin beetle, *Acrocinus longimanus*, but mates there as well (Zeh and Zeh 1992; Poinar, Curcic, and Cokendolpher 1998).

Driven by the ongoing confusion as to what actually constitutes a phoretic association, Walter and Proctor (1999) have put forward a looser interpretation of phoresy than that provided by Farish and Axtell. They define it simply as a form of temporary symbiosis in which a smaller individual (the phoretic) is transported on a larger individual (the carrier), irrespective of feeding or other activity. Based on the Walter-Proctor definition, all of the examples cited in the foregoing section may be considered phoretic.

Parasitic Mites

Acarine parasites of animals occur in most of the recognized acarine lineages (the order Opilioacarida is an exception), and many are of major importance to humans. The simple process of feeding by parasitic mites may cause damage to their animal hosts through weight loss, anemia, loss of reproductive potential, and tissue irritation, along with consequent allergic response. They may also expose damaged tissue to invasion by airborne microorganisms and other arthropods. Where mite numbers are unusually large or where host size is small, feeding can result in death of the host by exsanguination (Yunker 1973; Harwood and James 1979; Sonenshine 1993). Viruses, rickettsiae, spirochetes, and protozoans have been isolated from ectoparasitic mites, and many of these organisms cause virulent or debilitating diseases in humans and animals throughout the world. Some of these mites have altered the course of history through their impacts on human migration and settlement, world commerce, and military outcomes (Arthur 1962; Philip 1963; Audy 1968). Mites may also harbor and transmit animal parasites such as cestodes and helminths. For example, oribatid mite species representing several families are known or reputed intermediate hosts for a variety of anoplocephaline cestodes of domestic and wild animals, including the sheep tapeworm, *Moniezia expansa* (Allred 1954; Sengbusch 1977; Haq 2001).

Parasitic mites have a long and dark history as pests of humans and domestic animals. The many human and animal diseases that have been linked to attacks by these mites have inspired the production of an enormous body of literature on the mites themselves and on the etiology and immunology of the infectious diseases that they transmit. The effect of parasitic mites on host behavior (Moore 2002) has stimulated research on many fronts, including correlation of mate selection and genetic resistance to levels of parasitism (Møller 1994), host self-medication (Clayton and Wolfe 1993), and incorporation of parasiticidal plants into their nests by birds (Clark 1991).

Parasitic acarines are discussed below in broad terms of host type (vertebrate or invertebrate) and feeding site (external or internal). For purposes of this review, the respiratory, subcutaneous, and alimentary parasites are considered as internal forms. The few examples presented in the following subsections represent only a sampling of the known parasitic acarofauna. The reader is encouraged to seek additional information in subsequent chapters on specific taxa.

Ectoparasites of Vertebrates

The fur or feathers that cover the bodies of most vertebrates provide an ideal, protected setting for the evolution of a broad range of acarine symbionts. Virtually every vertebrate taxon has its complex of external mite parasites, some of which are specific to a particular host species (**monoxeny**) or to two or more hosts (**polyxeny**). Parasites may attack multiple hosts in the same genus (**oligoxeny**) or a variety of hosts representing different genera (**pleioxeny**). The members of some acarine families (e.g., the SPINTURNICIDAE (Mesostigmata) and the raphignathine family MYOBIIDAE) display a high level of host specificity, often extending to monoxeny (Dusbábek 2002). Few mite groups have been shown to be uniformly monoxenous (DEMODICIDAE, EPIMYODICIDAE), and species in a given genus may range from host-specific to pleioxenous (Radovsky 1967). Close correspondence between taxa of parasites and hosts implies a phylogenetic basis for the association, in which tracking of the host by the parasite over time consolidates the relationship between them and may lead to cospeciation. An understanding of the phylogeny of parasitic mite groups has occasionally contributed to a better understanding of the phylogeny of their hosts. For example, the recognition of the Mormoopidae as a separate family of bats is based in part on a clarification of the systematics of their spinturnicid mite parasites (Machado-Allison 1967; Smith 1972).

Although cospeciation may play a role in some observed host-parasite affiliations, many appear to be based on the availability of appropriate food resources or feeding sites on hosts that inhabit ecologically similar biotopes (**resource tracking**) rather than on recognition of phylogenetic cues (Kethley and Johnston 1975; Kim 1985b; Dusbábek 2002). An interesting example of resource tracking is found in the raphignathine family SYRINGOPHILIDAE, one of several families of prostigmatic and astigmatine sarcoptiform mites that live on or in the feathers of birds (**plumicoly** or **syringicoly**) or on the skin under the feathers (**dermicoly**) (Walter and Proctor 1999). Syringophilids invade the feather shafts and feed on their avian hosts through mite-created feeding ports in the feather calmus. Due to behavioral and morphological differences between the invading species, and to differences in feather morphologies and molting sequences, syringophilid genera occurring on a given host may be restricted to specific feather types. In a typical passerine bird, *Syringophilopsis* species may be found in the calmus of primary feathers, *Syringophiloidus* in the secondaries, and *Aulonastus* in the coverts (Kethley 1970). Resource tracking may well be an intermediate evolutionary step toward host tracking and, ultimately, to monoxeny.

Resource tracking is also considered a major strategy in the development of vertebrate parasitism by mite groups that make initial contact with prospective hosts through nidicoly. Mammal and bird nests provide attractive, long-term energy sources for microbivores and detritivores that in turn may serve as prey for predatory parasitiform and acariform mites from the surrounding soil and litter. The nest habitat offers a concentrated food source in a protected site, where feeding may proceed in a more or less uninterrupted fashion and where competitive distractions are minimized. Habituation to the nest habitat may provide these predators with a choice of food opportunities that includes not only microbes, detritus, and other arthropods but also host-produced substances such as scabs, lachrymal fluids, and other secretions. Certain nidicolous mesostigmatic mites, including some species of the genera *Androlaelaps*, *Eulaelaps*, and *Laelaps* (LAELAPIDAE), are also known to feed facultatively on nesting host animals by chewing or punching a hole in the skin (Zemskaya 1967; Radovsky 1994). The shift from casual to obligate nidicoly and the production of burgeoning mite populations in a confined niche allow males to find females quickly, obviating or reducing the need for males to feed. In the case of nidicolous Mesostigmata, male chelicerae have tended to become highly specialized for sperm transfer in facultatively parasitic species, often to the point where feeding is impossible. Nest habituation may also provide predators with opportunities for exploiting unusual host loci. For example, species of the astigmatine SYRINGOBIIDAE living in the quills of charadriform birds are subject to attack by the highly specialized, quill-invading predators of the cheyletid tribe Cheletosomatini (*Cheletopsis*, *Cheletosoma*, *Cheletoides*, and *Eucheletopsis*), a taxon with nidicolous roots (Volgin 1987; Bochkov, Fain, and Dabert 2002). Entire populations within a quill may be wiped out by these predators (B. M. OConnor, pers. obs.). Cheletosomatine CHEYLETIDAE may also share the quill habitat with species of SYRINGOPHILIDAE (see above), but their relationship is unclear. Finally, members of the astigmatine feather mite family APIONACARIDAE occur in parrot quills in company with syringophilids, but, again, the nature of their association has not been determined.

An evolutionary sequence from free-living predation to nidicoly to obligate nidicoly to parasitism has been postulated for the Dermanyssoidea (Mesostigmata) by Radovsky (1969, 1985, 1994), with the initial steps based largely on niche dynamics and morphological modifications of the male chelicerae. Development of obligate ecto- and endoparasitism in the Acari has led in some taxa to the manifestation of special ontogenetic features such as extreme sexual dimorphism or the repression of preimaginal instars or males. An example of ongoing intrageneric transition from nest parasites that feed intermittently on their hosts to full-time resident parasites is provided by the genus *Dermanyssus* (DERMANYSSIDAE). *D. gallinae* (DeGeer) and *D. hirundinus* (Hermann) are long-legged, ovate forms that pass their entire lives in the nest substrate, leaving it only intermittently to feed on their avian hosts (Zemskaya 1968, 1971; Moss 1978). Following a typically large blood meal, females may oviposit up to 20 eggs in the nest material. *D. grochovskae* Zemskaya and *D. quintus* Vitzthum, on the other hand, have shorter, thicker legs and a flatter body that suggest a closer physical association with their hosts. In fact, both species are more frequently collected on the host than in the nest, and smaller blood meals are consumed. The few eggs that are produced following each feeding episode are attached to the feathers of the host rather than being placed in the nest substrate (Radovsky 1994).

Dermanyssoid mites, exemplified by the DERMANYSSIDAE, LAELAPIDAE, MACRONYSSIDAE, IXODORHYNCHIDAE, and SPINTURNICIDAE, and the tick families IXODIDAE and ARGASIDAE, include ectoparasites found on virtually all terrestrial vertebrate taxa (Keirans 1992; Evans 1992; Radovsky 1985, 1994). They are joined by a host of acariform groups, including scab and fur mites of the astigmatine families PSOROPTIDAE, EPIDERMOPTIDAE, and LISTROPHORIDAE; the raphignathine CHEYLETIDAE and MYOBIIDAE; and a multitude of parasitengonine chiggers that are larval forms of the families TROMBICULIDAE and LEEUWENHOEKIIDAE (Wharton and Fuller 1952; Baker et al. 1956; Strandtmann and Wharton 1958; Yunker 1973; Fain and Hyland 1974; Fain and Lukoschus 1976b; Kim 1985a). Their feeding sites and methods of attack are varied and sometimes reflect the type of food ingested. Laelapid scavengers/facultative parasites such as *Haemogamasus reidi* (Ewing), *Androlaelaps fahrenholzi* (Berlese), and *Laelaps echidninus* (Berlese) range over their small mammal hosts and through their nests to feed on a variety of substances, including small arthropods and flea feces in the nest and ectoparasites on the host itself. They also feed on scabs and lachrymal secretions on the host skin and take fresh blood or lymph primarily from wounds gouged in the bare skin of nestlings (Furman 1959; Radovsky 1985). There is as yet no conclusive evidence that any of these facultatively parasitic mites transmits any human or animal disease in nature.

The IXODIDAE and ARGASIDAE, along with a number of dermanyssoid families, are obligatory hematophages that must cut through or pierce the host skin with modified chelicerae and probe deep enough to tap subsurface blood vessels (Evans 1992; Coons and Alberti 1999; Sonenshine 1991, 1993; Radovsky 1994). In like manner, the skin of mammals is pierced by members of the raphignathine families PSORERGATIDAE and MYOBIIDAE, and that of birds by species of the astigmatine family EPIDERMOPTIDAE. Astigmatine fur mites (LISTROPHORIDAE, ATOPOMELIDAE, and CHIRODISCIDAE) attach themselves to host hairs by means of a variety of leg and coxal field modifications and feed on hair surface scales (Hughes 1954a), or possibly on sebaceous secretions or detritus (see chapter 16). Psoroptid scab mites also penetrate the surface skin, but their feeding induces development of a weeping lesion that finally hardens into a protective scab under which the mites feed (Yunker 1973). Rather than inducing lesion formation, the ophioptine pit mites of snakes (HARPIRHYNCHIDAE) give rise to the formation of depressions under the posterior scale margins of their hosts, providing a protected site for feeding and development (Fain 1964d). Like many other larval parasitengonine parasites, trombiculid chiggers inject a digestive salivary enzyme into the feeding wound and suck up the lysed host tissues from the feeding channel, or **stylostome**, that is created (Wharton and Fuller 1952; Hase et al. 1978).

A number of mites are associated with freshwater fishes, and some of them are considered to be parasitic on their hosts. For example, *Hydronothrus aquariorum* Fain and Lambrechts (MALACONOTHRIDAE) was taken from necrotic lesions and associated mucus on young discus fishes (Fain and Lambrechts 1987) and was believed to have caused those symptoms. High mortality in a fish farm population of trout in Spain appeared to be associated with the presence of *Schwiebea estradai* Fain and Ferrando (ACARIDAE) on the gills, on the skin, and in the gut of their affected hosts (Fain and Ferrando 1990). *Histiostoma papillata* Halliday and Collins (HISTIOSTOMATIDAE) was likewise implicated as a parasite in the gills and fins of juvenile Murray cod, a large Australian freshwater fish (Halliday and Collins 2002). *H. papillata* appeared to inflict injuries that contributed to fish mortality, although B. M. OConnor (pers. comm.) considers it unlikely that *Histiostoma* or *Schwiebea* species are true parasites. They may instead feed on wound tissue or associated fungi (see also Walter and Kaplan 1990a).

Mention should be made here of the 30-plus astigmatine plumicolous feather mite families of the superfamilies Analgoidea and Pterolichoidea (the latter includes the formerly recognized Freyanoidea), although they are not generally considered to be parasitic (Gaud and Atyeo 1996a, b). Nevertheless, members of the family ASCOURACARIDAE and some plumicolous PYROGLYPHIDAE may consume the entire feather medulla. Limited observations on feeding by plumicolous feather mites indicate that they are commensals that subsist on uropygial gland fluid that coats the feather surfaces, and on detritus, pollen, and/or fungal spores that adhere to the feather barbs (Proctor and Owens 2000). Feather fragments or sloughed feather material may also be ingested (Dubinin 1951; OConnor 1982b), implying that mild damage may occur to the host bird in some cases. On the other hand, recent field studies showed a positive correlation between host condition and the number of plumicolous feather mites present, suggesting the possibility of a mutualistic relationship (Blanco, Tella, and Potti 1997).

Endoparasites of Vertebrates

The transition in mites from external to internal parasitism may be a gradual process, but many vertebrate endoparasites display remarkable morphological and behavioral traits that underline the antiquity and the ecological uniqueness of their cryptic habitats. The majority of species show a reduction in sclerotization and setae when compared to related external forms, many have reduced or modified appendages and mouthparts, and some assume bizarre shapes that allow them to invade and inhabit unique subcutaneous microniches. Like some ectoparasitic taxa, internal parasites often are monoxenous (Kim 1985a).

Many vertebrate endoparasites are associated with the respiratory systems of their hosts. For example, the dermanyssoid family RHINONYSSIDAE is largely confined to the nasal passages of birds, although *Sternostoma* species may invade the lungs and air sacs as well (Radovsky 1994). The nasal and lung habitats of birds have also been exploited by members of the prostigmatic family EREYNETIDAE and the astigmatine families TURBINOPTIDAE and CYTODITIDAE (Domrow 1969; Fain 1970c, d; Yunker 1973; Pence 1975), while members of at least five genera of the typically ectoparasitic TROMBICULIDAE occur in the nasal passages of bats and small mammals and in the tracheae and lungs of sea snakes and marine iguanas (Yunker and Jones 1961; Yunker and Brennan 1962; Vercammen-Grandjean 1965a; Nadchatram and Radovsky 1971; Nadchatram 2006). Like the trombiculids, the MACRONYSSIDAE are generally considered an ectoparasitic group, but the protonymphs of some species embed themselves in the oral mucosa of their phyllostomid bat hosts (Phillips, Jones, and Radovsky 1969; Radovsky, Jones, and Phillips 1971) and the adults of others in the nasal cavities of lizards (Yunker and Radovsky 1966). The nasal fossae of bats also are infested by members of the astigmatine family GASTRONYSSIDAE and those of primates by species of LEMURNYSSIDAE (Fain 1964c). Representatives of the astigmatine PNEUMOCOPTIDAE have been encountered in the lungs of rodents (Baker et al. 1956). Some species of the dermanyssoid family HALARACHNIDAE occur in the nasal passages of seals and walruses (Strandtmann and Wharton 1958; Furman and Smith 1973), while others may be embedded in the sinuses, bronchiae, tracheae, or lungs of apes, squirrels, springhares, and, occasionally, dogs (Fain 1962d; Clark and Clifford 1964; Yunker 1973) The dermanyssoid ENTONYSSIDAE are exclusively parasitic in the lungs and tracheae of reptiles (Fain 1961a).

The relative ease of entry into the vertebrate respiratory system, coupled with the protection that it provides from predation and desiccation, may explain the bewildering variety of acarines that have exploited it. The host epidermis, however, presents a considerably tougher barrier, one that restricts most acarine ectoparasites to simple probing and piercing of the host surface. Nevertheless, some parasitic astigmatine groups have successfully invaded and adapted to subcutaneous niches. For example, *Laminosioptes cysticola* (Vizioli) (LAMINOSIOPTIDAE) infests subsurface tissues of domestic fowl, where it causes the formation of nodules that may be felt through the skin, particularly in the pectoral area (Baker et al. 1956). Mange mites of the family SARCOPTIDAE burrow beneath the skin of their hosts and feed on subepidermal exudates produced at the site of injury. The human scabies mite, *Sarcoptes scabiei* (DeGeer), is a major parasite of mammals throughout the world, causing symptoms such as dermatitis, pruritis, and emaciation that may disable or kill the host (Yunker 1973; Fain 1978a; Fain and Hyland 1985). *Knemidocoptes mutans* (Robin and Lanquetin) (EPIDERMOPTIDAE) burrows beneath the epidermal scales on the legs of chickens, precipitating leg deformity and lameness. Similar related species tunnel into the skin beneath the feathers, into the base of the bill, or under the skin of the head of their various avian hosts (Fain 1962a, 1967c). Vermiform mites of the raphignathine family DEMODICIDAE occur in hair follicles, in gland ducts, and beneath the skin of a variety of mammals worldwide, including humans. Demodicids largely specialize in feeding on subcutaneous sebaceous gland products and have adapted morphologically in spectacular ways to life in the follicular niche. Symptoms of demodicid infestation may include dermatitis, scaling, pustule and subdermal nodule formation, and alopecia (Yunker 1973; Desch and Nutting 1972, 1973; Matthes 1996).

A peculiar relationship exists between species of the astigmatine genus *Myialges* (EPIDERMOPTIDAE) and their bird associates. *M.* (*Promyialges*) *macdonaldi* Evans, Fain, and Bafort is a skin parasite of the passeriform bird, *Parus caeruleus*, but females attach to hippoboscid flies to oviposit. Emerging larvae then move from the fly to attack the bird host (Evans, Fain, and Bafort 1963). Other bird-associated *Myialges* parasites use a variety of hippoboscid species and menoponid lice as sites for oviposition (Fain 1965d). Based on the broader interpretation of phoresy proposed by Walter and Proctor (1999; see also Phoresy, p. 73), *Myialges* species may be considered both parasitic and phoretic.

Accidental ingestion of live mites may lead to a condition known as acariasis, in which the mites survive and even reproduce in the alimentary or urinary tracts. Symptoms such as vomiting and diarrhea may occur in cattle that are fed grain heavily infested with astigmatine granary mites such as ACARIDAE or GLYCYPHAGIDAE (Hinman and Kampmeier 1934; Morini and Roveda 1974). It should be noted that a number of astigmatine mite species that live in human habitations are considered a major health problem because of the respiratory allergens that they generate (Bronswijk and Sinha 1971; Wharton 1976; Platts-Mills and de Weck 1988).

Ectoparasites of Invertebrates

The mite-associated fauna of invertebrates is vast, reflecting the antiquity and intimacy of the relationships between them (see chapter 2). Many of these mites use larger invertebrates simply for dispersal and for entry into host niches, where they may find food and shelter, but a significant number have evolved ectoparasitic associations with their carriers.

All of the active stages of ectoparasitic mites generally feed on their hosts, but in some cases only one stase is parasitic. In these instances, the remaining active forms typically retain more primitive predatory habits. For example, the larvae of six families of terrestrial Parasitengonina parasitize a wide range of arthropods worldwide (Welbourn 1983; Southcott 1961, 1987b, 1993b, 1999), but the remaining active instars (deutonymph and adult) are predaceous. This behavioral divergence is reflected in the striking morphological contrast between larval and postlarval instars. Similar behavioral and morphological dissimilarities are found in the parasitengonine water mites, whose larvae typically parasitize aquatic and semiaquatic insects and whose adults and active nymphs are predaceous (Prasad and Cook 1972; Mullen 1974; Smith and Oliver 1986). Larvae of the water mite *Hygrobates nigromaculatus* Lebert (HYGROBATIDAE) are unusual in that stream populations parasitize chironomid midges, while pool-inhabiting larvae simply remain in quiescent clutches until the deutonymphal molt occurs. Significant morphological differences may also be seen between the two populations (Martin and Davids 2002). The deutonymphs and adults of some hygrobatoid UNIONICOLIDAE are resident (and presumably parasitic according to Hevers 1980) in freshwater mussels, but the larvae parasitize chironomid flies (Mitchell 1955a; Baker 1991). Some postlarval water mites also are known to parasitize snails and crayfish (Walter and Proctor 1999). As in the Parasitengonina, the larva, deutonymph, and adult are the only active instars in the raphignathine family PTERYGOSOMATIDAE, but all three instars are parasitic (Bertrand 2002).

Embryonic development often occurs entirely within the physogastric female in parasitic Heterostigmatina (e.g., PYEMOTIDAE, CROTALOMORPHIDAE) (Fig. 13.84A, p. 410) and in the insect egg parasitoids of the family ACAROPHENACIDAE so that only the female is available to feed (Cross 1965; Lindquist 1969b; Kaliszewski, Athias-Binche, and Lindquist 1995; Lindquist and Krantz 2002). These females usually kill their egg or preadult hosts and are considered **parasitoids** in the sense of Lindquist (1983). Mite parasitoidism in the strict sense is recorded only for the Tarsonemina (Kaliszewski, Athias-Binche, and Lindquist 1995), but other invertebrate ectoparasites are known to disable or kill their hosts through feeding. For example, species of the pterygosomatid genus *Pimeliaphilus* parasitize triatomine Hemiptera, Coleoptera, cockroaches, and scorpions and may kill them when mite populations are high (Newell and Ryckman 1966). *Riccardoella limacum* (Schrank) (EREYNETIDAE) is a eupodine hematophage on the surface and in the mantle cavity of slugs and snails, and it sometimes occurs in large enough numbers to cause host death (Baker 1970a, b). Species of the astigmatine genus *Hemisarcoptes* (HEMISARCOPTIDAE) parasitize armored scales (Homoptera) and prey on their eggs, contributing significantly to scale mortality (Gerson, OConnor, and Houck 1990). The larva and nymphs of *Macrodinychus sellnicki* Hirschmann and Zirngiebel-Nicol, a member of the otherwise free-living mesostigmatic family DINYCHIDAE, parasitize and kill pupae of the crazy ant, *Paretrechina fulva* (Mayr), in tropical America. The free-living adult leaves the pupal carcass shortly after its emergence from the deutonymphal exuviae (Krantz, Gomez, and Gonzalez 2007).

The ascoid family OTOPHEIDOMENIDAE comprises three subfamilies and 10 recognized genera of mostly ectoparasitic insect associates that occur on Lepidoptera, Hemiptera, Orthoptera, and Isoptera (Treat 1975; Zhang 1995c; Mo 1996). Species of *Otopheidomenis* are especially likely to be found in the wing bases and the thoracoabdominal cleft of their sphingid and noctuid moth hosts. Unlike the moth ear mites of the laelapid genus *Dicrocheles* that invade the tympanic and countertympanic cavities to feed and lay eggs, lepidopterophilous otopheidomenids go no further than the external tympanic recesses, and then only to oviposit (Treat 1975).

The nests of social insects harbor many mite species, including predators, saprophages, fungivores, and pollenophages, and many of them use their insect hosts for dispersal to new nest sites. However, a small number of these mites are ectoparasites of adult bees or their brood, including some species of the obligatorily parasitic tarsonemoid genus *Acarapis* (TARSONEMIDAE) and the mesostigmatic genera *Varroa, Euvarroa* (VARROIDAE), and *Tropilaelaps* (LAELAPIDAE) (Eickwort 1990; Kaliszewski, Athias-Binche, and Lindquist 1995; Sammataro, Gerson, and Needham 2000). *A. dorsalis* Morgenthaler and *A. externus* Morgenthaler are hematophagous on the European honeybee (*Apis mellifera*) but apparently cause little harm to their host (DeJong, Morse, and Eickwort 1982). The introduction of *A. mellifera* into tropical Asia has exposed it to attack by species of *Varroa* and *Tropilaelaps*, which normally are associated with *Apis* species that are native to that area: *A. cerana, dorsata, florea,* and *koschevnikovi* (Delfinado and Baker 1961; Delfinado-Baker and Agrawal 1987; Woyke 1987; Guzman and Delfinado-Baker 1996). *V. destructor* Anderson and Trueman, identified from haplotypes previously included under the name *V. jacobsoni* (Oudemans), is a hematophagous ectoparasite originally described from *A. cerana* that has followed *A. mellifera* on its worldwide travels and is currently considered a major threat to international apiculture. Affected bees suffer reduced weight and abnormal body and wing development, and colonies often succumb to heavy infestations (DeJong, Morse, and Eickwort 1982; Anderson and Trueman 2000; Fernandez, Coineau, and Theron 2006).

Not all ectoparasites of invertebrates cause significant injury to their hosts. For example, all of the active instars of the heterostigmatine family HETEROCHEYLIDAE are parasitic on passalid beetles, but they apparently are so well adapted that they cause little injury to them. Heterocheylids are considered an ancient group, and they occur in every world region where the similarly ancient Passalidae are found. Clearly, an accommodation has been reached between host and parasite, one that is based on a relationship that dates back to mid-Jurassic times (Lindquist 1997).

Some ostensibly parasitic acarines on invertebrates should be classified as commensals or mutualists rather than parasites. For example, while the larviparous cockroach mite *Proctolaelaps nauphoetae* (Womersley) (ASCIDAE) is hematophagic and truly parasitic on its host (Egan and Hunter 1975), the laelapid *Gromphadorholaelaps schaeferi* Till, an associate of the Madagascar hissing cockroach, feeds only on roach saliva (Yoder 1996). The highly specialized genera of rhizoglyphine ACARIDAE exemplified by *Ewingia* species (Yunker1970a) attach to the gills or other body regions of coenobitid hermit crabs but are considered commensals rather than parasites (OConnor 1982a). Other "almost parasites" include the many mite inquilines of ants and other social insects that feed on bodily secretions or on bits of food during periods of trophalaxis and make themselves welcome in ant nests by producing exudates on which the insects feed, often camouflaging their presence with a chemical aura that is acceptable to their ant hosts (Eickwort 1990; Hölldobler and Wilson 1990). A singular relationship that might best be described as mutualistic exists between the macrochelid mite *Macrocheles rettenmeyeri* Krantz and its army ant host *Eciton dulcius* (Krantz 1962; Rettenmeyer 1962). The mite attaches to the membranous terminal tarsal lobes of legs III of large worker ants and presumably is carried in this fashion during swarming events. At the same time, the curved, inordinately long legs IV of *M. rettenmeyeri* serve as surrogate claws for clustering ants during bivouac or temporary nesting periods, although it is not clear why the ants would accept or even tolerate such an intrusion.

Endoparasites of Invertebrates

Few mite groups have adapted to endoparasitism in invertebrates, and most are found in the respiratory system. Many members of the morphologically unusual heterostigmatine family PODAPOLIPIDAE are ectoparasitic beneath the wings/elytra or in intersegmental membranous areas of orthopterans, beetles, hymenopterans, heteropterans, and cockroaches, but species of the genus *Locustacarus* are endoparasites in the tracheae and air sacs of acridid locusts and bombine bees, apparently piercing the tracheal walls to feed (Husband 1984a; Husband and Sinha 1970b). Carabid beetle parasites of the podapolipid genus *Ovacarus* occur in sacs associated with the withdrawn vaginal and aedeagal membranes and with the oviducts of their hosts (Stannard and Vaishampayan 1971; Husband 1974b). Unlike its ectoparasitic congeners, *Acarapis woodi* (Rennie) (TARSONEMIDAE) invades the prothoracic spiracles and occasionally the thoracic and abdominal air sacs of its honeybee host, piercing tracheal walls to feed and possibly impairing respiration and the ability of the bee to fly (DeJong, Morse, and Eickwort 1982; Lindquist 1986).

Otopheidomenids are typically ectoparasitic (see above), but the otopheidomenid species *Katydiseius nadchatrami* Fain and Lukoschus and *Orthopteroseius sinicus* Mo are exceptional in being endoparasitic in the tracheal systems of their katydid and grasshopper hosts (Fain and Lukoschus 1983b; Mo 1996). Also exceptional in terms of niche is the marine mite *Enterohalacarus minutipalpus* Viets (HALACARIDAE), an internal parasite of deep-sea urchins (Viets 1938). Members of other halacarid genera, including *Halixodes* and *Astacopsiphagus*, are gill and mantle parasites of chitons and freshwater decapods (Brucker and Trouessart 1900; Viets 1931; Bartsch 1986b, 1988), but most known halacarids are predatory or algivorous in oceanic and freshwater habitats (see chapter 13).

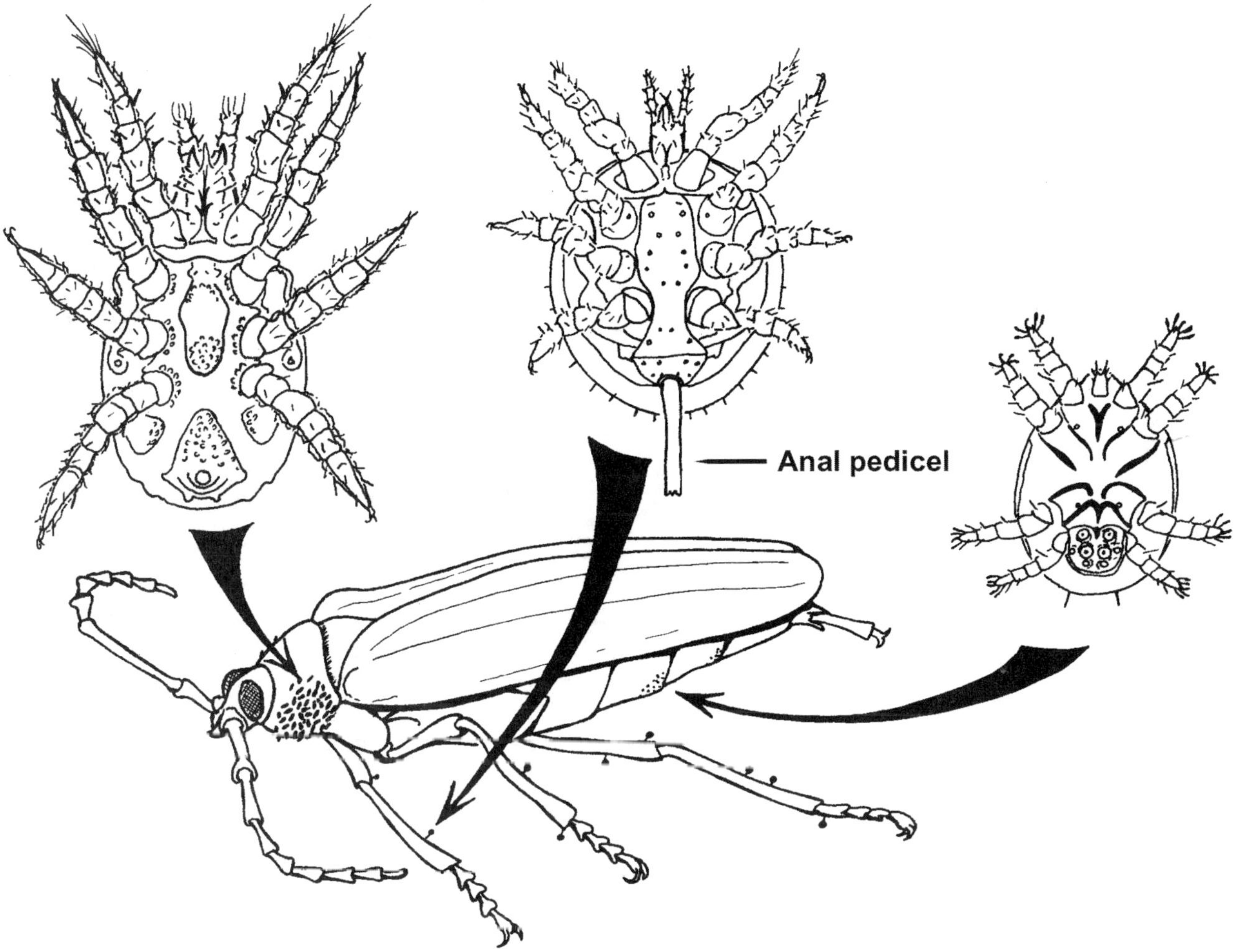

Fig. 6.1. Phoretic deutonymphs of (from left) POLYASPIDIDAE, UROPODIDAE, and ACARIDAE on a prionid beetle host. Arrows identify typical attachment sites.

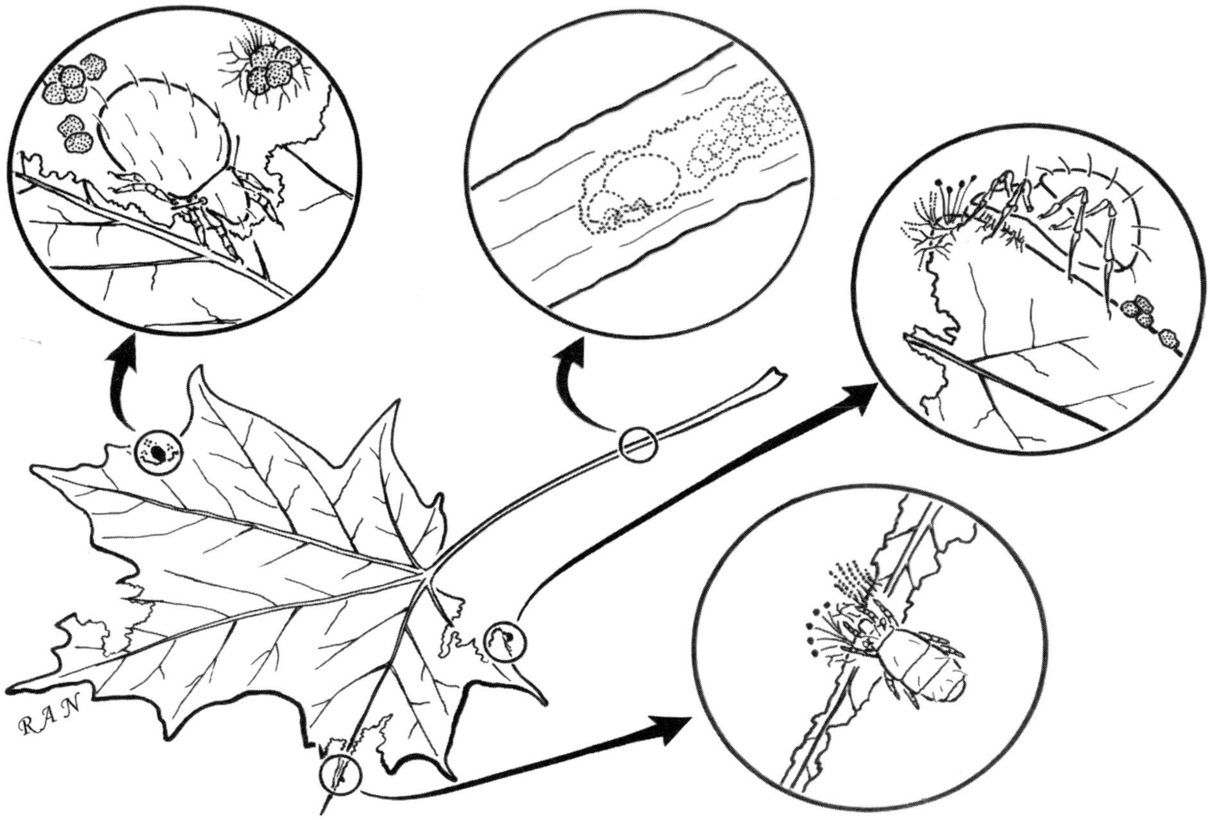

Fig. 6.2. Saprophagy and mycophagy by free-living acariform mites in soil. Clockwise from left: some saprophagous oribatid mites feed externally on softer dead leaf parts; others (especially as immatures) burrow through thicker, decaying woody tissue; many oribatid mites feed on saprophytic fungi. Pygmephorid mites also feed on fungi but, unlike oribatids, do not ingest particulate matter and do not produce fecal pellets. (Courtesy of R. A. Norton)

D.E. WALTER
G.W. KRANTZ

CHAPTER SEVEN

Collecting, Rearing, and Preparing Specimens

Mites are virtually everywhere, which means that finding mite habitats is a relatively simple matter. Extracting them from these habitats, however, may be a challenge. The first decision that must be made is whether the collection is to be qualitative or quantitative; that is, is the goal simply to determine what mite species are present, or is the sample to be subjected to statistical analyses of species, stage, or sex? If the latter, then the collector should first make enough qualitative collections to gain an understanding of the spatial dispersion of the mite species being studied and then become familiar with sampling techniques that will permit a valid quantitative analysis (e.g., Sabelis 1985b; Sumner 1999). The second decision to be made is whether the mites are to be collected as live or dead samples. Collecting live mites allows one to observe their natural appearance and behavior, establish cultures for life history information, and conduct molecular studies. Most standard collecting techniques are designed to retrieve mites in a preservative, so these techniques must be modified to obtain living mites.

Mites may be collected individually with a forceps or a fine brush, with the aid of a hand lens, from habitats such as vegetation, the underside of rocks or tree bark, or museum specimens of vertebrates or invertebrates. More typically, however, mites are extracted from representative substrate samples. Extraction methods may be either active or passive, with active methods based on mite response to physical or chemical perturbation of live samples and passive methods involving washing or centrifugation of sample concentrates (McSorely and Walter 1991). Active extraction may involve using light, heat, desiccation, or chemical stimulation to induce the mites to move down through the sample into an appended collecting container. The main drawback here is that nonmotile stages (e.g., eggs, quiescent molting individuals) are left behind. Furthermore, not all taxa react similarly to sample perturbation, and extraction efficiencies may vary, often from species to species. Passive extraction techniques supersede behavior and rely instead on the arthropod cuticle's affinity for hydrocarbons, on differences in specific gravity, or on sedimentation rates to separate the mites from their substrate.

Terrestrial Mites

Hand Collection

Because of the small size and delicacy of the subjects, the collector may encounter special problems when collecting individual mites in the field. An aspirator can be used to suck up larger mites on vegetation, rocks, logs, and so on, but the aspirator should be equipped with a fine mesh screen (50–100 µm Nitex or plankton netting is recommended) to prevent inhalation of the mites. Mechanical suction using a squeeze bulb (Fig. 7.1B) or a minivacuum cleaner is advisable if pathogenic fungal spores, toxic secretions, or solvents could be inhaled during suction. Mechanical suction devices are especially recommended if (1) collecting from carrion or dung or directly into a preservative or (2) the collector is prone to allergic reactions. Strips of moistened tissue paper in the aspirator jar provide both moisture and hiding places for live specimens, reducing mortality caused by desiccation and predation.

Funnel Extraction

The easiest way to collect large numbers of mites from organic substrates such as soil or nest material is to use a funnel (or canister) extractor (Fig. 7.1A). Essentially, the substrate sample is suspended in or over a screened funnel, and a humidity gradient is established by applying surface heat, which drives motile mites and other arthropods down through the drying sample, through the screen, and into a screw-top bottle or vial attached beneath the funnel. Antonio Berlese (1863–1927) developed the first extraction funnel using hot-water copper coils to establish a temperature-moisture gradient in the suspended sample (see Kevan 1985). Trägårdh and Forsslund (1932) later demonstrated the effectiveness of the Tullgren funnel, an extractor that utilizes an overhead incandescent light bulb rather than hot water for sample desiccation. This led to development of a modified

Tullgren apparatus by Haarløv (1947), variations of which have become standard for extracting mites from soil, litter, compost, and animal nest material. In recent years, all funnel extractors have generally come to be known as **Berlese funnels**. Many modifications of the Berlese funnel have been introduced (e.g., Macfadyen 1953; Southwood 1966; Merchant and Crossley 1970; Edwards and Fletcher 1971; Seastedt and Crossley 1978; Petersen 1978; Bieri, Delucchi, and Stadler 1986), primarily to establish strong temperature gradients in intact and inverted soil cores and improve the capacity of the funnel as a quantitative tool. It is recommended that anyone undertaking a quantitative sampling program review the recent literature on separation techniques (e.g., Robertson et al. 1999).

The wattage of light bulbs used for Berlese funnel separations depends in large part on the size and relative water content of the samples and on the distance of the bulb from the sample surface. Although high-wattage bulbs may be more effective for extracting large, damp samples, they often kill high percentages of lightly sclerotized, slow-moving acarines before they can work their way down through the sample and into the collecting jar (Fujikawa 1970). On the other hand, using less than 40 watts may greatly prolong the separation period and, in large samples, may even enhance continued feeding and reproduction in the substrate. Coineau (1974c) suggested that no heat be applied to the samples during the first day of extraction; if possible, heat should be increased progressively during the extraction period.

When electricity is not available, heavier-than-air volatiles—such as naphthalene crystals sprinkled on the surface of the sample—have been used to induce arthropods to move away from the surface and downward through the sample (McFadden 1961; Lauck 1965). Chemical extraction techniques require good ventilation and an understanding of the toxicity of the chemical being used. Extraction relying on the heat of the sun can be effective in field situations, but it tends to be slow and is impractical in wet climates. A portable funnel system for use in sun drying was described by Norton and Kethley (1988).

The walls of a Berlese funnel should be steep-sided, and the screen mesh should be large enough to allow passage of the largest mites. A layer of muslin under the sample can reduce the amount of dirt and debris that falls through the mesh during the sampling period (see below). Only a relatively small sample of litter or soil should be subjected to separation at a given time, because overly large samples often retain subsurface moisture, and mites may follow the humidity gradient into the center of the sample rather than continuing through it and into the collecting jar. Furthermore, large, wet organic samples can quickly become enmeshed in thick fungal mats that prevent mites from escaping. While negative reactions to heat and light play a role in the funnel extraction of mites, the major factor involved is their response to an established humidity gradient. Mesic materials such as moist soil, leaf litter, and compost are most appropriate for funnel extraction, but drier organic materials such as nests may also be extracted successfully if they are composed of relatively fine litter. Mites living in desert or dry grassland soils, deep mineral soils, or coarse woody material such as bark are much less efficiently collected with funnels. Their adaptation to naturally dry habitats may lead them to become quiescent rather than to flee, and some (e.g., NANORCHESTIDAE) appear able to tolerate very low humidity. Thus, funnel extractions from these habitats may be both inefficient and taxonomically biased (Walter, Kethley, and Moore 1987).

Usually only the larger, more active animals appear in the collecting jar during the first day or two of extraction. Relatively few animals are seen in the early phases of drying, but large numbers of mites will begin to appear as the critical level of dryness is reached. The length of time needed to extract most of the mites in a sample depends on sample size and its moisture content and on the weather. For example, a 500 cm^3 sample of rain-forest litter should remain in a funnel for four to five days in dry weather but may require a week or more of extraction during the rainy season.

Mites may be collected directly into a preservative (70%–80% ethanol with 1%–3% glycerin to keep the specimens moist if the alcohol evaporates) or collected live into a container with a floor of moistened charcoal plaster (see below). Alcohol is an adequate collecting fluid, but it is not an ideal preservative because the longer a specimen remains in alcohol, the more difficult it becomes to process for mounting. Polyethylene glycol or glycerin–glacial acetic acid solutions may be used (Krantz 1978), but carcinogenic and potentially explosive materials, such as picric acid, should be avoided.

When extraction is complete, the resulting preserved sample or **berleseate** often contains large amounts of soil and detritus that have been dislodged from the sample by escaping invertebrates. A layer of wide-mesh cloth (muslin or cheesecloth) or fine screening under the sample will help resolve this problem. Baffles or other autosegregators (e.g., Aoki 1984) that capture fallen soil under the sample also may be useful. Separation of mites from soil- or debris-contaminated berleseates may be facilitated by using one of the flotation techniques described below. Samples containing very sandy soils or dry, dusty clays will invariably contain many small, mite-sized particles, and a flotation technique may be the only efficient means for separating mites from this material.

Flotation Separation

Mites may be extracted from soil, house dust, and other particulate material by taking advantage of two properties common to mites and other arthropods: (1) their specific gravity differs from that of soil particles, and (2) their cuticle has a strong affinity for petroleum derivatives (McSorely and Walter 1991). **Specific gravity flotation** separates mites from their suspended substrate when the latter is mixed with a concentrated sugar or salt solution (e.g., Hart and Fain 1987; Pugh 1996c). Because of specific gravity differentials, mites will remain in suspension while soil particles settle out. The suspension may then be passed through a Buchner funnel apparatus (Fig. 7.1C) to collect the mites. **Hydrocarbon flotation** capitalizes on the tendency for some liquid hydrocarbons (e.g., heptane or kerosene) to adhere to the arthropod cuticle (Walter, Kethley, and Moore 1987; Geurs, Bongers, and Brussard 1991; Kethley 1991b). A sample chosen for extraction is poured into an Erlenmeyer flask, and a layer of an appropriate hydrocarbon, such as kerosene, is added. A magnetic stirrer may be used to agitate the sample to bring the mites into contact with the kerosene layer, or the flask may be gently shaken to mix the kerosene through the substrate (avoid vigorous shaking, which may result in a milky emulsion). The flask is then set aside until the kerosene droplets rise to the top of the flask, bringing the mites along with them. The kerosene layer is then decanted through a fine mesh sieve, and the mites are washed clean with 95%–100% ethanol. The decanting procedure should be repeated until mites cease to appear in the kerosene layer.

An advantage of hydrocarbon flotation is that soil samples may be preserved in ethanol and extracted at a later time. A major drawback is that large, heavy mites and those with encrusted soil particles, along with soil-inhabiting HISTIOSTOMATIDAE and NOTHRIDAE, are not efficiently extracted by this method. Hydrocarbon flotation should not be undertaken without adequate ventilation (a fume hood is advisable). Detailed instructions for a quantitative approach to flotation extraction may be found in Walter, Kethley, and Moore 1987.

Kethley (1991b) devised a method for collecting mites from mineral soil or beach sand that requires a large (30 L) bucket, a wooden stirring paddle, a scoop, a fine mesh sieve, a funnel, large collecting vials, ethanol in a squeeze bottle, and lots of water (fresh or salt). The bucket is filled with water, and a few liters of sand are added and gently stirred with the paddle. As the sand moves through the agitated liquid, lipids and other materials present in the soil form soapy bubbles that trap small mites. The resulting foam collects on the surface of the water and is scooped out onto the sieve. Ethanol is then squirted onto the sieve surface to cut the foam and to move the mites through an underlying funnel into the collecting vial. Because of the large amount of sand and organic material collected with the mites, these samples may benefit from subsequent kerosene flotation.

Trapping

Malaise traps, yellow pan traps, and other techniques used to collect flying insects will also collect their phoretic mites. Pitfall traps may be used to collect some of the larger surface-inhabiting soil mites such as ERYTHRAEIDAE, ANYSTIDAE, some of the larger Oribatida, and mites that are phoretic on beetles and other arthropods. Traps can be baited with carrion, rotting fruit, fungi, or dung to attract particular types of insects and their phoretic mites. Defaunated litter placed within open mesh bags or other containers that permit access to small invertebrates can be placed into holes bored in soil, caves, tree trunks, or other habitats, and any mites colonizing the litter may then be extracted using one of the methods detailed above. The litter can also be baited with prey items (e.g., nematodes or springtails from cultures) to make the samples more attractive to predators.

Arboreal Mites

Hand Collection

Sweep nets and beating sheets can be effective tools for separating and collecting live mites from vegetation (Steiner and Goodwin 1998). Alternately, leaves, flowers, fruit, stems, and galls can be collected with pruning shears for closer examination under a dissecting microscope. Mites hide wherever shelter is available, which means that clumps of detritus, dead insects, or plant structures such as leaf domatia should be separated with a scalpel or sharp probe for examination (Walter and O'Dowd 1995). Although manual collection of live mites from such samples can be tedious and time consuming, it allows for behavioral observations and provides a glimpse into protected sites from which mites are unlikely to be extracted using mechanical methods.

Mites may also be washed from plant parts or shaken loose into jars filled with alcohol or water to which a surfactant has been added (Zacharda, Pultar, and Muska 1988). Extraction efficiency is improved by alternating periods of agitation with periods of rest. The resulting supernatant may then be passed through a sieve or filter paper to collect the mites (Pratt and Croft 2000). Automated mite brushing machines have been used for collecting standardized mite samples from relatively tough, smooth leaves (Sabelis 1985b), but these appliances are less effective for separating mites from hairy or strongly veined leaf surfaces or from leaf domatia.

Eriophyoid mites that inhabit galls or blisters (Figs. 13.16D–G, p. 342) must be exposed by dissecting their habitat. Although they tend to be extremely small, sequestered eriophyoids usually occur in large enough numbers to be easily seen in infested tissue. Mites can even be retrieved from long-stored gall samples rehydrated by immersion in warm water (Amrine and Manson 1996). Mites in erineal patches on leaf undersides are difficult to recover from these dense patches of hairlike growth (see chapter 13). However, if the leaf is gradually dried by a heat source such as a warm light, the mites often respond by swarming to the surface of the erineum.

Mass Collection

Mass removal of mites from plant material may be expedited by placing infested vegetation directly into a Berlese funnel. Although separation efficiency may not be high, funnel processing allows large amounts of vegetation to be processed.

Vacuum samplers also are useful for removing mites from vegetation in the field, but the recovery of mites that produce dense webbing (some TETRANYCHIDAE) or that hide in crevices or plant domatia is likely to be low. Nevertheless, vacuum sampling has become a standard method for collecting arthropods, especially in grasslands and agricultural fields (Stewart and Wright 1995; Smith 1999).

Various trapping devices have been used to collect arboreal mites, especially those species that disperse on air currents. Plankton nets, tanglefoot-coated plates, and greased slides are widely used (Hoy 1982; Charles and White 1988), but other techniques have also been developed. For example, Franklin, Adis, and Woas (1997) used arboreal emergence traps (**photoeclectors**) and bark brushing to compare the trunk acarofauna of two floodplain forests in the Amazon basin. Trap plants can be used to attract both phytophagous mites and the predators that feed on them. For example, Walter, Azam et al. (1998) used spider mite–infested potted bean plants to sample the predatory mites attracted to the plants. James, Warren, and Whitney (1992) used a microwave oven to "excite" overwintering phytoseiid, tenuipalpid, bdellid, anystid, and tydeid mites out of stems and canes into a surrounding plastic bag. Perhaps the most unusual technique was developed by Zhao and Amrine (1997), who collected the top and bottom 2–3 cm of snow from the roof of a building and surrounding grounds in West Virginia and vacuum-filtered the meltwater through 4 µm nylon filters. They found an amazing array of mites (> 13 families), especially representatives of Eriophyoidea and TARSONEMIDAE.

Canopy fogging, which involves the fumigation of branches, individual trees, or sections of forest canopy with an insecticide (typically a pyrethrin) applied as a mist or fog, has been employed successfully in the mass collection of canopy arthropods, including mites (Majer and Recher 1988; Kitching, Bergelson, et al. 1993; Lowman, Kitching, and Carruthers 1996). The arthropods that fall from the canopy are collected on sheets and sorted under a dissecting microscope. Mites may also be collected from bark by attaching a collecting funnel with an appended vial of ethanol to the trunk being sampled and then spraying the area above the funnel with an insecticide. The sprayed area is swept later with a soft brush to dislodge any remaining arthropods (Walter, Seeman, et al. 1998).

Mushrooms and shelf fungi found on bark and other plant parts (see chapter 6) may be heavily infested by mites, with populations often changing with fungal age and state of decay. Fungi may be extracted in a Berlese funnel, or they may be placed on moist paper toweling in containers kept at room temperature for periodic examination as the fungi decay (Walter, Seeman, et al. 1998). A fine brush is a useful tool for removing mites from the paper surface. Fleshy basidiocarps (mushrooms) tend to be dominated by HISTIOSTOMATIDAE and other mites that favor wet, decaying substrates, but shelf fungi generally have a higher diversity of mites (O'Connell and Bolger 1997).

Freshwater and Marine Mites

Net Collection

Water mites (Hydrachnidiae) that frequent ponds or sluggish streams with little vegetation may often be recovered individually with a tea strainer, or they may be scooped up in numbers with a plankton net (50–150 µm mesh). Thicker aquatic vegetation or flowing water requires sturdier nets with coarser mesh (150–250 µm). These nets should have a relatively long bag because water flows slowly through the fine mesh, and a longer bag traps more water per sweep. Nets with a triangular or pentagonal opening are preferable to a round configuration, and the rim should be covered with stiff canvas to protect it from tearing. A few grommets in the canvas may aid in drainage. Water mites are best collected by sweeping aquatic vegetation or by dragging the net behind as the collector shuffles through bottom sediment. Stream gravel can be stirred up with a hand trowel and rocks overturned while holding the net downstream to capture displaced mites. Specimens may be collected from larger rocks, placed in the net, and scrubbed with a soft brush (Cook 1974b; Smith and Cook 1991; Proctor 2001). Efford (1965) extracted water mites from bottom mud by placing the sample in a concentrated magnesium sulfate solution and then decanting the floating mites.

An almost inevitable consequence of net sampling for aquatic mites is that they must first be separated from the muck, algae, and sediments that accompany them. The more active species can be collected by decanting the suspension into a white porcelain or plastic tray. As the sediments settle, mites (along with other organisms) may be seen swimming around the tray edges. An eyedropper or pipette can then be used to capture them. Mites that rise to the water surface (e.g., many water mite larvae, hydrophilic BLATTISOCIIDAE and HOMOCALIGIDAE) are best collected in a silk stocking net or tea strainer. Examining wet moss and aquatic macrophytes in tray samples may quickly yield a range of relatively large water mites, although this method is not efficient for recovering the more sedentary and less obvious hydrophilic Oribatida and freshwater Halacaroidea. Allowing the sample to sit for longer periods while periodically checking for trails in the sediment will sometimes yield results, but a better method may be to apply an anesthetizing agent such as chloroform to the sample, which causes the mites to relax their hold on the substrate and float free (see Filtration, below). Water-filled cavities in plants (**phytotelmata**) often have their own distinctive acarofauna (Kitching 2001), and collecting them can be as easy as grabbing a handful of wet leaves from a tree hole and sorting them under a dissecting microscope (Fashing 1998).

Filtration

Marine Halacaroidea are not easily collected by conventional means. These nonswimming, mostly nonparasitic forms frequent rocky or sandy bottom sediments or cling tightly to various algal or animal substrates in a variety of brackish and saltwater habitats. Because halacaroids are small and difficult to see, their recovery usually depends on the successful collection of mite-infested substrates. Littoral samples may be collected by conventional means (see above), but successful intertidal bottom collecting may require the initial separation of the mites from their sand or gravel substrates. Krantz (1976) recovered a variety of intertidal arenicolous marine mites from estuarine sand by rapidly digging a hole in the sand to a depth of 30–35 cm and then scooping out the water as it seeped into the hole from the surrounding sand interstices. Samples thus collected were concentrated by muslin bag filtration and stored in 95% ethanol for later study. Newell (1971a) extracted marine mites from subtidal bottom sediments by initially reducing sediment volume through differential screening and then removing silt with a 105 µm sieve. Recovering mites from deep subtidal and benthic substrates requires dredges, trawls, or grabs. A variety of deep-sea samplers have been described by Newell (1971a).

Mite extraction from algae, coral, or colonial invertebrates in the field is best accomplished by placing the substrate sample in a bucket filled with fresh seawater, to which chloroform (15 cm^3 in a 12 L bucket) has been added. The anesthetized mites relinquish their hold on the substrate, a process that is hastened by repeatedly pouring the mixture from one bucket to another. After 30 minutes, the substrate material is removed from the bucket, and the precipitate is separated by filtering through a muslin bag. The filtrate is then inverted into a petri dish for microscopic examination. Alternately, halacaroids can be extracted in the laboratory by washing the sample in a 2 mm mesh sieve over a 70 µm sieve using a jet of water (Green and Macquitty 1987). Staining sample precipitates with Rose Bengal imparts a bright red color to otherwise inconspicuous mites and may make their recovery simpler.

Flotation

Many freshwater Hydrachnidiae are large, active, and brightly colored; consequently, they tend to be singled out for attention by collectors. As a result, some of the less flamboyant water mites, along with the freshwater HALACARIDAE, HYADESIIDAE, and PEZIDAE and the aquatic representatives of the HYDROZETIDAE, LIMNOZETIDAE, TRHYPOCHTHONIIDAE, MALACONOTHRIDAE, ZETOMIMIDAE, ACARIDAE, HISTIOSTOMATIDAE, and BLATTISOCIIDAE, are often missed. These members of the aquatic acarofauna may be collected from stream and lake sediments by specific gravity flotation (e.g., Efford 1965; Pugh 1996c), but hydrocarbon flotation may be preferable. Combining traditional handpicking with kerosene flotation of alcohol-preserved residues can double the diversity of mites recovered from a sediment collection (Proctor 2001). A disadvantage is that the mites in alcohol-fixed samples become difficult to dissect if they are stored for long periods.

Interstitial Sampling

Interstitial and hyporheic zones of rivers and lakes often harbor great numbers of Hydrachnidiae, Halacaroidea, and Oribatida, but these sites have generally been ignored. An indication of the diversity of this cryptic fauna may be obtained by using variations on the method described for recovery of intertidal arenicolous forms (Krantz 1976)—that is, rapidly digging a hole in the sand or gravel bordering a body of water to a depth of 30–40 cm, quickly scooping up the outwash, and filtering it through a muslin bag or a fine net (Smith and Cook 1991). More thorough sampling involves standpipe corers (Williams 1984) or elaborate pipe and pump samplers (Boulton, Valett, and Fisher 1992). The latter methods are more quantitative and less susceptible to contamination from other habitats.

Trapping

Inverted funnel traps for collecting water mites have been devised by Pieczynski (1961), Conroy (1973), and others, and they may be baited with a chemiluminescent lure to attract mites at night (Barr 1979). Bricks, tiles, and synthetic sponges have all been used successfully as artificial aquatic substrates for colonization by mites. The exposed substrates are then removed to the laboratory for extraction. Pugh (1996c) successfully trapped a variety of intertidal mites in the subantarctic by using nylon pan scourers as surrogate algae.

Mite Symbionts of Animals

Hand Collection

Mite associates of both vertebrates and invertebrates may be recovered by collecting and examining the host habitat (nests, runway litter, animal manure, etc.) and the host itself. Careful examination of museum vertebrate specimens under a light microscope often reveals the presence of mites (e.g., feather mites on bird pelts, SPINTURNICIDAE on bats), and these may be removed from their hosts with a fine forceps or a brush without damaging the mites or the host. However, recovery of mites from museum collections of insects can be difficult because the mites, which are dry and brittle, often secrete themselves in deep, integumental recesses on their insect hosts. Breakage of both mites and host specimens may be kept to a minimum by first wetting the mites with a small droplet of water or alcohol and then carefully removing them with a small brush or pin and placing them in 70% alcohol for temporary storage prior to restoration (see Fixing, Preserving, and Preparing Specimens, this chapter). Access to subelytral and axillary sites may require that the host be relaxed in a humidifier before examination.

Robust insects such as beetles can be dipped in hot water to hasten the process.

Live vertebrates should be examined for mites with a jeweler's magnifier headset equipped with a light, which leaves both hands free for manipulating the animal. Fur and feathers, and the skin surface that underlies them, serve as habitat for a variety of mite species, as do the ears, eyes, and inguinal and anal regions. Fur mites are best removed from their mammal hosts by brushing the host over a white porcelain tray; feather mites can be recovered from surfaces of individual plucked feathers by gentle brushing or by hand picking, and then fixed in alcohol for later examination.

An alternative to hands-on examination involves placing the animal in a cloth collecting bag that has been treated with an insecticidal powder and then sorting the mites from the bag residue. Collecting ectoparasites from live hosts in the laboratory may be simplified by anesthetizing the host with an intramuscular injection of a 10% solution of phencyclidine hydrochloride (1 mg/kg) (Vercammen-Grandjean 1971). Furman et al. (1974) found that *Pneumonyssus simicola* Banks (HALARACHNIDAE), a lung mite of monkeys, was effectively recovered from anesthetized macaques by tracheobronchial swab irrigation. Mites also live inside feather quills and in the respiratory, gastrointestinal, and urogenital tracts of many vertebrates. Subdermal nodules, surface scabs, and scales also harbor parasitic mites. These mites are difficult to nearly impossible to collect without resorting to destructive sampling methods. Similarly, internal parasites of invertebrates (TARSONEMIDAE, PODAPOLIPODIDAE, and others) can be recovered only by careful dissection of the host (Husband 1969).

When mites are collected from live vertebrate hosts or from their nests, the potential for disease transmission from the host animal or its waste products, or from the bite of an associated mite parasite, should always be kept in the forefront. A bulb-operated aspirator (Singer 1964) rather than an oral vacuum aspirator is recommended for collecting nest Acari because it eliminates the possibility of accidentally inhaling potentially dangerous disease vectors or acariasis-producing mite species. Nest material lends itself well to Berlese funnel separation and should be considered the method of choice for larger samples.

Live invertebrates collected in the field may be examined for surface mite infestations using a 10X–16X hand lens before placing them in a killing jar or alcohol. Mites that are parasitic or phoretic on arthropods often die in situ when their hosts die, while others may drop onto the jar floor or into the alcohol where they may easily be retrieved. Larvae of parasitengonine mites that frequent aquatic or semiaquatic sites may parasitize mosquitoes and other insects in these habitats (Mullen 1974) and can be recovered with careful examination.

Washing

Mites may be removed from dead vertebrates by immersing the animal in a water-detergent solution (1%–2% soap) and shaking vigorously (Lipovsky 1951). This detaches many of the loosely attached ectoparasites, which may then be decanted into another container for collection with a brush, an eye dropper, or forceps. A more rigorous method involves mechanical shaking on a paint mixer for 5–10 minutes, which may also dislodge some nasal mites and most fur or feather mites into the supernatant (Clayton and Walther 1997). However, mechanical shaking is only moderately effective for removing attached ixodid ticks and subcutaneous parasites. The water-detergent solution is then poured into a Buchner funnel (Fig. 7.1C) in which filter paper has been inserted and is drawn through the filter paper with a sink vacuum. The paper may then be examined for mites, or the mites can be collected from the paper in soap suds cut with ethanol or by hydrocarbon flotation. Although some nasal mites are susceptible to separation by mechanical shaking, they are more easily removed from nasal passages using squirt bottle or hypodermic needle irrigation (Yunker 1961) or by probing sinuses with a needle. Where bird hosts are involved, splitting the bill between the nares may improve mite recovery.

Tissue Dissolution

Bits of skin, ears, and other soft tissues can be dissolved in a hot, saturated potassium hydroxide (KOH) solution, and the mites can be filtered and washed from the colloidal remains. Mites may be recovered from birds or small mammals by dissolving the whole animal in boiling KOH after incubation in trypsin. Exoskeletons of mites and other arthropods are recovered after passing the resulting slurry through a fine mesh screen and rinsing with ethanol and xylene to remove dissolved fats (Clayton and Walther 1997).

Trapping

Because many vertebrate parasites respond positively to carbon dioxide, subliming dry ice placed in animal runways or nest areas is a useful trapping tool. Chiggers (larval TROMBICULIDAE) are among those mite parasites attracted by CO_2, but foreign objects of almost any kind attract unattached trombiculid larvae. Porcelain or plastic disks, dishes, or squares of oilcloth dropped in areas of heavy infestation will almost always have chiggers on and under them within a few minutes (Wharton and Fuller 1952). A small (100 cm^2) square of black plastic dropped in an area thought to have chiggers or other parasitic mites (e.g., MACRONYSSIDAE, DERMANYSSIDAE) is an easy way to perform a semiquantitative survey. Inverted trays placed on the ground overnight often attract large numbers of chiggers, which can be collected when the tray is righted. Detecting mite ectoparasites in and around nests is often facilitated by placing a strip of plastic or cardboard in a contrasting color (black works well) on the ground or in the nest itself. Unattached ticks (Ixodoidea) may be collected by dragging strips of flannel cloth over suspected infestation sites (**flagging**). Ticks attach to the cloth as it passes over them.

Fixing, Preserving, and Preparing Specimens

Typically, mites are killed, fixed, and preserved in 70%–80% ethanol (concentrations of 95%–100% are best for specimens destined for DNA analysis). Prolonged preservation in high concentrations of ethanol, however, tends to harden internal tissues of small arthropods and often renders them unsatisfactory as whole mounts. Soft-bodied mites may become distorted, and the legs of many mites curl beneath their bodies. Lower alcohol concentrations, on the other hand, may allow microbial degradation to occur. Clearing and using gentle heating during mounting, or killing live mites in boiling water, sometimes circumvent these problems. Isopropyl alcohol also is acceptable for preservation, at least for short periods, but specimens tend to deform as they do in ethanol. Lactic acid (65%) is used as a preservative by some workers, although loss of cuticular mass (especially from setae) may occur over time. Various combinations of ethanol, methanol, and glacial acetic acid have been successfully used as preservatives. Saito and Osakabe (1992) tested a number of these formulations on tetranychid and phytoseiid mites and found that **MA80**, a 2:2:1 mixture of methanol (99.5%), glacial acetic acid (99.3%), and distilled water, fixed specimens with legs extended and with little disruption of the cuticle. Specimens stored in MA80 made good light microscope slides when mounted in Hoyer's medium (described on p. 90), but it is recommended that specimens destined for scanning electron microscopy (SEM) be transferred to ethanol after 20 minutes to avoid deformation (Y. Saito, pers. comm.).

Glycerin (1%–3%) added to ethanol fixatives provides protection from desiccation if evaporation occurs, and lactic acid is often used for long-term storage of acariform mites. **Oudemans' fluid**, a mixture of glycerin (5 parts), 70% ethanol (87 parts), and glacial acetic acid (8 parts), is one of the better-known formulations (Hughes 1976). Water mites are usually preserved in **Koenike's fluid**, a pungent but effective preservative. It is recommended that alcohol-preserved specimens be rinsed in water before transferring them to Koenike's fluid (and vice versa).

The formula for Koenike's fluid is as follows:

Glacial acetic acid, 10 mL
Distilled water, 40 mL
Glycerin, 50 mL

Light Microscopy

The study and identification of mites with a light microscope generally require magnifications of 100×–1,000×, but little can be learned even at the highest magnifications if the specimens under study are poorly prepared. The steps involved in slide preparation include maceration, dissection, rinsing, dehydration, mounting, drying, ringing, and labeling.

Maceration

Various macerating agents may be used to prepare mites for microscopic observation. The objective is to dissolve internal tissues (especially muscles), increase transparency, and soften the cuticle for possible dissection. The amount of time needed for maceration depends on the macerant being used, the length of time the mite has been preserved, and the size and type of mite being cleared. Longer preservation times require longer maceration periods. Treating preserved blood- or pigment-filled parasites is facilitated by first puncturing and gently expressing these substances from the body cavity. Freshly collected specimens may need little or no clearing, especially if they are soft bodied and unpigmented. Gentle warming accelerates the rate of maceration, while dilution of the macerant allows for extended periods of treatment (e.g., to accommodate a busy schedule or a weekend).

Lactic acid (60%–95% aqueous solution) is a relatively gentle and slow-acting macerant that is recommended for treating all but the oldest alcohol-preserved specimens. **Lactophenol**, commonly recommended in older texts, should be used with caution because of the known carcinogenic properties of phenol. Treating specimens that have hardened after long periods of alcohol preservation may require use of **Nesbitt's fluid** (see formula below) or a 5%–10% solution of KOH. Both of these substances are extremely corrosive and should be used with care, to protect both oneself (especially the eyes) and the mite specimens being treated. Excessive treatment in corrosives may weaken conjunctiva and cause specimens to fall apart. Gentle heating (45°C) in an oven, on a hot plate, or in a water bath can accelerate maceration, but excessive heating may be deleterious to the specimens.

The formula for Nesbitt's fluid is as follows:

Chloral hydrate, 40 g
Concentrated hydrochloric acid (HCl), 2.5 mL
Distilled water, 25 mL

Dissection

Heavily sclerotized or overly large mites often are difficult to study under the microscope, even when using macerants. Thick, ornamented shields may prevent easy observation of fine structures that are important in comparative studies and identification. A solution to this problem is to carefully separate the dorsum and venter with appropriate dissection tools and, in cases where the structure of the chelicerae and gnathotectum are of importance, to remove the chelicerae as well. The following basic tools are recommended (others are described in Norton and Saunders 1985):

- Fine watchmaker's forceps—straight shaft
- Minuten pin, finely sharpened, inserted in a wood matchstick or bamboo skewer base and fixed in place with paraffin wax or glue
- Microscalpel made of razor blade fragment embedded in wood handle

Specimens to be dissected should first be softened in a macerating agent, washed in several changes of water, and then transferred to 20%–40% ethanol on a spot plate or watch glass for dissection under a stereoscopic microscope. The specimen should be held firmly in the tips of the forceps while the minuten pin is inserted into the soft lateral integument near the vertex. If the specimen is properly softened, the pin may be pulled through the lateral integument posteriorly to a point behind coxae IV without harming the surrounding shields. The forceps should then be used to hold the dorsum or venter gently while the minuten pin is inserted into the opposite side of the specimen, where the tearing process is repeated. Because of their leathery integument, a microscalpel should be used instead of a minuten pin when dissecting ticks (Ixodida), both for the initial incision and for the separation of dorsum and venter. Internal tissues of the separated dorsum and venter should then be picked out with the forceps. The chelicerae may be removed from the gnathosoma by inserting the minuten pin into the open proximal end of the gnathosomatic foramen and forcing the chelicerae to protrude anteriorly, where they may be removed with the forceps. Where the proximal cheliceral segment is elongate and intact, the chelicerae may be removed by gently drawing the proximal segment back through the gnathosoma with the forceps and removing the entire structure from the body cavity.

Dissection leads almost inevitably to a loss of setae, and care should be taken to retain as many as possible. Ideally, dissections should be undertaken only when additional specimens of the same species are available for whole mount study.

Mounting Media

Three general classes of mounting media are used by acarologists: (1) water-miscible mixtures, (2) resin-based permanent mounts, and (3) polyvinyl alcohol (PVA) solutions (Upton 1991, 1993). Water-miscible media built around gum arabic may produce poor results when improperly formulated (Upton 1993), but these essentially semipermanent mountants are preferred because of their ease of use and their excellent optical properties. Resin-based mounts, on the other hand, have higher refractive indices and require extensive preparation with noxious or expensive solvents. Some, such as **Canada balsam** (refractive index = 1.535), are not miscible in alcohol or water and can be used only after passing the specimen through a number of intermediate fixations. Although balsam mounts are permanent, many turn red, brown, or black over time. Some PVA preparations have undesirable properties (Upton 1991), but more recent formulations show promise (Evans 1992) and are discussed below. Glycerin jelly is the preferred mounting medium for many water mite specialists (Newell 1947b; Cook 1974b).

Space limitations preclude mentioning many mountants and mounting techniques in the following discussion. Additional information may be found in a number of sources, including Clark and Morishita (1950); Evans, Sheals, and MacFarlane (1961); Barr (1973); Singer (1967); Jeppson, Keifer, and Baker (1975b); Vercammen-Grandjean (1973a); Krantz (1978); and Upton (1993).

Currently, the most widely used mounting medium for mites is an aqueous mixture containing **gum arabic** and known as **Hoyer's medium** (Singer 1967; Krantz 1978). The proportions of ingredients used in the formulation of Hoyer's medium vary widely among recipes (see one example below), and the lineage of the name is uncertain (Upton 1993). However, like most other aqueous mountants now in use, Hoyer's medium is a modification of Berlese's gum chloral fluid and contains distilled water, gum arabic, chloral hydrate, and glycerin mixed in sequence. Because crystalline gum arabic (i.e., unprocessed amorphic "rocks' tears") is difficult to obtain, the more easily available commercial preparations of powdered gum arabic (gum acacia) often must be substituted. Like any fine powder, some of the tiny gum particles usually fail to wet and dissolve when placed in water. Gentle heating of the gum arabic–water solution may correct the problem, and wetting the powder in ethanol before adding it to the water aliquot may predispose it to dissolve uniformly. Formulating Hoyer's medium is further complicated by the fact that chloral hydrate is a controlled substance in some countries, and a special permit may be required for its purchase.

Although Hoyer's medium has been the preferred choice of many acarologists because of its ease of use and excellent optical properties (refractive index = 1.2), Hoyer's medium slide preparations have a greater tendency to spoil than do slides made with nonaqueous media. Commonly, too little medium is used, and air bubbles invaginate under the coverslip as the slide dries. If the bubbles are not easily accessible for filling with additional Hoyer's medium, the preparation should be immersed in water in a petri dish, the coverslip floated off, and the specimen removed for remounting on another slide. Preparations that are not reprocessed may eventually crystallize or deteriorate, and the medium becomes filled with oily globules or takes on an opaque, granular appearance. The source of these latter abnormalities is not clear, but both presumably represent an interaction between the medium and some contaminant in the clearing agent or sealant, or in the mite itself. Since Hoyer's medium is hygroscopic, it must be ringed with a varnish or paint that prevents water from entering or leaving the mount. Insulating paints such as **Glyptal** tend to be flexible and chemically inert and are more effective vapor barriers than materials such as fingernail polish or asphaltum, which are susceptible to cracking or dissolution in solvents normally used to clean slides. Properly mounted and ringed Hoyer's medium slides may be expected to last for decades, while a poor preparation may spoil in a few months. When, for reasons of inadequate cleaning or poor ringing, deterioration of the preparation occurs, the specimen is easily retrieved by scoring the ring with a metal probe and soaking off the coverslip in water or rehydrating the mount in a humidifier. The freed specimen may be remounted directly onto a new slide.

One of the more widely used formulations of Hoyer's medium (Krantz 1978) is as follows:

Distilled water, 50 mL
Crystalline gum arabic (or gum acacia), 30 g
Chloral hydrate, 200 g
Glycerin, 20 mL

PVA preparations have been recommended by several acarologists over the years, although some earlier formulations often produced unsatisfactory results. **Heinze PVA** (HPVA), however, appears to be an adequate alternative to Hoyer's medium (Evans 1992). Good HPVA preparations require that enough medium be added to the slide to avoid invaginations under the coverslip during drying. Specimens can be recovered from PVA slides by soaking in warm glycerin, lactophenol (see warning, p. 89), or dilute Nesbitt's fluid. PVA preparations vary, and some may yield better results than others. The formula for HPVA (Evans 1992) is as follows:

Polyvinyl alcohol, 10 g
Chloral hydrate, 100 g
Glycerol, 10 g
Distilled water, 60 mL
85%–92% lactic acid, 35 mL

HPVA is prepared as follows:

1. Add water to PVA powder, stirring constantly in a water bath at just below boiling.
2. Add lactic acid and stir for a few minutes.
3. Add glycerol and stir until smooth.
4. Cool until only lukewarm.
5. Dissolve chloral hydrate in mixture.
6. Stir thoroughly, pass through filter paper in a suction funnel (or centrifuge).
7. Store in a brown bottle.

Water mites (Hydrachnidiae) have traditionally been prepared for study by first clearing them in 8%–10% KOH, expressing macerated material from the body cavity, and dissecting off the appendages and placing them adjacent to the body in glycerin jelly (Cook 1974b). Canada balsam, discussed below as a mountant, is the preferred ringing solution for these preparations.

Deterioration problems seen in aqueous mounts may be avoided by using more permanent, resin-based mountants such as **Euparal** (eucalyptus oil and paraldehyde) or Canada balsam. Specimens to be mounted in these media must first be thoroughly cleared to remove muscle tissue and internal organs and usually require an elaborate series of dehydrations and replacements (Rohde 1964; Saito et al. 1993). Canada balsam may be thinned with xylene, while Euparal is miscible in alcohol and can be thinned with absolute isopropyl alcohol or absolute ethanol (Upton 1991). Euparal is preferred over Canada balsam by some researchers (Halliday 1994) because of its superior optical properties (refractive index = 1.483), and it does not discolor over time. However, soft-bodied specimens may become deformed and distorted in Euparal as the medium sets, and the preparation may craze (become powdery) or polymerize. Some synthetic resins such as dimethyl hydrantoin formaldehyde (DMHF) and dimethyl hydrantoin formaldehyde resin (DMHFR) have excellent optical properties and do not require the dehydration of specimens (Steedman 1958; Bameul 1990), but data on how long these preparations will last are unavailable.

Slide Preparation

Making a good slide mount requires practice and more than a modicum of patience. Pitfalls to be avoided include improper centering of the medium droplet on the slide, using too little medium to accommodate the coverslip, and failure to rinse reactive macerants from the specimen prior to transfer to the mounting medium. Centering is made easier by drawing a slide outline on a 3 × 5 card and adding two intersecting diagonal lines from opposite corners to locate its center. The slide may then be placed on the card for precise application of the medium droplet. Using a droplet of the right size depends largely on the size of the coverslip, the thickness of the mite, and the viscosity of the medium. Aqueous media are hygroscopic and gain or lose water depending on relative humidity. Both Hoyer's medium and PVA contract slightly as the preparation dries, and using too much medium is better than using too little. When the slide has dried completely, and prior to ringing, excess media should be scraped off with a razor blade.

Mites may be mounted directly from water, alcohol, or Nesbitt's fluid into aqueous mounting media, but specimens that have been cleared in KOH or lactic acid must first be thoroughly rinsed in water. Mounting directly from KOH causes precipitation and degrading of the medium, while mounting directly from lactic acid eventually results in the growth of crystals that obscure the specimen. Improper rinsing of clearing agents may also result in degradation of Hoyer's medium preparations over time.

Large or heavily ornamented mites may need to be dissected before mounting to expose structures necessary for identification. Alternately, these mites can be studied in cavity slides in lactic acid (Grandjean 1949) or slide-mounted in a lactic acid pool surrounded by protective plastic rings to raise the coverslip (Foulkes 1983). In either case, however, much of the animal will not be available for observation. The carbon block technique developed by Grandjean provides an excellent temporary study platform for larger Oribatida (Grandjean 1949; Evans, Sheals, and MacFarlane 1961). The specimen is placed on the surface of a porous carbon block moistened by 75% ethanol that is delivered from a reservoir by a capillary tube. The mite is clearly observed under incident light because the moist carbon block provides a contrasting black background for it.

For most groups of Mesostigmata, a lateral view of the chelicerae (and, in many Phytoseioidea, the female sperm access system) often is necessary for proper species identification. Crush mounting (using tweezers to gently compress the mite between coverslip and slide) is sometimes a useful alternative to dissection with small mites, although the results often are less than aesthetically pleasing. With a well-cleared specimen (and some practice), it is possible to use crush mounting to quickly shear the dorsal from the ventral shield(s), expose the sperm access system, and flatten the chelicerae for easier observation.

Fresh specimens or those held for only a few months in alcohol may be mounted directly into aqueous media, but the preparation of permanent aqueous or resin-based slide mounts from preserved material should be attempted only after the mite has been taken through appropriate macerating, rinsing, and/or dehydration procedures as earlier described. The following steps are recommended for producing a slide mount:

1. Place a droplet of mounting medium in the center of a clean microscope slide.
2. Lift specimen from the spot plate or watch glass with a fine wire loop affixed to a matchstick base or gently grasp it with watchmaker's forceps.
3. Place the mite in the center of the medium droplet, gently press it to the surface of the slide so that friction will reduce movement when the coverslip is applied, and arrange it on a vertical axis with a minuten pin. If either dorsum or venter is known to be more important for identification (this will vary with the type of mite), orient that surface upward if only a single specimen is available. For dissected specimens, place parts in close enough proximity so that they may be easily located later.
4. Allow any air bubbles trapped in the medium to rise to the surface and dissipate.
5. Using a pair of forceps, apply a clean coverslip to the edge of the meniscus of the slide droplet and allow it to fall slowly into place. Last-second orientation of the specimen can be accomplished under the dissecting microscope with gentle pressure of the probe on the coverslip surface. Circular coverslips measuring 12–15 mm in diameter and #0 (0.08–0.13mm) thickness are appropriate for most mites, but larger coverslips should be used for specimens exceeding 3 mm in length.
6. If the legs or mouthparts are curled out of sight, they can sometimes be successfully extended in freshly made slides by applying light pressure to the coverslip or by gently warming over an alcohol flame or on a hot plate. However, this becomes difficult once the medium at the edge of the coverslip has begun to dry.
7. Turn the slide so that the gnathosoma of the mite is directed posteriorly. Mark the right side of the slide with an identifying number or letter. A diamond pen or indelible marker is recommended, since wax markings often become obliterated in succeeding steps.
8. Place the slide in an oven or slide warmer at 45°C–50°C until the medium around the rim is dry to the touch. Temperatures above 55°C may cause the medium to bubble at the coverslip edge or to darken. Drying times vary with the season and with the thickness of the mount. When the slide is dry, replace the temporary labeling with a permanent label that includes all relevant data (see Fig. 7.2 and step 11, below). Pinpoint the position of small or obscure specimens under the microscope and draw a circle of indelible ink on the underside of the slide to mark the location. This will save time in finding the specimen under the microscope.
9. Because oven heating causes contraction of the medium at the coverslip edges and consequent bowing of the coverslip, heat-treated slides should be held at room temperature for at least a week to allow the coverslips to return to their normal flat state. Pulling away of the medium from the coverslip edges often occurs at this time, and these imperfections should be filled with mounting medium before ringing is attempted.
10. A ring of insulating paint that provides a vapor barrier is recommended for sealing the coverslip of Hoyer's medium mounts (Travis 1968). Applying the sealant may be done with a no. 1 camel's hair brush or with a polyethylene bottle applicator (Wu 1986) on a slide turntable. Remove sealant from used brushes and from applicator bottle plugs in a solvent recommended by the manufacturer.
11. Unlabeled or poorly labeled slides are worthless for scientific research. Relevant information should be placed on two labels, one for collection data and one for identification information (Fig. 7.2). Systematists

tend to put taxonomic information on the left-hand label, while ecologists usually reserve that position for collection information. In any case, collection data should include the locality (preferably with coordinates), date of collection, habitat (including host ID if available), and the name of the collector. The other label should include identification, name of the identifier, year identified, and the mounting medium. Information should be printed in India ink or on professionally printed labels, but tentative identifications should be made in pencil. Additional detailed information on habitat or other aspects of the collection can be stored in a notebook or database and referenced by an accession number or bar code. Slides should be stored horizontally, coverslip up, in boxes or flat trays.

If the primary purpose of slide preparation is taxonomic research (or if a taxonomist is to examine the specimen), then the general rule is one mite per slide. However, mounting dorsal- and ventral-oriented specimens of the same species from the same collection on a single slide is sometimes a useful shortcut. For quantitative studies requiring species identifications, dozens of mites can be mounted on a single slide to save time and materials. Before doing so, however, the collection should be sorted under a dissecting microscope into what appear to be similar taxonomic entities, and representatives should be mounted singly to get a better understanding of what species may be present. Most taxonomists respond unfavorably to slides with numerous mites of different species, so be sure that you have voucher specimens for each species that follow the one mite per slide rule.

Digital Imaging

Digital imaging creates high-resolution descriptions of specimens that can be made available electronically. Such images may be further enhanced by the layering and filtering of focal plane series of slide-mounted mites (Valdecasas and Becerra 1997). Digital image capture is recommended as a backup for type material in the event that the type specimens deteriorate or are lost. Additionally, digital images in expert systems, as surrogate voucher specimens, in electronic identification tools (e.g., matrix-based computer interactive keys), and as vehicles for the rapid exchange of information make them especially important in modern systematic and ecological research (Behan-Pelletier and Newton 1999). Digital images may also be used as templates to produce accurate taxonomic drawings using standard tracing techniques or computer drawing programs.

Electron Microscopy

The study of mites can be greatly enhanced by use of electron microscopy. Transmission electron microscopy (**TEM**) is a valuable tool for understanding mite ultrastructure (Alberti and Coons 1999; Kennaway, Baker, and Ball 2004), but its use requires training and a special proficiency in fixation, embedding, and sectioning techniques (Glauert 1987; Glauert and Lewis 1998) or a generous budget for the professional preparation of specimens. However, **SEM** and a sputter coater (Alberti and Nuzzaci 1996) can produce images with a minimum of preparation time and expense. Ideally, mites to be subjected to SEM should be freshly killed in liquid nitrogen, 5% KOH, or boiling water (Evans and Loots 1972). Traditionally, mite specimens to be used for SEM studies have been fixed in gluteraldehyde or osmium tetroxide and dehydrated before critical point drying, but this approach provides numerous opportunities for losing or damaging specimens, and the results are often less than satisfactory (Achor et al. 2001). Furthermore, these elaborate techniques are not necessary for many mite taxa. In general, the more heavily sclerotized the mite, the less treatment will be required before SEM. For example, most brachypyline and ptychoid oribatid mites (see chapter 15) can be dehydrated in 100% ethanol, transferred to acetone, and dried under a fume hood in a few hours with little or no resulting deformation. Specimens may then be transferred to a SEM stub covered with double-sided sticky tape using a fine brush and sputter coated with gold, gold-palladium, platinum, or carbon.

An effective pre-SEM treatment regime that was originally developed for preparing internal tissues (Nation 1983) involves transferring the specimens from 100% ethanol into a solution of hexamethyldisilazane (**HMDS**). Because this solvent is highly toxic, preparations must be carried out under a fume hood while wearing rubber gloves. Mite transfers to HMDS should be done under a microscope set up under the hood. A microspatula may be used to move the specimens into an excavated block or porcelain spot plate with a few milliliters of HMDS. When processing large numbers of specimens, a small-bore (< 1 mm diameter) pipette is used for transferring the mites to an empty block, which is used again to remove excess ethanol before adding HMDS. The block is then covered with a glass plate and the specimens held for 2–4 hours (depending on the size of the mites) before removing most of the HDMS with a pipette and adding a few milliliters of fresh solution. The block should be covered again for an hour before the lid is removed to allow the HDMS to evaporate under the fume hood.

The HDMS technique is appropriate for many mites and traditional ambient temperature SEM procedures, but soft-bodied mites often collapse or deform. The best approach for dealing with soft-bodied mites, and especially mites on plant tissues, may be low-temperature electron microscopy (Achor et al. 2001). This technique has been used to produce striking images of PYEMOTIDAE, Eriophyoidea, TYDEIDAE, ACARIDAE, and other soft-bodied acarines that are rarely, if ever, free from deformities when prepared using other techniques. Environmental SEM allows the visualization of live specimens on leaf surfaces or specimens affixed to stubs under low vacuum (Ebrahim, Achor, and

Childers 1996). Chilling the stage of an environmental SEM reduces mite movement and lessens the chance that the specimen will escape from the stage.

Molecular Analysis

DNA may often be recovered from specimens stored in 70% ethanol, at least over the short term, but killing and storage in absolute alcohol produce more reliable results. Even more effective is flash freezing of live mites in small plastic tubes in a buffer solution at –70°C or lower. Standard Chelex DNA extraction technique (Hillis et al. 1996) works well with mites of moderate size using one to five specimens finely ground with a pestle in a micromortar. A variety of kits are commercially available that digest DNA from specimens and allow for the retrieval and mounting of the cuticle for use as voucher specimens.

Working with Live Mites

Soil-Litter Mites

Mites have a high surface/volume ratio, desiccate easily, and must be maintained at a relative humidity similar to that which they normally experience in nature. An atmosphere near saturation (> 95% relative humidity) is usually required for observing and rearing soil-litter mites, but some groups adapted to drier habitats may need a lower humidity. In either case, a mixture of plaster of paris, activated charcoal, and water, poured into a zipper vial or screw-cap bottle moistened to the degree appropriate for the mite being observed, can provide a suitable platform for maintaining these mites over time. Plaster normally sets as a crystalline matrix with pores large enough to admit water (and hence to maintain the desired humidity). The plaster by itself is white and tends to obscure small and light-colored mites. However, mixing one part powdered activated charcoal to every two parts of plaster (by volume) with enough water mixed in to make a thick slurry produces a platform of uniform light gray color that provides a contrasting background for mites, is easy on the eyes, and helps to absorb toxins (the original reason for using charcoal in this mixture). Adding too much water to the mixture yields a loose and crumbly plaster, while adding too little produces a virtually nonabsorbent, rock-hard surface that will not absorb much water later on. Observation vials and bottles must be small enough to fit under the microscope objective and must be straight sided (i.e., without necks) for optimal viewing. Shell vials and urine specimen containers work well. Modified Roberston cells drilled from Perspex (acrylic plastic) offer still another type of container for rearing and observing mites (Solomon and Cunnington 1964).

Plant Mites

Observing live plant mites in their native habitat may be as simple as putting a mite-infested leaf or stem under a microscope and using cool (fiber-optic) illumination. Leaves may be kept alive by placing them on water-saturated cotton or sponges, or mites can be restrained within small cells anchored to the leaves. There is an extensive literature on rearing spider mites (e.g., Helle and Overmeer 1985) and phytoseiid mites (Kostiainen and Hoy 1996). Most of the methods involve using water-saturated cotton or sticky barriers to keep mites confined to leaves, but hot wires also may be effective (Yao and Chant 1982). Techniques for rearing eriophyoid mites are reviewed in Oldfield and Perring (1996).

Water Mites

Hydracarina are easily maintained alive and may even be reared (especially those species with nonfeeding, i.e., nonparasitic, larvae) in small dishes, jars, and aquaria. Disposable plastic cell wells are an excellent tool for sorting live water mites during collection and for maintaining individual mites (Smith and Cook 1991).

Vertebrate Parasites

Special techniques often must be devised for the live observation of mite groups other than those discussed above, and this is especially true for parasites of vertebrates. Successful behavioral or life history studies on parasitic mites often require the presence of a host animal or its tissues or fluids, be it a cape buffalo (Sweatman, Walker, and Bindernagel 1969) or a five-lined skink (Durden et al. 2002). Because host animal requirements vary greatly, the techniques used to study their parasites often must be created *de novo* for each new situation. Regrettably, it is beyond the scope of this book to explore these techniques.

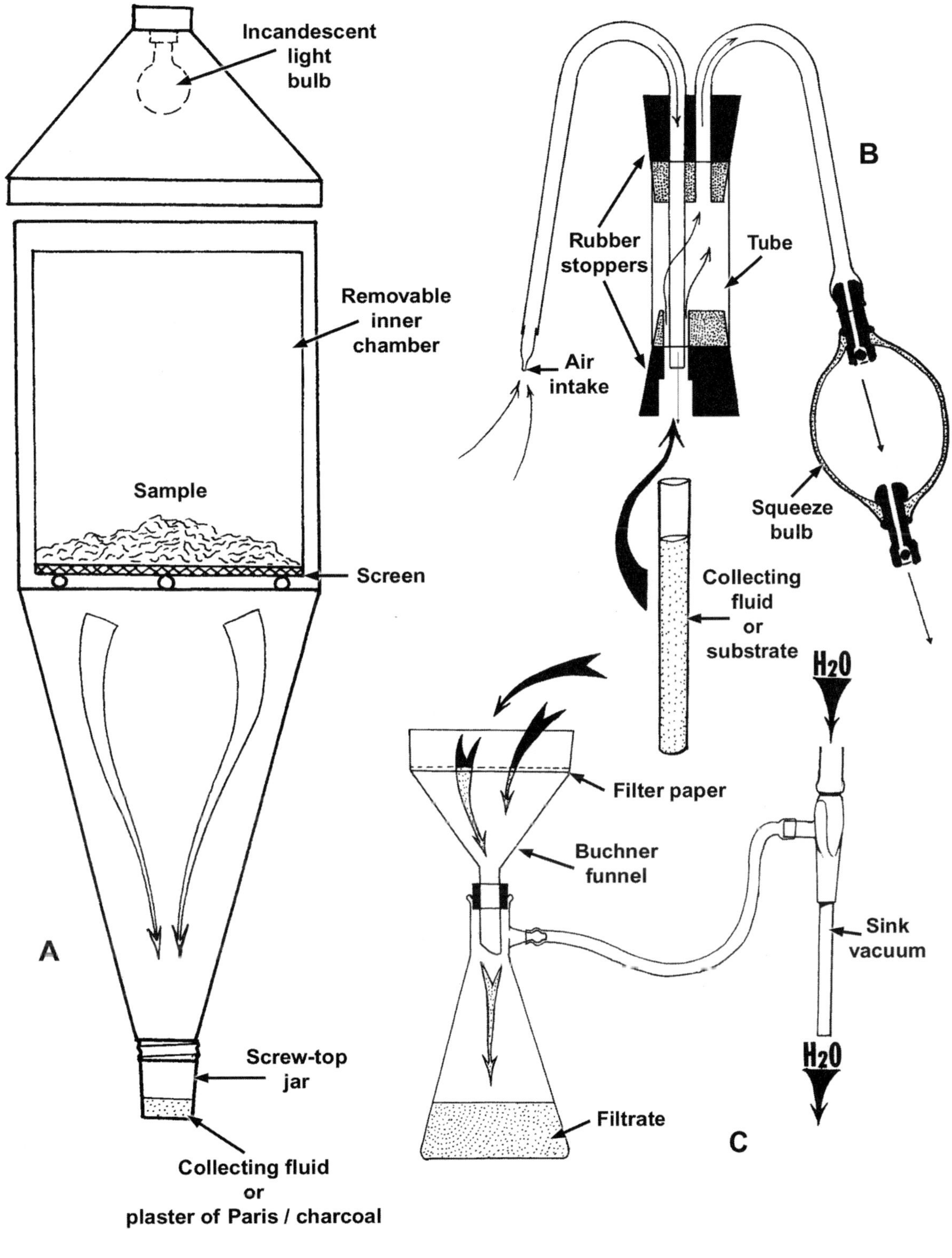

Fig. 7.1. Apparatus for concentrating and collecting acarines from sample substrates. **A**, modified Tullgren apparatus; **B**, Singer aspirator (after Singer 1964); **C**, Buchner funnel apparatus for vacuum filtration.

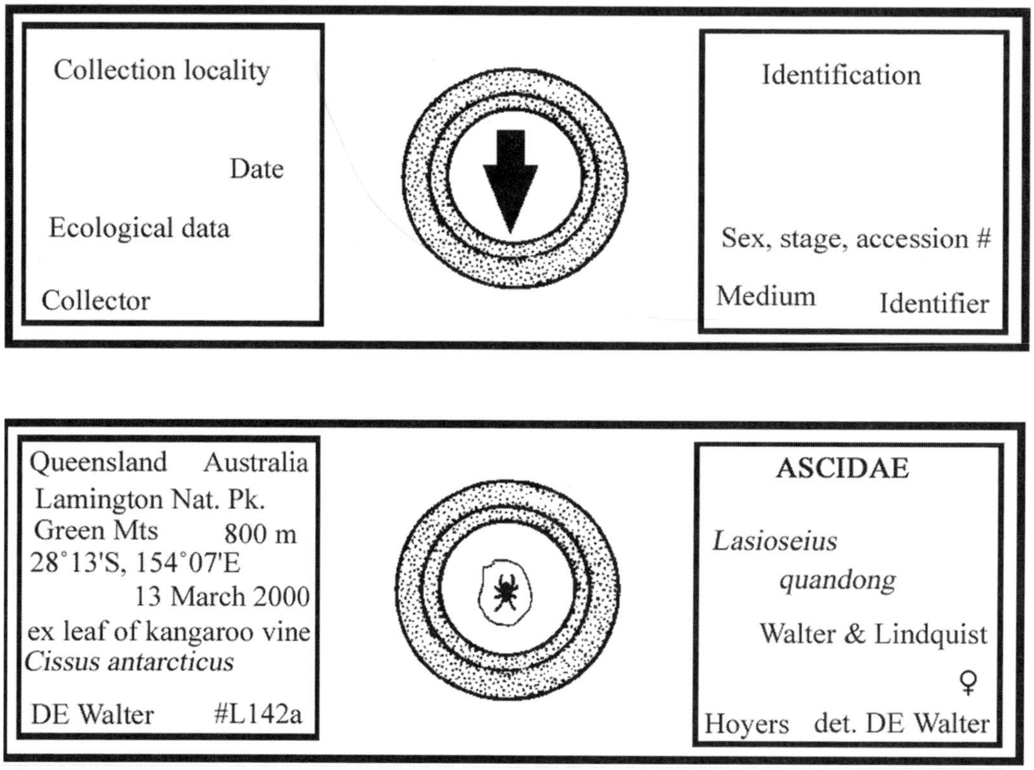

Fig. 7.2. Microscope slide preparation illustrating specimen orientation and an ecological-style label configuration, with identity label on the right.

CHAPTER EIGHT

E.E. LINDQUIST
G.W. KRANTZ
D.E. WALTER

Classification

Knowledge in the field of acarine systematics has increased dramatically over the last three decades. There have been major advances in our understanding of structural homologies, the anatomy of reproductive systems and their products, chromosomal patterns and genetic systems, and molecular markers that may serve as indicators of phylogenetic relationships among mite taxa. Discoveries of new mite species, genera, and even families have continued apace, and many of these findings have challenged existing familial- and generic-level concepts. Also, the widespread acceptance and application of the methods of phylogenetic systematics, or **cladistics**, to diagnoses of acarine groups have produced both new insights and major upheavals in acarine classification. Despite these advances, however, the higher classification of Acari, like that of the Arthropoda in general (Blaxter 2001), remains largely unresolved. The factors chiefly responsible for this disagreement are the application of different conceptual methodologies and the use of dissimilar character or data sets. Furthermore, current classifications are biased in that they are generally based on only a moderate knowledge of the acarofaunas of temperate Eurasia and North America and a severely limited understanding of those in other parts of the world, including the warm temperate to tropical habitats and the cool temperate regions of the Southern Hemisphere. A better resolution of the phylogenetic relationships and classification of mites must await a vastly improved knowledge of the world acarofauna, along with application of rigorous cladistic methodologies to a broader array of attributes, including the fine structure of reproductive and other organ systems, sensory receptors (especially of the palp and tarsus I), and genetic and molecular marker elements such as nuclear and mitochondrial DNA sequences.

The classification and keys to higher categories presented below reflect an ongoing attempt to bring our knowledge of the Acari up to date. The Acari is tentatively assumed to be monophyletic (1) because of the primitive sharing of a variety of presumably apomorphic traits by its two composite lineages, the Parasitiformes and the Acariformes, and (2) because there is no convincing evidence as yet that these two major lineages (superorders) are more closely related to any other group of Arachnida than they are to each other (Lindquist 1984; Norton et al. 1993; Alberti 2005). However, a possible relationship between Acariformes and Solifugae suggested by Grandjean (1954c) and more formally hypothesized by Van der Hammen (1989) is supported by similarities in the fine structure of sperm cells (Alberti and Peretti 2002). Thus, the scheme presented here is only one of many possible treatments that may differ at the highest levels, depending on whether the Acari (1) is considered a natural (monophyletic) or an artificial (di- or polyphyletic) assemblage or (2) represents a subclass or superorder of Arachnida. Although the fossil record of Acari is meager (see chapter 2), it clearly shows that diverse forms of early derivative acariform mites assignable to extant superfamilial groups existed during Devonian times (Norton, Bonamo, et al. 1988; Kethley et al. 1989). The ancestral acariform stock may therefore have arisen as one of the earliest lineages of terrestrial arachnids during the late Silurian period. Such a scenario would preclude further consideration of certain other arachnid orders as possible sister groups or out-groups of the Acari, if there is reasonable doubt that their lineages extend back into Silurian times. Admittedly, there is as yet no fossil evidence of such antiquity in the lineage of the Parasitiformes, which means that this same scenario could be used to argue against a sister relationship with their acariform counterparts.

Within each of the superorders Parasitiformes and Acariformes, some major classificatory changes have been made to reflect the results of current, widely accepted concepts in a cladistic context. For example, the Parasitiformes (or Anactinotrichida) here includes the order Opilioacarida (or Notostigmata), since the latter is more closely related to the other parasitiform orders than it is to any other higher taxon of Acari (Lindquist 1984; Evans 1992; Alberti 2005; M. A. Moraza and E. E. Lindquist, unpublished obs.). The Trigynaspidina is retained as a probable monophyletic group within the Mesostigmata, but the monogynaspidine counterpart is apomorphically definable only when divested of the early derivative Sejida. Within the Acariformes, some members of the paraphyletic endeostigmatan assemblage are treated here as primitive members of the Trombidiformes, while others are considered to be among the most primitive taxa of the Sarcoptiformes (OConnor 1984c; Norton et al. 1993). Recent findings in regard to the sarcoptiform Oribatida and "Astigmata" are especially interesting in terms of the traditional concepts behind these names. There is strong evidence that the Astigmata is a monophyletic group derived well within the Oribatida (Norton 1998), which means that the Oribatida is paraphyletic in the traditional sense of previous classifications. Therefore, if the Oribatida is to be treated as a natural (monophyletic) category, it must include the Astigmata as a subordinate group. These issues are considered further under the chapters for each of these major taxa.

From a phylogenetic perspective, certain acarine groups are now seen as having been overly ranked in previous classifications, and modifications are made in the present treatment to address this problem. For example, the Astigmata, previously ranked as a separate order, is lowered to cohort rank (Astigmatina) in the oribatid supercohort Desmonomatides, while the Parasitengona and Tarsonemina (or Heterostigmata), previously ranked as separate suborders in some earlier treatments (see relevant notes on pp. 102–3), are lowered to cohort rank as subsets of the supercohorts Anystides and Eleutherengonides, respectively. Although they may be justifiable on phylogenetic grounds, such lowered rankings create much compression of categories within the larger acarine suborders and orders. For example, cohorts like the Astigmatina and Parasitengonina, whose origins may extend back to at least the Triassic, are comparable to many groups of Insecta that are ranked as orders in entomological classifications.

When used as a higher category, the level of cohort is often placed between the levels of class and order (e.g., as in the Mammalia). However, with few exceptions (e.g., see Woolley 1988), the cohort has been used as a category between ordinal and family levels in acarine classifications since the time of Oudemans (1923). The same usage is retained in this treatment. Suffixes for supercohort and cohort names are standardized throughout the subclass (*-ides*, *-ina*, respectively). Where they are used, subcohort names are given the ending *-ae*. Due to constraints in the hierarchic system of zoological classification and in the number of formal categories that are traditionally recognized, the phyletic sequencing convention of Nelson (1974), as elaborated by Wiley (1979), is sometimes applied here at the superfamilial and higher levels, with an admission that not all taxa treated as superfamilies are equivalent in rank.

Key references to earlier literature on fossil evidence and higher classification are listed in chapter 2. Notes to alternative category names and rankings presented below are limited to major references dating from the publication of the second edition of this book (Krantz 1978).

Classification of the Higher Categories of the Subclass Acari

I. SUPERORDER PARASITIFORMES[1]
- A. Order Opilioacarida[2]
  - 1. Superfamily Opilioacaroidea
- B. Order Holothyrida[3]
  - 1. Superfamily Holothyroidea
- C. Order Ixodida[4]
  - 1. Superfamily Ixodoidea
- D. Order Mesostigmata[5]
  - 1. Suborder Sejida[6]
    - 1. Superfamily Sejoidea[7]
  - 2. Suborder Trigynaspida[8]
    - 1. Cohort Cercomegistina[9]
      - 1. Superfamily Cercomegistoidea
    - 2. Cohort Antennophorina[10]
      - 1. Superfamily Antennophoroidea
      - 2. Superfamily Celaenopsoidea
      - 3. Superfamily Fedrizzioidea
      - 4. Superfamily Megisthanoidea
      - 5. Superfamily Parantennuloidea
      - 6. Superfamily Aenictequoidea
  - 3. Suborder Monogynaspida[11]
    - 1. Cohort Microgyniina[12]
      - 1. Superfamily Microgynioidea
    - 2. Cohort Heatherellina[13]
      - 1. Superfamily Heatherelloidea
    - 3. Cohort Uropodina[14]
      - a. Subcohort Uropodiae[15]
        - 1. Superfamily Thinozerconoidea
        - 2. Superfamily Polyaspidoidea
        - 3. Superfamily Uropodoidea
        - 4. Superfamily Trachyuropodoidea
      - b. Subcohort Diarthrophalliae[16]
        - 1. Superfamily Diarthrophalloidea
    - 4. Cohort Heterozerconina[17]
      - 1. Superfamily Heterozerconoidea
    - 5. Cohort Gamasina[18]
      - a. Subcohort Epicriiae[19]
        - 1. Superfamily Epicrioidea
        - 2. Superfamily Zerconoidea
      - b. Subcohort Arctacariae[20]
        - 1. Superfamily Arctacaroidea

c. Subcohort Parasitiae[21] (Parasitina, Neotocospermata)
   1. Superfamily Parasitoidea
d. Subcohort Dermanyssiae[22] (Dermanyssina, Neopodospermata)
   1. Superfamily Veigaioidea
   2. Superfamily Rhodacaroidea
   3. Superfamily Eviphidoidea
   4. Superfamily Ascoidea
   5. Superfamily Phytoseioidea
   6. Superfamily Dermanyssoidea

II. SUPERORDER ACARIFORMES[23]
- A. Order Trombidiformes[24]
  - 1. Suborder Sphaerolichida[25]
    - 1. Superfamily Lordalycoidea
    - 2. Superfamily Sphaerolichoidea
  - 2. Suborder Prostigmata[26]
    - A. Supercohort Labidostomatides[27]
      - 1. Superfamily Labidostomatoidea
    - B. Supercohort Eupodides[28]
      - 1. Superfamily Bdelloidea
      - 2. Superfamily Halacaroidea
      - 3. Superfamily Eupodoidea
      - 4. Superfamily Tydeoidea
      - 5. Superfamily Eriophyoidea[29]
    - C. Supercohort Anystides
      - 1. Cohort Anystina[30]
        - 1. Superfamily Caeculoidea
        - 2. Superfamily Adamystoidea
        - 3. Superfamily Anystoidea
        - 4. Superfamily Paratydeoidea
        - 5. Superfamily Pomerantzioidea
      - 2. Cohort Parasitengonina[31]
        - a. Subcohort Erythraiae[32]
          - 1. Superfamily Calyptostomatoidea
          - 2. Superfamily Erythraeoidea
        - b. Subcohort Trombidiae[33]
          - 1. Superfamily Tanaupodoidea
          - 2. Superfamily Chyzerioidea
          - 3. Superfamily Trombidioidea
          - 4. Superfamily Trombiculoidea
        - c. Subcohort Hydrachnidiae[34]
          - 1. Superfamily Hydryphantoidea
          - 2. Superfamily Eylaoidea
          - 3. Superfamily Hydrovolzioidea
          - 4. Superfamily Hydrachnoidea
          - 5. Superfamily Lebertioidea
          - 6. Superfamily Hygrobatoidea
          - 7. Superfamily Arrenuroidea
        - d. Subcohort Stygothrombiae[35]
          - 1. Superfamily Stygothrombidioidea
    - D. Supercohort Eleutherengonides[36]
      - 1. Cohort Raphignathina[37]
        - 1. Superfamily Myobioidea
        - 2. Superfamily Pterygosomatoidea
        - 3. Superfamily Raphignathoidea
        - 4. Superfamily Tetranychoidea
        - 5. Superfamily Cheyletoidea
      - 2. Cohort Heterostigmatina[38]
        - 1. Superfamily Tarsocheyloidea
        - 2. Superfamily Heterocheyloidea
        - 3. Superfamily Dolichocyboidea
        - 4. Superfamily Trochometridioidea
        - 5. Superfamily Scutacaroidea
        - 6. Superfamily Pygmephoroidea
        - 7. Superfamily Pyemotoidea
        - 8. Superfamily Tarsonemoidea
- B. Order Sarcoptiformes[39]
  - 1. Suborder Endeostigmata
    - 1. Cohort Alycina[40]
      - 1. Superfamily Alycoidea
    - 2. Cohort Nematalycina[41]
      - 1. Superfamily Nematalycoidea
    - 3. Cohort Terpnacarina[42]
      - 1. Superfamily Oehserchestoidea
      - 2. Superfamily Terpnacaroidea
    - 4. Cohort Alicorhagiina[43]
      - 1. Superfamily Alicorhagioidea
  - 2. Suborder Oribatida[44]
    - A. Supercohort Palaeosomatides (Palaeosomata)[45]
      - 1. Superfamily Acaronychoidea
      - 2. Superfamily Palaeacaroidea
      - 3. Superfamily Ctenacaroidea
    - B. Supercohort Enarthronotides (Enarthronota)[46]
      - 1. Superfamily Brachychthonioidea
      - 2. Superfamily Atopochthonioidea
      - 3. Superfamily Hypochthonioidea
      - 4. Superfamily Protoplophoroidea
      - 5. Superfamily Heterochthonioidea
    - C. Supercohort Parhyposomatides (Parhyposomata)[47]
      - 1. Superfamily Parhypochthonioidea

D. Supercohort Mixonomatides (Mixonomata)[48]

1. Superfamily Nehypochthonioidea
2. Superfamily Eulohmannioidea
3. Superfamily Perlohmannioidea
4. Superfamily Epilohmannioidea
5. Superfamily Collohmannioidea
6. Superfamily Euphthiracaroidea
7. Superfamily Phthiracaroidea

E. Supercohort Desmonomatides (Desmonomata)[49]

1. Cohort Nothrina[50]
    1. Superfamily Crotonioidea
2. Cohort Brachypylina[51]
    1. Superfamily Hermannielloidea
    2. Superfamily Neoliodoidea
    3. Superfamily Plateremaeoidea
    4. Superfamily Damaeoidea
    5. Superfamily Cepheoidea
    6. Superfamily Polypterozetoidea
    7. Superfamily Microzetoidea
    8. Superfamily Ameroidea
    9. Superfamily Eremaeoidea
    10. Superfamily Gustavioidea
    11. Superfamily Carabodoidea
    12. Superfamily Oppioidea
    13. Superfamily Tectocepheoidea
    14. Superfamily Hydrozetoidea
    15. Superfamily Ameronothroidea
    16. Superfamily Cymbaeremaeoidea
    17. Superfamily Eremaeozetoidea
    18. Superfamily Licneremaeoidea
    19. Superfamily Phenopelopoidea
    20. Superfamily Achipterioidea
    21. Superfamily Oribatelloidea
    22. Superfamily Oripodoidea
    23. Superfamily Ceratozetoidea
    24. Superfamily Galumnoidea
3. Cohort Astigmatina (Astigmata)[52]
    1. Superfamily Schizoglyphoidea
    2. Superfamily Histiostomatoidea
    3. Superfamily Canestrinioidea
    4. Superfamily Hemisarcoptoidea
    5. Superfamily Glycyphagoidea
    6. Superfamily Acaroidea
    7. Superfamily Hypoderatoidea
    8. Superfamily Pterolichoidea
    9. Superfamily Analgoidea
    10. Superfamily Sarcoptoidea

Key 8.1. Key to the Superorders, Orders, and Suborders of Acari

1a. With 1–4 pairs of dorsolateral or ventrolateral stigmata posterior to coxae II; coxae of legs free, usually movable; tarsi of legs II–IV with peripodomeric fissure associated with slit organs; tarsus of leg I with dense dorsal cluster of solenidiform setae subdistally (these may be further elaborated into receptor organ complexes) **Superorder Parasitiformes—2**

1b. Without visible stigmata posterior to coxae II; coxae of legs integrated with venter of podosoma and often forming coxisterna; tarsi of legs II–IV without peripodomeric fissure and slit organs; tarsus of leg I with sparse pairings of dorsal setae distally and subdistally **Superorder Acariformes—7**

2a. Tarsus of palp with 1 or 2 terminal claws; adult opisthosoma with 4 pairs of dorsolateral stigmata posterior to level of coxae III; anus terminal; trochanters III and IV divided into 2 podomeres **Order Opilioacarida**

2b. Tarsus of palp without terminal claws, at most with a tined clawlike structure on inner basal surface; opisthosoma with 1 pair of ventrolateral stigmata in region lateral to coxae II–IV or posterior to coxa IV; anus ventral or ventro-subterminal; trochanters III and IV undivided **3**

3a. Subcapitulum with hypostome modified into a piercing organ with retrorse teeth but without corniculi; tarsus of palp without a tined apotelic claw; opisthosomatic stigmata without elongate peritremes; dorsum of tarsus I with a sensory complex (Haller's organ, Fig. 11.1A, p. 118) consisting of a deep posterior pit and a less well-defined anterior depression, both with sensory setae **Order Ixodida**

3b. Subcapitulum without hypostomatic modifications as above but with a pair of hornlike, setiform, or pliable corniculi; tarsus of palp usually with a tined apotelic claw; opisthosomatic stigmata usually with elongate peritremes extending anteriorly; dorsum of tarsus I without distinct Haller's organ, at most with a single pit or depression **4**

4a. Venter of subcapitulum with 5 or more pairs of setae (excluding the corniculi); tritosternum absent or, if present, consisting of a pair of flagellate laciniae on a small base; anal valves of adults with 2 or more pairs of setae; bases of chelicerae enclosed by pliable cuticle, epistome (gnathotectum) lacking **Order Holothyrida**

4b. Venter of subcapitulum with a maximum of 4 pairs of setae (excluding the corniculi); tritosternum usually present, with a distinctive base and 1–2 setulose laciniae (these structures reduced or absent in some parasitic taxa); anal valves of adults nude or at most with 1 pair of setae; bases of chelicerae enclosed by a sclerotized ring, usually with projecting epistome **Order Mesostigmata—5**

5a. Oviporus covered by a single plate (epigynal shield) with 0–1 pair of setae (rarely with 4 or 5 setae); tarsus of leg IV of deutonymphs and adults with a maximum of 18 setae, lacking setae *av4* and *pv4* and ventral intercalary sclerite **Suborder Monogynaspida**

5b. Oviporus covered by a single large shield (epigynal shield) usually bearing 6 to many setae (rarely 2 or 4 setae in some *Sejus*) or by a complex of 2–4 genital shields or their remnants; tarsus of leg IV of deutonymphs and adults with a minimum of 20 setae, setae *av4* and *pv4* present (absent in the paedomorphic millipede-associated NEOTENOGYNIIDAE, minute in some groups, e.g., PROMEGISTIDAE), usually on a ventral intercalary sclerite between the basitarsus and telotarsus **6**

6a. Oviporus covered by a single large shield (epigynal shield) bearing 6 (rarely 2–4) to many setae and often notched or excavated anteriorly; chelicerae without excrescences **Suborder Sejida**

6b. Oviporus covered by 2–4 shields (2 latigynals, a mesogynal, and rarely a sternogynal shield) that may be variously coalesced or reduced; mesogynal shield nude, usually subtriangular but often reduced or insensibly fused to other elements, latigynal shields each with one to many setae, free or fused posteriorly to ventral elements and/or medially to each other; movable cheliceral digit with medial or terminal dendritic, brushlike, or filamentous ventral excrescences **Suborder Trigynaspida**

7a. Chelicerae rarely chelate, fixed digit often regressed and movable digit usually a hook, knife, needle or stylet-like structure; cheliceral bases sometimes fused medially; palpi simple or modified into a thumb-claw process (Figs. 13.22A–C, p. 348), sometimes reduced; subcapitulum without rutella; ambulacra of at least legs II and III usually with 2 lateral claws and with or rarely without a median empodium that may be padlike or rayed and often armed with tenent hairs (Figs. 13.22A–C), or occasionally claw- or suckerlike; opisthosoma lacking paired lateral glands; opisthosomatic setal row *c* usually with 2 pairs of setae (*c1–c2*), rarely with 3 pairs or hypertrichous; tracheal system with 1 pair of stigmata opening between bases of chelicerae or on anterior prodorsum usually present and sometimes associated with peritremes dorsally on the cheliceral bases or on the anterior margin of prodorsum **Order Trombidiformes—8**

7b. Chelicerae typically chelate, usually dentate, rarely attenuate or styletlike; cheliceral bases always separate; palpi simple, never with thumb-claw process; subcapitulum usually with rutella or pseudorutella; ambulacra of legs I–IV usually with 1 or 3 claws, bidactyl condition rare, empodium clawlike or suckerlike, never padlike, rarely rayed; opisthosoma usually with pair of lateral glands (these are absent in more primitive taxa); opisthosomatic setal row *c* usually with 3–4 pairs of setae or hypertrichous; tracheal system absent or, when present, arising from bases of legs or as brachytracheae on various parts of the legs or idiosoma; stigmata and peritremes never present between cheliceral bases or on prodorsum **Order Sarcoptiformes—9**

8a. Tracheal system with 1 pair of stigmata opening between bases of chelicerae or on anterior prodorsum usually present (secondarily absent in Eriophyoidea, STIGMAEIDAE, some Dolichocyboidea); prodorsum usually with 4 or fewer pairs of setae or hypertrichous, sometimes including 1–2 pairs of bothridial sensilla; chelicerae rarely chelate (e.g., LABIDOSTOMATIDAE, RHAGIDIIDAE, some BDELLIDAE), usually with fixed digit sheathlike or completely regressed; coxal fields contiguous or II–III separated **Suborder Prostigmata**

8b. Tracheal system absent; prodorsum with 3 or 6 pairs of setae, including 2 pairs of filamentous bothridial sensilla, sometimes in a common pit (LORDALYCIDAE); chelicerae chelate and with elaborate dentition; coxal fields contiguous **Suborder Sphaerolichida**

9a. Typically small, soft-bodied mites with at most an indistinct prodorsal shield; genital valves present but not sclerotized in adults; leg tarsi with unpaired empodial claw usually present, simple or rayed (paired lateral claws present or absent); opisthosoma without lateral glands and with setae typically relatively short and often highly branched, wedge-shaped, dendritic or stellate. EITHER oval to globose mites with 5–6 pairs of prodorsal setae (or sometimes hypertrichous), of which 1–2 pairs usually are bothridial, OR elongate to wormlike mites (often with annulations), usually with reduced prodorsal setation, no bothridia, and sometimes with an unpaired rostral seta **Suborder Endeostigmata**

9b. Minute to large mites usually with a distinct prodorsal shield or fully sclerotized prodorsum and sometimes with extensive idiosomatic sclerotization; prodorsum with 6 or fewer pairs of setae, of which 0–1 pair is bothridial; genital valves absent or present and usually well sclerotized in adults; leg tarsi with unpaired empodial claw present and never rayed, or reduced and often integrated into a pulvillar pad, or absent (paired lateral claws present or absent); opisthosoma usually with lateral glands, and with setae sometimes elongate and variously ornamented, but rarely highly branched **10**

10a. Prodorsum without specialized sensory organs other than setiform setae; genital aperture exposed or partially covered by paragenital flaps, inversely V, U, or Y shaped, with usually 2 pairs of genital papillae in the adult or these variously reduced or modified; anal aperture without conspicuous paired plates; adult idiosoma usually weakly sclerotized, with epimeral plates undeveloped or weakly formed between bases of the legs; leg pretarsus with an empodial claw and fleshy pulvillus, or pretarsus suckerlike and true paired claws absent (Figs. 16.29F, 16.47D, pp. 630, 648); pretarsi III–IV often modified or absent in parasitic forms; palpi with only 2 segments, rarely 3; adult males with a sclerotized aedeagus, and often with a pair of copulatory adanal suckers **Suborder Oribatida, Cohort Astigmatina**

10b. Prodorsum usually with a pair of specialized setae arising from sensory pits or bothridia (pseudostigmatic organs); genital aperture usually covered by paired, bomb bay–like sclerotized valves, covering usually 3 pairs of genital papillae in the adult; anal aperture similarly covered by sclerotized valves; adult idiosoma usually well sclerotized, with well-developed epimeral plates between bases of the legs; leg pretarsi various but often with paired true claws with a median clawlike empodium (tridactyly), or with only the empodial claw present (monodactyly); occasionally bidactylous; no parasitic forms, and pretarsi III–IV usually not conspicuously modified; palpi usually with 5 segments, rarely 2–4; adult males without a sclerotized aedeagus and without adanal suckers **Suborder Oribatida (excluding Astigmatina)**

■ Notes

1. Order PARASITIFORMES of Krantz (1978); Superorder ANACTINOTRICHIDA of Van der Hammen (1989) and Evans (1992); Cohort PARASITIFORMES of Woolley (1988); Suborder PARASITIFORMES of Karg (1993b).
2. Suborder OPILIOACARIDA of Krantz (1978); Order OPILIOACARIFORMES of Johnston (1982) and Walter and Proctor (1999); Order OPILIOACARIDA of Woolley (1988) and Van der Hammen (1989); Order NOTOSTIGMATA of Evans (1992).
3. Suborder HOLOTHYRIDA of Krantz (1978) and Walter and Proctor (1999); Suborder HOLOTHYRINA of Johnston (1982); Order HOLOTHYRIDA of Woolley (1988), Van der Hammen (1989), and Evans (1992).
4. Suborder IXODIDA of Krantz (1978), Johnston (1982), and Walter and Proctor (1999); Order IXODIDA of Woolley (1988), Van der Hammen (1989), and Evans (1992).
5. Suborder GAMASIDA of Krantz (1978); Suborder MESOSTIGMATA of Johnston (1982) and Walter and Proctor (1999); Order GAMASIDA of Woolley (1988) and Van der Hammen (1989); Order MESOSTIGMATA of Evans (1992).
6. Cohort SEJINA of Krantz (1978).
7. Cohort SEJINA of Johnston (1982); Suborder SEJINA of Woolley (1988) and Evans (1992).
8. Supercohort TRIGYNASPIDES of Krantz (1978) and Krantz and Ainscough (1990).
9. Cohort CERCOMEGISTINA of Krantz (1978) and Johnston (1982); Suborder CERCOMEGISTINA of Woolley (1988) and Evans (1992).
10. Cohort ANTENNOPHORINA of Krantz (1978) and Johnston (1982); Suborder ANTENNOPHORINA of Woolley (1988) and Evans (1992).
11. Supercohort MONOGYNASPIDES of Krantz (1978) and Krantz and Ainscough (1990).
12. Cohort MICROGYNIINA of Johnston (1982); Suborder MICROGYNIINA of Woolley (1988) and Evans (1992).
13. Cohort HEATHERELLINA, **new cohort.**
14. Cohort UROPODINA of Krantz (1978) and Norton et al. (1993).
15. Cohort UROPODINA of Johnston (1982) and Karg (1989); Suborder UROPODINA of Woolley (1988) and Evans (1992).
16. Cohort DIARTHROPHALLINA of Johnston (1982); Suborder DIARTHROPHALLINA of Woolley (1988) and Evans (1992).
17. Cohort HETEROZERCONINA of Johnston (1982).
18. Cohort GAMASINA Krantz (1978) and Karg (1993b).
19. Cohort EPICRIINA of Johnston (1982); Suborder EPICRIINA of Woolley (1988) and Evans (1992).
20. Cohort ARCTACARINA of Johnston (1982); Suborder ARCTACARINA of Woolley (1988) and Evans (1992).
21. Cohort PARASITINA of Johnston (1982); Suborder PARASITINA of Woolley (1988) and Evans (1992).
22. Cohort DERMANYSSINA of Johnston (1982); Suborder DERMANYSSINA of Woolley (1988) and Evans (1992).
23. Order ACARIFORMES of Krantz (1978), Kethley (1982), and Walter and Proctor (1999); Cohort ACARIFORMES of Woolley (1988); Superorder ACTINOTRICHIDA of Van der Hammen (1989) and Evans (1992).
24. Suborder ACTINEDIDA of Krantz (1978); Suborder PROSTIGMATA of Kethley (1982); Suborder TROMBIDIFORMES of OConnor (1984c); Order ACTINEDIDA of Woolley (1988) and Van der Hammen (1989); Order PROSTIGMATA of Evans (1992).
25. Infraorder SPHAEROLICHIDA of OConnor (1984c).
26. Supercohort PROMATIDES of Krantz (1978); Infraorder PROSTIGMATA of OConnor (1984c).

27. Cohort LABIDOSTOMMATINA of Krantz (1978) and Kethley (1982); Suborder LABIDOSTOMMATINA of Woolley (1988) and Evans (1992); Suborder LABIDOSTOMMINA of Van der Hammen (1989).
28. Cohort EUPODINA of Krantz (1978) and Kethley (1982); Suborder EUPODINA of Woolley (1988) and Evans (1992); Suborder BDELLINA of Van der Hammen (1989).
29. Suborder TETRAPODILI of Woolley (1988); Suborder PHYTOPTINA of Van der Hammen (1989).
30. Subcohort ANYSTAE of Krantz (1978); Cohort ANYSTINA of Kethley (1982); Suborder ANYSTINA of Woolley (1988), Van der Hammen (1989), and Evans (1992).
31. Subcohort PARASITENGONAE of Krantz (1978); Cohort PARASITENGONA of Kethley (1982); Suborder PARASITENGONA of Woolley (1988) and Evans (1992); Suborder TROMBIDINA of Van der Hammen (1989).
32. Subcohort ERYTHRAEINA of Welbourn (1991).
33. Subcohort TROMBIDIINA of Welbourn (1991).
34. Phalanx HYDRACHNIDIA of Krantz (1978); Subcohort HYDRACARINA of Welbourn (1991).
35. Subcohort STYGOTHROMBIAE, **new subcohort.**
36. ELEUTHERENGONA of Lindquist (1976); Cohort ELEUTHERENGONINA of Krantz (1978); Suborder RAPHIGNATHINA of Van der Hammen (1989).
37. Subcohort RAPHIGNATHAE of Krantz (1978).
38. HETEROSTIGMATA of Lindquist (1976); Subcohort HETEROSTIGMAE of Krantz (1978); Suborder HETEROSTIGMATA of Woolley (1988); Suborder TARSONEMIDA of Van der Hammen (1989); Suborder TARSONEMINA of Evans (1992).
39. Suborders ACARIDIDA and ORIBATIDA of Krantz (1978); Suborders ASTIGMATA and ORIBATIDA of OConnor (1982a) and Johnston (1982); Suborder SARCOPTIFORMES of OConnor (1984c); Orders ASTIGMATA and ORIBATIDA of Woolley (1988), Van der Hammen (1989), and Evans (1992).
40. Infraorder BIMICHAELIIDA of OConnor (1984c).
41. Cohort NEMATALYCINA, **new cohort.**
42. Infraorder TERPNACARIDA of OConnor (1984c).
43. Infraorder ALICORHAGIIDA of OConnor (1984c).
44. Order ORIBATIDA of Woolley (1988), Van der Hammen (1989), and Evans (1992); Order CRYPTOSTIGMATA of Evans, Sheals, and MacFarlane (1961); Suborder CRYPTOSTIGMATA of Krantz (1970b); Supercohort ORIBATEI of Baker et al. (1958).
45. Suborder PALAEACARIFORMES of Trägårdh (1932); Suborder PALAEOSOMATA of Woolley (1988), Van der Hammen (1989), and Evans (1992); Infraorder PALAEOSOMATA of OConnor (1984c); Supercohort PALAECARI of Krantz (1970b); Cohort BIFEMORATINA of Krantz (1978); Cohort BIFEMORATA of Lee (1981); Cohort PALAEOSOMATA of Johnston (1982).
46. Suborder ENARTHRONOTA of Woolley (1988), Van der Hammen (1989), and Evans (1992); Cohort ENARTHRONOTA of Johnston (1982).
47. Suborder PARHYPOCHTHONATA of Woolley (1988); Suborder PARHYPOHTHONINA of Van der Hammen (1989); Suborder PARHYPOSOMATA of Evans (1992); Cohort PARHYPOCHTHONATA of Johnston (1982).
48. Suborder MIXONOMATA of Van der Hammen (1989); Suborders MIXONOMATA plus EUPTYCTIMA of Woolley (1988) and Evans (1992); Cohorts MIXONOMATA plus EUPTYCTIMA of Johnston (1982).
49. Suborder NOTHRONATA of Woolley (1988).
50. Suborder NOTHRINA of Van der Hammen (1989); Suborder DESMONOMATA of Evans (1992); Cohort NOTHRONATA of Johnston (1982).
51. Suborder CIRCUMDEHISCENTIAE of Van der Hammen (1989) and Evans (1992); Supercohort ORIBATEI SUPERIORES of Krantz (1970b); Supercohort BRACHYPYLIDES of Krantz (1978, *pars*); Cohort CIRCUMDEHISCENTIAE of Johnston (1982).
52. Suborder ACARIDIDA of Krantz (1978); Suborder ASTIGMATA of OConnor (1982a) and Johnston (1982); Order ASTIGMATA of Woolley (1988), Van der Hammen (1989), and Evans (1992).

CHAPTER NINE

D.E. WALTER
M.S. HARVEY

Order Opilioacarida

- Superfamily Opilioacaroidea (Fig. 9.1)

*DIAGNOSIS: Cuticle leathery, with purple or bluish markings and no distinct plates, body setae isotropic. Opisthosoma with relatively few setae (*Opilioacarus, Neocarus*) to many setae (*Panchaetes, Salfacarus*) and 4 pairs of dorsolateral spiracles (adults and tritonymphs), peritremes absent. Prodorsum with 2–3 pairs of lateral ocelli and numerous setae. With a pair of sternal verrucae between coxae II, genital region with paired setose verrucae, eugenital setae sometimes present, anus with 2 setose valves. Females with well-developed extrusible ovipositor terminating in 2–3 lobes. Gnathosoma always with more than 4 pairs of ventral setae, paired rutella, with discoid or biramous With's organs lateroventrally, and a large denticulate labrum; chelicerae 3-segmented, with 3–5 setae and 2 lyrifissures on middle article, palpi 5-segmented and with a well-developed, usually terminal, pair of claws. Tritosternum represented by unfused sternapophyses; leg coxae free, trochanters III–IV and tarsi I–IV divided in tritonymphs and adults.*

The monofamilial Opilioacarida is the smallest of the six recognized acarine orders, comprising the family OPILIOACARIDAE and 24 described species in nine genera. Opilioacarids are comparatively large (1,500–2,300 µm in length), typically long-legged, brownish gray, leathery acarines with distinctive blue or purple pigment patterns on the body and legs. They retain many ancestral features that have often been used to justify their status as the most primitive living mites (Grandjean 1936a; Van der Hammen 1970a). For example, the opisthosoma (or possibly hysterosoma; see Klompen 2000a) is thought to consist of 13 segments (Van der Hammen 1966), with remnants of primary segmentation still faintly demarcated (Fig. 9.1C; also see chapter 3). Some opilioacarids (*Paracarus, Siamacarus*) have three pairs of lateral ocelli (Kaiser and Alberti 1991) as compared to one pair in other Parasitiformes, and a maximum of two pairs in Acariformes. The tarsi of all legs are divided in opilioacarids, as in many nonacarine arachnids, while other Parasitiformes have at most an acrotarsus on legs I. The tarsi of Acariformes are entire (except for secondary subdivisions in some ANYSTIDAE). The opilioacarid palp bears a pair of distal claws that are presumed to be homologous with the clawlike palpal **apotele** (Camin, Clark, and Bourdeau 1958), a highly modified structure that is inserted at the base of the palptarsus in other parasitiform groups (acariform mites lack palptarsal claws). Most recent classifications place the Opilioacarida as the sister group of the order Parasitiformes (but see Murrell et al. 2005), although characters of uncertain polarity such as the rutella, isotropic setae, and particulate feeding could indicate a closer relationship to the Acariformes (Grandjean 1936a; Walter and Proctor 1998).

Opilioacarids have been collected in cave habitats, in litter, or under rocks in both semiarid environments and forest litter in tropical and warm temperate regions of the world. *Neocarus* species are restricted to the New World, ranging from the southern United States to Uruguay and Argentina (Chamberlin and Mulaik 1942; Van der Hammen 1966; Juvara-Bals and Baltac 1975; Lehtinen 1980; Palacios-Vargas and Vázquez 1988; Silvestri 1905; Vázquez and Klompen 2002). Species of *Opilioacarus* occur in the Mediterranean region, and *Adenacarus arabicus* (With) was described from Yemen (Van der Hammen 1969b; With 1904). Members of the genera *Panchaetes, Phalangiacarus*, and *Salfacarus* are found in Africa and/or Madagascar (Naudo 1963; Van der Hammen 1977b; Coineau and Van der Hammen 1979), those of *Siamacarus* and *Vanderhammenacarus* occur in forest litter and caves in Thailand (Leclerc 1989), and *Paracarus hexophthalmus* (Redikorzev) is restricted to central Asia (Redikorzev 1937; Van der Hammen 1968c). A fossil species of *Paracarus* was described from Eocene Baltic amber (Dunlop, Wunderlich, and Poinar 2004), suggesting an earlier, more northern distribution of the family in Europe. Undescribed species of uncertain generic position are known from various regions, including India and northern Australia (Harvey 1996c).

Although they are slow to react to disturbance, opilioacarids are capable of rapid movement over short distances, and one species has been reported to jump when threatened

(Vázquez and Palacios-Vargas 1988). They can readily autotomize one or more legs (up to four) when disturbed, and regeneration occurs during the subsequent molt. Opilioacarids have been observed to feed on pollen, fungal spores, and fragmented arthropod remains. Walter and Proctor (1998) found that the gut of an undescribed opilioacarid from northern Australia contained the nearly complete bodies of various prostigmatic mites, such as species of *Speleorchestes* (NANORCHESTIDAE) and the families EUPODIDAE, TENERIFFIIDAE, and TARSONEMIDAE, as well as fragments of oribatids and collembolans. One recently molted adult had ingested an entire opilioacarid cuticle (presumably its own). Laboratory observations revealed that food items are located with the palpi, picked up by the chelicerae, and ingested by short, rapid, alternating in-and-out movements of the chelicerae. While opilioacarids feed, the palpi constantly and rapidly palpate the food. The toothed rutella (Figs. 9.1A, B) project anterolaterally from the hypostome and may be used to cut off material held by the chelicerae. Bits of arthropod or fungus are snipped off and swallowed, while pollen is picked up by the chelicerae and swallowed whole. Ingested food forms discrete boluses in the gut.

Members of the family OPILIOACARIDAE have the full acarine ontogenetic sequence and are the only parasitiform mites known to have a prelarva. The large, oval eggs, which may be covered with dirt by the female (*Neocarus texanus* Chamberlin and Mulaik is an example), hatch into elattostatic prelarvae (see chapter 5). The prelarva is an incompletely formed instar in which the body and the legs lack setation. Only legs I–III are well formed, while legs IV are vestigial (Coineau 1973a; Coineau and Van der Hammen 1979; Klompen 2000a). Although it may be able to move, the prelarva cannot feed (Coineau and Van der Hammen 1979). Like the prelarva, the hexapod larval stage has functional legs I–III and vestiges of legs IV, and it also has well-developed chelicerae. Three octopod nymphal instars follow the larva. As in other Parasitiformes, legs I–III of the protonymph are formed within the old leg cuticle, allowing the pharate stage to move, but legs IV are formed outside the leg cuticle and curled under the body, in the same manner that regenerated limbs develop. At least some opilioacarids undergo adult molts. The prelarva and larva lack spiracles, but the protonymph has two pairs, the deutonymph three pairs, and the tritonymph and adult four pairs.

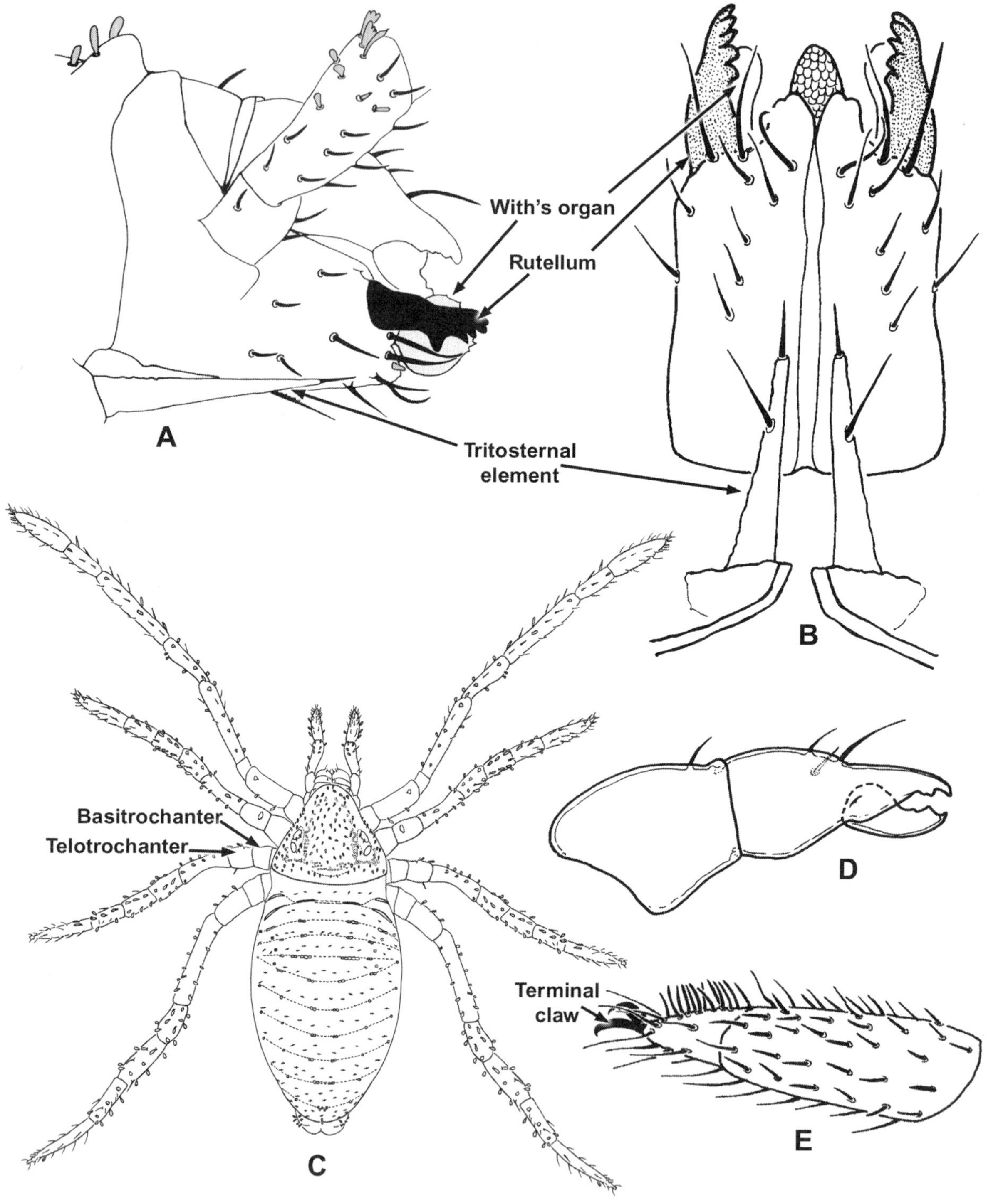

Fig. 9.1. Opilioacarida, generalized views. **A**, gnathosoma and tritosternum in lateral view; **B**, gnathosoma and tritosternum in ventral view; **C**, dorsal habitus; **D**, chelicera (antiaxial view); **E**, palptibia and tarsus with claws.

CHAPTER TEN

D.E. WALTER

Order Holothyrida (Holothyrina, Tetrastigmata)

■ Superfamily Holothyroidea (Figs. 10.1, 10.2)

DIAGNOSIS: Adults fully sclerotized with dome-shaped holodorsal shield with short, broad peritremes above legs III and with or without 1 pair of lateral ocelli. Holoventral shield encompasses a posterior pair of anal valves, each with ≥ 2 setae; male with a pair of transverse, setose genital valves in the intercoxal region, female with usually 4 setose genital shields. Tritosternum present and flagelliform or absent; hypostome with corniculi and ≥ 5 pairs of setae (including palpcoxal), sclerotized gnathotectum absent. Chelicerae 3-segmented, palpi 5-segmented. Family ALLOTHYRIDAE with 3 nymphal instars and a single hexapod larva (not yet described), ontogeny of other families unknown.

Adult holothyroids are large (2–7 mm), heavily sclerotized parasitiform mites that live in mesic litter habitats in South and Central America, on islands in the Caribbean Sea and Indian Ocean, and in Australasia. The name Holothyrida, derived from the Greek *holo* (whole) and *thyr* (shield), refers to their armored state. Holothyroids may be found in leaf litter, in mosses, and under stones in moist forests from near sea level to about 2,000 m in elevation. Members of the family ALLOTHYRIDAE (Figs. 10.1A–D, 10.2) and some HOLOTHYRIDAE (Figs. 10.1E–G) are known to scavenge, but none has been demonstrated to be a predator (Travé 1982; Walter and Proctor 1998).

The Holothyrida is currently considered to consist of three families, 13 genera, and approximately 22 described species, but numerous undescribed species and generic-level taxa occur in Australasia and the Neotropics. Only three New World holothyrans have been described to date (Lehtinen 1981, 1999; Kontschán and Mahunka 2004), and these have been placed in their own family, the NEOTHYRIDAE, as a sister group to the HOLOTHYRIDAE (Lehtinen 1991). The HOLOTHYRIDAE have a peculiar distribution, being found only on islands in the Indian Ocean (Mauritius, the Seychelles, Sri Lanka) and the Pacific Ocean (New Caledonia, Lord Howe, New Guinea). New Guinea appears to have the highest diversity of Holothyrida, with three endemic genera and 12 species; most other genera are restricted to single island groups. Allothyrids live in mesic forest litter and mosses (rain forest and wet sclerophyll) in southwest and eastern Australia and in New Zealand.

All holothyrids share a distinctive, dome-shaped holodorsal shield that incorporates the lateral stigmata and peritremes. The gnathosoma can be withdrawn under the anterior edge of the dorsum, and a holoventral shield protects the venter. The sternal region is often demarcated by lines or thickened areas of cuticle, and valves cover the genital and anal openings (Figs. 10.1D, E, 10.2A, B). Legs have the normal seven segments, and the coxae are free. All of the femora have a distinct basifemoral ring, and tarsi I of ALLOTHYRIDAE have separate acrotarsi. Just posterior to acrotarsus I of allothyrids is a pocketlike invagination with sensory setae that resembles the capsule of Haller's organ in ticks (Klompen and Oliver 1993a) and similar structures in many Mesostigmata.

Except for species of *Haplothyrus* from New Caledonia that have a single genital shield (Lehtinen 1995), the genital opening of adult females is covered by a complex of four shield elements: the anterior sternogynal shield, the posterior mesogynal shield, and the lateral latigynal shields (Figs. 10.1E, 10.2A, 12.20A). A similar and probably homologous arrangement is known in some trigynaspid Mesostigmata (Lehtinen 1991). Adult males (Figs. 10.1D, 10.2B) have a transverse genital opening between coxae IV that is covered by two small valves and often have large spurs ventrally on the femora of legs II. The male genital opening is within a large, concave region, the **epiandrum** (Figs. 10.1D, 10.2B), that is best developed in the HOLOTHYRIDAE.

Adult ALLOTHYRIDAE, *Neothyrus ana* Lehtinen (NEOTHYRIDAE), and some HOLOTHYRIDAE (e.g., *Lindothyrus*) are densely hairy throughout, but in most holothyrids the dorsum appears hairless and smooth (e.g., *Thonius*, *Hammenius*). The peritreme of *Allothyrus*, *Holothyrus*, and *Haplothyrus* is rather short and broad, but it extends much farther anteriorly in other genera. Its papillate structure is similar to that found in the Mesostigmata (see Pugh et al. 1991). A gland that opens on the lateral dorsal shield posterior to the peritreme (**Thon's organ**) was earlier confused with a second stigmatic opening and gave rise to the term *Tetrastigmata*, which was once applied to the Holothyrida. Thon's organ is considered by Travé (1983) to function as a defensive gland. Like some Mesostigmata, many allothyrids have a **peridium** (Fig. 10.2C), a deep depression at the base of coxa IV that contains numerous cuticular spines. Peridia resemble the lateral opisthosomatic organs seen in some sironid opilionids. The gnathosoma is withdrawn into the idiosoma except during feeding, and there is no sclerotized gnathotectum. The corniculi are either simple cones (ALLOTHYRIDAE) or toothed extensions (NEOTHYRIDAE, HOLOTHYRIDAE) (Fig. 10.1G) and are inserted dorsolaterally on the hypostome. In allothyrids, a long and a short seta arise dorsally above the corniculus, and the hypostome has six to many pairs of setae and a narrow deutosternal groove. The chelicerae are three-segmented and have either two (ALLOTHYRIDAE) (Figs. 10.1B, 10.2D) or a single (HOLOTHYRIDAE, NEOTHYRIDAE) dorsal seta. Species of the genera *Allothyrus*, *Australothyrus*, and *Leiothyrus* have a large tubercle on the palptrochanter that fits into a lateral hypostomatic notch, while neothyrids and holothyrids have a comb of stout setae on the palptibia (Fig. 10.1F). All known species of *Allothyrus* have both lateral ocelli (as in some ticks) and a biflagellate tritosternum (as in Mesostigmata), but the tritosternum is ostensibly absent in the allothyrid genus *Australothyrus* (Van der Hammen 1983b). No ocelli or tritosterna have been described in the HOLOTHYRIDAE or NEOTHYRIDAE.

Adult holothyrans are lethargic animals that play dead when disturbed, but nymphs tend to be more active. Nymphal *Allothyrus* have a pair of large dorsolateral idiosomatic glands (Fig. 10.1C) that excrete fluid when the mites are disturbed (Walter and Proctor 1998). Adults of *Holothyrus coccinella* Gervais produce a distasteful exudate that is reported to be fatal to fowl that eat them (Southcott 1976).

Adult female *Allothyrus* often carry one completely formed and one partially formed egg. A hexapod larval stage is known but has not been studied. The three octopod nymphal instars are recognizable on the basis of size increases and setal augmentations. Protonymphs of *Allothyrus* have a single pair of euanal setae, a single ventral palptrochanteral seta, and eight or nine pairs of ventral hypostomatic setae. Both the deuto- and tritonymph usually add setae in each of these body regions. Generally, there is far more variability between species and among individuals within a species (as well as symmetrically within an individual) than is considered normal for the Mesostigmata. A genital slit is present in the protonymph, and it increases in size in subsequent nymphs. New limbs are formed within the old during molting. Allothyrid nymphs are a bright white to light tan color, while adult holothyrans are red, reddish brown, nearly black, or brown with purplish stripes.

The development of sclerotized shields in *Allothyrus* is distinct from that of other parasitiform mites. While hard ticks (IXODIDAE) retain a scutum in each stage, and the sclerotized shields of Mesostigmata typically enlarge and coalesce through ontogeny (Fig. 12.1), the immature instars of *Allothyrus* have no discernible sclerotized shields. Instead, the body is covered with a leathery cuticle densely ornamented with tiny, rounded or toothlike projections and numerous setae (Fig. 10.1C), somewhat reminiscent of that of soft ticks (ARGASIDAE). The numerous integumental exocrine glands probably are involved in production of a secretion that spreads over the cuticle via excavation canals (Alberti and Seeman 2004). Similar glands occur in Thon's organs and in the peridia. Only the corniculi and cheliceral digits are strongly sclerotized, with the heavy armor of the idiosoma appearing only in the adult. Immature *Allothyrus* also lack ventral shields, although Johnston (1982) reported a sclerotized sternum in the nymph of an unidentified Holothyrida.

The relationship of the Holothyrida to the other orders of Parasitiformes (Opilioacarida, Ixodida, Mesostigmata) is unclear. Johnston (1982) suggested that the ALLOTHYRIDAE is the sister group of the Mesostigmata, but recent molecular studies (e.g., Dobson and Barker 1999; Murrell et al. 2005) and morphological studies (Lehtinen 1991; Moraza 2005c) suggest that ticks and holothyrans are sister groups and that this taxon is the sister group of the Mesostigmata. However, molecular studies, especially those based on a single gene system, are subject to problems such as saturation and long branch attraction that reduce their inference (Murrell et al. 2005). Morphological synapomorphies between ticks and holothyrans proposed by Lehtinen (1991) are likewise unconvincing (e.g., components of Haller's organ are present in the Mesostigmata and in the arachnid order Ricinulei). Some character states of uncertain polarity that may support a sister group relationship between Mesostigmata and Holothyrida include the tritosternal flagella, papillate peritremes, lateral stigmata, peridia (e.g., in ASTERNOSEIIDAE, p. 128), a sclerotized sternal shield, corniculi, fimbriated internal malae, and a subdistal, three-tined palpal apotele.

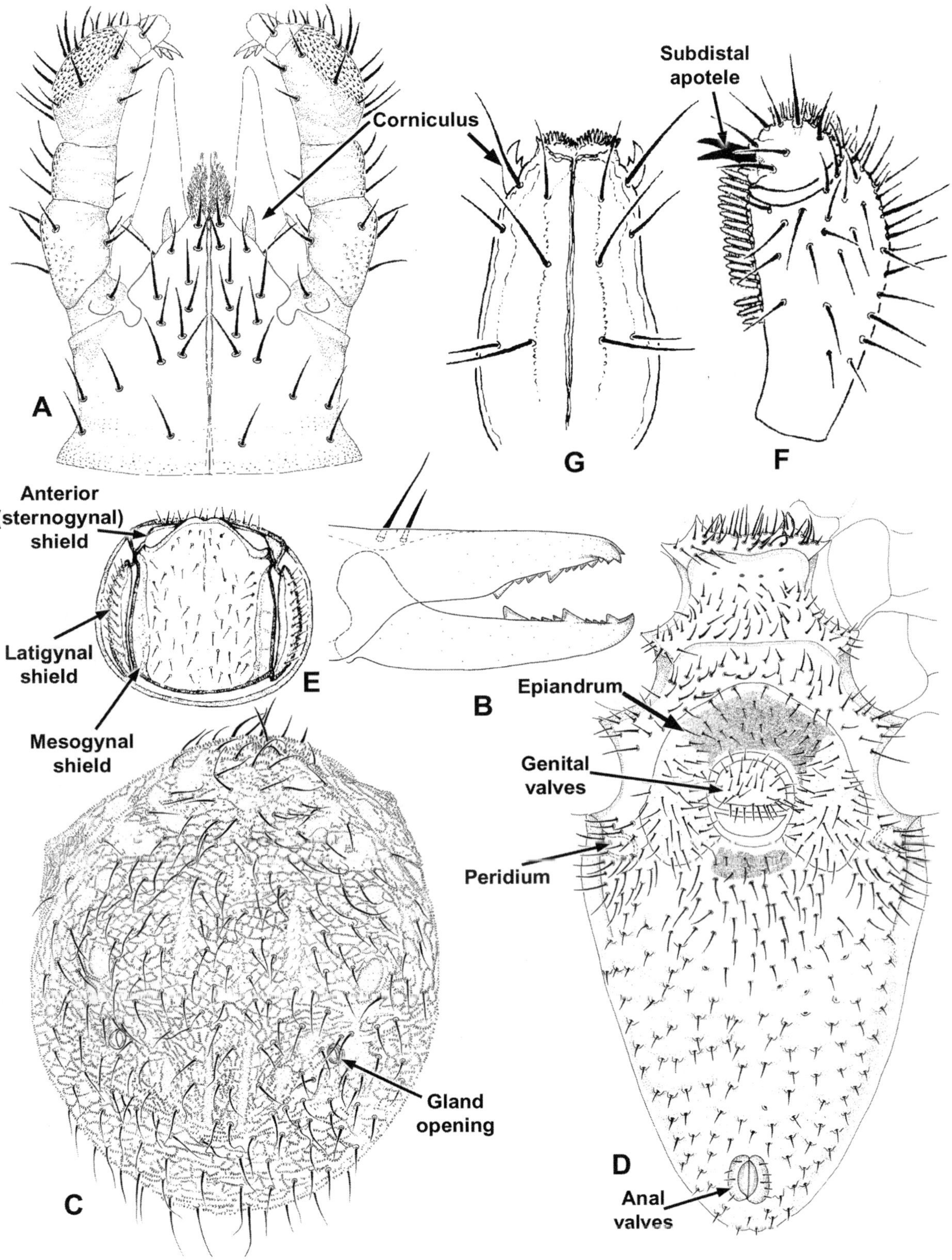

Fig. 10.1. ALLOTHYRIDAE. *Allothyrus* sp. (Queensland, Australia): **A**, venter of gnathosoma; **B**, chelicera; **C**, dorsum of nymph; **D**, venter of male.
HOLOTHYRIDAE. **E**, genital valves of female (New Guinea); **F**, palptibia and tarsus; **G**, subcapitulum.

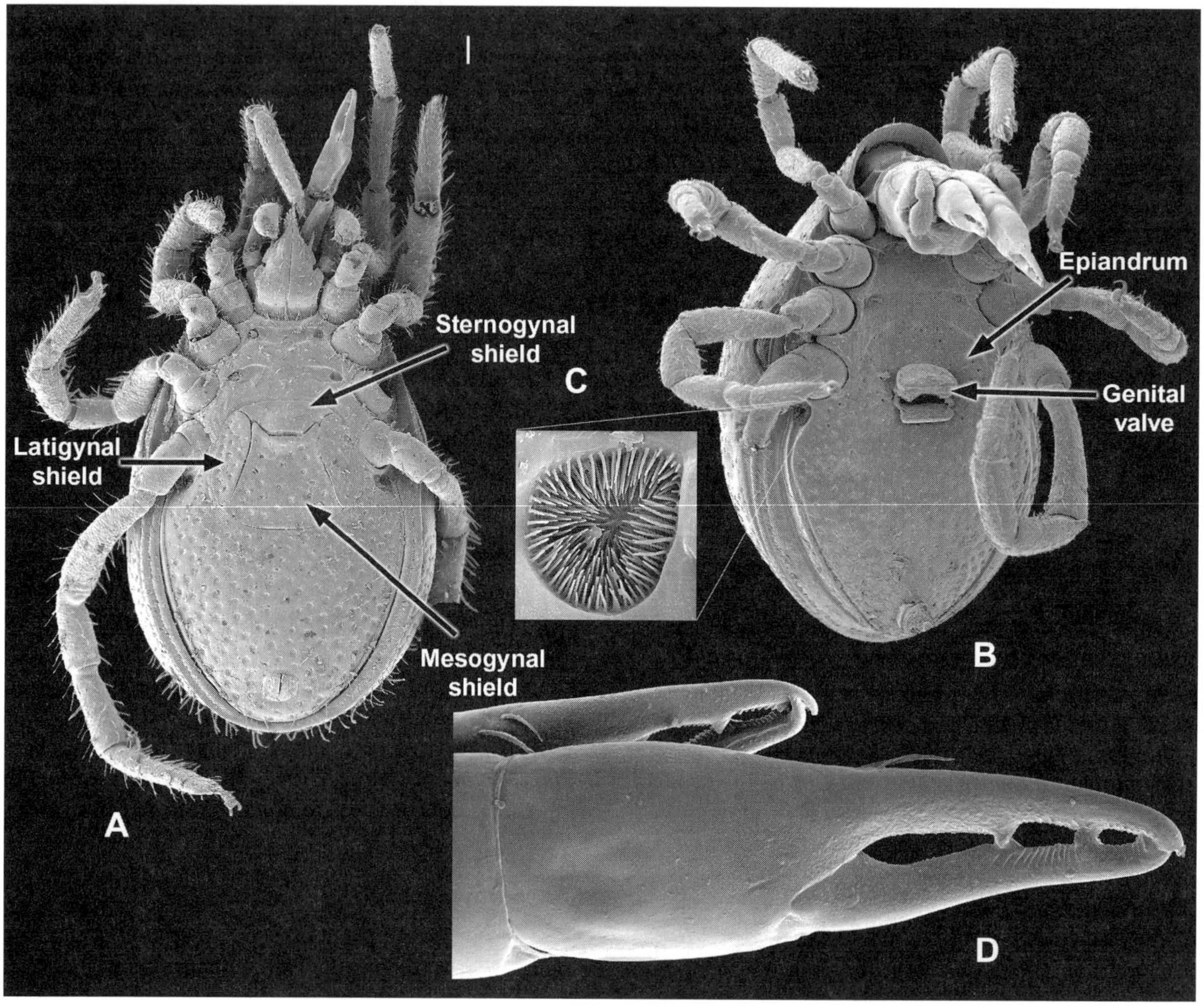

Fig. 10.2. ALLOTHYRIDAE. *Allothyrus* sp. (Queensland, Australia): **A**, venter of female. *Allothyrus* sp. (Tasmania, Australia): **B**, venter of male; **C**, detail of peridium; **D**, chelicerae of male.

CHAPTER ELEVEN

J.E. KEIRANS

Order Ixodida

The Ixodida, or ticks, comprise approximately 870 described species of parasitiform acarines that, in one or more of their life stages, are obligate hematophagous parasites on terrestrial vertebrates. Ticks, also known as the Metastigmata, occur throughout the world but are primarily tropical or subtropical in distribution (Oliver 1989; Keirans 1992; Horak, Camicas, and Keirans 2002). Nonetheless, they are often significant pests of humans and animals in temperate forest, steppe, and prairie habitats, as well as the harsher tundra of the arctic. Ticks may cause injury to their hosts through secondary infection or irritation at attachment sites, and when infestations are large, they can cause host decline or death through exsanguination. Certainly the pest status of ticks in many parts of the world is great, but their primary importance lies in their ability to transmit disease-causing agents to wildlife, domestic animals, and humans. Although mosquitoes may transmit disease agents of greater human significance than those transmitted by ticks (the malaria-causing *Plasmodium* spp. are examples), ticks transmit a wider variety of disease organisms to humans and animals than any other group of arthropods (Sonenshine 1991).

Their unusual morphology and great size (2,000 to over 30,000 µm) serve to distinguish the Ixodida from most other Acari (Hoogstraal 1973b; Coons and Alberti 1999). Ticks display considerable heteromorphy of developmental stages, but certain unique characteristics occur throughout ontogeny. Although the hypostome may be greatly reduced in some nonfeeding adults (i.e., *Otobius* spp. and some nidicolous male *Ixodes* spp.), it is typically produced as a distinctive, anteriorly projecting holdfast organ that bears a series of retrorse teeth designed to anchor the tick to its host during feeding (Figs. 11.3G, 11.5D). **Haller's organ**, a complex sensory field consisting of an anterior pit and a posterior capsule (Figs. 11.1A, 11.4B), is located dorsally on tarsi I of all postembryonic stages and provides sites for receiving chemical, olfactory, and microclimatic stimuli. Unlike other parasitiform orders, members of the Ixodida have palpi with four segments (only three distinct segments in the genus *Nuttalliella* (NUTTALLIELLIDAE)), and the chelicerae have two segments, with an inner (paraxial) digit on which a dorsal uncate appendage is articulated and a dentate outer (antiaxial) digit attached near the base of the inner element. A membranous mantle surrounds the inner digit, leaving only the cutting teeth of the digits exposed (Figs. 11.4F, 11.5D). Ticks also lack an epistome, corniculi, and a tritosternum. One or two pairs of simple eyes are sometimes present on the lateral margins of the scutum or in the pleural integument above the coxae. Most argasids lack eyes, however, and they are likewise absent in a number of ixodid genera, including *Ixodes* and *Haemaphysalis*, and the subgenus *Amblyomma* (*Aponomma*; synonymy established by Klompen, Dobson, and Barker 2002; see also Balashov 1972). Uniform, cell-like **festoons** (Fig. 11.4D) often are present on the posterior idiosomatic margin of ixodid ticks (absent in *Ixodes*).

Ontogenesis in the Ixodida is more or less typical for the order, but the number of nymphal instars may vary from only one in members of the family IXODIDAE to as many as eight in some ARGASIDAE. Ixodid larvae (Figs. 11.1B, C) possess a podonotal shield (**scutum**) that is retained in postlarval stadia and is often enlarged. Most known argasid larvae also have a scutum or dorsal plate (Fig. 11.2C), but it is lost at the larval molt. Larvae are hexapod and lack the paired ventrolateral spiracular plates that are common to succeeding nymphal and adult stages (Figs. 11.4B, 11.5B). Unlike most Parasitiformes, nymphal and adult Ixodida may not always exhibit obvious morphological differences, and sexual dimorphism is not clearly marked throughout the suborder. Nymphal IXODIDAE resemble adult females in that both retain the larval characteristic of an abbreviated podonotal scutum, a neoteny that allows for considerable idiosomatic expansion during feeding. Male ixodids have an entire scutum (Fig. 11.5A) and consequently are unable to consume large blood meals. Nymphal and adult soft ticks (ARGASIDAE) and the monotypic family NUTTALLIELLIDAE (Figs. 11.2A, B, 11.5A) have a leathery, mammillate or granular integument and lack a dorsal scutum. However, nymphs and adults of *Nothoaspis reddelli* Keirans and Clifford and *Nuttalliella namaqua* Bedford (males and larvae of

N. namaqua are unknown) are unique in having a leathery anterior scutum often referred to as a **pseudoscutum** (Fig. 11.2A) because it is only slightly differentiated from the integument that borders it (Keirans and Clifford 1975; Keirans et al. 1976; Coons and Alberti 1999). Except for differences in size, argasid nymphal instars are basically identical, and sexual dimorphism in the family is mostly expressed in internal structures (Balashov 1972).

Attraction and attachment responses of ticks are triggered, at least in part, by stimuli received through thin-walled contact and olfactory sensory setae at the apex of the palpal tibiotarsus, on the tarsi and pretarsi of legs I, and in Haller's organ (Balashov 1972; Ivanov and Leonovich 1983; Klompen and Oliver 1993a; Hess and Vlimant 1986; Coons and Alberti 1999). Mechanoreception (sudden vibration), photoreception (sudden shadows), and a perception of host warmth also appear to play a role in attraction and attachment, as does carbon dioxide emanating from a prospective host (Sonenshine 1991). The ocelli in ticks of the genus *Hyalomma* (Fig. 11.5C) may not only detect changes in light and shadow, but their elevated position may also allow them to perceive the host animal itself. For example, ocelli of *H. asiaticum* Schulze and Schlottke protrude far enough above the idiosomatic surface to aid the tick in orienting to its moving vertebrate host and to track it over considerable distances.

Tick attachment to a host is accomplished by piercing the host's skin with the chelicerae and anchoring to the site by inserting the toothed hypostome into the wound (Arthur 1957). A cementlike substance produced in the salivary glands of most ixodid ticks is deposited around the mouthparts and extends into the host tissue to secure the tick firmly in place (Chinery 1962, 1965, 1973). Argasid and ixodid ticks may produce a salivary anticoagulant that enhances blood flow during the feeding process (Bowman et al. 1997; see chapter 3 for additional information). Feeding ticks may engorge in a few days, while others may require considerably longer periods of attachment (Sonenshine 1993). Subcutaneous infestations of *Ixodes* and *Hyalomma* species have been observed on some hosts.

Lees (1948) observed that host finding and attachment are ensured by orientation responses that lead to favorable waiting postures at appropriate vegetation levels and direct response to an approaching host by assuming a "questing" posture. Chance of contact with a host is often enhanced by the tick's habit of climbing to the tops of grass blades or twigs in animal runways (Lees 1946). Desiccation of questing ticks in these exposed sites can occur within a few hours, although some tick genera (*Dermacentor* and *Hyalomma*, for example) are highly resistant to rapid water loss. Ticks that fail to contact a host may regain water balance by moving down to areas of higher relative humidity for short periods of time. Dropping off the host following engorgement appears to be subject to predictable periodicity in both ixodid and argasid ticks and may be synchronized with host behavior patterns to place the ticks at optimal sites for further development or reproduction (George 1971; Sonenshine 1993). Nidicolous ticks tend to detach when their hosts are inactive and in the nest, while nonnidicolous ticks are likely to do so only during periods of maximum host activity. Timing of the drop-off in all three stages of *Haemaphysalis leporispalustris* (Packard), and in larvae and nymphs of the argasid kangaroo tick, *Ornithodoros gurneyi* Warburton, is regulated by a circadian rhythm entrained by the daily light cycle (George 1971; Doube 1975). Circadian regulation of drop-off has also been suggested for species of *Argas*, *Hyalomma*, and *Rhipicephalus* (Hadani and Rechav 1969). Diapause expressed through delay in oviposition has been noted for *Argas arboreus* Kaiser, Hoogstraal, and Kohls (Khalil 1974).

Some ticks may feed on the hemolymph of engorged individuals of their own species (**homoparasitism**). An example of what appears to be obligate homoparasitism in the IXODIDAE was described by Moorhouse (1966), who observed male *Ixodes holocyclus* Neumann feeding on the hemolymph of partially engorged conspecific females. Males were never observed to feed directly on vertebrate hosts and may routinely procure their small blood meal from females, which apparently are unaffected by the male's feeding activities. Examples of cannibalism are common among nymphs and adults of the argasid genera *Ornithodoros* and *Argas* and are regarded by Beklemishev (1948) as vestigial manifestations of ancestral predation.

Ticks may be attacked by a variety of predators and parasites. Several bird and rodent species are known to feed on ticks (Wilkinson and Garvie 1975), a predilection that may explain the link between the bird–rabbit tick–lagomorph cycle and the rodent–dog tick–mammal cycle of spotted fever rickettsiae in North America. Lizards and amphibians also prey on ticks, as do many arthropods, including spiders, hemipterans, ants, and beetles (Samish and Rehacek 1999). Like other hematophagous acarines, ticks host numerous endosymbionts, including bacteria, protozoa, fungi, and nematodes, sometimes with lethal consequences (Burgdorfer, Brinton, and Hughes 1973; Benoit et al. 2005). Certain endosymbiotic bacteria apparently are capable of manipulating reproduction in ticks. Some obligate intracellular bacteria occur in the Malpighian tubules and ovaries and may supply essential nutrients to their tick hosts, while still others are inherited and feed on the mitochondria of their tick hosts (Perotti and Braig 1999). Ixodid ticks are parasitized by no fewer than seven species of the encyrtid wasp genus *Ixodiphagus* (equivalent to *Hunterellus pars*). The developing wasp larvae consume the entire tissue contents of their engorged nymphal and adult tick hosts, along with previously ingested blood (Oliver 1964; Davis 1986; Hu, Hyland, and Oliver 1998).

Recent molecular and morphological studies have suggested a sister group relationship between the Ixodida and the Holothyrida, based in part on the shared derived condition of the receptor complex of tarsus I (Haller's organ) and the presence of a postlarval acrotarsus (Moraza 2005c). The same study establishes the Ixodida + Holothyrida as the sister group of the order Mesostigmata, supporting the position held by Lehtinen (1991). However, there is evidence that suggests the possibility of a different phylogenetic scenario (see Holothyrida, p. 107).

Three families of Ixodida are presently recognized, and these are relegated to a single superfamily, the Ixodoidea. Because of space constraints, the following discussion should be considered only an introduction to this important group of acarine parasites. The literature on ticks and tick-borne diseases that has been cited here is likewise limited, but many of the more significant publications that are not included may be found in the bibliographies of works by Hoogstraal (1985), Oliver (1989), Camicas et al. (1998), Samish and Rehacek (1999), Sonenshine (1991, 1993), Coons and Alberti (1999), and Sonenshine, Lane, and Nicholson (2002).

▪ Superfamily Ixodoidea (Figs. 11.1–11.6)

DIAGNOSIS: Idiosoma with few or no sclerites but with a thick cuticle, with or without a podonotal scutum, occasionally with a leathery pseudoscutum, dorsolateral ocelli present or absent; integument folded or striate (IXODIDAE), leathery or mammillated/tuberculated (ARGASIDAE), or highly convoluted and pitted (NUTTALLIELLIDAE); posteromarginal festoons present or absent. Gnathosoma anterior or anteroventral in position, epistome and corniculi absent, with the hypostome typically modified as a holdfast organ in all postembryonic stages; palpi simple, telescoped or normally extended, with 3–4 segments, without palpal apotele. Postlarval stages with dorsolateral stigmatic plates between coxae III–IV (ARGASIDAE) or behind coxae IV, with or without coxal glands (see chapter 3), peritrematic elements absent. With a Haller's organ on the distidorsal aspect of tarsus I, all tarsi with claws.

The hard ticks, or IXODIDAE (Figs. 11.1, 11.3–11.6), number nearly 700 species that are accommodated in 12 genera. The family comprises two major phyletic lines: the Prostriata and the Metastriata (Sonenshine 1991). The Prostriata includes only the genus *Ixodes*, ostensibly the most primitive assemblage in the family. Most ixodids are **three-host ticks** that typically use a different host animal at each developmental stage (larva, nymph, and adult). Each stage feeds only once on each host, often for several days, and then drops off to molt to the next stage. Larvae and nymphs generally attach to birds or small mammals such as mice, voles, or shrews, while adults usually attach to large mammals. Some tick species, however, do not leave the original host animal until after adult emergence, feeding, and mating have taken place (**one-host ticks**). Others drop off the initial host following nymphal development and acquire a second host only after they have molted to the adult stage (**two-host ticks**). The Texas cattle fever tick, *Rhipicephalus* (*Boophilus*) *annulatus* (Say), is a one-host tick, while *Hyalomma marginatum* Koch is an example of a two-host tick.

While male ixodids may feed sparingly (or not at all), females require a large blood meal to produce an egg mass. They tend to be slow feeders and may engorge over a period of several days (Balashov 1972). Egg masses may contain up to 18,000 eggs but average 2,000–8,000. Production of the egg mass is a lengthy process in that each egg is coated with a waterproofing waxy material by the inflatable Gene's organ (see chapter 3) and then by an antioxidant produced by paired **porose areas** located on the dorsum of the capitulum (Sonenshine 1991). Mortality of hatching larvae and subsequent developmental stages is high because of both predation and the problems encountered in finding and successfully attaching to a suitable host (Hoogstraal 1973a).

Most ixodids show little specificity for particular hosts, but some demonstrate more than an incidental preference for particular animal groups. For example, *Dermacentor*, *Rhipicephalus*, *Margaropus*, *Rhipicentor*, *Nosomma*, and *Anomalohimalaya* species are primarily mammal parasites, while, with some exceptions (Klompen, Dobson, and Barker 2002), members of the subgenus *Aponomma* (genus *Amblyomma*) are restricted to reptiles. On the other hand, species of the primarily Old World genus *Haemaphysalis* occur commonly on both birds and mammals, while species of the Australian genus *Bothriocroton* occur on edentates and reptiles. The rare *Cosmiomma hippopotamensis* (Denny) belies its specific epithet, being found primarily on the black rhinoceros in southern Africa (Bezuidenhout and Schneider 1972).

As noted earlier, many ixodid species are vertebrate pests and transmitters of disease organisms. In addition to causing tissue damage and irritation through their feeding, ixodid ticks may serve as reservoirs and vectors for many infective viruses, rickettsiae, and bacteria, effecting transmission of these organisms through feeding or through contact of infective coxal fluids or excrement with the broken skin of the host (Hoogstraal 1973a; Sonenshine 1993). The salivary secretions of some ixodids can produce motor paralysis in humans and animals following an extended period of engorgement. Tick paralysis can result in death if feeding ticks are allowed to remain attached to the affected host (Gregson 1973; Sonenshine 1993).

The prostriate genus *Ixodes*, a worldwide group of roughly 250 species, commonly parasitize burrow- or den-inhabiting mammals and are mostly nonspecific in their choice of host, although a few species are highly specific to large artiodactyls or perissodactyls. Nonspecific *Ixodes* species rank among the most successful and medically important ticks in the world. Some notable examples are *I. scapularis* Say and *I. pacificus* Cooley and Kohls, which vector the spirochete that causes Lyme disease in North America. *I. ricinus* (L.) transmits Lyme disease in western Europe, as well as the viruses that cause louping ill, Crimean-Congo hemorrhagic fever, tick-borne encephalitis, and the causative agent of Q fever (Keirans, Needham, and Oliver 1999). Russian spring-summer encephalitis and Lyme disease are vectored in the eastern Palearctic by *I. persulcatus* Schulze (Sonenshine,

Lane, and Nicholson 2002), and *I. nipponensis* Kitaoka and Saito transmits Lyme disease in northeastern Russia and Japan. The genus *Ixodes* is notable in having retained, along with most members of the family ARGASIDAE, the primitive characteristic of producing spermatids as nymphs, and later copulating while off the host and transferring these spermatids without benefit of a blood meal (Oliver 1983).

The large metastriate genus *Amblyomma* includes over 100 species of large, brightly ornamented mammal parasites along with smaller, usually less ornamented forms that are found on reptiles and amphibians. *A. hebraeum* Koch and *A. variegatum* (Fabr.) are the primary transmitters of the organisms that cause heartwater disease in cattle and other ruminants in Africa. Recent molecular studies (Murrell, Campbell, and Barker 2000; Beati and Keirans 2001) have shown that the genus *Boophilus*, an assemblage of five relatively small one-host ticks, clusters as a subgenus within the large, primarily African genus *Rhipicephalus*. *Boophilus* species transmit cattle fever or babesiosis (*Babesia bigemina* (Smith and Kilbourne)) and other pathogens to domestic cattle in many parts of the world. *R.* (*B.*) *microplus* (Can.) is the most widespread member of the subgenus and is perhaps the world's most economically important tick (Sonenshine 1991). Among the 75 known species of *Rhipicephalus s. str.*, the brown dog tick, *R. sanguineus* (Latr.), is recognized as the world's most widely distributed tick species.

The approximately 30 known species of *Dermacentor* occur worldwide except in Australia and Antarctica, and some of them are important disease vectors. The American dog tick, *D. variabilis* (Say), and the Rocky Mountain wood tick, *D. andersoni* Stiles, are the primary vectors of the organism that causes Rocky Mountain spotted fever in eastern and western North America, respectively. *D. andersoni* occurs on a range of mammal hosts and is responsible for transmitting several disease organisms in addition to Rocky Mountain spotted fever, including Q fever, Colorado tick fever, and tularemia. James et al. (2006) provide detailed information on hosts and seasonality for *D. andersoni*, including the first ever county-level distribution map recording its presence in 14 states and 267 counties. Female *D. andersoni* and *D. variabilis* may also cause tick paralysis in humans and animals. Members of the large, predominantly Old World genus *Haemaphysalis* parasitize mammals and birds, and some species are known disease vectors. *H. leporipalustris* (Packard), one of only three known New World species, attacks a wide range of bird and mammal hosts and is an important vector of spotted fever and tularemia among wild animal populations. *H. leachi* (Audouin) is one of several ixodid vectors of the rickettsial organism that causes Mediterranean spotted fever (boutonneuse fever) in Europe, North Africa, and India (Sonenshine 1993). In addition to vectoring the virus that causes Crimean-Congo hemorrhagic fever in humans (Hoogstraal 1979), the large, highly active species of the genus *Hyalomma* can inflict significant feeding wounds on their hosts because of their inordinately long mouthparts. Lameness and toxicosis resulting in tick paralysis may also occur in host animals.

The soft ticks, or ARGASIDAE (Figs. 11.2, 11.5B, 11.6B), comprise a relatively small taxon (approximately 190 species in five genera). Most of them have retained various primitive biological characteristics, including diapause, production of spermatids in the last nymphal instar, multiple developmental stages, and great longevity (Oliver 1983). Argasids usually feed on hosts that return periodically to a permanent or semipermanent shelter. They are especially evident in arid semitropical and tropical regions, where they parasitize birds, small mammals, or reptiles in nests, caves, rookeries, and burrows (Cooley and Kohls 1944). Some species, however, may attack host animals outside the nest or burrow habitat, and members of a few genera (i.e., *Otobius* and *Ornithodoros*) parasitize larger mammals. Immature argasids usually feed only once in each developmental stage, although a nymphal instar may take two blood meals before molting to the next stage if the first feeding period is interrupted. Unlike the IXODIDAE, adult argasids may feed several times, with each meal followed by deposition of a relatively small mass of eggs by the female and the production of additional sperm by the male (Oliver 1989). As occurs in ixodids, each egg is waterproofed by Gene's organ secretions before deposition (porose areas are absent in argasids). The number of eggs per egg mass depends on blood meal volume, individual egg size, and the tick species involved (Sonenshine 1991). Arthur (1962) reported as few as 47 eggs per egg mass in *Argas persicus* (Oken), while *Otobius megnini* (Dugès) has been observed to produce from 800 to 1,500 eggs during a single oviposition episode.

Argasid ticks feed intermittently and rapidly (within minutes, in many cases), often at night and often on a single host. Many species have access to their hosts for only short periods of time (as little as 2–3 days for argasids parasitizing certain nesting birds and roosting bats) and have consequently adapted to long-term survival without feeding. For example, *Ornithodoros tholozani* Laboulbéne and Mègnin, a burrow-inhabiting rodent parasite, may survive for up to 11 years without a blood meal (Pavlovsky and Skrynnik 1960). Other argasids are similarly adapted (Balashov 1972).

Many argasids use birds and bats as their primary hosts, and at least one genus appears to be specific to a particular host group. Members of the New World genus *Antricola* are exclusively parasitic on bats, although only the larvae and nymphs actually feed. Adults of the approximately 14 known species often occur in great numbers in bat guano in Central and North American caves (Černý 1967; Hoogstraal 1973a), where the females oviposit without the benefit of a blood meal (**autogeny**; Sonenshine 1991). The bat cave habitat of *Antricola* species is shared by *Nothoaspis reddelli*, a monotypic cave-inhabiting argasid known only from southeastern Mexico (Keirans and Clifford 1975). One of the two known species of the genus *Otobius*, *O. megnini* (Dugès), is exceptional in that it leaves its hiding places in cracks and crevices or livestock shelters to parasitize large wild ungulates and domestic cattle, sheep, and goats from Canada southward into South America and in southern Africa and India. As in the genus *Antricola*, *Otobius* females are nonfeeding and autogenous (Oliver 1989; Sonenshine 1991, 1993). The cosmopolitan genus *Argas* includes some 60 species, more than 50 of which feed on bats and birds, with the remainder on lizards and small mammals. Virtually all species of *Argas* are found in desert or semidesert, steppe, or dry cave shelters in more humid areas, while members of the speciose genus *Ornithodoros* occur commonly on burrow- and ground-nesting rodents or birds and on bats, but rarely on reptiles, in a variety of habitats (Hoogstraal, Oliver, and Giurgis 1970).

Species of *Argas*, *Otobius*, and *Ornithodoros* may cause extreme irritation and trauma in their host animals and are sometimes responsible for host deaths (Hoogstraal 1973a, 1985). The spinose ear tick, *Otobius megnini*, is a noteworthy example in that the larvae and nymphs feed deep within the ears of their large mammal hosts, which gives rise in the host animal to scratching and head shaking and, in severe cases, to convulsions and death. As with ixodid ticks, argasids often attack humans when the opportunity arises, occasionally with significant consequences. For example, *Ornithodoros coriaceus* Koch uses deer and cattle as hosts but also attacks humans and is reported to have a painful and irritating bite that may ulcerate and persist for several weeks. The bite of *O. muesebecki* Hoogstraal, a bird parasite in the Middle East, causes fever, headache, itch, and erythema in human victims. A similar syndrome occurs in guano diggers attacked by *O. amblus* Chamberlain, a related Peruvian bird parasite (Hoogstraal, Oliver, and Giurgis 1970).

Because of their sedentary habits, the ARGASIDAE generally are considered less important than the more peripatetic IXODIDAE in terms of disease transmission capabilities. However, several species of *Ornithodoros* have been identified as vectors of tick-borne relapsing fever (Keirans 1992; Dworkin, Schwan, and Anderson 2002; Sonenshine, Lane, and Nicholson 2002), and *O. coriaceus* is believed to transmit epizootic bovine abortion in domestic cattle (Lane et al. 1985). In addition to causing severe tissue damage, *Otobius megnini* was found to harbor the rickettsial organism that causes Q fever (Jellison et al. 1948).

It should be noted here that Klompen and Oliver (1993b) established several new synonymies and combinations within the family ARGASIDAE based on phylogenetic analyses at the generic and subgeneric levels. Based on their findings, Klompen and Oliver recognized only four genera: *Argas*, *Carios* (which includes *Antricola* and *Nothaspis*, both recognized in this treatment as valid genera), *Ornithodoros*, and *Otobius*. Readers interested in argasid classification should consult Klompen and Oliver (1993b) and the publications of Pospelova-Shtrom (1969) and Camicas and Morel (1977), who have other approaches to generic classification in this family.

The family NUTTALLIELLIDAE (Figs. 11.2A, B, 11.5A) is represented by a single species, *Nuttalliella namaqua* Bedford, collected from nests of rock hyraxes (*Procavia*) and swallows in South Africa, Namibia, and Tanzania (Keirans et al. 1976). Less than 30 nymphal and female specimens of *N. namaqua* are known, and none has been collected from a host animal. *N. namaqua* shares a number of ixodid and argasid features but also displays a suite of characters that is unique to the species (i.e., pseudoscutum, a palp with 3 segments, and "ball and socket" leg articulations; Sonenshine 1991). Its true position in the suborder may not become clear until larvae and males can be examined.

The importance of ticks as human and animal pests has led to the creation of two major bibliographies that are indispensable tools for research on ticks and tick-borne diseases. A complete listing of these works may be found in Keirans and Robbins (1999). A current world list of valid tick names is available in Horak, Camicas, and Keirans (2002).

Key 11.1. Order Ixodida, Key to the Families and Genera

1a. Capitulum situated anteriorly in all known active stages, visible when viewed from above; a scutum or pseudoscutum present. Paired spiracular plates present in postlarval stages, situated dorsolaterally posterior to coxa IV **2**

1b. Capitulum situated anteriorly in larva only, situated ventrally and not visible from above in postlarval stages, dorsal plate present or absent in larvae; leathery pseudoscutum rarely present in nymphs and adults. Paired spiracular plates situated dorsolaterally between coxae III–IV **ARGASIDAE—3**

2a. With a leathery papillate podonotal pseudoscutum. Palpi apparently consisting of 3 segments; hypostome short with few denticles. Known only from a few nymphs and adults from South Africa, Namibia, and Tanzania (monotypic) **NUTTALLIELLIDAE, Genus *Nuttalliella***

2b. With a podonotal scutum present in larval and nymphal stages and in the adult female. A holonotal scutum is present in males. Scutal ornamentation present or absent. Palpi 4-segmented; in most species the 4th segment inserted into a pit ventrally on segment III. Hypostome usually well armed with denticles and usually with a specific dental formula. Larva often with large wax glands (absent in *Ixodes*) **IXODIDAE—7**

3a. With an obvious marginal sutural line separating dorsal and ventral body surfaces **Genus *Argas***

3b. Without an obvious marginal sutural line separating dorsal and ventral body surfaces **4**

4a. Anterior portion of dorsal surface covered by smooth, leathery integument **Genus *Nothoaspis***

4b. Anterior portion of dorsal surface tuberculated, mammillated or granular, not smooth, leathery **5**

5a. Nymphs with dorsal integument beset with spines; adults with granular integument and vestigial hypostomes (adults nonparasitic and infrequently collected) **Genus *Otobius***

5b. Integument of nymphs and adults similar; hypostome of various shapes but not vestigial **6**

6a. Dorsal integument tuberculated; hypostome scooplike. Associated with bats and bat caves **Genus *Antricola***

6b. Dorsal integument mammillated; hypostome usually with teeth but never scooplike. Associated with various animals, including bats **Genus *Ornithodoros***

7a. Festoons and eyes absent; anal groove distinct, extending anteriorly around anus **Genus *Ixodes***

7b. Festoons usually present, eyes present or absent; anal groove usually present and posterior to anus, may be indistinct or absent **8**

8a. Eyes absent **9**

8b. Eyes present but difficult to see in *Dermacentor* (*Anocentor*) *nitens* Neumann and some *Boophilus* species. Ornate or inornate ticks **12**

9a. Inornate ticks **10**

9b. Ornate ticks **11**

10a. Palpi short, with segment II as broad as long and usually much broader than long, obviously extended laterally, basis capituli dorsally rectangular. Primarily Old World in distribution. Three species in New World, common **Genus *Haemaphysalis***

10b. Palpi elongate, with segment II twice as long as broad and never extended laterally. Found in Nepal, China, and Russia. Three very rare species **Genus *Anomalohimalaya***

11a. Hypostomatic dentition 2/2 or 3/3, inner file of denticles reduced in size, bright iridescent ornamentation absent. Australasia, on reptiles and edentate mammals **Genus *Bothriocroton***

11b. Hypostomatic dentition usually 3/3 or higher, internal file of denticles not reduced in size; bright iridescent ornamentation usually present. Primarily Southeast Asia and Africa, on reptiles **Genus *Amblyomma* (*pars*, formerly, Genus *Aponomma*)**

12a. Festoons absent **13**

12b. Festoons present **14**

13a. Legs IV of males greatly enlarged. Africa, three uncommon species **Genus *Margaropus***

13b. Legs IV of males not greatly enlarged. Cosmopolitan, on cattle and goats **Genus *Rhipicephalus*, Subgenus *Boophilus***

14a. Basis capituli hexagonal dorsally, usually inornate species **15**

14b. Basis capituli rectangular or quadrangular, ornate or inornate species **16**

15a. Inornate, males without ventral plates. Africa, uncommon **Genus *Rhipicentor***

15b. Usually inornate (4 ornate species), males with ventral plates. Cosmopolitan, but most species in Africa, common **Genus *Rhipicephalus***

16a. With 9 festoons, spiracular plate and dorsum of basis capituli ornamented in ivory coloration. Africa, rare (monotypic) **Genus *Cosmiomma***

16b. Festoons of various numbers, but not 9, spiracular plate and dorsum of basis capituli not ornamented **17**

17a. Palpi much longer than basis capituli, palpal segment II at least twice as long as broad **18**

17b. Palpi about as long as basis capituli, palpal segment II about as broad as long **20**

18a. Scutum and palpi ornamented, palpal segment II with a dorsal and ventral flange, male with ventral plates. India, uncommon (monotypic) **Genus *Nosomma***

18b. Scutum ornate or inornate, palpi unornamented, palpal segment II without a dorsal and ventral flange, male with or without ventral plates **19**

19a. Scutum usually ornate, males without ventral plates, with 11 festoons, not coalesced, eyes usually flat or slightly raised, not bulging upward from distinct, deep sockets. Circumglobal in tropics and subtropics but rare in Palearctic; common **Genus *Amblyomma* (*pars*)**

19b. Scutum inornate, males with ventral plates, with 7–11 festoons, partially coalesced. Eyes large, bulging upward from distinct, deep sockets. Southern Europe, Africa, Middle East to India, common **Genus *Hyalomma***

20a. With 11 festoons, ornate. Cosmopolitan, common **Genus *Dermacentor***

20b. With 7 festoons, inornate. Neotropics into southern United States, relatively common **Genus *Dermacentor*, Subgenus *Anocentor***

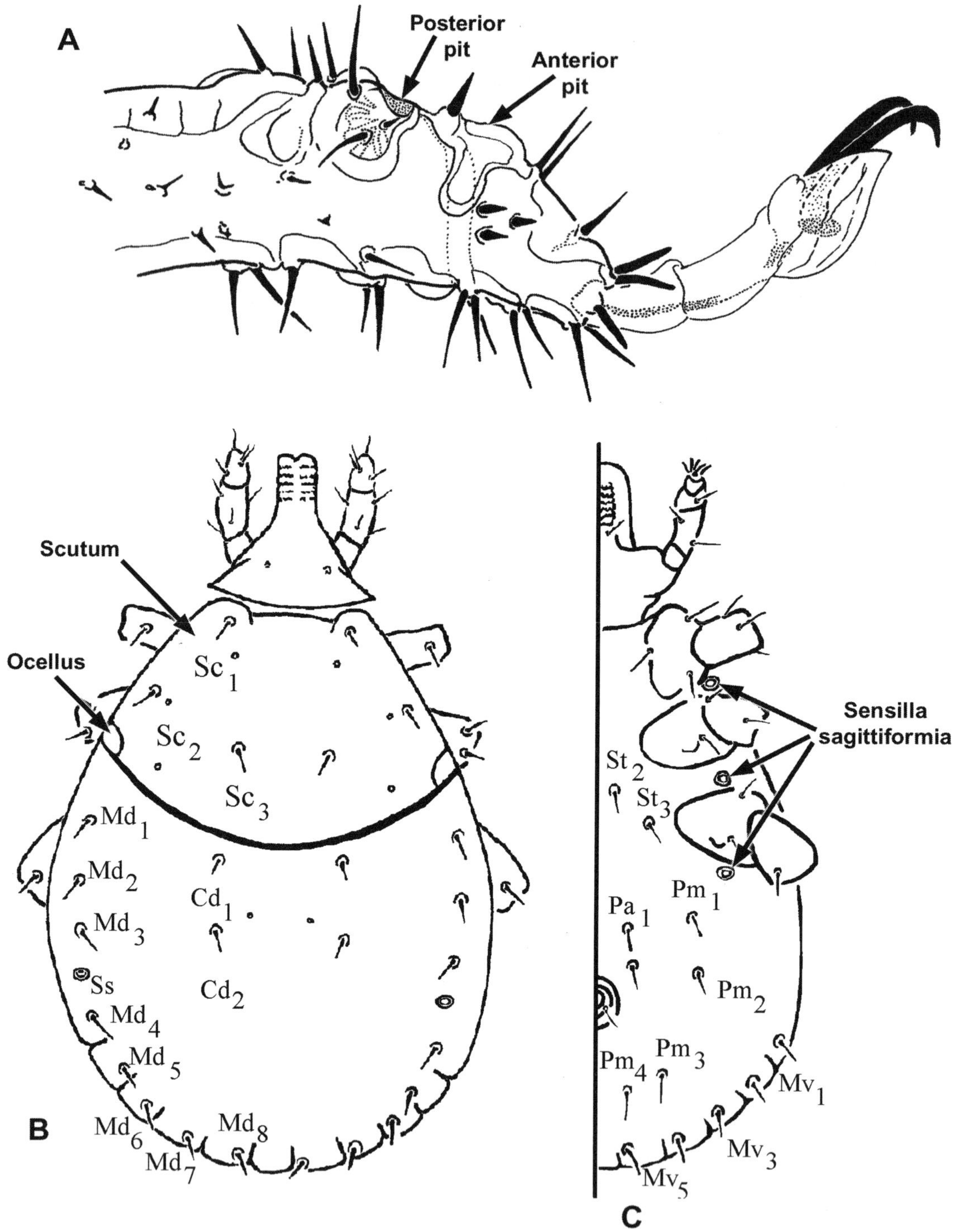

Fig. 11.1. IXODIDAE. *Dermacentor variabilis* (Say) (Massachusetts, USA): **A**, distal portion of tarsus I; **B**, dorsum of larva; **C**, venter of larva (setal notations after Clifford and Anastos 1960).

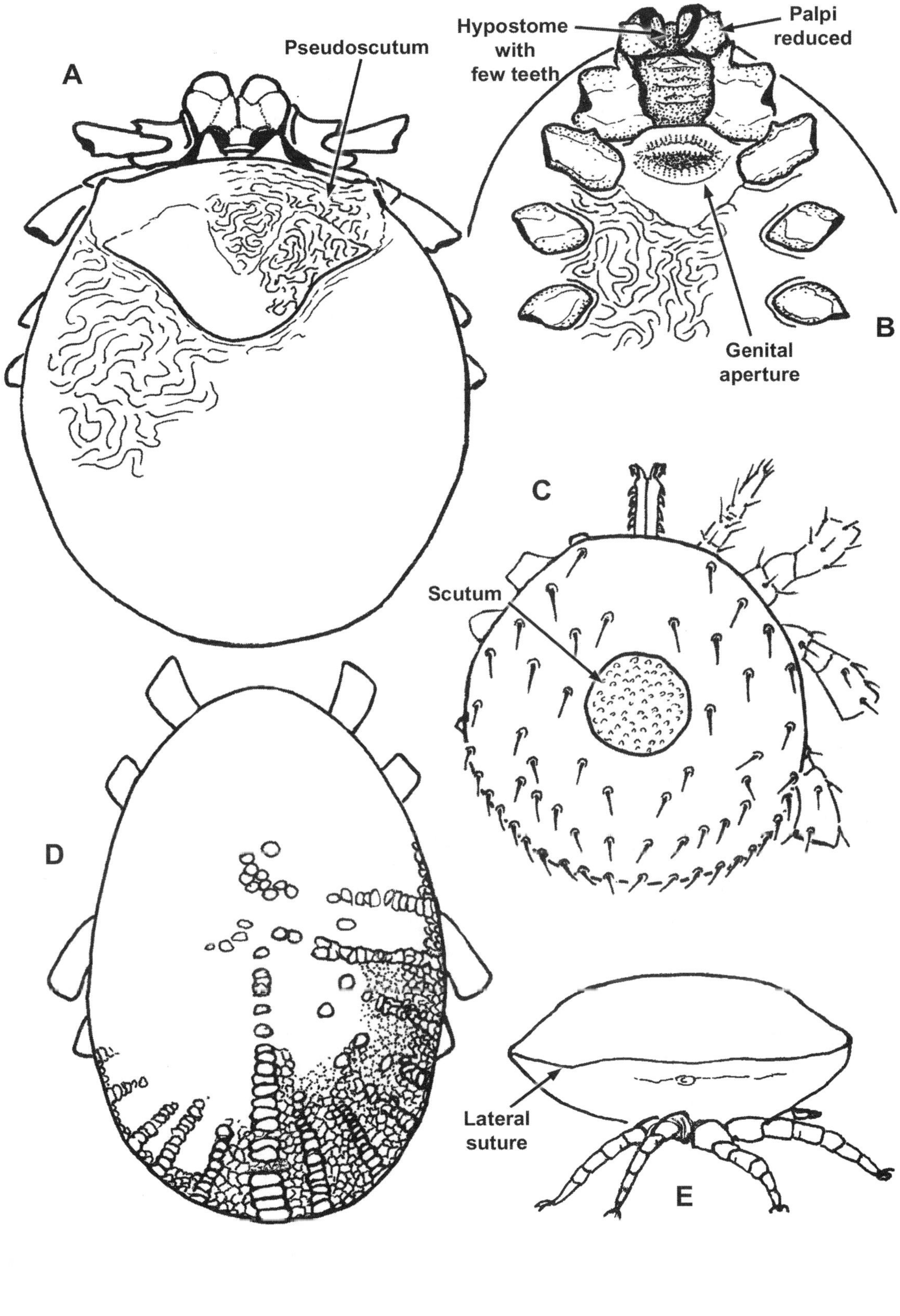

Fig. 11.2. NUTTALLIELLIDAE. *Nuttalliella namaqua* Bedford (Tanzania): **A**, dorsum of female; **B**, anteroventral aspect. ARGASIDAE. *Argas persicus* (Oken): **C**, dorsum of larva; **D**, dorsum of female; **E**, lateral aspect of partially engorged female.

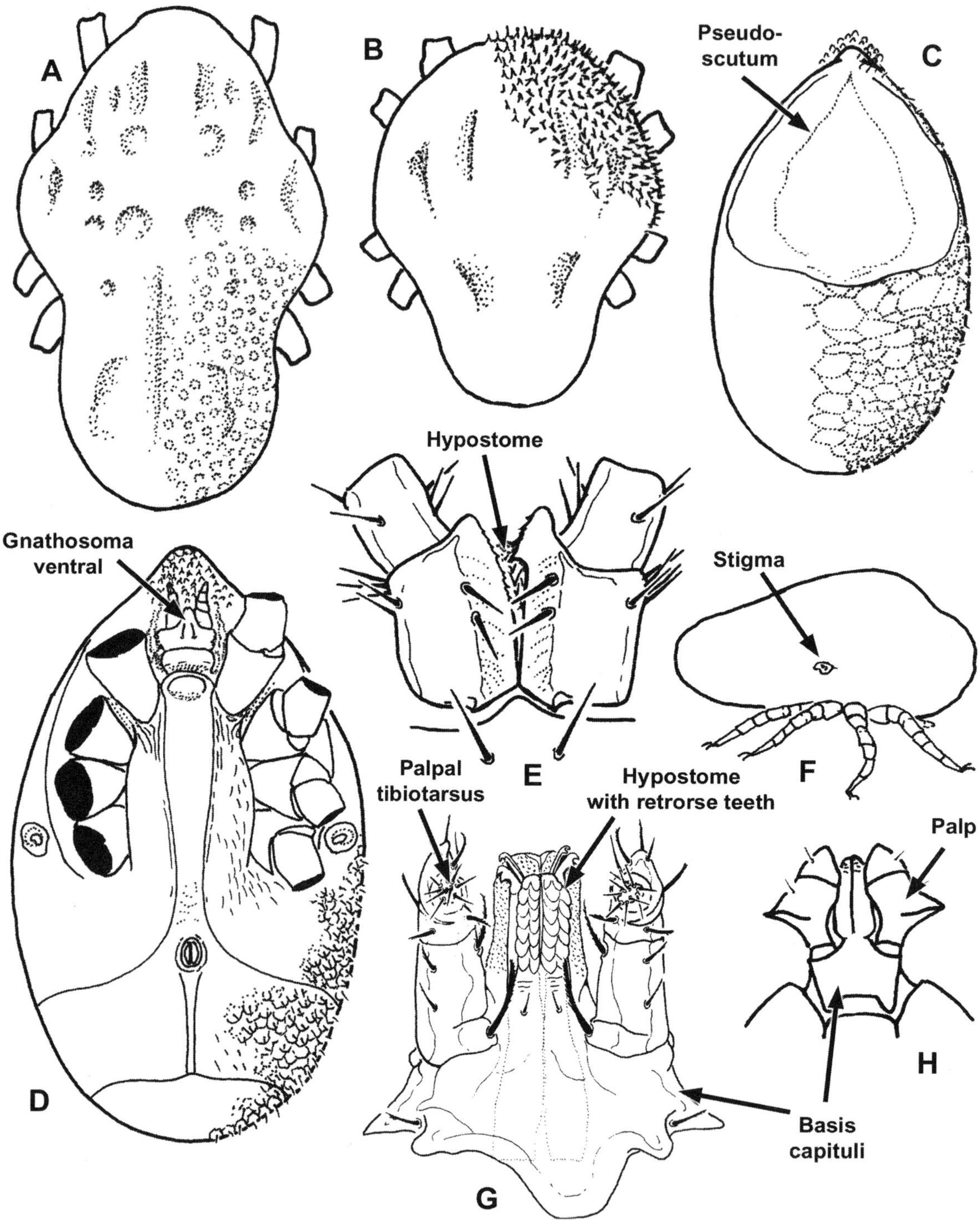

Fig. 11.3. ARGASIDAE. *Otobius lagophilus* Cooley and Kohls (after Johnston 1968): **A**, dorsum of male. *O. megnini* (Dugès): **B**, nymph. *Nothoaspis reddelli* Keirans and Clifford: **C**, dorsum of male; **D**, venter of male; **E**, ventral hypostomatic sheath formed by proximal palpal articles. *Ornithodorus* sp.: **F**, lateral aspect of partially engorged female.

IXODIDAE. **G**, ventral aspect of capitulum of ixodid tick (Republic of the Congo). *Haemaphysalis leporispalustris* (Packard): **H**, dorsal aspect of male capitulum.

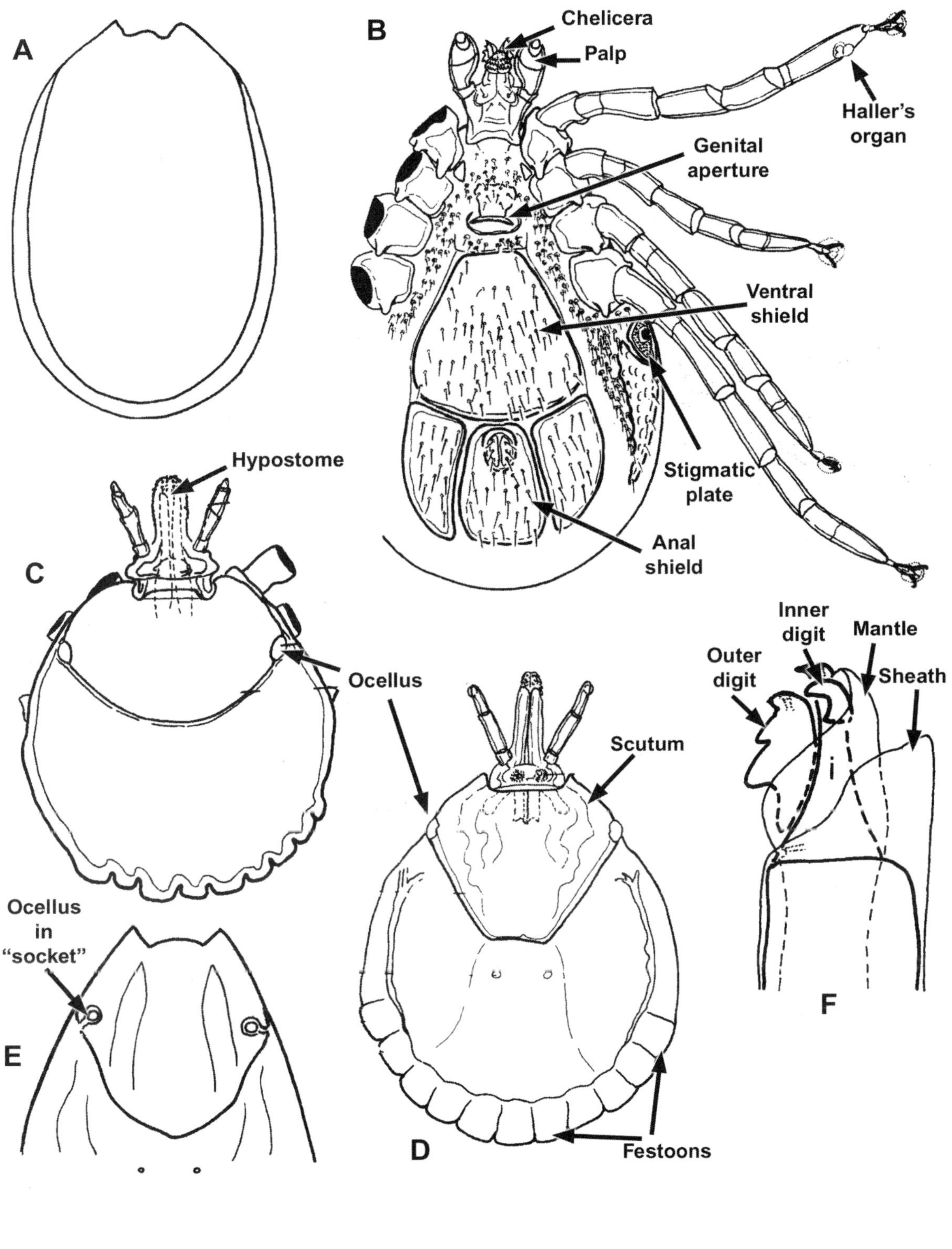

Fig. 11.4. IXODIDAE. *Ixodes* sp. (Oregon, USA): **A**, dorsum of male; **B**, venter of male. *Amblyomma cajennense* (Fabr.) (Mexico): **C**, dorsum of larva; **D**, dorsum of female. *Hyalomma asiaticum* P. and E. Schulze (after Balashov 1972): **E**, scutum of female; **F**, chelicera of female.

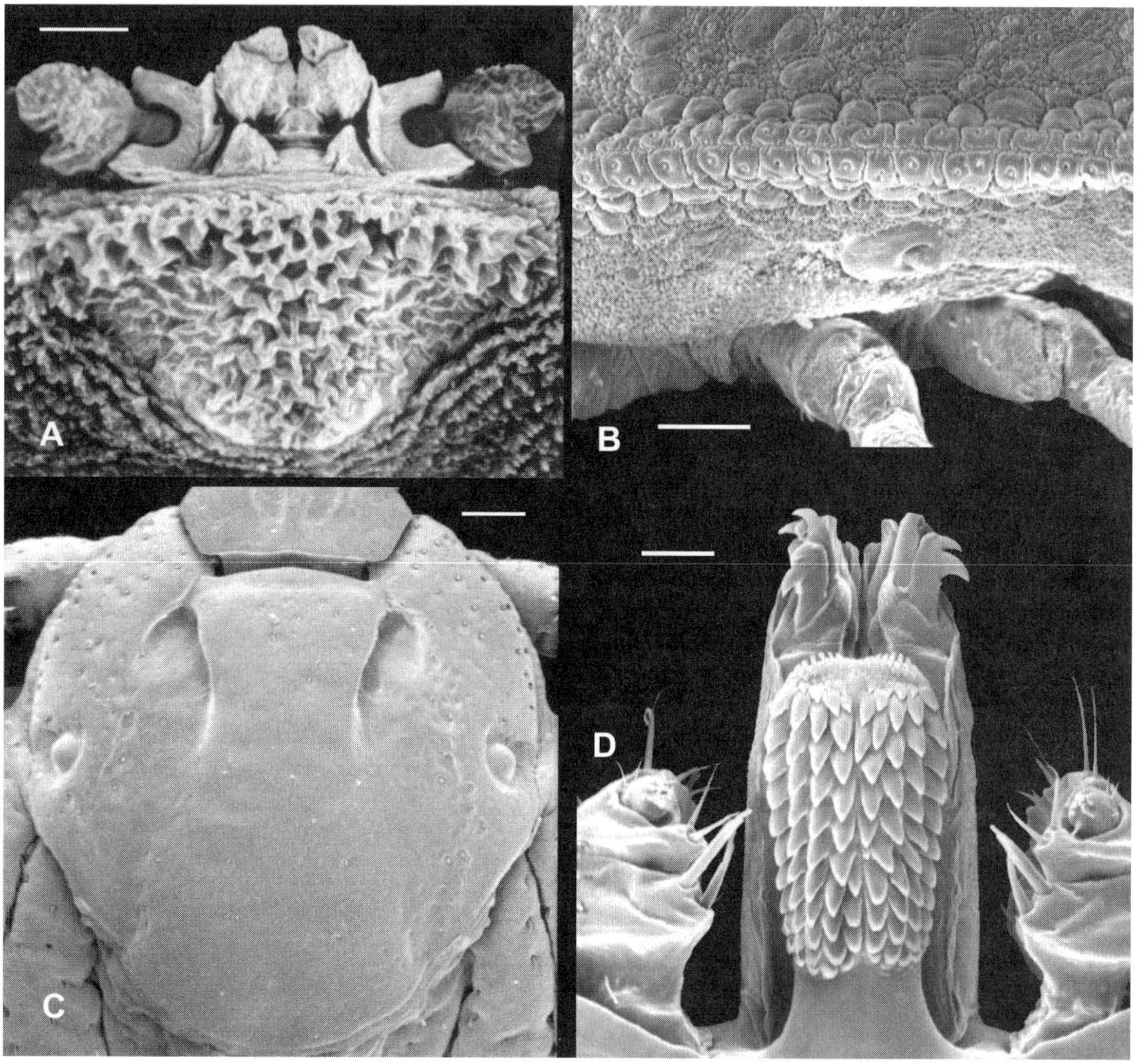

Fig. 11.5. NUTTALLIELLIDAE. *Nuttalliella namaqua* Bedford (Tanzania): **A**, scutum and dorsal gnathosoma of female.
ARGASIDAE. *Argas*: **B**, lateral view showing stigma.
IXODIDAE. *Hyalomma*: **C**, scutum. *Boophilus*: **D**, hypostome.

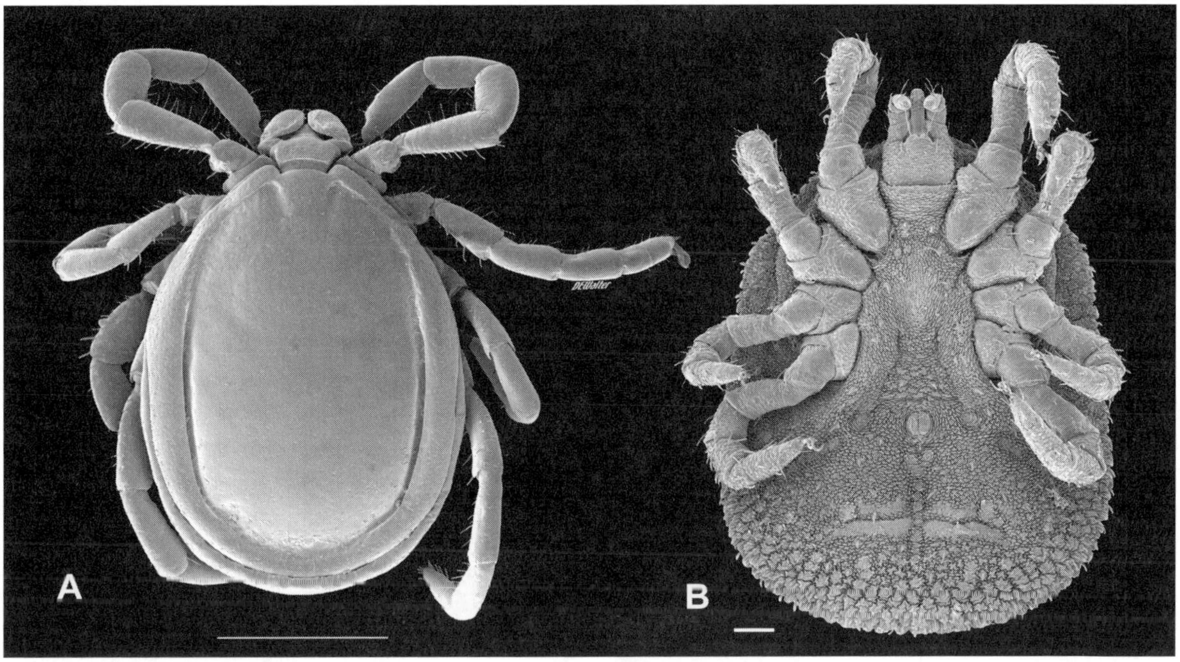

Fig. 11.6. IXODIDAE. *Ixodes holocyclus* Neumann (Queensland, Australia): **A**, dorsum of male. ARGASIDAE. *Carios* sp. (Queensland, Australia): **B**, venter of nymph.

CHAPTER TWELVE

E.E. LINDQUIST
G.W. KRANTZ
D.E. WALTER

Order Mesostigmata

The Mesostigmata (Gamasida) is a large, cosmopolitan assemblage of parasitiform mites that embrace an unusually diverse variety of lifestyles and habitats. The majority of species are free-living predators (Karg 1993b), while many others are parasites or symbionts of mammals, birds, reptiles, or arthropods (Strandtmann and Wharton 1958; Yunker 1973; Treat 1975; Walter and Proctor 1999). Relatively few feed on fungi, pollen, or nectar (Walter and Proctor 1999). Mesostigmatic mites may be found in association with soil, litter, rotting wood, compost, manure, carrion, nests, house dust and similar detritus-based niches, fungi, aerial niches on plants, and animals. Some species live in intertidal areas or along the margins of freshwater systems, where they may remain submerged for long periods of time.

Mesostigmatic mites range in size from 200 μm to 4,500 μm. They usually possess several distinctive sclerotized shields or plates on the idiosomatic dorsum and venter that illustrate characteristic incremental development, fusion, or fragmentation, beginning with the larval instar and ending with the adult molt (Fig. 12.1). Ontogenetic development in the Mesostigmata is limited to a larval and two nymphal instars prior to appearance of the adult (see chapter 5). Development signals the appearance of a pair of stigmata and elongation of accompanying peritremes in postlarval instars and, except where neoteny occurs, an increase in numbers of setae and porelike structures on the idiosoma and of setae on the appendages. The palpi of mesostigmatans provide a useful means for determining particular instars because certain of the five free segments acquire additional setae in a generally set pattern at the protonymphal and deutonymphal molts (Evans 1964). Exceptions occur in neotenous parasitic taxa and sometimes elsewhere. The following chart lists setal numbers on palpal segments in each instar of a typical mesostigmatic mite. Other features of ontogenetic development are illustrated in Fig. 12.1.

Ontogenetic Development

	Number of Setae on Palpal Segment				
Instar	Trochanter	Femur	Genu	Tibia	Tarsus
Larva	0	4	5	12	11
Protonymph	1	4	5	12	15
Deutonymph and Adult	2	5	6	14	15

There is pronounced sexual dimorphism in the intercoxal region of adult Mesostigmata, based in large part on an **epigynal shield** in the female (Fig. 12.2A) and, in many cases, on concomitant fragmentation of the surrounding ventral armature. The epigynal shield may be free or joined to or fused with adjacent sclerites. Males have a small genital opening usually set in a holoventral or sternitiventral shield complex (Fig. 12.3). In addition, cheliceral dimorphism occurs in groups where the male chelicerae are modified for sperm transfer to the female (see chapter 3 and below). Cheliceral modification in males often is accompanied by the appearance of spurlike structures or enlarged spinelike setae on various leg segments (especially legs II).

The idiosoma of mesostigmatic mites carries several distinctive and oftentimes diagnostic characters that serve to distinguish them from other parasitiform mites. The dorsal setae, along with those of the venter, often form predictable patterns that make possible a system of notation that often is diagnostic (Lindquist and Evans 1965 and others; Fig. 12.2B). In addition to setae, characteristic lyrifissures (cupules, poroids), gland pores, and muscle insertion scars (sigilla) also are found dorsally (Athias-Henriot 1969d, 1971, 1974; Krantz and Redmond 1987). Idiosomatic ocelli are absent. A pair of lateroventral or laterodorsal stigmata opens ventrally at the level of coxae II–IV, and these are usually associated with elongate peritremes in postlarval instars (Fig. 12.2A). Peritremes may be reduced or absent in certain families.

Female mesostigmatic mites have a transverse intercoxal

genital aperture that is covered by one to four shields. The male genital aperture is located either at or near the anterior margin of the shield in the intercoxal region (Fig. 12.3B) or well within it, usually at some level between coxae II–III (Fig. 12.3A). In the latter position, the aperture may be covered by one or two valves, the anterior of which may bear a pair of setae. The presence of an anterior aperture is linked to the presence of cheliceral spermatodactyli, which transfer sperm material to a female sperm induction pore or to her genital opening (see chapter 3). Mesostigmatans possess a pair of anal valves that are usually without setae, but a single pair (rarely two) may appear on the larval anal valves.

There is considerable diversity in gnathosomatic morphology in the Mesostigmata, but certain features remain more or less constant throughout the order. A pair of hornlike **corniculi** arise at the terminus of the hypostome (Fig. 12.2A), and they are usually well sclerotized and occasionally furcate or dentate. Corniculi are homologous with structures referred to as rutella in the parasitiform order Opilioacarida (chapter 9), but their homology with the rutella of sarcoptiform mites is problematical (Evans 1992). With's organs, the modified pair of setae normally associated with the opilioacarid rutella, are absent in the Mesostigmata. Three pairs of setae are commonly found on the ventral face of the hypostome, where their insertions on either side may describe a triangle (Fig. 12.2A) or a more or less longitudinal line (Fig. 12.31C). Only two pairs of hypostomatic setae occur in certain neotenous parasites and pollen feeders. Dorsally, a simple or variously shaped **epistome** (tectum capituli of Snodgrass 1948; gnathotectum of Evans 1992) projects anteriorly to form a roof over the gnathosoma (Fig. 12.14D). The epistome is absent or desclerotized in *Uropodella*, a genus of the supercohort Sejida. Cheliceral morphology varies greatly within the order, ranging from strongly developed dentate structures in many free-living forms to styletlike piercing organs in some parasitic groups (Fig. 12.62I). The fixed cheliceral digit usually carries a short, broad seta at its base and a subterminal pilus dentilis (cheliseta of Van der Hammen 1964). A subterminal pit sensillum may also be discerned, along with an antiaxial and a paraxial lyrifissure (Fig. 12.26B). The fixed digit may be reduced or absent in some groups (e.g., the DERMANYSSIDAE, the VARROIDAE, some BLATTISOCIIDAE, and the LAELAPIDAE). A tined claw, or apotele, is inserted at the base of the palptarsus (Fig. 12.2A). The apotele typically has two or three tines but can show considerable variation in some parasitic species (Pence and Casto 1976). It may be entirely absent in certain parasites.

An undivided tritosternum (sternapophysis of Van der Hammen 1966) arises ventrally behind the gnathosoma. Its base gives rise to one to three elongate laciniae, which may be secondarily anastomosed. The laciniae, and in some cases the entire tritosternum, may be vestigial or absent in certain parasitic families (e.g., SPINTURNICIDAE, RHINONYSSIDAE). As noted in chapter 3, the tritosternum has been shown to function in concert with the subcapitular groove of the gnathosoma as a fluid transport mechanism (Wernz and Krantz 1976).

Chaetotactic patterns on particular leg segments provide valuable taxonomic criteria for establishing relationships within the Mesostigmata (Evans 1963b, 1965, 1969b, 1972). A system of setal nomenclature and notation devised by Evans (1963b) permits characterization of a given segment by means of a simple formula. Setae inserted on the various faces of the segment can be recognized individually and are recorded in the following form for a leg on the right side of the body:

$$\text{anterolaterals (al)}\ \frac{\text{anterodorsals (ad)}}{\text{anteroventrals (av)}}\ \frac{\text{posterodorsals (pd)}}{\text{posteroventrals (pv)}}\ \text{posterolaterals (pl)}$$

Thus, a leg segment with two anterolateral and one posterolateral setae, two anterodorsals and two posterodorsals, one anteroventral and no posteroventrals, would be characterized as follows:

$$2\ \frac{2}{1}\ \frac{2}{0}\ 1 = 2\ 2/1,\ 2/0\ 1$$

A graphic example of this chaetotactic formula is shown in Fig. 12.4. Note that simply applying a setal count of 8 for this segment would provide only limited chaetotactic information because there are 64 possible permutations of an 8-setae configuration: 1 3/1, 2/1 0; 1 2/1, 2/1 1; 1 3/1, 2/0 1; 1 3/0, 3/0 1; and so on.

The dorsal terminus of tarsus I in the Mesostigmata carries a complex field of sensory setae (Haarløv 1943), many elements of which are also found on tarsus I of the related parasitiform orders Opilioacarida, Holothyrida, and Ixodida. Although they are homologous, the insertions of these shared setae may vary between orders.

The order Mesostigmata is divided here into three suborders, two of which are further subdivided into a number of cohorts. Approximately 70 families are included in the key to the order (p. 157), and these are grouped in 26 superfamilies. About 12,000 species have been described (Walter and Proctor 1999).

Suborder Sejida

- Superfamily Sejoidea (Figs. 12.5–12.7)

DIAGNOSIS: Body with moderate to extensive hypertrichy, numerous gland openings, and often densely tuberculate. Adult female dorsally with podonotal and pygidial shields (sometimes extensive) and usually with mesonotal or mesolateral shields; secondary hardening of the cuticle often obscures shield extent and number. Peritremes and peritrematic shields usually well developed, the latter fused anteriorly to the podonotal shield and sometimes included in a process over the gnathosoma (vertex); lateral setae often on small platelets. Venter of female with large ventrianal shield (sometimes developed dorsolaterally and fused to pygidial shield); metapodal shields usually free and well developed; epigynal shield large, usually bearing 6 to numerous setae (rarely 1–2 pairs) and covering the opening to a large, saclike spermatheca. Sternal area not fully sclerotized, but with seta-bearing platelets (sometimes obscured by genital shield) and often a shield bearing 2 pairs of setae. Males with more consolidated body shields; male genital aperture on an intercoxal shield, covered by a subcircular valve. Tarsus I often with an acrotarsus, with or without claws; tarsus IV with 20 or more setae, including setae av4 *and* pv4 *on a well-defined intercalary sclerite. Tritosternum with well-developed base and separate laciniae. Hypostomatic setae* h1 *often flattened, scalelike; corniculi entire or bifid. Palp apotele 2-tined. Chelicerae chelate-dentate, slender and with setiform pilus dentilis, to massive and without pilus dentilis; movable digit without arthrodial brush or excrescences. Phoretic deutonymph with a cylindrical anal process and 1–2 dorsal shields.*

The Sejida (Liroaspina, Trichopygidiina, Sejina) comprises densely setose Mesostigmata with fragmented dorsal and sternal shields. They often display striking plesiomorphies also found in some ticks, holothyrans, and opilioacarans. For example, tarsus I in deutonymphs and adults often has a terminal acrotarsus—also known in Heterozerconina, in LAELAPTONYSSIDAE, and in deutonymphs of some EVIPHIDIDAE, PARASITIDAE (Parasitiae), and OLOGAMASIDAE—and tarsus IV has an intercalary sclerite between the basitarsus-telotarsus that bears setae *av4*/*pv4* (also present in Trigynaspida). The single large female genital shield often has numerous, irregularly placed setae and an anterior notch as seen on the mesogynal shield of species of *Allothyrus*, a genus in the Holothyrida. The genetic systems in the Sejida have not been studied, but approximately the same number of males and females are encountered in field collections, suggesting diplodiploidy. The male chelicerae are not modified for sperm transfer, and the female is inseminated via the primary genital opening into a large, saclike spermatheca. One to four eggs are simultaneously matured in most species, but up to a dozen may be carried by individuals of large species. Some molecular analyses have grouped the cohort Heterozerconina with the Sejida (Lekveishvili and Klompen 2004a), but this relationship is not yet supported by clear morphological characters (Lekveishvili and Klompen 2004b).

Three families and nine genera of Sejida are currently recognized (Johnston 1982; Lekveishvili and Krantz 2004). The SEJIDAE (= Liroaspidae Trägårdh) includes the genera *Sejus* (= *Liroaspis*), *Epicroseius*, *Adenosejus*, *Africasejus*, and *Zuluacarus* (Lekveishvili and Klompen 2006); the UROPODELLIDAE contains the single genus *Uropodella*; and the ICHTHYOSTOMATOGASTERIDAE includes the genera *Asternolaelaps* and *Japanoasternolaelaps*. *Archaeopodella scopulifera* Athias-Henriot, a highly plesiomorphic sejid, currently is known only from nymphs (Athias-Henriot 1977). It is placed in the ICHTHYOSTOMATOGASTERIDAE in this treatment because of apparent synapomorphies with *Asternolaelaps*, including large, bifurcate corniculi (Fig. 12.7A); massive, grinding chelicerae; lack of a pilus dentilis; and an enlarged anal opening (Fig. 12.7B; see Lekveishvili and Klompen 2004a for a different interpretation). Athias-Henriot (1972, 1977) suggested that the families UROPODELLIDAE and ICHTHYOSTOMATOGASTERIDAE be synonymized, but her reasons for doing so were based on characters of uncertain polarity.

The SEJIDAE is the most diverse sejoid family, with about 30 described species of *Sejus* s. l., 16 of *Epicroseius*, and one of *Adenosejus*. The last genus was proposed by Lekveishvili and Krantz (2004) for a mite originally included in the genus *Sejus* in earlier editions of this volume (Krantz 1970b, 1978) and later designated a new species by Hirschmann (1991). An acrotarsus with or without claws may be present on leg I of sejids, and a pair of hornlike processes bearing distal setae is often present posterolaterally on the idiosoma. Adult females have one to six (mostly three) pairs of setae on the genital shield (Hirschmann 1991; Hirschmann and Wisniewski 1994). *Epicroseius* is well supported by derived characters (e.g., tarsus I without apical claws, epistome bifurcate, pygidial shield with median longitudinal keel), but *Sejus* probably contains more than one generic-level group (see Lekveishvili and Klompen 2006).

Sejids are common litter inhabitants in moist forests in the subtropics and tropics, especially in association with rotting wood, but are less common in the Holarctic region. Australian *Sejus* and *Epicroseius* are predatory on small arthropods and nematodes (Walter and Proctor 1998) and have been collected from decaying fungi on logs, wood mulch in gardens, litter around logs in rain forests, wet sclerophyll, and suspended soils in the canopy. The chelicerae have small, serrate teeth, and the anal opening is small—characteristics that relate to the fact that these mites digest their food externally and ingest only the resultant fluids. Males use their chelicerae to transfer a flask-shaped spermatophore to the anterior cleft of the female's genital shield. One to four eggs are carried by gravid females.

Deutonymphs of Sejida often are phoretic on beetles and flies and have a cylindrical extension of the anal opening that provides the means for attaching to their insect carriers (Kethley 1983b; Hirschmann 1991). At least some undescribed species of Australian *Epicroseius* produce two deutonymphal morphs in culture, one of which is a phoretomorph that has a perianal process similar to that described for *Uropodella lacinata* Berlese (Kethley 1983b) and a single dorsal shield that incorporates the podonotal, mesonotal, and pygidial elements of the nonphoretic nymph (D. E. Walter, unpublished).

Six species of *Uropodella* (UROPODELLIDAE) have been described from the Americas and Africa, mostly associated with woody habitats such as tree holes and bark beetle galleries (Camin 1955; Athias-Henriot 1972; Wisniewski and

Hirschmann 1991b). They are fluid feeders, but their biology is otherwise unknown. Species of *Asternolaelaps* (ICHTHYOSTOMATOGASTERIDAE), many of which are undescribed, are known from Europe, North and South America, Australia, and New Zealand. They have massive chelicerae and bifurcate corniculi that are used to "grind" food (Athias-Henriot 1972)—that is, to bite off and swallow pieces of fungi and small mites (Walter and Proctor 1998; D. E. Walter, unpublished), and enlarged anal openings for eliminating solid waste. Athias-Henriot (1972) created a new subgenus, *Parasternolaelaps*, to accommodate *A. castrii* Athias-Henriot, a species with pustulelike gland openings reminiscent of those seen on *Epicrius* (Gamasina). Hirschmann and Hiramatsu (1984) proposed a new genus, *Japanoasternolaelaps*, for species with trifurcate corniculi, a dense fringe of setae on the anterior margin of the female genital shield, and reduced dorsal and ventral plates in the adult.

As with other Sejida, *Asternolaelaps* species are often found in tree hollows or under bark and sometimes in nests, bat caves, and other habitats associated with mammals. Great numbers of an undescribed species of *Asternolaelaps* were recovered from nest balls of the wood rat, *Neotoma fuscipes*, in California, USA (G. W. Krantz and G. L. Parsons, unpublished). However, given the time and effort devoted over the past several decades to collecting mites associated with vertebrates, and the relatively few *Asternolaelaps* that have been found, a routine association with vertebrates seems unlikely.

Suborder Trigynaspida

Camin and Gorirossi (1955) proposed the suborder Trigynaspida, and the cohorts Cercomegistina and Antennophorina, to accommodate an assemblage of large and often striking mesostigmatic mites whose known species are mostly associated with arthropods or reptiles in warm temperate to tropical habitats. Trigynaspids are early derivative Mesostigmata distinguished primarily by a set of setal, cheliceral, and genital shield characters that are not necessarily unique to the group on an individual basis. For example, with rare exceptions, trigynaspids share the plesiomorphic retention of setae *av4* and *pv4* on tarsus IV (Figs. 12.16B, 12.19B) with the suborder Sejida. These setae are added on the deutonymph and usually occur on a free intercalary sclerite in the adult. At the same time, the Trigynaspida have an apparently unique set of setal characters on tarsi II–IV: an additional anterolateral seta (al_x) and a medioventral seta (Evans 1965). More striking are the often elaborate excrescences on the movable cheliceral digit (Figs. 12.8G, 12.9B, 12.11C, D, 12.13B, E, 12.15C, D, 12.16D, 12.17D). Similar structures also occur in the cohorts Microgyniina and Heterozerconina, although they arise on the fixed cheliceral digit in heterozerconines (see p. 137). Cheliceral excrescences should not be confused with the elongate arthrodial brushes (Figs. 12.44E, 12.46D, G) that originate in the membrane at the base of the cheliceral digits in some Dermanyssiae (e.g., PARHOLASPIDIDAE, MACROCHELIDAE and some Rhodacaroidea) and in some male Parasitiae (Fig. 12.38F).

Male trigynaspids lack dermanyssine spermatodactyli, but some Celaenopsoidea (e.g., *Cryptometasternum* spp.) have a sclerotized fingerlike process at the base of the movable digit that appears to be a modified sperm transfer organ. Additionally, the celaenopsoid male genital opening is located anteriorly at the base of the tritosternum, suggesting a modified role for the chelicerae in sperm transfer (see chapter 3). In most other Trigynaspida, the male genital opening is located more posteriorly in the intercoxal region.

The female oviporus is also the genital opening and is covered by a complex of shields (as in the Holothyrida, p. 107) or their remnants, not by the single shield that is characteristic of most Mesostigmata (Fig. 12.20). The name Trigynaspida derives from the presence, in many females, of three shield elements covering the oviporus (e.g., Fig. 12.8E), a posterior median mesogynal shield that is usually hinged posteriorly and is flanked by a pair of latigynal shields that form the walls of an egg trough (no ovipositor is present). Each latigynal shield bears one to many setae, and a lyrifissure often is present. The latigynals may be fused medially in cases where a discrete mesogynal shield is lost (e.g., Fig. 12.17A), or they may be contiguous with ventral elements (e.g., Fig. 12.14E). The mesogynal shield is usually without setae or pores, but it is sometimes insensibly fused to the latigynals (Figs. 12.14A, E, 12.18A) or to ventral seta-bearing elements and may be reduced to a tiny remnant. Some early derivative members of the cohort Uropodina (e.g., *Trachytes*) have genital shields that resemble those of some trigynaspid Cercomegistina, although more derived uropodines have latigynal elements that are fused to the endopodals and a single large, nude mesogynal shield. In contrast to the above examples, members of the cohort Gamasina have a single epigynal shield, usually with one pair of setae, that appears to represent fused latigynal-mesogynal shields.

In addition to the lateral (latigynal) and median-posterior (mesosgynal) shields covering the genital opening, some Cercomegistina (Figs. 12.8E, 12.9C, F, 12.10B, E) and Antennophorina (Figs. 12.16C, F, 12.18C, D) have an anterior genital shield or its remnant. A shield in the pregenital position is referred to as a **sternogynum** if it bears a pair of lyrifissures (e.g., Figs. 12.16A, F), and it may be divided (Fig. 12.18C) or entire as a **pseudosternogynum** (Fig. 12.17E) if pores are absent. Traditionally, pregenital shields have been considered sternal in origin (Camin and Gorirossi 1955; Kethley 1977a), but they may be homologous with the pregenital shield in the parasitiform order Holothyrida (Lehtinen 1991). In the FEDRIZZIIDAE (Fig. 12.16A), a sternogynum provides the only visible cover for the genital opening. Here, the remnants of both the mesogynal and latigynal shields are hidden under a plate that covers most of the venter. Such ventral plates, comprising elements of various shields, are common in the Trigynaspida and are especially characteristic of the Celaenopsoidea. The trigynaspid anal opening may be borne on the ventral plate or on free ventrianal or anal shields. Members of the superfamily Aenictequoidea have a pair of posterior sternal processes called **sternovaginal sclerites** (Fig. 12.12B) that may be homologous to the vaginal sclerites of the Celaenopsoidea (Fig. 12.14C) and the porose vaginal sclerites of the DAVACARIDAE (Fig. 12.9E). These structures may be derived from pregenital shields (Walter 2004).

The sternal region of trigynaspids usually bears four pairs of setae and three pairs of lyrifissures, but the sclerotization is often incomplete. The first pair of setae and lyrifissures may occur on a separate sclerite (**tetartosternum**) (Figs. 12.14C, 12.16A), on a pair of sclerites (**jugulars**) (Figs. 12.12E, 12.17A) or fleshy lobes (Fig. 12.17C), in apparently soft cuticle (Fig. 12.8E), or on a sternal shield(s) bearing other setae and covering part or all of the sternal region (Fig. 12.17E). Sclerites bearing the fourth sternal setae and the third pair of lyrifissures are sometimes present and may be referred to incorrectly as metasternal shields because of their similarity in placement to apparently analogous structures that occur in the Monogynaspida (see discussion under Celaenopsoidea).

Cohort Cercomegistina

■ Superfamily Cercomegistoidea (Figs. 12.8–12.11)

DIAGNOSIS: Body often with moderate hypertrichy. Adult female usually with separate podonotal and opisthonotal shields or holodorsal shield, opisthonotal region rarely with separate mesonotal and pygidial elements. Peritremes and peritrematic shields well developed, the latter fused anteriorly to podonotal shield. Lateral setae on platelets or on marginal shield. Venter usually with large ventrianal or genitoventral shield, rarely with small anal shield; genital region with 3 shields or their remnants, mesogynal shield often reduced, latigynal shields usually well developed; pregenital shield (sternogynum) free, fused to sternal elements, or absent; sternal shield well developed to fragmented. Spermatheca saclike, beneath oviporus; vaginal sclerites absent. Adult male genital opening on intercoxal shield, usually between coxae III, and covered by 1–2 transverse valves. Tarsus I with or without claws. Tarsus IV with setae av4 *and* pv4 *on a well-defined intercalary sclerite. Tritosternal laciniae usually fused for more than half their length. Unpaired postanal seta usually present in larva but absent in adult. Chelicerae chelate-dentate; movable digit with 1–3 brushlike excrescences; epistome (gnathotectum) denticulate, without ventral keel; palpgenu with 6 setae.*

The Cercomegistina was originally based on the family CERCOMEGISTIDAE, an assemblage of insect-associated trigynaspids with a gamasine-like epistome, median epistomatic apodeme, and small salivary processes (Camin and Gorirossi 1955). As currently defined (Kim 2004), the Cercomegistina includes the CERCOMEGISTIDAE and five additional families of free-living, soil-litter predators. Kethley (1977a) expanded the assemblage to include the ASTERNOSEIIDAE (Valle 1954) and two new families to accommodate previously described genera: the DAVACARIDAE for the genus *Davacarus* (Hunter 1970) and the SEIODIDAE for *Seiodes* (Berlese 1887; later redescribed by Athias-Henriot 1959). More recently, Lindquist and Moraza (1993) erected the PYROSEJIDAE, based on *Pyrosejus* and an undescribed genus from the Neotropics, and Walter (2000) erected the SALTISEIIDAE based on the new species *Saltiseius hunteri* Walter. Members of the CERCOMEGISTIDAE have been reported as phoretic on bark beetles and predatory in their galleries (Kinn 1971) and have been discovered in unusual situations, such as in webs of social spiders (Evans 1958b) and on pagurid crabs (André 1937).

The Cercomegistina is primarily Gondwanan in distribution (Walter 2000, 2004), with centers of diversity in the Neotropics, Africa, and Australia. The Neotropical region has a particularly rich, but largely undescribed, fauna of PYROSEJIDAE, ASTERNOSEIIDAE, and CERCOMEGISTIDAE. The latter two families are also diverse in Australia, along with the DAVACARIDAE, which extends from the tropics to subantarctic islands (Walter 2004), and SALTISEIIDAE. Asternoseiids are also found in Africa, along with the monotypic SEIODIDAE. Only a few cercomegistines occur in the north temperate zone, and these are mostly cercomegistids associated with beetles (Kinn 1967a, 1970, 1972).

Cohort Antennophorina

Unlike most Cercomegistina, antennophorines have fully separate tritosternal laciniae, always lack claws on legs I, and often have seven setae on the palpgenu, a mucronate gnathotectum with a ventral keel, and vaginal sclerites or sternovaginal processes. Six superfamilies and about 20 families of Antennophorina have been recognized, many of which are monogeneric (Kethley 1977; Kim 2004). Adult antennophorines are typically associated with arthropods (myriapods, beetles, ants, stingless bees, cockroaches, termites) or reptiles (snakes, lizards), but there are exceptions. Species of *Triplogynium* and *Funkotriplogynium* (TRIPLOGYNIIDAE) are free-living predators, *Celaenopsis* (CELAENOPSIDAE) can be found wandering on bark, and *Megacelaenopsis* (MEGACELAENOPSIDAE) on banana leaves. In contrast to adults, immature Antennophorina are only rarely found on arthropods; many are predatory in the galleries or nests of their adult hosts (Hunter and Davis 1965; Seeman 2000). Some Antennophorina are clearly parasitic (e.g., ANTENNOPHORIDAE) (Franks, Healy, and Byrom 1991).

■ Superfamily Aenictequoidea (Figs. 12.12A–F)

DIAGNOSIS: Mesogynal shield subtriangular, separate from ventrianal shield, and overlapped or covered by latigynal shields bearing 2–10 pairs of setae. Sternal shield with a pair of poste-

*rior projections lying under latigynal shields (**sternovaginal processes**); sternal setae* st1 *on separate shields or on entire sternal shield. Palptibia and tarsus insensibly fused; chelicerae edentate. Peritremes often short, sometimes associated with a peritreme-like groove.*

Aenictequoids are large, heavily sclerotized associates of ants. Two genera (*Messoracarus, Leptantennus*) of the family MESSORACARIDAE are described, but the other three families are monogeneric. *Aenicteques chapmani* Jacot (AENICTEQUIDAE) was described from the ant *Anictus martini* in the Philippines (Jacot 1939a), while the monotypic *Physalozercon raffrayi* (Wassmann) (PHYSALOZERCONIDAE) was found in the nest of the ant *Plagiolepis custodiens* in South Africa. Species of *Ptochacarus* (PTOCHACARIDAE) are found on ants (e.g., *Iridomyrmex, Camponotus*) in Australia (Womersley 1958).

- Superfamily Antennophoroidea (Figs. 12.13A, B)

DIAGNOSIS: Mesogynal shield reduced, with anterior pedicellate ventrianal element dividing the narrow latigynal shields; latigynals with more than 2 pairs of setae. Sternal setae st1 *free in integument, on separate anterior shield(s), or incorporated into entire sternal shield. Palptibia and tarsus distinct, unfused; chelicerae with minute teeth, movable digit with filamentous excrescences.*

The Antennophoroidea is represented by the family ANTENNOPHORIDAE (Figs. 12.13A, B) and the genus *Antennophorus*, an assemblage of large, circular ant associates found in Europe, North America, and the Neotropics. *A. grandis* Berlese is reported to be haplodiploid (Franks, Healy, and Byrom 1991), and all active stases are kleptoparasitic on their ant hosts. The mites position themselves under the ant's head and induce them to regurgitate a drop of fluid by stroking the ant's mouthparts with their hypertrophied legs I.

- Superfamily Celaenopsoidea (Figs. 12.13 C–F, 12.14A–E, 12.15A–E)

DIAGNOSIS: Deutonymph with separate podonotal, mesonotal, and pygidial shields; adults usually with holodorsal shield bearing an unpaired median seta at its anterior margin (adults of Neotenogynium *retain the free deutonymphal shields). Mesogynal shield small, triangular, and free or amalgamated with latigynal and/or ventral elements; latigynal shields free or fused to ventral plate; vaginal sclerites well developed. Sternal shield bearing 3–4 setae; free metasternal shield bearing* st4 *and* stp3 *often present. Venter covered by a single extensive plate, or plate flanked by free lateral and/or anal shields. Palptibia and tarsus distinct, unfused; palpgenu with 5–7 setae; chelicerae with robust, dentate digits, movable digit with large proximal tooth and dendritic or brushlike excrescences. Male genital opening on anterior margin of sternal shield.*

Eight families of Celaenopsoidea compose this most diverse group of Trigynaspida, and adults of its member species are associated with a great variety of arthropods and reptiles. Some of the hosts involved in these associations are unusual for the suborder: for example, *Celaenosthanus trigonophilus* Vitzthum and *Funkotriplogynium irapora* Flechtmann with stingless bees (Vitzthum 1930; Flechtmann 1983), and *Lobogyniella tragardhi* Krantz with termites (Krantz 1958b). Host associations are moderately predictable in some families: for example, all 11 genera of EUZERCONIDAE (Figs. 12.14D, E) are found on passalid beetles or millipedes in the Americas, Africa, India, and the Philippines (Funk 1980; Hunter and Rosario 1989, 1991). The poorly known members of the family COSTACARIDAE also are found on millipedes (Hunter 1993). Associations are more general in other groups, for example, species of the family SCHIZOGYNIIDAE (Fig. 12.15E) may be found on passalid beetles in South Africa and India (Trägårdh 1950; Ryke 1957) and in bark beetle galleries in western North America (Kinn 1966). A species of the schizogyniid genus *Indogynium* is associated with snakes (Sellnick 1954).

Members of the family DIPLOGYNIIDAE (Figs. 12.15A–D) are common on passalid and histerid beetles in the New World and Asia and on a variety of scarabaeine beetles in Africa. Adults of *Ophiocelaeno sellnicki* Johnston and Fain are found on snakes (Johnston and Fain 1964). Other collections include species from under bark (Trägårdh 1950) and from litter and tree holes. *Crenamargo binuseta* Hicks was described from the palm weevil, *Rhynchophorus palmarum*, in Nicaragua (Hicks 1958). Species of CELAENOPSIDAE (Fig. 12.13C) may be found on bark or in association with bark beetles either in Europe (*Celaenopsis* spp.), in Europe and North America (*Pleuronectocelaeno*), or in New Guinea (*Ceratocelaenopsis womersleyi* Trägårdh) (Vitzthum 1926; Trägårdh 1941a, 1950; Kinn 1968). *Brachycelaenopsis breviatus* (Banks) is an associate of ants (*Iridomyrmex*) in Tasmania (Trägårdh 1951).

The MEGACELAENOPSIDAE (Fig. 12.14A) comprises two genera of large celaenopsid mites that may be free living. *Megacelaenopsis oudemansi* Funk was found in association with banana leaves and debris from Central America, while *Pelorocelaenopsis camini* Funk was taken from unspecified materials in Zaire (Funk 1974). *Triplogynium krantzi* Funk (TRIPLOGYNIIDAE) was described from beetle galleries in Zaire (Funk 1977), *T. vallei* (Fox) from a rat trap in Puerto Rico (Fox 1959), and three additional species from unspecified sites in Indonesia and Africa. *Funkotriplogynium iagobadius* Seeman and Walter (Figs. 12.13D–F) is a free-living predator or rarely associated with passalid beetles in Australia (Seeman and Walter 1997). All stases of the paedomorphic *Neotenogynium malkini* Kethley (NEOTENOGYNIIDAE) (Figs. 12.14B, C) occur on spirostreptid millipedes in the Neotropics (Kethley 1974).

The variation in genital shield morphology of the Celaenopsoidea rivals their ecological diversity. As a result, the homology of the lateral genital shields in the Celaenopsoidea is often unclear. They are sometimes referred to as metasternal shields, especially in species where the seta they bear appears to be *st4* (e.g., Fig. 12.14E). A reduction in size and loss of the posterior sternal setae, however, is a consistent trend in the superfamily. In the less derived forms (e.g., Fig. 12.13D), similarly placed shields are clearly latigynal, and the setae they bear are the fifth or higher in the series. In the EUZERCONIDAE, however, the latigynals appear to be fused to the mesogynal shield, and the more anterior shields may be metasternal, separate anterior fragments of the latigynals, or possibly sternogynals. A separate anterior genital shield (pregenital shield, sternogynum, pseudosternogynum) appears to be absent in the Celaenopsoidea, but the vaginal sclerites (vaginal processes, sternovaginal shields) may be derived from the pregenital shield.

▪ Superfamily Fedrizzioidea (Fig. 12.16)

DIAGNOSIS: Holodorsal shield bearing numerous minute setae. Mesogynal and latigynal shields well developed or reduced and withdrawn under ventral plate. Sternal setae 1 on tetartosternum, sternal shield with 3 pairs of setae. Sternogynum entire, hinged along posterior margin of sternal shield and covering anterior genital opening. Endopodal shields fused to ventral plate (fused ventral, peritrematic, exopodal, and marginal elements) or free; pedofossae often present. Ventrianal shield obtriangular, meeting or partially fused to ventral plate. Palptibia and tarsus distinct, unfused; chelicerae with short digits and dense, moplike mass of filamentous and ribbonlike excrescences; corniculi broad, membranous.

Members of the Fedrizzioidea (*sensu* Kim 2004) are tropical in distribution and are found primarily on passalid beetles. The FEDRIZZIIDAE (*Fedrizzia, Neofedrizzia, Parafedrizzia*) and the KLINCKOWSTROEMIIDAE (*Klinckowstroemia, Klinckowstroemiella, Antennurella, Simantennurella*) comprise a closely related group of approximately 30 species whose adults are superficially uropodid-like. Klinckowstroemiids are known only from the Neotropics (Womersley 1959a; Hunter and Butler 1966), while fedrizziids occur in Australia, New Guinea, Indonesia, and the Philippines. Some species have not been found on or with passalid hosts (e.g., *Neofedrizzia vitzthumi* (Oudemans) and *N. camini* Womersley), but rather in locales that suggest the presence of passalids (rotting logs, fungi, etc.) (Womersley 1959b). In contrast to the well-developed genital-sternogynal shield complex found in adult female klinckowstroemiids, fedrizziid females have reduced and internalized genital sclerites and a greatly expanded sternogynal flap. Adult fedrizzioids are found on passalid beetles (rarely, and probably atypically, on carabids), and adults and immatures inhabit beetle pupation chambers and the tunnels that the beetles excavate in logs, where they feed on nematodes and scavenge dead arthropods (O. D. Seeman, pers. obs.).

▪ Superfamily Megisthanoidea (Figs. 12.17A–D, 12.19B–D)

DIAGNOSIS: Mesogynal shield obscure; latigynals insensibly fused to each other, free from, or fused to ventral shield; pregenital shield entire or divided, resembling a sternogynum but often bearing setae. Sternal setae 1 separate from other sternal elements on individual or common sclerite; sternal shield with 3 pairs of setae. Palptibia and tarsus separate; chelicerae strongly toothed, movable digit with dendritic excrescences, fixed digit often with denticulate, membranous excrescence.

Two monogeneric families are assigned to the Megisthanoidea: the MEGISTHANIDAE (Figs. 12.17B–D, 12.19B–D) and the HOPLOMEGISTIDAE (Fig. 12.17A). Adults of approximately 30 species of *Megisthanus* have been described from passalid beetles in the Americas, Australasia, and Africa, and numerous undescribed species are known. Megisthanids include the world's largest known Mesostigmata, with adult idiosomas of these species often exceeding 4 mm in length. Beetles commonly serve as hosts to a single male and several females, while immatures live in beetle galleries, where they feed on arthropods, nematodes, and perhaps fungi (Hunter and Davis 1965; Butler and Hunter 1968). *M. floridanus* Banks is a common associate of the passalid *Odontotaenius disjunctus* in the eastern United States. Adult *M. floridanus* are known to feed on Collembola, collembolan eggs, and their own young in the laboratory (Butler and Hunter 1968), but adult megisthanids are generally thought to be paraphagic on their beetle hosts, possibly feeding on glandular secretions. Larval and nymphal *M. floridanus* construct cells of frass in which they conceal themselves, possibly to avoid desiccation. Species of *Sternosternum* (= *Hoplomegistus*) (HOPLOMEGISTIDAE) are associated with passalid beetles in the Neotropics (Butler and Hunter 1966).

▪ Superfamily Parantennuloidea (Figs. 12.17E, F, 12.18A–G, 12.19A)

DIAGNOSIS: Oval, flattened mites often with reduced dorsal setae and a marginal fringe of enlarged, flattened setae. Mesogynal shield free from or fused to large ventral or ventrianal shield, sometimes with lateral setae; latigynal shields well developed or indistinct. Sternal shield variously fragmented. Palptibia and tarsus fused or separate; palpgenu with 6–7 setae; chelicerae edentate, digits reduced, movable digit with filamentous excrescences, fixed digit with membranous distal processes or fimbriate dorsal excrescence.

The Parantennuloidea is a variable group that might better be treated as two or more separate superfamilies (Kim 2004). Adults of two of the three described genera of PARANTENNULIDAE are known from myriapods in Europe (*Parantennulus* on centipedes and *Diplopodophilus* on millipedes), while all stases of the genus *Micromegistus* are associated with carabid beetles in the Americas and in Australia. *M. bakeri* Trägårdh, an associate of several carabid beetle species in the United States, is viviparous and gives birth while attached to its host. Larvae and nymphs of *M. bakeri* scavenge on dead arthropods (Nickel and Elzinga 1970a). *Philodana johnstoni* Kethley (PHILODANIDAE), a strongly hypertrichous mite found on the tenebrionid beetle *Neatus tenebrioides*, is unusual among Antennophorina in having a divided dorsal shield (Kethley 1977c).

The 26 described members of the parantennuloid family PARAMEGISTIDAE (Figs. 12.18C–F) are found on insects, myriapods, and reptiles. *Antennomegistus* and *Echinomegistus* species occur on carabid beetles, and *Paramegistus*, *Meristomegistus*, and *Neomegistus* are associated with millipedes (Kim and Klompen 2002). Species of *Ophiomegistus* are encountered on lizards and snakes in Australasia and appear to be obligate parasites on their reptilian hosts (Strandtmann and Voss 1967). Adult *Promegistus armstrongi* Womersley (PROMEGISTIDAE) (Figs. 12.17E, 12.19A) are found on carabid and passalid beetles in Australia (Womersley 1958).

Suborder Monogynaspida

The Monogynaspida is the largest and most speciose of the three suborders of Mesostigmata recognized in this treatment. It includes a multitude of free-living and parasitic species that occur in countless terrestrial and (occasionally) marine and freshwater habitats throughout the world. The suborder is represented by fewer than 12,000 described species, but a minimum estimate of the actual number of species is believed to exceed 94,000 (Walter and Proctor 1999). The suborder is considered to comprise 18 superfamilies, which are relegated to five cohorts. Unlike the Sejida and most of the Trigynaspida, members of the Monogynaspida have lost setae *av4*/*pv4* and lack a distinct intercalary sclerite on tarsus IV. Adults and deutonymphs usually carry five ventral setae in three verticils on tarsi II–IV (typically 2-1-2 or 1-2-2), but certain of these setae may be lost in some parasitic gamasine groups. Unlike other Mesostigmata, monogynaspid mites display an uncommonly broad array of dorsal and ventral shield configurations that range from virtually complete shielding in many Uropodina (e.g., Figs. 12.28A, D, 12.32C) to the greatly reduced sclerotization seen in highly specialized dermanyssoid parasites (Fig. 12.64). Female monogynaspids usually have a single epigynal shield that, in most cohorts, appears to be composed of fused latigynal and mesogynal elements and opens outwardly at its anterior margin. The shield may be hinged or fixed posteriorly in the ventral integument, or it may be fused or contiguous with sclerotized ventral/ventrianal/endopodal shield elements. A single pair of setae, *st5* (Fig. 12.20L), is normally inserted on the posterolateral shield margins or adjacent to them. Reduction or expansion of the epigynal shield occurs in several families, and, in certain parasitic taxa, the shield may be completely lost (Figs. 12.60D, 12.65C). Shield reductions often leave the insertions of setae *st5* in the soft cuticle, while expansion leads to the capture of one or more pairs of ventral setae in addition to *st5* (Fig. 12.20M). Free latigynal shields are present in some of the lower Uropodina (Fig. 12.20K), although these elements may be fused to the epigynal shield in some species (e.g., Fig. 12.23A). In most higher uropodines, however, the latigynals appear to have fused instead to the adjacent endopodal elements, leaving the epigynal shield as an independent entity (Fig. 12.20N). One or two simple arthrodial brushes may be present at the base of the movable cheliceral digit in some gamasine Monogynaspida (Figs. 12.44E, 12.46D, G), but the elaborate mid-digit excrescences common to the Trigynaspida occur only rarely in monogynaspid mites (see Microgyniina, below).

The development of tocospermy and podospermy in several monogynaspid subcohorts has given rise to specialized male cheliceral structures that either assist in placement of the sperm packet in the region of the genital opening (tocospermy) or transfer sperm to a more remote sperm induction pore or solenostome (podospermy; see chapter 3). These cheliceral modifications are largely absent in other mesostigmatic taxa, although tocospermy is known to occur in the mesostigmatic suborders Sejida and Trigynaspida and in the order Ixodida (see chapter 11); it is presumed to occur in the orders Holothyrida and Opilioacarida as well (Alberti 1991).

Cohort Microgyniina

- Superfamily Microgynioidea (Figs. 12.21A–E, 12.22A, B)

DIAGNOSIS: Monogynaspine Mesostigmata without a consolidated opisthonotal shield in the adults; mesonotal shield(s) or scutella and a pygidial shield present. Adult females with sternal shield divided between sternal setae st2 *and* st3*; epigynal shield with a single pair of setae. Adult males with sternitigenital shield bearing 4 pairs of setae; genital opening between coxae III; 5th pair of sternal setae on or near a separate plate between coxae IV; chelicerae not modified for sperm transfer. Intercalary sclerite of tarsus IV reduced to linear remnant; setae* av4 *and* pv4 *absent. Trochanter I of adult usually retaining larval complement of 4 setae (with 5 in* Microsejus*); femur, genu, and tibia I each with 10 setae; tibiae II–IV with 9 setae (2 1/1, 2/1 2). Hypostomatic setae in triangular array; palptibia and tarsus free or fused, palpal apotele 2-tined; movable digit of chelicerae often with excrescences. Larval idiosoma holotrichous (29 pairs of setae plus postanal seta), adults sometimes hypertrichous. Facultative, pedunculate, phoretic deutonymph known for* Microsejus.

The Microgyniina currently comprises two families, the MICROGYNIIDAE and the NOTHOGYNIIDAE. The three described species of MICROGYNIIDAE are relegated to the genera *Microgynium* (Figs. 12.21A–C, 12.22B) and *Microsejus*, and all of them are associated with rotting wood and litter in boreal forests (Trägårdh 1942; Krantz 1961; Gilyarov and Bregetova 1977; Evans and Till 1979; Wisniewski and Hirschmann 1991a). An undescribed microgyniid species probably representing a new genus was recently collected in the Tyndall Range of Tasmania in plant litter (O. D. Seeman, pers. comm.). The genus *Nothogynus* (NOTHOGYNIIDAE) (Figs. 12.21D, E, 12.22A) is known from two described species collected from tree holes and logs in rain forests in Queensland, Australia, which feed on nematodes and possibly other small invertebrates (Walter and Krantz 1999). Another undescribed nothogynid species is known from Lord Howe Island (D. E. Walter, unpublished).

Adult microgyniines resemble small Sejida in having separate mesonotal and pygidial shields (Figs. 12.21B, D), but this character also occurs in the DAVACARIDAE (Cercomegistina), some Uropodina, and in male *Heatherella* (HEATHERELLIDAE) (Fig. 12.22C) and female *Saprosecans* (Dermanyssiae, HALOLAELAPIDAE) (Walter 1997). In addition, species of *Nothogynus* and *Microgynium* have cheliceral excrescences (Figs. 12.21E) similar in form to those of Cercomegistina. However, microgyniines have a single genital shield (sometimes divided medially) that bears one pair of setae, and they have lost the setae of the intercalary sclerite of tarsus IV (*av4*, *pv4*). Thus, they cannot be accommodated in the Cercomegistina or Sejida as presently conceived, but may instead be a sister group of the latter suborder (see Johnston in Norton et al. 1993, Fig. 1.2, p. 15). In this connection, and based on integrated molecular and morphological analyses, Klompen (unpublished) has recognized four lineages within the Mesostigmata, one of which links the Microgyniina with the Uropodina. The Sejida is linked with the Heterozerconina (p. 137), while the Cercomegistina is recognized as a separate line of descent. The fourth lineage comprises the remaining Mesostigmata.

The Microgyniina share a number of derived characters; for example, deutonymphs and adults have 10 setae on femora I (Evans 1963b), and the intercoxal area in the adults is uniquely formed. The sternal shield in adult females is divided between sternal setae *st2* and *st3*, and *st5* in adult males are inserted on or near a plate between coxae IV, not on the sternitigenital shield. Microgyniines have a number of leg segment chaetomes that are otherwise found only in the cohort Uropodina (Evans 1972). For example, although *Microsejus truncicola* Trägårdh has five setae on trochanters I, species of *Microgynium* and *Nothogynus* have only four (i.e., no setae are added to the larval complement). In addition, all three described microgyniine genera display reductions from the normal gamasine setation on genua I (10), tibiae I (10), and femora II (9).

Cohort Heatherellina

- Superfamily Heatherelloidea (Figs. 12.3A, 12.21F–H, 12.22C, D)

DIAGNOSIS: Adults with gnathosoma hidden by the vertex of idiosoma, encased in a thick, waxy cerotegument and with marginally fragmented, sexually dimorphic dorsal shields: adult female dorsal shield flanked by narrow marginal shields and extending posteriorly to pygidial region on which there are 3 setose platelets, marginals fused anteriorly to dorsal and posterolaterally to circumventral shields; adult male with podonotal shield, 3 shields in the mesonotal region, and 3 setose pygidial shields as in female. Both sexes with opisthogastric, peritrematic and lateral regions covered by a circumventral shield carrying stalked lateral setae, 9 pairs of spoutlike gland openings, and stigmatic openings within raised, digitate structures; peritremes absent. Tritosternum with well-developed laciniae. Both sexes with sternal setae st1 *in soft cuticle; adult female with* st2 *on sternal shield,* st3 *and/or* st4 *on metasternal shields, and* st5 *on triangular-trapezoidal epigynal shield. Posterior margins of coxae IV without inguinal glands. Tarsus I without apotele, coxa I cylindrical, trochanter I with 6 setae, femur and genu I each with 13 setae, tibia I with 13 setae including 6 dorsals and 3 ventrals; tibia III with 9 setae (2 1/1, 2/1 2), and genu III (2 2/1, 2/1 2) and genu IV (2 2/1, 3/1 1) each with 10 setae. Peripodomeric suture of tarsi II–IV without intercalary sclerite bearing setae. Palptarsal claw 2-tined, chaetotaxy of palptrochanter-femur-genu-tibia 2-5-6-14. Corniculi well developed and hornlike, hypostomatic setae 2–3 transversely arranged, small paralaciniae present. Female tocospermous. Male genital opening between coxae III on sternitigenital shield, which carries* st2–st4 *and is separated from circumventral shield, opening covered by 2 subequal valves;* st5 *inserted on triangular region of circumventral plate behind coxae IV; chelicerae not modified for sperm transfer.*

The Heatherelloidea comprises one family, the HEATHERELLIDAE, represented by one known genus, *Heatherella*, and two described species from rain-forest litter in southeast Queensland, Australia (Walter 1997). Several additional undescribed species have been taken from forest litter in eastern Queensland and New South Wales. The taxonomic affinities of heatherellids are unclear. They superficially appear uropodine-like (e.g., Fig. 12.22C), but their morphological characters do not support a relationship with the Uropodina. Some morphological characters, such as the loss of the peritreme (Figs. 12.21F, 12.22D), thick cerotegument, hypertrophied cheliceral seta (Fig. 12.21H), and fragmented sternal shields (Fig. 12.21F), would seem to suggest a relationship to the Epicrioidea (p. 138), a group apparently absent from Australia.

Little is known about the life history of *Heatherella* species, but they do feed readily on nematodes in the laboratory. Individual males were observed to make venter-to-venter contact with females and to produce a large spermatophore, which was then applied to the female genital opening with the aid of the male chelicerae (Walter 1997).

Their greatly hypertrophied marginal gland openings (Fig. 12.21G) suggest that heatherellids may defend themselves chemically, or they may have other chemically mediated interactions with larger animals that share their habitat, such as some of the many ant species that are ubiquitous in Australian forest soils.

Cohort Uropodina (including Diarthrophallina and Trachyuropodina) (Figs. 12.23–32)

Except for the earliest derivative families, and aside from the highly specialized and phylogenetically enigmatic Diarthrophalloidea (p. 136), members of the Uropodina may be readily distinguished from other Mesostigmata by their characteristic reductions in leg and palpal chaetotaxy, elongate chelicerae usually with relatively diminutive terminal digits, and highly modified tritosterna. On the underside of the gnathosoma, setae *hyp3* typically lie well behind *hyp2*, describing nearly a straight line with *hyp1* (Figs. 12.25A, 12.31C). The corniculi may be simple and hornlike, distally toothed, or otherwise modified. The adult female genital shield is usually without setae and typically is intercoxal in position, although the shield is postcoxal in members of the genus *Metagynella* (Fig. 12.28F) and some species of passalobiine DIARTHROPHALLIDAE. The more highly derived uropodines have the dorsal and ventral shields coalesced into a tortoiselike armor, and the gnathosoma is withdrawn into a deep anteroventral recess (**camerostome**) that is protected from above by a projecting vertex. The camerostome may be bordered anteriorly by a scalelike **scabellum** (Fig. 12.29), a structure of uncertain function (Evans 1992; Krantz, Gomez, and Gonzalez 2007). The legs of derived forms may be withdrawn into ventral pitlike recesses (**pedofossae**). Early derivative uropodines display less fusion of dorsal and ventral shields, and these are often covered by a thick, ornamented cerotegument that is sometimes obscured by adherent soil and detritus. Numerous setae may be found in the soft marginal cuticle or on marginal platelets (scutella) that border the dorsal sclerites. Diarthrophalloids display major deficiencies in body and leg setation, have no palpal apotele, and often lack peritremes. Their inclusion in the Uropodina is questioned in some treatments (e.g., Trägårdh 1946; Schuster and Summers 1978), but Camin and Gorirossi (1955), Johnston (1961), and Norton et al. (1993) have been followed here in relegating this group to the Uropodina.

The biology of relatively few uropodines is known, but predation appears to be the plesiotypic feeding strategy for the cohort. Robust, raptorial cheliceral digits that suggest predatory behavior characterize early derivative uropodine families (e.g., THINOZERCONIDAE, POLYASPIDIDAE) (Fig. 12.24F), while the chelicerae of more highly derived groups are more typically elongate (usually more than half the length of the body) and often terminate in reduced, highly modified digits with unusual sensory structures (e.g., TRACHYUROPODIDAE, UROPODIDAE) (Figs. 12.24J, 12.26B, 12.29C). Attenuate chelicerae of this type are appropriate tools for extracting and feeding on nematodes that inhabit deep recesses in decaying organic matter, but nematophagy is often supplemented with feeding on maggots, slugs, fungi, or decaying organic matter (Faasch 1967; Ito 1971; Athias-Binche 1981; Karg 1986b, 1989). Sperm transfer is tocospermal (see chapter 3), with the male genital aperture usually located midsternally between coxae II–IV. As is typical for tocospermous forms, the chelicerae are not obviously modified for sperm transfer.

Uropodines reach their maximum diversity in tropical rain forests, where an impressive diversity of free-living taxa may be found in forest floor litter, suspended soils, and nests. Uropodines of temperate regions are more characteristic of ephemeral litter habitats such as compost, beach wrack, or dung. Like the related Microgyniina and Sejida, most uropodine deutonymphs are phoretic and typically attach to their carriers by an anal pedicel (see chapter 6). Beetles are common phoretic hosts, but other insects and even small vertebrates such as lizards may be recruited (Domrow 1981). Nonphoretic, or sedentary, deutonymphs have been demonstrated in some Uropodina (e.g., POLYASPIDIDAE) (Athias-Binche 1984). Relatively few uropodines are associated with insects in all stages, and a few are known to be parasites (Raut and Panigrahi 1991; Krantz, Gomez, and Gonzalez 2007). Some groups of litter-inhabiting Uropodina (e.g., members of the genus *Trachytes* (TRACHYTIDAE)) appear to have lost the phoretic form of the deutonymph.

Approximately 15–20 families of Uropodina in five superfamilies are currently recognized, although that number is likely to increase considerably as the little-studied Southern Hemisphere fauna becomes better known.

▪ Superfamily Protodinychoidea (Figs. 12.23A–D)

DIAGNOSIS: Coxae I widely separated, camerostome and pedofossae absent, vertex poorly developed; palpgenu with 6 setae. Adult with a prodorsal shield element on which the vertical setae are inserted, with a mediodorsal shield bearing 10–12 pairs of setae, marginal shields absent, with or without a pygidial shield. Tritosternal base narrow, laciniae free. Chelicerae elongate, with lyrifissure and small digits, distal sensory organ on fixed digit, pilus dentilis and 3 mushroomlike structures on fixed digit. Sternal, podal, and peritrematic elements fused, extending anteriorly to form the prodorsal element and posteriorly to a point well behind epigynal shield, peritremes linear and extending to anterior margin of coxae I. Femur I with 12 setae (2 5/3 2), genua III–IV each with 1 ventral seta, and tibia I with 4 dorsal and 3 ventral setae (2 2/1, 2/2 2). Epigynal shield hinged posteriorly, flanked by paragynal shields that each carry a seta and a pore. Male with holoventral shield; genital opening between coxae III–IV, anterior genital valve with a pair of setae.

The monogeneric PROTODINYCHIDAE (Figs. 12.23A–D) is the only family included in the Protodinychoidea. Its three known species appear to represent the earliest derivative living Uropodina (Athias-Binche and Evans 1981; Hutu and Calugar 2002). *Protodinychus punctatus* Evans was collected from floodwater debris in England and various grassland sites in Scotland and Ireland (Evans 1957a; Athias-Binche and Evans 1981), and *P. evansi* Hutu and Calugar was collected from a peat bog in Romania (Hutu and Calugar 2002). Adult and nymphal representatives of another species, recently described as *P. ainscoughi* Hutu and Calugar, were taken from beaver lodge debris collected in Ontario, Canada (Johnston 1961; Hutu and Calugar 2002). Deutonymphs of this species have also been found phoretic on adult leptinid, platypsyllid, and staphylinid beetles associated with the nests of beavers (Goulet 1965; Hutu and Calugar 2002).

- Superfamily Thinozerconoidea (Figs. 12.23E–G, 12.31D)

DIAGNOSIS: Coxae I widely separated, vertex well developed, camerostome present, pedofossae absent; palpgenu with 5 setae. Mediodorsal shield bearing > 20 pairs of setae, pygidial shield narrow and bearing 2 pairs of setae. Adult female with laterodorsal peritrematic shields fused to vertex and carrying vertical and numerous other setae. Chelicerae elongate, with lyrifissure and large dentate digits, pilus dentilis present on fixed digit. Peritremes well developed and recurved near stigma. Tritosternal base broad, laciniae free. Female with large jugular shields behind coxae I joined to endopodal elements and an otherwise weakly sclerotized sternal region. Femur I with 12 setae (2 5/4 1), genua III–IV each with 2 ventral setae, and tibia I with 5 dorsal and 4 ventral setae (2 2/2, 3/2 2). Epigynal shield tripartite, consisting of a tonguelike flap without setae and paired, elongate lateral elements widely separated by unsclerotized cuticle, each bearing a seta. Male with holodorsal shield; male genital opening between coxae III and with a separate ventrianal shield.

The Thinozerconoidea and the family THINOZERCONIDAE were proposed to accommodate a single species, *Thinozercon michaeli* Halbert, an unusual intertidal form recorded from three intertidal locations in Ireland (Evans 1957a; Athias-Binche 1982). Trägårdh (1941b, 1946) noted the presence of a pair of presternal setae in *T. michaeli*, but Athias-Binche (1982) corrected this misconception.

- Superfamily Polyaspidoidea (Figs. 12.24A–J, 12.31B, E, 12.32A, B)

*DIAGNOSIS: Coxae I flanking broad tritosternal base, vertex and camerostome well developed, distinct pedofossae absent; palpgenu with 5 setae. Body usually covered in thick, ornamented cerotegument; adults with a fragmented or holodorsal shield, peritrematic sclerites fused with mediodorsal shield anterodorsally and bearing vertical setae, without a separate vertex shield element; marginal shields usually present, often fragmented into smaller scutella or fused with mediodorsal shield, pygidial shield present, narrow to broad, sometimes fragmented, either nude or bearing 1 to several pairs of setae. Chelicerae elongate, without lyrifissure and with large to small digits, pilus dentilis absent. Peritremes well developed. Tritosternal base broader than long, laciniae fused basally, often subdivided distally. Femur I with 9 setae (typically 1 4/3 1), genua III–IV each with 1–2 ventral setae, and tibia I with 4 dorsal and 2 ventral setae (2 2/1, 2/1 2(1)). Female epigynal shield rounded or trapezoidal and with slender paragynal elements, each carrying a seta (*st5*) *and a pore, rarely fused to epigynal shield. Sternal region well sclerotized; large ventrianal shield present or incorporated into holoventral shield. Male with holoventral shield; genital opening between coxae III–IV, anterior genital valve sometimes with 1 pair of setae. Cerotegument usually thick, often reticulate, sometimes extended as wings on segments of legs I.*

Three families are currently recognized in the Polyaspidoidea: the TRACHYTIDAE, POLYASPIDIDAE, and DITHINOZERCONIDAE. Trachytids of the genus *Trachytes* (Figs. 12.20K, 12.24I, J, 12.32B) are known primarily from leaf litter, tree holes, moss, and decaying wood in moist temperate conifer and hardwood forest habitats in Europe, although some species have been described from North America and Japan (Pecina 1970; Hutu 1973; Bloszyk 1980a, 1999; Wisniewski and Hirschmann 1993). Parthenogenesis (see chapter 5) is widespread among *Trachytes* species, but thelytokous forms such as the ubiquitous Palearctic *T. aegrota* (C. L. Koch) and *T. pauperior* (Berlese) produce occasional males (Bloszyk and Szymkowiak 1996; Bloszyk 1999). Like *Trachytes*, members of the genus *Polyaspinus* (Figs. 12.24G, H, 12.32A) are denizens of forest litter and tree holes and are mostly restricted to the Northern Hemisphere. However, *P. tasmaniensis* Bloszyk and Halliday was taken from leaf litter in Tasmania, and *P. boliviensis* Hirschmann was collected from moss and lichens in Bolivia (Bloszyk and Halliday 2000; Hirschmann 1992). *P. cylindricus* Berlese is unique among the seven known members of the genus in being thelytokous, but Bloszyk and Halliday (2000) note that *P. quadrangularis* Athias-Binche, a closely related form that occurs within the range of *cylindricus*, may actually represent a bisexual population of *P. cylindricus* rather than a separate species (**geographical parthenogenesis** of Vandel 1969).

POLYASPIDIDAE (Figs. 12.24D–F) includes the widely distributed genus *Polyaspis*, many representatives of which are found in litter, moss, and tree holes in European forests (Athias 1975). A few species occur in tropical and subtropical America, Australia, and Java (Camin 1953b). *Dipolyaspis* shares the northern European forest habitat of some Palearctic *Polyaspis*, while the monotypic *Dyscritaspis whartoni* Camin ranges along the eastern seaboard of the United States north to Canada and west to Missouri. The rarely collected *Calotrachytes fimbriatipes* Michael has been found only in New Zealand (Michael 1908). The polyaspidid life cycle may include a phoretic deutonymph that is morphologically and behaviorally distinct from the typical sedentary deutonymphal form (Camin 1953b; Athias 1975).

Mites of the family DITHINOZERCONIDAE (Figs. 12.24A–C) are widespread in forest litter and moss in both the Old and New Worlds (Hirschmann and Hutu 1974; Karg 1989; Krantz and Ainscough 1990; Wisniewski and Hirschmann 1993). Like those of many other polyaspidoids, deutonymphs of *Dithinozercon* and *Uroseius* (the latter including Berlese's genus *Apionoseius*) are phoretic on or associated with insects. The deutonymphs of at least six species of *Uroseius* are known to be phoretic on phorid flies, suggesting a close behavioral relationship between these two taxa (Fain 1998a, b). Others occur virtually worldwide as phoretics of trogine and geotrupine Scarabaeidae, although *U. acuminatus* (C. L. Koch) was collected from honeybee hives in Europe. Members of the genus *Iphidinychus* are known from North America, Europe, and Japan (Johnston 1961; Wisniewski and Hirschmann 1993), while the monotypic *Caminella peraphora* Krantz and Ainscough is restricted to high altitude riparian moss habitats in the coastal mountains of the northwestern United States (Krantz and Ainscough 1960).

Little is known regarding the feeding habits of polyaspidoids, although *C. peraphora* was observed to feed readily on nematodes in laboratory cultures, often digging deep into the wet filter paper substrate with their elongate chelicerae to retrieve their prey (G. W. Krantz and B. D. Ainscough, pers. obs.). An Australian species of *Polyaspis* has been reared on springtails and on nematodes (D. E. Walter, pers. obs.).

- Superfamily Uropodoidea (Figs. 12.25–12.30, 12.31A, C, F, 12.32C–E)

DIAGNOSIS: Coxae I broad, covering or partially covering narrow tritosternal base, vertex and camerostome well developed, pedofossae often present; palpgenu with 4–5 setae. Dorsal shield of adults smooth or variously ornamented, sometimes with deep pits and ridges, entire or with a posterior pygidial element, free or contiguous with holoventral shield, fused anteriorly to peritrematic elements and often bordered by marginal scutella or coalesced marginal shields; pygidial shield, when present, may be narrow to broad, sometimes subdivided, with or without setae. Chelicerae elongate, sometimes appearing 4-segmented, without lyrifissure, terminal digits usually small and often bearing elaborate sensory structures, pilus dentilis present or absent, levator tendon with or without a sclerotized node. Peritremes well developed and often convoluted or serpentine, tritosternal laciniae fused basally and often subdivided distally. Femur I with 8–9 setae, genua III–IV without (or occasionally with 1) ventral setae, tibia I with only 3 dorsal and 2 ventral setae (typically 1 1/1, 2/1 1). Female genital opening covered by a single nude epigynal shield, paragynal elements absent. Male with holoventral shield, genital opening between coxae II–IV (rarely postcoxal), covered by 1–2 valves; protonymph with separate or fused podonotal and mesonotal elements; phoretic deutonymphs may attach to host via strand produced by anal pedicel.

The Uropodoidea, or "higher uropodines" (Evans 1972), is by far the largest of the uropodine superfamilies and comprises over 80 currently recognized genera, of which nearly half are member groups of the family UROPODIDAE (Figs. 12.25–12.27, 12.32D) (Wisniewski and Hirschmann 1993). Uropodoids were the subject of considerable systematic interest in the twentieth century, triggered largely by the discovery of countless new species in previously unexplored or unexamined tropical and subtropical habitats throughout the world. Numerous families and genera were proposed to accommodate these exotic forms, but many of them were subsequently relegated to lower rank or placed into synonymy with earlier named families or genera (Karg 1989; Wisniewski and Hirschmann 1993). Although the classification used here is based primarily on that of Karg (1989), it also recognizes the approach of Wisniewski and Hirschmann (1993), who used the *Gangmerkmale* principle of recognizing only characters that are similarly expressed throughout ontogeny in identifying uropodine families and genera (Hirschmann and Zirngiebl-Nicol 1964). It should be mentioned that *Gangmerkmale* classifications may prove less daunting for the user, but their disregard for important morphological or behavioral characters that may occur in only one or two developmental stages makes it debatable as to whether such systems can truly reflect natural relationships within the superfamily (Evans 1972).

Uropodoids are found in decaying organic substrates, including rotting wood and manure; in mammal, bird, and insect nests; and in humus, moss, beach wrack, and plant debris (Wisniewski and Hirschmann 1993). A few have been collected from the fur of mammals—for example, *Uroobovella marmorea* (Fox) (DINYCHIDAE) from a murine rodent (Fox 1948)—or from birds, but their presence on these animals is considered incidental to their primary *modus vivendi* as nest colonizers. Some uropodoid genera seem to prefer soil and humus substrates—for example, *Urodiaspis* and *Dinychus* (DINYCHIDAE), *Deraiophorus* and *Olodiscus* (UROPODIDAE)—while others, such as *Uropoda* spp. (UROPODIDAE), are more likely to be found in highly organic, insular deposits of manure and compost (Evans and Till 1979). Some species of the essentially European genus *Dinychus* (DINYCHIDAE) are known to have phoretic deutonymphs (Athias-Binche, Bloszyk, and Olszanowski 1989), and one species, *D. perforatus* Kramer, may feed on bacterial colonies under laboratory conditions. Generalized uropodoid nidicoles, including species of *Uroobovella* (DINYCHIDAE) and *Trichouropoda* (TREMATURIDAE) probably feed primarily on fungi and organic debris in nests and insect galleries. The mycetophagous habit may be shared by uropodoid species that colonize food storages (i.e., *Uroobovella marginata* (C. L. Koch) and *Trichouropoda orbicularis* (C. L. Koch)) (Sinha 1964), although *U. marginata* and *U. vegetans* (DeGeer) may also feed on nematodes, fly larvae, and other mites (Jalil and Rodriguez 1970a; Hughes 1976). *U. pyriformis* (Berlese) also preys on nematodes, frequenting habitats rich in decomposing compost and manure (Athias-Binche and Habersaat 1988). Members of the *Uropoda* subgenus *Phaulodinychus* (UROPODIDAE) are routinely encountered in litter, but some occur in intertidal beach wrack and algae (Halbert 1915; Willmann 1957; Hirschmann 1972). *P. mitis* (Leonard) is regularly submerged by tidal action on the northern Adriatic coast but is believed to be capable of plastron respiration during these periods (Krantz 1974) (see chapter 3).

Phoresy is a major behavioral trait in the Uropodoidea, with dispersing deutonymphs of nidicolous species often attaching to their host animals by means of an anal pedicel (Fig. 6.1, p. 81). Many insect groups serve as dispersal agents for uropodoid deutonymphs, especially the Scarabaeidae and the Formicidae. While uropodoid dispersal is largely a function of the deutonymph, members of the unusual Neotropical myrmecophile uropodid genus *Trichocylliba* (Fig. 12.27) disperse instead as adults and have evolved a range of unusual morphological modifications that prevent them from being dislodged from their nomadic doryline and ecitonine ant hosts (Elzinga 1978). Holdfast mechanisms in the form of strong claws and empodia on legs II–IV enable adults of *T.* (*Circocylliba*) species to attach tightly to the head, mandibles, thorax, or gaster of their ant carriers (Elzinga and Rettenmeyer 1974; Elzinga 1982a). *T.* (*Planodiscus*) *squamata* (Sellnick) and other members of the subgenus attach to a ventral tibial surface of their hosts and secure their position with the aid of toothed anterior idiosomatic lobes that curve ventrad to contact the tibial surface (Elzinga and Rettenmeyer 1970; Elzinga 1991). The entire idiosomatic margin of *T.* (*Coxequesoma*) species is curved ventrally to clasp a hind coxa or the alinotum of their ecitonine hosts (*Eciton* and *Labidus* spp.) (Elzinga 1982c). Ventral idiosomatic curvature is developed to an extreme in the subgenus *T.* (*Antennequesoma*), in which the entire body basically assumes a capsulate form that enfolds an antennal or leg segment of its host and presents a smooth hard exterior to the potential hazards of the ant nest environment (Sellnick 1926; Elzinga 1982b). Adults of the monotypic genus *Habeogula* are unique in having the idiosoma elongated posteriorly into a holdfast that clamps around the host ant's neck while the venter of the mite is appressed around the labial region of the host's head so that the mite's mouthparts are aligned directionally with those of the ant (Elzinga 1989).

Members of the uropodoid families TRACHYUROPODIDAE and OPLITIDAE (Fig. 12.30) have been described from litter, tree hole, and subcortical habitats, but most of the known species in both families are myrmecophilous, at least in the adult stage. However, two oplitids from South America, *Oplitis acinaca* (Sellnick) and *O. termitophila* Zirngiebl-Nicol and Hirschmann, were collected with termites, as was the South African *Trachyuropoda termitophila* Trägårdh (Trägårdh 1906; Zirngiebl-Nicol and Hirschmann 1973; Hirschmann 1976; Hunter and Farrier 1975, 1976). Many representatives of both families are tropical forms, although the genera *Trachyuropoda*, *Oplitis*, and *Urodiscella* are more widely distributed (Hirschmann and Hutu 1974). The OPLITIDAE are mostly small (< 750 µm) weakly sculptured species, while many trachyuropodids are quite large (> 1,000 µm) and heavily sclerotized, and adults often are ornamented with conspicuous dorsal and/or ventral depressions or ridges (e.g., Fig. 12.30B). These embellishments may be bordered by hypertrophied setae that entirely obscure the associated integumental structures (Hirschmann 1976). Unlike many uropodoids, phoretic trachyuropodid deutonymphs do not secrete an anal pedicel but rather attach themselves to their host by means of strong claws.

The association of at least one myrmecophilous uropodoid mite with its host goes well beyond phoresy. All of the immature stages of the dinychid mite, *Macrodinychus sellnicki* Hirschmann and Zirngiebl-Nicol (Fig. 12.28 E), are parasitic on the pupa of the Neotropical ant, *Paratrechina fulva*. The parasitized pupa slowly shrivels during mite development and finally dies just prior to ecdysis of the adult mite (Krantz, Gomez, and Gonzalez 2007).

Other uropodoid families, including the UROACTINIIDAE, TREMATURIDAE, and METAGYNURIDAE, essentially mirror the range of habitats and behaviors already noted for the UROPODIDAE, but each displays unique morphological characters not found in other uropodine taxa (Karg 1989). For example, members of the family UROACTINIIDAE (Figs. 12.29B, C) have an unusual radiating (actiniform) or mushroomlike sensory appendage on the terminus of the cheliceral fixed digit (Fig. 12.29C), and the epigynal shield of female metagynurids is posterior to coxae IV (Fig. 12.28D). *Uroactinia agitans* (Banks) (UROACTINIIDAE) may occur in great numbers in commercial fishworm beds, where they compete with the worms for nourishment (Stone and Ogles 1953).

▪ Superfamily Diarthrophalloidea (Figs. 12.31G–J)

DIAGNOSIS: Coxae I contiguous or widely separated, inserted well behind narrow tritosternal base, camerostome absent; idiosoma ovate or elongate, with entire, slightly abbreviated dorsal shield carrying 0–6 pairs of macrosetae and an indeterminate number of minute dorsomedian setae, with 4–7 pairs of long peripheral setae inserted in surrounding integument. Sternal shield fused or contiguous posteriorly with ventral or ventrianal shield; epigynal shield rounded anteriorly, free or fused posteriorly with ventral elements, located intercoxally or arising behind coxae IV. Paranal setae long, often more than half the length of body. Fixed cheliceral digit with a subterminal spur, epistome typically multibranched terminally, peritremes extremely short or absent. Leg setation highly deficient, without claws or empodia on tarsi I, palpal apotele absent. Known males with genital aperture medially located in holoventral shield.

The Diarthrophalloidea is represented by 18 genera and over 50 species of weakly sclerotized hypotrichous mites grouped into a single family, the DIARTHROPHALLIDAE. Diarthrophallids are associated exclusively with passalid beetles in the Old and New World tropics (the Nearctic *Diarthrophallus quercus* (Pearse and Wharton) is an exception) and may be found on the surface of the beetle or beneath the elytra (Womersley 1961; Hunter and Glover 1968a–c; Schuster and Summers 1978). Because diarthrophallid larvae, protonymphs, and deutonymphs often occur together with adult mites on their passalid hosts (Lombardini 1951; Hunter and Glover 1968b), and because diarthrophallids display a moderate degree of host specificity (Schuster and Summers 1978), it seems likely that the mite/beetle association is more intimate than simple commensalism.

Schuster and Summers (1978) considered the DIARTHROPHALLIDAE to comprise two subfamilies, the Diarthrophallinae and the Passalobiinae, which may be unambiguously differentiated based on the condition of coxae I (free in diarthrophallines and fused or contiguous in passalobiines) and the presence (in passalobiines) or absence of a minute dorsal **suranal seta** inserted above the terminal anal opening. Diarthrophallids have been taken from representatives of five passalid subfamilies, with the majority of collection records from the New World Proculinae and Passalinae and from the Australasian Aulacocyclinae. The diarthrophalline genera *Troctognathus* and *Brachytremella* were found on macrolinine passalids in New Guinea, and solenocyclines serve as hosts for species of the genera *Minyplax* and *Notoporus* in central Africa.

Cohort Heterozerconina

The Heterozerconina comprises two families of myriapod- or reptile-associated mites of uncertain relationship to other Monogynaspida. It may, in fact, be more closely related to the Trigynaspida (see concluding paragraph of the section that follows), or to the Sejida (Klompen 2000b; Lekveishvili and Klompen 2004b). Only the adults are found on myriapods, and these mites usually are readily distinguished by their paired ventral adhesive suckers. Such structures are rarely found in other Mesostigmata, although some bird nasal parasites in the genus *Rallinyssus* (RHINONYSSIDAE) have an unpaired anal sucker.

Regarding the possibility of a closer relationship of the Heterozerconina with the Sejida, recent morphological and molecular analyses by Lekveishvili and Klompen (2004b) have shown that heterozerconines consistently group phylogenetically with the sejoids, an outcome that led these authors to place the Heterozerconina as a subset within the Sejida. Pending the outcome of future studies by Lekveishvili and Klompen (G. W. Krantz, pers. corr.), the Heterozerconina is retained here as a valid entity.

- Superfamily Heterozerconoidea (Fig. 12.33)

DIAGNOSIS: Adult dorsal shield entire, hypertrichous, well sclerotized; female sternal shield entire or fragmented, often reduced to a medial remnant and a series of endopodal fragments that may carry sternal setae and lyrifissures, sternal setae 1 often on jugularia. Epigynal, ventral, and anal shield elements contiguous or fused behind coxae IV; with a pair of distinctive suckerlike adhesive organs arising behind coxae IV and flanking the anal valves, secondarily lost in some species. Male genital aperture at anterior margin of a holoventral shield or arising behind an entire sternal shield (DISCOZERCONIDAE, Discomegistus*), femur II with or without 1–2 ventral spurs; male sperm transfer process highly diverse, often elongated and convoluted, arising from base of fixed cheliceral digit (HETEROZERCONIDAE) or absent (DISCOZERCONIDAE). Tarsi I of adults and known immatures with discrete distal acrotarsi, tibia I with 4 dorsal setae and 1 or 2 posterolateral setae (2 2/1(2), 2/1 1(2)), femora III–IV without posterolateral setae. Palptarsal apotele 2-tined.*

The HETEROZERCONIDAE (Figs. 12.33A–C) comprises a bizarre assemblage of primarily Gondwanan acarines whose adults are associated with millipedes and, less commonly, snakes and lizards (Finnegan 1931; Lisazo 1979; Fain 1988, 1989; Flechtmann and Johnston 1990; Gerdeman, Klompen, and Tanigoshi 2000; Gerdeman and Klompen 2003). Adult heterozerconids attach to their hosts by means of paired ventral, suckerlike discs that are believed to achieve adhesion through alternate application of hemolymphatic pressure and dorsoventral muscle contraction (Evans 1992; Gerdeman and Alberti 2007). Some undescribed heterozerconids from South America do not have adhesive discs. *Heterozercon latus* Berlese and *Afroheterozercon ancoratus* Fain were described from termite nests in Paraguay and Zaire (Fain 1989), and earlier literature includes records of heterozerconids found on the termites themselves (Flechtmann and Johnston 1990). The latter collections were assumed to represent fortuitous invasions of millipede associates from nearby soil habitats, but it is questionable that such invasions would be tolerated in termite nests if the invaders lacked an appropriate chemical password.

The chelicerae and adjacent gnathosomatic processes of heterozerconids seem ideally adapted for gathering and feeding on host exudates. One or two basally expanded, ciliated processes arise from between the cheliceral digits, and the movable digit is bordered by a comb of strong fimbriae. The lateral lips of the gnathosoma of millipede-associated heterozerconids are pliable and may assist in directing host exudates to the mouthparts. The discovery of blood in the midgut of adult *Amheterozeron* (= *Zeterohercon*) *amphisbaenae* (Flechtmann and Johnston), a Neotropical associate of worm lizards (Amphisbaena), suggests that parasitism occurs in at least one heterozerconid species (Flechtmann and Johnston 1990). Other reptile-associated heterozerconids may also be hematophagous rather than paraphagic, but little information is available on the feeding habits in this group.

Although adult heterozerconids may be paraphagic or parasitic on their invertebrate or vertebrate hosts, there is evidence that the virtually unknown immature stages are predaceous. Immatures of *Narceoheterozercon ohioensis* Gerdeman and Klompen, an associate of the millipede *Narceus annularis* (Rafinesque) and the first heterozerconid species for which immatures have been identified, were observed to prey on Collembola and small mites in laboratory cultures (Gerdeman, Klompen, and Tanigoshi 2000; Gerdeman and Klompen 2003).

The DISCOZERCONIDAE (Figs. 12.33D, E) includes a small number of unusual and poorly known mite species found on centipedes in both the Old and New Worlds (Berlese 1888; Trägårdh 1911; Vitzthum 1925; Domrow 1956). Discozerconids are distinctive in that the idiosoma is virtually as wide as or wider than long and is bordered by a marginal shield on which there is a dense row of fusiform or platelike setae. The ventral adhesive discs are considerably larger than those of the HETEROZERCONIDAE and virtually fill the area between coxae IV and the anal shield. The two presently recognized genera are differentiated on the basis of sternal shield condition and form of the anal shield. Berlese (1910b) described *Discozercon mirabilis* Berlese as having only sternal shield remnants (jugular shields) and a fragmented anal shield consisting of two lateral and one posterior element, while *Discomegistus pectinatus* Trägårdh was characterized as having entire sternal and anal shields in both sexes (Trägårdh 1911). Males of a discozerconid species taken from a centipede in Arizona, USA, combine the attributes of a fragmented sternal shield and an entire anal shield, suggesting that the sternal/anal character suite may be an unreliable tool for generic separation. However, the entire sternal shield in male *D. pectinatus* is otherwise unknown for the family and for the Heterozerconoidea as a whole.

The presence of a sperm transfer process on the fixed cheliceral digit of male HETEROZERCONIDAE (Fig. 12.33B) is unique and differentiates the family from other podospermous Mesostigmata (i.e., the subcohort Dermanyssiae), in which a spermatodactyl arises on the movable cheliceral digit. Furthermore, the genesis and fine structure of heterozerconid sperm differ considerably from those of dermanyssine mites, a strong indication that podospermy in heterozerconids and in the Dermanyssiae may have arisen independently (Alberti, Seniczak, and Seniczak 2003). Male DISCOZERCONIDAE lack a sperm transfer process, and the genital aperture opens well behind the sternal area, which implies that discozerconids are tocospermous rather than podospermous. This basic divergence in sperm transfer strategies within the Heterozerconoidea provides reason enough to question whether its two member families share a common ancestry.

In addition to the question of relatedness within the Heterozerconoidea, there is presumptive and quantitative evidence that the superfamily may be more closely related to the Trigynaspida (see p. 127) than to the Monogynaspida. Trägårdh (1911) felt that the broad, spherical shape of the idiosoma of *D. pectinatus*, along with the presence of a marginal shield armed with a row of fusiform setae, an enlarged interdigital cheliceral membrane, and a male genital aperture posterior to the sternal shield (among other characters), pointed to an affinity of *Discomegistus* (and by inference Berlese's related genus *Discozercon*) with the Trigynaspida and more specifically with the Antennophorina. Based on cladistic analysis, Norton et al. (1993) concluded that the Heterozerconoidea *s. lat.* (= Heterozerconina) is the sister group of the supercohort Trigynaspida *s. lat.* (= Cercomegistina and Antennophorina), a finding that contradicts the conclusions of Lekveishvili and Klompen (2004b), who synonymized the Heterozerconina with the Sejida (see above). Pending further comparative morphological and behavioral studies, and for purposes of this review, the HETEROZERCONIDAE and DISCOZERCONIDAE are retained as member families of the Heterozerconoidea and placed in a separate cohort of the Monogynaspida as an assemblage with yet unresolved phylogenetic affinities.

Cohort Gamasina

The Gamasina comprises most of the described species of Mesostigmata and includes the most familiar families of soil predators, biocontrol agents, and vertebrate parasites. In this treatment, the Gamasina is considered to comprise four subcohorts and 10 superfamilies.

▪ Superfamily Epicrioidea (Fig. 12.34)

DIAGNOSIS: Female and male dorsal shield entire, usually expanded laterally and consolidated with peritrematic shields, which in turn are expanded posteriorly onto the ventrolateral surface behind coxae IV, ornamented with 11–19 pairs of dorsal glands and a network of large bi-, tri-, or quadrifurcate tubercles. Stigmata prominent, typically on protuberances, peritremes vestigial or lacking. Tritosternum reduced to stubby base and abbreviated laciniae. Sternal setae 1 on paired jugular plates, poroids may be absent. Female sternal shield with 2–3 pairs of setae, apparently without poroids; 4th sternal setae usually free on soft integument or rarely inserted on metasternal plates or on posterior edge of sternal shield; epigynal shield with 2 pairs (occasionally 1 pair) of setae and well separated posteriorly from an anal or small ventrianal shield. Males with sternitigenital shield separated from ventral and anal shields or from ventrianal or anal shield (the latter sometimes integrated with posterior margin of dorsal shield). Posterior margins of coxae IV with single or multiple inguinal glands. Leg I elongated, without tarsal claws and with 2 to several setae that are more or less elongated and minutely clubbed apically. Peripodomeric suture of tarsi II–IV usually with a dorsal intercalary sclerite bearing 2 setae. Femur I with 13 setae, including 4 ventral setae and 5 dorsal setae; tibia I with 4 ventral setae, 6 dorsal setae, and 2 anterolateral setae; tibia III with 9 setae (2 1/1, 2/1 2); genu III with 10 setae, including 2 ventral setae. Palptarsal claw 3-tined (occasionally 2-tined). Corniculi small, slender, much shorter than internal malae. Female tocospermous. Male with genital aperture located between coxae III and without spurs or modified spine-like setae on legs II–IV; male chelicerae not obviously elaborated for sperm transfer. Larva lacking pygidial shield; protonymphal pygidial shield with 8 pairs of setae, setae S4–S5 *on shield; dorsal shield of deutonymph entire.*

The Epicrioidea contains a single family, the EPICRIIDAE, which is represented by approximately 35 described species in four genera: *Epicrius*, *Berlesiana*, *Neoepicrius*, and *Adenoepicrius* (Evans 1955d; Balogh 1959; Athias-Henriot 1961a; Bregetova 1977a; Blaszak and Alberti 1989; Karg 1971, 1993b; Gwiazdowicz 2000a; Moraza and Johnston 2004; Moraza 2005a, b). Epicriids are found in moss (especially in seepage areas), humus, forest litter, decaying wood, and occasionally the nests of rodents and limestone caves (Bregetova 1977a; Evans and Till 1979; Ishikawa 1987; Evans 1992; Karg 1993b; Moraza 2005b). Their feeding habits have not been determined, but their cheliceral and gnathosomatic morphology and their reduced tritosternum suggest that they may feed on fungal fluids. All of the described species are known from the Northern Hemisphere, with *Berlesiana* species known only from the Palearctic region and those of *Adenoepicrius* only from the Nearctic (Moraza 2005a). Epicriids are clearly tocospermous, but no further information on their genetic system is available (Norton et al. 1993). An unusually slender, straight and pointed, fixed cheliceral digit has been noted in the males of some epicriids and is thought to be an adaptation for the transfer of spermatophores in the absence of a spermatodactyl (Moraza and Lindquist 1999). A similar cheliceral modification has been described by Di Palma et al. (2006) for the male of *Veigaia* species (VEIGAIIDAE) (see p. 143).

One of the few descriptions of an immature instar of an epicriid species (Solomon 1978) is notable in showing that the protonymph has a tritosternum with well-developed laciniae and stigmata with moderately well-developed peritremes. The deutonymph was not described, but the laciniae and peritremes are reduced in the adult. In other species, however, peritremes are vestigial, although longer than in adults, protonymphs, and deutonymphs (Moraza 2005b).

The presence of a dorsal intercalary sclerite bearing two setae (*ad3*, *pd3*) in the peripodomeric suture of tarsi II–IV is uniquely apomorphic for species of *Epicrius*, *Berlesiana*, and *Adenoepicrius* among all known taxa of Gamasina (Evans 1969b; Moraza 2005a, b; E. E. Lindquist, pers. obs.). Members of the recently described genus *Neoepicrius* (Figs. 12.34A–E) in western North America, however, lack intercalary sclerites on tarsi II–IV (Moraza and Johnston 2004).

▪ Superfamily Zerconoidea (Figs. 12.35, 12.37C)

*DIAGNOSIS: Female and male dorsal shield in ZERCONIDAE divided, rarely secondarily entire; podonotal shield often consolidated laterally with peritrematic shields; opisthonotal shield in COPROZERCONIDAE lacking on adult female but fully developed on deutonymph and adult male; opisthonotal shield in both families with 2 pairs of its gland pores (*gdZ3, gdZ5*) prominent, and with 1 pair of caudoventral setae (*JV5*) included in its setal complement; opisthonotal shield in ZERCONIDAE consolidated caudally with ventrianal shield and often with serrated lateral margins and a transverse row of usually 4 sclerotized fossae near its posterior margin. Stigmata with peritremes in adults reduced in length from normal, elongated state in deutonymph, sometimes vestigial. Tritosternum with well-developed laciniae. Female sternal shield with 3 pairs of setae or rarely fragmented such that any or all of* st1–st3 *on separate platelets; 4th sternal setae usually free on soft integument or rarely inserted on metasternal plates; sternal shield with a pair of gland pores on posterior margin and with sternal poroids present or rarely absent; epigynal shield with 1 pair of setae, shield either narrowly separated from an expansive ventrianal shield or (rarely) well separated from an anal shield. Males with sternitigenital shield separated from ventrianal shield. Posterior margins of coxae IV usually with single or multiple inguinal glands. Legs I not elongated, with claws and with tapered setae that are not markedly elongated. Peripodomeric suture of tarsi II–IV without intercalary sclerite bearing setae. Femur I with 13 setae, including 4 ventral setae and 5 dorsal setae; tibia I with 4 ventral setae, 6 dorsal setae, and 2 anterolateral setae; tibia III with 9 setae (2 1/1, 2/1 2); genu III with 10 setae, including 2 ventral setae. Palptarsal claw 2-tined. Corniculi well developed, hornlike. Female tocospermous. Male with genital aperture located between coxae III or (rarely) coxae II and without spurs or modified spinelike setae on legs II–IV; male chelicerae not obviously elaborated for sperm transfer. Larva with well-developed pygidial shield bearing several pairs of setae; protonymphal pygidial shield expanded into podonotal shield with about 15 pairs of setae, including* J1–J5, Z1–Z5, S2–S5, *and* JV5*; dorsal shield of deutonymph divided.*

The Zerconoidea includes the family ZERCONIDAE (Figs. 12.35A–C), with many species classified in some 37 currently recognized genera and subgenera (Halasková 1970; Blaszak 1976a, 1984; Lindquist and Moraza 1999), and the monobasic COPROZERCONIDAE (Moraza and Lindquist 1999; Fig. 12.35D). Both families are known only from the Northern Hemisphere. Zerconids are free-living mites found in forest humus and litter, grasslands, mosses, compost, decaying wood, and occasionally the nests of ants, small mammals, and birds (Petrova 1977; Evans 1992; Karg 1993b). Species of *Zercon* have been recorded frequently from both high altitudes and arctic latitudes. A species of *Mixozercon* may be unusual in being preferentially arboreal, as it has been found to be far more numerous in accumulations of suspended soil and litter in the canopy than in the floor litter of ancient coniferous rain forests in Pacific coastal Canada (N. N. Winchester and E. E. Lindquist, unpublished obs.).

Since the appearance of Sellnick's review of the ZERCONIDAE (1958), a number of works have been published on the zerconid fauna of various areas of the Palearctic region, including the British Isles (Skorupski and Luxton 1996), mainland Europe (e.g., Athias-Henriot 1961a; Halasková 1963, 1970; Blaszak 1974; Moraza 1990; Karg 1993b;

Urhan 2001), the southern Mediterranean coast of northern Africa (Athias-Henriot 1961a; Blaszak 1979b), northern Asia (e.g., Petrova 1977, 1978a, b, 1979), southern Asia (Athias-Henriot 1976; Blaszak 1979a), eastern Asia (Blaszak 1978), northern Africa (Blaszak 1979b), and the Korean Peninsula (Blaszak 1976b; Halasková 1979; Lim and Lee 2001). The comparatively sparse literature on the Nearctic zerconid fauna has been contributed primarily by European authors (e.g., Halasková 1969, 1977; Blaszak 1981a, b, 1984). *Coprozercon scopaeus* Moraza and Lindquist (COPROZERCONIDAE) may be unusual among zerconoids and epicrioids because it is cavernicolous. The only known specimens of this taxon were collected from a fecal depository of a wood rat some 20 m within a cave in Kentucky (Moraza and Lindquist 1999). Few other species of zerconids are recorded from caves (Mašán and Fenďa 2004). *Paleozercon cavernicolus* Blaszak et al. is known only from fossil specimens found embedded in calcium deposits of a stalagmite near a cave entrance (Blaszak, Cokendolpher, and Polyak 1995), and *Zercon triangularis* Koch is recorded from caves in Europe but is widespread in other habitats.

Although the keys and diagnoses to families of Gamasina by various authors routinely characterize the ZERCONIDAE as having a subtriangular idiosoma, dorsal shields with serrated lateral margins, and an opisthonotal shield with two pairs of large, sclerotized fossae or muscle insertion scars near the posterior margin, these attributes do not hold for a variety of genera. For example, adults of early derivative genera such as *Syskenozercon*, *Halozercon*, and *Aspar* have elliptical or oval idiosomas, dorsal shields with evenly rugose lateral margins, and undifferentiated muscle insertion scars posteriorly (Athias-Henriot 1976; Wisniewski, Karg, and Hirschmann 1992; Halasková 1977). Also, unlike species of most zerconid genera, neotrichy is evident among those of *Caurozercon* and *Echinozercon*, both of which have a symmetrically doubled number of setae along the margins of the opisthonotal shield of deutonymphs and adults. Species of *Syskenozercon* and *Halozercon* have more extensive dorsal shield neotrichy, with the marginal setae symmetrically multiple and the central and midlateral setae asymmetrical and haphazardly arranged (Lindquist and Moraza 1999).

Zerconid mites of diverse genera have remarkably similar chelicerae, and all are thought to be oligophagous predators (Evans and Till 1979). The only conclusive observations on their feeding were reported by Walter (1988a), who reared *Skeironozercon tricasus* Blaszak and an unidentified species of *Mixozercon* on nematodes, and by Martikainen and Huhta (1990), who observed *Parazercon radiatus* (Berlese) and *Zercon zelawaiensis* Sellnick preying on nematodes in microcosms simulating humus-litter layers of a spruce-birch forest.

Phoretic behavior has not been observed among zerconids. However, Moraza and Lindquist (1999) speculated that the coprozerconid *C. scopaeus* may be phoretic in that it has successfully adapted to the patchy transient habitats it occupies in cave systems. Furthermore, the deutonymph has more strongly sclerotized and ornamented dorsal shields than does the adult female, a trait that could offer protection against mechanical damage and desiccation during phoresy. In contrast, the dorsal shields of other zerconids are more weakly sclerotized and ornamented in deutonymphs than in adults. *C. scopaeus* is exceptional among Zerconoidea and Epicrioidea in that the opisthonotal shield is lost in the adult female but fully developed and well sclerotized in deutonymphs and the adult male, and one of the two pseudosymmetric pairs of ventral setae on tarsi II–IV is suppressed in the transformation from deutonymph to adult. Zerconids are clearly tocospermous, but no other information on the genetic system of zerconoids is available (Norton et al. 1993). The slender, straight, pointed form of the male fixed cheliceral digit has been noted for some species of zerconids and for *C. scopaeus* and is thought to be an adaptation for the transfer of spermatophores in the absence of a spermatodactyl (Moraza and Lindquist 1999).

Johnston and Moraza (1991) compared the idiosomatic glands and lyrifissures of zerconid mites with those of previously studied families of Gamasina and proposed a unified idionymic system of notation for these organules. Similarly, the ontogeny and homology of idiosomatic setae of zerconids and coprozerconids (Fig. 12.35D) were compared by Lindquist and Moraza (1999) and found to be fully homologous with those of other families of Gamasina. This facilitated a new interpretation of setal attributes used to characterize the superfamily Zerconoidea, including recognition of a caudodorsal displacement of a pair of opisthogastric setae, *JV5*, so as to be inserted on the opisthonotal shield in deutonymphs and adults of these families, a unique apomorphy of Zerconoidea. It also refutes an alternative interpretation of the ontogeny of idiosomatic setation by Balan (1996), which suggests that a limited anamorphosis occurs in zerconids. Deutonymphs of both families typically have long, well-developed peritremes that are subsequently shortened in adults.

▪ Superfamily Arctacaroidea (Figs. 12.36, 12.37A, B)

DIAGNOSIS: Female and male dorsal shield entire, heavily sclerotized but only lightly ornamented; female shield abbreviated, not expanded to consolidate with peritrematic shields laterally, and leaving posterior 1/4 to 1/2 of idiosoma with soft integument on which an unpaired pygidial remnant and sometimes mesonotal scutella are evident; male shield covering entire dorsum, expanded and consolidated laterally with peritrematic shields. Stigmata with peritremes well developed, generally extending to level of bases of legs I or II, peritrematic shields not expanded posteriorly onto the ventrolateral surface behind coxae IV. Tritosternum well developed, with long, unfused laciniae. Female sternal shield with broad endopodal extensions between bases of legs I–II and with 2 or (usually) 3 pairs of setae, 3 pairs of poroids, and a pair of gland pores near posterior margin; metasternal setae (st4) *free in soft integument; epigynal shield slender, without setae but flanked by 2 pairs of setae in soft integument, and with a rounded posterior margin well separated from anal shield. Male with sternitigenital shield entire or fragmented in region between coxae III–IV and separated from ventrianal shield, the latter free from posterior margin of dorsal shield. Opisthogaster usually with single or multiple inguinal glands behind coxae IV. Coxae of legs I slender, well removed anteriorly from those of legs II; legs I not elongated, with or without claws and with setae tapered, not markedly elongated. Peripodomeric suture of tarsi II–IV without intercalary sclerite bearing setae. Femur I with 13 setae, including 4 ventral setae and 5 dorsal setae; tibia I with 4 ventral setae, 6 dorsal setae, and 2 anterolateral setae; tibia III with 9 setae (2 1/1, 2/1 2); genu III with 10 setae, including 2 ventral setae. Palptarsal claw 3-tined. Corniculi well developed, hornlike, but shorter than internal malae. Female tocospermous. Male with genital aperture located between coxae II–III, sometimes with paired tubular protuberances of unknown function arising behind aperture, with spurs on legs I (trochanter), II, and IV; male chelicerae not obviously elaborated for sperm transfer. Chelicerae of protonymph and deutonymph multidentate; protonymph (and probably larva) lacking pygidial shield; dorsal shield of deutonymph divided, opisthonotal shield well developed, with 13–14 pairs of setae.*

The Arctacaroidea includes a single family, the ARCTACARIDAE, which is represented by two genera that each have three described species (Evans 1955b; Bregetova 1977d; Makarova 2003). Its phylogenetic affinities with other superfamilies are so uncertain that it remains segregated in its own higher category, pending further evidence of its affinities from sources other than external morphology (Moraza and Lindquist 1999). Recent ultrastructural studies on a species of *Proarctacarus*, however, suggest that the subcohort Arctacariae is indeed a valid lineage and that, based on the shared presence of lateral glands at the bases of coxae I (Alberti and Coons 1999), it may represent the sister group of the gamasine subcohorts Parasitiae and Dermanyssiae, as recognized in this treatment (Alberti and Krantz 2007). Apart from the morphological evidence of tocospermous sperm transfer, no information on the genetic system of arctacarids is available (Norton et al. 1993).

Arctacarids appear to have a circumberingian distribution in regions of high latitude or altitude in the more remote arctic and boreal regions of northeastern Asia and northwestern North America, and they are seldom collected. *Arctacarus rostratus* Evans has a widespread transberingian distribution in the arctic tundra (Evans 1955b; Watson, Davis, and Hanson 1966; MacLean, Behan, and Fjellberg 1978; Thomas and MacLean 1988). Two additional species of *Arctacarus* were described from material collected from forest litter and the nests and burrows of rodents (Bregetova 1977d). A species denoted as *Arctacarus* sp. in earlier works (Krantz 1978; Krantz and Ainscough 1990) has been described, along with two other species, in a newly proposed genus, *Proarctacarus*, by Makarova (2003), from montane coniferous forest litter sites in western North America (Figs. 12.36A–G).

Arctacarids are equipped with large, well-sclerotized, cheliceral digits, but their feeding habits are unknown. The male fixed chela in *A. rostratus* is thicker and less denticulate than in the female and the deutonymph (Evans 1955b), a subtle dimorphism that may indicate an adaptation in the male for transfer of spermatophores in the absence of a spermatodactyl (Moraza and Lindquist 1999). Similarly, the dentition on both cheliceral digits of the female of *Proarctacarus oregonensis* Makarova is highlighted by multiple small teeth, while that of the male is confined to only one median tooth on each digit. Diagnostic attributes noted for protonymphs and deutonymphs are based only on *A. rostratus* and *P. oregonensis* (E. E. Lindquist and G. W. Krantz, pers. obs.).

▪ Superfamily Parasitoidea (Fig. 12.38)

DIAGNOSIS: Adult dorsal shield entire or divided or with a pair of lateral incisions in female, entire with at most a transverse line of fusion of 2 shields in male. Female sternal shield with 3 pairs of setae; epigynal shield triangular, flanked by large discrete metasternal shields, and abutting or rarely coalescing with a ventrianal shield that is variously fused with podal, peritrematic, or, more rarely, dorsal shield elements; presternal platelets frequently present. Tibia I typically with 4 ventral setae, 6 dorsal setae, and 2 anterolateral setae (2 3/2, 3/2 2); tibia III with 8–9 setae (2 1/1, 2/1 1(2)); chaetotaxy of genu IV (2 2/1, 3/1 1). Palptarsal apotele 3-tined. Female tocospermous ("neotocospermous" sensu *Alberti 2002a). Male with genital aperture at anterior margin of a holoventral or sternitigenital shield and typically with highly developed spurs and apophyses on legs II; chelicerae with sperm-holding structure (spermatotreme) that is coalesced distally with movable digit. Larva without pygidial shield; protonymphal pygidial shield reduced laterally, generally with 4–5 pairs of setae (without setae S4–S5), or shield sometimes absent; dorsal shield of deutonymph divided in both subfamilies, and acrotarsus often present in subfamily Parasitinae.*

The Parasitoidea includes one speciose, cosmopolitan family, the PARASITIDAE. Its two constituent subfamilies, the Parasitinae and Pergamasinae, have been treated as separate families in some works (Woolley 1988). Parasitids are classified into relatively few genera by some authorities (e.g., Micherdzinski 1969; Tichomirov 1969; Karg 1993b) but into considerably more by others (e.g., Evans and Till 1979; Juvara-Bals 2002), including many of narrow scope proposed by Athias-Henriot (e.g., 1980, also references in 1982b). Parasitids are free-living predators found in a wide variety of primarily ground-level substrates that support microarthropod and nematode populations (Karg 1993b; Karg and Freier 1995).

Members of the subfamily Pergamasinae commonly inhabit relatively stable forest and grassland humus, moss, and soil ecosystems (Bhattacharyya 1963; Holzmann 1969; Karg

1993b, 1995). Their deutonymphs are not known to be phoretic (Evans 1992). Some members of the genus *Holoparasitus* frequent damp to wet habitats, such as fresh- and saltwater marshes and beach wrack (Hyatt 1987; Karg 1993b). Members of the subfamily Parasitinae commonly occur in temporary accumulations of organic debris, including manure, vegetable compost, and beach wrack, and in the nest habitats of mammals and insects (Hyatt 1980; Karg 1993b). Parasitine deutonymphs often are phoretic on insects (Rapp 1959; Binns 1982), using them for transport from one, often transient, feeding locus to another. The occurrence of these often conspicuous mites on insect carriers was misconstrued as parasitism by early observers and apparently led to their being named *Parasitus* and "parasitids" (Oudemans 1936a). Deutonymphs of species of *Parasitellus* are phoretic on bumblebees, while the other active instars are found free in bee nests, where they may feed on pollen and possibly on microarthropods (Richards and Richards 1976; Karg 1985a, 1993b). Some species of *Parasitus* are common phoretics of carabid and silphid beetles.

Mutualistic associations are evident between some parasitids and their insect hosts. For example, queen bumblebees that carry species of *Parasitellus* have been found to have relatively lower levels of parasitic nematodes (Schousboe 1987). The relationship between carrion-feeding silphid beetles and their ubiquitous mite associates, *Poecilochirus* spp., has been the subject of many studies (e.g., Springett 1968; Korn 1983; Müller and Schwarz 1990; Schwarz and Müller 1992; Blackman and Evans 1994; Brown and Wilson 1994). *Poecilochirus* deutonymphs (Fig. 12.38E) are phoretic on the beetles, from which they dismount to feed on fly larvae and nematodes that compete with the beetle's larvae for the carrion's nutrient resources. The beneficial effects of these mites, however, may be offset by negative effects of their predation on host beetle brood, a factor that requires further study (Beninger 1993).

Kinn (1971) cited previously published records of a *Eugamasus* species feeding on nematodes and mites in bark beetle galleries and of species of *Parasitus* preying on the bark beetles themselves (probably their brood). Those records probably refer to subsequently described species now placed in the genus *Schizosthetus* (Athias-Henriot 1982a), which are subcortical associates of bark beetles. *Parasitus coleoptratorum* (L.), *P. gregarius* Ito, and *Poecilochirus monospinosus* Wise et al. prey on pest fly larvae in manure (Wernz and Krantz 1976; Ito 1977; Axtell 1981; Wise, Hennessey, and Axtell 1988), but their impact on pest population levels is questionable (Krantz 1983a). Al-Amidi, Dunne, and Downes (1991) reported *Parasitus bituberosus* Karg as a potentially useful predator in biological control of sciarid gnat larvae in mushroom houses. The potential usefulness of targeted species of PARASITIDAE as biological indicators of soil ecosystem conditions was reviewed by Karg and Freier (1995).

In a series of ultrastructural investigations of the fine structure of sperm and genital systems in mesostigmatic mites, Alberti (2002a and previous papers) determined that the tocospermous Parasitoidea share several apomorphic attributes with, and are the probable sister group of, the podospermous taxa that compose the Dermanyssiae. The reproductive strategy of Parasitoidea, termed *neotocospermy* (also see chapter 3), is unique to this superfamily. The monobasic genus *Psilogamasus* Athias-Henriot (1969e) is notable in that all active instars have one of the setae on telotarsus IV modified as a trichobothrial seta. This modification is otherwise not known in the Mesostigmata, although Athias-Henriot (1969d) noted a somewhat similar condition for a seta denoted as a "trichocyst" on tarsus I of the deutonymph of some species of VEIGAIIDAE.

▪ Superfamily Veigaioidea (Fig. 12.39)

*DIAGNOSIS: Female dorsal shield divided or with a pair of deep lateral incisions, opisthonotal region commonly reduced; males with dorsal shield entire or divided in two, the posterior shield rarely subdivided medially into right and left halves. Female sternal shield large, with 3 pairs of setae; metasternal shields straplike to subtriangular, with 4th pair of sternal setae and 3rd pair of sternal lyrifissures; epigynal shield usually trapezoidal, with 1 pair of setae, and free from or partially to completely fused with ventral shield, or epigynal shield rarely reduced (*Gamasolaelaps*); a paired cluster of gland pores usually present behind posterior margins of legs IV; ventral shield well developed, free from or fused with peritrematic or epigynal shields; anal shield free, usually inversely subtriangular, bearing the 3 circumanal setae and rarely 1 pair of ventral setae; presternal plates usually present. Males with holoventral shield or separate sternitigenital and ventrianal shields or sternitigenital, ventral and anal shields. Dorsal subapical sensory field of tarsus I with lanceolate-tipped sensilla and with 2 adjacent sensilla fused basally and inserted together in a common alveolus. Tibia I typically with 4 ventral setae, 6 dorsal setae, and 2 anterolateral setae (2 3/2, 3/2 2); tibia III with 8 setae (2 1/1, 2/1 1); chaetotaxy of genu IV (2 2/1, 3/1 1). Palptarsal apotele 3- or 4-tined and with adjacent hyaline, scalelike process. Internal malae typically elaborated, fimbriated, often bilobed and moustachelike. Female podospermous, with sperm induction pores typically near base of coxae IV; sperm ducts lead to unpaired, central spermathecal sac (laelapid type). Males with genital aperture at anterior margin of sternal region and typically with spurs on legs II; spermatodactyl on movable cheliceral digit simple and fingerlike to extremely elongate, free distally or completely fused to the digit. Larva and protonymph without pygidial shield; protonymphal opisthonotum generally with 12 pairs of setae, lacking S5; dorsal shield of deutonymph divided or with a pair of deep lateral incisions.*

The Veigaioidea comprises a single family, the VEIGAIIDAE (Fig. 12.39), a rather small assemblage (less than 100 described species) of free-living mites. Veigaiids occur throughout much of the world from boreal to tropical regions, but they are best known in the Northern Hemisphere. Taxonomic treatments of the family have been done mostly by regions—for example, Evans (1955a, 1959) and Till (1988) primarily for the British Isles, Athias-Henriot (1961a) for the circum-Mediterranean region, Karg (1993b) for central Europe, Bregetova (1961, 1977g) for Russia, Ishikawa (1978) for Japan, Tseng (1994) for Taiwan, and Farrier (1957) and Hurlbutt (1965, 1984) for primarily the eastern regions of North America. Their seeming rarity in some regions of the Southern Hemisphere may be the result of a lack of collecting and reporting, as evidenced by their presence in collections made in eastern Africa (Hurlbutt 1983) and Brazil (G. W. Krantz, pers. obs.).

The VEIGAIIDAE contains three described genera, all of whose members are considered to be predaceous. Little is known of the biology of *Cyrthydrolaelaps* and *Gamasolaelaps*. Species of *Cyrthydrolaelaps* live in rocky fissures and beach wrack along seashores, and those of *Gamasolaelaps* inhabit moss and moist leaf litter. The genus *Veigaia* contains most of the described species. Along with members of the family PARASITIDAE (see p. 141), veigaiids are the dominant acarine predators in organic humus and litter habitats, especially those of temperate forests (Karg 1965; Evans and Till 1979; Hurlbutt 1984; Solomon 1985; Karg and Freier 1995). Feeding preferences of veigaiids are poorly known, but some species of *Veigaia* have been observed to prey on collembolans, proturans, pauropods, and soft-bodied mites (Karg 1961; Hurlbutt 1965). Hurlbutt (1964) noted a relationship between certain anatomical features of species of *Veigaia* and their vertical distribution in forest soils. Individuals of those species occurring predominantly in upper soil layers were dark in color and oval in shape and had long legs and long dorsal body setae; those found predominantly in deeper layers were a pale color, had slender bodies, and had relatively short legs and setae. The potential usefulness of targeted species of *Veigaia* as biological indicators of upper layer soil ecosystem conditions was reviewed by Karg and Freier (1995).

Thelytoky has been demonstrated only for one individual of VEIGAIIDAE through rearing of female progeny from eggs laid by a virgin female (Walter and Oliver 1990), but the continual absence of males in many repetitive field samplings is strong presumptive evidence that thelytoky occurs among at least some species of *Veigaia* and *Gamasolaelaps*. The apparently unisexual species of *Veigaia* are found more often in disturbed habitats and have broader distributions, particularly into more northern regions, than do bisexual forms (Hurlbutt 1979, 1984). Di Palma et al. (2006) described the fine structure and functional morphology of the male mouthparts of an undescribed species of *Veigaia* from hemlock litter in western North America, in which the spermatodactyl is totally fused with the greatly elongated movable cheliceral digit to form a modified sperm transfer structure.

There are no published records of phoretic relationships of veigaiids with insects, but phoretic deutonymphs of an undescribed veigaiid taxon were found on adult hemipterans of an undetermined taxon in the lowland tropical rain forest of Costa Rica (M. Paniagua and E. E. Lindquist, pers. obs.). Unlike deutonymphs of most other veigaiids, these phoretic forms retain an acrotarsus on legs I, an attribute that also is characteristic of phoretic PARASITIDAE and OLOGAMASIDAE (*Euryparasitus*). Whether phoretic or nonphoretic, deutonymphal veigaiids of most species have one subbasal seta on tarsus I (seta *pl* of the second-most basal verticil of five setae) modified somewhat like a trichobothrium. Denoted a "trichocyst" by Athias-Henriot (1969d), this modified structure is not known among other superfamilies of the Parasitiformes. The adult homologue of this seta is unmodified.

▪ Superfamily Rhodacaroidea (Figs. 12.40–12.43)

DIAGNOSIS: Female and male dorsal shield entire or more commonly divided, often with 2 pairs of sclerotic nodules (scleronoduli) on the anterior region. Female sternal shield usually with 4 pairs of setae (metasternal and sternal elements fused), but with 3 pairs of setae (metasternal elements free) and usually with separated endopodal fragments in HALOLAELAPIDAE; epigynal shield with 1 pair of setae, and free posteriorly from ventrianal or anal shield; microtrichia of postanal cribrum sometimes extending onto posterior margin of dorsal shield; presternal platelets present or absent. Males with holoventral shield or separate sternitigenital and ventrianal shields; when ventrianal shield separate, its anterolateral margins often incised or eroded and its posterolateral margins often fused with posterior dorsal shield, even when free from dorsal shield in female. Femur I usually with 13 setae, including usually 4 ventral setae; tibia I with 3–4 ventral setae, 5–6 dorsal setae, and 2 anterolateral setae (2 3/2, 2(3)/1 2 or 2 3/2, 3/2 2); tibia III with 7–8 (1 1/1, 2/1 1 or 2 1/1, 2/1 1); genu IV often lacking 1 anterolateral and with 0–1 posteroventral seta (1 2/1, 2/0 1 or 2 2/1, 3/1 1); leg chaetotaxy reduced and highly variable among species of LAELAPTONYSSIDAE and some DIGAMASELLIDAE. Palptarsal claw 2- or 3-tined. Female podospermous ("neopodospermous" sensu *Alberti 2002a); sperm induction pores on coxa, or trochanter, or femur of legs III, or on coxa or trochanter of legs IV, or on sternal region beside the bases of coxae III or IV; sperm ducts lead to unpaired, central spermathecal sac (laelapid type). Male with genital aperture at anterior margin of sternal region or rarely on a level with coxae II, and typically with well-developed spurs or modified spinelike setae on legs II and sometimes legs III–IV; spermatodactyl on movable cheliceral digit simple and fingerlike to variously modified, elongated or elaborated in shape. Larva with or without pygidial shield; protonymphal pygidial shield typically with 7–8 pairs of setae, setae* S4–S5 *may be inserted marginally on the shield or in adjacent soft cuticle; dorsal shield of deutonymph divided, even when entire in adult, except entire as in adult among genera of epiphidine OLOGAMASIDAE. Leg I of deutonymph occasionally with well-differentiated acrotarsus (i.e., some* Euryparasitus *(OLOGAMASIDAE) species).*

Rhodacaroids are primarily free-living forms that are especially common in the northern and southern temperate, tropical, and subantarctic zones (Athias-Henriot 1961a; Hunter 1970; Lee 1974; Lee and Hunter 1974; Karg 1993b, 1998c). They are widespread in surface and subsurface soils and in accumulations of decaying organic material such as compost, manure, and tidal debris. The potential usefulness of targeted species of this superfamily as biological indicators of soil ecosystem conditions has been reviewed primarily for northern temperate regions (e.g., Karg 1993c; Karg and Freier 1995; Koehler 1996) but also in a preliminary way for tropical and subtropical regions (Karg 1997b). The RHODACARIDAE (Fig. 12.40) is equivalent to the subfamily Rhodacarinae *sensu* Lee (1970) and Karg (1993b) and to the combined subfamilies Rhodacarinae and Rhodacarellinae *sensu* Shcherbak (1980). It includes mostly small, delicate forms frequently encountered in deeper soil layers, even into underlying groundwater habitats (Krantz 1986). Species of *Rhodacarus* and *Rhodacarellus* often are found in soil samples in both the Old and New Worlds (Evans, Sheals, and MacFarlane 1961; Price 1973; Bregetova and Shcherbak 1977). Rhodacarids prey on small collembolans, immature oribatid mites, and nematodes (Sardar and Murphy 1987; Karg 2000). Some species of the genera *Rhodacarus*, *Rhodacarellus*, and *Protogamasellopsis* (Fig. 12.40H) are thelytokous (Walter and Ikonen 1989; Walter and Kaplan 1990b). The taxonomically enigmatic genus *Tangaroellus*, for which Lee (1970) proposed a separate subfamily, has been tentatively regarded as a member of this family. Its one known species is associated with shore barnacles in intertidal rock fissures in New Zealand (Luxton 1968).

The family DIGAMASELLIDAE (Figs. 12.41, 12.42G) is equivalent to the subfamily Dendrolaelapinae *sensu* Shcherbak (1980) and Karg (1993b). Its members are classified into relatively few genera by some authors (e.g., Lindquist 1975b; Evans and Till 1979) but into considerably more taxa by others (e.g., Shcherbak 1980; Karg 1993b) to accommodate some of the many subgenera of narrow scope recognized by Hirschmann and Wisniewski (1982a, b). Digamasellids are widespread in surface and subsurface soils and in accumulations of decaying organic material such as compost, manure, and tidal debris, where they feed on collembolans, nematodes, arthropod eggs, and possibly fungi (Evans and Till 1979; Shcherbak 1980). Although many members of the speciose genus *Dendrolaelaps s. lat.* are soil or litter forms (Coineau 1974a; Karg 1993b), many others are found in decaying wood, bracket fungi, and the galleries of bark beetles (Hirschmann 1960; Hurlbutt 1967; McGraw and Farrier 1969; Moser and Roton 1971; Kinn 1971; Hirschmann and Wisniewski 1982a, b; Shcherbak 1980; Karg 1993b). Phoresy by deutonymphs is common among digamasellids. Those of the soil-dwelling nematophage *Dendrolaelaps fallax* (Leitner) are phoretic on sciarid flies (Binns 1973), while deutonymphs of *Digamasellus punctum* (Berlese), one of the few described species of the type genus, are phoretic on geotrupine dung beetles (Hirschmann 1960). Some species of *Dendrolaelaps s. lat.* are known only from deutonymphs collected from their insect carriers, especially bark- and wood-boring beetles (Hirschmann 1960; Hirschmann and Wisniewski 1982a, b). Deutonymphs and adults of some species of *Dendrolaelaps* prey on eggs and first-instar larvae of their bark beetle associates (Kielczewski and Balazy 1966; Moser 1975b), while those of other species feed readily on nematodes in beetle galleries (Hirschmann and Rühm 1953; Kinn 1967b). The relationship between the southern pine beetle and one of its common phoretic mites, *D. neodisetus* (Hurlbutt), may be mutualistic because beetles harboring this nematophagous mite are less affected by an endoparasitic nematode (Kinn 1980). In Japan, *D. fukikoae* Ishikawa and *D. unispinatus* Ishikawa have been observed to feed on pinewood nematodes in the pupal chambers and in the spiracles on the bodies of pine sawyer beetles (Enda and Tamura 1977). Their life cycles were completed when reared on nematodes in the laboratory (Tamura and Enda 1980a). Although mites of other recognized digamasellid genera occupy diverse habitats, the elongate-bodied forms of *Longoseius* species are exclusively subcortical, where they are found primarily in association with wood-boring beetles (Lindquist 1975b). Their feeding habits are unknown. The taxonomically enigmatic genus *Panteniphis*, which is provisionally included here in the DIGAMASELLIDAE, sometimes occurs in subaquatic habitats (Athias-Henriot 1969a). Specimens collected in litter and moss of alder forest stands in Poland provided the basis for description of the previously unknown deutonymph of *P. mirandus* Willmann (Fig. 12.42G), which is phoretic on carabid beetles (Gwiazdowicz 2000a).

The aberrant and monogeneric family LAELAPTONYSSIDAE (Figs. 12.42D–F) is based on seven known species of mites that are closely associated with termites and their nests (Krantz 2000; Halliday 2006). Their highly unusual morphology has led various authors to treat them as a subfamily or family with problematic affinities within the Rhodacaroidea and with related dermanyssine assemblages (Womersley 1956; Lee 1970; Halliday 1987; Krantz 2000). The disparity of derived attributes in laelaptonyssids is seen to have arisen as a consequence of their long-term association with the isolated arenas of their termite hosts (Krantz 2000). Their form and specialized association with a particular group of insects is reminiscent of the insect-associated DIGAMASELLIDAE, some of which also display setal deficiencies of the idiosoma and legs, reduction of peritremes, and abbreviation and weak ornamentation of body shields (Lindquist 1975b). *Teranyssus howardensis* Halliday, the type and only species of the termitophilous family TERANYSSIDAE, was found in the nests and on individuals of *Mastotermes darwiniensis* in Australia (Halliday 2006). *T. howardensis* shares many attributes with laelaptonyssids but differs from them in having an entire dorsal shield, 5-segmented palpi, and a straight rather than a recurved male spermatodactyl. The family is not keyed or illustrated in this treatment.

The OLOGAMASIDAE (Fig. 12.43), including the subfamily Epiphidinae *sensu* Kethley (1983a) and the families Gamasiphidae *sensu* Karg (1998c) and Euryparasitidae *sensu* Anthony (1980) and Hagele et al. (2005), is a large, widely distributed group of predators encountered in soil, humus, and compost. The family is much richer in genera and species in tropical to subantarctic regions of the Southern Hemisphere than in the Northern Hemisphere (Lee and Hunter 1974; Bregetova 1977g; Karg 1993b, 1998c). Species of *Cyrtolaelaps* and *Euryparasitus* commonly occur in nests of small mammals (Evans and Till 1979; Karg 1993a). Some species of *Hydrogamasus*, *Psammonsella*, and certain other ologamasid genera have adapted to intertidal marine habitats (Haq 1965; Lee 1970; Luxton 1990). *Hydrogamasus antarcticus* Trägårdh and *H. racovitzai* (Trouessart) have been collected from tussock grass and moss on the antarctic peninsula (Hunter 1967; Lee 1970). These and other of the most southerly distributed taxa of OLOGAMASIDAE appear to represent southward colonizations primarily from New Zealand (Lee and Hunter 1974). Electrophoretic analysis for prey traces in field populations, along with laboratory experiments, indicated that *H. racovitzai* preys primarily on collembolans and preferentially on *Cryptopygus antarcticus* in the maritime antarctic zone (Lister, Block, and Usher 1988). Prey specificity has also been observed in a number of ologamasid species kept in laboratory culture. Lee (1974) noted that *Gamasellus concinnus* (Womersley), *Euepicrius filamentosus* Womersley, and *Athiasella dentata* (Womersley) fed preferentially on collembolans, while *Gamasiphis fornicatus* Lee, *Acugamasus semipunctatus* (Womersley), and *Gamasiphoides propinquus* (Womersley) favored mites as prey. Various Australian ologamasids feed readily on nematodes as well (Beaulieu and Walter 2007).

Although phoresy is not a common phenomenon among ologamasids, deutonymphs of *Cyrtolaelaps* sp. and *Euryparasitus* sp. have been collected on the mammals whose nests they frequent (Bregetova 1977f; Kethley 1983a; Whitaker and French 1988). Other records of phoresy in this family, however, indicate a definite preference for carabid beetles. Deutonymphs of an undescribed species of *Epiphis* were taken from a Field Museum of Natural History (Chicago) collection of a carabid beetle in Tennessee (Kethley 1983a). Several collections of European carabid beetles yielded deutonymphs of *Iphidosoma fimetarium* (Müller) (Kethley 1983a), and additional deutonymphal specimens were found on carabid hosts taken from forest litter in eastern Europe (Bregetova 1977f). Contrary to other records of phoresy on insects by epiphidine ologamasids, deutonymphs of *I. razumovae* Bregetova were taken from a dipteran host (Bregetova 1977f). Without descriptive commentary, Bregetova illustrated the deutonymph of several species of *Iphidosoma* as having an acrotarsus on tarsus I. This was also noted by Kethley (1983a) and was confirmed by personal observations (E. E. Lindquist and D. E. Walter).

Members of the family HALOLAELAPIDAE (Figs. 12.42A–C) favor damp organic substrates, including compost, manure, and tidal debris (Hyatt 1956; Karg 1969, 1993c; Evans and Till 1979; Blaszak and Ehrnsberger 1993, 1995, 1997, 1998a–c, 1999, 2000a, b, 2002). Several species of *Halolaelaps* and *Halodarcia* have been found on marine algae and beach wrack in the intertidal zone. Deutonymphs of species of *Halolaelaps* in estuarine or intertidal habitats use amphipods for dispersal (Pugh et al. 1997), while those of species in other habitats use insects for dispersal (Evans and Till 1979). Those of some species of *Halodarcia* are subelytral phoretics of carabid beetles (Evans and Fain 1995). *Leitneria granulata* (Halbert), a thelytokous species occurring in forest soils in Europe, has only a single generation annually as a result of the unusually long aestivation period of the deutonymphal stage (Athias-Henriot 1976).

▪ Superfamily Eviphidoidea (Figs. 12.44–12.48)

DIAGNOSIS: Adult dorsal shield entire; sternal shield of female with 3–4 pairs of setae, metasternal setae on small shields that are amalgamated with adjacent sternal-endopodal elements or present as discrete shields, occasionally inserted free in the integument; preendopodal platelets present or absent, female with a ventrianal or anal shield. Tibia I with 1 posteroventral seta, 2–3 posterodorsal setae, and 1 or 2 anterolateral setae (1(2) 3/2, 2(3)/1 2), tibia III with 7–8 setae (1(2) 1/1, 2/1 1). Palptarsal apotele 2–3-tined, epistome variously produced. Female podospermous, sacculus of the "A," or laelapid type. Male with genital aperture at anterior margin of a holoventral or sternitigenital shield; spermatodactyl arises either from internal or external face of movable digit, free distally. Dorsal shield of deutonymph divided or entire; protonymph with 8–11 pairs of pygidial setae; larva with or without pygidial shield.

Although they are primarily humus- and litter-inhabiting predators, many eviphidoids have adapted to transient habitats such as dung, carrion, and beach wrack and have developed phoretic relationships with a variety of arthropods as a means of dispersal when these ephemeral substrates deteriorate. The PARHOLASPIDIDAE (see below) is exceptional in that none of its constituent species is known to be phoretic.

The family PARHOLASPIDIDAE (Fig. 12.44) comprises 11 genera of moderately to heavily sclerotized predatory mites that often are found in forest litter, soil, decaying vegetable matter, moss, and tree holes primarily in the Palearctic, Nearctic, and Oriental realms, often at higher altitudes (Evans 1956; Krantz 1960; Ishikawa 1979, 1980a–c, 1993, 1994; Datta and Bhattacharjee 1988). The single known female of *Neparholaspis crispus* Willmann (presumably conspecific with *N. spathulatus* Evans (Evans 1956)) was found in a cave in the Carnic Alps. Parholaspidids may attain high densities in some habitats, often surpassing other mite assemblages by virtue of sheer numbers. For example, according to Ishikawa (1980c), parholaspidids accounted for 26.8% (6,636 per square meter) of collected soil Mesostigmata from five mountain sites in Japan, followed by RHODACARIDAE (14.4%), UROPODIDAE (11.5%), ZERCONIDAE (10.6%), and PARASITIDAE (10.0%). Parholaspidids are distinctive in the superfamily in displaying at least 27 pairs of simple, spatulate, or whiplike idionotal setae, presternal platelets (absent in *Parholaspis kewensis* Evans), and corniculi that are more than three times longer than their basal width. Their wide geographical distribution in both the Northern and Southern hemispheres, coupled with their lack of reliance on phoresy, suggests that the PARHOLASPIDIDAE is an early derivative eviphidoid assemblage. Their common insular occurrence on mountaintop "islands" further suggests that they may represent a relict group whose distribution reflects early isolation brought about by geological upheavals in the late Mesozoic and the Cenozoic eras.

Although they are relatively common, little is known regarding the biology or behavior of parholaspidids. Ishikawa (1980c) found that the development of female *Parholaspulus ochraeus* (Ishikawa) from egg to adult may require as little as eight days and, as in all other studied Dermanyssiae, eggs are developed only one at a time. *P. ochraeus* was observed to feed on nematodes and microarthropods, a predilection that likely is shared by other members of the family.

The genera *Eviphis*, *Evimirus*, and *Scamiphis* of the family EVIPHIDIDAE (Fig. 12.45) include mostly nonphoretic forms that inhabit humus, moss, compost, grasslands, animal nests, and other relatively stable substrates. The monotypic genus *Rafaphis* is based on a single female taken from moss and may also be nonphoretic (Skorupski and Blaszak 1997). *Eviphis drepanogaster* Berlese is exceptional for the genus in that it is phoretic, dispersing on scarabaeid beetles that share its rodent nest habitat (Karg 1993b). Representatives of the remaining eight recognized eviphidid genera commonly occur as predators on nematodes and other microfauna in ephemeral habitats such as animal dung and beach wrack and are phoretic as deutonymphs and/or adults on arthropods that exploit these resources. Species of *Alliphis*, *Copriphis*, and *Scarabaspis* often are found on scarabaeid dung beetles (Evans 1957b; Karg 1993b), as is *Scarabacariphis grandisternalis* Mašán, a phoretic associate of *Onthophagus fracticornis* in central Europe (Mašán 1994). *Canestriniphis megalodacne* Potter and Johnston, an unusual species with cheliceral characteristics suggestive of insect parasitism, is phoretic on erotylid fungus beetles (Potter and Johnston 1976). Amphipods serve as dispersal agents for members of the genus *Thinoseius* in intertidal Nearctic, Palearctic, and Oriental habitats (Karg 1963, 1993b; Evans 1969a; Takaku 2000).

Members of the cosmopolitan family MACROCHELIDAE (Fig. 12.46), an assemblage of some 15 genera and approximately 400 described species, are predaceous on nematodes and the eggs and larvae of other microinvertebrates occurring in a wide range of organic substrates. Some macrochelids have retained the relatively simple soil- and litter-inhabiting, nonphoretic lifestyle of many other mesostigmatic predatory groups, feeding on small invertebrates in essentially permanent substrates such as forest litter and underlying soil and humus. The genera *Geholaspis* and *Dissoloncha*, and the early derivative species groups of the genus *Macrocheles*, exemplify these forms (Krantz 1998c). However, like the EVIPHIDIDAE, the great preponderance of known macrochelid species exploit ephemeral or specialized substrates such as animal dung or the nests or galleries of mammals, birds, and social insects (Berlese 1918; Evans and Browning 1956; Bregetova and Koroleva 1960; Filipponi 1964; Filipponi and Pegazzano 1960, 1962, 1963, 1967; Krantz and Filipponi 1964; Krantz 1965a, 1998b, c; Evans and Hyatt 1963; Costa 1966b, 1967, 1969b; Krauss 1970; Hyatt and Emberson 1988; Halliday 1990b, 2000; Takaku, Katakura, and Yoshida 1994; Takaku and Hartini 2001; Mašán 2003). Here, they feed on nematodes, oligochaetes, and insect eggs and larvae, often producing several generations of offspring before their habitats deteriorate. Because of the insular nature of many of these substrates, these mites commonly rely on phoretic associations with flying hosts to escape and colonize fresh substrates. Typically, only inseminated females disperse in this fashion, although males and immatures of some genera may also be phoretic—for example, all of the postlarval stases of *Neopodocinum* (Costa 1969b) and the protonymphs and deutonymphs of *dimidiatus* group species, genus *Macrocheles* (G. W. Krantz, pers. obs.). Members of the genera *Holocelaeno* and *Neopodocinum* and some species groups of *Macrocheles* exhibit varying degrees of specificity for their scarabaeid or muscid hosts, and an increasing commitment to host specificity often is correlated with an evolutionary progression from ancestral to highly derived forms (Krantz 1965b, 1991, 1998c; Costa 1967; Krantz and Mellott 1972). Positive phoretic responses in three phoretically specific species of *Macrocheles* were found to be triggered by a kairomone extracted from the cuticle of their beetle hosts (Krantz et al. 1991). *M. mycotrupetes* Krantz and Mellott, an unusual phoretic associate of psammophilous geotrupine dung beetles in Florida, USA, displays morphological and reproductive traits more typical of free-living edaphic macrochelids than of highly derived phoretics but nevertheless was capable in laboratory observations of perceiving the kairomonal signal of its beetle host and moving through a 15–20 mm sand barrier to reach it (Krantz and Royce 1994). An unusual mutualistic association between *M. rettenmeyeri* Krantz and its army ant host during swarming events (Krantz 1962) has already been discussed in chapter 6 under the heading Ectoparasites of Invertebrates.

Macrochelids have also been found in the fur of mammals and occasionally on birds (Evans and Browning 1956; Bregetova and Koroleva 1960; Krantz and Whitaker 1988; Karg 1993b). *M. lukoschusi* Krantz has established a remarkable relationship with the three-toed sloth, *Bradypus tridactylus*, in the forests of Amazonia. The mites live in the rectum of their sloth host, apparently feeding on nematodes in the dung stored in the rectal cavity. Weekly migration of the mites into the ventral pelage just prior to each host defecation episode is believed to be triggered by a positive phototaxic response to light penetrating into the opening rectal cavity (Krantz 1983b; Adis and Krantz 1985). Facultative parasitism is known or purported to occur in the *subbadius* and *muscaedomesticae* species groups of Macrocheles but only on adult drosophilid and muscoid Diptera (Jalil and Rodriguez 1970b; Polak 1996).

It should be noted that the basis for much of our present-day knowledge of the genetics, reproductive biology, and population ecology of coprophilous MACROCHELIDAE originated in the laboratory of Alessandro Filipponi and his coworkers who, among their many signal accomplishments during the 1950s and 1960s (see selected references in Krantz 1998c), verified the earlier observation by Pereira and de Castro (1945) that certain *Macrocheles* species can exert a controlling effect on synanthropic flies that breed in manure.

Species assigned to the seven currently recognized genera of PACHYLAELAPIDAE (Fig. 12.47) share the habits of macrochelids and eviphidids, preying on microfauna in litter, humus, moss, and the nests of mammals and social insects. Pachylaelapids are especially diverse in the Old and New World tropics, but they are also well represented in temperate North America and Europe. Members of the genus *Pachylaelaps* often disperse by attaching to arthropods, primarily scarabaeid beetles (Evans and Hyatt 1956; Hirschmann and Krauss 1965; Costa 1971, 1974b; Koroleva 1977; Karg 1993b, 1998c), although only one described species of the pachylaelapid genus *Zygoseius* (*Z. sarcinulus* Halliday) is known to be phoretic on scarabs (Halliday 1997b). However, other species of *Zygoseius* collected in dung and decomposing plant debris (Karg 1998a) would also seem likely candidates for phoretic associations with flying arthropods. Postlarval instars of *P. aegypticus* Hafez and Nasr, originally collected from soil and debris under *Lantana camara* in Giza, were observed to feed voraciously on housefly larvae in laboratory cultures (Gomaa et al. 1989). Ant nests are the preferred habitat for *Sphaerolaelaps* species (Berlese 1903; Koroleva and Sklyar 1982; Karg 1993b), while at least some members of the predominantly soil- and litter-inhabiting European genus *Olopachys* (*O. skalri* Koroleva and *O. vysotskaja* Koroleva) seem to prefer rodent nests (Koroleva 1976). Females of the edaphic genus *Pachyseius* differ from most other pachylaelapids in carrying sternal setae *st4* on free metasternal elements (Karg 1993b; Moraza 1993; Moraza and Peña 2005). As with the MACROCHELIDAE, many pachylaelapid species have been collected and recorded but remain to be described.

Mention should be made here of three eviphidoid taxa whose affinities remain largely unresolved. Evans and Till (1966) listed *Pseudolaelaps doderoi* (Berlese), a soil-inhabiting mite collected from various sites in Europe, as the sole British representative of a new dermanyssid subfamily Pseudolaelapinae. Karg (1993b) later moved *Pseudolaelaps* to the Eviphidoidea and elevated it to family level. While *P. doderoi* has morphological affinities with the PACHYLAELAPIDAE (Figs. 12.48A–C), the female displays characteristics that are not typical for the family as presently recognized (e.g., free metasternal shields, two ventral setae on genu and tibia I rather than three, and an epistome that is smooth and trispinate (Fig. 12.48C) rather than being forked or variously divided distally as in other pachylaelapids). Evans and Till (1966) described the palpal apotele of *P. doderoi* as being two tined, but Karg (1993b) noted that it is three tined as in PACHYLAELAPIDAE. Based on its general morphology, *Pseudolaelaps* is retained in the Eviphidoidea and is included provisionally as the sole representative genus of the pachylaelapid subfamily Pseudolaelapinae. Two additional species (*P. gamaselloides* (Berlese) and *P. paulseni* (Berlese)) are known (Gilyarov and Bregetova 1977).

In another work, Karg (1978a) elevated the Gondwanan laelapid genus *Leptolaelaps* to subfamilial status, concluding that it represented an intermediate form between the Eviphidoidea and the superfamily Dermanyssoidea. Later (1983), Karg assigned four additional genera to the subfamily (*Evansolaelaps*, *Ayersacarus*, *Cerambylaelaps*, and *Australocheles*) and placed the expanded taxon in the Eviphidoidea primarily because of the form and position of the male spermatodactyl and the presence of strong apophyses on legs II. In the same paper, Karg elevated the subfamily to family rank. Subsequently, he assigned two new genera (*Pulchraplaga* and *Indutolaelaps*) to this assemblage, along with the laelapid-like genera *Stevacarus*, *Hunteracarus*, *Cosmetolaelaps*, and *Paradoxiphis* (Karg 1997a). The considerable morphological diversity among these 11 genera (see Hunter 1964a, b, 1970; Marais and Loots 1969; Costa 1979; Costa and Allsopp 1979, 1981; Karg 1983, 1977a) (Figs. 12.48D–K) suggests that they may represent a polyphyletic grouping with mixed eviphidoid and dermanyssoid affinities. For this reason, they have not been formally included as a family entity in this treatment. An unidentified species of *Hunteracarus* and of another member of the generic grouping were observed to feed on nematodes and dead arthropods (Beaulieu and Walter 2007).

Finally, a family name was proposed for the genus *Bulbogamasus* and its sole representative species, *B. sinicus* Gu, Wang, and Duan, an eviphidoid-like gamasine mite whose adults are associated with scarab beetles in China (Gu, Wang, and Duan 1991). Although the original description omits certain morphological details, *B. sinicus* clearly has affinities with the Eviphidoidea and could represent a distinct family. Further clarification of its morphology, however, may reveal instead that it can be accommodated in an existing eviphidoid family. Because of its questionable status, the proposed family name Bulbogamasidae has not been formally recognized here.

■ Superfamily Ascoidea (Figs. 12.49–12.51)

DIAGNOSIS: Female and male dorsal shield entire, or with lateral incisions, or divided. Female sternal shield usually with 3 pairs of setae (metasternal setae on separate platelets or free in soft integument), but with 2 pairs or rarely 1 pair of setae (3rd and rarely 2nd sternal setae on separate platelets or soft integument) in some AMEROSEIIDAE; epigynal shield with 1 pair of setae, which may instead be in soft integument flanking shield and free posteriorly from ventrianal or anal shield. Males with sternitigenital shield usually separate from ventrianal shield, or with line of fusion usually evident when connected; ventrianal shield with anterolateral margins not incised or eroded and with posterolateral margins free from dorsal shield; ventrianal shield rarely divided into ventral and anal shields. Femur I usually with 12 setae, including 3 ventral setae; tibia I with 2–3 ventral setae, 5–6 dorsal setae, and 1–2 anterolateral setae; tibia III with 7–8 setae (1 1/1, 2/1 1 or 2 1/1, 2/1 1). Palptarsal claw usually 2-tined, appearing 3-tined when one tine split along mid-length. Female podospermous ("neopodospermous" sensu *Alberti 2002a); sperm induction pores on sternal region between bases of coxae III and IV; sperm ducts lead to unpaired, central spermathecal sac (laelapid type). Male with genital aperture at anterior margin of sternal region, and with or without spurs or modified spinelike setae on legs II and sometimes legs III–IV; spermatodactyl on movable cheliceral digit usually simple and fingerlike, sometimes elongated. Larva with or without pygidial shield; protonymphal pygidial shield with 5–8 pairs of setae, setae* S4–S5 *usually on shield; dorsal shield of deutonymph entire and with lateral incisions or divided if also divided in adult stage.*

The Ascoidea is richly represented in tropical, temperate, and arctic alpine regions, where many of its members are free-living predators of nematodes and microarthropods in soil or humus and suspended arboreal litter habitats. Others are adapted to more specialized niches, where they feed on fungi, pollen, or nectar. Members of the family AMEROSEIIDAE (Fig. 12.49) frequent a wide range of terrestrial and aboveground substrates, including moss, rotting straw, compost, manure, forest humus, decaying wood, bracket fungi, stored foods, and the nests of mammals, birds, and social insects; inflorescences may also be invaded (Evans and Till 1979; Bregetova 1977a; Karg 1993b; Halliday 1997a). Although many ameroseiids have stout rutella-like corniculi and coarsely toothed chelicerae adapted for feeding on fungi and ingesting spores and hyphal fragments, others have less robust mouthparts more suitable for feeding on pollen and nectar (Evans 1963a). No observations have confirmed a predatory feeding behavior for any ameroseiid species. Many members of the genera *Ameroseius* and *Epicriopsis* occur in soil and humus of grasslands and forests and in the middens of rodents (Bregetova 1977c; Evans and Till 1979; Karg 1993b). Some *Ameroseius* species are fungivorous associates of bracket fungi and the decaying wood of trees that have succumbed to attacks by bark- and wood-boring beetles (Westerboer and Bernhard 1963; De Leon 1964a; Flechtmann 1985). *Kleemannia plumosa* (Oudemans) is a cosmopolitan inhabitant of stored food and the nests of small mammals and social insects, where it feeds on fungi (Hughes 1976; Halliday 1997a). *K. plumosa* and several other species of *Kleemannia* were reported as nuisances in new houses and office blocks in Germany, where vegetable material used as insulation during construction absorbed moisture from drying brick or concrete and became moldy, thus permitting large-scale development of the mites. The infestations disappeared as the new structures gradually dried out (Rack 1963, 1972b). Phoretic associations with other animals have been established for many genera of ameroseiids, predominantly as adult females. However, records of a few phoretic males and immatures also exist (Ramanan and Ghai 1984; Haitlinger 1987f; Seeman and Walter 1995; Seeman 1996).

Species of *Sertitympanum* are commonly phoretic as adult females on rodents and likely are fungivores in the nests of their carriers, similar to some species of the closely related genus *Kleemannia* (Rack 1972b; Domrow 1979a; Elsen and Whitaker 1985; Bassols, Vargas, and Polaco 1993). Other ameroseiids have established phoretic associations with insects. For example, adult females of several *Ameroseius* species are found on a variety of sphecid wasps (Elsen 1973), while females of other species occur on beetles inhabiting bracket fungi and wood (Westerboer and Bernhard 1963; Karg 1993b). Adult females of *Brontispalaelaps leveri* Womersley were collected on hispine chrysomelid beetles in southeast Asia (Womersley 1956), while those of *B. marianneae* Halliday were found attached to the wings of a phycitine pyralid moth in Australia (Halliday 1997a).

Adult females of the ameroseiid genera *Neocypholaelaps*, *Afrocypholaelaps*, and *Hattena* occur on bees, wasps, butterflies, and flower-visiting flies and scarab beetles in the Old World tropics and feed on nectar and probably on pollen of flowers visited by their insect carriers (Elsen 1972, 1974; Mo 1972; Manson 1974; Treat 1975; Delfinado-Baker, Baker, and Phoon 1989; Seeman and Walter 1995; Seeman 1996). Adult female *N. indica* (Evans) may attach to the petite workers of *Apis cerana*, the Indian honeybee, in great numbers. Up to 400 mites per bee were counted by Ramanan and Ghai (1984), and Kshirsagar (1968) found an average of 47 mites per bee. A maximum of 247 *A. africana* (Evans) was noted by Seeman and Walter (1995) on one worker honeybee. All instars of *N. indica* may be found in flowers normally favored by their insect carriers (Evans 1963a; Treat 1975). Evans (1955e) presented morphological notes on a possible stridulating mechanism in male *Neocyphlolaelaps stridulans* (Evans), but no evidence of vibratory production or perception was provided. Adult females of three species of the genus *Hattena* have been recorded, along with a few males and nymphs, as running in and out of the nares and along the beak of honeyeaters (family Meliphagidae) and sunbirds (Nectariniidae) (Domrow 1979a; Noske 1993; Seeman 1996; Halliday 1997a). This fascinating mite-bird-flower association may parallel that of hummingbird flower mites of the family MELICHARIDAE (see below). The first known Neotropical species of *Hattena* was described by Faraji and Cornejo (2006).

The MELICHARIDAE (Fig. 12.50) has been classified as a subfamily or a tribe in previous accounts (Hirschmann 1962; Lindquist and Evans 1965), but it is treated here as a separate family and probable sister group of the AMEROSEIIDAE based on a uniquely shared modification of the cheliceral pilus dentilis. This diverse group has adapted to a broad spectrum of terrestrial and aboveground habitats similar to those of ameroseiids (Lindquist and Evans 1965; Bregetova 1977b; Karg 1993b; Halliday, Walter, and Lindquist 1998; Gwiazdowicz 2007). Although many species of melicharids are predatory, others have adapted to feeding on fungi, pollen, and nectar, and one is reported to be parasitic on cockroaches (Egan and Hunter 1975). *Melichares agilis* Hering and *Proctolaelaps pomorum* (Oudemans) are cosmopolitan in cultivated soil and arthropod-infested stored foods, including dried fruit, where they may prey on eggs and immatures of acarid mites (Hughes 1976). *P. regalis* De Leon is one of several species that undergo their life cycle in freshly rotting fruit. *P. regalis* preys on larvae and pupae of mixed populations of drosophilid flies and may be involved in the horizontal gene transfer of P-elements from one host species to another (Houck et al. 1991). *P. pygmaeus* (Müller), also known as *P. hypudaei* (Oudemans), is a ubiquitous species with varied feeding behavior. Although observed to prey on a variety of mites by some authors (Nesbitt 1951; Mathys and Tencalla 1959), *P. pygmaeus* has been reported to feed and reproduce on fungal mycelia by others (Mathys and Tencalla 1959; Muma 1961; Listemann 1984) and occasionally may cause skin irritation in humans (Andrews and Ramsay 1982). Adults and immatures of the fungivorous genus *Mycolaelaps* are unusual in having elongated bodies adapted to living in the pore tubes of bracket fungi (Lindquist 1995; Makarova 2004).

Phoresy by adult females is widespread in the genus *Proctolaelaps* (Karg 1985b). Species whose females are phoretic on bark beetles or their subcortical beetle associates undergo development in the gallery systems of their hosts, either as predators of nematodes or the eggs and larvae of the beetles or as fungivores (Hirschmann and Rühm 1953; Samšinák 1960; Lindquist and Hunter 1965; Moser and Roton 1971; Kinn 1983). Those of several other species of *Proctolaelaps* (also recorded under the name *Garmaniella*) are phoretic on bumblebees and undergo their life histories in the bees' nests, where they probably feed on pollen (Westerboer 1963; Karg 1993b; Schwarz, Huck, and Schmid-Hempel 1997). Adult females of the genera *Rhinoseius*, *Tropicoseius*, and of some species of *Proctolaelaps* are phoretic on hummingbirds, using them for transport from flower to flower, where the mites feed either on pollen and nectar or on other arthropods (Fain, Hyland, and Aitken 1977; Colwell 1979; Naskrecki and Colwell 1998). Sexually dimorphic deutonymphs and heteromorphic adult males are evident among some species of *Rhinoseius* and *Tropicoseius* (Wiese and Fain 1996). A similar relationship may exist between the African *Proctolaelaps vandenbergi* (Ryke) and its honey bird carrier (Lindquist and Evans 1965). All instars of this mite occurred in large numbers in flowers of *Protea*, where they were thought to prey on small phytophagous invertebrates (Ryke 1964). Adults were found on the heads of honey birds, which were suggested as a means of transport to fresh flowers. Many adult females and a few males of *P. spencerae* Domrow were found phoretic on the honey possum, *Tarsipes spencerae*, in Western Australia (Domrow 1979a). Two species of melicharine flower mites of an as yet undescribed genus undergo their life cycles on inflorescences of a *Calyptrogyne* palm and a *Spathiphyllum* aroid, respectively, in lowland Neotropical rain forests. One fully developed inflorescence may harbor hundreds of the quickly developing mites in all instars (Lindquist 2001). Female mites are transferred by *Artibeus* bat pollinators and *Lagochile* scarab beetle visitors to the palm host (Tschapka and Cunningham 2004) and by *Trigona* bee pollinators on the aroid host (Lindquist 2001). Two species of *Xanthippe* are known only from the inflorescences of a Neotropical palm, where they are thought to have a phoretic association with nitidulid beetles that pollinate the plant (Naskrecki and Colwell 1995). Other species of *Proctolaelaps* are known only from tropical flowers (De Leon 1963). Adult females of several species of *Proctolaelaps* use the well-protected interpalpal region of butterflies and moths for shelter during phoretic transfer (Treat 1975). *P. lobata* De Leon uses a variety of butterfly species for dispersal but is specific to *Lantana* flowers during its development (Boggs and Gilbert 1987). A more complex relationship was noted between the palm flower weevil, *Elaedobius kamerunicus* (Faust), a major pollinator of oil palms in Malaysian plantations, and an undescribed species of *Proctolaelaps*. The mite is predaceous on phoretic nematodes during the transport of both organisms by adult weevils from senescent to new inflorescences (Krantz and Poinar 2004).

Adult females of the genus *Mucroseius* are found in the thoracic spiracular atria and under the elytra of wood-boring cerambycid beetles in North America and Japan (Lindquist and Wu 1991). Although a predatory or parasitic relationship with the beetles has not been ruled out, these mites (misidentified as *P. hystrix* in papers by Enda and Tamura) are thought to prey on nematodes, including the pinewood nematode (Enda and Tamura 1980; Tamura and Enda 1980b; Lindquist and Wu 1991), or on coexisting digamasellid mites (Kinn 1987). Adult females of *P. cossi* (Dugès) are phoretic on the wood-boring goat moth, *Cossus cossus*, in Eurasia; they are abundant in the galleries, where immatures and adults feed on the moth's larvae and ingest hemolymph exuding from wounds (Andreev 1985). As noted earlier, all active instars of *P. nauphoetae* (Womersley) feed on the blood of their cockroach hosts and thus may be considered parasitic (Egan and Hunter 1975) (see also chapter 6).

As treated here, the family ASCIDAE (Fig. 12.51) includes two subfamilies, the Arctoseiinae and the Ascinae, and excludes the melicharines (see above) along with the blattisociines and platyseiines (classified in the Phytoseioidea in this treatment). In this stricter sense, ascids are relatively unspecialized, free-living inhabitants of both soil and arboreally suspended litter and humus layers. Although ascids are generally most diverse in the tropics, the arctoseiine genus *Arctoseius* is especially speciose in arctic regions (Lindquist 1961; Makarova 2000a, b 2002). However, *A. cetratus* (Sellnick) and *A. semiscissus* (Berlese) have widespread distributions in grassland and cultivated soils and in a variety of other synanthropic habitats, where they prey on nematodes, microarthropods, and sciarid fly eggs (Weis-Fogh 1948; Karg 1961; Binns 1975; Walter 1988a; Rudzinska 1998). The potential usefulness of targeted species of *Arctoseius* as biological indicators of soil ecosystem conditions was reviewed for northern temperate regions (Karg 1993c; Karg and Freier 1995; Koehler 1999) and, in a preliminary way, for a Neotropical region (Karg 1998b). The arctoseiine genera *Iphidozercon*, *Zerconopsis*, and *Xenoseius* are known primarily from species inhabiting dry to moist forest litter, seepages, compost, manure, bracket fungi, and loose bark in the Northern Hemisphere (Evans 1958a; Evans and Hyatt 1960; Ishikawa 1969; Bregetova 1977b; Karg 1993a; Gwiazdowicz 2007). However, a few indigenous species of these genera and *Arctoseius* are known from warm temperate and subtropical regions of Australia and Ecuador (Halliday, Walter, and Lindquist 1998; Karg 1998b); a few others are undescribed from lowland and tropical rain-forest habitats in Costa Rica (E. E. Lindquist, unpublished). *Iphidozercon corticalis* Evans appears to be another synanthropic species with a cosmopolitan distribution; it feeds readily on various nematodes under laboratory conditions (Halliday, Walter, and Lindquist 1998). Most species in all arctoseiine genera appear to be bisexual, with obligatory mating and no indications of thelytoky (Walter and Lindquist 1995). Some species are adapted to living in wet or inundated substrates. For example, *Zerconopsis muestairi* (Schweizer) may survive continuous submergence on reservoir flood plains for up to 128 days in the egg stage (Tamm 1984). Phoretic dispersal by adult female mites using nematoceran Diptera that frequent the same habitats is known for species of *Arctoseius* on sciarids, chironomids, cecidomyiids, and trichocerids; *Iphidozercon* on psychodids; and *Zerconopsis* on tipulids (references cited in Walter and Lindquist 1995).

The subfamily Ascinae appears to be the least morphologically cohesive taxon of the superfamily (Walter and Lindquist 1995). It is well represented throughout the world, both in ground-level and suspended soil habitats, except at high latitudes and high altitudes. The monotypic genus *Ectoantennoseius* in Australia (Walter 1998a) and several members of the speciose genus *Asca*, ubiquitous in subtropical and tropical regions, clearly prefer arboreal niches (De Leon 1967; Karg 1979; Walter, Halliday, and Lindquist 1993). *Asca* species prey on other mites, collembolans, small insects, and nematodes (Hurlbutt 1963; Walter, Halliday, and Lindquist 1993). Arboreal species may play a role in the natural control of herbivorous eriophyoid and tenuipalpid mites (Walter, Halliday, and Lindquist 1993). Several species of *Leioseius* and the one known species of *Arctoseiodes* occur in salt marsh and seashore habitats (Evans 1958a; Luxton 1989; Karg 1993b). Members of the genus *Neojordensia* frequent damp to wet substrates in grasslands and meadows and along water margins (Athias-Henriot 1973; Karg 1993b). Species of *Antennoseius* are found in moss, humus, meadows, fields, salt marshes, and the ground-level shelters and nests of rodents and birds, where they may prey on a variety of nematodes and small arthropods (Ryke 1962; Lindquist and Walter 1989; Karg 1993b). Adult females of some species of *Antennoseius* and of the one known species of the closely related genus *Anystipalpus* are phoretic under the elytra of carabid beetles (Costa 1969a; Lindquist and Walter 1989; Fain, Noti, and Dufrêne 1995; Eidelberg 2000). Adult female dimorphism has been documented for a species of *Antennoseius* in which one morph is phoretic and the other free living (Lindquist and Walter 1989). Most other ascine genera comprise free-living species that occur in relatively stable habitats where strong dispersal abilities are not required, and phoretic associations are unknown (Walter and Lindquist 1995). Males do occur in most species of Ascinae, and mating may be obligatory, but arrhenotoky has been experimentally demonstrated in some populations of *Protogamasellus* and *Gamasellodes* (Afifi, Hassan, and Nawar 1986; Walter 1987b, 2003; Walter and Lindquist 1995). Males are unknown in some widely distributed species of *Asca*, including *A. aphidioides* (Linnaeus) and *A. garmani* Hurlbutt, and thelytoky has been demonstrated in several species of *Protogamasellus* and in populations of *Gamasellodes bicolor* (Berlese) (Hurlbutt 1979; Norton et al. 1993; Walter and Lindquist 1995). Species of *Leioseius*, *Arctoseiodes*, *Ectoantennoseius*, and *Anephiasca* appear to be bisexual, but their mode of reproduction is unknown. Members of the genus *Anephiasca* are unusual among the ASCIDAE in probably being fungivorous, as indicated by their gut content and the rutella-like form of their corniculi (Athias-Henriot 1969b). Although species of *Protogamasellus* and *Gamasellodes* are generally euedaphic and prey on nematodes and microarthropods (Genis, Loots, and, Ryke 1967; Afifi, Hassan, and Nawar 1986; Walter 1987b), large numbers of several species have been encountered in termite nests at sites where microflora is abundant. The mycophagous habits of *P. ascleronodulus* Shcherbak and Petrova is clearly indicated by the regular presence of fungal spores in their gut content (Shcherbak and Petrova 1987), and *P. mica* (Athias-Henriot) has been reared on fungi (Walter and Lindquist 1989).

■ Superfamily Phytoseioidea (Figs. 12.52–12.55)

DIAGNOSIS: Female and male dorsal shield usually entire but divided in some OTOPHEIDOMENIDAE and a few PHYTOSEIIDAE. Female sternal shield usually with 3 pairs of setae (metasternal setae on separate platelets or free in soft integument) but rarely with 2 or fewer pairs of setae; epigynal shield usually with 1 pair of setae, or these on soft integument flanking shield, shield free posteriorly from ventrianal or, rarely, anal shield. Male with sternitigenital shield usually separate from ventrianal shield or with line of fusion evident when connected; ventrianal shield with anterolateral margins not incised or eroded and with posterolateral margins free from dorsal shield; ventrianal shield rarely divided into ventral and anal shields. Femur I usually with 11–12 setae, including 3 ventral setae and 4–5 dorsal setae; tibia I with 2–3 ventral setae, 4–6 dorsal setae, and 1–2 anterolateral setae; tibia III with 7–8 setae (1 1/1, 2/1 1 or 2 1/1, 2/1 1), rarely 9 when pl2 *present. Palptarsal claw usually 2-tined (3-tined in PODOCINIDAE). Female podospermous ("neopodospermous"* sensu *Alberti 2002a, b); sperm induction pores on sternal region between the bases of coxae III–IV; each major sperm duct leads to an often sclerotized spermathecal calyx, with an associated saclike vesicle, atrium, and minor duct (phytoseiid type; see chapter 3). Male with genital aperture at anterior margin of sternal region, and with or without spurs or modified spinelike setae on legs II and sometimes legs III–IV; spermatodactyl on movable cheliceral digit simple and fingerlike to variously modified, elongated, or elaborated in shape. Larva with or without pygidial shield; protonymphal pygidial shield with 5–8 pairs of setae, setae* S4–S5 *usually on shield; dorsal shield of deutonymph entire, with lateral incisions or divided if divided in adult.*

The Phytoseioidea includes predatory, parasitic, fungivorous, and pollenophagous species found in both ground and aerial habitats. Its constituent families share a uniquely derived combination of sperm morphology in the males and sperm access system in the females, the complexity of which renders the option of convergence highly improbable (Walter and Lindquist 1997; Alberti 2002a, b).

Most of the 1,800-plus described species of PHYTOSEIIDAE (Fig. 12.52), about 15% of the known diversity of the Mesostigmata, have been described in the last 50 years as a direct result of their role as biological control agents (Kostianinen and Hoy 1996). Typically, they inhabit vegetation and feed on small arthropods, pollen, honeydew, plant exudates or leaf cell contents, and, to a certain extent, fungi (Zemek and Prenerova 1997). Phytoseiids also are known to be common inhabitants of leaf domatia (Walter 1996). A few species are found in soil (e.g., some *Neoseiulus*, *Proprioseiopsis*, and *Amblyseius*) or in animal nests (usually species of *Neoseiulus*). Although those that specialize on spider mites (e.g., species of *Phytoseiulus*, *Galendromus*, and some *Neoseiulus*) are the best known, most PHYTOSEIIDAE are polyphagous and feed on a variety of plant products and small arthropods, including pestiferous thrips, whiteflies, scale crawlers, and eriophyoid mites (McMurtry and Croft 1997). Most phytoseiids are multivoltine, and the adult female is the overwintering (or aestivating) stage (Veermann 1992). The life cycle of phytoseiid mites from egg to egg usually takes 8–10 days at 22°C–24°C (Helle and Sabelis 1985b) but varies with subfamily and with genera within subfamilies (Zhang 1995d). Some species, for example, *Phytoseiulus persimilis* Athias-Henriot, have generation times as short as five days at summer temperatures. Given enough food, females can lay three to five eggs per day for several weeks. Except for a few known thelytokous species, phytoseiids are parahaploid (pseudoarrhenotokous) (Hoy 1985).

Phytoseiid subfamilial categories (Phytoseiinae, Typhlodrominae, Amblyseiinae) and some genera (e.g., *Phytoseius* and *Phytoseiulus*) have remained relatively stable over time. However, generic concepts have varied widely over the years, with as few as eight genera recognized by some authors, and as many as four dozen recognized by others. This has resulted in considerable confusion in the literature. For example, the western predatory mite, *Galendromus occidentalis* (Nesbitt), is often listed as a species of *Metaseiulus* or *Typhlodromus*. Similar confusion exists for other economically useful species that have been moved from *Amblyseius* to *Euseius* or to *Neoseiulus* or even to *Typhlodromus*, or from the latter to *Neoseiulella*. Kostianinen and Hoy's (1996) bibliography of the PHYTOSEIIDAE is an excellent reference for sorting out the confusion in generic identities, as are the catalogue by de Moraes, McMurty, and Denmark (1986) and the series of subfamily-level revisions by Chant and McMurtry (1994, 2003, 2004, 2005a, b, 2006).

The OTOPHEIDOMENIDAE (Fig. 12.53) comprises a distinctive group of phytoseioid mites, all active instars of which are parasites of insects in temperate and tropical regions throughout the world. The family may be divided into three subfamilies based on host preferences and morphological attributes (Fain and Lukoschus 1983b; Zhang 1995c). Members of the Otopheidomeninae, with the genera *Otopheidomenis*, *Noctuiseius*, and *Prasadiseius*, are found in the thoracoabdominal and wing axillary areas of noctuiid and sphingid moths (Prasad 1968; Wainstein 1972; Treat 1975). Eggs of *Otopheidomenis* and *Noctuiseius* generally are oviposited bilaterally in the tympanic recesses of moth hosts, but feeding appears to occur in the tegular region and thus does not impair the auditory function of the tympani. Aside from a single observation of a live female of *O. zalelestes* Treat, all records of otopheidomenid-moth associations are based on preserved material (Treat 1975). Members of the Treatiinae, with the genera *Hemipteroseius*, *Treatia*, *Entomoseius*, and *Nabiseius*, are found on the dorsum of the thorax and abdomen under the hemielytra of pyrrhocorid, lygaeid, and nabid bugs (Krantz and Khot 1962; Evans 1963c; Treat 1965; Chant 1965; Chant and Lindquist 1965; Costa 1968). Heavy infestations of *Hemipteroseius indicus* (Krantz and Khot) may paralyze and kill red cotton bug hosts, *Dysdercus koenigii* (Banerjee and Datta 1980; Shahi and Krishna 1981). The Katydiseiinae includes the monobasic genera *Katydiseius* and *Eickwortius*. All stases of *Katydiseius* were collected in the "thoracic tracheae" (perhaps the large prothoracic spiracular atria and their tracheal trunks) of a Malaysian katydid (Fain and Lukoschus 1983b), and adult females of *Eickwortius* were found on African termites (Zhang 1995c). According to the earlier observations of Wainstein (1972) and Fain and Lukoschus (1983b), the family concept of Zhang (1995c) as presented here may be diphyletic because members of the subfamilies Otopheidomeninae and Katydiseiinae do not appear to have a phytoseiid-type spermathecal system.

The free-living members of the family PODOCINIDAE (Fig. 12.54) inhabit humus and litter habitats throughout the tropics and into the mild temperate zones of Eurasia, Africa, and the Americas (Evans and Hyatt 1957; Bhattacharyya 1994). *Podocinum pacificum* Berlese, however, is a widely distributed species of higher latitudes, for example, in North America (De Leon 1964b; Lindquist and Wu 1987), Eurasia (Blaszak and Alberti 1985; Gilyarov and Bregetova 1977), Japan (Ishikawa 1970), and Australia (Halliday 1990a). In addition to litter habitats, *P. pacificum* has been recorded from rodent nests, ant nests, and grain field debris. It preys readily on collembolans in laboratory cultures and reproduces by thelytokous parthenogenesis (Wong 1967). *P. sagax* (Berlese), a widely distributed tropical species, has been collected in the Boboli Gardens in Florence, Italy, where it may have been accidentally introduced with ornamental plants (Evans and Hyatt 1957, recollected in Florence by G. W. Krantz in 1963). In culture, *P. sagax* is a voracious predator of collembolans and, to some extent, other mites. Males are known, and mating is required before females can reproduce (Wong 1967). However, Wong's observation that some virgin females of *P. sagax* became gravid but subsequently could not oviposit eggs without being mated remains unexplained. Placement of the enigmatic genus *Andregamasus* in the PODOCINIDAE by Karg (1986a, 1993b) is considered questionable in the absence of information confirming that these unusual associates of terrestrial hermit crabs (Costa 1965) actually have a phytoseiid-type spermathecal system.

The BLATTISOCIIDAE (Fig. 12.55) is a diverse group that has adapted to a broad spectrum of terrestrial, arboreal, and subaquatic habits. As treated here, the family includes two subfamilies, the Platyseiinae and the Blattisociinae. Some platyseiine species, particularly of the genera *Platyseius* and *Cheiroseius*, inhabit various subaquatic habitats, including sod in and around marshes, mosses in springs and streams, spray zones of waterfalls, sewage filter beds, and, in a few cases, salt marshes or seashore habitats (Lindquist and Evans 1965; Evans and Till 1979; Karg 1981, 1993b). Platyseiines typically have several enlarged and acutely pointed structures on the apices of legs II–IV, including the pretarsal paradactyli and a median projection of the pulvillus (Fig. 12.55G); these may be adaptations for movement in moist substrates (Lindquist 2003). The peritremes of adult *Platyseius* and some *Cheiroseius* species are large in surface area, functioning as plastrons when submerged (Hinton 1971). Some of those species have potential usefulness as biological indicators of ecosystem soil or water conditions in both temperate and tropical regions (Karg and Freier 1995; Karg 1998b). Other species inhabit drier habitats, such as grassland soils and forest litter, or more temporary or patchy substrates such as decaying vegetation, dung, and bracket fungi (Evans and Hyatt 1960; Karg 1981, 1993b). Adults of the one known species of *Cheiroseiulus* were collected from water pools on epiphytes in Jamaica (Evans and Baker 1991). The chelicerae of most species of Platyseiinae are similar in size and dentition (Evans and Till 1979), suggesting generally similar feeding habits at the generic level (Karg 1982) on vermiform prey such as nematodes and dipteran larvae (Karg 1993b; Halliday, Walter, and Lindquist 1998). Adult females of *Cheiroseius* are phoretic on nematoceran flies, including Tipulidae and Culicidae (Whitsel and Schoeppner 1973; Walter and Lindquist 1995). Although most species of Platyseiinae appear to be bisexual and require mating for reproduction, a few species of *Cheiroseius* are thelytokous (Walter and Lindquist 1995).

Members of the subfamily Blattisociinae frequent a diverse range of terrestrial and aboveground habitats similar to that noted for the AMEROSEIIDAE and the MELICHARIDAE (Evans and Till 1979; Bregetova 1977b; Karg 1993b; Halliday, Walter, and Lindquist 1998). Their feeding habits include predation on nematodes and microarthropods, fungivory, pollenophagy, and parasitism of insects. Some widespread species of *Blattisocius* are common predators of acarid mites and insect eggs and larvae in stored food and the nests of small mammals and birds (Evans 1958a; Chant 1963; Hughes 1976; Haines 1978; Karg 1993b). Although these species may occasionally annoy people handling arthropod-infested stored food products (Domrow 1979a) and are sometimes regarded as uninvited pests in laboratory cultures of arthropods (Halliday, Walter, and Lindquist 1998), *B. tarsalis* (Berlese) is considered to have beneficial potential in the control of stored grain pests in closed systems (Haines 1981). Adult blattisociine females attach phoretically to moths and may feed subsequently on the eggs and larvae of these moths, among other insects (Treat 1975). *B. patagiorum* Treat has slender edentate movable chelae and appears to be parasitic on adult noctuid moths. Adults of both sexes, along with nymphs, may occur on the moths and can reproduce without other sources of food (Treat 1966, 1975). Two species of *Blattisocius* have been described from adult females on honeybees or their brood, but the nature of these associations is unknown (Chinniah and Mohanasundaram 1995; Basha and Yousef 2000). Several other species of *Blattisocius* are found with ripe, dried, and rotting vegetables and fruit in soil and amid foliage on orchard trees, where they may prey on acarid mites (Chant 1963; Nasr, Nawar, and Afifi 1988; Karg 1993b).

Over 150 species of *Lasioseius*, the most speciose genus of BLATTISOCIIDAE, have been described (Christian and Karg 2006), but the great majority remain to be characterized (E. E. Lindquist, pers. obs.). *Lasioseius* is also the most disparate blattisociid genus in terms of cheliceral and other morphological structures (Walter and Lindquist 1997), feeding habits, and habitat preferences. Some species frequent soil habitats, including grasslands, forest litter, moldering straw, compost, and the terrestrial nests of small mammals and birds (Chant 1963; Bregetova 1977b; Karg 1980, 1993b). Some of these species have been observed to prey voraciously on nematodes and microarthropods, and they also feed readily on fungi (Walter and Lindquist 1989). A few species prefer wet substrates, and others are found associated with beetles under tree bark or on bracket fungi. *L. safroi* (Ewing) (= *L. dendroctoni* Chant) was observed to prey on nematodes under the bark of pine infested with *Ips* beetles but ignored opportunities to prey on mites encountered there (Lindquist 1964). Several species of *Lasioseius*, including *L. allii* Chant (= *L. bispinosus* Evans and *L. martini* Tjying) may prey on rhizoglyphine acarid mites that frequent belowground soil habitats in association with fleshy plant structures such as bulbs, corms, and tubers (Poe 1971; Tjying 1971; Lesna et al. 1995). Three species of *Lasioseius* known only from the adult stase are unusual in having body shapes that conform to the spiracular depressions of the millipedes on which they are phoretic (Evans and Sheals 1959). Certain species of *Lasioseius* are among the most numerous predatory mites on foliage in the tropical rain forests of Australia (Walter and Lindquist 1997). Still others occur in flowers, with their adult females phoretic on flower-visiting moths, beetles, and birds. Moth-associated *Lasioseius* species use the well-protected interpalpal region for shelter during phoretic transfer (Treat 1975). Adult females of a few species of *Lasioseius* are phoretic on hummingbirds and coinhabit hummingbird-pollinated inflorescences with species of the melicharid genera *Rhinoseius* and *Proctolaelaps* (Naeem, Dobkin, and OConnor 1985; Ohmer, Fain, and Schuchmann 1991). Their chelicerae are typical of predatory blattisociids rather than the nectarivorous melicharids mentioned above.

Among other blattisociine genera, all instars of the monobasic *Adhaerenseius* are found in the flower capsules of *Poellnitzia rubriflora* in South Africa, where they are thought to feed on pollen (Loots and Theron 1992). Members of the genus *Hoploseius* are strictly associated with bracket fungi, and their chelicerae are modified for rasping, shredding, and feeding on the cellular content of live fungal tissue (Lindquist 1963; Walter 1998b; Mašán and Walter 2004). Adult females of some species are phoretic on drosophilid flies (Lindquist 1963). All active instars of *H. tenuis* Lindquist in North America, along with an as yet undescribed species of the same genus in Russia, are unusual in having elongated bodies adapted to living inside the pore tubes of bracket fungi (Lindquist 1965; Makarova 2004). Nothing is known about the behavior of the two described species of the genus *Orthadenella*, *O. lawrencei* (Evans) and *O. tennesseensis* (De Leon), which frequent decaying wood and moss on tree trunks in forested regions of Europe and North America (Evans 1958a; De Leon 1963; Athias-Henriot 1973). Adult females of the genus *Aceodromus* have been found in the large prothoracic spiracular atria or at the bases of the wings of orthopterans (Lindquist and Chant 1964; Lindquist and Evans 1965), although a parasitic relationship has not been demonstrated. Adults and immatures of other *Aceodromus* have been found living freely in tropical grassy areas (E. E. Lindquist, unpublished obs.) or infesting phytoseiid colonies in Australian glasshouses (D. E. Walter, unpublished obs.).

Many species of Blattisociinae appear to be bisexual, with mating obligatory for reproduction, but a few species of *Lasioseius* and some populations of *Zercoseius spathuliger* (Leonardi) have been demonstrated to be thelytokous (Walter and Lindquist 1989, 1995; Athias-Henriot 1967). Males generally appear to be haploid, and pseudoarrhenotoky (parahaploidy) has been suspected but not conclusively demonstrated for the family as a whole (Norton et al. 1993). However, the one known species of the Australian genus *Arrhenoseius* has been determined to be arrhenotokous rather than pseudoarrhenotokous (Walter and Lindquist 2001). *A. gloriosus* Walter and Lindquist occurs in both subtropical forest floor litter and litter suspended in rain-forest canopies, and all active instars prey voraciously on nematodes in culture.

▪ Superfamily Dermanyssoidea (Figs. 12.56–12.66)

DIAGNOSIS: Adult dorsal shield entire, divided, or variously fragmented, sometimes greatly reduced or absent; sternal shield of female normally developed, reduced, or absent, sternal setae or setal remnants generally present; metasternal setae present or absent. Chaetotaxy of legs highly variable except in free-living forms; femur I commonly with 13 setae, including 4 ventral setae and 5 dorsal setae; tibia I typically with 6 dorsal setae (4 or 5 in certain obligate parasites), genu IV almost always with 2 anterolateral setae and with 0–1 posteroventral seta; palptarsal apotele 2-(rarely 1- or 3-) tined. Male genital aperture anterior, spermatodactyl arises externally on the movable digit, fused to digit basally and free distally. Female podospermous, with a "type A" spermathecal complex. Larva without pygidial shield; protonymphal pygidial shield either broad, narrowed, or represented by mesonotal scutella; entire pygidial shield with 1–8 pairs of setae (J3, S4, and S5 inserted on the shield or on the soft adjacent integument where the shield is reduced); dorsal shield of deutonymph entire, with lateral incisions or with a separate small pygidial element posteriorly.

The Dermanyssoidea is a large and diverse assemblage of free-living and parasitic taxa that, despite their morphological and behavioral heterogeneity, appear to be derivative groups of a basic free-living stock exemplified by the subfamily Hypoaspidinae (LAELAPIDAE) (Radovsky 1969). Many classification schemes have been proposed in past years for this complex group, including those of Berlese (1892–1903, 1913), Vitzthum (1940–43), Baker and Wharton (1952), Evans (1957a), and Karg (1965, 1993b). A major reorganization of the dermanyssoids was put forward by Evans (1957a) and Evans and Till (1966), who revised and significantly broadened Vitzthum's earlier concept of the LAELAPIDAE to include 16 subfamilies comprising free-living forms, paraphages, and facultative and obligate vertebrate and invertebrate ecto- and endoparasites. Based on rules of priority, Evans and Till (1966) applied the family name DERMANYSSIDAE to this assemblage, relegating the LAELAPIDAE *s. str.* to subfamily status. In the present treatment, the DERMANYSSIDAE is considered in the narrower sense of Evans and Till's Dermanyssinae, and the LAELAPIDAE is retained essentially as outlined by Radovsky (1994). Recognition of the Macronyssinae and the Dermanyssinae as valid familial entities (Radovsky 1967), and the subsequent elevation to family level of several other dermanyssid subfamilies (Johnston 1982; Evans 1992), made necessary the recognition of a superfamilial category to accommodate these taxa. Rules of priority were invoked in using the name Dermanyssoidea (rather than Laelapoidea *sensu* Evans 1957a) for this grouping. Due to limitations of space, the following paragraphs contain only a sampling of this species-rich and remarkably diverse group.

The LAELAPIDAE (Figs. 12.56, 12.57, 12.58A, B, 12.66A, B) comprises a multitude of morphologically and behaviorally diverse dermanyssoid mites that are free living or associated with arthropods, mammals, or birds. The family is here considered to comprise nine subfamilies, based in part on a phylogenetic study of free-living and arthropod-associated laelapid taxa by Casanueva (1993), and includes the recently characterized subfamily Acanthochelinae (Radovsky and Gettinger 1999). Among these subfamilies are predators living in the soil-litter column or in the nests of vertebrates or invertebrates, paraphages of arthropods, and facultative or obligatory parasites of mammals.

Members of the subfamily Hypoaspidinae (Hypoaspididae of Karg 1993b) often are collected in litter or soil substrates (the genera *Pseudoparasitus*, *Ololaelaps* (Figs. 12.56D, E), and those often included under *Hypoaspis*, such as *Gaeolaelaps* (Figs. 12.56A, B), *Cosmolaelaps,* and *Stratiolaelaps*, are examples), while species of these and other genera are routinely or occasionally encountered in the nests of mammals or arthropods or directly associated with insects (Evans and Till 1966, 1979; Hunter and Rosario 1988; Karg 1978b, 1993b; Krantz and Platnick 1995). Some species of *Gaeolaelaps* and *Stratiolaelaps* show potential for use as inoculative biological control agents against fungus gnats and flower thrips in greenhouses (Gillespie and Quiring 1990; Chambers, Wright, and Lind 1993; Enkegaard, Brodsgaard, and Sardar 1995; Walter and Campbell 2003; Zhang 2003; Beaulieu and Weeks 2007). *Coleolaelaps* and some *Hypoaspis* species are found beneath the elytra of beetles; *Dicrocheles* species are parasitic in the tympanic air sacs of noctuid moths, and *Blaberolaelaps* and *Gromphadorholaelaps* are paraphagic on cockroaches (Till 1969; Costa and Hunter 1970; Treat 1975; 1988; Costa 1980). Yoder (1996, 1997) established that *Gromphadorholaelaps schaeferi* Till, an associate of the Madagascar hissing cockroach, feeds on the saliva of its host. *G. schaeferi* was also observed to cannibalize its own young and attack smaller parasitic mite associates of another roach species in the laboratory. Based on their cheliceral morphology, members of the Australasian genus *Raymentia* are thought to feed on pollenkitt in the nests of their halictid bee hosts (Walter et al. 2002). *Cosmolaelaps ungeri* Karg and *C. brevipedstra* Karg were taken from termite nests in Cuba (Karg 1985c), although most members of this cosmopolitan genus have been collected from soil and litter.

The subfamily Melittiphidinae *sensu* Casanueva (1993) includes genera that have established close associations with ants and bees, and they often appear in considerable numbers in host nests and hives. Examples may be found in the genus *Pneumolaelaps*, which is restricted to the nests and bodies of bumblebees, *Melittiphis alvearius* (Berlese) (Fig. 12.58A), which is phoretic on honeybees and feeds on pollen in their nests, and in the genera *Stigmatolaelaps* and *Xylocolaelaps*, whose members occur only in association with carpenter bees (Costa 1966a; Hunter and Husband 1973; Krantz 1998a; Royce and Krantz 2003). The latter two genera are unique for the subfamily in retaining seta *ad2* on tibia II and are unusual in having seven setae on femora III–IV, a character that they share with some other bee-associated melittiphidine genera (see below). Members of the Neotropical genera *Bisternalis* and *Melittiphisoides* are denizens of stingless bee nests, although *B. hunteri* Baker, Delfinado-Baker, and Ordaz inexplicably was recovered from a wasp nest in Panama (Baker, Flechtmann, and Delfindo-Baker 1984; Delfinado-Baker, Baker, and Flechtmann 1984). Species of *Gymnolaelaps*, *Laelaspis*, and *Myrmozercon* (= *Myrmonyssus*) (Fig. 12.58B) are nest myrmecophiles in both the Old and New World (Hunter 1963, 1964c; Hunter and Hunter 1963; Evans and Till 1966; Walter 2003).

Adults of the melittiphidine tribe Varroini, considered in the present treatment as the dermanyssoid family VARROIDAE, were characterized by Casanueva (1993) as lacking hypostomatic setae 3 on the gnathosoma. However, both the genus *Dinogamasus*, whose members are paraphagic on carpenter bees, and the monotypic trigonid bee associate *Stevelus amiculus* Hunter, which was taken from the nest of a stingless bee, have a full complement of hypostomatic setae (Hunter 1963; Lundqvist 1998). In that connection, omission of femoral chaetotactic characters in Casanueva's analysis may have forfeited an important data source for defining the Varroini and identifying its member taxa. For example, the honeybee-associated varroinine genus *Tropilaelaps* has a normal gamasine complement of six setae on femora III–IV, but *Dinogamasus* and members of the genus *Urozercon*

(= *Uronyssus* Halliday 2006), bizarre termite nest associates, have seven setae on both femora III–IV. *Varroa* (Fig. 12.58C) and *Euvarroa* species (Figs. 12.58D, E) may have seven to nine setae on femur III and six to seven on femur IV (Akimov et al. 1988; Guzman and Delfinado-Baker 1996; Anderson and Trueman 2000). Varroids invade honeybee colonies (*Apis* spp.), where they feed on and parasitize both the immature and adult bees, often causing death of affected bees through both feeding injury and transmission of viral pathogens. Decline and eventual death of infested colonies occur in apiaries where control practices are inadequate (DeJong, Morse, and Eickwort 1982; Fernandez and Coineau 2002, 2007; Fernandez, Coineau, and Theron 2006). *Tropilaelaps clarae* Delfinado and Baker is a major pest of honeybees in southeast Asia and India and is both phoretic and parasitic on adult apid bees. Adult female *T. clarae* feeding on larvae of *Apis mellifera* in the laboratory were observed to create a fold in the soft integument of the host by means of palpal manipulation, after which the relatively short chelicerae were pressed into the tissue fold. The terminal hook of the movable cheliceral digit appeared to provide an anchor for the acuminate tip and contiguous subterminal tooth of the fixed digit to produce a feeding wound in the integument (G. W. Krantz, pers. obs.).

Morphologically, the four member genera of the fundamentally Holarctic subfamily Haemogamasinae (*Haemogamasus*, *Eulaelaps*, *Brevisterna*, and *Ischyropoda*) have diverged little from the basic hypoaspidine pattern and often share the nidicolous habits of hypoaspidine and melittiphidine laelapids. Some have developed obligatory parasitic relationships with small mammals and (occasionally) with birds, but many have retained a predilection for predation and saprophagy, along with hematophagy (Evans and Till 1966; Radovsky 1967, 1985). *Haemogamasus pontiger* (Berlese) appears to be entirely free living and has only loose connections to nest habitats, but other members of the genus are obligatory nest dwellers and either are exclusively hematophagous (*H. liponyssoides* Ewing is an example) or polyphagous, accepting a range of food substances in addition to blood. *H. reidi* Ewing and other members of the *reidi* species group can complete their development on an exclusive diet of blood but will also feed readily on flea feces or arthropods. *Eulaelaps stabularis* (Koch) and *E. cricetuli* Vitzthum are mostly predaceous and saprophagous, while *Brevisterna utahensis* (Ewing) is primarily hematophagous (Radovsky 1985). Haemogamasines share a number of morphological similarities with both the nest-inhabiting Australasian subfamily Mesolaelapinae, a monogeneric assemblage of marsupial and rodent associates, and the Neotropical Acanthochelinae, which currently comprises two monotypic genera: *Acanthochela chilensis* Ewing from marsupials and *Lukoschus maresi* Radovsky and Gettinger from rodents (Tenorio and Radovsky 1974; Radovsky and Gettinger 1999).

The unusually long second segment of the male chelicera is a major distinguishing characteristic of *Alphalaelaps aplodontiae* (Jellison), the sole representative of the subfamily Alphalaelapinae (Tipton 1960; Radovsky 1985). *A. aplodontiae* is closely associated with the mountain beaver, *Aplodontia rufa*, the last known surviving species of a once extensive lineage of sciuromorph rodents that arose during the Paleocene (OConnor 1982b). *A. aplodontiae* is considered a relict form that has coevolved with its reclusive host and probably is parasitic.

The Laelapinae includes a large number of genera that are primarily facultative nest parasites of rodents, although some have associations with bats or primates (Radovsky 1967, 1969, 1985). A morphological departure in some laelapines from the generalized short chelate-dentate chelicerae to a more slender, elongate piercing type (as occurs also in the subfamily Hirstionyssinae) is indicative of an adaptation for penetrating the skin and regular feeding on blood and/or lymph (Radovsky 1985). Among these species are *Hymenolaelaps princeps* Furman, a parasite of murine rodents (Furman 1972), and the highly specialized bat-associated genus *Neolaelaps*. Members of the obligatorily nidicolous genus *Androlaelaps* (Figs. 12.57F, G) (placed in the Hypoaspidinae by Casanueva 1993) are able to feed with varying degrees of facility on the blood of their rodent or bird hosts (Reytblat 1965) and may penetrate intact skin of nestlings. *Laelaps* species (Figs. 12.56C, 12.57A) may feed on lymphatic secretions of their mammal hosts, in addition to imbibing blood. Feeding on lymph is a trait shared by *Myonyssus*, the sole representative genus of the laelapine subfamily Myonyssinae. Many laelapine genera favor hosts that share particular habitats. For example, a number of *Laelaps* species appear to prefer myomorph rodents and, moreover, myomorphs that occur in moist habitats. Species of *Mysolaelaps*, a Neotropical genus, show a similar preference, while members of the genus *Steptolaelaps* occur on rodent hosts that favor arid regions (Strandtmann and Wharton 1958; Furman 1972). Further information on feeding strategies in the Laelapinae may be found in chapter 6.

Based on phylogenetic evidence, Casanueva (1993) elevated the laelapid subfamily Iphiopsidinae *sensu* Evans (1955c) to family level. The IPHIOPSIDIDAE includes three subfamilies and 10 genera of mites that are associated with millipedes, centipedes, spiders, and terrestrial crustaceans. Iphiopsidids display a variety of striking morphological reductions in sclerotization, chaetotaxy, peritremes, cheliceral size and dentition, and ambulacral claws (Evans 1955c; Womersley 1956; Ryke 1959; Domrow 1975a; Maes 1983; Ishikawa 1986, 1988). The loss of seta *pl2* on genu IV in iphiopsidids phylogenetically defines the family as an entity separate from the LAELAPIDAE (Casanueva 1993), but its laelapid roots may clearly be seen in the genus *Julolaelaps* (Fig. 12.59C), an assemblage of iphiopsidine millipede associates that had long been considered a subgenus of the broadly defined laelapid genus *Hypoaspis* (Ryke 1959; Ishikawa 1986). Little is known regarding the biology of iphiopsidids, although the regressive nature of the characters that define it suggests a paraphagic or parasitic mode of life on their terrestrial arthropod hosts. *Trichoaspis julus* Gu, Wang, and Li, an incompletely described millipede associate from China, appears to be an iphiopsidid, although the describers have proposed a new family, the Trichoaspididae, to accommodate it (Gu, Wang, and Li 1991). Pending additional information, the proposed family name is not formally recognized in this treatment.

The remarkable grublike species of the monogeneric family LARVAMIMIDAE (Figs. 12.59A, B) are found in the nests and bivouacs of Neotropical army ants (Elzinga 1993). Both males and females are moderately to greatly elongated (675–2,500 µm), unsclerotized, hypertrichous forms that closely resemble the ant larvae that they mimic. The four species of *Larvamima*—*marianae* Elzinga, *schneirlai* Elz., *carli* Elz., and *cristata* Elz.—display considerable reductions in or loss of morphological structures; that is, both the ambulacral claws and tritosternum are absent, the legs are shortened, and the corniculi, peritremes, and cheliceral digits are reduced. Leg setation suggests the retention of immature chaetotaxy (immatures are unknown), and the adult idiosoma is neotenic. Males possess a cheliceral spermatodactyl that is more or less typical of many other dermanyssoid mites, although the fixed digit is greatly reduced. Mimicry of ant larvae by adult larvamimids extends to the presence in females of idiosomatic pseudosegments that resemble the true segmentation of the ant larva. Rettenmeyer (1961) has observed worker army ants carrying adult mites in the same way that ant larvae are transported during the daily peregrinations of the colony. The mites apparently are not distinguished from larvae by the ants because of their overall morphological resemblance to the ant larva and, in all likelihood, because they emit a chemical signal that mirrors the pheromonal bouquet of their ant hosts.

The remaining 11 dermanyssoid families comprise vertebrate parasites, and many are specific to particular animal groups. For example, the families HALARACHNIDAE (Fig. 12.60), DASYPONYSSIDAE (Figs. 12.61B–E), and MANITHERIONYSSIDAE (Fig. 12.61A) are restricted to mammals (Newell 1947a; Radovsky 1969; Radovsky and Yunker 1971; Yunker 1973; Furman 1979; Domrow 1980); the SPINTURNICIDAE (Figs. 12.62A–G, 12.66E) to bats (Rudnick 1960; Furman 1966; Domrow 1972; Radovsky 1994); the RHINONYSSIDAE (Fig. 12.64) to birds (Domrow 1969; Pence 1975); and the IXODORHYNCHIDAE (Figs. 12.59D–F), ENTONYSSIDAE (Figs. 12.61F–H), and OMENTOLAELAPIDAE (Figs. 12.65A, B) to reptiles (Fain 1961a, b, 1962a). The genus *Dermanyssus*, one of two currently recognized genera of the family DERMANYSSIDAE (Figs. 12.62H–L), is exclusively parasitic on birds, but some species of *Liponyssoides* parasitize rodents instead (Moss 1968, 1978; Domrow 1979b; Radovsky 1985). Many members of the MACRONYSSIDAE (Figs. 12.63D–G, 12.66C, D) are bat parasites (Radovsky 1967), but several genera occur on birds (some *Ornithonyssus*, *Pellonyssus*), rodents (*Lepronyssoides*), and reptiles (*Ophionyssus*, *Draconyssus*) (Radovsky 1985; Fain and Bannert 2000). Macronyssids are generally considered to be exclusively parasitic, but *Mitonyssoides stercoralis* Yunker, Lukoschus, and Giesen, an associate of molossid and vespertilionid bats, was found to be predaceous on the rosensteiniid mites that feed on bat droppings (Radovsky, Krantz, and Whitaker 1997). Finally, *Hystrichonyssus turneri* Keegan, Yunker, and Baker (HYSTRICHONYSSIDAE) (Figs. 12.63A–C), a highly derived dermanyssoid species with questionable affinities, was taken from porcupines and (possibly) from a snake in Malaysia (Keegan, Yunker, and Baker 1960).

Dermanyssoid mites can cause primary injury to host tissues by piercing the skin, as occurs with *Dermanyssus* (DERMANYSSIDAE), *Haemogamasus* (LAELAPIDAE), *Raillietia* (HALARACHNIDAE, Raillietinae), and *Ornithonyssus* (MACRONYSSIDAE), and by producing allergic reactions in sensitive host animals. Attacks by species of *Ornithonyssus* and *Ophionyssus* on their bird or reptile hosts may involve mass feeding that can result in anemia or death of the host through exsanguination (Yunker 1973; Southcott 1976; Møller 1990). Lung congestion, rhinitis, and sinusitis are common in birds and mammals invaded by respiratory parasites of the families RHINONYSSIDAE, ENTONYSSIDAE, and HALARACHNIDAE (Baker et al. 1956; Yunker 1973; Domrow 1987). *Pneumonyssus simicola* Banks, the halarachnid lung mite of monkeys, causes few overt symptoms, but the lung parenchyma of autopsied infested animals have characteristic spotty lesions similar to those caused by tuberculosis. *Pneumonyssoides caninum* (Chandler and Ruhe) infests the sinuses and nasal passages of dogs, where symptoms generally are limited to excessive mucus production and rhinitis (Baker et al. 1956; Yunker 1973). Members of the families LAELAPIDAE, DERMANYSSIDAE, and MACRONYSSIDAE are proven vectors of disease organisms in birds, reptiles, and mammals—including humans (see also chapter 6).

Mention should be made here of the remarkable, rarely collected mites of the Neotropical family SPELAEORHYNCHIDAE (Figs. 12.65C–E), a taxon of bat parasites long considered to be a member group of the Dermanyssoidea. Spelaeorhynchids were earlier classified as a subfamily of ticks (Neumann 1911), based largely on their great size (up to 1,800 µm) and their general appearance. Their mesostigmatic origins were recognized by Banks (1917), and most subsequent authors are in agreement (Vitzthum 1940–43; Baker and Wharton 1952; Strandtmann and Wharton 1958; Fain et al. 1967). Radovsky (1985) reviewed the history of the family, noting that its modern placement in the Dermanyssoidea is unjustified. Female spelaeorhynchids display a suite of unique apomorphic characters including short and stout, heavily toothed chelicerae; a ventrally oriented camerostome; a genital slit that is unprotected and located just anterior to the anus; and a bizarre life cycle that includes the abscission of ambulacra in older females. The male is unknown.

Radovsky concluded that spelaeorhynchids may have closer ties to other mesostigmatic groups than they do to the Dermanyssoidea and excluded them from his 1985 review. Its unique morphological characters suggest that the SPELAEORHYNCHIDAE merits separate superfamilial status, but its proper placement will likely hinge on a critical examination of male cheliceral and sperm morphology.

Key 12.1. Order Mesostigmata, Key to Families (Figs. 12.1–12.66)

Note: References to the female or the male indicate the adult female or adult male.

1a. Oviporus covered by a single large shield usually bearing 6 to many setae (rarely 2 or 4 setae in some *Sejus*) or by a complex of 2–3 genital shields or their remnants; tarsus of leg IV of deutonymphs and adults with a minimum of 20 setae, setae *av4* and *pv4* present (absent in the paedomorphic, millipede-associated NEOTENGYNIIDAE, minute in some groups, e.g., PROMEGISTIDAE), usually on a ventral intercalary sclerite between the basitarsus and telotarsus **2**

1b. Oviporus covered by a single shield, with 0–1 pair of setae (sometimes expanded and with 4–5 pairs), or shield absent; tarsus of leg IV of deutonymphs and adults with a maximum of 18 setae but lacking setae *av4* and *pv4* and ventral intercalary sclerite. A highly diverse, primarily free-living assemblage that includes predators, fungivores, pollenophages, symbionts, and a broad spectrum of parasites associated with both invertebrate and vertebrate hosts **Suborder Monogynaspida—32**

2a. Oviporus covered by a single large epigynal shield bearing 6 (rarely 2–4) to many setae and often notched or excavated anteriorly (Figs. 12.5D, 12.6B); chelicerae without excrescences; femur IV with 7 setae. Found in diverse soil/litter and forest habitats, deutonymphs phoretic on beetles **Suborder Sejida—3**

2b. Oviporus covered by 2–3 shields (2 latigynal and 1 mesogynal) that may be variously coalesced or reduced (Figs. 12.20D–J); median element (mesogynal shield) nude, subtriangular but often reduced or insensibly fused to other elements, latigynal shields each with 1 to many setae, free or fused posteriorly to ventral elements and/or medially to each other; movable cheliceral digit with medial or terminal dendritic, brushlike, or filamentous ventral excrescences (Figs. 12.15C, 12.16D); femur IV with 4–11 setae. A diverse assemblage of mostly large (up to 5,000 μm) predatory, fungivorous, paraphagic, phoretic, and saprophagous mites associated with arthropods, reptiles, and mammals **Suborder Trigynaspida—5**

3a. Corniculi large, bifurcate; chelicerae massive, with few coarse teeth and lacking pilus dentilis; idiosomatic shields with scalelike ornamentation; anal opening enlarged (Figs.12.7A, B) **ICHTHYOSTOMATOGASTERIDAE**

3b. Corniculi small, horn shaped; chelicerae moderate in size, denticulate, and with setiform pilus dentilis; idiosomatic shields tuberculate to coarsely reticulate; anal opening small **4**

4a. Apex of podonotum produced into a spiky knob; cheliceral digits with few teeth; female epigynal shield subrectangular, with numerous setae irregularly inserted over most of surface; opisthonotum of female with a pair of lateral shields and a pygidial shield that extends onto the mesonotal region and lacking hornlike processes posteriorly (Figs. 12.5A–D) **UROPODELLIDAE**

4b. Apex of podonotum without spiky knob; cheliceral digits serrate; female epigynal shield with 1–6 pairs of setae inserted laterally; opisthonotum of female without lateral shields but with 1 or more mesonotal shields and a pygidial shield that often has a pair of posterior, hornlike processes bearing setae (Figs. 12.6, 12.7C, D) **SEJIDAE**

5a. Adult female with separate podonotal, mesonotal, and pygidial shields (sometimes obscured by leathery, secondary sclerotization) **6**

5b. Adult female with holonotal (entire) shield or with separate podonotal and opisthonotal shields **7**

6a. Mesonotum with a single, large shield; pygidial shield small, on tail-like posterior process; metapodal shields and peritremes absent; female with well-developed and tapering vaginal sclerites. On Neotropical millipedes (Figs. 12.14B, C) **Superfamily Celaenopsoidea, NEOTENOGYNIIDAE**

6b. With a pair of mesonotal shields and a well-developed pygidial shield; free metapodal shields present, distinct in teneral or lightly sclerotized adults but often obscured by secondary leathery cuticle; peritremes present and well developed; female with a pair of straplike internal genital plates with numerous large pores. Free living (Figs. 12.9A–E, 12.11C, D) **Superfamily Cercomegistoidea, DAVACARIDAE**

7a. Often yellowish to light brown in color; weakly to strongly sclerotized and with a holodorsal shield or separate podonotal and opisthonotal shields; palpgenu with 6 setae; gnathotectum denticulate, usually not coming to point and without median keel; tritosternal laciniae usually fused for half or more of length, often rodlike; chelicerae toothed; large internal sclerites in genital region absent; tarsus I with or without claws. Free living or associated with arthropods **Cohort Cercomegistina *s. str.*—8**

7b. Often dark reddish brown in color; strongly sclerotized and with a holodorsal shield; palpgenu with 5–7 setae; gnathotectum usually smooth, with median spur and ventral keel; tritosternal laciniae usually free; chelicerae toothed or edentate; vaginal sclerites or sternovaginal processes often present; tarsus I without claws. Associated with arthropods or reptiles **Cohort Antennophorina *s. str.*—13**

8a. Adults with a holodorsal shield **10**

8b. Adults with separate podonotal and opisthonotal shields **9**

9a. Dorsal shields with relatively few setae; lateral idiosomal setae on marginal shields, not on platelets in soft cuticle; ventrianal shield fused posteriorly to opisthonotal and marginal shields; tarsus I with or without claws (Figs. 12.10D, E) **PYROSEJIDAE**

9b. Dorsal shields strongly hypertrichous; lateral idiosomal setae on platelets in soft cuticle; ventrianal shield free posteriorly; tarsus I without claws (Figs. 12.8A–D) **CERCOMEGISTIDAE**

10a. Female with separate latigynal and mesogynal shields (Fig. 12.20E) **12**

10b. Latigynal and mesogynal shields fused, contiguous, or coalesced (Figs. 12.20A, B) **11**

11a. Latigynal and mesogynal shields insensibly fused into intercoxal shield; ventrianal shield fused to holodorsal shield posteriorly; plicate ventral cuticle without setae; tarsus I with claws (Fig. 12.9F) **SEIODIDAE**

11b. Latigynal shields free medially but fused posterolaterally to mesogynal-opisthoventrianal shield, which is free from holodorsal shield posteriorly; plicate ventral cuticle with setae; tarsus I without claws. Associated with crabs **Genus *Vitzthumegistus*, CERCOMEGISTIDAE?**

12a. Anterior sternal area occupied by sclerotized sternal shield bearing *st1* and *st2*; anal opening in small shield bearing only paranal setae; legs IV hypertrophied, modified for jumping; peritrematic shields with longitudinal groove behind coxae IV (Figs. 12.10A–C, 12.11B) **SALTISEIIDAE**

12b. Anterior sternal region desclerotized, fragmented, *st1* in soft or lightly sclerotized cuticle; anal opening in large ventrianal shield; legs IV normal; peritrematic shields without longitudinal groove behind coxae IV (Figs. 12.11A, 12.8E–G) **ASTERNOSEIIDAE**

13a. Chelicerae edentate or with minute serrations (Fig. 12.13B) or baleenlike comb **14**

13b. Chelicerae with well-developed, sclerotized teeth (Fig. 12.19D) **22**

14a. Fixed digit of chelicera without excrescence or membranous process; sternal shield with or without internal sternovaginal processes; latigynal shields usually extensively overlapping mesogynal region. Associated with ants **18**

14b. Fixed digit with fringed excrescence or distally membranous-denticulate or with row of small teeth; sternal shield without sternovaginal sclerites; latigynal shields various but not overlapping mesogynal shield. Associated with reptiles, beetles, or myriapods **Superfamily Parantennuloidea—15**

15a. Fixed digit with dorsal excrescence; anal opening in small shield with 1–3 pairs of setae; mesogynal shield bearing lateral setae. Associated with carabid beetles, centipedes, or millipedes (Figs. 12.18A, B) **PARANTENNULIDAE**

15b. Fixed digit without dorsal excrescence, but terminating distally in denticulate, membranous process; anal opening in large ventral or ventrianal shield with numerous setae; mesogynal shield nude or fused to setose ventral elements. Associated with beetles, myriapods, or reptiles **16**

16a. Dorsal shield entire; anal opening in large ventral shield; mesogynal shield free or fused to ventral elements; setae *av4*, *pv4* reduced in size, often minute. Associated with beetles, myriapods, or reptiles **17**

16b. Dorsal shield divided; anal opening in straplike ventrianal shield; mesogynal shield absent; setae *av4, pv4* normally developed. Associated with tenebrionid beetles (Fig. 12.18G) **PHILODANIDAE**

17a. Sternal shield entire, bearing *st1–st3* and sternal pores 1–2; pregenital shield fusiform and without sternal pores 3 (pseudosternogynum); palpgenu with 6 setae. Associated with carabid or passalid beetles (Figs. 12.17E, 12.19A) **PROMEGISTIDAE**

17b. Sternal shield variously divided; pregenital shield divided into two subtriangular shields, sometimes narrowly joined medially and bearing sternal pores 3 (sternogynum); palpgenu with 7 setae. Associated with carabid beetles, millipedes, or reptiles (Figs. 12.18C–F) **PARAMEGISTIDAE**

18a. Sternal shield with a pair of sternovaginal sclerites; ventrianal shield without anterior process **Superfamily Aenictequoidea—19**

18b. Sternal shield without sternovaginal sclerites; ventrianal shield with anterior process (Figs. 12.13A, B) **Superfamily Antennophoroidea, ANTENNOPHORIDAE**

19a. Latigynal shields subrectangular, with parallel, approximate mesal margins; mesogynal shield obscured or absent (Figs. 12.12C, D) **MESSORACARIDAE**

19b. Latigynal shields subtriangular, margins diverging posteriorly; mesogynal shield well developed, subtriangular **20**

20a. Latigynal shields each with 2 pairs of setae; pseudoperitreme present lateral and posterior to true peritreme (Fig. 12.12A) **AENICTEQUIDAE**

20b. Latigynal shields each with 10 or more pairs of setae; pseudoperitreme absent **21**

21a. Setae *st1* and associated pores on separate anterior platelets (jugularia) (Figs. 12.12E, F) **PTOCHACARIDAE**

21b. Setae *st1–st3* (*4*) and pores on an entire sternal shield (Fig. 12.12B) **PHYSALOZERCONIDAE**

22a. Setae *st1* on separate shield from *st2–st3*; well-developed sternogynum (entire or divided, and sometimes setose) present; movable digit of chelicera without enlarged proximal tooth; movable digit of male without sclerotized process. Associated with passalid beetles (rarely on carabid beetles) **23**

22b. Setae *st1* on same shield as *st2–st3*; sternogynum absent (free metasternal plates sometimes present); movable digit of chelicera with enlarged proximal tooth; movable digit of male with sclerotized excrescence. Free living or associated with a variety of arthropods or reptiles **Superfamily Celaenopsoidea (*pars*)—26**

23a. Cheliceral digits short, with moplike mass of long, filamentous, and ribbonlike excrescences (Fig. 12.16D); uropodid-like mites often with pedofossae **Superfamily Fedrizzioidea—24**

23b. Cheliceral digits robust, with dendritic or brushlike excrescences; very large, long-legged mites without pedofossae **Superfamily Megisthanoidea—25**

24a. Latigynal and mesogynal shields well developed; male genital aperture ellipsoidal, between coxae III (Figs. 12.16C–F) **KLINKOWSTROEMIDAE**

24b. Latigynal and mesogynal shields reduced and hidden by ventral plate; male genital aperture subcircular, between coxae II III (Figs. 12.16A, B) **FEDRIZZIIDAE**

25a. Sternogynum narrow, divided; fused latigynal shields free from ventral shield. Neotropical (Fig. 12.17A) **HOPLOMEGISTIDAE**

25b. Sternogynum broad, divided, or entire and surrounded by straplike, fused sternal-latigynal-ventral shield. Neotropical, African, and Australasian (Figs. 12.17B–D, 12.19B–D) **MEGISTHANIDAE**

26a. Mesogynal shield usually small, triangular, overlapped marginally by latigynals, and free from or fused to ventral plate; latigynal shields well developed and freely hinged to ventral plate **27**

26b. Mesogynal shield fused to ventral plate, free from or fused to latigynal elements; latigynals insensibly fused to ventral plate or extended posteriorly to coxae IV and narrowly joined to ventral plate **29**

27a. Anal opening on small plate separate from ventral plate. Free living or associated with insects or myriapods **28**

27b. Anal opening on large ventral plate. Adults associated with arthropods or snakes (Figs. 12.15A–D) **DIPLOGYNIIDAE**

28a. Free ventromarginal plates present. Free living in soil litter, on bark, in bark beetle galleries, or in the nests of stingless bees; rarely on beetles (Figs. 12.13D–F) **TRIPLOGYNIIDAE**

28b. Free ventromarginal plates absent. Associated with millipedes (Fig. 12.15F) **COSTACARIDAE**

29a. Latigynal-mesogynal complex insensibly fused, with at most a small median notch present on anterior margin of ventral-genital plate; metasternal setae on sternal shield or on well-developed metasternal plates. Free living or associated with bark beetles **30**

29b. Latigynal and mesogynal elements free from each other mesally; metasternal setae on sternal shield or on narrow, straplike plate or plates. Associated with a variety of arthropods or snakes **31**

30a. Setae *st4* on large, well-developed metasternal plates; endopodal-peritrematic shields fused to ventral plate. Associated with bark and bark beetles (Fig. 12.13C) **CELAENOPSIDAE**

30b. Setae *st4* on sternal shield; endopodal-peritrematic shields fused but free from ventral plate. Free living (Fig. 12.14A) **MEGACELAENOPSIDAE**

31a. Median separations between mesogynal and latigynal elements not extending past level of coxae III. Associated with passalid beetles and millipedes (Figs. 12.14D, E) **EUZERCONIDAE**

31b. Latigynal elements elongate, separate from mesogynal element to or beyond level of coxae IV. Associated with beetles or snakes (Fig. 12.15E) **SCHIZOGYNIIDAE**

32a. Adult opisthogaster with a pair of distinctive, suckerlike adhesive organs (secondarily lost in some species); tarsus I of adults with acrotarsus; sternal shield of female often fragmented and reduced, represented by lateral remnants that are contiguous with endopodal plates; sternal pores 2–3 usually absent; male chelicerae with sperm transfer process arising on the fixed or the movable digit **Superfamily Heterozerconoidea—33**

32b. Adult opisthogaster without adhesive organs; tarsus I of adults usually without acrotarsus; sternal shield of female usually well developed or, if reduced, not as described above; sternal pores 2–3 usually present; male sperm transfer process, when present, always arising on the movable digit **34**

33a. Genua III–IV each with 10 setae, including 3 posterodorsals (2 2/1, 3/1 1); male with spermatostylus arising on fixed cheliceral digit. Associated with millipedes and snakes (Figs. 12.33A–C) **HETEROZERCONIDAE**

33b. Genua III–IV each with 8–9 setae, including two posterodorsals (2 2/1, 2/1 1 or 2 2/1, 2/1 0); male with spermatodactyl arising on movable cheliceral digit. Associated with scolopendrine centipedes (Figs. 12.33D, E) **DISCOZERCONIDAE**

34a. Female and male wormlike, hypertrichous, pseudosegmented, without idiosomatic sclerites; fixed cheliceral digit greatly reduced, male spermatodactyl entirely fused to reduced movable digit. Associated with army ants (Figs. 12.59A, B) **Subcohort Dermanyssiae, Superfamily Dermanyssoidea (*pars*), LARVAMIMIDAE**

34b. Females and males without above combination of characters **35**

35a. Female with sternal setae *st1–2* and *st3–4* on separate shields, or all sternal setae on separate shields; male with sternitigenital shield bearing 4 pairs of sternal setae and a mid-sternal genital opening, and with sternal setae *st5* on or near a separate plate between coxae IV, which is rarely fused posteriorly with expansive circumventral shield; idiosomatic dorsum of adults with podonotal, mesonotal, pygidial, and rarely marginal shields. Free living in litter, rotten wood **36**

35b. Female with arrangement of sternal setae and shields not as above; male with sternitigenital or holoventral shield usually bearing all 5 or more pairs of sternal setae (rarely without *st4*) and a presternal or mid-sternal genital opening, or, if sternal setae *st5* on or near a separate plate, then genital opening presternal; idiosomatic dorsum of adults with 1–2 shields or, if mesonotal and pygidial elements present, then male with genital opening presternal **38**

36a. Idiosomatic venter of adult with expansive circumventral shield incorporating ventrianal, peritrematic, and exopodal shields and bearing laterally on either side a dense row of 20–26 ventrolateral setae, 6–9 spoutlike glandular openings, and a sparser row of 12–13 setae above these glands. Idiosomatic dorsum of female covered by large anteromedian shield incorporating podonotal and mesonotal elements and flanked by a marginal plate on either side and 3 pygidial plates posteriorly. Dorsum of male with fused podonotal-marginal shield anteriorly, separate mesonotal and broad marginal shields medially, and 3 pygidial plates posteriorly; stigmata lacking peritremes; legs I with 13 setae on each of femur, genu, and tibia, including 5 dorsals on femur and 6 dorsals on genu and tibia (Figs. 12.3A, 12.21F–H, 12.22C, D) **Cohort Heatherellina, HEATHERELLIDAE**

36b. Idiosomatic venter of adult with ventrianal shield surrounded by soft cuticle separating it from peritrematic and expodal shields, without a dense row of ventrolateral setae and spoutlike glandular openings. Idiosomatic dorsum of female and male with podonotal, 1–2 mesonotal shields and 1 pygidial shield, lacking marginal shields; stigmata with peritremes; legs I with 10 setae, including 4 dorsals, on each of femur, genu, and tibia **Cohort Microgyniina—37**

37a. Podonotal shield with 16–22 pairs of setae; mesonotal region with 1–2 plates bearing setae; pygidial shield with setae; tarsus I with claws; palpi with 5 free segments, tibia and tarsus separate (Figs. 12.21A–C, 12.22B) **MICROGYNIIDAE**

37b. Podonotal shield with 9–10 pairs of setae; mesonotal region with 4 suboval platelets that lack setae; pygidial shield without setae; tarsus I without claws; palpi with 4 free segments, tibia and tarsus fused (Figs. 12.21D, E, 12.22A) **NOTHOGYNIIDAE**

38a. Epigynal shield of female flask or wedge shaped or subtriangular, often extending into the opisthogastric region and bearing setae *st5* (Fig. 12.20M) or flanked by those setae in the postcoxal region; epigynal shield reduced or obliterated in some endoparasitic groups; shield usually not surrounded by fused sternal-endopodal-ventral elements; adults with 1–2 dorsal shields, lacking marginal shields or their platelets; peritreme typically linear, sometimes reflexed anteriorly or posteriorly, rarely vestigial; tritosternal base cylindrical or flattened, never subrectangular or columnar; femur IV of deutonymph and adult typically with 6 setae **Cohort Gamasina—39**

38b. Epigynal shield of female oval, subtriangular, or tongue shaped, usually nude and partially or entirely enclosed by fused sternal-endopodal-ventral shielding bearing sternal setae 1–5 (*st5* rarely on separate shields that may be partially fused to epigynal shield) and usually confined to the podosomatic region, rarely displaced posteriorly or fused with ventral elements; adults with 1 to several dorsal shields, marginal shields or platelets generally present; peritreme often sinuous, sometimes on hornlike projections, rarely vestigial; tritosternal base often enlarged, subrectangular or columnar; femur IV of deutonymph and adult with 7–8 setae (if only 6 setae are present, then hypostomatic setae *h2* and *h3* longitudinally aligned **Cohort Uropodina—75**

39a. Genua III and IV each with 10 setae, including 2 posterolaterals on genu III (2 2/1, 2/1 2) and 1 posteroventral on genu IV (2 2/1, 3/1 1); female epigynal shield lacking expanded hyaline rim anteriorly; male lacking cheliceral sperm-transferring appendages and with mid-sternal genital opening; peritremes often short or vestigial on adults **40**

39b. Genua III and IV each commonly with 9 or fewer setae, including 1 posterolateral on genu III (2 2/1, 2/1 1) and no posteroventral on genu IV (2 2/1, 3/0 1) but with 10 setae, including 2 posterolaterals on III and a posteroventral on IV, in PARASITIDAE and VEIGAIIDAE; female epigynal shield with hyaline rim usually expanded anteriorly to often reach or overlap posterior edge of sternal shield; male chelicerae with a sperm transfer appendage on movable digit and with presternal genital opening; adult peritremes variously developed, often extending anteriorly to or beyond insertions of coxae II **43**

40a. Tritosternum with rudimentary laciniae; peritremes vestigial or absent in deutonymph and adult; adult with dorsal shield entire, distinctively tuberculate-reticulate; legs I elongated, their tarsi with several long minutely clubbed setae and lacking ambulacrum and paired claws; peripodomeric suture of tarsi II–IV with or without a dorsal intercalary sclerite bearing 2 setae. Free living, Northern Hemisphere (Figs. 12.34A–H) **Superfamily Epicrioidea, EPICRIIDAE**

40b. Tritosternum with well-developed laciniae; peritremes well developed in deutonymph, often reduced in length in adult; adult with dorsal shield entire or more commonly divided into well-developed podonotal and opisthonotal shields, the latter rarely absent in female; legs I not elongated, without long, minutely clubbed setae, and with or without ambulacrum and paired claws; peripodomeric suture of tarsi II–IV without intercalary sclerite **41**

41a. Adult dorsal shield abbreviated and leaving posterior one-fourth to one-half of idiosoma uncovered in female but covering entire dorsum in male; female epigynal shield without setae but flanked by 2 pairs of setae in soft integument; male ventrianal shield free caudally from posterior margin of dorsal shield; peritremes well developed in deutonymph and adult, extending to level of bases of legs I or II; palptarsal apotele 3-tined. Free living in high northern latitudes and altitudes (Figs. 12.36A–G, 12.37A, B) **Superfamily Arctacaroidea, ARCTACARIDAE**

41b. Adult with separate, well-developed podonotal and opisthonotal shields, the latter rarely absent in female; female epigynal shield with 1 pair of setae and not flanked by setae in soft integument; male ventrianal shield usually fused caudally with posterior margin of dorsal shield; peritremes well developed in deutonymph, usually reduced in length in adult and usually extending at most to level of bases of legs III, rarely to legs II; palptarsal apotele 2-tined **Superfamily Zerconoidea—42**

42a. Female, male, and deutonymph with podonotal and opisthonotal shields well developed, those of adult ornamented and expanded laterally so as to bear the marginal setae along their usually serrated margins; opisthonotal shield of deutonymph and adult fused caudally with ventrianal shield and often with a transverse row of 4 (sometimes coalesced to 3 or 2) clearly defined fossae near posterior margin. Male genital opening between coxae II–III; female and male with a broad ventrianal shield; adult peritremes often short but rarely reduced to size of stigmata. Free-living predators, Northern Hemisphere (Figs. 12.35A–C; 12.37C) **ZERCONIDAE**

42b. Female with podonotal shield unornamented, reduced in extent so as not to bear the marginal setae, and lacking shield on opisthonotum, leaving setae inserted on soft cuticle; deutonymph and male with podonotal and opisthonotal shields well developed and ornamented but free caudally from ventral shields, those of male expanded laterally so as to bear the marginal setae along their smooth margins; opisthonotum lacking well-defined muscle-insertion fossae. Male genital opening between coxae II; female with an anal shield and male with a narrow ventrianal shield; adult peritremes reduced to size of stigmata. Cavernicoles (Fig. 12.35D) **COPROZERCONIDAE**

43a. Female with sternal setae *st4* inserted on large metasternal plates flanking an anteriorly acuminate epigynal shield anterolaterally; female sperm access system opening via an unpaired, usually strongly developed endogynum beneath epigynal shield, often with ancillary sclerotized processes. Male chelicera with a spermatotreme, a sperm-holding process that is coalesced distally with movable digit; males usually with highly developed, sexually dimorphic, spinelike setae, spurs, and apophyses on legs II. Predators, free living or associated with insects (Figs. 12.38A–F) **Subcohort Parasitiae, PARASITIDAE**

43b. Female with sternal setae *st4* inserted on small metasternal plates or on soft integument or on posterolateral corners of sternal shield; female sperm access system opening by pair of small solenostomes in region of coxae III or IV. Male chelicera with a spermatodactyl, a sperm-holding process that projects distally, usually free from movable digit; males with or without sexually dimorphic structures on legs II **Subcohort Dermanyssiae—44**

44a. Palptarsal apotele 3- or 4-tined and with adjacent hyaline scalelike process; hypostome with internal malae typically elaborated, fimbriated, often bilobed, and moustachelike; inguinal area usually with a cluster of gland pores behind posterior margin of legs IV; dorsal subapical sensory field of tarsus I with 2 adjacent sensilla fused basally and inserted together in common alveolus; genu IV with 10 setae, including 2 ventrals. Free-living predators (Figs. 12.39A–F) **Superfamily Veigaioidea, VEIGAIIDAE**

44b. Palptarsal apotele 2- or 3-tined and lacking adjacent hyaline process, or rarely if with such a process, then apotele 2-tined; hypostome with internal malae usually unmodified and lightly fringed; inguinal area with or without a single gland pore behind posterior margin of legs IV; dorsal subapical sensory field of tarsus I lacking fusion of any two adjacent sensilla; genu IV with maximally 9 or rarely 10 setae, including 1 or rarely 2 ventrals **45**

45a. Female sternal shield usually fused with metasternal plates so as to include 4th pair of sternal setae but with metasternal elements free and with separated endopodal fragments in HALOLAELAPIDAE; female epigynal shield separate from ventrianal or anal shield; male with holoventral shield or separate sternitigenital and ventrianal shields, when separate, anterolateral margins of ventrianal shield often incised or eroded, and posterolateral margins often fused caudally with posterior margin of dorsal shielding; microtrichia of postanal cribrum sometimes extending onto posterior margin of dorsal shield **Superfamily Rhodacaroidea—46**

45b. Female sternal shield usually not fused with metasternal plates, leaving 4th pair of sternal setae free on soft integument or on metasternal plates that are free or attached to endopodal plates; if sternal setae *st4* on endopodal-metapodal plates that are incorporated with sternal shield (as in PACHYLAELAPIDAE), then epigynal shield expanded into a genitiventral shield; male with holoventral shield or separate sternitigenital and ventrianal shields; when separate, anterolateral margins of ventrianal shield intact, and posterolateral margins free caudally from posterior margin of dorsal shielding; microtrichia of cribrum confined to postanal region of anal or ventrianal shield **50**

46a. Fixed and movable digits of chelicerae elongate and attenuate, each with 1–3 retrorse teeth; corniculi narrow, acicular; stigmata with vestigial peritremes; palpi 4-segmented, with fused tibia and tarsus; ambulacra of tarsi I–IV without claws; tarsus I with ventrally delineated acrotarsus. Associated with termites (Figs. 12.42D–F) **LAELAPTONYSSIDAE**

46b. Fixed and movable cheliceral digits typically robust, not attenuate, with dentition variable but usually not retrorse; corniculi typically robust, hornlike; stigmata with peritremes usually well developed, rarely reduced; palpi 5-segmented, with separate tibia and tarsus; ambulacra of tarsi I–IV with claws, or leg I with ambulacra and claws reduced or absent; tarsus I without acrotarsus **47**

47a. Dorsal shield of adults divided, the anterior shield usually with setal pair *j2* nearly in transverse alignment with vertical and paravertical pairs *j1* and *z1*, and usually with 2–4 distinctive sclerotic nodules in mid-posterior region (Figs. 12.40E, 12.41C) (absent in *Digamasellus* and *Longoseius*); anterior portion of female sternal shield and of male sternitigenital shield weakly defined, usually carrying sternal setae *st1*; spermatodactyl of male often recurved basally **48**

47b. Dorsal shield of adults divided or entire, the anterior shield with setal pair *j2* in usual position well behind vertical and paravertical pairs *j1* and *z1*, *and* without sclerotic nodules; anterior portion of female sternal shield and male sternitigenital shield undifferentiated from more posterior portions; spermatodactyl of male not recurved basally **49**

48a. Palptarsal apotele 2-tined; legs IV usually with 7 setae on genu (1 2/1, 2/0 1) and 7 on tibia (1 1/1, 2/1 1); male with sternal setae *st5* on separate platelets. Predators, insect associates, and a few fungivores (Figs. 12.41A–H, 12.42G) **DIGAMASELLIDAE**

48b. Palptarsal apotele 3-tined; legs IV with 10 setae on genu (2 2/1, 3/1 1) and 10 on tibia (2 1/1, 3/1 2); male with sternal setae *st5* on sternitigenital shield. Free-living soil predators (Figs. 12.40A–H) **RHODACARIDAE**

49a. Exopodal elements generally distinct as sclerotized strips alongside coxae II–IV, endopodal elements usually present and fused with sternal shield in female and with sternitigenital shield in male; tibia IV typically with 10 setae (2 1/1, 3/1 2); tarsi II–IV without elongate dorsodistal setae. Free-living predators (Figs. 12.43A–F) **OLOGAMASIDAE**

49b. Exopodal elements not developed, and endopodal elements absent or present as small pieces separate from sternal shield in female and from sternitigenital shield in male (except *Halodarcia*); tibia IV typically with 8 setae (2 1/1, 2/1 1); tarsi II–IV often with a pair of elongate, distally acuminate, dorsodistal setae that extend well beyond tarsal claws. Free-living and arthropod associates (Figs. 12.42A–C) **HALOLAELAPIDAE**

50a. Arthrodial envelope at base of movable digit of chelicerae with 1 or 2 plumose or filamentous processes or with a brush and an adjacent fringed coronet; female typically with an epigynal and a ventrianal shield (if anal shield is present, it is widely separated from epigynal shield); male typically with a holoventral shield **51**

50b. Arthrodial processes, if present, at base of movable cheliceral digit with at most a fringed coronet but without distinct filamentous or brushlike processes; female with an epigynal and an anal or a ventrianal shield, or rarely with a genitiventral shield narrowly separated from or fused with an anal shield; male with separate sternitigenital and ventrianal or anal shields or with a holoventral shield **52**

51a. Peritremes generally looped proximally, joining the stigmata posteriorly; genu I usually with 2 ventral setae; tarsus I usually without claws; paired paradactyli of pretarsi II–IV usually broad and extending to apex of claws and divided or deeply serrated distally when observed obliquely; female with a pair of conspicuous accessory sclerites beneath the lateral margins of epigynal shield. Predators, free living or associated with insects (Figs. 12.46A–H) **MACROCHELIDAE**

51b. Peritremes normal, joining the stigmata anteriorly; genu I usually with 3 ventral setae; tarsus I with or without claws; paired paradactyli of pretarsi II–IV usually setiform, undivided distally; female with lateral accessory sclerites faint or absent beneath the lateral margins of epigynal shield. Free-living predators (Figs. 12.44A–E) **PARHOLASPIDIDAE**

52a. Tarsus II in female and male usually with 1 or 2 of the distal setae enlarged into stout spines; if lacking such spines, female with a genitiventral shield closely bordered by an anal shield or with peritrematic-exopodal shields extending posteriorly beyond coxae IV to fuse with metapodal plates; tibia III and genu and tibia IV each with 1 anterolateral seta. Predators, free living, or associated with insects (Figs. 12.47A–C, 12.48A–C) **PACHYLAELAPIDAE**

52b. Tarsus II usually without enlarged distal spinose structures in female but sometimes with such attributes dimorphically in male; female with an epigynal shield separate from a ventrianal or anal shield or, rarely, with an expanded genitiventral or genitiventrianal shield in some LAELAPIDAE; female with peritrematic-exopodal shields usually not extending posteriorly to fuse with metapodal plates, but rarely with such extensions in some BLATTISOCIIDAE, LAELAPIDAE, and EVIPHIDIDAE; tibia III and genu and tibia IV each with 1 or more commonly 2 anterolateral setae. Predators, fungivores, nidicoles, insect associates, and ecto- and endoparasites of vertebrates **53**

53a. Female sternal shield with 1 or typically 2 pairs of setae (sternal setae 1–2); sternal setae 3 often on adjacent platelets or on soft cuticle, rarely on sternal shield; opisthonotum without caudal setae *J5* and marginal *R* setae; posterior row of deutosternal denticles extending laterally beyond insertions of capitular setae; corniculi often divided distally or entire; femur II with 10 setae, including 4 dorsals. Fungal, algal, pollen, or nectar feeders; free living, nidicoles, or associates of insects and birds (Figs. 12.49A–H) **AMEROSEIIDAE**

53b. Female with 0 to typically 3 pairs of sternal setae on the sternal shield; opisthonotum usually with caudal setae *J5* and usually with one or more marginal *R* setae (*R* setae absent in PODOCINIDAE, some PHYTOSEIIDAE, and some OTOPHEIDOMENIDAE); posterior row of deutosternal denticles not extending laterally beyond insertions of capitular setae; corniculi usually entire, rarely divided distally; femur II with 10–11 setae, including 5 dorsals **54**

54a. Female with epigynal shield truncate or weakly convex posteriorly and either narrowly separated from or abutting a ventrianal shield or widely separated from an anal shield that is round or oval but usually not inversely subtriangular; male with sternitigenital shield usually delineated from, though often abutting, a ventrianal shield or separate ventral and anal elements **55**

54b. Female with epigynal shield broadly or narrowly rounded posteriorly, usually widely separated from inversely subtriangular anal shield, or epigynal shield expanded into a genitiventral or genitiventrianal shield in some LAELAPIDAE; male with either a consolidated holoventral shield or a sternitigenital shield widely separated from an anal shield **60**

55a. Deutonymphs and adults with less than 20 pairs of dorsal shield setae (the podocinid species *Podocinella plumosa*, with 23 pairs, is an exception), setae *J1* absent, and female with less than 4 pairs of marginal setae on soft integument; female with phytoseiid-type sperm access system, including a usually sclerotized calyx and often an associated minor duct (Fig. 12.55C). Free living or phoretically associated with, or parasitic on, insects **56**

55b. Deutonymphs and adults with more than 20 pairs of dorsal shield setae (the blattisociid genus *Aceodromus*, with 18–21 pairs, is an exception); setae *J1* present; female usually with more than 4 pairs of marginal setae on soft integument; female with phytoseiid-type or laelapid-type sperm access system, with or without a sclerotized calyx and an associated minor duct. Free living or phoretically associated with arthropods or birds **58**

56a. Legs I greatly elongated, with genual, tibial, and tarsal segments attenuate, subequal; tarsus I without claws and with 1–2 apical, whiplike setae; tibia II with 10 setae, including 2 anterolaterals and 4 dorsals, and tibia IV with 10 setae, including 2 anterolaterals and 2 ventrals; dorsal shield with a pair of large, distinctive gland pores between setae *J3* and *Z3*. Free living (Figs. 12.54A–F) **PODOCINIDAE**

56b. Legs I usually not elongated, with genual, tibial, and tarsal segments unequal in length; tarsus I usually with claws and lacking whiplike setae apically; typically, tibia II with 7 setae, including 1 anterolateral and 3 dorsals; tibia IV with 6 setae, including 1 anterolateral and 1 ventral; dorsal shield with porelike structures but without an enlarged pair between the *J* and *Z* setal series **57**

57a. Fixed and movable cheliceral digits normally developed, similar in length and apposed; tritosternum present, with well-developed laciniae; anal opening posteroventral, usually in a ventrianal shield. Free-living predators, fungivores, and pollen feeders in aerial and soil habitats (Figs. 12.52A–I) **PHYTOSEIIDAE**

57b. Fixed cheliceral digit absent or reduced to less than 1/4 the length of the slender, pointed movable digit; tritosternum commonly absent or reduced to a basal remnant; anal opening terminal or occasionally subterminal, usually in an anal shield. Parasites of insects (Figs. 12.53A–J) **OTOPHEIDOMENIDAE**

58a. Female with 3rd pair of sternal poroids on posterolateral corners of sternal shield and with sternal setae *st4* usually free on soft cuticle; male with endopodal strips beside coxae III–IV usually free or narrowly connected to sternitigenital shield (except genus *Antennoseius*); movable cheliceral digit usually bidentate (Figs. 12.51A–H) **ASCIDAE**

58b. Female with 3rd pair of sternal poroids off sternal shield and associated with sternal setae *st4* usually on metasternal plates or on soft cuticle; male with endopodal strips beside coxae III–IV fully integrated with sternitigenital shield; movable cheliceral digit with 0 to many teeth, often tridentate **59**

59a. Fixed cheliceral digit with pilus dentilis modified to a hyaline flap; movable cheliceral digit usually with a pointed process (mucro) on its mid-ventral face; peritrematic shield of adults free posteriorly from, or narrowly attached to, exopodal plate beside coxa IV; female with laelapid-type sperm access system, lacking a sclerotized spermathecal calyx and associated minor duct; female with epigynal shield gently rounded posteriorly, and usually with an oval or elliptical anal shield bearing only the 3 anal setae (or rarely expanded to capture the nearest pair of opisthogastric setae (Figs. 12.50A–G) **MELICHARIDAE**

59b. Fixed cheliceral digit with setiform pilus dentilis; movable cheliceral digit lacking a ventral mucro; peritrematic shield of adults broadly fused posteriorly to exopodal plate curving behind coxa IV; female with phytoseiid-type sperm access system, including a sclerotized calyx and an associated minor duct; female with epigynal shield usually truncate posteriorly, and usually with a ventrianal shield bearing 2–7 of the opisthogastric setae in addition to the anal setae (Figs. 12.55A–I) **BLATTISOCIIDAE**

60a. Tibia and genu I each with 1 anterolateral seta (1 3/2, 2/1 2 or 1 2/1, 2/1 1); tectum usually produced into an elongated, lancelike process, rarely multidenticulate; dorsal shield entire; female and male with anal shield. Free living or associated with insects or amphipods (Figs. 12.45A–F) **EVIPHIDIDAE**

60b. Tibia and genu I each with 2 anterolateral setae (2 3/2, 2/1 2 or 2 3/2, 3/1 2 or 2 3/1, 2/1 2); tectum with smooth or denticulate anterior margin, usually not produced into an elongated process; dorsal shield entire or with lateral incisions or divided; female with anal or, rarely, genitiventrianal shield, male usually with holoventral shield **Superfamily Dermanyssoidea (*pars*)—61**

61a. Sternal shield over 6 times wider than long at its widest point; epigynal membrane broad, convoluted, with epigynal setae flanking the shield remnant, with separate anal shield. Opisthosoma considerably broader than long, with a fringe of spatulate setae. Parasites of snakes (Figs. 12.65A, B) **OMENTOLAELAPIDAE**

61b. Without the above combination of characters **62**

62a. Chelicerae massive, hooked; with a broad, heavy epistome overlying the gnathosoma. Parasites of Neotropical bats (Figs. 12.65C–E) **SPELAEORHYNCHIDAE**

62b. Chelicerae and epistome not as above **63**

63a. Peritremes absent or greatly reduced. Respiratory tract parasites of mammals, reptiles, or birds **64**

63b. Peritremes rarely absent, occasionally reduced. Free living or ectoparasites of vertebrates or insects **66**

64a. Epigynal shield absent or rudimentary (distinct in *Zumptiella bakeri*), sternal shield usually present; stigmata ventral or lateroventral. Respiratory tract parasites of terrestrial and marine mammals (Figs. 12.60A–F) **HALARACHNIDAE**

64b. Epigynal shield distinct but sometimes reduced, shield setae reduced or absent; sternal shield present or absent, stigmata lateral or dorsal. Parasites of snakes or birds **65**

65a. Sternal and epigynal shields well developed but often weakly sclerotized, sternal setae minute or absent, stigmata lateral. Respiratory tract parasites of snakes (Figs. 12.61F–H) **ENTONYSSIDAE**

65b. Sternal shield reduced or absent but with distinct, and often distinctive, sternal setae, epigynal shield well developed or reduced, stigmata dorsal. Respiratory tract parasites of birds (Figs. 12.64A–H) **RHINONYSSIDAE**

66a. Legs I extremely stout, with heavy sessile claws; legs II–IV slender, with long pretarsi and small claws, leg coxae widely separated. Sternal shield absent or barely visible as an interruption in the ventral integumentary striae; opisthogaster with distinctive spurlike or broad flattened setae. Parasites of armadillos (Figs. 12.61B–E) **DASYPONYSSIDAE**

66b. Legs I–IV of comparable thickness or coxae I–IV contiguous; sternal shield well developed or reduced but distinctly tanned and easily visible. Opisthogaster without spurlike or flattened setae **67**

67a. Tritosternum absent or represented by tritosternal base remnant (if tritosternal base is well developed, then the peritremes extend only to level of anterior edge of coxae III). Sternal setae inserted at the margins of reduced sternal shield or in integument bordering it; epigynal shield narrowed or otherwise reduced, with or without setae. Parasites of bats (Figs. 12.62A–G, 12.66E) **SPINTURNICIDAE**

67b. Tritosternum well developed, with laciniae **68**

68a. Sternal shield subrectangular, reduced laterally, carrying only sternal setae *st1* and associated pores, sternal setae *st2* and *st3* in adjacent integument. Anal shield elongate, narrowly produced posterior to postanal seta; opisthogastric margin with 2 pairs of long, flagellate setae. Parasites of edentates (Fig. 12.61A) **MANITHERIONYSSIDAE**

68b. Sternal and anal shields variously developed, often reduced or expanded but not as above **69**

69a. Chelicerae of female whiplike, styliform; cheliceral digits minute, chelate; corniculi membranous, indistinct **70**

69b. Chelicerae variously produced but not styliform; corniculi variously developed **71**

70a. Elongate second cheliceral segment of female far exceeding basal cheliceral segment 1 in length, male chelicerae with second segment of normal length. Idiosoma broadly rounded posteriorly. Parasites of mammals and birds (Figs. 12.62H–L) **DERMANYSSIDAE**

70b. Second cheliceral segment normally developed, considerably shorter than greatly elongated basal cheliceral segment. Idiosoma with strongly narrowed opisthosoma. Parasites of porcupines and snakes (Figs. 12.63A–C) **HYSTRICHONYSSIDAE**

71a. Chelicerae elongate, edentate; corniculi membranous, usually lobate. Palptrochanter often with a raised medioventral keel; with a large anterior nonsetigerous spur on leg coxa II (rarely minute or absent), other coxae without spurs but occasionally with small ridges (e.g., *Chirocetes*); genu IV typically with 2 ventral setae. Parasites of mammals, birds, and reptiles (Figs. 12.63D–G, 12.66C, D) **MACRONYSSIDAE**

71b. Chelicerae various, dentate or edentate, corniculi strongly sclerotized or membranous, hornlike, barbed, or lobate. Palptrochanter without raised medioventral keel; generally with more than 1 large, nonsetigerous coxal spur (e.g., *Hirstionyssus* (LAELAPIDAE)), or coxal spurs absent (a single anterior spur is present on coxa II of *Scutanolaelaps* (IXODORHYNCHIDAE)); genu IV typically with 1 ventral seta **72**

72a. Corniculi attenuate-acuate, often barbed (barbs absent in *Hemilaelaps*, *Scutanolaelaps*, *Strandtibbetsia*, and *Asiatolaelaps*); with spurlike setae on coxae II, I–II, or I–III; palptarsal claw greatly reduced, generally with a single tine. External parasites of snakes (Figs. 12.59D–F) **IXODORHYNCHIDAE**

72b. Corniculi not as above, spurlike setae on coxae present or absent; palptarsal claw normally produced, with 2–3 tines **73**

73a. Fixed cheliceral digit absent; with only 2 pairs of hypostomatic setae; peritremes of female short, looped medially or apically, confined to level of coxae III or coxae III–IV; males with peritremes very short and crooked or vestigial. Parasites of bees (Figs. 12.58C–E) **VARROIDAE**

73b. Fixed cheliceral digit present, chelicerae dentate or edentate; with 3 pairs of hypostomatic setae; peritremes of adults variously produced, typically well developed and elongate, occasionally absent **74**

74a. Tibia I usually with 1 ventral seta, lacking seta *av2*; genu IV usually with 1 posterolateral seta, lacking seta *pl2*; subcapitulum with internal malae usually weakly developed, with nearly smooth lateral margins and shorter than corniculi. Paraphages of chilopods, diplopods, spiders, and crustaceans (Fig. 12.59C) **IPHIOPSIDIDAE**

74b. Tibia I usually with 2 ventral setae, including *av2*; genu IV usually with 2 posterolateral setae, including *pl2*; subcapitulum with internal malae usually well developed, with fimbriate lateral margins, and equal to or longer than corniculi. A heterogenous group comprising free-living forms and facultative and obligate parasites of arthropods, birds, and mammals (Figs. 12.56A–F, 12.57A–G, 12.58A, B, 12.66A, B) **LAELAPIDAE**

75a. Deutonymph and adults with a single dorsal shield that is never flanked by marginal shields; dorsal shield with several pairs of greatly elongated marginal setae, and paranal setae also markedly elongated; palptarsal claw absent, palpgenu with 4 setae; tibiae I–IV with only 1 dorsal seta (1 0/1, 1/1 1); femur IV with 8 setae (1 2/1, 2/1 1). Parasites of passalid beetles (Figs. 12.31G–J) **Subcohort Diarthrophalliae, DIARTHROPHALLIDAE**

75b. Deutonymph and adults with more than 1 shield dorsally, adults usually with 1 or more pygidial shields or marginal shields in addition to dorsal shield; dorsal shield without differentiated pairs of greatly elongated marginal setae and paranal setae not elongated, although 1 pair of ventrocaudal setae sometimes so elongated; palptarsal claw present, palpgenu with 5–6 setae; tibiae I–IV with at least 3 dorsal setae (1 1/1, 2/1 1); femur IV with 6 setae in various configurations. Free-living predators, fungivores, and scavengers, or associated with arthropods **Subcohort Uropodiae—76**

76a. Coxae of legs I narrow, cylindrical, each in its own acetabulum; tritosternum with 2 long, feathered laciniae; female epigynal shield loosely tripartite (Fig. 12.23G); chelicerae with lateral slit organs; palpgenu with 6 setae; tibia of leg I with 3 or 4 ventral setae **Superfamily Thinozerconoidea—77**

76b. Coxae of legs I inserted together in a gnathopodal cavity and often widened so that the tritosternum is partly covered; tritosternum with variously shaped laciniae; epigynal shield of female usually unipartite, occasionally trigynaspid-like (Fig. 12.20K); chelicerae without lateral slit organs; palpgenu usually with 4–5 setae; tibia of leg I with 2 ventral setae **78**

77a. Peritremes and stigmata dorsal; coxa of legs IV with 2 setae; epistome with a narrow middle spine (Figs. 12.23E–G) **THINOZERCONIDAE**

77b. Peritremes and stigmata ventrolateral; coxa of legs IV with 1 seta; epistome without differentiated middle spine (Figs. 12.23A–D) **PROTODINYCHIDAE**

78a. Coxae of legs I normally produced, not covering base of tritosternum, tritosternal base about twice as wide as long; ventral leg cavities (pedofossae) absent **Superfamily Polyaspidoidea—79**

78b. Coxae of legs I usually widened and flattened so that the tritosternum is entirely or partially covered; base of the tritosternum usually not wider than long; pedofossae (Fig. 12.25E) usually well developed, sometimes absent **Superfamily Uropodoidea—81**

79a. Genu I with 2 anterolateral and 2 posterolateral setae; femur IV with 2 ventral setae **80**

79b. Genu I usually with 2 anterolateral and 1 posterolateral seta, femur IV with 2 ventral setae (if only 1 is present, then genu I with 2 posterolateral setae (Figs. 12.24A–C) **DITHINOZERCONIDAE**

80a. Idiosoma typically elongate, subtriangular, or pyriform, marginal shields entire or fragmented into numerous platelets, pygidial shield entire or variously subdivided. Genu IV with 1 ventral seta (Figs. 12.24G–J, 12.32A, B) **TRACHYTIDAE**

80b. Idiosoma broadly ovate, marginal shields absent or variously coalesced, pygidial shield entire. Genu IV with 2 ventral setae (Figs. 12.24D–F) **POLYASPIDIDAE**

81a. Apex of fixed cheliceral digit with a large flower- or mushroom-shaped sensory organ; corniculi with terminal teeth (Figs. 12.29B, C) **UROACTINIIDAE**

81b. Apex of fixed cheliceral digit not as above, corniculi generally smooth terminally **82**

82a. Chelicerae with a distinctive internal sclerotized node associated with the levator tendon **84**

82b. Chelicerae without internal sclerotized node **83**

83a. Hypostomatic setae *h1* short, spiniform; genital shield of female located behind coxae IV. Small mites, with adults under 400 μm in length (Fig. 12.28F) **METAGYNURIDAE**

83b. Hypostomatic setae *h1* typically long and setiform; female genital shield located between coxae II–IV. Adults generally over 400 μm in length (Figs. 12.25A–E, 12.26A–E, 12.27A–F, 12.32D) **UROPODIDAE**

84a. Internal malae of hypostome simple, without marginal fimbriations or distal moustachelike excrescences; dorsal shield of adults often notched marginally (Fig. 12.29A) **TREMATURIDAE**

84b. Internal malae with short marginal fimbriations and/or with elaborated distal moustachelike excrescences **85**

85a. Fixed cheliceral digit typically with a rounded or acuminate apical "finger" that extends well beyond the movable digit (absent in *Caminella*). Peritremes well developed and often elaborated to form distinctive serpentine patterns, pygidial shield sometimes present in adults (Figs. 12.28A–E, 12.32E) **DINYCHIDAE**

85b. Fixed cheliceral digit subequal to or little longer than movable digit, sometimes with short rounded extension. Peritremes variously developed but not as above, pygidial shield absent in adults **86**

86a. Dorsal shield of adults strongly ornamented with pits and ridges, often with numerous dorsal T-shaped setae. Movable cheliceral digit more than twice the length of its basal width (Figs. 12.30B, 12.32C) **TRACHYUROPODIDAE**

86b. Dorsal shield of adults without conspicuous ornamentation or T-shaped dorsal setae. Movable cheliceral digit shorter than twice its basal width (Fig. 12.30A) **OPLITIDAE**

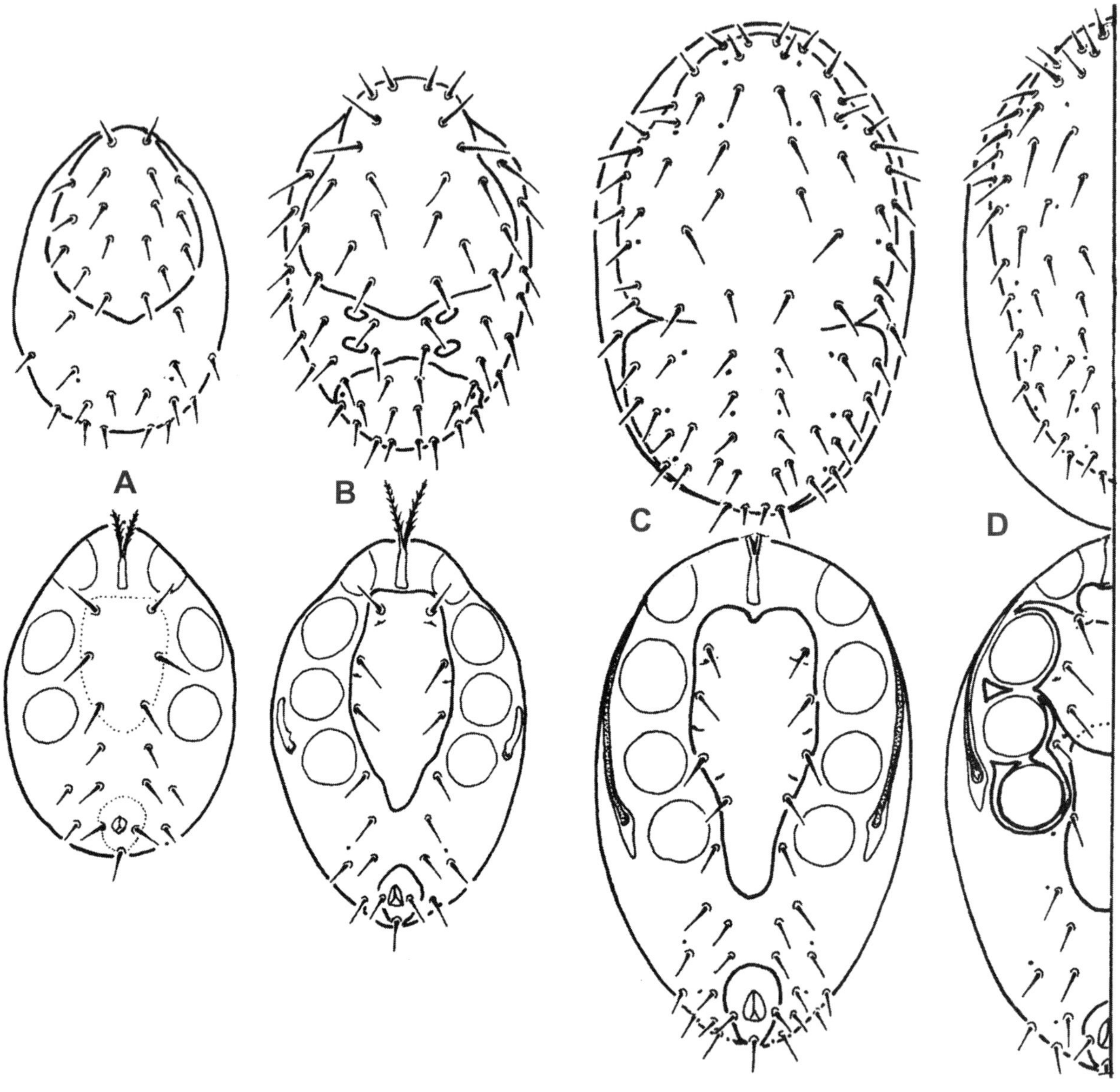

Fig. 12.1. Ontogenetic development of *Gaeolaelaps aculeifer* (Canestrini) (LAELAPIDAE), showing dorsal (top) and ventral aspects of each instar. **A**, larva; **B**, protonymph; **C**, deutonymph; **D**, adult female

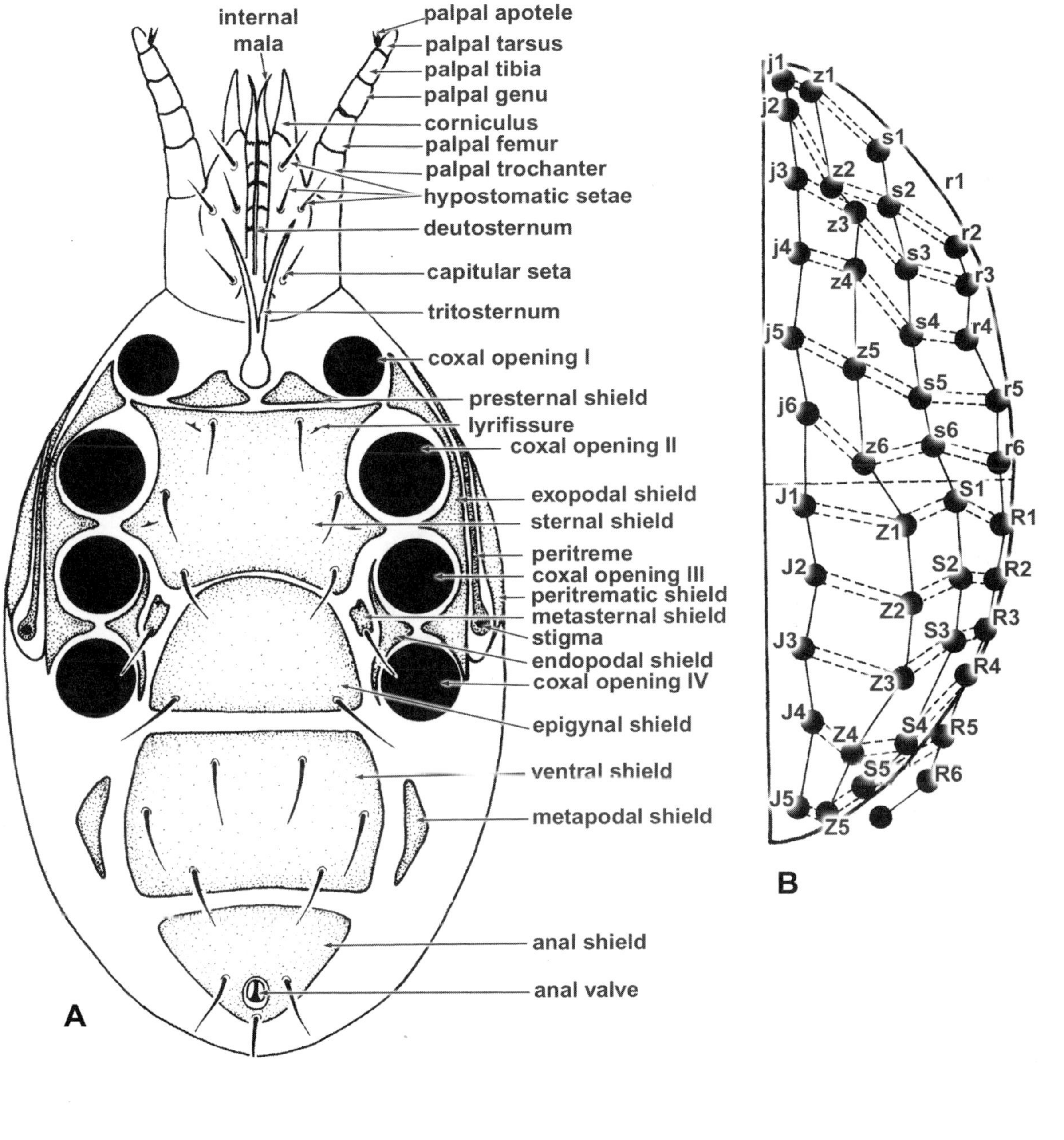

Fig. 12.2. Generalized mesostigmatic mite. **A**, venter; **B**, dorsum. Dorsal chaetotactic system is that of Lindquist and Evans (1965); dotted and solid lines denote assumed vertical and horizontal relationships of setal bases so connected.

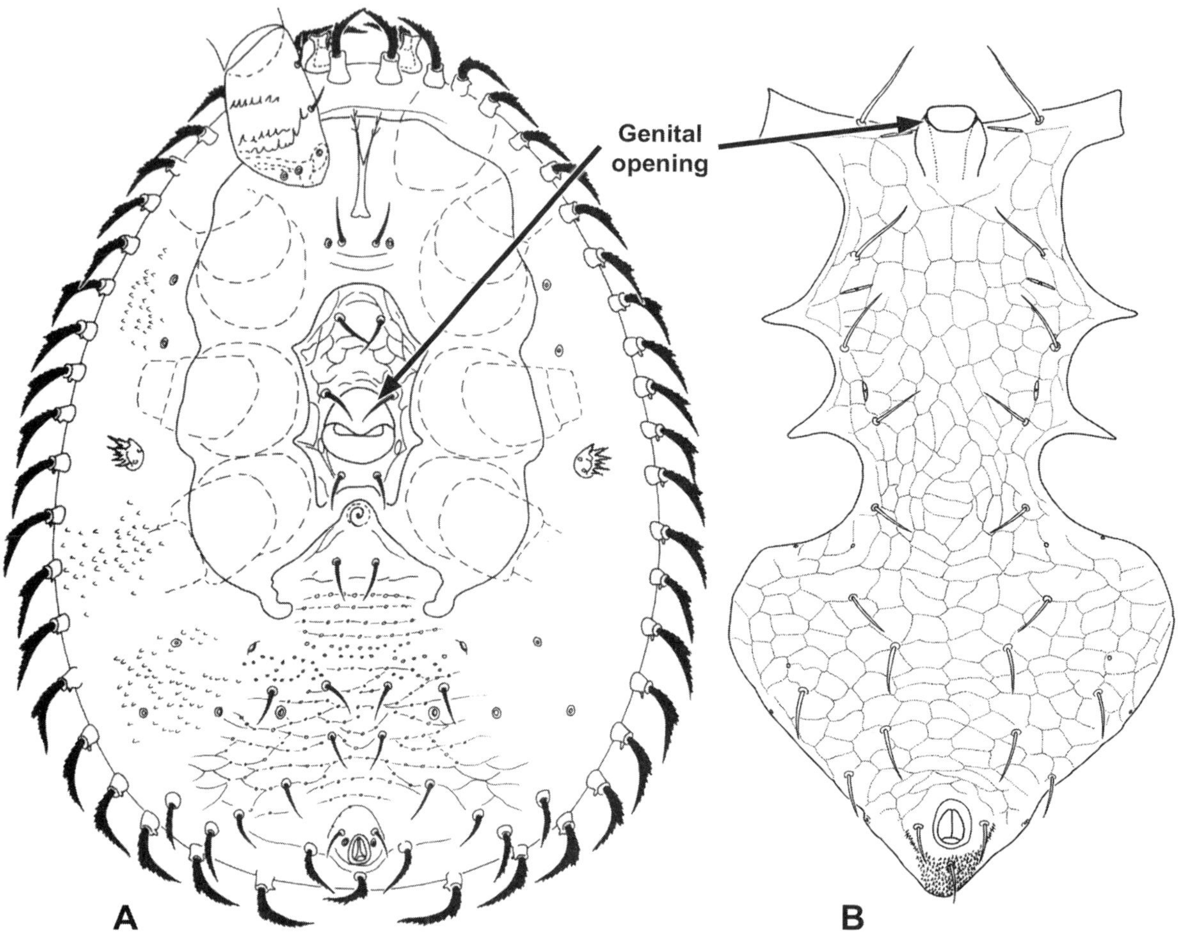

Fig. 12.3. Venters of adult male Mesostigmata. **A**, *Heatherella acanthocharis* Walter (HEATHERELLIDAE) (Australia), with genital opening between the bases of legs III–IV and covered by two valves; **B**, *Stratiolaelaps lamington* Walter and Campbell (LAELAPIDAE) (Australia), holoventral shield with genital opening on anterior margin and covered by a single valve.

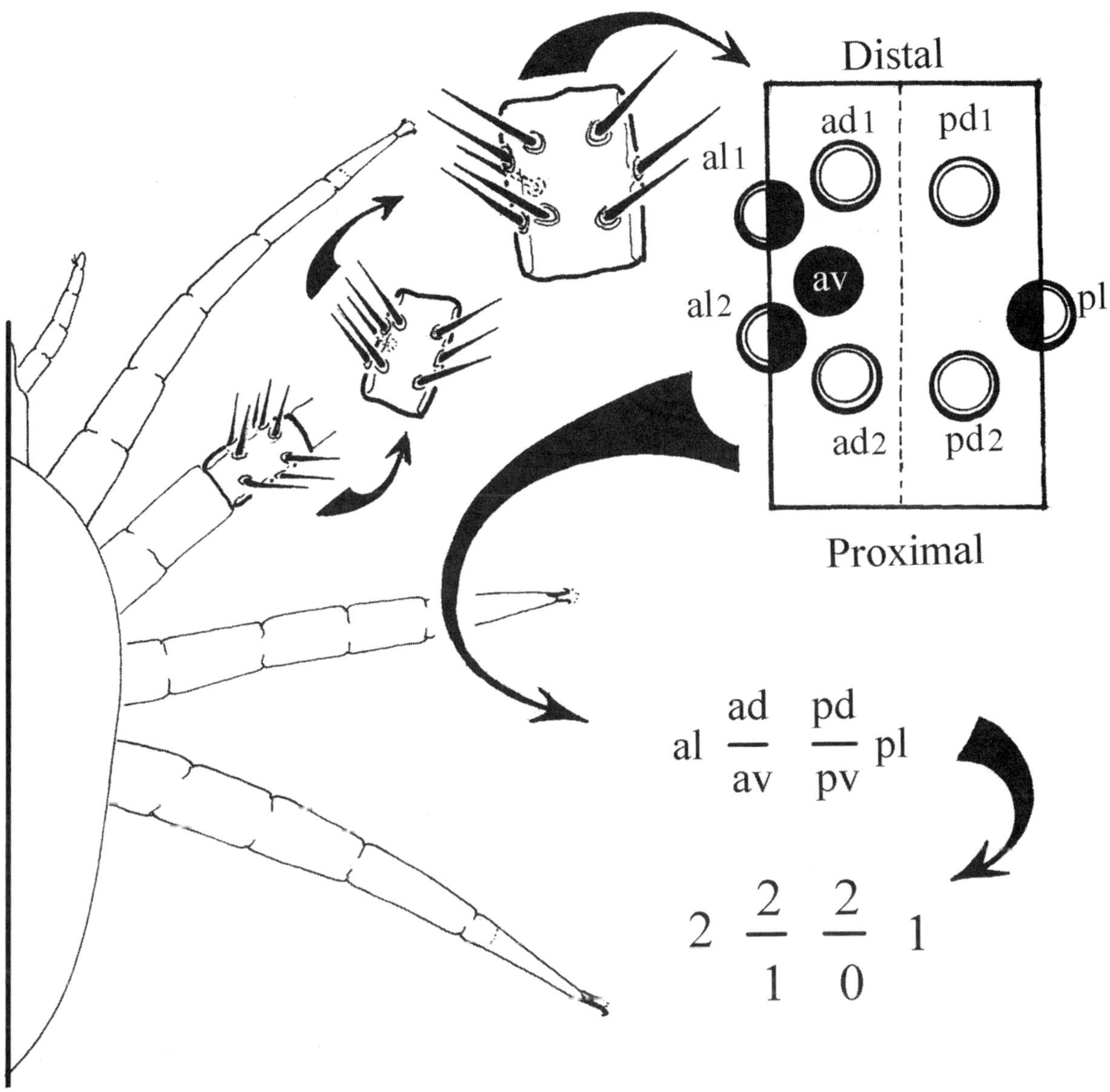

Fig. 12.4. Leg segment chaetotaxy in the Mesostigmata. Genu II has been separated from leg II and oriented so that its leading edge (the anterolateral face, or *al*) corresponds to the vertical left-hand margin in the segment diagram. Genu II is viewed in dorsal aspect, and its top margin corresponds to the distal extremity of the segment. Dorsal setal insertions are indicated by open circles and ventral insertions by solid circles.

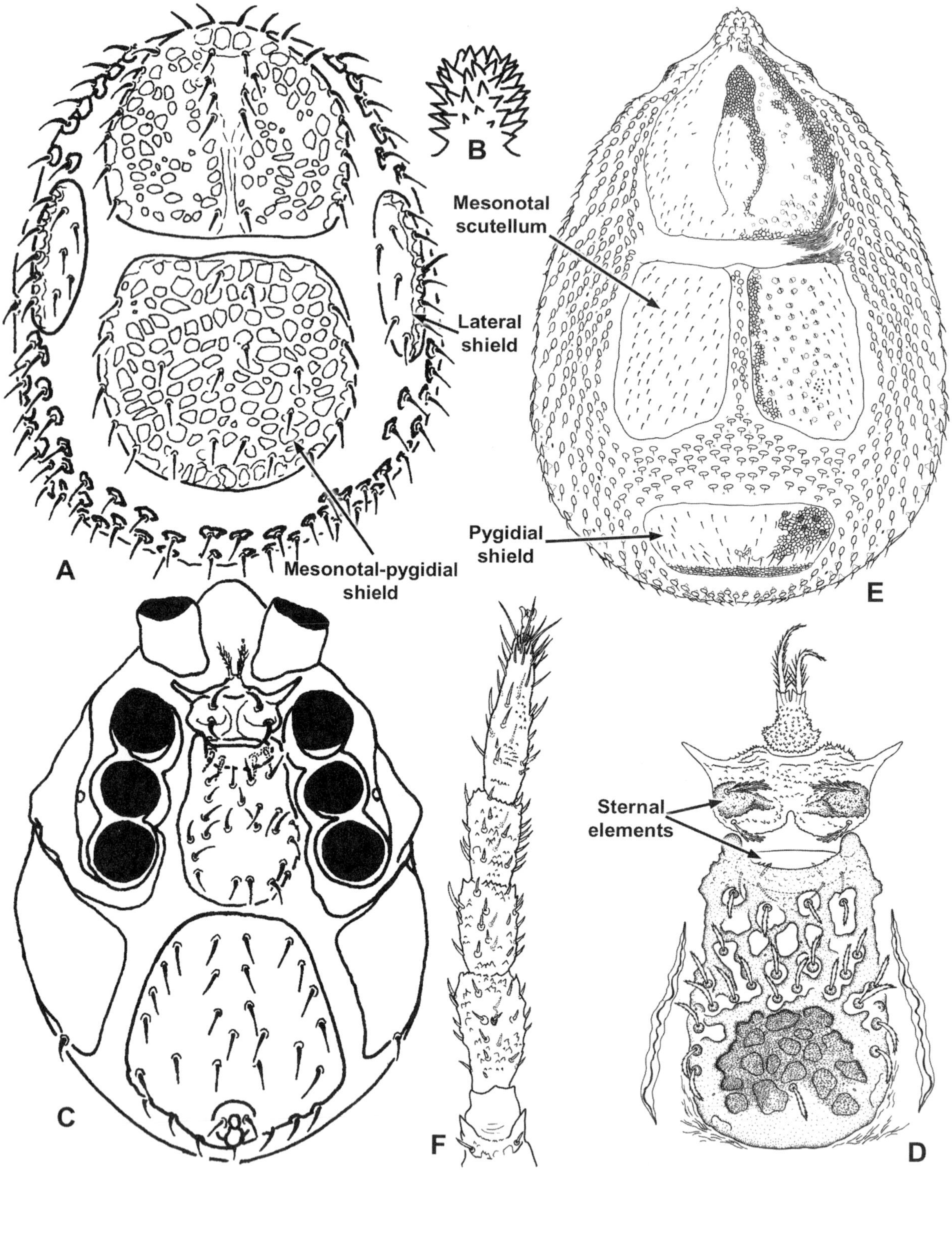

Fig. 12.5. UROPODELLIDAE. *Uropodella* sp. (Ohio, USA) female: **A**, dorsum; **B**, spiky knob on vertex; **C**, venter; **D**, sternitigenital region.

ICHTHYOSTOMATOGASTERIDAE? Undescribed taxon (Australia): **E**, dorsum of female; **F**, left leg I of same.

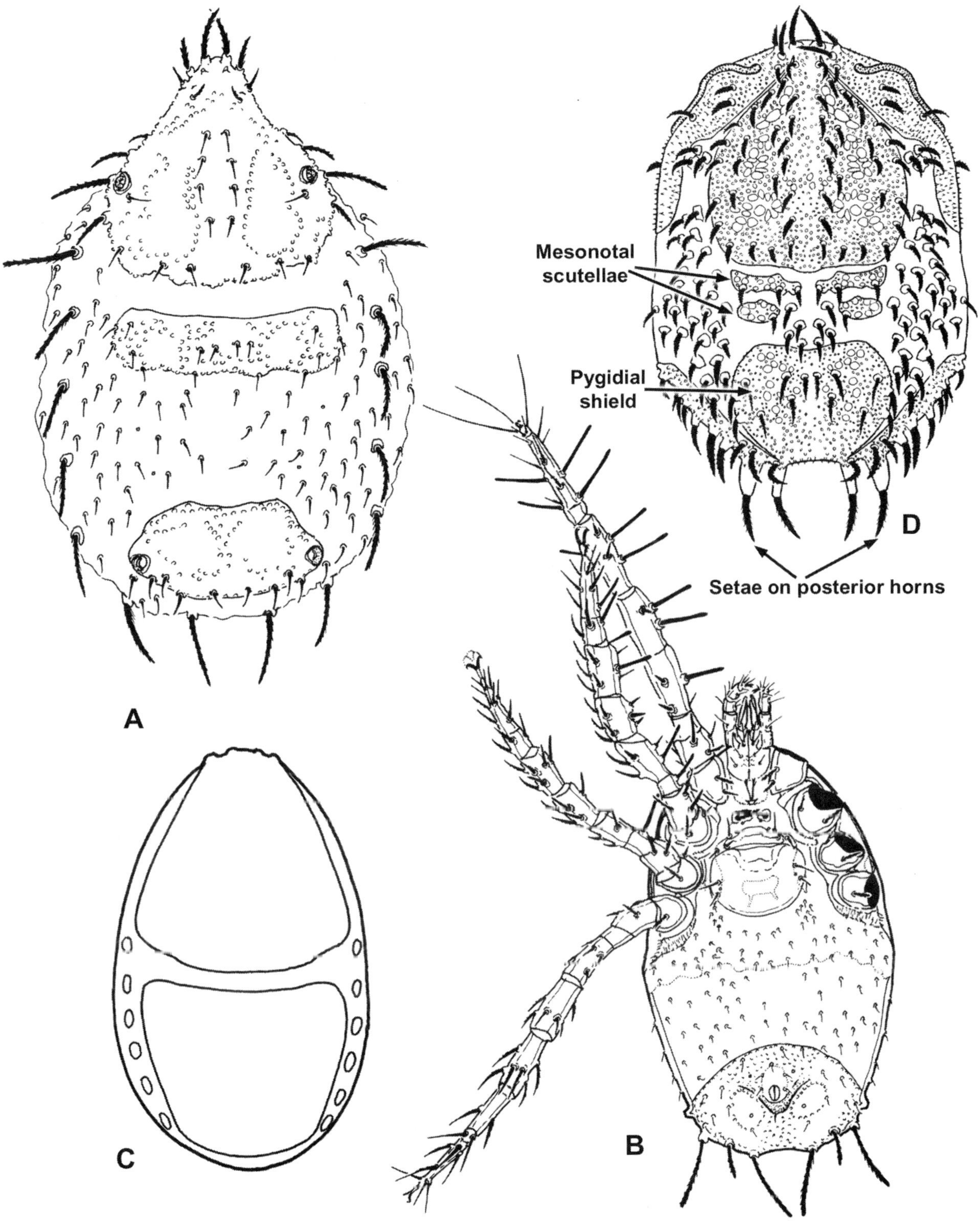

Fig. 12.6. SEJIDAE. *Adenosejus krantzi* (Hirschmann) (Oregon, USA): **A**, dorsum of female; **B**, venter of female; **C**, dorsum of male. *Sejus baloghi* (Athias-Henriot): **D**, dorsum of male (after Athias-Henriot 1960).

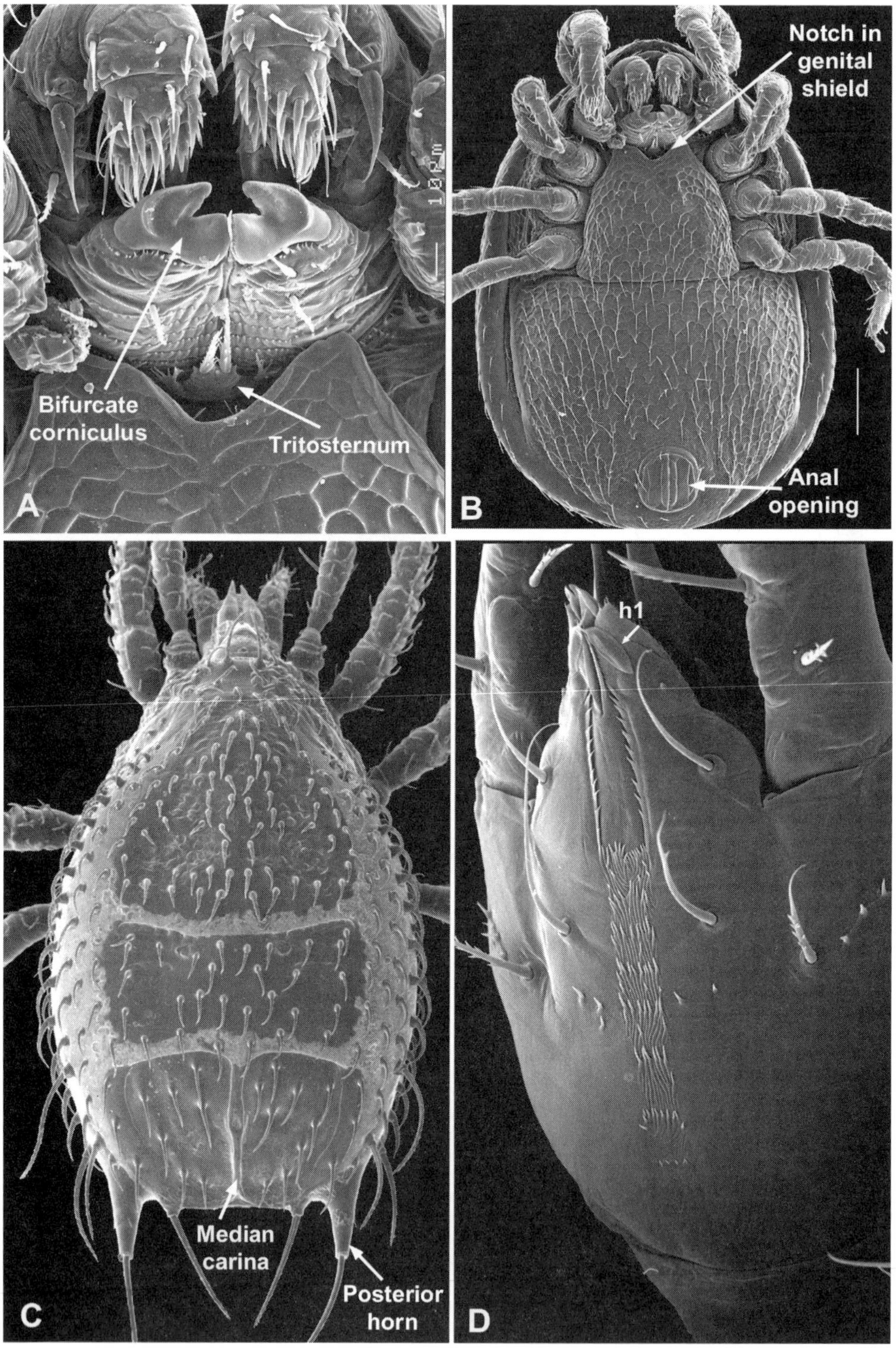

Fig. 12.7. ICHTHYOSTOMATOGASTERIDAE. *Asternolaelaps* sp. (Australia) female: **A**, subcapitulum, tritosternum, and anterior margin of genital shield; **B**, venter.

SEJIDAE. *Epicroseius* sp. (Australia): **C**, dorsum of female; **D**, subcapitulum showing scalelike hypostomatic seta *h1*.

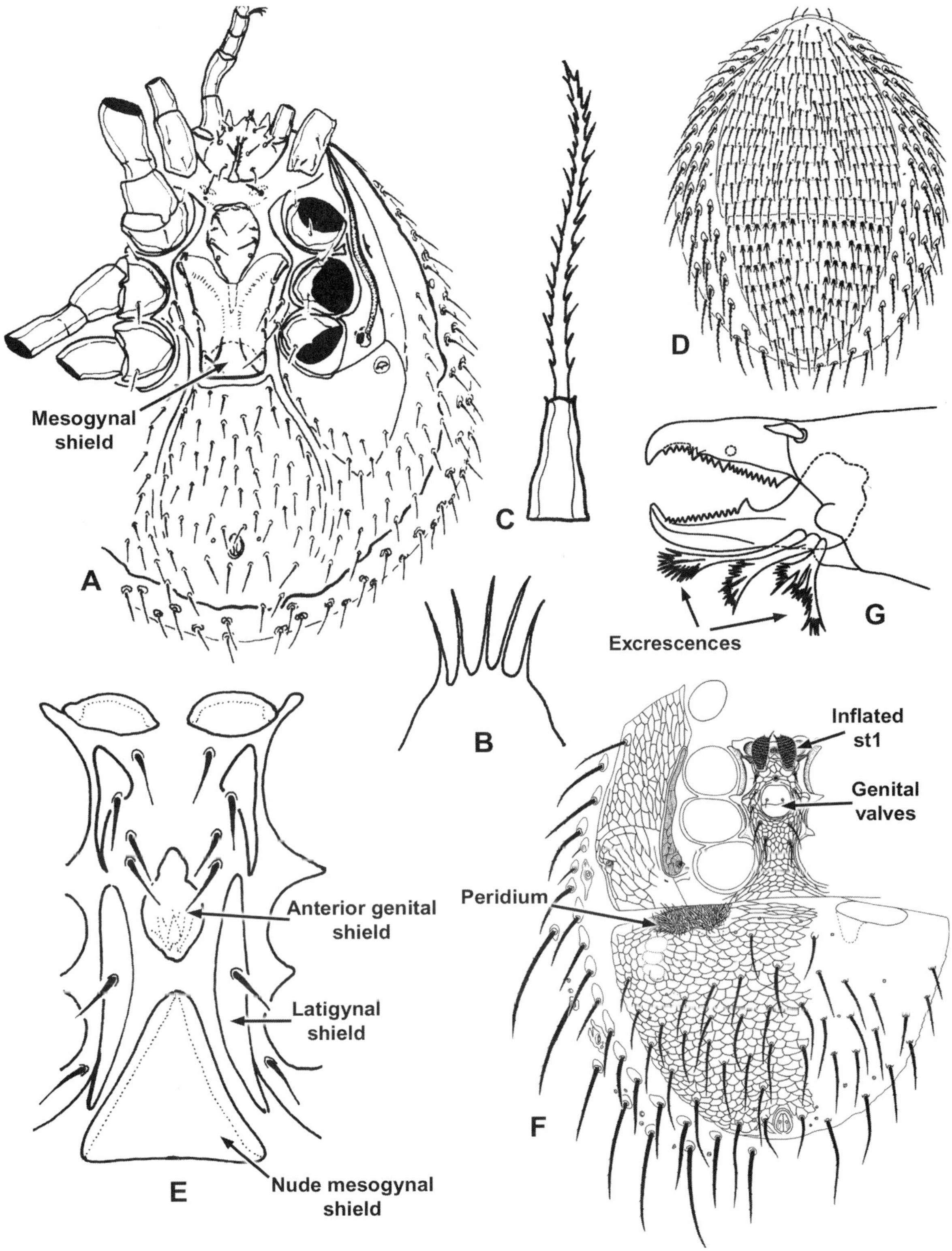

Fig. 12.8. CERCOMEGISTIDAE. *Cercomegistus evonicus* Kinn (Oregon, USA) female: **A**, dorsum; **B**, epistome; **C**, tritosternum. *Cercomegistus* sp. female: **D**, dorsum.

ASTERNOSEIIDAE. *Asternoseius ciliatus* Berlese (Italy): **E**, sternitigenital region of female *Asternoseius* sp. (Australia); **F**, venter of male; **G**, chelicera of female.

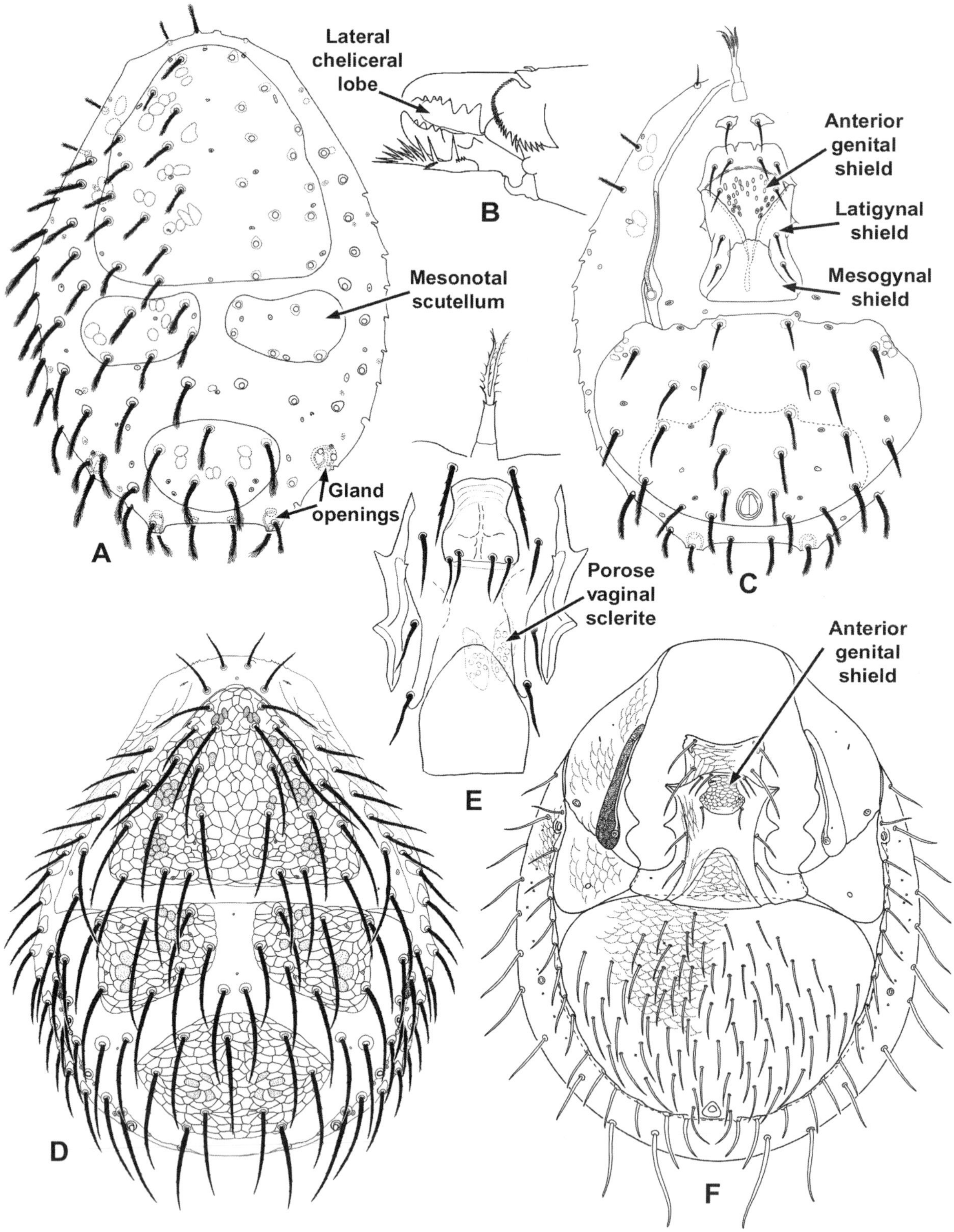

Fig. 12.9. DAVACARIDAE. *Acanthodavacarus klompeni* Walter (Australia) female: **A**, dorsum; **B**, chelicera; **C**, venter. *Davacarus reginaldi* Walter (Australia): **D**, male dorsum; **E**, female sternitigenital region (from Walter 2004).
SEIODIDAE. *Seiodes* sp. (Africa): **F**, female venter.

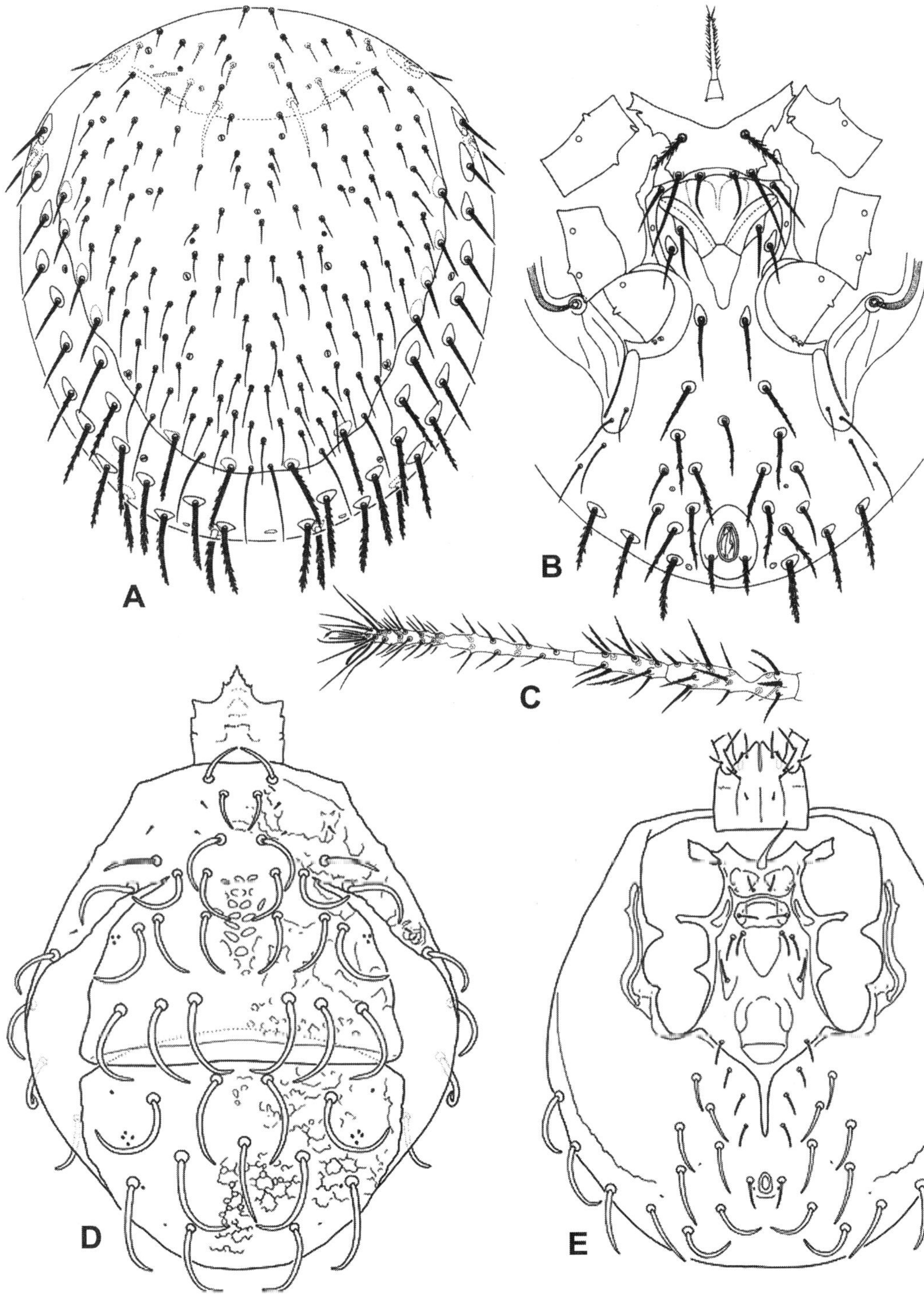

Fig. 12.10. SALTISEIIDAE. *Saltiseius hunteri* Walter (Australia) female: **A**, dorsum; **B**, venter; **C**, leg I. PYROSEJIDAE. Female (Costa Rica): **D**, dorsum; **E**, venter.

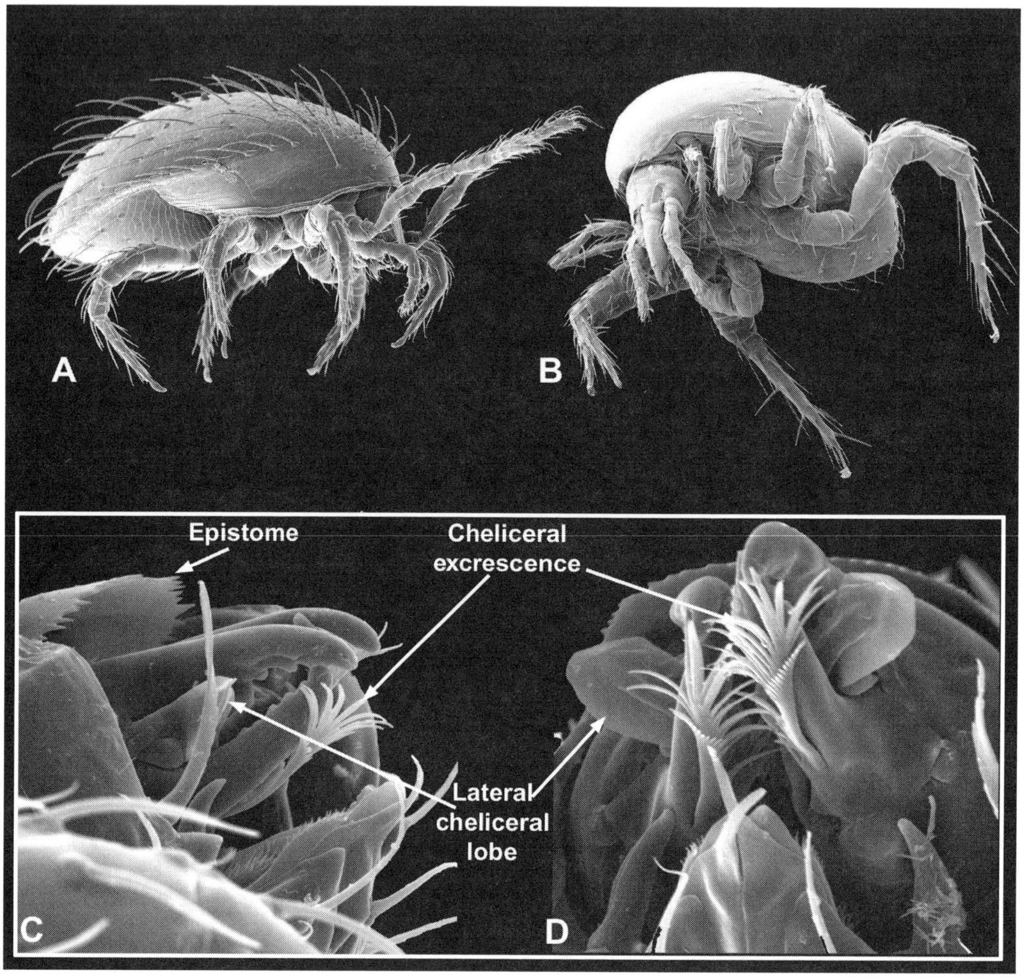

Fig. 12.11. ASTERNOSEIIDAE. *Asternoseius* sp. (Australia): **A**, lateral view of female.
SALTISEIIDAE. *Saltiseius hunteri* Walter (Australia): **B**, fronto-ventral view of male.
DAVACARIDAE. *Acanthodavacarus klompeni* Walter (Australia): **C**, lateral view of apex of gnathosoma; **D**, fronto-ventral view of gnathosoma.

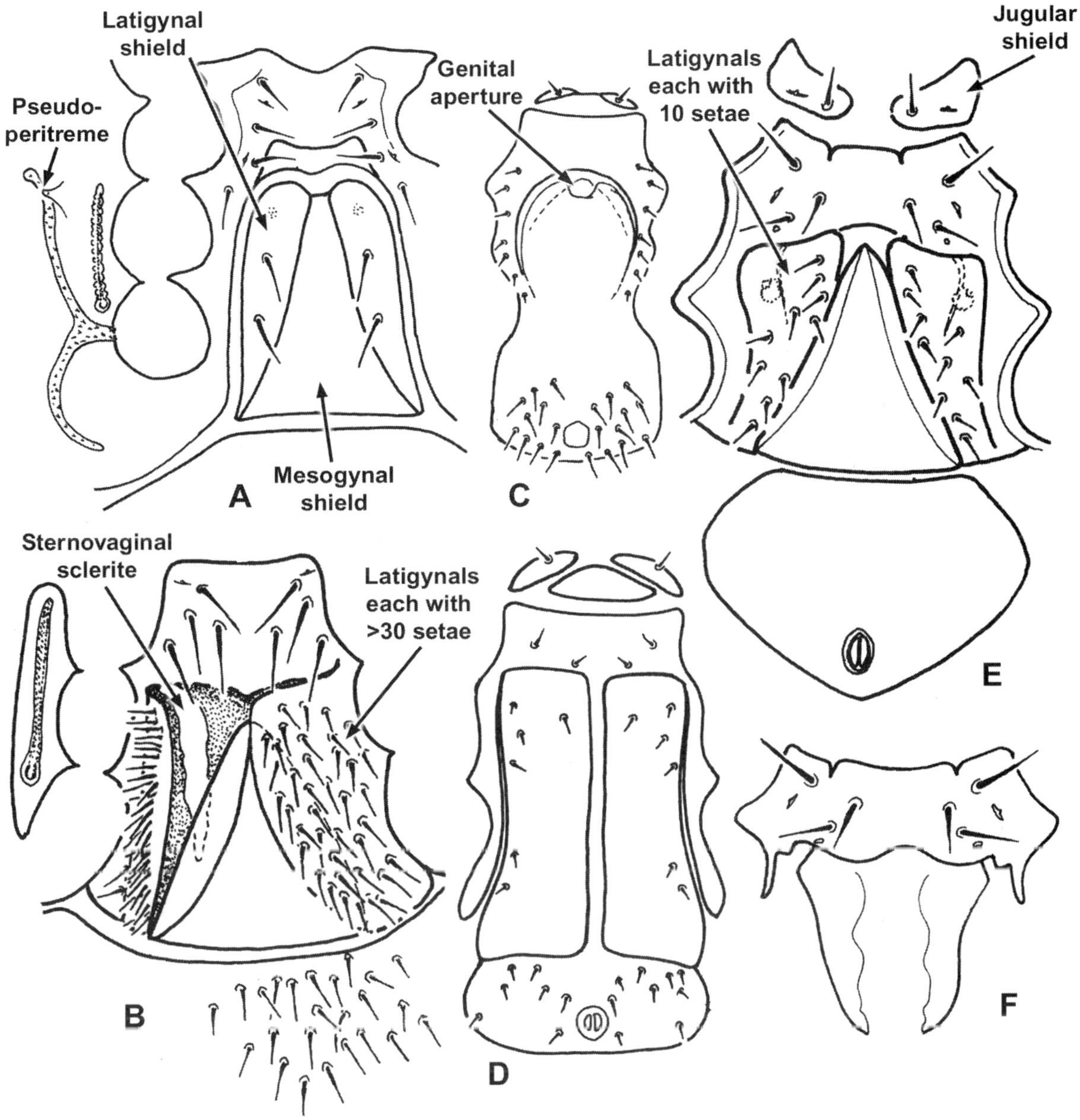

Fig. 12.12. AENICTEQUIDAE. *Aenicteques chapmani* Jacot (Philippines): **A**, sternitigenital region of female.

PHYSALOZERCONIDAE. *Physalozercon raffrayi* Wassmann: **B**, sternitigenital region of female with one latigynal shield folded back to expose a sternovaginal shield.

MESSORACARIDAE. *Messoracarus mirandus* Silvestri: **C**, venter of male; **D**, venter of female.

PTOCHACARIDAE. *Ptochacarus silvestri* Womersley (Australia): **E**, venter of female; **F**, sternal shield with posterolateral sternovaginal shields exposed.

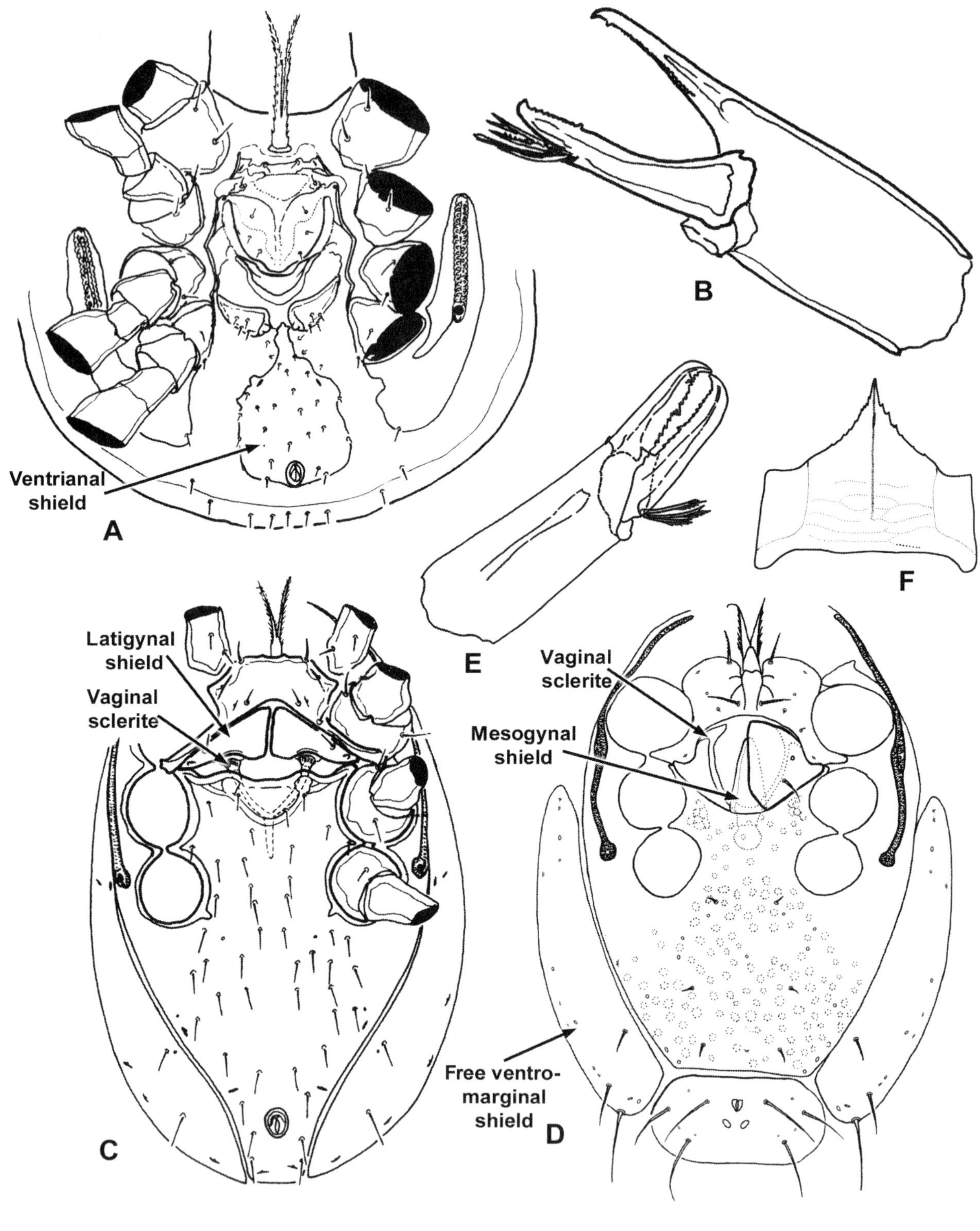

Fig. 12.13. ANTENNOPHORIDAE. *Antennophorus* sp. (British Columbia) female: **A**, venter; **B**, chelicera.
CELAENOPSIDAE. *Pleuronectocelaeno* sp. (Colorado, USA): **C**, venter of female.
TRIPLOGYNIIDAE. *Funkotriplogynium iagobadius* Seeman and Walter (Australia) female: **D**, venter with right latigynal shield removed to show vaginal sclerite and hidden part of mesogynal shield; **E**, chelicera; **F**, epistome.

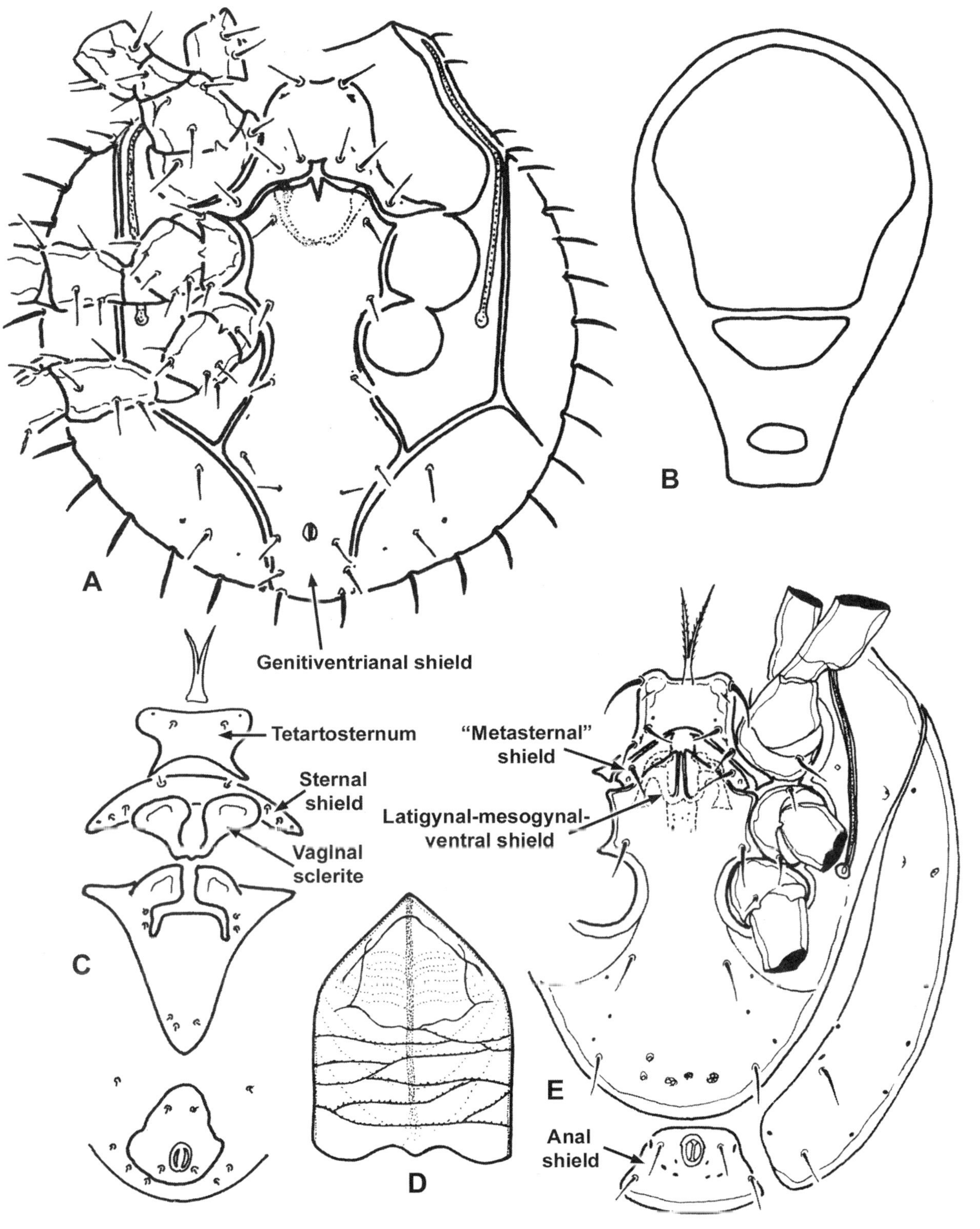

Fig. 12.14. MEGACELAENOPSIDAE. *Megalocelaenopsis oudemansi* Funk (Panama) female: **A**, venter. NEOTENOGYNIIDAE. *Neotenogynium malkini* Kethley (Ecuador) female: **B**, dorsum; **C**, venter. EUZERCONIDAE. *Euzercon* spp.: **D**, epistome (Mexico); **E**, venter of female (Arkansas, USA).

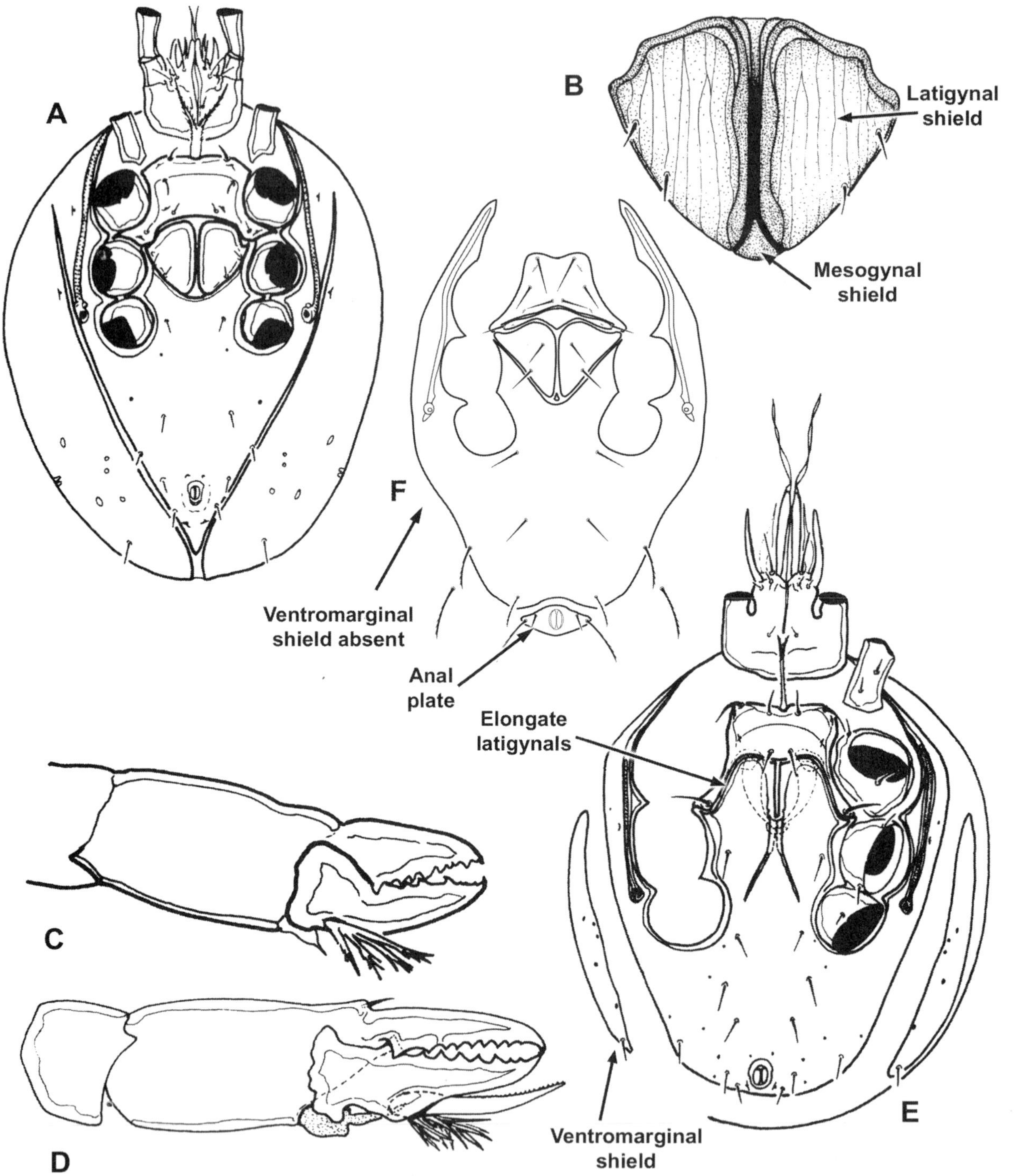

Fig. 12.15. DIPLOGYNIIDAE. Undescribed sp. (Ohio, USA) female: **A**, venter; **B**, latigynal and mesogynal shields; **C**, chelicera. *Lobogyniella tragardhi* Krantz (Oregon, USA): **D**, chelicera of female.

SCHIZOGYNIIDAE. *Schizogynum* sp. (Panama) female: **E**, venter.

COSTACARIDAE. *Costacarus reyesi* Hunter (Mexico): **F**, venter of female (after Hunter 1993).

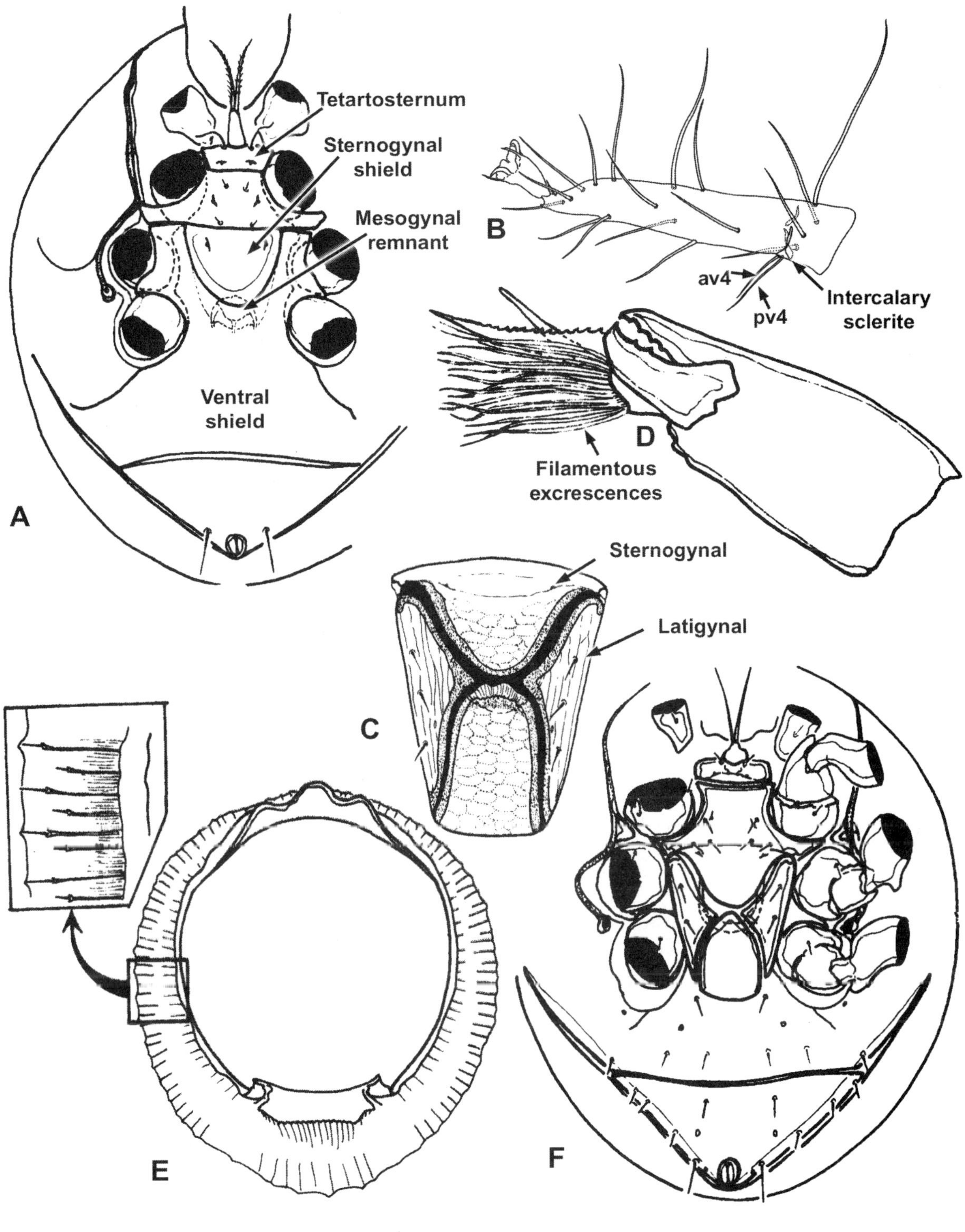

Fig. 12.16. FEDRIZZIIDAE. *Fedrizzia* spp.: **A**, venter of female (Papua New Guinea); **B**, tarsus IV (Australia). KLINKOWSTROEMIDAE. *Klinckowstroemia* sp. (Brazil) female: **C**, genital shields; **D**, chelicera; **E**, dorsum with detail of marginal ornamentation; **F**, venter.

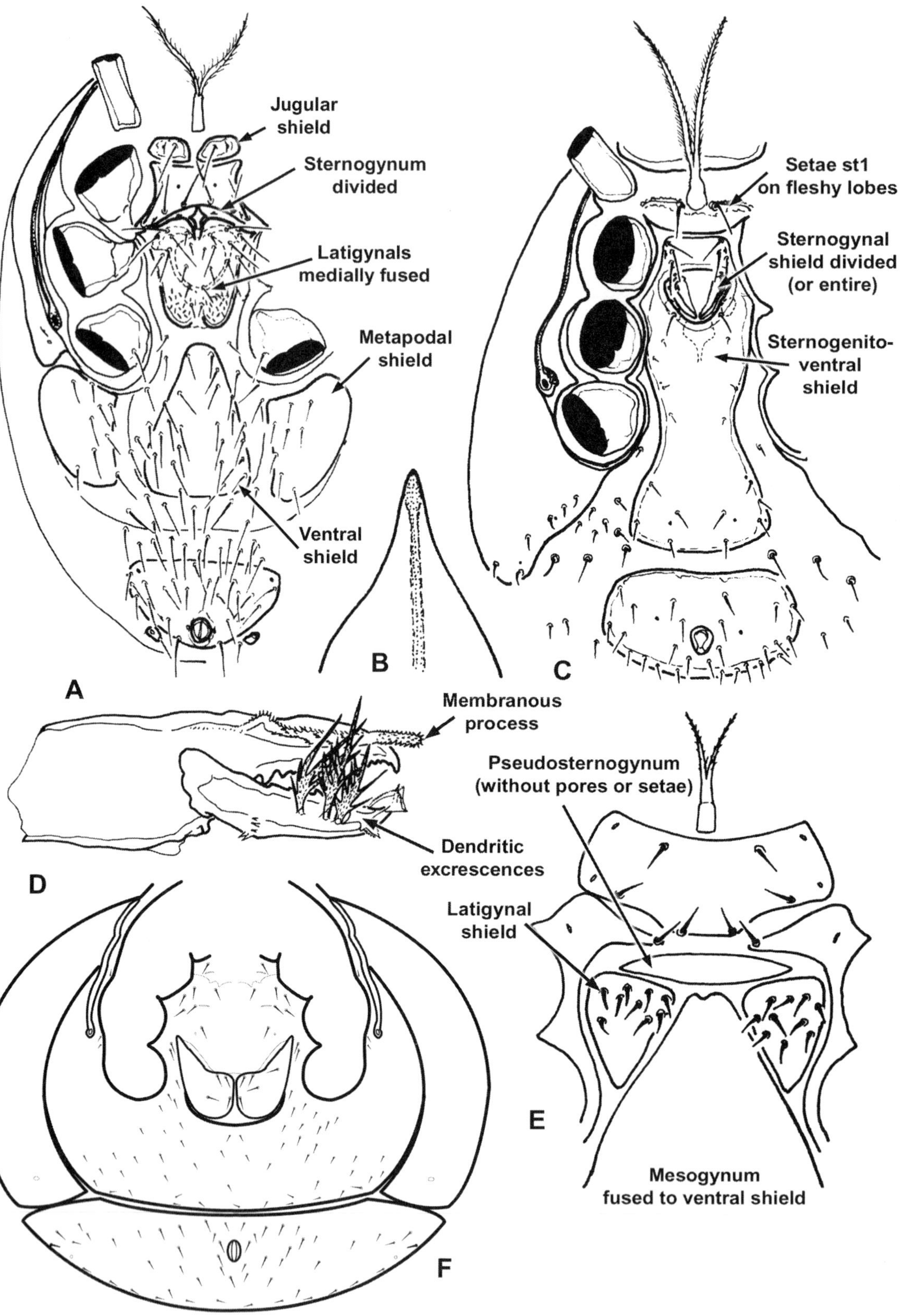

Fig. 12.17. HOPLOMEGISTIDAE. *Hoplomegistus* sp. (Brazil): **A**, venter of female.
MEGISTHANIDAE. *Megisthanus floridanus* Banks (Florida, USA) female: **B**, epistome; **C**, venter; **D**, chelicera.
PROMEGISTIDAE. *Promegistus armstrongi* Womersley (Australia): **E**, sternitigenital region of female.
PHILODANIDAE. *Philodana johnstoni* Kethley (Michigan, USA): **F**, venter of female.

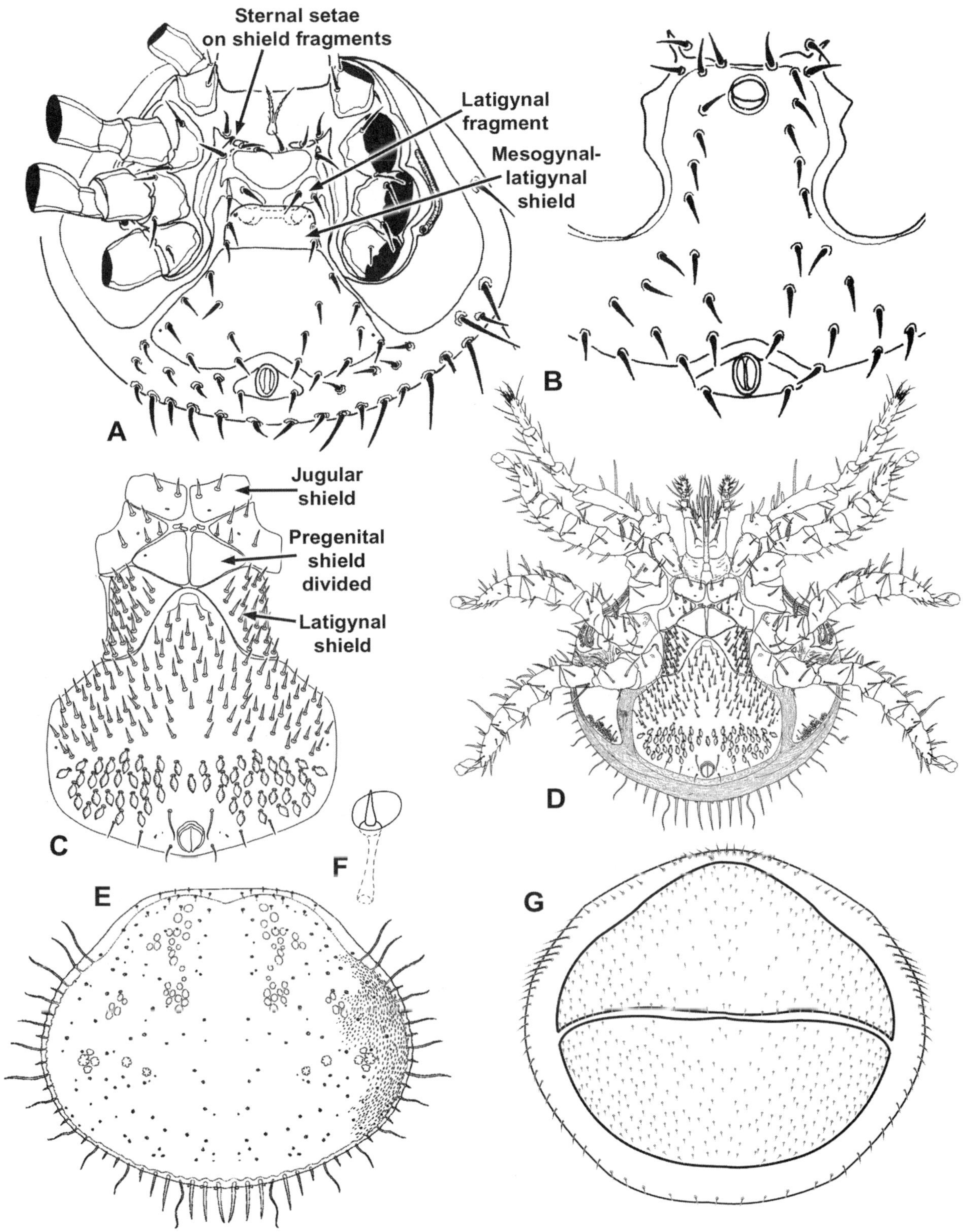

Fig. 12.18. PARANTENNULIDAE. *Micromegistus bakeri* Trägårdh (Ohio, USA): **A**, venter of female; **B**, venter of male.
PARAMEGISTIDAE. *Ophiomegistus* sp. (Australia) female: **C**, ventral complex of plates; **D**, venter; **E**, dorsal shield; **F**, detail of recessed dorsal seta.
PHILODANIDAE. *Philodana johnstoni* Kethley (Michigan, USA): **G**, dorsum of female.

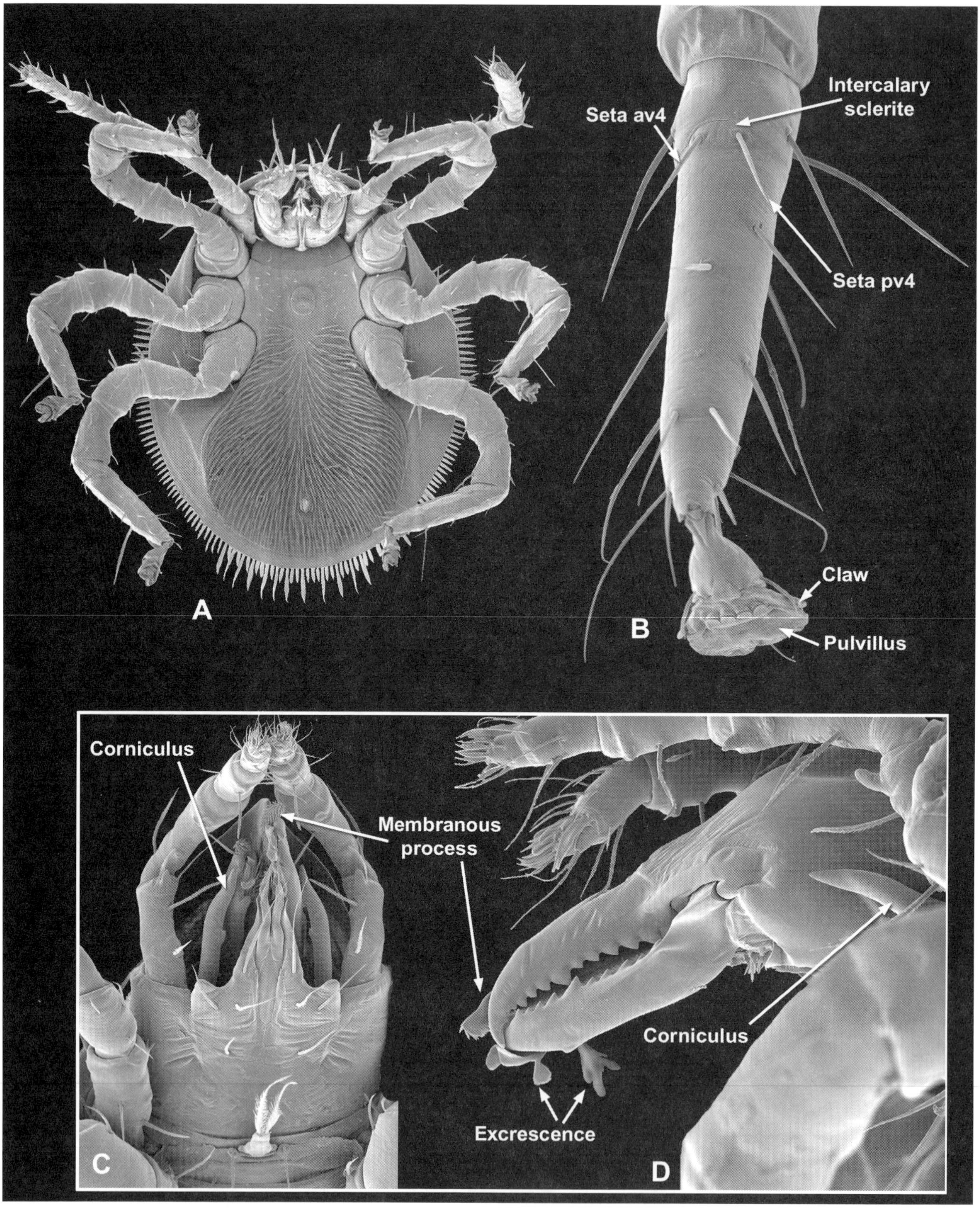

Fig. 12.19. PROMEGISTIDAE. *Promegistus armstrongi* Womersley (Australia): **A**, venter of male.
MEGISTHANIDAE. *Megisthanus* sp. (Australia) female: **B**, venter of tarsus IV; **C**, subcapitulum; **D**, lateral view of chelicera.

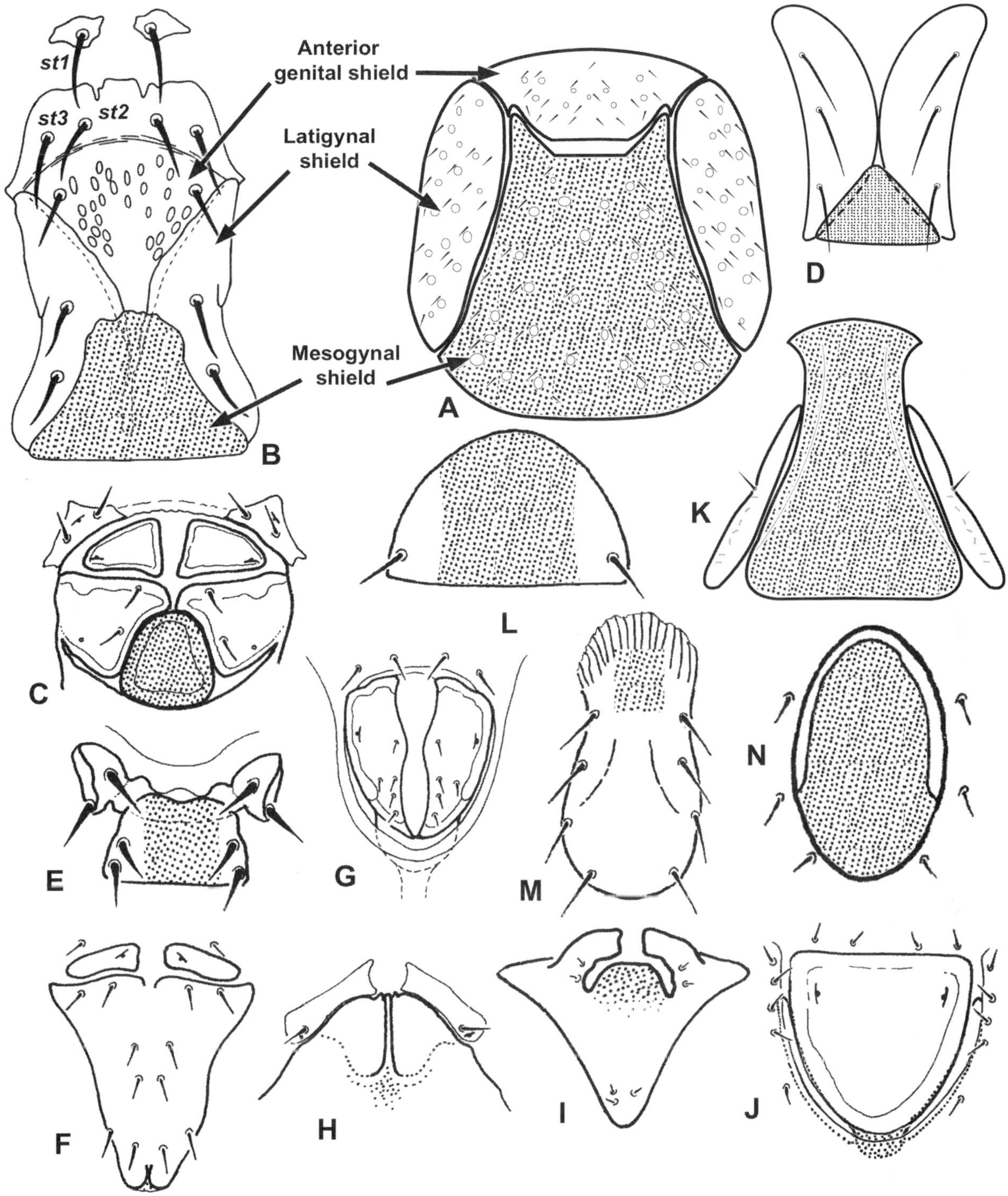

Fig. 12.20. Representative genital shields in adult female Parasitiformes. Stippled areas represent the posterior or median genital element or its hypothesized remnant. Holothyrida, quadrigynaspid condition (four genital shields): **A**, ALLOTHYRIDAE.

Trigynaspida, quadrigynaspid condition: **B**, DAVACARIDAE; **C**, PARAMEGISTIDAE.

Trigynaspida, trigynaspid condition (three genital shields) and various modifications: **D**, CERCOMEGISTIDAE; **E**, PARANTENNULIDAE; **F**, HOPLOMEGISTIDAE; **G**, MEGISTHANIDAE; **H**, EUZERCONIDAE; **I**, NEOTENOGYNIIDAE; **J**, FEDRIZZIIDAE.

Monogynaspida, trigynaspid condition (three genital shields): **K**, TRACHYTIDAE.

Monogynaspida, monogynaspid condition (one genital shield): **L**, OLOGAMASIDAE (Gamasiphinae), free genital shield; **M**, LAELAPIDAE (Hypoaspidinae), fused genitiventral shields; **N**, DIARTHROPHALLIDAE.

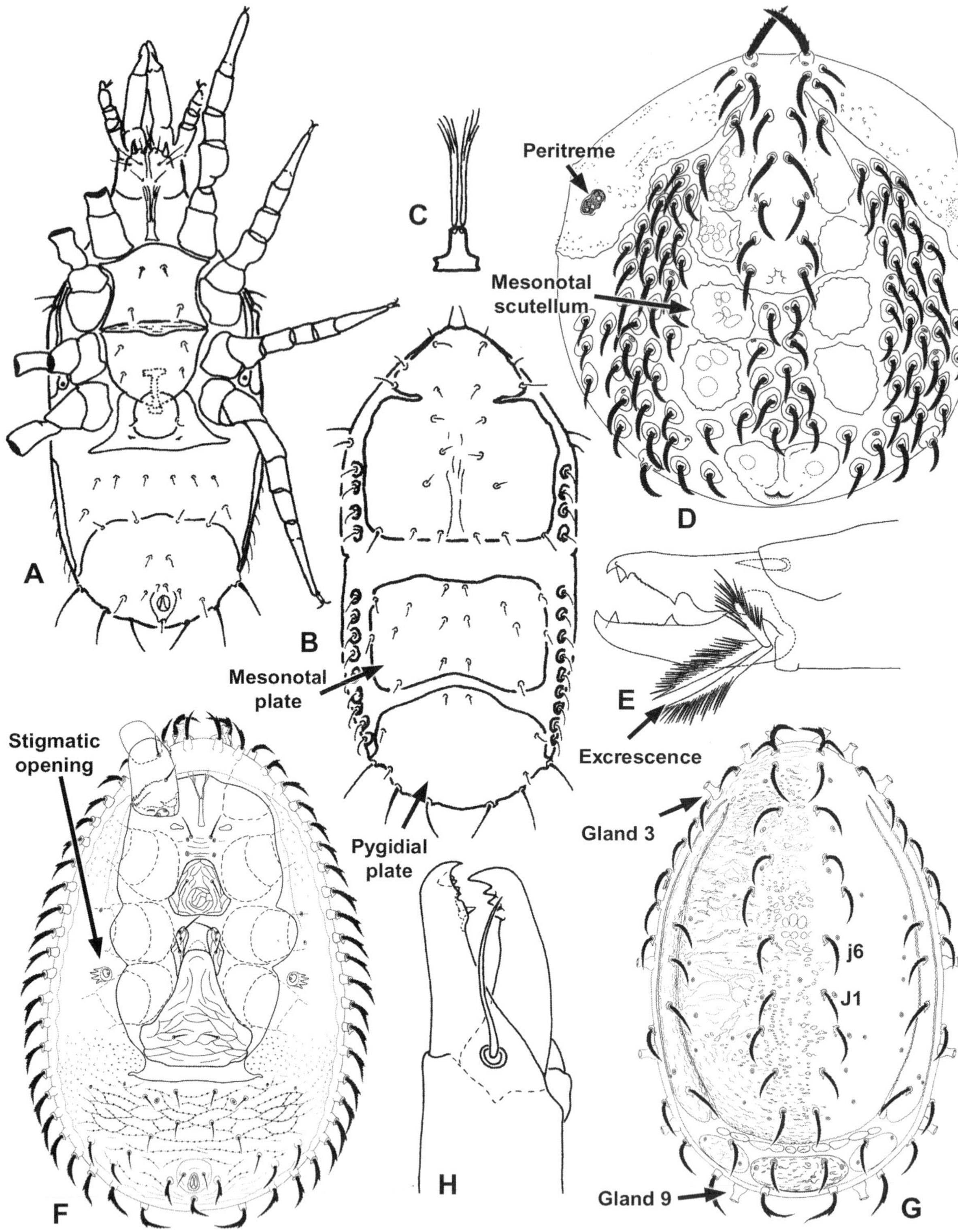

Fig. 12.21. MICROGYNIIDAE. *Microgynium incisum* Krantz (Oregon, USA): **A**, venter of female; **B**, dorsum of female; **C**, tritosternum.

NOTHOGYNIIDAE. *Nothogynus klompeni* Walter and Krantz (Queensland, Australia): **D**, dorsum of female; **E**, chelicera with excrescences.

HEATHERELLIDAE. *Heatherella acanthocharis* Walter (Australia): **F**, venter of female; **G**, dorsum of female. *H. callimaulos* Walter (Australia): **H**, chelicera of female.

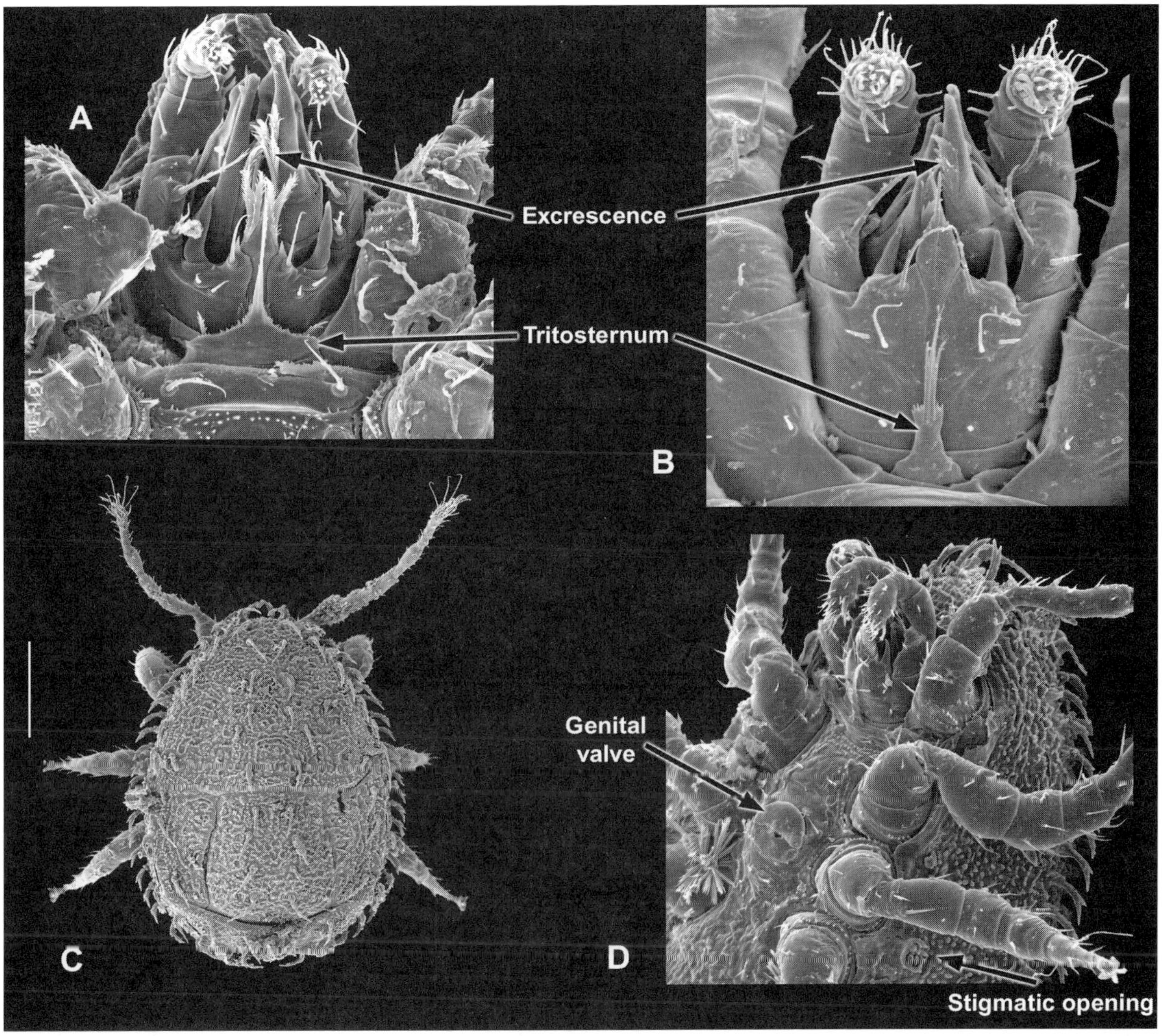

Fig. 12.22. NOTHOGYNIIDAE. *Nothogynus klompeni* Walter and Krantz (Queensland, Australia): **A**, tritosternum and venter of gnathosoma.

MICROGYNIIDAE. *Microgynium incisum* Krantz (Oregon, USA): **B**, tritosternum and venter of gnathosoma.

HEATHERELLIDAE. *Heatherella acanthocharis* Walter (Australia): **C**, dorsum of male; **D**, ventrolateral view of male showing genital opening.

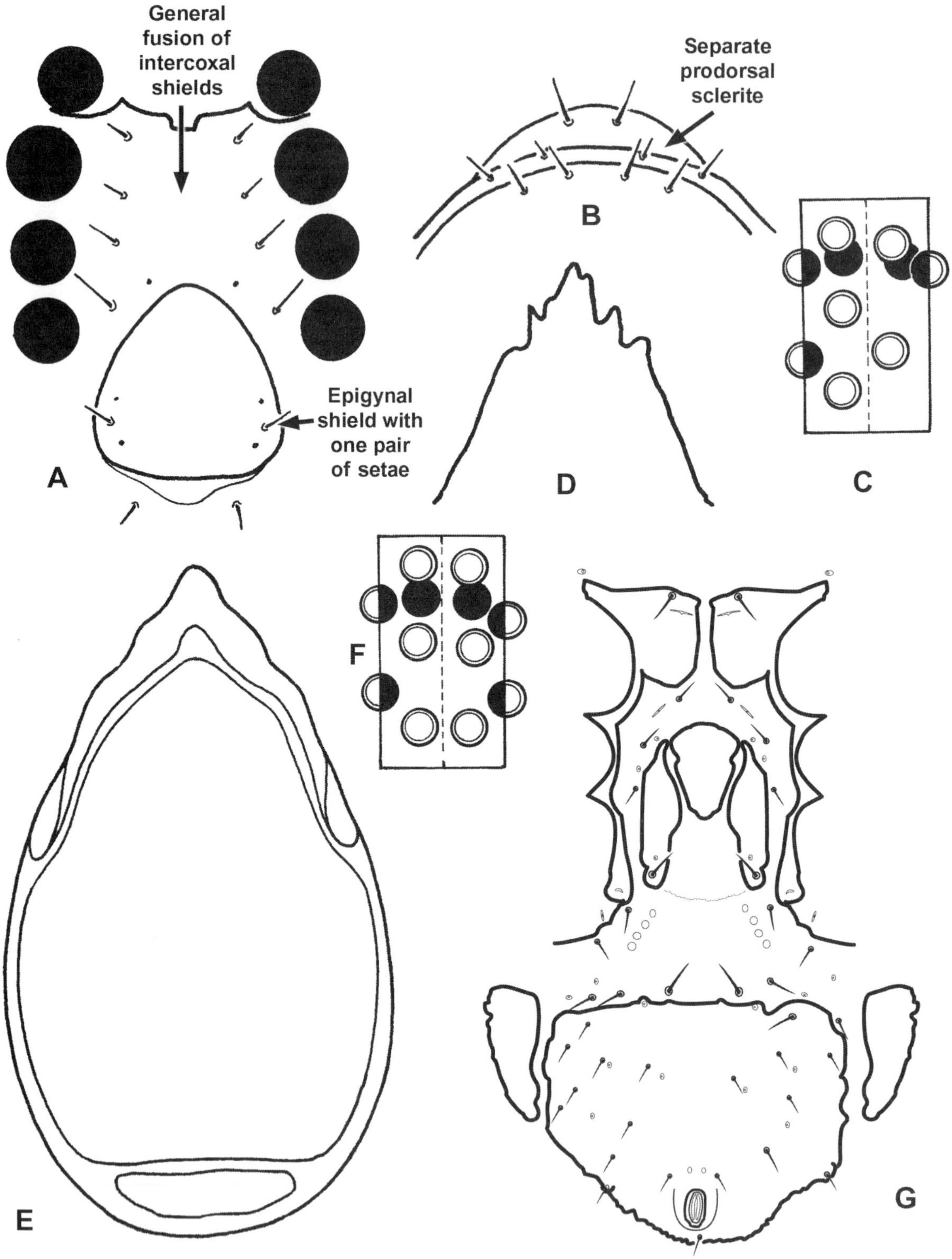

Fig. 12.23. PROTODINYCHIDAE. *Protodinychus* sp. (Canada): **A**, sternal and genital region of female. *P. punctatus* Evans (England): **B**, anterodorsal portion of idiosoma; **C**, chaetotaxy of genu I; **D**, epistome.

THINOZERCONIDAE. *Thinozercon michaeli* Halbert (Ireland): **E**, dorsum of female; **F**, chaetotaxy of genu I; **G**, venter of female (after Athias-Binche 1982).

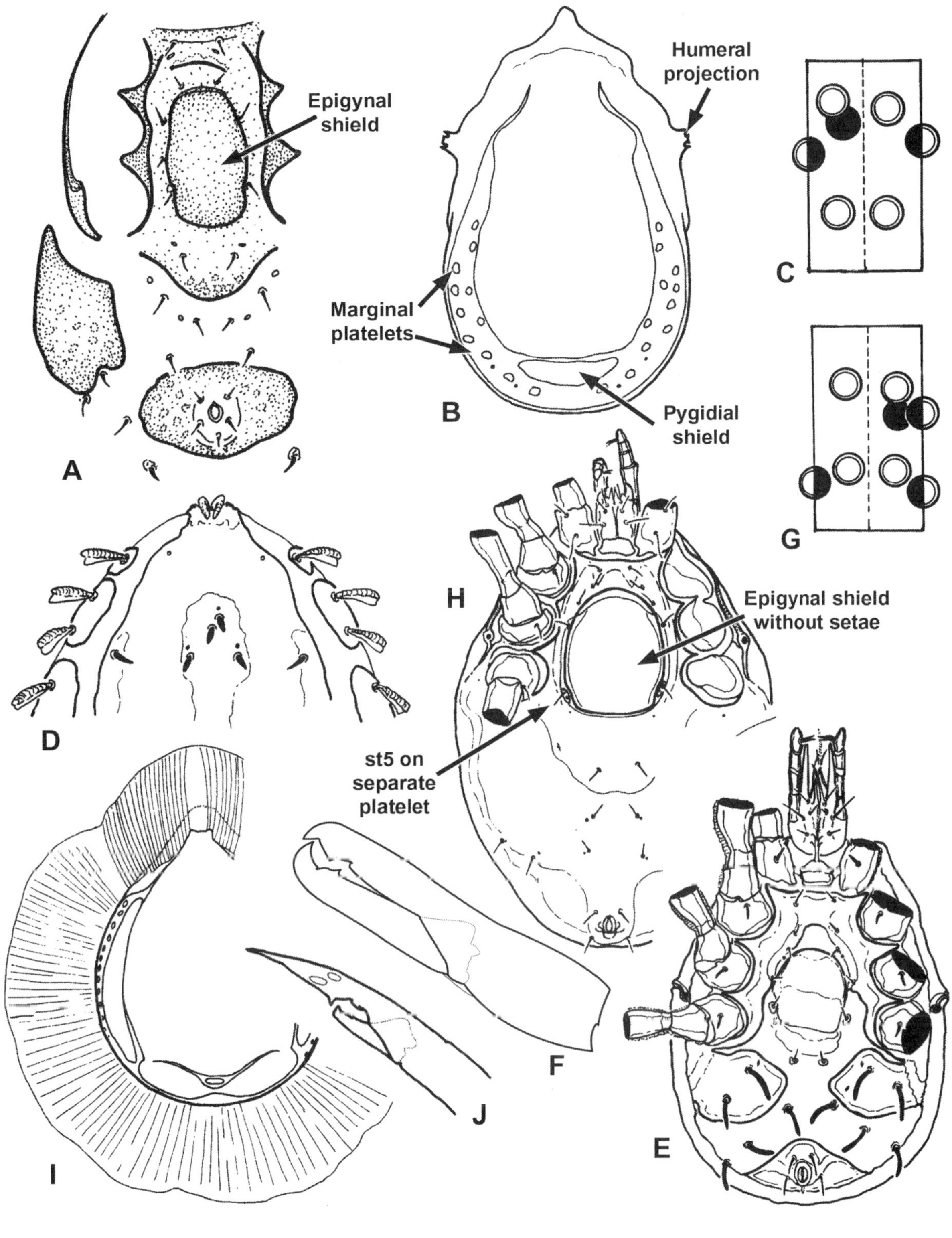

Fig. 12.24. DITHINOZERCONIDAE. *Apionoseius castrii* Hirschmann (Chile): **A**, venter of female. *Caminella peraphora* Krantz and Ainscough (Oregon, USA): **B**, dorsum of female; **C**, chaetotaxy of genu IV of many dithinozerconids.

POLYASPIDIDAE. *Polyaspis* sp. (Oregon, USA): **D**, anterodorsal aspect of idiosoma of female. *Polyaspis* sp. (Brazil): **E**, venter of female. *Polyaspis* sp. (Lord Howe Island): **F**, chelicera of female.

TRACHYTIDAE. *Polyaspinus* sp. (Oregon, USA): **G**, chaetotaxy of genu IV; **H**, venter of female. *Trachytes* sp. (Oregon, USA): **I**, dorsum of female; **J**, distal chelicera of female.

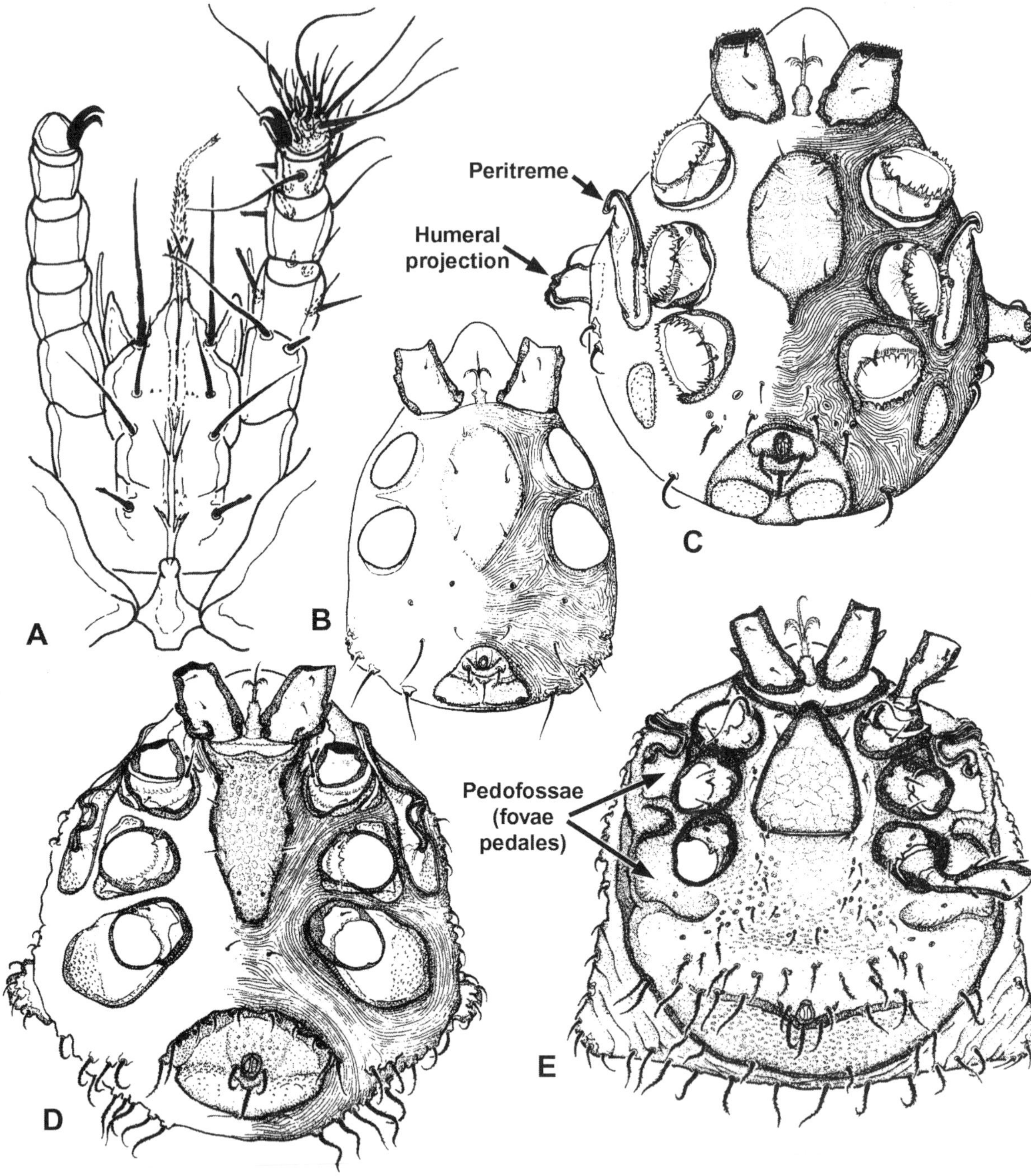

Fig. 12.25. UROPODIDAE. *Uropoda* (*Phaulodinychus*) *mitis* Leonardi (Italy), ventral aspect: **A**, venter of gnathosoma. *Deraiophorus maya* (Krantz): **B**, larva; **C**, protonymph; **D**, deutonymph; **E**, female.

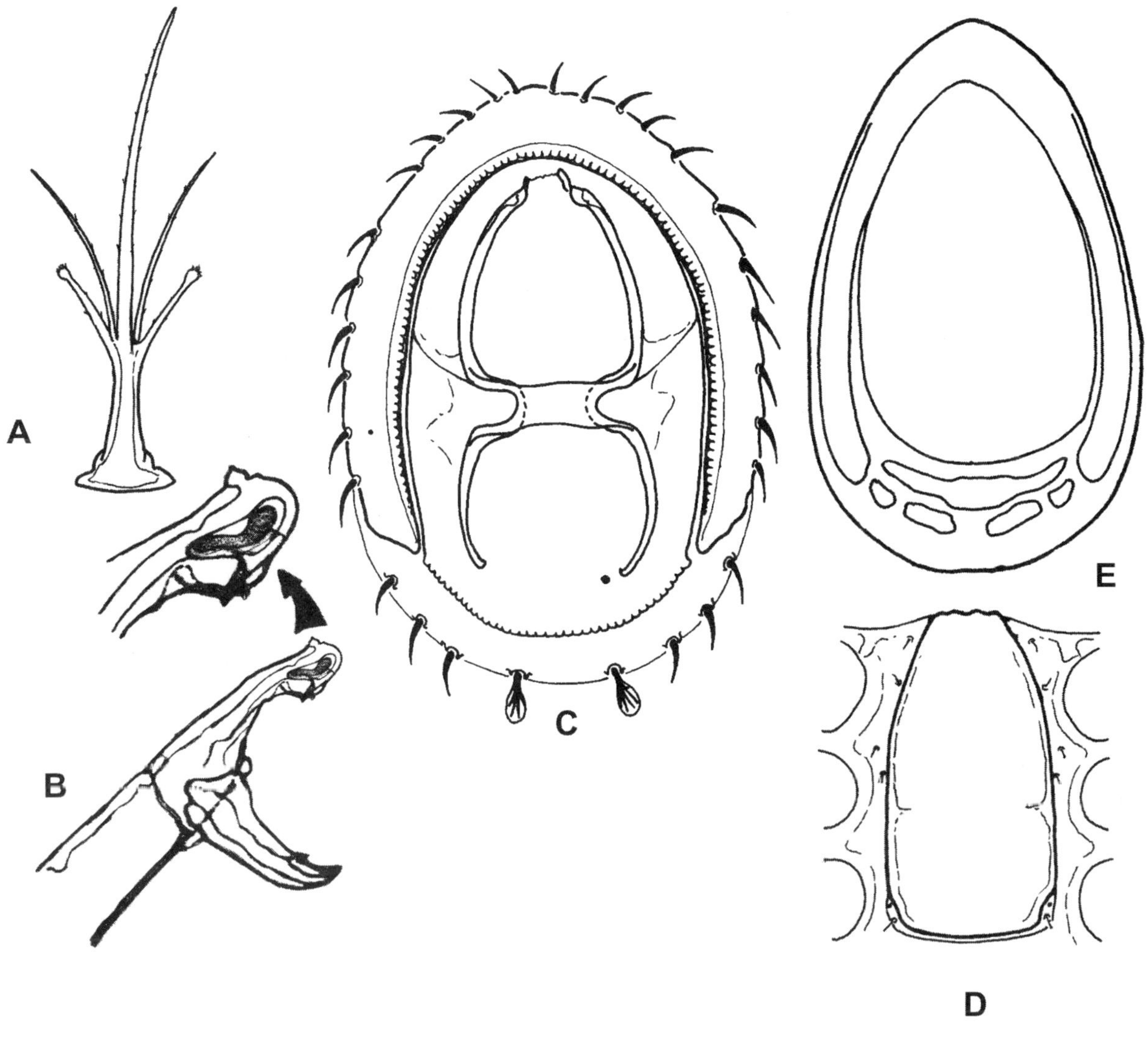

Fig. 12.26. UROPODIDAE. **A**, typical uropodid tritosternum. *Uropoda* (*Phaulodinychus*) *mitis* Leonardi (Italy): **B**, chelicera of female. *U.* (*Phaulodinychus*) *krantzi* Hirschmann (Zaire): **C**, dorsum. *Discourella* sp. (North America): **D**, sternitigenital region of female; **E**, dorsum of female (after Johnston 1961).

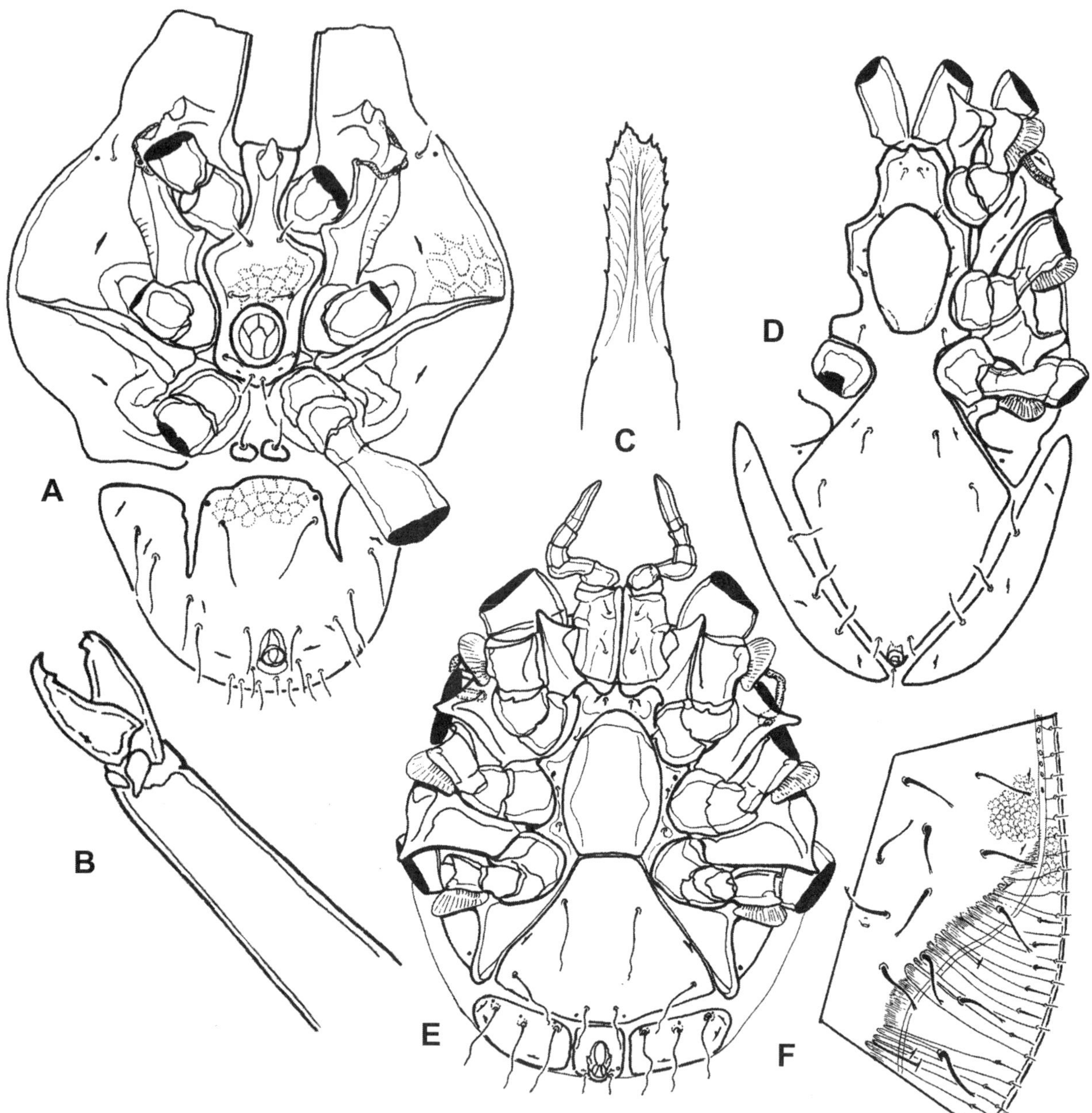

Fig. 12.27. UROPODIDAE. *Trichocylliba* subgenera. *T.* (*Coxequesoma*) *panamensis* (Hirschmann) (Panama): **A**, venter of male; **B**, chelicera of male. *T.* (*Planodiscus*) *burchelli* Elzinga and Rettenemeyer (Panama): **C**, epistome; **D**, venter of female. *T.* (*Circocylliba*) *krantzi* (Hirschmann) (Panama): **E**, venter of female; **F**, dorsal ornamentation and setation.

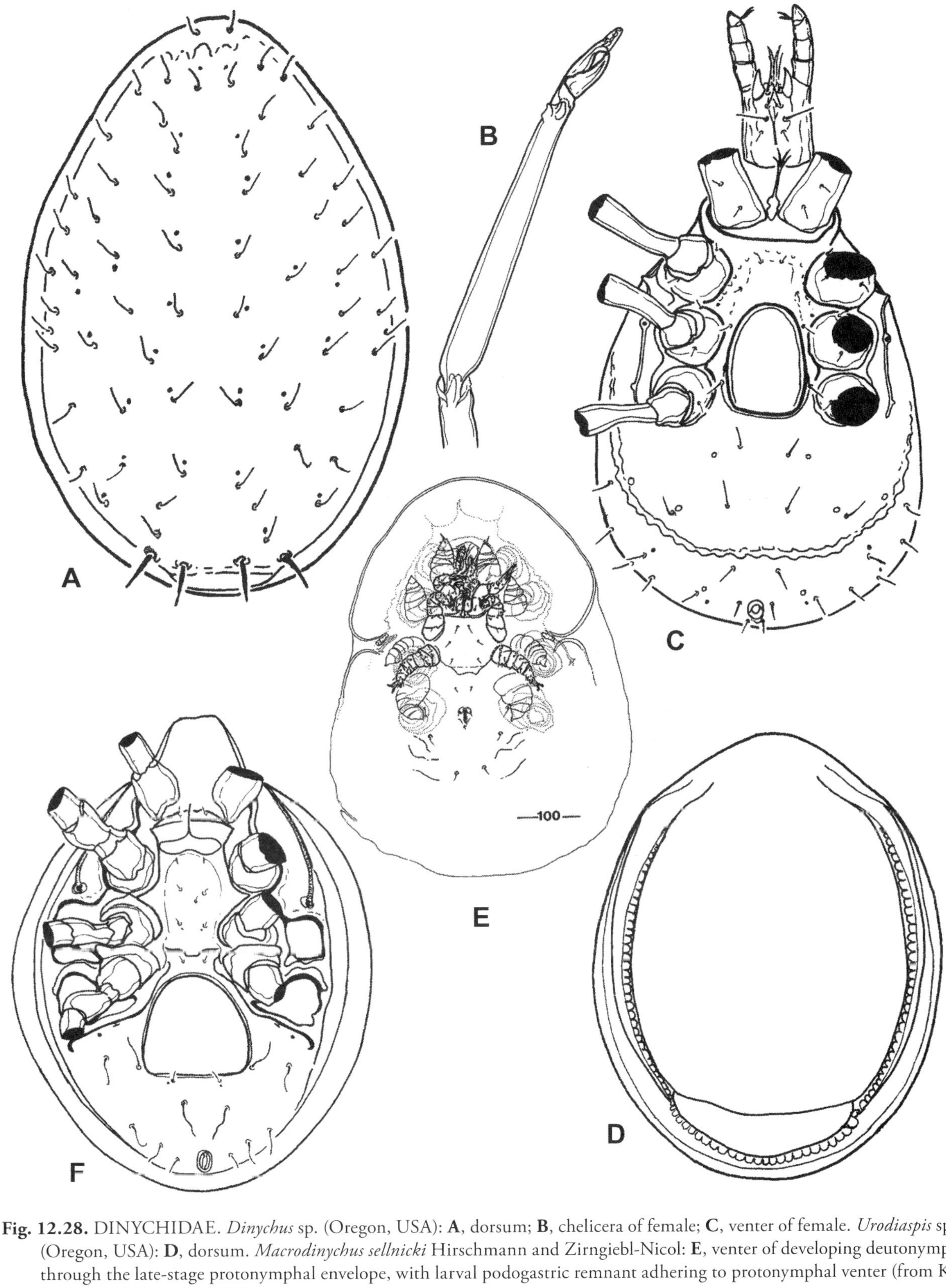

Fig. 12.28. DINYCHIDAE. *Dinychus* sp. (Oregon, USA): **A**, dorsum; **B**, chelicera of female; **C**, venter of female. *Urodiaspis* sp. (Oregon, USA): **D**, dorsum. *Macrodinychus sellnicki* Hirschmann and Zirngiebl-Nicol: **E**, venter of developing deutonymph viewed through the late-stage protonymphal envelope, with larval podogastric remnant adhering to protonymphal venter (from Krantz, Gomez and Gonzales 2007).

METAGYNURIDAE. *Metagynella* sp. (South Africa): **F**, venter of female.

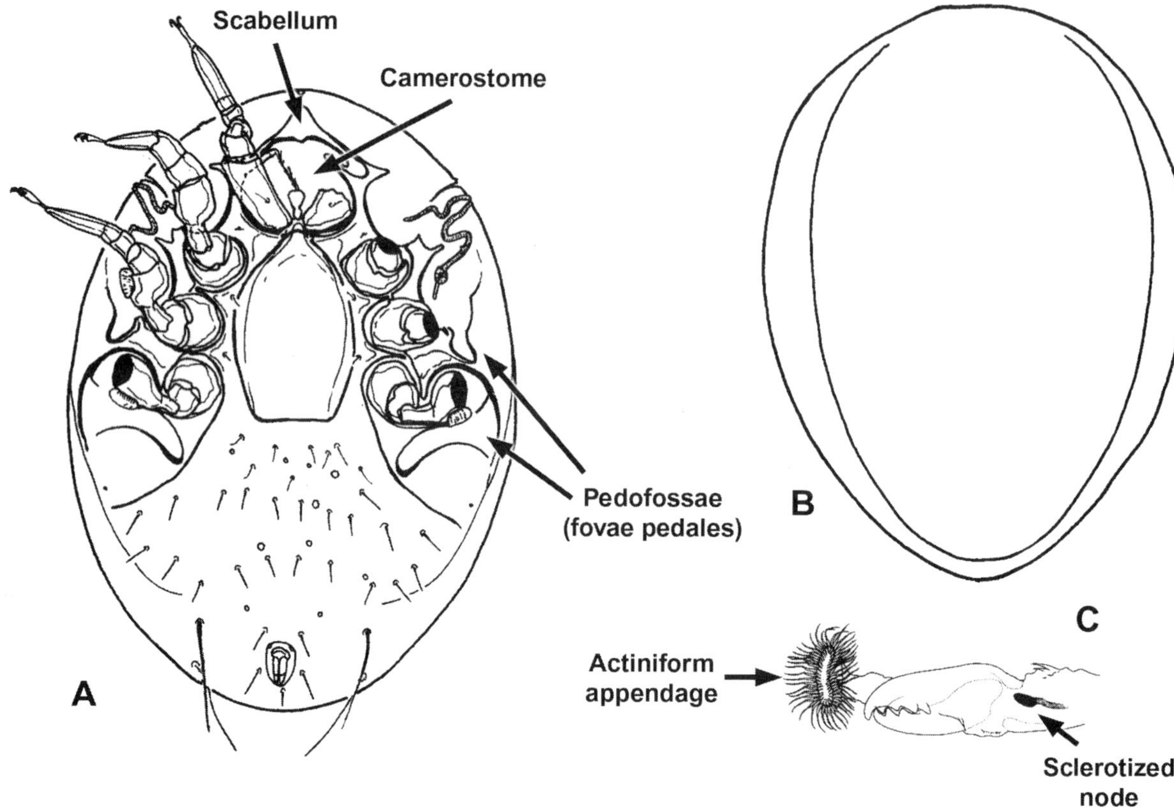

Fig. 12.29. TREMATURIDAE. *Trichouropoda krantzi* Hirschmann (Zaire): **A**, venter of female. UROACTINIIDAE. *Uroactinia agitans* (Banks): **B**, dorsum; **C**, chelicera of female.

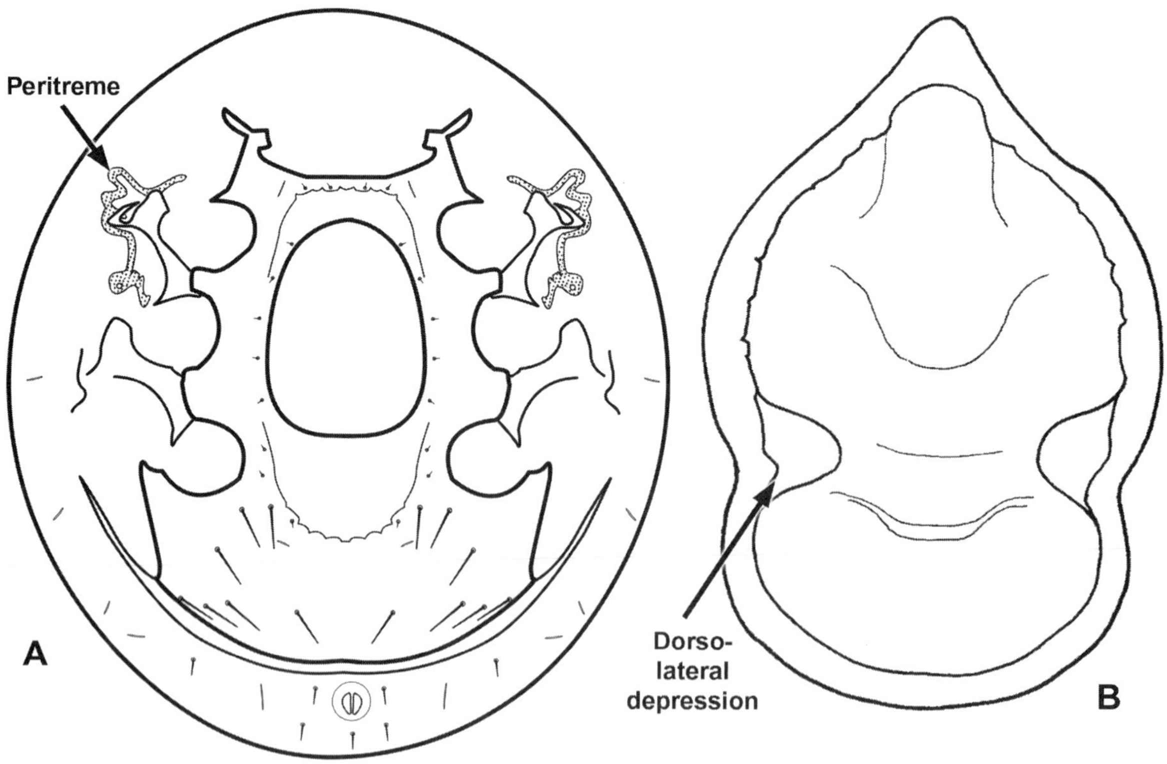

Fig. 12.30. OPLITIDAE. *Oplitis aktia* Hunter and Farrier (Southeast, USA): **A**, venter of female (after Hunter and Farrier, 1976). TRACHYUROPODIDAE. *Trachyuropoda.* sp. (Zaire): **B**, dorsum.

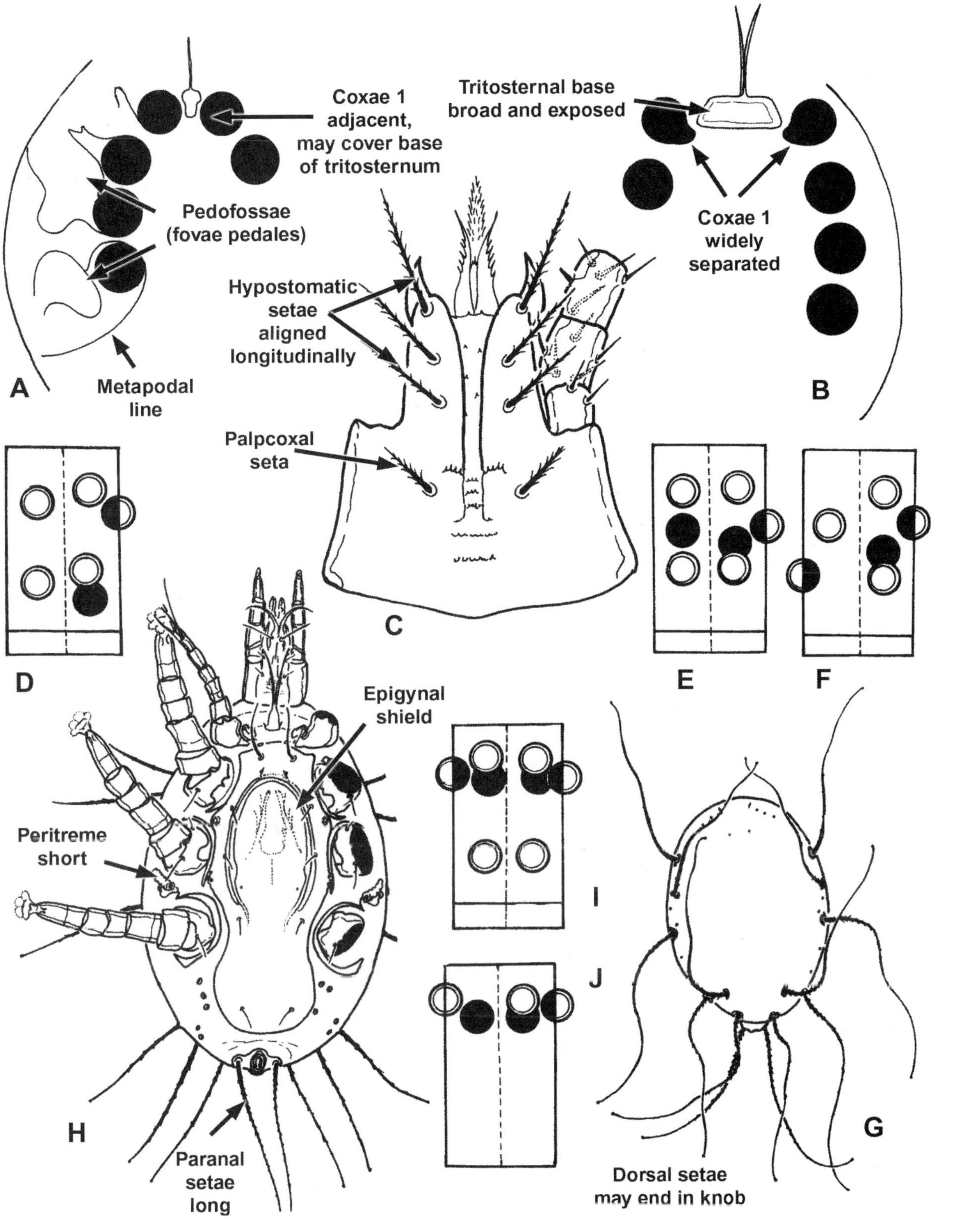

Fig. 12.31. Uropodina, venters of major uropodine types: **A**, Uropodoidea; **B**, Polyaspidoidea. **C**, typical uropodine gnathosoma. Chaetotaxies of femur IV in Uropodina: **D**, Thinozerconoidea; **E**, Polyaspidoidea; **F**, Uropodoidea.

DIARTHROPHALLIDAE. *Diarthrophallus quercus* (Pearse and Wharton) (Mexico): **G**, dorsum of female; **H**, venter of female; **I**, chaetotaxy of femur IV; **J**, chaetotaxy of tibia I.

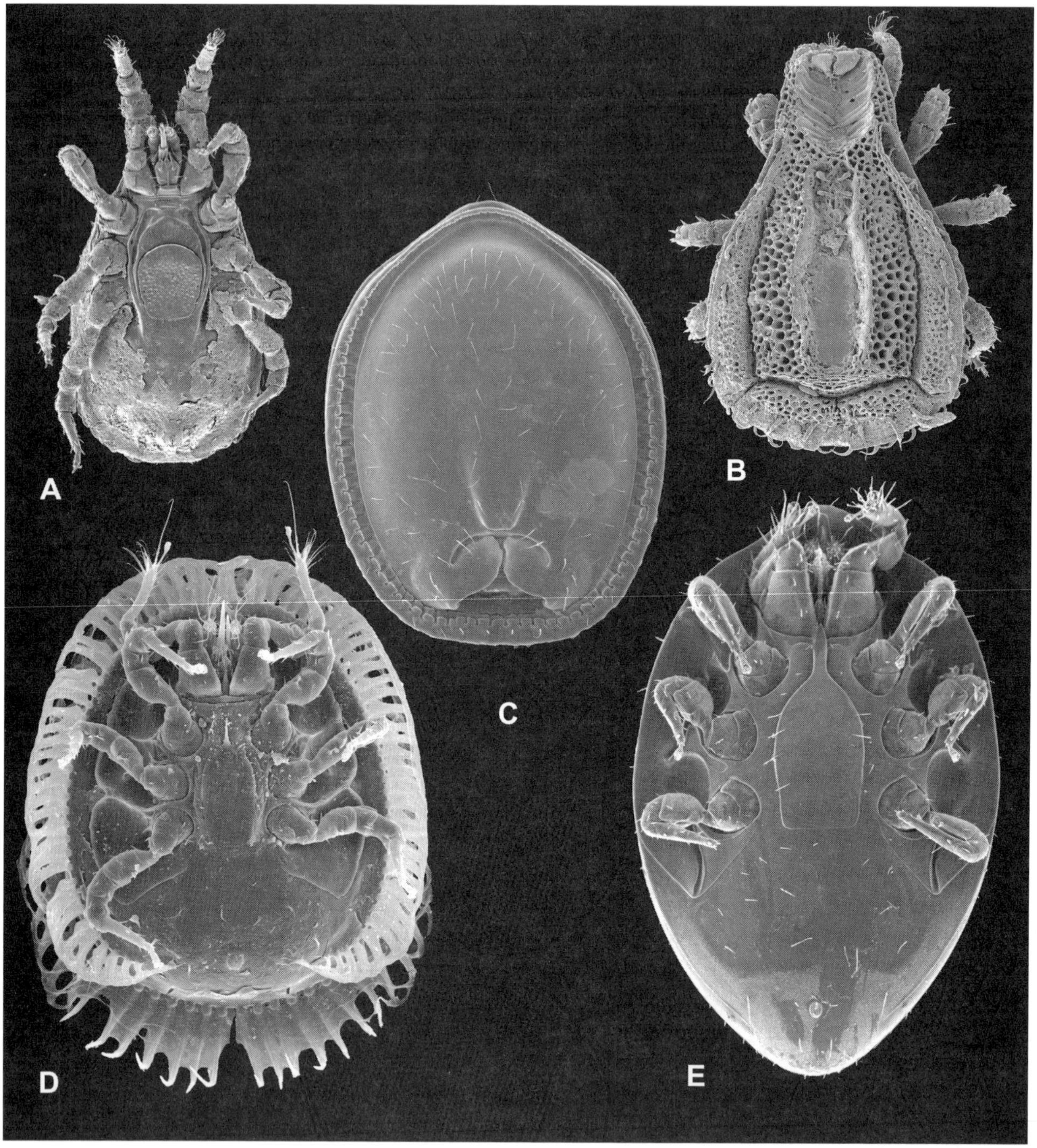

Fig. 12.32. TRACHYTIDAE. *Polyaspinus* sp. (Queensland, Australia): **A**, venter of female. *Trachytes* sp. (Oregon, USA): **B**, dorsum.
TRACHYUROPODIDAE? **C**, dorsum (British Columbia, Canada).
UROPODIDAE. *Clausiadinychus*? sp.: **D**, venter of female (Queensland, Australia).
DINYCHIDAE. *Uroobovella marginata* (Koch): **E**, venter of female (Queensland, Australia).

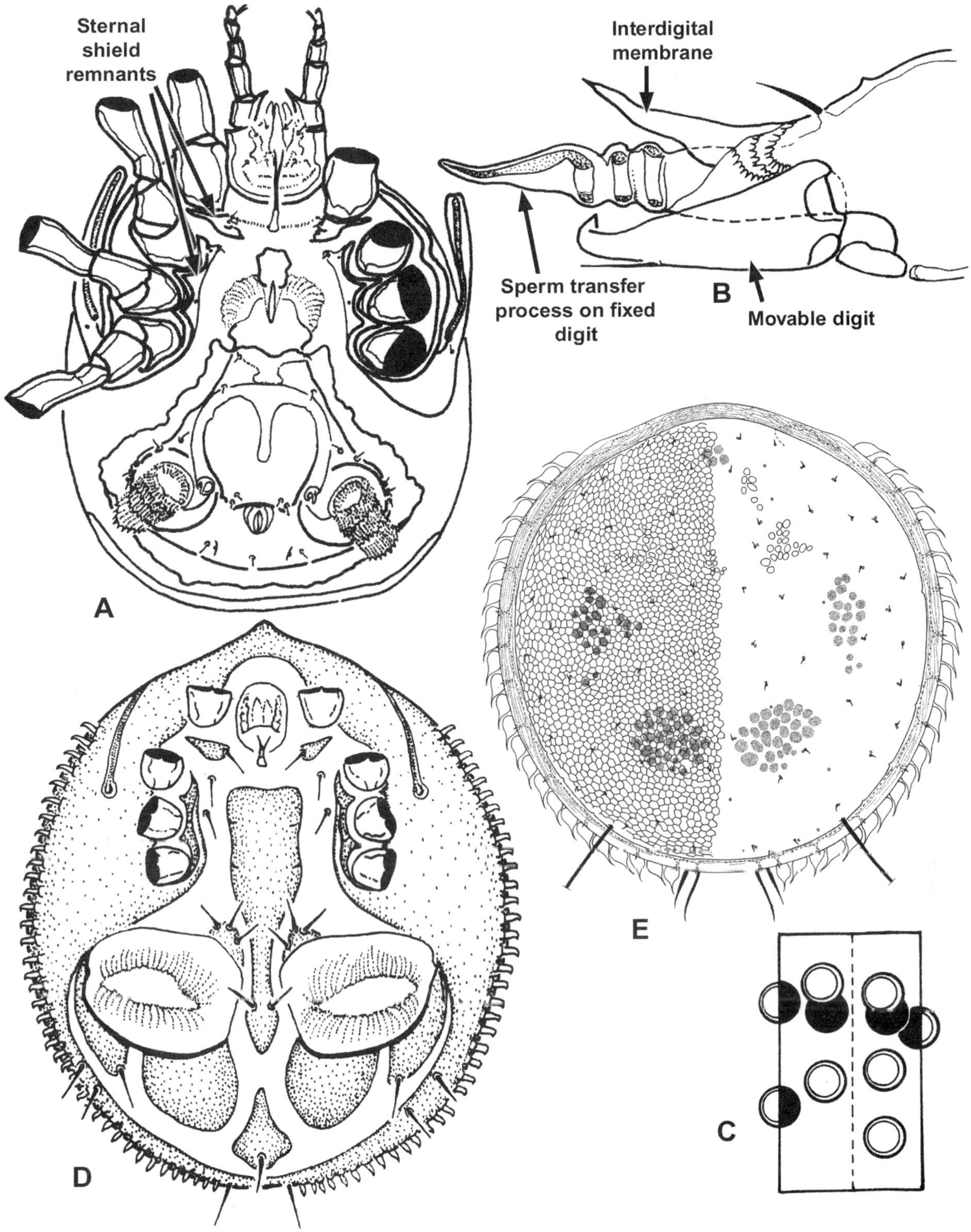

Fig. 12.33. HETEROZERCONIDAE. *Heterozercon* spp.: **A**, venter of female (Florida, USA); **B**, chelicera of male (Tanzania); **C**, chaetotaxy of genua III and IV (Tanzania).
DISCOZERCONIDAE. *Discozercon* spp.: **D**, venter of female (after Baker et al. 1958); **E**, dorsum (Queensland, Australia).

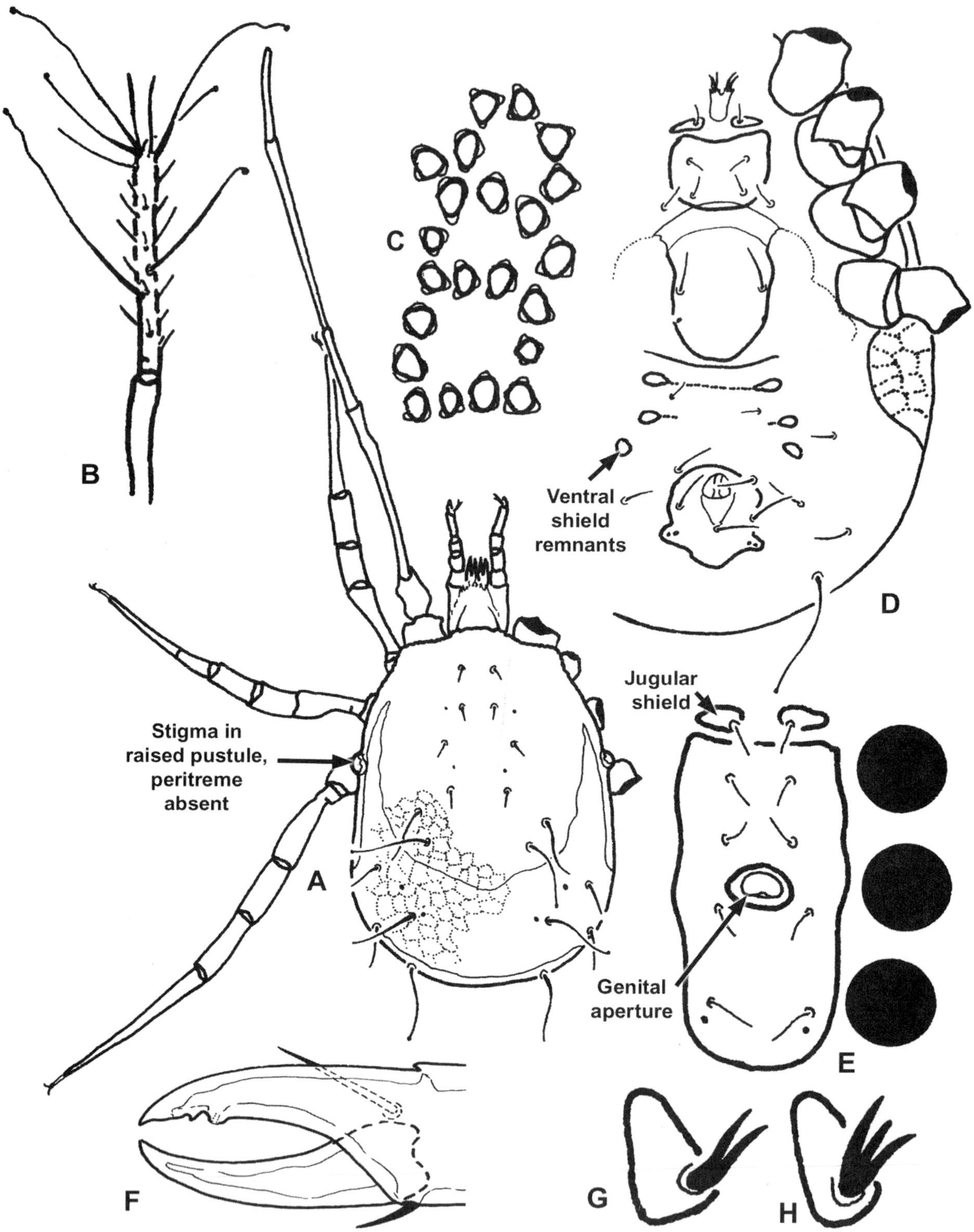

Fig. 12.34. EPICRIIDAE. *Neoepicrius* sp. (Oregon, USA): **A**, venter of female; **B**, tarsus I of female; **C**, dorsal ornamentation; **D**, venter of female; **E**, sternitigenital region of male. *Epicrius mollis* (Kramer): **F**, chelicera of female. **G**, **H**, epicriid palptarsal claw types.

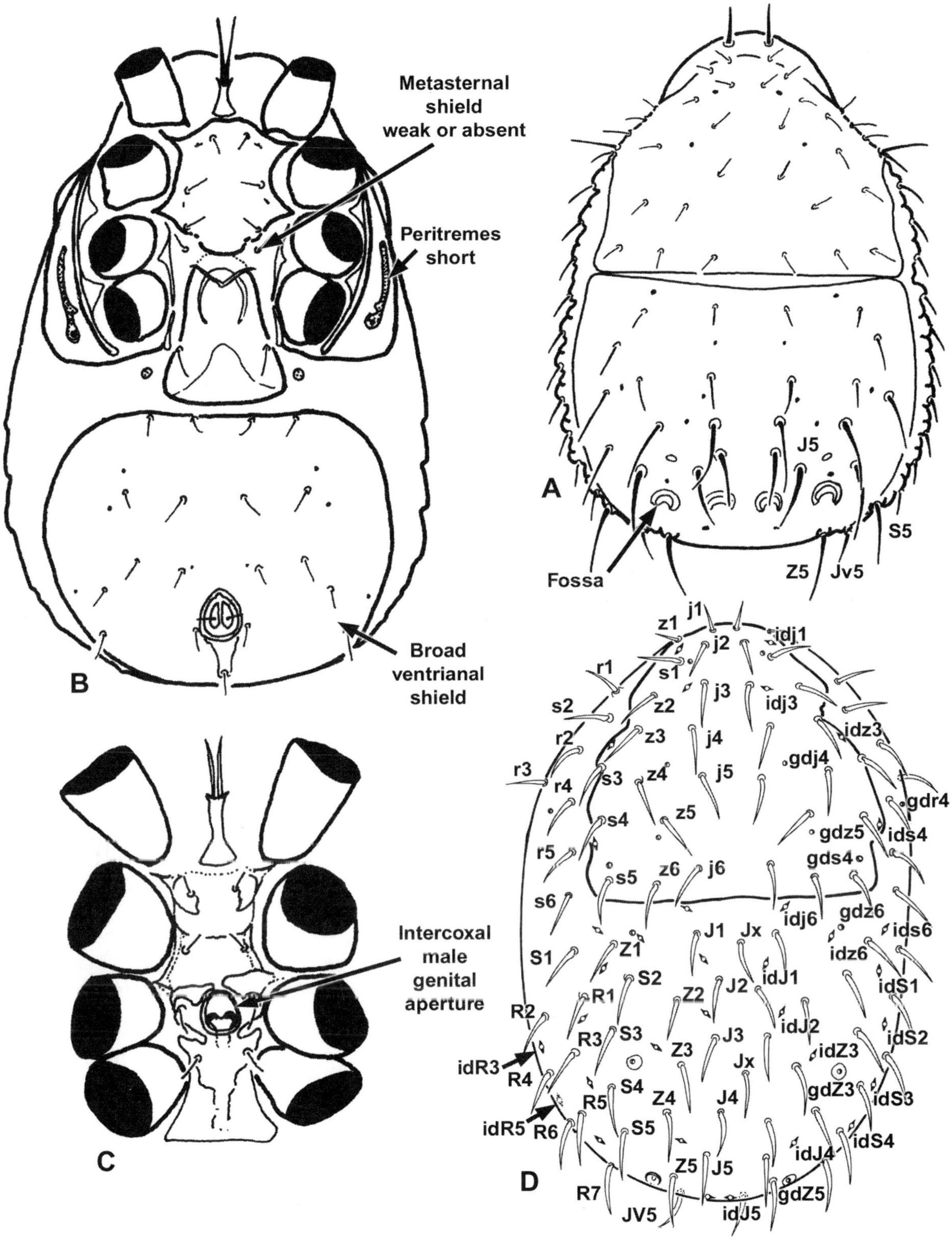

Fig. 12.35. ZERCONIDAE. *Zercon* sp. (Norway): **A**, dorsum of female. *Zercon* sp. (Oregon, USA): **B**, venter of female; **C**, podogastral region of male.

COPROZERCONIDAE. *Coprozercon scopaeus* Moraza and Lindquist: **D**, dorsum of adult female (after Moraza and Lindquist 1999).

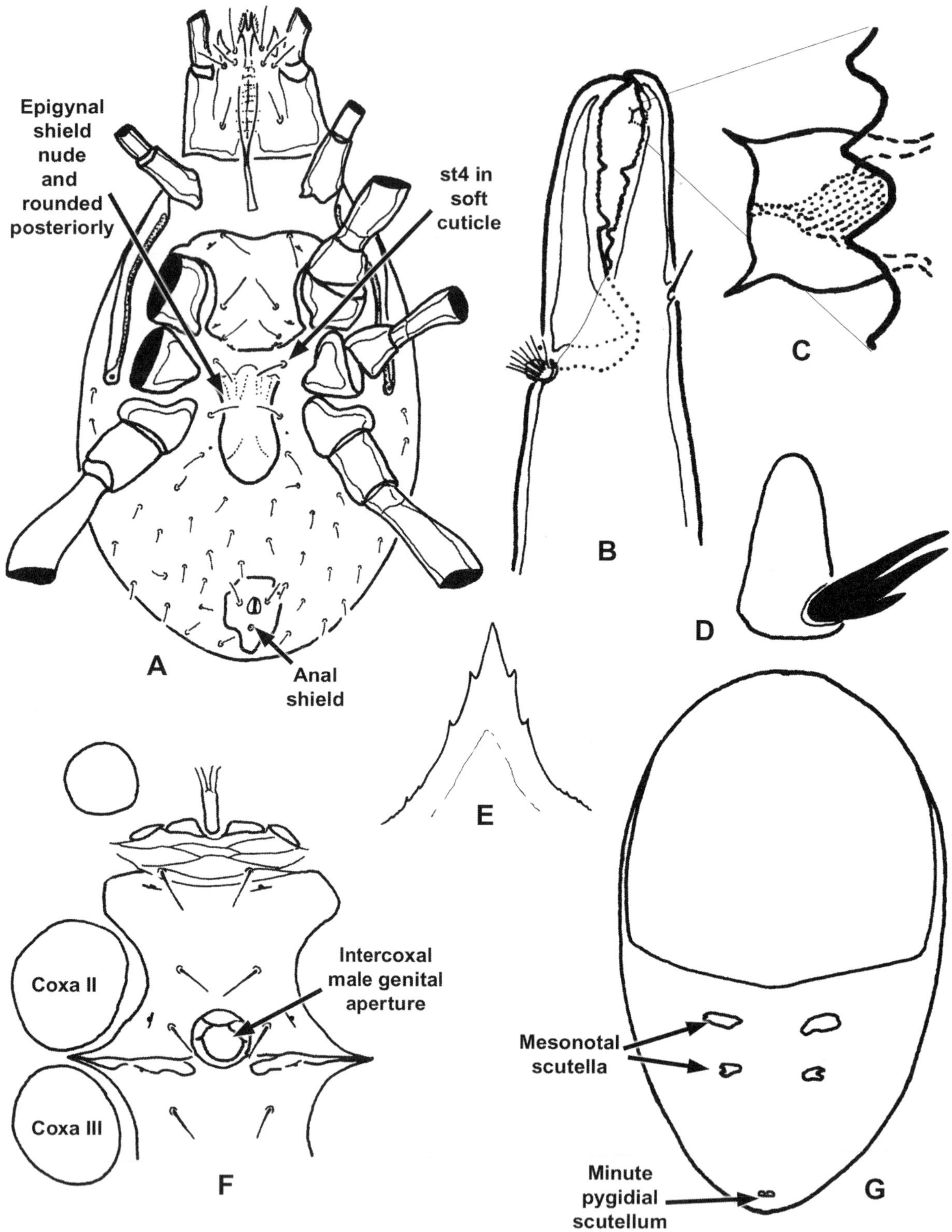

Fig. 12.36. ARCTACARIDAE. *Proarctacarus oregonensis* Makarova (Oregon, USA): **A**, venter of female; **B**, chelicera of female; **C**, detail of pilus dentilis; **D**, palptarsal claw; **E**, epistome of female; **F**, sternitigenital region of male; **G**, dorsum of female.

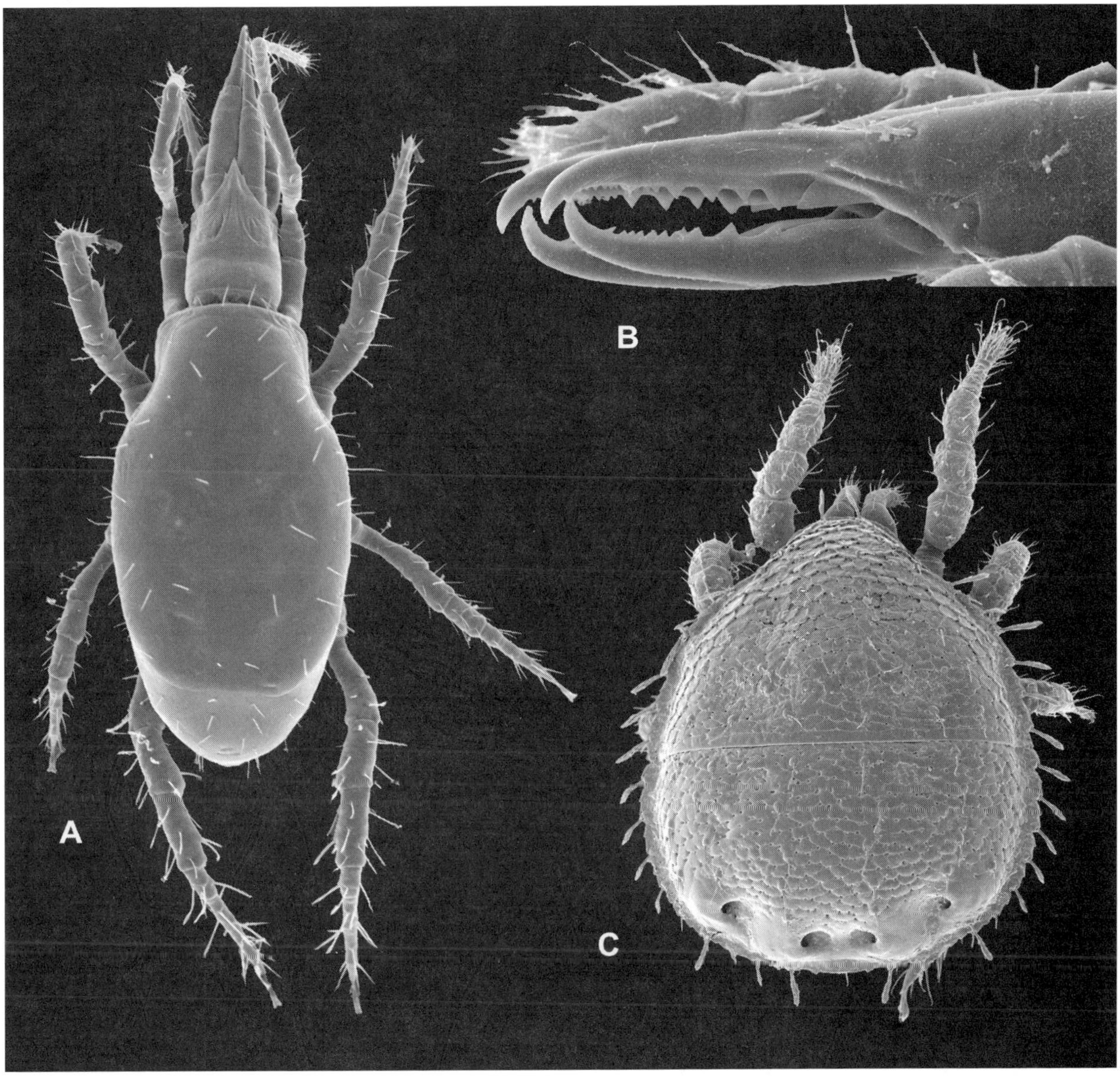

Fig. 12.37. ARCTACARIDAE. *Arctacarus rostratus* Evans (Yukon, Canada): **A**, dorsum of female; **B**, chelicera of female. ZERCONIDAE. **C**, dorsum of female.

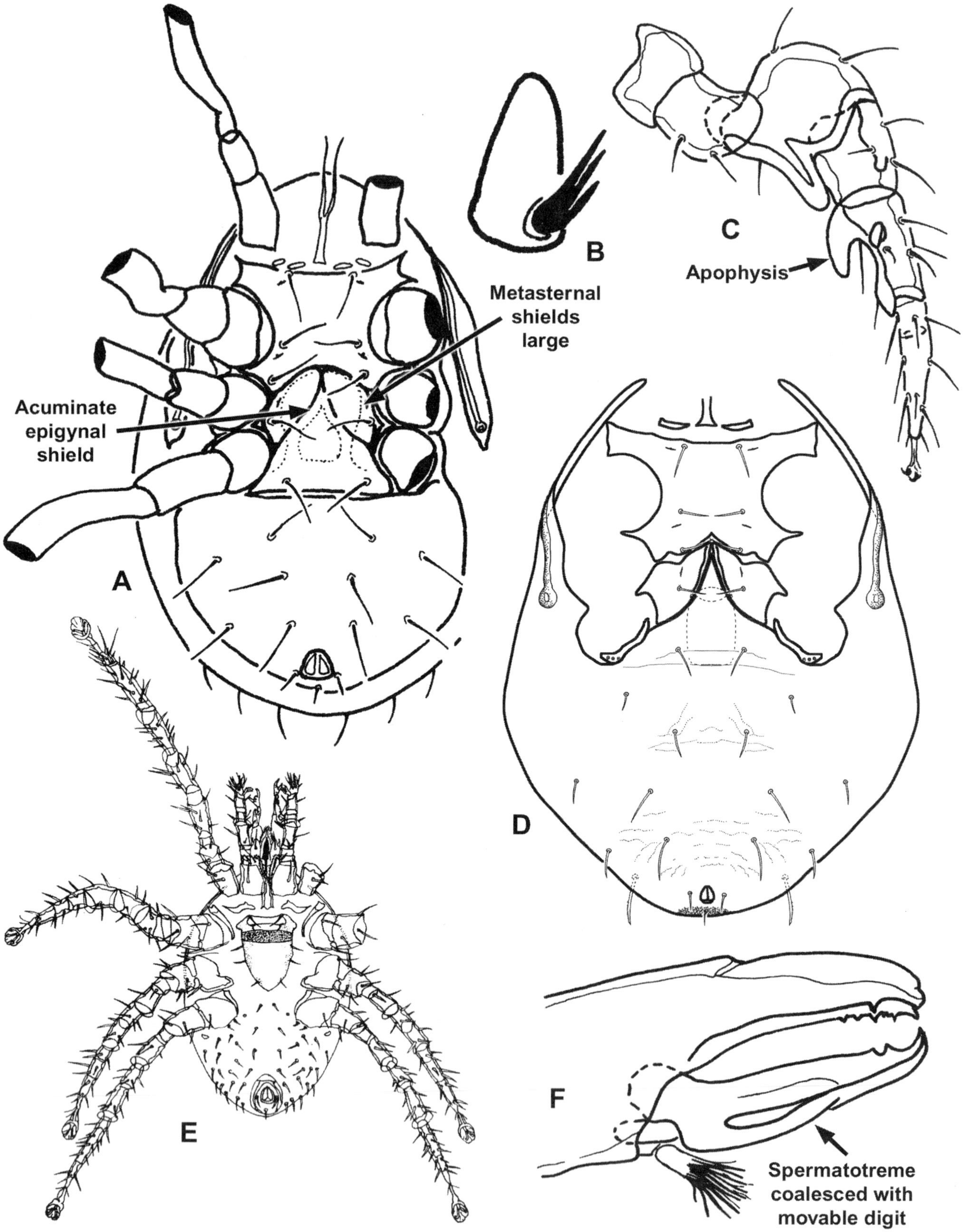

Fig. 12.38. PARASITIDAE. *Parasitus* sp. (Oregon, USA): **A**, venter of female; **B**, palptarsal claw; **C**, leg II of male. *Gamasodes queenslandicus* Halliday and Walter: **D**, venter of female. *Poecilochirus necrophori* Vitzthum (Oregon, USA): **E**, venter of deutonymph. *Parasitus diversus* Halbert: **F**, chelicera of male (after Karg 1971).

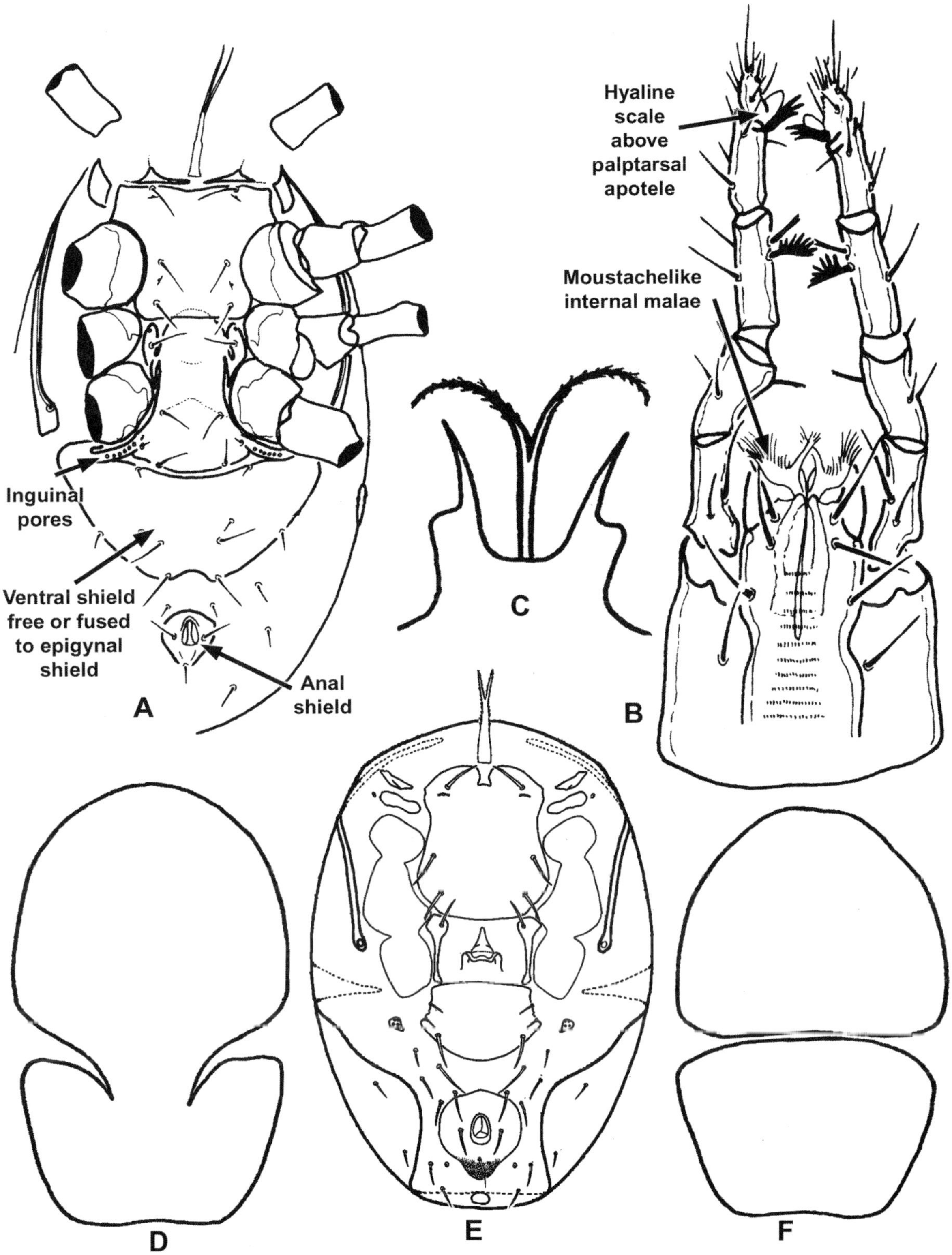

Fig. 12.39. VEIGAIIDAE. *Veigaia* sp. (Oregon, USA): **A**, venter of female; **B**, venter of female gnathosoma. *Gamasolaelaps* sp. (Queensland, Australia): **C**, epistome; **E**, venter of female; **D**, **F**, typical veigaiid dorsal shield configurations.

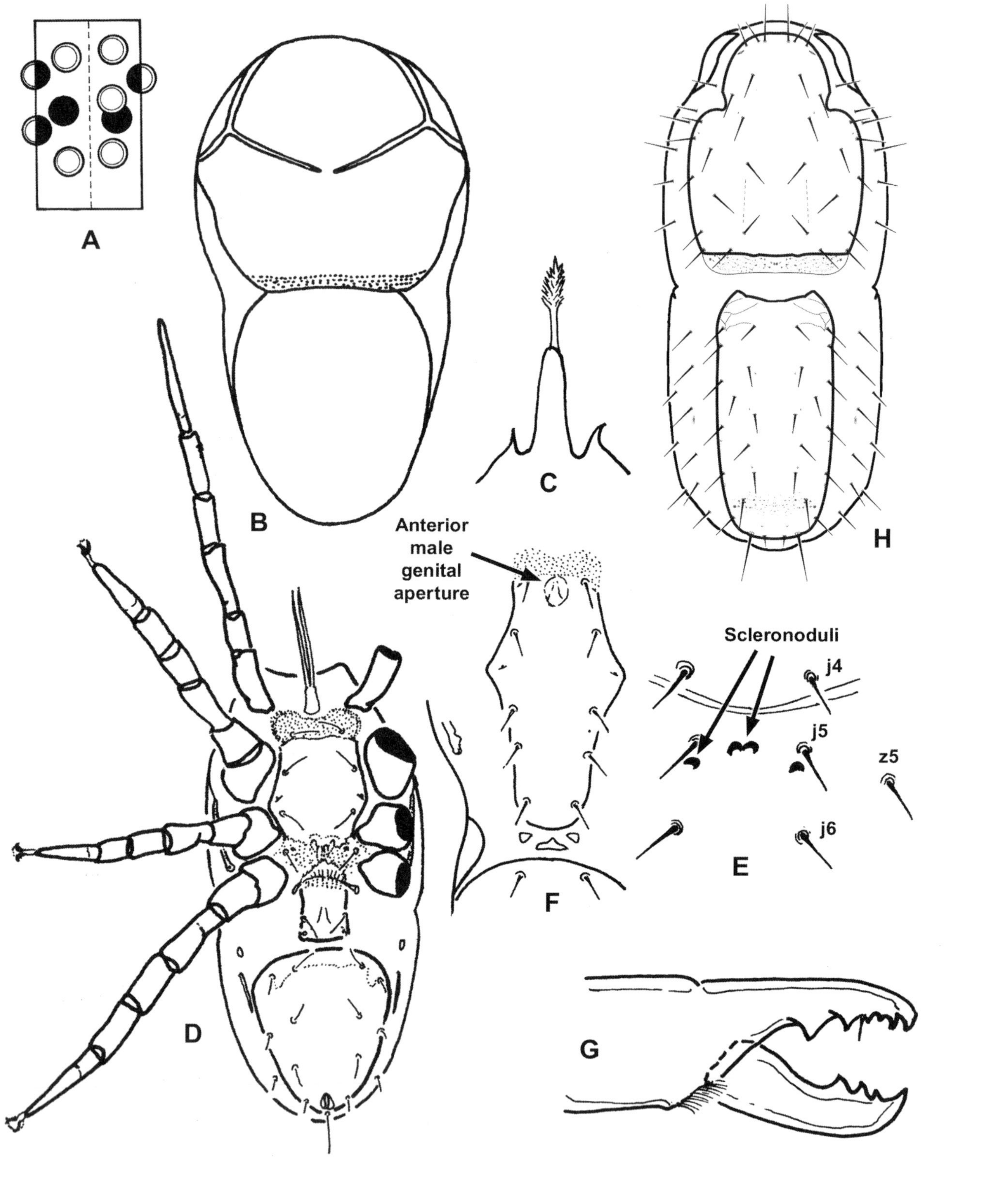

Fig. 12.40. RHODACARIDAE. *Rhodacarus roseus* Oudemans (Oregon, USA): **A**, chaetotaxy of genu IV; **B**, dorsum of female; **C**, epistome; **D**, venter of female; **E**, position of scleronoduli; **F**, sternitigenital shield of male; **G**, rhodacarid female chelicera. *Protogamasellopsis corticalis* Evans and Purvis: **H**, dorsum of female.

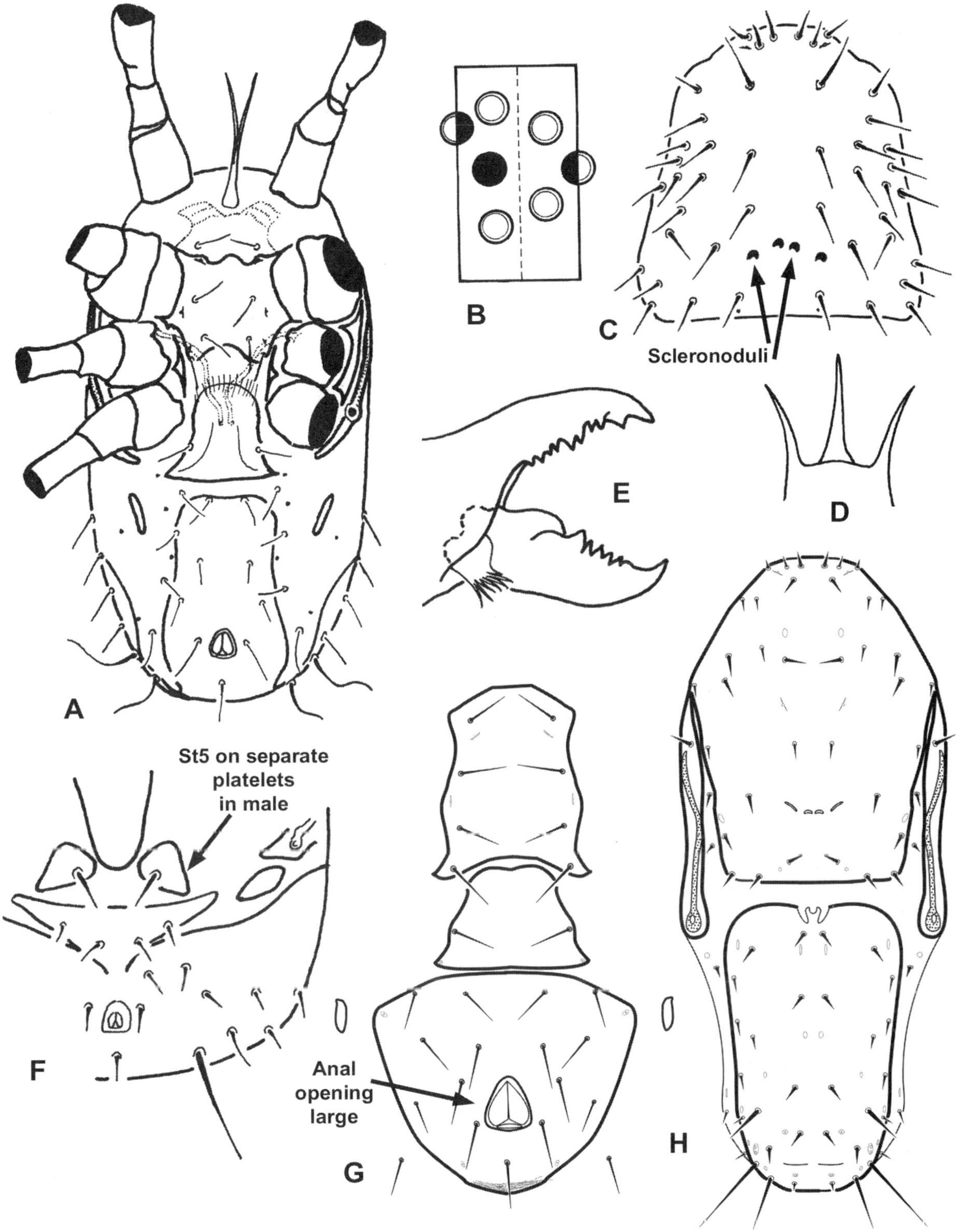

Fig. 12.41. DIGAMASELLIDAE. *Dendrolaelaps* sp. (Oregon, USA): **A**, venter of female; **B**, chaetotaxy of genu IV; **C**, podonotal shield of female; **D**, epistome of female; **E**, chelicera of female. *D. ulmi* Hirschmann: **F**, posteroventral aspect of male. *Digamasellus* sp. (Queensland, Australia): **G**, ventral plates of female. *Longoseius brachypoda* (Hurlbutt): **H**, dorsum of female.

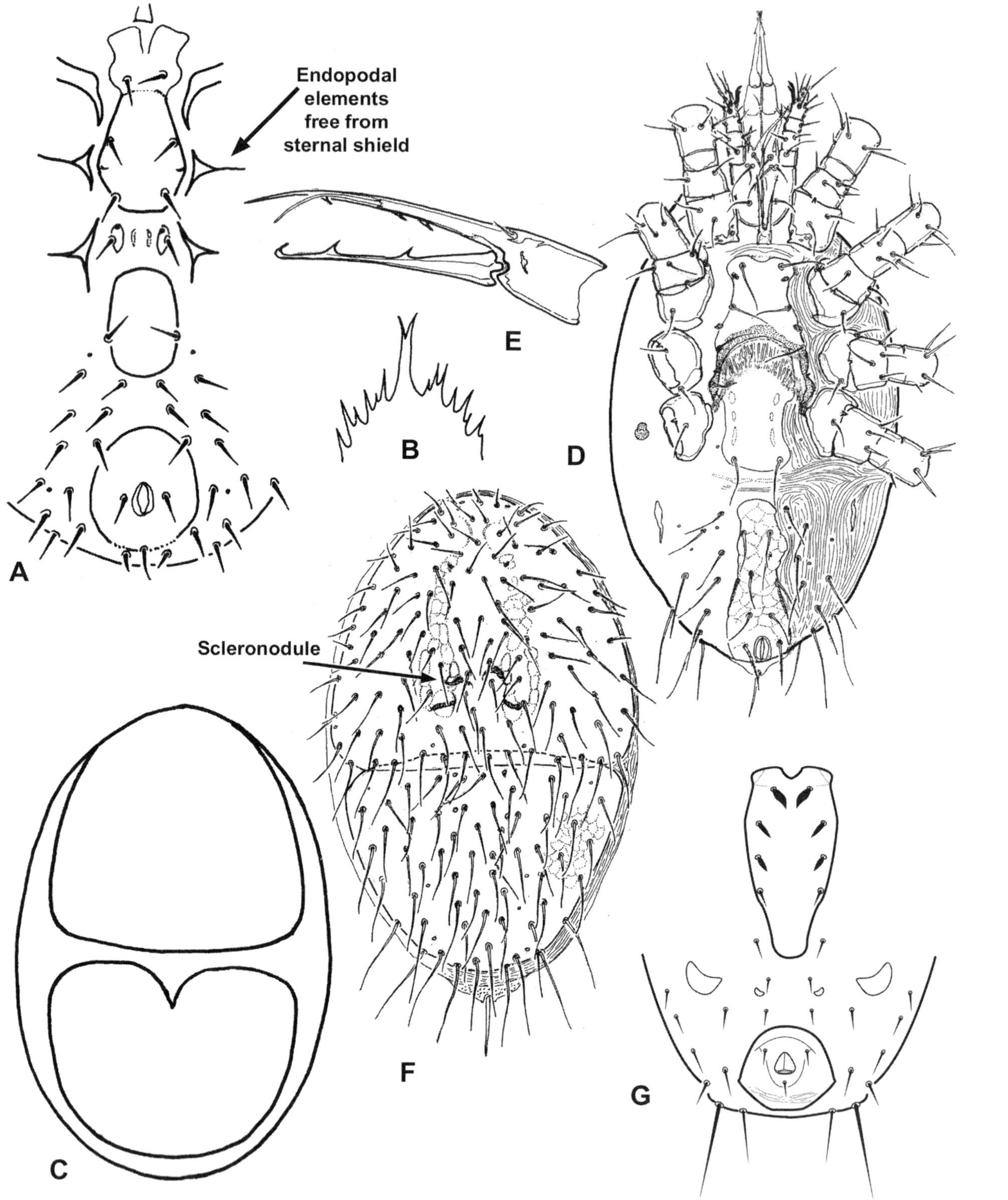

Fig. 12.42. HALOLAELAPIDAE. *Halolaelaps* sp. (Oregon, USA): **A**, venter of female; **B**, epistome of female; **C**, typical halolaelapid dorsal shield conformation.

LAELAPTONYSSIDAE. *Laelaptonyssus setosus* Krantz (Mississippi, USA): **D**, venter of female; **E**, chelicera of female; **F**, venter of female (from Krantz 2000).

DIGAMASELLIDAE. *Panteniphis mirandus* Willmann (Poland): **G**, venter of deutonymph (after Gwiazdowicz 2000a).

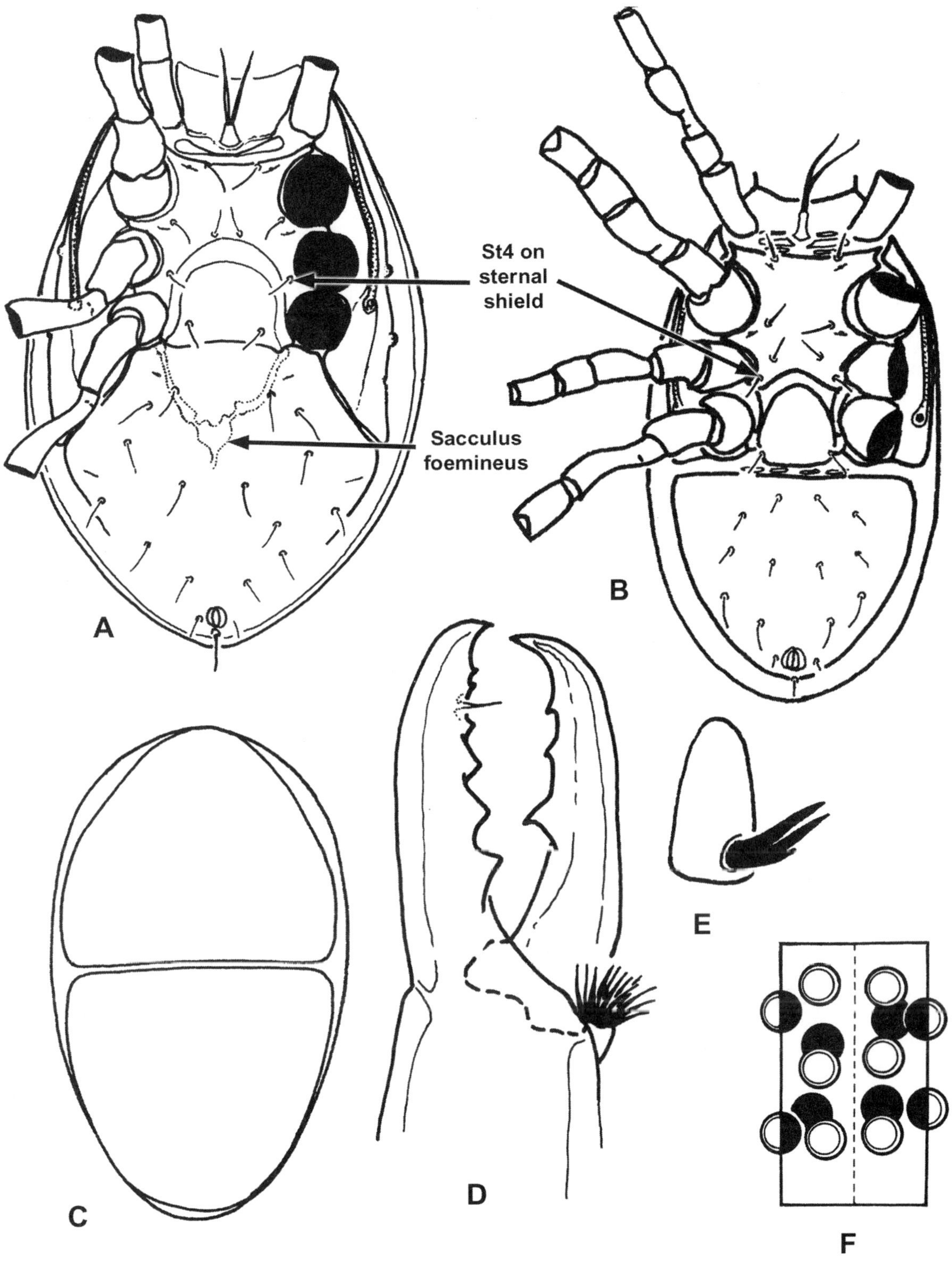

Fig. 12.43. OLOGAMASIDAE. *Gamasiphis* sp. (India): **A**, venter of female. *Gamasellus* sp. (Oregon, USA): **B**, venter of female; **C**, dorsal shield conformation of female. *Euryparasitus emarginatus* (C. L. Koch): **D**, female chelicera. **E**, typical ologamasid palptarsus; **F**, typical ologamasid chaetotaxy of tibia I.

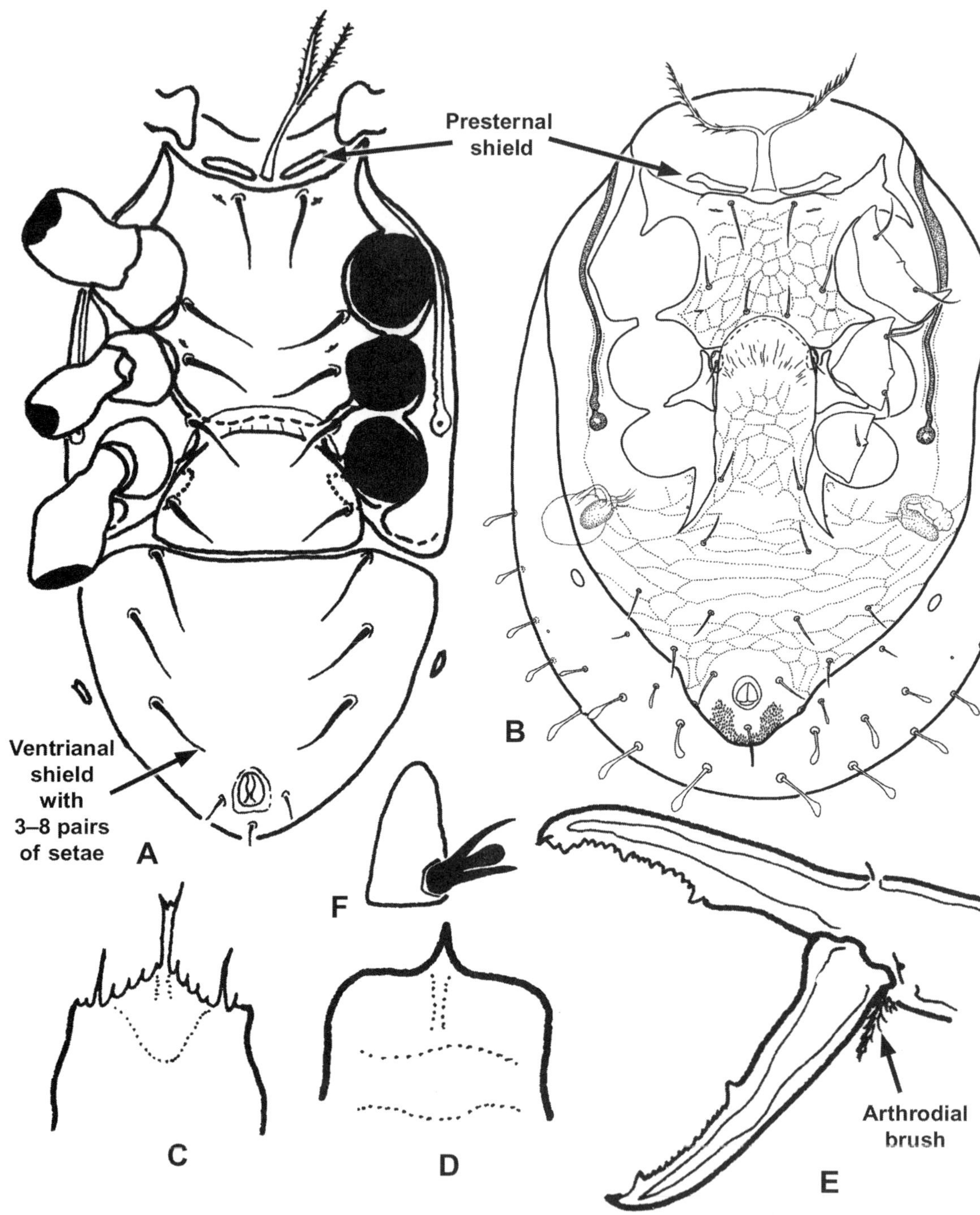

Fig. 12.44. PARHOLASPIDIDAE. *Snaveolaspis parvilobatus* (Krantz) (Oregon, USA): **A**, venter of female. *Holaspulus* sp. (Queensland, Australia): **B**, venter of female. *Holaspina coalescens* (Krantz) (Louisiana, USA): **C**, epistome. *Calholaspis berlesei* Krantz (Maryland, USA): **D**, epistome. *P. lobatus* Krantz (Oregon, USA): **E**, chelicera of female; **F**, palptarsus of typical parholaspidid.

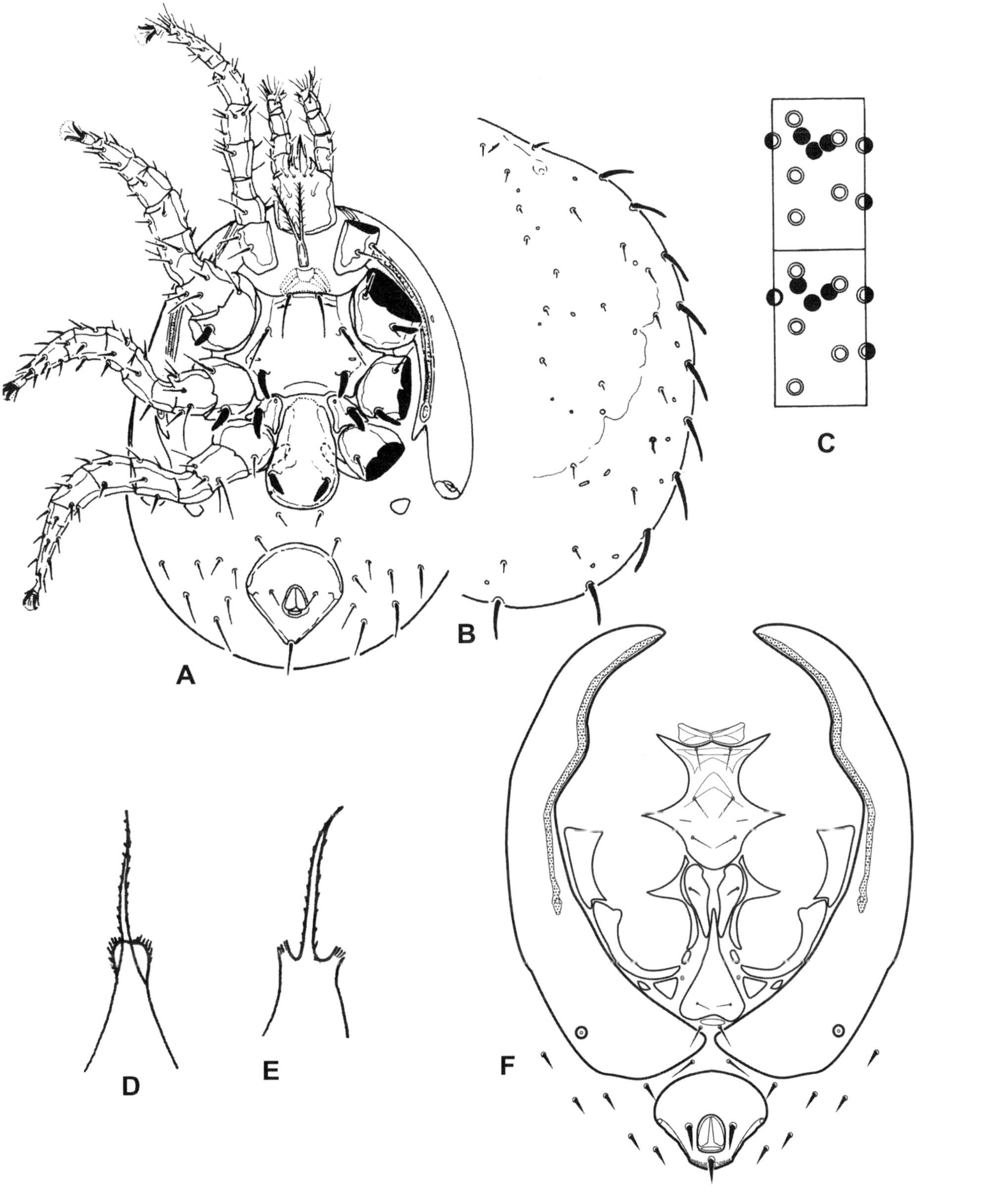

Fig. 12.45. EVIPHIDIDAE. *Eviphis* sp. (Africa): **A**, venter of female; **B**, dorsum of female; **C**, chaetotaxy of genu and tibia I of an eviphidid. *E. stephaniniarus* Berlese (Africa): **D**, epistome. *Scarabaspis rykei* Shoemake and Krantz (Zaire): **E**, epistome. *Evimirus* sp. (Queensland, Australia): **F**, venter of female.

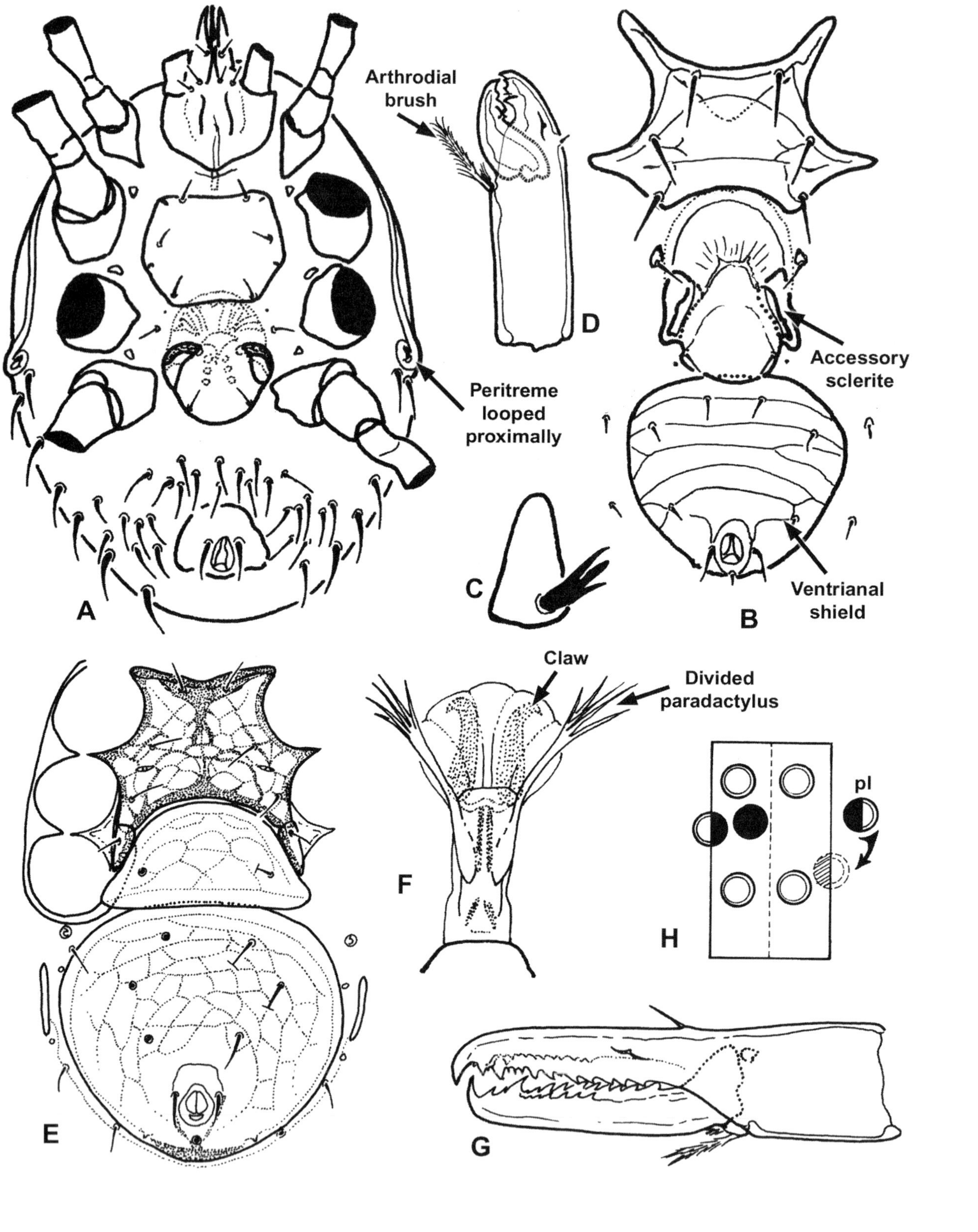

Fig. 12.46. MACROCHELIDAE. *Lordocheles rykei* Krantz (Zaire): **A**, venter of female. *Holocelaeno melisi* Krantz (Brazil): **B**, venter of female; **C**, typical macrochelid palpal tarsus. *Holostaspella bifoliata* (Trägårdh) (Oregon, USA): **D**, chelicera of female. *H. punctata* Krantz (Germany): **E**, venter of female. *Macrocheles* sp. (Trinidad): **F**, tarsus III. *Geholaspis* (*Longicheles*) *mandibularis* (Berlese) (Germany): **G**, chelicera of female; **H**, chaetotaxy of genu IV in Macrochelidae; an additional seta (*pl*) may occur in some species.

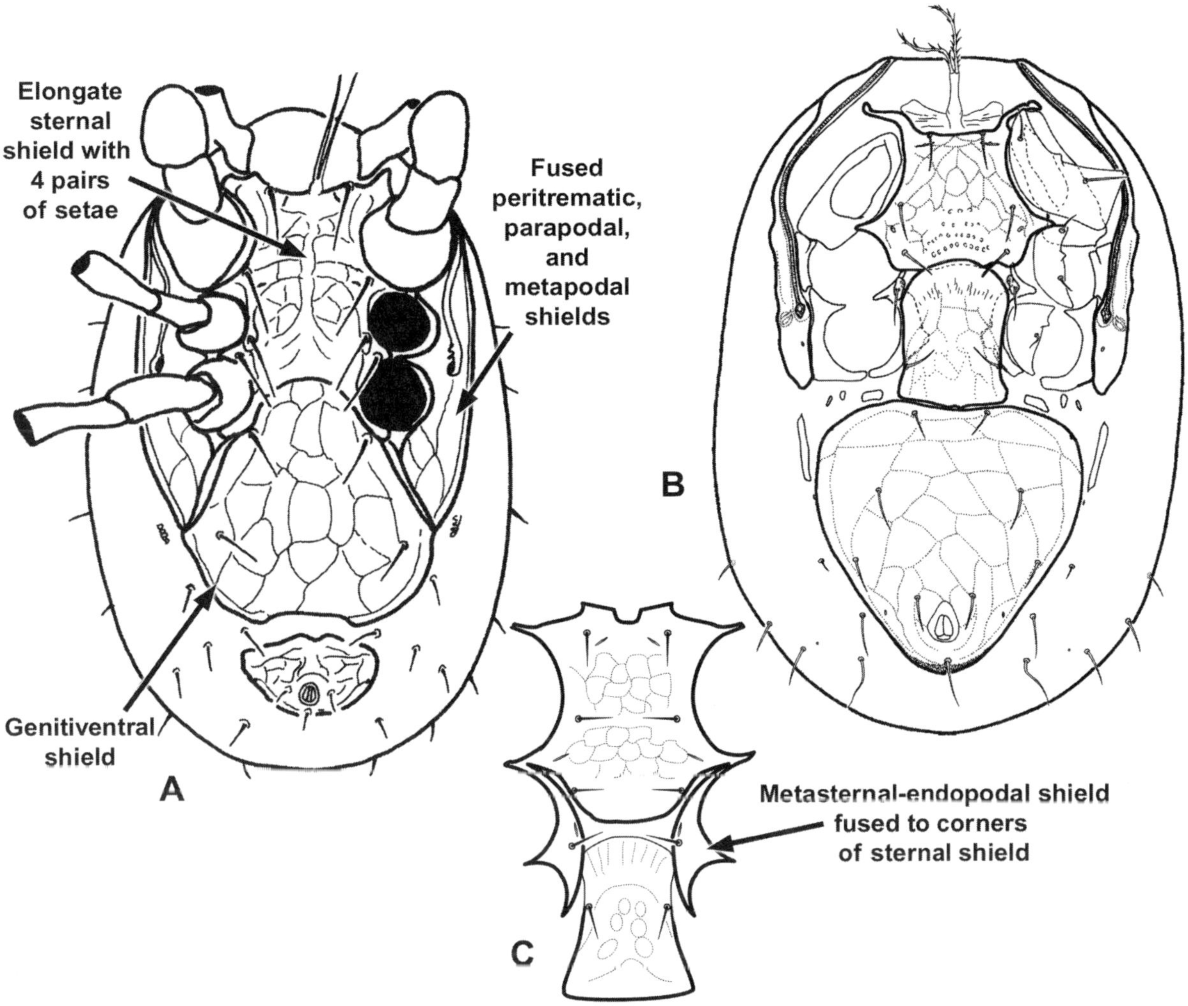

Fig. 12.47. PACHYLAELAPIDAE. *Pachylaelaps* sp. (Oregon, USA): **A**, venter of female. *Pachyseius* sp. (Northern Territory, Australia): **B**, venter of female. *Zygoseius furciger* Berlese (Colorado, USA): **C**, sternal and genital shields of female.

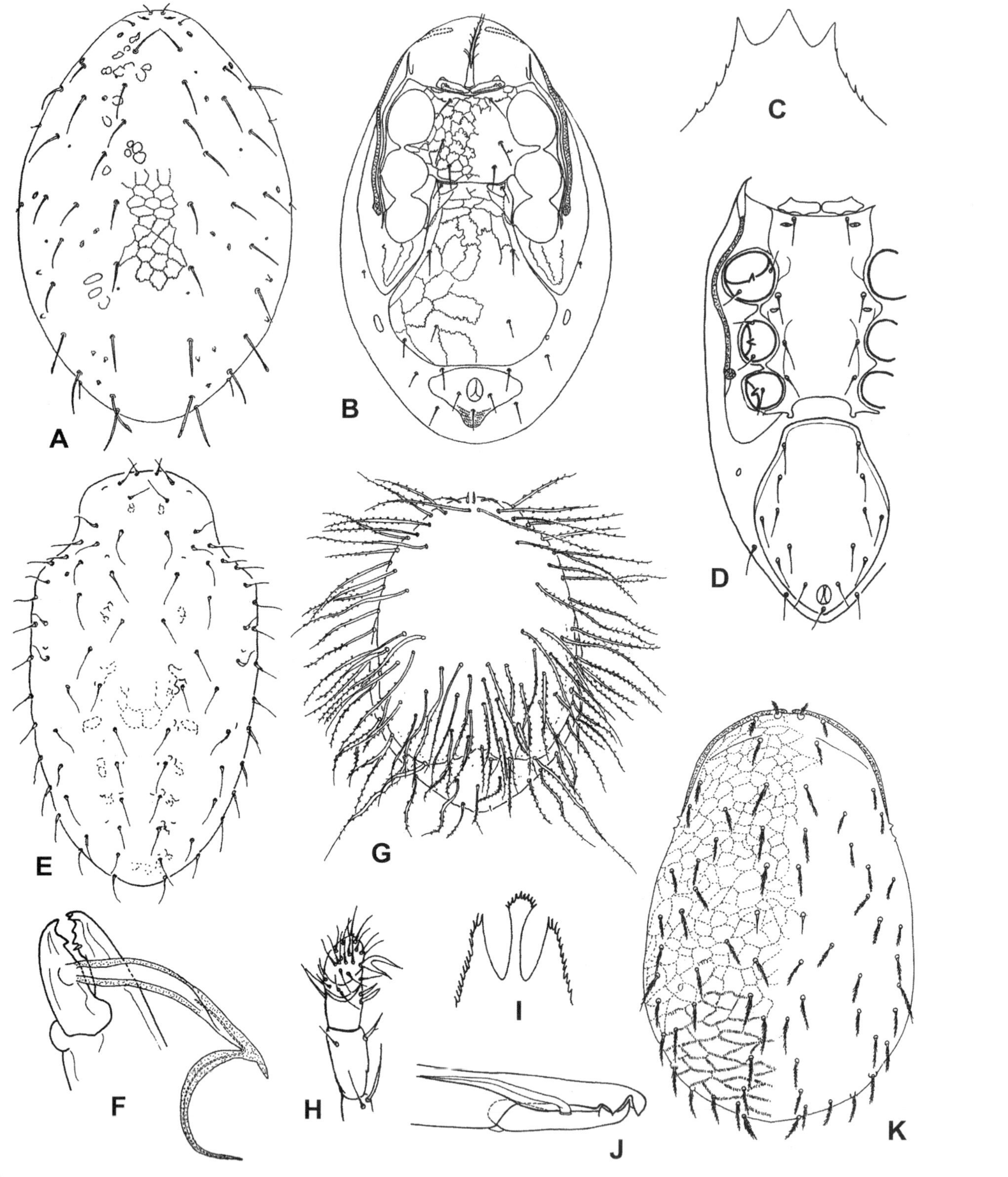

Fig. 12.48. PACHYLAELAPIDAE. Subfamily Pseudolaelapinae, *Pseudolaelaps doderoi* (Berlese): **A**, dorsum; **B**, venter; **C**, epistome (from Evans and Till 1966, by permission, British Natural History Museum, London).

Leptolaelapidae (*sensu* Karg 1983, 1997a). *Indutolaelaps squamosus* Karg (New Caledonia): **D**, venter of female. *Leptolaelaps capensis* Evans (South Africa): **E**, dorsum of female; **F**, chelicera of male. *Paradoxiphis longisetosus* Costa and Allsop (Australia): **G**, dorsum of adult. *Pulchraplaga caledonia* Karg (New Caledonia): **H**, palp of male; **I**, epistome of male; **J**, chelicera of male; **K**, dorsum of male.

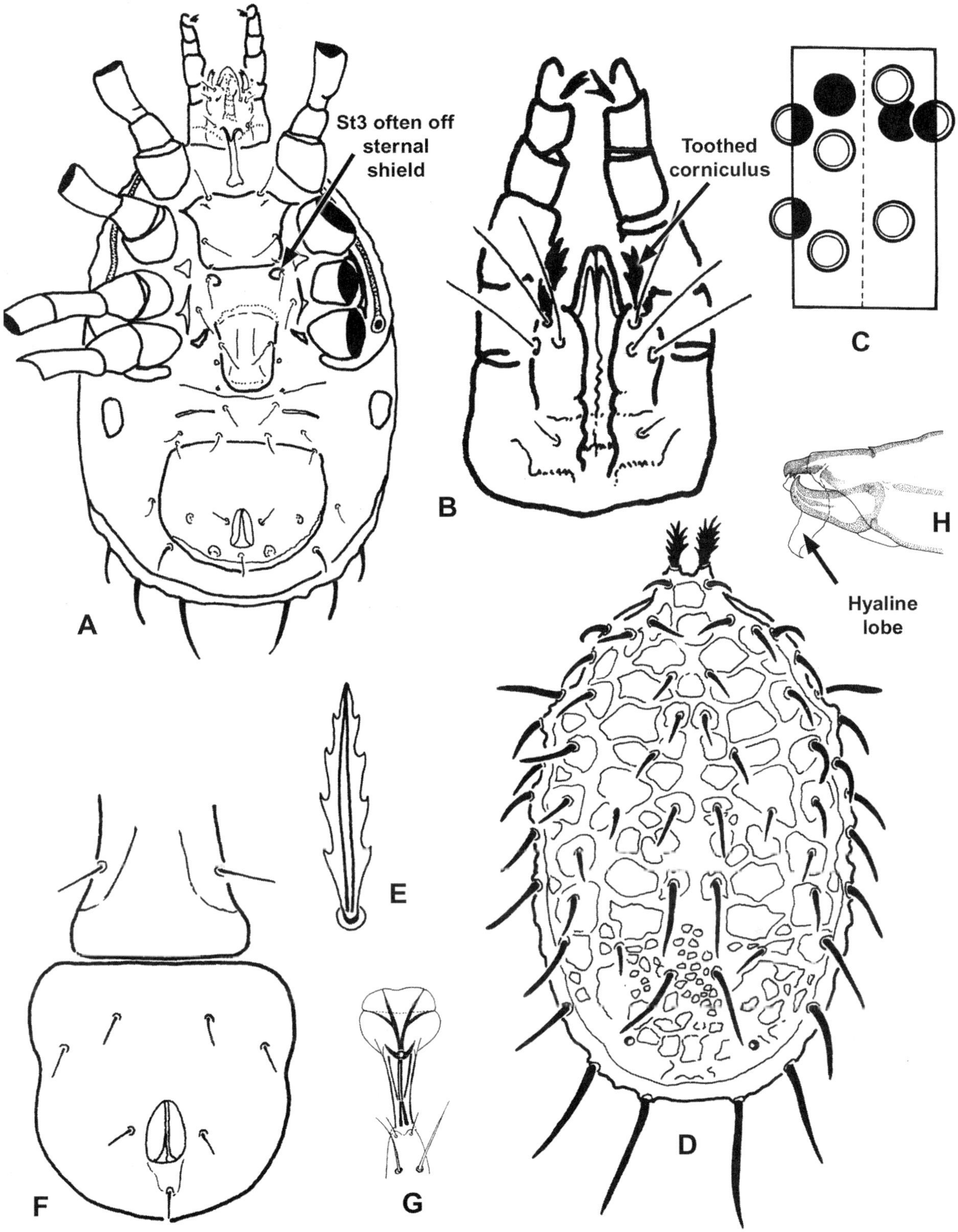

Fig. 12.49. AMEROSEIIDAE. *Ameroseius* sp. (Oregon, USA): **A**, venter of female; **B**, venter of gnathosoma; **C**, chaetotaxy of tibia IV; **D**, dorsum of female; **E**, typical dorsal seta. *Kleemannia* sp. (Oregon, USA): **F**, epigynal and ventrianal shields of female. *Afrocypholaelaps africana* (Evans) (Queensland, Australia): **G**, padlike tarsal apotele. *Neocypholaelaps* sp. (Queensland, Australia): **H**, chelicera of female.

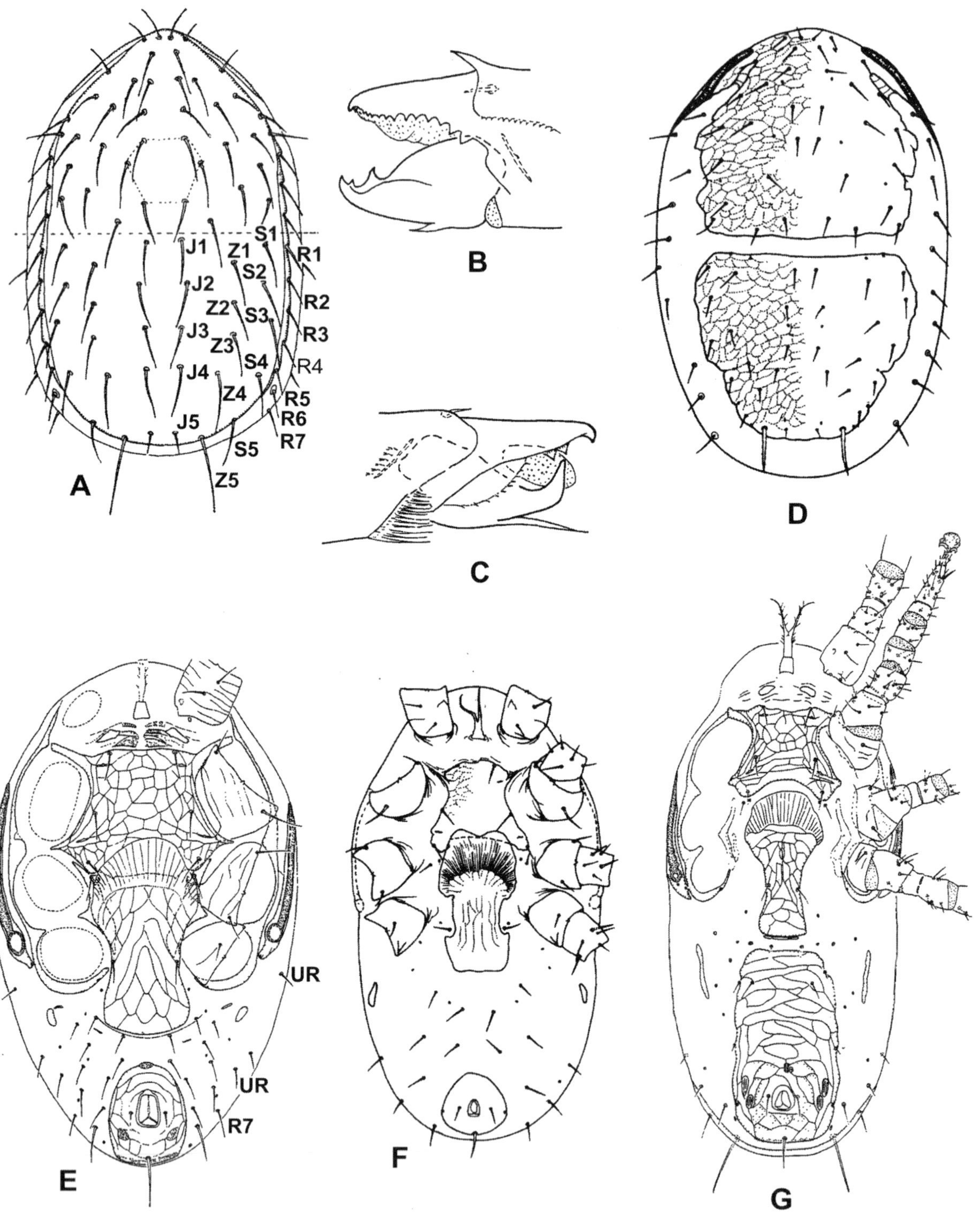

Fig. 12.50. MELICHARIDAE. *Proctolaelaps pygmaeus* (Müller): **A**, dorsum of female (after Evans 1958a); **B**, chelicera of female (after Lindquist and Evans 1965). *Tropicoseius wetmorei* Baker and Yunker: **C**, chelicera of female (after Lindquist and Evans 1965). *T. fuentesi* Naskrecki and Colwell: **D**, dorsum of female (after Naskrecki and Colwell 1998). *P. subcorticalis* Lindquist: **E**, venter of female (after Lindquist 1971). *Rhinoseius richardsoni* Hunter: **F**, venter of female (after Naskrecki and Colwell 1998). *Mycolaelaps maxinae* Lindquist: **G**, venter of female (after Lindquist 1995).

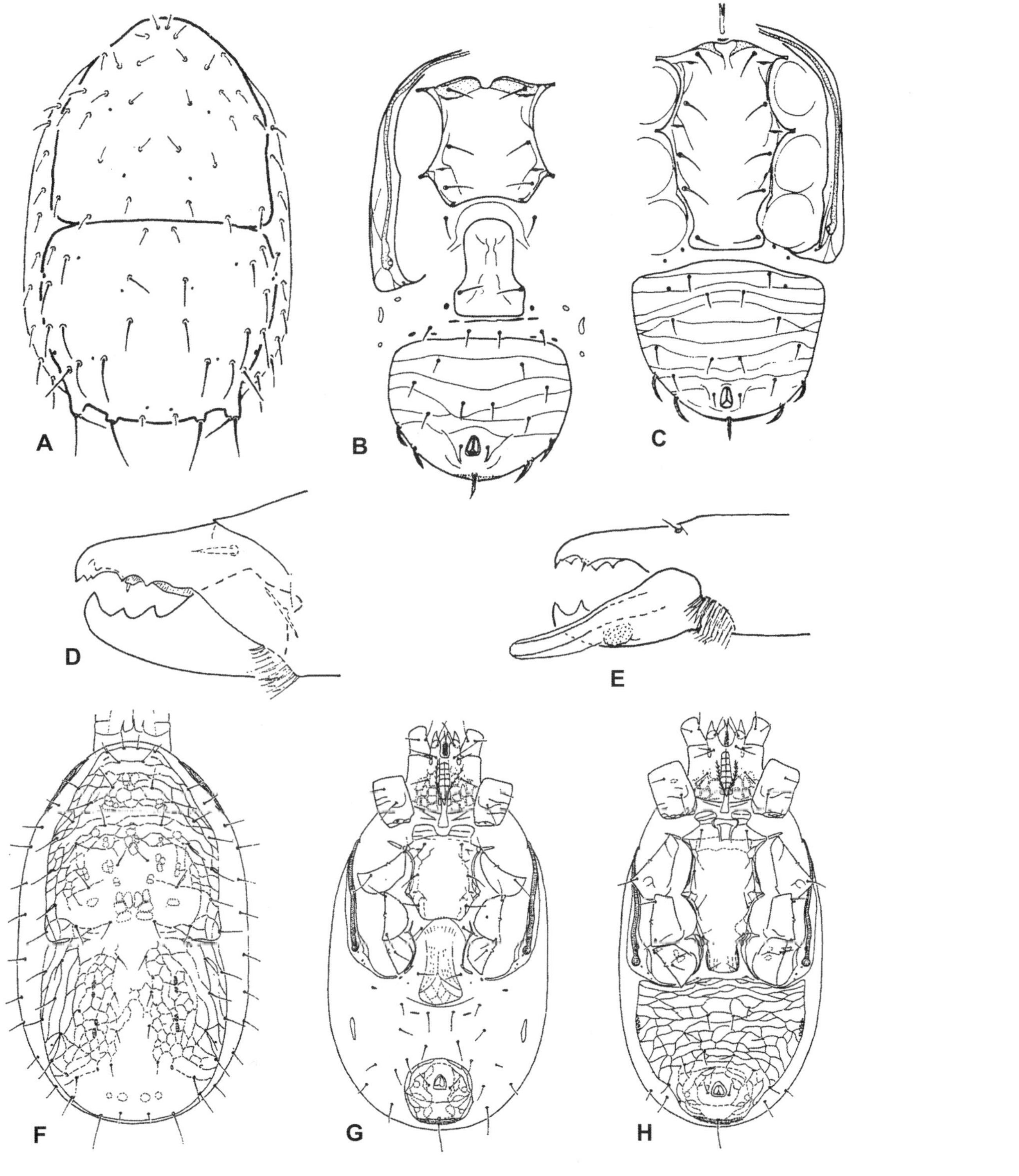

Fig. 12.51. ASCIDAE. *Asca* sp. (Oregon, USA): **A**, dorsum of female. *A. grostali* Walter, Halliday, and Lindquist (Queensland, Australia) (after Walter, Halliday, and Lindquist 1993): **B**, venter of female; **C**, venter of male. *Arctoseius confusus* Lindquist: **D**, chelicera of female (after Lindquist and Evans 1965). *Arctoseius minor* Lindquist: **E**, chelicera of male (after Lindquist 1961). *A. idiodactylus* Lindquist (after Lindquist 1961): **F**, dorsum of female; **G**, venter of female; **H**, venter of male.

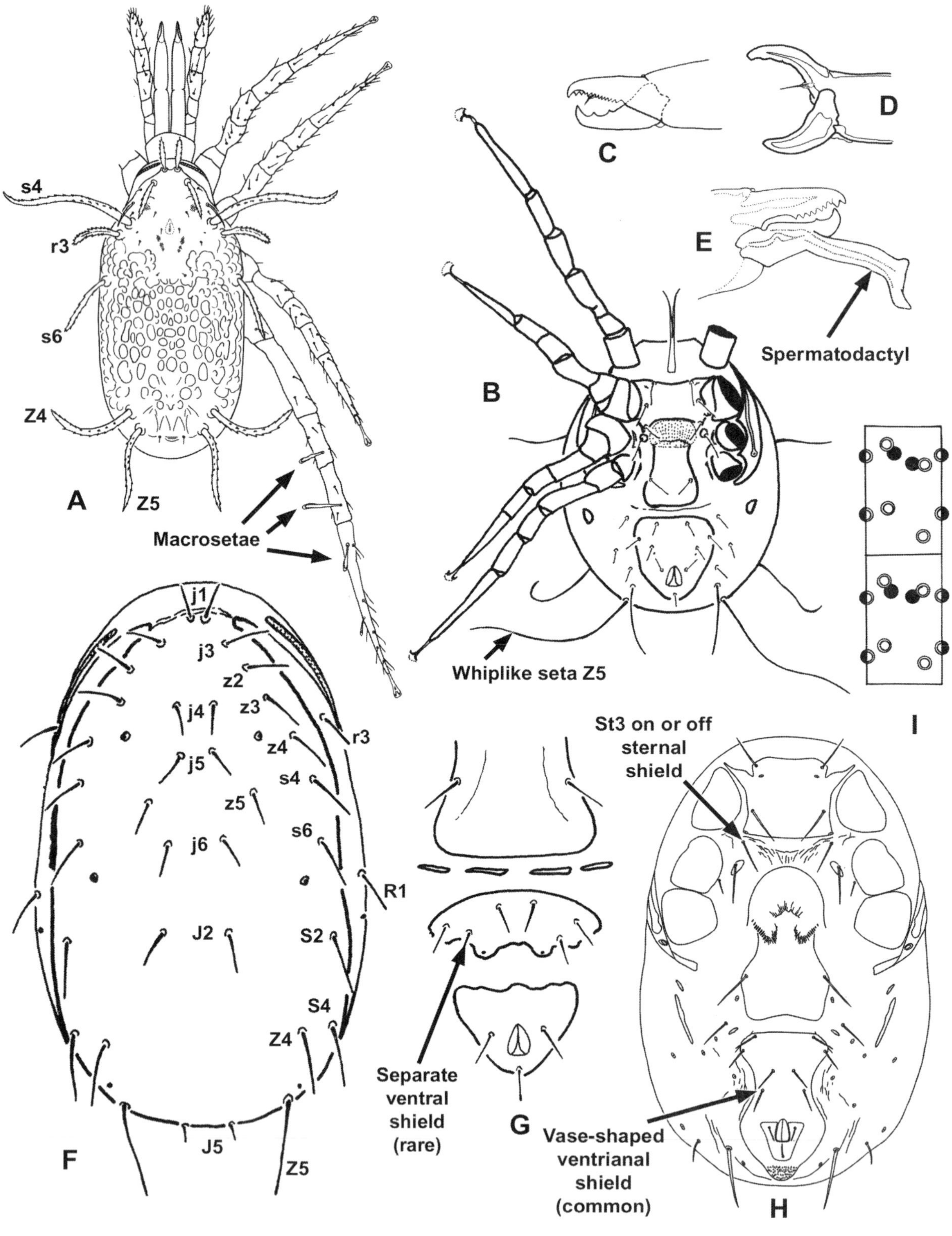

Fig. 12.52. PHYTOSEIIDAE. *Phytoseius* sp.: **A**, dorsum of female. *Amblyseius* sp. (Oregon, USA): **B**, venter of female. Amblyseiinae chelicerae: **C**, **D**, examples of female chelicerae; **E**, male chelicera. *Typhlodromus pyri* Scheuten (Oregon, USA): **F**, dorsum of female. *Iphiseius degenerans* (Berlese) (Italy): **G**, epigynal and ventrianal region of female. *Metaseiulus* sp.: **H**, venter of female; **I**, typical chaetotaxy of genu (bottom) and tibia (top).

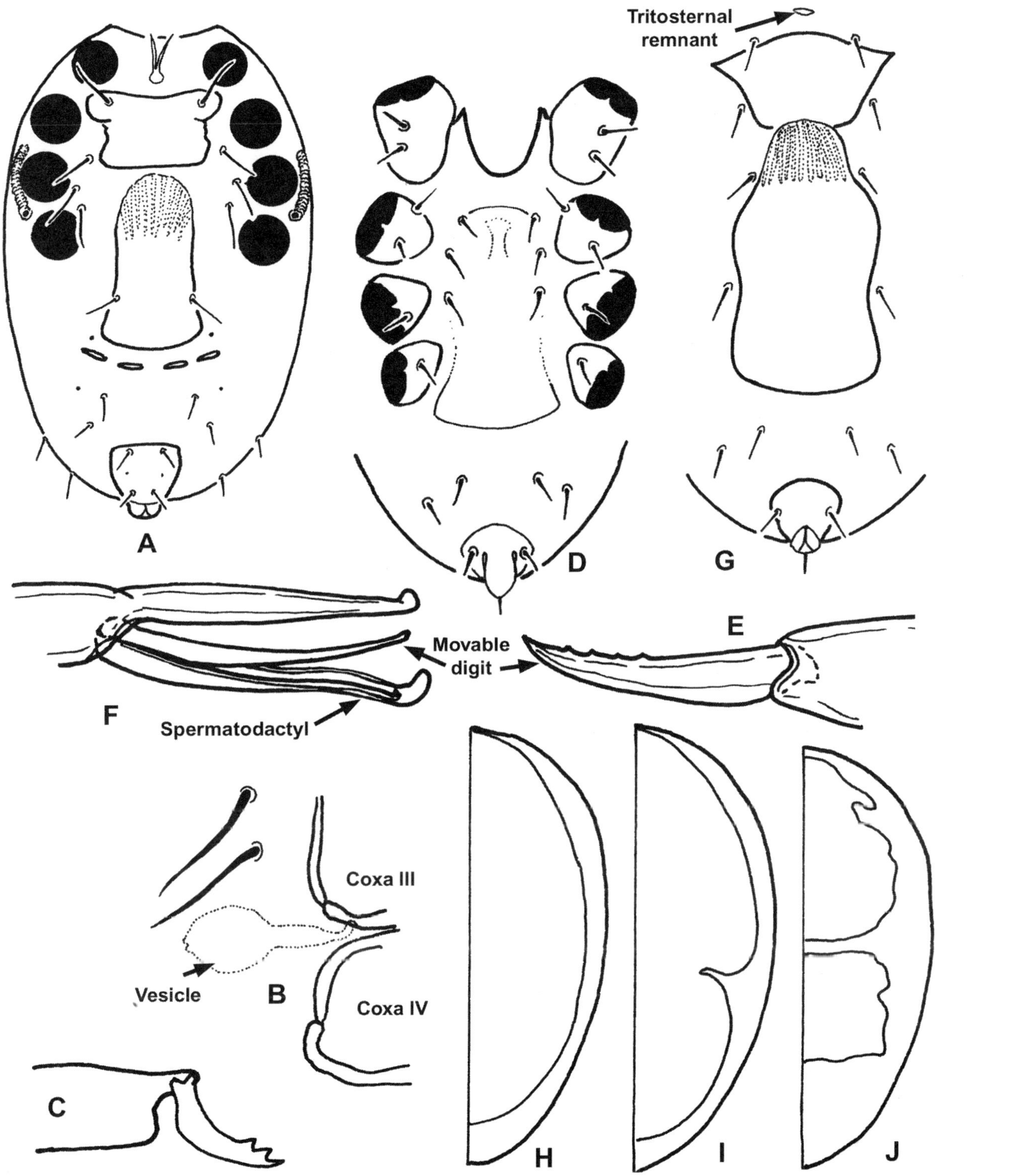

Fig. 12.53. OTOPHEIDOMENIDAE. *Hemipteroseius indicus* (Krantz and Khot) (India): **A**, venter of female; **B**, spermatheca; **C**, chelicera of female. *Otopheidomenis zalelestes* Treat: **D**, venter of female (after Treat 1975); **E**, chelicera of female; **F**, chelicera of male. *Otopheidomenis* sp.: **G**, venter of female. *Nabiseius* sp.: **H**, dorsum of female. *Otopheidomenis* sp.: **I**, dorsum of female. *Hemipteroseius* sp.: **J**, dorsum of female.

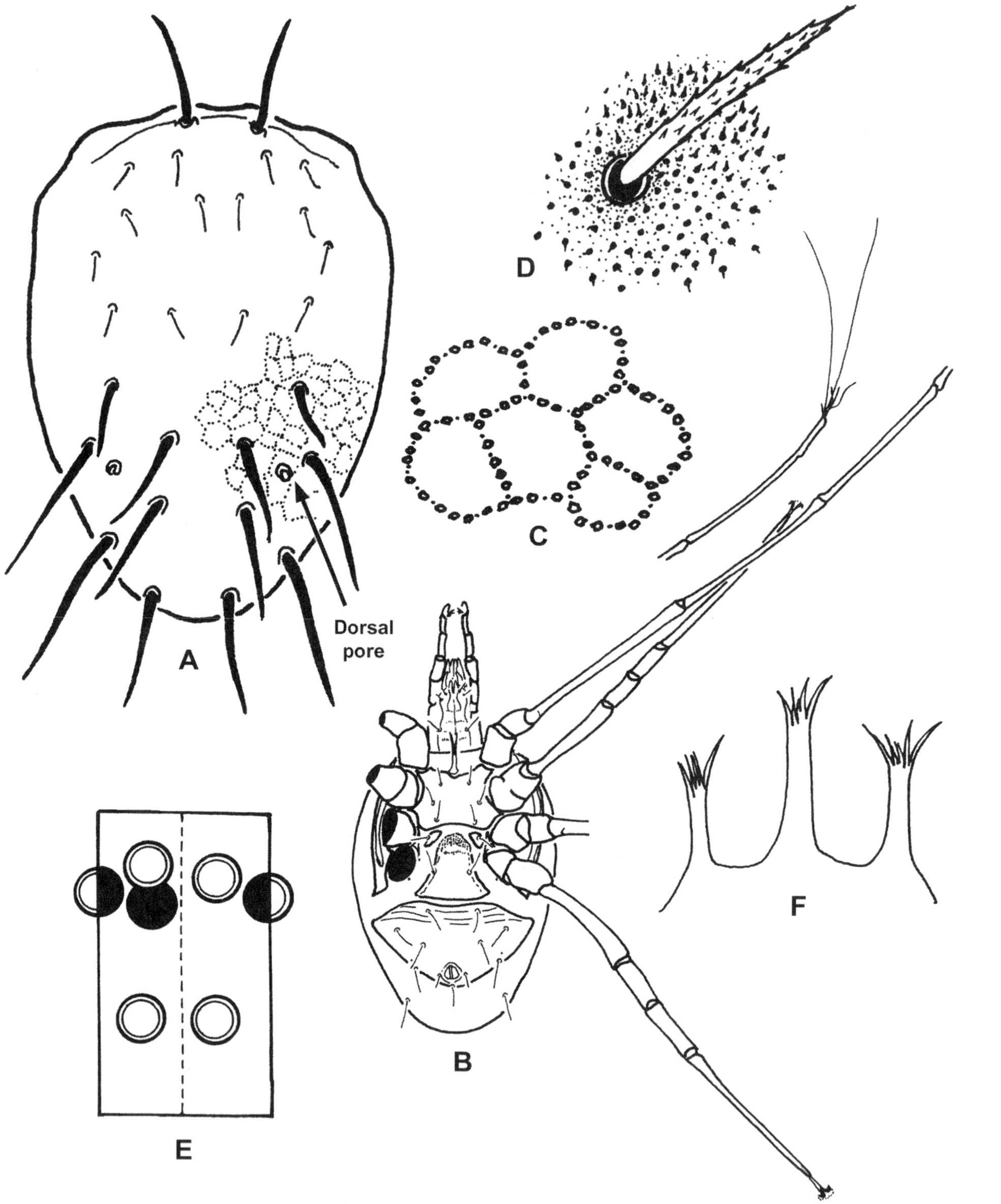

Fig. 12.54. PODOCINIDAE. *Podocinum* sp. (Kansas, USA): **A**, dorsum of female; **B**, venter of female; **C**, ornamentation of dorsal shield. *Podocinella* sp. (Mexico): **D**, dorsal seta and surrounding ornamentation; **E**, chaetotaxy of genu II; **F**, epistome.

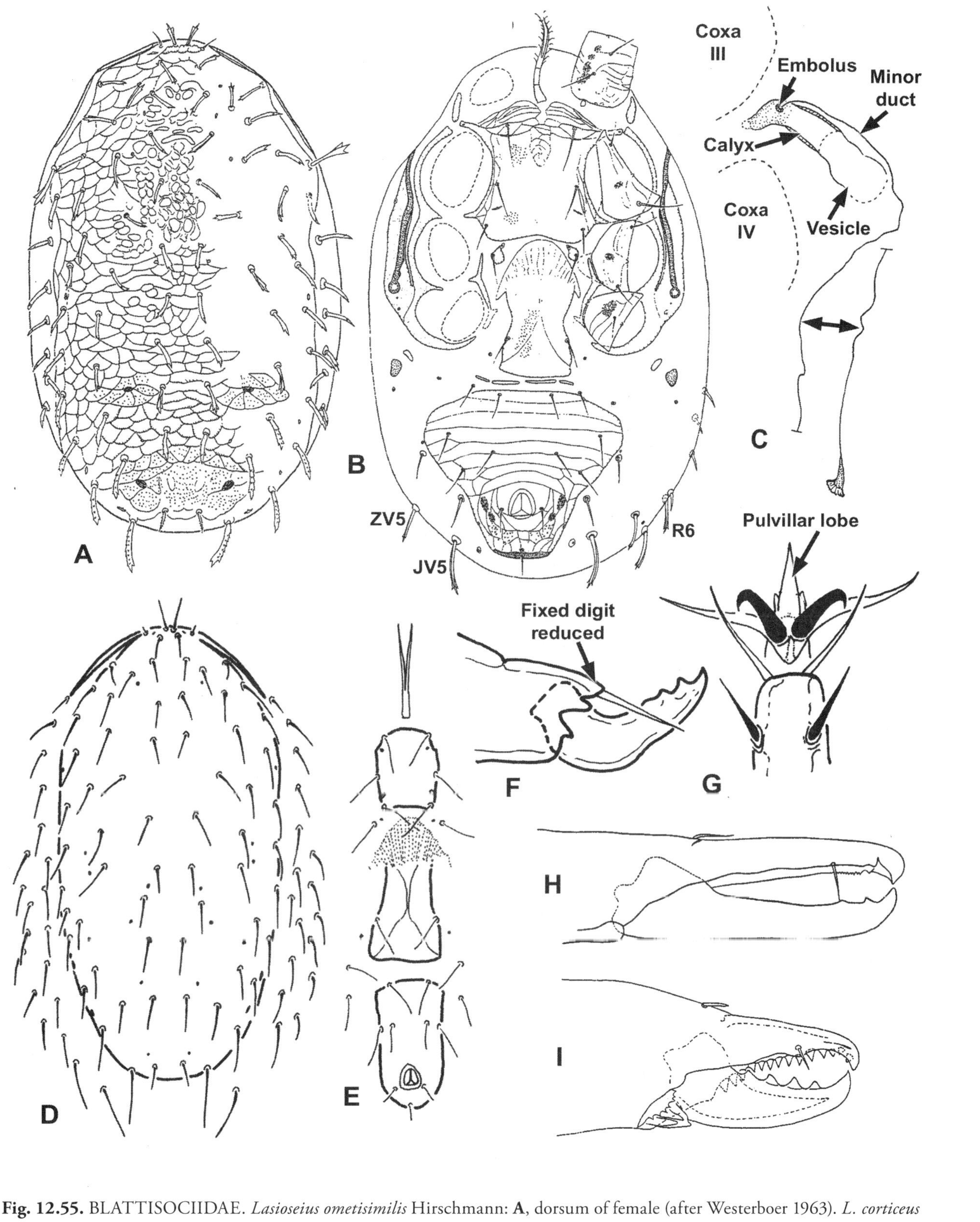

Fig. 12.55. BLATTISOCIIDAE. *Lasioseius ometisimilis* Hirschmann: **A**, dorsum of female (after Westerboer 1963). *L. corticeus* Lindquist: **B**, venter of female (after Lindquist 1971). *Hoploseius tenuis* Lindquist: **C**, sperm-reception structures of female (after Lindquist 1995). *Blattisocius tarsalis* (Berlese): **D**, dorsum of female; **E**, venter of female; **F**, chelicera of female. *Cheiroseius* sp.: **G**, tarsus II of female, dorsal aspect. *Platyseius major* (Halbert): **H**, chelicera of female (after Westerboer 1963). *L. wondjina* Walter and Lindquist (Queensland, Australia): **I**, chelicera of female (after Walter and Lindquist 1997).

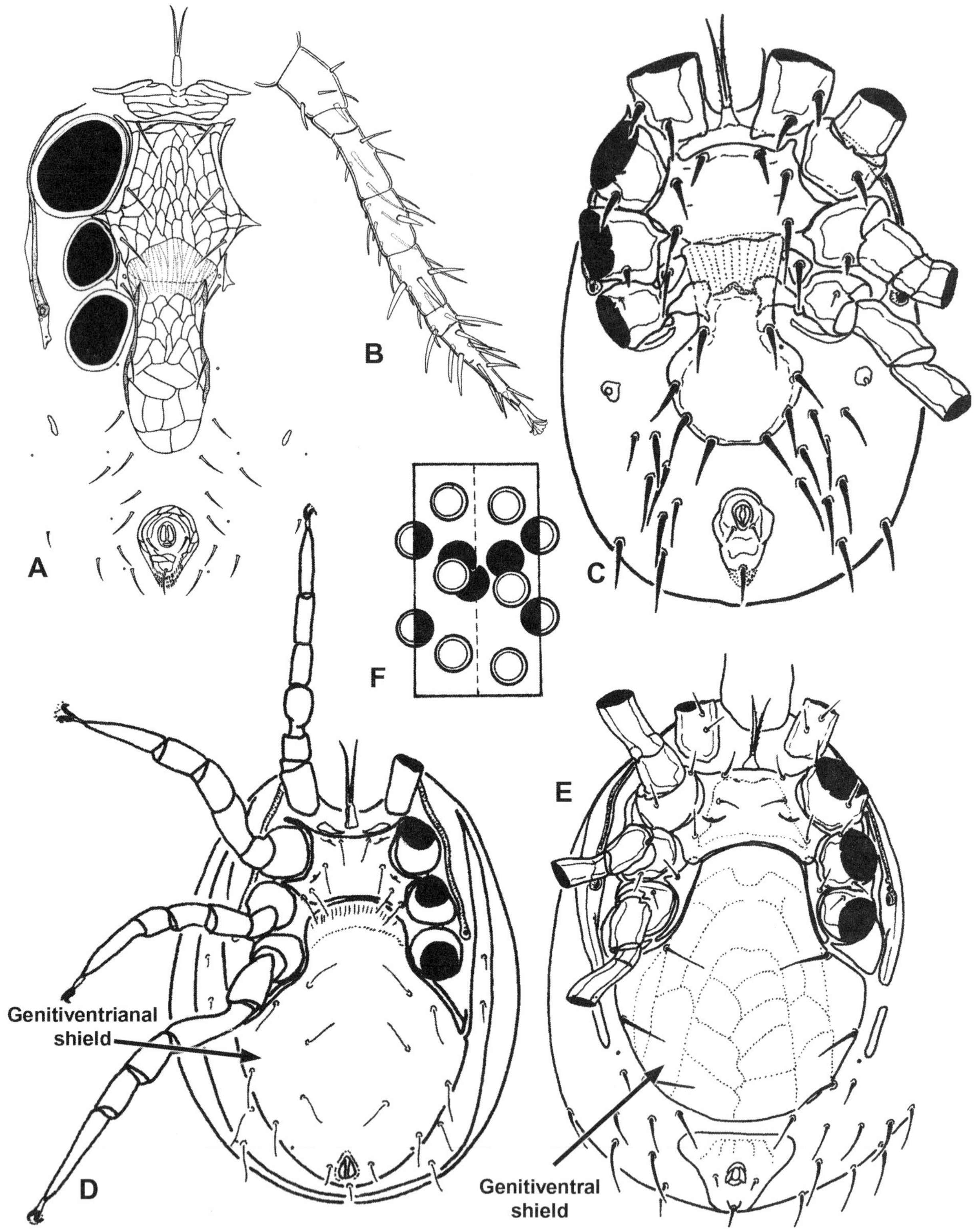

Fig. 12.56. LAELAPIDAE. *Gaeolaelaps oreithyiae* Walter and Oliver (Colorado, USA): **A**, venter of female; **B**, leg IV of female. *Laelaps* sp. (Brazil): **C**, venter of female. *Ololaelaps* sp. (Oregon, USA): **D**, venter of female. *Pseudoparasitus*? sp. (Oregon, USA): **E**, venter of female; **F**, typical laelapid chaetotaxy of tibia I.

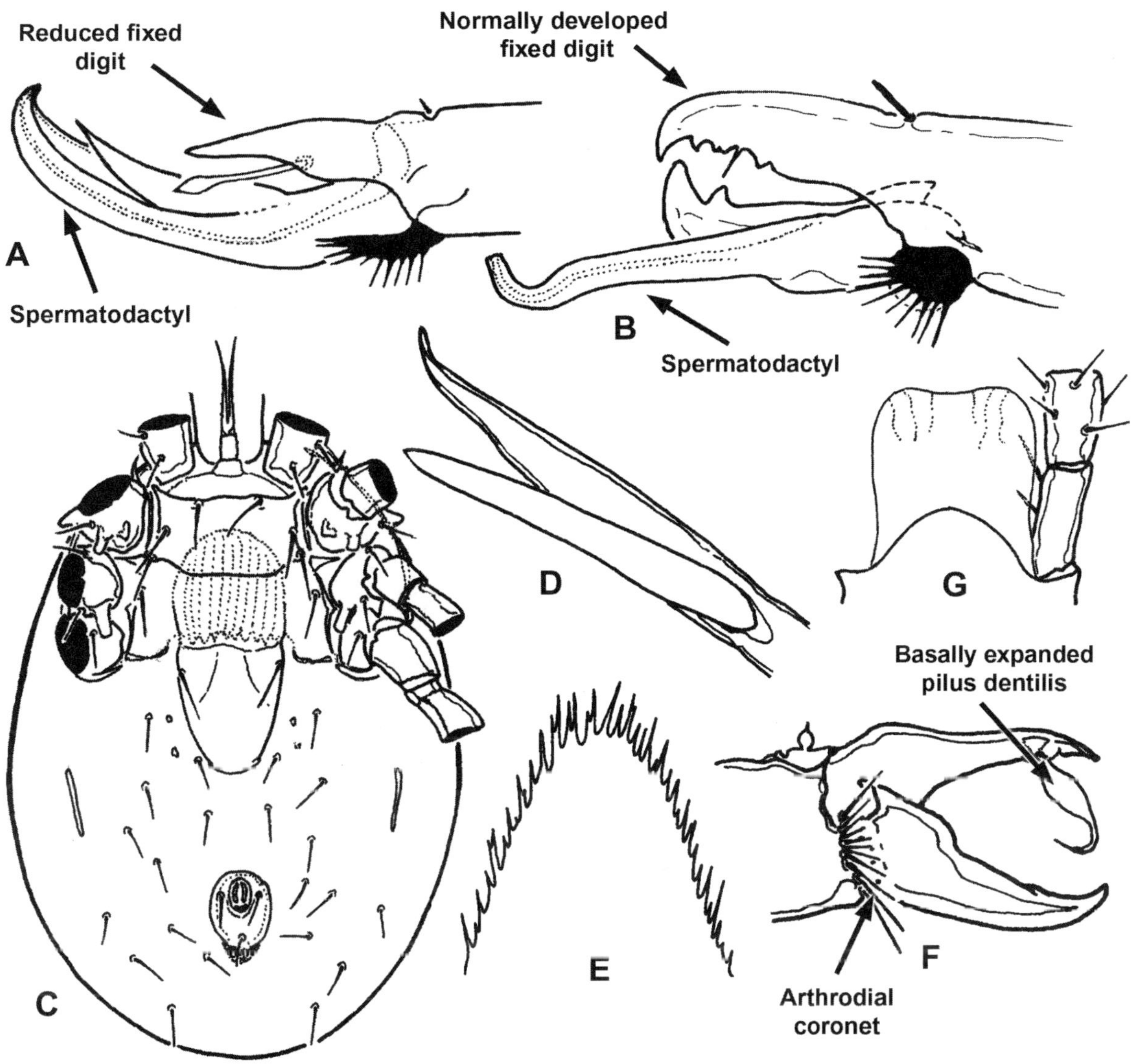

Fig. 12.57. LAELAPIDAE. *Laelaps hilaris* Koch: **A**, chelicera of male (after Evans and Till 1965). *Hypoaspis krameri* (Canestrini): **B**, chelicera of male (after Evans and Till 1965). *Hirstionyssus* sp. (Oregon, USA): **C**, venter of female; **D**, chelicera of female. *Haemogamasus pontiger* (Berlese) (Oregon, USA): **E**, epistome. *Androlaelaps fahrenholtzi* (Berlese) (Oregon, USA): **F**, chelicera of female; **G**, epistome.

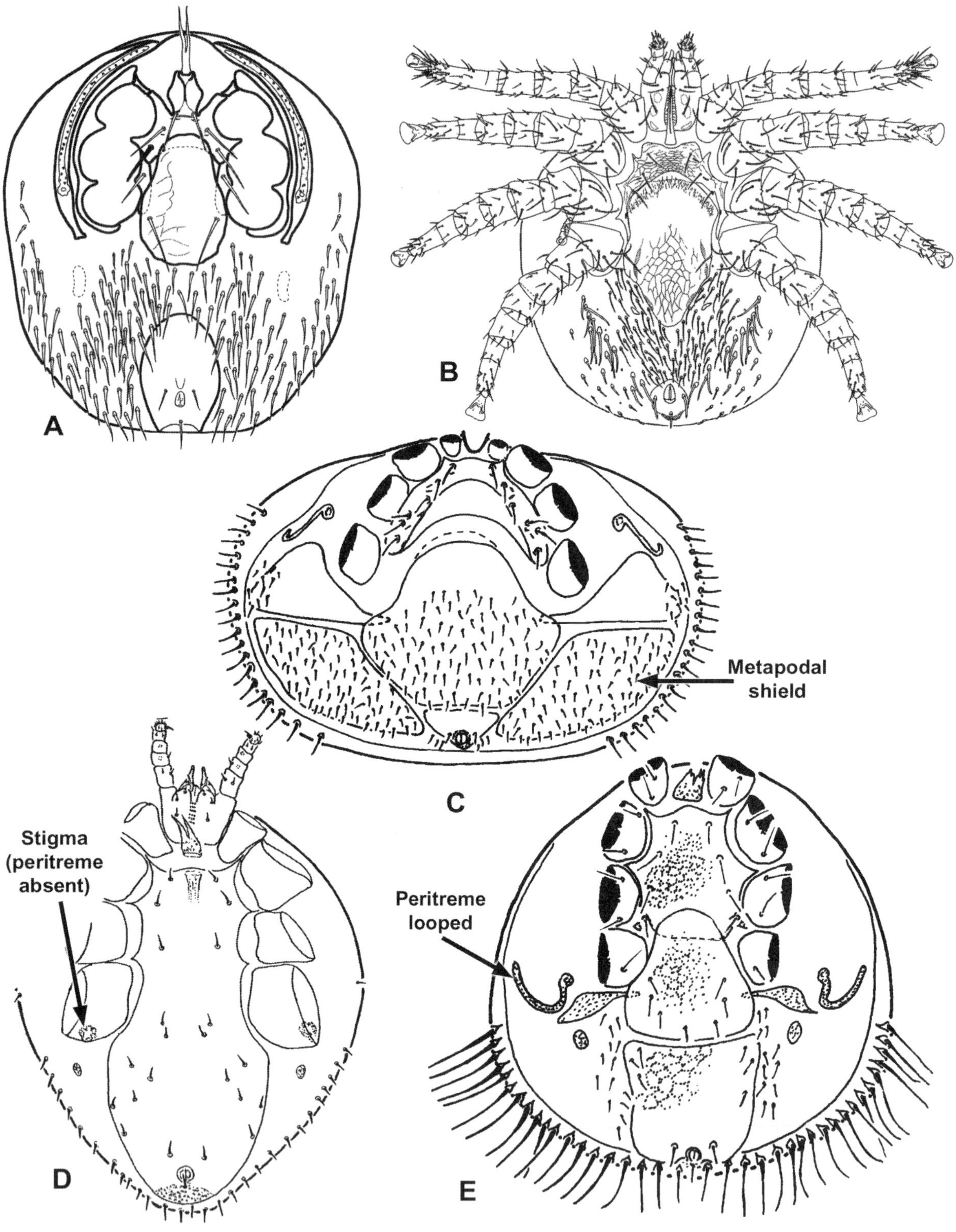

Fig. 12.58. LAELAPIDAE. *Melittiphis alvearius* (Berlese): **A**, venter of female (from Walter et al. 2002). *Myrmozercon iainkayi* Walter (Queensland, Australia):
B, venter of female.
VARROIDAE. *Varroa* sp.: **C**, venter of female. *Euvarroa sinhai* Delfinado and Baker: **D**, venter of male; **E**, venter of female.

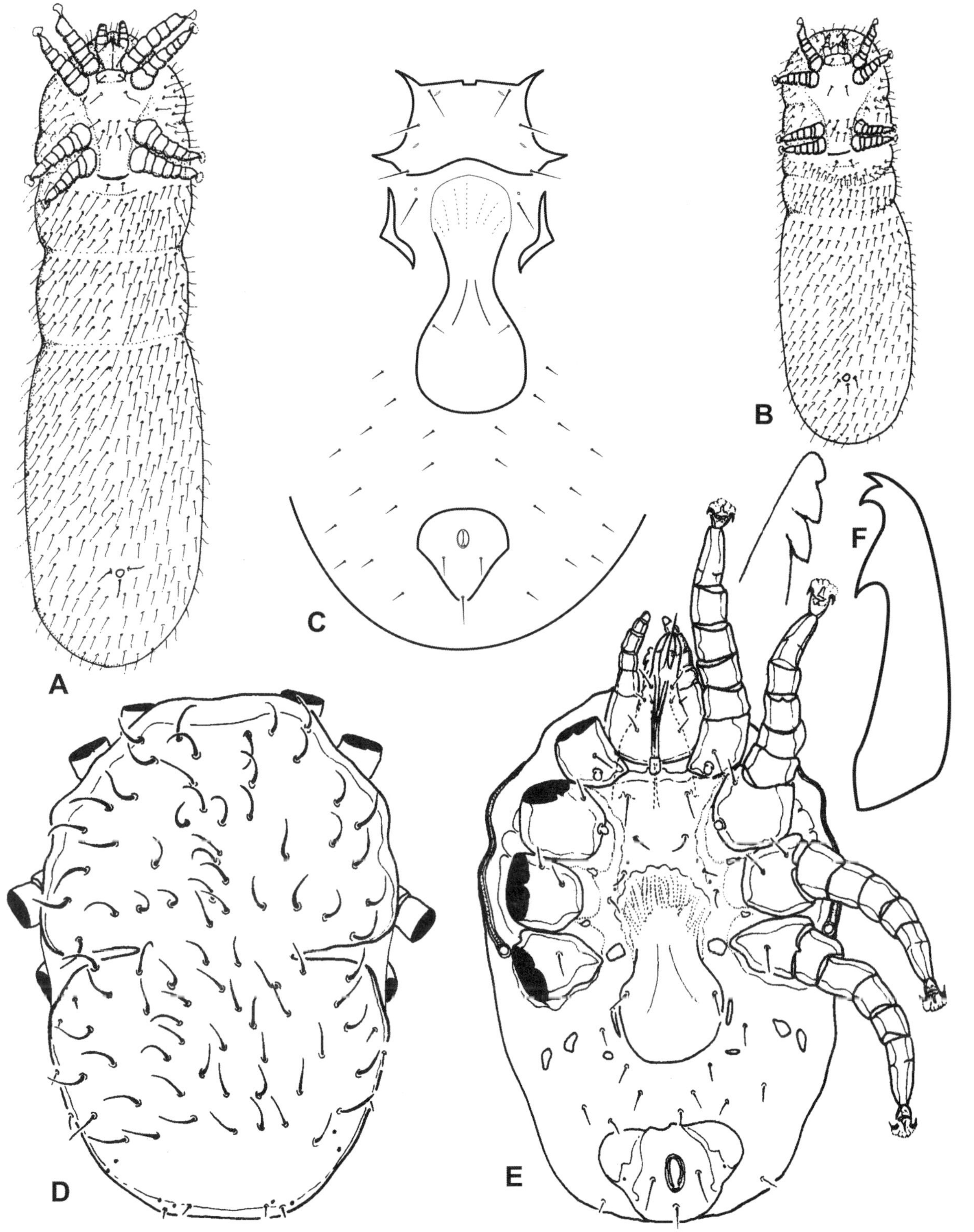

Fig. 12.59. LARVAMIMIDAE. *Larvamima carli* Elzinga (Panama) (from Elzinga 1993): **A**, venter of female. *L. schneirlai* Elzinga (Panama): **B**, venter of female.

IPHIOPSIDIDAE. *Julolaelaps* sp.: **C**, venter of female.

IXODORHYNCHIDAE. *Ixodorhynchus liponyssoides* Ewing: **D**, dorsum of female; **E**, venter of female; **F**, distal tip of corniculus (right) and chelicera (left) of female.

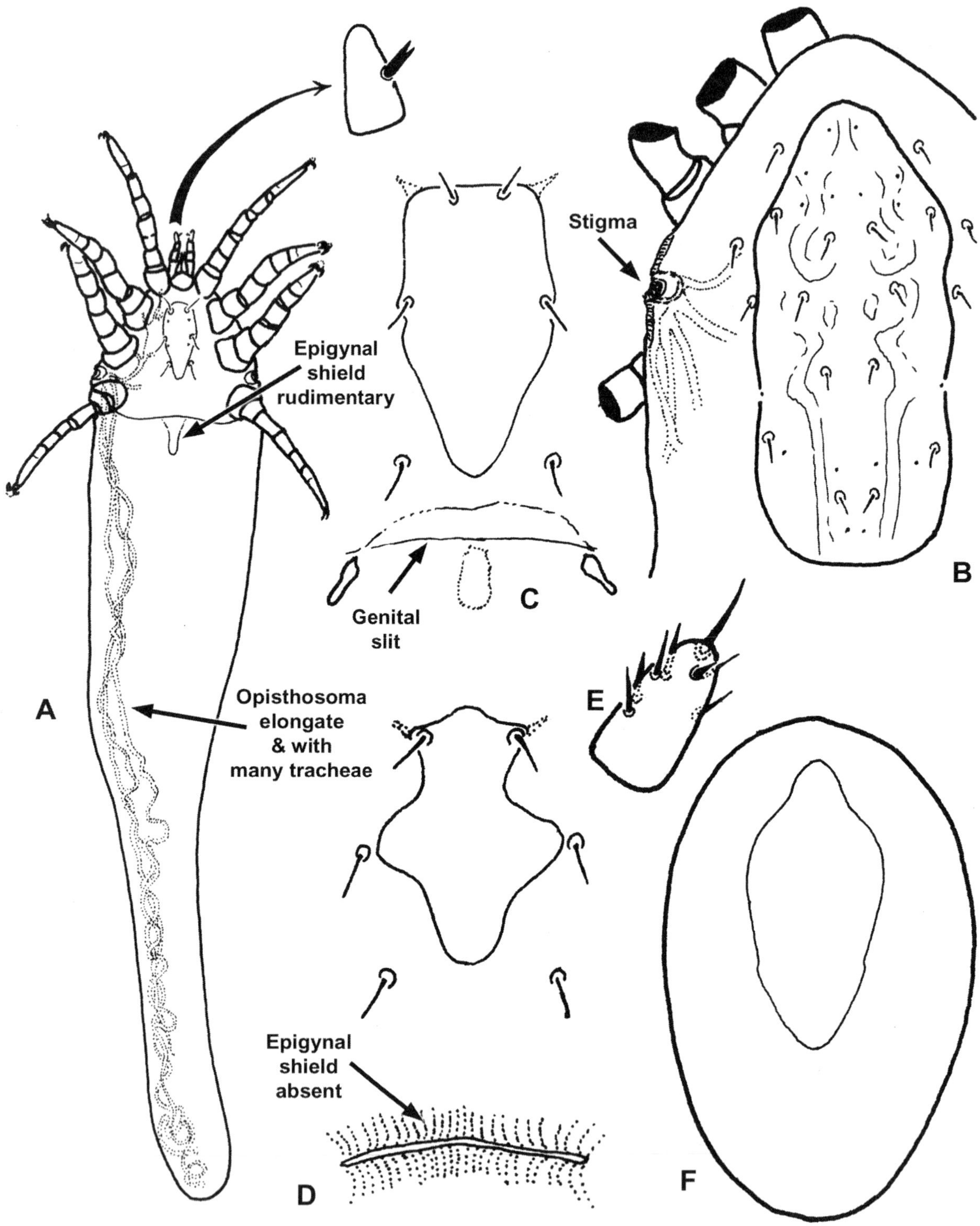

Fig. 12.60. HALARACHNIDAE. *Orthohalarachne attenuata* (Banks) (North Pacific): **A**, venter of female; **B**, anterior dorsal aspect of female. *Halarachne americana* Banks: **C**, sternal and genital region of female. *Pneumonyssoides caninum* (Chandler and Ruhe): **D**, sternal and genital region of female; **E**, palptarsus; **F**, dorsum of female.

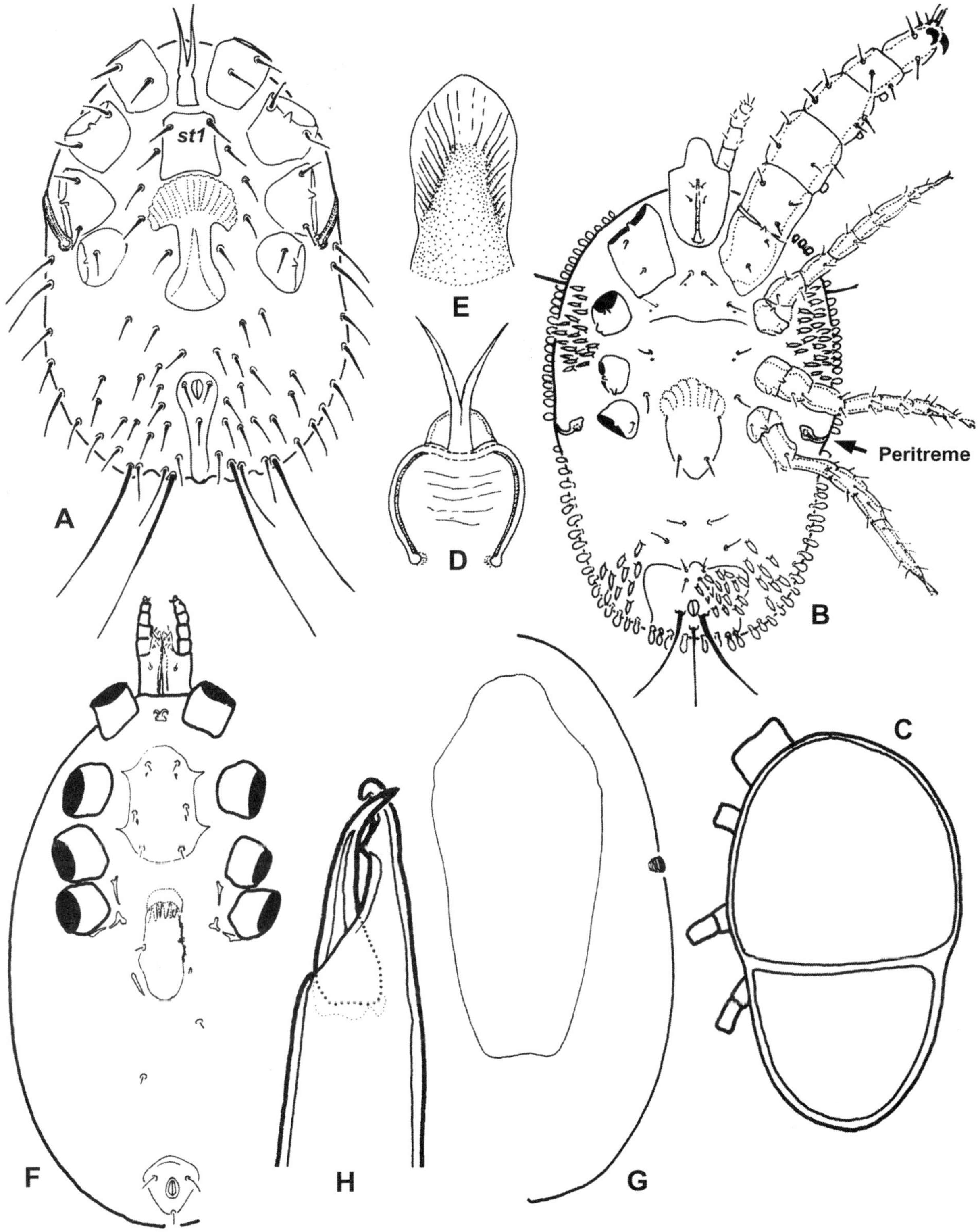

Fig. 12.61. MANITHERIONYSSIDAE. *Manitherionyssus heterotarsus* Vitzthum: **A**, venter of female (after Strandtmann and Wharton 1958).

DASYPONYSSIDAE. *Dasyponyssus neivai* Fonesca: **B**, venter of female; **C**, dorsum (diagrammatic). *Xenarthronyssus furmani furmani* Radovsky and Yunker: **D**, tritosternum and presternal region; **E**, epistome (after Radovsky and Yunker 1971).

ENTONYSSIDAE. *Ophiopneumicola* sp. (Washington, USA): **F**, venter of female; **G**, dorsum of female; **H**, chelicera of female.

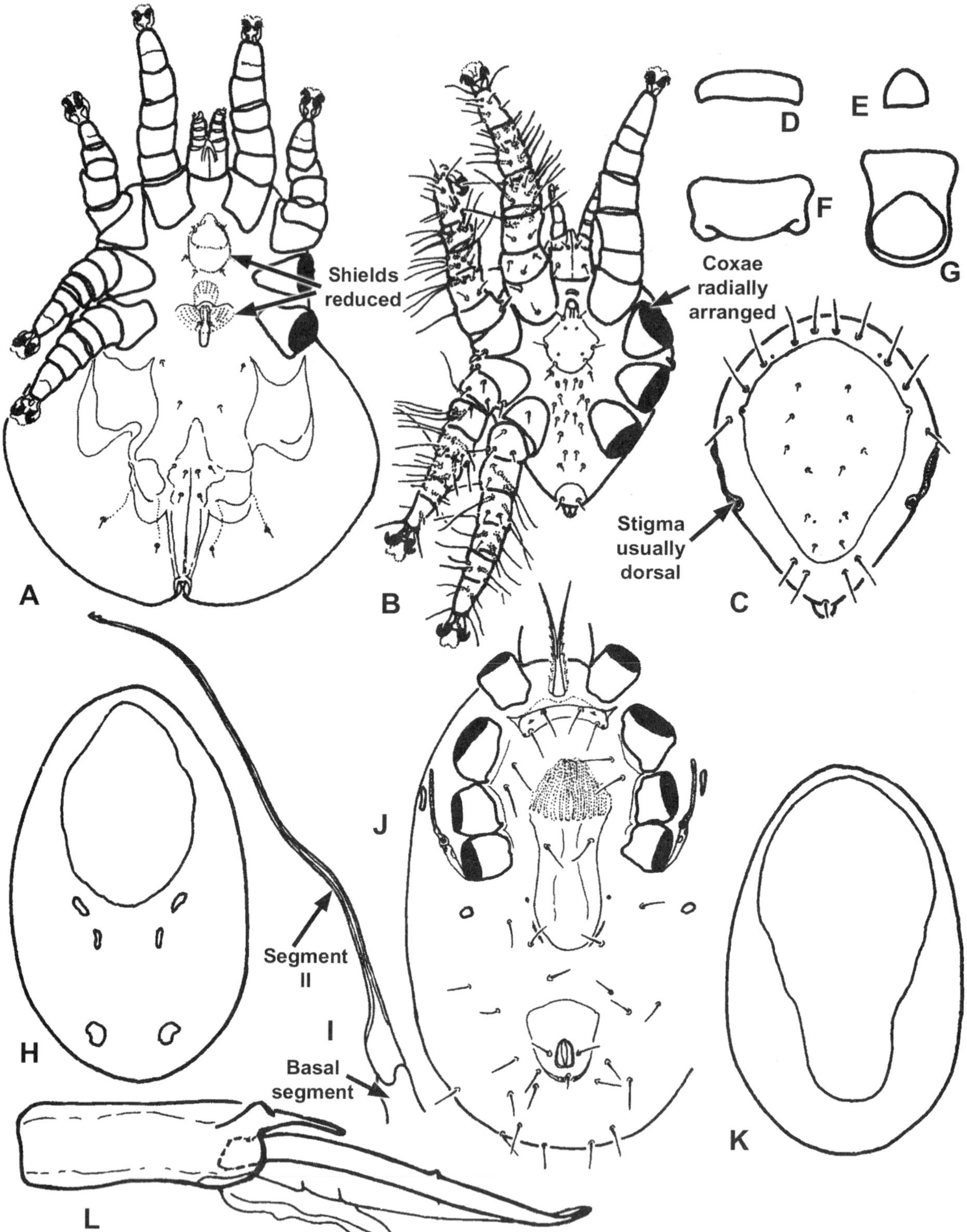

Fig. 12.62. SPINTURNICIDAE. *Periglischrus* sp. (Mexico): **A**, venter of female. *Spinturnix* sp. (Indiana, USA): **B**, venter of male; **C**, dorsum of male; **D–G**, tritosternal bases of various spinturnicids (after Rudnick 1960).

DERMANYSSIDAE. *Dermanyssus triscutatus* Krantz (Alaska, USA): **H**, dorsum of female; **I**, chelicera of female; **J**, venter of female. *D. gallinae* (DeGeer) (Oregon, USA): **K**, dorsum of female; **L**, chelicera of male (after Evans and Till 1965).

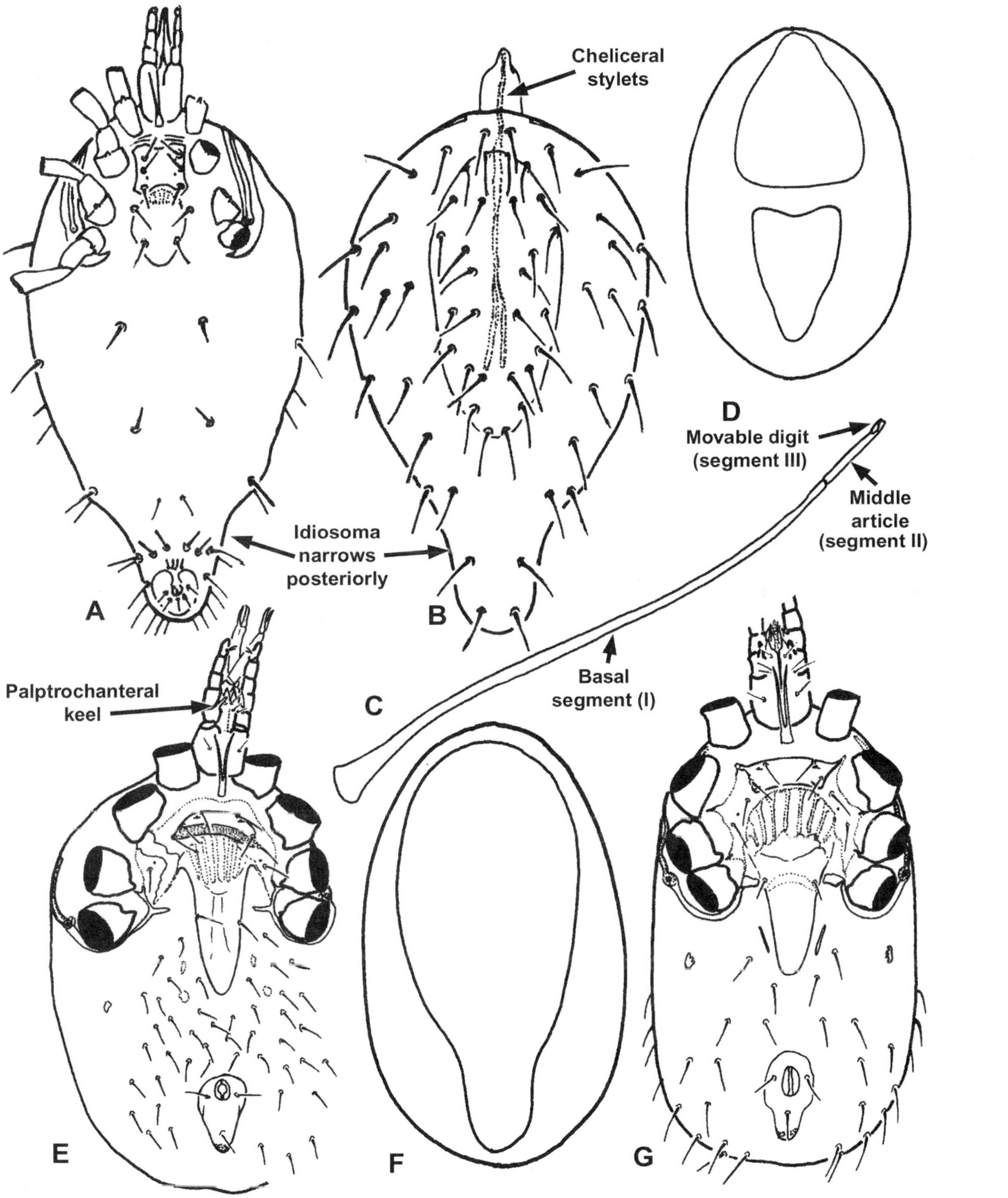

Fig. 12.63. HYSTRICHONYSSIDAE. *Hystrichonyssus turneri* Keegan, Yunker, and Baker (Malaysia): **A**, venter of female; **B**, dorsum of female; **C**, chelicera of female.

MACRONYSSIDAE. *Steatonyssus* sp. (California, USA): **D**, dorsum of female; **E**, venter of female. *Ornithonyssus sylviarum* (Canestrini and Fanzago) (Oregon, USA): **F**, dorsum of female; **G**, venter of female.

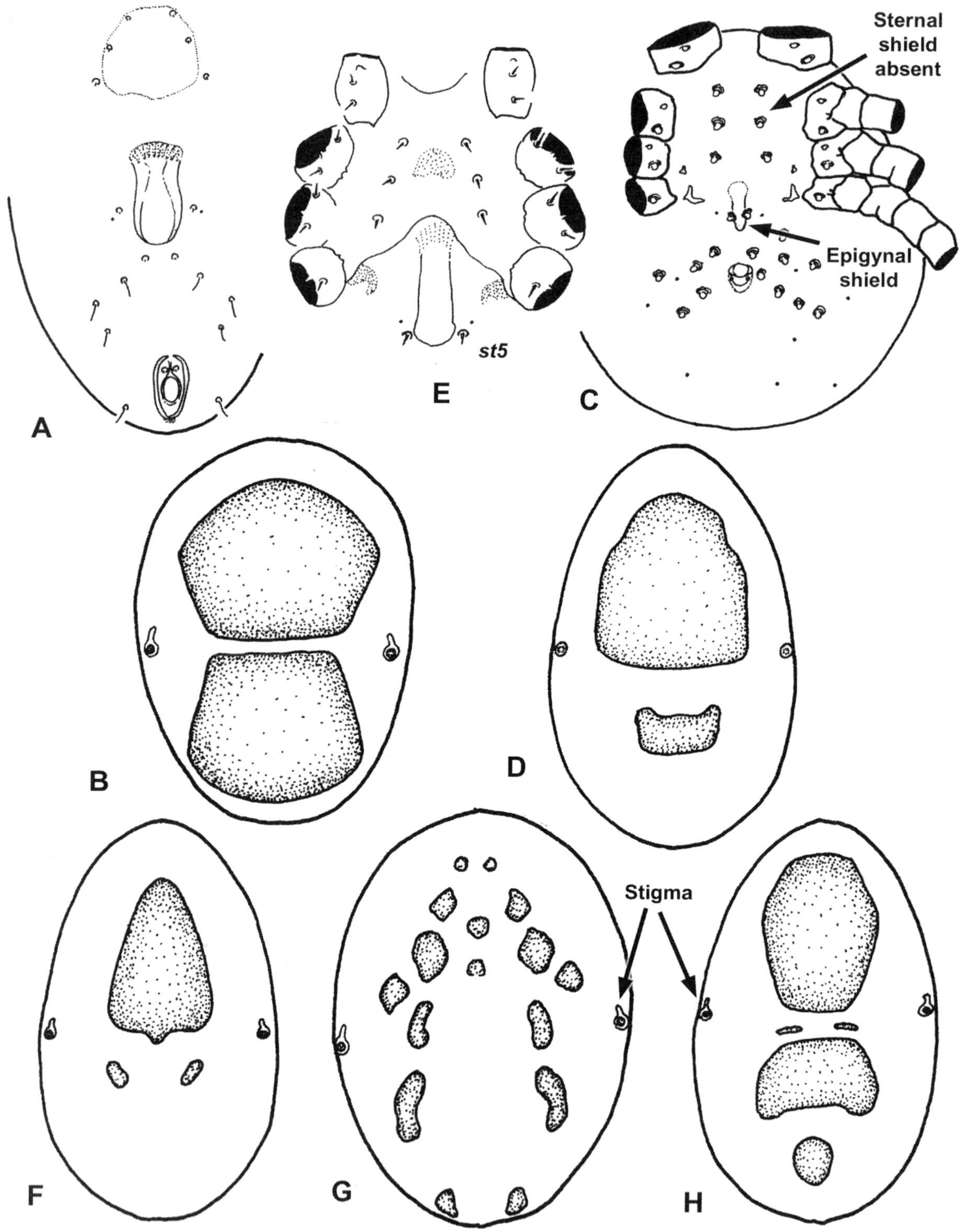

Fig. 12.64. RHINONYSSIDAE. *Neonyssus columbae* Crossley (Texas, USA): **A**, venter of female; **B**, dorsum of female. *Cas angrensis* Castro (Texas, USA): **C**, venter of female; **D**, dorsum of female. *Ptilonyssus ohioensis* Fain and Johnston (Ohio, USA): **E**, sternal and genital region of female (after Fain and Johnston 1966). *Rhinoecius* sp.: **F**, female dorsum. *Larinyssus* sp.: **G**, female dorsum. *Ptilonyssoides* sp.: **H**, female dorsum.

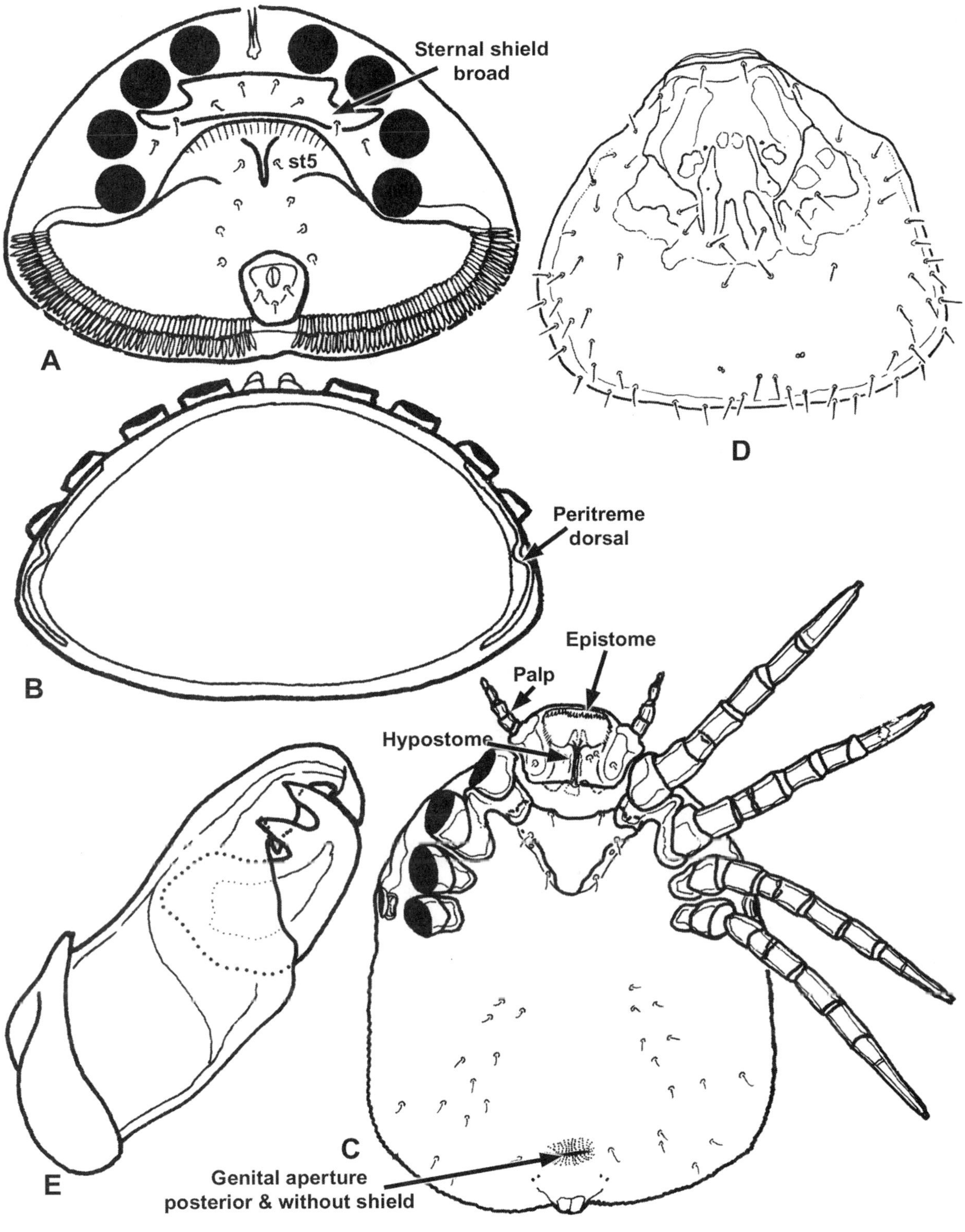

Fig. 12.65. OMENTOLAELAPIDAE. *Omentolaelaps mehelyae* Fain (Zaire): **A**, venter of female; **B**, dorsum of female (after Fain 1961a, b).
SPELAEORHYNCHIDAE. *Spelaeorhynchus praecursor* Neumann (South America): **C**, venter of female; **D**, dorsum of female; **E**, chelicera of female.

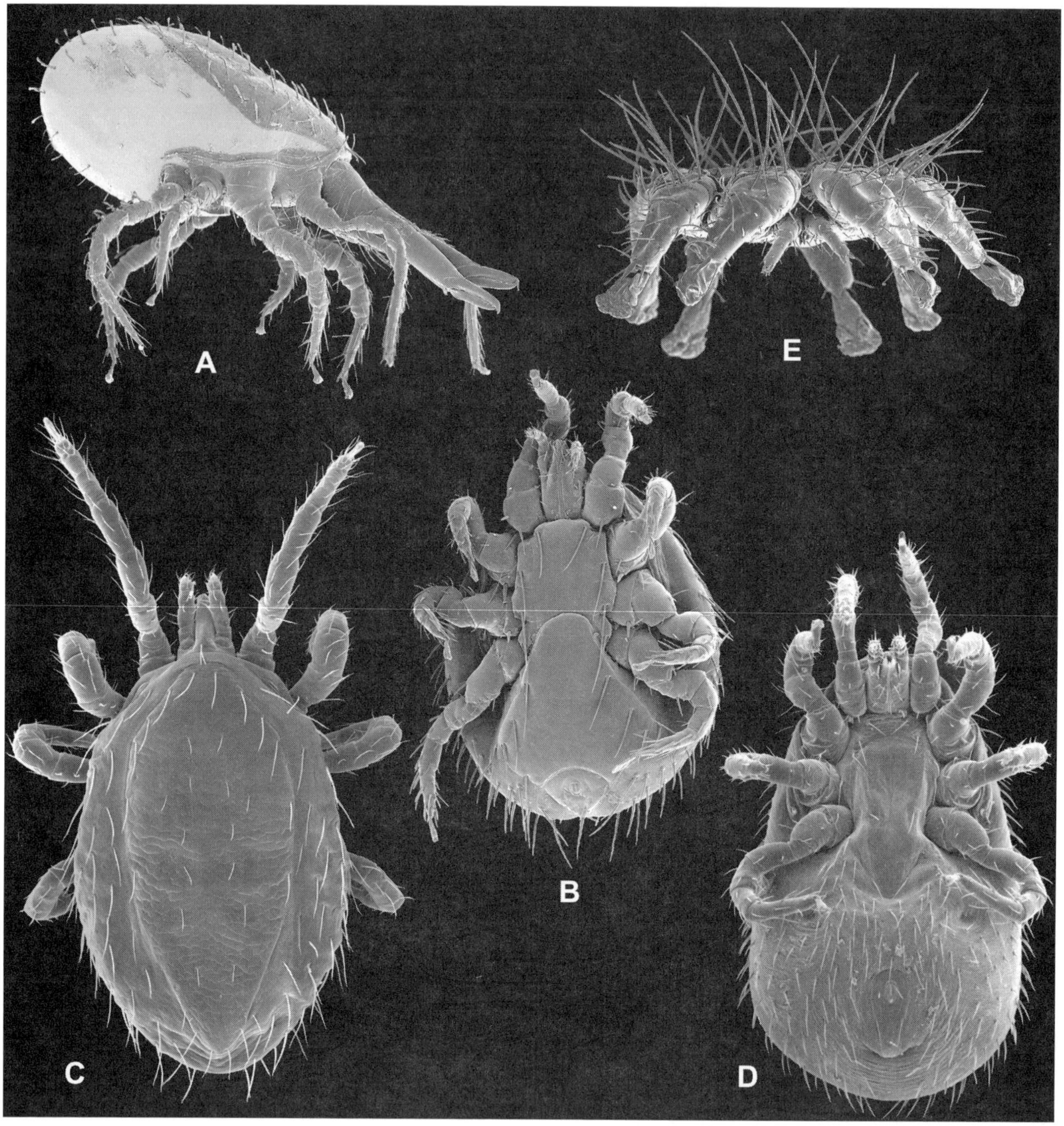

Fig. 12.66. LAELAPIDAE. *Stratiolaelaps scimitus* (Womersley) (Queensland, Australia): **A**, adult female in lateral view. *Laelaps echindninus* Berlese (Queensland, Australia): **B**, venter of female.

MACRONYSSIDAE. *Ornithonyssus bursa* (Berlese) (Queensland, Australia): **C**, dorsum of female. *O. bacoti* (Hirst) (Northern Territory, Australia): **D**, venter of female.

SPINTURNICIDAE. *Spinturnix loricata* Domrow (Victoria, Australia): **E**, anterior view of male.

D.E. WALTER
E.E. LINDQUIST
I.M. SMITH
D.R. COOK
G.W. KRANTZ

CHAPTER THIRTEEN

Order Trombidiformes

The order Trombidiformes is a diverse assemblage of acariform mites that currently comprise two suborders: the Prostigmata and the Sphaerolichida. There are approximately 38 currently recognized trombidiform superfamilies, all but two of which are classified under the Prostigmata (see below). The sphaerolichid superfamilies Sphaerolichoidea and Lordalycoidea are each represented by a single family and a few described species of small, weakly ornamented mites that (1) lack a tracheal system and associated stigmatic openings, (2) tend to display a strong posteroventral curvature of the idiosoma or caudal bend (referred to in chapter 3), and (3) often retain the primitive character of six pairs of prodorsal setae (including two pairs of bothridial setae). Tracheal systems may also be absent in some Prostigmata (i.e., Halacaroidea, Eriophyoidea, and some Raphignathoidea and Dolichocyboidea), but these absences represent secondary losses. Unlike the Sphaerolichida, prostigmatic taxa generally lack a distinct idiosomatic caudal bend (some Eupodoidea, Tydeoidea, Eriophyoidea, and Raphignathoidea are exceptions). Although the Labidostomatoidea retain the plesiomorphic condition of six pairs of primary prodorsal setae and members of the eupodoid family STRANDTMANNIIDAE have five pairs, other Prostigmata have no more than four primary prodorsal pairs, including the bothridial setae. Like many Prostigmata, row *c* of the opisthosoma of Sphaerolichida is deficient (setae *c3–4* are absent in both superfamilies), but, unlike some Prostigmata, they display no dorsal hypertrichy. Tarsal claws may be present or absent in both suborders, and the tarsal empodia of at least legs II–III typically are padlike, membranous, or rayed, or they may be absent. Clawlike (e.g., Figs. 13.44C, H) or suckerlike (e.g., Figs. 13.82A, B, 13.92B, F) empodia occur in some trombidiform taxa.

Suborder Sphaerolichida

*DIAGNOSIS: Prodorsum with 2 pairs of filamentous trichobothria (*si, vi*) and 1 (*se*) to 4 (*in, ve, exp*) additional pairs of setae; naso nude (anterior pair of trichobothria may be inserted near its base) or absent; median eye present on underside of naso or absent, 0–2 pairs of lateral eyes. Body chaetome reduced, setae* c3–c4 *absent, segments AN and PA not added during ontogeny; deutonymphs and adults with 2 pairs of genital papillae. Rutella absent, palpi with 4–5 free segments, chelicerae each with 2 setae. Tracheal openings between coxae I–II have been reported in* Sphaerolichus.

The Sphaerolichida contains two monogeneric families and approximately 16 described species. They are considered to be closely related to the Prostigmata (OConnor 1984c). *Lordalycus* Grandjean (LORDALYCIDAE) is a junior synonym of *Hybalicus* Berlese, which Theron (1974) designated as the type genus for his family Hybalicidae. The species of *Hybalicus* described in that work are referable instead to the endeostigmatic genus *Oehserchestes* Jacot (OESERCHESTIDAE) (Kethley 1977b; see p. 421).

Members of the LORDALYCIDAE (Figs. 13.1H–K, 13.2A, B) (Lordalychidae in Theron and Ryke 1975a; Krantz 1978; and Kethley 1982) are distinctive mites with globular bodies and strongly ornamented cuticle and are common worldwide. Little is known about their biology, but some species have been observed to carry fungal materials in their guts (Theron 1979; Walter 1988b). The prelarva splits the eggshell but remains inactive after doing so. The prelarval dorsum is protected by well-developed dorsal setae, much as in some Anystina.

The genus *Sphaerolichus* (SPHAEROLICHIDAE) (Figs. 13.1A–G) is common in many types of litter worldwide (Shiba 1968; Theron and Ryke 1975b). Australian species of *Sphaerolichus* are bright pink in life and move slowly while waving their highly modified front legs (Fig. 13.1D). They climb readily and may be ambush predators (D. E. Walter, unpublished), although the guts of some South African species have been reported to contain fungal material (Theron 1979).

Key 13.1. Suborder Sphaerolichida, Key to the Families (Figs. 13.1, 13.2, pp. 327–28)

1a. Anterior pair of trichobothria (*vi*) associated with well-developed naso (Fig. 13.1A), often with median eye on its underside; 0–2 pairs of lateral eyes present; pretarsus I without empodium (Figs. 13.1D–E), other empodia clawlike (Fig. 13.1F); chelicerae chelate, with lobelike teeth (Fig. 13.1C) **SPHAEROLICHIDAE (*Sphaerolichus*)**

1b. Anterior pair of trichobothria (*vi*) in communal depression (Figs. 13.1H, 13.2A); naso absent; eyes absent; all pretarsi with padlike, setulate empodia (Fig. 13.1I); fixed digit of chelicera truncate, movable digit serrulate (Figs. 13.1J, 13.2B) **LORDALYCIDAE (*Hybalicus*)**

Suborder Prostigmata

The 36 superfamilies of Prostigmata include a bewildering variety of terrestrial, aquatic, and marine predators, phytophages, saprophages, paraphages, and parasites. Their range of morphological, ontogenetic, and behavioral attributes far exceed those of any other higher category of Acari. Although there are exceptions, prostigmatans tend to be weakly or incompletely sclerotized forms that only rarely undergo a formulized, incremental ontogenetic development of idiosomatic plates. Many fall within expected size limits of 300–500 µm, but some parasitic species of Heterostigmata may measure less than 100 µm in length, while certain members of the parasitengonine family TROMBIDIIDAE can exceed 12,000 µm. Where an internal respiratory system occurs, it may debouch via paired stigmata at or near the cheliceral bases (Figs. 13.62B, 13.74C) or at the humeral angles of the propodosoma (Figs. 13.81B, 13.86A). Similar diversity exists in the structure of the chelicerae, which may be completely external or largely concealed within the gnathosomatic capsule (Figs. 13.77D, 13.82A, B). They may be chelate in form (Figs. 13.5C, 13.6E, 13.21B), but more often the fixed cheliceral digits are modified as a fused stylophore (Figs. 13.66A, 13.74C) or they may be reduced or absent, with the movable digit assuming a styliform, hooklike, or spearlike form (Figs. 13.13C, 13.22D, 13.24C, 13.68I). The palpi also show considerable variation, often diverging in remarkable ways from the simple segmented appendages common to early derivative, free-living prostigmatic taxa. They may be grasping or fanglike (Figs. 13.6G, H, 13.7F), truncated (Figs. 13.18A, E), profoundly reduced (Figs. 13.84B, C, 13.86C), or absent (Fig. 13.87C), or they may also be well developed with distinctive palptibial claws (Figs. 13.20B, 13.22A–C, F). The leg coxae are immovably joined or insensibly fused to the podogastric surface. A pair of urstigmata, or organs of Claparède, often is found on or in the area of coxae I–II of larvae. With some exceptions, the genital and anal openings are approximate or contiguous on the opisthogaster, with the anal aperture commonly being terminal in position. Two or three pairs of genital papillae (Figs. 13.3F–H) may flank the genital opening of the adult, although some adult Hydrachnidiae (water mites) may have well over 100 pairs (Fig. 13.34). Sperm transfer generally is accomplished by means of stalked spermatophores, but an intromittent aedeagus may mediate sperm transfer in certain prostigmatic groups (the Raphignathina and Heterostigmatina provide many examples).

There is considerable range in idiosomatic sclerotization, from virtually complete plate coverage both dorsally and ventrally (e.g., the LABIDOSTOMATIDAE) (Figs. 13.2C, D, 13.5A, B, E), to little more than a weak prodorsal sclerite (e.g., terrestrial Parasitengonina) (Figs. 13.40C, F, I). Many prostigmatic taxa have no discernible sclerotization, either dorsally or ventrally. One or two pairs of bothridial organs may be present on the prodorsum, but a posterior opisthosomatic pair, along with a single prodorsal pair, occurs in certain Tydeoidea (Fig. 13.14C). Additional bothridial setae (trichobothria) may be found on the posttrochanteral segments of the legs and on the terminal palpal segments. Ocelli and dorsal cupules may be present or absent.

Other than in Heterostigmatina and some Raphignathina, sexual dimorphism rarely is pronounced in the suborder. Ontogenetic development (Fig. 13.3) in most taxa is gradual and subtle and, aside from the unique hexapod condition of the larva, all stases tend to be similar in appearance as well as habitat. However, transformation from the larval to postlarval instars in the cohort Parasitengonina is accompanied by dramatic changes in appearance and behavior. The heteromorphic parasitengonine larva is parasitic, the protonymphs and tritonymphs are nonfeeding calyptostases, and the active deutonymph and adult are predaceous. Other character state changes may occur during ontogenetic development of prostigmatic mites. For example, in many families that have genital papillae, the number of papillar pairs increases from one pair in the protonymph to two in the deutonymph and finally to three in the tritonymph and adult (Figs. 13.3F–H). Incremental papillar augmentation of this type exemplifies **Oudeman's rule**. Similar incremental increases may be seen in the number of solenidia on tarsus I (Fig. 13.4) and in the number of setae that make up the genital chaetome.

Supercohort Labidostomatides

The Labidostomatides are unusual in having an undivided, heavily armored dorsal shield (Figs. 13.2C, D) and a complete ventral armature (Figs. 13.5A, B, E). Both the dorsum and venter are embellished with distinctive alveolar ornamentation that may vary from taxon to taxon. Ventrally, the coxal fields are expanded and mesally contiguous and, along with the genital and anal valves, share the sclerotization of the surrounding venter. Members of this supercohort have two pairs of prodorsal setae (*ro*, *bo*) produced as bothridial sensilla and are unusual in having a total of six pairs of prodorsal setae (Fig. 13.5B), a plesiomorphic characteristic that is shared only by some members of the superfamily Sphaerolichoidea (Sphaerolichida, see above). The prodorsum also carries a pair of lateral ocelli (rarely absent) and an unpaired, naso-associated anteromedian ocellus that may be distinct or obscure. In most taxa (the genus *Sellnickiella*, Fig. 13.2D, is an exception), one or more pairs of lateral **pustules** (Seitenhöckern of Thor 1931) lie on the idiosomatic margin posterior to the lateral ocelli (Figs. 13.2C, 13.5A, B). These enigmatic organs, which Grandjean (1942b) considered homologues of cuticular pores, may be uniporous or multiporous and may number over 50 pairs in some taxa (Coineau 1964b; Robaux 1977a; Bertrand 1990a). There is sexual dimorphism in the arrangement of genital and anal valves, with those of females (Fig. 13.5F) being contiguous and those of males (Fig. 13.5E) clearly separated (the male of *Sellnickiella brasiliense* Feider and Vasilu may be an exception). Two pairs of genital papillae are present. The strongly chelate chelicerae carry both setae *cha* and *chb* on the ornamented fixed digit (Fig. 13.5C), a primitive trait that lends credence to the assumption of antiquity in the supercohort. A pair of stigmata opens at the cheliceral bases. The palpi have four segments, with an unusual fusion of the genu and tibia, and they lack a thumb-claw complex. The femora of legs I–IV are subdivided into basi-, meso-, and telofemora, although there is only a basifemoral division on leg I in the genus *Eunicolina*.

- Superfamily Labidostomatoidea (Figs. 13.2C, D, 13.5A–F)

DIAGNOSIS: With the characteristics of the supercohort; medium to large, heavily sclerotized mites (300–1500 µm) with distinctive dorsal and ventral ornamentation, sometimes neotrichous, appearing yellow, orange, or dark olive-green to greenish black in color often even in preserved material. Chelicerae dentate or edentate, cheliceral bases not fused to rostrum or to each other, terminal palpal segment with an erect solenidion; leg tarsus I with a simple, bifid, or variously branched dorsal famulus and a pair of erect solenidia (Fig. 13.5D), ambulacra I bidactylous or absent, ambulacra II–IV tridactylous, empodium clawlike. With characteristic dorsal chaetotaxy, sometimes obscured by neotrichy. With an elattostatic larva and 3 active nymphal instars.

The Labidostomatoidea is represented by the family LABIDOSTOMATIDAE (Labidostomidae, Labidostommidae, Labidostommatidae, and Nicoletiellidae of authors), a cosmopolitan assemblage of approximately 50 described species of predatory mites considered by Kethley (1982) to represent nine genera and three subfamilies. Bertrand (1990a), however, recognized four broadly defined labidostomatid genera (*Eunicolina*, *Akrostomma*, *Sellnickiella*, and *Labidostoma*) and reduced several of Kethley's generic entities to subgeneric rank. Labidostomatids occur in surface or deep soil niches and in overlying organic material, lichens, and moss, where they prey on collembolans and other small arthropods. Specimens have also been recovered from caves in Europe. While labidostomatids have achieved their greatest diversity in the Southern Hemisphere, the seven species composing the genera *Eunicolina* and *Akrostomma* occur only in southern Europe and North America (Greenberg 1952; Robaux 1977a; Bertrand and Coineau 1979; Bertrand 1990a). In addition, there are some European representatives of the primarily tropical subgenus *L.* (*Labidostoma*) (Grandjean 1942b; Coineau 1964a; Bertrand 1981; Bertrand and Corpus-Raros 1997), and the subgenus *L.* (*Nicoletiella*) is exclusively Holarctic (Feider and Vasilu 1969, 1972; Shiba 1969; Bloszyk 1980b; Vistorin 1980b). Species of *L.* (*Atyeonella*) have been described only from Australia, although a South American form also has been identified (Atyeo and Crossley 1961b; Bertrand 1990a).

The genus *Sellnickiella* (Fig. 13.2D) appears to have achieved its greatest radiation in the Southern Hemisphere (Atyeo and Crossley 1961a, b; Feider and Vasilu 1970; Feider, Vasiliu, and Calugar 1974; Bertrand 1990b; Bertrand and Theron 1990, 1992) and is considered by Bertrand (1990a) to represent an early derivative Gondwanan lineage. Representatives of the genus have been collected in South America, Australia, New Zealand, and South Africa. The five known species of *S.* (*Dicastriella*) have been found only in Chile and may compose a relict assemblage that arose in isolation following the Andean uplift during the Miocene. Their small size (idiosomatic length generally less than 350 µm), reductions in ocular and bothridial setal development, and relatively short legs are thought to reflect the confining nature of their deep soil microhabitat.

There is little published information on the behavior of labidostomatids. Schuster and Schuster (1969) observed live *L.* (*Nicoletiella*) *cornuta* (Canestrini and Fanzago) collected in Austria and found that males produced and deposited numbers of stalked spermatophores when females were present, but few in their absence. One male in the company of a female, for example, produced a total of 21 spermatophores in 10 days. Labidostomatids move about fairly rapidly on legs II–IV, using legs I as surrogate antennae and for capturing, holding, and manipulating prey. The tarsi of legs II may also be used while feeding, leaving the predators to stand on only legs III–IV (Vistorin 1980a). Adult and nymphal specimens of *L.* (*Nicoletiella*) *pacifica* (Ewing), a familiar faunal component of forest litter in western Oregon, USA, seemed indifferent to confinement in laboratory observation chambers and actively stalked collembolans provided as prey, using legs I as probes and pincers. They continued to move about the chamber even after a successful capture, discarding the shriveled remains of their prey after some minutes and then seeking new targets (G. W. Krantz, pers. obs.). Vistorin (1980a) observed that adults and nymphs of *L.* (*N.*) *cornuta* and *L.* (*N.*) *denticulata* (Schrank) prey on a wide range of arthropods, including hard-shelled adult oribatids and soft-bodied mites and collembolans. These and a few other species were observed to prey only on vagile individuals, but *L.* (*N.*) *jaquemarti* (Coineau) fed also on morbid hosts.

Bertrand (1989) examined the development times and distributions of three European species of *Labidostoma* and found that only the widely distributed *L.* (*Nicoletiella*) *luteum* Kramer has been successful in surmounting mountain barriers and climatic constraints and invading the more northerly portions of the continent following the last glaciation. Unlike *L.* (*N.*) *cornuta* C. and F. and *L.* (*N.*) *denticulata* (Schrank), *L.* (*N.*) *luteum* has a short development time (one year) and is parthenogenetic in its northern range. Thelytokous populations of *L. luteum* in southern Finland show no obvious preference for any of the wide range of soil habitats in which they are found, thriving in microclimates that range from dry barren stone beds to wet sphagnum. Many records of these populations are connected with an abundance of ants, especially with nests of *Lasius* in soil, under stones, or under the bark of tree stumps (Lehtinen and Niemi 1995).

Supercohort Eupodides

The Eupodides is a large, diverse, and probably heterogeneous assemblage of predators, parasites, fungivores, and phytophages. One superfamily, the Halacaroidea, contains primarily marine and estuarine forms. Most eupodines are soft-bodied forms with striated integument, weak subcutaneous sclerites, and a reduced opisthosomatic chaetotaxy. The disjugal furrow may sometimes be obscure or absent. Except in cases where neotrichy occurs (e.g., PENTHALEIDAE, STRANDTMANNIIDAE), at most two pairs of setae are found in setal rows *c*, *f*, and *h*, and adanal and anal setae (*ad*, *an*) may be present or absent. One pair of prodorsal bothridia and associated trichobothria occurs in most eupodidid groups (two pairs in Bdelloidea, none in Eriophyoidea), and a naso usually is present (absent in the Tydeoidea and Eriophyoidea and rarely present in the EUPODIDAE). Eupodidids often have one or two pairs of lateral ocelli, and an anteromedian "eye" occurs in some taxa (e.g., some Bdelloidea and Halacaroidea). Most of the known species have two or three pairs of genital papillae (absent in Eriophyoidea and reduced or absent in Tydeoidea and Halacaroidea). As in most other prostigmatic mites, eupodidids have postcheliceral stigmata, although these structures have been lost, apparently secondarily, in some groups (e.g., Eriophyoidea, Halacaroidea). Chelicerae often are elongate and may be chelate-edentate, with the fixed digit subequal to the movable digit or greatly reduced; alternatively, the movable digit may take the form of a needle- or sicklelike extension of the fused or contiguous fixed digits. One or both digits may be twisted or otherwise distorted (e.g., some EUPODIDAE), and the fixed digit may take the form of a membranous extension. The palpi have one to five segments and may be simple or raptorial; a thumb-claw complex is lacking. Leg tarsus I usually has at least one erect, recumbent, or occasionally recessed dorsal solenidion, along with a famulus that often is reduced and terminally branched (often referred to as *stellate* in literature). The famulus may be absent in some groups (e.g., the Tydeoidea and Eriophyoidea). Solenidia also may be found on genu and tibia I and on other leg segments, including tarsus II, genu and tibia III, genu IV, and femur III. The pretarsi usually are armed with a pair of claws and an empodium (claws absent in Eriophyoidea). Leg trichobothria are present or absent on tibia IV (absent in Tydeoidea and Eriophyoidea) but only occur on other leg segments among members of the superfamily Bdelloidea. Larvae have urstigmata, or Claparède organs, in the region of coxal fields I–II (reduced in Tydeoidea but absent in Eriophyoidea; see chapter 3).

The five superfamilies included in the Eupodides—the Bdelloidea, Halacaroidea, Eupodoidea, Tydeoidea, and Eriophyoidea—are characterized in the sections that follow.

▪ Superfamily Bdelloidea (Figs. 13.6A–J)

DIAGNOSIS: Soft-bodied to well-sclerotized forms, with 2 pairs of prodorsal sensilla and with 12–14 additional pairs of dorsal, lateral, and caudal setae, lateral ocelli present or absent. With 1–3 pairs of dorsal cupules (3 pairs in BDELLIDAE, 1–3 pairs in CUNAXIDAE) and 1 pair posteroventrally (lyrifissure ih *or* ips*); rudimentary caudal segment PS present, along with setae* ps1 *and* ps2, *segment AD and associated setae* ad1 *may be present (BDELLIDAE). Ventrally with 3 pairs of genital papillae plus 2 or more eugenital setae (BDELLIDAE) or with 2 pairs of genital papillae and typically lacking eugenital setae (CUNAXIDAE); hypostome either with 2 pairs of macrosetae (CUNAXIDAE, subfamilies Bdellinae and Odontoscirinae of BDELLIDAE) or 6 pairs (bdellid subfamilies Cytinae and Spinibdellinae). Cheliceral bases separated and hinged to rostrum, chelicerae usually elongate and capable of lateral scissors-like movement, both cheliceral setae usually present (BDELLIDAE), or seta* cha *absent (CUNAXIDAE); cheliceral digits either well developed and chelate (some BDELLIDAE) or fixed digit reduced and movable digit hooklike; palpi with 3–5 segments, sometimes armed with strong paraxial spines, palpal tibiotarsus terminating in a pair of long tactile setae (1 seta in* Monotrichobdella*) (BDELLIDAE) or clawlike and raptorial (CUNAXIDAE). Podocephalic canals often apparent internally or externally above coxal fields I. Tarsi I–IV with claws, empodia pad- or clawlike; always with a leg trichobothrium on tibia IV, trichobothria present or absent on tibia I–II and on tarsi III–IV, always absent on tarsus I. Eggs globular or flat and lenticular, often tuberculate or ornamented with spiny projections; male sperm transfer indirect in BDELLIDAE and most CUNAXIDAE, but a sclerotized aedeagus present in some cunaxine* *genera. With five postembryonic noncalyptostatic instars: larva, protonymph, deutonymph, tritonymph, and adult; calyptostatic prelarva present in BDELLIDAE.*

The Bdelloidea comprises a moderately large assemblage of predatory species grouped in two cosmopolitan families: the BDELLIDAE and the CUNAXIDAE. The BDELLIDAE (Figs. 13.6A–F) includes 16 genera, five subfamilies, and approximately 135 species of active eupodidid mites that prey on small arthropods and their eggs in soil and litter (Atyeo 1960, 1963a; Wallace and Mahon 1972, 1976; Wainstein et al. 1978; Tseng 1978; Michocka 1987; Swift and Goff 1987; Van der Schyff, Theron, and Ueckermann 2003, 2004). The elongate gnathosoma characteristic of most bdellid genera has led to their being referred to as "snout mites" in some treatments. Bdellids are mostly weakly sclerotized (*Trachymolgus* is an exception); are predominantly red, green, purple, or yellow in life; and tend to be active and moderately fast moving. Some are of great size (> 3,000 µm), but most measure between 500 and 2,400 µm. A prelarval calyptostase has been described for a variety of bdellid genera (Alberti 1975).

Bdellid species often are found in soil and litter in a variety of situations, ranging from dry exposed desert to cool moist forest habitats. Mosses, lichens, and algae on rocks and on tree bark also are common habitats. Some species are found under tree bark or on tree foliage, in low herbs and grasses, under rocks, or in littoral habitats in association with algae. As an example of the latter, *Neomolgus littoralis* (L.), a large, bright red circumpolar species, sometimes occurs in great numbers in algae on intertidal rock outcrops on the northern Oregon, USA, coast (G. W. Krantz, pers. obs.). Similarly, *Bdellodes hygrotes* Swift and Goff was described from algae-covered rocks at Cape Kinau on Oahu, Hawaii (Swift and Goff 1987). The activity level of *Bdella septentrionalis* Atyeo and Tuxen, another littoral species that inhabits rocky seashores and estuaries in the Baltic basin, clearly follows a tidal rhythm. Mites become active during daylight hours but only after their substrate has been dampened by high tides or sea sprays. Estuarine mites kept in a laboratory seawater aquarium under constant artificial light displayed a 12-hour circatidal periodicity in activity that is believed to have been endogenous. However, mites collected from the seashore, where there is a relatively low tidal range compared to that of the estuarine habitat, did not establish a recognizable activity rhythm in the same aquarium setting (Ernst 1995).

Some bdellids may be effective in regulating populations of economically important arthropods (Gerson and Smiley 1990). Examples are *Bdella depressa* Ewing on the clover mite, *Bryobia praetiosa* Koch in the midwestern and western United States (Snetsinger 1956; Jeppson, Keifer, and Baker 1975b), and *Bdellodes lapidaria* on the collembolan *Sminthurus viridis* in Australia (Wallace and Walters 1974). *Odontoscirus virgulus* (C. and F.) was found in considerable numbers with the red-legged earth mite, *Halotydeus destructor* (Tucker) (PENTHALEIDAE) in South Africa, but a predator-prey relationship was not clearly established between the two organisms (Womersley 1933; Meyer and Ryke 1959a). Large numbers of the spinibdelline *Biscirus uncinatus* (Kramer) were collected from balsam fir trees that were heavily infested by the balsam woolly aphid, which probably provided a major food source for these large predators.

Members of the BDELLIDAE secrete silk from the buccal cavity via the podocephalic canals (see chapter 3) and use this silk to secure their prey to the substrate and construct molting chambers (Wallace and Mahon 1972, 1976; Alberti 1973; Alberti and Ehrnsberger 1977). Bdellids also have been observed to utilize a stringy exudate originating in the supraesophageal organ to capture prey (Alberti 1973). Some bdellids have been reported to feed on arthropod eggs and on nematodes (Sorenson, Kinn, and Doutt 1983; Pugh and King 1985a).

Male bdellids produce uniquely shaped stalked spermatophores to receive their sperm packets (see Fig. 3.20A, p. 46). Their genital tract is quite complex, as is the female ovipositor (Atyeo 1960; Michocka 1987). Bdellid eggs are reddish in color and often covered with spiny projections (Wallace and Mahon 1972; Alberti 1974). Whether bdellids have a diplodiploid or haplodiploid genetic system remains undetermined (Norton et al. 1993).

The family CUNAXIDAE (Figs. 13.6G–J) comprises small to moderately large (400–1,000 µm), active, mostly weakly sclerotized predators (*Scutascirus* is an exception), many of which feed on microarthropods in soil, plant debris, moss, or straw (Baker and Hoffmann 1948; Smiley 1975, 1992; Den Heyer 1976, 1977a, b, 1978a, b, 1979a–d, 1980a, b, 1981; Kuznetsov and Livshits 1978; Tseng 1980; Sionti and Papadoulis 2003, 2004). Cunaxids may be whitish or translucent, but many are predominantly red, orange, or yellow in color. Their palpi are often raptorial with a clawlike terminal palptarsus and armed internally with spines and spurs that better enable them to grasp their prey (Figs. 13.7G, H). However, the palpi in species of *Parabonzia* terminate instead in two long sensory setae and a solenidion, much as in the BDELLIDAE (Fig. 13.6A). Because they share this and a variety of other attributes with bdellids, *Parabonzia* has been considered the earliest derivative cunaxid genus (Den Heyer 1978a).

Two types of prey capture strategies are known in the family, and these are reflected in the morphology and use of the palpi (Walter and Proctor 1999). The Cunaxinae (e.g., species of *Armascirus*, *Cunaxa*, *Dactyloscirus*, and *Rubroscirus*) are ambush predators that have very long prodorsal trichobothria and elongate palpi that are held reflexed over the body. When prey (e.g., springtails, soft-bodied mites, and small insects) pass, cunaxines pounce and grab the prey in the palpi, impaling them on their chelicerae. The prey is often elevated off the ground as it is digested and drained of fluids. In contrast, members of the Coleoscirinae (e.g., species of *Neoscirula* and *Coleoscirus*) and Cunaxoidinae (e.g., species of *Pulaeus*, *Cunaxoides*, and *Neocunaxoides*) are cruise predators that search out arthropod prey but also take less active forms such as nematodes. These mites have relatively short palpi that are held parallel to their chelicerae as they search out their prey (Walter, Hunt, and Elliott 1988; Walter and Kaplan 1991; Walter and Proctor 1999).

Many cunaxids are aerial forms that prey on phytophagous insects and mites (Meyer and Ryke 1959a; Muma 1975; Gerson and Smiley 1990; Walter 1999; Walter and Proctor 1999). Their effect on prey populations is unclear, although Ewing and Webster (1912) referred to *Cunaxoides parvus* (Ewing) as an enemy of the oyster-shell scale on apples. *C. oliveri* (Schruft) was observed to prey on the eriophyid mite *Calepitrimerus vitis* (Can.) (ERIOPHYIDAE) infesting grape vines (Schruft 1971; Gerson and Smiley 1990). Cunaxids are commonly found on the bark of fruit trees infested by spider mites (TETRANYCHIDAE) in western Oregon, USA, but their effect on prey populations has not been determined. *Cunaxa capreolus* Berlese, a common inhabitant of citrus trees in Florida, fed preferentially on active stases of prey (the citrus brown mite and booklice) under experimental conditions in Egypt (Zaher, Soliman, and El-Bishlawy 1975). *C. setirostris* (Hermann), a common denizen in litter habitats worldwide, has also been collected from stored barley meal and wheat in Europe (Hughes 1976). Under laboratory conditions, *Rubroscirus africanus* Den Heyer (*C. setirostris* of Meyer and Ryke 1959a and Den Heyer and Ryke 1970) was observed to prey on the citrus red mite, *Panonychus citri*, and a tydeoid mite, *Pronematus* sp., in South Africa (Den Heyer 1979d). An undescribed Australian species of *Rubroscirus* was found to feed on mites and small insects such as thrips, as well as drinking drops of honeydew (Walter and Proctor 1999).

Like the BDELLIDAE, cunaxids are known to spin silk for a variety of purposes. In addition to employing silk strands as a means for securing prey, *C. setirostris* constructs an irregular silken net and waits beneath it for prey to become entrapped, a strategy similar to that used by theridiid spiders to capture prey (Foelix 1982). Nymphs of *C. taurus* (Kramer), *Dactyloscirus inermis* Trägårdh, *Coleoscirus simplex* (Ewing), and an undescribed species of *Pulaeus* are known to construct a silken chamber around themselves for protection during molting (Alberti and Ehrnsberger 1977; Walter and Kaplan 1991).

Little is known about reproductive behavior in cunaxids, although precopulatory guarding of the quiescent tritonymphal female has been reported in the subfamilies Cunaxinae, Coleoscirinae, and Cunaxoidinae (Walter and Kaplan 1991). A well-developed, sclerotized aedeagus has been observed in males of *Dactyloscirus* and a less well-formed, poorly sclerotized one in *Rubroscirus* and *Armascirus* (Den Heyer 1978b, 1979c, d, 1981). A haplodiploid genetic system has been confirmed for *Cunaxa capreolus*, whose male lacks an aedeagus (Helle et al. 1984). The eggs of cunaxids are often the same color as the adults and may be highly ornamented with reticulate sheaths or spines.

▪ Superfamily Halacaroidea (Figs. 13.7, 13.8)

DIAGNOSIS: Soft-bodied or leathery mites; typically with 1–4 variously ornamented dorsal shields in a striated integument (anterior, posterior, and paired lateral ocular plates) and (sometimes) associated smaller platelets, shields occasionally overlapping (HALACARIDAE, Mictognathinae) or dorsal plates fragmented into 12–16 poorly defined platelets (PEZIDAE), dorsum usually with 5–10 pairs of setae (16–20 pairs in PEZIDAE), typically with 1–2 pairs of dorsolateral ocelli (occasionally absent), sometimes with an anteromedian pigmented photosensitive field usually visible only in live or freshly prepared specimens, bothridial sensilla and naso absent. With 3–5 pairs of dorsal pores (5 pairs clearly discernible in Halacarus *and* Lohmannella*), some of which likely represent cupules* ia, im, *and* ip *of other Eupodides, and "rosette pores" that form part of the dorsal shield ornamentation in some genera (e.g.,* Copidognathus, Agauopsis, Pelacarus*). Venter generally with anterior and posterior epimeral plates on which legs I–II and III–IV are articulated, often expressed laterodorsally, sometimes reduced or absent (e.g., some Rhombognathinae), anterior epimerals typically fused ventrally into single plate and carrying 3 pairs of setae, with paired epimeral glands (occasionally absent) opening internal to trochanters I–II, larvae (rarely postlarval instars) may have epimeral urstigmata; posterior epimerals not fused, usually with 4 pairs of setae. Genital apparatus often within a plate that is contiguous with and partially encircling terminal anal papilla and sclerites or may encompass them to form a genito-anal plate, eugenital setae often present. Adult marine forms generally with 2–3 pairs of genital papillae located beneath genital valves and often obscure, fresh- and brackish water forms with 2 to several external pairs of papillae on or near genital valves, occasionally internal (*Peregrinacarus reticulatus *Bartsch and PEZIDAE) or with both internal and external papillae (*Caspihalacarus hyrcanus *Viets). Gnathosomatic rostrum normally produced or strongly curved upward to accommodate overlying curved cheliceral shafts; movable cheliceral digit often hooklike or elongate, sometimes wedge shaped and toothed internally (e.g.,* Halixodes*), fixed digit absent or represented by a membranous extension. Palpi 2–4-segmented, simple, or with terminal segment curved and pointed, occasionally greatly elongated (*Corallihalacarus chilcottensis *Otto), attached laterally or dorsally to the subcapitulum. Podocephalic canals entirely internal, occasionally visible in clear specimens; stigmata and peritremes absent. Tarsi I–IV with variously ornamented paired claws, with or without basilar sclerites, often with a comblike internal fringe and a dorsal accessory tooth that may be expanded into a fan-shaped comb (e.g., some Rhombognathinae), a median uni- or bidentate empodial claw often present; tarsi I–II each with an erect solenidion, tarsus I with a famulus that may be obscure or reduced to a pore, terminal eupathidia and other parambulacral setae may be present on all tarsi. In uni- or mul-*

*tivoltine forms, sperm transfer primarily through spermatophore deposition; parthenogenesis known for only one marine species (*Isobactrus setosus *(Lohmann)), but probably common in freshwater forms. Life stages include larva, 1–3 nymphs, and adult.*

The Halacaroidea is a cosmopolitan assemblage of over 1,000 described species and more than 50 genera of marine and freshwater mites considered here to represent two families: the HALACARIDAE (Figs. 13.7A–H) and the monogeneric PEZIDAE (Figs. 13.8A–D) (Bartsch 1996, 2004; Harvey 1990a). Two additional family categories (Rhombognathidae, Lohmannellidae) have been recognized by some authors (e.g., Newell 1984; Harvey 1990e), and *Halixodes* was earlier considered to represent yet another family (Viets 1927). Although there may be enough morphological homogeneity within the HALACARIDAE to warrant retention of a single family category, within-family diversity has led to the recognition of a number of subfamilial categories over the past 75 years. Currently, the family consists of 14 subfamilies: the Halacarinae Viets, Limnohalacarinae Viets, Copidognathinae Bartsch, Mictognathinae Otto, Actacarinae Viets, Acarochelopodinae Bartsch, Halixodinae Viets, Anomalohalacarinae Bartsch, Enterohalacarinae Viets, Astacopsiphaginae Viets, Simognathinae Viets, Ropohalacarinae Bartsch, Lohmannellinae Viets, and Rhombognathinae Viets.

Halacarids tend to be small in size, with most species measuring 200–800 µm in length. They do not swim, but their strong curved tarsal claws probably facilitate limited movement, even in the face of tidal surges, and serve as anchors against being dislodged. Species of *Halixodes*, *Parhalixodes*, *Astacopsiphagus*, and *Enterohalacarus* are known or suspected parasites or obligate associates of other invertebrate animals, including gastropods, decapod crustaceans, nemerteans, sponges, and echinoids (Brucker 1897; Viets 1931, 1938; Newell 1956; Stout and Viets 1959; Laubier 1960; Bartsch 1986b, 1987, 1988, 1995). Extreme palpal reduction in *Spongihalacarus longiscutis* Otto, an Australian halacarid found associated with "sponge weed" (a symbiotic pairing of a sponge, *Haliclona cymaeformis*, with a red alga, *Ceratodictyon spongiosum*) suggests that it too may be a parasitic form (Otto 2006). The great preponderance of known halacarids, however, are considered to be predaceous or algivorous. Crustaceans, annelids, and other halacarids have been verified as food items for members of the halacarid genera *Thalassarachna* and *Agauopsis* (Lohmann 1893; Teschner 1963; Kirchner 1969; Krantz 1970a; Bartsch 1974a, b; MacQuitty 1984). Members of the marine subfamily Rhombognathinae are commonly found in association with diatoms and other algae on which they are thought to feed (André 1946; Newell 1947b; Bartsch 1977a, 1982b, 1983, 1989b, 2003; Green and MacQuitty 1987). Their typically green color would seem to corroborate this supposition, but Abé (1998) observed *Rhombognathus* species attacking bryozoans on an algal substrate. Individuals of the freshwater halacarine genus *Porohalacarus* often have green gut contents, suggesting that they also feed on plant material (Bartsch 1996).

Marine halacarids have been collected from intertidal and shallow subtidal sand, sediments, salt marshes, tidal flats, and rocky surfaces and from a wide variety of niches in the plant and animal communities that thrive in these zones (Lohmann 1893; Viets 1940b; André 1946, 1959; Newell 1947b; Angelier 1953; Krantz 1971, 1973a, 1976; Green and MacQuitty 1987; Bartsch 1977b, 1989b, 1992, 2004; Pugh and King 1985c; Abé 1998; Otto 1999a–c; Chatterjee and DeTroch 2000; Chatterjee and Chang 2005, 2006). At the same time, a considerable number of species have been taken from deeper waters, occasionally from depths exceeding 1,000 m (Newell 1967b; Krantz 1982; Bartsch 1982a, 1989a, 1994, 2004), and some have been recovered from bottom samples collected at depths exceeding 4,000 m (Newell 1967a; Sokolov and Yankowskaya 1969). Approximately 50 marine halacarid species have been recorded from bathyal and hadal collections (Bartsch 1994), including a small number taken from volcanic hydrothermal vent habitats in both the Atlantic and Pacific basins. Although most of the latter species are thought to be deep-sea forms with no special affinity for the hydrothermal vent milieu, two morphologically similar species of *Copidognathus*—*C. papillatus* Krantz and *C. nautilei* Bartsch—have thus far been collected only from hydrothermal sites (Krantz 1982; Bartsch 1991, 1997). *C. nautilei* is represented in only one collection from the Mid-Atlantic Ridge, but *C. papillatus* has been recovered from the Galapagos Rift, the East Pacific Ridge, and a number of other hydrothermal sites in the Pacific basin. Members of the arenicolous halacarid genera *Actacarus* and *Arenihalacarus* are commonly present in culture beds of the Japanese scallop, *Patinopecten yessoensis*, in shallow sands off Hokkaido, Japan, and may prove to be suitable indicators for identifying future high-productivity sites for culturing this economically valuable scallop species (Abé, Sasaki, and Hiromi 2001).

Members of several halacarid genera occur in the brackish waters of estuaries and bays, while approximately 60 known species have adapted entirely to freshwater habitats, including the benthos of rivers and lakes, lava caves, subterranean pools, hyporheic zones, springs, and driven wells (Viets 1940a, c; Imamura 1970a, b, 1971, 1981; Petrova 1971; Bartsch 1973, 1989c, 1996, 1999: Schwoerbel 1986c). Some species may have gained access to continental waters through a combination of lateral subterranean migration (Delamare Deboutteville 1960) and invasion via benthic interstices from surface water habitats. Adaptation to freshwater habitats by halacarids is thought to have taken place on a number of occasions over time, beginning with the colonization of Pangaean shores from adjacent marine habitats by the ancestors of the freshwater subfamily Limnohalacarinae and the halacarine genera *Lobohalacarus* and *Porohalacarus*, the lohmannelline genus *Porolohmannella*, and the ropohalacarine genus *Ropohalacarus* (Bartsch 1996). Modern-day representatives of these taxa are found in freshwater habitats that are far removed from the sea, a circumstance that—along with their morphological adaptations to freshwater habitats (external genital papilla) and their broad geographical and ecological distribution—may further attest to the antiquity of these forms (Petrova 1971). That the movement of present-day halacarid taxa from marine into freshwater habitats is an ongoing process has been documented by Petrova (1971, 1974b), who conducted extensive sampling of coastal Black Sea and adjacent continental habitats. Petrova noted that populations of several species of the marine genus *Halacarellus* followed a gradient that suggested a gradual migration from coastal waters into subterranean continental habitats, and some species displayed a greater tendency to invade continental waters than did others. Similarly, a member of the marine genus *Copidognathus* (*C. tectiporus* Viets) is found in Black Sea sands and adjacent brackish lakes, while a closely related species has colonized a freshwater spring approximately 30 km inland. Although representatives of some marine taxa have succeeded in invading continental waters, the freshwater taxa mentioned earlier are, in general, farthest removed from the seashore.

The HALACARIDAE has a worldwide distribution, but its diversity is highest in the tropical and temperate provinces and decreases significantly in the polar regions. Based on available data, diversity also is highest in the upper 200 m, while relatively fewer species are known from depths exceeding 1,000 m. However, aside from fresh- and brackish water taxa and algivorous forms that have special nutritional ties to littoral habitats (e.g., the freshwater Limnohalacarinae, the algivorous Rhombognathinae, and the brackish water genera *Peregrinacarus*, *Caspihalacarus*, and *Acarothrix*), halacarid mite groups found in littoral habitats may also be expected to occur beyond the continental shelf (Bartsch 2004).

Marine halacarids are likely descendents of terrestrial prostigmatic mite antecedents that first invaded marine habitats during the Palaeozoic or early Mesozoic eras. Newell (1984) observed that both halacarine and lohmannelline HALACARIDAE display morphological affinities with the terrestrial eupodidid superfamily Bdelloidea (p. 236). Using particular genera as exemplars, Newell noted that *Halacarus* could well be placed in or near the BDELLIDAE, while *Lohmannella* displays strong affinities with the CUNAXIDAE. He also felt that rhombognathines, exemplified by the genus *Rhombognathus*, are unique in terms of other halacarids and were independently derived. Abé (1998) believes that the early derivative rhombognathine genera *Rhombognathus* and *Isobactrus* arose during the late Palaeozoic era (220–280 mya) somewhere on the coast of Pangaea, while *Rhombognathides* and *Metarhombognathus* were derived from a common rhombognathine ancestor on the evolving northern Atlantic coastline during the Paleocene or Eocene. Bartsch (2004) considers the HALACARIDAE *s. lat.* to have Mesozoic or possibly pre-Mesozoic roots. Based on their biology and broad geographical distribution, representative species of some halacarine and rhombognathine genera (e.g., *Halacarellus*, *Thalassarachna*, *Isobactrus*, *Metarhombognathus*, and *Rhombognathides*) probably have survived virtually unchanged for more than 50,000,000 years.

Structures identified as urstigmata, or organs of Claparède (see chapter 3), have been described from the larva of *Thalassarachna rhaphidochela* Krantz, a mussel bed inhabitant of the Oregon, USA, coast (Krantz 1977). These papillate structures occur adjacent to the epimeral pores on the anterior epimeral plate but, unlike the pores, are absent in subsequent nymphal and adult stages. Although urstigmata may be widespread in larval HALACARIDAE, there remains some question as to their identity in that they occupy a position closely adjacent to the epimeral pores and may be difficult to discern. Bartsch (1973, 1974c) described epimeral organs in larval rhombognathine and halacarine HALACARIDAE that appear to coincide in morphology with the urstigmata described for *T. raphidochela*. Further study of these enigmatic structures is clearly indicated.

The halacaroid family PEZIDAE comprises a single genus and two species of freshwater mites, one of which (*Peza ops* Harvey) occurs widely in slow-moving river systems and pools in southeastern Australia, while the other (*P. daps* Harvey) was recovered from the gill chamber of a freshwater crayfish, where it was presumed to be parasitic (Harvey 1990e). Pezids differ from halacarids in having 12–16 dorsal platelets (1–4 plates in HALACARIDAE) that carry 10–14 pairs of long, serrate setae in raised insertions, and in having an additional six pairs of dorsal "microsetae" (Fig. 13.8A). Bartsch (1996) feels that these setae may be homologous with some of the dorsal pores (cupules, rosette pores) found in most marine and freshwater halacarids. The legs of pezids are directed radially rather than having legs I–II directed forward and legs III–IV directed posteriorly as in HALACARIDAE (Figs. 13.7A, H). Legs IV are widely separated from legs III and are inserted near the posterior extremities of the opisthogaster (Fig. 13.8B). While there are notable similarities between pezids and lohmannelline halacarids, there are major differences in palpal setation (many more setae in pezids) and in the attachment sites of the palpi to the subcapitulum (more widely separated in pezids) (Figs. 13.8C, D). Females of both species of *Peza* and of *Limnohalacarus wackeri* (Walter), a freshwater lohmannelline mite collected in Italy (Ramazzotti and Nocentini 1960), have been reported to carry eggs on their legs. However, at least in the case of the *Peza* species, these "eggs" have since been identified as protozoans of the subclass Peritrichia (M. S. Harvey, pers. comm.).

▪ Superfamily Eupodoidea (Figs. 13.9–13.12A–D)

DIAGNOSIS: Mostly soft-bodied, weakly sclerotized forms (PENTHALODIDAE is an exception), typically with 1 pair of prodorsal bothridial setae (setae vi *may represent a second pair (Baker 1990a)), with a pair of opisthonotal trichobothria also present in eupodid genera* Benoinyssus *and* Aethosolenia, *one pair of lateral ocelli present (rarely absent); usually with an anteromedian naso (absent in* Claveupodes delicatus *Strandtmann and Prasse), or with a nasolike epirostral projection bearing paired (rarely unpaired)* ro *setae. Opisthosoma typically with 8 pairs of dorsal setae, rarely lacking setal pair* f *and one of the* h *series, or sometimes with postlarval neotrichy; cupules (lyrifissures) usually present, sometimes obscure, or absent. Ventrally with 2 (rarely 3) pairs of genital papillae; anal opening ventral or terminal, occasionally dorsoterminal or dorsal (i.e., PENTHALEIDAE); elements of opisthosomatic segments PS and AD present, usually including 2–3* ps *setae and sometimes* ad1. *Cheliceral bases separate and free from rostrum; with opposable cheliceral digits, or movable digits hook- or styletlike and inserted subapically on remnant of fixed chela, fixed digit sometimes deformed, usually with 1–2 dorsal cheliceral setae (*cha, chb; *absent in* Penthalodes *and* Hawaiieupodes*). Palpi 4- or rarely 5-segmented, palptarsus with a typically recumbent antiaxial solenidion. Tarsi I–IV with claws, and each usually with setulose and padlike, or rarely nude and clawlike, empodium. Tarsus I usually with 1 or more recumbent solenidia lying flush with tarsal surface in dorsal depressions, often with similar solenidia on tarsus II (rarely on tarsus III) and on tibiae I–II; solenidion of tibia II, when present, either completely recessed (RHAGIDIIDAE, STRANDTMANNIIDAE) or inserted on the surface. With 5 active postembryonic instars: larva, protonymph, deutonymph, tritonymph, and adult; a calyptostatic prelarva may be present.*

The Eupodoidea includes seven families: the EUPODIDAE, RHAGIDIIDAE, STRANDTMANNIIDAE, PENTAPALPIDAE, PENTHALODIDAE, PENTHALEIDAE, and ERIORHYNCHIDAE. Eupodoids are worldwide in distribution and often are found in extreme environments, including high alpine and polar habitats. They may be predaceous, fungivorous, bryophagous, or phytophagous and are not known to be symbiotic on other animals. Where sperm transfer occurs, it is via spermatophore deposition, although parthenogenesis appears to be common in some groups (e.g., RHAGIDIIDAE). Coineau (1976b) and Ehrnsberger (1974) have developed descriptions of the prelarval calyptostases of the eupodoid families EUPODIDAE and RHAGIDIIDAE.

The EUPODIDAE (Fig. 13.9) comprises at least 10 genera and over 80 species of mites that are commonly encountered in soil, litter, low-growing vegetation, lichens, or moss in temperate, boreal, and polar latitudes. Eupodids have been recovered from habitats that range in elevation from sea level to high montane (> 2,200 m). Some occur in coastal intertidal zones, and others have been found in caves, under stones, and in animal nests. *Hawaiieupodes thermophilus* Strandtmann and Goff was collected on the island of Hawaii from volcanic steam vent microhabitats where temperatures of 41°C were recorded (Strandtmann and Goff 1978). Other collections from the Hawaiian Islands suggest that at least one eupodid species, *Eupodes sigmoidensis* Strandtmann and Goff, was austral in origin and probably spread into the Pacific oceanic region during ancient times (Strandtmann and Goff 1978). However, other species, including the circumpolar *E. voxencollinus* Thor, probably were introduced into Hawaii by colonizing human populations.

Many of the known species of the type genus *Eupodes* have greatly swollen femora IV (Fig. 13.9A) (Thor and Willmann 1941; Strandtmann 1971) and are accomplished jumpers. Others, such as *E. minutus* (Strandtmann), *E. tottanfjella* Str., and *E. angardi* Str. and Sømme have normally developed femora on all legs (Strandtmann 1967, 1970; Strandtmann and Sømme 1977). Swollen femora IV also are characteristic of the small, soft-bodied species of the eupodid genus *Cocceupodes*, an assemblage common to boreal, northern temperate, and arctic regions but also found on subantarctic islands and Hawaii (Strandtmann and Tilbrook 1968; Shiba 1969; Strandtmann and Goff 1978). Similarly, most known species of the genus *Linopodes* have swollen femora IV and are capable of rapid backward movement when threatened. Legs I of eupodids are typically elongate (e.g., *Eupodes longisetatus* Strandtmann), but those of *Linopodes* species are unusually long (Fig. 13.9G) and ideally suited to serve as surrogate antennae as the mites explore their surroundings. Some *Linopodes* species may be encountered in mushroom houses, where they are considered economic pests. However, Hussey, Read, and Hesling (1969) suggested that *L. antennapes* Banks, a common inhabitant of outdoor and greenhouse mushroom beds, is not a fungivore as surmised in earlier literature but rather is a predator on fungivorous astigmatine and heterostigmatine mites that feed on mushroom tissue. Species formerly referred to *Protereunetes*, now referred to *Eupodes* (Baker 1987), occur both in arctic and subantarctic habitats and probably feed preferentially on algae. Maritime antarctic species of *Eupodes* favor mossy habitats over lichen-encrusted rocky sites (Booth, Edwards, and Usher 1985). All life stages of one species, *E. boerneri* (Thor), were recovered from old and new nests of the brown lemming, *Lemmus trimucronatus*, in Alaska (Strandtmann 1971). Larval, protonymphal, and deutonymphal stages of *E. paulinae* (Gless) observed in culture appeared to feed on filamentous algae (Gless 1972).

Soil- and litter-dwelling species of the genus *Benoinyssus* are unique among eupodids (and other families of Eupodoidea) in having opisthonotal setae *f1* modified as trichobothria. However, the homologous pair of setae in some species of *Eupodes* assumes a distinctive form with more deeply inserted alveoli than other dorsal opisthosomatic setae and have been tentatively considered as trichobothria (Grandjean 1939b; Coineau 1976b). *Benoinyssus najae* Fain, the type species of the genus, was first found in the nasal fossae of the cobra *Naja melanoleuca* in Central Africa (Fain 1958), an association that is now considered to have been accidental in that large numbers of *B. najae* (along with seven additional species of the genus) have since been collected in soil and leaf litter samples (Olivier and Theron 1997). The monobasic, leaf litter–dwelling eupodid genus *Aethosolenia* also is characterized in part as having a pair of opisthonotal trichobothria, but, unlike *B. najae*, these are setae *h1* instead of *f1* (Baker and Lindquist 2002).

Members of the predaceous cosmopolitan family RHAGIDIIDAE (Figs. 13.10A–D) are soft bodied, rapidly moving mites that occupy many of the same habitats as the EUPODIDAE. Many of the species in this relatively large family (approximately 21 genera and more than 125 species) are opaque white in life, but some are yellow, orange, or red in color (see color photos in Ehrnsberger 1981). Rhagidiids have been collected in alpine/high montane situations as well as at lower elevations. They have also been found in arctic and antarctic tundra, caves, and littoral habitats on the seashore (Strandtmann 1967, 1971; Schuster 1958; Zacharda 1980, 1983). Rhagidiids commonly occur in the uppermost layers of the soil-litter profile or within the soil column and seem to prefer moist, dark, cool niches. They move erratically in a zigzag manner using legs II–IV, with legs I extended anteriorly as tactile organs (Ehrnsberger 1981). Some species may be multivoltine, and their developmental stases, unlike those of univoltine species, may appear at any season (Zacharda 1979a). Thelytokous parthenogenesis has been documented for several species of rhagidiids and is suspected in others (Ehrnsberger 1977; Zacharda 1980). In all rhagidiids that have been studied, the prelarva is a calyptostase, and the larval instar, although mobile, is an elattostase and does not feed (Ehrnsberger 1974, 1981). Larval and nymphal stases of rhagidiids use silk from the podocephalic gland system to spin a web around themselves prior to molting (Ehrnsberger 1979). Adults of *Rhagidia longisensilla* Shiba also spin silk for catching prey, for communal refuge and protection of the eggs, and for constructing molting nests (Ehrnsberger 1979).

A number of rhagidiid species occur both in the Palearctic and Nearctic realms (*Robustocheles mucronata* (Willmann), *Shibaia longisensilla* (Shiba), *Poecilophysis weyerensis* Packard, and *Coccorhagidia pittardi* Strandtmann are examples), while others (e.g. *Tuberostoma keithi* (Str.), *Troglocheles gineti* Rack, and *Krantzia quadriseta* Zacharda) appear to be more restricted in distribution. Although relatively few rhagidiid species have been found in the Southern Hemisphere (Schuster 1958; Meyer and Ryke 1960b; Zacharda 1980), many have been described from the Nearctic region and are believed to represent only a hint of the diversity that occurs there (Zacharda 1997). A high degree of endemism is indicated from the wide variety of cave-dwelling species that have been described (Zacharda and Elliot 1981; Zacharda 1987; Zacharda and Pugsley 1988). The discovery of *Rhagidia occulta* Zacharda and Pugsley in the innermost passages of a cave system where no other arthropods except troglobitic crustaceans were found suggests that *R. occulta* may have adapted to using these crustaceans as prey.

Rhagidiids are unusual in having fields of modified recumbent dorsal solenidia on tarsi I–II that are distinctive enough to be diagnostic at the generic level (Zacharda 1980). These **rhagidial organs** (Fig. 13.10A) usually are inserted in individual or communal grooves or troughs and typically are modified so that each solenidion takes the form of a pick hammer, with the offset stem representing the shortened handle. Similar but less strongly developed rhagidial organ complexes also occur in other eupodoid families. Zacharda (1980) considers rhagidial organs to be chemosensory in function. In that connection, it is interesting that the cave-inhabiting species of the genus *Troglocheles* have as many as nine rhagidial organs on tarsus I rather than the usual number (three to five), suggesting that troglobitic species may have special needs for detecting the presence of prey in this challenging habitat. In contrast, edaphic forms such as *Brevipalpia* and *Arhagidia* have few rhagidial organs that no longer are recumbent (Zacharda 1980; Lindquist and Zacharda 1987).

The family STRANDTMANNIIDAE is based on the single genus *Pilorhagidia* (= *Strandtmannia*) and its two species. *P. hirsuta* Strandtmann and Goff was found in *Metrosideros* scrub in the Hawaiian Islands (Strandtmann and Goff 1978), while *P. celtarum* (Zacharda) was collected from oak forest litter in the Czech Republic (Zacharda 1979b). Other than having an additional pair of prodorsal setae, a neotrichous opisthosoma, and an attenuated palptarsus, strandtmanniids are morphologically similar to rhagidiids (Kethley 1982) (see Figs. 13.10H–K). The monobasic family PENTAPALPIDAE (Figs. 13.11A–D) also is morphologically similar to rhagidiids, differing primarily in retaining apparently five-segmented palpi (with separate femur and genu) and in having a nude, clawlike empodium between the paired claws on the tarsi of all legs (Olivier and Theron 2000). *Pentapalpus unguempodius* Olivier and Theron is known from a variety of soil and litter habitats in South Africa. Strandtmanniids, pentapalpids, and rhagidiids are noteworthy among eupodoids in having strongly developed chelate chelicerae (Figs. 13.10D, J, 13.11C). Although their feeding habits are unknown, it seems likely that strandtmanniids and pentapalpids share the predaceous habits of the RHAGIDIIDAE (Halliday 1991).

Like rhagidiids, the often colorful, ornamented species of the family PENTHALODIDAE (Figs. 13.10E–G) have been considered as possible predators (Baker 1946; Baker and Wharton 1952), although *Stereotydeus mollis* Womersley and Strandtmann observed in culture was thought to feed on various species of algae and lichens (Pittard 1971). Penthalodids often are found in moss and litter in temperate and subtropical habitats (antarctic *Stereotydeus* species are exceptions) and generally are strongly sclerotized and ornamented with reticulate or punctate patterns. Of the two genera of penthalodids, *Penthalodes* is Holarctic in distribution, while *Stereotydeus* is worldwide. In a comparative study of the life stages of *S. mollis*, Pittard (1971) observed that, along with additions of genital papillae and setae and of tactile leg setae through ontogeny, there is also an increase in rhagidial organ number. Thus, while the larva and protonymph have only one rhagidial organ on tarsi I–II, the number increases to two on both tarsi at the deutonymphal molt and to three on tarsus I at the tritonymphal molt (Fig. 13.4). Similar incremental additions occur in the RHAGIDIIDAE (Zacharda 1980) and other eupodoids as well.

The PENTHALEIDAE (Fig. 13.12) is distinctive among eupodoid families in that some of its species are obligately phytophagous, and two are of considerable economic importance (Jeppson, Keifer, and Baker 1975a). The strong, styletlike form of the movable cheliceral digits among plant-feeding species appears to be an adaptation for phytophagous behavior (Nuzzaci and de Lillo 1991; Di Palma 1995; de Lillo, Di Palma, and Nuzzaci 2001). *Penthaleus major* (Dugès), the winter grain mite (also known as the blue oat or pea mite), is an important phytophage distributed throughout temperate and tropical realms (Wallace and Mahon 1971; Jeppson, Keifer, and Baker 1975a; Halliday 2005), where it attacks a variety of plants including clover (Swan 1934), peas (Campbell 1941), and grain (Chada 1956). Cotton, peanuts, and ornamental flowers may also serve as hosts. The idiosoma of *P. major* appears black in life, but the body contents of crushed specimens are, in fact, dark green in color. In contrast, the legs are red, and the dorsomedian anal opening is bordered by a red perianal zone. Females are thelytokous, and males are unknown. Other species of *Penthaleus* include *P. minor* (Canestrini), which may be distinguished from *P. major* on the basis of its relatively small chelicerae (Jeppson, Keifer, and Baker 1975a), and *P. tectus* Halliday, a recently described pest of grain crops in eastern Australia (Halliday 2005). A fourth species, *P. crinitis* Narayan, was recovered from under rocks in south-central Mexico (Narayan 1957).

Halotydeus destructor (Tucker), the red-legged earth mite or black sand mite, is polyphagous, feeding chiefly on annual broad-leaved plants. It is a common pest of vegetable and legume crops, tobacco, and grasses in Australia, New Zealand, South Africa, Zimbabwe, and Malawi (Meyer and Ryke 1960b; Wallace and Mahon 1971; Jeppson, Keifer, and Baker 1975a; Halliday 1991). Like *P. major*, *H. destructor* appears black in color and, as its name implies, has red legs. Unlike *P. major*, however, the anal opening of *H. destructor* and other species of the genus is terminal rather than dorsal in position. A distinctive behavioral feature of *H. destructor* is its group feeding in pasture foliage canopies, with many individuals often aggregating on the upper leaf surfaces. At the same time, feeding individuals tend to disperse quickly to the soil surface when disturbed (Gaull and Ridsdill-Smith 1996). Other species of *Halotydeus* include *H. hydrodromus* Berlese and Trouessart from littoral and intertidal rocks in France, Italy, and Ireland; *H. signiensis* Strandtmann and Tilbrook from moss in the South Orkney Islands; and *H. mollis* Luxton from a mangrove beach in Hong Kong (Halliday 1991). *H. anthroopus* Qin and Halliday was collected from clover, grass, and cape weed in South Africa, and three other species of the genus occur in native vegetation in Australia (Qin and Halliday 1996b). *Chromotydaeus* Berlese is now recognized as a genus distinct from *Penthaleus* and is represented by *C. egregius* (Berlese) from mainly dry soil habitats in Italy, and *C. quartus* Qin and Halliday from native shrub and pasture habitats in Australia (Qin and Halliday 1996a).

Both *P. major* and *H. destructor* produce winter and summer eggs, which permit the mites to develop populations under a variety of environmental conditions. *P. major* produces bright pink to orange-red eggs that either hatch after a short incubation period (winter eggs) or shrivel and lose their color in response to desiccation and as the prelude to a period of aestivation that may exceed 110 days (Jeppson, Keifer, and Baker 1975a). These eggs eventually split when moisture becomes adequate, revealing a calyptostatic prelarva that finally emerges as an active larva. The bright orange winter eggs of *H. destructor* are deposited in masses, mainly on the undersurfaces of plant leaves in damp localities (Meyer 1981). When temperatures rise above 18°C, females produce summer eggs that are retained within the female's body until she dies, which provides early protection from desiccation. Liberated summer eggs lose moisture and become shrunken and flattened but regain moisture and assume the appearance of winter eggs when exposed to wet conditions. These factors assume special importance when considering the potential of these two plant pests for invading and causing damage in previously uninfested agricultural settings (Wallace 1970; Wallace and Mahon 1971).

Adults of the family ERIORHYNCHIDAE (Figs. 13.11E–H) are large (idiosomatic lengths 1,100–1,800 μm), soft-bodied, long-legged mites, described as being dark in color, red legged, and hairy in appearance (Qin and Halliday 1997). Feeding behavior is unknown, but the dark green pigment within their bodies suggests a phytophagous or algophagous feeding habit. Eriorhynchids are distinctive in having neotrichous palpi and subcapitula, small opposable cheliceral digits, and strongly neotrichous bodies and leg segments. In addition, the prodorsal naso is uniquely incorporated into an enlarged, neotrichous naso-prodorsal process that is delineated from the remainder of the prodorsal surface and bears all the prodorsal setae. The single genus, *Eriorhynchus*, contains five species known only from a variety of natural habitats in Australia, including rotting wood, native pasture grasses, rain-forest leaf litter, and moss (Qin and Halliday 1997).

Several enigmatic monobasic eupodoid taxa have unclear familial affinities. *Linopenthaloides novaezealandicus* Strandtmann, a large, cave-dwelling mite, was originally placed in the PENTHALEIDAE (Strandtmann 1981), but the anal opening is terminal rather than dorsal or dorsoterminal as in other penthaleids (Kethley 1990), and the solenidia of tarsi I–II are erect rather than recumbent. *L. novaezealandicus* also is unusual in being strongly hypertrichous and in having contiguous coxae rather than the usual wide separation between coxae II–III and in having inordinately long legs. Its position is unclear, but it may well represent a separate family. *Protopenthalodes coniunctus* Jesionowska, a presumably primitive taxon recovered from mossy turf in Poland, was originally placed in the PENTHALODIDAE but is exceptional for that family in having a soft, poorly sclerotized idiosoma that lacks prominent dorsal furrows or a prodorsal projection over the gnathosoma, and rhagidial organs on tarsi I–II that represent two (tarsus I) or three (tarsus II) fused solenidia in a single, elongated groove (Jesionowska 1989). Because *Protopenthalodes* does not have the basic apomorphic attributes generally used to define PENTHALODIDAE, it seems more appropriate to place it in the EUPODIDAE (Qin 1997). The neotrichous, short-legged *Turanopenthalodes polytrichus* Barilo, collected from under rocks and on soil in meadow vegetation in Samarqand, Uzbekistan, was placed in the PENTHALODIDAE but is exceptional in having a weakly sclerotized, finely tuberculate idiosoma and a dorsal anal opening, the latter condition thought to be convergent with some penthaleids. *T. polytrichus* is also unusual in having erect rhagidial solenidia on tarsus I (Barilo 1988). The body of the mite was described as black, again suggesting a concentration of dark green pigment similar to that seen in phytophagous species of *Penthaleus* and *Halotydeus*. *Callipenthalodes pennyae* Qin, collected from moss on *Nothofagus* logs in New Zealand, also was placed in the PENTHALODIDAE but is unique in that it has a soft, neotrichous idiosoma and erect rhagidial solenidia on tarsi I–II. In addition, *C. pennyae* has a ventroterminal anal opening and is notably long legged (Qin 1998). Both *Turanopenthalodes* and *Callipenthalodes* are characterized by the presence of a well-developed epirostral projection of the prodorsum over the gnathosoma, possibly the only attribute common to all penthalodids except for *Protopenthalodes*.

▪ Superfamily Tydeoidea (Figs. 13.12E–G, 13.14)

DIAGNOSIS: Soft bodied or ornamentally sclerotized dorsally. Gnathosoma with postcheliceral stigmata and tracheae distinct or obscure; cheliceral bases fused or closely contiguous along their mesal edges, lacking dorsal cheliceral setae; fixed cheliceral digit reduced, lobelike, movable cheliceral digit styletlike or needlelike, rarely elongated, whiplike in some IOLINIDAE; subcapitulum characteristically with inversely T-shaped, sclerotized delineation of medial walls of lateral lips; palpi simple, with 1–4, (rarely 5) segments. Prodorsum lacking anteromedian naso, with 4 pairs (sometimes as few as 1 pair) of setae, including 1 pair of bothridial sensilla; eyes present, with either surface cornea or subdermal pigmented areas, or absent. Opisthosoma with 7–11 pairs of setae dorsally, laterally, and caudally, lacking postlarval setal additions or hypertrichy dorsally, but including a second pair of bothridial sensilla on the posterodorsal aspect in certain EREYNETIDAE and TYDEIDAE; cupules (lyrifissures) present, sometimes obscure or absent; elements of opisthosomatic segments PS and AD present, including ps1 *and, rarely,* ps2 *or* ad1. *Adult genital aperture positioned behind legs III–IV, usually longitudinal but transverse in some IOLINIDAE, and usually with 2 pairs of genital papillae (reduced to 1 pair or absent in IOLINIDAE). Males usually lacking an aedeagus (present in some IOLINIDAE). Leg setation lacking bothridial setae; tarsus I with 1–2 usually erect solenidia (with 1 recumbent solenidion in a rhagidial-like depression in certain EREYNETIDAE) and with or without claws, occasionally with only a median empodium; tarsi II–IV generally with paired claws and a padlike setulose, but not rayed, median empodium. Generally with 5 active postembryonic instars: larva, protonymph, deutonymph, tritonymph, and adult (tritonymph suppressed in some EREYNETIDAE, TYDEIDAE, and IOLINIDAE, and all nymphal instars reduced to calyptostases in speleognathine EREYNETIDAE).*

The superfamily Tydeoidea is worldwide in distribution and is represented by a diversity of species that have colonized habitats ranging from soil and arboreal niches to cavities on insects and mollusks and the nasal cavities of vertebrates. Free-living members of Tydeoidea are abundant in such climatically extreme regions as the arctic tundra and the antarctic maritime (Thor 1933; André 1980; Usher and Edwards 1986). The superfamily includes omnivorous species that feed on fungi, pollen, and leaf tissues; predators that feed on arthropod eggs and other mites and nematodes; and highly specialized hematophagous parasites. Tydeoid mites are distinctive among the Trombidiformes in having a reduced form of the typically larval Claparède organs, or urstigmata, associated with coxae I–II, retained into the postlarval instars (André 1991). Classification of the Tydeoidea, particularly of the family TYDEIDAE as noted by Kethley (1982), has long been in a state of flux, owing to the apparent paraphyly of groups recognized as families or subfamilies. In their cladistic reappraisal and systematic reorganization of the superfamily, André and Fain (2000) recognize and redefine four families, the TRIOPHTYDEIDAE (previously Meyerellidae), EREYNETIDAE, TYDEIDAE, and IOLINIDAE.

The TRIOPHTYDEIDAE is an early-derivative group of free-living soil, cortical, and foliar forms with poorly understood feeding habits, characterized by the presence of three sets of prodorsal eyespots (André and Fain 2000). It includes approximately 40 species placed in two subfamilies: the Triophtydeinae (André 1979) and the Edbakerellinae (André 2004, previously the Meyerellinae of André 1979), which were formerly placed in the TYDEIDAE. It should be noted here that, in proposing *Edbakerella* and Edbakerellinae as new replacement names for *Meyerella* Baker 1968 and Meyerellinae André 1979, André (2004) also applied the family name Edbakerellidae for the taxon that includes the subfamilies Triophtydeinae André 1979 and Edbakerellinae André 2004. In accordance with International Code of Zoological Nomenclature rules, therefore, the oldest available name for the family is TRIOPHTYDEIDAE rather than Edbakerellidae.

Nutritional experiments by Brickhill (1958) indicated that *Triophtydeus bakeri* (Brickhill) would not develop in the presence of leaf tissue alone or on eriophyid mites or the eggs of coccid insects or European red mites, but it developed readily on two-spotted spider mite eggs and even more successfully on scale-produced honeydew.

With one known exception, all members of the diverse family EREYNETIDAE, with some 180 described species relegated to three subfamilies, are characterized by a peculiar sensory complex of unknown function, the **ereynetal organ** (Figs. 13.14E, F), on the tibia of legs I (Fain 1962c, 1985a). This organ typically consists of three parts: a deeply inserted solenidion enclosed in a sclerotized pouch, a fine canal connecting the pouch to the exterior face of the tibia, and a famulus whose alveolus is contiguous with the solenidial pouch. Ereynetids seem to be restricted to humid or wet habitats, whether they are cavities of hosts colonized by parasitic species or soils of temperate or tropical regions (André and Fain 2000). They are rarely found in exposed foliar habitats or arid regions. Some members of the subfamily Ereynetinae are common under bark and in dung, leaf litter, moss, and lichens, while others may be intimately associated with terrestrial arthropods or gastropods. Snail and slug mites of the ereynetine genus *Riccardoella* are usually found on the surface and within the mantle cavity of their hosts (Baker 1970b; Fain and Van Goethem 1986). They may temporarily wander away from their hosts and move about the immediate vicinity and then follow the slime trail of their hosts to relocate them (Turk and Phillips 1946). Under normal conditions, *Riccardoella* species have been thought to have no effect on their hosts (André and Lamy 1930; Turk and Phillips 1946). However, slug mites have been demonstrated to be hematophagous, forming a food canal or stylostome in the host tissue for injection of saliva and ingestion of food material; heavy infestations on slugs under confined laboratory conditions may result in considerable loss of blood and even death of the hosts (Baker 1970a, 1973). Long thought to be confined to the Holarctic region, these gastropod parasites have been confirmed to occur naturally in New Zealand and elsewhere in the southwestern Pacific (Fain and Barker 2004). Curiously, two species of *Riccardoella* and the monobasic genus *Hanriccardoella* seem to be free living and restricted to karstic and other calcareous-rich sites, including cave substrates. This may point to an association of the entire *Riccardoella* complex with high calcium levels in that parasitic members of the complex occur on hosts that are good reservoirs of calcium (André, Ducarme, and Lebrun 2004). Other, presumably predaceous, members of the subfamily Ereynetinae are free-living or close associates of arthropods. Several species of *Ereynetes* are subcortical and phoretic associates of bark beetles (Hunter, Rosario, and Moser 1989), nidicolous associates of native bees (Hunter and Cross 1968; Flechtmann, Fain, and Leal 1985), or phoretic associates of dung-inhabiting beetles and flies. Two species of *E.* (*Anereynetes*) are associated with pagurid land crabs (Hunter and Poe 1971; Fain 1985; Fain and Camerik 1994). The genus *Hydranetes* is known only from adult females of two species collected under the elytra of hydrophilid water beetles (Kethley 1971a). The enigmatic genus *Pseudotydeus*, originally based on a few tritonymphs (mistaken for adult females) of a single species collected from a woodlot in Ohio, USA (Baker and Delfinado 1974), and placed in its own subfamily in the TYDEIDAE, was subsequently transferred to the EREYNETIDAE (André and Fain 2000). An adult female of a second species has since been described by André and Ducarme (2003) from a Belgian cave.

Ereynetids of the subfamily Lawrencarinae are parasitic in the nasal passages of amphibians (Fain 1962c). *Lawrencarus eweri* (Fain) and *Xenopacarus africanus* Fain, Baker, and Tinsley are African species that appear to be hematophagous in the nasal cavities of their frog and toad hosts (Lawrence 1952; Baker 1973). Other species of these genera, as well as members of the genus *Batracarus*, may share the hematophagous habit. Members of the ereynetid subfamily Speleognathinae inhabit the nasal mucosa of a wide variety of birds and mammals throughout the world (Clark 1960; Fain 1963b, 1985b; Pence 1973; Fain and Hyland 1975). Mammalian hosts include such early derivative groups as marsupials and bats (Domrow 1975b; Fain and Lukoschus 1971). Birds of a wide variety of orders have been recorded as hosts of the speciose genus *Boydaia* (Fain 1971a) as well as of 15 other speleognathine genera in the tribes Boydaiini and Trispeleognathini (Fain 1985b). *Boydaia sturnellae* Clark, a North American speleognathine, appears to be capable of feeding on whole blood in the turbinates of its meadowlark host (Clark 1960), as does *B. nigra* Fain (Baker 1973). Speleognathines have no active nymphal instars in their ontogenetic development and transform directly from the hexapod larval form to the adult. Fain (1972a) has observed the vestigial immobile nymphal forms developing as calyptostases within the larval skin.

The TYDEIDAE (Figs. 13.13A–D) is a large, cosmopolitan family of weakly to heavily sclerotized striate or reticulate mites. Earlier classifications recognized relatively few genera (i.e., the many papers by E. W. Baker in the United States and N. N. Kuznetsov and I. Z. Livshits in the former Soviet Union during the 1960s and 1970s (fully cited in relatively recent papers, for example, Sepasgosarian 1997; Kazmierski 1998b), compared to later schemes that recognized a large number (over 50 genera in seven subfamilies (André 1979, 1980, 1981a, b; Kazmierski 1996a, b, 1998b)). Some of these taxa are based on permutations and combinations of setal attributes that are not demonstrably apomorphic. Recently, however, André and Fain (2000) redefined the TYDEIDAE to include only the three subfamilies Tydeinae, Pretydeinae, and Australotydeinae. However, the family is large, with some 30 genera and 340 described species, and may be paraphyletic.

Tydeids have been recorded as predators; as fungus, pollen, and plant feeders; and as scavengers (Oatman 1963; Baker 1965; Gerson 1968a). Phoresy is a rarely noted habit among tydeids, but a few specimens of the tydeine genera *Lorryia* and *Tydeus* have been noted as phoretic on noctuid moths (Treat 1970) and as associates of beehives. Specimens of *T. xylocopae* Oudemans were recorded from carpenter bees (Oudemans 1926a), and others of the monobasic genus *Melissotydeus* were identified as nest associates of meliponine bees (André 1985). *L. superba* Oudemans may be unique in having been collected from a mammal, although Oudemans (1928) was doubtful as to the validity of the association. *T. molestus* (Moniez) has been reported to bite humans and domestic animals in Europe and cause dermal irritation (Thor's 1933 reference to earlier records by Moniez), but no additional records are available to confirm that tydeids may be irritants of vertebrates. Tydeids are often encountered in moss, litter, straw, soil or humus, fungi, bird nests, stored food products, and arboreally on plants (Marshall 1970; Kazmierski 1998b). They are notably resistant to desiccation, which contributes to their abundance in exposed foliar habitats as well as to their occurrence in desert habitats (André and Fain 2000). A six-month overwintering dormancy as a tritonymphal calyptostase by the corticolous tydeid mite *Venilia liberta* (Livshits) may be an adaptive strategy for survival during extended dry periods (Kuznetsov 1980).

Although a number of tydeid species have been categorized as to general lifestyle, little is actually known about the interactions of tydeids with their habitats. The ultrastructure and functional anatomy of the tydeid gnathosoma and its mouthparts are similar to those of penthaleids, although more consolidated and compact (Nuzzaci and Di Palma 2002). Only one case of silk production in a tydeid has been recorded. Adult *L. schusteri* (André and Naudo) produce fine silken strands from the buccal region to form a loosely constructed cocoon that may be used both as a refuge and as a thigmotactic indicator of the immediate environs by the mite (Schuster 1972). Many tydeids may be unspecialized feeders, which could account for the ongoing difficulty in analyzing their behavior. *T. californicus* (Banks) certainly fits the unspecialized category, having been considered a predator by Baker and Wharton (1952) and Rizk, Soliman, and Ali (1979) and a phytophage by Fleschner and Arakawa (1953) and Zaher and Shehata (1963). Bayan (1986) reported the larvae and protonymphs of *T. californicus* to feed primarily on aphid honeydew, while other instars fed and developed normally on aphid-free apple leaves. This mite is also thought to serve as alternative prey for phytoseiid mites in the absence of other types of food in Dutch apple orchards (Calis, Overmeer, and Van der Geest 1988). Although *L. formosa* Cooreman has been considered fungivorous (Jeppson, Keifer, and Baker 1975a), various authors noted by Aguilar, Childers, and Welbourn (2001) have reported it to cause significant damage to green branches and young citrus fruit tissue. However, the same mite is thought to be beneficial on citrus trees in Israel, where it feeds on sooty mold and scale-excreted honeydew that contaminate the leaves and fruit (Mendel and Gerson 1982). Similar dietary omnivory and flexibility are evident among other tydeine species reared experimentally. For example, *L. ferula* Baker develops readily on spider mite eggs but is even more successful when scale-produced honeydew is available (Brickhill 1958). *T. kochi* Oudemans is a facultative mycophage that develops more readily on a diet of eriophyid mites or spider mite eggs (Kul'chitskii 1993). The remarkable geographical and habital ranges reported for some species of *Lorryia* leave it unclear as to whether species complexes with discrete patterns and habits actually exist, or whether some species are ubiquitous and omnivorous (Kethley 1982). For example, the geographical and host ranges of *L. reticulata* (Oudemans) (= *L. bedfordiensis* Evans)—Europe, North and South America, and Japan on plants and in plant litter, fungi, stored grain, and moss (Shiba 1969; Marshall 1970)—suggest the existence of an *L. reticulata* species complex. In a recent paper by André (2005), the genus *Lorryia* is restricted to its type species, *L. superba* Oudemans, based on an apomorphic attribute of having asymmetrically unpaired claws on tarsi I–IV, while the many other *Lorryia* species with paired claws are relegated to the previously obscure genus *Brachytydeus*. For purposes of this treatment, the current generic concepts of earlier works are retained.

The IOLINIDAE (Figs. 13.13E–I) has been expanded by André and Fain (2000) to include the subfamilies Tydaeolinae and Pronematinae. Along with the Iolininae, the family includes some 125 described species in 36 genera and is characterized by the palplike form and function of legs I, the loss of genital papilla, and, within the Iolininae, the acquisition of direct copulation. Iolinid species are free living in soil, on plants, and in beehives or otherwise associated with, or phoretic on, insects; some are known to occur on citrus plants in southern Africa (Ueckermann and Grout 2007). An active tritonymphal instar is suppressed in the tydaeoline genera *Microtydeus* and *Tydaeolus* (Kuznetsov 1980). Most of the known species of the cosmopolitan pronematine genus *Proctotydeus* are insect associates (Kazmierski 1998a). For example, adult females and a few immatures of *P.* (*Proctotydulus*) *pyrrohippeus* (Treat) have been recovered from the tympana of noctuid moths (Treat 1970, 1975). Adult females of three species of *Proctotydaeus* (*Neotydeolus*) are nest symbionts of meliponine bees (Rosa, André, and Flechtmann 1985), and those of another species of *Proctotydaeus* (*Oriola*) and *P.* (*Proctotydulus*) are associates of bark beetle galleries (Baker 1968; Khaustov 1997). Among freely living pronematines, *Pronematus ubiquitus* (McGregor) is an apparently omnivorous species, feeding on honeydew, fungi, and dead arthropods in citrus trees (Jeppson, Keifer, and Baker 1975a) but also preying on eriophyoid mites on figs (Abou-Awad, El-Sawaf, and Abdel Kader 1999). Repeated observations have shown *Homeopronematus anconai* (Baker) serving as alternative prey for phytoseiid mites in San Joaquin Valley, California, vineyards (Flaherty and Hoy 1971; Calvert and Huffaker 1974; Knop and Hoy 1983). The same mite was reported to be a significant predator in reducing populations of the tomato russet mite *Aculops lycopersici* (Massee) in the same region (Hessein and Perring 1986). *Iolina nana* Pritchard is known from specimens representing all active stases removed from the wing bases of the cockroaches *Blaberus craniifer* and *Diplotera dytiscoides* collected in the continental United States and Hawaii (Pritchard 1956; André 1984b). André discerned a regressive calyptostatic prelarva within the eggs of *I. nana*. All active stases of a second monobasic genus, *Idiolina*, were retrieved from material collected in a light trap in a cavern of Batu Caves in Malaysia (André 1984b), an improbable site for another association with orthopterans.

Although indirect sperm transfer is the basic and prevalent mode of reproduction among tydeoid mites (Schuster and Schuster 1970), the presence of a sclerotized aedeagus and of copulatory sperm transfer by males has been observed among some pronematine and iolinine genera of IOLINIDAE (Price 1972; Baker and Delfinado 1974; Knop 1985; André 1984b). Direct sperm transfer may be an evolutionary adaptation to full association of all instars of some iolinids with insect hosts, perhaps as paraphages or parasites—for example, species of *Proctotydaeus* with orthopterans (Fain and Evans 1966) and other pronematines with moths (Treat 1970). However, as evidenced by its presence in free-living forms such as *H. anconai*, this behavior clearly precedes such specialized associations.

▪ Superfamily Eriophyoidea (Figs. 13.15–13.19)

DIAGNOSIS: Soft bodied or armored dorsally, annulate, and wormlike. Gnathosoma with an open dorsomedian infracapitular gutter that ensheaths 7–9 styletlike structures, flanked by simple, weakly 4-segmented palpi; cheliceral shafts styletlike (imperceptibly split into fixed and movable parts apically), accompanied by a pair of auxiliary stylets, an unpaired oral stylet (the labrum), and a lateral pair of guidelike structures that sometimes end in freely projecting styletlike apices; cheliceral bases not enlarged, not forming a stylophore; without stigmata and tracheae. Prodorsum covered by a prodorsal shield bearing maximally 5, minimally 0, simple setae, none of which is inserted in bothridia; prodorsum usually lacking eyes, or rarely with a pair of smooth, somewhat convex, eyelike areas laterally. Opisthosoma either vermiform, with numerous narrow annuli differentiated little from dorsum to venter, or fusiform, with dorsal annuli consolidated into a series of fewer, thicker, less flexible "tergites," which delineated laterally from more numerous narrow ventral annuli; opisthosoma lacking all cupules (lyrifissures), and with setation reduced to maximally 7 pairs of setae, of which 3–4 pairs displaced ventrolaterally to resemble ventral setae; caudally, opisthosoma terminating with adhesive structure flanked dorsally by pair of elongate caudal setae. Genital opening transverse, positioned more or less closely behind legs II; genital opening exposed in male but covered by an anteriorly hinged flap in female. With only 2 pairs of anteriorly inserted legs, lacking pairs III–IV; legs lacking true (paired) claws, but with a well-developed, unpaired empodial featherclaw. Leg setation reduced to maximum of 6 setae each, none bothridial; tarsi I–II consistently with a solenidion, which is usually inserted dorsodistally and often clawlike in form. Three active postembryonic instars: larva, nymph, and adult.

The Eriophyoidea, also known as the Tetrapodili or four-legged mites, are worldwide in distribution, occurring on a multitude of primarily perennial hosts. The entire lineage consists of obligately phytophagous mites that are extremely small to moderate in adult body length (typically 100–150 µm but with a range of 80–500 µm) and that have diversified into a large number of taxa. Nearly 3,500 species in about 300 genera are recognized (Amrine, Stasny, and Flechtmann 2003; de Lillo, Amrine, and Stasny, unpublished comm., December 2003), but the eriophyoid fauna remains unknown for vast regions of the world, particularly the subtropics and tropics from which perhaps fewer than 5% of the species have been described (Amrine and Stasny 1994). Although previously thought to have phylogenetic affinities with the Tetranychoidea (p. 297), the Eriophyoidea are currently believed to be a much more ancient lineage, possibly representing the sister group of the Tydeoidea (Norton et al. 1993; Lindquist 1996d). An even more ancient origin and relationship with the ALYCIDAE (p. 421) has also been considered (Lindquist 1998). Either scenario would have provided ample evolutionary time for the eriophyoid lineage to have become as morphologically distinctive and specialized for phytophagy as it is at present. Based on their morphology, ontogeny, life cycle, host and site specificity; stimulation of host plant growth; the nondestructive nature of their parasitic associations; and a capacity to transmit viral or similar disease agents of plants, the Eriophyoidea clearly is the most highly adapted plant-feeding lineage among the Acari (Lindquist and Amrine 1996).

The literature presents a variety of reflections on general evolutionary trends within the Eriophyoidea (e.g., Farkas 1966, 1969; Shevchenko 1970, 1976; Hong and Zhang 1996; Bagnyuk, Sukhareva, and Shevchenko 1997; Lindquist and Oldfield 1996; Sabelis and Bruin 1996). Following Keifer (1975); Amrine and Stasny (1994); Lindquist and Amrine (1996); and Amrine, Stasny, and Flechtmann (2003), three families of Eriophyoidea are recognized here: (1) PHYTOPTIDAE (including the subfamily Nalepellinae with tribes Nalepellini, Trisetacini, and Pentasetacini, all of which have been treated as separate families by some authors (e.g., Shevchenko, Bagnyuk, and Sukhareva 1991; Bagnyuk, Sukhareva, and Shevchenko 1997); (2) the ERIOPHYIDAE (including the subfamily Ashieldophyinae, which was originally proposed as a separate family by Mohanasundaram (1984)); and (3) the DIPTILOMIOPIDAE.

Details of the external morphology (Fig. 13.17) and the

internal anatomy and physiology of eriophyoid mites were given by Lindquist (1996b) and Nuzzaci and Alberti (1996), respectively. Updated keys to the world genera of Eriophyoidea were presented by Amrine, Stasny, and Flechtmann (2003), and keys restricted to the genera with species occurring on plants of economic importance, including notes on individual species in each genus, were given by Lindquist and Amrine (1996). Amrine and Stasny (1994) provided a catalog of the world species of Eriophyoidea, which continues to be updated electronically (de Lillo, Amrine, and Stasny, unpublished comm., December 2003). A guide to the eriophyoid mites of the United States by Baker, Andrews, and Fox (1996) includes figures and keys to the known species of that country. A preliminary account of a computerized database for Eriophyoidea, in which all references to descriptions, illustrations, hosts, symptoms, distributions, and so on, are linked to individually named species, is given by de Lillo and Amrine (1998).

Indirect sperm transfer by spermatophores, and sex determination by haplodiploidy, are ancestral to the eriophyoid lineage (Oldfield 1973; Oldfield and Michalska 1996; Helle and Wysoki 1984, 1996). Although thelytoky has not been demonstrated for any species of Eriophyoidea (Keifer 1975), arrhenotokous parthenogenesis is found throughout the lineage, with uninseminated females capable of producing all-male progeny that have the potential of mating back with their mothers (Oldfield and Newell 1973; Oldfield and Michalska 1996). Adult females store sperm consistently in one or the other of their paired spermathecae or consistently in both of them (Oldfield 1973; Oldfield and Michalska 1996). A trend toward sperm storage in only one spermatheca has evolved at least twice from the ancestral condition of bilateral storage as retained in PHYTOPTIDAE, and patterns of bilateral versus unilateral sperm storage may provide clues for resolution of some taxonomic and phylogenetic problems among some taxa of ERIOPHYIDAE (Lindquist and Oldfield 1996; Oldfield 1999).

Eriophyoid mites generally rely on passive dispersal by wind currents for transfer from one patch of host plants to another (Lindquist and Oldfield 1996; Lindquist 1998; see also chapter 6). An analysis of snowborne arthropods as evidence of potential for long distance aerial dispersal indicated that the diversity of species and numbers of individuals of eriophyoids far exceeded other mites (Zhao and Amrine 1997). Because passive dispersal is independent of host site, Sabelis and Bruin (1996) anticipated that phoretic transport by insects may play a more important role than previously thought for relatively host-specific mites like eriophyoids. However, observations of eriophyoids found on insects are exceedingly few (e.g., Shvanderov 1975; Waite and McAlpine 1992) and perhaps fortuitous, as the insect carriers were often wingless or winged forms with dissimilar or unspecialized host plant preferences (Lindquist 1998).

Although their injury may be economically significant, eriophyoids are highly adapted to their natural hosts and rarely kill them. Exceptionally devastating effects, like those caused by *Aculops lycopersici* (Tryon) on tomatoes, reflect the fact that tomatoes evidently are not one of the natural solanaceous hosts of this mite (Sabelis and Bruin 1996). Similarly, the original host of *Aceria guerreronis* Keifer, now one of the most important pests of coconut (see below) is a non-coconut palm of the Neotropics, a zone greatly removed from the Indo-Pacific where the coconut originated (Navia et al. 2005). Like the Tetranychoidea, eriophyoids feed by inserting their cheliceral stylets into plant cells and sucking up their liquid contents (Keifer 1975; Nuzzaci and Alberti 1996; Westphal and Manson 1996; Figs. 13.16A, B). In contrast with most tetranychoids, however, eriophyoids illustrate a high degree of host specificity, in most cases at least to the generic level of their hosts (Oldfield 1996a). Host reaction to eriophyoid feeding often is bizarre, involving the formation of galls and other teratosities that the mites then utilize as protected feeding niches (Westphal and Manson 1996) (Figs. 13.16D–G). Many eriophyoids are referred to as **gall mites**, **blister mites**, **bud mites**, or **rust mites**, depending on host reaction to their feeding (Oldfield 1996b; see also chapter 6). Those that live continuously on exposed surfaces and cause little or no observable host injury are referred to as **leaf vagrants**. Symptomatology may be useful, at least to some extent, in defining family categories in the Eriophyoidea. For example, the family DIPTILOMIOPIDAE consists entirely of leaf vagrants that do little or no damage to their hosts, although a few species are known to cause rusting or browning on leaves of orchard or ornamental plants (Jeppson and Keifer 1975; Lindquist and Amrine 1996). The ERIOPHYIDAE, on the other hand, commonly initiate gall formation and other host abnormalities, as do some members of the early-derivative family PHYTOPTIDAE.

Eriophyoids in temperate and subarctic latitudes often produce special overwintering females called **deutogynes**. This is primarily an evolutionary adaptation for survival on deciduous plants in regions with well-defined winters, although it is found secondarily in milder regions and infrequently in tropical regions and on evergreen hosts (Manson and Oldfield 1996). Deutogynes differ morphologically from normal females or **protogynes** (which generally resemble males of the same species), and they generally are difficult to identify. Deutogynes may appear early or late in the growing season, depending on host physiology, but always seek an aestivation or hibernation niche, sometimes in host plant tissue below the soil surface (Krantz and Ehrensing 1990). Deuterogyny is known to occur in all three families of Eriophyoidea, but it is rarely found among phytoptids and is thought to have arisen independently within each family and perhaps repeatedly within ERIOPHYIDAE (Keifer 1975). Fertilization of deutogynes by spermatophores is known to occur prior to hibernation in *Eriophyes emarginatae* Keifer and *Aculus fockeui* (Nalepa and Trouessart) (Oldfield 1969; Oldfield and Newell 1973) and probably occurs in other species as well (Oldfield and Michalska 1996). Detailed information on deuterogyny is presented by Keifer (1942, 1952, 1975) and Manson and Oldfield (1996).

Several species of ERIOPHYIDAE, including three of the subfamily Aberoptinae, apparently are capable of producing weblike coatings on leaves, but no silk-producing organs have been confirmed in these forms (Manson and Gerson 1996). Lindquist and Oldfield (1996) have speculated that interspecific coexistence or intraspecific polymorphism may occur among some of the forms that live under such patches of weblike or waxlike coatings. Production of waxlike coatings has been rarely noted in the PHYTOPTIDAE, but it occurs more frequently in the more derivative ERIOPHYIDAE and DIPTILOMIOPIDAE (Manson and Gerson 1996; Lindquist and Oldfield 1996). Liquid secretion is known only in two diptilomiopid genera, suggesting that this capability is correlated with the larger, deeper-penetrating stylets characteristic of this family (Manson and Gerson 1996).

The PHYTOPTIDAE includes 21 genera and some 145 species that primarily infest ancient hosts such as Araucariaceae, Coniferaceae, Palmaceae, Cyperaceae, Fagaceae, Corylaceae, and Betulaceae (Lindquist and Amrine 1996). Generally recognized as the earliest derivative eriophyoid lineage, phytoptids retain a greater array of plesiomorphic attributes than the other families and are difficult to define apomorphically. A preliminary cladistic analysis, although incomplete and based on some homoplastic characters, indicates that the family may be paraphyletic (Hong and Zhang 1996). While no phytoptid species are known to initiate leaf erineal growth, a few cause galls or similar deformations such as "big buds" on their hosts, in which the inner walls of affected tissues are often densely matted with papillae (Keifer 1975; Castagnoli 1996). Leaf and twig galls are caused by *Pentasetacus araucaria* Schliesske on araucaria (Schliesske 1985), and *Trisetacus pini* (Nalepa) is unique in inducing twig and bark swellings on pine that may be continuously occupied for up to 10 years (Shevchenko, Bagnyuk, and Rinne 1993). Comparatively few phytoptids are of economic significance, although several feed on economically important hosts. *Phytoptus avellanae* Nalepa (Fig. 13.15) produces big bud symptoms in hazelnut and commercial filberts, often causing significant bud losses (Krantz 1979). *P. hedericola* Keifer produces leaf stunting and deformation of English ivy, while *Retracrus johnstoni* Keifer and *R. elaeis* Keifer cause frond discoloration in ornamental and oil palms (Keifer et al. 1982). Species of *Trisetacus* live on coniferous hosts of a wide variety of genera of Cupressaceae, Pinaceae, and Taxodiaceae (Smith 1984b; Bagnyuk, Sukhareva, and Shevchenko 1995), where their feeding may cause bud proliferation or abortion, rosette galls, and stunted or discolored needles (Jeppson and Keifer 1975; Keifer et al. 1982; Castagnoli 1996). The juniper bud mite, *T. quadrisetus* (Thomas), lives and feeds inside the berries of its host, transforming them to gall-like structures (Keifer et al. 1982); three other species of *Trisetacus* have also been reported to inhabit cupressaceous berries or cones (Smith 1978b; Oganezova and Pogosova 1994). *Nalepella tsugifoliae* Keifer and *N. haarlovi* Boczek may be pests of conifers, primarily in nurseries and greenhouses, where they cause needle chlorosis and browning, sometimes leading to the death of young trees (Keifer 1953; Boczek 1962; Löyttyniemi 1969). Other species of *Nalepella* that occasionally damage coniferous trees of a variety of pinaceous genera other than *Pinus* were reviewed by Marshall and Lindquist (1972) and Castagnoli (1996).

The DIPTILOMIOPIDAE includes 53 genera and approximately 315 species that are characterized by having an enlarged gnathosoma that is abruptly bent ventrally. The cheliceral and associated stylets also are relatively elongate and abruptly bent (Figs. 13.18E–G), which suggests that these mites may be capable of feeding on deeper tissues than do other eriophyoids (Keifer 1975). However, the only noticeable symptoms produced by diptilomiopids involve surface leaf cells (Royalty and Perring 1996). Many diptilomiopid immatures and adults produce a flocculent, waxy exudate that covers the mite's dorsum (Manson and Gerson 1996), presumably to protect the exposed vagrant mites from excessive water loss. *Catarhinus tricholaenae* Keifer, a grass-inhabiting diptilomiopid from Brazil, causes leaf rust on corn (Jeppson and Keifer 1975; Perring 1996). A species of *Cheiracus* identified as *C. sulcatus* Keifer by Hong et al. (2005) causes rust stripes and serious withering damage to leaves of rice in China. Other rust-producing diptilomiopids are found in the genera *Rhynchaphytopus*, *Diptilomiopus*, and *Diptacus* (Jeppson and Keifer 1975; Lindquist and Amrine 1996).

The ERIOPHYIDAE is by far the largest family of Eriophyoidea, with approximately 227 genera and over 3,000 species, of which many are of economic importance. Some eriophyid genera such as *Aceria* are not only large (over 800 species) but are vaguely defined (Amrine and Stasny 1994). Injury and distortions to leaves and other organs of their plant hosts produced by eriophyids may be of several types and may involve surface or mesophyll tissues (Keifer 1952, 1975; Westphal and Manson 1996; Oldfield 1996b; Royalty and Perring 1996). *Aculops lycopersici*, the tomato russet mite, is an important pest on solanaceous crops in many parts of the world. It severely damages the leaves and stems of tomato plants, often killing its host. Heavy infestations of *A. lycopersici* may also injure leaves of tobacco, potato, pepper, and petunia (Jeppson and Keifer 1975; Perring and Farrar 1986; Perring 1996). The citrus rust mite, *Phyllocoptruta oleivora* (Ashmead), and the pink citrus rust mite, *Aculops pelekassi* Keifer, are among the more serious pests of citrus worldwide, causing russeting and discoloration of fruit and leaves, distortion of new growth, and leaf drop; similar damage may be produced by several other vagrant species found on citrus (McCoy 1996). *Aculus fockeui*, the plum nursery mite, reproduces on many Palearctic species of *Prunus*, including the commercially important fruit and nut species (Castagnoli and Oldfield 1996). This mite causes leaf spotting on peach, plum, and cherry in the spring, followed by leaf curling and silvering or browning later in the season. *A. fockeui* may cause more severe injury to plums than to other hosts, including curling of terminal shoots and dwarfism of foliage. Available evidence indicates that mites formerly treated under the name *A. cornutus* (Banks), the peach silver mite, are conspecific with *A. fockeui* (Oldfield 1984).

Several species of *Aceria* may cause serious damage to other kinds of fruit trees. For example, *A. oleae* (Nalepa) seriously deforms the leaves and fruit of olive trees (Castagnoli and Oldfield 1996), and *A. guerreronis* Keifer severely damages floral bracts and may distort, stunt, and cause premature dropping of nuts of coconut palm (Moore and Howard 1996). *A. guerreronis* also causes meristem necrosis, leaf distortion, and death in young queen palms (Ansaloni and Perring 2004). *A. mangiferae* Sayed causes stunting and bud proliferation on mango, and *Metaculus mangiferae* (Attiah) produces russeting and distortion of mango leaves, buds, and inflorescences (Jeppson and Keifer 1975). Bud and leaf injuries on fig trees infested by *A. ficus* (Cotte) result from both the feeding effects of this mite and the fig mosaic pathogen that is transmitted by it (Jeppson and Keifer 1975). Severe leaf edge-rolling is caused by *A. granati* (Canestrini and Massalongo) on pomegranate (Meyer 1996), and several species of *Aceria* cause distortion and discoloration of leaves on walnut, pecan, and other nut trees (Castagnoli and Oldfield 1996). The effects of these mites usually do not influence crop production, although they may adversely affect the appearance of ornamental plants. The pink tea mite, *Acaphylla theae* (Watt and Mann), and the purple tea mite, *Calacarus carinatus* (Green), are important pests of tea in southern Asia, where they may severely damage the leaves of their hosts (ChannaBasavanna 1996). *Aculus comatus* (Nalepa) causes rusting of filbert leaves in Europe and North America but may also give rise to leaf edge-rolling (Krantz 1973b; Castagnoli and Oldfield 1996).

Some eriophyid species induce a secondary development of papillar outgrowths on leaves or petioles, which afford shelter for the mites. Papillar development may be patchy or generalized and may appear on either leaf surface. Open papillar masses are known as **erinea** (Fig. 13.16F) and constitute "open galls" in which pocketing is minimal (Westphal and Manson 1996; Royalty and Perring 1996). Various species of *Aceria*, *Eriophyes*, *Acalitus*, and *Colomerus* induce erineal formations on their hosts that may sometimes be remarkably colorful, such as the glossy crimson erineal patches caused by *Aceria elongatus* (Hodgkiss) on leaves of sugar maple (Keifer et al. 1982). The injurious grape erineum mite, *C. vitis* (Pagenstecher), appears to comprise three strains or races, only one of which precipitates development of erinea (Smith and Stafford 1948; Stafford and Doutt 1974). The other two strains feed on bud tissues or cause leaf curl in susceptible grape varieties (Duso and de Lillo 1996).

True galls are produced by host reactions to feeding by numerous species of eriophyid mites. Galls may assume a variety of shapes on leaves (nail, bead, finger, bladder, etc.), but all are modifications of localized, virtually closed leaf puckerings or **purse galls**, in which the abnormal growth of papillar material provides a substrate in which mites may feed with little danger from predators or desiccation (Figs. 13.16D, E). Galls may be produced on the upper sides of leaves (with an opening on the lower side) or on flowers, stems, buds, or petioles. All are formed prior to maturation of the affected plant part. Keifer (1975) and Westphal and Manson (1996) note a variety of gall-inducing ERIOPHYIDAE, and Felt (1940), Liro and Roivainen (1951), Mani (1964), and Keifer et al. (1982) provide illustrations of many gall types. Gall distortions induced by these mites on leaves usually do not adversely affect plant productivity.

Pears, apples, quince, and other pomaceous plants may be affected by *Eriophyes pyri* (Pagenstecher) or another of the *E. pyri* complex of blister mites. Unlike galls, **blisters** are formed within the parenchymal leaf tissues (Fig. 13.16G). Ingress occurs early in the spring following hypertrophy and necrosis of surface cells previously fed on by early emerging mites. The collapse of necrotic cells leaves tiny openings through which mites invade and develop subsequent populations within these shelters. *E. pyri* may also damage bud tissues in the fall and winter (Jeppson and Keifer 1975; Easterbook 1996). Injury by other eriophyids may be caused by feeding in buds or by bud gall formation. The citrus bud mite, *Aceria sheldoni* (Ewing), feeds in buds and fruit buttons, particularly of lemons, and may severely distort buds, leaves, flowers, and fruit (Boyce and Korsmeier 1941; Jeppson and Keifer 1975; McCoy 1996). Members of the genus *Cecidophyopsis* induce bud gall formation in a variety of hosts, including yew, currant, and filbert (Fig. 13.15) (Jeppson and Keifer 1975; Krantz 1979; Castagnoli 1996; Castagnoli and Oldfield 1996; de Lillo and Duso 1996). Careful biological and morphological studies have revealed that five species of *Cecidophyopsis*, which have long been confused with one another, occur only on species of gooseberries and currants of the genus *Ribes*. Four of these species are each restricted to a different species of host and are important or potentially important as vectors of currant reversion disease (Amrine et al. 1994).

In addition to causing leaf and bud injury, some eriophyid mites are responsible for the development of "witches broom" (adventitious twig development), flower galls, shortening of internodes, and plant stunting (Westphal and Manson 1996). A diversity of eriophyoids, nearly all in the family ERIOPHYIDAE, infest and sometimes damage cereal and other grasses, including forage and pasture grasses and turfgrass. Those that infest cereals were reviewed by Sukhareva (1992), and those on other grasses by Frost and Ridland (1996). Wheat streak mosaic is the only serious problem of grain plants associated with eriophyid mites (Styer and Nault 1996). Eriophyid pests of forage legumes, especially the lucerne bud mite, *Aceria medicaginis* (Keifer), were reviewed by Ridland (1996). Meyer (1996) reviewed a variety of eriophyoid species, all in the family ERIOPHYIDAE, that cause damage to ornamental plants.

Many eriophyids that do not cause galls or other distortions of host tissues may affect the appearance of their hosts either by feeding on epidermal cells or by their salivary toxins, which can cause altering underlying tissue. Such **toxemias** and other nondistortive feeding effects were reviewed by Oldfield (1996b). Examples of toxemias include such abnormalities as red berry on various berries of the genus *Rubus*, caused by *Acalitus essigi* (Hassan). Feeding by this mite prevents ripening of drupelets, parts or all of which remain bright red and offer an extended protective niche for the mites through autumn and winter (Keifer et al. 1982; de Lillo and Duso 1996). Kernel red streak on corn is caused by *Aceria tulipae* (Keifer), which produces reddened stripes of discoloring in kernels along part or all of the pericarp (Keifer et al. 1982; Styer and Nault 1996). *A. tulipae* also infests bulbs of tulip and species of *Allium*, including garlic, onion, shallot, and ornamentals, and causes the outer scales to become dull and discolored instead of glossy and white. Heavily infested bulbs dry out, and root and shoot development is retarded or inhibited (Conijn, Van Aartrijk, and Lesna 1996). Foliage that grows from infested bulbs becomes stunted, curled, and discolored, and ornamental varieties may become permanently disfigured (Jeppson and Keifer 1975; Perring 1996). The various effects of feeding by *A. tulipae* on different parts of its host plants exemplify a species whose feeding may cause distortive or nondistortive effects on its host, depending on which organ it attacks (Oldfield 1996b).

With few exceptions, eriophyoids are the only mites that have been demonstrated to transmit important plant viruses and other infectious agents to herbaceous and woody crop or ornamental plants (Slykhuis 1969, 1972, 1980; Oldfield 1970; Jeppson 1975; Oldfield and Proeseler 1996). Eriophyoids known to be vectors of plant disease agents are members of only six genera, representing the subfamilies Eriophyinae, Phyllocoptinae, and Cecidophyinae, all in the family ERIOPHYIDAE (Lindquist and Oldfield 1996). Considering the small number of such diseases that seem to be transmitted by eriophyoids, and the few taxa that have been shown to transmit them, the conditions that must be satisfied for these mites to be competent vectors must be many and complex and perhaps do not often lead to compatible systems that result in a vector-pathogen-host relationship (Sabelis and Bruin 1996). The relationship between an eriophyid vector and a transmitted agent is highly specific, to the point that no plant pathogen is known to be transmitted by more than one species of eriophyid (Oldfield and Proeseler 1996). For example, viral pathogens of cereal grass diseases such as ryegrass mosaic are transmitted by *Abacarus hystrix* (Nalepa), while wheat streak and wheat spot mosaics are transmitted by members of the *Aceria tulipae* species complex including *A. tosichella* Keifer (= *A. tritici* Shevchenko), but not *A. tulipae* as widely recorded in previous literature (see Shevchenko et al. 1970; Sukhareva 1992; Amrine and Stasny 1994; Halliday and Knihinicki 2004). The pathogen of currant reversion, a serious disease of black currants, is transmitted by *Cecidophyopsis ribis* (Westwood), but whether the agent is viral remains in doubt (Slykhuis 1980). Other woody perennials are affected by probable viral pathogens transmitted by *Aceria ficus* (fig mosaic), *Eriophyes insidiosus* Keifer and Wilson (peach mosaic), *E. inaequalis* Wilson and Oldfield (cherry mottle leaf), and *Phyllocoptes fructiphilus* Keifer (rose rosette). Pigeon pea sterility mosaic, an important disease of *Cajanus cajan* in India, is thought to be caused by a virus transmitted by *A. cajani* ChannaBasavanna (Reddy et al. 1989). Slykhuis (1980) commented on a variety of other diseases that may be caused by eriophyid mites in the absence of an infectious agent or that involve doubtful reports of transmission by eriophyids, as well as others that are thought to be transmitted by eriophyids. The association between the eriophyid *Calacarus citrifolii* Keifer and a bacterial spiroplasma-like organism in lesions of concentric ring blotch on citrus remains unclear (Kotzé et al. 1987).

The potential beneficial effects of some eriophyoid mite species have received considerable attention in recent years. Dunley and Croft (1996) reviewed the role of eriophyoids as competitors of other, more harmful, phytophagous mites, especially in apple orchards. The potential value of eriophyoids as alternative prey for beneficial predatory mites in orchards and vineyards was reviewed by Sabelis and Van Rijn (1996). A number of species of ERIOPHYIDAE have been investigated, at least in a preliminary way, for possible use in the biological control of weeds (Cromroy 1983; Rosenthal 1996; Amrine 1996; Craemer et al. 1999; Goolsby, Zonneveld, and Bourne 2004). These include *Aceria chondrillae* (Can.) on skeletonweed, *A. malherbae* Nuzzaci on field bindweed, *A. acroptiloni* Kovalov and Shevchenko on Russian knapweed, *A. drabae* (Nalepa) on hoary cress, *A. centaureae* (Nalepa) and *A. thessalonicae* Castagnoli on diffuse knapweed, *Epitrimerus taraxaci* Liro on dandelion, *Aculops hyperici* (Liro) on St. John's wort, *Phyllocoptes fructiphilus* Keifer on multiflora rose, *Cecidophyes rouhollahi* Craemer et al. on false cleavers, and *Floracarus perrepae* Knihinicki and Boczek on Old World climbing fern. The potential of certain other eriophyid species for weed control was noted by Cromroy (1983) and by Boczek and Petanovic (1996).

Supercohort Anystides

Cohort Anystina

The Anystina includes eight families grouped in five superfamilies. Members of the cohort have short, often sicklelike movable cheliceral digits arising terminally from broad unfused bases, and the fixed digits are greatly reduced or absent. A well-defined palpal thumb-claw process is present throughout the subcohort, but it may be reduced in some groups and is absent in the PARATYDEIDAE. Prodorsal ocelli and postcheliceral stigmata and peritremes usually are distinct, and a naso and one or two pairs of bothridial organs may be present. Larval urstigmata and postlarval genital papillae are retained in most taxa except for the PSEUDOCHEYLIDAE, and anamorphic additions of adanal and, rarely, anal setae usually occur in postlarval instars (the PSEUDOCHEYLIDAE and POMERANTZIIDAE are exceptions). Males lack an aedeagus, and copulation involving direct sperm transfer is unknown in the cohort. Many anystines are xerophilic, and most are considered predatory, ranging over arid and often exposed substrates in their search for prey or lying in wait to ambush them.

■ Superfamily Caeculoidea (Figs. 13.20A–D)

DIAGNOSIS: Heavily sclerotized forms, usually with 8 dorsal shields arranged in a characteristic pattern. Cheliceral bases unfused, apparently capable of limited lateral motion; palptibia always with a single, well-defined terminal claw and ancillary subterminal spines; with 2 pairs of lateral eyes and an anteromedian prodorsal eye associated with a naso. Setal elements of opisthosomatic segment AD present; with 3 pairs of genital papillae. Legs I with strong, internal spinose setae, tibia, and tarsus I each with a solenidion and a vestigial actinopilous seta in an integumental sink. Paired claws on legs usually unequal in size, empodia absent. With prelarval, larval, and 3 nymphal stases.

The Caeculoidea includes only one family, the CAECULIDAE (Figs. 13.20A–D), also known as rake-legged mites. Caeculids are large (1,000–3,000 μm), heavily armored, wrinkled, slow-moving forms that typically inhabit dry exposed habitats such as beaches and beach vegetation, tree bark, and rocky niches in desert and mountain habitats (Enns 1958; Coineau 1974b; Coineau and Haupt 1978; Hagan 1985; Otto 1993). Despite their large size, their slow movement and dull coloration, coupled with the camouflage provided by a rough cerotegument and adhering organic matter, often make caeculids difficult to discern. Fossil caeculids are known from Cretaceous and early Tertiary amber (Coineau and Magowski 1994; Coineau and Poinar 2001; Grimaldi and Engel 2005).

Most caeculids appear to be ambush predators of collembolans, small insects such as bark lice, and soft-bodied mites (Coineau 1973b, 1974b; Walter and Proctor 1999). Immature and adult individuals of an Australian *Neocaeculus* sp. have been observed to remain motionless for days with their heavily spined legs I held at a 45° angle to the substrate, waiting for an unsuspecting prey to wander into range (Walter and Proctor 1999). *Microcaeculus pica* Otto, a psammophilous species, catches its collembolan prey in much the same way, using the strong internal setae of legs I to help prevent prey escape (Otto 1993). These same leg setae in *M. pica* and related beach-dwelling *Microcaeculus* species tend to be clavate rather than spinose and may have evolved for the dual purpose of securing prey and digging in the sand substrate (Otto 1993). *M. pica* has a tolerance for long-term immersion in seawater, similar to that reported for other littoral mite species by Schuster (1979). Specimens in a laboratory setting survived submersion for six months, and one lived for 294 days (Otto 1993). Laboratory observations by Crossley and Merchant (1971) on a species of *Caeculus* suggested that some caeculids also may feed on fungi. The mites never were observed to feed on collembolans or their eggs but reproduced in cultures where only fungal hyphae were available as a food source. Feeding in fungal cultures was corroborated by recovery of ^{134}Cs from mites confined on fungi treated with this radioactive material.

The biology, morphology, and behavior of the CAECULIDAE have been reviewed in depth by Coineau (1963, 1964c, 1966, 1970, 1973b, 1974b).

■ Superfamily Adamystoidea (Figs. 13.21A–D)

DIAGNOSIS: With 1–2 shields partially or completely covering dorsum; 4 pairs of setae in the prodorsal region, including bothridial setae vi *on a well-developed naso and bothridial setae* si *posterior to the naso; 1–2 pairs of eyes; dorsosejugal furrow absent, but idiosoma often weakly constricted posterior to legs IV. Setal elements of opisthosomatic segments AD and AN added developmentally; anal valves with 2–4 pairs of* ad *setae; 2 pairs of genital papillae in adult. Stigmatic openings at base of chelicerae, often produced into short peritremes. Chelicerae weakly chelate to hooklike, base bulbous and freely articulated, capable of scissorslike movement over the rostrum, fixed digit produced as slender hook or reduced, movable digit weakly toothed and hooklike. Palpi linear, with 5 free segments, palptibial claw complex absent. Coxal bases weakly radiate, tarsal claws weakly pectinate, empodium clawlike. Prelarval (at least some of which are elattostases), larval, and 3 nymphal stases are present.*

The Adamystoidea includes only one family, the ADAMYSTIDAE (Figs. 13.21A–D), in the present treatment. Members of this assemblage are medium-sized, reddish mites with a sclerotized plate covering some or most of the dorsum; a ventral plate may also be present. The cheliceral bases are bulbous and separate, the fixed digit is reduced but apparent, and the movable digit is hooked. The palpi are linear and lack a thumb-claw process, and the tarsal claws may be bipectinate. Adamystids have two pairs of genital papillae, although they are neither mentioned nor illustrated for *Saxidromus anahoffmannae* López-Campos (López-Campos 1996). The more anterior pair of prodorsal bothridial organs is inserted on a well-developed naso. Stigmata and simple peritremes are present at the cheliceral bases (Kethley 1982; Walter and Proctor 2001).

Adamystids are thought to be free-living predators and are found mostly in dry soils and litter over much of the world (Cunliffe 1957; Hunter and Crossley 1968; Coineau 1974d, 1979; Barilo 1987; Ueckermann 1989; Rafalski 1982; McDaniel and Bolen 1983; Palacios-Vargas and Prieto-Trueba 1995; López-Campos 1996; Swift 2000). Species of

Adamystis also are found on the leaves of rain-forest trees in Australia (Walter and O'Dowd 1995). The prelarva of *S. delamarei* is an active elattostase, and adult mating behavior is complex (Coineau 1976b). The male carries the female on his dorsum until he has deposited a stalked spermatophore on the substrate, at which time he manipulates the female onto the spermatophore to achieve insemination. *S. delamarei* and other adamystids are plesiotypic in that they retain postlarval opisthosomatic additions of adanal and anal setae on the protonymph and deutonymph, respectively (Coineau and Naudo 1986).

■ Superfamily Anystoidea (Figs. 13.22A–G, 13.23A, B)

DIAGNOSIS: Soft bodied, often with a weakly developed prodorsal shield; dorsal setal bases may be sclerotized; posterior and coxal field hypertrichy often present. Prodorsum with 4 pairs of setae, including bothridial setae si, *setae* vi *normal, or bothridial and usually present on a distinct naso in ANYSTIDAE, naso weakly expressed and nude in TENERIFFIIDAE and absent from PSEUDOCHEYLIDAE; 1 or 2 pair of eyes; dorsosejugal furrow absent or present (PSEUDOCHEYLIDAE). Setal elements of opisthosomatic segment AD added (ANYSTIDAE) or absent (PSEUDOCHEYLIDAE, TENERIFFIIDAE); with 3 pairs of small genital papillae or papillae absent (PSEUDOCHEYLIDAE). Stigmatic openings at base of chelicerae, produced into linear to sinuous peritremes, sometimes emergent distally. Cheliceral bases freely articulated, capable of scissorslike movement over the rostrum, fixed digit reduced or absent; palptibia with 1–3 clawlike setae, palptarsus strongly developed or reduced. Leg tarsi entire or subdivided distally into several distinct subsegments; pretarsi padlike and produced on annulate stalks in PSEUDOCHEYLIDAE, claws obscure or absent; tarsal claws often pectinate in other families, empodia variously developed. Prelarval (sometimes elattostatic), larval, and three nymphal stases are present in ANYSTIDAE and TENERIFFIIDAE; a prelarva has not been reported in the PSEUDOCHEYLIDAE.*

Three families are included in the Anystoidea in the present treatment: ANYSTIDAE (Figs. 13.22A–E), TENERIFFIIDAE (Figs. 13.22F, G), and PSEUDOCHEYLIDAE (Figs. 13.23A, B). The ANYSTIDAE includes soft-bodied, moderately large (500–1,500 μm), long-legged, mostly bright orange or red, rapidly moving mites that often frequent plants and bare rocky surfaces on which they search for prey. The family is currently considered to consist of two subfamilies: the Anystinae, which comprises mostly broader-bodied species with two pairs of ocelli and three claws on the palptibia; and the Erythracarinae, which includes species that are typically longer than broad, have one or two pairs of ocelli, fewer than three palptibial claws, and flexible or secondarily subsegmented leg tarsi (Oudemans 1936b; Meyer and Ryke 1960a; Otto 2000a). There is evidence suggesting that the Anystinae may be more closely related to the TENERIFFIIDAE than to the Erythracarinae (Otto 2000a), but both anystid subfamilies are plesiotypic relative to teneriffiids in retaining the anamorphic expression of adanal setae on the protonymph (Grandjean 1943b; E. E. Lindquist, pers. obs.). *Anystis baccarum* (L.) and other species of the genus are aerial predators of phytophagous insects and mites in Europe, Australia, Africa, and North America, but their efficacy as control agents on economic crops may be offset by their slow rate of increase, lack of prey preference, cannibalistic tendencies, and sensitivity to pesticides (Jeppson, Keifer, and Baker 1975a; Gerson and Smiley 1990). *A. wallacei* Otto is partial to pasture situations in southern Europe, while its congener *A. salicinus* (L.) occurs north of the Alps on trees and in surrounding soil (Otto 1992). The preference of *A. wallacei* for a Mediterranean climate and grassland habitats has led to its introduction into Australia and South Africa (under the name *A. salicinus*) for the control of mite pests of pasture grasses (Meyer and Ueckermann 1987; Otto and Halliday 1991). *A. agilis* (Banks), the common whirligig mite, feeds on a variety of invertebrate pests in citrus and apple orchards, vineyards, and alfalfa plantings in North America (Mostafa, de Bach, and Fisher 1975; Sorenson et al. 1976; Frazer and Nelson 1981). *A. agilis* preferred the citrus thrips over the citrus red mite, *Panonychus citri* McGregor (TETRANYCHIDAE), in the laboratory and produced progeny only with the former prey (Mostafa, de Bach, and Fisher 1975). Because of the obstacle posed by their webbing, spider mites tend to be immune from attack by *A. agilis* (Sorenson et al. 1976). Grobler (1962) observed *A. agilis* associated with over 50% of the overwintering eggs of the woolly pine needle aphid, an important pest of *Pinus* in Canada. Laboratory trials have shown that some anystine species confined to small spaces can attack and kill cattle tick larvae when offered no alternative prey and perhaps merit further investigation of their potential as natural enemies of ticks (Holm and Wallace 1989).

Lee (1961) and Southcott (1976) noted that *Anystis* sp. (probably *baccarum*) pierces human skin and has no difficulty in feeding to repletion on subdermal fluids. Southcott (1976) described the effect of an anystid bite as consisting of a stinging sensation followed in some cases by swelling and localized hemorrhaging.

Members of the anystid subfamily Erythracarinae are predators in soil and litter, on or under bare rock, and on low-lying plants and have been found on all continents except Antarctica (Meyer and Ryke 1960b; Meyer and Ueckermann 1987; Otto 1999d–f). Their tarsi are subdivided apically into a series of flexible pseudosegments, which enables them to run rapidly over irregular terrain (Grandjean 1952; Otto 1999f). In Europe, *Chaussieria venustissima* (Berlese) may be found in rocky areas from sea level on Crete to altitudes of 2,800 m in Switzerland (Hutterer 1978). *Tarsotomus alticolus* (Hirst) was collected on Mount Everest at 5,200 m (Hirst 1927; Meyer and Ueckermann 1987), although Otto (1999f) noted the apparent preference of *Tarsotomus* species for hot dry surfaces and warmer climates. Otto (1999d–f) has reviewed the erythracarine genera *Erythracarus*, *Tarsotomus*, *Paratarsotomus*, and *Chaussieria*, relegating a number of previously recognized genera into synonymy with them (i.e., *Chabrieria*, Walzia, *Paranandia*, *Absoloniana*, *Parabsolonia*, and *Bechsteinia*). Males of some erythracarines stack their spermatophores, an unusual behavior that may indicate sperm competition in these taxa (Otto 1999e). The prelarva of *C. venustissima* is a motile, nonfeeding elattostase that is negatively phototaxic (Otto and Olomski 1994). In contrast, those of *Erythracarus* spp. may be able to move their legs and change their position slightly, while others are entirely immobile (Otto 1997).

Members of the family TENERIFFIIDAE (Figs. 13.22F, G) are red, yellow, or brownish long-legged predators that are capable of rapid movement. Teneriffiids have two pairs of eyes and one pair of filiform bothridial setae on the prodorsum, the bases of which have a rosette pattern. Coxal fields are hypertrichous, tarsi III–IV have two or more pseudosegments, and the tarsal claws of at least legs I are strongly bipectinate. Although Kethley (1990) cited Ehara (1965) in indicating that teneriffiids have adanal setae, there is no indication of anamorphic additions of these setae during postlarval development (E. E. Lindquist, pers. obs.). Adults have three pairs of genital papillae. Interestingly, Otto (1997) observed the prelarva of a species of *Parateneriffia* to remain partly enclosed by the eggshell and completely immobile (Otto 1997).

Teneriffiids have been collected in habitats ranging from intertidal sand banks to the undersides of rocks in the Tyrol at altitudes exceeding 1,000 m (Irk 1939b; Eller and Strandtmann 1963). *Parateneriffia uta* (Tibbetts), *P. luxoriensis* (Hirst), and other members of the genus seem to share a predilection for arid desert habitats (McDaniel, Morihara, and Lewis 1976). One species was collected in sage and saltbrush litter in the sandy high desert country of central Oregon, USA, a locale where abundant marine fossils may attest to its having been covered by a shallow sea during the Mesozoic era (G. W. Krantz, pers. obs.). It is possible that *Parateneriffia* spp., or its antecedents, preyed on littoral invertebrates on its ancient shores. A preference for coastal habitats is apparent in the genus *Teneriffia*, several species of which are encountered only in seashore situations. For example, *T. quadripapillata* Thor, the type species for the family, was collected on a beach in Tenerife, *T. marina* (Hirst) is an intertidal predator in Malaysia, and *T. mexicana* McDaniel, Morihara, and Lewis occurs in a variety of intertidal substrates in Baja California (McDaniel, Morihara, and Lewis 1976). Members of other teneriffiid genera occur in seashore habitats of Japan (Shiba and Furukawa 1975). In contrast, *Austroteneriffia* species primarily inhabit trees in Australia (Walter and Proctor 1999), and a species of *Neoteneriffiola* was found on pine trees near the seashore in Japan (Shiba and Furukawa 1975).

Members of the family PSEUDOCHEYLIDAE (Figs. 13.23A, B) are poorly studied mites that have a weakly defined prodorsal plate and four-segmented palpi with laterally expanded basal segments. The palptibia usually has one to three internal, spinelike or expanded setae, and the palptarsus apparently is absent. One pair of filiform or apically swollen bothridia and one pair of eyes occur on the prodorsum, and well-developed peritremes are present on the anterior margin of the prosoma. Larval urstigmata and postlarval genital papillae are absent. The insertions of legs II are widely separated from those of legs III, and the tarsal claws are produced on elongate, annulate pretarsi (Fig. 13.23B). Pseudocheylids lack anamorphic additions of adanal and anal setal elements during ontogeny (E. E. Lindquist, pers. obs.).

Representatives of the three recognized pseudocheylid genera (*Pseudocheylus*, *Anoplocheylus*, and *Neocheylus*) have been collected primarily from moss, but they also have been found in soil and dry leaf litter. One species, *N. nidicolus* Lawrence, was collected in Uganda from a weaverbird nest that had been occupied by *Galago senegalensis*, a nocturnal lorisid primate (Lawrence 1954). Pseudocheylids are known mostly from southern Europe, Australia, South America, and South Africa (Van Dis and Ueckermann 1991; Walter and Proctor 2001), but this may reflect inadequate collecting. Records from outside the regions noted above include *N. americanus* (Ewing) and *A. transiens* Delfinado and Baker, taken, respectively, from under maple bark in Illinois, USA, and leaf litter in New York, USA (Baker and Atyeo 1964; Delfinado and Baker 1974); *A. tauricus* Livshits and Mitrofanov from *Euphorbia* in Crimean Ukraine (Livshits and Mitrofanov 1973); and an undetermined species from southern Ontario, Canada (E. E. Lindquist and I. M. Smith, pers. obs.).

▪ Superfamily Paratydeoidea (Figs. 13.23C–I)

DIAGNOSIS: Small, soft-bodied elongate mites that are pale white to orange or violet in life. Legs I–II widely separated from legs III–IV and with distinct constrictions posterior to segment C and often segment D (postpedal furrows). Prodorsum with or without poorly defined shield bearing 1 pair of bothridial setae si *and normally developed setae* ve *and* se *(minute* vi *present in STIGMOCHEYLIDAE) and 0–2 pairs of eyes; paired stigmata opening between cheliceral bases, peritremes produced at anterior margin of prodorsum; naso absent. Cheliceral bases separate, movable digit small and hooked; palpi with (STIGMOCHEYLIDAE) or without (PARATYDEIDAE) tibial claw complex. Genital and anal opening approximate or separate, with 2 or 3 pairs of genital papillae; setal elements of opisthosomatic segment AD (and sometimes AN) present. Femora of legs subdivided, pretarsi with claws (except for legs I in STIGMOCHEYLIDAE); empodia minute (PARATYDEIDAE) or absent (STIGMOCHEYLIDAE). Known ontogeny includes egg, larva, 3 nymphal stases, and the adult.*

The two paratydeoid families PARATYDEIDAE and STIGMOCHEYLIDAE were relegated to the Eleutherengonides as early derivative relatives of the Heterostigmata in a cladistic hypothesis put forward by Kethley (in Norton et al. 1993). We tentatively group them here in a superfamily based on their characteristic elongate bodies with widely separate coxae II–III and postpedal constriction.

The PARATYDEIDAE (Figs. 13.23C–G) comprises about seven genera and more than a dozen described species, along with numerous undescribed forms. Seeman and Walter (2000) recognized two main clusters within the family: one group is defined by the apomorphic loss of eyes, conjunction of the anal and genital regions, and presence of a constriction at the posterior margin of segment D (e.g., *Hexatydeus*, *Neotydeus*, *Tanytydeus*, and *Walytydeus*), and the remaining genera usually have eye lenses, separate anal and genital regions, and a postpedal furrow only behind segment C (e.g., *Scolotydeus*, *Sacotydeus*, and *Paratydeus*). Some paratydeids have been described with only two pairs of genital papillae, but these may have been based on deutonymphs rather than adults. Paratydeids have been collected worldwide from an array of habitats, including soil, litter, moss, tree branches, and bird nests (Baker 1949a, 1950; Theron, Meyer, and Ryke 1969; Price 1973; Kuznetsov 1973; Flechtmann 1992; Kandeel 1992; Seeman and Walter 2000). They are particularly abundant in dry soils and are assumed to be predatory.

Members of the family STIGMOCHEYLIDAE (Figs. 13.23H, I) are small, elongate mites with a poorly defined, rectangular prodorsal shield and paratydeid-like chelicerae. Their palpi, however, are well developed, with five segments and a distal tibial claw and preapical thumblike tarsus. The prodorsal shield bears one pair of club-shaped bothridial setae, and eyes are absent or obscure. Very short, two to four chambered peritremes emerge on the anterior margins of the prodorsum. Unlike paratydeids, stigmocheylids lack claws on tarsi I, and tarsi III–IV lack empodia. Setal elements of segments AD and AN are added during postlarval development, and two pairs of genital papillae are present in adults (only adanal setae are added during ontogeny, and three pairs of genital papillae are present in adult PARATYDEIDAE). Stigmocheylids are known from Australia, Europe, North America, and North Africa (Baker and Atyeo 1964; Soliman and Zaher 1975; Norton and Kinnear 1999), where they have been taken from soil and dry leaves.

▪ Superfamily Pomerantzioidea (Figs. 13.24A–D)

DIAGNOSIS: Elongate mites with a series of dorsal sclerites. Legs I–II widely separated from legs III–IV. Prodorsum with large shield bearing normally developed setae ve, si, *and* se; *eyes and trichobothria absent; paired stigmata with short peritremes opening at bases of chelicerae. Chelicerae with bases separate, each with a subterminal seta* (cha); *movable digit small and hooked; fixed digit vestigial; palpi stout, thumb-claw process well developed. Dorsal hysterosoma of tritonymph and adult with 5 plates corresponding to segments C, D, E, F, and H. Genital and anal openings separate, with 3 pairs of genital papillae in adult and tritonymph and well-developed ovipositor in adult female; larva with urstigmata. Femora of legs subdivided, pretarsi with claws, empodia absent. Developmental stages include egg, prelarva, 3 nymphal stases, and the adult female; males are unknown.*

Members of the enigmatic family POMERANTZIIDAE (Figs. 13.24A–D) belong to the genera *Pomerantzia* and *Apomerantzia*, which comprise small, elongate mites known mostly from deeper soil horizons in the United States and China (Baker 1949b; Price 1974; Price and Benham 1976; Kethley 1989a; Fan and Chen 2005). Their euedaphic way of life may have led to secondary losses of structures such as the prodorsal naso and bothridia and protruding peritremes. Price (1971) argued against a relationship between pomerantziids and the Raphignathoidea or heterostigmatic superfamilies. Fan and Chen (2005) accept the hypothesis of Kethley (in Norton et al. 1993) in allying the Pomerantzioidea with the supercohort Eleutherengonides instead of with the Anystides, but Fan and Chen's conclusions are based on attributes that are subject to widespread homoplasy among families of Prostigmata, for example, losses of bothridia on the prodorsum and legs, lack of a naso (which also occurs in the anystine families PSEUDOCHEYLIDAE and PARATYDEIDAE), and suppression of anamorphosis during postlarval development (which also occurs in the PSEUDOCHEYLIDAE). Moreover, pomerantziids are characterized by attributes typical of taxa of Anystides rather than the Eleutherengonides. The absence of an empodial structure in the presence of paired claws on tarsi I–IV is found in various anystines but not in eleutherengonid taxa. Furthermore, female pomerantziids retain three pairs of genital papillae and a well-developed ovipositor equipped with eugenital setae, and larvae retain distinct urstigmata (Price 1974), as does the prelarva of *P. philippina* Bochkov and Walter, the only known prelarva in the family (Bochkov and Walter 2007). The absence of males in samples where entire series of immatures have been found (Price 1974) may indicate that at least some pomerantziid species are thelytokous (Kethley, in Norton et al. 1993).

Cohort Parasitengonina

The cohort Parasitengonina is a very large and diverse clade comprising four subcohorts and is well represented in all regions of the world. Members of the group share an apotypical life history pattern that includes egg, inactive prelarva, active larva, inactive protonymph, active deutonymph, inactive tritonymph, and active adult stages (Smith, Cook, and Smith 2001; Wohltmann 2001). The larva is typically parasitic on an invertebrate or vertebrate host and is often strongly heteromorphic and so different from the postlarval instars in appearance and behavior that associations can be established only by rearing. The inactive postlarval stases, corresponding to the protonymph and tritonymph, are calyptostases (see chapter 5) during which major morphological and anatomical reorganization takes place. Dispersion occurs mainly during the parasitic larval stage, when the mites are transported by hosts, and may result in dispersal when the engorged mites drop off the host in a habitat suitable for postlarval development (Smith and Oliver 1986; Smith and Cook 1999a; Smith, Cook, and Smith 2001; Wohltmann 2001).

Parasitengonine mites are diplodiploid, and all species for which observations are available reproduce sexually, although there may be circumstantial evidence of thelytoky in certain species of ERYTHRAEIDAE, in which males appear to be rare or absent (Norton et al. 1993). Spermatophore transfer is usually indirect, with females using pheromonal cues to find and pick up spermatophores deposited on the substrate by males (Witte 1984; see also chapter 6). Many species of terrestrial parasitengonines appear to have little contact between sexes during spermatophore transfer, but others show varying degrees of interaction, and some exhibit complex courtship displays (Witte 1984). The widest range of courtship behavior is found among the Hydrachnidiae, in which members of basal groups show minimal sexual interaction and those of numerous derivative lineages exhibit various forms of elaborate courtship culminating in copulation in many clades and direct transfer of spermatophores in a few of them (Proctor 1991b, 1992a; Smith, Cook, and Smith 2001; also see chapter 6 and below).

Classification of many families and superfamilies in the major subcohorts of Parasitengonina remains provisional, reflecting the present incomplete state of knowledge of taxonomic diversity and distribution coupled with tentative understanding of phylogenetic relationships. The practice of describing species and proposing genera based on adults or larvae that have not been associated with other instars has led to proliferation of available species-level and genus-level names in many families of terrestrial Parasitengonina. Many of these names have been used inconsistently and often incorrectly by later authors, resulting in considerable confusion in nomenclature and classification of the group, much of which remains unresolved. Efforts to revise and stabilize the superfamily classification of terrestrial subcohorts have been largely unsuccessful for several reasons, including overreliance on knowledge of a few arbitrarily chosen species used as surrogates for large and diverse taxonomic groups, analysis of small and incomplete character sets, selective and uncritical acceptance of previously published data, adoption of untested taxonomic concepts, and problematic interpretations of character polarities based on within-group comparisons. The relatively comprehensive and critical recent studies by Southcott and the more rigorous phylogenetic approaches advocated by Welbourn, Wohltmann, and Zhang have laid some of the groundwork for substantial progress in the classification of terrestrial families. However, much more carefully designed and rigorously conducted observational, descriptive, experimental, and analytical research will be necessary to improve the stability and predictiveness of the higher classification of the Parasitengonina. The following account generally reflects the superfamily and family arrangements for aquatic groups proposed by Cook (1974b) and for terrestrial groups proposed by Welbourn (1991), supplemented and modified as necessary based on more recently published information.

Subcohort Hydrachnidiae

The subcohort Hydrachnidiae or water mites (also frequently referred to as Hydrachnida, Hydrachnidia, or Hydrachnellae) is a highly diverse group of Parasitengonina comprising well over 6,000 described species and represented in all regions of the world except Antarctica (Cook 1974b; Viets 1987). At least as many additional species await discovery and description, and new genera and families continue to be reported regularly, even from relatively well-known areas such as the United States and Canada. We estimate that only about half of the species living in North America have been named thus far. Larval water mites can be distinguished morphologically from those of other parasitengonines in having two setae, rather than one, on the genu of the palp (as in Fig. 13.52D). Deutonymphal and adult water mites differ from all other Acari in having a series of paired **glandularia** (Figs. 13.28E, 13.31D) on the idiosoma, each typically consisting of a small gland that opens through the integument on a platelet bearing an associated seta.

The most comprehensive inventories documenting the nomenclature, classification, and geographic occurrence of water mites are found in the monumental bibliographies and species catalogs by Karl Viets (1955a, 1956) and Kurt Viets (1982, 1987) and the synopsis at the generic level by Cook (1974b). Of course, these comprehensive reference works were based on the accumulated efforts of many workers over the previous 150 years that also resulted in the publication of several highly useful regional monographs and major compilations. Perhaps the most notable among these are for the faunas of Switzerland (Walter 1907, 1922a, b; Bader 1975a, b, 1994), the British Isles (Soar and Williamson 1925, 1927, 1929), Sweden (Lundblad 1927a, 1962, 1968), France (Motaş 1928), Spain (Viets 1930), Germany (Viets 1936), Russia (Sokolow 1940), Romania (Şoarec 1942; Motaş and Tanasachi 1946; Motaş et al. 1947, 1958), Hungary (Szalay 1964), the Netherlands (Besseling 1964; Smit and Van der Hammen 2000), West Africa (Viets 1913/14, 1916, 1925; Cook 1966), Central Africa (Viets 1953; Bader 1968; Viets and Böttger 1974a, b); East Africa (Lundblad 1927b, 1952; Walter and Bader 1952), India (Cook 1967), tropical South America (Lundblad 1941, 1942, 1943a, b, 1944; Viets 1954a, b; Cook 1980), austral South America (Cook 1988), Southeast Asia (Walter 1929; Viets 1935; Lundblad 1969, 1971), New Zealand (Cook 1983, 1991, 1992), and Australia (Cook 1986; Harvey 1988a). A new initiative by a consortium of acarologists aims to provide comprehensive treatment of the

Hydrachnidiae and Stygothrombiae, along with semiaquatic groups of Trombidiae and Erythraiae, in central Europe (Gerecke 2007).

An overview of information on the morphology, systematics, evolution, ecology, and behavior of water mites, with an extensive review of the recent literature, was published by Smith, Cook, and Smith (2001). Knowledge of the parasitic associations of water mite larvae with their insect hosts was reviewed by Smith and Oliver (1986) and Smith (1988). Relatively recent comprehensive reviews of water mite biogeography (Cook 1984; Smith and Cook 1999a), courtship and mating (Proctor 1991b, 1992a), and feeding behavior (Proctor and Pritchard 1989) also are available. A wealth of information, along with many references to earlier literature, can be found in these sources. The primary focus of the following discussion is to summarize these accounts and supplement them based on more recent publications. In the taxonomic review presented below, we generally refer to Cook (1974b) for a synthesis of previous information and essential references on classification and distribution, supplemented by records listed in Viets (1987) when appropriate. We attempt to provide more intensive citation of literature only for the period since 1987 but do include references from between 1974 and 1987 that were not listed by Viets or contain new information on life history, phylogeny, or classification. The terminology and notations used in the text and keys follow Smith, Cook, and Smith (2001).

There is general agreement that members of the seven currently recognized superfamilies of water mites belong to a monophyletic clade (Cook 1974b; Viets 1987; Smith, Cook, and Smith 2001), although relationships with terrestrial parasitengonines remain uncertain (see below). An eighth superfamily, the enigmatic aquatic Stygothrombidioidea, appear to be more closely related to water mites than to terrestrial superfamilies, but their phylogenetic status remains unclear. The fossil record of water mites is sparse, comprising two published accounts of larval specimens representing highly derived taxa from Tertiary deposits (Cook 1957; Poinar 1985). Attempts to reconstruct the phylogeny of water mites, therefore, depend on interpretations of transformations in morphological and behavioral characters within the context of biogeographical hypotheses. Knowledge of water mite phylogeny has steadily improved recently as a result of extensive taxonomic work on the fauna of the Southern Hemisphere and cladistic synthesis of information on the various life history instars.

Water mites evolved from terrestrial ancestors, probably at a formative stage of the divergence of Parasitengonina from anystoid stock. They are often assumed to have descended from an unspecified terrestrial group that already possessed the defining attributes of Parasitengonina (Mitchell 1957b; Davids and Belier 1979; Witte 1991b). However, certain extant members of the superfamily Hydryphantoidea exhibit morphological and behavioral attributes that appear to be extremely plesiotypical within Parasitengonina and probably not derived from homologous character states observed in any extant terrestrial parasitengonine group (Smith and Cook 1999a). These attributes apparently represent primitive conditions not only for water mites but also for all Parasitengonina, suggesting that the earliest parasitengonines may have been recognizable water mites. Ancestral water mites may have evolved the characteristic life history pattern of Parasitengonina as a set of adaptations for exploiting spatially discontinuous and temporally intermittent aquatic habitats, such as temporary pools and, possibly, springs (see Wiggins, MacKay, and Smith 1980; Smith and Oliver 1986; Smith and Cook 1999a; Smith, Cook, and Smith 2001). The phylogenetic relationships of water mites with other Parasitengonina require comprehensive reassessment through cladistic analyses of morphological, behavioral, and life history attributes incorporating data for all instars, with reference to Anystoidea as the probable sister group (Smith and Cook 1999a; Wohltmann 2001).

Host associations and distributional data appear to support an ancient origin for water mites. Larvae of the earliest derivative superfamilies parasitize hosts belonging to insect orders that originated and flourished initially during Paleozoic times, and those of more recently derived clades are associated with host groups that differentiated and diversified during the Mesozoic. All superfamilies, except Hydrovolzioidea, appear to be endemic to all zoogeographical regions. We conclude that water mites probably originated during mid-Paleozoic time, and that extant superfamilies had differentiated by the Triassic period. However, the relationships of many family group taxa remain uncertain. As mentioned above, discovery of additional new species and genera will likely continue for some time, and the new information that results will necessitate substantial revision of most families. Knowledge of relationships at all taxonomic levels will improve greatly as larvae and deutonymphs of more taxa are discovered and described.

The sequence of evolutionary events leading to the obligate association of larval parasitengonine mites with a host is a subject for speculation. The plesiotypical condition may have involved facultative, opportunistic feeding by larval mites on arthropods that occasionally resulted in incidental transport of the mites with the host between habitats (Wiggins, MacKay, and Smith 1980; Smith, Cook, and Smith 2001; Wohltmann 2001). For mites adapting to spatially discrete and possibly temporally intermittent habitats, the selective advantage of phoretic transport by a winged host from one suitable site to another would have strongly favored evolution of an obligate relationship. It is interesting to speculate that the early evolution of Parasitengonina may well have occurred during the Permian and early Triassic, when warm, dry conditions prevailed globally, and adaptations promoting reliable, efficient transport between moist habitats would have been an essential prerequisite for survival. Whatever the relationship of water mites to ancestral terrestrial parasitengonine stock may have been, larvae of all extant water mite clades exhibit elegant adaptive mechanisms that integrate their parasitic and dispersal functions (see Smith and Cook 1999a; Smith, Cook, and Smith 2001).

Larval water mites differ markedly in morphology from other instars (Prasad and Cook 1972; Wainstein 1980; Smith 1976b, 1978a, 1979, 1982, 1983f–i, 1984a, 1989d, e, 1992a; I. M. Smith 1989b; Tuzovskij 1977, 1978, 1979, 1981, 1982, 1984b, 1985, 1986c, 1987a, b, 1988b; Wiles 1993a; Martin 2000, 2003; Smith, Cook, and Smith 2001). They are so heteromorphic that reliable association with deutonymphs and adults of the same species can be established only by rearing. With very few exceptions (see Smith 1998), water mite larvae seek out an appropriate insect host and become ectoparasites immediately after emerging from egg membranes. The larva then engorges on host fluids for several hours to a few days, often inducing formation of characteristic feeding tubes, or stylostomes (Smith 2003). During the parasitic phase, the larva is passively transported and typically leaves the parental aquatic habitat with the host (see Smith and Oliver 1986). The fully fed larva drops off the host when the latter returns to water and transforms to a quiescent **nymphochrysalis**, which represents the protonymphal instar. Extensive structural reorganization occurs during this stage to produce the deutonymph, which resembles the adult in being octopod and predaceous. However, the deutonymph is sexually immature and exhibits incomplete sclerotization and chaetotaxy (Tuzovskij 1990). It typically feeds and grows substantially before entering a second quiescent stage, the **imagochrysalis**, representing the tritonymphal instar. The mite completes its metamorphosis during this stage before emerging as an adult. Deutonymphs and adults are active and often are capable of rapid, agile, and protracted locomotion directed toward prey capture and, in adults, location of mates. Shape and degree of sclerotization of the body in these forms are typically modified to promote efficient locomotion in particular habitats. However, movement is limited to the confines of the aquatic habitat, and capabilities for dispersion and potential for dispersal between water bodies are negligible. The dispersal function in water mites resides virtually exclusively in the parasitic phase of the larval stage. All populations of species that have retained parasitic larvae experience dispersal, or the opportunity to disperse, between different habitats during each generation.

With the exception of those parasitic in mollusks, deutonymphs and adults are active predators of aquatic invertebrates, particularly crustaceans and insects (Proctor and Pritchard 1989; Smith, Cook, and Smith 2001), and those of many taxa exhibit both morphological and behavioral adaptations for exploiting particular groups and life stages of prey. Adult water mites display an amazing range of courtship behavior associated with spermatophore transfer (Proctor 1991b, 1992a; Smith, Cook, and Smith 2001). In basal clades of most families, males deposit spermatophores that are subsequently detected by females using pheromonal cues, with no direct interaction between the sexes. Various grades of association of the sexes during spermatophore transfer have evolved repeatedly and independently in derivative clades of many families, including numerous examples of indirect transfer without copulation, indirect transfer with copulation, and direct transfer with intromission. Males of species in which copulation occurs often have the idiosoma and appendages highly modified to assist in grasping and positioning females during elaborate courtship rituals (also see chapter 6).

All water mite superfamilies except Hydrovolzioidea, and many of their early derivative families and subfamilies, are richly represented in all zoogeographical regions. Knowledge of global distribution patterns of water mite taxa has been substantially improved by recent studies of the fauna of austral regions, especially those by Viets, Cook, and Harvey. Basic distribution patterns are the result of vicariance due to plate tectonics (Cook 1986, 1988; Smith and Cook 1999a). Dispersal of mites and their hosts between adjacent land masses (e.g., Southeast Asia and Australia) or across continents (e.g., Africa) significantly altered these patterns as more recent groups progressively displaced ancient ones.

Distributions of water mite taxa were dramatically influenced by climatic changes associated with plate tectonics that resulted in cooling and glaciation at high latitudes and desertification in subtropical latitudes during the late Tertiary and Pleistocene. Distributional limits are frequently correlated with habitat specializations. For example, many interstitial taxa in the Northern Hemisphere have relict Tertiary distributions because their habitats were destroyed in most northern regions of North America and Eurasia by Pleistocene glaciation and will require a long period of climatic stability to become reestablished. As a result, recently diverged sister species may exhibit strikingly discontinuous distributions in temperate areas of North America and Eurasia. In contrast, species adapted to cold stenothermic habitats that have been quickly reestablished in glaciated areas during the Holocene have essentially circumpolar or boreal Holarctic distributions.

Water mites are among the most numerous and diverse groups of arthropods in freshwater habitats throughout the world and have significant impact on the organization and dynamics of freshwater communities as both parasites and predators. They have great potential for use as indicators of water quality and the health of aquatic ecosystems (Schwoerbel 1964; Gerecke and Schwoerbel 1991; Smit and Van der Hammen 1992; Steenbergen 1993; Cantonati, Gerecke, and Bertuzzi 2006).

Preliminary assessment of sister groupings within the Hydrachnidiae suggests that the Hydryphantoidea, Hydrovolzioidea, Eylaoidea, and Hydrachnoidea represent relatively early derivative lineages. The Hydryphantoidea, Hydrovolzioidea, and Eylaoidea may have diverged successively from a persistent "protohydryphantoid" ancestral lineage or originated through a series of dichotomies involving undiscovered sister groups, some of which may be extinct. Species of these three superfamilies retain plesiotypical behavioral attributes, such as aerial larvae that parasitize various ancient insect groups. Their deutonymphs and adults typically inhabit temporary pools, marginal aquatic situations such as wet detritus and mosses in springs, or, in a few cases, interstitial habitats.

Based on tradition and current usage among water mite specialists, the term **genital acetabula** has been used in the following section to identify the ventral osmoregulatory structures usually referred to as genital papillae (see chapter 3).

■ Superfamily Hydryphantoidea (Figs. 13.25I, J, 13.26I, J, 13.27A, 13.28G, H, 13.29A–G, 13.43A, H, I, 13.44A–C, I–P)

*DIAGNOSIS: Adult gnathosoma with palpi typically chelate, with dorsodistal tibial "claw" extending well beyond base of tarsus; tibia longer than genu. Chelicerae 2-segmented. Idiosoma with lateral eye capsules absent or present, small and not borne on dorsal plate; body highly variable in both shape, ranging from ovoid to dorsoventrally flattened or vermiform, and degree of sclerotization, ranging from soft with dorsum bearing only glandularia platelets to heavily armored with dorsum entirely covered by a dorsal shield. Genital field bearing genital acetabula, which are often borne on or beneath movable genital flaps flanking gonopore. Larvae essentially terrestrial. Gnathosoma lacking elaborate camerostome. Idiosoma with lateral eyes borne on separate platelets; integument soft with dorsum either lacking dorsal plate or with small dorsal plate confined to anterior third of body and bearing at least 3 pairs of setae—*ve *and* vi *anteriorly and* si *posterolaterally. Venter with coxal plates all separate; with paired urstigmata laterally between coxal plates I–II; coxal plates III located near midlength of body and laterally directed; excretory pore plate lacking setae or bearing 1–2 pairs of setae (or their alveoli). Legs with 6 movable segments, with basifemur and telofemur separated; tarsus with unmodified claws and empodium.*

The Hydryphantoidea includes the large and diverse family HYDRYPHANTIDAE and five relatively small families of highly modified species. The HYDRYPHANTIDAE (Figs. 13.26I, J, 13.43A, H, I, 13.44A–C, K, M, N) is an ancient assemblage comprising 12 subfamilies, including some of the earliest derivative extant groups of water mites (Smith and Cook 1999a). Deutonymphs and adults of most of these subfamilies are sluggish mites adapted for walking or crawling in aquatic vegetation and detritus. The subfamily Hydryphantinae includes the large genus *Hydryphantes*, with numerous species inhabiting ponds and temporary pools worldwide (Cook 1974b; Özkan 1981, 1982, 1988a; Viets 1987; Gerecke 1997b, 2004c; Smit 1998c; Boyaci and Özkan 2001; Tuzovskij 2003b) and *Hydrobaderia ilicaensis* Özkan from Turkey (Özkan 1985). Deutonymphs and adults are active but awkward swimmers. Those of *Hydryphantes tenuabilis* Marshall fed exclusively on dragonfly eggs in successful laboratory rearings (Lanciani 1979), suggesting that members of the genus utilize aquatic insect eggs as food under natural conditions. Larvae of the marsh inhabiting species *H. tenuabilis* parasitize hydrometrid bugs (Lanciani 1971), while those of *H. ruber* (DeGeer), and possibly other temporary pool species, use ephydrid flies as hosts (Smith and Oliver 1986). The Wandesiinae is a subfamily of vermiform, mainly hyporheic mites, with the subgenus *Wandesia* (*Pseudowandesia*) appearing to be the most basal clade with representatives in Australia, austral South America, and throughout the Northern Hemisphere (Cook 1974b, 1986; Orghidan and Georgesco 1977; Viets 1987; Gerecke 1997b). Larvae of many species of *Wandesia* are parasites of stone flies (Plecoptera) (Smith and Oliver 1986). Members of the subgenus *Partnuniella* are known from both North and South America, and some species, such as *W. thermalis* Viets from the western United States and *W. chechoi* Cook from Chile, live in hot spring habitats (Cook 1974b, 1988; Viets 1987). In a hot spring system in Yellowstone National Park (Wyoming, USA), larvae of *W. thermalis* parasitize adult brine flies (Ephydridae), and the adult mites feed on the eggs of these flies (Collins 1975). Larval success in locating hosts is apparently only 25%, primarily because the distribution of hosts is correlated with sporadic and highly clustered blooms of the algal-bacterial mats on which their larvae feed (Collins, Mitchell, and Wiegert 1976). Species of other subgenera of *Wandesia* have relatively restricted distributions in Eurasia and North America (Cook 1974b; Viets 1987; Tuzovskij 1987b, 1988b; Gerecke 1997b) suggesting Laurasian origin, while species of the Gondwanan genus *Euwandesia* occur in Australia, New Zealand, and southern South America (Cook 1974b, 1986, 1988; Viets 1987). The Tartarothyadinae includes only the ancient genus *Tartarothyas*, comprising several species living in springs and streams in temperate Eurasia and North America and Australia (Cook 1974b; Viets 1987; Harvey 1987; Gerecke 1997b; Smith and Cook 1999d). The small subfamily Pseudohydryphantinae includes the genera *Pseudohydryphantes*, with a few described species known from ponds and lakes in northern Eurasia and North America (Cook 1974b; Viets 1987; Gerecke 1997b; Tuzovskij 2001d), Australia (Cook 1986; Harvey 1987, 1988a; Smit 1998c), and New Zealand (Cook 1983), and *Cyclohydryphantes*, known only from Australia (Cook 1974b, 1986; Harvey 1987; Smit 1998c). Deutonymphs and adults of this subfamily are unusual among hydryphantid mites in that they are active and agile swimmers.

The hydryphantid subfamily Protziinae comprises numerous stream-inhabiting species in three genera (Cook 1974b). Members of *Protzia* occur throughout temperate areas of Eurasia and North America as far south as Costa Rica (Cook 1974b, 1980; Özkan 1982; Viets 1987; Gerecke 1997a, b, 1999a; Goldschmidt and Gerecke 2003), species of the closely related *Partnunia* are distributed in Eurasia and western North America (Cook 1974b; Viets 1987; Gerecke 1992b; Gerecke and Smit 1994; Panesar and Gerecke 1994;

Gerecke 1997b), and those of *Neocalonyx* live in austral and alpine areas of South America and as far north as Costa Rica (Cook 1974b, 1980; Viets 1987; Goldschmidt and Gerecke 2003). The Eupatrellinae is a monobasic subfamily including two described species from Africa: one from northern South America and one from Costa Rica (Cook 1974b; Viets 1987; Goldschmidt and Gerecke 2003). Another monobasic subfamily, the Diplodontinae, is represented by only a few species of the genus *Diplodontus* inhabiting ponds in southern Eurasia, Africa, and Australia (Cook 1974b; Viets 1980, 1987; Gerecke 1997b, 2004a, c). The Mamersinae is another monobasic subfamily comprising several species of a single genus, *Mamersa*, that live in ponds in tropical areas of Eurasia, Africa, and Australia (Cook 1974b; Viets 1987; Harvey 1988a; Gerecke 2004c). The subfamily Ankelothyadinae is based on a single specimen of *Ankelothyas emydoides* Besch, with distinctively modified dorsal plates and anterior legs, collected in a stream in Chile (Cook 1974b, 1988; Viets 1987). The Cyclothyadinae includes a small number of species of the genus *Cyclothyas* reported from springs and streams in highly disjunct areas of southeastern Eurasia and southwestern North America (Cook 1974b; Smith 1983b; Viets 1987), while the monobasic subfamily Cowichaniinae is based on the highly unusual species *Cowichania interstitialis* Smith, which lives in springs on the west coast of North America (Smith 1983e; Viets 1987). The Chimerothyadinae contains only the recently discovered species *Chimerathyas cooki* Mitchell, a highly specialized resident of temporary pools in Ohio, USA, in which the larval instar is suppressed (Mitchell 2003).

The subfamily Euthyadinae is a large and probably paraphyletic subfamily that includes numerous species in over 20 genera living in springs, streams, and temporary pools throughout Eurasia and North America, with a few species each in Africa, Central and South America, and Australia (Cook 1974b, 1986, 1988; Özkan and Bader 1988; Viets 1987; Harvey 1988b, 1996b; Smith and Cook 1998b, 1999b, c; Gerecke 1997b, 1999a; Mitchell 2003). Özdikmen (2006) has determined that the generic name *Thyas* Koch, 1835 (Acari: Hydryphantidae) is a junior homonym of *Thyas* Hübner, 1824 (Lepidoptera: Noctuidae). He has proposed the replacement names *Acerbitas* for *Thyas* Koch, 1835, and Euthyadinae Viets, 1931, for Thyadinae Viets, 1926. Species of the *Acerbitas*-like genera and the *Panisus*-like genera have boreal distributions in Eurasia and North America, suggesting Laurasian origin (Cook 1974b; Viets 1987; Tuzovskij 1994a, 2000a, 2005, 2007; Gerecke 1997b; Smith and Cook 1999a). Members of several other Laurasian clades, such the *Tadjikothyas* + *Japonothyas* + *Columbiathyas* group (Cook 1974b; Viets 1987; Özkan 1988b; Gerecke 1997b; Smith and Cook 1998b) and the *Thyasella* + *Panisellus* + *Parathyasella* + *Amerothyasella* group (Cook 1974b; Smith and Cook 1999b), exhibit highly disjunct distributions in southern areas of Eurasia and North America resulting from fragmentation of more continuous distributions during the Tertiary by Pleistocene glaciation (Smith and Cook 1999a). Certain genera of the *Trichothyas*-like group also have southern Holarctic distributions, although some of them are also represented in Africa (Cook 1974b; Özkan 1982; Viets 1987; Özkan and Boyaci 1992d; Gerecke 1997b). Although superficially similar to basal euthyadine genera, the genus *Notopanisus* has an anomalous distribution for this subfamily, with species reported from mountain streams in austral South America (Cook 1974b, 1980), Australia (Harvey 1988a), and New Zealand (Smit 1996b), suggesting Gondwanan origin. This genus may not be a euthyadine, and its systematic placement remains uncertain. Deutonymphs and adults of both *Acerbitas barbigera* (Viets) and *A. stolli* (Koenike) were observed to feed on mosquito eggs under laboratory conditions (Mullen 1975b), suggesting that mites of this subfamily may use insect eggs as food in their natural habitats. Adults of species inhabiting vernal temporary pools, such as *A. barbigera*, endure drought in the pool basin for up to 10 months during the summer, autumn, and winter (Wiggins, MacKay, and Smith 1980). Mitchell (2003) reported that deutonymphs and two cohorts of adults are represented in overwintering populations of this species in Ohio, USA. *Trichothyas muscicola* Mitchell has large and thick idiosomatic sclerites to anchor the massive leg muscles needed for crawling through mats of wet vegetation (Mitchell 1953, 1957a), as do many other euthyadine species living in springs and seepages. Euthyadine larvae parasitize various aquatic Hemiptera, Diptera, and Trichoptera and also opportunistically use terrestrial hosts of the orders Collembola, Thysanoptera, and Homoptera that regularly visit wet vegetation and litter at the edges of aquatic habitats (Smith and Oliver 1986). Mitchell (1953) observed clusters of more than 100 recently emerged larvae of *T. muscicola* attending pupae of the tipulid fly *Limonia humidicola* before attaching to the adults at ecdysis.

The family HYDRODROMIDAE (Figs. 13.27A, 13.44L, O, P) is a monobasic hydryphantoid family represented by the genus *Hydrodroma*, with several species inhabiting streams, ponds, and lakes worldwide (Cook 1974b, 1986; Özkan 1982; Wiles 1985, 1986; Viets 1987; Gerecke 2004c). Deutonymphs and adults are active but awkward swimmers, and those of the widespread species *H. despiciens* (Müller) reportedly feed on chironomid eggs (Wiles 1982). Larvae parasitize nematocerous Diptera, especially Chironomidae (Meyer 1985; Smith and Oliver 1986; Wiles 1987).

The TERATOTHYADIDAE (Figs. 13.28G, H) is a small family comprising the genera *Teratothyas* and *Teratothyasides*, which are known from streams in Africa and Sumatra (Cook 1974b; Viets 1987). The RHYNCHOHYDRACARIDAE (Figs. 13.29D, E) includes three subfamilies: Rhynchohydracarinae, with a few species of *Rhynchohydracarus* from streams in South America (Cook 1974b; Viets 1987; Valdecasas 2001a); Clathrosperchontinae, comprising several species of *Clathrosperchon* from streams in South and North America and a few species of *Clathrosperchonella* from streams in tropical South America (Cook 1974b, 1980; Viets 1987); and Santiagocarinae, including members of the genera *Santiagocarus* and *Gledhillia* recently described from Panama (Valdecasas 2001a).

The CTENOTHYADIDAE (Figs. 13.29A–C) is a small family of highly unusual mites including *Ctenothyas verrucosa* Lundblad from Java (Cook 1974b; Viets 1987) and two species of *Stellulathyas* from New Zealand (Cook and Hopkins 1998). Adults of *S. magnifica* Cook and Hopkins and *S. lundbladi* Cook and Hopkins were collected in mosses and grasses in essentially terrestrial habitats and have not thus far been found in water. They appear to be more adapted for terrestrial living than any other water mites. The peculiar large, stellate sclerites with recumbent spines associated with their glandularia may function to retain moisture (Cook and Hopkins 1998).

The THERMACARIDAE (Figs. 13.29F, G, 13.44I, J) is a monobasic family containing a few large, heavily sclerotized species of the unusual genus *Thermacarus* inhabiting hot springs in central Eurasia, western North America, and austral South America (Cook 1974b; Viets 1987; Schwoerbel 1987; Martin and Schwoerbel 2002). Adults of this genus are atypical among hydryphantoid mites in exhibiting striking sexual dimorphism, as illustrated by Baker (1985) for *T. nevadensis* Marshall. Deutonymphs of *T. nevadensis* grow to five times their original size during their development, involving differential expansion of the dorsal and ventral shields that results in substantial reorganization of the body musculature (Mitchell 1962a). Larvae of the South American species *T. andinus* Martin and Schwoerbel are highly unusual in parasitizing a species of toad, *Bufo spinulosus* (Schwoerbel 1987; Martin and Schwoerbel 2002). The hosts of larvae of other species of *Thermacarus* are unknown, possibly because amphibians breeding in their hot spring habitats have not been considered as likely candidates.

Zelandothyas and *Australiothyas*, two austral genera of the hydryphantoid family ZELANDOTHYADIDAE (Figs. 13.25I, J) discovered in hyporheic habitats, have been tentatively placed in the subfamilies Zelandothyadinae and Australiothyadinae, respectively (Cook 1983, 1986; Harvey 1998a). They exhibit combinations of attributes found in Hydryphantoidea and Eylaoidea and have not yet been assigned to a superfamily.

- Superfamily Eylaoidea (Figs. 13.26A, B, D–H, 13.44D–H, 13.45A–F)

DIAGNOSIS: Adult gnathosoma with palpi not chelate, usually with 5 movable segments and tibia longer than genu, rarely with distal segments fused into a single movable segment flexing medially to oppose anterior edge of capitulum. Chelicerae 2-segmented. Idiosoma with lateral eye capsules usually borne on medial prodorsal plate, rarely on rugose paired platelets separated by a triangular medial prodorsal platelet; body variable in shape, ranging from roughly ovoid to vermiform, and degree of sclerotization, ranging from soft with dorsum bearing only prodorsal plate and glandularia platelets to heavily armored with dorsum entirely covered by a series of paired rugose platelets. Genital field bearing genital acetabula borne on plates or directly on ventral integument. Larva usually essentially terrestrial, rarely aquatic. Gnathosoma lacking elaborate camerostome. Idiosoma with lateral eyes borne on fused platelets; dorsum with moderately large to large dorsal plate covering well over anterior third of body and bearing setae si *near midlength. Venter with coxal plates all separate; with paired urstigmata laterally between coxal plates I–II; coxal plates III located near midlength of body and laterally directed; excretory pore plate bearing 1 pair of setae. Legs with 6 movable segments, with basifemur and telofemur separated; tarsus with claws modified or reduced.*

The Eylaoidea is an ancient and possibly polyphyletic group comprising the EYLAIDAE (Figs. 13.26E, H, 13.44D, H, 13.45A, B) and three additional families of relatively large water mites. Numerous species of the eylaid genus *Eylais* inhabit ponds and temporary pools worldwide (Cook 1974b; Özkan 1982; Viets 1987; Gerecke 2004c; Davids et al. 2005). Deutonymphs and adults feed on ostracod and cladoceran crustaceans, and larvae are subelytral parasites of various aquatic Coleoptera and Hemiptera (Smith 1986; Smith and Oliver 1986). Females of *Eylais* produce large numbers of eggs, with those of *E. discreta* Koenike typically laying between 10,000 and 13,000 each (Davids 1973b). Larvae of species that breed in vernal temporary pools in temperate latitudes spend up to 10 months in diapause, attached to their hosts in permanent water habitats, while the vernal pools are dry during the summer, autumn, and winter (Lanciani 1970; Wiggins, MacKay, and Smith 1980; Smith, Cook, and Smith 2001). The water boatman species *Cenocorixa expleta* thrives in highly saline lakes in British Columbia but is excluded from lakes with lower salinity because it is unable to survive parasitism by larvae of the species of *Eylais* that breed in these habitats (Scudder 1983; Smith 1988).

The subfamily Limnocharinae (LIMNOCHARIDAE; Figs. 13.26F, 13.44E, G, 13.45D) contains the genera *Limnochares* and *Neolimnochares*, both of which have species inhabiting streams, ponds, and lakes in various parts of the world, including temperate Eurasia and North America, Africa, and Australia (Cook 1974b, 1986; Viets 1987; Harvey 1990c; Gerecke 2004c; Smith and Cook 2005a, b; Davids et al. 2005), with one species of the latter genus reported from Cuba (Orghidan and Georgesco 1977). Adult *Limnochares* feed on larval Chironomidae, and larvae parasitize surface-dwelling aquatic Hemiptera, especially Gerridae, and various Odonata (Smith and Oliver 1986). Larval parasitism by *L. aquatica* (Linnaeus) has significant impact on the survival of juvenile water striders of the genus *Gerris*, reducing recruitment and increasing variation in age structure within populations of host species (B. P. Smith 1989b). The subfamily Rhyncholimnocharinae includes several species of *Rhyncholimnochares* living in streams in South America and eastern North America (Cook 1974b; Viets 1987) and *Austrolimnochares womersleyi* (Lundblad) from streams in Australia (Cook 1974b, 1986; Harvey 1998b). Larvae of *Rhyncholimnochares kittitanniana* Habeeb are unusual among eylaoid mites in being fully aquatic parasites of riffle beetles (Elmidae) (B. P. Smith 1989a).

The eylaoid family PIERSIGIIDAE (Figs. 13.26D, G, 13.44F, 13.45C) is a small assemblage of species grouped in two monobasic subfamilies. Species of *Piersigia* (Piersigiinae) live mainly in temporary pools and springs in temperate Eurasia and North America (Cook 1974b; Viets 1987; Tuzovskij 1986c), and *Austrapiersigia montana* Smit was collected from interstitial water in Australia (Smit 1998a). The unusual species *P. crusta* Mitchell lives in interstitial habitats in the central United States and was first collected from the outflow of tile drains in Illinois (Mitchell 1955b). Adults of *Piersigia* are relatively well sclerotized compared to other eylaoid mites, and larvae are subelytral parasites of Hydrophilidae (Coleoptera) (Imamura and Mitchell 1967; Smith and Oliver 1986). The subfamily Stygolimnocharinae includes only two species of *Stygolimnochares*, known from hyporheic interstitial habitats in India and Australia (Cook 1974b, 1986; Viets 1987), and *Parawandesia chappuisi* Angelier from similar situations in southern Europe (Gerecke and Cook 1995). Adults of these genera are soft bodied and vermiform. The three recently described species of *Apheviderulix* (APHEVIDERULICIDAE; (Figs. 13.26A, B, 13.45E, F) occur in springs and streams in widely separated areas of southern Eurasia and North America (Gerecke, Smith, and Cook 1999). Only the highly distinctive larvae and heavily sclerotized deutonymphs have been collected thus far, suggesting that adults may live only in deep interstitial habitats. Larvae are subelytral parasites of Dytiscidae (Coleoptera).

- Superfamily Hydrovolzioidea (Figs. 13.25C, D, 13.43C–G, J)

*DIAGNOSIS: Adult gnathosoma with palpi not chelate; tibia longer than genu. Chelicerae 2-segmented. Idiosoma with lateral eye capsules absent or present and not borne on dorsal plate; body oval or fusiform, dorsoventrally flattened; well sclerotized, with dorsum bearing large anterior and posterior unpaired plates surrounded by 5 or 10 pairs of paired platelets. Genital field with movable genital flaps flanking gonopore; genital acetabula numerous and borne on posterior coxal plates. Larvae essentially terrestrial or aquatic. Gnathosoma with elaborate camerostome enclosing chelicerae and with palpi inserted ventrally. Idiosoma with lateral eyes borne on fused platelets; integument soft with dorsum bearing small dorsal plate confined to anterior third of body and bearing only two pairs of setae—*ve *and* vi *anteriorly. Venter with paired rows of small urstigmata between coxal plates I–II or with urstigmata absent; coxal plates III located posteriorly on body and posteriorly directed. Legs with 6 movable segments, with basifemur and telofemur separated; tarsus with unmodified claws and empodium.*

The Hydrovolzioidea comprises two small families of highly distinctive, well-sclerotized mites (Tuzovskij, Benefatti, and Gerecke 2001; Smith, Cook, and Smith 2001). The HYDROVOLZIIDAE (Figs. 13.25D, 13.43D, E, J) includes several species of *Hydrovolzia* that live in springs and cold brooks in southern areas of Eurasia (including India and Indochina) and North America (Cook 1974b; Özkan 1982; Viets 1987), *Hydrovolziella lata* Walter from a waterfall in West Africa (Cook 1974b; Viets 1987), and *Stygovolzia uenoi* Imamura from interstitial waters in Japan (Cook 1974b; Viets 1987). The cryptically placed genital acetabula of *Hydrovolzia* species were overlooked until recently discovered on the posterior pairs of coxal plates of *H. placophora* (Monti) using scanning electron microscopy (Alberti and Bader 1990). Larvae of *Hydrovolzia* are terrestrial. Those of *H. gerhardi* Mitchell reportedly parasitize Mesoveliidae (Hemiptera) (Mitchell 1954), while those of other North American species commonly occur on Empididae (Diptera) (I. M. Smith, pers. obs.). Members of the two described genera of ACHERONTACARIDAE (Figs. 13.25C, 13.43C, F, G) inhabit interstitial and seepage habitats. Several species of *Acherontacarus* are known from southern Europe (Cook 1974b; Viets 1987; Gerecke 1999b; Tuzovskij, Benefatti, and Gerecke 2001), and an undescribed species has recently been collected in central North America. Known larvae are aquatic and emerge from eggs nearly as large as adults, suggesting that the parasitic phase may be suppressed. The two known species of *Bharatovolzia* occur in India (Cook 1967, 1974b) and Iran (Schwoerbel and Sepasgosarian 1980).

- Superfamily Hydrachnoidea (Figs. 13.25E–G, 13.43B, 13.45G–I)

*DIAGNOSIS: Adult gnathosoma typically with long, curved, pointed rostrum. Palpi chelate, with dorsodistal tibial claw extending well beyond base of tarsus; genu longer than tibia. Chelicerae 1-segmented and styletiform. Idiosoma with lateral eye capsules present and not borne on dorsal plate; body nearly spherical, variable in degree of sclerotization, ranging from soft with dorsum bearing only glandularia platelets to heavily armored with dorsum entirely covered by a dorsal shield. Genital field heart shaped in males and rounded in females, with gonopore located posteriorly; bearing numerous genital acetabula borne on plates flanking gonopore; with conspicuous tubular ovipositor in females. Larvae aquatic. Gnathosoma lacking elaborate camerostome. Idiosoma with lateral eyes borne on dorsal plate; dorsum with large dorsal plate bearing 8 pairs of setae, including verticils, scapulars, *c3*, and 3 additional pairs of hysterosomatic setae. Venter with coxal plates all separate; with paired urstigmata laterally between coxal plates I–II; coxal plates III located near midlength of body and laterally directed; excretory pore plate lacking setae. Legs with 5 movable segments,* *with basifemur and telofemur fused; tarsus with claws lacking but empodium present.*

The large, monobasic family HYDRACHNIDAE (Figs. 13.25E–G, 13.43B, 13.45G–I) includes many species of *Hydrachna* inhabiting temporary pools, ponds, and lakes worldwide (Cook 1974b; Özkan 1982; Viets 1987; Gerecke 2004c; Davids et al. 2005). Adults of *H. conjecta* Koenike feed on eggs of the water boatman species *Sigara striata* laid in submerged stems of aquatic plants (Davids 1973b). Females of *Hydrachna* also oviposit in plant stems. Those of *H. conjecta* use their styletiform chelicerae to penetrate the epithelium and their uniquely developed ovipositor to lay up to 1,500 eggs in intracellular air spaces (Davids 1973b). Larval *Hydrachna* are parasites of various aquatic Hemiptera and Coleoptera (Smith and Oliver 1986; Smith 1987). Parasitic larvae of *H. conjecta* reduce fecundity and survival of *S. striata* and other corixid hosts (Davids 1973b; Davids and Schoots 1975). Those of *H. virella* Lanciani have similar effects on the notonectid *Buenoa scimitra* (Lanciani 1982) and also retard growth and development of this host (Lanciani 1980). As in the EYLAIDAE, larvae of *Hydrachna* species that breed in vernal temporary pools in temperate latitudes spend up to 10 months in diapause attached to their hosts in permanent water habitats while the vernal pools are dry during the summer, autumn, and winter (Wiggins, MacKay, and Smith 1980; Smith, Cook, and Smith 2001).

The remaining three hydrachnid superfamilies (Lebertioidea, Hygrobatoidea, and Arrenuroidea) appear to represent a monophyletic lineage that diverged from hydryphantoid stock. Species of these clades have aquatic larvae that predominantly parasitize Diptera (Smith and Oliver 1986) and exploit a remarkably broad array of freshwater habitats.

- Superfamily Lebertioidea (Figs. 13.26C, 13.27B, D–J, 13.28A–F, 13.46A–H, 13.47A–F, 13.48A–H, 13.49A–C)

DIAGNOSIS: Adult gnathosoma with palpi not chelate; tibia usually longer than genu. Chelicerae 2-segmented. Idiosoma with lateral eyes absent or present, small and not borne on dorsal plate; body highly variable in both shape, ranging from ovoid to dorsoventrally flattened or laterally compressed, and degree of sclerotization, ranging from soft with dorsum bearing only glandularia platelets to heavily armored with dorsum entirely covered by a dorsal shield. Genital field bearing genital acetabula borne in gonopore beneath movable genital flaps. Larvae aquatic. Gnathosoma lacking elaborate camerostome. Idiosoma with lateral eyes borne on separate platelets; dorsum covered by large dorsal plate bearing at least 4 pairs of setae including verticils and scapulars and rarely 1 or more pairs of hysterosomatic setae. Venter with coxal plates all separate or variously fused; with paired urstigmata laterally between coxal plates I–II; coxal plates III with insertions of legs III located near midlength of body and laterally directed; excretory pore plate bearing 2–3 pairs of setae. Legs with 5 movable segments, with basifemur and telofemur fused; tarsus with unmodified claws and empodium.

Based on morphology, the families STYGOTONIIDAE, SPERCHONTIDAE, RUTRIPALPIDAE, and TEUTONIIDAE appear to be early derivative clades of Lebertioidea. Their relatively soft-bodied adults (as well as known larvae and deutonymphs) exhibit traits that appear to be intermediate between the plesiomorphic conditions found in Hydryphantoidea and the striking synapomorphies that characterize major derivative clades such as LEBERTIIDAE, OXIDAE, TORRENTICOLIDAE, and the various lineages within ANISITSIELLIDAE (see figure citations below).

The monobasic family STYGOTONIIDAE (Fig. 13.26C) is represented by the highly unusual species *Stygatonia ambigua* Cook living in interstitial habitats in New Zealand (Cook 1992). The SPERCHONTIDAE (Figs. 13.27B, 13.28E, 13.46A, B, D, F, G) includes two subfamilies, one of which is highly speciose. The Sperchontinae includes many species of *Sperchon* and *Sperchonopsis* living mainly in springs and streams throughout Eurasia and North America, with a few species of *Sperchon* known from Africa and tropical South America (Cook 1974b, 1980; Özkan 1982, 1989; Viets 1987; Smit 1995; Tuzovskij 2002b, e). Larvae are parasites of various nematocerous Diptera and Trichoptera (Smith and Oliver 1986). Some species of *Sperchon* commonly parasitize black flies (Simuliidae) and may affect their fecundity (Gledhill, Cowley, and Gunn 1982). The subfamily also includes several species of *Notosperchonopsis* and *Illesiella* from springs and streams in austral South America (Cook 1974b, 1988; Viets 1987). Two species of Apeltosperchontinae are known from stream habitats: one from Chile (Cook 1988) and the other from New Zealand (Cook 1983). This disjunct distribution is probably an example of vicariance following the breakup of Gondwanaland (Smith and Cook 1999a).

The RUTRIPALPIDAE (Figs. 13.27D–F) is a small monobasic family with two described species of *Rutripalpus* known from springs in southern Europe and eastern North America, respectively (Cook 1974b; Smith 1991a; Gerecke and Tuzovskij 2001). A recent study of all active instars of the Palearctic species *R. limicola* Sokolow suggests that this genus may be related to both TEUTONIIDAE and certain genera currently placed in ANISITSIELLIDAE. Unengorged larvae of this species are phoretic on larval Ptychopteridae (Diptera), suggesting that they may parasitize adults of this group (Gerecke and Tuzovskij 2001). The TEUTONIIDAE (Figs. 13.27G, 13.47A, B) contains several species of *Teutonia* from springs and streams in northern areas of Eurasia and North America (Cook 1974b; Viets 1987) and *Limnolegeria longiseta* Motaş known only from a lake in southern France and streams in Sicily (Cook 1974b; Gerecke 1988a). Adults of several species of *Teutonia* are unusual among lebertioid water mites in being rapid and agile swimmers. Larvae of *T. lunata* Marshall are parasites of tanypodine Chironomidae (Diptera) (Smith 1982).

The family LEBERTIIDAE (Figs. 13.28B–D, 13.47C–E) comprises several genera, including *Lebertia*, which is represented by many species throughout Eurasia and North America and a few in Africa and tropical South America (Cook 1974b; Özkan 1982; Viets 1987; Tuzovskij 2006a). Deutonymphs and adults of the various species of *Lebertia* inhabit an impressive array of surface water habitats, including springs, streams, ponds, and lakes. Adults of most species are walkers, but those of a few are competent swimmers. Larvae of this genus parasitize Chironomidae (Diptera) (Smith and Oliver 1986). The family also includes a few species of laterally compressed mites in the genus *Estelloxus* (Cook 1974b; Smith 1982; Viets 1987) and the dorsoventrally flattened and relatively heavily sclerotized species *Scutolebertia trinitensis* Smith (Smith 1991a), all from springs and streams in western North America. Numerous species of the genera *Oxus* and *Frontipoda* (OXIDAE; Figs. 13.28A, 13.47F, 13.48A, B) inhabit pools, ponds, and lakes worldwide (Cook 1974b; Viets 1987; Tuzovskij 1984b, 2002c, d, 2006c), and several species of the austral oxid genus *Flabellifrontipoda* live in similar habitats in South America, New Zealand, and Australia (Cook 1974b, 1986, 1988; Viets 1987). Adult oxids are slightly to greatly compressed laterally and are typically strong swimmers. Known larvae are parasites of Chironomidae (Smith and Oliver 1986). The unusual species *Acucapito naso* Wiles described from Brunei (Wiles 1996) is mentioned here pending further study to clarify its relationship to members of the families LEBERTIIDAE and OXIDAE.

The lebertioid family TORRENTICOLIDAE (Figs. 13.27H–J, 13.48C–G) includes three subfamilies of heavily sclerotized, dorsoventrally flattened mites living mainly in springs and streams, although a few species occur in lakes and interstitial habitats (Cook 1974b; Viets 1987). Known larvae of this family parasitize Chironomidae (Smith and Oliver 1986). The subfamily Torrenticolinae includes many species of the genera *Torrenticola* and *Monatractides*, along with a few species of *Pseudotorrenticola* (Cook 1974b). Most torrenticoline species are found in temperate areas of Eurasia and North America (Cook 1974b; Özkan 1982; Viets 1987; Wiles 1989b, 1991b, 1997a; Smith 1991a; Cramer 1992b; Özkan and Boyaci 1992c; Gerecke 1996, 2004b; Gerecke and Di Sabatino 1996; Tuzovskij 2000b; Cramer and Cook 2000), but some are known from Africa (Cook 1966; Viets 1987), the Americas (Cook 1980; Viets 1987; Goldschmidt 2007), and Australia (Cook 1986). The subfamily Testudacarinae includes numerous species of *Testudacarus* from temperate areas of Eurasia and North America (Cook 1974b; Viets 1987; Cramer 1992b; Wiles 1997a). The Neoatractidinae, comprising several species of *Neoatractides*, was recently substantially revised by Wiles (1997a). Species of the nominate subgenus are distributed from tropical South America to the southwestern United States (Cook 1974b; Viets 1987; Cramer 1992b; Cramer and Cook 2000), while members of the subgenus *Heteratractides* occur in South America (Cook 1974b; Viets 1987) and Southeast Asia (Wiles 1997a), and species of *Allotorrenticola* are known from India (Cook 1967) and Southeast Asia (Wiles 1997a).

The family ANISITSIELLIDAE (Figs. 13.28F, 13.46C, E, H, 13.48H, 13.49A–C) is a large paraphyletic (and possibly polyphyletic) group of derivative lebertioid taxa, including some that appear to represent the sister groups of clades currently placed in other families and superfamilies (Cramer and Smith 1993; Smith and Cook 1999a). This was recognized by Panesar (2004), but his proposals for revision of the group were not well supported by rigorous and comprehensive analysis of phylogenetic relationships among anisitsiellid taxa and their supposed relatives in other lebertioid families and the hygrobatoid family LIMNESIIDAE. We consider many of his conclusions to be problematic and follow a more conservative family-level classification here pending more extensive and inclusive revisionary studies. Known hosts of larvae are various nematocerous Diptera (Smith 1982; Smith and Oliver 1986). There are currently two subfamilies, with the Anisitsiellinae comprising a mixed assemblage of genera representing several distinct lineages worldwide (Cook 1974b; Viets 1987; Orghidan and Gruia 1987; Cramer and Smith 1993; Harvey 1990b; Gerecke 2004c). One possibly holophyletic group of well sclerotized, dorsoventrally flattened anisitsielline mites includes the genera *Bandakia*, *Utaxatax*, and *Bandakiopsis*, each with groups of related species inhabiting springs and both surface and hyporheic waters of streams in highly disjunct areas of Eurasia and North America (Cook 1974b; Smith 1979, 1982; Viets 1987; Cramer and Smith 1991; Gerecke 1988b, 1991b, 1994; Panesar 2004), and the monobasic genera *Cookacarus* and *Oregonacarus* currently known only from similar habitats in western North America (Barr 1977; Smith 1979; I. M. Smith 1989a). Other probably monophyletic anisitsielline clades have austral distributions exhibiting vicariance resulting from the fragmentation of Gondwanaland. For example, the stream-inhabiting genus *Anisitsiellides*, with species in austral South America, New Zealand, and Australia, is probably the sister genus of *Australiotonia*, from interstitial habitats in Australia (Cook 1983, 1986, 1988). The recently described *Rubicundula placibilis* Gerecke from Ecuador is unusual for this family in having large numbers of genital acetabula (Gerecke 1995). The anisitsielline genus *Mamersellides*, which occurs in tropical South America and southern North America, appears to be most closely related to certain genera in the subfamily Nilotoniinae (Cramer and Smith 1993; Gerecke and Smith 1993), and species of *Psammotorrenticola* and *Stygomamersopsis* from interstitial habitats in southwestern Eurasia (Cook 1974b; Viets 1987; Gerecke 1994) appear to be allied to early derivative groups of the hygrobatoid family LIMNESIIDAE (see below). The Nilotoniinae includes several species of *Nilotonia* and related genera from a wide range of habitats throughout tropical and warm temperate areas of the world (Cook 1974b; Özkan 1982; Viets 1987; Orghidan and Gruia 1987;

Özkan and Bader 1988; Özkan and Soysal 1989; Özkan and Boyaci 1990; Gerecke 1992a, 1994, 2004b; Özkan and Küçükönder 1993). These taxa also appear to have close affinities with LIMNESIIDAE. Comprehensive revision of the ANISITSIELLIDAE and related taxa based on phylogenetic proposals derived from analysis of morphological and behavioral characters of all active instars remains a priority.

- Superfamily Hygrobatoidea (Figs. 13.25A, B, H, 13.27C, 13.29H, I, 13.30A, B, G–I, 13.31B–G, 13.32A–C, 13.36E–I, 13.37A–K, 13.52A–I, 13.53A–J, 13.54A–H, 13.55A–L)

DIAGNOSIS: Adult gnathosoma with palpi not chelate; tibia longer than genu. Chelicerae 2-segmented. Idiosoma with lateral eyes absent or present, small and not borne on dorsal plate; body highly variable in both shape, ranging from ovoid to dorsoventrally flattened or laterally compressed, and degree of sclerotization, ranging from soft with dorsum bearing only glandularia platelets to heavily armored with dorsum entirely covered by a dorsal shield. Genital field bearing genital acetabula on plates flanking gonpore. Larvae aquatic. Gnathosoma lacking elaborate camerostome. Idiosoma with lateral eyes borne on separate platelets; dorsum covered by large dorsal plate bearing at least 4 pairs of setae including verticils and scapulars and rarely 1 or more pairs of hysterosomatic setae. Venter with coxal plates variously fused, with plates II–III on each side at least fused medially; with paired urstigmata laterally between coxal plates I–II; coxal plates III with insertions of legs III located near midlength of body and laterally directed; excretory pore plate bearing 2–4 pairs of setae. Legs with 5 movable segments, with basifemur and telofemur fused; tarsus with unmodified claws and empodium.

The PONTARACHNIDAE (Figs. 13.25A, B) is a small family with several species in two genera: *Pontarachna* and *Litarachna* (Cook 1974b; Viets 1987). These mites live mostly in littoral marine habitats worldwide, but two species are known from freshwater (Cook 1996). The family LIMNESIIDAE (Figs. 13.27C, 13.30A, B, 13.52A, B) is an enigmatic assemblage that probably represents a paraphyletic group with close affinities to certain genera of ANISITSIELLIDAE. A world revision of this family and its close relatives is a high priority in water mite systematics. Eleven subfamilies, most of which are monobasic, are currently recognized. The Limnesiinae includes many species of *Limnesia* and several closely allied genera from streams, ponds, and lakes worldwide (Cook 1974b, 1980, 1986, 1988; Özkan 1982; Viets 1987; Fernandez 1992; Smit 1994, 1998d; Tuzovskij 1997; Wiles 1997c; Gerecke 2004b, c). Known larvae are parasites of Chironomidae (Smith and Oliver 1986). Species of *Limnesia* are currently arranged in 11 subgenera that do not reflect natural species groupings and tend to obscure relationships (Smith and Cook 1999a). A number of limnesiine genera are endemic to Australia, including *Heterolimnesia*, *Physolimnesia*, and *Timmsilimnesia* (Cook 1986; Viets 1987; Harvey 1998a), while all known species of *Fijilimnesia* are endemic to Viti Levu, Fiji (Smit 2002b). Members of the austral genus *Tubophorella* exhibit vicariance, with one group of related species living in Australia and another in tropical areas of South and North America (Cook 1980, 1986; Viets 1987; Fernandez 1992). Most species of *Centrolimnesia* are confined to South America (Cook 1980; Viets 1987; Gerecke, Fisher-Hartig, and Steinitz-Kannan 1996), but *C. bondi* Lundblad is distributed from Brazil north to the West Indies and even the extreme southeastern United States (Cook 1980; Viets 1987; Smith and Cook 1996). *Kawamuracarus*, the sole representative genus of the subfamily Kawamuracarinae, includes several large, soft-bodied species from interstitial habitats throughout southern areas of Eurasia and North America (Cook 1974b, 1980; Viets 1987; Cramer 1987; Smit 1992a, 2003a; Smith and Cook 2000).

Deutonymphs and adults of the following six subfamilies are small, elongate, well-sclerotized mites that appear to be derived from common ancestral stock. The small subfamily Neomamersinae contains numerous species of *Neomamersa* and *Meramecia* and two species of *Arizonacarus*, found in hyporheic habitats in temperate North and South America, including the West Indies (Cook 1980, 1981a, 1988; Orghidan and Gruia 1983b; Viets 1987; Cramer 1987; Smith and Cook 1994; Fernandez 2001, 2002), while the two known species of *Psammolimnesia* (Psammolimnesiinae) live in hyporheic habitats in Mexico and Costa Rica (Cook 1974b, 1980). The Nicalimnesiinae is represented by the genus *Nicalimnesia*, known from hyporheic gravels in India (Cook 1967, 1974b); its apparent sister group, Stygolimnesiinae, includes only *Stygolimnesia japonica* Imamura from interstitial water in Japan (Imamura 1979b). The small subfamily Protolimnesiinae includes species of *Protolimnesia* and three closely related genera mainly from springs and streams in tropical South America (Cook 1974b; Viets 1987; Orghidan and Gruia 1987; Fernandez 1987b, 1994; Smit 2000), although *P.* (*Protolimnesella*) *ventriplacophora* Smith and Cook is known from hyporheic habitats in the southwestern United States (Smith and Cook 1996), and a few species of *Mixdea* occur on Caribbean islands (Orghidan and Gruia 1980b; Cook 1981a; Viets 1987). The Mixolimnesiinae includes several species of *Mixolimnesia* from hyporheic habitats in Chile (Cook 1988) and Argentina (Smit 2000).

The subfamily Neotorrenticolinae includes several large, distinctive species of *Neotorrenticola* from streams in tropical South America and Mexico (Cook 1974b; Viets 1987; Cramer 1987). *Epallagopus tecticoxalis* Viets, the only species of the subfamily Epallagopodinae, was taken from a spring in El Salvador (Cook 1974b). The Tyrrellinae is a small subfamily with two genera, the largest of which is *Tyrrellia*. Included in this genus are several species from South America and a few from North America as far north as southern Canada (Cook 1974b, 1980, 1988; Viets 1987; Smith 1987). Deutonymphs and adults of *Tyrrellia* are unusual among water mites in that they live in the film of water that coats plants and litter at the edges of springs, seepage areas, streams, and ponds and are typically found above the water line (Mitchell and Mitchell 1958). Their larvae are parasites of Ceratopogonidae (Diptera) (Smith and Oliver 1986). Most of the known species of *Neotyrrellia* occupy marginal habitats associated with streams in South America (Cook 1974b; Viets 1987), but one species, *N. anitahoffmannae* Smith and Cook, occurs in the southwestern United States (Smith and Cook 1996). Adults of this genus have unusual acetabula-like structures on the fourth coxal plates (Goldschmidt, Alberti, and Meyer 1999).

The small family OMARTACARIDAE (Fig. 13.29I) appears to be a basal group of Hygrobatoidea that may have diverged either from LIMNESIIDAE or from ancestral lebertioid stock. It includes two monobasic subfamilies, the Omartacarinae and the Maharashtracarinae (Cook 1980). Several species of *Omartacarus* are known from hyporheic habitats in southern North America and South America (Cook 1974b, 1980, 1988; Viets 1987; Fernandez 1987a, 1993, 2002; Fernandez and Grosso 1991) and Australia (Cook 1986), while the two maharashtracarine species are found in interstitial water in India (Cook 1967, 1974b) and Costa Rica (Cook 1980), respectively.

The families WETTINIDAE, FRONTIPODOPSIDAE, and LETHAXONIDAE also appear to be early derivative hygrobatoid clades (Cook, Smith, and Harvey 2000). The WETTINIDAE (Figs. 13.31B, E–G, 13.53F, G) includes species living in streams and oligotrophic lakes representing the genera *Wettina* in northern Eurasia and North America; *Wheenyella*, *Wheenyoides*, and *Tasmanaxona* in Australia; *Stormaxonella* in South Africa (Cook 1974b, 1986; Smith 1976a; Harvey 1990e, 1996a; Cook, Smith, and Harvey 2000). Adults of *Wettina* are rapid and agile swimmers, and their larvae are parasites of Chironomidae (Smith and Oliver 1986). The monobasic family FRONTIPODOPSIDAE (Figs. 13.32A–C) comprises the genus *Frontipodopsis* (Cook, Smith, and Harvey 2000), with several species of the nominate subgenus known from hyporheic habitats in South America (Cook 1974b, 1980, 1988) and Australia (Cook 1986), and members of the subgenus *Frontipodopsella* reported from similar situations in southern Eurasia and North America (Cook 1974b, 1986; Viets 1987; Valdecasas 2001b). The interstitial family LETHAXONIDAE (Figs. 13.31C, D) includes a few species of *Lethaxona* from highly disjunct areas of southern Eurasia, eastern Africa, and western North America (Cook 1974b, 1981b; Viets 1987; Valdecasas 2001b) and two described species of *Lethaxonella* from southern North America and South America (Cook 1974b, 1980).

Although the remaining six families of Hygrobatoidea probably represent a monophyletic group, phylogenetic relationships within most and among all of them are poorly understood and need to be resolved by rigorous cladistic analysis of characters of all active instars. One of these families, the monobasic FERRADASIIDAE (Fig. 13.29H), is represented by a single known species, *Ferradasia musicola* Cook, from streams in Argentina (Cook 1980).

The HYGROBATIDAE (Figs. 13.25H, 13.36G–I, 13.52C–F) is a large and diverse family found in a wide variety of habitats worldwide (Cook 1974b; Viets 1987). Although no subfamilies are currently recognized, the family contains several distinctive clades. Species of the large genera *Hygrobates* and *Atractides*, both of which have numerous subgenera, are among the dominant water mites in flowing water habitats in the Northern Hemisphere (Cook 1974b; Özkan 1982; Viets 1987; Gerecke 2003, 2004b). A few species of both genera are found in similar habitats in Africa (Cook 1974b; Viets 1987) and South America (Cook 1974b, 1980), several species of *Hygrobates* are known from Australia (Cook 1974b, 1986; Viets 1987), and *Atractides schwoerbeli* Orghidan and Gruia was reported from Cuba (Orghidan and Gruia 1983b). Members of *Mixobates*, recently elevated from a subgenus of *Hygrobates* to generic rank, occur widely in streams in Eurasia and North America (Cook 1974b; Viets 1987; Tuzovskij and Gerecke 2003). Known hygrobatid larvae are parasites of Chironomidae (Smith and Oliver 1986). Predation by *Hygrobates nigromaculatus* Lebert and *H. trigonicus* Koenike has been shown to be an important factor limiting populations of chironomid larvae in lakes (Ten Winkel 1985; Ten Winkel, Davids, and de Nobel 1989). Females of the highly unusual species *Hygrobates salamandrarum* Goldschmidt, Gerecke, and Alberti are the only known adult water mites to parasitize a vertebrate, the newt *Pachytriton labiatus* (Unterstein) (Goldschmidt, Gerecke, and Alberti 2002). The family exhibits much greater diversity at the generic level in the Southern Hemisphere, where many clades exhibit endemism and vicariance belying Gondwanan origin (Cook 1974b, 1986, 1988; Viets 1987; Orghidan and Gruia 1987; Harvey and Cook 1988; Harvey 1989; Wiles 1997b, c; Smit 2000, 2001, 2002a, c; Smith and Cook 1999a; Gerecke 2004c; Cook 2005). For example, members of the rheobiontic genera *Aciculacarus* from New Zealand (Cook 1983), *Rhynchaustrobates* from Australia (Cook 1986), and *Rhynchaturus* and *Brevaturus* from austral South America (Cook 1988) represent a clade of closely related species with obvious Gondwanan connections (Smith and Cook 1999a). Similar patterns of austral vicariance are evident among species of other genera, such as *Aspidiobates* (New Zealand, New Caledonia, Australia, and southern South America) and *Corticacarus* (South America, southern North America, Australia, and New Zealand) (Cook 1974b, 1980, 1983, 1986, 1988; Wiles 1991a; Smit 2002c; Smith and Cook 1999a). Members of the distinctive genus *Diamphidaxona* illustrate an atypical distribution pattern extending from Argentina to southern Canada, with greatest species diversity in Mexico and the southwestern United States (Cook 1974b, 1980; Viets 1987; Orghidan and Gruia 1987; Fernandez 1987b, 1994; Cramer-Hemkes and Letechipia-Torres 1996; Smith and Cook 2006), and one species, *D. margaretae* Orghidan and Gruia, reported from the West Indies (Orghidan and Gruia 1983a).

The unusual, recently described hygrobatid species from Nigeria, *Dockovdia coolearum* Gledhill and *D. oruensis* Gledhill and Agbolade, inhabit the mantle cavities of the prosobranch gastropods *Lanistes libyeus* and *Potadoma moerchi*, respectively (Gledhill 2003; Gledhill and Agbolade 2006).

The ATURIDAE (Figs. 13.37J, K, 13.53H–J, 13.54A–G) is a large and highly diverse family of well-sclerotized and typically dorsoventrally flattened mites grouped in four subfamilies (Cook 1974b). Adults of many aturid genera are strongly sexually dimorphic, implying elaborate courtship behavior involving copulation (Proctor 1991b). Numerous species in many genera of the subfamily Axonopsinae live in a wide variety of habitats worldwide (Cook 1974b; Viets 1987). Known larvae are parasites of Chironomidae (Smith 1984a; Smith and Oliver 1986). The *Axonopsella*-like mites comprise a large group of genera, with many species groups exhibiting vicariance in the Southern Hemisphere (Cook 1974b; Viets 1987; Smit 1996d), probably reflecting Gondwanan origin (Smith and Cook 1999a). For example, most species groups of *Axonopsella* are represented in Australia and New Guinea (Cook 1986; Wiles 1997c) and in South America (Cook 1980, 1988; Viets 1987; Orghidan and Gruia 1987). A few species of *Axonopsella*, *Neoaxonopsella*, and *Polyaxonopsella* have been reported from Cuba (Orghidan and Gruia 1981), and a single species, *A. bakeri* Smith and Cook, is known to occur in the southeastern United States (Smith and Cook 1998a), all evidently resulting from recent dispersal of the group from South America. Species of several other genera, such as *Stygalbiella*, *Miraxona*, *Miraxonides*, and *Submiraxona*, are distributed from South America to southern North America (Cook 1974b, 1980, 1981b; Viets 1987; Cramer 1992a; Cramer and Cook 1991; Smith and Cook 1998a), suggesting similar northward dispersal during the Pliocene and Pleistocene (Smith and Cook 1999a). Species groups in many genera of *Axonopsis*-like mites, such as *Axonopsis*, *Erebaxonopsis*, *Ljania*, *Brachypoda*, and *Woolastookia*, illustrate strongly disjunct distributions mainly in temperate areas of Eurasia and North America (Cook 1974a, b, 1975, 1981b; Valdecasas 1981; Viets 1987; Wiles 1989a; Gerecke 1991a; Özkan and Boyaci 1992b) and, in the case of *Axonopsis*, the West Indies (Orghidan, Gruia, and Iavorschi 1981). These groups were probably distributed throughout Laurasia until the Miocene (Smith and Cook 1999a). Members of two monobasic genera of this group have boreal distributions. One of them, the Nearctic species *Estellacarus unguitarsus* (Habeeb) lives in cold streams and lakes from New Brunswick north to the Canadian arctic and west to coastal Alaska, British Columbia, and Oregon (Cook 1974b; Smith 1984a, 1991d). The other, *Neobrachypoda eckmani* Lundblad, has been reported from tundra lakes in arctic areas of Europe and Canada (Cook 1974b; Viets 1987) and occurs in cold springs and alpine pools as far south as Oregon in western North America (Smith 1984a, 1991d). In contrast, most genera of the large group of *Axonopsalbia*-like mites are confined to Eurasia and Africa (Cook 1974b, 1999; Viets 1987; Tuzovskij 1986b; Gerecke and Meyer 1989; Gerecke 1994), with only a few species such as *Albaxona (Vietsaxona) nearctica* Cook and an undescribed species of *Javalbia* known from temperate North America (Cook 1974b; Viets 1987). These taxa may represent clades of southeastern Laurasian origin that only began to diversify extensively shortly before the separation of Eurasia and North America (Smith and Cook 1999a).

The subfamily Albiinae includes numerous species of *Albia* found in streams, ponds, and lakes throughout most of the world and the highly unusual species *Parasitalbia sumatrensis* Viets, adult females of which parasitize larval mayflies (Ephemeroptera), known only from a lake in Sumatra (Cook 1974b, 1986). Known larvae of *Albia* are parasites of adult Trichoptera (Smith and Oliver 1986). Species of the large subgenus *Albiella* live in streams, ponds, and lakes in southern Eurasia, Africa, North America, South America, and Australia (Cook 1974b, 1986; Viets 1987; Wiles 1991c, 1997c; Smit 1996c, d). In contrast, members of the subgenus *Albia* are known from streams in southern regions of Eurasia and North America and those of the subgenera *Dentalbia* and *Spinalbia* from similar habitats in tropical Eurasia and Africa (Cook 1974b; Viets 1987), with one species of the latter subgenus also reported from Australia (Cook 1986; Smit 1996d).

The remaining two aturid subfamilies, the Aturinae and the Notoaturinae, appear to be ancient sister groups that diversified in Laurasia and Gondwanaland, respectively (Cook 1974b, 1984, 1986, 1988; Smith and Cook 1999a). Several genera of the Aturinae are represented in streams and interstitial habitats, mainly in the Northern Hemisphere (Cook 1974b; Viets 1987). Known larvae parasitize Chironomidae (Smith and Oliver 1986). Members of the large genera *Aturus* and *Kongsbergia* are among the dominant mites in stream and spring habitats throughout temperate Eurasia and North America, with a few species in each genus occurring in Africa and South America (Cook 1974b; Özkan 1982; Valdecasas and Avilés 1983; Viets 1987; Tuzovskij 1994b; Cramer 2000). The unusual *K. (Crocokongsbergia) cooki* Orghidan and Gruia, with elongate and recurved anterior projections of the first coxal plates, was recently described from Cuba (Orghidan and Gruia 1980a, 1983c). Highly developed courtship behavior has been reported in adults of the Palearctic species *A. scaber* Kramer (Lundblad 1929a). Species of *Phreatobrachypoda* and *Bharatalbia* are found mainly in hyporheic habitats in widely disjunct southern areas of Eurasia and North America (Cook 1967, 1974b; Imamura 1977, 1984; Viets 1987; Smith 1991c). Other genera have more limited distributions, with species of *Neoaturus* and *Aturides* occurring from tropical South America to southern North America (Cook 1974b, 1980; Viets 1987) and those of *Subalbia* and *Subaturus* confined to tropical Africa (Cook 1966, 1974b; Viets 1987). Members of numerous genera of the Notoaturinae occupy stream and interstitial habitats in the Southern Hemisphere (Cook 1974b, 1983, 1986, 1991; Viets

1987). Species of *Notoaxona* living in austral South America seem to represent a basal clade. Members of the other two genera found in southern South America, *Notoaturus* and *Noesaturus*, show affinities with taxa in New Zealand and Australia, respectively. Species of *Masabania* from East Africa (Cook 1974b, 1998) and those of *Amperaturus* and *Valsaturus* from South Africa (Cook 1998) appear to be early derivative taxa not closely related to other known notoaturine genera. By far the largest diversity of notoaturine mites is found in New Zealand, with about 40 known species in 19 endemic genera (Cook 1983, 1991), and Australia, with more than 30 species in 7 endemic genera (Cook 1986).

The monobasic FELTRIIDAE (Figs. 13.36E, 13.37F, 13.52G–I) includes many species of *Feltria* inhabiting springs and streams, including hyporheic habitats, throughout Eurasia and North America (Cook 1974b; Özkan 1982; Viets 1987; Tuzovskij 1988b, 1998, 1999, 2002a; Valdecasas 2001b). The most southerly record for the family in the New World is of the species *F. anahoffmannae* Cramer from a stream on Popocatépetl volcano near Mexico City (Cramer 1986). Larvae of *Feltria* are parasites of Chironomidae (Smith and Oliver 1986).

The family UNIONICOLIDAE (Figs. 13.36F, 13.37C–E, G, 13.53A–E) is a large and diverse family with many genera assigned to four subfamilies. The Pionatacinae includes numerous species living in ponds, lakes, and streams throughout the world, most of which belong to three large genera (Cook 1974b; Viets 1987). Species of the basal subgenus *Koenikea* (*Notomideopsis*) are abundant in Australia (Cook 1986; Wiles 1997c) and South America (Cook 1980, 1988; Viets 1987), with a few species in the West Indies (Orghidan and Gruia 1980b) and southern North America (Cook 1980). On the other hand, species of the more recently evolved nominate subgenus are well represented in South America and southern North America (Cook 1976b, 1980; Viets 1987). Members of *Recifella* also show Gondwanan connections among the numerous species in Australia (Cook 1986; Wiles 1997c) and South America (Cook 1980, 1988; Viets 1987). In contrast, *Neumania* shows greatest diversity in the Northern Hemisphere, with many species in North America and Eurasia and relatively modest numbers in South America, Africa, and Australia (Cook 1974b; Viets 1987; Wiles 1997c). Known pionatacine larvae are parasites of Chironomidae or Trichoptera (Smith and Oliver 1986). Proctor (1991a) observed that adults of *N. papillator* Marshall pose with their strong and highly setose anterior legs extended to ambush crustacean prey and demonstrated that males capitalize on this adaptation to attract females to spermatophore masses during courtship. The Unionicolinae includes numerous species of *Unionicola* from streams, ponds, and lakes worldwide and several smaller genera with more restricted distributions that apparently represent derivative species groups of *Unionicola* (Cook 1974b; Hevers 1977, 1984; Vidrine 1986; Viets 1987; Gerecke 2004c). Free-living species of *Unionicola* are essentially planktonic (Riessen 1982). Some species are commensals or parasites in sponges (Crowell and Davids 1979a, b; Vidrine 1984; Crowell 1984, 1990; Davids, Crowell, and de Groot 1985; Konopacka and Sicinski 1985; Mari and Morselli 1987; Proctor and Pritchard 1990). Many others have obligate relationships with mollusks (Davids 1973a; Cook 1974b; Ferradas 1974; Hevers 1978b, 1980; Gordon, Swan, and Paterson 1979; Özkan 1982; Gledhill 1985; Fashuyi 1990; Edwards and Dimock 1997; Gledhill and Vidrine 2002; Vidrine, Borsari, and Bastian-Stanford 2005) that may result in damage to host tissue (Baker 1976a, b, 1977). Species living in the mantle cavity of mussels may have highly developed courtship rituals and population structure promoting successful reproduction in these enclosed habitats (Hevers 1978c, 1980; Dimock 1983, 1985; Davids, Crowell, and de Groot 1985; Edwards and Dimock 1988, 1991b; Edwards and Vidrine 1994). Mussel associates also exhibit behavioral adaptations for recognizing appropriate hosts, some of which are chemically mediated (Dimock and LaRochelle 1980; Hevers 1980; Edwards and Dimock 1991a, 1995). Members of some species have been shown to be remarkably sensitive and responsive to subtle differences in the intensity and frequency of light (Dimock and Davids 1985; Dimock 1988). Morphologically similar species of the subgenus *Parasitatax* show both genetic and host selection differences (Edwards and Dimock 1997; Edwards, Bogardus, and Wilhite 1998). DNA data may prove useful for diagnosing closely related *Parasitatax* species and may provide insight into their phylogenetic relationships (Edwards, Deatherage, and Ernsting 2004; Ernsting et al. 2006). Known larvae of *Unionicola* parasitize Chironomidae or Trichoptera (Hevers 1978a; Smith and Oliver 1986) and may affect the survival and fecundity of their host (Weiberg and Edwards 1997). Species representing two monobasic subfamilies, Pollicipalpinae from streams in tropical West Africa and Australia (Cook 1974b, 1986; Viets 1987) and Encentridophorinae from ponds in tropical Africa, Asia, and Australia (Cook 1974b, 1986; Viets 1987; Gerecke 2004c), exhibit Gondwzanan connections. Finally, the monobasic subfamily Omanohydracarinae was recently proposed for *Omanohydracarus arabicus* Gerecke, a distinctive species from groundwater habitats in Oman (Gerecke 2004b).

The large and diverse family PIONIDAE (Figs. 13.37A, B, H, I, 13.54H, 13.55A–L) includes seven subfamilies with many species in springs, streams, and standing water habitats worldwide (Cook 1974b; Viets 1987). Larval pionids are parasites of Chironomidae (Smith and Oliver 1986). Adults of many genera of this family exhibit marked sexual dimorphism, and elaborate courtship involving copulation has been observed for several species (Mitchell 1957c; Smith 1976b; Proctor 1991b). The basal subfamily Huitfeldtiinae is represented by lake-inhabiting species belonging to two monobasic genera, *Huitfeldtia* in northern Eurasia and North America (Cook 1974b; Smith 1976b; Viets 1987) and *Larri* in Australia (Harvey 1996a). Another ancient subfamily, the Tiphyinae, includes several species of *Tiphys* and *Pionopsis* from temporary pools and ponds in Eurasia and North America, a few species of *Neotiphys* from streams in North America, and one species of *Pionides* from pools in Eurasia (Cook 1974b; Smith 1976b; Özkan 1982; Viets 1987; Tuzovskij 2003a), along with two species each of *Acercella* (ponds) and *Australiotiphys* (streams) in Australia (Cook 1974b; Harvey 1996a). Species of *Tiphys* and *Piona* (see below) inhabiting temporary pools survive the dry period during summer, autumn, and winter in the pond basin as quiescent protonymphs attached by the chelicerae of the larval exoskelton at leaf axils of damp mosses (Smith 1976b; Wiggins, MacKay, and Smith 1980). The monobasic subfamily Hydrochoreutinae is represented by several species of *Hydrochoreutes* found in stream pools and lakes in North America and Eurasia (Cook 1974b; Smith 1976b; Viets 1987; Tuzovskij 2001e, 2003c), while the Najadicolinae contains only the unusual species *Najadicola ingens* Koenike that is a commensal in the mantle cavity of mussels in rivers and lakes in eastern North America and southeastern Eurasia (Cook 1974b; Vidrine, Vidrine, and Vidrine 1976; Simmons and Smith 1984; Viets 1987). The Foreliinae is a moderately large subfamily comprising a few species in each of three widespread Holarctic genera, namely *Pionacercus* in pools and ponds, *Pseudofeltria* in springs, and *Forelia* in stream pools, ponds, and lakes (Cook 1974b; Smith 1976b; Viets 1987; Özkan and Boyaci 1992a; Tuzovskij 2001a, b). The Pioninae comprises a few species of *Nautarachna* from stream pools and springs in Eurasia and North America (Smith 1972, 1976b; Cook 1974b; Viets 1987) and many species of the genus *Piona* from a wide variety of standing water habitats worldwide (Cook 1974b, 1980, 1986; Özkan 1982; Viets 1987; Jin and Gua 1992b; Harvey 1996a; Wiles 1997c; Gerecke 2004c). In a few cases, members of this genus have become established on oceanic islands remote from their region of origin. The recently discovered Hawaiian species *P. lapointei* Smith and Cook has its nearest relatives living in tropical South America (Smith and Cook 2004). Deutonymphs and adults of several species of *Piona* are essentially pelagic in ponds and lakes, where they are voracious predators, primarily on planktonic Cladocera. For example, in Madden Lake, Panama, members of *P. limnetica* Biesiadka reportedly consume cladocerans at a sufficiently high rate to reduce the standing crop of their prey by 50% per week (Gliwicz and Biesiadka 1975). Finally, the subfamily Schminkeinae is a small group of highly distinctive mites representing the genera *Schminkea* and *Twinforksella* from interstitial habitats in New Zealand (Schwoerbel 1984; Cook 1992).

The family ASTACOCROTONIDAE (Figs. 13.30G–I) includes only one species, the highly unusual *Astacocroton molle* Haswell, known only from gill chambers of the freshwater decapod crustacean *Astacopsis serratus* in southeastern Australia (Haswell 1922; Cook 1974b). Little is known of the biology of this mite, and its phylogenetic relationship to other hygrobatoid water mites is unclear.

- Superfamily Arrenuroidea (Figs. 13.26K, 13.30C–F, 13.31A, 13.32D–H, 13.33A–J, 13.34A–H, 13.35A–K, 13.36A–D, 13.49D–J, 13.50A–I, 13.51A–I)

DIAGNOSIS: Adult gnathosoma with palpi not chelate but often uncate; tibia longer than genu. Chelicerae 2-segmented. Idiosoma with lateral eyes absent or present, small and not borne on dorsal plate; body highly variable in both shape, ranging from ovoid to dorsoventrally flattened, and degree of sclerotization, ranging from soft with dorsum bearing only glandularia platelets to heavily armored with dorsum entirely covered by a dorsal shield. Genital field bearing genital acetabula lying in gonopore, on acetabular plates flanking gonopore, or both. Larvae aquatic. Gnathosoma lacking elaborate camerostome. Idiosoma with lateral eyes borne on separate platelets; dorsum covered by large dorsal plate bearing at least 4 pairs of setae including verticils and scapulars and rarely 1 or more pairs of hysterosomatic setae. Venter with coxal plates usually all separate, rarely variously fused; with paired urstigmata laterally between coxal plates I–II; coxal plates III with insertions of legs III located near midlength of body and laterally directed; excretory pore plate bearing 2 pairs of setae. Legs with 5 movable segments, with basifemur and telofemur fused; tarsus with unmodified claws and empodium.

The superfamily Arrenuroidea comprises at least three major lineages, including the family MOMONIIDAE, the mideopsiform families, and the arrenuriform families (*sensu* Smith 1990b), respectively. The MOMONIIDAE (Figs. 13.31A, 13.49G–J) is a moderately large, ancient family of moderately to well-sclerotized mites, including numerous genera exhibiting endemism in various parts of the world (Cook 1974b, 1983, 1986, 1988; Orghidan et al. 1977; Viets 1987; Smith 1989c–e, 1991b; Harvey 1990d, 1998b; Smith 1998b). Deutonymphs and adults of all species have short, modified tarsi on the first pair of legs bearing a single stout claw that flexes proximad to oppose the dorsal surface of the segment. Known larvae of MOMONIIDAE are parasites of Trichoptera (Smith and Oliver 1986). Members of the austral genera *Momoniella* (all southern continents), *Partidomomonia* (Australia, New Zealand), *Neomomonia* (New Zealand), and *Notomomonia* (Chile), along with species of the recently described Australian genera *Hesperomomonia* and *Australomomonia*, are primarily interstitial mites illustrating vicariance reflecting Gondwanan origins (Cook 1974b, 1980, 1983, 1986, 1988; Viets 1987; Harvey 1990d, 1998b; Wiles 1997c; Smith and Cook 1999a). Species of the genera *Momonia* (in streams and lakes) and *Stygomomonia* (in hyporheic habitats) have Tertiary relict distributions in southern Eurasia and North America (Cook 1974b; Viets 1987; Smith 1989d, e, 1991d; Valdecasas 2001b) consistent with Laurasian origin for these genera. The unusual species *Momonisia phreatica* Petrova from interstitial water in southeastern Europe (Petrova 1974; Gerecke 1994) seems allied to species of

Momonia, while the distinctive *Cyclomomonia andrewi* Smith from hyporheic habitats in western North America (I. M. Smith 1989b) may be related to those of *Stygomomonia*. The three known species of *Momonides* from streams in southern Eurasia (Cook 1974b; Viets 1987; Gerecke 1994) and *Xenomomonia subcentrata* Orghidan and Gruia from hyporheic deposits in Cuba (Orghidan et al. 1977) appear to represent divergent clades not closely related to other genera in the family (Smith and Cook 1999a). The unusual species *Cladomomonia mirifica* Orghidan and Gruia described from Cuba and placed in the subfamily Cladomomoniinae (Orghidan and Gruia 1980a, 1983c) is not a member of this family and appears to be a highly divergent member of the superfamily Hygrobatoidea.

The mideopsiform group of Arrenuroidea includes an array of moderately to well-sclerotized mites representing a few ancient families of Pangaean origin (NUDOMIDEOPSIDAE, MIDEOPSIDAE, and ATHIENEMANNIIDAE), a number of Laurasian families (CHAPPUISIDIDAE, NEOACARIDAE, MIDEIDAE, and ACALYPTONOTIDAE), and a few families of recent origin in Eurasia (KANTACARIDAE, NIPPONACARIDAE, and possibly HARPAGOPALPIDAE) or North America (ARENOHYDRACARIDAE, AMOENACARIDAE, and LAVERSIIDAE) (see figure citations below).

The NUDOMIDEOPSIDAE (Figs. 13.34A, 13.50H, I, 13.51A–C), a small family of mites inhabiting springs, hyporheic gravels, and deep interstitial waters, may be one of the most ancient basal groups of Arrenuroidea (Smith 1990b). The few known species of *Nudomideopsis* from southern Eurasia, North America, and New Zealand (Smith 1977, 1983h, 1990b; Cook 1992), *Paramideopsis* from western North America and New Zealand (Smith 1983f; Cook 1992; Cook and Hopkins 1998), and *Neomideopsis* from western North America (Smith 1983a) are found in poorly sampled habitats and may represent only a small proportion of the world fauna of this family. Larvae are parasites of Chironomidae (Smith 1983h; Smith and Oliver 1986). The monobasic MIDEIDAE (Figs. 13.34B, C, 13.51D, E) includes only a few species of the genus *Midea* living in pools, ponds, and lakes in northern Eurasia and North America (Cook 1974b; Smith 1978a; Viets 1987; Tuzovskij 1984a). Adults of the European species *M. orbiculata* (Müller) exhibit elaborate courtship involving copulation and direct transfer of spermatophores (Lundblad 1929a). Known larvae of this family are chironomid parasites (Smith 1978a; Smith and Oliver 1986). The MIDEOPSIDAE (Figs. 13.35I–K, 13.50F, G) is a moderately large family of heavily sclerotized mites grouped in five subfamilies (Cook 1974b, 1983, 1986, 1992; Imamura 1983), although separation of the Mideopsinae and Mideopsellinae is not well supported by phylogenetic evidence (Smith and Cook 1999a). Numerous species of *Mideopsis* live in a wide variety of habitats throughout North and South America, Eurasia, and Africa (Cook 1974b, 1976c, 1980, 1986, 1988; Viets 1987; Özkan and Erman 1990b; Jin 1992; Tuzovskij 1984a, 2001c, 2002f, 2006b; Valdecasas 2001b). Known larvae parasitize Chironomidae (Smith and Oliver 1986). Although previously placed together in Mideopsellinae, the genera *Mideopsella* and *Tiramideopsis* appear to be disparate derivative species groups of *Mideopsis* that have invaded streams in South America and interstitial habitats in India and the Arabian peninsula, respectively (Cook 1974b; Viets 1987; Gerecke 2004b). Interstitial species of the genera *Penemideopsis* from Australia (Cook 1986) and *Phreatomideopsis* from austral South America (Schwoerbel 1986c) may be related to basal members of *Mideopsis* (*Xystonotus*), a large group of species living in hyporheic habitats in temperate areas of North America and Europe (Cook 1974b, 1976c; Viets 1987). The species *Mideopsellides geae* Viets from streams in South Africa and *Tilia davisae* Harvey from a brook in Australia (Harvey 1990e) appear to represent distinct austral clades. Members of the monobasic austral subfamilies Plaumanniinae, comprising species of *Plaumannia* from streams in tropical South America (Cook 1974b; Viets 1987), and Gretacarinae, with numerous species of *Gretacarus* from streams and ponds in Australia (Cook 1986; Viets 1987; Wiles 1997c; Smit 1998b), appear to form a sister group exhibiting vicariance reflecting Gondwanan origin (Cook 1986; Smith and Cook 1999a). The austral, interstitial subfamily Guineaxonopsinae includes species of *Guineaxonopsis* in New Guinea (Imamura 1983; Wiles 1997c), New Zealand (Cook 1983), and Australia (Cook 1986) and *Kuschelacarus ovalis* Cook in New Zealand (Cook 1992), with similarities to members of both the MIDEOPSIDAE and CHAPPUISIDIDAE (see below).

The ATHIENEMANNIIDAE (Figs. 13.32F, G, 13.51F, G) includes four subfamilies of heavily sclerotized and dorsoventrally flattened mites (Cook 1974b, 1986, 2000; Smith 1990a; Harvey 1988b, 2003). The Athienemanniinae is the most diverse subfamily with nine genera. Members of *Chelomideopsis* inhabit springs, with four species in North America and one species in western Eurasia (Cook 1974b; Viets 1987; Smith 1992a), while two species of *Platyhydracarus* (Smith 1989b) and one species of *Chelohydracarus* (Smith 1998) occur in hyporheic deposits in western North America. Known larvae of these genera parasitize Chironomidae (Smith and Oliver 1986; Smith 1989b, 1992a). Five small genera are confined to the Old World, with species of *Mundamella* living in streams in western Eurasia (Cook 1974b; Viets 1987), those of *Penemundamella* and *Bleptomundamella* inhabiting streams in South Africa (Cook 2000), species of *Phreatohydracarus* found in interstitial habitats in western Eurasia (Cook 1974b; Viets 1987) and those of *Stygohydracarus* known from interstitial waters of both western Eurasia and South Africa (Cook 1974b, 2000; Viets 1987). The distinctive species *Anamundamella zelandica* Cook inhabits interstitial waters in New Zealand (Cook 1992). The remaining three subfamilies are relatively small and restricted in distribution. The monobasic Stygameracarinae includes two species of *Stygameracarus* known from hyporheic gravels in North America (Smith 1990a); the monobasic Africasiinae comprises several species of *Africasia* living in streams in southeastern Eurasia and tropical Africa (Cook 1967, 1974b, 2000; Viets 1987); and the Notomundamellinae contains species assigned to the monobasic austral genera *Notomundamella*, *Mellamunda*, and *Davecookia* found in hyporheic gravels in Australia (Cook 1986; Harvey 1988b, 2003).

Members of the CHAPPUISIDIDAE (Figs. 13.33D–J), a moderately diverse family of well-sclerotized and dorsoventrally flattened mites, are confined to interstitial habitats in the Northern Hemisphere (Cook 1974b; Viets 1987; Smith 1992b). A number of genera include species groups exhibiting striking Tertiary relict distributions in disjunct areas of southern Eurasia and North America (Smith and Cook 1999a). There are four distinctive subfamilies that apparently represent disparate holphyletic clades of uncertain relationship to one another. A family reassessment will be necessary when larvae and deutonymphs of more species are discovered and studied morphologically. The Chappuisidinae comprises several species of *Chappuisides* in widely separated areas of temperate Eurasia and North America (Cook 1974b; Imamura 1979a; Viets 1987; Smith 1992b; Valdecasas 2001b), while the Uchidastygacarinae includes numerous species of *Uchidastygacarus* in eastern Asia and temperate North America (Cook 1974b; Viets 1987; Smith 1983d, 1992b). The Morimotacarinae contains the genera *Morimotacarus* and *Yachatsia*, both of which comprise species pairs living in eastern Asia and western North America (Cook 1974b; Tuzovskij 1986a; Viets 1987; Smith 1992b). The Tsushimacarinae is represented by *Tsushimacarus uenoi* Imamura reported from hyporheic gravel in Japan (Cook 1974b; Viets 1987). It should be noted here that the mite described as *Ussuriaxona elliptica* Tuzovskij (1986a) and placed in the ATURIDAE is not a member of that family but appears to belong to the chappuisidid genus *Yachatsia* that is otherwise known only from western North America (Smith 1992b).

The family NEOACARIDAE (Figs. 13.33A–C, 13.51H) includes several species of well-sclerotized and dorsoventrally flattened mites with strongly uncate palpi and sexually dimorphic third legs, living in hyporheic habitats. Thirteen species of *Neoacarus* are known from North America, and one species occurs in western Eurasia (Cook 1974b; Viets 1987; Smith 2003b). Known larvae parasitize Chironomidae (Smith 1983g; Smith and Oliver 1986). Two described species of *Volsellacarus* are found in temperate North America (Cook 1974b). The ACALYPTONOTIDAE (Figs. 13.32H, 13.49D–F) includes a few weakly to moderately well sclerotized species of *Acalyptonotus* living in cold springs, pools, and lakes in arctic and alpine areas of Europe and North America (Cook 1974b; Smith 1983i, Viets 1987) and *Paenecalyptonotus fontinalis* Smith from springs in eastern North America (Smith 1976a). Known larvae are parasites of Chironomidae (Smith 1983i; Smith and Oliver 1986).

Members of the monobasic families KANTACARIDAE (Fig. 13.35A), represented by *Kantacarus matsumotoi* Imamura, and NIPPONACARIDAE (Figs. 13.30C, E), with three known species of *Nipponacarus*, are well-sclerotized and dorsoventrally flattened mites found in interstitial habitats in Japan (Cook 1974b; Viets 1987). The monobasic HARPAGOPALPIDAE (Figs. 13.32D, E) includes six known species of *Harpagopalpus* living in streams in West Africa, Madagascar, and India (Cook 1966, 1967, 1974b; Gerecke 2006). Adults of this family exhibit striking idiosomatic neotrichy and have strongly uncate palpi.

The three small, monobasic families ARENOHYDRACARIDAE, AMOENACARIDAE, and LAVERSIIDAE are confined to the New World. The ARENOHYDRACARIDAE (Figs. 13.35B–D) includes several tiny, elongate, heavily sclerotized species of *Arenohydracarus* inhabiting interstitial waters in the southwestern United States, Mexico, and Cuba (Cook 1974b, 1980; Orghidan and Gruia 1983b). The AMOENACARIDAE (Figs. 13.35F–H) includes the highly distinctive and strongly sexually dimorphic species *Amoenacarus dixiensis* Smith and Cook from streams in southeastern North America (Smith and Cook 1997). The LAVERSIIDAE (Figs. 13.34D, E, 13.51I) is represented by *Laversia berulophila* Cook from cold springs, streams, and deep oligotrophic lakes throughout temperate North America (Cook 1974b; Smith 1978a). Larvae of *Laversia* are parasites of Chironomidae (Smith 1978a; Smith and Oliver 1986).

The ancient families KRENDOWSKIIDAE and ARRENURIDAE, both of Pangaean origin, and (provisionally) the Laurasian families BOGATIIDAE and HUNGAROHYDRACARIDAE comprise an assemblage referred to as the arrenuriform group (Smith 1990b). The KRENDOWSKIIDAE (Figs. 13.35E, 13.50D, E) is a moderately large family of heavily armored, nearly spherical mites (Cook 1974b). Known larvae of this family are parasites of tanypodine Chironomidae (Smith 1978a; Smith and Oliver 1986). Numerous species of *Krendowskia* live in streams and lakes from tropical South America to southern Canada, and a few species are known from similar habitats in the southern Palearctic (Cook 1974b, 1980; Viets 1987; Smit 1997b). Known species of *Allokrendowskia* are restricted to South Africa and Madagascar, and this genus seems to be only distantly related to others in the family (Cook 1974b; Viets 1987). Many species of *Geayia* are known from streams and ponds in South America, and several species occur in eastern North America as far north as southern Canada (Cook 1974b, 1980; Smith 1983c). Species of *Roqueella* are confined to tropical areas of the New World (Cook 1974b, 1980).

The ARRENURIDAE (Figs. 13.26K, 13.36A–D, 13.50A–C) includes a multitude of species of the hyperdiverse genus *Arrenurus*, which occur in virtually all types of freshwater habitats worldwide, and species of a few small genera in disjunct tropical areas that probably represent divergent species groups of *Arrenurus* (Cook 1974b, 1976a, 1980, 1986, 1988; Viets 1987; Orghidan and Gruia 1987; Orghidan et al. 1977; Wiles 1988, 1997c; Smith and Harvey 1989; Özkan and Boyaci 1991; Özkan and Erman 1990a, 1991; Cramer and Cook 1992a, b; Erman 1992, 1993; Jin and Gua 1992a; Özkan, Erman, and Boyaci 1993; Wiles 1993b; Anderson and Smith 1996; Jin and Wiles 1996; Erman and Özkan 1997; Smit 1992b, 1996a, c, 1997a, 1999, 2003b; Smit, Gerecke, and Di Sabatino 2000b; Gerecke 2004c). Adults of *Arrenurus* are strongly sclerotized, and those of species living in standing water are typically strong swimmers. Many species feed on ostracods, but some prey on other invertebrates. Adults of most species exhibit striking sexual dimorphism, with the idiosoma of females ovoid or nearly spherical and that of males elaborately and characteristically modified posteriorly as an adaptation for spermatophore transfer. Studies of the elaborate courtship behavior of several species have been published (Lundblad 1929b; Böttger 1962, 1965; Böttger and Schaller 1961; Proctor 2003b; Proctor and Smith 1994). Known larvae of *Arrenurus* species are parasites of various nematocerous Diptera or Odonata (Smith and Oliver 1986; Böttger 1981; Mitchell 1998). The associations of several species of larval *Arrenurus* and their insect hosts have been intensively studied. Larvae of most species exhibit strong host specificity and use spatial, temporal, behavioral, and chemical cues to locate, attach to, and detach from hosts (Mitchell 1959, 1961, 1964b; Mullen 1976; Stechmann 1977, 1978, 1980; Åbro 1979, 1986; Münchberg 1982; Smith and MacIver 1984a, b; Forbes and Baker 1990; Sarkar et al. 1990; Smith and Laughland 1990; Smith and Cook 1991; Forbes, Muma, and Smith 1999; Anderson 2003). They typically feed by forming a well-defined stylostome through which they extract fluids from the host (Åbro 1979, 1984, 1991, 1992; Redmond and Hochberg 1981; Smith 2003). Their feeding activities can affect the survival, longevity, reproductive success, and fecundity of hosts (Åbro 1982, 1990; Forbes and Baker 1991; Lanciani 1986; Lanciani and Boyt 1977; Reinhardt 1996; Smith and MacIver 1984c). Larvae of the European species *A. globator* (Müller) have an exceptionally broad spectrum of hosts and appear able to engorge and complete their development on both aerial and aquatic hosts (Böttger and Martin 2003).

The family BOGATIIDAE (Figs. 13.30D, F) contains two species of heavily sclerotized, interstitial mites with relict distributions in the Holarctic region, namely *Bogatia maxillaris* Motaş and Tanasachi in southeastern Europe and *Horreolanus orphanus* Mitchell in central North America (Cook 1974b; Viets 1987). The HUNGAROHYDRACARIDAE (Figs. 13.34F–H) is another small, Holarctic family of strongly armored mites found in hyporheic habitats. The subfamily Hungarohydracarinae includes a few species of *Hungarohydracarus* and *Bharatohydracarus* distributed widely in southern Eurasia (Cook 1974b; Viets 1987; Valdecasas 1984; Gerecke 2004b) and *Stygarrenurus* found in Costa Rica and Mexico (Cook 1980; Cramer and Cook 1996). The monobasic subfamilies Balcanohydracarinae and Cubanohydracarinae are represented by two species of the genus *Balcanohydracarus* in southern Europe (Cook 1974b; Viets 1987) and *Cubanohydracarus elegans* Orghidan and Gruia in Cuba (Orghidan and Gruia 1980b, respectively).

Subcohort Stygothrombiae

The Stygothrombiae comprises only the superfamily Stygothrombidioidea.

- Superfamily Stygothrombidioidea (Figs. 13.38A–E, 13.56A–D)

DIAGNOSIS: Adult gnathosoma retractable into idiosoma, chelicerae with movable digit bladelike and curved; palpal femur and genu fused, tibia bearing 1 or more enlarged spinelike setae adjacent to tibial claw and tarsus bearing a long terminal solenidion; prodorsal area with slender prodorsal plate bearing unpaired anteromedial seta (representing vi*), 1 pair of bothridia (*si*) and 2 pairs of posterior setae (*ve *and* se*); eyes absent; cuticle smooth and bearing relatively sparse complement of short, simple setae and several pairs of modified stomatoid lyrifissures that resemble the glandularia of true water mites, without setiferous papillae; genital field bearing 3 pairs of stalked acetabula, pregenital tubercle absent; legs each with postaxial stomatoid lyrifissures on femur, genu, and tibia, and with tarsus bearing pectinate claws and simple empodium. Larva moderately heteromorphic; gnathosoma with genu bearing 2 setae and with tibial claw 4-pronged; with prodorsal plate bearing unpaired anteromedial seta (representing* vi*), 1 pair of trichobothria (*si*), and 2 pairs of posterior setae (*ve *and* se*); venter bearing stalked urstigmata and well-defined anus, lacking setiferous anal plate; coxal plates I–III bearing 2, 1, and 1 pairs of setae, respectively, plates I–II contiguous; legs each with postaxial lyrifissures present on femur, genu, and tibia, genu with more than 4 setae, and tarsi each with pectinate claws and simple empodium.*

The Stygothrombidioidea includes only the enigmatic STYGOTHROMBIDIIDAE (Figs. 13.38A–E, 13.56A–D). Deutonymphs and adults of this family live in interstitial aquatic habitats, presumably as predators of small invertebrates, throughout the Northern Hemisphere (Robaux 1967a; Mullen and Vercammen-Grandjean 1980; Smith, Cook, and Smith 2001). Larvae are also aquatic and parasitize both nymphs and adults of Plecoptera (Mullen 1979; Mullen and Vercammen-Grandjean 1980; Smith, Cook, and Smith 2001; Yasick, Simmons, and Earle 2003). Stygothrombidioid mites exhibit unique combinations of character states in both larval and postlarval instars (Mullen and Vercammen-Grandjean 1980; Vercammen-Grandjean 1980). Some of these attributes, such as the presence of two setae on the larval palpgenu and the modified glandularia-like lyrifissures on the idiosoma of deutonymphs and adults, may indicate relationship with Hydrachnidiae. However, until these and other characters are analyzed more rigorously to determine their variability and polarity among species of Stygothrombidioidea, Hydrachnidiae, and terrestrial groups of Parasitengonina, the systematic position of this superfamily will remain unclear. Several stygothrombidioid species and genera have been proposed either on the basis of adults (Viets 1932, 1934, 1955b, 1959; Sokolow 1944, 1948; Motaş and Tanasachi 1946; Walter 1947; Habeeb 1955, 1973, 1974a–c, 1979; Robaux 1967a, 1969b, c) or larvae (Motaş 1928; Imamura 1956; Vercammen-Grandjean 1976, 1980), but the taxonomy of the group remains poorly understood. All currently recognized species have been grouped together in a single genus, *Stygothrombium* (Mullen and Vercammen-Grandjean 1980).

Subcohort Erythraiae

The Erythraiae includes two superfamilies, the Calyptostomatoidea and Erythraeoidea.

- Superfamily Calyptostomatoidea (Figs. 13.38F, G, 13.56E–G)

DIAGNOSIS: Adult gnathosoma entirely retractable into idiosoma; chelicerae with movable digit styletiform, palpi lacking enlarged spinelike setae adjacent to tibial claw; prodorsal area bearing pair of trichobothria on small medial platelet and pair of conspicuous ocular pits; cuticle covered by stellate reticular pattern and bearing moderately dense pelage of short, simple setae, without setiferous papillae; genital field with 2 pairs of papillae and lacking pregenital tubercle. Larva homeomorphic; gnathosoma with palpgenu bearing 2–4 setae; prodorsal plate absent but with 1 pair of trichobothria; venter bearing sessile urstigmata and well-defined anus, lacking setiferous anal plate; coxal plates each with 15 or more setae; plates I–II contiguous; legs with genua each bearing more than 4 setae.

Deutonymphs and adults of the monobasic family CALYPTOSTOMATIDAE (Figs. 13.38F, G, 13.56E–G) are large, sluggish mites found in wet litter and moss habitats in Europe (Irk 1939b; Turk 1945; Schweizer and Bader 1963; Vistorin-Theis 1976; Wainstein 1977; Gabryś and Haitlinger 1986; Haitlinger 2003b), Japan (Shiba 1969), Australia (Womersley 1942a), Africa (Lawrence 1944), and North America (Smith and Lindquist 1979). They apparently are predators of larval Diptera (Vistorin-Theis 1977). Larvae of *Calyptostoma* parasitize adult Tipulidae and Limoniidae (Diptera) (Shiba 1969; Vistorin-Theis 1975; Rack 1976).

- Superfamily Erythraeoidea (Figs. 13.38H, 13.39A–C, 13.56H–J, 13.57A–E)

DIAGNOSIS: Adult gnathosoma entirely retractable into idiosoma or with only chelicerae retractable; chelicerae with movable digit styletiform, palpi lacking enlarged spinelike setae adjacent to tibial claw; prodorsal area bearing 2 pairs of bothridia on prodorsal plate and 1 pair of lateral eyes; cuticle smooth and bearing sparse to extremely dense pelage of setae that vary from short and simple to long and highly modified in shape, setiferous papillae absent; without genital papillae and pregenital tubercle. Larva moderately to strongly heteromorphic; gnathosoma with palpgenu bearing fewer than 2 setae; prodorsal plate bearing 2 pairs of trichobothria; venter lacking urstigmata, well-defined anus, and setiferous anal plate; coxal plates usually bearing 1 or 2 setae each, plates I–II widely separated; legs with genua each with more than 4 setae.

The Erythraeoidea includes two families. Deutonymphs and adults of the SMARIDIDAE (Figs. 13.38H, 13.56H–J) are moderately large, active predators of small arthropods primarily in grassland and litter habitats. Members of both subfamilies, Smaridinae and Hirstiosomatinae, are widely distributed globally (Southcott 1961, 1963, 1995). Larvae of *Smaris prominens* (Banks) were reported as parasites of psocids in Australia (Womerseley and Southcott 1941) and those of other species have subsequently been found on a range of insect hosts, including scale insects in Cyprus (Zhang 1995b) and leafhoppers in India (Zhang 1995a).

The ERYTHRAEIDAE (Figs. 13.39A–C, 13.57A–E) is a large and diverse family, with eight included subfamilies of moderately to very large, often reddish mites found in a broad range of edaphic and arboreal habitats worldwide. Larvae are typically parasitic on other arthropods, although those of some species are parasites of vertebrates, while others are free living and nonparasitic. Erythraeid mites are unusually well represented in the fossil record, and larval specimens of a number of species have been reported and, in some cases, described and named. These include a larva of *Proterythraeus southcotti* Vercammen-Grandjean in Cretaceous amber from Cedar Lake, Manitoba, Canada (Vercammen-Grandjean 1971), unnamed larvae attached to tineid and gracillarid moths in Dominican amber (Poinar, Treat, and Southcott 1991), unnamed larvae attached to ceratopogonid flies in Lebanese amber (Poinar, Acra, and Acra 1994), and an unnamed larva attached to a chironomid fly in Cretaceous amber from the Foremost Formation, Canada (Poinar et al. 1997). Four of the eight extant subfamilies (Erythraeinae, Callidosomatinae, Abrolophinae, and Balaustiinae) are diverse assemblages of genera, while the Phanolophinae, Myrmicotrombidiinae, Leptinae, and Augustsonellinae are monobasic taxa.

Deutonymphs and adults of species of the worldwide subfamily Erythraeinae are often abundant in open woodland and grassland habitats, where they actively hunt and feed on small invertebrates. Numerous species of *Erythraeus* are distributed in Eurasia (Southcott 1961; Beron 1982; Gabryś 1989; Saboori and Akrami 2001; Saboori, Cobanoglu, and Bayram 2004; Haitlinger 1987b, 2004a) and North America. Adults of *E. ghilarovi* Schuster and Turk were collected in large numbers running on the soil surface in a puna grass barren at 4,700 m in the Peruvian Andes (Schuster and Turk 1987). Those of *Lasioerythraeus whitcombi* (Smiley) feed on eggs of the cotton bollworm *Heliothis zea* (Boddie) in Arkansas (Smiley 1964, 1966), and those of *L. johnstoni* Welbourn and Young appear to be economically significant predators of nymphs of the tarnished plant bug, *Lygus lineolaris*, in Mississippi (Welbourn and Young 1987; Young and Welbourn 1987). Adults of *Erythrites reginae* (Hirst) occur commonly under loose eucalyptus bark in riparian parkland in South Australia (Southcott 1961), and *Erythrites* larvae are known to parasitize various insects on that continent (Greenslade and Southcott 1980). Larvae of *E. jacksoni* Southcott were reported as parasites of brachypterous females of the grasshopper *Brachaspis collinus* (Hutton) in New Zealand (Southcott 1988d), while those of *Erythraeus* (= *Bochartia*) species parasitize Homoptera (Pussard and André 1929; André 1951; Yano and Ehara 1982; Goldarazena-Lafuente and Zhang 1998a). Young and Welbourn (1987, 1988) reported *Lygus lineolaris* as the primary host of *Lasioerythraeus johnstoni* larvae and listed several alternate hosts in the orders Hemiptera and Thysanoptera.

Postlarval instars of mites belonging to the Callidosomatinae, another large cosmopolitan erythraeoid subfamily, are active predators that may be recognized by the conspicuous enlargement of their palptarsi. Larvae of several species of the mainly austral genus *Caeculisoma* are known to parasitize acridid grasshoppers (Southcott 1961, 1972; Key 1990) or geometrid moths (Southcott 1972), while those of *C. mouldsi* Southcott were reported from two cicada species in Queensland, Australia (Southcott 1988d). Postlarval instars of *Callidosoma metzi* Sharma, Drooz, and Treat prey on insect eggs, and larvae of this mite parasitize several species of moths (Sharma, Farrier, and Drooz 1983; Treat 1985; Adler and Brown 1986). Larvae of other species of *Callidosoma* are associated with Lepidoptera (Treat 1985) and other insect groups (Southcott 1972). As discussed by Southcott (1991b), the synonymy of the long-standing generic names *Charletonia*, based on larvae, and *Sphaerolophus*, based on adults, was established through rearing experiments conducted independently by Ishii (1954), Southcott (1961), Treat (1980), and Rosa and Flechtmann (1980). The amalgamation of these two assemblages (under the name *Charletonia*) has produced a highly speciose and cosmopolitan genus. Larvae of both *C. venus* Southcott and *C. krendowskyi* (Feider) parasitize *Mantis religiosa* in southern Europe (Southcott 1961; Beron 1975), and those of several other species of the genus are parasites of grasshoppers in Australia (Southcott 1966, 1972, 1983, 1991b; Key 1991) and of a broad range of arthropod hosts worldwide (Welbourn 1983; Treat 1975; Southcott 1991b). Larvae of another callidosomatine mite, *Carastrum ferrari* Southcott, select ventral abdominal attachment sites on several host species of calyptrate Diptera in Australia (Southcott 1988a).

Deutonymphs and adults of *Abrolophus* (Abrolophinae) are moderately large mites frequently found in moist forest litter habitats throughout Eurasia (Southcott 1961) and North America (Yao, Snider, and Snider 2000). Larvae assigned to the genus *Hauptmannia* have been reported as parasites of Thysanoptera from many areas of the Holarctic (Southcott 1961, 1994b; Haitlinger 1986a, 1987a, d, e). Grandjean's (1947a) conclusion that members of *Hauptmannia* are larvae of *Abrolophus* was not fully accepted by Southcott (1961), and confusion over use of these names has continued to the present time despite efforts by Welbourn and Young (1987) and Goldarazena-Lafuente and Zhang (1998b) to resolve generic concepts by reinterpreting previously published and unpublished evidence. As pointed out by Zhang and Goldarezena-Lafuente (1996), rearing experiments to associate larvae with deutonymphs and adults of the various species currently placed in these two genera are badly needed.

Postlarval instars of species of Balaustiinae, moderately large mites with distinctive prodorsal, pit-shaped urnulae (see Lindquist 2001; Yoder et al. 2006), are found in arboreal and litter habitats worldwide (Southcott 1961; Smiley 1966, 1968; Welbourn and Jennings 1991; Gabryś 2000; Halliday 2001; Saboori 2001). Deutonymphs and adults of *Balaustium* species are general predators of small arthropods, although they also feed on pollen grains. *Balaustium* larvae are exceptional among erythraeoid mites in being free-living predators (Newell 1963; Cadogan and Laing 1977; Childers and Enns 1975; Childers and Rock 1981; Welbourn 1995). All active instars have also been observed feeding on pollen grains (Grandjean 1946b). Predation by large numbers of an undetermined species of *Balaustium* markedly increased egg mortality in *Colias alexandra* (Edwards) in the Rocky Mountains of Colorado, USA, but without overall significant impact on the butterfly population (Hayes 1985). Members of the North American species *B. putmani* Smiley are considered to be economically significant predators of pest insects and mites in commercial fruit orchards (Putman and Herne 1966; Putman 1970; Cadogan and Laing 1982). Several species of *Balaustium* have been reported to bite humans, in some cases causing mild to severe dermatitis (Newell 1963; Rack 1973b; Southcott 1976; Ido et al. 2004).

The Leptinae includes many species, all of which are placed in the widespread genus *Leptus* and nearly all of which are based either on larvae or adults (Fain 1987c; Fain and Gummer 1987; Haitlinger 1987c, 1990b, 1991, 1992b, 1998a, 1999c, 2001c, 2002b, 2004c; Southcott 1988c, 1991a, 1992b, 1993a, 1994b, 1999; Chinniah and Mohanasundarum 1998; Saboori and Atamehr 1999; Zhang 2001b; Saboori and Masoud 2003). Postlarval instars are large predators in litter and herbaceous plant habitats and are recognized by their bushy, usually darkly colored idiosomatic setae and their prominent eyes, which are located anterior to the middle of the prodorsal plate. Larvae of *Leptus* parasitize a wide range of arthropod hosts, including various insects (Houseweart et al. 1980; Baker 1982; Haitlinger 1987g, 1990b, c, e; Quintero 1987; Dhiman and Singh 1988; Southcott 1988b, 1989a, b, 1991a, 1992b, 1993a, 1999; Wilson, Rubink, and Collins 1990; Fain 1991; Welbourn and Jennings 1991; Baker and Selden 1997; Saboori and Ostovan 2000; Beenan 2001; Zheng 2003) and arachnids (Åbro 1988; Norton, Welbourn, and Cave 1988; Fain and Jocque 1996b; McAloon and Durden 2000; Stasiov 2003). The life history and phenology of the Palearctic *L. trimaculatus* (Hermann) was described and discussed by Wendt et al. (1992).

Little is known regarding the biology of postlarval instars of *Phanolophus oedipodarum* (Frauenfeld) (Phanolophinae), an uncommon erythraeid mite known from southern Europe and northern Africa and characterized by cone-shaped setae on the idiosoma and appendages (Southcott 1961). Larvae of this species parasitize Orthoptera (Southcott 1961). *Myrmicotrombium brevicristatum* Womersley (Myrmicotrombiinae) is associated with ant nests in Australia (Womersley 1934; Southcott 1957b), while *Augustsonella tuberculata* (Augustson) (Augustsonellinae) inhabits dune and desert habitats in western North America (Augustson 1940; Southcott 1961; Hoffmann and Mendez 1973). Adults of *A. tuberculata* are large, brown, densely setose mites with numerous large, conical setae on the genu and tibia of the palp.

Subcohort Trombidiae

The Trombidiae is a large and diverse group that includes the four superfamilies Tanaupodoidea, Chyzerioidea, Trombiculoidea, and Trombidioidea.

- Superfamily Tanaupodoidea (Figs. 13.39D–J, 13.57F, G)

DIAGNOSIS: Adult gnathosoma not retractable into idiosoma; chelicerae with movable digit bladelike and curved, palp bearing enlarged spinelike setae adjacent to tibial claw; prodorsal area bearing 1 pair of trichobothria on prodorsal plate and 2 pairs of lateral eyes; cuticle lacking setiferous papillae, smooth and bearing relatively sparse pelage of simple setae inserted on small sclerites; with 3 pairs of genital papillae and bearing a conspicuous pregenital tubercle. Larva heteromorphic; gnathosoma with palpgenu bearing 1 seta; idionotum with prodorsal plate (scutum) bearing 4 pairs of setae, including 2 pairs of trichobothria but lacking scutellum, with setae c1 *inserted on paired platelets; venter bearing sessile urstigmata and well-defined anus, with paired anal plates bearing 2 pairs of setae; coxal plates I–III bearing 2, 2, and 4 or 5 setae, respectively, plates I–II contiguous; legs with genua each bearing more than 4 setae.*

The Tanaupodoidea includes only the family TANAUPODIDAE (Figs. 13.39D–J, 13.57F, G). Deutonymphs and adults of this family are moderately large mites inhabiting moist litter habitats in the Northern Hemisphere. Known larvae are parasites of Diptera (Newell 1957). Welbourn (1991) recognized three subfamilies, namely, the Tanaupodinae, Lasseniinae, and Polydisciinae, and speculated that the TANAUPODIDAE may be the sister group of Hydrachnidiae. However, this relationship is neither reflected in his dendrogram nor well supported by available evidence. *Eothrombium nemoricola* (Berlese) has been reported from several parts of Europe (Berlese 1910b; Feider 1955a; Robaux 1966c, 1967a). Other species of the genus, including those previously placed in *Rhinothrombium* and *Lassenia*, are known from Poland (Haitlinger 1995a), China (Zhang 1988, 1993), and North America (Newell 1957). The unusual *Amphotrombium jenseni* (Zhang), described from a single larval specimen associated with an aphid host (Zhang 1998), is mentioned here pending further study on its association with the adult stage.

- Superfamily Chyzerioidea (Figs. 13.40A–C, 13.57H, I)

DIAGNOSIS: Adult gnathosoma not retractable into idiosoma; chelicerae with movable digit bladelike and curved, palp bearing enlarged spinelike setae adjacent to tibial claw; prodorsal area bearing 1 pair of trichobothria on prodorsal plate and 2 pairs of lateral eyes; dorsal cuticle peripherally with several pairs of long papillae densely covered with plumose setae, strongly reticulate medially and bearing moderately dense pelage of simple setae; with 3 pairs of genital papillae, pregenital tubercle absent. Larva heteromorphic; palpgenu with 1 seta; idionotum with prodorsal plate (scutum) bearing 4 pairs of setae including 1 pair of trichobothria but lacking scutellum, setae c1 *located on paired platelets; ventrally with sessile urstigmata and well-defined anus, anal plate bearing 2 pairs of setae; coxal plates I–III bearing 2, 1, and 1 setae, respectively, plates I–II contiguous; legs with genua each bearing 4 setae.*

The Chyzerioidea includes only the family CHYZERIIDAE (Figs. 13.40A–C, 13.57H, I). Welbourn (1991) recognized three subfamilies, namely, the Chyzeriinae, Pteridopodinae, and Ralphaudyninae. Adults of *Chyzeria* (Chyzeriinae) are large mites found in various edaphic habitats throughout the Southern Hemisphere (Robaux 1969d; Southcott 1987a). Larvae of *C. derricki* Southcott reportedly parasitize tettigoniid grasshoppers in Australia, and those of *C. hirsti* Womersley apparently feed on adults of their own species (Southcott 1982). Pteridopodines and ralphaudynines are known only from larvae. Those of *Pteridopus auditor* Newell and Vercammen-Grandjean and *Ralphaudyna amamiensis* Vercammen-Grandjean et al. were found as parasites of crickets (Gryllidae) in southern Africa and Japan, respectively (Newell and Vercammen-Grandjean 1964; Southcott 1987a), and the larva of *R. iranensis* was recently described from Iran (Zhang and Saboori 1995). Larvae of three other species, all associated with orthopteroid hosts, have also been assigned to CHYZERIIDAE (Haitlinger 1999d). Two of these, *Cretessenia leoni* Haitlinger from Crete and *Napassenia aginapaica* Haitlinger from Cyprus, appear related to *Pteridopus* species. The third species, *Perumaropta mirsadi* Haitlinger, described from Peru, was placed in a separate subfamily, the Perumaroptinae.

- Superfamily Trombiculoidea (Figs. 13.40D–I, 13.41A–D, 13.58A–J, 13.59A–C)

DIAGNOSIS: Adult gnathosoma not retractable into idiosoma; chelicerae with movable digit bladelike and curved, palp bearing enlarged spinelike setae adjacent to tibial claw; prodorsal area bearing 1 pair of trichobothria on prodorsal plate and 2 pairs of lateral eyes; dorsal cuticle without long papillae, bearing moderately to extremely dense pelage of setae; with 2–3 pairs of genital papillae, pregenital tubercle absent. Larva heteromorphic; gnathosoma with palpgenu bearing 1 seta; with prodorsal plate (scutum) bearing 4 pairs of setae, including 1 pair of trichobothria, or an unpaired anteromedial seta and 3 pairs of setae, including 1 pair of trichobothria but always lacking scutellum, setae c1 *located on paired platelets; ventrally with sessile urstigmata and well-defined anus, without setiferous anal plate; coxal plates I–III bearing 2, 1, and 1 setae, respectively, plates I–II contiguous; legs with genua each bearing 4 setae.*

The Trombiculoidea is a large and diverse group that includes five families. Deutonymphs and adults of the JOHNSTONIANIDAE (Figs. 13.40E–H, 13.58A–C) are moderately large mites found in moist litter habitats. Known larvae are parasites of various insects. Welbourn (1991) placed the family in Trombiculoidea and recognized the subfamilies Johnstonianinae and Charadacarinae, an arrangement that was not accepted by Wohltmann, Mąkol, and Gabryś (2004), who followed earlier authors in including the JOHNSTONIANIDAE in the Trombidioidea. Several species of Johnstonianinae representing the genera *Johnstoniana*, *Centrotrombidium*, and *Diplothrombium* are known from both North America (Newell 1957; Robaux 1978) and Europe (Robaux 1967a, 1970; Wendt and Eggers 1994; Haitlinger 2001b; Wohltmann, Mąkol, and Gabryś 2004), and a few species of the latter two genera have been reported from Australia (Hirst 1928; Womersley 1939) and Chile (Robaux 1968c). Newell (1957) reported that larvae of *Diplothrombium monoense* parasitizing pupae of aquatic Coleoptera immediately detached from the host when preserved in alcohol and interpreted this as evidence of self-detaching behavior that permits larvae to transfer between hosts. Other reports of parasitism by larval Johnstonianinae mention only adult semiaquatic Diptera as hosts (Vercammen-Grandjean and Feider 1973; Vercammen-Grandjean and Cochrane 1974; Haitlinger 1993d; Wohltmann et al. 1994; Wohltmann 1996; Wohltmann and Wendt 1996; Baker 1999). Species of the small subfamily Charadacarinae are known from North America and Europe (Newell 1960).

Welbourn (1991) suggested that larval mites allocated to the genus *Tetrathrombium* (subfamily Tetrathrombiinae, family JOHNSTONIANIDAE) by Feider (1955b) and Feider and Suciu (1956) were instead euthyadine water mites. These species, along with others subsequently placed in the Tetrathrombiinae by Haitlinger (1993c) and Fain (1997), are, in fact, members of several genera representing various subfamilies of the water mite family HYDRYPHANTIDAE (Hydrachnidiae). All will probably require new combinations, and some will no doubt prove to be synonyms.

Most species of TROMBICULIDAE (Figs. 13.40D, 13.58D, F) belong to the highly diverse, cosmopolitan subfamily Trombiculinae and the relatively small Gahrliepiinae, an assemblage that shares their habits and distribution. Members of the family LEEUWENHOEKIIDAE (Fig. 13.58E), comprising the subfamilies Leeuwenhoekiinae and Apoloniinae, are closely related to the TROMBICULIDAE and have often been considered a trombiculid subfamily. Identification and classification of mites in these two families are based largely on the larvae, which are commonly known as chiggers (Vercammen-Grandjean 1973b; Brennan and Goff 1977). Deutonymphs and adults are common predators in various edaphic habitats, including nests, but taxonomic diagnoses and biological observations of these instars are limited to relatively few species. Insect eggs appear to be a main source of food (Lipovsky 1954). The anatomy of the mouthparts and blind gut of the trombiculid *Blankaartia acuscutellaris* (Walch) were described in the detailed studies by Mitchell (1962b, 1964a, 1970) to illustrate how the structure, arrangement, and functioning of the digestive organs permit efficient digestion in the absence of a closed circulatory system. The remarkable method used by these mites to eliminate solid waste, termed schizeckenosy, has been investigated by Mitchell and Nadchatram (1969) and Shatrov (1993); see also chapter 3.

The relatively well-studied larvae of trombiculoid families are typically parasites of a wide range of vertebrate hosts, including wildlife, domesticated species, and humans (see bibliographies in Fuller 1952; Audy 1968; Traub and Wisseman 1974; Domrow and Lester 1985; Goff, Loomis, and Wrenn 1986; and Hoffmann 1990). Intensive study of the larval trombiculoids has led to the development of highly specialized terminology, which only recently has been harmonized with more general systems used for members of related groups (Goff et al. 1982).

Chiggers induce formation of a characteristic feeding tube, or **stylostome**, at the site of attachment to the host (André 1927; Cross 1964; Voigt 1970; Hase et al. 1978). Chigger bites often result in a condition known as **trombidiosis** (or trombiculosis), which is characterized by intensely itchy lesions in the skin of the host that may become inflamed or infected (Wrenn 1996). Members of several genera, notably *Eutrombicula*, *Neotrombicula*, and *Schöngastia*, regularly cause trombidiosis in humans. Larvae of *N. autumnalis* (Shaw), the harvest mite, are the most frequently reported causative agents in Europe (Vater 1982; Daniel 1983; Kampen 2002), and those of *E. alfreddugesi* (Oudemans) and *E. splendens* Ewing in North America (Mallow et al. 1984; Clopton and Gold 1993), although many of the North American records may in fact be based on misidentified specimens of other species in the genus (Loomis and Wrenn 1984; Wrenn and Loomis 1984). The response to chigger bites in humans depends to some extent on the normal host associations of the mite. The most severe reactions appear to be caused by chigger species that normally parasitize birds or reptiles and are ill adapted for feeding on mammalian hosts (Audy 1951; Traub and Wisseman 1974; Wrenn 1996). Some chiggers act as vectors for zoonotic diseases, the most important being scrub typhus, a rickettsial disease of eastern Asia that is spread among rodents and transmitted to humans primarily by members of the genus *Leptotrombidium* (Traub and Wisseman 1974; Traub, Wisseman, and Farhang-Azad

1977; Brown 1978; Nadchatram 1970a, 1979, 1984; Philip 1980; Burgdorfer 1988; Uchikawa and Kumada 1989; Wu and Liu 1991; Takahashi et al. 1993; Uchikawa et al. 1994; Takahashi and Tanaka 1995; Shi, Zhou, and Chen 2000; Huang, Jia, and Li 2003; Takahashi et al. 2004).

Most chiggers attach to exposed areas of skin, but many species in various genera typically infest the nasal passages of their hosts (Vercammen-Grandjean 1965a, 1975b; Nadchatram 1970b). Larvae of some apparently exceptional species, such as *Microtrombicula eltoni* Audy (Audy 1956), *Neotrombicula scorpionis* Nadchatram and Dohany (Nadchatram and Dohany 1974), and a few others (see André 1943 and Lourenço 1982), parasitize scorpion hosts. Phenological studies on various chigger species have demonstrated marked temporal fluctuations in wild populations that appear to be influenced by temperature and moisture (Loomis 1956; Cunningham, Kunz, and Price 1976; Shatrov 1992; Takahashi et al. 1993, 1994, 1995).

The small family NEOTROMBIDIIDAE (Figs. 13.40I, 13.58G–J) includes four genera in two subfamilies, the Neotrombidiinae and Anomalothrombiinae (Vercammen-Grandjean 1973c). Neotrombidiid deutonymphs and adults are moderately large mites that are typically found in forest litter and tree bark habitats in many parts of the world (Lindquist and Vercammen-Grandjean 1971). Singer (1971b) successfully reared *Neotrombidium beeri* Singer in the laboratory on a diet consisting of eggs of the beetle *Tenebrio molitor* L. The unusual *Discotrombidium villasiense* Feider is a member of the bat guano community living in Cuban caves (Feider 1977), and postlarval instars of an undescribed species of *Monunguis* have been found in a similar habitat in a cave in Missouri. Larvae of various species of *Neotrombidium* are parasites of wood-boring or subcortical beetles, and those of *Monunguis streblida* Wharton are hyperparasites of streblid flies parasitizing bats in the Caribbean region (Borland 1956; Lindquist and Vercammen-Grandjean 1971; Singer 1971a).

The TROMBELLIDAE (Figs. 13.41A, B, 13.59A–C) is a widely distributed family with several genera in four subfamilies (Welbourn 1991). Deutonymphs and adults are large, often colorful mites found in forest litter habitats. *Nothrotrombidium otiorum* (Berlese) (Trombellinae) occurs throughout southern Europe (Feider 1958; Robaux 1967a), and the striking *Trombella lusitanica* André (Trombellinae), notable for its conspicuous dorsal sclerotized depressions, is known from Spain and Portugal (Robaux 1967a). Adults of *Moyanella gigax* Boshell and Kerr (Moyanellinae) from South America are over 3 mm in length, are bright orange in color, and have a prominent sclerotized band that completely encircles the idiosoma (Robaux 1967b). Members of the related genera *Typhlothrombium*, *Neotyphlothrombium*, and *Paratyphlothrombium* (Spelaeothrombiinae) are also very large mites distributed through much of South America (Robaux 1968b). Adults of still another spelaeothrombiine mite, *Speothrombium monoculata* Robaux, were found in a cave in Cuba (Robaux 1972b). Larvae of *Nothrotrombidium treati* Southcott (Trombellinae) have been reported as parasites of Lepidoptera (Southcott 1987a), while those of several species of *Trombella* are associated with various orthopteran hosts (Southcott 1986a, b; Key and Southcott 1986; Key 1990). Parasitism by larval *Trombella alpha* Southcott (Trombellinae) on the cricket *Teleogryllus commodus* (Walker) in South Australia shows marked seasonality, peaking in early summer and declining by autumn (Southcott 1986b). Larvae of *Durenia singaporensis* Vercammen-Grandjean and Audy (Trombellinae) and several other species of this genus are parasites of mosquitoes (Vercammen-Grandjean 1955, Vercammen-Grandjean and Audy 1959; Mullen 1975a; Fain and Grootaert 1996). Those of *Taraxithrombium gordoni* Robaux were not collected from an arthropod host but rather from decomposing wood in the trunk of a downed redwood tree in California (Robaux 1977b). Robaux believed that the larvae may have fed prior to being collected. *Austrotrombella leprosa* Southcott, an unusual species described from a swamp in South Australia and placed in the TROMBELLIDAE by Southcott (1991d), was subsequently reassigned to the water mite family HYDRYPHANTIDAE (Harvey 1996b).

Larvae of *Audyana thompsoni* Womersley, the only member of the monobasic family AUDYANIDAE (Figs. 13.41C, D), parasitize scorpions in Malaysia (Womersley 1954a). Although a deutonymphal specimen has been reared from engorged larvae (Womersley 1954b), the systematic placement of *Audyana* will remain problematic until adults are known.

■ Superfamily Trombidioidea (Figs. 13.41E–L, 13.42A–O, 13.60A–J)

DIAGNOSIS: Adult gnathosoma not retractable into idiosoma; chelicerae with movable digit bladelike and curved, palp bearing enlarged spinelike setae adjacent to tibial claw; prodorsal area bearing 1 pair of trichobothria on prodorsal plate and 2 pairs of lateral eyes; dorsal cuticle lacking long papillae, with a moderately to extremely dense pelage of setae; genital field bearing 2–3 pairs of acetabula, pregenital tubercle absent. Larva heteromorphic; palpgenu bearing 1 seta; with prodorsal plate (scutum) bearing 4 pairs of setae including 1 pair of trichobothria and at least 1 other unpaired sclerite (scutellum) bearing setae c1; ventrally with sessile urstigmata and well-defined anus, setiferous anal plate absent; coxal plates I–III bearing 2, 1, and 1 setae, respectively, plates I–II contiguous; legs with genua each bearing 4 setae.

The Trombidioidea is a large and diverse group containing four families. The TROMBIDIIDAE (Figs. 13.41E–L, 13.59D–I) is a large worldwide family of about 20 genera grouped in three subfamilies (Welbourn 1991). Four of these genera have recently been placed in synonymy under *Allothrombium* and *Trombidium* by Mąkol (2007), who also established six new trombidiid genera in her recent world review of the family. Trombidiid deutonymphs and adults are large, densely setose mites that are often bright red in color and are commonly referred to as velvet mites. They are active predators of small invertebrates in habitats that range from subaquatic situations to deserts, while their larvae are parasites of arthropods. Postlarval instars of species of the widespread genus *Trombidium* (Trombidiinae) are common in edaphic habitats, and their larvae parasitize many groups of insects and arachnids (Vercammen-Grandjean and Popp 1967; Vercammen-Grandjean, Van Driesche, and Gyrisco 1977; Southcott 1986c; Welbourn and Young 1988; Pugh, King, and Fordy 1991; Peterson et al. 1992; Benfekih and Doumandji-Mitiche 1997; Mąkol and Wohltmann 2000; Conrad et al. 2002). The familiar *Trombidium holosericeus* L. occurs throughout much of Europe (Robaux 1967a; Mąkol and Wohltmann 2000); several other species have more restricted distributions in the Palearctic (Robaux 1967a; Southcott 1986c; Gabryś 1986; Zhang and Saboori 1996; Wohltmann 2000; Mąkol 2001) and North America. Adults of *Parathrombium*, another cosmopolitan trombidiine genus, are usually found in moist litter (Newell 1958) or in moss on and around the base of tree trunks. Larvae of *Parathrombium*, which are characterized by the highly modified comblike seta on the first coxal plate (see Robaux 1969a), frequently parasitize Hymenoptera but also may be found on various Hemiptera and Homoptera (Southcott 1997; Goldarazena-Lafuente and Zhang 1997; Goldarazena-Lafuente, Jordana, and Zhang 1999).

The large and strikingly colored deutonymphs and adults of the worldwide trombidiid subfamily Allothrombiinae are among the most conspicuous and readily recognized free-living Acari. They are voracious predators of small invertebrates, and some species are known to attack important agricultural pests. For example, *Allothrombium lerouxi* Moss feeds on aphids and caterpillars in apple orchards in eastern North America (Moss 1960; Mathur and LeRoux 1965), *A. pulvinum* Ewing is an important predator of spider mites in peach orchards (Zhang 1992c) and cotton fields (Chen and Zhang 1991; Zhang 1992b) in China, and *A. mitchelli* Davis feeds on balsam woolly aphid in the southeastern United States (Davis 1961). Based on available evidence, Zhang (1991a) concluded that species of *Allothrombium* have considerable potential as economically significant biological control agents. *Allothrombium* larvae may use other arthropods as hosts (Zhang 1992d; Zhang and Norbakhsh 1995), including opilionids, spiders, beetle larvae, and chironomid flies (Moss 1960; Robaux 1972a). *A. pulvinum* larvae are important early season parasites of *Aphis gossypii* in cotton fields in China (Zhang and Chen 1993; Zhang et al. 1993; Chen et al. 1994) and significantly reduced survivorship of pea aphids, *Acyrthosiphon pisum*, in controlled experiments (Zhang 1991b). Zhang (1992a) investigated the phenomenon of **superparasitism**—the presence of several larvae on the same host individual—in *A. pulvinum* and hypothesized that it could be an adaptation to enhance the probability of successfully locating a mate following the parasitic stage.

The vividly colored adults of the allothrombine genus *Dinothrombium* are among the largest known mites, with some individuals of the African species *D. tinctorum* (L.) exceeding 12 mm in length. Deutonymphs and adults of *D. pandorae* (Newell and Tevis) inhabit deserts of the southwestern United States, where they spend nearly all of their lives in the soil, emerging on the surface only after significant rains (Newell and Tevis 1960; Tevis and Newell 1962). Mass emergences following heavy summer downpours may be spectacular, with dense aggregations of up to 50 mites per square meter covering several acres of the desert floor (Newell and Tevis 1960). Postlarval stases of *D. pandorae* apparently are predators of termites, while the larvae are parasites of grasshoppers (Tevis and Newell 1962). Several species of the related genus *Austrothrombium* are distributed throughout Australia (Southcott 1986c), and larvae of the austral genus *Clinotrombium* parasitize spiders and moths (Southcott 1986c).

Deutonymphs and adults of the widespread subfamily Podothrombiinae (the family Podothrombiidae of Mąkol 2007) are moderately large mites found in soil and litter habitats. Species of *Podothrombium* have been reported to occur throughout Eurasia (Robaux 1967a, 1982; Zhang and Xin 1989; Haitlinger 1994b, 1995a, 2001d, 2003b; Fain and Ripka 1998; Mąkol and Marusik 1999; Mąkol 2000, 2007; Zhang 2001a) and in North America (Robaux 1977b; Zhang and Jensen 1995). *Podothrombium* larvae are parasites of aphids (Zhang and Xin 1989; Zhang and Jensen 1995). The South American species *Variathrombium thibaudi* Robaux appears to be closely related to *Podothrombium* (Robaux 1969d; Southcott 1986c), and the Australian mite *Pollicothrombium miniatum* (Womersley) may also belong in this subfamily.

The NEOTHROMBIIDAE (Figs. 13.60A–D, F, G), a small but widespread family of approximately 15 species in 11 genera, was reviewed by Zhang (1994). Deutonymphs and adults occur in various edaphic habitats, and the larvae of most species are parasites of Orthoptera (Zhang 1994; Haitlinger 1998b, 2000; Saboori 2002). The Palearctic genus *Neothrombium* includes the type species *N. neglectum* (Bruyant) in Europe, two species in China (Robaux 1972a; Zhang 1994), and one in Vietnam (Haitlinger 1998b). Some neothrombidiids, such as *Calothrombium paolii* (Berlese), have been collected only on rare occasions and presumably have secretive habits (Robaux 1968b). Larvae of *Ceuthothrombium cavaticum* Robaux, Webb, and Campbell parasitize cave crickets in the southwestern United States (Robaux, Webb, and Campbell 1976; Welbourn 1983). All active instars of this species are white and translucent, a trait often found in cavernicoles (Webb, Robaux, and Campbell 1977). Published examples of parasitic associations between larval neothrombiid mites and other host groups include *Paputrombidium grootaerti* Fain parasitizing dolichopodid Diptera in

Papua New Guinea (Fain 1992a), *Silphitrombium furculigerum* Fain and *Neosilphitrombium gratum* Fain on silphid beetles in Europe (Fain 1992b), and the unusual *Wondeclia centipedae* Southcott on the otostigmid centipede *Rhysida nuda* in Queensland, Australia (Southcott 1987b).

Deutonymphs and adults of the EUTROMBIDIIDAE (Figs. 13.42A–F, 13.60E) are large mites that are found in a wide variety of edaphic habitats worldwide. Eutrombiids often display a conspicuous plate posteriorly on the idionotum. Species of the widespread genus *Eutrombidium* (Eutrombidiinae) have been reported from Europe, North America, and the Australian region (Robaux 1967a; Southcott 1992a, 1993c; Key 1994; Haitlinger 2004b). Postlarval instars of *Eutrombidium locustarum* (Walsh) are predators of grasshopper eggs (Severin 1944). Some species of Eutrombidiinae, such as *Bruyantella smithi* Southcott from eastern North America, are eyeless (Southcott 1991c); others, notably the Palearctic species *Leptothrombium oblongum* (Trägårdh), are associated with ant nests (Robaux 1968a). Larvae of several species of *Eutrombidium* are parasites of various Orthoptera (Southcott 1993c; Wohltmann, Wendt, and Waubke 1996; Saboori and Nemati 2001; Haitlinger 2003a), and those of *Verdunella lockleii* (Welbourn and Young) and *Spinnitrombium kenyense* Fain and Jocqué parasitize spiders (Welbourn and Young 1988; Fain and Jocqué 1996a). The subfamily Hexathrombiinae is known only from larvae (Southcott 1993c). The larva of *Hoplothrombium quinquescutatum* Ewing was described from a single specimen collected in association with an oribatid mite from the stomach of a toad (Ewing 1925; Vercammen-Grandjean 1967). Larvae of two of the known species of *Hexathrombium* parasitize carabid Coleoptera (Cooreman 1944; Southcott 1993c), and those of other species are associated with beetles of the families Staphylinidae (Fain and Drugmand 1993) or Erotylidae (Haitlinger 1997). Larvae of *Beronium coiffati* (Beron) were found on carabid beetles in Morocco (Beron 1973; Southcott 1986c), and those of four other species assigned to this genus have been reported from carabid hosts in Chile, China, Sumatra, Madagascar, and the Canary Islands (Haitlinger 1994a). The two known species of Milliotrombiinae, *Milliotrombidium milliopodum* Shiba and *Kepongia malayensis* Southcott, are both based on larvae removed from myriapod hosts (Shiba 1976; Southcott 1993c).

The cosmopolitan family MICROTROMBIDIIDAE (Figs. 13.42G–O, 13.60H–J) is a large and diverse group of about 60 genera (Gabryś 1999) arranged in three subfamilies. Their densely setose deutonymphs and adults are among the most commonly encountered parasitengonine mites in soil, litter, and moss habitats (Robaux 1967a; Gabryś 1996). Adults of the Palearctic species *Microtrombidium pusillum* (Hermann) live in forest litter near bodies of water (Gabryś and Wohltmann 2001), and those of *Sucidothrombium sucidum* (Koch) are found in fens and alpine meadows (Mąkol and Gabryś 2002). The idiosomatic setae of microtrombidiids often are elaborately modified, as in members of *Dendrothrombidium betschi* Robaux (Feideriinae) from Madagascar (Robaux 1966a), *Willmanniella johnstoni* (Robaux) (Microtrombidiinae) from eastern North America (Robaux 1976), and *Echinothrombium rhodinum* (Koch) from Europe (Gabryś and Wohltmann 2003). Larvae of the subfamily Microtrombidiinae were critically reviewed by Southcott (1994a). Reported hosts of larval microtrombidiid mites include Diptera (Vercammen-Grandjean and Cochrane 1974; Shiba 1976; Mullen and Vercammen-Grandjean 1978; Fain and Baker 1993; Southcott 1994a; Fain and Grootaert 1993, 1994, 2001; Zhang and Saboori 1997; Baker 1999; Metz and Irwin 2001; Gabryś and Wohltmann 2001; Wohltmann 2001) and, in a few cases, Coleoptera (Moser and Vercammen-Grandjean 1979).

Key 13.2. Cohort Parasitengonina, Adults, Key to the Families (Figs. 13.25–13.42)

1a. Idiosoma and appendages usually not hypertrichous (Figs. 13.25A, C, 13.38A) (exception: CTENOTHYADIDAE); idiosoma with series of stomatoid lyrifissures or glandularia comprising glandlike structures that each open on small platelet, often adjacent to associated seta (Fig. 13.36E, arrows). Aquatic forms **2**

1b. Idiosoma and appendages hypertrichous (Figs. 13.38F, 13.39G, 13.40A, B); idiosoma without glandularia. Terrestrial forms **3**

2a. Idiosoma variously shaped but rarely vermiform (exception: HYDRYPHANTIDAE, subfamily Wandesiinae); palpfemur and genu separate, and palptarsus lacking long, terminal solenidia (Figs. 13.25F, 13.26J, K, 13.30C, 13.33I, J); prodorsal area various but never with elongate prodorsal plate bearing unpaired *vi* seta anteromedially; legs with claws usually simple or with a few distal prongs (Figs. 13.31B, 13.37E, F), rarely pectinate **Subcohort Hydrachnidiae 4**

2b. Idiosoma vermiform, greatly elongated posterior to podosoma (Figs. 13.38A, D); prodorsal area with elongate prodorsal plate bearing unpaired *vi* seta anteromedially (Fig. 13.38B); palpfemur and genu fused, and palptarsus bearing long, terminal solenidia (Fig. 13.38E); legs with claws strongly pectinate (Fig. 13.38C)
Subcohort Stygothrombiae, Superfamily Stygothrombidioidea, STYGOTHROMBIDIIDAE

3a. Chelicerae with movable digit usually bladelike and not retractable (Figs. 13.39G, 13.40A, E, I, 13.41A, L, 13.42A, H) **Subcohort Trombidiae—59**

3b. Chelicerae with movable digit styletiform and retractable, either independently (Figs. 13.39A, C) or with entire gnathosoma (Figs. 13.38F, H) **Subcohort Erythraiae—71**

4a. Mostly marine; 4th coxal plates with 2 pairs of long posterior apodemes (Fig. 13.25A, arrows); genital acetabula wheel-like in appearance (Fig. 13.25B) and usually not in gonopore region (Fig. 13.25A, arrow) **PONTARACHNIDAE**

4b. Living in fresh water and without above combination of characters **5**

5a. Genital acetabula small and inconspicuous, lying on coxal plates IV **6**

5b. Genital acetabula conspicuous, either closely associated with gonopore or scattered in ventral integument **7**

6a. Dorsum with large central plate surrounded by platelet anteriorly and many smaller but similar platelets laterally and posteriorly (Fig. 13.25C); glandularia represented by setae, with gland portion absent **ACHERONTACARIDAE**

6b. Dorsum with large central plate surrounded by platelet anteriorly and several smaller dissimilar platelets laterally and posteriorly (Fig. 13.25D); glandularia with gland portion present **HYDROVOLZIIDAE**

7a. Idiosoma nearly spherical (Fig. 13.25E); palptibia much shorter than genu and bearing dorsodistal projection that extends beyond origin of tarsus (Fig. 13.25F); chelicera styletiform and 1-segmented (Fig. 13.25G) **HYDRACHNIDAE**

7b. Idiosoma variously shaped but rarely spherical; palptibia usually longer than genu, but when shorter, then lacking distodorsal projection; chelicera 2-segmented (Fig. 13.25H) **8**

8a. Genital flaps present and partially or completely covering 3 pairs of acetabula (Fig. 13.25I); palptibia with 2 thick dorsodistal setae (Fig. 13.25J, arrow); known only from Australia and New Zealand **ZELANDOTHYADIDAE**

8b. Genital flaps present or absent; palptibia lacking 2 thick dorsodistal setae as described above **9**

9a. Capitulum broadly widened distally, palpi with all segments fused and flexing medially to oppose anterior edge of capitulum (Fig. 13.26A, arrow); dorsum and venter with numerous rugose glandularia platelets (Figs. 13.26A, B) **APHEVIDERULICIDAE**

9b. Without above combination of characters **10**

10a. Soft bodied; all coxal plates fused on their respective sides and narrowly joined at anterior end (Fig. 13.26C); movable genital flaps covering 3 pairs of genital acetabula. Known only from interstitial waters in New Zealand **STYGOTONIIDAE**

10b. Without above combination of characters **11**

11a. Mouth opening surrounded by a distinct membranous fringe (Fig. 13.26D, arrow); lateral eyes usually present and located on common prodorsal plate (Figs. 13.26E–G) **12**

11b. Mouth opening with indistinct or no membranous ring; lateral eyes, if present, not on common prodorsal plate **14**

12a. Prodorsal plate noticeably wider than long, constricted near middle and bearing 1 pair of setae (Fig. 13.26E); coxal plates as indicated in Fig. 13.26H **EYLAIDAE**

12b. Prodorsal plate either longer than wide or only slightly wider than long, not constricted near middle and bearing 4 pairs of setae **13**

13a. Prodorsal plate much longer than wide (Fig. 13.26F); genital acetabula scattered in ventral integument **LIMNOCHARIDAE**

13b. Prodorsal plate approximately as long as wide (Fig. 13.26G); genital acetabula located on acetabular plates **PIERSIGIIDAE**

14a. Palpi chelate with dorsodistal portion of tibia extending beyond base of tarsus, either as a spur (Fig. 13.27A) or a thick seta (Fig. 13.26J). (Not to be confused with uncate palp, in which tarsus is able to fold against produced ventrodistal portion of tibia to form a grasping organ as in Fig. 13.26K.) Capitulum lacking an anchoral process **15**

14b. Palpi rarely chelate; when appearing chelate (some PIONIDAE), then capitulum with an anchoral process (Fig. 13.37I, arrow) **16**

15a. Dorsodistal extension of palptibia an extremely long spur (Fig. 13.27A, arrow); idiosoma without platelets dorsally **HYDRODROMIDAE**

15b. Dorsodistal extension of palptibia a relatively short, thick seta (Fig. 13.26I, J); idiosoma usually with varying degrees of plate or platelet development dorsally **HYDRYPHANTIDAE**

16a. Genital field with movable genital flaps flanking gonopore and either partially or completely covering gonopore when closed; genital acetabula lying free in gonopore, not on flaps or acetabular plates (Fig. 13.27B) **17**

16b. Genital field usually without movable genital flaps, when flaps present then genital acetabula lie on flaps rather than under them (Fig. 13.27C) **23**

17a. Coxal plates IV with median margins reduced to median angles and bearing a pair of glandularia near tip of these angles (Fig. 13.27D); palptarsus bearing terminal pad-shaped setae and appears spatulate (Figs. 13.27E, F) **RUTRIPALPIDAE**

17b. Coxal plates IV with median margins usually well developed, when reduced to median angles, then lacking glandularia in position illustrated above **18**

18a. Integument of idiosoma soft; coxal plates IV with pair of glandularia located in small area of soft integument anteromedially, partially (Fig. 13.27G, arrow) or completely surrounded by sclerotization of coxal plates IV; gonopore bearing 3 pairs of genital acetabula **TEUTONIIDAE**

18b. Without above combination of characters **19**

19a. Dorsal and ventral shields present; either with gonopore bearing 6 pairs of genital acetabula (Fig. 13.27H) and dorsal shield consisting of a large posterior plate and 1 or 2 pairs of anterior platelets (as in Fig. 13.27I) (in rare cases anterior platelets are fused with posterior plate but suture lines are evident) or with gonopore bearing 3 pairs of genital acetabula and dorsal shield consisting of a central plate surrounded by several pairs of platelets (as shown in Fig. 13.27J) **TORRENTICOLIDAE**

19b. Without above combination of characters **20**

20a. Dorsal shield absent, dorsal integument soft and with or without scattered platelets; insertions of all pairs of legs near anterior end of idiosoma (Fig. 13.28A); median coxal suture line usually absent but present in some species in Southern Hemisphere **OXIDAE**

20b. Dorsal shield present or absent; insertions of at least legs IV near middle of idiosoma; median coxal suture line or lines present **21**

21a. Venter with suture lines between coxal plates II–III incomplete (Fig. 13.28B, arrow) but touching genital field area posteriorly; palp typically with at least 5 long setae on medial surface of genu (Fig. 13.28C), but members of North American genus *Scutolebertia* have 4 or 5 long setae on dorsal surface of femur (Fig. 13.28D) **LEBERTIIDAE**

21b. Venter with suture lines between coxal plates II–III complete; palp without long setae on medial surface of genu **22**

22a. Legs IV bearing claws; lateral eyes in capsules lying on anterodorsal surface of idiosoma (Fig. 13.28E, arrow) **SPERCHONTIDAE**

22b. Claws either present or absent on legs IV (claws commonly present in Holarctic species but generally absent in species from other parts of world); when claws present, then lateral eyes located on platelets associated with ventral shield (Fig. 13.28F, arrow) (exception: Asian genus *Bharatonia*) **ANISITSIELLIDAE**

23a. Dorsum and venter with series of closely fitting reticulate platelets (as shown in Figs. 13.28G and 13.28H); genital acetabula varying from 3 pairs to numerous. Known from Africa and Indonesia **TERATOTHYADIDAE**

23b. Dorsum and venter not as illustrated above **24**

24a. Genital field with 3–4 pairs of genital acetabula located on outer edges of movable genital flaps (Fig. 13.29A); legs and palpi with numerous, short, bladelike setae, with those of legs nearly completely covering segments (Figs. 13.29B, C) **CTENOTHYADIDAE**

24b. Without above combination of characters **25**

25a. Numerous genital acetabula on movable genital flaps; dorsum covered either with closely fitting reticulate platelets (Fig. 13.29D) (North and South America) or with closely fitting porous platelets (Fig. 13.29E) (South America) **RHYNCHOHYDRACARIDAE**

25b. Without above combination of characters **26**

26a. Dorsal and ventral shields present; coxal plates I with 2 rows of large setae extending posteriorly from capitular bay (Fig. 13.29F); palpfemur with 2 long medial setae (Fig. 13.29G); known only from hot springs **THERMACARIDAE**

26b. Without above combination of characters **27**

27a. Coxal plates III extending far posteriorly and excluding coxal plates IV from midline; coxal plates IV lacking glandularia; male with gonopore placed far forward and extending beyond base of capitulum (Fig. 13.29H); known only from Argentina **FERRADASIIDAE**

27b. Without above combination of characters **28**

28a. Integument soft; idiosoma somewhat elongated; all coxal plates grouped together, with medial edges of anterior coxal group noticeably longer than those of posterior group; suture lines between coxal plates III–IV extending at right angles to midline or slightly posterolaterally (Fig. 13.29I); interstitial **OMARTACARIDAE**

28b. Without above combination of characters **29**

29a. Palpfemur, with extremely rare exceptions, bearing a ventral seta (Fig. 13.30A); all females, and males of many genera, with genital acetabula borne on movable genital flaps (Fig. 13.30B) **LIMNESIIDAE**

29b. Without above combination of characters **30**

30a. Palpfemur bearing 2–4 ventral setae (Figs. 13.30C, D) and not uncate **31**

30b. Palpfemur usually not bearing ventral setae, but when bearing ventral setae, then palp uncate (Fig. 13.34F) **32**

31a. Dorsal and ventral shields present; a pair of glandularia located in indentations at posterior end of ventral shield (Fig. 13.30E); palptibia and tarsus long, slender, and functionally fused, genu much higher than long (Fig. 13.30C). Known only from interstitial waters in Japan **NIPPONACARIDAE**

31b. Dorsal and ventral shields present but without free glandularia at posterior end of ventral shield (Fig. 13.30F); palpgenu and tibia extremely short (Fig. 13.30D) or fused together. Known from widely scattered areas of North America and Europe **BOGATIIDAE**

32a. Integument soft; all coxal plates fused on their respective sides; coxal plates I and IV with short, peglike setae (Fig. 13.30G); male with a long petiole (Fig. 13.30H); female with small acetabular plates bearing 4 pairs of acetabula and flanking a long slitlike gonopore (Fig. 13.30I). Known only from gill cavities of freshwater crayfish in Australia **ASTACOCROTONIDAE**

32b. Without above combination of characters **33**

33a. Tibia I much longer than tarsus (Fig. 13.31A); claw of leg I flexing proximally rather than distally (Fig. 13.31A) **MOMONIIDAE**

33b. Without above combination of characters **34**

34a. Legs I with claw socket at least half as long as tarsus and with tarsus longer than tibia (Fig. 13.31B); idiosoma variously sclerotized but never with a bipartite dorsal shield **35**

34b. Legs I with claw socket usually less than half as long as tarsus, when more than half as long as tarsus, then either tarsus shorter than tibia or idiosoma with a bipartite dorsal shield **36**

35a. Idiosoma flattened dorsoventrally; legs IV inserted medially near midline (Fig. 13.31C); dorsal shield consisting of a central plate surrounded by 8 or 9 paired platelets (Fig. 13.31D) **LETHAXONIDAE**

35b. Idiosoma usually somewhat laterally compressed (Figs. 13.31E, F); legs IV inserted laterally; dorsal shield usually absent, but South African genus *Stormaxonella* has dorsal shield consisting of a central plate surrounded by 5 pairs of platelets (Fig. 13.31G) **WETTINIDAE**

36a. Idiosoma strongly laterally compressed (Figs. 13.32A, B); segments of legs IV dorsoventrally expanded and laterally flattened (Fig. 13.32C) **FRONTIPODOPSIDAE**

36b. Without above combination of characters **37**

37a. Dorsal and ventral shields present; genital acetabula scattered laterally in integument as well as in gonopore region (Fig. 13.32D); palp uncate, palptibia greatly expanded ventrally and tarsus elongate and sharply pointed (Fig. 13.32E). Known from West Africa and India **HARPAGOPALPIDAE**

37b. Without above combination of characters **38**

38a. Dorsal and ventral shields present; palp uncate with palptibia rotated approximately 90° relative to genu (Fig. 13.32F); capitulum with anterior end pointed in lateral view but wide in dorsal view and bearing a pair of long subterminal setae (Figs. 13.32F, G) **ATHIENEMANNIIDAE**

38b. Without above combination of characters **39**

39a. Medial margins of combined coxal plates III–IV reduced to sharp angles, with suture line between coxal plates III–IV incomplete (Fig. 13.32H); all coxal plates close together with no body pores separating them; no glandularia on coxal plates IV **ACALYPTONOTIDAE**

39b. Without above combination of characters **40**

40a. Dorsal and ventral shields present; ventral shield with suture lines between coxal plates III–IV extending posteromedially to genital field region and well separated from each other medially (Fig. 13.33A); no glandularia on coxal plates IV; genital acetabula in single rows on each side (Fig. 13.33A), 3–5 pairs lying in gonopore in male, 5–9 pairs flanking gonopore in females; palpi uncate (Figs. 13.33B, C) **NEOACARIDAE**

40b. Without above combination of characters **41**

41a. Dorsal and ventral shields present; dorsal shield with 4 pairs of glandularia, including a pair anterior to postocular setae (at or near anterior edge of shield) (Fig. 13.33D); ventral shield variously modified, with either margins of first 3 pairs of coxal plates rounded and without a capitular bay (Fig. 13.33E), median margins of coxal plates IV long, and suture line between coxal plates III–IV extending at right angles to long axis of body (Fig. 13.33F) or with coxoglandularia 2 located at anteromedial corner of coxal plates IV (Fig. 13.33G, arrow); palpi various, from slightly modified (Fig. 13.33H) to uncate (Fig. 13.33I), or highly modified (Fig. 13.33J) **CHAPPUISIDIDAE**

41b. Without above combination of characters **42**

42a. Dorsal and ventral shields present; coxoglandularia 1 shifted far forward on coxal plates II (Figs. 13.34A, B, arrows); genital acetabula of female either lying in gonopore (Fig. 13.34A) or on acetabular plates closely flanking the gonopore (Fig. 13.34B) **43**

42b. Dorsal and ventral shields present or absent; coxoglandularia 1 not shifted far forward on coxal plates II; genital acetabula variously arranged **44**

43a. Suture lines between coxal plates III–IV ending far anterior to genital field region (Fig. 13.34A); genital field region of male not highly modified **NUDOMIDEOPSIDAE**

43b. Suture lines between coxal plates III–IV extending to genital field region (Fig. 13.34B); male genital field region highly modified (Fig. 13.34C) **MIDEIDAE**

44a. Dorsal and ventral shields present; coxoglandularia 1 located slightly anterior to suture line between coxal plates II–III (Figs. 13.34D, E); male with acetabula both on acetabular plates and in gonopore and with genital field extending anteriorly to level of coxal plates II (Fig. 13.34E); females with genital acetabula on acetabular plates extending well lateral to gonopore (Fig. 13.34D) **LAVERSIIDAE**

44b. Without above combination of characters **45**

45a. Dorsal and ventral shields present; palpi uncate with palpfemur bearing 2 medioventral setae (Fig. 13.34F) in species from Old World but with patch of medial setae in a genus known from Mexico and Costa Rica; genital acetabula numerous, lying on acetabular plates flanking gonopore in females (Fig. 13.34G) but confined to gonopore in males (Fig. 13.34H) **HUNGAROHYDRACARIDAE**

45b. Without above combination of characters **46**

46a. Dorsal and ventral shields present; openings for insertion of legs IV covered by lobed extensions of coxal plates IV (Fig. 13.35A); 3 pairs of genital acetabula closely flanking gonopore; dorsal shield bearing 6 pairs of glandularia. Known only from a female specimen taken in interstitial waters in Japan **KANTACARIDAE**

46b. Without above combination of characters **47**

47a. Dorsal and ventral shields present; coxal plates IV with posterior suture lines obliterated and lacking projections associated with insertions of legs IV (Figs. 13.35B, C); males with 1–2 pairs of genital acetabula lying in gonopore, as well as additional acetabula present on integument posteriorly (Fig. 13.35B); females with all acetabula lying free in integument flanking gonopore (Fig. 13.35C); palptibia with slight distoventral bulge but not forming a true uncate palp (Fig. 13.35D). Known only from interstitial waters in North and South America **ARENOHYDRACARIDAE**

47b. Without above combination of characters **48**

48a. Dorsal and ventral shields present; palpi uncate; median margins of coxal plates IV reduced to medial angles and bearing a pair of glandularia at angles (Fig. 13.35E, arrow); genital acetabula located in gonopore in both sexes; openings for insertion of legs IV covered by extensions of coxal plates IV **KRENDOWSKIIDAE**

48b. Without above combination of characters **49**

49a. Dorsal and ventral shields present; male genital field located at extreme posterior end of idiosoma and with 2 pairs of acetabula in gonopore and several pairs flanking it (Fig. 13.35F); openings for insertion of legs IV covered by projections of coxal plates IV, which extend well lateral to sides of idiosoma; males with tarsi IV greatly expanded (Fig. 13.35G); females with numerous genital acetabula lying on plates that flank gonopore but are separate from remainder of ventral shield; openings for insertion of legs IV with associated projections (Fig. 13.35H). Known only from southeastern United States **AMOENACARIDAE**

49b. Without above combination of characters **50**

50a. Dorsal and ventral shields present; typically with 3 pairs of genital acetabula but in rare instances up to 7 pairs; acetabula confined to gonopore in both sexes (Fig. 13.35I); openings for insertion of legs IV with either no or only very small projections (Fig. 13.35I); palpi typically as shown in Figs. 13.35J and 13.35K, but, in rare instances (5 known species from scattered areas in Southern Hemisphere), the palpi are uncate **MIDEOPSIDAE**

50b. Without above combination of characters **51**

51a. Dorsal and ventral shields present; palpi uncate (with exception of a species of *Micruracaropsis*, known only from trapped water in an epiphytic bromeliad in Surinam); genital acetabula numerous and lying on acetabular plates extending laterally from gonopore with none in gonopore (exception: a species of *Thoracophoracarus* from Chile in which there is 1 pair of acetabula in male gonopore); idiosoma of female relatively unmodified (Fig. 13.36A) but posterior end of idiosoma in males moderately to extremely modified (Figs. 13.36B–D) **ARRENURIDAE**

51b. Without above combination of characters **52**

52a. Sclerotization of dorsum varying from scattered platelets of varying size up to a complete dorsal shield; integument between coxal plates IV and genital field with 2 pairs of glandularia arranged more or less in a row (Fig. 13.36E, arrows); palptibia lacking a peglike seta **FELTRIIDAE**

52b. Sclerotization of dorsum and venter variable; typically without 2 pairs of glandularia in a row between coxal plates IV and genital field; when glandularia arranged as above, then palptibia bearing a peglike seta (Fig. 13.36F, arrow) **53**

(The following four families—HYGROBATIDAE, ATURIDAE, PIONIDAE, and UNIONICOLIDAE—contain a large number of genera assigned to diverse subfamilies of varying relationship to other members of the family in which they have been placed. Because the larval and nymphal stages are unknown for so many genera, especially from the Southern Hemisphere, their assignment to particular families has often been completely arbitrary. Until a thorough analysis and revision of these four families can be undertaken, the following couplets are offered for identification purposes. They should key out known Holarctic species but may not be suitable for all species from other areas.)

53a. A pair of glandularia present on coxal plates IV (Fig. 13.36G, arrow), occasionally extending far forward on anteriorly directed loops of coxal suture line (Fig. 13.36H) **54**

53b. Glandularia absent on coxal plates IV **55**

54a. Usually soft bodied, but a few scattered dorsal or ventral platelets may be present; if complete dorsal and ventral shields present (usually only in male), then with a downturned seta on tibia I (Fig. 13.36I, arrow) **HYGROBATIDAE**

54b. Dorsal and ventral shields present in both sexes; without a downturned seta on tibia I **ATURIDAE (pars)**

55a. Dorsal and ventral shields present **56**

55b. Soft bodied or with scattered platelets but without distinct dorsal and ventral shields **58**

56a. Genital acetabula numerous (more than 10 pairs); genu of leg IV either concave on one side and with numerous peglike setae (Fig. 13.37A) or with median surface of palptibia bearing a peglike seta at distal end (Fig. 13.37B) and coxal plates lacking well-developed projections associated with insertions of legs IV **PIONIDAE (*pars*)**

56b. Without above combination of characters **57**

57a. Openings for insertions of legs IV with large associated projections that extend laterally or only slightly posteriorly, or, when projections extend decidedly posteriorly (Fig. 13.37C), then palptibia with lateral projections as shown in Fig. 13.37D (species with the latter set of characters are known only from West Africa and Australia) **UNIONICOLIDAE (*pars*)**

57b. Openings for insertions of legs IV with either small or no associated projections (Fig. 13.37K), or, when large projections present they are directed posteriorly or decidedly posterolaterally (Fig. 13.37J) and palptibia lacks lateral projection shown in Fig. 13.37D **ATURIDAE (*pars*)**

58a. Claws simple (Fig. 13.37E) or with clawlets (Fig. 13.37F); legs I usually but not always with numerous long thickened setae or *"Rillborsten"* (Fig. 13.37G); posterior margins of coxal plates IV truncate or only slightly rounded; chelicera either fused or separate medially; most species are commensals in freshwater Mollusca (usually clams) or sponges, but there are free-living species **UNIONICOLIDAE (*pars*)**

58b. Claws always with clawlets; legs I never with long thickened setae as illustrated above; posterior margins of coxal plates IV more or less strongly angled as indicated in Figs. 13.37H and 13.37I; chelicerae separate medially; typically free living as adults, but members of the Holarctic genus *Najadicola* live as commensals in freshwater Mollusca **PIONIDAE (most species)**

59a. Pregenital tubercle present (Figs. 13.39F, H, J, arrows) **Superfamily Tanaupodioidea, TANAUPODIDAE**

59b. Pregenital tubercle absent (Figs. 13.40H, 13.42I) **60**

60a. Idionotum usually bearing numerous setiferous, fingerlike projections and with central area densely covered by long setae (Figs. 13.40A, B) **Superfamily Chyzerioidea, CHYZERIIDAE**

60b. Idionotum lacking setiferous, fingerlike projections **Superfamilies Trombiculoidea + Trombidioidea—61**

61a. Idionotal setae trifurcate or spoon shaped (Fig. 13.40I, arrow) **NEOTROMBIDIIDAE**

61b. Idionotal setae variously shaped but not trifurcate or spoon shaped (Fig. 13.41D, E, H, 13.42L–N) **62**

62a. Idionotum either entirely surrounded by peripheral sclerotized band or bearing 2–16 circular depressions with peripheral sclerotized bands (Figs. 13.41A, B) **TROMBELLIDAE (*pars*)**

62b. Idionotum lacking both peripheral sclerotized band and circular depressions **63**

63a. Idionotal setae borne on raised papillae (Figs. 13.41C, D) **64**

63b. Idionotal setae not borne on raised papillae **65**

64a. Idionotal setae borne on basal sclerites (Figs. 13.41C, D) **AUDYANIDAE**

64b. Idionotal setae not borne on basal sclerites **TROMBELLIDAE (*pars*)**

65a. Palptibia bearing large spine ventrally or laterally (Fig. 13.40G, arrow); idionotal setae nude or sparsely barbed (Fig. 13.40F) **JOHNSTONIANIDAE**

65b. Palptibia lacking large spine ventrally or laterally (although large dorsal spine may be present adjacent to tibial claw); idionotal setae usually branched (Figs. 13.41E, H, 13.42B, E, L–N) **66**

66a. Idiosoma usually conspicuously constricted between legs II–III; palptibia bearing group of 2–3 lateral spinose setae in tandem (Fig. 13.40D, arrow) **67**

66b. Idiosoma not constricted between legs II–III; palptibia lacking group of lateral spinose setae in tandem **68**

67a. Prodorsal plate bearing 2 *vi* setae anteriorly; palpfemur bearing more setae than genu **LEEUWENHOEKIIDAE**

67b. Prodorsal plate bearing 0–1 *vi* setae anteriorly; palpfemur bearing about same number of setae as genu **TROMBICULIDAE**

68a. Idionotum bearing large plate posteriorly (Figs. 13.42A, F) **EUTROMBIDIIDAE**

68b. Idionotum lacking plate posteriorly (Fig. 13.42H) **69**

69a. Palptibia bearing 1 or more large, spinose setae adjacent to claw and with other spinose setae arranged in conspicuous rows (Figs. 13.42G, K, O) **70**

69b. Palptibia usually lacking large spinose setae adjacent to claw and either lacking any spinose setae or bearing spinose setae that are not arranged in rows (Figs. 13.41G, K); when bearing large spinose setae arranged in rows, then anterior edge of prodorsal plate concave (Fig. 13.41I) **TROMBIDIIDAE**

70a. Prodorsal plate bearing 1–2 setae anteriorly; eyes either absent or incorporated into prodorsal plate; idionotal setae with basal spines or bifurcate **NEOTHROMBIIDAE**

70b. Prodorsal plate usually bearing more than 2 setae anteriorly (Fig. 13.42J); eyes conspicuously present and not incorporated into prodorsal plate (Fig. 13.42J); idionotal setae setiform or highly modified (Figs. 13.42H, L–N) **MICROTROMBIDIIDAE**

71a. Prodorsal area bearing 1 pair of bothridial organs (Fig. 13.38F) **Superfamily Calyptostomatoidea, CALYPTOSTOMATIDAE**

71b. Prodorsal area bearing 2 pairs of bothridial organs (Fig. 13.39A) **Superfamily Erythraeoidea—72**

72a. Gnathosoma entirely retractable into idiosoma (Fig. 13.38H); idionotal setae usually flattened, broad, and serrate (Fig. 13.38H) **SMARIDIDAE**

72b. Gnathosoma with only chelicerae retractable into idiosoma (Fig. 13.39A); idionotal setae usually setiform (Fig. 13.39A) **ERYTHRAEIDAE**

Key 13.3. Cohort Parasitengonina, Larvae, Key to the Families (Figs. 13.43–13.60)

1a. Palpgenu bearing 2 setae (Figs. 13.44E, 13.46A, 13.48A, 13.56A). Active in or on water and parasitic on aquatic insects **2**

1b. Palpgenu usually bearing 1 seta (Figs. 13.56H, 13.58I), rarely 2–4 setae (Fig. 13.56F). Active on terrestrial substrata and parasitic on terrestrial hosts **3**

2a. Prodorsum bearing paired *vi* setae anterolaterally (Figs. 13.43D, 13.44B, E, 13.46C); venter with urstigmata sessile (Figs. 13.43E, 13.44A, 13.45C, 13.49H, 13.52C, 13.55G); palptibial claw simple or bisected (Figs. 13.46E, G, 13.49F); legs with claws not pectinate (Figs. 13.49J, 13.51F) **Subcohort Hydrachnidiae—4**

2b. Prodorsal plate bearing an unpaired *vi* seta anteromedially (Fig. 13.56A, arrow); venter with urstigmata stalked (Figs. 13.56A, C); palptibial claw 4-pronged (Fig.13.56B, arrow); legs with claws pectinate (Fig. 13.56D) **Subcohort Stygothrombiae, Superfamily Stygothrombidioidea, STYGOTHROMBIDIIDAE**

3a. Urstigmata present between coxal plates I–II (Figs. 13.57G, I, 13.58I, 13.59B, E, 13.60H); anus usually present (Figs. 13.57G, I, 13.59B); coxal plates I–II contiguous or nearly so (Figs. 13.57G, I, 13.58I, 13.59B, E, 13.60H) **Subcohort Trombidiae—41**

3b. Urstigmata usually absent between coxal plates I–II (Figs. 13.56J, 13.57B) (exception: CALYPTOSTOMATIDAE; Fig. 13.56G); anus usually absent (Figs. 13.56J, 13.57B) (exception: CALYPTOSTOMATIDAE; Fig. 13.56G); coxal plates I–II usually widely separated (Figs. 13.56J, 13.57B) (exception: CALYPTOSTOMATIDAE; Fig. 13.56G) **Subcohort Erythraiae—41**

4a. Legs with 6 movable segments, with basifemur and telofemur separated (Fig. 13.43A) **5**

4b. Legs with 5 movable segments, with basifemur and telofemur fused (Fig. 13.43B) **13**

5a. Gnathosoma with elaborate camerostome enclosing chelicerae, with palpi inserted ventrally (Fig. 13.43C); dorsal plate present and bearing only 2 pairs of setae (verticils—*ve* and *vi*) near anterior edge (Figs. 13.43D, F); coxal plates III located posteriorly on idiosoma with insertions of legs III located at posterolateral edges (posterior to level of excretory pore) and legs III directed posteriorly (Figs. 13.43E, G) **Superfamily Hydrovolzioidea—6**

5b. Gnathosoma lacking elaborate camerostome, palpi inserted anteriorly; dorsal plate absent (Fig. 13.43H), or present and bearing more than 2 pairs of setae with verticils (*ve* and *vi*) and at least internal scapulars (*si*) (Fig. 13.43I); coxal plates III usually located near midlength of idiosoma with insertions of legs III located anterolaterally (anterior to level of excretory pore) and legs III directed laterally (Fig. 13.44A) (exception: APHEVIDERULICIDAE) 7

6a. Dorsal plate relatively small (Fig. 13.43D); venter with rows of numerous small urstigmata between coxal plates I–II (Fig. 13.43E, arrow); coxal plates I–II separated from one another on each side and from opposite members medially (Fig. 13.43E); palpi massive, with tibia bearing 4 long, thick, curved setae (Fig. 13.43J); setae *c3* often foliate (Fig. 13.43D) **HYDROVOLZIIDAE**

6b. Dorsal plate relatively large (Fig. 13.43F); venter with urstigmata inconspicuous and apparently absent (Fig. 13.43G); coxal plates I–II fused together to form single anterior plate (Fig. 13.43G); palpi moderate in size, with tibia bearing 2 slightly thickened, straight setae (Fig. 13.43C) **ACHERONTACARIDAE**

7a. Dorsal plate absent (Fig. 13.43H) or small, covering less than 1/3 length of idiosoma and usually with internal scapular setae (*si*) at posterior edge (Fig. 13.44B); lateral eyes borne on separate platelets (Fig. 13.44B); leg tarsi with unmodified claws and empodium (emp) (Fig. 13.44C) **Superfamily Hydryphantoidea—8**

7b. Dorsal plate large, covering well over 1/3 length of idiosoma and with internal scapular setae (*si*) near midlength (Figs. 13.44D–F); lateral eyes borne on single eye plates (Fig. 13.44E); leg tarsi with claws modified or reduced (Figs. 13.44G, H) **Superfamily Eylaoidea—10**

8a. Claw (movable digit) of chelicerae over 1/2 length of basal segment (Fig. 13.44I); excretory pore setae (*ps1* and *ps2*) and their alveoli absent (Fig. 13.44J) **THERMACARIDAE**

8b. Claw (movable digit) of chelicerae less than 1/3 length of the basal segment (Fig. 13.44K); excretory pore setae (*ps1* and *ps2*), or at least their alveoli, present (Figs. 13.44L, M) **9**

9a. Palptarsus with all setae slender; solenidia on leg tarsi slender (Fig. 13.44N, arrow) **HYDRYPHANTIDAE**

9b. Palptarsus with 2 thickened, bladelike setae (Fig. 13.44O, arrows); solenidia on tarsi I–II very thick (Fig. 13.44P, arrow) **HYDRODROMIDAE**

10a. Coxal plates I–III on each side all bearing 2 or more setae (Figs. 13.45A, C, E); tarsi of legs bearing paired claws and claw-like empodium (Fig. 13.44H) **11**

10b. Coxal plates I–III on each side bearing 2 or 1, 1, and 1 setae respectively (Fig. 13.45D); tarsi of legs bearing 2 dissimilar clawlike structures terminally (Fig. 13.44G) **LIMNOCHARIDAE**

11a. Coxal plates I–III on each side bearing about 25, 16, and 13 blunt, conical setae, respectively (Fig. 13.45E); palptrochanter, femur, genu, and tibia fused into single segment bearing 6 setae, palptarsus small and bearing only 2 setae + a solenidion (Fig. 13.45F) **APHEVIDERULICIDAE**

11b. Coxal plates I–III on each side bearing 2, 2, and 2 setae, respectively (Fig. 13.45C); palpi with 5 movable segments (Fig. 13.45A) and palptarsus bearing more than 2 setae + a solenidion (Fig 13.45C) **12**

12a. Idionotum nearly covered by single, elongate dorsal plate in unengorged larvae, bearing 7 pairs of setae (or their alveoli) (Fig. 13.44D); coxal plates with setae all simple (Fig. 13.45A); excretory pore plate elongate (Fig. 13.45B) **EYLAIDAE**

12b. Idionotum with 3 separate plates bearing 5, 1, and 1 pairs of setae, respectively (Fig. 13.44F); coxal plates with setae *1a*, *1b*, *2b*, and *3b* blunt and conical in shape (Fig. 13.45C); excretory pore plate obcordate (Fig. 13.45C) **PIERSIGIIDAE**

13a. Dorsal plate bearing 8 pairs of setae, including verticils, scapulars, *c3*, and 3 additional pairs of hysterosomatic setae (Fig. 13.45G); leg tarsi lacking paired claws (Fig. 13.45I); excretory pore plate tiny and bearing neither setae nor their alveoli (Fig. 13.45H) **Superfamily Hydrachnoidea, HYDRACHNIDAE**

13b. Dorsal plate bearing 4–6 pairs of setae, including only verticils and scapulars (Fig. 13.46A) or, in some cases, verticils, scapulars, and 1–2 pairs of hysterosomatic setae (Fig. 13.46C); leg tarsi bearing paired claws (Fig. 13.53D); excretory pore plate small to large, always bearing setae *ps1* and *ps2* or at least their alveoli (Figs. 13.46B, 13.53E) **14**

14a. Coxal plates I–III on each side all separate (Figs. 13.46F, H), or all fused with plates of the two sides fused medially (Fig. 13.49E) and dorsal plate round and bearing setae *c1* laterally (Fig. 13.49D) **15**

14b. Coxal plates I–II on each side separate and plates II–III fused at least medially (Fig. 13.47C), or all plates on each side fused (Fig. 13.48E) with plates of two sides separate medially and dorsal plate elliptical and not bearing setae *c1* (Fig. 13.48C) **27**

15a. Tarsus III bearing 12 or more setae, including *Ta8* (Fig. 13.46D, arrow); dorsal plate usually elongate and elliptical and bearing only 4 pairs of setae (verticils and scapulars) (Fig. 13.46A); when dorsal plate round and bearing 5 pairs of setae, including *c1* laterally (Fig. 13.46C), then coxal plates III bearing setae *pa* and *h2* in addition to *3a* (Fig. 13.46H) and all setae on palptarsi short and bladelike (Fig. 13.46E) **Superfamily Lebertioidea (*pars*)—16**

15b. Tarsus III usually bearing 11 or fewer setae, lacking at least *Ta8* (Figs. 13.50C, E, 13.51C, E); dorsal plate usually nearly round (Figs. 13.50F, 13.51A, G, H), rarely elongate and elliptical (Fig. 13.49G); when tarsus III bears 12 setae, then dorsal plate round and bearing 5 pairs of setae, including *c1* laterally (Fig. 13.50A) *and* coxal plates III bearing only setae *3a* (Fig. 13.50B) *and* at least 1 seta on palptarsi long and whiplike (similar to Fig. 13.50I) **Superfamily Arrenuroidea—18**

16a. Palptarsi with no setae as long as palp (Fig. 13.46E); tarsi I–III bearing 13 or 14 (+1ω), 13 or 14 (+1ω), and 12 setae, respectively, lacking *Ta15* **ANISITSIELLIDAE (*pars*)**

16b. Palptarsi with at least 1 seta as long as palp (Fig. 13.46G); tarsi I–III bearing 15 (+1ω), 15 (+1ω), and 13 or 14 setae, respectively, including *Ta15* (Fig. 13.46D) **17**

17a. Dorsal plate longitudinally striate (Fig. 13.46A); coxal plates III bearing only setae *3a* (Fig. 13.46F); excretory pore plate small and usually nearly quadrangular (Fig. 13.46B) **SPERCHONTIDAE**

17b. Dorsal plate with reticulate sculpturing (Fig. 13.47A); coxal plates III bearing 2 pairs of setae, *3a* and supernumerary setae *3b* (Fig. 13.47B); excretory pore plate relatively large and diamond shaped (Fig. 13.47B) **TEUTONIIDAE**

18a. Coxal plates I–III on each side all fused and coxal plates of two sides fused medially, with setae *1a*, *2b*, and *3a* reduced to vestiges (Fig. 13.49E); palptarsi with no setae as long as palp (Fig. 13.49F); dorsal plate with setae *si* reduced to vestiges (Fig. 13.49D) **ACALYPTONOTIDAE**

18b. Coxal plates I–III on each side all separated by complete suture lines and coxal plates of two sides separate (Figs. 13.49H, 13.50G); palptarsi with at least 1 seta as long as palp (Fig. 13.50I) **19**

19a. Idiosoma moderately flattened dorsoventrally; gnathosoma projecting beyond anterior edge of dorsal plate, entirely exposed in dorsal view (Fig. 13.49G); excretory pore plate variously shaped but usually not attenuate anteriorly (Fig. 13.49H); palptarsi with long, thick, distal seta straight or only slightly bowed basally; leg genua with setae *Ge5* borne on tubercles that usually are prominent (Fig. 13.49I, arrow) **20**

19b. Idiosoma extremely flattened dorsoventrally; gnathosoma recessed beneath protruding anterior edge of dorsal plate, partially or entirely concealed in dorsal view (in unmounted specimens) (Fig. 13.50F); excretory pore plate triangular with anterior apex attenuate (Fig. 13.50H); palptarsi with long, thick, distal seta strongly bowed, and usually lobed, basally (Fig. 13.50I, arrow); leg genua with setae *Ge5* not borne on tubercles **22**

20a. Dorsal plate bearing 4 pairs of setae (verticils and scapulars), with setae *c1* on lateral membranous integument (Fig. 13.49G); setae *c2* and *d2* thick and long relative to other hysterosomatic setae (Fig. 13.49G); excretory pore setae *ps1* and *ps2* absent, represented by their alveoli on excretory pore plate (Fig. 13.49H); tibia I bearing 8 setae (+2ϕ), including *Ti11*; leg tarsi lacking setae *Ta14* (Figs. 13.49I, J) **MOMONIIDAE**

20b. Dorsal plate bearing 5 pairs of setae, including the verticils, scapulars, and setae *c1* laterally (Fig. 13.50A); setae *c2* and *d2* similar to other hysterosomatic setae (Fig. 13.50A); excretory pore setae *ps1* and *ps2* present on plate (Figs. 13.50B, D), tibia I bearing 7 setae (+2ϕ), lacking *Ti11*; leg tarsi bearing setae *Ta14* (Figs. 13.50C, E, arrows) **21**

21a. Tarsus III bearing 9 setae, lacking *Ta12* (Fig. 13.50E); tarsi II–III with setae *Ta4* and *Ta6* long (usually as long as respective segments) (Fig. 13.50E); coxal plates II (and in some species coxal plates I) with conspicuous denticulate projections on posterior edges (Fig. 13.50D, arrows), coxal plates III with transverse muscle attachment scars; excretory pore plate subtriangular (Fig. 13.50D) **KRENDOWSKIIDAE**

21b. Tarsus III bearing at least 11 setae, including *Ta12* (Fig. 13.50C); tarsi II–III with setae *Ta4* and *Ta6* usually short (conspicuously shorter than respective segments) (Fig. 13.50C); coxal plates I and II lacking denticulate projections on posterior edges, coxal plates III lacking transverse muscle attachment scars; excretory pore plate variously shaped, but rarely triangular (Fig. 13.50B) **ARRENURIDAE**

22a. Dorsal plate bearing 5 pairs of setae, including verticils, scapulars and setae *c1* laterally (Fig. 13.50F) **MIDEOPSIDAE**

22b. Dorsal plate bearing 4 pairs of setae (verticils and scapulars), with setae *c1* on lateral membranous integument (Figs. 13.51A, G, H) **23**

23a. Tibiae II–III bearing 9 setae (+2ϕ and +1ϕ, respectively), including both *Ti10* and *Ti11* (Figs. 13.51C, E) **24**

23b. Tibiae II–III bearing fewer than 9 setae (+2ϕ and +1ϕ, respectively), lacking either *Ti10* or *Ti11*, or both (Figs. 13.51F, I) **25**

24a. Palptarsus with long, thick, distal seta bowed, but not deeply lobed or fringed basally, and with most medial seta moderately thick but not fringed (Fig. 13.51D); legs II–III with setae *Ge5*, *Ti9*, and *Ti11* much longer than respective segments and plumose, and setae *Ta4* and *Ta6* much longer than respective segments (Fig. 13.51E) **MIDEIDAE**

24b. Palptarsus with long, thick, distal seta bowed, deeply lobed and fringed basally, and most medial seta thick and fringed (Fig. 13.50I); legs II–III with setae *Ge5*, *Ti9*, and *Ti11* shorter than respective segments and simple, and setae *Ta4* and *Ta6* shorter than respective segments (Fig. 13.51C) **NUDOMIDEOPSIDAE**

25a. Tibia II bearing 7 setae (+2ϕ), lacking *Ti10* and *Ti11* (Fig. 13.51I); tibia III bearing 8 setae (+1ϕ), lacking *Ti10* **LAVERSIIDAE**

25b. Tibiae II and III bearing 8 setae (+2ϕ and +1ϕ, respectively), including *Ti10* but lacking *Ti11* (Fig. 13.51F) **26**

26a. Dorsum with setae *si* located well in anterior half of plate and setae *c1* located in lateral integument near midlength of plate, posterior to level of setae *si* (Fig. 13.51H) **NEOACARIDAE**

26b. Dorsum with setae *si* near midlength of plate and setae *c1* located in lateral integument near setae *c3*, anterior to level of setae *si* (Fig. 13.51G) **ATHIENEMANNIIDAE**

27a. Coxal plates III bearing setae *3a* and at least 1 other pair of setae (*pa* or *3b*) and excretory pore plate bearing only setae *ps1* and *ps2* (Figs. 13.47C, 13.48B, E, F) **Superfamily Lebertioidea (*pars*)—28**

27b. Coxal plates III usually bearing only 1 pair of setae, *3a* (Figs. 13.49B, 13.52A, C, 13.54C, 13.55A), when coxal plates III also bearing setae *pa*, then excretory pore plate bearing setae *h2* in addition to *ps1* and *ps2* (Figs. 13.52E, 13.54A, D, F) **30**

28a. Coxal plates III truncate posteriorly, bearing 3 pairs of setae, including *3a* and both *pa* and *h2* on posterior edges (Fig. 13.47C); tibiae of legs I–III bearing 9 setae (+2ϕ, +2ϕ, and +1ϕ, respectively), including both *Ti9* and *Ti10* (Fig. 13.47D); tarsi I–II bearing 14 setae (+1ω), including *Ta14* and *Ta15* (Fig. 13.47D); cheliceral bases separate (Fig. 13.47E, arrow) **LEBERTIIDAE**

28b. Coxal plates III rounded or pointed posteriorly, bearing 2 pairs of setae, including setae *3a* and either *3b* laterally (Fig. 13.48B) or *pa* posteromedially (Figs. 13.48E, F); tibiae I–III bearing 8 setae (+2ϕ, +2ϕ, and +1ϕ, respectively), lacking either *Ti9* or *Ti10* (Figs. 13.47F, 13.48G); tarsi I–II bearing 13 setae (+1ω), lacking *Ta15* (Fig. 13.48G), or 12 setae (+1ω), lacking both *Ta14* and *Ta15* (Fig. 13.47F); cheliceral bases fused (Figs. 13.48C, D) **29**

29a. Dorsal plate bearing 5 pairs of setae including verticils, scapulars, and setae *c1* laterally; integument beneath posterior edge of dorsal plate intricately folded (Fig. 13.48A, arrow); coxal plates I elongate, extending posteriorly nearly to level of excretory pore plate; coxal plates III bearing setae *3b* laterally (Fig. 13.48B); tarsi I–II bearing 12 setae (+1ω), lacking *Ta14* (Fig. 13.47F) **OXIDAE**

29b. Dorsal plate bearing 4 pairs of setae (verticils and scapulars) (Fig. 13.48C) or 5 pairs including verticils, scapulars, and setae *d1* laterally (Fig. 13.48D); integument beneath posterior edge of dorsal plate not intricately folded (Fig. 13.48C); coxal plates I not elongate, extending posteriorly only to level of insertion of legs III (Fig. 13.48F), or fused with plates II (Fig. 13.48E); plates III bearing setae *pa* posteromedially (Figs. 13.48E, F); tarsi I–II bearing 13 setae (+1ω), including *Ta14* (Fig. 13.48G) **TORRENTICOLIDAE**

30a. Cheliceral bases separate (Fig. 13.49A); tarsus III bearing 12 setae, including *Ta9* (Fig. 13.48H, arrow); excretory pore plate very large (equal in width to coxal plates of one side), bearing setae *ps1* and *ps2* anteromedially and occasionally also setae *h2* at posterolateral angles, and with excretory pore near anterior edge (Fig. 13.49C); coxal plates I separate from posterior coxal groups, coxal plates III bearing only setae *3a* (Fig. 13.49B) **Superfamily Lebertioidea (*pars*), ANISITSIELLIDAE (*pars*)**

30b. Cheliceral bases fused (Figs. 13.52D, I); tarsus III usually bearing 11 or fewer setae, lacking *Ta9* (Fig. 13.54H) (exception: LIMNESIIDAE); excretory pore plate small to very large, when equal in width to coxal plates of one side (measured from midline to insertion of leg III) (Figs. 13.52C, E), then either excretory pore located near posterior edge of plate (Fig. 13.54C, arrow), or coxal plates I fused to posterior coxal groups (Fig. 13.52C), or coxal plates III bearing setae *pa* in addition to setae *3a* (Fig. 13.54A) **Superfamily Hygrobatoidea—31**

31a. Two pairs of urstigmata borne distally between coxal plates I–II (Fig. 13.52A, arrows); tibiae I–II bearing 7–8 setae (+2ϕ), including only 3 ventral setae, lacking either *Ti8* and *Ti9* or *Ti10* and *Ti11* (Fig. 13.52B) **LIMNESIIDAE**

31b. One pair of urstigmata borne distally between coxal plates I and II (Fig. 13.52C, arrow); tibiae of legs I and II usually bearing 9 or more setae (+2ϕ), including 5 ventral setae (*Ti7, Ti8, Ti9, Ti10*, and *Ti11*) (Fig. 13.52F) (exception: FELTRIIDAE) **32**

32a. Coxal plates I–III on each side all fused (Fig. 13.52C) **HYGROBATIDAE**

32b. Coxal plates I separate from posterior coxal group on each side (Fig. 13.52H) **33**

33a. Tarsi I–II bearing 10 setae (+1ω), lacking *Ta14* and either *Ta12* (Fig. 13.52G) or *Ta9* (Fig. 13.53B) **34**

33b. Tarsi I–II bearing 11 or more setae (+1ω), including *Ta9* (Figs. 13.53I, 13.54B, 13.55E, F, arrows) **35**

34a. Dorsal and coxal plates, and leg sclerites, conspicuously longitudinally striate (as in Fig. 13.52I); coxal plates III rounded posteriorly (Fig. 13.52H); tibia I bearing 7 setae (+2ϕ), lacking *Ti10* and *Ti11* (Fig. 13.52G) **FELTRIIDAE**

34b. Dorsal and coxal plates reticulate (Fig. 13.53A); coxal plates III bearing pointed projections posteriorly (Fig. 13.53C, arrow); tibia I bearing 9 setae (+2ϕ), including *Ti10* and *Ti11* (Fig. 13.53B) **UNIONICOLIDAE**

35a. Tibiae I–III bearing 7 setae (+2ϕ, +2ϕ, and +1ϕ, respectively), lacking *Ti9* and *Ti11* (Fig. 13.53I) **ATURIDAE (*pars*)**

35b. Tibiae I–III bearing 8 or more setae (+2ϕ, +2ϕ, and +1ϕ, respectively), including *Ti9* and *Ti11* (Fig. 13.54B) **36**

36a. Coxal plates III with pointed or lobed projections posteriorly and bearing 2 pairs of setae, *3a* anteromedially and *pa* posteromedially (Fig. 13.54A); excretory pore plate large, bearing 3 pairs of setae, *ps1* and *ps2*, and *h2* posterolaterally (Figs. 13.54D, F) **ATURIDAE (*pars*)**

36b. Coxal plates III without projections (Figs. 13.54C, E, G, 13.55A, G, L), or with small projections (Fig. 13.53F), posteriorly and bearing only setae *3a*; excretory pore plate small or large, bearing only setae *ps1* and *ps2* (Figs. 13.53G, 13.55B) **37**

37a. Coxal plates III with small projections posteriorly (Fig. 13.53F, arrow); excretory pore plate small, little larger than combined area of excretory pore and bases of setae *ps1* and *ps2* (Fig. 13.53G) **WETTINIDAE**

37b. Coxal plates III without projections posteriorly (Figs. 13.54C, G, 13.55A, G, L); excretory pore plate considerably larger than combined area of excretory pore and bases of setae *ps1* and *ps2* (Fig. 13.55B) **38**

38a. Tarsi I bearing 13 setae (+1ω), including *Ta8* (Figs. 13.54B, 13.55E, arrows); excretory pore plate large and usually triangular or obcordate (Figs. 13.54C, E, 13.55B–D) **39**

38b. Tarsi I bearing 12 setae (+1ω), lacking *Ta8* (Fig. 13.55F); excretory pore plate small to large, and variously shaped (Figs. 13.54G, 13.55H–K) **40**

39a. Excretory pore plate with setae *ps1* close together adjacent to excretory pore in posterior half of plate (Figs. 13.55B–D) **PIONIDAE (*pars*)**

39b. Excretory pore plate with setae *ps1* widely separated and removed from excretory pore in anterior half of plate (Figs. 13.54C, E, arrow) **ATURIDAE (*pars*)**

40a. Excretory pore plate small and subtriangular, with setae *ps2* borne near posterolateral angles of plate (Fig. 13.54G); suture lines between coxal plates II–III parallel to anterior edge of plate II, coxal plates without distinct lateral coxal apodemes and coxal plates III lacking medial coxal apodemes and transverse muscle attachment scars (Fig. 13.54G) **ATURIDAE (*pars*)**

40b. Excretory pore plate variously shaped (Figs. 13.55H–K), when subtriangular with setae *ps2* borne near posterolateral angles of plate then suture lines between coxal plates II–III terminating in distinct lateral coxal apodemes that are usually nearly transverse and coxal plates III bearing at least medial coxal apodemes (Fig. 13.55L) **PIONIDAE (*pars*)**

41a. Urstigmata present between coxal plates I–II (Figs. 13.56G, 13.57G, I, 13.58I, 13.59B, E, 13.60H); anus usually present (Figs. 13.56G, 13.57G, I, 13.59B); prodorsal region with 1 or 2 pairs of trichobothria (Figs. 13.56E, 13.57F, I, 13.58B, D, E, G, 13.59A, 13.60B, E, F, I); coxal plates I–II contiguous or nearly so (Figs. 13.56G, 13.57G, I, 13.58I, 13.59B, E, 13.60H) **42**

41b. Urstigmata absent (Figs. 13.56J, 13.57B); anus usually absent (Figs. 13.56J, 13.57B); prodorsal region usually with 2 pairs of trichobothria (Figs. 13.56H, 13.57A, D); coxal plates I–II widely separated (Figs. 13.56J, 13.57B) **Superfamily Erythraeoidea—54**

42a. Prodorsal plate (scutum) present (Figs. 13.57F, H, 13.58B, D, E, G, 13.60B, D, F, I); coxal plates usually bearing 1–2 (rarely many) setae each (Figs. 13.57G, I, 13.58I, 13.59B, E, 13.60H); palpgenu usually bearing fewer than 2 setae (Figs. 13.57I, 13.58I, 13.59A, 13.60H) **43**

42b. Prodorsal plate absent (Fig. 13.56E); coxal plates bearing 15 or more setae each (Fig. 13.56G); palpgenu bearing 2–4 setae (Fig. 13.56F) **Superfamily Calyptostomatoidea, CALYPTOSTOMATIDAE**

43a. Idionotum with unpaired prodorsal plate, and at least 1 other unpaired sclerite (scutellum) bearing at least setae *c1* (Figs. 13.59D, G, 13.60B, D, F, I) **Superfamily Trombidioidea—44**

43b. Idionotum with unpaired prodorsal sclerite but lacking scutellum, with setae *c1* located on paired platelets (Figs. 13.57F, I, 13.58A, F, I, 13.59A) **47**

44a. Femur I bearing 5 setae (Fig. 13.59I); femur II usually bearing 4 (rarely 5) setae; genua of legs II–III each bearing 3–4 setae **TROMBIDIIDAE**

44b. Femur I bearing 6 setae; femur II bearing 4 or 5 setae (Fig. 13.60C); genua of legs II–III each usually bearing 2 (rarely 4) setae **45**

45a. Femur II bearing 4 branched setae; genu of leg II lacking microseta κ **NEOTHROMBIIDAE**

45b. Femur II with 5 branched setae; genu II usually bearing microseta κ **46**

46a. Coxal plates with setae *1b*, *3b*, and usually *2b*, thick and bifid (Fig. 13.60G); tarsus I with famulus usually distal to solenidia ω; parasitic on Orthoptera **EUTROMBIDIIDAE**

46b. Coxal plates with setae *1b*, *2b*, and *3b* setiform (Fig. 13.60H); tarsus I with famulus proximal or distal to solenidia ω; parasitic on Diptera **MICROTROMBIDIIDAE**

47a. Femora I–III usually lacking solenidia θ; parasitic on arthropods or vertebrates **48**

47b. Femur I bearing 1 or more solenidia θ; parasitic on arthropods **Superfamily Trombiculoidea (*pars*)—52**

48a. Anal plate present (Figs. 13.57G, I); parasitic on arthropods **49**

48b. Anal plate absent; usually parasitic on vertebrates (rarely arthropods) **Superfamily Trombiculoidea (*pars*)—50**

49a. Coxal plates II–III each bearing 2–4 setae (Fig. 13.57G) **Superfamily Tanaupodoidea, TANAUPODIDAE**

49b. Coxal plates II–III each bearing 1 seta (Fig. 13.57I) **Superfamily Chyzerioidea, CHYZERIIDAE**

50a. Pretarsi I–II each bearing paired claws and clawlike empodium (as in Fig. 13.59F). Parasitic on vertebrates **51**

50b. Pretarsi I–II each with paired claws but lacking empodium. Parasitic on scorpions **AUDYANIDAE**

51a. Genu II bearing 4 barbed setae; prodorsal plate usually bearing 2 (rarely 1) *vi* setae (Fig. 13.58E) **LEEUWENHOEKIIDAE**

51b. Genu II bearing 3 barbed setae; prodorsal plate bearing 0–1 *vi* setae (Fig. 13.58D) **TROMBICULIDAE**

52a. Femora II–III each bearing 2 or more solenidia θ; pretarsus with empodium **53**

52b. Femora II–III each bearing 0–1 solenidia θ (Fig. 13.58C); pretarsus lacking empodium **JOHNSTONIANIDAE**

53a. Femur I bearing 7 branched setae (Fig. 13.58H); coxal plates distinctly reticulate (Fig. 13.58J) **NEOTROMBIDIIDAE**

53b. Femur I bearing 5–6 branched setae (Fig. 13.59C); coxal plates smooth (Fig. 13.59B) **TROMBELLIDAE**

54a. Leg I bearing 1 or more trichobothria on genu, tibia, and/or tarsus (Fig. 13.56I, arrows); tibia I bearing 3 or more solenidia ϕ (Fig. 13.56I) **SMARIDIDAE**

54b. Legs lacking trichobothria; tibia I usually bearing 2 (rarely 3) solenidia ϕ (Fig. 13.57C) **ERYTHRAEIDAE**

Supercohort Eleutherengonides

This assemblage consists of two sister cohorts that include 13 superfamilies of Prostigmata. Few of these superfamilial aggregations are entirely limited to predation, and most of them illustrate highly specialized lifestyles in at least some of their constituent families. Phytophagy is a major evolutionary trend in the cohort Raphignathina, although members of several raphignathine families are parasitic or predaceous. Many members of the cohort Heterostigmatina have established intimate phoretic, parasitic, or other symbiotic relationships with insects and other animals. Some have evolved as fungivores or phytophages, while others are predators, often in singular habitats.

In contrast with members of the supercohorts Eupodides and Labidostomatides, but in common with the Parasitengonina, those of the supercohorts Anystides and Eleutherengonides typically have a distal palptibial seta that is modified into a clawlike element (e.g., Fig. 13.62B) that often extends beyond the reduced, laterally displaced tarsus so that it is terminal on the palp. This palpal apotele or **thumb-claw process**, however, is secondarily reduced or lost in some members of the cohort Raphignathina and in many of the Heterostigmatina. The fixed cheliceral digits are reduced or lacking, and the movable cheliceral digits are hooklike or styletlike and are inserted apically or subapically on the closely contiguous or fused cheliceral bases. When present, solenidia on tarsi I are erect rather than closely appressed to the segment.

The supercohort Eleutherengonides is considered in this treatment to include two cohorts, the Raphignathina with five superfamilies and the Heterostigmatina with eight superfamilies. The hierarchic scheme used here is based in part on that suggested by Lindquist (1998) and by Norton et al. (1993).

Cohort Raphignathina

The Raphignathina comprises five superfamilies with about 4,000 described species accommodated in approximately 360 genera of small to moderate-sized mites. About half of the known species are obligate plant parasites in the superfamily Tetranychoidea, including most of the major acarine pests of agriculture. More than 20% of the described raphignathine species are parasites of mammals (Myobioidea, Cheyletoidea), birds (Cheyletoidea), and, to a lesser extent, reptiles (Pterygosomatoidea, Cheyletoidea). Relatively few species are parasites of arthropods (species of *Pimeliaphilus* in PTERYGOSOMATIDAE and a few CHEYLETIDAE) or phoretic on insects (CHEYLETIDAE, STIGMAEIDAE, DASYTHYREIDAE). As far as is known, most of the remaining described species are free-living predators in soil, on vegetation, on tree trunks or logs, in moss and nests, and, rarely, on the surface of freshwater (HOMOCALIGIDAE, STIGMAEIDAE).

Raphignathines range from soft bodied to enclosed in a series of sclerotized shields or single dorsal and ventral plates (Figs. 13.71A–E). The prodorsum often has a prodorsal plate bearing three to four pairs of setae, none of which are bothridial. Prodorsal and opisthosomatic neotrichy occurs in the PTERYGOSOMATIDAE (Fig. 13.62C) and the DASYTHYREIDAE (Fig. 13.65B) and on the opisthosoma of some Tetranychoidea as well. Genital papillae are absent. Stigmata typically are present and open between the bases of the chelicerae. They are connected to a short set of chambered peritremes on the leading edge of the prodorsum, emerge from it as hornlike structures (Figs. 13.62B, D), or sometimes are elaborated on the dorsal surface of the chelicerae (Figs. 13.65A, E, 13.72A, B). Primitively, the chelicerae are separate with highly regressed fixed digits, each of which forms a sheath for a usually styletlike movable digit (Figs. 13.68A, F, I, 13.72B–D). However, the chelicerae are fused into a stylophore in the Tetranychoidea (Fig. 13.74C) and many Raphignathoidea (Figs. 13.66A, B, D, 13.68D) and are fully integrated with the subcapitulum into a gnathosomatic capsule in the Myobioidea and Cheyletoidea (Figs. 13.77A, D). The palpi usually have four free segments and often have a well-developed thumb-claw process, but they may be highly regressed in vertebrate parasites. The pretarsal empodia often have tenent hairs or other modifications but are sometimes regressed; claws may also have tenent hairs or rarely are lost (e.g., Fig. 13.68E).

Adult males are typically smaller than females and have a sclerotized aedeagus that can be extruded from a terminal or, more rarely, dorsal opening. Arrhenotoky may be the basic genetic system in the cohort, and thelytoky is common in some families. Those of Myobioidea and Pterygosomatoidea have three nymphal stases, although the larva and tritonymph are calyptostases in the Pterygosmatoidea. In most other raphignathine families, the tritonymph appears to be suppressed (the XENOCALIGONELLIDIDAE in the Raphignathoidea and the TUCKERELLIDAE in the Tetranychoidea are exceptions).

▪ Superfamily Myobioidea (Fig. 13.61)

DIAGNOSIS: Soft-bodied, dorsoventrally flattened mites with legs I with or without claws, shortened and modified for grasping hairs, immatures with ventral unpaired clasping organ on tarsus I; legs II–IV with claws, but without empodia or tenent hairs; immatures with femora and genua fused on all legs; coxal seta 2a *present; coxal fields vestigial. Prodorsum with four pairs of nontrichobothrial setae (*vi, ve, si, se*), idiosomatic shield and eyes absent. Chelicerae styliform, retractile within the gnathosomatic capsule, fused basally with each other and with the subcapitulum except at hypostomatic apex, remnant cheliceral seta (*chb*) or its alveolus present dorsally on capsule; palpi 2-segmented in adults, absent in immatures; gnathosoma of immatures proboscidiform. Female anal opening dorsal, separate from genital opening; male genito-anal opening dorsal, aedeagus typically elongate and generally directed forward. Dorsal lyrifissures (cupules) absent. With 5 active postembryonic instars: larva, protonymph, deutonymph, tritonymph, and adult.*

The Myobioidea is represented by a single cosmopolitan family, the MYOBIIDAE (Figs. 13.61A–E), which comprises 50 genera and more than 450 species of ectoparasites that cling to the fur of their marsupial, rodent, bat, elephant shrew, and eulipotyphlan insectivore hosts by means of legs I, which are modified as clasping organs (Bochkov and Fain 2003). The family consists of five subfamilies (Fain 1994; Bochkov 1997), the most generalized of which (Xenomyobiinae) is represented by one species, *Xenomyobia hirsuta* Fain and Lukoschus, that parasitizes the primitive Peruvian marsupial *Lestoros inca* (Fain and Lukoschus 1976a). Legs I of *X. hirsuta* have shortened tarsi with small claws but lack attachment organs; legs III and IV are used to cling to hairs. Females of the subfamily Archemyobiinae comprise two genera whose members are parasitic on South American marsupials (Fig. 13.61C). Archomyobiines have paired ventral clasping organs on the genua of legs I, while members of the subfamilies Australomyobiinae (two genera parasitic on Australian marsupials) and Myobiinae have an unpaired attachment organ dorsally on genu I. The Myobiinae consists of four tribes. The tibia and tarsus of legs I are fused in derived members of the myobiine tribes Acanthophthiriini (parasites of bats) and Protomyobiini (parasites of eulipotyphlan insectivores) and in all members of the Elephantulobiini (parasites of elephant shrews (Macroscelididae)). In mites of the tribe Myobiini (parasites of rodents), however, the three apical segments of legs I are fused.

Although MYOBIIDAE occur throughout the world, their origin and center of dispersion are thought to have been in South America, where the most primitive marsupials also are found (Carroll 1988). Myobiids may have arisen there as mammalian parasites before divergence occurred between marsupials and placental mammals, later coevolving with these lineages (Dusbábek 1969; Carroll 1988). Myobiids and their mammalian hosts show high levels of phylogenetic parallelism and are commonly correlated in degree of primitiveness or progressiveness with their hosts (Dusbábek 1969, 1973; Fain 1975b, 1994; Uchikawa 1988; Bochkov 1997, 2001).

The MYOBIIDAE appears to represent an early derivative lineage of parasitic mites within the cohort Raphignathina. They display certain morphological convergences with the Cheyletoidea, including styletlike chelicerae and a stylophore fused with the subcapitulum to form a gnathosomatic capsule (Bochkov 2002). Basal species in the family have relatively rich leg chaetomes that are more similar to the those of Tetranychoidea than to the Cheyletoidea, but more derived species show a clear tendency toward reduction of leg setae (Fain 1994; Bochkov 1997). Based primarily on cheliceral and subcapitular features, a number of earlier authors, including Cunliffe (1955), Dubinin (1957), and Kethley (1990), considered the MYOBIIDAE to comprise a member taxon of the Cheyletoidea, but the compelling cladistic and phylogenetic evidence presented by Bochkov and coworkers points to their representing a superfamily separate from the Cheyletoidea. Bochkov's conclusions are followed in this treatment.

Bochkov and Labrzycka (2003) concluded that the tritonymphal stage is retained in myobiids, although the tritonymph has been suppressed in most raphignathine groups that have been studied (Raphignathoidea except XENOCALIGONELLIDIDAE, Cheyletoidea, and all Tetranychoidea except TUCKERELLIDAE). Further study of myobiid postlarval development would be useful for verifying this unusual ontogeny.

Feeding in myobiids is more or less confined to the hair follicle bases and often involves hematophagy. Female myobiids ingest blood during reproductive periods (F. S. Lukoschus, pers. comm.) but may feed on lymph and other extracellular tissue fluids as well (Yunker 1973; Bochkov and Fain 2003). *Myobia murismusculi* (Schrank), *Radfordia* (*Radfordia*) *ensifera* (Poppe) and *R.* (*R.*) *affinis* (P.) are common myobiid parasites of laboratory rodents and cause dermatitis, alopecia, and trauma in their hosts (Baker et al. 1956; Yunker 1973). Their cosmopolitan distribution parallels that of their widely distributed hosts.

▪ Superfamily Pterygosomatoidea (Figs. 13.62A–G)

*DIAGNOSIS: Soft bodied with a prodorsal shield with 2 to many pairs of setae and a pair of eyes (or eyes on separate sclerites); trichobothria absent; moderate to extensive idiosomatic hypertrichy common; 2 pairs of cupules (*ia *and* im *absent). Chelicerae separate, bases expanded, extremities sheathed in troughlike to tubular subcapitulum; movable digit curved outward, hooklike or with 1 or more teeth; fixed digit reduced to membranous sheath. Peritremes emergent, running from stigmata at base of chelicerae to tips of fingerlike processes. Femora I–IV entire, pretarsi with paired claws bearing tenent hairs, empodia absent. Palpi with 5 free segments, including a well-developed thumb-claw process. Female anal opening terminal; genital papillae, eugenital setae, and ovipositor absent. Male with coalesced genito-anal opening in a terminal to dorsal position and an eversible aedeagus. Full ontogenetic sequence present in females, but prelarva, protonymph, and tritonymph calyptostases; larva, deutonymph, and adult parasitic on reptiles or arthropods.*

The Pterygosomatoidea includes a single family, the PTERYGOSOMATIDAE (Figs. 13.62A–G), comprising more than 120 described species of bright red mites found primarily on lizards, tortoises, and arthropods in Africa, Asia, Australia, and the Americas. Most of the described species are parasites of lizards (Agamidae, Gekkonidae, Iguanidae, Eublepharidae, Xantusiidae, Zonuridae) and are members of the genera *Cyclurobia*, *Geckobia*, *Geckobiella*, *Hirstiella*, *Ixodiderma*, *Pterygosoma*, *Scaphothrix*, *Tequisistlana*, and *Zonurobia* (Lawrence 1935, 1936; Cunliffe 1949a; Davidson 1958; Jack 1961b; Hoffmann and Sanchez 1979; Domrow 1983; de la Cruz 1984; Haitlinger 1988; Baker 1998; Bertrand, Paperna, and Finkelman 1999). They attach under scales, between the toes, or in areas known as mite pockets (Bertrand and Modry 2004) and often are confused with chiggers. Pterygosomatids feed on body fluids of their host, upon which their impact varies from benign to severe pathological disorders such as anemia and intense skin irritation (Baker 1998; Walter and Proctor 1999; Walter and Shaw 2002). Apparently, some species are vectors of protozoan diseases of lizards (Goodwin 1954; Newell and Ryckman 1964; Walter and Shaw 2002).

The broad, short species of the genus *Pterygosoma* are common on agamid lizards in North and East Africa and India (Jack 1961a) and also parasitize iguanid lizards in the Americas (de la Cruz, Morando, and Avila 2004). The 50 or so known species that compose the genus *Geckobia* (Figs. 13.62C–E) are mostly ectoparasites of geckos (Gekkonidae, Eublepharidae) (Bertrand, Paperna, and Finkelman 1999), but *Geckobia enigmatica* Bertrand and Pedrono (1999) was collected from captive-reared tortoises after their reintroduction into native areas in Madagascar. Other *Geckobia* species are common on geckos in southeast Australia, Asia, and the Americas (Rivera et al. 2003). These mites are frequently found in the periocular skin folds, in the tympanum, under scales, in interdigital spaces, and in axillae and appear to have species-specific attachment sites (Womersley 1942b). Apparently, parthenogenetic geckos are especially susceptible and may endure heavy *Geckobia* infestations involving hundreds of mites (Moritz 1991). *Geckobiella texana* (Banks) has been recovered from iguanids of the genus *Sceloporus* in Texas and California. Species of *Hirstiella* and *Cyclurobia* primarily infest iguanas, including the chuckwalla (Cunliffe 1949a; André 1961; Newell and Ryckman 1964; Werman 1983; de la Cruz 1984; Baker 1998; Bertrand, Paperna, and Finkelman 1999). *Hirstiella diolii* Baker (Figs. 13.62A, B) is a pest of iguanas in zoos and is associated with the transmission of a protozoan blood parasite of the lizards (Baker 1998; Walter and Shaw 2002, 2003).

Although most pterygosomatids are parasites of reptiles, some may be associated with other host taxa. Species of *Pimeliaphilus* Trägårdh (1905) are sometimes found on geckos but usually infest triatomine bugs (Reduviidae: Triatominae), cockroaches, or scorpions (Jack 1964). Two species attach to the conjunctival tissues of scorpions, where they feed on the blood of their hosts (Beer 1960; Cunliffe 1949b), while *P. podapolipophagus* is an ectoparasite of cockroaches (Cunliffe 1952) and often proves troublesome in laboratory rearing operations. Heavy infestations of this mite can cause roach mortality within a few hours after initial exposure. Several species of triatomine kissing bugs (Reduviidae) are attacked by species of *Pimeliaphilus* (Newell and Ryckman 1966, 1969). Under laboratory conditions, *P. triatomae* Cunliffe causes nymphal mortality, increased developmental time, and reduced oviposition in its triatomine host (Zumaquero et al. 2004). Prasad (1975) described *Bharatoliaphilus punjabensis* from a dove (*Streptopelia decaocta*) in India, but the description was based on a single specimen, and its host relationships are unclear.

Pterygosomatids have a life cycle that involves **alternating calyptostasy** (Kethley 1991b), which results in the alternating appearance of active and quiescent (regressive) instars. Jack and Girot (1965) determined that adult female *Hirstiella insignis* emerged from the calyptostatic tritonymph that was sequestered in the deutonymphal exuviae. Baker (1998) observed that males of *H. diolii* Baker were enclosed in a larval exoskeleton, and that only females had deutonymphal exuviae. Goodwin (1954) made similar observations for *Geckobiella texana* (Banks). However, it is not yet clear if the males pass through the deutonymphal and tritonymphal stases or emerge directly from the protonymph.

- Superfamily Raphignathoidea (Figs. 13.63–13.72)

DIAGNOSIS: Soft bodied or well sclerotized, with a holonotal shield or a series of smaller weakly tanned, reticulate, or dimpled dorsal sclerites, rarely without shields (i.e., BARBUTIIDAE, many CALIGONELLIDAE, some STIGMAEIDAE); idiosomatic neotrichy present in some families (CAMEROBIIDAE, DASYTHYREIDAE, XENOCALIGONELLIDIDAE); trichobothria absent; prodorsum usually with 3–4 pairs of setae (si may be absent), rarely neotrichous; with 0–2 pairs of ocelli, 2nd pair (postocular organ) sometimes stalked. Cheliceral bases usually swollen (CRYPTOGNATHIDAE and some STIGMAEIDAE are exceptions), free, adnate, partially fused, or amalgamated into a stylophore, each with a styliform movable digit encased in the sheathlike remnant of the fixed digit; chelicerae not fused to underlying rostral elements, retractile in CRYPTOGNATHIDAE. Palpi with 5 free segments, palptibial claw (thumb-claw process) usually present but often weakly developed or obsolete, occasionally absent. Stigmatic openings present at base of chelicerae or absent; peritremes absent or simple to highly convoluted on cheliceral bases, or located on anterior margin of prodorsum and distally emergent (RAPHIGNATHIDAE). Empodial elements of tarsi commonly with tenent hairs, claws usually present on all legs (replaced by padlike structures in some STIGMAEIDAE), tenent hairs lacking on claws except in BARBUTIIDAE. Males with aedeagus, genital opening typically terminal, rarely dorsal; typically with solenidia on all tarsi. Generally with 4 active postembryonic instars: larva, protonymph, deutonymph, and adult (tritonymph present in XENOCALIGONELLIDIDAE).

Although members of the family HOMOCALIGIDAE (Figs. 13.70A–C, 13.71C) and some STIGMAEIDAE (e.g., Fig. 13.68D) are associated with wet habitats such as ponds, streams, and salt marshes, most raphignathoids are found in a variety of xeric habitats throughout the world. The MECOGNATHIDAE (Figs. 13.69A, B), EUPALOPSELLIDAE (Figs. 13.69C, D), and many genera of STIGMAEIDAE (e.g., Figs. 13.67C–F, 13.68H, 13.71E, 13.72E, F) are typically arboreal (epiphytes, leaf surfaces, twigs, and bark), although others are usually found in the soil (e.g., Figs. 13.68A, B, F, 13.71D). Members of the RAPHIGNATHIDAE (Figs. 13.63D–F, 13.71A, 13.72C, D), CAMEROBIIDAE (Figs. 13.64A, B), XENOCALIGONELLIDIDAE (Figs. 13.65D, E), and DASYTHYREIDAE (Figs. 13.65A–C, 13.72A) are associated with low plants, tree trunks, or logs, while the BARBUTIIDAE (Figs. 13.63A–C), RAPHIGNATHIDAE, CALIGONELLIDAE (Figs. 13.63G, 13.66A–G, 13.72B), and CRYPTOGNATHIDAE (Figs. 13.64C, D, 13.71B) are characteristic of dry soils and moss. Although the feeding habits of most species of Raphignathoidea are unclear, predation appears to be the dominant mode, except in some moss-feeding STIGMAEIDAE and possibly in the CRYPTOGNATHIDAE. Attachment to insects has been observed in some STIGMAEIDAE and DASYTHYREIDAE, but most raphignathoids appear to disperse on their own.

Systematic relationships among the 11 recognized raphignathoid families are poorly understood and in need of study. Members of the superfamily display considerable diversity in attributes such as idiosomatic sclerotization, which may vary from fully armored (HOMOCALIGIDAE,

CRYPTOGNATHIDAE, some STIGMAEIDAE) to essentially nonexistent (BARBUTIIDAE, CALIGONELLIDAE, some STIGMAEIDAE, some RAPHIGNATHIDAE). In most Raphignathoidea, the chelicerae are fused and function as a single unit, the stylophore, which bears the peritremes on its dorsal surface (Figs. 13.65A, E, 13.66A, B, D, G, 13.72A, B). Stigmata and peritremes, however, are absent in the STIGMAEIDAE, HOMOCALIGIDAE, EUPALOPSELLIDAE, and MECOGNATHIDAE. In the latter three families and in some STIGMAEIDAE, the chelicerae are coalesced into a stylophore. The chelicerae in many stigmaeids may instead be closely aligned while operating more or less independently (Figs. 13.67C, 13.68A, F, 13.72E, F), a condition often referred to as *adnate* in the literature. The CRYPTOGNATHIDAE have separate chelicerae that are deeply retractile (Figs. 13.64C, D), while the BARBUTIIDAE have chelicerae that are fused at their bases (Fig. 13.63A).

The four described species of the monogeneric family BARBUTIIDAE (Figs. 13.63A–C) are found in seasonally arid soils (Fan, Walter, and Proctor 2003a). *Barbutia perretae* Robaux occurs in forest litter in the western United States (Robaux 1975a), while *B. anguinea* (Berlese) has been collected in Italy and California (Summers 1964). Two species are known from dry soils in Australia. Adult female barbutiids are minute, elongate, soft-bodied mites with the cheliceral bases mesally fused at their proximal extremities and with minute peritremes arising on their dorsal surfaces. The palptibial claw has a ventral tooth. Each empodium has three pairs of tenent hairs, and each claw bears two pairs of tenent hairs, the latter character being unique within the Raphignathoidea. The African genus *Namibarbutia*, originally assigned to the BARBUTIIDAE, has been determined to be a member of the CALIGONELLIDAE (André 1996; Fan, Walter, and Proctor 2003a).

The families XENOCALIGONELLIDIDAE (Figs. 13.65D, E) and DASYTHYREIDAE (Figs. 13.65A–C, 13.72A) are unusual among raphignathoids in that they are neotrichous (see also CAMEROBIIDAE), and many setae project in a pincushion-like array to a distance of several body diameters around their bodies (Walter and Gerson 1998). Species in both families are commonly found on twigs, tree trunks, and logs. Xenocaligonellidid mites have holodorsal shields bearing 15 or more pairs of setae, of which the marginal pairs are long and whiplike and those along the midline and posterior are shorter. *Xenocaligonellidus ovaerialis* De Leon was described from the twigs and bark of several tree species from Florida and Mexico. These mites are slow moving, produce eggs on long stalks and aggregate them in clusters, and appear to produce droplets of fluid at the spiny tips of their median and posterior dorsal setae (De Leon 1959). Gonzalez (1978) described *X. galapagus* from a plant on an island in the Galápagos, and Meyer and Ueckermann (1989) described *X. teclae* from a plant in Natal, South Africa. Hu and Liang (1995) identified a fourth species, *X. smileyi*, from China, and Fan (2000) described its ontogeny, including the tritonymph. Fan and Chen (1996) described *Echinopsis* (= *Apocaligonellidus*) *fukiensis*, a new xenocaligonellidid genus and species with 22 pairs of dorsal setae, from under logan bark in Fujian Province, China. Members of the related family DASYTHYREIDAE have a divided dorsal shield with 19 or more pairs of mostly long, whiplike setae. *Xanthodasythyreus toohey* Walter and Gerson (1998), a slow-moving beige-colored inhabitant of soil and litter, often occurs at the base of grass trees in the open forests of eastern Australia. The bright red *Dasythyreus hirsutus* Atyeo bears 179 whiplike setae and was described from tree bark in Arkansas, USA (Atyeo 1961). Undescribed species of *Dasythyreus* have been found on logs in boreal forest in Alberta, Canada (D. E. Walter, unpublished), and phoretic as adult females on elater beetles (Coleoptera: Elateridae) (E. E. Lindquist, unpublished).

More than 40 species of the family RAPHIGNATHIDAE Kramer (Figs. 13.63D–F, 13.71A, 13.72C, D) have been described from around the world, and all but one of them has been assigned to the genus *Raphignathus*. Species of this genus have well-developed dorsal shields that are more extensive in males than females, while the female of the monotypic *Neoraphignathus howei* Smiley and Moser, a mite collected from a pine trunk bolt infested with bark beetles, has no distinct dorsal shields (the male is unknown) (Smiley and Moser 1968). Raphignathids are often found on or beneath tree bark, in moss, and in dry organic litter (Atyeo 1963b; Atyeo, Baker, and Crossley 1961) but may also be collected on low-growing herbs and shrubs and in synanthropic habitats, such as house dust or stored grain. *R. gracilis* (Rack) and R. *collegiatus* Atyeo, Baker, and Crossley occur in the deeper soil layers in California pine forests (Price 1973). Raphignathids appear to be especially diverse in Africa (Meyer and Ueckermann 1989), China (Fan and Yin 2000), and Australia, although most of the Australian fauna is undescribed (D. E. Walter, unpublished).

The CRYPTOGNATHIDAE (Figs. 13.64C, D, 13.71B) is a well-defined family of two recognized genera and more than 40 described species of small (idiosoma < 0.4 mm in length), armored, highly ornamented mites found in leaf litter, tree bark, boulders, moss, and lichens in many parts of the world (Krantz 1958a; Luxton 1973, 1993; Fan 1997b; Krisper and Schneider 1998). Adult cryptognathids are easily recognized by their holodorsal shield bearing 11 pairs of setae, an elongate and retractile gnathosoma, and a distinctive prodorsal hood that may be hyaline or deeply sculptured (Summers and Chaudri 1965; Luxton 1973). The two cryptognathid genera (*Cryptognathus* and *Favognathus*) may be distinguished by the condition of the presternal "apron" at the distal margin of the ventral plate; it is subtriangular and ornamented with punctae in *Favognathus* and crescentic and lightly sclerotized in *Cryptognathus*. Luxton (1973) noted a green material in the hindgut of these mites, which suggests that they may be phytophagous or algivorous.

Members of the EUPALOPSELLIDAE (Figs. 13.69C,

D) and MECOGNATHIDAE (Figs. 13.69A, B) are plant-inhabiting predaceous mites that have distinctive elongate cheliceral stylets and lack peritremes and stigmata, as in the related STIGMAEIDAE (Gerson and Walter 1998). Mecognathids have three pairs of prodorsal setae (*si* appear to be absent), the empodia have three pairs of tenent hairs, and the palptibia is four to five times longer than the palptarsus. The family comprises three described species of *Paraeupalopsellus* from North America, South Africa, and the Philippines (Rimando and Corpuz-Raros 1996) and two species of *Mecognatha* (= *Acaciacarus*) from Australia and New Zealand (Wood 1967; Gerson, Frost, and Swift 1997; Gerson and Walter 1998).

Eupallopsellids are cosmopolitan and are most commonly encountered on twigs and bark. They have four pairs of prodorsal setae, the empodia have one (rarely two) pairs of tenent hairs, and the palptarsi are elongate. Five genera are currently placed in the family (Meyer and Ueckermann 1989; Gerson and Walter 1998). The eight described species of *Eupalopsellus* are especially common in very dry regions (sage, heather, brushland habitats), although *E. olandicus* Sellnick has been collected from coastal areas in northern Europe (Meyer and Ueckermann 1984, 1989). The genus *Eupalopsis* consists of three described species from the Mediterranean region, South Africa, and eastern Australia (Meyer and Ueckermann 1984; Gerson 1994), while *Peltasellus* is monotypic and known only from South Africa (Meyer and Ueckermann 1984). The five described species of *Saniosulus* often are associated with scale insects (Fan, Zhang, and Liu 2000), and *S. nudus* Summers can be a pest in commercial production of scale parasitoids. The genus *Exothorhis* is the most diverse and widespread genus in the family, with 15 described species known from around the world (Rimando and Corpuz-Raros 1996; Hu and Chen 1996). *E. caudata* Summers and *S. nudus* are predators in citrus crops (Summers 1960a), and both *S. nudus* and *Eupalopsis maseriensis* (Canestrini and Fanzago) prey on the citrus chaff scale, *Parlatoria pergandii*, in Israel (Gerson 1966, 1967; Gerson and Blumberg 1969). *E. pinicola* Oudemans is found on apple trees in eastern Canada (Baker and Wharton 1952) and is a scale insect predator in Europe. Other eupalopsellids have been taken in moss or leaf mold.

The cosmopolitan family STIGMAEIDAE (Figs. 13.67A–F, 13.68A–I, 13.71D, E, 13.72E, F) is the most diverse, and problematical, assemblage in the Raphignathoidea, with about 30 recognized genera and well over 300 described species. The two largest genera, *Stigmaeus* and *Eustigmaeus* Berlese, are especially in need of revision. Stigmaeid genera often may be recognized on the basis of their distinctive dorsal shield configurations (e.g., see Figs. 13.67, 13.68) (Summers 1966a). At least three contain species that are predators of phytophagous mites in orchard crops. *Zetzellia mali* (Ewing) feeds on several species of spider mites in North America, Europe, and Israel (Jeppson, Keifer, and Baker 1975a) and is regarded as a beneficial species. Several representatives of the stigmaeid genera *Agistemus* and *Mediolata* also prey on spider mites and other orchard-associated arthropods in various parts of the world (Gonzalez-Rodriguez 1965; Summers 1966b; Jeppson, Keifer, and Baker 1975a).

Eustigmaeus s. lat., the largest genus in the STIGMAEIDAE, is composed of globate, strongly ornamented species (Figs. 13.67C, 13.71E) that may be encountered in soil and litter habitats (Summers and Price 1961; Price 1973). Some species have been observed to feed on mosses (Gerson 1972a). *E. frigida* (Habeeb), for example, completed its entire life cycle on a diet of *Heterophyllium.* The adult females of some *Eustigmaeus* and *Stigmaeus* species have been found attached to and apparently feeding on adult phlebotomine sandflies (Diptera: Psychodidae) in several regions of the world (Swift 1987; Zhang and Gerson 1995; Shehata and Baker 1996). It may be that only particular lineages of *Eustigmaeus* are moss feeders, or that these mites are more omnivorous than has been assumed. Some recent workers have assigned some *Eustigmaeus* species to genera based on previously recognized categories (e.g., *Ledermuelleria* Oudemans), and have proposed new genera (*Chaudhria*, *Wooderia*) for others (Rimando and Corpuz-Raros 1997). *Ledermuelleriopsis*, a related genus of more than 20 described species, appears to be basal to a clade that includes *Eustigmaeus s. lat.* and *Cheylostigmaeus* (Fan, Walter, and Proctor 2003b).

In contrast to the strongly sclerotized species discussed above, many stigmaeids have minimal dorsal plating. For example, the five known species of *Storchia* (Fan and Yan 1997) and the approximately 20 species of *Eryngiopus* have little more than vague prodorsal shields, while *Summersiella ancydactyla* Gonzalez, a predator in leaf domatia in New Zealand, is practically unsclerotized. Most stigmaeids have paired claws (Fig. 13.68C) and an empodium bearing paired tenent hairs (capitate raylets), but the arboreal genera *Mullederia*, *Neilstigmaeus* (Fig. 13.68E), and *Pilonychiopus* lack claws and have a membranous pouch (arolium) bearing the tenent hairs (Gerson and Meyer 1995).

Two genera currently placed in the STIGMAEIDAE share a fusion of the cheliceral bases, and an ecological association with aquatic habitats characteristic of the fully aquatic family HOMOCALIGIDAE. Species of *Cheylostigmaeus* are known mostly from marshy habitats in North America, Europe, and Asia (Summers and Ehara 1965), while the two known species of *Caligohomus* were described from swampy habitats in the eastern United States (Habeeb 1966; Fan and Walter 2004). The HOMOCALIGIDAE (Figs. 13.70A–C, 13.71C) are capable of skating across the surface of water, and the adults have what appear to be accessory sacs or tubes apparently used in respiration or flotation. *Homocaligus muscorum* Habeeb was collected from a pasture swamp in eastern North America (Habeeb 1962), *H. scapularis* (Koch) from a variety of aquatic habitats in Europe (Wood 1969), and *H. crassipes* Fan from water hyacinth in China (Fan 1997b). A second genus, *Annerossella*, is known from swampy and other aquatic habitats in South Africa

(Meyer and Ryke 1959b), Malaysia (Wood 1969), China (Fan 1997b), and Australia (Halliday 1998).

The CALIGONELLIDAE (Figs. 13.63G, 13.66A–G) currently consists of five genera of small, mostly soft-bodied orange to red mites that are primarily associated with dry soils, but they are also found on bark, in bird nests, on low-growing plants, and occasionally in litter overlying wet soil (Summers and Schlinger 1955; Smiley and Moser 1968; Robaux 1975a; Meyer and Ueckermann 1989). *Molothrognathus fulgidus* Summers and Schlinger occurs under the bark of prune and almond trees in California, and *M. rosei* Smiley and Moser was collected in association with bark beetles. In addition to the approximately 18 known *Molothrognathus* species, unidentified species have been collected from cherry tree twigs in the Pacific Northwest and from grain storages in Canada. *Neognathus terrestris* (Summers and Schlinger) inhabits the soil substrata in pine forest soils of California, and seven additional species have been described from similar dryland soils in Africa (Summers and Schlinger 1955; Soliman 1971; Meyer and Ueckermann 1989) and Asia (Chaudri, Akbar, and Rasool 1979; Fan 2000). Numerous undescribed species of *Neognathus* are known from dry soils in western North America and Australia (D. E. Walter, unpublished), while five additional species are accommodated in the closely related genus *Paraneognathus* (= *Sinognathus*) (Fan and Li 1995; Fan 2000).

Approximately seven genera and more than 130 species of stilt-legged mites are accommodated in the family CAMEROBIIDAE (Figs. 13.64A, B). Camerobiids are found in leaf mold, on or under tree bark, or on vegetation or epiphytes in temperate and tropical realms. They feed on plant pests such as scale insects and plant-parasitic mites (ERIOPHYIDAE, TENUIPALPIDAE) and on other small mites that scavenge or feed on fungi on plant surfaces (e.g., TARSONEMIDAE, TYDEIDAE) (McGregor 1950b; De Leon 1967; Bolland 1983; Gerson, Smiley, and Ochoa 2003). One species of *Camerobia* is common in sage and salt grass litter in central and eastern Oregon, USA, and another has been recorded from mite-infested farm grain storages. *C. australis* Southcott was found beneath *Eucalyptus* bark in Queensland and South Australia (Southcott 1957a), and *C. fissus* (De Leon) was collected on a cocoa tree that was heavily infested by a false spider mite, *Tenuipalpus orilloi* Rimando, in Trinidad (De Leon 1967).

▪ Superfamily Tetranychoidea (Figs. 13.73–13.76)

DIAGNOSIS: Soft bodied to leathery and with striate or reticulate patterning, with or without a distinct sejugal furrow and weakly defined prodorsal shield, or discrete reticulate to rugose dorsal shields. Gnathosoma with postcheliceral stigmata and peritremes, peritremes emergent onto soft pliable anterior margin of prodorsum in some genera; cheliceral bases fused, lacking dorsal cheliceral setae and forming deeply retractable stylophore moving separately from subcapitulum; fixed cheliceral digits vestigial, lobelike, movable cheliceral digits greatly elongated, whiplike, retractile, inserted within stylophore, grooved medially and their apices juxtaposed when protruded to form a hollow puncturing-sucking organ for feeding on plants; palpi primitively with 5 segments and a thumb-claw process, but with palptarsal segment terminal, lacking the tibial claw, often with reduced segmentation in TENUIPALPIDAE; palptarsus in TETRANYCHIDAE with or without enlarged eupathidial spinneret. Prodorsum lacking anteromedian naso, with 3–4 pairs of setae, none bothridial; 2 pairs of lateral eyes present or absent. Opisthosoma with 6–26 pairs of setae dorsally, laterally, and caudally, often with some larval neotrichy including c3, d3, e3, h3 *but usually lacking postlarval setal additions; cupules present, sometimes obscure, or absent; setal elements of opisthosomatic segment PS present, those of AD absent. Adult female genital aperture transverse, positioned behind legs III–IV, without genital papillae. Adult males with a caudal sclerotized aedeagal apparatus. Leg setation lacking bothridial setae; tarsi I–II with erect solenidia that often form part of a duplex seta set (see below); all legs with paired claws and empodia, claws with tenent hairs and padlike, clawlike, or short rodlike empodia with or without tenent hairs and variable in form. Generally with 5 active postembryonic instars: larva, protonymph, deutonymph, and adult; tritonymph apparently retained in TUCKERELLIDAE; postlarval stases may be hexapod in TENUIPALPIDAE.*

The superfamily Tetranychoidea includes approximately 2,000 described species of obligately phytophagous mites assigned to five recognized families. Most of the species that are of major economic importance are contained in two families, the TETRANYCHIDAE or spider mites and the TENUIPALPIDAE or false spider mites. The families TUCKERELLIDAE, LINOTETRANIDAE, and ALLOCHAETOPHORIDAE are not as taxonomically diverse and contain relatively few described species. Members of the superfamily are unique in possessing an independently movable stylophore composed of the fused cheliceral bases that may be deeply retracted into the propodosoma, and in having greatly elongated movable cheliceral stylets that are strongly recurved basally within the stylophore. This arrangement allows for a greater degree of retraction and protraction of both the stylophore and the stylets than in mites of other superfamilies. The protruded stylets interlock to form a single hollow probe and a salivary channel during feeding. Following stylet retraction, plant juices appear to be sucked up via a hermetically sealed food channel formed by the labrum and preoral groove (Nuzzaci and de Lillo 1989, 1991). A haplodiploid arrhenotokous genetic system along with copulatory behavior and direct sperm transfer are basic to the superfamily, but female populations and species with diploid thelytoky have arisen independently between, and within, the families (Gutierrez, Bolland, and Helle 1979; Bolland and Helle 1981; Bolland, Gutierrez, and Helle 1981; Gutierrez and Helle 1985; Helle and Pijnacker 1985). Spider mites and false spider mites run the gamut of plant host specificity, from species that prefer just one species or genus of host plants to others that are known from many hundreds of species of hosts in diverse families.

The TETRANYCHIDAE (Figs. 13.73–13.74) comprises over 1,200 described species in about 70 genera and is the most economically important family and one of the most speciose (second only to the ERIOPHYIDAE) of phytophagous mites (Bolland, Gutierrez, and Flechtmann 1998). Spider mites occur throughout the world where higher plants flourish and are found on virtually every major food crop and ornamental plant, often causing serious injury to, or death of, the host. Members of the family are green, yellow, orange, or red in color; those of the subfamily Tetranychinae may commonly be found associated with fine silk webbing that they spin from large unicellular glands located in the palpi (Grandjean 1948b; Alberti and Storch 1974; Mothes and Seitz 1981a; Alberti and Crooker 1985). Tetranychids feed by inserting their whiplike cheliceral stylets through or between the palisade cells and into the underlying parenchyma (Jeppson, Keifer, and Baker 1975b; André and Remacle 1984). Feeding symptoms may be confined to simple stippling of injured tissue, or the host plant may suffer leaf and fruit drop or twig dieback. Overall plant vigor, growth, and physiology may be affected as well (Tomczyk and Kropczynska 1985). In contrast to the phytophagous eriophyoid mites (p. 247), and with rare exceptions (e.g., transmission of a barley virus by *Petrobia latens* Müller) (Robertson and Carroll 1988), tetranychids are generally not effective vectors of plant viruses (Oldfield 1970; Krantz and Lindquist 1979).

The association of an elongated tapered solenidion with a shorter seta to form a duplex set (Figs. 13.73B, 13.74B) on tarsi I and II of the larval and postlarval instars, followed by a second such set on tarsus I in postlarval instars, is a uniquely apomorphic trait in the TETRANYCHIDAE. Separation of the elements of these sets occurs secondarily in a few genera of the subfamily Tetranychinae, while further reduction of the setal element of these sets to a vestige occurs secondarily in a few species of Bryobiinae (Lindquist 1985a, b). Thelytokous parthenogenesis appears to be common among members of the Bryobiinae, but it is rare in the Tetranychinae (Gutierrez and Helle 1985). There is much diversity and specialization in the reproductive, diapausing, and chemical communicative behavior among the tetranychids (Cone 1985; Sabelis 1985a; Veerman 1985), which would seem to support a hypothetical evolutionary scheme that ranks them, along with the TENUIPALPIDAE, as one of the most recently derived taxa of Tetranychoidea (Mitrofanov 1983). At the same time, there is considerable morphological evidence that the TETRANYCHIDAE may be one of the more early derivative lineages among the families of Tetranychoidea, in contrast to the TENUIPALPIDAE (Lindquist 1985a).

The subfamily Bryobiinae is readily distinguished by the terminal placement of duplex sets on tarsi I–II and their greatly reduced (sometimes vestigial) setal elements, the presence of tenent hairs on the leg empodia, and the unmodified form of all three palptarsal eupathidia, one of which is modified as a spinneret in tetranychines to produce silk (Lindquist 1985a, c). Observations of silk spinning by larval *Petrobia apicalis* (Banks) (Smith and Weber 1954; Zein-Elden 1956) need further confirmation, and the means by which silk production is accomplished in this species remains to be clarified.

Grasses and other low-growing plants commonly are infested by bryobiines. The clover mite, *Bryobia praetiosa* Koch, is found on grasses in many parts of the world (Pritchard and Baker 1955; Jeppson, Keifer, and Baker 1975b; Meyer 1974, 1987). It completes several generations during the growing season and may overwinter in any stage. Adults and nymphs often invade buildings in great numbers prior to the onset of cold weather. *B. praetiosa* is regarded by some specialists as a complex of closely related races that differ somewhat in host plant preference, life history, and behavior. The complex includes *B. kissophila* Van Eyndhoven, a multivoltine species on ivy that overwinters in all stages; *B. ribis* Thomas, a univoltine form on gooseberry that overwinters only in the egg stage; and *B. graminum* Schrank, a multivoltine inhabitant of grasses and fruit trees that overwinters primarily in the egg stage. The brown almond mite, *B. rubrioculus* (Scheuten), is an important pest of pome and stone fruits in both the Old and New Worlds (Morgan 1960; Jeppson, Keifer, and Baker 1975b; Meyer 1987). Other species of *Bryobia* of occasional or regional economic importance are reviewed by Jeppson, Keifer, and Baker (1975b) and Zhang (2003).

Bryobiines of the genus *Monoceronychus* are mostly confined to grasses, while members of the genera *Petrobia*, *Aplonobia*, *Hystrichonyssus*, and *Pseudobryobia* (among others) occur on a variety of low-growing plants. The brown wheat mite, *Petrobia latens* (Müller), and the legume mite, *P. apicalis* (Banks), are serious economic pests in many parts of the world. An infestation of *Tetranycopsis horridus* (Canestrini and Fanzago), a pest of filberts in Europe, persists in a filbert planting in California (Jeppson, Keifer, and Baker 1975b).

The subfamily Tetranychinae is distinguished apormorphically on the basis of several attributes, including the absence of tenent hairs on the leg empodia and the enlarged form of one of the three palptarsal eupathidia, which is modified as a spinneret and correlated with the capacity to produce silk (Grandjean 1948b; Alberti and Crooker 1985; Lindquist 1985a). Patterns of web-spinning behavior vary greatly among tetranychine spider mites, from those that spin little silk and tend to live on upper surfaces of leaves to those that spin copiously and often live exclusively on leaf undersides (Gutierrez and Helle 1985). Production of webbing serves a variety of functions, including protecting the eggs, locating members of the opposite sex, dispersal, and protecting populations against adverse environmental factors and natural enemies (Gerson 1985b). Some species of *Schizotetranychus* (including a group segregated into a separate genus, *Stigmaeopsis*, by Saito et al. 2004) exhibit a subsocial behavior, living gregariously and cooperating in spinning nestlike structures that protect largely inbreeding colonies of mites (Saito 1985, 1990). Adult males and females may effectively defend their nest and offspring by counterattacks against intrusive immature phytoseiid predators (Saito 1986). Although tetranychines are not known to play out strands of silk to disperse on the wind (ballooning) in the manner of spiders, many species do dangle from silken threads and are borne aloft when these threads snap (Fleschner et al. 1956; Kennedy and Smitley 1985; Bell et al. 2005; see also chapter 6).

A number of tetranychine species, especially in the genera *Tetranychus*, *Panonychus*, and *Mononychellus*, represent either species complexes or genetically distinct populations that may be recognized by body color or strial microstructure, diapausing forms, host preferences, or other behavioral or life history traits (Oloo, Ogol, and Kambona 1987; Gotoh, Suwa, and Kitashima 1993; Hance, Neuberg, and Noèl-Lastelle 1998). Molecular techniques have been useful in better understanding these pliable complexes (e.g., Navajas and Gutierrez 1997; Navajas et al. 1994, 1996; Navajas, Gutierrez, and Gotoh 1997; Gotoh and Tokioka 1996; Gotoh, Gutierrez, and Navajas 1998; Navajas and Fenton 2000; Gotoh, Kitashima, and Goka 2007). Still, the systematic status of the green and the red forms of *T. urticae* Koch, including the form known as *T. cinnabarinus* (Boisduval), is unclear because the ability of the two forms to interbreed varies considerably depending on the populations used (Navajas et al. 2000; Sugasawa, Kitashima, and Gotoh 2002). A wide range of genetic divergence, host preferences, and reproductive incompatibility is also known among populations of *T. kanzawai* Kishida (Gotoh, Gomi, and Nagata 1999).

All higher plant groups are infested by tetranychines, many of which are economically important species. Most of the examples noted here are also mentioned in the review by Jeppson, Keifer, and Baker (1975b). The two-spotted spider mite, *Tetranychus urticae*, is perhaps the single most economically important species of spider mites worldwide and is a major pest of deciduous fruit trees, while the carmine spider mite, *T. cinnabarinus*, causes serious damage to cotton, vegetable crops, strawberries, and, under greenhouse conditions, tomatoes (Van de Vrie 1985; Hussey and Scopes 1985; Zhang 2003). *T. kanzawai* (= *T. hydrangeae* Pritchard and Baker) (Navajas et al. 2001) and *T. mcdanieli* McGregor also are important fruit tree pests, the former especially in eastern Asia (where it is also a major pest of tea) and the latter in northwestern North America. Cotton and low-growing vegetable crops and greenhouse plants often are infested by the desert spider mite, *T. desertorum* Banks; the strawberry spider mite, *T. turkestani* (Ugarov and Nikolski) (= *T. atlanticus* McGregor); the tumid spider mite, *T. tumidus* Banks; and *T. ludeni* Zacher. Ironically, *T. lintearius* Dufour, a species that is morphologically similar to but reproductively isolated from *T. urticae*, has been found to be so host specific to *Ulex* species that it has been introduced from Europe as a beneficial biological control agent against gorse, a seriously invasive weed of pastoral land and forest areas in New Zealand, western North America, and several other North American sites (Stone 1986; Hill and O'Donnell 1991). A related species, *T. pueraricola* Ehara and Gotoh, infests kudzu, *Pueraria lobata*, an Asian endemic that has become a major economic problem in the southern United States (Baskin 2002). Although *T. pueraricola* may have potential as a biological control agent against kudzu, it also has a high reproductive capacity, reproductive incompatibility between strains, and a potentially broad host range, all of which may make it a questionable choice for weed control in agricultural settings (Ehara and Gotoh 1996; Akiyuki, Kitashima, and Rezk 2004; Gotoh, Suwa, and Kitashima 2004; Suwa, Kitashima, and Gotoh 2006).

Among the relatively few described species of *Panonychus* (Ehara and Gotoh 1992), several are serious pests. The European red mite, *P. ulmi* (Koch), may be found in association with *Tetranychus* species on deciduous fruit trees and is considered a major economic problem in European, North American, and eastern Asian orchards (Groves and Massee 1951; Ehara 1975b; Lee 1989; Tseng 1990). The citrus red mite, *P. citri* (McGregor), attacks citrus trees and other fruit and ornamental plants in nurseries and outdoors in North America, South Africa, and eastern Asia and has been highly injurious in California orchards, particularly in coastal and intermediate citrus-growing areas (Jeppson, Keifer, and Baker 1975b; Zhang 2003). *P. citri* and *P. mori* Yokoyama are important pests of deciduous fruit trees in Japan (Gotoh and Kubota 1997; Gotoh, Ishikawa, and Kitashima 2003). The aptly named *P. thelytokus* Ehara and Gotoh, living on elms in Japan, is a rare example of a thelytokous species of the subfamily Tetranychinae. Its males are rarely produced and evidently are sterile (Ehara and Gotoh 1992).

Several species of the genus *Mononychellus* feed preferentially or exclusively on cassava and other species of *Manihot*. Cassava-producing areas in Africa were greatly altered by the introduction and rapid dissemination from South America of the green cassava mite, *M. tanajoa* (Bondar), which resulted in major tuber yield losses (Lyon 1973). Two other species, *M. caribbeanae* (McGregor) and *M. progresivus* (Doreste), are also frequently reported to cause damage on cassava (Bellotti 1985).

A number of species of *Oligonychus* are considered to be of major economic importance. The Banks grass mite, *O. pratensis* (Banks), has proven to be a serious pest of grasses in North and Central America, Hawaii, and Africa (Jeppson, Keifer, and Baker 1975b; Meyer 1987). Wheat, corn, bluegrass, sorghum, sugarcane, and rice compose a partial list of host plants for this species. Coniferous and deciduous trees are major hosts for species groups of *Oligonychus*. The spruce spider mite, *O. ununguis* (Jacobi), is a major pest of conifers throughout the world. An epidemic outbreak of *O. ununguis* in Montana and Idaho in 1957 severely damaged over 800,000 acres of Douglas fir forests (Furniss and Carolin 1977). *O. subnudus* (McGregor) and *O. milleri* (McGregor) may also cause severe injury in ornamental plantings of conifers. The oak mite, *O. bicolor* (Banks), is a serious pest of oak and other ornamental trees in widespread areas of North America. The avocado brown mite, *O. punicae* (Hirst), sometimes attains sufficiently high population densities to cause partial defoliation and lessened fruit production of avocado trees in southern California (McMurtry 1985). Similar symptoms are associated with the avocado red mite, *O. yothersi* (McGregor), in Florida and south into South America (McMurtry 1985). The tea red spider mite, *O. coffeae* (Nietner), is widely distributed and is considered the most serious of all tea pests in some Asian countries (Banerjee and Cranham 1985; Meyer 1987). The southern red mite, *O. ilicis* (McGregor), is injurious on holly, rhododendron, and other hosts in the eastern United States (Weidhaas and Reeves 1963). It is also a widely distributed pest of coffee in Brazil. As among some species of *Tetranychus*, divergence in host range and reproductive compatibility has been observed among strains of *Oligonychus gotohi* Ehara (Gotoh et al. 2007).

Members of the genus *Eotetranychus* feed primarily on trees and shrubs, and some species are considered to be regional or periodic economic pests of ornamental trees, citrus and deciduous fruit trees, grapes, berries, and other plants in greenhouses. For example, *E. carpini carpini* (Oudemans) may become a problem in unsprayed apple orchards in Europe and North America, although natural enemies tend to keep it in check. The yellow spider mite, *E. carpini borealis* (Ewing), may be injurious to apple and pear trees in the northwestern United States; the yellow vine mite, *E. carpini vitis* (Oudemans), to grape vines in Italy and France; and the fig spider mite, *E. hirsti* Pritchard and Baker, to figs in India and Pakistan. The six-spotted spider mite, *E. sexmaculatus* (Riley), is a periodic pest of citrus and avocado in California, Florida, and Formosa and grape vines in Arizona and New Zealand. The Lewis spider mite, *E. lewisi* (McGregor), feeds on many plant species, including citrus in California and papaya in Mexico, and is a significant pest of poinsettia in greenhouses. The conclusion that males of *E. lewisi* have only one nymphal stage (Zhang 2003) needs confirmation. Finally, *E. pruni* (Oudemans) and *E. uncatus* Garman may be pests of pomaceous and prunaceous fruit trees and grape vines in Europe and eastern North America.

Although the Willamette spider mite, *E. willamettei* (McGregor), may occasionally cause extensive injury of grape vines elsewhere in California, it tends to have low economic impact in California's Central Valley, where its antagonistic interaction with the more damaging Pacific spider mite, *Tetranychus pacificus* McGregor, has led to its being introduced experimentally as a "vaccination" to prevent outbreaks of *T. pacificus* on grapes (Karban and English-Loeb 1988, 1990).

A comprehensive review of injurious tetranychid mites may be found in Jeppson, Keifer, and Baker (1975b). Excellent surveys of ecological factors affecting host injury by important tetranychid species are presented by Huffaker, Van de Vrie, and McMurty (1970); McMurtry, Huffaker, and Van de Vrie (1970); and Van de Vrie, McMurty, and Huffaker (1972). A two-volume work edited by Helle and Sabelis (1985a, b) presents accumulated knowledge on anatomy, phylogeny and systematics, reproduction and development, physiology and genetics, and ecology of spider mites in the first volume, followed by treatment of their natural enemies and biological and integrated control in the second volume. In recent decades, many guides with keys and illustrations to species have been published to identify spider mites in a single country or region, including Africa (Meyer 1987), Taiwan (Tseng 1990), the United States (Baker and Tuttle 1994), and Japan (Ehara 1999). These and other such surveys are noted by way of introduction in a catalog to the species of TETRANYCHIDAE of the world by Bolland, Gutierrez, and Flechtmann (1998), which includes hosts and distribution and keys to subfamilies, tribes, and genera.

The false spider mites or TENUIPALPIDAE (Figs. 13.76A–C) comprise about 30 genera and 800 described species (Sepasgosarian 1983, 1990; Smiley, Frost, and Gerson 1996; Zhang 2003) of small, somewhat flattened, slow-moving, pale, beige, red, or green mites that appear to be best adapted to, and most diverse in, subtropical and tropical climates. The ranges of some species, however, extend into temperate areas via greenhouse infestations (Zhang 2003), while others occur naturally in temperate zones, occasionally under extreme climatic conditions. For example, *Aegyptobia montana* Bury and Krantz was described from soil ridge patches of *Penstemon* in a snowfield at an elevation of 2,377 m on Mt. Hood, Oregon, USA, where conditions are so harsh that mites were confined to basal or subterranean portions of their host plant, rather than venturing onto exposed stems and leaves (Bury and Krantz 1978). Species occurring in extreme temperate climates generally overwinter as adult females.

Tenuipalpids may feed on stems, fruit, or either leaf surface but occur most commonly on lower leaf surfaces near the midrib or veins (Baker and Pritchard 1960). Some species feed in floral heads or under sheaths of leaves or grass blades, or even through the thin bark of plant stems, and the most specialized species form plant galls (Meyer 1979). The genera *Brevipalpus* and *Tenuipalpus* are highly speciose, the former with more than 300 and the latter around 200 described species (Meyer 1979, 1993; Welbourn et al. 2003), and are of particular importance as plant pests, as indicated by the examples presented below. The dispersal behavior of tenuipalpids has not received attention but is thought to be passive via air currents and human transportation of infested plants (Zhang 2003).

The suppression of the palptibial thumb-claw complex is the most distinctive synapomorphy for the TENUIPALPIDAE and is one of the indicators of its relatively highly derivative phylogenetic position among the families of Tetranychoidea (Lindquist 1985a). Evolutionary plasticity also is remarkable within the family. The number of free palpal segments in tenuipalpids often is reduced from the plesiomorphic five-segmented state characteristic of other families of Tetranychoidea to four, three, two, one, or the vestige of a segment. Extreme palpal reduction appears to be homoplastic and occurs in *Larvacarus transitans* (Ewing) and *Obdulia tamaricis* Pritchard and Baker, which induce twig galls on jujube trees in India and on *Tamarix* trees in Israel, respectively (Pritchard and Baker 1958; Gerling, Kugler, and Lupo 1976). *Obuloides rajamohani* Baker and Tuttle, an inhabitant of eriophyid galls in India (Baker and Tuttle 1975a), also has greatly reduced palpi, as does *Capedulia maritima* Gerson and Smith Meyer, collected from the roots of *Limonium meyeri* on a coastal plain occasionally immersed by seawater in Israel (Gerson and Meyer 1980). Another independently reductive and perhaps homoplastic trend is the suppression of legs IV in postlarval instars. Taxa with hexapodan adult females include *Larvacarus transitans* (Ewing) and *Raoiellana allium* Baker and Tuttle on *Allium* bulbs from Turkey and species of *Phytoptipalpus* on *Acacia* and *Albizzia* in Africa and India (Pritchard and Baker 1958; Baker and Tuttle 1972; Meyer 1979).

A classification of the TENUIPALPIDAE proposed by Mitrofanov (1973a, b) comprised two subfamilies, eight tribes, and over 30 genera, but subsequent authors (Meyer 1979; Baker and Tuttle 1987) argued that some of the characters used were variable or arbitrary, that the differences between some genera were insignificant, and that Mitrofanov's genera were not easily placed into natural suprageneric groups (tribes). Nevertheless, descriptions of new genera based on arbitrary attributes continue to persist in the literature (e.g., Smiley and Gerson 1995; Smiley, Frost, and Gerson 1996).

There is considerable genetic and developmental evidence for evolutionary plasticity within the TENUIPALPIDAE. Arrhenotoky has been confirmed in a variety of genera, including numerous species of *Brevipalpus*, and diploid thelytoky has been demonstrated in a few species of *Aegyptobia* and *Tenuipalpus* (Bolland and Helle 1981). More remarkable is the occurrence of haploid female parthenogenesis in populations of *Brevipalpus obovatus* Donnadieu and evidently also in two closely related species, *B. phoenicis* (Geijskes) and *B. californicus* (Banks), a phenomenon otherwise unknown among animals (Pijnacker et al. 1980; Bolland and Helle 1981; Weeks, Marec, and Breeuwer 2001). Low frequencies of adult males and intersexes persist among such populations, and males are able to copulate but insemination does not take place (Pijnacker, Ferwerda, and Helle 1981). Paedogenesis, with eggs developing and emerging from protonymphs and deutonymphs has been reported in a species of *Brevipalpus* (Baker 1979). Sexual dimorphism occurs in the deutonymph, protonymph, and sometimes in the larva of some species of *Tenuipalpus* (Collyer 1973b). A high degree of variation in length and form of setae has been noted on same instar nymphs in some species of *Tenuipalpus* and *Brevipalpus* (Collyer 1973b; Evans, Cromroy, and Ochoa 1993).

In terms of destructive potential, tenuipalpids are considered the third most important family of phytophagous mites. Citrus trees in commercial orchards are but one of a multitude of hosts for the widely distributed *Brevipalpus californicus* (Banks), *B. lewisi* McGregor, *B. obovatus* Donnadieu, and *B. phoenicis* (Geijskes) (Childers, Rodrigues, and Welbourn 2003). Symptomatology includes foliar chlorosis and necrosis, gall formation, and cracking, blistering, and deformation of fruit. Viral leprosis and other viral or virus-like disorders may also be caused by association of these mites with plant pathogens (Jeppson 1975; Evans, Cromroy, and Ochoa 1993; Childers, French, and Rodrigues 2003; Kitajima, Chagas, and Rodrigues 2003). For example, *B. phoenicis* is a confirmed vector of coffee ringspot and passion fruit green spot viruses in Brazil and Costa Rica (Chagas, Kitajima, and Rodrigues 2003; Kitajima, Rezende, and Rodrigues 2003b), in addition to being a major pest of tea throughout much of Indonesia (Oomen 1982). *B. californicus* is a known vector of orchid fleck virus in many parts of the world (Kondo, Maeda, and Tamada 2003). *B. chilensis* Baker causes serious damage to grapes in Chile, and *B. russulus* (Boisduval) is a widely spread pest of ornamental cacti and other succulents (Chiavegato 1976; Denmark 1978; Ashley and Manson 1987). Ornamental coniferous trees and hedges, excluding members of the Pinaceae, may be disfigured by several species of *Pentamerismus*, while species of *Brevipalpus* replace them on pines (Pritchard and Baker 1958; Pegazzano 1971). Other hosts for *Brevipalpus* species include ornamentals, berry crops, grapes, olives, orchids, and other plants grown in greenhouses (Zhang 2003).

Members of the genus *Tenuipalpus* infest grasses, tea, orchids, and pomegranates, among other hosts. *T. pacificus* Baker attacks orchids in European greenhouses (Dosse 1954) and was noted to cause damage to cultivated ferns in Brazil (Veiga and Flechtmann 1980). Other species of *Tenuipalpus* are known to feed on ferns in China and New Zealand (Gerson and Collyer 1984). *Cenopalpus lanceolatisetae* (Attiah) and *C. pulcher* (Canestrini and Fanzago) are pests of deciduous fruit trees in Europe, Asia, and North Africa (Jeppson, Keifer, and Baker 1975b). More recently, *C. pulcher* was found infesting apple and pear foliage in Oregon, USA, the first report of this species from the New World (Bajwa, Krantz, and Kogan 2001). Feeding by *Larvacarus transitans* on *Ziziphus* fruit trees in India not only induces distorted twig galls but also causes the setting of undersized and crumpled fruit (Sharma 1991).

The elongate bright red tenuipalpid species of the genus *Dolichotetranychus* are found primarily on monocotyledonous plants, especially under the leaf sheaths or in the floral heads of grasses (Pritchard and Baker 1958). *D. australianus* (Womersley) damages lawn grasses in parks, golf courses, and rugby fields, and *D. floridanus* (Banks) may cause injury to pineapples wherever they are cultivated (Meyer 1979). *D. vandergooti* is only known to occur on orchids (Collyer 1973a).

The remaining three families of Tetranychoidea—the TUCKERELLIDAE (Figs. 13.75D, E), ALLOCHAETOPHORIDAE (Figs. 13.75A, B), and LINOTETRANIDAE (Figs. 13.76D, E)—include few species of agricultural importance. Tuckerellids are slow moving and typically bright red with conspicuous white setae and occur in soil and on stems, leaves, and fruit of plants. They are often referred to as peacock mites because of their long plumose caudal setae. According to Ochoa (1989a), these setae are swept over the body to deter predators. Nearly 30 species of TUCKERELLIDAE are currently described, and all are included in the genus *Tuckerella.* Some species occur in soil, probably in association with underground plant parts. These include *T. knorri* Baker and Tuttle, collected from the roots of *Pandanus* and *Achras* in Thailand; *T. hypoterra* McDaniel, Morihara, and Lewis, from pasture soil in South Dakota; and *T. coleogynis* Jorgensen, from foliage and debris at the base of grasses in Nevada (Baker and Tuttle 1975b; McDaniel, Morihara, and Lewis 1975; Jorgensen 1967). An unidentified species of *Tuckerella* was collected from plant roots in California, and *T. litoralis* Collyer was found in moss on boulders bordering a saltwater harbor (Collyer 1969). Other species of the genus are aerial forms, including several from conifers (Miller 1964; Zaher and Rasmy 1969; Chaudhri 1971; McDaniel, Morihara, and Lewis 1975; Ehara 1975b; Servin and Otero 1989; Ochoa 1989a).

Most species of TUCKERELLIDAE appear to have restricted distributions, about half of them having type localities from the Oriental region (Corpus-Raros 2001). However, the type species *T. ornata* (Tucker), *T. pavoniformus* (Ewing), and *T. knorri* Baker and Tuttle, originally described from South Africa, Hawaii, and Thailand, respectively, are each reported from several continents (Ehara 1975a; Corpuz-Raros 1989; Meyer and Ueckermann 1997; Lin and Fu 1997; 2000; Beard and Walter 2005). *T. knorri* is economically important as a causal associate of leprosis on *Citrus* in Costa Rica (Ochoa 1989a). Species of *Tuckerella* may be unique in the superfamily in that, based on genital and leg setae, the tritonymphal stage is evidently retained (Miller 1964; Jorgensen 1967; Quiros-Gonzalez and Baker 1984; Rasmy and Abou-Awad 1984; Servin and Otero 1989; Ochoa 1989b; Corpuz-Raros 2001; Beard and Walter 2005). However, rearing studies that track the number of postlarval molts or observations of pharate females and males in nondimorphic tritonymphal skins will be needed to confirm this morphological evidence and eliminate the alternative possibilities of sexually dimorphic deutonymphs (Lindquist, in Servin and Otero 1989), or a male developmental sequence that lacks a tritonymphal instar (Beard and Walter 2005). Paedogenesis, with eggs developing and emerging from tritonymphs, has been reported in one species (Ochoa 1989b).

Mites of the family LINOTETRANIDAE are inconspicuous forms with slender, reticulated, colorless, and eyeless idiosomas measuring approximately 300–400 μm in length. They have been reported from moss, soil, and rotting trees, but most collections indicate an association with the crown and roots of grasses or sedges in dryland habitats (Beard and Walter 2004). The family has been known for nearly a century, but for nearly half that time from a single species, *Linotetranus cylindricus* Berlese, described from moss in Panormitanti, Italy (Berlese 1910a). Currently, there are four described linotetranid genera and 12 species. Baker and Pritchard (1953) described *L. achrous* associated with salt grass, *Distichlis spicata*, in California. In the same paper, they redescribed *L. cylindricus* from two female specimens collected from rotting alpine fir in Utah, basing their identification on a comparison made by G. Lombardini of their illustrations with the Berlese holotype. However, the identity of the Utah specimens remains uncertain. Subsequent papers have reported linotetranids from North America (Smith and Lindquist 1979; Walter 1987a), South America (Arganaraz and de-Alzuet 1991), Africa (Athias-Henriot 1961b; André 1996; Meyer and Ueckermann 1997), and Australia (Kinnear 1991; Noble, Whitford, and Kaliszewski 1996; Beard and Walter 2004). Prior to recognition of the genera *Afrolinotus* and *Anoplopalpus* by Meyer and Ueckermann (1997) and of *Austrolinus* by Beard and Walter (2004), all previous reports of unspecified linotetranids from Australia (and probably elsewhere) referred to *Linotetranus* sp. (e.g., Kinnear 1991). Because only one species, *L. achrous*, has been reported on a specific host plant, nothing can be said here regarding host preferences. However, the elongated form of these and other unrelated mites that are obligately phytophagous has been noted as a probable adaptation to living in sheaths of graminaceous hosts (E. E. Lindquist, unpublished obs.).

Long the mystery mite of the Tetranychoidea, *Allochaetophora californica* McGregor (ALLOCHAETOPHORIDAE) was described from a single deutonymph collected from a tuft of Bermuda grass (*Cynodon dactylon*) in California (McGregor 1950a). Meyer and Ueckermann (1997) described adults of both sexes and the larva and deutonymph of a new species, *Allochaetophora africana*, from grass and soil in South Africa. Lindquist (1985a) suggested that ALLOCHAETOPHORIDAE was the likely sister group of the TENUIPALPIDAE, and that the LINOTETRANIDAE was the sister group to both families. The recent discovery of adult allochaetophorids opens the door to reexamining this hypothesis. As in linotetranids, the elongated form of allochaetophorids may reflect adaptation to sequestered habitats in the sheaths of graminaceous plants.

▪ Superfamily Cheyletoidea (Figs. 13.77–13.80)

DIAGNOSIS: Soft bodied but often with weak dorsal shields. Gnathosoma well defined, composed of cheliceral stylophore either free from the subcapitulum or fused into a gnathosomatic capsule; movable cheliceral digits styliform, needlelike, retractile, approximate terminally; palpi variously developed, often reduced with fused tibiotarsus and femorogenu, or telescoped, sometimes large and armed with a distinctive tibial claw; post-cheliceral stigmata and peritremes present in free-living and some parasitic forms but absent in many parasites; when present, peritremes are at the base of the chelicerae or on the dorsal surface of the gnathosomatic capsule, dorsal cheliceral seta rd *absent in larva. Prodorsum lacking anteromedian naso, typically with 4 pairs of setae, none inserted in bothridia; eyes present or absent. Opisthosoma with 0 to many setae dorsally and laterally, with or without postlarval setal additions or hypertrichy dorsally; cupules (lyrifissures) absent; elements of opisthosomatic segments AD absent. Larval urstigma and postlarval genital papilla absent. Claws always present on tarsi II, and usually on tarsi III–IV unless legs are reduced or absent, claws present or absent on tarsus I; empodia present or absent, often with tenent hairs; genua and tibiae of all legs with a maximum of 3 setae; solenidion of genu I present or absent. Female genital opening ventral, terminal, or dorsal, free from or fused to anal opening; oviporus adjacent to or fused with genital opening. Male with a terminal or dorsal genito-anal opening. With 4 active postembryonic instars including larva, 2 nymphal instars (only 1 occurs in male of* Nodele calamondin *Muma), and adult. Developmental stases may be apodous, or postlarval stases may be hexapod.*

The Cheyletoidea comprises seven families, of which only one—the CHEYLETIDAE—includes arthropod associates. About a quarter of the described species, along with those of the remaining six cheyletoid families, are parasites of reptiles, birds, and mammals, including humans. As in the mesostigmatan Dermanyssoidea (p. 153), the Cheyletoidea illustrates what appears to be a series of evolutionary stepping stones from free-living predation in vertebrate nests to ecto- and endoparasitism on the nesting hosts and can be considered a model system for study of the evolution of nest-associated parasitism (see Radovsky 1994).

The more than 130 genera and 600 described species of Cheyletoidea are characterized by the close approximation or fusion of the oviporus with the genital opening of the female, the reduction of setation of genua and tibiae I–IV to three or fewer setae, and the suppression of larval seta *rd* (Bochkov, Mironov, and Fain 1999; Bochkov 2002). All cheyletoids have a stylophore composed of the fused cheliceral bases (Summers and Witt 1971; Gorgol' and Yastrebtsov 1989). The stylophore in the CHEYLETIDAE (free-living, skin and quill parasites, and predators) and SYRINGOPHILIDAE (quill parasites) is fused to the subcapitulum to form a **gnathosomatic capsule** (rostral shield of some authors), and the peritremes are elaborated on its dorsal surface (Figs. 13.77D, 13.78D). The peritremes are located at the base of the chelicerae in the bird-parasitic members of the HARPIRHYNCHIDAE (Fig. 13.79A), but they are reduced to stigmatic openings in the Ophioptinae, a harpirhynchid subfamily parasitic on snakes. Stigmata and peritremes are absent in the highly reduced intradermal and internal parasites of the PSORERGATIDAE, DEMODICIDAE, CLOACARIDAE, and EPIMYODICIDAE. Paedomorphosis is known to occur in members of the cheyletid genus *Metacheyletia*, which live in the feathers of parrots. These mites have highly modified larviform adults in which legs IV are vestigial or absent (Atyeo, Kethley, and Perez 1984). Silken retreats or capsulelike structures are spun by members of the CHEYLETIDAE, HARPIRHYNCHIDAE, DEMODICIDAE, PSORERGATIDAE, and possibly members of other families.

The CHEYLETIDAE (Figs. 13.77A–G) includes approximately 74 genera and more than 370 described species that occupy a broad range of habitats, including soil, litter, tree bark, vegetation, debris, and grain storages. Cheyletids have also been collected from arthropod and bird nests, caves, and bat roosts in association with beetles, bugs, bees, flies, scorpions, millipedes, lizards, birds, and mammals (Smiley 1970; Summers and Price 1970; Volgin 1987; Gerson, Fain, and Smiley 1999; Bochkov, Fain, and Dabert 2002; Bochkov and OConnor 2003). Subsocial behavior has been reported in some cheyletids, including defense of eggs (Mori, Saito, and Tho 1999; Walter and Proctor 1999). Approximately three-fourths of the described genera and species are considered predators.

Haplodiploid genetic systems have been reported in the CHEYLETIDAE, as well as in the cheyletoid families SYRINGOPHILIDAE, HARPIRHYNCHIDAE, and DEMODICIDAE (Norton et al. 1993). Although observations on reproductive modes in cheyletids are relatively few in number, both arrhenotoky and thelytoky have been observed in the family (Beer and Dailey 1956; Summers, Witt, and Regev 1972; Barker 1992). Most of the published records deal with species of the type genus *Cheyletus*. Thelytoky has been demonstrated in *C. eruditus* (Schrank) and *C. cacahuamilpensis* Baker, and arrhenotoky in *C. trouessarti* Oudemans and *Nodele calamondin* Muma (Barker 1992). Males of *N. calamondin* have only one nymphal stase, while females have two (Barker 1992). Male polymorphism, especially in relation to relative palpal length, has been noted among species of *Cheyletus* and *Eutogenes* (Regev 1974; Smiley and Moser 1975).

Species of *Cheyletus*, *Cheletomorpha*, *Cheyletia*, *Nodele*, and *Acaropsis* often occur as predators in grain storages infested by graminivorous mites (Hughes 1976; Barker 1992). *Cheyletus eruditus* is considered by some observers to be effective in controlling grain mite populations (Pulpan and Verner 1965; Zdarakova and Pulpan 1973; Zdarakova 1986), but it may be seasonal in its effectiveness (Solomon 1946a). The mite was noted to reduce laboratory populations of the granary mite, *Acarus siro* L., when prey and predator populations were equal or imbalanced in favor of the predator, at 20°C and 80% relative humidity (Solomon 1969). *A. siro* populations quickly outnumbered *C. eruditus* at lower temperatures. House dust mites (Astigmatina, PYROGLYPHIDAE) are effectively controlled by *C. eruditus* in laboratory populations (F. S. Lukoschus, pers. comm.). *C. eruditus* also has been observed to prey on eggs of the cat flea, *Ctenocephalides felis* (Williams and Hallas 1988), and on a variety of microarthropods in the laboratory that were small or weak enough to be overpowered (Beer and Dailey 1956). Ambush behavior by *C. eruditus* was described by Beer and Dailey (1956), and that of *C. aversor* Rohdendorf was studied by means of closed-circuit television by Wharton and Arlian (1972).

Many cheyletid species are found on vegetation, and some may be useful predators of crop pests (Avidov, Blumberg, and Gerson 1968; de Moraes, Neto, and Pinto 1989). Phytophilous species include members of the genera *Cheletacarus, Cheletogenes, Cheletomimus, Hemicheyletia, Oudemansicheyla, Philippicheyla, Prosocheyla,* and *Tutacheyla* (Volgin 1987). The elongate, cylindrical species of the genus *Neochelacheles* are mycetophiles that live in the spore tubes of polypore fungi and disperse on mycetophilous tenebrionid beetles (OConnor 1984a; Bochkov and OConnor 2004a). *Bak deleoni* Yunker also lives in the spore tubes of fungi (De Leon 1962).

Although many cheyletids are assumed to be free-living predators, some of them have established unusual associations with other animals. For example, *Chelacaropsis moorei* Baker and species of *Hemicheyletus* are sometimes found in the fur of mammals, where they prey on ectoparasitic arthropods sharing the same habitat (Lawrence 1954; Volgin 1960). Species of *Cheyletus, Eucheyletia,* and *Cheletonella* may also be associated with mammals (Whitaker and Wilson 1974; Reisen et al. 1976), while members of the genus *Cheletopsis* live in the quills of charadriiform birds, where they are known to prey on syringophilid quill mites (Bochkov, Fain, and Dabert 2002). *Acaropsis docta* Berlese and a species of *Cheletonella* were observed to feed on the skin of house martins in England (Woodroffe 1956), although neither genus is considered truly parasitic. On the other hand, *Pavlovskicheyla platydemae* Thewke and Enns appears to be a true ectoparasite of the tenebrionid beetle *Platydema ruficorne* (Thewke and Enns 1975), with the mites attaching beneath the elytra of the beetle host. Members of the genus *Cheletophyes* that are phoretic among the hairs or in the acarinaria of carpenter bees of the genus *Xylocopa* in Asia and Australia (also reported in Panama; B. M. OConnor, pers. comm.) appear to be mutualists that feed on other mites in bee nests (OConnor 1993c; Walter et al. 2002).

As noted earlier, some cheyletids are parasitic on birds and small mammals. Although these mites have sometimes been treated as a separate subfamily (Cheyletiellinae) or family (Cheyletiellidae), it now appears that parasitism in this assemblage has arisen independently several times (Bochkov and Fain 2001). Approximately six genera and 60 species are known to be parasitic on birds, including both skin and quill parasites (Bochkov and OConnor 2003). For example, *Bakericheyla chanayi* (Berlese and Trouessart) feeds on the blood of several species of wild sparrows, including the white-crowned sparrow, *Zonotrichia leucophrys.* Migrating sparrows and those held in captivity may develop heavy infestations. In extreme cases, captive birds are enmeshed in large amounts of silk webbing and become emaciated and listless (Furman and Sousa 1969). Another cheyletielline, *Ornithocheyletia* sp., can cause severe hyperkeratinosis and consequent secondary fungal infection in the common pigeon, *Columba livia* (Haarløv and Mørch 1975). Members of the genera *Nihelia, Eucheyletiella, Criokeron,* and *Sciurocheyla* and some *Cheyletiella* are mammal parasites whose ancestors were free-living predators that spun silken nests to protect their eggs and themselves during molting (Walter and Proctor 1999; Bochkov and OConnor 2004b). In the case of birds, silken webs are spun on the host skin, while on mammals the webs are found at the bases of hairs on which eggs may be attached.

Several species of *Cheyletiella* are host specialists that live on the epidermis of small to medium-sized mammals and may cause a type of mange referred to as **cheyletiellosis** (Bronswijk and Kreek 1976). *C. yasguri* Smiley is associated with dogs (Rack 1971; Fain, Gerrits, and Lukoschus 1982) and is also a hyperparasite of dog-infesting louse flies in Iran (Vercammen-Grandjean and Rak 1968). *C. blakei* Smiley infests cats (Keh 1975; Pfeiffer 1979), while rabbits are host to *C. parasitivorax* (Mégnin). Infestations of pets and laboratory animals can be benign, but pruritis, hair loss, and crustose lesions can develop in severe cases. Like their free-living ancestors, species of *Cheyletiella* are highly active mites whose pale color and mobility have led to their being referred to as "walking dandruff." They are able to move readily from pets or laboratory animals to pet owners and laboratory workers, where they may cause a transient dermatitis characterized by raised red wheals with blisterlike centers. Cheyletiellosis is probably common in many parts of the world but often is misdiagnosed because of the elusive nature of the mites that cause it (Yunker 1973; Lee 1991). Although domestic cheyletid species usually feed on house dust mites or stored product mites, they may come into contact with humans, especially when their population densities become high. Skin lesions have been induced experimentally in human subjects by exposure to some of these species—for example, *Cheyletus malaccensis* Oudemans, *C. fortis* Oudemans, and *Chelacaropsis* spp. (Yoshikawa 1985; Hatsushika, Okino, and Miyoshi 1989).

The HARPIRHYNCHIDAE (Figs. 13.79A, B) comprises some 14 genera whose species are parasitic in the skin (sometimes forming intradermal nodules), scales, and feathers of their hosts (Fritsch 1954; Zumpt 1961; Fain 1972d, 1976b; Filipich and Domrow 1985; Bochkov, Mironov, and Fain 1999). The subfamilies Harpirhynchinae and Harpypalpinae are confined to bird hosts, while members of the Ophioptinae (Fig. 13.79B) are found under or between the scales of colubrid and elaphid snakes (Bochkov 2002), typically in small pits. Sambon (1928) reported that *Ophioptes parkeri* Sambon was responsible for the formation of these pits in the scales of colubrids, a conclusion also reached by Ewing (1933) for *O. tropicalis* on another colubrid species. It has since been determined that the female makes the pit and deposits an egg therein. The apodous larvae feed, expand, and molt to apodous nymphs, which then give rise to males and females (Fain 1964d). Harpirhynchine harpirhynchids are known to occur in the skin and in intradermal nodules of 16 orders of birds, while harpipalpine species (also with apodous juveniles) are confined to intradermal nodules on the wings of passerine hosts (Bochkov 2002). Species of North American passerine birds representing four families are known to be parasitized by *Harpyrhynchus novoplumaris* Moss, Oliver, and Nelson, a harpirhynchine species that usually is found on the head and neck of its hosts. The female feeds at a feather base and forms a silken sheath that covers both her and her eggs. Reproduction is synchronized with the breeding season of the host, and females die after oviposition (Moss, Oliver, and Nelson 1968).

Members of the family SYRINGOPHILIDAE (Figs. 13.78C, D) are permanent monoxenous or oligoxenous parasites that feed on fluids of their bird hosts through the walls of the feather quills in which they reside. More than 150 species representing 33 genera have been described to date, but it has been estimated that the family may comprise upward of 5,000 species (Kethley 1970, 1973; Johnston and Kethley 1973; Bochkov, Fain, and Skoracki 2004). Syringophilids are highly specific to host and feather type (primaries, secondaries, coverts, contour feathers), with the thickness and size of a given quill largely determining the mite species complex inhabiting it (Kethley 1971b). Stylet length is a critical factor for survival in a chosen feeding site, as is maximizing development of female dispersants at host molting time in the autumn and nestling development in the spring. While members of the subfamily Syringophilinae tend to be restricted to particular host orders or closely related orders, the approximately 20 known species of Picobiinae (*Picobia* and *Calamincola*) are recorded from the body feather quills of at least six orders (Skoracki, Bochkov, and Wauthy 2004). The SYRINGOPHILIDAE is the sister group of the CHEYLETIDAE and differs from them mostly on the basis of autapomorphic characters (Kethley 1970; Bochkov 2002; also see Ectoparasites of Vertebrates in chapter 6, p. 75).

Members of the PSORERGATIDAE (Figs. 13.78A, B) are minute (130–200 μm), round mites that live in superficial skin layers of their mammal hosts. Three genera are currently recognized (*Psorergates*, *Psorergatoides*, and *Psororbia*), with the most important pest species in the last genus (Giesen 1990). Psorergatids often cause dermatitis, mange, or follicular infection, which is aggravated by rubbing or biting by the host. The sheep itch mite, *Psorobia ovis* Womersley, is a major pest of sheep in the United States and Australia (Baker et al. 1956). *P. bos* (Johnston), an itch mite of cattle in the southwestern United States and Europe, precipitates symptoms in its host that are similar to those caused by *P. ovis* on sheep (Johnston 1964; Roberts and Meleney 1965). *P. foinae* Fain and Lukoschus and *P. mustelae* Lukoschus are unusual for the genus in being parasites of carnivores rather than ungulates (Fain and Lukoschus 1968). Other psorergatids have been collected from insectivores, bats, rodents (including laboratory animals), and primates in Africa, Asia, North and South America, Europe, and Australia (Fain, Lukoschus, and Hallmann 1966; Lukoschus, Fain, and Beaujean 1967; Yunker 1973; Lukoschus, Louppen, and Maa 1974; Giesen 1990; Domrow 1991). Feeding by *Psorergates glaucomys* Ah, Peckham, and Atyeo on flying squirrels induces production of dermal cysts containing aggregations of these mites (Ah, Peckham, and Atyeo 1973).

The minute, usually wormlike species of the family DEMODICIDAE (Figs. 13.79C–F) are parasites that may cause epithelial destruction, hyperplasia, and granuloma in their mammal hosts. Demodicids are known to invade hair follicles and meibomian (holocrine) glands or ducts, and may penetrate the epidermis. Records also exist of invasion of auricular tissues, subdermal fascia, the digestive tract, and even the circulatory system (Nutting 1975, 1985). *Demodex caprae* Railliet may cause metaplasia of follicular epithelial cells of infested goats, followed by hyperkeratinization and eventual cratering of the skin after extrusion of the mites (Lebel and Nutting 1971). Severe mange in canine puppies can be caused by *D. canis* Leydig (Yunker 1973). *D. equi* Railliet causes pruritis and alopecia in horses, and subsurface skin pustules containing demodicid mites are routinely found in cattle, sheep, and goats. Pathology due to *Demodex* is limited in most wild animals (Nutting 1985), but massive infestations that induce hyperkeratosis and alopecia may occur for reasons that are not entirely understood. Captive wild animals (Nutting 1986; Holz 1998; Vogelnest, Vogelnest, and Mueller 2000) are especially prone to such infestations, although wild animal populations of black bears (*Ursus americanus*) may sometimes suffer similar outbreaks (Forrester et al. 1993).

Bats seem particularly susceptible to demodicid-derived pathology (Nutting 1985; Vargas et al. 1995). The greatly elongated *D. longissimus* Desch and Nutting, and the short, squat *Opthalmodex molossi* (Lukoschus, Woeltjes, Desch, and Nutting) (Lukoschus et al. 1980; Veal et al. 1984) attach inside the large posterior Meibomian gland of Neotropical bats in such a way that the opisthosoma hangs out over the eyeball of the host (Desch, Nutting, and Lukoschus 1972). Immobile larvae and nymphs have greatly enlarged hooked posterior legs with which they secure themselves to the host tissue (Fig. 13.79E).

Prosimian bushbabies and galagos are parasitized by species of *Stomatodex* and *Rhinodex* (Fain 1959), and humans serve as the natural host for *D. folliculorum* (Simon) and *D. brevis* Akbulatova (Desch and Nutting 1972; Sengbusch and Hauswirth 1986). Both of the latter *Demodex* species may be collected from the skin of the forehead, nose, and alar regions of the face, but each exploits a different niche. *D. folliculorum* is found in simple hair follicles above the level of the sebaceous glands, while *D. brevis* colonizes sebaceous glands of the deeper vellus hairs (Desch and Nutting 1978). Population infestation levels increase with age, and the mites are especially common on the cheeks, eyelashes, and foreheads of adult males. The mites feed on epithelial and sebaceous gland cells, which can lead to swelling and keratinization that appears to be nonpathogenic in most individuals. However, heavy infestations of *Demodex* mites may be associated with severe cases of acne rosacea, and they may contribute to the inflammatory reaction (e.g., by blocking pores or inducing allergic reaction or by vectoring bacteria on the skin). Cases of pityriasis, chalazion, granuloma of the eyelid, and blepharitis also are associated with high mite populations (English and Nutting 1981; Bonnar, Eustace, and Powell 1993; Roihu and Kariniemi 1998), but whether the mites are causative agents or merely responding to reduced immune response or increased food availability (or both) is not known.

Species of the highly modified family CLOACARIDAE (Figs. 13.80A–C) are commonly found in the mucosa of the cloacae of turtles (Camin et al. 1967; Fain 1968c). However, like the closely related EPIMYODICIDAE (see below), they may normally reside in the subcutaneous fascia, but only the females migrate to the cloaca to disperse (B. M. OConnor, pers. obs.). One species, *Pneumophagus bubonis* Fain and Smiley, was found in the connective tissue surrounding the lungs and bronchi of the great horned owl, *Bubo virginianus* (Fain and Smiley 1989). The peculiar find of an avian host in a family primarily associated with reptiles led Bochkov (2002) to suggest that the CLOACARIDAE is an ancient group that may have infested the ancestors of both turtles and birds before the birds split off from the Reptilia. Cloacarids are thought to be venereally transmitted and are strongly host specific. *Caminacarus theodori* Fain and *Theodoracarus testudinis* Fain both tend to invade the mucosa itself, while other species attach at the mucosal surface by means of their peculiar hooked, fanglike palpi.

Members of the genus *Epimyodex* (EPIMYODICIDAE) (Figs. 13.80D, E) are parasites of shrews, moles, and mice (Fain and Orts 1969; Fain et al. 1982; Fain and Bochkov 2001; Bochkov 2002). They occur widely in the tissues of their hosts and migrate to the reproductive system to disperse, resembling cloacarids in this and other respects. Based on morphological and ecological evidence, Bochkov (2002) considers the EPIMYODICIDAE and the CLOACARIDAE to be sister groups. Furthermore, he feels that the common ancestor of both families may well have made the transition from ectoparasitism to endoparasitism in the cloacae of late Palaeozoic and early Mesozoic synapsid reptiles.

Cohort Heterostigmatina

The Heterostigmatina comprises eight superfamilies with over 2,000 described species of small to moderate-sized mites (80–675 μm; adult females of the CROTALOMORPHIDAE exceptionally much larger) with a longitudinal series of dorsal opisthosomatic plates (generally four in number), with the tergites of segments E and F fused and the caudal pseudanal block covered, thereby giving the impression of only four segments that are sometimes weakly sclerotized or rarely fragmented or desclerotized. Adult females—and rarely males—usually have a pair of anterolateral prodorsal stigmata (**heterostigmata**) and associated tracheae, but peritremes are absent. A pair of bothridia with (usually) capitate sensilla arise on the prodorsum of most adult females and rarely of males, but not of immature instars. Eyes are absent. The palpi often are greatly reduced, comprising only one or two indistinct segments, and are closely appressed to a capsulate gnathosoma. In earlier derivative groups, however, the palpi may be conspicuous three-segmented appendages with a well-developed terminal claw. The movable cheliceral digits are curved basally, being either bladelike and partly retractable in more early derivative superfamilies or styletlike and fully retractable in more highly derivative superfamilies. The gnathosomatic capsule and its components are reduced and trophically nonfunctional in the males of two superfamilies. Typically, tarsi II–III each have a stalked flexible empodium that usually is flanked by paired claws. These structures may be present or lost on tarsus IV and, more rarely, on tarsi II–III. Tarsus I usually has no empodium, but paired claws or more commonly a single claw may be present or absent. Genital papillae are absent. Sexual dimorphism is pronounced throughout the cohort, and sperm transfer is direct in all taxa for which mating behavior is known.

The system of superfamilial classification followed here for the Heterostigmatina is based largely on the concepts elaborated by Lindquist (1976, 1986) and Kaliszewski, Athias-Binche, and Lindquist (1995).

▪ Superfamily Tarsocheyloidea (Fig. 13.81)

DIAGNOSIS: Gnathosoma elongate, subconical, easily seen from above, with 1–2 pairs of dorsal setae; palpi well developed, usually projecting well beyond gnathosomatic apex, with 3 segments and a distinctive terminal tibiotarsal claw, palptarsal "thumb" absent; movable cheliceral digits bladelike or styletlike, partially retractile. Prodorsal shield of adult females and males with a pair of anterolateral stigmata with associated tracheae, and with 3 pairs of setiform setae and a pair of anterolateral bothridia with capitate sensilla, the latter represented by normal scapular setae in immature instars. Opisthosoma of females and males with dorsal plates not overlapping each other posteriorly; first dorsal plate C divided into median plate (bearing seta c1*) and a lateral plate (bearing seta* c2*) on each side, and separated from second dorsal plate D. Adults and nymphs with a pair of platelets between coxae IV, each with 2 setae; coxisternal sclerites I–II separated at midline by soft striated integument. Female genital opening exposed, flanked by pair of elongate valves, each with 4 setae; flanking pair of aggenital plates each with 2 setae. Legs I–IV similarly developed on both sexes, each with paired tarsal claws, those on tarsus I sessile; tarsi II–IV with flexible pretarsus bearing hyaline empodium between claws. Adult coxisternal plates I–II together with 7 pairs of setae; trochanters I–IV each with a seta, except with usually 2 on trochanter III; genu I* *with 5 setae; genu II with 4 setae. With 4 active postembryonic instars: larva, protonymph, deutonymph, and adult.*

The Tarsocheyloidea includes a single cosmopolitan family, the TARSOCHEYLIDAE, comprising approximately 12 described species in two genera. The type genus *Tarsocheylus* is represented by a single species, *T. paradoxus* Berlese, while other species are accommodated in the genus *Hoplocheylus* (Atyeo and Baker 1964). Tarsocheylids are slender mites, with adult body lengths (including gnathosoma) of 275–575 µm. They are found in soil, sand, humus, forest, rodent nest litter, and rotting wood or trunk cavities of deciduous trees or under the elytra of passalid beetles (Lindquist 1976). *H. pickardi* Smiley and Moser was taken from under pine bark infested with bark beetles (*Dendroctonus*) in Louisiana (Smiley and Moser 1968). Characteristics distinguishing each of the immature instars are given in Lindquist (1987). Mating behavior and the method of sperm transfer are unknown.

■ Superfamily Heterocheyloidea (Fig. 13.82)

DIAGNOSIS: Gnathosoma elongate, subconical, easily seen from above, without dorsal setae; palpi well developed, projecting well beyond gnathosomatic apex, with 2–3 segments that may not be freely articulated; tibiotarsal claw reduced, rounded apically, palptarsal "thumb" absent; movable cheliceral digits bladelike, hooked. Prodorsal shield of adult female with a pair of anterolateral stigmata with associated tracheae, and with 2 pairs of setiform setae and a pair of posteromarginal ampulliform organs, the latter represented by a pair of normal scapular setae in immature instars and males. Opisthosoma of females and males with first dorsal plate C entire, and partially or completely fused with second dorsal plate D. Adults with an elongate, rectangular single shield medially between fused coxisternal plates III–IV, bearing a single pair of setae; coxisternal plates I–II united medially. Female genital opening exposed, flanked by pair of elongate valves, each with 1 seta; flanking pair of aggenital plates each with 1 seta. Legs I somewhat less robust than legs II–IV and lacking empodium and claws; legs II–IV similarly developed on both sexes, lacking claws but with a stalked discoid empodium and a heavy spinose ventrodistal subunguinal seta. Adult coxisternal plates I–II together with 2 pairs of setae; trochanters I–IV lacking setae; femur I with 2 setae; genu I with 1 seta; genu II lacking setae. With 4 active postembryonic instars: larva, protonymph, deutonymph, and adult.

The Heterocheyloidea consists of a single monogeneric family, the HETEROCHEYLIDAE. About 25 species of *Heterocheylus* have been described, and all are subelytral symbionts of passalid beetles. Heterocheylids are elongate fusiform mites that usually are distinctly broadened in the region between legs II–III. Adult body lengths (including gnathosoma) range from 250 to 675 µm, with males 10%–20% shorter than females. Their known distribution and habitats overlap those of their beetle hosts: decaying wood in tropical and temperate forests throughout the world except in Europe and northern and western North America (Lindquist and Kethley 1975). The relationship between *Heterocheylus* species and passalids may involve parasitism, although only females are routinely found on the beetles. A fairly high degree of host specificity has been noted by Schuster and Lavoipierre (1970), at least at the host generic level. Inasmuch as passalids are one of the more ancient families of scarabaeoid beetles, with a fossil record dating back to Lower Jurassic times, the ubiquitous occurrence of heterocheylid mites with passalids indicates an association that may date back to the mid-Jurassic, some 175 mya (Lindquist 1996a). Characteristics distinguishing each of the immature instars are noted by Schuster and Lavoipierre (1970). Mating behavior and method of sperm transfer are unknown, but the presence of an extrusible copulatory apparatus in males is suspected (Lindquist and Kethley 1975). Furthermore, the elongated anal valves on deutonymphs resemble claspers that possibly may be used in some sort of precopulatory behavior.

Heterocheyloids have been considered to have close phylogenetic ties with the Tarsocheyloidea (Lindquist and Kethley 1975), and both taxa were previously accommodated in the same family (Atyeo and Baker 1964). Both occur on passalid beetles (Lindquist 1976), although the tarsocheyloids have a much less specialized body plan and use nonpassalid habitats as well. More recent phylogenetic studies suggest that HETEROCHEYLIDAE is the sister group of the entire assemblage of superfamilies that constitute the Tarsonemina (Lindquist 1986; Kaliszewski, Athias-Binche, and Lindquist 1995).

■ Superfamily Dolichocyboidea (Figs. 13.83, 13.84)

DIAGNOSIS: Female gnathosoma capsulate, visible from above, usually with 1–2 pairs of dorsal setae; palpi moderately well developed, usually freely projecting somewhat beyond gnathosomatic apex, with 2 segments and rudimentary tibiotarsal claw; movable cheliceral digits bladelike or styletlike, partially retractile; pharynx muscular, undivided. Male gnathosomatic capsule and its components reduced in size, its cheliceral stylets probably not functional trophically in the absence of pharyngeal structures. Prodorsal shield of females lacking a pair of anterolateral stigmata and associated tracheae, and with 2–3 pairs of setiform setae and with or without a pair of anterolateral bothridia with capitate sensilla, the latter represented by a pair of normal scapular setae in males. Opisthosoma of females with first dorsal plate C entire, divided into median plate and lateral plate on each side, or fragmented into 4 platelets. Adults with coxisternal plates I–II usually united medially, but coxisternal plates III–IV usually separated medially by narrow strip of soft cuticle. Female genital opening exposed, flanked by 1–4 pairs of genital setae, these sometimes rudimentary or indiscernible; with 1–2 pairs of aggenital setae on unsclerotized cuticle or on plates either paired and flanking genital opening or consolidated with each other and with plate H and pseudanal plate to form ventrocaudal plate. Males with plates C and D separate, and with plates H, Ps, Ag variably consolidated but not forming a genital capsule. Legs I–IV similarly developed in both sexes, each with symmetrically or asymmetrically paired tarsal claws; trochanter IV similar in shape to trochanters I–III, usually subtriangular. Adult coxisternal plates I–II together with 2–6 pairs of setae; trochanters I–IV lacking setae, except 1 rarely present on legs III; femur I with 1–4 setae; genu I with 1–5 setae; genu II with 1–4 setae. With only 1 active postembryonic instar (the adult); larval and single nymphal instars inactive calyptostases.

Until recently, only the DOLICHOCYBIDAE, a widely distributed family with five genera, had been included in the Dolichocyboidea (Lindquist 1986; Kaliszewski, Athias-Binche, and Lindquist 1995). However, the enigmatic monobasic family CROTALOMORPHIDAE is tentatively included here as a member family, although information on instars other than the adult female is still lacking (Lindquist and Krantz 2002). The Dolichocyboidea is considered the most primitive superfamily of the Tarsonemina *sensu* Lindquist (1986). Adult females in both families retain an exposed genital opening that is usually flanked by one to several pairs of genital setae that are in turn flanked by one or two pairs of aggenital setae. The aggenitals are sometimes inserted on a pair of small plates, but the plates are united, and aggenital setae may be absent in the dolichocybid subfamily Formicomotinae (Magowski 1988). Members of both families are plesiomorphic in having paired claws on legs I, and their coxisternal plates III–IV usually are separated from each other medially by a strip of soft cuticle. However, the families are singularly derived in other respects, including the secondary losses of prodorsal stigmata and associated tracheal respiratory system, and of trochanter seta *v'* on all legs.

Adults of the approximately 25 described species of DOLICHOCYBIDAE (Figs. 13.83A–I) are small, with body lengths of 100–280 μm. *Dolichocybe indica* Mahunka was recovered from rotting straw in India, and *Acanthomastix spinipes* Mahunka from primary tropical forest litter in New Guinea (Mahunka 1971, 1972a). *Pavania tahanae* Sevast'yanov and Abo-Korach was collected in soil under cotton plants in Egypt (Sevast'yanov and Abo-Korach 1985), and several species of *Dolichocybe* and *Dolichomotes* have been found under tree bark in Europe, North America, and Hawaii (Rack 1967, 1973c). These and other members of the family appear to be associated with subsocial and social insects, particularly beetles and ants, on which they are phoretic as adult females. For example, two species of *Acanthomastix* are associates of subcortical weevils in conifers in North America, and another species is associated with ants in dead trees in Europe (Magowski and Moser 1993). *Pavania fusiformis* Lombardini, *P. riparia* Sevast'yanov, and *P. tadjikistanica* Sevast'yanov are associated with scarabeid beetles in Eurasia (Rack 1973c; Sevast'yanov 1980), and several other species of *Pavania* were recovered from unidentified insects in Ghana (Mahunka 1973a, 1974a, 1975b). The monobasic genus *Rafamastix* is known only from a few females collected from an unknown substrate in a valley at 3,600 m in Ecuador (Magowski 1997).

Although little is known about their ways of life, dolichocybids do not appear to be parasites of their hosts, but seem instead to be fungivorous and live in habitats conducive to fungal growth (Kaliszewski, Athias-Binche, and Lindquist 1995). However, the possibility of facultative parasitoidism, as is known in the related family TROCHOMETRIDIDAE, cannot be ruled out. *Dolichocybe hippocastani* Rack and *Formicomotes heteromorphus* Magowski are confirmed fungivores that undergo extensive physogastry during feeding and give rise directly to as many as 200 adult male and female progeny (Rack 1967; Magowski 1988). Adults emerge through the skins of a larval and a nymphal apoderm as well as the egg chorion (Rack 1967). Adult female dimorphism is expressed in *F. heteromorphus* and probably also in a species of *Acanthomastix* associated with ants (Magowski 1988; Magowski and Moser 1993). In Thailand, *F. heteromorphus* may occur in large numbers on the cultivated mushroom, *Auricularia polytricha*, on which it causes retardation of mycelial growth (Kantaratanakul, Chaiwat, and Prayoonrat 1989). Zou and Gao (1996) described two species of *Dolichocybe* from edible fungi in China.

The crotalomorphid genus *Crotalomorpha* (Figs. 13.84A–I) was included, with a habitus figure of a greatly physogastric and gravid female, in the second edition of *A Manual of Acarology* and ascribed to a manuscript being prepared by J. H. Camin (Krantz 1978) but never published. The family and genus were described by Lindquist and Krantz (2002) to validate these names, with *C. camini* Lindquist and Krantz as the type species. Their description was based on the few known, greatly engorged, gravid adult females found under the elytra of carabid beetles of the subgenus *Agonoderus*, genus *Stenolophus*, in the midwestern United States. The large size (idiosomatic length approximately 3 mm) of these apparent parasitoids, even accounting for their being engorged, is atypical of Heterostigmatina. A few additional specimens from north central to mid-southern United States, found in varying degrees of engorgement on carabids of the same subgenus, have been noted by Husband and Husband (2003, 2004).

▪ Superfamily Trochometridioidea (Fig. 13.85)

DIAGNOSIS: Female gnathosoma capsulate, visible from above, with 1–2 pairs of dorsal setae; palpi moderately well developed, freely projecting somewhat beyond gnathosomatic apex, with 2 segments and rudimentary tibiotarsal claw; movable cheliceral digits bladelike, partially retractile; pharynx muscular, undivided. Male gnathosomatic capsule and its components reduced in size, its cheliceral stylets probably not functional trophically in absence of pharyngeal structures. Prodorsal shield of females with pair of anterolateral stigmata and associated tra-

cheae, and with 2–3 pairs of setiform setae and with or without a pair of anterolateral bothridia with capitate sensilla, the latter represented by a pair of normal scapular setae in males. Opisthosoma of females with first dorsal plate C usually divided into median plate and lateral plate on each side, or rarely entire. Adults with coxisternal plates I–II united medially, forming prominent prosternal apodeme; female coxisternal plates III–IV separated from each other medially by soft cuticle that bears a separate triangular sternal plate with a pair of setae; male coxisternal plates III–IV fused medially but not forming a poststernal apodeme. Female genital opening small, covered by consolidated aggenital plate, and flanked by 1–3 pairs of tiny or rudimentary genital setae or these setae absent; 1–3 pairs aggenital setae. Males with plates C and D consolidated, and with plates H and PS not forming a caudal genital capsule (setigenous processes left exposed). Females with legs I thicker than legs II–IV and with single claw, usually enlarged for grasping; legs II–IV similarly developed, each with paired tarsal claws and stalked empodium; trochanter IV subtriangular or subquadrangular, similar to subtriangular trochanters I–III in being not longer than wide. Adult coxisternal plates I–II together with usually 6 pairs of setae; trochanters I–IV each with a seta; femur I with 5 setae; genu I with 5 setae; genu II with 3 setae. Male legs IV similar in form to legs II–III, with paired claws and stalked empodium. With only 1 active postembryonic instar (the adult); larval and single nymphal instars inactive calyptostases.

Two families are recognized in this superfamily, the TROCHOMETRIDIIDAE (Figs. 13.85A–F), based on a single genus of mites associated with a variety of ground-nesting bees and their hymenopterous associates, and a broadened concept of the ATHYREACARIDAE (Figs. 13.85G–J) that includes four genera associated with various ground-dwelling beetles. A third family, the CARABOACARIDAE, was included in this superfamily by Kaliszewski, Athias-Binche, and Lindquist (1995), but it is tentatively treated here as a member of the Pyemotoidea (see p. 314). Adult females of the two described species of *Trochometridium*—*T. tribulatum* Cross and *T. chinensis* (Mahunka)—have body lengths of 210–300 µm and are characterized in part by having a pair of invaginated, saclike sporothecae between the bases of legs III and IV, an adaptation for carrying fungal spores (Lindquist 1985b). The life cycle and close association of *T. tribulatum* with ground-nesting bees has been extensively studied (Cross and Bohart 1969, 1978). The female mite apparently kills the egg or larva of the bee in the cell in which she will raise her brood, thereby assuming the role of a kleptoparasite. This allows development of fungal growth in the provisioned brood cell, providing food for the mite. After undergoing physogastry, she produces a mass of eggs that hatch into nonfeeding larvae, most of which develop into females. The few emergent males have reduced gnathosomas and are nonfeeding (Kaliszewski, Athias-Binche, and Lindquist 1995). Adult female mites are sometimes found phoretically on kleptoparasitic bee and predatory wasp associates of their bee hosts (Cross and Bohart 1969).

As originally described, *Athyreacarus* and ATHYREACARIDAE were based on a single described genus and eight undescribed species, all of which are associated with Neotropical scarab beetles of the genus *Neoathyreus* (Lindquist, Kaliszewski, and Rack 1990). Discovery of a closely related undescribed genus from scarabs of another tribe of Geotrupinae in Australia indicates that the general association of athyreacarid mites with geotrupine beetles predates the breakup of continental elements of Gondwanaland, around 145 mya (Lindquist 1996a). Athyreacarids evidently live symbiotically in the burrows of their beetle hosts; only adult phoretic females are known, and these display a variable degree of host specificity for their scarab carriers. All of the females are well-sclerotized, medium-sized mites with body lengths of 470–650 µm. Although feeding habits and other aspects of their reproductive behavior are unknown, the cheliceral and pharyngeal structures of athyreacarids closely resemble those of trochometridiids. Extensive areas of highly plicated soft cuticle indicate a probable physogastric capability.

To avoid inclusion of a monogeneric family-group taxon based on few derivative (apomorphic) and homoplastic attributes, the family ATHYREACARIDAE is further broadened here and in the key to include the monobasic genus *Bembidiacarus*, for which Khaustov (2000b) proposed the family Bembidiacaridae. Another undescribed monobasic genus noted earlier as a trochometridiid by Lindquist (1985b) also is included here. Members of both genera are known solely from adult females that are phoretic under the elytra of beetles, the former on bembidiine carabids, the latter on trogid scarabaeoids.

▪ Superfamily Pygmephoroidea (Figs. 13.86, 13.87)

DIAGNOSIS: Female gnathosoma capsulate, circular, oval or elongate, visible or hidden from above, with 1 or usually 2 pairs of dorsal setae; palpi closely appressed, reduced; movable cheliceral digits styletlike, partially retractile; pharynx divided into 2–3 muscular sections. Male gnathosoma reduced, tubiform, nonfunctional trophically. Prodorsal shield of females with pair of anterolateral stigmata (extended into troughlike peritremes in some taxa) and associated tracheae, and with 2 or usually 3 pairs of setiform setae and with or without a pair of anterolateral bothridia with capitate sensilla, the latter represented by a pair of normal scapular setae in larvae and males. Opisthosoma of females with first dorsal plate C entire, not expanded anteriorly over prodorsum. Adults with coxisternal plates I–II united medially, forming prominent prosternal apodeme; female coxisternal plates III–IV united medially, usually forming poststernal apodeme. Female genital opening small, covered by consolidated aggenital plate, genital and aggenital setae absent (exceptionally, 2 pairs of aggenital setae present in Meristoplaxa*). Males with plates C and D consolidated, and with aedeagus tubiform, usually short and stiff, enclosed in caudal genital capsule formed by* *consolidated plates H, PS, Ag. Females with legs I similar to or thicker than legs II–IV and with single claw sometimes enlarged for grasping; legs II–IV dissimilar in form, each with paired claws and stalked empodium or these absent on legs IV; trochanter IV quadrangular, longer than wide, in contrast to subtriangular trochanters I–III. Adult coxisternal plates I–II together with 4–6 pairs of setae; trochanters I–IV each with a seta; femur I with usually 4 (exceptionally 3 in* Asensilla, *5 in* Meristoplaxa*) setae; genu I with 2–4 setae; genu II with 2–3 setae. Male leg IV stout, its tarsus variably formed, with or without a single claw, with or without a broad, short empodium, sometimes with 1–2 apical setae elongated and 1–2 other apical setae short, spinelike. With 2 active postembryonic instars (larva and adult); single nymphal instar an inactive calyptostatic apoderm.*

The two families of Pygmephoroidea, SITEROPTIDAE (Figs. 13.86A–E) and PYGMEPHORIDAE (Figs. 13.87A–F), consist primarily of free-living species, some of which have both phoretic and nonphoretic female forms. Female body lengths range from 110 to 520 µm. Some species are common on plants, but feeding behavior is generally fungivorous, and no suspected cases of parasitism or parasitoidism are known (Kaliszewski, Athias-Binche, and Lindquist 1995). Species that are associated with bark beetles and fungus-inhabiting flies may display a phoretic and nonphoretic female dimorphism referred to as phoretomorphy (Moser and Cross 1975; Smiley and Moser 1976; Martin 1978; Cross and Kaliszewski 1988; also see chapter 6). Recognition of this phenomenon clarified and corrected previous interpretations that some species of *Siteroptes* had active nymphal females and adult females, either of which could undergo physogastry and reproduce. However, discovery of phoretomorphy also led to problems in the recognition of certain genera. For example, two species associated with bark beetles and described under the genus *Siteroptes* produce two distinct female types, the phoretic form of which was referable to the pygmephorid genus *Pediculaster* (Moser and Cross 1975; Smiley and Moser 1976). These and related findings by Gurney and Hussey (1967), Rack (1974), and other authors raised questions as to the validity of existing generic concepts and placement of various species and pointed to a confusion in literature citations of these taxa. *Siteroptes*, *Pediculaster*, and *Pygmephorellus* were synonymized by Smiley and Moser (1976), but they were reestablished as distinct genera based on other characteristics by Mahunka (1979). Both Livshits, Mitofanov, and Sharonov (1986) and Kaliszewski (1987) further defined *Siteroptes* as distinct and in a separate family from *Pediculaster*. However, their generic concepts are discordant enough to make problematic the definition and generic content of SITEROPTIDAE in distinction to PYGMEPHORIDAE. Also, the placement of the monobasic genus *Meristoplaxa*, for which Mahunka (1970c) proposed a separate subfamily, Meristoplaxinae, remains unresolved in the absence of information on female pharyngeal structures and on males and larvae.

The family SITEROPTIDAE in a restricted sense (based in part on Kaliszewski 1988a, b; E. E. Lindquist, pers. comm.) includes about eight genera (several more genera are recognized by Livshits, Mitrofanov, and Sharonov (1986)) and approximately 80 species. Among its genera, members of the monobasic *Asensilla* occur in soil (Rack 1974), while those of *Diroptes* and *Sevastianovella* (in part synonyms) occur in litter, rotten plants, grasses, and soil (Kaliszewski 1988a; Livshits, Mitrofanov, and Sharonov 1986). Species of *Pseudobakerdania* have been found in forest litter and soil (Kaliszewski 1988b), those of *Dudichiana* in ants' nests (Mahunka 1970b), and those of *Metasiteroptes* either on or in sugarcane stems or as phoretics on aradid bugs (Cross 1965). *Siteroptes* species have been collected in a wide variety of habitats, including foliage of graminaceous plants (Rack 1972a; Livshits, Mitrofanov, and Sharonov 1986). The monobasic genus *Meristoplaxa*, which is provisionally placed in this family, is known from a single female recovered from a pitfall trap in sandy soil along a lakeshore in Mongolia (Mahunka 1970c). Another enigmatic monobasic genus, *Zambedania*, known only from two females found in soil probes in Zimbabwe (Mahunka 1972b), may also belong in this family.

Siteroptes cerealium Kirchner (= *S. graminum* (Reuter)) is a known vector of *Fusarium poae*, the fungal pathogen that causes carnation bud rot (Cooper 1940), and it may also be implicated in silver top of grasses in North America and Eurasia (Hardison 1959; Suski 1973, 1984). Females of *S. cerealium* and of some other species of *Siteroptes* carry fungal spores on the external surfaces of their bodies, while those of still other species possess paired eversible sporothecae on the ventral body surface that are adapted for carrying only the spores of fungi preferred for food (Lindquist 1985c). In *S. avenae* (Müller) and closely related species, the sporothecae are deeply invaginated, located between coxisternal plates II and III, and designed to carry several conidiospores of particular species of *Fusarium*, *Botrytis*, and *Harzia* (Suski 1973; Su, Beng, and Chen 1979; Su, Chen, and Beng 1981; de Lillo 1996), including the pathogen *F. poae*, noted above. Sporothecae in the *S. reniformis* species group are shallower pouches that are located behind coxisternal plates IV and are designed to accommodate just one conidiospore each of *Nigrospora* fungi, including the causal agents of lint rot of cotton, stalk rot of corn and sorghum, and similar rots of spikes of wheat and other grains (Alfaro 1946; Laemmlen and Hall 1973). The mutualistic association between these mites and *Nigrospora* is further elaborated by Laemmlen and Hall (1973) and Su, Chen, and Beng (1981).

Adult females of some species of *Siteroptes* undergo extreme hysterosomatic physogastry while feeding, with resultant high fecundity. In *S. cerealium*, up to 500 progeny per female may be produced, of which about 95% are female (Cooper 1937; Krczal 1959). The progeny pass through an active larval instar and usually attain adulthood within the maternal hysterosomatic sphere, where copulation often takes place before birth. A mass birth usually is coincident with rupture of the hysterosomatic sphere and death of the mother.

The PYGMEPHORIDAE is a large assemblage of mites that occur in soil, litter, and organic substrates, in nests of birds and mammals, and in association with subsocial and social insects (Krczal 1959; Cross 1965; Mahunka 1969b, 1970a; Savulkina 1981). At least 350 species in about 30 genera are described, the great majority of which are based solely on adult females whose body lengths range from 110 to 520 µm. Members of many pygmephorid genera with one or few species, along with some species of *Elattoma*, *Pediculaster*, and *Bakerdania*, are found in a wide variety of forest and cultivated soils and other organic and herbaceous substrates. *Pediculitopsis tubiphorus* Mahunka evidently is an example of a species displaying adult female dimorphism, although one of its forms was tentatively thought to be a nymph (Mahunka 1970d). Both forms were extracted together from a soil sample, and the one having legs I incrassate and with fused tibiotarsi appears to be a phoretic form (Savulkina 1981). Adult females of various genera use insects or mammals for phoretic dispersal. Life history and behavioral studies of a few species scattered in various genera indicate a general pattern of an active, fungivorous larva, a nonfeeding adult male, and a fungivorous female that undergoes a greater or lesser degree of physogastry. This pattern has been confirmed for two phoretomorphic species of *Pediculaster* common in commercial mushroom houses (Košir 1975; Cross and Kaliszewski 1988), a species of *Luciaphorus* that infests cultivated mushrooms (Rack 1983; Kantaratanakul and Jitrat 1984; Zou, Gao, and Ma 1993), two phoretomorphic species described under the name *Siteroptes* associated with bark beetles (Smiley and Moser 1976), a halictid bee–associated species of *Parapygmephorus* (Rack and Eickwort 1980), and a myrmecophilous species of *Petalomium* (Ebermann and Rack 1982). Košir (1975) found that *Pediculaster mesembrinae* (Canestrini) could develop on a wide variety of fungi, whereas *Bakerdania quadrata* (Ewing) developed readily on only one fungal genus. The life history of *Luciaphorus perniciosus* Rack (also reported under the names *L. hauseri* Mahunka and *L. auriculariae* Gao et al.) (see Kaliszewski, Athias-Binche, and Lindquist 1995) on commercially grown fungi is similar to that of species of *Siteroptes*. The female undergoes extensive physogastry, with development of her progeny to adulthood (and sometimes accompanied by progeny copulation) taking place within her body. The female of *Xenaster longiabdominalis* Mohanasundaram, whose true generic placement is currently uncertain (*Brasilopsis*?), has a similar life history; the female feeds and undergoes physogastry on fungal growth amid fallen coconut buttons. Newly emergent male progeny carry away virgin females as they emerge from the greatly enlarged maternal hysterosoma (Mohanasundaram 1994).

Pygmephorids associated with arthropods generally are free-living commensals on their typically subsocial or social insect hosts, and female dimorphism is known or thought to occur in many of them (Rack and Kaliszewski 1985). Among the pygmephorid associates of Coleoptera, females of *Geotrupophorus* and *Spatulaphorus* are phoretic exclusively on geotrupine scarabs (Mahunka 1970a; Dastych and Rack 1993; Khaustov 2005b). Females of a species of *Acarothorectes* are phoretic on curculionid beetles (Cross 1965), and those of several species of *Elattoma* and *Pygmephorellus* are associated with a variety of subcortical and dung-inhabiting beetles (Cross and Moser 1971; Savulkina 1981; Khaustov 2007a). Large numbers of females of the monobasic genus *Sasadania* were found under the elytra of several species of platynine carabid beetles in Japan (Kurosa 1989a). Among the pygmephorids found with Hymenoptera, females of *Acinogaster* and many of *Petalomium* are found on ants and in their nests (Cross 1965; Savulkina 1981; Ebermann and Rack 1982; Khaustov 2005a). Females of *Sicilipes* and some of *Parapygmephorus* are phoretic on halictid bees (Cross 1965; Mahunka 1974b, 1980). An underwing moth, *Amphipyra pyramidoides*, was found to carry females of *Propygmephorus treati* Cross (Treat 1975), while females of *Termitacarus ishiharai* Kurosa and of an undescribed species of *Mahunkania* and of *Luciaphorus* were collected on termites or in their nests in Japan (Kurosa 1994). Other unusual phoretic associations of females include a species of *Pseudopygmephorus* with a fossorial mygalomorph spider (Vincent and Rack 1982), one of the speciose genus *Bakerdania* with terrestrial isopods (Colloff and Hopkin 1986), and two species of *Allopygmephorus* with aquatic hydrophilid beetles (Cross 1965). A series of 25 phoretic females of *Siteroptoides trombidiphilus* (Rack and Vercammen-Grandjean) were found as phoretics only on large red velvet mites, *Camerotrombidium* sp., even though more likely arthropod hosts were present in the same compost heap (Rack and Vercammen-Grandjean 1979). Phoretic females of some dung-inhabiting species of *Pediculaster* appear to exhibit strong preferences for certain kinds of flies as carriers and for certain kinds of dung or shoreline debris as habitats (Camerik and Ueckermann 1995; Camerik 2005). Members of the speciose genus *Pygmephorus* commonly occur in the nests of small mammals and are phoretic in the hosts' fur. They also occur, although less frequently, in the nests of birds. Although some 50 species of *Pygmephorus* have been described, males of only two of them are known (Kaliszewski and Rack 1985). Information on their life history and behavior, and on the occurrence of nonphoretic females, remains unavailable (Dastych, Rack, and Wilson 1992). Members of the genera *Cerattoma* and *Rackia* are also found in the burrows and nests of small mammals (Mahunka 1970a, 1975c).

- Superfamily Scutacaroidea (Figs. 13.88, 13.89A, B, 13.90)

DIAGNOSIS: Female gnathosoma capsulate, circular, oval or elongate, generally not visible from above, with 2 pairs of dorsal setae; palpi closely appressed, usually reduced, rarely elongated and bladelike; movable cheliceral digits styletlike, partially retractile, rarely vestigial; pharynx divided into 2–3 muscular sections. Male gnathosoma reduced, tubiform, nonfunctional trophically. Prodorsal shield of females with pair of anterolateral stigmata and associated tracheae and with 1–2 pairs of setiform setae and usually 1 pair of anterolateral bothridia with capitate sensilla, the latter represented by normal scapular setae in larvae and males. Opisthosoma of females with first dorsal plate C entire, sometimes expanded anteriorly to overlap prodorsum partially (MICRODISPIDAE) or with a free anterior margin greatly expanded to cover prodorsum entirely (SCUTACARIDAE). Adults with coxisternal plates I–II united medially, forming prominent prosternal apodeme; female coxisternal plates III–IV united medially, usually forming poststernal apodeme. Female genital opening small, covered by consolidated aggenital plate; genital and aggenital setae absent. Males with plates C and D consolidated, and with aedeagus tubiform, elongate, and curved, enclosed in caudal genital capsule formed by consolidated plates H, PS, Ag. Females with legs I thicker than legs II–IV and with single grasping claw, or thinner than other legs and without ambulacrum and claws; legs II–IV dissimilar in form, each with paired claws and stalked empodium or these absent on legs IV; trochanter IV quadrangular, longer than wide, in contrast to subtriangular trochanters I–III. Adult coxisternal plates I–II together with maximally 4 pairs of setae; trochanters I–IV with a seta; femur I with 3 setae; genu I with 3 setae; genu II usually with 3 setae. Male leg IV stout, its tarsus usually ending with broad, short, padlike empodium flanked by 1–2 elongated setae and 1–2 short, spinelike setae but with or without a small single claw. With 2 active postembryonic instars, (larva and adult); single nymphal instar an inactive calyptostatic apoderm.

As in the Pygmephoroidea, the two recognized families of Scutacaroidea, SCUTACARIDAE and MICRODISPIDAE, comprise primarily free-living species, some with phoretic and nonphoretic female forms that often are found in association with insects and, to a lesser extent, in the nests of birds and small mammals. The SCUTACARIDAE (Figs. 13.88A–E, 13.89A, B) is a large and ubiquitous family of mites represented in soil, forest litter, humus, manure, compost, the nests of small animals, and on a variety of insects on which they are phoretic (Karafiat 1959; Mahunka 1965; Ebermann 1991a). About 750 species in some 30 genera and subgenera have been described, and nearly all of them are based on adult females. These generally have broad, tortoise-shaped bodies ranging in length from 120 to 370 μm. The larva and adult male instars of a scutacarid species were first described by Michael (1884b), but relatively few have been characterized for other species since that time (e.g., Norton and Ide 1974; Rack 1975; Delfinado and Baker 1976; Elbadry, Yousef, and Metwally 1976; Ebermann 1982b, 1983, 1994). However, males probably exist for most species, inasmuch as thelytoky has not been confirmed for any members of the family (Norton et al. 1993). Nonfeeding males display a precopulatory behavior of carrying immobile pharate females (Norton and Ide 1974; Eickwort 1979; Binns 1979; Cross and Bohart 1992), sometimes for periods up to 37 hours prior to final female ecdysis (Ebermann 1982a) (Figs. 13.89A, B). Ebermann (1982a) observed that copulation in *Imparipes* (*I.*) *histricinus* Berlese involves a "retroconjugate" or rear-to-rear position (Fig. 13.89B) and that unfertilized females give rise to all-male progeny.

Many members of the speciose genus *Scutacarus*, as well as those of smaller genera, are found in various of the soil and organic substrates noted above. Many other scutacarids are associated with insects (especially social ones), either as phoretic females on the insects themselves or as nest associates (e.g., Karafiat 1959; Mahunka 1969a, 1975e; Delfinado and Baker 1976; Delfinado, Baker, and Abbatiello 1976; Kurosa 1976b, 1990; Binns 1979). Many species of *Archidispus* show preferential associations with carabid beetles (e.g., Rack 1973a; Kurosa 1989b), while others, along with many species of *Imparipes* and *Scutacarus*, are found on ants or in their nests (e.g., Karafiat 1959; Rack 1973a; Mahunka 1970f, 1977a–c; Ebermann 1981). Species of the genera *Lophodispus*, *Rettenmeyerella*, and some of *Thaumatopelvis* also are ant associates (Kurosa 1972; Mahunka 1977a; Ebermann 1980, 1982b), while those of the genera *Nasutiscutacarus* and *Parascutacarus* are associated with halictid bees and bumblebees, respectively (Beer and Cross 1965; Baker and Delfinado 1975).

Females of a few scutacarid species are phoretic on arthropods other than insects. For example, *Imparipes caledonicus* Mahunka occurs on millipedes (Mahunka 1992), *I. tocatlphilus* Ebermann and Palacios-Vargas on ricinuleids (Ebermann and Palacios-Vargas 1988), two species of *Scutacarus* on nemesiid trapdoor spiders (E. Ebermann, pers. comm.), and three species of *Scutacarus* on deutonymphal parasitid mites (Ebermann 1991a). The latter are examples of **hyperphoresy**, where the parasitid mite carriers are in turn phoretic on bumblebees. Oudemans' (1914) observation of five females of *Scutacarus* (= *Disparipes*) *talpae* (Oudemans) phoretic on representatives of "several dozen" deutonymphs of *Haemogamasus nidi* Michael (= *H. michaeli* Oudemans) in a rodent nest is peculiar, because females rather than deutonymphs are the dispersing instar among haemogamasine LAELAPIDAE. However, Oudemans' search through enormous numbers of *H. nidi* and *H. hirsutus* Berlese (and of various parasitids) failed to turn up more *S. talpae*.

Dimorphism among adult female scutacarids may be frequent and is correlated fully with phoretic behavior in this family (Ebermann 1991a). This phenomenon may be so extreme as to involve the coincident absence of a claw on legs I of nonphoretic females and the presence of an enlarged claw on legs I of phoretic females of the same species, a difference that previously was considered sufficient to assign such morphs to separate subgenera (Ebermann 1990). However, some species that are phoretic have been shown to have only one female morph (Ebermann 1991a). Whether this indicates a suppression of the nonphoretic morph is uncertain.

Although the feeding behavior of scutacarids is poorly known, they generally appear to be fungivorous, without indication of parasitoidism or parasitism (Ebermann 1991a; Kaliszewski, Athias-Binche, and Lindquist 1995). Females of the subgenus *Imparipes* (*Sporichneuthes*) are characterized by having a broadened gnathosoma with structures adapted to feeding on the contents of conidiospores, in contrast to the general scutacarid behavior of feeding on the contents of hyphae (Ebermann 1998). Adult females of three species of the subgenus *I.* (*Imparipes*) that are phoretic on wood- and soil-dwelling sphecid wasps and wild bees carry fungal spores in a pair of shallow sporothecae (Ebermann and Hall 2003; Hall and Ebermann 2005). The cavities are located on either side of the genital sclerite in the fissure at the base of the aggenital plate, where it is overlapped by the extended posterior margin of the consolidated coxisternal plates IV, much as in siteroptids of the *Siteroptes reniformis* group (Lindquist 1985b). The relationship between the mites, the fungi, the bees, and their nests remains to be clarified. Females of species of *Rhynchodispus*, *Nasutiscutacarus*, and *Imparipes* (*Parimparipes*) have unusually elongated gnathosomas, but their feeding behavior and its possible relationship to gnathosomatic morphology remains unstudied (Ebermann 1998). A limited degree of swelling of the body of gravid females has been noted in a few fungivorous scutacarids (Binns 1981; Cross and Bohart 1992), but this is not equivalent to the well-developed physogastry that occurs in some other families of Tarsonemina. Ebermann (1991b) noted **thanatosis**, or feigning of death, in females of several species of three genera of scutacarids that, after being touched, retract their legs into body recesses and remain motionless for up to 6 minutes. In contrast, females of *I. dispar* (Rack) employ an escape response to danger. They are capable of spectacular jumps of at least 75 times their own body length, a phenomenon otherwise unknown among heterostigmatine mites (Ebermann 1995).

The MICRODISPIDAE (Figs. 13.90A–F) has fewer described species than the SCUTACARIDAE and includes mostly soil and litter forms, some of which possess greatly enlarged and/or ornate dorsal setae (Cross 1965; Mahunka 1969b, 1972a, 1975a). Microdispids are small, with adult females (the only instar known for all but one species) ranging in body length from 80 to 315 μm, and males of at least one species, *Microdispus* (*Premicrodispus*) *lambi* (Krczal), averaging about 35% smaller than females (Kaliszewski and Rack 1986). The systematic position of this family is controversial, with various researchers considering them to be a subset either of the PYGMEPHORIDAE (Cross 1965; Savulkina 1981) or of the SCUTACARIDAE (Mahunka and Mahunka-Papp 1994). Others consider them a separate group of uncertain relationship between the PYGMEPHORIDAE and SCUTACARIDAE (Mahunka 1970a; Rack and Kaliszewski 1985; Ebermann 1986; Kaliszewski and Wrensch 1993), or a sister family of SCUTACARIDAE (Kaliszewski, Athias-Binche, and Lindquist 1995). The majority of known species (approximately 100 in 15 genera), occur in the tropics. Many species of the genera *Microdispus*, *Brennandania*, *Phyllodispus*, *Cochlodispus*, *Tubulodispus*, *Vietodispus*, *Scutodispus*, and *Punicodoxa* inhabit soil, forest litter, decaying vegetation, rotting wood, moss beds, and similar substrates (Cross 1965; Mahunka 1975a; Savulkina 1981). *M.* (*Premicrodispus*) *lambi* is a pest of cultivated mushrooms in Australia (Clift and Toffolon 1981). Species of the genera *Glyphidomastax*, *Myrmecodispus*, *Unguidispus*, *Caesarodispus*, *Reductodispus*, and some of *Microdispus* are associated with, and phoretic on, ants (Cross 1965; Mahunka 1977a–c; Savulkina 1981), but some of *Premicrodispus* are found with termites (Cross 1965). *Brennandania scolopendrae* Rack is unusual in being phoretic on centipedes (Rack 1979). Although fungivory appears to be the primitive feeding behavior in this family, there is evidence of several parasitoid associations with insects. The monotypic genera *Glyphidomastax* and *Perperipes* are associated with army ants, and their females are phoretic on the adult ants. Although their chelicerae are vestigial, the females have long, movable bladelike palpi that appear analogous to the chelicerae in other parasitic or parasitoid Heterostigmatina used for piercing host cuticle. These females are thought to feed on ant larvae or eggs (Cross 1965). Gravid females are physogastric, are greatly elongated (Figs. 13.90E, F), and seem to mimic the form of young ant larvae with which they are found. Females of the monobasic genus *Peponacara* are associated with the polyphagine roach *Arenivaga apacha* and are also thought to be parasites or parasitoids. Although they lack modified palpi, these mites have enlarged bladelike chelicerae with which they attach to their hosts. Many of the attached females were noted by Cross (1965) to be gravid.

▪ Superfamily Pyemotoidea (Figs. 13.91, 13.92)

DIAGNOSIS: Female gnathosoma capsulate, circular, oval, or somewhat elongate, visible or hidden from above, with 1–2 pairs of dorsal setae; palpi closely appressed, often weakly delineated or reduced; movable cheliceral digits usually strongly styletlike, partially retractile; pharynx muscular, undivided, usually prominent. Male gnathosoma similarly developed, though sometimes less robust than in female. Prodorsal shield of females with pair of anterolateral stigmata and associated tracheae, and with 2–3 pairs of setiform setae and with or without a pair of anterolateral bothridia with capitate sensilla, the latter represented by normal scapular setae in males. Opisthosoma of females with first dorsal plate C entire, not expanded anteriorly over prodorsum. Adults with coxisternal plates I–II united medially, forming prominent prosternal apodeme; female coxisternal plates III–IV separated from each other medially by soft cuticle that bears a separate triangular sternal plate with a pair of setae, or united medially and not forming a poststernal apodeme; male coxisternal plates III–IV fused medially, with or without a poststernal apodeme. Female genital opening small, covered by consolidated aggenital plate; genital setae absent, 1 pair of aggenital setae present or absent. Males with plates C and D consolidated, and with caudal or ventrocaudal genital capsule formed by consolidated plates H, PS, Ag, with short sclerotized aedeagus flanked by a pair of blunt or hooked setigenous processes and surrounded by adhesive flange or disc. Females with legs I similar to or thicker than legs II–IV and with single claw sometimes enlarged for grasping; legs II–IV similar in form, each usually with paired claws and stalked empodium; trochanter IV subtriangular, similar to trochanters I–III. On adults, coxisternal plates I–II together with 1–4 pairs of seta; trochanters I–IV with or without a seta; femur I with 3–4 setae; genu I with 3 or usually 4 setae; genu II with 2–3 setae. Male legs IV stout, with a single sessile claw in PYEMOTIDAE; similar in form to legs III, with pretarsus and symmetrically paired claws (if claws present on legs III) in ACAROPHENACIDAE; and unmodified, but legs III with large, spurlike tarsal process in RESINACARIDAE. With only 1 active postembryonic instar (the adult); larva and single nymphal instars inactive calyptostatic apoderms, except a free-living larval instar apparently present in RESINACARIDAE.

The Pyemotoidea includes the families PYEMOTIDAE (Figs. 13.91A–E) and ACAROPHENACIDAE (Figs. 13.92A–E), adult females of which are parasitoids on the eggs and immature instars of insects including Thysanoptera, Homoptera, Lepidoptera, Coleoptera, Diptera, and Hymenoptera (Krczal 1959; Rack 1959; Cross and Krantz 1964; Cross 1965; Cross and Moser 1971; Goldarazena et al. 2001). High fecundities are achieved by relatively short life cycles (suppression of the larval instar), no intermediate hosts, and development of numerous progeny at approximately the same time within the mother's physogastric body on a single host. Furthermore, all offspring are born as sexually mature adults, 95%–99% of which usually are female (Bruce and Wrensch 1990). Adults are small (80–325 µm in unfed individuals), and males are considerably smaller than females. Physogastry in adult female PYEMOTIDAE may be extensive but nevertheless is limited to the opisthosoma. It is less extreme in female ACAROPHENACIDAE, but it affects the entire idiosoma. The families RESINACARIDAE (Figs. 13.91F, G; Mahunka 1975a) and CARABOACARIDAE (Figs. 13.92F–H; Mahunka 1970a) also are placed in the Pyemotoidea, albeit tentatively. *Resinacarus resinatus* Vitzthum is plesiomorphic relative to pyemotids and acarophenacids in retaining a free-living larval instar. Its larva and male have a fully developed, apparently functional gnathosoma (a slender, undivided pharyngeal pump is evident), and its female can undergo physogastry much as seen in acarophenacids. The male aedeagus is flanked by a pair of setigerous hooks, similar to those found in acarophenacid males, but the male of *R. resinatus* differs in having legs III rather than legs IV modified for copulation.

Only females of the family CARABOACARIDAE (Figs. 13.92F–H) are known, and the consequent lack of information on male characteristics (i.e., degree of gnathosomatic development and form of the genital capsule and its components) and on whether a free-living larva exists, makes placement of the family in the Pyemotoidea uncertain. Adult female *Caraboacarus* clearly resemble those of other pyemotoids in the form of their pharynx and prodorsal structures. However, they have a greatly widened gnathosomatic capsule in which the palpi are completely subsumed, and a pair of discoid structures that apparently are adhesive modifications of setae on coxisternal plates I. In addition, legs I are reduced in size relative to legs II–IV and lack ambulacrum and claws, and tibia and tarsus IV are at least partially fused and have vestigial claws on otherwise well-developed pretarsi.

The PYEMOTIDAE (Figs. 13.91A–E) currently includes only one genus, *Pyemotes*, with about 20 identifiably described species. Phoretic females of undescribed species that may represent another genus have been found under the wings of grain-infesting lathridiid beetles in Canada (Barker 1993). Adult *Pyemotes* are small (85–325 µm), with males considerably smaller than females. Females of *Pyemotes* are divided into two species groups based on their toxicity to hosts and their degree of host specificity (Cross and Moser 1975; Moser 1975a; Moser, Smiley, and Otvos 1987). Members of the *P. scolyti* group tend to be restricted in their host relationships and habitats, but they are usually associated with bark beetles and feed on the eggs, larvae, and pupae of their hosts. Heteromorphy occurs in both sexes, and females are apparently nonvenomous. Phoretic females, or phoretomorphs (also see chapter 6), are shorter, broader, and have thicker and more heavily clawed legs I than nonphoretic females (Cross and Moser 1975). *P. parviscoltyi* Cross and Moser is phoretic on a pine bark beetle, *Pityophthorus bisulcatus*, but on no other pine bark insects (Moser, Cross, and Roton 1971). *P. scolyti* (Oudemans) may be an assemblage of cryptic species in that there seem to be ecological barriers between morphologically similar populations associated with *Scolytus multistriatus* on elms and *S. ventralis* on conifers (Moser and Roton 1972). In contrast to the *P. scolyti* group, species of the *P. ventricosus* group have wide host ranges and occur in a wide range of habitats; their females are monomorphic, highly venomous, and attack all instars of their hosts. Members of this assemblage are frequently encountered as pests in laboratory insect cultures and stored foods. The toxin injected by biting females of the *ventricosus* group (e.g., *P. ventricosus* (Newport), *P. tritici* (Lagrèze-Fossat and Montagné), and *P. herfsi* (Oudemans)) causes paralysis and eventual death of their insect hosts. The same toxin may cause "grain itch," a pruriginous dermatitis in individuals who are routinely exposed to these mites, such as food handlers and granary workers. Itching may be accompanied by asthma, nausea, and other symptoms, including fever and local lymphadenopathy in some cases (Baker et al. 1956; Krczal 1959; Cross 1965; Moser 1975a; Treat 1975; Southcott 1976; Le Fichoux et al. 1980). The ubiquitous *P. tritici* has one of the highest reproductive rates yet documented for any animal (Bruce and Wrensch 1990) and is a common mortality factor for larval and pupal Lepidoptera and other stored products insects (Tawfik, El-Husscini, and Awadallah 1981). Populations of *P. tritici* are easily cultured in the laboratory, and the species is routinely commercialized (Moser, Smiley, and Otvos 1987; Moser et al. 1987; Bruce 1989). It has been used (under the name *P. herfsi*) in the biological control of insects such as the pink bollworm, *Pectinophora gossypiella*, in Egypt (Tawfik and Awadallah 1971), and it has also been introduced for biological control of the imported fire ant, *Solenopsis invicta*, in the United States with inconsistent results (Bruce and LeCato 1980; Jouvenaz and Lofgren 1986; Thorvilson and Phillips 1987). Females of *P. barbara* Moser, Smiley, and Otvos readily attack immature instars of a variety of forest insects offered in the laboratory, and the species was suggested as a candidate for biological control of cone moths and other seed insects (Moser, Smiley, and Otvos 1987; Moser et al. 1987).

Details concerning ovoviviparous development, male behavior during the birth of adult progeny, precopulatory versus postcopulatory searching behavior of females for host carriers, highly skewed sex ratios, and exposure of the maternal mite's diploid genome in the haploid condition of her sons are presented by Bruce and Wrensch (1990), Kaliszewski and Wrensch (1993), and Kaliszewski, Athias-Binche, and Lindquist (1995). Athias-Binche (1990) reviewed host-phoretic and host-parasitoid specificity as limiting factors in considering various species of *Pyemotes* for potential use as biological control agents.

The ACAROPHENACIDAE (Figs. 13.92A–E) includes seven genera and approximately 35 species whose adults somewhat resemble tarsonemids in body shape and size and range in length from 100 to 320 µm. Adult females are parasitoids of insect eggs, and the sex ratio is maximally biased such that only one son usually appears in a set of progeny produced by a fertilized female (Newstead and Duvall 1918; Rack 1959; Elbadry and Tawfik 1966; Steinkraus and Cross 1993). The male may fertilize his sisters inside the engorged body of their physogastric mother just before birth or outside immediately thereafter. In the former case, inbreeding seems unavoidable (Kaliszewski, Athias-Binche, and Lindquist 1995). Certain members of the genus *Acarophenax* may be beneficial in that they are parasitoids of graminivorous beetles of the genera *Tribolium*, *Cryptolestes*, and *Alphitobius* (Newstead and Duvall 1918; Cross and Krantz 1964; Livshits and Mitrofanov 1974; Steinkraus and Cross 1993; Faroni, Guedes, and Matioli 2000). Their fecundity is generally an order of magnitude less than in *Pyemotes* species, but, in view of their host specificity and because they feed only on the eggs of insect hosts and have no toxic effects on people or domesticated animals, Steinkraus and Cross (1993) considered them potentially useful as biological control agents in closed systems such as poultry houses or stored products situations. *A. rackae* Mahunka and Zaki was collected from sheep dung (Mahunka and Zaki 1990), in which it could be associated with any of a variety of insects, and females of *A. lukoschusi* Mahunka and Fain were found phoretic on a sphecid wasp (Mahunka and Fain 1989). Species of *Paracarophenax* are associated with graminivorous beetles and other Coleoptera (Rack 1959; Cross 1965; Mahunka 1975e) and also are found in debris under decaying fruit (Cross 1965) and in rotting wood (Mahunka and Rack 1977). The 15 known species of *Adactylidium* are associates of phlaeothripid thrips (Cross 1965; Elbadry and Tawfik 1966; Goldarazena, Jordana, and Zhang 1997; Goldarazena 2001; Khaustov 2007b), and *Aethiophenax ipidarius* (Redikortsev) occurs with bark beetles in Eurasia (Redikortsev 1947; Magowski 1994). Females of *Paradactylidium* have been encountered in soil litter samples in Mongolia, the Ukraine, and the eastern United States, *Aegiptophenax* occurs in chicken manure and pigeon nests in Egypt, and females of *Aethiophenax* turned up in a stream bank soil probe in Ethiopia, all without reported insect associates (Mahunka 1973b, 1981; Goldarazena, Ochoa, and Jordana 1999; Rady 1992; Khaustov 2007b). *Protophenax kotejii* Magowski was described from three females attached to a male coccid insect, *Inka minuta* Koteja (Magowski 1994), in Cretaceous Siberian amber.

Little is known about the way of life of *Resinacarus resinatus*. Adult females are dimorphic, with one form having a separate tibia and tarsus with a moderate-sized claw on legs I and the other having a fused tibiotarsus with a larger claw on legs I (Mahunka 1975f). Physogastric females with up to 10–12 eggs were found in galleries bored in spruce resin by larvae of syrphid flies of the genus *Myolepta* and another unidentified form (Vitzthum 1927).

The known species of *Caraboacarus* have been described from adult females found under the elytra of various species of carabid beetles (Krczal 1959; Cross 1965; Nickel and Elzinga 1969; Husband and Husband 1984; Husband 1986b; Haitlinger 1990a). Some have been observed with their chelicerae inserted into the veins of their hosts' hind wings (Nickel and Elzinga 1969), but none has appeared engorged or gravid (Cross 1965; Husband 1986b). A parasitic way of life seems doubtful in that males, eggs, and immature instars are not found on their host insects, although parasitoidism remains a possibility (Kaliszewski, Athias-Binche, and Lindquist 1995). In his monograph on the fauna of Tarsonemina in Hungary, Mahunka (1972b) included *Pyemotes aradii* Krczal in *Resinacarus*, but this species was not mentioned in his subsequent paper on the genus (Mahunka 1975f).

▪ Superfamily Tarsonemoidea (Figs. 13.89C, D, 13.93, 13.94)

DIAGNOSIS: Female gnathosoma capsulate, circular, subquadrate, ovoid, or elongate, visible or hidden from above, with 1 pair of dorsal setae; palpi closely appressed, reduced; movable cheliceral digits styletlike, partially to deeply retractile; pharynx undivided, variably muscular or sclerotized. Male gnathosoma usually similarly developed, though sometimes less robust than in female. Prodorsal shield of females with a pair of anterolateral stigmata and associated tracheae, and with 1 or usually 2 pairs of setiform setae and with or without a pair of anterolateral bothridia with usually capitate sensilla, the latter represented by normal scapular setae in larvae and males. Opisthosoma of females with first dorsal plate C entire, not expanded anteriorly over prodorsum. Adults with coxisternal plates I–II united medially, usually forming prominent prosternal apodeme; female and male coxisternal plates II–IV united medially, sometimes with a set of lateral and poststernal apodemes. Female genital opening small, covered by consolidated aggenital plate; genital setae absent, 1 pair of aggenital setae present or absent. Males with plates C and D consolidated, and with caudal or dorsocaudal genital capsule formed by consolidated plates H, PS, Ag, with a pair of terminal setigenous stylets flanking aedeagus that sometimes is surrounded by adhesive flange or disc. Legs I of adults with single claw or paired claws or lacking claws; legs II–III similar in form, or trochanter on legs III more elongate and rigid than on legs II, or these legs absent in females though present in males. Legs IV of female either absent or small, narrow, with fused femorogenu, lacking claws, and ending with 2–3 attenuated setae. On adults, coxisternal plates I–II together with maximally 2 pairs of setae; trochanters I–IV lacking a seta except 1 sometimes present on trochanter of legs IV of males; femur I with 1–4 setae; genu I with 0–4 setae; genu II with 0–3 setae. Legs IV of male either absent or stout, usually with a single sessile claw. With 2 active postembryonic instars (larva, which consistently retains its 3 pairs of legs, and adult). Only 1 active, sexually reproductive instar in males of some PODAPOLIPIDAE; single nymphal instar is an inactive calyptostatic apoderm or lacking.

The Tarsonemoidea is considered to comprise two families, both of which have evolved complex relationships with insects, other mites, and plants. Body lengths of tarsonemoid larvae and adults range from 90 to 400 µm, although engorged individuals may be larger. The TARSONEMIDAE (Figs. 13.93A–E) is a large family of insectophilous, nidicolous, algivorous, fungivorous, and phytophagous mites grouped in about 40 genera including some 530 species, about half of which are relegated to the genus *Tarsonemus* (Lindquist 1986; Lin and Zhang 2002). A key to Palearctic species of the genus *Tarsonemus* based on adult females (Kaliszewski 1993) treats 77 species, most of which are fungivores, and many of which have a wide distribution. Species of *Polyphagotarsonemus*, *Hemitarsonemus*, *Phytonemus*, and *Steneotarsonemus* are important phytophages, with the broad mite, *Polyphagotarsonemus latus* (Banks), and the cyclamen mite, *Phytonemus pallidus* (Banks), being of prime economic significance (Moznette 1917; Gadd 1946; Schaarschmidt 1959; Karl 1965a, b; Jeppson, Keifer, and Baker 1975c; Crop Protection Compendium 1999; Zhang 2003). Both species may produce teratosities in the plant tissues on which they feed, presumably through injection of salivary toxins. Similar tissue toxemia has been reported in other tarsonemid mite-plant relationships, including one involving the monobasic genus *Suskia* on myrsine, and others involving species of *Hemitarsonemus* on ferns and some of *Steneotarsonemus* on monocots (Jeppson, Keifer, and Baker 1975c; Lo and Ho 1979; Lindquist 1986; Ochoa, Aquilar, and Vargas 1994b; Lin and Zhang 1999; Zhang 2003). As associates of pathogenic fungi, several species of *Steneotarsonemus* may further harm their host plants. For examples, *S. ananas* (Tryon) initiates fungal infection in pineapple (Tryon 1898), the feeding injury of *S. bancrofti* (Michael) allows entry of spores of red rot in sugarcane (Butani 1959), and S. *spinki* Smiley is a vector of conidia that cause sheath rot in rice (Chow et al. 1980). Adult females of the tarsonemelline genera *Tarsonemella*, *Paratarsonemella*, and *Ficotarsonemus* are closely associated with agaonid fig wasps, whose winged females are used by the mites for dispersal. Pollinating wasps carry the mites through the ostiole into the fig fruit, where the mites initiate gall formation among the fig ovules and then develop on these galls, often in dense aggregations (Compton 1993; Lindquist 1998). Adult females of a related monobasic genus, *Alkithoenemus*, are phoretic on a species of fruit bat that may frequent figs in which these mites are thought to undergo their life history (Ochoa, OConnor, and Klompen 1996).

Several taxa of possibly fungivorous tarsonemids are recorded by Attiah (1970) from citrus trees in Florida. Females of *Fungitarsonemus pulvirosus* Attiah, an inhabitant of citrus orchards, carry debris on their dorsa that obscures the mites and perhaps offers a degree of protection from predators (Muma 1975). Adult females of undescribed Australian rainforest *Fungitarsonemus* also carry debris on their dorsa and attach debris in patches on leaf surfaces in which eggs are laid and larvae become quiescent and molt (D. E. Walter, pers. obs.). Larvae, males, and elaborately ornamented females of the genera *Daidalotarsonemus*, *Ceratotarsonemus*, and *Excelsotarsonemus* frequent lichen growth and bryophytes on bark and foliage of trees, where they are thought to feed on algae or fungi (Lindquist 1986; Ochoa et al. 1997; Ochoa and OConnor 1998) or on foliage itself (Lofego, Ochoa, and Moraes 2005). Other phytophagous, algophagous, and fungivorous tarsonemids are noted by Hussey, Reed, and Hesling (1969); Jeppson, Keifer, and Baker (1975c); Lindquist (1986); Ochoa, Aquilar, and Vargas (1994b); Goldarazena et al. (2002); and Zhang (2003). For example, *Xenotarsonemus belemnitoides* (Weis-Fogh), *Tarsonemus confusus* Ewing, and *T. waitei* Banks are widespread and may occur together in soil and litter and on many plants, including food crops and ornamentals. Although thought to be mostly fungivores, these mites have been reported as foliage-feeding pests of some ornamentals in greenhouses in Europe and of citrus and tomatoes in North America. Tarsonemids that usually are not closely associated with particular kinds of winged insects may normally depend on wind for passive dispersal, although females of *P. latus* have been found phoretic on *Bemisia* whiteflies (Gerson 1992). Based on an analysis of snowborne arthropods as evidence of potential for long distance aerial dispersal, tarsonemids were second only to eriophyids in diversity of species and numbers of individuals recovered (Zhao and Amrine 1997). The absence of phoretic behavior is thought to be a secondary loss among more derivative taxa of tarsonemids (Lindquist 1998).

Many tarsonemid associates of insects are phoretic as adult females on the insects with which they otherwise appear to coexist independently. Species of *Pseudotarsonemoides*, *Ununguitarsonemus*, *Heterotarsonemus*, *Pseudotarsonemus*, and some *Tarsonemus* are phoretic on bark beetles and undergo their life cycles in the subcortical galleries of their hosts (Lindquist 1969a, 1970b, 1986; Smiley and Moser 1974; Khaustov 1998, 2001; Magowski and Khaustov 2006a, b). *T. ips* Lindquist and *T. krantzi* Smiley are involved along with species of *Dendroctonus* as vectors of pathogenic fungi causing blue stain of coniferous trees (Bridges and Moser 1983; Moser and Bridges 1986). The genera *Nasutitarsonemus* and *Tarsanonychus* are known only from adult females phoretic on hispine beetles, and they often occur under the elytra of their hosts (Lindquist 1986). Two species of *Tarsonemus* live in the nests of carpenter bees (Magowski 1986), while female *T. apis* (Rennie), and *Pseudacarapis indoapis* (Lindquist) and *P. trispicula* Ochoa and Pettis are phoretic on honeybees, with all instars living as fungus- or pollen-feeding commensals with the bees in their nests (Lindquist 1968, 1986; Kaliszewski, Athias-Binche, and Lindquist 1995; Sumangala 1999; Ochoa, Pettis, and Mireles 2003). Females of *P. indoapis* and *P. trispicula* evidently prefer to locate in the region of the neck and tentorial pits of the bee's head. The genus *Crossacarapis* is known only from females of one species phoretic on an orchid bee of the genus *Euglossa* (Ochoa and OConnor 1996).

All instars of the monobasic taxon *Tarsonemus* (*Chaetotarsonemus*) *dispar* Beer are found under the elytra of the passalid beetle, *Odontotaenius disjunctus*, in eastern North America (Beer 1954; Lindquist 1986). They are clearly symbionts of passalids in that they undergo their entire life history on their hosts, but their feeding behavior is unknown. Similarly, all instars of the monobasic genus *Flechtmannus* were found in Brazil under the elytra of the curculionid beetle, *Homalinotus coriaceus* (Gyllenhal), where they were thought to feed either mutualistically on fluid exudates or parasitically on hemolymph (de Moraes, Lindquist, and Lofego 2002).

Other tarsonemid genera are documented as parasitoids, parasites, and predators. Adult females of *Iponemus* are exclusively parasitoids that are specifically associated with ipine bark beetles. Of the 18 *Iponemus* species associated with pine-feeding ipines in North America, 16 are monospecific (Lindquist 1969b). Adult female mites feed exclusively on the beetles' eggs and use emerging adult beetles for dispersal. While feeding, the mites undergo idiosomatic physogastry, but eggs are laid externally to the idiosoma. Although fully mobile, larvae, adult male, and young adult female mites remain clustered together and do not feed, using instead the energy reserves derived through the egg from their engorged mother's reserves (Terry 1966; Lindquist 1969b). Male adults emerge first and usually represent only about 5% of the progeny. Because of the clustering behavior of the progeny, few males are sufficient for mating with the numerous emerging female adults.

Larvae and adults of *Acaronemus* prey on eggs of tetranychid and tenuipalpid mites on the foliage of coniferous trees (Smiley and Landwehr 1976; Lindquist and Smiley 1978). There is evidence that some members of the related genus *Dendroptus* may prey on gall-producing eriophyid mites or on eggs of arboreal tydeid mites (Lindquist 1986). Some species of *Dendroptus* (previously reported under the genus *Steneotarsonemus*) have been thought to be "social parasites" of certain gall mites of the family ERIOPHYIDAE. *D. fulgens* (Beer) and *D. nitidus* (Beer) invade the galls of eriophyids and feed on the galligenous tissues of leaves of *Populus* and *Betula*. This precipitates a reduction in the number of eriophyids that apparently abandon their galls (Beer 1963). However, these observations do not exclude the possibility of facultative predation by the tarsonemids on their eriophyid associates (Lindquist 1986).

Larvae and adults of the genus *Acarapis* live as parasites only on adult honeybees, *Apis* spp. Of the three recognized species, *Acarapis dorsalis* Morgenthaler and *A. externus* Morgenthaler occur on external body surfaces of honeybees, apparently causing no injury to their hosts (Morgenthaler 1934; Eckert 1961; Bailey 1963; Delfinado-Baker and Baker 1982). However, parasitism by the honeybee tracheal mite, *A. woodi* (Rennie), hastens the physiological aging of bees both through the physical damage caused by piercing the bee's tracheal walls and feeding on hemolymph and by reducing air exchange as a result of increasing mite numbers in the tracheae, which finally results in hindering of flight (Hirst 1921; Hirschfelder and Sachs 1952; Maki et al. 1988). *A. woodi* and its pathogenicity, commonly known as acarine disease, acariosis, or Isle of Wight disease, are now known wherever honeybees naturally occur or have been introduced (Australia, New Zealand, the Hawaiian Islands, and Scandinavia are exceptions) (Sammataro, Gerson, and Needham 2000). Larvae and adults of the coreitarsonemine genera *Coreitarsonemus*, *Amcotarsonemus*, and *Asiocortarsonemus* have been found only in the atria of the odoriferous thoracic glands of coreid bugs (Fain 1970a, 1971d, 1978c, 1980a, b). Patterns of host specificity in coreitarsonemines are not yet clear, but at least one species, *C. anoplocnemis* Fain, has been recorded from hosts representing 10 genera of Coreidae (Fain 1978c). Evidence that these mites are parasites is reviewed by Kaliszewski, Athias-Binche, and Lindquist (1995).

Certain tarsonemids have been implicated in disorders of humans and other mammals, early cases of which were reviewed by Beer (1954). *Daidalotarsonemus hewitti* Mahunka was shaken from the clothing of a person afflicted with a skin rash (Mahunka 1974c), and a few species of *Tarsonemus* (also under the name *Tarsonemoides*) have been isolated from the surface and subsurface skin tissues of persons with dermal abnormalities (Hewitt et al. 1973). In one case, a person was working with laboratory mice that had similar symptoms (Fagasinski 1986). These associations may be the result of accidental or incidental contamination (Beer 1954; Samšinák et al. 1976; Lindquist 1986), but a Danish study found *Tarsonemus* mites to be second in frequency only to PYROGLYPHIDAE in house dust and implicated them as one of the causal agents of human allergies (Korsgaard and Hallas 1979).

As mentioned for the Pygmephoroidea and Scutacaroidea, male tarsonemids often display the precopulatory behavior of carrying a pharate adult female on their posterior dorsum, using the caudal genital papillae and sometimes their modified legs IV (Figs. 13.89C, D) to hold the female in place and to guard it prior to copulation (Schaarschmidt 1959; Nucifora 1963; Karl 1965a; Suski 1972; Lindquist 1986; Sumangala and Haq 2002). Similar to the groups mentioned above, copulation between tarsonemid adults involves a retroconjugate position (Fig. 13.89D).

The PODAPOLIPIDAE comprises about 30 genera and some 200 described species of internal and external arthropod parasites whose adult females often display bizarre reductions in numbers of legs and whose males exhibit modifications of the genital apparatus (Figs. 13.94A–M). Podapolipids occur under the elytra and in the vaginal conjunctiva of beetles, in the tracheae and air sacs of orthopterans and hymenopterans, and externally on terrestrial insects of the orders Blattaria, Orthoptera, Heteroptera, Hymenoptera, and, especially, Coleoptera (Regenfuss 1968; Husband 1991, 1998). Members of the genera *Bakerpolipus*, *Panesthipolipus*, and *Peripolipus*, and some *Tetrapolipus* and *Podapolipus* feed externally on cockroaches (Husband and Sinha 1970a; Husband 1984a), while those of *Orthapolipus* do so on tettigoniids (Husband, OConnor, and Ochoa 2005). Members of the monobasic genus *Cydnipolipus* are parasites of cydnid bugs of the genus *Aethus* and are the only known podapolipids routinely associated with Heteroptera (Kurosa and Husband 1994). A number of genera of PODAPOLIPIDAE appear to be restricted to one family of host insects. On beetles, for example, *Dorsipes*, *Eutarsopolipus*, *Ovacarus*, and *Regenpolipus* are found only on Carabidae (Regenfuss 1968, 1972; Husband 1986c, 1998, 1999; Husband, Camerik, and Stals 2006); *Coccipolipus* on Coccinellidae (Husband 1972a, 1989c); *Chrysomelobia* and *Parobia* on Chrysomelidae (Husband and OConnor 2004; Seeman and Nahrung 2005); *Tarsopolipus*, *Archipolipus*, *Scarabapolipus*, *Dilopolipus*, and *Stenopolipus* on Scarabaeidae (Husband 1984c, 1989b; Husband and Kurosa 1994); *Tetrapolipus* and *Kurosapolipus* on Cerambycidae (Husband and Li 1993); *Buprestapolipus* on Buprestidae (Husband, Ochoa, and Naskrecki 1995); *Stigmacarus* and *Rhynchopolipus* on Curculionidae (Feldman-Muhsam and Havivi 1977; Husband 1985; Husband and OConnor 1999); and *Silphopolipus* on Silphidae (Kurosa, Khaustov, and Husband 2004). However, species of *Podapolipus* parasitize hosts of three different orders of insects: the Blattaria, Orthoptera, and Coleoptera (Husband 1986a, 1989b). Species of *Podapolipoides* (including *Locustipolipus* Lo 1990) and some of *Podapolipus* feed externally on locusts and grasshoppers (Husband 1972b, 1992), while *Locustacarus tracheali*s Ewing and *L. masoni* Husband occur in the respiratory systems of their locust hosts, apparently piercing the tracheal walls to feed (Wehrle and Welch 1925; Husband and Sinha 1970b; Husband 1974a). *L. buchneri* (Stammer) also is a tracheal parasite but occurs in apid bees rather than in orthopterans (Stammer 1951). Among the approximately 25 species of *Dorsipes* and *Eutarsopolipus* associated with carabid beetles in Europe, many show highly specific preferences among seven genera of hosts (Regenfuss 1968). In some cases, the same host is shared by two or three species of these mites, which are adapted to different microspaces having different integumentary thicknesses to penetrate under the host's elytra (Regenfuss 1972; Husband and Husband 2007). *Ovacarus clivinae* Stannard and Vailshampayan and *O. peellei* Husband also parasitize carabids, but they appear to be restricted to the infolded vaginal conjunctiva and oviduct sacs of their hosts (Stannard and Vaishampayan 1971; Husband 1974b).

The unusual life history of *L. buchneri* (Stammer) probably reflects that of other podapolipids, at least in basic features. Female mites overwinter in the tracheae of queen bees (*Bombus bimaculatus*) and in the spring begin to feed on bee hemolymph through the tracheal walls. Within a week, the greatly enlarged, immobile female deposits up to 50 eggs that soon hatch into diminutive nonfeeding hexapod adult males and motile, hexapod larval females. Mating occurs immediately and the mated larval females seek worker bees. They attach to the tracheal walls, apparently molt to the adult female instar, and begin to feed and enlarge. Adult females are saclike, have only one pair of legs, and are considerably larger than the tiny males and larval females. Several generations may thus be passed in the bee's nest, with larval females dispersing and infesting successive worker broods. New queens leaving the nest in late summer carry with them enough larval females to initiate a new cycle following winter hibernation (Stammer 1951).

Small and large adult males of the same species have been observed among different genera of podapolipids, but the triggering mechanism and adaptive value of this phenomenon are unknown (Husband and Dastych 1998). Males of most podapolipid mites evidently lack a mobile and sexually inactive larval instar (i.e., one capable of movement but lacking copulatory structures), although such larval males have been observed for some species of *Scarabapolipus*, *Chrysomelobia*, and *Tarsopolipus* (Husband 1989a, 1991; Husband and Kurosa 1994). Lack of evidence of pharate male adults bearing genitalia within a larval cuticle may indicate suppression of an active asexual larval stase (as suggested by Seeman and Nahrung 2003) or of a postlarval stase, leaving a sexually reproductive larva instead. The behavioral characteristic of males mating with apparently preadult females is peculiar to members of the PODAPOLIPIDAE.

Members of the genera *Chrysomelobia* (Fig. 13.94G) and *Parobia* are subelytral parasites of leaf beetles in Europe, North America, and Australia and appear to represent the most primitive known podapolipid genera. Adult males and females have four pairs of legs, resembling members of the family TARSONEMIDAE (Figs. 13.93A–E) in this and certain other respects. However, leg segmentation and setation closely approximate those of other podapolipid genera (Eickwort 1975; Lindquist 1986; Seeman and Nahrung 2003; Husband and OConnor 2004). Clues to evolutionary trends in the PODAPOLIPIDAE may lie in the morphological variation and position of the male aedeagal apparatus and, consequently, the mating stance of representative taxa (Regenfuss 1973; Eickwort 1975; Husband 1991). The aedeagus ranges from a seemingly primitive, terminal, posteriorly directed organ (e.g., *Eutarsopolipus* (Fig. 13.94K), *Archipolipus*, and *Ovacarus*), to a bizarre structure that arises anterodorsally and extends to or beyond the gnathosoma (e.g., *Rhynchopolipus*, *Locustacarus*, and *Podapolipus* (Fig. 13.94M)). Retroconjugate or rear-to-rear mating as practiced by the former group of genera is similar to that found in the sister family TARSONEMIDAE and has been thought to indicate a lower phylogenetic level of development than that attained by the latter group of genera, which mate in an inverted proconjugate position. However, based on evidence from other character states, a posterior position of the aedeagus in the three genera noted above may be a reversal in evolutionary direction and, therefore, a derived condition (Husband 1991).

Several species of podapolipid mites have been noted as potential biological control agents against insect pests, especially *Coccipolipus epilachnae* Smiley on the Mexican bean beetle, *Epilachna varivestis* (Husband 1984b). The host range of *C. epilachnae* is limited to phytophagous members of the coccinellid subfamily Epilachninae, which makes it suitable for release as a biological control agent. *C. epilachnae* was initially found to reduce fecundity and increase mortality of bean beetles (Schroder 1979, 1982), but subsequent studies found mite infestations to have little or no significant effect in suppressing populations of its host (Cantwell, Cantelo, and Cantwell 1985; Hochmuth et al. 1987). Similar investigations on the impact of *Chrysomelobia labidomerae* Eickwort on populations of the Colorado potato beetle, *Leptinotarsa decemlineata*, and the milkweed leaf beetle, *Labidomera clivicollis*, indicated variable results in affecting fecundity and mortality of infested beetles and limited potential to affect host population dynamics (Eickwort and Eickwort 1986; Drummond, Casagrande, and Logan 1989, 1992).

Key 13.4. Suborder Prostigmata, Key to the Families (excluding Parasitengonina) (Figs. 13.2–13.24 and 13.61–13.94)

1a. Adults and deutonymphs often densely hypertrichous, typically free-living predators; larvae usually strongly heteromorphic, parasitic on arthropods or vertebrates, protonymphs and tritonymphs calyptostatic (see chapter 5). Terrestrial forms with 1 or 2 pairs of bothridial setae on a prodorsal sclerite, aquatic forms highly diverse, without bothridial setae, usually with multiple modified idiosomatic lyrifissures (glandularia) that open on small platelets or amalgamated shields **COHORT PARASITENGONINA** (keys 13.2 and 13.3)

1b. Adults and nymphs with or without dense hypertrichy, bothridial setae and prodorsal sclerite present or absent, without glandularia. Larvae more or less similar to other stages, free living or parasitic. Alternating calyptostasis rare **2**

2a. With undivided, heavily armored dorsal and ventral shields, coxal fields expanded and mesally contiguous. Prodorsum with 2 pairs of trichobothria (bothridial setae *vi*, *si*) and 4 additional pairs of setae; a pair of lateral ocelli (rarely absent) and an unpaired anteromedian ocellus; one or more pairs of lateral pustules. Genital and anal valves of females contiguous, those of males usually separated; 2 pairs of genital papillae. Chelicerae separate and strongly chelate, each with 2 setae (*cha*, *chb*) (Figs. 13.2C, D, 13.5) **Supercohort Labidostomatides, LABIDOSTOMATIDAE**

2b. Variously armored or soft bodied. Prodorsum with a maximum of 5 pairs of setae, including 0–2 pairs of bothridial setae, or hypertrichous; ocelli present or absent, lateral pustules absent. 0–3 pairs of genital papillae in adults. Chelicerae various **3**

3a. Females usually with a pair of stigmata and associated tracheae anterolaterally on prodorsum and commonly with a pair of capitate bothridial setae (both usually absent in immatures and males); peritremes absent. Opisthosoma usually with a longitudinal series of 4 dorsal plates (tergites), often overlapping. Tarsi of legs II–III with stalked, smooth, padlike empodium usually flanked by paired claws, or rarely without empodium but with paired claws. Sexual dimorphism pronounced, males often with a genital capsule (Figs. 13.81–13.94) **Cohort Heterostigmatina—56**

3b. Females and males with a pair of stigmata between cheliceral bases or on posterodorsal margin of gnathosoma and often opening into a simple to complex peritreme, or absent; prodorsal trichobothria, if present, of various forms and on immatures and adults of both sexes. Opisthosoma variously armored or soft bodied. Tarsi of legs II–III generally with setulose, rayed, or clawlike empodium, often flanked by paired claws. Sexual dimorphism strong or weak, but males without genital capsule **4**

4a. With 1–2 pairs of variously shaped prodorsal bothridial setae usually inserted in distinctly cup-shaped bothridia **10**

4b. Without prodorsal bothridial setae **5**

5a. Legs III–IV absent in all stages; legs I–II ending in a highly branched empodial featherclaw; opisthosoma elongated, annulated, vermiform, or fusiform. Obligate plant parasites (Figs. 13.15–13.19) **Supercohort Eupodides (*pars*), Superfamily Eriophyoidea—6**

5b. Legs III present and legs IV usually present in postlarval instars; one or more pairs of legs typically ending in an empodium of various forms and often with a pair of lateral claws; opisthosoma variously shaped. Free living or aquatic, or associated with a variety of plant and animal hosts **8**

6a. Prodorsal shield with 1, 3, 4, or 5 setae; spermathecal tubes usually long; tibia I with or without solenidion **PHYTOPTIDAE**

6b. Prodorsal shield nude or with 1 pair of setae; spermathecal tubes short; tibia I without solenidion 7

7a. Gnathosoma large compared to body and stylets abruptly curving ventrally near base; featherclaw deeply divided medially or entire **DIPTILOMIOPIDAE**

7b. Gnathosoma relatively small compared to body and stylets more or less straight; featherclaw entire, rarely divided medially **ERIOPHYIDAE**

8a. Marine, brackish, and freshwater mites usually found attached to substrates (including algae and marine invertebrates) or in interstitial samples; tarsi I–IV with strongly developed lateral claws, usually with serrate margins, and often with a clawlike or bifid empodium; adults with 2 to several pairs of genital papillae on, near, or under a pair of genital valves (Figs. 13.7, 13.8) **Supercohort Eupodides (*pars*), Superfamily Halacaroidea—9**

8b. Mostly terrestrial mites, occasionally intertidal or on the surface of freshwater, or parasitic on terrestrial plants and animals; tarsal apoteles various but not modified into clawlike holdfasts with serrate margins; adults with 3 pairs of genital papillae or genital papillae absent **29**

9a. With 5–10 pairs of dorsal idiosomatic setae; marine and freshwater forms (Fig. 13.7) **HALACARIDAE**

9b. With 16–20 pairs of dorsal setae; freshwater forms (Fig. 13.8) **PEZIDAE**

10a. Prodorsal peritremes present (sometimes emergent), palptibia usually with at least 1 clawlike seta distally and often apposed to a reduced or thumblike palptarsus; if thumb-claw process absent, then body elongate and with 1 or more postpedal furrows or oval with a postpedal constriction **Supercohort Anystides (*pars*)—23**

10b. Peritremes absent, palptibia without clawlike setae or thumblike palptarsus; postpedal furrows absent **Supercohort Eupodides (*pars*)—11**

11a. Leg trichobothria absent; chelicerae fused or free but with little potential for lateral motion, inserted on or fused with short subcapitulum **13**

11b. Trichobothria on tibia IV, and sometimes on tibia I, tarsus III, and tarsus IV; chelicerae free and capable of considerable lateral motion, inserted on elongate subcapitular "snout" (Fig. 13.6) **Superfamily Bdelloidea—12**

12a. Adults with 3 pairs of genital papillae and 2 or more eugenital setae; palpi usually ending in 1–2 elongate setae and often elbowed, without strong spines or knobs; tarsi III–IV and tibia I typically with trichobothria (Figs. 13.6A–F) **BDELLIDAE**

12b. Adults with 2 pairs of genital papillae and usually without eugenital setae; palpi usually ending in a stout spine and sometimes with strong spines or knobs; tarsi III–IV and tibia I without trichobothria (Figs. 13.6G–J) **CUNAXIDAE**

13a. Chelicerae freely articulating, digits variously formed; tarsus I with a famulus and at least 1 recumbent solenidion; idiosomatic chaetome sometimes with postlarval hypertrichy; naso usually present (Figs. 13.9–13.12) **Superfamily Eupodoidea—14**

13b. Chelicerae closely contiguous or fused medially, fixed digit regressed and movable digit styletlike to whiplike; tarsus I without famulus, solenidia erect; idiosoma never hypertrichous, naso absent (Figs. 13.12E–G, 13.14) **Superfamily Tydeoidea—20**

14a. Body fully sclerotized, dorsum with V- or Y-shaped furrow and an anterior epirostrum projecting over the gnathosoma (Figs. 13.10E–G) **PENTHALODIDAE**

14b. Soft bodied, without V- or Y-shaped furrow or epirostrum **15**

15a. Palptibia with 6–7 setae; subcapitulum with more than 15 setae; naso-prodorsal process present and hypertrichous (Figs. 13.11E–H) **ERIORHYNCHIDAE**

15b. Palptibia with 3 or fewer setae; subcapitulum typically with 2 or 4 pairs of setae; naso, when present, not enlarged but usually bearing one pair of setae **16**

16a. Chelicerae strongly chelate, shearlike, with well-developed fixed digit opposed to large, often toothed, movable digit **17**

16b. Chelicerae weakly chelate, hooked, or styletlike, not shearlike **19**

17a. Palpi with 5 free segments; tarsus I with 2 solenidia; empodia clawlike (Figs. 13.11A–D) **PENTAPALPIDAE**

17b. Palpi with 4 free segments; tarsus I with 3 or more solenidia; empodia padlike **18**

18a. Palptibia with 3 setae; tibia I with 2 or more solenidia, prodorsum with 1 pair of bothridia and 3 pairs of setae (Figs. 13.10A–D) **RHAGIDIIDAE**

18b. Palptibia with 1 seta; tibia I with 1 solenidion, prodorsum with 1 pair of bothridia and 4 pairs of setae (Figs. 13.10H–K) **STRANDTMANNIIDAE**

19a. Anal opening dorsal or dorsoterminal; intercoxal region with 15 or more pairs of setae; femora IV similar to others (Figs. 13.12A–D) **PENTHALEIDAE**

19b. Anal opening ventral or terminal; intercoxal region with fewer than 10 pairs of setae; femora IV normal or strongly swollen, used in jumping (Fig. 13.9) **EUPODIDAE**

20a. Tibia I with deeply recessed solenidion in saclike structure connecting to surface by narrow duct; legs and prodorsum often with netlike ornamentation; adults with 2 pairs of genital papillae (Fig. 13.14) **EREYNETIDAE**

20b. Solenidion of tibia I on surface or absent; with plicate ornamentations or sclerotized plates; adults with 0–2 pairs of genital papillae **21**

21a. Adults with only 1 pair of genital papillae; or papillae absent; movable digit of chelicerae styletlike or whiplike; palptarsus usually with 3 or 4 elongated setae; males typically with an aedeagus (Figs. 13.13E–I) **IOLINIDAE**

21b. Adults with 2 pairs of genital papillae; movable digit of chelicerae styletlike; palptarsus with no markedly elongated setae; males without aedeagus **22**

22a. Legs I antenniform, apotele vestigial or absent; median eyespot consisting of subcuticular silver granules present; epimeral seta *3d* absent (Figs. 13.12E–G) **TRIOPHTYDEIDAE**

22b. Legs I with claws; median eyespot absent; epimeral seta *3d* usually present (Figs. 13.13A–D) **TYDEIDAE**

23a. Genital papillae absent; palptarsus reduced; tarsal apoteles padlike, produced on annulate stalks (Figs. 13.23A, B) **PSEUDOCHEYLIDAE**

23b. With 2–3 pairs of genital papillae; palptarsus usually well developed; tarsal apoteles sessile **24**

24a. Elongate mites with 1 or more postpedal furrows; legs I–II widely separated from legs III–IV; with 1 pair of prodorsal bothridial setae (Not always elongate) **25**

24b. Oval to subrectangular mites without postpedal furrows, although sometimes with a postpedal constriction; legs I–IV closely inserted; with 2 pairs of prodorsal bothridial setae **26**

25a. Palpi linear, without thumb-claw process; legs I ending in paired claws; prodorsal bothridial setae filiform (Figs. 13.23C–F) **PARATYDEIDAE**

25b. Palps with well-developed thumb-claw process; legs I without apoteles; bothridial setae clubbed (Figs. 13.23H, I) **STIGMOCHEYLIDAE**

26a. Moderately to strongly sclerotized mites with 1–2 large dorsal shields or a complex of 8 dorsal plates **27**

26b. Soft-bodied mites with at most a weakly sclerotized prodorsal shield **28**

27a. Adults leathery, with 8 heavily sclerotized dorsal shields and 3 pairs of genital papillae; legs I with strongly spinose anterolateral setae (rake-legged mites) (Fig. 13.20) **CAECULIDAE**

27b. Adults with 1–2 large, moderately sclerotized dorsal shields and 2 pairs of genital papillae; legs I without spinose setae as above (Fig. 13.21) **ADAMYSTIDAE**

28a. Naso weakly developed and nude; peritremes sinuous but not emergent distally; prodorsal bothridia usually with rosette pattern; palptarsus reduced, not extending beyond palptibial claw; pretarsal empodia I–II absent (Figs. 13.22F, G) **TENERIFFIIDAE**

28b. Naso usually present and bearing setae *vi*; peritremes linear and often emergent distally; prodorsal bothridia simple; palptarsus usually extending beyond palptibial claw; pretarsal empodia I–II padlike or clawlike (Figs. 13.22A–E) **ANYSTIDAE**

29a. Adult females with 3 pairs of genital papillae, an elongate, extrusible ovipositor, a well-developed prodorsal shield, and a series of 5 dorsal hysterosomatic shields (Fig. 13.24) **Supercohort Anystides (*pars*), POMERANTZIIDAE**

29b. Genital papillae absent in all stages; ovipositor absent; dorsal shields various **Cohort Raphignathina—30**

30a. Pretarsal claws well developed or reduced but always with tenent hairs; empodium, if present, with or without tenent hairs; palpal thumb-claw process usually well developed (absent in TENUIPALPIDAE) **31**

30b. Pretarsal claws lacking tenent hairs; empodia, if present, usually with tenent hairs; palpal thumb-claw process well developed, reduced, or absent **37**

31a. Chelicerae independently movable, elongate, with swollen bases and hooklike movable digit with 1 or more teeth; empodia absent; peritremes emergent. Parasites of lizards, cockroaches, triatomines, and scorpions (Fig. 13.62) **Superfamily Pterygosomatoidea, PTERYGOSOMATIDAE**

31b. Chelicerae fused basally and with short, needlelike movable digits or fully integrated into a retractable stylophore with long, whiplike digits; empodia usually present, rarely strongly reduced or seemingly absent; peritremes sessile or emergent **32**

32a. Chelicerae fused basally and with short, needlelike movable digits and bearing short, 3–5 chambered peritremes on their paraxial faces; palptibial claw with a ventral tooth. Free living in soil and litter (Figs. 13.63A–C) **Superfamily Raphignathoidea (*pars*), BARBUTIIDAE**

32b. Cheliceral bases fused into a movable stylophore that is deeply retractable into idiosoma; movable cheliceral digits elongated, whiplike, recurved basally within stylophore. Obligate plant parasites **Superfamily Tetranychoidea—33**

33a. Setae of posterior margin of idiosoma (row *h*) forming a fanlike transverse row of 5 or more pairs of flagellelate or bipectinate setae; dorsal and lateral opisthosomatic setae expanded, fanlike, the anterior row (*c*) comprising 5 or more pairs (Figs. 13.75D, E) **TUCKERELLIDAE**

33b. Setae of posterior margin of idiosoma slender, simple, not forming a fanlike row; dorsal and lateral opisthosomatic setae variable in form, the anterior row (*c*) comprising at most 4 pairs **34**

34a. Tarsi I–II without peg-shaped or bulbous solenidia, and with 1–2 long, slender, tapered solenidia usually closely associated with a short or minute seta to form duplex sets; stylophore attachment to idiosoma without ribbed collar; palpi 5-segmented, with thumb-claw process; palptarsus with or without one of distal eupathidia enlarged as a spinneret (Figs. 13.73, 13.74) **TETRANYCHIDAE**

34b. Tarsi I–II with distal, peg-shaped or bulbous solenidia, and with no solenidia closely associated with a seta to form duplex sets; stylophore attached to idiosoma by retractable ribbed collar; palpi with 5 or fewer segments and with or without thumb-claw process; palptarsus without eupathidial modification as above **35**

35a. Prodorsum with 4 pairs of setae (*v1* present) and without eyes; anterior row (*c*) of opisthosomatic setae comprising 4 pairs (*c1–4* present); Figs. 13.76D, E **LINOTETRANIDAE**

35b. Prodorsum with 3 pairs of setae (*v1* absent) and usually with 2 pairs of eyes; anterior row (*c*) of opisthosomatic setae comprising at most 3 pairs (*c1* or *c4* absent) **36**

36a. Palpi linear, with 5 or fewer segments and lacking a thumb-claw process; opisthosoma with at most 13 pairs of setae dorsally and laterally (*c1* usually present, *c4* absent, 2 pairs *h* setae); anal valves with 1–3 pairs of simple pseudanal setae (Figs. 13.76A–C) **TENUIPALPIDAE**

36b. Palpi 5-segmented, with a thumb-claw process; opisthosoma with 14 pairs of setae dorsally and laterally (*c1* absent, *c4* present, 3 pairs *h* setae); anal valves with 3 pairs of pseudanal setae, some of which are bifurcate or dendritic (Figs. 13.75A–C) **ALLOCHAETOPHORIDAE**

37a. Chelicerae and subcapitulum fused into gnathosomatic capsule that is retractile within the idiosoma; legs I modified for clasping mammalian hairs; immatures with a ventral, unpaired clasper on tarsus I and femora fused to genua on all legs; female anal opening dorsal. Ectoparasites in the fur of bats, rodents, and various insectivores (Fig. 13.61) **Superfamily Myobioidea, MYOBIIDAE**

37b. Chelicerae separate, fused into stylophore or incorporated into gnathosomatic capsule (if the latter, then exerted and not retractile); legs I not modified for clasping hairs; immatures without clasper on tarsus I or femoral-genual fusions as above, female anal opening typically ventral, sometimes terminal. Free-living predators, or feeding on mosses, or ectoparasitic on mammals, birds, and reptiles (rarely insects), or living in the skin pores or mucosa of mammal and reptiles (rarely birds) **38**

38a. Chelicerae separate, partially fused, or completely coalesced into a stylophore; tibiae and often genua of some legs with more than 3 setae; palpi with 5 free segments, thumb-claw process usually well developed, but palptarsus sometimes reduced and palptibial claw sometimes reduced or absent; legs well developed, ambulatory, and each with a pair of claws and an empodium with tenent hairs. Free-living predators, moss associates, or insect phoretics **Superfamily Raphignathoidea (*pars*)—39**

38b. Chelicerae fused into a stylophore, or fully integrated into a gnathosomatic capsule; all tibiae and genua with 3 or fewer setae; palpi often with fewer than 5 segments, palptibial claw various, typically absent in parasitic groups. Capsulate forms with well-developed legs, usually with paired claws and with tenent hairs on empodia (sometimes absent on legs I) and well-developed palpi. Free-living predators, insect associates, or parasites of vertebrates. Stylophoric forms with legs flattened, telescoped, or otherwise reduced, often with highly modified empodia; claws may be absent. Parasitic on vertebrates **Superfamily Cheyletoidea—50**

39a. Fully armored mites with a holodorsal shield produced anteriorly into a hyaline or ornamented collar into which the narrow, elongate gnathosoma can be retracted (Figs. 13.64C, D) **CRYPTOGNATHIDAE**

39b. Soft bodied or with an array of dorsal shields separated by striate cuticle, or with 1–2 dorsal shields; prodorsal collar absent, gnathosoma not deeply retractable **40**

40a. Stigmata opening at bases of chelicerae; peritremes simple to highly convoluted, produced on cheliceral bases or on anterior edge of prodorsum **46**

40b. Peritremes and stigmata absent **41**

41a. Cheliceral bases often closely associated but apparently freely movable **42**

41b. Cheliceral bases fused at their bases or coalesced into a distinct stylophore **43**

42a. Fully armored and strongly hemispherical mites with a holodorsal shield bearing an arched groove posterior to the large eyes and internal flasklike or funnel-shaped structures. On surface of standing freshwater and associated vegetation or in subaquatic soils (Fig. 13.70) **HOMOCALIGIDAE**

42b. Mostly soft bodied, or with one or more dorsal plates separated by striate cuticle, rarely with a holodorsal shield but, if present, lacking arched groove and internal structures as above; found in many habitats including mosses, vegetation, bark, soil, marshes, and freshwater habitats **43**

43a. Cheliceral bases insensibly fused into stylophore **44**

43b. Cheliceral bases distinct and fused medially, at least in basal 1/3 (Figs. 13.67C, 13.72E) **STIGMAEIDAE (*pars*)**

44a. Palpi slender, much longer than chelicerae **45**

44b. Palpi robust, subequal to chelicerae in length (Fig. 13.68D) **STIGMAEIDAE (*pars*)**

45a. Palptarsus longer than palptibia; empodia on all tarsi with 1–2 pairs of tenent hairs (Figs. 13.69C D) **EUPALOPSELLIDAE**

45b. Palptarsus shorter than palptibia; empodia on all tarsi with 3 pairs of tenent hairs (Figs. 13.69A, B) **MECOGNATHIDAE**

46a. Chambered peritremes embedded in dorsal surface of stylophore; coxae II narrowly to widely separated from coxae III **47**

46b. Peritremes linear on prodorsum, usually emergent: coxae II–III contiguous (Figs. 13.63D–F) **RAPHIGNATHIDAE**

47a. Leg tibiae and tarsi subequal in length; body shapes various **48**

47b. Leg tibiae about 3 times longer than respective tarsi; oval to subcircular mites with very long legs (Figs. 13.64A, B) **CAMEROBIIDAE**

48a. With 1 or 2 plates covering most of dorsum and bearing 15 or more pairs of setae, of which at least 6 pairs are much longer than idiosoma; palptibial claw peglike or absent **49**

48b. Soft bodied, with at most a small prodorsal shield and/or small sclerites around the setal bases; all 11–12 pairs of dorsal setae considerably shorter than idiosoma; palptibial claw usually well developed (Fig. 13.66) **CALIGONELLIDAE**

49a. With a single holodorsal shield; palptarsus elongate and tapering; palptibial claw absent (Figs. 13.65D, E) **XENOCALIGONELLIDIDAE**

49b. With 2 dorsal shields; palptarsus short and mound shaped; palptibial claw peglike (Figs. 13.65A–C) **DASYTHYREIDAE**

50a. Cheliceral stylophore fused to subcapitulum to form a gnathosomatic capsule; peritremes elaborated on dorsal surface of capsule; genu I with solenidion **51**

50b. Cheliceral stylophore separate from subcapitulum; peritremes at base of chelicerae or absent; genu I without solenidion **52**

51a. Body oval to rounded, occasionally elongate; palptibial claw present and well developed. Free living, predators, insect associates, or parasites of vertebrates (Fig. 13.77) **CHEYLETIDAE**

51b. Body elongate; palptibial claw absent. Parasites within the quills of birds (Figs. 13.78C, D) **SYRINGOPHILIDAE**

52a. Gnathosoma relatively well developed, chelicerae present, movable digit styletlike; bodies oval, flattened, or wormlike. Parasites on, in, or (rarely) under the skin of vertebrates **54**

52b. Gnathosoma highly reduced, chelicerae absent. Internal parasites of reptiles (rarely birds) or small mammals **53**

53a. Weakly sclerotized mites with striate opisthosomatic cuticle; gnathosoma reduced, but distinct; palpi with apical bifid claw; legs with lateral claws. Internal parasites in the tissues of shrews, moles, and mice (Figs. 13.80D, E) **EPIMYODICIDAE**

53b. Well-sclerotized mites without opisthosomatic striae; gnathosoma reduced to single-segmented palpi, each with a long, fanglike stylet apically; legs without distinct claws. Internal parasites in the tissues of reptiles and birds (Figs. 13.80A–C) **CLOACARIDAE**

54a. Body wormlike with numerous annuli; legs strongly telescoped, highly reduced, or modified into hooks. Parasites in the hair follicles, dermal glands, or subdermal regions of mammals (Figs. 13.79C–F) **DEMODICIDAE**

54b. Body oval, flattened; legs I–II well developed, legs III–IV well developed or reduced. Skin parasites of mammals, birds, and reptiles **55**

55a. All legs ending in paired lateral claws and padlike or bilobed empodia; femora I–IV each with a strong ventral spur. Parasites on the skin of mammals (Figs. 13.78A, B) **PSORERGATIDAE**

55b. Legs I–II with empodial tenent hairs and with or without lateral claws, legs III–IV without lateral claws or legs IV absent; femora I–IV without strong ventral spurs. Skin parasites under the feathers of birds or under the scales of reptiles (Figs. 13.79A, B) **HARPIRHYNCHIDAE**

56a. Cheliceral bases fused into a subconical stylophore that is not consolidated with the subcapitulum; palpi pronounced, 3-segmented, extending well beyond other gnathosomatic extremities, and with well-developed apical claw. Female genital opening large, flanked by conspicuous pair of genital valves, and functioning in oviposition **57**

56b. Cheliceral stylophore consolidated with subcapitulum to form a capsulate gnathosoma; palpi reduced, usually with no more than 2 readily distinguishable nonarticulating segments, extending to or slightly beyond other gnathosomatic extremities, and with reduced or vestigial apical claw. Female genital opening usually small and not flanked by genital valves, usually functioning in insemination but rarely in oviposition **58**

57a. Prodorsum of female and male with a pair of stigmata and associated tracheae, and a pair of capitate bothridial sensilla laterally. Tarsi II–IV with paired claws and a stalked empodium, tarsus I with paired claws and no empodium. Gnathosoma with 2 pairs of dorsal setae. Free living or associated with passalid beetles (Fig. 13.81) **Superfamily Tarsocheyloidea, TARSOCHEYLIDAE**

57b. Prodorsum of female only with a pair of stigmata and associated tracheae laterally and a pair of ampulliform organs posteromedially; male prodorsum lacking stigmata and tracheae, and with a pair of normal setae instead of ampulliform organs. Tarsi II–IV without claws but with large discoid empodium, tarsus I without claws or empodium. Gnathosoma without dorsal setae. Subelytral associates of passalid beetles (Fig. 13.82) **Superfamily Heterocheyloidea, HETEROCHEYLIDAE**

58a. Legs IV of female with separate femur and genu, with total of 5–12 setae on tibia and tarsus, and often with paired claws and empodium; legs IV of male 5-segmented, with symmetrically or asymmetrically paired claws or single or vestigial claw **59**

58b. Legs IV of female, when present, typically with fused femur and genu, with total of 2–3 setae on tibia and tarsus, and without claws and empodium; legs IV of male, when present, usually 4-segmented, with a single sessile or vestigial claw **Superfamily Tarsonemoidea—70**

59a. Tarsus I with paired claws; trochanters of legs I, II, IV, and usually III, lacking setae. Females lacking the pair of stigmata and associated tracheae anterolaterally on prodorsum; female genital aperture exposed, either on genital plate and flanked by paired aggenital plates or flanked by paired valves bearing genital setae. Male legs IV with pretarsus, empodium, and paired or reduced claws like those on legs II–III **Superfamily Dolichocyboidea—60**

59b. Tarsus I with single claw or without claws; trochanters of legs I–IV usually each with a seta. Females with the pair of stigmata and associated tracheae anterolaterally on prodorsum; female genital aperture covered by posterior margin of consolidated aggenital plate. Male legs IV with claws variously shaped but often different in form from those on legs II–III **61**

60a. Prodorsum of females with a pair of expansive winglike processes that overlap with similar expansions of tergite C. Female genital aperture elongate, flanked by 4 pairs of readily visible genital setae. Tarsi II–IV lacking empodium between claws. Trochanter III with a seta; setation of genua I-II-III-IV, 5-4-3-1. Associates of carabid beetles (Fig. 13.84) **CROTALOMORPHIDAE**

60b. Prodorsum and tergite C of females normal in form, without winglike expansions. Female genital aperture small, flanked by 1–3 pairs of minute or vestigial genital setae, or these absent. Tarsi II–IV with smooth flexible empodium between claws. Trochanter III without a seta; setation of genua I-II-III-IV, maximally 2-1-1-1. Free living or associated with insects (Fig. 13.83) **DOLICHOCYBIDAE**

61a. Trochanter IV quadrangular, generally longer than wide, different in form from subtriangular trochanter III. Female pharynx divided into 2 or 3 muscular sections. Male gnathosoma reduced, tubiform, lacking chelicerae and palpi. Larval instar active **62**

61b. Trochanter IV subtriangular, similar in form to trochanter III, or, if compressed and subquadrangular, then as wide as or wider than long. Female pharynx large, muscular, undivided. Male gnathosoma not reduced to tubiform appendage, chelicerae and palpi present. Larval instar usually inactive, calyptostatic **65**

62a. Prodorsum of females not covered posteriorly by tergite C and usually with 3 pairs of setiform setae (*v1* and usually *v2* present). Coxisternal plates I–II together with 4–6 pairs of setae. Femur I usually with 4 setae. Male aedeagus usually relatively short, stiff **Superfamily Pygmephoroidea—63**

62b. Prodorsum of females usually covered to greater or lesser extent by tergite C and with 1–2 pairs of setiform setae (*v1* and sometimes *v2* absent). Coxisternal plates I–II together with maximum of 4 pairs of setae. Femur I with 3 setae. Male aedeagus relatively long, flexible, curved, or looped **Superfamily Scutacaroidea—64**

63a. Femur I of females with all setae slender, attenuate (seta *d* setiform). Coxisternal plates I–II together usually with 4 pairs of setae. Prodorsum of females with stigmata often extended into longitudinal grooves (peritremes). Plant, fungus associates (Fig. 13.86) **SITEROPTIDAE**

63b. Femur I of females with seta *d* thickened, rigid, and bladelike, rodlike, or hooked. Coxisternal plates I–II together with 4–6 pairs of setae. Prodorsum of females with stigmata round or oval, not extended into elongate grooves. Free living or associated with fungi, insects, and animal nests (Fig. 13.87) **PYGMEPHORIDAE**

64a. Tergite C of females expansive, covering prodorsum completely, forming a roof over gnathosoma; prodorsum usually with 2 pairs of short setiform setae (covered, often difficult to discern). Idiosoma broadly oval or rounded, compact, with distance between insertions of legs II–III usually similar to that between legs III–IV. Femur I with seta *d* short, stout, usually branched. Male aedeagal tube apparently annulated or striated. Free living or associated with insects and animal nests (Figs. 13.88, 13.89A, B) **SCUTACARIDAE**

64b. Tergite C of females covering posterior region of prodorsum to variable extent, not covering entire prodorsum and gnathosoma; prodorsum with usually 1 pair of setiform setae, a 2nd pair (*v2*) rudimentary or absent. Idiosoma oval, distance between insertions of legs II and III at least twice that between legs III and IV. Femur I with seta *d* attenuated, smooth or barbed. Male aedeagal tube apparently smooth. Free living or associated with insects and animal nests (Fig. 13.90) **MICRODISPIDAE**

65a. Coxisternal plates I–II together with usually 6 pairs of setae. Femur I and genu I each with 5 setae. Female gnathosomatic capsule with palpi prominent, extending slightly beyond apex of stylophore. Males with incompletely formed caudal genital capsule, with exposed setigenous processes **Superfamily Trochometridioidea—66**

65b. Coxisternal plates I–II together with maximally 4 pairs of setae. Femur I and genu I each with 3 or usually 4 setae. Female gnathosomatic capsule with palpi usually somewhat reduced or poorly delineated, usually not extending beyond apex of stylophore. Males with fully formed caudal or ventrocaudal genital capsule, with retractable setigenous processes **Superfamily Pyemotoidea—67**

66a. Prodorsum of females with anterior pair of vertical setae inserted posterior to stigmata; aggenital plate with 1 pair of setae; tergite EF with setae *e* and *f* aligned transversely; caudal plate PS rounded posteriorly, without a caudal protrusion. Male gnathosoma reduced, lacking chelicerae and delineated palpi; male leg I lacking claws. Associated with bees (Figs. 13.85A–F) **TROCHOMETRIDIIDAE**

66b. Prodorsum of females with anterior pair of vertical setae inserted anterior or medial to stigmata; aggenital plate with 2 or 3 pairs of setae; tergite EF with setae *e* and *f* aligned longitudinally; caudal plate PS with a caudally protruding process. Males unknown. Associated with beetles (Figs. 13.85G–J) **ATHYREACARIDAE (includes the family Bembidiacaridae of Khaustov 2000)**

67a. Females with gnathosomatic capsule large, clearly wider than long. Female coxisternal plates I with a pair of discoid adhesive structures. Female legs I lacking claws, smaller than legs II–IV; legs IV with tibia and tarsus at least partially fused, and with reduced claws. Associated with carabid beetles (Figs. 13.92F–H) **CARABOACARIDAE**

67b. Females with gnathosomatic capsule of normal or reduced size, no wider than long. Female coxisternal plates I lacking discoid structures. Female legs I with or without a single claw, similar in size to legs II–IV; legs IV with tibia and tarsus separate, and with paired claws if these present on legs II–III **68**

68a. Prodorsum of females lacking bothridia and bothridial setae. Coxisternal plates I–II together with maximally 2 pairs of setae. Gnathosoma somewhat hypognathous, enclosed and partly hidden in propodosomatic camerostome; palpi not differentiated from lateral walls of gnathosoma. Male genital capsule ventrocaudal in position. Parasitoids of insect eggs (Figs. 13.92A–E) **ACAROPHENACIDAE**

68b. Prodorsum of females with pair of bothridia and capitate bothridial setae. Coxisternal plates I and II together with 4 pairs of setae. Gnathosoma prognathous, well exposed from propodosoma; palpi delineated from lateral walls of gnathosoma. Male genital capsule caudal in position **69**

69a. Female coxisternal plates III–IV separated from each other medially by soft cuticle that bears a separate triangular sternal plate. Male legs IV differentiated from legs II–III in being somewhat stouter and ending with a single sessile claw. Parasitoids of immature insects (Figs. 13.91A–E) **PYEMOTIDAE**

69b. Female coxisternal plates III–IV united medially, effacing separate medial plate. Male legs III differentiated from legs II and IV in being somewhat stouter and having a large, spurlike tarsal process. Associated with insects (Figs. 13.91F–I) **RESINACARIDAE**

70a. Females always with 4 pairs of legs; legs IV 3-segmented, usually much more slender than legs II–III. Males always with 4 pairs of legs; legs IV 3- or 4-segmented, always inserted ventrally; male genital capsule caudal in position. Free living or plant, fungus, or insect associates (Figs. 13.89C, D, 13.93) **TARSONEMIDAE**

70b. Females typically with 1–3 pairs of legs; legs IV, if present, 5-segmented and more slender than legs II–III. Males with 3 or 4 pairs of legs; legs IV, if present, 5-segmented and inserted ventrally or reduced and inserted dorsally; male genital capsule caudal or dorsal in position. Parasites of insects (Fig. 13.94) **PODAPOLIPIDAE**

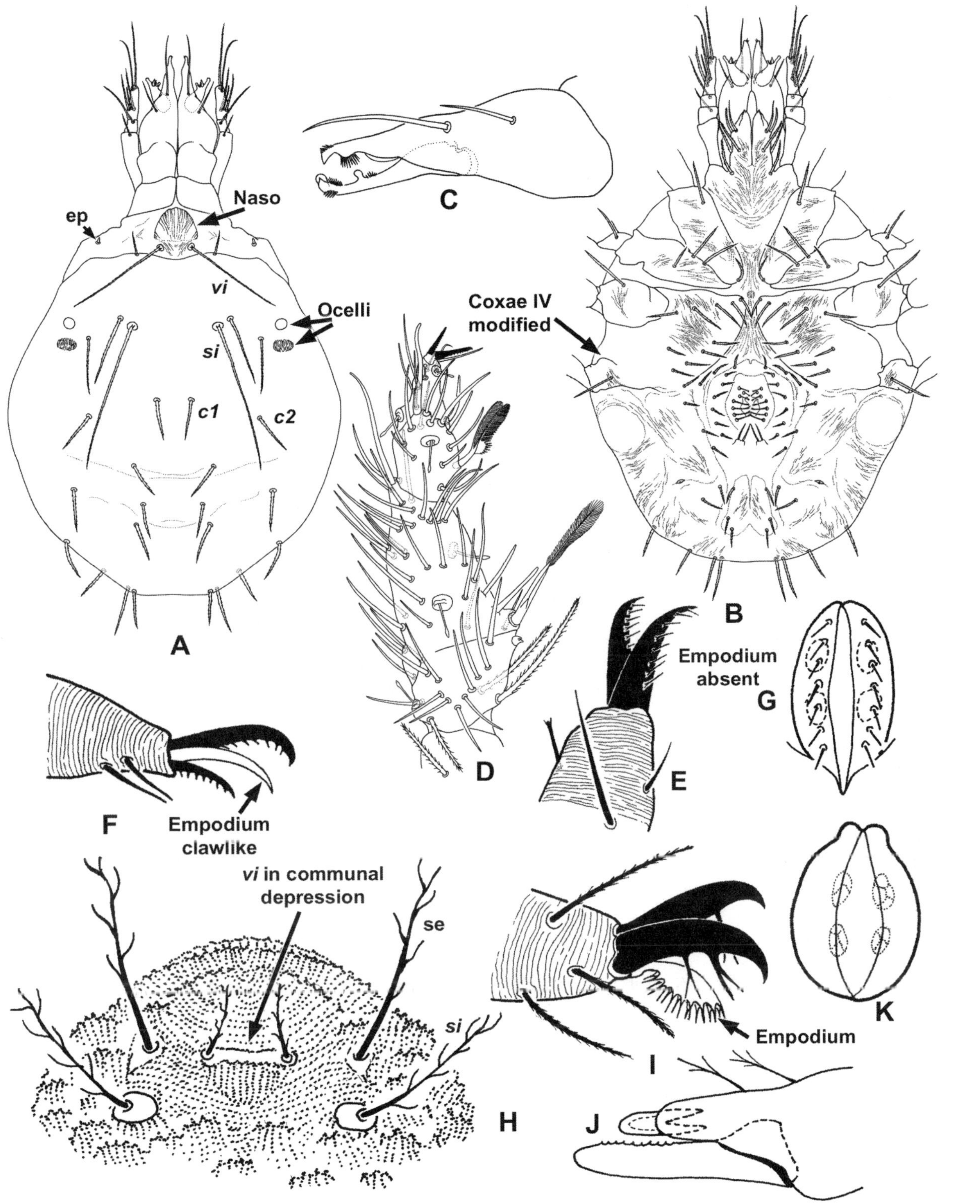

Fig. 13.1. SPHAEROLICHIDAE. *Sphaerolichus* sp. (Queensland, Australia): **A**, dorsum; **B**, ventral aspect; **C**, lateral view of chelicera; **D**, distal segments of leg I. *S. barbatus* Grandjean (France): **E**, pretarsus I claws; **F**, pretarsus III claws and empodium; **G**, genital region of female.

LORDALYCIDAE. *Hybalicus peraltus* (Grandjean): **H**, propodosoma of female; **I**, pretarsus I; **J**, cheliceral extremities; **K**, genital region of female.

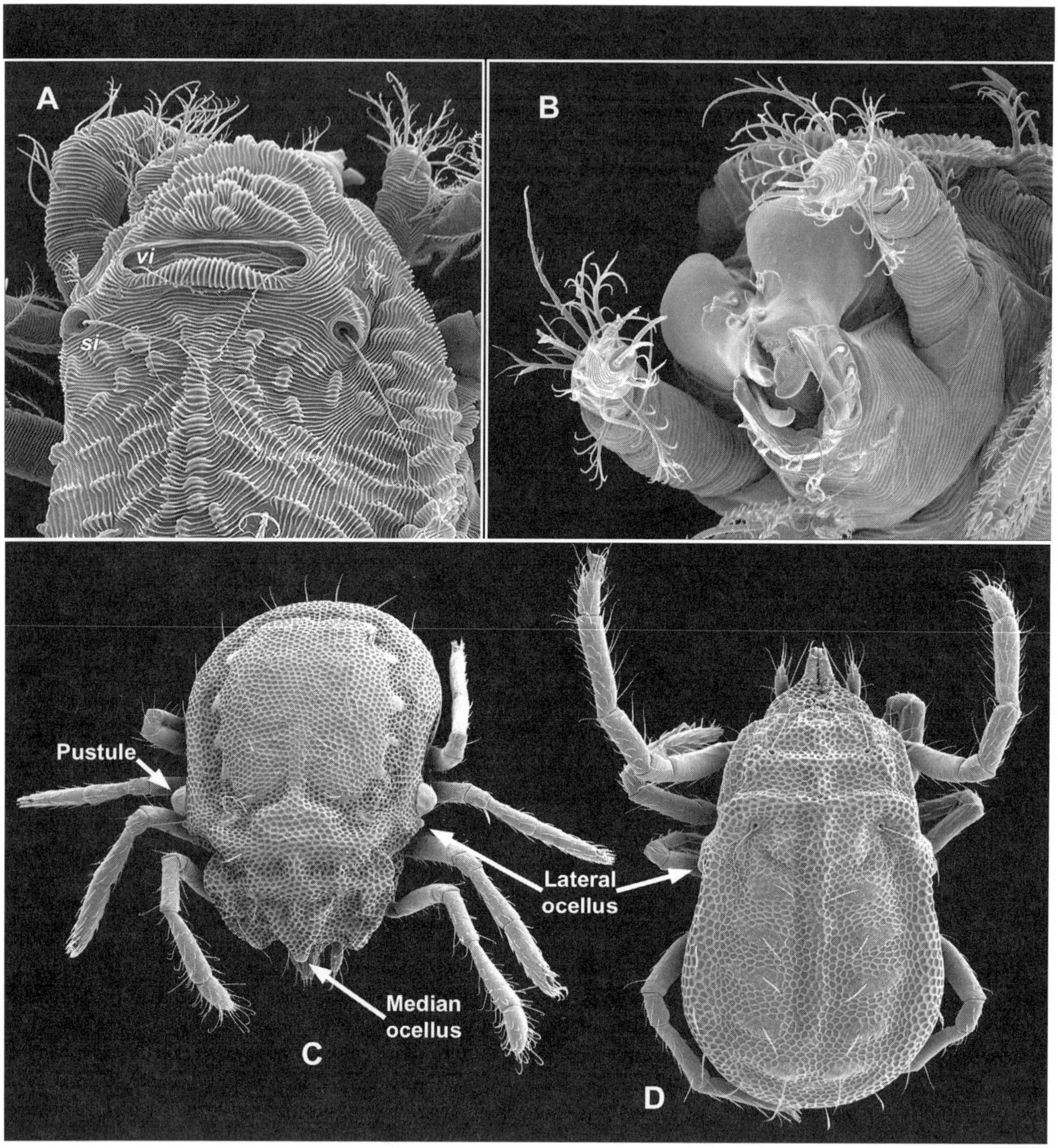

Fig. 13.2. LORDALYCIDAE. *Hybalicus* sp. (Queensland, Australia): **A**, anterodorsal region showing bothridial sensilla *vi* in communal depression; **B**, anteroventral aspect of capitulum.

LABIDOSTOMATIDAE. *Labidostoma* sp. (Queensland, Australia): **C**, anterodorsal aspect. *Sellnickiella* sp. (Queensland, Australia): **D**, dorsal aspect.

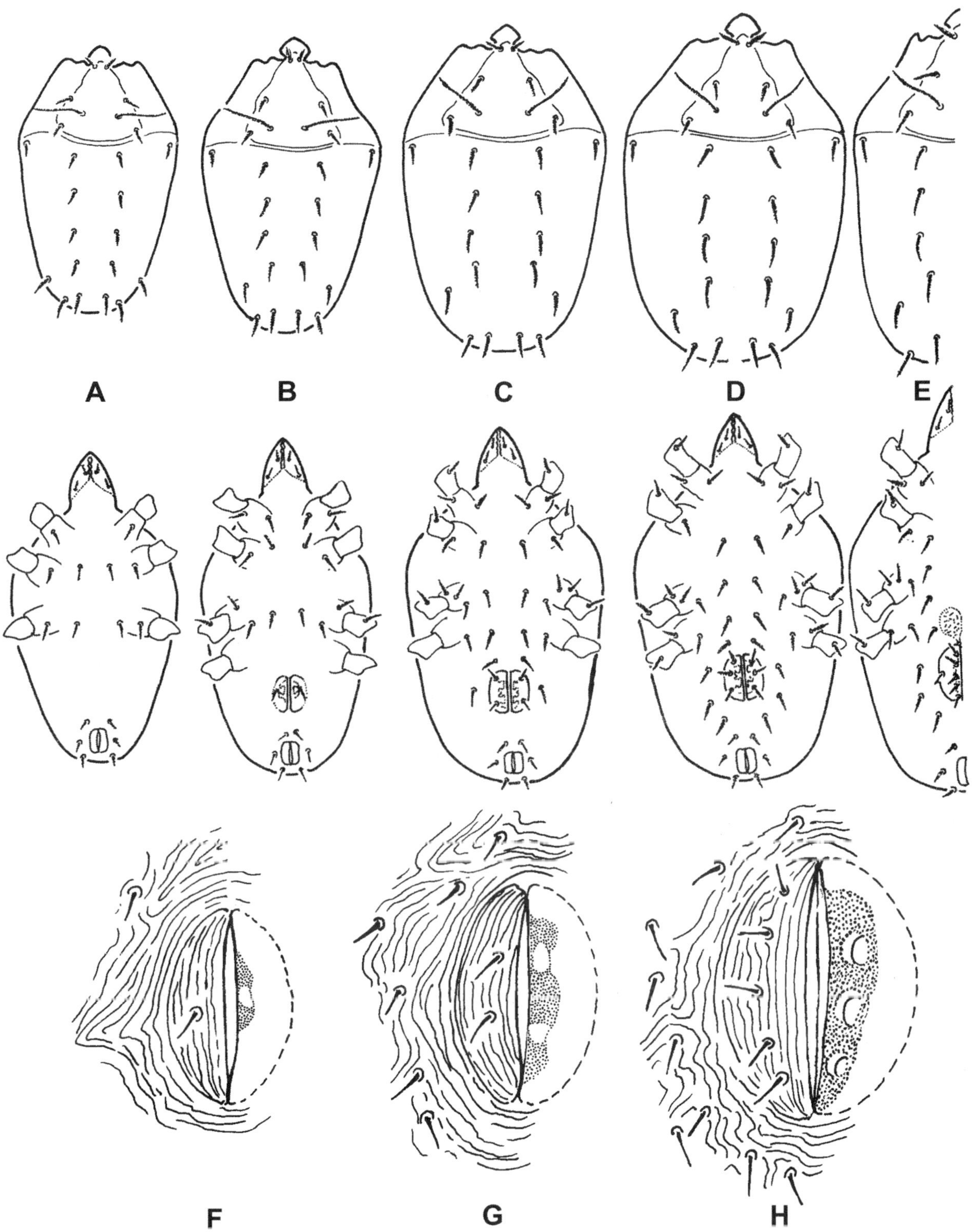

Fig. 13.3. PENTHALODIDAE. *Stereotydeus mollis* Womersley and Strandtmann, dorsal and ventral aspects during development: **A**, larva; **B**, protonymph; **C**, deutonymph; **D**, tritonymph; **E**, adult male.

BDELLIDAE. *Bdella semiscutata* Sig Thor, genital field showing additions of setae and genital papillae during development: **F**, protonymph; **G**, deutonymph; **H**, tritonymph (after Grandjean 1938b).

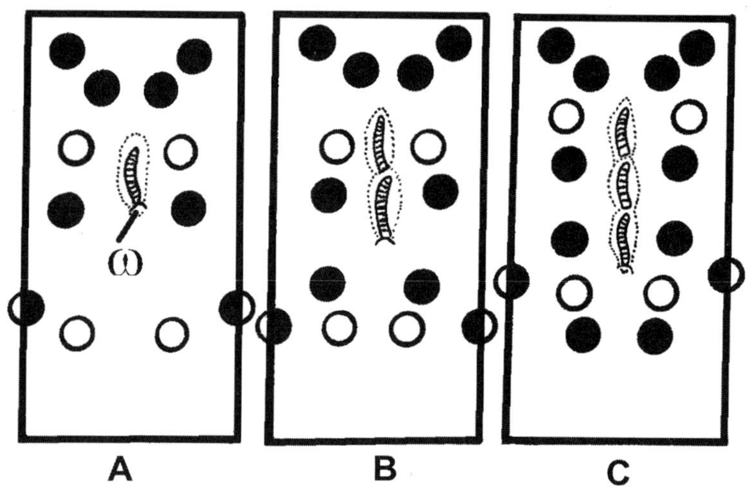

Fig. 13.4. PENTHALODIDAE. *Stereotydeus mollis* Womersley and Strandtmann, tarsus I showing ontogenetic additions of setae and solenidia (solid circles = ventral setae): **A**, protonymph; **B**, deutonymph; **C**, tritonymph (after Pittard 1971).

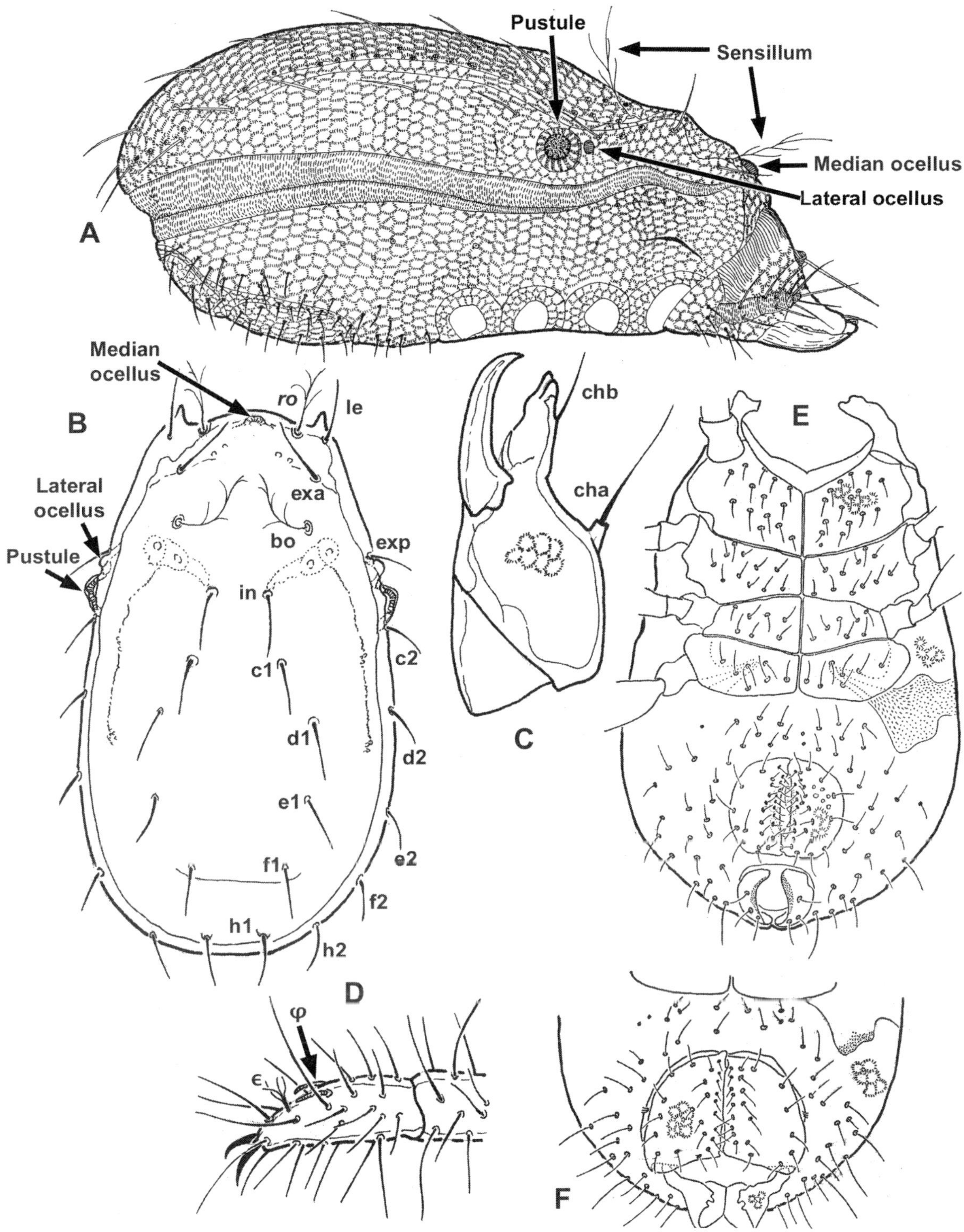

Fig. 13.5. LABIDOSTOMATIDAE. *Labidostoma nepalense* Feider and Vasilu (Nepal): **A**, lateral aspect of female (from Feider and Vasilu 1968). *L.* (*Nicoletiella*) *pacifica* (Ewing) (Oregon, USA): **B**, dorsum of female; **C**, chelicera of female; **D**, tarsus I of female. *Eunicolina tuberculata* Berlese (Europe) (from Vistorin 1979): **E**, venter of male; **F**, opisthogaster of female.

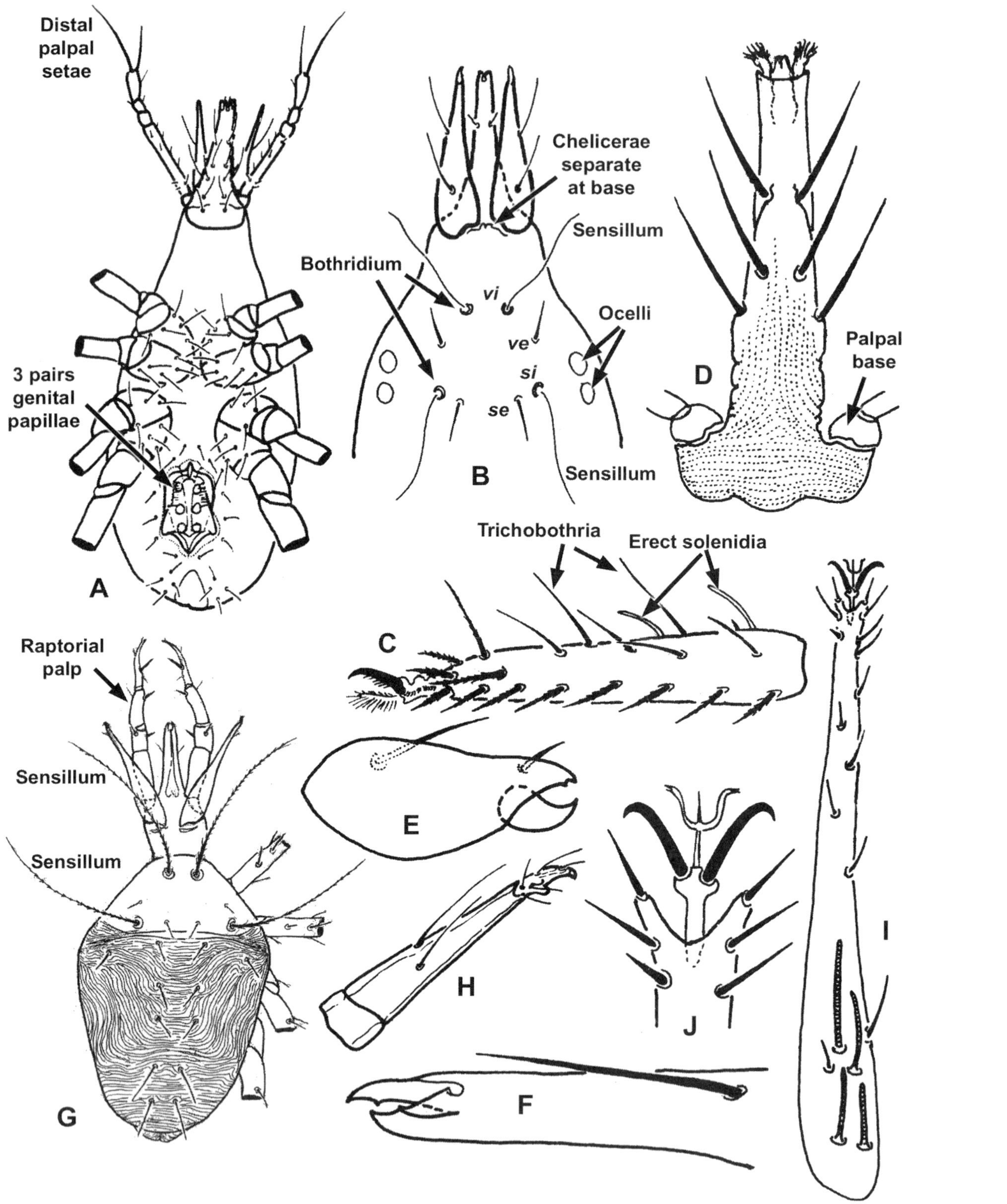

Fig. 13.6. BDELLIDAE. *Bdella* sp. (Oregon, USA): **A**, venter of male; **B**, prodorsum; **C**, tarsus I. *B. longicornis* (L.) (Japan): **D**, venter of gnathosoma (after Shiba 1969). *Cyta latirostris* (Hermann): **E**, chelicera. *Bdellodes longirostris* (Hermann): **F**, chelicera (after Shiba 1969).

CUNAXIDAE. *Cunaxa* sp.: **G**, dorsum. *Cunaxoides* sp. (Central Africa): **H**, palp. *Cunaxa setirostris* (Hermann): **I**, tarsus I (after Smiley 1975); **J**, detail of tarsus I.

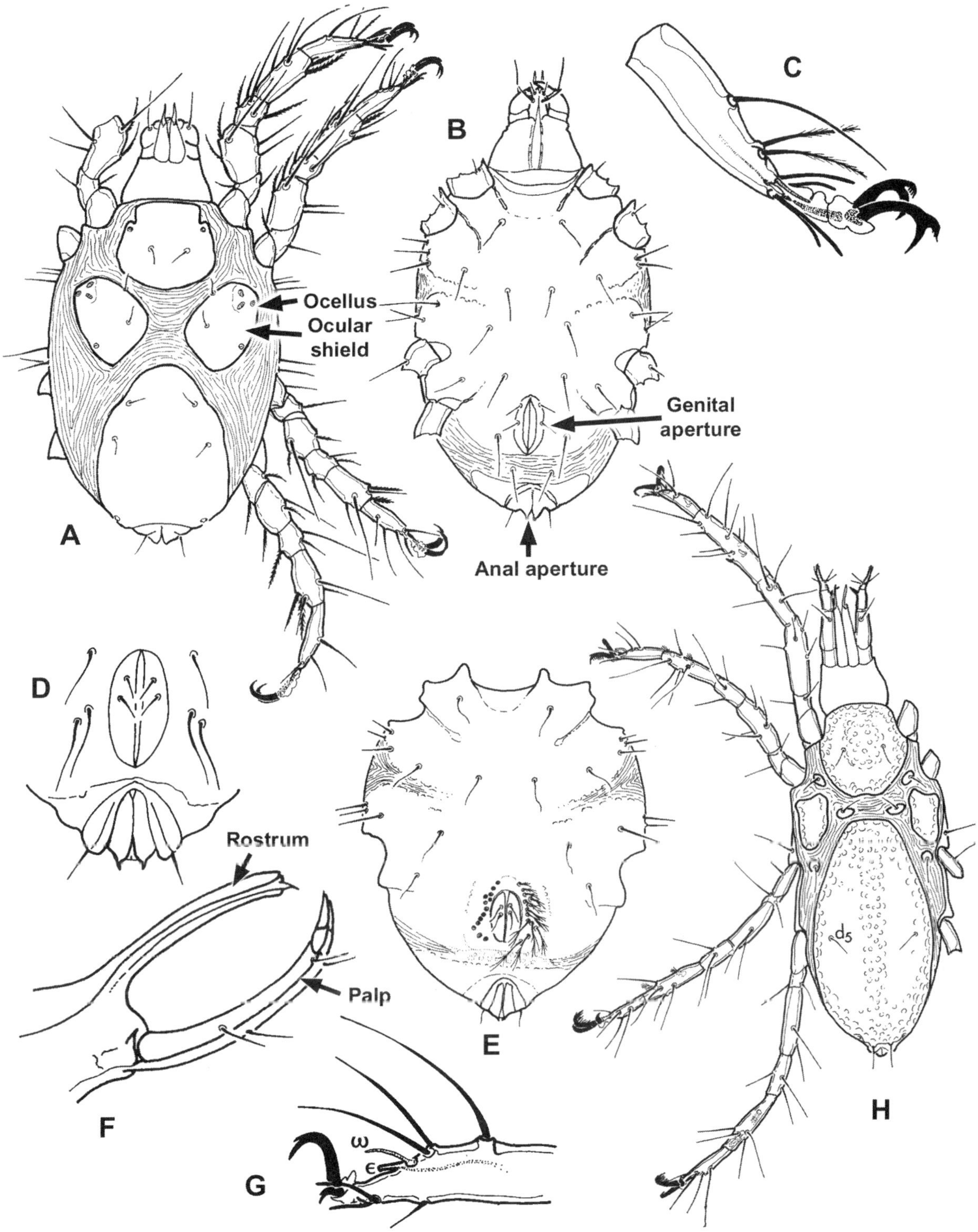

Fig. 13.7. HALACARIDAE. *Rhombognathus leurodactylus* Krantz (Oregon, USA): **A**, dorsum of female; **B**, venter of female. *R. reticulatus* Krantz (Oregon, USA): **C**, tarsus I of male; **D**, genito-anal region of female; **E**, venter of male. *Lohmannella* sp.: **F**, lateral aspect of gnathosoma. *Thalassarachna psammophila* Krantz (Oregon, USA): **G**, terminal portion of tarsus I of male showing famulus (ε) and solenidion (ω); **H**, dorsum of female.

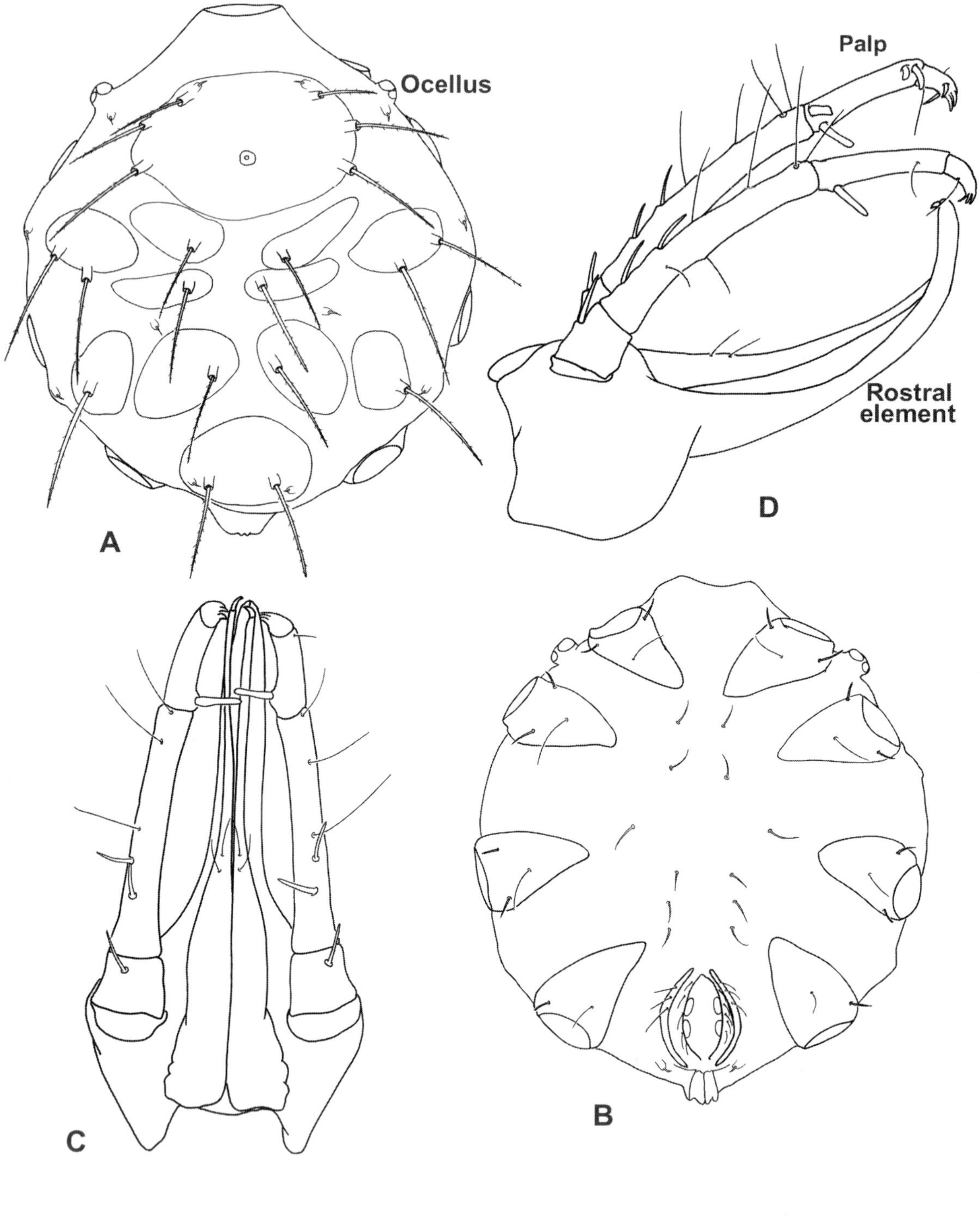

Fig. 13.8. PEZIDAE. *Peza ops* Harvey (Australia): **A**, dorsum of female; **B**, venter of female; **C**, dorsal aspect of female gnathosoma; **D**, lateral aspect of female gnathosoma (all from Harvey 1990a).

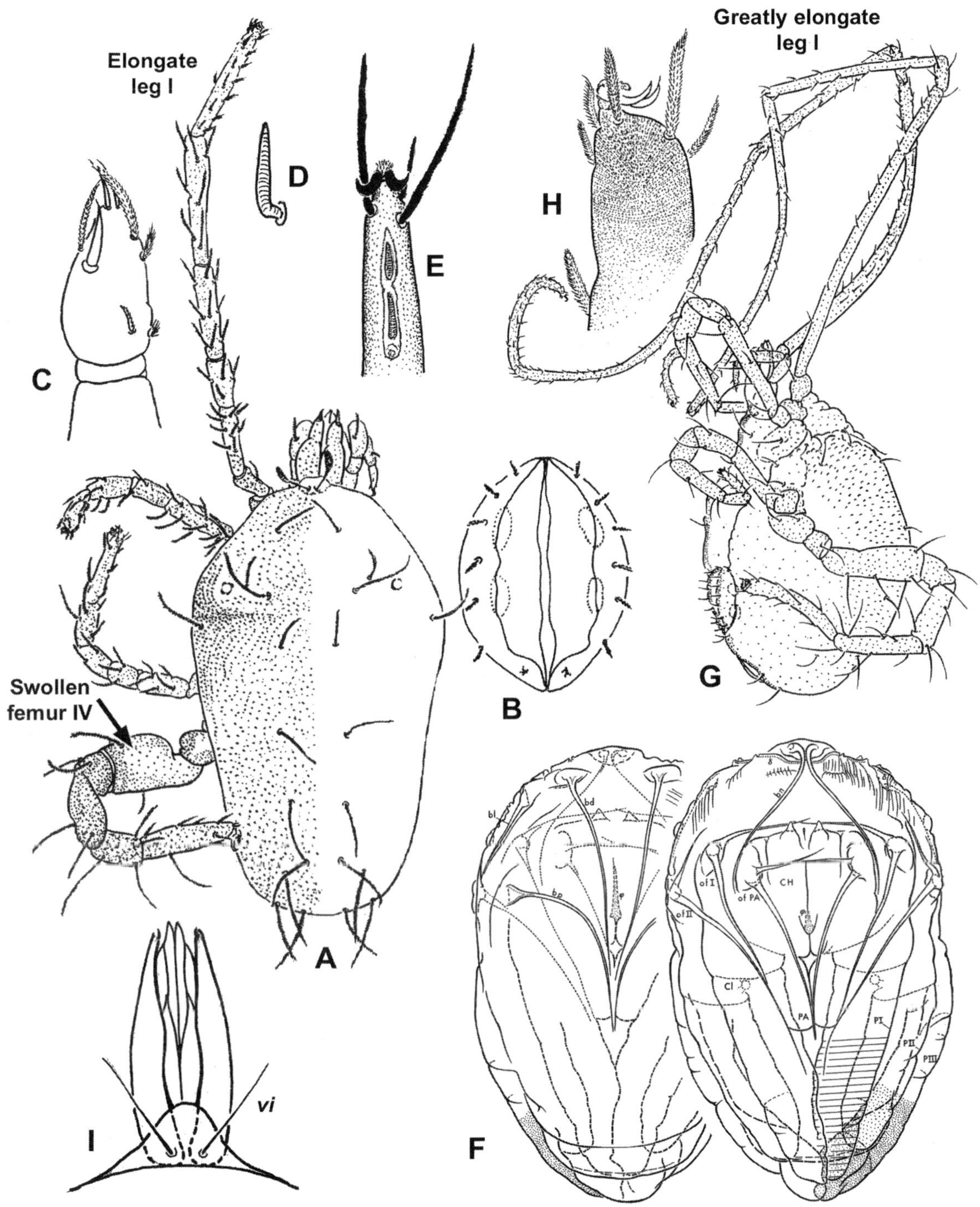

Fig. 13.9. EUPODIDAE. *Cocceupodes communis* Shiba (Japan): **A**, dorsum of female; **B**, genital region of female; **C**, detail of palp; **D**, tarsal rhagidial organ (i.e., recumbent solenidion) (from Shiba 1969, with permission, National Science Museum, Tokyo). *Eupodes longisetatus* Strandtmann (Japan): **E**, tarsus I rhagidial organs. *E. strandtmanni* Coineau (France): **F**, dorsal and ventral aspects of prelarva (from Coineau 1976b). *Linopodes iwatensis* Morikawa (Japan): **G**, lateral aspect of female; **H**, terminus of tarsus I (from Shiba 1969, with permission, National Science Museum, Tokyo). *Linopodes* sp. (Oregon, USA): **I**, dorsum of gnathosoma and naso bearing setae *vi*.

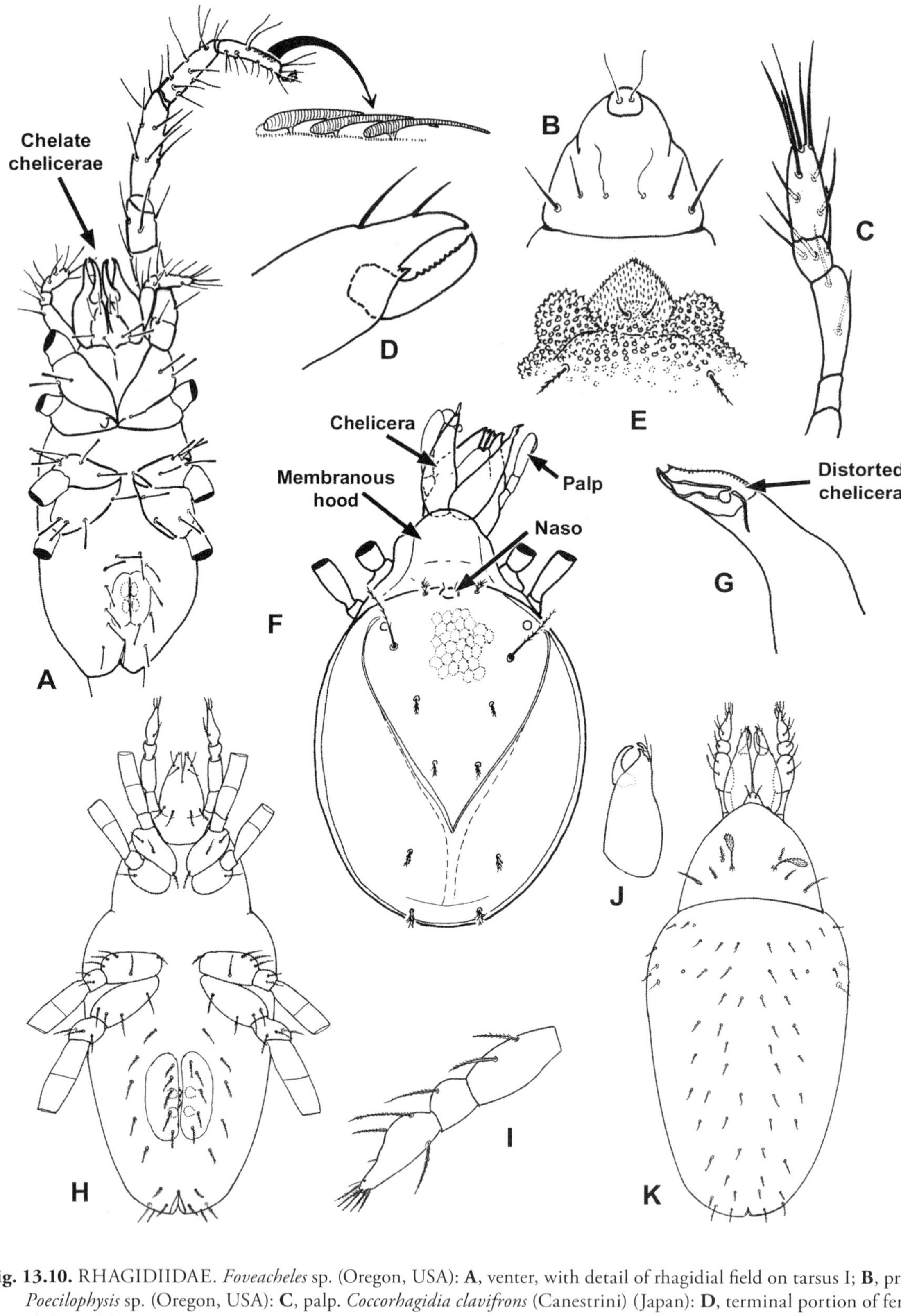

Fig. 13.10. RHAGIDIIDAE. *Foveacheles* sp. (Oregon, USA): **A**, venter, with detail of rhagidial field on tarsus I; **B**, prodorsum. *Poecilophysis* sp. (Oregon, USA): **C**, palp. *Coccorhagidia clavifrons* (Canestrini) (Japan): **D**, terminal portion of female chelicera (after Shiba 1969).

PENTHALODIDAE. *Stereotydeus longipes* Strandtmann: **E**, anterior aspect of male propodosoma (after Strandtmann 1970). *Penthalodes* sp. (Oregon, USA), female: **F**, dorsum; **G**, chelicera.

STRANDTMANNIIDAE. *Pilorhagidia hirsuta* Strandtmann and Goff: **H**, female venter; **I**, palp; **J**, chelicera; **K**, dorsum (from Strandtmann and Goff 1968, with permission, Bishop Museum Press, Honolulu).

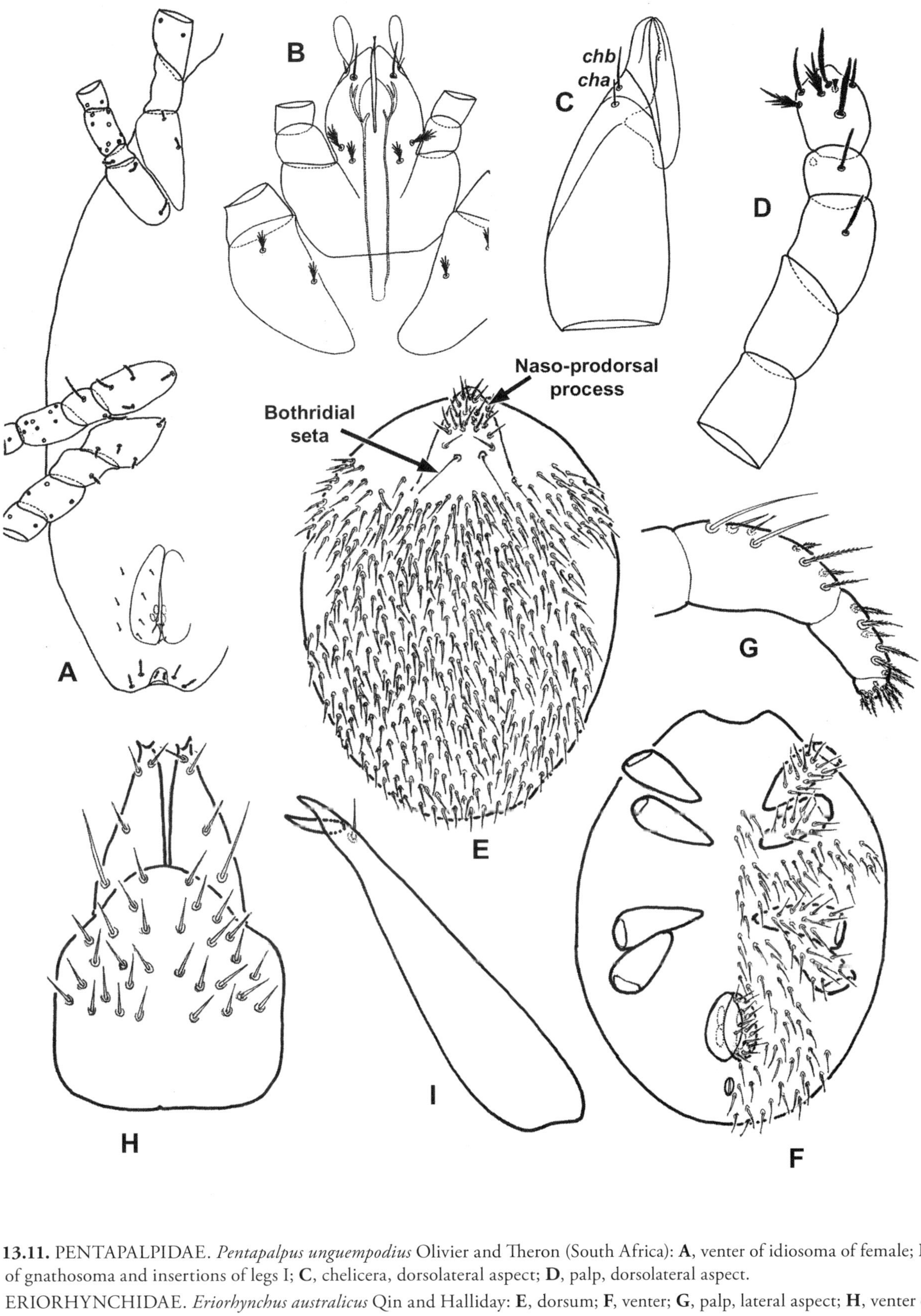

Fig. 13.11. PENTAPALPIDAE. *Pentapalpus unguempodius* Olivier and Theron (South Africa): **A**, venter of idiosoma of female; **B**, venter of gnathosoma and insertions of legs I; **C**, chelicera, dorsolateral aspect; **D**, palp, dorsolateral aspect.

ERIORHYNCHIDAE. *Eriorhynchus australicus* Qin and Halliday: **E**, dorsum; **F**, venter; **G**, palp, lateral aspect; **H**, venter of gnathosoma; **I**, chelicera.

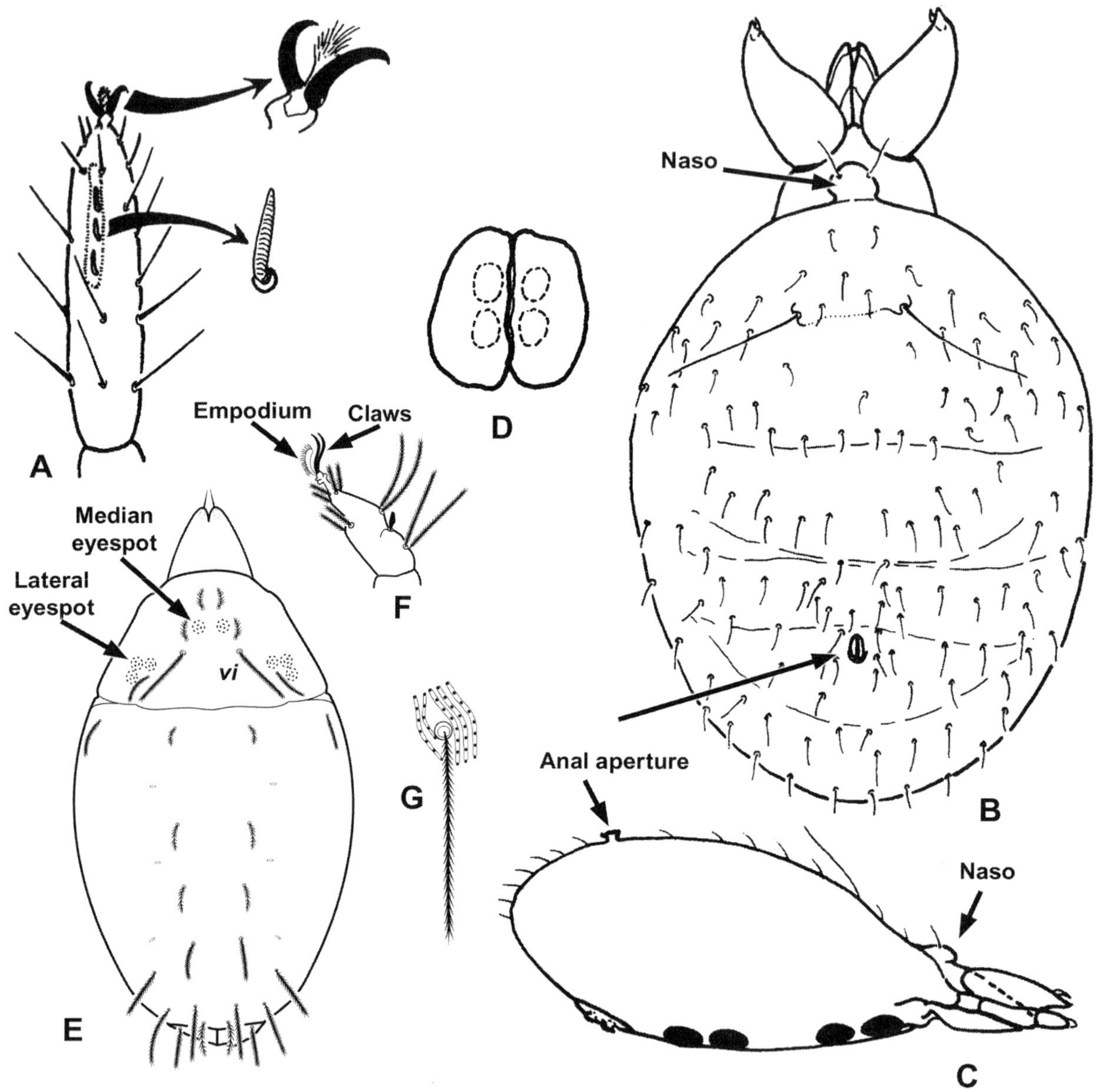

Fig. 13.12. PENTHALEIDAE. *Penthaleus major* (Dugès) (Oregon, USA): **A**, tarsus I with details of solenidion and pretarsus; **B**, dorsum of female; **C**, lateral aspect; **D**, genital region.

TRIOPHTYDEIDAE. *Triophtydeus immanis* Kuznetzov (Crimea, Ukraine): **E**, dorsum; **F**, tarsus and pretarsus I; **G**, bothridial seta *vi* with details of integument.

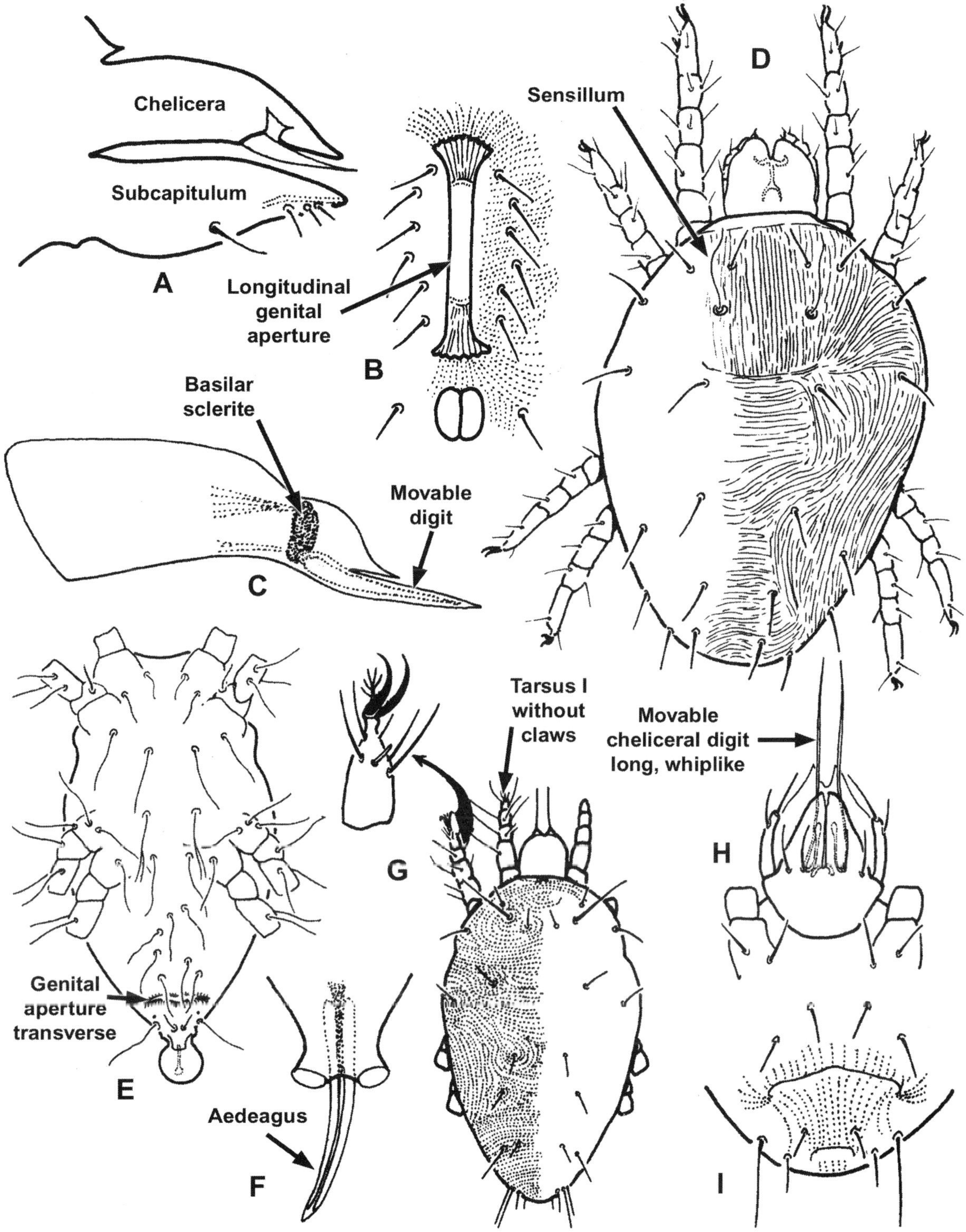

Fig. 13.13. TYDEIDAE. *Lorryia reticulata* (Oudemans) (Canada): **A**, lateral aspect of gnathosoma; **B**, genital region of female (after Marshall 1970). *L. montrealensis* Marshall (Canada): **C**, chelicera, antiaxial view. *Paralorryia* sp. (Oregon, USA): **D**, dorsum of female.

IOLINIDAE. *Proctotydeus schistocercae* Fain and Evans (Galápagos Islands): **E**, venter of female (after Fain and Evans 1966); **F**, male genitalia. *Iolina nana* Pritchard (Massachusetts, USA): **G**, dorsum of female with detail of tarsus II (after Pritchard 1956); **H**, venter of gnathosoma; **I**, genito-anal region of female.

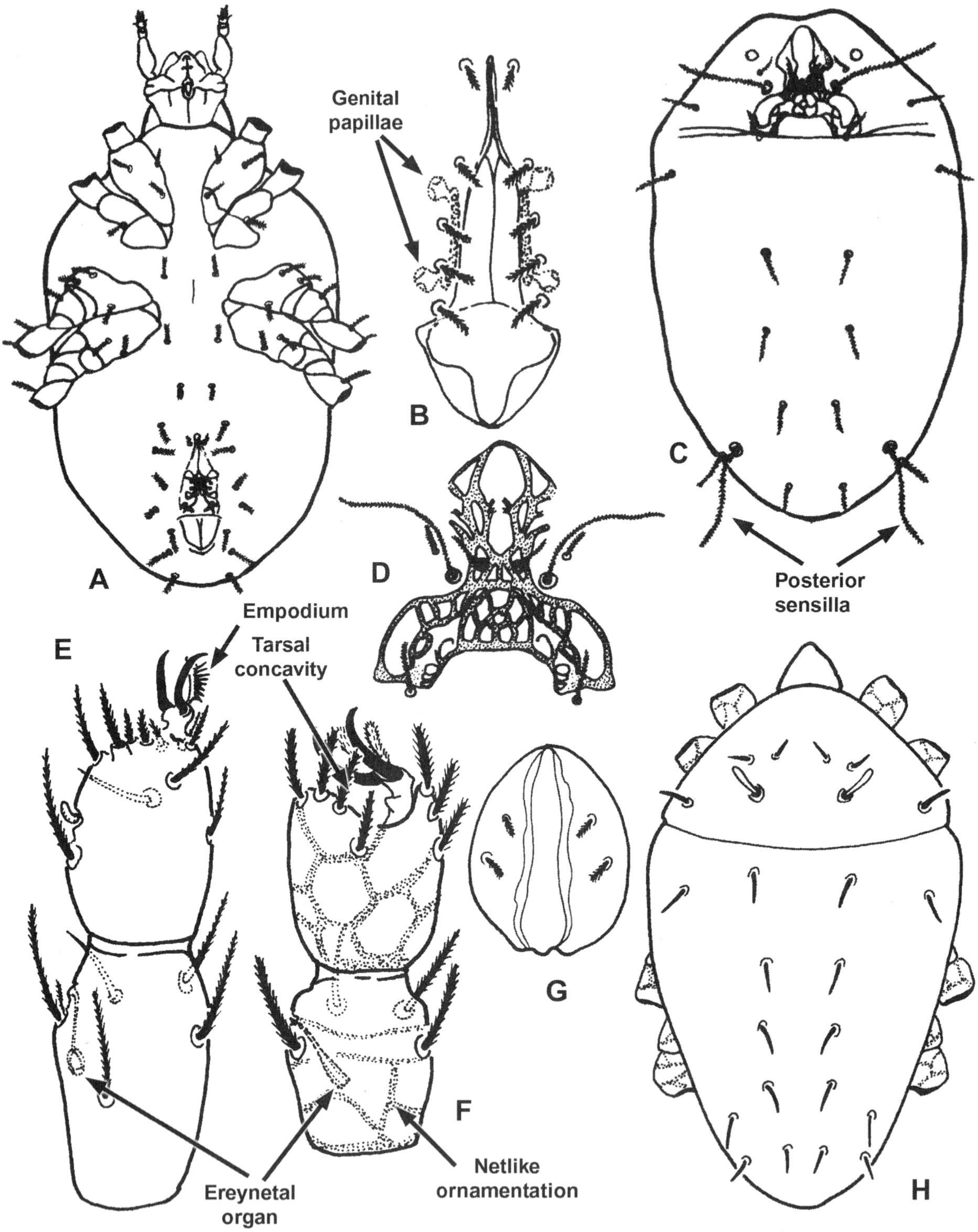

Fig. 13.14. EREYNETIDAE. *Ereynetes* sp. (Oregon, USA): **A**, venter of female; **B**, genito-anal region; **C**, dorsum of female; **D**, detail of propodosomatic ornamentation; **E**, tibia and tarsus I. *Boydaia sturni* Womersley: **F**, tibia and tarsus I. *Speleognathus* sp.: **G**, genital region; **H**, dorsum (diagrammatic).

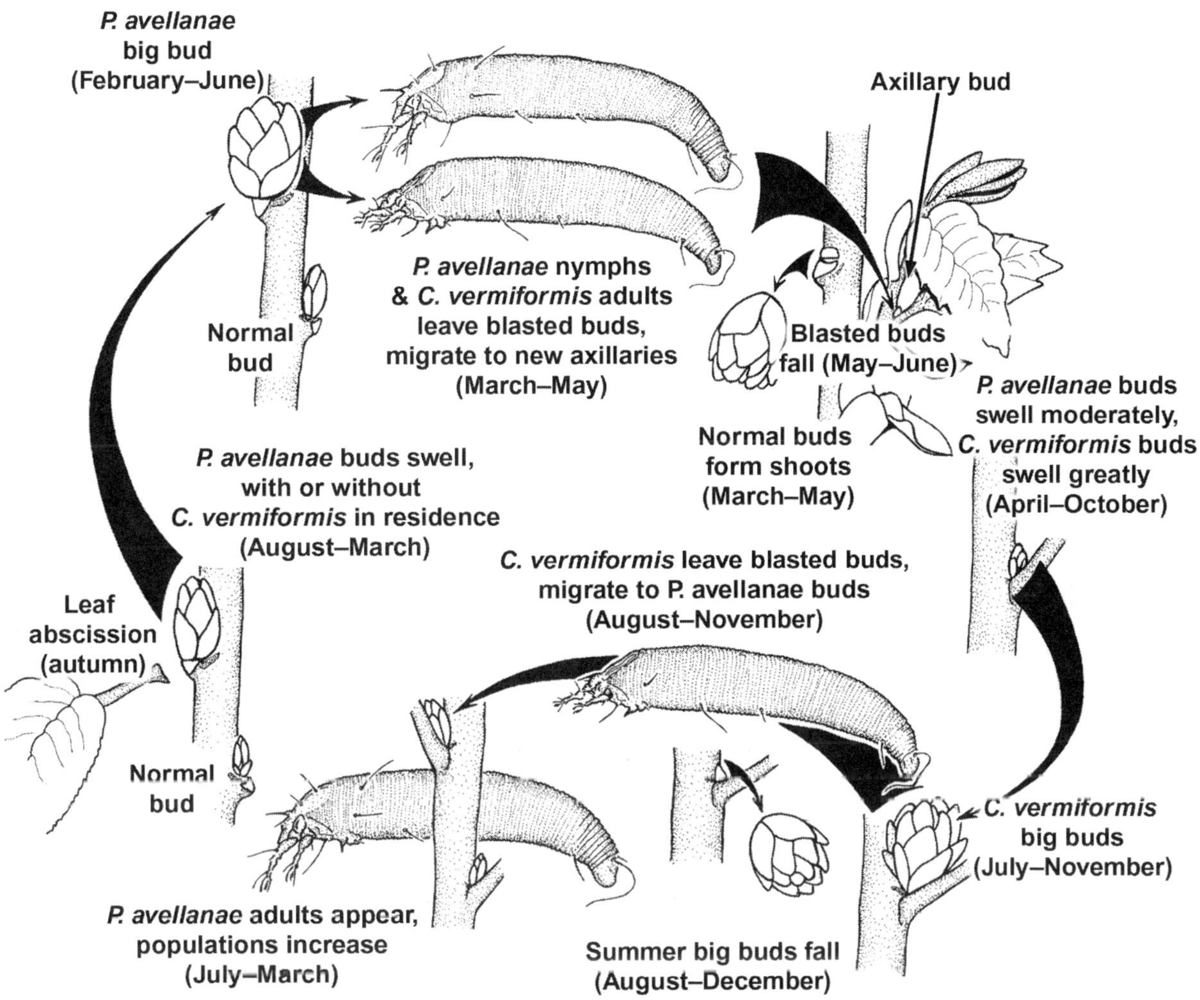

Fig. 13.15. Dynamics of the filbert bud mites *Phytoptus avellanae* Nalepa (PHYTOPTIDAE) and *Cecidophyopsis vermiformis* (Nalepa) (ERIOPHYIDAE) in commercial filbert trees (Oregon, USA) (after Krantz 1974).

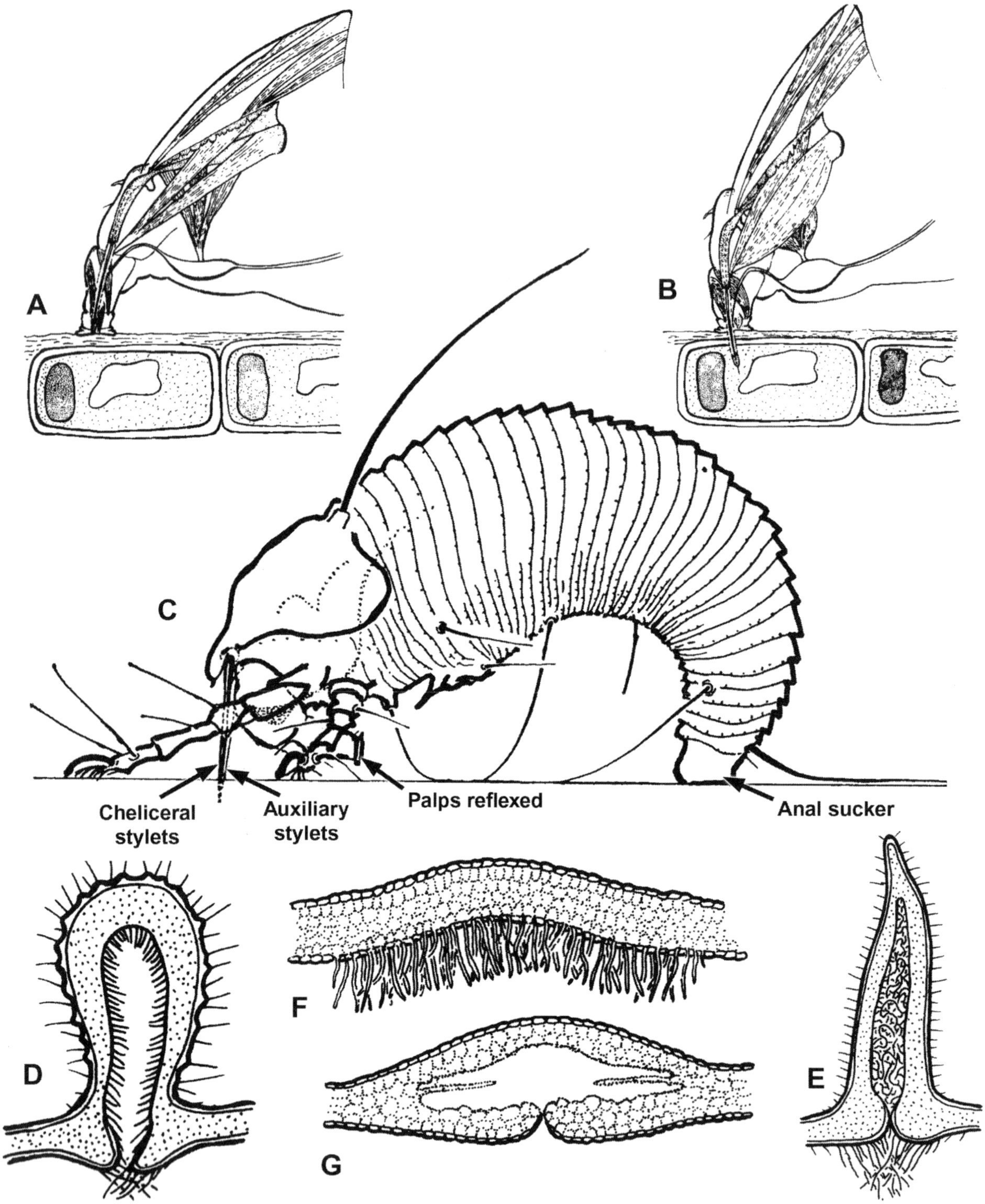

Fig. 13.16. Superfamily Eriophyoidea, diagrammatic feeding stance of an eriophyid mite, showing function of palpal suction pad and retractor muscles in forcing the auxiliary stylets (partly removed to show other stylets) into a plant cell: **A**, suction pad affixed to plant surface; **B**, intrinsic muscles of palp contracted, telescoping the palp and forcing stylets into plant cell, and pharyngeal dilatory muscles contracted, enlarging lumen of pharynx for feeding (after Nuzzaci and Alberti 1996); **C**, feeding stance of *Aculus comatus* (Nalepa), showing an alternative posturing with body strongly arched and palpi bent backward to initiate feeding.

Plant host leaf reactions to eriophyid feeding: **D**, **E**, closed gall types lined inside with papillar outgrowths and with tiny opening on ventral leaf surface; **F**, open erineal patch with papillar outgrowths on ventral leaf surface; **G**, mesophyll blister.

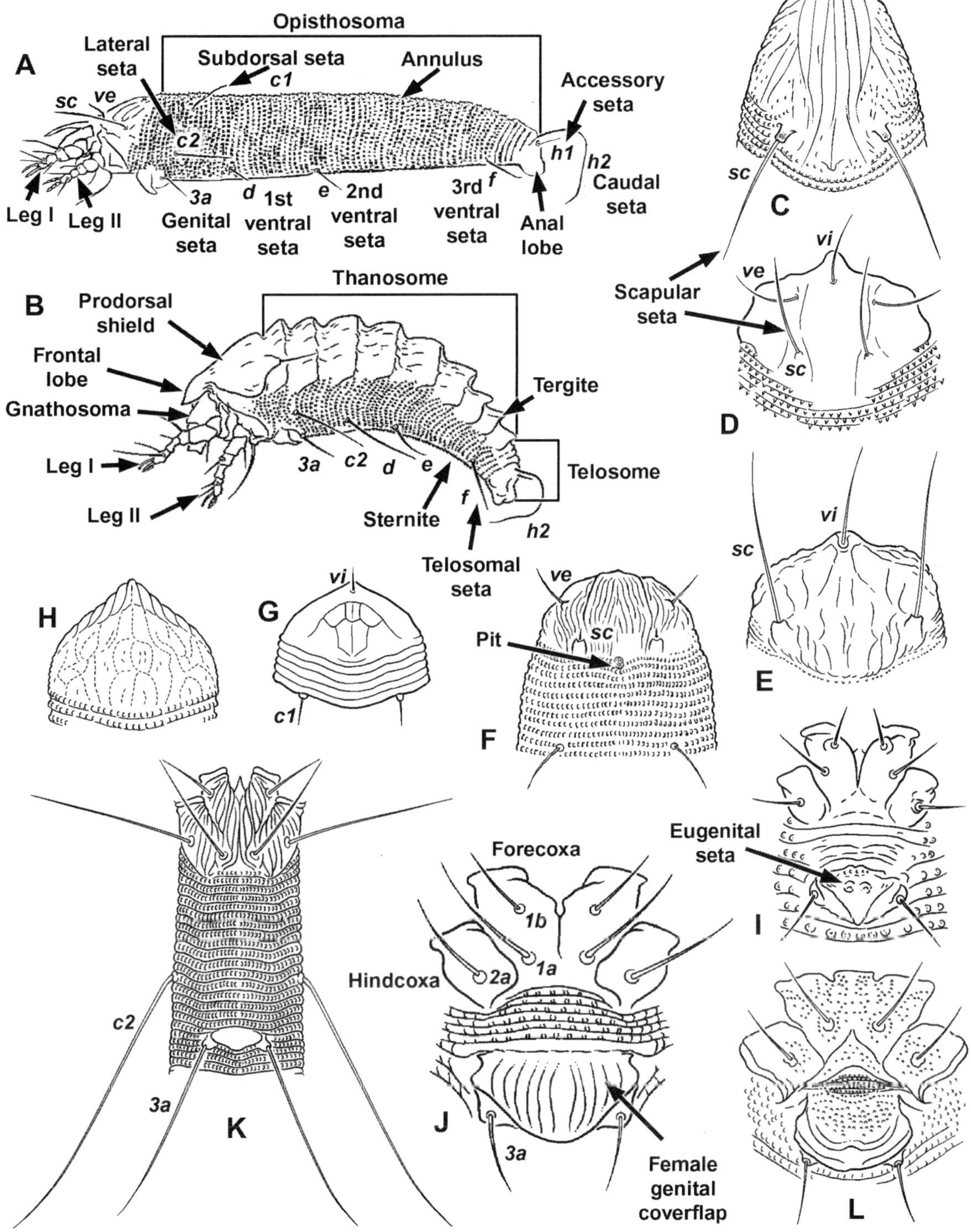

Fig. 13.17. Superfamily Eriophyoidea (after Lindquist 1996b), lateral views of the two major body forms: **A**, *Phytoptus leucothonius* Keifer, a vermiform mite; **B**, *Anthocoptes helianthella* Keifer, a fusiform mite.

Prodorsa of various genera, showing diverse shapes and setation: **C**, *Aceria*; **D**, *Pentasetacus*; **E**, *Nalepella*; **F**, *Acathrix*; **G**, *Boczekella*; **H**, *Cecidophyes*.

Ventral views of coxisternal and genital regions of various genera, showing diverse shapes and distances between structures: **I**, *Aceria*, male; **J**, *Aceria*, female, with genital aperture normally spaced from coxisternal plates; **K**, *Novophytoptus*, female, with genital aperture distant from coxisternal plates; **L**, *Floracarus*, female, with genital aperture appressed to coxisternal plates.

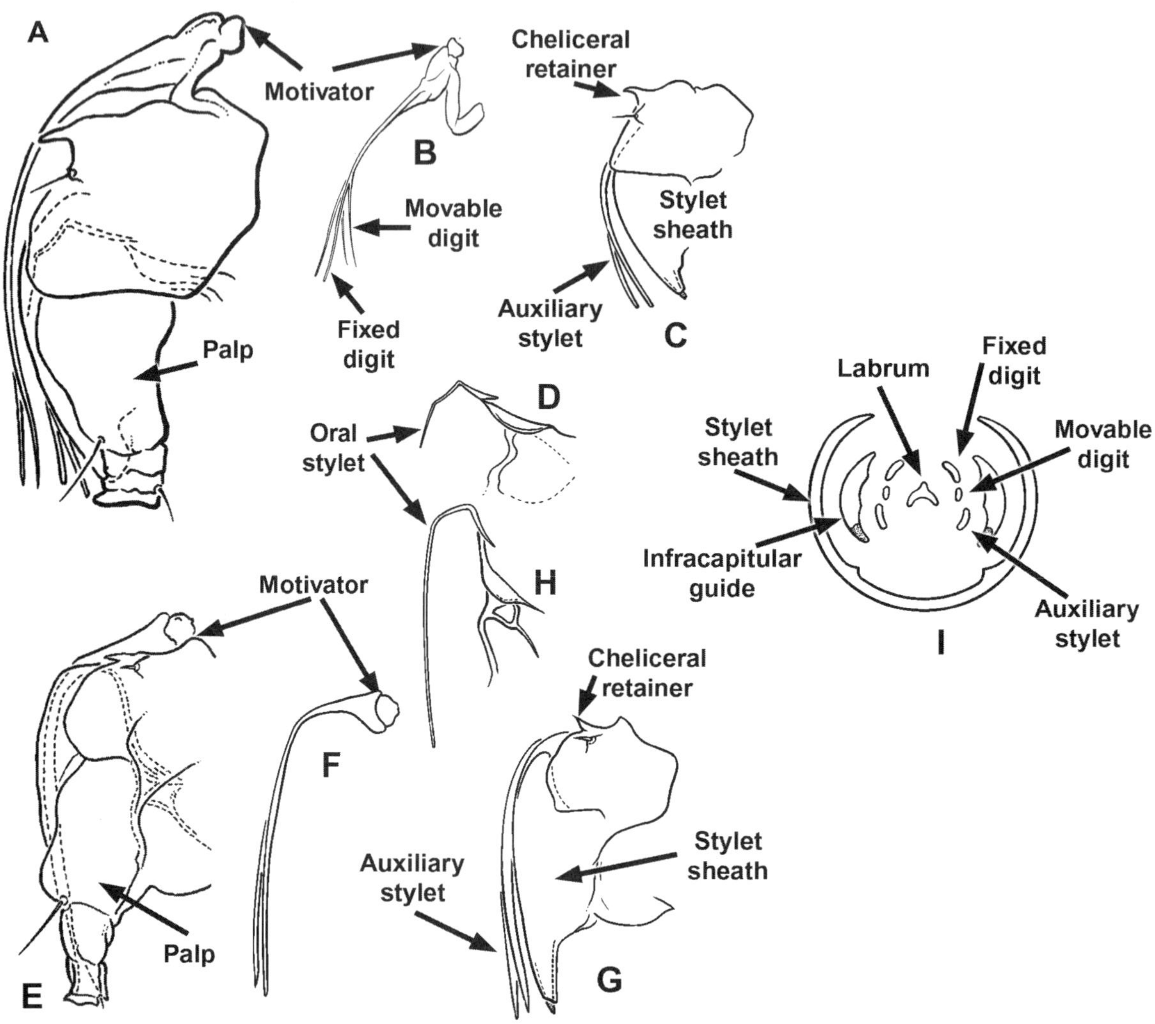

Fig. 13.18. Superfamily Eriophyoidea (after Lindquist 1996b). Exploded diagrammatic lateral views of gnathosomatic structures.

ERIOPHYIDAE. **A**, composite; **B**, cheliceral stylets apart from other structures; **C**, palpcoxal base, infracapitulum, and auxiliary stylets apart from other structures; **D**, labrum (= oral stylet) and pharynx apart from other structures.

DIPTILOMIOPIDAE. **E**, composite; **F**, cheliceral stylets apart from other structures; **G**, palpcoxal base, infracapitulum, and auxiliary stylets apart from other structures; **H**, labrum (= oral stylet) and pharynx apart from other structures; **I**, diagrammatic transverse section of cheliceral and associated structures at level near apices of stylets.

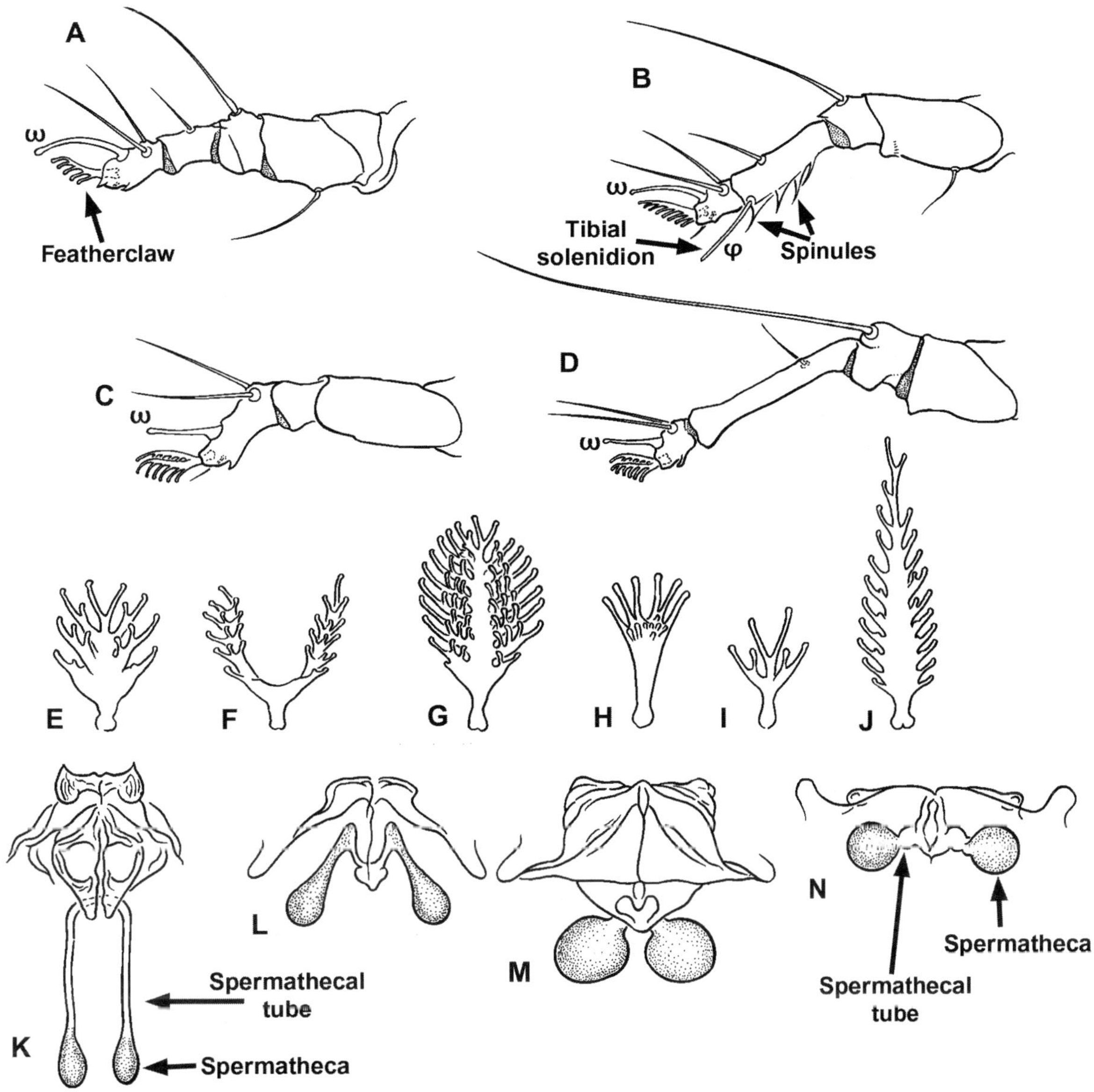

Fig. 13.19. Superfamily Eriophyoidea (after Lindquist 1996b). Lateral views of leg I of various genera showing diversity in length, fusion, and setation of segments: **A**, *Aceria*; **B**, *Nalepella*; **C**, *Diptilomiopus*; **D**, *Dialox*.

Ventral views of diverse forms of empodial featherclaws: **E**, *Acathrix*; **F**, *Diptilomiopus*; **G**, *Nalepella*; **H**, *Acrinotus*; **I**, *Tetra*; **J**, *Trisetacus*.

Dorsoventral views of spermathecal and associated structures in adult females of various genera showing diverse lengths and orientations of spermathecal tubes: **K**, *Trisetacus*; **L**, *Acathrix*; **M**, *Anthocoptes*; **N**, *Cecidophyes*.

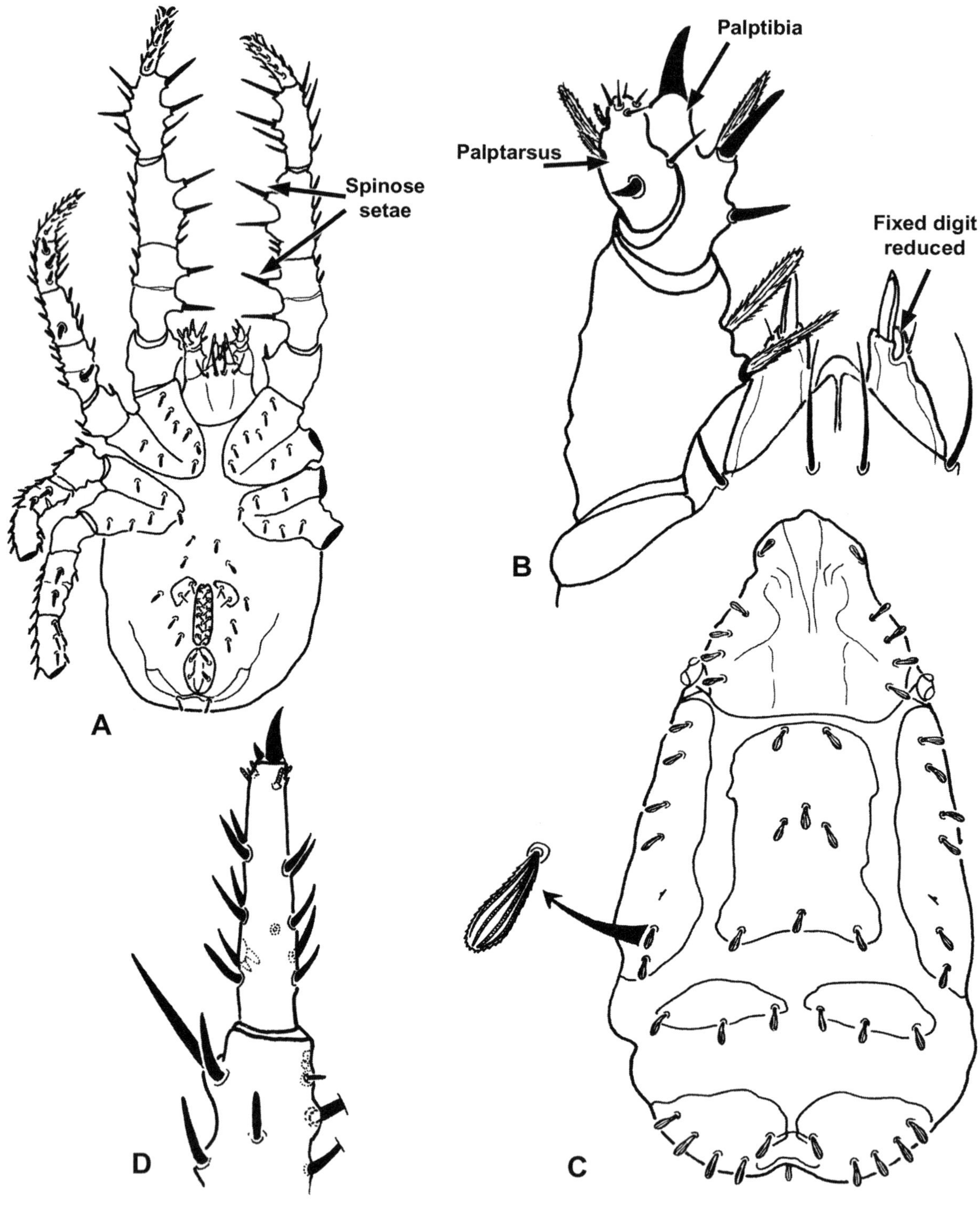

Fig. 13.20. CAECULIDAE. *Caeculus* sp. (Oregon, USA): **A**, venter; **B**, venter of gnathosoma including the right palp; **C**, dorsum with detail of seta. *C. liguricus*? Vitzthum (France): **D**, tibia and tarsus I (after Grandjean 1944a). (See Fig. 3.13B for details regarding sensory structures on these segments.)

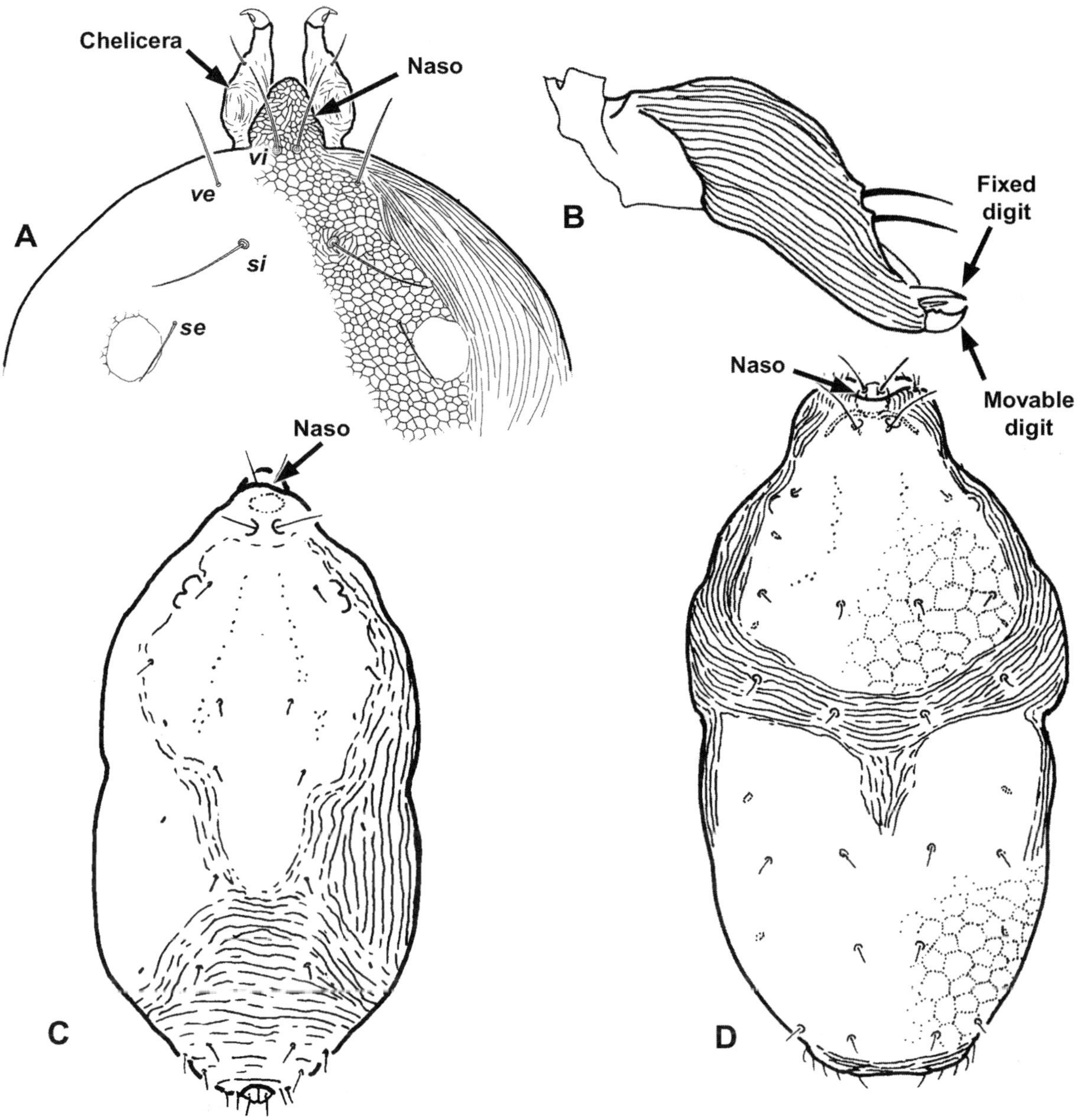

Fig. 13.21. ADAMYSTIDAE. *Adamystis* sp. (Victoria, Australia): **A**, anterior prodorsum and chelicerae. *A. fonsi* Coineau (France) (after Coineau 1974d): **B**, chelicera in antiaxial view; **C**, dorsum. *Saxidromus delamarei* Coineau (France) (after Coineau 1974d): **D**, dorsum.

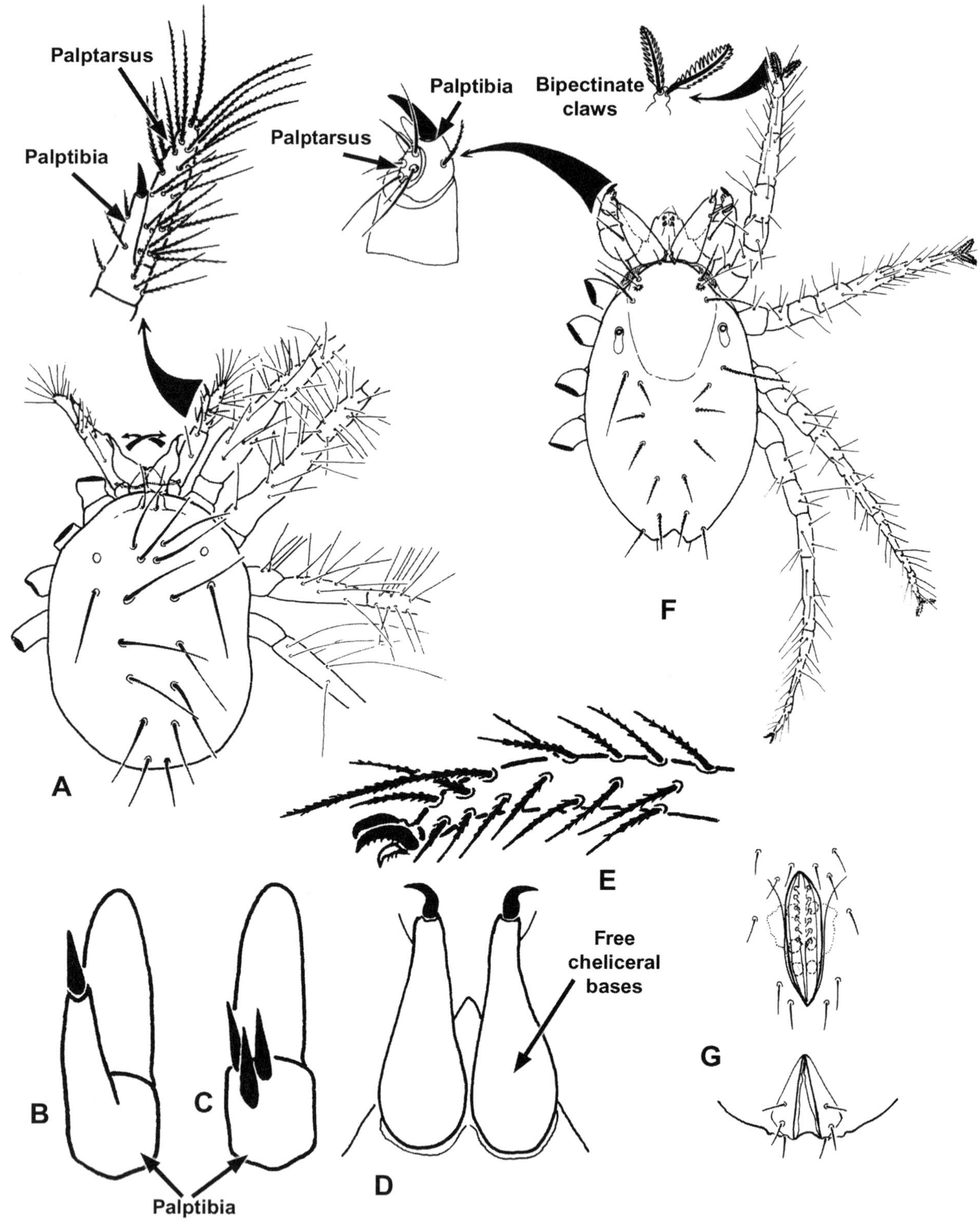

Fig. 13.22. ANYSTIDAE. *Erythracarus* sp. (Oregon, USA): **A**, dorsum with detail of palpal thumb-claw process; **B**, typical thumb-claw process in Erythracarinae; **C**, typical thumb-claw process in Anystinae; **D**, chelicerae typical of ANYSTIDAE. *Anystis* sp. (Oregon, USA): **E**, distal portion of tarsus I.

TENERIFFIIDAE. *Teneriffia* sp. (Oregon, USA): **F**, dorsum of female with detail of palpal thumb-claw process and pretarsus I; **G**, genito-anal region of female.

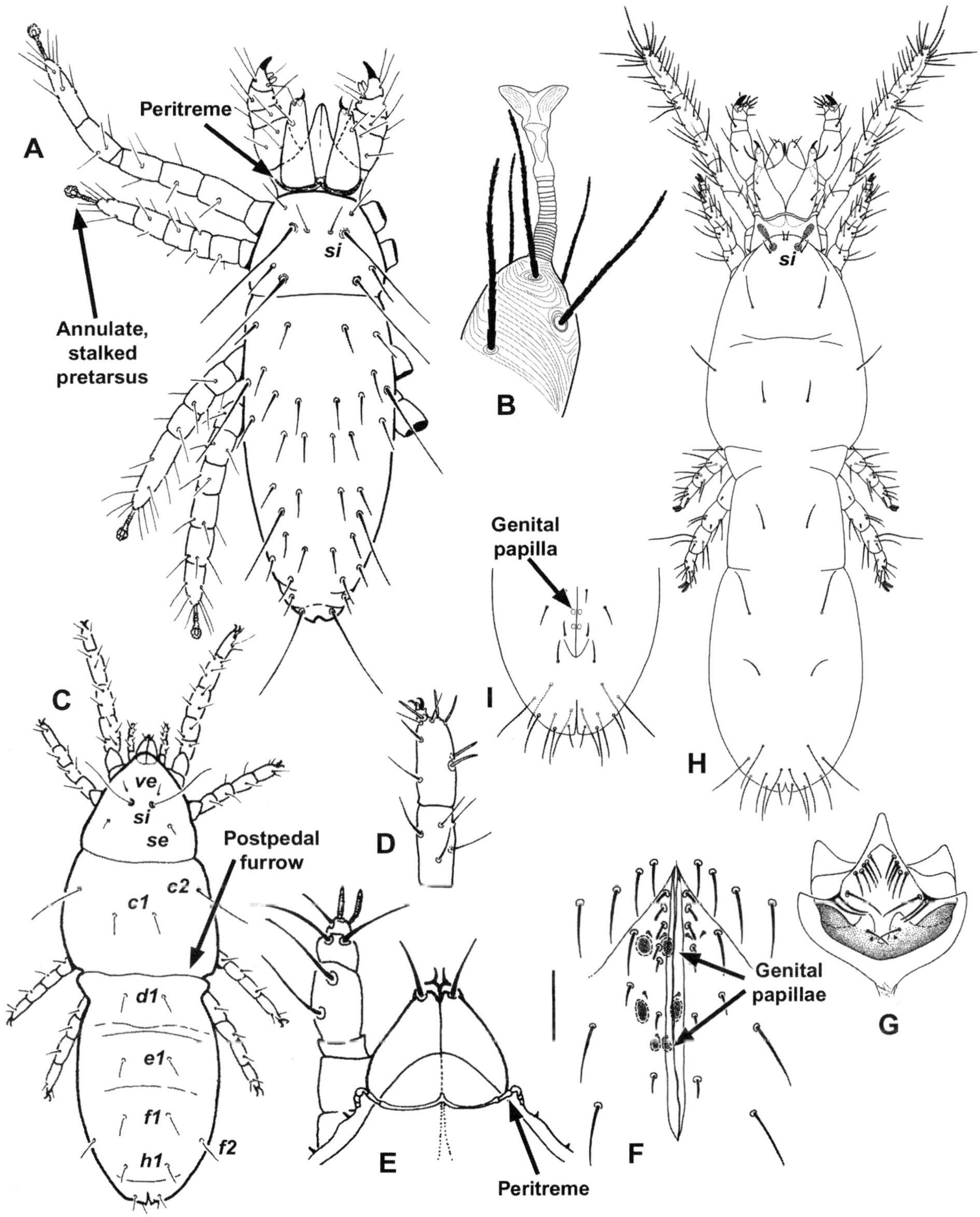

Fig. 13.23. PSEUDOCHEYLIDAE. *Pseudocheylus* sp.: **A**, dorsum. Undescribed pseudocheylid species (Queensland, Australia): **B**, distal tarsus and pretarsus III.

PARATYDEIDAE. *Paratydeus* sp. (Arizona, USA): **C**, dorsum of female; **D**, tibia and tarsus I of female; **E**, anterodorsal aspect of female. *Tanytydeus lamington* Seeman and Walter (Queensland, Australia): **F**, genital region of male; **G**, internal spermatophoric organ of male (after Seeman and Walter 2000).

STIGMOCHEYLIDAE. *Stigmocheylus* sp.: **H**, dorsum of female; **I**, genito-anal region of female.

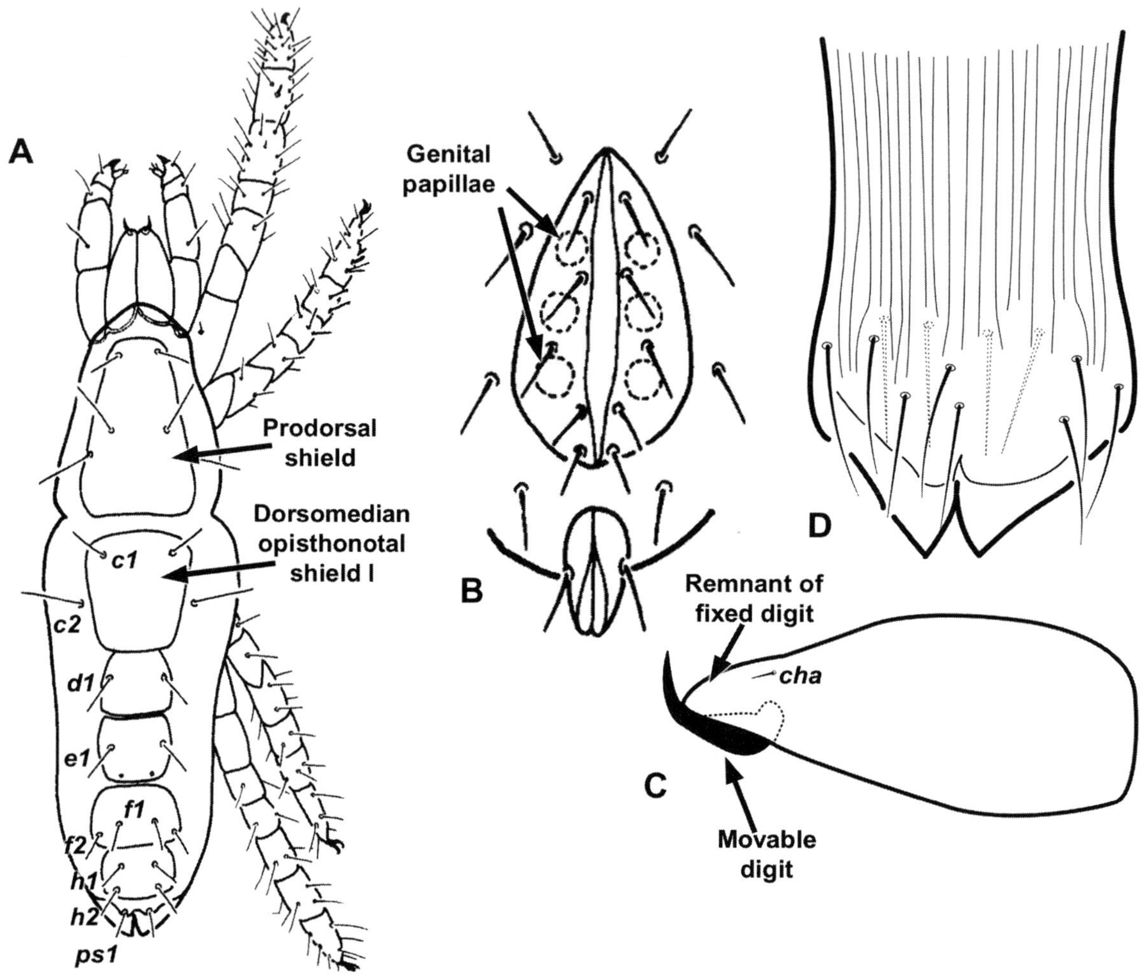

Fig. 13.24. POMERANTZIIDAE. *Pomerantzia charlesi* Baker (Georgia, USA): **A**, dorsum of female; **B**, genito-anal region of female (after Baker 1949b). *P. subterranea* Fan and Chen (Fujian, China): **C**, chelicera of female; **D**, ovipositor of female (after Fan and Chen 2005).

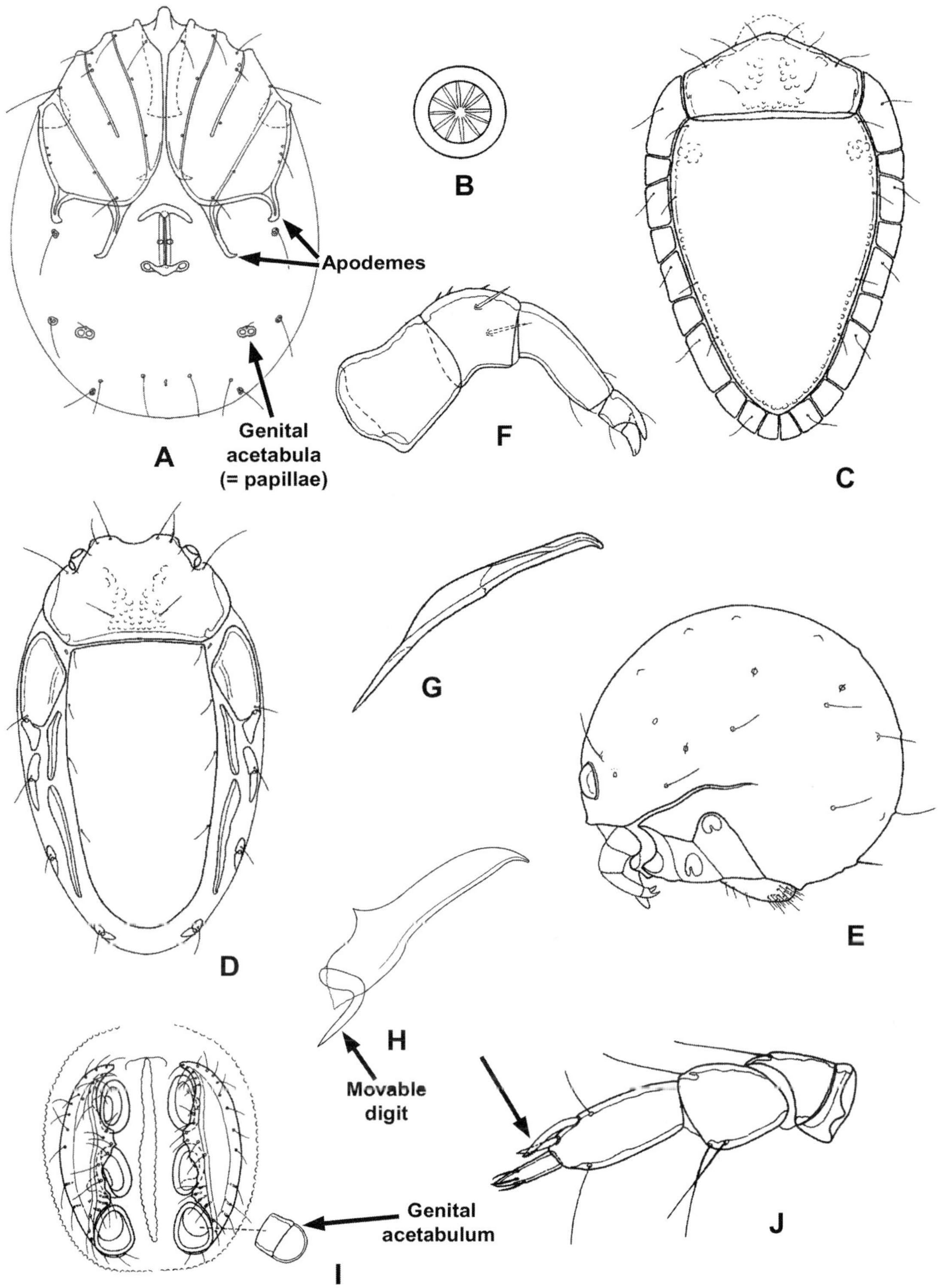

Fig. 13.25. PONTARACHNIDAE. *Pontarachna hoffmannae* Cook: **A**, venter of idiosoma; **B**, genital acetabulum (= papilla).
ACHERONTACARIDAE. *Acherontacarus bicornis* Cook: **C**, dorsal shield.
HYDROVOLZIIDAE. *Hydrovolzia marshallae* Cook: **D**, dorsum of idiosoma.
HYDRACHNIDAE. *Hydrachna testudinata* Cook: **E**, lateral aspect; **F**, palp. *H. perplexa* Cook: **G**, chelicera.
HYGROBATIDAE. *Hygrobates* sp.: **H**, chelicera.
ZELANDOTHYADIDAE. *Zelandothyas diamphida* Cook: **I**, genital field; **J**, palp.

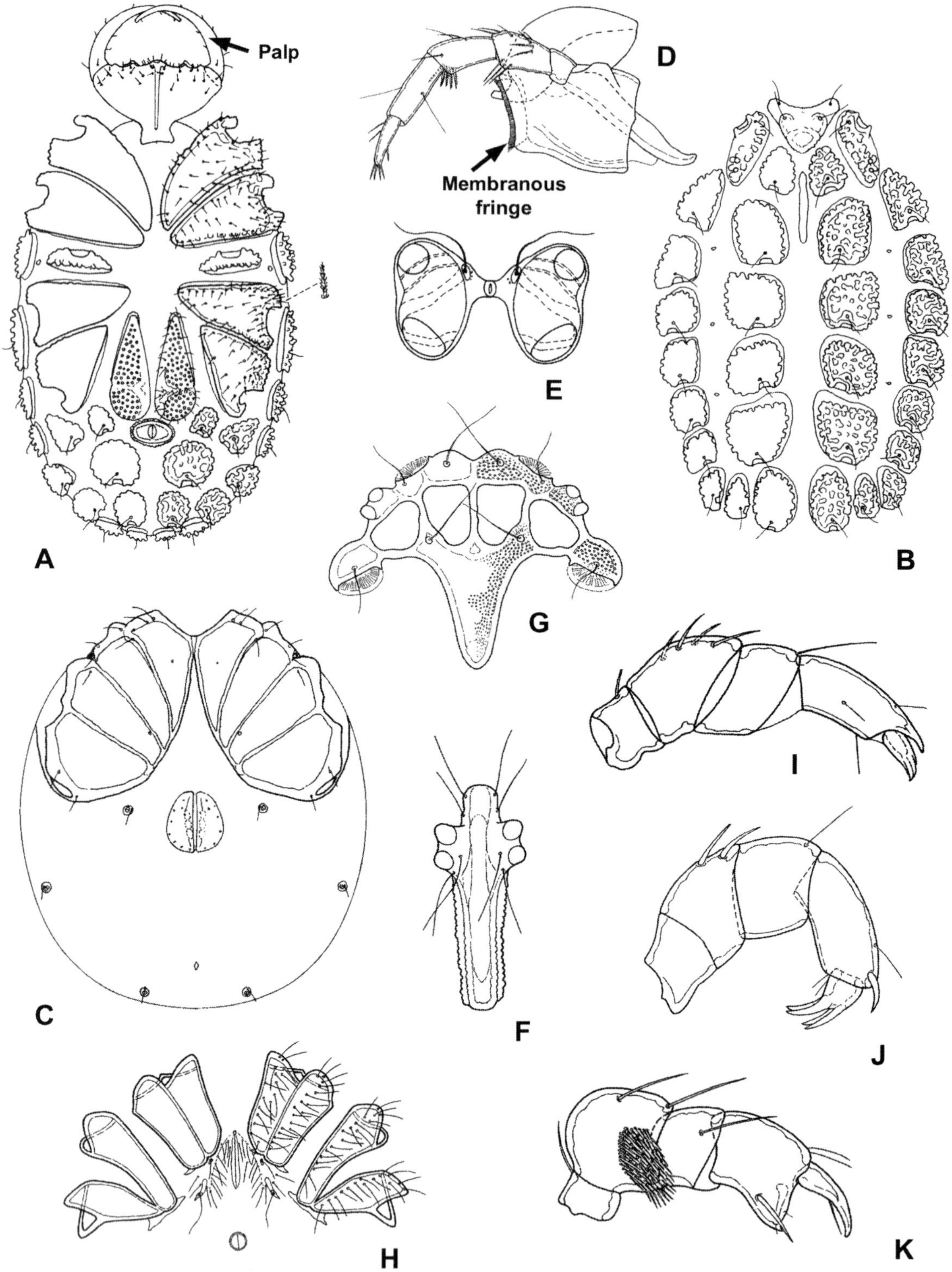

Fig. 13.26. APHEVIDERULICIDAE. *Apheviderulix santana* Gerecke, Smith, and Cook, deutonymph: **A**, venter; **B**, dorsum of idiosoma.
STYGOTONIIDAE. *Stygotonia ambigua* Cook: **C**, venter of idiosoma.
PIERSIGIIDAE. *Piersigia limnophila* Protz: **D**, gnathosoma, lateral view; **G**, prodorsal region.
EYLAIDAE. *Eylais montana* Ribaga: **E**, eye plate. *Eylais* sp.: **H**, ventral sclerites of idiosoma.
LIMNOCHARIDAE. *Limnochares crinita* Koenike: **F**, eye plate.
HYDRYPHANTIDAE. *Wandesia vermiformis* Cook: **I**, palp. *Euwandesia vietsiella* Besch: **J**, palp.
ARRENURIDAE. *Arrenurus ceylonicus* Daday: **K**, palp.

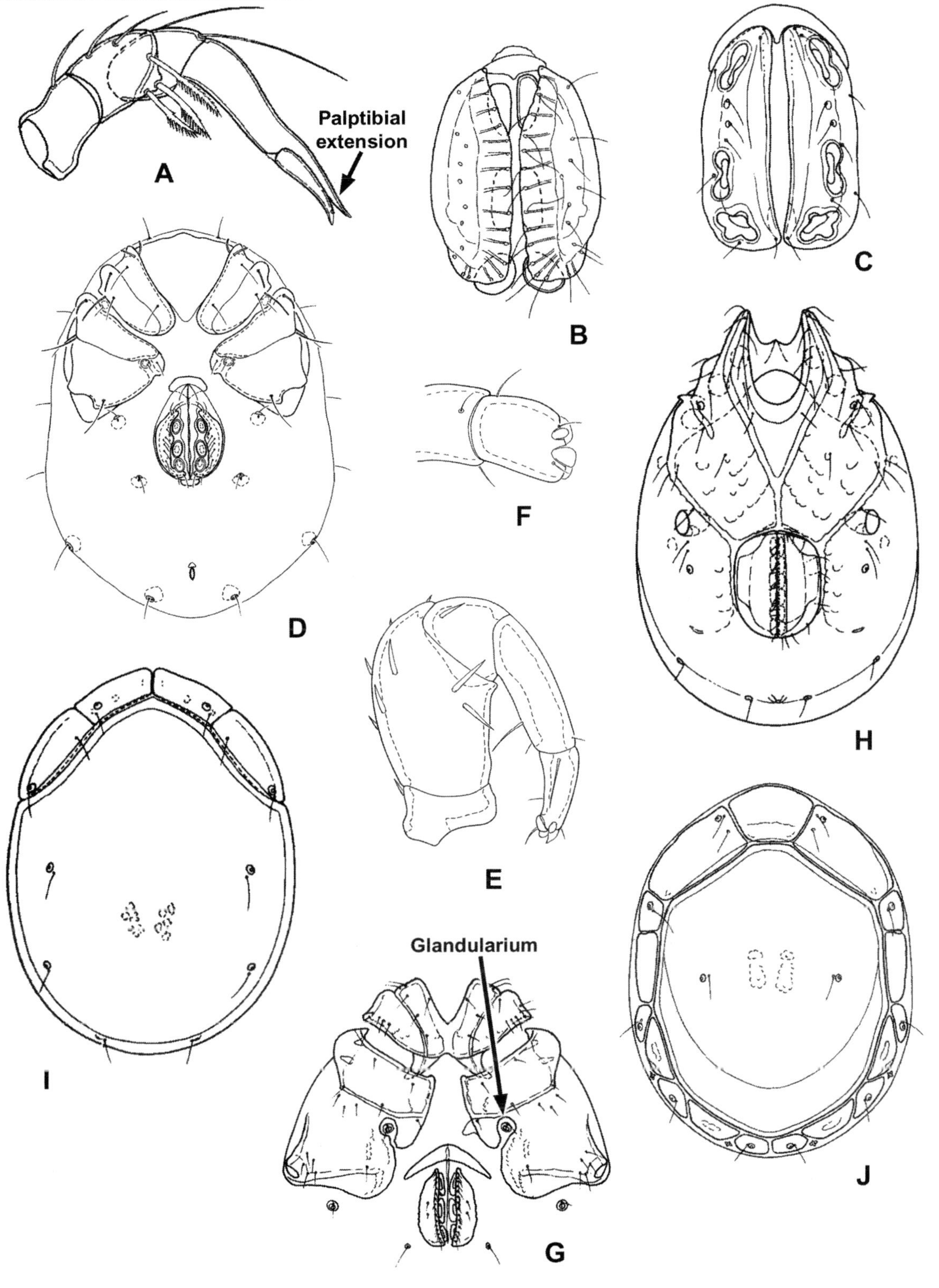

Fig. 13.27. HYDRODROMIDAE. *Hydrodroma monticola* Piersig: **A**, palp.
SPERCHONTIDAE. *Sperchonopsis nova* Prasad and Cook: **B**, genital field.
LIMNESIIDAE. *Protolimnesia interstitialis* Cook: **C**, genital field.
RUTRIPALPIDAE. *Rutripalpus canadensis* Smith: **D**, venter of idiosoma; **E**, palp; **F**, palptarsus.
TEUTONIIDAE. *Teutonia* sp.: **G**, venter of idiosoma.
TORRENTICOLIDAE. *Torrenticola amala* Cook: **H**, ventral shield; **I**, dorsal shield. *Testudacarus americanus* Marshall: **J**, dorsal shield.

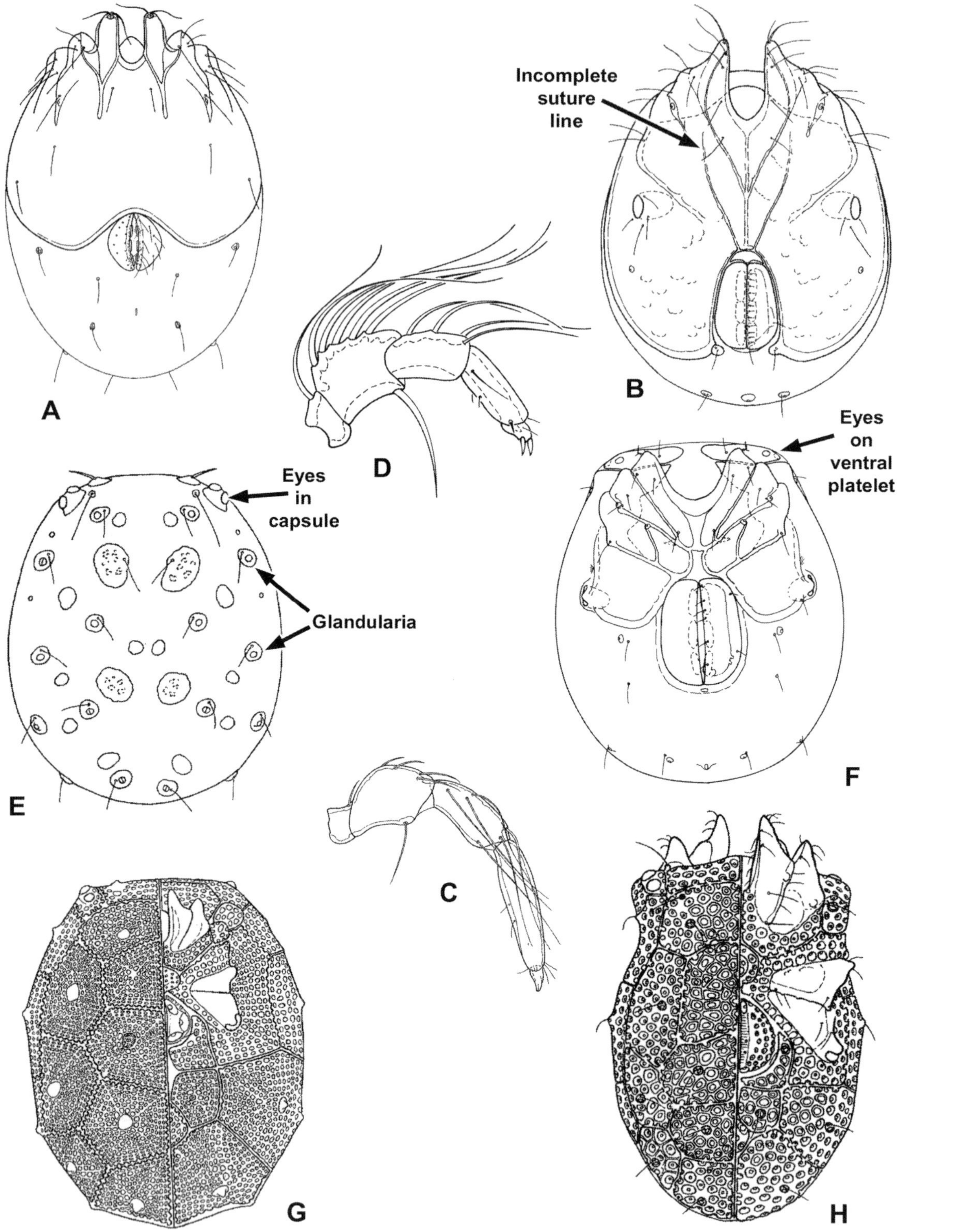

Fig. 13.28. OXIDAE. *Oxus pictus* Daday: **A**, venter of idiosoma.

LEBERTIIDAE. *Lebertia ventriscutata* Cook: **B**, venter of idiosoma. *L. michiganensis* Cook: **C**, palp. *Scutolebertia trinitensis* Smith: **D**, palp.

SPERCHONTIDAE. *Sperchon nilgiris* Cook: **E**, dorsum of idiosoma.

ANISITSIELLIDAE. *Bandakia fragilis* Smith: **F**, ventral shield.

TERATOTHYADIDAE. *Teratothyasides undulatus* Cook: **G**, split view of dorsal and ventral shields. *T. sagariphorus* Cook: **H**, split view of dorsal and ventral shields.

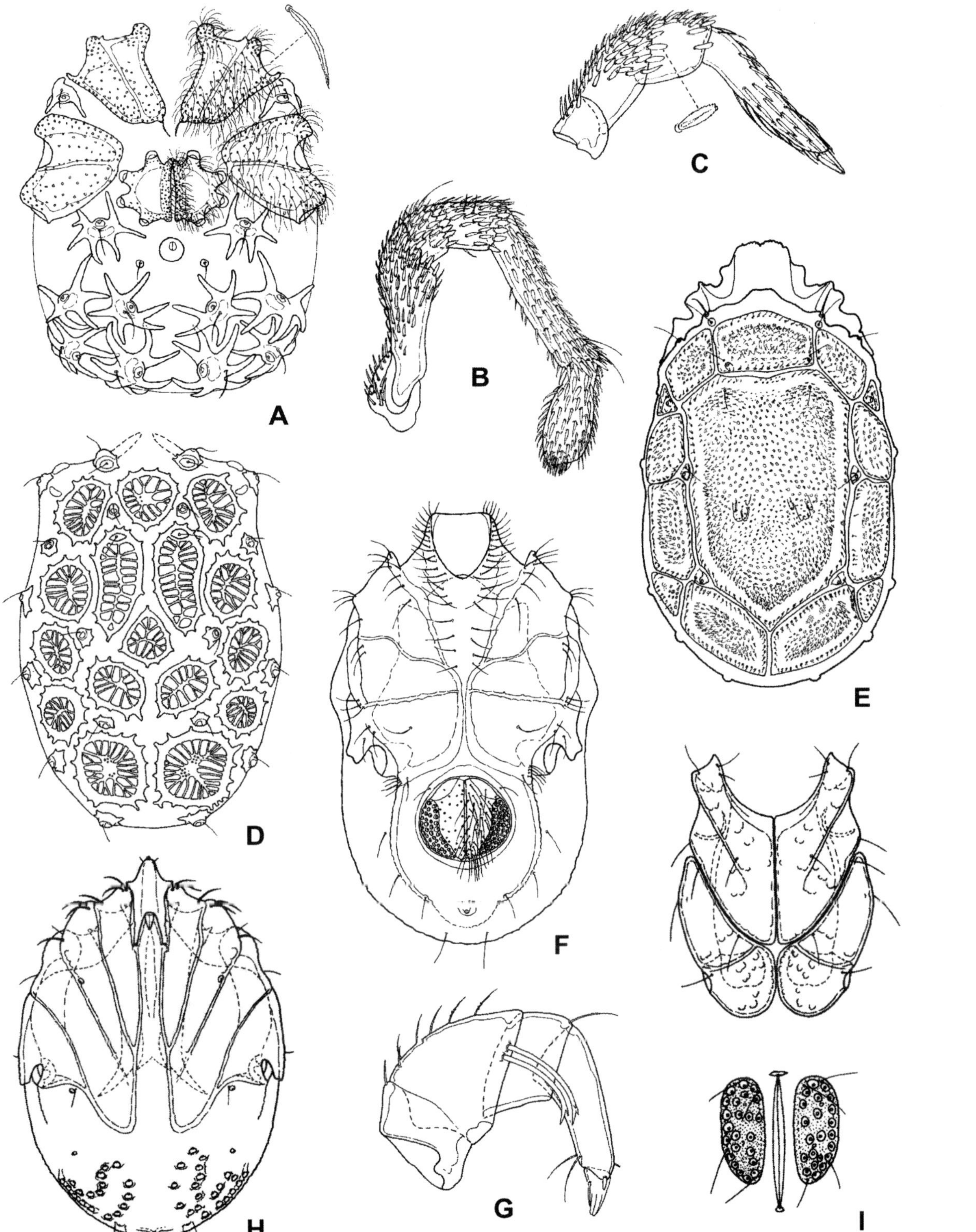

Fig. 13.29. CTENOTHYADIDAE. *Stellulathyas lundbladi* Cook and Hopkins: **A**, venter of idiosoma; **B**, leg I; **C**, palp. RHYNCHOHYDRACARIDAE. *Clathrosperchon ornatus* Cook: **D**, dorsum of idiosoma. *Rhynchohydracarus testudo* Lundblad: **E**, dorsum of idiosoma.
THERMACARIDAE. *Thermacarus nevadensis* Marshall: **F**, venter of idiosoma; **G**, palp.
FERRADASIIDAE. *Ferradasia musicola* Cook: **H**, venter of idiosoma, male.
OMARTACARIDAE. *Omartacarus brevipalpus* Cook: **I**, venter of idiosoma.

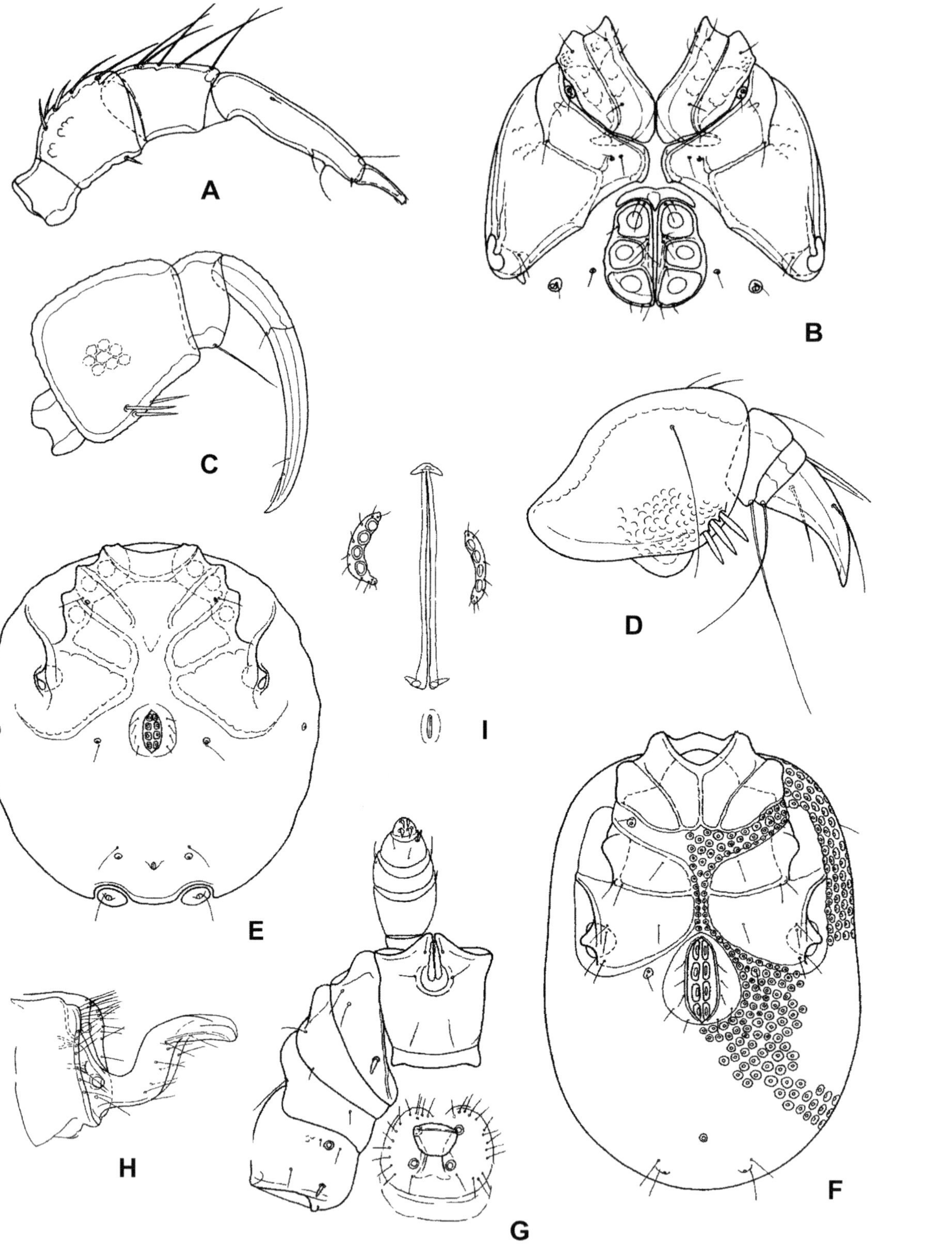

Fig. 13.30. LIMNESIIDAE. *Limnesia tucumanensis* Cook: **A**, palp. *L. patagonica* Lundblad: **B**, venter of idiosoma.
NIPPONACARIDAE. *Nipponacarus matsumotoi* Imamura: **C**, palp; **E**, venter of idiosoma.
BOGATIIDAE. *Bogatia maxillaris* Motaş and Tanasachi: **D**, palp. *Horreolanus orphanus* Mitchell: **F**, ventral shield, male.
ASTACOCROTONIDAE. *Astacocroton molle* Haswell: **G**, venter of idiosoma; **H**, lateral view of genital field, male; **I**, genital field, female.

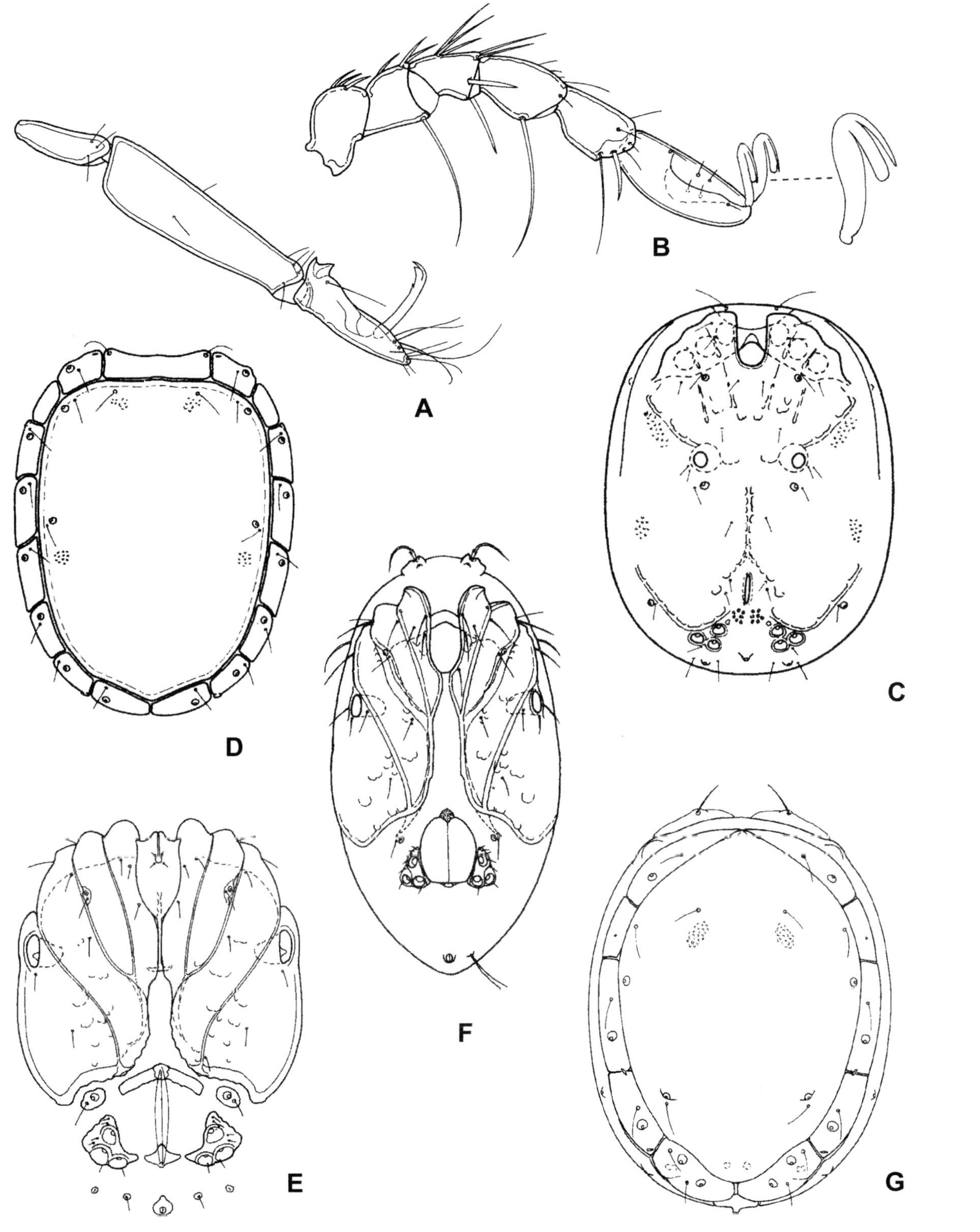

Fig. 13.31. MOMONIIDAE. *Stygomomonia separata* Cook: **A**, distal segments of leg I.
WETTINIDAE. *Wettina occidentalis* Cook, Smith, and Harvey: **B**, leg I; **E**, venter of idiosoma, female. *Wheenyella smithi* Cook: **F**, venter of idiosoma, female. *Stormaxonella scutulata* K. O. Viets: **G**, dorsal shield, male.
LETHAXONIDAE. *Lethaxona oregonensis* Cook: **C**, ventral shield, male. *Lethaxonella parva* Cook: **D**, dorsal shield.

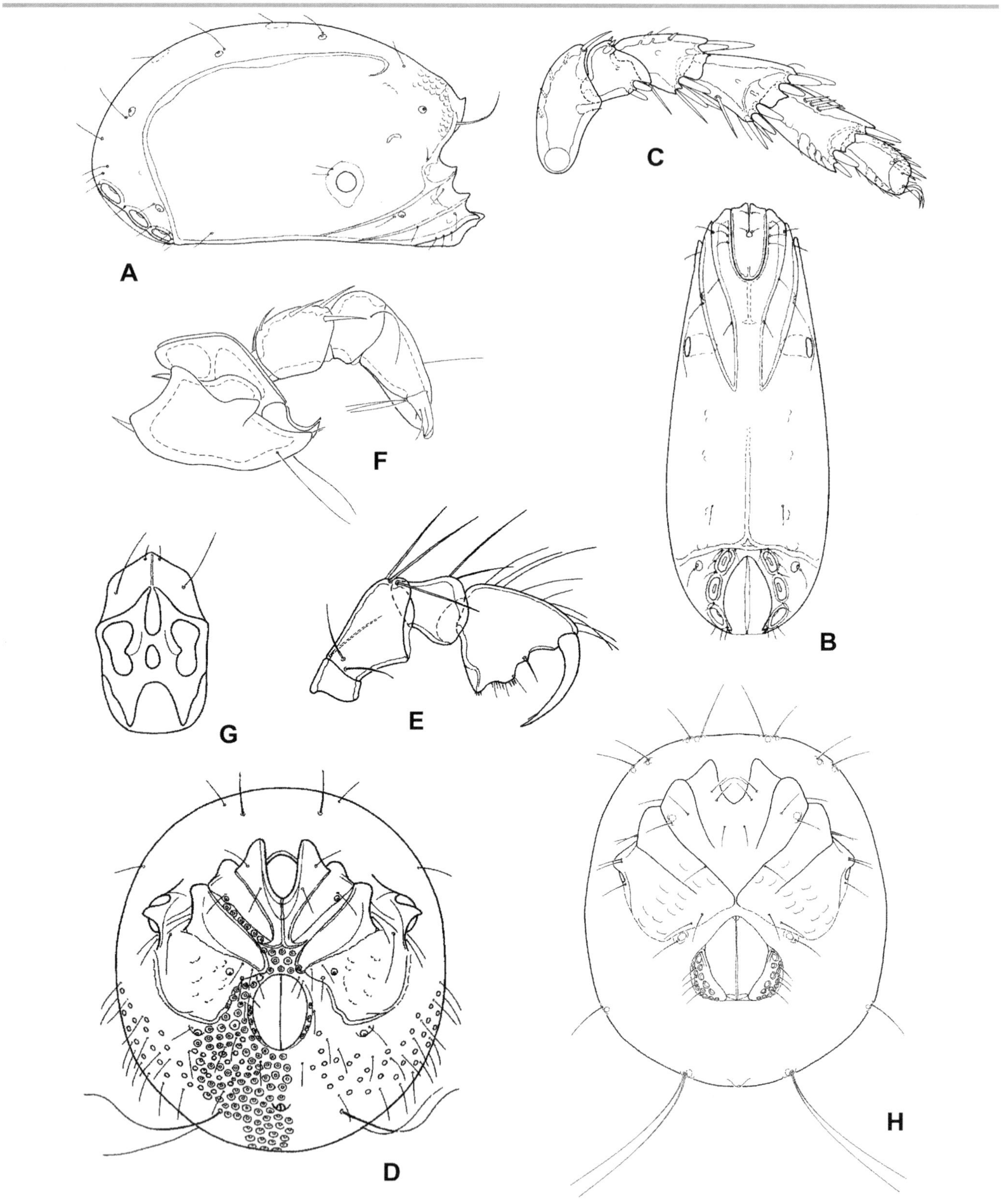

Fig. 13.32. FRONTIPODOPSIDAE. *Frontipodopsis sudafricanus* Cook, Smith, and Harvey: **A**, lateral view of idiosoma; **B**, venter of idiosoma; **C**, leg IV.

HARPAGOPALPIDAE. *Harpagopalpus indicus* Cook: **D**, ventral shield, female; **E**, palp.

ATHIENEMANNIIDAE. *Chelomideopsis besselingi* (Cook): **F**, lateral view of gnathosoma. *C. annemiae* Romijn: **G**, dorsal view of capitulum.

ACALYPTONOTIDAE. *Acalyptonotus neoviolaceus* Smith: **H**, venter of idiosoma, female.

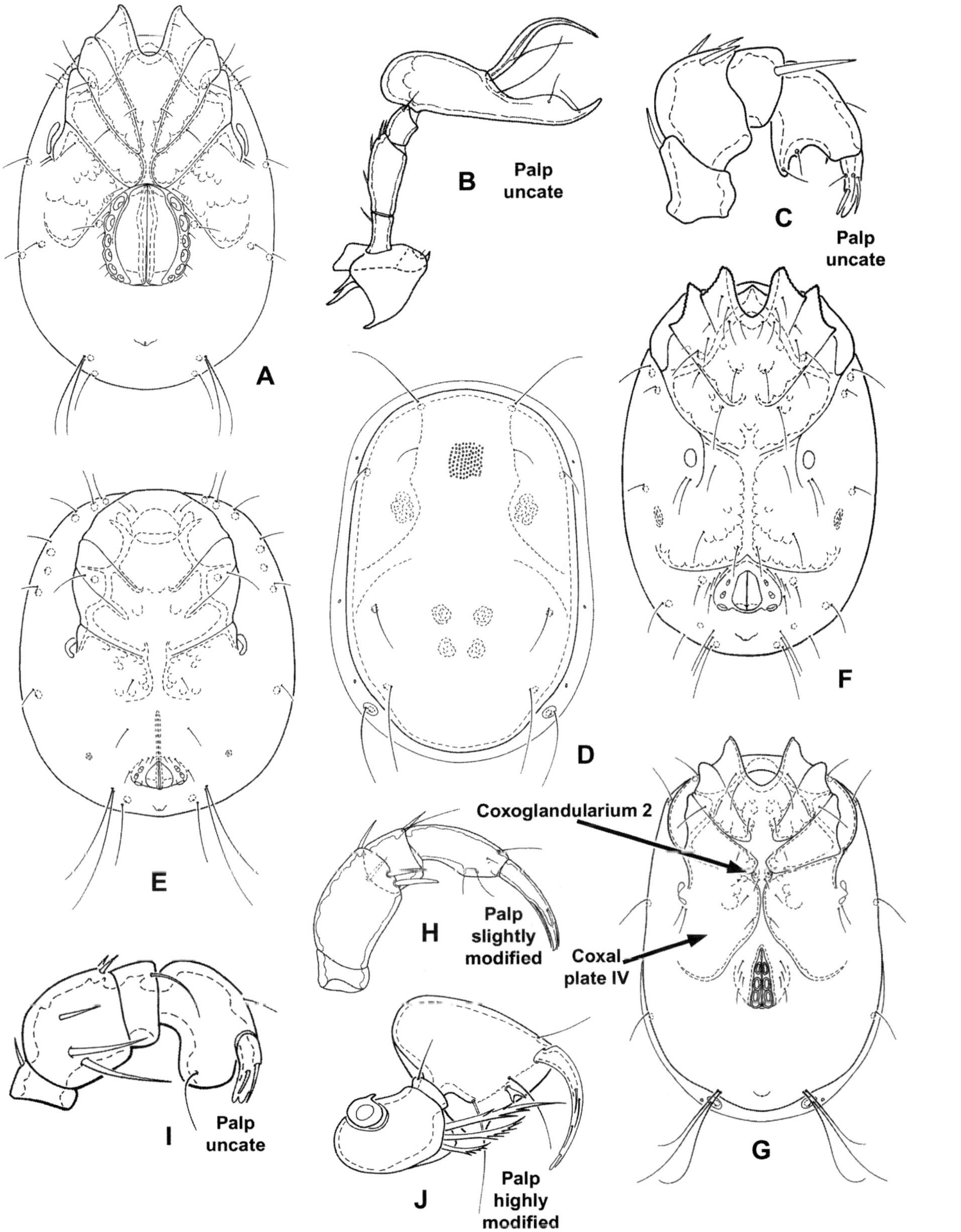

Fig. 13.33. NEOACARIDAE. *Neoacarus davecooki* Smith: **A**, ventral shield, female. *Volsellacarus sabulonis* Cook: **B**, capitulum and palp. *N. occidentalis* Cook: **C**, palp.

CHAPPUISIDIDAE. *Chappuisides acadianus* Smith: **D**, dorsal shield. *Uchidastygacarus imamurai* Cook: **E**, ventral shield, female. *Yachatsia mideopsoides* Cook: **F**, ventral shield, female. *C. acadianus* Smith: **G**, venter of idiosoma, male; **I**, palp. *Tsushimacarus uenoi* Imamura: **H**, palp. *U. acadiensis* Smith: **J**, palp.

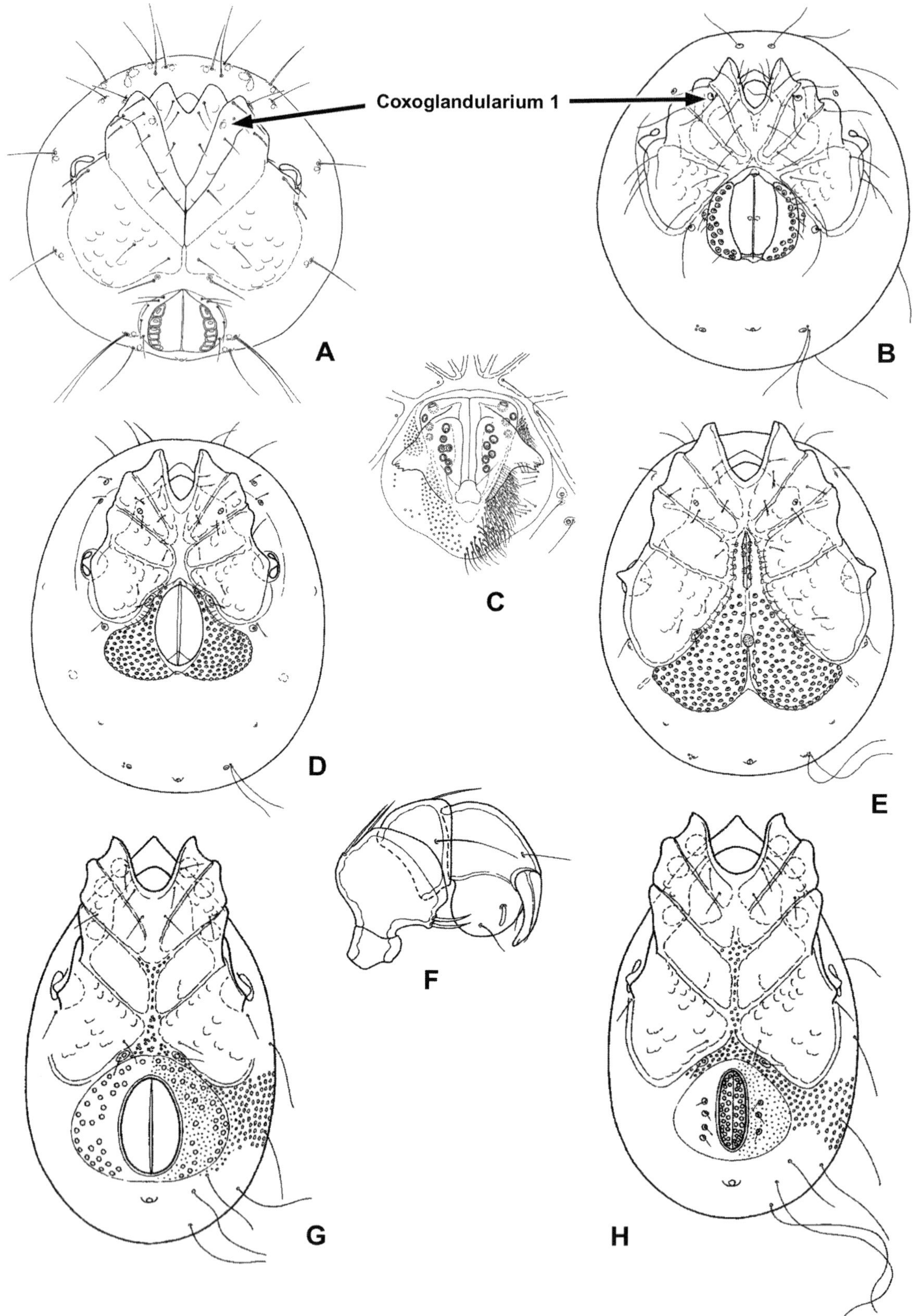

Fig. 13.34. NUDOMIDEOPSIDAE. *Paramideopsis susanae* Smith: **A**, ventral shield, female.
MIDEIDAE. *Midea expansa* Marshall: **B**, ventral shield, female. *M. orbiculata* (Müller): **C**, genital field, male.
LAVERSIIDAE. *Laversia berulophila* Cook: **D**, ventral shield, female; **E**, ventral shield, male.
HUNGAROHYDRACARIDAE. *Hungarohydracarus szalayi* Cook: **F**, palp; **G**, ventral shield, female; **H**, ventral shield, male.

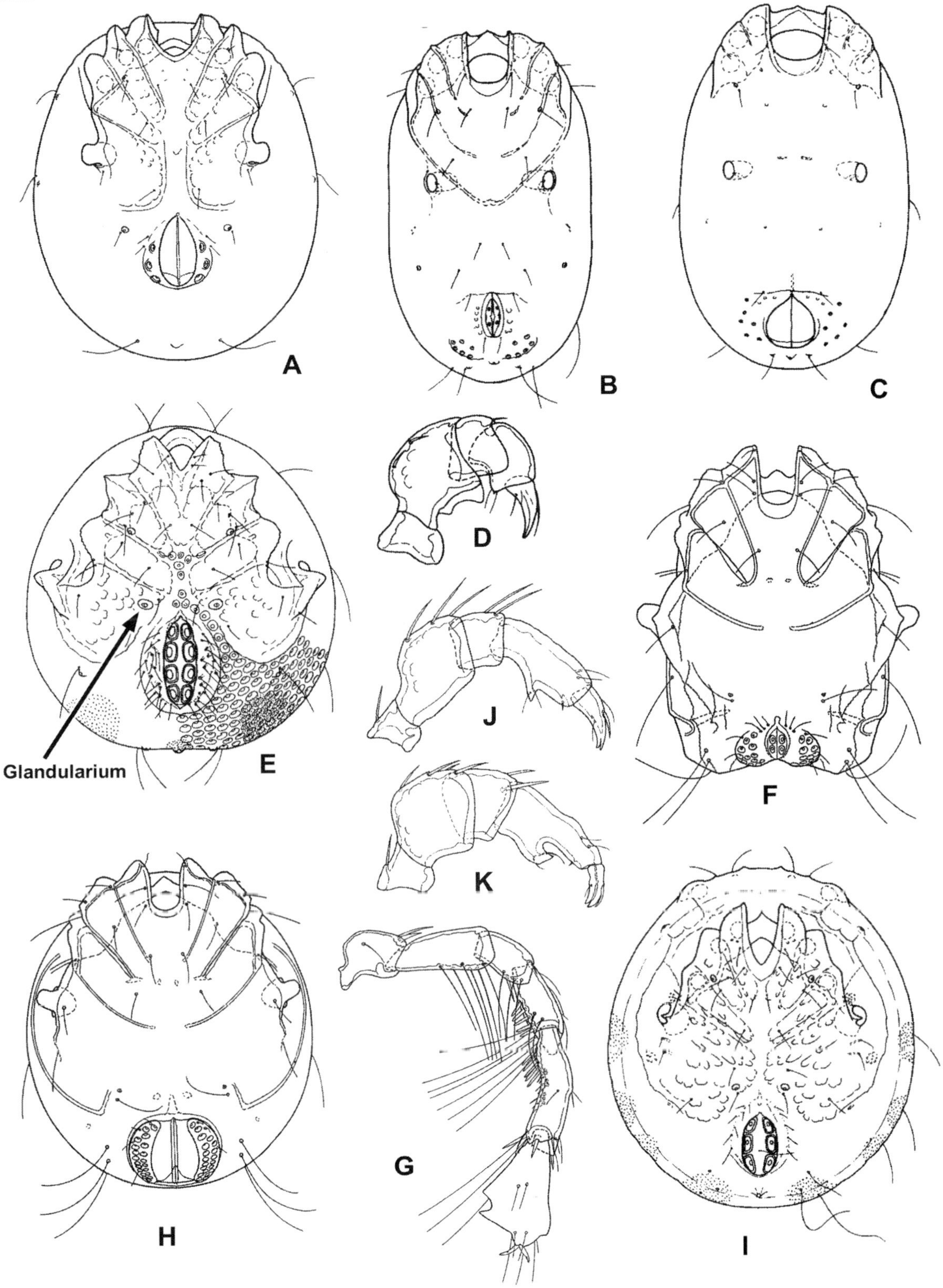

Fig. 13.35. KANTACARIDAE. *Kantacarus matsumotoi* Imamura: **A**, ventral shield, female.
ARENOHYDRACARIDAE. *Arenohydracarus minimus* Cook: **B**, ventral shield, male; **C**, ventral shield, female; **D**, palp.
KRENDOWSKIIDAE. *Geayia mitchelli* Cook: **E**, ventral shield, male.
AMOENACARIDAE. *Amoenacarus dixiensis* Smith and Cook: **F**, ventral shield, male; **G**, leg IV, male; **H**, ventral shield, female.
MIDEOPSIDAE. *Mideopsis sucaba* Cook: **I**, ventral shield, female; **J**, palp. *M. mexicana* Cook: **K**, palp.

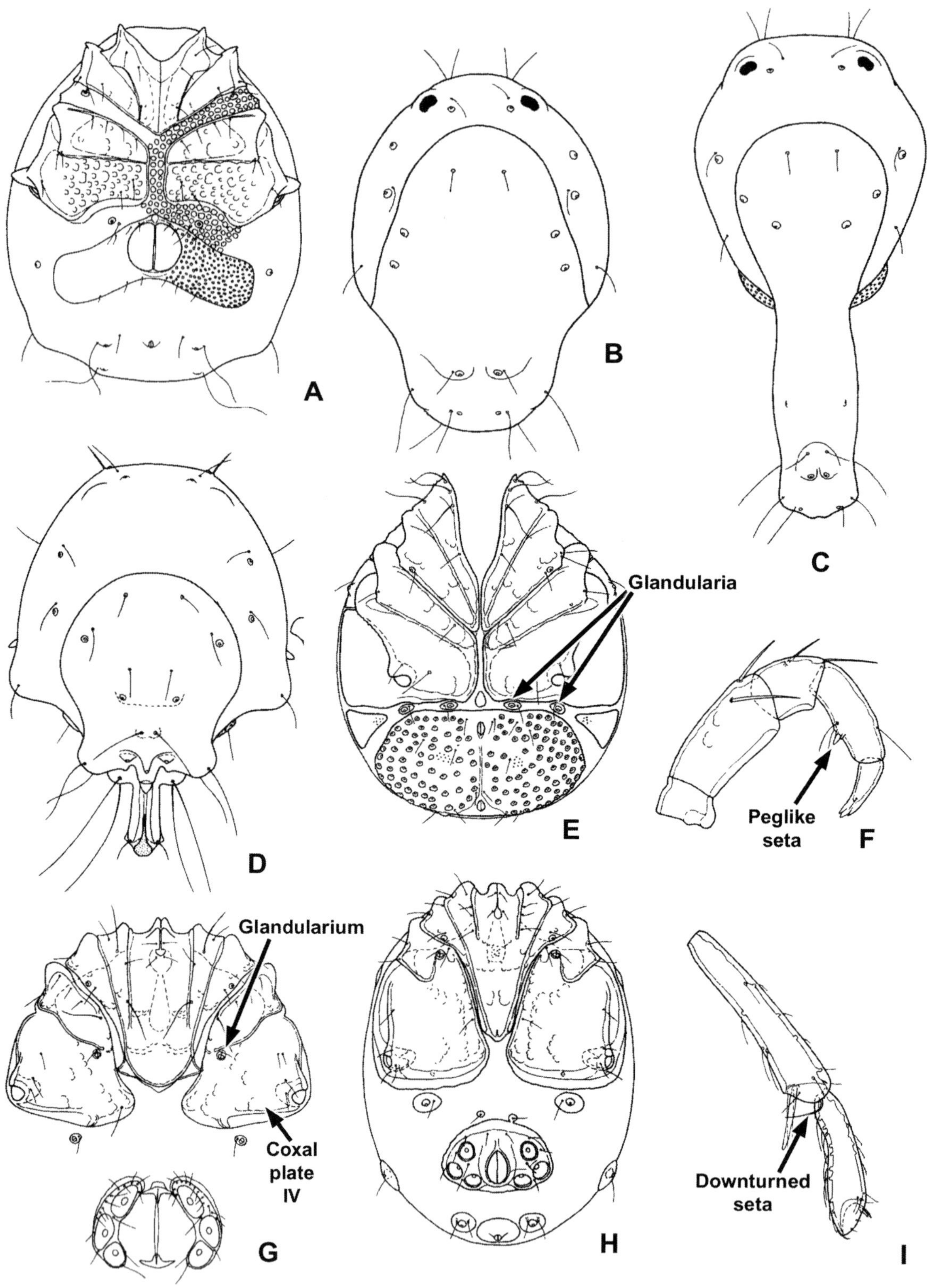

Fig. 13.36. ARRENURIDAE. *Arrenurus valdiviensis* K. O. Viets: **A**, ventral shield, female. *A. dinotoformis* Cook: **B**, dorsum of idiosoma, male. *A. longicaudatus* Marshall: **C**, dorsum of idiosoma, male. *A. munovus* Cook: **D**, dorsum of idiosoma, male.

FELTRIIDAE. *Feltria wyomingensis* Cook: **E**, venter of idiosoma, male.

UNIONICOLIDAE. *Koenikea bispina* Cook: **F**, palp.

HYGROBATIDAE. *Hygrobates ampliatus* Viets: **G**, venter of idiosoma, female. *H. calvotus* Cook: **H**, venter of idiosoma, male. *Atractides tembolus* Cook: **I**, tibia and tarsus of leg I.

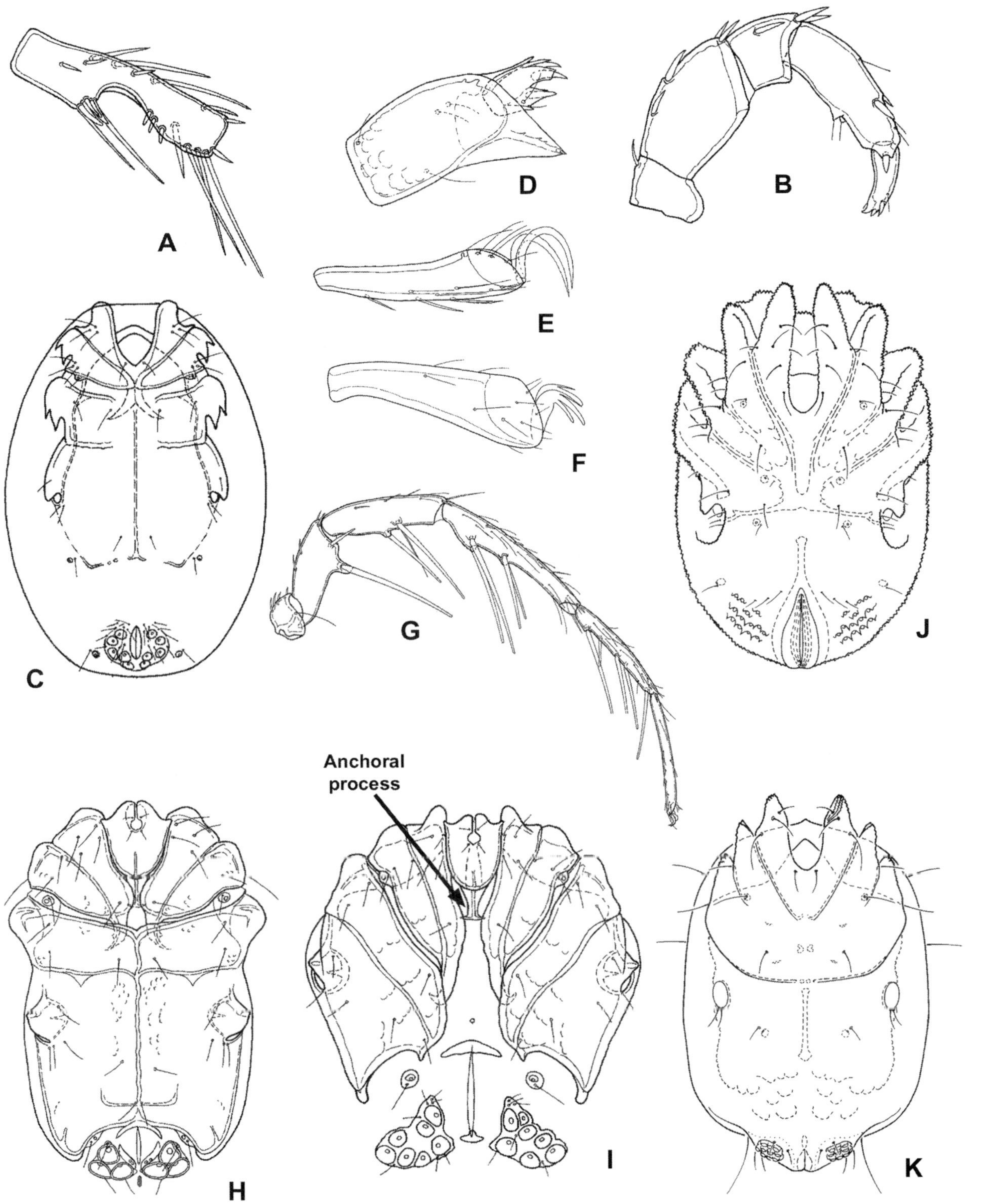

Fig. 13.37. PIONIDAE. *Piona indica* Cook: **A**, genu of leg IV. *Pseudofeltria laversi* Cook: **B**, palp. *Tiphys gracilipes* Cook: **H**, venter of idiosoma, male. *Forelia borealis* (Habeeb): **I**, venter of idiosoma, female.

UNIONICOLIDAE. *Pollicipalpus scutatus* Viets: **C**, ventral shield, male. *P. australicus* Cook: **D**, tibia and tarsus of palp. *Koenikea sorpresa* Cook: **E**, tarsus of leg IV. *Unionicola neoaffinis* Cook: **G**, leg I.

FELTRIIDAE. *Feltria virginiensis* Cook: **F**, tarsus of leg III.

ATURIDAE. *Bharatalbia cooki* Smith: **J**, venter of idiosoma, female. *Phreatobrachypoda oregonenis* Smith: **K**, venter of idiosoma, male.

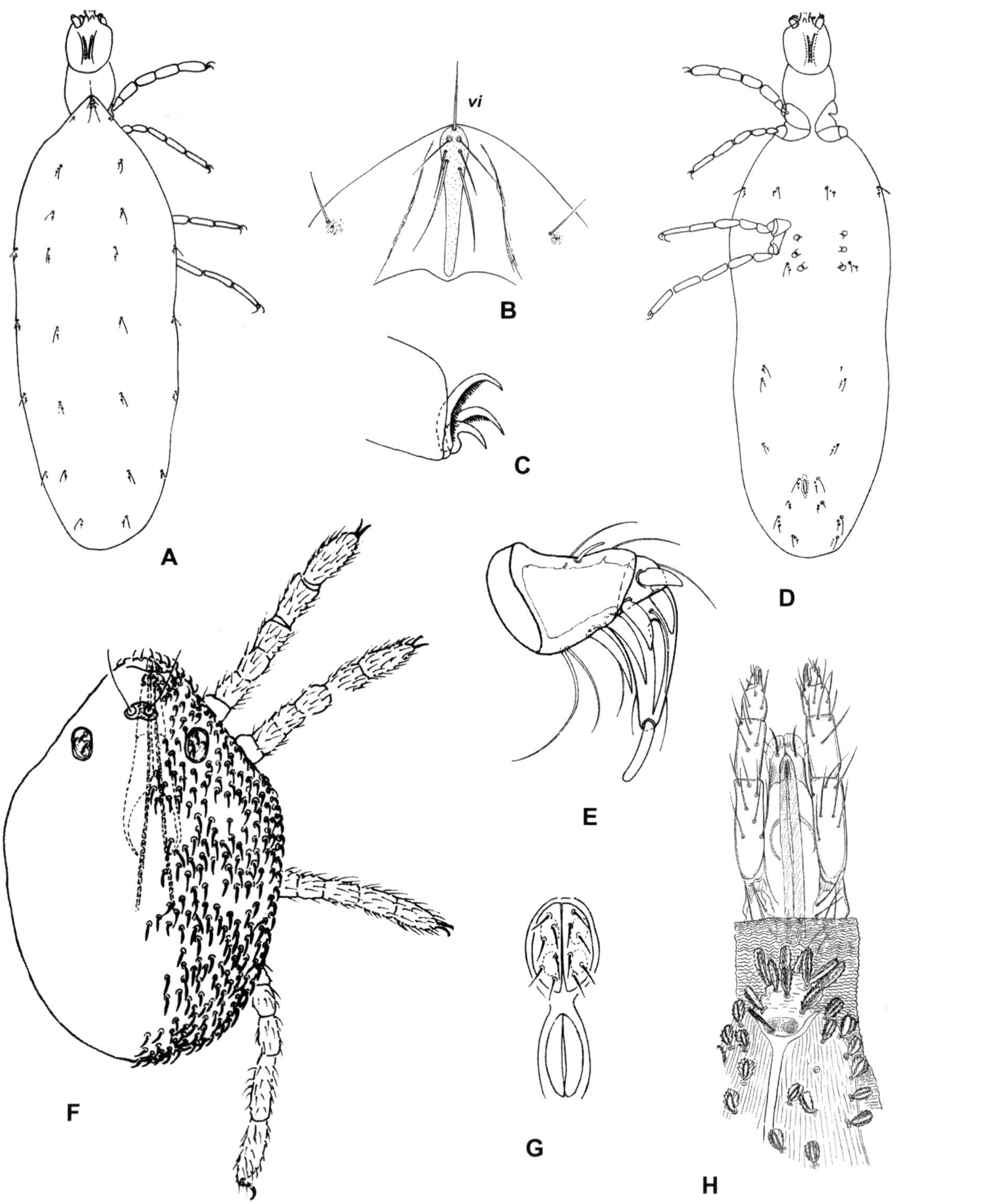

Fig. 13.38. STYGOTHROMBIDIIDAE. *Stygothrombium cooki* (Robaux): **A**, dorsum; **B**, prodorsal area; **D**, venter (all after Robaux 1969c). *Stygothrombium* sp.: **C**, claws and empodium of leg I; **E**, palp.

CALYPTOSTOMATIDAE. *Calyptostoma* sp.: **F**, dorsum with gnathosoma retracted; **G**, genito-anal field.

SMARIDIDAE. *Trichosmaris jacoti* (Southcott): **H**, prodorsal area with gnathosoma extruded (after Southcott 1963).

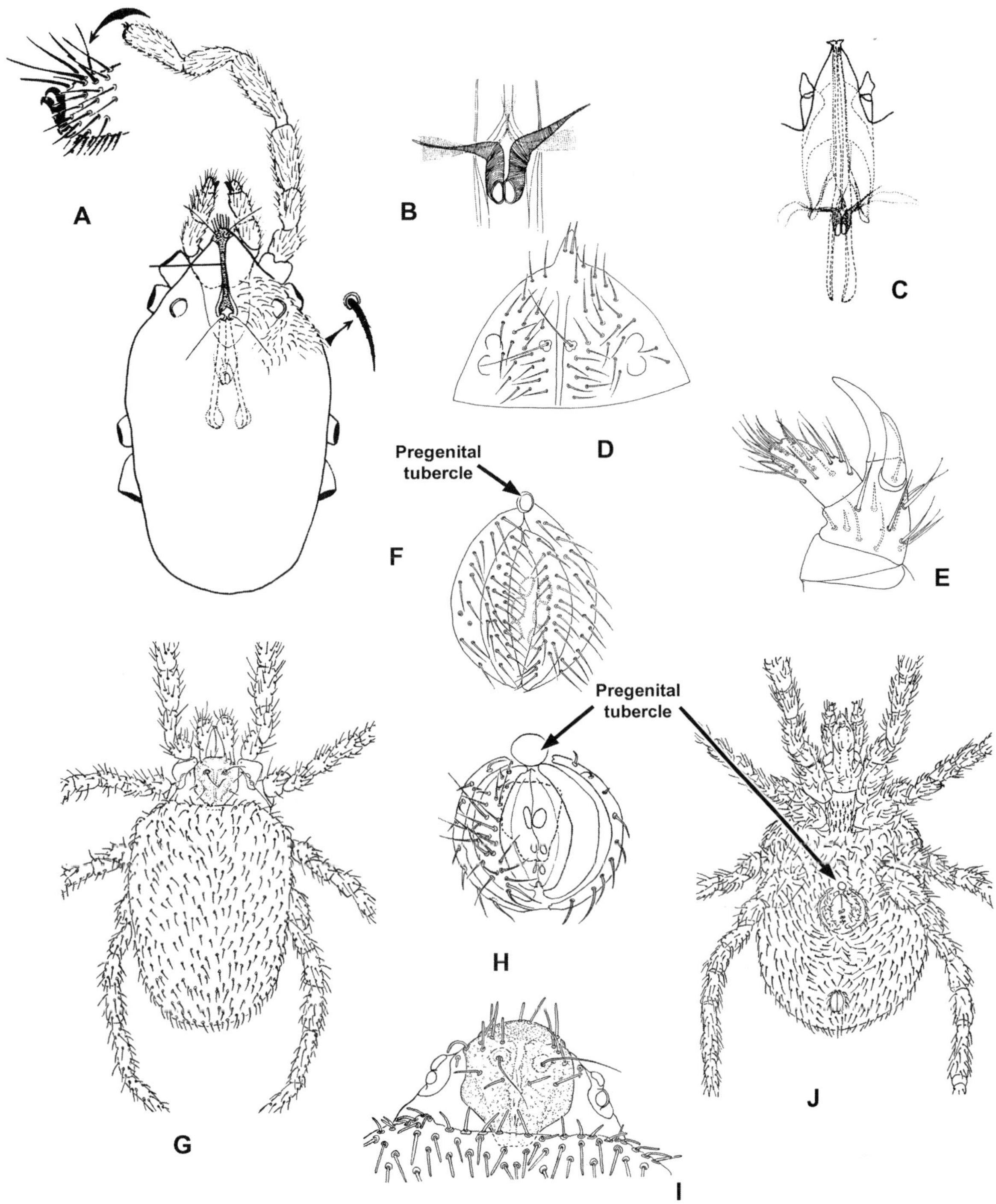

Fig. 13.39. ERYTHRAEIDAE. *Balaustium* sp.: **A**, dorsum with chelicerae retracted and showing detail of tarsus of leg I; **B**, stigmata and tracheal trunks; **C**, rostrum and associated respiratory structures.

TANAUPODIDAE. *Eothrombium nemoricola* (Berlese): **D**, prodorsal area; **E**, distal segments of palp; **F**, genital field (all after Robaux 1966c). *E. lasseni* (Newell): **G**, dorsum; **H**, genital field; **I**, prodorsal area; **J**, venter (all after Newell 1957).

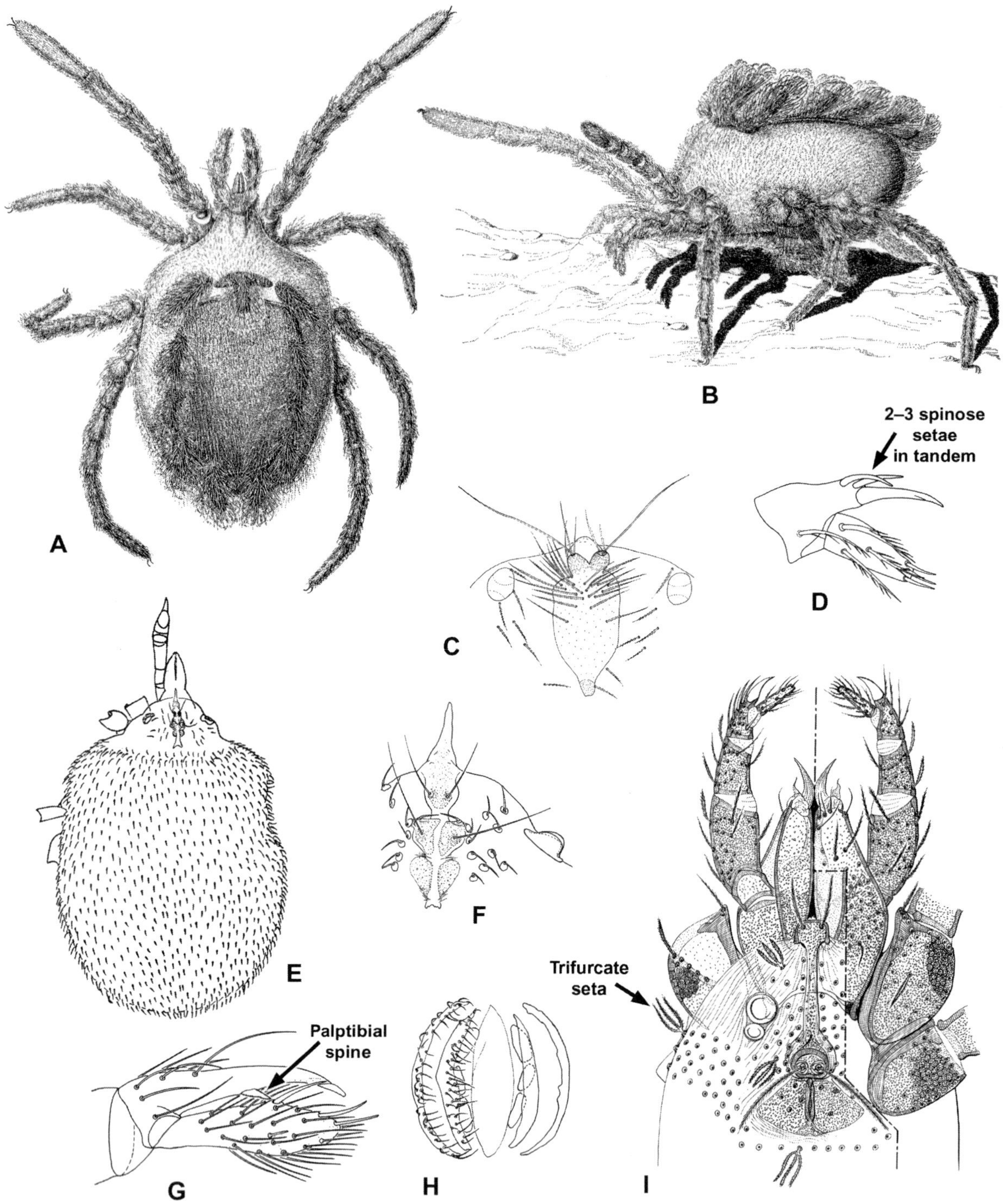

Fig. 13.40. CHYZERIIDAE. *Chyzeria poivrei* Robaux: **A**, dorsal habitus; **B**, lateral habitus; **C**, prodorsal area (all after Robaux 1969d).
TROMBICULIDAE. *Euschongastia jonesi* Lipovsky and Loomis: **D**, distal segments of palp of deutonymph (after Crossley 1960).
JOHNSTONIANIDAE. *Diplothrombium monoense* Newell: **E**, dorsum; **F**, prodorsal area; **G**, distal segments of palp; **H**, genital field (all after Newell 1957).
NEOTROMBIDIIDAE. *Neotrombidium barringunense* Hirst: **I**, split view of dorsum (left) and venter (right) of gnathosoma and propodosomatic area (after Lindquist and Vercammen-Grandjean 1971).

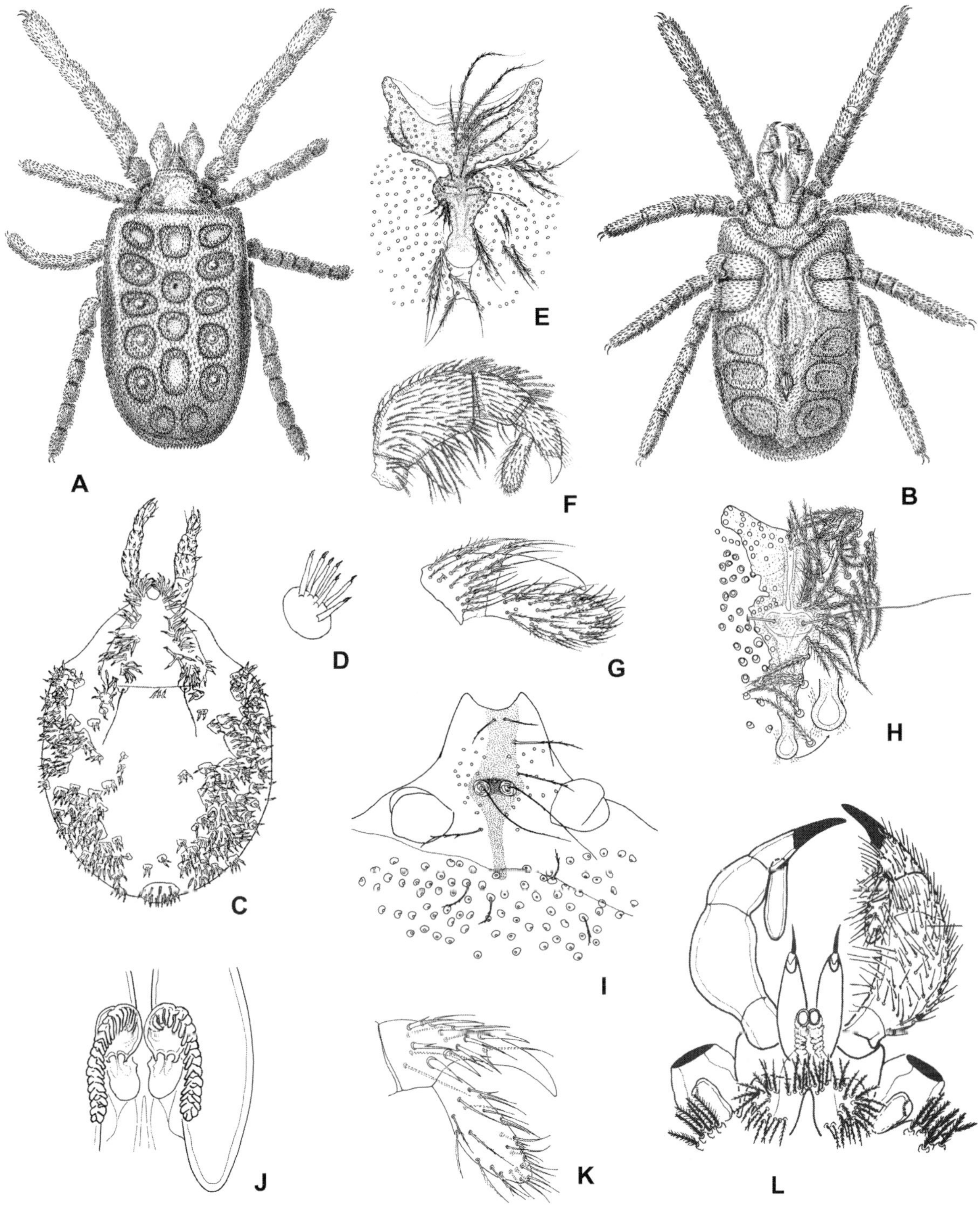

Fig. 13.41. TROMBELLIDAE. *Trombella lusitanica* André: **A**, dorsal habitus; **B**, ventral habitus (both after Robaux 1967a, with permission, Muséum National d'Histoire Naturelle, Paris).

AUDYANIDAE. *Audyana thompsoni* Womersley: **C**, dorsum; **D**, dorsal seta (both after Womersley 1954b).

TROMBIDIIDAE. *Dinothrombium pandorae* (Newell and Tevis): **E**, prodorsal area (after Newell and Tevis, 1960). *Austrothrombium hirsutum* (Womersley): **F**, palp (after Southcott 1986c). *Parathrombium quadriseta* Newell: **G**, distal segments of palp (after Newell 1958). *P. bidactylus* Newell: **H**, prodorsal area (after Newell 1958). *Podothrombium filiforme* Mąkol: **I**, prodorsal area (after Mąkol and Marusik 1999). *Trombidium* sp.: **J**, internal air sacs and peritremes; **L**, gnathosoma and prodorsal area. *P. dubium* Robaux: **K**, distal segments of palp (after Robaux 1966c).

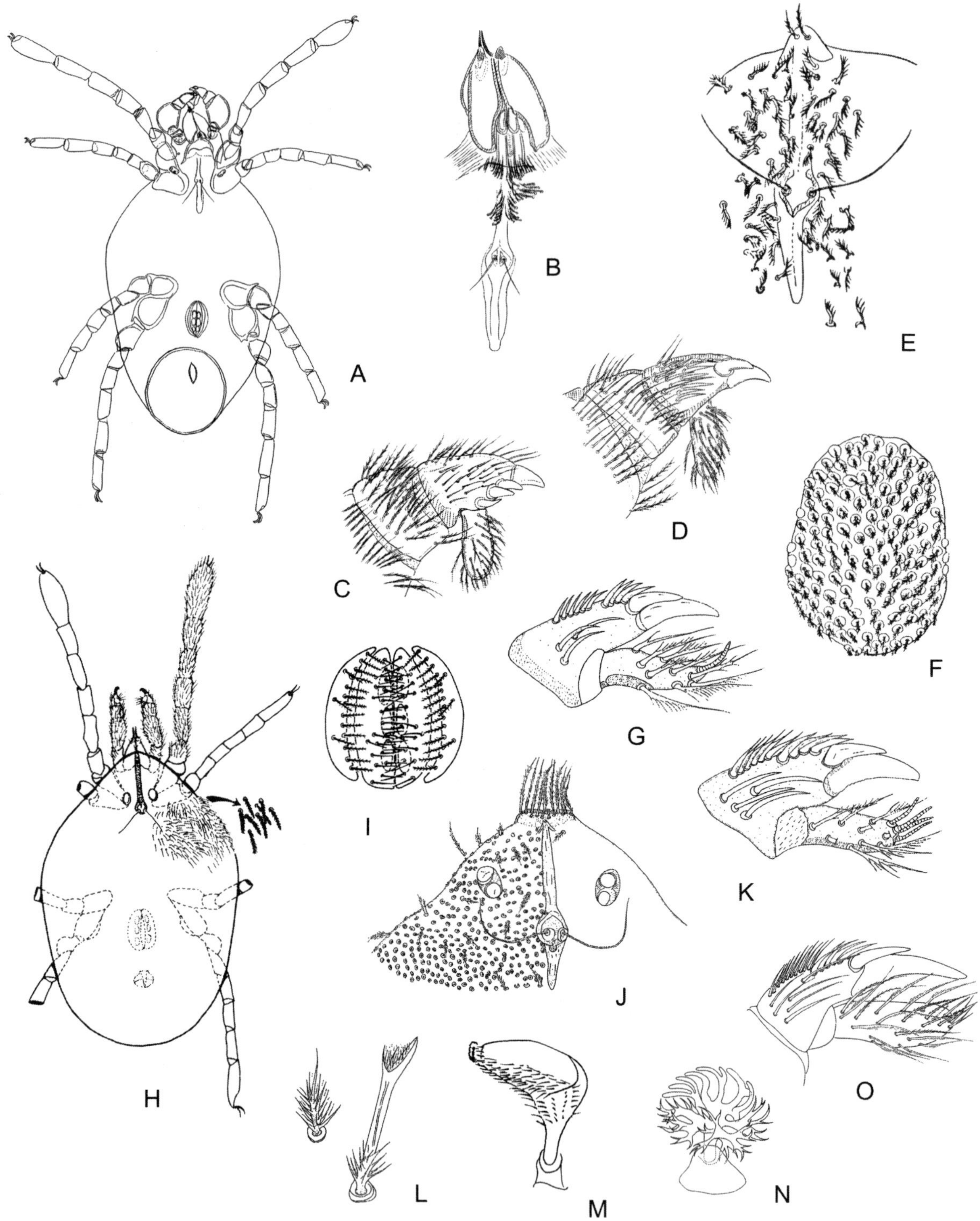

Fig. 13.42. EUTROMBIDIIDAE. *Eutrombidium lebaroni* Southcott: **A**, combination transparent view of dorsum and venter; **B**, prodorsal area; **C**, distal segments of palp, lateral view; **D**, distal segments of palp, medial view (all after Southcott 1992a). *Bruyantella smithi* Southcott: **E**, prodorsal area; **F**, posterior dorsal plate (both after Southcott 1991c).

MICROTROMBIDIIDAE. *Microtrombidium pusillum* (Hermann): **G**, distal segments of palp, medial view (after Gabryś and Wohltmann 2001). *Microtrombidium* sp.: **H**, dorsum; **I**, genital field. *Echinothrombium rhodinum* (Koch): **J**, prodorsal area; **K**, distal segments of palp, medial view; **L**, dorsal setae (all after Gabryś and Wohltmann 2003). *Willmanniella johnstoni* (Robaux): **M**, dorsal seta (after Robaux 1976). *Dendrothrombidium betschi* Robaux: **N**, dorsal seta (after Robaux 1966b, with permission, Muséum National d'Histoire Naturelle, Paris). *Campylothrombium barbarum* (Lucas): **O**, distal segments of palp, medial view (after Robaux 1966c).

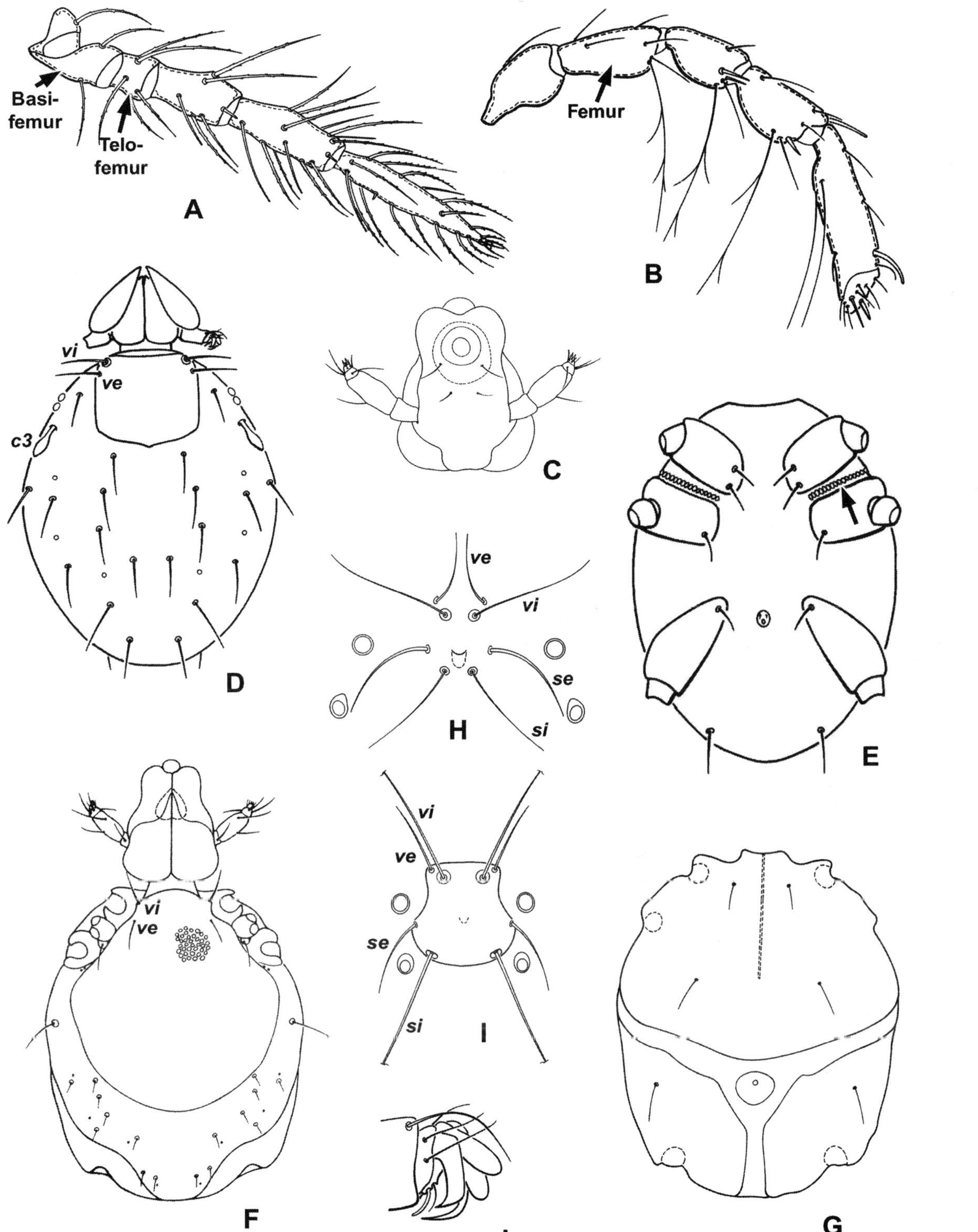

Fig. 13.43. HYDRYPHANTIDAE. *Acerbitas stolli* (Koenike): **A**, leg I. *Protzia* sp.: **H**, prodorsal region. *Tartarothyas* sp.: **I**, dorsal plate region.
HYDRACHNIDAE. *Hydrachna magniscutata* Marshall: **B**, leg I.
ACHERONTACARIDAE. *Acherontacarus* sp.: **C**, venter of gnathosoma; **F**, dorsum of idiosoma and gnathosoma; **G**, venter of idiosoma.
HYDROVOLZIIDAE. *Hydrovolzia gerhardi* Mitchell: **D**, dorsum of idiosoma and gnathosoma; **E**, venter of idiosoma; **J**, distal segments of palp.

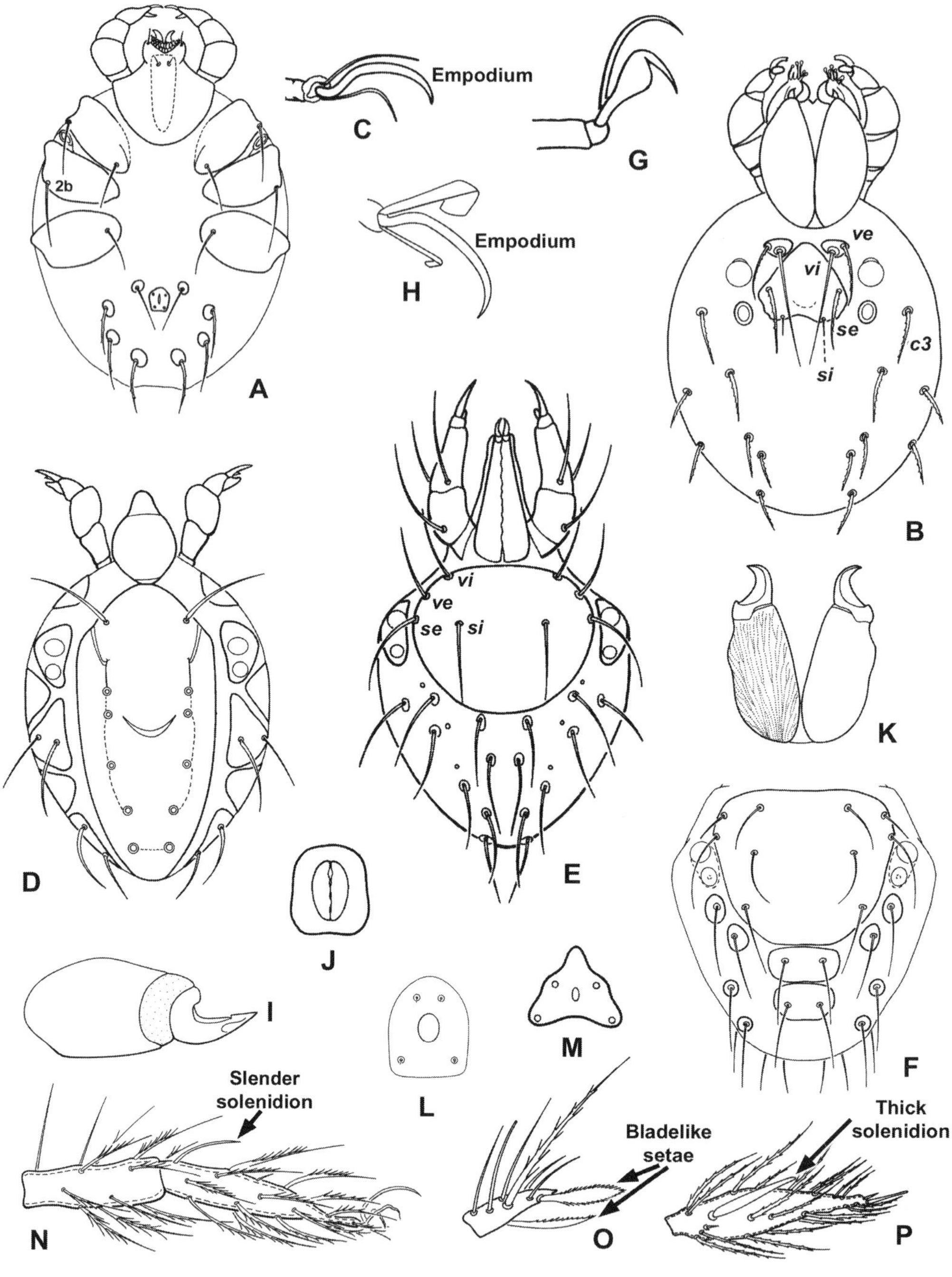

Fig. 13.44. HYDRYPHANTIDAE. *Acerbitas stolli* (Koenike): **A**, venter of idiosoma and gnathosoma; **C**, claws and empodium of leg I. *Hydryphantes ruber* De Geer: **B**, dorsum of idiosoma and gnathosoma; **K**, chelicerae. *Protzia* sp.: **M**, excretory pore plate; **N**, tibia and tarsus of leg I.

EYLAIDAE. *Eylais major* Lanciani: **D**, dorsum of idiosoma and gnathosoma; **H**, claws and empodium of leg I.

LIMNOCHARIDAE. *Limnochares americana* Lundblad: **E**, dorsum of idiosoma and gnathosoma. *Neolimnochares* sp.: **G**, claw and empodium of leg I.

PIERSIGIIDAE. *Piersigia* sp.: **F**, dorsum of idiosoma.

THERMACARIDAE. *Thermacarus nevadensis* Marshall: **I**, chelicera; **J**, excretory pore plate.

HYDRODROMIDAE. *Hydrodroma despiciens* (Müller): **L**, excretory pore plate; **O**, palptarsus; **P**, tarsus of leg I.

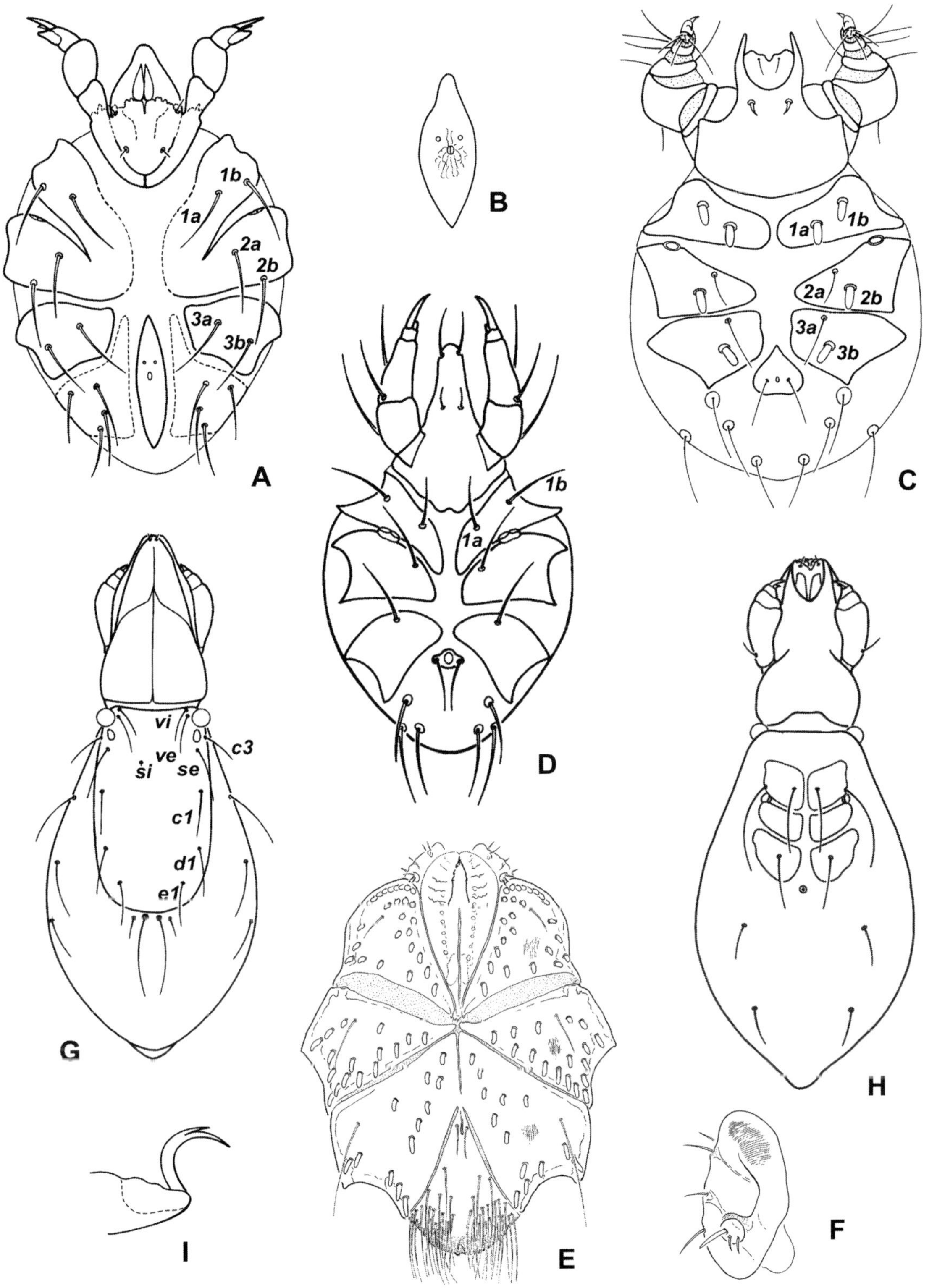

Fig. 13.45. EYLAIDAE. *Eylais major* Lanciani: **A**, venter of idiosoma and gnathosoma; **B**, excretory pore plate.
PIERSIGIIDAE. *Piersigia* sp.: **C**, venter of idiosoma and gnathosoma.
LIMNOCHARIDAE. *Limnochares americana* Lundblad: **D**, venter of idiosoma and gnathosoma.
APHEVIDERULICIDAE. *Apheviderulix welwitchioides* Gerecke, Smith, and Cook: **E**, venter of idiosoma and gnathosoma; **F**, palp.
HYDRACHNIDAE. *Hydrachna magniscutata* Marshall: **G**, dorsum of idiosoma and gnathosoma; **H**, venter of idiosoma and gnathosoma; **I**, empodium of leg I.

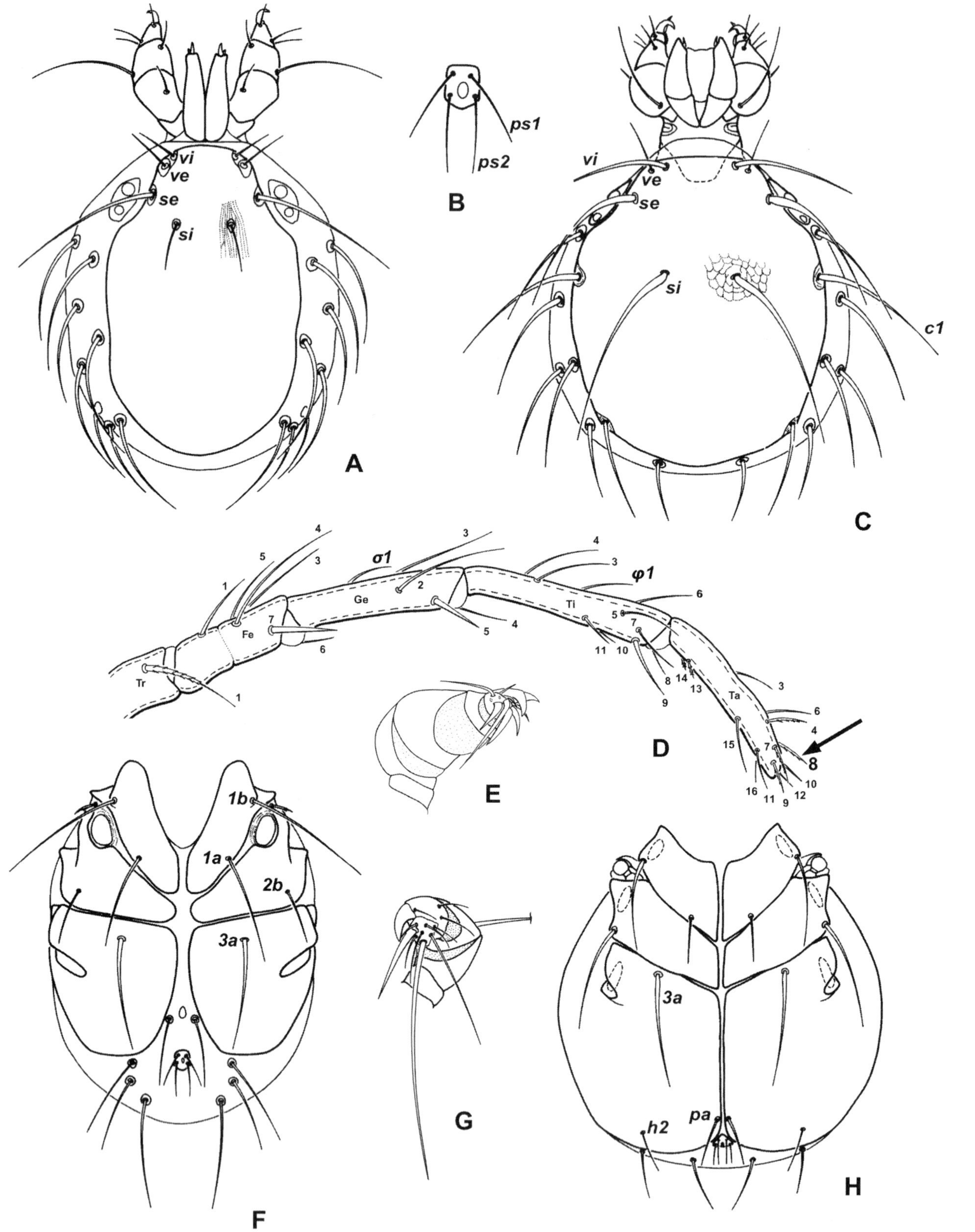

Fig. 13.46. SPERCHONTIDAE. *Sperchon glandulosus* Koenike: **A**, dorsum of idiosoma and gnathosoma; **B**, excretory pore plate; **F**, venter of idiosoma; **G**, palp. *Sperchonopsis ecphyma* Prasad and Cook: **D**, leg III.

ANISITSIELLIDAE. *Bandakiopsis fonticola* Smith: **C**, dorsum of idiosoma and gnathosoma; **E**, palp; **H**, venter of idiosoma.

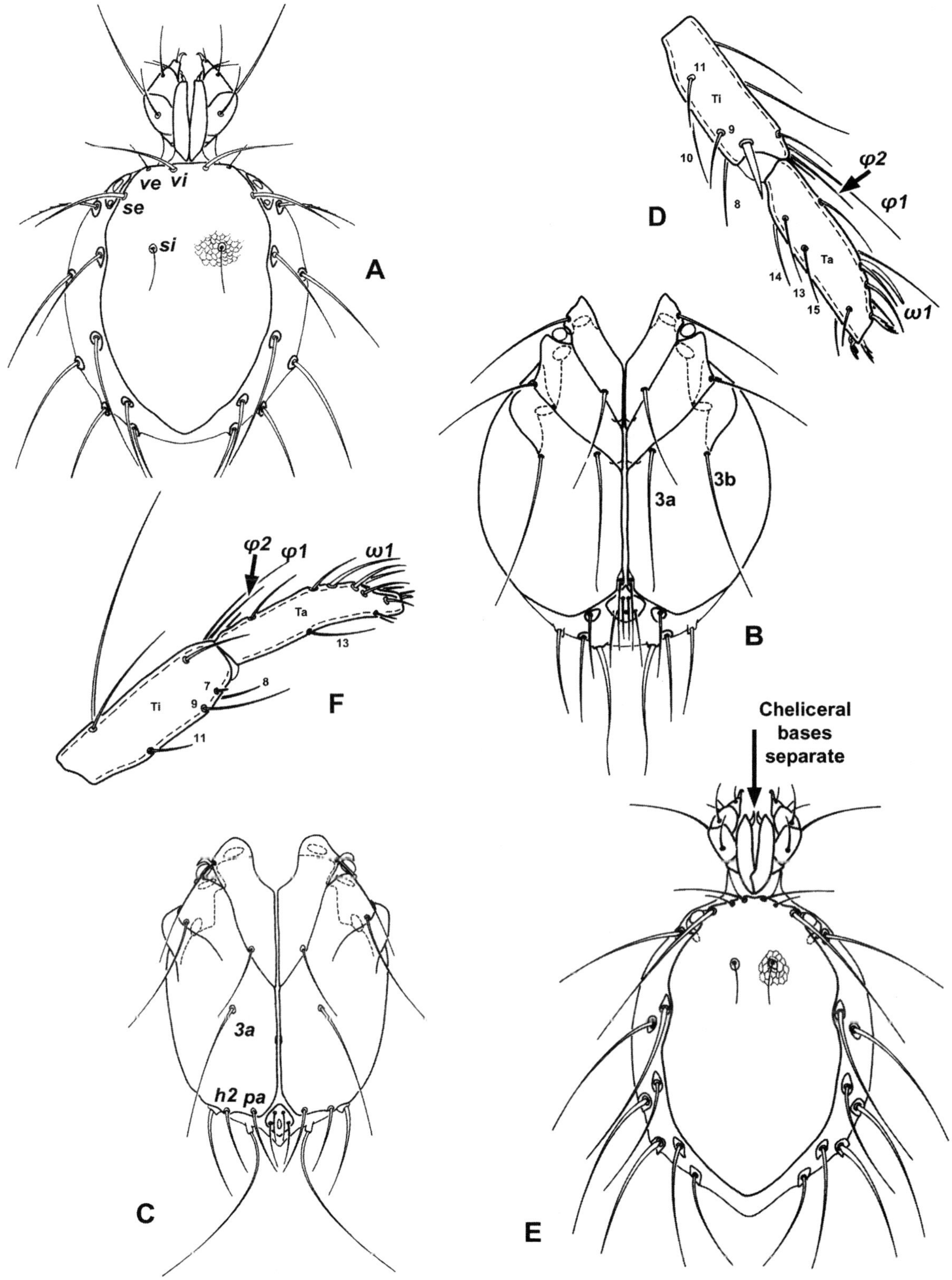

Fig. 13.47. TEUTONIIDAE. *Teutonia lunata* Marshall: **A**, dorsum of idiosoma and gnathosoma; **B**, venter of idiosoma. LEBERTIIDAE. *Lebertia* sp.: **C**, venter of idiosoma; **D**, tibia and tarsus of leg I; **E**, dorsum of idiosoma and gnathosoma. OXIDAE. *Oxus elongatus* Marshall: **F**, tibia and tarsus of leg I.

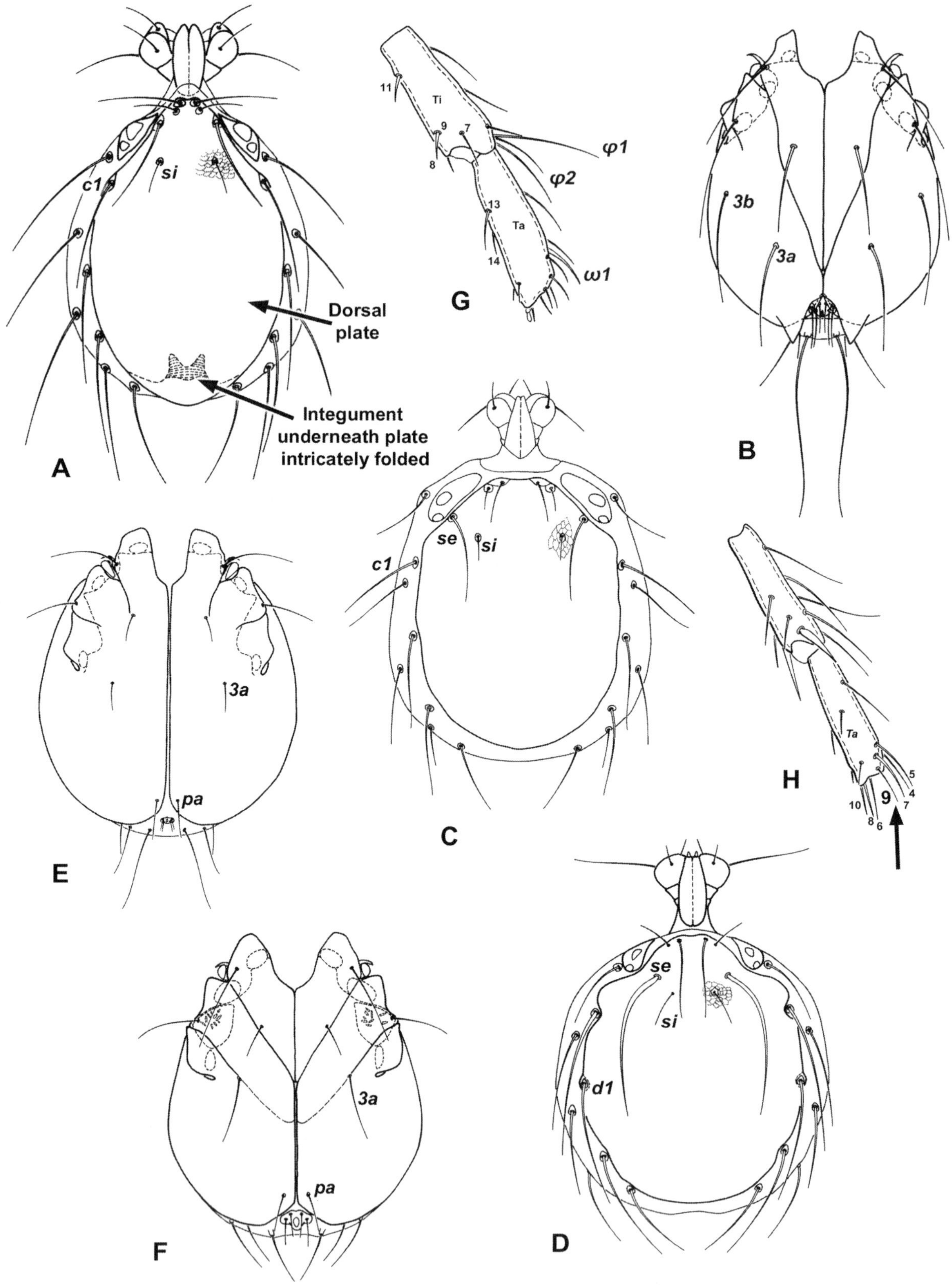

Fig. 13.48. OXIDAE. *Oxus elongatus* Marshall: **A**, dorsum of idiosoma and gnathosoma; **B**, venter of idiosoma. TORRENTICOLIDAE. *Torrenticola* sp.: **C**, dorsum of idiosoma and gnathosoma; **E**, venter of idiosoma. *Testudacarus* sp.: **D**, dorsum of idiosoma and gnathosoma; **F**, venter of idiosoma; **G**, tibia and tarsus of leg I. ANISITSIELLIDAE. *Bandakia phreatica* Cook: **H**, tibia and tarsus of leg III.

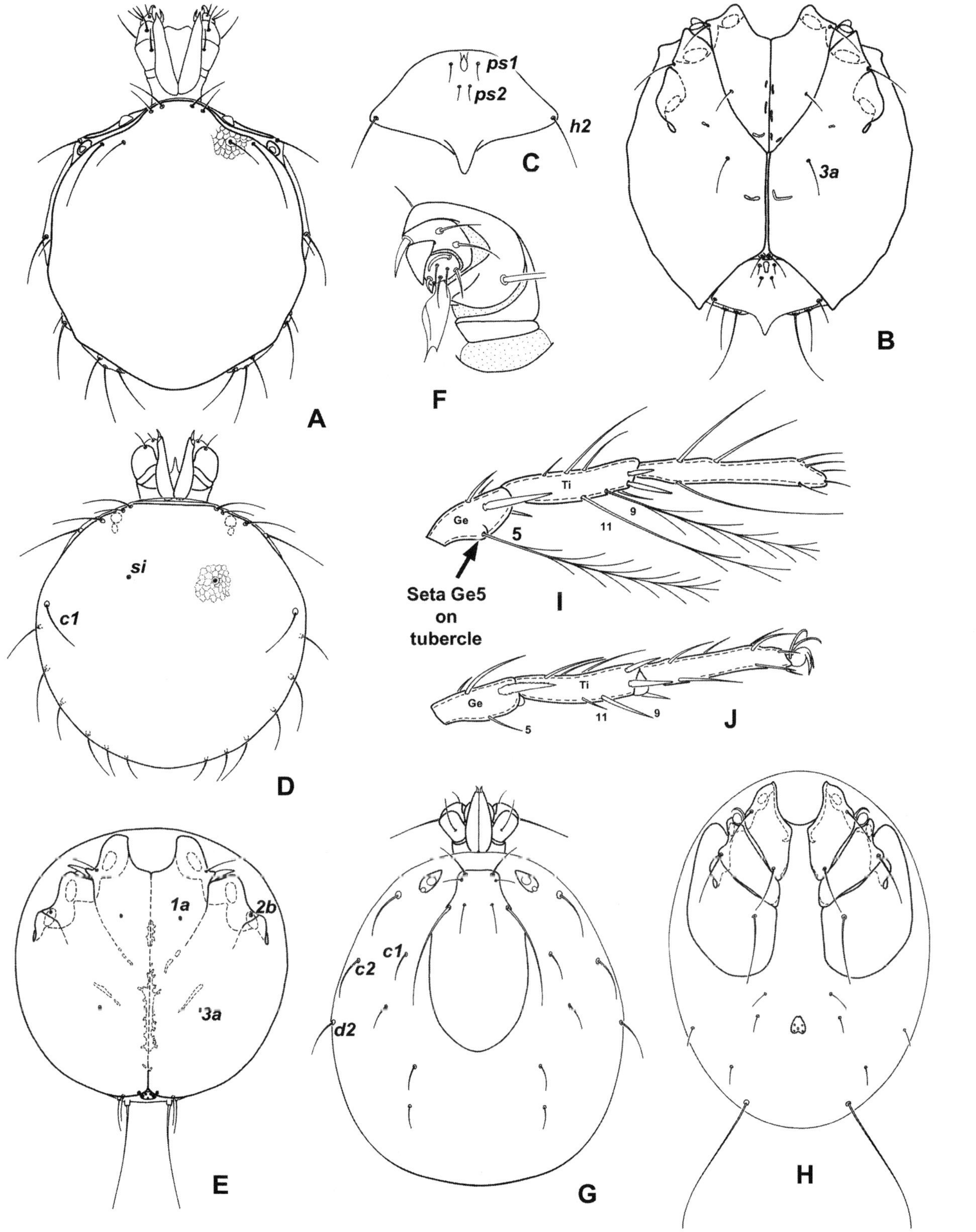

Fig. 13.49. ANISITSIELLIDAE. *Bandakia phreatica* Cook: **A**, dorsum of idiosoma and gnathosoma; **B**, venter of idiosoma; **C**, excretory pore plate.

ACALYPTONOTIDAE. *Acalyptonotus neoviolaceus* Smith: **D**, dorsum of idiosoma and gnathosoma; **E**, venter of idiosoma; **F**, palp.

MOMONIIDAE. *Stygomomonia mitchelli* Smith: **G**, dorsum of idiosoma and gnathosoma (engorged specimen); **H**, venter of idiosoma (engorged specimen); **J**, distal segments of leg III. *Momonia campylotibia* Smith: **I**, distal segments of leg III.

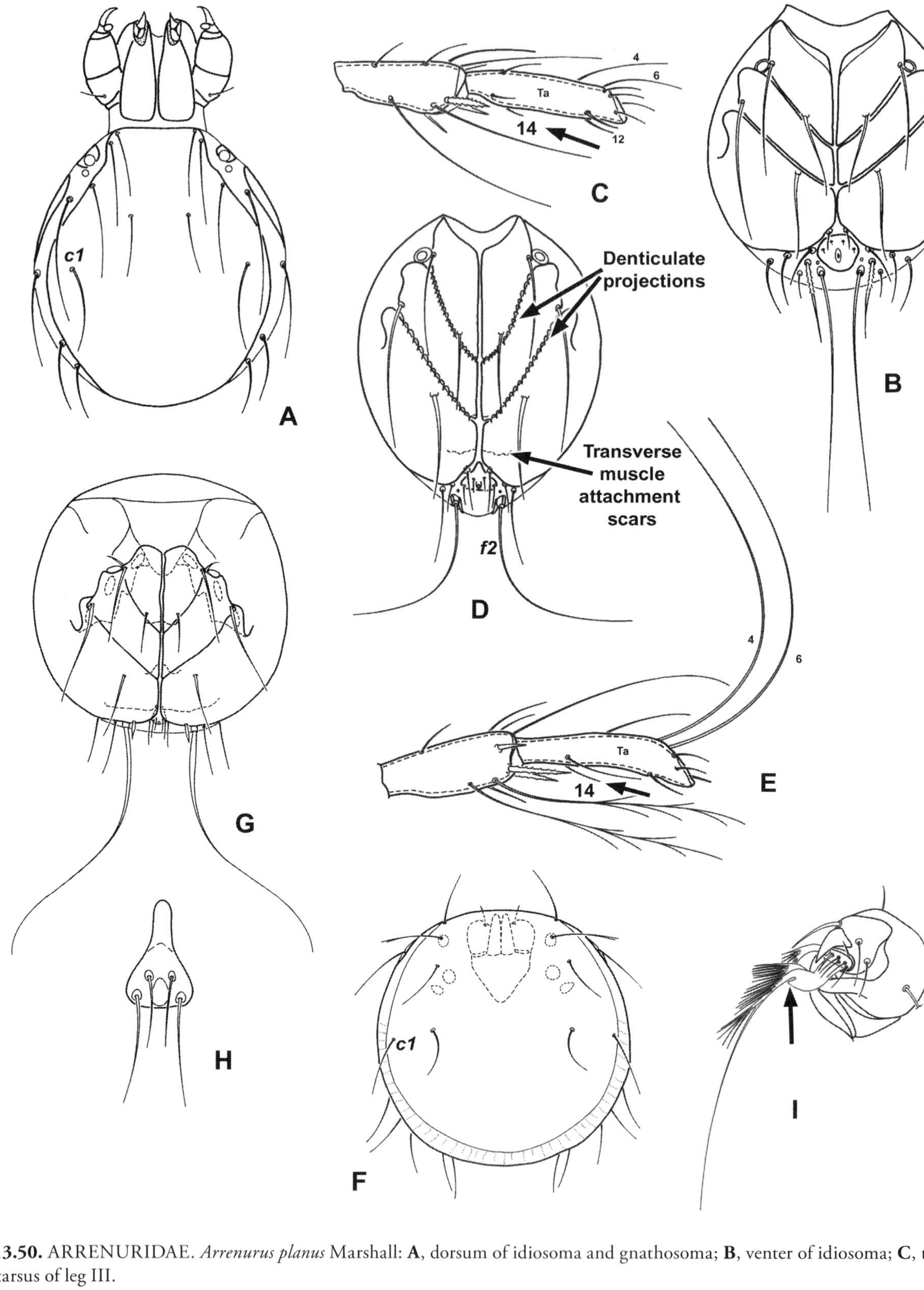

Fig. 13.50. ARRENURIDAE. *Arrenurus planus* Marshall: **A**, dorsum of idiosoma and gnathosoma; **B**, venter of idiosoma; **C**, tibia and tarsus of leg III.

KRENDOWSKIIDAE. *Krendowskia similis* Viets: **D**, venter of idiosoma; **E**, tibia and tarsus of leg III.

MIDEOPSIDAE. *Mideopsis borealis* Habeeb: **F**, dorsum of idiosoma and gnathosoma; **G**, venter of idiosoma.

NUDOMIDEOPSIDAE. *Nudomideopsis magnacetabula* (Smith): **H**, excretory pore plate. *Paramideopsis susanae* Smith: **I**, palp.

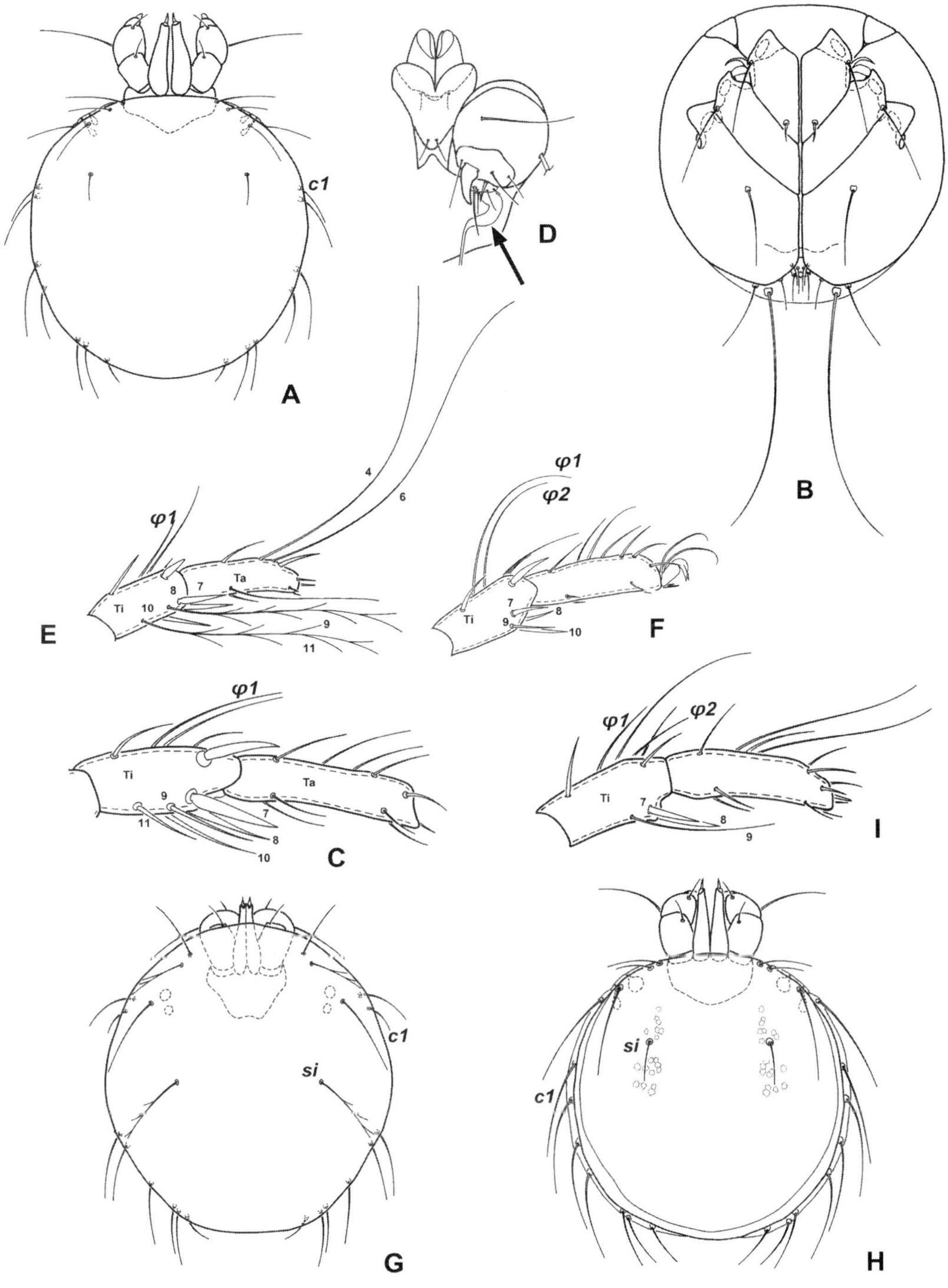

Fig. 13.51. NUDOMIDEOPSIDAE. *Nudomideopsis magnacetabula* (Smith): **A**, dorsum of idiosoma and gnathosoma; **B**, venter of idiosoma. *Paramideopsis susanae* Smith: **C**, tibia and tarsus of leg III.

MIDEIDAE. *Midea expansa* Marshall: **D**, venter of gnathosoma; **E**, tibia and tarsus of leg III.

ATHIENEMANNIIDAE. *Platyhydracarus juliani* Smith: **F**, tibia and tarsus of leg II; **G**, dorsum of idiosoma and gnathosoma.

NEOACARIDAE. *Neoacarus occidentalis* Cook: **H**, dorsum of idiosoma and gnathosoma.

LAVERSIIDAE. *Laversia berulophila* Cook: **I**, tibia and tarsus of leg II.

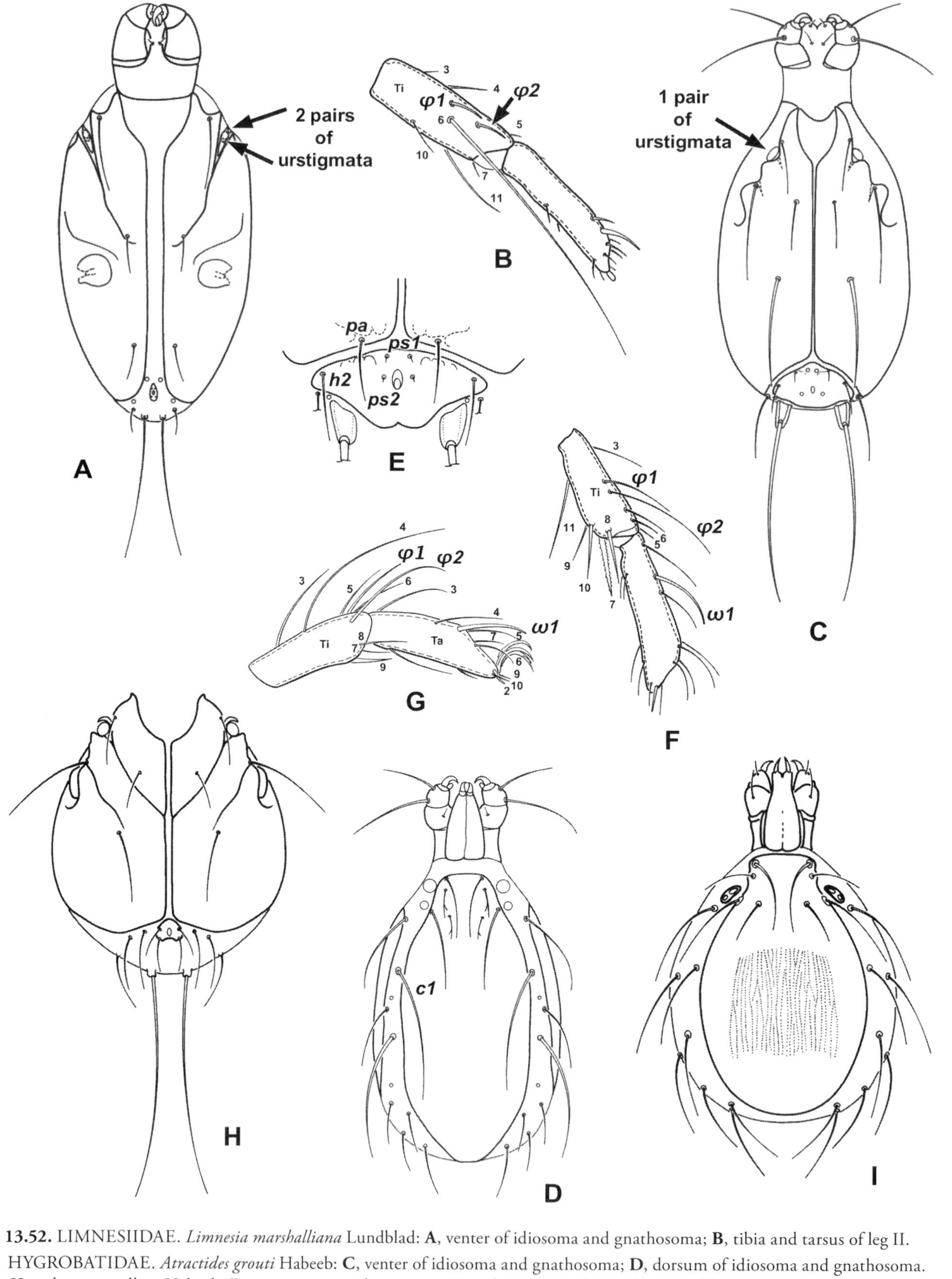

Fig. 13.52. LIMNESIIDAE. *Limnesia marshalliana* Lundblad: **A**, venter of idiosoma and gnathosoma; **B**, tibia and tarsus of leg II. HYGROBATIDAE. *Atractides grouti* Habeeb: **C**, venter of idiosoma and gnathosoma; **D**, dorsum of idiosoma and gnathosoma. *Hygrobates neocalliger* Habeeb: **E**, excretory pore plate region. *Hygrobates* sp.: **F**, tibia and tarsus of leg II. FELTRIIDAE. *Feltria* sp.: **G**, tibia and tarsus of leg I; **H**, venter of idiosoma; **I**, dorsum of idiosoma and gnathosoma.

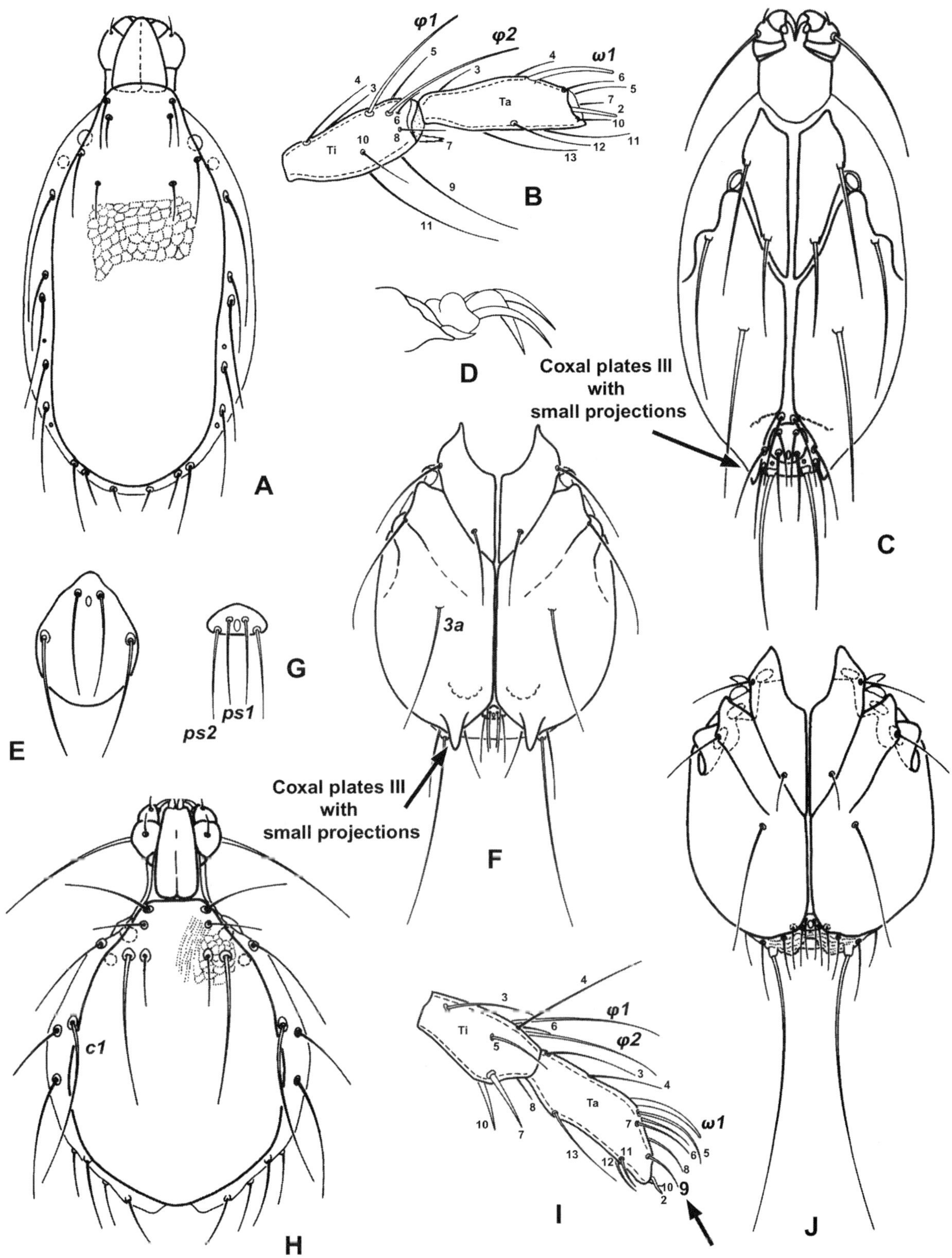

Fig. 13.53. UNIONICOLIDAE. *Unionicola* sp.: **A**, dorsum of idiosoma and gnathosoma; **B**, tibia and tarsus of leg I; **C**, venter of idiosoma and gnathosoma. *Neumannia punctata* Marshall: **D**, claws and empodium of leg I; **E**, excretory pore plate.
WETTINIDAE. *Wettina ontario* Smith: **F**, venter of idiosoma; **G**, excretory pore plate.
ATURIDAE. *Aturus* sp.: **H**, dorsum of idiosoma and gnathosoma; **I**, tibia and tarsus of leg I; **J**, venter of idiosoma.

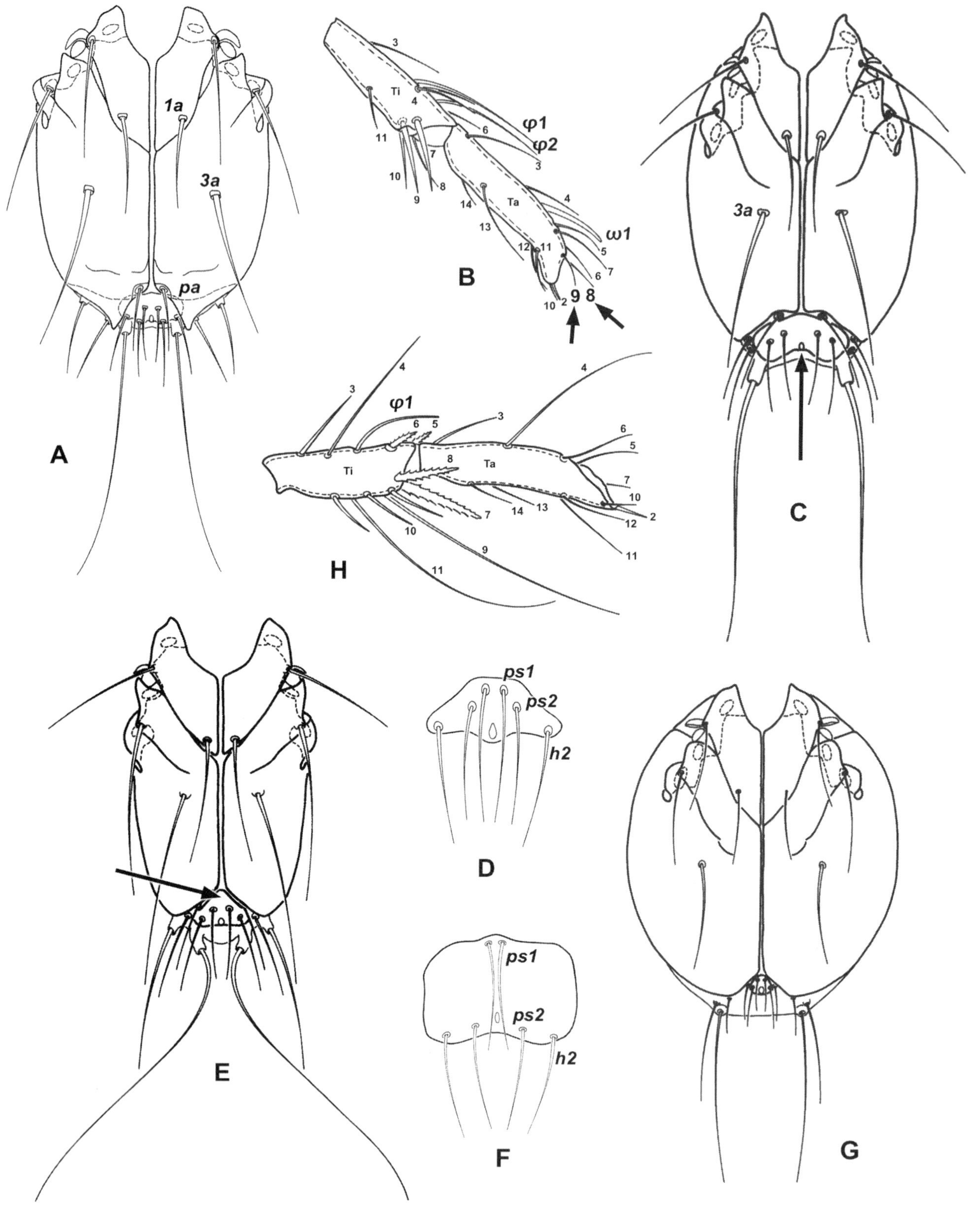

Fig. 13.54. ATURIDAE. *Axonopsis setonensis* Habeeb: **A**, venter of idiosoma. *Ljania bipapillata* Thor: **B**, tibia and tarsus of leg I; **C**, venter of idiosoma. *Estellacarus unguitarsus* (Habeeb): **D**, excretory pore plate. *Albia neogaea* Habeeb: **E**, venter of idiosoma. *Woolastookia pilositarsa* (Habeeb): **F**, excretory pore plate. *Neobrachypoda ekmani* (Walter): **G**, venter of idiosoma. PIONIDAE. *Forelia ovalis* Marshall: **H**, tibia and tarsus of leg III.

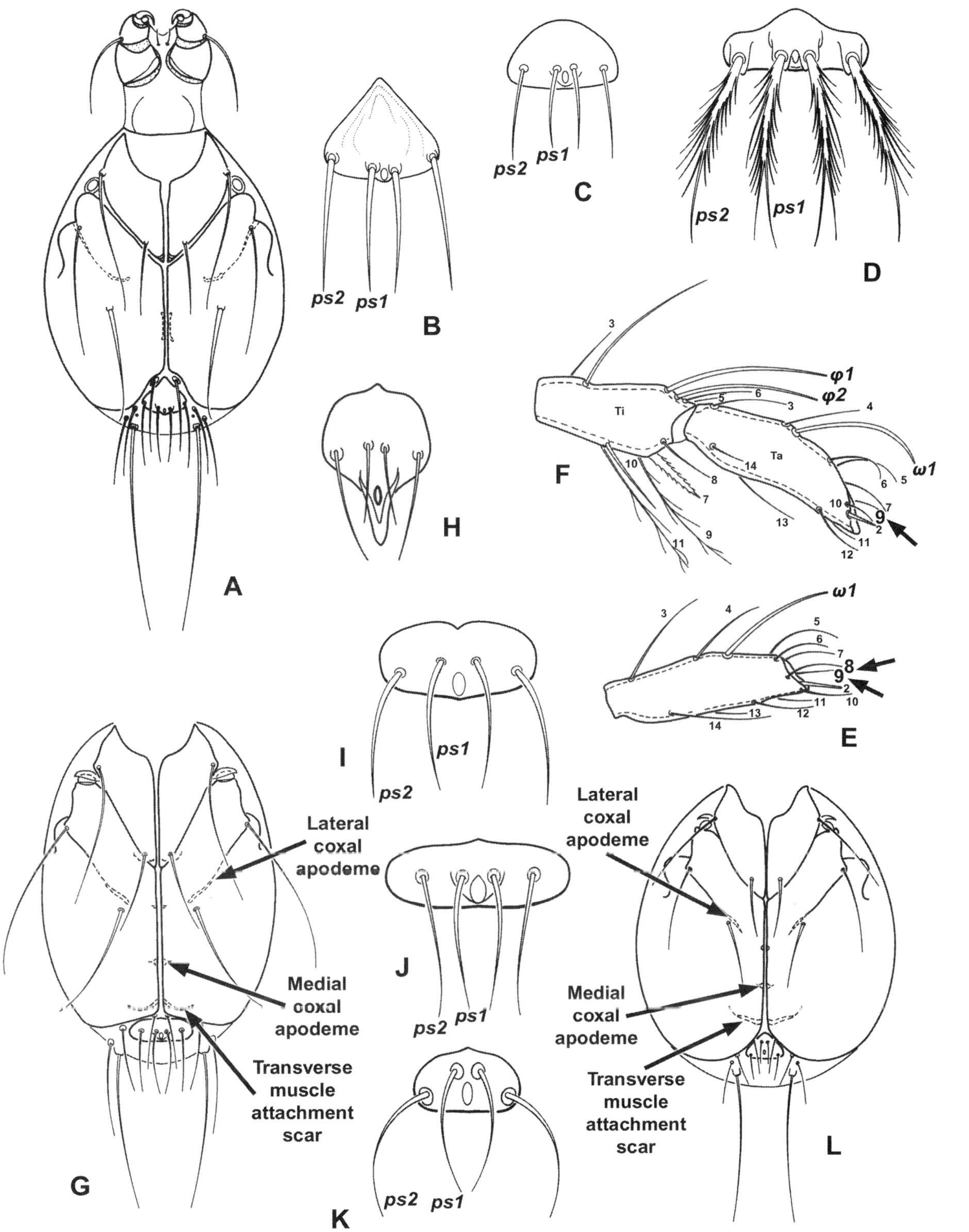

Fig. 13.55. PIONIDAE. *Piona interrupta* Marshall: **A**, venter of idiosoma and gnathosoma; **B**, excretory pore plate. *Najadicola ingens* Koenike: **C**, excretory pore plate. *Nautarachna muskoka* Smith: **D**, excretory pore plate. *P. mitchelli* Cook: **E**, tarsus of leg I. *Tiphys ornatus* (Koch): **F**, tibia and tarsus of leg I. *Neotiphys pionoidellus* (Habeeb): **G**, venter of idiosoma; **J**, excretory pore plate. *Hydrochoreutes minor* Cook: **H**, excretory pore plate. *Huitfeldtia rectipes* Thor: **I**, excretory pore plate. *Forelia onondaga* Habeeb: **K**, excretory pore plate. *Pseudofeltria multipora* Cook: **L**, venter of idiosoma.

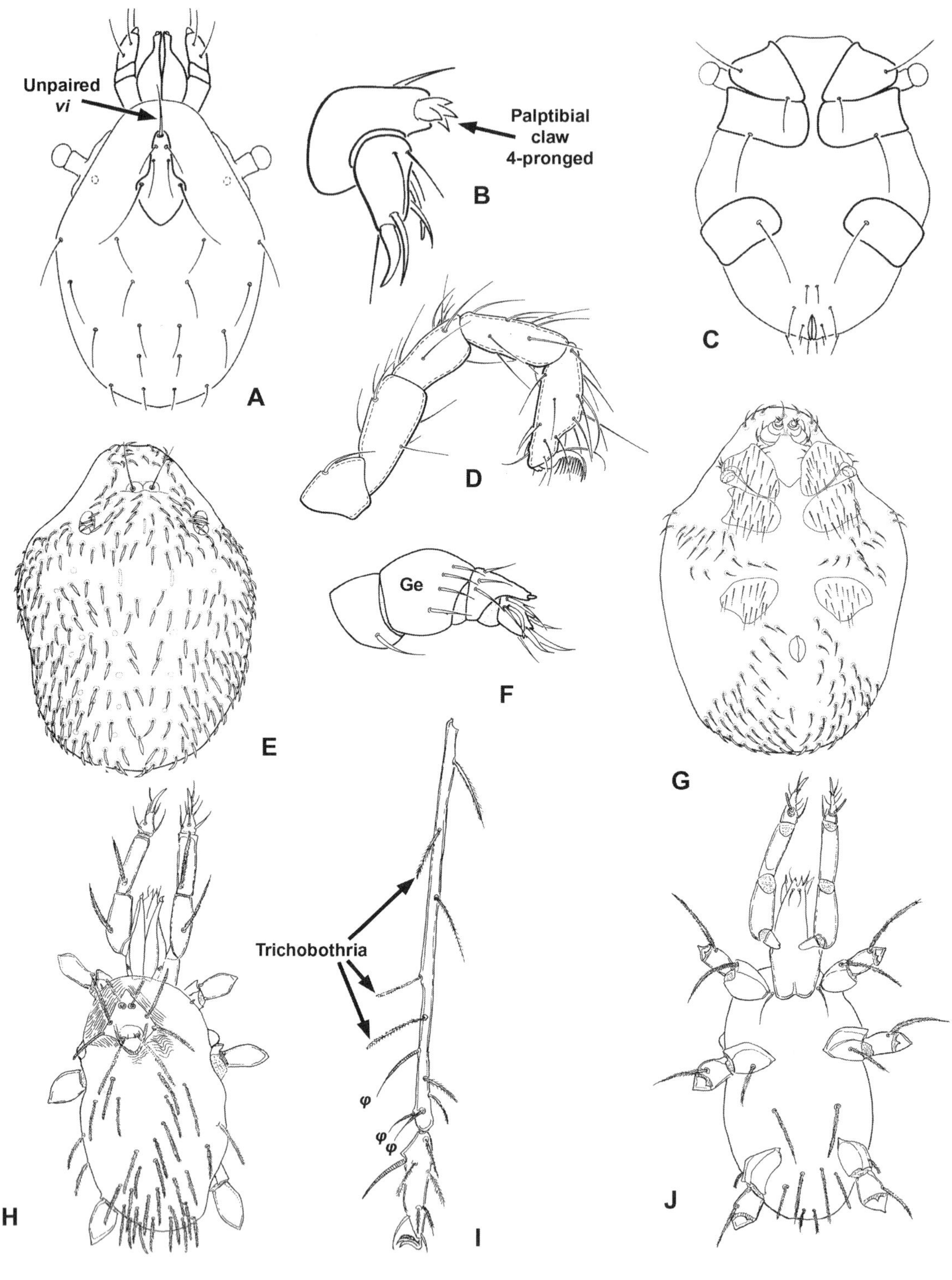

Fig. 13.56. STYGOTHROMBIDIIDAE. *Stygothrombium* sp.: **A**, dorsum; **B**, distal segments of palp; **C**, venter of idiosoma; **D**, leg I.
CALYPTOSTOMATIDAE. *Calyptostoma velutinus* (Müller): **E**, dorsum of idiosoma; **F**, palp; **G**, venter (all redrawn after Vistorin-Theis 1978).
SMARIDIDAE. *Surasmaris longirostris* Southcott: **H**, dorsum; **I**, tibia and tarsus of leg I; **J**, venter (all after Southcott 1995).

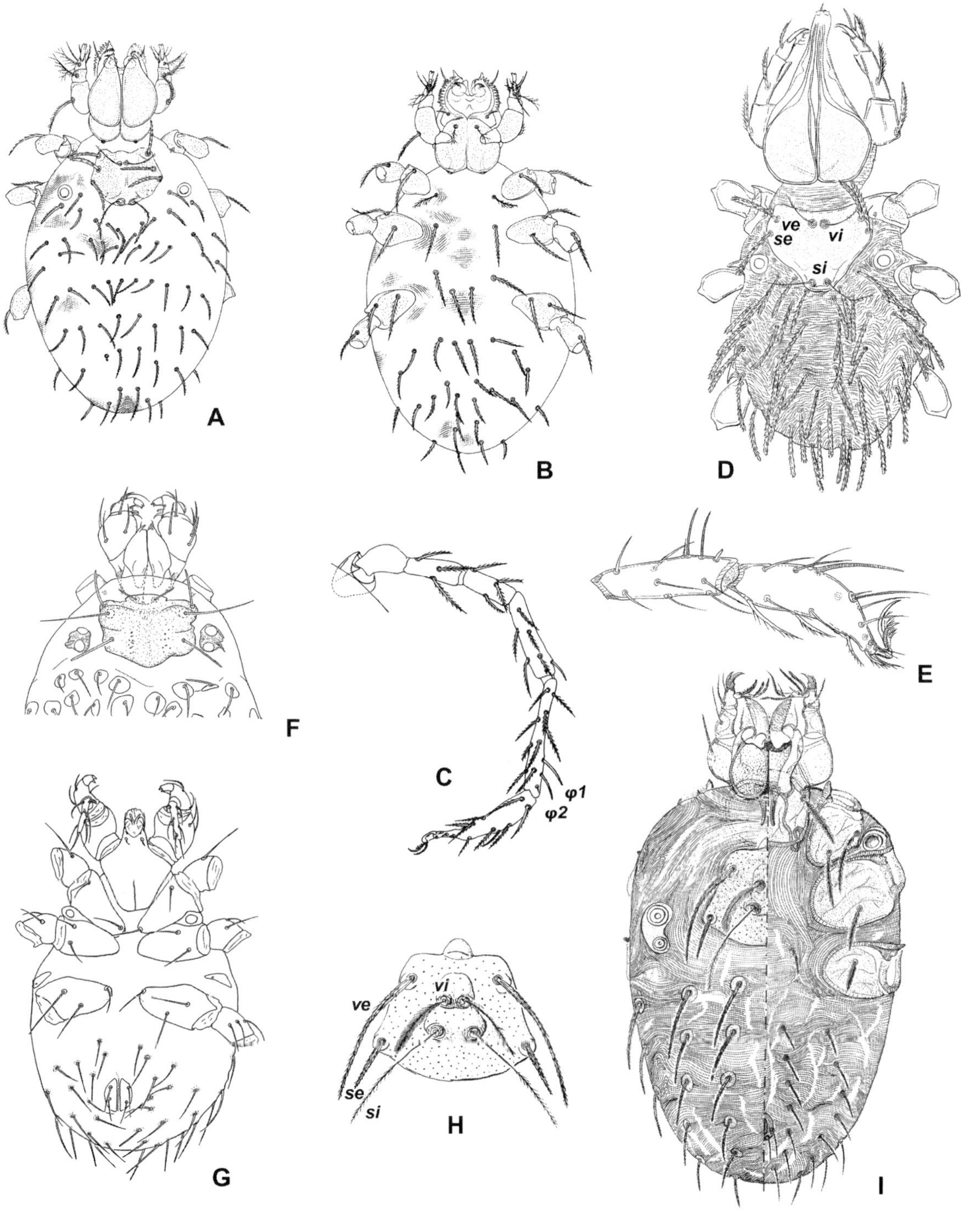

Fig. 13.57. ERYTHRAEIDAE. *Charletonia rocciai* Treat and Flechtmann: **A**, dorsum; **B**, venter; **C**, leg I, posterior view (all after Treat and Flechtmann 1979). *Leptus ariel* Southcott: **D**, dorsum (after Southcott 1989b). *Charletonia blascoi* Southcott: **E**, tibia and tarsus of leg I, anterior view (after Southcott 1993b).

TANAUPODIDAE. *Eothrombium lasseni* (Newell): **F**, prodorsal area; **G**, venter (both after Newell 1957).

CHYZERIIDAE. *Pteridopus pseudhannemania* Newell and Vercammen-Grandjean: **H**, prodorsal plate; **I**, split view of dorsum (left) and venter (right) of gnathosoma and idiosoma (both after Newell and Vercammen-Grandjean 1964).

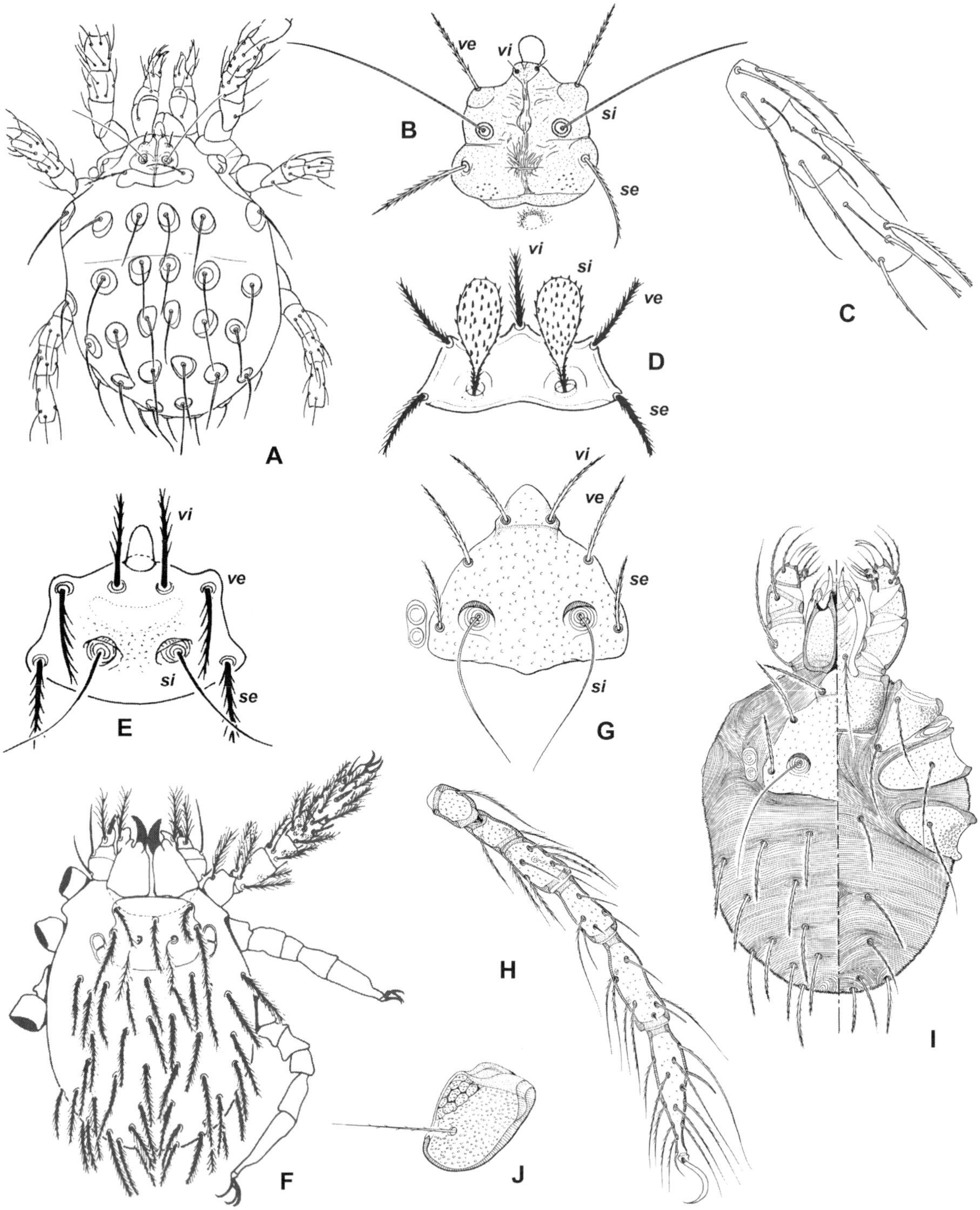

Fig. 13.58. JOHNSTONIANIDAE. *Diplothrombium micidium* Newell: **A**, dorsum. *D. monoense* Newell: **B**, prodorsal plate; **C**, telofemur, genu, and tibia of leg III (all after Newell 1957).

TROMBICULIDAE. *Euschongastia* sp.: **D**, prodorsal plate. *Trombicula* sp.: **F**, dorsum.

LEEUWENHOEKIIDAE. *Leeuwenhoekia americana* (Ewing): **E**, prodorsal plate.

NEOTROMBIDIIDAE. *Neotrombidium samsinaki* (Daniel): **G**, prodorsal plate; **H**, leg I, anterolateral view; **I**, split view of dorsum (left) and venter (right) of gnathosoma and idiosoma (all after Lindquist and Vercammen-Grandjean 1971). *N. anuroporum* Lindquist and Vercammen-Grandjean: **J**, coxal plate III (after Lindquist and Vercammen-Grandjean 1971).

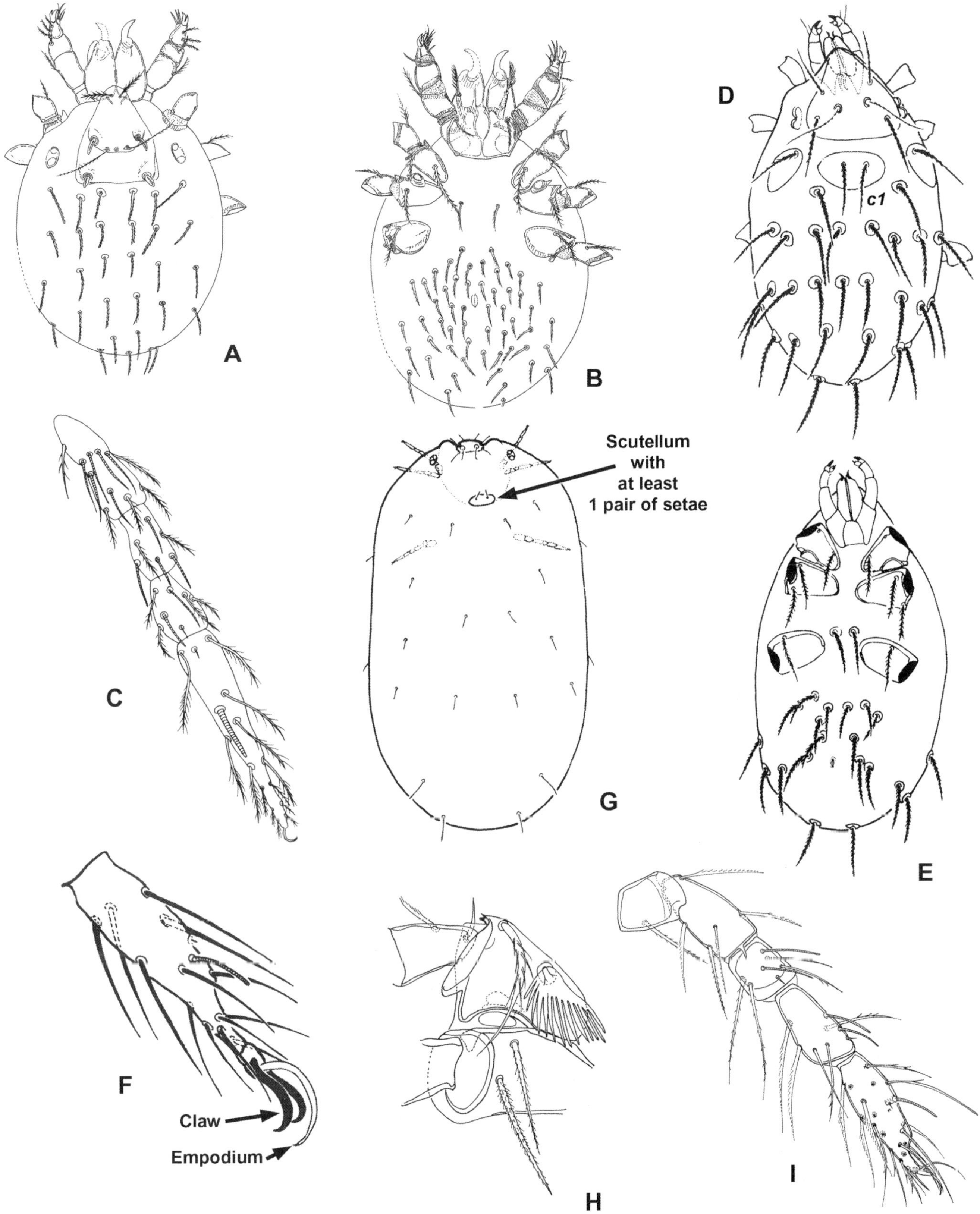

Fig. 13.59. TROMBELLIDAE. *Trombella cucumifera* Southcott: **A**, dorsum; **B**, venter (both after Southcott 1986a). *Durenia papuana* Fain and Grootaert: **C**, distal segments of leg I (after Fain and Grootaert 1996).

TROMBIDIIDAE. *Allothrombium* sp.: **D**, dorsum; **E**, venter; **F**, tarsus of leg I. Genus and species undetermined: **G**, engorged larva. *Parathrombium megalochirum* Robaux: **H**, coxal plates I and II; **I**, leg I, anterolateral view (both after Robaux 1969a).

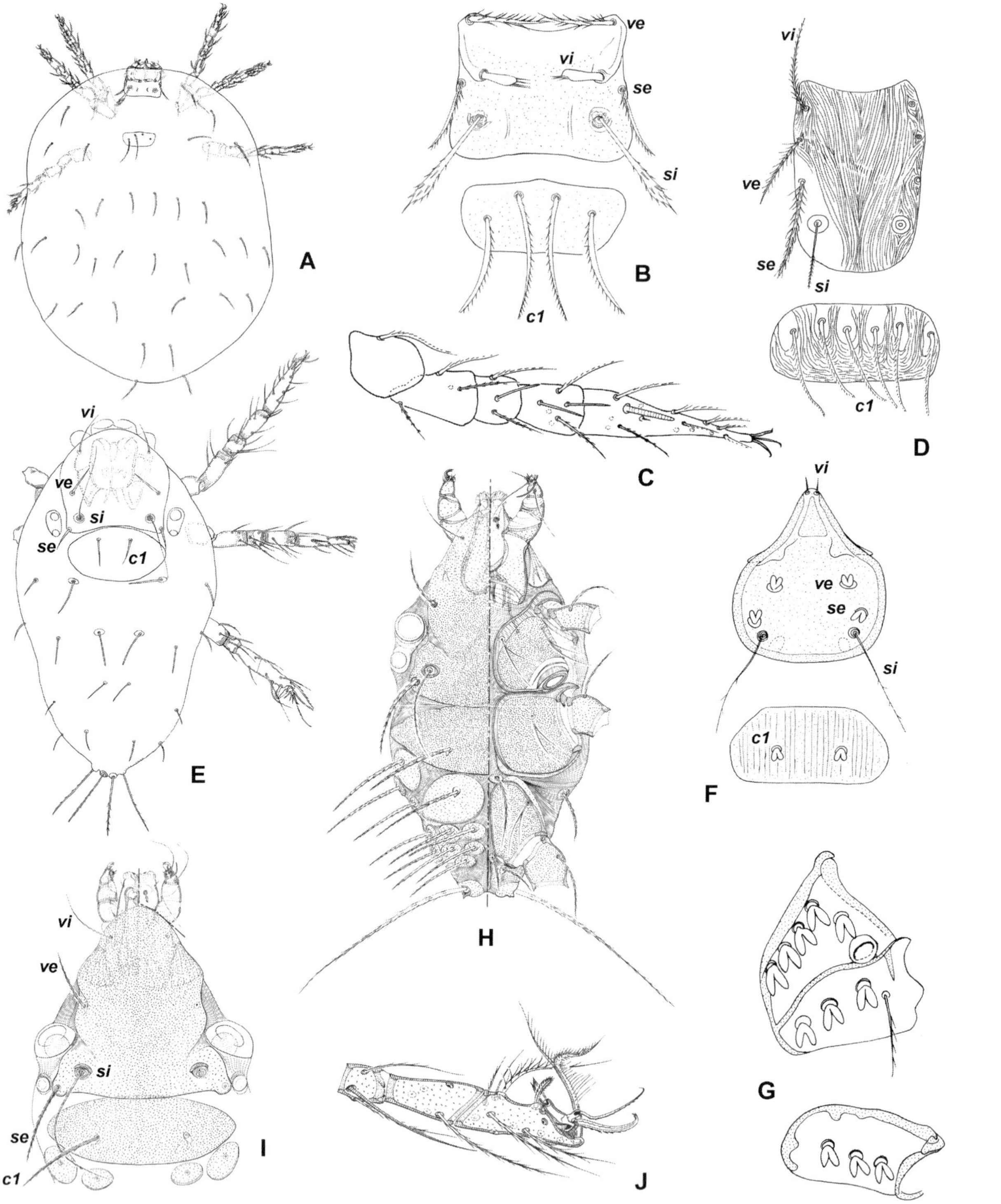

Fig. 13.60. NEOTHROMBIIDAE. *Wondeclia centipedae* Southcott: **A**, combination transparent view of dorsum and venter; **B**, prodorsal plate (scutum) and scutellum (both redrawn after Southcott 1987b). *Ceuthothrombium cavaticum* Robaux, Webb, and Campbell: **C**, leg II (after Robaux, Webb, and Campbell 1976). *Neothrombium neglectum* (Bruyant): **D**, prodorsal plate (scutum) and scutellum (after Robaux 1972a). *Silphitrombium furculigerum* Fain: **F**, prodorsal (scutum) and scutellum; **G**, coxal plates (after Fain 1992b).

EUTROMBIDIIDAE. *Eutrombidium orientale* Southcott: **E**, dorsum (after Southcott 1993c).

MICROTROMBIDIIDAE. *Microtrombidium wellesIyi* (Mullen and Vercammen-Grandjean): **H**, split view of dorsum (left) and venter (right) of gnathosoma and idiosoma. *M. danbyi* (Mullen and Vercammen-Grandjean): **I**, prodorsal area; **J**, distal segments of leg III (all after Mullen and Vercammen-Grandjean 1978).

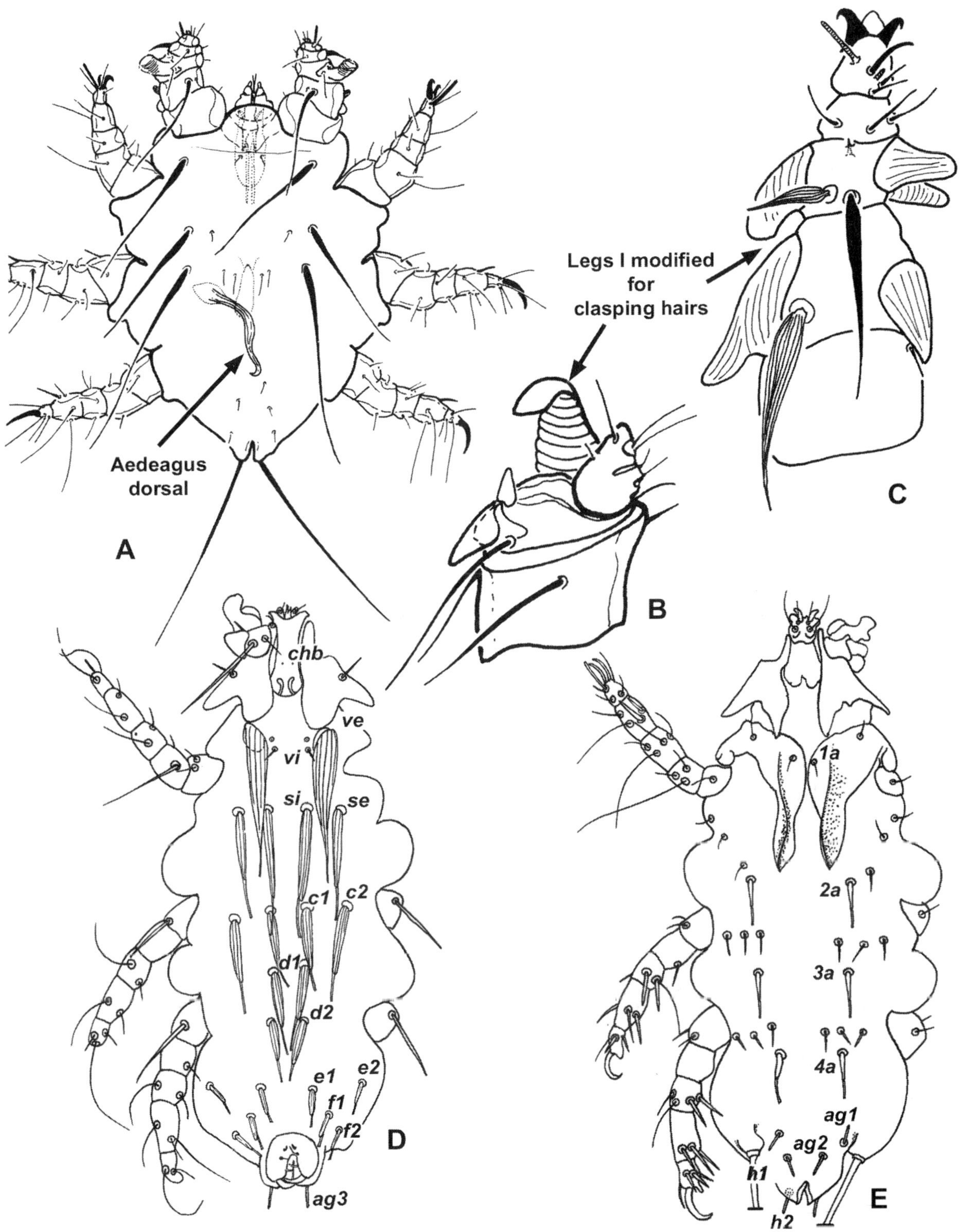

Fig. 13.61. MYOBIIDAE. *Protomyobia* sp. (Oregon, USA): **A**, dorsum of male. *Radfordia* sp. (Oregon, USA): **B**, leg I of female. *Archemyobia philander* Lukoschus, Dusbábek, and Jameson (Netherlands): **C**, leg I of female (dorsal) (after Lukoschus, Faint, and Driessen 1972). Diagrammatic representation of generalized myobiid: **D**, dorsum; **E**, venter (after A. Bochkov, with permission).

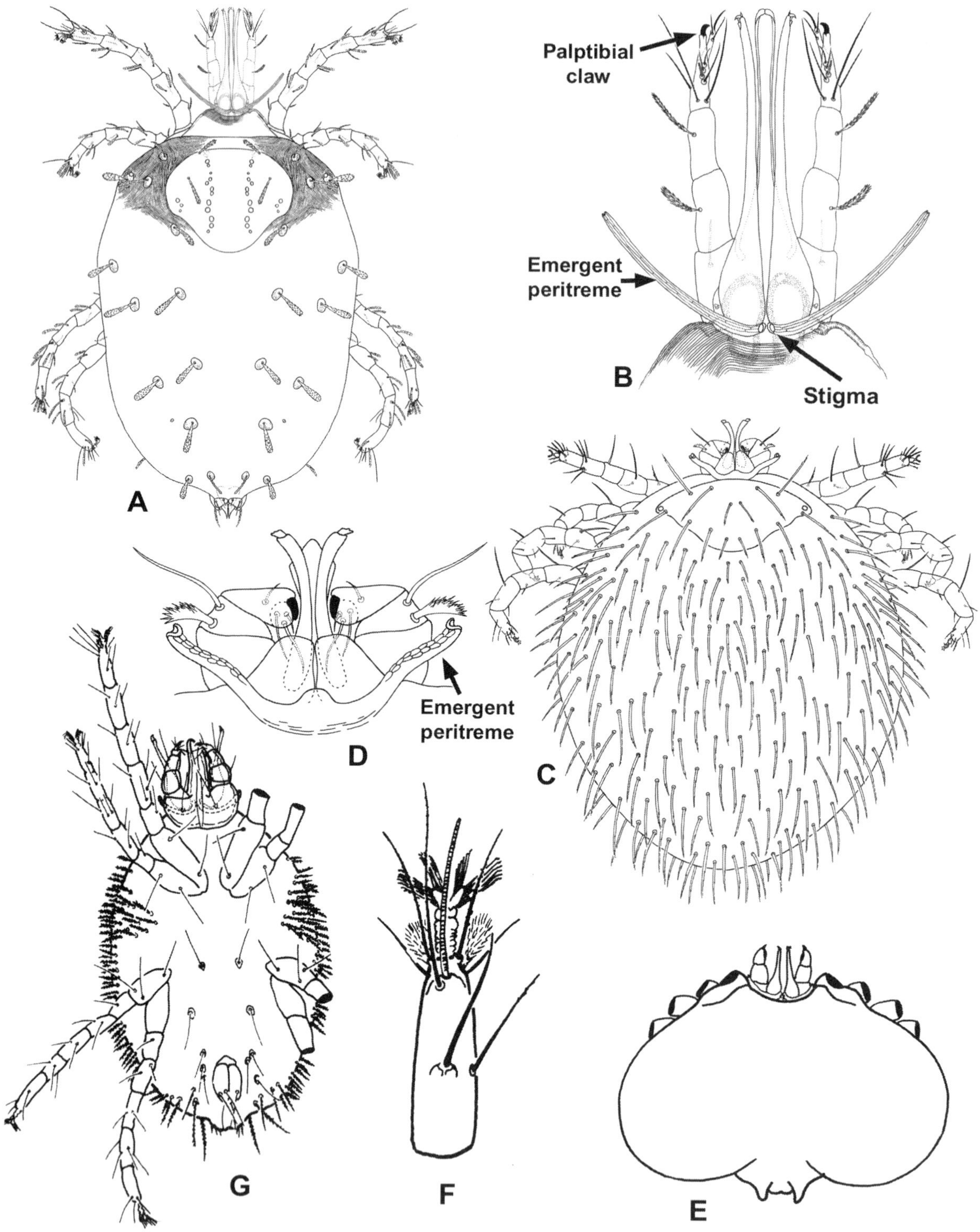

Fig. 13.62. PTERYGOSOMATIDAE. *Hirstiella diolii* Baker (New South Wales, Australia): **A**, dorsal habitus; **B**, details of dorsal gnathosoma. *Geckobia* sp. (Queensland, Australia): **C**, dorsal habitus; **D**, details of dorsal gnathosoma; **E**, diagrammatic of broad-bodied form. *Geckobiella texana* Banks (California, USA): **F**, tarsus I dorsum; **G**, venter of female.

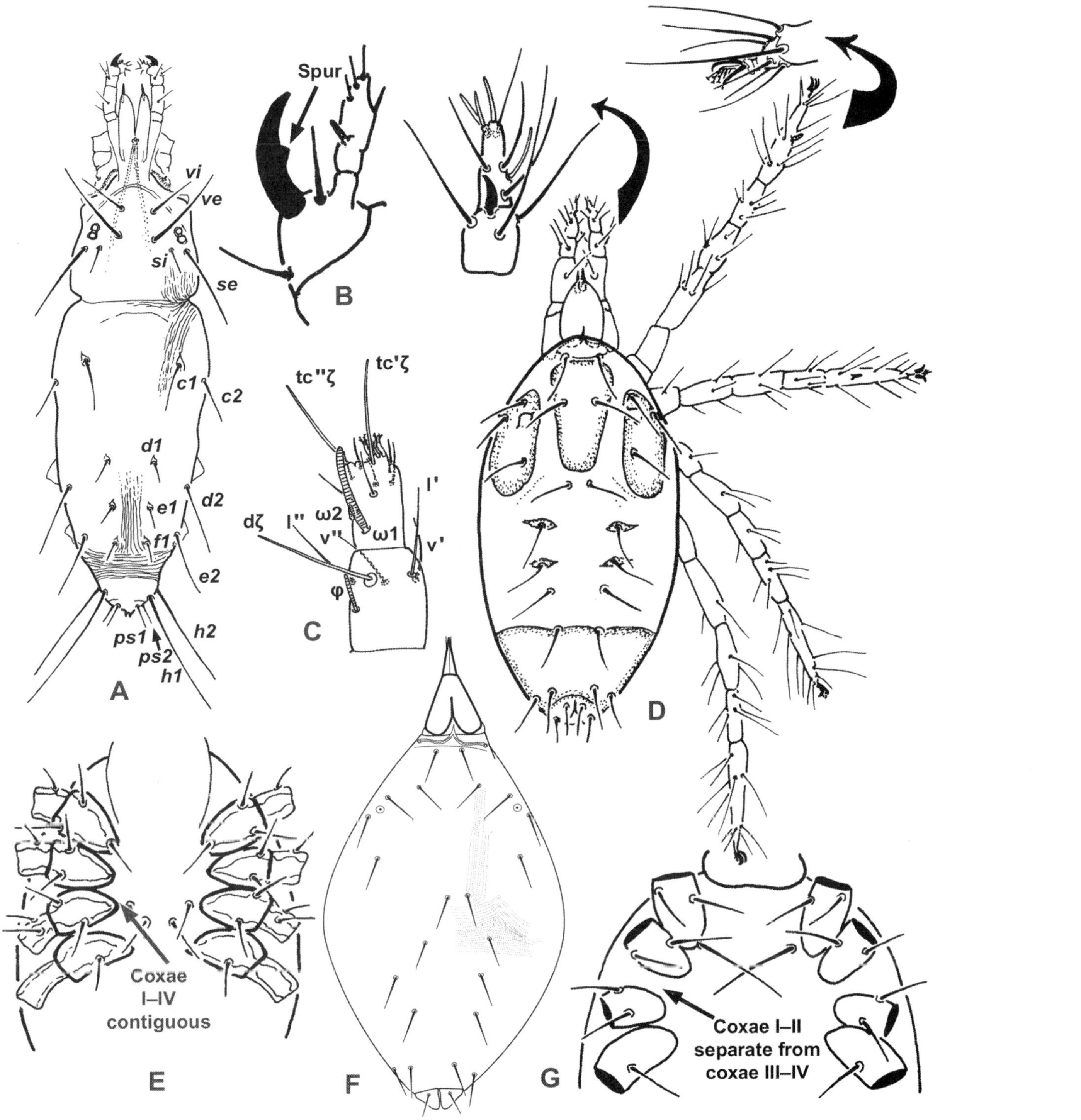

Fig. 13.63. BARBUTIIDAE. *Barbutia perretae* Robaux (California, USA): **A**, dorsum of female. *B. anguinea* (Berlese): **B**, palpal thumb-claw process (both after Robaux 1975b). *B. australis* Fan et al. (Queensland, Australia): **C**, tarsus I of female (after Fan, Walter, and Proctor 2003a).

RAPHIGNATHIDAE. *Raphignathus* sp. (Oregon, USA): **D**, dorsum of female with details of the palptarsus and distal tarsus I; **E**, coxal arrangement on venter. *Neoraphignathus howei* Smiley and Moser: **F**, dorsal idiosoma and chelicerae of female (after Smiley and Moser 1968).

CALIGONELLIDAE. *Coptocheles* sp. (Oregon, USA): **G**, coxal arrangement on venter.

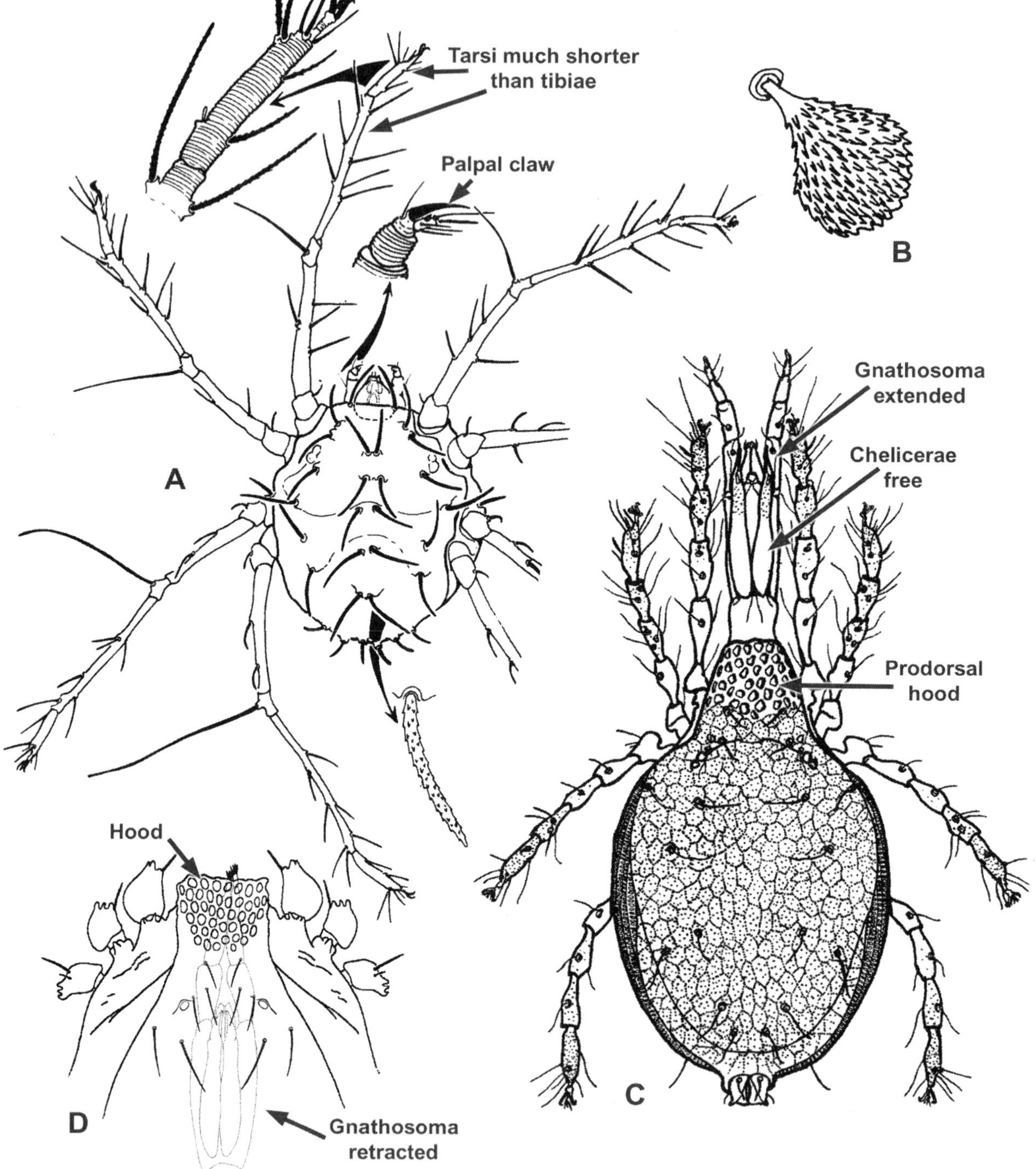

Fig. 13.64. CAMEROBIIDAE. *Camerobia* sp. (Oregon, USA): **A**, dorsum with (top to bottom) detail of tarsus I, the palpal tibia and tarsus, and a dorsal seta. *C. australis* Southcott (Australia): **B**, body seta (after Southcott 1956).
CRYPTOGNATHIDAE. *Cryptognathus lagena* Kramer (Oregon, USA): **C**, dorsum with gnathosoma extended (from Krantz 1958a). *Favognathus* sp. (Queensland, Australia): **D**, prodorsal region with gnathosoma retracted.

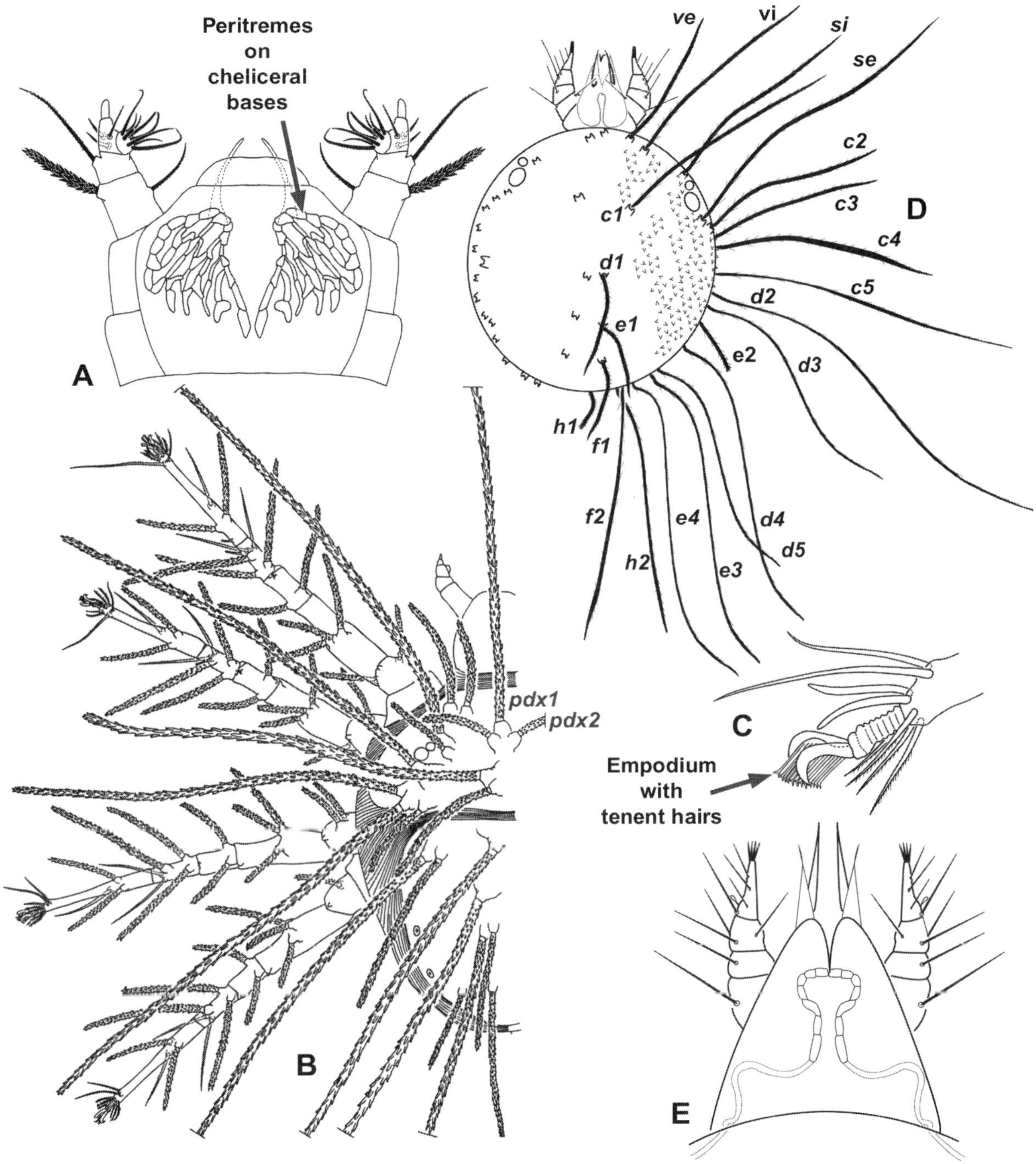

Fig. 13.65. DASYTHYREIDAE. *Xanthodasythyreus toohey* Walter and Gerson (Queensland, Australia): **A**, dorsal gnathosoma; **B**, dorsum of female; **C**, tarsus I of female.

XENOCALIGONELLIDIDAE. *Echinopsis fukiensis* Fan and Chen (Fujian, China): **D**, dorsum of adult female (after Fan and Chen 1996). *Xenocaligonellidus galapagus* Gonzalez: **E**, dorsal gnathosoma (after Gonzalez 1978).

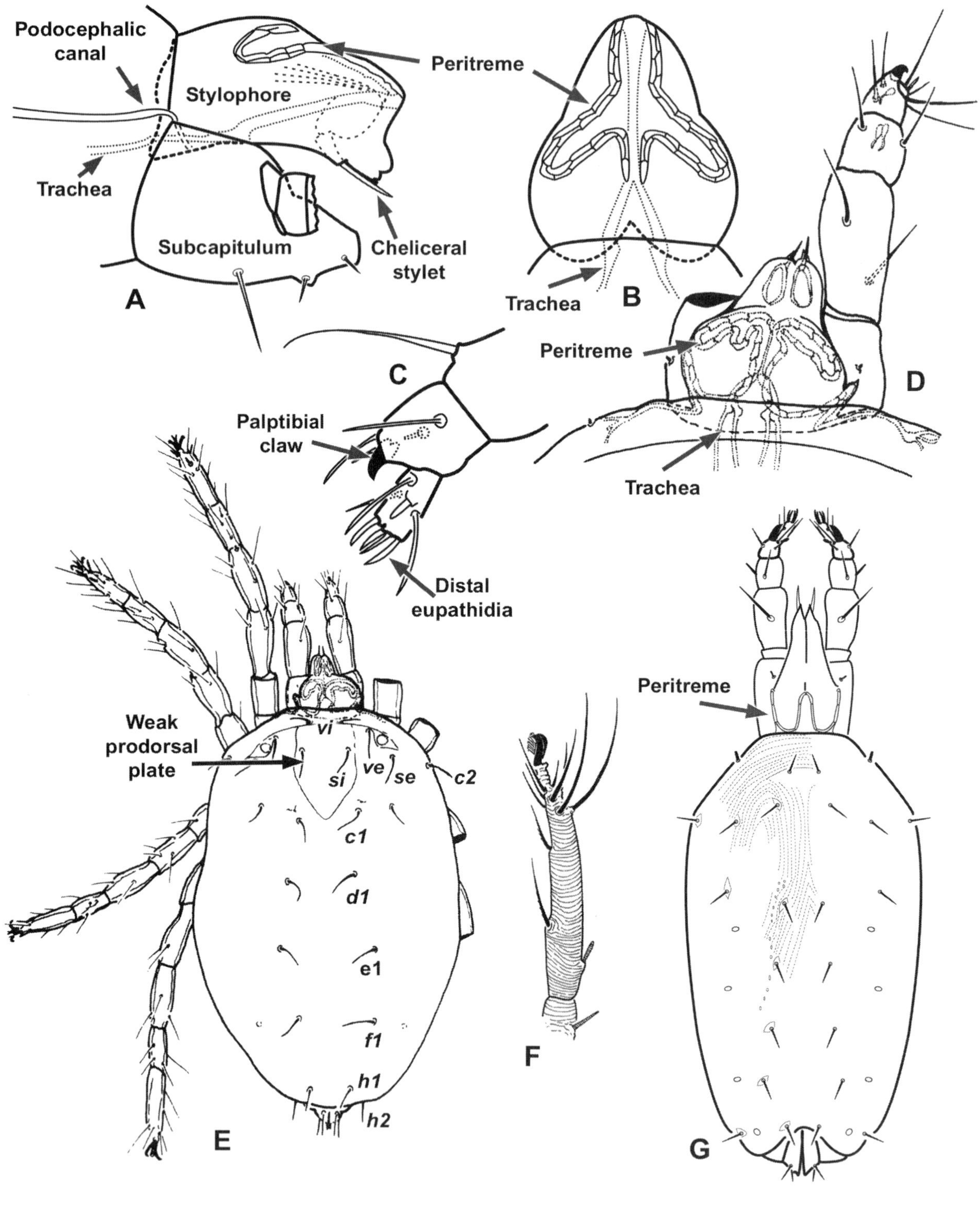

Fig. 13.66. CALIGONELLIDAE. *Caligonella humilis* (Koch) (after Grandjean 1946a): **A**, lateral view of gnathosoma; **B**, dorsal view of stylophore; **C**, lateral view of palp. *Coptocheles* sp. (Oregon, USA): **D**, dorsal gnathosoma; **E**, dorsum of female; **F**, tarsus I of caligonellid type. *Neognathus terrestris* (Summers and Schlinger): **G**, dorsal idiosoma and gnathosoma of female (after Summers and Schlinger 1955).

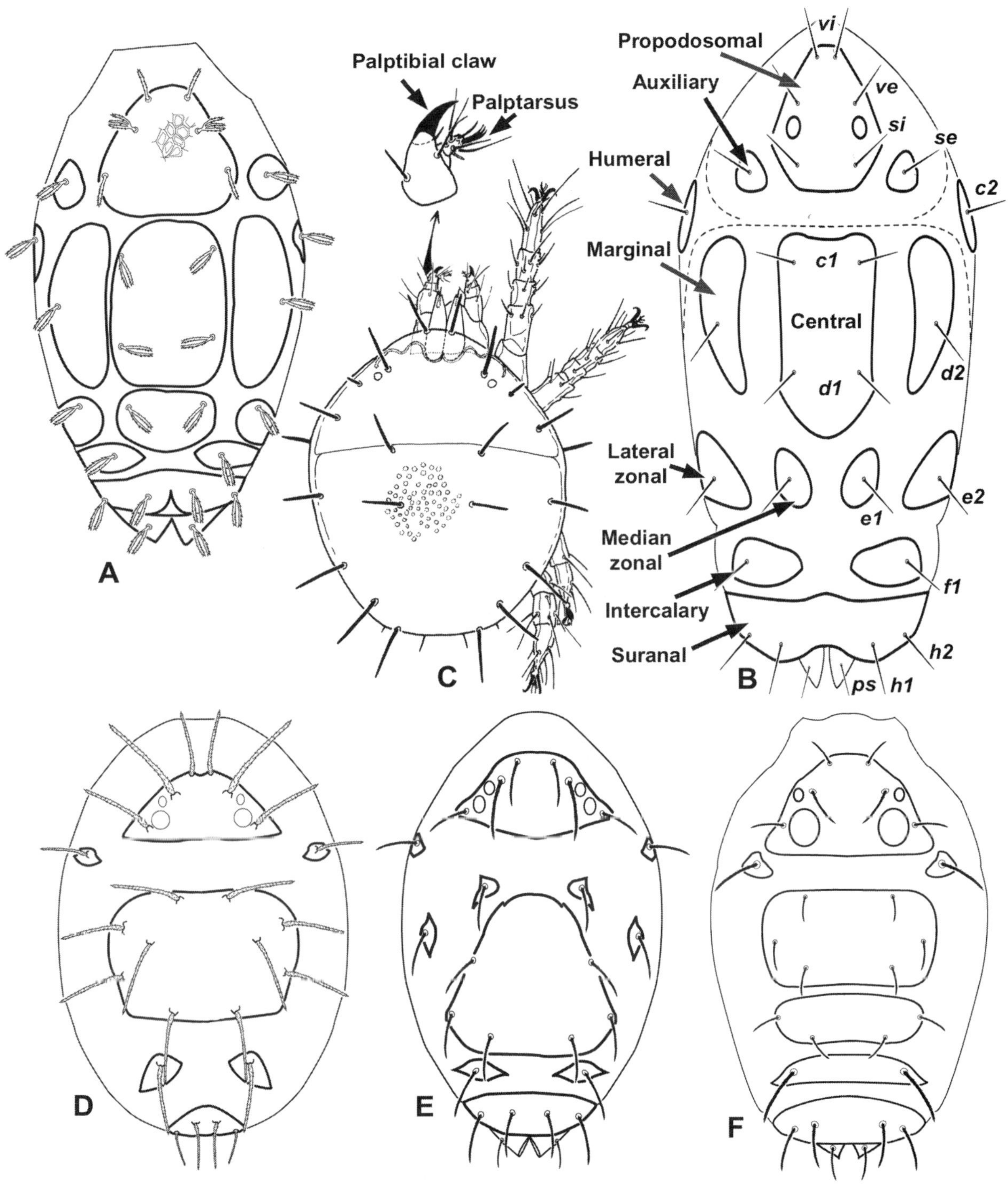

Fig. 13.67. STIGMAEIDAE. *Stigmaeus sinai* Swift (Egypt): **A**, dorsum of female (after Swift 1987). **B**, diagrammatic representation of the idiosomatic plating in stigmaeids (after Summers 1962). *Eustigmaeus* sp. (Oregon, USA): **C**, dorsum with details of palpal thumb-claw complex. Dorsal plates in adult female stigmaeids: **D**, *Agistemus* sp.; **E**, *Zetzellia mali* (Ewing); **F**, *Mediolata* sp.

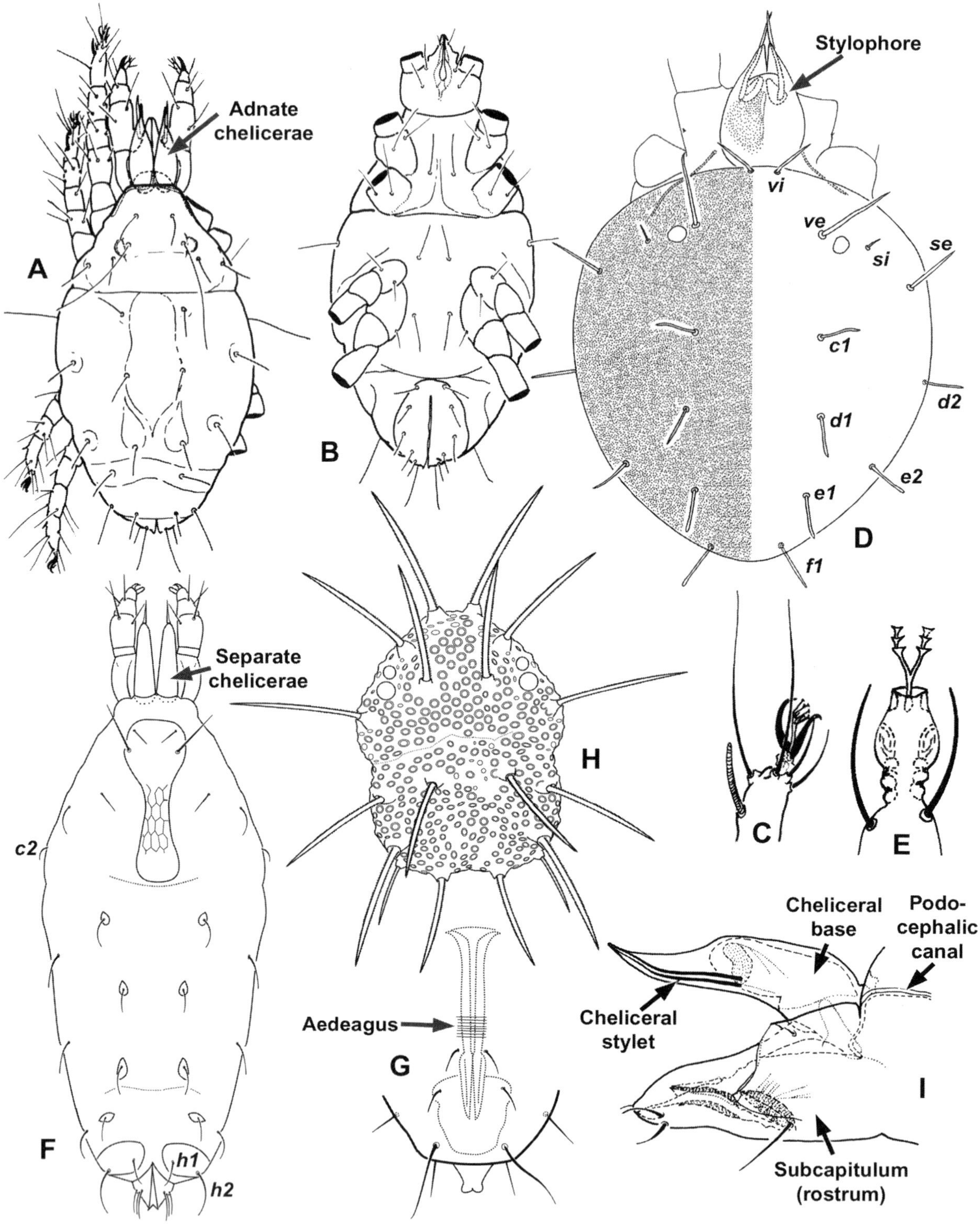

Fig. 13.68. STIGMAEIDAE. *Stigmaeus* sp. (Oregon, USA): **A**, dorsum of female; **B**, venter; **C**, tarsus I. *Caligohomus durus* Fan and Walter (Florida, USA): **D**, dorsum of female (after Fan and Walter 2004). *Neilstigmaeus lamingtoni* Gerson and Meyer (Queensland, Australia) (after Gerson and Meyer 1995): **E**, tarsal arolium, leg IV. *Storchia robusta* (Berlese): **F**, dorsum of female; **G**, male aedeagus ventral view (after Grandjean 1944a). *Mullederia* sp. (Queensland, Australia): **H**, dorsum of idiosoma. *Mediolata mariaefrancae* André (Belgium): **I**, lateral aspect of gnathosoma (after André 1977).

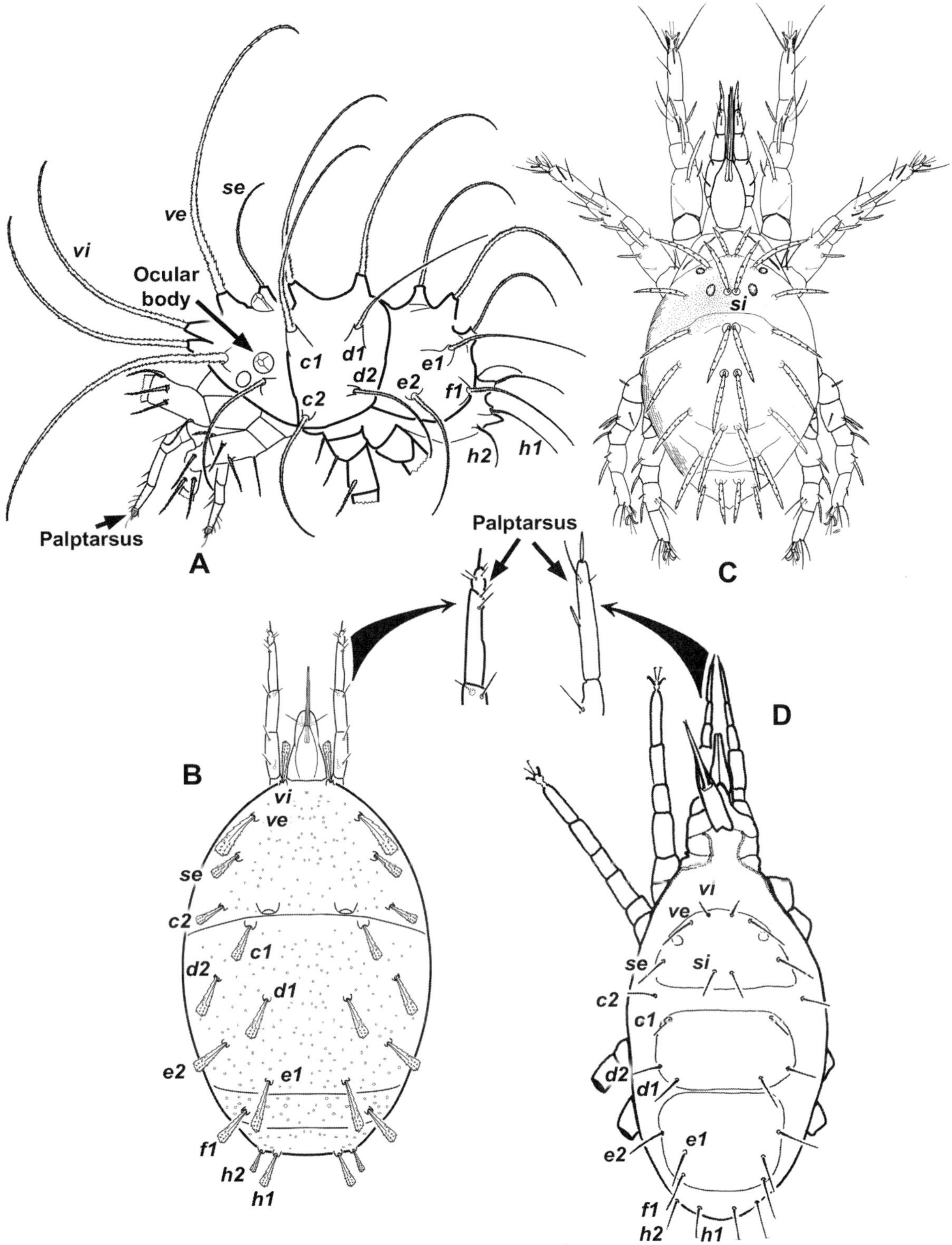

Fig. 13.69. MECOGNATHIDAE. *Mecognatha* sp. (Queensland, Australia): **A**, dorsolateral view with distal leg segments removed. *Paraeupalopsellus clareae* (Philippines): **B**, dorsum of female with detail of palp (after Rimando and Corpuz-Raros 1996). EUPALOPSELLIDAE. *Exothorhis* sp. (Queensland, Australia): **C**, dorsal habitus. *Eupalopsellus* sp. (Oregon, USA): **D**, dorsum with detail of palp.

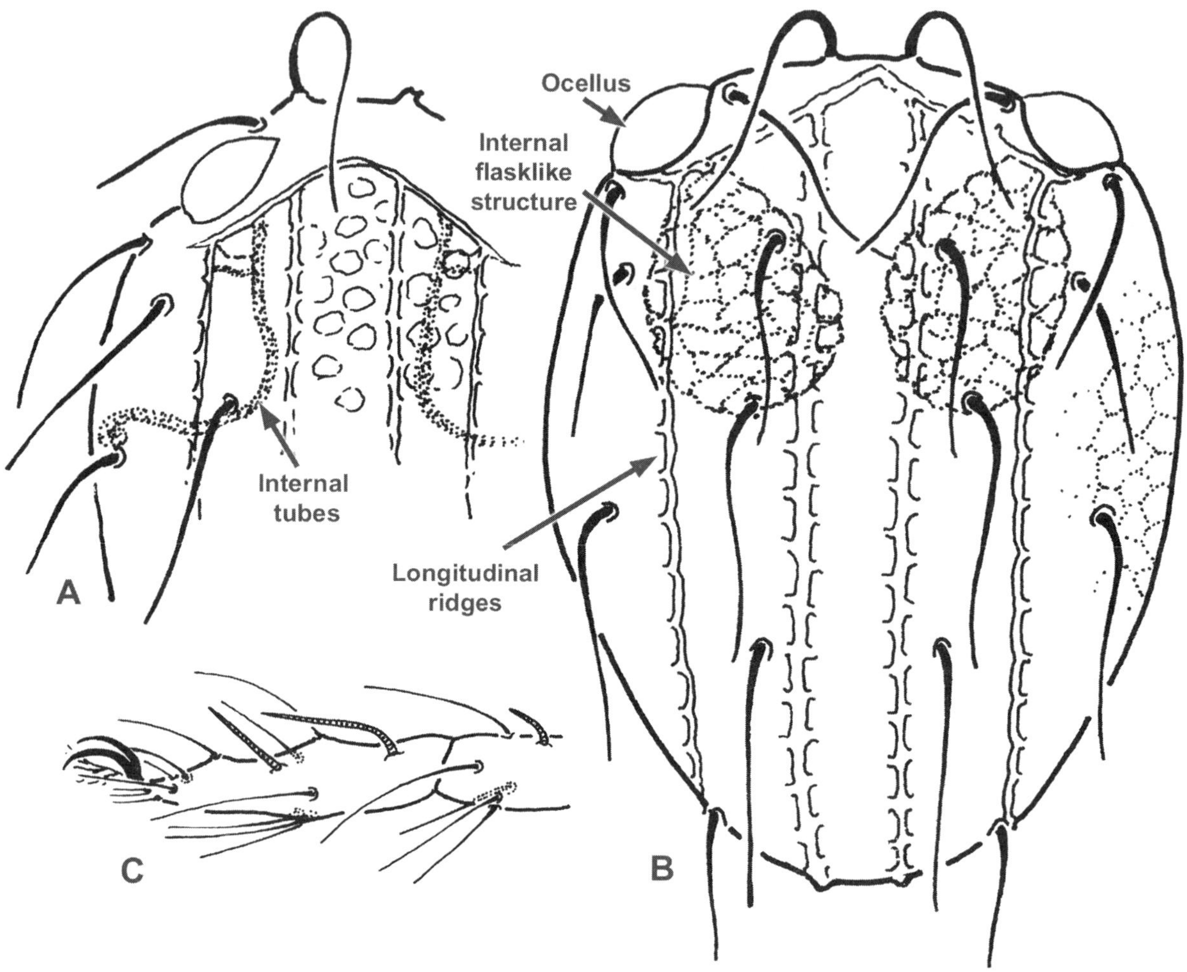

Fig. 13.70. HOMOCALIGIDAE. *Annerossella pacifica* Wood (Malaysia): **A**, portion of male dorsum showing internal tubes; **B**, dorsum of female; **C**, distal segments of leg I.

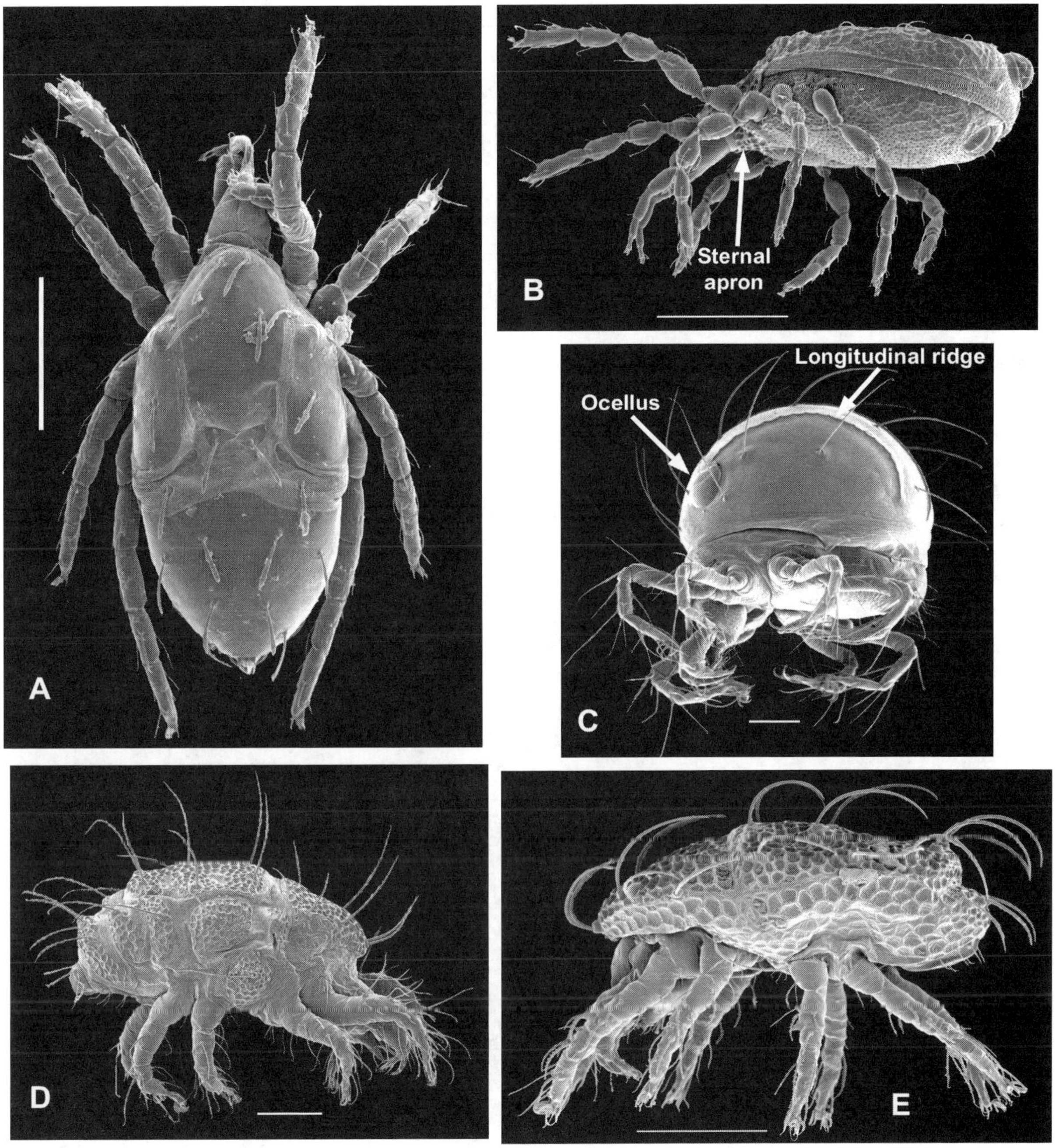

Fig. 13.71. RAPHIGNATHIDAE. *Raphignathus* sp.: **A**, dorsal view of female.
CRYPTOGNATHIDAE. *Favognathus* sp.: **B**, ventrolateral view.
HOMOCALIGIDAE. *Annerossella* sp.: **C**, lateral view.
STIGMAEIDAE. *Stigmaeus* sp.: **D**, lateral view of female. *Eustigmaeus s. lat.* sp.: **E**, lateral view of female. (All from Queensland, Australia). Scale bars = 0.1 mm.

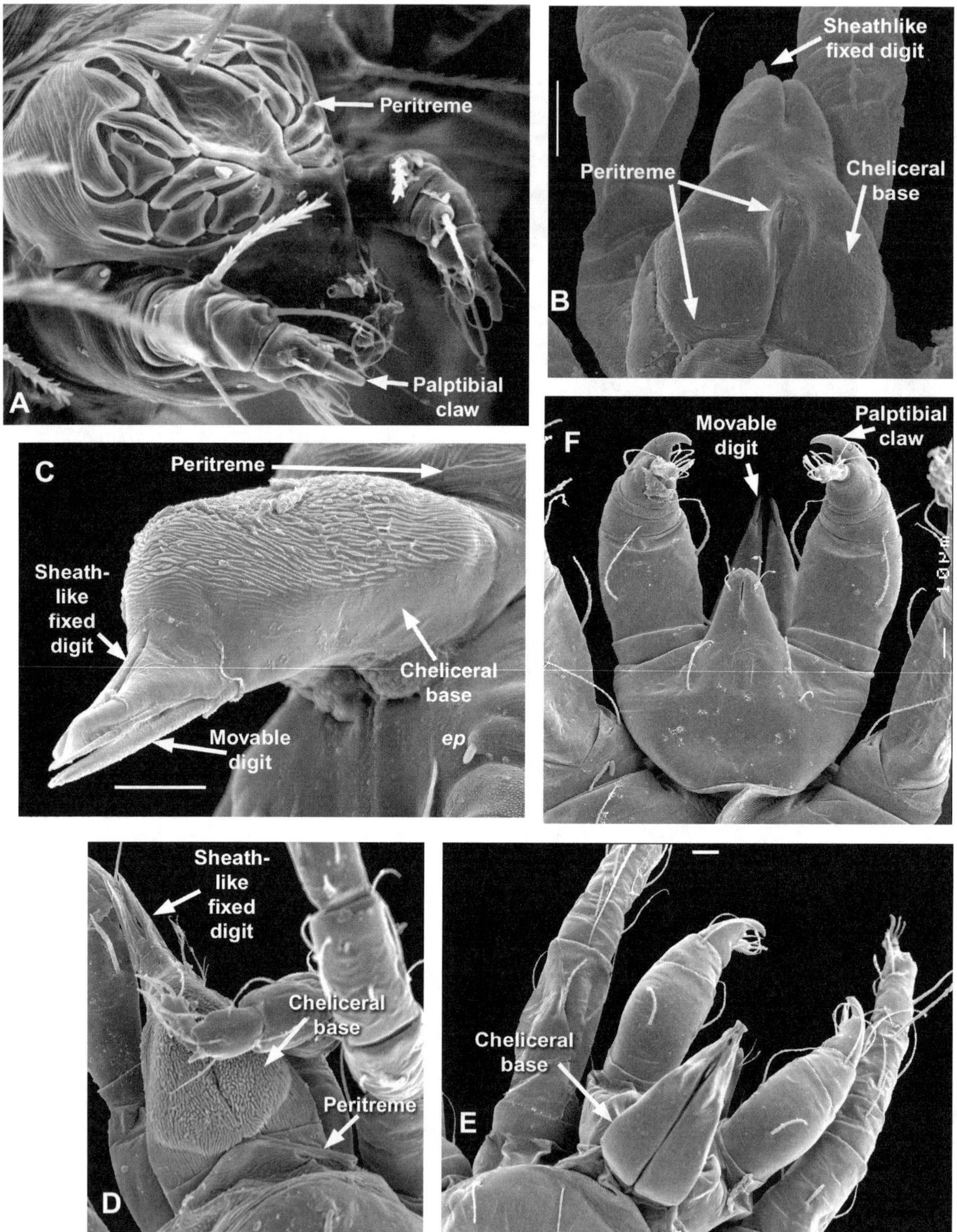

Fig. 13.72. DASYTHYREIDAE. *Xanthodasythyreus toohey* Walter and Gerson: **A**, anterodorsal aspect of gnathosoma. CALIGONELLIDAE. *Neognathus* sp.: **B**, dorsal gnathosoma. RAPHIGNATHIDAE. *Raphignathus* sp.: **C**, lateral aspect of stylophore; **D**, dorsal aspect of gnathosoma. STIGMAEIDAE. *Agistemus* sp.: **E**, dorsal gnathosoma; **F**, ventral gnathosoma. (All from Queensland, Australia). Scale bars = 0.01 mm.

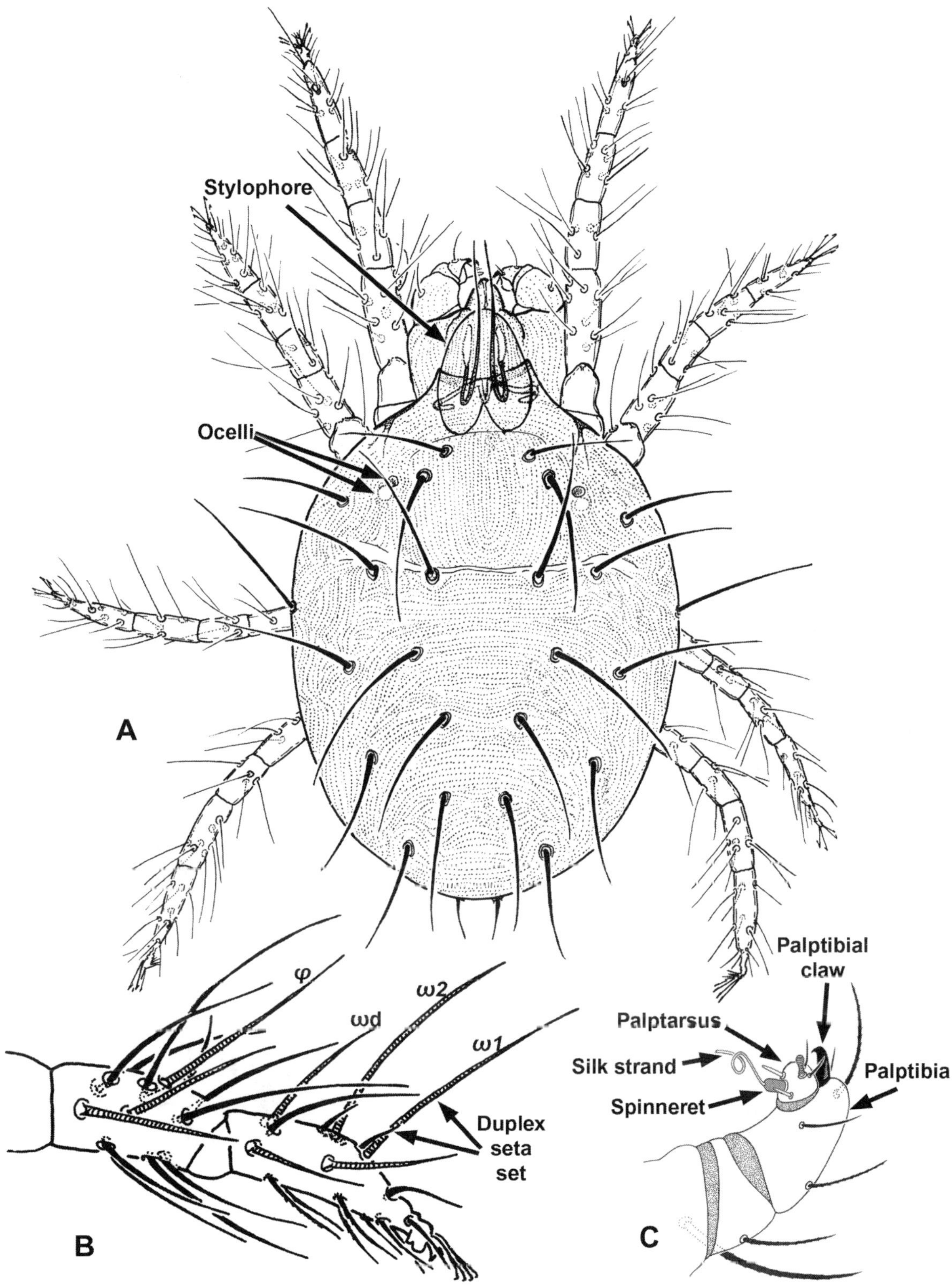

Fig. 13.73. TETRANYCHIDAE. *Tetranychus* (*Armenychus*) *mcdanieli* McGregor (Oregon, USA): **A**, dorsum of female. *T. urticae* Koch: **B**, tibia and tarsus of leg I of male (antiaxial), showing position of setae and solenidia (ϕ, ω), including two duplex seta sets composed of an erect solenidion ω and a shorter seta (after Grandjean 1948b). *Tetranychus* sp.: **C**, distal palpal segments.

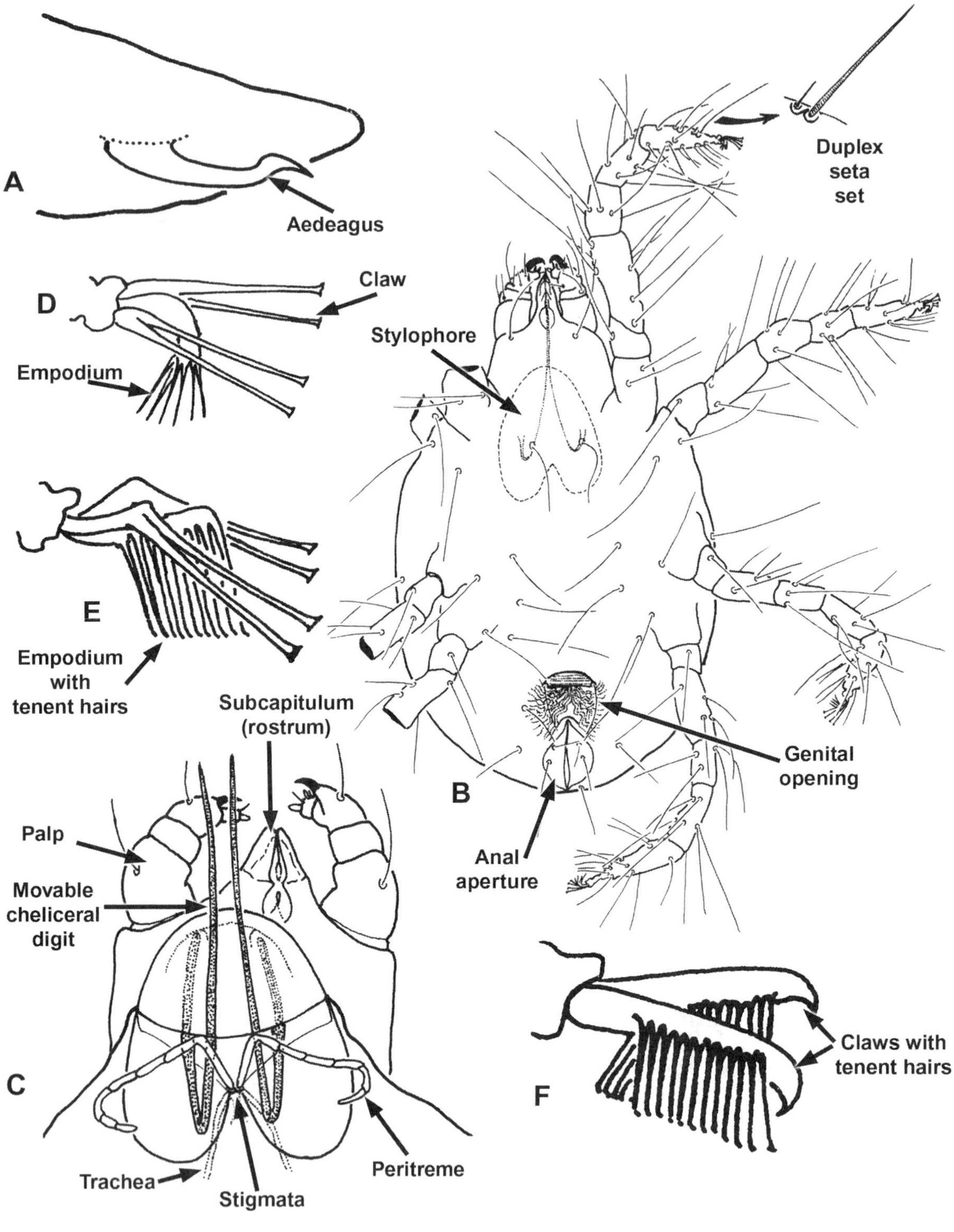

Fig. 13.74. TETRANYCHIDAE. *Tetranychus* sp. (Oregon, USA): **A**, aedeagus of male, lateral aspect; **B**, venter of female with detail of duplex seta set; **C**, dorsal aspect of gnathosoma; **D**, pretarsus. *Petrobia* sp.: **E**, pretarsus. *Tetranycopsis* sp.: **F**, pretarsus.

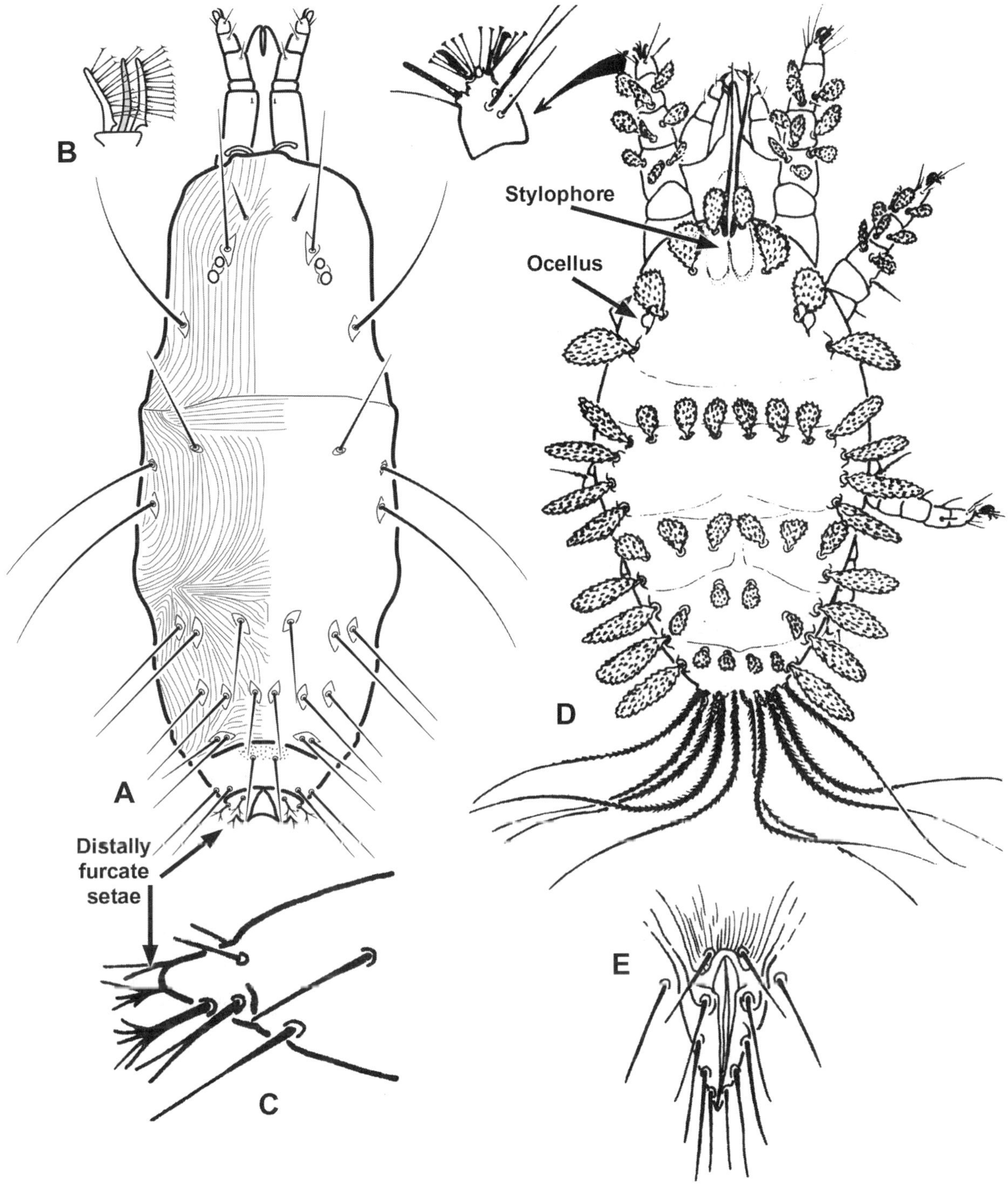

Fig. 13.75. ALLOCHAETOPHORIDAE. *Allochaetophora africana* Meyer and Ueckermann (South Africa) (after Meyer and Ueckermann 1997): **A**, dorsum of female; **B**, pretarsus. *A. californica* McGregor (California, USA): **C**, caudal aspect of nymph. TUCKERELLIDAE. *Tuckerella* sp. (Hawaii, USA): **D**, dorsum of male with detail of tarsus I; **E**, anogenital area of male.

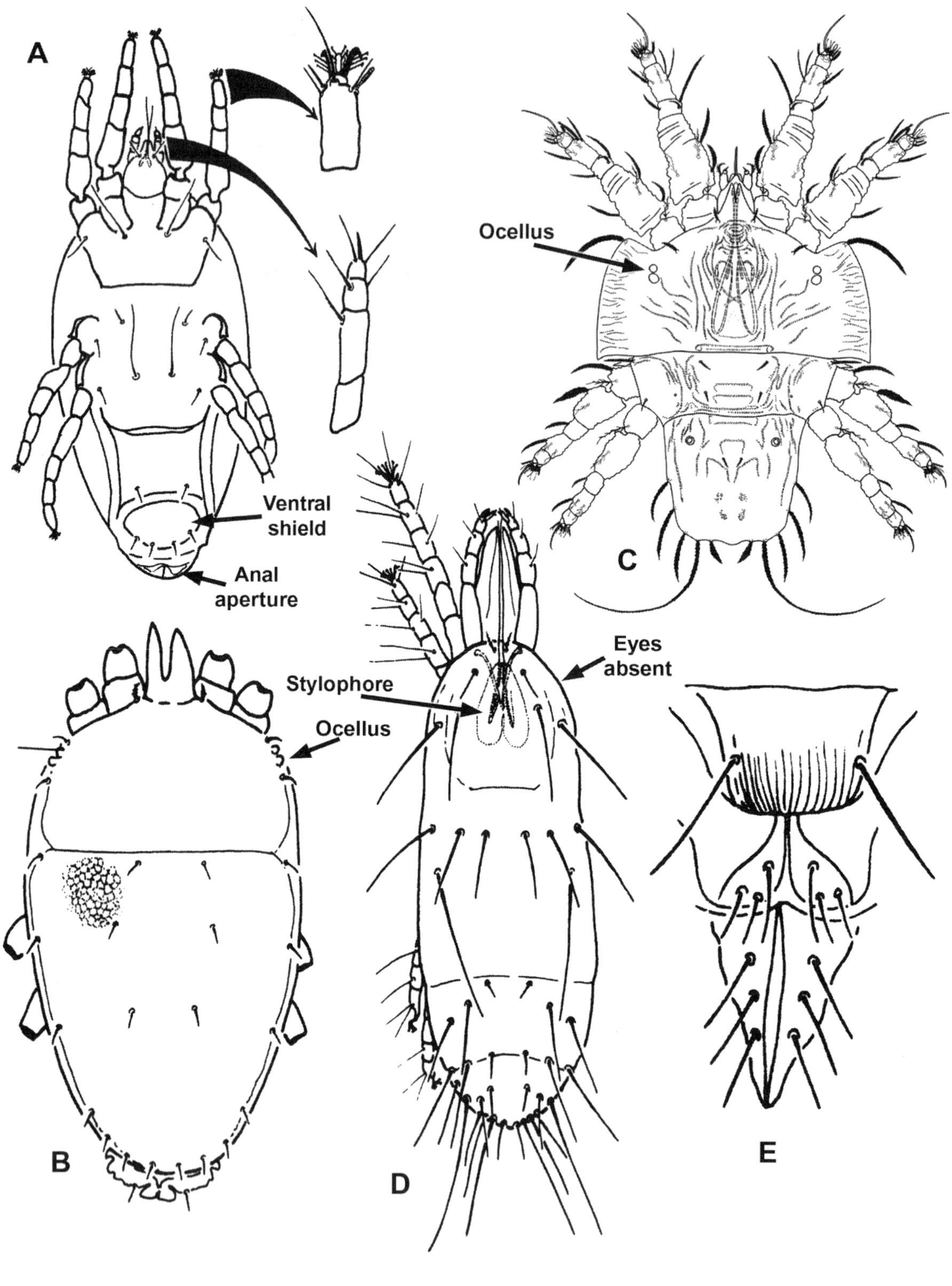

Fig. 13.76. TENUIPALPIDAE. *Brevipalpus essigi* Baker (Oregon, USA): **A**, venter of female with details of tarsus II and palp (bottom); **B**, dorsum. *Tenuipalpus* sp. (Queensland, Australia): **C**, dorsum of female.
LINOTETRANIDAE. *Linotetranus* sp.: **D**, dorsum of female; **E**, anogenital area of female.

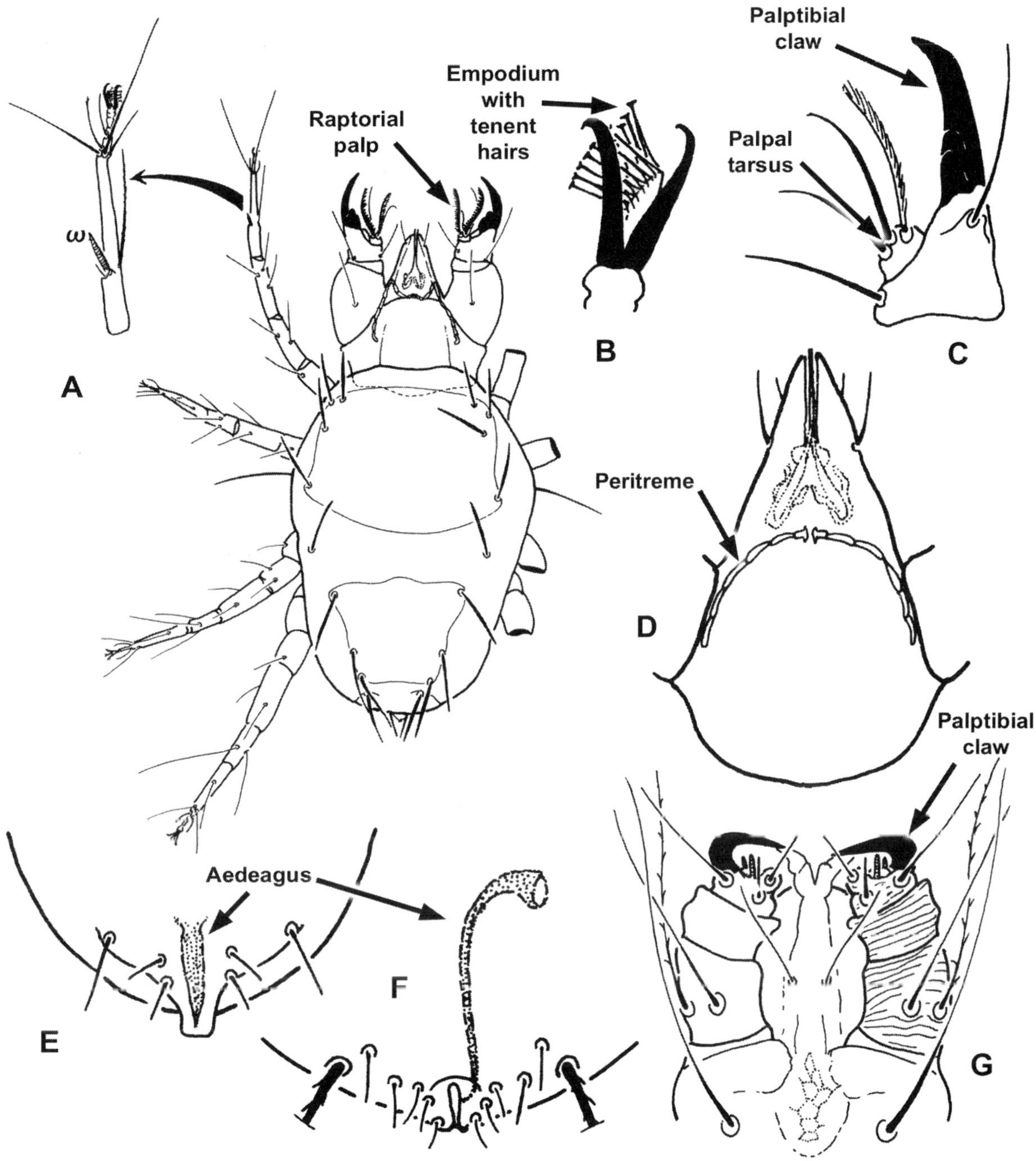

Fig. 13.77. CHEYLETIDAE. *Cheyletus malaccensis* Oudemans. **A**, dorsum with detail of tarsus I; **B**, pretarsus I; **C**, palpal thumb-claw process. *Acaropsis docta* Berlese: **D**, dorsomedian aspect of gnathosomatic capsule; **E**, aedeagal area of male. *Cheyletiella parasitivorax* (Mégnin): **F**, posterodorsal aedeagal area of male. *Ornithocheyletia* sp.: **G**, venter of gnathosoma (all from Oregon, USA).

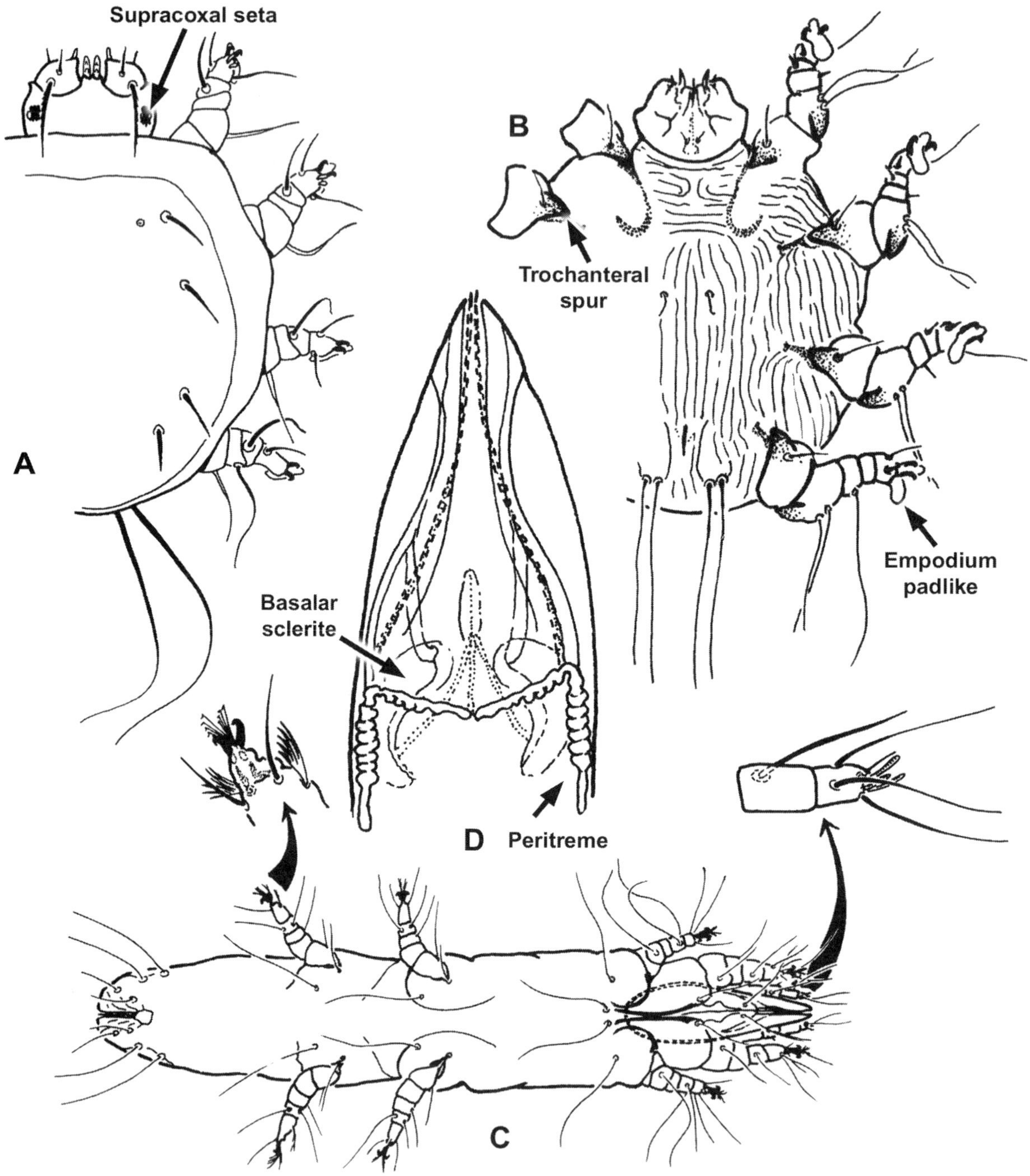

Fig. 13.78. PSORERGATIDAE. *Psorobia bos* (Johnston) (New Mexico, USA): **A**, dorsum of female. *Psorergates talpae* Lukoschus (Netherlands): **B**, venter of female.

SYRINGOPHILIDAE. Syringophylid (Delaware, USA): **C**, venter of female with detail of palp and of distal aspect of tarsus IV; **D**, dorsal aspect of gnathosomatic capsule of female showing position of chelicerae and reniform basalar sclerites.

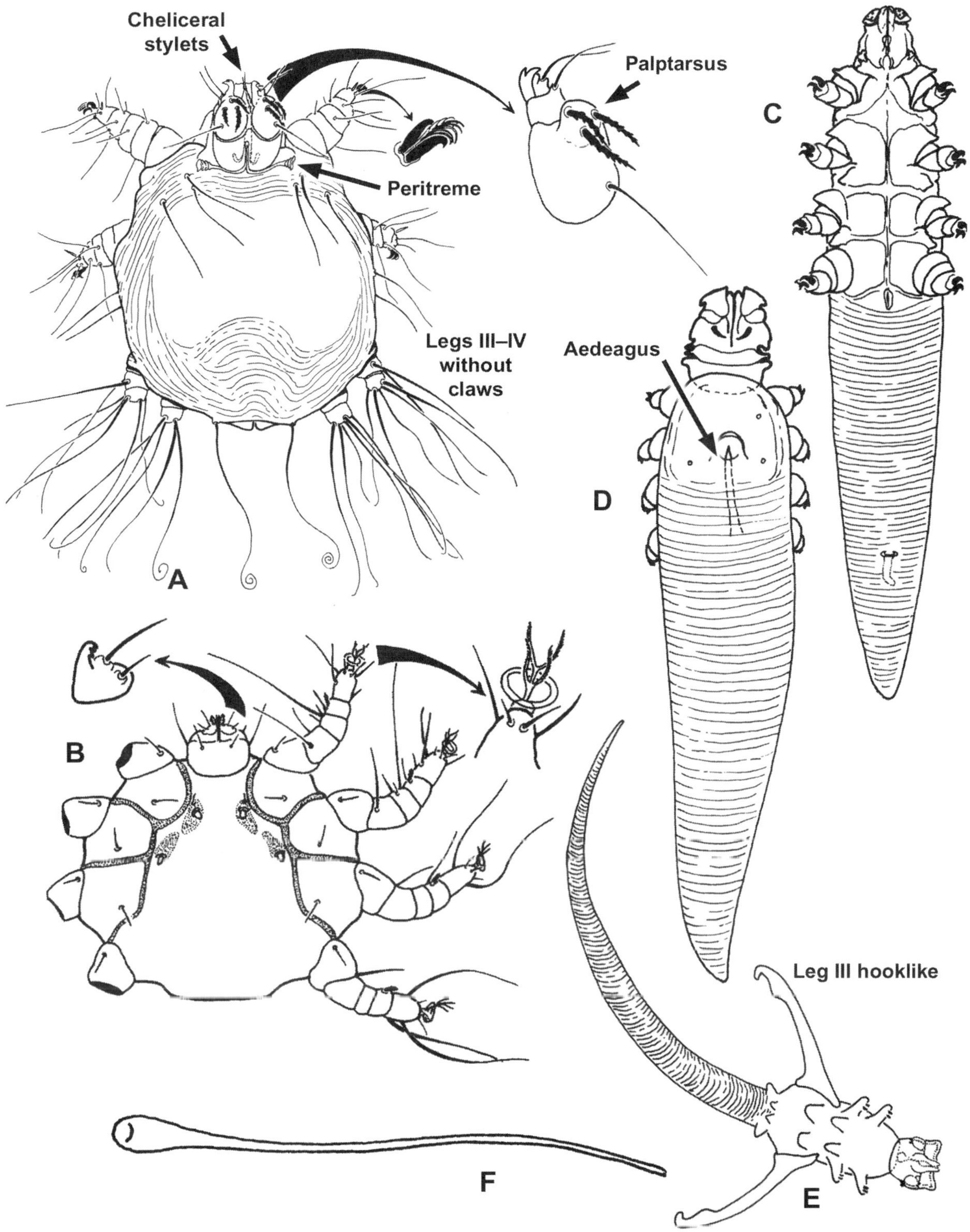

Fig. 13.79. HARPIRHYNCHIDAE. *Harpyrhynchus* sp. (California, USA): **A**, dorsum of female with details of pretarsus I and palp. *Ophioptes southcotti* Fain: **B**, venter of female with details of palp and empodium I (after Fain 1964d).
DEMODICIDAE. *Demodex* sp. (Oregon, USA): **C**, venter of female; **D**, dorsum of male. *D. longissimus* Desch and Nutting (Surinam): **E**, venter of nymph (opisthosoma normally twice length illustrated); **F**, ovum (after Desch and Nutting 1972).

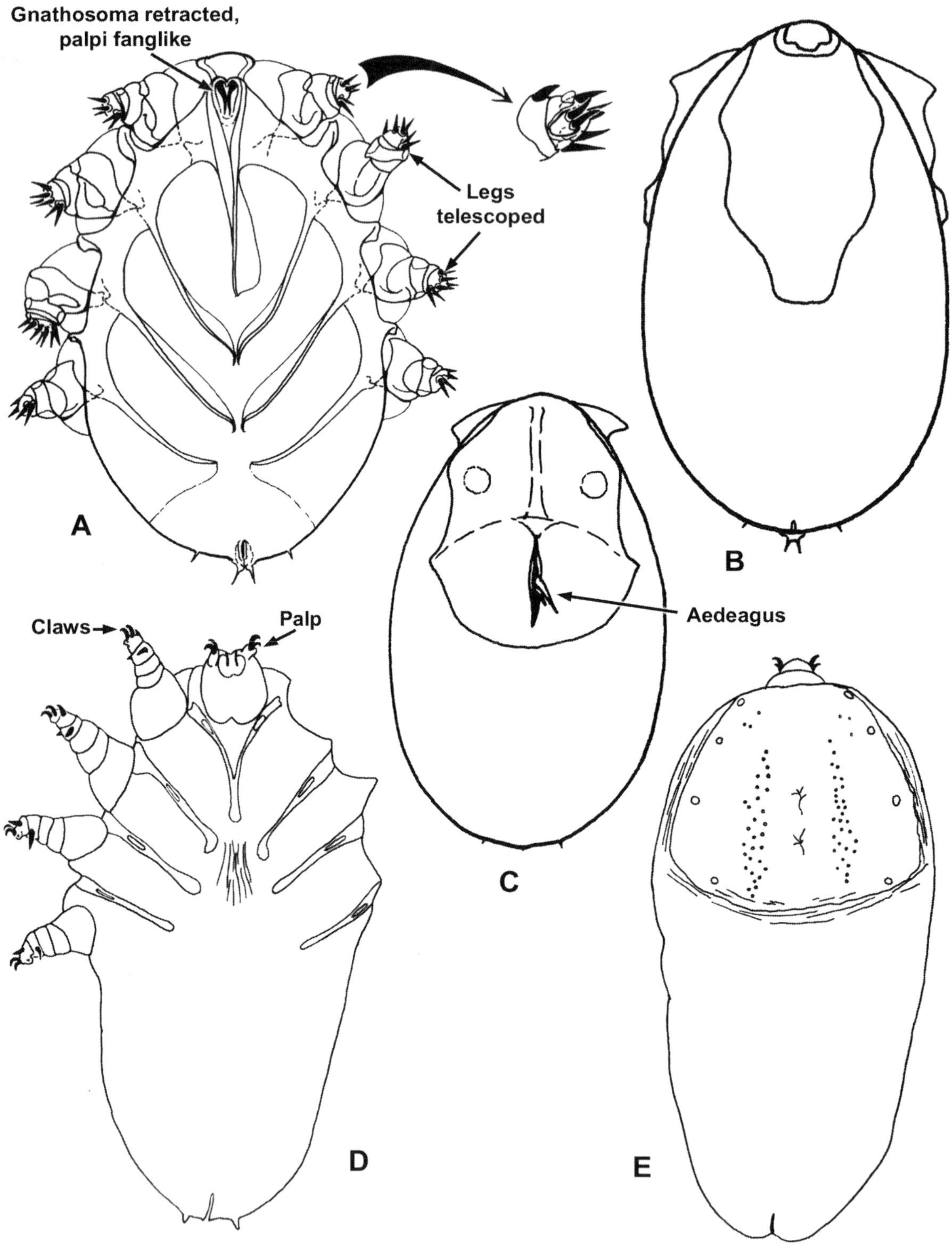

Fig. 13.80. CLOACARIDAE. *Cloacarus faini* Moss, Oliver, and Singer (Missouri, USA): **A**, venter of female; **B**, dorsum of female; **C**, dorsum of male (after Camin et al. 1967).

EPIMYODICIDAE. *Epimyodex soricis* Fain and Bochkov adult female (Oregon, USA): **D**, venter; **E**, dorsum (after Fain and Bochkov 2001).

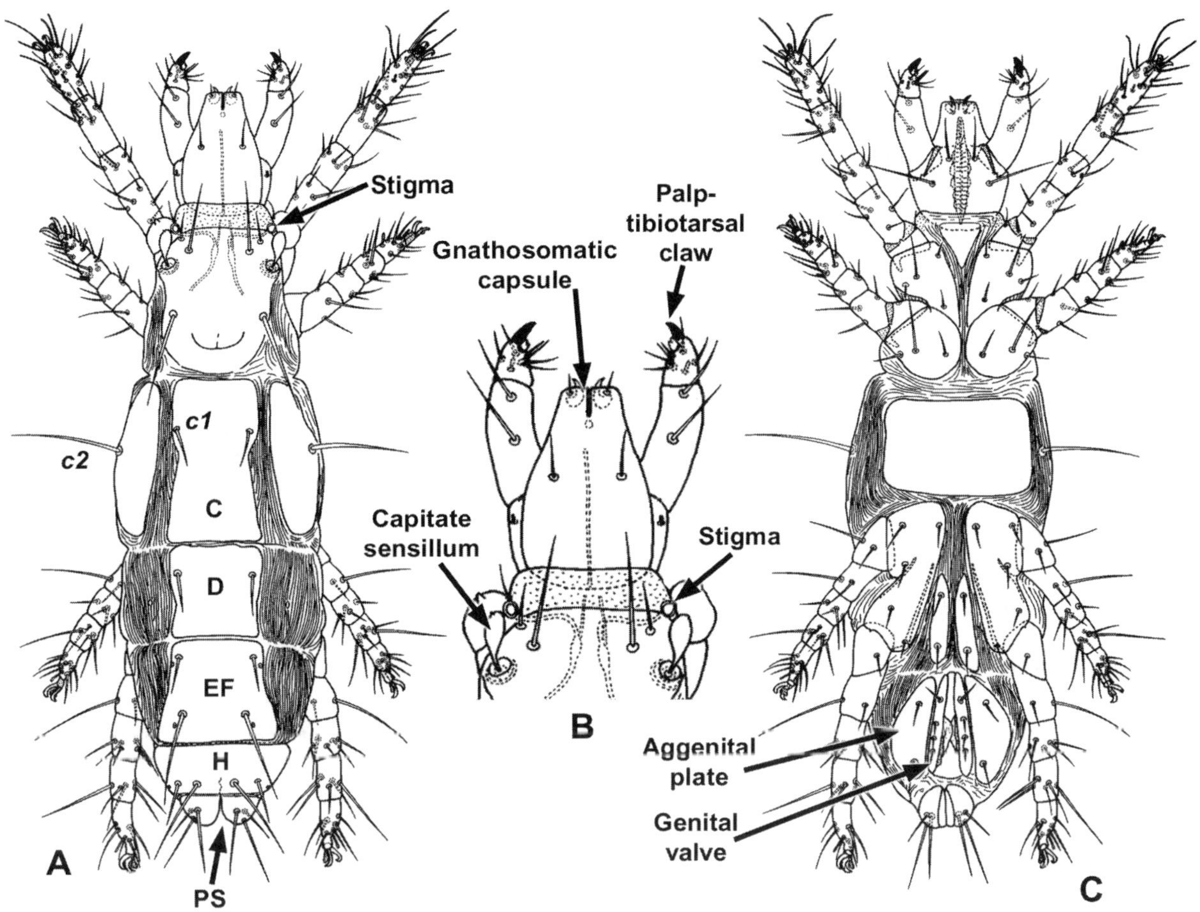

Fig. 13.81. TARSOCHEYLIDAE. *Hoplocheylus* sp. (after Lindquist 1976): **A**, dorsum of female; **B**, details of anterior; **C**, venter of female.

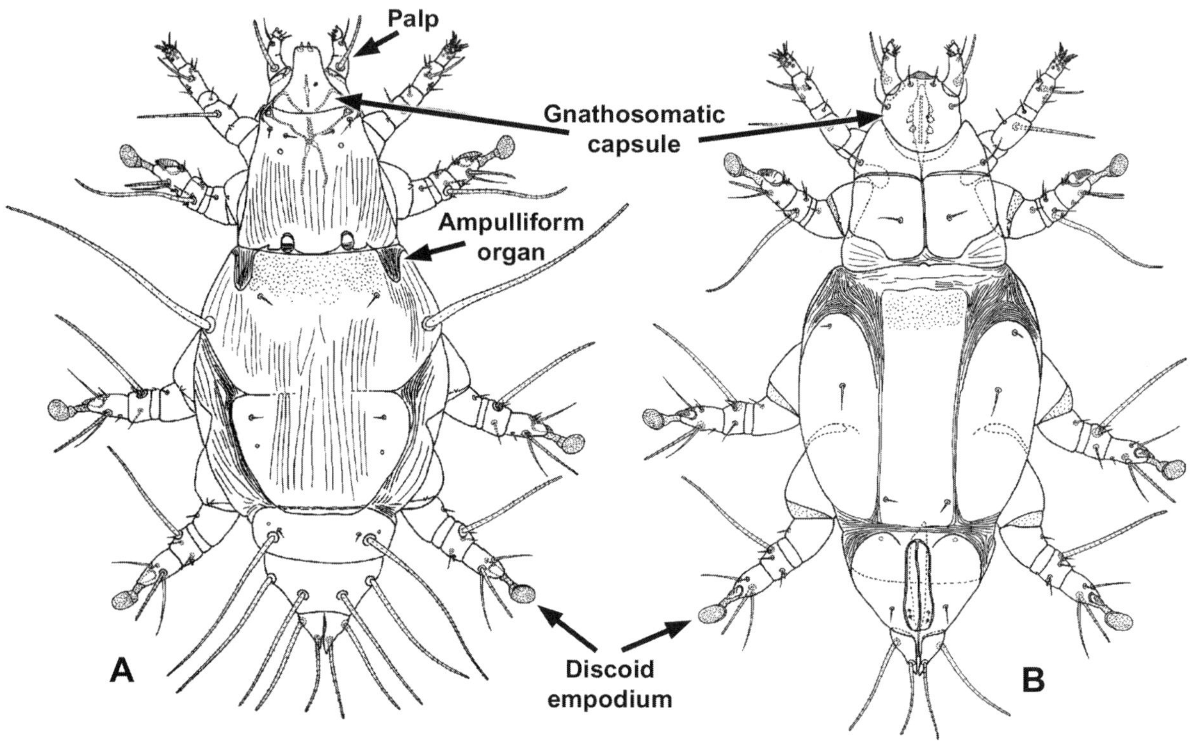

Fig. 13.82. HETEROCHEYLIDAE. *Heterocheylus* sp. (after Lindquist and Kethley 1975): **A**, dorsum of female; **B**, venter of female.

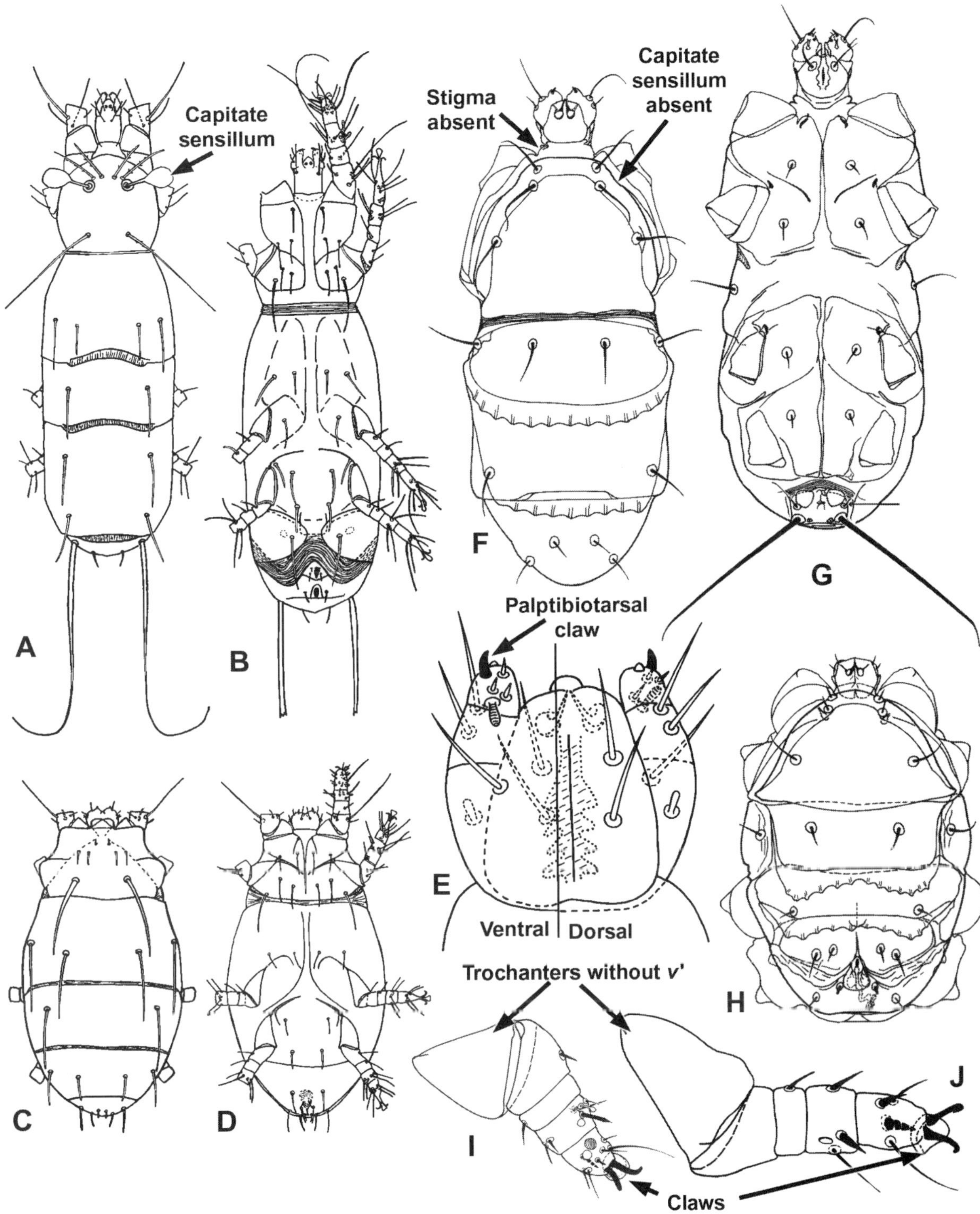

Fig. 13.83. DOLICHOCYBIDAE. *Dolichocybe keiferi* Krantz (after Cross 1965): **A**, dorsum of female; **B**, venter of female; **C**, dorsum of male; **D**, venter of male. *Pavania* sp.: **E**, gnathosoma (after Lindquist 1976). *Formicomotes heteromorphus* Magowski (after Magowski 1988): **F**, dorsum of female; **G**, venter of female; **H**, dorsum of male; **I**, leg I of male; **J**, leg IV of male.

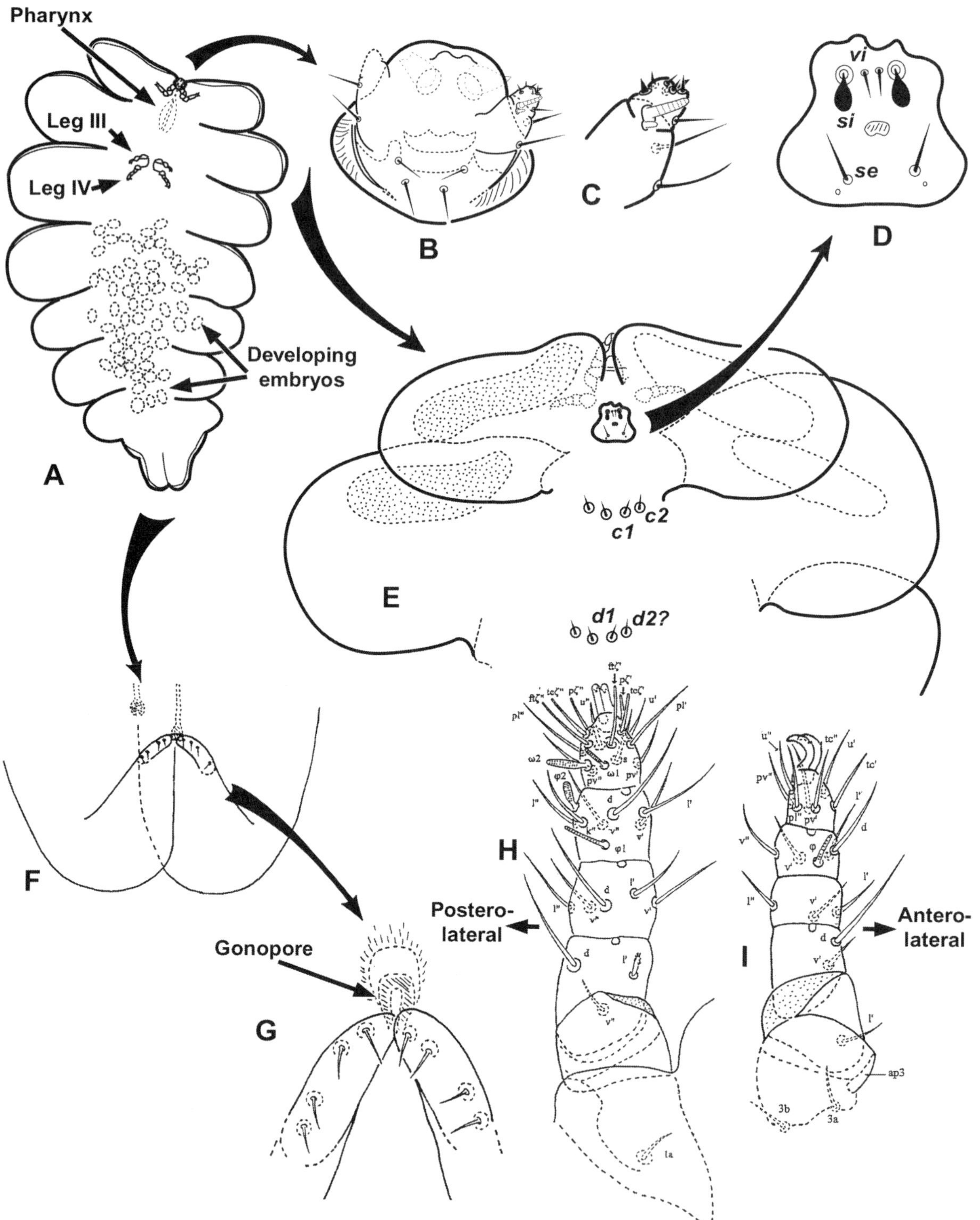

Fig. 13.84. CROTALOMORPHIDAE. *Crotalomorpha camini* Lindquist and Krantz, female (after Lindquist and Krantz 2002): **A**, ventral habitus of gravid female; **B**, ventral view of subcapitular structures; **C**, detail of palp; **D**, prodorsal shield; **E**, dorsal view of anterior section of idiosoma showing tergitelike expansions flanking prodorsum and remnants of opisthosomatic segments C and D; **F**, ventral face of bilobed caudum of idiosoma; **G**, detail of genital valves, genital setae, and gonopore; **H**, leg I, dorsal aspect; **I**, leg III, dorsolateral aspect.

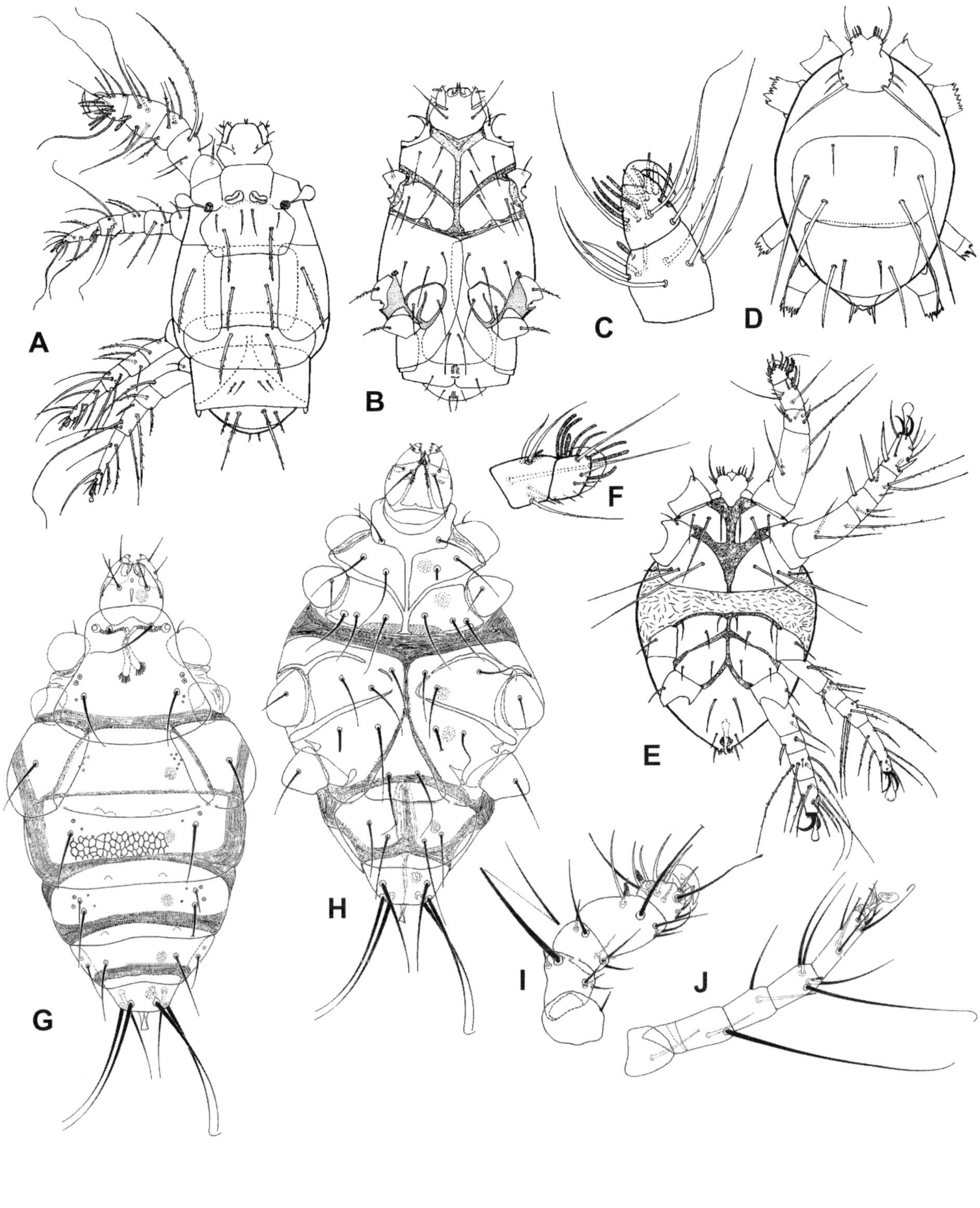

Fig. 13.85. TROCHOMETRIDIIDAE. *Trochometridium tribulatum* Cross (after Cross 1965): **A**, dorsum of female; **B**, venter of female; **C**, detail of female tibia and tarsus I; **D**, dorsum of male; **E**, venter of male; **F**, detail of male tibia and tarsus I.

ATHYREACARIDAE. *Athyreacarus pleiotretus* Lindquist, Kaliszewski, and Rack (after Lindquist, Kaliszewski, and Rack 1990): **G**, dorsum of female; **H**, venter of female; **I**, leg I of female; **J**, leg IV of female.

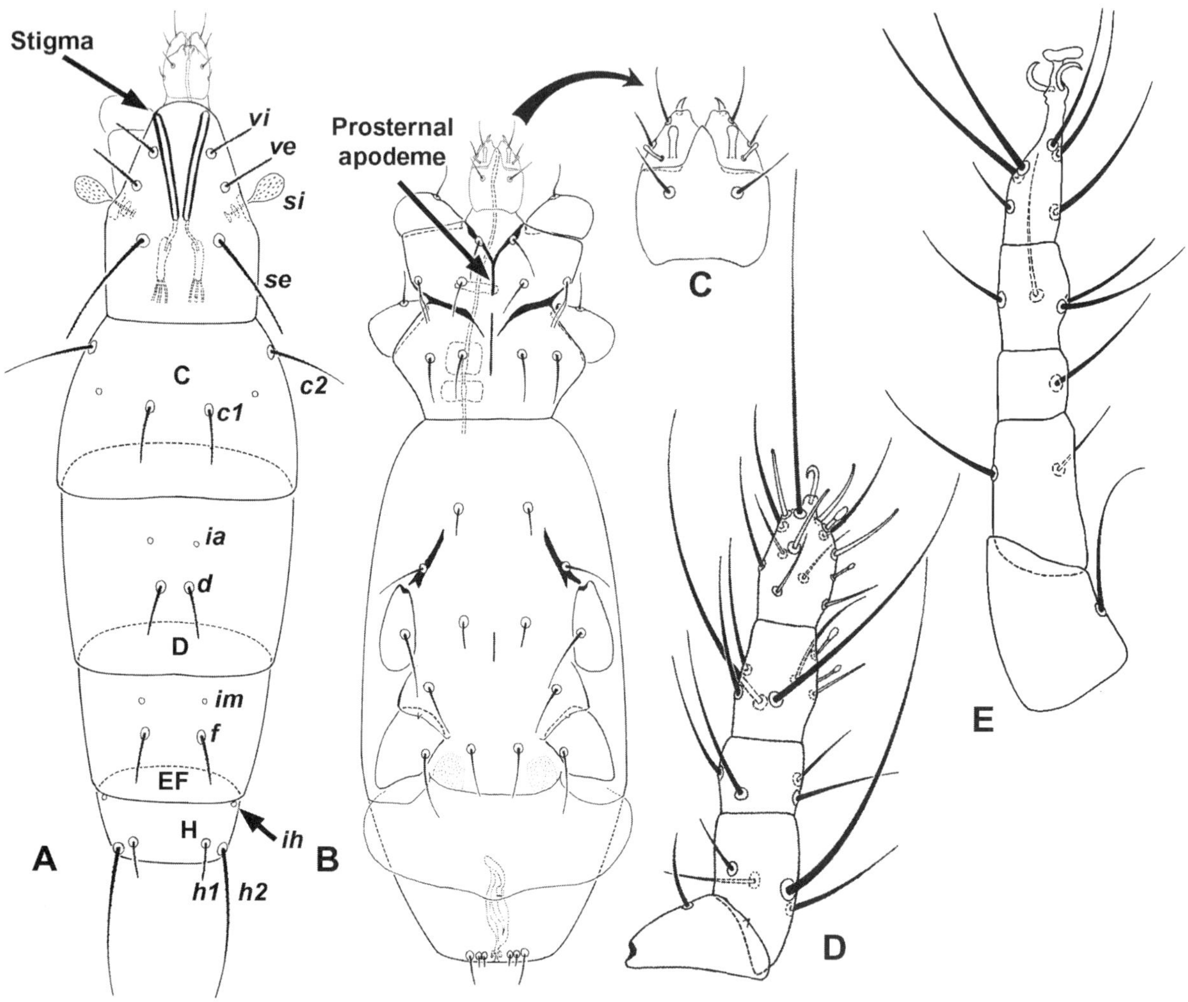

Fig. 13.86. SITEROPTIDAE. *Siteroptes longisomus* Kaliszewski (after Kaliszewski 1987): **A**, dorsum of female; **B**, venter of female; **C**, venter of gnathosoma; **D**, leg I of female; **E**, leg IV of female.

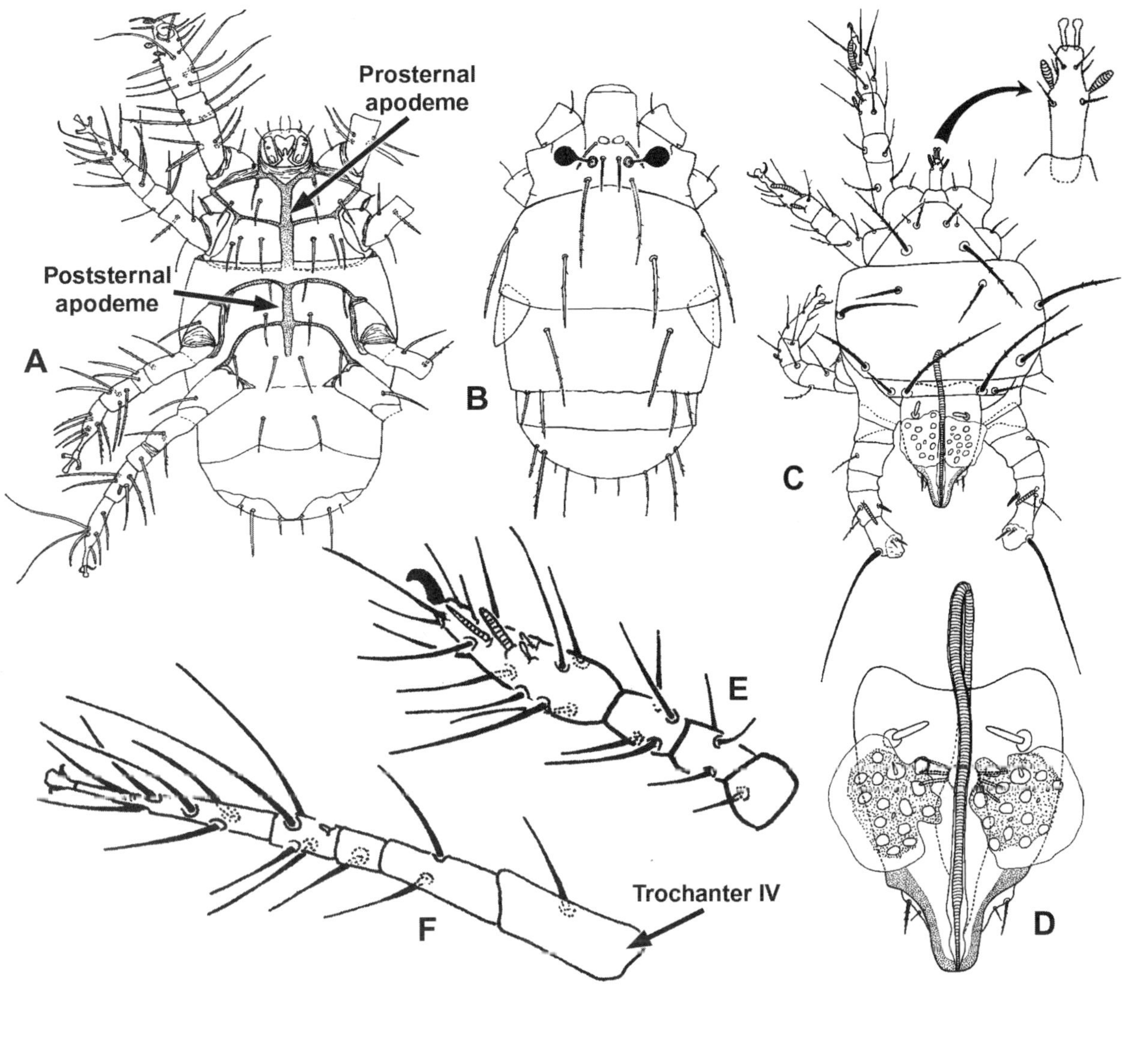

Fig. 13.87. PYGMEPHORIDAE. *Pediculaster americanus* (Banks) (after Cross 1965): **A**, venter of female; **B**, dorsum of female. *Bakerdania exigua* (Mahunka) (after Rack 1974): **C**, dorsum of male, with detail of gnathosoma; **D**, dorsum of male genital capsule. *Bakerdania* sp.: **E**, leg I; **F**, leg IV of female of a typical pygmephorid.

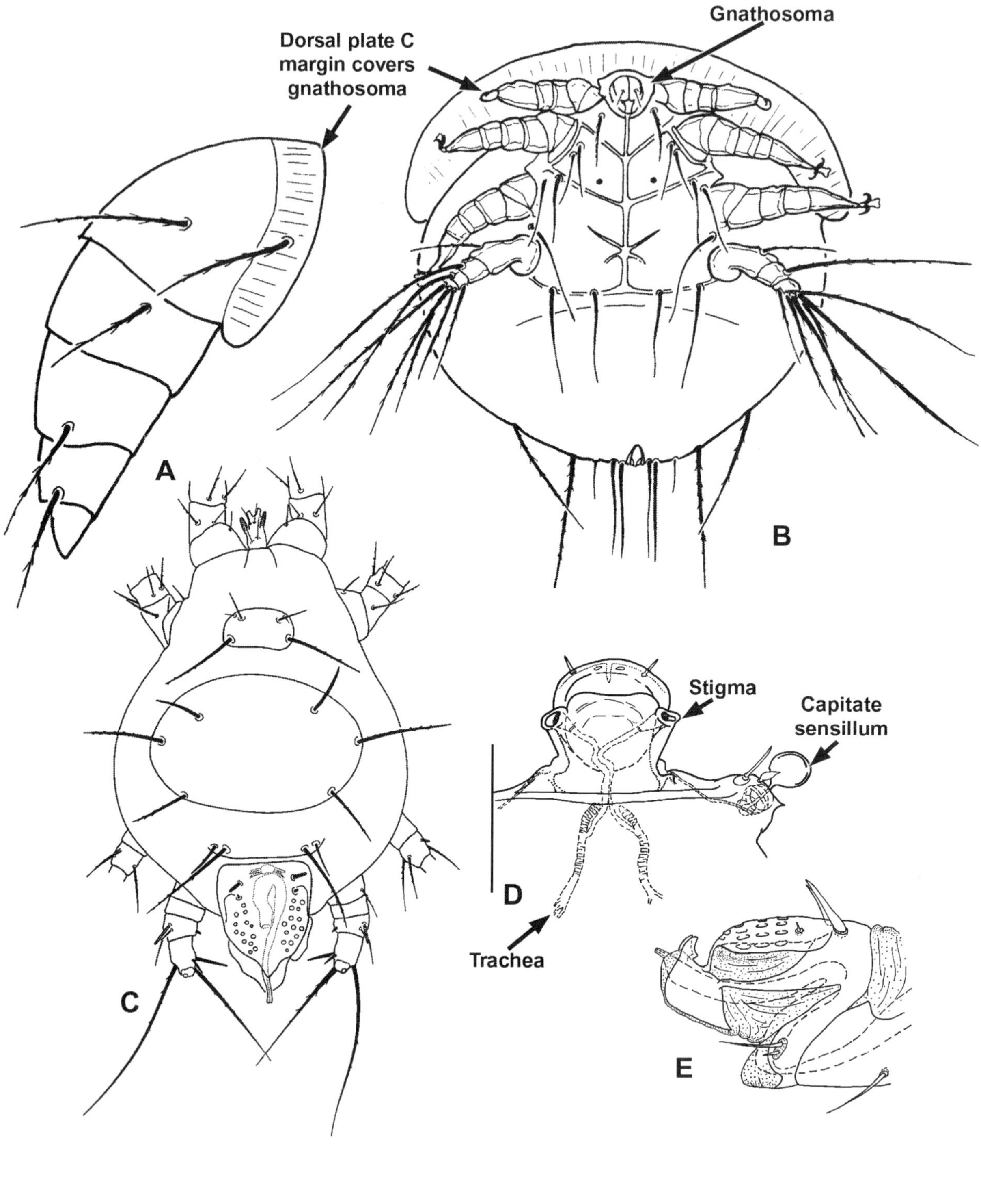

Fig. 13.88. SCUTACARIDAE. *Scutacarus* sp., female: **A**, lateral aspect of dorsal plates; **B**, venter. *Lophodispus irregularis* Ebermann: **C**, dorsum of male (after Ebermann 1982). *Scutacarus mediotarsus* Athias: **D**, prodorsum of female (after Athias 1973). *S. baculitarsus* Mahunka: **E**, lateral aspect of male genital capsule (after Norton and Ide 1974).

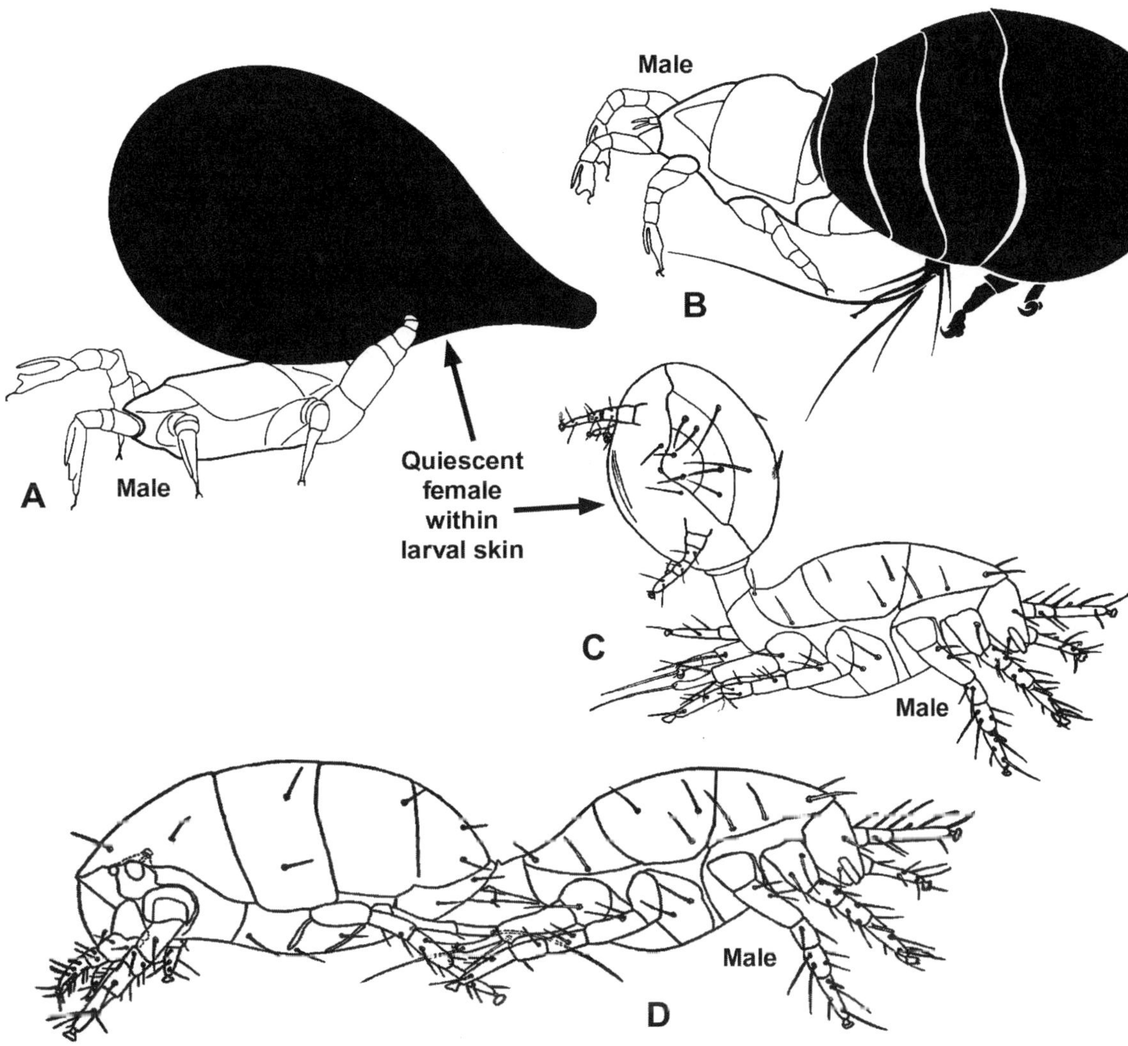

Fig. 13.89. Precopulatory and copulatory behavior of *Imparipes histricinus* Berlese (SCUTACARIDAE) (**A**, **B** after Ebermann 1982b) and *Polyphagotarsonemus latus* (Banks) (TARSONEMIDAE) (**C**, **D** after Nucifora 1963): **A**, **C**, male precopulatory behavior, carrying a quiescent female not yet emerged from larval skin (legs of larval skin not shown in **A**; longitudinal axis of female's body at right angles to that of male in **C**); **B**, **D**, retroconjugate mating position between male and emerged female adult. (Female indicated in black in **A**, **B**.)

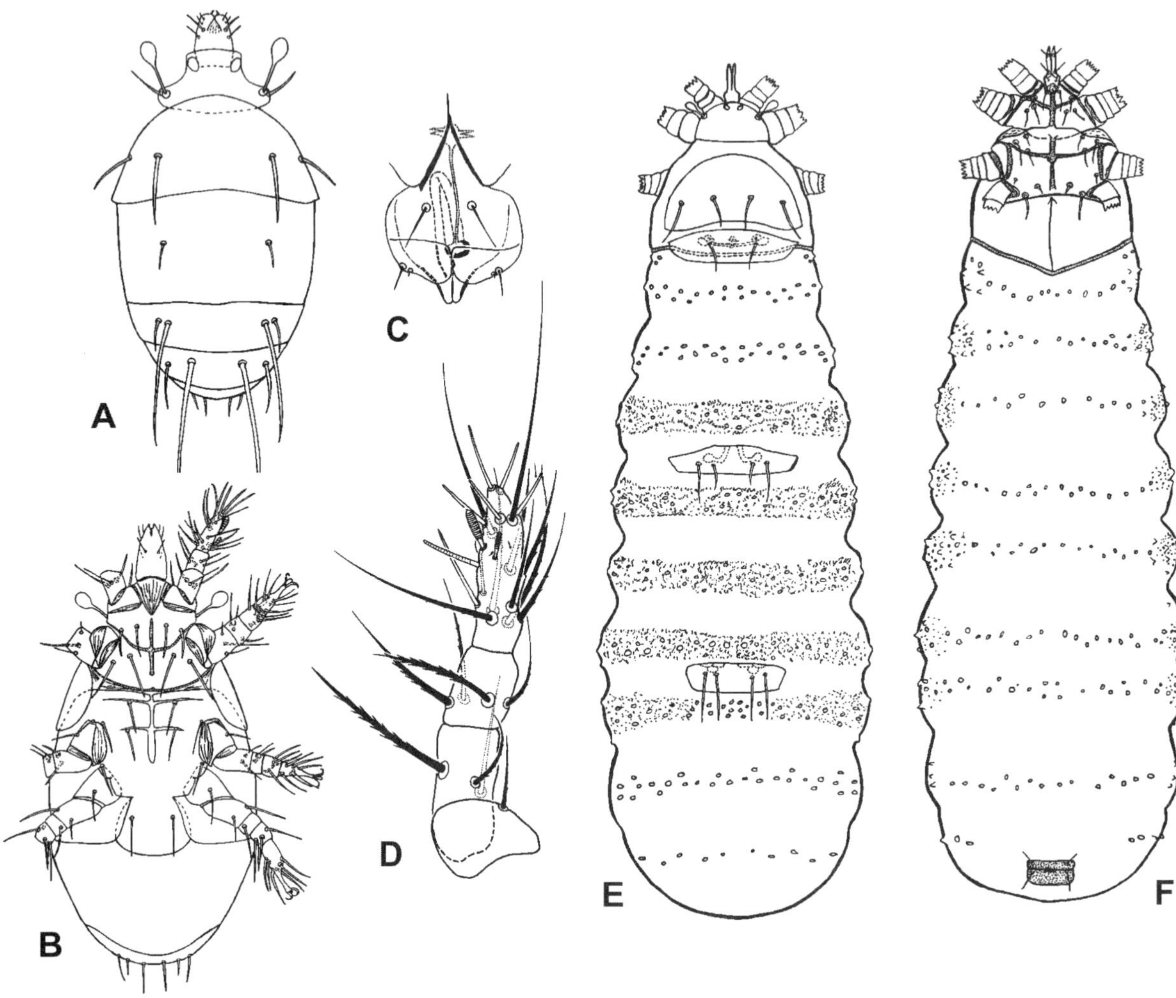

Fig. 13.90. MICRODISPIDAE. *Myrmecodispus dorylini* Cross (after Cross 1965): **A**, dorsum of female. *Microdispus* (*Premicrodispus*) *chandleri* Cross (after Cross 1965): **B**, venter of female. *M.* (*Premicrodispus*) *lambi* (Krczal) (after Kaliszewski and Rack 1986): **C**, venter of male genital capsule; **D**, leg I of female. *Perperipes ornithocephala* Cross, gravid female (after Cross 1965): **E**, dorsum; **F**, venter.

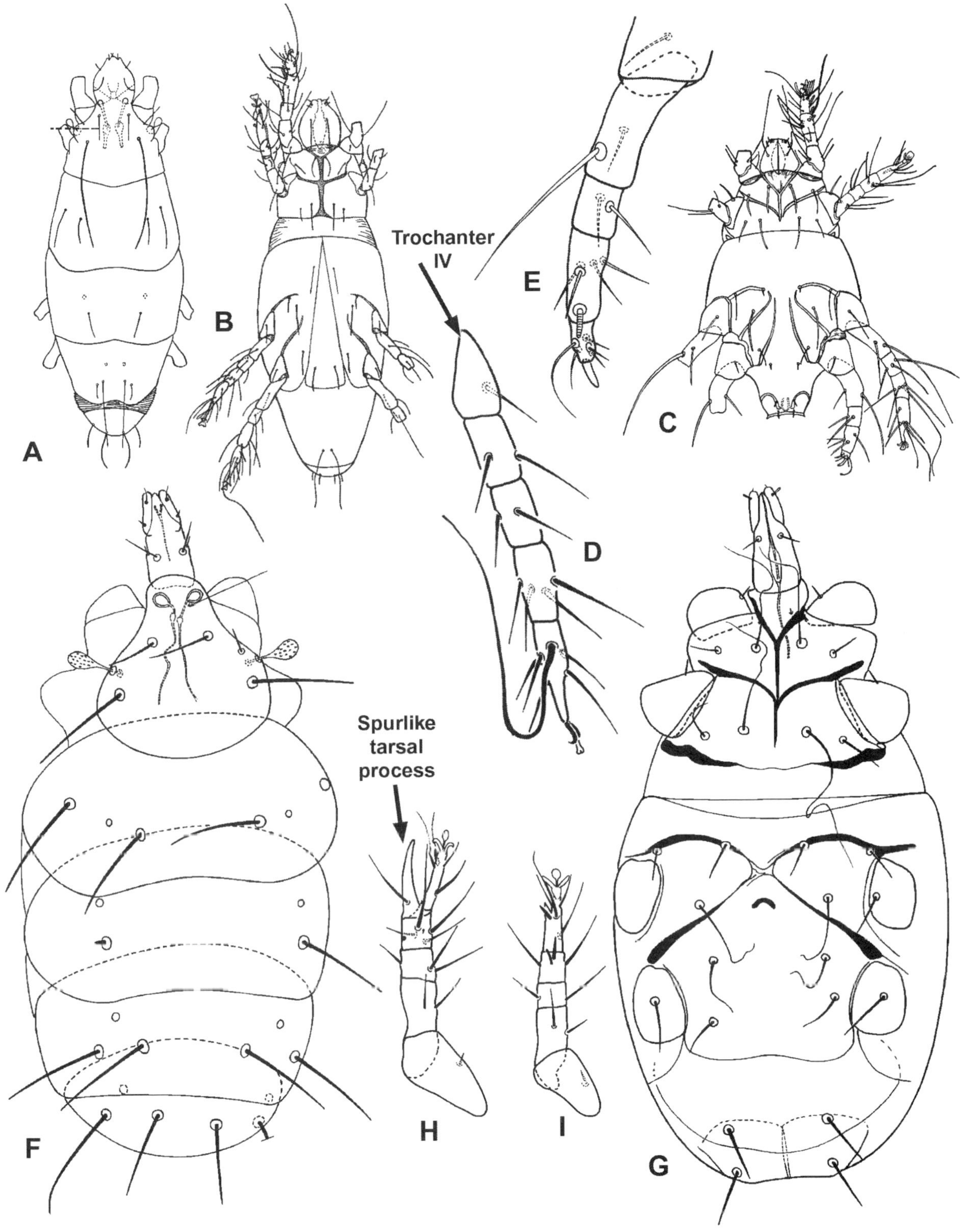

Fig. 13.91. PYEMOTIDAE. *Pyemotes* sp. (after Cross 1965): **A**, dorsum of female; **B**, venter of female; **C**, venter of male; **D**, leg IV of female (after Krantz 1978); **E**, leg IV of male (after Lindquist 1986).

RESINACARIDAE. *Resinacarus resinatus* Vitzthum (after Mahunka 1975f): **F**, dorsum of female; **G**, venter of female; **H**, leg III of male; **I**, leg IV of male.

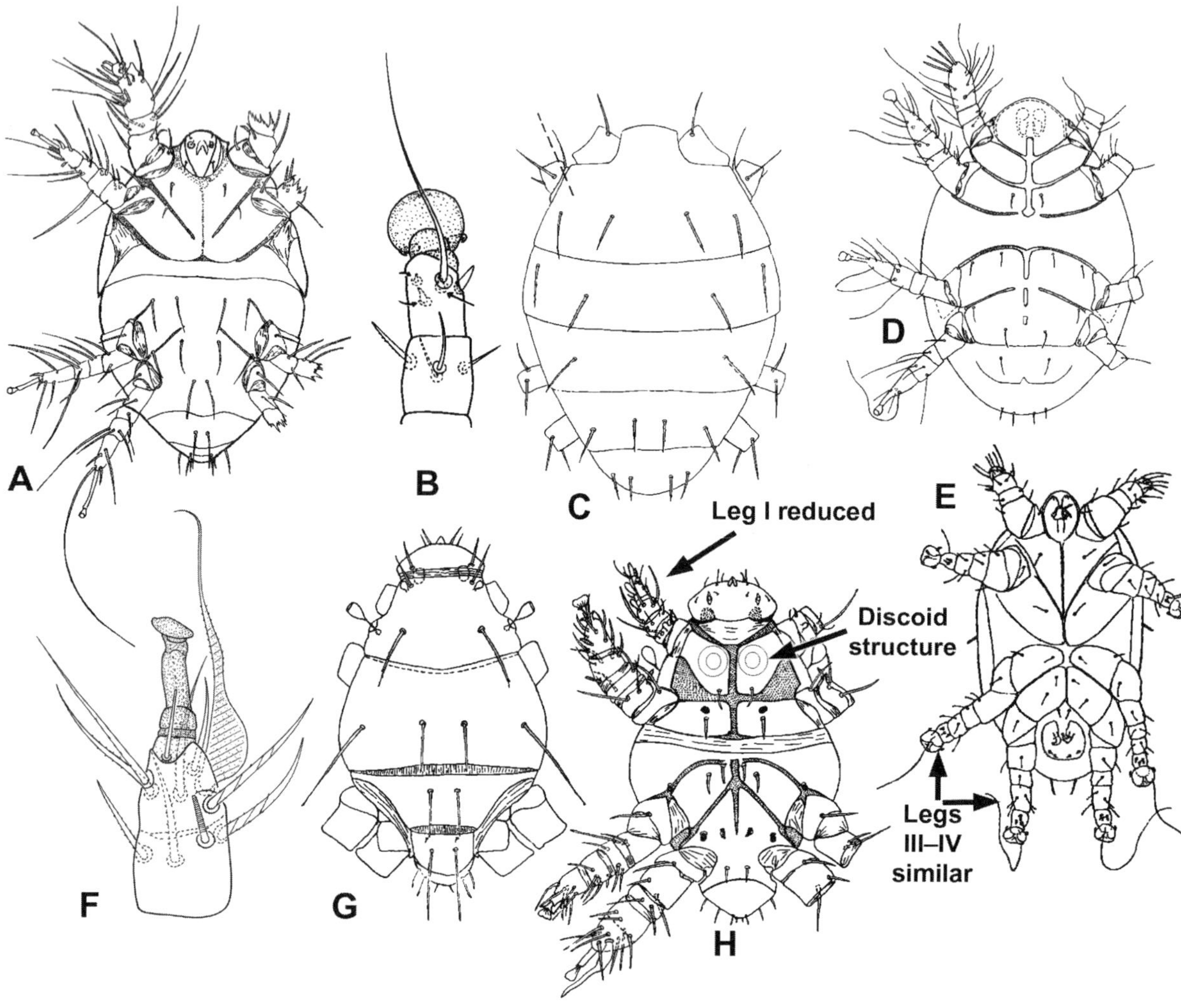

Fig. 13.92. ACAROPHENACIDAE. *Acarophenax tribolii* Newstead and Duvall (after Cross 1965): **A**, venter of female. *Acarophenax* sp. (after Lindquist 1986): **B**, tibia and tarsus IV of male. *Paracarophenax dybasi* Cross (after Cross 1965): **C**, dorsum of female. *Adactylidium beeri* Cross (after Cross 1965): **D**, venter of female. *P. dermestidarium* (Rack) (after Rack 1959): **E**, venter of male. CARABOACARIDAE. *Caraboacarus* sp. female (after Cross 1965): **F**, tibiotarsus of leg IV; **G**, dorsum; **H**, venter.

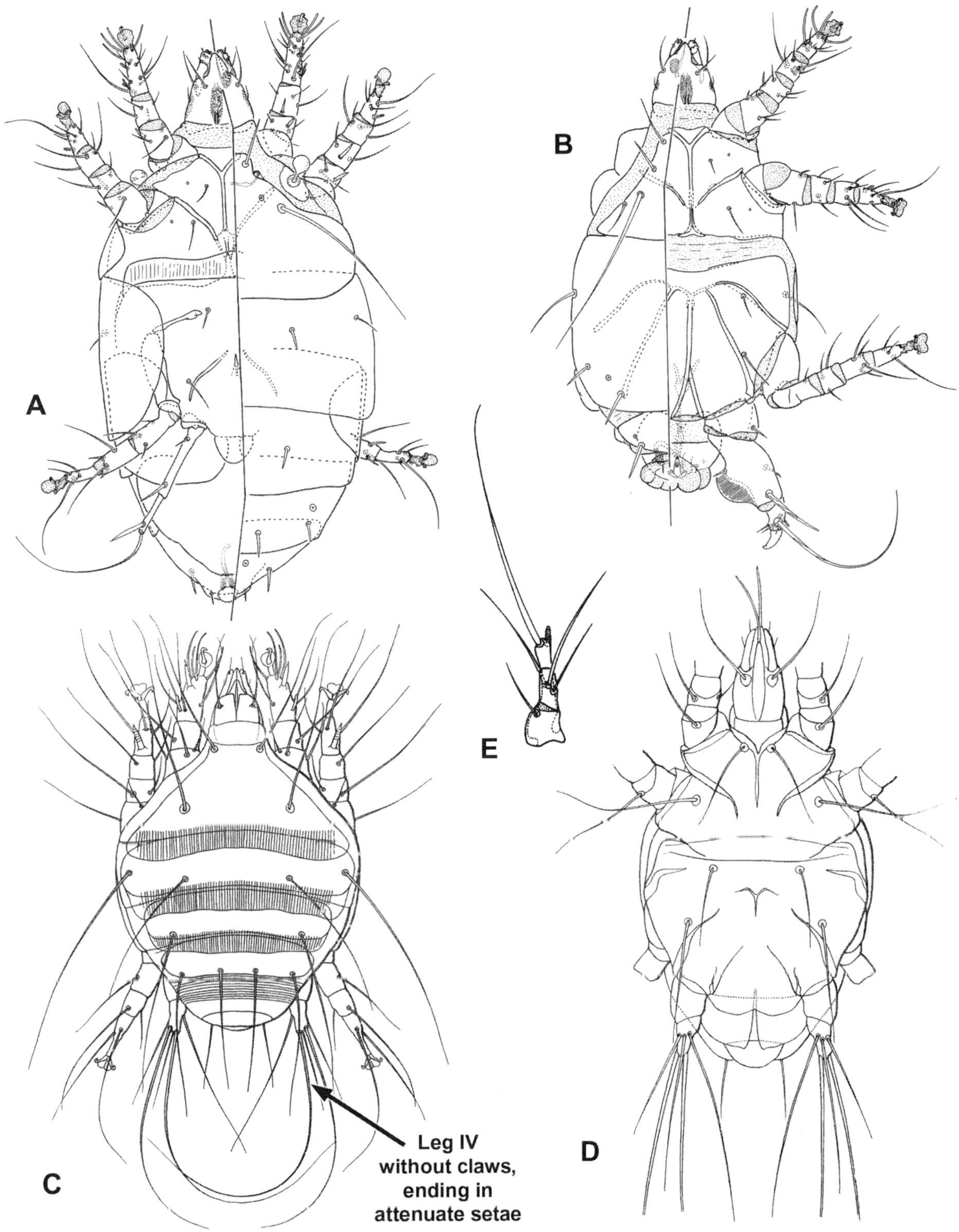

Fig. 13.93. TARSONEMIDAE. *Phytonemus pallidus* (Banks) (after Lindquist 1986): **A**, habitus of female, ventral aspect on left side, dorsal aspect on right; **B**, habitus of male, dorsal aspect on left side, ventral aspect on right. *Acarapis woodi* (Rennie), female (after Hirst 1921): **C**, dorsum; **D**, venter. *Acarapis* sp. (after Lindquist 1986). **E**, leg IV of male.

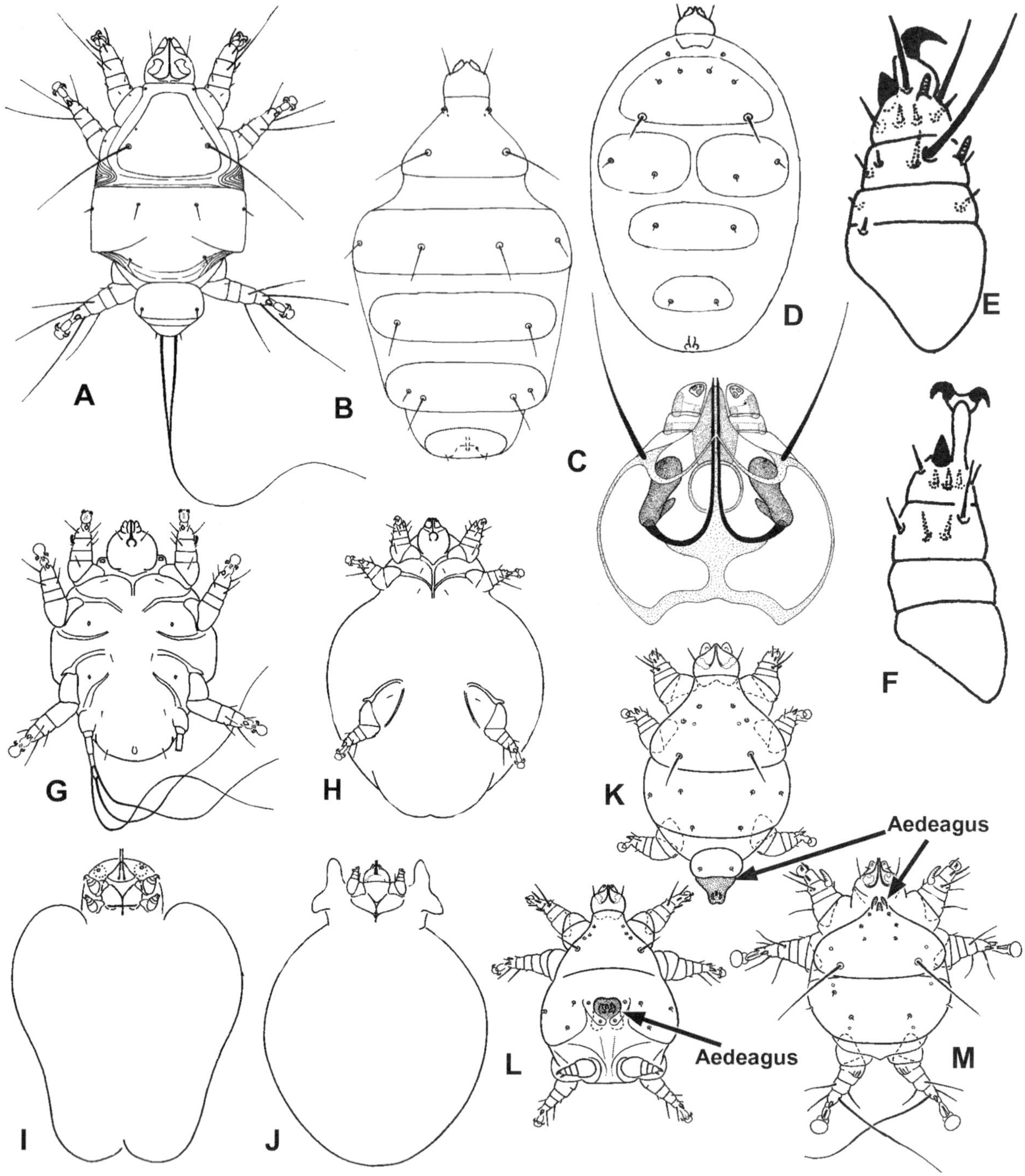

Fig. 13.94. PODAPOLIPIDAE. *Dorsipes dorsipes* Regenfuss (after Regenfuss 1968): **A**, dorsum of larval female; **B**, dorsum of adult female; **C**, gnathosoma. *Eutarsopolipus agonobius* Regenfuss (after Regenfuss 1968): **D**, dorsum of adult female. *Eutarsopolipus* sp., adult female: **E**, leg I; **F**, leg III.

Venters of adult females of genera showing different numbers of legs retained (after Regenfuss 1973): **G**, *Chrysomelobia mahunkai* Regenfuss; **H**, *D. inflatus* Regenfuss; **I**, *Tetrapolipus rhynchophori* Ewing; **J**, *Podapolipoides* sp.

Dorsa of males of genera showing differences in position of aedeagus and numbers of legs retained (after Regenfuss 1973): **K**, *E. acanthomus* Regenfuss; **L**, *Dorsipes* sp.; **M**, *Podapolipus* sp.

CHAPTER FOURTEEN D.E. WALTER

Suborder Endeostigmata

*DIAGNOSIS: Paraphyletic group of early derivative sarcoptiform mites with numerous plesiomorphies. Body often with constrictions reflecting primary opisthosomatic segmentation, and either hypertrichous or bearing most of the setal pairs (usually 28–35 pairs in extant species) of a hypothesized holotrichous ancestor; segment PA sometimes suppressed. Prodorsum unsclerotized or with weakly defined plate and usually with 6 pairs of setae (*ro, le, bo, exa, in, exp*); prodorsal setae* bo *(except in Nematalycoidea) and sometimes* ro *or* le *formed as trichobothria, setae* in *and/or* exp *sometimes suppressed; dorsosejugal furrow usually distinct; naso and/or median eye often present, lenslike lateral ocelli and postocular body often present. Subcapitulum with lateral lips bearing 0–3 pairs of adoral setae, usually with narrow rutella with distal blade or teeth; 2–3 or more pairs of setae in the genal area and usually with a pair of peglike epicoxal setae (*ep*); palpi with 4–6 apparent segments (palpfemur sometimes subdivided) and without a palptibial claw. Chelicerae 2–3 segmented, chelate-dentate to attenuate-edentate and each bearing 0–2 setae. Genital shields unsclerotized, bearing numerous genital setae, and covering 3 (rarely 2) pairs of genital papillae in the adult. Legs well supplied with solenidia and other setae but without typical trichobothria; epicoxal seta* epl *sometimes present; pretarsi usually with empodium and often with claws; legs IV sometimes modified for jumping. Body cuticle usually plicate, sometimes with secondary ornamentation; colorless (white) to pale lilac, bright pink, or yellowish brown in color. Tracheal systems have been reported in some families.*

The Endeostigmata (cohort Pachygnathina of Krantz 1978) was proposed by Grandjean (1937d) for a cluster of early derivative acariform mites that he believed linked the Oribatida to the Prostigmata. Over the next few years, Grandjean studied several genera of PACHYGNATHIDAE (now ALYCIDAE) in detail and proposed three of the currently recognized families (ALICORHAGIIDAE, TERPNACARIDAE, NANORCHESTIDAE). Subsequent descriptions by Kethley (1977c) (OEHSERCHESTIDAE, GRANDJEANICIDAE), Coineau and Theron (1983) (MICROPSAMMIDAE), Kethley (1989b) (PROTEONEMATALYCIDAE), and Lindquist and Palacios-Vargas (1991) (PROTERORHAGIDIIDAE) raised the number of endeostigmatic families to nine. Additionally, the NEMATALYCIDAE Strenzke (1954) is currently considered a member family of the Endeostigmata (Evans 1992). The families of Endeostigmata may be separated by the key at the end of this chapter.

Although the Endeostigmata has traditionally been treated as a primitive group of Prostigmata, its affinities lie with the Oribatida (see chapter 15). However, two of Grandjean's families, the SPHAEROLICHIDAE and LORDALYCIDAE, are exceptions in that they have affinities instead with the Prostigmata (see chapter 13). They presently compose the suborder Sphaerolichida, a sister group to the Prostigmata (OConnor 1984c). In addition to problems with higher-level placement, the Endeostigmata has been subject to confusing nomenclatorial problems. For example, the first nominate family, the PACHYGNATHIDAE—raised to accommodate members of the genus *Pachygnathus* Dugês, a genus studied in depth by Grandjean (1936b, 1937a–d)—is a junior homonym of a spider family of the same name (based on *Pachygnatha*; the spider family itself is a junior synonym of Tetragnathidae). The family name ALYCIDAE Canestrini and Fanzago (based on *Alycus* C. L. Koch) is now preferred over the BIMICHAELIIDAE of some authors (Judson 2000). Additionally, *Lordalycus* Grandjean (LORDALYCIDAE) has been identified as a junior synonym of *Hybalicus* Berlese, which was designated by Theron (1974) as the type genus for his family Hybalicidae. The species of *Hybalicus* described in that work are referable instead to the endeostigmatic genus *Oehserchestes* Jacot (OEHSERCHESTIDAE) (Kethley 1977b). Finally, the PEDICULOCHELIDAE, an enarthronote oribatid assemblage, often has been misplaced in the Endeostigmata.

The Endeostigmata, in the strict sense used in this treatment, includes mostly tiny, globular, or bizarrely elongate sarcoptiform mites that display numerous primitive morphological characters (Grandjean 1937d, 1939d, 1943b, c) and are often found in extreme soil habitats (e.g., cold and hot deserts, microbial crusts, seashores, sandy soils, and deep soil layers). Fossil endeostigmatans have been described from some of the earliest known terrestrial faunas (Hirst 1923; Dubinin 1962; Kethley et al. 1989), and their lifestyles tend to reflect their primordial roots. For example, most of them display the fundamental acariform ontogeny (see chapter 5): an egg, an inactive hexapod prelarva, an active hexapod larva, and active octopod protonymph, deutonymph, tritonymph, and adult forms (Walter and Proctor 1999; but see Kethley 1990, 1991a). The active, feeding larva has urstigmata between legs I–II but lacks genital papillae and the posterior opisthosomatic segments AN, AD, and PA (see Fig. 3.4, p. 30). In some taxa the prelarva is retained within the egg, but in others it expands and ruptures the eggshell and then remains quiescent (Walter 1988b; Kethley 1992). At least some prelarvae of NANORCHESTIDAE are able to move, but they cannot feed (Schuster and Pötsch 1989).

As noted in the subordinal diagnosis above, the full complement of prodorsal setae usually is present in endeostigmatic mites. However, it should be mentioned that various systems for designation of these setae occur in the literature and that setal interpretations may differ, especially between *exa* and *le*. When the prodorsal chaetome is reduced, it is usually setae *in* and/or *exp* that are absent. Folds indicating remnants of opisthosomatic segmentation are present in many TERPNACARIDAE and ALYCIDAE, to the point that some Russian researchers have referred to members of these families as "segmented mites." These folds are best seen in species of the terpnacarid genera *Terpnacarus* (Figs. 14.1A, 14.5A) and *Alycosmesis*, the latter of which have expanded and leaflike to palmate dorsal setae (Grandjean 1939d). Body region *C* (possibly representing body segments VII + VIII) often carries 4 pairs of setae (*c1–c4*). Other putative segments may have 4 or more pairs of apparently normal setae in TERPNACARIDAE, and the holotrichous condition (28–35 pairs in extant species) appears to be unusually prevalent. Hysterosomatic hypertrichy (apparently neotrichy) occurs in the ALYCIDAE and NANORCHESTIDAE, and the former may also exhibit prodorsal neotrichy (Fig. 14.6C).

As in oribatid mites, the posterior body segments and pairs of genital papillae are added during development (anamorphosis); that is, segment AN + one pair of genital papillae in the protonymph, segment AD + the second pair of genital papillae in the deutonymph, and segment PA + the third pair of genital papillae in the rarely suppressed tritonymph (see Oudemans' rule, chapter 3). The stage between molts is quiescent and unable to move. The number of eggs matured at one time by a female seems to vary with size, with small species tending to develop a single egg. A short ovipositor (Fig. 14.1A) is primitively present but is reduced or absent in many groups. Males usually have a well-developed organ for producing stalked spermatophores, but parthenogenesis is common and males are unknown for many taxa. Members of the genus *Alicorhagia* (ALICORHAGIIDAE) spin silken cocoons for the deposition of eggs and before molts (Walter 1988b); cocoons of a variety of endeostigmatans can be found in soil extractions using flotation methods (see chapter 7).

Larvae and other active stages of ALICORHAGIIDAE, TERPNACARIDAE, OEHSERCHESTIDAE, GRANDJEANICIDAE, and MICROPSAMMIDAE ingest solid food fragments (i.e., particulate feeding) that form discrete gut boluses, as in many other sarcoptiform mites and in the Opilioacarida (Walter and Proctor 1998). Fungi and sclerotized bits of small invertebrates often can be identified in gut boluses (Schuster 1979; Theron 1979; Walter 1988b). *Grandjeanicus theroni* Walter feeds on immature oribatids and other mites as well as on fungi (Walter 2001). Some ALYCIDAE (e.g., *Alycus roseus* Koch) appear to be exclusively predatory on nematodes (Walter 1988b), but others (e.g., species of *Bimichaelia*) have highly modified, elongate, needlelike cheliceral digits of unknown function. Theron (1979) speculated that species of *Bimichaelia* might suck juices from plant roots, and Lindquist (1998) expanded on the possible evolution of phytophagy in the ALYCIDAE and the similarity of their mouthparts to those in Eriophyoidea. Alternatively, as species of *Bimichaelia* are abundant in forest litter and not obviously associated with plant roots, some other feeding strategy seems a more likely scenario. Members of the NANORCHESTIDAE are fluid feeders and have a sclerotized tubular labrum that may play a role in the feeding process. *Nanorchestes amphibius* Topsent and Trouessart was reported to feed on green algae (Schuster and Schuster 1977).

The PROTERORHAGIIDAE are probably predatory, based on their massive raptorial chelicerae (Fig. 14.4A), but the feeding habits of PROTEONEMATALYCIDAE and NEMATALYCIDAE are unknown. The NEMATALYCIDAE comprises four genera of highly derived acariform mites that resemble worms and live in deep, sandy soils. Although they lack plesiomorphies characteristic of other endeostigmatic mites (e.g., trichobothria, rutella, naso, and apparent segmental remnants), the family has been referred to the Endeostigmata since the discovery of two "missing link" taxa that lack trichobothria but that have other endeostigmatan symplesiomorphies. *Micropsammus littoralis* Coineau and Theron, an elongate mite with a strong constriction behind legs IV, is the least aberrant of these forms and retains three pairs of prodorsal setae, a naso bearing setae *ro*, and well-developed rutella (Coineau and Theron 1983). *Proteonematalycus wagneri* Kethley, an even more wormlike mite, has an unpaired median seta *ro* and minute prodorsal setae *exa*, *in*, and *exp* (Kethley 1989b). It also retains median and lateral eyes, as well as rutella and three pairs of genital papillae.

Key 14.1. Suborder Endeostigmata, Key to the Families (Figs. 14.1–14.6)

1a. Prodorsum with 1–2 pairs of trichobothria (bothridial setae) (Figs. 14.1A, B, F, 14.2C, G, 14.3A, D, F), one of which may be clavate or capitate (Fig. 14.3F); body shape various but never wormlike; lateral and/or median eyes present (Fig. 14.1A) or absent **2**

1b. Prodorsum without bothridia or clavate-capitate setae (Figs. 14.4C–E); body shape elongate, often wormlike (Fig. 14.4B); eyes absent **Nematalycoidea—8**

2a. Prodorsum with 2 pairs of trichobothria (Figs. 14.3A, D, F, 14.4A), one of which may be clavate or capitate; naso absent or weakly developed and nude; rutella present or absent **3**

2b. Prodorsum with 1 pair of filiform trichobothria (*bo*) (Figs. 14.1A, B, F, 14.2A, C, G); naso well developed and bearing setiform setae *ro*; rutella present and dentate **5**

3a. Chelicerae slender, weakly chelate-dentate, needlelike or with short movable digit (Fig. 14.3C); opisthosomatic chaetome usually hypertrichous **4**

3b. Chelicerae massive, chelate-serrulate, and much larger than prodorsum (Fig. 14.4A); opisthosomatic chaetome hypotrichous **PROTERORHAGIIDAE**

4a. All tarsi with paired claws and rayed empodium (Fig. 14.3A, detail); coxae IV not modified for jumping; chelicera chelate or needlelike **ALYCIDAE**

4b. Tarsi with only rayed empodium (Fig. 14.3E), lateral claws absent; coxae IV modified for jumping; chelicera chelate **NANORCHESTIDAE**

5a. Pretarsi II–IV with paired claws and empodium (Fig. 14.1D); each chelicera with 2 setae (Fig. 14.1G); 3 pairs of genital papillae in adult (Fig. 14.1A); body setae simple, dendritic, globose, rodlike, or leaflike **6**

5b. All pretarsi with a single, simple, clawlike empodium and without lateral claws (Fig. 14.2E); each chelicera with 1 seta (Fig. 14.2D); 2 (Fig. 14.2F) or 3 pairs of genital papillae in adult; body setae simple or branched-plumose **ALICORHAGIIDAE**

6a. Prodorsum with 5 pairs of setae (including *bo*), external bothridial seta *exp* absent (Figs. 14.1F, 14.2A); tarsus I without lateral claws, with (Fig. 14.1E) or without (Fig. 14.2B) an empodium **7**

6b. Prodorsum with 6 pairs of setae, external bothridial seta *exp* present (Figs. 14.1A, B); tarsus I with both claws and empodium (Fig. 14.1D) **TERPNACARIDAE**

7a. Tarsus I (Fig. 14.1E) ending in small, clawlike empodium; without pilose flagellum (distal, daggerlike seta sometimes present) **OEHSERCHESTIDAE**

7b. Tarsus I without claws or empodium but with a pilose flagellum (Fig. 14.2B) several times longer than the tarsus **GRANDJEANICIDAE**

8a. Body elongate but not wormlike; rutella present (Figs. 14.4D, E); genital and anal openings situated near each other; prodorsum with naso and 11–12 setae (posterior pairs may be minute); eyes present or absent **9**

8b. Body wormlike (Fig. 14.4B), > 10× as long as wide; rutella absent; genital and anal openings widely separated, prodorsum without naso and with ≤ 7 setae (Fig. 14.4C); eyes absent **NEMATALYCIDAE**

9a. Prodorsal seta *ro* paired; eyes absent; chelicerae massive (Fig. 14.4D) **MICROPSAMMIDAE**

9b. Prodorsal seta *ro* unpaired; eyes present; chelicerae not hypertrophied (Fig. 14.4E) **PROTEONEMATALYCIDAE**

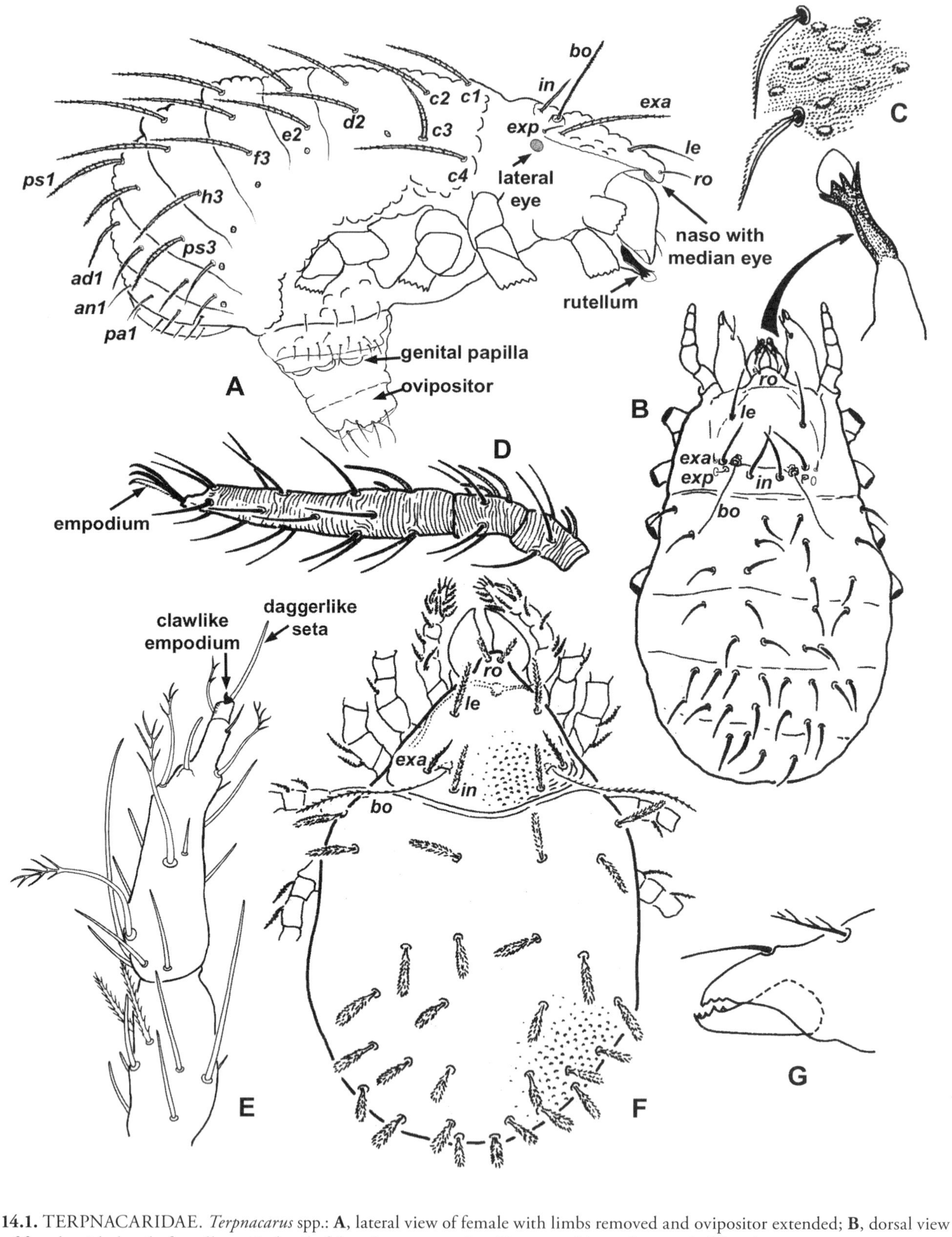

Fig. 14.1. TERPNACARIDAE. *Terpnacarus* spp.: **A**, lateral view of female with limbs removed and ovipositor extended; **B**, dorsal view of female with detail of rutellum; **C**, detail of dorsal ornamentation; **D**, genu, tibia, and tarsus I of female.

OEHSERCHESTIDAE. *Oehserchestes arboriger* (Theron) (Australia): **E**, tibia and tarsus I of female. *O. dorysetatus* (Theron) (South Africa): **F**, dorsum of female; **G**, chelicera of female.

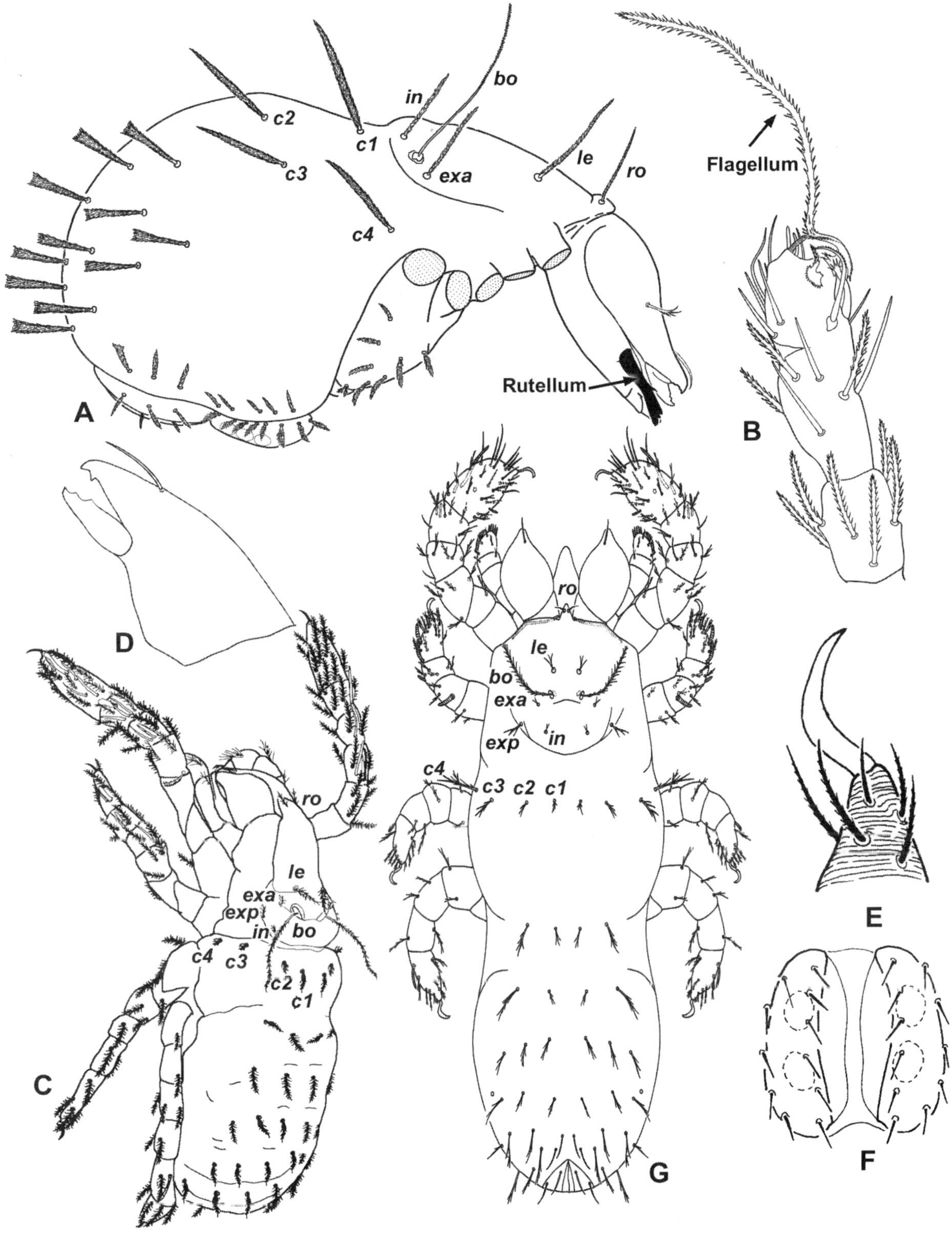

Fig. 14.2. GRANDJEANICIDAE. *Grandjeanicus theroni* Walter (Australia): **A**, lateral view of body of female; **B**, genu, tibia, and tarsus I.

ALICORHAGIIDAE. *Alicorhagia* sp.: **C**, dorsolateral view of female; **D**, chelicera of female; **E**, apex of tarsus I; **F**, genital region of female. *Stigmalychus veretrum* Theron, Ryke, and Meyer (Australia): **G**, dorsum of female.

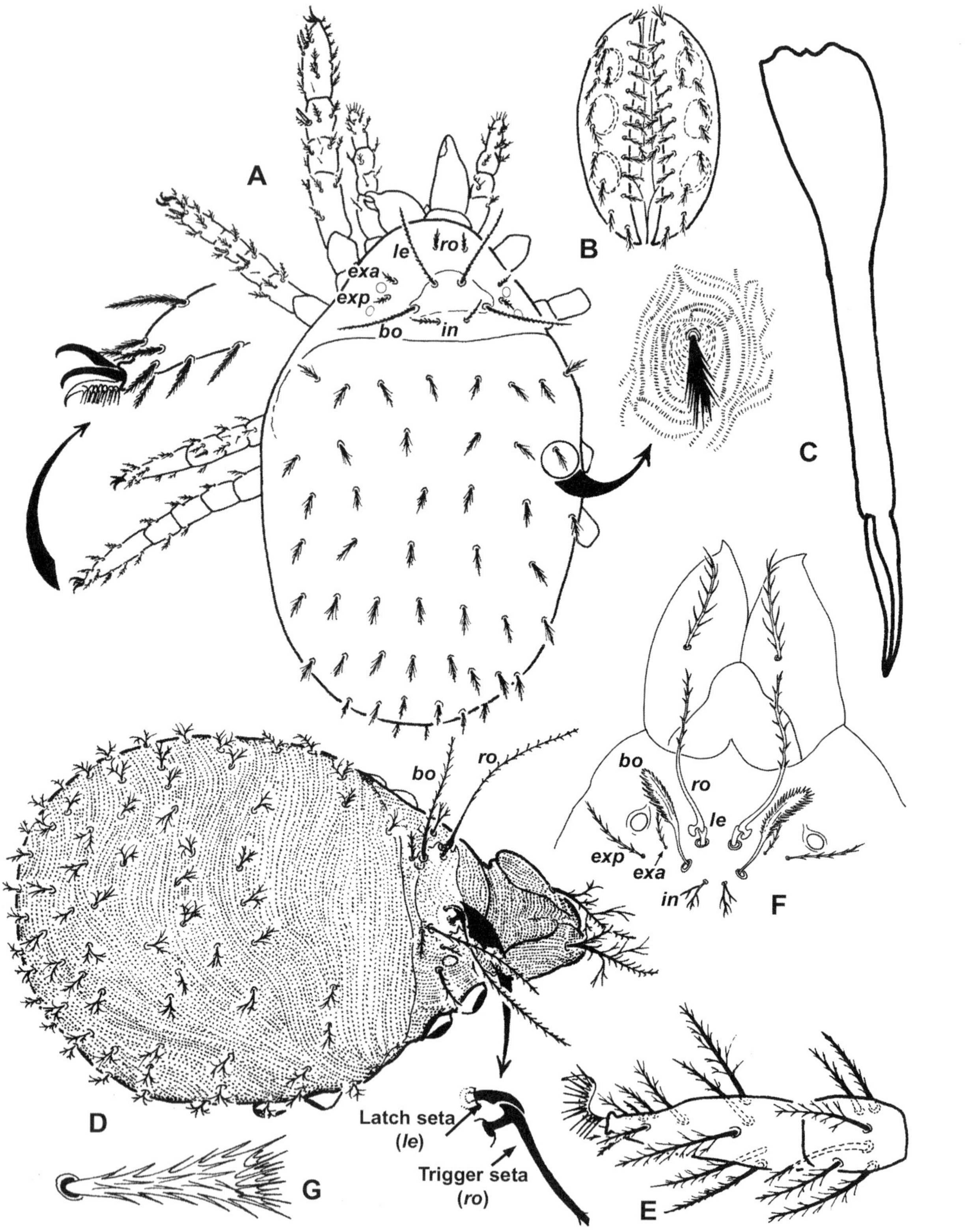

Fig. 14.3. ALYCIDAE. *Alycus* sp. (Oregon, USA): **A**, dorsum of female with detail of tarsus IV and dorsal ornamentation; **B**, genital region of female. *Bimichaelia diadema* Grandjean: **C**, chelicera of female.

NANORCHESTIDAE. *Nanorchestes* sp. (Oregon, USA): **D**, dorsolateral aspect of female with detail of sensillary apparatus; **E**, tibia and tarsus I. *Neonanorchestes*? sp. (Australia): **F**, prodorsum and chelicerae. *Speleorchestes* sp. (Oregon, USA): **G**, idiosomatic seta.

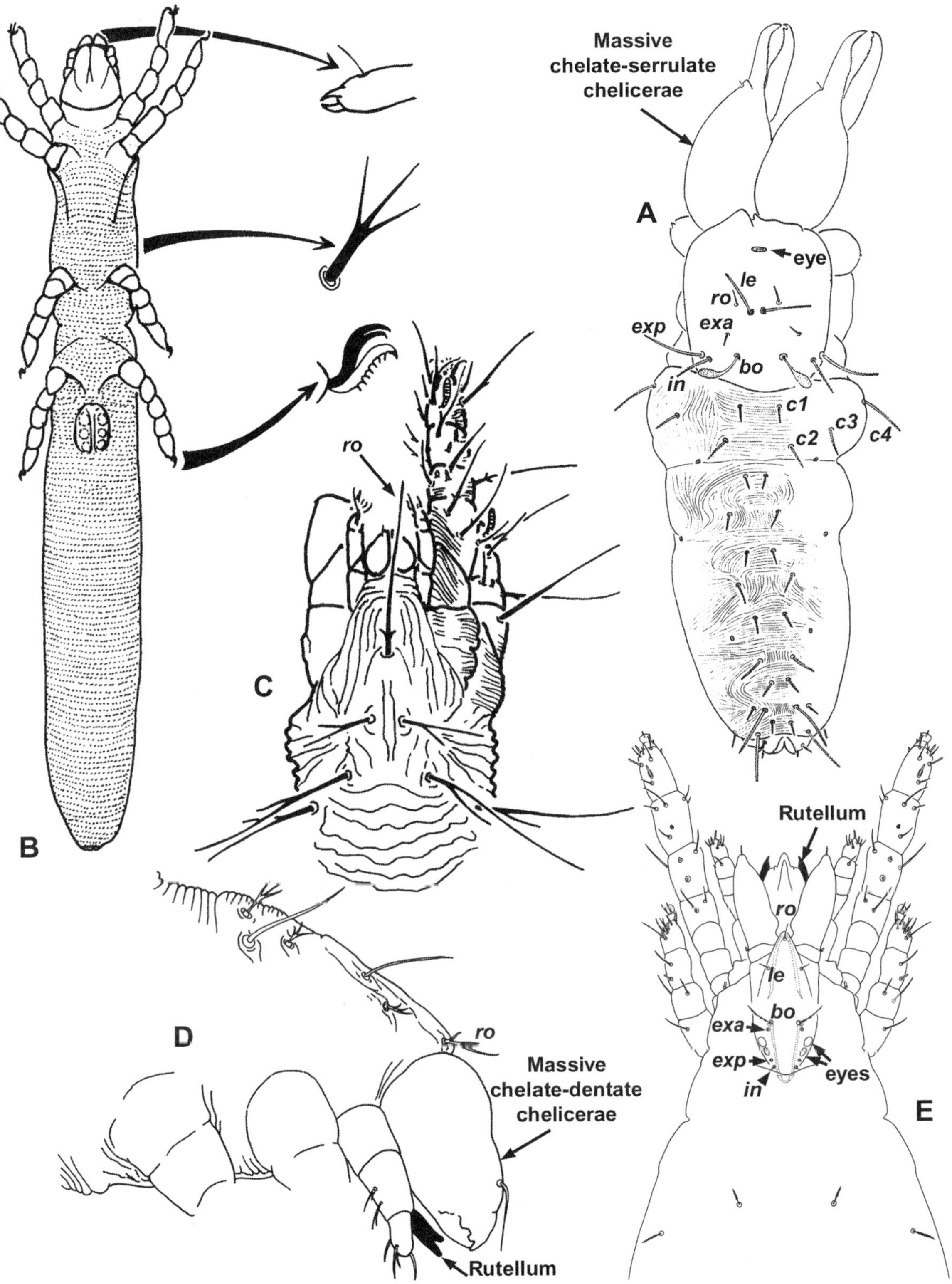

Fig. 14.4. PROTERORHAGIIDAE. *Proterorhagia oztotloica* Lindquist and Palacios-Vargas (Mexico): **A**, dorsum of female.
NEMATALYCIDAE. *Nematalycus* (diagrammatic): **B**, venter of female with details of chelicera, dorsal seta, and pretarsus IV. *Psammolychus delamarei* (Schubart): **C**, dorsum of prosoma.
MICROPSAMMIDAE. Undescribed genus (USA): **D**, lateral view of anterior.
PROTEONEMATALYCIDAE. *Proteonematalycus wagneri* Kethley (USA): **E**, dorsal view of anterior.

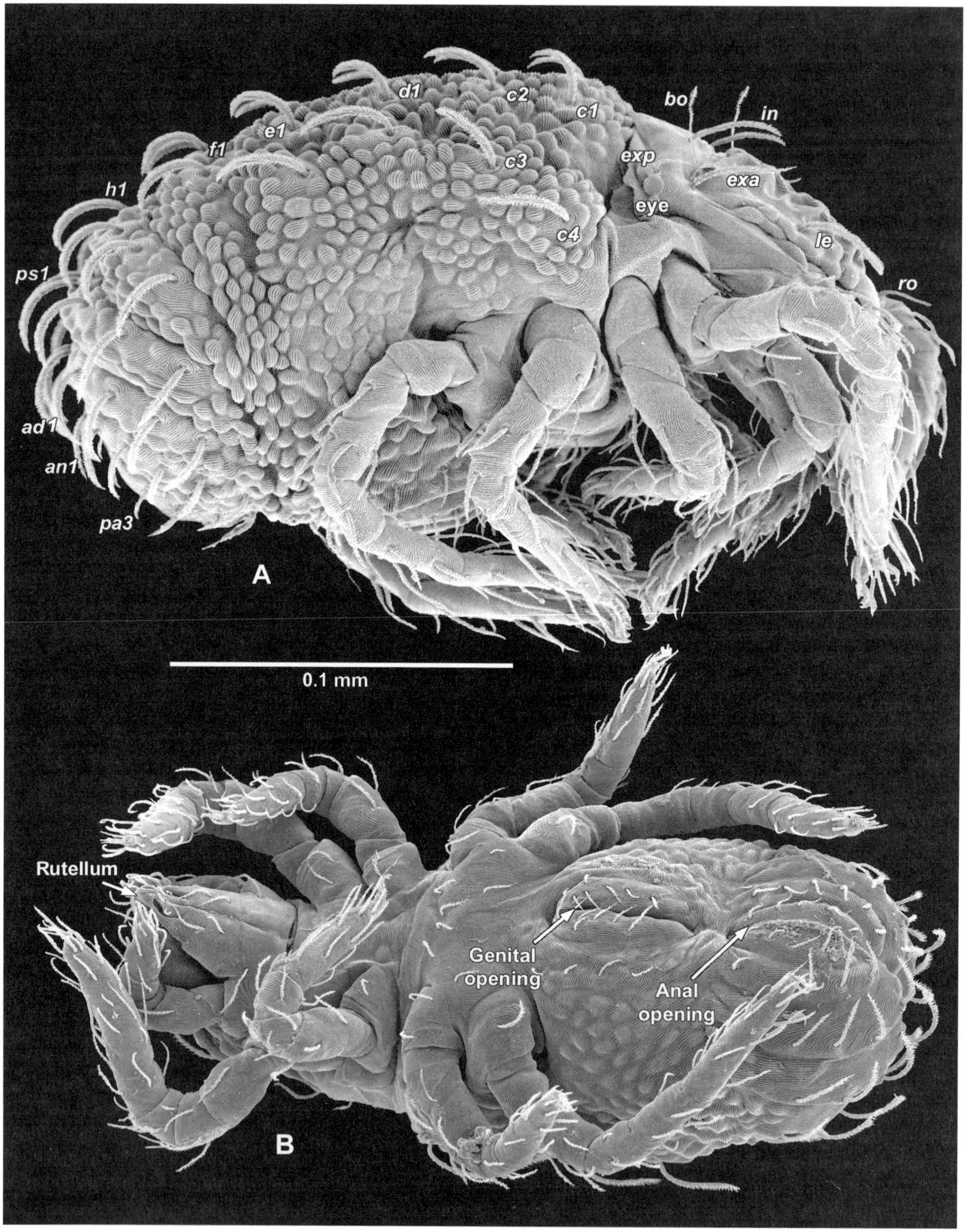

Fig. 14.5. TERPNACARIDAE. *Terpnacarus gibbosus* (Womersley) (Australia): **A**, lateral view of adult female; **B**, ventral view of female tritonymph.

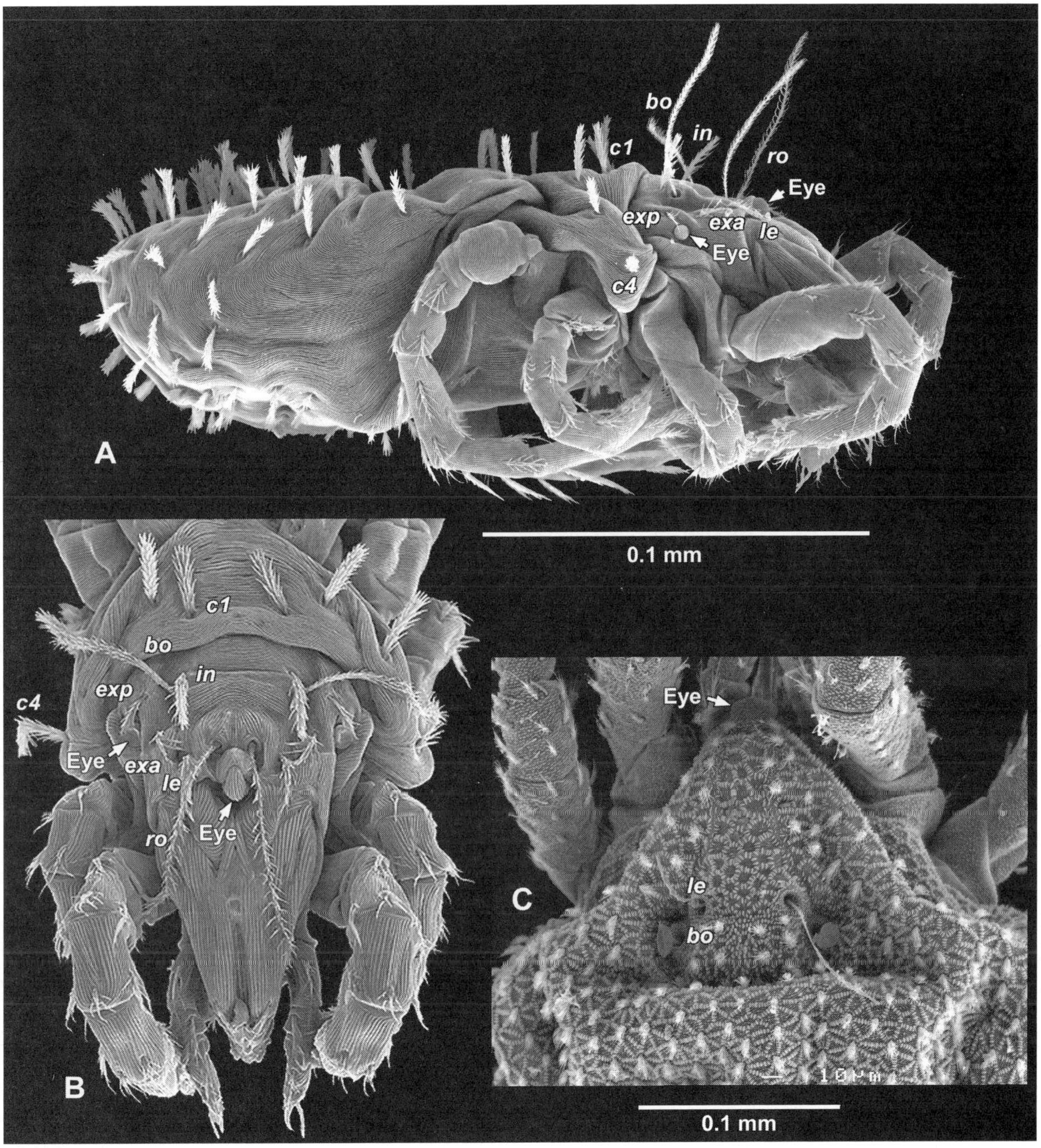

Fig. 14.6. NANORCHESTIDAE. *Speleorchestes* sp. (Australia): **A**, lateral view of adult female; **B**, dorsal view of anterior of same. ALYCIDAE. *Bimichaelia* sp. (Australia): **C**, prodorsum and anterior hysterosoma showing hypertrichy (apparent neotrichy) and globose (*bo*) and filiform (*le*) bothridial sensilla.

CHAPTER FIFTEEN

R.A. NORTON
V.M. BEHAN-PELLETIER

Suborder Oribatida

In the classic sense of the group, oribatid mites (also called beetle mites, armored mites, or moss mites) comprise more than 9,000 named species (Schatz 2002, 2005; Subías 2004) representing 172 families. These numbers do not include members of the large cohort Astigmatina, which are mentioned in this chapter only in passing. Because of their radically different biology, many morphological and life cycle innovations, and traditional distinction, astigmatine mites are treated separately (chapter 16) and, unless otherwise noted, are not included in generalizations about oribatid mites.

Although many are arboreal and a few are aquatic, most oribatid mites inhabit the soil-litter system. They are often the dominant arthropod group in highly organic soils of temperate forests, where 100–150 species may have collective densities exceeding 100,000 m^{-2}. Oribatid mites are mostly known as particle-feeding saprophages and mycophages (Fig. 6.2, p. 82), but opportunistic predation on nematodes and other microfauna and scavenging on small dead arthropods (necrophagy) are probably underestimated (Schuster 1956; Luxton 1972; Schneider et al. 2004). As adults, most are medium to dark brown as a result of melanization, but they may be colorless or have yellow to red pigment. Adult body length is typically 300–700 µm, but collectively they span an order of magnitude in size, from about 150 to 2,000 µm.

In all known instances, oribatid mites have a plesiotypic acariform life cycle including a calyptostatic prelarva, mobile hexapod larva, three nymphs, and adult. Sexual dimorphism is minimal and usually restricted to slightly smaller size and proportionally smaller genital plates in males. Fertilization is usually indirect, by stalked spermatophore, without direct association of male and female, but several exceptions are known (see also chapters 3 and 4). Oviposition is the rule, but embryogenesis is sometimes completed internally prior to oviposition; in some species (particularly aquatic or semi-aquatic forms), the second instar may be reached (larviposition) (Norton 1994). The Oribatida displays the greatest known concentration of female parthenogenesis (thelytoky) in the Acari, and much of it exists in taxonomic clusters (Norton and Palmer 1991; Norton et al. 1993). There are isolated parthenogenetic species in otherwise sexual genera, and there are wholly parthenogenetic genera, families, and even superfamilies; several of the families include more than 100 species.

Oribatid mites usually exhibit K-style life history traits, most notably low reproductive output and long life cycles. Life spans typically are one to two years in temperate to boreal regions but can reach four or five. Rather than being an adaptive "strategy," these traits probably represent constraints on secondary production (Norton 1994). The result is that, after taking a long time to reach maturity, adults must live a relatively long time to accrue enough resources for reproduction. This ensures strong selection pressure for defensive mechanisms, which abound in oribatid mites and particularly in adults. Among these are various forms of protective setae, camouflage, waxy exudates, defensive glands (see chapter 3), cuticular hardening, and a striking array of body forms and protective structures (Sanders and Norton 2004; Norton 2007). While these defenses seem effective against small predators (Peschel et al. 2006), certain beetles and ants can overcome them (Schmidt 1988; Molleman and Walter 2001; Wilson 2005). A highly specialized terminology has been developed to identify and describe these various defensive devices.

The following overview of adult morphology is focused on features used in the family key and diagnoses or on their structural context. A computer-based interactive glossary of these and other terms is available (Hunt et al. 1998). Some of the more general terms also appear earlier in this text (chapter 3). Most oribatid terminology was coined by Francois Grandjean in the 1930s, 1940s, and 1950s (see Travé and Vachon 1975 and Hammen 1980 for references).

Integument and Body Form

Oribatid mite cuticle has the general features described in chapter 3, but, compared to other mites, there is a greater frequency of regions with dense, conspicuous pore canals (Norton et al. 1997). When extensive (as is common on the coxisternum and prodorsum), these regions usually represent respiratory surfaces. By contrast, small, localized **porose organs** may be either secretory, with large epidermal cells that are probably involved in cuticular maintenance, or respiratory, with a very thin epidermis to maximize gas exchange (Alberti et al. 1997). The cuticular part is called a porose area (= area porosa) if it is not invaginated (Fig. 15.3F). If invaginated, porose areas have a variety of names: small pouches are **saccules** (Fig. 15.3G); large, flat, lamelliform pouches are **platytracheae**; thick, relatively short tubes are **brachytracheae** (Fig. 15.2E); and extremely long, filamentous tubes are tracheae.

Extensive sclerotization and the associated melanization are typical of oribatid mites and are the basis for the common names "beetle mites" and "armored mites." However, hard cuticular plates may derive from mineralization rather than sclerotization in some cases, which means that some rather hard species can be light colored. Known minerals involved in this process include calcium carbonate, hydrated calcium oxalate (whewellite), and calcium phosphate, with at least the latter two being deposited in epicuticular chambers (Norton and Behan-Pelletier 1991b; Alberti, Norton, and Kashbohm 2001). Articulations between certain plates are covered by a rooflike extension from the edge of one plate, called a **tectum** (Figs. 15.2B, 15.3C, 15.5C); such structures allow mobility while protecting the vulnerable soft cuticle of the articulation from predators.

Although the cuticle of oribatid mites may be clean and smooth, various materials can make it appear dull and irregular. The cerotegument, or secretion layer, is often highly developed in oribatid mites. It appears as a white or gray coating in live mites and has consistent form within species. It may comprise a generally distributed thin coating; often there are excrescences in the form of small tubercles, cylindrical or conical projections, long, cottony filaments, or reticulations. In some cases, the cerotegument is solid and birefringent in polarized light, either platelike (MALACONOTHRIDAE), blocky (e.g., PHENOPELOPIDAE, some GYMNODAMAEIDAE), or in an amorphous mass (Fig. 15.11G). Various groups of oribatid mites incorporate organic or mineral debris with cerotegument or pack such debris in solid dorsal masses (e.g., CAMISIIDAE, CROTONIIDAE, DAMAEIDAE). Partial exuviae of immature instars are also carried by adults of some species. Such exuviae may represent only the tritonymph (HERMANNIELLIDAE, some CROTONIIDAE) or all previous instars (many Brachypylina), with successive **scalps** from the opisthonotum stacked like a pagoda (Figs. 15.11E, 15.13A, D).

Some primitive oribatid mites, particularly the Archeonothroidea, have a body form much like that of endeostigmatic mites, from which they differ in having numerous small plates at the base of many body setae (Fig. 15.1A). However, in the large majority of oribatid mites, most visible cuticle is hardened, forming two expansive dorsal plates and various arrangements of ventral plates. The dorsal plate anterior to the sejugal furrow is called a **prodorsum** (or **aspis**, when it is isolated from ventral plates). The prodorsum is fused to the ventral plate in all Brachypylina and in a few other groups, making its boundaries imprecise. The dorsal and lateral cuticle of the hysterosoma (i.e., that portion bearing setal rows *c* through *ps*) is called the **notogaster** if it is sclerotized and the **opisthonotum** (= gastronotic region) if it is mostly unsclerotized (as in the adults of some early derivative taxa and in the immatures of most oribatid mites).

Several body forms are created by various arrangements of plates and connecting articulations. Oribatid mites are **dichoid** if the sejugal furrow remains soft, such that the proterosoma and hysterosoma articulate freely (Fig. 15.1B). In some dichoid mites (e.g., Enarthronotides) the notogaster comprises several components separated by transverse articulations, or **scissures** (see below); in others (many Mixonomatides), the notogaster is entire. If a soft, postpedal furrow also occurs so that the body has two primary articulations (Fig. 15.1G), the mite is **trichoid** (e.g., most Parhyposomatides). Several groups (Enarthronotides: MESOPLOPHORIDAE, PROTOPLOPHORIDAE; and Mixonomatides: Euphthiracaroidea, Phthiracaroidea) have a **ptychoid** body form (Figs. 15.1E, F, 15.10E, F), a defensive adaptation that allows the mite to close like a seed when disturbed. The common name "box mite" is sometimes applied to these forms. Ptychoidy involves use of a set of large muscles that pulls the coxisternum and legs into the opisthosoma, which is possible because all of the podosomatic cuticle other than the coxisternum is soft (Sanders and Norton 2004). Another set of muscles deflects the aspis and pulls it toward the opisthosoma so that the legs are captured inside a secondary chamber. In the **holoid** body form, which characterizes most Nothrina (Fig. 15.1C) and all Brachypylina (Fig. 15.1D), epimera II and III have fused, eliminating the sejugal furrow ventrally. The result is an idiosoma comprising a single functional unit that lacks a major transverse articulation. In holoid mites, the principal body articulation is the more or less horizontal **circumgastric scissure** (Figs. 15.1D, 15.3D, E) that separates the notogaster from ventral plates. In the NANHERMANNIIDAE (Figs. 15.27A, B), the circumgastric scissure does not exist, and the notogaster and ventral regions merge.

Prodorsum

In most oribatid mites, the prodorsum extends far anteriorly as a rostral tectum, or simply **rostrum**; such a prodorsum is called **stegasime** (Fig. 15.2B). The rostrum usually curves ventrally, creating a protected secondary vestibule (camerostome) within which the chelicerae operate. When the mite is disturbed, the chelicerae are fully retracted and the subcapitulum is levated to close the vestibule. The edge of a rostrum may be smooth and simple or variously modified with a reflexed edge, projecting teeth, or emarginations of various types, some of which are narrow indentations that create toothlike remnants (Fig. 15.12E). One such emargination, typical of several brachypyline groups, is the **genal notch** (Fig. 15.2J) at the lateral corner of the rostrum, which delimits the **genal tooth** (Fig. 15.12B). The ancestral condition of the prodorsum in oribatid mites is essentially like that of the Endeostigmata in that there is an inconspicuous aspis that ends abruptly at the articulation with the chelicerae, which are fully exposed even when retracted (Fig. 15.2A). This **astegasime** condition is characteristic of the cohort Palaeosomatides but occurs in some Enarthronotides, Parhyposomatides, and Mixonomatides as well. A naso is retained in some Palaeosomatides (ARCHEONOTHRIDAE) and Enarthronotides (BRACHYCHTHONIIDAE). True eyes are restricted to a few Palaeosomatides and Enarthronotides (see chapter 3). *Galapagacarus* (HERMANNIIDAE) has an unpaired, eyelike structure on the prodorsum that, although little studied, may represent a transitional organ linking prodorsal eyes and the light-receptive notogastral lenticulus (below) of many Brachypylina.

The surface of the prodorsum is smooth in the basal supercohorts Palaeosomatides, Enarthronotides, and Parhyposomatides, but it may be microtuberculate, areolate, reticulate, or have other surface relief in more derived cohorts. In particular, mites in the cohort Brachypylina are commonly provided with tubercles, carinae, ridges, or other projecting prodorsal structures. Many have a pair of longitudinal structures, medially or laterally, with a spectrum of forms. If these are simple, low ridges, they are called **costulae** (Figs. 15.2G, 15.11C); if they are bladelike, with at least one free edge, they are called **lamellae** (Figs. 15.2H–J, 15.11A, B, 15.12B). Costulae and lamellae are probably not homologous structures, but the distinction can be unclear, and both terms are used imprecisely in the literature. Unlike most costulae, a lamella runs forward from the bothridium (see below) and, when of sufficient size, it is a protective structure under which leg I is retracted when the mite is disturbed. A lamella commonly extends anteriorly as a projecting tooth- or knifelike **lamellar cusp** of various sizes and forms on which the lamellar seta (below) typically inserts. The lamellae may be independent, may appear connected by a transverse ridge- or bladelike **translamella** (Fig. 15.2I), or may fuse for part or all of their length (Fig. 15.11B). A narrow carina, the **prolamella**, may run distally from each lamella toward the rostral margin as in HUMEROBATIDAE and some SCHELORIBATIDAE (Fig. 15.2H). The latter family may also have lamellae that fork posteriorly, with the more ventral branch—the **sublamella**—running below the bothridium. Some Brachypylina have a ridge- or bladelike longitudinal **tutorium** (Figs. 15.2J, 15.11A) on each lateral face of the prodorsum. Like the lamella, the tutorium is of diverse size and shape and may have a distal cusp; if bladelike, the free edge is dorsal, and the distal part of leg I lies in the resulting valley when retracted. In Brachypylina, the insertions of legs I and II are often protected by **pedotecta**, scale- or earlike (auriculiform) structures that project from the body wall immediately posterior to the respective acetabulum (Figs. 15.2J, 15.6E). In contrast, a tooth- or hornlike lateral projection between legs I and II that is not closely adjacent to an acetabulum is called a **propodolateral apophysis** (P; = **pedotectal tooth)** (Fig. 15.6F), and such projections do not coexist with pedotecta; a scalelike (tectiform) projection in a similar position is a **patronium,** which may coexist with pedotecta (Fig. 15.50D).

An **enantiophysis** may occur at several places on the prodorsum of brachypyline mites (Figs. 15.2F, 15.6F). It comprises a set of two tubercles that oppose each other across an articulation or furrow. Typically, the orientation of these tubercles is such that one (*a*, or 1) is anterior and the other (*p*, or 2), posterior. When present, the **prodorsal enantiophysis** (*Aa*, *Ap*) spans a transverse or paired groove at midlength of the prodorsum. The **lateral enantiophysis** (*La*, *Lp*) spans the sejugal furrow laterally, and the **humeral enantiophysis** spans the same furrow dorsally, with tubercles on the bothridial wall and the humeral angle of the notogaster.

Insertions of the paired cheliceral retractor muscles are often conspicuous on the prodorsum due to their associated sigilla ("muscle scars"), a large group of which is often present in the region between setae *in* and *le* (see below); another group is located on the lateral face (Fig. 15.2G). These sigillary aggregations have been referred to in some publications as "maculae" or "spots." The prodorsum is sometimes modified in some way above the more dorsal insertions, such as by paired domes. In many Brachypylina, particularly in poronotic families, cheliceral muscles insert instead on two paired apodemes that project internally from the posterior margin of the prodorsum (Fig. 15.2I). The **dorsophragmata** (= dorsophragmatic apophyses) are in the dorsosejugal region, and in transmitted light may seem to come from the anterior margin of the notogaster. The **pleurophragmata** are more laterally positioned.

Several porose organs (all of which are probably dermal glands) are found on the prodorsum in poronotic superfamilies of Brachypylina (Norton et al. 1997; Alberti et al. 1997). The paired **sublamellar** porose area (*Al*) may be present below the lamella (Fig. 15.2H). A humerosejugal series may include one or more of the following: the **dorsosejugal** (*Ad*) and **humerosejugal** (*Aj*) porose areas on the prodorsum and one (*Ah*) or two (also *Am*) **humeral** porose areas in the subhumeral region of the ventral plate (Figs. 15.2J, 15.6G).

Ancestrally, six pairs of mechanoreceptive setae insert on the prodorsum (Figs. 15.1B, 15.2J). These are homologs of setae present in Endeostigmata, but different names typically are given to them in oribatid mites—names that relate to structures not present in endeostigmatids. Three of the pairs typically have a somewhat medial position. **Rostral setae** (*ro*) are anteriormost; when the rostral tectum is present, they insert on it. Posterior to the rostrals are the **lamellar setae** (*le*), which insert at or near the anterior end of the lamellae if the latter are present. The most proximal of the three pairs is the **interlamellar setae** (*in*), which lie in the space between lamellae. A trichobothrium usually inserts in each posterolateral corner of the prodorsum. It consists of a bothridial seta (*bo*), also called a pseudostigmatic organ or a **sensillus**, and a deep, cuplike **bothridium** in which the seta inserts. The bothridium is a simple, inverted cone in primitive oribatid mites (Palaeosomatides and some Enarthronotides) (Fig. 15.2C), but it has a strong curve (usually S shaped) at its base in most oribatid mites (Fig. 15.2D); the shape is mimicked by the inserted basal part of the bothridial seta. Externally, the bothridium may have a simple rim that is flush with the prodorsal surface, or it may be variously projected, sometimes like a funnel. Inside the bothridium, the cuticle may invaginate further to form porose saccules (Fig. 15.27E), brachytracheae (Fig. 15.2E), or short tracheae (Grandjean 1934d; Norton et al. 1997); most such structures are probably respiratory surfaces (Alberti et al. 1997). Bothridial setae vary greatly in shape among oribatid mites, but generally they differ from other prodorsal setae. Ventrolateral to each bothridium are two pairs of **exobothridial setae**; these have been given several designations, including *exa/exp*, *ex1/ex2*, *xs/xi*, and *xa/xp*, but their correspondence is unclear. Among prodorsal setae, the exobothridials are most susceptible to loss. Members of Nothrina and Brachypylina always lack at least one pair (the remaining pair is labeled *ex*), and in some cases both are lost. In some groups, lost exobothridial setae are represented by a porelike alveolar vestige.

Notogaster

The notogaster in Brachypylina is a single, caplike sclerite, separated by the circumgastric scissure from the prodorsum anteriorly and from the ventral region posteriorly. The anterior separation is variously called the dorsosejugal furrow, groove, or (incorrectly) suture (Fig. 15.1C). In some Nothrina and Brachypylina, it is effaced by a fusion of the prodorsum and notogaster, either imperceptibly or marked by a true, thickened suture indicating the ancestral line of contact. Aside from the Brachypylina, there are many variations in notogastral structure. A paired, longitudinal **suprapleural scissure** often separates a dorsal **notaspis** from a lateral **pleuraspis** (Fig. 15.3K), while in Enarthronotides the notaspis itself is usually divided by 1–3 transverse scissures. These may demarcate an anterior **pronotaspis** and a posterior **pygidium** (Fig. 15.1G), both of which vary in segmental composition. Three types of transverse scissures are known (Figs. 15.5A–C), and these differ in both structure and function (Grandjean 1947b; Norton 2001). **Type E scissures** are simple articulations between plates; edges of the two adjacent plates are unmodified, and the intervening band of soft cuticle is usually narrow (Fig. 15.10D). A **type L scissure** is a defensive specialization that permits significant telescoping of two plates while protecting a broad articulating cuticle. The posterior edge of the more anterior plate is hypertrophied as a tectum that is broad enough that no soft cuticle is exposed, even at maximum plate separation. A **type S scissure** is a more compound structure. In its typical form, the space between two major plates is occupied by a transverse series of four closely adjacent, intercalary sclerites, each bearing a seta. These small sclerites may be variously combined, depending on the taxon. Ancestrally, setae of type S scissures are large and erectile and serve a defensive role (Fig. 15.10G), but they are regressed in some groups (see below). The occurrence and distribution of these scissure types vary among taxa (Figs. 15.5D–G). Large fixed setae may also play a defensive role (Fig. 15.10C, 15.11H).

The notogaster may be hemispherical, flattened, concave, or laterally compressed, according to group. Like that of the prodorsum, the notogastral cuticle may be smooth or it may have various forms of small-scale relief. Conspicuous tubercles or spines may project toward or across the dorsosejugal furrow, and longitudinal ridges may be present, either freestanding or connected to anterior tubercles; such a tubercle-ridge complex in Brachypylina is sometimes called a **crista** (Fig. 15.11F). The humeral region may be simple or it may present various projecting structures, from simple tubercles to anteriorly directed knifelike processes or laterally directed tecta. A humeral tectum that is well defined and large enough to conceal all or part of the retracted legs is called a **pteromorph** (Figs. 15.3A, B, 15.11A, B) and is known only in Brachypylina. In some taxa, the base of a pteromorph is completely or partially desclerotized to form a linear hinge, and some of the dorsoventral musculature is modified to pull the resulting movable pteromorph against the body when the mite is disturbed (Figs. 15.3I, 15.11D). Ramsay and Wallwork (1972) speculated that larger pteromorphs such as those in the GALUMNIDAE and the PARAKALUMMIDAE might also restrict airflow to tracheae and thereby control water loss. The posterior margin of the notogaster may also project as a tectum, overhanging the circumgastric scissure (Figs. 15.3B, C). This tectum is incomplete medially in some poronotic oribatids where lobelike projections overlap (Fig. 15.62H). In some groups, a raised central region is delineated by a circular or U-shaped submarginal depression, the **circummarginal furrow** (Fig. 15.11C, arrowhead).

Other than in the Brachypylina, the most general complement of mechanoreceptive setae on the oribatid notogaster is 16 pairs, a condition known as **holotrichy**. Grandjean (1934a) gave these setae notations according to their presumed segmental origin (Figs. 15.4A, B). Except for HERMANNIELLIDAE, brachypylines have fewer than 16 setal pairs. The basic number is 15, but uncertainty about which seta in the middle of the notogaster was lost prompted Grandjean (1934a) to create a different chaetotaxy, the **unideficiency nomenclature** (Figs. 15.4C, D). The principal distinction is that the six pairs of setae in the middle of the notogaster are given notations unrelated to the holotrichous nomenclature; three pairs of dorsocentral setae (*da*, *dm*, *dp*) are flanked by three pairs of dorsolateral setae (*la*, *lm*, *lp*). In most cases, adult Brachypylina have either 15 pairs or a lesser number that can be easily identified with homologous setae of the unideficiency nomenclature. Grandjean (1951e) also developed a special chaetotaxy for species with 10 or fewer pairs—the **multideficiency**, or "Dometorina nomenclature" (Figs. 15.4E, F). He soon abandoned this system after resolving certain homologies, but it still lingers in the literature. Rarely, the notogaster of some species is neotrichous, sometimes with dozens of supernumerary setae. Notogastral setae take many forms and may be highly specialized. For example, certain setae in Palaeosomatides and Enarthronotides are large and movable and are erected when the mite is disturbed (Figs. 15.5B, 15.10G). In the enarthronote group Atopochthonioidea, platelike setae serve as defensive armor (Figs. 15.10B, C). Particular notogastral setae may be represented only by vestiges, in which only the porelike alveolar canal remains or there is a minute setal remnant. One or both *f*-row setae often are vestigial in Mixonomatides and Nothrina, and in some poronotic Brachypylina all or most notogastral setae are vestigial.

Other structures of the notogaster include lyrifissures, an opisthonotal gland opening, porose organs, a lenticulus, and muscle sigilla. There are five pairs of notogastral lyrifissures in most oribatid mites: *ia*, *im*, *ip*, *ih*, and *ips* (Fig. 15.1B). In the PARHYPOCHTHONIIDAE, the adanal segment becomes incorporated into the notogaster, along with its setae and lyrifissure *iad*. Lyrifissure *im* is modified as a **piriform organ** of unknown function in some ZETOMOTRICHIDAE (Fig. 15.62F). Except for the Palaeosomatides and Enarthronotides, most oribatid mites possess **opisthonotal glands** (also called opisthosomatic or oil glands) that open in the mid-lateral region of the notogaster (see chapter 3). Usually the opening is inconspicuous, but it may be on a tubercular or funnel-like projection (PARHYPOCHTHONIIDAE, HERMANNIELLIDAE, PLASMOBATIDAE) (Figs. 15.3J, 15.10A). In some early-derivative Brachypylina (HERMANNIELLIDAE, NEOLIODIDAE), numerous small respiratory saccules (**macropores**) may be present (Alberti et al. 1997; Norton et al. 1997) (Fig. 15.3H). The notogaster of some LOHMANNIIDAE (Enarthronotides) and many Brachypylina have well-circumscribed secretory porose organs. Usually the cuticular component of such dermal glands is a porose area (= area porosa), but sometimes the porose cuticle is invaginated as a saccule or trachealike **tubule**.

When dermal glands occur on the notogaster of Brachypylina, they are typically arranged in four pairs, forming the **octotaxic system** of porose organs (Fig. 15.3A). They are usually conspicuous when formed as porose areas, but less so if formed as saccules and very difficult to see if represented by tubules. The anteriormost porose area is the **adalar** (*Aa*), with the others being **mesonotic** (*A1, A2, A3*). Saccules and tubules have similar designations, using *S* or *T* instead of *A*. These structures differ considerably in size and shape among species and are often used as diagnostic characters. In rare cases, the octotaxic system may include a mixture of porose areas and saccules. Only two examples are known: *Drymobates silvicola* Grandjean (DRYMOBATIDAE) and *Glanderemaeus hammerae* Balogh and Csiszar (in males only: family affiliation uncertain) (Norton et al. 1997). Although the typical complement is four pairs of octotaxic organs, higher or lower numbers are not uncommon (Norton and Alberti 1997).

The lenticulus (including the less well demarcated "clear spot" of authors) is an unpaired light receptor organ situated near the anterior margin of the notogaster in some Brachypylina (Alberti and Fernandez 1990; see chapter 3) (Figs. 15.40A, 15.64A). Muscle sigilla, excavations on the internal face of the cuticle where muscles attach, may be very conspicuous on the notogaster, especially if the cuticle is well sclerotized; the thinner cuticle of the sigillum shows a lighter color in transmitted light (thus being sometimes mistaken for porose areas). In Brachypylina, most notogastral muscle sigilla are arranged in a U-shaped curve just inside the margin, indicating the origins of the circumgastric muscle band that controls hemolymph pressure (Figs. 15.3A, D).

Ventral Structures

The venter of an oribatid mite exhibits three regions that are usually easily distinguished by sclerotization patterns. Anteriormost is the subcapitulum, a part of the gnathosoma that is discussed below. Behind this, and usually separated from it by a narrow articulation, is the **coxisternum** (or coxisternal region), which forms the floor of the podosoma and serves to support the legs. Behind the coxisternum is the **anogenital region**, in which the anal and genital openings are situated.

The four epimera (I–IV) that compose the coxisternum (Fig. 15.6B) always exhibit some level of sclerotization. Minimally, each exhibits a pair of epimeral plates (epimeral halves or demi-epimera), medially separated by a soft longitudinal band and is independent of other epimera (Fig. 15.6A). Maximally, all are fully sclerotized and collectively fused as a single unit (Fig. 15.6E), which characterizes the holoid body form. In dichoid body forms there are usually two connected groups (I–II and III–IV) separated by the sejugal articulation, but various levels of fusion exist. The anterior edge of epimere I may form a tectum (**mentotectum**) that overhangs the base of the subcapitulum (Fig. 15.6C). Bladelike apodemes may form at the juncture of two fused epimera, but dark borders may also delineate epimera in a fused coxisternum (Fig. 15.6C). These borders are made conspicuous when the middle area of epimera is internally excavated by many muscle sigilla. The coxisternal setation is usually represented as a formula, with the number of pairs on epimera I–IV in succession. The formula 3-1-3-4 is most common. Setae are numbered according to epimere and are given letters according to their distance from the midline in the larva (e.g., *1a*, *3b*; Fig. 15.6E); setae added subsequently during ontogeny are lettered in order of appearance. Coxisternal setation is neotrichous in a few groups. Legs attach at the lateral extent of the epimera, either by a simple articulation with the trochanter, or (in Brachypylina) by a ball and socket joint in which the epimere is invaginated laterally to form the socket, or **acetabulum** (Figs. 15.6C, D). These are not to be confused with genital acetabula (papilla) of water mites (see chapter 13). This difference is also expressed in the direction in which legs leave the body (Fig. 15.6B, C).

The anogenital region ancestrally comprises four pairs of plates: the **genital plates**, flanked by the **aggenital plates**, and the **anal plates**, flanked by the **adanal plates** (Fig. 15.6A). Each plate bears various numbers of setae having the same respective names: aggenital (*ag*) and genital (*g*) setae are numbered from anterior to posterior; anal (*an*) and adanal (*ad*) setae are numbered in the opposite direction (Fig. 15.1B) due to the paraproctal origin of their plates and the caudal bend (see chapter 5). Adanal plates usually bear a pair of lyrifissures (*iad*), and anal plates may bear another (*ian*); neither pair exists in any member of Palaeosomatides or Enarthronotides. The genital plates (which are sometimes transversely divided) close to form the genital vestibule, into which the paired genital papillae and unpaired genital organ are retracted. With rare exception, females have a tubular ovipositor (Figs. 15.7C, D), and males have a homologous but much smaller spermatopositor (also called male organ, male genital sclerite, or penis; Figs. 15.7E, F). Eugenital setae insert on these structures. Three pairs of similar genital papillae (*Va*, *Vm*, *Vp* from anterior to posterior) generally are present (Fig. 15.7D); in some groups one is larger or smaller than others, rarely they are all minute, and rarely one or all are absent. The anal plates close to form the anal vestibule, or rectum. An unpaired preanal plate often lies in front of them (Fig. 15.6A) or may be incorporated in the anal vestibule as a **preanal organ** (Brachypylina), which projects internally as an apodeme (Figs. 15.7A, B); both forms serve as the origin for muscles that insert on the genital plates. Rarely, the **peranal segment** (PA) is added in the tritonymph (some Enarthronotides, some Parhyposomatides). If so, a **peranal plate** usually forms a narrow band at the medial edge of the anal plate (Fig. 15.22G).

From the complex arrangement described above evolved a variety of plate fusions, particularly diverse in Enarthro notides, Mixonomatides, and Nothrina. Various combined terms can be applied to these fusions, including anogenital plates and aggenito-adanal plates. In several groups, particularly Euphthiracaroidea, a pair of large, intercalary **plicature plates**, bearing neither setae nor lyrifissures, lies between the venter and the notogaster proper (Fig. 15.10F). In some Nothrina (HERMANNIIDAE) and all Brachypylina, the adanal and aggenital plates are hypertrophied and imperceptibly fused; they also fuse with the coxisternum and usually fuse in front of and behind the anal plates (Fig. 15.1D). The result is a strongly constructed ventral plate in which the relatively small genital and anal plates open like bomb-bay doors (Fig. 15.6E). This is called the **brachypyline** condition, in contrast to the **macropyline** venter of other, supposedly more primitive oribatid mites. The normal setation of a brachypyline venter includes one pair of aggenital and three pairs of adanal setae, but both losses and neotrichy can occur.

Several ventral or lateroventral structures are restricted to (but not universal in) the Brachypylina. A spine- or ridge-like **discidium** often occurs between acetabula III–IV, and a paired **circumpedal carina** may curve anteromedially from the edge of the ventral plate, running behind leg IV and forward across epimera IV and III (Figs. 15.6H, 15.12A, 15.71). The circumpedal carina may merge with the discidium, and in some cases the anterior end projects as an inconspicuous, knifelike **custodium**. Enantiophyses may span any furrow in the coxisternal region (Fig. 15.6F), with anterior and posterior tubercles designated like those on the prodorsum. Most common are the **parastigmatic enantiophysis** (S), which spans the sejugal furrow ventrolaterally between legs II and III, the **ventrosejugal enantiophysis** (V), which spans it ventrally, and **epimeral enantiophysis IV** (E4). In some groups a small canal (taenidium) of unknown function runs laterally from each anteriolateral corner of the genital vestibule to acetabulum IV; a narrow tectum (**minitectum**) overhangs the taenidium from the posterior side (Fig. 15.7A).

Various porose organs occur on the ventral plate of some Brachypylina, the more obvious of which are porose areas that probably represent dermal glands. A common one is the unpaired **postanal porose area**, a typically elongated area immediately behind the anal plates (Fig. 15.58H). A marginoventral series of porose areas may be present near the circumgastric scissure (Fig. 15.6G), resembling an extension of the more common humerosejugal series (above). The main respiratory surfaces of Brachypylina are internalized porose organs (usually tracheae), but their stigmata on the ventral plate are difficult to see. The name Cryptostigmata, a synonym of Oribatida, derives from this trait. There are three pairs of stigmata in most oribatid families, one within acetabulum I, another in the sejugal furrow, and a third within acetabulum III (Fig. 15.6G). The first two typically bifurcate close to the stigmata (Fig. 15.6D) so that five long pairs of tracheae usually aerate the body. Variations of this **apodematic-acetabular tracheal system** are characteristic of particular families or superfamilies.

Gnathosoma

Oribatid mites show considerable variation in the structure of all three gnathosomatic components: the unpaired subcapitulum (= infracapitulum), the palpi, and the chelicerae. In ventral aspect (Fig. 15.8C) the subcapitulum usually exhibits a large proximal base, the **mentum** (with one or two pairs of setae); paired anterior lobes, or **genae** (with several pairs of setae); and, most distally, a pair of **lateral lips** that partially define the mouth (with up to three pairs of **adoral setae**, *or1–or3*). Rarely, an unpaired **ventral lip** (= inferior lip) is present (ARCHEONOTHRIDAE) (Fig. 15.8A). Grandjean (1957b) recognized three basic subcapitular forms based on the nature of the **labiogenal articulation**, which separates mentum from genae and allows deformation during feeding. An **anarthric** subcapitulum (Figs. 15.8B, 15.12D) lacks this articulation and is typical of many Enarthronotides, along with members of other groups with highly modified mouthparts. The labiogenal articulation in a **stenarthric** subcapitulum is Λ shaped and runs obliquely from a central point in the midline to each posterolateral corner. This produces a triangular mentum and is typical of most early- to middle-derivative groups (Fig. 15.8C). On a **diarthric** subcapitulum (Fig. 15.8D), the labiogenal articulation runs transversely at a level just posterior to the palpi and demarcates a quadrangular mentum (= hypostome) typical of most Brachypylina. Diarthric subcapitula may have a **mental tectum** (= hypostomatic tectum) that projects forward to cover the articulation (e.g., GALUMNIDAE) (Fig. 15.8E). Distally on each anterior lobe (gena) of the subcapitulum is a modified seta, the rutellum (Fig. 15.8A). The rutella (also see chapter 3) may be setiform in some Enarthronotides, but in most species they are enlarged and have distal teeth or cutting blades for scraping or cutting that are used in conjunction with the chelicerae to cut food into small particles (Fig. 15.12C). If chelicerae are elongated (see below), the rutellum may be modified to form half a tube (Fig. 15.8F) that, when paired, serves to guide the chelicerae (Fig. 15.12G). Some Brachypylina have a small **axillary saccule** that opens at the base of each palp (Fig. 15.8G); Alberti et al. (1997) showed a filamentous form of this structure in the PHENOPELOPIDAE to be a secretory organ.

The palpi lack a pretarsus and ancestrally have five segments (Fig. 15.9A). This condition remains the typical form in all major groups, but there are various fusions to form four, three, or only two functional segments. Palpi may be proportionally large in Palaeosomatides, but in other groups they are small and inconspicuous and probably are used mostly as sense organs. The palptarsus has a proximodorsal lyrifissure and one solenidion, various normal mechanoreceptive setae, and several distal eupathidial setae (eupathidia). More proximal segments have few setae, with a typical formula for the basal four segments being 0-2-1-3 (setal nomenclature indicated in Fig. 15.9A). Most setiform organs on the palp are independent, but in some cases distal eupathidia are fused at the base to form forked or trifurcate structures. Many Brachypylina have a eupathidium (*acm*) fused to the solenidion to form what is often called a **double horn** (Fig. 15.9B).

Chelicerae are two-segmented, but this is obvious only in primitive groups such as Palaeosomatides and Enarthronotides (Fig. 15.8H). The trochanter usually is greatly regressed, being little more than a proximoventral vestige (Fig. 15.8I). This regression is concomitant with a change in operating direction from near vertical to near horizontal. In Nothrina and Brachypylina, as well as in the Hypochthonioidea (Enarthronotides), the body wall attaches rather distally on the chelicerae so that about one-third of the appendage is internalized as an apodeme. In the former two groups, a movable, fingerlike **Trägårdh's organ** (function unknown) extends distally from near this point on the paraxial face (Fig. 15.8I). Most oribatid chelicerae have two setae, one dorsal or slightly paraxial (*cha*) and the other usually lower on the antiaxial face (*chb*); one or both may be absent. Chelicerae are typically chelate-dentate (i.e., robust, with strong, conspicuously dentate chelae), but many modifications exist. The greatest diversity of cheliceral form is found in Enarthronotides and Brachypylina, but there has been no serious attempt to learn their function or even to classify and name them. One exception is **pelopsiform** (= peloptoid) chelicerae, which are greatly elongated structures with small but well-formed and dentate chelae (Fig. 15.8J). Pelopsiform chelicerae evolved independently in various brachypyline families. In the family PHENOPELOPIDAE they are used to cut small holes in the cuticle of dead leaves and remove parenchymous tissue (Riha 1951). Attenuate-edentate chelicerae (sometimes called "suctorial") are best known in the SUCTOBELBIDAE (Figs. 15.8K, 15.12E) but are found in other groups as well. Those of GUSTAVIIDAE lack the fixed digit and have a styliform movable digit with distal serration (Figs. 15.12F, 15.51C). The structure of some other forms suggests that they are modified to scrape or filter soft substrates.

Legs

The legs of adult Palaeosomatides have six free segments, not including the pretarsus (Fig. 15.9C). Those of all other oribatid mites have five segments, due to a fusion of basifemur and telofemur (Fig. 15.9E) that, even in the Palaeosomatides (also called Bifemorata), do not fully articulate. As in other acariform mites, there is no free coxa. The pretarsus is **monodactylous** (with only an empodial claw), **tridactylous** (with empodial and paired lateral claws), or **bidactylous** (with either only lateral claws or with empodial and one lateral claw). In some basal groups (Palaeosomatides, Enarthronotides, Parhyposomatides), a tridactylous pretarsus may have only a minute, hooklike empodial claw (Fig. 15.9J). The tibia and tarsus are sometimes functionally fused by having a very narrow articulation (e.g., CARABODIDAE). Rarely, leg I has a truly fused tibiotarsus (e.g., some SELENORIBATIDAE), or there is no discernable pretarsus (e.g., some STAUROBATIDAE). Some arboreal species have a padlike pulvillus below the claw (e.g., ADHAESOZETIDAE). The most aberrant pretarsus is found in the PEDICULOCHELIDAE, where it is represented by only a bilobed terminal pad that incorporates the remnant of an empodial claw.

In most early-derivative groups, leg segments are simple tubular structures, with the genu and tibia having similar form and size; in some cases (e.g., LOHMANNIIDAE) femora have broad ventral keels associated with protecting retracted legs. However, the legs of Brachypylina have much more diverse structure. The genu is much shorter (and usually thinner) than the tibia and lacks its own musculature, serving primarily as a point of flexion (a knee) and as a conduit for tendons (Fig. 15.9F). Because legs insert in acetabula, each has a right-angle bend near its base (Fig. 15.6C). On legs I–II the trochanters are very small, mostly hidden within their respective acetabula, and the bend is at the base of the femur (Fig. 15.9F). On legs III–IV the trochanters are large and possess the bend (Fig. 15.9E). If muscles are concentrated in swollen parts of segments, legs have a **moniliform** appearance, but if segments are extremely thin and elongated, the legs appear **filiform**. In some Brachypylina the proximal part of some segments can have a **retrotectum**, a sclerotized, gauntletlike covering of the articulation that produces a socket joint (Fig. 15.9D). Tarsi I–IV each have a dorsal lyrifissure just distal to the articulation with the tibiae (Fig. 15.9H).

Respiratory porose areas are commonly found on the femora and on trochanters III and IV (Fig. 15.9G), and small areas are occasionally found on the tibia and tarsus (reviewed by Norton et al. 1997). Leg porose areas are internalized as saccules or tracheae in some families of Brachypylina, primarily in groups inhabiting dry or alternating wet and dry habitats (Norton and Alberti 1997).

Ontogeny

Oribatid mites are conservative in that most maintain the basic attributes of acariform mite development, as overviewed in chapter 5. However, several developmental traits vary among groups, and three of these are used below in taxonomic diagnoses: anamorphic development, metamorphosis, and molting. Other than the cohorts Enarthronotides and Parhyposomatides, all oribatid mites cease to add segments at the deutonymphal molt, so the anal segment is paraproctal in adults. A few primitive enarthronotid families add the peranal segment, as does the parhyposomatid family PARHYPOCHTHONIIDAE. The latter is unique in that the peranal plate is well developed, and the adanal segment merges with the opisthonotum.

Many oribatid mites exhibit a metamorphosis between tritonymph and adult that is striking enough to make the association of adults and immatures difficult. This is the rule in brachypyline mites (Figs. 15.13A–G, 15.14A–F; see details below) and in all ptychoid mites in the Enarthronotides and Mixonomatides. In most other groups, adults differ from immatures mostly in the degree of sclerotization and the addition of genitalia.

Oribatid mites exhibit several modes of ecdysis that differ in the position of the **line of dehiscence** (δ), along which the old cuticle splits, and in the direction in which the mite emerges (Norton and Kethley 1994). One of the three more common modes is the ancestral type, **prodehiscence** (Fig. 15.15A), in which the U-shaped dehiscence line is anterior, and the mite leaves anteriorly. In **transdehiscence** (Figs. 15.15B, C), the line runs across the opisthonotum (δt), sometimes accompanied by a sagittal branch (δs), and the mite leaves posteriorly. In **circumdehiscence** (Fig. 15.15D), the line passes around the opisthonotum posteriorly, and the mite leaves posteriorly in the same direction.

Classification and Identification

The higher classification used here (see chapter 8) is one of many, and, like all others in the literature, it recognizes a mixture of natural (monophyletic) and unresolved paraphyletic groupings. We follow Grandjean (1954b, 1969a) and Marshall, Reeves, and Norton (1987) in abandoning two paraphyletic groupings that are widespread in the literature. One, the "lower" oribatid mites (Macropylina, Archoribatida, Oribatei Inferiores, or variants of these), merges all early-derivative taxa that complement the diverse cohort Brachypylina. The other grouping is a paraphyletic assemblage of Brachypylina that includes the bulk of those superfamilies lacking an octotaxic system of notogastral dermal glands and pteromorphs. It is most often referred to by one of two roughly equivalent names, or variants of them: the Pycnonota (a complement to the monophyletic group Poronota, members of which have the octotaxic system) and the Apterogasterina (a complement to the Pterogasterina, which have notogastral pteromorphs). Neither name is used in this treatment (see below). Unlike the above cited authors, we include the extremely diverse cohort Astigmatina (= Astigmata, Acaridei, and Acaridida) in the Oribatida, thereby recognizing their phylogenetic origin. This placement is strongly supported on both morphological (OConnor 1984c; Norton 1998) and biochemical (Sakata and Norton 2001) grounds, but it is not supported by molecular studies (Domes et al. 2007a). The Astigmatina are treated in chapter 16.

Various monographs, catalogs, and checklists are available to further identify specimens or check known distributions of Oribatida. Foremost among them are a series of books authored by János and Peter Balogh and by Sándor Mahunka, each with keys and illustrations. World genera are keyed directly (not to family) in Balogh and Balogh (1992) and in an earlier version by Balogh (1972). Other works treat species of the Neotropical region (Balogh and Balogh 1988, 1990), non-Holarctic species (Balogh and Balogh 2002), and nonbrachypyline species of the Palearctic region (Balogh and Mahunka 1983). Species-level monographs are also available for the former Soviet Union (Gilyarov and Krivolutskij 1975), Australia (Hunt et al. 1998), and New Zealand (Luxton 1985). Three books collectively monograph the cohort Brachypylina for the Iberian Peninsula (Pérez-Íñigo 1993a, 1997; Subías and Arillo 2001). Another helpful book focuses on species described by the prolific Antonio Berlese, which originate from many areas of the world (Mahunka and Mahunka-Papp 1995).

Catalogs and checklists exist for the world (Subías 2004) and for many countries or regions. For example, the Nearctic region of the United States and Canada is covered by Marshall, Reeves, and Norton (1987), and the Yukon is addressed by Behan-Pelletier (1997a). The northern Palearctic region is treated by Karppinen and Krivolutsky (1982); Golosova, Karppinen, and Krivolutsky (1983); Karppinen, Krivolutsky, and Poltavskaja (1986); and Karppinen et al. (1987, 1992). Treatments for individual countries include Denmark (Gjelstrup 1978), Svalbard (Coulson 2007), Sweden (Lundqvist 1987), Finland (Niemi, Karppinen, and Uusitalo 1997), the Netherlands (Van der Hammen 1952), the British Isles (Luxton 1996), Austria (Schatz 1983), Germany (Weigmann and Kratz 1981; Weigmann 2006), Hungary (Mahunka and Mahunka-Papp 2000, 2004), Poland (Niedbała and Olszanowski 1997; Olszanowski, Rajski, and Niedbała 1996), Lithuania (Ejtminaviciute 2003), the Czech Republic (Stary 2000), Ukraine (Sergienko 1994, with key), the former Yugoslavia (Tarman 1983), and Romania (Vasiliu, Ivan, and Vasiliu 1993). The western Mediterranean fauna is treated by Subías and Gil-Martin (1997); that of Italy by Bernini, Castagnoli, and Nanelli (1995); and that of Greece by Flogaitis (1992). The fauna of China is reviewed by Wang, Wen, and Chen (2002, 2003) and Aoki, Yamamoto, and Hu (2000); that of Korea by Choi (1997); that of Vietnam by Vu and Thi (1995); and that of Japan by Fujikawa, Fujita, and Aoki (1993). Neotropical reviews include those for Argentina (Martinez and Velis 2000), Chile (Covarrubias 1986), the Galápagos Islands (Schatz 1991), Mexico (Palacios-Vargas 1994), Central America (Schatz 2006), and Amazonia (Woas 2002). Reviews of Ethiopian fauna include those for Algeria (Doumandji 1985), Egypt (Abd-El-Hamid, Bayoumi, and Mohamad 1980), Ghana (Wallwork 1961), Zambia (Mwase and Baker 2006), and Morocco and Western Sahara (Subías, Arillo, and Gil-Martin 1994). The fauna of India is treated by Sanyal and Bhaduri (1986) and of the Philippines by Corpuz-Raros (1979). In the Austro-Pacific region, Colloff and Halliday (1998) review the oribatid fauna of Australia, and Swift and Norton (1998) that of the Hawaiian Islands. Antarctic and subantarctic faunas are treated by Pugh (1993) and by Stary and Block (1998).

Supercohort Palaeosomatides

This group (also called Palaeosomata, Bifemorata, and Bifemoratina) usually is assumed to comprise the most primitive Oribatida, although some Enarthronotides are also rich in ancestral traits (Grandjean 1969a). They are small to relatively large (150–700 μm), pale, and lack the strong cuticular sclerotization that characterizes most oribatid mites. However, all have inconspicuous, weak sclerites that are small to extensive, according to group, and these create several recognizable body forms. For example, members of the Acaronychoidea lack large plates and have an endeostigmatid-like body. The PALAEACARIDAE and CTENACARIDAE are essentially dichoid, while the APHELACARIDAE and ADELPHACARIDAE are essentially trichoid. Some have an **aesthenic** zone, a wide dorsal region of soft, malleable cuticle between prodorsum and pronotaspis that Grandjean (1954c) considered a possible remnant of regressed segmental elements. All palaeosomatid setae are unusual in being birefringent only at their base. The prodorsum is astegasime, although there is a narrow rostral tectum, and a naso may also be present. The bothridial seta is simple, never with barbs or other ornamentation, and is straight at its base, like the bothridium. Lyrifissures assume the form of cupules; *ian* and *iad* are always absent, and others may also be absent. If a notogaster seems to be present (rarely), it is transversely divided. With rare exception, there are at least 17 pairs of opisthonotal setae (*p4* present, pygidial neotrichy may occur), with some often much larger than others and conspicuously darkened. Setae *d2* and *e1* are erectile in some groups and, unlike those of enarthronotid mites (see below), are individually movable (Norton 2001). The coxisternum has a richer setation than that of any other group (except for neotrichous taxa), with three pairs occurring on epimere II. Ventral plates are poorly defined or absent, and there is no preanal plate. Legs have six free segments beyond the coxisternum, and all femora are divided, at least in the adult. Most species have a rich complement of solenidia, but these are never flagelliform or coupled to normal setae. Opisthonotal glands are never present. The subcapitulum may be stenarthric or anarthric and may have a ventral lip under the mouth, an ancestral state unknown in other oribatid mite groups. The chelicerae are always chelate-dentate, and sometimes have elongated digits. Chelicerae are oriented vertically or obliquely, they have no Trägårdh's organ, and their bases are not inserted into the body as apodemes. Palpi may be relatively large. Where studied, palaeosomatid immatures molt by prodehiscence. A peranal segment never forms.

Grandjean (1954c and cited papers) provided the foundation for the study of morphology and classification of the Palaeosomatides. Few faunistic works exist; the fauna of Russia (and the former USSR) was summarized by Bulanova-Zachvatkina (1980).

The relatively quick movements of these primitive oribatid mites often surprise observers, but little else is known of their biology. Males are unknown in the PALAEACARIDAE (Grandjean 1954c; Norton and Palmer 1991), but all other families appear to be bisexual.

▪ Superfamily Acaronychoidea (Figs. 15.18A–E)

DIAGNOSIS: Naso present, with ventral protuberance. Bothridial seta filiform. Larger exobothridial seta (xi) *barbed, positioned below the other* (xs) *and similar in length to seta* in. *Aesthenic zone small or absent. Hysterosoma relatively short. Opisthonotum without notogaster but with series of small, weak, mostly paired sclerites, each bearing 4 or fewer setae; with 17 pairs of setae, some large and darkly pigmented but none erectile. Cupules present. Pretarsi tridactylous. Famulus large, erect, ciliated; most basally positioned setiform organ on tarsus I, adjacent to lyrifissure. Tibia II with 2 solenidia. Without modifications of genital papillae or genital setae (eugenital setae may be modified). Subcapitulum with large or small unpaired ventral lip. Anterior adoral seta* (or1) *smaller than others. Cheliceral trochanter complete, main part of chelicerae oriented almost vertically, with narrow, elongated digits. Palptarsus short, little longer than tibia; eupathidia smooth.*

This superfamily, often wrongly called Archeonothroidea, includes the ACARONYCHIDAE (Figs. 15.18A–C), with two genera, and the ARCHEONOTHRIDAE (Figs. 15.18D, E), with five genera. Some are rather large, reaching 700 µm. Archeonothroids are perhaps the most plesiomorphic Oribatida in that they retain vertically orientated chelicerae with a well-formed trochanter, have a ventral subcapitular lip, and lack a notogaster. The acaronychid *Acaronychus traegardhi* Grandjean has been collected from dry habitats throughout the Holarctic (Marshall, Reeves, and Norton 1987). The archeonothrid genera *Loftacarus* and *Stomacarus* are widely distributed in the Southern Hemisphere, with the latter restricted to tropical soils (Colloff and Halliday 1998). In addition to fungal hyphae and spores, Walter and Proctor (1998) found remains of small animals in the guts of some Australian archeonothrids. *Andacarus ligamentifer* Hammer is arboreal, living on lichens and moss on trunks (Hammer 1972). *Zachvatkinella* is widespread in the Northern Hemisphere.

- Superfamily Palaeacaroidea (Figs. 15.18F–H)

*DIAGNOSIS: Naso absent. Bothridial seta filiform. Larger exobothridial seta (*xa*) positioned anterior to the other (*xp*) and usually about half length of seta* in. *Hysterosoma moderately elongated, with large aesthenic zone. Opisthonotum with distinct pygidial shield bearing all dorsal setae posterior to row* e. *Several pairs of opisthonotal setae large, darkly pigmented, with* d2 *and* e1 *largest, erectile;* h1 *and* p1 *unusually short, swollen, or spinelike. Cupules present. Most anterior pair of genital setae strongly modified. All genital papillae of normal size. Pretarsi of all legs bidactylous. Famulus of tarsus I large, approximately equal in length to solenidia or only slightly shorter; inserted on conspicuous tubercle in proximal quarter or third of tarsus. Tibia II with 1 solenidion. Subcapitulum without unpaired ventral lip. Chelicerae chelate-dentate, oriented obliquely, anteroventrally; digits not narrowed. Anterior adoral seta (*or1*) simple, smaller than others. Palptarsus elongated, at least twice tibial length; distal eupathidia barbed.*

The single family, PALAEACARIDAE (Figs. 15.18F–H) comprises two genera that are represented throughout the Holarctic and in western Australia (Colloff and Halliday 1998). Although widespread, palaeacarids are infrequently encountered and always at low densities. *Palaeacarus hystricinus* Trägårdh is found in temperate-forest litter where it apparently feeds on fungi.

- Superfamily Ctenacaroidea (Fig. 15.19)

*DIAGNOSIS: Naso absent. Bothridial seta slightly to greatly expanded (rarely filiform). Larger exobothridial seta (*xa*) positioned anterior to the other (*xp*) and usually about half the length of seta* in *or less. Hysterosoma moderately to much elongated, with aesthenic zone. Opisthonotum with weakly definable notogaster; posterior shield bears all setae posterior to row* c, *with latter on separate anterior sclerite; separation between sclerites narrow or broad. Opisthonotum usually neotrichous in pygidial region; rarely with 16 pairs (*p4 *absent). Setae of various form, large and darkly pigmented or not,* d2 *and* e1 *erectile or not; setae* h1 *and* p1 *not unusually short. Cupules not evident. Anterior pair of genital papillae greatly reduced or absent; most anterior pair of genital setae normal or strongly modified. Famulus of tarsus I much smaller than solenidia; inserted at midpoint or in distal half of tarsus. Tibia II with 1 solenidion. Leg pretarsi II–IV tridactylous, with empodial claw well developed or much reduced. Adoral seta* or1 *strongly barbed or pectinate, usually larger than other 2 setae. Chelicerae chelate-dentate, oriented obliquely, anteroventrally; digits not narrowed. Palptarsus elongated, at least twice tibial length; eupathidia smooth.*

The Ctenacaroidea includes the families CTENACARIDAE (Figs. 15.19A–C), with four extant genera, and ADELPHACARIDAE (Figs. 15.19G–I) and APHELACARIDAE (Figs. 15.19D–F), each with one extant genus. Erectile setae occur only in ctenacarids, where they insert on poorly defined sclerites. Mites of the ctenacarid genus *Ctenacarus* are common and often abundant components of the soil fauna in deserts and dry coniferous forests. The widespread aphelacarid *Aphelacarus acarinus* (Berlese) lives in similar habitats, where its elongated, trichoid body seems adapted to moving among soil pores. Incongruously, it also occurs in buildings in some more northerly regions (Grandjean 1954c). The reduction in Ctenacaroidea of the size or number of genital papillae, which are osmoregulatory organs (see chapter 3), parallels that in the enarthronotid superfamily Protoplophoroidea, most of which also inhabit dry environments (Norton, OConnor, and Johnston 1983). Recently, three new genera of extinct ctenacaroids were described from fossils in Upper Devonian deposits in New York and Lower Carboniferous deposits in Northern Ireland (Subías and Arillo 2002).

Supercohort Enarthronotides

This group of early-derivative oribatid mites (= Enarthronota, Arthronota, Arthronotina) is arguably the most morphologically diverse supercohort of oribatid mites, by virtue of extensive plasticity in body form (Norton 2001). Enarthronotid mites range widely in size (100–1,000 µm), but most are small. Although some are pale, many are tanned to various degrees, and various forms of mineralization are known. The supercohort includes the only oribatid mites with strong yellow, orange, or red pigmentation. Enarthronotids usually are recognized by having a multipartite notogaster that is subdivided by transverse scissures (Grandjean 1947b, 1969a). They have fully capitalized on the adaptive potential of such a serial arrangement in a way reminiscent of early arthropod tagmosis. Plates have variously regressed, hypertrophied, or fused in different ways, and the articulations have been functionally modified (Norton 2001). Among the more striking examples, all transverse articulations have fused in all members of the family LOHMANNIIDAE and in some HYPOCHTHONIIDAE (*Nothrolohmannia*), while a ptychoid body form has evolved independently in the families MESOPLOPHORIDAE and PROTOPLOPHORIDAE. Paedomorphic loss of most sclerotization, including all notogastral plates, characterizes the PEDICULOCHELIDAE (Figs. 15.22A–C). When combined with an emphasis on differences rather than similarities, major evolutionary shifts in form as noted above have tended to play havoc with attempts at enarthronotid classification. For example, mesoplophorids and protoplophorids have sometimes been included with ptychoid members of the supercohort Mixonomatides in a polyphyletic taxon called Ptyctima, rather than being grouped with Enarthronotides. Alternately, they have been considered a diphyletic assemblage, Arthropyctima (Balogh and Mahunka 1979, 1983; Fujikawa 1991), with nonptychoid families paraphyletically grouped as Euarthronota. Grandjean (1969a) included LOHMANNIIDAE with the Mixonomatides (see also Subías 2004) but had long known of similarities with enarthronotids (Grandjean 1950b). Norton (2001) suggested the closest relatives of the LOHMANNIIDAE were in the Hypochthonioidea, where Woas (2002) subsequently included them. The hypochthoniid genus *Nothrolohmannia* was considered a separate family (Nothrolohmanniidae) and linked at different times to several nonenarthronotid taxa before being recognized as an enarthronotid (Norton 2001). The PEDICULOCHELIDAE had been variously placed in the Astigmatina, Endeostigmata, and Prostigmata before being recognized as a paedomorphic enarthronotid (Norton, OConnor, and Johnston 1983). Herein we recognize five enarthronotid superfamilies, of which the Brachychthonioidea, Hypochthonioidea, and Protoplophoroidea are monophyletic. The Atopochthonioidea and Heterochthonioidea each contain superficially similar early-derivative families and probably are not monophyletic.

As was the case for the Palaeosomatides, Grandjean (1947b, 1948a, 1950a, 1954a) provided a morphological foundation for the study of Enarthronotides. A dichoid body form is typical, although two families are ptychoid (see above). The prodorsum usually has a well-developed rostral tectum but may be astegasime, either plesiomorphically so (e.g., some Atopochthonioidea) or reflecting a derived paedomorphosis (some Protoplophoroidea). Eyes are weakly developed or absent. The bothridium may be simple and straight throughout, or it may have the strong proximal curve typical of most oribatid mites. The bothridial seta varies greatly, ranging from filiform to capitate and from smooth to having a wide range of barbs or pectination. A longitudinal suprapleural scissure on the notogaster usually partly or fully isolates a paired pleuraspis from the dorsal notaspis.

The notaspis comprises two to four dorsal plates separated by 1–3 transverse scissures of three possible types (see above) in a variety of combinations (Norton 2001, his fig. 5). In the simplest arrangement, all scissures are type E: simple, narrow articulations (BRACHYCHTHONIIDAE, HAPLOCHTHONIIDAE). Type S scissures ancestrally carry hypertrophied erectile setae, but setae are reduced or vestigial in HYPOCHTHONIIDAE. Erectile setae are present in one (ATOPOCHTHONIIDAE) or, usually, two rows (*e*, *f*). They are directed posteriorly when at rest but quickly raised in concert when the mite is disturbed. The presence of hypertrophied setae *e* and *f* in oribatid mite fossils from the Devonian in New York is evidence that erectile setae appeared early in enarthronotid evolution (Norton, Shear, et al. 1988). The mechanism of setal erection is unclear; Grandjean (1948a) thought the force was provided by distension of the hysterosoma, but Alberti (R. A. Norton, pers. comm.) has seen direct muscular attachments to the intercalary sclerites in *Heterochthonius gibbus* (Berlese). Telescoping, type L scissures are found in two nonptychoid families, the ENIOCHTHONIIDAE and the SPHAEROCHTHONIIDAE, each of which is the sister group of a ptychoid family (MESOPLOPHORIDAE and PROTOPLOPHORIDAE, respectively). The large volumetric changes that are possible with type L scissures, along with cuticular mineralization that provides an efficient hydraulic envelope, may have been a precondition for the evolution of ptychoidy (Norton 2001). Most of these mites have 16 pairs of notogastral setae, although some may be vestigial; neotrichy is known only in the LOHMANNIIDAE. Opisthonotal glands are never present, but notogastral lyrifissures are well formed in all groups.

Enarthronotid epimera may fuse in various ways, but the coxisternum is never fused across the sejugal furrow into a united sclerite. Ventral plates assume a variety of shapes, and like the notogastral plates they have fused in various combinations. A preanal plate is present in some groups, although it is often inconspicuous. The peranal segment, along with its paired paraproctal plate and seta, is present or absent. Lyrifissures *ian* and *iad* are always absent. Legs have five free segments, with flagellate solenidia and solenidion-seta coupling occurring in some groups. The subcapitulum is anarthric, with few exceptions (LOHMANNIIDAE, some TRICHTHONIIDAE), and both palpi and chelicerae show a wide variety of forms; chelicerae never have Trägårdh's organ. Immatures of studied enarthronotids are prodehiscent, except for those of LOHMANNIIDAE, which have a form of circumdehiscence (Norton and Kethley 1994; Norton, Florian, and Manning 2001).

Few generalities can be made about the biology of enarthronotids. Fungi and minute organic fragments appear to be their most common food items, but necrophagy is also reported (see below). Except for the LOHMANNIIDAE, enarthronotids do not appear to feed on higher plant remains. Most enarthronotid families are entirely parthenogenetic, including the speciose BRACHYCHTHONIIDAE and LOHMANNIIDAE. Sexuality is known only in the MESOPLOPHORIDAE (most species) and in members of the Protoplophoroidea (Norton and Palmer 1991; Norton et al. 1993).

■ Superfamily Brachychthonioidea (Figs. 15.22F, G)

DIAGNOSIS: Body dichoid, pale or pigmented, often yellow, orange, or red. Prodorsum stegasime. Naso not evident, but remnant may exist in form of eyes underneath rostrum; paired lateral eye present or absent in exobothridial region. Bothridium and base of bothridial seta sharply bent. One pair of exobothridial setae present. Notogaster with 16 pairs of setae, none erectile or otherwise strikingly different. With 2 transverse type E scissures between setal rows d–e *and* e–f, *respectively, such that pronotaspis bears 6 pairs of setae; pleural region comprised of multiple plates. Epimere II with 1 pair of setae; epimere III fused to IV but medially divided. Genital plates and combined anal/peranal plates subequal in size, with 3 pairs of genital papillae. With 1 pair of aggenital setae and 2 pairs of anal setae; narrow peranal plates with 1 pair of setae. Preanal plate absent. Leg pretarsi monodactylous. Trochanters I–II without setae. Tarsal solenidia* ω2 *and* ω3 *coupled, fully attached; tibia I solenidion flagellate; genu I with 2 solenidia; seta* d *coupled to solenidion on most genua and tibiae; tibia IV with 1 solenidion. Chelicerae of various form but not strongly built, with teeth absent or very small; cheliceral base not inserted into body as apodeme.*

This superfamily includes only the BRACHYCHTHONIIDAE (Figs. 15.22F, G), an assemblage of 170 species in 12 genera. Brachychthoniids are small mites, usually less than 250 μm in length. Geographically and ecologically, brachychthoniids are perhaps the most widely distributed family of oribatid mites. In addition to being diverse and abundant in moist forest soil and litter, they are common in disclimax communities and newly created habitat (Norton and Palmer 1991) and are known from the arctic and subarctic (Behan-Pelletier 1999b). They are also the dominant oribatid family in fescue prairie of North America (Clapperton, Kanashiro, and Behan-Pelletier 2002). Almost nothing is known of their food habits, although several species have been cultured on unicellular algae (R. A. Norton, unpublished).

European species of BRACHYCHTHONIIDAE were studied by Moritz (1976, 1977), and many of the species described in these works are also found in North America (Marshall, Reeves, and Norton 1987). Japanese species are reviewed by Chinone (1974) and Chinone and Aoki (1972). Ontogeny was studied by Grandjean (1963).

■ Superfamily Atopochthonioidea (Fig. 15.23)

DIAGNOSIS: Body pale, without noticeable sclerotization; dichoid, though not obviously so due to generally weak sclerotization. Prodorsum astegasime or weakly stegasime; naso not evident, but small rostral tectum projects medially in some. Bothridium and bothridial seta with sharp bend at base. With 2 pairs of exobothridial setae. Notogaster with 2–3 transverse scissures; most notogastral setae broad, phylliform, or platelike, with distinct pattern of foveae or reticulation and no barbs or cilia, 1 row of setae may be erectile (ATOPOCHTHONIIDAE) or permanently erect (PHYLLOCHTHONIIDAE). Epimere II with 1 pair of setae. With 6 or 8 pairs of genital setae and 1 pair of aggenital setae; eugenital setae of ovipositor normal or enlarged, hornlike. Anal plates with 2 pairs of setae; peranal plates present or absent, preanal plate absent. Leg pretarsi mono- or bidactylous; tibia I solenidion flagellate or not; other tibial solenidia coupled with setae or not. Genua with or without solenidia. Chelicerae of various form but never strongly developed or with base inserted in body as apodeme. Palptarsus either terminating in long, ribbon-shaped eupathidia, about equal to length of distal 3 palpal segments combined, or with forked distal eupathid, similar in length to the solenidion.

The three monogeneric families included in the Atopochthonioidea—ATOPOCHTHONIIDAE (Figs. 15.23E, F), PTEROCHTHONIIDAE (Figs. 15.23A–D), and PHYLLOCHTHONIIDAE (Figs. 15.23G–I)—appear to be rather early-derivative, phylogenetic relicts and are apparently parthenogenetic (Norton et al. 1993). They share the characters of small size (200–350 μm), pale color, squat form, and expanded, foveolate setae, but otherwise their morphology is highly divergent, and it seems unlikely that they constitute a monophyletic group (Norton 2001). *Pterochthonius angelus* (Berlese) and *Atopochthonius artiodactylus* Grandjean are found throughout the Holarctic in mesic to dry forest habitats, and the latter species is common in sphagnum bogs in northeastern North America. Although rarely reported, phyllochthoniids seem to have a somewhat complementary distribution in warmer climates. *Phyllochthonius aoutii* Travé is known only from the Ivory Coast, but the genus is also reported from Mexico (Vazquez 1999) and Brazil (Oliveira and de Moraes 2001) and is represented along the Gulf Coast of North America (R. A. Norton, unpublished). Virtually nothing is known of their feeding biology.

■ Superfamily Hypochthonioidea (Figs. 15.16D, E, 15.21B–F, 15.25A–D)

DIAGNOSIS: Small to large mites, with body moderately sclerotized or mineralized, dichoid or ptychoid; color ranging from pale yellow to reddish brown, rarely almost black. Prodorsum stegasime; naso not evident. Bothridium and base of bothridial seta with sharp bend. With 2 pairs of exobothridial setae. Notogaster with a united notaspis or with 1 transverse scissure that may be of type L or type S. Never with erectile setae; if with type S scissure, the setae (row e*) are regressed or vestigial, with intercalary sclerites united as single piece; setal row* d *not reduced in size. Lyrifissure* im *inserted on pronotaspis, rather than pleuraspis. If ptychoid, type L scissure is terminal, such that hysterosoma* *covered dorsally by an apparent notogaster with only 6–8 pairs of setae (rows* c, d, *and* e). *If dichoid, posterior coxisternal plate with sternal apodeme. Epimere II with 1 pair of setae. Genital plates transversely divided, or not; genital plates fuse broadly and imperceptibly with epimere IV; otherwise, ventral plates separate or variously combined. Preanal plate present, but may be inconspicuous, occupying vertical cuticle anterior to anal plates. Peranal segment not developed. Leg pretarsi monodactylous; trochanters I and II without setae. Solenidia and setae* d *of tibiae and genua either coupled or* d *absent. With 0–2 pairs of anal setae. Subcapitulum anarthric, rarely (LOHMANNIIDAE) stenarthric; with some adoral setae large, modified. Chelicerae of various form, but with basal third inserted into body as apodeme.*

The Hypochthonioidea is one of two groups of enarthronotid mites that show extreme body form divergence. Two of the dichoid families have limited diversity: the HYPOCHTHONIIDAE (Figs. 15.21E, F, including Malacoangeliidae and Nothrolohmanniidae of authors) comprises four genera, while the ENIOCHTHONIIDAE (Figs. 15.21B–D) is monogeneric; together, there are fewer than two dozen valid species. The ptychoid MESOPLOPHORIDAE (Figs. 15.16D, E, including Archoplophoridae and Apoplophoridae of authors), has four genera and about 30 species, including the only known sexual species in the superfamily. The greatest diversity is in the family LOHMANNIIDAE (Figs. 15.25A–D), with 25 genera and over 190 species. Monophyly of the superfamily is well supported, and its internal relationships are comparatively well known (Norton 1984, 2001). Evolution of body form seems closely linked with mineralization of the cuticle. In *Hypochthonius rufulus* C. Koch, epicuticular chambers are filled with calcium phosphate in localized patches (Alberti, Norton, and Kasbohm 2001) that lie above insertions of larger muscle groups, which suggests that they increase rigidity at these points of stress. In LOHMANNIIDAE, mineralized chambers form in narrow transverse bands—sometimes confused with body segmentation—that may aid in support of the arched notogaster. In other HYPOCHTHONIIDAE, the distribution of mineral-filled chambers has expanded to cover all exposed plates and probably impedes penetration by predators (Norton 2001). The same is true of the ENIOCHTHONIIDAE and MESOPLOPHORIDAE, but in these cases the strengthening mineral is calcium oxalate (Norton and Behan-Pelletier 1991a).

The HYPOCHTHONIIDAE includes species with cosmopolitan distributions and others that are highly endemic. The common *Hypochthonius rufulus* is widely distributed in forests and peatlands throughout the Holarctic (Marshall, Reeves, and Norton 1987; Behan-Pelletier and Bissett 1994). *H. rufulus* is acidophilous (Van Straalen and Verhoef 1997) and intolerant of heat and drought extremes (Siepel 1996). It is known to feed on fungi (Maraun et al. 1998) and perhaps on bacteria (Luxton 1972), but some laboratory observations and gut content analyses of this species and of *H. luteus* Oudemans suggest that these mites are necrophages, scavenging small dead soil invertebrates (Riha 1951; Schuster 1960a; Schneider et al. 2004). *Eohypochthonius* is widespread in warm-temperate and tropical regions; *Malacoangelia* is pantropical, but its sister genus *Nothrolohmannia*, long placed in a monobasic family, is endemic to Papua New Guinea (Norton 2001). A fossil species of HYPOCHTHONIIDAE was described from Lower Carboniferous deposits in Northern Ireland (Subías and Arillo 2002).

The few known species of ENIOCHTHONIIDAE are pale, heavily mineralized mites. *Eniochthonius minutissimus* (Berlese) is found worldwide in forest soil and is one of the most common oribatids in peatlands (Behan-Pelletier and Bissett 1994). The calcium oxalate that hardens both adults and immatures, even in calcium-poor environments, probably is mobilized from their fungal food (Norton and Behan-Pelletier 1991a).

Unlike other ptychoid families of oribatid mites, species of MESOPLOPHORIDAE have the ptychoid body form as both adults and immatures. *Mesopolophora* is virtually cosmopolitan, *Archoplophora* (which includes the only known asexual mesoplophorid; Norton et al. 1993) is widespread in the Holarctic, and species of *Apoplophora* are found throughout Southeast Asia. Most mesoplophorids are associated with soil and litter, but some *Mesoplophora* species inhabit highly decayed wood and are among the few oribatid mites outside the Astigmatina to be phoretic on insects. They attach to hairs on wood-inhabiting beetles and cockroaches using their ptychoid closing mechanism, aided by small projections on the genital plate (Norton 1980). Recent identification keys to mesoplophorid genera and species are available for the Ethiopian (Niedbała 2001) and Nearctic regions (Niedbała 2002b). Without supportive discussion, Subías (2004) recently seems to have rejected the inclusion of this family (but not the other ptychoid enarthronotid family, PROTOPLOPHORIDAE; see below) in Enarthronotides, listing it instead among the Euptyctima (Mixonomatides).

A few species of LOHMANNIIDAE inhabit warm- to mid-temperate climatic zones, but most are tropical. Where abundant, they appear to be ecologically important decomposers of higher-plant residues and are the only enarthronotids with this feeding style. Mites in the genera *Meristacarus*, *Torpacarus*, and *Heptacarus* create tunnels in wood that are occupied by single individuals or are communal feeding areas occupied by a mixture of adults and immatures (Grandjean 1934a; Haq 1982; Ramani and Haq 1991); wood feeding by *H. hirsutus* Wallwork depends on the presence of microbes (Haq 1984). Some species have been observed to burrow in plant roots (Pérez-Íñigo 1967; Shereef 1976). Adults of the genera *Annectacarus*, *Javacarus*, *Lohmannia*, *Meristacarus*, *Mixacarus*, and *Paulianacarus* have numerous secretory porose areas distributed over the body surface. In one studied species, these were found to have nervous-system control (Alberti et al. 1997; Norton and Alberti 1997), but their role is unknown. Lohmanniid mites are well protected from predators by an anterior notogastral tectum that extends over the sejugal furrow dorsally and by having ventrolateral pedofossae into which legs are retracted when the mite is disturbed. A key to lohmannid genera and species of the world was published by Balogh and Balogh (1987b).

- Superfamily Protoplophoroidea (Figs. 15.16A–C, 15.21G, 15.22A–E, 15.24E–G)

DIAGNOSIS: Small to medium-size mites, pale to light brown; body dichoid, ptychoid, or without large sclerites. Prodorsum usually stegasime, rarely (secondarily) astegasime. Narrow, naso-like lobe rarely evident (PEDICULOCHELIDAE). Bothridium and base of bothridial seta with sharp bend, with the seta usually expanded distally. With 2 pairs of exobothridial setae. Notogaster either with 3 transverse scissures of type E or type S; or with 3 simple grooves; or with 1 large, telescoping type L scissure posterior to setal row c *and posterior 2 scissures vestigial or absent. Type S scissures with multiple small intercalary sclerites bearing barbed, pectinate or phylliform erectile setae. Lyrifissure* im *inserted on pleuraspis or in soft lateral cuticle. Coxisternum with 2 pairs of epimere II setae; without sternal apodeme. Aggenital plates, if present, not fused to epimere IV; genital, aggenital plates never transversely divided. With 2–3 pairs of genital papillae, often small and inconspicuous, rarely absent. Preanal plate present or absent, peranal segment not formed. Pretarsi with 1–3 claws (sometimes differing among legs) or with highly modified, heart-shaped hyaline flap (PEDICULOCHELIDAE). Subcapitulum with unique pharyngeal complex; large dorsal muscles originate on cupola-like capitular apodeme and insert on small sclerotized dorsal region of pharynx. Chelicerae highly divergent among groups, base not inserted into body as apodeme.*

Five families compose the Protoplophoroidea: the COSMOCHTHONIIDAE (Figs. 15.24E–G), HAPLOCHTHONIIDAE (Figs. 15.22D, E), SPHAEROCHTHONIIDAE (Fig. 15.21G), PROTOPLOPHORIDAE (Figs. 15.16A–C), and PEDICULOCHELIDAE (Figs. 15.22A–C). Monophyly of the group is well established (Norton, OConnor, and Johnston 1983), with the most striking synapomorphy being the unique pharyngeal complex present in all members. Collectively, protoplophoroid mites exhibit as much body form diversity as does the Hypochthonioidea and share with it at least one evolutionary path (Norton 2001). Mirroring the relationship between the hypochthonoid ENIOCHTHONIIDAE and the MESOPLOPHORIDAE, members of the two genera of SPHAEROCHTHONIIDAE possess a telescoping type L scissure, and the 12 genera of its sister family, the PROTOPOLOPHORIDAE, are ptychoid. However, the two ptychoid families are quite distinct, mostly because the type L scissure is positioned dorsally between setal rows *c* and *d* in protoplophorids and is inconspicuously terminal in mesoplophorids. Also, there is analogous variation in the associations and fusions of ventral plates (Grandjean 1954a). Unlike the Hypochthonioidea, no known protoplophoroid mites have a unified notaspis, but a paedomorphic trend is apparent in the three genera of HAPLOCHTHONIIDAE and culminates in the monogeneric PEDICULOCHELIDAE (Norton, OConnor, and Johnston 1983; Norton 2001).

Cuticular mineralization occurs in some Protoplophoroidea, but it is less well studied than in Hypochthonioidea. Among the genera of PROTOPLOPHORIDAE, only *Prototritia* has been studied chemically, and, as in mesoplophorids, it deposits calcium oxalate (Norton and Behan-Pelletier 1991a). Phylogenetic patterns of mineralization are not yet clear. For example, all members of the cosmochthoniid genus *Phyllozetes* have large, thin, epicuticular plaques that are strikingly birefringent (Norton and Behan-Pelletier 1991b), although none of the other 50+ species of this family is known to be mineralized. One Australian species of *Sphaerochthonius* has similar plaques, but most species do not (Alberti, Norton, and Kasbohm 2001).

Body form plasticity has affected the classification of the Protoplophoroidea in several ways, including the diverse positioning of the PEDICULOCHELIDAE as discussed above. The superfamily is often considered to comprise only the PROTOPLOPHORIDAE, with most other families either similarly elevated (e.g., Balogh and Mahunka 1979) or grouped as Cosmochthonioidea (e.g., Grandjean 1969a). Relictual plesiomorphic enarthronotids with erectile setae have routinely been included in COSMOCHTHONIIDAE in some earlier classifications. These forms have been grouped instead as a paraphyletic superfamily, the Heterochthonioidea, in this treatment (see below).

Most members of the Protoplophoroidea inhabit arid or semiarid regions or periodically dry microhabitats in mesic or humid regions. Like ctenacaroid palaeosomatids, which also prefer dry habitats, protoplophoroids have regressive osmoregulatory organs (e.g., genital papillae), or they have lost them altogether. Unique internalized genital tracheae may occur in the protected genital vestibule, or they may be represented by vestigial pockets (Grandjean 1947b, 1948a, 1962c; Norton, OConnor, and Johnston 1983). Cosmochthoniid and haplochthoniid species are common in house and barn debris and in periodically dry soils. Three species of *Haplochthonius*, purportedly from Antarctica, were proposed by Sanyal, Basak, and Barman (2002), but the substrate was carried to India before being extracted in Berlese funnels. Consequently, there is some possibility that these mites originated from the laboratory of the authors. A species of *Cosmochthonius* is found at the top of rain-forest trees in Costa Rica, a microhabitat of highly varying moisture (R. A. Norton and V. B. Pelletier, unpublished). Members of the family PEDICULOCHELIDAE are naturally associated with dry soils, probably worldwide, but they occur synanthropically in nests of bees, rats, and chickens (Norton, OConnor, and Johnston 1983).

The feeding habits of protoplophoroid mites are virtually

unstudied. Nothing is known about how their unique pharyngeal complex is employed in feeding, although its structure suggests that very powerful aspiration can be generated (Grandjean 1947b).

Recent identification keys to genera and species are available for PROTOPLOPHORIDAE of the Ethiopian region (Niedbała 2001), the Nearctic (Niedbała 2002b), the Neotropical region (Niedbała 2003, 2004a), and the world fauna (Niedbała 2004b), and each of these treatments includes a biogeographical analysis. Schatz (2003) provided a key to Neotropical *Sphaerochthonius* (SPHAEROCHTHONIIDAE) species. An extinct species of COSMOCHTHONIIDAE and another of PROTOPLOPHORIDAE were collected from Lower Carboniferous deposits in Northern Ireland (Subías and Arillo 2002).

- Superfamily Heterochthonioidea (Figs. 15.21A, 15.24A–D)

DIAGNOSIS: Small to medium sized (200–800 μm), pale to light brown; body dichoid. Prodorsum stegasime; naso not evident; medial and lateral eyes present (HETEROCHTHONIIDAE) or absent. Bothridium and base of bothridial seta straight (TRICHTHONIIDAE) or with proximal bend. With 2 pairs of exobothridial setae. Notogaster with 2–3 transverse scissures: type E between setal rows c *and* d *present or absent; 2 type S scissures with large erectile setae of rows* e *and* f*. Erectile setae of various forms, simple and smooth to broadly phylliform, with surface barbs and other ornamentation. Intercalary sclerite of each seta separate (usually) or fused with others in row. Lyrifissure* im *usually located on pleuraspis or in soft lateral cuticle, rarely on notaspis or apparently absent. Coxisternum with 1 pair of epimere II setae; without sternal apodeme, epimeral halves fused medially or not. Aggenital plates, if present, not fused to epimere IV; genital plates transversely divided or not. Preanal plate present or absent, peranal segment formed or not. Leg pretarsi monodactylous. Subcapitulum anarthric or stenarthric; capitular apodeme tonguelike or not evident, but never an enlarged cupola; muscle insertion on pharynx not highly localized. Chelicerae of diverse form, but not strongly chelate-dentate; base never inserted into body as apodeme.*

A small number of plesiomorphic, relictual, and apparently asexual enarthronotid mites are included in this paraphyletic assemblage, pending further analysis of their relationships. These mites have conspicuous erectile setae and therefore present the facies of the COSMOCHTHONIIDAE, but they lack the derived traits of Protoplophoroidea. When first recognized (Grandjean 1969a), this superfamily included only the HETEROCHTHONIIDAE (Figs. 15.24B–D), comprising three genera and about a dozen species. Added herein are the TRICHTHONIIDAE (Fig. 15.24A) and the monobasic ARBORICHTHONIIDAE (Fig. 15.21A).

The family TRICHTHONIIDAE is a small paraphyletic assemblage of three genera of primitive enarthronotids with broad, but mostly relictual, distributions. *Gozmanyina* is known from temperate eastern regions of the Nearctic and Palearctic, as well as from the Andes Mountains, and *G. majesta* (Marshall and Reeves) can be locally abundant in acid forest soil or sphagnum bogs in eastern North America. *Trichthonius pulcherrimus* (Hammer) has been collected from the Andes and southern Australia and *Nipponiella* from Japan and California. *Arborichthonius* species are known only from southeastern Canada and eastern China. The HETEROCHTHONIIDAE is represented throughout the temperate Palearctic but in no other region. The feeding habits of these mites have not been studied.

Supercohort Parhyposomatides

This group (also called Parhyposomata, Parhypochthonata, or Monofissurae) comprises the single superfamily Parhypochthonioidea, which in some classifications is included with superficially similar enarthronotid mites under the name Arthronota (or variants thereof). Due to the absence of males, all members are presumed to be parthenogenetic (Norton and Palmer 1991; Norton et al. 1993). Immatures exhibit prodehiscence in molting, although slight variations occur (Norton and Kethley 1994).

- Superfamily Parhypochthonioidea (Figs. 15.20A–F)

DIAGNOSIS: Rather weakly sclerotized mites, white to pale yellow; cuticle smooth, rarely weakly punctate (ELLIPTOCHTHONIIDAE). Body trichoid, or nearly so, with postpedal furrow weakly or strongly developed. Prodorsum astegasime or stegasime. Bothridium and bothridial seta with sharp proximal bend, seta pectinate or clavate with barbs. With 2 pairs of exobothridial setae. Notogaster with 1 simple transverse scissure or groove, lying between setal rows d *and* e*; row* e *inserted well posterior to scissure. Seta* f2 *either well developed, reduced, or absent but represented by vestige. Epimere III fused to IV or not; pairs fused medially or not; 1–2 pairs of setae on epimere II. Lyrifissures* iad *and* ian *present. Opisthonotal gland present. Peranal segment developed (PARHYPOCHTHONIIDAE) or not; if developed, adanal segment incorporated into notogaster. Pretarsi with well-developed lateral claws; empodial claw reduced, hooklike, or absent. Solenidia never flagellate or coupled to setae. Subcapitulum stenarthric, rarely anarthric (ELLIPTOCHTHONIIDAE); rutella strongly developed, distally broad and toothed; 1 or 3 pairs of adoral setae. Palpi with 5 segments (rarely 4, with trochanter and femur fused). Chelicerae normal, chelate-dentate, with base not inserted into body as apodeme.*

The Parhypochthonoidea includes the families PARHYPOCHTHONIIDAE (Figs. 15.20A, B), ELLIPTOCHTHONIIDAE (Figs. 15.20C, D), and GEHYPOCHTHONIIDAE (Figs. 15.20E, F), each with a single extant genus.

Composition of the superfamily varies according to author; for example, Haumann (1991) in effect excluded ELLIPTOCHTHONIIDAE, and Woas (2002) included the mixonomatid family NEHYPOCHTHONIIDAE.

Species of PARHYPOCHTHONIIDAE are plesiomorphic in many regards. Most striking is the strong development of the peranal segment and plate, accompanied by the unique incorporation of the adanal segment into the notogaster. The family is widely distributed throughout the Holarctic and is known from Australia (Colloff and Halliday 1998). *Parhypochthonius aphidinus* Berlese is widespread both geographically and ecologically. It is common in North American forest soils, and large numbers may be found in bogs of eastern Canada (Behan-Pelletier and Bissett 1994) and the northeastern United States. *P. aphidinus* also inhabits decaying logs and tree holes (Grandjean 1969a; R. A. Norton, unpublished). Its feeding is not well studied, but large food boluses are easily observed through the translucent cuticle. The opisthonotal gland opens on a flared process, and its secretions are readily exhausted when the mite is disturbed, suggesting that they constitute a predator defense mechanism (Sakata and Norton 2001). The main secretion component is 3-ethylphenol, and the phenol odor is easily detectable, even from a single mite.

Like the parhypochthoniids, the GEHYPOCHTHONIIDAE is a widespread assemblage, with species being known throughout the Holarctic and the Southern Hemisphere (Lee 1982; Martinez and Bernava-Laborde 2000). *Gehypochthonius rhadamanthus* Jacot and similar species are abundant in unincorporated sand or sandy soils. They are small and strongly trichoid, traits that seem adaptive for moving among sand grains. *G. urticinus* (Berlese) is abundant in sphagnum bogs in North America (Marshall, Reeves, and Norton 1987). Unlike *P. aphidinus*, the opisthonotal gland of *G. urticinus* opens on a flat plate, and the main component of its secretion seems to be an isomer of 1-methyl-2-naphthol (Sakata and Norton 2001). An extinct species of GEHYPOCHTHONIIDAE was described from Lower Carboniferous deposits in Northern Ireland (Subías and Arillo 2002).

The single species of ELLIPTOCHTHONIIDAE, *Elliptochthonius profundus* Norton, is known only from its original discovery in deep mineral soil in California (Norton 1975). As with small *Gehypochthonius* species, the elongated form and strongly trichoid body of *E. profundus* probably facilitate movement within the soil pore structure.

Supercohort Mixonomatides

This supercohort (= Mixonomata) includes dichoid and ptychoid mites in seven superfamilies: Nehypochthonioidea, Eulohmannioidea, Perlohmannioidea, Epilohmannioidea, Collohmannioidea, Phthiracaroidea, and Euphthiracaroidea. Most are stegasime, all have paired bothridial setae and bothridia with a sharp proximal bend and an undivided notogaster, and most have opisthonotal glands. Lyrifissure *iad* is generally present (*ian* present or absent), the peranal segment never forms, and the cheliceral base does not insert noticeably into the body as an apodeme. In its present configuration, the Mixonomatides are paraphyletic in that the closest relatives of the supercohort Desmonomatides almost certainly lie within this group (Haumann 1991). However, precise relationships remain uncertain (Norton 1998). Species in the Phthiracaroidea and Euphthiracaroidea are ptychoid; these are sometimes grouped under the name Euptyctima, with the residual set of superfamilies (Dichosomata) given equivalent rank (Balogh and Mahunka 1979). In some classifications, the Euptyctima is removed from Mixonomatides and given equal rank (Johnston 1982; Evans 1992).

▪ Superfamily Nehypochthonioidea (Figs. 15.25G, H)

DIAGNOSIS: White, weakly sclerotized mites, 670–770 µm; body form vaguely dichoid, sejugal furrow poorly defined laterally. Prodorsum astegasime, rostral tectum very weakly developed, chelicerae mostly exposed. Bothridium and bothridial seta with strong basal bend, seta pectinate. Notogaster with vague anterior limits due to light sclerotization, posteriorly slightly more sclerotized, weakly punctate; without suprapleural scissure; opisthonotal gland present. Notogastral setae f1 *and* f2 *both strongly regressed, unlike larger neighboring setae; narrow, elongated porose area between setae* f1 *and* h1. *Epimera not fused medially. Genital plates poorly defined; anal plates narrow, elongated; preanal plate absent; lyrifissure* iad *present,* ian *absent. Leg pretarsi tridactylous, with empodial claw minute, hooklike; several solenidia subflagellate, some on genua and tibiae coupled to respective seta* d. *Subcapitulum stenarthric, rutella large, distally toothed; palpi 5-segmented. Chelicerae robust, chelate-dentate, Trägårdh's organ absent.*

This superfamily includes only the monogeneric NEHYPOCHTHONIIDAE (Figs. 15.25G, H). Since its original description, the type species *Nehypochthonius porosus* Norton and Metz has been collected in litter and moss in pine forests throughout the southeastern and Gulf coast regions of the United States, but local populations also exist in mixed forests of central New York state (R. A. Norton, unpublished). *N. porosus* also is common and abundant on the six main Hawaiian Islands (Swift and Norton 1998), but even though records are from natural environments (higher elevations in wet, humid forests), these populations probably represent one or more human-mediated introductions to the island chain. A second species, *N. yanoi* Aoki, is known only from Japan.

Little is known about nehypochthoniid biology. *N. porosus* appears to be parthenogenetic, based on the absence of males from large populations (Norton and Metz 1980). Gut contents appear to be mostly fragments from higher plants, with occasional fungal hyphae (R. A. Norton, unpublished). Secretions of the opisthonotal gland include mostly the hydrocarbon tridecane and an unknown substance, possibly a monoterpene (Sakata and Norton 2001).

■ Superfamily Eulohmannioidea (Figs. 15.26D–F)

DIAGNOSIS: Body dichoid, light yellow, elongated (600–700 μm); proterosoma noticeably constricted behind level of bothridium. Prodorsum stegasime; with 2 pairs of exobothridial setae. Bothridial seta pectinate. Notogaster without suprapleural scissure; setae f1, f2 *both present, not reduced. Opisthonotal gland absent. Posterior coxisternal plate uniquely formed: epimera III–IV fused without demarcation, also fused seamlessly to hypertrophied, neotrichous aggenital region and to notogaster. Genital plates unusually small. With 4 pairs of anal and 4 pairs of adanal setae. Subcapitulum stenarthric; rutella strongly dentate; palpi 5-segmented. Chelicerae robust, chelate-dentate, without Trägårdh's organ. Leg pretarsi apparently bidactylous, but with minute empodial vestige present; solenidia long, finely tapered, but not flagellate, none coupled to setae.*

The Eulohmannioidea includes a single family, the EULOHMANNIIDAE (Figs. 15.26D–F), with one, widespread asexual species, *Eulohmannia ribagai* Berlese. The unique fusion of the posterior coxisternal plate of *E. ribagai* to the notogaster and aggenital region results in a U-shaped scissure projecting posteriorly from base of legs IV and meeting behind genital plates. This configuration is sometimes contrasted with that of the NANHERMANNIIDAE (Desmonomatides, below) and is called **pseudodiagastry**. *E. ribagai* is known from forest habitats throughout the Holarctic (Marshall, Reeves, and Norton 1987), where it is associated with humus rather than with the litter layer (Grandjean 1939c; Lions 1978).

■ Superfamily Perlohmannioidea (Figs. 15.25E, F)

DIAGNOSIS: Body dichoid, moderately sclerotized; medium to large (620–1,400 μm). Prodorsum stegasime; with 2 pairs of exobothridial setae; bothridial seta pectinate, longer than seta le. *Hysterosoma subrectangular in dorsal aspect and dorsoventrally flattened. Notogaster with 15 pairs of similar, short setae* (f1 *absent but represented by alveolar vestige): 8 setae of rows* c, d, e *usually in 2 longitudinal rows, with lateral row on carina; suprapleural scissure developed under carina or not. Opisthonotal gland present. Epimera III–IV fused but well delineated, each separated medially by unsclerotized cuticle; not fused to notogaster or adanal plates; with 1 pair of epimere II setae. Genital plate divided transversely or not, posterior half with 2 setae; paired aggenital plates distinct, with 2 pairs of setae. Anal and adanal plates long, narrow, each with 3 pairs of setae; lyrifissures* iad *and* ian *present. Preanal plate absent, unpaired postanal plate present; paired plicature plates weakly developed between adanal plates and notogaster. Pretarsi monodactylous (usually) or tridactylous; tarsus I rich in ventral setae. Tibia I solenidion thin and tapered but not flagellate; seta* d *of various genua and tibiae fully developed and independent from solenidion, or reduced, closely adjacent to solenidion. Subcapitulum stenarthric, with strong, toothed rutella. Palpi with 5 free segments. Chelicerae without Trägårdh's organ.*

The single family, PERLOHMANNIIDAE (Figs. 15.25E, F) has a Holarctic, but often relictual, distribution and is represented by three similar genera and nine species, all of which are probably sexual. The family is represented in North America by one named species, *Hololohmania alaskensis* Kubota and Aoki, and several undescribed species from the Appalachian Mountains and the Pacific Northwest (R. A. Norton and V. B. Pelletier, unpublished). Grandjean (1958c) did a comprehensive morphological study of the Palearctic species *Perlohmannia dissimilis* (Hewitt). Taxonomic treatments include those of Schuster (1960b), Bulanova-Zachvatkina (1960), and Fujikawa and Aoki (1969). Most species are associated with forest litter, but *P. dissimilis* is known to damage the root systems of potato, strawberry, and tulip plants (Evans, Sheals, and MacFarlane 1961). The opisthonotal gland of an undescribed American *Perlohmannia* species produces a complex secretion containing pentadecene, the monoterpenes neral and geranial, and the aromatic γ-acaridial, components that are widely distributed in Astigmatina (Sakata and Norton 2001).

■ Superfamily Epilohmannioidea (Figs. 15.26G, H)

DIAGNOSIS: Body dichoid, with unusually broad sejugal articulation; somewhat cylindrical, moderately elongated; proterosoma without noticeable posterior constriction; small to large (320–800 μm). Color yellowish to dark reddish brown, moderately sclerotized. Prodorsum stegasime, but rostrum sometimes strongly emarginated laterally; with 2 pairs of exobothridial setae. Bothridial seta filiform to clavate, never pectinate; bothridium with or without brachytracheae or porose pockets. Notogastral setae f1, f2 *absent, represented only by inconspicuous alveoli. Opisthonotal gland present. Hysterosoma ventrally with much fusion: epimera III and IV fused medially and to each other but outlined by borders and apodemes; epimere IV fused to hypertrophied aggenital plates, which themselves broadly fuse anterior to genital plates; adanal plates fuse anterior to anal plates. Aggenital and adanal region separated by transverse scissure (sometimes called schizogastry) or indistinguishably fused; often with 3 pairs of aggenital setae, or neotrichous; with 3 pairs each of anal and adanal setae. Subcapitulum stenarthric; mentum fused with coxisternum or not. Palpi with 2–3 free segments, with at least palptrochanter, femur, and genu fused (tibia may be distinct, but not independently movable). Chelicerae robust, chelate-dentate, without Trägårdh's organ; rutella strongly developed, distally toothed. Leg pretarsi monodactylous. Tibia I solenidion flagellate; seta* d *of genua and tibiae not independent, absent or minute, inconspicuous, and coupled to solenidion.*

The Epilohmannioidea is represented by the widespread family EPILOHMANNIIDAE (Figs. 15.26G, H), which comprises four named genera and about 55 species. The world fauna was reviewed by Shtanchaeva (1996) and a key to world species of *Epilohmannia* was presented by Pérez-Íñigo and Baggio (1985). Several small species of *Epilohmannoides* and *Epilohmannia*, including *E. cylindrica* (Berlese) and *E. pallida* Wallwork, probably are asexual (Norton and Palmer 1991; Norton et al. 1993) and are widely distributed (Marshall, Reeves, and Norton 1987). Epilohmanniids are common in many soil types, from poor grassland to rich forests, and the smaller species are often found in lower organic horizons or even in mineral soil. At least two undescribed large North American species of *Epilohmannia* feed on decaying wood or bark as adults (Seastedt, Reddy, and Cline 1989; R. A. Norton unpublished). Their immatures tunnel in decaying wood, in a manner similar to those of PHTHIRACARIDAE (see below).

■ Superfamily Collohmannioidea (Figs. 15.26A–C)

DIAGNOSIS: Body dichoid; slightly compressed laterally or not; proterosoma without posterior constriction. Large mites (1,400–2,000 µm); color medium to dark brown, well sclerotized. Prodorsum stegasime; bothridial seta subfiliform, much shorter than seta le*; with 2 pairs of exobothridial setae. Notogaster convex, without lateral carina or suprapleural scissure. With 15 pairs of notogastral setae:* f1 *absent, represented by alveolar vestige; setae* d2, h2, p1 *flagellate, much longer than others. Epimera III–IV fused to each other but divided medially by unsclerotized cuticle; not fused to notogaster or to anogenital region. Genital plate entire. With large, paired aggenital plates, each with 2 setae, mostly separate from adanal plates (may be small area of fusion anteriorly); paired plicature plates weakly developed between adanal plates and notogaster. With 3 pairs each of anal and adanal setae, preanal and postanal plates absent. Subcapitulum stenarthric, with large, distally toothed rutella. Palpi with 5 free segments. Chelicerae robust, chelate-dentate, Trägårdh's organ present. Leg pretarsi tridactylous, with empodial claw equal to others; tarsus I unusually rich in setae, in both ventral and lateral regions; tibia I solenidion flagellate; seta* d *of genua and tibiae absent or apparently so (coupled to solenidion but minute, inconspicuous).*

The Collohmannioidea comprises a single monogeneric family, the COLLOHMANNIIDAE (Figs. 15.26A–C). The three named species inhabit rich forest litter in Europe and western Asia, and one undescribed species is known from West Virginia, USA. Although their geographical distribution is relictual and highly localized, collohmanniids can be abundant. Gut content analysis and laboratory observations suggest that their primary food is decaying leaves (R. A. Norton, unpublished). Despite their rarity, these mites are interesting and important for a number of reasons. Studies of the opisthonotal gland of *Collohmannia gigantea* Sellnick by Raspotnig et al. (2001) and Raspotnig, Schuster, and Krisper (2003) have revealed that the saclike reservoir comprises a single layer of flat cells lined with cuticle, and that it probably does not itself make the secretion. A flap that closes the gland aperture is opened by small muscles. The gland secretion includes the strongly lemon-scented monoterpene aldehydes neral and geranial, the scented monoterpene ester neryl formate, a scented aromatic aldehyde, and two nonscented hydrocarbons. This set of compounds is also known from many Astigmatina and some Nothrina (below).

Unlike most oribatid mites, collohmanniids are modestly dimorphic. Males are not only smaller and proportioned differently from females, but they have an unusually large spermatopositor and have various modifications of genu IV, traits that relate to a unique associative mating behavior. Males of the European species *C. gigantea* trail females closely prior to mating, and then raise the posterior of the body, project legs IV backward in parallel, and use the long spermatopositor to deposit a drop of nuptial food at the conjoined genua, which are modified to hold it (Schuster 1962b). The female then moves her proterosoma underneath his raised hysterosoma and eats the food. Spermatophores are never produced, so insemination probably occurs underneath the pair at this time, but it has not been observed. The unnamed species from West Virginia has a similar behavior (R. A. Norton, unpublished).

Grandjean (1969a) noted many similarities between *Collohmannia* and the ptychoid family ORIBOTRITIIDAE, the most basal family of Euphthiracaroidea. The COLLOHMANNIIDAE probably represents the sister taxon of that superfamily (Haumann 1991) and therefore has a phylogenetic position analogous to those of the dichoid families ENIOCHTHONIIDAE and SPHAEROCHTHONIIDAE, each of which is sister taxon to a ptychoid enarthronotid family.

■ Superfamily Euphthiracaroidea (Figs. 15.17A–F)

DIAGNOSIS: Body ptychoid, pale to dark brown; medium to large mites (about 400–2,000 µm when fully extended). Prodorsum stegasime; usually with paired ***manubrium*** *(narrow, projecting posterolateral apodeme); often with sagittal apodeme; with 0–2 pairs of exobothridial setae. Bothridial seta various, never pectinate; bothridium with or without respiratory saccules or fine trachea, but never with 3 brachytracheae. Notogaster laterally compressed; only anterior margin extended as collarlike tectum. Usually with 14 pairs of notogastral setae (rarely more);* f1 *and* f2 *absent, represented only by alveolar vestiges. Opisthonotal gland usually present. Coxisternum reduced in size, with little fusion among epimera; breadth of epimera decreases I–IV; 1 pair of epimere II setae. Combined anogenital region narrow, length usually greater than 3 times maximum width; component plates separate or variously fused; narrow plicature plates usually present. Anterior genital papillae never reduced; posterior pair rarely reduced. Subcapitulum stenarthric;*

rutella broad, massive; palp 3–5-segmented. Pretarsi mono- or tridactylous. Tibia I solenidion (and others) flagellate; seta d *of genua and tibiae absent or apparently so (coupled to solenidion but minute, inconspicuous).*

The Euphthiracaroidea comprises the three families ORIBOTRITIIDAE (Figs. 15.17A, B), with 15 genera and about 150 species; EUPHTHIRACARIDAE (Figs. 15.17C, D), with six genera and about 125 species; and SYNICHOTRITIIDAE (Figs. 15.17E, F) with four genera and 13 species. Regions for which there are species-level monographs or biogeographical analyses of the Euphthiracaroidea include Hawaii (Niedbała 1994, 1998a), northern California (Walker 1964), the Oriental and Australian regions (Niedbała 2000), the Ethiopian region (Niedbała 1998b, 2001, 2006), the Nearctic (Niedbała 2002b), the Neotropics (Balogh and Balogh 1987a; Niedbała 2004a), Costa Rica (Niedbała 2003), New Zealand (Niedbała 1993), and Bermuda (Niedbała 2002a). Mahunka (1990) provided a world overview of the superfamily.

Euphthiracaroid mites show trends toward fusion of ventral plates that probably relate to biomechanical factors (Sanders and Norton 2004). During normal activity in these ptychoid mites, muscles effect a lateral compression that creates the hydrostatic pressure needed to support the legs (also see chapter 3). When the mite is disturbed, the muscles relax (collapsing the high pressure), the legs are pulled into the opisthosoma, and the prodorsum deflects and retracts to cover the secondary space. Ventral plates that are not composed of multiple articulated sclerites are probably more efficient in containing hydrostatic pressure. Consequently, in most EUPHTHIRACARIDAE, the original four pairs of plates are fused into a single large pair of holoventral plates. The same seems to have occurred in the genera *Synichotritia* and *Sabahtritia* (SYNICHOTRITIIDAE), which evolved ventral plate fusion independently from euphthiracarid mites. These two genera have taken the trend to its zenith with the fusion of the paired plates into a single unit, anterior to the anal aperture. Their holoventral plates articulate medially but do not open. In other synichotritiids (*Apotritia, Temburongia*), fusion across the genital region is complete and results in an unpaired genital sclerite, although their ano-adanal plates have remained separate (Norton and Lions 1992). Earlier stages in this trend toward plate fusions are seen in the ORIBOTRITIIDAE, a paraphyletic family containing species that retain some degree of ventral plate separation (Haumann 1991). Both hydraulic efficiency and resistance to penetration are improved, at least in some groups, by mineralization of the cuticle (Norton and Behan-Pelletier 1991a; Sanders and Norton 2004); unlike the ptychoid enarthronotids, the mineral involved is calcium carbonate.

Hydraulic pressure is probably also responsible for the jumping ability of some ORIBOTRITIIDAE. Adults of *Indotritia heterotrichia* Mahunka from New Guinea can jump backward when stimulated with a fine brush, and the animal spins 12–18 times before landing and bouncing (Wauthy et al. 1998). The only obvious morphological adaptation is a spine on legs IV that may operate as a catch mechanism (see also chapter 3).

Euphthiracaroid mites are considered primarily macrophytophagous, feeding on the decaying parts of higher plants. All studied immatures are endophagic, creating individual burrows in dead wood or woody structures of plants that they leave only after maturing (e.g., Michael 1888). Rohde (1955) found that eggs of *Rhysotritia* sp. were frequently laid on the surface of wood, from which the hatchlings would begin burrowing. He also noted that eggs of *Euphthiracarus flavus* Ewing and *Oribotritia* sp. were deposited in the feeding passages of the adult, where they were inserted in holes formed by large wood tracheae. Adult *Rhysotritia duplicata* (Grandjean) showed resistance to contamination with cadmium and lead at levels far exceeding natural conditions, possibly storing some of the accumulated metal in their thick cuticle (Ludwig, Kratzmann, and Alberti 1991). *R. duplicata* is tolerant of drought but not of heat extremes (Siepel 1996). Gregarine and microsporidian protozoans are abundant parasites in examined adults of this species, whereas Microsporidia were more abundant than gregarines in *R. ardua* (C. Koch) (Purrini and Bukva 1984).

▪ Superfamily Phthiracaroidea (Figs. 15.16F, G)

DIAGNOSIS: Body ptychoid, pale to nearly black, sometimes opalescent; medium to large mites (about 400–1,500 μm when fully extended). Prodorsum stegasime; without manubrium; often with sagittal apodeme, with 0–1 pair of exobothridial setae. Bothridial seta various, never pectinate, sometimes short, inconspicuous; 3 respiratory tubes (brachytracheae) open into each bothridium. Notogaster convex, never laterally compressed; margins extended as deep, nearly continuous tectum. Usually with 15 pairs of notogastral setae (seta p4 *present;* f1 *and* f2 *absent, represented only by alveolar vestiges); sometimes with weak to strong neotrichy. Opisthonotal gland absent. Coxisternum reduced in size, with little fusion among epimera; breadth of epimera decreases from I to IV; with 1 pair of epimere II setae. Combined anogenital region broad, length approximately twice width; with imperceptibly fused genital/aggenital plates, and fused anal/adanal plates), without plicature plates. Anterior genital papillae (*Va*) reduced. Subcapitulum stenarthric; rutella broad, massive. Chelicerae robust, chelate-dentate. Palpi with 3 segments (trochanter, femur, tibia fused). Pretarsi monodactylous. Tibia I solenidion (and others) flagellate; seta* d *of genua and tibiae absent or apparently so (coupled to solenidion but minute, inconspicuous).*

The Phthiracaroidea is among the more species-rich oribatid mite superfamilies, with over 600 species in 20 genera in the single cosmopolitan family PHTHIRACARIDAE (Figs. 15.16F, G). Some authors recognize a second family, the Steganacaridae, which contains most phthiracaroid genera. However, the grouping is clearly paraphyletic, and we follow Woas (2002) in recognizing a single, well-defined traditional family. The world fauna has been treated by Niedbała (1992), and more recent regional monographs exist for the Oriental and Australian regions (Niedbała 2000), the Ethiopian region (Niedbała 2001), the Nearctic (Niedbała 2002b), and the Neotropics (Niedbała 2003, 2004a). Each of these recent treatments includes a biogeographical analysis.

Like ptychoid Enarthronotides, phthiracarid mites have a cuticle that is hardened by mineralization to a greater or lesser degree. Where identified, the mineral has proven to be calcium carbonate rather than calcium oxalate (Norton and Behan-Pelletier 1991a). Many species are mineralized to the point of having brittle cuticle, sometimes with a pale color and an opalescent sheen in reflected light. In all known instances, eggs are retained by females to the completion of embryogenesis; that is, they deposit prelarvae rather than eggs.

Immature phthiracarid mites are endophages in partially decayed woody tissues of logs, twigs, seed-bearing structures, and similar substrates or in coniferous needles (Jacot 1939b; Edsberg and Hågvar 1999). In wood, they create characteristic linear, frass-filled burrows that usually align with the wood grain. Immatures are rarely found outside such materials, but adults are common in leaf litter and have more catholic tastes, consuming also leaf litter and fungi (Anderson 1975). Phthiracarid mites are considered important decomposers, especially in coniferous forests, where they can contribute substantially to respiratory metabolism of the oribatid community (Luxton 1981). For example, *Phthiracarus borealis* Trägårdh and three other oribatid species consumed 0.22% of the annual primary production of an alpine meadow in Europe (Reutimann 1987). Phthiracarids may be arboreal in Tasmania, where they are a component of the epicorticolous fauna (Colloff and Niedbała 1996).

Steganacarus magnus (Nicolet), the best-studied member of the family, is tolerant of drought and heat extremes (Siepel 1996) and has a demonstrated supercooling point between −7° C and −38°C (Block 1979). This is made possible by the accumulation of **cryoprotectants**, low molecular weight solutes such as sugars and polyols (Webb and Block 1993). Life history traits of *S. magnus* were summarized by Lebrun et al. (1991). Adult females chew shallow pits in woody substrates such as pinecones and the cupules of beech (*Fagus* sp.) to deposit prelarvae, from which larvae burrow into the substrate (Harding and Easton 1984; Webb 1991). This large species takes more than a year to become mature and lives for more than three years (Webb 1977, 1989). Other PHTHIRACARIDAE may take up to five years for a generation (Soma 1990).

Supercohort Desmonomatides

The Desmonomatides encompasses most of the diversity in Oribatida, focused in two of its three cohorts. As previously indicated, one of these cohorts, the Astigmatina, is dealt with separately in this treatment (chapter 16) because of its different evolutionary trajectory and historical and traditional separation. The cohort Brachypylina comprises the bulk of species and family-group taxa of traditional oribatid mites. An assemblage of seven early-derivative families, with as-yet uncertain interrelationships, are included here in the cohort Nothrina. Because the monophyletic taxa Astigmatina and Brachypylina probably both originated among these families (chapter 8), the Nothrina may be considered doubly paraphyletic.

With few exceptions, desmonomatid mites have opisthonotal glands. Well-sclerotized representatives (i.e., Nothrina and Brachypylina) have a stegasime prodorsum, the bothridial seta and bothridium have a strong proximal bend, and they have a holoid body form (except as noted) and either a macropyline or brachypyline venter. The subcapitulum may be stenarthric, diarthric, or secondarily anarthric. The chelicerae have only a vestige of the trochanter, and the basal portion (often the proximal third) is inserted into the body as an apodeme that is paraxially emarginated to increase the angles of muscle insertion. Trägårdh's organ is absent in the Astigmatina but usually present in the other groups. The palptibia articulates fully with the tarsus in Brachypylina, but there is dorsal fusion in the other groups. Pretarsal claws are usually supported by two rodlike struts (condylophores). Molting is either transdehiscent or circumdehiscent. These and other characters are more fully discussed by Norton (1998).

Cohort Nothrina

The Nothrina is a paraphyletic assemblage that is also known as Nothronata (Johnston 1982), Desmonomata (Woolley 1973), Holosomata (Balogh and Mahunka 1979, 1983; Fujikawa 1991), and Nothroidea *s. lat.* (Grandjean 1954b; Travé et al. 1996; Woas 2002). The seven relevant families are sufficiently well defined, but their arrangement in superfamilies has varied and has sometimes been contradictory. As with the cohort itself, these superfamilies have little justification other than temporary convenience, pending further study.

Most commonly, three superfamilies of Nothrina have been recognized. Two families with significant ventral plate fusions (NANHERMANNIIDAE and HERMANNIIDAE) have each been given superfamily rank, and the remaining five, more plesiomorphic families have generally been grouped as the Crotonioidea (= Nothroidea *s. str.*). Some authors (e.g., Knülle 1957; Weigmann 1997a, b) group two of these families, the TRHYPOCHTHONIIDAE and the MALACONOTHRIDAE, under the superfamily Malaconothroidea, separate from the Crotonioidea. This has been a convenient arrangement for purposes of discussion because, in the view of Norton (1998, 2007), a paedomorphic trend began in the Malaconothroidea and culminated in the Astigmatina. Malaconothroids have a narrow ventrosejugal articulation instead of the holoid venter typical of other Nothrina, a feature that Norton interpreted as retention of an immature trait. Conversely, Woas (2002) considered the articulation to be ancestral, making the TRHYPOCHTHONIIDAE and MALACONOTHRIDAE basal in Desmonomatides. Lee (1985) thought similarly, but included the two families with mixonomatids as the subgroup Neomixosomatae. Because relationships among the plesiomorphic families remain obscure, we follow Grandjean (1954b) and Woas (2002) in recognizing a single superfamily that, considering the rules of nomenclature, must carry the name Crotonioidea.

■ Superfamily Crotonioidea (Figs. 15.27A–G, 15.28A–J)

*DIAGNOSIS: Body usually holoid; rarely (TRHYPOCHTHONIIDAE, MALACONOTHRIDAE) with narrow ventrosejugal articulation. Prodorsum stegasime; fused to coxisternum or not. Bothridial seta various but never pectinate, rarely highly regressed or absent; bothridium with or without respiratory saccules or tubules. Ventral region usually macropyline, with large plates and little or no distance between genital and anal apertures; sometimes diagastric (see below; NANHERMANNIIDAE) or brachypyline (HERMANNIIDAE). Notogaster with opisthonotal gland or its vestige; usually with 16 or 15 pairs of notogastral setae (*f1 *present or absent, represented by vestige). Epimere II with 1 or with 3 or more pairs of setae. Aggenital setae present or absent; genital plate with 4–24 pairs of setae; preanal plate distinct, inconspicuous, or absent. Subcapitulum usually stenarthric, rarely anarthric or diarthric; rutella well developed, with distal teeth or carina. Chelicerae usually robust, chelate-dentate; usually with Trägårdh's organ. Palptibia fused with tarsus dorsally. Lyrifissure* iad *present,* ian *usually present.*

The Crotonioidea includes the families CAMISIIDAE (Figs. 15.28D, E), NOTHRIDAE (Figs. 15.27C–E), CROTONIIDAE (Figs. 15.28A–C), NANHERMANNIIDAE (Figs. 15.27A, B), TRHYPOCHTHONIIDAE (Figs. 15.28I, J), MALACONOTHRIDAE (Figs. 15.28F–H) and HERMANNIIDAE (Figs. 15.27F–H). Pending further study, we do not recognize the Galapagacaridae in this treatment but instead follow the author of the monotypic genus *Galapagacarus* (P. Balogh 1985) and Woas (2002) in considering it a member of the HERMANNIIDAE. The Crotonioidea is cosmopolitan and well represented in most terrestrial habitats.

Most species of Crotonioidea are parthenogenetic. Sexual species are unknown among the over 140 species of MALACONOTHRIDAE, the 70 species of TRHYPOCHTHONIIDAE, the 84 species of CAMISIIDAE, and the 56 species of NANHERMANNIIDAE, nor do they occur in the moderately large genus *Nothrus* (NOTHRIDAE) (Norton et al. 1993). Some asexual crotonioid mites inhabit intermittently wet and permanent freshwater habitats, leading Norton and Palmer (1991) to postulate that these plesiotypically parthenogenetic nothrines might be preadapted to exploit such environments, where the typical oribatid mite style of fertilization might be maladaptive. All examined species of CROTONIIDAE and HERMANNIIDAE are sexual.

Members of the family CAMISIIDAE are found worldwide and include common and abundant mites, many of which are camouflaged by adherent organic debris. Classical and recent monographic treatments include those of Sellnick and Forsslund (1955), Colloff (1993), and Olzsanowski (1996). Species of *Camisia* are primarily arboreal, living on the trunks of trees or in the canopy or among mosses and lichens on rock surfaces (Travé 1963). They are only rarely found in soil. *C. carrolli* André inhabits twigs of certain age on Douglas fir trees in the northwestern United States, where fungal food is sufficient and where fir needles retain surface moisture that the mites require for active feeding (André and Voegtlin 1982). Two other common genera, *Platynothrus* and *Heminothrus*, are common in forest soil but also are found in peatlands, mosses, and other habitats. The best-known camisiid mite is *Platynothrus peltifer* (Koch), an asexual species that is widespread both geographically and ecologically and that, genetic studies suggest, has existed for perhaps 100 million years (Heethoff et al. 2007). *P. peltifer* inhabits forest soil and litter, mosses, peatlands, various freshwater habitats, and has even been found in benthic habitats (Schatz and Gerecke 1996). Adults are relatively tolerant of drought and heat extremes (Siepel 1996). Molecular studies of the ribosomal internal transcribed spacer region 1 (ITS 1) also indicate that *P. peltifer* has a general-purpose genotype adapted to a wide range of habitats (Heethoff 2000; Heethoff, Maraun, and Scheu 2000). All instars feed on decaying plant material and fungi (Hartenstein 1962b; Siepel 1990), and the generation time in temperate European forests is probably at least one year (Weigmann 1975; Schenker 1986). Because of its availability *P. peltifer* has been widely used in ecotoxicological studies; for example, those of Dennemann and Van Straalen (1991), Crommentuijn et al. (1995), Van Straalen and Verhoef (1997), and Lebrun and Van Straalen (1995). *P. peltifer* has been found to have good tolerance to metal contamination from smelters (Zaitsev and Van Straalen 2001).

Like the CAMISIIDAE, the speciose family NOTHRIDAE is cosmopolitan, although its two known sexual genera, *Trichonothrus* and *Novonothrus*, are restricted to the Southern Hemisphere (Casanueva and Norton 1997). Species of the third genus, *Nothrus*, are general feeders on decaying plant material and fungi (Siepel 1990). Monographic treatments include those of Sellnick and Forsslund (1955) and Olzsanowski (1996). *N. palustris* Koch is probably the best-known member of the genus, thanks to numerous studies on its population dynamics, feeding habits, and influence of temperature on development (e.g., Lebrun et al. 1991; Grishina 1993). Akimov and Yastrebtsov (1989) described its musculature. The opisthonotal gland secretions of adult *N. palustris* are composed mainly of a monoterpene, dehydrocineole, and a hydrocarbon. In contrast, nymphs produce neral and geranial, the latter of which functions as an alarm pheromone (Shimano et al. 2002). *N. palustris* is quite tolerant of metal contamination from smelters (Zaitsev and Van Straalen 2001).

The CROTONIIDAE is principally a Southern Hemisphere family, with most of the 50 described species being from Australia, New Zealand, South America, and Oceania but with a few records from Central America, the Antilles, and southern Mexico (Norton and Olszanowski 1989; Olszanowski 1999). Species of *Crotonia*, the most common of the three genera that compose the family, are large and carry much tightly adhered organic matter on their bodies. Although they are found in leaf litter, many seem to be primarily arboreal. Their prodorsal trichobothrium takes the unusual form of a short, capitate bothridial seta lying within an enclosing bothridium. Molecular studies suggest that the CROTONIIDAE evolved from within the CAMISIIDAE and that a sexual reproductive mode was re-evolved in this family (Domes et al. 2007b).

The cosmopolitan family NANHERMANNIIDAE includes over 50 very heavily sclerotized, slow-moving species that share a unique facies (Figs. 15.27A, B). The body is subcylindrical, and the notogaster is imperceptibly fused to the venter in its posterior half, such that in ventral view a pair of crescentlike scissures is directed between widely spaced genital and anal plates. **Diagastry** is a term sometimes used to denote this unique sclerotization pattern. *Nanhermannia*, the largest of eight recognized genera, is diverse in the Holarctic but also is found in Australia (Colloff and Halliday 1998) and the temperate Neotropics. *N. elegantula* Berlese, one of the most abundant oribatid generalists in beech litter in France (Cancela da Fonseca 1994), is tolerant of heat extremes but not of drought (Siepel 1996). Other genera are tropical or subtropical.

The TRHYPOCHTHONIIDAE, an assemblage of 12 genera and more than 70 species, is treated here in a broad sense to include Trhypochthoniellidae, Allonothridae, and Mucronothridae of authors. Also included here is the Parallonothridae of Badejo, Woas, and Beck (2002), which is itself a junior synonym of Allonothridae Lee 1985 (*n. syn.*). Norton (1998, 2007) recognized a paedomorphic trend within the family and believed that the closest relative of the cohort Astigmatina was likely to be found either within this family or the MALACONOTHRIDAE. Weigmann (1997a, b) has also studied relationships in this group. Some species of the cosmopolitan genus *Trhypochthonius*, 1 of 12 genera in the family, inhabit peatlands, while others are found in litter and moss in mesic to arid habitats. Meier, Scherrer, and Honegger (2002) found that fecal pellets of *T. tectorum* (C. Koch) that have fed on lichen thalli contain both viable ascospores and photobiont cells and may be involved in vegetative short- and long-distance dispersal of lichenforming organisms. *Archegozetes longisetosus* Aoki is among the best studied of Oribatida. It is a good colonizer, lives in disclimax habitats, and has high fecundity and short generation time (Norton and Palmer 1991; Honciuc 1996). *A. longisetosus* is becoming a model oribatid mite for studies of genetics and development (Palmer and Norton 1992; Telford and Thomas 1998a; Thomas and Telford 1999; Thomas 2002), ecotoxicology (Sarkar, Pramanik, and Joy 2000; Seniczak and Seniczak 2002), and morphology (Alberti, Seniczak, and Seniczak 2003). Like *Allonothrus*, *Archegozetes* species are tropical and subtropical in distribution.

Many of the 70+ species of TRHYPOCHTHONIIDAE are associated with moist habitats, such as peatlands (Kehl 1997). *Mainothrus badius* (Berlese) is widespread in the Palearctic and Nearctic regions (Seniczak, Norton, and Wang 1998), while *Trhypochthoniellus setosus* Willmann is the dominant mite in sphagnum moor in Japan, where densities can reach 318,150 m^{-2} (Kuriki 1996). Other trhypochthoniids are truly aquatic. For example, Fain and Lambrechts (1987) considered *T. crassus* to be an external parasite of young *Discus* fish in aquaria, but the mites may have been grazing microbes on the surface of wounds (Seniczak, Norton, and Wang 1998). The two described species of *Mucronothrus* are found in streams, in springs, and at the bottom of very cold lakes (Norton et al. 1988; Norton, Behan-Pelletier, and Wang 1996a; Schatz and Gerecke 1996). *M. nasalis* (Willmann) is an unspecialized feeder with low fecundity and probably a multiyear generation time (Norton and Palmer 1991). Movement of *M. nasalis* outside free water is slow and awkward, whereas in water it moves well, especially on irregular surfaces (Norton, Palmer, et al. 1988). Legs I–II have unusual rotation ability and are commonly splayed outward. Restriction of the species to oligotrophic freshwater, along with worldwide but discontinuous distribution, led Hammer and Wallwork (1979) to postulate that *M. nasalis* existed at the breakup of Pangaea, some 200 mya.

Study of opisthonotal gland chemistry of several species of TRHYPOCHTHONIIDAE has resulted in identification of a suite of compounds, most of which are also common in Astigmatina. However, the exact composition of the compounds is diagnostic for each species (Sakata, Tagami, and Kuwahara 1995; Sakata, Shimano, and Kuwahara 2003; Sakata and Norton 2003; Raspotnig, Krisper, and Schuster 2004).

The MALACONOTHRIDAE is a cosmopolitan family of four genera that may be closely related to the TRHYPOCHTHONIIDAE (Knülle 1957). Two unique features of the family are the presence of a nearly continuous film of birefringent cerotegument and the lack of a prodorsal trichobothrium. The opisthonotal gland appears to be undeveloped, although a vestige of the opening is present. Species of *Malaconothrus* and *Trimalaconothrus* have been recorded from peatlands, streams, springs, and lakes worldwide (e.g., Tarras-Walhlberg 1985; Gjelstrup, Hansen, Warneke 1991; Kehl 1997; Yamamoto and Aoki 1998). Little is known of their biology.

With its brachypyline venter, the HERMANNIIDAE, a family comprising three genera and over 80 species, often is considered transitional between other Crotonioidea and the cohort Brachypylina. Accordingly, some authors have extended the Brachypylina to include this family (e.g., Balogh 1972; Krantz 1978; Haumann 1991). Woas (1981 and cited works) reviewed the taxonomy and phylogeny of the family. The best-known species, *Hermannia gibba* (C. L. Koch), is common throughout the temperate Palearctic and is found in tundra habitats in the western North American arctic (Behan-Pelletier 1999b). Its life cycle in Bavarian pine and spruce forests can require two years (Bäumler 1970). *H. gibba* is a common, unspecialized feeder on decaying litter and fungi (Nanelli 1990). Bäumler (1970) recorded bacteria as its most important food, but this may have been related to an abundant gut flora. *H. gibba* displays longer survival times at all relative humidities at lower temperatures (5°C, 10°C) than at higher temperatures (15°C–25°C) (Atalla and Hobart 1964; Madge 1964). Yastrebsov (1987) described its musculature; Fernandez, Alberti, and Kümmel (1991) characterized its sperm and spermatophores; and Waitzbauer (1983) provided details on spermatogenesis in the species. *H. gibba* is one of the few oribatid mites for which the effects of parasites have been studied (Bäumler 1970; Purrini and Bäumler 1977).

Cohort Brachypylina

The Brachypylina (= Circumdehiscentiae) is the largest, most taxonomically rich group of traditional oribatid mites. This monophyletic lineage is defined by the combination of a holoid body type, brachypyline venter, distinct acetabula with trochanters I–II almost totally contained within them, and genua of legs I–III (and usually IV) shorter than tibiae and lacking intrinsic musculature. Most have a typical apodemato-acetabular system of tracheae, but all have some type of internalized respiratory surface (Norton et al. 1997). Lyrifissure *ian* is usually absent or reduced in size. The basic chelate-dentate chelicerae seen in many Oribatida (Fig. 15.18H–I) has evolved into elongated or filtering forms in many families of Polypterozetoidea, Ameroidea, Gustavioidea, and Oppioidea, and the subcapitulum is often modified accordingly.

Two special traits relate to immature brachypyline instars. As noted above, there usually is a striking metamorphosis between tritonymph and adult that is otherwise found only in ptychoid taxa. This prevents easy association of adult with immature specimens based on morphology alone (compare immatures in Figs. 15.13 and 15.14 with corresponding adult). Also, brachypyline mites are circumdehiscent when molting (with minor variations); posteriorly, δ passes just above the anal region and is directed toward the dorsosejugal region anteriorly. Depending on the group, δ may be circular—that is, complete in the dorsosejugal region—or it may stop short on each side, assuming a U shape (Fig. 15.15D). In the latter case the exuviae remains intact. If δ is circular, a dorsal cap, or scalp, breaks away as the mite ecloses. In most such cases, the emerging nymphs (and adults in some genera) retain the scalps from all previous instars, which accumulate in a pagoda-like fashion.

Since immature Brachypylina generally seem to have a more conservative morphology than do adults, Grandjean (1954b) recognized their potential value as indicators of natural taxonomic groups. Several characters have been given priority, one of which is the nature of dehiscence and scalp retention. Some superfamilies are **apheredermous**, which means that nymphs (and adults) do not retain scalps regardless of whether or not line δ is complete (Figs. 15.13C, E–G). Others are **eupheredermous**, meaning that δ is complete and that nymphs retain scalps in a pagoda-like fashion, tightly pressed to their bodies (Figs. 15.13A, 15.14F), but adults may or may not retain them (Fig. 15.11E). These two types encompass the majority of Brachypylina, but two others are recognized. Members of the ORIBATELLIDAE are **apopheredermous**; that is, δ is complete and nymphs retain scalps, but the scalps are held away from the body by setae (Fig. 15.13D). On the other hand, nymphs of the HERMANNIELLIDAE do not retain scalps; that is, they are **opsiopheredermous**. However, the adult retains the tritonymphal scalp, although it is pressed tightly to the notogaster and is difficult to discern.

Another important trait relates to opisthonotal setation (i.e., setae in the region of the presumptive notogaster). The number of setae that are missing from the ancestral oribatid mite complement of 16 pairs (13 in larva, in which segment H does not yet contribute to the opisthonotal region) is counted as a deficiency. **Uni-**, **bi-**, **quadri-**, and **quinquedeficient** nymphs would therefore have 15, 14, 12, and 11 pairs of setae, respectively. Greater losses are usually indicated as multideficiency. Unideficiency is most common, with the missing seta perhaps being *f1*, but this is uncertain (see above). Adults of species that have unideficient immatures may retain the 15 pairs, but often there are further losses (setal regressions). Eupheredermous nymphs are generally quadrideficient, lacking the dorsocentral pairs *da*, *dm*, and *dp*; scalps are located directly over the central region normally occupied by these mechanoreceptive setae (Figs. 15.13A, B). Adults in eupheredermous taxa never reform these setae. Rarely, apheredermous taxa are similarly quadrideficient. Gustavioid families like TENUIALIDAE and LIACARIDAE are examples, and Norton (1983) suggested that these might have had eupheredermous ancestors. Members of the apopheredermous ORIBATELLIDAE are unideficient, with the dorsocentral setae being well formed underneath the perched scalps. One of these setal pairs (*da*) is modified as an attachment device (Grandjean 1953a). Members of the opsiopheredermous family HERMANNIELLIDAE are unique for the Brachypylina in being holotrichous.

A third important immature character in the Brachypylina relates to properties of the hysterosomatic cuticle, particularly in the opisthonotal region. One widespread nymphal condition is for the cuticle to be somewhat tanned and leathery and provided with many folds (Figs. 15.13F, 15.14B). These forms are referred to as **plicate nymphs**, and they are found in both apheredermous and eupheredermous taxa.

This seems a highly plesiomorphic, if not ancestral, trait in the Brachypylina and is found even in some poronotic groups. Large sclerites are present on the opisthonotal cuticle in some assemblages, including some with plicate nymphs (e.g., some Ameronothroidea). When these sclerites occupy considerable parts of the surface, they are called **macrosclerites** (Fig. 15.13E). Macrosclerites are typical of the poronotic groups Ceratozetoidea (where there is at least a large pygidial sclerite) and GALUMNIDAE (where a sclerite covers nearly the whole hysterosomatic dorsum). Some or all of the opisthonotal setae of immature Oripodoidea have a dermal gland at their base (Norton et al. 1997), and both the setal insertion and the porose area associated with the gland are on small **excentrosclerites** (= microsclerites; Fig. 15.13G). Rarely, porose areas are similarly adjacent to setae on a macrosclerite, as in HUMEROBATIDAE (Ceratozetoidea).

Based in part on Grandjean (1954b, 1965a), and Woas (2002), we recognize 24 brachypyline superfamilies: Hermannielloidea, Neoliodoidea, Plateremaeoidea, Damaeoidea, Cepheoidea, Polypterozetoidea, Microzetoidea, Ameroidea, Eremaeoidea, Gustavioidea, Carabodoidea, Oppioidea, Tectocepheoidea, Hydrozetoidea, Ameronothroidea, Cymbaeremaeoidea, Eremaeozetoidea, Licneremaeoidea, Phenopelopoidea, Achipterioidea, Oribatelloidea, Oripodoidea, Ceratozetoidea, and Galumnoidea. The last seven of these are often considered a natural, monophyletic subgroup based on their possession of the octotaxic system in adults (Grandjean 1954b). However, the limits of this group—often named Poronota—are equivocal because of much within-taxon variation in gland expression, particularly in more basal families. Herein, we avoid the formal name Poronota and refer to any taxon that possesses the octotaxic system, or some part of it, as being **poronotic**. We exclude Nanhermannioidea and Hermannioidea from the Brachypylina (compare Balogh 1972; Krantz 1978). The former is an earlier derivative group, probably related to the CAMISIIDAE, in which an unusual amount of plate fusion occurred. The latter is usually considered the closest outgroup to Brachypylina, with which it shares a brachypyline venter but not most other traits of the cohort.

Considerable uncertainty remains in regard to the internal classification of Brachypylina. Phylogenetic relationships among the superfamilies are poorly understood, although several are addressed by Woas (2002). Immatures of some families, such as the HUMEROBATIDAE and the EUZETIDAE (Ceratozetoidea), show intermediate character states that suggest close but equivocal relationships among the Oripodoidea, Ceratozetoidea, and Galumnoidea. Because superfamilial diagnoses include adult and immature characters, the placement of many brachypyline families remains ambiguous for those whose ontogenies are unknown.

Relationships of the enigmatic, monobasic family ARIBATIDAE (Figs. 15.30A–D) are unclear, and, like the original authors (Aoki, Takaku, and Ito 1994), we leave it unplaced at the superfamily level; the association with Eremaeoidea, tentatively proposed by Subías (2004), seems weak. The only known species, *Aribates javensis* Aoki, Takaku, and Ito, is an obligate myrmecophile in nests of a myrmecine ant, *Myrmecina* sp., in the Oriental tropics (Aoki, Takaku, and Ito 1994). The ants care for all stages of the mite, which cannot survive without this attention. Ants feed on dead mites and only eat live individuals when there is a shortage of food in the colony (Ito and Takaku 1994; also see chapter 6). Adults of *A. javensis* have a vestigial bothridium, and the sensillum is absent. As adults, they have two rather than three pairs of genital papillae (Aoki, Takaku, and Ito 1994), which may be a paedomorphic condition.

▪ Superfamily Hermannielloidea (Figs. 15.29A–G)

*DIAGNOSIS: Adults: Usually delicately ornamented; paired opisthonotal glands opening on distinct, funnel-shaped tubes or on large apophyses laterally on notogaster. Dorsophragmata and pleurophragmata absent. Subcapitulum diarthric, chelicerae chelate-dentate or pelopsiform. Notogaster without tectum posteriorly; with macropores. Notogaster with scalps, or retaining tritonymphal scalp only; with 16 pairs of notogastral setae (*f1 *present), or 14 pairs. Apodemato-acetabular system abnormal. Palpal eupathidium* acm *separate from solenidion. Rostrum may have deep medial incision. Ventral and anal plates may have series of porose areas or saccules. Seta* d *present on tibiae and genua when respective solenidion is present. Trochanters and femora I–II with or without retrotecta.*

Immatures: With platytracheae or brachytracheae; with trichobothrium vestigial or normal. Seta d *present on tibiae and genua when respective solenidion is present. Larva holotrichous or unideficient. Nymphs holotrichous or multideficient. Tritonymphal line of dehiscence may be abnormal, with setae* h2, h3, *and* p1 *below it, not on resulting scalp. Setation of protonymphal leg IV is 0-0-0-0-5 or 0-0-0-0-7.*

The Hermannielloidea includes the families HERMANNIELLIDAE (Figs. 15.29F, G) and PLASMOBATIDAE (Figs. 15.29A–E). Although they are placed in a single superfamily, there is no evidence that they share an immediate common ancestry (Grandjean 1962a). Hermanniellids are uniquely opsiopheredermous in Grandjean's system of classification (1954b, 1965), with only adults bearing an exuviae (that of the tritonymph). In contrast, plasmobatids are eupheredermous, with nymphs and adults bearing exuviae.

The 55 known hermanniellid species are primarily litter inhabitants in temperate and tropical forests and coastal habitats where adults appear to be general macrophytophages, feeding on moist fallen leaves (Riha 1951). The endophagous immatures burrow in decaying plant materials such as dead wood and foliar needles (Michael 1882; Luxton 1972; Nanelli 1979). All nine of the currently described hermanniellid genera are represented in the Neotropics, where some species are canopy dwellers.

The three genera of PLASMOBATIDAE are primarily

tropical, although *Plasmobates* sp. has been found at 48°N in eastern Canada (Marshall, Reeves, and Norton 1987). *P. pagoda* Grandjean is one of the best-known arboricolous brachypylines, living on epiphytic plants on tree trunks and in tree canopies (Grandjean 1929). Four arboricolous species of *Plasmobates* are found in lowland rain forests of Peru (Wunderle 1992).

▪ Superfamily Neoliodoidea (Figs. 15.29H, I)

DIAGNOSIS: Adults: Large mites (1,000–2,000 μm), heavily sclerotized, bearing concentric nymphal exuviae. Apodemato-acetabular system abnormal and variable at the genus level. Dorsophragmata and pleurophragmata absent. Leg segments with saccules or brachytracheae. Genital plates with transverse scissure, with 7–8 pairs of setae. Preanal organ with brachytrachea or platytrachea. Lyrifissure ian *present. Palpal eupathidium* acm *separate from solenidion; seta* d *present on tibiae and genua when respective solenidion is present. Tibiae of legs with* (Platyliodes) *or without retrotecta.*

Immatures: With platytracheae or brachytrachea. Larva unideficient. Nymphs eupheredermous, multideficient, with maximum of 6 pairs of opisthonotal setae. Setation of protonymphal leg IV is normal (0-0-0-0-7) or variable at genus and species level.

The single family NEOLIODIDAE (Figs. 15.29H, I) is found worldwide in temperate to tropical regions. We follow the rationale of Pérez-Íñigo (1997) and consider *Neoliodes* the valid name for the type genus. The 55 species, accommodated in four genera, appear to favor dry niches and inhabit soil and leaf litter under vegetation (Riha 1951) or are arboreal in bark habitats. *Neoliodes theleproctus* (Hermann), a common neoliodid in southern Europe and northern Africa, is often arboricolous (Grandjean 1936e), as are species of *Platyliodes* and some other members of the family. At least one member of the tropical genus *Telioliodes*, *T. zikani* (Sellnick), appears to favor arboreal habitats. It is twice as common in the canopy of lowland tropical rain forests as in underlying soil and litter (Wunderle 1992).

The relationship of Neoliodoidea to other Brachypylina is unclear. Species of HERMANNIELLIDAE, along with the neoliodid genera *Poroliodes* and *Platyliodes*, have macropores on the notogaster (Grandjean 1934d) and minute saccules that appear to be respiratory organs (Norton et al. 1997), so a relationship between the groups is possible (Norton and Alberti 1997). The porose organ associated with the preanal organ in Neoliodoidea is especially large and, in *Poroliodes farinosus* (Koch), is a platytrachea (Grandjean 1934d; the anal sac of Michael 1884a). Norton and Alberti (1997) speculated on the evolution of changes in respiratory morphology in Oribatida to accommodate prevention of water loss, protection from predators, and procurement of high-quality resources, of which the porose structures in both Hermannielloidea and Neoliodoidea are excellent examples.

Neoliodes brevitarsus (Woolley) has been found in amber in Oligocene-Miocene deposits in Mexico (Norton and Poinar 1993). *Platyliodes ensigerus* Sellnick and *Neoliodes quadrisetatus* (Sellnick) were recovered from Baltic amber (Krivolutsky et al. 1990).

▪ Superfamily Plateremaeoidea (Figs. 15.30G, H, 15.31A–I, 15.32E, 15.33A–H)

DIAGNOSIS: Adults: With subnormal apodemato-acetabular system, lacking trachea I; with or without porose sac opening into acetabulum. Well-developed cerotegument covering body and setae. Prodorsum usually with deep transverse furrow. Prodorsal lamellae and costulae absent. Parastigmatic enantiophysis, dorsophragmata, and pleurophragmata absent. Notogaster flattish in lateral aspect; usually concave inside distinct margin; 4, 5, or 6 pairs of notogastral setae. Genal notch absent. Subcapitulum diarthric; palpal eupathidium acm *separate from solenidion. Tibia I with long apophysis distally, bearing solenidia* φ1 *and* φ2*; femora III–IV with retrotecta; tibiae with or without retrotecta. Tarsus II with 1–2 solenidia. Leg segments with or without internalized porose organs.*

Immatures: Larva plicate, unideficient. Nymphs eupheredermous, quadrideficient. Platytracheae or brachytracheae present. Small porose sclerites in adanal or aggenital region present or absent.

The Plateremaeoidea, a senior synonym of Gymnodamaeoidea (Marshall, Reeves, and Norton 1987; Woas 2002), includes 10 families. We follow the rationale of Hunt (1966a, b) and consider HAMMERIELLIDAE and NOOLIODIDAE as junior synonyms of PHEROLIODIDAE (Figs. 15.31G–I). There is some ambiguity in the diagnoses of the monogeneric families LYRIFISSELLIDAE (Figs. 15.33A–C) and IDIODAMAEIDAE (Figs. 15.30G, H) (Woas 2002). For example, Woas (1992) considered *Idiodamaeus illecebrosus* Paschoal, the type species of the genus and family, to belong to ALEURODAMAEIDAE (Figs. 15.31D–F). Furthermore, the monobasic *Nacunansella* (NACUNANSELLIDAE) (Fig. 15.32A) is a possible synonym of *Licnodamaeolus* (LICNODAMAEIDAE) (Figs. 15.33G, H), which is itself a heterogeneous assemblage of species in three genera Woas (2002). Clearly, these groups are in need of revision.

Fossils of PLATEREMAEIDAE (Figs. 15.32B, C) and GYMNODAMAEIDAE (Figs. 15.31A–C) are known from the Cretaceous (Labandeira, Phillips, and Norton 1997), and representatives of both families are found worldwide today. Species of the two genera of LICNOBELBIDAE (Figs. 15.33D–F) seem restricted to the Northern Hemisphere. The GYMNODAMAEIDAE, comprising five genera, are cosmopolitan, with species well represented in tundra habitats. However, they are most diverse in central and southeastern Europe (Woas 2002) and in dry habitats in western North America (Paschoal 1989). The family was revised by Woas

(1992). Other plateremaeoid families are more restricted to warm temperate and tropical regions worldwide.

The family PEDROCORTESELLIDAE (Figs. 15.32D, E) includes about 30 species and is particularly well represented in Australia, Africa, and South America, with species of *Pedrocortesella* showing an essentially Gondwanan distribution extending into Japan and the eastern Palearctic (Hunt 1996d; Bayartogtokh and Smelyansky 2002, 2004). In Australia, species of *Pedrocortesella* are most common in litter and soil of dry sclerophyll forests and semiarid woodland. However, they are more common on tree trunks than in litter in rain-forest habitats. Species of *Hexachaetoniella* and *Novazelandiella* are primarily arboreal and show morphological adaptations for life in the overstory, including short, ovoid bothridial setae and long, ventrally directed barbs on the proral and unguinal tarsal setae (Aoki 1973; Hunt 1996c). Like pedrocortesellids, the 70+ species of PHEROLIODIDAE are primarily Gondwanan in distribution and are especially diverse in Argentina (e.g., Fernandez, Martinez, and Eguaras 1991) and Chile (Covarrubias 1968a). However, outliers occur in the southeastern United States of America and the Mediterranean region (Hunt 1996a). As in the arboreal pedrocortesellids, tree-inhabiting species of the pheroliodid genera *Labiogena* and *Darthvaderum* have modified bothridial setae (Hunt 1996b).

Certain plateremaeoid and neoliodid genera have internalized porose organs in certain leg segments. In *Pheroliodes*, where they are best known, each femoral organ opens via a dorsal stigma, which on femora III and IV lies underneath the large retrotectum (Norton et al. 1997). Most members of Plateremaeoidea and other taxa with internalized leg porose organs are arboricoles or saxicoles that often inhabit mosses, lichens, or thin accumulated humus found in trees or on rocks in dry soils (Norton and Alberti 1997). Although mites in these groups are occasionally found in samples of moist organic soil, and in some cases may be known only from this microhabitat as rare species (presumably accidentals), their primary ecological affiliations seem to be aboveground in microhabitats that are subject to significant variations in moisture. These mites, and the lichens and mosses that they often inhabit, are both known to base their activity cycles on moisture availability (Norton and Alberti 1997).

▪ Superfamily Damaeoidea (Figs. 15.39C–F)

DIAGNOSIS: Adults: Scalps present or absent, sometimes with mass of debris on notogaster. Cerotegument often dense and conspicuous, covering body and setae. Prodorsal lamellae and costulae absent. Tutorium absent. Well-developed parastigmatic enantiophysis present. Dorsophragmata and pleurophragmata absent. Subcapitulum diarthric; palpal eupathidium acm *separate from solenidion. Genal notch absent. Notogaster with 11 pairs of setae, of which 9 pairs in longitudinal row. Genital setation 6 pairs. Legs moniliform; seta* d *present or absent on tibiae and genua when respective solenidion is expressed. Tarsus II with 2 solenidia.*

Immatures: Larva unideficient; nymphs eupheredermous, quadrideficient. Seta d *present on tibiae and genua when respective solenidion is present.*

This superfamily includes the single family DAMAEIDAE (Figs. 15.39C–F), a large assemblage comprising 280 species in 31 genera. Some species are found in the tropics, but the family is most diverse in temperate to boreal regions of the Holarctic (Marshall, Reeves, and Norton 1987; Behan-Pelletier 1997a; Bayartogtokh 2001; Bayartogtokh and Norton 2007). None is recorded from Australia (Colloff and Halliday 1998). Damaeids are known from both Baltic and Sicilian amber (Labandeira, Phillips, and Norton 1997).

Most damaeids are humus and litter inhabitants, but a number of species are arboreal or associated with a range of microhabitats. Debris carried on the notogaster (Michael 1884a) may offer a degree of protective camouflage. All known damaeids appear to be mycophagous, browsing on fungi, algae, and bacterial films. Three species of *Damaeus* (under *Belba*) were reared on bark with *Pleurococcus* algae (Pauly 1956), *Belba corynopus* (Hermann) eats *Penicillium* (Krantz 1978), and *Damaeus clavipes* (Hermann) feeds on *Protococcus* algae and mold in rearing containers, keeping fungal growth in check. *D. clavipes* requires 84 days to go from egg hatch to adult (Wunderle 1992a).

▪ Superfamily Cepheoidea (Figs. 15.30E, F, 15.42F–H, 15.52G–K, 15.53F, G, 15.55A–C)

DIAGNOSIS: Adults: Prodorsum with or without lamellae and lamellar cusps. Tutorium absent. Parastigmatic enantiophysis absent. Genal notch absent. Dorsophragmata and pleurophragmata absent. Subcapitulum diarthric, chelicerae chelate dentate or pelopsiform; palpal eupathidium acm *associated with solenidion, fusion often incomplete. Pedotectum I deeply incised; pedotectum II present. Notogaster with or without well-developed humeral apophysis; 9–10 pairs of notogastral setae, all in marginal or posteromarginal position.*

Immatures: Larva unideficient. Nymphs eupheredermous, with large, dorsoventrally flattened setae; quadrideficient. Seta d *present on tibiae and genua when respective solenidion is present.*

The Cepheoidea is a cosmopolitan assemblage of seven families, including families previously placed in the Eutegaeoidea by Luxton (1988) and Balogh and Balogh (1992). The latter authors overlooked the transfer of EUTEGAEIDAE (Figs. 15.53F, G), a family with 12 genera, to the Cepheoidea by Marshall, Reeves, and Norton (1987). In this treatment, we consider the Pterozetidae Luxton 1988 and Compactozetidae Luxton 1988 as junior synonyms of EUTEGAEIDAE and CEPHEIDAE (Figs. 15.55A–C), respectively, because they lack character state differences that are diagnostic at the family level.

Species of *Cepheus* are common in moss and litter, but they may also be found in rotting wood in association with other arthropods. *C. latus* (Nicolet) is a true wood borer in yellow birch twigs (Wallwork 1958). Immature cepheids are among the most beautiful oribatids known, especially those of *Conoppia palmicinctum* (Michael), collected from under lichens (Michael 1884a). Fossil *Cepheus* is known from Baltic amber (Labandeira, Phillips, and Norton 1997).

The cepheoid genera *Eupterotegaeus* and *Ommatocepheus* are considered here to be members of CEPHEIDAE, although Woas (2002) provisionally reassigned them to the EUTEGAEIDAE. Species of *Ommatocepheus* are primarily arboreal in suspended soil and bark habitats, while *Eupterotegaeus* species are primarily litter associates. Adults and immatures of the arboreal *O. ocellatus* (Michael) may remain inactive for days in lichen mats under dry conditions, with immatures displaying more drought resistance than adults (Travé 1963). The cepheid *Conoppia microptera* (Berlese) was collected in vegetation at an elevation of 3,000 m in Spain (Pérez-Íñigo 1970).

Species in the two genera of MICROTEGEIDAE (Figs. 15.52G–I) are common in litter and soil of tropical rain forests throughout Central and South America (Balogh and Balogh 1988). Microtegeids have pelopsiform mouthparts and may be closely related to the EUTEGAEIDAE (Woas 2002). Members of the three genera of CEROCEPHEIDAE (Figs. 15.52J, K), a family established by Mahunka (1986), not Subías (2004), also have pelopsiform chelicerae, as do species of NOSYBEIDAE (*Nosybea* and *Topalia*), a family that was inadvertently omitted from the key to families (see Mahunka 1993). *Pterobates incertus* Balogh and Mahunka, of the monobasic family PTEROBATIDAE (Figs. 15.30E, F), is known only from Brazil (Balogh and Mahunka 1977), while species in the genera *Anderemaeus*, *Christeremaeus*, and *Yungaseremaeus* (ANDEREMAEIDAE) (Figs. 15.42F–H) are more broadly distributed but are restricted to high elevations or southern latitudes in South America. Pterobatids and anderemaeids may be better accommodated in the Ameroidea (Woas 2002) but are retained in the Cepheoidea in this treatment, pending revision of genera and families in both superfamilies.

A new species of *Eupterotegaeus* was described recently from Lower Cretaceous amber (approximately 113 million years old) from northern Spain (Arillo and Subías 2002). *Cepheus cepheiformis* (Nicolet) is an intermediate host of the tapeworms *Cittotaenia ctenoides* and *C. denticulata* in Germany (Denegri 1993).

▪ Superfamily Polypterozetoidea (Figs. 15.50C–K, 15.51D–K)

DIAGNOSIS: Adults: Scalps present or absent. Prodorsum with broad lamellae, translamella present or absent, without genal notch. Tutorium present. Pedotectum I present; patronium present or absent; pedotectum II present or absent. Dorsophragmata and pleurophragmata absent. Subcapitulum diarthric or anarthric, with modified chelicerae and rutella; subcapitular mentum with tectum; palpal eupathidium acm *separate from recumbent solenidion. Notogaster often obscured by adherent dirt attached to cerotegument; with humeral apophysis that may appear like pteromorph; with or without medial apophysis. Genital setation 6 pairs.*

Immatures: Only single species of POLYPTEROZETIDAE known: eupheredermous; hysterosoma usually obscured by covering of adherent dirt attached to cerotegument; quadrideficient; seta d *present to tritonymph on tibiae and genua when respective solenidia are present.*

Balogh (1961) first recognized this superfamily based on Grandjean's (1959d) description of the monospecific POLYPTEROZETIDAE (Figs. 15.50C–G). The superfamily diagnosis was subsequently modified by Piffl (1972), who included PODOPTEROTEGAEIDAE (Figs. 15.50H–K), TUMEROZETIDAE (Figs. 15.51D–H), NODOCEPHEIDAE (Figs. 15.51I–K), and EREMAEOZETIDAE (see Eremaeozetoidea, below). The placement of TUMEROZETIDAE and NODOCEPHEIDAE is problematic; they were included in Cepheoidea by Subías (2004), but we retain them in the Polypterozetoidea pending discovery of their immatures.

Polypterozetoids are found at high elevations or in regions with cool temperatures. Each family has a single genus with at most a few species. The POLYPTEROZETIDAE and PODOPTEROTEGAEIDAE have Holarctic distributions. Immatures and adults of *Polypterozetes cherubin* Berlese produce a cementlike secretion with which they affix minute mineral and debris fragments to their cerotegument, so that when they are motionless, they are entirely camouflaged (Grandjean 1959d). The NODOCEPHEIDAE is the most diverse polypterozetoid family, with seven species described in the genus *Nodocepheus*. Nodocepheids are found at high elevations in Costa Rica, South America, Pakistan, and New Zealand. The monobasic *Nemacepheus*, described from Japan, was tentatively considered a nodocepheid by Fujikawa (2001) and removed from the TECTOCEPHEIDAE. Species of the monogeneric TUMEROZETIDAE have unusual anterior extensions on the notogaster and are known only from New Zealand (Luxton 1985). The mouthparts of all polypterozetoid species are modified to some extent, but their feeding habits are unknown.

▪ Superfamily Microzetoidea (Figs. 15.59F–I)

DIAGNOSIS: Adults: Proterosoma subequal to hysterosoma in length, so that notogaster is as wide as or wider than long; epimeral border IV clearly evident, extending transversely anterior to genital plates, with tectum. Prodorsum with broad lamellae, translamella present or absent, without genal notch. Tutorium present. Dorsophragmata and pleurophragmata absent. Subcapitulum diarthric, eupathidium acm *separate from solenidion; chelicerae dorsally with bacilliform tubercle. Notogaster free or fused to prodorsum; posterior notogastral tectum and pteromorphs present. Notogastral setation 9 pairs. Circumpedal carina present. Pedotectum I well developed; pedotectum II present. Epimeral setation 3-1-3-3. Genital setation 6 pairs. Tarsus II with 2 solenidia; proral setae of tarsi II–IV short and thick.*

Immatures: Larva unideficient; with bothridial regression. Nymphs eupheredermous, scalps without attachment mechanism; quadrideficient. Seta d *present on tibiae and genua when respective solenidia are present.*

Species in the single family MICROZETIDAE (Figs. 15.59F–I) are most diverse in the tropics but are found in temperate regions to latitude 48°N in eastern North America. Over 50 genera and 180 species are known from soil, litter, and arboreal habitats in the Neotropics, Africa, and the southwest Pacific (Balogh and Balogh 1992). Microzetids are particularly diverse in rain forests of the Neotropical region (Woas 2002), and new species have recently been recorded from the Galápagos Islands (Schatz and Palacios-Vargas 1999).

Microzetids have an unusual suite of apomorphies for eupheredermous Brachypylina: posterior notogastral tectum, pteromorphs, dorsal projection of the chelicerae (Grandjean 1954b), and a porose organ in the area of Ah (Norton et al. 1997). The relationships of this superfamily are unclear, but they are usually placed close to the Polypterozetoidea and Ameroidea (Marshall, Reeves, and Norton 1987; Balogh and Balogh 1992; Colloff and Halliday 1998; Woas 2002). A relationship with the ORIBATELLIDAE was suggested by Woas (1991), implying that the octotaxic system was lost in microzetid taxa.

- Superfamily Ameroidea (Figs. 15.34E–I, 15.35I–L, 15.36E–K, 15.37A–L, 15.38A–I, 15.39A, B, G, H, 15.41F–H, 15.43A–C)

DIAGNOSIS: Adults: Normal brachypyline tracheal system present, or trachea I expressed as brachytrachea. Prodorsal, bothridial, lateral, and sejugal apophyses present or absent. Prodorsum with or without costulae, without genal notch. Tutorium absent. Dorsophragmata and pleurophragmata absent. Subcapitulum diarthric, eupathidium acm *separate from recumbent solenidion; chelicerae and rutella modified or not; 0–2 cheliceral setae. Axillary saccule of the subcapitulum occasionally present. With constriction in sejugal region or in region of epimere IV. Prodorsum and notogaster fused or not. Notogastral setation 8–11 pairs. Posterior notogastral tectum absent; humeral apophysis present or absent. Circumpedal carina, custodium, and discidium absent. Pedotectum I usually present, may be reduced; pedotectum II present or absent; propodolateral apophysis present or absent. Coxisternal setation 3-1-3-3. Genital setation 6 pairs. Aggenital and adanal neotrichy present or absent. Preanal organ without caecum. Proral setae of tarsi II–IV short and spiniform. Seta* d *occasionally present on tibiae and genua when respective solenidion is present.*

Immatures: Larva unideficient or bideficient. Nymphs usually eupheredermous, scalps with or without attachment mechanism (apheredermous with complete δ *in* Gymnodampia*); generally quadrideficient or quinquedeficient, unideficient if apheredermous. According to genus, seta* d *may be present on tibiae and genua of immatures when respective solenidion is present.*

The Ameroidea (= Amerobelboidea, = Eremuloidea) comprises a number of families, none of which is rich in genera or species. Diagnoses for most included families are clear (Grandjean 1965a; Miko and Travé 1996; Avanzati et al. 2003), but recent studies of immatures and adults of species of *Gymnodampia* (AMERIDAE) have expanded both family and superfamily diagnoses (Chen et al. 2004). We follow Woas (2002) in including RHYNCHORIBATIDAE (Figs. 15.35I–L) and SPINOZETIDAE (Figs. 15.41F–H), each with two genera, and the monogeneric OXYAMERIDAE (Figs. 15.36E–H) in this superfamily pending study of their immatures. Similarly, the CALEREMAEIDAE (Figs. 15.43A–C) is included here, based on the eupheredermous immatures of its member species (Subías and Arillo 2001) and on adult characters that they share with the HUNGAROBELBIDAE (Figs. 15.39G, H) (Miko and Travé 1996), although Weigmann (2006) placed it in its own superfamily (Caleremaeoidea).

Species of Ameroidea are found in forest and prairie litter and in soil in temperate and tropical regions of the world. Some are known to occur at depths of 10–15 cm (Grandjean 1961a). European species of AMERIDAE (Figs. 15.37J–L) have been subjected to biochemical analyses that have helped clarify their biogeographical range (Avanzati et al. 2003). The genera *Caleremaeus* and *Veloppia* (CALEREMAEIDAE) are Holarctic in distribution, and the former has been found in Baltic amber (Labandeira, Phillips, and Norton 1997). Members of the families BASILOBELBIDAE (Figs. 15.36I–K), HETEROBELBIDAE (Figs. 15.37A–D), EREMULIDAE (Figs. 15.39A, B), and EREMOBELBIDAE (Figs. 15.38F–I) are primarily circumtropical and subtropical in distribution. However, the eremulid genus *Eremulus* and the eremobelbid genus *Eremobelba* are diverse in warm temperate regions of the Holarctic and have been found as far north as southern Canada (Marshall, Reeves, and Norton 1987).

Ameroid species have some of the most unusual and highly varied mouthpart configurations known for Oribatida. For example, species of OXYAMERIDAE have pelopsiform chelicerae (Figs. 15.36E, F) (Aoki 1965), those of RHYNCHORIBATIDAE have attenuate-edentate chelicerae, and those of the six genera of DAMAEOLIDAE (Figs. 15.34E–I) have reduced palpal segmentation (fused trochanter, femur, and genu), enlarged, membranous rutella, and broadly toothed chelicerae (Grandjean 1965b). Their feeding habits are mostly unknown, although, based on stable isotope studies, *Amerus troisi* (Berlese) may live predominantly on an animal diet, possibly of nematodes (Schneider et al. 2004).

Some species in the eight genera of AMEROBELBIDAE (Figs. 15.38A–C) are sexually dimorphic. For example, tarsus II and lamellar setae are dimorphic in *Mongaillardia* spp. (Grandjean 1961a), males of *Hellenamerus ionicus* have both modified hairs and unique ventrolateral porose areas (Norton and Alberti 1997), and male *Amerobelba* sp. have anal plate porose areas (Behan-Pelletier, unpublished). Species of STAUROBATIDAE (Figs. 15.37E–I) found in subtropical Brazilian forests, the widely distributed monogeneric family CTENOBELBIDAE (Figs. 15.38D, E), and the amerid *Hymenobelba ypsilon* Balogh from Madagascar have tactile legs I that may play a role in sexual behavior.

■ Superfamily Eremaeoidea (Figs. 15.34A–D, 15.42A–C, 15.54A, B, F–I)

DIAGNOSIS: Adults: Prodorsum usually with costulae. Genal notch absent. Dorsophragmata and pleurophragmata absent. Pedotecta I–II present; discidium and circumpedal carina absent. Notogastral setation 7–11 pairs, or neotrichous. Notogastral cerotegument usually with large tubercles (globules). Palpal eupathidium acm *separate from recumbent solenidion. Axillary saccule of the subcapitulum occasionally present. Epimere II with 1–3 pairs of setae. Anal and/or adanal neotrichy present or absent. Tarsi I–IV often with porose areas present ventrally; tarsus I with 2–3 solenidia; legs IV may be modified for jumping.*

Immatures: Larva unideficient. Nymphs with platytracheae or brachytracheae; eupheredermous; opisthonotal setation quadrideficient, quinquedeficient, or neotrichous. Preanal organ hollowed as caecum.

The Eremaeoidea includes families whose immatures have an apodemato-acetabular tracheal system in the form of brachytracheae (Grandjean 1967a). Adults of these families are distinct and include the EREMAEIDAE (Figs. 15.34C, D) with nine genera, the monogeneric MEGEREMAEIDAE (Figs. 15.34A, B), and the ZETORCHESTIDAE (Figs. 15.54F–I), although the latter has been included in its own superfamily by Marshall, Reeves, and Norton (1987); Balogh and Balogh (1992); and Subías (2004). The monogeneric family NIPHOCEPHEIDAE (Figs. 15.54A, B), placed in Cepheoidea by Subías (2004) and in its own superfamily (Niphocepheoidea) by Weigmann (2006), also is included here, but with reservations. It shares precocious development of the apodemato-acetabular tracheal system with other Eremaeoidea as well as with the Neoliodoidea (Grandjean 1967a), Hermannielloidea, and Plateremaeoidea (Norton et al. 1997), and Woas (2002) suggested a relationship with the EREMAEIDAE and MEGEREMAEIDAE. Based primarily on adult characters, Woas (2002) also considered genera in the ARCEREMAEIDAE (Figs. 15.42A–C) to be related to the Eremaeoidea. We tentatively concur, pending discovery of immatures. Arceremaeids are tropical, although a species of *Tecteremaeus* is found in subtropical Florida, USA.

Members of the six genera of ZETORCHESTIDAE are found in plant debris, moss, and lichens and occasionally among rocks or on tree trunks or foliage (Grandjean 1951b). They have a temperate to tropical distribution but are most diverse in warm temperate regions. *Zetorchestes micronychus* (Berlese) is considered to be a saprophage and a mycophage (Schuster 1956). Grandjean (1951b) noted fungal mycelia and pollen in the guts of *Saxicolestes auratus* Grandjean, *Litholestes altitudinus* Grandjean, and *Z. flabrarius* Grandjean, species that live on exposed rocks where wind-blown pollen may collect. Adaptations to this lifestyle include reflective cuticle and an enlarged pharynx (Travé 1963). Some zetorchestids may jump when disturbed, often for considerable distances (Krisper 1990, 1991). Using structurally modified legs IV (Fig. 15.54G), they can jump in any direction, most often sideways, and rotate during the jump. The longest recorded jump was 14.6 cm, about 290 times their body length (also see chapter 6).

Species of both EREMAEIDAE and MEGEREMAEIDAE are mycophages found primarily in dry habitats in arctic, alpine, boreal, and temperate regions of the Holarctic. Collectively, their generic diversity is greater in the Palearctic (seven genera) than in the Nearctic (three genera). Eremaeid fossils are known from Baltic amber (Labandeira, Phillips, and Norton 1997), and there are fossil records of *Proteremaeus* from Tertiary northern North America, although distribution of extant species is temperate to subarctic in the Palearctic (Behan-Pelletier and Ryabinin 1991). Eremaeoid species richness in North America is especially striking in the west and southwest (Behan-Pelletier 1990, 1993). Eremaeids are found primarily in the litter layer or in moss and lichens on the soil surface. As many as five species may occur in a given habitat, although nothing is known about resource partitioning among these mycophages (Mitchell 1978). The eremaeid *Eueremaeus oblongus* (Koch) is a major inhabitant of the epiphytic lichens *Hypogymnia physodes* and *Parmelia sulcata* in Europe (Smrź and Kocourkova 1999), and other species in this genus also are common in arboreal habitats. Behan-Pelletier (1990, 1993) revised the EREMAEIDAE and MEGEREMAEIDAE of North America.

■ Superfamily Gustavioidea (Figs. 15.51A–C, 15.53B–E, 15.55D–G, 15.56A–G)

DIAGNOSIS: Adults: Prodorsum with lamellae, tutorium, with or without translamella, without genal notch. Dorsophragmata, pleurophragmata present (LIACARIDAE, GUSTAVIIDAE, TENUIALIDAE) or absent. Subcapitulum diarthric, chelicerae chelate-dentate, pelopsiform, or styliform. Palpal eupathidium acm *and solenidion attached or separate. Axillary saccule of subcapitulum absent. Pedotectum I with or without deep incision. Notogaster with or without knifelike humeral pro-*

cess; pteromorph absent. Notogastral setation 8, 10, or 11 pairs, or glabrous, but with multiple alveoli. Genital setation 5–6 pairs. Seta d *occasionally present on tibiae and genua when respective solenidion is present.*

Immatures: Larva unideficient. Nymphs apheredermous or eupheredermous; opisthonotal setation quadrideficient or quinquedeficient. According to genus, seta d *may be present on tibiae and genua, when respective solenidion is present.*

This superfamily, once widely known as the Liacaroidea, comprises a heterogeneous, cosmopolitan group of families (Balogh and Balogh 1992; Subías 2004). Of these, inclusion of TENUIALIDAE (Figs. 15.53B–E) is most equivocal in that it may be more closely related to ameroid or eremaeoid genera (Grandjean 1965a). The six described tenuialid genera are found primarily in rich forest litter in the temperate Holarctic. The family was revised for North America by Norton (1983). Inclusion of KODIAKELLIDAE has also been questioned; Subías (2004) considered it more closely related to eremaeoid genera.

The cosmopolitan LIACARIDAE (Figs. 15.55D–G) is a large assemblage, comprising approximately 220 species in 12 genera. We consider the XENILLIDAE to be a junior synonym of LIACARIDAE, since the character state that differentiates these families (sculptured integument) is not diagnostic at the family level. The family is highly diverse in temperate regions of the Northern and Southern hemispheres and in the New and Old World tropics. The taxonomically diverse family PELOPPIIDAE (Figs. 15.52E, F) (the Ceratoppiidae and Metrioppiidae of older literature and of Subías (2004)) comprises almost 80 species in 22 genera and is recorded from arctic to tropical regions of the world, while the ASTEGISTIDAE (Figs. 15.56F, G), a family of 10 genera, is cosmopolitan but is most diverse in warm temperate regions and includes the monospecific *Maorizetes* from Australia and New Zealand. Both of the monogeneric families GUSTAVIIDAE (Figs. 15.51A–C) and KODIAKELLIDAE (Figs. 15.56C–E) are primarily Holarctic, although 3 of 13 known gustaviid species occur in Africa. The 2 described species of *Kodiakella* are found on Kodiak Island, Alaska, and in Spain, respectively. Member species of the 2 genera of the Gondwanan family MULTORIBULIDAE (Figs. 15.56A, B) occur in Chile and South Africa. *Ceratoppia* sp. (PELOPPIIDAE), *Cultroribula* sp. (ASTEGISTIDAE), and a species of *Xenillus* (LIACARIDAE) are known from Baltic amber (Krivolutsky et al. 1990).

Many gustavioid species are endophagous in decaying wood or in conifer needles (e.g., Riha 1951). Immatures and adults of *Xenillus punctulatus* (Banks) create feeding chambers in untreated wood in the southeastern United States, even in hard locust-wood fence posts (R. A. Norton, unpublished). Immatures and adults of *Liacarus* spp. may each feed on different substrates (Trávníček 1989), reflecting the often-different habitats of adult and immature Oribatida (Siepel 1990). For example, adults of the astegistid *Adoristes ovatus* (Koch) have specific preferences for acid conditions (Van Straalen and Verhoef 1997) and feed in spruce litter, while the immatures feed entirely within spruce needles after hatching from eggs that are usually deposited in the petiole (Gourbière, Lions, and Pepin 1985). In the spruce forest studied, 50% of needles were colonized by one or two immatures per needle (Lions and Gourbière 1988). The average density of juvenile mites developing inside needles at any one time may be as high as 40,000–60,000 m^{-2} (Edsberg and Hågvar 1999).

The chelicerae of most gustavioids are chelate-dentate, but those of some PELOPPIIDAE (e.g., *Metrioppia*), are pelopsiform, and those of GUSTAVIIDAE are styliform, with serrations distally. It has been suggested that gustaviid chelicerae may be modified to scrape bacterial films from damp substrates (Krantz 1978).

Species of *Liacarus*, *Xenillus*, and *Ceratoppia* have been implicated as intermediate hosts of tapeworms (Denegri 1993). Adult *C. bipilis* (Hermann) was observed to jump 2–3 cm when the long notogastral setae were touched with a pin (Suzuki 1976).

- Superfamily Carabodoidea (Figs. 15.36A–D, 15.43D–H, 15.53A, 15.57D–I)

DIAGNOSIS: Adults: Integumental birefringence present or absent. Prodorsum with or without lamellae, positioned laterally, and tutorium. Genal notch absent. Dorsophragmata and pleurophragmata absent. Subcapitulum diarthric, eupathidium acm *separate from recumbent solenidion; chelicerae and rutella normal, or rutella tubular and chelicerae pelopsiform; 1–2 cheliceral setae. Prodorsum and notogaster fused or not; dorsosejugal depression present or absent. Notogastral setation 10–14 pairs; deep depressions on notogaster present or absent. Humeral apophysis present or absent. Circumpedal carina, custodium, and discidium absent. Pedotecta I–II present. Coxisternal setation 3-1-3-3, with occasional loss. Genital setation 3–6 pairs.*

Immatures: Apheredermous, with smooth integument; unideficient. Legs short; anus subterminal; trichobothrial reduction present. Seta d *absent on tibiae and genua I–III, present to tritonymph on tibia and genu IV. Incomplete retrotecta may be present on femora III–IV.*

The Carabodoidea includes the speciose families CARABODIDAE (Figs. 15.57D–F) and OTOCEPHEIDAE (Figs. 15.43D, E), the smaller families NIPPOBODIDAE (Fig. 15.53A) and DAMPFIELLIDAE (Figs. 15.36A–D)—each with two genera—and the monogeneric TOKUNOCEPHEIDAE (Figs. 15.43F–H) and CARABOCEPHEIDAE (Figs. 15.57G–I). Most of the described genera are exclusively tropical (Mahunka 1986, 1987), but the carabodid genera *Carabodes*, *Odontocepheus*, *Austrocarabodes*, *Yoshiobodes*, and *Klapperiches* (Reeves 1998); the otocepheid *Dolicheremaeus*; and the dampfiellid *Dampfiella* are more widespread. Various authors have recognized a superfamily

Otocepheoidea that includes all the above families except Carabodidae (e.g., Balogh 1972; Subías 2004).

The CARABODIDAE, a family of 48 genera, was revised by Mahunka (1986, 1987). In North America, both *Carabodes* and the CARABODIDAE as a whole are most diverse in the southern Appalachian Mountains (Reeves 1998), and only 3 of the 28 species of *Carabodes* known from North America have a Holarctic distribution (Reeves and Behan-Pelletier 1998). *C. brevis* Banks and *C. radiatus* Berlese prefer arboreal habitats; and *C. granulatus* Banks, *C. labyrinthicus* (Michael), *C. niger* Banks, and *C. willmanni* Bernini may also be arboreal where lichens, moss, or fungi are present. The widely distributed *C. labyrinthicus* is known from arctic to temperate habitats in North America, from European forest soils, and from cave debris in Spain. It accounts for more than 80% of oribatid mites on beech bark in Germany, occupying a dominant position in the bark synusium studied by Wunderle (1992a). Wunderle found *C. labyrinthicus* to have the longest development period of all the oribatid bark associates present, requiring an average of 168 days from larva to adult. Species of *Otocepheus* (OTOCEPHEIDAE) and *Carabodes*, including *C. labyrinthicus*, are known from Baltic amber (Labandeira, Phillips, and Norton 1997).

Although they range widely as adults, all carabodids probably are endophages as immatures, burrowing in a variety of substrates, including decaying wood (e.g., *Odontocepheus* in Michael 1882). Immatures of most *Carabodes* species feed within fungi (especially bracket fungi) and lichens (Bellido 1979a; Reeves 1988; Materna 2000). *C. labyrinthicus* is one of the major inhabitants of epiphytic lichens (Smrž and Kocourkova 1999), and grazing by adults and immatures on *Evernia prunastri* can induce variability in the growth of lichen thalli (Prinzing and Wirtz 1997). Schneider, Renker, and Maraun (2005) found *C. femoralis* (Nicolet) to feed on a range of ectomycorrhizal fungi. In contrast to most brachypyline oribatid mites, cuticular birefringence is common in *Carabodes* species and some otocepheid genera (Reeves and Behan-Pelletier 1998), possibly caused by calcium carbonate deposited in the cuticle.

The large family OTOCEPHEIDAE (39 genera) and the TOKUNOCEPHEIDAE include elongated species that are especially common in moist litter, moss, and wet decaying wood in tropical regions. Adults of the otocepheid genus *Dolicheremaeus* live in spongy, very wet decaying wood and have respiratory taenidia of a type more commonly found in Oribatida from aquatic and semiaquatic habitats. Adults can survive immersion of 2–3 hr, and immatures dig galleries into decaying wood (Travé 1986). Similarly, taenidia are found in the primarily tropical dampfiellid genera *Beckiella* and *Dampfiella* (Grandjean 1964b). Both genera are diverse in lowland tropical rain forest in the Neotropics, Asia, and Africa. The Chinese NIPPOBODIDAE were revised by Chen and Wang (2007).

- Superfamily Oppioidea (Figs. 15.35A–H, 15.41D, E, 15.42D, E, 15.44A–I, 15.45D–F, 15.46A–I, 15.47A, C–F, 15.48A–G, 15.50A, B)

DIAGNOSIS: Adults: Prodorsum with or without costula; large, paired flat regions (tectopedial fields) present or absent; without genal notch. Tutorium absent. Dorsophragmata and pleurophragmata absent. Subcapitulum diarthric, eupathidium acm *separate from recumbent solenidion; chelicerae and rutella normal or modified (suctobelbid type), with chelicerae attenuate and edentate or with fine teeth distally; 2 cheliceral setae; palptarsus with 8–9 setae. Body with constriction in sejugal region. Prodorsum and notogaster separate. Notogaster often with tubercles or crista anteriorly; notogastral setation 9–14 pairs. Posterior notogastral tectum usually absent; humeral apophysis present or absent. Circumpedal carina, custodium, and discidium absent. Pedotectum I usually present, small, scalelike; pedotectum II absent. Epimera III and IV fused; coxisternal setation 3-1-3-3, or epimera III and IV neotrichous; 4–7 pairs of genital setae; ventral neotrichy absent. Preanal organ without caecum. Legs moniliform. Trochanter III with 1–2 setae; tarsus II with 2 solenidia; proral setae of tarsi II–IV short and spinelike; pretarsi monodactylous.*

Immatures: Apheredermous, unideficient, or nymphs bideficient. According to genus, seta d *may be present on tibiae and genua when respective solenidion exists.*

The Oppioidea (including part of Eremelloidea and Trizetoidea of Subías 2004) is the largest superfamily in the Oribatida and includes its most species-rich family, the OPPIIDAE (Figs. 15.49F–H), comprising more than 1,000 species in approximately 170 genera. The families of Oppioidea (*sensu* Balogh and Balogh 1992) are in need of comprehensive phylogenetic analysis. Recent major works by Woas (1986, 2002) and Subías and Arillo (2001) may provide the necessary foundation. A number of families often included in the Oppioidea have been assigned in this treatment to other superfamilies: OXYAMERIDAE, CALEREMAEIDAE, RHYNCHORIBATIDAE, and SPINOZETIDAE to the Ameroidea; DAMPFIELLIDAE to the Carabodoidea; and EREMELLIDAE to the Licneremaeoidea. The family CUNEOPPIIDAE is possibly a junior synonym of the widely distributed SUCTOBELBIDAE (Figs. 15.35A–E), from which it differs primarily in nondiagnostic characters. The monobasic families ENANTIOPPIIDAE (Figs. 15.48A, B), HEXOPPIIDAE (Figs. 15.50A, B), LUXTONIIDAE (Figs. 15.49A–E), NOSYBELBIDAE (Figs. 15.44D–I), PLATYAMERIDAE (Figs. 15.46F, G), PAPILLONOTIDAE (Figs. 15.48F, G), MACHADOBELBIDAE (Figs. 15.42D, E), and CHAVINIIDAE (Figs. 15.48C–E) are retained in Oppioidea, pending discovery of immatures.

Comprehensive keys to Oppioidea (*sensu* Balogh and Balogh 1992) of the Iberian Peninsula are presented in Subías and Arillo (2001). Woas (2002) included substantive family diagnoses for the OPPIIDAE, SUCTOBELBIDAE, and QUADROPPIIDAE (Figs. 15.49I–L); and Woas (1986) redescribed many common oppioid species. Keys to oppiid genera of the world are found in Subías and Balogh (1989).

The diagnosis of Oppioidea used herein is based on the concepts of Grandjean (1954b) and Woas (2002) and includes two informal groups erected by Woas (2002). A "Protoppiid" group includes the MACHUELLIDAE (Figs. 15.46H, I), the TERATOPPIIDAE (Figs. 15.46A–E), the monobasic STERNOPPIIDAE (Figs. 15.47A, B), the AUTOGNETIDAE (Figs. 15.45D–F), and the EPIMERELLIDAE (Figs. 15.47C, D), plus selected genera previously included in other oppioid families (Woas 2002). The monobasic Australasian family TUPAREZETIDAE (Figs. 15.44A–C), might also be included in this group. Woas' "Archoppiid" group includes the TRIZETIDAE (Figs. 15.35F–H) and GRANULOPPIIDAE (Figs. 15.47E, F). Woas (2002) considered the THYRISOMIDAE (Figs. 15.41D, E), which was included in the Oppioidea by Balogh (1961 and subsequent references), to be related to the Gustavioidea. However, thyrisomid nymphs are unideficient, like those of OPPIIDAE and SUCTOBELBIDAE (Grandjean 1954b), and the traditional placement is retained herein pending further study. The family THYRISOMIDAE, with seven genera, was revised by Fujikawa (1978) under the synonym Banksinomidae.

Oppia (OPPIIDAE), *Suctobelbella* (SUCTOBELBIDAE), and *Autogneta* (AUTOGNETIDAE) are known from Baltic amber, and *Oppia* is also known from Dominican and Chiapas amber (Labandeira, Phillips, and Norton 1997). Extant oppioids are most diverse in the tropics, although there are a great number of temperate species. They are common inhabitants of moss, humus, litter, and pasture sod in both moist and dry situations. The oppiid genus *Moritziella* is one of only five oribatid genera found in polar desert habitats (Makarova 2002). Populations of *Lauroppia translamellata* Willmann were unaffected by an experimentally induced, thick surface ice layer at a high arctic site (Coulson et al. 2000). Some oppiid species have worldwide distributions, for example, the thelytokous species *Microppia minus* (Paoli) and *Oppiella nova* (Oudemans). The latter is among the most intensively studied oribatid mites (Marshall, Reeves, and Norton 1987; Kaneko 1988). High population densities of *O. nova* have been associated with high litter C/N ratios (McLean and Parkinson 2000) but not with litter species diversity or litter structural diversity (Hansen 2000). *O. nova* and *M. minus* were consistently found to depths of 64 cm in forest soil (Moskacheva 1967).

Feeding studies indicate that fungi make up the bulk of oppiid diets. Masses of fungal spores have been identified from the gut of *Oppia neerlandica* (Oudemans) and *O. nitens* C. L. Koch (Luxton 1972). Feeding habits of the over 290 species of SUCTOBELBIDAE in 25 genera are unclear. Many species are found worldwide and have been the subject of recent revisions, for example, Mahunka and Mahunka-Papp (2001) and Chinone (2003).

Certain species of the beetle family Scydmaenidae appear to be specialized predators of Oppioidea (Molleman and Walter 2001).

■ Superfamily Tectocepheoidea (Figs. 15.57A–C)

DIAGNOSIS: Adults: Granular cerotegument present. Prodorsum with lamellae and lamellar cusps, translamella present or absent. Genal notch present. Dorsophragmata and pleurophragmata present. Chelicerae chelate-dentate; palp eupathidium acm *and solenidion separate, or incompletely coupled; subcapitular mentum without tectum. Axillary saccule of subcapitulum usually absent. Discidium, custodium, and postanal porose area absent. Pedotectum I large, extending to insertion of exobothridial seta, or small, barely covering acetabulum I; pedotectum II present. Notogastral setation 10 pairs. Pteromorphs present, without hinge, or absent. Genital setation 6 pairs.*

Immatures: Plicate, apheredermous, unideficient. Seta d *present to tritonymph on tibiae and genua I to IV when respective solenidion exists.*

The Tectocepheoidea was first recognized by Marshall, Reeves, and Norton (1987) to accommodate the TECTOCEPHEIDAE (Figs. 15.57A–C), a family whose relationships are unclear. The family was included in the Carabodoidea by Balogh and Balogh (1992), but, unlike the smooth immatures of Carabodoidea, immature tectocepheids are plicate. The placement of other genera in this family by Balogh and Balogh (1992) is dubious. Nübel-Reidelbach and Woas (1992) questioned the inclusion of *Tegeozetes*, but they did not give an alternative placement. The genus *Lamellocepheus*, included by Balogh and Balogh (1992), is considered invalid (Kok 1968) because there is no generic description, and the type species is, in fact, a species of *Tectocepheus* (Woas 2002). *Tectocepheus* and *Tegeozetes* are two of the taxa that share immature character states with the Cymbaeremaeoidea, Licneremaeoidea, and Ameronothroidea.

Members of the TECTOCEPHEIDAE are widespread inhabitants of soil and litter in arctic to tropical regions of the world. Species of *Tectocepheus* are common in all disturbed or early successional habitats, regardless of where these habitats occur or whether the disturbance is natural (e.g., frost boils in the high arctic) or anthropogenic (cultivated sites, demolition sites, mine tailings, etc.). Despite the number of species and varieties of *Tectocepheus* described in the literature, Nübel-Reidelbach (1994) recognized only the type species, *T. velatus* (Michael) and *T. minor* Berlese in her revision of the genus. However, the combined morphological and molecular study of Laumann et al. (2007) indicates that this may be an excessively conservative view of species in this wholly parthenogenetic genus. Morphological variability in *T. velatus* was further reviewed by Caballero and Iturronbeitia (2000) and Weigmann (2002), and additional *Tectocepheus* species have since been described from Mongolia (Bayartogtokh 1998) and Japan (Fujikawa 2001).

T. velatus is a cosmopolitan species that occurs from arctic tundra to tropical forests. It is also the most frequently encountered oribatid species in the plumage of birds, having been recovered from 12 host species (Krivolutsky, Lebedeva, and Matyukhin 2001; Krivolutsky and Lebedeva 2003). Populations of *T. velatus* from forest, grassland, and crop fields show variations in morphological characters and ecological features, with those in forest being smaller than crop field populations (Fujikawa 1995). Populations of both *T. velatus* and *T. minor* can be maintained under no-till field conditions (Fujita and Fujiyama 2001), and both deutonymphs and adults are moderately resistant to drying (50%–85% relative humidity) (Iordansky and Stein-Margolina 1993). *T. velatus* goes through two generations per year in European broadleaf forest (Schuster 1986); its life cycle in the field was analyzed by Totschnig and Schatz (1998).

T. velatus can be reared on algae, and it has been variously considered a mycophage (Wunderle 1992a), an opportunistic herbofungivore (Siepel and de Ruiter-Dijkman 1993), or a particle-feeding saprophage and mycophage (Bellido and Deleporte 1994). Interestingly, a specimen was found lodged in the conjunctiva of a human eyelid (Matsubara, Chinzei, and Aoki 1996). This species is among the many oribatids that can be infected with anoplocephalid cysticercoids (Fritz 1995). All studied species of *Tectocepheus* are thelytokous (Norton et al. 1993), and all specimens of *T. velatus* that Fujikawa (1995) examined (over 4,000) were female, although Grandjean (1941b) recorded 3 males in a population with 430 females.

Tegeozetes tunicatus Berlese is a very common tectocephoid in Neotropical rain-forest litter (Nübel-Reidelbach 1994).

- Superfamily Hydrozetoidea (Figs. 15.40A–C, 15.60A–C)

DIAGNOSIS: Adults: Prodorsum with or without lamellae, adjacent medially or separated, and lamellar cusps. Tutorium present or absent. Genal notch present or absent. Dorsophragmata and pleurophragmata present. Subcapitulum diarthric. Palpal eupathidium associated with solenidion. Chelicerae and rutella normal. Axillary saccule of subcapitulum absent. Bothridial seta reduced or absent (frequently broken). Notogaster with 10–11 or 15–17 pairs of notogastral setae (if more than 15, several are neotrichous h *setae); pteromorphs present or absent; lenticulus present or absent; octotaxic system absent. Epimera II–IV clearly delineated; epimeral borders II–III often fused anterolateral of genital plates, and meeting epimeral border IV at, or anterior to, genital aperture. Seta* d *inserted on proximal fifth of femora I–III, proximal to other femoral setae.*

Immatures: Apheredermous, plicate, integument tuberculate or not; opisthonotal setae normal or very long; unideficient. Seta d *absent from tibiae and genua when respective solenidion is present.*

All members of the two monogeneric hydrozetoid families, HYDROZETIDAE (Figs. 15.40A–C) and LIMNOZETIDAE (Figs. 15.60A–C), are semiaquatic. Although they recently have been assigned to separate superfamilies by Subías (2004), we believe that the similarities noted by Grandjean (1951a) and Behan-Pelletier (1989) convincingly link the two families. *Limnozetes* is the most species-rich oribatid genus in sphagnum moss of Holarctic bogs, and its members often are the only microarthropods in wet sphagnum peat layers (Bocard 1991) and bog pools (Behan-Pelletier and Bissett 1994). Species of *Hydrozetes* are found throughout the world in aquatic situations, generally in association with plants. Individuals are found on the exposed surfaces or underwater on various plant parts and have been observed in subsurface roots of anchored aquatic plants (Krantz and Baker 1982). The type species, *H. lemnae* (Coggi), has been collected from several meters below the waterline to several centimeters above it (Buford 1976). Fossils of *Hydrozetes* are known from the Jurassic, the Tertiary, and the Quarternary (Labandeira, Phillips, and Norton 1997); and fossil *Limnozetes* are known from the Quarternary (Solhøy and Solhøy 2000).

All known *Limnozetes* species are parthenogenetic, as are some species of *Hydrozetes*. However, other *Hydrozetes* are sexual (*H. confervae* Oudemans and *H. incisus* Grandjean) and have modified setae on tarsus I that may be associated with mating (Norton and Palmer 1991). Females of *Hydrozetes* deposit eggs in canals beneath the exposed surfaces of decomposing logs or in crevices and other small depressions in aquatic plants. Aggregations of ovigerous females averaging 140 cm^{-2} have been recorded for the parthenogenetic *H. lemnae* (Buford 1976), and large numbers have also been found on the gills of dying young fish (*Discus* sp.) in aquaria (Fain, Lambrechts, and Wauthy 1988; also see chapter 6).

Adult and immature *H. lemnae* reared in laboratory aquaria feed on leaves of aquatic plants that are in the early stage of decomposition (Covarrubias and Mellado 1998). In an Argentinian study, immature *H. lemnae* restricted their feeding to *Lemna gibba* L., which the mites can hollow and kill by their feeding, whereas adults fed on a variety of duckweed species (Athias-Binche and Fernandez 1986). All stages of *H. lemnae* are detritivores under natural conditions, but they also feed on decaying mosses, algae, plants, cyanobacteria, bacterial films, and decaying bryozoans (Buford 1976). Adults have been observed on the damaged or deteriorating leaves of the aquatic plant *Salvinia rotundifolia* in Florida (Krantz 1978). *Hydrozetes* and *Limnozetes* have lifestyles that change during development, with adults moving over the surface of aquatic vegetation or being free from vegetation, and immatures living in the fronds (Fernandez and Athias-Binche 1986; V. M. Behan-Pelletier, unpublished). Correlated with this behavior, adults are more exposed to predator pressure than are immatures and have been found in the gut of mayflies and damselflies (Behan-Pelletier and Bissett 1994). Predation is especially heavy in spring, when aggregations of adults on stable, submerged surfaces can reach densities of 60 to 394 mites cm^{-2} (Buford 1976).

Adults of *Limnozetes* and *Hydrozetes* have taenidia and an extensive plastron mechanism. This allows for gas interchange across an air-water interface by diffusion (Krantz and Baker 1982), and the airspace is continuous with that of the tracheae. Furthermore, the cerotegument of *Hydrozetes* contains up to 66% airspace (Pugh, King, and Fordy 1987b). The plastron of *Hydrozetes* proved to be resistant to low and high pressures, maintaining itself through a pressure range of -98kPa to +550kPa (Messner, Adis, and Franklin Ribeiro 1992). Immatures of some *Limnozetes* and *Hydrozetes* species have a papillate integument that retains an air layer when individuals are submersed. *Hydrozetes* species avoid anoxic conditions by forming a gas bubble in the midgut, through which they become buoyant and rise in the water column (Newell 1945). A bubble also forms when mites are agitated and dislodged or when light changes from high to low intensity (Newell 1945; Fernandez and Athias-Binche 1986).

The lenticulus of *H. lemnae* is among the simplest but, relative to body size, the largest photoreceptor organ in the animal kingdom (Alberti and Fernandez 1988; Alberti 1998). Adaptations and constraints of aquatic Oribatida, particularly members of the Hydrozetoidea, were reviewed by Behan-Pelletier and Eamer (2007).

- Superfamily Ameronothroidea (Figs. 15.40D–H, 15.41A–C, 15.54C–E)

DIAGNOSIS: Adults: Integument often weakly sclerotized and easily deformed. Prodorsum with or without lamellae, adjacent medially or separated, and lamellar cusps. Tutorium and translamella present or absent. Genal notch absent. Bothridial seta normally developed, or extremely small, or absent. Chelicerae chelate-dentate; palpal eupathidium acm *separate from solenidion; subcapitular mentum without tectum. Axillary saccule of subcapitulum absent. Dorsophragmata and pleurophragmata present or absent. Discidium present or absent, custodium absent. Pedotectum I large, extending to base of bothridium, subdivided, or absent. Pedotectum II present or absent. With or without tecta laterally to maintain plastron around sejugal region and acetabulum III. Coxisternal setation various (3-1-3-2, 3-1-2-2, 3-1-3-2, 2-1-2-3, or 1-1-1-1). Genital setation 3, 5, or 6 pairs. Ovipositor and spermatopositor with 11–12 setae. Notogaster overhanging ventral plate. Lenticulus present or absent.*

Immatures: Apheredermous, plicate, opisthonotum often smooth and convex centrodorsally; unideficient or bideficient. Prodorsal, adanal, and latero-hysterosomal porose regions (or their saccule homologues) present or absent. Seta d *variably present on tibiae and genua when respective solenidion is present.*

The four families of Ameronothroidea (AMERONOTHRIDAE (Figs. 15.40F–H), with nine genera; FORTUYNIIDAE (Figs. 15.41A–C), with three genera; SELENORIBATIDAE (Figs. 15.40D, E) with five genera; and the monogeneric TEGEOCRANELLIDAE (Figs. 15.54C–E)) are primarily halophilous and hygrophilous. They may have a cerotegumental plastron and taenidia and tecta in the pleural region that allow gas exchange during periodic immersion (Pugh, King, and Fordy 1990). Littoral species of AMERONOTHRIDAE, FORTUYNIIDAE, and SELENORIBATIDAE tolerate submergence in salt- or freshwater but are not active when submersed. Individuals either tolerate the daily tidal cycle or avoid it by vertical migrations (Bücking, Ernst, and Siemer 1998).

Several species of *Ameronothrus* (AMERONOTHRIDAE) occur in intertidal algae and lichens and in salt marsh vegetation on temperate to arctic coastlines (Schubart 1975; Schulte, Schuster, and Schubart 1975; Schuster 1979; Bücking, Ernst, and Siemer 1998). *Ameronothrus marinus* (Banks) is among the few species of mites inhabiting the artificial rocky shores of estuaries in northern Germany (Bücking, Ernst, and Siemer 1998), while *Alaskozetes* and *Podacarus* species occur in algae and other substrates in the antarctic (Wallwork 1967). The three species in the ameronothrid genera *Aquanothrus* and *Chudalupia* require water for activity but can withstand dry periods. They inhabit geographically disjunct, ephemeral rock pool habitats in semiarid or arid regions (Norton, Graham, and Alberti 1996). *Aquanothrus montanus* Engelbrecht occurs in vegetated pools on sandstone at 1,700–2,000 m in mountains in Orange Free State, South Africa (Engelbrecht 1975), while *Chudalupia meridionalis* Wallwork inhabits pools on granite in southwestern Australia (Wallwork 1981). A species of *Aquanothrus* lives in shallow depressions in Navajo sandstone in deserts of western North America, where its broad diet includes orange bdelloid rotifers (Norton, Graham, and Alberti 1996).

Species of *Tegeocranellus* (TEGEOCRANELLIDAE) live in temporary swamps and streams, and immatures have been collected only during periods of inundation (Behan-Pelletier 1997b). The family was revised for North and Central America by Behan-Pelletier (1997b) and placed in Ameronothroidea. The FORTUYNIIDAE are intertidal from temperate to tropical regions, and new species have recently been recorded from rocky shores and mangrove swamps (Marshall and Pugh 2002). Similarly, species of SELENORIBATIDAE are intertidal from temperate to tropical regions (Schuster 1977). *Selenoribates mediterraneus* Grandjean is frequently found in the intertidal area of the Mediterranean (Grandjean 1966b).

The ameronothrid species *Alaskozetes antarcticus* (Michael) and *Ameronothrus lineatus* (Thorell) live in extreme environments of the antarctic and high arctic, respectively. *Alaskozetes antarcticus* displays low enzyme activation energies and elevated metabolic rates at low temperatures and has associated low optimum temperatures for activity, feeding, and growth when compared to temperate species (Block and Convey 1995); it has a life cycle of five to six years (Convey 1994). Although considered general herbivores or detritivores (Covarrubias 1968b), these mites can be selective in the type of microalgae they choose (Worland and Lukesova 2000). The gamasine mite *Gamasellus racovitzai* (Trouessart) is a known predator of *A. antarcticus* where populations of both species occur (Lister 1984). *Ameronothrus lineatus* (Thorell), an inhabitant of temperate estuaries of northern Germany, has a two-year life cycle (Bücking, Ernst, and Siemer 1998), whereas populations living on cyanobacterial mats or in salt marshes in Svalbard have a minimum five-year life cycle (Søvik et al. 2003). Synchronous larviposition and molting has been observed in *A. lineatus* (Søvik et al. 2003). Serial sections and computer-assisted reconstructions were used for establishing the three-dimensional structure of the gut of this species (Bücking 2002).

▪ Superfamily Cymbaeremaeoidea (Figs. 15.45A–C)

DIAGNOSIS: Adults: Prodorsum with or without lamellae and lamellar cusps. Interlamellar setae present or absent. Pedotecta I–II present; genal notch absent; tutorium present or absent. Dorsophragmata and pleurophragmata present. Chelicerae chelate-dentate; palpal eupathidium acm *inserted on large tubercle, separate from solenidion. Subcapitular mentum with tectum. Axillary saccule of subcapitulum absent. Notogaster with or without lenticulus and humeral projection; without posterior tectum. Pteromorphs absent. Octotaxic system absent. Notogastral setation 10 or 13 pairs. All leg femora, tibiae, and tarsi generally with saccules. Seta* s *on tarsus I noneupathidial. Seta* d *present on tibiae and genua when respective solenidion is present (few species).*

Immatures: Apheredermous, with plicate integument, unideficient. Seta d *present to tritonymph on tibiae and genua when respective solenidion is present. Prodorsal and opisthonotal porose areas or sacculi usually present. Tarsi with or without bilobed subunguinal pulvillus.*

The Cymbaeremaeoidea includes a single family, the CYMBAEREMAEIDAE (Figs. 15.45A–C), and eight genera. Cymbaeremaeids are cosmopolitan but are found primarily in temperate regions. Fossils are known from the Jurassic, and a few species have been recovered from Tertiary Baltic amber (Labandeira, Phillips, and Norton 1997). Wunderle (1992a) found *Cymbaeremaeus cymba* (Nicolet) to be the dominant corticolous oribatid mite at mid-height in beech trees in Europe, and immatures were often found in the cast skins of scale insects. Wunderle successfully cultured this species on *Protococcus* algae. Species of the genus *Scapheremaeus* prefer epilithic and epiphytic habitats (Woas 2002) and are among the dominant corticolous mites in temperate and tropical regions of the world, including the Holarctic, Australia, Mexico, Central and South America, India, and the Galápagos Islands. Woas (2002) suggested the family name Scapheremaeidae for *Scapheremaeus*, but the name was not proposed according to nomenclatural rules.

Ametroproctus is widely distributed in western North America, Japan, the Russian Far East, and Switzerland, where species are found in dry, primarily subalpine or alpine habitats and the canopy of conifers (Behan-Pelletier 1987; Lindo and Stevenson 2007). The species are probably mycophages, and in some habitats they can be the numerically dominant oribatid mite. Subías (2004) recently proposed the family Ametroproctidae to include *Ametroproctus* and *Scapuleremaeus*, but we retain both in the CYMBAEREMAEIDAE.

▪ Superfamily Eremaeozetoidea (new status; Figs. 15.60D–I)

DIAGNOSIS: Adults: Integument often tuberculate, nodulate, or reticulate. Cerotegument well developed, covering all of body except lenticulus. Prodorsum with lamellae meeting medially. Tutorium and pedotecta I–II present; genal notch absent. Dorsophragmata and pleurophragmata present. Subcapitulum diarthric, eupathidium acm *separate from solenidion. Chelicerae chelate-dentate. Axillary saccule of subcapitulum absent. Prodorsum and notogaster separate or partially fused. Notogastral setation 10 pairs. Notogaster with octotaxic system present as 4 pairs of saccules (IDIOZETIDAE), or absent; lenticulus, pteromorphs, and posterior notogastral tectum present. Circumpedal carina present.*

Immatures: Apheredermous, plicate, unideficient. Coupled seta d *present to tritonymph or absent from immatures.*

The families EREMAEOZETIDAE and IDIOZETIDAE are included in this newly recognized taxon. Eremaeozetids (Figs. 15.60G–I) are known from the Oriental, Ethiopian, Neotropical, and Oceanian regions. The family is comparatively species rich, with 33 known species in three genera. It is an enigmatic assemblage that, for superficial reasons, has been included either in the Polypterozetoidea (Balogh 1972; Marshall, Reeves, and Norton 1987; Colloff and Halliday 1998), the Cepheoidea (Balogh and Balogh 1992), or, most recently, in the Unduloribatoidea (Subías 2004), a superfamily that is not recognized in this treatment. However, Schatz (2000) noted the close similarity between immatures of *Eremaeozetes irenae* Schatz and those of *Scapheremaeus* and postulated a close relationship between EREMAEOZETIDAE and CYMBAEREMAEIDAE.

Although most diverse in warmer climates, eremaeozetids extend north into southern Georgia, USA, and at high elevations in the Himalayas (Piffl 1972). They have recently been recorded from South Africa (Schatz 2001a), and fossils are known from Dominican amber (Norton and Poinar 1993). Eremaeozetids are known from arid habitats (Schatz 2001a, b) and from epiphytic moss, suspended soils, and organic debris in tropical forests. Two species are endemic to the Galápagos Islands (Schatz 2001b). Adults often have a thick layer of dark cerotegument and can be almost indistinguishable from humus particles.

The IDIOZETIDAE (Figs. 15.60D–F) includes only the type genus, *Idiozetes*, represented by two species from Java and West Malaysia forests. Their morphology is so unusual that *Idiozetes* was relegated to its own family (Aoki 1976) and superfamily (Balogh and Balogh 1992). Noting the morphological similarity between *Idiozetes* and *Eremaeozetes*, Woas (2002) included *Idiozetes* in the EREMAEOZETIDAE, but we retain both families, pending discovery of idiozetid immatures. Because EREMAEOZETIDAE is the older name, the rules of nomenclature require the superfamily name to be Eremaeozetoidea.

■ Superfamily Licneremaeoidea (Figs. 15.44J–L, 15.45G, H, 15.52A–D, 15.63A–F, 15.64A–K)

DIAGNOSIS: Adults: Prodorsum with or without lamellae or costulae; lamellar cusps and translamella present or absent; tutorium absent. Genal notch present or absent, according to family. Dorsophragmata and pleurophragmata present. Chelicerae chelate-dentate; palpal eupathidium acm *attached to, or separate from, solenidion; subcapitular mentum with or without complete tectum. Axillary saccule of the subcapitulum present or absent, according to family. Discidium, custodium, and postanal porose area absent. Octotaxic system present—as 1–4 pairs of porose areas or saccules—or absent. Integument often tuberculate, nodulate, or reticulate. Pedotectum I present; pedotectum II present or absent. Notogaster with or without lenticulus, posterior notogastral tectum present or absent, if present, overlapping medially or not; pteromorphs absent. Notogastral setation 10–15 pairs. Coxisternal setation either 3-1-3-3 or 3-1-2-2. Porose organs of legs variously developed as porose areas, saccules, or tracheae. Genital setation 4–6 pairs. Genital plate proportionally large, subtriangular, with posterior margin little removed from anal plate and anterior margin aligned with posterior border of epimere IV. Anal setation 1–2 pairs. Solenidion* σ1 *of tibia I inserted on distinct, cylindrical tubercle.*

Immatures: Apheredermous, plicate. Larva unideficient; nymphs unideficient, bideficient, or trideficient. ***Humeral organ*** *(small, usually papilliform porose organ located anterolateral to seta* c2*) absent. Coupled seta* d *present to the tritonymph or absent from immatures.*

The Licneremaeoidea comprises nine families, including the MICREREMIDAE (Figs. 15.44J–L) and LAMELLAREIDAE (Figs. 15.63D–F), which were placed here primarily because at least one species in each family has a partial octotaxic system in the form of notogastral porose areas (Kok 1968; Norton and Alberti 1997). The EREMELLIDAE (Figs. 15.45G, H) is included in Licneremaeoidea following the rationale of Woas (2002). Woas subsumed the EREMELLIDAE under MICREREMIDAE, but both families are retained here. However, we agree with Subías (2004) that FENICHELIIDAE, which usually has been included in Oripodoidea, should be subsumed under MICREREMIDAE.

The relationships of CHARASSOBATIDAE (Figs. 15.52A–D) are problematic; they have been relegated to their own monobasic superfamily (Marshall, Reeves, and Norton 1987; Balogh and Balogh 1992; Subías 2004), and Woas (2002) argued for a closer relationship to the TEGEOCRANELLIDAE (Ameronothroidea). However, based on Grandjean's (1958b) arguments, we include the family in the Licneremaeoidea because he considered *Charassobates cavernosus* Grandjean to be close to *Licneremaeus*, with which it shares a large pedotectum I, absence of pedotectum II, unusual properties of the bothridial seta (it corrodes in lactic acid), anal plates smaller than genital plates, similar preanal organ, and genital setal ontogeny of 1-2-4-5. Also included in the Licneremaeoidea is the genus *Glanderemaeus*, whose familial placement is unclear. While the octotaxic system of females includes the typical four pairs of minute but normal porose areas, the most posterior pair in males instead takes the form of a large saccule (Norton and Alberti 1997).

Licneremaeoid mites are mycophages that occur in periodically or continuously xeric habitats such as desert, shortgrass prairie, and tree canopies. Most are confined to the tropics or warm temperate regions, but both genera of PASSALOZETIDAE (Figs. 15.64A–C), *Passalozetes* and *Bipassalozetes*, are diverse in Mongolia (Bayartogtokh and Aoki 1997). Shtanchaeva (1986) revised the Central Asian passalozetid fauna.

Species of *Micreremus*, one of five genera in MICREREMIDAE, are small oribatids that occur in arboreal habitats in Europe (Balogh and Balogh 1998). *M. brevipes* (Michael), one of the most common arboreal mites in Europe (Grandjean 1954e), was found living in holes and crevices in beech canopy in Germany by Wunderle (1992a). She found this species to have a one year life cycle, with protonymphs being the main overwintering stage. Females produced eggs only during the summer. Immatures in culture were often found in the cast skins of older instars. Wunderle also found *M. brevipes* sheltered in old coccid scales, presumably for protection from desiccation and predators while molting and ovipositing.

The arboreal family ADHAESOZETIDAE (Figs. 15.64H–K) comprises two genera known only from Australia, New Zealand, Tonga, and Japan. *Adhaesozetes polyphyllos* Walter and Behan-Pelletier is found in arboreal habitats in relictual rain forests of eastern Australia, where it is abundant on leaves with smooth surfaces or with hairs closely appressed to the leaf surface. Both immatures and adults have expanded tarsal pulvilli that possibly help maintain their grip on smooth leaf surfaces. Immatures often molt in insect punctures, wounds on the surface of leaves, and within exuviae. As in *Micreremus*, eggs are often laid in exuviae (Walter and Behan-Pelletier 1993). Species of *Phylleremus* share the same general habitat as *A. polyphyllos* but are usually found on plants with densely tomentous hairs (Behan-Pelletier and Walter 2007).

The monobasic family DENDROEREMAEIDAE (not included in the key to families) comprises two species from arboreal habitats in western North America. Adults share character states with Oripodoidea, but immatures confirm relationships with other Licneremaeoidea (Behan-Pelletier, Eamer, and Clayton 2005).

The arboreal family LICNEREMAEIDAE (Figs. 15.63A–C) includes *Licneremaeus* and the monobasic genus *Huilicheremaeus*, which was tentatively placed in this family by Fernandez, Marcangeli, and Eguaras (1997). These are small mites (150–255 μm) with flabelliform bothridial setae and one or two pairs of notogastral porose areas. Most species are known from the dry tropics, but they do occur in temperate regions (Marshall, Reeves, and Norton 1987).

Species of the three genera of LAMELLAREIDAE have been collected from moist microhabitats, including soil, moss, and plant litter in North America and South Africa (Coetzee 1987; Kok 1968; Norton and Alberti 1997).

The SCUTOVERTICIDAE (Figs. 15.64D–G) includes 11 genera and approximately 64 species found in xeric and arboreal habitats. Most genera lack evidence of the octotaxic system, but *Exochocepheus* and *Scutovertex* have one and three pairs of sacculi, respectively. The world fauna of SCUTOVERTICIDAE was reviewed recently by Shtanchaeva and Netuzhilin (2003), who also provided a key to the genera. *Scutovertex minutus* (C. L. Koch) has been experimentally infected with the eggs of the tapeworm *Bertiella studeri* (Stunkard 1940) and is an intermediate host for other anoplocephalid species (Denegri 1993).

■ Superfamily Phenopelopoidea (Figs. 15.61A–G)

DIAGNOSIS: Adults: Cerotegument well developed, with thick, blocky structure on notogaster and ventral plate; strongly birefringent in polarized light. Prodorsum with lamellae, translamella, tutorium, and genal notch. Discidium and custodium present. Pedotecta I–II present. Dorsophragmata and pleurophragmata present. Chelicerae chelate-dentate or pelopsiform. Palpal eupathidium acm *attached to solenidion, at least distally, but their insertions well separated. Axillary saccule of subcapitulum present. Notogaster with lenticulus, well or poorly circumscribed; without tectum posteriorly. Pteromorph well developed, with or without hinge. Octotaxic system represented by small porose areas, usually adjacent to notogastral setae; sometimes absent, or with fewer than 4 pairs. Notogastral setation 10 pairs; pair* h1 *(and often others) thick, often somewhat clavate. Tibia IV with dorsal carina and/or grooved or flattened dorso-antiaxial surface, and with thick spinelike seta* l''. Seta d *present on tibia IV when respective solenidion is present.*

Immatures: Apheredermous, with plicate integument and marginally flattened, carinate opisthonotum; unideficient. Humeral organ absent. Seta d *present on tibiae IV when respective solenidion is present.*

The Phenopelopoidea includes the cosmopolitan families PHENOPELOPIDAE (Figs. 15.61D–G) and UNDULORIBATIDAE (Figs. 15.61A–C). The PHENOPELOPIDAE (Pelopidae of most older literature) is known primarily from forest litter in temperate regions of the Northern and Southern hemispheres. Some species of the four phenopelopid genera occur at high elevations in the Andes and in montane forest in the Neotropics; others have been collected in Central Africa, South Africa, and Australia. The three genera in the subfamily Phenopelopinae have pelopsiform mouthparts, suggesting that they are mycophages (e.g., Luxton 1972; Behan-Pelletier and Hill 1983). However, Riha (1951) observed members of an unidentified species of *Eupelops* using their long chelicerae to cut a hole in the underside of dead leaves and reach into the parenchyma, where they excavated the central area and left only the two layers of leaf cuticle. While feeding in this way, they probably also ingested fungal material nonselectively. There is no intraspecific genetic variation in the D3 region and its flanking regions of 28S rRNA in the closely related sexual species *Eupelops hirtus* (Berlese) and *E. torulosus* (C. L. Koch) (Maraun et al. 2003, 2004).

Species of UNDULORIBATIDAE, a small group comprising two genera, are known only from northern latitudes (Marshall, Reeves, and Norton 1987) or high-elevation habitats in the Himalayas (e.g., Nübel-Reidelbach and Woas 1992). Little is known of their biology.

The phenopelopid *E. punctulatus* (Sellnick) and the unduloribatid *Unduloribates parvus* (Sellnick) were recorded from Baltic amber (Krivolutsky et al. 1990).

■ Superfamily Achipterioidea (Figs. 15.59A, B, 15.61H, I, 15.62A–C)

DIAGNOSIS: Adults: Prodorsum with lamellae, adjacent medially or fused medially or separated; lamellar cusps, tutorium, translamella present or absent. Genal notch present or absent. Dorsophragmata and pleurophragmata present. Chelicerae chelate-dentate; palpal eupathidium acm *attached to solenidion; subcapitular mentum with or without complete tectum. Axillary saccule of subcapitulum usually absent. Discidium and custodium present, postanal porose area present or absent. Pedotectum I large, extending to base of bothridium; pedotectum II present. Notogaster with or without knifelike humeral process; with tectum posteriorly, overlapping medially or not; pteromorphs present, with or without hinge; lenticulus present. Notogastral setation 10 pairs. Octotaxic system developed as porose areas or saccules, sometimes minute. Genital setation 6 pairs. Bothridium with or without internal, spiral thickenings. Genu IV often concave dorsally, subequal in length or longer than tibia IV; tibia IV with or without solenidion, with or without carina on dorsal surface.*

Immatures: Plicate, apheredermous or eupheredermous, integument at least partially tuberculate or not; unideficient. Humeral organ absent. Integument surrounding opening of opistho-

notal gland porose or not. Seta d *present to tritonymph on tibiae and genua I–IV, or only IV, when respective solenidion exists. Protonymphal legs IV setation 0-0-0-0-7, 0-1-0-0-7, or 0-0-2-0-7. Retrotecta on tibiae and tarsi present or absent.*

The Achipterioidea includes the two widely distributed families ACHIPTERIIDAE (Figs. 15.61H, I), an assemblage of 9 genera, and TEGORIBATIDAE (Figs. 15.62A–C), with 10 genera. Like the Licneremaeoidea and Phenopelopoidea, achipterioids have immatures with a plicate cuticle, and these three superfamilies are informally considered together as *higher plicates* in contrast to the so-called lower plicates, which include taxa with plicate immatures but without the octotaxic system. Grandjean (1954b) envisioned all of these taxa as forming a natural (monophyletic) group, but more likely the plicate condition is a plesiomorphy that is carried into the poronotic clade. A third family, the EPACTOZETIDAE (Figs. 15.59A, B), is tentatively included in this superfamily. Its relationships are unclear (Nübel-Reidelbach and Woas 1992; Woas 2002; Subías 2004), but epactozetid immatures are plicate (Norton and Alberti 1997) and lack the macrosclerites characteristic of immature Galumnoidea, the superfamily in which the family usually is placed (Balogh and Balogh 1992).

Species of ACHIPTERIIDAE and TEGORIBATIDAE are diverse in temperate soil and litter, and in mosses and liverworts in both the Northern and Southern hemispheres. *Achipteria coleoptrata* (L.) and *Parachipteria willmanni* Van der Hammen, are particularly common in forest soils in Europe; the immatures of these species were described by Seniczak (1978). Generation time for *A. coleoptrata* is about one year (Luxton 1981), similar to that of *A. holomonensis* Cancela da Fonseca and Stamou, which is an important component of forest litter in Greece (Stamou 1986). Achipteriid species in the tropics are mainly found at higher elevations, for example, in cloud forest litter. The tegoribatid genera *Lepidozetes* and *Scutozetes* are diverse at high elevations or high latitudes. For example, *L. singularis* Berlese and *S. lanceolatus* Hammer are widely distributed throughout the arctic and subarctic (Behan-Pelletier 1999b) and are found in Mongolia (Bayartogtokh and Aoki 1999). In contrast, the tegoribatid *Hypozetes laysanensis* Aoki has been collected in sand washings from stabilized dunes among grasses and mesquite seedlings in Hawaii (Behan-Pelletier 2001). Species of EPACTOZETIDAE are most common in rain-forest litter in the Neotropics (Woas 2002). Fossils of ACHIPTERIIDAE are known from the Jurassic and from Baltic amber (Labandeira, Phillips, and Norton 1997).

Achipterioid mites whose feeding habits have been studied are saprophages and mycophages that apparently feed opportunistically on available resources ranging from fungi to algae to decaying plant material. *Lepidozetes singularis* feeds on lichens at high elevations in Europe (Smrž and Kocourkova 1999), while the achipteriid *Achipteria coleoptrata* was reared successfully under laboratory conditions on the green bark alga *Desmococcus vulgaris*, on grass and herb litter, and on fungi growing on the litter (Hubert, Zilova, and Pekar 2001).

■ Superfamily Oribatelloidea (Figs. 15.69F–H)

DIAGNOSIS: Adults: Prodorsum with large, broad lamellae, fused at base of lamellar cusps; tutorium usually rectangular and dentate distally. Genal notch present. Dorsophragmata and pleurophragmata present. Chelicerae chelate-dentate; palpal eupathidium acm *attached to solenidion; subcapitular mentum without tectum. Axillary saccule of subcapitulum present. Discidium and custodium present; postanal porose area present or absent. Pedotectum I large, extending to base of bothridium; pedotectum II present. Notogastral setation 10–15 pairs; pteromorphs present, without hinge. Octotaxic system developed as 4 pairs of porose areas. Genital setation 6 pairs. Bothridium without internal, ringlike thickenings, cup shaped, or with medial dens. Genu IV shorter than tibia IV; tibia IV with solenidion.*

Immatures: Unideficient; nymphs apopheredermous, hysterosoma with lateral carina, without plicae, microsclerites or macrosclerites. Humeral organ present. Seta d *lost on tibiae and genua when respective solenidion is present.*

The single oribatelloid family ORIBATELLIDAE (Figs. 15.69F–H) is cosmopolitan in distribution but is particularly diverse in warm temperate regions of the Holarctic (e.g., Cistjakov 1979). In the past, the Oribatelloidea has been the repository of a mixture of unrelated taxa. Even recently, Pérez-Íñigo (1993a) included the ACHIPTERIIDAE and the TEGORIBATIDAE (Achipterioidea) in the ORIBATELLIDAE because they share large lamellae in the adult. However, as noted by Grandjean (1954d), the apheredermous plicate immatures of achipterioid mites contrast strongly with those of ORIBATELLIDAE. Several of the 14 genera included in ORIBATELLIDAE by Balogh and Balogh (1992) are misplaced. For example, *Latilamellobates*, known from the temperate Holarctic, belongs in the CERATOZETIDAE (Behan-Pelletier 2000). *Plakoribates*, known from the Neotropics, Java, Sri Lanka, and India, has affinities with known phenopelopoid taxa (Woas 2002), and *Scutozetes* belongs in the TEGORIBATIDAE (Marshall, Reeves, and Norton 1987). The genus *Novoribatella* may be related to the achipteriid genus *Tectoribates* (Norton and Alberti 1997), and the placement of *Adoribatella* in the ORIBATELLIDAE is questionable (Marshall, Reeves, and Norton 1987).

The genus *Oribatella* is comparatively diverse, with over 100 species distributed throughout the Nearctic, Palearctic, and Neotropical regions. It has been particularly well studied in Europe (e.g., Bernini and Avanzati 1983) but needs revision in other regions. Species of *Oribatella* are primary decomposers (Schneider et al. 2004) and may feed on fungal hyphae (Behan-Pelletier and Hill 1983) or algae (Wunderle 1992a).

- Superfamily Oripodoidea (Figs. 15.58I, 15.62D–H, 15.63G, H, 15.65A–I, 15.66A–H, 15.67A–I, 15.68A–K, 15.69A–E)

DIAGNOSIS: Adults: Prodorsum with lamellae well developed, reduced to a narrow carina, or absent; tutorium and translamella present or absent. Bothridium often covered by anterior margin of notogaster. Genal notch absent. Prodorsum and notogaster separate or fused. Dorsophragmata and pleurophragmata present. Chelicerae chelate-dentate or pelopsiform; palpal eupathidium acm *attached to solenidion; subcapitular mentum without tectum. Axillary saccule of subcapitulum absent. Discidium and custodium present or absent; postanal porose area absent. Pedotectum I small, usually not covering acetabulum I, pedotectum II small or absent. Circumpedal carina present. Notogaster with or without lenticulus, without tectum posteriorly; pteromorphs present or absent, if present, with or without hinge. Notogastral setation 9–15 pairs. Octotaxic system diverse, often with more or fewer than 4 pairs, rarely absent. Genital setation 1–6 pairs. Genital plates well removed from anal plates and virtually bisecting epimere IV. Usually with porose organs on tibiae and tarsi. Genu IV shorter than tibia IV; tibia IV with solenidion.*

Immatures: Apheredermous, unideficient; some or all opisthonotal setae usually with excentrosclerites. Humeral organ absent. Seta d *absent from tibiae and genua when respective solenidion is present.*

This large superfamily, also known in the literature as the Excentrosclerosae, includes 18 families. Woas (1992) questioned the inclusion of a number of these families because their immatures are unknown. These include the CRASSORIBATULIDAE (Figs. 15.69A, B) with four genera; the CALOPPIIDAE (Figs. 15.68E–G) with three genera; and the monogeneric CAMPBELLOBATIDAE (Figs. 15.66D–F), NASOBATIDAE (Figs. 15.67A–C), NESOZETIDAE (Figs. 15.63G, H), STELECHOBATIDAE (Figs. 15.67F–I), and TUBULOZETIDAE (Figs. 15.65A, B). Similarly, immatures are unknown for the three genera of BIROBATIDAE, which we consider a junior synonym of ORIPODIDAE (Figs. 15.65E–I), following Subías (2004). Immatures also remain to be identified for the two known species of *Areozetes* (AREOZETIDAE), which may prove to be members of the family MOCHLOZETIDAE (Figs. 15.67D, E). Lee (1991) divided the Oripodoidea into nine groups on the basis of the octotaxic system, shape of the preanal organ, and the chaetotaxy of femur II, a system followed by Colloff and Halliday (1998). The adult diagnosis for the superfamily is in need of revision (Badejo, Woas, and Beck 2002), but the defining autapomorphy for the superfamily (excentrosclerotic immatures) is unknown for most included species and genera.

A number of taxonomic problems at the family group level in the Oripodoidea remain unresolved. For example, *Reticuloppia* (in ORIBATULIDAE (Figs. 15.68H–K) according to Balogh and Balogh (1992) and Subías (2004)) was included in CRASSORIBATULIDAE by Colloff and Halliday (1998) but is related to *Chaunoproctus* according to Woas (2002). Therefore, CRASSORIBATULIDAE may be a junior synonym of CALOPPIIDAE. Most members of these latter two families (other than *Chaunoproctus*, which is circumtropical) are known from India, tropical Asia, Australia, and New Zealand. Their immatures are unknown, and little is known of their ecology. A more important problem is the use and context of the family names HAPLOZETIDAE, XYLOBATIDAE, and PROTORIBATIDAE. The name HAPLOZETIDAE (Figs. 15.65C, D) is used here in the original broad sense of Grandjean (1936d), which includes the Xylobatidae of Balogh and Balogh (1984). The latter authors distinguished these families only on the nature of the octotaxic system (saccules and porose areas, respectively), but this highly plastic character exhibits many independent transitions in poronotic mites (Norton and Alberti 1997). Balogh and Balogh (1984) also proposed the family Protoribatidae, which is not recognized here. This taxon is technically synonymous with Xylobatidae in that the respective type genera are synonyms (Weigmann, Miko, and Nanelli 1993; Weigmann and Miko 2002; Weigmann 2006). The other genera included in Protoribatidae by Balogh and Balogh (1984), including some species of *Protoribates* itself, are in need of reassignment, but this is beyond the scope of the present treatment. Although each case needs detailed attention, most would probably fit into a redefined SCHELORIBATIDAE (Figs. 15.68A–D).

Multiplication of the octotaxic system is common in genera of MOCHLOZETIDAE and HAPLOZETIDAE, with up to 50 pairs being noted in the mochlozetid genera *Rykella* and *Drymobatoides* (Norton et al. 1997).

The six genera of PARAKALUMMIDAE (Fig. 15.58I) are represented throughout temperate and tropical regions, usually in forest litter. *Neoribates auranticus* (Oudemans) is common in decaying wood and bracket fungi and is found from temperate to arctic habitats in North America (Behan-Pelletier 1999b). This family is often confused with the GALUMNIDAE, with which it shares large, movable, auriculate pteromorphs. However, galumnids have a large subcapitular (mental) tectum as adults and large opisthonotal sclerites as immatures. The 12 genera of ZETOMOTRICHIDAE (Figs. 15.62F–H) occur in temperate to dry areas of the world, including Chile (Covarrubias 1969) and Australia (Colloff and Halliday 1998). They were placed in their own superfamily, Zetomotrichoidea, by Subías (2004). Based on their morphology, adults of *Zetomotrichus* species are probably saltatory (Grandjean 1955b). They have large, spinelike setae on legs IV, similar to those in zetorchestid genera. Adults do not have a defined octotaxic system per se, but they do have supernumerary macropores on the prodorsum that possibly represent dispersed porose areas (Norton et al. 1997).

The 12 genera of MOCHLOZETIDAE, the monogeneric NEOTRICHOZETIDAE (Figs. 15.62D, E), the monobasic family DRYMOBATIDAE (Figs. 15.66G, H), and most of the 24 genera of ORIPODIDAE are unusual in that they inhabit living plants, most often trees. The occasional collection of specimens in soil and litter probably results from accidental dislodgement from vegetation or as a result of dispersal behavior (Norton 1983). The two species of NEOTRICHOZETIDAE are found in arboreal habitats or associated with grasses in Australia, New Zealand, and the southwestern part of South America (Behan-Pelletier and Walter 2000). They seem to prefer humid environments and in South America have been found in a *Nothofagus* forest (Woas 2002). Immatures have not been described. The MOCHLOZETIDAE is most diverse in temperate to tropical regions of the Americas, although species of *Podoribates* are widespread in the Holarctic, where they are associated with grasses and wet pastures. All mochlozetids, drymobatids, and certain other arboreal oripodoids have a ventral subterminal tooth on each lateral claw of the leg tarsi that is thought to enhance grasping efficiency on plants (Norton 1983). Some oripodids are saxicolous, living in lichens on rocks (Travé 1963).

The families HAPLOZETIDAE, ORIBATULIDAE, and SCHELORIBATIDAE are the most species-rich families of Oripodoidea, with each having over 300 species in more than 40 genera. All three families are cosmopolitan, but they are most diverse in the tropics. The latter two families were revised by Grandjean (1958d) and by Wunderle, Beck, and Woas (1990). The parthenogenetic haplozetid mite *Rostrozetes ovulum* (Berlese) is a frequent inhabitant of bogs in eastern North America (Norton and Palmer 1991), bird's nest fern in southwestern Japan (Karasawa and Hijii 2006), and various edaphic and plant habitats in southeast Brazil (Oliveira, Norton, and de Moraes 2005). It is also the dominant oribatid species (as *R. foveolatus*) in Amazonian floodplains (Franklin et al. 2001). Schatz (1998) postulated that *Polyxylobates diversiporosus* Hammer dispersed from Tonga to the Galápagos Islands on floating mangrove seedlings. The oribatulid *Oribatula tibialis* (Nicolet) is a virtually cosmopolitan species found in many environments, even in association with some insectivorous plants (Antor and Garcia 1995). The scheloribatid *Dometorina plantivaga* (Berlese) lives exclusively among epiphytic lichens and algae, where densities can reach 250 individuals per 25 cm^2 (Travé 1963). It creates cavities in lichen thallus and uses feces as building material to close openings (Grandjean 1951d). The oribatulid genus *Phauloppia* is arboreal or found in lichen patches throughout the temperate Holarctic (e.g., Prinzing and Wirtz 1997). The two species of *Phauloppia* that inhabit lichens on boulder tops in Norway show remarkable tolerance for temperature and moisture fluctuations, surviving losses of up to 90% of body water content (Sjursen and Sømme 2000). Species of *Scheloribates* (SCHELORIBATIDAE) are among the most common and abundant mites in soil and litter, even in anthropogenic habitats. One of these, *S. laevigatus* (C. Koch), is common in North American lawns and may have been introduced from Europe. Larviparity is known in some European populations, and females die soon after egg laying, based on observed sex ratios (Hubert 2000).

Oripodoid mites are primarily mycophagous, feeding on algae and fungi. *Oribatula tibialis* is a selective feeder on ectomycorrhizal fungi (Schneider, Renker, and Maraun 2005). However, some species of *Haplozetes* and *Peloribates* feed on nematodes (Walter 1988a), and arthropod fragments were noted in the guts of *Dometorina plantivaga* (Berlese) and *Liebstadia humerata* Sellnick (Wunderle 1992a). For those taxa living on plants, the host itself is not usually a direct source of nutrition. Most of these mites feed on fungal spores and hyphae, although *Dynatozetes obesus* Grandjean and *Terrazetes mauritius* Grandjean (MOCHLOZETIDAE) are known to ingest plant material (Grandjean 1960d). The saprophagous and mycophagous *S. laevigatus* has been shown to have a significant impact on decomposition of grass litter through feeding on fungi (Hubert 2001). The Ecuadoran species *Tubulozetes rostratus* P. Balogh (TUBULOZETIDAE) is the only species in this superfamily with pelopsiform mouthparts (Balogh 1989), but its food is unknown.

Sexual dimorphism, especially hypertrophied or novel porose organs of the male octotaxic system, is expressed in a number of oripodoid genera, including the scheloribatid genus *Parapirnodus* and the mochlozetid genera *Mochlozetes* and *Mochloribatula*. Norton and Alberti (1997) postulated that these organs play a role in intraspecific communication, possibly producing semiochemicals that have a direct or indirect role in reproductive biology. Oliveira et al. (2007) observed courtship behavior in a species of *Mochloribatula* that is consistent with this idea. Such secretions, perhaps sex-attractant or general aggregation pheromones, either may be produced in larger quantities by the dimorphic males or may be absent from females altogether. Adults of *Sellnickia caudata* (Michael) (SELLNICKIIDAE) (Figs. 15.69C–E), a species known from New Zealand and Australia, have a dimorphic rostral lobe with a porose leading edge (Grandjean 1958e). The male lobe is strongly curved distally and has the appropriate size and shape to fit over the pygidial tubercle of the female. The mating biology of this species is as yet unknown, but should the male be found to "nuzzle" the tubercle of the female, its rostral porose area could well be sensory (Norton and Alberti 1997). Sexual dimorphism is also expressed in the monobasic SYMBIORIBATIDAE (Figs. 15.66A–C), with male *Symbioribates papuensis* Aoki and *S. aokii* Karasawa and Behan-Pelletier having both hypertrophied porose organs and unusually enlarged rostral setae. All stages of *S. papuensis* have been collected only from cryptogamic plants, where they live on the elytra of arboreal curculionid beetles (Aoki 1966a). *S. aokii*, however, is known only from bark.

A number of oripodoids may be of potential concern to humans in that they feed directly on economically important plants. For example, *Scheloribates* sp. damages maize roots in Hawaii (Jacot 1934), *S. pallidulus* (C. Koch) feeds on greenhouse cucumbers (Karg 1984), *S. curvialatus* Hammer and *Rostrozetes ovulum* attack the apical portion of roots of pineapple (Sanyal and Das 1989), and *S. oryzae* Wu et al. damages rice shoots in China (Lan, Xin, and Aoki 1986). *Oribatula sakamori* Aoki attacks prince melon in greenhouses (Suzuki 1971), *Xylobates pembertoni* Jacot makes minute holes in cane roots (Jacot 1934), and *Zygoribatula agaveae* Aoki and Wang eats the leaf tissue of the sisal plant *Agave sisalana* in China, breaking the fibers and rendering them economically worthless. The corticolous oribatulid species *Liebstadia humerata* and *Z. laubieri meridionalis* Travé, along with the scheloribatid *S. latipes* (C. L. Koch), are potential vectors of chestnut canker *Cryphonectria parasitica* (Murr.) Barr, including hypovirulent strains. RNA evidence shows unaffected passage of the blight through the digestive system of these species (Nanelli, Turchetti, and Maresi 1998).

Several oripodoid species are important vectors of tapeworms. *S. laevigatus* can transmit eight different cestode species (Allred 1954). Fully formed cysticercoids developed in *S. latipes* experimentally infected with eggs that were in turn able to infect sheep (Polec and Moskwa 1994). *L. similis* (Michael) is an intermediate host for the rabbit tapeworm, and *O. minuta* (Banks) has been found to carry *Monoecocestus sigmodontis* and the sheep tapeworm, *Moniezia expansa* (Krantz 1978).

There are relatively few reports of beneficial activities by oripodoids. *S. azumaensis* Yamamoto et al. prevents radish root rot through feeding on the pest fungi *Rhizoctonia* spp. (Enami and Nakamura 1996). Lan, Xin, and Aoki (1986) report that the scheloribatid *Dometorina predatoria* Wu, Xin, and Aoki eats gall mites on orange trees. There is strong evidence that oribatid mites, in particular members of the SCHELORIBATIDAE and HAPLOZETIDAE, are a major dietary source for the alkaloids present in the poison frog *Dendrobates pumilio* (Anura: Dendrobatidae) (Takada et al. 2005, Saporito et al. 2007) (see also chapter 3).

Species of *Metaleius* and *Paraleius* (ORIBATULIDAE), along with *Euscheloribates* and *Scheloribates* (SCHELORIBATIDAE) and the haplozetid genus *Protoribates* (= *Xylobates*), are phoretic on passalid, scarabaeid, cerambycid, and scolytid beetles. A species of *Metaleius* is recorded as phoretic on the dipteran *Pantophthalmus heydeni* (Norton 1980), and *Paraleius leontonycha* (Berlese) was found as a commensal in the brood galleries of a bark beetle (Vitzthum 1926).

The Oripodoidea is well represented in the fossil record (Labandiera, Phillips, and Norton 1997). *Neoribates* (PARAKALUMMIDAE) is known from Baltic amber, as are *Eporibatula*, *Liebstadia* (ORIBATULIDAE), *Oripoda* (ORIPODIDAE), and *Scheloribates* (SCHELORIBATIDAE). *Oripoda* is also known from Dominican amber, and *Benoibates* (ORIPODIDAE), *Parapirnodus*, and *Scheloribates* (SCHELORIBATIDAE) from Chiapas amber. *Mochlozetes* and *Mochloribatula* (MOCHLOZETIDAE) are represented in Dominican and Chiapas amber, respectively. Oribatulids are also represented in Manitoban amber of the Cretaceous and Sakalin amber of the Tertiary (Labandiera, Phillips, and Norton 1997).

▪ Superfamily Ceratozetoidea (Figs. 15.59C–E, 15.70A–G, 15.71A–F, 15.72A–H)

DIAGNOSIS: Adults: Prodorsum with lamellae, adjacent medially or separated; tutorium, lamellar cusps, translamella present or absent. Genal notch present. Dorsophragmata and pleurophragmata present. Chelicerae chelate-dentate; palpal eupathidium acm *attached to solenidion; subcapitular mentum with or without complete tectum. Axillary saccule of subcapitulum generally present. Discidium and custodium present, postanal porose area generally present. Pedotectum I large, extending to base of bothridium. Pedotectum II present. Notogaster with or without lenticulus, posterior tectum present or absent, if present, overlapping medially or not; pteromorphs usually present, with or without hinge. Notogastral setation 10–15 pairs. Octotaxic system developed as 1–7 pairs (usually 4) of porose areas, saccules, or tubules. Genital setation 4–6 pairs. Bothridium without internal, spiral thickenings, cup shaped, or with medial and lateral scales. Genu IV shorter than tibia IV; tibia IV with solenidion.*

Immatures: Opisthonotum with macrosclerites, apheredermous, integument smooth or partially tuberculate; unideficient. Humeral organ generally present. Opisthonotal setae without excentric porose area at base (exception HUMEROBATIDAE). Integument surrounding opening of opisthonotal gland porose. Single circumgenital sclerite present (EUZETIDAE) or absent. Seta d *absent from tibiae and genua when respective solenidion is present.*

The superfamily Ceratozetoidea includes nine families of detritus feeders, mycophages, and necrophages found in terrestrial soil and litter, arboreal, and semiaquatic habitats. Although represented in the tropics, the superfamily is most diverse in temperate to boreal regions (Pavlitshenko 1994). These mites have been found to altitudes of 4,500 m in the Himalayas (Wang et al. 2001) and are diverse in the arctic (Behan-Pelletier 1999b). The superfamily was revised for the Ukraine (Pavlitshenko 1994), and Shaldybina (1977) described immatures of representatives of most families.

The CERATOZETIDAE (Figs. 15.72E–H) is a large, cosmopolitan family of approximately 310 species in 50 genera that are common in forest, grassland, and tundra habitats from the tropics to warm-temperate to high arctic and subantarctic regions (Behan-Pelletier 1999b). Although found mainly in litter and soil, ceratozetid mites may occur in large numbers in grass (Michael 1884a), canopy habitats (Behan-Pelletier 2000), mosses and lichens (Materna 2000), or wet, mesotrophic bogs (Weigmann 1991). Certain species, such as *Trichoribates incisellus* (Kramer), are early colonizers of litter (Serralheiro and Madeira 1991), and *Ceratozetes gracilis* (Michael) is common in deciduous litter throughout the temperate Holarctic (e.g., Mitchell 1978). The feeding habits of ceratozetids are as diverse as their habitats. Based on gut content analysis and feeding experiments, species of *Ceratozetes* and *Fuscozetes* are considered saprophages and mycophages (Luxton 1972). *C. gracilis* feeds primarily on pigmented fungal hyphae (Mitchell and Parkinson 1976), *Fuscozetes* spp. readily eat dead Collembola and other arthropods (Wallwork 1958; Sakata and Norton, unpublished), and *Ceratozetes* sp. feeds on nematodes (Walter 1988b). *Dentizetes ledensis* Behan-Pelletier uses phytopathogenic fungi as food (Behan-Pelletier 2000), and *Oromurcia sudetica* Willmann, the dominant oribatid at timberline in central Europe, is a mycophage (Schatz 1979, 1985). *Diapterobates humeralis* (Hermann) feeds on the woolly filaments that envelop the ovisacs of the hemlock woolly adelgid and dislodges their eggs in the process (McClure 1995). Meier, Scherrer, and Honegger (2002) found that fecal pellets of *Trichoribates trimaculatus* (C. L. Koch) that have fed on thalli of the bright yellow wall lichen, *Xanthoria parietina*, contain both viable ascospores and the photobiont cells (*Trebouxia arboricola*). These authors speculated that passage of viable elements in feces might provide a common and successful mode of vegetative short- and long-distance dispersal of lichen-forming ascomycetes and their photobionts, as suggested by Stubbs (1995) for fungal soridia.

The widespread ceratozetid genera *Melanozetes* and *Sphaerozetes* have been identified from Baltic amber and *Diapterobates* from Australian amber. Both *Diapterobates* and *Svalbardia* are known from the Tertiary (Labandeira, Phillips, and Norton 1997).

The families CHAMOBATIDAE (Fig. 15.70G) and MAUDHEIMIIDAE (Figs. 15.72A–C) each include two genera, whereas the EUZETIDAE (Fig. 15.72D) is monogeneric. *Chamobates cuspidatus* (Michael) is one of the most commonly encountered oribatid mites in central European pine forests (Sylwestrowicz and Kostecka 1991). *Maudheimia* and *Multimaudheimia* are the only genera of Acari known from continental (= East) Antarctica that are endemic. A Gondwanan origin is proposed for these genera based on their antiquity (which was inferred from endemism), their widespread distribution throughout continental Antarctica, and their limited dispersal capacity (Marshall and Coetzee 2000). Species of *Ramsayellus* are known only from New Zealand and Southern Australia (Colloff and Halliday 1998). The RAMSAYELLIDAE may be a junior synonym of MYCOBATIDAE (see below), inasmuch as the type species, *Ramsayellus grandis* (Hammer) has 4 pairs of small porose areas (V. M. Behan-Pelletier, pers. obs.). Their presence goes counter to the key diagnostic character separating this family from MYCOBATIDAE in Luxton (1985).

Species of *Euzetes*, including *E. seminulum* (Mueller), live on the soil surface in woods with heavy leaf fall (Pauly 1956). Adults of *E. globulus* (Nicolet) have been found in the hair of small mammals in numbers that are sometimes six times greater than in surrounding natural habitats, suggesting active dispersal on these mammals (Miko and Stanko 1991). Chamobatids occurring in the former USSR were revised by Shaldybina (1989).

Species of MYCOBATIDAE (Figs. 15.71D–F), a family comprising 22 genera, are found in habitats similar to those of CERATOZETIDAE and, like ceratozetids, are more diverse in temperate to arctic regions than in the tropics (Behan-Pelletier 1999b). Behan-Pelletier (1999a) considered character states of Ceresellidae (Pavlichenko 1991) to fall within the diagnosis of MYCOBATIDAE. Many species of *Mycobates* are associated with lichens (Materna 2000), and one species, *M. parmeliae* (Michael), is named for the lichen on which it often feeds (Michael 1884a). Immatures of *Minunthozetes semirufus* (Koch) are presumed to burrow in the stems of grasses (Evans, Sheals, and MacFarlane 1961). Shaldybina (1987) has revised the mycobatid fauna of the former USSR. A species of *Punctoribates* has been identified from Baltic amber (Labandeira, Phillips, and Norton 1997).

Sexually dimorphic notogastral porose areas are found in species of the mycobatid genus *Zachvatkinibates* that inhabit the marine littoral. Some species in the semiaquatic family ZETOMIMIDAE (Figs. 15.70C–F) also are sexually dimorphic, but the type of dimorphism varies from porose anal and ventral plates in the male (species of *Zetomimus* and *Heterozetes*), to a modified rostrum in *Naiazetes* (Behan-Pelletier and Eamer 2003). All male zetomimids have a large spermatopositor relative to those in other Ceratozetoidea, suggesting modified sexual behavior associated with their semiaquatic habitat. The placement of zetomimids in Ceratozetoidea was questioned by Woas (2002) because immatures lack macrosclerites, a key apomorphy of the Ceratozetoidea. Zetomimids have a tuberculate integument (Shaldybina 1977; Behan-Pelletier and Eamer 2003), but whether this is an adaptation to a semiaquatic habitat is unclear. The ZETOMIMIDAE was revised for North America by Behan-Pelletier and Eamer (2003).

The family ONYCHOBATIDAE (Figs. 15.70A, B) consists of a single species, *Onychobates nidicola* Hammer, known only from a nest of a grey warbler, *Gerygone igata*, in New Zealand (Hammer 1967). Woas (2002) subsumed the family into the ZETOMIMIDAE based on claw numbers on the legs. However, the anterior position of the adalar porose area (Aa) in *O. nidicola* is unique for Ceratozetoidea, and the family is retained here pending further study. Species of the monogeneric HUMEROBATIDAE (Fig. 15.71A–C), are found worldwide in temperate regions in association with plants. Adults and immatures feed on fungi and algae on trees and are virtually unknown from litter samples. Adults of *Humerobates rostrolamellatus* Grandjean move from the trunk to terminal shoots of fruit trees in spring and early summer where they deposit eggs and where hatched larvae and adults feed. The population moves back to the trunk in autumn for overwintering (Murphy and Balla 1973).

Species in the nine genera of CERATOKALUMMI-

DAE (Figs. 15.59C–E) are found throughout the world in temperate and tropical habitats. This family, placed in the Oribatelloidea by Subías (2004), is tentatively included here in the Ceratozetoidea because adults share ceratozetoid character states such as a genal notch, axillary saccule of the subcapitulum, and large pedotectum I. The relationships of the enigmatic genera *Lamellobates*, *Paralamellobates*, and *Sacculozetes* will likely remain unresolved until immatures are described (Behan-Pelletier 1999a). Woas (2002) suggested the family LAMELLOBATIDAE for these genera, but the name is unavailable according to nomenclatural rules.

A number of ceratozetoids may feed directly on economically important plants, although such relationships are not obligatory. For example, *Chamobates lapidarius* (Lucas) damages azaleas, *Ceratozetes* sp. feeds on peaches, and *Humerobates rostrolamellatus* attacks split cherries in Japan (Suzuki 1971). *H. rostrolamellatus* has also been implicated as an invasive nuisance pest of buildings (Scott 1958; Aoki 1963). *Lamellobates palustris* (Hammer) and *Paralamellobates bengalensis* (Bhaduri and Raychaudhuri) feed on the apical portion of roots of pineapple (Sanyal and Das 1989), and the latter species also feeds on the green leaves of *Dioscorea alata* (Ramani and Haq 1984). Finally, *Punctoribates manzanoensis* (Hammer) causes heavy damage to rice shoots in China (Lan, Xin, and Aoki 1986).

Species of *Ceratozetes*, *Trichoribates*, and *Fuscozetes* are known intermediate hosts of tapeworms (Denegri 1993; Denegri et al. 1998). Tapeworm cysticercoids were recovered from species of the ceratozetid genus *Kilimabates* in an infection trial using *Moniezia expansa* eggs (Schuster, Coetzee, and Putterill 2000).

▪ Superfamily Galumnoidea (Figs. 15.58A–H)

DIAGNOSIS: Adults: Prodorsum with lamellae absent, at most narrow ridges present (line L); tutorium reduced to ridge (line S) or absent. Genal notch absent. Dorsophragmata and pleurophragmata present. Chelicerae chelate-dentate or pelopsiform; palpal eupathidium acm *attached to solenidion; subcapitular mentum with complete tectum. Axillary saccule of subcapitulum present. Discidium present; custodium absent; postanal porose area present or absent. Pedotectum I small, not covering acetabulum I; pedotectum II present. Notogaster with hinged, auriculate pteromorphs, with* ***alary furrow*** *(appearing bilobed); with or without lenticulus; with tectum posteriorly, not overlapping medially. Notogastral setation 10–15 pairs. Octotaxic system developed as 1–5 pairs of porose areas (occasionally porose areas multiplied), saccules or tubules, rarely absent; some species or genera with medial porose area or scattered pores. Genital setation 6 pairs. Bothridium without internal, spiral thickenings, cup shaped. Genu IV shorter than tibia IV; tibia IV with solenidion.*

Immatures: Opisthonotum with macrosclerites, often bearing octotaxic system of porose areas (GALUMNIDAE); apheredermous; unideficient. Humeral organ present. Opisthonotal setae without excentric porose area at base. Integument surrounding opening of opisthonotal gland porose. Single circumgenital sclerite present. Seta d *absent from tibiae and genua when respective solenidion is present.*

The Galumnoidea includes the families GALUMNIDAE (Figs. 15.58E–H), a large assemblage of 450 species in 50 genera, and the GALUMNELLIDAE (Figs. 15.58A–D), with seven genera. Galumnoids occur from polar regions to the tropics and are particularly diverse in the Neotropics, especially in soil and arboreal habitats in rain forests. They are also recorded from many other habitats, including moss, pasture sod, forest litter, and rotting wood. Galumnoids appear to prefer moist niches but occasionally are found on tree bark (Hammer 1973) or in nest material (Evans et al. 1961). Fossil galumnoids are known from Baltic and Dominican amber (Labandeira, Phillips, and Norton et al. 1997).

The feeding habits of galumnoid mites are diverse. Species of GALUMNELLIDAE have pelopsiform mouthparts and probably are mycophages. A few GALUMNIDAE feed on economically important plants. For example, *Orthogalumna terebrantis* Wallwork feeds on water hyacinth (*Eichornia crassipes*), an important aquatic pest in tropical and subtropical waterways (Wallwork 1965). Adult *O. terebrantis* consume previously damaged plant tissue and oviposit into the leaflike lamina, where the immatures burrow and develop to adults. Such damage attracts certain beetles that are more effective than *O. terebrantis* in controlling the plant (del Fosse, Cromroy, and Habek 1975). A *Galumna* sp. in Japan was reported to damage seeds of winter wheat after sowing (Suzuki 1971), while another species, *G. formicarius* (Berlese), may be an obligate coprophage on fecal material in the galleries of wood-boring insects (Wallwork 1958). Some galumnid species have been observed to prey on nematodes and may contribute to their control. Both *Pergalumna omniphagous* Rockett and Woodring and *Pilogalumna cozadensis* Nevin were found to be nematophagous in laboratory cultures, and the latter would also capture and feed on slowly moving or injured springtails (Rockett and Woodring 1966; Walter 1988a; Walter, Hunt, and Elliott 1988). A species of *Galumna* fed on Enchytraeidae and Collembola in a laboratory culture (Wunderle 1992a).

Sexual dimorphism is expressed in a number of galumnid genera. For example, Norton and Alberti (1997) postulated that hypertrophied porose areas of *Dicatozetes uropygium* (Grandjean) males might produce semiochemicals that are involved directly or indirectly in reproductive biology. As in similarly dimorphic oripodoids and ceratozetoids (see appropriate sections above), such secretions either may be produced in larger quantities by males than by females or may be produced only by males. The mediodorsal porose cluster in males of *Acrogalumna longipluma* (Berlese) is innervated (Alberti et al. 1997) and is considered a likely source for sex pheromones (Norton and Alberti 1997).

The reproductive biology of GALUMNIDAE appears

more diverse than that of other Brachypylina that have been studied. Spermatophore deposition in a species of *Pergalumna* is dependent on the presence of the female, and the male deposits signaling structures as guides to the spermatophore (Oppedisano, Eguaras, and Fernandez 1995). Mate pairing has been noted in a few galumnid species. Males of *Erogalumna zeucta* Grandjean press the prodorsum against the anal valves of the female and rest the first pair of legs on her flanks (Grandjean 1964a). Males of *Pilogalumna* sp. forcefully turn teneral females on their side, assume a venter-to-venter positioning, deposit a stalkless spermatophore in the coxisternal region, and quickly disengage. Females later lie on their backs and push the spermatophore mass toward the genital opening with their legs (Estrada-Venegas, Norton, and Moldenke 1996). Starvation seems to induce ovovivipary or larvipary in *Galumna elimata* (Hubert and Sustr 2001), but the fate of the progeny is unknown.

G. virginiensis Jacot is a major carrier of the sheep tapeworm, *Moniezia expansa*, and both *G. nigra* (Ewing) and *Pergalumna emarginata* (Banks) serve as important intermediate hosts (Denegri 1993). In separate infection trials using *M. expansa* eggs or oncospheres, tapeworm cysticercoids were recovered from *Pergalumna nervosa* (Berlese) (Polec and Moskwa 1994) and *G. gracilis* (Schuster, Coetzee, and Putterill 2000).

Key 15.1. Suborder Oribatida (Excluding Astigmatina), Key to Families (Figs. 15.1–15.72, pp. 493–64)

Note: MICROTEGEIDAE, MULTORIBULIDAE, DAMPFIELLIDAE, and PELOPPIIDAE are keyed more than once because these families include genera with both diarthric and anarthric subcapitula. THYRISOMIDAE is keyed more than once because the presence or absence of apodeme 4 is genus specific in this family. CYMBAEREMAEIDAE, HYDROZETIDAE, TECTOCEPHEIDAE, and MICREREMIDAE are keyed more than once because their respective coxisternal region can be misinterpreted.

1a. Genua of legs noticeably shorter than tibiae, differently shaped, functioning as knee. Legs articulate with body in deep pockets (acetabula), as a "ball and socket" joint; trochanters I–II almost totally contained within acetabula. Venter of the brachypyline type, comprising a unified rigid plate in which only the subcapitulum and the paired genital valves and anal valves are distinct and movable (i.e., coxisternal, aggenital, and adanal regions are fused into a single unit that carries their respective setae). Subcapitulum usually diarthric; if (secondarily) without labiogenal articulation, then rutella and chelicerae are highly modified. (The "higher" oribatid mites) (Figs. 15.29–15.72) **Cohort BRACHYPYLINA—43**

1b. Genua of legs similar to tibiae in size and shape, not reduced to kneelike segment. Leg articulation not in deep acetabula, at most in shallow depressions; trochanters I–II small but clearly external. Venter of various form but rarely as above (exception: HERMANNIIDAE); paired aggenital and adanal plates often distinguishable; coxisternal region often transversely divided by sejugal articulation and usually not fused to aggenital region. Subcapitulum usually stenarthric (if anarthric then rutella and chelicerae not noticeably modified; diarthric in several TRHYPOCHTHONIIDAE). The "lower" or "macropyline" oribatid mites **2**

2a. Body form ptychoid; legs can be fully withdrawn as a group and aspis rotated ventrad such that it touches the notogaster and ventral plate (thus fully covering the retracted legs). Legs relatively short, those on each side adjacent, attached to a rather inconspicuous, narrow coxisternal region that connects to a voluminous, soft, deformable pleural region (seen on distended specimens) **3**

2b. Body form not ptychoid; legs longer or, if short, clearly not capable of being withdrawn and covered by prodorsum. Coxisternal region conspicuous, wide; pleural region sclerotized, or, if not, noticeable sclerotization generally absent from whole body **8**

3a. Notogaster clearly of multiple parts; dorsal plate (notaspis) subdivided transversely by a telescoping (type L) articulation; only 4 pairs of setae (row *c*) on anterior plate (pronotaspis); posterior plate (pygidium) with setae of rows *e* and *f* closely associated with 2 transverse carinae; setae of row *d* minute or absent; seta *p1* and row *h* on ventral margin; setae *p2* and *p3* inserted posteriorly on pair of large lateral plates (pleuraspis). Epimere II with 2 pairs of setae. Genua of all legs without solenidia (Figs. 15.16A–C) **Superfamily Protoplophoroidea (*pars*), PROTOPLOPHORIDAE**

3b. Hysterosoma covered dorsally and laterally by single, undivided sclerite that bears at least 6 pairs of setae. One pair of epimere II setae. Genua I–III with 1 or 2 solenidia **4**

4a. Hysterosoma covered dorsally by an apparent notogaster with only 6–8 pairs of setae (rows *c*, *d*, and *e*); body region bearing setae of rows *f*, *h*, *p* (and *ad* in *Mesoplophora*), shifted ventrally, incorporated into a ventral plate in which genital and anal (and sometimes adanal) plates insert. Subcapitulum anarthric; rutella narrow. Genu I with 1 solenidion (Figs. 15.16D, E) **Superfamily Hypochthonioidea (*pars*), MESOPLOPHORIDAE**

4b. Hysterosoma covered with notogaster of typical structure, bearing 14 or 15 (rarely more) pairs of setae (rows *c*, *d*, *e*, *h*, *p* present; row *f* represented only by inconspicuous alveoli). No separate, seta-bearing ventral plate in which anal or genital valves insert. Subcapitulum stenarthric; rutella broad, massive. Genu I with 2 solenidia **Supercohort Mixonomatides (*pars*)—5**

5a. Combined anogenital region broad, length approximately twice width, composed of 2 fully independent pairs of plates (fused genital/aggenital plates and fused anal/adanal plates). Without plicature plates. Pretarsi monodactylous. Anterior genital papillae reduced (Figs. 15.16F, G) **Superfamily Phthiracaroidea, PHTHIRACARIDAE**

5b. Combined anogenital region narrow, length usually greater than 3 times maximum width; structure of ventral plates otherwise. Paired, narrow plicature plates (without setae) present between notogaster and aggenital/adnanal plates; rarely absent (some SYNICHOTRITIIDAE). Pretarsi monodactylous or tridactylous. Three pairs of genital papillae well formed, or only posterior pair reduced **Superfamily Euphthiracaroidea—6**

6a. Anal plates (bearing anal setae) discrete, although they may be very narrow posteriorly, not fused to adanal plate or to adanal portion of aggenito-adanal plate; genital plate of various form, fused to aggenital region or not. Notogastral integument usually smooth, uniformly shiny in reflected light (Figs. 15.17A, B) **ORIBOTRITIIDAE**

6b. Neither anal nor genital plates discernable; that is, genital plates fused to aggenitals, and anal plates fused to adanals. Notogastral integument often pitted (at least anteriorly), usually not shiny 7

7a. Paired ventral plates connected across midline in preanal position by a short region of interdigitating ridges (in transmitted light appearing as dark interlocking triangle). Genital plates not attached medially, genital setae spread the length of genital region, located on medial edge, all genital papillae of approximately equal size. Palp 3-segmented (trochanter/femur/genu fused) (Figs. 15.17C, D) **EUPHTHIRACARIDAE**

7b. Preanal region without interlocking triangle. Paired plates attached to each other medially in genital region, either fused medially as a single plate or articulating by narrow scissure; genital setae concentrated anteriorly, not restricted to medial region, with one pair of genital papillae greatly reduced, 1/2 the size of other two. Palp 4- (femur/genu fusion) or 5- (no fusions) segmented (Figs. 15.17E, F) **SYNICHOTRITIIDAE**

8a. Legs with 6 free segments past coxisternum, all femora divided. Prodorsum astegasime, with no rostral tectum. Bothridium and base of bothridial seta simple, without obvious bend, bothridial seta never barbed, ciliate, or pectinate. Body white, at least in anterior half; usually with inconspicuous, thin sclerites on hysterosoma. Several pairs of notogastral setae often large and conspicuously darkened. Coxisternum II with 3 pairs of setae. Solenidia of tibiae and genua clearly separate from respective setae *d* **Supercohort Palaeosomatides—9**

8b. Legs with 5 free segments past coxisternum; femora undivided. Prodorsum with or without rostral tectum. Bothridium and base of bothridial seta with conspicuous sharp proximal *d* bend (except TRICHTHONIIDAE), bothridial seta of diverse form, smooth or barbed. Body (adults) usually tan or brown, with noticeable sclerotization; rarely white (Parhyposomatides) or brightly colored (some BRACHYCHTHONIIDAE). Notogastral setae not darkened, although some may be enlarged. Epimere II usually with 1 pair of setae, rarely 2 or 3. Solenidia and setae *d* of tibiae and genua usually either adjacent (coupled), or *d* absent **14**

9a. With 2 exobothridial setae oriented almost vertically; ventral (*xi*) similar in size and shape to seta *in*, and much larger than dorsal (*xs*). Famulus large, erect, ciliate, the most proximodorsal setiform organ on tarsus I. Main part of chelicerae oriented almost vertically, with narrow, elongated digits. Palptarsus short, little longer than tibia. Genital papillae and genital setae unmodified. Tibia II with 2 solenidia **Superfamily Archeonothroidea—10**

9b. With 2 exobothridial setae oriented almost horizontally; anterodorsal one (*xa*) much larger than other (*xp*), but both clearly smaller than *in*. Famulus large or small, but smooth and inserted more distally. Chelicerae normal, oriented anteriorly or anteroventrally and digits shorter, thicker. Palptarsus at least twice length of tibia. Anterior pair of genital setae strongly modified and/or anterior genital papillae strongly reduced in size (or absent). Tibia II with 1 solenidion **11**

10a. Famulus not broadened distally. Inferior lip of mouth large, easily distinguished. Setae of pairs *c1* and *c2* all on single unpaired sclerite; pair *e1* usually on single unpaired sclerite. Prodorsum in midregion often with vague, internal sclerotized thickenings forming a U or square shape (Figs. 15.18D, E) **ARCHEONOTHRIDAE**

10b. Famulus broadened distally, somewhat spatulate. Inferior lip minute, difficult to distinguish. Setae of pair *c2* on individual small sclerites; setae of pair *e1* on individual sclerites. Prodorsum without thickenings (Figs. 15.18A–C) **ACARONYCHIDAE**

11a. Leg pretarsi II–IV bidactylous. Famulus large, approximately equal in length to solenidia or only slightly shorter, inserted on conspicuous tubercle in proximal quarter or third of tarsus. All genital papillae of normal size. Anterior adoral seta (*or1*) simple, smaller than others. Bothridial seta filiform. Opisthonotal setae *h1* and *p1* unusually short, swollen or spinelike; seta *p4* present (17 pairs of opisthonotal setae; Figs. 15.18F–H) **Superfamily Palaeacaroidea, PALAEACARIDAE**

11b. Leg pretarsi II–IV tridactylous, with empodial claw well developed or much reduced. Famulus much smaller than solenidia, inserted at midpoint or in anterior half of tarsus. Anterior pair of genital papillae greatly reduced or absent. Adoral seta *or1* usually strongly barbed or pectinate, larger than other 2. Bothridial seta slightly to greatly expanded (rarely filiform). Opisthonotal setae *h1* and *p1* not unusually short; with either 16 (*p4* absent) or more than 17 (neotrichy) pairs of opisthonotal setae **Superfamily Ctenacaroidea—12**

12a. Opisthonotal setae heteromorphic, 2 pairs (*d2* and *e1*) enlarged, the longest of dorsal setae, darkly pigmented and erectile, 1 pair near posterior margin broadly vaned or heavily barbed. Body form normal, hysterosoma about 1.5 times longer than broad. Anterior genital papillae reduced, middle and posterior papillae normally developed. Leg solenidia normally developed (Figs. 15.19A–C) **CTENACARIDAE**

12b. Opisthonotal setae similar in structure, with various lengths but all thin and smooth. Body elongated, hysterosoma about twice as long as broad. All genital papillae noticeably reduced in size, anterior papillae present or absent. Most solenidia of genua and tibiae II–IV unusually small or vestigial **13**

13a. Bothridial seta conspicuously and abruptly broadened, flattened distally, straight to strongly curved, shorter than seta *in*. With only a groove (no striations) between setal rows *c* and *d*; *c1* much longer than *c2*. Anterior genital papillae absent. Tarsus III with 1 solenidion. Palpfemur with 1 seta (*sup* absent) (Figs. 15.19D–F) **APHELACARIDAE**

13b. Bothridial seta straight, thin, very slightly and gradually broadened distally, much longer than *in*. Broad region of soft, transversely striated cuticle between setal rows *c* and *d*; *c1* much shorter than *c2*. Anterior genital papillae present. Tarsus III with 2 solenidia. Palpfemur with 2 setae (*sup* present) (Figs. 15.19G–I) **ADELPHACARIDAE**

14a. Opisthonotum subdivided transversely by 1–3 complete scissures (if dorsum is otherwise sclerotized) or grooves (if dorsum is rather soft) **15**

14b. Opisthonotum represented by a single notogaster, not subdivided transversely, usually well sclerotized; if superficial transverse grooves appear in sclerotized cuticle, then they number more than 3 and are usually incomplete medially (some LOHMANNIIDAE; see Figs. 15.25A–D) **30**

15a. Opisthonotal gland present. Notogaster with a single transverse scissure or groove, lying between setal rows *d* and *e*; row *e* inserted well posterior to scissure. Rather weakly sclerotized mites, colorless to pale yellow. Rutella strongly developed, distally broad and toothed. Pretarsi with well-developed lateral claws; empodial claw reduced, hooklike, or absent **Supercohort Parhyposomatides—16**

15b. Opisthonotal gland absent. Notogaster with 1–3 transverse scissures or grooves (if only 1, then the following is true: mites more strongly pigmented, not soft bodied; scissure runs either anterior to row *d* or posterior to *e* or actually contains row *e* on narrow intercalary sclerite). Rutella various, but rarely as above. Pretarsi various, usually monodactylous **Supercohort Enarthronotides (*pars*)—18**

16a. Prodorsum astegasime; rostral tectum absent or poorly developed, much of chelicerae visible in dorsal aspect. With 3 pairs of adoral setae. Palp 5-segmented, trochanter and femur freely articulated. Notogastral seta *f2* much smaller than *f1* or absent **17**

16b. Prodorsum stegasime; rostral tectum covering most of chelicerae in dorsal aspect. With 1 pair of adoral setae. Palp 4-segmented, trochanter and femur fused. Notogastral seta *f2* well developed, similar to *f1* (Figs. 15.20C, D) **ELLIPTOCHTHONIIDAE**

17a. Opisthonotal gland opens on flared, funnel-shaped protuberance bearing seta *f2*. With 4 pairs of adanal, 4 pairs of anal setae; peranal segment present, bearing 1 pair of setae. With 1 pair of epimere II setae; genu II and tarsus III with 2 and 1 solenidia, respectively (Figs. 15.20A, B) **PARHYPOCHTHONIIDAE**

17b. Opisthonotal gland opens without conspicuous protuberance; seta *f2* absent, represented only by inconspicuous alveolus near gland opening. With 3 pairs of adanal, 2 pairs of anal setae; peranal segment absent. With 2 pairs of epimere II setae; genu II and tarsus III with 1 and 0 solenidia, respectively (Figs. 15.20E, F) **GEHYPOCHTHONIIDAE**

18a. Notogaster with a single transverse scissure, which lies posterior to setal row *c*, *d*, or *e*. (Telescoping, type L scissures can appear as 2 in transmitted light, due to overlapping structures; some SPHAEROCHTHONIIDAE have vestiges of 1–2 additional posterior scissures, but they appear simply as thin ridges, fused to surrounding cuticle; ENIOCHTHONIIDAE have a transverse groove near setal row *d*, that can be confused with a scissure.) **19**

18b. Notogaster with 2–3 transverse scissures or grooves, separating 3–4 dorsal plates or regions **22**

19a. Single transverse scissure of type S, with intercalary sclerite containing large, erectile setae of row *e*. Setae *f1*, *f2* also erectile, stout, with long branches, inserted on intercalary sclerites in pair of circular unsclerotized depressions; setae *p1*, *p2* phylliform. Prodorsum astegasime. Leg pretarsi bidactylous, empodial claw absent (Fig. 15.21A) **ARBORICHTHONIIDAE**

19b. Single transverse scissure of other type; either telescoping type L or, if type S then setae of row *e* (on intercalating sclerite) smaller than nearby notogastral setae, often vestigial. No opisthonotal setae conspicuously enlarged or erectile. Prodorsum stegasime. Leg pretarsi monodactylous to tridactylous; empodial claw well developed **20**

20a. Integument distinctly patterned (foveolate/reticulate); at least some dorsal setae of notogaster and/or prodorsum heavily ciliate, T shaped. Transverse scissure type L, posterior to setal row *c*; 4 pairs of setae insert on anterior notogastral plate (pronotaspis). With 4–10 pairs of anal setae in linear arrangement. Trochanters I–II each with 1 seta (Fig. 15.21G) **SPHAEROCHTHONIIDAE**

20b. Integument smooth or sculpted but without foveae or reticulation; dorsal setae simple, setiform, or flattened, lanceolate. Transverse scissure of either type (L, S), located posterior to setal row *d*, such that 6–8 pairs of setae borne on pronotaspis. With 0–2 pairs of anal setae. Trochanters I–II without setae **21**

21a. Transverse scissure of type L, located behind setal row *e*; pronotaspis with 8 pairs of setae (setae *e* similar to others) and with transverse, medially incomplete sulcus at level of setal row d. Setae *h2*, *h3* on separate, longitudinally elongated lateral plate. Aggenital setae present, on separate plate. Femur I with 3 setae; genu I with 2 solenidia and 5 setae (Fig. 15.21B–D) **ENIOCHTHONIIDAE**

21b. Single transverse scissure a modified type S, with undivided intercalary plate bearing small or vestigial setae *e1*, *e2*. Pronotaspis with 6 pairs of setae, without sulcus. Without isolated lateral plate bearing setae. Neither separate aggenital plate nor aggenital seta present. Femur I with 5 setae, genu I with 1 solenidion and 3 setae (Figs. 15.21E, F) **HYPOCHTHONIIDAE (*pars*)**

22a. All scissures (2 or 3) simple, type E, or only grooves. Notogastral setae generally homogeneous, similar in shape and size; none extremely broad (platelike) or unusually large or erectile; bothridial seta clavate **23**

22b. At least 1 (of 2 or 3) scissure of type S, with intercalary sclerite and large, erectile setae; or notogastral setae extremely broad, platelike, covering body like shields. Bothridial seta of various forms **25**

23a. Body more than 3 times longer than wide; legs II–III widely separated. Pretarsi without distinct claw; empodium highly modified, with hyaline flap on either side. Prodorsum astegasime, with a small, narrow anterior lobe bearing setae *ro*; bothridial seta smooth (Figs. 15.22A–C) **PEDICULOCHELIDAE**

23b. Body little more than twice maximum width; legs II–III almost adjacent. Pretarsi with empodium developed as typical claw. Prodorsum stegasime, without obvious, narrow naso; bothridial seta barbed or ciliate **24**

24a. Notogaster with 3 transverse type E scissures; 4 pairs of setae (*c* row only) on anteriormost plate. With cupule immediately lateral to each mid-dorsal seta (*c1*, *d1*, *e1*, *f1*), in addition to normal complement. Genital plates twice the size of anal plates. Aggenital seta absent, anal plates with 4 pairs of setae; peranal plate and seta absent (Figs. 15.22D, E) **HAPLOCHTHONIIDAE**

24b. Notogaster with 2 transverse type E scissures; 6 pairs of setae on anteriormost plate (*c* and *d* rows). Without extra cupules immediately lateral to mid-dorsal setae. Genital plates and combined anal/peranal plates subequal in size. With 1 pair of aggenital setae, anal plates with 2 pairs of setae; peranal plates present, with 1 pair of setae (Figs. 15.22F, G) **Superfamily Brachychthonioidea, BRACHYCHTHONIIDAE**

25a. Some or all setae of prodorsum and notogaster broad, shieldlike, sometimes covering all or most of dorsal surface; setal surface with distinct pattern of foveae or reticulation; without barbs or cilia. With or without erectile setae. Pale mites, with little sclerotization **Superfamily Atopochthonioidea—26**

25b. Dorsal setae of various size and ornamentation—smooth, ciliate, or pectinate—but not foveate or reticulate. At least 4 pairs (*e1*, *e2*, *f1*, *f2*) much longer than others, inserted on intercalary sclerites in 2 type S scissures. Cuticle usually with noticeable sclerotization, light brown color **28**

26a. Leg pretarsi monodactylous. All dorsal setae shieldlike, similar in shape and mostly of similar size, none erectile or inserted on large tubercles. Palptarsus terminating in long, ribbon-shaped eupathidia. Rutellum minute, conical, smooth. Leg genua apparently without solenidia (vestige may be present on genu I). Setae of ovipositor enlarged, hornlike eupathidia (Figs. 15.23A–D) **PTEROCHTHONIIDAE**

26b. Leg pretarsi bidactylous, with strong lateral claws but without empodium. One or 2 pairs of notogastral setae different from others, in shape and orientation. Palptarsus terminating in short, forked eupathid. Rutellum elongated, with narrow base, broad and toothed distally. Leg genua with 2-1-1-1 solenidia (I–IV). Setae of ovipositor normal, not hornlike **27**

27a. With 2 pairs of erectile setae inserted on large, independent tubercles; these setae narrowed distally, phylliform, distinctly longer than all other dorsal setae. Dorsal setae clearly independent, well spaced; interlamellar setae shieldlike, as broad as long, foveate. With 6 pairs of genital setae. Tarsus III without solenidion (Figs. 15.23E, F) **ATOPOCHTHONIIDAE**

27b. With 1 pair of notogastral setae (*e1*) erect, in fixed position; these setae broad, not tapering. Remaining notogastral setae shieldlike, overlapping and appearing fused together, covering entire hysterosoma; interlamellar seta much longer than broad, without foveae. With 8 pairs of genital setae. Tarsus III with solenidion (Figs. 15.23G–I) **PHYLLOCHTHONIIDAE**

28a. Notogaster with only 2 transverse scissures, both of type S, bearing erectile setae *e* and *f*. Setal rows *c* and *d* similar and on same plate; bothridial seta straight at base, with simple insertion, bothridium without noticeable curve (Fig. 15.24A) **TRICHTHONIIDAE**

28b. Notogaster with 3 transverse scissures, such that setal rows *c* and *d* similar or different in shape but not on same plate; bothridial seta and bothridium strongly curved at base **29**

29a. Notogaster and prodorsum without noticeable sculpturing, rostrum not fenestrate. Setae of rows *c–f* compressed into anterior 1/3 of notogaster, leaving middle 1/3 devoid of setae; setal row *d* much enlarged, similar to rows *e* and *f* but not erectile. Genital plate transversely divided; aggenital plates large, broadly fused to coxisternum; adanal plate pair not fused posterior to anal aperture. Pretarsi of legs monodactylous (Figs. 15.24B–D) **HETEROCHTHONIIDAE**

29b. Notogaster and prodorsum usually with reticulate or foveate sculpturing; rostrum distinctly fenestrate. Setae of rows *c–f* occupy anterior 1/2 of notogaster; no large area without setae; setal row *d* not enlarged, similar to row *c*. Genital plate not divided; aggenital plates small, not fused to coxisternum; adanal plates fused posteriorly to form a U shape. Pretarsi of legs bi- or tridactylous (Figs. 15.24E–G) **COSMOCHTHONIIDAE**

30a. Dichoid mites: with full sejugal articulation, such that prodorsum is capable of horizontal and lateral movement relative to hysterosoma, often slightly retractable into it; epimera II–III therefore not fused, separated by band of unsclerotized cuticle. With 2 pairs of exobothridial setae **31**

30b. Holoid mites: epimera II–III usually entirely fused, such that sejugal articulation present only in dorsal and sometimes lateral region; proterosoma not retractable into hysterosoma (exception: a narrow band of unsclerotized cuticle connects epimera II–III in some Malaconothroidea, allowing slight dorsoventral movement of proterosoma). With 1 or 0 pair of exobothridial setae **Superfamily Crotonioidea—37**

31a. With 10 pairs of genital setae, some in lateral row. Aggenital setae absent; aggenital plates small, inconspicuous, or absent; usually with 2 pairs of anal setae (rarely 0 or 1). Opisthonotal gland and lyrifissures *iad*, *ian* absent **32**

31b. With fewer than 10 pairs of genital setae, usually all in medial row. With 2–3 pairs of aggenital setae, aggenital region variously formed; with 3–4 pairs of anal setae. Opisthonotal gland, lyrifissure *iad*, and often *ian* present **Supercohort Mixonomatides (*pars*)—33**

32a. With characteristic body shape: ventral surface flat, dorsal surface strongly arched, prodorsum posteriorly equally as broad as notogaster. Anterior margin of notogaster with large, thin tectum overhanging posterior region of prodorsum. Preanal plate conspicuous. Subcapitulum stenarthric; adoral setae all broadest near base. Femora of at least legs I–II with ventral keels (Figs. 15.25A–D) **LOHMANNIIDAE**

32b. Differently shaped; prodorsum only 2/3 width of notogaster. Anterior margin of notogaster without tectum. Preanal plate minute, usually hidden in ventral aspect. Subcapitulum anarthric; adoral seta *or2* broadest distally, with conspicuous tines. Femora with small ventral spurs, but without keels **HYPOCHTHONIIDAE (*pars*, *Nothrolohmannia*)**

33a. Colorless, weakly sclerotized mites. Rostral tectum weakly developed, chelicerae mostly exposed. Leg pretarsi tridactylous, with empodial claw minute, hooklike. Notogastral setae *f1* and *f2* present, but equally small, unlike larger neighboring setae (Figs. 15.25G, H) **NEHYPOCHTHONIIDAE**

33b. More sclerotized forms, with distinct yellow or brown pigmentation. Rostral tectum covers retracted chelicerae, at least dorsally. Leg pretarsi various, if tridactylous, empodial claw similar in size to lateral claws. Notogastral seta of row *f* not as above: one or both absent, or both present but similar to other notogastral setae **34**

34a. Large mites: total length over 800 µm. Epimeral halves III–IV separated medially by unsclerotized cuticle. Genital plates immediately posterior to epimere IV, located in anterior 1/2 of hysterosoma; with 2 pairs of aggenital setae. Palp with 5 free segments **35**

34b. Smaller mites, total length less than 700 µm, usually much less. Epimeral halves III–IV fused medially, demarcated with suture or not. Genital plates well posterior to epimere IV, in posterior half of hysterosoma; with 3 (rarely more) pairs of aggenital setae. Palp with 2–4 free segments **36**

35a. Hysterosoma rather rectangular in dorsal aspect and dorsoventally flattened. All notogastral setae short; *c1*, *d1*, *d2*, *e1* in longitudinal row; *c3*, *c2*, *cp*, *e2* in another row on lateral carina. With unpaired postanal plate; 3 pairs of adanal setae (Figs. 15.25E, F) **PERLOHMANNIIDAE**

35b. Hysterosoma convex, both in dorsal and lateral aspects. Notogastral setae *d2*, *h2*, *p1* flagellate, much longer than others; setae not in rows as above, no lateral carina. Without postanal plate; 2 pairs of adanal setae (Figs. 15.26A–C) **COLLOHMANNIIDAE**

36a. Body light yellow, cylindrical, proterosoma constricted behind level of bothridium. Epimera III–IV and aggenital region fused without delineation; U-shaped scissure passes between adanal, anal, and genital plates. With 4 pairs of anal, 4 pairs of adanal, and 9 pairs of genital setae. Leg pretarsi bidactylous. Palp with 4 free segments (Figs. 15.26D–F) **EULOHMANNIIDAE**

36b. Body light to dark brown, moderately elongated, without constriction as above. Epimera III–IV clearly outlined by borders and apodemes; without U-shaped scissure. With 3 pairs of anal, 3 pairs of adanal, and 7–8 pairs of genital setae. Leg pretarsi monodactylous. Palp with 2–3 free segments (Figs. 15.26G, H) **EPILOHMANNIIDAE**

37a. Body cylindrical, somewhat elongated. Prodorsum posteriorly with tubercles or ridges directed over dorsosejugal furrow. Notogaster indistinguishably fused to ventral region in posterior 1/2, giving appearance of pair of crescent-shaped scissures directed between widely separated genital and anal plates (Figs. 15.27A, B) **NANHERMANNIIDAE**

37b. Body not cylindrical, often globose or dorsally flattened. Prodorsum posteriorly without tubercles or ridges. Notogaster distinct from ventral region throughout its length; genital and anal plates adjacent or only narrowly separated **38**

38a. Ventral region of brachypyline type, similar to that of cohort BRACHYPYLINA: adanal and aggenital regions broad, completely fused, and aggenital region fused with epimere IV. Hysterosoma rather globose, may have granules, but without carinae or other strong surface topography (Figs. 15.27F, G) **HERMANNIIDAE**

38b. Ventral region of macropyline type: adanal and aggenital plates (when latter is distinct) relatively narrow; if sclerotized, aggenital region never fully fused to epimere IV. Hysterosoma not globose; usually somewhat dorsoventrally flattened, often with distinct surface topography of carinae, broad ridges or depressions **39**

39a. Epimere II with 3 or more pairs of setae. With 9 pairs of genital setae, 1 posterior pair far from medial edge of plate. Rostrum with short medial incision. Bothridial seta setiform, longer than seta *in*; bothridium basally with many saccules or short brachytracheae. Integument distinctly foveolate (Figs. 15.27C–E) **NOTHRIDAE**

39b. Epimere II with 1 (rarely 2 or 0) pair of setae. With 4–24 pairs of genital setae, all near medial edge. Rostrum without medial incision. Bothridium at most with single saccule, usually none; bothridial seta often clavate, length usually equal to or less than that of seta *in*. Integument various, but rarely foveolate **40**

40a. Prodorsum and notogaster fused dorsally. Bothridial seta with globular head, extremely short, entirely within closed bothridial chamber. Coxisternum uniformly fused; epimeral halves not delineated by sagittal features. Often with much adherent organic debris and with posterior setae on long apophyses. Mostly in Southern Hemisphere (Figs. 15.28A–C) **CROTONIIDAE**

40b. Prodorsum separated from notogaster by band of soft cuticle. Bothridial seta, if present, of various forms, but clearly projecting from normal bothridium (with rare exception). Coxisternum with some type of sagittal feature (e.g., longitudinal groove, soft cuticle, apodeme) that delineates at least some epimeral halves. Cosmopolitan families **41**

41a. Anal plate wider or only slightly narrower than adanal plate, not greatly reduced. Medial margin of genital plates, bearing genital setae, delimited from rest of plate by distinct line or carina; aggenital setae present. Body often with adherent organic debris (Figs. 15.28D, E) **CAMISIIDAE**

41b. Anal plate usually reduced (exception: *Allonothrus*), much narrower than adanal plate, sometimes inconspicuous. Genital plate without carina that delimits marginal seta-bearing region; aggenital setae absent. Body usually without conspicuous organic debris **42**

42a. Bothridial seta and bothridium lost without trace. Subcapitulum without labiogenal articulation (anarthric). Palpfemur without setae. Cerotegument present in form of large, thin plates that appear waxy, without excrescences, strongly birefringent in polarized light (Figs. 15.28F–H) **MALACONOTHRIDAE**

42b. Bothridial seta present, usually with well-developed bothridium (vestigial or absent in some *Trhypochthoniellus*). Subcapitulum with labiogenal articulation (usually stenarthric). Palpfemur with 1 seta. Cerotegument, if present, of different form, not birefringent (Figs. 15.28 I, J) **TRHYPOCHTHONIIDAE**

43a. Notogaster without octotaxic system of porose areas or saccules, pteromorphs absent, dorsophragmata and pleurophragmata usually absent (present in LIACARIDAE, GUSTAVIIDAE, TENUIALIDAE). Epimera II–IV usually rectangular in shape; apodemes 2 and sejugal apodeme transverse. Discidium and circumpedal carina usually absent. Humerosejugal porose organ Am absent, Ah usually absent **44**

43b. Notogaster usually with octotaxic system of porose areas or saccules, pteromorphs present or absent, dorsophragmata and pleurophragmata present. Epimera II–IV never rectangular in shape; apodemes 2 and sejugal apodeme angled toward genital plate. Discidium and circumpedal carina usually present (absent in PASSALOZETIDAE). Humerosejugal porose organs Am and Ah present **137**

44a. Paired opisthonotal glands opening on distinct, funnel-shaped tubes or on large apophyses laterally on notogaster **45**

44b. Paired opisthonotal glands opening directly on notogaster, without funnel-shaped tubes **46**

45a. Gnathosoma modified, chelicerae pelopsiform. Paired opisthonotal glands opening on large anteriorly directed apophyses laterally on notogaster. Rostrum with deep medial incision; lamellar setae small, hardly visible. Trochanters and femora I–II with retrotecta. Adults with or without scalps, scalps when present concentrically arranged and very evident. Length: small to medium, 330–440 µm (Figs. 15.29A–E) **PLASMOBATIDAE**

45b. Gnathosoma normal, chelicerae chelate-dentate. Paired opisthonotal glands opening on distinct, funnel-shaped tubes laterally on notogaster. Rostrum without incision; lamellar setae well developed. Trochanters and femora I–II without retrotecta. Adults with only inconspicuous tritonymphal scalp adhering closely to notogaster. Length: 550–800 µm (Figs. 15.29F, G) **HERMANNIELLIDAE**

46a. Genital plates with transverse scissure; 7–8 (exceptionally up to 18) pairs of genital setae. Adults with scalps, scalps strongly sclerotized. Femora I–IV and trochanters III–IV with brachytrachea. Length: 800–1,100 µm (Figs. 15.29H, I) **NEOLIODIDAE**

46b. Genital plates without transverse scissure; usually with 3–6 pairs of genital setae. Adults with or without scalps, scalps if present weakly sclerotized. Femora I–IV and trochanters III–IV generally with porose areas; rarely with filamentous tracheae **47**

47a. Prodorsum without true lamellae; with or without costulae **48**

47b. Prodorsum with true lamellae **111**

48a. Bothridium vestigial, bothridial seta absent; with 2 pairs of genital papillae, 7 pairs of genital setae, and 3 or 4 pairs of anal setae. Single species is a myrmecophile in Java. Length: 650–720 µm (Figs. 15.30A–D) **ARIBATIDAE**

48b. Bothridium developed, bothridial seta normal or reduced. Usually with 3 pairs of genital papillae and 3–6 pairs of genital setae, with 2–9 pairs of anal setae **49**

49a. Notogaster with large, triangular humeral projections and with concave impressions, anterior notogastral margin concave. Costulae positioned laterally on prodorsum. Setae of epimera I–II spiniform. Length: 340–350 µm (Figs. 15.30E, F) **PTEROBATIDAE**

49b. Without combination of above characters. Notogaster with or without triangular humeral projections, anterior notogastral margin usually straight to convex, notogaster with or without concave impressions. Costulae variously positioned or absent. Setae of epimera I–II usually setiform **50**

50a. Notogaster usually flat-topped and elliptical (convex in ALEURODAMAEIDAE), with 2–9 pairs of posteromarginal setae, dorsocentral setae absent or only 1 pair present. Usually with thick, ornamented cerotegument **Superfamily Plateremaeoidea—51**

50b. Notogaster usually convex, with 9–15 pairs of setae or setal alveoli, only 3 pairs of setae positioned posteromarginally. Ornamented cerotegument usually absent **60**

51a. Pedotecta I–II present, pedotectum I auriculate, propodolateral apophysis absent. Legs thin, filiform. Notogastral setae positioned posterior to lyrifissure *im* **52**

51b. Pedotectum I–II absent; propodolateral apophysis present or absent. Legs never filiform. Notogastral setae variously positioned relative to lyrifissure *im* **54**

52a. With 3 pairs of anal setae. Tarsus II with 1 solenidion. Length: approximately 500 µm (Figs. 15.30G, H) **IDIODAMAEIDAE**

52b. With 2 pairs of anal setae. Tarsus II with 2 solenidia **53**

53a. Notogaster concave in lateral aspect, caudal notogastral setae long or not but never twisted and corkscrewlike. Length: 480–700 µm (Figs. 15.31A–C) **GYMNODAMAEIDAE**

53b. Notogaster convex in lateral aspect, caudal notogastral setae long and twisted, corkscrewlike. Length: 430–450 µm (Figs. 15.31D–F) **ALEURODAMAEIDAE**

54a. Prodorsum with enantiophysis A bordering deep, transverse furrow. Femora with trachea. Length: 300–500 µm (Figs. 15.31G–I) **PHEROLIODIDAE**

54b. Prodorsum without enantiophysis A, transverse furrow present or absent. Femora with porose areas **55**

55a. Notogaster with circummarginal furrow; notogaster extending anteromedially to level of bothridium. Genua, tibiae, and tarsi I–IV with retrotecta. Length: 350–370 µm (Fig. 15.32A) **NACUNANSELLIDAE**

55b. Notogaster without circummarginal furrow; notogaster not extending anteromedially to level of bothridium. Genua, tibiae, and tarsi I–IV with or without retrotecta **56**

56a. Femur I with minimum of 8 setae, femur IV with minimum of 5 setae, trochanter IV with 3 setae. Bothridial seta filiform. Coxisternal region neotrichous. Length: 600–800 µm (Figs. 15.32B, C) **PLATEREMAEIDAE**

56b. Femur I with 5 setae, femur IV with 3 setae, trochanter IV with 1 setae. Head of bothridial seta flat, leaf shaped, fusiform, or club shaped. Coxisternal setation is 3-1-3-3 **57**

57a. Lamellar setae inserted close to rostral margin, anterior to rostral setae. Bothridial seta fusiform or club shaped. Exobothridial seta absent; setae *p2* and *p3* positioned dorsally to dorsolaterally on notogaster. Genital plates with 6–7 pairs of setae **58**

57b. Lamellar setae removed from rostral margin, at same level or posterior to rostral setae. Bothridial seta flat, leaf shaped. Exobothridial seta present; setae *p2* and *p3* positioned ventrally on notogaster. Genital plates with 5–6 pairs of setae **59**

58a. Genital and anal apertures partially joined; aggenital setae lateral to genital plates. Notogaster slightly invaginated posteriorly, ovate. Lateral claws of pretarsi slender, weak. Length: 500–600 µm (Figs. 15.32D, E) **PEDROCORTESELLIDAE**

58b. Genital and anal apertures completely joined, aggenital setae posterior to genital plates. Notogaster rounded posteriorly, round also in general shape. Three claws of pretarsi strongly developed. Length: 500–800 µm (Figs. 15.33A–C) **LYRIFISSELLIDAE**

59a. Scalps absent. Notogaster with 9–10 pairs of setae; notogastral lyrifissures long, integument reticulate or smooth, without pair of tubercles lateroposteriorly. Femora with porose area. Length: 190–420 µm (Figs. 15.33G, H) **LICNODAMAEIDAE**

59b. Scalps present, with reticulate sculpture, tritonymphal scalp attached anteriorly to notogaster. Notogaster with 4–5 pairs of notogastral setae, of which *p* series and pair posterior to lateral tubercles distinct; notogastral lyrifissures short; integument of adult notogaster shiny, smooth; notogaster with pair of tubercles lateroposteriorly. Femora with dorsal saccule. Length: 250–300 µm (Figs. 15.33D–F) **LICNOBELBIDAE**

60a. Anal plates with 2–9 pairs of setae. Porose areas distoventrally on tibiae I–IV and proximoventrally on tarsi I–IV. Notogaster without humeral projections. **Superfamily Eremaeoidea (*pars*)—61**

60b. Anal plates with 2 (exceptionally 3) pairs of setae. Tibiae and tarsi without porose areas. Notogaster with or without humeral projections **62**

61a. Oblique carina present lateral to each costula. Rostral carina present. Notogaster with 2 pairs of tubercles anteriorly; with 10 pairs of setae. With 3 pairs of anal setae. Length: 550–1,100 µm (Figs. 15.34A, B) **MEGEREMAEIDAE**

61b. Rostral and oblique lateral carinae absent. Notogaster without tubercles anteriorly; usually with 10–11 pairs of setae, occasionally up to 21 pairs. With 2–9 pairs of anal setae. Length: 400–800 µm (Figs. 15.34C, D) **EREMAEIDAE**

62a. Gnathosoma modified: either rutella very large, thin and leaflike, or paired rutella forming tube; chelicerae with unusually large teeth, or pelopsiform, or digits tapered, elongate. Chelicerae with 0–1 seta **63**

62b. Gnathosoma normal, chelicerae with dentate chelae; rutella normally developed. Chelicerae with 2 setae **68**

63a. Subcapitulum diarthric; rutella very large, thin, and leaflike. Chelicerae with unusually large teeth. Palpgenu without seta, fused or not to femur; palptarsus with 3–5 setae. Length: 215–250 µm (Figs. 15.34E–I) **DAMAEOLIDAE**

63b. Subcapitulum anarthric; rutella forming tube. Chelicerae pelopsiform, or attenuate-edentate. Palpgenu with 1 seta, not fused to femur; palptarsus with 8 or 9 setae **64**

64a. Chelicerae attenuate-edentate. Distal eupathids on palptarsus fused basally, appearing like forked seta. Palptarsus with 8 setae **65**

64b. Chelicerae pelopsiform. Distal eupathids on palptarsus separate, not appearing like forked seta. Palptarsus with 9 setae **67**

65a. Prodorsum with or without paired flat regions (tectopedial fields) and short costulae or knoblike tubercle. Tarsus IV with 2 ventral setae more strongly developed, clavate, terminating bluntly, often pinnate, trochanter III with 1 seta. Prodorsum centrally granular or not; and/or rostral setae geniculate. Length: 170–300 µm (Figs. 15.35A–E) **SUCTOBELBIDAE**

65b. Prodorsum without paired tectopedial fields or knoblike costulae. Tarsus IV without strongly developed, clavate, or pinnate ventral setae, trochanter III with 2 setae. Prodorsum centrally not granular; rostral setae curved gently, not geniculate **66**

66a. Fused epimera III–IV very elongate; with 5 pairs of genital setae. Prodorsum large, broad, triangular; bothridium positioned well anterior of dorsosejugal furrow. Length: 430–450 µm (Figs. 15.35F–H) **TRIZETIDAE**

66b. Fused epimera III–IV not elongate; with 7 pairs of genital setae. Prodorsum elongate, with rostrum gradually or abruptly attenuated; bothridium positioned only slightly anterior of dorsosejugal furrow. Length: 500–1,000 µm (Figs. 15.35I–L) **RHYNCHORIBATIDAE**

67a. Notogaster narrow about twice as long as wide; with strong depression laterally between setae *c* and *la*; porelike alveolar vestige of second exobothridial seta present, costulae present. With 3–4 pairs of genital setae. Tarsi and tibiae functionally fused, move as single segment. Length: 600–1,400 µm (Figs. 15.36A–D) **DAMPFIELLIDAE (*pars*)**

67b. Notogaster not elongated, slightly longer than wide; without strong depression laterally between setae *c* and *la*; alveolar vestige of second exobothridial seta absent, costulae absent. With 6 pairs of genital setae. Tarsi and tibiae with normal articulation. Length: 270–515 µm (Figs. 15.36E–H) **OXYAMERIDAE**

68a. Notogaster with scalps (at least tritonymphal scalp) tightly affixed by conspicuous anterior attachment structure **69**

68b. Notogaster without scalps **70**

69a. Scalps affixed on medial tubercle at anterior of notogaster. Bothridial seta with ringlike thickenings proximally. Genital plates with 6 pairs of setae; circumventral carina present curving posterior to anal plates. Trochanter II with broad ventral keel that mimics pedotectum II, setae (*p*) absent from tarsi II to IV. Length: 400–750 µm (Figs. 15.36I–K) **BASILOBELBIDAE**

69b. Scalps affixed on pair of humeral tubercles at anterior of notogaster. Bothridial seta without ringlike thickenings. Genital plates with 7 pairs of setae; circumventral carina absent. Trochanter II without broad ventral keel, setae (*p*) present on tarsi II–IV. Length: 300–400 µm (Figs. 15.37A–D) **HETEROBELBIDAE**

70a. Prodorsum subequal in length to notogaster; strongly constricted anterior to legs I. Legs I palplike; claw present or absent; eupathidial setae (*p*) of tarsus I very long, almost twice length of claw, when claw present, tarsus I is thickest segment; leg segments without porose areas. Subcapitular mentum with tectum. Length: 400–500 µm (Figs. 15.37E–I) **STAUROBATIDAE**

70b. Prodorsum clearly shorter in length than notogaster; prodorsum not constricted anterior to legs I. Legs I normal, leglike; claw(s) present; eupathidial setae (*p*) of tarsus I shorter than or subequal to length of claw, tarsus I not thickest segment; at least femora with porose areas. Subcapitular mentum with or without tectum **71**

71a. Posterior of prodorsum and anterior of notogaster strongly flattened and usually fused; longer notogastral setae often showing two distinct patterns: *lm*, *lp*, and *h3* arch toward midline, and *h1* and *h2* arch posteriorly; rostrum usually with pair of incisions. Axillary saccule present at base of palp. Length: 600–1,050 µm (Figs. 15.37J–L) **AMERIDAE**

71b. Posterior of prodorsum and anterior of notogaster convex, fused or not; notogastral setae of similar size and shape, usually none arching medially; rostrum with single incision or incisions absent. Axillary saccule absent **72**

72a. Aggenital neotrichy present, that is, 3 or more pairs of aggenital setae; ventral plate setae branched or not **73**

72b. Aggenital neotrichy absent, 0–2 pairs of aggenital setae; coxisternal and ventral plate setae not branched **75**

73a. Adanal neotrichy present (more than 3 pairs of setae). Tibiae and tarsi I–IV with retrotecta, femora I–IV and trochanters III–IV with saccules. Length: 300–550 µm (Figs. 15.38F–I) **EREMOBELBIDAE**

73b. Adanal setation 3 pairs. Tibiae and tarsi I–IV without retrotecta, femora I–IV and trochanters III–IV with porose areas **74**

74a. Ventral plate setae branched. Costulae present. Pedotectum I present; propodolateral apophysis absent. Length: 275–500 µm (Figs. 15.39A, B) **EREMULIDAE**

74b. Ventral plate setae not branched. Costulae absent. Pedotectum I absent; propodolateral apopysis present. Length: 335–415 µm (Figs. 15.39G, H) **HUNGAROBELBIDAE**

75a. Notogaster with 10 pairs of setae (1 seta anterior to lyrifissure *ia*). Pedotectum II present. Proral setae of tarsi spiniform **76**

75b. Notogaster with 10–11 pairs of setae (1–2 setae anterior to lyrifissure *ia*, or notogastral setae (other than *p* series) arranged in 2 conspicuous longitudinal rows). Pedotectum II present or absent. Proral setae of tarsi setiform **77**

76a. Prodorsum without costulae, usually fused to notogaster. Length: 500–800 µm (Figs. 15.38A–C) **AMEROBELBIDAE**

76b. Prodorsum with parallel costulae; prodorsum usually separate from notogaster. Length: 400–500 µm (Figs. 15.38D, E) **CTENOBELBIDAE**

77a. Notogastral setae (other than *p* series) arranged in 2 conspicuous longitudinal rows. Well-developed parastigmatic enantiophysis present. Legs moniliform, tarsal segments swollen proximally, tapered distally; femora and tibiae narrow proximally, swollen distally. Length: 450–1,300 µm (Figs. 15.39C–F) **Superfamily Damaeoidea, DAMAEIDAE**

77b. Notogastral setae not arranged in 2 longitudinal rows. Parastigmatic enantiophysis absent. Legs moniliform (Oppioidea) or not **78**

78a. Epimera III–IV distinctly delineated **79**

78b. Epimera III–IV not distinctly delineated; border of epimere III not visible **89**

79a. Seta *d* inserted on proximal 1/5 of femora I–III, proximal to other femoral setae; with 13 or 15–17 pairs of notogastral setae (secondary *h* setae). Bothridial seta reduced or absent (frequently broken). Lenticulus present, circular, strongly convex. Length: 400–700 µm (Figs. 15.40A–C) **HYDROZETIDAE (*pars*)**

79b. Seta *d* not most proximal seta on femora I–III; 10–15 pairs of notogastral setae, (only 3 pairs of *h* setae). Bothridial seta normal, reduced, or absent. Lenticulus, if present, not strongly convex **80**

80a. Bothridial seta very small or absent **81**

80b. Bothridial seta normally developed **83**

81a. Genital plates with 3 pairs of setae. Coxisternal setation 1-0-1-1, tarsus II with 1 solenidion. Length: 370–420 µm (Figs. 15.40D, E) **SELENORIBATIDAE**

81b. Genital plates with 5 or 6 pairs of setae. Coxisternal setation 3-1-2-2, 3-1-2-3, or 3-1-3-2, tarsus II with 2 solenidia **82**

82a. Genital plates with 5 pairs of setae. Lyrifissure *iad* positioned anterior to *ad3*. Lateral plastron (Van der Hammen's organ) absent. Leg tibiae and tarsi completely separate. Length: 550–800 µm (Figs. 15.40F–H) **AMERONOTHRIDAE**

82b. Genital plates with 6 pairs of setae. Lyrifissure *iad* positioned beside or posterior to *ad3*. Lateral plastron (Van der Hammen's organ) present. Tibiae and tarsi partially fused. Length: 400–500 µm (Figs. 15.41A–C) **FORTUYNIIDAE**

83a. Notogaster narrow, elongated, about twice as long as wide, prodorsum and notogaster fused; vestigial alveolus of second exobothridial seta present. Leg tarsi and tibiae functionally fused, move as single segment. Genital plates with 3–4 pairs of setae. Length: 300–850 µm (Figs. 15.36A–D) **DAMPFIELLIDAE (*pars*)**

83b. Notogaster oval, length to width ratio less than 1.5:1.0, prodorsum and notogaster usually separate; vestigial alveolus of second exobothridial seta absent. Leg tarsi and tibiae articulate normally. Genital plates with 4–6 pairs of setae **84**

84a. Border of epimere IV with taenidium and minitectum. Length: 300–500 µm (Figs. 15.41D, E) **THYRISOMIDAE (*pars*)**

84b. Border of epimere IV generally without taenidium and minitectum **85**

85a. Notogaster with pair of deep depressions between setae *c* and *la*, lamellar setae arising from medial tubercle. Body elongated, prodorsum nearly as long as notogaster. Costulae with long, thin cusps. Prodorsum usually with spine(s) between interlamellar and lamellar setae. Bothridial seta with ringlike thickenings proximally. Rostrum with pair of incisions. Length: 300–400 µm (Figs. 15.41F–H) **SPINOZETIDAE**

85b. Notogaster without paired depressions between setae *c* and *la*. Lamellar setae not arising from medial tubercle. Body not elongated. Costulae without cusps, or costulae absent. Prodorsum without spine(s). Bothridial seta without ringlike thickenings. Rostrum rounded or dentate **86**

86a. Femora I–IV with very broad ventral keel, trochanters III–IV without porose areas. Costulae absent. Propodolateral apophysis present. Enantiophyses L and E4 present. Length: 450–600 µm (Figs. 15.42A–C) **ARCEREMAEIDAE**

86b. Femora I–IV without broad ventral keel, trochanters III–IV with porose areas. Costulae present. Propodolateral apophysis absent. Enantiophyses L and E4 absent **87**

87a. Femora III–IV with retrotecta. Bothridial seta long, bifurcated. Tutorium, humeral apophysis, parastigmatic enantiophysis S, and enantiophysis V absent. Notogaster with 2 pairs of cristae. Length: 400–500 µm (Figs. 15.42D, E) **MACHADOBELBIDAE**

87b. Femora III–IV without retrotecta. Bothridial seta not bifurcated. Tutorium, humeral apophysis, parastigmatic enantiophysis S, and enantiophysis V present. Notogaster with or without crista **88**

88a. Bothridium with internal spiral thickening. Notogaster without tubercles on anterior margin, depressions absent between notogastral setae *c* and *la*; pedotectum II present, enantiophysis A absent. Fixed digit of chelicerae with unsclerotized lobe with 7–10 small teeth. Tarsus II with 2 solenidia. Length: 280–850 µm (Figs. 15.42F–H) **ANDEREMAEIDAE**

88b. Bothridium without internal thickening. Notogaster with tubercles on anterior margin, with 1 to several pairs of small, circular depressions between notogastral setae *c* and *la*; pedotectum II absent, enantiophysis A present. Fixed digit of chelicerae without unsclerotized lobe. Tarsus II with 1 solenidion. Length: 300–475 µm (Figs. 15.43A–C) **CALEREMAEIDAE**

89a. Body elongated. Pedotectum II large, well developed, auriculate or not. Distance between genital and anal plates greater than twice length of genital plate **90**

89b. Body not elongated. Pedotectum II weakly developed, never auriculate. Distance between genital and anal plates less than twice length genital plate **91**

90a. Costulae present. Pedotectum II auriculate. Border of epimere IV without minitectum. Length: 500–1,600 µm (Figs. 15.43D, E) **OTOCEPHEIDAE**

90b. Costulae absent. Pedotectum II not auriculate. Border of epimere IV with minitectum. Length: 650–750 µm (Figs. 15.43F–H) **TOKUNOCEPHEIDAE**

91a. Tarsi I–II shorter than tibiae I–II, respectively. Chelicerae with edentate or dentate chelae **92**

91b. Tarsi I–II longer than or subequal to tibiae I–II, respectively. Chelicerae with dentate chelae **93**

92a. Notogaster and posterior of prodorsum covered with very thick, white cerotegument, held between long, thick, and heavily barbed interlamellar setae and 5 pairs of long, thick, and heavily barbed notogastral setae. Subcapitulum diarthric. Chelicerae with dentate chelae. Length: ca. 425 µm (Figs. 15.44A–C; 15.11G) **TUPAREZETIDAE**

92b. Thick, white cerotegument absent. Notogastral and interlamellar setae setiform. Subcapitulum anarthric. Chelicerae with edentate chelae. Length: 370–400 µm (Figs. 15.44D–I) **NOSYBELBIDAE**

93a. Lenticulus present or absent. Tubercle present bearing palptarsal eupathidium *acm*, associated with solenidion or solenidion positioned at base of tubercle **94**

93b. Lenticulus absent. Tubercle bearing palptarsal eupathidium *acm* absent; *acm* not closely associated with or fused to solenidion **95**

94a. Aggenital setae absent. Notogaster with 14 pairs of setae. Tarsal setation I–IV reduced, 13-12-9-8, respectively (not including solenidia). Femora III–IV with 2 porose areas each, 1 proximally on femoral bulge and 1 paraxially on segment. Subcapitular mentum without tectum. Solenidion σ absent from genu III, porose areas or saccules absent from tibiae and tarsi. Palpal eupathidium *acm* forming imperfect "double horn" with solenidion. Length: 240–315 µm (Figs. 15.44J–L) **MICREREMIDAE (*pars*)**

94b. Aggenital setae present. Notogaster usually with 10 or 13 pairs of setae. Tarsal setation not reduced, usually (17–20)-15-15-12, respectively (not including solenidia), femora III–IV with 1 porose area, on paraxial face. Subcapitular mentum with or without tectum. Solenidion σ present on genu III, tibiae and tarsi with porose areas or saccules. Palpal eupathidium *acm* free from solenidion. Length: 350–800 µm (Figs. 15.45A–C) **CYMBAEREMAEIDAE (*pars*)**

95a. Border of epimere IV with minitectum. Genital and anal plates very large, almost touching, separated by distance of less than 1/2 length genital plate. Length: 290–500 µm (Figs. 15.41D, E) **THYRISOMIDAE (*pars*)**

95b. Border of epimere IV generally without minitectum. Genital and anal plates separated by distance greater than 1/2 length of genital plate **96**

96a. Rostrum with straight, deep, medial incision. Medially positioned parallel costulae present; humeral enantiophysis present. Tibia I with large dorsodistal tubercle overhanging tarsus I. Length: 250–350 µm (Figs. 15.45D–F) **AUTOGNETIDAE**

96b. Rostrum without medial incision. Costulae not parallel; humeral enantiophysis absent. Tibia I without large dorsodistal tubercle overhanging tarsus I **97**

97a. Costulae H shaped. Seven pairs of notogastral setae arranged in submarginal longitudinal row. Length: 250–350 µm (Figs. 15.45G, H) **EREMELLIDAE**

97b. Costulae, if present, never H shaped. Notogastral setae arranged otherwise **98**

98a. Tibiae I–II expanded laterally, with apophyses. Genital plates with 6 pairs of setae. Length: 300–450 µm (Figs. 15.46A–E) **TERATOPPIIDAE**

98b. Tibiae I–II without apophyses. Genital plates with 3–6 pairs of setae **99**

99a. Genital plates with 3 pairs of setae. Prodorsum flattened; rostrum very broad. Lamellar setae originating near anterior margin of rostrum in transverse alignment with rostral setae; interlamellar setae positioned midway between bothridia and rostrum. Length: 450–900 µm (Figs. 15.46F, G) **PLATYAMERIDAE**

99b. Genital plates with 4–6 pairs of setae. Prodorsum not flattened; rostrum tapered. Lamellar setae inserted posterior to rostral setae; interlamellar setae positioned in interbothridial region **100**

100a. Coxisternal setal formula 2-2-4-5 or up to 17 pairs of coxisternal setae; coxisternal setae directed medially toward ventrosejugal region, forming a "basket" within a thick layer of cerotegument. Anterior margin of epimere I with medial tooth. Distance between genital and anal plates less than length of genital plate. Length: 150–250 µm (Figs. 15.46H, I) **MACHUELLIDAE**

100b. Coxisternal setal formula usually 3-1-3-3 or setation reduced; coxisternal setae not directed medially toward ventrosejugal region, not forming a "basket" in thick layer of cerotegument. Anterior margin of epimere I without tooth. Distance between genital and anal plates greater than length of genital plate **101**

101a. Coxisternal setae large, plumose; coxisternum covered by 2 overlapping plates. Length: 250–350 µm (Figs. 15.47A, B) **STERNOPPIIDAE**

101b. Coxisternal setae setose; coxisternum not covered by 2 overlapping plates **102**

102a. Borders of epimera III–IV separated medially by area subequal in width to genital aperture. Length: 300–500 µm (Figs. 15.47C, D) **EPIMERELLIDAE**

102b. Borders of epimera III–IV meet medially **103**

103a. Epimeral border IV absent **104**

103b. Epimeral border IV present **106**

104a. Setae *ad1* (most posterior pair) in adanal position, far removed from anal plates; setae *ad3* in preanal position; costulae present. Length: ca. 450–680 µm (Figs. 15.47E, F) **GRANULOPPIIDAE**

104b. Setae *ad1* (most posterior pair) in postanal position (if in adanal position, costulae absent); setae *ad3* in preanal or adanal position **105**

105a. Epimera III–IV long, extending well posterior of genital plates. Notogaster without anteromedial sclerotized elevation between 2 oval concavities. Length: 120–550 µm (Figs. 15.49F–H) **OPPIIDAE (*pars*)**

105b. Epimera III–IV normal in length, not extending posterior of genital plates. Notogaster with anteromedial sclerotized elevation between 2 oval concavities. Length: 400–500 µm (Figs. 15.48A, B) **ENANTIOPPIIDAE**

106a. Epimeral border IV straight, transverse, posterior to genital plates. Papillate cerotegument covering body. Costulae strongly developed, with anterior cusp; notogastral setae papillate. Length: 220–250 µm (Figs. 15.48F, G) **PAPILLONOTIDAE**

106b. Epimeral border IV either slightly convex or concave, never straight transverse. Cerotegument if present, not papillate. Costulae, if present, without anterior cusp; notogastral setae not papillate **107**

107a. Prodorsum with lateral paired flat regions (tectopedial fields), notogaster without crista. Epimera appear separated by broad sternal region. Length: 260–290 µm (Figs. 15.48C–E) **CHAVINIIDAE**

107b. Prodorsum without paired tectopedial fields, notogaster with or without crista. Epimera usually not appearing separated by broad sternal region **108**

108a. Ventral plate with postanal saccule. Posterior margin of notogaster divided medially, with margins overlapping. Solenidion ω*2* on tarsus I arising well proximal of famulus; *ω1* on tarsus II positioned proximally on segment. Length: 230–260 µm (Figs. 15.49A–E) **LUXTONIIDAE**

108b. Ventral plate without postanal saccule. Posterior margin of notogaster undivided. Solenidion ω*2* on tarsus I arising close to famulus; ω*1* on tarsus II positioned medially on segment **109**

109a. Costulae, if present, shorter than 1/2 length of prodorsum. Crista, if present, not longer than 1/3 length of notogaster. Length: 120–550 µm (Figs. 15.49F–H) **OPPIIDAE (*pars*)**

109b. Costulae longer than 1/2 length of prodorsum. Crista 1/3 length of notogaster or longer **110**

110a. Notogaster with 9 pairs of setae, similar in size and shape. Length: 150–230 µm (Figs. 15.49I–L) **QUADROPPIIDAE**

110b. Notogaster with 10 pairs of setae, of which 3 pairs are long and densely pilose. Length: 180–200 µm (Figs. 15.50A, B) **HEXOPPIIDAE**

111a. Scalps present, obscured by thick cerotegument bearing mass of dirt. Tutorium present. Femur IV and trochanters III–IV with large ventral keel **112**

111b. Scalps absent. Tutorium present or absent. Femur IV and trochanters III–IV with or without large ventral keel **113**

112a. Cerotegument very thick and bearing dirt mass. Lamellae very large and touching medially, medially thickened and with thin lateral extensions. Patronium present, pedotectum II absent. Subcapitulum anarthric; mentum and gena extending anteriorly over rutella like a tectum; rutella in form of large bifurcate seta. Chelicerae highly modified with strainerlike teeth; cheliceral setae setiform. Length: 400–560 µm (Figs. 15.50C–G) **POLYPTEROZETIDAE**

112b. Cerotegument thin, without adherent dirt. Lamellae very large and fused medially, patronium absent. Small pedotectum II present. Subcapitulum diarthric; mentum and gena not extending anteriorly over rutella; rutella normal. Chelicerae chelate-dentate; cheliceral seta *cha* curved and coiled many times back over digit. Length: 350–400 µm (Figs. 15.50H–K) **PODOPTEROTEGAEIDAE**

113a. Subcapitulum anarthric. Cheliceral digits highly modified, or chelicerae pelopsiform or styliform (movable digit serrate distally) **114**

113b. Subcapitulum diarthric. Chelicerae chelate-dentate **120**

114a. Notogaster with or without large fossae. Ventral plate with gutterlike tectum, accommodating edge of notogaster. Chelicerae pelopsiform. Length: 340–420 µm (Figs. 15.52A–D) **CHARASSOBATIDAE**

114b. Notogaster without fossae. Ventral plate without tectum. Chelicerae highly modified, or pelopsiform or styliform (movable digit serrate distally) **115**

115a. Chelicerae styliform, movable digit serrate distally, fixed digit absent. Border of epimere IV with minitectum. Length: ca. 600 µm (Figs. 15.51A–C) **GUSTAVIIDAE**

115b. Chelicerae highly modified or pelopsiform. Border of epimere IV without minitectum **116**

116a. Humeral processes large, subrectangular, curving dorsally, covering seta *c* or not. Gnathosoma not strongly tapered; cheliceral digits highly modified. Rutella in form of large bifurcate setae; mentum and gena fused and extended as tectum that covers base of rutella **117**

116b. Humeral processes, if present, small, subtriangular, not curving dorsally, not covering seta *c*, or humeral process absent. Gnathosoma strongly tapered. Chelicerae pelopsiform; digits not highly modified. Rutella tubular; mentum and gena fused but not extended as tectum that covers base of rutella **118**

117a. Notogaster with large projection anteromedially, curving above prodorsum and forming large open depression with anterior of notogaster; prodorsum with broad converging lamella-like tubercles, bearing interlamellar setae, these tubercles with connecting transverse ridge. With 10 pairs of notogastral setae; seta *c2* absent. Length: ca. 500 µm (Figs. 15.51D–H) **TUMEROZETIDAE**

117b. Notogaster without projection anteromedially; prodorsum with large medial hump, bearing interlamellar setae. With 11 pairs of notogastral setae; seta *c2* hidden by dorsally curved humeral process. Length: 300–350 µm (Figs. 15.51I–K) **NODOCEPHEIDAE**

118a. Integument shiny; cerotegument weakly developed, mainly in acetabular region. Notogaster without humeral processes or tubercles. Length: 400–800 µm (Figs. 15.52E, F) **PELOPPIIDAE (*pars*)**

118b. Integument roughened; cerotegument well developed. Notogaster with humeral processes or tubercles **119**

119a. Rostrum without medial incision. Bothridium cup shaped, without internal thickening. Enantiophysis E4 present on either side of genital plates; deep pit present medially over sejugal apodeme. Tutorium absent. Length: 210–380 µm (Figs. 15.52G–I) **MICROTEGEIDAE (*pars*)**

119b. Rostrum with long, deep medial incision. Bothridium trumpet shaped, with internal thickening. Enantiophysis E4 absent; pit absent medially over sejugal apodeme. Tutorium present. Length: approximately 400 µm (Figs. 15.52J, K) **CEROCEPHEIDAE**

120a. Notogaster with long, triangular or subrectangular humeral processes, projecting anteriorly to bothridia, or projecting dorsally over bothridium. Genital neotrichy absent **121**

120b. Notogaster without long, triangular humeral processes, if humeral tubercles present, not reaching level of bothridium. Genital neotrichy present or absent **124**

121a. Subcapitular mentum with tectum. Palptarsal eupathidium *acm* on tubercle, associated with solenidion, or solenidion positioned at base of tubercle. Length: 350–800 µm (Figs. 15.45A–C) **CYMBAEREMAEIDAE (*pars*)**

121b. Subcapitular mentum without tectum. Palptarsal eupathidium not on tubercle, not associated with or fused to solenidion **122**

122a. Notogaster with large subtriangular or subrectangular humeral processes projecting dorsally over bothridium; prodorsum with large processes, interlocked or not with humeral processes. Length: 600–750 µm (Fig. 15.53A) **NIPPOBODIDAE**

122b. Notogaster with long, triangular or subrectangular humeral processes, projecting anterolaterally to bothridium; prodorsum without processes **123**

123a. Prodorsum and notogaster smooth, shiny; bothridium without internal spiral thickenings. Podosomal region with depressions for reception of retracted legs. Epimeral border IV with minitectum. Trochanters III–IV and femora III–IV without ventral carina. Length: 800–1,200 µm (Figs. 15.53B–E) **TENUIALIDAE**

123b. Prodorsum and notogaster sculptured, not shiny; bothridium with internal spiral thickenings. Podosomal region without concavities for reception of retracted legs. Epimeral border IV without minitectum. Trochanter III–IV and femora III–IV with well-developed ventral carina. Length: 700–1,050 µm (Figs. 15.53F, G) **EUTEGAEIDAE**

124a. Genital neotrichy present; 11–18 pairs of genital setae in 2 longitudinal rows. Epimere IV with 4 setae. Length: 850–1,150 µm (Figs. 15.54A, B) **NIPHOCEPHEIDAE**

124b. Genital neotrichy absent; 4–7 pairs of genital setae. Epimere IV with at most 3 setae **125**

125a. Integument of genital area usually darker than surrounding area. Epimeral setae *4a* and *4b* closely adjacent. Lenticulus usually present. Pedotectum I in two parts (narrow dorsal part and wider ventral part). Length: 250–500 µm (Figs. 15.54C–E) **TEGEOCRANELLIDAE**

125b. Integument of genital area same color as anal region. Epimeral setae *4a* and *4b* not closely adjacent. Lenticulus absent. Pedotectum I not in two separate parts **126**

126a. Subcapitular mentum with tectum. Eupathidium *acm* of palptarsus on tubercle. Lamellar cusps touching medially, fused or not, covering rostrum medially; lamellar setae arising anteroventrally on lamellar cusps. Length: 350–800 µm (Figs. 15.45A–C) **CYMBAEREMAEIDAE (*pars*)**

126b. Subcapitular mentum without tectum. Eupathidium *acm* of palptarsus not on tubercle. Lamellar cusps, if touching medially, not covering rostrum; lamellar setae arising anteriorly or dorsally on lamellae or lamellar cusps **127**

127a. Tarsus I with 3 solenidia, pretarsus with very small claw, tarsal pulvillus present; legs IV modified for jumping or not. Cerotegument with large globules. Adanal setation 1 or 2 pairs, positioned postanally. Length: 330–550 µm (Figs. 15.54F–I) **ZETORCHESTIDAE**

127b. Tarsus I with 2 solenidia, pretarsus with normal size claw, tarsal pulvillus absent; legs IV not modified for jumping. Cerotegument without globules. Adanal setation 3 pairs, positioned adanally and postanally **128**

128a. Notogaster with 0–1 pair of setae positioned centrodorsally; all other notogastral setae in marginal and posteromarginal rows and clearly visible; enantiophysis E4 present or absent **129**

129b. Notogaster with 2 pairs of setae positioned centrodorsally, or notogastral setae other than *p* series absent, with or without alveoli; enantiophysis E4 absent **130**

129a. Notogaster with 1 pair of setae centrodorsally; other notogastral setae in marginal or posteromarginal position; enantiophysis E4 present on either side of genital aperture. Integument sculptured. Length: 210–380 µm (Figs. 15.52G–I) **MICROTEGEIDAE (*pars*)**

129b. Notogaster without setae centrodorsally; all notogastral setae in marginal or posteromarginal position; enantiophysis E4 absent. Integument usually sculptured (smooth in *Conoppia*). Length: 500–1,200 µm (Figs. 15.55A–C) **CEPHEIDAE**

130a. Humeral region with 2 pairs of closely adjacent setae *c* (*c1* and *c2*). Length: 500–1,160 µm (Figs. 15.55D–G) **LIACARIDAE**

130b. Humeral region with 0–1 pair of setae *c* **131**

131a. Integument smooth. Lamellae converging or parallel, if parallel, not positioned close to lateral margin of prodorsum **132**

131b. Integument distinctly sculptured with coarse ridges, tubercles, or areoles. Lamellae usually broad, almost parallel, close to lateral margin of prodorsum **135**

132a. Notogaster with 50–55 pairs of setal alveoli. Length: 600–750 µm (Figs. 15.56A, B) **MULTORIBULIDAE**

132b. Notogaster with 9–13 pairs of setae or alveoli **133**

133a. Anal plates with 3 pairs of setae. Length: 750–900 µm (Figs. 15.56C–E) **KODIAKELLIDAE**

133b. Anal plates with 2 pairs of setae **134**

134a. Genital plates large; distance between genital and anal plates usually less than length of anal plates. Notogaster glabrous or with 10 pairs of notogastral setae. Length: 250–600 µm (Figs. 15.56F, G) **ASTEGISTIDAE**

134b. Genital plates medium in size; distance between genital and anal plates usually greater than length of anal plates. Eight pairs of notogastral setae present or vestigial. Length: 400–800 µm (Figs. 15.52E, F) **PELOPPIIDAE (*pars*)**

135a. Genal notch present. Circumpedal carina developed caudally. Anal plate triangular. Leg tarsi and tibiae articulate normally. Interlamellar setae positioned in interbothridial region. Length: 225–300 µm (Figs. 15.57A–C) **TECTOCEPHEIDAE (*pars*)**

135b. Genal notch absent. Circumpedal carina absent. Anal plate rectangular. Leg tarsi functionally fused to tibiae. Interlamellar setae usually positioned well anterior of bothridia or on lamellae **136**

136a. Rostrum rounded; lamellar cusps present. Cerotegument of notogaster with foveae or tubercles. Tutorium without free cusp. Length: 350–600 µm (Figs. 15.57D–F) **CARABODIDAE**

136b. Rostrum elongate; lamellar cusps absent, lamellae merging with prodorsum anteriorly. Cerotegument of notogaster granular. Tutorium with free cusp. Length: ca. 600 µm (Figs. 15.57G–I) **CARABOCEPHEIDAE**

137a. Pteromorphs movable, auriculate, that is, extending anteriorly and posteriorly past articulation with notogaster such that all legs are covered when pteromorphs are closed. Seta *c* positioned on pteromorph or body of notogaster **138**

137b. Pteromorphs present or absent; if present, fixed or movable but not auriculate. Seta *c* positioned on body of notogaster **142**

138a. Lamellae absent, at most a narrow carina present (line L). Pteromorph with alary furrow, appearing bilobed. Subcapitular mentum with tectum; tutorium reduced to line (S) or absent **Superfamily Galumnoidea—139**

138b. Lamellae or broad carina present. Pteromorph without alary furrow. Subcapitular mentum without tectum; tutorium present, lamelliform **140**

139a. Rostrum acuminate. Chelicerae pelopsiform. Length: 440–500 µm (Figs. 15.58A–D) **GALUMNELLIDAE**

139b. Rostrum rounded. Chelicerae chelate-dentate. Length: 300–1100 µm (Figs. 15.58E–H) **GALUMNIDAE**

140a. Lamella usually narrow, directed anteriorly from bothridium as simple carina. Seta *c* positioned on body of notogaster, notogaster without posterior tectum. Tibiae and tarsi I–IV with porose areas ventrally. Marginoventral porose areas present. Adalar porose organ of octotaxic system expressed as porose area or saccule, other 3 (mesonotic porose organs) always expressed as saccules. Length: 300–820 µm (Fig. 15.58I) **PARAKALUMMIDAE**

140b. Lamellae broader, usually converging and with distinct cusps. Seta *c* positioned on pteromorph; notogaster with posterior tectum. Tibiae and tarsi I–IV without porose areas ventrally. Marginoventral porose areas absent. Octotaxic system expressed as porose areas or saccules **141**

141a. Notogaster and pteromorphs foveate. Genital plates with 6 pairs of setae; postanal porose area absent. Length: 210–300 µm (Figs. 15.59A, B) **EPACTOZETIDAE**

141b. Notogaster and pteromorphs smooth. Genital plates with 5 pairs of setae; postanal porose area present. Length: 250–350 µm (Figs. 15.59C–E) **CERATOKALUMMIDAE**

142a. Epimeral border IV clearly evident, extending transversely anterior to genital plates; border of epimere IV with tectum. Prodorsum often longer than notogaster; the latter as wide as long, or wider than long. Chelicerae dorsally with bacilliform tubercle. Tarsi II–IV with short, thick setae (*p*). Length: 150–550 µm (Figs. 15.59F–I) **MICROZETIDAE**

142b. Epimeral border IV not clearly evident, or, if present, not transverse but angled anteriorly to meet genital plates. Border of epimere IV without tectum. Prodorsum shorter than notogaster; the latter longer than wide. Chelicerae dorsally without bacilliform tubercle. Tarsi II–IV with setiform setae (*p*) **143**

143a. Bothridial seta weakly developed, reduced, or absent. Seta *d* inserted on proximal fifth of femora I–III, proximal to other femoral setae; epimera II–IV clearly delineated **144**

143b. Bothridial seta well developed. Seta *d* not most proximal seta on femora; epimera II–IV weakly delineated **145**

144a. Lenticulus absent. Notogaster with or without pteromorphs, with 10 pairs of notogastral setae, lamellae and tutoria well developed. Length: 260–380 µm (Figs. 15.60A–C) **LIMNOZETIDAE**

144b. Lenticulus present, circular, strongly convex. Notogaster without pteromorphs, with 13 or 15–17 pairs of notogastral setae (because of secondary *h* setae), prodorsal carina and tutoria weakly expressed. Length: 400–700 µm (Figs. 15.40A–C) **HYDROZETIDAE (*pars*)**

145a. Cerotegument well developed, very thick, with or without blocky structure, sometimes birefringent in polarized light. Pteromorphs present **146**

145b. Cerotegument developed as thin layer, not blocky in structure, never birefringent. Pteromorphs present or absent **149**

146a. Lenticulus distinct convex, shiny, oval structure, free of cerotegument. Interlamellar setae present or absent. Cerotegument not birefringent in polarized light. Chelicerae normal **147**

146b. Lenticulus, if present not distinctly convex or oval (exception: some PHENOPELOPIDAE with pelopsiform chelicerae), not shiny. Interlamellar setae present. Cerotegument birefringent in polarized light. Chelicerae pelopsiform or normal **148**

147a. Notogaster shorter than prodorsum, strongly convex, with anterior process bearing lenticulus; apophyses lateral to process bearing setae *c*; humeral processes shaped as prominent discal projections; with pteromorphs long, attenuated, strongly bent downward. Interlamellar setae present. Length: 260–405 µm (Figs. 15.60D–F) **IDIOZETIDAE**

147b. Notogaster longer than prodorsum, weakly convex, without anterior process; without apophyses laterally; humeral processes usually subrectangular or round; with pteromorphs small to long, attenuated or not, often strongly bent downward. Interlamellar setae usually absent. Length: 260–600 µm (Figs. 15.60G–I) **EREMAEOZETIDAE**

148a. Genital plates with 8 or more pairs of setae. Octotaxic system absent. Notogastral setal pair *h1* divergent, posteromarginal region of notogaster undulate. Pedotectum I without transverse carina. Interlamellar seta setose. Chelicerae normal. Length: 550–670 µm (Figs. 15.61A–C) **UNDULORIBATIDAE**

148b. Genital plates with 6 pairs of setae. Octotaxic system of porose areas, if present, all closely associated with setae. Posteromarginal region of notogaster smooth, notogastral setal pair *h1* convergent or directed posteriorly. Pedotectum I with transverse carina. Interlamellar seta large, spatulate or short, setose. Chelicerae pelopsiform or normal. Length: 400–1,000 µm (Figs. 15.61D–G) **PHENOPELOPIDAE**

149a. Bothridium with internal, spiral thickenings. Genu IV long thin segment, concave dorsally, longer than genu III, usually longer than tibia IV. Pteromorph present, with or without long, knifelike anterior projection. Length: 390–850 µm (Figs. 15.61H, I) **ACHIPTERIIDAE**

149b. Bothridium without internal, spiral thickenings. Genu IV straight dorsally, short segment, subequal in length to genu III, shorter than tibia IV. Pteromorph, if present, without knifelike projection **150**

150a. Lamellae completely fused medially (exception: separate in *Hypozetes*). Tibia IV usually without solenidion. Length: 340–480 µm (Figs. 15.62A–C) **TEGORIBATIDAE**

150b. Lamellae, if present, touching medially or separate, never completely fused. Tibia IV with solenidion **151**

151a. Lenticulus sharply defined, oval, at center of concavity on notogaster. Prodorsum without lamellae; with or without costulae; lamellar setae positioned closely adjacent to rostral setae, not associated with costulae. Tibiae and tarsi with retrotecta. Pteromorph absent. Length: 230–460 µm (Figs. 15.64A–C) **PASSALOZETIDAE**

151b. Lenticulus, if present, not sharply defined, rectangular to subtriangular in shape. Prodorsum with or without lamellae, lamellar setae borne on lamellae or lamellar cusps or on adjacent prodorsum, lamellar setae not closely adjacent to rostral setae. Tibiae and tarsi without retrotecta. Pteromorph present or absent **152**

152a. Circumpedal carina partially developed close to lateral margin of ventral plate. Octotaxic system absent. Sejugal apodeme and apodemes 3–4 meet medially anterior to genital plate. Genal notch present. Length: 250–280 µm (Figs. 15.57A–C) **TECTOCEPHEIDAE (*pars*)**

152b. Circumpedal carina complete. Octotaxic system present or absent. Sejugal apodeme and apodemes 3–4 not meeting medially anterior to genital plate. Genal notch present or absent **153**

153a. Notogastral neotrichy present, with 30–35 pairs of long, dark, notogastral setae. Length: 700–750 µm (Figs. 15.62D, E) **NEOTRICHOZETIDAE**

153b. Notogastral neotrichy absent, with at most 15 pairs of notogastral setae **154**

154a. Pedotectum I at most weakly developed as small lamina. Prodorsum without genal notch. Axillary sacculus of subcapitulum absent. Postanal porose area absent. **155**

154b. Pedotectum I well developed as large lamina. Prodorsum with genal notch. Axillary sacculus of subcapitulum usually present. Postanal porose area usually present (absent in MAUDHEIMIIDAE and CHAMOBATIDAE). **175**

155a. Notogaster with triangular humeral process bearing large seta *c* apically, seta *c* larger than other notogastral setae; rostrum denticulate. Legs IV sometimes modified as jumping legs. Trochanter IV without seta. Lyrifissure *im* often modified as piriform organ. Large humeral saccule present or absent. Length: 350–550 µm (Figs. 15.62F–H) **ZETOMOTRICHIDAE**

155b. Notogaster without triangular humeral process; seta *c* subequal in size and length to other notogastral setae; rostrum never denticulate. Legs IV never modified as jumping leg. Trochanter IV with seta. Lyrifissure *im* not modified as piriform organ. Humeral saccule absent **156**

156a. Pedotectum II usually absent. Genital and anal plates closely adjacent, separated by less than half length of genital plates. Notogaster with small porose areas *Aa*, *A2* only **157**

156b. Pedotectum II present. Genital and anal plates separated by more than half length of genital plate. Notogaster with 0–many pairs of porose areas or saccules **158**

157a. Bothridial seta large, strongly widened distally, leaf shaped; prodorsum with costulae. Notogaster extending medially between bothridia and setae *in*, forming acute angle; with large circular to reticulate sculpturing, independent of sculpturing formed by cerotegument. Length: 200–250 µm (Figs. 15.63A–C) **LICNEREMAEIDAE**

157b. Bothridial seta not strongly widened distally, not leaf shaped; prodorsum with lamellae and translamella; notogaster curved to flat anteriorly, not extending medially between bothridia and setae *in*. Notogaster and cerotegument without large circular to reticulate sculpturing. Length: 200–300 µm (Figs. 15.63D–F) **LAMELLAREIDAE**

158a. Prodorsum laterally with broad hyaline expansions. Aggenital setae absent, with 2 pairs of adanal setae. Length: ca. 270 µm (Figs. 15.63G, H) **NESOZETIDAE**

158b. Prodorsum laterally without hyaline expansions. Aggenital setae usually present, with 2–3 pairs of adanal setae **159**

159a. Aggenital setae absent. Femora III–IV each with 2 porose areas, 1 proximally on femoral bulge and 1 paraxially on segment, solenidion σ absent from genu III. Length: 240–315 μm (Figs. 15.44J–L) **MICREREMIDAE (*pars*)**

159b. Aggenital setae present. Femora III–IV each with 1 porose area, solenidion σ present on genu III **160**

160a. Posterior notogastral tectum present **161**

160b. Posterior notogastral tectum absent **162**

161a. Tutorium present. Posterior notogastral tectum without overlapping lobes. Tarsal pulvillus absent, brachytracheae or saccules present on tibiae I–IV. Length: 350–500 μm (Figs. 15.64D–G) **SCUTOVERTICIDAE**

161b. Tutorium absent. Posterior notogastral tectum with overlapping lobes. Tarsal pulvillus present; porose areas, brachytracheae or saccules absent from tibiae I–IV. Length: 400–450 μm (Figs. 15.64H–K) **ADHAESOZETIDAE**

162a. Pteromorphs well developed, with hinge; insertions of adductor muscles for pteromorphs evident on notogaster **163**

162b. Pteromorphs if present, without hinge; pteromorph adductor muscles absent **164**

163a. Subcapitulum anarthric. Chelicerae pelopsiform. Notogastral setae reduced to length of alveolus, octotaxic system of 4 pairs of saccules. Length: approximately 670 μm (Figs. 15.65A, B) **TUBULOZETIDAE**

163b. Subcapitulum diarthric. Chelicerae chelate-dentate. Notogastral setae developed, octotaxic system of porose areas or saccules. Length: 250–650 μm (Figs. 15.65C, D) **HAPLOZETIDAE**

164a. Tarsi, particularly tarsus I, strongly truncate. Prodorsum not curved over gnathosoma, and rostral tectum not completely covering chelicerae dorsally. Bothridium inserted lateroposteriorly, posterior to level of anterior notogastral margin, thus anterolateral border of notogaster partially or completely covering bothridial seta. Chelicerae with 1–2 setae. Anal, and/or adanal setae often long, flagelliform **165**

164b. Tarsi not truncate. Prodorsum curving over gnathosoma, covering chelicerae dorsally. Bothridium inserted anterior to notogastral margin. Chelicerae with 2 setae. Anal and adanal setae never long and flagelliform **167**

165a. Pretarsi tridactylous. With 4 pairs of saccules. With 1–3 pairs of genital setae. Length: 250–550 μm (Figs. 15.65E–I) **ORIPODIDAE**

165b. Pretarsi monodactylous. With 4 pairs of porose areas or octotaxic system absent. With 4 pairs of genital setae **166**

166a. Octotaxic system present as porose areas; sexual dimorphism in octotaxic system such that porose areas of male differ from those of female. Aggenital setae present. Setae (*u*) of tarsi I–IV setiform. Sexually dimorphic rostral setae present, those of female setose, directed anteromedially, those of male robust, tapered projections directed anteriorly. Length: 220–260 μm (Figs. 15.66A–C) **SYMBIORIBATIDAE**

166b. Octotaxic system absent, or 1–2 pairs of saccules present, sexual dimorphism in octotaxic system absent. Aggenital setae absent. Setae (*u*) of tarsi I–IV short, dark colored and terminally bifid. Sexually dimorphic rostral setae absent. Length: 230–335 μm (Figs. 15.66D–F) **CAMPBELLOBATIDAE**

167a. Notogaster with both porose areas and saccules, Sa and S1 saccules, remainder of octotaxic system porose areas. Length: 800–1,020 μm (Figs. 15.66G, H) **DRYMOBATIDAE**

167b. Notogaster with either porose areas or saccules but not both **168**

168a. Rostrum with medial projection, almost tubelike. Granular cerotegument present over entire body. Notogastral setae lanceolate; tutorium present. Length: ca. 430 μm (Figs. 15.67A–C) **NASOBATIDAE**

168b. Rostrum rounded, no tubelike projection. Granular cerotegument present at most in region between pteromorph, pedotectum I and lateral body wall. Notogastral setae not lanceolate; tutorium present or absent **169**

169a. Notogaster with 4 or more pairs of porose areas, their limits often well defined by internally thickened border; porose areas sexually dimorphic, or not; notogaster without conspicuous setae; 3 pairs of short setae in *p* row and 7 pairs of alveoli; prodorsum fused to notogaster; tutorium present. Distinct notch present antiaxially on chelicerae. Length: 500–800 μm (Figs. 15.67D, E) **MOCHLOZETIDAE**

169b. Notogaster usually with 4 (or fewer) pairs of porose areas or saccules, without thickened border; notogaster with 10–14 pairs of well-developed setae; prodorsum fused or not to notogaster. Tutorium present or absent. Without notch antiaxially on chelicerae **170**

170a. Notogaster with 4 pairs of curved, sausage-shaped saccules, notogastral setae unusually long. Tutorium and pleural carinae present. Length: 380–500 μm (Figs. 15.67F–I) **STELECHOBATIDAE**

170b. Notogaster with porose areas or saccules; if saccules present, never curved like a sausage; notogastral setae short to long; tutorium and pleural carinae absent **171**

171a. Pteromorphs or humeral projections usually present (if absent, octotaxic system as saccules). Distal eupathidia on palptarsus in same plane (Fig. 15.68D). Notogaster with 2–4 pairs of porose areas or saccules. With 1–4 pairs of genital setae. Prodorsum with prolamella often present; sublamella present. Sternal furrow extending from genital plates anteriorly on coxisternum often present. Length: 200–650 µm (Figs. 15.68A–D) **SCHELORIBATIDAE**

171b. Pteromorphs or humeral projections absent. Distal eupathidia on palptarsus not in same plane (Fig. 15.68H). Notogaster with 4–5 pairs of porose areas; 4–5 pairs of genital setae. Prodorsum without prolamella and sublamella. Sternal furrow absent **172**

172a. Adanal setation 2 pairs. Octotaxic system with 5 pairs of saccules, integument strongly foveate, notogastral setae strongly barbed to plumose, translamella present. Length 350–850 µm (Figs. 15.68E–G) **CALOPPIIDAE**

172b. Adanal setation 3 pairs. Octotaxic system comprised of porose areas or saccules; if more than 4 pairs, they are porose areas, integument usually smooth. Notogastral setae not strongly barbed to plumose **173**

173a. Marginoventral porose areas present. Tarsus I without dorsal porose area. Length: 250–900 µm (Figs. 15.68H–K) **ORIBATULIDAE**

173b. Marginoventral porose areas absent. Tarsus I with dorsal porose area between solenidia **174**

174a. Octotaxic system of 5 pairs of small porose areas. Femur II with 4 or fewer setae. Notogaster without caudal protuberance. Circumpedal carina incomplete posterior to acetabulum IV. Tarsus II with 2 solenidia, without porose area at base. Length: 300–400 µm (Figs. 15.69A, B) **CRASSORIBATULIDAE**

174b. Octotaxic system of 4 pairs of porose areas. Femur II with 5 setae. Notogaster with caudal protuberance bearing porose area and seta *p1*. Circumpedal carina complete posterior to acetabulum IV. Tarsus II with 1 solenidion, with porose area at base. Length: 575–625 µm (Figs. 15.69C–E) **SELLNICKIIDAE**

175a. Lamellae well developed, fused at base of cusps; lamellar cusps almost covering entire prodorsum, usually deeply incised, forming large medial and lateral tooth, or large lateral tooth; lamellar setae inserted at base of incision. Tutorium usually rectangular, dentate distally, or if tapered, notogaster with indistinct hexagonal pattern anteriorly. Pteromorph without hinge. Length: 240–750 µm (Figs. 15.69F–H) **ORIBATELLIDAE**

175b. Lamellae well or weakly developed, usually not fused at level of base of cusps; lamellar cusps not covering entire prodorsum; lamellae not deeply incised; medial and lateral tooth, if present, small. Tutorium not rectangular, usually tapered, dentate or not distally. Pteromorph with or without hinge **176**

176a. Pretarsi I and IV usually with different number of claws. Notogaster with posterior tectum **177**

176b. Pretarsi I and IV with same number of claws. Notogaster with or without posterior tectum **178**

177a. Notogaster with porose areas *Aa* and *A3* only, *Aa* positioned anteriorly, just posterior or medial to seta *c*. Claw of leg I expanded medially. With 10 pairs of notogastral setae. Length: ca. 315 µm (Figs. 15.70A, B) **ONYCHOBATIDAE**

177b. Notogaster with 4 pairs of porose areas, *Aa* positioned well posteriorly of seta *c*, or poros areas absent. Claw of leg I not expanded medially. With 10–11 pairs of notogastral setae or their alveoli; if 11 pairs, then 2 pairs of setae *c* (or their alveoli) closely adjacent on anterior of notogaster. Length. 250–750 µm (Figs. 15.70C–F) **ZETOMIMIDAE**

178a. Notogaster with posterior tectum. With 10 pairs of notogastral setae **179**

178b. Notogaster without posterior tectum. With 10–15 pairs of notogastral setae **181**

179a. Lamellar setae inserted on prodorsal surface, lamellar cusps absent. Postanal porose area absent. Length: 290–700 µm (Fig. 15.70G) **CHAMOBATIDAE**

179a. Lamellar setae inserted on lamellae or on lamellar cusps, which usually are present. Postanal porose area present **180**

180a. Prolamella extending from tip of lamellae to rostral margin, positioned medial to rostral seta. Tibiae and tarsi I–IV with distoventral and proximoventral porose areas. Tutorium without distinct cusp. Subcapitular mentum without tectum. Length: 600–936 µm (Figs. 15.71A–C) **HUMEROBATIDAE**

180b. Prolamella absent. Tibiae and tarsi I–IV without porose areas. Tutorium with distinct cusp. Subcapitular mentum with or without tectum. Length: 400–650 µm (Figs. 15.71D–F) **MYCOBATIDAE**

181a. Pteromorph small, not projecting ventrally. Custodium absent. Postanal porose area absent. Length: 500–650 µm (Figs. 15.72A–C) **MAUDHEIMIIDAE**

181b. Pteromorph usually well developed, projecting ventrally (exception *Jugatala*). Custodium present. Postanal porose area present **182**

182a. Subcapitular mentum with tectum. Epimere IV neotrichous (4 or more pairs of setae). With 10 pairs of minute notogastral setae or setae reduced to alveoli. Length: 1,000–1,200 µm (Fig. 15.72D) **EUZETIDAE**

182b. Subcapitular mentum without tectum. Epimere IV with 2 or 3 pairs of setae. With 10–15 pairs of minute to long notogastral setae. Length: 270–990 µm (Figs. 15.72E–H) **CERATOZETIDAE**

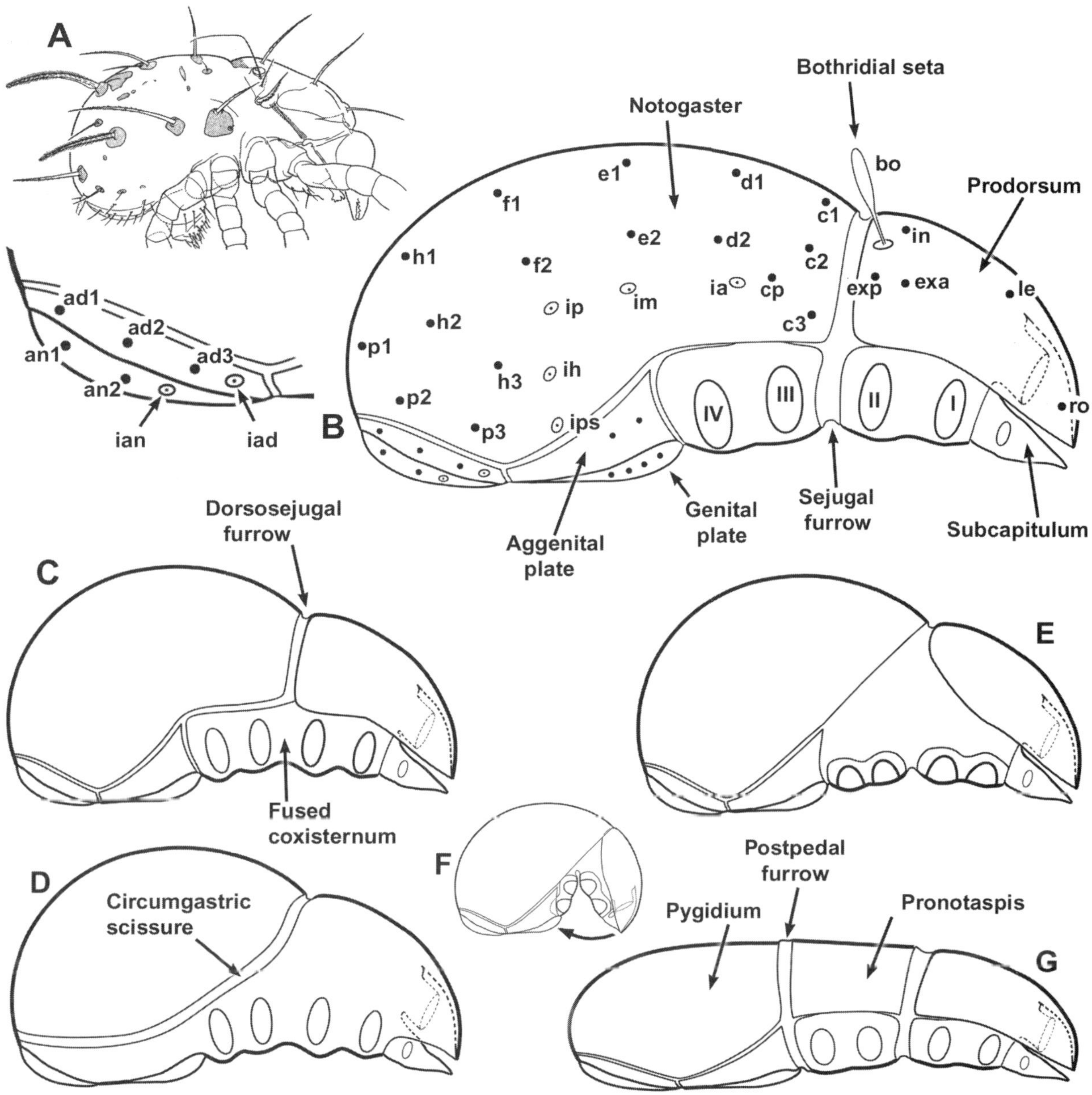

Fig. 15.1. General body forms of oribatid mites (in lateral aspect; ellipses indicate base of appendages). **A**, *Acaronychus traegardhi* Grandjean, illustrating primitive acariform mite body without major sclerites (after Grandjean 1954c); **B**, schematic dichoid mite showing general cuticular features (enlargement of anal region to left); **C**, schematic holoid mite, macropyline type; **D**, same, but brachypyline type; **E**, schematic ptychoid mite in active posture; **F**, same, partially closed; **G**, schematic trichoid mite. (**C–E, G** after Norton 2001; **F** after Sanders and Norton 2004).

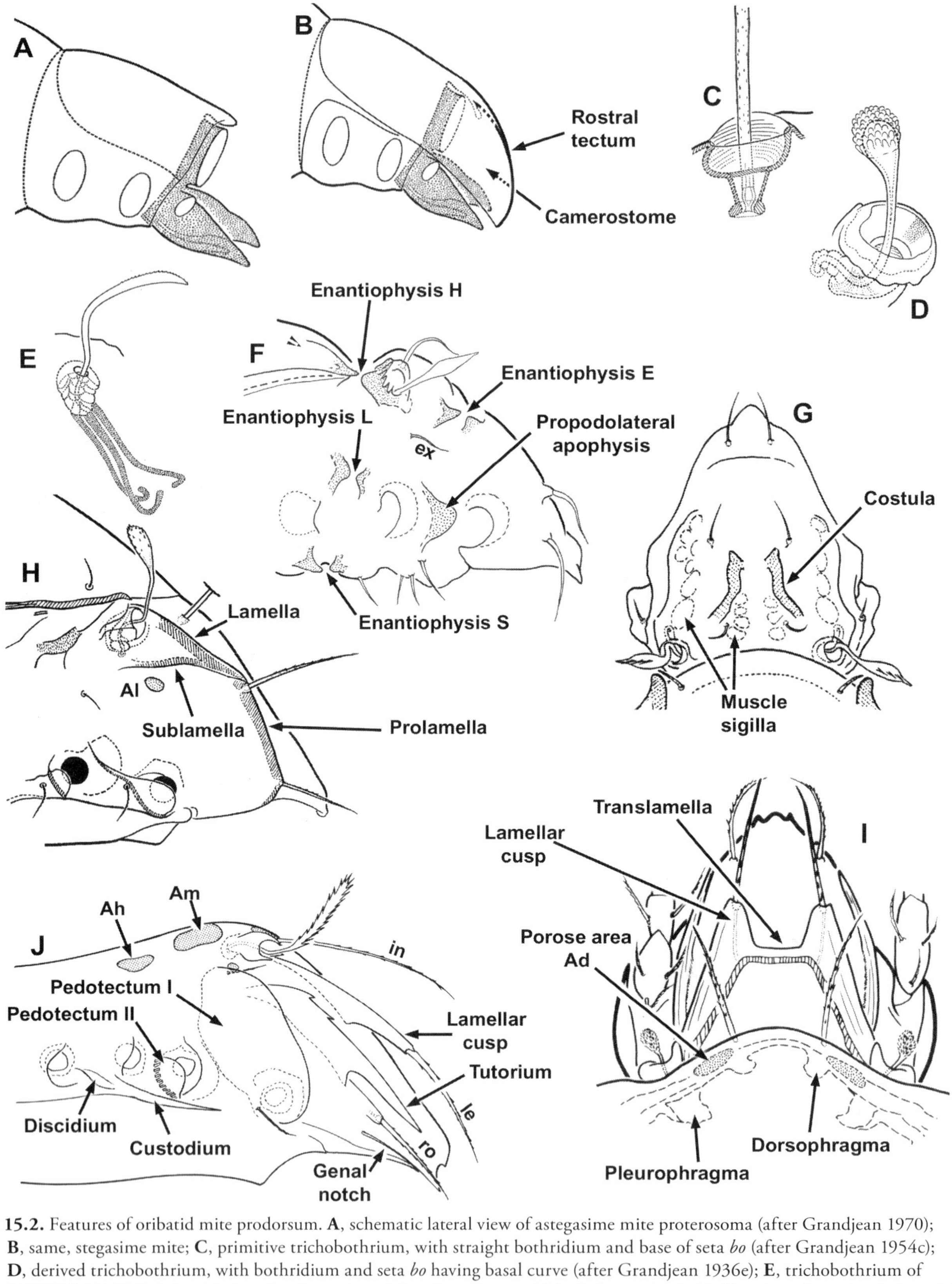

Fig. 15.2. Features of oribatid mite prodorsum. **A**, schematic lateral view of astegasime mite proterosoma (after Grandjean 1970); **B**, same, stegasime mite; **C**, primitive trichobothrium, with straight bothridium and base of seta *bo* (after Grandjean 1954c); **D**, derived trichobothrium, with bothridium and seta *bo* having basal curve (after Grandjean 1936e); **E**, trichobothrium of PHTHIRACARIDAE, with three brachytracheae (after Grandjean 1934d); **F–I**, anterior half of brachypyline mites in lateral (**F**, **H**, **J**) and dorsal (**G**, **I**) views; **F**, enantiophyses and other cuticular projections (R. A. Norton, unpubl.); **G**, OPPIIDAE (modified from Aoki 1983); **H**, SCHELORIBATIDAE (after Grandjean 1953a); **I**, CERATOZETIDAE (after Behan-Pelletier 1984); **J**, CERATOZETIDAE, prodorsum (after Behan-Pelletier 1986).

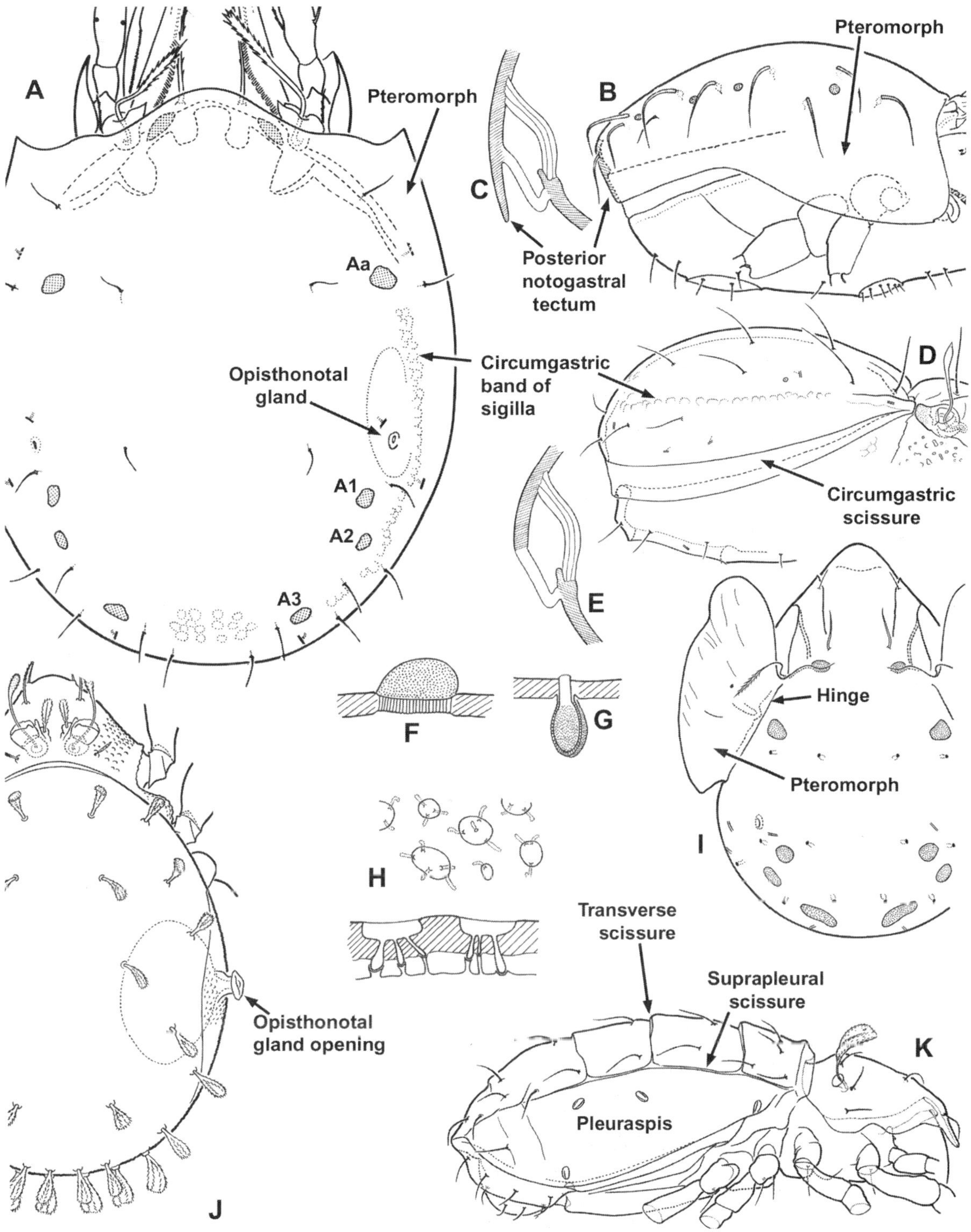

Fig. 15.3. Features of oribatid mite notogaster. **A**, CERATOZETIDAE, dorsal view (from Behan-Pelletier 1984); **B**, ORIBATELLIDAE, lateral view (from Grandjean 1956c); **C**, schematic cross section of posterior margin with tectum (after Grandjean 1959b); **D**, AUTOGNETIDAE, lateral view (after Grandjean 1960c); **E**, schematic cross section of posterior margin without tectum (after Grandjean 1959b); **F**, schematic section of cuticular part of porose area (after Norton et al. 1997); **G**, schematic section of cuticular part of saccule (after Norton et al. 1997); **H**, macropores in NEOLIODIDAE, top, dorsal view (after Grandjean 1934d) and HERMANNIELLIDAE, bottom, cross section (after Grandjean 1962a); **I**, GALUMNIDAE, dorsal view (after Grandjean 1956b); **J**, HERMANNIELLIDAE, dorsal view (after Grandjean 1962a); **K**, HAPLOCHTHONIIDAE, lateral view (after Grandjean 1947b).

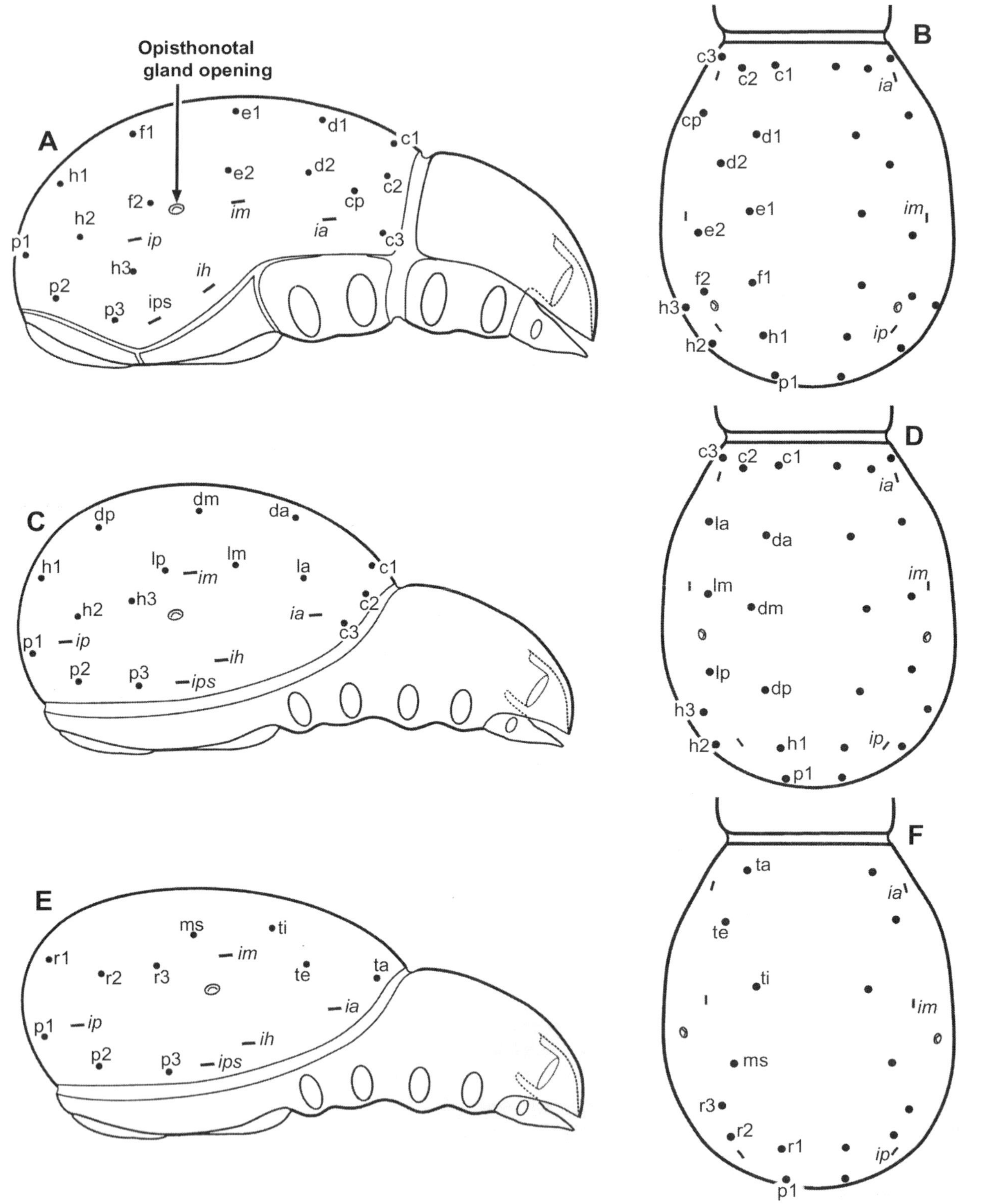

Fig. 15.4. Comparison of chaetotaxies used for notogastral setae of oribatid mites, shown in lateral (**A**, **C**, **E**) and dorsal (**B**, **D**, **F**) views. **A**, **B**, holotrichy; **C**, **D**, unideficiency; **E**, **F**, multideficiency, or "Dometorina nomenclature."

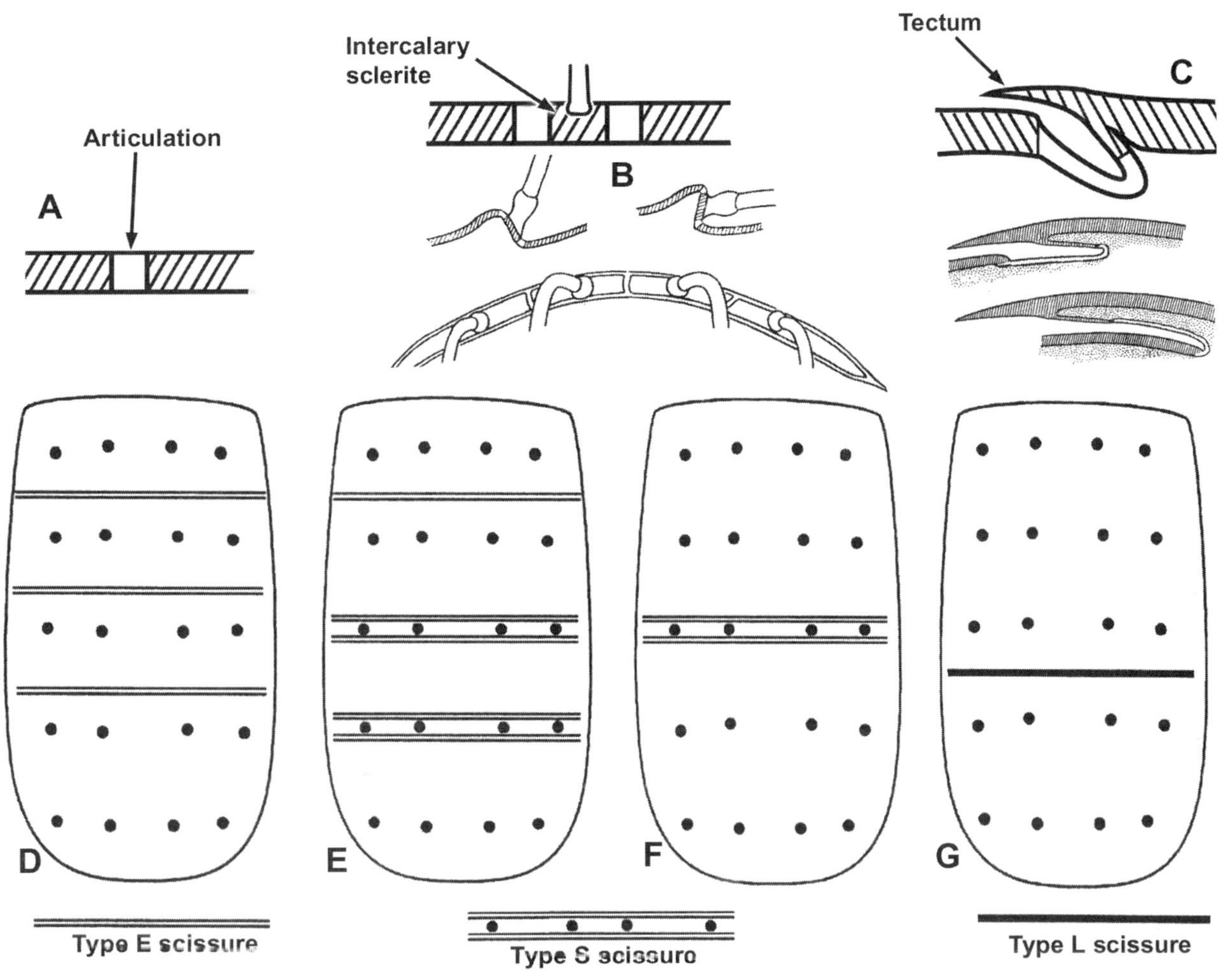

Fig. 15.5. Structure and distribution of transverse notogastral scissures in Enarthronotides. **A**, schematic sagittal section of type E scissure; **B**, schematic sagittal sections of type S scissure showing erect (top) and reclined (middle) positions of erectile setae and dorsal view of same (bottom); **C**, schematic sagittal section of type L scissure (top), and the same showing the range of motion (middle is extended; bottom is contracted); **D–G**, common distribution patterns of scissures in Enarthronotides, using key at bottom. (After Grandjean 1931b, 1934b; Norton 2001).

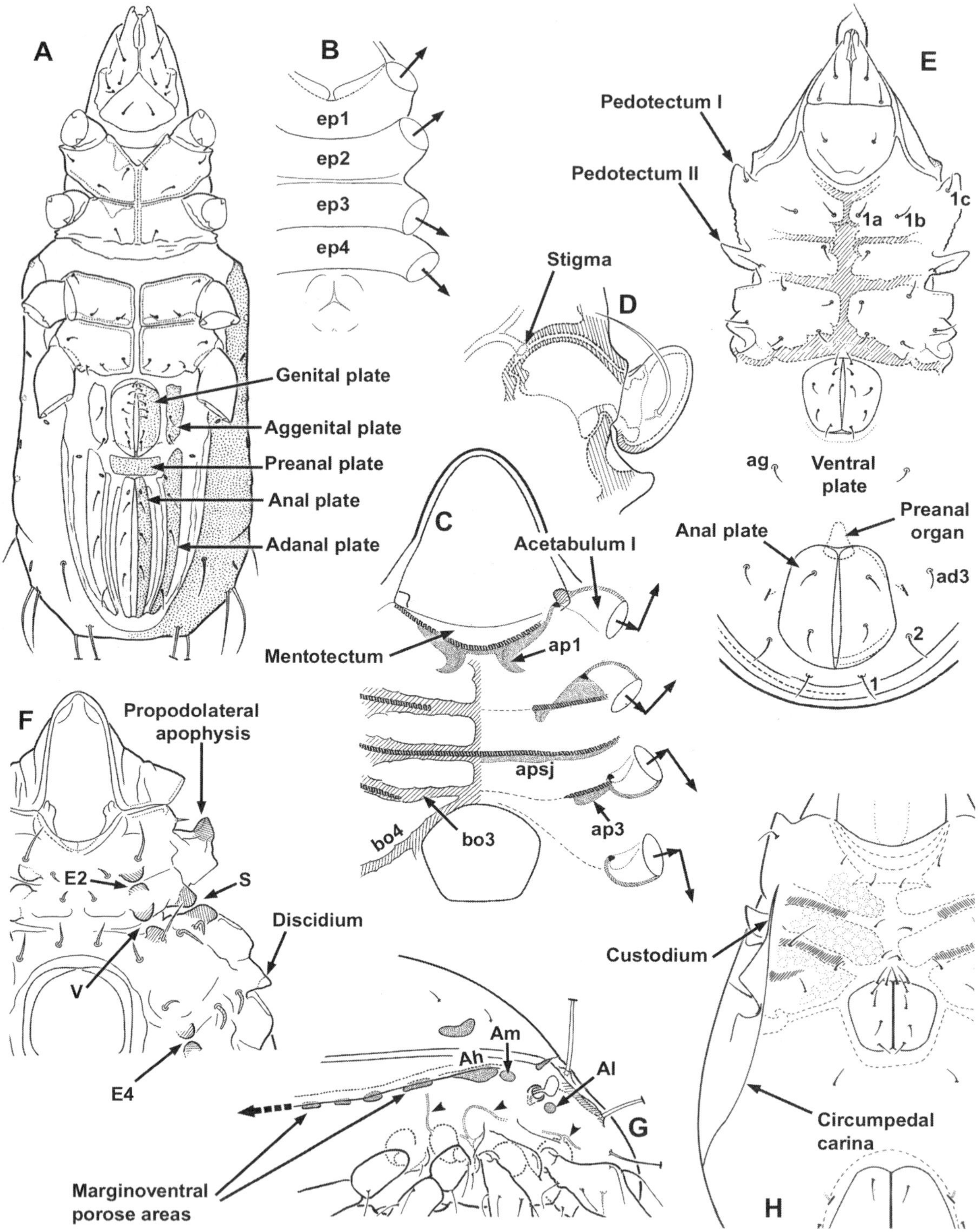

Fig. 15.6. Ventral and lateral structures of oribatid mites. **A**, macropyline type of ventral structure (after Grandjean 1958c); **B**, schematic structure of primitive coxisternum, with arrows indicating direction of legs as they leave respective epimera (1–4) (after Grandjean 1952a); **C**, schematic structure of brachypyline coxisternum, showing acetabula, apodemes (*ap*), and epimeral borders (*bo*), with arrows indicating direction of legs as they leave respective acetabula (after Grandjean 1952a); **D**, acetabulum I of brachypyline mite, showing opening (stigma) of two-branched trachea (after Grandjean 1959a); **E**, brachypyline type of ventral structure (from Grandjean 1960b); **F**, DAMAEIDAE, partial venter showing enantiophyses (*S*, *V*, *E2*, *E4*) and other cuticular projections (after Grandjean 1960c); **G**, lateral features of poronotic brachypyline mite, showing porose areas and tracheae (arrowheads) (after Grandjean 1971); **H**, CERATOZETIDAE, partial venter (after Behan-Pelletier 1984).

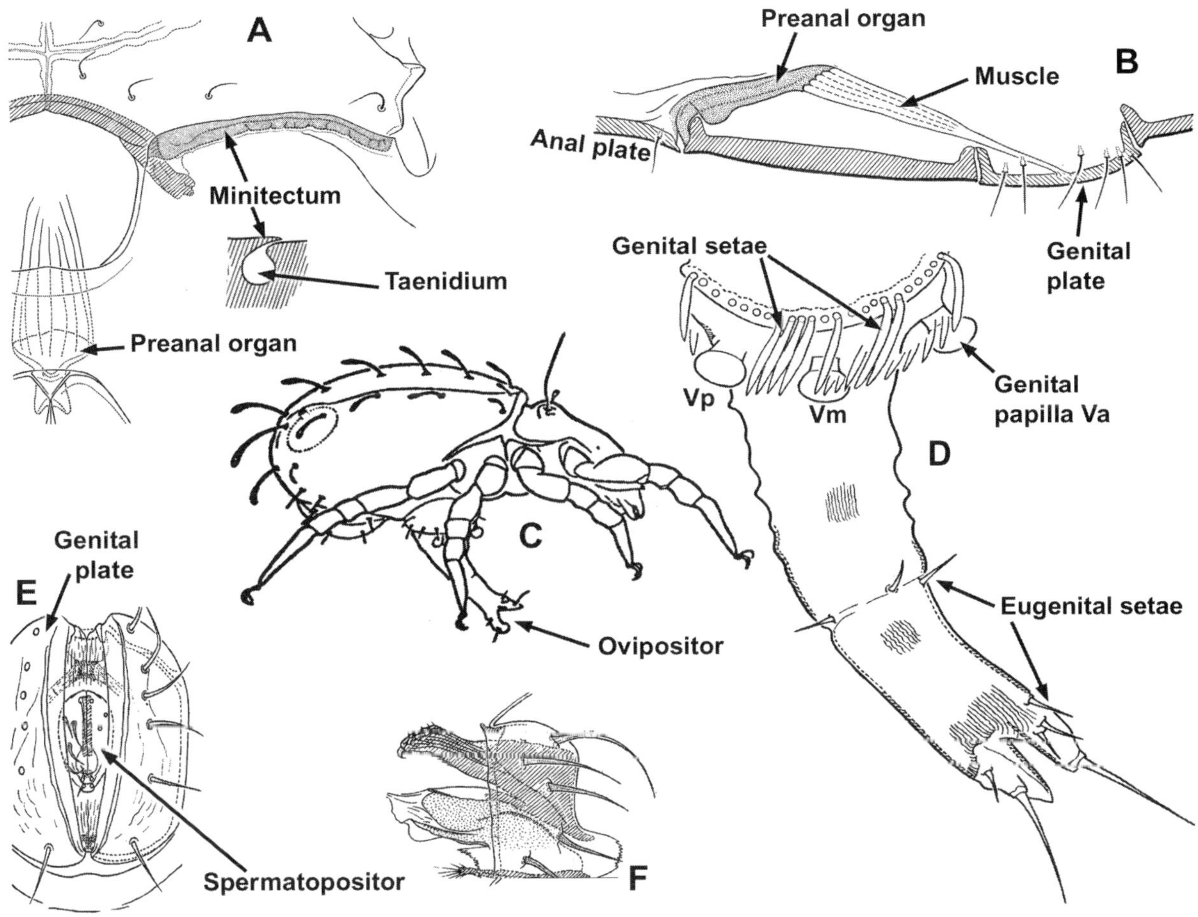

Fig. 15.7. Further ventral features of oribatid mites. **A**, partial venter of brachypyline mite showing taenidium and minitectum on epimeral border IV, along with preanal organ; **B**, sagittal section of brachypyline venter, showing preanal organ and muscles inserting on genital plate; **C**, NOTHRIDAE, female, lateral aspect, showing extended ovipositor; **D**, same, ovipositor in lateral view; **E**, DAMAEIDAE, genital region of male, showing spermatopositor retracted within genital vestibule; **F**, same, spermatopositor removed, lateral view. (**A** after Grandjean 1968; **B** after Grandjean 1969b; **C**, **D** after Grandjean 1956a; **E**, **F** after Grandjean 1956).

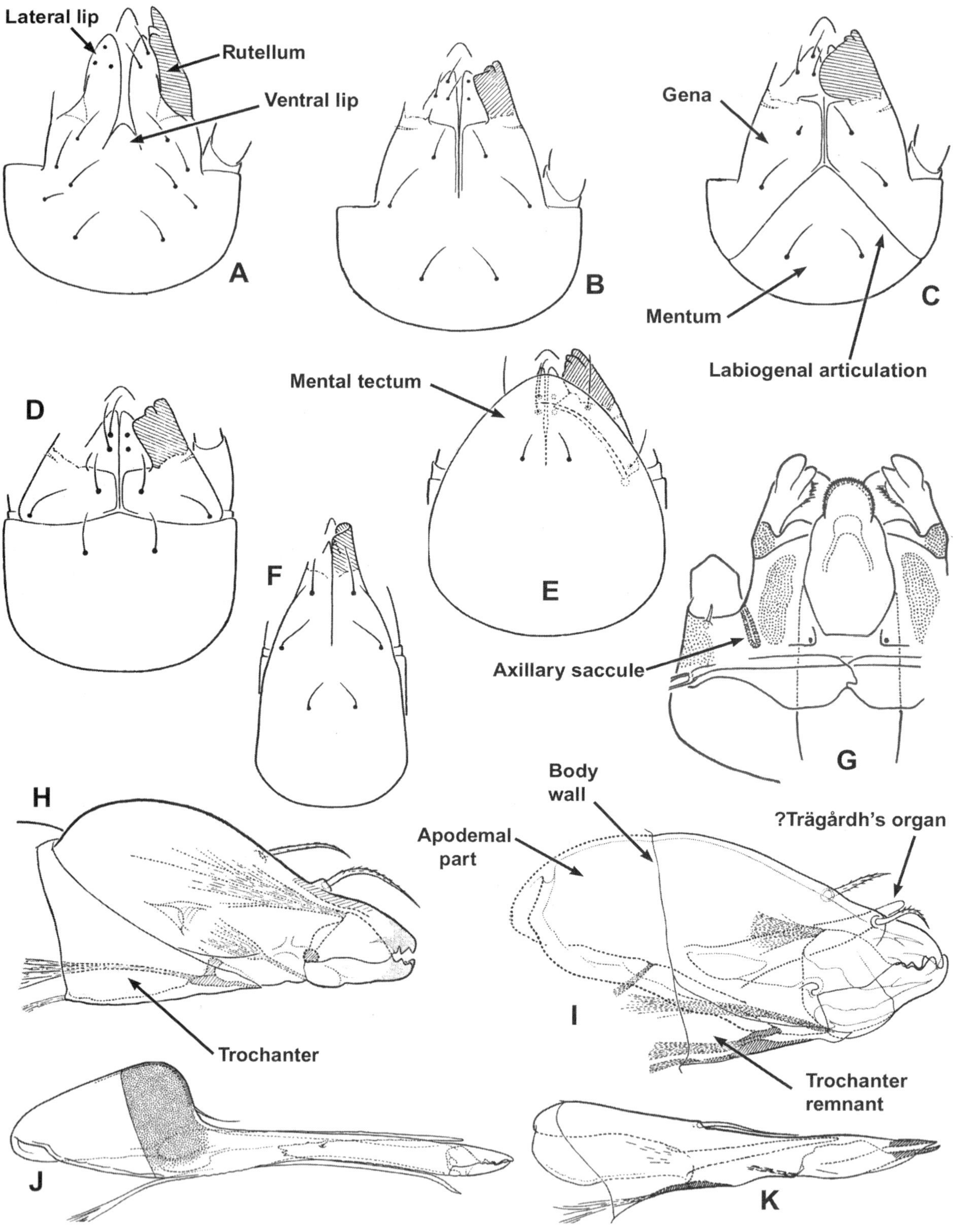

Fig. 15.8. Subcapitula (**A–F** ventral, **G** dorsal) and chelicerae (**H–K**, antiaxial) of oribatid mites. **A**, anarthric, with ventral lip; **B**, anarthric, without ventral lip; **C**, stenarthric; **D**, diarthric; **E**, diarthric, with mental tectum; **F**, secondarily anarthric, pelopsiform; **G**, composite, showing axillary saccule at base of palp; **H**, primitive form, with complete trochanter; **I**, typical chelate-dentate form; **J**, pelopsiform type; **K**, attenuate-edentate type. (**A–F** after Grandjean 1957b; **G** after Van der Hammen 1968a; **H**, **I**, **J**, **K** after Grandjean 1954c, 1947e, 1936e, 1951c, respectively).

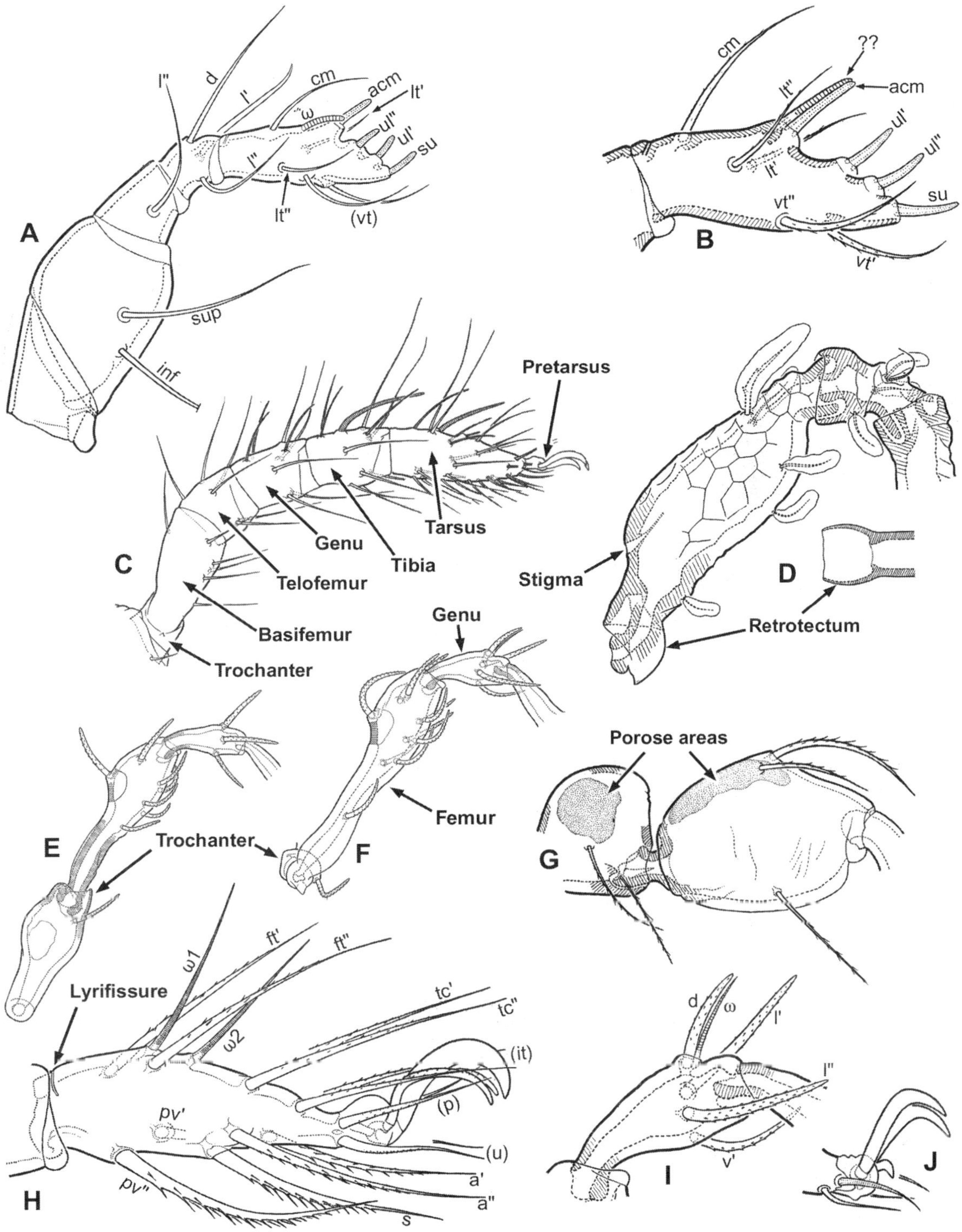

Fig. 15.9. Palps (**A**, **B**) and legs (**C–J**) of oribatid mites. **A**, palp of brachypyline mite showing typical setation; **B**, tarsus of poronotic brachypyline mite showing fused solenidion ω and eupathidium *acm* ("double horn"); **C**, palaeosomatid-type leg, with subdivided femur; **D**, brachypyline leg with articulations in sockets formed by retrotecta (insert is schematic); **E**, proximal half of leg IV of brachypyline mite showing large trochanter; **F**, same, but leg I, showing small trochanter; **G**, trochanter and femur III of brachypyline mite in antiaxial aspect, showing porose areas on opposite face; **H**, tarsus and tridactylous pretarsus III of brachypyline mite in antiaxial aspect, showing typical setation and normal, large empodial claw; **I**, genu of brachypyline mite showing coupled solenidion σ and seta *d*; **J**, tridactylous pretarsus with reduced, hooklike empodial claw. (**A**, **D** after Grandjean 1964c; **B** after Grandjean 1960d; **C**, **J** after Grandjean 1954c; **E**, **F**, **I** after Grandjean 1960c; **G** after Grandjean 1953b; **H** after Grandjean 1940).

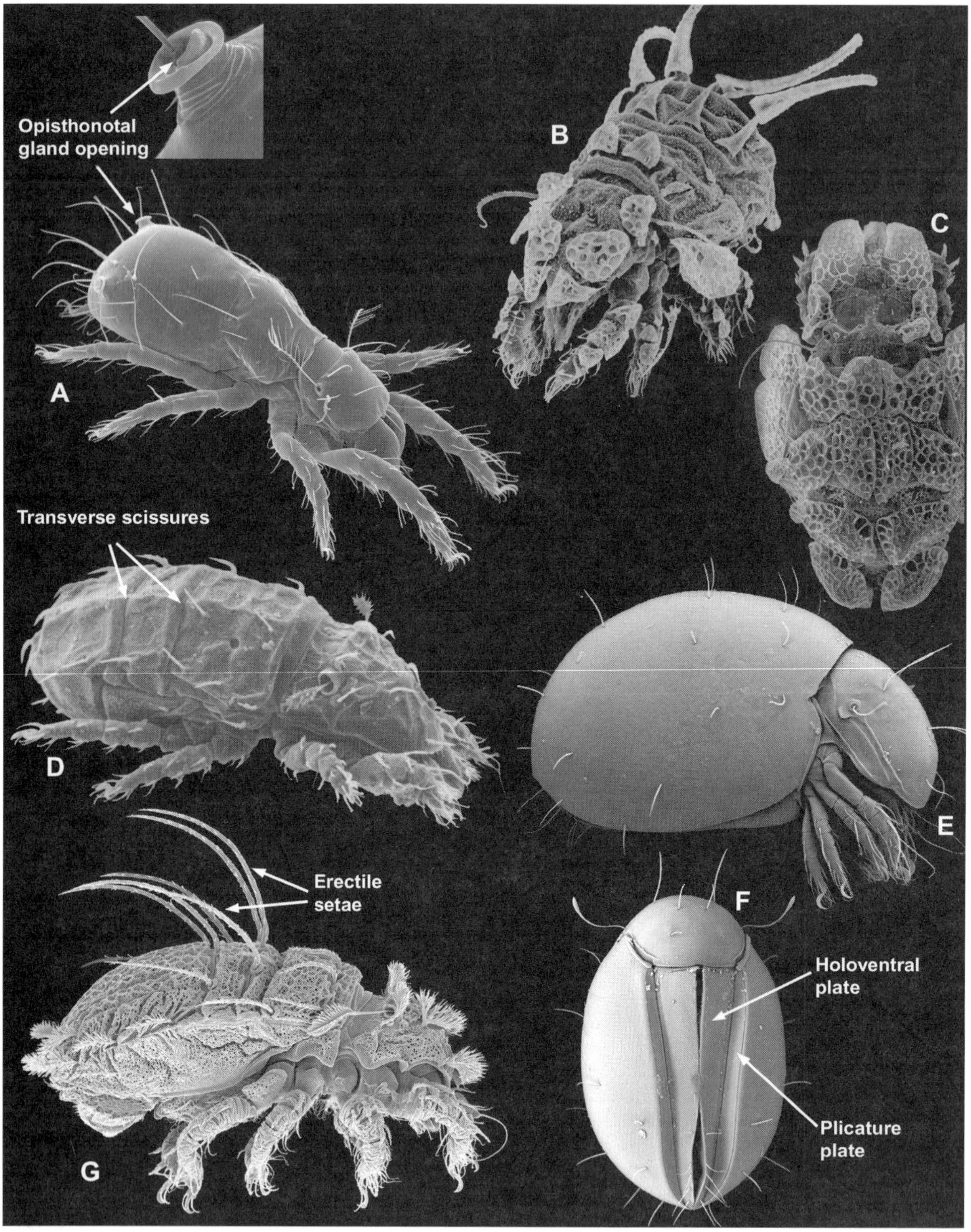

Fig. 15.10. Macropyline oribatid mites. **A**, *Parhypochthonius aphidinus* Berlese (PARHYPOCHTHONIIDAE), dorsolateral (insert, enlargement of opening to opisthonotal gland); **B**, *Atopochthonius artiodactylus* Grandjean (ATOPOCHTHONIIDAE), anterolateral; **C**, *Pterochthonius angelus* (Berlese) (PTEROCHTHONIIDAE), dorsal; **D**, *Brachychthonius* sp. (BRACHYCHTHONIIDAE), dorsolateral; **E**, *Euphthiracarus cooki* Norton et al. (EUPHTHIRACARIDAE), lateral, in active posture; **F**, same, but ventral, in defensive posture; **G**, *Cosmochthonius* sp. (COSMOCHTHONIIDAE), defensive setae erected (**A**, **G** by D. E. Walter; **E**, **F**, and **A** insert courtesy of Maria Minor; **B** anonymous, courtesy of the Ministry of Agriculture, Fisheries and Food, London; **C** courtesy of Günther Krisper; **D** courtesy of Marilyn Clayton).

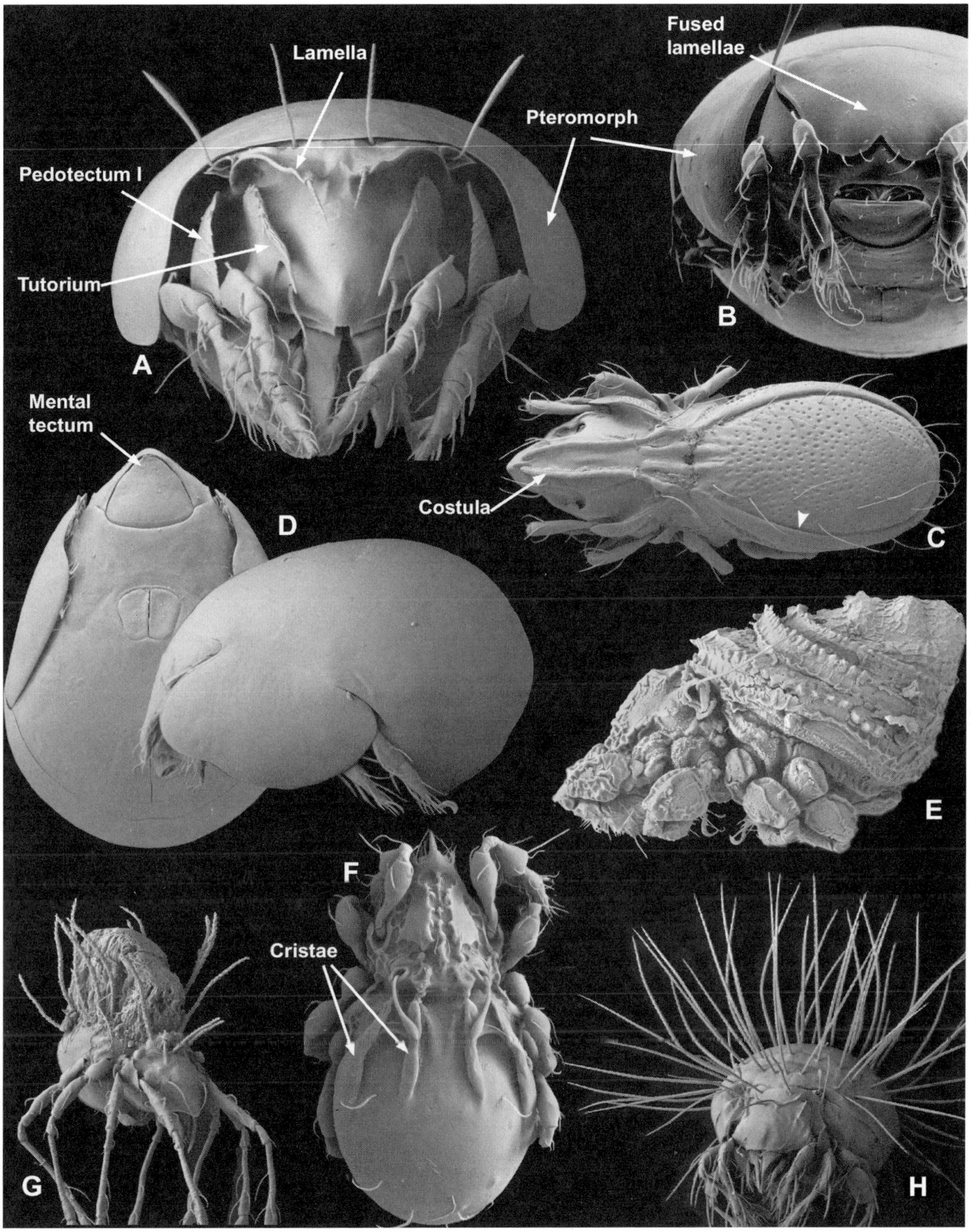

Fig. 15.11. Brachypyline oribatid mites. **A**, *Ceratozetes* sp. (CERATOZETIDAE), anterior; **B**, *Lepidozetes* sp. (TEGORIBATIDAE), anterior; **C**, OTOCEPHEIDAE (undetermined), dorsal; **D**, GALUMNIDAE (undetermined), lateral in active posture, ventral in defensive posture; **E**, *Plasmobates* sp. (PLASMOBATIDAE), lateral, showing exuvial scalps; **F**, *Suctobelbella* sp. (SUCTOBELBIDAE), dorsal; **G**, *Tuparezetes* sp. (TUPAREZETIDAE), anterolateral, showing amorphous cerotegument on notogaster; **H**, *Neotrichozetes* sp. (NEOTRICHOZETIDAE), anterolateral. (**B** courtesy of Deborah Sillman; remainder courtesy of Sue Lindsay).

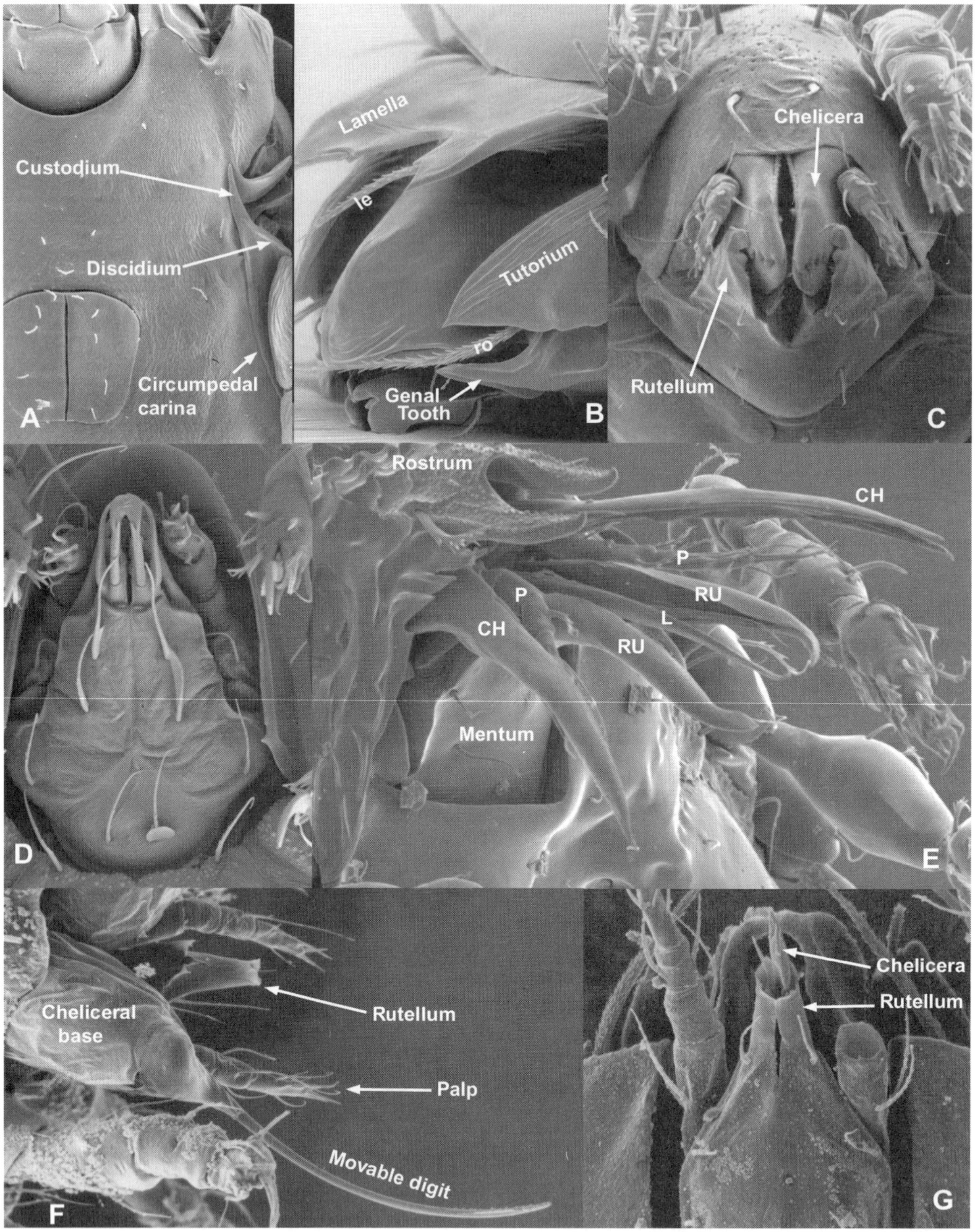

Fig. 15.12. A, *Latilamellobates baloghi* (Mahunka) (CERATOZETIDAE), coxisternal and genital regions; **B**, same, prodorsum, anterolateral; **C**, *Melanozetes crossleyi* Behan-Pelletier (CERATOZETIDAE), anterior view of tritonymph showing gnathosoma; **D**, anarthric subcapitulum of *Eniochthonius* sp. (ENIOCHTHONIIDAE); **E**, *Suctobelbella falcata* (Forsslund) (SUCTOBELBIDAE), anteroventral view showing extended gnathosoma (*CH*, chelicera; *L*, labrum; *P*, palp; *RU*, rutellum); **F**, *Gustavia* sp. (GUSTAVIIDAE), extended gnathosoma of nymph, with chelicerae pulled away from rutellar tube; **G**, *Peloptulus* sp. (PHENOPELOPIDAE), ventral view of gnathosomatic region, with one pelopsiform chelicera partly extending from rutellar tube. (**A–C** from Behan-Pelletier 2000; **D** courtesy of Sue Lindsay; **E** courtesy of Maria Minor; **F** courtesy of Marilyn Clayton; **G** after Norton and Behan-Pelletier 1986).

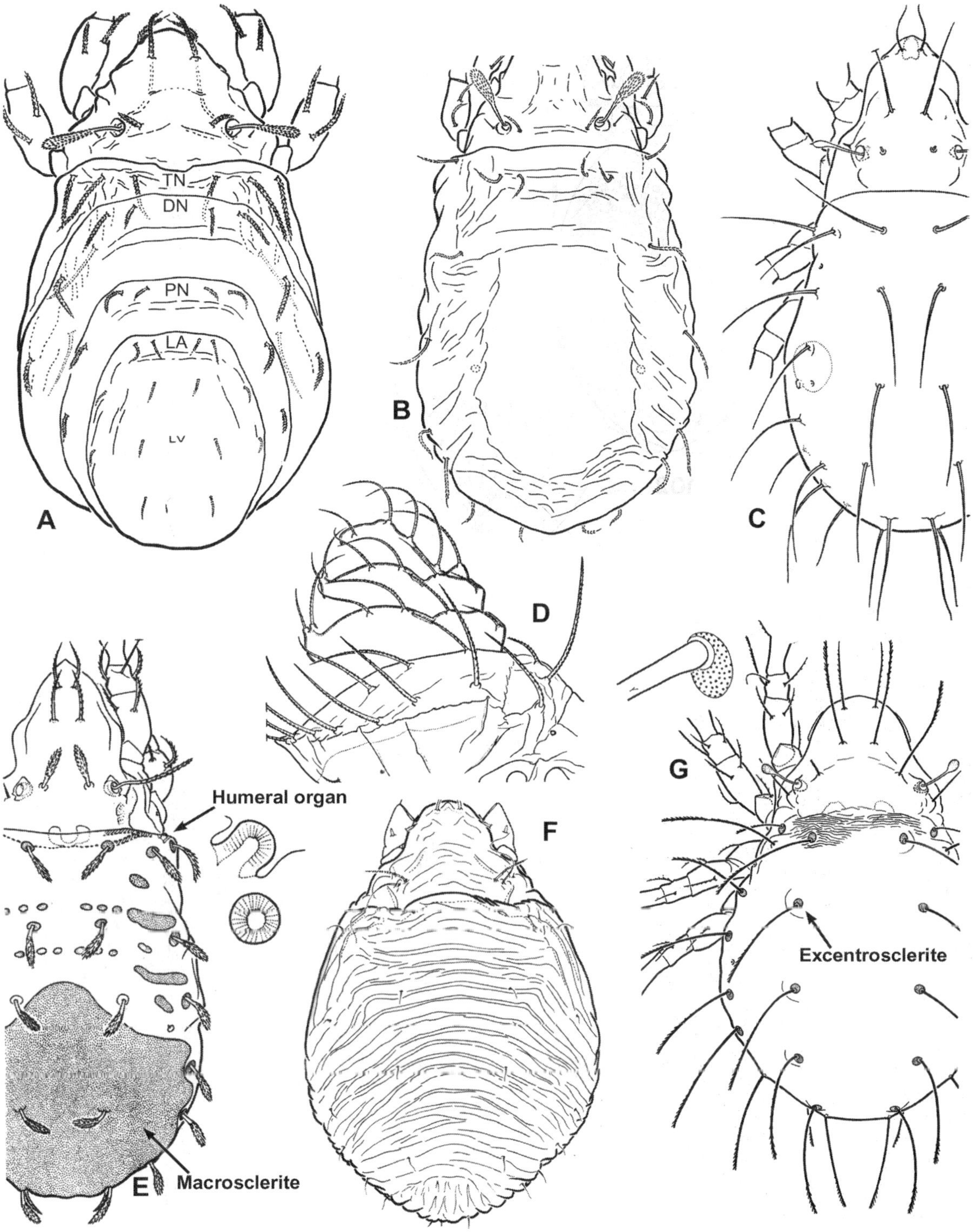

Fig. 15.13. Nymphs of brachypyline oribatid mites (dorsal views, except **D**, partial lateral). **A**, eupheredermous nymph of *Eueremaeus stiktos* (Higgins) with exuvial scalps attached (cuticle of tritonymph, deutonymph, protonymph, and larva indicated by labels); **B**, eupheredermous nymph of *E. higginsi* Behan-Pelletier with scalps removed, showing absence of dorsocentral opisthonotal setae; **C**, apheredermous nymph of *Raphigneta numidiana* Grandjean (AUTOGNETIDAE); **D**, apopheredermous nymph of *Oribatella calcarata* (C. Koch) (ORIBATELLIDAE); **E**, apheredermous nymph of *Ceratozetes peritus* Grandjean (CERATOZETIDAE), showing macrosclerites (insert of enlarged humeral organ); **F**, plicate apheredermous tritonymph of *Limnozetes borealis* Behan-Pelletier; **G**, apheredermous nymph of *Podoribates longipes* (Berlese) (MOCHLOZETIDAE), showing setae inserted on excentrosclerites (insert with enlargement). (**F**, **A**, **B** after Behan-Pelletier 1989 (**F**), 1993 (**A**, **B**), respectively; **E**, **D**, **C**, **G** after Grandjean 1951d (**E**), 1953a (**D**), 1960b (**C**), 1963b (**G**), respectively).

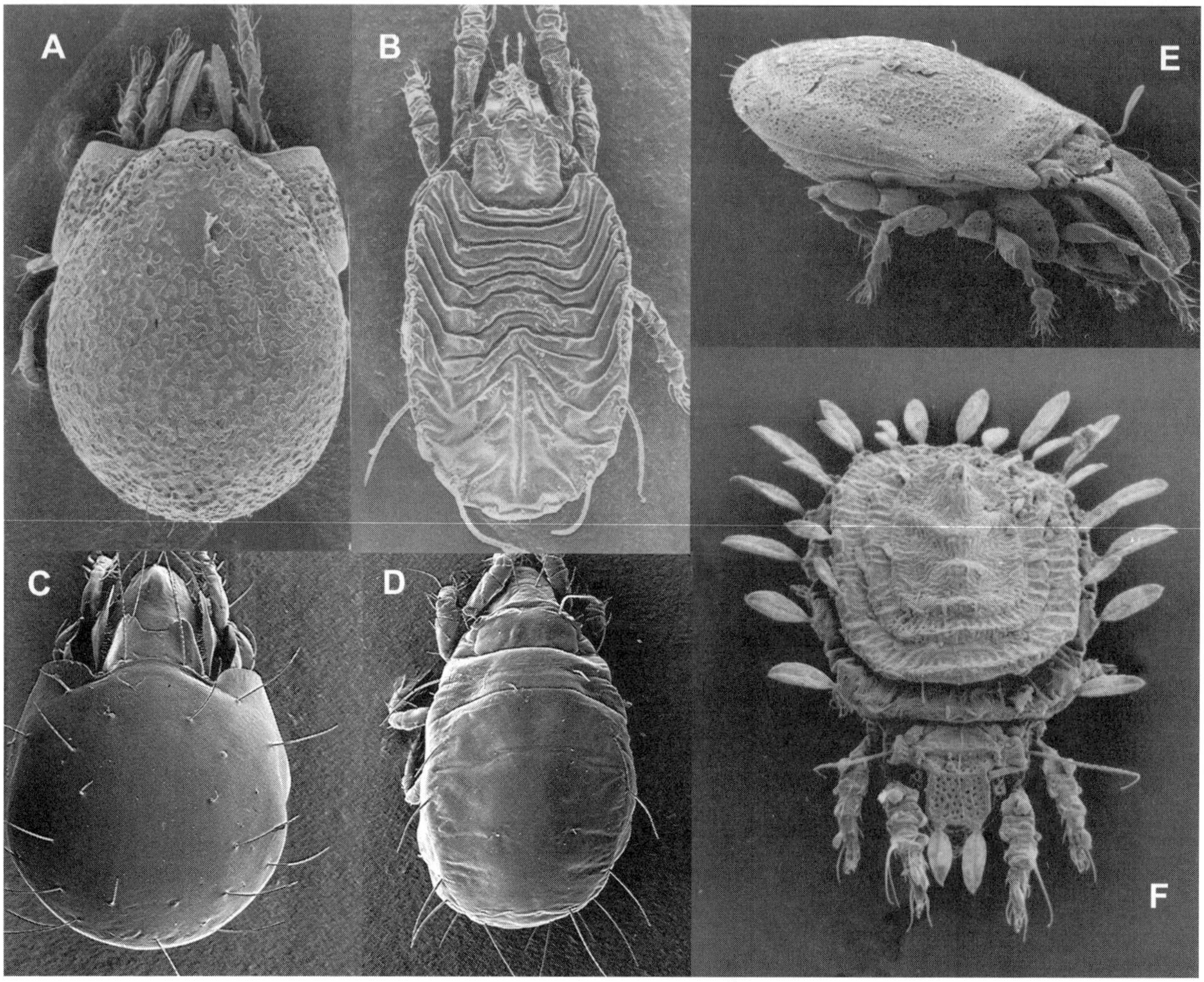

Fig. 15.14. Comparison of adult and immature brachypyline oribatid mites (all dorsal except **F**, dorsolateral). **A**, *Eupelops* sp. (PHENOPELOPIDAE) adult; **B**, nymph of same; **C**, *Diapterobates humeralis* (Hermann) adult; **D**, nymph of same; **E**, *Eupterotegaeus rhamphosus* Higgins and Woolley, adult; **F**, nymph of same, with exuvial scalps attached. (**A**, **B** after Norton and Behan-Pelletier 1986; **C**, **D** by V. M. Behan-Pelletier; **E**, **F** courtesy of Marilyn Clayton).

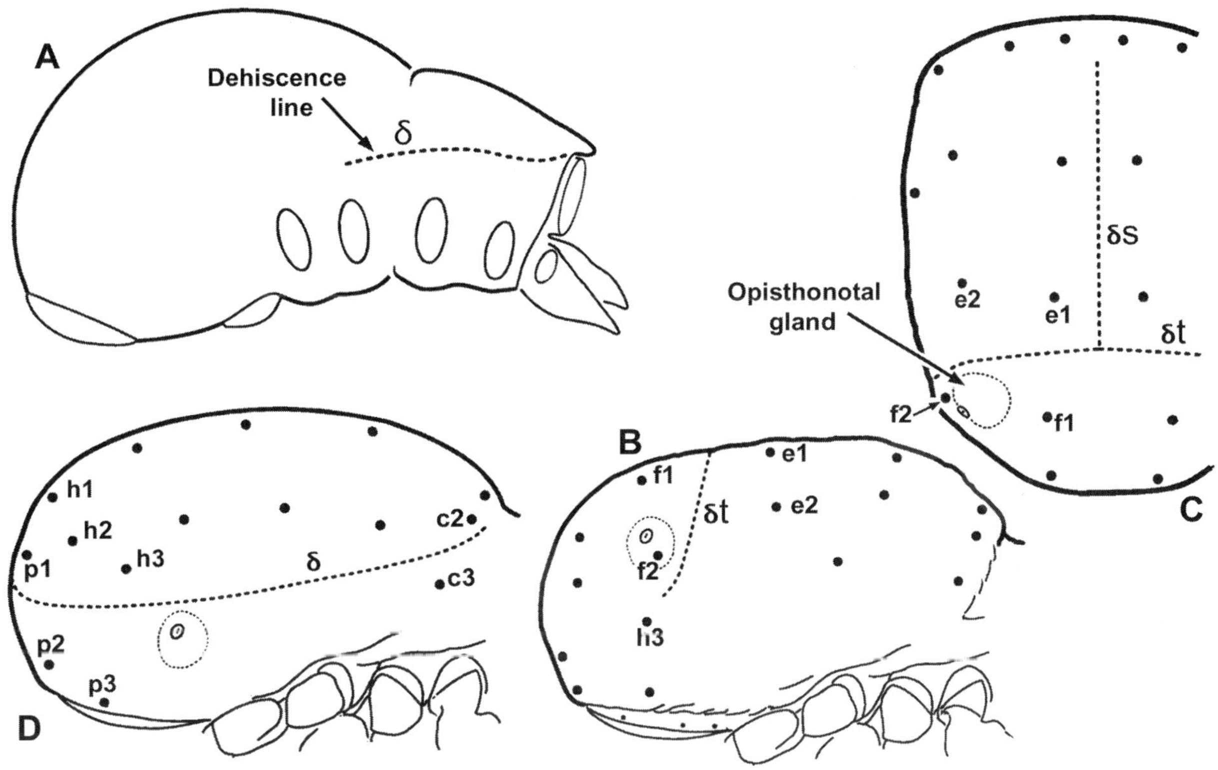

Fig. 15.15. Schematic representation of three principal patterns of ecdysial dehiscence in oribatid mites, indicated by the dehiscence line (δ, δ*s*, δ*t*). **A**, prodehiscence, lateral view; **B**, transdehiscence, lateral view; **C**, same, dorsal view; **D**, circumdehiscence, lateral view. (After Norton and Kethley 1994).

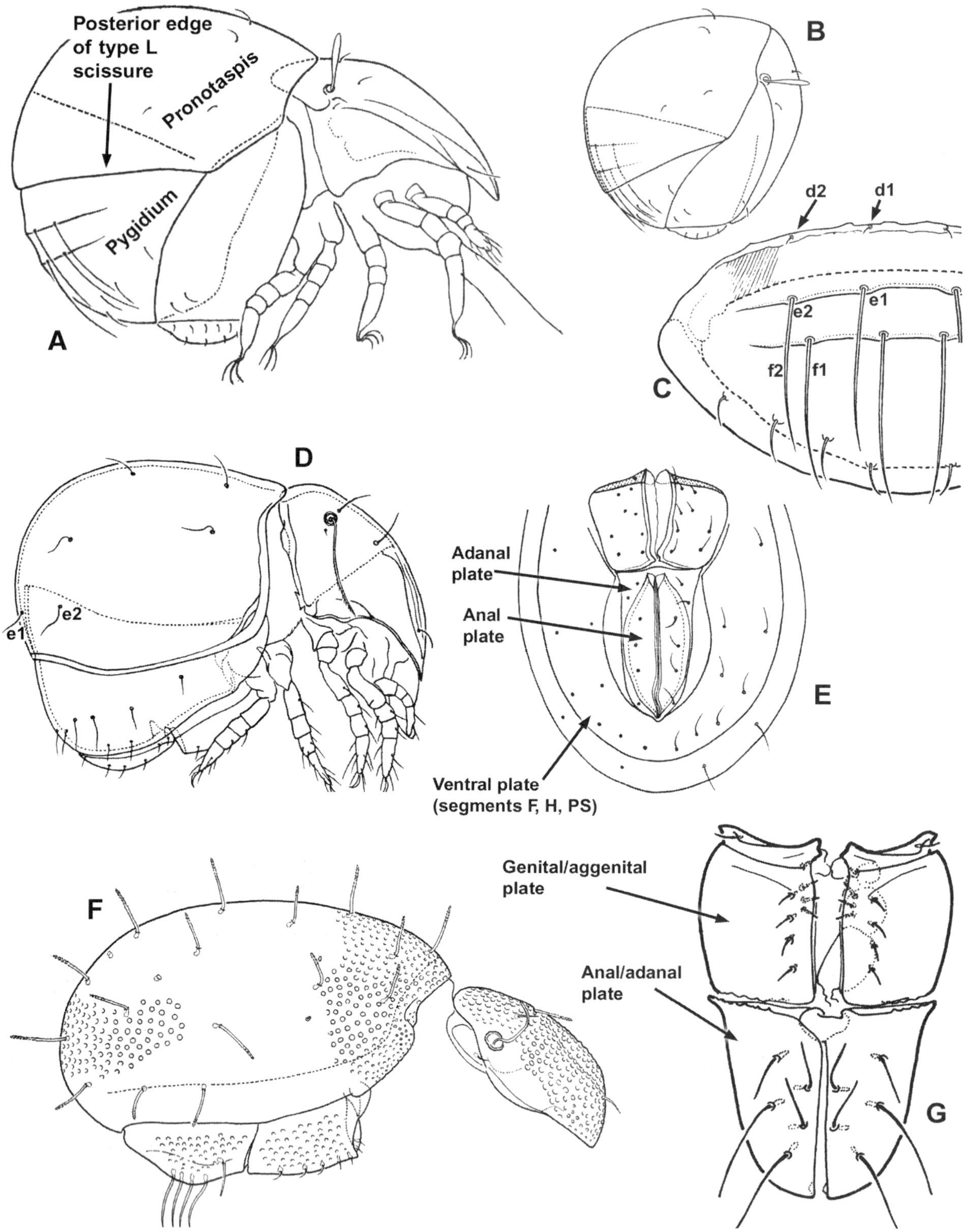

Fig. 15.16. PROTOPLOPHORIDAE. *Aedoplophora glomerata* Grandjean (Venezuela): **A**, open lateral; **B**, closed lateral; **C**, pygidium, posterior aspect. (**A**, **B** after Grandjean 1932; **C** after Grandjean 1954a).

MESOPLOPHORIDAE. *Archoplophora rostralis* (Willmann) (Japan): **D**, lateral; **E**, ventral (after Aoki 1980c).

PHTHIRACARIDAE. **F**, *Atropacarus striculus* (C. L. Koch) (Holarctic), lateral (after Aoki 1980a); **G**, *Phthiracarus murphyi* Harding (England), plates (after Harding 1976).

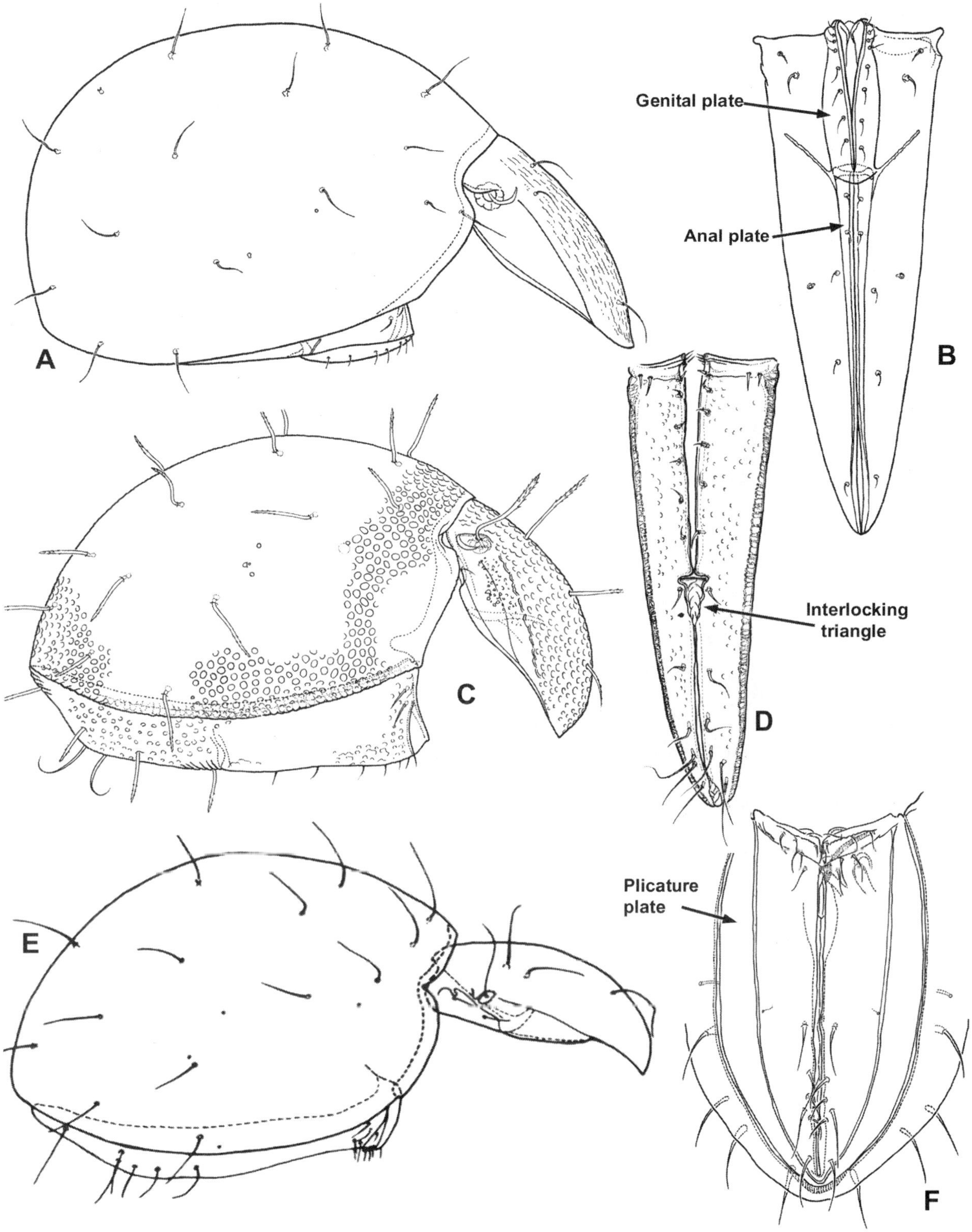

Fig. 15.17. ORIBOTRITIIDAE. *Oribotritia chichijimensis* Aoki (Japan): **A**, lateral; **B**, ventral plates (after Aoki 1980a).
EUPHTHIRACARIDAE. *Euphthiracarus foveolatus* Aoki (Japan): **C**, lateral (after Aoki 1980b); **D**, ventral plates (after Shimano and Norton 2003).
SYNICHOTRITIIDAE. *Synichotritia caroli* Walker (California, USA): **E**, dorsal (after Walker 1964); **F**, ventral (after Lions and Norton 1998).

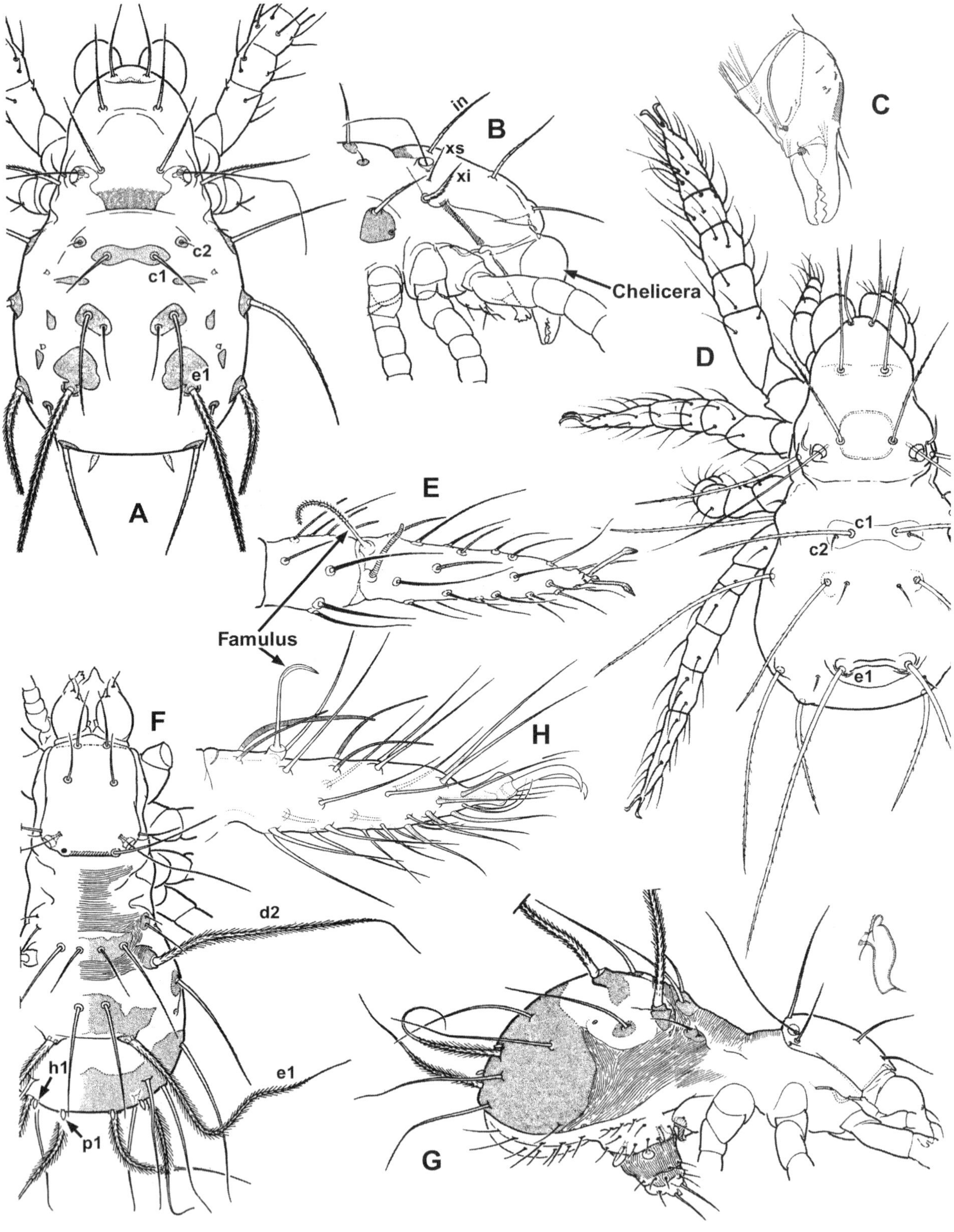

Fig. 15.18. ACARONYCHIDAE. *Acaronychus traegardhi* Grandjean (Morocco): **A**, dorsal, **B**, lateral; **C**, chelicera (after Grandjean 1954c).

ARCHEONOTHRIDAE. **D**, *Zachvatkinella nipponica* Aoki (Japan), dorsal (after Aoki 1980); **E**, *Z. belbiformes* Lange (Siberia), tibia, tarsus I (after Lange 1954).

PALAEACARIDAE. *Palaeacarus hystricinus* Trägårdh (Holarctic): **F**, dorsal; **G**, lateral, with enlarged anterior genital seta to right; **H**, tarsus I (after Grandjean 1954c).

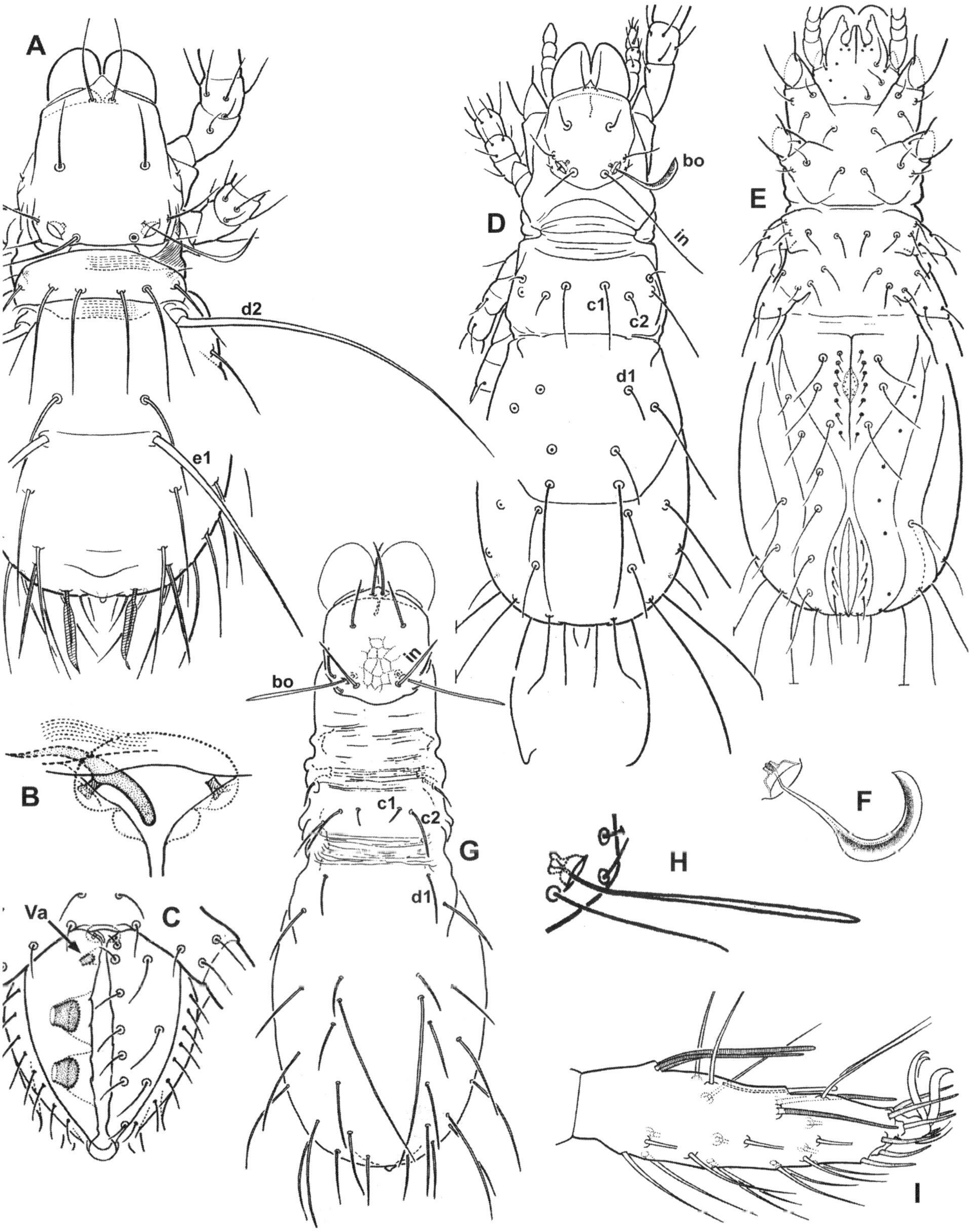

Fig. 15.19. CTENACARIDAE. *Ctenacarus araneola* Grandjean (Morocco): **A**, dorsal; **B**, modified anterior genital setae; **C**, genital region (after Grandjean 1954c).

APHELACARIDAE. *Aphelacarus acarinus* (Berlese) (Holarctic): **D**, dorsal; **E**, ventral; **F**, trichobothrium (after Grandjean 1954c).

ADELPHACARIDAE. *Adelphacarus sellnicki* Grandjean (Europe): **G**, dorsal; **H**, trichobothrium and nearby setae; **I**, tarsus I. (**G** after Beck and Woas 1991; **H**, **I** after Grandjean 1952c).

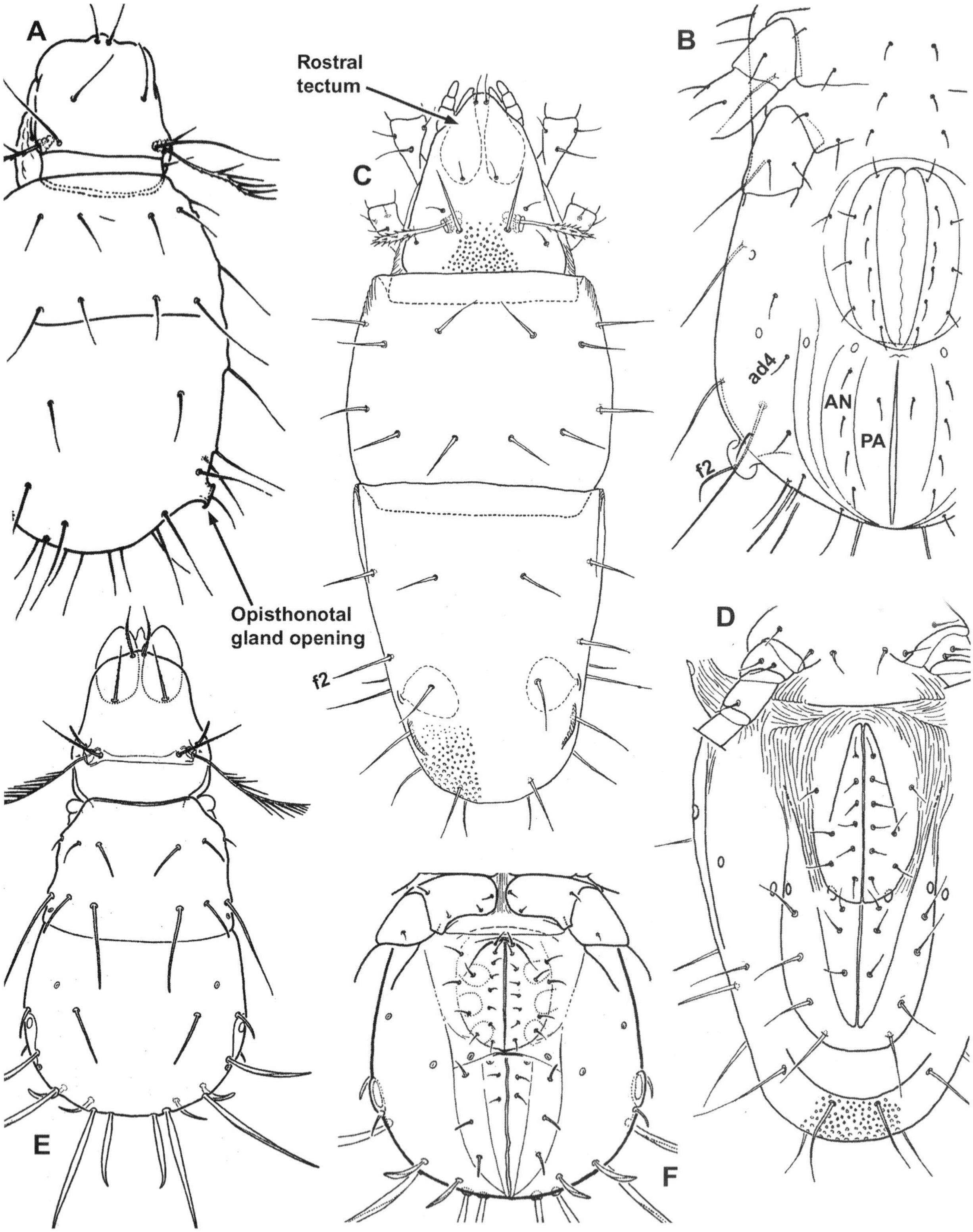

Fig. 15.20. PARHYPOCHTHONIIDAE. **A**, *Parhypochthonius aphidinus* Berlese (Holarctic), dorsal (after Aoki 1969); **B**, *Parhypochthonius* sp. ventral (after Grandjean 1934c).
ELLIPTOCHTHONIIDAE. *Elliptochthonius profundus* Norton (California, USA): **C**, dorsal; **D**, ventral (after Norton 1975).
GEHYPOCHTHONIIDAE. *Gehypochthonius xarifae* Strenzke (Maldives): **E**, dorsal; **F**, ventral (after Strenzke 1963).

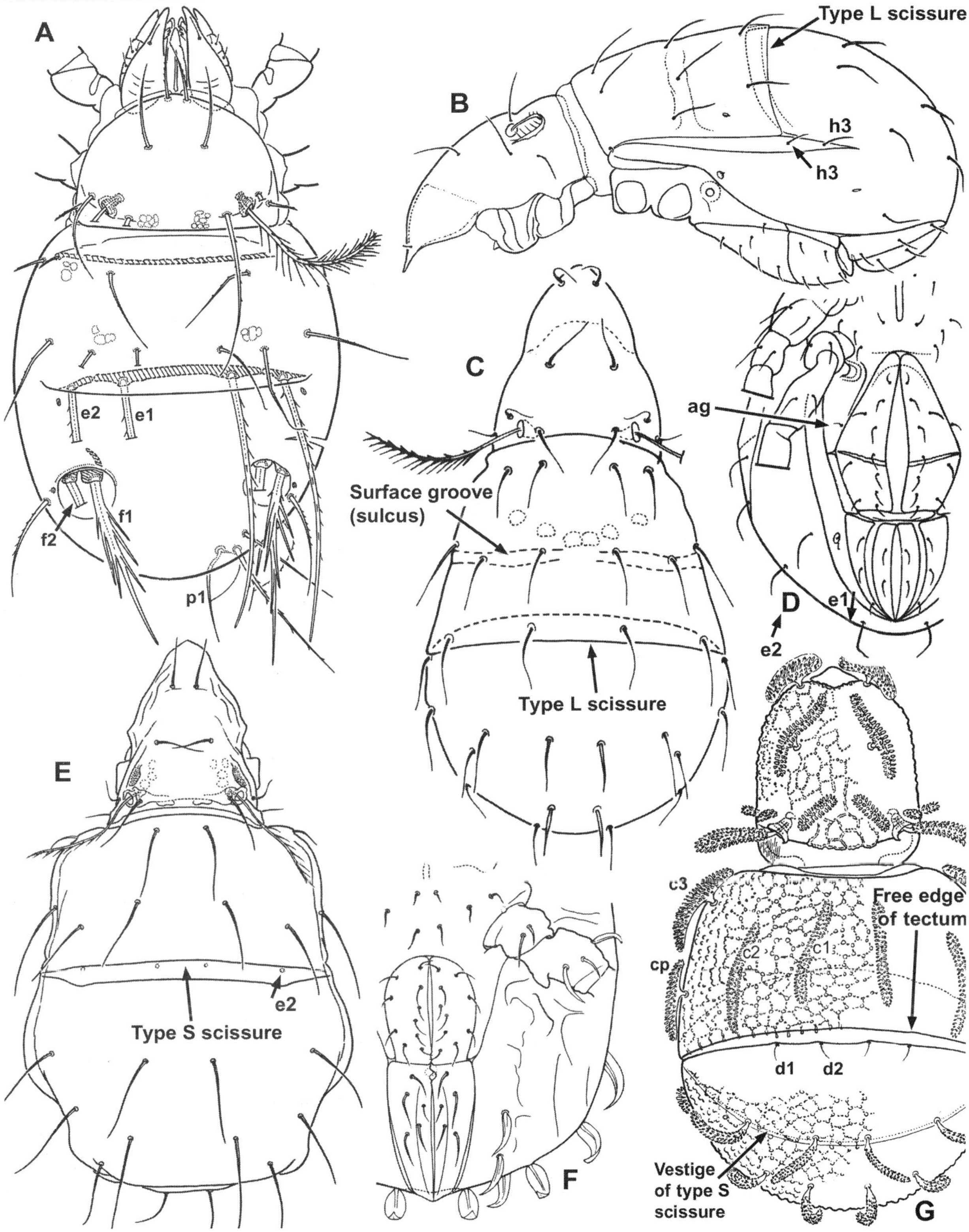

Fig. 15.21. ARBORICHTHONIIDAE. *Arborichthonius styosetosus* Norton (Canada): **A**, dorsal (after Norton 1982).
ENIOCHTHONIIDAE. *Eniochthonius minutissimus* (Berlese) (Holarctic): **B**, lateral (after Grandjean 1934a); **C**, dorsal (after Balogh and Mahunka 1983); **D**, ventral (after Grandjean 1933).
HYPOCHTHONIIDAE. **E**, *Hypochthonius rufulus* C. Koch (Holarctic), dorsal (after Aoki 1965b); **F**, *Malacoangelia remigera* Berlese (Java), ventral (after Grandjean 1935c).
SPHAEROCHTHONIIDAE. *Sphaerochthonius suzukii* Aoki (Japan): **G**, dorsal (after Aoki 1977b).

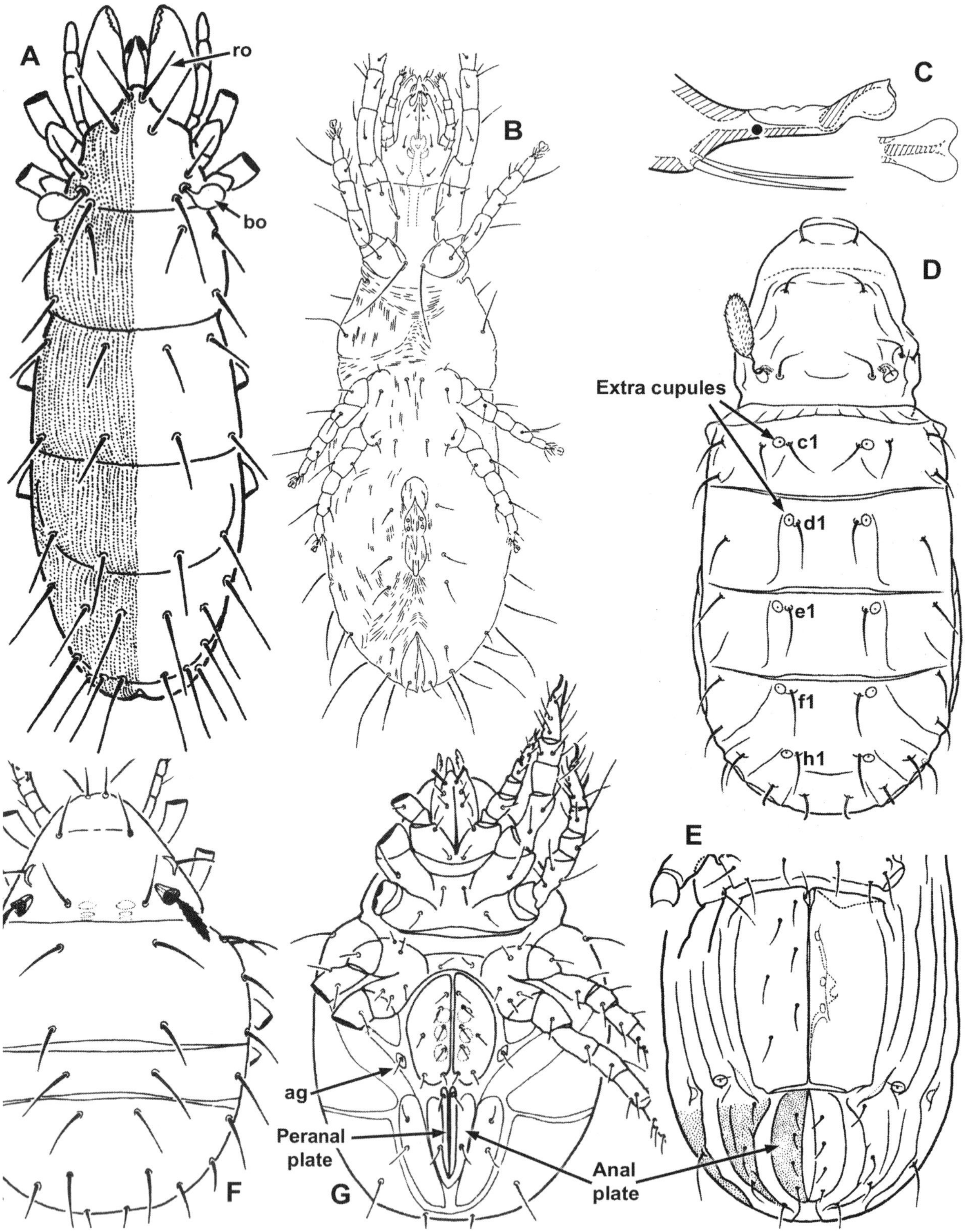

Fig. 15.22. PEDICULOCHELIDAE. **A**, *Paralycus* sp., dorsal; **B**, **C**, *P. lavoipierrei* (Price) (California, USA): **B**, ventral; **C**, distal part of tarsus III, lateral (left) and pretarsus (dorsal). (**A** after Krantz 1978; **B** after Price 1973b; **C** after Norton, OConnor, and Johnston 1983).

HAPLOCHTHONIIDAE. *Haplochthonius simplex* Willmann (Europe): **D**, dorsal; **E**, hysterosomatic ventral (after Grandjean 1947b).

BRACHYCHTHONIIDAE. *Eobrachychthonius* sp. (Oregon, USA): **F**, dorsal; **G**, ventral (after Krantz 1978).

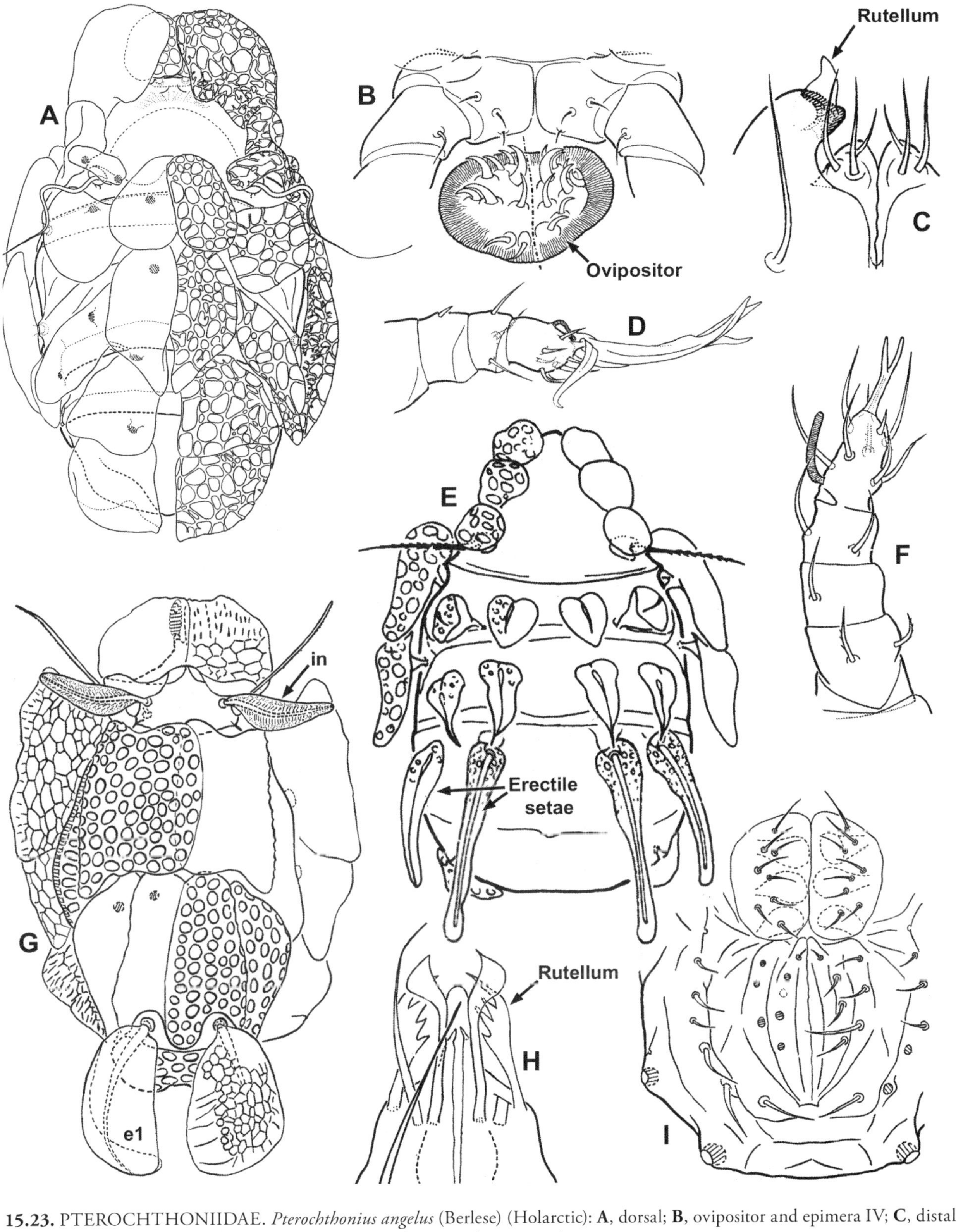

Fig. 15.23. PTEROCHTHONIIDAE. *Pterochthonius angelus* (Berlese) (Holarctic): **A**, dorsal; **B**, ovipositor and epimera IV; **C**, distal subcapitulum; **D**, palp. (**A** after Suzuki 1977; **B–D** after Grandjean 1950a).

ATOPOCHTHONIIDAE. *Atopochthonius artiodactylus* Grandjean (Europe): **E**, dorsal (after Krantz 1978); **F**, palp (after Grandjean 1948a).

PHYLLOCHTHONIIDAE. *Phyllochthonius aoutii* Travé (Ivory Coast): **G**, dorsal; **H**, distal subcapitulum; **I**, ventral hysterosoma (after Travé 1967).

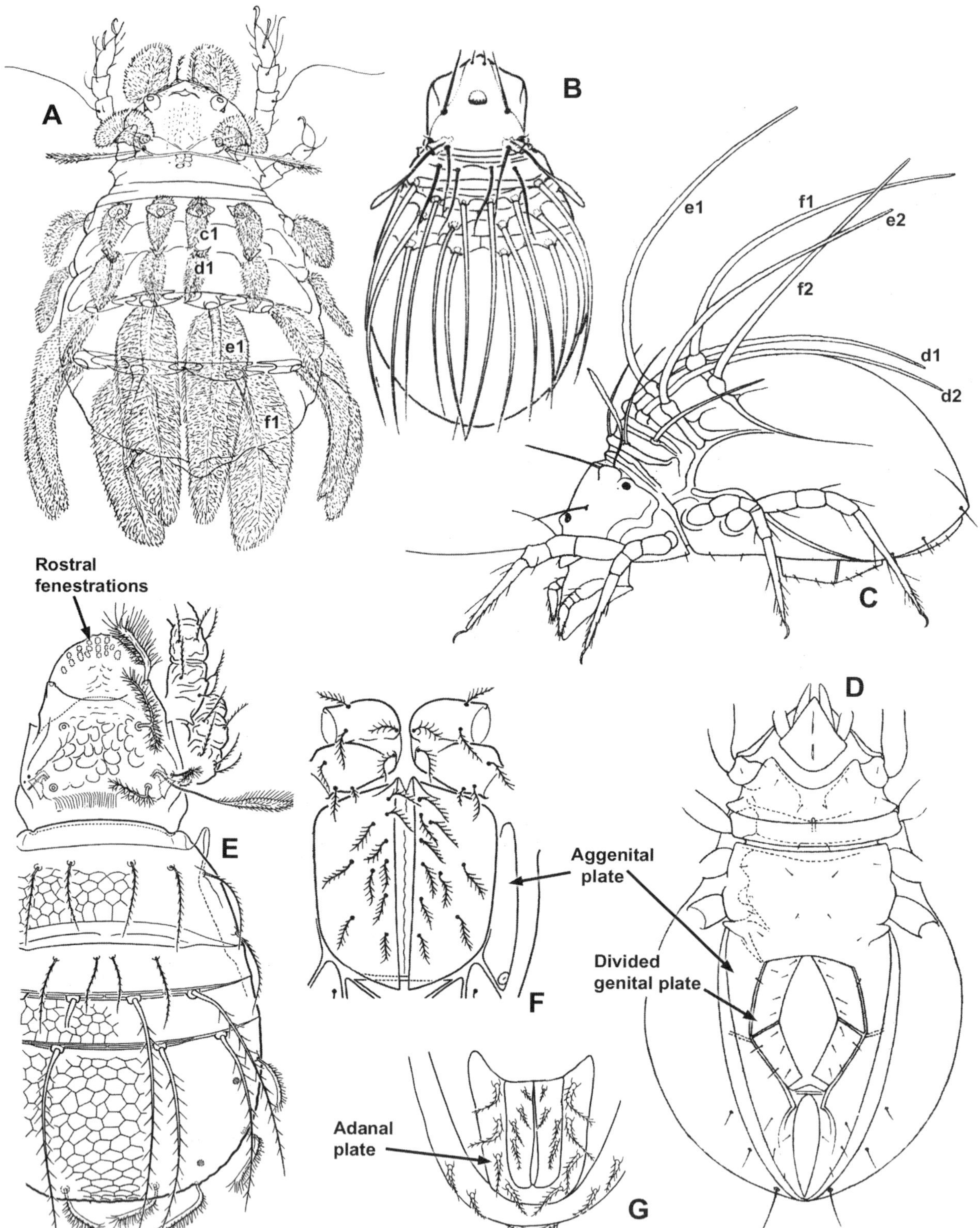

Fig. 15.24. TRICHTHONIIDAE. *Trichthonius pulcherrimus* (Hammer) (Argentina): **A**, dorsal (after Hammer 1958). HETEROCHTHONIIDAE. *Heterochthonius gibbus* (Berlese) (Italy): **B**, dorsal; **C**, lateral; **D**, ventral. (**B** after Kunst 1971; **C**, **D** after Grandjean 1928). COSMOCHTHONIIDAE. *Cosmochthonius reticulatus* Grandjean (France): **E**, dorsal; **F**, genital region (after Grandjean 1962c). *Phyllozetes emmae* (Berlese) (New Mexico, USA): **G**, anal region (after Kamill, Wallwork, and MacQuitty 1986).

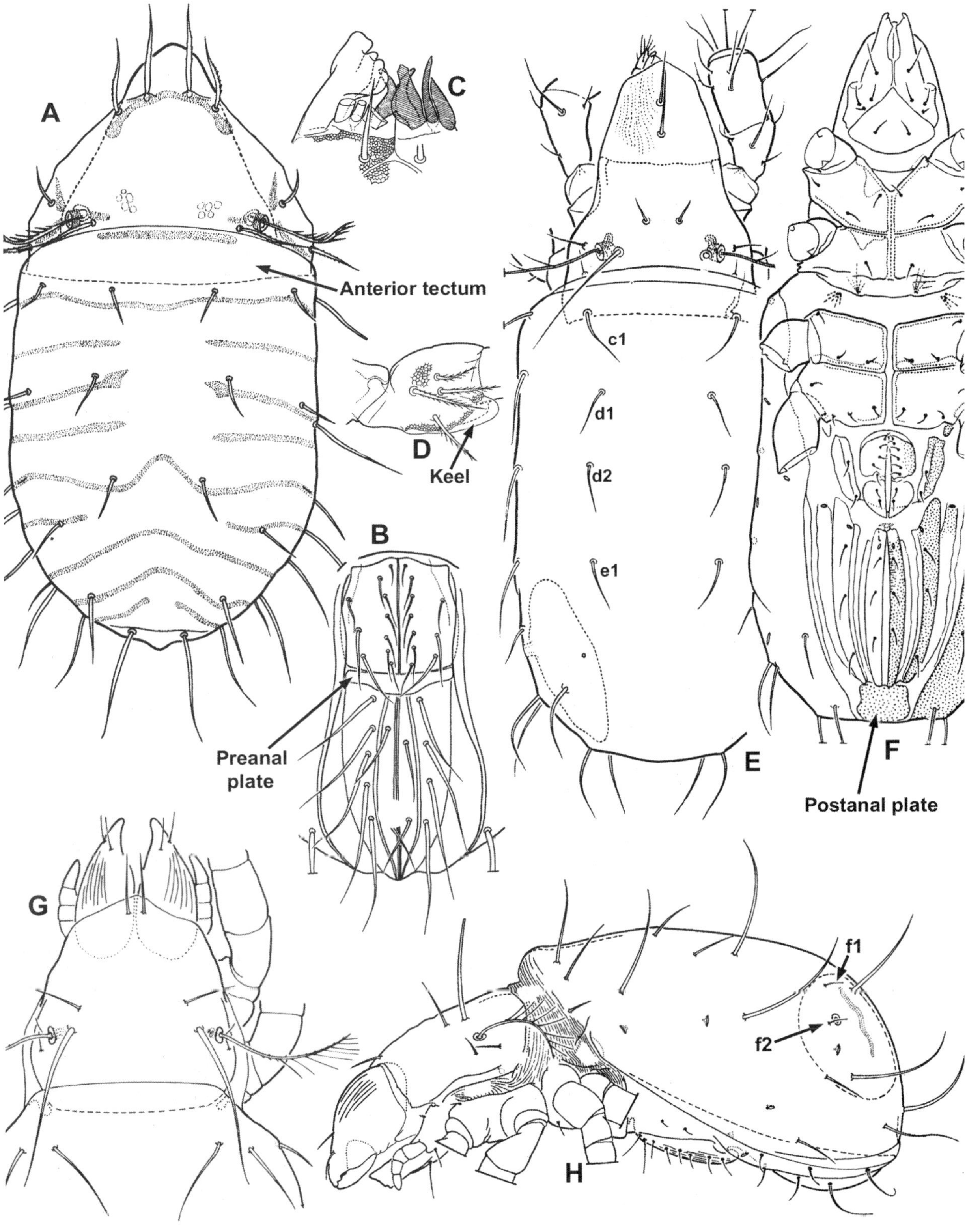

Fig. 15.25. LOHMANNIIDAE. *Mixacarus exilis* Aoki (Japan): **A**, dorsal; **B**, hysterosomatic venter (after Aoki 1970). *Annectacarus mucronatus* Grandjean (Venezuela): **C**, distal subcapitulum; **D**, femur III (after Grandjean 1950b).
PERLOHMANNIIDAE. *Perlohmannia dissimilis* (Hewitt) (England): **E**, dorsal; **F**, ventral (after Grandjean 1958c).
NEHYPOCHTHONIIDAE. *Nehypochthonius porosus* Norton and Metz (Florida, USA): **G**, dorsal; **H**, lateral (after Norton and Metz 1980).

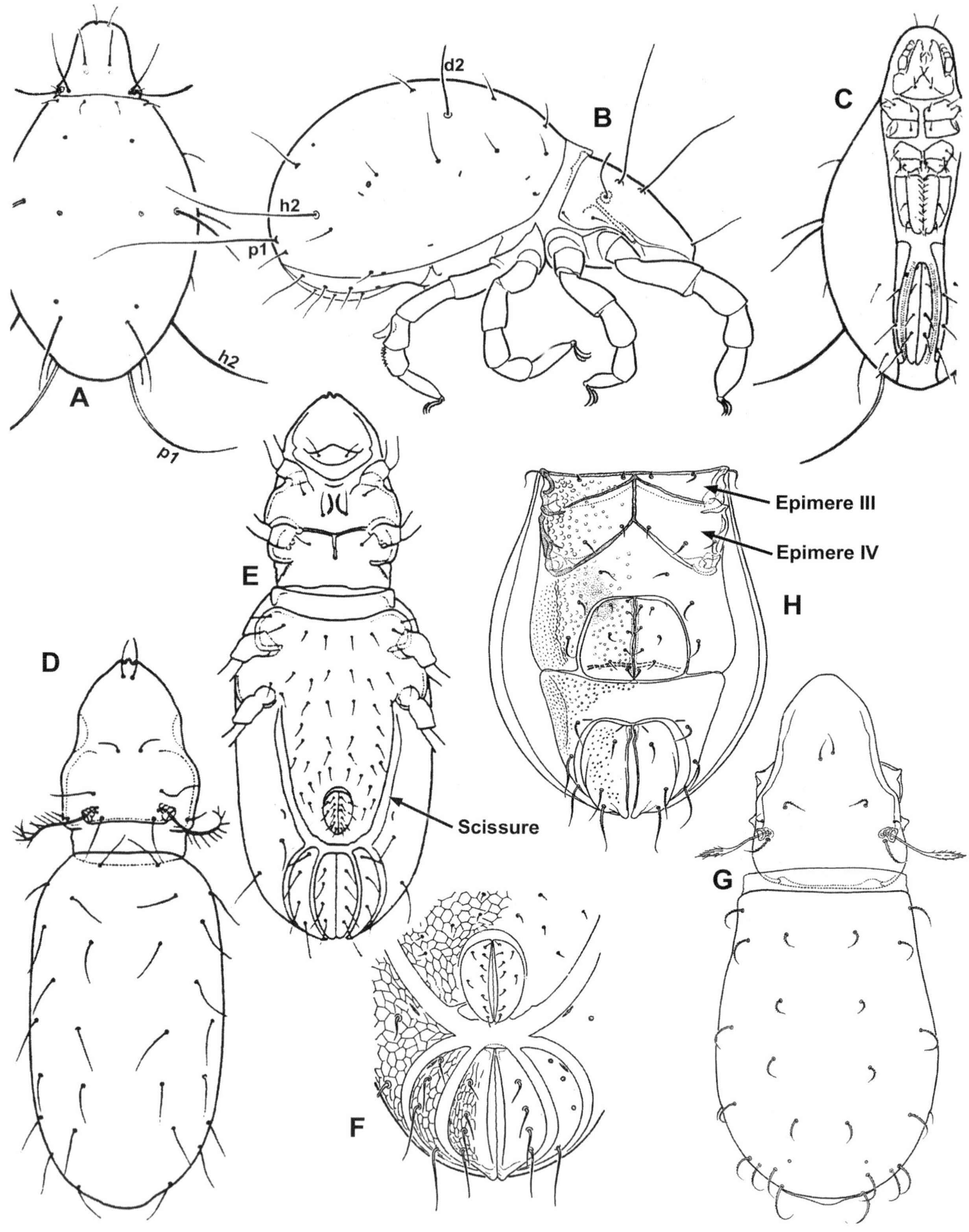

Fig. 15.26. COLLOHMANNIIDAE. *Collohmannia gigantea* Sellnick (Europe): **A**, dorsal; **B**, lateral; **C**, ventral. (**A**, **C** after Kunst 1971; **B** after Grandjean 1966a).

EULOHMANNIIDAE. *Eulohmannia ribagai* Berlese (Holarctic): **D**, dorsal; **E**, ventral (after Aoki 1975); **F**, posterior ventral (after Grandjean 1956d).

EPILOHMANNIIDAE. *Epilohmannia pallida pacifica* Aoki (Hawaii): **G**, dorsal; **H**, ventral hysterosoma (after Aoki 1965a).

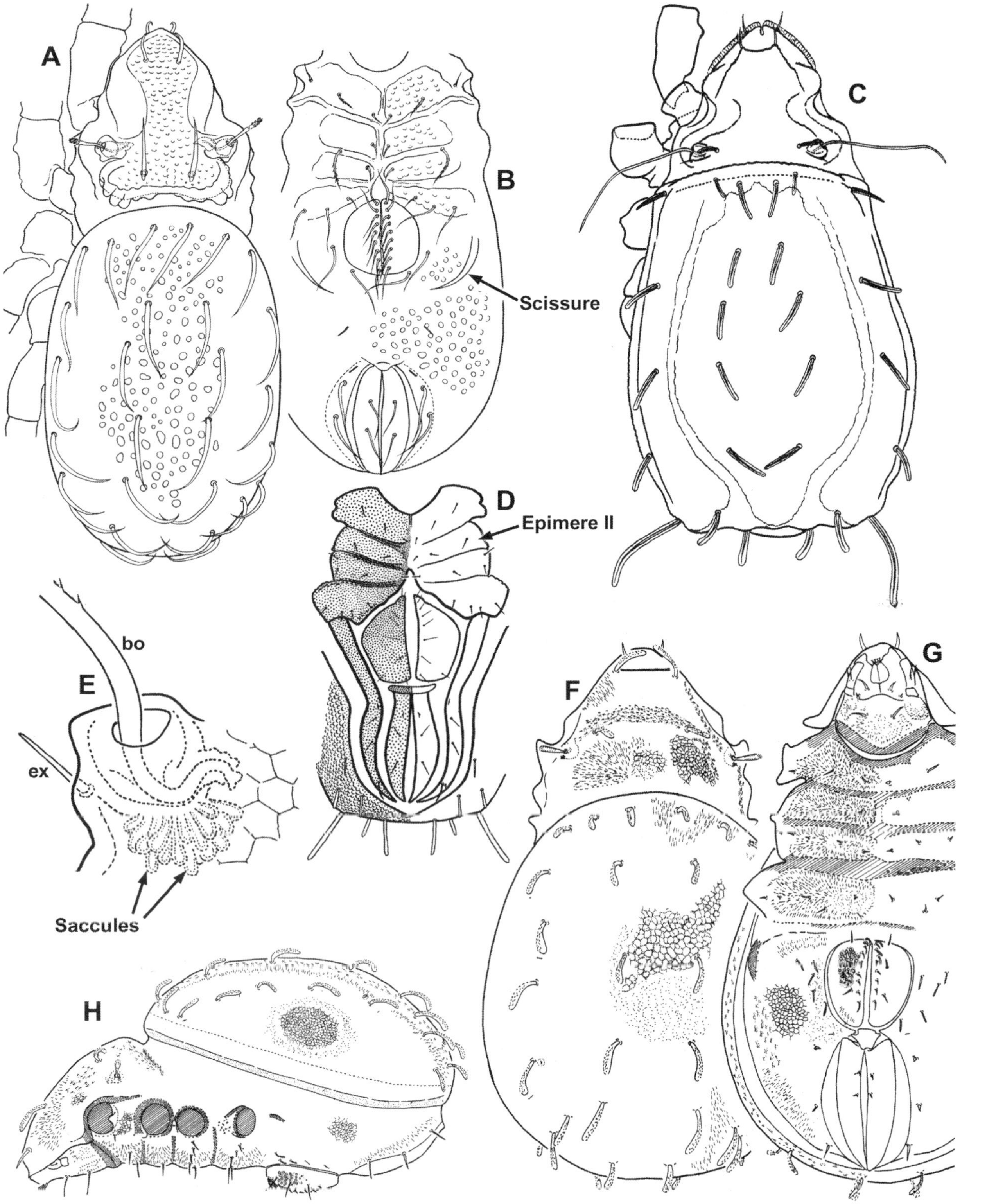

Fig. 15.27. NANHERMANNIIDAE. *Nanhermannia thaiensis* Aoki (Thailand): **A**, dorsal; **B**, ventral (after Aoki 1965c).
NOTHRIDAE. *Nothrus silvestris* Nicolet (Holarctic): **C**, dorsal (after Aoki 1969); **D**, schematic ventral hysterosoma (after Sellnick and Forsslund 1955). *Nothrus* sp.: **E**, bothridium and surrounding structures (after Grandjean 1934d).
HERMANNIIDAE. *Hermannia subglabra* Berlese (Holarctic): **F**, dorsal; **G**, ventral; **H**, lateral (after Woas 1981).

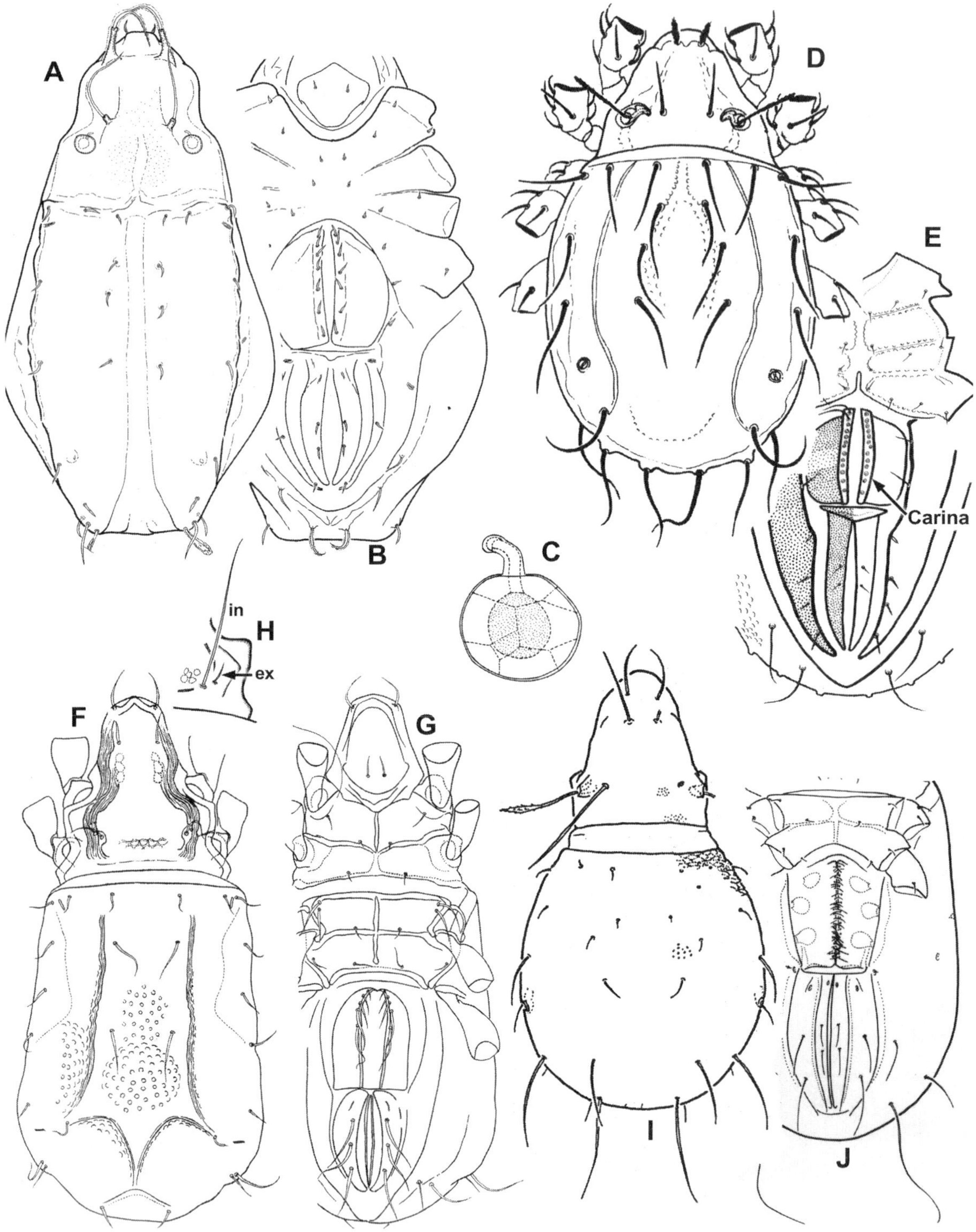

Fig. 15.28. CROTONIIDAE. *Holonothrus venetiolanus* Olszanowski (Venezuela): **A**, dorsal; **B**, ventral (after Olszanowski 1999). *H. foliatus* Wallwork (Macquarie Island): **C**, trichobothrium (after Wallwork 1963).

CAMISIIDAE. *Platynothrus banksi* (Michael) (Oregon, USA): **D**, dorsal (after Krantz 1978). *Platynothrus* sp.: **E**, schematic ventral hysterosoma (after Sellnick and Forsslund 1955).

MALACONOTHRIDAE. *Malaconothrus robustus asiaticus* Aoki (Thailand): **F**, dorsal; **G**, ventral (after Aoki 1967). *Trimalaconothrus yachidairaensis* Yamamoto, Kuriki, and Aoki (Japan): **H**, posterolateral corner of prodorsum, showing absence of trichobothrium (after Yamamoto, Kuriki, and Aoki 1993).

TRHYPOCHTHONIIDAE. *Mainothrus badius* (Berlese) (Holarctic): **I**, dorsal (after Weigmann 1997a). *Mucronothrus nasalis* (Willmann) (global): **J**, ventral hysterosoma (after Norton, Behan-Pelletier, and Wang 1996).

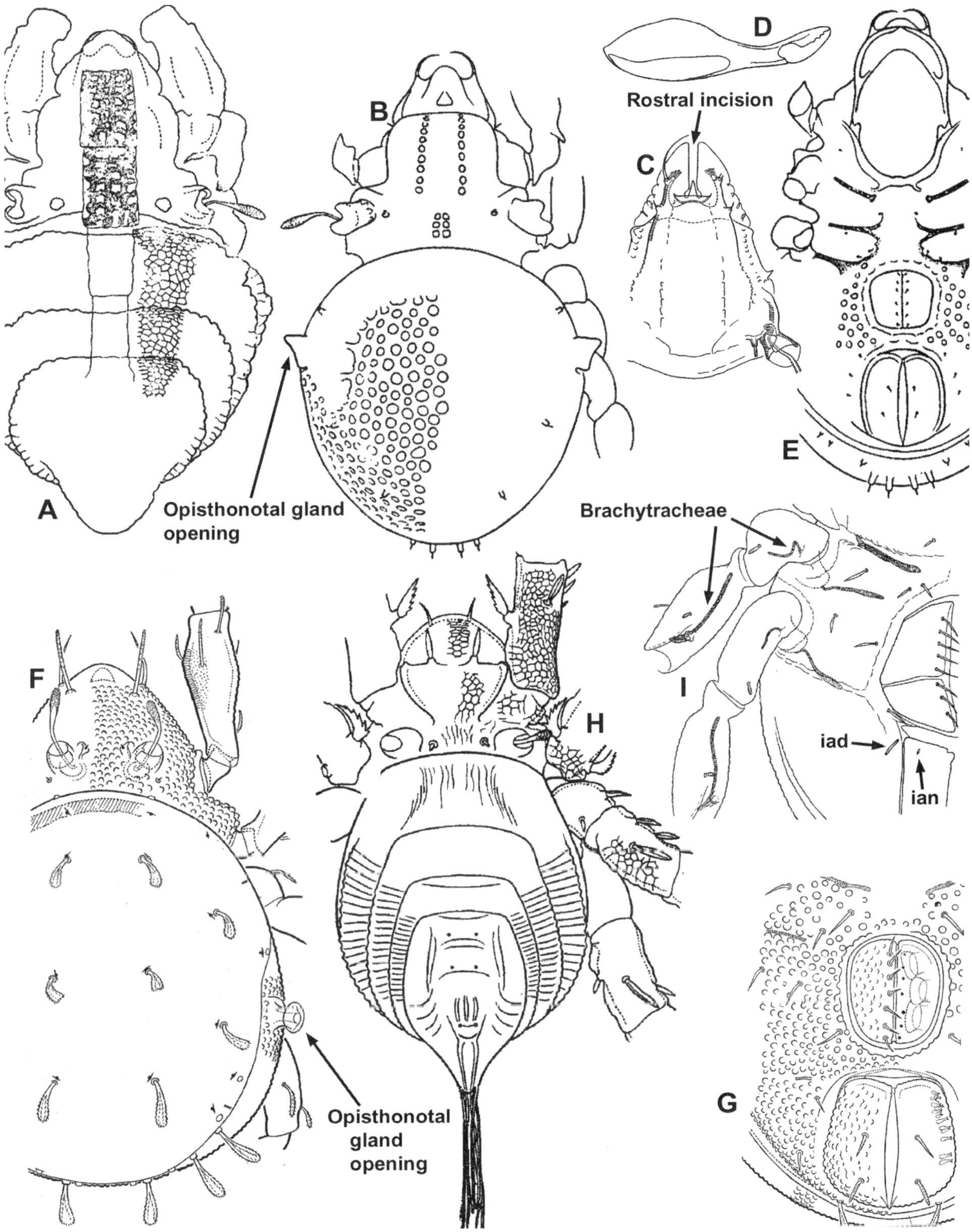

Fig. 15.29. PLASMOBATIDAE. *Plasmobates pagoda* Grandjean (Venezuela): **A**, dorsal; **B**, dorsal (scalps removed); **C**, prodorsum; **D**, chelicerae (after Grandjean 1929); **E**, ventral (**A**, **B**, **D**, **E** after Grandjean 1929; **C** after Grandjean 1961c). HERMANNIELLIDAE. *Ampullobates nigriclavatus* Grandjean (Panama): **F**, dorsal; **G**, ventral (after Grandjean 1962a). NEOLIODIDAE. *Telioliodes madininensis* Grandjean (Martinique): **H**, dorsal. *Neoliodes theleproctus* (Hermann) (Europe): **I**, ventral (partial) (after Grandjean 1934d).

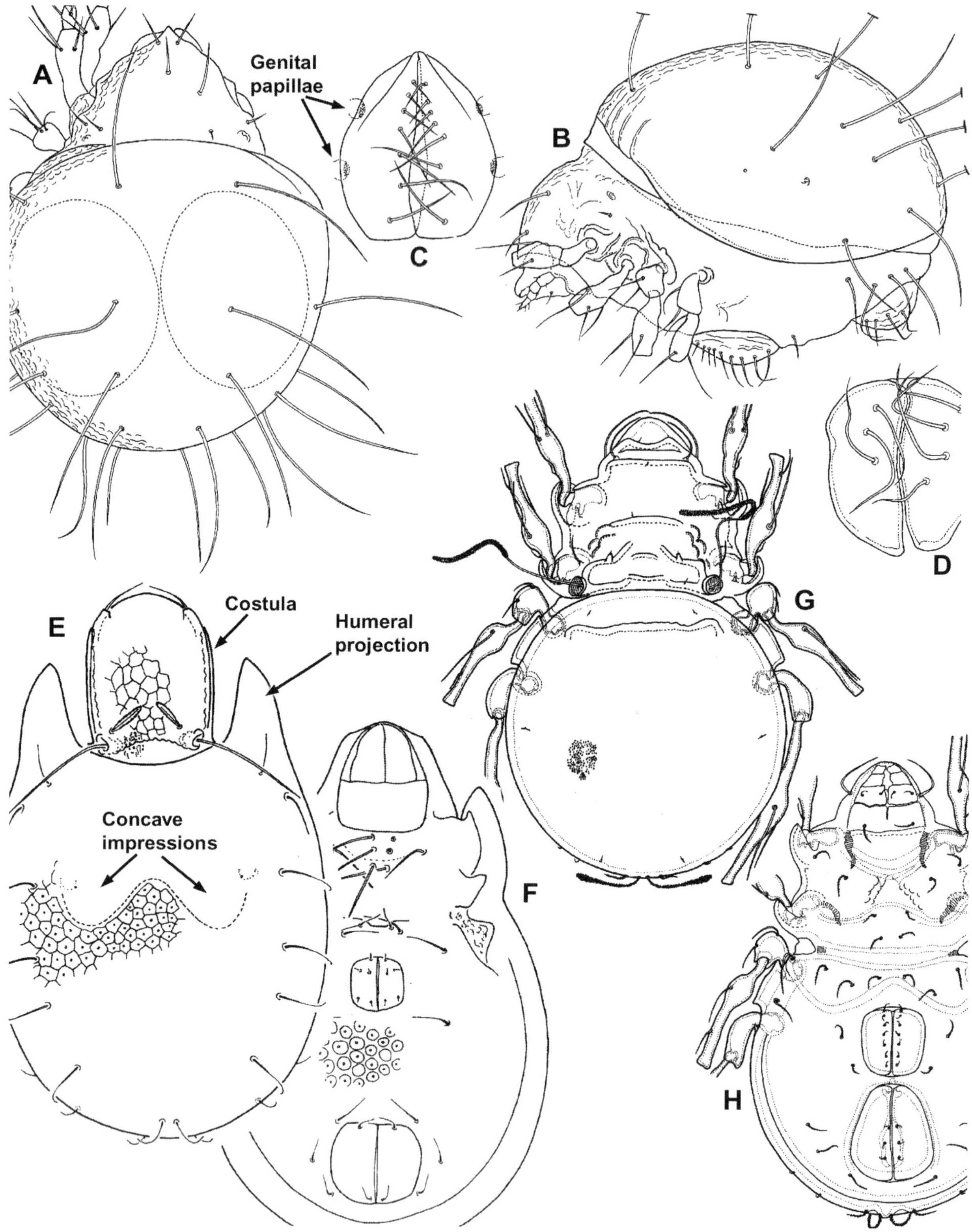

Fig. 15.30. ARIBATIDAE. *Aribates javensis* Aoki, Takaku, and Ito (Java): **A**, dorsal; **B**, lateral; **C**, genital plates; **D**, anal plates (after Aoki, Takaku, and Ito 1994).

PTEROBATIDAE. *Pterobates incertus* Balogh and Mahunka (Brazil): **E**, dorsal; **F**, ventral (after Balogh and Mahunka 1977).

IDIODAMAEIDAE. *Idiodamaeus illecebrosus* Paschoal: **G**, dorsal; **H**, ventral (after Paschoal 1984).

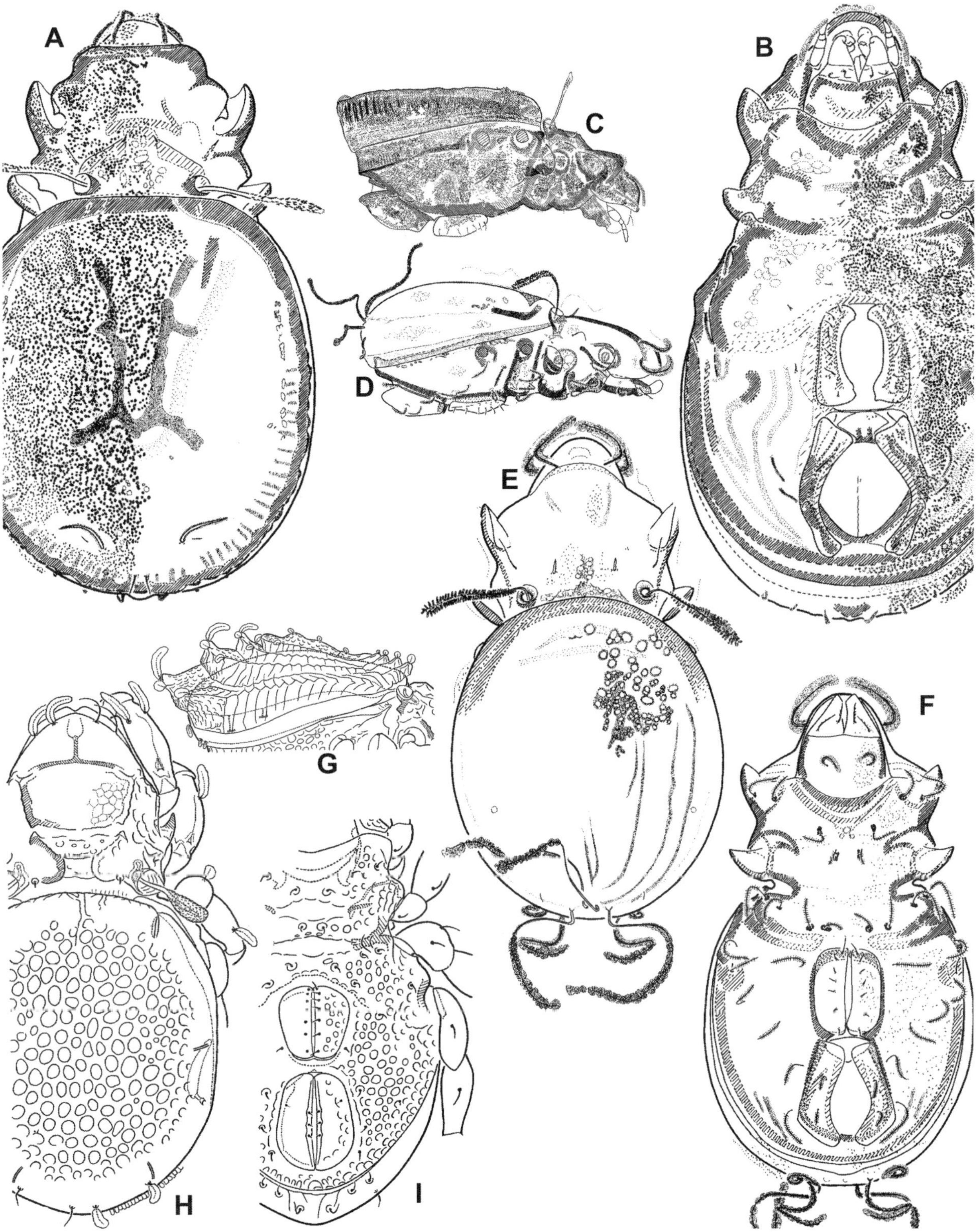

Fig. 15.31. GYMNODAMAEIDAE. *Gymnodamaeus helveticus* Woas (Switzerland): **A**, dorsal; **B**, ventral; **C**, lateral (after Woas 1992). ALEURODAMAEIDAE. *Aleurodamaeus australis* Woas (Australia): **D**, lateral; **E**, dorsal; **F**, ventral (after Woas 1992). PHEROLIODIDAE. *Pheroliodes wehnckei* (Willmann) (Venezuela): **G**, lateral of notogaster; **H**, dorsal; **I**, ventral (after Grandjean 1964c).

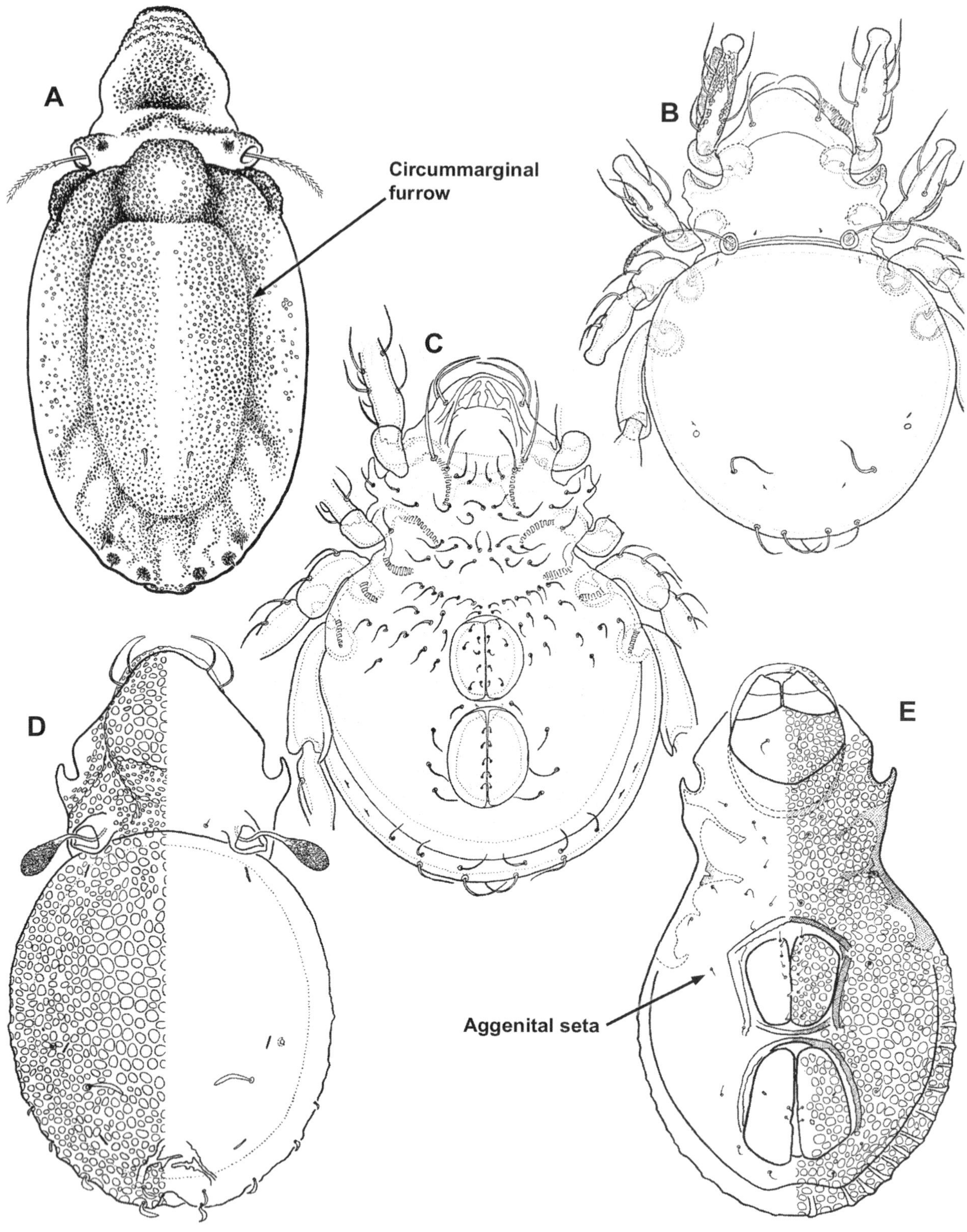

Fig. 15.32. NACUNANSELLIDAE. *Nacunansella diminuta* Fernandez and Cleva (Argentina): **A**, dorsal (after Fernandez and Cleva 1997).
PLATEREMAEIDAE. *Allodamaeus ewingi* Banks (North Carolina, USA): **B**, dorsal; **C**, ventral (after Paschoal 1988).
PEDROCORTESELLIDAE. *Pedrocortesella japonica* Aoki and Suzuki (Japan): **D**, dorsal; **E**, ventral (after Aoki and Suzuki 1970).

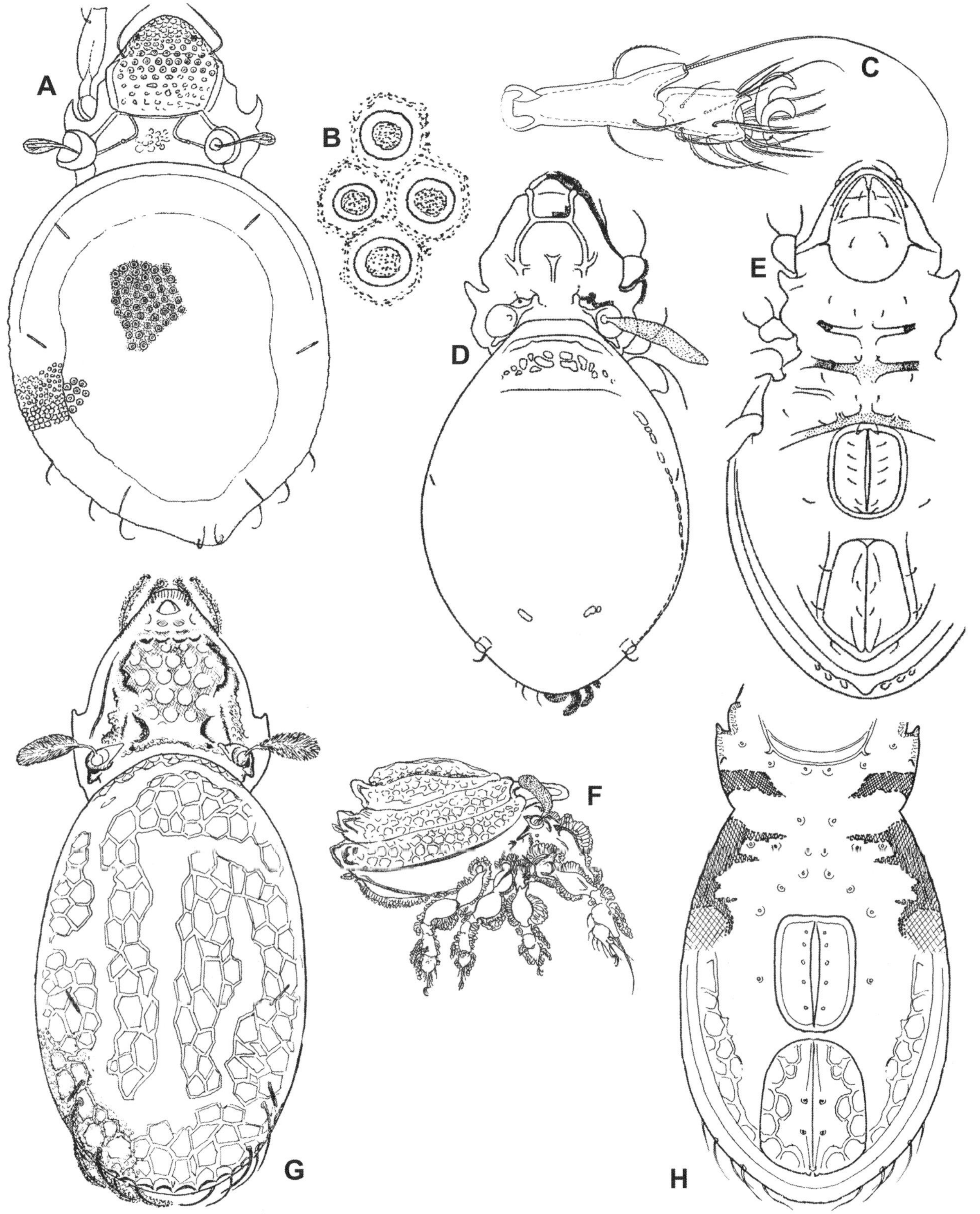

Fig. 15.33. LYRIFISSELLIDAE. *Lyrifissella latoclava* (Hammer) (New Zealand): **A**, dorsal; **B**, pattern on notogaster; **C**, tibia, tarsus I (from Hammer 1966).

LICNOBELBIDAE. *Licnobelba alestensis* Grandjean (France, Switzerland): **D**, dorsal; **E**, ventral (after Grandjean 1931a). *L. latiflabellata* (Paoli) (Italy): **F**, lateral (after Pérez-Íñigo 1993b).

LICNODAMAEIDAE. *Licnodamaeus costula* Grandjean (Spain): **G**, dorsal; **H**, ventral (after Pérez-Íñigo 1970).

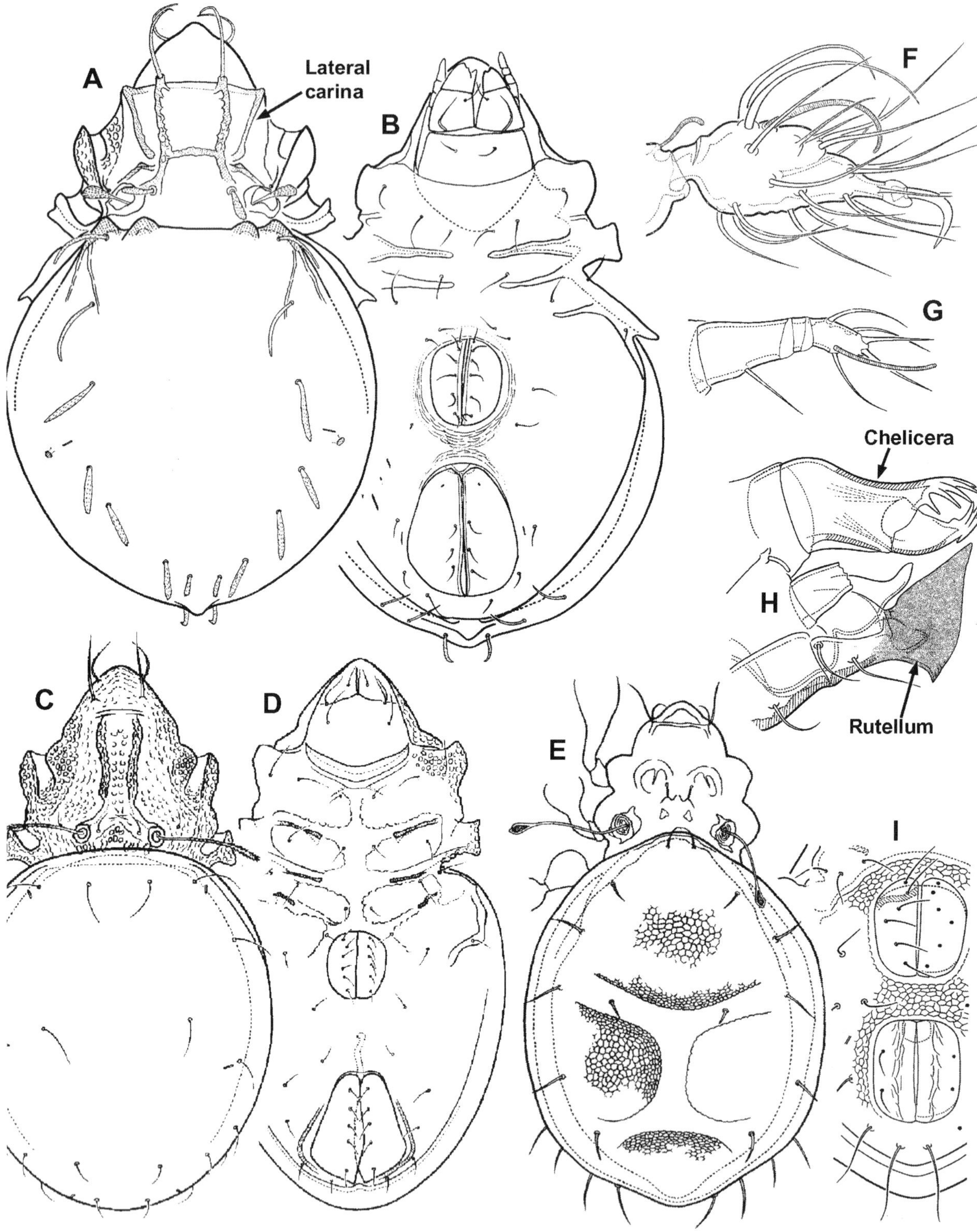

Fig. 15.34. MEGEREMAEIDAE. *Megeremaeus expansus* Aoki and Fujikawa (Japan): **A**, dorsal; **B**, ventral (after Aoki and Fujikawa 1971).

EREMAEIDAE. *Eremaeus tenuisetiger* Aoki (Japan): **C**, dorsal; **D**, ventral (after Aoki 1970).

DAMAEOLIDAE. *Fosseremus quadripertitus* (Paoli) (Italy): **E**, dorsal; **F**, tarsus I; **G**, palp; **H**, gnathosoma (lateral); **I**, ventral (after Grandjean 1965b).

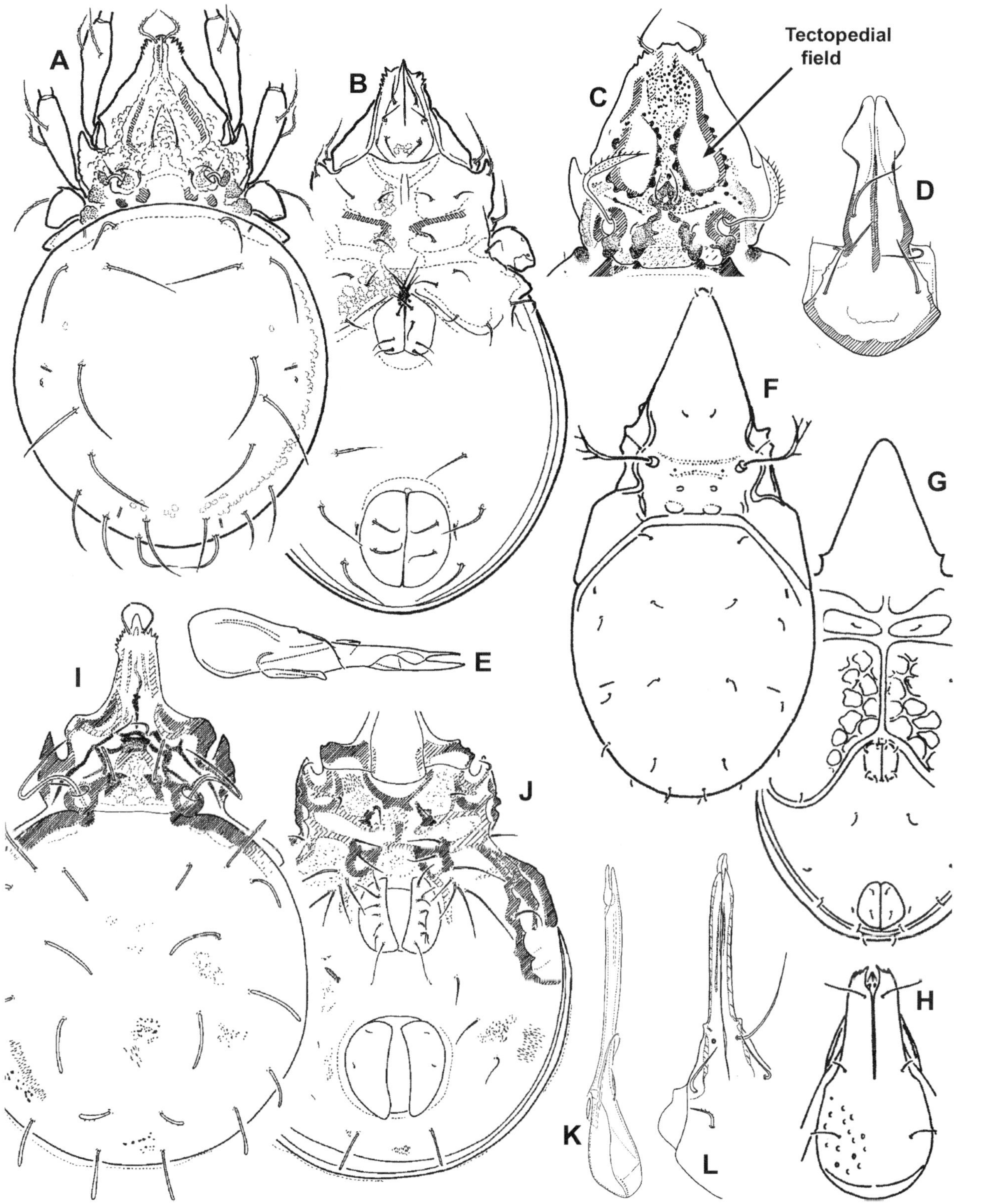

Fig. 15.35. SUCTOBELBIDAE. *Suctobelba* sp. (Florida, USA): **A**, dorsal; **B**, ventral. *S. carcharodon* Moritz (Italy): **C**, prodorsum. *S. centroamericana* Woas (El Salvador): **D**, subcapitulum; **E**, chelicera. (**C–E** after Woas 1986).

TRIZETIDAE. *Trizetes pyramidalis* Berlese (Italy): **F**, dorsal; **G**, ventral; **H**, subcapitulum (after Balogh and Balogh 1992).

RHYNCHORIBATIDAE. *Rhynchoribates brasiliensis* Woas (Brazil): **I**, dorsal; **J**, ventral; **K**, chelicera; **L**, subcapitulum (after Woas 1986).

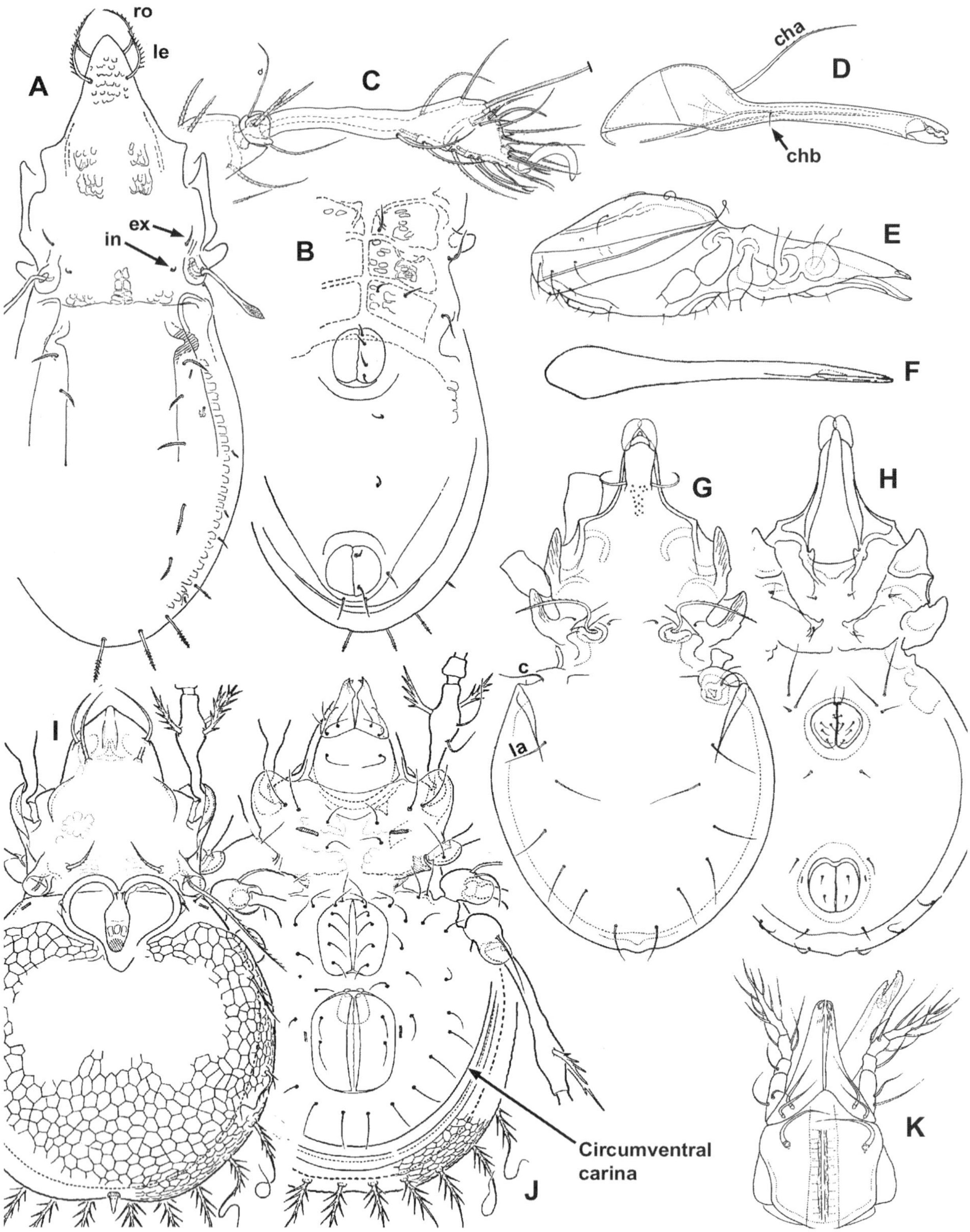

Fig. 15.36. DAMPFIELLIDAE. *Dampfiella cejansis* Beck (Peru): **A**, dorsal; **B**, ventral; **C**, leg I; **D**, chelicera (after Beck 1962b). OXYAMERIDAE. *Oxyamerus spathulatus* Aoki (Thailand): **E**, lateral; **F**, chelicera; **G**, dorsal; **H**, ventral (after Aoki 1965c). BASILOBELBIDAE. *Basilobelba sollertius* (Grandjean) (Australia): **I**, dorsal; **J**, ventral (after Grandjean 1959a). *Xiphobelba setosa* Aoki (New Britain Island): **K**, gnathosoma (after Aoki 1968).

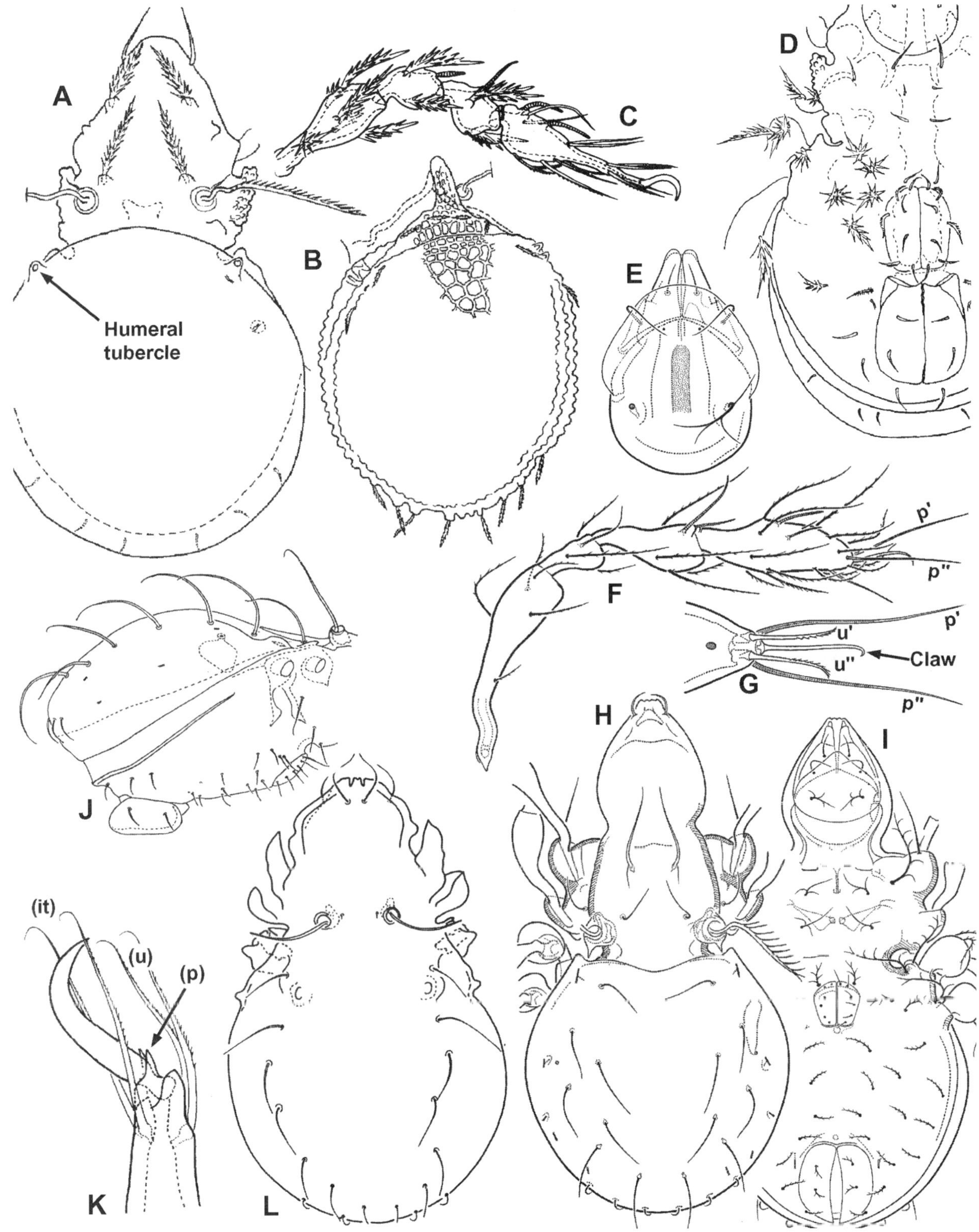

Fig. 15.37. HETEROBELBIDAE. *Heterobelba barbata* Beck (Peru): **A**, dorsal; **B**, cerotegument; **C**, leg II; **D**, ventral (after Beck 1962a). STAUROBATIDAE. *Staurobates schusteri* Grandjean (Brazil): **E**, subcapitulum; **F**, leg I; **G**, tip of tarsus I. *Stauroma cephalotum* Grandjean (Brazil): **H**, dorsal; **I**, ventral (after Grandjean 1967b).

AMERIDAE. *Amerus troisi* (Berlese) (Europe): **J**, lateral (after Avanzati et al. 2003). *Caenosamerus spatiosus* Aoki (Japan): **K**, apical part of tarsus II (after Aoki 1977a). *Amerus laticephalus* Berlese (Italy): **L**, dorsal (after Mahunka and Mahunka-Papp 1995).

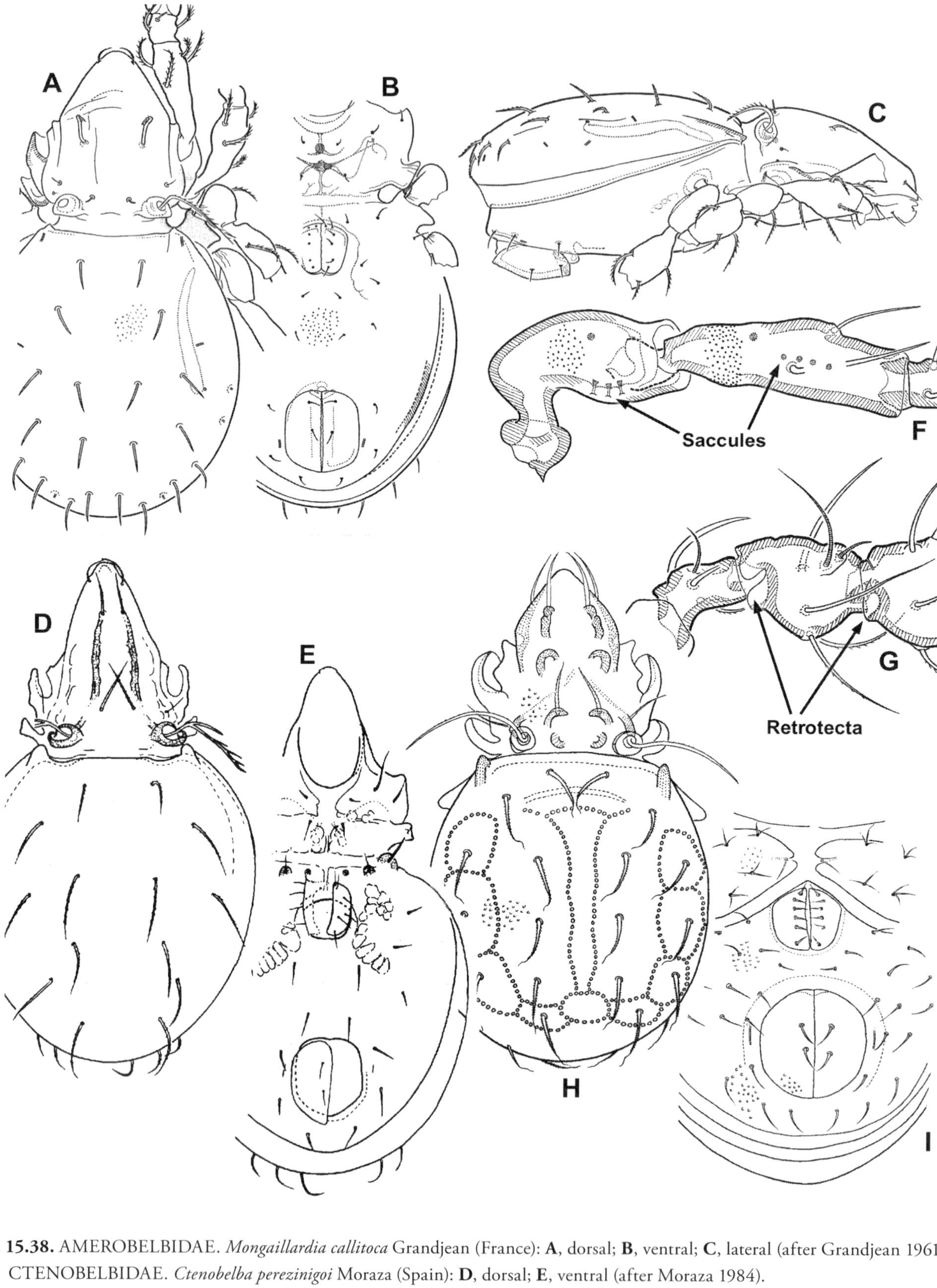

Fig. 15.38. AMEROBELBIDAE. *Mongaillardia callitoca* Grandjean (France): **A**, dorsal; **B**, ventral; **C**, lateral (after Grandjean 1961a). CTENOBELBIDAE. *Ctenobelba perezinigoi* Moraza (Spain): **D**, dorsal; **E**, ventral (after Moraza 1984). EREMOBELBIDAE. *Eremobelba geographica* Berlese (Italy): **F**, genu, tibia leg I; **G**, trochanter, femur leg IV (after Grandjean 1965a). *E. minuta* Aoki and Wen (Japan): **H**, dorsal; **I**, ventral (after Aoki and Wen 1983).

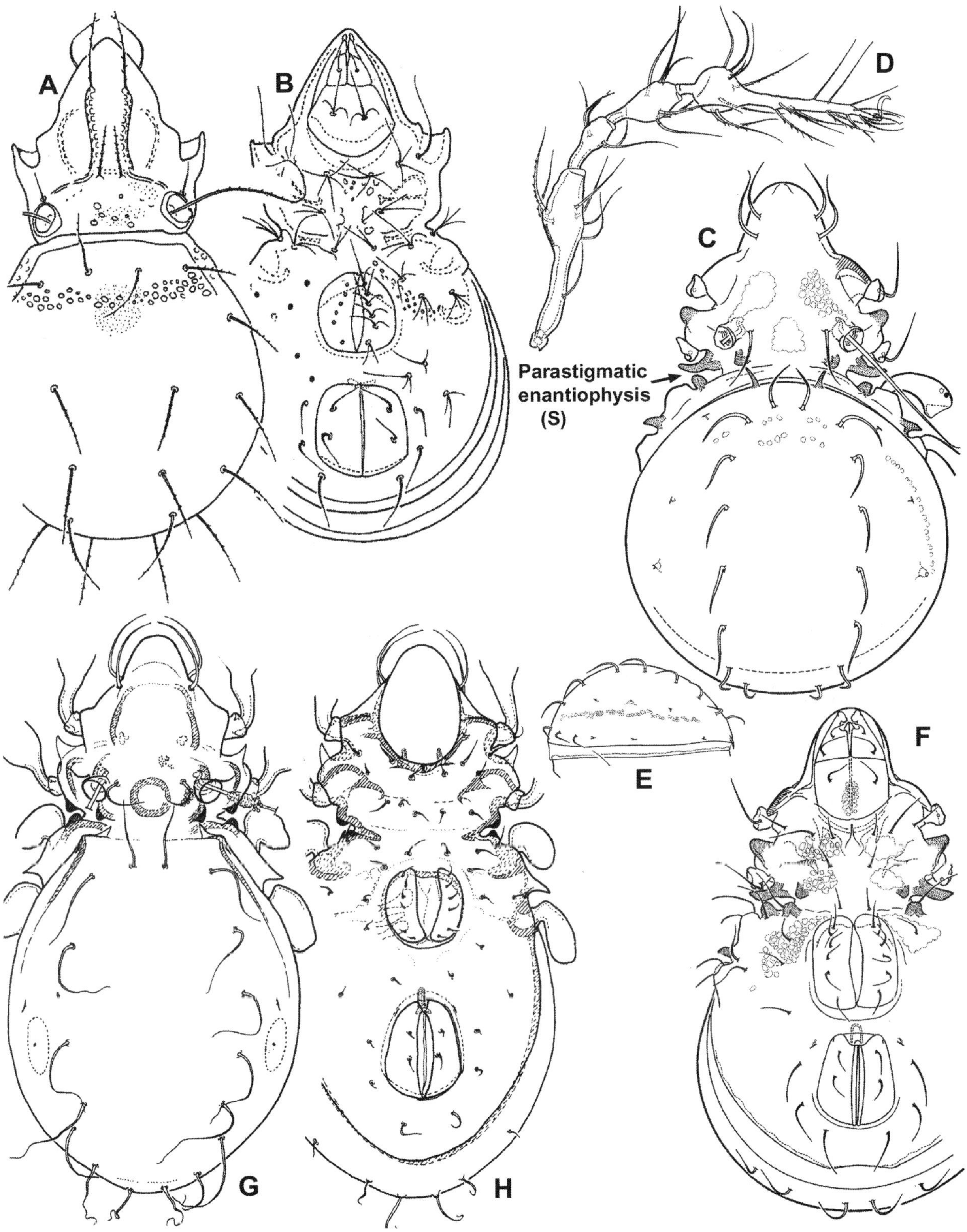

Fig. 15.39. EREMULIDAE. *Eremulus rigidisetosus* Balogh and Mahunka (Bolivia): **A**, dorsal; **B**, ventral (after Balogh and Mahunka 1969).

DAMAEIDAE. *Epidamaeus bakeri* (Hammer) (arctic North America): **C**, dorsal; **D**, leg I; **E**, lateral; **F**, ventral (after Behan-Pelletier and Norton 1983).

HUNGAROBELBIDAE. *Hungarobelba pyrenaica* Miko and Travé (France): **G**, dorsal; **H**, ventral (after Miko and Travé 1996).

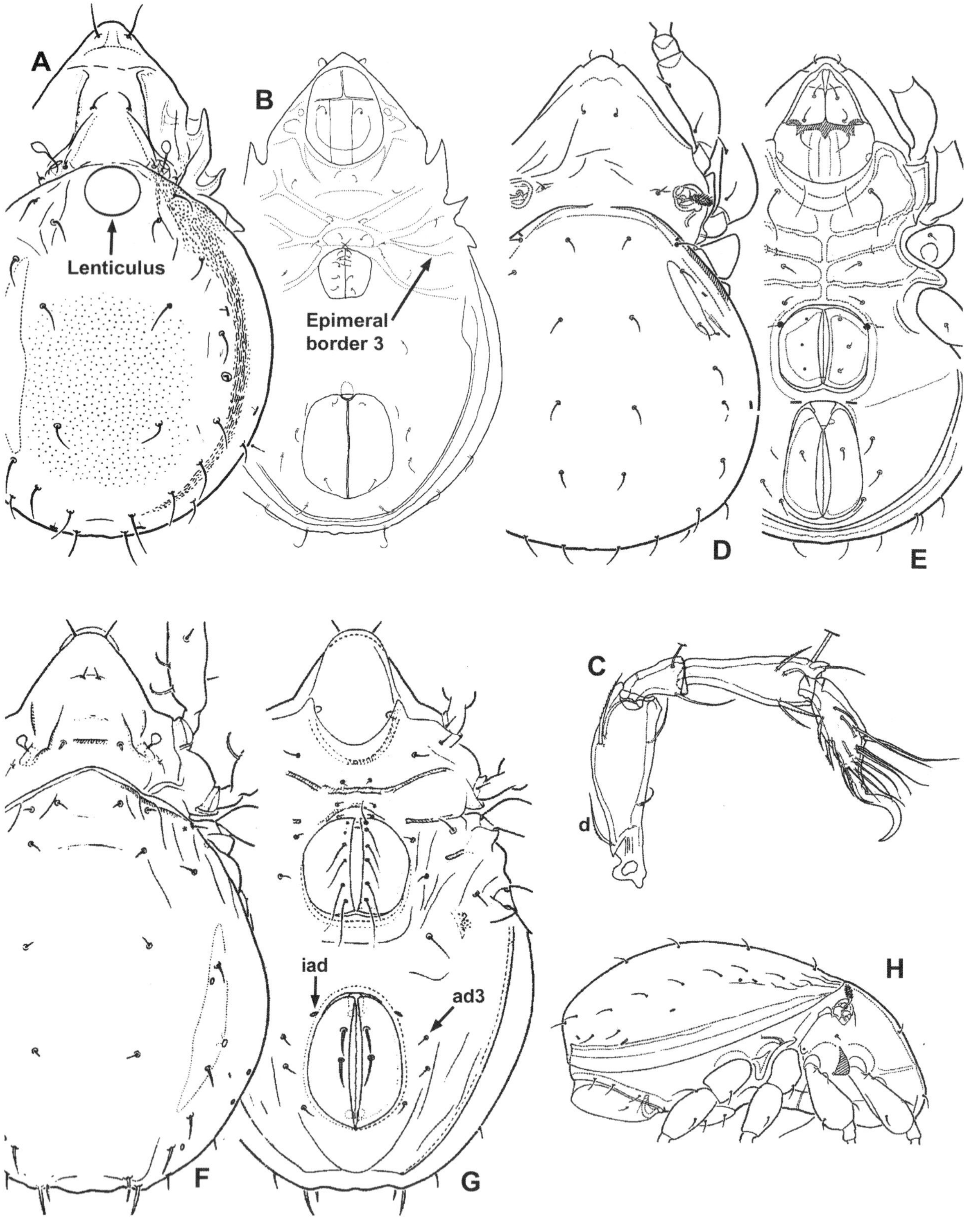

Fig. 15.40. HYDROZETIDAE. *Hydrozetes incisus* Grandjean (western Europe): **A**, dorsal (after Grandjean 1948c). *H. lemnae* Coggi (worldwide): **B**, ventral; **C**, leg I (after Engelbrecht 1974).
SELENORIBATIDAE. *Selenoribates mediterraneus* Grandjean (Mediterranean): **D**, dorsal; **E**, ventral (after Grandjean 1966b).
AMERONOTHRIDAE. *Podacarus auberti* Grandjean (Kerguelen Islands): **F**, dorsal; **G**, ventral; **H**, lateral (after Grandjean 1955a).

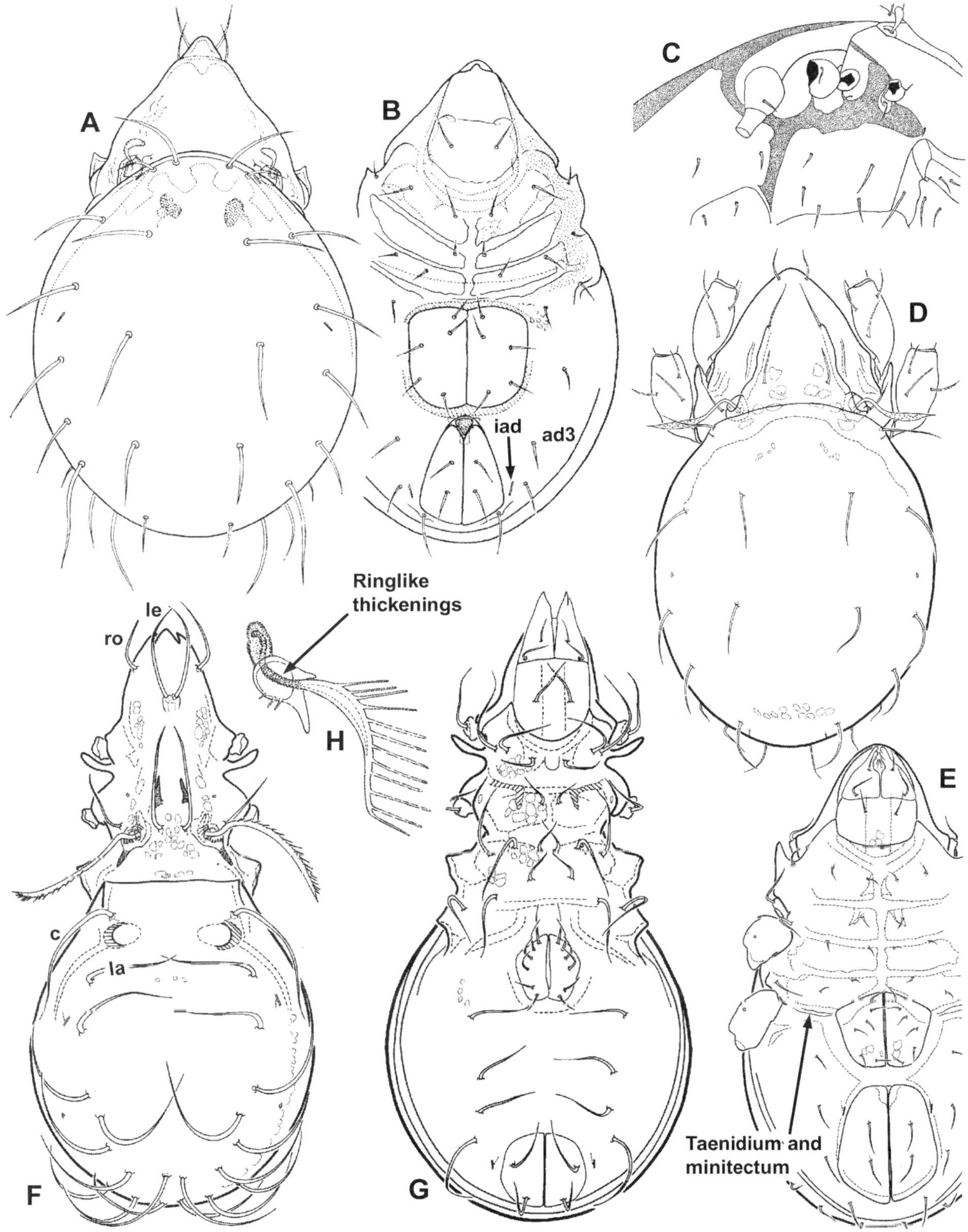

Fig. 15.41. FORTUYNIIDAE. *Fortuynia elamellata shibai* Aoki (Japan): **A**, dorsal; **B**, ventral (after Aoki 1974). *F. yunkeri* Van der Hammen (Panama): **C**, Van der Hammen's organ (after Pugh, King, and Fordy 1990).
THYRISOMIDAE. *Gemmazetes crosbyi* (Berlese) (eastern USA): **D**, dorsal; **E** ventral (V. M. Behan-Pelletier, unpubl.).
SPINOZETIDAE. *Spinozetes* sp. (California, USA): **F**, dorsal; **G**, ventral (V. M. Behan-Pelletier, unpubl.). *S. inexpectatus* Piffl: **H**, detail of bothridium and bothridial seta (after Piffl 1966).

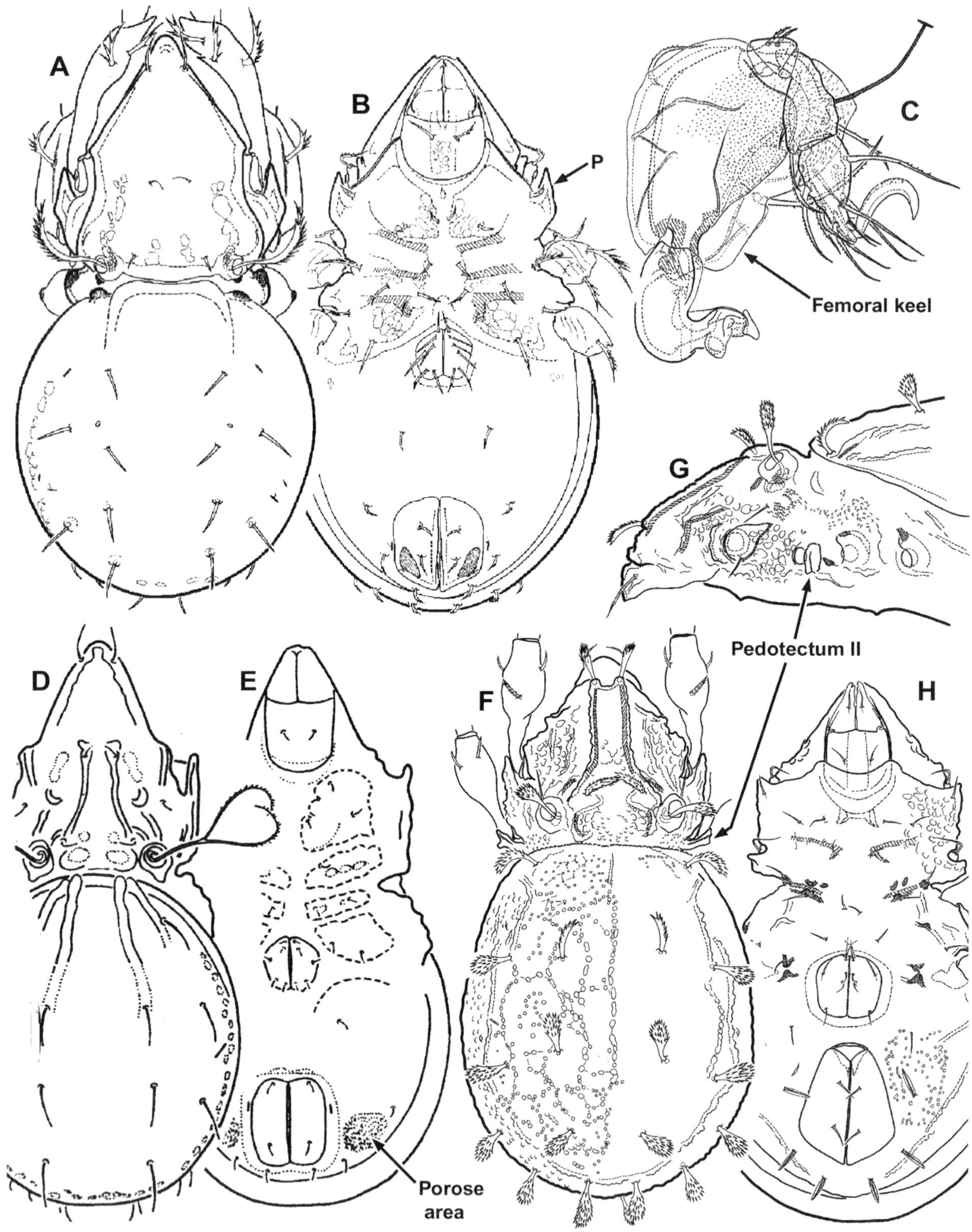

Fig. 15.42. ARCEREMAEIDAE. *Tecteremaeus* sp. (Florida, USA): **A**, dorsal; **B**, ventral (V. M. Behan-Pelletier, unpubl.). *T. cachoeirensis* Franklin and Woas (Brazil): **C**, leg IV (after Franklin and Woas 1992).

MACHADOBELBIDAE. *Machadobelba symmetrica* Balogh (Zaire): **D**, dorsal; **E**, ventral (after Balogh 1958).

ANDEREMAEIDAE. *Epieremulus* sp. (Alabama, USA): **F**, dorsal; **G**, lateral; **H**, ventral (V. M. Behan-Pelletier, unpubl.).

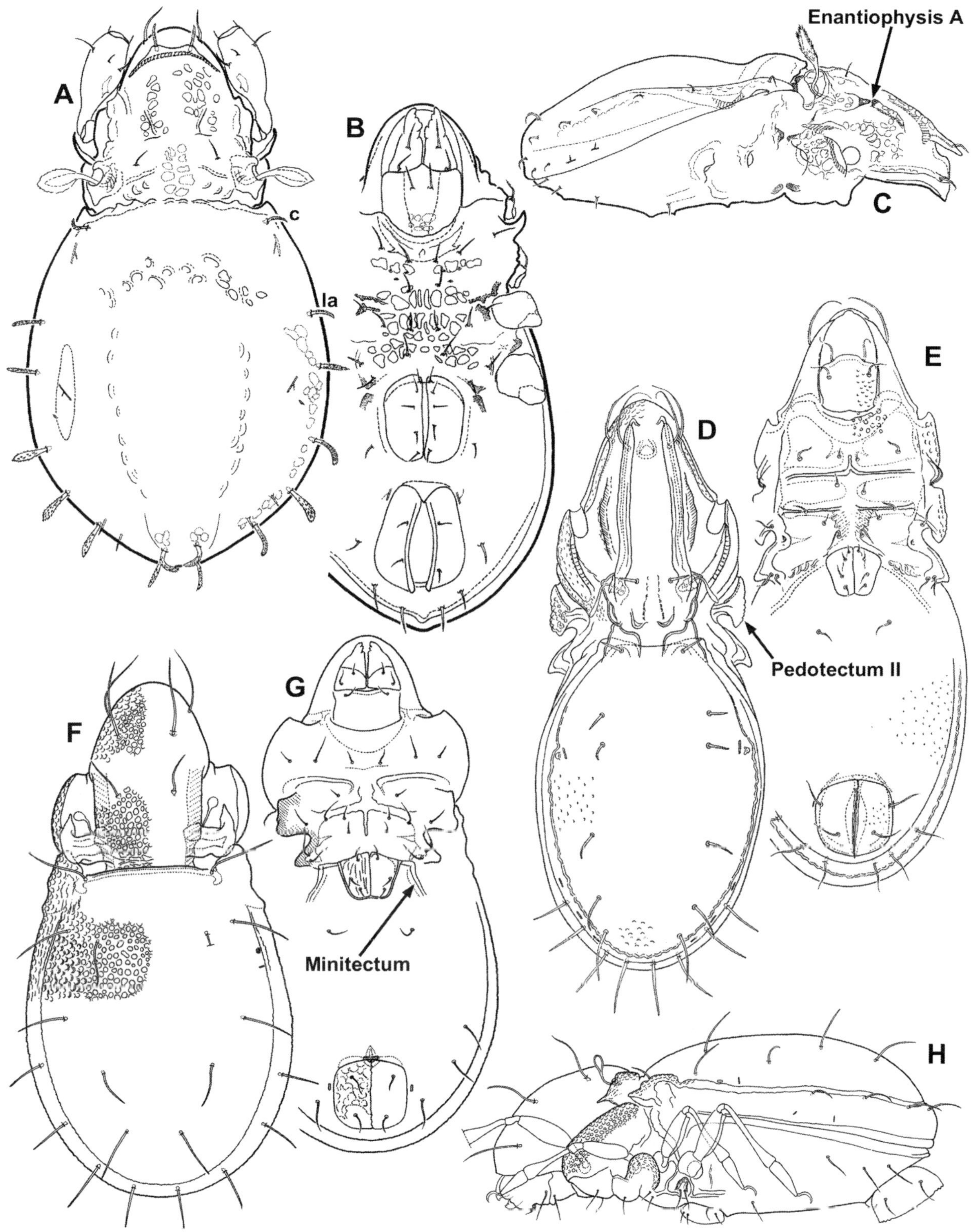

Fig. 15.43. CALEREMAEIDAE. *Caleremaeus* sp. (Georgia, USA): **A**, dorsal; **B**, ventral; **C**, lateral (V. M. Behan-Pelletier, unpubl.). OTOCEPHEIDAE. *Otocepheus* (*Acrotocepheus*) *excelsus* Aoki (Japan): **D**, dorsal; **E**, ventral (after Aoki 1965d). TOKUNOCEPHEIDAE. *Tokunocepheus mizusawai* Aoki (Japan): **F**, dorsal; **G**, ventral; **H**, lateral (after Aoki 1966b).

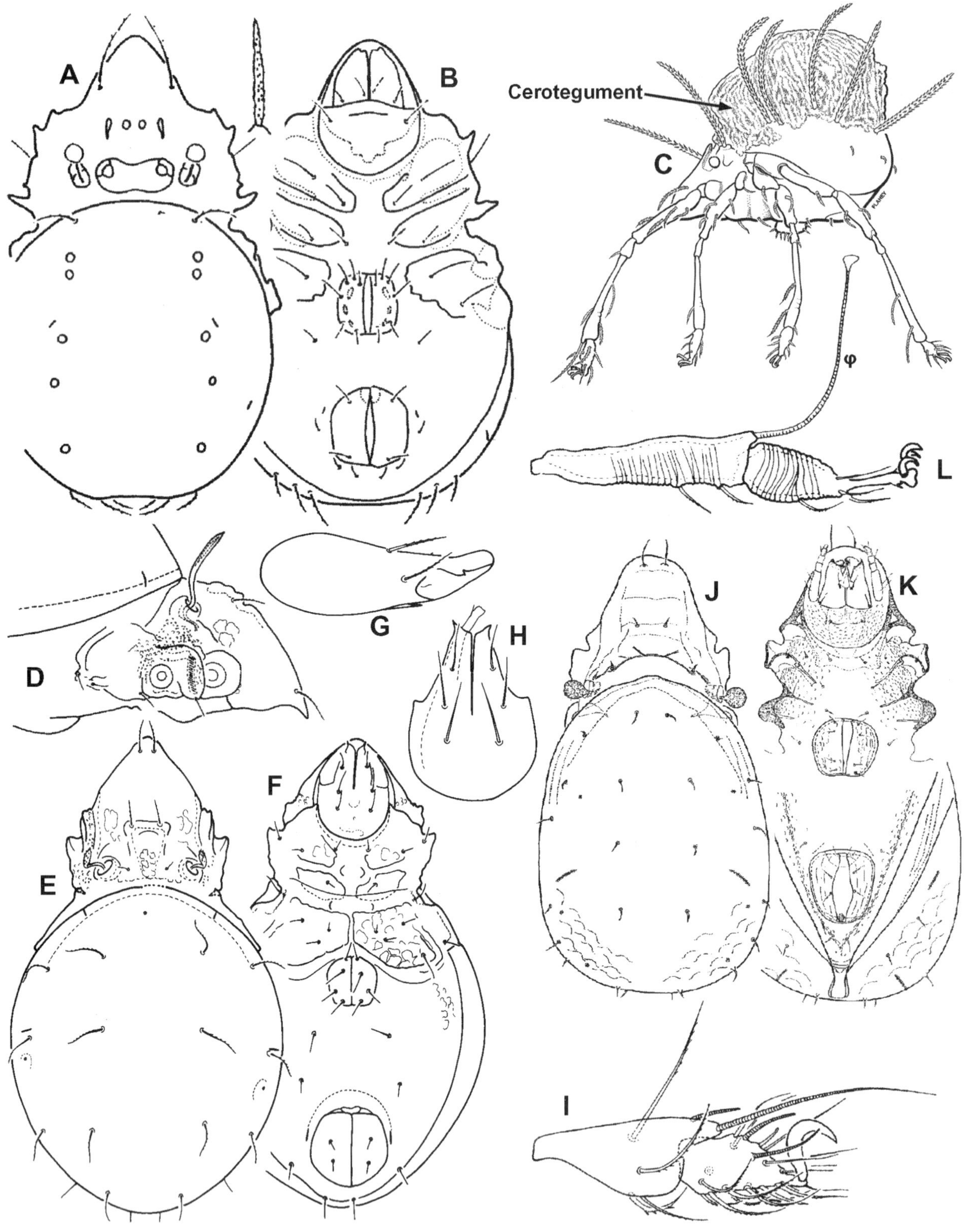

Fig. 15.44. TUPAREZETIDAE. *Tuparezetes christineae* Spain (New Zealand): **A**, dorsal; **B**, ventral (after Spain 1969); **C**, lateral (after Hunt et al. 1998).

NOSYBELBIDAE. *Nosybelba oppiana* Mahunka (Madagascar): **D**, lateral; **E**, dorsal; **F**, ventral; **G**, chelicera; **H**, subcapitulum; **I**, tibia and tarsus I (after Mahunka 1994).

MICREREMIDAE. *Micreremus subglaber* Ito (Japan): **J**, dorsal, **K**, ventral (after Ito 1982). *Porofenichelia porosa* Mahunka (South Africa): **L**, tibia, tarsus IV (after Mahunka 1985).

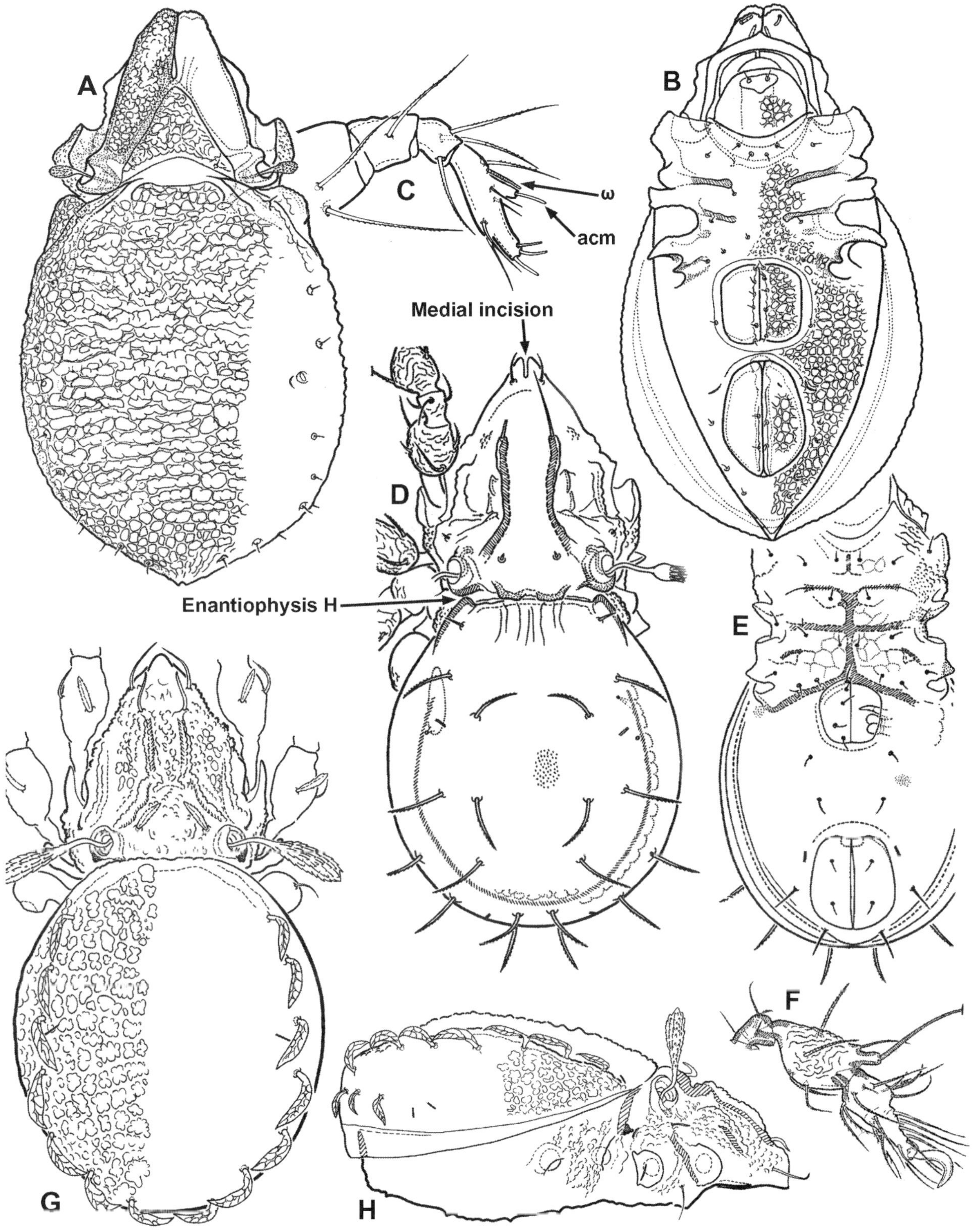

Fig. 15.45. CYMBAEREMAEIDAE. *Ametroproctus reticulatus* (Aoki and Fujikawa) (Japan, North America): **A**, dorsal; **B**, ventral (after Aoki and Fujikawa 1972): *A. oresbios* Higgins and Woolley (western North America): **C**, palp (after Behan-Pelletier 1987). AUTOGNETIDAE. *Autogneta penicillum* Grandjean (France): **D**, dorsal; **E**, ventral; **F**, tibia and tarsus I (after Grandjean 1960c). EREMELLIDAE. *Eremella* sp. (Florida, USA): **G**, dorsal; **H**, lateral (V. M. Behan-Pelletier, unpubl.).

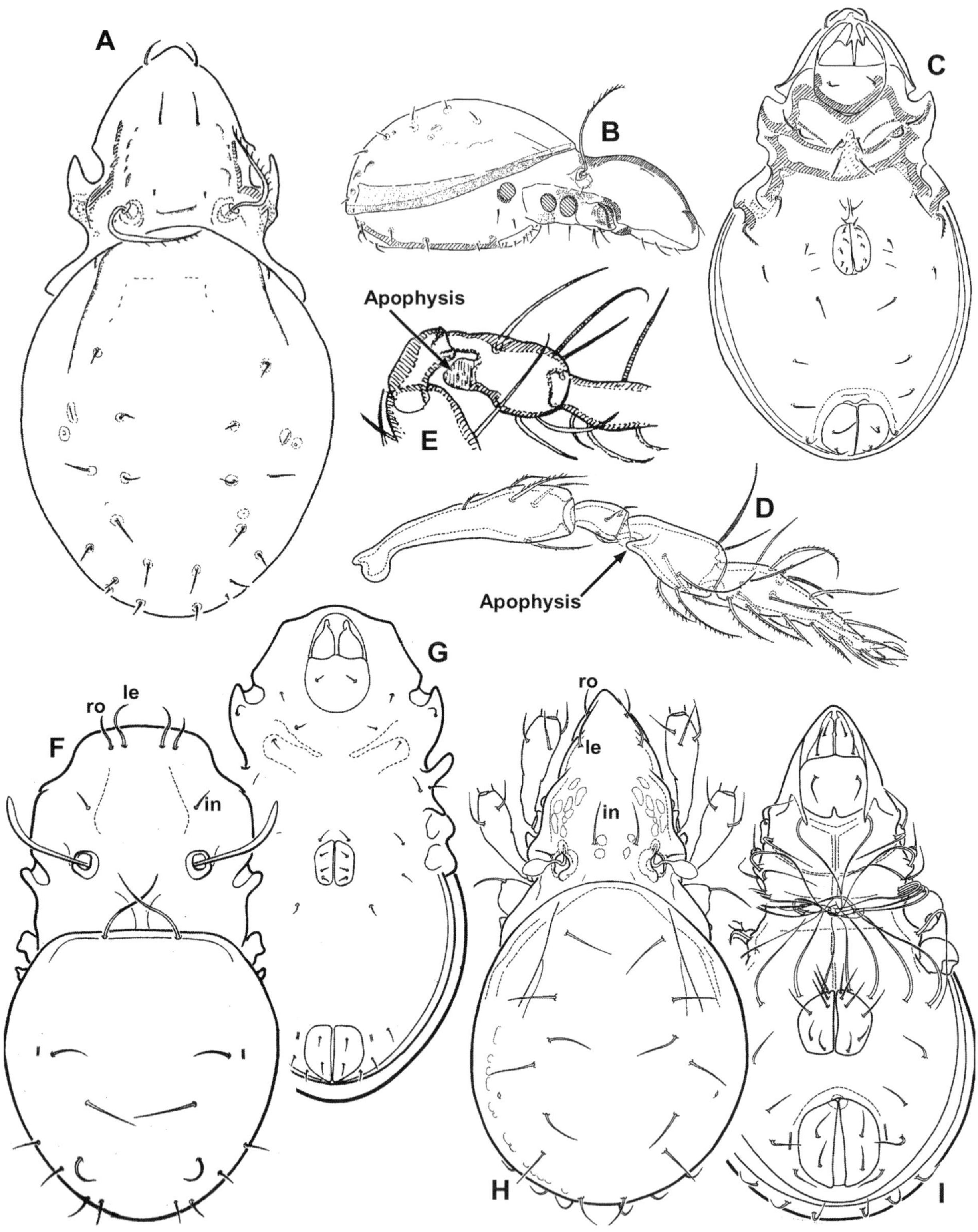

Fig. 15.46. TERATOPPIIDAE. *Teratoppia centroamericana* Woas (El Salvador): **A**, dorsal; **B**, lateral; **C**, ventral; **D**, leg I (after Woas 1986): *T. uspiensis* Pérez-Íñigo and Baggio (Brazil): **E**, tibia I (after Pérez-Íñigo and Baggio 1980).
PLATYAMERIDAE. *Platyamerus peculiarus* J. and P. Balogh (Australia): **F**, dorsal; **G**, ventral (after Hunt et al. 1998).
MACHUELLIDAE. *Machuella* sp. (southern Ontario, Canada): **H**, dorsal; **I**, ventral (V. M. Behan-Pelletier, unpubl.).

Fig. 15.47. STERNOPPIIDAE. *Sternoppia brasiliensis* Franklin and Woas (Brazil): **A**, dorsal; **B**, ventral (after Franklin and Woas 1992). EPIMERELLIDAE. *Enisella turcicus* Ayyildiz and Luxton (Turkey): **C**, dorsal; **D**, ventral (after Ayyildiz and Luxton 1989). GRANULOPPIIDAE. *Senectoppia rugosa* Aoki (West Malaysia): **E**, dorsal; **F**, ventral (after Aoki 1976).

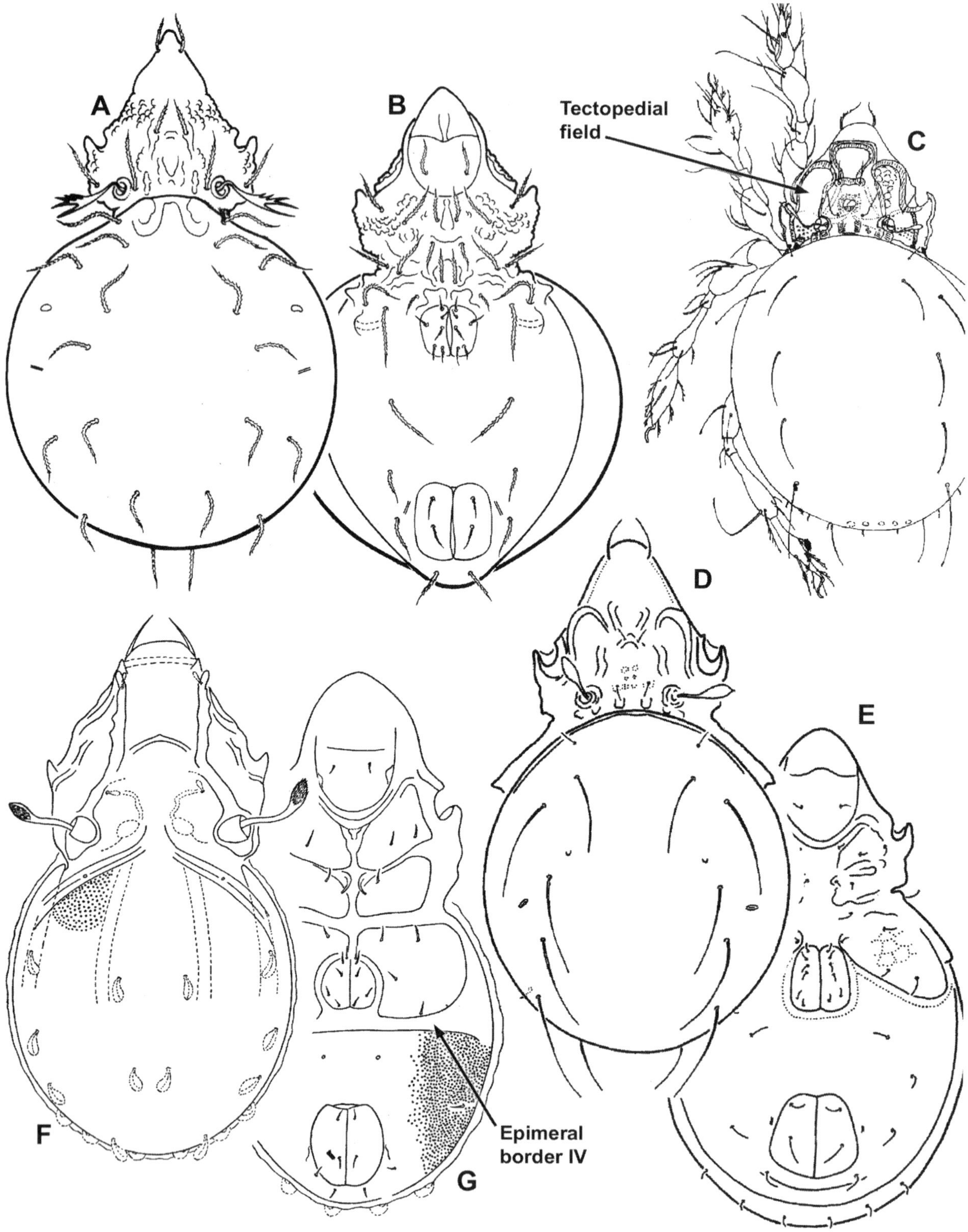

Fig. 15.48. ENANTIOPPIIDAE. *Enantioppia multituberculata* Balogh and Mahunka (Bolivia): **A**, dorsal; **B**, ventral (after Balogh and Mahunka 1969).

CHAVINIIDAE. *Chavinia paradoxa* Hammer (Peru): **C**, dorsal (after Hammer 1962). *C. similis* P. Balogh (Colombia): **D**, dorsal; **E**, ventral (after P. Balogh 1984).

PAPILLONOTIDAE. *Papillonotus maculatus* Wallwork (Ghana): **F**, dorsal; **G**, ventral (after Wallwork 1961).

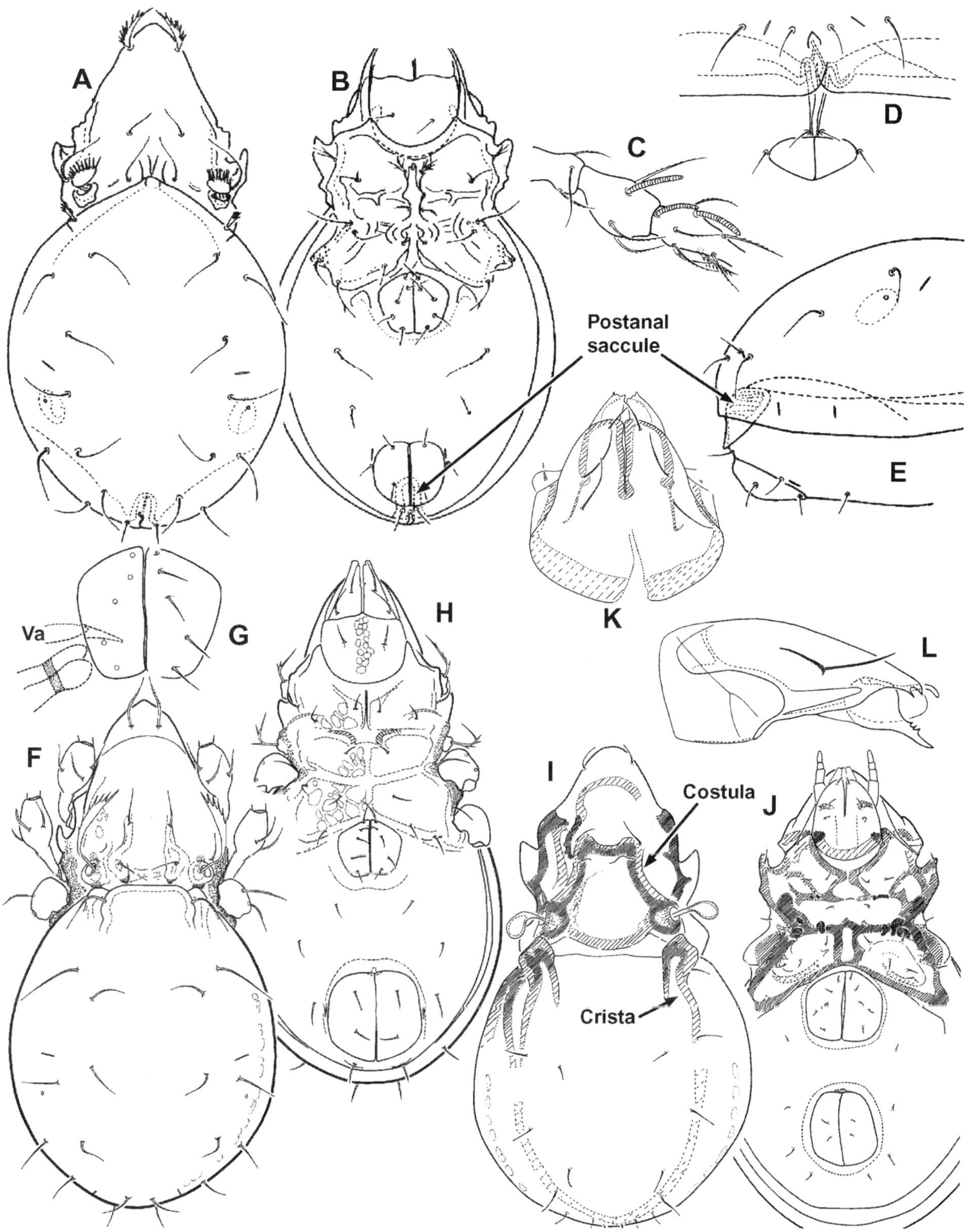

Fig. 15.49. LUXTONIIDAE. *Luxtonia hauseri* Mahunka (Brunei): **A**, dorsal; **B**, ventral; **C**, tibia and tarsus II; **D**, posterior view; **E**, posterior of notogaster in lateral view (after Mahunka 2001).

OPPIIDAE. *Oppiella nova* (Oudemans) (cosmopolitan): **F**, dorsal; **G**, genital papillae; **H**, ventral (V. M. Behan-Pelletier, unpubl.).

QUADROPPIIDAE. *Quadroppia quadricarinata* (Michael) (cosmopolitan): **I**, dorsal; **J**, ventral; **K**, subcapitulum; **L**, chelicera (after Woas 1986).

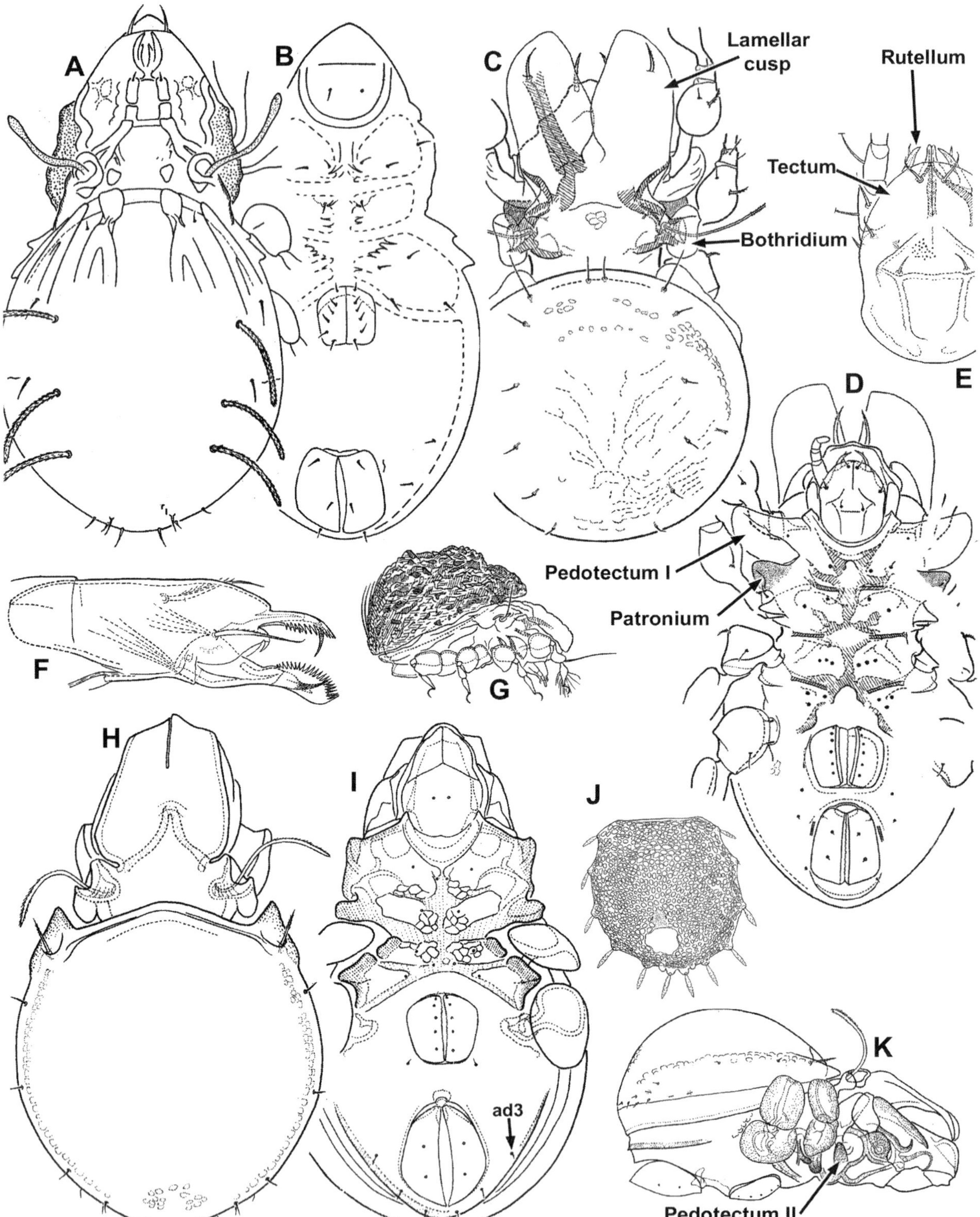

Fig. 15.50. HEXOPPIIDAE. *Hexoppia heterotricha* Balogh (Africa): **A**, dorsal; **B**, ventral (after Wallwork 1961).
POLYPTEROZETIDAE. *Polypterozetes cherubin* Berlese (Italy): **C**, dorsal; **D**, ventral; **E**, subcapitulum; **F**, chelicera; **G**, lateral (after Grandjean 1959d).
PODOPTEROTEGAEIDAE. *Podopterotegaeus tectus* Aoki (Japan): **H**, dorsal; **I**, ventral; **J**, scalp; **K**, lateral (after Aoki 1969).

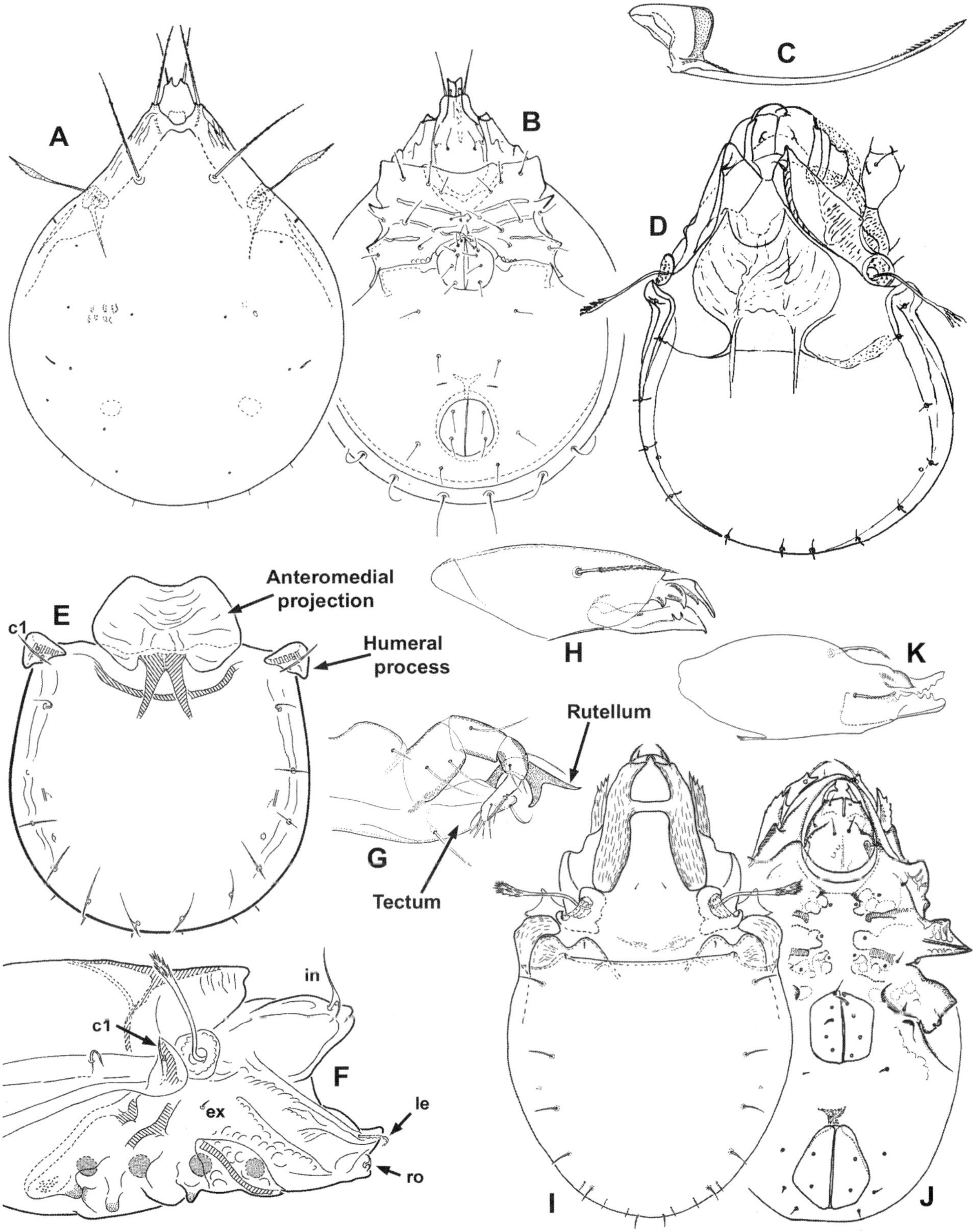

Fig. 15.51. GUSTAVIIDAE. *Gustavia aethiopica* Mahunka (Ethiopia): **A**, dorsal; **B**, ventral (after Mahunka 1982). *Gustavia* sp. (New York, USA): **C**, chelicera (R. A. Norton, unpubl.).

TUMEROZETIDAE. *Tumerozetes bifurcatus* Hammer (New Zealand): **D**, dorsal (after Hammer 1966); **E**, notogaster; **F**, lateral (V. M. Behan-Pelletier, unpubl.); **G**, subcapitulum (lateral); **H**, chelicera (after Piffl 1972).

NODOCEPHEIDAE. *Nodocepheus laterodentatus* Piffl (Costa Rica): **I**, dorsal (H. Schatz, unpubl.); **J**, ventral (after Piffl 1972). *N. dentatus* Hammer (Argentina): **K**, chelicera (after Piffl 1972).

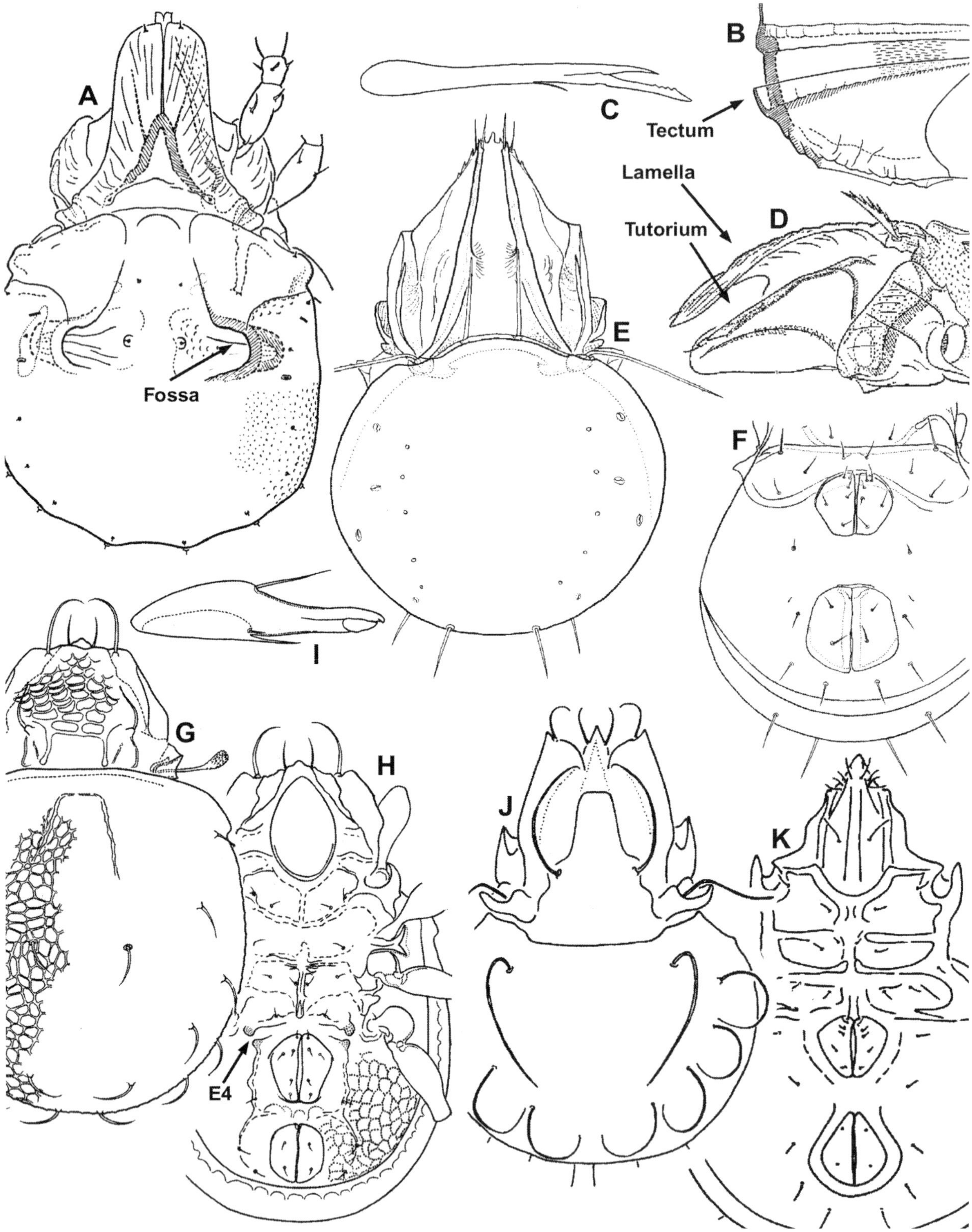

Fig. 15.52. CHARASSOBATIDAE. *Charassobates cavernosus* Grandjean (Venezuela): **A**, dorsal; **B**, lateral (posterior); **C**, chelicera; **D**, lateral (anterior) (after Grandjean 1929, 1958b).
PELOPPIIDAE. *Ceratoppia incisa* Kaneko and Aoki (Japan): **E**, dorsal; **F**, ventral (after Kaneko and Aoki 1982).
MICROTEGEIDAE. *Microtegeus reticulatus* Aoki (Japan, Nepal): **G**, dorsal; **H**, ventral; **I**, chelicera (after Aoki 1965c).
CEROCEPHEIDAE. *Cerocepheus mirabilis* Trägårdh (Juan Fernández Islands): **J**, dorsal; **K**, ventral (after Trägårdh 1931).

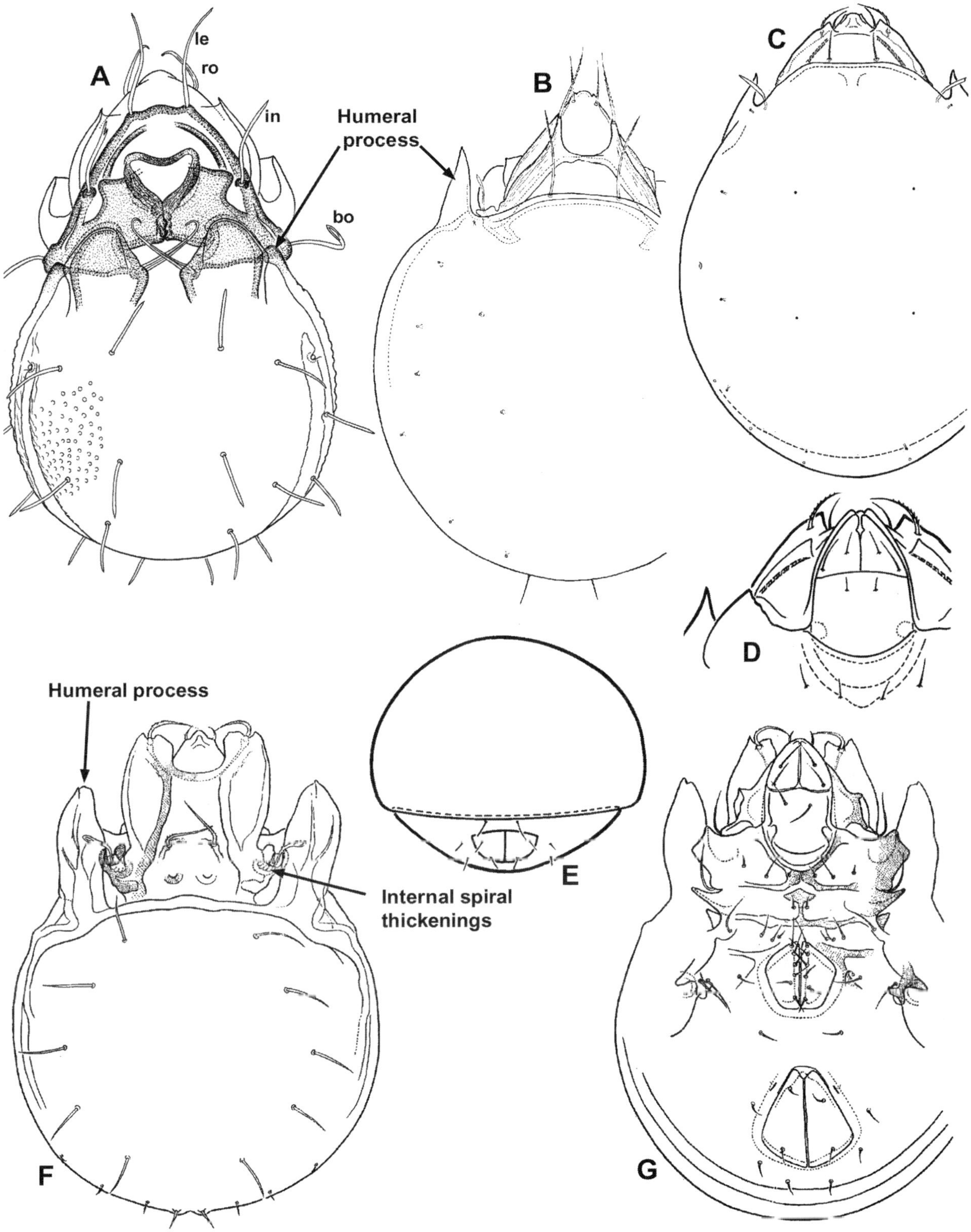

Fig. 15.53. NIPPOBODIDAE. *Leobodes anulatus* Aoki (Himalayas): **A**, dorsal (after Aoki 1965b).
TENUIALIDAE. **B**, *Tenuialoides translamellatus* (Aoki and Fujikawa) (Japan): dorsal (after Aoki and Fujikawa 1969). *Hafenferreffia nitidula* (Banks) (New Hampshire, USA): **C**, dorsal; **D**, partial ventral; **E**, posterior (after Norton 1983).
EUTEGAEIDAE. *Eutegaeus papuensis* Aoki (Papua New Guinea): **F**, dorsal; **G**, ventral (after Aoki 1964).

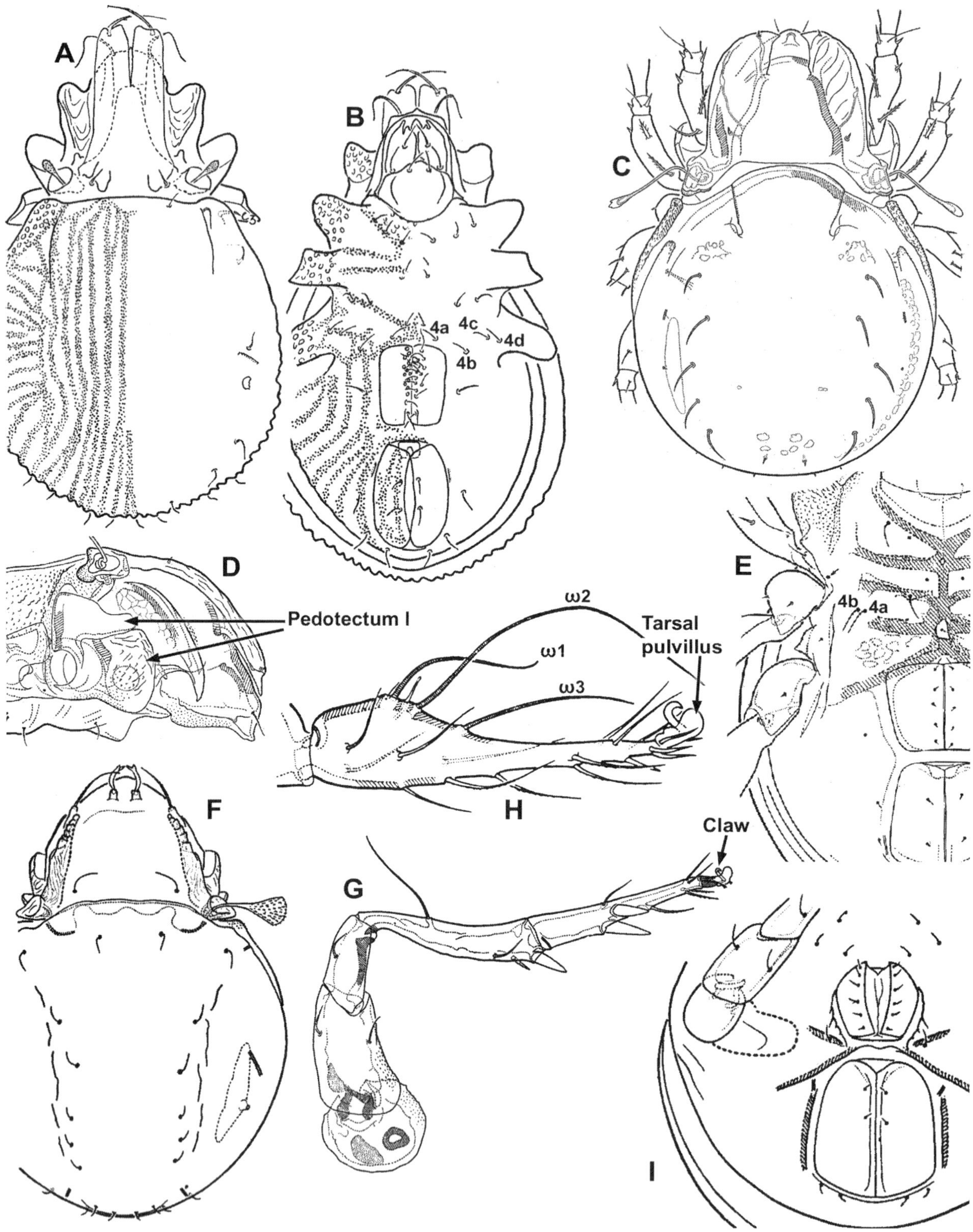

Fig. 15.54. NIPHOCEPHEIDAE. *Niphocepheus nivalis delmarei* Travé (France): **A**, dorsal; **B**, ventral (after Travé 1959). TEGEOCRANELLIDAE. *Tegeocranellus laevis* (Berlese) (Italy): **C**, dorsal; **D**, lateral; **E**, ventral (after Grandjean 1962b). ZETORCHESTIDAE. *Zetorchestes flabrarius* Grandjean (France): **F**, dorsal; **G**, leg IV; **H**, tarsus I; **I**, partial ventral (after Grandjean 1951b).

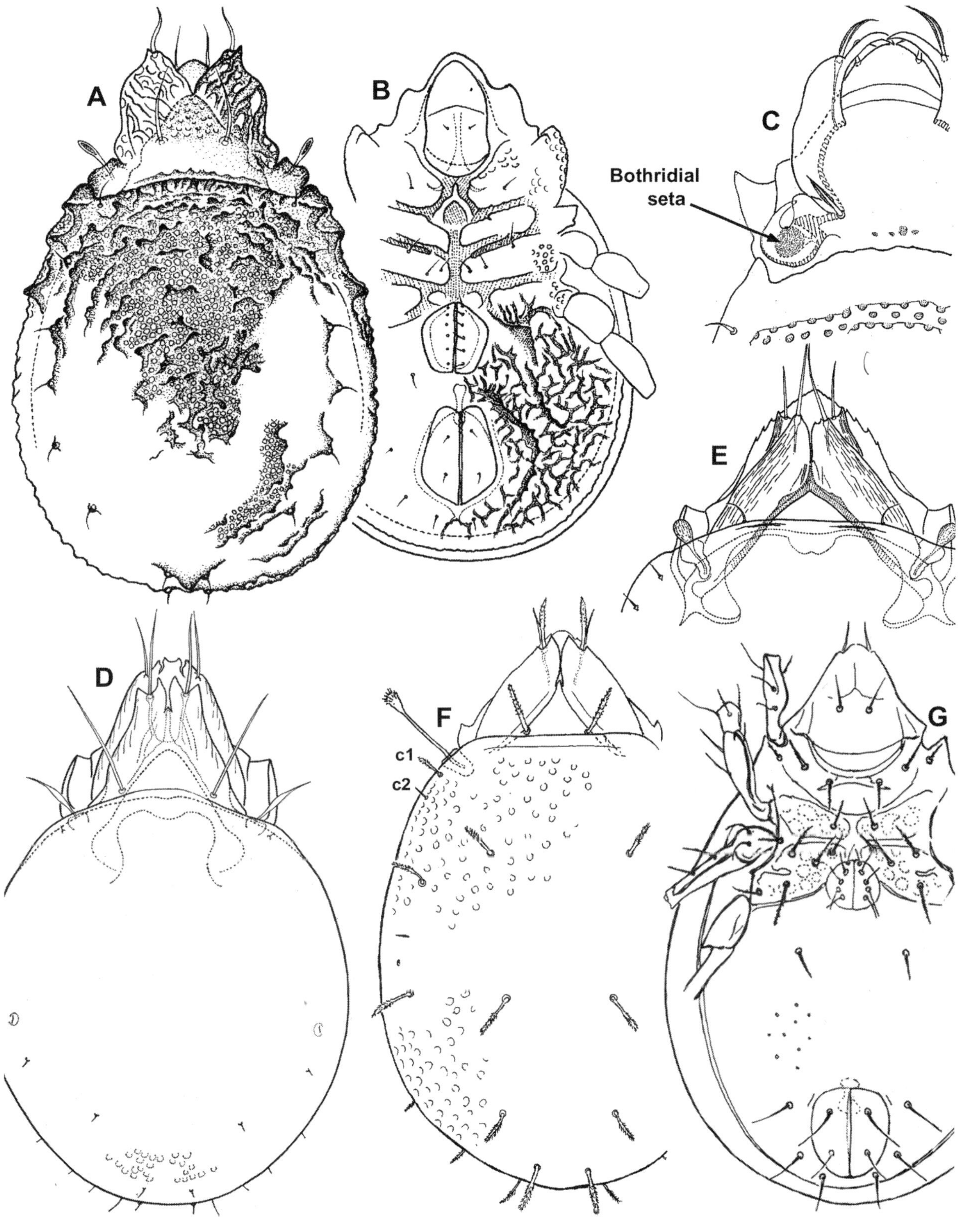

Fig. 15.55. CEPHEIDAE. *Cepheus takasago* Aoki (Taiwan): **A**, dorsal; **B**, ventral (after Aoki 1991). *Ommatocepheus ocellatus* (Michael) (Europe): **C**, prodorsum (after Travé 1963).

LIACARIDAE. *Liacarus yayeyamensis* Aoki (Japan): **D**, dorsal (after Aoki 1973). *Dorycranosus clavatus* Fujikawa and Aoki (Japan): **E**, prodorsum and anterior of notogaster (after Fujikawa and Aoki 1970). *Xenillus imitator* Pérez-Íñigo and Baggio (Brazil): **F**, dorsal (after Pérez-Íñigo and Baggio 1993). *X. moyae* Pérez-Íñigo and Peña (Canary Islands): **G**, ventral (after Pérez-Íñigo and Peña 1994).

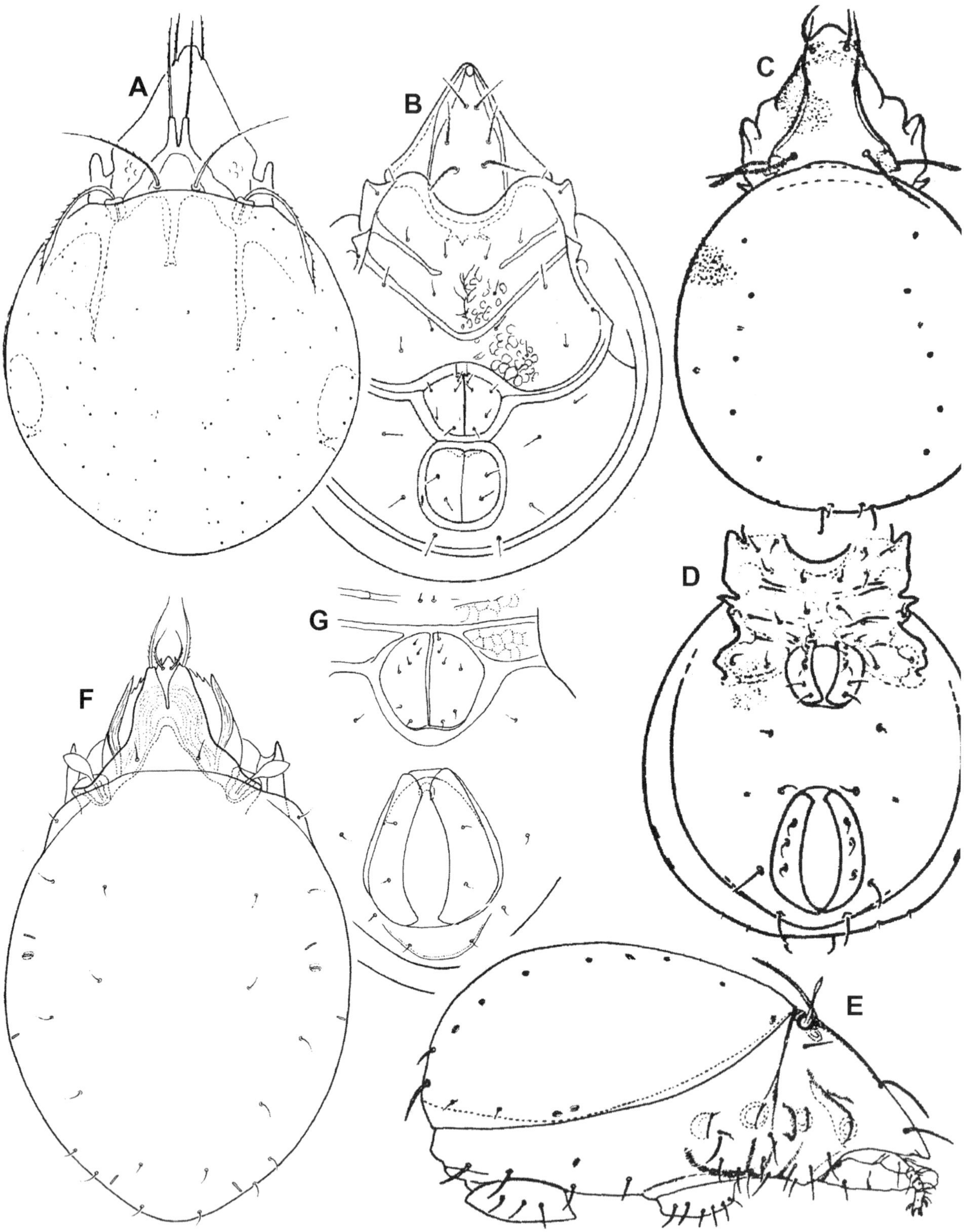

Fig. 15.56. MULTORIBULIDAE. *Peloptoribula spinulosa* Mahunka (South Africa): **A**, dorsal; **B**, ventral (after Mahunka 1984).
KODIAKELLIDAE. *Kodiakella dimorpha* Pérez-Íñigo and Subías (Spain): **C**, dorsal; **D**, ventral; **E**, lateral (after Pérez-Íñigo and Subías 1978).
ASTEGISTIDAE. *Cultroribula breviclavata* Aoki (Japan): **F**, dorsal; **G**, ventral (after Aoki 1984b).

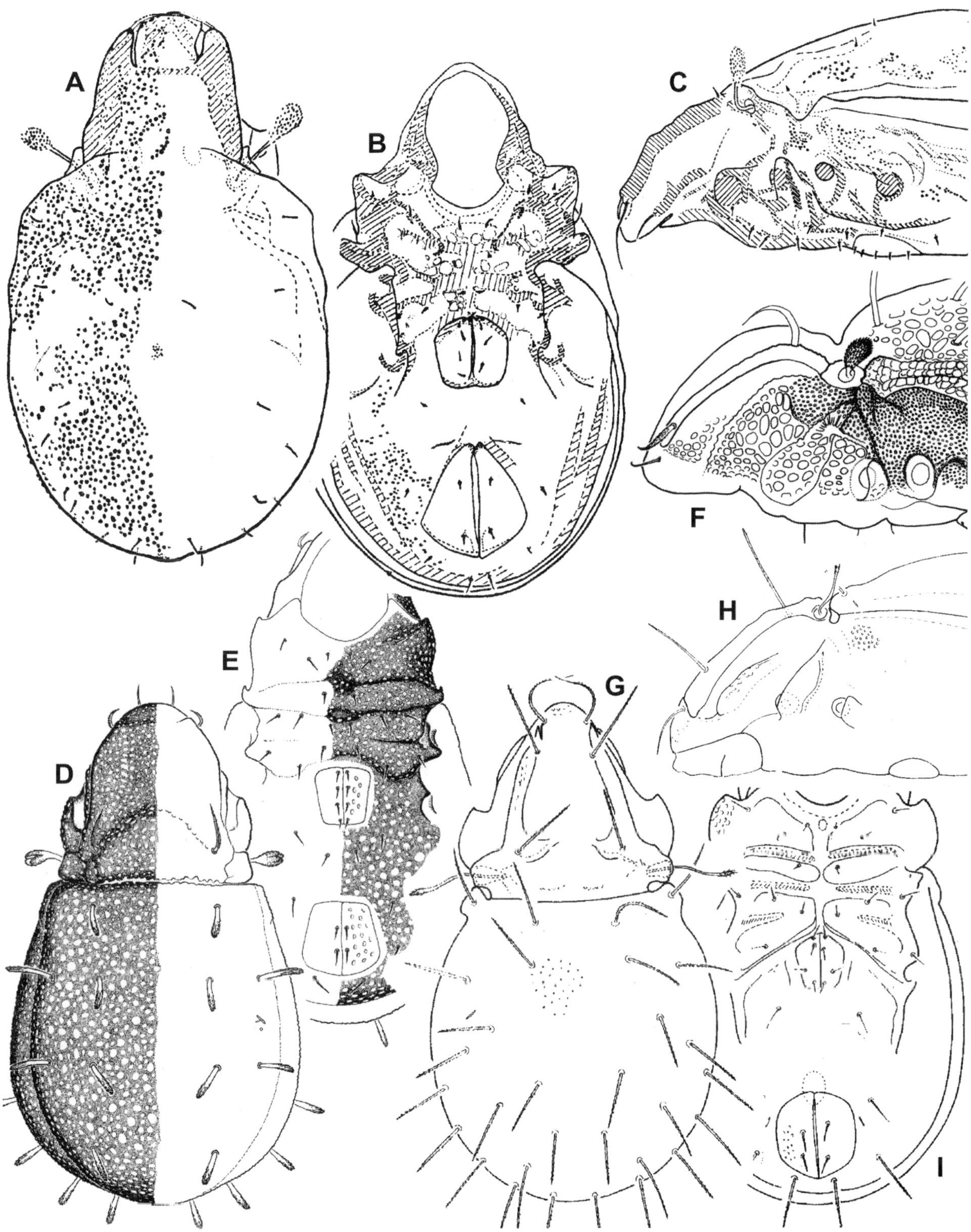

Fig. 15.57. TECTOCEPHEIDAE. *Tectocepheus velatus* Michael (cosmopolitan): **A**, dorsal; **B**, ventral; **C**, lateral (after Nübel-Reidelbach 1994).

CARABODIDAE. *Carabodes chandleri* Reeves (eastern USA): **D**, dorsal; **E**, ventral; **F**, lateral (after Reeves 1992).

CARABOCEPHEIDAE. *Carabocepheus lounsburyi* Berlese (South Africa): **G**, dorsal; **H**, lateral; **I**, ventral (after Mahunka 1986).

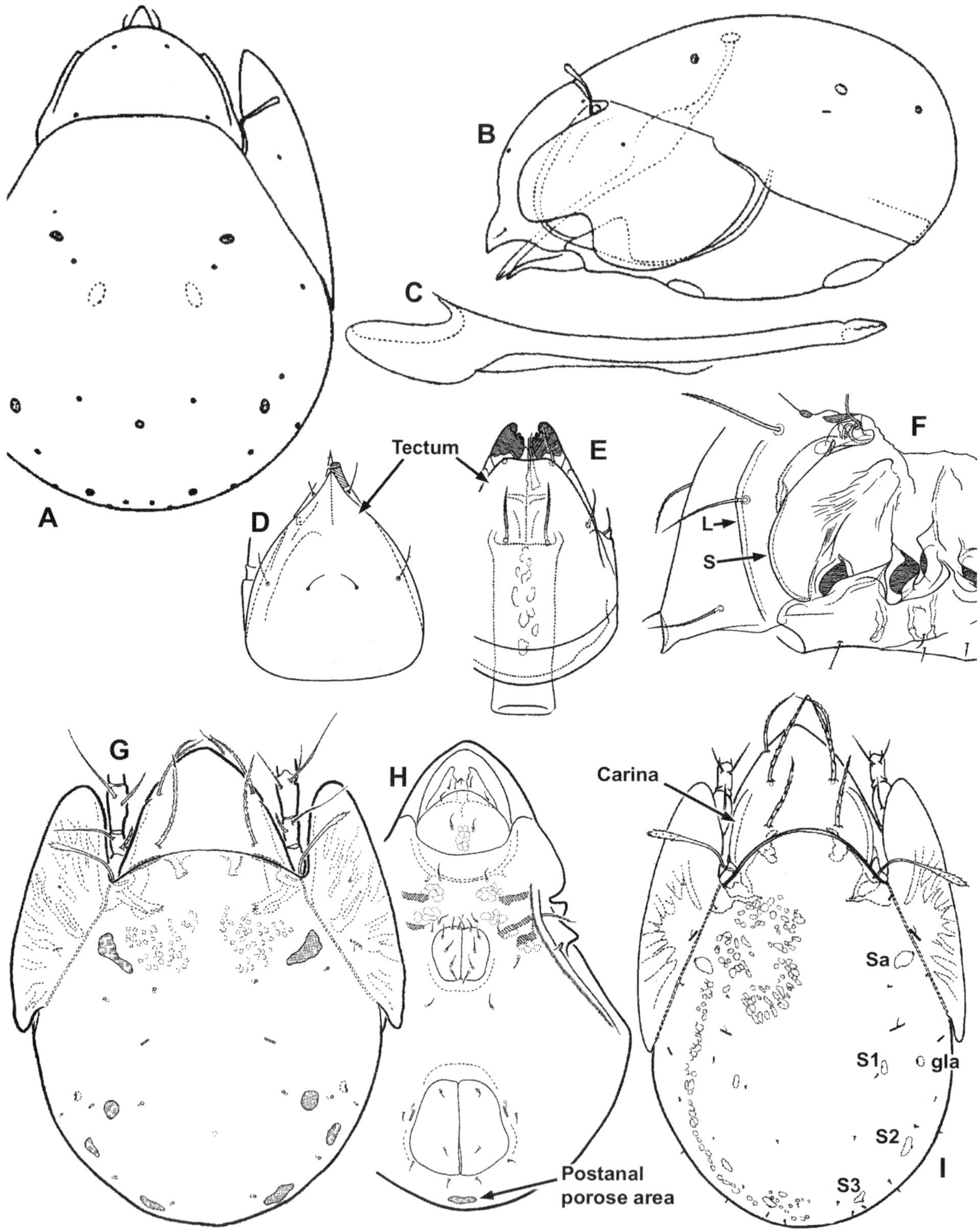

Fig. 15.58. GALUMNELLIDAE. *Galumnopsis holoscripta* Grandjean (Brazil): **A**, dorsal; **B**, lateral; **C**, chelicera; **D**, subcapitulum (after Grandjean 1931b, 1957b).

GALUMNIDAE. *Galumna alatus* (Hermann) (Germany): **E**, subcapitulum; **F**, lateral (partial) (after Grandjean 1936e). *Pergalumna* sp. (Florida, USA): **G**, dorsal; **H**, ventral (V. M. Behan-Pelletier unpubl.).

PARAKALUMMIDAE. *Neoribates auranticus* (Oudemans) (Holarctic): **I**, dorsal (V. M. Behan-Pelletier, unpubl.).

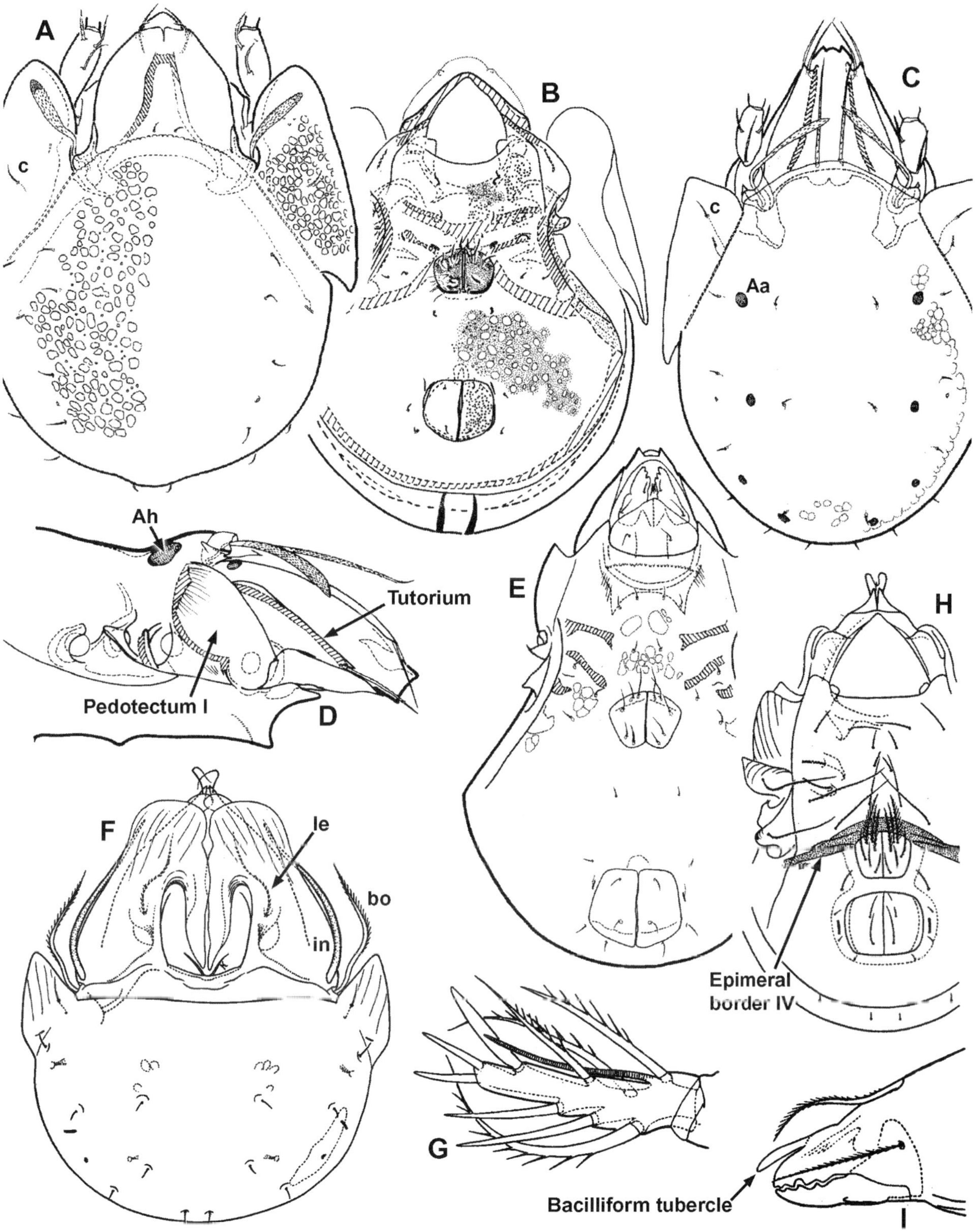

Fig. 15.59. EPACTOZETIDAE. *Epactozetes* sp. (Florida, USA): **A**, dorsal (V. M. Behan-Pelletier, unpubl.). *E. setosus* Balogh and Mahunka (Amazonia): **B**, ventral (after Nübel-Reidelbach and Woas 1992).

CERATOKALUMMIDAE. *Cultrobates* sp. (Florida, USA): **C**, dorsal; **D**, lateral; **E**, ventral (V. M. Behan-Pelletier, unpubl.).

MICROZETIDAE. *Phylacozetes membranulifer* Grandjean (Venezuela): **F**, dorsal; **G**, palptarsus; **H**, ventral; **I**, chelicera (after Grandjean 1936c).

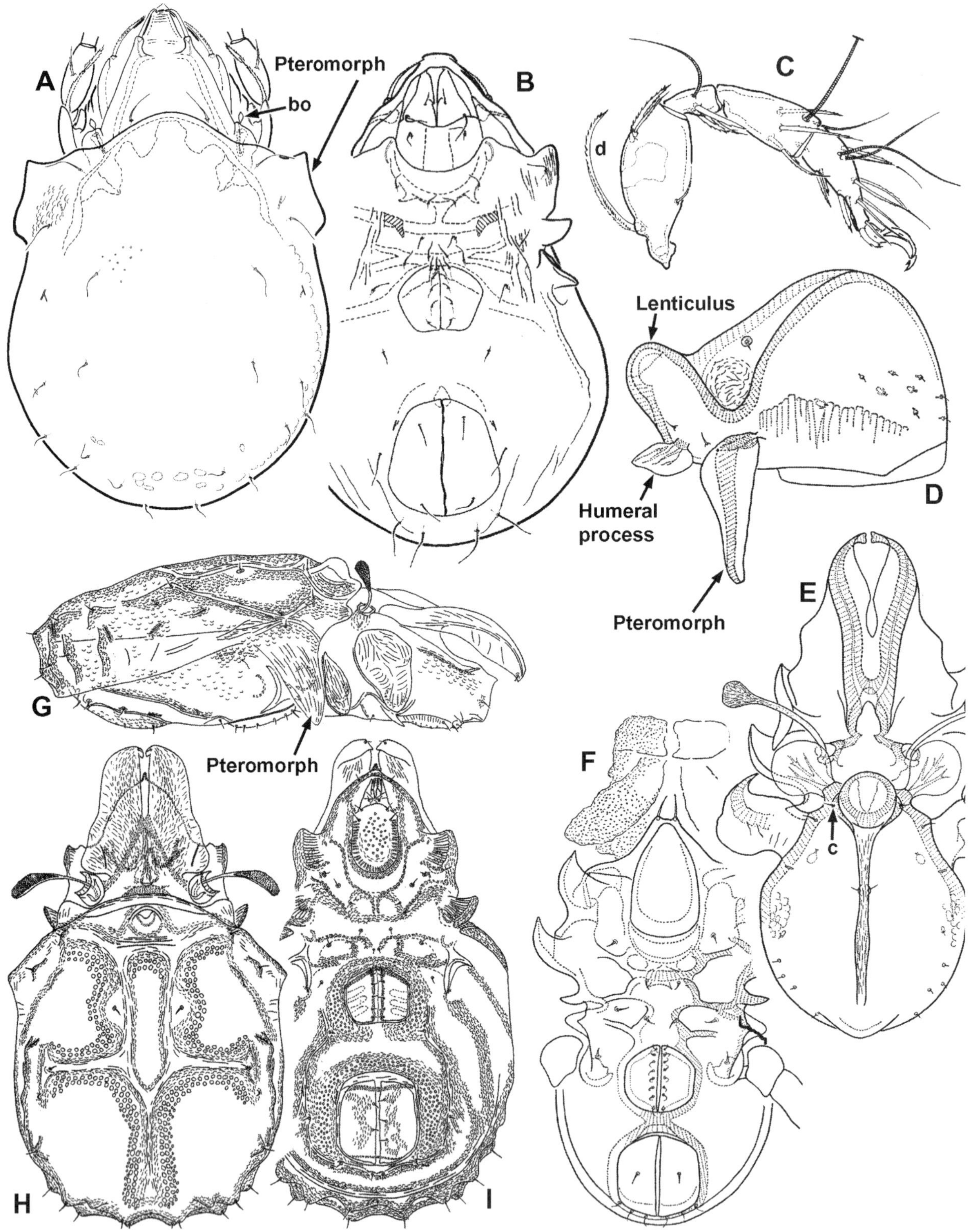

Fig. 15.60. LIMNOZETIDAE. *Limnozetes atmetos* Behan-Pelletier (North America): **A**, dorsal; **B**, ventral. *L. onondaga* Behan-Pelletier (North America): **C**, leg I (after Behan-Pelletier 1989).

IDIOZETIDAE. *Idiozetes erectus* Aoki (West Malaysia): **D**, lateral view of notogaster; **E**, dorsal; **F**, ventral (after Aoki 1976).

EREMAEOZETIDAE. *Eremaeozetes irenae* Schatz (Galápagos Islands): **G**, lateral; **H**, dorsal; **I**, ventral (after Schatz 2001b).

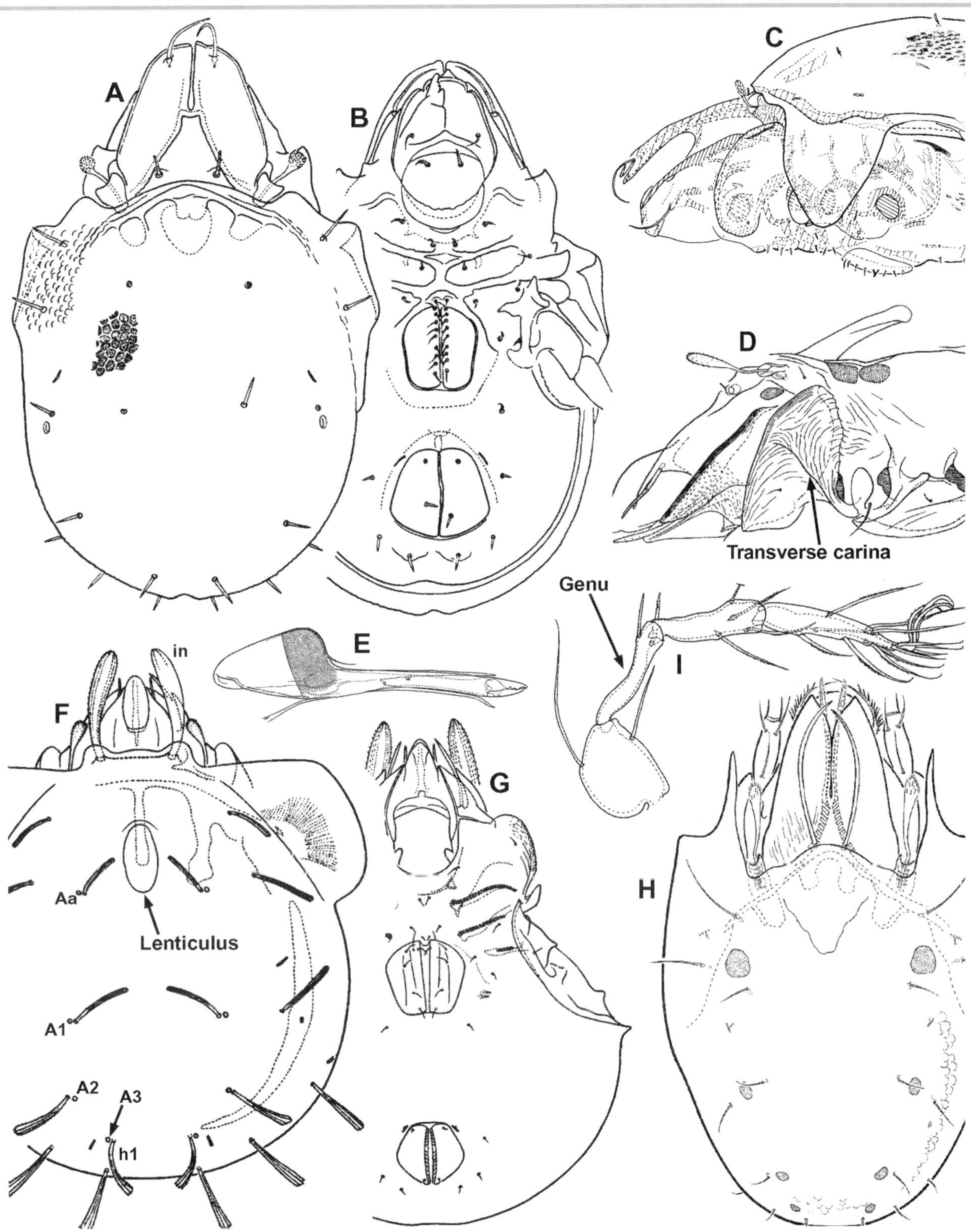

Fig. 15.61. UNDULORIBATIDAE. *Unduloribates hebes* Aoki (Himalaya): **A**, dorsal; **B**, ventral (after Aoki 1965b). *U. brevisetosus* Nübel-Reidelbach and Woas (Nepal): **C**, lateral (after Nübel-Reidelbach and Woas 1992).

PHENOPELOPIDAE. *Eupelops acromios* (Hermann) (Holarctic): **D**, lateral; **E**, chelicera; **F**, dorsal; **G**, ventral (after Grandjean 1936e).

ACHIPTERIIDAE. *Parachipteria* sp. (North America): **H**, dorsal (V. M. Behan-Pelletier, unpubl.). *Achipteria serrata* Hirauchi and Aoki (Japan): **I**, leg IV paraxial (after Hirauchi and Aoki 1997).

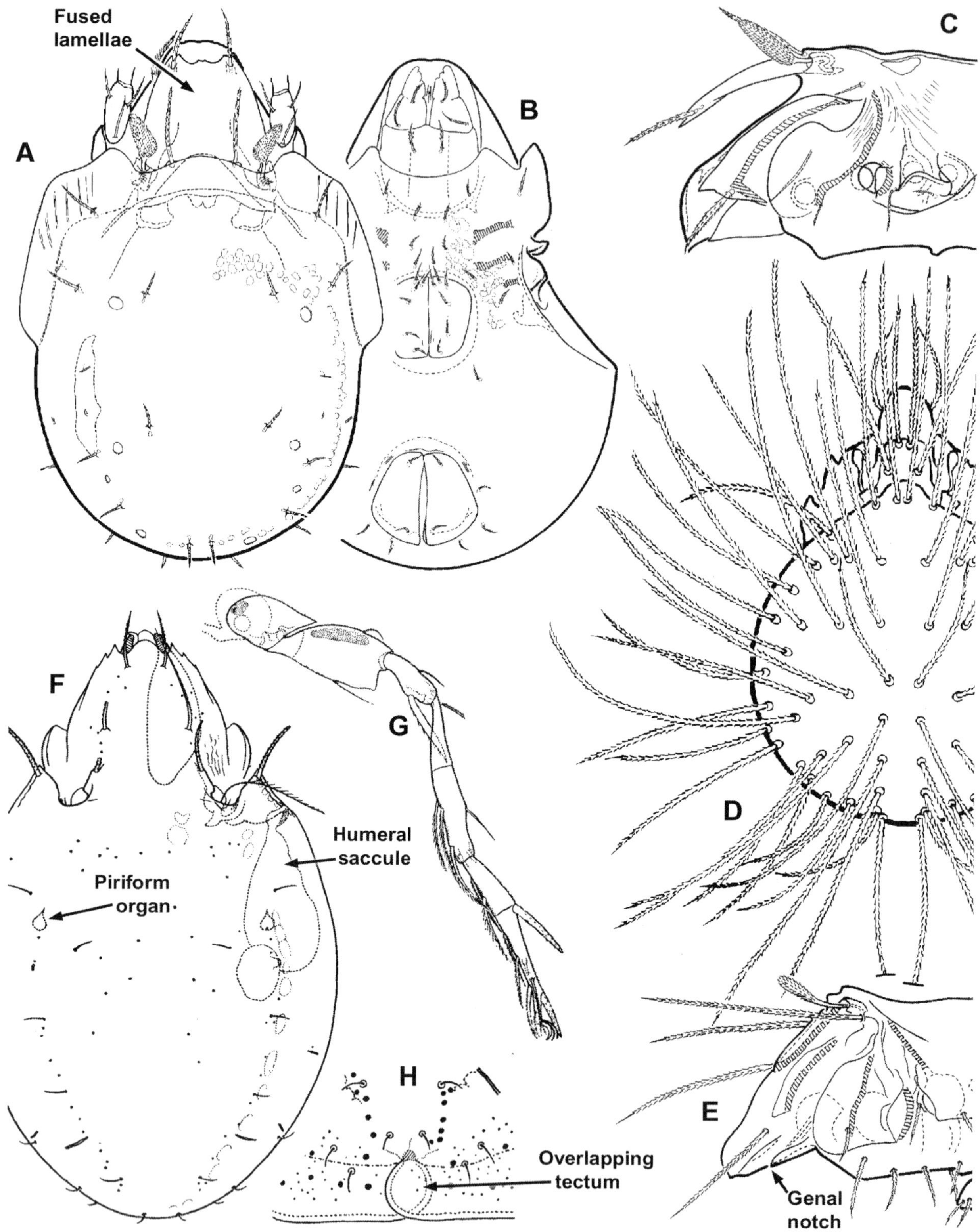

Fig. 15.62. TEGORIBATIDAE. *Lepidozetes latipilosus* Hammer (arctic; subarctic): **A**, dorsal; **B**, ventral; **C**, lateral (V. M. Behan-Pelletier, unpubl.).

NEOTRICHOZETIDAE. *Neotrichozetes spinulosa germaineae* Travé (Argentina): **D**, dorsal; **E**, lateral of prodorsum (after Travé 1961).

ZETOMOTRICHIDAE. *Zetomotrichus lacrimans* Grandjean (Algeria): **F**, dorsal; **G**, leg IV; **H**, posterior of notogaster (after Grandjean 1934e, 1955b).

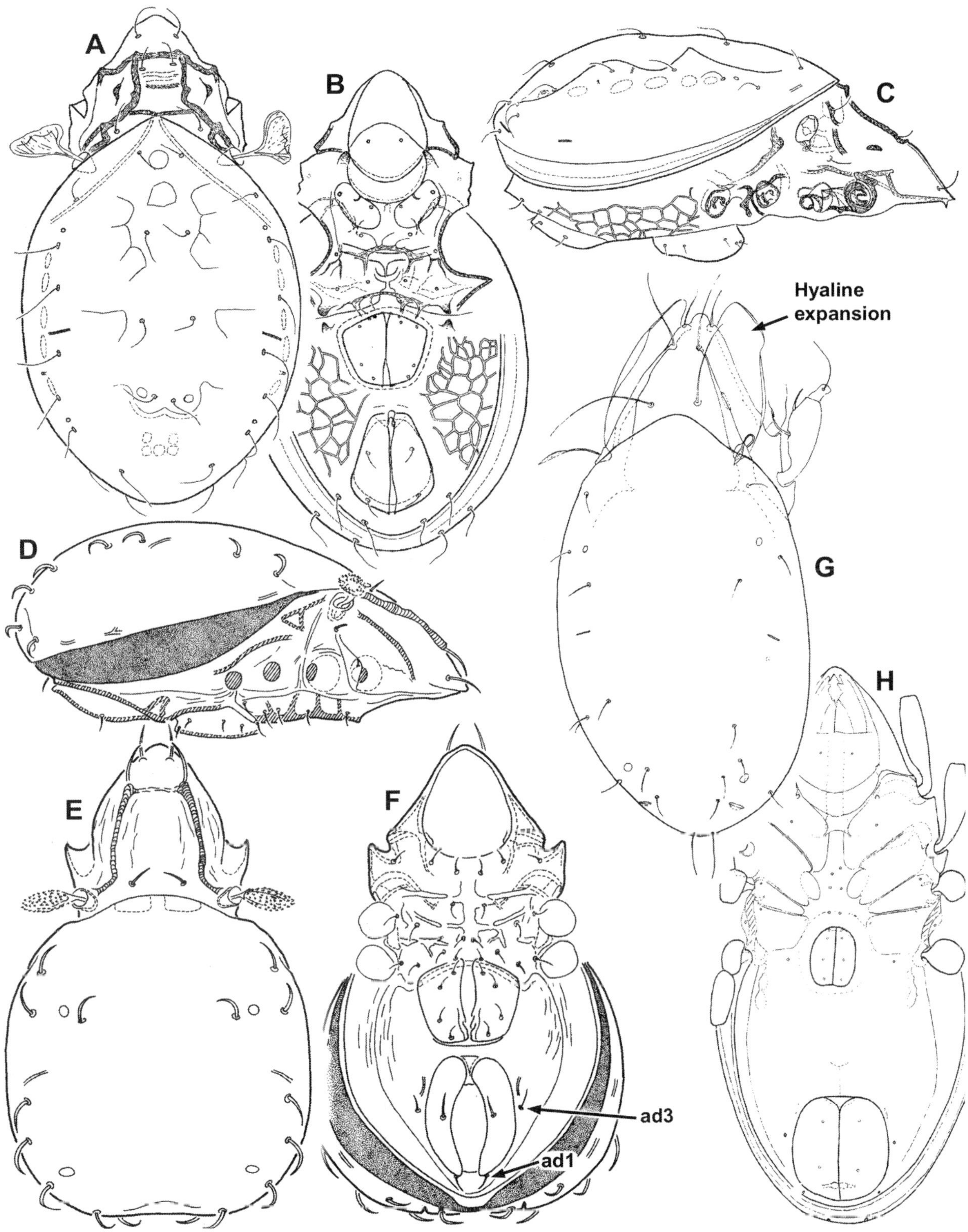

Fig. 15.63. LICNEREMAEIDAE. *Licneremaeus licnophorus* (Michael) (Europe, Japan): **A**, dorsal; **B**, ventral; **C**, lateral (after Ito 1982).
LAMELLAREIDAE. *Tenuelamellarea argentinensis* Martinez et al. (Argentina): **D**, lateral; **E**, dorsal; **F**, ventral (after Martinez et al. 1995).
NESOZETIDAE. *Nesozetes rostropterus* Hammer (Fiji): **G**, dorsal; **H**, ventral (after Hammer 1971).

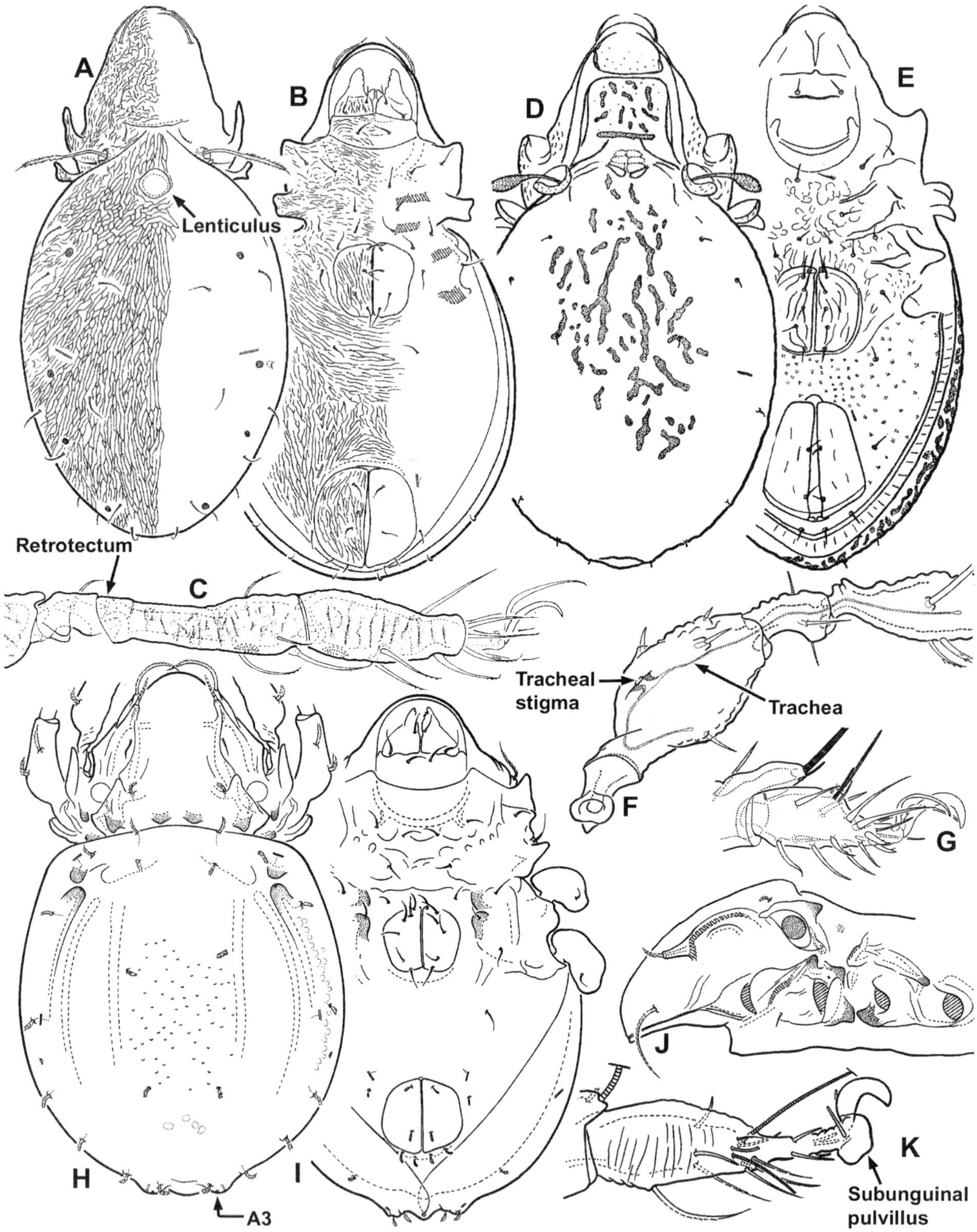

Fig. 15.64. PASSALOZETIDAE. *Passalozetes* sp. (British Columbia, Canada): **A**, dorsal; **B**, ventral (V. M. Behan-Pelletier, unpubl.). *P. perforatus* (Berlese) (Italy): **C**, leg IV (after Woas 1998).

SCUTOVERTICIDAE. *Scutovertex pictus* Kunst (Europe): **D**, dorsal; **E**, ventral (after Kunst 1959). *S. sculptus* Michael (UK): **F**, femur I; **G**, tarsus I (after Woas 1998).

ADHAESOZETIDAE. *Adhaesozetes polyphyllos* Walter and Behan-Pelletier (Australia): **H**, dorsal; **I**, ventral; **J**, lateral; **K**, tarsus I (after Walter and Behan-Pelletier 1993).

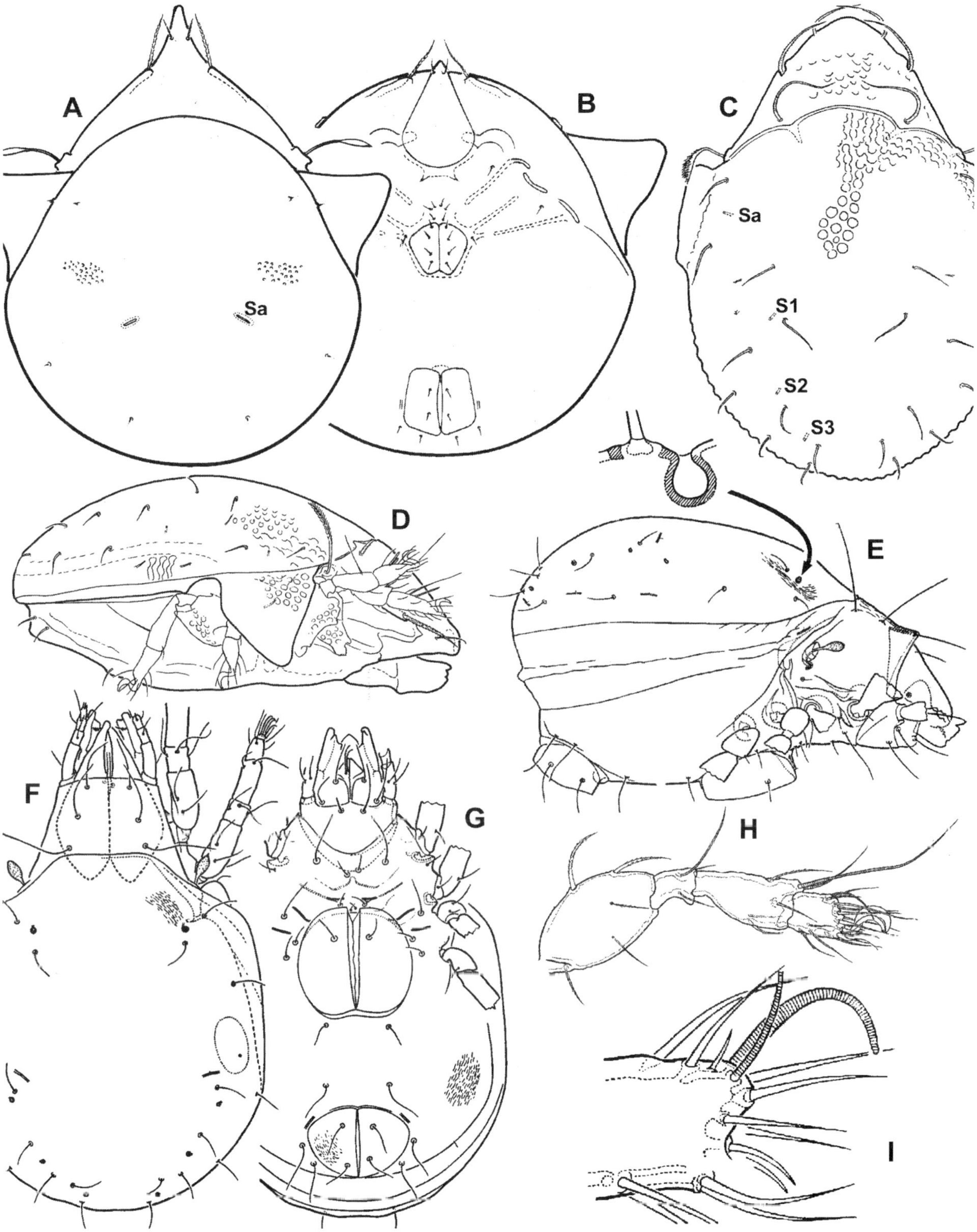

Fig. 15.65. TUBULOZETIDAE. *Tubulozetes rostratus* P. Balogh (Ecuador): **A**, dorsal; **B**, ventral (after Balogh 1989). HAPLOZETIDAE. *Rostrozetes ovulum* (Berlese) (Nearctic and Neotropics): **C**, dorsal; **D**, lateral (after Beck 1965). ORIPODIDAE. *Pirnodus detectidens* Grandjean (France): **E**, lateral with detail of saccule, in section; **F**, dorsal; **G**, ventral; **H**, leg I; **I**, extremity of tarsus I (after Grandjean 1956e).

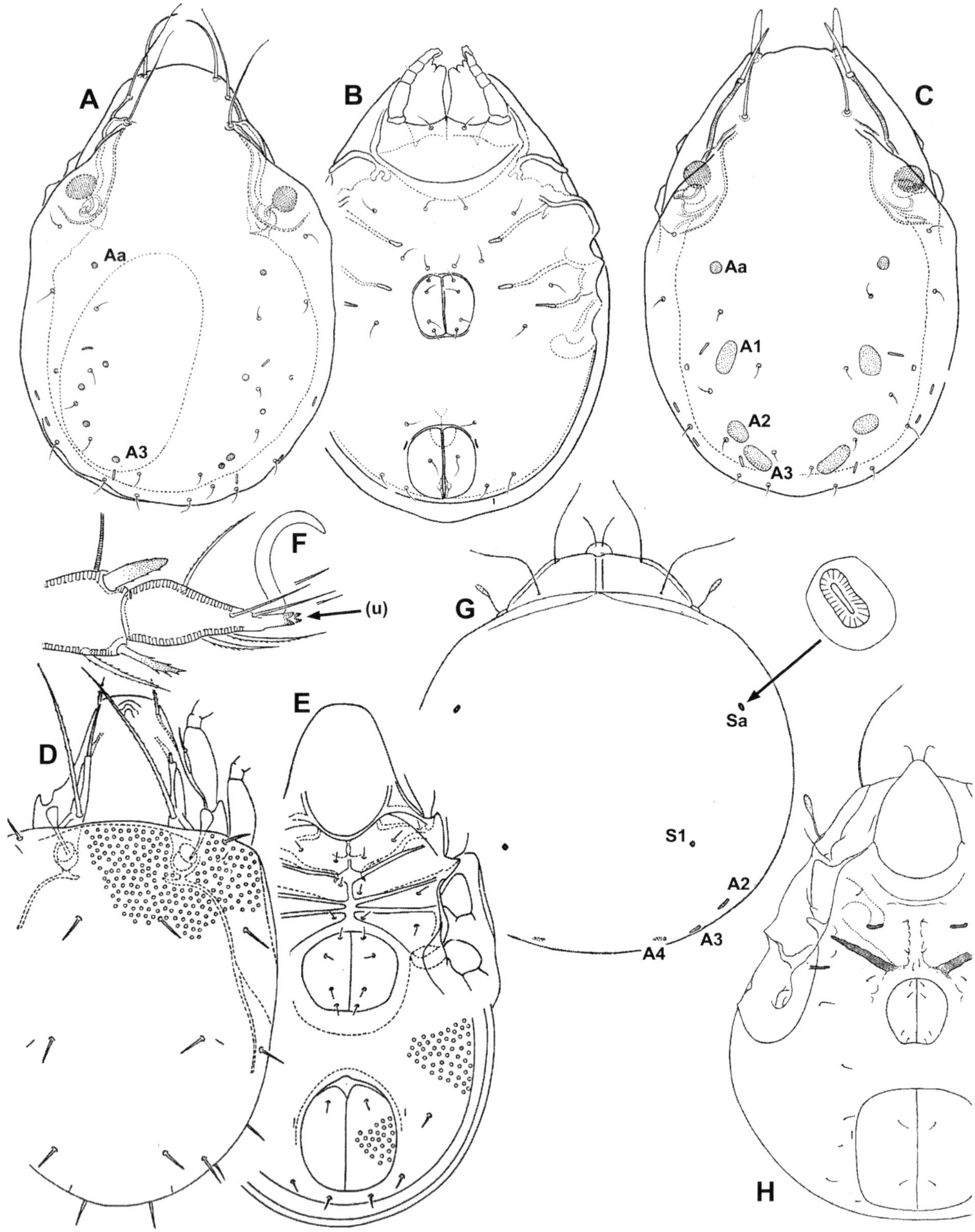

Fig. 15.66. SYMBIORIBATIDAE. *Symbioribates papuensis* Aoki (Papua): **A**, dorsal female; **B**, ventral male; **C**, dorsal male (after Aoki 1966a).

CAMPBELLOBATIDAE. *Campbellobates acanthus* Wallwork (Campbell Island): **D**, dorsal; **E**, ventral; **F**, tibia and tarsus IV (after Wallwork 1964).

DRYMOBATIDAE. *Drymobates silvicola* Grandjean (Martinique): **G**, dorsal with detail of adalar saccule; **H**, ventral (after Grandjean 1930).

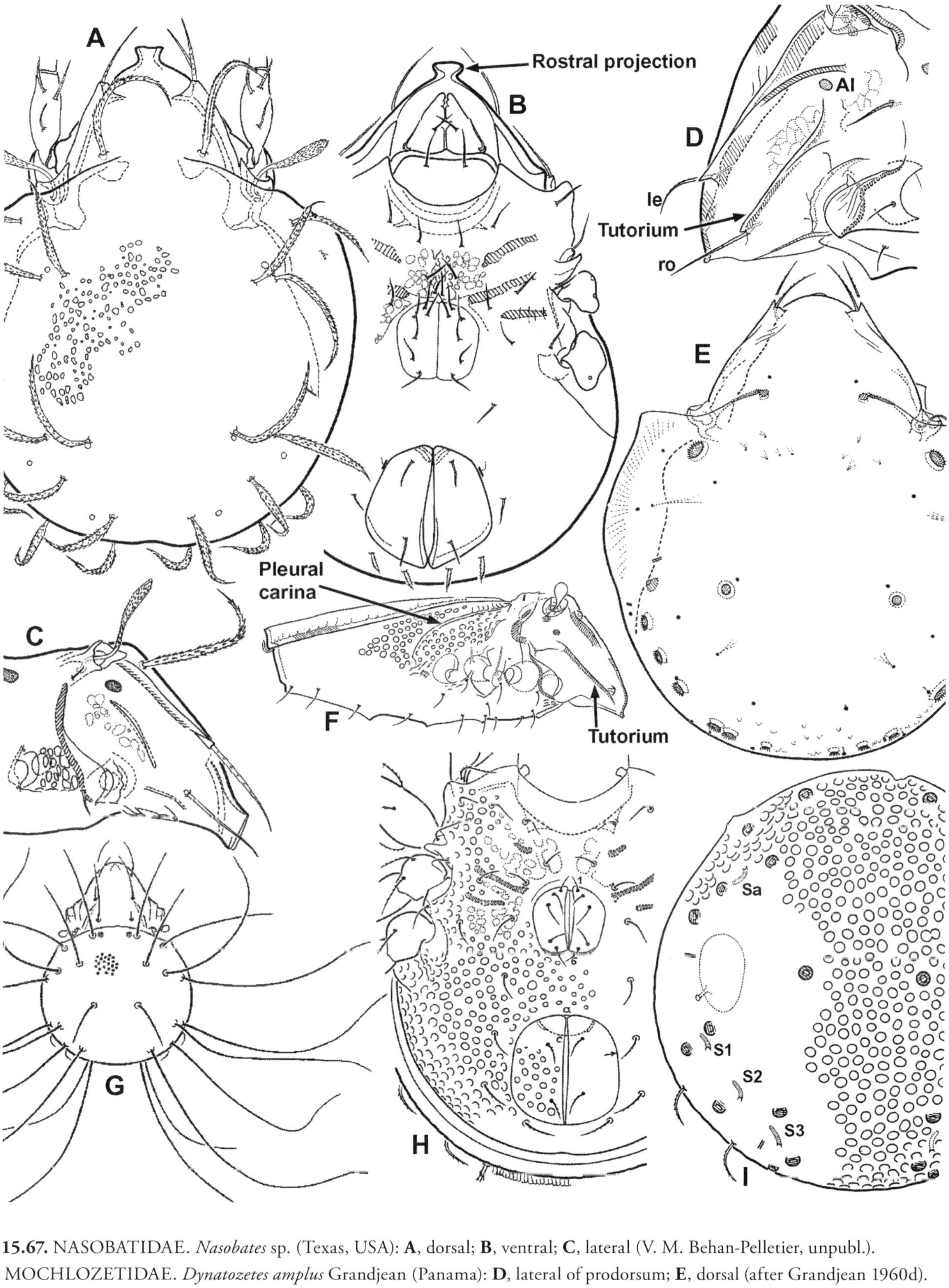

Fig. 15.67. NASOBATIDAE. *Nasobates* sp. (Texas, USA): **A**, dorsal; **B**, ventral; **C**, lateral (V. M. Behan-Pelletier, unpubl.). MOCHLOZETIDAE. *Dynatozetes amplus* Grandjean (Panama): **D**, lateral of prodorsum; **E**, dorsal (after Grandjean 1960d). STELECHOBATIDAE. *Stelechobates megalotrichus* Grandjean (France): **F**, lateral; **G**, dorsal; **H**, ventral; **I**, notogaster (after Grandjean 1965c).

Fig. 15.68. SCHELORIBATIDAE. *Scheloribates laevigatus* (C. Koch) (Holarctic): **A**, dorsal; **B**, ventral; **C**, lateral (after Wunderle, Beck, and Woas 1990); **D**, palptarsus (after Grandjean 1958d).

CALOPPIIDAE. *Chaunoproctellus rugosus* Mahunka (Senegal): **E**, dorsal; **F**, lateral; **G**, ventral (after Mahunka 1992).

ORIBATULIDAE. *Oribatula tibialis* (Nicolet) (Holarctic): **H**, palptarsus (after Grandjean 1958d); **I**, dorsal; **J**, ventral; **K**, lateral (after Wunderle, Beck, and Woas 1990).

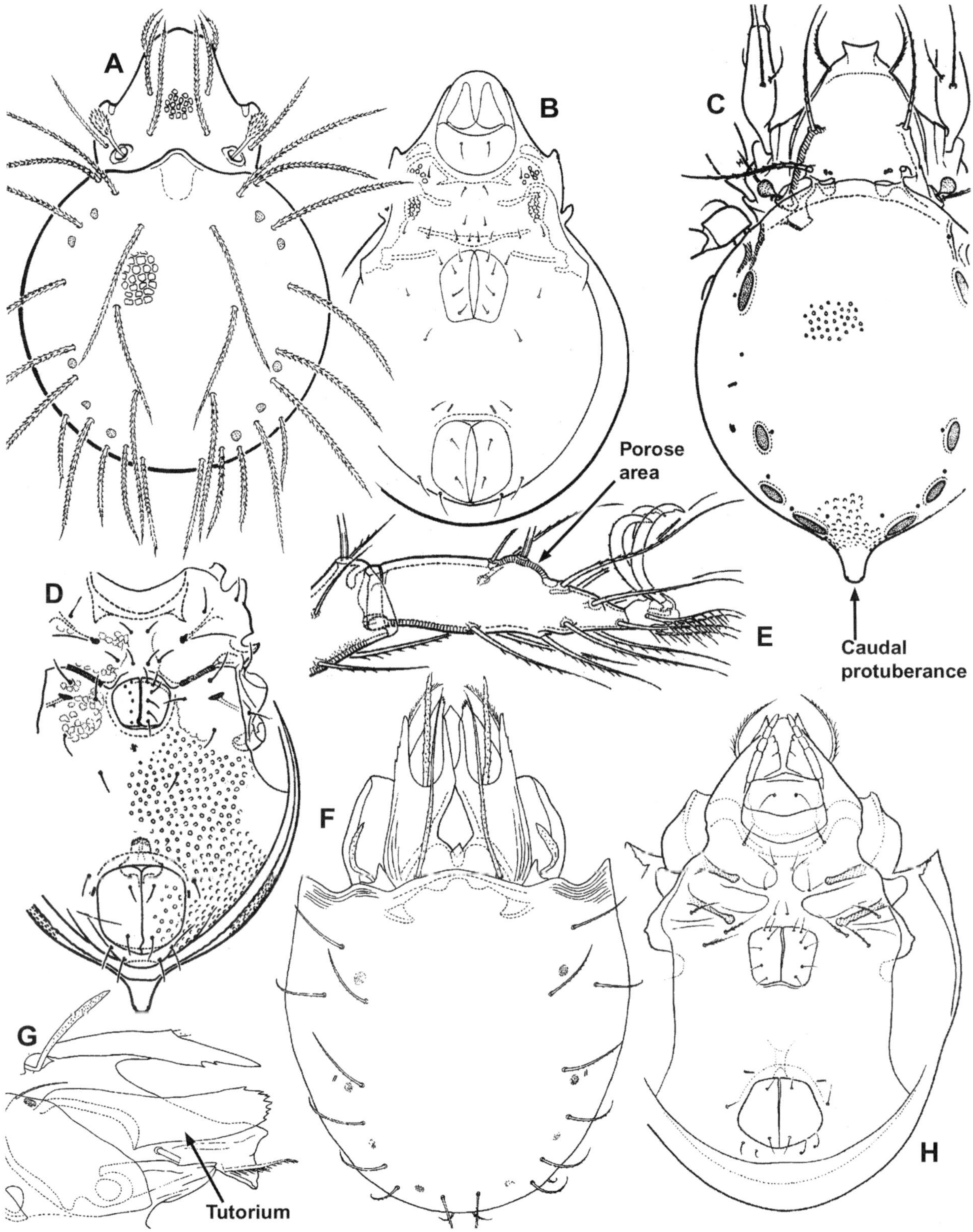

Fig. 15.69. CRASSORIBATULIDAE. *Reticuloppia reticulata* Balogh and Mahunka (Australia): **A**, dorsal, **B**, ventral (after Hunt et al. 1998).

SELLNICKIIDAE. *Sellnickia caudata* (Michael) (New Zealand): **C**, dorsal male; **D**, ventral male; **E**, tarsus II (after Grandjean 1958e).

ORIBATELLIDAE. *Oribatella berlesei* (Michael) (Europe): **F**, dorsal; **G**, lateral. *O. meridionalis* Berlese (Palearctic): **H**, ventral (after Aoki 1970).

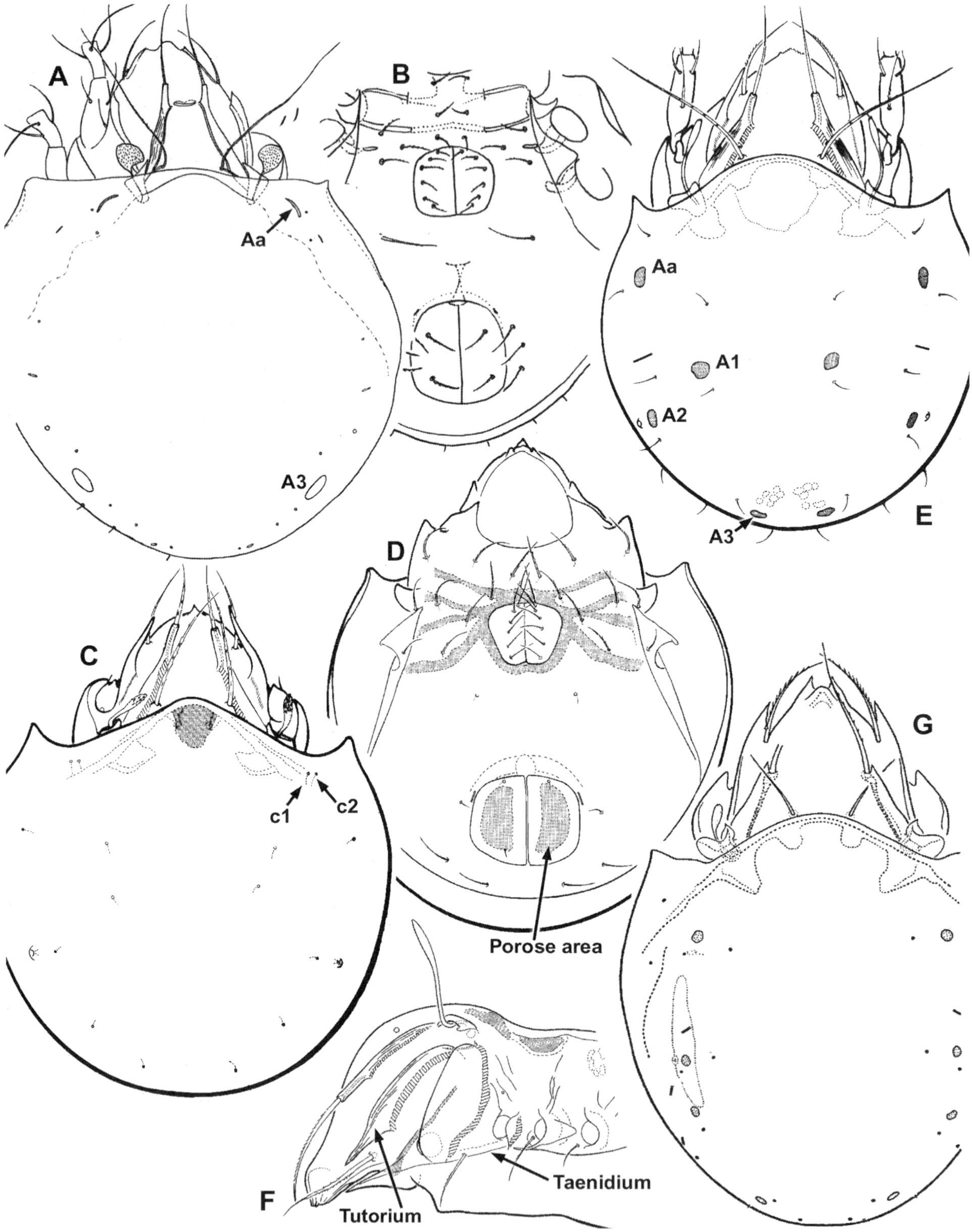

Fig. 15.70. ONYCHOBATIDAE. *Onychobates nidicola* Hammer (New Zealand): **A**, dorsal; **B**, ventral (after Hammer 1967). ZETOMIMIDAE. *Heterozetes minnesotensis* (Ewing) (North America): **C**, dorsal; **D**, ventral of male. *Zetomimus setosus* (Banks) (North America): **E**, dorsal; **F**, lateral (after Behan-Pelletier and Eamer 2003). CHAMOBATIDAE. *Chamobates* sp. (Italy): **G**, dorsal (after Grandjean 1962c).

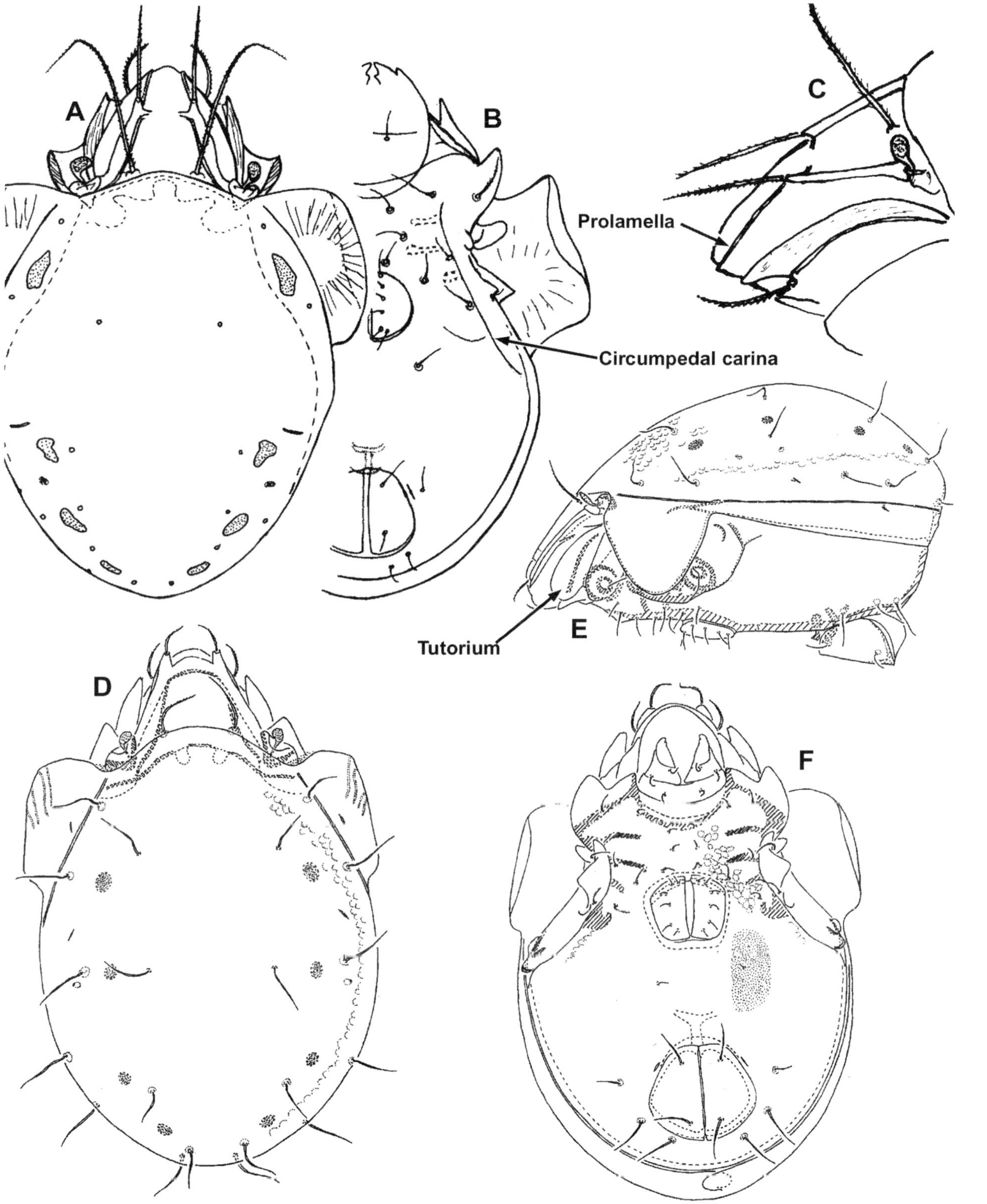

Fig. 15.71. HUMEROBATIDAE. *Humerobates flechtmanni* Pérez-Íñigo and Pérez-Íñigo Jr. (Brazil): **A**, dorsal; **B**, ventral; **C**, lateral of prodorsum (after Pérez-Íñigo and Pérez-Íñigo Jr. 1993).
MYCOBATIDAE. *Mycobates sarekensis* (Trägårdh) (Holarctic): **D**, dorsal; **E**, lateral; **F**, ventral (after Solhøy 1997).

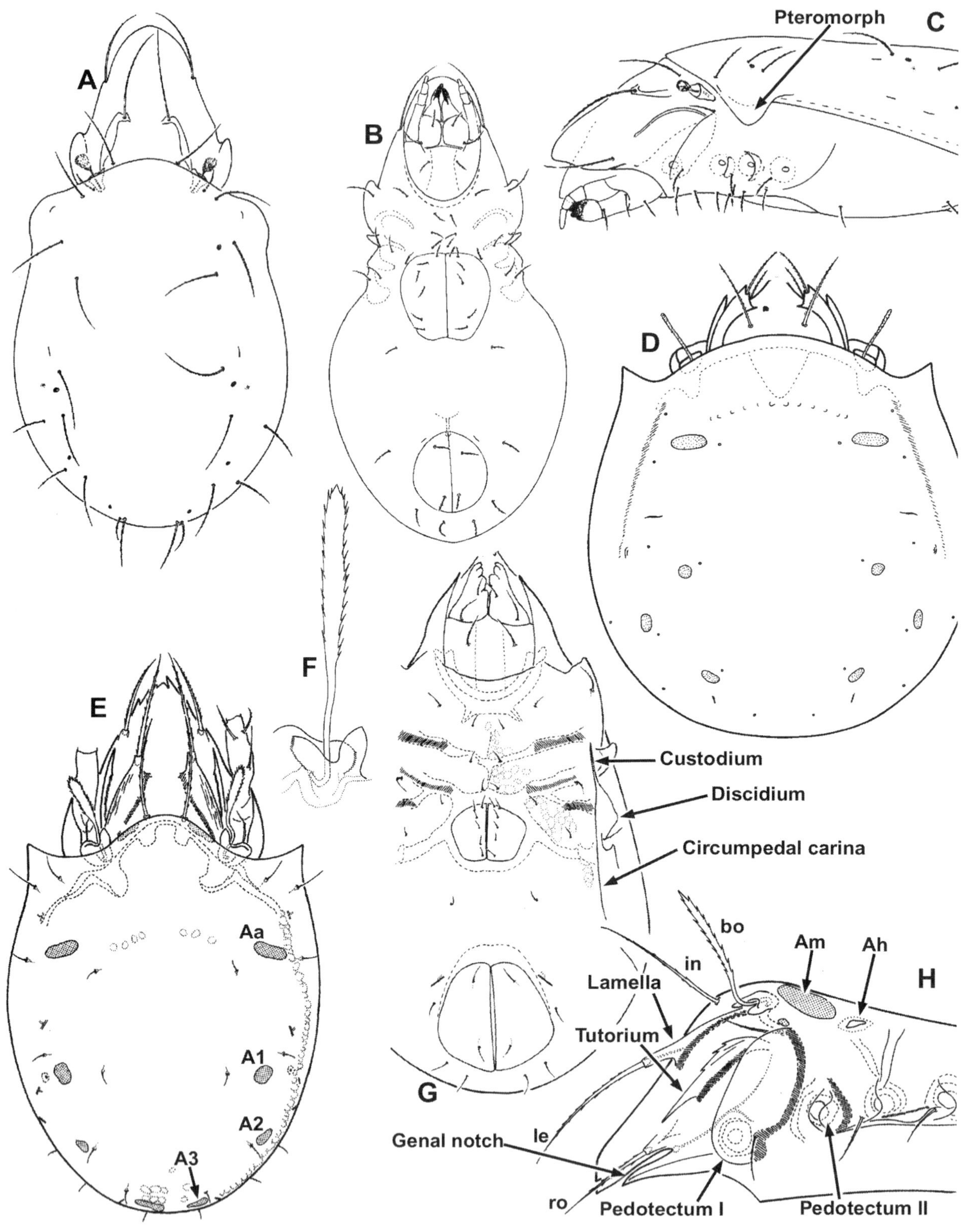

Fig. 15.72. MAUDHEIMIIDAE. *Maudheimia petronia* Wallwork (Antarctica): **A**, dorsal; **B**, ventral; **C**, lateral (after Coetzee 1997). EUZETIDAE. *Euzetes globulus* (Nicolet) (Europe): **D**, dorsal (after Pérez-Íñigo 1993a). CERATOZETIDAE. *Ceratozetes mediocris* Berlese (Holarctic): **E**, dorsal; **F**, bothridium and bothridial seta; **G**, ventral; **H**, lateral (after Behan-Pelletier 1984).

CHAPTER SIXTEEN

B.M. OCONNOR

Cohort Astigmatina

The Astigmatina (= Astigmata, Acaridiae, Acaridida) has traditionally been ranked at a level coordinate with the Prostigmata and Oribatida. However, evidence has accumulated supporting its inclusion within the Oribatida, as first postulated in 1953 by Zakhvatkin (OConnor 1984c; Norton 1998). Norton (1998) cited 14 derived character states that place the Astigmatina within the oribatid supercohort Desmonomatides, a clade that includes the cohorts Nothrina and Brachypylina. Another 13 characters more specifically support a relationship with the nothrine Crotonioidea, including such distinctive traits as the presence of opisthonotal glands, paraproctal atrichosity, and transdehiscent ecdysis (also see chapter 15). From these origins the astigmatine lineage has diversified to the point where 71 families comprising 960 genera and more than 5,000 species are currently recognized. Adaptations based on three key evolutionary processes—(1) reproductive mode, (2) neoteny, and (3) ontogenetic reorganization—appear to have played a major role in achieving this diversity. Regarding reproductive mode, most crotonoid Oribatida reproduce by thelytokus parthenogenesis, in which males are rarely formed and are generally thought to be unnecessary for reproduction. The ancestral astigmatine mite reverted to the more ancient reliance on sexual reproduction and developed direct insemination via a male aedeagus. Regarding neoteny, ancestral patterns of metamorphosis involving distinct reorganization of the body prior to the final molt were suppressed, leading to the generally paedomorphic appearance of adult astigmatine mites. Finally, perhaps coupled with failure of this ancestral form to metamorphose beyond the deutonymphal level of organization, a new and more striking metamorphosis has evolved. Prior to the molt from feeding protonymph to nonfeeding deutonymph, the body is completely reorganized, acquiring an entirely new habitus. The deutonymph is highly specialized for dispersal or for withstanding adverse environmental conditions, which allows the species to exploit patchy or temporary habitats normally unavailable to mites without such specialization. Perhaps more remarkably, once conditions become conducive for resuming development, the nonfeeding deutonymph (or hypopus) molts into a feeding tritonymph, a *retrograde metamorphosis* that may also be seen in certain Mesostigmata (see chapter 12), where loss and recovery of a well-developed tritosternum and chelicerae through ontogeny are correlated with loss and recovery of the ability to feed (Wernz and Krantz 1976).

Astigmatine mites often compose the dominant acarofaunal component of naturally occurring patchy or ephemeral habitats such as decaying logs, fungal fruiting bodies, dung, carrion, sap flows, tree holes, phytotelmata, and caves. In these habitats mites feed as saprophages on decaying organic materials, fungi, or bacteria (also see chapter 6). Some have adapted to consuming seeds or other specialized plant tissues such as bulbs and tubers. Fully aquatic species may consume algae, and rare instances of predation on nematodes or insect eggs have been recorded (Muraoka and Ishibashi 1976). Most free-living astigmatine mites are classic *r-strategists* having a short developmental time, high fecundity, and excellent dispersal capacity (also see chapter 3). Deutonymphs often disperse among habitat patches through phoretic association with insects or other arthropods that also use these resources. The phoretic association is ancestrally commensalistic, in which the arthropod host is not burdened by the mite's presence. Having adapted to this type of relationship, many astigmatine mites have developed closer associations with their insect carriers, specializing in resources created by the tunneling of wood-boring beetles or the building of complex nests by social insects, and in resources provided by these insects in the form of nest provisions. They also exploit resources in the nests of birds and mammals, and many have adapted to dispersal on the bodies of their vertebrate hosts. A final transition has led to temporary or permanent reliance of these astigmatines on their arthropod or vertebrate hosts for food resources, which in turn set the stage for the evolution of parasitism (also see chapter 6).

As a consequence of the neotenic and metamorphic processes noted above, astigmatine mites have evolved two distinct morphological types that differ from those of their desmonomatid ancestors. Early derivative astigmatine mites tend to resemble the juvenile rather than adult stages of related Oribatida. Feeding stages generally have little sclerotization except for a prodorsal sclerite. The sejugal furrow is retained ancestrally but is lost in numerous derived groups. The cuticle may be smooth, striate, or covered with various forms of ornamentation, such as tiny microtrichia, scales, or triangular or rounded protuberances. Some HISTIOSTOMATIDAE have thin cuticular extensions that trap and hold debris on the surface, providing camouflage or protection from predators. Many species in the superfamilies Histiostomatoidea and Hemisarcoptoidea sequester guanine crystals in the fat body underlying the epidermis, giving the mites a shiny, white appearance. A more heavily sclerotized body has evolved independently in several more highly derived lineages. For example, the GUANOLICHIDAE (Fig. 16.8E), some ALGOPHAGIDAE (*Fusohericia*) (Fig. 16.16B), the GLYCYPHAGIDAE (*Fusacarus*), the ACARIDAE (*Stereoglyphus*), and the HYPODERATIDAE (*Tytodectes*) are essentially completely sclerotized, while discrete dorsal hysterosomatic sclerites have developed in adults of HETEROCOPTIDAE (Fig. 16.10C) and in the ancestral psoroptidian lineage. Many taxa have ventrally sclerotized coxal fields.

Respiration in the Astigmatina is presumed to be integumental in that no obvious stigmata or extensive tracheal systems are present. Paired trachealike structures do occur in scattered taxa such as the "genital tracheae" of *Gohieria fusca* (Oudemans) (GLYCYPHAGIDAE) (Hammen 1968b, 1972), the "pseudotracheae" opening above the bases of legs II in *Grammolichus* species (GLYCYPHAGIDAE) and some PYROGLYPHIDAE (Fain 1965a, 1968a), and the long tubular structures extending posteriorly from the gnathosoma in *Rosensteinia* and *Mydopholeus* (ROSENSTEINIIDAE) (B. M. OConnor, pers. obs.). However, most of these structures may be glandular rather than respiratory in function.

The central portion of the prodorsum of feeding astigmatine stases may be completely unsclerotized (CARPOGLYPHIDAE, ECHIMYOPODIDAE, CHORTOGLYPHIDAE); more commonly, a rectangular, shieldlike sclerite is present (e.g., ACARIDAE) (Fig. 16.29B). The shield is more extensive in many HISTIOSTOMATIDAE and psoroptidian groups, and incorporates the bases of the rostral and lamellar setae. In some nidicolous groups, the sclerotization takes the form of two parallel bands with soft cuticle between (e.g., EUGLYCYPHAGIDAE) (Fig. 16.17B). In some species of *Glycyphagus* (GLYCYPHAGIDAE), there is a thin median sclerite associated with the rostral setae that has been termed a **crista metopica**. The supracoxal seta of legs I is highly variable in form, ranging from a short spine or simple seta in species living in high humidity conditions to a highly elaborated, many-branched structure (Fig. 16.29D). The latter form covers the gland opening and minimizes desiccation in species living in drier habitats. An elaborated **Grandjean's organ** may be found at the apex of the podocephalic sclerite in some taxa. Its structure ranges from a small lobe to an elongate, fingerlike process; it may also be broad and strongly fimbriate as in some ACARIDAE (Fig. 16.29D). The function of Grandjean's organ is unknown. In a few GLYCYPHAGIDAE (*Lophuromyopus*, *Xenoryctes*) and PYROGLYPHIDAE (e.g., *Gymnoglyphus*), the entire adult body may be sclerotized, and the propodosoma may bear costulae as in many other oribatid taxa. The propodosoma bears a maximum of four pairs of setae that have been homologized with similar setae in juvenile and some adult Crotonioidea. Prior treatments of astigmatine chaetotaxy have referred to the anterior two pairs as external and internal vertical setae and the posterior two pairs as external and internal scapular setae. To standardize chaetotactic signatures across the suborder, we adopt here the terminology used in other oribatid taxa: the anterior two pairs become the rostral (*ro*) and lamellar (*le*) setae, and the posterior two pairs become the interlamellar (*in*) and exobothridial (*ex*) setae (Fig. 16.1). As in many crotonoid oribatids, astigmatine mites lack prodorsal trichobothria and bothridial setae and the second pair of exobothridial setae.

The notogastral region of the adult posterior to the sejugal furrow bears 12 pairs of setae in rows *c* through *h*. Several schemes of setal notation have been used for astigmatine mites, but the system used here is the one developed for oribatid mites by Grandjean and applied to the Astigmatina by Griffiths et al. (1990) (Fig. 16.1). Setae *f1* are always lacking, although a vestigial alveolus in this position appears in some HISTIOSTOMATIDAE. The setae of row *c* are arranged dorsally as in related oribatids in this family as well, but in all other Astigmatina, seta *c3* has shifted into a ventrolateral position. The idiosomatic setae have a great variety of forms in different species, ranging from short and setiform to very long and densely barbed. Others may be expanded and fanlike, flattened and bladelike, or strongly bipectinate. Four pairs of cupules are retained: *ia*, *im*, *ip*, and *ih*—the last ventrally positioned lateral to the anal opening. The opisthonotal glands open dorsolaterally between setal rows *d* and *e* and may be more or less conspicuous when viewed externally, depending on their contents. A wide variety of hydrocarbon compounds has been identified in various species of Astigmatina, including straight-chain hydrocarbons and aromatics. Most common among the straight-chain hydrocarbons are the monoterpenes neryl formate, neral, and geranial, and among the aromatics are 2,6-HMBD (2-hydroxy-6-methylbenzaldehyde) and α-acaridial (Kuwahara 1999; Sakata and Norton 2001). Some of these compounds function as antipredator or antifungal allomones, while others serve as alarm, aggregation, or sex pheromones (also see chapter 3).

The ventral body surface is characterized by distinct coxal apodemes. In the HISTIOSTOMATIDAE, the epimeres or ventral coxal surfaces are arranged in a more or less transverse manner somewhat similar to the ancestral oribatid form. In more derived families, the coxal fields are more oblique, and the division between the bases of legs II–III is wider. The anterior apodemes of coxal fields I in many Astigmatina fuse together to form a sternal apodeme, but they remain free in some parasitic taxa. Coxal field setation is strongly reduced from that of ancestral Oribatida, with coxal fields I and III typically bearing only one pair of setae (a setal pair also may occur on coxal fields II in some ALGOPHAGIDAE), and coxal fields IV bearing two pairs, the inner pair of which may be associated with the female genital valves (Norton 1998). The anus of most astigmatine mites is located on the posterior ventral surface, but it may shift more posteriorly to a terminal position (some HEMISARCOPTIDAE, SARCOPTIDAE) or to a mid-dorsal position in most female SARCOPTIDAE (Fig. 16.52D).

The position of the adult genital opening is highly variable, ranging from far posterior and almost confluent with the anus (HEMISARCOPTIDAE) (Fig. 16.12C), to far anterior between coxal fields I (some GLYCYPHAGIDAE, e.g., *Pycnoglycyphagus*; AEROGLYPHIDAE, e.g., *Aeroglyphus* (Fig. 16.20A); and some psoroptidian groups such as RECTIJANUIDAE (Fig. 16.35B)). The position of the genital atrium in homeomorphic nymphs is fairly constant, with the position between or slightly posterior to coxal fields IV, reflecting the ancestral position. The adult genital valves are unsclerotized except in totally sclerotized taxa such as *Fusacarus* (GLYCYPHAGIDAE). The female valves are fused to the body posteriorly and open anteriorly and medially in the GUANOLICHIDAE and are fused together medially in the HISTIOSTOMATIDAE, although the band of fusion is somewhat flexible. In all other Astigmatina, the valves are fused to the body anteriorly but free posteriorly and medially, leaving the genital opening in the form of an inverted V. In some psoroptidians, the genital valves themselves are indistinct in females, which results in a simple and inverted U-shaped or transverse opening (some PSOROPTIDAE, CHIRORHYNCHOBIIDAE, GAUDOGLYPHIDAE, FALCULIFERIDAE, and others). Two pairs of genital papillae often are associated with the genital valve elements in postprotonymphal stases, but they may be indistinct or absent in parasitic, hygrophilous, and halophilous taxa.

In keeping with their neotenous development, the true ovipositor of the female and the homologous spermatopositor in the male are not added ontogenetically. The terminal portion of the female reproductive tract may be eversible, however, and serves as a pseudovipositor in most Acaroidea and a few other groups. The female astigmatine reproductive system is a complete tube, with sperm introduced through a median copulatory opening that leads to a spermatheca. In many groups, the copulatory opening is placed at the end of a copulatory tube, or spermaduct, that enters a duct leading from the primary male genital opening to meet the aedeagus at its terminus (e.g., some EUSTATHIIDAE, CRYPTUROPTIDAE) (Fig. 16.36C). Alternately, the spermaduct may slide along a groove in an elaborately sclerotized male genital apparatus (some TROUESSARTIIDAE) (Fig. 16.43C). Males typically possess a sclerotized aedeagus that may vary widely in form, even within a single genus (also see chapter 3). Male secondary sexual characters will be discussed below.

The gnathosoma of astigmatine mites is relatively simple compared to the ancestral form (see chapter 15). The subcapitulum is usually smooth but bears a distinct pattern of transverse ventral ridges in the GLYCYPHAGIDAE (Fig. 16.22B) and may bear transverse striae in hair-clasping psoroptidians (some LOBALGIDAE, ATOPOMELIDAE). The subcapitulum carries a single pair of ventral setae that are ancestrally associated with the genal element of the capitular base. Ventrodistally, the rutella cover most of the ventral apex of the subcapitulum. Ancestrally, the rutella have a lateral element that may bear one or two sclerotized distal teeth in many free-living forms and a broad lobe that extends ventromedially. Many taxa bear membranous processes extending from the rutellar surface. Such processes are well developed in many ROSENSTEINIIDAE (Fig. 16.20D) and in some groups of psoroptidians inhabiting bird feathers, such as the PROCTOPHYLLODIDAE and TROUESSARTIIDAE (Fig. 16.43A). A pair of small lateral lips is present between the median rutellar lobes. These lack the ancestral adoral setae in all known feeding forms, although the heteromorphic (probable) deutonymph of SCHIZOGLYPHIDAE bears two pairs of tiny setae in this position (Fig. 16.6B).

The chelicerae of most astigmatine mites are chelate-dentate and may either crush solid food materials or act as scrapers, removing food material from the substrate surface. The chelicerae are highly modified in the Histiostomatoidea, with the movable digit fused basally to the fixed digit and the entire structure flattened laterally and typically bearing numerous fine teeth (Fig. 16.8D). These teeth are used to filter particles from the water films inhabited by these mites (also see chapter 3). The chelicerae of females of the genus *Hypodectes* (HYPODERATIDAE) are reduced and nonfunctional, while those of the male are greatly enlarged and probably function in agonistic behavior toward other males (Figs. 16.31E, F). The chelicerae of parasitic psoroptidians tend to be small, but otherwise they are not particularly modified. An exception is found in the family CYTODITIDAE, members of which are endoparasites in the respiratory tract of birds. Here, the chelicerae are very small, and the fixed digits or the entire chelicera are highly reduced (Figs. 16.40E, F).

Astigmatine palpi are typically small and reduced to two apparent segments, which probably represent a partially fused tibia and tarsus. A third basal segment persists in the putative deutonymph of SCHIZOGLYPHIDAE and probably represents the genu (Norton 1998) (Fig. 16.6B). Ancestrally, the palpi bear three tactile setae in the more proximal portion and a short solenidion and one or two eupathidial setae at the tip. The palpi of some HISTIOSTOMATIDAE are strongly modified, with the distal portion often reflexed outward and bearing a very long solenidion and sometimes a long eupathidial seta as well (Fig. 16.7B). The basal setae are strongly reduced and difficult to resolve with the light microscope. In their excellent SEM study of a histiostomatid palp, Wurst and Kovac (2003) regarded both of the long apical structures as solenidia because neither was birefringent under polarized light. This would be a highly abnormal condition, however, since no sensillum was regarded as a eupathidium.

The legs of most free-living and parasitic Astigmatina are relatively unspecialized and, with rare exceptions, have freely articulating segments (members of the glycyphagid genus *Fusacarus* have fused femora/genua). Fusion of segments is more common among parasitic psoroptidians; for example, species of the feather mite families ALLOPTIDAE and XOLALGIDAE have fused femora and genua on all legs (Fig. 16.42), while femur and genu III–IV are fused in the TROUESSARTIIDAE (Fig. 16.43A). Among mammalian fur mites, tibia and tarsus III–IV are fused in ATOPOMELIDAE, a family in which some taxa also exhibit fusion of the femur and genu (Fig. 16.55A). The same variation is seen in the skin-inhabiting families SARCOPTIDAE and RHYNCOPTIDAE. Extreme fusion of most segments of the anterior legs characterizes most CHIRODISCIDAE, where the flattened legs clasp mammalian hair (Figs. 16.54C, D). One or both pairs of hind legs may be reduced in size and segmentation in some PSOROPTIDAE and RHYNCOPTIDAE (Fig. 16.51C), while legs IV may be completely absent in highly paedomorphic species of the families SARCOPTIDAE (Figs. 16.51D, 16.52D), EPIDERMOPTIDAE, and APIONACARIDAE (Fig. 16.41A). The legs of some free-living species may be strongly sclerotized and heavily ridged (SCATOGLYPHIDAE and some GLYCYPHAGIDAE). The setation of the legs is strongly paedomorphic, with few setae added to the fundamental complement during ontogeny. A set of signatures for leg setae developed by Grandjean (1939a) has been in general use, but in the interest of standardizing terminology among all sarcoptiform groups, we adopt here a modified set of setal homologies somewhat different from those hypothesized by Van der Hammen (1982; see Fig. 16.3). An explanation of the rationale behind these proposed homologies will be presented in a separate publication. The pretarsus in the Astigmatina is highly variable and has been studied in detail by Atyeo (1979), whose terminology we adopt here (Fig. 16.4). Ancestrally, the pretarsus is similar to the juvenile pretarsus of related oribatid groups, with a small membranous ambulacrum and a well-developed, clawlike empodium. In some free-living taxa, the empodium is lost (HEMISARCOPTIDAE, MELIPONOCOPTIDAE, HYPODERATIDAE), and the condylophores are thin in many others. The separation between ambulacral stalk and an expanded ambulacral disc is obvious in psoroptidians and insect paraphages of the families CANESTRINIIDAE, HETEROCOPTIDAE, and ROSENSTEINIIDAE. The empodium often is flattened and incorporated into the disc as a central sclerite in these mites, and additional sclerites may be arranged around the central empodial sclerite.

Sexual dimorphism in primary genitalic structures was discussed earlier (see chapter 3). With the need for maintaining position during copulation, and also with precopulatory mate guarding and agonistic behavior toward other males being common in Astigmatina, males have developed an amazing array of modifications for these functions. In most groups, males are either similar in size to females or are smaller (e.g., some PSOROPTIDAE, SARCOPTIDAE, GLYCYPHAGIDAE, and others), but males may sometimes be considerably larger than females, as in some feather-inhabiting ALLOPTIDAE (*Laminalloptes*) (Fig. 16.42F) and AVENZOARIIDAE (*Zachvatkinia*). Two mating positions are exhibited in the Astigmatina, one with the male above and facing parallel to the female, and the other with the male above and facing away (reversed position). Secondary sexual dimorphism often reflects the mating position, with taxa that use the parallel position typically having male modifications on the anterior body and legs, and those adopting the reversed position having modifications posteriorly. Males ancestrally do not add the three pairs of setae of paraproctal segment AD that are added in the adult female (some GLYCYPHAGIDAE are exceptional), but a pair of suckers is present in some Histiostomatoidea and retained in many other more derived taxa (Fig. 16.2C). Witalinski (1990) and Witalinski, Dabert, and Walzl (1992) describe a median central papilla in the suckers that takes the form of a thin-walled seta in an undescribed species of *Bonomoia* (HISTIOSTOMATIDAE) (Fig. 16.7E) and in *Donnelafontia calostegis* Lavoipierre (CANESTRINIIDAE) (B. M. OConnor, pers. obs.). A thin seta without the sucker appears in a similar position in *Lemanniella minotauri* Wurst (LEMANNIELLIDAE), and a fully developed seta is present in *Scatoglyphus polytrematus* Berlese (SCATOGLYPHIDAE). These observations suggest that the paranal suckers are derived from alveoli of one of the *ad* setae. Families in which the mating position is reversed tend to retain the suckers, while those adopting the parallel position generally lose them. The HISTIOSTOMATIDAE and GUANOLICHIDAE are unusual in that some early derivative genera retain the suckers despite the parallel mating position, but in these taxa the suckers may be displaced anterior to the anus and lie beside or anterior to the aedeagus (Fig. 16.7E). Males of the family HEMISARCOPTIDAE have a median ventral sucker anterior to the aedeagus formed from the fused alveoli of setae *4b* (Fig. 16.12A). In many psoroptidians with the reversed mating posture, the posterior end of the male is strongly bilobed or excavated ventrally, and the posterior setae are often elongated, fanlike, or otherwise modified (Figs. 16.35F, 16.47B). Precopulatory guarding is taken to the extreme in most taxa of the fur mite family CHIRODISCIDAE, where juvenile females lack functional legs and depend on adult males to attach to them with the aid of a highly modified posterior portion of the idiosoma (Fig. 16.54D).

The legs of male astigmatines may also be modified for guarding, copulating, or fighting. Taxa with the reversed mating position may have legs III and/or IV enlarged. Legs III are modified in the free-living GLYCACARIDAE (Fig. 16.26B), LARDOGLYPHIDAE (Fig. 16.27B), and in some ACARIDAE, as well as in many analgoid feather mites (Fig. 16.46A) and some mammal parasites (PSOROPTIDAE, RHYNCOPTIDAE) (Fig. 16.51C). Legs IV are enlarged in many fur mites of the family ATOPOMELIDAE (Fig. 16.55D). Although mites in these groups use the enlarged legs merely for holding juvenile or adult females, legs III of some ACARIDAE (*Sancassania*, *Cosmoglyphus*, and perhaps others) are used for fighting, with the winning male impaling his rival with the strongly pointed ends of the enlarged legs. Some setae of the hind tarsus may be modified as suckers to help hold the lateral sides of the female during mating. Setae *ft*" and *tc*" take this form in most Acaroidea, Hypoderatoidea, and many psoroptidian lineages (Fig. 16.2C), while more basal setae are similarly modified in some ALGOPHAGIDAE (*Hericia*, *Fusohericia*) (Fig. 16.16B). Males of New World ECHIMYOPODIDAE (*Blomia*, *Echimyopus*) have an unusual twist in legs IV and a suckerlike modification of the articulation between tibia and tarsus (Fig. 16.19B).

Males in taxa that employ a parallel mating posture may have modified legs I or II, as has been observed in some HISTIOSTOMATIDAE (some *Histiostoma*), HYADESIIDAE (Fig. 16.13A), GLYCYPHAGIDAE (*Baloghella*), and ROSENSTEINIIDAE (*Chiroptoglyphus*, *Micronychites*) (Fig. 16.21A). Legs I also are enlarged in the genus *Acarus* (ACARIDAE) and in some PYROGLYPHIDAE (Fig. 16.45D), despite their reversed mating posture. Less obvious modifications involve the anterior pretarsi. In most WINTERSCHMIDTIIDAE, for example, the ventral apices of tarsi I–II are excavated, and the pretarsal ambulacral stalks form suckers (Fig. 16.11D). The ambulacral discs of pretarsi I–II in CARPOGLYPHIDAE are asymmetrical, with a posterior lobe supported by a shifted condylophore (Fig. 16.15C). In CHAETODACTYLIDAE, the proximal parts of the condylophores themselves fuse and emerge as hooks from the base of the ambulacral stalk, while in the glycyphagid subfamily Labidophorinae the entire ambulacral stalk is expanded and fimbriated.

Much of the diversity in secondary sexual characteristics is complicated by the phenomenon of male polymorphism, which may be expressed discretely or continuously in a variety of characters. Discrete polymorphism occurs in some genera of ACARIDAE, where males either have enlarged legs III for fighting, or legs III are similar to those of the female. Males of *Sancassania* (= *Caloglyphus*) may also vary discretely in dorsal setal length. These traits vary independently, resulting in the formation of four discrete morphs (Türk and Türk 1957; Woodring 1969a). Timms, Ferro, and Emberson (1982) demonstrated that the male morph type was determined environmentally, with fighting morphs produced in low-density colonies at a much higher rate than in high-density colonies. Fighting morphs would have a fitness advantage in low-density colonies because they could eliminate their homeomorphic rivals, whereas in high-density populations, fighting would take time away from mating and not increase a male's chance to mate (Radwan 1993). In some psoroptidian feather mite families, two discrete male morphs are produced that vary in a variety of traits, including body size and shape of the posterior setae. Polymorphism also has been observed in the genera *Freyana* (FREYANIDAE) and *Vexillaria* (VEXILLARIIDAE), but its implications are unknown.

More common are continuous polymorphisms that may or may not involve allometric effects on particular parts of the male body. Size-related continuous polymorphism has been documented in the feather mite genus *Analges* (ANALGIDAE), where smaller males have only slightly enlarged legs III and body setae, while larger males show greatly enlarged legs III and much longer body setae (Fig. 16.46A). Legs I vary allometrically with body size in the rosensteiniid genus *Chiroptoglyphus*, with a small increase in body size yielding a much greater increase in leg size (OConnor and Reisen 1978) (Fig. 16.21A). Similar allometries may affect legs I in males of *Acarus* (ACARIDAE) and *Dermatophagoides* (PYROGLYPHIDAE). Another allometric effect involves inordinate enlargement of the chelicerae, subcapitulum, and legs I (and sometimes legs II) in several distantly related taxa. The phenomenon is best documented in the feather mite genera *Falculifer* and *Cheiloceras* (FALCULIFERIDAE) (Fig. 16.39C), *Bdellorhynchus* (AVENZOARIIDAE), and *Suladectes* (FREYANIDAE), in which the chelicerae and legs I of the largest males are monstrously enlarged. The enlarged chelicerae of male *Bdellorhynchus* are used in combat with rival males. A single male was observed to defend an entire flight feather on a duck host against rivals, and the male with the largest chelicerae would drive the losers from the feather (B. M. OConnor, unpublished obs.). The sex ratio appeared to be strongly female biased, which means that males holding territories had access to numerous potential mates. Similar allometric enlargement of chelicerae occurs in nonpsoroptidian groups such as *Horstia* (= *Ceroglyphus*) (ACARIDAE). Other males may exhibit asymmetrical development of the legs (ALLOPTIDAE) (Fig. 16.2D) or body setae (FREYANIDAE, KRAMERELLIDAE) (Fig. 16.2E).

The modified deutonymphs of astigmatine mites (also referred to as hypopi or hypopodes by some authors) are surely the key to the success of the group. Formation of the deutonymph in a particular individual's ontogeny is facultative in many groups of free-living Astigmatina. In many cases protonymphs sense factors of their environment such as food quality and quantity and mite population density, and if these factors favor the mite's ability to survive and reproduce in place, the protonymph will molt directly to the tritonymph. In the opposite situation, the deutonymph will be formed, and the mite will attempt to disperse (Griffiths 1966). In some cases, the availability of a potential dispersal agent is a key factor in mediating the appearance of a deutonymphal stage. Fashing (1975) found that deutonymphs were produced by *Naiadacarus arboricola* Fashing (ACARIDAE) only when the adults of their syrphid fly host were emerging from the tree-hole habitat, despite the abundance of food at the time. In a few taxa, formation of deutonymphs seems obligatory. In species of WINTERSCHMIDTIIDAE inhabiting the nests of solitary vespid wasps, the life cycle of mite and host are closely timed so that deutonymphs can attach to the new adult wasp as it emerges from its cocoon (Cooper 1955; Okabe and Shun'ichi 2003). Deutonymphs of the bivoltine mite *Winterschmidtia nataliae* (Zakhvatkin) (= *Calvolia fraxini*), a predator of bark beetle eggs, are produced only by the summer generation, and they disperse or overwinter in the host galleries. Mites in the spring generation molt directly from protonymph to tritonymph (Kielczewski and Seniczak 1972). The ability to form deutonymphs in some species varies with different populations, as is the case with *Carpoglyphus lactis* (L.) (Chmielewski 1971), while in others this ability has been lost. In taxa occupying more continuous habitats, such as the marine intertidal HYADESIIDAE and most species of the widespread, polyphagous acarid genus *Tyrophagus*, mites may disperse in the absence of phoresy. The same is true for permanent parasites that disperse directly from host to host, such as the beetle-parasitizing CANESTRINIIDAE and HETEROCOPTIDAE and the vertebrate-parasitic psoroptidian groups. Harder to explain is the loss of the deutonymph from the glycyphagoid lineage comprising the AEROGLYPHIDAE and ROSENSTEINIIDAE, most of whose species are free-living inhabitants of bat roosts.

The specialized astigmatine deutonymph functions not only in dispersal but also in withstanding adverse environmental conditions. Deutonymphs can survive much greater extremes of temperature and humidity than feeding stages, and may live for a long period without molting (OConnor 1994). Many species of astigmatine mites exploit both deutonymphal functions through a polymorphism, with one morph specialized for phoretic dispersal and the other for sitting and waiting. Different terminology has been applied to these morphs; for example, nonphoretic morphs are often referred to as "inert" because many have a highly regressive morphology. These forms typically have the legs and attachment organ vestigial or absent and persist as a calyptostase within the protonymphal cuticle. Deutonymphal polymorphism involving an "inert" sedentary form has been documented in the HISTIOSTOMATIDAE (*Tensiostoma veliaphilum* Wurst and Kovac), CHAETODACTYLIDAE (*Chaetodactylus*), GLYCYPHAGIDAE (*Glycyphagus*, *Baloghella*), CHORTOGLYPHIDAE (*Alabidopus*) (Fig. 16.18E), and ACARIDAE (*Acarus*, *Stereoglyphus*). Two active forms are produced in some ALGOPHAGIDAE (*Hericia*), but only one has a functional attachment organ and disperses through phoresy (Fashing 1991). Similar deutonymphs with normal morphology but a vestigial attachment organ are known in the HISTIOSTOMATIDAE and may actually represent alternate morphs of other species.

Deutonymphal astigmatine mites exhibit numerous morphological modifications for their dual role in dispersal and survival. These modifications have led to a classification of deutonymphal types (Fain 1971b; Volgin 1973), but these tend to obscure the diversity and evolutionary origin of different forms. The body of phoretic forms is generally well sclerotized and dorsoventrally flattened (Fig. 16.2A), the chelicerae are absent, and the remaining gnathosomatic structures are reduced in size or eliminated. The leg tarsi often bear specialized foliate setae that assist in keeping contact with a host, but the major modification for attachment is the development of a posterior ventral adhesive organ. Ancestrally, this organ includes setae *p1* and *p2* modified as rounded conoids, the alveoli of setae *ad1* and *ad2* fused on each side and expanded to form large median suckers, the alveoli of setae *ad3* forming large, stalked, movable, anterior suckers, and the *de novo* development of two pairs of lateral suckers and a posterior median sucker (Fig. 16.2B). The adhesive organ facilitates attachment to smooth arthropod cuticle and probably is ancestral in the Astigmatina (the attachment organ of the enigmatic SCHIZOGLYPHIDAE is somewhat different in structure) (Fig. 16.6A). The suckers are reduced in many GLYCYPHAGIDAE, and in the histiostomatid genus *Fibulanoetus,* the conoidal setae are elongated and ridged and a pair of cuticular flaps extends over the organ (Figs. 16.8A, 16.23C–E). This modified attachment organ serves to attach the glycyphagid mite to the hair of a mammal, or the histiostomatid mite to the seta of a beetle.

Deutonymphs of many groups of glycyphagoid mites and those of the family HYPODERATIDAE actually enter the tissues of a vertebrate host, either into a hair follicle or into the subcutaneous tissues. Most of these deutonymphs have a strongly reduced attachment organ, or the organ may be completely absent (Figs. 16.23B, 16.31D). Many grow while in the host tissue, despite the lack of a mouth or functional anterior gut. Although this phenomenon of deutonymphal parasitism has been known for almost 200 years in some HYPODERATIDAE, the mechanism for nutrient intake has not been verified. A few insect-associated deutonymphs are known or are suspected to be parasitic. *Hemisarcoptes cooremani* (Thomas) was shown conclusively to take host hemolymph into its hindgut through the anus (Houck and Cohen 1995), thus providing a possible mechanism for the extreme engorgement seen in some deutonymphs parasitizing vertebrates (also see chapter 6).

Classification

The classification used here is a modification of that presented by OConnor (1982a), which was based on a phylogenetic analysis of the nonpsoroptidian groups. Prior classifications (e.g., Yunker 1955; Krantz 1978) that divided the Astigmatina into two major groups recognized a paraphyletic taxon containing primarily free-living forms (Acaridia,

Acaridides) and a hypothetically monophyletic grouping primarily of vertebrate parasites (Psoroptidia or Psoroptides). The present classification recognizes a number of superfamilies that diverged before the origin of the Psoroptidia. A cladogram of hypothesized relationships is presented in Fig. 16.5.

The Astigmatina has a poor fossil record limited in its usefulness in determining the age of the group. A deutonymph described from Mexican amber of Miocene age, *Amphicalvolia hurdi* Türk, is virtually indistinguishable from the modern genus *Winterschmidtia* associated with subcortical beetles (Türk 1963; B. M. OConnor, pers. obs.). A fossil psoroptidian preserved along with mammalian hair in Dominican amber of lower Miocene age was placed in the family LISTROPHORIDAE (Poinar 1988, 1992; Poinar and Poinar 1999), but the habitus of the mite is strongly reminiscent of modern fur mites of the family ATOPOMELIDAE that occur on extant mammals in the Dominican Republic. Klompen's (1992, 1993) coevolutionary analyses of mammalian skin-inhabiting mites of the families SARCOPTIDAE and RHYNCOPTIDAE show a basal dichotomy in the phylogeny of each of these sister families between parasites of marsupials and parasites of placental mammals. The timing of the divergence of these host groups was in the early Cretaceous at the latest, which means that the most derived lineage of Astigmatina (the superfamily Sarcoptoidea) was already present in the Mesozoic era. This, coupled with the cosmopolitan distribution of many astigmatine genera, suggests a Palaeozoic origin for the lineage.

■ Superfamily Schizoglyphoidea (Figs. 16.6A–F)

DIAGNOSIS: Deutonymph (?) with gnathosomatic remnant well developed; hypostome with 2 pairs of tiny adoral setae; palpi 3-segmented with 1 seta on tibia, 4 setae and 1 solenidion on tarsus. Dorsal idiosoma sclerotized, punctate; sejugal furrow well developed. Propodosomatic chaetome complete; notogastral chaetome lacking setae f2. *Coxisternal region with normal chaetome. Genital valves each with 4 setae; 3 pairs of genital papillae present. Posterior ventral attachment organ with anterior (*ad3*) and median (*ad1 + 2*) suckers well developed; structures derived from* ps *setae vestigial; with a pair of large, ovoid, anterior cuticular suckers encompassing setal vestiges, and a single, very broad posterior sucker curving around posterior half of attachment organ. Legs normally segmented, chaetome complete, including unguinal setae, but* ft' *absent on tarsus II,* s *absent on tarsi I–IV; setae* ft' *present on tarsi III–IV, and solenidion* ω *present on tarsus II; some tarsal setae foliate. Pretarsus without membranous ambulacrum.*

The single enigmatic species, *Schizoglyphus biroi* Mahunka, of the monogeneric family SCHIZOGLYPHIDAE, was described from two individuals collected from the tenebrionid beetle, *Dioedus tibialis* Kaszab, in western New Guinea (Mahunka 1978). The absence of chelicerae and the presence of a posterior ventral attachment organ and foliate tarsal setae are characteristic only of deutonymphal Astigmatina (Fig. 16.6A). However, the presence of three pairs of genital papillae (Fig. 16.6D) indicates that either the third pair is accelerated in the deutonymph, or that these individuals represent heteromorphic tritonymphs. The post-deutonymphal stages remain unknown.

■ Superfamily Histiostomatoidea (Figs. 16.7, 16.8)

DIAGNOSIS: Adults with gnathosoma highly modified for filter feeding. Chelicerae laterally flattened, fixed digit typically comblike, with numerous, finely spaced teeth (rarely with few, bladelike teeth), movable digit short, fused to base of fixed digit, with few short, apical teeth. Palpi short, simple or more often enlarged, reflexed outward, and with tarsal solenidion and eupathidial seta elongate; palpi often with complex, membranous processes. Idiosoma rarely completely sclerotized, more often with discrete propodosomatic sclerite and unsclerotized hysterosoma. Sejugal furrow present or absent. Idiosomatic chaetome complete or with reduced paraproctal setation. Coxisternal skeleton with apodemes largely transverse, chaetome usually complete, but seta 4b *always absent. Female with genital valves fused to body posteriorly and fused to each other medially in most, oviporus functionally transverse. Male with genital and anal regions usually closely associated, often on ventrally directed cone. Genital papillae always ringlike, never papilliform, often disassociated from genital region; homeomorphic nymphs without genital atrium, papillar rings on ventral integument. Male with or without paranal suckers; suckers, when present, may be anterior to aedeagus. Mating position parallel. Legs with normal chaetome, lacking* a'' *on tarsus I, many leg setae spinelike. Males often more heavily sclerotized than females, and often with modified legs I, II, or (rarely) IV.*

Deutonymphs: Known only in HISTIOSTOMATIDAE. *Gnathosomatic remnant well developed or reduced. Idiosoma well sclerotized; propodosomatic and hysterosomatic sclerites separated by sejugal furrow. Idiosomatic chaetome complete, but anterior propodosomatic setae (*ro, le*) always absent. Coxisternal skeleton well developed, usually with posterior median apodeme (sometimes reduced); chaetome complete except* 4b *absent. Attachment organ usually well developed, with all structures usually present; organ modified in some for clasping insect setae or leg segments, mammal hair, or for pinching membranous host cuticle; organ occasionally highly reduced. Anterior legs normally developed; legs III–IV often with some segments fused and with secondary articulation in midtarsus; articulations of trochanter-femur III–IV anteriorly directed in repose. Pretarsi with small membranous ambulacrum and empodial claw; empodium IV sometimes absent.*

The Histiostomatoidea includes the family GUANOLICHIDAE, a small taxon comprising two genera and three species inhabiting bat guano in the Afrotropical and Neotropical regions, and the cosmopolitan, speciose HISTIOS-

TOMATIDAE (= Anoetidae) with 58 genera and over 500 species. Adult female guanolichids retain short, relatively unmodified palpi, and genital valves that are unfused medially. Small remnants of both pairs of genital papillae are closely associated with the genital valves (Fain 1968b, 1978b; Fashing 1979) (Fig. 16.8E). The genital valves of female histiostomatids are fused medially, creating a transverse oviporus. The two pairs of ringlike genital papillae are not closely associated, and the posterior pair typically is located on the ventral opisthosoma (Fig. 16.7A).

Histiostomatoid mites inhabit wet substrates, where the highly modified mouthparts are normally used to filter organic material and microorganisms from water films. Their enlarged, highly modified genital papillae appear to be an adaptation for maintaining osmotic balance in aquatic habitats (Witalinski, Liana, and Alberti 2002). Species that live submerged in freshwater habitat patches tend to have enlarged papillae, while those of species living in hyperosmotic environments such as beach wrack and bat guano are quite reduced. Preferred habitats for histiostomatids are typically patchy or ephemeral, and consequently, some species have the shortest known generation times among the Astigmatina. Some populations develop from founding deutonymph through next generation deutonymph in as little as six days (Hill and Deahl 1978; Hughes and Jackson 1958). Species in the genera *Histiostoma*, *Myianoetus*, *Glyphanoetus*, *Rhopalanoetus*, *Aphodanoetus*, *Copronomoia*, and others are typical inhabitants of vertebrate dung, with deutonymphs often specialized for phoresy on particular groups of dung-associated insects (Bongers, OConnor, and Lukoschus 1985). Vertebrate carrion is inhabited by species of *Spinanoetus*, *Pelzneria*, and *Peripatetes*, and the deutonymphs are typically phoretic on silphid and staphylinid beetles (Scheucher 1957). Fungal fruiting bodies are exploited by species of *Histiostoma*, which feed on necrotic tissue and may spread pathogenic microorganisms in commercial mushroom operations (Hill and Deahl 1978). Species in many other genera have specific associations with habitats created by insect hosts. For example, wet subcortical habitats are colonized by species of *Histiostoma* and *Bonomoia* associated with scolytid beetles, while galleries of Passalidae are home to species of *Scolianoetus*, *Kanoetus*, and *Histiostoma* (Woodring and Moser 1970; B. M. OConnor, pers. obs.). Galleries of palm weevils of the genus *Rhynchophorus* are inhabited by species of *Curculanoetus*, while those of African pycnocerine Tenebrionidae are utilized by species of *Chiloanoetus*. Vertebrate nests harbor species of *Hexanoetus*, *Prowichmannia*, *Myianoetus*, and *Histiostoma* (Fain and Philips 1979; Scheucher 1957).

A great diversity of histiostomatids may be found in naturally occurring "container habitats," with water-filled plant cavities, or phytotelmata, being especially attractive. Many species living in such fully aquatic environments have elongate legs and/or long, bladelike empodial claws. Species of *Zwickia*, *Nepenthanoetus*, *Creutzeria*, and some undescribed taxa inhabit the leaf pitchers of Old World *Nepenthes* pitcher plants, while species of *Sarraceniopus* inhabit the pitchers of New World Sarraceniaceae (Fashing 2002; Fashing and OConnor 1984). *Creutzeria* species are unique among astigmatine mites in having the ability to swim. These mites have fusiform bodies, anteriorly positioned legs, and swimming hairs on legs II–IV and are remarkably convergent with certain aquatic Prostigmata (Fashing, OConnor, and Kitching 1996). Species of *Hormosianoetus* inhabit water-filled tree holes and the leaf axils of Bromeliaceae, while many undescribed taxa have been collected from water-filled leaf axils and flower bracts of Araceae, Heliconiaceae, Marantaceae, and other plants (Fashing 1994; B. M. OConnor, pers. obs.). Species inhabiting phytotelmata often use Hemiptera, Diptera, or Coleoptera as dispersing agents (Wurst and Kovac 2003; B. M. OConnor, pers. obs.).

A few histiostomatids have developed more intimate associations with animal hosts. Species of *Loxanoetus* and *Otanoetus* inhabit the ears of large mammals such as elephants and antelope (Domrow and Ladds 1984; Fain and Zumpt 1974), but their feeding ecology is unknown. *Histiostoma murchiei* Hughes and Jackson has bladelike chelicerae and feeds as a predator in earthworm egg cocoons (Oliver 1962). Species of *Anoetus* (*s. str.*) live in mutualistic association with halictid bees, with feeding stages filtering potentially pathogenic microorganisms from the pollen/nectar provisions in the nest cells. Female bees carry mite deutonymphs in a specialized acarinarium on the first metasomatic tergite (Eickwort 1979, 1994).

Most histiostomatid deutonymphs have a normally developed attachment organ that serves to attach the mite to smooth arthropod cuticle (Fig. 16.7C). More variation has evolved in this group, however, than in any other family of Astigmatina. The attachment organ in species of *Prowichmannia* is elongate, and the conoidal setae *p1*, *p2*, and *4a* are strongly ridged to hold the hair of the mammalian host (Fig. 16.8B). Species of *Fibulanoetus* have similarly ridged conoidal setae and also have paired cuticular flaps laterally, similar to many deutonymphs of GLYCYPHAGIDAE (Fig. 16.8A). In contrast to the mammalian phoretic glycyphagids, however, *Fibulanoetus* deutonymphs attach to the body setae of their scarabaeid beetle hosts (Fain et al. 1980). Species of *Xenanoetus* have enlarged, hooklike conoidal setae *p1* and *p2* that are positioned behind the median suckers (Fig. 16.8C). These mites pinch a fold of intersegmental membrane of their dipteran host's abdomen between the conoids and central suckers to form a tight attachment (B. M. OConnor, pers. obs.). Some histiostomatid deutonymphs, such as those of *Amyzanoetus* and *Comyianoetus* species, have very small attachment organs although they may still disperse through phoretic attachment to insects. A highly regressive, calyptostatic deutonymphal morph has been described in *Tensiostoma veliaphilum* Wurst and Kovac (Wurst and Kovac 2003).

A few histiostomatids are of some economic importance. *Histiostoma feroniarum* (Dufour) is a virtually cosmopolitan species feeding in wet, damaged areas of bulbs, tubers, and fruits (Hughes and Jackson 1958; Qin and Rohitha 1996), and *H. heinemanni* Hill and Deahl can form large populations in commercial mushroom operations. Although not directly damaging, these mites can be difficult to remove from the mushrooms prior to marketing. Finally, *Hormosianoetus laboratorium* (Hughes) is a pest in laboratory *Drosophila* colonies.

▪ Superfamily Canestrinioidea (Figs. 16.9, 16.10)

DIAGNOSIS: Adults with body modified for insect paraphagy. Gnathosoma normally developed, palpi and chelicerae occasionally elongated; chelicerae typically chelate, rarely elongate and edentate. Idiosoma with cuticle smooth, striate, reticulate, scaly, or sometimes well sclerotized; sejugal furrow inconspicuous; chaetome complete or with paraproctal setation reduced; setae simple, barbed, flattened, or otherwise modified; supracoxal seta of legs I typically displaced laterally from supracoxal gland opening. Coxisternal skeleton normally developed or with apodemes III–IV reduced; chaetome complete, female often with setae 4b *anterior to oviporus. Female with genital valves fused to body anteriorly, forming inverted* V*-shaped oviporus. Male with or without paranal suckers; often with posterior opisthosoma highly modified. Mating position reversed. Legs normally formed, cylindrical, or rarely strongly flattened; chaetome complete or with some reductions, unguinal setae usually absent, usually with seta* v' *present on trochanter IV. Pretarsi typically with enlarged membranous ambulacrum; ambulacral disc often with secondary internal sclerotization; empodial claw small or incorporated into ambulacral disc as flattened sclerite.*

Deutonymphs: Known only in genus Megacanestrinia *(CANESTRINIIDAE). Body very broadly teardrop shaped. Gnathosomatic remnant well developed. Idiosoma well sclerotized, sejugal furrow well developed. Coxisternal skeleton well developed; apodemes III–IV fused medially to form coxal fields III; chaetome complete. Attachment organ very large, all structures present. Legs normally developed, deutonymphal chaetome complete; pretarsal ambulacra absent, empodial claws I–IV well developed.*

All species included in the Canestrinioidea are obligate external or subelytral paraphages of adult Coleoptera. *Diplopodocoptes transkeiensis* Fain, described as a canestriniid from millipedes (Fain 1987a), is here considered a member of the family Chetochelacaridae, superfamily Acaroidea. The grouping of the families CANESTRINIIDAE and HETEROCOPTIDAE may be a matter of convenience because characteristic modifications for permanent insect paraphagy such as laterally inserted legs and enlarged ambulacral discs may be the result of convergence in the two families. The CANESTRINIIDAE is a large family (300 species in 95 genera) whose members are moderate to large in size and very weakly sclerotized. Early derivative taxa in the family are African in distribution. Species of *Megacanestrinia* (= *Coleoglyphus*) are unique in retaining the deutonymph in the life cycle, a stage lost in all other canestriniids (Fig. 16.10A). Species in this genus are normally associated with large African carabid beetles of the genus *Tefflus*, where feeding stages occur in the subelytral space and deutonymphs congregate on the ventral thorax (Haitlinger 1990d; Lombardini 1939; B. M. OConnor, pers. obs.). All other canestriniids have abandoned deutonymphal dispersal and presumably transfer between hosts during host mating, since mites are never associated with host larvae. Other deutonymphs described as belonging to this family (e.g., Nesbitt 1976) were not positively correlated with other life stages and clearly belong to other groups. The feeding behavior of canestriniids has not been described, but most species retain chelate chelicerae that may be used to scrape exudates from the host cuticle. Some species commonly occur on or around host mouthparts and may feed on masticated or regurgitated material. *Vereoxia bogeri* Haitlinger lives on Asian Lucanidae and possesses extremely elongate, nonchelate chelicerae and elongate palpi that suggest a different feeding strategy (Haitlinger 1995b; B. M. OConnor, pers. obs.). The body and legs of *Passalophagus georgei* Nesbitt, associated with a passalid beetle in New Guinea, are extremely flattened, and the lateral body setae are large and fanlike, giving the entire mite a disclike appearance (Nesbitt 1976).

Old World canestriniid mites are commonly associated with beetles of the families Carabidae, Tenebrionidae, Scarabaeidae, Lucanidae, and Passalidae and less frequently with the Chrysomelidae, Cerambycidae, Erotylidae, and a few other host groups. Canestriniid mites have radiated extensively in the Old World from their African roots and have also reached the Neotropical region, where their greatest diversity is in association with the Chrysomelidae. A single, undescribed species occurs in southern Florida (USA), the only Nearctic representative of the family (B. M. OConnor, pers. obs.). Host specificity varies from strict to weak, with the same mite species sometimes occurring on different genera of hosts (Haitlinger 1992b, 1993a, b; Samšinák 1971).

The family HETEROCOPTIDAE is a comparatively small assemblage, with 13 genera and 29 species. Heterocoptids have a round body outline and possess well-defined sclerites dorsally in both adult and immature stages (Fig. 16.10C). Males may or may not retain paranal suckers and often have elaborate posterior projections of the idiosoma (Fig. 16.10D) or flattened posterior legs. The mating position in heterocoptids has not been described, but, based on the posterior modifications of the males, it is presumed to be reversed.

Heterocoptids occupy niches similar to those of canestriniids on Chrysomelidae, Erotylidae, and Passandridae in the Afrotropical and Oriental regions (Fain 1987b; Haitlinger 1996, 1999b, 2002b). Few New World species are known from Trogositidae and Passalidae, although the latter host association requires verification (Haitlinger 2001a; B. M. OConnor, pers. obs.). An undescribed species has been collected in Australia but without verified host associations

(B. M. OConnor, pers. obs.). Haitlinger (2002a) provides a key to genera.

The enigmatic family LEMANNIELLIDAE (Figs. 16.28A–C) includes one Holarctic genus and two species associated with ants (Wurst 2001). It presents a combination of ancestral and derived states that makes assignment to the Canestrinioidea (or any other recognized superfamily) problematic. The body of feeding stages is strongly ornamented with protuberances and lacks a sejugal furrow (Fig. 16.28A). Adults possess a large, tuberculate, and pitted median sclerite on the idiosoma, positionally reminiscent of early derivative HETEROCOPTIDAE. The ambulacral discs are expanded, but the empodial claws are well developed. Males are abbreviated posteriorly and lack paranal suckers, but they retain a thin-walled seta in the position of *ad1*. The mating position is reversed. Deutonymphs are flattened, with only the palpal solenidia retained on the highly reduced gnathosoma (Fig. 16.28C). The dorsal propodosoma is strongly abbreviated and bears only one pair of setae. Legs III–IV are reduced, have fused tibiotarsi, and lack pretarsi. The attachment organ is of the ancestral form.

Lemanniella minotauri Wurst is associated with the ant *Lasius brunneus* (Latreille) in central Europe, where feeding stages consume fungi in the nest (Wurst 2001). Deutonymphs of *L. reducta* Mahunka were described from *Myrmica sabuleti* Meinert in France, and an undescribed species has been collected from ants in western North America (B. M. OConnor, pers. obs.).

▪ Superfamily Hemisarcoptoidea (Figs. 16.11–16.16)

DIAGNOSIS: Adults with gnathosoma normally formed; chelicerae chelate-dentate, or rarely fixed digit reduced. Subcapitulum and palpi normally formed, rarely elongated. Idiosoma with cuticle smooth, mammillate, microtrichous, or occasionally strongly sclerotized. Propodosomatic sclerite shieldlike, sometimes elongate, shortened, or occasionally absent. Ocelli sometimes present on propodosoma. Sejugal furrow present or absent. Dorsal chaetome complete except setae le *always absent; paraproctal chaetome complete to strongly reduced. Coxisternal skeleton normally developed; chaetome complete or setae* g *and/or* 4b *modified or absent in males. Female with oviporus in form of inverted* V*, epigynal apodeme often present. Male without paranal suckers; often with modified legs or pretarsi I–II; legs IV strongly modified in some male ALGOPHAGIDAE; males polymorphic in some CHAETODACTYLIDAE. Mating position parallel. Leg chaetome always lacking* a'' *and usually* ft' *from tarsus I, other reductions common. Pretarsal ambulacra well developed, empodial claws normally developed, enlarged, or absent.*

Deutonymphs: Known in 5 of 7 families. Gnathosomatic remnant variously developed or absent. Idiosoma fusiform or round in outline, well sclerotized, typically with strong pattern of striations, occasionally smooth; propodosomatic ocelli often present; sejugal furrow usually present, rarely absent. Coxisternal skeleton normally developed or reduced in some (CARPOGLYPHIDAE); chaetome complete, or with some or all coxal setae reduced to alveolar vestiges. Attachment organ with ancestral complement of suckers and conoidal setae, strongly reduced in regressive forms. Legs normally developed, or legs IV reduced, lacking pretarsus and with very long tarsal setae; tibia-tarsus IV fused in some; chaetome complete except always lacking a'' *and* ft' *from tarsus I; setal reductions common on tarsus IV. Pretarsi with or without membranous ambulacrum; empodial claws normally present, rarely absent. Some taxa with morphologically regressive deutonymphs.*

Hemisarcoptoid mites are common inhabitants of various arboreal or aquatic habitats, with many species specialized as inquilines in nests of Hymenoptera. The genera *Hyadesia* and *Amhyadesia*, and 47 species of the family HYADESIIDAE (Figs. 16.13A, B) cataloged in Proches and Marshall (2001), inhabit marine intertidal to littoral habitats, occurring on virtually all coastlines and often associated with barnacles, algal mats, or kelp holdfasts (Pugh and King 1985a, b; Schuster 1979). In keeping with their saline environment, genital papillae of hyadesiids are quite reduced, and the empodial claws of the posterior legs are greatly enlarged for locomotion in their aquatic environment (Pugh, King, and Fordy 1987a). Hyadesiids are also unique in possessing a pair of propodosomatic lyrifissures (Fig. 16.13B). They feed primarily on marine algae, but other organic debris may also be consumed (Pugh and King 1985a). *Hyadesia fusca* (Lohmann), a common species in northern Europe and northeastern North America, has only two generations per year, and the tritonymph overwinters (Bücking 1999). Deutonymphs are never formed in the family, and dispersal is accomplished by feeding forms.

Members of the small family ALGOPHAGIDAE (seven genera, 18 species) (Fig. 16.16) also are aquatic, with various species inhabiting sap flows on trees (*Hericia*, *Fusohericia*), water-filled tree holes (*Lamingtonacarus*), rivers and lakes (*Algophagopsis*), or other aquatic or subaquatic habitats (Vitzthum 1931b; Türk and Türk 1957; Fain and Johnston 1975; Fashing, OConnor, and Kitching 2000). Tree-hole mites feed by skeletonizing leaves or scraping fungi from leaf surfaces, while *A. pneumatica* Fain and Johnston ingests primarily diatoms, which it crushes with massive chelicerae (Fashing 1998; B. M. OConnor, pers. obs.). An undescribed species of *Algophagopsis* was taken in great numbers from submerged rock surfaces covered with mosses (*Drepanocladus*), cyanobacteria (*Nostoc*), algae (*Cladophora*), lichens, and epiphytic diatoms in Crater Lake (Oregon, USA) at depths up to 100 m (Anderson and Krantz 1993). Species of *Algophagus* and *Neohyadesia* are generally restricted to subantarctic islands, where they range from coastal brackish pools to somewhat inland tussock grass soils, where they feed primarily on algae (Goddard 1982; Marshall, OConnor, and Pugh 2003). Some algophagid mites are soft bodied, but most show some degree of idiosomatic sclerotization. Most species also have relatively long legs and well-developed empodial claws, traits that are common in aquatic mites. As in the HYADESIIDAE, genital papillae are strongly reduced in algophagids, but the uniquely developed axillary organs between legs I–II that characterize the family (Figs. 16.16A, B) are likely osmoregulatory organs (Fashing 1984; Fashing and Marcuson 1996).

The posterior legs of male *Hericia* and *Fusohericia* are strongly modified for precopulatory mate guarding, with some tarsal setae modified to facilitate attachment (Fashing 2001) (Fig. 16.16B). These modifications are clearly convergent with those seen in more derived Astigmatina, since the modified setae are not homologous. The sap-inhabiting taxa of the algophagid subfamily Hericiinae form deutonymphs for dispersal on sap-frequenting Coleoptera (Nitidulidae, Nosodendridae) and Lepidoptera (Fashing 1991; Türk and Türk 1957). Both dispersing and sedentary deutonymphal morphs are known in *Hericia*, with the latter having a highly reduced attachment organ and unmodified setae *tc"* of tarsi I–II (Fashing 1991). Of the remaining algophagine genera, only *Lamingtonacarus oreillyorum* Fashing, OConnor, and Kitching is known to form deutonymphs, but these have not been collected from any insect host. The normal dispersal mode of algophagine species is unknown. A generic phylogeny was proposed by OConnor and Moser (1985).

The CHAETODACTYLIDAE (Fig. 16.14) includes three genera and 112 species of obligate nest associates of solitary or communally nesting bees. Feeding stages (known for relatively few species) have soft, globose bodies with a distinctly mammillate cuticle, lack a sejugal furrow, and have greatly enlarged empodial claws that allow for movement on host pollen/nectar stores. Phoretic deutonymphs bear 1–2 dorsal sclerites, and, in most species, pretarsi I–III bear enlarged, twisted empodial claws that may be used to attach to host setae (Fig. 16.14A).

Species of the chaetodactylid genus *Sennertia* occur in association with carpenter bees (Apidae: *Xylocopa* and *Ceratina*). Feeding stages of *S. cerambycina* (Scopoli) infest relatively few nests of the European *Xylocopa violacea* (L.) and few cells within infested nests. *S. cerambycina* is a kleptoparasite, destroying host eggs or larvae and feeding on pollen and nectar stores (Vicidomini 1996). In contrast, *S. splendidulae* Alzuet and Abrahamovich is a harmless commensal in nests of *Xylocopa splendidula* in Argentina (Abrahamovich and Alzuet 1990). Species of *Chaetodactylus*, an assemblage primarily associated with cavity-nesting hosts, are kleptoparasites of various species of megachilid and apid bees. *C. osmiae* (Dufour) in Europe and *C. krombeini* Baker in North America can cause economic losses in populations of mason bees (*Osmia* spp.) managed for crop pollination (Krombein 1962b; Fain 1966; Bosch 1992). A highly reduced, sedentary deutonymphal morph may remain in the nest cavity to reinfest subsequent host generations, while the phoretic morph disperses on surviving adult bees (Fain 1966). A generic phylogeny was proposed by OConnor (1993a).

Members of the two genera of MELIPONOCOPTIDAE (Figs. 16.13C, D) inhabit the nests of stingless bees (Apidae: Meliponini) in Brazil. Additional undescribed species have been collected in Thailand. Only the feeding stages are described, and their biology is unknown (Fain and Rosa 1983b; B. M. OConnor, pers. obs.). Adults are similar to those of the HEMISARCOPTIDAE in lacking empodial claws and a number of leg setae, but these mites are relatively well sclerotized, and males lack the midventral sucker characteristic of hemisarcoptids.

The best-known representative of the CARPOGLYPHIDAE is the dried-fruit mite, *Carpoglyphus lactis* (L.). This species is a cosmopolitan pest of stored materials with a high sugar content, such as dried fruit, powdered milk, and honey, as well as of other stored food substrates (Chmielewski 1971; Hughes 1976). *C. lactis* is a common natural inhabitant of honeybee hives and may also occur in sap flows (Baker and Delfinado 1978). Formation of deutonymphs is genetically determined, with some populations having lost the ability to disperse in this fashion (Knülle 1995). Deutonymphs have been collected from Lepidoptera and Coleoptera (Türk and Türk 1957). Four other species of *Carpoglyphus* have been described from composite flower inflorescences, vertebrate nests, and peridomestic areas (Fain and Rack 1987; Hubard and Fashing 1996).

The 11 genera and 31 described species of HEMISARCOPTIDAE represent a small fraction of this diverse assemblage. The family was found to be the most diverse group of Astigmatina in an old growth, mixed boreal forest in northern Michigan (USA), with most deutonymphs being nonspecific associates of cortical and subcortical insects (OConnor and Houck 1989; OConnor 1991). Species of *Nanacarus* and *Nanacaroides* often inhabit the spore tubes of polypore fungi (Pielou and Varma 1968; Volgin and Mironov 1979), and *Nanacarus minutus* Oudemans is a common inhabitant of house dust and stored food products. *Linobia coccinellae* (Scopoli) is a rare astigmatine example of a hematophagous parasite. Its host is a European chrysomelid beetle, *Chrysomela populi* (L.) (Haitlinger 1999a). Feeding stages of *L. coccinellae* lack a fixed cheliceral digit and use the movable digit to pierce the dorsal intersegmental membranes of the host's abdomen (B. M. OConnor, pers. obs.).

Species of *Hemisarcoptes* have a unique, doubly parasitic ecology. Feeding stages are obligate parasites of diaspidid scale insects in most parts of the world but are nonspecific as to host; more than one adult mite feeding on a host typically results in mortality. Some species have been employed in biological control efforts directed against several scale-insect pests, with mixed results (Gerson, OConnor, and Houke 1990). The association between *Hemisarcoptes* deutonymphs and coccinellid beetles of the genus *Chilocorus* is highly unusual. Originally assumed to be a simple phoretic association, it has been demonstrated that some deutonymphs actually are parasitic. Detailed studies have shown that individuals of *H. cooremani* (Thomas) produce holes in the cuticle of the underside of the beetle's elytra and take host hemolymph into the mite hindgut through the anus (Houck and Cohen 1995). Parasitism of a host beetle was necessary for deutonymphs to continue development in laboratory colonies (Houck 1994).

The 25 genera and 138 species of WINTERSCHMIDTIIDAE (= Saproglyphidae) (Fig. 16.11) make it the largest family in the Hemisarcoptoidea. Winterschmidtiids display a wide range of ecological associations. Species of *Saproglyphus* are fungivores in decaying materials, bark crevices, and woody fungi (Türk and Türk 1957; Volgin 1980), and deutonymphs are dispersed by a wide variety of insect hosts. Other species are common fungivores on leaf surfaces; *Oulenzia* species are common in warmer regions while *Czenspinskia transversostriata* (Oudemans) (= *Calvolia lordi* Nesbitt) (Fig. 16.11C) is virtually cosmopolitan. The latter species is unusual in being thelytokus and in not forming dispersing deutonymphs. In temperate regions, the tritonymph is specialized as a resistant overwintering stage (Dosse and Schneider 1957). Species of *Winterschmidtia* are also unusual in being predators of the eggs of bark beetles (Scolytidae). *W. nataliae* (Zakhvatkin) (= *Calvolia fraxini* Türk and Türk) undergoes two generations/year in galleries of the European ash beetle, *Leperisinus fraxini*, with overwintering deutonymphs produced only in the second generation (Kielczewski and Seniczak 1972; Khaustov 2000b). Related taxa are known from deutonymphs associated with various families of subcortical Coleoptera. Other winterschmidtiid genera are found in various types of decaying material such as dung, carrion, and beach wrack and in vertebrate nests. *Procalvolia zacheri* (Oudemans) and *Acalvolia squamata* (Oudemans) occasionally infest stored food products (Rack 1980; Fain and Knülle 1981), and a mite resembling *P. zacheri* was associated with severe dermatitis in a horse (Carelle et al. 2000). Most of these mites are generalist phoretics, while some associations are more specialized, such as those between nidicolous species of *Psylloglyphus* and fleas (Fain and Beaucournu 1993).

The largest lineage of WINTERSCHMIDTIIDAE, the subfamily Ensliniellinae, includes obligate associates of Hymenoptera. Species of *Vidia* are associated with leafcutter bees of the genus *Megachile*, where they feed on fungi growing on the leaf pieces lining the host nest cells (OConnor and Eickwort 1988). Other taxa are associated with sphecid wasps (*Crabrovidia* and one unnamed genus), but most species are specific associates of Vespidae. *Sphexicozela* species are found in nests of social *Polistes* species in the Holarctic region, with deutonymphs dispersing in the genital capsule of reproductives (Mahunka 1970e; B. M. OConnor, pers. obs.). Other genera are associated with solitary wasps of the subfamilies Masarinae (pollen wasps) and Eumeninae. A partial list of associations and a phylogeny for the subfamily was provided by OConnor and Klompen (1999). Although few life cycles are known, some species such as *Monobiacarus quadridens* Baker and *Kurosaia jiju* OConnor and Okabe feed on hemolymph of paralyzed arthropod prey provisioned by the female wasp. Other species, in the genera *Ensliniella*, *Kennethiella*, and *Vespacarus* feed directly on hemolymph of the wasp host larva or pupa (Cooper 1955; Krombein 1967; Klompen, Lukoschus, and OConnor 1987; Okabe and Shun'ichi 2003), but no host mortality has been associated with this type of parasitism. The latter three genera are unusual in having two male morphs—a large type that develops ultimately from founding deutonymphs in the host cell, and a small type that quickly develops parthenogenetically from females in the same cell (Klompen, Lukoschus, and OConnor 1987). Many female (or rarely male) wasp hosts have more or less elaborated acarinaria in different locations on the body that harbor dispersing deutonymphs. Based on their phylogeny, OConnor and Klompen (1999) concluded that acarinaria have evolved independently in many discrete host groups.

▪ Superfamily Glycyphagoidea (Figs. 16.17–16.23)

DIAGNOSIS: Adults with gnathosoma normally formed, chelicerae chelate-dentate. Subcapitulum and palpi normally formed; palpal supracoxal seta always elongate and strongly barbed; palptarsus often with dorsal and ventral eupathidial setae somewhat elongate. Ventral subcapitular ridges prominent in GLYCYPHAGIDAE. Idiosoma with cuticle typically microtrichous, sometimes with striations, scales, or irregular protuberances, rarely smooth or entirely sclerotized. Prodorsal sclerotization in the form of paired parallel sclerites, single "crista metopica" or widened shield. Sejugal furrow present in early derivative groups, absent in most. Idiosomatic chaetome complete, or setae le *absent; paraproctal setation complete or reduced; dorsal setae often very elongate and strongly barbed, sometimes flattened, spinelike, or rarely simply filiform. Coxisternal skeleton normally developed or apodemes III and/or IV reduced, chaetome complete. Female oviporus in the form of an inverted V; epigynal apodeme often present; copulatory opening terminal, often with external copulatory tube. Male with (CHORTOGLYPHIDAE) or without (other families) paranal suckers; other secondary sexual characters variable. Mating position reversed in early derivative taxa, parallel in GLYCYPHAGIDAE, AEROGLYPHIDAE, and ROSENSTEINIIDAE. Legs, especially tarsi, typically elongate, sometimes relatively short. Chaetome complete except* a" *always absent from tarsus I, unguinal and sometimes proral setae of tarsi reduced or absent. Some leg setae often strongly barbed. Pretarsi typically reduced, with membranous ambulacrum, empodial claws small or absent.*

*Deutonymphs: Known in 5 of 7 families. Wide variation in body form. Early derivative groups (EUGLYCYPHAGIDAE) with typical insecticolous form, attachment organ with ancestral set of suckers and conoidal setae. Other families with body specialized for mammal association; all with anterior and central suckers (*ad3, ad1+2*) reduced or absent. Some GLYCYPHAGIDAE with attachment organ setae* p1–p2 *elongate and strongly ridged, and with posterior idiosomatic flaps for holding hair; other GLYCYPHAGIDAE, PEDETOPODIDAE, CHORTO-*

GLYPHIDAE, and ECHIMYOPODIDAE with deutonymphs parasitic in hair follicles or phoretic in host ears, these with most or all attachment organ structures vestigial or absent. Some with modified setae on tibiae or tarsi II–IV. Morphologically regressive forms known in GLYCYPHAGIDAE and CHORTOGLYPHIDAE.

Species in the Glycyphagoidea are inhabitants of vertebrate nests ancestrally and share a suite of characters common to early derivative nidicolous taxa in the Acaroidea and Hypoderatoidea that suggest a common ancestry. These include a generally soft, rounded body; long, barbed idiosomatic setae that may be defensive in function; long legs; and prodorsal sclerotization in the form of paired parallel sclerites. Traditional classifications (e.g., Zakhvatkin 1941; Fain 1986) included most taxa in a broadly conceived family GLYCYPHAGIDAE along with a few groups that belong elsewhere, but some recognized CHORTOGLYPHIDAE and ROSENSTEINIIDAE as distinct families. Based on phylogenetic analysis, OConnor (1982a) recognized several groups previously ranked as subfamilies as being distinct families. Deutonymphs have evolved quite different associations, beginning with typical insect phoresy and progressing theoretically through mammal phoresy to endofollicular parasitism in several different groups. Many glycyphagoid species from several families have moved from their ancestral nest habitats to colonize human habitations, where they form an important component of the acarofauna of house dust and stored food products (Hughes 1976; OConnor 1979).

Euglycyphagus intercalatus Fain and Philips (EUGLYCYPHAGIDAE) (Figs. 16.17A–D) inhabits the nests of falconiform birds in North America. Adults are typically glycyphagoid, but the anus is anteriorly displaced to a position directly behind the genital region. Deutonymphs retain well-developed suckers and small conoidal setae on the attach ment organ. Those of *E. intercalatus* are phoretic on trogid beetles that inhabit raptor nests (Fain and Philips 1977). Adults of the second genus, *Lomelacarus*, share the characteristic of an anteriorly displaced anus, but the dorsal setae are short. The two described species of *Lomelacarus* are known only from house dust and livestock enclosures in Africa and China. Deutonymphs are unknown in this genus.

The single species in the family PEDETOPODIDAE, *Pedetopus zumpti* Fain (Figs. 16.17E, F), is known only from deutonymphs parasitic in the hair follicles of *Pedetes capensis* (Forster), a relictual rodent restricted to southern Africa. Unlike many parasitic deutonymphs, the attachment organ is retained (although small) and bears remnants of the suckers and conoidal setae characteristic of the ancestral condition (Fain 1969a).

Members of the family CHORTOGLYPHIDAE (Fig. 16.18) are naturally associated with the nests of sciurid, aplodontid, and murid rodents in North and Central America (*Chortoglyphus* and *Aplodontopus*) and with sciurid and murid rodents in Australasia and a mouse lemur in Madagascar (*Alabidopus*). Adult chortoglyphids have highly reduced coxal apodemes III–IV, and the males are the only glycyphagoids to possess paranal suckers. Some, but not all, males have modified suckerlike or spinelike setae on tarsus IV, as in males of more derivative superfamilies. Deutonymphs of all genera are endofollicular parasites and possess modified, comblike setae *v′* on tibiae III–IV that engage the lining of the host hair follicle to provide anchorage. Dorsal setae are mostly spinelike, bifurcate apically, and anteriorly directed, which also serves to maintain the mite's position in the follicle. Unlike superficially similar parasitic deutonymphs in the family GLYCYPHAGIDAE, chortoglyphids retain all coxal setae, and setae *p1* and *p2* of the attachment organ take the form of tiny spines at the opisthosomatic terminus. Growth of some deutonymphs is extreme, with body length increasing 10-fold during engorgement. *Alabidopus asiaticus* Lukoschus et al. forms both well-developed parasitic deutonymphs and highly regressive, sedentary forms (Lukoschus et al. 1981). Deutonymphs of *Aplodontopus marmotae* Krantz et al. were reported to cause severe dermatitis in their marmot host (Krantz et al. 2003). A single species, *Chortoglyphus arcuatus* (Troupeau) (Figs. 16.18A, B), has adapted to human associations and commonly infests stored grain and flour throughout the world (Hughes 1976).

The family ECHIMYOPODIDAE (Fig. 16.19) is a small assemblage (five genera, 31 species) whose members occur naturally in North and South America and Australia, where species are associated with marsupials, rodents, and one xenarthran. Adults are soft bodied; bear long, barbed dorsal setae; and have regressive coxal apodemes III (Fig. 16.19A). Males of the native New World genera *Echimyopus* and *Blomia* have highly modified legs IV for maintaining position during mating (Bronswijk and Cock 1973). Deutonymphs are endofollicular parasites characterized by a complete lack of attachment organ structures and a circular apodeme surrounding a primordial genital region (Figs. 16.19C, D). Deutonymphs of the Australian genus *Marsupiopus* are somewhat elongate and lack obvious modifications for maintaining position in the follicle, while New World deutonymphs have strongly hooked empodial claws that articulate with the tarsal apex for grasping host tissue (Fain 1969a). Those of *Echimyopus dasypus* Fain et al. are unusual in occurring in the subcutaneous tissues of their armadillo host, where they cause tissue damage and hyperkeratosis (Fain, Lukoschus, and Nadchatram 1981). Deutonymphs of *Marsupiopus zyzomys* Lukoschus et al. are pathogenic, causing tissue necrosis leading to progressive loss of the host's tail (Lukoschus, Janssen Duijghuijsen, and Fain 1979). Species of the genus *Blomia* are known only from postdeutonymphal stages occurring in human habitations and stored food products in many parts of the world (Bronswijk and Cock 1973). *B. tropicalis* is the most common house dust mite in South America and is an important agent of house dust allergy (Rosa and Flechtmann 1979; Caraballo 1999).

Species in the closely related families AEROGLYPHI-

DAE (Figs. 16.20A–C) and ROSENSTEINIIDAE (Figs. 16.20D–F, 16.21) typically have body cuticle that is at least partially striated or covered in scalelike protrusions dorsally. They lack tectal setae on tarsus IV and do not form deutonymphs for dispersal. Species in both families are commonly associated with bat colonies, where rosensteiniids are often the dominant element of a diverse mite community in guano. Many of the 19 known species of AEROGLYPHIDAE are associated with guano, but some have been collected from bird, mammal, and bee nests or peridomestic habitats. In contrast with earlier derivative glycyphagoid families, where the mating position is reversed, aeroglyphids share the parallel mating position with the ROSENSTEINIIDAE and GLYCYPHAGIDAE. The genus *Aeroglyphus* is Holarctic and Afrotropical in distribution, while *Glycycometus* (= *Austroglycyphagus*) is cosmopolitan. *Glycycometus* species are unusual in having highly attenuated tarsi, with the ventral setae *pv'* basally positioned and elongate and wrapping around the length of the tarsus, a condition convergent with some true GLYCYPHAGIDAE. *Aeroglyphus robustus* (Banks) is a common species in stored food products, where its ability to tolerate relatively lower humidity conditions allows it to survive when other mite pests cannot (Sinha 1966). Several species of *Glycycometus* have been collected from house dust in tropical areas and may be responsible for allergic reactions in humans (Miranda, Quintero, and Almanza 2002).

Despite being ancestrally restricted to bat guano, the 15 genera and 46 species of ROSENSTEINIIDAE have diversified into different morphological forms and have adopted new types of associations. Dorsal hysterosomatic setae are often flattened or strongly modified as fans, barbed blades, or strong spines and, curiously, the setae of genua I–II are often modified in the same fashion. Males of *Chiroptoglyphus americanus* OConnor and Reisen have greatly enlarged legs I that grasp the base of the female's anterior leg during mating. Strong size variation in leg development is allometrically related to body size (OConnor and Reisen 1978). *Micronychites postverrucosus* Fain and Lukoschus exhibits a similar allometrically related enlargement but of legs II (Fain and Lukoschus 1983a). Males in other genera show little secondary sexual dimorphism.

Although most rosensteiniids are guano inhabitants, Neotropical species of *Rosensteinia* (Fig. 16.21C) and an unnamed Afrotropical genus are external commensals of large cockroaches. These mites are strongly convergent with many CANESTRINIIDAE in having flattened bodies, laterally positioned legs, and greatly enlarged ambulacral discs (Haitlinger 1999e). Species of several other genera have been similarly collected from the bodies of specialized Dermaptera (suborder Arixenina) associated with the molossid bat, *Cheiromeles torquatus*, in Southeast Asia (Fain, Lukoschus, and Nadchatram 1981; Fain and Lukoschus 1983a). The same bat is host to *Cheiromelichus trochanteralis* Fain, Lukoschus, and Nadchatram, the eggs of which also were found attached to the skin of the bat host. *C. trochanteralis* may be parasitic on its host (Fain, Lukoschus, and Nadchatram 1981).

Troglotacarus hauseri Fain, described from bat guano in Kenya (Fain 1977a), and *Lophonotacarus minutus* Fain, described from millipedes in South Africa (Fain 1987a), were originally described as the sole members of the families Troglotacaridae and Lophonotacaridae. Both species have small, rounded, or elongate and well-sclerotized bodies, and the pretarsal ambulacral discs are small in *T. hauseri*. However, both possess the anteriorly positioned genital region, dorsally shifted supracoxal gland openings, and modified genual setae characteristic of many ROSENSTEINIIDAE, and they are treated here as part of that family.

The family GLYCYPHAGIDAE (Figs. 16.22, 16.23) is the largest in the superfamily, comprising 41 genera and 192 species. Glycyphagids generally are associated with the nests of rodents, insectivores, and New World didelphimorph marsupials, but a few are associated with species in other mammalian orders. The family exhibits a diversity of body and setal forms, but virtually all bear a pattern of external ridges on the venter of the subcapitulum (the strongly sclerotized *Fusacarus* is exceptional in having all cuticular surfaces smooth). Most glycyphagids are soft bodied, with prodorsal sclerotization either shieldlike, reduced to a median crista metopica, or absent. These species generally retain the long, barbed dorsal idiosomatic setae, the cuticle is covered with pointed microtrichia or lobate protrusions, and the tarsi generally are elongate. Species in the subgenus *Glycyphagus* (*Lepidoglyphus*) have attenuate tarsi with a supporting scalelike *pv'* seta, as in the aeroglyphid genus *Glycycometus*.

While some glycyphagid taxa retain the same general morphology throughout the feeding stages, others exhibit a striking metamorphosis at the last molt. For example, adults of species of *Xenoryctes*, *Gohieria*, *Lophuromyopus*, *Ctenoglyphus*, and others become more heavily sclerotized, the idiosomatic setae change form, and the legs may develop strong cuticular ridges. *Xenoryctes* species develop propodosomatic costulae reminiscent of those in more derived oribatid families (Fain 1969b). Males are typically smaller than females and show a variety of secondary modifications for grasping the female during mating. Some *Glycyphagus* species have modified, comblike anterior tibial setae that lock around the bases of the female's dorsal setae (Volgin 1961; Fain and Lukoschus 1974), while males of *Baloghella melis* Mahunka (= *Melesodectes auricularis*) have greatly enlarged legs II for clasping the legs of the female or for precopulatory guarding of female tritonymphs (Lukoschus, de Cock, and Fain 1971; Wurst and Pfister 1990). Males in the glycyphagid subfamily Labidophorinae have enlarged ambulacral stalks on pretarsi I–II and have enlarged legs III–IV that clasp the underside of the female's body (Fain 1969b). Species of *Ctenoglyphus* and *Rodentopus* have strongly divergent forms of the dorsal idiosomatic setae in the two sexes.

Deutonymphs in the subfamilies Marsupialichinae, Glycyphaginae, Labidophorinae, and Fusacarinae (here including *Xenoryctes*) attach to a hair shaft of their mammalian host by means of a clasping organ comprising the elongate, ridged *p1* and *p2* setae, large lateroventral flaps of the body, and the posterior two pairs of legs (Figs. 16.3C–E, 16.23A). Most species attach at the base of the hair, but species of Glycyphaginae attach farther up on the hair shaft. Different species in the subfamily Metalabidophorinae show a transition to endofollicular parasitism, with *Paralabidophorus* species attaching at the base of the hair as simple phoretics, *Metalabidophorus* species embedding partially within the follicle, and species of *Neolabidophorus* and several other genera becoming fully endofollicular and undergoing significant growth (Fain 1969a; Fain and Lukoschus 1978b). All metalabidophorine deutonymphs retain the attachment organ structures, although the organ is small in the fully endofollicular parasites. Lophuromyopodine (Fig. 16.23B) and ctenoglyphine deutonymphs are endofollicular parasites with totally regressed attachment organs and modified setae on tibiae III–IV, like those of the CHORTOGLYPHIDAE (Fain 1969a).

Lophioglyphus liciosus Volgin (= *Apodemopus apodemi*) causes host tissue degeneration around the mite in the follicles of the European wood mouse, *Apodemus sylvaticus* (Lukoschus, Fain, and Driessen 1972). *Baloghella melis* Mahunka, the sole species in the subfamily Melesodectinae, has all attachment organ elements regressed but apparently is not parasitic. Its deutonymphs occur in waxy exudates in the ears of its host, the European badger, *Meles meles*. Deutonymphal polymorphism is known in the genera *Baloghella* and *Glycyphagus*, where both phoretic and sedentary morphs are formed (Wurst and Pfister 1990; Knülle 2003). Regressive deutonymphs of *Glycyphagus domesticus* (DeGeer) and *G. destructor* (Schrank) remain within the protonymphal cuticle and use airborne dispersal methods (Hughes 1976).

Little is known of the natural ecology of nidicolous GLYCYPHAGIDAE. Host specificity varies from strict to very loose, with some species collected from the nests of birds as well as mammals (Fain 1968a), and some species of *Glycyphagus* collected from foliage or litter (Luxton 1995; B. M. OConnor, pers. obs.). Best known are the numerous species that invade peridomestic habitats, stored food products, and house dust. Species in several lineages ancestrally associated with rodents have colonized grain stores and habitations (Hughes 1976; OConnor 1979). *G.* (*Lepidoglyphus*) *destructor* (Schrank) is one of the most common and widespread stored product mites, while *G. domesticus* has been reported from house dust in many countries. Other widespread synanthropic glycyphagids are *Gohieria fusca* (Oudemans), *Ctenoglyphus plumiger* (Koch), and *Diamesoglyphus intermedius* (Canestrini), the latter not yet reported outside the Palearctic region. Several of these species, especially those with long body setae, have been implicated as causal agents of topical dermatitis ("grocer's itch") in individuals handling stored grain and straw (Hughes 1976).

■ Superfamily Acaroidea (Figs. 16.24–16.27, 16.28A–C, 16.29, 16.30)

DIAGNOSIS: Adults with gnathosoma normally formed; chelicerae chelate-dentate. Subcapitulum and palpi normally formed. Idiosoma with cuticle smooth, mammillate, tuberculate, rarely entirely sclerotized. Propodosomatic sclerotization typically shieldlike, rarely in the form of two parallel sclerites, occasionally absent. Sejugal furrow usually present, sometimes absent. Dorsal and paraproctal chaetome complete or strongly reduced. Coxisternal skeleton normally developed with addition of paired apodematic elements in ventral sejugal region; chaetome ancestrally complete. Female oviporus in the form of an inverted V*; internally with strongly folded pseudovipositor; epigynal apodeme rarely apparent. Male normally with paranal suckers, occasionally reduced or absent; with modified, suckerlike setae on tarsus IV; sometimes with legs III variously enlarged, or rarely legs I enlarged. Mating position reversed. Legs normally developed; chaetome complete to strongly reduced. Pretarsi with or without membranous ambulacrum, empodial claws usually present, rarely absent.*

Deutonymphs: Known in 5 of 8 families. Gnathosomatic remnant variously developed or absent. Idiosoma typically ovoid to rounded in outline, usually well sclerotized, smooth or with pattern of striations or punctations dorsally; propodosomatic ocelli rarely present (some ACARIDAE). Sejugal furrow always present. Coxisternal skeleton normally developed or strongly reduced and without posterior median apodeme; chaetome ancestrally complete or some or all coxal setae absent. Attachment organ well developed, with all structures present. Legs normally developed, or legs IV reduced and lacking pretarsus; chaetome complete or some setae absent. Pretarsi with or without membranous ambulacrum; empodial claws normally present, occasionally absent (some SUIDASIIDAE). Some taxa with morphologically regressive deutonymphs.

The Acaroidea is a diverse lineage united by the presence of ventral sejugal apodemes in feeding stages and a strongly folded pseudovipositor that the female uses for egg deposition. Males retain paranal suckers in most taxa and share the modification of setae *ft''* and *tc''* as suckers on tarsus IV with the remaining, more derived groups. Deutonymphs are typical insect phoretics, generally retaining the ancestral form of the attachment organ. Ecological association with the nests of vertebrates or Hymenoptera is common in basal lineages, but the large family ACARIDAE has radiated into many habitats. Earlier taxonomic treatments included taxa here treated in the families LARDOGLYPHIDAE, SCATOGLYPHIDAE, and SUIDASIIDAE within an expanded, paraphyletic concept of ACARIDAE (Zakhvatkin 1941; Hughes 1976).

The monobasic families GLYCACARIDAE (Figs. 16.27A, B) and SCATOGLYPHIDAE (Figs. 16.27C, D) are associated with bird nests. *Glycacarus combinatus* Griffiths was collected in the nest burrow of a seabird, *Procellaria aequinoctialis* on subantarctic islands (Griffiths 1977), while *Scatoglyphus polytrematus* Berlese has been found in poultry litter (Zakhvatkin 1941) and in the nest of an owl (B. M. OConnor, pers. obs.). Both are described only from feeding stages. The mating position has not been described for either family.

Species in the small family LARDOGLYPHIDAE (Fig. 16.26) are best known as pests of stored meat and seafood products (Hughes 1976) but are also reported from other types of stored food products, bird nests (Philips and Norton 1978; Asmat and Kader 1990), and vertebrate carrion, including human mummies (Early and Goff 1986; Baker 1990b; Iverson et al. 1996). Legs III of males are enlarged, the tarsus terminates in a spinelike process, seta *s* is spinelike and may be fused to the segment, and the pretarsus is lacking (Fig. 16.26B). Unlike some male ACARIDAE, there is no polymorphic enlargement of legs III in lardoglyphid males. Deutonymphs of *Lardoglyphus* are normally associated with beetles of the genus *Dermestes* and with their nest and carrion habitats. Deutonymphs commonly attach to the intersegmental membranes of the abdominal dorsum of adult beetles and appear swollen before detachment. Most deutonymphs of *L. konoi* Sasa and Asanuma in laboratory culture required attachment to a beetle host prior to "swelling" and molting, suggesting a possible parasitic relationship (Okamoto, Matsumoto, and Shirasaka 1991). Deutonymphs also attach to larval *Dermestes* and may kill larvae and cause decline in museum colonies of *Dermestes* used to clean skeletons (Iverson et al. 1996). *Lardoglyphus zacheri* Oudemans and *L. konoi* can be serious pests of dried meat and fish products, respectively.

The GAUDIELLIDAE (including Partamonacoptidae and Platyglyphidae) (Fig. 16.24) inhabit the nests of bumblebees and meliponine stingless bees. Adults have a circular body outline and lack a sejugal furrow, and most have at least some stout, strongly barbed dorsal setae. (*Platyglyphus* is exceptional in having all short, simple setae.) *Cerophagus* and *Partamonacoptes* have a strongly mammillate cuticle, while *Gaudiella* (= *Trigonacoptes*) and *Platyglyphus* are smooth. Sexual dimorphism is minimal, and the mating position is undescribed (Atyeo, Baker, and Delfinado 1974; Kurosa 1976a; Fain and Rosa 1983a; OConnor 1992). Deutonymphs are known only for the genus *Cerophagus* (Figs. 16.24A, B), and these are associated with *Bombus* in the Holarctic region. Genera associated with meliponine bees are known from the Neotropical and Oriental regions. Nothing is known of their ecology.

The 18 known species of the family SUIDASIIDAE (including Sapracaridae) (Fig. 16.25) may be encountered in a variety of habitats. Most are similar to lardoglyphids in possessing clawlike proral setae on the tarsi and lacking a pretarsus IV in the deutonymph, but feeding stages differ in possessing a mammillate cuticle (smooth in some *Sapracarus*), in retaining the sejugal furrow, and in exhibiting little sexual dimorphism. Deutonymphs are known only in the genera *Tortonia* and *Sapracarus*. Species of *Suidasia*, *Neosuidasia*, and *Sinosuidasia* are best known from stored food products and house dust. *Suidasia pontifica* Oudemans (= *S. medanensis*) is restricted to warmer climates, while *S. nesbitti* Hughes is more widespread (Hughes 1976; Fain and Philips 1978b; Balmes-Pacia and Corpus-Raros 1998). *Tortonia* species are restricted to the nests of solitary Hymenoptera, where they may be kleptoparasites (Krombein 1962a, 1967; Qu et al. 2003). Species of *Namibacarus* and *Sapracarus* are among the smallest astigmatine mites and are unusual in inhabiting deep soil environments (Fain et al. 1993). *S. tuberculatus* Fain and Philips has also been collected in bird nests, bat roosts, and peridomestic habitats (Fain and Philips 1978a; B. M. OConnor, pers. obs.).

The family CHETOCHELACARIDAE (Figs. 16.28D–F) was proposed for a single female specimen of *Chetochelacarus mamillatus* Fain, collected from a millipede in Congo (Fain 1987a). The family has tentatively been placed here in the Acaroidea. *C. mamillatus* shares similarities with the female of *Diplopodocoptes transkeiensis* Fain, also described from millipedes in Africa and originally assigned to the CANESTRINIIDAE, but here placed with *C. mamillatus*. The male of *D. transkeiensis* (Fig. 16.28E) is strongly bilobed posteriorly, and legs IV are enlarged, with tarsus IV bearing suckerlike setae *ft"* and *tc"*. Adults and tritonymphs of *D. transkeiensis* were collected from several species of millipede hosts (Fain 1987a). Their modified body form and enlarged pretarsal ambulacra suggest that they may be permanent paraphages. Deutonymphs are unknown.

With 80 genera and over 500 species, the family ACARIDAE (Figs. 16.29, 16.30) is one of the largest in the Astigmatina. The family is relatively conservative morphologically, but some taxa are highly divergent. The body is generally soft and smooth, but some are completely sclerotized (*Stereoglyphus*, some *Schwiebea*) or are ornamented with cuticular tubercles (*Fagacarus*). Propodosomatic sclerotization is shieldlike in soft-bodied species but indistinct in sclerotized forms. Idiosomatic setae may be long or short, simple or weakly barbed (strongly barbed in *Fagacarus*). The idiosomatic chaetome is ancestrally complete, but regression of many dorsal and paraproctal setae occurs in derived taxa (e.g., *Schwiebea*). Legs range from very short in burrowing forms (*Schwiebea*) to quite long in aquatic species (*Naiadacarus*). A phylogenetic framework for the family was presented by OConnor (2001).

The ACARIDAE is ecologically diverse, with early derivative taxa (*Acarus* and related genera) associated with vertebrate nests. *Acarus* deutonymphs are ancestrally phoretic on fleas (Fain and Beaucournu 1993), but some *Acarus* species are best known as serious pests of stored food products. *A. siro* L. is the most damaging of all described species, infesting many types of grains, flour, and cheese and feeding on both the substrate and associated fungi (Griffiths 1964, 1970; Hughes 1976). Species in the subfamily Horstiinae inhabit the nests of solitary bees, in which some may be kleptoparasites (Krombein 1962a; OConnor 2001). Other horstiine species are nidicoles of social insects, especially of ants and termites, while species of *Kuzinia* are obligate associates of bumblebees and feed on pollen in the nest (Chmielewski 1969; Volgin and Kadzhaya 1973). The genus *Tyrophagus* is exceptional in that, while the early derivative species *T. formicitorum* Volgin remains in ant nests and forms dispersing deutonymphs (Fain and Chmielewski 1987), other *Tyrophagus* species colonize a wide variety of habitats and have abandoned deutonymphal dispersal. *T. putrescentiae* (Schrank), perhaps the most ubiquitous astigmatine mite, lives in many peridomestic situations and also on plant tissue and is a common pest in laboratory insect and fungus cultures (Robertson 1959; Hughes 1976). Other species such as *T. longior* (Gervais) and *T. similis* Volgin inhabit grassland litter and contaminate hay and straw (Samšinák 1962), while *T. neiswanderi* Johnston and Bruce feeds directly on plant tissue and may cause damage in greenhouse cultivation (Czajkowska and Kropczynska 1991). *T. putrescentiae*, *T. similis*, and *T. zachvatkini* Volgin have been observed to feed voraciously on nematodes in laboratory cultures, and the latter two species were successfully reared on a diet composed exclusively of nematodes (Walter, Hudgens, and Freckman 1986; Walter, Hunt, and Elliot 1988).

More highly derived acarid taxa colonize moister habitats than do early derivative forms and use a variety of insects for dispersal. Species of *Sancassania* (= *Caloglyphus*) occur in manure, compost, and decaying fungi and often use scarabaeoid beetles for dispersal (Klimov and OConnor 2003; Klimov et al. 2004). Species of *Rhizoglyphus* inhabit bulbs, tubers, and other plant tissues; *R. robini* Claparède and *R. echinopus* (Fumouze and Robin) are widespread pests in commercial cultivation of tulips, onions, and other root crops (Diaz et al. 2000; Fan and Zhang 2004). Species in other genera are common inhabitants of decaying wood (*Schwiebea*, *Histiogaster*, *Thyreophagus*, *Reckiacarus*, *Fagacarus*), polypore fungi (*Boletoglyphus*), or phytotelmata (*Naiadacarus*). Perhaps most unusual are highly morphologically modified acarids that live as commensals of crustaceans. These species, formerly placed in a separate family Ewingiidae, were transferred to the ACARIDAE by OConnor (1982a) because they share the diagnostic features of the family. Species of *Ewingia* and *Hoogstraalacarus* live on the gills of hermit crabs (Coenobitidae) and use highly modified legs III–IV to attach to their crab hosts (Yunker 1970a). *Askinasia* species use differently modified posterior legs to attach to setae near the back of the carapace of the same hosts, while *Kanekobia potamona* (Kaneko and Kadosaka) attaches to the gills of its freshwater crab host using only modified legs IV (Kaneko and Kadosaka 1978; Fain, Lukoschus, and Rosmalen 1982). All of these species have similarly modified chelicerae, which are flattened and may serve to filter particles from water currents in and around the hosts' gill chambers.

▪ Superfamily Hypoderatoidea (Fig. 16.31)

DIAGNOSIS: Adults with gnathosoma normally formed, or male chelicerae, subcapitulum and palpi greatly enlarged and female gnathosomatic structures vestigial. Idiosoma with cuticle smooth, scaly, or strongly sclerotized, all or partially striate in homeomorphic immatures. Propodosomatic sclerotization in the form of parallel sclerites, or propodosoma without distinct sclerite. Sejugal furrow present or absent. Hysterosoma often with a system of cuticular grooves connecting idiosomatic cupules on each side; dorsal chaetome complete; paraproctal chaetome lacking two pairs of ad *setae. Coxisternal skeleton normally developed; chaetome complete. Female with oviporus in the form of an inverted* V*, epigynal apodeme large, in the form of an inverted* U*, or very reduced or absent. Male typically with posterior opisthosoma bilobed, paranal suckers well developed; with modified, suckerlike setae* ft" *and* tc" *on tarsus IV; legs III and/or IV often with enlarged, spinelike setae. Mating position reversed. Legs normally developed, often elongate; chaetome complete, tarsus I with solenidion ω2 distally displaced. Pretarsi with well-developed, elongate ambulacra; empodial claws absent.*

Deutonymphs: highly modified for subcutaneous parasitism of birds or (rarely) rodents. Gnathosomatic remnant absent; idiosoma rounded or fusiform, moderately sclerotized to poorly sclerotized and capable of great enlargement in many taxa; coxisternal skeleton weakly developed, capable of neosomic growth in some; chaetome reduced. Attachment organ completely absent; legs normally developed; chaetome complete or with some regressions; pretarsi with small membranous ambulacra and well-developed empodial claws; pretarsus IV absent.

The 80 known species of the family HYPODERATIDAE are obligate vertebrate nidicoles, the majority of which are known only from the parasitic deutonymphal stage. Most deutonymphs are found in the subcutaneous tissues of altricial birds, but species of *Acotylopus* (= *Muridectes*) and *Dipodomydectes* have been collected from several different groups of desert-adapted rodents (Fain 1967b, 1969a; Fain and Lukoschus 1978a). Some hypoderatids such as *Tytodectes strigis* (Gené) have a normal ontogeny, with all stages active and feeding (Wurst and Havelka 1997), while others exhibit morphological regression of immature stages. Larvae and protonymphs of *T. falconis* Fain have regressive chelicerae, and legs IV of the protonymph are strongly reduced (Wurst and Havelka 1998). Species of *Hypodectes* and *Phalacrodectes* are most extreme in having completely calyptostatic larval and homeomorphic nymphal stages (where known) (Fain and Bafort 1967; Fain and Lukoschus 1986). Adult males of *Hypodectes* species have hypertrophied chelicerae that are probably used in agonistic behavior toward other males, while females have vestigial cheliceral digits (Figs. 16.31E, F). It is clear from their gnathosomatic morphology that adults of these species do not feed. This leads to the paradoxical situation where the stage that acquires all the nutrition for the entire life cycle and reproductive output (the deutonymph) has no mouth or normal gut. These deutonymphs undergo enormous engorgement within the host tissues, increasing body volume up to 1,000 times and sometimes undergoing neosomic reinforcement of coxal apodemes and dorsal muscle attachments (Fain and Bafort 1967) (see also chapter 3).

The Psoroptidia

The Psoroptidia, an unranked but hypothetically monophyletic group that includes most of the permanent astigmatine symbionts of vertebrates, was first proposed as Psoroptoidae by Thor (1929) on ecological grounds. The group was formally diagnosed by Yunker in 1955 under the name Psoroptidia, but none of the character states he listed was actually a synapomorphy. A monophyletic Psoroptidia was accepted by Fain (1962b), who suggested that birds were the original hosts of the group. Krantz (1978) gave a better diagnosis of the group (as Psoroptides), listing (among other states) the regression or absence of genital papillae, loss of one seta (*c*″) from tibiae I–II, and loss of the deutonymph from the life cycle. OConnor (1982a) hypothesized that the group was most closely related to the Hypoderatoidea, sharing with that lineage the striate cuticle, regression of propodosomatic setae *le*, reduction of the sejugal furrow, reduction of Grandjean's organ, loss of at least one pair of paraproctal setae from segment AD, reduction of the genital papillae, ancestral presence of a large, inverted, U-shaped epigynal apodeme in the female, and presence of sclerotization on the male hysterosoma. Synapomorphies diagnosing a monophyletic Psoroptidia with respect to the Hypoderatoidea are the reduction in size of the genital valves and at least partial fusion of the valves to the body wall in females, loss of seta *c*″ from tibiae I–II and *v*′ from femur IV, loss of unguinal setae from all tarsi and seta *a*″ from tarsus I, regression of the empodial claws (either small and still clawlike or fused into an expanded ambulacral disc), and the elimination of the deutonymph from the life cycle. An alternate opinion suggesting that certain groups associated with bird feathers were derived from different free-living ancestors was presented by Atyeo and Gaud (1979a) and seemingly accepted by Dabert and Mironov (1999) and Proctor (2003a), but supporting evidence has not been provided. Mironov (2003) referred again to a presumably monophyletic Psoroptidia.

Psoroptidian mites exploit bird and mammal hosts in many ways, but the only observations that have been made on the interaction of these mites with their hosts have been on detrimental parasites of domestic or laboratory animals. The majority of psoroptidian diversity comprises "feather mites" and "fur mites," species living on the integumentary covering of birds or mammals. These species are assumed to feed primarily on oils from avian uropygial glands (Dubinin 1951; Proctor 2003a) or on mammalian sebaceous secretions (OConnor 1982b), and occasionally on other materials such as fungi or algae that contaminate plumage or pelage (reviewed in Proctor 2003a; also see chapter 6). Species living permanently on a host typically transfer from adult hosts to juveniles by direct contact (Dubinin 1951), but transfer between adult hosts may also occur, particularly among bird species that are brood parasites or otherwise do not raise their own young (Atyeo and Gaud 1983; Proctor 2003a). More intimate associations have evolved in numerous lineages, with mites adapting to life on or in the skin of the host, in feather or hair follicles, subdermal tissues, the respiratory tract, and occasionally in other body cavities.

The great morphological variation that accompanied adaptation to this variety of habitats on host bodies has made subdivision of the Psoroptidia difficult. Some classifications have emphasized traits that are habitat related, such as grouping together all mites that attach to hair shafts, live on feathers, or that inhabit respiratory tracts. Recognizing that such classifications often group taxa based on convergently acquired characters, Gaud and Atyeo (1978) revised the classification of most of the psoroptidian groups associated with birds from the more ecologically based classification developed by Dubinin (1954) to one based primarily on the structure of the pretarsus and the setation of the tarsus, traits they believed were less prone to convergent evolution. OConnor (1982a) slightly modified Gaud and Atyeo's concept and also grouped all of the psoroptidian families associated with mammals in a single superfamily, but noted that the morphological distinctness of the Analgoidea versus the Sarcoptoidea (= Psoroptoidea of OConnor 1982a) was not well supported. That classification, with the return of OConnor's Pyroglyphoidea to the Analgoidea, is followed here.

▪ Superfamily Pterolichoidea (Figs. 16.32–16.40A–C)

DIAGNOSIS: Tarsi with proral setae usually present, typically flattened and bifurcate apically, rarely filiform or absent; condylophores typically L shaped, sometimes rectangular; ambulacral discs expanded, typically with lateral and unguiform sclerites, condylophore guide absent, empodium flattened as central sclerite, discs sometimes relatively small in quill-inhabiting forms. Paraproctal chaetome with 0–3 pairs of p *setae and 0–2 pairs of* ad *setae. Paraphages or parasites of birds.*

The Pterolichoidea includes 18 families of mites living permanently on the bodies of birds, either externally on feathers (almost exclusively on the vanes of wing and tail feathers) or within the quills of these feathers. Gaud and Atyeo (1978, 1996a, b) excluded the families FREYANIDAE, CAUDIFERIDAE, and VEXILLARIIDAE from the Pterolichoidea on the basis of the purported absence of proral setae and the rectangular form of the condylophores, placing these families in a superfamily Freyanoidea. OConnor (1982a) and Ehrnsberger, Mironov, and Dabert (2001) regarded these families as belonging within Pterolichoidea, noting that many FREYANIDAE actually retain proral setae (Fig. 16.33D).

Fourteen pterolichoid families include species inhabiting feather vanes (the genus *Petitota* in the KRAMERELLIDAE is exceptional in living on plumaceous feathers). The family PTEROLICHIDAE (Figs. 16.40A–C) includes most of the diversity of pterolichoid feather mites. This assemblage of 92 recognized genera and hundreds of species probably is paraphyletic with respect to many of the other families and is diagnosed only on the basis of retained ancestral character states (Ehrnsberger, Mironov, and Dabert 2001). Species occur on a wide variety of bird hosts, with the greatest diversity on Galliformes (including *Pterolichus obtusus* Robin, the common wing mite of domestic poultry) (Figs. 16.40A–C), Psittaciformes, Ciconiiformes, and Gruiformes, with fewer taxa on Struthioniformes, Falconiformes, Cuculiformes, Charadriiformes, and Coraciiformes. Most genera are restricted to a single host family. Gaud and Atyeo (1996a, b) provide a key to subfamilies and genera.

The remaining pterolichoid families can be diagnosed as monophyletic groups and tend to be more restricted in host range. An exception is the family GABUCINIIDAE (Figs. 16.35D–F), the members of which occur on the wing and tail feather vanes of Falconiformes, Caprimulgiformes, Coraciiformes, Culiciformes, Piciformes, and Passeriformes (Corvidae). The family is easily characterized by the posterior displacement of the genital papillar remnants with respect to the oviporus in the female (Fig. 16.35D). Males always exhibit a posteriorly bilobed opisthosoma and show neither enlargement of posterior legs nor any polymorphism. The family was reviewed by Gaud and Atyeo (1975).

The three families sometimes grouped as the Freyanoidea share the simple form or complete absence of the proral setae and rectangular (not L shaped) pretarsal condylophores. The family FREYANIDAE (Figs. 16.33A–D) includes a large number of species grouped in 17 genera associated with the wing and, occasionally, tail feathers of aquatic birds in the orders Anseriformes, Gruiformes, Charadriiformes, Pelecaniformes, and Procellariiformes. Many freyanid mites are broad and have ventrally inserted posterior legs and inhabit feathers with overlapping tegmina extending from the barbs. Microhabitat specificity is strong in the family, with two different species of *Freyana* often living between the same pair of barbs, but with one near the rachis where barb tegmina overlap, and the other more distally located in the exposed space (Dubinin 1953). Male polymorphism is pronounced in *Freyana* species infesting anseriform birds. *Diomedacaraus gigas* (Trouessart), one of the largest species of Astigmatina, inhabits the wings of albatrosses and is unusual in retaining two pairs of *ad* setae in the paraproctal region of the female. The genera and many species of FREYANIDAE have been reviewed by Gaud and Atyeo (1982a, b; 1985), and a phylogeny was proposed by Ehrnsberger, Mironov, and Dabert (2001). The latter authors suggested that the small family CAUDIFERIDAE (Figs. 16.32D–G), which is associated with ibises (Ciconiiformes: Threskiornithidae), may have diverged from within the FREYANIDAE. The African species of CAUDIFERIDAE were reviewed by Gaud (1982). The 12 genera of VEXILLARIIDAE (Figs. 16.33E, F) are primarily associated with hornbills (Coraciiformes: Bucerotidae), but two species of *Calaobia* have colonized African estrildid finches (Passeriformes). Species of *Vexillaria* exhibit an unusual male polymorphism where the two distinct morphs are so unlike each other that they were originally described as separate genera. The polymorphism was resolved and the African species reviewed by Gaud (1980b). The subfamily Calaobiinae was reviewed by Gaud and Atyeo (1990).

Species of KRAMERELLIDAE (Figs. 16.32A–C) also occur on a variety of host groups, including Strigiformes, Ciconiiformes, Pelecaniformes, Falconiformes, and Gruiformes. Many species exhibit a strongly widened and flattened body and ventrally positioned legs that are reminiscent of the FREYANIDAE, in which they formally had been placed (Dubinin 1953). Also, as in freyanids, kramerellids live in the space created by overlapping tegmina of adjacent feather barbs in owls, hawks, and eagles. Species of *Petitota* are unusual among pterolichoid mites in living on plumaceous feathers of owls; males have enlarged legs III, a trait convergent with many analgoid mites living in similar habitats. Males of the genus *Freyanella* on ibis hosts may be strongly asymmetrical (Atyeo, Gaud, and Humphreys 1972) (Fig. 16.2E), while males of *Parabdellorhynchus* from pelicans exhibit an allometrically related polymorphism of the gnathosoma and anterior legs (Dubinin 1956).

The remaining pterolichoid families inhabiting feather vanes are much more restricted to particular host groups. Species in the three genera of PTILOXENIDAE (Figs. 16.37A–C) are found on flight feathers of Charadriiformes, Podicepediformes, and Ciconiiformes (Phoenicopteridae—flamingoes), respectively. Females are bilobed posteriorly and retain one pair of *ad* setae. The family was reviewed and a phylogeny presented by Dabert and Ehrnsberger (1998). Species of FALCULIFERIDAE (Figs. 16.39A–C) are characterized by having the female oviporus transverse or in the form of an inverted U, and the epigynal apodeme is absent. Males of some genera are polymorphic and display an allometrically related enlargement of the gnathosoma and anterior legs (Fig. 16.39C). Species representing the 14 described genera are restricted to the Columbiformes. The African falculiferids were reviewed by Gaud (1976).

Adults of the family EUSTATHIIDAE (Figs. 16.34A–C) are somewhat paedomorphic, with the genital and trochanteral setae not added ontogenetically and the male genitalia posteriorly displaced in most species. All of the 65 known species are restricted to birds in the order Apodiformes, families Apodidae and Hemiprocnidae (Peterson, Atyeo, and Moss 1980). The CRYPTUROPTIDAE (Figs. 16.36A–C) includes nine genera and 16 species restricted to the Neotropical order Tinamiformes. Females are characterized by a well-developed external spermaduct or copulatory tube (Gaud, Atyeo, and Berla 1972).

Members of the family CHEYLABIDIDAE (Figs. 16.39D, E) are accommodated in two genera found on falconiform birds and one on sandgrouse (Columbiformes: Pteroclididae). They are unusual in retaining two pairs of *ad* setae in the paraproctal region of the female (Gaud and Atyeo 1984; Perez and Atyeo 1991). The OCHROLICHIDAE (Figs. 16.34D–F) includes three monotypic genera that are unusual among pterolichoid mites in occurring on passeriform birds. The mites, which are found on African larks (Alaudidae) and warblers (Sylviidae), are small (< 400 µm) and do not add setae on the trochanters. They also lack solenidia on all genua, along with idiosomatic seta *d1* (Gaud and Atyeo 1996a, b). The monogeneric family RECTIJANUIDAE (Figs. 16.35A–C) includes 12 species associated with waterfowl (Anseriformes, Anatidae). These very small, elongated mites have a longitudinally opening oviporus surrounded by a long, narrow epigynal apodeme in the female and are unusual in living on neck feathers and wing coverts rather than on flight feather vanes (Atyeo and Peterson 1976). Finally, the family THORACOSATHESIDAE (Figs. 16.36D, E) comprises two species in a single genus associated with Australasian megapodes (Galliformes: Megapodiidae). Females possess a long copulatory tube, and femur and genu IV of the male are fused (Atyeo 1992).

The remaining four pterolichoid families inhabit the quills of the wing and tail feathers of numerous orders of birds. Of these, the family SYRINGOBIIDAE (Figs. 16.38E, F) is the least specialized and most diverse, with 14 genera and 67 species inhabiting the quills of charadriiform birds (Dabert 2003a, b). Adults are comparatively well sclerotized for quill inhabitants, and some species move regularly from the lower vane surface into and out of the quill through the superior umbilicus of the feather. All species feed on keratinaceous material, with most feeding on pulp caps in the lower part of the feather. Species of *Plutarchusia* and *Paidoplutarchusia*, however, consume the pithy feather medulla and can cause considerable damage and weakening of the feathers. A phylogeny proposed by Dabert (2003b) suggests a complex pattern of evolution involving cospeciation, host switching, and sorting events. Species of ASCOURACARIDAE (Figs. 16.38A, B) are very large mites with massive chelicerae. The propodosoma is strongly sclerotized, while the hysterosoma of females and some males is unsclerotized. Ascouracarid females retain one pair of *ad* setae as in those of the SYRINGOBIIDAE, and some males lack paranal suckers. The seven genera and 21 known ascouracarid species occur on birds of numerous orders (Falconiformes, Galliformes, Caprimulgiformes, Apodiformes, Psittaciformes, Piciformes, and Passeriformes). A single adult female and male typically inhabit a feather, where they feed by tunneling in the medulla and often cause significant damage. The very elongate, sclerotized larvae disperse by chewing a hole in the wall of the rachis and entering a new feather through the superior umbilicus (Dabert and Ehrnsberger 1992).

The two species of the monogeneric pterolichoid family KIWILICHIDAE (Figs. 16.38C, D) inhabit the vestigial flight quills of kiwis (Apterygiformes) in New Zealand. Adults have squat bodies and bear two sclerites on the dorsal hysterosoma (Dabert 1994). The OCONNORIIDAE (Figs 16.37D–F) is represented by a single species collected from the flight quills of an owl (Strigiformes) in the Philippines. *Oconnoria inexpectata* Gaud, Atyeo, and Klompen is a small, poorly sclerotized mite of uncertain affinities (Gaud, Atyeo, and Klompen 1989).

▪ Superfamily Analgoidea (Figs. 16.40D–F, 16.49)

DIAGNOSIS: All tarsi lacking proral setae, tarsus IV lacking seta s; *genu I with 0–2 solenidia; pretarsus with empodium clawlike in center of ambulacral disc or flattened as central sclerite; condylophore guide typically present at junction of ambulacral stalk and disc; pretarsus often retractable. Paraproctal chaetome with 0–3 pairs of* p *setae and 0* ad *setae. Paraphages or parasites of birds, rarely nidicolous or synanthropic (some PYROGLYPHIDAE).*

The Analgoidea shares most diagnostic characters with the Sarcoptoidea living on mammals, and relationships among these lineages are uncertain. Analgoid species exhibit a much broader range of habitats on their avian hosts than is seen in the Pterolichoidea. They occur not only on and in feathers but also in feather follicles, upper epidermal layers, subcutaneous tissue, and the respiratory tract. Five families of Analgoidea include species typically inhabiting the vanes of flight, tail, covert, and head feathers. They exhibit a general morphology similar to that of the Pterolichoidea, having flattened, well-sclerotized bodies, regressive setae on the central dorsal hysterosoma, expanded ambulacral discs with internal sclerotization, and a posteriorly bilobed opisthosoma in most males.

The 23 recognized genera of ALLOPTIDAE (Figs. 16.42D–F) are associated with aquatic bird hosts (Gaviiformes, Procellariiformes, Pelecaniformes, Ciconiiformes, Anseriformes, Gruiformes, and Charadriiformes). Although the family shares the characteristic fusion of the femur/genu on all legs with the XOLALGIDAE and THYSANOCERCIDAE, alloptids can be distinguished by a female hysteronotal shield that extends to the posterior terminus of the body; setae *ro* are often but not always present. Small alloptid species live on covert feathers between overlapping barbule fields or wedged at the juncture between barb and rachis. Spectacular asymmetry involving legs on either side of the body is observed in males of *Dinalloptes* and *Hyperpedalloptes* (Fig. 16.2D) (Dubinin 1955; Atyeo and Peterson 1966; Peterson and Atyeo 1968; Peterson 1971). The monogeneric family THYSANOCERCIDAE (Fig. 16.43D) was previously included in the ALLOPTIDAE based on the fused femur/genu of all legs, but females of *Thysanocercus* have elaborated posterior lobes bearing setae *h2* on their apices, and the hysteronotal sclerites are highly ornamented (Gaud and Atyeo 1996a, b). Species inhabit the wings of Apodiformes (Apodidae). The African species were reviewed by Gaud and Peterson (1987).

The closely related families AVENZOARIIDAE and PTERONYSSIDAE were formerly united under the first cited name but were separated by Mironov (2001a). These mites are characterized by strong enlargement of the pretarsal ambulacral discs, which are positioned at an angle to the ambulacral stalks. The modified empodial claw is large, and the tip extends past the margin of the disc (Figs. 16.47D, G). Males are bilobed posteriorly, and legs III and/or IV may be enlarged. Species of the 21 genera of PTERONYSSIDAE occur on the avian orders Coraciiformes, Piciformes, and Passeriformes and are characterized by the proximal enlargement of the ambulacral stalk as a membrane along the ventral side of the tarsus and by the presence of obvious lacunae in the lateral sclerites of the ambulacral disc (Fig. 16.47G). The redefined AVENZOARIIDAE includes 18 genera whose members occur on a variety of aquatic bird orders. Avenzoariids differ from pteronyssids in lacking modified ambulacral stalks and lateral sclerite lacunae. Males of the avenzoariid genus *Bdellorhynchus* are notable in that they exhibit allometric polymorphism of the chelicerae and anterior legs. A phylogeny of the subfamily Avenzoariinae based on morphological data was presented by Mironov and Dabert (1999); Dabert, Dabert, and Mironov (2001) added molecular sequence data that largely corroborated earlier morphological findings.

The family PROCTOPHYLLODIDAE (Figs. 16.41E–H) is one of the largest in the Astigmatina (36 genera and several hundred species) and contains the dominant wing vane mites of passeriform birds and hummingbirds (Apodiformes). Most species are elongate and narrow, and all lack propodosomatic setae *ro* and the solenidion on genu II. Females and many males are bilobed posteriorly, with females possessing elongate projections from the posterior lobes. The male paranal suckers are often modified as cylinders that fit over a pair of short, dorsally directed posterior protuberances of juvenile females, facilitating precopulatory guarding. Two genera occurring on hummingbirds are unusual in that they inhabit the interior of the flight feather quills. Species of *Allodectes* are little modified and enter and exit the quills through an enlarged superior umbilicus, while species of *Rhamphocaulus* are permanent quill inhabitants with cylindrical bodies. Most species occur on the two above-mentioned host groups, but rare colonization events have led to permanent associations with host species in the orders Charadriiformes, Coraciiformes, Gruiformes, and Piciformes. The large genus *Proctophyllodes* was revised by Atyeo and Braasch (1966), and the genera of the subfamily Pterodectinae were diagnosed by Park and Atyeo (1971). Park and Atyeo (1975) summarized the associations of this group with hummingbirds.

Species of TROUESSARTIIDAE (Figs 16.43A–C) are typically larger mites with broad anterior bodies tapering to pointed lobes in females and some males or to scalloped lobes in other males. At least one *ro* seta is retained on the propodosoma. Trouessartiids are unusual in that they live on the dorsal surface of the wing coverts and head feathers in areas where the surfaces of adjacent feathers overlap. Most of the species diversity in the family is in the large genus *Trouessartia*, which occurs primarily on passeriform birds. The remaining 10 genera are associated with Caprimulgiformes, Cuculiformes, Coraciiformes, and Piciformes. A general taxonomic framework for the family has been developed by Orwig (1968) and Santana (1976).

Species in three large families of Analgoidea have adapted to life on plumaceous barbs of body contour feathers or the basal areas of wing and tail feathers. Adaptations to life in a more three-dimensional space than is occupied by vane-inhabiting mites include a rounded or cylindrical body form, longer dorsal setae, and various types of projections on the anterior leg segments for moving among the barbs. Typically, legs III or IV of males are enlarged for precopulatory mate guarding. The family ANALGIDAE (Fig. 16.46) includes 33 genera occurring on a variety of host groups, mostly the Passeriformes. The family is characterized by the form of the pretarsi, with the ambulacral stalk inserted on the ventral tarsal apex to form a distinctly convex surface (Fig. 16.46D). The ambulacral discs are relatively small (completely reduced in *Ancyralges*), and the anterior legs bear apophyses, spines, or modified setae that aid the mite in maneuvering among downy barbs. Males of many genera exhibit a continuous polymorphism, with the enlargement of legs III and elongation of the idiosomatic setae correlated with body size (Gaud and Atyeo 1996a, b). Large populations of *Megninia* species on domestic poultry have been associated with weight loss and a decline in egg production (Gaud and Atyeo 1996a).

The family XOLALGIDAE (Figs. 16.42A–C) includes 21 genera whose species exhibit a variety of body forms. The family shares the fusion of femur and genu on all legs with the ALLOPTIDAE and THYSANOCERCIDAE, but its species are much less sclerotized and inhabit plumaceous feathers of many bird orders, both aquatic and terrestrial. Male xolalgids often have enlarged legs III or IV, and those of the subfamily Xolalginae have a strongly modified tibia and tarsus IV for grasping the strongly reduced hind leg of immature females during precopulatory guarding. Both sexes in species of *Beaucournuella* from rallid hosts (Gruiformes) have enlarged, hooklike legs III–IV that lack pretarsi and function to clasp the base of the rachis of body contour feathers (Gaud 1974). Extreme site specificity has been reported for seven species of *Fainalges* associated with the green conure, *Aratinga holochlora* (Psittaciformes) (Perez 1995).

The family PSOROPTOIDIDAE (Figs. 16.45E–G) was formerly included within the ANALGIDAE but retains the more ancestral form of the pretarsus, with a narrow ambulacral stalk arising directly from the tarsal apex (Fig. 16.45G) (Gaud and Atyeo 1982c). Females typically have a very large, inverted U-shaped epigynal apodeme and always bear a dorsal hysterosomatic sclerite. Small apophyses may be present on the anterior legs for moving through their plumaceous feather habitat. Psoroptoidids are associated with many avian orders, with the genus *Mesalgoides* being especially diverse on passeriform hosts (Gaud and Atyeo 1982c, 1996a, b).

The space within feather quills has been colonized by species in a number of families, four of which are exclusively syringicolous. The DERMOGLYPHIDAE (Figs. 16.48A–C) includes elongate, cylindrical inhabitants of the quills of many avian orders. The 10 described species are characterized by the terminal position of the anus, absence of the female epigynal apodeme and male paranal suckers, and the ventrally expanded ambulacral stalk of the pretarsus, similar to that of the ANALGIDAE (Fig. 16.48C) (Gaud and Atyeo 1996a, b). The two genera that occur on Tinamiformes are less elongate and retain dorsal hysterosomatic sclerotization (Gaud et al. 1973). Legs III or IV may be enlarged in males. Dermoglyphids may feed on keratinaceous material within the quill or on fluids, including blood, from the basal feather papilla (B. M. OConnor, pers. obs.). The single species of GAUDOGLYPHIDAE, *Gaudoglyphus minor* (Nörner) (Figs. 16.48D–F) inhabits the flight quills of the domestic chicken, *Gallus gallus*. *G. minor* is similar to some other quill mites in being small and cylindrical in form, but it is unique in that the female oviporus is transverse. The ambulacral stalk is thin and apically positioned, in contrast to the analgid-like pretarsus of the DERMOGLYPHIDAE. The tritonymph also is unique among the Psoroptidia in being calyptostatic and in having extremely reduced appendages (Bruce and Johnston 1976).

The APIONACARIDAE (Figs. 16.41A–D) is a small, poorly known family whose four genera are associated with charadriiform, galliform, and passeriform birds. Adults are globose and have no hysterosomatic sclerotization. They are strongly paedomorphic, adding nothing except genital structures beyond the protonymphal level of development; setae *4a*, *4b*, and all trochanter setae are omitted, legs IV (when present) bear a single seta on the tarsus, and only one pair of remnant genital papillae is present. Some males fail to add legs IV and remain hexapod (Fig. 16.41A) (Gaud and Atyeo 1976; Mironov 2001b). The larvae are elongate and serve as the dispersing stage. Apionacarids typically cohabit quills with mites of the trombidiform family SYRINGOPHILIDAE, but the nature of the association is unknown (Gaud and Atyeo 1976; B. M. OConnor pers. obs.).

Three families of Analgoidea occur on or in the skin of their avian host rather than on feathers. In earlier classifications, the DERMATIONIDAE was considered an element of a broader family EPIDERMOPTIDAE, but Gaud and Atyeo (1996a) gave family rank to the assemblage. The Knemidokoptidae was formerly treated as a distinct family, but phylogenetic analysis indicates that this lineage is derived from within the EPIDERMOPTIDAE (*s. str.*) (Mironov, Bochkov, and Fain 2005), and it is so treated here. Both newly diagnosed families are generally characterized by relatively small, round, flattened, or globose bodies and the loss of dorsal idiosomatic setae *ro* and *c1* and the solenidion from genu III. Members of the DERMATIONIDAE (Figs. 16.43E–G) are the least specialized of the three skin-inhabiting families and live on the skin surface of their hosts, attaching via retrograde apophyses on legs III and/or IV (Figs. 16.43E, F). Species in the 12 genera are distributed on hosts of the orders Procellariiformes, Pelecaniformes, Anseriformes, Coraciiformes, Strigiformes, Apodiformes, and Passeriformes. Dermationids are relatively nonpathogenic, but high population densities have been associated with skin damage in several different hosts (Fain 1965d). The redefined EPIDERMOPTIDAE (Figs. 16.44A–C) includes 12 genera of mites that are more specialized for a life in the upper layers of the host epidermis; that is, the body is more sclerotized, the legs are shorter and stouter, and the anterior tarsi bear apical clawlike projections. Most epidermoptid mites burrow in the superficial layers of the epidermis, a level of contact that is intimate enough to lead to noticeable effects on their hosts. For example, *Epidermoptes bilobatus* Rivolta may cause mangelike lesions on the skin of domestic poultry (Fain 1965d).

Some species of EPIDERMOPTIDAE are unusual in that they have developed hyperparasitic relationships with parasitic flies of the family Hippoboscidae. Adult females of species of *Myialges* and *Promyialges* attach to the abdomen of the dipteran host by means of highly modified processes of tarsus I, where they feed on hemolymph and often become greatly distended. Juveniles and males remain on the avian host in epidermal burrows. The relative lack of host specificity of hippoboscid flies to their avian hosts leads to a much greater dispersal and colonization potential for these mites, and host specificity in these genera is extremely weak. Thus, the same mite species may parasitize birds belonging to several different orders. Other analgoid mites occasionally have been collected from hippoboscid flies (Philips and Fain 1991), but Jovani et al. (2001) dismissed such associations as unimportant in the dispersal of nonepidermoptid families. Due to their close contact with avian hosts and their lack of specificity, epidermoptid mites can be quite pathogenic. Gilardi et al. (2001) reported severe dermatitis leading to mortality in nestling Laysan albatrosses, *Phoebastria immutabilis* (Rothschild), caused by *Myialges nudus* Fain.

The six genera and 17 species of the epidermoptid subfamily Knemidokoptinae (Fig. 16.44C) are highly specialized for burrowing into host tissue (Dubinin 1953; Fain and Elsen 1967). Knemidocoptine females and nymphs are characterized by a globose body, an even greater shortening of the legs than in other epidermoptids, and by the presence of a pair of strong dorsal apodemes underlying the propodosomatic sclerite. These sclerites presumably serve as origins for strong muscles that allow the mites to burrow in host tissue.

Females of *Evansacarus* are highly paedomorphic and remain larviform throughout ontogeny, but males develop normally (Fain 1962b). Species of *Knemidokoptes* are well-known pathogens of many groups of birds. *K. mutans* (Robin and Lanquetin) and *K. jamaicensis* Turk cause *scaly leg* in poultry and passerine birds, respectively, while *K. pilae* Lavoipierre and Griffiths damages the cere and beak of psittaciform birds (Fain and Elsen 1967). *K. jamaicensis* has been implicated as a mortality factor among migratory passerine birds (Latta and OConnor 2001).

The 28 species that compose the family LAMINOSIOPTIDAE (Figs. 16.49D, E) are unusual in that they inhabit feather follicles and subcutaneous cysts and have been found embedded in the matrix of developing feather quills. The family is characterized by an elongated, flattened body, very short legs I–II, and pretarsi with elongate ambulacral stalks and small discs. Follicle- and quill-inhabiting species of the subfamily Fainocoptinae lack hysterosomatic sclerotization and often bear large membranous lateral projections posterior to legs II (Fig. 16.49D). Subcutaneous species in the nominate subfamily are smaller, are relatively shorter, bear well-developed hysterosomatic sclerites, and lack membranous projections laterally (Lombert and Lukoschus 1979; Fain 1981a). *Laminosioptes cysticola* (Vizioli) (Fig. 16.49E), a common parasite of domestic fowl, may cause disease by invading internal organs (Smith, Quist, and Crum 1997).

Members of the endoparasitic family CYTODITIDAE (Figs. 16.40D–F) inhabit the respiratory tract of birds in the orders Galliformes, Psittaciformes, Columbiformes, Gruiformes, Coraciiformes, Cuculiformes, and Passeriformes. Cytoditids resemble laminosioptids in the regression of their central dorsal setae, shortened segments of the anterior legs, and loss of paranal suckers in the male. However, cytoditid mites are even more regressive and are characterized by a rounded body, strong reduction of sclerotization, loss of all paraproctal setae, strong reduction in size of the remaining setae, and strongly regressed chelicerae that apparently are absent in some species (Fain 1960, 1964a). Species of *Cytonyssus*, one of the two described genera, exhibit somewhat better developed dorsal sclerotization and inhabit the upper respiratory tract of their hosts, while the completely unsclerotized species of *Cytodites* inhabit the lungs, bronchi, and air sacs. Some pathology has been noted in domestic poultry with heavy infestations of *C. nudus* (Vizioli) (Baker et al. 1956; Yunker 1973).

A final lineage of analgoid mites includes the PYROGLYPHIDAE, PTYSSALGIDAE, and TURBINOPTIDAE, grouped by OConnor (1982b) as a separate superfamily Pyroglyphoidea but treated here in the Analgoidea following Gaud and Atyeo (1996a, b). Species share the distal migration of solenidion *ω1* on tarsus I to a position near solenidion *ω3* at the apex of the segment (Fig. 16.45D). Most species bear a clawlike process at the tip of the tarsus, a feature shared with the EPIDERMOPTIDAE and one that led to their inclusion in the latter family by earlier workers (e.g., Dubinin 1953). The female epigynum usually is well developed, although typically reduced in endoparasites. Legs III of the male are often enlarged, but less so in the endoparasitic TURBINOPTIDAE.

The family PYROGLYPHIDAE (Figs. 16.45A–D) demonstrates the greatest ecological breadth among the Psoroptidia. Representatives of the 19 genera range from being largely unsclerotized dorsally except for a propodosomatic sclerite (e.g., *Dermatophagoides*) to being virtually completely sclerotized (e.g., *Pyroglyphus*). The coxal fields are well sclerotized in most taxa. The female epigynal apodeme ranges from a large inverted U to a smaller, almost transverse form, and is completely absent in *Malayoglyphus*. The pretarsal ambulacral disc may be relatively large and sclerotized internally or reduced to a bell-shaped disc with a clawlike empodium centrally. Males of some *Dermatophagoides* and *Sturnophagoides* species have enlarged legs I, with the degree of enlargement allometrically related to body size. Some genera remain obligate parasites of avian hosts; for example, species of *Onychalges* live externally on the feathers or skin of passeriform birds (Gaud 1968), while *Paralgopsis* species inhabit the quills of Psittaciformes, where they feed on the keratinaceous pulp caps (Gaud and Atyeo 1996a; B. M. OConnor, pers. obs.). Both genera closely resemble other analgoid groups, notably the PSOROPTOIDIDAE, in their general appearance. Most other pyroglyphid species are bird nest inhabitants, but some are associated with mammal nests where they feed on sloughed skin or hair residues and on other organic materials. Species of *Dermatophagoides*, *Euroglyphus*, and *Gymnoglyphus* are best known because several species, notably *D. pteronyssinus* (Trouessart), *D. farinae* Hughes, *E. maynei* (Cooreman), and *G. longior* (Trouessart), have adapted to peridomestic situations and are the dominant house dust mites in many parts of the world, in addition to infesting stored food products. House dust mites, along with their shed exuviae and feces, are primary sources for allergens in house dust, and they cause respiratory allergy and, occasionally, dermatitis in sensitive individuals (Wharton 1976; Fain, Guérin, and Hart 1990). The taxonomy of the family was reviewed by Fain, Guérin, and Hart (1990), who recognized five subfamilies, and by Atyeo and Gaud (1996a), who recognized only three.

The monogeneric family PTYSSALGIDAE (Figs. 16.44D, E) includes species that inhabit the flight feather quills of hummingbirds (Apodiformes: Trochilidae). Ptyssalgids are similar to the quill-inhabiting pyroglyphid species of the genus *Paralgopsis* in retaining *ro* setae (one in PTYSSALGIDAE, two in *Paralgopsis*) and in having enlarged legs III in the male (Atyeo and Gaud 1979b). A single species is described, but others have been collected (Gaud and Atyeo 1996a, b; B. M. OConnor, pers. obs.).

Species of TURBINOPTIDAE (Figs. 16.49A–C) inhabit the upper respiratory tract of many avian orders, living in the relatively dry area just inside the external nares of the host (Fain 1970d; Hyland 1979). The nine genera and 38 species are characterized by very short tarsi, at least on legs I–II, and by a particularly well-developed clawlike apical process. Species of *Congocoptes* are primarily parasites of piciform birds and strongly resemble species of PYROGLYPHIDAE. Those of other genera exhibit regressive trends involving reduction of the female epigynal apodeme and widening of the oviporus, weaker sclerotization of the dorsum and coxal fields, shortening of the pretarsal ambulacral stalks, and reduction or loss of paranal suckers in the male. A catalog of species that includes hosts and distributions was prepared by Fain (1977b).

■ Superfamily Sarcoptoidea (Figs. 16.50–16.56)

DIAGNOSIS: All tarsi lacking proral setae, tarsus IV lacking seta s*; genu I with 0–1 solenidion (2 in* Hyracoptes*, PSOROPTIDAE); pretarsus with empodium clawlike in center of ambulacral disc or may be flattened as central sclerite in fur mites, empodium rarely absent; if ambulacral disc well developed, condylophore guide typically present at junction of ambulacral stalk and disc; pretarsus not retractable. Paraproctal chaetome with 0–3* p *setae and 0* ad *setae. Paraphages or parasites of mammals.*

The Sarcoptoidea includes 12 families of mites that have adapted to numerous habitats on and in the bodies of mammals. Like the closely related Analgoidea on birds, some sarcoptoid species live on the integumentary covering (fur mites); others live on the skin surface, in hair follicles or intradermal tissues, or in the respiratory tract. Some highly specialized bat parasites (GASTRONYSSIDAE) inhabit the eye orbits or even the stomach of their hosts. The fur mite families LISTROPHORIDAE, ATOPOMELIDAE, and CHIRODISCIDAE contain the greatest diversity of sarcoptoid mites. They are specialized for life on hair shafts, where they feed on sebaceous materials and generally cause no damage to the hair. The three families were formerly united with one or more additional families within the family LISTROPHORIDAE (e.g., Dubinin 1954) but were later recognized as separate, monophyletic groups (McDaniel 1968; Fain 1971c). Species in all three families have relatively elongate bodies, and legs I–II are modified to a greater or lesser extent for clasping host hair. Early derivative taxa in the three fur mite families are dorsoventrally flattened, while more derived taxa are cylindrical or laterally compressed. Some taxa retain three pairs of pseudanal setae in the paraproctal region, while most other sarcoptoid mites retain at most two pairs.

The family LISTROPHORIDAE (Figs. 16.54A, B) includes 21 genera and over 150 species characterized by the presence of a projecting tegmen that extends over the gnathosoma to contact the host hair (Fig. 16.54B). The propodosoma is sclerotized in adults, and the hysterosoma may bear sclerites as well. Males are usually more sclerotized than females. The main hair-clasping structure consists of a pair of striated flaps that extend anteriorly and ventrally from coxal fields I and fold over the hair. These flaps were earlier thought to be modified palpi (Krantz 1978) and are complemented by legs I–II, which are somewhat modified for hair clasping but do not bear expanded projections. In some taxa (e.g., *Geomylichus*), coxal fields II are strongly striated and engage the hair scales as well. All leg segments are freely articulated, and the ambulacral discs are expanded and bear internal sclerotization. Males are usually smaller than females and may have stouter legs III–IV, their opisthosoma is generally bilobed and somewhat flattened posteriorly, and paranal suckers are retained. However, setae *ft"* and *tc"* of tarsus IV are not modified. Females of *Listrophorus caudatus* Fain, Smith and Whitaker bear a long caudal projection that is not an external spermaduct since the actual copulatory opening is at the base of the projection (Fain, Smith, and Whitaker 1986). The genus *Aplodontochirus*, associated with the mountain beaver, *Aplodontia rufa*, is dorsoventrally flattened, while all other listrophorid genera are more or less cylindrical. Each genus is generally associated with a particular host order, including Rodentia, Eulipotyphla, Lagomorpha, and Carnivora, and host specificity is pronounced. Several species are associated with domestic and laboratory animals, including *Lynxacarus radovskyi* Tenorio on domestic cats, *Leporacarus gibbus* (Pagenstecher) on rabbits, and *Afrolistrophorus maculatus* Fain and *A. musculus* (Wilson and Lawrence) on commensal rats and mice, respectively. The last two species have not yet been reported from laboratory populations of their hosts. Although *L. radovskyi* has been associated with skin and fur abnormalities in feline hosts, a causal relationship has not been established (Tenorio 1974; Bowman 1978).

An interesting case of multiple colonization of a novel host involves six species of the normally rodent-associated listrophorid genus *Aprolistrophorus* found on the Andean caenolestid marsupial, *Lestoros inca* (Thomas) (Fain and Lukoschus 1984). Synhospitaly also occurs with six species of *Listrophorus* occurring together on the muskrat, *Ondatra zibethicus* (Fain and Hyland 1974). Listrophorids are widespread, but they do not occur on native hosts in Australia or Madagascar. Regional monographs are available for the Nearctic (Fain and Hyland 1974), Neotropical (Fain 1973), and Afrotropical (Fain 1971c) regions, and a key to genera is provided by Fain (1981b).

The ATOPOMELIDAE (Fig. 16.55) is closely related to the LISTROPHORIDAE, sharing the striated coxal fields that engage the scales on the host hair shaft when the hair is clasped by legs I–II. However, the coxal fields in atopomelids do not extend outward as clasping plates, and there is no projecting tegmen over the gnathosoma. The tibia and tarsus of legs III–IV are fused, and the tibial solenidion sometimes migrates distad onto the tarsal portion of the fused segment. The pretarsi are relatively small and are retained on all legs, as in the LISTROPHORIDAE. Body sclerotization varies from a remnant propodosomatic sclerite (e.g., *Listrocarpus*,

Koalachirus), to almost complete sclerotization of the body (e.g., *Oryzomysia*). The female usually retains a small epigynal apodeme, and males may or may not retain paranal suckers. Sexual dimorphism is generally more pronounced than in the LISTROPHORIDAE, and legs IV of the male often are enlarged. The family includes 47 genera and approximately 400 species that are found almost exclusively in the Southern Hemisphere and the Asian tropics. Hosts include most orders of marsupials in Australia and South America and the placental orders Eulipotyphla, Afrosoricida, Rodentia, Carnivora, and Primates. An extreme case of synhospitaly occurs with the potoroo, *Potorus tridactylus*, in Australia, from which 21 species of *Cytostethum* have been described across its range (Fain and Domrow 1974). *Chirodiscoides caviae* Hirst is a common fur mite of laboratory guinea pigs, and *Listrophoroides cucullatus* (Trouessart) is a widespread fur mite of commensal rats in tropical regions. Regional monographs have been prepared for the Neotropical (Fain 1979), Afrotropical (Fain 1972c), and Australian (Fain 1972b) regions and for Madagascar (Fain 1976a).

All of the 28 genera and approximately 250 species of CHIRODISCIDAE (Figs. 16.54C–E) have strongly modified legs I–II to clasp host hairs. Early derivative genera (*Chirodiscus, Lemuroecius, Schizocoptes*) are dorsoventrally flattened, but lateral compression of the body characterizes most other genera. Compared with that of the preceding two families, the body of chirodiscids is poorly sclerotized, and the anterior legs always possess striated processes for clasping host hairs. Some early derivative genera retain free segments on the anterior legs, but the majority of species (subfamily Labidocarpinae) have free articulation only between trochanter and coxa, with all distal segments fused and expanded as clasping plates. The pretarsus is absent on the anterior legs and may be reduced or lost on legs III or IV as well. Males of the Labidocarpinae, most of whose species occur on bats, are strongly modified for precopulatory guarding. The paranal suckers are typically well developed and situated in a posterior concavity of the body, modifications that presumably aid the male in securing attachment to the immature female, which has only rudimentary legs II–IV (Fig. 16.54D). Females acquire functional legs only at their final ecdysis.

The CHIRODISCIDAE is cosmopolitan, with virtually all groups of bats harboring one to several genera. Nonchiropteran hosts include Afrosoricida (Chrysochloridae), Eulipotyphla (African Soricidae), Primates (Cheirogaleidae), Carnivora (Mustelidae), and Rodentia (Castoridae). Synhospitaly of the genus *Schizocarpus* on beavers (Castoridae) is spectacular, with 50 species of this genus described from the two extant beaver species. The North American *Castor canadensis* harbors 17 species, with up to 9 occurring on the same host individual (Fain and Whitaker 1988); the European *Castor fiber* hosts 33 distinct species (Fain and Lukoschus 1985). A combination of habitat partitioning on the host and geographical isolation of host and mite populations has been invoked to explain this diversity (Fain and Whitaker 1988). Regional monographs are available for the Neotropical (Fain 1973) and Afrotropical (Fain 1971c) regions.

The small family LOBALGIDAE (Figs. 16.50D, E) includes highly flattened mites that live on specialized hairs of Neotropical mammals. The seven described species resemble some Neotropical mites classified in the subfamily Psoralginae (PSOROPTIDAE) in retaining a well-developed epigynal apodeme and dorsal hysterosomatic sclerotization in the female, and in lacking hysterosomatic setae *c1* and leg apophyses. Lobalgids may be distinguished by their more elongate body form, a bilobed opisthosoma with well-sclerotized lobes in both sexes, the broadening of the subcapitulum basally, and well-developed pretarsal ambulacral discs with internal sclerotization. Species of *Echimytricalges* are strongly flattened and bear several hysterosomatic sclerites in the female. Pretarsal ambulacra are very large, and hysterosomatic setae *cp* are lanceolate and laterally positioned. Overall, their appearance is reminiscent of some vane-inhabiting feather mites, a clear case of convergence given that these mites live between longitudinal ridges at the edges of the flattened, spiny hairs of echimyid rodent hosts (Fain and Lukoschus 1970). *Lobalges trouessarti* Fonseca, the remaining species in the family, lives in the fur of sloths of the genus *Bradypus* (Fonseca 1954). The genus *Coendalges*, represented by *C. panamaensis* Fain and Mendez collected from a Neotropical porcupine and assigned to the LOBALGIDAE, is here considered a member of the PSOROPTIDAE (see below).

The family PSOROPTIDAE (Figs. 16.50A–C) includes a diversity of taxa living on the skin or in the ears of many groups of mammals. These mites live on the skin surface, at the base of hairs, or occasionally are embedded in superficial tissue. They feed by scraping the skin with their chelicerae and ingesting tissue fluid. Their morphology is generalized, with the female body usually rounded and weakly sclerotized, but species of the subfamily Listropsoralginae may be strongly sclerotized dorsally. Setae *ro* may be present or absent, and some taxa have lost additional pairs of hysterosomatic setae. The oviporus ranges from an inverted V shape to being almost transverse, and the epigynal apodeme ranges from a large inverted U to absent. Retrograde apophyses, which assist the mite in attaching to the host skin, occur on some leg segments in some taxa, but they are absent in others. Females of some primate parasites bear strong, hooklike apophyses at the posterior end of the body. Pretarsi are always present on the anterior legs but may be lost on legs III or IV in either sex. The ambulacral disc is typically rounded or bell shaped with a small empodial claw remnant, and the ambulacral stalk is usually elongate (subdivided in the genus *Psoroptes*). Sexual dimorphism is pronounced, with males generally being bilobed posteriorly. Legs III or IV may be enlarged or otherwise modified, or, alternatively, legs IV may be very short. The 33 genera and 61 species composing the family are found on didelphimorph and diprotodont marsupials and the placental orders Xenarthra, Eulipotyphla (Erinaceidae), Rodentia (some Neotropical caviomorph families),

Carnivora, Primates, Hyracoidea, Artiodactyla, and Perissodactyla. A phylogenetic analysis conducted by OConnor (1984b) supported a hypothesis of coevolution between psoroptid parasites of Primates and their hosts.

The PSOROPTIDAE includes the only astigmatine mites that live on large eutherian mammals, and some species are of considerable importance as parasites of domestic animals. *Psoroptes* mites live on the skin or in the ears of horses, sheep, other livestock, and domestic rabbits, as well as on wild sheep and goats. Allergic reactions to feeding by these mites can lead to severe lesions and significant economic losses (Mullen and OConnor 2002). Species concepts in *Psoroptes* have been controversial; recent studies using both morphological and molecular evidence provide no support for the idea of multiple species. However, some local adaptation of populations is suggested by the difficulty in cross-infesting other host species (Bates 1999; Zahler et al. 2000). *Chorioptes* species occur on cattle and sheep, as well as wild hosts, and *Otodectes cynotis* (Hering) is the common ear mite of domestic cats and dogs. All of these species irritate the host through their feeding and through formation of scabs on the skin or ears (Mullen and OConnor 2002). Keys to some genera are found in Fain (1963a, 1965b, 1975a).

The MYOCOPTIDAE (Figs. 16.53A–D) is a small family that probably is related to the psoroptid subfamily Listropsoralginae. Early derivative taxa on marsupial hosts in both groups are well sclerotized (Fig. 16.53C) and have well-developed apophyses on the leg segments for attaching against the host skin. Myocoptid mites are notable in having legs III–IV in juveniles and females, and legs III in males, enlarged and specialized for grasping host hairs. Unlike true fur mites, myocoptid mites emulate psoroptids in feeding on tissue fluid from the skin. Myocoptid parasites of placental mammals are less sclerotized, and the leg apophyses are weak or absent; most are dorsoventrally flattened and grasp one hair in legs III–IV on each side. Species of the genus *Trichoecius* are unusual in having cylindrical females, and legs III–IV are ventrally positioned (Fig. 16.53D). These mites grasp a single host hair using all four posterior legs. The genus *Dromiciocoptes* occurs on didelphid, caenolestid, and microbiotheriid marsupials in South America (Fain 1970e; B. M. OConnor, pers. obs.), while the remaining genera parasitize rodents in all parts of the world and, occasionally, Australian marsupials. *Myocoptes musculinus* (Koch) (Figs. 16.53A, B) can be a problem in laboratory mouse colonies, where high population densities may lead to skin irritation and hair loss. The myocoptid fauna of Europe has been reviewed by Fain, Munting, and Lukoschus (1970) and Haitlinger (1986b) and that of the Afrotropical region by Fain (1970b).

Two closely related families of sarcoptoid mites—the RHYNCOPTIDAE (including Audycoptidae) (Figs. 16.51A–C) and the SARCOPTIDAE (including Teinocoptidae and Bakerocoptidae of the authors) (Figs. 16.51D, E, 16.52)—have adapted to living in the skin of mammals. Dorsal hysterosomatic setae *c1* and *c2* in these families are ancestrally modified as strong spines that may be strongly ridged or layered. Rhyncoptids and sarcoptids also share the loss of both fastigal setae on tarsi I–II, and leg segments may bear retrograde apophyses as in some PSOROPTIDAE and MYOCOPTIDAE. Rhyncoptids are globose or elongate mites that inhabit the hair follicles of species in several mammalian orders. Femora and genua III–IV are fused, and most species bear retrograde apophyses on the palpi. *Caenolestocoptes inca* Fain and Lukoschus is globose, exhibits relatively little sexual dimorphism, and occurs in the hair follicles of the tail of caenolestid marsupials (Fain and Lukoschus 1976c). Species of *Audycoptes* and *Saimirioptes*, on the other hand, are elongate mites that occur in the follicles of the facial vibrissae of cebid primates. *Rhyncoptes* females have greatly enlarged legs I–II that anchor the mite in the follicle on various areas of the host body, while the saclike posterior body with its greatly reduced legs III–IV projects from it (Fig. 16.51A). Sexual dimorphism is extreme in *Rhyncoptes*, with males being much smaller and having enlarged legs III (Fig. 16.51C). *Rhyncoptes* species inhabit the follicles of the body and appendages of cebid, callitrichid, and cercopithecid monkeys and also occur on African porcupines (Hystricidae). The two species of *Ursicoptes* inhabit the hair follicles of bears (Ursidae) and raccoons (Procyonidae), respectively. Heavy infestations can cause skin thickening and hair loss (mange) (Yunker et al. 1980; B. M. OConnor, pers. obs.). A phylogeny for the family has been proposed by Klompen (1993).

The 17 genera and 117 species of SARCOPTIDAE burrow in the upper epidermal layers of many mammalian hosts. Females are typically globose, but some are elongated dorsoventrally. The family is distinguished from the RHYNCOPTIDAE by the transverse female oviporus, loss of paranal suckers in the male, and loss of paraproctal seta *p3*. Sexual dimorphism is apparent, with males being smaller than females and typically displaying some sclerotization of the hysterosoma (Fig. 16.52C). In addition, males often have a more complex coxisternal skeleton than do females, and legs III are sometimes somewhat enlarged. The family was revised and a phylogeny developed by Klompen (1992), who recognized three subfamilies: the Diabolicoptinae, which comprises two genera associated with Australian dasyuromorph marsupials; the Sarcoptinae, which includes four genera associated with Rodentia and Primates; and the Teinocoptinae, which contains nine genera and the bulk of the species and is ancestrally associated with Chiroptera. According to Klompen (1992), cospeciation was the primary process producing the diversity of species, although some host switching was apparent.

The human scabies mite, *Sarcoptes scabiei* (L.) (Figs. 16.52A, B) is an ancestral parasite of humans that forms subdermal burrows parallel to the skin surface in susceptible individuals. Normal population density is small, with generally fewer than 100 mites per host. Symptoms of papular scabies are erythematous patches at the site of the mite burrows and intense itching in other areas caused by a generalized immune response. Transmission is typically by direct contact, although mites may survive for two or three days away from a host under suitable temperature and humidity conditions (Arlian 1989; Mullen and OConnor 2002). A condition known as crusted scabies may occur in affected individuals whose immune system is compromised by age, drug therapy, or HIV infection. In these cases, thousands of mites may be present in hyperkeratotic lesions, and the condition is highly contagious. Humans have transmitted *S. scabiei* to many species of domestic animals, and subsequent transmission to wild hosts also has occurred. Nonhuman hosts typically suffer a condition similar to crusted scabies termed **sarcoptic mange**, which is characterized by severe hyperkeratosis and hair loss and typically is complicated by bacterial infections. The condition can lead to host population declines, particularly in the case of wild canid species in North America (Stone et al. 1972) and wombats in Australia (Skerratt, Martin, and Handasyde 1998). Other genera of Sarcoptinae may be somewhat pathogenic in their normal hosts. *Kutzerocoptes gruenbergi* Lavoipierre caused a scabieslike condition in captive capuchin monkeys, and species of *Prosarcoptes* cause similar problems in captive macaques and baboons (Lavoipierre 1970; Smiley and OConnor 1980). *Trixacarus* species can cause problems in laboratory rats and guinea pigs.

Most species of the sarcoptid subfamily Teinocoptinae cause little damage to the skin of their chiropteran hosts. Females of most species merely embed the ventral body surface in the epidermis and remain in one position. Species of *Nycteridocoptes* form intradermal cysts in the skin, where the adult females become physogastric (see chapter 5 for more on physogastry). *Rousettocoptes*, *Teinocoptes*, and *Chirobia* species are unusual in that the females are paedomorphic, remaining larviform in structure while growing through a normal ontogeny. Some species of *Teinocoptes* undergo extreme dorsoventral elongation (Fig. 16.52D). A number of *Notoedres* species have colonized nonchiropteran hosts and may cause mange in rodents, rabbits, and carnivores. *N. cati* (Hering) causes crusting of the skin of the head, ears, and feet of domestic cats.

The remaining four families of Sarcoptoidea share a strong reduction in dorsal idiosomatic setation, retaining at most propodosomatic setae *in* and *ex* and hysteronotal setae *cp* and *h2*. Males generally are similar in form to females and lack paranal suckers. The two species composing the family CHIRORHYNCHOBIIDAE (Figs. 16.53F, G) have an enlarged gnathosoma, with the strongly toothed chelicerae and heavily sclerotized palpi and subcapitulum forming an attachment organ that anchors the mite on the terminal edge of the wings of Neotropical phyllostomid bat hosts. The opisthosoma is extremely abbreviated, the anus is dorsal, and the female oviporus is a small and longitudinal posteroventral opening. The legs are well developed, and the pretarsi are large, reminiscent of some fur mites. Nothing is known of the immature stages (Fain 1968d; Yunker 1970b).

Like chirorhynchobiids, most of the seven genera and 28 species of GASTRONYSSIDAE (Figs. 16.56A–E) parasitize bats, but all are endoparasitic. A propodosomatic sclerite is generally retained, but the body is otherwise unsclerotized and striated. Most species are found in the nasal passages of bats of most families, but *Opsonyssus brutsaerti* Fain commonly occurs in the eye orbits of its pteropodid hosts (Fain 1957). *Gastronyssus bakeri* Fain is unique in living in the stomach of several species of pteropodid fruit bats. *G. bakeri* is extremely elongate and attaches to the stomach lining, where it feeds on blood (Fig. 16.56A) (Fain 1956; Stiller and Sullivan 1977). Species of *Yunkeracarus* and *Sciuracarus* inhabit the nasal passages of murid and sciurid rodents, respectively. A key to the species of *Yunkeracarus* is given by Zabludovskaya (1990).

Members of the LEMURNYSSIDAE and PNEUMOCOPTIDAE are endoparasites in the respiratory tract of Primates and Rodentia and share the characteristics of well-developed hysterosomatic sclerites and loss of propodosomatic seta *in*. Lemurnyssids (Figs. 16.56G–I) are elongate parasites in the nasal cavities of African bushbabies (Galagonidae) and of South American monkeys and marmosets (Cebidae and Callitrichidae) (Fain 1964c). The monogeneric PNEUMOCOPTIDAE (Fig. 16.56F) includes four species that have a rounded body and strongly abbreviated opisthosoma, legs IV situated near the posterior end of the body, and a well-developed sclerite that covers most of the dorsal hysterosoma. In contrast to the other respiratory endoparasites, pneumocoptids inhabit the lungs of their rodent hosts. *Pneumocoptes penrosei* (Weidman) and *P. banksi* (Wellman and Wherry) parasitize ground squirrels and prairie dogs in western North America, *P. jellisoni* Baker was described from species of *Peromyscus* and *Onychomys*, and *Pneumocoptes tiollaisi* Doby occurs in bank voles in western Europe (Baker 1951; Kouchakji and Loos-Frank 1984). The ontogeny of the viviparous *P. tiollaisi* was described by Loos-Frank and Abel (1983), who also documented transmission from mother to young in laboratory nests.

Key 16.1. Cohort Astigmatina (Adults), Key to the Families (Figs. 16.1–16.56)

1a. With the following combination of characters: body cuticle usually at least partially striate; genital papillae greatly reduced or absent; tibiae I–II with a dorsal solenidion and 1 (*v'*) or rarely 0 ventral setae; unguinal setae absent from all tarsi (Figs. 16.3D, E); pretarsi often enlarged, with ambulacral stalk and disc well developed, empodial claws usually reduced or incorporated into ambulacral disc as central sclerite (Figs. 16.33D, E). Parasites of birds or mammals, rarely insects (some EPIDERMOPTIDAE) with a few free-living in nests, stored products/house dust (some PYROGLYPHIDAE) **Psoroptidia—28**

1b. Without all of the above characteristics; body cuticle smooth or striate; genital papillae normal in form, strongly enlarged and ringlike (Histiostomatidae) or reduced; tibiae I–II with dorsal solenidion and 0–2 ventral setae; unguinal setae present or absent; pretarsi variously formed, sometimes enlarged in arthropod parasites; empodial claws variously formed or absent. Free living, parasites of insects or crustaceans, rarely parasites or external commensals of mammals (some ROSENSTEINIIDAE, HISTIOSTOMATIDAE) **2**

2a. Chelicerae laterally compressed, often attenuate, usually with numerous finely divided teeth, nonchelate (Fig. 16.8D); palpi with a terminal solenidion and usually at least 1 elongate, eupathidial seta, normal setae strongly reduced, palp usually reflexed outward (Fig. 16.7B); female genital valves fused to body posteriorly, open anteriorly forming a transverse oviporus; genital papillae of both sexes often in the form of large rings on ventral body surface (Fig. 16.7A) **3**

2b. Chelicerae usually chelate, with few teeth, rarely chelicerae vestigial or fixed digit absent; palpi reflexed inward (Fig. 16.22B), usually bearing 3 filiform setae and terminal solenidion; female oviporus longitudinal with genital valves fused to body anteriorly, free posteriorly; genital papillae always associated with genital opening and never in the form of large rings **4**

3a. Genital papillae small, arising from a common furrow on either side of genital opening or apparently absent in males; body strongly sclerotized, dorsoventrally flattened and with legs laterally positioned (Fig. 16.8E); very small species. In bat guano; Afrotropical, Neotropical **GUANOLICHIDAE**

3b. Genital papillae ringlike, usually large, not arising from a common furrow and not directly associated with the genital opening (Fig. 16.7A); body relatively larger and not dorsoventrally flattened. In a wide variety of wet habitats; cosmopolitan **HISTIOSTOMATIDAE (= Anoetidae)**

4a. Pretarsal ambulacra usually greatly expanded (Figs. 16.10B, 16.21C); legs laterally positioned. Commensals or parasites of adult insects or myriapods, or nidicoles in social insect nests **5**

4b. Pretarsal ambulacra not greatly expanded; legs ventrally positioned or rarely dorsal or lateral; not associated with adult insects in all instars (*Linobia coccinellae*, a hemisarcoptid parasite of chrysomelid beetles is exceptional; this genus has reduced ambulacra and lacks the fixed cheliceral digit) **9**

5a. Dorsum almost entirely covered by 1–2 large sclerites **6**

5b. Dorsum with only a rectangular propodosomatic sclerite; supracoxal sclerites and rarely a small, median, opisthosomatic sclerite **7**

6a. Body rounded, dorsum bearing a large domed sclerite with deep punctations (Fig. 16.28A); legs tapering; tarsi with proral setae foliate, partially fused to pretarsal ambulacrum. Monogeneric, in ant nests; Holarctic **LEMANNIELLIDAE**

6b. Body flattened, dorsum with 1–2 large sclerites, smooth or with linear or transverse grooves, never with deep punctations (Fig. 16.10C); legs cylindrical; tarsi with proral setae simple or absent (Fig. 16.10B). External commensals or parasites of Coleoptera; Afrotropical, Oriental, Australian, Nearctic, Neotropical **HETEROCOPTIDAE**

7a. Supracoxal seta of legs I displaced laterally away from supracoxal gland opening (Fig. 16.9A); tibiae I–II with 0–1 ventral seta (rarely 2), femur IV usually without seta. Associates of many beetle families; cosmopolitan (except Nearctic) **CANESTRINIIDAE**

7b. Supracoxal seta of legs I closely associated with supracoxal gland opening; tibiae I–II usually with 2 ventral setae; femur IV usually with 1 ventral seta **8**

8a. Body cuticle smooth, striate or scaly; solenidion $\omega 2$ of tarsus I apical or absent; male without paranal suckers or suckerlike setae on tarsus IV. Associates of Blattaria, Dermaptera, and possibly Diplopoda; Afrotropical, Neotropical, Nearctic **ROSENSTEINIIDAE (*pars*; including Lophonotacaridae, Troglotacaridae)**

8b. Body cuticle mammillate, at least in female (Fig. 16.28A); solenidion ω*2* of tarsus I basal in position; male with paranal suckers and suckerlike setae on tarsus IV (Fig. 16.28B). Associates of Diplopoda; Afrotropical **CHETOCHELACARIDAE**

9a. Prodorsum with lamellar (*le*) setae absent; pretarsi with long, thin condylophores or condylophores fused or absent; empodial claws present or absent; dorsal setae may be elongate, but never heavily barbed; males without paranal suckers or suckerlike setae on tarsus IV **22**

9b. Prodorsum with lamellar (*le*) setae present, or if absent, then pretarsi with short, strong condylophores (some ACARIDAE) (Fig. 16.29F), or some dorsal setae long and heavily barbed (some GLYCYPHAGIDAE) (Fig. 16.22A), or empodial claws absent and opisthosoma bilobed posteriorly (HYPODERATIDAE) (Fig. 16.31A); males with or without paranal suckers and suckerlike setae on tarsus IV **10**

10a. Anus positioned directly behind genital opening, displaced from the posterior edge of the body by a distance greater than the length of the anus (Figs. 16.17A, D); without ventral ridges on subcapitulum; male with paranal suckers vestigial or absent **11**

10b. Anus positioned near posterior margin of body, or if more anterior, then subcapitulum with a distinct pattern of ventral ridges (some GLYCYPHAGIDAE) (Fig. 16.22B); male with or without paranal suckers **12**

11a. Dorsal setae relatively short, heavily barbed; leg segments with strong ridges; female with copulatory opening surrounded by a large, round sclerite (Fig. 16.27C); male with a pair of modified, suckerlike setae on tarsus IV; male with 4 pairs of setae in anal region (3 *p*, 1 *ad*; Fig. 16.27D); Monobasic (*Scatoglyphus*), in bird nests; Holarctic **SCATOGLYPHIDAE**

11b. Dorsal setae very long and barbed (*Euglycyphagus*) (Fig. 16.17A) or short and nude (*Lomelacarus*); female copulatory opening not surrounded by a large sclerite; male without suckerlike setae on tarsus IV and with only 3 pairs of setae in anal region (3 *p*). In bird nests and house dust; Holarctic, Afrotropical **EUGLYCYPHAGIDAE**

12a. Female and sometimes male with empodial claws bifurcate (Fig. 16.26C); male with legs III enlarged, terminating in a large, straight, empodial claw and an enlarged, straight spinelike seta similar in length and form to empodial claw (Fig. 16.26B); male with paranal suckers and suckerlike setae on tarsus IV. In vertebrate nests, carrion, and stored products; cosmopolitan **LARDOGLYPHIDAE**

12b. Both sexes with empodial claws simple or absent; male with legs III similar to legs IV, or if legs III enlarged, then empodial claw shorter and curved and without an enlarged seta similar in form to empodial claw **13**

13a. Ventral subcapitulum with a prominent pattern of external transverse and oblique ridges (Fig. 16.22B) not to be confused with internal pharyngeal sclerotization (ridges reduced in heavily sclerotized *Fusacarus*); empodial claws small or absent; condylophores very thin or absent (Figs. 16.22A, E); female usually with an epigynal apodeme and often with an external copulatory tube; male without paranal suckers or modified setae on tarsus IV. In nests of vertebrates, stored products, house dust, or occasionally in soil, litter, or plant foliage; cosmopolitan **GLYCYPHAGIDAE**

13b. Ventral subcapitulum without external ridges; other characters variable **14**

14a. Body cuticle at least partially striated, often in dorsal propodosomatic region or posteroventrally, or with a pattern of striations broken into scalelike structures (Fig. 16.20B); if striations or scales absent, then with all dorsal setae elongate and heavily barbed except *c1*, which is long and nude; solenidion ω*2* of tarsus I more distal than ω*1*; tarsus IV of both sexes with tectal setae absent **15**

14b. Body cuticle without striations; small rounded protuberances or small triangular microtrichia may be present; solenidion ω*2* variable in position, usually more basal than ω*1*; tectal setae present on all tarsi or one member of pair may be absent on all tarsi **16**

15a. Opisthosoma with a row of small, closely spaced microtrichia extending between the bases of most lateral body setae; most dorsal setae elongate and heavily barbed (Fig. 16.20B); pretarsal ambulacra short, relatively simple; empodial claws reduced or absent. In bat roosts, vertebrate nests, stored products, house dust; cosmopolitan **AEROGLYPHIDAE**

15b. Opisthosoma without microtrichia running between the bases of lateral body setae; body often ornamented with scalelike protuberances; dorsal setae often highly modified, rarely elongate and heavily barbed; gnathosoma usually with a pair of enlarged rutellar processes (Fig. 16.20D); pretarsal ambulacrum usually large, divided into 3 distinct regions (Fig. 16.20E; reduced in *Troglotacarus*); empodial claws well developed. In bat roosts or in fur of bats; cosmopolitan **ROSENSTEINIIDAE (*pars*)**

16a. Discrete coxal apodemes III and sometimes IV absent (Fig. 16.18A); discrete propodosomatic sclerite absent **17**

16b. Discrete coxal apodemes III–IV present, projecting obliquely from bases of trochanters; propodosomatic sclerite usually present **18**

17a. Body cuticle covered by small microtrichia (Fig. 16.19A); male without paranal suckers or modified suckerlike setae on tarsus IV (a basal sucker may be present in tibio-tarsal articulation, Fig. 16.19B). In nests of mammals, stored products, and house dust; Australian, Nearctic, Neotropical; 1 genus (*Blomia*) introduced cosmopolitan **ECHIMYOPODIDAE**

17b. Body without microtrichia; male with paranal suckers and often with suckerlike setae on tarsus IV (Fig. 16.18C). In nests of mammals, stored products, and house dust; Afrotropical, Oriental, Australian, Nearctic; 1 species (*Chortoglyphus arcuatus*) introduced cosmopolitan **CHORTOGLYPHIDAE**

18a. Tarsi with tectal setae asymmetrical, with *tc″* usually spinelike, rarely filiform; unguinal setae usually larger and stouter than proral setae (Fig. 16.30A) **21**

18b. Tarsi with both tectal setae filiform, similar in length; proral setae spinelike or enlarged and clawlike; unguinal setae spinelike, reduced or absent, never distinctly larger than proral setae **19**

19a. Propodosoma with lamellar (*le*) setae absent; empodial claws absent, tibiae I and II with two ventral setae (Fig. 16.31A); body cuticle usually smooth or with scalelike ornamentation; opisthosoma usually bilobed posteriorly (Fig. 16.31B), if rounded, then chelicerae vestigial or greatly enlarged (*Hypodectes*) (Figs. 16.31E, F). In nests of birds or, rarely, mammals; cosmopolitan with few genera described as adults **HYPODERATIDAE**

19b. Propodosoma with lamellar (*le*) setae present; empodial claws present, or if absent, then tibiae I–II with only l ventral seta; body cuticle often with rounded protuberances; opisthosoma not bilobed posteriorly **20**

20a. Body outline round, length similar to width (Figs. 16.24A, B); some dorsal setae often heavily pectinate or plumose (Fig. 16.24B), occasionally all short and simple (*Platyglyphus*); tarsi with proral and unguinal setae similar in form. In nests of social bees; Holarctic, Neotropical, Oriental **GAUDIELLIDAE**

20b. Body distinctly longer than wide; dorsal setae filiform, unbarbed; tarsi with proral setae enlarged and clawlike, unguinal setae reduced or absent (Fig. 16.25D). In vertebrate or insect nests, stored products, house dust, rarely in soil; cosmopolitan **SUIDASIIDAE (including Sapracaridae)**

21a. Prodorsal sclerotization in the form of 2 thin, elongate, parallel sclerites (Fig. 16.27A); dorsal setae elongate and heavily barbed; cuticle ornamented with small, triangular microtrichia; male with legs III much larger than legs IV but otherwise unmodified (Fig. 16.27B). Monobasic (*Glycacarus*), in bird nests; subantarctic islands **GLYCACARIDAE**

21b. Prodorsal sclerotization in the form of a rectangular shield (Fig. 16.29B) that may be incised posteriorly, or prodorsal sclerotization absent; dorsal setae variable but not usually both elongate and heavily barbed; cuticle smooth or rarely with small rounded protuberances; male with legs III similar to legs IV, or legs III enlarged and bearing a large empodial claw and reduced tarsal setation. In a wide variety of habitats; cosmopolitan **ACARIDAE**

22a. Tarsi I–II very short, bearing a greatly enlarged, apical, clawlike seta (Figs. 16.13A, B); pretarsi I–II elongate, bearing a very small empodial claw; pretarsi III–IV shorter, empodial claws very large (Fig. 16.13A); supracoxal setae greatly elongate and heavily barbed; with a pair of lyriform organs posterior to the prodorsal sclerite (Fig. 16.13B). Marine intertidal to subtidal, cosmopolitan in coastal areas **HYADESIIDAE**

22b. Tarsi I–II more elongate, without a greatly enlarged apical seta; pretarsi variable in form but similar on all legs; supracoxal setae short; without lyriform organs on prodorsum although ocelli may be present **23**

23a. Supracoxal gland opening on, or adjacent to, a large, sclerotized region directly above legs I (Figs. 16.14C, 16.16A) **24**

23b. Supracoxal gland opening not associated with a large sclerotized region **25**

24a. Supracoxal gland opening on an elongate, oval, sclerotized region, restricted to dorsal area above legs I (Fig. 16.14C); female pretarsus with paired elongate condylophores; male with condylophores fused along their length and emerging ventrobasally as a hooklike structure in the base of the pretarsus. In bee nests; cosmopolitan **CHAETODACTYLIDAE**

24b. Supracoxal gland opening hidden by a large, straplike sclerotized region (axillary organ) that usually extends ventrally between legs I–II (Fig. 16.16A) and may extend posteriorly over ventral surface (Fig. 16.16B); condylophores normally developed or absent in both sexes. In sap fluxes, phytotelmata, or fully aquatic habitats; Holarctic, Neotropical, Oriental, subantarctic islands **ALGOPHAGIDAE**

25a. Prodorsal sclerite absent; rostral (*ro*) setae situated about halfway between anterior edge of propodosoma and interlamellar (*in*) setae; with a pair of large ocelli in propodosomatic region (Fig. 16.15A); coxal apodemes I fused medially with coxal apodemes II closing coxal fields I in both sexes (Fig. 16.15B); condylophores elongate and separate in female, asymmetrical in male (Fig. 16.15C); male with genital setae (*g*) and coxal setae *4b* present. Monogeneric (*Carpoglyphus*), in stored products, vertebrate nests, bee nests, flowers; cosmopolitan **CARPOGLYPHIDAE**

25b. Prodorsal sclerite well developed; rostral (*ro*) setae at anterior edge of propodosoma or absent; ocelli present or absent; apodemes I not fused medially with apodemes II, coxal fields I open; condylophores fused to each other or absent; male with genital setae vestigial or absent, setae *4b* filiform, or absent but with alveoli modified into a sucker **26**

26a. Empodial claws present; condylophores usually fused into a V-shaped sclerite or a more elongate sclerite in base of ambulacral stalk (Fig. 16.11C); tibiae I–II with 1–2 ventral setae; male with all traces of genital setae and coxal setae *4b* absent; male usually with pretarsal ambulacra I–II arising from ventral apex of tarsus, often modified as a sucker (Fig. 16.11D). In a wide variety of habitats; cosmopolitan **WINTERSCHMIDTIIDAE (=Saproglyphidae)**

26b. Empodial claws absent, condylophores apparently absent; tibiae I–II with 0–1 ventral seta **27**

27a. Both sexes with genital opening between or posterior to coxal fields IV, female oviporus often confluent with anal opening (Fig. 16.12C); male with alveoli of setae *4b* fused medially forming a sucker anterior to aedeagus (Fig. 16.12A). In a variety of habitats; cosmopolitan **HEMISARCOPTIDAE**

27b. Both sexes with genital opening between coxal fields III–IV (Fig. 16.13C); male with setae *4b* filiform, alveoli not fused to form a sucker (Fig. 16.13D). In nests of stingless bees (Apidae: Meliponini), Neotropical, Oriental **MELIPONOCOPTIDAE**

28a. Parasitic or paraphagic on or in mammals; genu I with 0–1 solenidion (2 in *Hyracoptes* parasitic on hyraxes) **29**

28b. Parasitic or paraphagic on birds, hyperparasitic on hippoboscid flies or lice (some EPIDERMOPTIDAE), nidicolous or free living (some PYROGLYPHIDAE); genu I with 0–2 solenidia **40**

29a. Dorsal hysterosoma with at least 1 pair of setae between interlamellar (*in*) setae and posterior terminus (Fig. 16.50C), or if setae absent, then subcapitulum and palpi greatly enlarged (Fig. 16.53F). External parasites, on hair or skin or in hair follicles, subcutaneous tissues, mouth, or ears **32**

29b. Interlamellar and/or exobothridial setae and posterior terminal setae *h2* may be present, other dorsal hysterosomatic setae absent (Fig. 16.56D); palpi not greatly enlarged. Endoparasites in respiratory tracts of Chiroptera, Rodentia, and Primates or in stomach or eye orbits of Chiroptera **30**

30a. Dorsal hysterosoma without sclerites, although scales may be present (Fig. 16.56D); tarsi very short, equal in length or shorter than tibiae. In nasal passages of Chiroptera and Rodentia or eye orbits or stomach of Chiroptera; cosmopolitan **GASTRONYSSIDAE (including Yunkeracaridae)**

30b. Dorsal hysterosoma with one or more sclerites; tarsi usually longer than tibiae **31**

31a. Body rounded; legs IV inserted in posterior 1/4 of body; opisthosoma covered by a single large dorsal sclerite (Fig. 16.56F). In lungs of Rodentia (Sciuridae, Muridae); Holarctic **PNEUMOCOPTIDAE**

31b. Body very elongate; legs IV inserted more anteriorly; opisthosoma with 1–3 dorsal sclerites and a distinct posterior ventral sclerite (Fig. 16.56G). In nasal passages of Primates (Galagonidae, Cebidae); Afrotropical, Neotropical **LEMURNYSSIDAE**

32a. Female with oviporus in the form of a transverse slit situated between coxal fields II–III (Fig.16.52A); genital apodemes very small or absent; idiosoma globular or elongate, anus usually dorsal; all legs very short. In skin of many mammalian orders; cosmopolitan **SARCOPTIDAE (including Teinocoptidae, Bakerocoptidae)**

32b. Female with oviporus in the form of an inverted U or V, rarely transverse (if transverse, then legs I–II elongate or oviporus situated between coxal fields III–IV **33**

33a. Females with pretarsal ambulacral stalks I–II elongate, usually longer than the tarsi, pretarsi III–IV absent; legs III–IV of female short, with reduced segmentation; both sexes usually with gnathosoma elongate, bearing retrorse spines on subcapitulum and/or palpi (Fig. 16.51A). In hair follicles of Paucituberculata (Caenolestidae), Primates (Cercopithecidae), Carnivora (Ursidae, Procyonidae), and Rodentia (Hystricidae); Afrotropical, Neotropical, Nearctic **RHYNCOPTIDAE (including Audycoptidae)**

33b. Female with pretarsal ambulacral stalks of all legs similar, or tarsi III and/or IV without pretarsi but bearing more than 1 long seta; gnathosoma with at most 1 pair of retrorse spines on subcapitulum, none on palpi. On skin or hair or in ears **34**

34a. Gnathosoma greatly enlarged, strongly sclerotized; palpi pointed, unsegmented, forming a holdfast apparatus with chelicerae (Fig. 16.53F); dorsal hysterosoma with at most 1 pair of setae. Monogeneric (*Chirorhynchobia*), on wings of Chiroptera (Phyllostomidae); Neotropical **CHIRORHYNCHOBIIDAE**

34b. Gnathosoma not greatly enlarged; dorsal hysterosoma with more than 1 pair of setae **35**

35a. At least 1 pair of legs modified for clasping hair or skin; body usually cylindrical or laterally flattened, occasionally dorsoventrally flattened; on hairs of hosts **37**

35b. Legs not modified for clasping hairs; body globose or dorsoventrally flattened; on skin or hairs **36**

36a. Anterior apodemes of coxal fields I fused to form a sternum; opisthosoma elongate, bilobed posteriorly (Fig. 16.50D). On hairs of Xenarthra (Bradypodidae) and Rodentia (Echimyidae); Neotropical **LOBALGIDAE**

36b. Anterior apodemes of coxal fields I not fused to form a sternum (Fig. 16.50A); body rounded; female opisthosoma rarely bilobed, male with or without bilobed opisthosoma. On skin or in ears of many mammal orders **PSOROPTIDAE**

37a. Legs III–IV of female and III of male strongly modified for clasping hairs (Fig. 16.53A), rarely legs IV of female reduced; body usually dorsoventrally flattened, occasionally females cylindrical (*Trichoecius*) (Fig. 16.53D). Parasites of Didelphimorphia, Paucituberculata, Microbiotheria, and Rodentia **MYOCOPTIDAE**

37b. Legs III–IV not modified for clasping hairs **38**

38a. Legs I–II terminating in flat, platelike attachment organs (Fig. 16.54E), or entire legs I–II flattened and platelike (Figs. 16.54C, D); pretarsal elements absent from legs I–II. Parasites of Primates, Afrosoricida, Eulipotyphla, Carnivora (Mustelidae), Rodentia (Castoridae), and most species of Chiroptera **CHIRODISCIDAE**

38b. Legs I–II not terminating in platelike attachment organs; pretarsi present **39**

39a. Gnathosoma covered dorsally by a projecting tegmen (Fig. 16.54A); tibiae-tarsi III–IV freely articulated. Parasites of many mammalian orders **LISTROPHORIDAE**

39b. Gnathosoma not covered by a projecting tegmen; tibiae-tarsi III–IV completely fused (Figs. 16.59A, B). Parasites of many mammalian orders; primarily Southern Hemisphere **ATOPOMELIDAE**

40a. Pretarsal ambulacral discs well developed, without condylophore guide (Fig. 16.3D); tarsus IV with 3 ventral setae (*s* present); all tarsi with proral setae usually present, typically flattened and bifurcate apically (Fig. 16.3E), rarely filiform or absent **41**

40b. Pretarsal ambulacral discs variously formed, if large, then condylophore guide present (Fig. 16.4A); rarely pretarsi absent; tarsus IV with fewer than 3 ventral setae (*s* always absent) (Fig. 16.3E); proral setae always absent **58**

41a. Tarsus III with 1 ventral seta (*s*; Fig. 16.32C) **42**

41b. Tarsus III with 3 ventral setae (Fig. 16.34C) **43**

42a. Tarsi with proral setae present; 4 ventral setae on tarsi I–II positioned in middle of segment (Fig. 16.32B). Parasites of Ciconiiformes, Strigiformes **KRAMERELLIDAE**

42b. Tarsi with proral setae absent; 2 ventral setae on tarsi I–II, both positioned subapically (Fig. 16.33F). Parasites of Coraciiformes (Bucerotidae) or, rarely, Passeriformes (Emberizidae) **VEXILLARIIDAE**

43a. Proral setae present on all tarsi, flattened and typically bifurcate apically; condylophores L shaped (Fig. 16.32C) **45**

43b. Proral setae filiform, spinelike or absent on some or all tarsi; condylophores flattened and pointed apically, not L shaped (Figs. 16.33C, D, F) **44**

44a. Prodorsum with a single rostral (*ro*) seta (Fig. 16.32E); posterior legs inserted marginally. Parasites of Ciconiiformes (Threskiornithidae) **CAUDIFERIDAE**

44b. Prodorsum with 2 rostral setae or none; posterior legs ventrally inserted (Figs. 16.33A, B). Parasites of many orders of aquatic birds **FREYANIDAE**

45a. Trochanters I–III without setae; no seta on tarsus I longer than solenidion $\omega 3$ **46**

45b. Trochanters I–III with 1 seta each, or if setae absent, then 2 setae on tarsus I longer than solenidion $\omega 3$ **47**

46a. Solenidia σ absent from genua I–III; genital setae (*g*) present (Figs. 16.34D, F). Parasites of Passeriformes (Alaudidae, Sylviidae); Afrotropical **OCHROLICHIDAE**

46b. Solenidia σ present on genua I–III; genital setae (*g*) absent (Figs. 16.34A–C). Parasites of Apodiformes (Apodidae) **EUSTATHIIDAE**

47a. Female oviporus in the form of a longitudinal slit (Fig. 16.35B); ambulacral disc with long point in axis of disc. Parasites of Anseriformes (Anatidae) **RECTIJANUIDAE**

47b. Oviporus in the form of an inverted U, V, or Y, or transverse; ambulacral discs otherwise **48**

48a. Female with genital papillae posterior to genital setae (*g*) (Fig. 16.35D); genital papillae of male relatively distant from aedeagus that originates posterior to bases of legs IV (Fig. 16.35F); tibia IV without seta *v'*; male anterior coxal apodemes I ending freely, not fused into a V shape. On many bird orders **GABUCINIIDAE**

48b. Female with genital papillae anterior to genital setae (*g*); tibia IV with seta *v'* usually present (if *v'* absent, males with coxal apodemes I V shaped (some PTEROLICHIDAE) or male genital papillae very close to aedeagus (some FALCULIFERIDAE and PTEROLICHIDAE) **49**

49a. Female with external median copulatory tube at least as long as 1/2 length of tarsus IV (Fig. 16.36C); male with femur and genu IV fused and/or aedeagus positioned between coxal apodemes I **50**

49b. Female with external copulatory tube small or absent; male with femur and genu IV freely articulated and with aedeagus positioned posterior to coxal apodemes I **51**

50a. Female copulatory tube rigid, heavily sclerotized, and very thick (Fig. 16.36E). Parasites of Galliformes (Megapodiidae); Oriental, Australian **THORACOSATHESIDAE**

50b. Female copulatory tube thin, flexible, length variable (Fig. 16.36C). Parasites of Tinamiformes (Tinamidae); Neotropical **CRYPTUROPTIDAE**

51a. Propodosoma with interlamellar (*in*) setae distant from each other and inserted posteromesal to exobothridial (*ex*) setae; lines drawn through *in* and *ex* of each side will connect at the meson in a right angle posterior to these setae **52**

51b. Setae *in* inserted closer to *ex*; lines drawn through setae on each side will connect in a more oblique angle **53**

52a. Dorsal idiosoma largely covered by well-developed propodosomatic and hysterosomatic shields (Fig. 16.37B); female with 3 *p* and 1 *ad* setae in paraproctal region. External parasites of Podicipediformes, Charadriiformes, and Ciconiiformes (Phoenicopteridae) **PTILOXENIDAE**

52b. Propodosomatic shield small and narrow; dorsal hysterosoma without shield (Fig. 16.37E); paraproctal region of female with 2 *p* and 0 *ad* setae. In quills of Strigiformes (Strigidae); Oriental **OCONNORIIDAE**

53a. Posterior legs inserted ventrally; seta *c3* filiform; male with well-sclerotized coxal fields; female with 3 *p* and 1 *ad* seta in paraproctal region. In feather quills **54**

53b. Posterior legs inserted laterally, rarely ventrally (CHEYLABIDIDAE); seta *c3* typically spinelike, rarely filiform (FALCULIFERIDAE); female typically with 2–3 *p* and 0 *ad* setae (2 *ad* present in CHEYLABIDIDAE). On feather vanes **56**

54a. Coxal apodemes I ending freely or fused in a V shape; female oviporus in the form of an inverted Y, with sclerotized lateral borders (Fig. 16.38B); female idiosoma globose, body often longer than 1 mm. In quills of Psittaciformes, Falconiformes, Galliformes, Apodiformes, Caprimulgiformes, Piciformes, and Passeriformes **ASCOURACARIDAE**

54b. Coxal apodemes fused in a Y shape forming a sternum; female oviporus Ω shaped, with anterolateral border strongly folded; idiosoma elongate **55**

55a. Dorsal hysterosoma with a single, undivided shield (Fig. 16.38F). In quills or rarely externally on feathers of Charadriiformes; cosmopolitan **SYRINGOBIIDAE**

55b. Dorsal hysterosoma with 2 longitudinal shields in female (Fig. 16.38D); shields of male fused, but with obvious suture line medially. Monogeneric (*Kiwilichus*), in quills of Apterygiformes; New Zealand **KIWILICHIDAE**

56a. Female without epigynal apodeme; oviporus in the form of an inverted U or a transverse slit (Fig. 16.39A); setae *c3* long, filiform. On feathers of Columbiformes and Psittaciformes; cosmopolitan **FALCULIFERIDAE**

56b. Female with epigynal apodeme present, or if absent, then oviporus an inverted Y and setae *c3* short and bladelike **57**

57a. Posterior legs inserted ventrally; anterior tarsi 2 times longer than corresponding tibiae; female with 3 *p* and 2 *ad* setae in paraproctal region (Fig. 16.39D). On feathers of Falconiformes, Afrotropical **CHEYLABIDIDAE**

57b. Female with either legs inserted laterally, or anterior tarsi less than twice the length of tibiae; female without *ad* setae (Fig. 16.40A). On feathers of many bird orders; cosmopolitan **PTEROLICHIDAE**

58a. Chelicerae with fixed digit absent or entire chelicerae vestigial; leg and body setae highly reduced (Figs. 16.40D–F). In respiratory passages, lungs, and air sacs of numerous avian orders **CYTODITIDAE**

58b. Chelicerae well developed; leg and body setae variously formed. In a variety of microhabitats, including respiratory passages **59**

59a. Body saclike or elongate; prodorsal sclerite with a strongly developed internal apodeme on each side (Fig. 16.44C); all legs very short, similar in form; in feather follicles or skin lesions of numerous avian orders **EPIDERMOPTIDAE (*pars*; former Knemidokoptidae)**

59b. Body variously formed; prodorsal sclerite without lateral apodemes, or if apodemes present, then legs III–IV different in form from legs I–II **60**

60a. Pretarsi I–II with ambulacral stalks as long as or longer than the tarsi; rostral (*ro*) setae absent. Endoparasites in respiratory passages, feather follicles, quill walls, or subcutaneous tissues **76**

60b. Pretarsi I–II with ambulacral stalks shorter than the tarsi; rostral (*ro*) setae present or absent. On feathers or skin or in lumen of quills **61**

61a. Tarsus IV with only 1 seta or legs IV absent (Fig. 16.41D). In quills of several avian orders, associated with syringophilid mites **APIONACARIDAE**

61b. Tarsus IV with more than 1 seta **62**

62a. Tibia IV without ventral seta **63**

62b. Tibia IV with 1 ventral seta **67**

63a. Genu II without dorsal solenidion σ (Fig. 16.41F). Parasites of Passeriformes, Apodiformes (Trochilidae), rarely other groups **PROCTOPHYLLODIDAE**

63b. Genu II with a dorsal solenidion σ **64**

64a. Genua and femora fused in all legs (Fig. 16.42B) **65**

64b. Genua and femora of legs I–II not fused (Fig. 16.43A). On many host groups **TROUESSARTIIDAE**

65a. Rostral setae (*ro*) always absent (Fig. 16.42C); tibia III with seta *v′* present, generally elongate (Figs. 16.42A, B); males with setae *h2* simple, never dilated; female with hysterosomatic sclerite always interrupted anterior to posterior terminus of body. On many bird groups, usually on body feathers **XOLALGIDAE**

65b. Rostral setae present or absent; tibia III with seta *v′* very reduced or absent (if present, 1 *ro* seta present or males with setae *h2* leaflike); female with hysterosomatic sclerite entire, not interrupted anterior to posterior terminus **66**

66a. Female with opisthosomatic lobes each prolonged by bipectinate appendage bearing an apical seta (Fig. 16.43D); male with 2 pairs of setae (*g*, *p3*) immediately anterior to paranal suckers and well posterior to aedeagus in most species. Monogeneric (*Thysanocercus*), on wings of Apodiformes **THYSANOCERCIDAE**

66b. Female terminal appendages variable in form or absent but without apical seta; male with seta *p3* immediately anterior to paranal suckers, genital setae always closer to aedeagus than to paranal suckers (Fig. 16.42F). On aquatic birds **ALLOPTIDAE**

67a. Genu III with solenidion σ absent. On skin and in feather follicles **68**

67b. Genu III with solenidion σ present **69**

68a. Anterior legs conical with short tarsi bearing apicodorsal clawlike process (Fig. 16.44B). Parasites of many avian orders or hyperparasites of Hippoboscidae **EPIDERMOPTIDAE (*pars*)**

68b. Anterior legs cylindrical; tarsi without apicodorsal clawlike process (Fig. 16.43G). Parasites of several avian orders **DERMATIONIDAE**

69a. Tarsus I with solenidion ω*1* inserted subapically, very near solenidion ω*3* **70**

69b. Tarsus I with solenidion ω*1* inserted basally **71**

70a. Tarsus II with solenidion ω inserted subapically (Fig. 16.44E). In quills of Apodiformes (Trochilidae) **PTYSSALGIDAE**

70b. Tarsus II with solenidion ω inserted in basal half of the segment (Fig. 16.45C). On feathers and skin or in quills of many avian orders, bird or mammal nests, house dust, or stored food products; cosmopolitan **PYROGLYPHIDAE**

71a. Pretarsal ambulacral stalk thickened, asymmetrical, with ventral surface convex, distally rounded at point of contact with ambulacral disc, arising ventrally from tarsal apex (Figs. 16.46D–F) **72**

71b. Ambulacral stalk with parallel sides, arising from apex of tarsus; stalk subdivided, with basal portion apically truncated, spherical distally **73**

72a. Both sexes with legs III and IV ventrally inserted; female without epigynum (Fig. 16.48A). In quills of many avian orders **DERMOGLYPHIDAE**

72b. Legs III and IV inserted laterally or, if ventrally, then an epigynum is present in females (Figs. 16.46A, B). Usually on body feathers of many avian orders **ANALGIDAE**

73a. Pretarsi with ambulacral discs narrowing apically, with central sclerite almost as wide as disc (Fig. 16.47D). On wing or tail feather vanes **74**

73b. Ambulacral discs rounded or convex apically, central sclerite narrow (Fig. 16.45G). On feather vanes, plumaceous feathers or in quills **75**

74a. Ventral surface of tarsi I–II with extensive unsclerotized longitudinal membrane; lateral sclerites of ambulacral disc with distinct clear areas (lacunae). On Coraciiformes, Piciformes, and Passeriformes **PTERONYSSIDAE**

74b. Ventral surface of tarsi I–II completely sclerotized; lateral sclerites of ambulacral disc evenly sclerotized, without lacunae. On aquatic birds and osprey **AVENZOARIIDAE**

75a. Posterior legs ventrally inserted; female oviporus transverse (Fig. 16.48D). In quills of Galliformes (Phasianidae) **GAUDOGLYPHIDAE**

75b. Posterior legs laterally inserted; female oviporus shaped like an inverted Y (Fig. 16.45F). On downy feathers of many avian orders **PSOROPTOIDIDAE**

76a. Body elongate and flattened, often with elongate lateral processes (Fig. 16.49D); some lateral body setae very elongate; tarsi without large, hooked processes; male without paranal suckers. In feather follicles, quill walls, or subcutaneous cysts of many avian orders **LAMINOSIOPTIDAE**

76b. Body rounded; lateral body setae very short (Fig. 16.49A); tarsi usually with large, hooked processes (Fig. 16.49C); male with paranal suckers. In nasal passages of many avian orders **TURBINOPTIDAE**

Key 16.2. Cohort Astigmatina (Deutonymphs), Key to the Families (Figs. 16.1–16.31

Deutonymphs are not formed in any family of Psoroptidia. All deutonymphs belong to nonpsoroptidid groups.

1a. Genital valves with 4 pairs of minute setae (Fig. 16.6C); palpi with 2 freely articulating segments and vestiges of a third, more basal segment (Fig. 16.6B); attachment organ with conoidal setae *p1* and *p2* vestigial (Fig. 16.6A); tarsus II with 2 solenidia (Fig. 16.6E). Monogeneric (*Schizoglyphus*), with 1 species from a tenebrionid beetle; New Guinea **SCHIZOGLYPHIDAE**

1b. Genital valves with at most a single pair of setae; palpi without articulating segments, often vestigial or absent; attachment organ variously developed; tarsus II with 0–1 solenidion **2**

2a. Legs III–IV with femur-genu-tibia-tarsus partially or completely fused; legs III–IV flexed at trochanter-femur articulation and directed anteriorly in repose (Fig. 16.7C); rostral and lamellar pairs of setae absent; setae *4b* absent. Cosmopolitan on a variety of hosts **HISTIOSTOMATIDAE (Amoetidae)**

2b. Legs III–IV usually without above degree of fusion of segments; legs III–IV posteriorly or laterally directed in repose, or if anteriorly directed, then attachment organ vestigial or absent and living in hair follicles of mammals; at least rostral setae usually present; setae *4b* usually present **3**

3a. Attachment organ with anterior and median suckers vestigial or absent; setae *p1–p2* in the form of ribbed claspers (Figs. 16.23C–E), small vestigial conoids, or absent **4**

3b. Attachment organ with anterior and median suckers (*ad3* and *ad1+2*) well developed and setae *p1–p2* in the form of conoids (Fig. 16.10A); if attachment organ is reduced, at least 1 pair of suckers remains **8**

4a. Coxal setae *3a* and *4a* present, *1a* present or absent; attachment organ vestigial or absent (Fig. 16.18D) **5**

4b. Coxal setae *1a*, *3a*, and *4a* absent (occasionally a vestigial *1a* present); attachment organ with *p1–p2* as claspers or vestigial **7**

5a. Genital region terminal or subterminal, with genital papillae projecting posteriorly (Fig. 16.18D); ventral setae of tibiae III–IV in the form of strongly toothed combs; dorsal opisthosomatic setae spinelike, often bifurcate apically. Endofollicular parasites of Rodentia, Eulipotyphla, and Primates; Nearctic, Afrotropical, Australasian **CHORTOGLYPHIDAE**

5b. Genital region ventral with papillae in normal position; genital region often surrounded by a sclerotized ring (Fig. 16.19C); tibial setae of legs III–IV unmodified; dorsal setae variously formed **6**

6a. Coxal fields II closed; rostral (*ro*) setae situated on a strong, hooked, anterior protuberance (Fig. 16.17E); setae *p1–p2* present as very small conoids (Fig. 16.17F). Monobasic (*Pedetopus*), in hair follicles of *Pedetes*; Afrotropical **PEDETOPODIDAE**

6b. Coxal fields II open (closed in one undescribed species; Fig. 16.19C); rostral setae near anterior apex of propodosoma; setae *p1* and *p2* absent; body elongate or rounded. In hair follicles and subcutaneous tissues of various marsupial orders, Rodentia, and Xenarthra; Neotropical, Nearctic, Australian **ECHIMYOPODIDAE**

7a. Attachment organ with setae *p1–p2* in the form of ribbed claspers (Fig. 16.23C), or if absent, then vestiges of their alveoli present as small sclerotized rings (*Baloghella*) (Fig. 16.23F), or attachment organ reduced and tibiae III–IV with large ventral, comblike setae (Fig. 16.23B); posterior coxal apodemes with distinct edges. On hairs or in hair follicles of Didelphimorphia, Paucituberculata, Xenarthra, Eulipotyphla, Afrosoricida, and Rodentia, 1 species in ears of Carnivora; sedentary forms wind dispersed; cosmopolitan **GLYCYPHAGIDAE**

7b. All traces of attachment organ completely absent; legs well developed; tibiae III–IV without comblike setae (Fig. 16.31D); posterior coxal apodemes often with indistinct edges. In subcutaneous tissues of birds and rodents; cosmopolitan **HYPODERATIDAE**

8a. Legs I–II without pretarsi; coxal apodemes II–IV end freely **9**

8b. Pretarsal ambulacrum and/or empodial claw present on legs I–II; coxal apodemes II–IV end freely or fuse with other apodematal elements **10**

9a. Dorsal sclerites smooth; propodosomatic sclerite very short, interlamellar (*in*) and exobothridial (*ex*) setae apparently absent (Fig. 16.28C); genua I–II without dorsal solenidion. Monogeneric (*Lemanniella*), associated with ants; Holarctic **LEMANNIELLIDAE**

9b. Dorsal sclerites smooth or with rounded protuberances; propodosomatic sclerite normally developed, bearing interlamellar and exobothridial setae; genu I–II each with dorsal solenidion **SUIDASIIDAE (*pars*; genus *Sapracarus*)**

10a. Legs IV with empodial claw absent; legs IV often shorter than legs III, with some fusion of segments and usually with 1 or more long apical setae **11**

10b. Legs IV with empodial claw present, or if absent, then empodial claw III also absent, but claws I–II present; legs IV generally similar in form to legs III **15**

11a. Empodial claws I–III borne on membranous ambulacra (small in the CHAETODACTYLIDAE where claws are very large and hooked) **12**

11b. Empodial claws I–III arising directly from tarsal apices, or some or all claws absent (Figs. 16.25B). Phoretic on Coleoptera and Hymenoptera; cosmopolitan **SUIDASIIDAE (*pars*)**

12a. Pretarsal ambulacra with a pair of very thin, symmetrical condylophores extending from tarsal apices to empodial claws (Fig. 16.26F); dorsal hysterosomatic sclerite usually reticulate anteriorly, smooth or striate posteriorly (Fig. 16.26E). One genus (*Lardoglyphus*) phoretic on *Dermestes* beetles; cosmopolitan **LARDOGLYPHIDAE**

12b. Condylophores either restricted to tarsal apices, or if extending through ambulacral stalk, then very strong and asymmetrical **13**

13a. Empodial claws greatly enlarged and hooklike, ocelli absent (Fig. 16.14A); body relatively circular in outline and with 2 dorsal sclerites, or ellipsoid and with 1 dorsal sclerite. Phoretic on bees: Apidae *s. lat.* and Megachilidae; cosmopolitan **CHAETODACTYLIDAE**

13b. Empodial claws usually small and simple, if enlarged and hooklike, then body fusiform and with 2 dorsal sclerites; ocelli often present **14**

14a. Tarsus I with solenidia widely separated, solenidion $\omega 3$ apical; coxal apodemes III usually free apically, rarely weakly fused to median ventral apodeme; ocelli present on apex of propodosoma, pigment spots underlying ocelli fused medially; coxal setae *4a* well developed; tibiae I–II with 0–1 ventral seta (Fig. 16.12D). On a wide variety of insects; cosmopolitan **HEMISARCOPTIDAE**

14b. Tarsus I with solenidia closely associated, usually in the basal half of the tarsus; coxal apodemes III either fused together, to apodemes of coxal fields IV or to median apodeme, rarely free; ocelli present on apex of propodosoma with separate pigment spots (Fig. 16.11A), rarely widely separated, or occasionally absent; tibiae I–II with 1–2 ventral setae. On a variety of insect groups; cosmopolitan **WINTERSCHMIDTIIDAE**

15a. Pretarsi with membranous ambulacra **16**

15b. Pretarsi without membranous ambulacra, empodial claws arising from tarsal apices **17**

16a. Coxal fields III closed but not fused to each other medially; posterior median ventral apodeme absent; ocelli present, widely separated on propodosoma (Fig. 16.15D). Monogeneric (*Carpoglyphus*), phoretic on Lepidoptera and Coleoptera; cosmopolitan **CARPOGLYPHIDAE**

16b. Coxal fields III open or closed, but coxal apodemes III always fused medially; posterior median ventral apodeme present; ocelli, when present, at anterior apex of propodosoma (Figs. 16.16D, E). Phoretic on Diptera, Coleoptera, Lepidoptera; Holarctic, Neotropical, Oriental Australian **ALGOPHAGIDAE**

17a. Coxal apodemes III–IV ending freely; all coxal fields completely open (Fig. 16.24C); propodosomatic ocelli absent. Associates of social bees; Holarctic, Neotropical, Oriental **GAUDIELLIDAE**

17b. Coxal apodemes III–IV usually fused medially; if unfused (some ACARIDAE), then posterior apodemes II contiguous or fused with apodemes III, or propodosomatic ocelli present **18**

18a. Body very large, broadly teardrop shaped, coxal fields III narrowing medially, connected by a short apodeme (Fig. 16.10A). One genus (*Megacanestrinia*) on *Tefflus* spp. (Coleoptera: Carabidae), Afrotropical **CANESTRINIIDAE**

18b. Body not greatly widened posteriorly; coxal fields variable but coxal fields III usually quadrate or longer than wide **19**

19a. Rostral (*ro*) setae situated on a ventral tubercle between gnathosoma and overhanging propodosoma (Fig. 16.17C). One genus (*Euglycyphagus*) on trogid beetles; Holarctic **EUGLYCYPHAGIDAE**

19b. Rostral setae otherwise; if propodosoma overhangs gnathosoma anteriorly, then rostral setae inserted on its ventral surface (Figs. 16.30B, C); cosmopolitan **ACARIDAE**

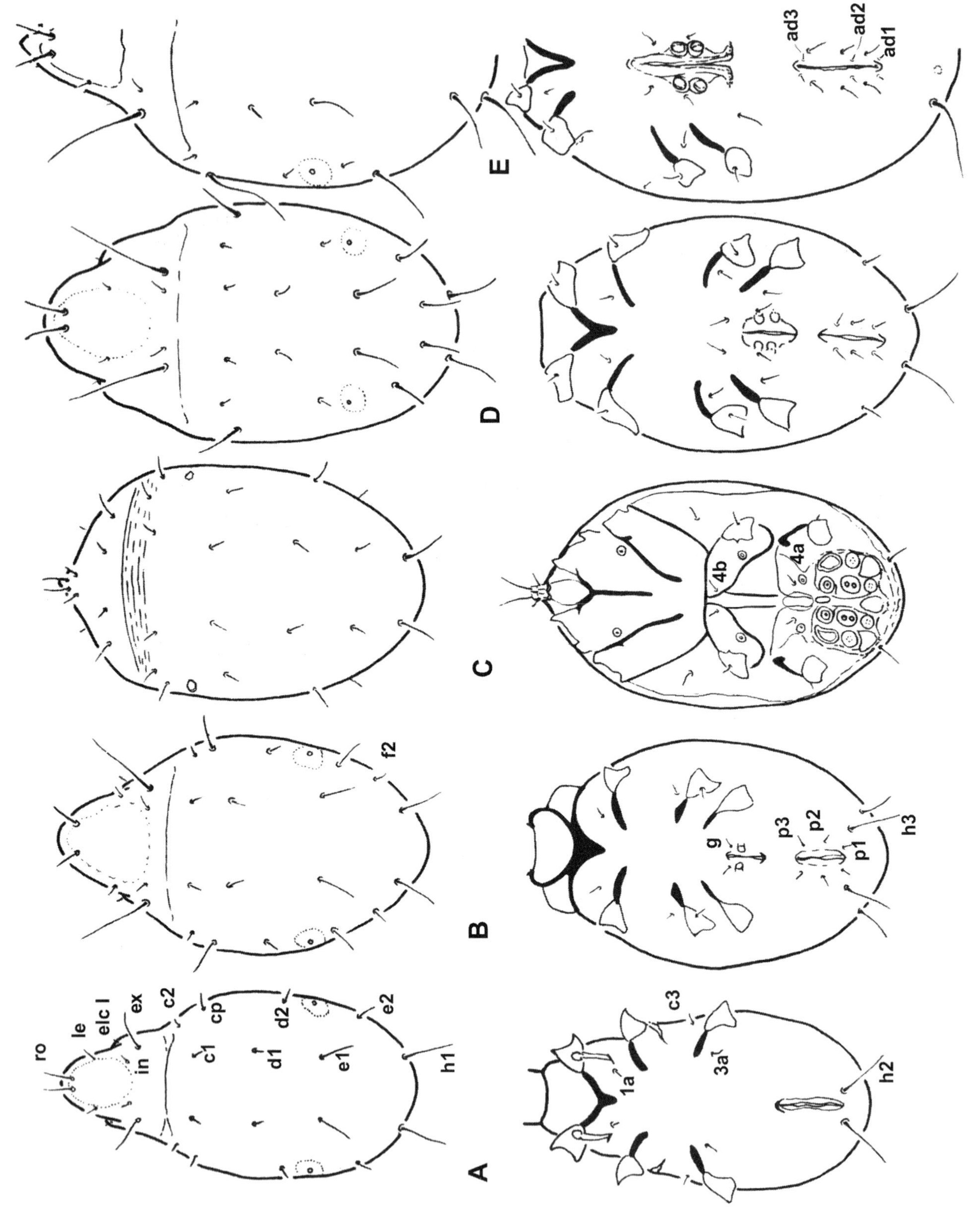

Fig. 16.1. Ontogenetic development of idiosomatic setae of *Sancassania* sp. showing dorsal (top) and ventral aspects of each instar. **A**, larva; **B**, protonymph; **C**, deutonymph; **D**, tritonymph; **E**, adult female (modified from Oseto and Mayo 1975).

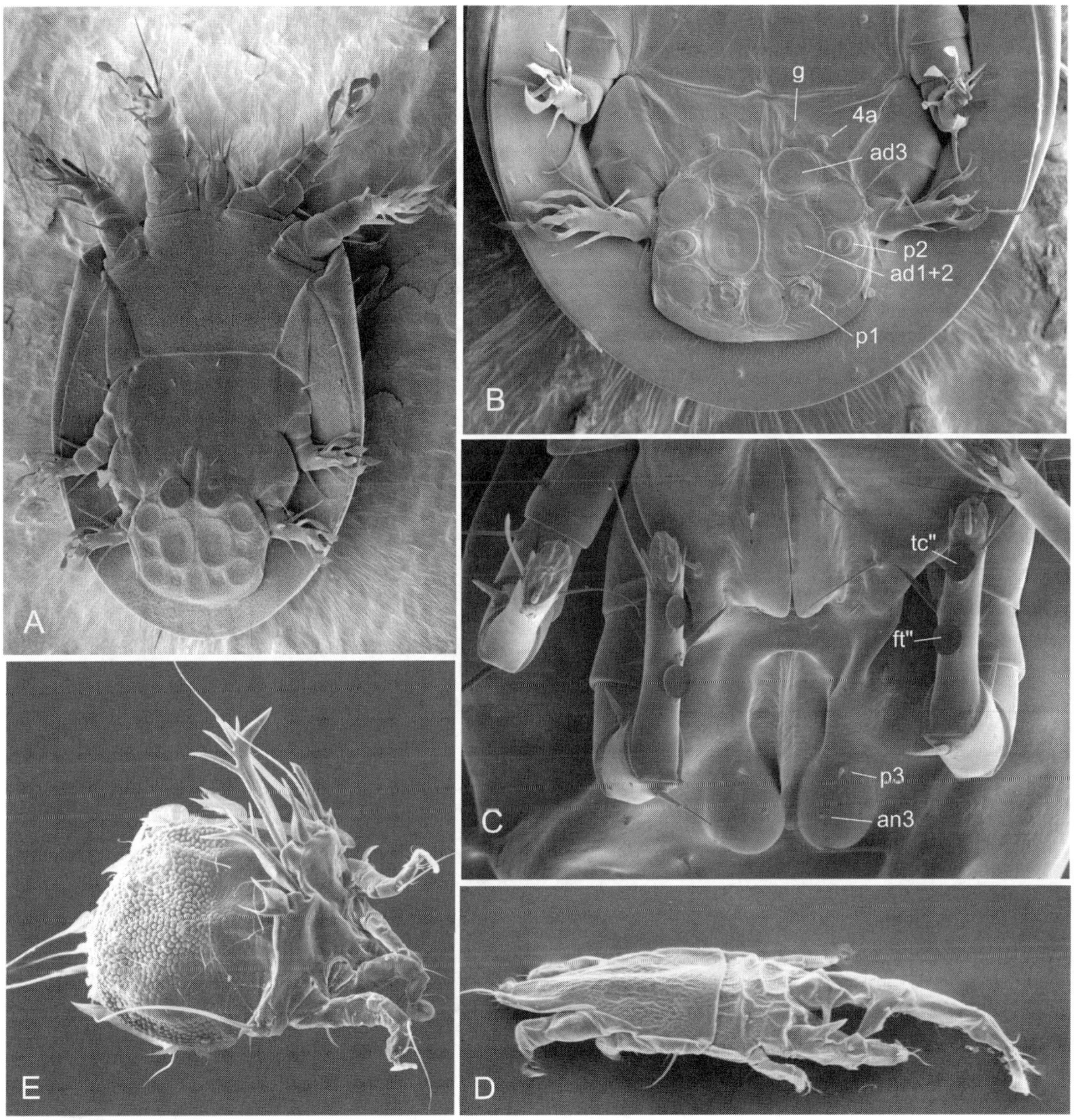

Fig. 16.2. *Sancassania* sp. (Maryland, USA): **A**, venter of deutonymph; **B**, detail of attachment organ; **C**, genito-anal region and leg IV of male showing paranal and tarsal suckers. *Hyperpedalloptes titanopedes* Dubinin: **D**, dorsolateral aspect of male (from Gaud and Atyeo 1996). *Freyanella tarandus* (Trouessart and Neumann): **E**, dorsolateral aspect of male.

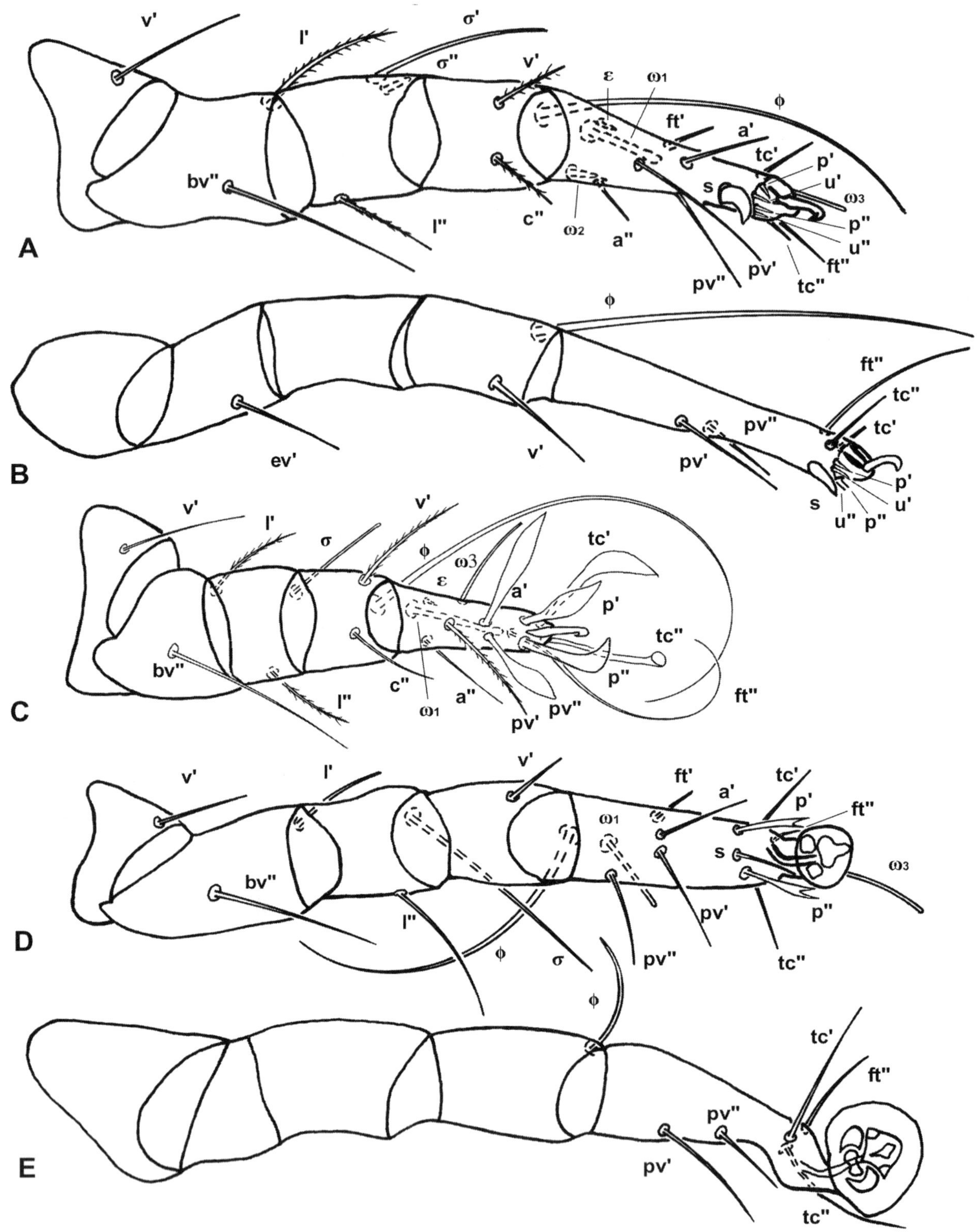

Fig. 16.3. Leg setation of representative Astigmatina.
ACARIDAE. *Acarus siro* L.: **A**, leg I of female; **B**, leg IV of female; **C**, leg I of deutonymph.
OCONNORIIDAE. *Oconnoria inexpectata* Gaud, Atyeo, and Klompen: **D**, leg I of female.
PROCTOPHYLLODIDAE. *Proctophyllodes glandarinus* (Koch): **E**, leg IV of female.

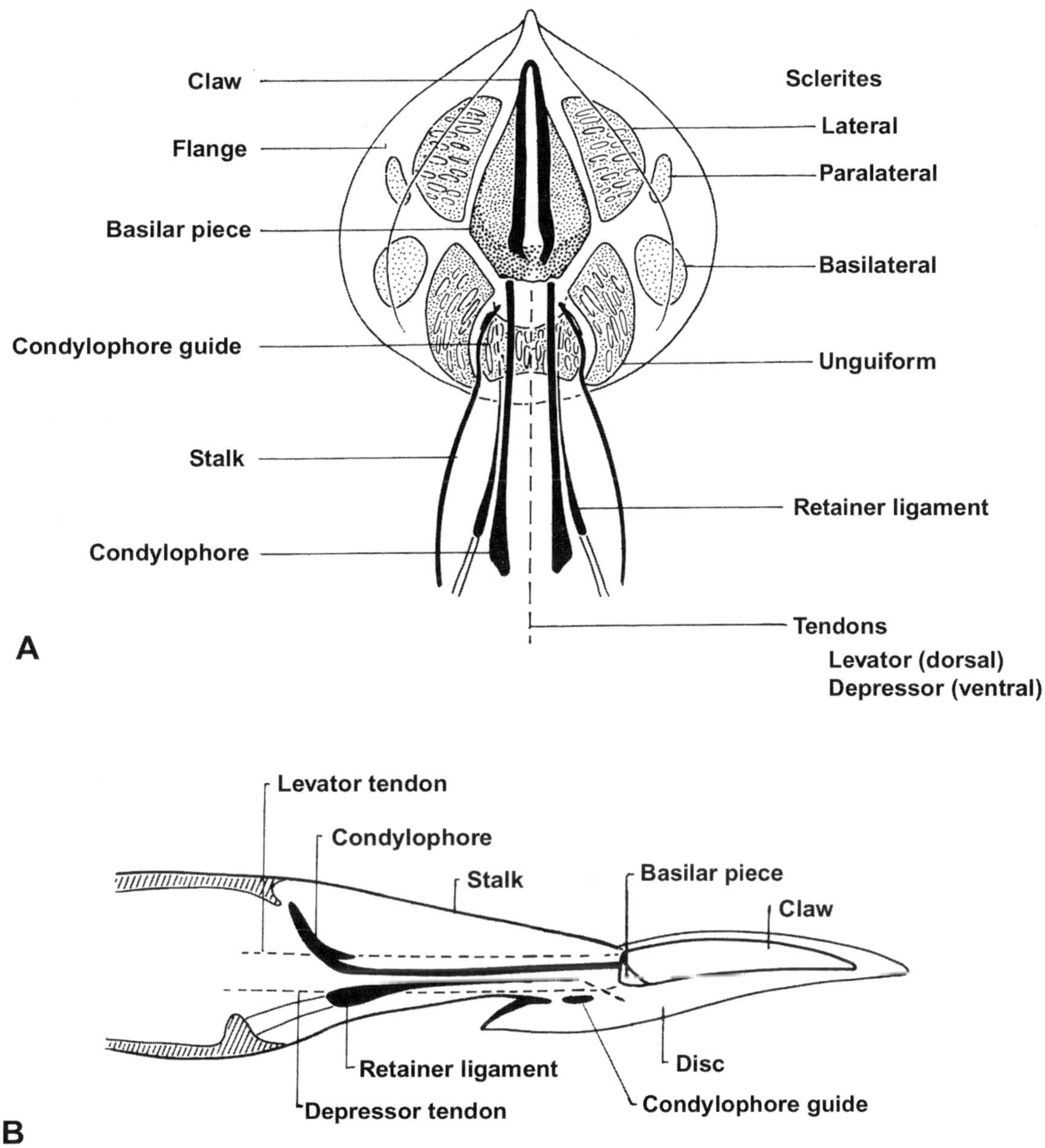

Fig. 16.4. Hypothetical psoroptidian mite pretarsus. **A**, dorsal aspect; **B**, lateral aspect (from Atyeo 1979).

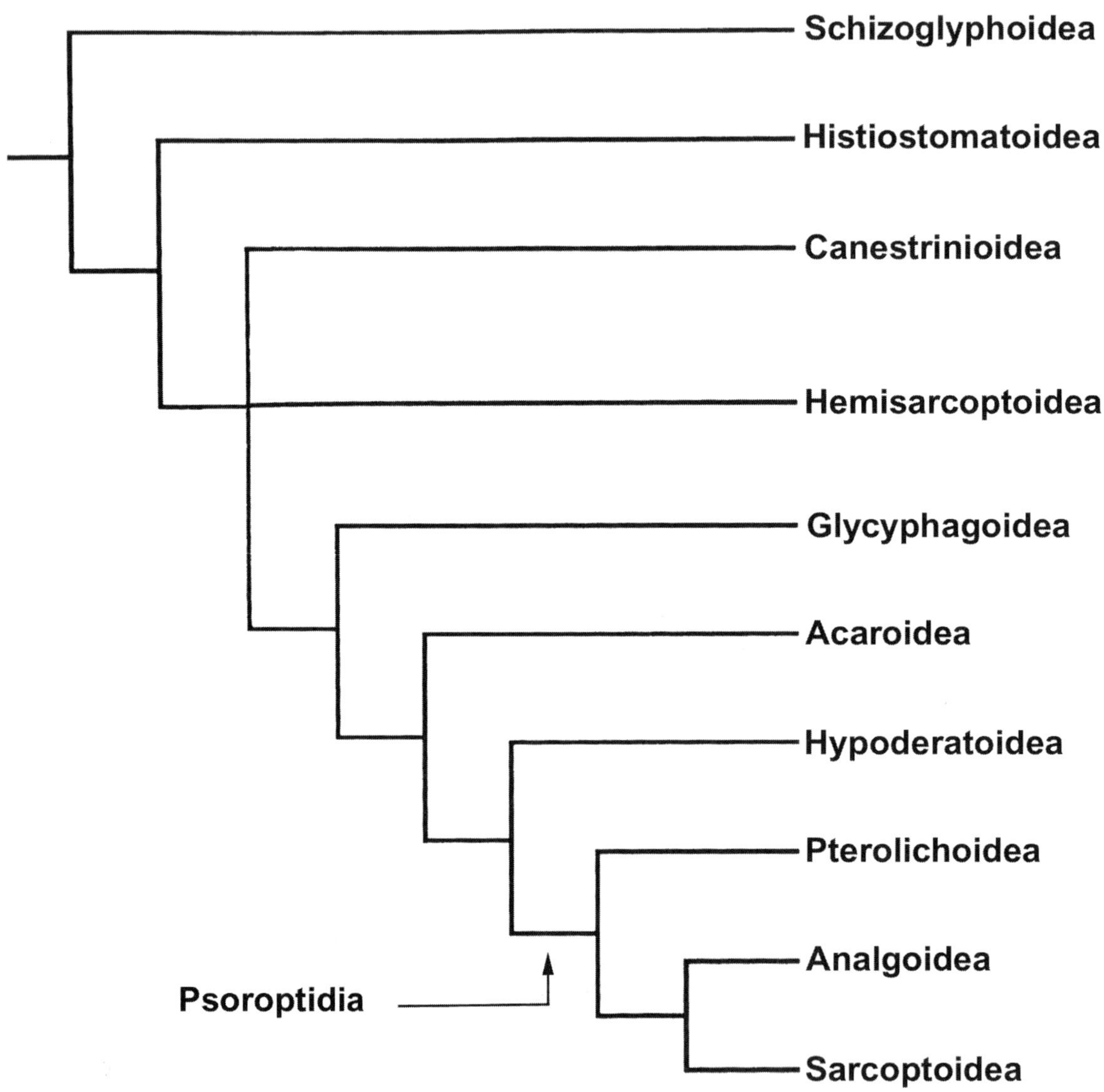

Fig. 16.5. Phylogeny of superfamilies of Astigmatina (from OConnor in Norton et al. 1993).

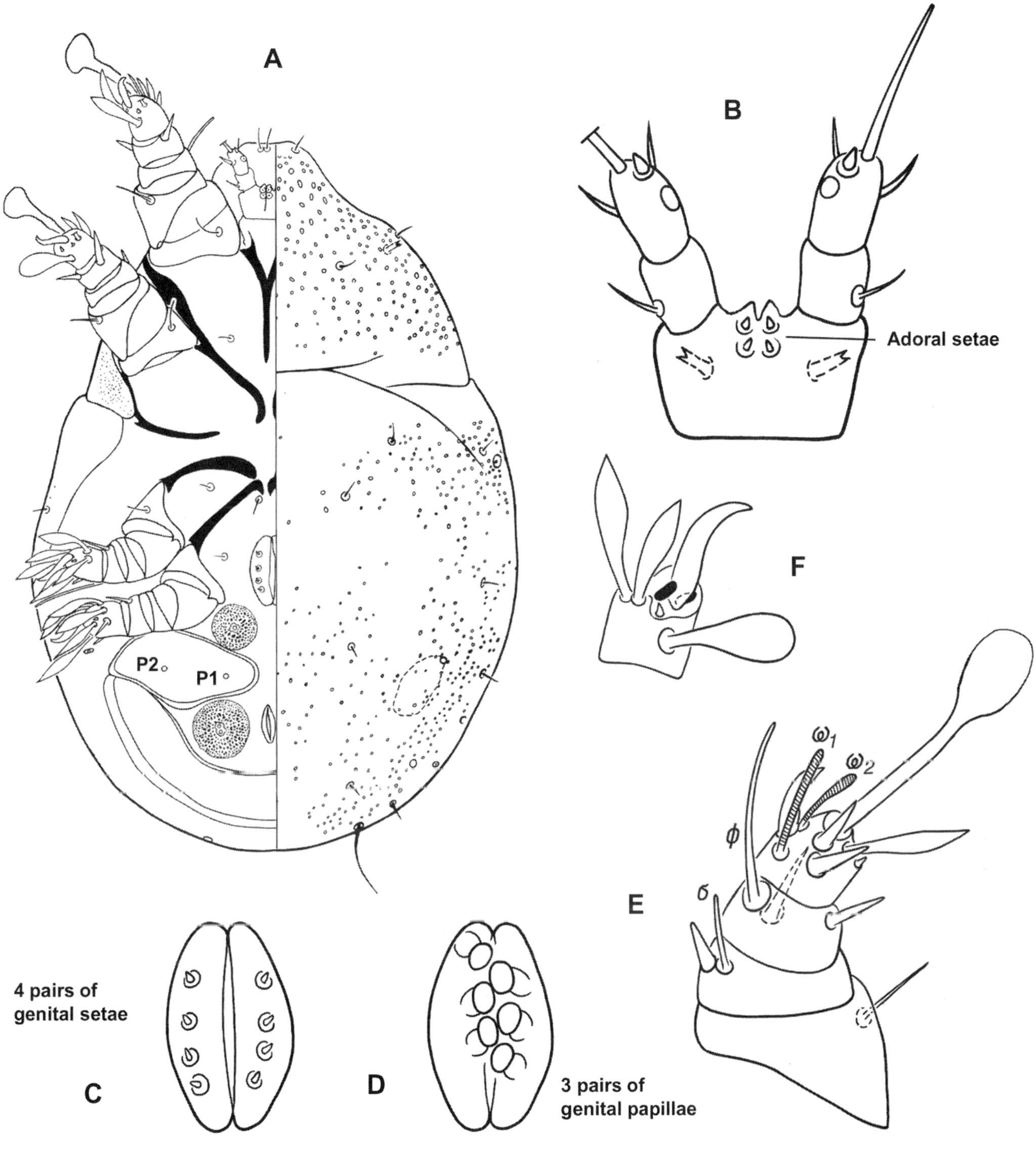

Fig. 16.6. SCHIZOGLYPHIDAE. *Schizoglyphus biroi* Mahunka (Papua, Indonesia): **A**, composite dorsal and ventral view of deutonymph; **B**, gnathosomatic remnant; **C**, genital valves; **D**, genital papillae; **E**, dorsal aspect of leg II with solenidia indicated; **F**, venter of tarsus II.

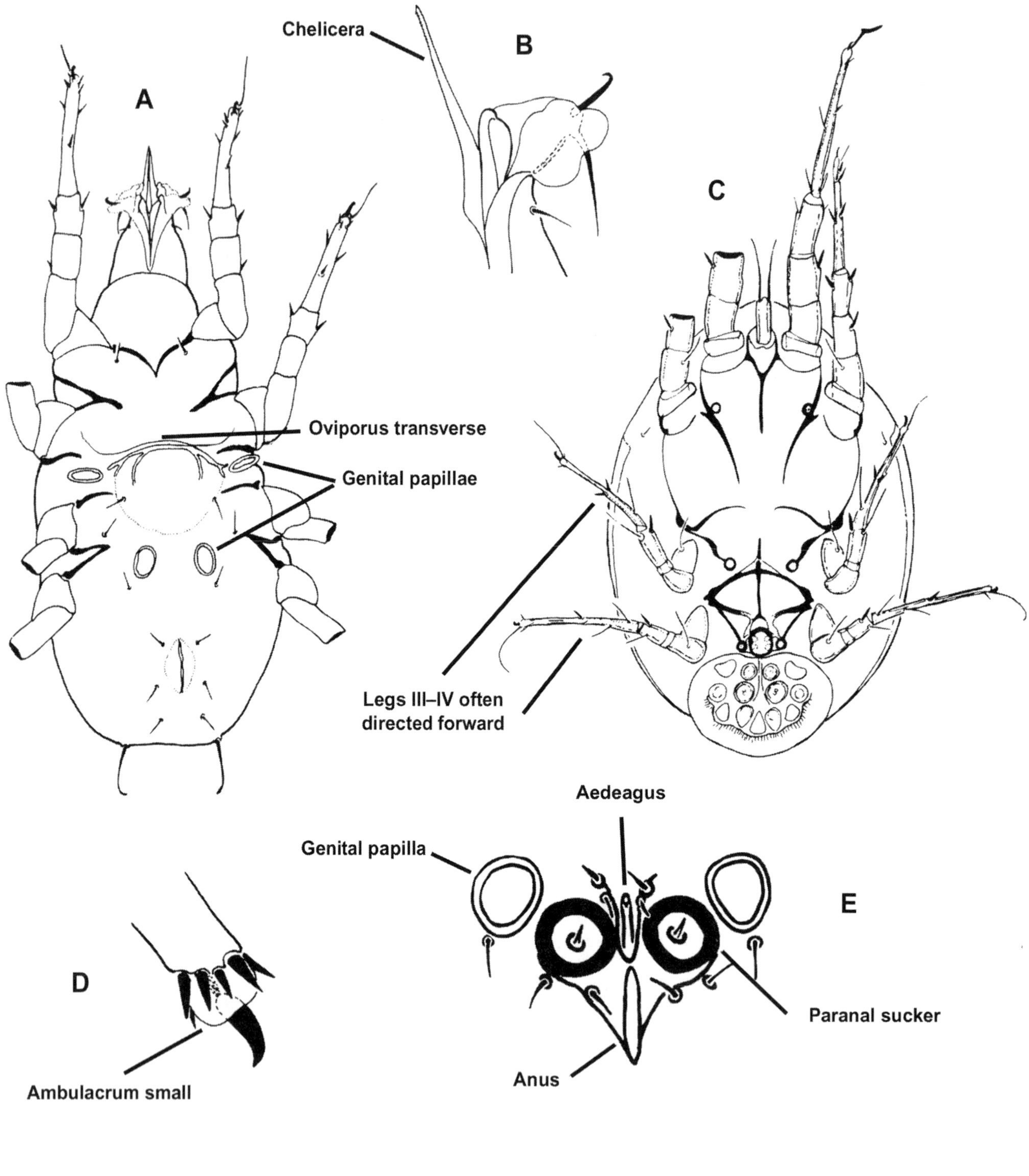

Fig. 16.7. HISTIOSTOMATIDAE. *Histiostoma* sp. (Oregon, USA): **A**, venter of female; **B**, modified palp and portion of chelicera; **C**, venter of deutonymph; **D**, pretarsus III of female. *Bonomoia* sp. (Mexico): **E**, genito-anal region of male.

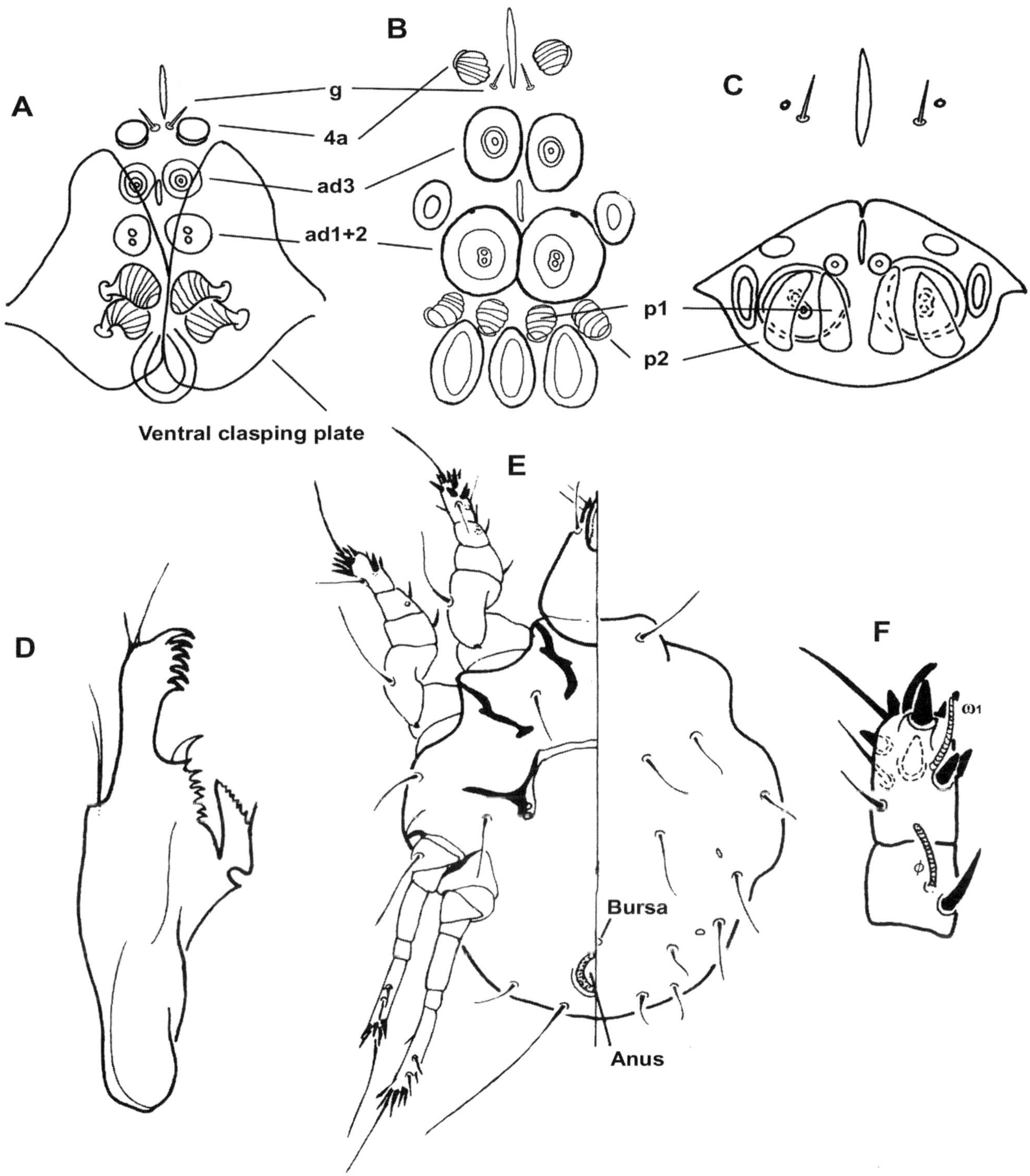

Fig. 16.8. HISTIOSTOMATIDAE. *Fibulanoetus* sp. (Kenya): **A**, deutonymphal attachment organ. *Prowichmannia* sp. (New York, USA): **B**, deutonymphal attachment organ. *Xenanoetus kofrankova* Samšinák (Florida, USA): **C**, deutonymphal attachment organ. GUANOLICHIDAE. *Guanolichus gabonensis* Fain (Gabon) (after Fain 1968b): **D**, chelicera of female; **E**, composite dorsum and venter of female; **F**, tibia and tarsus I of male.

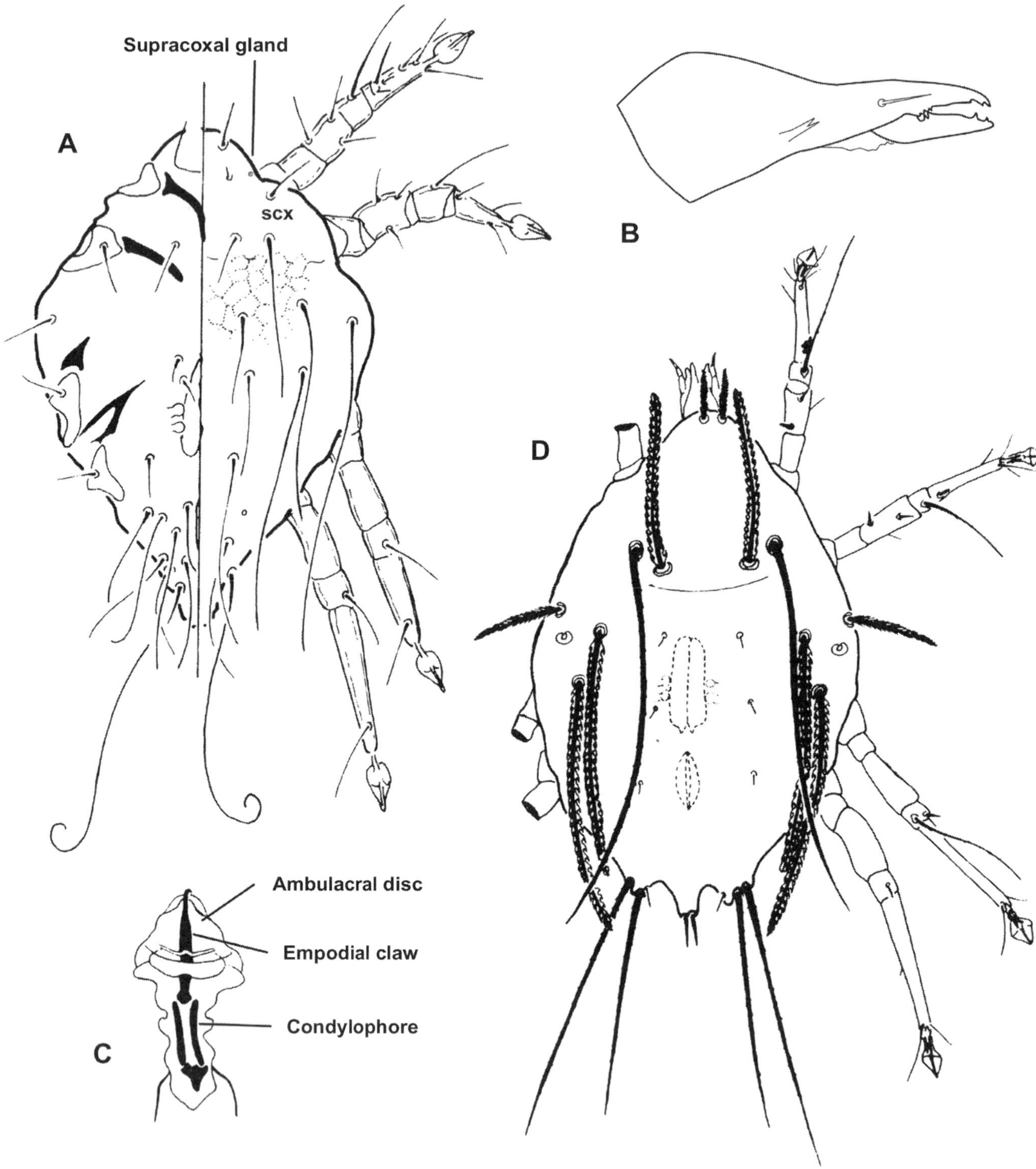

Fig. 16.9. CANESTRINIIDAE. *Mesophotia penicillata similis* Samšinák (Spain): **A**, composite dorsum and venter of female (adapted from Samšinák 1971). *Procericola ichthyoides scabricola* Samšinák (Turkey): **B**, chelicera of male. *Melisia* sp. (India): **C**, pretarsus II. *Melisia* sp.: **D**, dorsum of female.

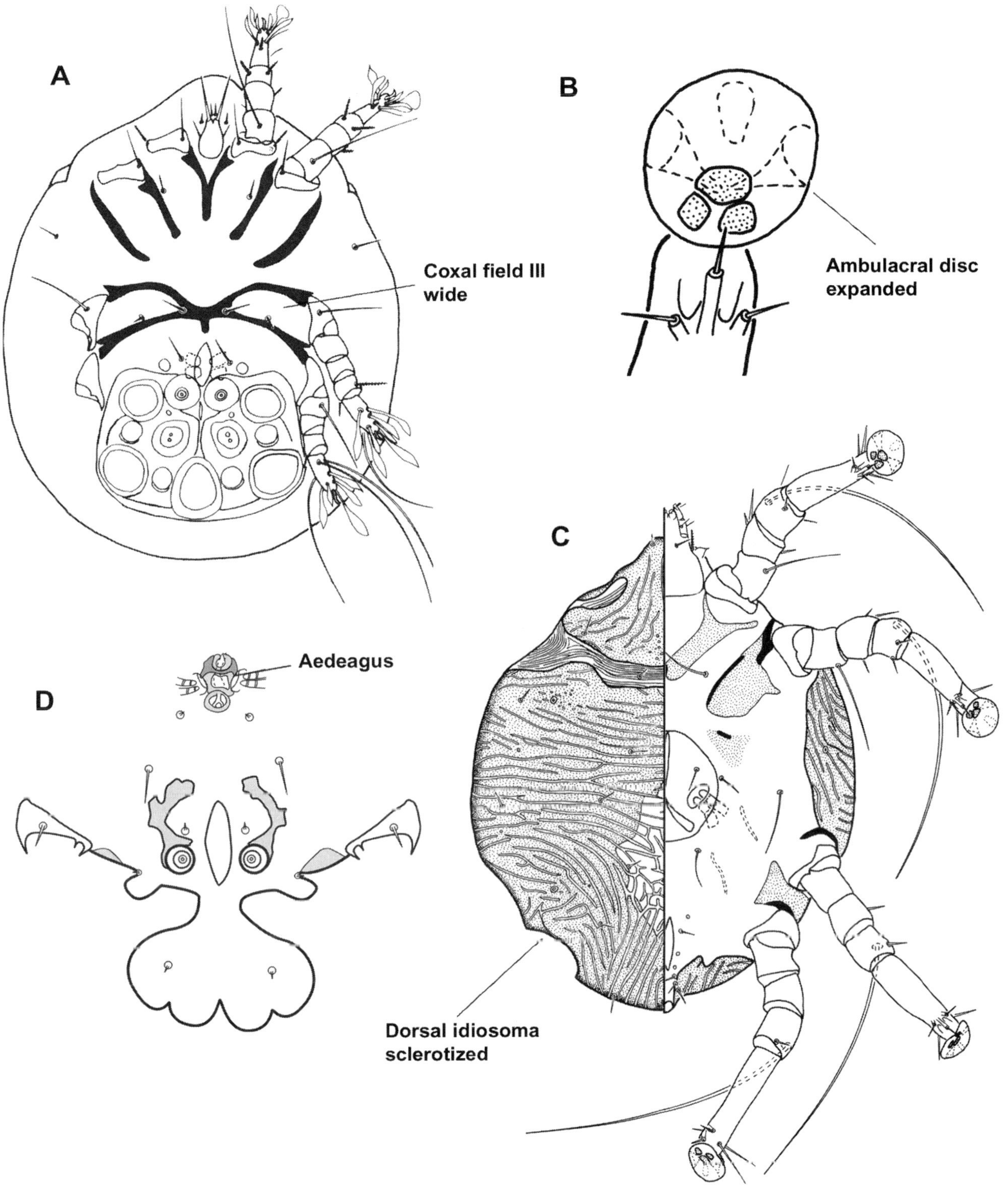

Fig. 16.10. CANESTRINIIDAE. *Megacanestrinia* sp. (Tanzania): **A**, venter of deutonymph.
HETEROCOPTIDAE. *Heterocoptes tarsii* Fain (India): **B**, distal tarsus and pretarsus I; **C**, composite dorsum and venter of female. *Erotylocoptes helenae* Haitlinger (South Africa): **D**, genito-anal region of male.

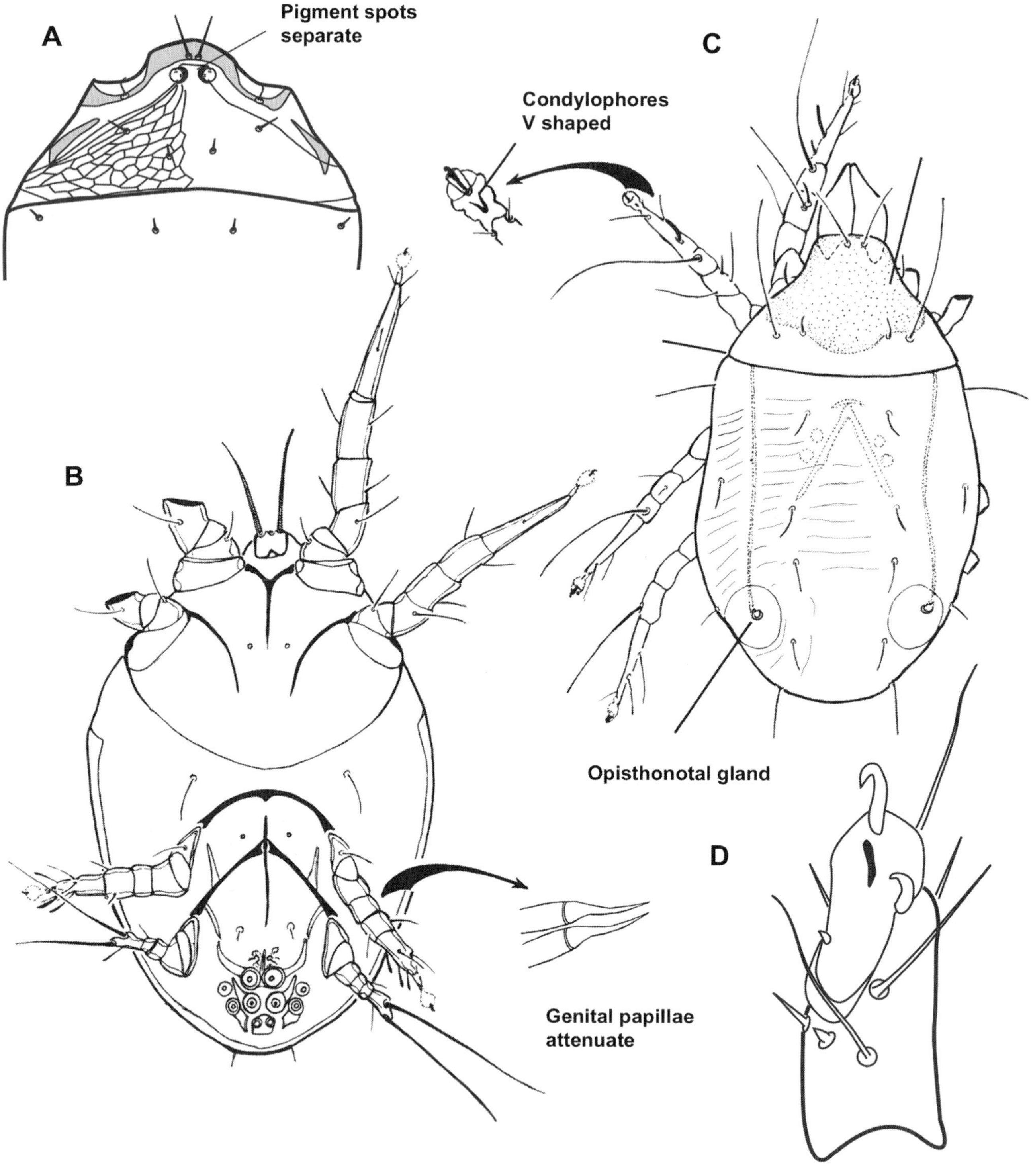

Fig. 16.11. WINTERSCHMIDTIIDAE. *Acalvolia squamata* (Oudemans) (Oregon, USA), deutonymph: **A**, propodosoma; **B**, venter. *Czenspinskia transversostriata* (Oudemans) (Oregon, USA): **C**, dorsum of female with detail of pretarsus. *Saproglyphus* sp. (New York, USA): **D**, tarsus and pretarsus of male.

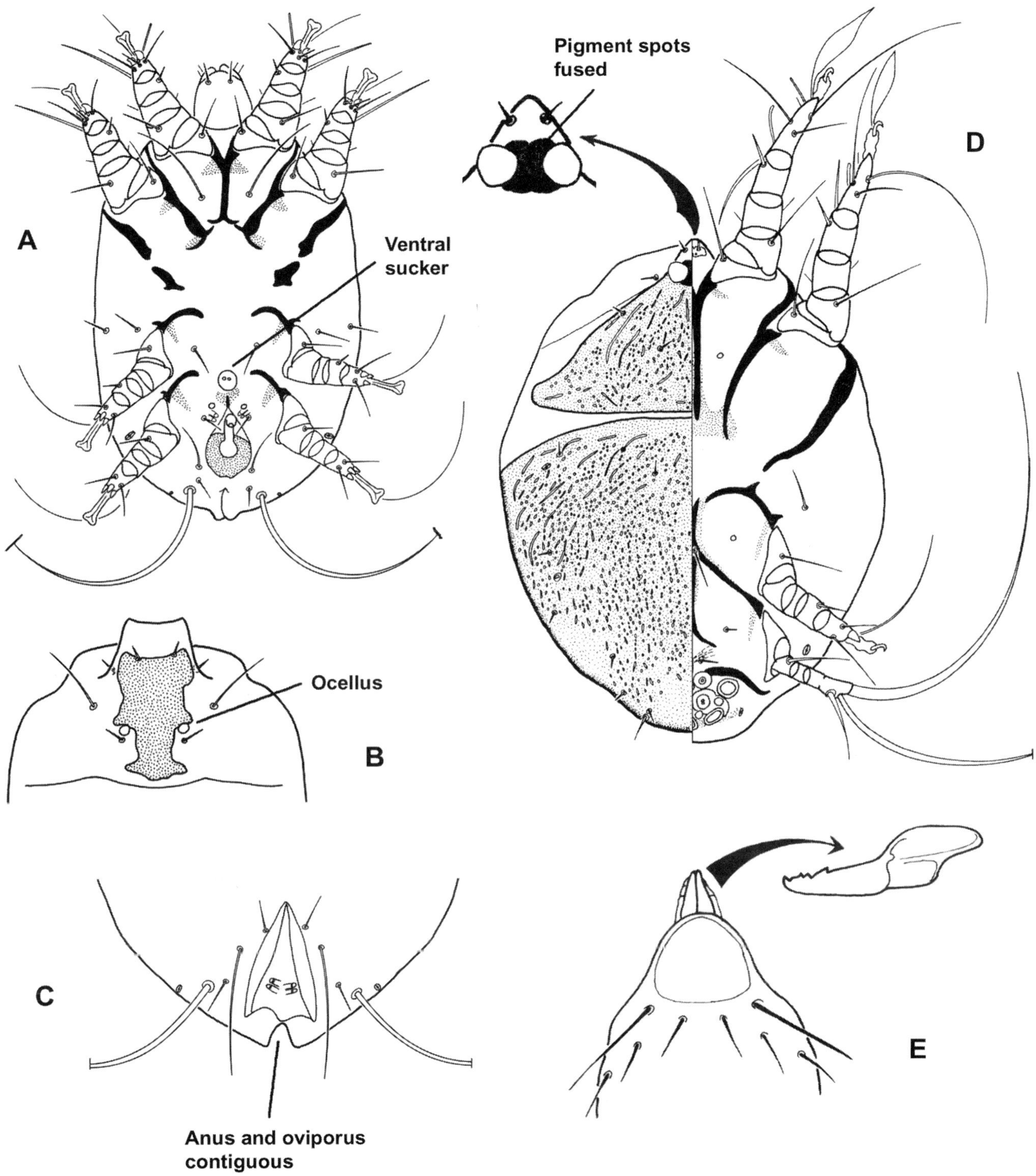

Fig. 16.12. HEMISARCOPTIDAE. *Hemisarcoptes cooremani* (Thomas) (Texas, USA): **A**, venter of male; **B**, propodosoma of male; **C**, genito-anal region of female; **D**, composite dorsum and venter of deutonymph with detail of propodosomatic ocelli (**A–D** from Houck and OConnor, 1990). *Linobia coccinellae* (Scopoli): **E**, anterodorsal aspect of female with detail of chelicera (after Baker et al. 1958).

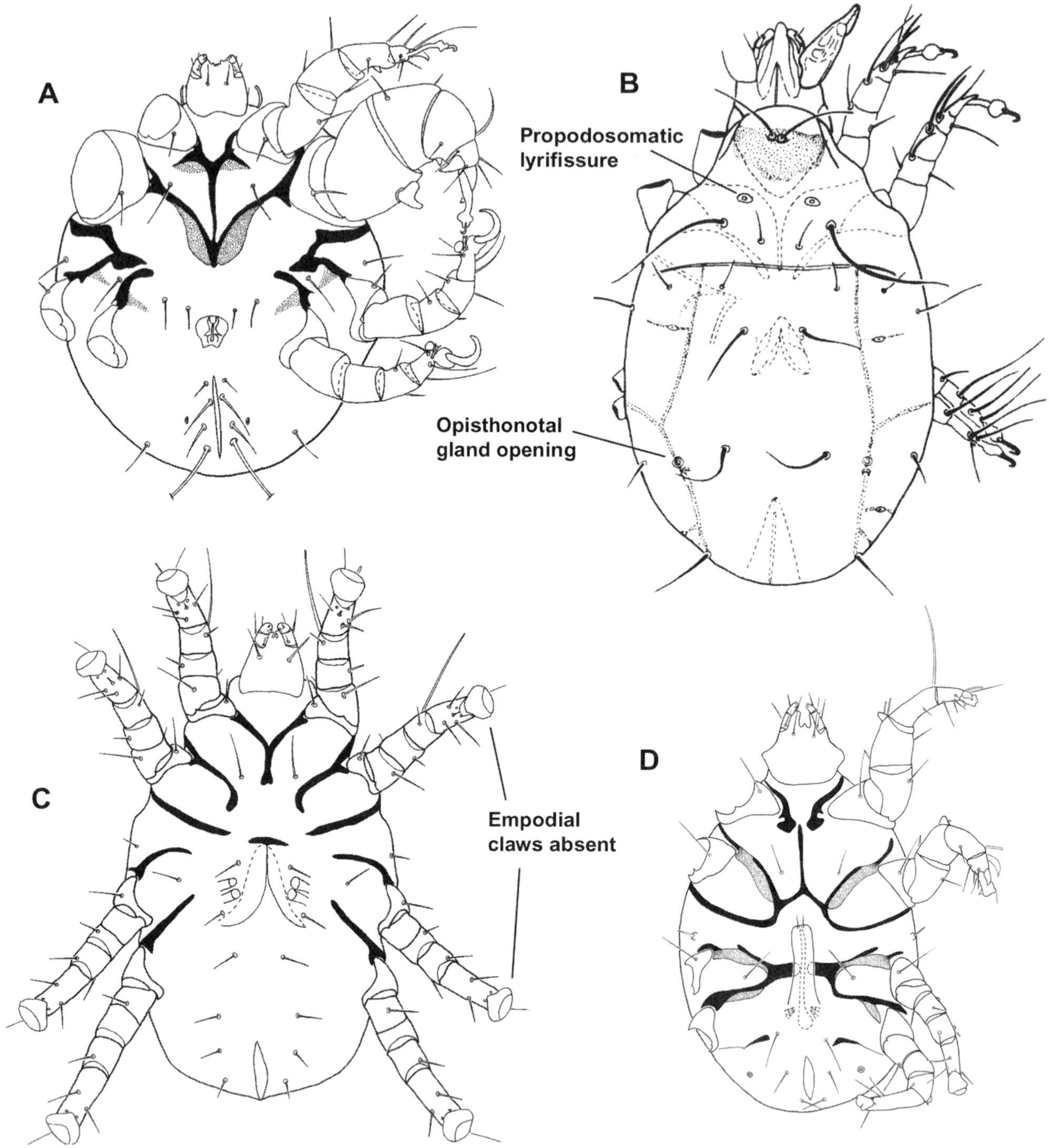

Fig. 16.13. HYADESIIDAE. *Hyadesia sanjuanensis* Fain and Ganning (British Columbia, Canada): **A**, venter of male. *Hyadesia* sp. (Oregon, USA): **B**, dorsum of female.

MELIPONOCOPTIDAE. Unnamed genus (Thailand): **C**, venter of female. *Meliponocoptes orphanus* Fain and Rosa (Brazil): **D**, venter of male.

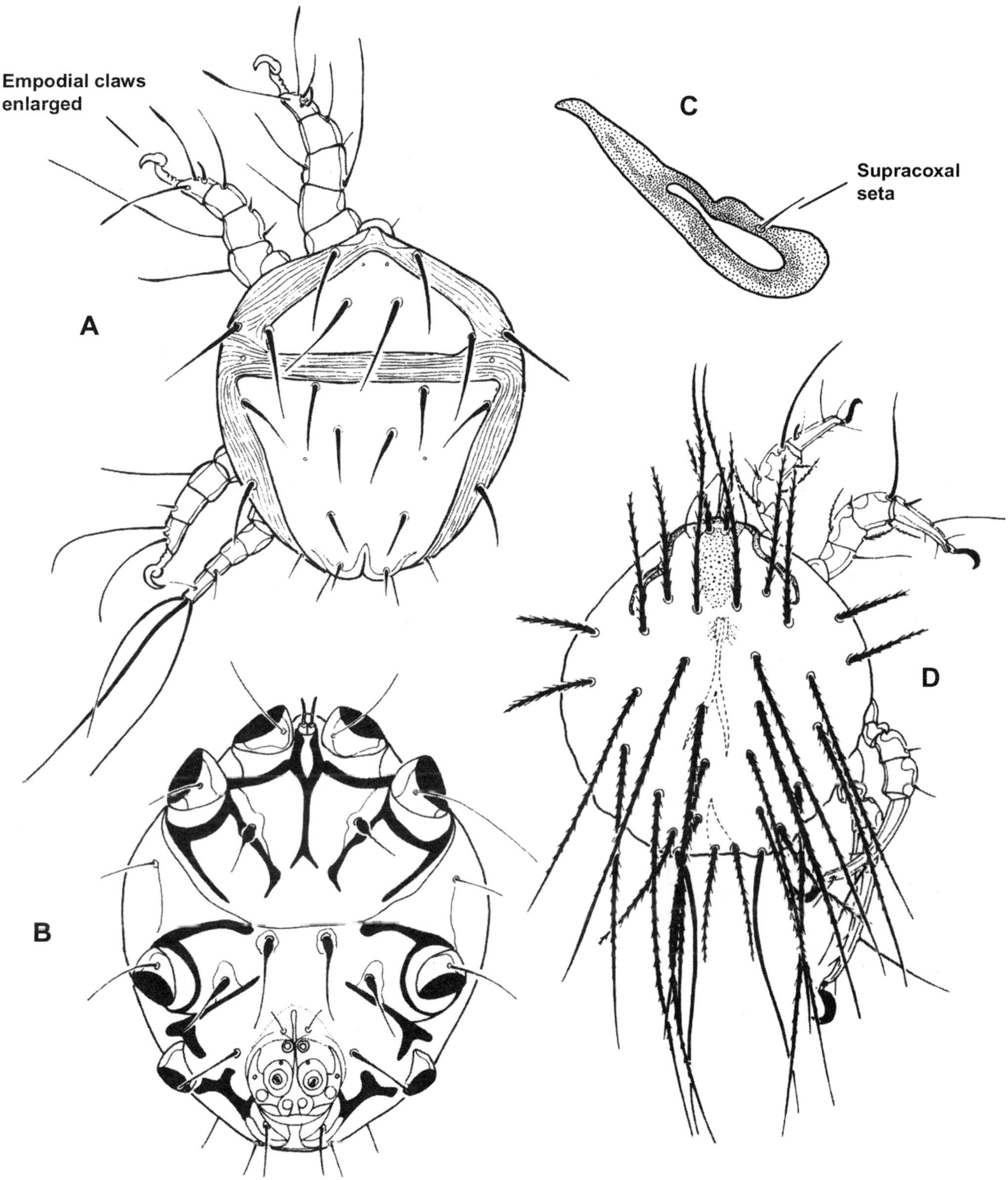

Fig. 16.14. CHAETODACTYLIDAE. *Chaetodactylus krombeini* Baker (Oregon, USA), deutonymph: **A**, dorsum; **B**, venter. *C. krombeini* (Washington, USA): **C**, supracoxal gland region of male; **D**, *Sennertia* sp. (Congo), dorsum of male.

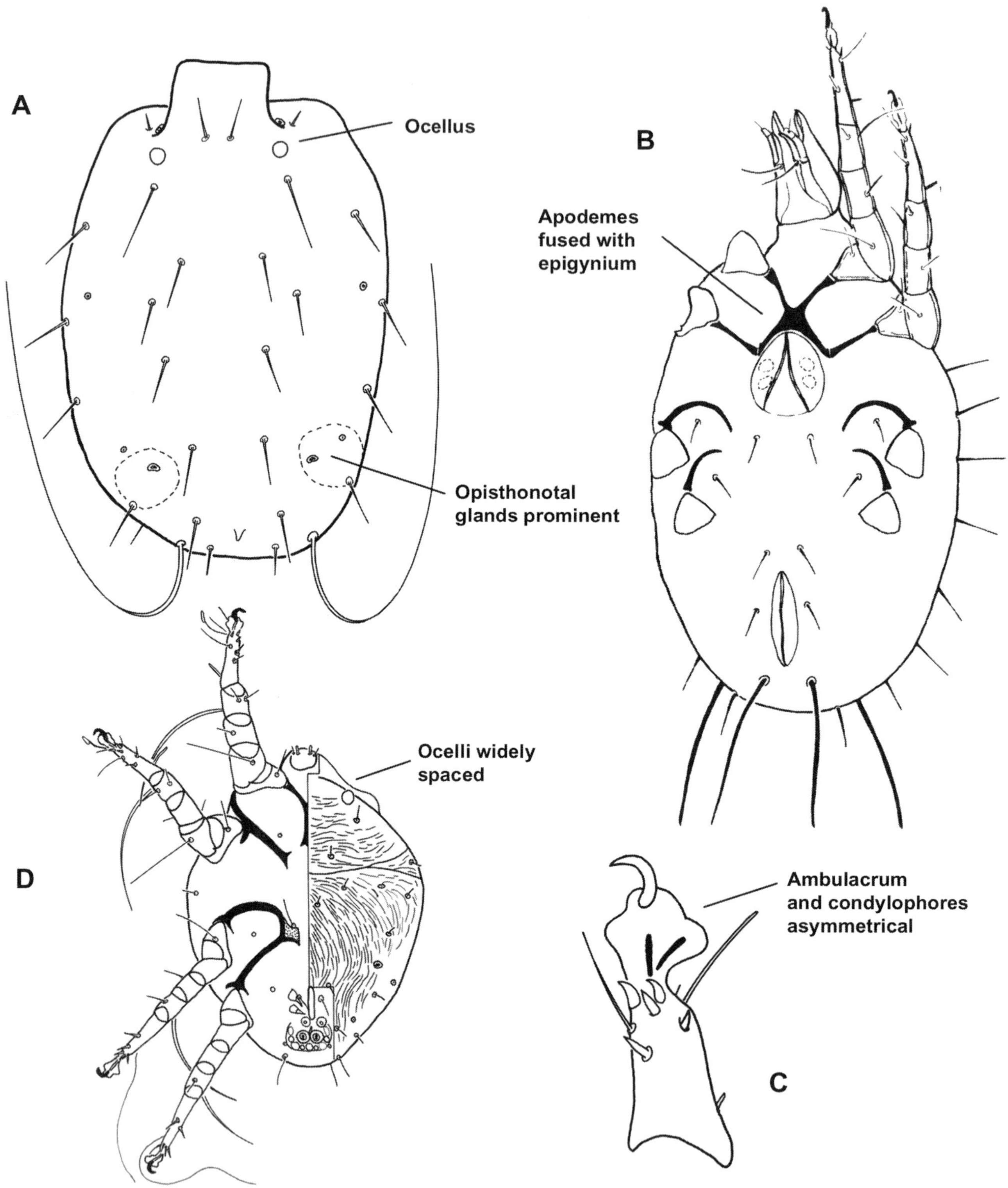

Fig. 16.15. CARPOGLYPHIDAE. *Carpoglyphus lactis* (L.) (Oregon, USA): **A**, dorsum of female; **B**, venter of female; **C**, tarsus and pretarsus I of male. *Carpoglyphus* sp. (Illinois, USA): **D**, composite dorsum and venter of deutonymph.

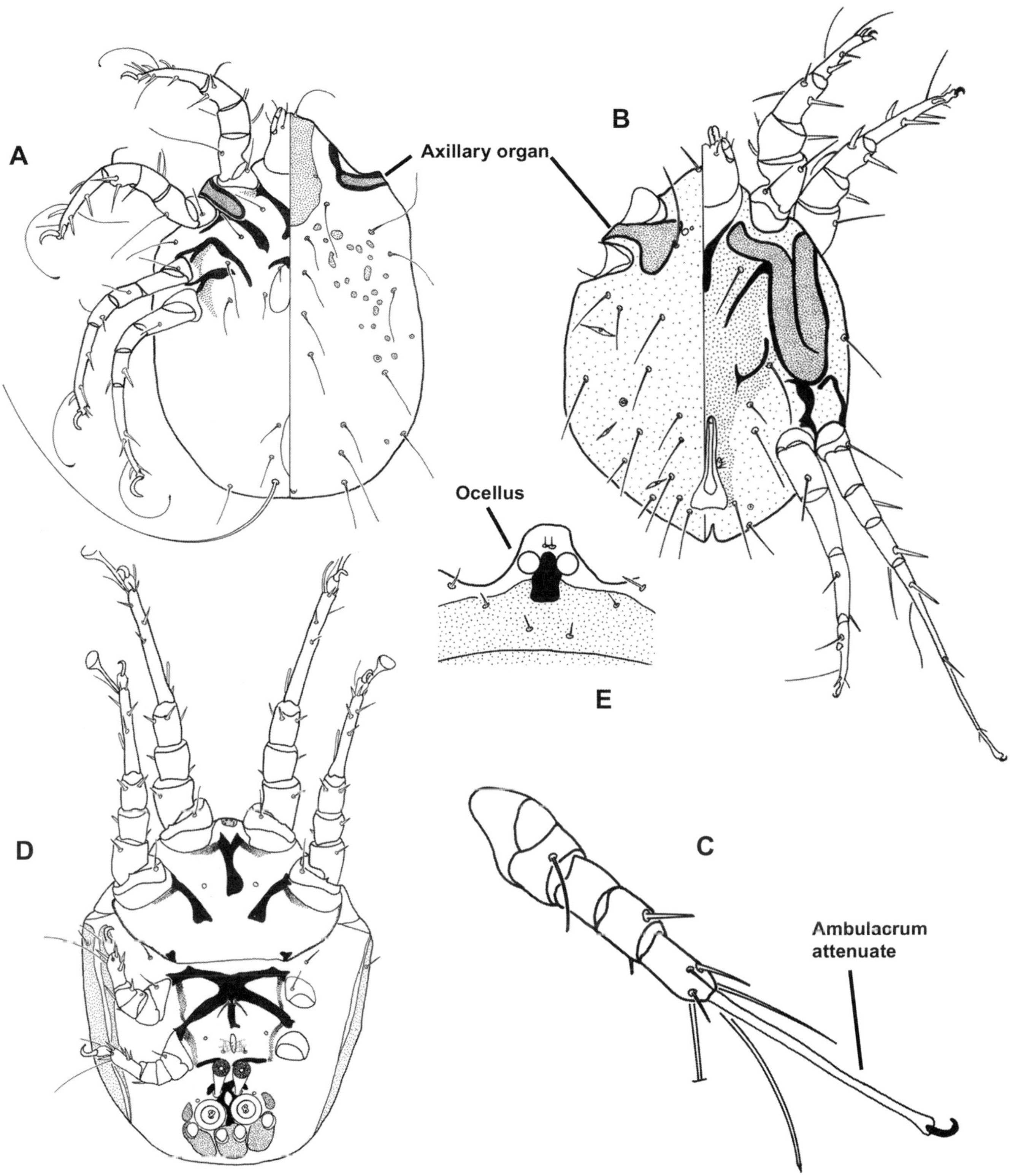

Fig. 16.16. ALGOPHAGIDAE. *Algophagus antarcticus* Hughes (Prince Edward Island, South Indian Ocean): **A**, composite dorsum and venter of female. *Fusohericia lawrencei* Baker and Crossley (Michigan, USA): **B**, composite dorsum and venter of male; **C**, leg IV of deutonymph. *Prohericia longipes* OConnor and Moser (Germany), deutonymph: **D**, venter; **E**, dorsal propodosoma (from OConnor and Moser 1985).

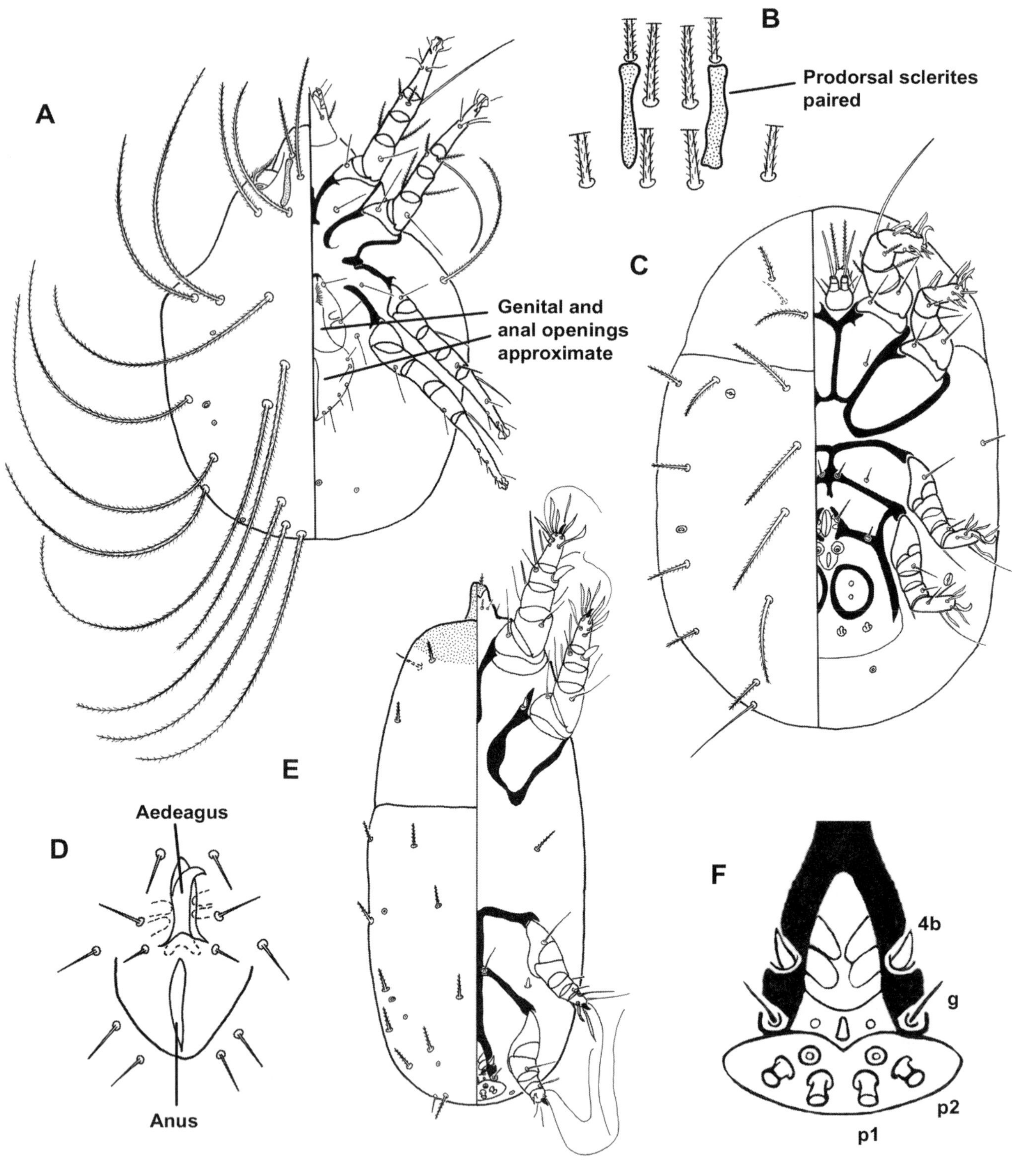

Fig. 16.17. EUGLYCYPHAGIDAE. *Euglycyphagus intercalatus* Fain and Philips (Michigan, USA): **A**, composite dorsum and venter of female; **B**, prodorsal region of female; **C**, composite dorsum and venter of deutonymph; **D**, genito-anal region of male. PEDETOPODIDAE. *Pedetopus zumpti* Fain (South Africa): **E**, composite dorsum and venter of deutonymph; **F**, attachment organ of deutonymph.

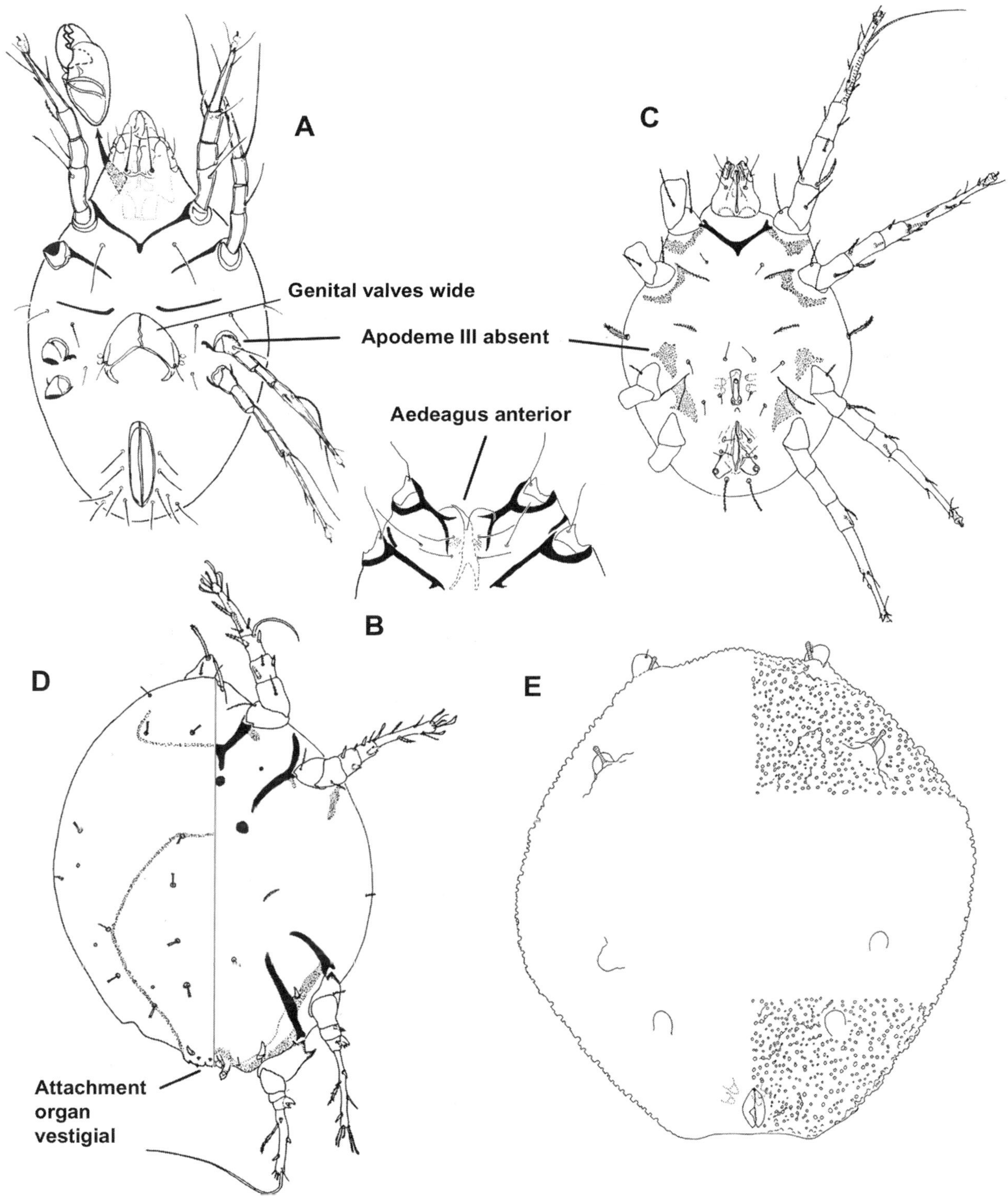

Fig. 16.18. CHORTOGLYPHIDAE. *Chortoglyphus arcuatus* (Troupeau) (Oregon, USA): **A**, venter of female; **B**, genital region of male. *Alabidopus asiaticus* Lukoschus et al. (Malaysia): **C**, venter of male; **D**, composite dorsum and venter of parasitic deutonymph; **E**, venter of sedentary deutonymph (from Lukoschus et al. 1981).

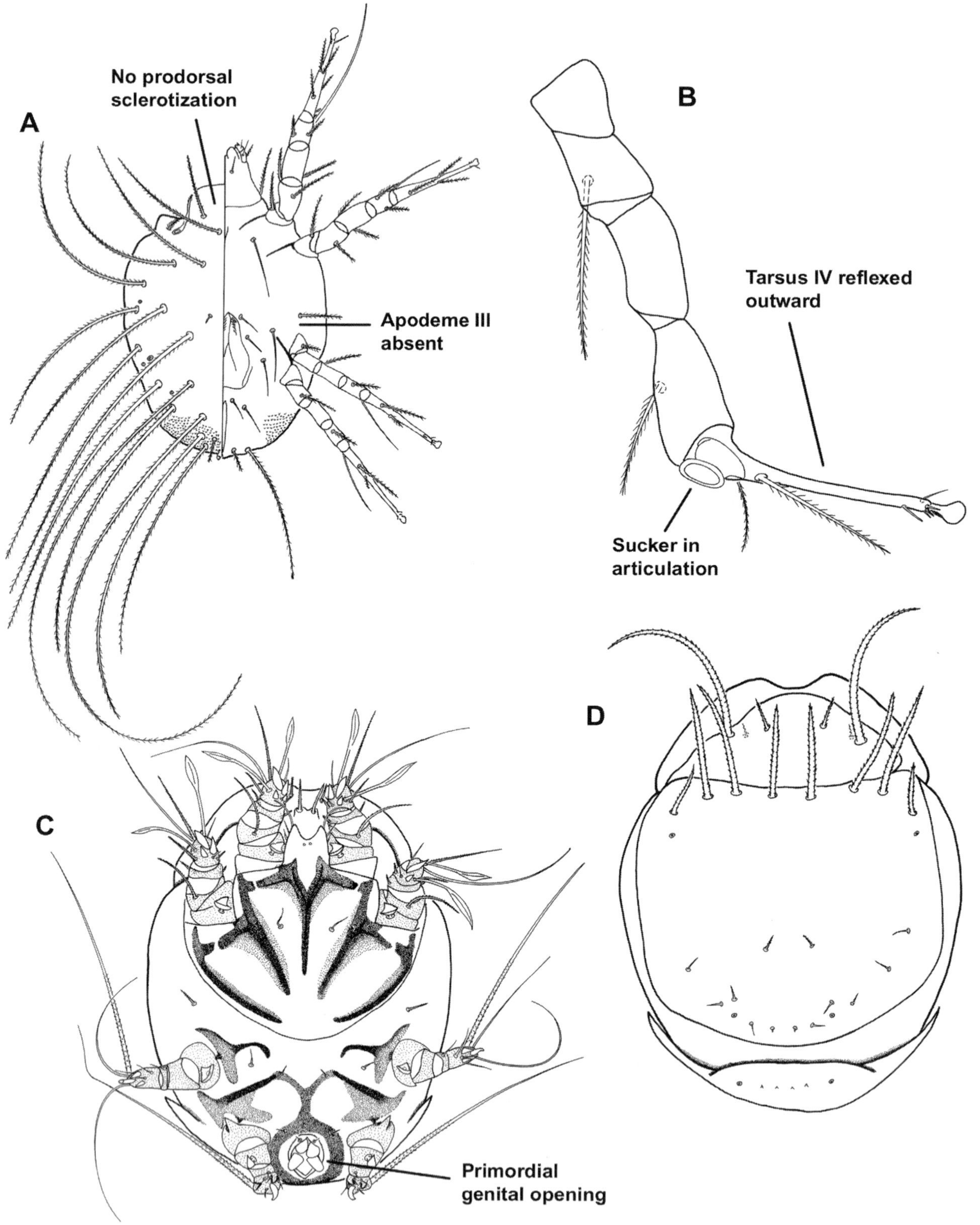

Fig. 16.19. ECHIMYOPODIDAE. *Echimyopus* sp. (Philippines): **A**, composite dorsum and venter of female; **B**, leg IV of male. *Marmosopus mesoamericanus* OConnor, Lukoschus, and Giesen (Costa Rica): **C**, venter of deutonymph; **D**, dorsum of deutonymph (from OConnor, Lukoschus, and Giesen 1982).

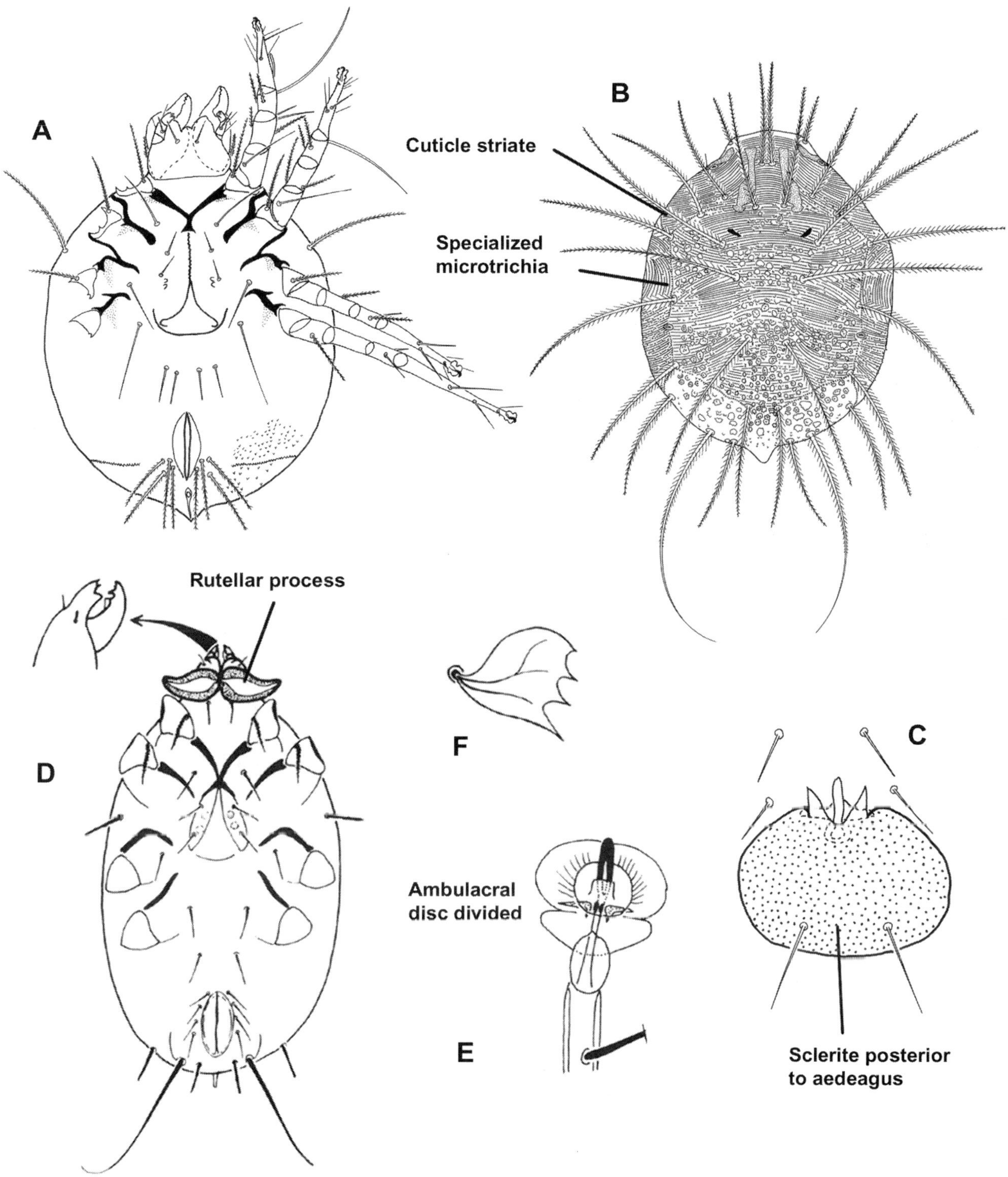

Fig. 16.20. AEROGLYPHIDAE. *Aeroglyphus robustus* (Banks) (Michigan, USA): **A**, venter of female; **B**, dorsum of female. *Glycycometus lukoschusi* (Fain) (Brazil): **C**, genital region of male.

ROSENSTEINIIDAE. *Nycteriglyphus vespertilio* Ah and Hunter (South Korea): **D**, venter of female with detail of chelicera (after Ah and Hunter 1968). *N. bifolium* Strandtmann: **E**, pretarsus (from Atyeo 1979). *N. pterophorus* (Berlese): **F**, body seta (after Zakhvatkin 1941).

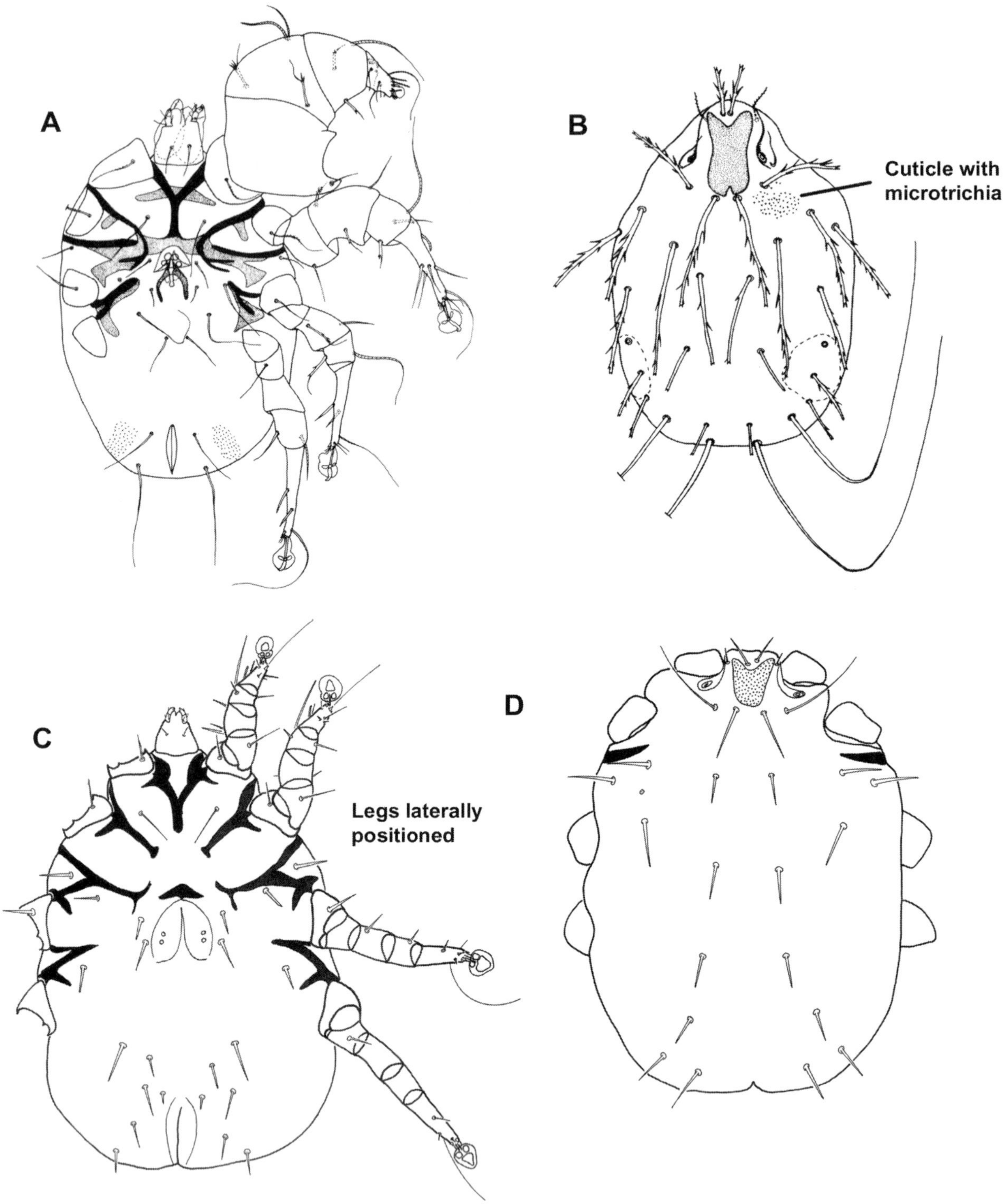

Fig. 16.21. ROSENSTEINIIDAE. *Chiroptoglyphus americanus* OConnor and Reisen (Oklahoma, USA): **A**, venter of male; **B**, dorsum of male (from OConnor and Reisen 1978). *Rosensteinia sieversi* Oudemans (Nicaragua): **C**, venter of female; **D**, dorsum of female.

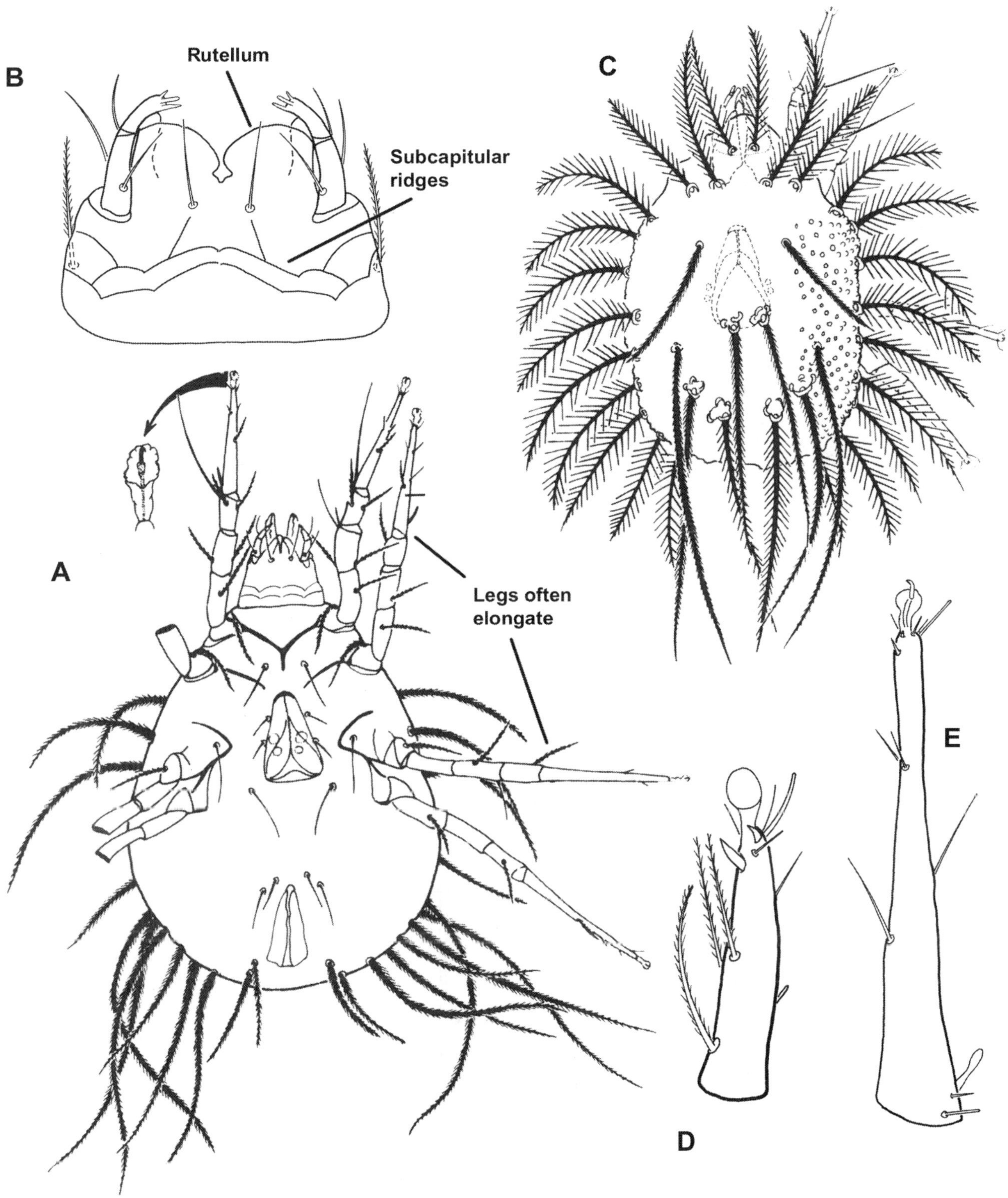

Fig. 16.22. GLYCYPHAGIDAE. *Glycyphagus domesticus* (DeGeer) (Oregon, USA): **A**, venter of female with detail of pretarsus I. *G. destructor* (Schrank) (Michigan, USA): **B**, subcapitulum and palpi of female. *Ctenoglyphus plumiger* (Koch) (Oregon, USA): **C**, dorsum of female. *Grammolichus malukuensis* (Fain) (Costa Rica): **D**, tarsus I of female. *Marsupialichus johnstoni* Fain (Louisiana, USA): **E**, tarsus I of female.

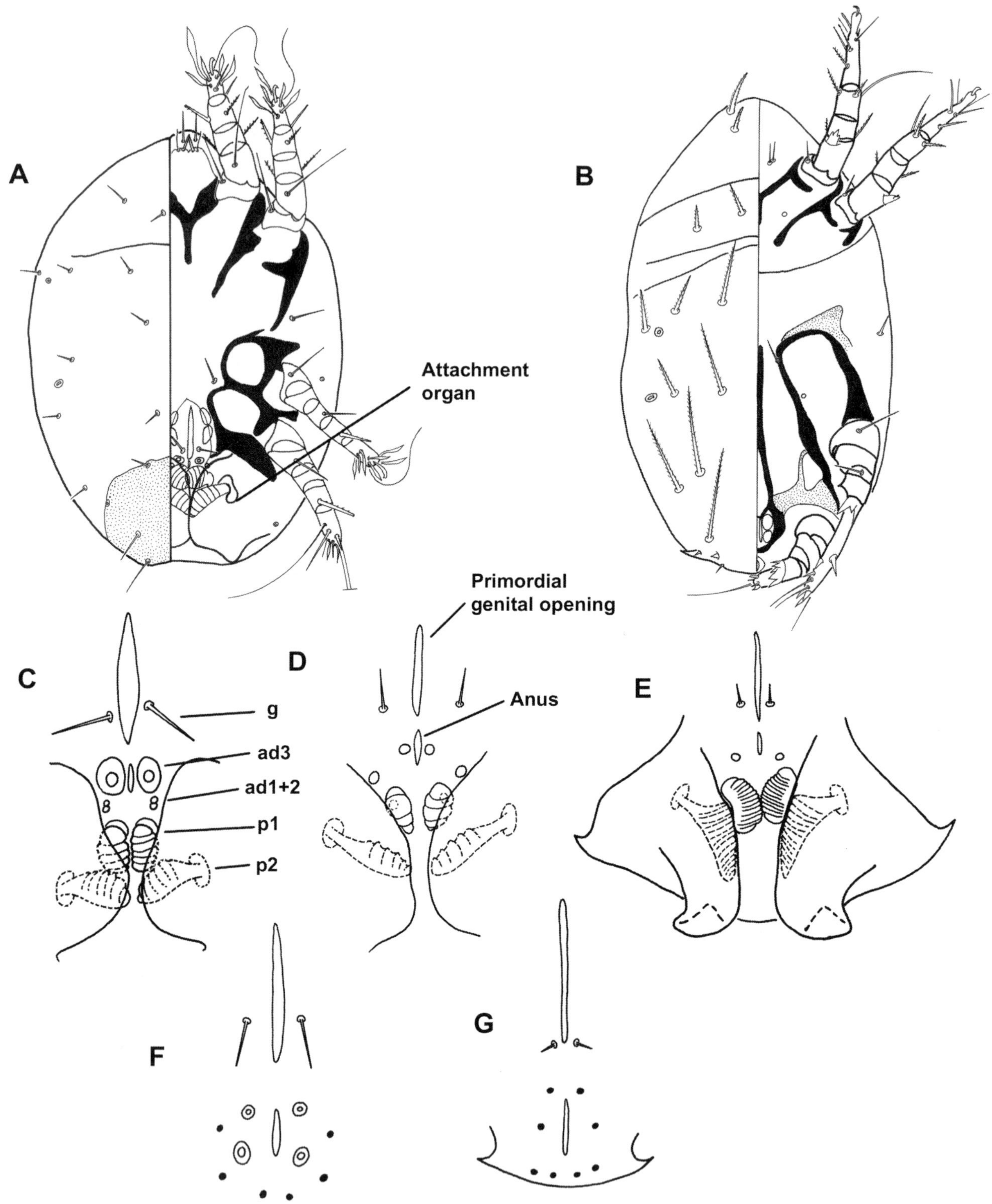

Fig. 16.23. GLYCYPHAGIDAE. *Dermacarus tamiasciuri* Rupes, Yunker, and Wilson (Michigan, USA): **A**, composite dorsum and venter of deutonymph. *Lophuromyopus schoutedeni* Fain (Tanzania): **B**, composite dorsum and venter of deutonymph. *Marsupialichus johnstoni* Fain (Louisiana, USA): **C**, attachment organ of deutonymph. *Glycyphagus hypudaei* (Koch) (Michigan, USA): **D**, attachment organ of deutonymph. *Orycteroxenus soricis* (Oudemans) (Netherlands): **E**, attachment organ of deutonymph. *Baloghella melis* Mahunka (Germany): **F**, posterior ventral region of deutonymph. *Cryptomyopus zachvatkini* Fain (South Africa): **G**, posterior ventral region of deutonymph.

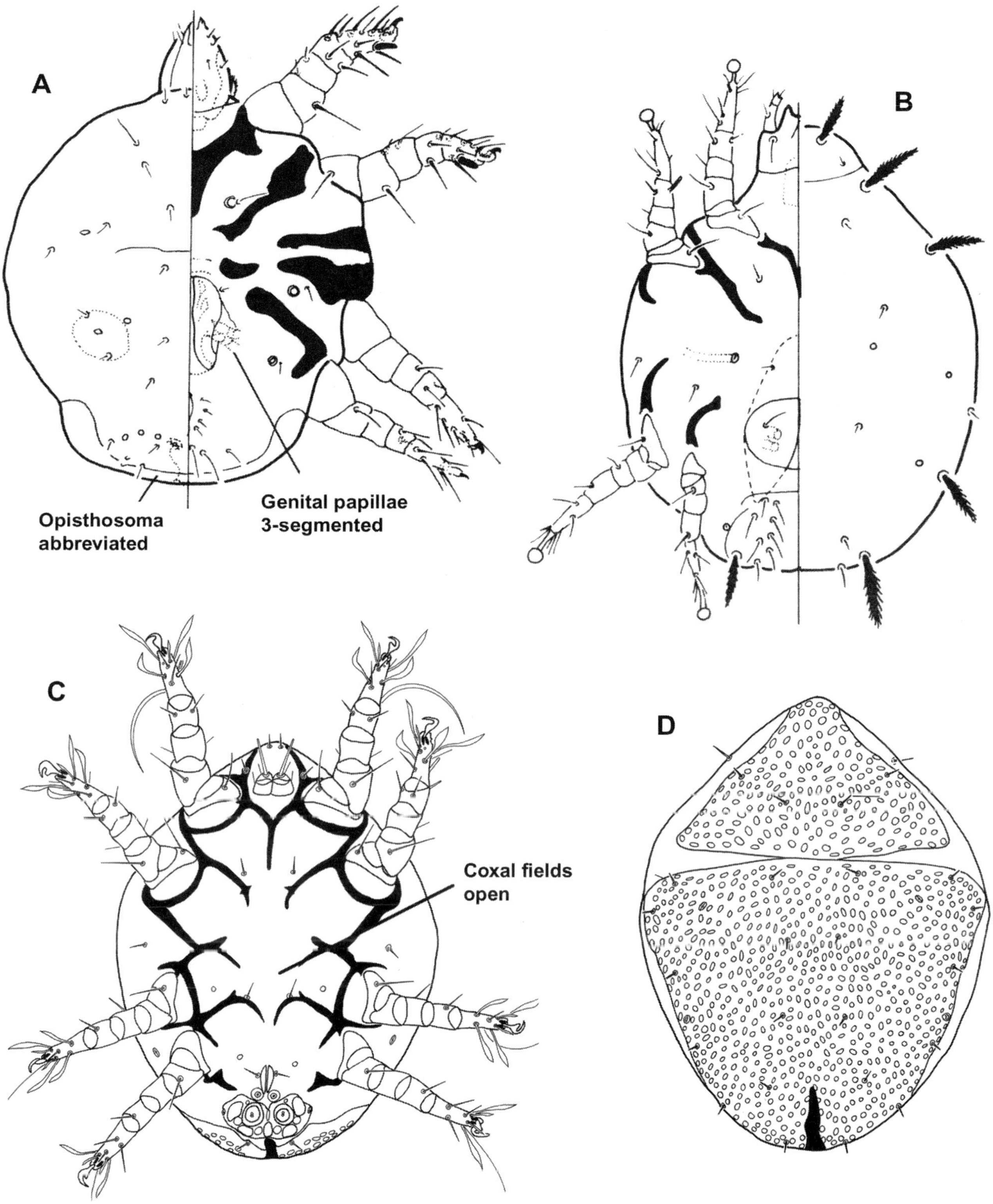

Fig. 16.24. GAUDIELLIDAE. *Platyglyphus malayanus* Kurosa (Malaysia): **A**, composite dorsum and venter of female (adapted from Kurosa 1976). *Gaudiella minuta* Atyeo, Baker, and Delfinado (Brazil): **B**, composite dorsum and venter of female (adapted from Atyeo, Baker, and Delfinado 1974). *Cerophagus nearcticus* OConnor (Michigan, USA): **C**, venter of deutonymph; **D**, dorsum of deutonymph (from OConnor 1992).

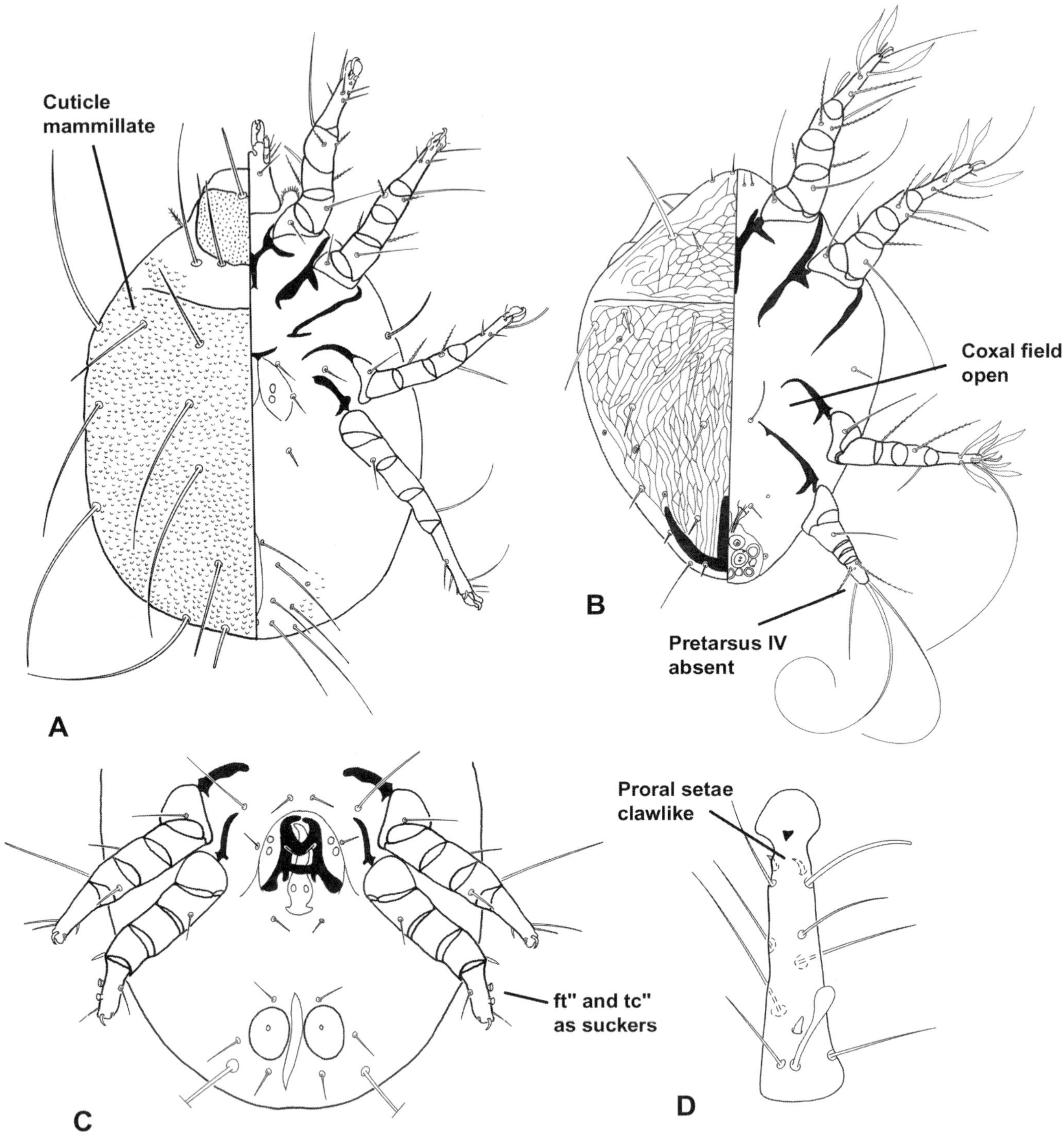

Fig. 16.25. SUIDASIIDAE. *Tortonia quadridens* Baker (New York, USA): **A**, composite dorsum and venter of female; **B**, composite dorsum and venter of deutonymph. *Suidasia pontifica* (Oudemans) (Surinam): **C**, venter of male. *Sapracarus tuberculatus* Fain and Philips (Michigan, USA): **D**, leg I of female.

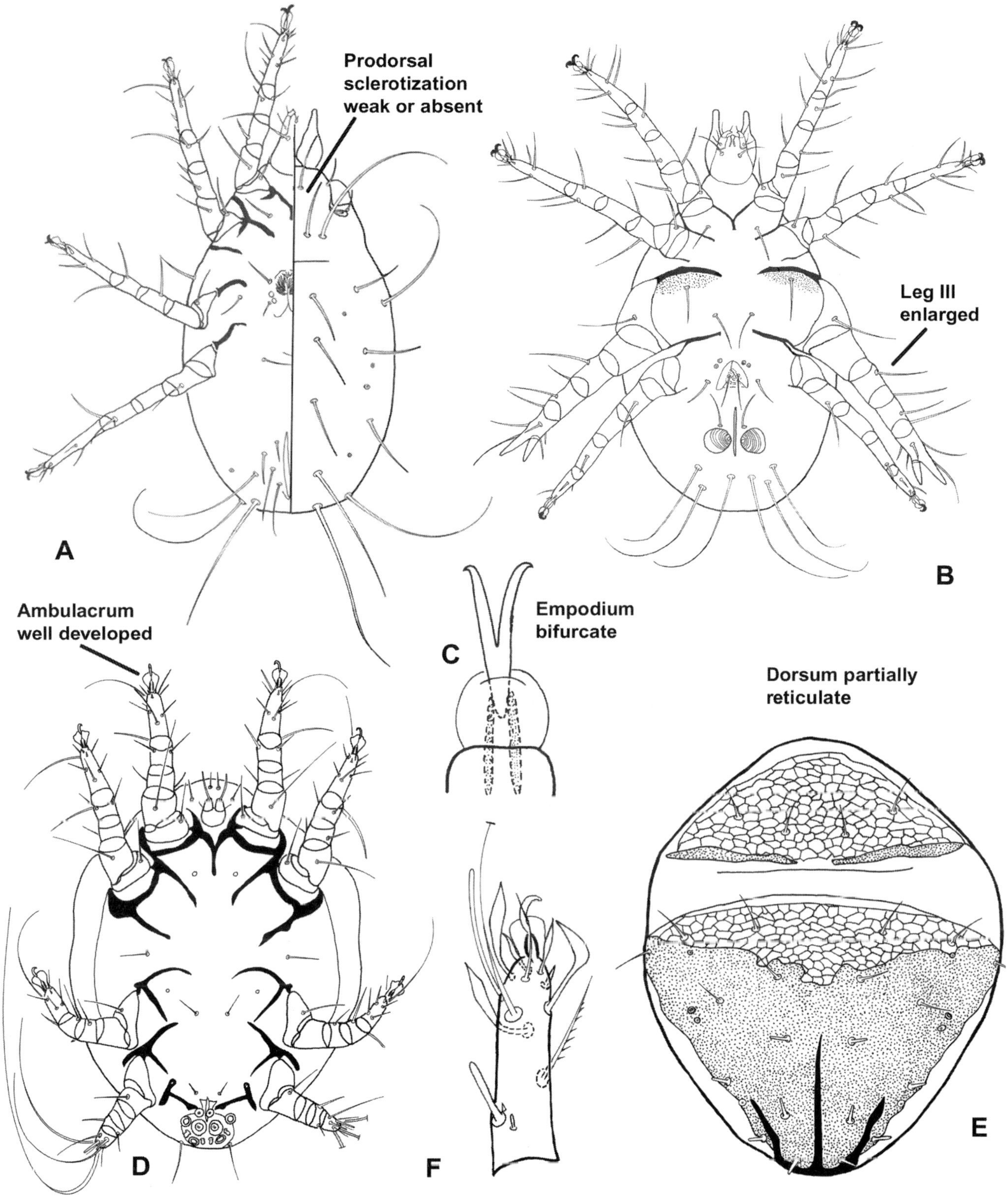

Fig. 16.26. LARDOGLYPHIDAE. *Lardoglyphus zacheri* Oudemans (New York, USA): **A**, composite dorsum and venter of female; **B**, venter of male; **C**, pretarsus of female; **D**, venter of deutonymph; **E**, dorsum of deutonymph. *L. falconidus* Philips and Norton (New York, USA): **F**, tarsus I of deutonymph.

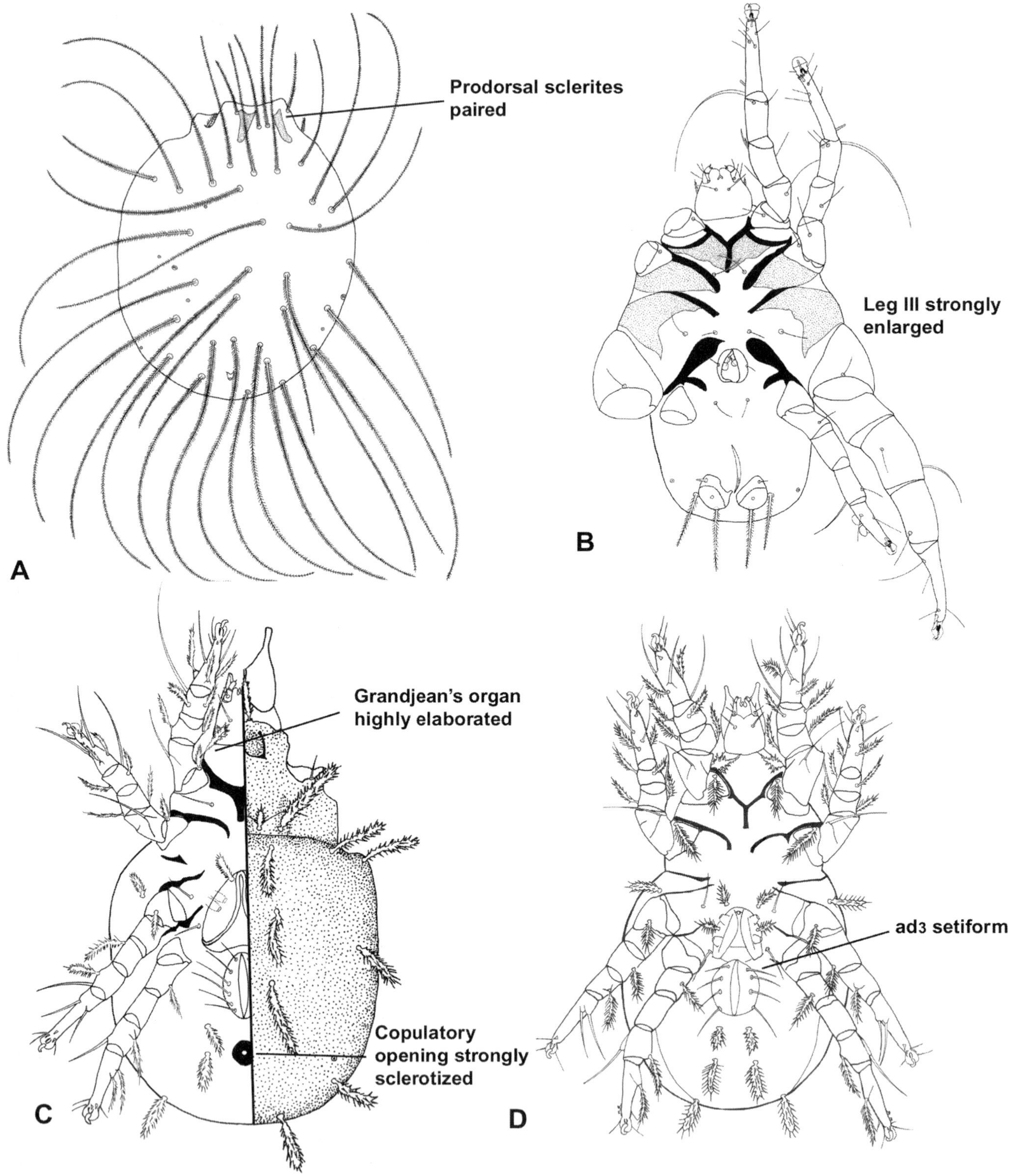

Fig. 16.27. GLYCACARIDAE. *Glycacarus combinatus* Griffiths (Prince Edward Islands, South Indian Ocean): **A**, dorsum of female, **B**, venter of male.

SCATOGLYPHIDAE. *Scatoglyphus polytrematus* Berlese (Illinois, USA): **C**, composite dorsum and venter of female; **D**, venter of male.

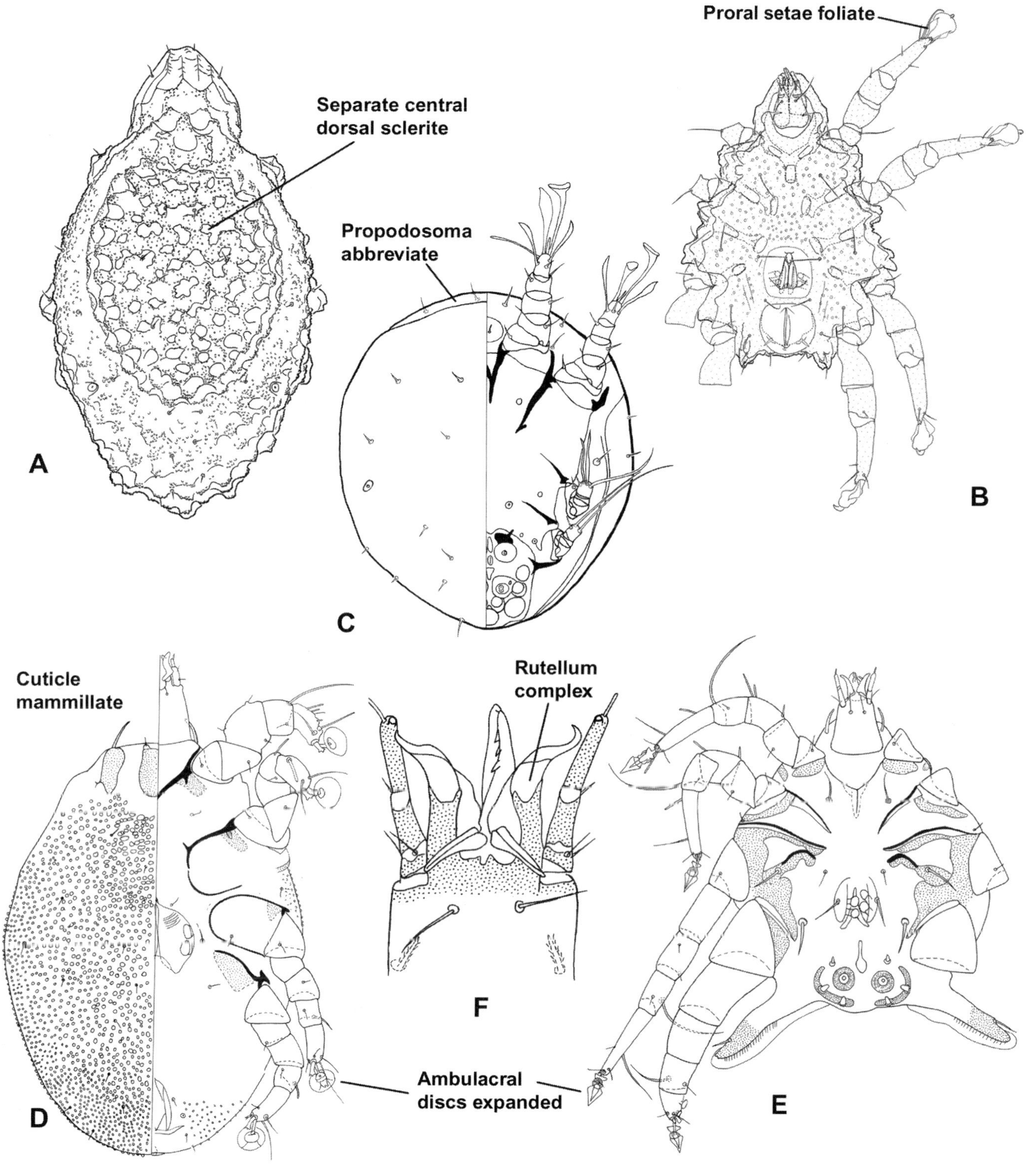

Fig. 16.28. LEMANNIELLIDAE. *Lemanniella minotauri* Wurst (Germany): **A**, dorsum of female; **B**, venter of male (from Wurst 2001). *L. reducta* Mahunka (France); **C**, composite dorsum and venter of deutonymph.

CHETOCHELACARIDAE. *Chetochelacarus mamillatus* Fain (Congo); **D**, composite dorsum and venter of female (modified from Fain 1987a). *Diplopodocoptes transkeiensis* Fain (South Africa): **E**, venter of male; **F**, gnathosoma of male (**E–F** from Fain 1987a).

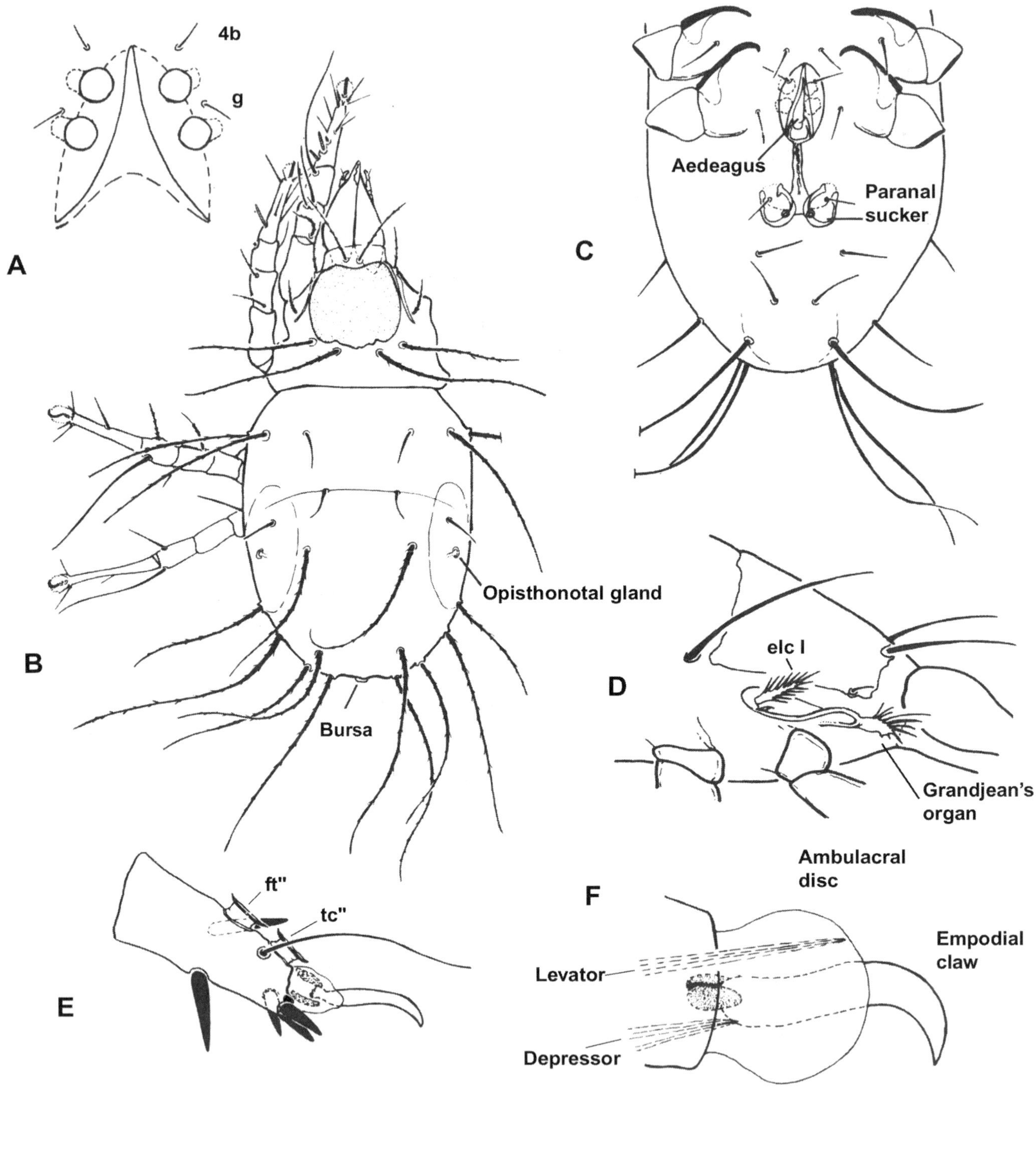

Fig. 16.29. ACARIDAE. *Tyrophagus* sp. (Oregon, USA): **A**, female genital region; **B**, dorsum of female; **C**, *Sancassania* sp. (Oregon, USA), venter of male; **D**, *Acarus siro* L., anterolateral aspect of female; **E**, *Rhizoglyphus echinopus* (Fumouze and Robin), tarsus IV of male; **F**, typical pretarsus of an acarid mite.

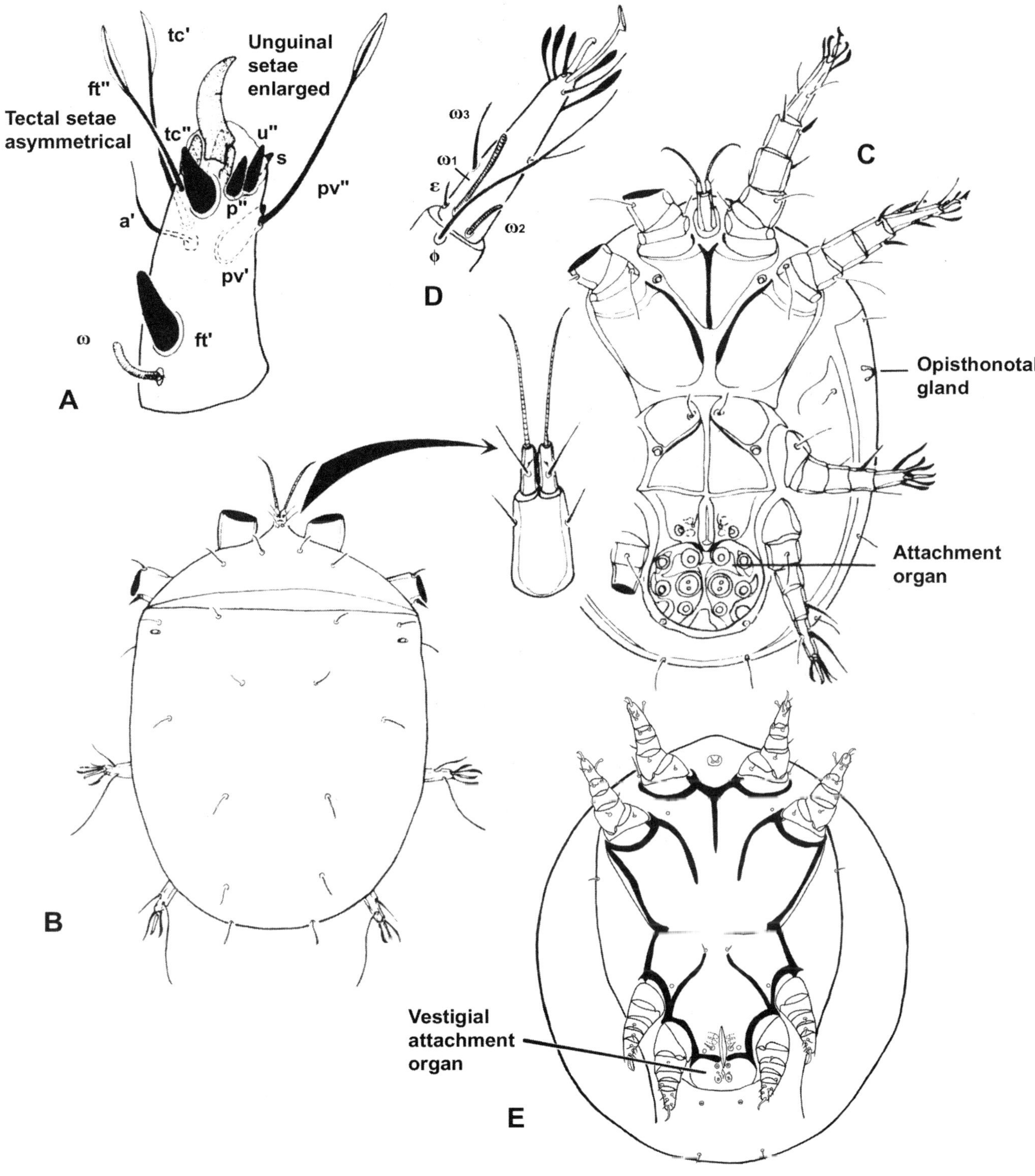

Fig. 16.30. ACARIDAE. *Rhizoglyphus echinopus* (Fumouze and Robin): **A**, tarsus II of female; **B**, dorsum of deutonymph with detail of gnathosomatic remnant; **C**, venter of deutonymph; **D**, tarsus I of deutonymph (dorsal). *Acarus immobilis* Griffiths (Michigan, USA): **E**, venter of deutonymph.

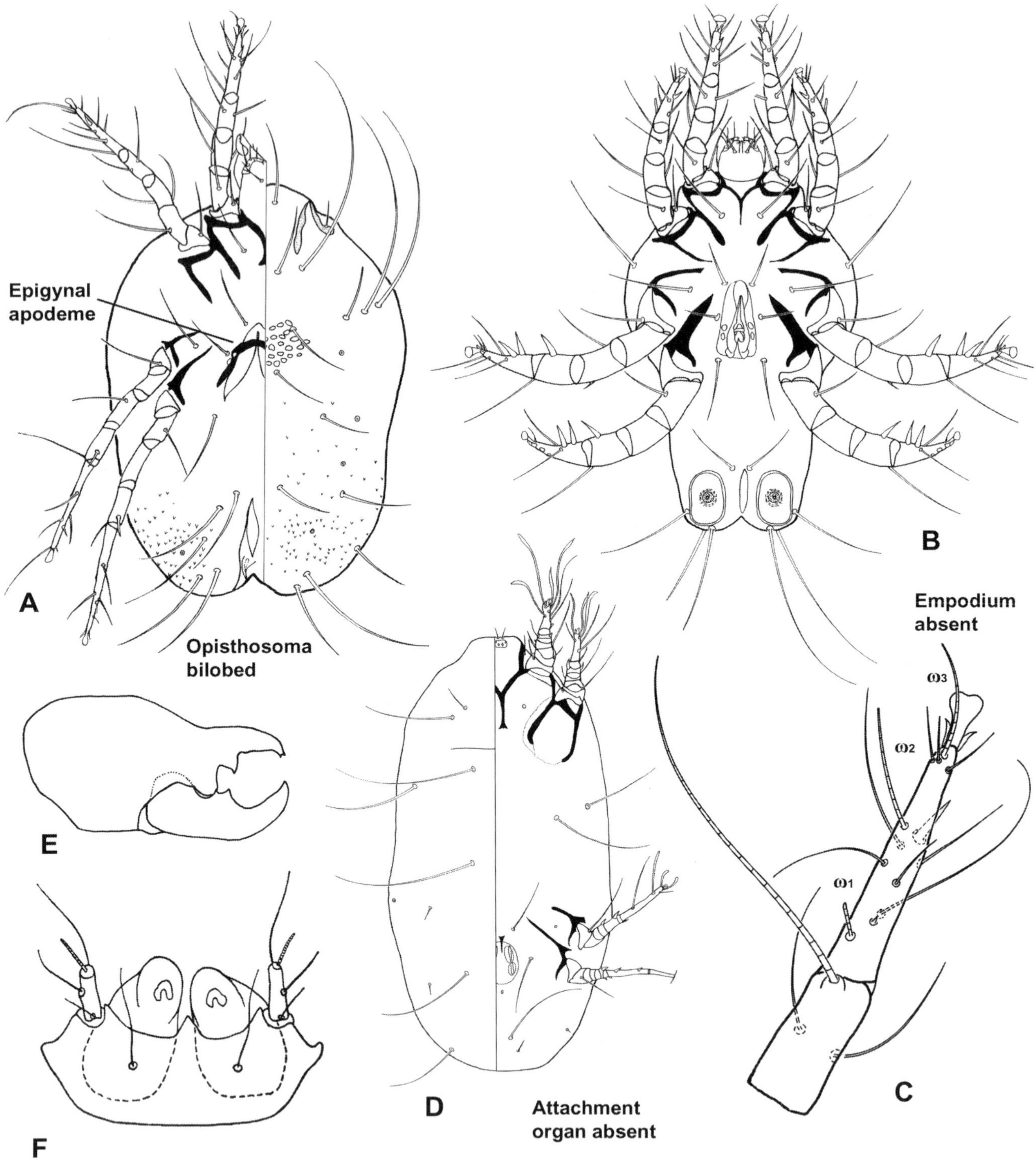

Fig. 16.31. HYPODERATIDAE. *Neottialges plegadicola* Fain (Maine, USA): **A**, composite dorsum and venter of female; **B**, venter of male; **C**, tibia-tarsus I of female; **D**, composite dorsum and venter of deutonymph. *Hypodectes propus* Nitzsch (Georgia, USA); **E**, chelicera of male; **F**, gnathosoma of female (from Fain and Bafort 1966).

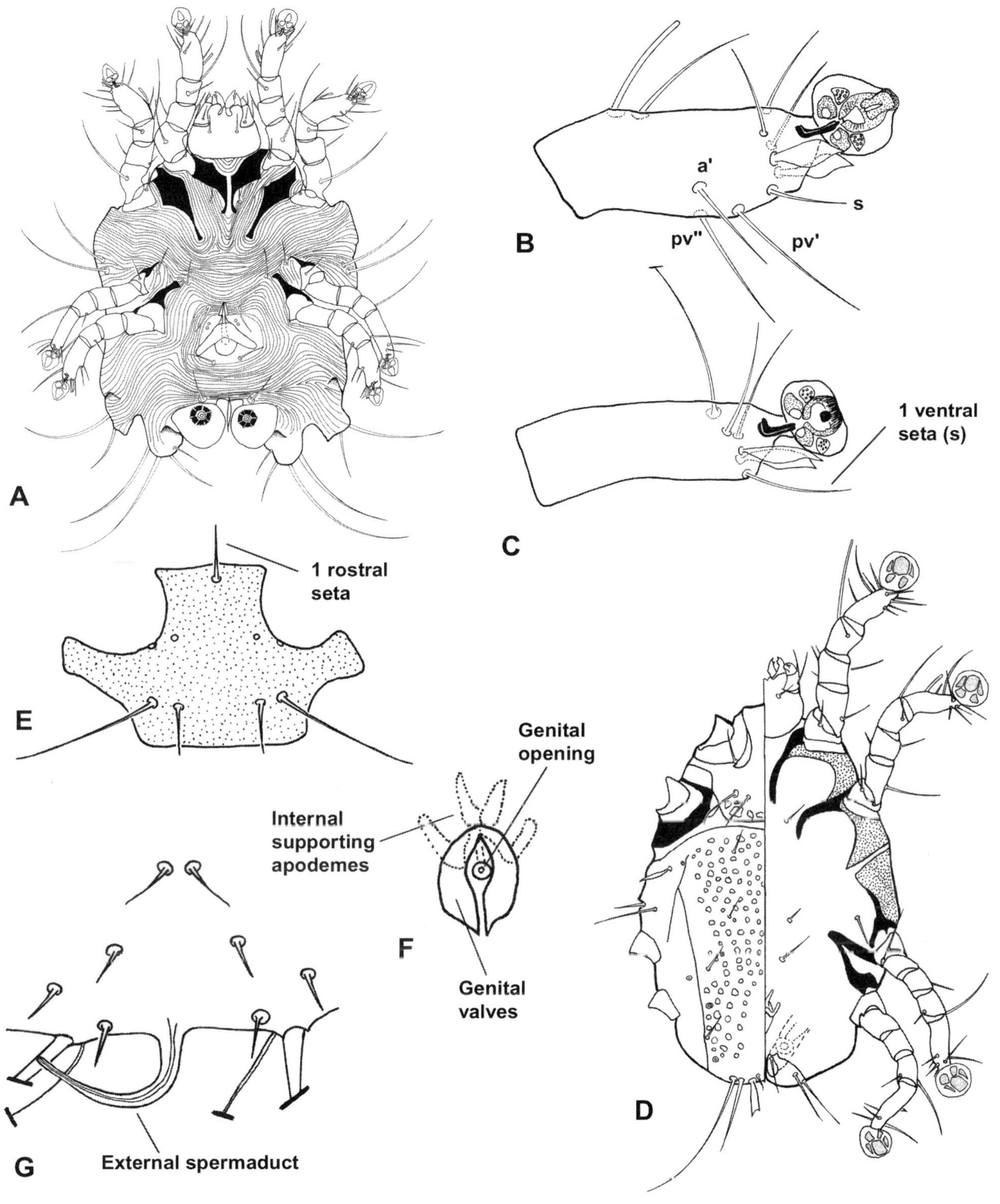

Fig. 16.32. KRAMERELLIDAE. *Kramerella maculata* Gaud (Uganda): **A**, venter of male; **B**, tarsus I of male; **C**, tarsus III of male. CAUDIFERIDAE. *Caudifera marginata* (Trouessart) (Congo): **D**, composite dorsum and venter of male; **E**, dorsal propodosoma of male; **F**, genital region of male; **G**, posterior dorsal region of female.

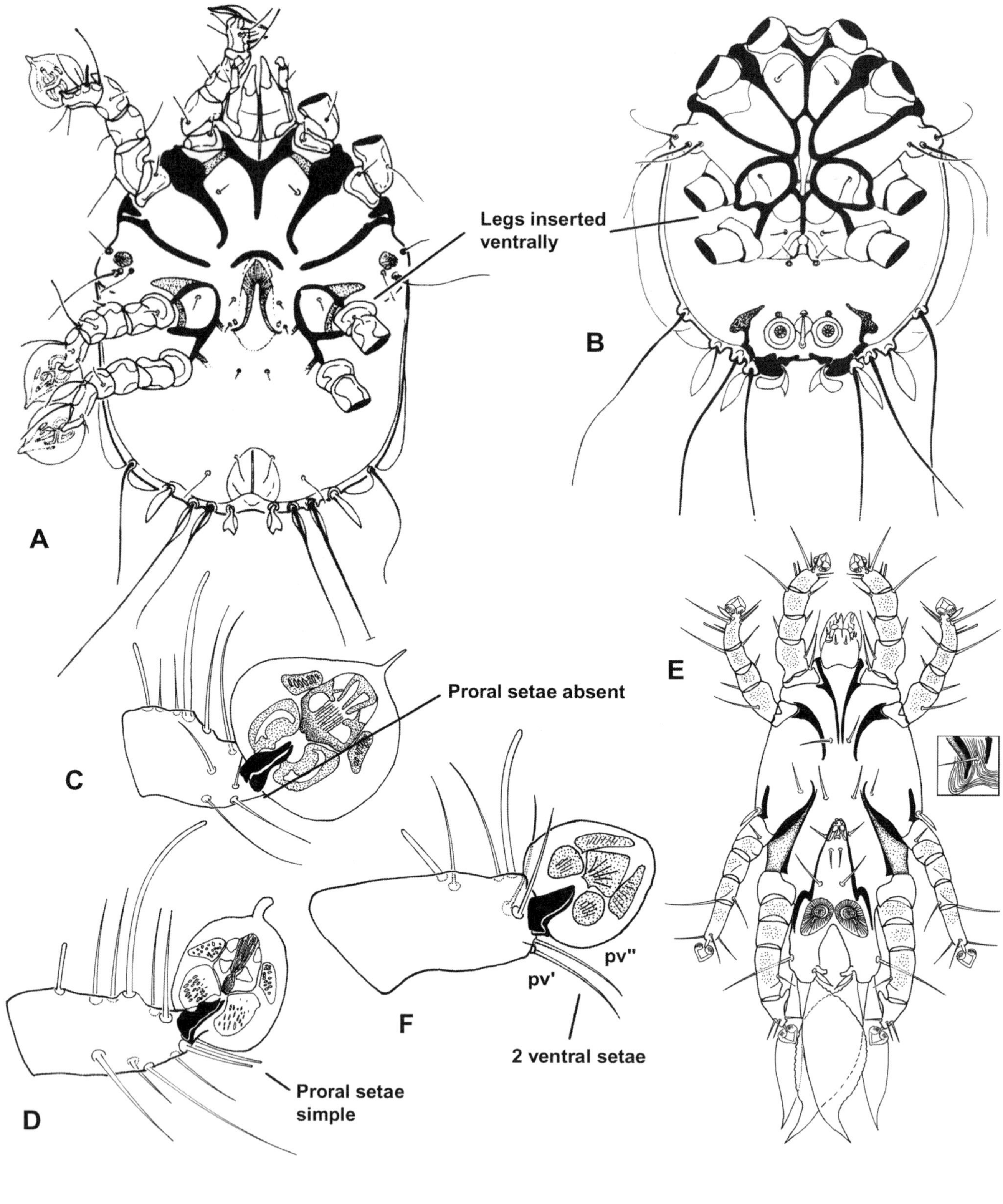

Fig. 16.33. FREYANIDAE. *Freyana* sp. (Texas, USA): **A**, venter of female; **B**, venter of male; **C**, tarsus I of female. *Dobyella longipalpa* Gaud and Atyeo (Australia): **D**, tarsus I of female.

VEXILLARIIDAE. *Vexillaria intermedia* Gaud and Mouchet (Cameroon): **E**, venter of male; **F**, tarsus I of male.

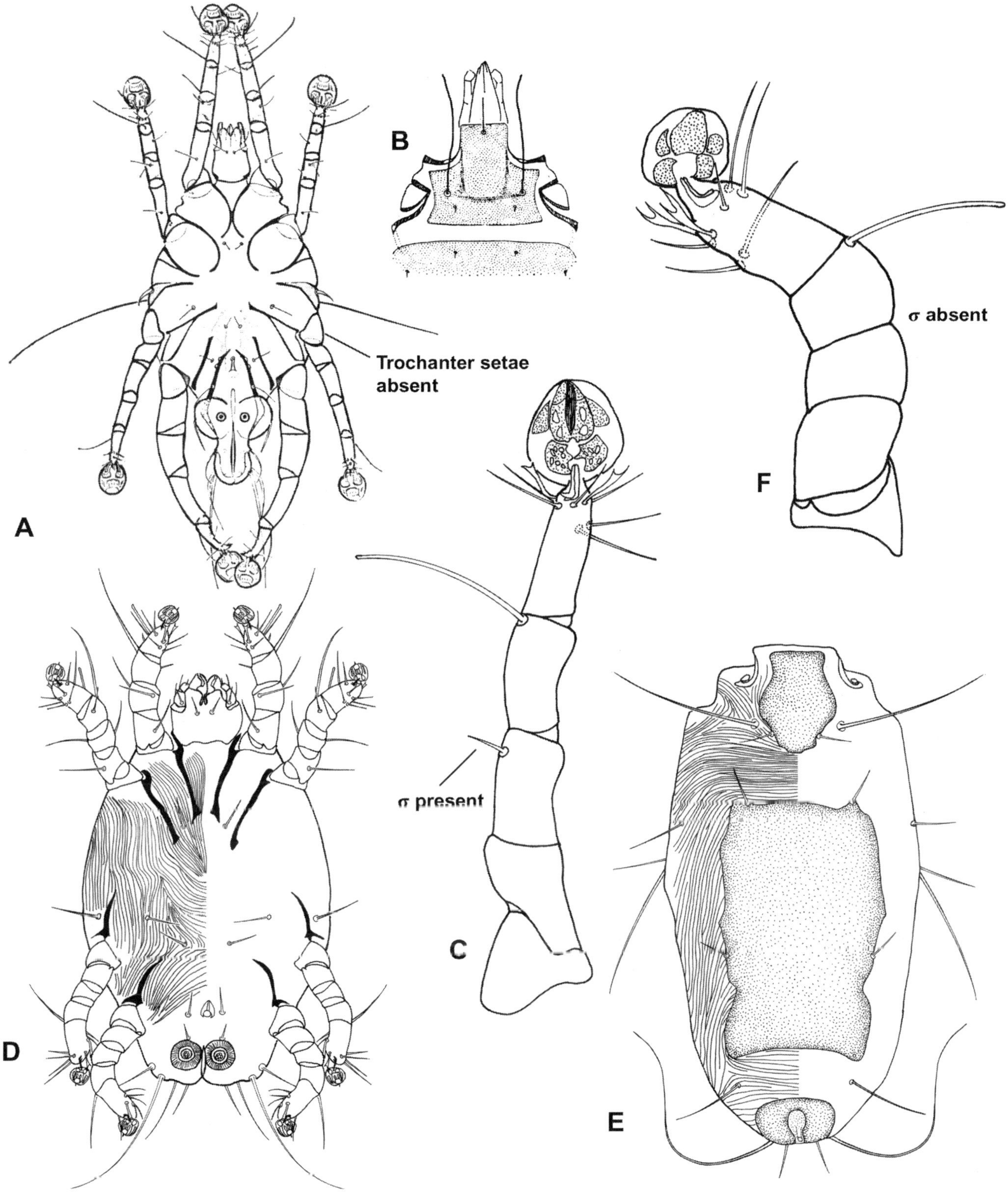

Fig. 16.34. EUSTATHIIDAE. *Alleustathia ungulata* Gaud and Atyeo (Rennell Island): **A**, venter of male; **B**, dorsal propodosoma of female; **C**, leg III of female (after Gaud and Atyeo 1967).
OCHROLICHIDAE. *Psilolichus brachyurus* Gaud and Atyeo (South Africa): **D**, venter of male; **E**, dorsum of female; **F**, leg III of female.

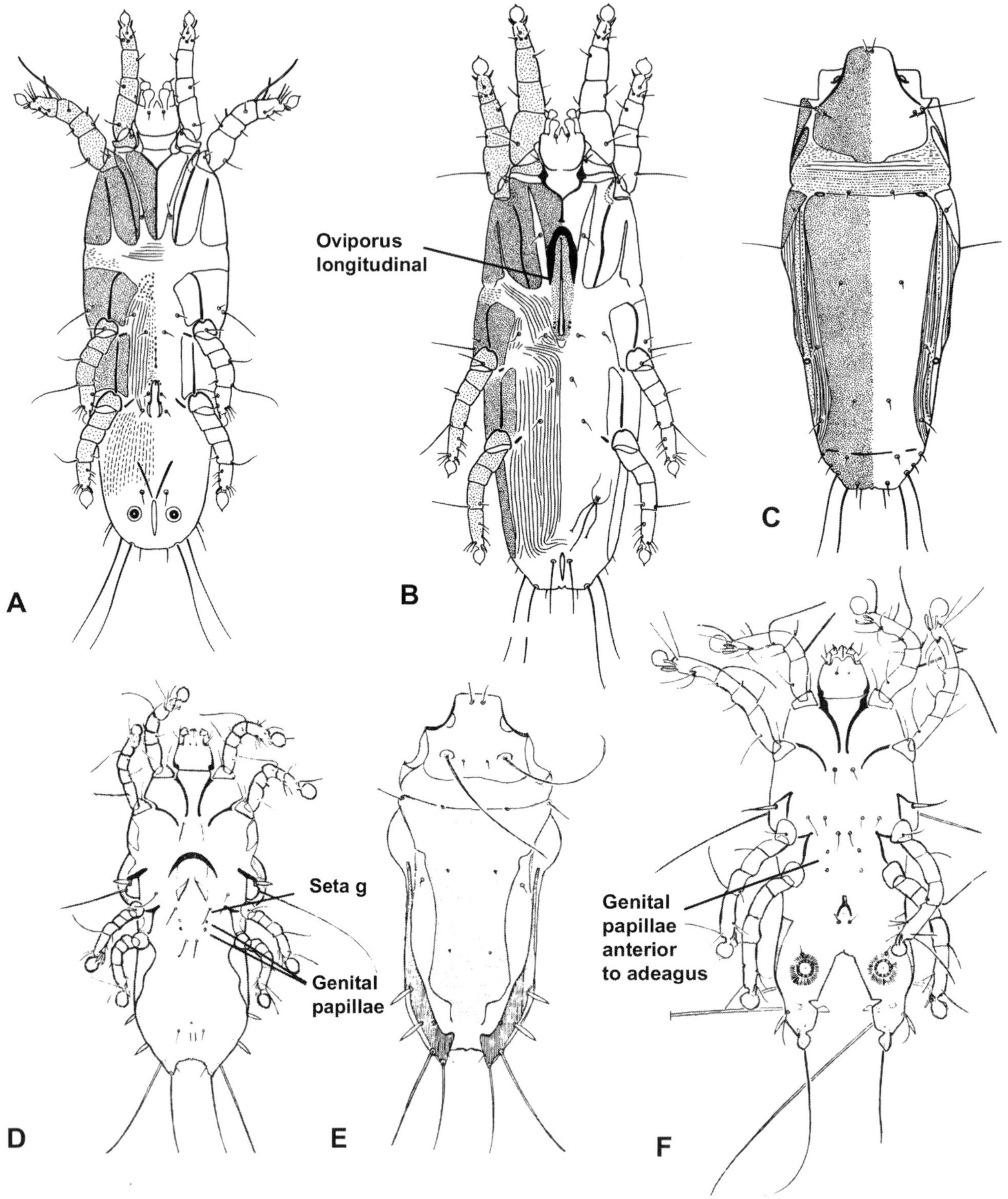

Fig. 16.35. RECTIJANUIDAE. *Rectijanua striata* Atyeo and Peterson (Thailand, Malaysia): **A**, venter of male; **B**, venter of female; **C**, dorsum of female (from Atyeo and Peterson 1976).

GABUCINIIDAE. *Capitolichus tetroplourus* Gaud and Atyeo (Colombia): **D**, venter of female; **E**, dorsum of female (from Gaud and Atyeo 1974). *Aetacarus phylloproctus* (Megnin and Trouessart) (New Guinea): **F**, venter of male (from Gaud and Atyeo 1974).

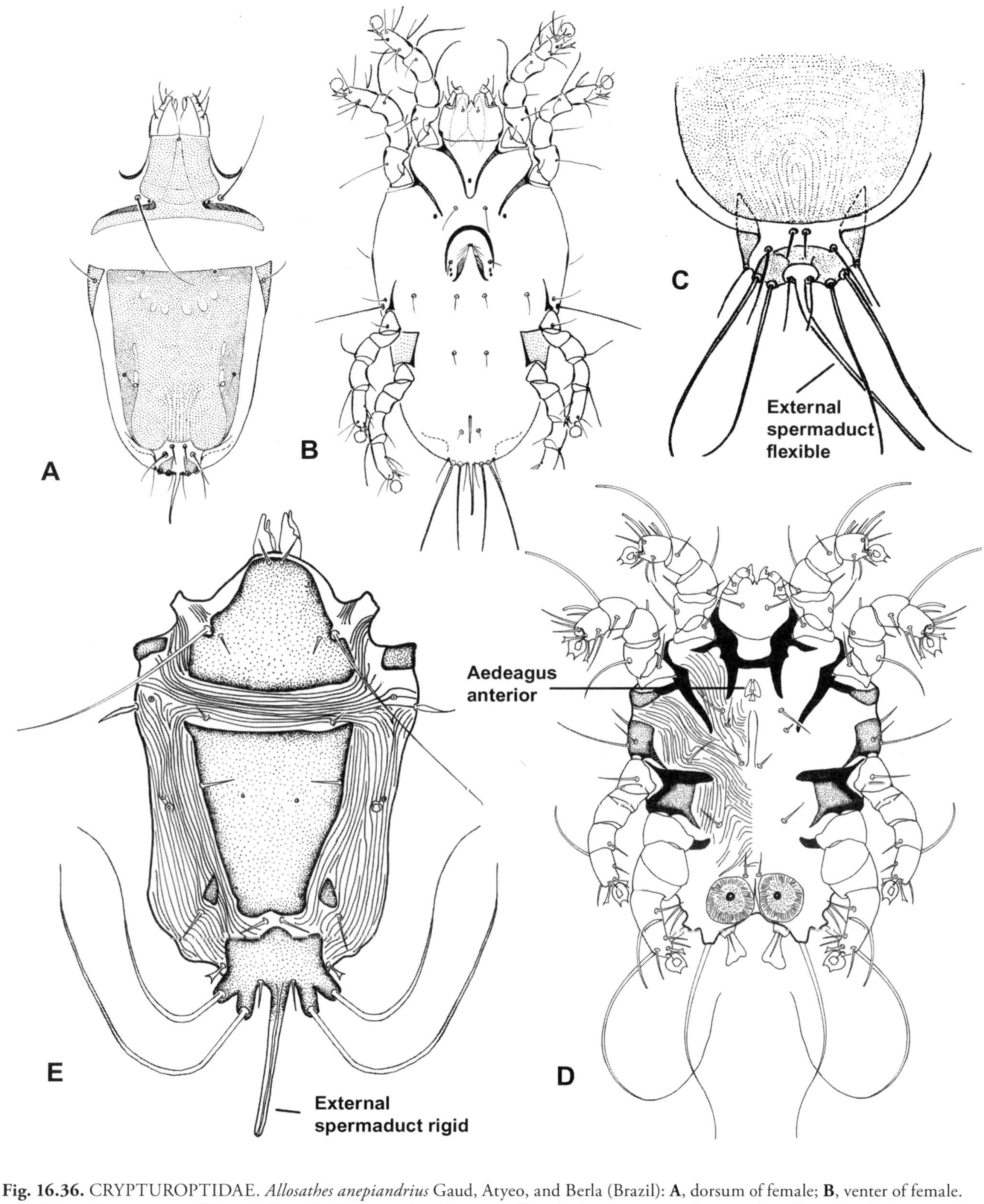

Fig. 16.36. CRYPTUROPTIDAE. *Allosathes anepiandrius* Gaud, Atyeo, and Berla (Brazil): **A**, dorsum of female; **B**, venter of female. *Mesosathes tetrasetosus* (Novaes) (Brazil): **C**, posterodorsal aspect of female showing terminal spermaduct (from Gaud, Atyeo, and Berla 1972).

THORACOSATHESIDAE. *Thoracosathes thoracosathes* (Trouessart and Neumann) (Papua, Indonesia): **D**, venter of male; **E**, dorsum of female.

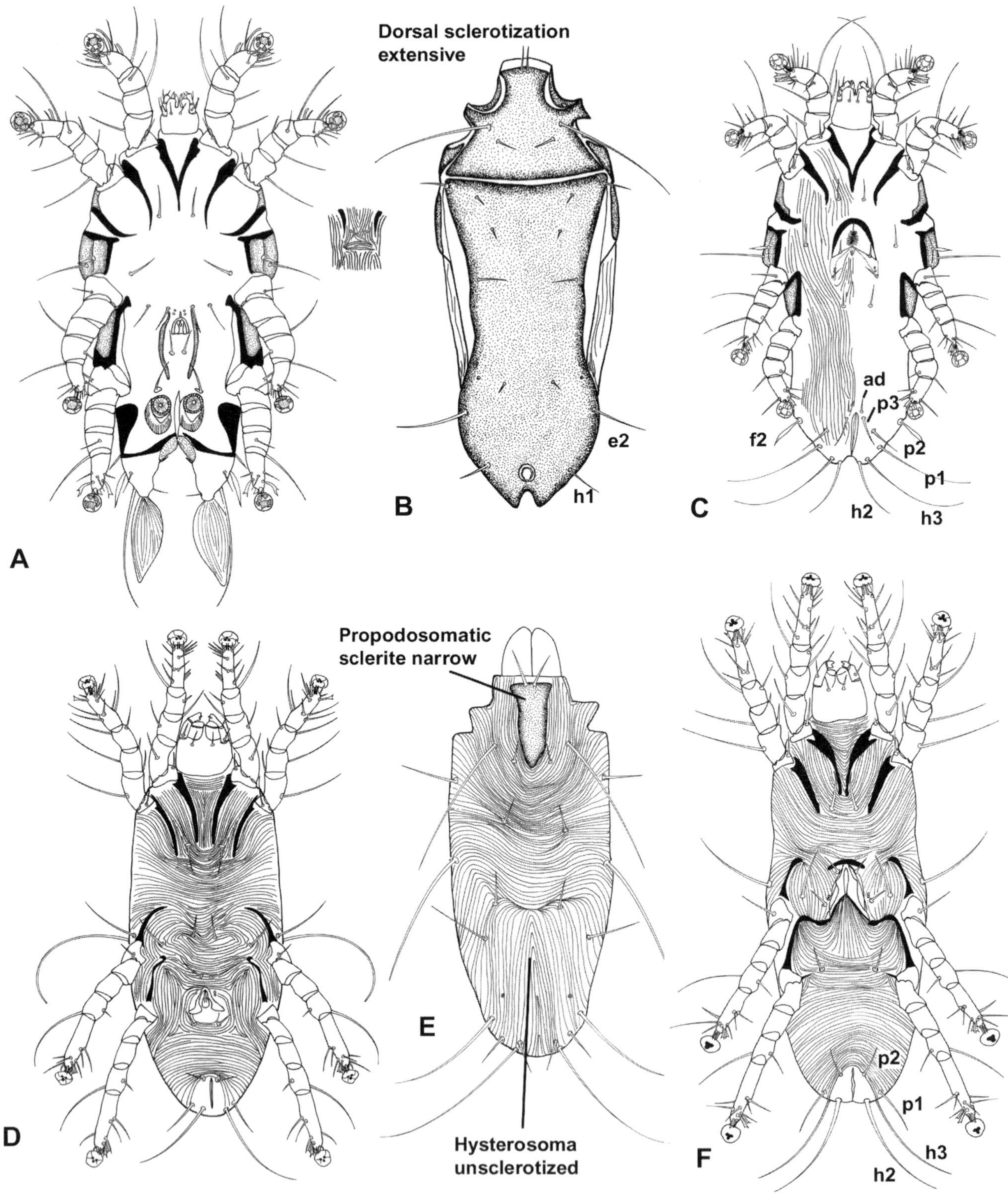

Fig. 16.37. PTILOXENIDAE. *Ptiloxenus major* (Megnin and Trouessart) (Canada): **A**, venter of male; **B**, dorsum of male; **C**, venter of female.

OCONNORIIDAE. *Oconnoria inexpectata* Gaud, Atyeo, and Klompen (Philippines): **D**, venter of male; **E**, dorsum of male; **F**, venter of female.

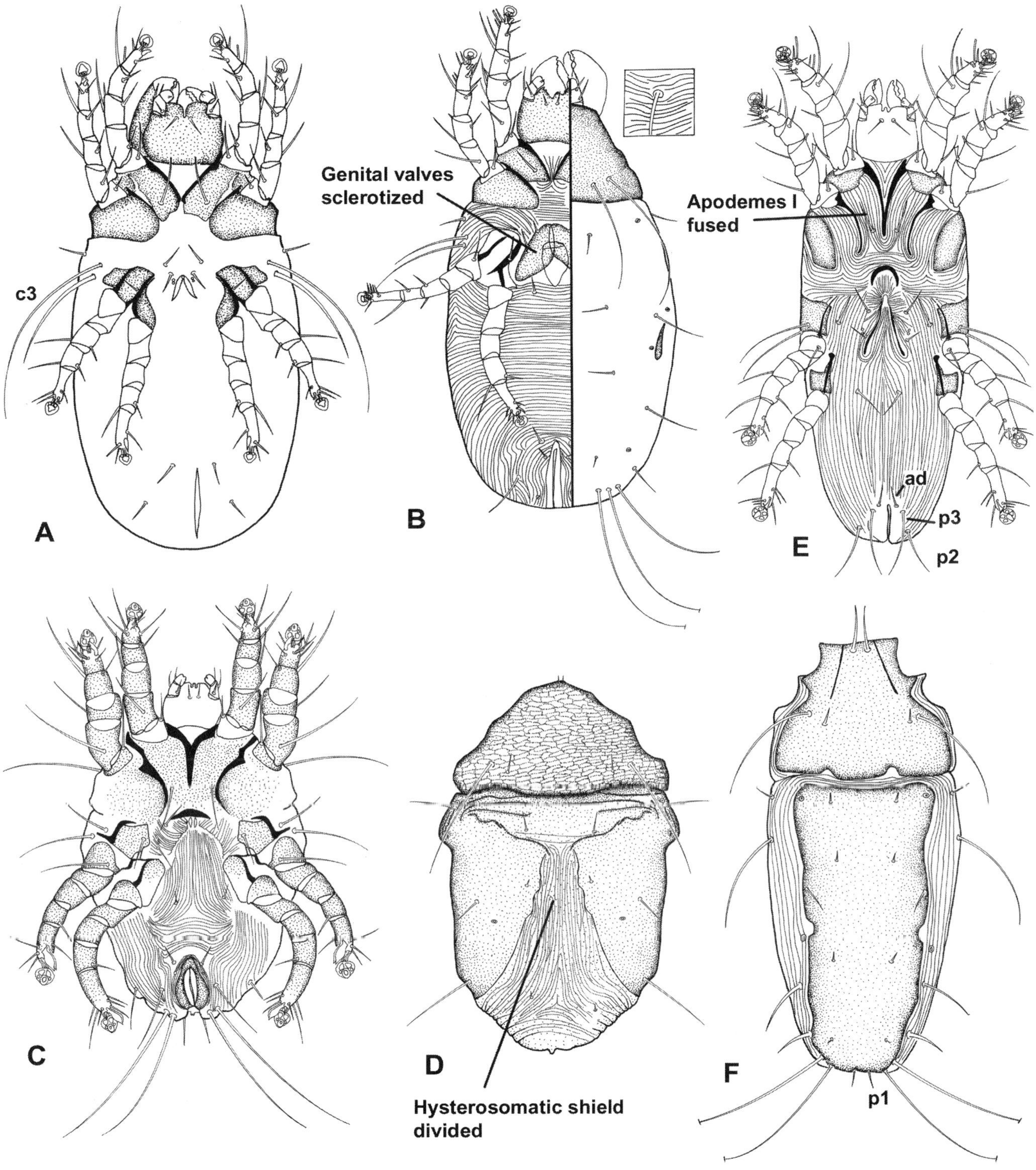

Fig. 16.38. ASCOURACARIDAE. *Ascouracarus michigani* Dabert and Ehrnsberger (Michigan, USA): **A**, venter of male; **B**, composite dorsum and venter of female.
KIWILICHIDAE. *Kiwilichus delosikyus* Gaud and Atyeo (New Zealand): **C**, venter of female; **D**, dorsum of female.
SYRINGOBIIDAE. *Phyllochaeta tenuiseta* Dabert and Atyeo (Michigan, USA): **E**, venter of female; **F**, dorsum of female.

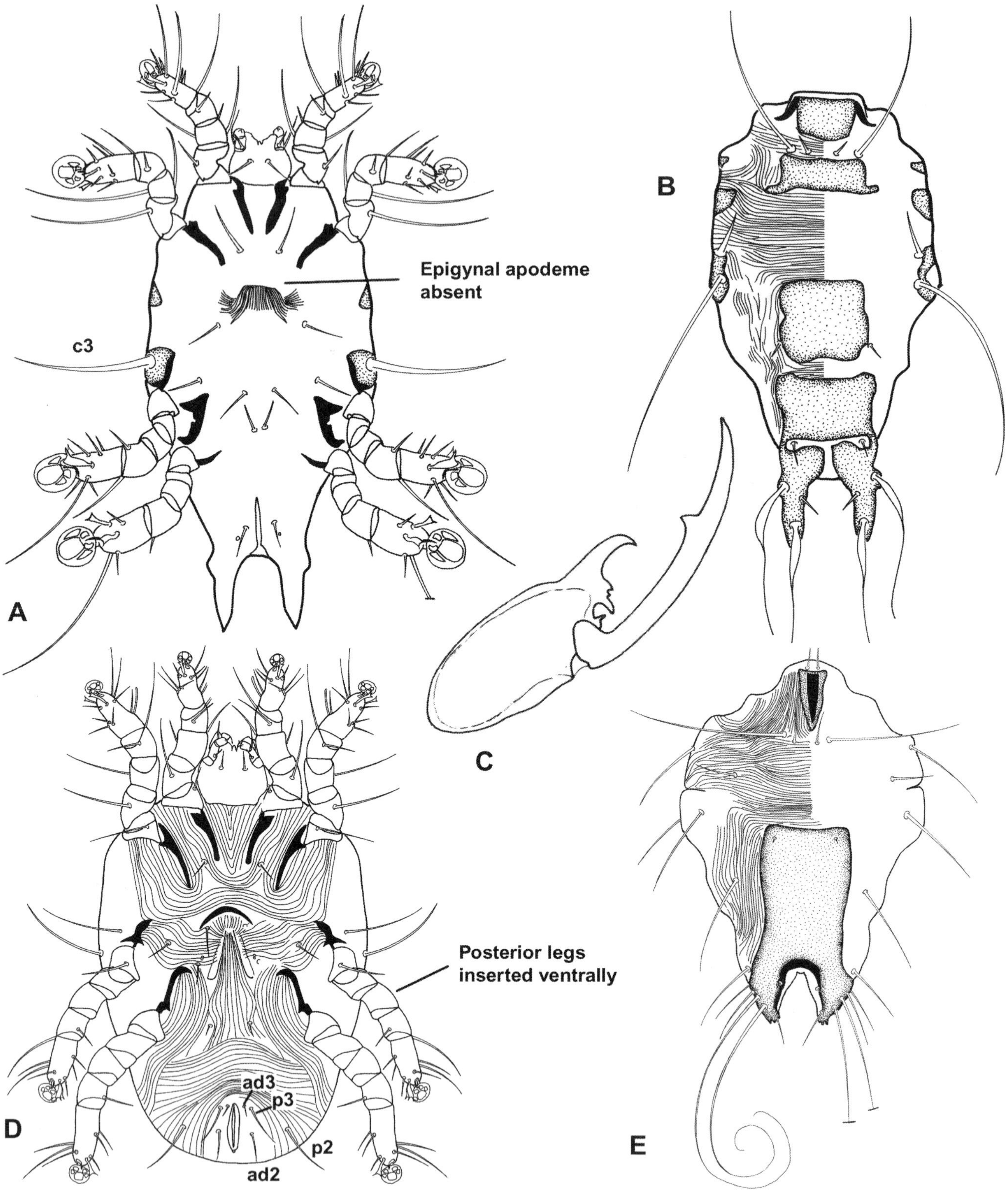

Fig. 16.39. FALCULIFERIDAE. *Atyeonia bifurcata* Gaud (Trinidad): **A**, venter of female; **B**, dorsum of female. *Falculifer* sp. (Cuba): **C**, chelicera of heteromorphic male.

CHEYLABIDIDAE. *Gangolichus prionurus* Perez and Atyeo (Somalia): **D**, venter of female; **E**, dorsum of male.

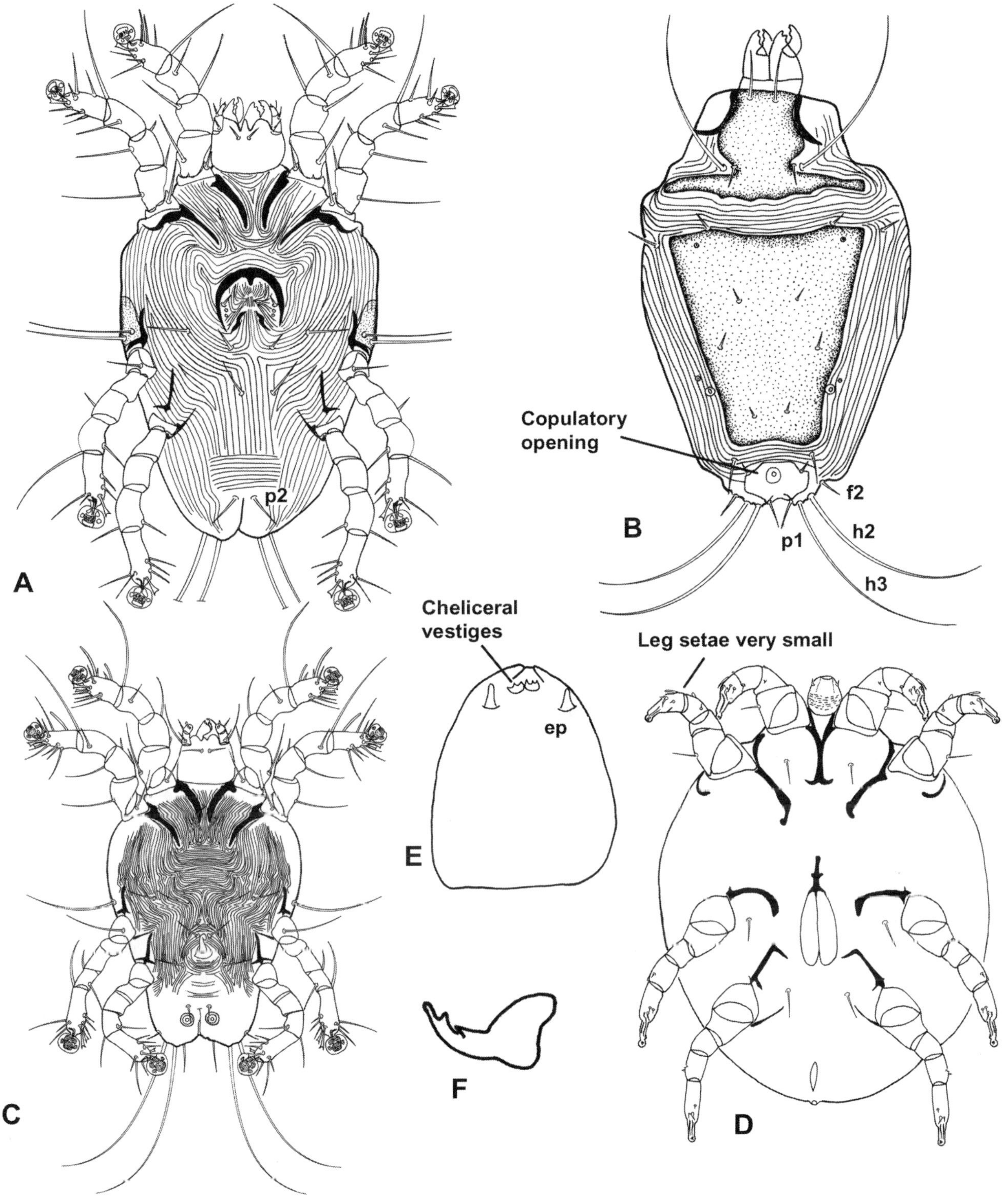

Fig. 16.40. PTEROLICHIDAE. *Pterolichus obtusus* Robin (Georgia, USA): **A**, venter of female; **B**, dorsum of female; **C**, venter of male. CYTODITIDAE. *Cytodites nudus* (Vizioli) (Nebraska, USA): **D**, venter of female. *Cytodites* sp. (Peru): **E**, gnathosoma of female in dorsal view. *Cytonyssus temenuchi* Fain (Kenya): **F**, chelicera of female in lateral view.

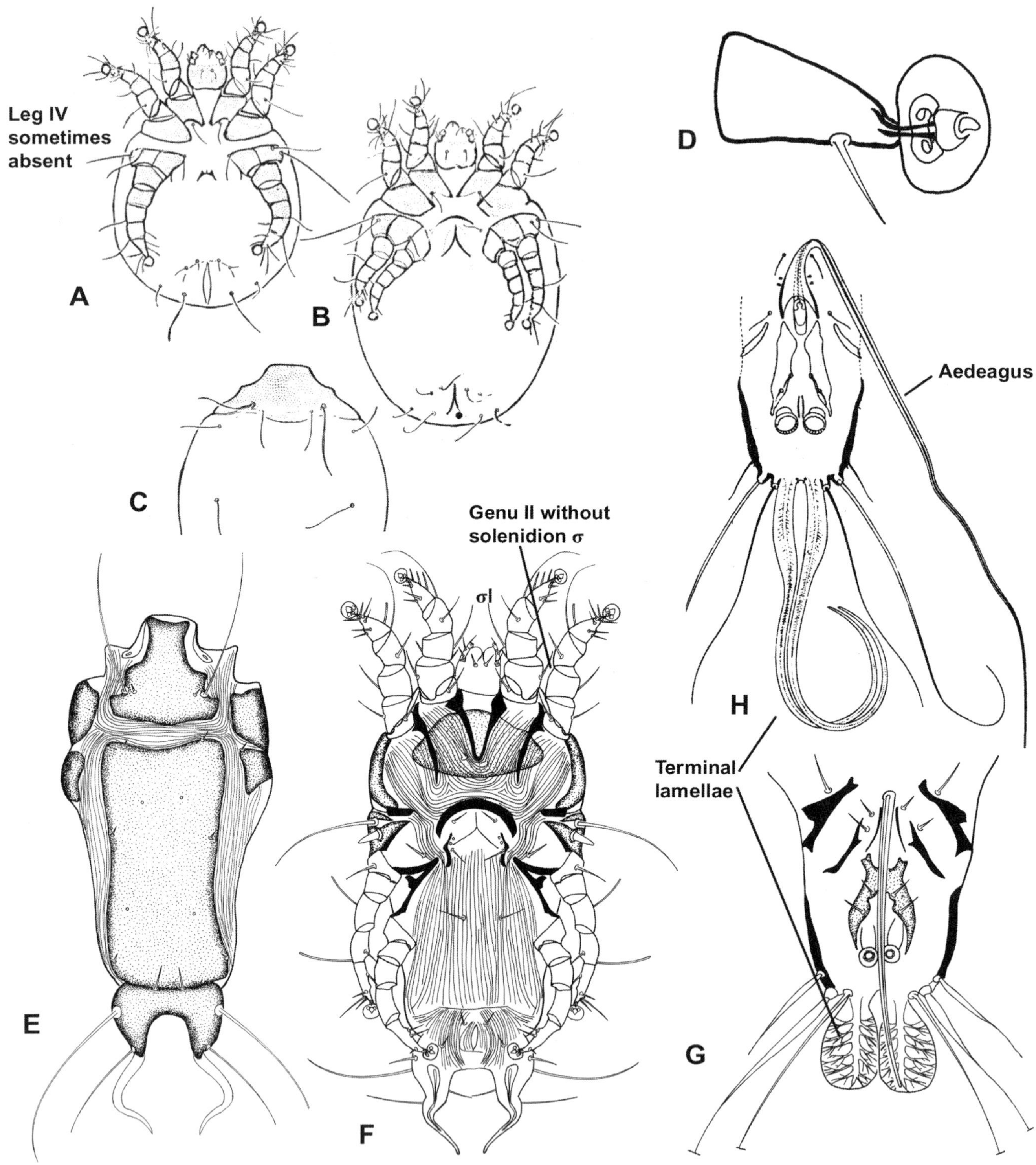

Fig. 16.41. APIONACARIDAE. *Atelespoda crena* (McDaniel) (Texas, USA): **A**, venter of male; **B**, venter of female; **C**, dorsum of male (after Gaud and Atyeo 1975). *Atelespoda* sp. (Egypt): **D**, tarsus IV of female.

PROCTOPHYLLODIDAE. *Proctophyllodes glandarinus* (Koch) (Bulgaria): **E**, dorsum of female; **F**, venter of female; **G**, posterior ventral aspect of male. *P. longiphyllus* Atyeo and Braasch (Texas, USA): **H**, posterior ventral aspect of male.

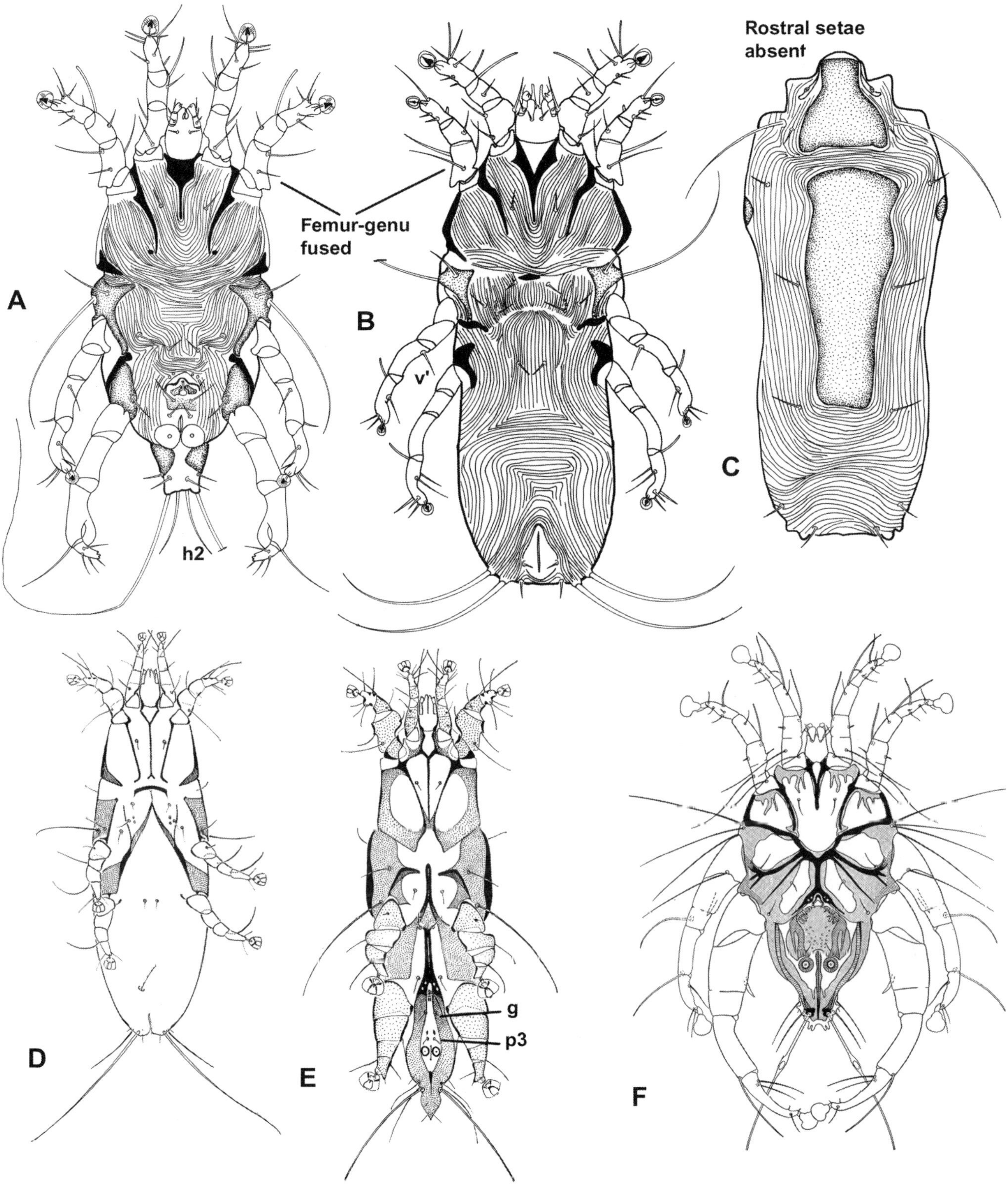

Fig. 16.42. XOLALGIDAE. *Xolalgoides palmai* Mironov and Perez (Galápagos Islands): **A**, venter of male; **B**, venter of female; **C**, dorsum of female.

ALLOPTIDAE. *Oxyalges cardiurus* Gaud and Atyeo (Falkland Islands): **D**, venter of female; **E**, venter of male (from Gaud and Atyeo 1966). *Laminalloptes phaethontis* (Fabr.): **F**, venter of male (from Atyeo and Peterson 1967).

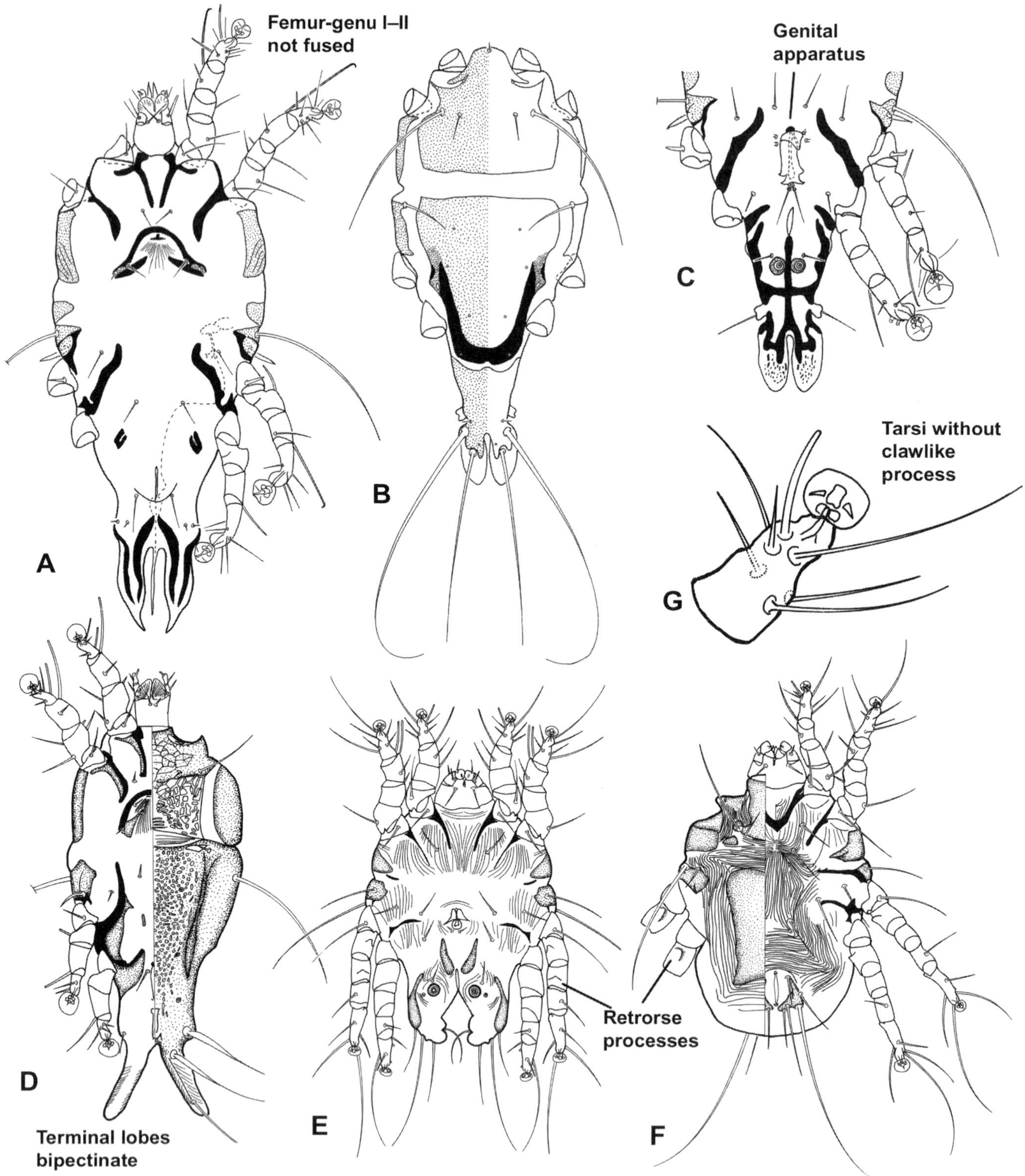

Fig. 16.43. TROUESSARTIIDAE. *Trouessartia geospiza* OConnor, Foufopoulos, and Lipton (Galápagos Islands): **A**, venter of female; **B**, dorsum of male; **C**, posteroventral aspect of male (from OConnor et al. 2005).

THYSANOCERCIDAE. *Thysanocerus affinis* Gaud and Peterson (Kenya): **D**, composite dorsum and venter of female.

DERMATIONIDAE. *Passeroptes dermicola* (Trouessart) (Philippines): **E**, venter of male; **F**, composite dorsum and venter of female; **G**, tarsus I of male.

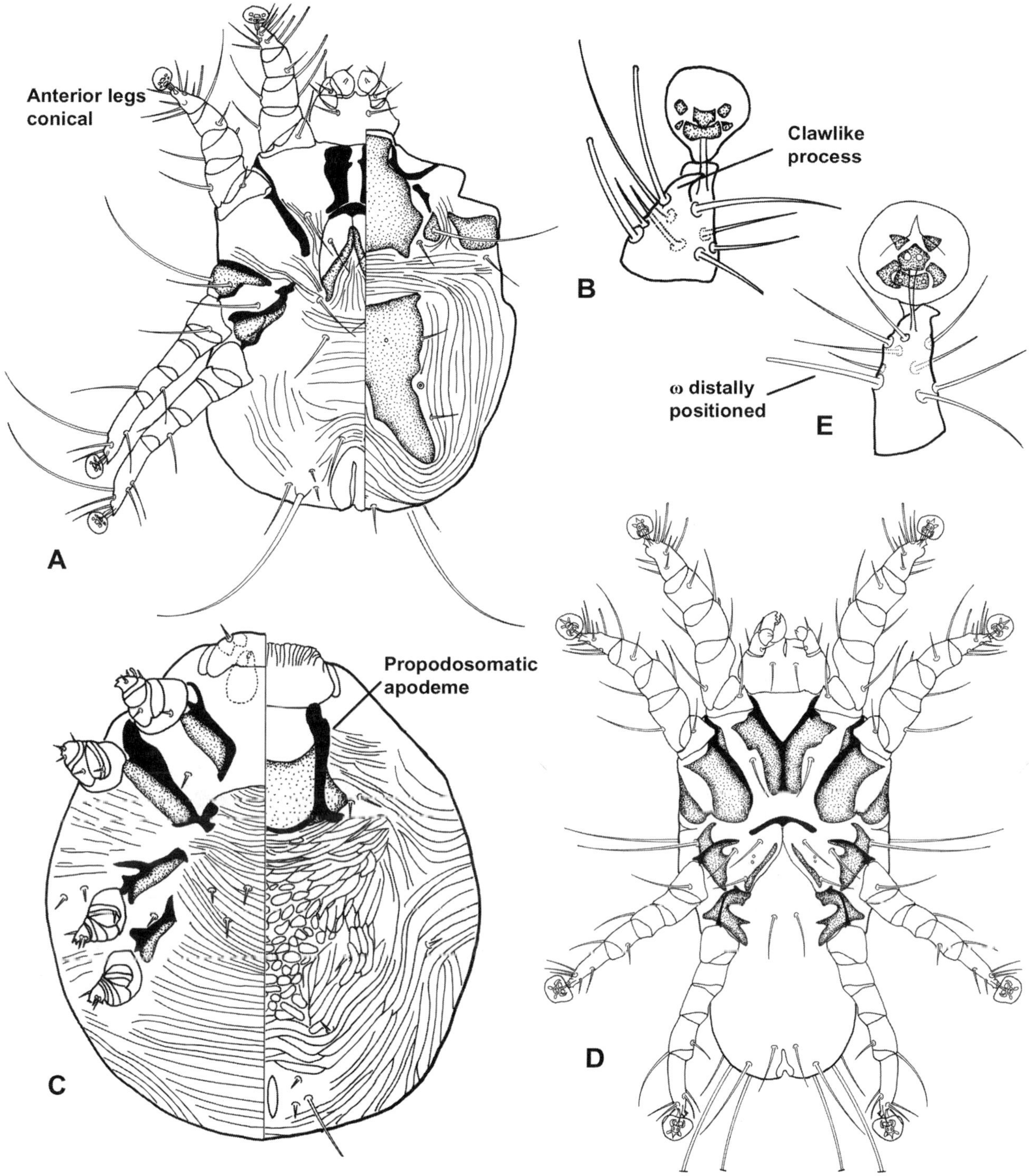

Fig. 16.44. EPIDERMOPTIDAE. *Rallepidermoptes fulicae* Fain (Michigan, USA): **A**, composite dorsum and venter of female; **B**, tarsus I of female. *Knemidokoptes mutans* (Robin and Lanquetin) (Netherlands): **C**, composite dorsum and venter of female. PTYSSALGIDAE. *Ptyssalges* sp. (Peru): **D**, venter of female; **E**, tarsus II of female.

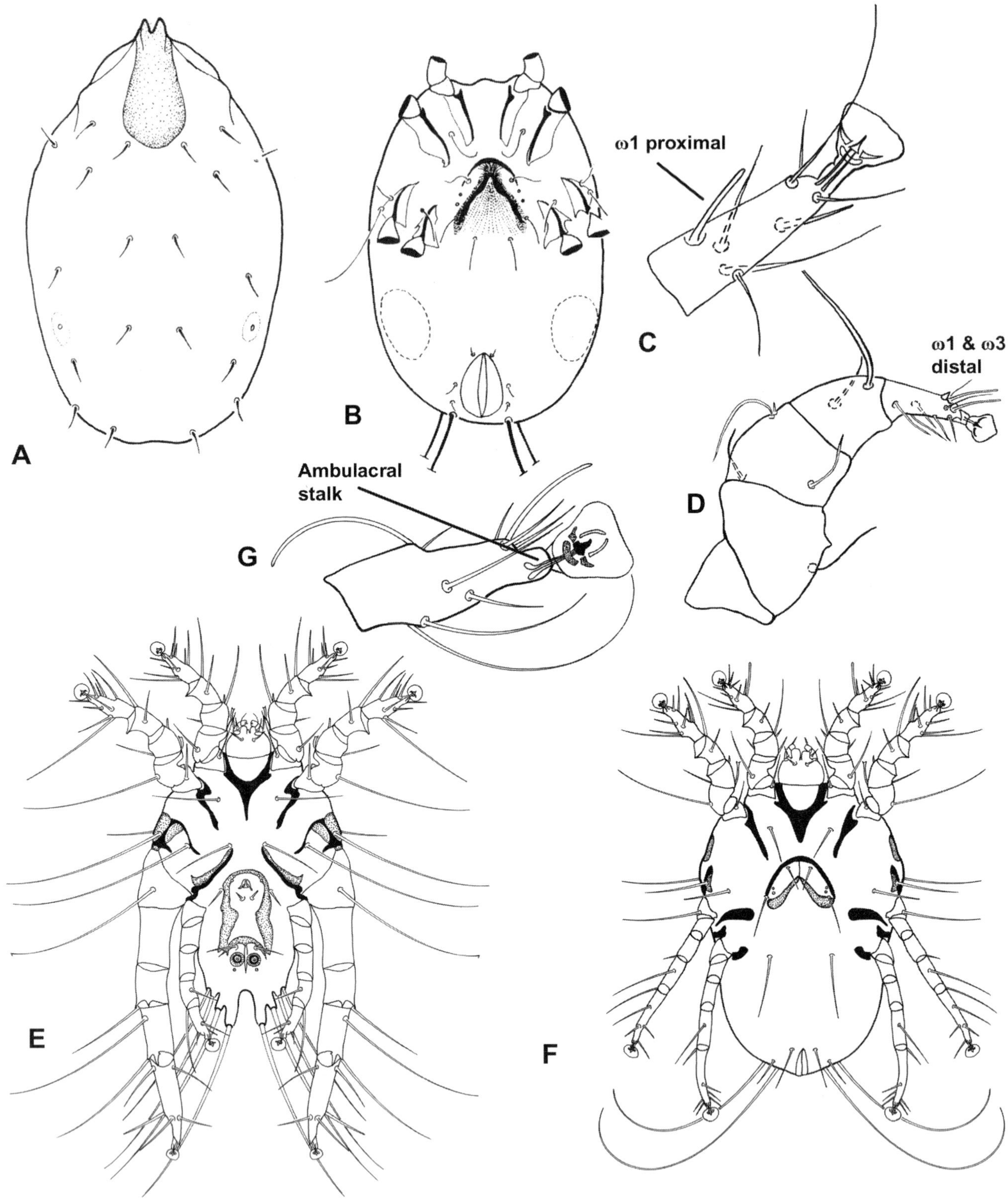

Fig. 16.45. PYROGLYPHIDAE. *Euroglyphus longior* (Trouessart) (England): **A**, dorsum of female. *Dermatophagoides pteronyssinus* (Trouessart) (Oregon, USA): **B**, venter of female; **C**, tarsus II of female; **D**, leg I of male.

PSOROPTOIDIDAE. *Pandalura strigisoti* (Buchholz) (Netherlands): **E**, venter of male; **F**, venter of female; **G**, tarsus and pretarsus I of female.

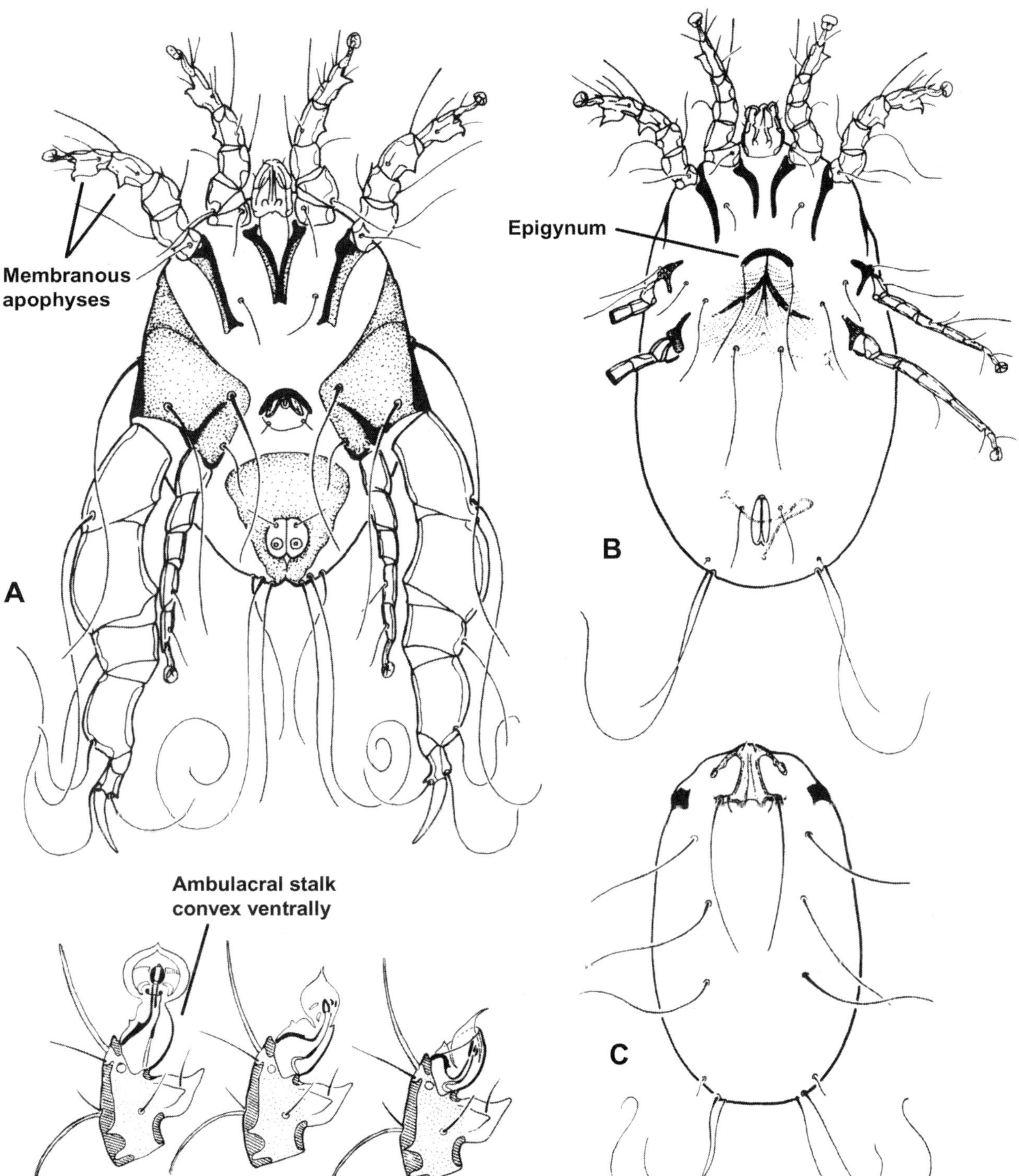

Fig. 16.46. ANALGIDAE. *Analges* sp. (Oregon, USA): **A**, venter of male; **B**, venter of female; **C**, dorsum of female. *Anhemialges* sp.: **D–F**, retraction sequence of pretarsus (from Gaud and Atyeo 1977).

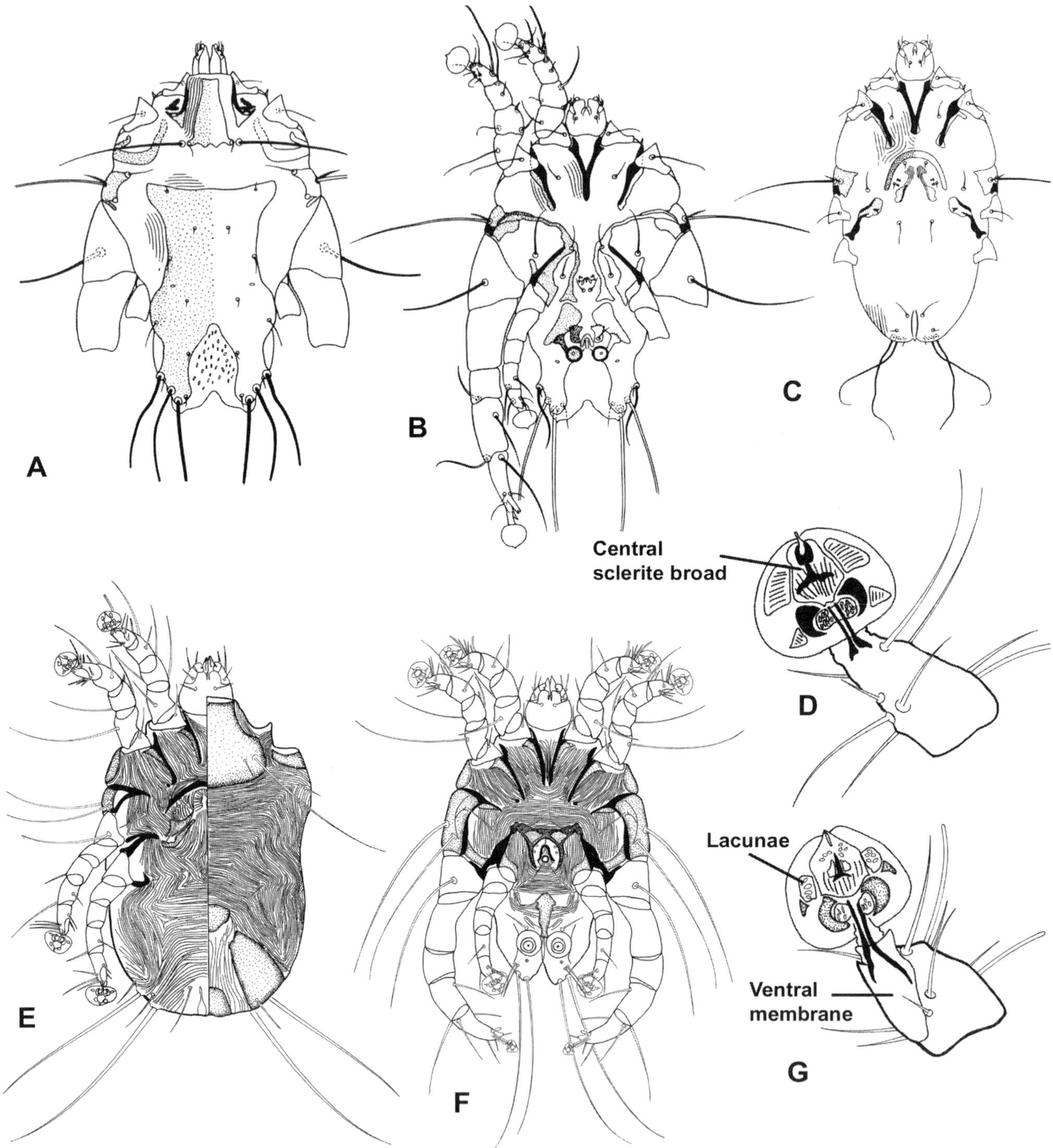

Fig. 16.47. AVENZOARIIDAE. *Scutomegninia subantarctica* Mironov (Antarctica): **A**, dorsum of male; **B**, venter of male; **C**, venter of female (from Atyeo and Peterson 1967). *S. phalacrocoracis* (Dubinin and Dubinina): **D**, pretarsus I of female.
PTERONYSSIDAE. *Pteronyssoides passeris* (Gaud) (Madagascar): **E**, composite dorsum and venter of female; **F**, venter of male.
Dicrurobius monacrotrichus (Gaud) (Madagascar): **G**, tarsus and pretarsus I of male.

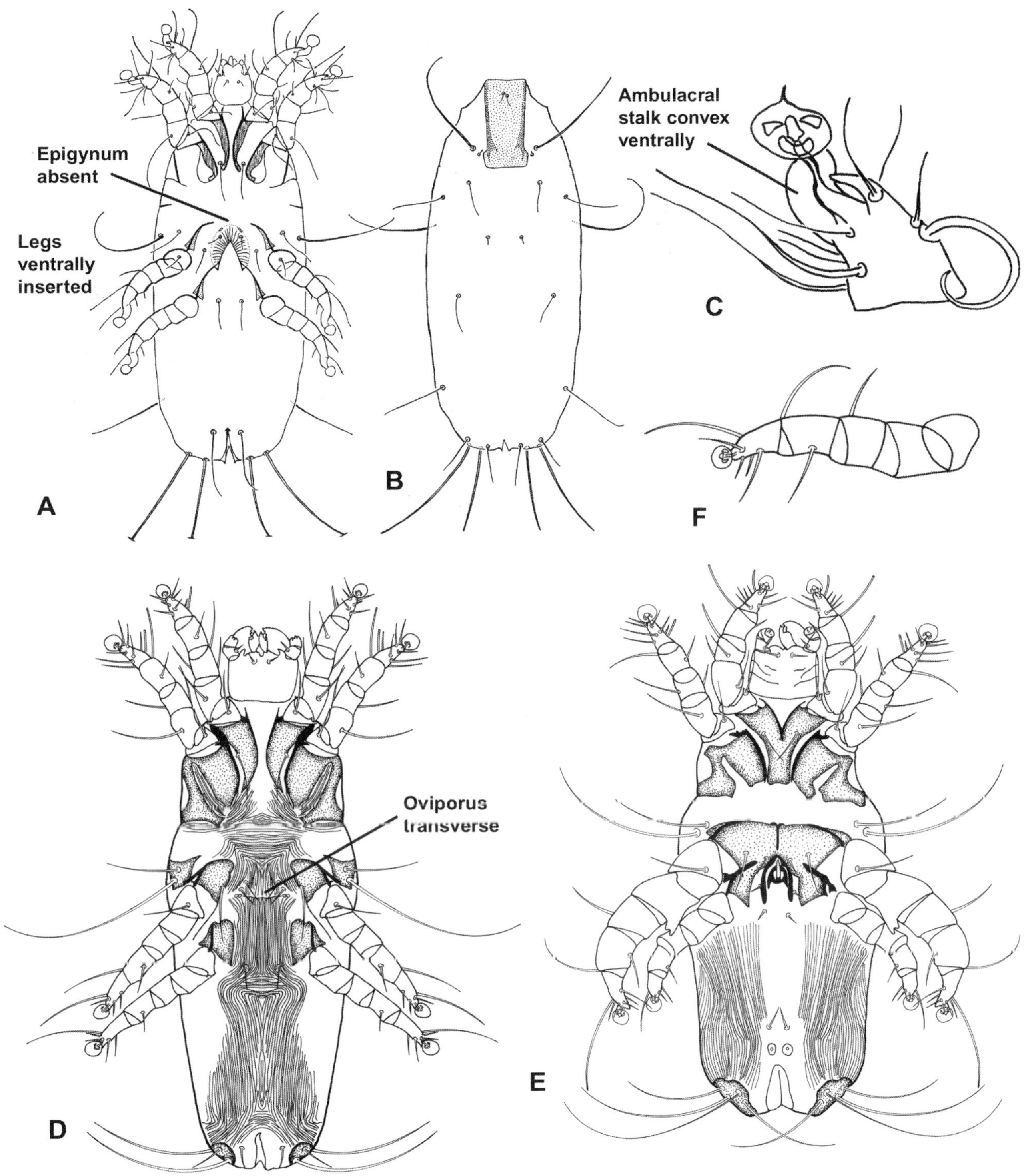

Fig. 16.48. DERMOGLYPHIDAE. *Paralges microtrichus* Gaud (Morocco): **A**, venter of female; **B**, dorsum of female; **C**, tarsus and pretarsus I (from Gaud 1973).
GAUDOGLYPHIDAE. *Gaudoglyphus minor* (Nörner) (Tanzania): **D**, venter of female; **E**, venter of heteromorphic male; **F**, leg III of homeomorphic male.

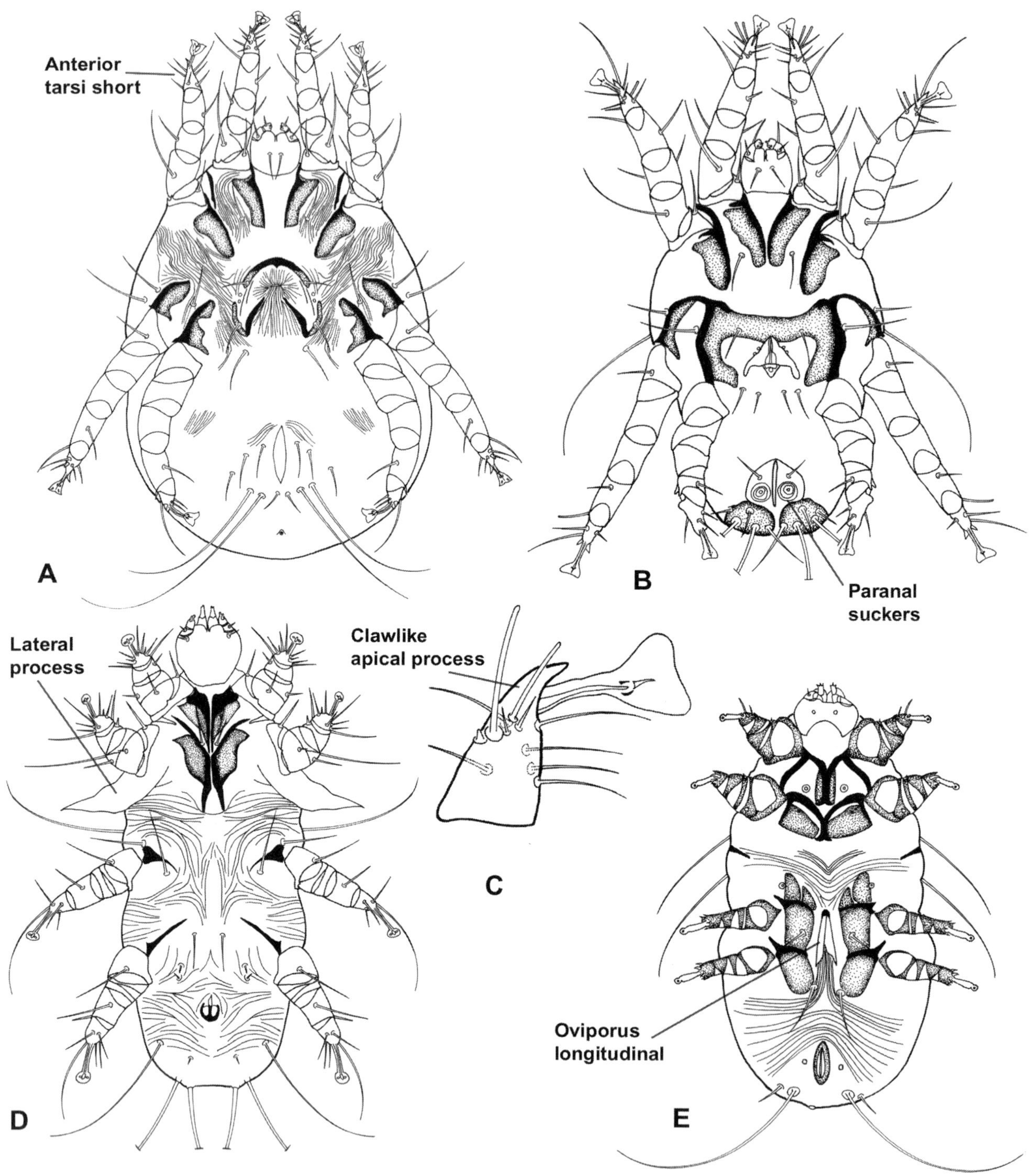

Fig. 16.49. TURBINOPTIDAE. *Congocoptes sphyrapicicola* Pence (Michigan, USA): **A**, venter of female; **B**, venter of male; **C**, tarsus I of female.

LAMINOSIOPTIDAE. *Fainocoptes columbicola* Lukoschus and Lombert (Netherlands): **D**, venter of male. *Laminosioptes cysticola* (Vizioli) (Mexico): **E**, venter of female.

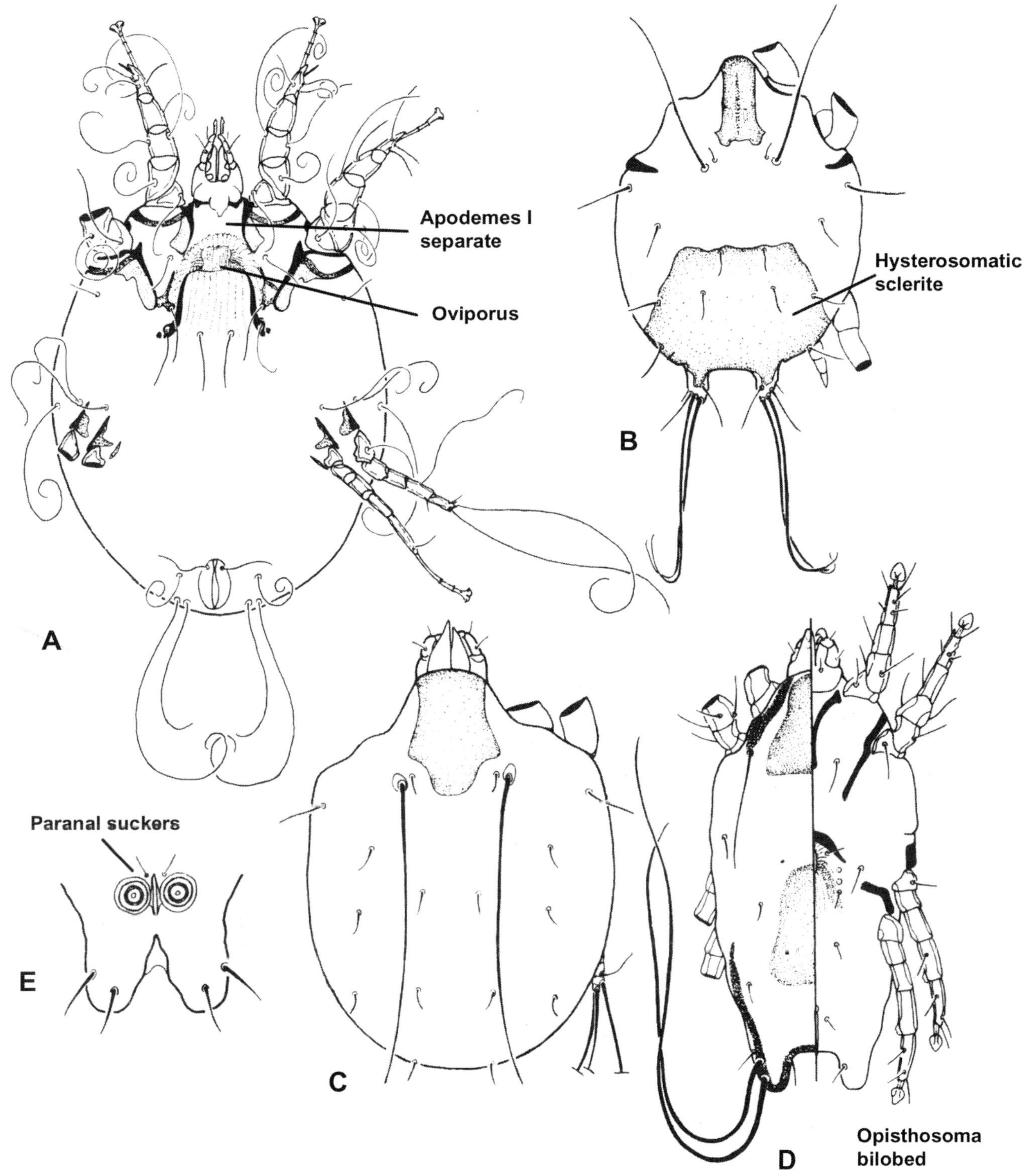

Fig. 16.50. PSOROPTIDAE. *Psoroptes cunniculi* (Delafond) (Oregon, USA): **A**, venter of female; **B**, dorsum of male. *Caparinia tripilis* (Michael) (New Zealand): **C**, dorsum of female.

LOBALGIDAE. *Lobalges trouessarti* Fonseca (Brazil): **D**, composite dorsum and venter of female (after Fain 1965b); **E**, posteroventral aspect of male.

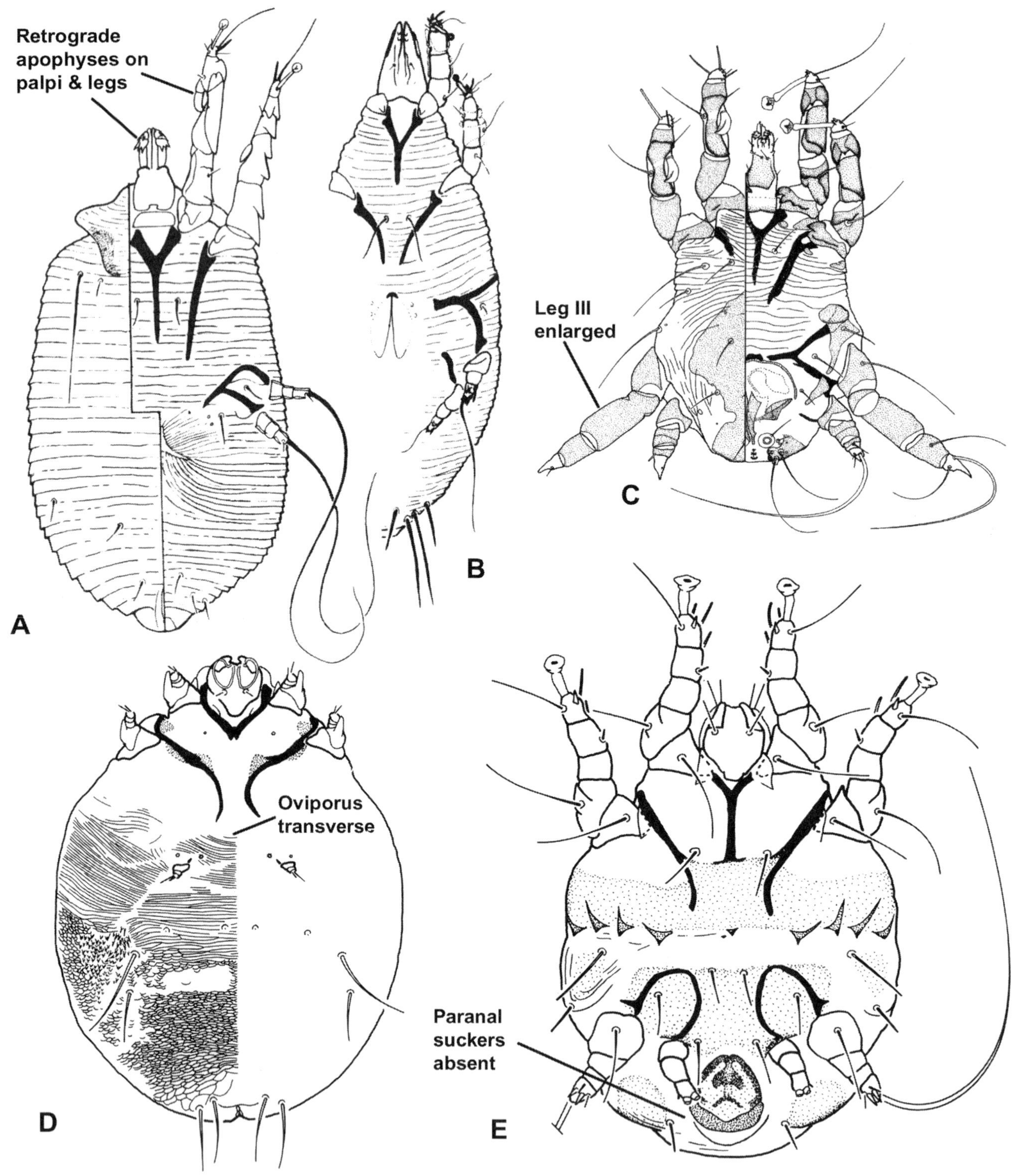

Fig. 16.51. RHYNCOPTIDAE. *Rhyncoptes anastosi* Fain (South America): **A**, composite dorsum and venter of female (adapted from Fain 1965e). *Saimirioptes paradoxus* Fain (South America): **B**, venter of female (after Fain 1968d). *R. grabberi* Klompen (Michigan, USA): **C**, composite dorsum and venter of male.

SARCOPTIDAE. *Chirobia haplonycteris* Klompen (Philippines): **D**, venter of female. *Teinocoptes vandeuseni* Mitchell and Fain (Philippines): **E**, venter of male (**D**, **E** from Klompen 1992).

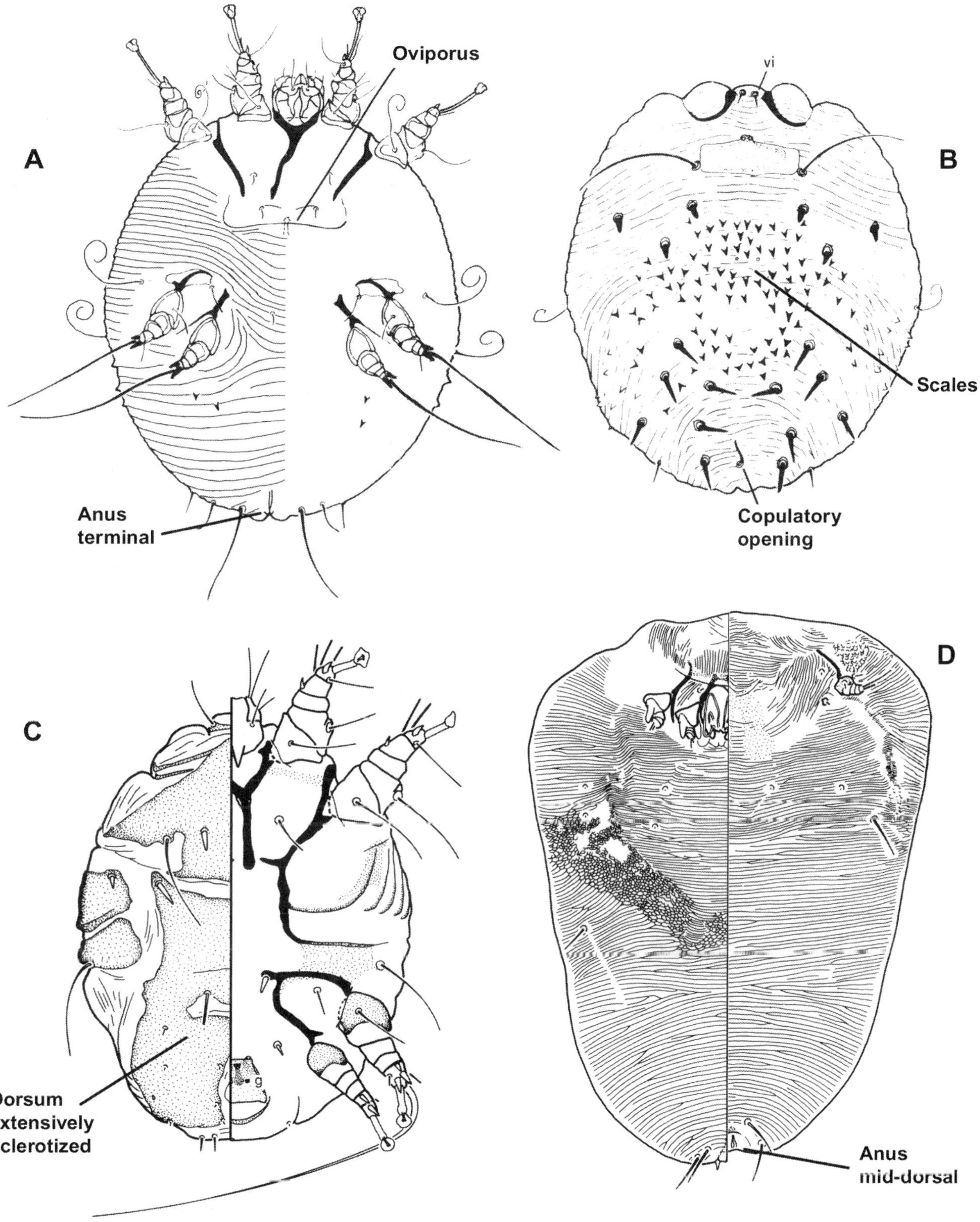

Fig. 16.52. SARCOPTIDAE. *Sarcoptes scabiei* (L.) (Oregon, USA): **A**, venter of female; **B**, dorsum of female. *Nycteridocoptes heidemanni* Klompen (Philippines): **C**, composite dorsal and ventral view of male. *Teinocoptes vandeuseni* Mitchell and Fain (Philippines): **D**, composite anterior and posterior view of female. (**C**, **D** after Klompen, 1992).

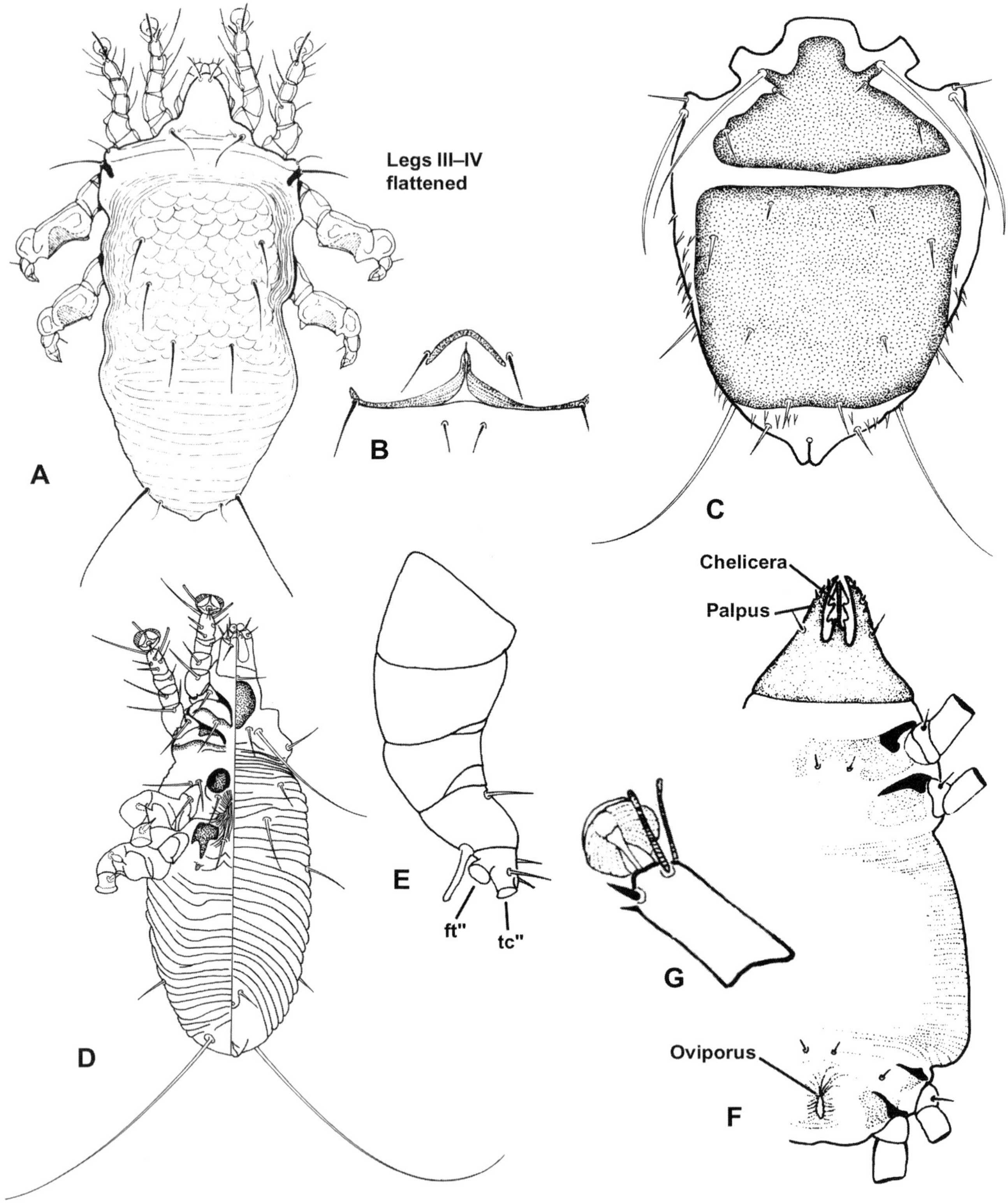

Fig. 16.53. MYOCOPTIDAE. *Myocoptes musculinus* (Koch) (Oregon, USA): **A**, dorsum of female; **B**, oviporus region of female. *Dromiciocoptes caenolestes* Bochkov and OConnor (Ecuador): **C**, dorsum of female. *Trichoecius tenax* (Michael) (New York, USA): **D**, composite dorsum and venter of female; **E**, leg IV of male.

CHIRORHYNCHOBIIDAE. *Chirorhynchobia urodermae* Fain: **F**, venter of female; **G**, tarsus I of female (after Fain 1968d).

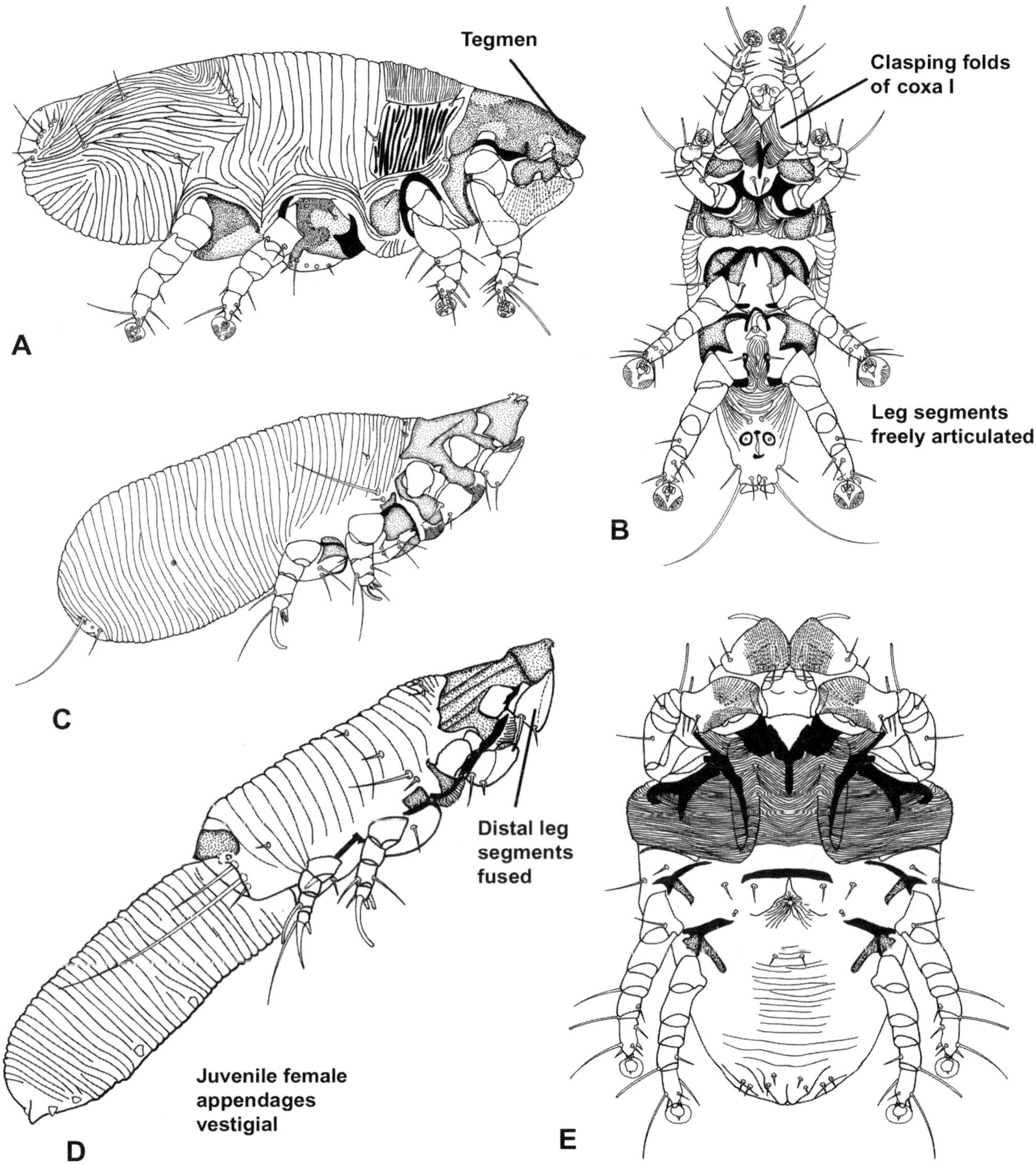

Fig. 16.54. LISTROPHORIDAE. *Listrophorus americanus* Radford (Michigan, USA): **A**, lateral aspect of female; **B**, venter of male. CHIRODISCIDAE. *Alabidocarpus chalinolobi* Fain and Lukoschus (Australia): **C**, lateral aspect of female. *A. vietnamensis* Fain (Philippines): **D**, lateral aspect of male and female nymph in precopulatory guarding posture. *Schizocoptes conjugatus* Lawrence (Congo): **E**, venter of female.

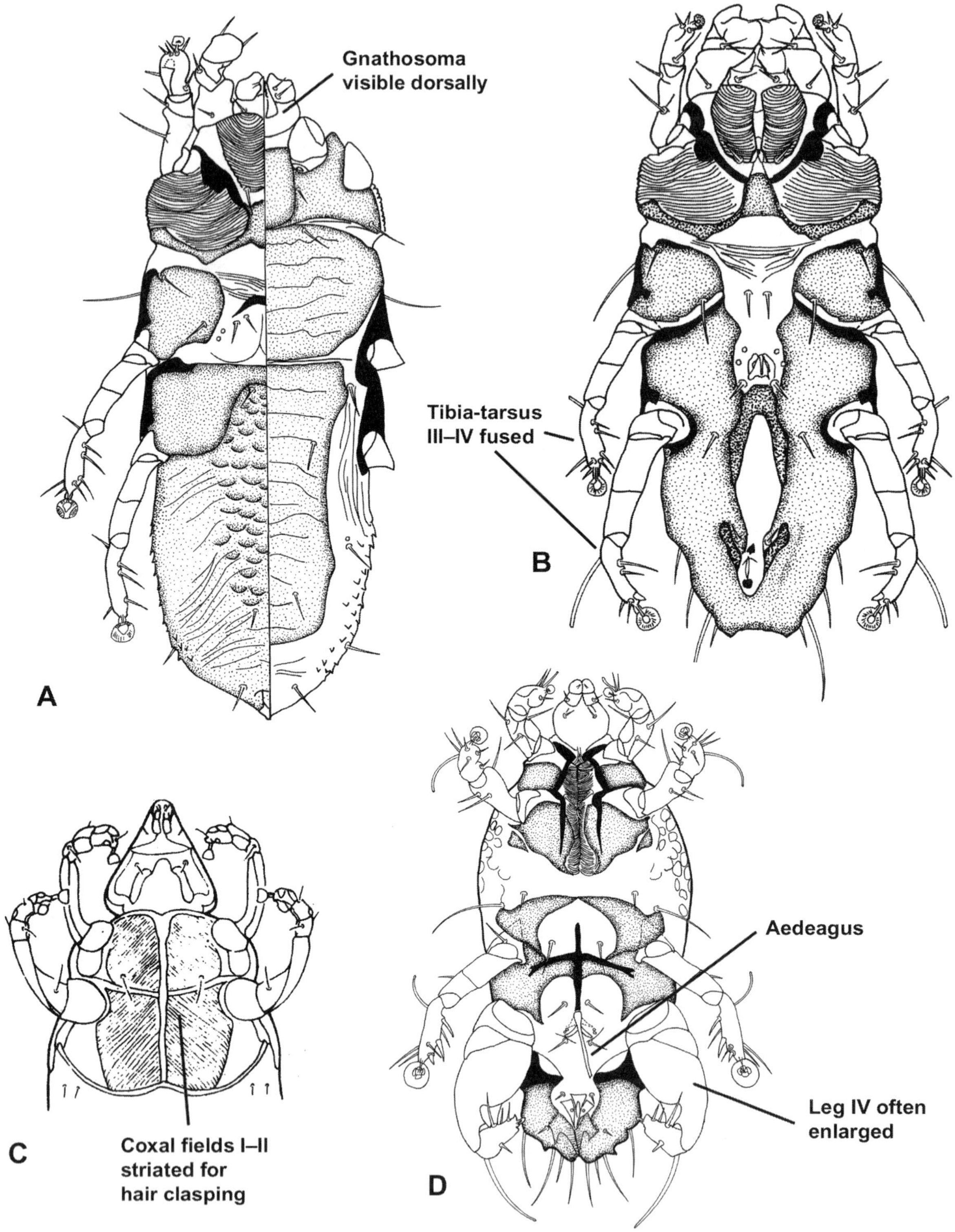

Fig. 16.55. ATOPOMELIDAE. *Listrophoroides cucullatus* (Trouessart) (Indonesia): **A**, composite dorsum and venter of female; **B**, venter of male. *Chirodiscoides caviae* Hirst: **C**, coxisternal aspect of female. *Didelphilichus serrifer* Fain (New York, USA): **D**, venter of male.

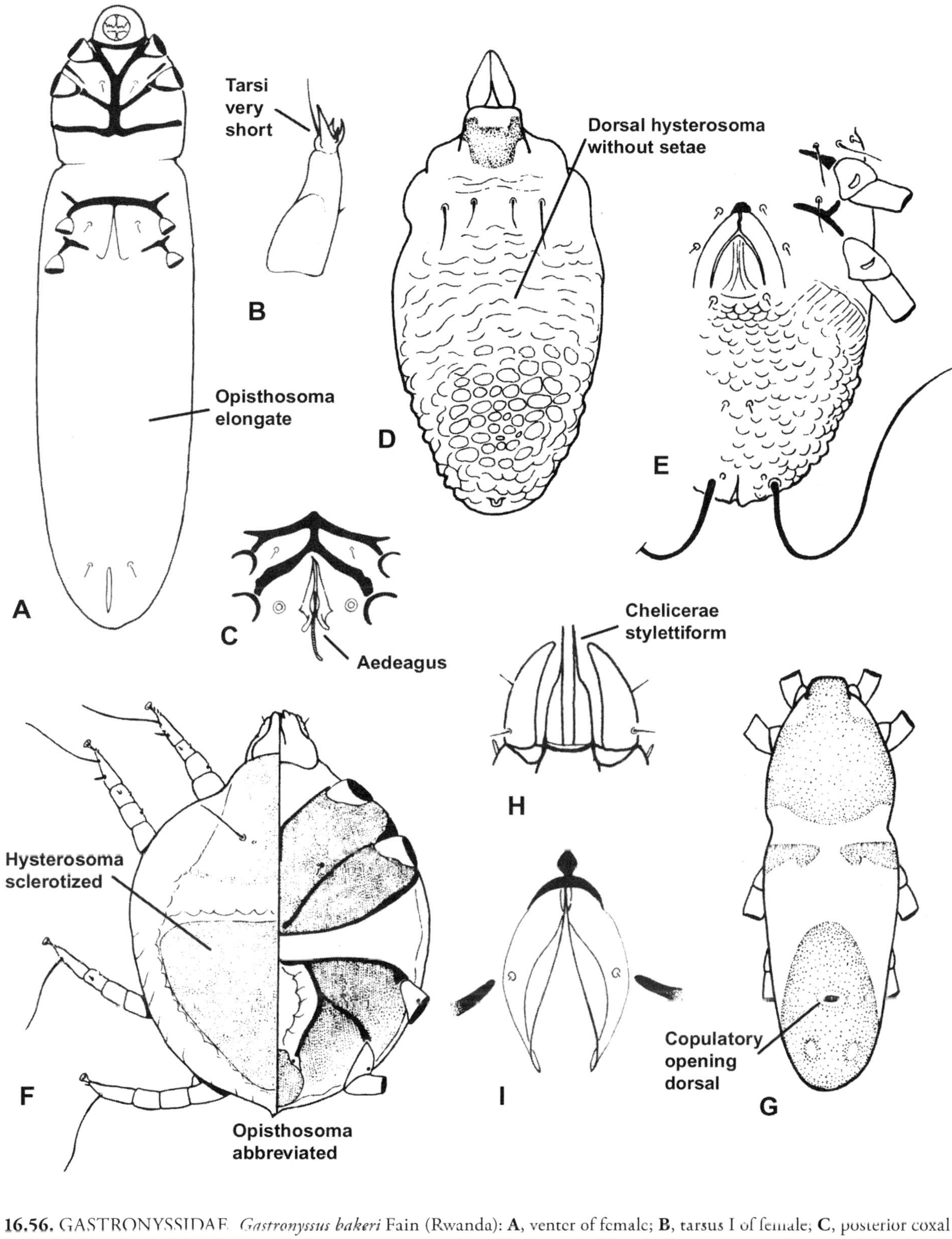

Fig. 16.56. GASTRONYSSIDAE. *Gastronyssus bakeri* Fain (Rwanda): **A**, venter of female; **B**, tarsus I of female; **C**, posterior coxal and genital fields of male (after Fain 1959). *Sciuracarus paraxeri* Fain: **D**, dorsum of female; **E**, posteroventral aspect of female (after Fain 1964b).

PNEUMOCOPTIDAE. *Pneumocoptes jellisoni* Baker (Idaho, USA): **F**, composite dorsum and venter of female (adapted from Baker 1951).

LEMURNYSSIDAE. *Lemurnyssus galagoensis* Fain (Congo): **G**, dorsum of female; **H**, gnathosoma of female; **I**, oviporus of female (after Fain 1957).

Bibliography

Abd-El-Hamid, M. E., B. M. Bayoumi, and A. I. Mohamad. 1980. Checklist of the oribatid mites (Acari: Oribatei). *Egypt. Bull. Fac. Sci.* (Assiut Univ.) 9:139–57.

Abdussalam, M. 1941. Pterygosomid mites from two north Indian lizards. *Indian J. Entomol.* 3:65–72.

Abé, H. 1998. *Rhombognathine mites: Taxonomy, phylogeny, and biogeography.* Sapporo, Japan: Hokkaido Univ. Press.

Abé, H., T. Sasaki, and J. Hiromi. 2001. Halacarid mites (Acari: Halacaridae) as possible indicators of preferable culture beds of Japanese scallop *Patinopecten yessoensis* (Jay) (Pterioida: Pectinidae). *Internat. J. Acarol.* 27:91–96.

Abou-Awad, B. A., B. M. El-Sawaf, and A. A. Abdel Kader. 1999. Life history and life table of *Pronematus ubiquitus* (McGregor) as a predator of eriophyoid mites in Egypt (Acari: Tydeidae). *Acarologia* 40:29–32. [IOLINIDAE]

Aboul-Nasr, A. E., and T. T. M. Bassal. 1971. Biochemical and physiological studies of certain ticks (Ixodoidea). The sugar content and concentration in *Argas* and *Hyalomma* biological fluids. *J. Med. Entomol.* 8:521–24.

Abrahamovich, A. H., and A. B. de Alzuet. 1990. Tipos de asociacion entre acaros e hymenopteros. 1, *Sennertia (A.) splendidulae* Alzuet y Abrahamovich, 1988 (Acarina: Chaetodactylidae) comensal y foretico de *Xylocopa (S.) splendidula splendidula* Lepeletier, 1841 (Hymenoptera: Anthophoridae). *Turrialba* 40:319–22.

Åbro, A. 1979. Attachment and feeding devices of water-mite larvae (*Arrenurus* spp.) parasitic on damselflies (Odonata, Zygoptera). *Zool. Scripta* 8:221–34.

———. 1982. The effects of parasitic water mite larvae (*Arrenurus* spp.) on zygopteran imagoes (Odonata). *J. Invertebr. Pathol.* 39:373–81.

———. 1984. The initial stylostome formation by parasitic larvae of the water-mite genus *Arrenurus* on zygopteran imagines. *Acarologia* 24:33–45.

———. 1986. On the phoretic phase of *Arrenurus* larvae (Acari, Hydrachnellae) associated with zygopteran nymphs. *Acarologia* 27:251–54.

———. 1988. The mode of attachment of mite larvae *Leptus* spp. to harvestmen (Opiliones). *J. Nat. Hist.* 22:123–30. [ERYTHRAEIDAE]

———. 1990. The impact of parasites in adult populations of Zygoptera. *Odonatologica* 19:223–34.

———. 1991. The incipient stylostome of parasitic water mite larvae (*Arrenurus* spp.). *J. Parasitol.* 77:313–14.

———. 1992. On the feeding and stylostome composition of parasitic water mite larvae (*Arrenurus* spp.) on damselflies (Zygoptera, Odonata). *Zool. Beitr.* 34:241–48.

Achor, D. S., R. Ochoa, E. F. Erbe, H. Aguilar, W. P. Wergin, and C. C. Childers. 2001. Relative advantages of low temperature versus ambient temperature scanning electron microscopy in the study of mite morphology. *Internat. J. Acarol.* 27:3–12.

Adis, J., and G. W. Krantz. 1985. Notes on the natural history of *Macrocheles* (Acari: Gamasida: Macrochelidae) associated with 3-toed sloths *Bradypus* spp. (Edentata: Bradypodidae) in the Central Amazon. *Zool. Anz.*, Jena 3/4:222–24.

Adler, C. R. L., and M. W. Brown. 1986. Attachment and dispersion of *Callidosoma metzi* (Acari: Erythraeidae) parasitizing *Platynota idaeusalis* (Lepidoptera: Tortricidae). *Ann. Entomol. Soc. Amer.* 79:56–59.

Aeschlimann, A. 1958. Développement embryonnaire d'*Ornithodoros moubata* (Murray) et transmission transovarienne de *Borrelia duttoni*. *Acta Trop.* 15:15–64.

Aeschlimann, A., and E. Hess. 1984. What is our current knowledge of acarine embryology? In Griffiths and Bowman, 1:90–99.

Afifi, A. M., M. F. Hassan, and M. S. Nawar. 1986. Notes on the biology feeding habits of *Protogamasellus minutus* Hafez, El-Badry & Nasr (Acari: Gamasida: Ascidae). *Bull. Soc. Entomol. Egypt* 66:251–59.

Agbede, R. I. S., and D. H. Kemp. 1985. Digestion in the cattle-tick *Boophilus microplus* (Canestrini). Light microscope study of the gut cells in nymphs and females. *Internat. J. Parasitol.* 15:147–57.

Aguilar, H., C. C. Childers, and W. C. Welbourn. 2001. Relative abundance and seasonal occurrence of mites in the family Tydeidae on citrus in Florida. In Halliday et al. (2001), 376–80.

Agwaral, A. A., and R. Karban. 1997. Domatia mediate plant-arthropod mutualism. *Nature* 387:562–63.

Ah, H.-S., and P. E. Hunter. 1968. *Nycteriglyphus vespertilio* n. sp., a new acarid mite associated with bats from Korea (Acarina: Rosensteiniidae). *Acarologia* 10:269–75.

Ah, H.-S., J. C. Peckham, and W. T. Atyeo. 1973. *Psorergates glaucomys* sp. n. (Acari: Psorergatidae), a cystogenous mite from the southern flying squirrel (*Glaucomys v. volans*), with histopathologic notes on a mite-induced dermal cyst. *J. Parasitol.* 59:369–74.

Ainscough, B. D. 1960. The internal anatomy of *Caminella peraphora* Krantz and Ainscough, with descriptions of the immature stages. Thesis, Oregon State Univ., Corvallis, LD 4330.

Akimov, I. A. 1974. Morphological and functional characteristics of the mouthparts of the Acaridae mites (Acaridae Ewing et Nesbitt, 1942). In Piffl (1979), 569–74.

Akimov, I. A., I. V. Badanin, and I. Wita. 2002. Some aspects of trophic specialization in both free-living and parasitic prostigmatid mites (Acari: Actinotrichida) from a morpho-functional viewpoint. *Acarina* 10:65–73.

Akimov, I. A., I. S. Starovir, A. V. Yastrebstov, and V. T. Gorgal. 1988. *The Varroa mite—the causative agent of varroatosis of bees. A morphological outline*. Kiev, Ukraine: Naukova Dumka. [VARROIDAE]

Akimov, I. A., and A. V. Yastrebtsov. 1984. Reproductive system of *Varroa jacobsoni*. I, Female reproductive system and oogenesis. *Vestn. Zool.* 6:61–68.

———. 1989. Muscular system and elements of the skeleton of *Nothrus palustris* (Acariformes, Oribatida). *Zool. Zh.* 48:57–66. [NOTHRIDAE]

Akov, S. 1982. Blood digestion in ticks. *Current Themes Trop. Sci.* 1:197–211.

Al-Amidi, A. H. K., R. Dunne, and M. J. Downes. 1991. *Parasitus bituberosus* (Acari: Parasitidae): An agent for control of *Lycoriella solani* (Diptera: Sciaridae) in mushroom crops. *Exp. Appl. Acarol.* 11:159–66.

Alberti, G. 1973. Ernährungsbiologie und Spinnvermögen der Schnabelmilben (Bdellidae, Trombidiformes). *Z. Morphol. Tiere* 76:285–338.

———. 1974. Fortpflanzungsverhalten und Fortpflanzungsorgane der Schnabelmilben (Acarina: Bdellidae, Trombidiformes). *Z. Morphol. Tiere* 78:111–57.

———. 1975. Prälarven und Phylogenie der Schnabelmilben (Bdellidae, Trombidiformes). *Z. Zool. Syst. Evol. Forsch.* 13:44–62.

———. 1979. Fine structure and probable function of genital papillae and Claparède organs of Actinotrichida. In Rodriguez, 2:501–7.

———. 1980a. Zur Feinstruktur der Spermien und Spermiocytogenese der Milben (Acari). I, Anactinotrichida. *Zool. Jb. Anat.* 104:77–138.

———. 1980b. Zur Feinstruktur der Spermien und Spermiocytogenese der Milben (Acari). II, Actinotrichida. *Zool. Jb. Anat.* 194:144–203.

———. 1984. The contribution of comparative spermatology to problems of acarine systematics. In Griffiths and Bowman, 1:479–90.

———. 1988. Genital system of Gamasida and its bearing on phylogeny. In ChannaBasavanna and Viraktamath, 1:197–204.

———. 1991. Spermatology in the Acari: Systematic and functional implications. In Schuster and Murphy, 77–105.

———. 1998. Fine structure of receptor organs in oribatid mites (Acari). In Ebermann (1998a), 27–77.

———. 2000. Chelicerata. In *Reproductive biology of invertebrates, 9B*, ed. B. G. M. Jamieson, 311–88. New Delhi: Oxford & IBH Publishing.

———. 2002a. Reproductive systems of gamasid mites (Acari, Anactinotrichida) reconsidered. In Bernini et al. (2002), 125–39.

———. 2002b. Ultrastructural investigations of sperm and genital systems in Gamasida (Acari: Anactinotrichida): Current state and perspectives for future research. *Acarologia* 42:107–26.

———. 2005. On some fundamental characteristics in acarine morphology. In Morphologia funzionale degli Acari. Tavole Rotonde sui maggiore problemi riguardante l'entomologia agraria in Italia XII, coord. G. Nuzzaci. *Atti Accad. Naz. Ital. Entomol.* 53:315–60.

Alberti, G., and C. Bader. 1990. Fine structure of external "genital" papillae in the freshwater mite *Hydrovolzia placophora* (Hydrovolziidae, Actinedida, Actinotichida, Acari). *Exp. Appl. Acarol.* 8:115–24. [HYDROVOLZIIDAE]

Alberti, G., and L. B. Coons. 1999. Mites. In Harrison and Foelix, 8C:xviii, 515–1265.

Alberti, G., and A. R. Crooker. 1985. Internal anatomy. In Helle and Sabelis (1985a), 1A:29–62. [TETRANYCHIDAE]

Alberti, G., and A. Di Palma. 2002. Fine structure of the phytoseiid-type sperm access system (Acari, Gamasida, Phytoseiidae). In Bernini et al. (2002), 241–52.

Alberti, G., A. Di Palma, and G. W. Krantz. 2005. Preliminary observations of the fine structure of spermatodactyls in males of three species of Gamasida. In Weigmann et al. (2005), 123–28.

Alberti, G., and R. Ehrnsberger. 1977. Rasterelektronenmicroscopische untersuchungen Zum Spinnvermögen der Bdelliden und Cunaxiden (Acari, Prostigmata). *Acarologia* 19:55–61.

Alberti, G., and N. A. Fernandez. 1988. Fine structure of a secondarily developed eye in the fresh water moss mite, *Hydrozetes lemnae* (Coggi 1899) (Acari: Oribatida). *Protoplasma* 146:106–17. [HYDROZETIDAE]

———. 1990. Aspects concerning the structure and function of the lenticulus and clear spot of certain oribatids (Acari, Oribatida). *Acarologia* 31:65–72.

Alberti, G., N. A. Fernandez, and G. Kümmel. 1991. Spermatophores and spermatozoa of oribatid mites (Acari: Oribatida). Part II, Functional and systematical considerations. *Acarologia* 32:435–49.

Alberti, G., A. Gegner, and W. Witalinski. 2000. Fine structure of the spermatophore and spermatozoa in inseminated females of *Pergamasus* mites (Acari: Gamasida: Pergamasidae). *J. Morphol.* 245:1–18.

Alberti, G., B. Gerdeman, and H. Klompen. 2007. Fine structure of spermiogenesis and sperm in a heterozerconid mite (Heterozerconidae; Heterozerconina; Gamasida). In Morales-Malacara et al., 557–60.

Alberti, G., and H. Hänel. 1986. Fine structure of the genital system in the bee parasite, *Varroa jacobsoni* (Gamasida: Dermanyssina) with remarks on spermiogenesis, spermatozoa and capacitation. *Exp. Appl. Acarol.* 2:63–104.

Alberti, G., T. Kaiser, and N. A. Fernandez. 1991. Fine structure of photoreceptor organs in Acari. In Dusbábek and Bukva, 2:343–48.

Alberti, G., T. Kaiser, and A.-K. Klauer. 1996. New ultrastructural observations of the coxal glands (nephridia) of Acari. In Mitchell et al. (1996), 1:309–18.

Alberti, G., and G. W. Krantz. 2007. Some ultrastructural observations on a species of Arctacaridae (Arctacarina: Gamasida), with remarks on their phylogenetic significance. In Morales-Malacara et al., 571–74.

Alberti, G., A. I. Moreno, and M. Kratzmann. 1994. The fine structure of trichobothria in moss mites with special emphasis on *Acrogalumna longipluma* (Berlese, 1904) (Oribatida, Acari, Arachnida). *Acta Zool. (Stockholm)* 75:57–74.

Alberti, G., R. A. Norton, J. Adis, N. A. Fernandez, E. Franklin,

M. Kratzmann, A. I. Moreno, E. F. Ribeiro, G. Weigmann, and S. Woas. 1996. Porose areas and related organs in oribatid mites (Oribatida). In Mitchell et al. (1996), 1:277–83.
Alberti, G., R. A. Norton, J. Adis, N. A. Fernandez, E. Franklin, M. Kratzmann, A. I. Moreno, G. Weigmann, and S. Woas. 1997. Porose integumental organs of oribatid mites (Acari, Oribatida). 2, Fine structure. *Zoologica (Stuttgart)* 146:33–114.
Alberti, G., R. A. Norton, and J. Kasbohm. 2001. Fine structure and mineralization of cuticle in Hypochthonioidea and Lohmannioidea (Acari: Oribatida). In Halliday et al. (2001), 230–41.
Alberti, G., and G. Nuzzaci. 1996. SEM and TEM techniques. In Lindquist, Sabelis, and Bruin, 6:399–410.
Alberti, G., and A. V. Peretti. 2002. Fine structure of male genital system and sperm in Solifugae does not support a sister-group relationship with Pseudoscorpiones (Arachnida). *J. Arachnol.* 30:268–74.
Alberti, G., and R. Schuster. 2005. Behavioural and ultrastructural peculiarities of reproduction in *Collohmannia gigantea* (Oribatida: Mixonomata). In Weigmann et al. (2005), 129–40.
Alberti, G., and O. Seeman. 2005. Ultrastructural observations on Holothyridaa (Acari: Anactinotrichida). In Weigmann et al. (2005), 103–11.
Alberti, G., A. Seniczak, and S. Seniczak. 2003. The digestive system and fat body of an early-derivative oribatid mite, *Archegozetes longisetosus* Aoki (Acari: Oribatida, Trhypochthoniidae). *Acarologia* 43:149–219.
Alberti, G., and V. Storch. 1973. Zur Feinstruktur der "Munddrüsen" von Schnabelmilben (Bdellidae, Trombidiformes). *Z. Wiss. Zool. (Leipzig)* 186:149–60.
———. 1974. Über Bau und Funktion der Prosoma-Drüsen von Spinnmilben (Tetranychidae, Trombidiformes). *Z. Morphol. Tiere* 79:133–53.
———. 1976. Ultrastruktur-Untersuchungen am männlichen Genitaltrakt und am Spermien von *Tetranychus urticae* (Tetranychidae, Acari). *Zoomorphology* 83:283–96.
Alberti, G., V. Storch, and H. Renner. 1981. Über den feinstrukturellen Aufbau der Milbencuticula (Acari, Arachnida). *Zool. Jb. Anat.* 105:183–236.
Alekseev, A. N., L. A. Burenkova, V. M. Podboronov, and S. P. Chunikhin. 1995. Bacteriocidal qualities of ixodid ticks (Acarina: Ixodidae) salivary cement plugs and their changes under the influence of a viral tick-borne pathogen. *J. Med. Entomol.* 32:578–82.
Alfaro, A. 1946. El ácaro *Pediculopsis graminum* Reuter y el hongo *Nigrospora oryzae* (Berk. & Br.), en asociación parasitaria sobre trigos aragonenses. *Bol. Patol. Veg. Entomol. Agricola Madrid* 14:321–34. [SITEROPTIDAE]
Allred, D. 1954. Mites as intermediate hosts of tapeworms. *Proc. Utah Acad. Sci. Arts Letters* 31:44–51.
———. 1970. New ameroseiid mites from birds of New Guinea. *J. Med. Entomol.* 7:99–102.
Amrine, J. W., Jr. 1996. *Phyllocoptes fructiphilus* and biological control of multiflora rose. In Lindquist, Sabelis, and Bruin, 6:741–49.
Amrine, J. W., Jr., G. H. Duncan, A. T. Jones, S. C. Gordon, and L. A. Roberts. 1994. *Cecidophyopsis* mites (Acari: Eriophyidae) on *Ribes* spp. (Grossulariaceae). *Internat. J. Acarol.* 20:139–68.
Amrine, J. W., Jr., and D. C. M. Manson. 1996. Preparation, mounting and descriptive study of eriophyoid mites. In Lindquist, Sabelis, and Bruin, 6:383–96.
Amrine, J. W., Jr., and T. A. Stasny. 1994. *Catalog of the Eriophyoidea (Acarina: Prostigmata) of the world.* West Bloomfield, MI: Indira Publishing House.
Amrine, J. W., Jr., T. A. Stasny, and C. H. W. Flechtmann. 2003. *Revised keys to world genera of Eriophyoidea (Acari: Prostigmata).* West Bloomfield, MI: Indira Publishing House.
Anderson, D. L., and J. W. H. Trueman. 2000. *Varroa jacobsoni* (Acari: Varroidae) is more than one species. *Exp. Appl. Acarol.* 24:165–89.
Anderson, D. T. 1973. *Embryology and phylogeny in annelids and arthropods.* Oxford, UK: Pergamon Press.
Anderson, J. M. 1975. Succession, diversity and trophic relationships of some soil animals in decomposing leaf litter. *J. Anim. Ecol.* 44:475–95.
Anderson, N. H., and G. W. Krantz. 1993. Biological notes on a new species of aquatic mite from Crater Lake. In *Crater Lake limnological studies final report. Tech. Rept. NPS/PNROSU/NRTR-93/03*, ed. G. L. Larson, C. D. McIntire, and R. W. Jacobs, 653–60. Seattle, WA: National Park Service, Northwest Region. [ALGOPHAGIIDAE]
Anderson, R. C. 1968. Ecological observations on three species of *Pimeliaphilus* parasites of Triatominae in the United States (Acarina: Pterygosomidae) (Hemiptera: Reduviidae). *J. Med. Entomol.* 5:459–64.
Anderson, T. M. 2003. Detachment of *Arrenurus hamrumi* larvae (Hydrachnida: Arrenuridae) from *Argia vivida* (Odonata: Coenagrionidae). In I. M. Smith (2003a), 1–10.
Anderson, T. M., and I. M. Smith. 1996. *Arrenurus hamrumi* (Hydrachnida: Arrenuridae), a new species of water mite from rangeland springs in central Oregon. *Internat. J. Acarol.* 22:285–90.
André, H. M. 1977. Note sur le genre *Mediolata* (Actinedida: Stigmaeidae) et description d'une nouvelle espèce corticole. *Acarologia* 18:463–74.
———. 1979. A generic revision of the family Tydeidae (Acari: Actinedida). I, Introduction, paradigms and general classification. *Ann. Soc. Roy. Zool. Belg.* 108:189–208. [TRIOPHTYDEIDAE, TYDEIDAE, IOLINIDAE]
———. 1980. A generic revision of the family Tydeidae (Acari: Actinedida). IV, Generic descriptions, keys and conclusions. *Bull. Ann. Soc. Roy. Belg. Entomol.* 116:103–30, 139–68. [TRIOPHTYDEIDAE, TYDEIDAE, IOLINIDAE]
———. 1981a. A generic revision of the family Tydeidae (Acari: Actinedida). II, Organotaxy of the idiosoma and gnathosoma. *Acarologia* 22:31–46. [TRIOPHTYDEIDAE, TYDEIDAE, IOLINIDAE]
———. 1981b. A generic revision of the family Tydeidae (Acari: Actinedida). III, Organotaxy of the legs. *Acarologia* 22:165–78. [TRIOPHTYDEIDAE, TYDEIDAE, IOLINIDAE]
———. 1984a. Notes on the ecology of corticolous epiphyte dwellers. 3, Oribatida. *Acarologia* 25:385–96.
———. 1984b. Redefinition of the Iolinidae (Acari: Actinedida) with a discussion of their familial and superfamilial status. In Griffiths and Bowman, 1:180–85.
———. 1985. Acari domun meliponinarum brasiliensium habitantes. 10, *Melissotydeus macrosolenus* gen. n., sp. n. (Acari: Tydeidae). *Bull. Ann. Soc. Roy. Belg. Entomol.* 121:243–46.
———. 1986. Notes on the ecology of corticolous epiphyte dwellers. 4, Actinedida (especially Tydeidae) and Gamasida (especially Phytoseiidae). *Acarologia* 27:107–15.

———. 1991. The Tydeoidea: A striking exception to the Oudemans-Grandjean rule. In Dusbábek and Bukva, 2:293–96.

———. 1996. Two new mites of the rare families Barbutiidae and Linotetranidae (Acari), from the Namib Desert. *Acarologia* 36:97–106.

———. 2004. Revalidation of *Oriola* and replacement name for *Meyerella* (Acari: Tydeoidea). *Internat. J. Acarol.* 30:279–80. [IOLINIDAE]

———. 2005. In search of the true *Tydeus* (Acari, Tydeidae). *J. Nat. Hist.* 39 (13): 975–1001.

André, H. M., and X. Ducarme. 2003. Rediscovery of the genus *Pseudotydeus* (Acari: Tydeoidea), with description of the adult using digital imaging. *Insect Syst. Evol.* 34:373–79.

André, H. M., X. Ducarme, and P. Lebrun. 2004. New ereynetid mites (Acari: Tydeoidea) from karstic areas: True association or sampling bias? *J. Cave Karst Stud.* 66:81–88.

André, H. M., and A. Fain. 2000. Phylogeny, ontogeny and adaptive radiation in the superfamily Tydeoidea (Acari: Actinedida), with a reappraisal of morphological characters. *Zool. J. Linn. Soc.* 130:405–48.

André, H. M., and M.-I. Noti. 1993. Extracting sand microarthropods: A carbon tetrachloride flotation method. *Eur. J. Soil Biol.* 29:91–96.

André, H. M., M.-I. Noti, and P. Lebrun. 1994. The soil fauna: the other last biotic frontier. *Biodivers. Conserv.* 3:45–56.

André, H. M., and C. Remacle. 1984. Comparative and functional morphology of the gnathosoma of *Tetranychus urticae* (Acari: Tetranychidae). *Acarologia* 25:179–90.

André, H. M., and D. J. Voegtlin. 1982. Some observations on the biology of *Camisia carrolli* (Acari: Oribatida). *Acarologia* 23: 81–89. [CAMISIIDAE]

André, M. 1927. Digestion "extra-intestinale" chez le Rouget (*Leptus autumnalis* Shaw). *Bull. Mus. Nat. Hist. Natur. Paris* 33:509–16. [ERYTHRAEIDAE]

———. 1934. Acariens terrestres adaptés à la vie marine. Paper #17, 67th meeting, Section Science Congrès Soc. Savantes 17: 134–56.

———. 1937. Trois espèces d'acariens (Gamasoidea) Pagurophiles. *Bull. Soc. Zool. Fr.* 62:45–70. [CERCOMEGISTIDAE]

———. 1943. Une espèce nouvelle de *Leeuwenhoekia* parasite de scorpions. *Bull. Mus. Nat. Hist. Natur. Paris*, ser. 2, 15:294–98. [LEEUWENHOEKIIDAE]

———. 1946. Halacariens marins. *Faune Fr.* 46:1–152.

———. 1951. Nouvelles observations sur le *Bochartia kuyperi* Oudemans (Acarien). *Bull. Mus. Hist. Nat. Paris*, ser. 2, 25:253–55. [ERYTHRAEIDAE]

———. 1959. XXVI—Acari. I, Contribution a l'étude des halacariens de la Mer Rouge. *Rés. Sci. Miss. R. P. Dollfussen Egypte* 1927–29, Paris, no. 26:93–119.

———. 1961. Observations sur *Hirstiella insignis* (Berlese). *Acarologia* 3:159–64.

André, M., and E. Lamy. 1930. Les Acariens parasites des mollusques. *J. Conchyliol.* 74:199–221. [EREYNETIDAE]

Andreev, E. A. 1985. Mites associated with caterpillar galleries of the common goat moth (*Cossus cossus* L.), an important trunk pest in southern Turkmenistan. *Biol. Nauk.* 1985 (8): 34–38. [MELICHARIDAE] [in Russian]

Andrews, J. R. H., and G. W. Ramsay. 1982. A case of papular dermatosis in man attributed to an ascid mite (Acari). *J. Med. Entomol.* 19:111–12. [MELICHARIDAE]

Angelier, E. 1953. Halacariens des sables littoraux méditerranéens. *Vie Milieu* 4:281–89.

Ansaloni, T., and T. M. Perring. 2004. Biology of *Aceria guerreronis* (Acari: Eriophyidae) on queen palm, *Syagrus romanzoffiana* (Arecaceae). *Internat. J. Acarol.* 30:63–70.

Anthony, L. M. M. K. d. A. 1980. A phylogenetic analysis of the Rhodacaroidea (Acari: Mesostigmata). Ph.D. diss., Ohio State Univ., Columbus.

Antor, R. J., and M. L. Garcia. 1995. A new mite-plant association: mites living amidst the adhesive traps of a carnivorous plant. *Oecologia* 101:51–54.

Aoki, J. 1963. Untersuchungen über die Zönosen der Oribatiden in Nikko in Beziehung zu Pflanzendeckung und Boden. IV, Pflanzendecken und Oribatidenzönosen. *Jpn. J. Ecol.* 13:139–51.

———. 1964. Studies on Oribatei (Acarina) from the South Pacific. II, *Eutegaeus papuensis* n. sp. from New Guinea with a note on the genus. *Pac. Insects* 7 (2): 301–8. [EUTEGAEIDAE]

———. 1965a. Notes on the species of the genus *Epilohmannia* from the Hawaiian Islands (Acarina: Oribatei). *Pac. Insects* 7 (2): 309–15. [EPILOHMANNIIDAE]

———. 1965b. Oribatid mites (Acarina: Oribatei) from Himalaya with descriptions of several new species. *J. Coll. Arts Sci., Chiba* 4 (3): 289–302.

———. 1965c. Oribatiden (Acarina) Thailands. I. *Nat. Life S. E. Asia, Kyoto* 4:129–93.

———. 1965d. A preliminary revision of the family Otocepheidae (Acari: Cryptostigmata). I, Subfamily Otocepheinae. *Bull. Nat. Sci. Mus., Tokyo* 8 (3): 259–341.

———. 1966a. Epizoic symbiosis: An oribatid mite, *Symbioribates papuensis*, representing a new family, from cryptogamic plants growing on backs of Papuan weevils (Acari: Cryptostigmata). *Pac. Insects* 8 (1): 281–89. [SYMBIORIBATIDAE]

———. 1966b. A remarkable new oribatid mite from South Japan (Cryptostigmata: Tokunocepheidae, fam. nov.). *Acarologia* 8 (2): 358–64.

———. 1967. Oribatiden (Acarina) Thailands. II. *Nat. Life S. E. Asia, Kyoto* 5:189–207.

———. 1968. A new species of the genus *Xiphobelba* from New Britain Island (Acari, Cryptostigmata). *Bull. Nat. Sci. Mus., Tokyo* 11 (3): 269–74. [BASILOBELBIDAE]

———. 1969. Taxonomic investigations on free living mites in the subalpine forest on Shiga heights IBP area—III. Cryptostigmata. *Bull. Nat. Sci. Mus., Tokyo* 12 (1): 117–41.

———. 1970. The oribatid mites of the Island of Tsushima. *Bull. Nat. Sci. Mus., Tokyo* 13 (3): 395–442.

———. 1973. Soil mites (oribatids) climbing trees. In Daniel and Rosický, 59–65.

———. 1974. The first record of the intertidal oribatid genus, *Fortuynia*, from Asia. *Annot. Zool. Jpn.* 47 (3): 170–74. [FORTUYNIIDAE]

———. 1975. Taxonomic notes on some little known oribatid mites of Japan. *Bull. Nat. Sci. Mus., Ser. A (Zool.), Tokyo* 1 (1): 57–65.

———. 1976. Oribatid mites from IBP study area, Pasoh Forest Reserve, West Malaysia. *Nat. Life S. E. Asia* 7:39–59.

———. 1977a. Discovery of a mite of the genus *Caenosamerus* from Japan. *Annot. Zool. Jpn.* 50 (4): 255–58.

———. 1977b. New and interesting species of oribatid mites from Kakeroma Island, Southwest Japan. *Acta Arachnol.* 27:85–93.

———. 1980a. A revision of the oribatid mites of Japan. I. The families Phthiracaridae and Oribotritiidae. *Bull. Inst. Environ. Sci. Technol., Yokohama* 6 (2): 1–88.

———. 1980b. A revision of the oribatid mites of Japan. II, The family Euphthiracaridae. *Acta Arachnol.* 29 (1): 9–24.

———. 1980c. A revision of the oribatid mites of Japan. III, Families Protoplophoridae, Archoplophoridae and Mesoplophoridae. *Proc. Jpn. Soc. Syst. Zool.* 18:5–16.

———. 1983. Some new species of oppiid mites from South Japan (Oribatida: Oppiidae). *Internat. J. Acarol.* 9:165–72.

———. 1984a. A modified Tullgren funnel with double porous discs for preventing soil particles from dropping into collecting tube. *Bull. Inst. Environ. Sci. Technol.* 11:103–5.

———. 1984b. New and unrecorded oribatid mites from Kanagawa, Central Japan (I). *Bull. Inst. Environ. Sci. Technol., Yokohama* 11 (1): 107–18.

———. 1991. Oribatid mites of high altitude forests of Taiwan: I, Mt. Pei-ta-wu Shan. *Acta Arachnol.* 40 (2): 75–84.

Aoki, J., and T. Fujikawa. 1969. Description of a new species of the genus *Hafenferrefia* Jacot (Acari, Tenuialidae). Taxonomic notes on oribatid mites of Hokkaido. I. *Annot. Zool. Jpn.* 42 (4): 216–19.

———. 1971. Occurrence of a Japanese representative of the North American genus *Megeremaeus* (Acari, Megeremaeidae). Taxonomic notes on oribatid mites of Hokkaido. IV. *Annot. Zool. Jpn.* 44 (2): 109–12.

———. 1972. A new genus of oribatid mites, exhibiting both the characteristic features of the families Charassobatidae and Cymbaeremaeidae (Acari: Cryptostigmata). *Acarologia* 14: 258–67.

Aoki, J., and K. Suzuki. 1970. A new species of the genus *Pedrocortesella* from Japan (Acari: Cryptostigmata). *Bull. Nat. Sci. Mus., Tokyo* 13 (2): 117–20. [PEDROCORTESELLIDAE]

Aoki, J., G. Takaku, and F. Ito. 1994. Aribatidae: a new myrmecophilous oribatid mite family from Java. *Internat. J. Acarol.* 20:1–8.

Aoki, J., and Z. Wen. 1983. Two new species of oribatid mites from Shikine islet. *Bull. Inst. Environ. Sci. Technol., Yokohama* 9 (1): 165–69.

Aoki, J., Y. Yamamoto, and S. Hu. 2000. A checklist of oribatid mites of Yunnan Province in China (Acari: Oribatida). In *Taxonomical studies on the soil fauna of Yunnan Province in Southwest China*, ed. J. Aoki, W. Yin, and G. Imadate, 33–41. Tokyo: Tokai Univ. Press.

Araman, S. F., and A. Said. 1972. Biochemical and physiological studies of certain ticks (Ixodoidea). The ionic regulatory role of the coxal organs of *Argas* (*Persicargas*) *persicus* (Oken) and *A.* (*P.*) *arboreus* Kaiser, Hoogstraal and Kohls. *J. Parasitol.* 58:348–53.

Arganaraz, M. S., and A. B. de-Alzuet. 1991. Linotetranidae: una familia de Acaros (Prostigmata) nueva para la fauna Argentina. *Neotropica* 36:99–100.

Arillo, A., and L. S. Subías. 2002. Second fossil oribatid mite from the Spanish Lower Cretaceous amber. *Acarologia* 42:403–6. [EUTEGAEIDAE]

Arlian, L. G. 1989. Biology, epidemiology and host relations of *Sarcoptes scabiei*. *Ann. Rev. Entomol.* 34:139–61. [SARCOPTIDAE]

Arthur, D. 1957. The capitulum and feeding mechanism of *Dermacentor parumapertus*. *Parasitology* 47:169–84. [IXODIDAE]

———. 1962. *Ticks and disease.* Oxford, UK: Pergamon Press.

Ascerno, M. E., F. L. Pfleger, and H. F. Wilkins. 1981. Effect of root rot and *Rhizoglyphus robini* on greenhouse-forced Easter lily development. *Environ. Entomol.* 10:947–49.

Ashley, E., and D. C. M. Manson. 1987. The occurrence of *Brevipalpus essigi* and *B. russulus* (Acari: Tenuipalpidae) in New Zealand. *New Zeal. Entomol.* 10:131–33.

Asmat, G. S. M., and B. Kader. 1990. A preliminary survey of mites from the nest of certain birds in the Chittagong University campus. *Bangladesh J. Zool.* 18:177–83. [LARDOGLYPHIDAE]

Atalla, E. A. R., and J. Hobart. 1964. The survival of some soil mites at different humidities and their reaction to humidity gradients. *Entomol. Exp. Appl.* 7:215–28.

Athias, F. 1973. Scutacaridae de la savane de Lamto (Côte d'Ivoire) (Acariens: Tarsonemida). I, Données quantitatives sur le peuplement, liste des espèces de Scutacarinae, description de trois espèces nouvelles. *Acarologia* 14:638–56.

———. 1975. Observations morphologique sur *Polyaspis patavinus* Berlese 1881 (Acariens: Uropodides). I, Morphologie de l'idiosoma au cours du développement postembryonnaire. *Acarologia* 17:410–35.

Athias-Binche, F. 1978. Développement postembryonnaire d'*Allonychus flagellifer* (Berlese 1910) (Gamasida: Uropodina). *Acarologia* 20:44–57.

———. 1981. Écologie des uropodides édaphiques (Arachnides: Parasitiformes) de trois écosystemes forestiers. I, Introduction, matérial, biologie. *Vie Milieu* 31:137–47.

———. 1982. A redescription of *Thinozercon michaeli* Halbert, 1915 (Uropodina: Thinozerconoidea) with notes on its systematic status. *Proc. Roy. Irish Acad.* 82:261–76.

———. 1984. La phorésie chez les acariens uropodides (Anactinotriches), une stratégie écologique originale. *Acta Oecol. Gener.* 5:119–33.

———. 1990. Sur le concept de symbiose. L'example de la phorésie chez les acariens et son évolution vers le parasitisme ou le mutualisme. *Bull. Soc. Zool. Fr.* 115:77–98. [PYEMOTIDAE]

———. 1994. *La phoresie chez les acariens. Aspects adaptatifs et évolutifs.* Perpignan, France: Editions du Castillet.

Athias-Binche, F., J. Bloszyk, and Z. Olszanowski. 1989. *Dinychus ruseki* n.sp. (Acari: Uropodina) from Canada, with remarks on the habitats and distribution of the members of the genus *Dinychus*. *Can. J. Zool.* 67:1482–88.

Athias-Binche, F., and G. O. Evans. 1981. Observations on the genus *Protodinychus* Evans, 1957 (Acari: Mesostigmata) with descriptions of the male and phoretic deuteronymph. *Proc. Roy. Irish Acad.* B 81:1–36.

Athias-Binche, F., and N. A. Fernandez. 1986. Analyse demographique d'une population d'*Hydrozetes lemnae* Coggi, Acarien Oribate infeode a la lentille d'eau *Lemna gibba* L. en Argentine. 2, Les relations predateur/proie. *Zool. Jb. Syst.* 113:229–49. [HYDROZETIDAE]

Athias-Binche, F., and U. Habersaat. 1988. An ecological study of *Janetiella pyriformis* (Berlese, 1920), a phoretic Uropodina from decomposing organic matter (Acari: Anactinotrichida). *Mitt. Schweiz. Entomol. Gesell.* 61:377–90. [DINYCHIDAE]

Athias-Henriot, C. 1959. Redescription du stade adulte de *Seiodes ursinus* Berlese, 1887 (Parasitiformes, Antennophorina). *Zool. Anz.* 163:11–25. [SEIODIDAE]

———. 1960. Contribution aux mésostigmates d'Algerie (Parasitiformes: Liroaspidae, Veigaiidae). *Acarologia* 2:159–74.

———. 1961a. Mésostigmates (Urop. excl.) édaphiques méditerranéens

(Acaromorpha, Anactinotrichida) (Collect. Prof. H. Franz et C. Athias-Henriot). Première Série. *Acarologia* 3:381–509.

———. 1961b. Nouveaux acariens phytophages d'Algerie (Actinotrichida, Tetranychoidea: Tetranychidae, Linotetranidae). *Ann. École Nat. Agr. Alger.* 3 (3): 1–10.

———. 1967. Observations sur les *Lasioseius spathuliger* méditerranéens (Parasitiformes, Laelapoidea). *Rev. Ecol. Biol. Sol* 4:123–49. [BLATTISOCIIDAE]

———. 1968. L'appareil d'insémination laelapoïde (Acariens Anactinotrichides: Laelapoidea, ♀♀). Premières observations. Possibilité d'emploi à des fins taxonomiques. *Bull. Sci. Bourgogne* 25:229–74.

———. 1969a. Contribution aux peuplements acarologiques des biotopes terrestres humides médioeuropéens: Observations sur le genre *Panteniphis* Willm. (Mésostigmates, Rhodacaridae). *Rev. Ecol. Biol. Sol* 6:69–78. [DIGAMASELLIDAE]

———. 1969b. Gamasides Chiliens (Acariens Anactinotriches) 1, Genre *Anephiasca* n. g. (Ascidae). *Rev. Biol.* 7:123–49.

———. 1969c. Notes sur la morphologie externe des gamsides (Acariens Anactinotriches). [Notes on the external morphology of the gamasid mites (Anactinotrichous Acari).] *Acarologia* 11:609–29.

———. 1969d. Les organes cuticulaires sensoriels et glandulaires des Gamasides. Poroidotaxie et adenotaxie. *Bull. Soc. Entomol. Fr.* 94:485–92.

———. 1969e. *Psilogamasus hurlbutti*, n. g., n. sp., gamaside nouveau de Tanzanie [Acariens Anactinotriches, Parasitidae]. *Ann. Soc. Entomol. Fr.*, n.s., 5:439–49.

———. 1970. Observations sur la morphologie des gamasides, mises au point terminologiques. *Acarologia* 12:25–27.

———. 1971. La divergence neotaxique des Gamasides (Arachnides). Deuxième partie. *Bull. Sci. Bourgogne* 28:93–106.

———. 1972. Gamasides Chiliens (Arachnides) II, Revision de la famille Ichthyostomatogasteridae Sellnick 1953 (= Uropodellidae Camin, 1955). *Arq. Zool. Sao Paulo* 22:113–91.

———. 1973. Observations sur les genres *Neojordensia* et *Orthadenella* n. g. en Europe occidentale (Gamasides, Dermanyssina, Ascidae). *Acarologia* 15:18–32. [includes BLATTISOCIIDAE]

———. 1975a. The idiosomatic euneotaxy and epineotaxy in gamasids (Arachnida, Parasitiformes). *Z. Zool. Syst. Evol. Forsch.* 13:97–109.

———. 1975b. Nouvelle notes sur les Amblyseiini. II, Le rélève organotaxique de la face dorsal adulte (Gamasides protoadeniques, Phytoseiidae). *Acarologia* 17:20–29.

———. 1976. Sur la bioécologie en France tempérée de Leitneria granulata, gamaside édaphique forestier (Arachnides, Parasitiformes). *Pedobiologia* 16:151–60. [HALOLAELAPIDAE]

———. 1977. A new Australian mite, *Archaeopodella scopulifera* gen. et sp. n. (Gamasida: Liroaspidae). *J. Aust. Entomol. Soc.* 16: 225–35.

———. 1980. Parasitidae nord-coréens (Parasitiformes): Apparentés au genre *Neogamasus* Tikh., avec création de trois nouveaux genres. *Ann. Hist.-Nat. Mus. Nat. Hung.* 72:285–94.

———. 1982a. *Schizosthetus* n. g. (type *Eugamasus lyriformis* McGr. & Farr., 1969) avec deux espèces nouvelles (Parasitiformes, Parasitidae). *Acarologia* 23:207–14.

———. 1982b. Sur le genre monotypique *Parasitus* Latr. (Parasitiformes, Parasitidae). Diagnose, définition, distribution. *Rev. Suisse Zool.* 89:3–6.

Atkinson, P. W., and K. C. Binnington. 1973. New evidence on the function of the porose areas of ixodid ticks. *Experientia* 29: 799–801.

Attiah, H. H. 1970. New tarsonemid mites associated with citrus in Florida (Acarina: Tarsonemidae). *Fla. Entomol.* 53:179–201.

Atyeo, W. T. 1960. A revision of the mite family Bdellidae in North and Central America (Acarina, Prostigmata). *Univ. Kans. Sci. Bull.* 40:345–499.

———. 1961. The taxonomic position of the genus *Neophyllobius* Berlese, 1886 (Acarina, Caligonellidae) with the description of a new genus and species. *Acarologia* 3:153–58.

———. 1963a. The Bdellidae (Acarina) of the Australian realm. *Bull. Univ. Nebraska St. Mus.* 4:113–210.

———. 1963b. New and redescribed species of Raphignathidae (Acarina) and a discussion of the chaetotaxy of the Raphignathoidea. *J. Kans. Entomol. Soc.* 36:172–86.

———. 1979. The pretarsi of astigmatid mites. *Acarologia* 20: 244–69.

———. 1992. The pterolichoid feather mites (Acarina, Astigmata) of the Megapodiidae. *Zool. Scripta* 21:265–305. [PTEROLICHIDAE, THORACOSATHESIDAE]

Atyeo, W. T., and E. W. Baker. 1964. Tarsocheylidae, a new family of prostigmatic mites (Acarina). *Bull. Univ. Nebraska St. Mus.* 4:243–56.

Atyeo, W. T., E. W. Baker, and D. A. Crossley, Jr. 1961. The genus *Raphignathus* Dugès (Acarina, Raphignathidae) in the United States with notes on the Old World species. *Acarologia* 3:14–20.

Atyeo, W. T., E. W. Baker, and M. D. Delfinado. 1974. *Gaudiella minuta*, a new genus and species of mite (Acarina: Acaridia) belonging to the new family Gaudiellidae. *J. Wash. Acad. Sci.* 64:295–98.

Atyeo, W. T., and N. L. Braasch. 1966. The feather mite genus *Proctophyllodes* (Sarcoptiformes: Proctophyllodidae). *Bull. Univ. Nebraska St. Mus.* 5:1–354. Lincoln.

Atyeo, W. T., and D. A. Crossley. 1961a. Labidostomidae from Australia (Acarina, Prostigmata) with the description of a new species. *Trans. Roy. Soc. Aust.* 84:83–86.

———. 1961b. The Labidostomidae of New Zealand. *Rec. Dominion Mus. (Wellington)* 4:28–48.

Atyeo, W. T., and J. Gaud. 1979a. Feather mites and their hosts. In Rodriguez, 2:355–61.

———. 1979b. Ptyssalgidae, a new family of analgoid feather mites (Acarina, Acaridida). *J. Med. Entomol.* 16:306–8.

———. 1983. Feather mites of obligate brood parasites. *J. Parasitol.* 69:455–58.

Atyeo, W. T., J. Gaud, and W. J. Humphreys. 1972. The feather mite genus *Freyanella* Dubinin, 1953 (Analgoidea: Pterolichidae). *Acarologia* 13:383–409. [KRAMERELLIDAE]

Atyeo, W. T., J. B. Kethley, and T. M. Perez. 1984. Paedomorphosis in *Metacheyletia* (Acari: Cheyletidae), with the description of a new species. *J. Med. Entomol.* 21:125–31.

Atyeo, W. T., and P. C. Peterson. 1966. The feather mite genus *Dinalloptes* (Acarina, Proctophyllodidae). *Acarologia* 8:470–74. [ALLOPTIDAE]

———. 1967. The feather mite genus *Laminalloptes* (Proctophyllodidae: Alloptinae). *J. Kans. Entomol. Soc.* 40:447–58. [ALLOPTIDAE]

———. 1976. The species of the feather mite family Rectijanuidae (Acarina: Analgoidea). *J. Georgia Entomol. Soc.* 11:349–66.

Audy, J. R. 1951. Trombiculid mites and scrub itch. *Aust. J. Sci.* 14:94–96.
———. 1956. Trombiculid mites infesting birds, reptiles and arthropods in Malaya, with a taxonomic revision and descriptions of a new genus, two new subgenera, and six new species. *Bull. Raffles Mus., Singapore* 28:27–80.
———. 1968. *Red mites and typhus.* London: Univ. London, Athlone Press. [TROMBICULIDAE]
Audy, J. R., F. J. Radovsky, and P.-H. Vercammen-Grandjean. 1972. Neosomy: Radical intrastadial metamorphosis associated with arthropod symbioses. *J. Med. Entomol.* 9:487–94.
Augustson, G. F. 1940. The fauna and flora of the El Segundo sand dunes. 9, Some of the mites (Acarina) of the dunes. *Bull. S. Calif. Acad. Sci.* 38:191–97. [ERYTHRAEIDAE]
Avanzati, A. M., N. Salomone, M. Baratti, and F. Bernini. 2003. Taxonomic revision of *Amerus troisi* (Berlese, 1883) (Acari, Oribatida, Ameridae) using morphological and biochemical characters. *J. Nat. Hist.* 37:797–819. [AMERIDAE]
Avidov, Z., D. Blumberg, and U. Gerson. 1968. *Cheletogenes ornatus* (Acarina: Cheyletidae), a predator of the chaff scale on citrus in Israel. *Israel J. Entomol.* 3:77–94.
Ayyildiz, N., and M. Luxton. 1989. Epimerellidae (Acari, Oribatida), a new mite family. *J. Nat. Hist.* 23 (6): 1381–86.
Axtell, R. C. 1963. Effect of Macrochelidae (Acarina: Mesostigmata) on house fly production from dairy cattle manure. *J. Econ. Entomol.* 56:317–21.
———. 1969. Macrochelidae (Acarina: Mesostigmata) as biological control agents for synanthropic flies. In Evans (1969c), 401–16.
———. 1981. Use of predators and parasites in filth fly IPM programs in poultry housing. In *Proceedings of Workshop "Status of Biological Control of Filth Flies,"* Univ. Florida, Gainesville, Feb. 4–5, 26–43.
Badejo, M. A., S. Woas, and L. Beck 2002a. Description of six species of nothroid mites from Nigeria and Brazil (Acari: Oribatida: Nothroidea). *Genus (Warsaw)* 13:505–48.
———. 2002b. *New pterogasterine mites from Nigeria and Brazil. I,* Scheloribates, Muliercula *and* Peloribates. Syst. Appl. Acarol. Soc. Spec. Publ. 12. Turveren, Belgium. [SCHELORIBATIDAE, HAPLOZETIDAE]
Bader, C. 1968. *Wassermilben aus Zentralafrika.* Spec. Publ. Ann. Mus. Roy. Afr. Centr. 163. Tervuren, Belgium: Musée Royale de l'Afrique Centrale.
———. 1975a. Die Wassermilben des Schweizerischen Nationalparks. 1, Systematisch-Faunistischer Teil. *Ergebn. Wiss. Untersuch. Schweiz. Nationalpark* 14:1–270.
———. 1975b. Die Wassermilben des Schweizerischen Nationalparks. 2, Nachtrag zum systematisch-faunistischen Teil. *Ergebn. Wiss. Untersuch. Schweiz. Nationalpark* 14:373–97.
———. 1994. Die Wassermilben des Schweizerischen Nationalparks. 4, Zweiter Nachtrag zum systematisch-faunistischen Teil. *Ergebn. Wiss. Untersuch. Schweiz. Nationalpark* 16:223–87.
Bagnyuk, I. G., S. I. Sukhareva, and V. G. Shevchenko. 1995. Analiz roda *Trisetacus* Keifer kak klyuchevogo v semeistve Nalepellidae Roiv. (Acari, Tetrapodili) [Analysis of the genus *Trisetacus* Keifer as a key of the family Nalepellidae Roiv. (Acari, Tetrapodili)]. *Acarina, Russian J. Acarol.* 3:31–65. [in Russian, with briefer English version]
———. 1997. Osnovnye napravleniya zvolyutsii chetyrekhnogikh kleshchei kak spetsializirovannoi gruppy organizmov na primere semeistv Pentasetacidae Shev., Nalepellidae Roiv. i Phytoptidae Murray (Acari: Tetrapodili) [Major trends in evolution of four-legged mites as a specialized group, using families Pentasetacidae Shev., Nalepellidae Roiv. and Phytoptidae Murray (Acari: Tetrapodili) as examples]. *Acarina, Russian J. Acarol.* 6:59–76. [in Russian, with briefer English version]
Bailey, L. 1963. *Infectious diseases of the honey-bee.* London: Land Books. [TARSONEMIDAE]
Bajwa, W. I., G. W. Krantz, and M. Kogan. 2001. Discovery of *Cenopalpus pulcher* (C.& F.) (Acari: Tenuipalpidae) in the New World. *Proc. Entomol. Soc. Wash.* 103:754–56.
Baker, A. S. 1987. *Caleupodes*, a new genus of eupodoid mite (Acari: Acariformes) showing primary opisthosomal segmentation. *Bull. Brit. Mus. (Nat. Hist.), Zool.* 53:103–13. [EUPODIDAE]
———. 1990a. A survey of external morphology of mites of the superfamily Eupodoidea Banks, 1894 (Acari: Acariformes). *J. Nat. Hist.* 24:1227–61.
———. 1990b. Two new species of *Lardoglyphus* Oudemans (Acari: Lardoglyphidae) found in the gut contents of human mummies. *J. Stored Prod. Res.* 26:139–47.
———. 1996. A redescription of *Halotydeus destructor* (Tucker) (Prostigmata: Penthaleidae), with a survey of ontogenetic setal development in the superfamily Eupodoidea. *Internat. J. Acarol.* 21:261–82.
———. 1998. A new species of *Hirstiella* Berlese (Acari: Pterygosomatidae) from captive rhinoceros iguanas, *Cyclura cornuta* Bonnaterre (Reptilia: Iguanidae). *Syst. Appl. Acarol.* 3:183–92.
———. 1999. Two new species of larval mites (Acari: Trombidioidea: Microtrombidiidae and Johnstonianidae) parasitizing *Culicoides impunctatus*, the highland midge (Insecta: Ceratopogonidae), in Scotland. *Syst. Parasitol.* 44:37–47.
Baker, A. S., A. H. Andrews, and M. T. Fox. 1996. First record of the cattle itch mite, *Psorobia bos* (Prostigmata: Psorergatidae), in the Palaearctic region. *Syst. Appl. Acarol.* 1:213–16.
Baker, A. S., and E. E. Lindquist. 2002. Aethosolenia laselvensis *gen. nov., sp. nov., a new eupodoid mite from Costa Rica (Acari: Prostigmata).* Syst. Appl. Acarol. Spec. Publ. 11. London. [EUPODIDAE]
Baker, A. S., and P. A. Selden. 1997. New morphological and host data for the ectoparasitic larva of *Leptus hidakai* Kawashima (Acari, Acariformes, Erythraeidae). *Syst. Parasitol.* 36:183–91.
Baker, E. W. 1946. New species of North and Central American mites of the family Penthaleidae (Acarina). *J. Wash. Acad. Sci.* 36:421–25. [PENTHALODIDAE]
———. 1949a. Paratydeidae, a new family of mites. *Proc. Entomol. Soc. Wash.* 51:119–22.
———. 1949b. Pomerantziidae, a new family of prostigmatic mites. *J. Wash. Acad. Sci.* 39:269–71.
———. 1950. Further notes on the family Paratydeidae (Acarina), with a description of another new genus and species. *J. Wash. Acad. Sci.* 40:289–91.
———. 1951. *Pneumocoptes*, a new genus of lung-inhabiting mite from rodents (Acarina: Epidermoptidae). *J. Parasitol.* 37:583–86. [PNEUMOCOPTIDAE]
———. 1965. A review of the genera of the family Tydeidae (Acarina). *Adv. Acarol.* 2:95–133.
———. 1968. Two new genera of Tydeidae (Acarina). *Ann. Entomol. Soc. Amer.* 61:968–70.
———. 1979. A note on paedogenesis in *Brevipalpus* sp. (Acari: Tenuipalpidae), the first such record for a mite. *Internat. J. Acarol.* 5:355–56.

Baker, E. W., and W. T. Atyeo. 1964. A review of the mites of the family Pseudocheylidae Oudemans, 1909 (Acarina, Prostigmata). *Bull. Univ. Nebraska St. Mus.* 4:257–72.

Baker, E. W., J. H. Camin, F. Cunliffe, T. A. Woolley, and C. E. Yunker. 1958. *Guide to the families of mites.* College Park, MD: Institute of Acarology.

Baker, E. W., and M. D. Delfinado. 1974. Pseudotydeinae, a new subfamily of Tydeidae (Acarina). *Proc. Entomol. Soc. Wash.* 76:444–47. [EREYNETIDAE]

———. 1975. *A new genus of Scutacaridae (Acarina) on a bumble bee from India.* USDA Coop. Econ. Insect Rpt. 25, no. 19. Washington, DC: USDA.

———. 1978. Notes on the driedfruit mite *Carpoglyphus lactis* (Acarina: Carpoglyphidae) infesting honeybee combs. *J. Apicult. Res.* 17:52–54.

Baker, E. W., T. M. Evans, D. J. Gould, W. B. Hull, and H. L. Keegan. 1956. *A manual of parasitic mites of medical or economic importance.* Nat. Pest Control Assoc. Tech. Publ. New York: Nat. Pest Control Assoc.

Baker, E. W., C. H. W. Flechtmann, and M. Delfinado-Baker. 1984. Acari domum meliponinarum brasiliensium habitantes. VI, New species of *Bisternalis* Hunter (Laelapidae: Acari). *Internat. J. Acarol.* 10:181–89.

Baker, E. W., and A. Hoffmann. 1948. Acaros de la familia Cunaxidae. *An. Esc. Nac. Cienc. Biol.* 5:229–75.

Baker, E. W., T. Kono, J. W. Amrine, Jr., M. D. Delfinado-Baker, and T. A. Stasny. 1996. *Eriophyoid mites of the United States.* West Bloomfield, MI: Indira Publishing House.

Baker, E. W., and A. E. Pritchard. 1953. The family categories of tetranychoid mites, with a review of the new families Linotetranidae and Tuckerellidae. *Ann. Entomol. Soc. Amer.* 46:243–58.

———. 1960. The tetranychoid mites of Africa. *Hilgardia* 29 (11): 455–74.

Baker, E. W., and D. M. Tuttle. 1972. New species and further notes on the Tetranychoidea mostly from the southwestern United States (Acarina: Tetranychidae and Tenuipalpidae). *Smithsonian Contr. Zool.* 116:37.

———. 1975a. *A new genus of Tenuipalpidae (Acarina) from India.* USDA Coop. Econ. Insect Rpt., vol. 25, no. 22, 453–55. Washington, DC: USDA.

———. 1975b. *A new species of* Tuckerella *(Acarina: Tuckerellidae) from Thailand.* USDA Coop. Econ. Insect Rpt., vol. 25, no. 17, 337–40. Washington, DC: USDA.

———. 1987. *The false spider mites of Mexico (Tenuipalpidae: Acari).* USDA Tech. Bull. no. 1706. Washington, DC: USDA.

———. 1994. *A guide to the spider mites (Tetranychidae) of the United States.* West Bloomfield, MI: Indira Publishing House.

Baker, E. W., and G. W. Wharton. 1952. *An introduction to acarology.* New York: Macmillan.

Baker, G. T. 1982. Site attachment of a protelean parasite (Erythraeidae: *Leptus* sp.). *Experientia* 38:923.

———. 1985. External morphology of *Thermacarus nevadensis* Marshall (Hydracarina: Hydryphantoidea: Thermacaridae). *Zool. Anz.* 215:391–98.

Baker, R. A. 1970a. The food of *Riccardoella limacum* (Schrank)—Acari: Trombidiformes and its relationship with pulmonate molluscs. *J. Nat. Hist.* 4:521–30. [EREYNETIDAE]

———. 1970b. Studies on the life history of *Ricardoella limacum* (Schrank) (Acari-Trombidiformes). *J. Nat. Hist.* 4:511–19.

———. 1973. Notes on the internal anatomy, the food requirements and development in the family Ereynetidae (Trombidiformes). *Acarologia* 15:43–52.

———. 1976a. Cellular responses in the freshwater mussel *Anodonta anatina* produced by the mite *Unionicola intermedia*. *Parasitology* 73:33.

———. 1976b. Tissue damage and leucocytic infiltration attachment of the mite *Unionicola intermedia* to the gills of the bivalve mollusc *Anodonta anatina*. *J. Invertebr. Pathol.* 27:371–76.

———. 1977. Nutrition of the mite *Unionicola intermedia* Koenike and its relationship to the inflammatory response induced in its molluscan host *Anodonta anatina* L. *Parasitology* 75:301–8.

———. 1991. Development and life-history strategies in mussel mites (Hydrachnellae: Unionicolidae). In Schuster and Murphy, 65–73.

Balan, P. G. 1996. On postembryonic development of zerconine mites (Acari, Mesostigmata, Zerconina). Vestnik Zoologii 1996 (3): 43-47. [ZERCONIDAE] [in Russian]

Balashov, Y. S. 1961. The structure of digestive organs and blood digestion in argasid ticks. *Parazitol. Sb.* (Leningrad) 23:8–18.

———. 1972. Bloodsucking ticks (Ixodoidea): Vectors of diseases of man and animals [originally published in Russian in 1968 by Nauka, Leningrad]. *Misc. Publ. Entomol. Soc. Amer.* 8 (5): 161–376.

Balmes-Pacia, J. S., and L. A. Corpuz-Raros. 1998. Biology of *Suidasia pontifica* Oudemans (Acari: Acaridida: Suidasiidae). *Philipp. Entomol.* 12:137–53.

Balogh, J. 1958. Oribatides nouvelles de l'Afrique tropicale. *Rev. Zool. Bot. Afr.* 58 (1–2): 1–34.

———. 1959. Neue Epicriiden aus Bulgarien (Acari, Mesostigmata). *Acta Zool. Acad. Sci. Hung.* 4:115–30.

———. 1961. Identification keys of world Oribatid (Acari) families and genera. *Acta Zool. Acad. Sci. Hung.* 7:243–344.

———. 1972. *The oribatid genera of the world.* Budapest: Akad. Kiadó.

Balogh, J., and P. Balogh. 1984. A review of the Oribatuloidea Thor, 1929 (Acari: Oribatei). *Acta Zool. Acad. Sci. Hung.* 30:257–313. [ORIPODOIDEA]

———. 1987a. Identification keys of the ptychoid Mixonomata of the Neotropical region (Acari, Oribatei). *Acta Zool. Acad. Sci. Hung.* 33:1–36.

———. 1987b. A new outline of the family Lohmanniidae Berlese, 1916 (Acari, Oribatei). *Acta Zool. Acad. Sci. Hung.* 33:327–98.

———. 1988. *Oribatid mites of the Neotropical region I.* Amsterdam: Elsevier.

———. 1990. *Oribatid mites of the Neotropical region II.* Amsterdam: Elsevier.

———. 1992. *The oribatid mites genera of the world.* Vol. 1. Budapest: Hung. Nat. Mus. Press.

———. 1998. On the family Micreremidae Grandjean, 1954 (Acari, Oribatei). *Opusc. Zool.* 31:17–23.

———. 2002. *Identification keys to the oribatid mites of the extra-Holarctic regions.* Vols. 1–2. Budapest: Well Press.

Balogh, J., and S. Mahunka. 1969. The scientific results of the Hungarian Soil Zoological expedition to South America. 12, Acari: Oribatids from the materials of the second expedition. III. *Acta Zool. Acad. Sci. Hung.* 15 (3–4): 255–75.

———. 1977. New data to the knowledge of the oribatid fauna of Neogaea (Acari). II, *Acta Zool. Acad. Sci. Hung.* 23:247–65. [EUTEGAEIDAE, MICROZETIDAE, LIACARIDAE, MULTIORIBATULIDAE, and others]

———. 1979. New taxa in the system of the Oribatida (Acari). *Acta Zool. Acad. Sci. Hung.* 71:279–90. [BRACHYCHTHONIIDAE, PHTHIRACARIDAE]

———. 1983. *Primitive oribatids of the Palaearctic region.* Vol. I, *The soil mites of the world*, ed. J. Balogh and S. Mahunka. Amsterdam: Elsevier.

Balogh, P. 1984. Oribatid mites from Colombia (Acari). *Acta Zool. Acad. Sci. Hung.* 30 (1–2): 29–51.

———. 1985. *Galapagacarus schatzi* gen. et sp. n. (Acari, Oribatei). *Opusc. Zool., Budapest* 21:31–34. [HERMANNIIDAE]

———. 1989. Oribatid mites from Ecuador (Acari, Oribatida). *Acta Zool. Hung.* 35:17–28. [OTOCEPHEIDAE, MICROZETIDAE, TUBULOZETIDAE, ORIBATELLIDAE]

Bameul, F. 1990. Le DMHF: Un excellent milieu de montage en entomologie. *L'Entomologie* 46:233–39.

Banerjee, B., and J. E. Cranham. 1985. Tea. In Helle and Sabelis (1985b), 1B:371–74. [TETRANYCHIDAE]

Banerjee, P., and S. Datta. 1980. Biological control of red cotton bug, *Dysdercus koenigii* Fabricius by mite, *Hemipteroseius indicus* (Krantz and Khot). *Indian J. Entomol.* 42:265–67. [OTOPHEIDOMENIDAE]

Banks, N. 1904. *A treatise on the Acarina, or mites.* Proc. U.S. Nat. Mus., vol. 28, no. 1382. Washington, DC: GPO.

———. 1917. New mites, mostly economic. *Entomol. News* 28: 193–99. [SPELEORHYNCHIDAE]

Barilo, A. B. 1987. On the mites of the family Adamystidae (Acariformes, Prostigmata) found in Middle Asia. *Zool. Zh.* 66: 1414–16.

———. 1988. A new genus of mite of the family Penthalodidae (Acariformes) from central Asia. *Zool. Zh.* 67:617–19.

Barker, P. S. 1992. Bionomics of *Nodele calamondin* Muma (Acarina: Cheyletidae) fed on *Lepidoglyphus destructor* (Schrank) (Acarina: Glycyphagidae) at two constant temperatures. *Can. J. Zool.* 70:2333–37.

———. 1993. Phoretic mites found on beetles associated with stored grain in Manitoba. *Can. Entomol.* 125:715–19. [PYEMOTIDAE]

Barker, P. S., and L. B. Smith. 1968. Locomotion in two species of Acarina. *Manitoba Entomol.* 2:66–69.

Barr, D. 1973. *Methods for the collection, preservation, and study of water mites (Acari: Parasitengona).* Life Sci. Misc. Publ. Roy. Ontario Mus.

———. 1977. A new water mite genus from western Canada (Acari: Parasitengona: Anisitsiellidae). *Can. J. Zool.* 55:877–81.

———. 1979. Water mites (Acari, Parasitengona) sampled with chemoluminescent bait in underwater traps. *Internat. J. Acarol.* 5:187–94.

Bartsch, I. 1972. Ein Beitrag zur Systematik, Biologie und Ökologie der Halacaridae (Acari) aus dem Litoral der Nord- und Ostsee. 1, Systematik und Biologie. *Abh. Verh. Naturwiss. Ver. Hamb. (N. F.)* 16:155–230.

———. 1973. *Porohalacarus alpinus* (Thor) (Halacaridae, Acari), ein morphologischer Vergleich mit marinen Halacariden nebst Bemerkungen zur Biologie dieser Art. *Entomol. Tidskr.* 94:116–23.

———. 1974a. Ein Beitrag zur Systematik, Biologie, und Öklogie der Halacaridae (Acari) aus dem Litoral der Nord- und Ostsee. II, Ökologische Analyse der Halacaridenfauna. *Abh. Verh. Naturwiss. Ver. Hamb.* 17:9–53.

———. 1974b. *Simognathus minutus* (Hodge) (Halacaridae, Acari) eine Wiederbeschreibung nebst Bemerkungen zur Unterfamilie der Simognathinae. *Cah. Biol. Mar.* 15:275–84.

———. 1974c. Über das Auftreten von Epimeralpore besonders bei den Rhombognathinae (Halacaridae, Acari). *Zool. Anz.* 193:266–68.

———. 1977a. *Einführung in Morphologie, Anatomie, Biologie und Verbreitung der Halacaridae, (Acari).* Publ. Razi Univ. 2. Kermanshah, Iran.

———. 1977b. *Interstitielle Fauna von Galapagos. XX, Halacaridae (Acari).* Mikrofauna Meeresbodens 65. Mainz: Akademie der Wissenschaften und der Literatur.

———. 1982a. Zur Gattung *Bathyhalacarus* (Acari, Halacaridae), nebst Beschreibung fünf neuer Arten. *Zool. Scripta* 11:209–20.

———. 1982b. Zur Gattung *Rhombognathus* (Acari, Halacaridae). Übersicht über alle Arten. Deren Verbreitung und eine Bestimmungstabelle. *Zool. Jb. Syst.* 109:83–97.

———. 1983. Zur Halacaridenfauna der Philippinen Beschreibung von fünf Arten der Gattung *Rhombognathus* (Acari, Halacaridae). *Entomol. Mitt. Zool. Mus. Hamb.* 7:399–416.

———. 1986a. Acari: Halacaridae. In Botosaneanu, 638–42.

———. 1986b. A new species of *Halixodes* (Halacaridae, Acari) and a review of the New Zealand species. *J. Roy. Soc. New Zeal.* 16:51–56.

———. 1987. *Australacarus inexpectatus* gen. et spec. nov. (Halacaroidea, Acari), mit einer Übersicht über parasitisch lebende Halacariden. *Zool. Anz.* 218:17–24.

———. 1988. *Thalassophthirius auster* gen. et spec. nov., a halacarid mite (Acari) suspected of being a parasite. *Polar Res.* 6:181–84.

———. 1989a. Deep-sea mites (Halacaridae, Acari), from the southwestern Pacific. *Cah. Biol. Mar.* 30:455–71.

———. 1989b. Marine mites (Halacaroidea: Acari): A geographical and ecological survey. *Hydrobiologia* 178:21–42.

———. 1989c. Süsswasserbewohnende Halacariden und ihre Einordnung in das System der Halacaroidea (Acari). *Acarologia* 30:217–39.

———. 1991. *Copidognathus papillatus* Krantz (Acari, Halacaridae), a hydrothermal vent mite in the Pacific Ocean. *Zool. Sci.* 6: 789–92.

———. 1992. *Phacarus flavellus* gen. et spec. nov. (Copidognathinae, Halacaroidea, Acari), eine Meeresmilbe auf Kalkalgen. *Zool. Anz.* 228:212–19.

———. 1994. Halacarid mites (Acari) from hydrothermal deep sea sites. New records. *Cah. Biol. Mar.* 35:479–90.

———. 1995. *Halacarellus obsoletus* n. sp. (Arachnida: Acari: Halacaridae), an associate with antarctic hexactinellid sponges. *Mitt. Zool. Mus. Berl.* 71:121–28.

———. 1996. Halacarids (Halacaroidea: Acari) in freshwater. Multiple invasions from the Paleozoic onwards? *J. Nat. Hist.* 30: 67–99.

———. 1997. *Copidognathus nautilei* (Acari: Halacaridae), a new mite from a hydrothermal vent field in the Mid-Atlantic Ridge. *Cah. Biol. Mar.* 38:29–33.

———. 1999. *Peregrinacarus reticulatus* gen. nov., spec. nov., a freshwater halacarid mite (Acari, Halacaridae) from Marion Island. *Hydrobiologia* 392:225–32.

———. 2003. The subfamily Rhombognathinae: developmental pattern and re-evaluation of the phylogeny (Arachnida, Acari, Halacaridae). *Senckenb. Biol.* 82:15–57.

———. 2004. Geographical and ecological distribution of marine halacarid genera and species (Acari: Halacaridae). *Exp. Appl. Acarol.* 34:37–58.

Basha, A-A., and A-T. Yousef. 2000. New species of Laelapidae and Ascidae from Egypt: genera *Androlaelaps* and *Blattisocius* (Acari: Gamasida). *Acarologia* 41:395–402. [BLATTISOCIIDAE]

Baskin, Y. 2002. *A plague of rats and rubbervines: The growing threat of species invasions.* Washington, DC: Island Press/Shearwater Books.

Bassols, B., I. M. Vargas, and O. J. Polaco. 1993. Algunas consideraciones sobre *Sertitympanum contiguum* Elsen y Whitaker, 1985 (Mesostigmata, Ameroseiidae) y descripción del macho. *Folia Entomol. Mex.* 88:1–7.

Bates, P. G. 1999. Inter- and intra-specific variation within the genus *Psoroptes* (Acari: Psoroptidae). *Vet. Parasitol.* 83:201–17.

Bäumler, W. 1970. Zur Morphologie, Biologie und Ökologie von *Hermannia gibba* (C. L. Koch) (Acarina: Oribatei) unter Berücksichtigung einiger Begleitarten. I. Z. *Angew. Entomologie*, 66:257–77. [HERMANNIIDAE]

Bayan, A. 1986. Tydeid mites associated with apples in Lebanon (Acari: Actinedida: Tydeidae). *Acarologia* 27:311–16.

Bayartogtokh, B. 1998. Three species of oribatid mites of the genus *Tectocepheus* (Acari: Oribatida: Tectocepheidae) from Mongolia. *Edaphologia* 63:55–73.

———. 2001. Three new soil mites of the genus *Epidamaeus* (Acari, Oribatida, Damaeidae) from Mongolia. *Zoosystema, Paris* 23:29–49.

Bayartogtokh, B., and J. Aoki. 1997. Oribatid mites of the genus *Bipassalozetes* Mihelcic, 1957 (Acari: Oribatei: Passalozetidae) from Mongolia. *Acta Arachnol.* 46 (2): 87–99.

———. 1999. Oribatid mites of the family Tegoribatidae (Acari: Oribatida) from Mongolia. *Acta Arachnol.* 2:107–25.

Bayartogtokh, B., and R. A. Norton. 2007. The *Dyobelba tectopediosa* species-group (Acari: Oribatida: Damaeidae) from the southeastern USA, with a key to world species of *Dyobelba* and notes on their distribution. *Zootaxa* 1591:39–66.

Bayartogtokh, B., and I. E. Smelyansky. 2002. Oribatid mites of the superfamilies Gymnodamaeoidea and Plateremaeoidea (Acari, Oribatida) from East Kazakhstan. *Mitt. Mus. Nat. Kd. Berl., Zool. Reihe* 78 (1): 71–86.

———. 2004. Oribatid mites of the superfamilies Gymnodamaeoidea and Plateremaeoidea (Acari, Oribatida) from steppe of Russia. *Mongolian J. Biol. Sci.* 2 (1): 3–17.

Beament, J. W. L. 1951. The structure and formation of the egg of the fruit tree spider mite, *Metatetranychus ulmi* Koch. *Ann. Appl. Biol.* 38:1–24.

Beard, J. J., and D. E. Walter. 2004. Cryptic false spider mites: A review of the family Linotetranidae (Acari: Prostigmata: Tetranychoidea) with descriptions of a new genus, *Austrolinus*, and 2 new species. *Ann. Appl. Biol.* 18:593–606.

———. 2005. A new species of *Tuckerella* (Prostigmata: Tetranychoidea: Tuckerellidae) from Australia with descriptions of all stages and a discussion of the tritonymphal stage. *Acarologia* 45:49–60.

Beati, L., and J. E. Keirans. 2001. Analysis of the systematic relationships among ticks of the genera *Rhipicephalus* and *Boophilus* (Acari: Ixodidae) based on mitochondrial 12S ribosomal DNA gene sequences and morphological characters. *J. Parasitol.* 87:32–48.

Beaulieu, F., and D. E. Walter. 2007. Predation in suspended and forest floor soils: Observations on Australian mesostigmatic mites. *Acarologia* 47:43–54.

Beaulieu, F., and A. R. Weeks. 2007. Free-living mesostigmatic mites in Australia: Their roles in biological control and bioindication. *Aust. J. Exp. Agr.* 47:460–78.

Beck, L. 1962a. Beiträge zur Kenntnis der neotropischen Oribatidenfauna. 2, Nothridae, Camisiidae, Heterobelbidae (Arach., Acari). *Senckenb. Biol., Frankfurt* 43 (5): 385–407.

———. 1962b. Beiträge zur Kenntnis der neotropischen Oribatidenfauna. 3, *Dampfiella* (Arach. Acari). *Senckenb. Biol., Frankfurt* 43 (6): 475–87. [DAMPFIELLIDAE]

———. 1965. Über Variabilität und Wertigheit morphologischer Merkmale bei adulten Oribatiden (Arachnida, Acari) am Beispiel der Gattung *Rostrozetes* Sellnick 1925. *Abh. Senckenb. Naturforsch. Ges.* 508:5–64. [HAPLOZETIDAE]

Beck, L., and S. Woas. 1991. Die Oribatiden-Arten (Acari) eines südwestdeutschen Buchenwaldes I. *Carolinea (Karlsruhe)* 49:37–82.

Becker, E. C., ed. 1956. *Proc. X Internat. Congr. Entomol.* Vol. 1. Ottawa: Mortimer.

Beenan, R. 2001. *Leptus mariae* (Acari: Erythraeidae) paraiterend op bladkevers (Coleoptera: Chrysomelidae). *Entomol. Ber.* 61:201–2.

Beer, R. E. 1954. A revision of the Tarsonemidae of the western hemisphere. *Univ. Kans. Sci. Bull.* 36:1091–1387.

———. 1960. A new species of *Pimeliaphilus* (Acarina: Pterygosomidae) parasitic on scorpions, with discussion of its postembryonic development. *J. Parasitol.* 46:433–40.

———. 1963. Social parasitism in the Tarsonemidae, with description of a new species of tarsonemid mite involved. *Ann. Entomol. Soc. Amer.* 56:153–60.

Beer, R. E., and E. A. Cross. 1965. A new genus and two new species of scutacarid mites (Acarina) phoretic upon bees. *J. Kans. Entomol. Soc.* 33:49–57.

Beer, R. E., and D. T. Dailey. 1956. Biological and systematic studies of two species of cheyletid mites, with a description of a new species (Acarina, Cheyletidae). *Univ. Kans. Sci. Bull.* 38: 393–437.

Behan-Pelletier, V. M. 1984. *Ceratozetes* (Acari: Ceratozetidae) of Canada and Alaska. *Can. Entomol.* 116:1449–1517.

———. 1986. Ceratozetidae (Acari: Oribatei) of the western North American subarctic. *Can. Entomol.* 118:991–1057.

———. 1987. Redefinition of *Ametroproctus* (Acari: Oribatida) with descriptions of new species. *Can. Entomol.* 119:505–35. [CYMBAEREMAEIDAE]

———. 1989. *Limnozetes* (Acari: Oribatida: Limnozetidae) of northeastern North America. *Can. Entomol.* 121:453–506.

———. 1990. Redefinition of *Megeremaeus* (Acari: Megeremaeidae) with description of new species and nymphs of *M. montanus* Higgins and Woolley. *Can. Entomol.* 122:875–900.

———. 1993. Eremaeidae (Acari: Oribatida) of North America. *Mem. Entomol. Soc. Can.* 168:1–193.

———. 1997a. Oribatid mites (Acari: Oribatida) of the Yukon. In *Insects of the Yukon*, ed. H. V. Danks and J. A. Downes, 116–49. Ottawa: Biological Survey of Canada [Terrestrial Arthropods].

———. 1997b. The semi-aquatic genus *Tegeocranellus* (Acari: Oribatida: Ameronothroidea) of North and Central America. *Can. Entomol.* 129:537–77. [TEGEOCRANELLIDAE]

———. 1999a. Ceratozetoidea (Acari: Oribatida) of lowland tropical rainforest, La Selva, Costa Rica. *Acarologia* 39:349–81.

———. 1999b. Oribatid mite fauna of northern ecosystems: A product of evolutionary adaptations or physiological restraints? In Needham, Mitchell, et al., 2:87–106.

———. 2000. Ceratozetidae (Acari: Oribatida) of arboreal habitats. *Can. Entomol.* 132:153–82.

———. 2001. Phylogenetic relationships of *Hypozetes* (Acari: Tegoribatidae). In Halliday et al. (2001), 50–57.

Behan-Pelletier, V. M., and B. Bissett. 1994. Oribatida of Canadian peatlands. *Mem. Entomol. Soc. Can.* 169:73–88.

Behan-Pelletier, V. M., and B. Eamer. 2003. Zetomimidae (Acari: Oribatida) of North America. In I. M. Smith, 21–56.

———. 2005. *Zachvatkinibates* (Acari: Oribatida: Mycobatidae) of North America, with descriptions of sexually dimorphic species. *Can. Entomol.* 137:631–47.

———. 2007. Aquatic Oribatida: Adaptations, constraints, distribution and ecology. In *Acarology XI: Proceedings of the International Congress*, ed. J. B. Morales-Malacara, V. Behan-Pelletier, E. Ueckermann, T. M. Pérez, E. G. Estrada-Venegas, and M. Badii, 71–82. Mexico City: Instituto de Biología and Facultad de Ciencias, Universidad Nacional Autónoma de México; Sociedad Latinoamericana de Acarología.

Behan-Pelletier, V. M., B. Eamer, and M. Clayton. 2005. Dendroeremaeidae n. fam., from forest trees in western North America (Acari: Oribatida: Licneremaeidae). *Acarologia* 46 (4): 321–39.

Behan-Pelletier, V. M., and S. B. Hill. 1978. Feeding habits and spore dispersal of oribatid mites in the North American arctic. *Rev. Ecol. Biol. Sol* 15:497–516.

———. 1983. Feeding habits of sixteen species of Oribatei (Acari) from an acid peat bog, Glenamoy, Ireland. *Rev. Ecol. Biol. Sol* 20:221–67.

Behan-Pelletier, V. M., and G. Newton. 1999. Linking soil biodiversity and ecosystem function—the taxonomic dilemma. *Bioscience* 49:149–53.

Behan-Pelletier, V. M., and R. A. Norton. 1983. *Epidamaeus* (Acari: Damaeidae) of arctic western North America and extreme northeast U.S.S.R. *Can. Entomol.* 115:1253–89.

Behan-Pelletier, V. M., and N. A. Ryabinin. 1991. Taxonomy and biogeography of *Proteremaeus* (Acari: Oribatida: Eremaeidae). *Can. Entomol.* 123:559–65.

Behan-Pelletier, V. M., and D. E. Walter. 2000. Biodiversity of oribatid mites (Acari: Oribatida) in tree canopies and litter. In *Invertebrates as webmasters in ecosystems*, ed. D. C. Coleman and P. F. Hendrix, 187–202. Wallingford, Oxon, UK: CAB International.

——— 2007. *Phylleremus* n. gen. from leaves of deciduous trees in eastern Australia (Oribatida. Licneremaeoidea). *Zootaxa* 1386:1–17. [ADHAESOZETIDAE]

Bei, N.-X., and S.-G.Yin. 1995. A new species and a new record of the genus *Ledermuelleria* (Acari: Stigmaeidae). *Acta Zootax. Sinica* 20:185–88.

Beklemishev, W. N. 1948. On interrelationships between the systematic position of the agent and the vector of transmissible diseases of terrestrial vertebrates and man. *Med. Parasitol. Parazit. Bolezni* 17:385–400.

———. 1969. *Principles of comparative anatomy of invertebrates.* Vol. 2, *Organology.* Chicago: Univ. Chicago Press.

Bell, J. R., D. A. Bohan, E. M. Shaw, and G. S. Weyman. 2005. Ballooning dispersal using silk: World fauna, phylogenies, genetics and models. *Bull. Entomol. Res.* 95:69–114.

Bellido, A. 1979a. Développement postembryonnaire de *Carabodes willmanni* Bernini, 1975 (Acari, Oribatei). *Acarologia* 20: 419–32. [CARABODIDAE]

———. 1979b. Ecologie de *Carabodes willmanni* Bernini 1975 (Acari, Oribatei) dans les formations pionierres de la lande armoricaine. *Rev. Ecol. Biol. Sol* 16:195–218. [CARABODIDAE]

Bellido, A., and S. Deleporte. 1994. Oribatid mites Diptera interactions in a deciduous forest leaf litter—an application of multivariate analysis under linear constraints. *Pedobiologia* 38: 429–47.

Bellotti, A. C. 1985. Cassava. In Helle and Sabelis (1985b), 1B: 333–38. [TETRANYCHIDAE]

Benfekih, L., and B. Doumandji-Mitiche. 1997. Observations sur le parasitisme de stauronautes marocains solitaires *D. maroccanus* (Orthoptere, Gomphocerinae) par un acarien *Trombidium* sp. dans le sud de Medea (Algerie). *Medd. Fac. Landbouwk. Toeg. Biol. Wetens., Rijksuniv. Gent.* 62:437–47. [TROMBIDIIDAE]

Beninger, C. W. 1993. Egg predation by *Poecilochirus carabi* (Mesostigmata: Parasitidae) and its effect on reproduction of *Nicrophorus vespilloides* (Coleoptera: Silphidae). *Environ. Entomol.* 22:766–69.

Benoit, J. B., J. A. Yoder, J. T. Ark, and E. J. Rellinger. 2005. Fungal fauna of *Ixodes scapularis* Say and *Rhipicephalus sanguineus* (Latreille) (Acari: Ixodida) with special reference to species-associated internal mycoflora. *Internat. J. Acarol.* 31:417–22.

Berger, R. S. 1972. 2,6-dichlorophenol, sex pheromone of the lone star tick. *Science* 177:704–5.

Berger, R. S., J. C. Dukes, and Y. S. Chow. 1971. Demonstration of a sex pheromone in three species of hard ticks. *J. Med. Entomol.* 8:84–86.

Bergstrom, J. 1979. Morphology of fossil arthropods as a guide to phylogenetic relationships. In *Arthropod Phylogeny*, ed. A. P. Gupta, 1–56. New York: Van Nostrand Reinhold.

Berlese, A. 1882–1903. Acari, Myriapoda, et Scorpiones hucusque in Italia reperta. Fasc. 1–101. Padova.

———. 1887. Acari, Myriopoda et Scorpiones hucusque in Italia reperta. Ordo Mesostigmata (Gamasidae). Fasc. 38, 70.

———. 1888. Acari Austro-Americani quos collegit Aloysius Balzan Manipulus 1. *Bull. Soc. Entomol. Ital.* 20:171–222. [DISCOZ ERCONIDAE]

———. 1892. Acari, Myriopoda et Scorpiones hucusque in Italia reperta. Fasc. 66, no. 1.

———. 1903. Acari nuovi. *Redia* 1:235–52. [PACHYLAELAPIDAE]

———. 1904. Acari mirmecofili. *Redia* 1:299–474.

———. 1910a. Acari nuovi—Manipulus V. *Redia* 6:199–234. [BARBUTIIDAE]

———. 1910b. Brevi diagnosi di generi e specie nuovi di Acari. *Redia* 6:346–88. [DISCOZERCONIDAE, TANAUPODIDAE]

———. 1913. Acarotheca Italica, Firenze. Fasc. 1, 2.

———. 1918. Centuria quarta di Acari nuovi. *Redia* 13:113–90.

———. 1920. Centuria quinta di Acari nuovi. *Redia* 14:143–95.

Bernhard, F. 1963. Die Familie Ascaidae (Oudemans 1905) Bernhard nov. comb. In H. J. Stammer, 33–177.

Bernini, F. 1991. Fossil Acarida: Contribution of paleontological data to acarid evolutionary history. In *The early evolution of metazoa and the significance of problematic taxa*, ed. A. M. Simonetta and S. Conway Morris, 253–62. Cambridge, UK: Cambridge Univ. Press.

Bernini, F., and A. M. Avanzati. 1983. Notulae Oribatologicae

XXIX. Le Oribatellidae (Acarida, Oribatida) viventi sul massiccio sardo-corso. *Lav. Soc. Ital. Biogeogr.*, n.s. 8:347–400.

Bernini, F., M. Castagnoli, and R. Nannelli. 1995. Arachnida, Acari. In *Checklist delle specie della fauna italiana,* 24, ed. A. Minelli, S. Rufo, and S. La Posta, 1–131. Bologna: Calderini.

Bernini F., R. Nannelli, G. Nuzzaci, and E. de Lillo, eds. 2002. *Acarid phylogeny and evolution. Adaptations in mites and ticks.* Proceedings IV EURAAC Symposium, Siena. Dordrecht: Kluwer Academic Publishers.

Beron, P. K. 1973. Une nouvelle larve d'acarien (*Hoplothrombium coiffati* sp. n., Trombidiidae), parasite d'un coléoptère cavernicole du Maroc. *Ann. Spéléol.* 28:413–16. [EUTROMBIDIIDAE]

———. 1975. Erythraeidae (Acariformes) larvaires de Bulgarie. *Acta Zool. Bulg.* 1:45–75.

———. 1982. Deuxième contribution à l'étude des Erythraeidae (Acariformes) larvaires de Bulgarie. *Acta Zool. Bulg.* 19:46–47.

Berry, R. E. 1973. Biology of the predaceous mite *Pergamasus quisquiliarum* on the garden symphylan, *Scutigerella immaculata*, in the laboratory. *Ann. Entomol. Soc. Amer.* 66:1354–56.

Bertrand, M. 1981. New species of Labidostomidae de Corse. Intérêt biogéographique. *Acarologia* 23:27–38.

———. 1989. La durée de développement, un facteur limitant de l'extension de l'aire de repartition: Le cas de trios espèces européennes de Labidostomidae (Acari, Actinedida). In *Ontogeny and the concept of stase in arthropods*, ed. H. M. André and J.-Cl. Lions, 123–28. Wavre, Belgium: Agar Publ.

———. 1990a. La famille des Labidostomidae Oudemans, 1904 (Acari: Actinedida). Révision des genres et sous-genres et catalogue des espèces décrites. *Acarologia* 31:31–38.

———. 1990b. Nouvelles espèces de Labidostomidae d'Amérique australe. *Acarologia* 31:177–85.

———. 2002. Morphologic adaptations to parasitism on reptiles: Pterygosomatidae (Prostigmata: Raphignathina). In Bernini et al. (2002), 233–40.

Bertrand, M., and Y. Coineau. 1979. Une nouvelle forme biologique eu-édaphique d'acariens: *Akrostomma coralloides* n. sp. Labidostomidae aveugle. *Vie Milieu*, ser. C, 28/29:101–10.

Bertrand, M., and L. A. Corpuz-Raros. 1997. Diversité du genre *Labidostomma* Kramer: nouvelles données apportées par l'étude de *Labidostoma (Labidostoma) guadalupense* n. sp. des Philippines (Acari, Actnedida, Labidostomatidae). *Acarologia* 38: 361–68.

Bertrand, M., and I. Ineich. 1987. Contribution a la connaissance des Pterygosomatidae du pacifique sud. *Acarologia* 28:241–50.

———. 1989. Repartition des Pterygosomatidae du genre *Geckobia* Megnin, 1878. Ectoparasites du gecko *Gehyra oceanica* (Lesson, 1826), en Polynesie Française. *Acarologia* 30:365–71.

Bertrand, M., and D. Modry. 2004. The role of mite pocket like structures on *Agama caudospinosa* (Agamidae) infested by *Pterygosoma livingstonei* sp. n. (Acari: Prostigmata: Pterygosomatidae). *Folia Parasitol.* 51:61–66.

Bertrand, M., I. Paperna, and S. Finkelman. 1999. Pterygosomatidae: Descriptions et observations sur les genres *Pterygosoma*, *Geckobia*, *Zonurobia* et *Hirstiella* (Acari: Actinedida). *Acarologia* 40:275–304.

Bertrand, M., and M. Pedrono. 1999. Euryxenie et stenoxenie du genre *Geckobia* Megnin (Actinedida: Pterygosomatidae): Recolte de *Geckobia enigmatica* n. sp. sur une tortue terrestre Malgache (*Geochelone yniphora*). *Acarologia* 40:147–53.

Bertrand, M., and P. Theron. 1990. Nouveaux Labidostomidae d'Afrique australe: le genre *Labidostoma* Kramer 1879. *Acarologia* 31:247–57.

———. 1992. Nouveaux Labidostomidae d'Afrique australe: Le genre *Sellnickiella* Feider et Vasiliu, 1969. *Acarologia* 33:45–52.

Besseling, A. J. 1964. *De Nederlandse Watermijten (Hydrachnellae Latreille, 1802).* Monogr. Nederl. Entomol. Ver. 1. Amsterdam: Netherlands Entomological Society.

Bezuidenhout, J. D., and H. P. Schneider. 1972. Studies on the biology of *Cosmiomma hippopotamensis* Denny, 1843 in South West Africa. *J. S. Afr. Vet. Med. Assoc.* 43:301–4. [IXODIDAE]

Bhattacharyya, S. K. 1963. A revision of the British mites of the genus *Pergamasus* Berlese s. lat. (Acari: Mesostigmata). *Bull. Brit. Mus. (Nat. Hist.), Zool.* 11 (2): 131–242 + 8 plates. [PARASITIDAE]

———. 1994. On a collection of Podocinidae (Acarina: Mesostigmata). *Rec. Zool. Surv. India* 94:55–61.

Bieri, M., V. Delucchi, and M. Stadler. 1986. Optimization of extraction conditions in the air-conditioned Macfadyen funnel extractor for soil arthropods. *Pedobiologia* 30:127–35.

Binnington, K. C. 1972. The distribution and morphology of probable photoreceptors in eight species of ticks (Ixodoidea). *Zeitschr. Parasitenk.* 40:321–32.

———. 1978. Sequential changes in salivary gland structure during attachment and feeding of the cattle tick *Boophilus microplus. Internat. J. Parasitol.* 8:97–115.

Binnington, K. C., and F. D. Obenchain. 1982. Structure and function of the circulatory, nervous and neuroendocrine systems in ticks. *Current Themes Trop. Sci.* 1:351–98.

Binns, E. S. 1973. *Digamasellus fallax* Leitner (Mesostigmata: Digamasellidae) phoretic on mushroom sciarid flies. *Acarologia* 15:10–17.

———. 1975. Notes on the biology of *Arctoseius cetratus* (Sellnick) (Mesostigmata: Ascidae). *Acarologia* 16:577–82.

———. 1979. *Scutacarus baculitarsus* Mahunka (Acarina: Scutacaridae) phoretic on the mushroom phorid fly *Megaselia halterata* (Wood). *Acarologia* 21:91–107.

———. 1981. Physogastry in *Scutacarus baculitarsus* Mahunka (Acarina, Scutacaridae). *Entomol. Monthly Mag.* 116:157–59.

———. 1982. Phoresy as migration—some functional aspects of phoresy in mites. *Biol. Rev.* 57:571–620.

Blackman, S. W. M., and G. O. Evans. 1994. Observations on a mite (*Poecilochirus davydovae*) predatory on the eggs of burying beetles (*Nicrophorus vespilloides*) with a review of its taxonomic status. *J. Zool. Lond.* 234:217–27. [PARASITIDAE]

Blanco, G., J. L. Tella, and J. Potti. 1997. Feather mites on group-living red-billed choughs: a non-parasitic interaction? *J. Avian Biol.* 28:197–206.

Blaszak, C. 1974. *Zerconidae (Acari, Mesostigmata) polski.* Monografie fauny polski 3. Pol. Akad. Nauk, Zakl. Syst. Doswiad. (In Polish)

———. 1976a. Systematic studies on family Zerconidae I. A revision of the family Zerconidae (Acari, Mesostigmata). *Acarologia* 17:553–69.

———. 1976b. Systematic studies on family Zerconidae II. North Korean Zerconidae (Acari, Mesostigmata). *Acta Zool. Cracov.* 21:527–52.

———. 1978. Mongolian Zerconidae (Acari: Mesostigmata). *Acta Zool. Acad. Sci. Hung.* 24:301–20.

———. 1979a. Systematic studies on the family Zerconidae IV.

Asian Zerconidae (Acari, Mesostigmata). *Acta Zool. Cracov.* 24:3–112.
———. 1979b. Tunisian Zerconidae (Acari: Gamasida). *Folia Entomol. Hung.* 32:13–26.
———. 1981a. Two new genera of mites (Acari, Gamasida: Zerconidae) from the United States of America. *Bull. Acad. Pol. Sci.*, Cl. II, Ser. Sci. Biol. 28:637–45.
———. 1981b. Three new genera of zerconid mites (Acari, Gamasida: Zerconidae) from the United States of America. *Can. J. Zool.* 59:2038–47.
———. 1984. Three new genera of zerconid mites (Acari, Mesostigmata, Zerconidae) from the United States of America with a key to the American genera. *Polskie Pismo Entomol.* 53:587–601.
Blaszak, C., J. C. Cokendolpher, and V. J. Polyak. 1995. *Paleozercon cavernicolus*, n. gen., n. sp., fossil mite from a cave in the southwestern U.S.A. (Acari, Gamasida: Zerconidae), with a key to Nearctic genera of Zerconidae. *Internat. J. Acarol.* 21:253–59.
Blaszak, C., and G. Alberti. 1985. Podocinidae Berlese, 1913: Eine neue Milbenfamilie (Acari: Mesostigmata) in der Bundesrepublik Deutschland. *Zool. Anz.* 215:168–76.
———. 1989. Eine neue *Epicrius*-Art aus Österreich (Acari: Gamasida: Epicriidae) mit einem Schlüssel zur Bestimmung europäischer Arten. *Verh. Zool.-Bot. Ges. Öster.* 126:67–75.
Blaszak, C., and R. Ehrnsberger. 1993. Beitrag zur Kenntnis von *Halolaelaps* (*Saprogamasellus*) Gotz, 1952 (Acari: Gamasida: Halolaelapidae). *Genus (Warsaw)* 4:143–267.
———. 1995. Beitrag zur Kenntnis von *Halolaelaps* (*Halogamasellus* subgen. nov.) (Acari: Gamasida: Halolaelapidae). *Osnabrück. Naturwiss. Mitt.* 20–21:25–94.
———. 1997. *Halolaelaps* (*Halolaelaps*) *rafaljanus* sp. nov. eine neue Art von der Nordseeküste Deutschlands (Acari: Gamasida: Halolaelapidae). *Genus (Warsaw)* 8:3–7.
———. 1998a. Beitrage zur Kenntnis von *Halolaelaps* (*Halolaelaps* s. str.) (Acari: Gamasida: Halolaelapidae). *Osnabrück. Naturwiss. Mitt.* 24:159–81.
———. 1998b. *Halolaelaps* (*Saprogamasellus*) *giganteus* sp. nov. (Acari, Gamasida: Halolaelapidae), eine neue Art von der Ostseeküste Deutschlands. In Ebermann (1998a), 79–92.
———. 1998c. Eine neue Untergattung *Haloseius* subgen. nov. in der Gattung *Halolaelaps* Berlese & Trouessart, 1889 (Acari: Gamasida: Halolaelapidae). *Genus (Warsaw)* 9:421–29.
———. 1999. A new species of the genus *Halolaelaps* Berlese & Trouessart, 1889 (Acari, Gamasida, Halolaelapidae) from the Spanish Mediterranean coast. *Graellsia* 55:3–7.
———. 2000a. Eine neue Raubmilbenart *Saprolaelaps claudiae* sp. nov., aus der Gattung *Saprolaelaps* Leitner, 1946 (Acari: Gamasida: Halolaelapidae). *Genus (Warsaw)* 11:613–18.
———. 2000b. *Saprolaelaps reticulates* nov. sp., eine neue Milbenart der Gattung *Saprolaelaps* Leitner, 1946 (Acari: Gamasida: Halolaelapidae). *Osnabrück. Naturwiss. Mitt.* 26:135–38.
———. 2002. Beiträge zur Kenntnis von *Saprolaelaps* Leitner, 1946 in Europa (Acari: Gamasida: Halolaelapidae). *Osnabrück. Naturwiss. Mitt.* 28:159–97.
Blauvelt, W. E. 1945. *The internal anatomy of the common red spider mite (*Tetranychus telarius *Linn.).* Mem. Cornell Univ. Agr. Exp. Sta. 270. Ithaca, NY.
Blaxter, M. 2001. Sum of the arthropod parts. *Nature* 413 (Sept. 13): 121–22.
Block, W. 1979. Cold tolerance of micro-arthropods from Alaskan taiga. *Ecol. Entomol.* 4:103–10.
Block, W., and P. Convey. 1995. The biology, life-cycle and ecophysiology of the antarctic mite *Alaskozetes antarcticus*. *J. Zool. Soc. London* 236:431–49. [AMERONOTHRIDAE]
Bloszyk, J. 1980a. Mites of the genus *Trachytes* Michael, 1894 (Acari, Mesostigmata) in Poland. *Poznan. Towarz. Przy. Nauk* 54:5–52. [in Polish with English summary]
———. 1980b. Study on the family Nicoletiellidae (Acari, Prostigmata) in Poland. Materialy do Fauny Roztoczy (Acari) Polski. *Poznan. Towarz. Przy. Nauk* 54:53–85 [LABIDOSTOMATIDAE] [in Polish]
———. 1999. *Geographical and ecological variability of mites of the cohort Uropodina (Acari: Mesostigmata) in Poland.* 1, *Uropodine mites of oak-hornbeam forests* (Carpinion betuli). Poznan: Wydawnictwo Kontekst. [in Polish, with English summary]
Bloszyk, J., and R. B. Halliday. 2000. Observations on the genus *Polyaspinus* Berlese 1916 (Acari: Trachytidae). *Syst. Appl. Acarol.* 5:47–64.
Bloszyk, J., and P. Szymkowiak. 1996. *Trachytes kaliszewskii* n. sp. (Acari: Uropodina), from the Great Basin (Utah, USA), with remarks on the habitats and distribution of the members of the genus *Trachytes*. *Great Basin Nat.* 56:59–72.
Bochkov, A. V. 1997. New classification of myobiid mites (Acari, Acariformes). *Entomol. Obozr.* 76:938–951. [in Russian]
———. 2001. Parallel evolution of mites of the family Myobiidae (Acari: Prostigmata) and jerboas (Rodentia: Dipodoidea). *Parazitologiya* 35:9–18. [in Russian]
———. 2002. The classification and phylogeny of the mite superfamily Cheyletoidea (Acari, Prostigmata). *Entomol. Obozr.* 81:488–513. [in Russian]
Bochkov, A. V., and A. Fain. 2001. Phylogeny and system of the Cheyletidae (Acari: Prostigmata) with special reference to their host-parasite associations. *Bull. Inst. Roy. Sci. Nat. Belg., Entomol.* 71:5–36.
———. 2003. New observations on mites of the family Myobiidae MEGNIN, 1877 (Acari: Prostigmata) with special reference to their host-parasite relationships. *Bull. Inst. Roy. Sci. Nat. Belg., Entomol.* 73:5–50.
Bochkov, A. V., A. Fain, and J. Dabert. 2002. A revision of the genus *Cheletopsis* (Acari Cheyletidae). *Bull. Inst. R. Sci. Nat. Belg., Entomologie* 72:5–26.
Bochkov, A. V., A. Fain, and M. Skoracki. 2004. New quill mites of the family Syringophilidae (Acari: Cheyletoidea). *Syst. Parasitol.* 57:135–50.
Bochkov, A. V., and A. Labrzycka. 2003. A revision of the European species of the genus *Myobia* von Heyden, 1826 (Acari, Myobiidae). *Acta Parasitol.* 48:24–40.
Bochkov, A. V., S. V. Mironov, and A. Fain. 1999. Phylogeny and host-parasite relationships of the family Harpirhynchidae (Acari, Prostigmata). *Acarina* 7:69–87.
Bochkov, A. V., and B. M. OConnor. 2003. New cheyletid mites (Acari: Cheyletidae) associated with birds. *Acta Parasitol.* 48:265–79.
———. 2004a. Phylogeny, taxonomy and biology of mites of the genera *Chelacheles* and *Neochelacheles* (Acari: Cheyletidae). *Invertebr. Syst.* 18:547–92.
———. 2004b. Systematics of the genus *Sciurocheyla* (Acari: Cheyletidae), parasites of tropical squirrels. *J. Med. Entomol.* 41:308–20.
Bochkov, A. V., and D. E. Walter. 2007. The life cycle of *Pomerantzia*

philippina sp. n. (Prostigmata: Pomerantziidae) described from the Philippines. *Acarina* 15:159–70.

Boctor, F. N. 1972. Biochemical and physiological studies of certain ticks (Ixodoidea). Free amino acids in female *Argas* (*Persicargas*) *arboreus* Kaiser, Hoogstraal and Kohls (Argasidae) analyzed by gas-liquid chromatography. *J. Med. Entomol.* 9:201–4.

Boczek, J. 1962. *Nalelpella haarlovi* n. sp. (Acarina, Eriophyidae). *Entomol. Meddr.* 31:195–97.

Boczek, J., and R. Petanovic. 1996. Eriophyid mites as agents for the biological control of weeds. In *Proc. IX Internat. Symposium on Biological Control of Weeds, Jan. 19–26*, ed. V. C. Moran and J. H. Hoffmann, 127–31. Stellenbosch, Univ. S. Africa.

Boggs, C. L., and L. E. Gilbert. 1987. Spatial and temporal distribution of *Lantana* mites phoretic on butterflies. *Biotropica* 19:301–5. [MELICHARIDAE]

Bolland, H. R. 1983. A description of *Neophyllobius aesculi* n. sp. and its developmental stages (Acari: Camerobiidae). *Entomol. Ber.* 43:42–47.

———. 1986. Review of the systematics of the family Camerobiidae (Acari: Raphignathoidea). I, The genera *Camerobia, Decaphyllobius, Tillandsobius*, and *Tychenobius*. *Tijdschr. Entomol.* 129:191–215.

Bolland, H. R., J. Gutierrez, and C. H. W. Flechtmann. 1998. *World catalogue of the spider mite family (Acari: Tetranychidae).* Leiden: K. Brill.

Bolland, H. R., J. Gutierrez, and W. Helle. 1981. Chromosomes in spider mites (Tetranychidae: Acari). *Acarologia* 22:271–75.

Bolland, H. R., and W. Helle. 1981. A survey of chromosome complements in the Tenuipalpidae. *Internat. J. Acarol.* 7:157–60.

Bongers, M. G. H., B. M. OConnor, and F. S. Lukoschus. 1985. Morphology and ontogeny of histiostomatid mites (Acari: Astigmata) associated with cattle dung in the Netherlands. *Zool. Verh., Leiden* 223:1–56. [HISTIOSTOMATIDAE]

Bonnar E., P. Eustace, and F. C. Powell. 1993. The *Demodex* mite population in rosacea. *J. Amer. Acad. Dermatol.* 28:443–48. [DEMODICIDAE]

Booth, R. G., M. Edwards, and M. B. Usher. 1985. Mites of the genus *Eupodes* (Acari, Prostigmata) from maritime Antarctica: a biometrical and taxonomic study. *J. Zool. Lond.* A, 207: 381–406. [EUPODIDAE]

Booth, T. F., D. J. Beadle, and R. J. Hart. 1984. The ultrastructure of Gene's organ in the cattle tick, *Boophilus microplus* Canestrini. In Griffiths and Bowman, 1:261–67.

Borcard, D. 1991. The oribatid mites of the Swiss Jura peat bogs (Acari, Oribatei). 1, Aspects of the oribatid community of the Cachot peat bog. *Rev. Suisse Zool.* 98:303–17.

Borland, J. G. 1956. The genus *Neotrombidium* (Acarina: Trombidioidea) in the United States. *J. Kans. Entomol. Soc.* 29:29–35. [NEOTROMBIDIIDAE]

Bosch, J. 1992. Parasitism in wild and managed populations of the almond pollinator *Osmia cornuta* Latr. (Hymenoptera: Megachilidae). *J. Apicult. Res.* 31:77–82. [CHAETODACTYLIDAE]

Botosaneanu, L., ed. 1986. *Stygofauna mundi: A faunistic, distributional, and ecological synthesis of the world fauna inhabiting subterranean waters (including the marine interstitial).* Leiden: E. J. Brill/Dr. W. Backhuys.

Böttger, K. 1962. Zur Biologie und Ethologie der einheimischen Wassermilben *Arrenurus (Megaluracarus) globator* (Müll.), 1776, *Piona nodata nodata* (Müll.), 1776 und *Eylais infundibulifera meridionalis* (Thon), 1899. (Hydrachnellae, Acari). *Zool. Jb. Syst., Ökol. Geogr. Tiere* 89:501–84. [ARRENURIDAE].

———. 1965. Zur Ökologie und Fortpflanzungsbiologie von *Arrenurus valdiviensis* K. O. Viets 1964 (Hydrachnellae, Acari). *Zeitschr. Morphol. Ökol. Tiere* 55:115–41.

———. 1981. Zur Parasitologie und postparasitischen Entwicklung der neotropischen Wassermilbe *Arrenurus valencius* (Hydrachnellae, Acari), nebst einigen faunistischen Angaben zum See "Laguna Chichoj" in Guatemala. *Stud. Neotrop. Fauna Environ.* 15:155–66.

Böttger, K., and P. Martin. 2003. On the morphology and parasitism of *Arrenurus globator* (O. F. Müller, 1776) (Hydrachnidia, Acari), a water mite with an unusually extensive host spectrum. *Acarologia* 43:49–57.

Böttger, K., and F. Schaller. 1961. Biologische und ethologische Beobachtungen an einheimischen Wassermilben. *Zool. Anz.* 167:46–50. [ARRENURIDAE].

Boulton, A. J., H. M. Valett, and S. G. Fisher. 1992. Spatial distribution and taxonomic composition of the hyporheos of several Sonoran Desert streams. *Arch. Hydrobiol.* 125:37–61.

Bowman, A. S., L. B. Coons, G. R. Needham, and J. R. Sauer. 1997. Tick saliva: Recent advances and implications for vector competence. *Med. Vet. Entomol.* 11:277–85.

Bowman, C. E. 1984. Some aspects of feeding and digestion in the soil predatory mite *Pergamasus longicornis* (Berlese) (Parasitidae: Mesostigmata). In Griffiths and Bowman, 1:316–22.

Bowman, W. L. 1978. The cat fur-mite (*Lynxacarus radovskyi*) in Australia. *Aust. Vet. J.* 54:403–4. [LISTROPHORIDAE]

Boyaci, Y. Ö., and M. Özkan. 2001. A new species of *Hydryphantes* C. L. Koch, 1841 (Hydryphantidae, Hydrachnellae, Acari) for the Turkish fauna. *Turk. J. Zool.* 25:369–73.

Boyce, A. M., and R. B. Korsmeier. 1941. The citrus bud mite, *Eriophyes sheldoni* Ewing. *J. Econ. Entomol.* 34:745–56. [ERIOPHYIDAE]

Bregetova, N. G. 1961. Mites of the family Veigaiaidae Oudemans, 1939, of fauna USSR. *Parazit. Sbornik Zool., Inst. Akad. Nauk SSSR* 20:10–106. [in Russian]

———. 1977a. Cohort Epicriina. In Gilyarov and Bregetova, 28–39. [in Russian]

———. 1977b. Family Aceosejidae Baker et Wharton 1952 (sensu Evans, 1958). In Gilyarov and Bregetova, 169–226. [includes BLATTISOCIIDAE, ASCIDAE, MELICHARIDAE] [in Russian]

———. 1977c. Family Ameroseiidae (Berlese, 1919) Evans, 1961. In Gilyarov and Bregetova, 148–69. [in Russian]

———. 1977d. Family Arctacaridae Evans, 1955. In Gilyarov and Bregetova, 145–48. [in Russian]

———. 1977e. Family Laelaptidae Berlese, 1892. In Gilyarov and Bregetova, 483–554. [in Russian]

———. 1977f. Family Ologamasidae Ryke, 1962. In Gilyarov and Bregetova, 308–15. [in Russian]

———. 1977g. Family Veigaiaidae Oudemans, 1939. In Gilyarov and Bregetova, 108–45. [in Russian]

Bregetova, N. G., and E. G. Koroleva. 1960. The macrochelid mites (Gamasoidea, Macrochelidae) in the USSR. *Parazit. Sbornik Zool., Inst. Akad. Nauk SSSR* 19:32–154. [in Russian]

Bregetova, N. G., and G. I. Shcherbak. 1977. Sem. Rhodacaridae Oudemans, 1902. In Gilyarov and Bregetova, 256–307. [in Russian]

Brennan, J. M., and M. L. Goff. 1977. Keys to the genera of chiggers of the Western Hemisphere (Acarina: Trombiculidae). *J. Parsitol.* 63:554–66.

Brennan, J. M., and E. K. Jones. 1959. Keys to the chiggers of North America with synonymic notes and descriptions of two new genera (Acarina: Trombiculidae). *Ann. Entomol. Soc. Amer.* 52:7–16.

Brickhill, C. D. 1958. Biological studies of two species of tydeid mites from California. *Hilgardia* 27 (20): 601–20.

Bridges, J. R., and J. C. Moser. 1983. Role of two phoretic mites in transmission of bluestain fungus, *Ceratocystis minor. Ecol. Entomol.* 8:9–12. [TARSONEMIDAE]

Brinton, L. P., and J. H. Oliver, Jr. 1971. Fine structure of oogonial and oocyte development in *Dermacentor andersoni* Stiles (Acari: Ixodidae). *J. Parasitol.* 57:720–47.

Brody, A. R. 1970. Observations on the fine structure of the developing cuticle of a soil mite *Oppia coloradensis* (Acarina: Cryptostigmata). *Acarologia* 12:421–31.

Brody, A. R., J. C. McGrath, and G. W. Wharton. 1972. *Dermatophagoides farinae*: The digestive system. *J. New York Entomol. Soc.* 80:152–77.

———. 1976. *Dermatophagoides farinae*: The supracoxal glands. *J. New York Entomol. Soc.* 84:34–47.

Bronswijk, J. E. M. H. van, A.W.A. M. de Cock, and S. Oshima. 1973. The genus *Blomia* Oudemans (Acari: Glycyphagidae). II, Comparison of its species. *Acarologia* 15:490–504. [ECHIMYOPODIDAE]

Bronswijk, J.E. M. H. van, and E. J. de Kreek. 1976. *Cheyletiella* (Acari: Cheyletiellidae) of dog, cat and domesticated rabbit: A review. *J. Med. Entomol.* 13:315–27. [CHEYLETIDAE]

Bronswijk, J. E. M. H. van, and R. N. Sinha. 1971. Pyroglyphid mites (Acari) and house dust allergy. *J. Allergy* 47:31–52.

Brown, G. W. 1978. Recent studies in scrub typhus: a review. *J. Roy. Soc. Med.* 71:507–10. [TROMBICULIDAE]

Brown, J. M., and D. S. Wilson. 1994. *Poecilochirus carabi*: Behavioral and life-history adaptations to different hosts and the consequences of geographical shifts in host communities. In Houck (1994b), contribution no. 1, 1–22. [PARASITIDAE]

Bruce, W. A. 1989. Artificial diet for the parasitic mite *Pyemotes tritici* (Acari: Pyemotidae). *Exp. Appl. Acarol.* 6:11–18.

Bruce, W. A., and D. E. Johnston. 1976. *Gaudoglyphus* n. gen., based on *Analges minor* Nörner (Acari: Gaudoglyphidae n. fam.). *Internat. J. Acarol.* 2:29–33.

Bruce, W. A., and G. L. LeCato. 1980. *Pyemotes tritici*: A potential new agent for biological control of the red imported fire ant, *Solenopsis invicta* (Acari: Pyemotidae). *Internat. J. Acarol.* 6:271–74.

Bruce, W. A., and D. L. Wrensch. 1990. Reproductive potential, sex ratio, and mating efficiency of the straw itch mite (Acari: Pyemotidae). *J. Econ. Entomol.* 83:384–91.

Brucker, E. A. 1897. Sur un nouvel Acarien marin. *C.R. Soc. Biol.* 4:632–33.

Brucker, E. A., and E. L. Trouessart. 1900. Descriptions d'espèces nouvelles d'Halacaridae (par le Dr. Trouessart) et description d'un genre nouveau (par Brucker et Trouessart). *Bull. Étud. Sci. Angers* 29:209–34.

Bruin, J., L. P. S. van der Geest, and M. W. Sabelis, eds. 1999. *Ecology and evolution of the Acari*. Dordrecht: Kluwer Academic Publishers.

Brust, G. E., and G. J. House. 1988. A study of *Tyrophagus putrescentiae* (Acari: Acaridae) as a facultative predator of southern corn rootworm eggs. *Exp. Appl. Acarol.* 4:335–44.

Bücking, J. 1999. Population biology of a phytophagous mite of marine and estuarine rocky shores (Astigmata, Hyadesiidae). In Bruin, van der Geest, and Sabelis, 441–50.

———. 2002. Three-dimensional structure of the gut system of the mites *Ameronothrus lineatus* (Oribatida: Ameronothridae) and *Hyadesia fusca* (Astigmata: Hyadesiidae). In Bernini et al. (2002), 217–25.

Bücking, J., H. Ernst, and F. Siemer. 1998. Population dynamics of phytophagous mites inhabiting rocky shores: K-strategists in an extreme environment? In Ebermann (1998a), 93–143.

Buford, D. R. 1976. Morphology and life history of *Hydrozetes bushnelli* n. sp. (Oribatei, Hydrozetidae). Ph.D. diss., Univ. Colorado, Boulder. [HYDROZETIDAE]

Bulanova-Zachvatkina, E. M. 1960. New representatives of primitive oribatid mites from USSR. Perlohmannoidea Grandj. 1958. (Acariformes, Oribatei). *Zool. Zh.* 39:1835–48.

———. 1980. On the Palaeosomata (Acariformes, Oribatei) of the U.S.S.R. *Entomol. Obozr.* 59:679–88.

Bull, J. J. 1983. *Evolution of sex determining mechanisms*. London: Benjamin Cummings.

Burgdorfer, W. 1988. Ecological and epidemiological considerations of Rocky Mountain spotted fever and scrub typhus. *Biol. Rickettsial Dis.* 1:33–50. [TROMBICULIDAE]

Burgdorfer, W., L. Brinton, and L. Hughes. 1973. Isolation and characterization of symbionts from the Rocky Mountain wood tick, *Dermacentor andersoni*. *J. Invertebr. Pathol.* 22:424–34. [IXODIDAE]

Bury, R. J., and G. W. Krantz. 1978. A new species of *Aegyptobia* (Acari: Tenuipalpidae) from a high altitude habitat in Oregon. *Acarologia* 19:417–21.

Butani, D. K. 1959. Sugarcane mites—a review. *Proc. Indian Acad. Sci.* B 49:99–102. [TARSONEMIDAE]

Butler L., and P. E. Hunter. 1966. A new species of *Hoplomegistus* from Costa Rica (Acarina: Hoplomegistidae). *Acarologia* 8:525–31. [HOPLOMEGISTIDAE]

———. 1968. Redescription of *Megisthanus floridanus* with observations on its biology (Acarina: Megisthanidae). *Fla. Entomol.* 51:187–97.

Caballero, A. I., and J. C. Iturrondobeitia. 2000. Aportaciones al conocimiento de la especie *Tectocepheus velatus* (Michael, 1880) (Acari: Oribatida: Tectocepheidae). *Boln. Asoc. Esp. Entomol.* 24:9–13.

Cadogan, B. L., and J. E. Laing. 1977. A technique for rearing the predaceous mite, *Balaustium putmani* (Acarina: Erythraeidae), with notes on its biology and life history. *Can. Entomol.* 109:1535–44.

———. 1982. A study of *Balaustium putmani* (Acari: Erythraeidae) in apple ordchards in southern Ontario. *Proc. Entomol. Soc. Ont.* 112:13–22.

Calis, J. N. M., W. P. J. Overmeer, and L. P. S. van der Geest. 1988. Tydeids as alternative prey for phytoseiid mites in apple orchards. *Medd. Fac. Landbouwk. Rijksuniv. Gent* 53:793–98.

Calvert, D. J., and C. B. Huffaker. 1974. Predator (*Metaseiulus occidentalis*)-prey (*Pronematus* spp.) interactions under sulfur and cattail pollen applications in a noncommercial vineyard. *Entomopohaga* 19:361–69. [IOLINIDAE]

Camerik, A. M. 2005. Two new *Pediculaster* (Acari: Siteroptidae) species from Thailand and Mexico. *Bull. Ann. Soc. Roy. Belg. Entomol.* 141:63–71.

Camerik, A. M., and E. A. Uekermann. 1995. *Pediculaster norrbomialis* sp. n. and *P. gracilis* sp. n. (Acari, Heterostigmata: Pygmephoridae) from South Africa, with notes on host and dung preference of their phoretic females. *Mitt. Hamb. Zool. Mus. Inst.* 92:73–86.

Camicas, J.-L., J.-P. Hervy, F. Acam, and P. C. Morel. 1998. *Le Tiques du monde (Acarida: Ixodida). Nomenclature, stades décrits, hôtes, rèpartition*. Paris: Editions de l'ORSTOM.

Camicas, J.-L., and P. C. Morel. 1977. Position systématique et classification des tiques (Acarida: Ixodida). *Acarologia* 18:410–20.

Camin, J. H. 1953a. *Observations on the life history and sensory behavior of the snake mite,* Ophionyssus natricis *(Gervais) (Acarina: Macronyssidae).* Chicago Acad. Sci. Spec. Publ. 10.

———. 1953b. A revision of the cohort Trachytina, 1938, with the description of *Dyscritaspis whartoni*, a new genus and species of polyaspid mite from tree holes. *Bull. Chicago Acad. Sci.* 9: 335–85.

———. 1955. Uropodellidae, a new family of mesostigmatid mites based on *Uropodella lacinata* Berlese, 1988 (Acarina, Liroaspina). *Bull. Chicago Acad. Sci.* 10:65–81.

Camin, J. H., G. M. Clark, and F. G. Bourdeau. 1958. The palpal "tined seta" in the Mesostigmata, a homologue of the palpal claw in the Onychopalpida (Acarina). In Becker, 1:903–8.

Camin, J. H., and F. E. Gorirossi. 1955. *A revision of the suborder Mesostigmata (Acarina) based on new interpretations of comparative morphological data*. Chicago Acad. Sci. Spec. Publ. 11. [DIARTHROPHALLIDAE]

Camin, J. H., W. W. Moss, J. H. Oliver, Jr., and G. Singer. 1967. Cloacaridae, a new family of cheyletoid mites from the cloaca of aquatic turtles (Acari: Acariformes: Eleutherengona). *J. Med. Entomol.* 4:261–72.

Campbell, R. E. 1941. Further notes on the blue oat or pea mite *Penthaleus major* (Dugès) in California. *Bull. Calif. Dept. Agr.* 30:312–14. [PENTHALEIDAE]

Cancela da Fonseca, J. P. 1969. Le spermatophore de *Damaeus quadrihastatus* Markel et Meyer (Acarien, Oribate). In Evans (1969c), 227–32.

———. 1994. Ecological niches of decomposer oribatid mites. *Cairo Univ. Afr. Stud. Rev.* 15:1–20.

Canestrini, G. 1891. Abbozzo del sistema Acarologico. *Atti Ist. Ven.* 38:699–725.

Cangialosi, K. R. 1997. Foraging versatility and the influence of host availability in *Argyrodes trigonum* (Araneae, Theridiidae). *J. Arachnol.* 25:182–93.

Cantonati, M., R. Gerecke, and E. Bertuzzi. 2006. Springs of the Alps—sensitive ecosystems to environmental change: From biodiversity assessments to long-term studies. *Hydrobiologia* 562:59–96.

Cantwell, G. E., W. W. Cantelo, and M. A. Cantwell. 1985. Effect of a parasitic mite, *Coccipolipus epilachnae*, on fecundity, food consumption and longevity of the Mexican bean beetle. *J. Entomol. Sci.* 20:199–203. [PODAPOLIPIDAE]

Caraballo, L. 1999. Caracterizacion inmunoquimica y molecular de los alergenos de *Blomia tropicalis*, un acaro causante de asma en el tropico. *Rev. Acad. Colomb. Cien. Exact. Fisic. Natur.* 23 (88): 433–43. [ECHIMYOPODIDAE]

Carroll, R. L. 1988. *Vertebrate paleontology and evolution*. Vol. 3. New York: Freeman.

Casanueva, M. E. 1993. Phylogenetic studies of the free-living and arthropod associated Laelapidae (Acari: Mesostigmata). *Guyana Zool.* 57:21–46.

Casanueva, M. E., and R. A. Norton. 1997. New nothroid mites from Chile: *Novonothrus covarrubiasi* n. sp. and Novonothrus puyehue n. sp. (Acari: Oribatida). *Rev. Chilena Hist. Nat.* 70:435–45. [NOTHRIDAE]

Castagnoli, M. 1996. Ornamental coniferous and shade trees. In Lindquist, Sabelis, and Bruin, 6:661–71.

Castagnoli, M., and G. N. Oldfield. 1996. Other fruit trees and nut trees. In Lindquist, Sabelis, and Bruin, 6:543–59.

Cerný, V. 1967. Two new species of argasid ticks (Ixodoidea, Argasidae). *Folia Parasitol.* 14:141–48.

Chada, H. L. 1956. Biology of the winter grain mite and its control in small grains. *J. Econ. Entomol.* 49:515–20. [PENTHALEIDAE]

Chagas, C. M., E. W. Kitajima, and J. C. V. Rodrigues. 2003. Coffee ringspot virus vectored by *Brevipalpus phoenicis* (Acari: Tenuipalpidae) in coffee. *Exp. Appl. Acarol.* 30:203–13.

Chamberlin, R. V., and S. Mulaik. 1942. On a new family in the Notostigmata. *Proc. Biol. Soc. Wash.* 55:125–32.

Chambers, R. J., E. M. Wright, and R. J. Lind. 1993. Biological control of glasshouse sciarid flies (*Bradysia* spp.) with the predatory mite, *Hypoaspis miles*, on cyclamen and poinsettia. *Biocontrol Sci. Technol.* 3:285–93. [LAELAPIDAE]

ChannaBasavanna, G. P. 1996. Sugarcane, coffee and tea. In Lindquist, Sabelis, and Bruin, 6:631–40.

ChannaBassavanna, G. P., and C. A. Viraktamath, eds. 1988. *Progress in acarology*. Vol. 1. New Delhi: Oxford & IBH Publishing.

Chant, D. A. 1959. Phytoseiid mites (Acarina: Phytoseiidae). Part I, Bionomics of seven species in southeastern England. *Can. Entomol.* (Suppl. 12) 91:5–44.

———. 1963. The subfamily Blattisocinae Garman (= Aceosejinae Evans) (Acarina: Blattisocidae Garman) (= Aceosejidae Baker and Wharton) in North America, with descriptions of new species. *Can. J. Zool.* 41:243–305. [includes ASCIDAE, MELICHARIDAE]

———. 1965. Generic concepts in the family Phytoseiidae (Acarina: Mesostigmata). *Can. Entomol.* 97:351–74.

———. 1993. Paedomorphosis in the family Phytoseiidae (Acari: Gamasina). *Can. J. Zool.* 71:1334–49.

Chant, D. A., and E. E. Lindquist. 1965. *Nabiseius duplicisetus*, a new genus and species of Otopheidomeninae (Acarina: Phytoseiidae) from nabid bugs. *Can. Entomol.* 97:515–21.

Chant, D. A., and J. A. McMurtry. 1994. A review of the subfamilies Phytoseiinae and Typhlodrominae (Acari: Phytoseiidae). *Internat. J. Acarol.* 20:223–310.

———. 2003. A review of the subfamily Amblyseiinae Muma (Acari: Phytoseiidae). Part I, Neoseiulini new tribe. *Internat. J. Acarol.* 29:3–46.

———. 2004. A review of the subfamily Amblyseiinae Muma (Acari: Phytoseiidae). Part III, The tribe Amblyseiini Wainstein, subtribe Amblyseiina n. subtribe. *Internat. J. Acarol.* 30: 171–228.

Chant, D. A., and J. A. McMurtry. 2005a. A review of the subfamily Amblyseiinae Muma (Acari: Phytoseiidae). Part VII, Typhlodromipsini n. tribe. *Internat. J. Acarol.* 31:315–40.

———. 2005b. A review of the subfamily Amblyseiinae Muma (Acari: Phytoseiidae). Part VIII, The tribes Macroseiini Chant, Denmark and Baker, Phytoseiulini n. tribe, Afroseiulini n. tribe and Indoseiulini Ehara and Amano. *Internat. J. Acarol.* 32: 13–25.

Chapman, R. F. 1982. *The insects: Structure and function*. 3d ed. Cambridge, MA: Harvard Univ. Press.

Charles, J. G., and V. White. 1988. Airborne dispersal of *Phytoseiulus persimilis* (Acarina: Phytoseiidae) from a raspberry garden in New Zealand. *Exp. Appl. Acarol.* 5:47–54.

Chatterjee, T., and C. Y. Chang. 2005. A new species of *Limnohalacarus* (Acari: Halacaridae) from India. *Bull. Inst. Roy. Sci. Nat. Belg.* 75:23–27.

———. 2006. Descriptions of two new species of the genus *Copidognathus* and females of *Copidognathus jejuensis* (Acari: Halacaridae) from Korea. *J. Nat. Hist.* 40:1761–74.

Chatterjee, T., and M. DeTroch. 2000. Halacaridae (Acari) from Gazi Bay (Kenya): Description and biogeography of three new and two known species. *Hydrobiologia* 427:177–94.

Chaudhri, W. M. 1971. Four new species of the genus *Tuckerella* from Pakistan. *Pakistan J. Zool.* 3:79–85. [TUCKERELLIDAE]

Chaudhri, W. M., S. Akbar, and A. Rasool. 1979. *Studies on the predatory leaf inhabiting mites of Pakistan*. Faisalbad, Pakistan: Univ. Agr. Faisalabad.

Chen, J., R. A. Norton, V. M. Behan-Pelletier, and H. F. Wang. 2004. Analysis of the genus *Gymnodampia* (Acari: Oribatida) with redescription of *G. setata* and description of two new species from North America. *Can. Entomol.* 136:793–821. [AMERIDAE]

Chen, J., and H. F. Wang. 2007. Taxonomic study on the family Nippobodidae (Acari: Oribatida) from China. *Zootaxa* 1464:45–63.

Chen, P.-R., and Z.-Q. Zhang. 1991. Biology of *Allothrombium pulvinum* Ewing (Acari, Trombidiidae) and its impact on twospotted spider mite (Acari, Tetranychindae) in cotton fields. *J. Appl. Entomol.* 112:31–37.

Chen, P.-R., Z.-Q. Zhang, K. Wang, X.-Y. Wang, W.-L. Xu, and Z.-L. Gao. 1994. *Allothrombium pulvinum* Ewing (Acari, Trombidiidae), an important early-season natural enemy of *Aphis gossypii* Glover (Hom., Aphididae) in cotton. *J. Appl. Entomol.* 117:113–21.

Cheng, L. 1976. Insects in marine environments. In *Marine insects*, ed. L. Cheng, 1–4. Amsterdam: North-Holland.

Chiavegato, L. G. 1976. *Brevipalpus russulus* (Boisduval) Oudemans, 1938 (Acari: Tenuipalpidae) severa praga de cactáceas ornamentais. *Cienc. Cult.* 28:342–43.

Childers, C. C., and W. R. Enns. 1975. Predaceous arthropods associated with spider mites in Missouri apple orchards. *J. Kans. Entomol. Soc.* 48:453–71. [ERYTHRAEIDAE]

Childers, C. C., J. V. French, and J. C. V. Rodrigues. 2003. *Brevipalpus californicus, B. obovatus, B. phoenicis*, and *B. lewisi* (Acari: Tenuipalpidae): a review of their biology, feeding injury and economic importance. *Exp. Appl. Acarol.* 30:5–28.

Childers, C. C., and G. C. Rock. 1981. Observations on the occurrence and feeding habits of *Balaustium putmani* (Acari, Erythraeidae) in North Carolina apple orchards. *Internat. J. Acarol.* 7:63–68.

Childers, C. C., J. C. V. Rodrigues, and W. C. Welbourn. 2003. Host plants of *Brevipalpus californicus, B. obovatus*, and *B. phoenicis* (Acari: Tenuipalpidae). *Exp. Appl. Acarol.* 30:29–105.

Chinery, W. A. 1962. Cytochemical observation on the salivary glands of an ixodid tick. *Trans. Roy. Soc. Trop. Med. Hyg.* 56:268.

———. 1965. Studies on the various glands of the tick *Haemaphysalis spinigera* Neumann 1897. III, The salivary glands. *Acta Trop.* 22:321–49. [IXODIDAE]

———. 1973. The nature and origin of the "cement" substance at the site of attachment and feeding of adult *Haemaphysalis spinigera* (Ixodidae). *J. Med. Entomol.* 10:355–62. [IXODIDAE]

Chinniah, C., and M. Mohanasundaram. 1995. Record of new ascid mites (Ascidae: Acari) infesting insects in Tamil Nadu, India. *Entomon* 20:233–36. [includes BLATTISOCIIDAE]

———. 1998. Two new species of genus *Leptus* Latreille (Acari: Erythraeidae: Leptinae) from Tamil Nadu, India. *J. Insect Sci.* 11:1–4.

Chinone, S. 1974. Further contribution to the knowledge of the family Brachychthoniidae from Japan. *Bull. Biogeogr. Soc. Japan* 30:1–29.

———. 2003. Classification of the soil mites of the family Suctobelbidae (Oribatida) of Japan. *Edaphologia* 72:1–110.

Chinone, S., and J. Aoki. 1972. Soil mites of the family Brachychthoniidae from Japan. *Bull. Nat. Sci. Mus., Tokyo* 15:217–51.

Chmielewski, W. 1969. Obserwacje nad biologia nowego dla akarofauny polskiej gatunku *Kuzinia laevis* (Dujardin, 1849) (Acarina, Acaridae). *Polsk. Pismo Entomol.* 39:603–17.

———. 1971. Morfoligia, Biologia i Ekologia *Carpoglyphus lactis* (L., 1758) (Glycyphagidae, Acarina). *Prace Nauk. Instytut. Ochr. Roslin.* 13:63–166. [CARPOGLYPHIDAE]

Choi, S. S. 1996. Newly recorded oribatid mites (Acari: Oribatei) from Korea: I. *Korean Arachnol.* 12:103–10.

———. 1997. Checklist of oribatid mites of Korea. *Korean Arachnol.* 13:83–104.

Chow, Y. S., S. S. Tzean, C. S. Chang, and C. H. Wang. 1980. A morphological approach of the tarsonemid mite *Steneotarsonemus spinki* Smiley (Tarsonemidae) as a rice plant pest. *Acta Arachnol.* 29:25–41.

Christian, A., and W. Karg. 2006. The predatory mite genus *Lasioseius* Berlese, 1916 (Acari, Gamasina). *Abh. Ber. Naturk. Görlitz* 77:99–250.

Cicolani, B. 1979. The intrinsic rate of natural increase in dung macrochelid mites, predators of Musca domestica eggs. *Boll. Zool.* 46:171–78.

Cistjakov, M. P. 1979. Zur Oribatidenfauna (Oribatei) der Überfamilie Oribatelloidea (Woolley, 1956) des Oblast Gorki. *Fauna, Sist., Biol. Ekol. Gelmintov Promezutoc. Chozjaev, Gorky*: 83–87. [in Russian]

Clapperton, M. J., D. A. Kanashiro, and V. M. Behan-Pelletier. 2002. Changes in abundance and diversity of microarthropods associated with fescue prairie grazing regimes. *Pedobiologia* 46:496–511.

Clark, E. W., and F. Morishita. 1950. C-M medium: A mounting medium for small insects, mites and other whole mounts. *Science* 112:789–90.

Clark, G. M. 1960. Three new nasal mites (Acarina: Speleognathidae) from the gray squirrel, the common grackle, and the meadowlark in the United States. *Proc. Helminthol. Soc.* 27:103–10. [EREYNETIDAE]

Clark, G. M., and C. M. Clifford. 1964. *Zumptiella citelli* sp. n. (Acarina: Halarachnidae) from the Columbian ground squirrel, Citellus columbianus. *J. Parasitol.* 50:471–73.

Clark, L. 1991. The nest protection hypothesis: The adaptive use of plant secondary compounds by European starlings. In *Bird-parasite interactions: Ecology, evolution, and behaviour*, ed. J. E. Loye and M. Zuk, 205–21. New York: Oxford Univ. Press.

Clayton, D. H., and B. A.Walther. 1997. Collection and quantification of arthropod parasites of birds. In *Host-parasite evolution. General principles and avian models*, ed. D. H. Clayton and J. Moore, 419–40. New York: Oxford Univ. Press.

Clayton, D. H., and N. D. Wolfe. 1993. The adaptive significance of self-medication. *TREE* 8:60–63.

Clifford, C. M., and G. Anastos. 1960. The use of chaetotaxy in the identification of larval ticks (Acarina: Ixodidae). *J. Parasitol.* 46:567–78.

Clift, A. D., and S. F. Larsson. 1987. Phoretic dispersal of *Brennandania lambi* (Krczal [*sic*]) (Acari: Tarsonemida: Pygmephoridae) by mushroom flies (Diptera: Sciaridae and Phoridae) in New South Wales, Australia. *Exp. Appl. Acarol.* 3:11–20.

Clift, A. D., and R. B. Toffolon. 1981. Biology, fungal host preferences and economic significance of two pygmephorid mites (Acarina: Pygmephoridae) in cultivated mushrooms, N. S. W., Australia. *Mushroom Sci.* 11 (2): 245–53. [MICRODISPIDAE]

Clopton, R. E., and R. E. Gold. 1993. Distribution and seasonal and diurnal activity patterns of *Eutrombicula alfreddugesi* (Acari: Trombiculidae) in a forest edge ecosystem. *J. Med. Entomol.* 30:47–53.

Coetzee, L. 1987. The South African Lamellaraeidae Balogh, 1972 (Acari, Oribatei). *Navors. Nas. Mus., Bloemfontein* 5:326–53.

———. 1997. The antarctic mite genus *Maudheimia* (Acari, Oribatida). *Navors. Nas. Mus., Bloemfontein* 13 (3): 101–35. [MAUDHEIMIIDAE]

Coineau, Y. 1963. Contribution à l'étude des Caeculidae. Première série: Développement postlarvaire de *Allocaeculus catalanus* Franz 1954. Première partie: La chétotaxie du corps. *Acarologia* 5:189–212.

———. 1964a. Contribution à l'étude des Caeculidae. Première série: Développement postlarvaire de *Allocaeculus catalanus* Franz 1954. Deuxième partie: La chétotaxie des pattes. *Acarologia* 6:47–72.

———. 1964b. Un nouveau *Labidostoma* à pustules multiples: *Labidostoma jacquemarti* n. sp. (Labidostommidae Acari Prostigmata). *Rev. Ecol. Biol. Sol* 1:543–52.

———. 1964c. Une nouvelle espèce française de Labidostomidae (Acariens, Prostigmates). Eunicolina travei sp. n. *Vie Milieu* 15:153–75.

———. 1966. Contribution a l'étude des Caeculidae. Deuxième série: Développement postlarvaire de *Microcaeculus hispanicus* Franz 1952. *Acarologia* 8:23–44.

———. 1970. A propos de l'oeil antérieur du naso des Caeculidae. *Acarologia* 12:109–18.

———. 1973a. A propos de quelques caracteres particulierement primitifs de la prélarve et de la larve d'un Opilioacaridae du Gabon (Acariens). *C.R. Hebdomadaires Séanc. Acad. Sci. Paris*, ser. D, 276:1181–84.

———. 1973b. Les Caeculidae (Acariens Prostigmatides) quelques aspects de leurs particularités éco-éthologiques. *Bull. Ecol.* 4:329–37.

———. 1974a. Introduction a l'étude des microarthropodes du sol. In *Documents pour l'enseignement pratique de l'ecologie*, 57–83. Paris: DOIN, éditeurs.

———. 1974b. *Éléments pour une monographie morphologique, écologique et biologique des Caeculidae (Acariens)*. Mem. Mus. Nat. Hist. Natur., n.s. A, Zool., 81. Paris: Muséum national d'Histoire naturelle.

———. 1974c. Nouvelles techniques pour l'etude de la morphologie des formations chitineuses des acariens. *Acarologia* 16:4–10.

———. 1974d. Un type nouveau d'Acariens Prostigmates libres: les Saxidromoidea nouvelle super-famille. *C.R. Acad. Sci. Paris*, ser. D, 278:1059–62. [ADAMYSTIDAE]

———. 1976a. Les pariades sexuelles des Saxidrominae Coineau 1974 (Acariens Prostigmates, Adamystidae). *Acarologia* 28: 234–40.

———. 1976b. La première prélarve connue du genre *Eupodes, Eupodes strandtmanni*, n. sp. *Acarologia* 18:56–64. [EUPODIDAE]

———. 1979. Les Adamystidae, une étonnante famille d'acariens prostigmates primitifs. In Piffl (1979), 4:431–35.

———. 1980. Un exemple d'Actinedida libres—les Caeculidae. Gnathosoma et region dorsale du corps, morphologie et chétotaxie evolutives. *Salfia* 1:1–45.

Coineau, Y., and J. Haupt. 1978. Biogeographie, ecologie et chétotaxie evolutive a propos de *Microcaeculus stephani* n. sp. Caeculidae Sabulicole (Acari, Actinedida). *Acarologia* 19:622–32.

Coineau, Y., and R. Legendre. 1975. Sur un mode de régénération appendiculaire inédit chez les Arthropodes: La régénération des pattes marcheuses chez les Opilioacariens (Acari: Notostigmata). *C.R. Acad. Sci. Paris* 280:41–43.

Coineau, Y., and W. L. Magowski. 1994. Caeculidae in amber. *Acarologia* 35:243–46.

Coineau, Y., and M. H. Naudo. 1986. Contribution a l'étude de la morphologie et du développement postprélarvaire de *Saxidromus delamarei*, Y. Coineau, 1974. Première partie: Chétotaxie du corps et région buccale. *Acarologia* 27:303–9. [ADAMYSTIDAE]

Coineau, Y., and G. Poinar. 2001. Un Caeculidae de l'ambre de la Republique Dominicaine. *Acarologia* 41:141–44.

Coineau, Y., and P. Theron. 1983. Les Micropsammidae, n. fam. d'acariens Endeostigmata des sables fin. *Acarologia* 24:275–80.

Coineau, Y., P. D. Theron, and C. Valette. 1997. Une association de phanères, constituant probablement un système de communication sonore original chez les acariens. *Acarologia* 38:111–16.

Coineau, Y., and L. van der Hammen. 1979. The postembryonic development of Opilioacarida, with notes on new taxa and on a general model for the evolution. In Piffl (1979), 4:437–41.

Collins, N. C. 1975. Tactics of host exploitation by a thermophilic water mite. *Misc. Publ. Entomol. Soc. Amer.* 9:250–54. [HYDRYPHANTIDAE]

Collins, N. C., R. D. Mitchell, and R. G. Wiegert. 1976. Functional analysis of a thermal spring ecosystem, with an evaluation of the role of consumers. *Ecology* 57:1221–32. [HYDRYPHANTIDAE]

Colloff, M. J. 1993. A taxonomic revision of the oribatid mite genus *Camisia* (Acari: Oribatida). *J. Nat. Hist.* 27:1325–1408. [CAMISIIDAE]

Colloff, M. J., and R. B. Halliday. 1998. *Oribatid mites: A catalogue of Australian genera and species*. Monographs on Invertebrate Taxonomy, vol. 6. Collingswood, Victoria, Australia: CSIRO Publishing.

Colloff, M. J., and S. P. Hopkin. 1986. The ecology, morphology and behaviour of *Bakerdania elliptica* (Acari: Prostigmata: Pygmephoridae), a mite associated with terrestrial isopods. *J. Zool. Lond.*, ser. A, 208:109–23.

Colloff, M. J., and W. Niedbała. 1996. Arboreal and terrestrial habi-

tats of phthiracaroid mites (Oribatida) in Tasmanian rainforests. In Needham, Mitchell, et al., 2:607–11.
Collyer, E. 1969. Two species of *Tuckerella* (Acarina: Tuckerellidae) from New Zealand. *New Zeal. J. Sci.* 12:811–14.
———. 1973a. A new species of the genus *Dolichotetranychus* (Acari: Tenuipalpidae) from New Zealand. *New Zeal. J. Sci.* 16:747–49.
———. 1973b. New species of the genus *Tenuipalpus* (Acari: Tenuipalpidae) from New Zealand, with a key to the world fauna. *New Zeal. J. Sci.* 16:915–55.
Colwell, R. K. 1979. The geographical ecology of hummingbird flower mites in relation to their host plants and carriers. *Rec. Adv. Acarol.* 2:461–68. [MELICHARIDAE]
———. 1989. Hummingbirds of the Juan Fernandez Islands: Natural history, evolution and population status. *Ibis* 131:548–66.
———. 1995. Effects of nectar consumption by the hummingbird flower mite *Proctolaelaps kirmesi* on nectar availability in *Hamelia patens. Biotropica* 27:206–17.
Colwell, R. K., and S. Naeem. 1994. Life-history patterns of hummingbird flower mites in relation to host phenology and morphology. In Houck (1994b), 2:23–44.
Compton, G. L., and G. W. Krantz. 1978. Mating behavior and related morphological specialization in the uropodine mite, *Caminella peraphora. Science* 200:1300–1301.
Compton, S. F. 1993. One way to be a fig. *Afr. Entomol.* 1:151–58. [TARSONEMIDAE]
Cone, W. W. 1985. Mating and chemical communication. In Helle and Sabelis (1985a), 1A:243–51. [TETRANYCHIDAE]
Conijn, C. G. M., J. van Aartrijk, and I. Lesna. 1996. Flower bulbs. In Lindquist, Sabelis, and Bruin, 6:651–59.
Conrad, L., S. A. Corbet, T. J. Roper, and E. J. Bodsworth. 2002. Parasitism by the mite *Trombidium breei* on four U.K. butterfly species. *Ecol. Entomol.* 27:651–59. [TROMBIDIIDAE]
Convey, P. 1994. Growth and survival strategy of antarctic mite *Alaskozetes antarcticus. Ecography* 17:97–107. [AMERONOTHRIDAE]
Cook, D. R. 1957. Order Acarina. Suborder Hydracarina. Genus *Protoarrenurus* Cook, n. gen. In *Miocene arthropods from the Mojave Desert, California.* Professional Papers, Geol. Survey No. 294-G, ed. A. R. Palmer, 248–49. [ARRENURIDAE]
———. 1966. *The water mites of Liberia.* Mem. Amer. Entomol. Inst. 6. Gainesville, FL.
———. 1967. *Water mites from India.* Mem. Amer. Entomol. Inst. 9. Gainesvilla, FL.
———. 1974a. North American species of the genus *Axonopsis* (Acarina: Aturidae: Axonopsinae). *Great Lakes Entomol.* 7:55–79.
———. 1974b. *Water mite genera and subgenera.* Mem. Amer. Entomol. Inst. 21. Gainesville, FL.
———. 1975. North American species of the genus *Brachypoda* (Acarina: Aturidae: Axonopsinae). *Proc. Entomol. Soc. Wash.* 77:278–89.
———. 1976a. Contributions to the water mite fauna of North America. New North American species of the genus *Arrenurus*, mostly from Florida (Acarina: Arrenuridae). *Contr. Amer. Entomol. Inst.* 11:1–58.
———. 1976b. Contributions to the water mite fauna of North America. North American species of the genus *Koenikea* (Acarina: Unionicolidae). *Contr. Amer. Entomol. Inst.* 11:59–100.
———. 1976c. Contributions to the water mite fauna of North America. North American species of the genus *Mideopsis* (Acarina: Mideopsidae). *Contr. Amer. Entomol. Inst.* 11:101–48.
———. 1980. *Studies on Neotropical water mites.* Mem. Amer. Entomol. Inst. 31. Gainesville, FL.
———. 1981a. Amsterdam expeditions to the West Indian Islands, Report 11. New hyporheic water mites from Haiti. *Bijdragen Dierkunde* 51:135–44. [LIMNESIIDAE].
———. 1981b. New or incompletely known axonopsin water mites (Acarina: Aturidae) from western North America. *Internat. J. Acarol.* 7:91–104.
———. 1983. Rheophilic and hyporheic water mites of New Zealand. *Contr. Amer. Entomol. Inst.* 21:1–224.
———. 1984. Preliminary review of the relationships of the water mite fauna of South America, Australia and New Zealand. In Griffiths and Bowman, 2:959–64.
———. 1986. *Water mites from Australia.* Mem. Amer. Entomol. Inst. 40. Gainesville, FL.
———. 1988. *Water mites from Chile.* Mem. Amer. Entomol. Inst. 42. Gainesville, FL.
———. 1991. Water mites from driven wells in New Zealand, the subfamily Notoaturinae Besch (Acarina, Aturidae). *Stygologia* 6:235–53.
———. 1992. Water mites (Hydracarina) from driven wells in New Zealand: Taxa other than Notoaturinae Besch. *Stygologia* 7: 43–62. [STYGOTONIIDAE, PIONIDAE, MIDEOPSIDAE, ATHIENEMANNIIDAE]
———. 1996. A freshwater species of *Pontarachna* (Acari, Pontarachnidae) from South Africa, with a discussion of genital acetabula in the family. *An. Inst. Biol., Univ. Nac. Aut. Méx., Ser. Zool.* 67:259–64.
———. 1998. New genera and species of Notoaturinae (Acari, Aturidae) from South Africa. *Acarologia* 39:335–48.
———. 1999. *Axonopsalbia*-like water mites from South Africa (Acari, Hydrachnida, Aturidae). *Internat. J. Acarol.* 25:317–25.
———. 2000. New and incompletely known genera and species of Arrenuroidea (Acari, Hydracarina) from South Africa. *Acarologia* 41:233–48.
———. 2005. A new subgenus and two new species of *Hygrobatopsis* (Acari: Hydrachnida: Hygrobatidae) from South Africa. *Internat. J Acarol.* 31:249–54.
Cook, D. R., and C. L. Hopkins. 1998. New water mite species (Acari: Hydracarina) from New Zealand. *Acarologia* 39:257–63.
Cook, D. R., I. M. Smith, and M. S. Harvey. 2000. Assessment of lateral compression of the idiosoma in adult water mites as a taxonomic character and reclassification of *Frontipodopsis* Walter, *Wettina* Piersig and some other basal Hygrobatoidea (Acari: Hydrachnida). *Invertebr. Taxon.* 14:433–48.
Cooley, R. A. 1932. *The Rocky Mountain wood tick.* Montana St. Coll. Bull. 268. Bozeman.
Cooley, R. A., and G. M. Kohls. 1944. *The Argasidae of North America, Central America, and Cuba.* Amer. Midl. Nat. Monogr. 1. Notre Dame, IN: Univ. Notre Dame.
Coons, L. B., and G. Alberti. 1999. Acari: Ticks. In Harrison and Foelix, 8B:267–514.
Coons, L. B., and R. C. Axtell. 1971. Cellular organization in the synganglion of the mite *Macrocheles muscaedomesticae* (Acarina: Macrochelidae). *Z. Zellforsch.* 119:309–20.
———. 1973. Sensory setae of the first tarsi and palps of the mite *Macrocheles muscaedomesticae. Ann. Entomol. Soc. Amer.* 66:539–44.
Coons, L. B., and M. A. Roshdy. 1973. Fine structure of the salivary

glands of unfed *Dermacentor variabilis* (Say) (Ixodoidea, Ixodidae). *J. Parasitol.* 59:900–912.

Coons, L. B., M. A. Roshdy, and R. C. Axtell. 1974. Fine structure of the central nervous system of *Dermacentor variabilis* (Say), *Amblyomma americanum* (L.), and *Argas arboreus* Kaiser, Hoogstraal, and Kohls (Ixodoidea). *J. Parasitol.* 60:687–98.

Cooper, K. W. 1937. Reproductive behaviour and haploid parthenogenesis in the grass mite, *Pediculopsis graminum* (Reut.) (Acarina, Tarsonemidae). *Proc. Nat. Acad. Sci.* 23:41–44. [SITEROPTIDAE]

———. 1940. Relations of *Pediculopsis graminum* and *Fusarium poae* to bud rot of carnation. *Phytopathology* 30:853–59. [SITEROPTIDAE]

———. 1955. Venereal transmission of mites by wasps, and some evolutionary problems arising from the remarkable association of *Ensliniella trisetosa* with the wasp *Ancestrocerus antilope.* Biology of eumenine wasps II. *Trans. Amer. Entomol. Soc.* 80:119–74. [WINTERSCHMIDTIIDAE]

Cooreman, J. 1944. Notes et observations sur les acariens. III. *Bull. Inst. Roy. Sci. Nat. Belg.* 20:1–16. [EUTROMBIDIIDAE]

Corente, Ch., and W. Knülle. 1996. Induction of the hypopus stage by ingestion of different grain products in the astigmatic mite, *Lepidoglyphus destructor* (Schrank) (Glycyphagidae). In Mitchell et al. (1996), 657–62.

———. 2003. Trophic determinants of hypopus induction in the stored-product mite *Lepidoglyphus destructor* (Acari: Astigmata). *Exp. Appl. Acarol.* 29:89–107.

Corpuz-Raros, L. A. 1979. Philippine Oribatei (Acarina). I, Preliminary list of species and descriptions of forty new species. *Philipp. Agr.* 62:1–82.

———. 1989. A review of Philippine *Tuckerella* (Tuckerellidae, Acari). *Philipp. Entomol.* 7:557–60.

———. 2001. *Tuckerella filipina*, a new species of Tuckerellidae (Acari) from the Philippines. *Internat. J. Acarol.* 27:71–74.

Costa, M. 1961. Mites for the nests of the mole-rat (*Spalax ehrenbergi*) in Israel. *Ann. Mag. Nat. Hist.*, ser. 13:481–503. [ASTERNOSEIIDAE]

———. 1965. *Andregamasus* n. gen., a new genus of mesostigmatic mites associated with terrestrial hermit crabs. Israel South Red Sea Expedition, 1962, Report no. 6. Sea Fish. *Res. Sta. Haifa Bull.* 38:6–14.

———. 1966a. The biology and development of *Hypoaspis* (*Penumolaelaps*) *hyatti* (Acari: Mesostigmata). *J. Zool.* 148:191–200.

———. 1966b. Notes on macrochelids associated with manure and coprid beetles in Israel. I, *Macrocheles robustulus* (Berlese, 1904), development and biology. *Acarologia* 8:532–48.

———. 1967. Notes on macrochelids associated with manure and coprid beetles in Israel. II, Three new species of the *Macrocheles pisentii* complex with notes on their biology. *Acarologia* 9:304–29.

———. 1968. Notes on the genus *Hemipteroseius* Evans (Acari: Mesostigmata) with the description of a new species from Israel. *J. Nat. Hist.* 2:1–15. [OTOPHEIDOMENIDAE]

———. 1969a. *Antennoseius bytinskii* sp. nov., with notes on the genus *Antennoseius* Berlese (Acari: Mesostigmata) in Israel. *Israel J. Entomol.* 4:217–26. [ASCIDAE]

———. 1969b. The association between mesostigmatic mites and coprid beetles. *Acarologia* 11:411–28.

———. 1971. Mites of the genus *Pachylaelaps* Berlese (Acari: Mesostigmata, Pachylaelapidae) from litter in Israel. *Israel J. Zool.* 20:253–77.

———. 1974a. Mesostigmatic mites (Acari: Mesostigmata) from the Mediterranean shores of Israel. I, The genus *Hypoaspis* Canestrini, 1884. *Israel J. Entomol.* 9:219–28.

———. 1974b. A note on mites associated with *Heliocopris japetus* Klug from Kenya. *Entomol. Mitt. Zool. Mus. Hamb.* 4 (86): 481–97. [PACHYLAELAPIDAE]

———. 1979. *Cerambylaelaps nadchatrami*, n. gen., n. sp., an unusual mesostigmatic mite (Acari) associated with a cerambycid beetle in Malaysia. *Acarologia* 20:188–95.

———. 1980. *Blaberolaelaps matthiesensis* gen. n, sp. n., a new laelapine mite associated with the cockroach *Blaberus piracicabensis* in Brazil. *Rev. Bras. Biol.* 40:547–51. [LAELAPIDAE]

Costa, M., and P. G. Allsopp. 1979. Gamasine mites associated with Australian scarabaeid beetles. I, The genus *Paradoxiphis* Berlese, symbionts of the Bolboceratini. *Aust. J. Zool.* 27:825–65.

———. 1981. Gamasine mites associated with Australian scarabaeid beetles. II, The genus *Cosmetolaelaps* Womersley, symbionts of *Onthophagus* Latreille. *Aust. J. Zool.* 29:269–87.

Costa, M., and P. E. Hunter. 1970. The genus *Coleolaelaps* Berlese, 1914 (Acarina: Mesostigmata). *Redia* 52:323–60. [LAELAPIDAE]

Coulson, S. J. 2007. Terrestrial and freshwater invertebrate fauna of the high arctic archipelago of Svalbard. *Zootaxa* 1448:41–58.

Coulson, S. J., H. P. Leinaas, R. A. Ims, and G. Søvik. 2000. Experimental manipulation of the winter surface ice layer: The effects on a high arctic soil microarthropod community. *Ecography* 23:299–306.

Covarrubias, R. 1968a. Observations sur le genre Pheroliodes. I, *Pheroliodes roblensis* n. sp. (Acarina, Oribatei). *Acarologia* 10:657–95. [PLATEREMAEIDAE]

———. 1968b. Some observations on antarctic Oribatei (Acarina). *Liochthonius australis* sp. n., and two *Oppia* ssp. n. *Acarologia* 10:313–56. [BRACHYCHTHONIIDAE, OPPIIDAE]

———. 1969. Observations sur le genre *Mikizetes* (Oribates). *Acarologia* 11:828–46. [ZETOMOTRICHIDAE]

———. 1986. Estado actual de nuestros conocimientos sobre los Acaros Oribatida de Chile. *Acta Entomol. Chilena* 13:167–75.

Covarrubias, R., and I. Mellado. 1998. Oribátidos de Chile (Acarina: Oribatida). II, Especies asociadas a plantas acuaticas. *Acta Entomol. Chilena* 22:37–44.

Cowan, D. P. 1985. Life history and male dimorphism in the mite *Kennethiella trisetosa* (Acarina: Winterschmidtiidae), and its symbiotic relationship with the wasp *Ancistrocerus antilope*. *Ann. Entomol. Soc. Amer.* 77:725–32.

Craemer, C., R. Sobhian, A. S. McClay, and J. W. Amrine, Jr. 1999. A new species of *Cecidophyes* (Acari: Eriophyidae) from *Galium aparine* (Rubiaceae) with notes on its biology and potential as a biological control agent for *Galium spurium*. *Internat. J. Acarol.* 25:255–63.

Cramer, C. 1986. Primer registro de la familia Feltriidae (Acarida: Hydrachnellae) para México y descripción de una especie nueva. *Folia Entomol. Mex.* 67:37–43.

———. 1987. Nuevas especies Mexicanas de hidracáridos, pertenecientes a los generos *Neomamersa, Kawamuracarus* y *Neotorrenticola* (Acarida: Limnesiidae). *Folia Entomol. Mex.* 73:185–203.

———. 1992a. Estudios sobre hidracáridos Mexicanos. Especies nuevas de la familia Aturidae. II, *Aturus, Stygalbiella. Folia Entomol. Mex.* 84:113–29.

———. 1992b. Estudios sobre hidracáridos Mexicanos, familia Torrenticolidae. I, Cinco especies nuevas de *Neoatractides* y *Torrenticola* y primer registro de *Testudacarus* para México. *An. Inst. Biol., Univ. Nac. Aut. Méx., Ser. Zool.* 63:13–27.

———. 2000. New species of the water mite genus *Konigsbergia* (Acari: Hydrachnida: Aturidae) from Mexico. *Internat. J. Acarol.* 26:271–78.

Cramer, C., and D. R. Cook. 1991. A new species of *Miraxonides* (Aturidae, Axonopsinae) from Mexico. *Internat. J. Acarol.* 17:13–16.

———. 1992a. New species of *Arrenurus* (Acari, Arrenuridae) from Mexican lakes. *Acarologia* 33:349–66.

———. 1992b. New species of *Arrenurus* (*Dadayella*) new rank (Acari, Arrenuridae) from Mexico with a discussion of the latter's relationships. *Internat. J. Acarol.* 18:221–29.

———. 1996. A new species of *Stygarrenurus* (Acari, Hungarohydracaridae) and a discussion of its systematic position. *Internat. J. Acarol.* 22:29–32.

———. 2000. Water mites of the genus *Neoatractides* Lundblad and *Pseudotorrenticola* Walter (Acari, Hydrachnida, Torrenticolidae) from Mexico. *Internat. J. Acarol.* 26:51–61.

Cramer, C., and I. M. Smith. 1991. New species of water mites (Acari: Hydrachnida) of the genera *Bandakia* (Anisitsiellidae), *Chappuisides* (Chappuisididae), and *Neoacarus* (Neoacaridae) from Mexico. *Can. Entomol.* 123:795–809.

———. 1993. A new species of the genus *Mamersellides* Lundblad, 1937 (Acari: Hydrachnida) with remarks on the family Anisitsiellidae. *Can. Entomol.* 125:769–83.

Cramer-Hemkes, C., and M. del C. Letechipia-Torres. 1996. Especie nueva de ácaro acuático (Acari: Hydracarina) del género *Diamphidaxona*, de arroyos Mexicanos. *An. Inst. Biol., Univ. Nac. Aut. Méx., Ser. Zool.* 67:279–85. [HYGROBATIDAE]

Croft, B. A. 1990. *Arthropod biological control agents and pesticides.* New York: Wiley Interscience.

Crommentuijn, T., C. J. A. M. Doodeman, J. J. C. van der Pol, A. Doornekamp, M. C. J. Rademaker, and C. A. M. van Gestel. 1995. Sublethal sensitivity index as an ecotoxicity parameter measuring energy allocation under toxicant stress: Application to cadmium in soil arthropods. *Ecotox. Environ. Safe.* 31: 192–200.

Cromroy, H. L. 1983. Potential use of mites in biological control of terrestrial and aquatic weeds. In Hoy, Cunningham, and Knutson, 61–66.

Crop protection compendium, global module. 1999. CAB International. CD-ROM. [TARSONEMIDAE]

Cross, E. A. 1965. The generic relationships of the family Pyemotidae (Acarina: Trombidiformes). *Univ. Kans. Sci. Bull.* 45:29–275.

Cross, E. A., and G. E. Bohart. 1969. Phoretic behavior of four species of alkali bee mites as influenced by season and host sex. *J. Kans. Ent. Soc.* 42:195–219. [TROCHOMETRIDIIDAE]

———. 1978. Some observations of the habits and distribution of *Trochometridium* Cross, 1965 (Acarina: Pyemotidae). *Acarologia* 20:286–93. [TROCHOMETRIDIIDAE]

———. 1992. The biology of *Imparipes apicola* (Acari: Scutacaridae) and its relationships to the alkali bee, *Nomia melanderi* (Hymenoptera: Halictidae), and to certain fungi in the bee cell ecosystem. *J. Kans. Entomol. Soc.* 65:157–73.

Cross, E. A., and M. J. Kaliszewski. 1988. The life history of a mushroom pest mite, *Pediculaster flechtmanni* (Wicht) (Acari: Pygmephoroidea), with studies of alternate morph formation. *Environ. Entomol.* 17:309–15. [PYGMEPHORIDAE]

Cross, E. A., and G. W. Krantz. 1964. Two new species of the genus *Acarophenax* Newstead and Duvall 1918 (Acarina: Pyemotidae). *Acarologia* 6:287–95. [ACAROPHENACIDAE]

Cross, E. A., and J. C. Moser. 1971. Taxonomy and biology of some Pyemotidae (Acarina: Tarsonemoidea) inhabiting bark beetle galleries in North American conifers. *Acarologia* 13:47–64.

———. 1975. A new dimorphic species of *Pyemotes* and a key to previously described forms (Acarina: Tarsonemoidea). *Ann. Entomol. Soc. Amer.* 68:723–32. [PYEMOTIDAE]

Cross, H. F. 1964. Observations on the formation of the feeding tube by *Trombicula splendens* larvae. *Acarologia* 6 (fasc. h.s.): 255–61. [TROMBICULIDAE]

Crossley, D. A. 1960. Comparative external morphology and taxonomy of nymphs of the Trombiculidae (Acarina). *Univ. Kans. Sci. Bull.* 40:135–321.

———. 1977. Oribatid mites and nutrient cycling. In Dindal (1977), 71–85.

Crossley, D. A., Jr., and V. A. Merchant. 1971. Feeding by caeculid mites on fungus demonstrated with radioactive tracers. *Ann. Entomol. Soc. Amer.* 64:760–62.

Crowe, J. H., and C. G. Camara. 1973. Studies on acarine cuticles. I, Cuticular pores in a marine mite. *Comp. Biochem. Physiol.* 45A:757–66.

Crowell, R. M. 1984. Scanning electron-microscopic distinctions among the sponge-associated water mites (Acarina: Unionicolidae). In Griffiths and Bowman, 2:965–70.

———. 1990. *Unionicola billieaehonore* new species, a sponge-associated Hydracarina (Acari, Unionicolidae) from New Zealand. *New Zeal. J. Zool.* 17:265–70.

Crowell, R. M., and C. Davids. 1979a. The developmental cycle of sponge-associated mites. In Rodriguez, 1:563–65. [UNIONICOLIDAE].

———. 1979b. Systematics of *Unionicola laurentiana*, n. sp., and *U. nearctica*, n. sp., sponge-associated Hydracarina (Parasitengona: Unionicolidae) from North America. *Ohio J. Sci.* 79:178–86.

Crozier R. H., and P. Pamilo. 1996. *Evolution of social insect colonies: Sex allocation and kin selection.* Oxford, UK: Oxford University Press.

Cruickshank, R. H., and R. H. Thomas. 1999. Evolution of haplodiploidy in dermanyssine mites (Acari: Mesostigmata). *Evolution* 53:1796–1803.

Cummins, K. W. 1974. Structure and function of stream ecosystems. *Bioscience* 24:631–41.

Cunliffe, F. 1949a. *Hirstiella pelaezi*, a new lizard parasite from Mexico (Acarina, Pterygosomidae). *Proc. Entomol. Soc. Wash.* 51: 25–26.

———. 1949b. *Pimeliaphilus isometri*, a new scorpion parasite from Manila, P.I. *Proc. Entomol. Soc. Wash.* 51:123–24. [PTERYGOSOMATIDAE]

———. 1952. Biology of the cockroach parasite, *Pimeliaphilus podapolipophagus* Tragardh, with a discussion of the genera *Pimeliaphilus* and *Hirstiella*. *Proc. Entomol. Soc. Wash.* 54:153–69. [PTERYGOSOMATIDAE]

———. 1955. A proposed classification of the trombidiforme mites. *Proc. Entomol. Soc. Wash.* 57:209–18.

———. 1956. A new species of *Nematalycus* Strenzke with notes on the family (Acarina, Nematalycidae). *Proc. Entomol. Soc. Wash.* 58:353–55.

———. 1957. Notes on the Anystidae with a description of a new genus and species, *Adamystris donnae*, and a new subfamily Adamystinae (Acarina). *Proc. Entomol. Soc. Wash.* 59:173–75.

Cunningham, J. R., S. E. Kunz, and M. A. Price. 1976. Effects of selected environmental factors on damaging populations of *Neoschongastia americana*. *J. Econ. Entomol.* 69:161–64. [TROMBICULIDAE]

Cuy, L. 1979. Synopsis of Philippine Pterygosomidae (Acarina: Prostigmata). *Kalikasan* 8:155–61.

Czajkowska, B., and D. Kropczynska. 1991. The influence of different host plants on the reproductive potential of *Tyrophagus putrescentiae* Schrank and *Tyrophagus neiswanderi* Johnston and Bruce (Acaridae). In Schuster and Murphy, 313–18.

Dabert, J. 1994. Kiwilichidae fam. nov. eine neue Federmilbenfamilie (Astigmata, Pterolichoidea). *Entomol. Mitt. Zool. Mus. Hamb.* 11 (149): 101–10.

———. 2003a. The feather mite family Syringobiidae Trouessart, 1896 (Acari, Astigmata, Pterolichoidea). I, Systematics of the family and descriptions of new taxa. *Acta Parasitol. Warsaw* 48 (Suppl.): 1–184.

———. 2003b. The feather mite family Syringobiidae Trouessart, 1896 (Acari, Astigmata, Pterolichoidea). II, Phylogeny and host-parasite evolutionary relationships. *Acta Parasitol. Warsaw* 48 (Suppl): 185–233.

Dabert, J., M. Dabert, and S. V. Mironov. 2001. Phylogeny of feather mite subfamily Avenzoariinae (Acari: Analgoidea: Avenzoariidae) inferred from combined analyses of molecular and morphological data. *Mol. Phylo. Evol.* 20:124–35.

Dabert, J., and R. Ehrnsberger. 1992. Neue Arten bei der Federmilbenfamilie Ascouracaridae Gaud & Atyeo, 1976. *Osnabrück. Naturwiss. Mitt.* 18:109–50.

———. 1998. Phylogeny of the feather mite family Ptiloxenidae Gaud, 1982 (Acari: Pterolichoidea). *Biosyst. Ecol. Ser.* 14: 145–78.

Dabert, J., and S. V. Mironov. 1999. Origin and evolution of feather mites (Astigmata). *Exp. Appl. Acarol.* 23:437–54. (Also appears in Bruin, van der Geest, and Sabelis, 89–103.)

Daniel, M. 1983. Trombiculosis—a serious problem of urbanized areas. *Med. Parazitol. Parazit. Bolezni* 5:52–57. [TROMBICULIDAE]

Daniel, M., and B. Rosický, eds. 1973. *Proceedings 3rd International Congress of Acarology, Prague*. The Hague: W. Junk b.v.

Dastych, H., and G. Rack. 1993. *Spatulaphorus* Rack gen. n. and three new species of pygmephorid mites associated with scarab beetles in Botswana and Vietnam (Acari: Heterostigmata). *Mitt. Hamb. Zool. Mus. Inst.* 90:265–84.

Dastych, H., G. Rack, and N. Wilson. 1992. Notes on mites of the genus *Pygmephorus* (Acari: Heterostigmata) associated with North American mammals (Part II). *Mitt. Hamb. Zool. Mus. Inst.* 89:141–56. [PYGMEPHORIDAE]

Datta, A. K., and P. C. Bhattacharjee. 1988. *Krantzholaspina rebatii*, a new genus and a new species (Acari: Mesostigmata: Parholaspididae) from Dibrugarh, Assam, India. In ChannaBassavana and C. A. Viraktamath, 1:411–14.

Davids, C. 1973a. The relations between mites of the genus *Unionicola* and mussels *Anodonta* and *Unio*. *Hydrobiologia* 41:37–44. [UNIONICOLIDAE]

———. 1973b. The water mite *Hydrachna conjecta* Koenike, 1895 (Acari, Hydrachnellae), bionomics and relation to species of Corixidae (Hemiptera). *Neth. J. Zool.* 23:363–429. [HYDRACHNIDAE]

Davids, C., and R. Belier. 1979. The spermatophores of *Hydrachna conjecta* Koenike and the life history of the land-living ancestors of this water mite. In Piffl (1979), 147–51. [HYDRACHNIDAE]

Davids, C., R. M. Crowell, and C. J. de Groot. 1985. The developmental cycles of two co-occurring sponge mites *Unionicola crassipes* (Müller) and *Unionicola minor* (Soar) (Acari, Hydrachnellae). *Hydrobiologia* 122:199–205. [UNIONICOLIDAE]

Davids, C., J. Holtslag, and R. V. Dimock, Jr. 1988. Competitive exclusion, harem behaviour and host specificity of the water mite *Unionicola ypsilophora* (Hydrachnellae, Acari) inhabiting *Anodonta cygnea* (Unionidae). *Internat. Rev. Hydrobiol.* 73:651–57. [UNIONICOLIDAE]

Davids, C., and C. J. Schoots. 1975. The influence of the water mite species *Hydrachna conjecta* and *H. cruenta* (Acari, Hydrachnellae) on the egg production of the Corixidae *Sigara striata* and *Cymatia coleoptrata* (Hemiptera). *Verh. Internat. Verein. Limnol.* 19:3079–82. [HYDRACHNIDAE]

Davids, K., A. Di Sabatino, R. Gerecke, T. Gledhill, and H. Smit. 2005. On the taxonomy of water mites (Acari: Hydrachnidia) described from the Palearctic. Part 1, Hydrachnidae, Limnocharidae and Eylaidae. *Zootaxa* 1061:36–64.

Davidson, J. A. 1958. A new species of lizard mite and a generic key to the family Pterygosomidae (Acarina, Anystoidea). *Proc. Entomol. Soc. Wash.* 60:75–78.

Davis, A. J. 1986. Bibliography of the Ixodiphagini (Hymenoptera, Chalcidoidea, Encyrtidae), parasites of ticks (Acari, Ixodidae), with notes on their biology. *Tijdschr. Entomol.* 129:181–90.

Davis, R. 1961. A mite, *Allothrombium mitchelli*, new to science, predator on the balsam woolly aphid. *Proc. Entomol. Soc. Wash.* 63:269–72. [TROMBIDIIDAE]

de Guzman, L. I., and M. Delfinado-Baker. 1996. A new species of *Varroa* (Acari: Varroidae) associated with *Apis koschevnikovi* (Apidae: Hymenoptera) in Borneo. *Internat. J. Acarol.* 22: 23–27.

DeGeer, C. 1778. *Memoires pour servir a l'histoire des insectes*. Stockholm.

DeJong, D., R. A. Morse, and G. C. Eickwort. 1982. Mite pests of honey bees. *Ann. Rev. Entomol.* 27:229–52. [VARROIDAE]

de la Cruz, J. 1984. Sistemática de la familia Pterygosomidae (Acarina: Prostigmata), con la descripción de un nuevo género y especie. *Poeyana* 278:1–22.

de la Cruz, K-D., M. Morando, and I. L. Avila. 2004. Description of a new pterygosomatid mite (Acari, Actinedida: Pterygosomatidae) parasitic on *Liolaemus* spp. (Iguania: Liolaemini) from Argentina. *Zootaxa* 521:1–6.

Delamare Deboutteville, C. 1960. *Biologie des eaux souterraines littorales et continentales*. Chapter XIII, *Arthropodes autres que les crustacés: Tardigrades, pycnogonides, acariens, collemboles*. Paris: Hermann.

De Leon, D. 1959. A new genus of mites occurring in Florida and Mexico (Acarina: Caligonellidae). *Fla. Entomol.* 42:17–19.

———. 1962. Three new genera and seven new species of cheyletids (Acarina: Cheyletidae). *Fla. Entomol.* 45:129–37.

———. 1963. A new genus and twelve new species of mites from Mexico and southeast United States (Acarina: Blattisocidae). *Fla. Entomol.* 46:197–207. [includes MELICHARIDAE]

———. 1964a. Four new *Ameroseius* Berlese 1903 from the United States (Acarina: Ameroseiidae). *Fla. Entomol.* 47:213–17.

———. 1964b. Two new *Podocinum* from the United States with distribution notes on three described species (Acarina: Podocinidae). *Fla. Entomol.* 47:39–41.

———. 1967. *Some mites of the Caribbean area*. Lawrence, KS: Allen Press. [ASCIDAE, PHYTOSEIIDAE, CAMEROBIIDAE, TENUIPALPIDAE, CHEYLETIDAE]

Del Fosse, E. S., H. L. Cromroy, and D. H. Habek. 1975. Determination of the feeding mechanism of the waterhyacinth mite. *Hyacinth Contr. J.* 13:53–55. [GALUMNIDAE]

Delfinado, M. D., and E. W. Baker. 1961. *Tropilaelaps*, a new genus of mite from the Philippines (Laelaptidae s. lat.: Acarina). *Fieldiana, Zool.* 44:53–65.

———. 1974. Terrestrial mites of New York (Acarina: Prostigmata). I, Tarsocheylidae, Paratydeidae, and Pseudocheylidae. *J. New York Entomol. Soc.* 82: 202–11.

———. 1976. New species of Scutacaridae (Acarina) associated with insects. *Acarologia* 18:264–301.

———. 1982. Notes on honey bee mites of the genus *Acarapis* Hirst (Acari: Tarsonemidae). *Internat. J. Acarol.* 8:211–26.

Delfinado, M. D., E. W. Baker, and M. J. Abbatiello. 1976. Terrestrial mites of New York. III, The family Scutacaridae. *J. New York Entomol. Soc.* 84:106–45.

Delfinado-Baker, M., and K. Aggarwal. 1987. A new *Varroa* (Acari: Varroidae) from the nest of *Apis cerana* (Apidae). *Internat. J. Acarol.* 13:233–37.

Delfinado-Baker, M. D., E. W. Baker, and C. H. W. Flechtmann. 1984. Acari domum meliponinarum brasiliensium habitantes. V, Two new genera and species of Laelapidae (Mesostigmata: Acari) from stingless bee nests. *Internat. J. Acarol.* 10:3–10.

Delfinado-Baker, M. D., E. W. Baker, and A. C. G. Phoon. 1989. Mites (Acari) associated with bees (Apidae) in Asia, with description of a new species. *Amer. Bee J.* 129:609–10, 612–13. [AMEROSEIIDAE]

de Lillo, E. 1996. *Siteroptes avenae* (Müller) (Acari Siteroptidae) e *Fusarium* spp. (Hyphomycetes Tuberculariaceae). *Entomologica, Bari* 30:7–18.

de Lillo, E., and P. Aldini. 2002. Fine morphology of the bothridial receptor organs of *Pediculaster mesembrinae* (Acari: Siteroptidae) adult female. In Bernini et al. (2002), 253–61.

de Lillo, E., and J. W. Amrine, Jr. 1998. Eriophyoidea (Acari) on a computer database. *Entomologica* 32:7–21.

de Lillo, E., A. Di Palma, and G. Nuzzaci. 2001. Morphological adaptations of mite chelicerae to different trophic activities (Acari). *Entomologica, Bari* 35:125–80.

de Lillo, E., and C. Duso. 1996. Currants and berries. In Lindquist, Sabelis, and Bruin, 6:583–91.

de Lillo, E., and G. Nuzzaci. 2005. Sensorial structures in mites and perspectives of research. In Weigmann et al. (2005), 59–81.

de Moraes, G. J., E. E. Lindquist, and A. C. Lofego. 2002. A new genus and species of tarsonemid mite (Acari: Tarsonemidae) associated with a Neotropical curculionid beetle (Coleoptera). *Invertebr. Syst.* 16:687–95.

de Moraes, G. J., J. A. McMurtry, and H. A. Denmark. 1986. *A catalog of the mite family Phytoseiidae: References to taxonomy, synonymy, distribution and habitat.* Brasilia: EMBRAPA–DDT.

de Moraes, G. J., R. S. Neto, and H. C. S. Pinto. 1989. Morphology, biology and pesticide tolerance of *Chelotogenes ornatus* [Acari: Cheyletidae]. *Entomophaga* 34:477–84.

Den Heyer, J. 1974. A new genus *Cunabdella* (Prostigmata: Acari) with a description of a new species from the Ethiopian region. *Acarologia* 16:664–70.

———. 1976. *Scutascirus*, a new cunaxid genus (Prostigmata: Acari) from South Africa. *Wetensk. Bydraes P.U. vir C.H.O.*, ser. B, 92:1–10.

———. 1977a. A new genus *Neoscirula* (Cunaxidae: Prostigmata: Acari) from the Ethiopian region. *J. Entomol. Soc. S. Afr.* 40:73–86.

———. 1977b. Six new species of *Pseudobonzia* Smiley, 1975 (Prostigmata: Acari) from the Ethiopian region. *J. Entomol. Soc. S. Afr.* 40:171–94.

———. 1978a. Bonzinae, a new subfamily of the Cunaxidae (Prostigmata: Acari). *Acarologia* 19:601–18.

———. 1978b. Four new species of *Armascirus* gen. nov. (Prostigmata: Acari) from the Ethiopian region. *J. Entomol. Soc. S. Afr.* 41:217–39.

———. 1979a. Coleoscirinae, a new cunaxid subfamily and two new South African species of *Coleoscirus* Berlese, 1916 (Prostigmata: Acari). *Acarologia* 20:522–41.

———. 1979b. Descriptions of seven African species of *Cunaxa* von Heyden, 1826 (Actinedida: Acarida), with remarks on the genus. *Phytophylactica* 11:23–42.

———. 1979c. Notes on the cunaxid genus *Dactyloscirus* (Actinedida: Acarida) with descriptions of two new species from the Ethiopian region. *Phytophylactica* 11:87–98.

———. 1979d. *Rubroscirus*, a new cunaxid genus (Prostigmata: Acari) with three new species from the Ethiopian region. *Acarologia* 20:70–92.

———. 1980a. A classification system for the family Cunaxidae (Actinedida: Acarida). *Publ. Univ. of the North,* ser. A23:1–12.

———. 1980b. Six new species of the subfamily Coleoscirinae (Cunaxidae: Actinedida: Acarida). *Phytophylactica* 12:105–28.

———. 1981. Systematics of the family Cunaxidae Thor, 1902 (Actinedida: Acarida). *Publ. Univ. of the North*, ser. A24:1–19.

Den Heyer, J., and P. A. J. Ryke. 1970. The mite complex and associated insects on citrus trees at Zebediela Estates. *Wetensk. Bydraes P.U. vir C.H.O., Reeks B: Natuurwetensk.* 32:1–23.

Denegri, G. M. 1993. Review of oribatid mites as intermediate hosts of the Anoplocephalidae. *Exp. Appl. Acarol.* 17:567–80.

Denegri, G. M., W. Bernadina, J. Perez-Serrano, and F. Rodriguez-Caabeiro. 1998. Anoplocephalid cestodes of veterinary and medical significance: A review. *Folia Parasitol.* 45:1–8.

Denegri, G. M., M. C. Elissondo, and M. C. Dopchiz. 2002. Oribatid mites as intermediate hosts of *Thysanosoma actinioides* (Cestoda: Anoplocephalidae): A preliminary study. *Vet. Parasitol.* 103:267–71.

Denmark, H. A. 1978. *A mite,* Brevipalpus russulus *(Boisduval) in Florida (Acarina: Tenuipalpidae)*. Fla. Dept. Agr. Consumer Serv., Div. Plant Indust., Entomol. Circ. 188. Gainesville: Fla. Dept. Agr.

Denneman, C. A. J., and N. M. van Straalen. 1991. The toxicity of lead and copper in reproduction tests using the oribatid mite *Platynothrus peltifer*. *Pedobiologia* 35:305–11. [CAMISIIDAE]

Desch, C. E., Jr. 1984. The reproductive anatomy of *Demodex folliculorum* (Simon). In Griffiths and Bowman, 1:464–69.

Desch, C. E., and W. B. Nutting. 1972. *Demodex folliculorum* (Simon) and *D. brevis* Akbulatova of man: Redescription and reevaluation. *J. Parasitol.* 58:169–77. [DEMODICIDAE]

———. 1973. Demodicids (Trombidiformes: Demodicidae) of

medical and veterinary importance. In Daniel and Rosický, 499–505.

———. 1978. Morphology and functional anatomy of *Demodex folliculorum* (Simon) of man. *Acarologia* 19:422–62. [DEMODICIDAE]

Desch, C. E., W. B. Nutting, and F. Lukoschus. 1972. Parasitic mites of Surinam. VII, *Demodex longissimus* n. sp. from *Carollia perspicillata* and *D. molossi* n. sp. from *Molossus molossus* (Demodicidae: Trombidiformes): Meibomian complex inhabitants of Neotropical bats (Chiroptera). *Acarologia* 14:35–53.

Dhiman, S. C., and R. K. Singh. 1988. Studies of the parasitism of *Leptocoris augur* Fabr. (Heteroptera: Rhopalidae) by *Leptus* sp. of mite (Acarina: Erythraeidae). *Uttar Pradesh J. Zool.* 8:180–85.

Di Palma, A. 1995. Morfologia funzionale delle parti boccali di *Penthaleus major* (Dugès) (Eupodoidea: Penthaleidae). *Entomologica, Bari* 29:69–86.

Di Palma, A., and G. Alberti. 2001. Fine structure of the female genital system in phytoseiid mites with remarks on egg nutrimentary development, sperm-access system, sperm transfer, and capacitation (Acari, Gamasida, Phytoseiidae). *Exp. Appl. Acarol.* 25:525–91.

Di Palma, A., G. Alberti, G. Nuzzaci, and G. W. Krantz. 2006. Fine structure and functional morphology of the mouthparts of a male *Veigaia* sp. (Gamasida: Veigaiidae) with remarks on the spermatodactyl and related sensory structures. *J. Morphol.* 267:208–20.

Díaz, A., K. Okabe, C. J. Eckenrode, M. G. Villani, and B. M. OConnor. 2000. Biology, ecology, and management of the bulb mites of the genus *Rhizoglyphus* (Claparède) (Acari: Acaridae). *Exp. Appl. Acarol.* 24:85–113.

Dicke, M., and P. Grostal. 2001. Chemical detection of natural enemies by arthropods: An ecological perspective. *Ann. Rev. Ecol. Syst.* 32:1–23.

Dimock, R. V., Jr. 1983. In defense of the harem: Intraspecific aggression by male water mites (Acari: Unionicolidae). *Ann. Entomol. Soc. Amer.* 76:463–65.

———. 1985. Population dynamics of *Unionicola formosa* (Acari: Unionicolidae), a water mite with a harem. *Amer. Midl. Nat.* 114:168–79.

———. 1988. Photobiology and behavior of the water mite genus *Unionicola*: A minireview. *Comp. Biochem. Physiol.* 91:193–98.

Dimock, R. V., Jr., and C. Davids. 1985. Spectral sensitivity and photo-behaviour of the water mite genus *Unionicola*. *J. Exper. Biol.* 119:349–63.

Dimock, R. V., Jr., and P. B. LaRochelle. 1980. Chemically mediated host recognition: a behavioral basis for the specificity of water mite symbioses. *Amer. Zool.* 20:922. [UNIONICOLIDAE]

Dindal, D. L., ed. 1977. *Biology of oribatid mites.* Syracuse: SUNY.

———, ed. 1990. *Soil biology guide.* New York: John Wiley & Sons.

Dobson, S. J., and S. C. Barker. 1999. Phylogeny of the hard ticks (Ixodidae) inferred from 18S rRNA indicates the genus *Aponomma* is paraphyletic. *Mol. Phylogen. Evol.* 11:288–95.

Domes, K., M. Althammer, R. A. Norton, S. Scheu, and M. Maraun. 2007. The phylogenetic relationship between Astigmata and Oribatida (Acari) as indicated by molecular markers. *Exp. Appl. Acarol.* 42:159–171.

Domes, K., R. A. Norton, M. Maraun, and S. Scheu. 2007. Reevolution of sexuality breaks Dollo's law. *Proc. Nat. Acad. Sci* 104:7139–44.

Domrow, R. 1955. A second species of *Holothyrus* (Acarina: Holothyroidea) from Australia. *Proc. Linn. Soc. N. S. W.* 74:159–62.

———. 1956. The family Discozerconidae (Acarina, Mesostigmata) in Australia. *Proc. Linn. Soc. N. S. W.* 81:193–96.

———. 1957. Some Acarina Mesostigmata from the Great Barrier Reef. *Proc. Linn. Soc. N. S. W.* 81:197–216.

———. 1969. The nasal mites of Queensland birds (Acari: Dermanyssidae, Ereynetidae and Epidermoptidae). *Proc. Linn. Soc. N. S. W.* 93:297–426.

———. 1972. Acari Spinturnicidae from Australia and New Guinea. *Acarologia* 13:552–84.

———. 1975a. *Ljunghia* Oudemans (Acari: Dermanyssidae), a genus parasitic on mygalomorph spiders. *Rec. S. Aust. Mus.* 17:31–39. [IPHIOPSIDIDAE]

———. 1975b. A new species of *Paraspeleognathopsis* Fain (Acari: Ereynetidae) from an Australian dasyurid marsupial. *J. Aust. Entomol. Soc.* 14:97–99.

———. 1976. Cercomegistina in Australia. *J. Austral. Entomol. Soc.* 15:201–5. [CERCOMEGISTIDAE]

———. 1979a. Ascid and ameroseiid mites phoretic on Australian mammals and birds. *Rec. W. Aust. Mus.* 8:97–116.

———. 1979b. Dermanyssine mites from Australian birds. *Rec. W. Aust. Mus.* 7:403–13. [DERMANYSSIDAE]

———. 1980. The genus *Raillietia* Trouessart in Australia (Acari: Dermanyssidae). *Proc. Linn. Soc. N. S. W.* 104:183–93. [IPHIOPSIDIDAE]

———. 1981. A small lizard stifled by phoretic deutonymphal mites (Uropodina). *Acarologia* 22:247–52.

———. 1983. Acari from Operation Drake in New Guinea. 1, Pterygosomatidae. *Acarologia* 24:393–402.

———. 1987. Acari Mesostigmata parasitic on Australian vertebrates: an annotated checklist, keys and bibliography. *Invertebr. Taxon.* 1:817–948.

———. 1991. Acari Prostigmata (excluding Trombiculidae) parasitic on Australian vertebrates: An annotated checklist, keys and bibliography. *Invertebr. Taxon.* 4:1283–1376.

Domrow, R., and P. W. Ladds. 1984. A new ear mite from the Indian elephant (Acari: Anoetidae). *J. Nat. Hist.* 18:759–64. [HISTIOSTOMATIDAE]

Domrow, R., and L. N. Lester. 1985. *Chiggers of Australia (Acari: Trombiculidae): An annotated checklist, keys and bibliography.* Aust. J. Zool., suppl. ser. 33 (114). Collingwood, Australia: CSIRO. [TROMBICULIDAE]

Doreste, S. E. 1984. *Acarologia.* San Jose, Costa Rica: Inst. Interamericano de Cooperacion para la Agricultura.

Dosse, G. 1954. *Tenuipalpus orchidarum* Parfitt nun auch in deutschen Gewachshäusern. *Zeitschr. Angew. Entomol.* 36: 304–15. [TENUIPALPIDAE]

———. 1958. Sie Spermathecae, ein zusätzliches Bestimmungsmerkmal beiRaubmilben (Acar., Phytoseiidae). *Pflanzenschutz-Berichte* 20:1–11.

Dosse, G., and I. Schneider. 1957. Biologie und Lebensweise von *Czenspinskia lordi* Nesbitt (Acar., Sarcoptiformes). *Zeit. Angew. Entomol.* 44:403–18. [WINTERSCHMIDTIIDAE]

Doube, R. M. 1975. Regulation of the circadian rhythm of detachment of engorged larvae and nymphs of the argasid kangaroo tick, *Ornithodoros gurneyi*. *J. Med. Entomol.* 12:15–22.

Doumandji, B. 1985. Catalogue des Acariens terrestres d'Algerie. *Bull. Zool. Agr.*, No. spec.:1–31.

Drummond, F. A., R. A. Casagrande, and P. A. Logan. 1989. Population dynamics of *Chrysomelobia labidomerae* Eickwort, a parasite of the Colorado potato beetle. *Internat. J. Acarol.* 15:31–45. [PODAPOLIPIDAE]

———. 1992. Impact of the parasite, *Chrysomelobia labidomerae* Eickwort, on the Colorado potato beetle. *Internat. J. Acarol.* 18:107–15. [PODAPOLIPIDAE]

Drummond, R. O., T. M. Whetstone, and W. J. Gladney. 1971. Oviposition of the lone star tick. *Ann. Entomol. Soc. Amer.* 64:191–94.

Du Toit, B. J., P. D. Theron, and E. A.Ueckermann. 1998. A new genus and four new species of the family Camerobiidae (Acari: Raphignathoidea) from South Africa. *Internat. J. Acarol.* 24: 3–19.

Dubinin, V. B. 1951. *Feather mites (Analgesoidea).* Part I, *Introduction to their study.* Fauna SSSR, Pauk. 6 (5).

———. 1953. *Per'evie kleshchi (Analgesoidea).* Part II, *Semeistva Epidermoptidae i Freyanidae.* Fauna SSSR, Pauk. 6 (6).

———. 1954. Novaya klassifikatsiya per'evikh kleshchei nadsemeistva Analgesoidea i polozhenie ego v sisteme otryada Acariformes A. Zachv. 1952. *Izvest. Akad. Nauk SSSR Ser. Biol.* 4: 59–75.

———. 1955. Novie rodi per'evikh kleshchei (Analgesoidea). *Trudy Zool. Inst. Akad. Nauk SSSR* 18:248–87. [ALLOPTIDAE]

———. 1956. *Per'evie kleshchi (Analgesoidea).* III, *Semeistvo Pterolichidae.* Fauna SSSR, Pauk. 6 (7).

———. 1957. A new classification of mites of the superfamilies Cheyletoidea W. Dub. and Demodicoidea W. Dub. (Acariformes, Trombidifiormes). *Parazitol. Sb. (Leningrad)* 17: 71–136.

———. 1962. Class Acaromorpha: Mites, or gnathosomic chelicerate arthropods. In *Fundamentals of palaeontology*, ed. B. B. Rodendorf, 447–73. Moscow: Acad. Sci. USSR. [in Russian]

Dugès, A. 1839. Recherches sur l'ordre des Acariens. *Ann. Sci. Nat. Zool.* 1:18–63.

Dunley, J. E., and B. A. Croft. 1996. Eriophyoids as competitors of other phytophagous mites. In Lindquist, Sabelis, and Bruin, 6:751–55.

Dunlop, J. 1995. Are the fossil phalangiotarbids just big opilioacarid mites? *Newsletter Brit. Arachnol. Soc.* 74:8–9.

Dunlop, J., J. Wunderlich, and G. O. Poinar, Jr. 2004. The first fossil opilioacariform mite (Acari: Opilioacariformes) and the first Baltic amber camel spider (Solifugae). *Trans. Roy. Soc. Edin., Earth Sci.* 94:261–73.

Dunlop, J. A., and G. Alberti. 2007. The affinities of mites and ticks: A review. *J. Zool. Syst. Evol. Res.* 10:1–18.

Durden, L. A., and J. E. Keirans. 1996. Host-parasite coextinction and the plight of tick conservation. *Amer. Entomol.* (Summer): 87–91.

Durden, L. A., J. H. Oliver, Jr., C. W. Banks, and G. N. Vogel. 2002. Parasitism of lizards by immature stages of the black-legged tick, *Ixodes scapularis* (Acari, Ixodidae). *Exp. Appl. Acarol.* 26:257–66.

Dusbábek, F. 1969. To the phylogeny of genera of the family Myobiidae (Acarina). *Acarologia* 11:537–74.

———. 1973. A systematic review of the genus *Pteracarus* (Acariformes:Myobiidae). *Acarologia* 15:240–88.

———. 2002. Adaptation of mites and ticks to parasitism. Medical and veterinary aspects. In Bernini et al. (2002), 399–418.

Dusbábek, F., and V. Bukva, eds. 1991. *Modern acarology: Proceedings VIII International Congress of Acarology.* 2 vols. Prague: Academia; The Hague, SPB Publ. b.v.

Duso, C., and E. de Lillo. 1996. Grape. In Lindquist, Sabelis, and Bruin, 6:571–82.

Dworkin, M. S., T. G. Schwan, and D. E. Anderson. 2002. Tick-borne relapsing fever in North America. *Med. Clin. N. Amer.* 86:417–33. [ARGASIDAE]

Early, M., and M. L. Goff. 1986. Arthropod succession patterns in exposed carrion on the island of O'ahu, Hawaiian Islands, USA. *J. Med. Entomol.* 23:520–31. [LARDOGLYPHIDAE]

Easterbrook, M. A. 1996. Damage and control of eriophyoid mites in apple and pear. In Lindquist, Sabelis, and Bruin, 6:527–41.

Ebermann, E. 1980. Zur Zoogeographie, Taxonomie und Larvalentwicklung einiger Scutacariden-Arten aus Nordamerika (Acari, Trombidiformes). *Zool. Anz., Jena* 205:123–35.

———. 1981. *Imparipes* (*Imparipes*) *brevitarsus* n. sp. (Acari, Scutacaridae), eine mit Ameisen vergesellschaftete neue Milbenart aus Österreich. *Mitt. Naturwiss. Ver. Steiermark* 111:183–92.

———. 1982a. Fortpflanzungsbiologische Studien an Scutacariden (Acari, Trombidiformes). *Zool. Jb. Syst.* 109:98–16.

———. 1982b. Zuchtversuche und morphologische Untersuchungen an heimischen Milben (Acari, Scutacaridae). *Mitt. Naturwiss. Ver. Steiermark* 112:155–65.

———. 1983. Zur Morphologie der Milbenart *Scutacarus peractus* Karafiat, 1959 (Acari, Trombidiformes). *Mitt. Naturwiss. Ver. Steiermark* 113:91–98. [SCUTACARIDAE]

———. 1986. To the knowledge of the tarsonemid fauna of Italian islands (Acari: Microdispidae, Scutacaridae). *Redia* 69:469–80.

———. 1990. Taxonomic consequences of the polymorphism found in scutacarids (Acari: Scutacaridae). *Entomol. Mitt. Zool. Mus. Hamb.* 10 (139/140): 29–42.

———. 1991a. Das Phänomen Polymorphismus in der Milbenfamilie Scutacaridae (Acari, Heterostigmata, Tarsonemina, Scutacaridae). *Zoologica (Stuttgart)* 141.

———. 1991b. Thanatosis or feigning death in mites of the family Scutacaridae. In Schuster and Murphy, 399–401.

———. 1994. First demonstration of males in species of the genus *Archidispus* Karafiat (Acari, Tarsonemina, Scutacaridae). *Internat. J. Acarol.* 20:169–82.

———. 1995. Indication of jumping ability in the mite family Scutacaridae (Acari, Tarsonemina). *Entomol. Mitt. Zool. Mus. Hamb.* 11 (152): 205–9.

———, ed. 1998a. *Arthropod biology: Contributions to morphology, ecology and systematics.* Austrian Acad. Sci. Biosystematics and Ecology Series 14. Vienna: Austrian Acad. Sci. Press.

———. 1998b. *Imparipes* (*Sporichneuthes* nov. subgen.), a remarkable new taxon in the mite family Scutacaridae (Acari, Heterostigmata). In Ebermann (1998a), ed., 179–214.

Ebermann, E., and M. Hall. 2003. First record of sporothecae within the mite family Scutacaridae (Acari, Tarsonemina). *Zool. Anz.* 242:367–75.

Ebermann, E., and J. G. Palacios-Vargas. 1988. *Imparipes* (*Imparipes*) *tocatlphilus* n. sp. (Acari, Tarsonemina, Scutacaridae) from Mexico and Brazil: first record of ricinuleids as phoresy hosts for scutacarid mites. *Acarologia* 29:347–54.

Ebermann, E., and G. Rack. 1982. Zur Biologie einer neuen myrmecophilen Art der Gattung *Petalomium* (Acari, Pygmephoridae). *Entomol. Mitt. Zool. Mus. Hamb.* 7 (115): 175–92.

Ebrahim, H. M., D. S. Achor, and C. C. Childers. 1996. A new method for preparation of eriophyid mites (Eriophyidae) for

scanning electron microscopy using sputter coat-vacuum drying. In Mitchell et al. (1996), 1:269–72.
Eckert, J. E. 1961. Acarapis mites of the honey bee, *Apis mellifera* Linnaeus. *J. Insect Pathol.* 3:409–25. [TARSONEMIDAE]
Edsberg, E., and S. Hågvar. 1999. Vertical distribution, abundance, and biology of oribatid mites (Acari) developing inside decomposing spruce needles in a podsol soil profile. *Pedobiologia* 43:413–21.
Edwards, C. A., and K. E. Fletcher. 1971. A comparison of extraction methods for terrestrial arthropods. In *Methods of study in quantitative soil ecology*, IBP Handbook No. 18, ed. J. Phillipson, 150–85. Oxford, UK: Blackwell.
Edwards, D. D., R. Bogardus, and N. Wilhite. 1998. Geographic differences in host specialization between the symbiotic water mites *Unionicola formosa* and *U. foili* (Acari: Unionicolidae). In Bruin, van der Geest, and Sabelis, 195–206.
Edwards, D. D., E. Deatherage, and B. R. Ernsting. 2004. Random amplified polymorphic DNA analysis of kinship within host-associated populations of the symbiotic water mite *Unionicola foili (Acari: Unionicolidae)*. *Exp. Appl. Acarol.* 34:67–77.
Edwards, D. D., and R. V. Dimock, Jr. 1988. A comparison of the population dynamics of *Unionicola formosa* from two anodontine bivalves in a North Carolina farm pond. *J. Elisha Mitchell Sci. Soc.* 104:90–98. [UNIONICOLIDAE]
———. 1991a. The ontogeny of host recognition behavior by the symbiotic water mite *Unionicola formosa*. *Amer. Zool.* 31:145A. [UNIONICOLIDAE]
———. 1991b. Relative importance of size versus territorial residency in intraspecific aggression by symbiontic male water mites (Acari: Unionicolidae). *Exp. Appl. Acarol.* 12:61–65.
———. 1995. Specificity of the host recognition behaviour of larval *Unionicola* (Acari: Unionicolidae): The effects of larval ontogeny and early larval experience. *Anim. Behav.* 50:343–52.
———. 1997. Genetic differentiation between *Unionicola formosa* and *U. foili* (Acari: Unionicolidae): Cryptic species of molluscan symbionts. *Invertebr. Biol.* 116:124–33. [UNIONICOLIDAE]
Edwards, D. D., and M. F. Vidrine. 1994. A new species in the water mite subgenus *Parasitatax* (Acari: Unionicolidae: *Unionicola*) from the North American freshwater mussel, *Utterbackia imbecillis* (Bivalvia: Unionidae). *J. Elisha Mitchell Sci. Soc.* 110:1–6.
Edwards, M. A. 1975. The chaetotaxy of the pedipalps and legs of some larval ticks (Acari: Argasidae). *Trans. Zool. Soc. Lond.* 33:1–76.
Edwards, M. A., and G. O. Evans. 1967. Some observations on the chaetotaxy of the legs of larval Ixodidae (Acari: Metastigmata). *J. Nat. Hist.* 4:595–601.
Efford, I. E. 1965. The ecology of the watermite *Feltria romijni* Besseling. *J. Anim. Ecol.* 34:233–51.
———. 1966. Observations on the life history of three stream-dwelling watermites. *Acarologia* 8:86–93.
Egan, M. K., and P. E. Hunter. 1975. Redescription of a cockroach mite, *Proctolaelaps nauphoetae*, with notes on its biology. *Ann. Entomol. Soc. Amer.* 68:361–64. [MELICHARIDAE]
Ehara, S. 1965. Two new species of Teneriffiidae from Japan, with notes on the genera *Heteroteneriffia* and *Neoteneriffiola* (Acarina: Prostigmata). *Seto Marine Biol. Lab.* 13:221–29.
———. 1975a. Description of a new species of *Tuckerella* from Japan (Acarina: Tuckerellidae). *Internat. J. Acarol.* 1:1–5.
———. 1975b. A guide to the tetranychid mites of agricultural importance in Japan. In *Approaches to biological control*, ed. K. Yasumatsu and H. Mori, 7:15–23. Tokyo: JIBP Synthesis.
———. 1999. Revision of the spider mite family Tetranychidae of Japan (Acari, Prostigmata). *Species Diversity* 4:63–141.
Ehara, S., and T. Gotoh. 1992. Descriptions of two *Panonychus* spider mites from Japan, with a key to species of the genus in the world (Acari: Tetranychidae). *Appl. Entomol. Zool.* 27:107–15.
———. 1996. Two new species of spider mites occurring in Japan (Acari: Tetranychidae). *J. Acarol. Soc. Jpn.* 5:17–25.
Ehrnsberger, R. 1974. Prälarval und Larvalentwicklung bei Rhagidien (Acari: Prostigmata). *Osnabrück. Naturwiss. Mitt.* 3: 85–117.
———. 1977. Zwei neue Rhagidien aus Norddeutschland: *Rhagidia danica* n. sp. und *Rhagidia clavicrinita* n. sp. (Acari: Prostigmata). *Osnabrück. Naturwiss. Mitt.* 5:95–107.
———. 1979. Spinnvermögen bei Rhagidiidae (Acari, Prostigmata). *Osnabrück. Naturwiss. Mitt.* 6:45–72.
———. 1981. Ernährungsbiologie der bodenbewohnenden Milbe *Rhagidia* (Trombidiformes). *Osnabrück. Naturwiss. Mitt.* 8: 127–32. [RHAGIDIIDAE]
———. 1988. Mating behaviour of *Linopodes* sp. (Acariformes: Eupodoidea). In ChannaBasavanna and Viraktamath, 1:211–18.
Ehrnsberger, R., S. V. Mironov, and J. Dabert. 2001. A preliminary analysis of phylogenetic relationships of the feather mite family Freyanidae Dubinin, 1953 (Acari: Astigmata). *Biol. Bull. Poznan* 38:182–201.
Eickwort, G. C. 1975. Development and bionomics of *Chrysomelobia labidomerae* (Acari: Tarsonemina: Podapolipidae), a parasite of the milkweed leaf beetle (Coleoptera: Chrysomelidae). *Can. Entomol.* 107:627–38.
———. 1979. Mites associated with sweat bees (Halictidae). In Rodriguez, 1:575–81. [HISTIOSTOMATIDAE, SCUTACARIDAE]
———. 1990. Associations of mites with social insects. *Ann. Rev. Entomol.* 35:469–88.
———. 1994. Evolution and life-history patterns of mites associated with bees. In Houck (1994b), contribution no. 9, 218–51.
Eickwort, R. C., and G. C. Eickwort. 1986. Effects of parasitism by the mite *Chrysomelobia labidomerae* (Acari: Podapolipidae) on the longevity and fecundity of its host beetle, *Labidomera clivicollis* (Coleoptera: Chrysomelidae). *Internat. J. Acarol.* 12:223–27.
Eidelberg, M. M. 2000. Three new species of carabidophilic mites of the family Antennoseiidae (Parasitiformes, Gamasina). *Entomol. Rev.* 80:704–9. [ASCIDAE]
Ejtminaviciute, I. S. 2003. *Soil mites of Lithuania*. (Acari) Catalogue. Vilnius: Institute of Ecology of Vilnius Univ.
Elbadry, E. A., and M. S. F. Tawfik. 1966. Life cycle of the mite *Adactylidium* sp. (Acarina: Pyemotidae), a predator of thrips eggs in the United Arab Republic. *Ann. Entomol. Soc. Amer.* 59:458–61. [ACAROPHENACIDAE]
Elbadry, E. A., A. A. Yousef, and S. H. Metwally. 1976. Life history studies on the soil mite *Scutacarus aegypticus* with description of the immature stages (Acarina: Scutacaridae). *Pediobiologia* 16:167–70.
Eller, R., and R. W. Strandtmann. 1963. Notes on Teneriffiidae (Acarina: Prostigmata). *Southwest. Nat.* 8:23–31.
Elsen, P. 1972. Sept nouvelles espèces de *Neocypholaelaps* Vitzthum, 1941 (Mesostigmata: Ameroseiidae) phorétiques d'Hyménoptères africains. *Rev. Zool. Bot. Afr.* 86:17–30.

———. 1973. Nouvelles espèces d'*Ameroseius* Berlese, 1903 (Acari, Mesostigmata: Ameroseiidae) phorétiques d'Hyménoptères et de Coléoptères. *Rev. Zool. Bot. Afr.* 87:729–57.

———. 1974. Une nouvelle espèce du genre *Afrocypholaelaps* Elsen, 1972 et redéfinition du genre (Meostigmata: Ameroseiidae). *Rev. Zool. Afr.* 88:159–64.

Elsen, P., and J. O. Whitaker, Jr. 1985. *Sertitympanum*, a new genus of Ameroseiidae (Acarina, Mesostigmata) taken from rodents in the United States: Including descriptions of three new species in the genus. *Acarologia* 26:117–22.

Elzinga, R. J. 1978. Holdfast mechanisms in certain uropodine mites (Acarina: Uropodina). *Ann. Entomol. Soc. Amer.* 71: 896–900. [EUROPODIDAE]

———. 1982a. The generic status and six new species of *Trichocylliba* (Acari: Uropodina). *Acarologia* 23:3–18.

———. 1982b. The genus *Antennequesoma* (Acari: Uropodina) and descriptions of four new species. *Acarologia* 23:319–25. [EUROPODIDAE]

———. 1982c. The genus *Coxequesoma* (Acari: Uropodina) and descriptions of four new species. *Acarologia* 23:215–24.

———. 1989. *Habeogula cauda* (Acari: Uropodina), new genus and species of mite from the army ant *Labidus praedator* (F. Smith). *Acarologia* 30:341–44. [EUROPODIDAE]

———. 1991. Two new elongate species of *Planodiscus* (Acari: Uropodina) with a key to the known species. *Acarologia* 32:109–14. [EUROPODIDAE]

———. 1993. Larvamimidae, a new family of mites (Acari: Dermanyssoidea) associated with army ants. *Acarologia* 34:95–103.

Elzinga, R. J., and C. W. Rettenmeyer. 1970. Five new species of *Planodiscus* (Acarina: Uropodina). *Acarologia* 12:59–70.

———. 1974. Seven new species of *Circocylliba* (Acarina: Uropodina). *Acarologia* 16:595–611.

Emmanuel, N. G. 1982. Greece. In Prasad (1982), 195–210.

Enami, Y., and Y. Nakamura. 1996. Influence of *Scheloribates azumaensis* (Scheloribatidae) on *Rhizoctonia solani*, the cause of the radish-root rot. *Pedobiologia* 40:251–54.

Enda, N., and H. Tamura. 1977. Matsunomadara kamikiri seichu kara kenshutsu sareta senchu hoshokusei dani-rui [Nematophagous mites found on adults of the Japanese pine sawyer]. *Shinrin Boeki* [Forest Pests] 26:188–90. [DIGAMASELLIDAE] [in Japanese]

———. 1980. Nematode-feeding mites found on adults of the Japanese pine sawyer (continued). *Shinrin Boeki* [Forest Pests] 29:17–20. [MELICHARIDAE] [in Japanese]

Engelbrecht, C. M. 1974. The genus *Hydrozetes* (Oribatei: Acari) in South Africa. *Navors. Nas. Mus., Bloemfontein* 3 (3): 41–49. [HYDROZETIDAE]

English, F. P., and W. B. Nutting. 1981. Demodicosis of ophthalmic concern. *Amer. J. Ophthalmol.* 91:362–72. [DEMODICIDAE]

Enkegaard, A., M. A. Sardar, and H. F. Brodsgaard. 1997. The predatory mite *Hypoaspis miles*: Biological and demographic characteristics on two prey species, the mushroom sciarid fly, *Lycoriella solani*, and the mould mite, *Tyrophagus putrescentiae*. *Entomol. Exp. Appl.* 82:135–46. [LAELAPIDAE]

Enkegaard, H. F. Brodsgaard, and M. A. Sardar. 1995. *Hypoaspis miles*—a polyphagous predatory mite for control of soil-dwelling pests: Biology and food preference. *S P Rapport Statens Planteavlsforsog* 4:247–55. [LAELAPIDAE]

Enns, W. R. 1958. A new species of rake-legged mite from Missouri (Acarina, Caeculidae). *J. Kans. Entomol. Soc.* 31:107–13.

Entrekin, D. L., and J. H. Oliver, Jr. 1982. Aggregation of the chicken mite, *Dermanyssus gallinae* (Acari: Dermanyssidae). *Med. Entomol.* 19:671–78.

Epstein, A., and J. H. Hill, eds. 1994. Rose rosette and other eriophyid mite-transmitted plant disease agents of uncertain etiology. Proc. Internat. Symposium, May 19–21, Iowa State Univ., Ames.

Erman, O. 1992. Türkiye Faunası İçin Yeni *Arrenurus* (s. str.) Dugès, 1834 (Arrenuridae, Hydrachnellae, Acari) Türleri. *Doğa TU Zool., Derg.* 16:193–208.

———. 1993. Three new species of water mites in the subgenus *Arrenurus* from Turkey. *Acarologia* 34:223–30.

Erman, O., and M. Özkan. 1997. Some new and unrecorded *Arrenurus* Dugès 1834 (Acari: Hydrachnellae) species from Turkey. *J. Nat. Hist.* 31:1417–28. [ARRENURIDAE]

Ernst, H. 1995. Endogenous circatidal activity rhythm in a littoral mite, *Bdella septentrionalis* (Acarina, Bdellidae). *Acarologia* 36:315–24.

———. 1996. The distribution and zonation of terrestrial mites inhabiting artificial rocky shores of estuaries in northern Germany. In Mitchell et al. (1996), 1:529–32.

Ernsting, B. R., D. Edwards, M. F. Vidrine, K. S. Myers, and C. M. Harmon. 2006. Phylogenetic relationships among species of the subgenus *Parasitatax* (Acari: Unionicolidae: *Unionicola*) based on DNA sequence of the mitochondrial Cytochrome Oxidase I gene. *Internat. J. Acarol.* 32:195–202.

Erwin, T. L. 1991. How many species are there?: revisited. *Conserv. Biol.* 5:330–33.

Estrada-Venegas, E., R. A. Norton, and A. R. Moldenke. 1996. Unusual sperm transfer in *Pilogalumna* sp. (Galumnidae). In Mitchell et al. (1996), 1:565–67.

Evans, G. A., H. L. Cromroy, and R. Ochoa. 1993. The Tenuipalpidae of Honduras (Tenuipalpidae: Acari). *Fla. Entomol.* 76:126–55.

Evans, G. O. 1954. On the genus *Asternolaelaps* Berlese, 1923 (Acarina-Mesostigmata). *Entomol. Monthly Mag.* 90:88–90.

———. 1955a. British mites of the genus *Veigaia* Oudemans (Mesostigmata-Veigaiaidae). *Proc. Zool. Soc. Lond.* 125:569–86.

———. 1955b. A collection of mesostigmatid mites from Alaska. *Bull. Brit. Mus. (Nat. Hist.), Zool.* 2:285–307. [ARCTACARIDAE]

———. 1955c. A review of the laelaptid paraphages of the Myriopoda with descriptions of three new species (Acarina: Laelaptidae) *Parasitology* 45:352–68. [IPHIOPSIDIDAE]

———. 1955d. A revision of the family Epicriidae (Acarina-Mesostigmata). *Bull. Brit. Mus. (Nat. Hist.), Zool.* 3:169–200 + 2 pl.

———. 1955e. A stridulating organ in the Acarina. *Bull. Natl. Inst. Sci. India* 7:107–9. [AMEROSEIIDAE]

———. 1956. On the classification of the family Macrochelidae with particular reference to the subfamily Parholaspinae (Acarina-Mesostigmata). *Proc. Zool. Soc. Lond.* 127:345–77. [PARHOLASPIDIDAE]

———. 1957a. An introduction to the British Mesostigmata (Acarina) with keys to families and genera. *J. Linn. Soc. (Zool.)* 43:203–59. [PROTODINYCHIDAE]

———. 1957b. On the genus *Scarabaspis* Womersley (Acarina-Mesostigmata). *Ann. Mag. Nat. Hist.* 10 (12): 409–16. [EVIPHIDIDAE]

———. 1958a. A revision of the British Aceosejinae (Acarina:

Mesostigmata). *Proc. Zool. Soc. Lond.* 131:177–229. [ASCIDAE, MELICHARIDAE, BLATTISOCIIDAE]
———. 1958b. Some mesostigmatid mites from a nest of social spiders in Uganda. *Ann. Mag. Nat. Hist.* 13:580–90. [CERCOMEGISTIDAE]
———. 1959. The genera *Cyrthydrolaelaps* Berlese and *Gamasolaelaps* Berlese (Acarina-Mesostigmata). *Acarologia* 1:201–15. [VEIGAIIDAE]
———. 1963a. The genus *Neocypholaelaps* Vitzthum (Acari: Mesostigmata). *Ann. Mag. Nat. Hist.*, ser. 13, 6:209–30. [AMEROSEIIDAE]
———. 1963b. Observations on the chaetotaxy of the legs in the free-living Gamasina (Acari: Mesostigmata). *Bull. Brit. Mus. (Nat. Hist), Zool.* 10:277–303.
———. 1963c. Observations on the classification of the family Otopheidomenidae (Acari: Mesostigmata) with descriptions of two new species. *Ann. Mag. Nat. Hist.*, ser. 13, 5:609–20.
———. 1963d. Some observations on the chaetotaxy of the pedipalps in the Mesostigmata (Acari). *Ann. Mag. Nat. Hist.* 6:513–27.
———. 1964. Some observations on the chaetotaxy of the pedipalps in the Mesostigmata (Acari). *Ann. Mag. Nat. Hist.*, ser. 13, 6:513–27.
———. 1965. The ontogenetic development of the chaetotaxy of the tarsi of legs II–IV in the Antennophorina (Acari: Mesostigmata). *Ann. Mag. Nat. Hist*, ser. 13, 8:81–83.
———. 1968. The external morphology of the postembryonic developmental stages of *Spinturnix myoti* Kol. (Acari: Mesostigmata). *Acarologia* 10:589–608.
———. 1969a. A new mite of the genus *Thinoseius* Halbt. (Gamasina-Eviphididae) from the Chatham Islands New Zealand. *Acarologia* 11:505–14.
———. 1969b. Observations on the ontogenetic development of the chaetotaxy of the tarsi of legs II–IV in the Mesostigmata (Acari). In Evans (1969c), 195–200.
———, ed. 1969c. *Proceedings 2nd International Congress of Acarology, Sutton Bonnington, UK.* Budapest: Akad. Kiadó.
———. 1972. Leg chaetotaxy and the classification of the Uropodina (Acari: Mesostigmata). *J. Zool. Lond.* 167:193–206.
———. 1992. *Principles of acarology.* Wallingford, Oxon, UK: CAB International.
Evans, G. O., and A. S. Baker. 1991. A new genus of the mite subfamily Platyseiinae from Jamaica (Parasitiformes: Ascidae). *J. Zool. Lond.* 224:121–26. [BLATTISOCIIDAE]
Evans, G. O., and E. Browning. 1956. British mites of the subfamily Macrochelinae Trägårdh (Mesostigmata-Macrochelidae). *Bull. Brit. Mus. (Nat. Hist.), Zool.* 4:1–54.
Evans, G. O., and A. Fain. 1995. A new phoretic deuteronymph of *Halodarcia* Karg (Acari: Mesostigmata: Halolaelapidae) associated with carabid beetles in Belgium with a review of the genus. *Bull. Inst. Roy. Sci. Nat. Belg., Entomol.* 65:63–71.
Evans, G. O., A. Fain, and J. Bafort. 1963. Découverte du cycle évolutif du genre *Myialges* avec description d'un espèce nouvelle (Myialgesidae: Sarcoptiformes). *Bull. Ann. Soc. Roy. Belg. Entomol.* 99:486–500.
Evans, G. O., and K. H. Hyatt. 1956. British mites of the genus *Pachylaelaps* Berlese (Gamasina-Pachylaelapidae). *Entomol. Monthly Mag.* 42:118–29.
———. 1957. The genera *Podocinum* and *Podocinella*. *Ann. Mag. Nat. Hist.*, ser. 12, 10:913–32. [PODOCINIDAE]
———. 1960. A revision of the Platyseiinae (Mesostigmata: Aceosejidae) based on material in the collections of the British Museum (Natural History). *Bull. Brit. Mus. (Nat. Hist.), Zool.* 6:27–101. [BLATTISOCIIDAE]
———. 1963. Mites of the genus *Macrocheles* Latr. (Mesostigmata) associated with coprid beetles in the collections of the British Museum (Natural History). *Bull. Br. Mus. (Nat. Hist.), Zool.* 9:327–401.
Evans, G. O., and G. C. Loots. 1972. Scanning electron microscopy in the study of the gnathosoma of the Acari. *Wetensk. Bydr. Potchefstroomse Univ. B (Natuur.)* 49.
———. 1975. Scanning electron microscope study of the structureof the hypostome of *Phityogamasus*, *Laelaps* and *Ornithonyssus* (Acari: Mesostigmata). *J. Zool. Lond.* 176:425–36.
Evans, G. O., and J. G. Sheals. 1959. Three new mesostigmatic mites associated with millipedes in Indonesia. *Entomol. Ber.* 19:107–11. [BLATTISOCIIDAE]
Evans, G. O., J. G. Sheals, and D. MacFarlane. 1961. *The terrestrial Acari of the British Isles: An introduction to their morphology, biology, and classification.* Vol. 1, *Introduction and biology.* London: British Museum (Natural History).
Evans, G. O., and W. M. Till. 1965. Studies on the British Dermanyssidae (Acari: Mesostigmata). I, External morphology. *Bull. Brit. Mus. (Nat. Hist.), Zool. Lond.* 13:249–94.
———. 1966. Studies on the British Dermanyssidae (Acari: Mesostigmata). II, Classification. *Bull. Brit. Mus. (Nat. Hist.), Zool. Lond.* 14:109–370. [LAELAPIDAE, PACHYLAELAPIDAE]
———. 1979. Mesostigmatic mites of Britain and Ireland (Chelicerata: Acari-Parasitiformes). An introduction to their external morphology and classification. *Trans. Zool. Soc. Lond.* 35: 139–270.
Eveleigh, E. S., and W. Threlfall. 1975. The biology of *Ixodes* (*Ceratixodes*) *uriae* White, 1852 in Newfoundland. *Acarologia* 16:621–35.
Ewing, H. E. 1925. A contribution to our knowledge of the taxonomy of chiggers including the description of a new genus, six new species, and a new variety. *Amer. J. Trop. Med.* 5:251–65. [TROMBICULIDAE]
———. 1933. A new pit-producing mite from the scales of a South American snake. *J. Parasitol.* 20:53–56. [HARPIRHYNCHIDAE]
Ewing, H., and R. L. Webster. 1912. Mites associated with oystershell scale (*Lepidosaphes ulmi* Linne). *Psyche* 19:121–34.
Faasch, H. 1967. Beitrag zur Biologie der einheimischen Uropodiden *Uroobovella marginata* (C. L. Koch, 1839) und *Uropoda orbicularis* (O. F. Müller, 1776) und experimentelle Analyse ihres Phoresieverhaltens. *Zool. Jb. Syst.* 94:521–608.
Fagasinski, A. 1986. Inwazja myszy przez *Tarsonemoides noxius* [Infestation in mice of *T. noxius*]. *Wiad. Parazytol.* 32:373–75. [TARSONEMIDAE] [in Polish]
Fain, A. 1956. Une nouvelle famille d'acariens endoparasites des chauves-souris: Gastronyssidae fam. nov. (Acarina: Sarcoptiformes). *Ann. Soc. Belg. Med. Trop.* 36:87–98.
———. 1957. Notes sur quelques acariens parasites. *Rev. Zool. Bot. Afr.* 55:253–54. [GASTRONYSSIDAE]
———. 1958. Un nouvel acarien trombidiforme parasitant les fosses nasals d'un serpent au Ruanda-Urundi. *Rev. Zool. Bot. Afr.* 57:177–83. [EUPODIDAE]
———. 1959. Deux nouveaux genres d'acariens vivant dans

l'épaisseur des muqueuses nasale et buccale chez un lémurien (Trombidiformes: Demodicidae). *Ann. Soc. Belg. Med. Trop.* 95:263–73.

———. 1960. Révision du genre *Cytodites* (Megnin) et description de deux espèces et un genre nouveaux dans la famille Cytoditidae Oudemans (Acarina: Sarcoptiformes). *Acarologia* 2:238–49. [CYTODITIDAE]

———. 1961a. Les acariens parasites endopulmonaires des serpentes. *Bull. Inst. Roy. Sci. Nat. Belg.* 37 (6): 1–135. [ENTONYSSIDAE]

———. 1961b. Une nouvelle famille d'acariens. *Rev. Zool. Bot. Afr.* 66:283–96. [OMENTOLAELAPIDAE]

———. 1962a. Un acarien remarquable combinant les caracteres de plusiers familles: *Evansacarus lari* n. g., n. sp. (Evansacaridae nov. fam.: Sarcoptiformes). *Bull. Ann. Soc. Roy. Belg. Entomol.* 98:125–45. [EPIDERMOPTIDAE]

———. 1962b. Les acariens mésostigmatiques ectoparasites des serpentes. *Bull. Inst. Roy. Sci. Nat. Belg.* 38 (18): 1–149. [IXODORHYNCHIDAE]

———. 1962c. Les acariens parasites nasicoles des Batraciens. Revision des Lawrencarinae Fain, 1957 (Ereynetidae: Trombidiformes). *Bull. Inst. Roy. Sci. Nat. Belg.* 38 (25): 1–69.

———. 1962d. Un curieux acarien parasite du Lièvre-sauteur, *Pedetes cafer*, d'Afrique du Sud. *Rev. Zool. Bot. Afr.* 65:357–64.

———. 1963a. Les acariens producteurs de gale chez les lémuriens et les singes avec une étude des Psoroptidae (Sarcoptiformes). *Bull. Inst. Roy. Sci. Nat. Belg.* 39 (32): 1–125.

———. 1963b. Chaetotaxie et classification des Speleognathinae. *Bull. Inst. Roy. Sci. Nat. Belg.* 39 (9): 1–80. [EREYNETIDAE]

———. 1964a. Les acariens de la famille Cytoditidae (Sarcoptiformes): Description de sept espèces nouvelles. *Acarologia* 6:504–28.

———. 1964b. Chaetotaxie et classification des Gastronyssidae avec description d'un nouveau genre parasite nasicole d'un ecureuil sudafricain (Acarina: Sarcoptiformes). *Rev. Zool. Bot. Afr.* 70:40–52.

———. 1964c. Les Lemurnyssidae parasites nasicoles des Lorisidae africains et des Cebidae sud-américains. Description d'une espèce nouvelle (Acarina: Sarcoptiformes). *Ann. Soc. Belg. Med. Trop.* 44:453–58.

———. 1964d. Les Ophioptidae, acariens parasites des écailles des serpents (Trombidiformes). *Bull. Inst. Roy. Sci. Nat. Belg.* 40 (15): 1–57. [HARPIRHYNCHIDAE]

———. 1965a. Les acariens nidicoles et détriticoles de la famille Pyroglyphidae Cunliffe. *Rev. Zool. Bot. Afr.* 72:257–88.

———. 1965b. Les acariens producteurs de gale chez les edentés et les marsupiaux (Psoroptidae et Lobalgidae: Sarcoptiformes). *Bull. Inst. Roy. Sci. Nat. Belg.* 41 (17): 1–141.

———. 1965c. Un nouveau type d'hypope, parasite cuticole de rongeurs africains (Acari: Sarcoptiformes). *Zeitschr. Parasitenk.* 26:82–90.

———. 1965d. *A review of the family Epidermoptidae Trouessart parasitic on the skin of birds (Acarina: Sarcoptiformes).* Vol. I, pts. 1 and 2. Konink. Acad. Wetensch. Let. schone Kunst. 84. Brussels: Paleis der Academiën. [EPIDERMOPTIDAE, DERMATIONIDAE]

———. 1965e. A review of the family Rhyncoptidae Lawrence parasitic on porcupines and monkeys (Acarina: Sarcoptiformes). In *Advances in acarology*, ed. J. A. Naegele, 2:135–59. Ithaca, NY: Cornell Univ. Press.

———. 1966. Notes sur la biologie des acariens du genre *Chaetodactylus* et en particulier de *C. osmiae*, parasite des abeilles solitaires *Osmia rufa* et *O. cornuta* en Belgique (Sarcoptiformes: Chaetodactylidae). *Bull. Ann. Soc. Roy. Belg. Entomol.* 102:249–61.

———. 1967a. Les hypopes parasites des tissus cellulaires des oiseaux (Hypodectidae: Sarcoptiformes). *Bull. Inst. Roy. Sci. Nat. Belg.* 43 (4): 1–139. [HYPODERATIDAE]

———. 1967b. Observations sur les Rodhainyssinae acariens parasites des voies respiratoires des chauves-souris (Gastronyssidae: Sarcoptiformes). *Acta Zool. Pathol. Antverp.* 44:3–35.

———. 1968a. Acariens nidicoles et détriticoles en Afrique au sud du Sahara. III, Espèces et genres nouveaux dans les sous-familles Labidophorinae et Grammolichinae (Glycyphagidae: Sarcoptiformes). *Acarologia* 10:86–110.

———. 1968b. Deux nouveaux acariens cavernicoles du Gabon (Sarcoptiformes). *Biol. Gabon* 4:195–205. [GUANOLICHIDAE, ROSENSTEINIIDAE]

———. 1968c. Notes sur les acariens de la famille Cloacaridae Camin et al. parasites du cloaque et des tissus profonds des tortues (Cheyletoidea: Trombidiformes). *Bull. Inst. Roy. Sci. Nat. Belg.* 44 (15): 1–33.

———. 1968d. Notes sur trois acariens remarquables (Sarcoptiformes). *Acarologia* 10:76–291. [CHIRORHYNCHOBIIDAE, CHIRODISCIDAE, RHYNCOPTIDAE]

———. 1969a. Les deutonymphes hypopiales vivant en association phoretique sur les mammiferes (Acarina: Sarcoptiformes). *Bull. Inst. Roy. Sci. Nat. Belg.* 45 (33): 1–262. [GLYCYPHAGIDAE, CHORTOGLYPHIDAE, ECHIMYOPODIDAE, PEDETOPODIDAE, HYPODERATIDAE]

———. 1969b. Morphologie et cycle évolutif des Glycyphagidae commensaux de la taupe Talpa europaea (Sarcoptiformes). *Acarologia* 11:750–95.

———. 1970a. *Coreitarsonemus* un nouveau genre d'acariens parasitant la glande odoriférante des Hémiptères Coreidae (Tarsonemidae: Trombidiformes). *Rev. Zool. Bot. Afr.* 82:315–44.

———. 1970b. *Les Myocoptidae en Afrique au sud du Sahara (Acarina: Sarcoptiformes).* Ann. Mus. Roy. Afr. Cent. (Sci. Zool.) 179. Tervuren, Belgium: Musée Royale de l'Afrique Centrale.

———. 1970c. Nomenclature des poils idiosomaux et description de trois espèces nouvelles dans la famille Ereynetidae (Trombidiformes). *Acarologia* 12:313–25.

———. 1970d. Nouveaux acarines nasicoles de la famille Turbinoptidae (Sarcoptiformes). *Bull. Ann. Soc. Roy. Belg. Entomol.* 106:28–36.

———. 1970e. Un nouvel acarien pilicole du Chile (Myocoptidae: Sarcoptiformes). *Bull. Ann. Soc. Roy. Belg. Entomol.* 106: 181–88.

———. 1971a. Clé et liste des espèces du genre *Boydaia* Womersley (Ereynetidae: Trombidiformes). *Acarologia* 13:98–112.

———. 1971b. Évolution de certains groupes d'hypopes en fonction du parasitisme (Acarina, Sarcoptiformes). *Acarologia* 13:171–75.

———. 1971c. Les listrophorides en Africa au sud du Sahara (Acarina: Sarcoptiformes). II, Familles Listrophoridae et Chirodiscidae. *Acta Zool. Pathol. Antverp.* 54:1–231.

———. 1971d. Note sur la répartition géographique des Coreitarsoneminae parasites des Hémiptères coreides avec description de taxa nouveaux (Acarina: Trombidiformes). *Bull. Ann. Soc. Roy. Belg., Entomol.* 107:81–88. [TARSONEMIDAE]

———. 1972a. Développement postembryonnaire chez les acariens de la sous-famille Speleognathinae (Ereynetidae: Trombidiformes). *Acarologia* 13:607–14.

———. 1972b. *Les listrophorides en Afrique au sud du Sahara (Acarina: Sarcoptiformes). III, Famille Atopomelidae.* Ann. Mus. Roy. Afr. Cent. (Sci. Zool.) 197. Tervuren, Belgium: Musée Royale de l'Afrique Centrale.

———. 1972c. Les listrophorides d'Australie et de Nouvelle-Guinée (Acarina: Sarcoptiformes). *Bull. Inst. Roy. Sci. Nat. Belg.* 48 (5): 1–196. [ATOPOMELIDAE, CHIODISCIDAE]

———. 1972d. Notes sur les Acariens des familles Cheyletidae et Harpyrhynchidae producteurs de gale chez les oiseaux ou les mammiferes. *Acta Zool. Pathol. Antverp.* 56:37–60. [HARPIRHYNCHIDAE]

———. 1973. Les listrophoridés d'Amérique néotropicale (Acarina: Sarcoptiformes). I, Familles Listrophoridae et Chirodiscidae. *Bull. Inst. Roy. Sci. Nat. Belg.* 49 (6): 1–149.

———. 1975a. Nouveaux taxa dans les Psoroptinae hypothese sur l'origine de ce groupe (Acarina, Sarcoptiformes, Psoroptidae). *Acta Zool. Pathol. Antverp.* 61:57–84.

———. 1975b. Observations sur les Myobiidae parasites des rongeurs. Evolution parallele hôtes-parasites (Acariens: Trombidiformes). *Acarologia* 16:441–75.

———. 1976a. *Arachnides, Acariens, Astigmata, Listrophoroidea.* Faune de Madagascar 42. Paris: ORSTOM, CNRS. [ATOPOMELIDAE, CHIRODISCIDAE]

———. 1976b. Notes sur les Harpyrhynchidae. Description de quatre espèces nouvelles (Acarina: Prostigmates). *Acarologia* 18:124–32. [HARPIRHYNCHIDAE]

———. 1977a. Nouveaux acariens astigmates cavernicoles du Kenya. *Rev. Suisse Zool.* 84:565–81. [ROSENSTEINIIDAE]

———. 1977b. Observations sur les Turbinoptidae d'Afrique au sud du Sahara. *Rev. Zool. Afr.* 91:83–116.

———. 1978a. Epidemiological problems of scabies. *Dermatology* 17:20–30.

———. 1978b. Notes sur les acariens astigmates cavernicoles. Description de deux genres nouveaux et de quatre espèces nouvelles. *Acarologia* 20:116–27. [GUANOLICHIDAE, ROSENSTEINIIDAE]

———. 1978c. Notes sur les Coreitarsoneminae (Acari, Prostigmata, Tarsonemidae) parasites de la glande odoriférante d'Hémiptères Coreidae en Afrique centrale. *Rev. Zool. Afr.* 92:959–70.

———. 1979. Les listrophorides d'Amérique néotropicale (Acarina: Astigmates). II, Famille Atopomelidae. *Bull. Inst. Roy. Sci. Nat. Belg.* 51 (7): 1–158.

———. 1980a. Les Coreitarsoneminae (Prostigmata, Tarsonemidae) parasites de la gland odoriférante d'Hémiptères de la famille Coreidae. I, Espèces d'Asie et d'Australie. *Acarologia* 21:279–90.

———. 1980b. Les Coreitarsoneminae (Prostigmata, Tarsonemidae) parasites de la gland odoriférante d'Hémiptères de la famille Coreidae. II, Espèces américaines. *Acarologia* 21:346–60.

———. 1981a. Notes on the genus *Laminosioptes* Megnin, 1880 (Acari, Astigmata) with description of three new species. *Syst. Parasitol.* 2:123–32. [LAMINOSIOPTIDAE]

———. 1981b. Notes sur les Listrophoridae (Acari, Astigmata). I, Distribution géographique, caractères morphologiques de clé des genres. *Acarologia* 22:305–12.

———. 1985a. Nouvelles observations sur l'organe ereynetal et les solenidions chez les Ereynetidae (Acari, Prostigmata). *Bull. Ann. Soc. Roy. Belg. Entomol.* 121:247–60.

———. 1985b. Systematic notes on the Speleognathinae (Acari, Ereynetidae) with description of new taxa and a key to the Trispeleognathini. *Bull. Ann. Soc. Roy. Belg. Entomol.* 121:143–52.

———. 1986. Observations sur les Glycyphagidae (Acari, Astigmata) avec description de deux espèces nouvelles. *Bull. Ann. Soc. Roy. Belg. Entomol.* 122:155–69.

———. 1987a. Notes on mites associated with Myriapoda. 1, Three new astigmatic mites from Afrotropical Myriapoda (Acari, Astigmata). *Bull. Inst. Roy. Sci. Nat. Belg., Entomol.* 57:161–72. [CHETOCHELACARIDAE]

———. 1987b. Nouveaux taxa dans la famille Heterocoptidae Fain, 1967 (Acari, Astigmata). *Bull. Ann. Soc. Roy. Belg. Entomol.* 123:243–53.

———. 1987c. Observations sur les larves du genre *Leptus* Latreille, 1795 (Acari, Erythraeidae) d'Afrique centrale. *Rev. Zool. Afr.* 101:103–23.

———. 1988. Notes on mites associated with Myriapoda. III, Two new species of the genus *Heterozercon* Berlese, 1888 (Acari, Mesostigmata) from Afrotropical myriapods. *Bull. Ann. Soc. Roy. Belg. Entomol.* 124:237–42.

———. 1989. Notes on mites associated with Myriapoda. IV, New taxa in the Heterozerconidae (Acari, Mesostigmata). *Bull. Inst. Roy. Sci. Nat. Belg.* 59:145–56.

———. 1991. A new larva of *Leptus* Latreille, 1796 (Acari, Erythraeidae) parasitic on Lepidoptera from Belgium. *Acarologia* 32:145–49.

———. 1992a. A new larval trombidiid, *Paputrombidium grootaerti* n. g. and n. sp. (Acari, Trombidiidae), parasitic on *Cymatopus* spp. (Diptera) from Papua New Guinea. *Bull. Inst. Roy. Sci. Nat. Belg., Entomol.* 62:105–8. [NEOTHROMBIIDAE]

———. 1992b. Two new genera and species of larval Trombidiinae (Acari, Prostigmata) from a carrion-beetle (Coleoptera, Silphidae) from Belgium. *Bull. Inst. Roy. Sci. Nat. Belg., Entomol.* 62:137–42. [NEOTHROMBIIDAE]

———. 1994. Adaptation, specificity and host-parasite coevolution in mites (Acari). *Internat. J. Parasitol.* 24:1273–78.

———. 1997. Description of a new genus and three new species of Tetrathrombiinae (Acari: Johnstonianidae) from their larval stages. *Bull. Inst. Roy. Sci. Nat. Belg., Entomol.* 67:57–64. [HYDRYPHANTIDAE]

———. 1998a. Description of mites (Acari) phoretic on Phoridae (Insecta: Diptera) with description of four new species of the genus *Uroseius* Berlese (Parasitiformes, Uropodina, Polyaspididae). *Internat. J. Acarol.* 24:213–20. [TRACHYTIDAE]

———. 1998b. New mites (Acari) phoretic on Phoridae and Ephydridae (Diptera) from Thailand. *Bull. Inst. Roy. Sci. Nat. Belg.* 68:53–61. [TRACHYTIDAE]

Fain, A., G. Anastos, J. Camin, and D. Johnston. 1967. Notes on the genus *Speleorhynchus.* Description of *S. praecursor* Neumann and of two new species. *Acarologia* 9:535–56.

Fain, A., and J. Bafort. 1966. Les hypopes parasitant les tissues cellulaires des pigeons sont les deutonymphes d'un acarien libre et pas celles d'un acarien plumicole (Note préliminaire). *Rev. Zool. Bot. Afr.* 74:313–16. [HYPODERATIDAE]

———. 1967. Cycle evolutif et morphologie de *Hypodectes (Hypodectoides) propus* (Nitzsch) acarien nidicole à deutonymphe parasite tissulaire des pigeons. *Bull. Acad. Roy. Belg. Classe Sci.*, ser. 5, 53:501–33. [HYPODERATIDAE]

Fain, A., and A. Baker. 1993. On some larval Microtrombidiidae (Acari, Prostigmata) parasitic on phlebotomine sandflies (Diptera, Psychodidae). *Bull. Ann. Soc. Roy. Belg. Entomol.* 129: 325–39.

Fain, A., and B. Bannert. 2000. Two new species of *Ophionyssus* Mégnin (Acari: Macronyssidae) parasitic on lizards of the genus *Gallotia* Boulenger (Reptilia: Lacertidae) from the Canary Islands. *Internat. J. Acarol.* 26:41–50.

Fain, A., and G. M. Barker. 2004. A new species of the genus *Riccardoella* Berlese, 1923 (Acari: Ereynetidae) occurring as a parasite in the pallial cavity of Athoracophoridae (Gastropoda) in New Zealand. *Bull. Soc. Roy. Belg. Entomol.* 40:43–48.

Fain, A., and J. C. Beaucournu. 1993. Les hypopes des astigmates (Acari) phorétiques sur les puces (Siphonaptera) de mammifères de d'oiseaux. *Bull. Inst. Roy. Sci. Nat. Belg., Entomol.* 63:77–93. [ACARIDAE, HISTIOSTOMATIDAE, WINTERSCHMIDTIIDAE]

Fain, A., and A. V. Bochkov. 2001. A new species of the genus *Epimyodex* Fain and Orts, 1969 (Acari: Cloacaridae: Epimyodicinae) parasitizing *Sorex trowbridgii* (Soricidae). *Internat. J. Acarol.* 27:221–23. [EPIMYODICIDAE]

———. 2002. On some little known and a new species of Myobiidae (Acari) associated with rodents. *Bull. Soc. Roy. Belg. Entomol.* 138:95–105.

Fain, A., and A. M. Camerik. 1994. Notes on the mites of the genus *Ereynetes* Berlese (Acari: Ereynetidae), with description of five new species from South Africa. *Bull. Inst. Roy. Sci. Nat. Belg., Entomol.* 64:145–64.

Fain, A., A. M. Camerik, F. S. Lukoschus, and F. M. Kniest. 1980. Notes on the hypopi of *Fibulanoetus* Mahunka, 1973, an anoetid genus with a pilicolous clasping organ. *Internat. J. Acarol.* 6:39–44. [HISTIOSTOMATIDAE]

Fain, A., and W. Chmielewski. 1987. The phoretic hypopi of two acarid mites described from ants' nest: *Tyrophagus formicetorum* Volgin, 1948, and *Lasioacarus nidicolus* Kadzhaja and Sevastianov, 1967. *Acarologia* 28:53–61. [ACARIDAE]

Fain, A., and J. M. Clark. 1994. Description and life-cycle of *Suladectes hughesae antipodus* subspec. nov. (Acari: Hypoderatidae) associated with *Sula bassana serrator* Gray (Aves: Pelecaniformes) in New Zealand. *Acarologia* 35:361–71.

Fain, A., Y. Coineau, and H. M. Andre. 1993. Description de deux espèces nouvelles d'astigmates sabulicoles (Acari). *Acarologia* 34:149–58. [SUIDASIIDAE]

Fain, A., and R. Domrow. 1974. The subgenus *Cytostethum* Domrow (Acari: Atopomelidae): Multiple speciation on the marsupial *Potorus tridactylus* (Kerr). *Aust. J. Zool.* 22:549–72.

Fain, A., and D. Drugmand. 1993. Notes on the genus *Hexathrombium* Cooreman, 1944 (Acari, Trombidiidae) with description of a new tribe and species from Afrotropical Staphylinidae (Coleoptera). *Bull. Ann. Soc. Roy. Belg. Entomol.* 129:121–28. [EUTROMBIDIIDAE]

Fain, A., and P. Elsen. 1967. Les acariens de la famille Knemidokoptidae producteurs de gale chez les oiseaux (Sarcoptiformes). *Acta Zool. Pathol. Antverp.* 45:3–145.

Fain, A., and G. O. Evans. 1966. The genus *Proctotydaeus* Berl. (Acari: Iolinidae) with descriptions of two new species. *Ann. Mag. Nat. Hist.*, ser. 13, 9:149–57.

Fain, A., and M. Ferrando. 1990. A new species of *Schwiebea* Oudemans (Acari: Acaridae) parasitizing the trout *Salmo trutta fario* in Spain. *Rev. Iber. Parasitol.* 50:67–71.

Fain, A., P. Gerrits, and F. S. Lukoschus. 1982. *Teinocheylus gundii* spec. nov. from *Ctenodactylus gundi* (Acari, Cheyletidae). *Rev. Zool. Afr.* 96:448–56.

Fain, A., and P. Grootaert. 1993. A new larval trombidiid of the genus *Microtrombidium* (Acari, Trombidiidae) parasitic on shore flies (Diptera: Canacidae) in the Galapagos Is. *Bull. Ann. Soc. Roy. Belg. Entomol.* 129:151–57.

———. 1994. Four new species of larval Microtrombidiidae (Acari) parasitic on Diptera (Insecta) from Papua New Guinea and Brunei (Borneo). *Bull. Inst. Roy. Sci. Nat. Belg., Entomol.* 64:165–75.

———. 1996. The genus *Durenia* Vercammen-Grandjean, 1955 (Acari: Trombellidae) in Papua New Guinea. Description of a new species. *Bull. Inst. Roy. Sci. Nat. Belg.* 66:51–55.

———. 2001. A new larval Microtrombidiidae (Acari) phoretic on a ceratopogonid fly (Insecta, Diptera) from southern China. *Belg. J. Entomol.* 3:189–94.

Fain, A., B. Guérin, and B. J. Hart. 1990. *Mites and allergic disease.* Varennes en Argonne, France: Allerbio.

Fain, A., and S. L. Gummer. 1987. Two new species of *Leptus* Latreille, 1796 (Acari: Erythraeidae) from the USA. *Internat. J. Acarol.* 13:135–40.

Fain, A., and K. Hyland. 1974. The listrophoroid mites in North America. II, The family Listrophoridae Megnin and Trouessart (Acarina: Sarcoptiformes). *Bull. Inst. Roy. Sci. Nat. Belg.* 50: 1–69.

———. 1975. Speleognathinae collected from birds in North America (Acarina: Ereynetidae). *J. New York Entomol. Soc.* 83:203–8.

———. 1985. Evolution of astigmatid mites on mammals. In Kim (1985a), 641–58.

Fain, A., K. E. Hyland, and T. H. G. Aitken. 1977. Flower mites of the family Ascidae phoretic in nasal cavities of birds (Acarina: Mesostigmata). *Acta Zool. Pathol. Antverp.* 69:99–154. [MELICHARIDAE]

Fain, A., and R. Jocqué. 1996a. A new genus and species of larval Eutrombidiinae Thor, 1935 (Acari: Microtrombidiidae) from an Afrotropical spider. *Internat. J. Acarol.* 22:11–16.

———. 1996b. A new larva of the genus *Leptus* Latreille, 1796 (Acari: Erythraeidae) parasitic on a spider from Rwanda. *Internat. J. Acarol.* 22:101–8.

Fain, A., and D. E. Johnston. 1966. Nouveaux acariens nasicoles d'Oiseaux nord-Americains (Acari: Rhinonyssidae). *Bull. Soc. R. Zool. Anvers* 38:25–41.

———. 1975. A new algophagin mite, *Algophagopsis pneumatica* gen. n., sp. n. living in a river (Astigmata: Hyadesiidae). *Bull. Ann. Soc. Roy. Belg. Entomol.* 111:66–70. [ALGOPHAGIDAE]

Fain, A., and W. Knülle. 1981. The life-cycle of *Acalvolia squamata* (Oudemans, 1909) (Acari, Astigmata, Saproglyphidae). *Internat. J. Acarol.* 7:139–42. [WINTERSCHMIDTIIDAE]

Fain, A., and L. Lambrechts. 1987. Observations on the acarofauna of fish aquariums. I, Mites associated with *Discus* fish. *Bull. Ann. Soc. Roy. Belg. Entomol.* 123:87–102.

Fain, A., L. Lambrechts, and G. Wauthy. 1988. Observations on the acarofauna of fish aquariums. III, Presence of *Hydrozetes lemnae* (Coggi, 1899) in aquariums in Belgium. *Bull. Ann. Soc. Roy. Belg. Entomol.* 124:255–56. [HYDROZETIDAE]

Fain, A., and F. Lukoschus. 1968. *Psorergates* (*Psorobia*) *foinae* sp. n., acarien producteur de gale chez la fouine en Belgique. *Bull. Inst. Roy. Sci. Nat. Belg.* 44 (14): 1–6. [PSORERGATIDAE]

———. 1970. Parasitic mites of Surinam. II, Skin and fur mites of

the families Psoroptidae and Lobalgidae. *Acta Zool. Pathol. Antverp.* 51:49–60.
———. 1971. Parasitic mites of Surinam. XV, Nasal ereynetid mites of bats with a key of the known species (Trombidiformes). *Bull. Ann. Soc. Roy. Belg. Entomol.* 107:284–97.
———. 1974. Observations sur le développement postembryonnaire des acariens de la famille Glycyphagidae à hypopes pilicoles ou endofolliculaires (Acarina: Astigmata). *Bull. Acad. Roy. Belg. (Cl. Sci.)*, ser. 5, 60:1137–59.
———. 1976a. A new genus and species of Myobiidae from the marsupial *Lestoros inca* (Acarina: Prostigmata). *Acarologia* 18:489–95.
———. 1976b. Observations sur les Myobiidae d'insectivores avec description de taxa nouveaux (Acarina: Prostigmates). *Acta Zool. Pathol. Antverp.* 66:121–88.
———. 1976c. Two new species and two new genera of Sarcoptidae from a marsupial and an insectivore. *Internat. J. Acarol.* 2:1–8. [RHYNCOPTIDAE, SARCOPTIDAE]
———. 1978a. *Dipodomydectes americanus* gen. and sp. n. (Acari: Hypoderidae) from the kangaroo rat. *J. Parasitol.* 64:137–38. [HYPODERATIDAE]
———. 1978b. New endofollicular or subcutaneous hypopi from mammals (Acarina: Astigmata). *Acarologia* 19:484–93. [CHORTOGLYPHIDAE, GLYCYPHAGIDAE, HYPODERATIDAE]
———. 1983a. Five new species of Rosensteiniidae (Acarina, Astigmata) from Indonesia, associated with bats or with the earwig *Xeniaria jacobsoni*. *Zool. Meded. Leiden.* 57:31–42.
———. 1983b. *Katydiseius nadchatrami* n. g., n. sp. (Acari, Otopheidomenidae) from the tracheae of a Malaysian katydid *Chloracris brullei* Pictet & Saussure, 1892 (Orthoptera, Pseudophyllidae). *Internat. J. Acarol.* 9:173–78.
———. 1984. New observations on the genus *Prolistrophorus* Fain, 1970 (Acari: Astigmata: Listrophoridae). *Syst. Parasitol.* 6: 161–85.
———. 1985. The genus *Schizocarpus* Trouessart, 1896 (Acari, Chirodiscidae) from the beaver *Castor fiber* L.: an example of multiple speciation. *Entomol. Ab. Staat. Mus. Tierk. Dresden* 49: 35–68.
———. 1986. Observations on the life cycle of *Neottialges* (*Pelecanectes*) *evansi* Fain, 1966 and *Phalacrodectes whartoni* Fain, 1967 with descriptions of new taxa (Acari, Hypoderatidae). *Syst. Parasitol.* 8:291–316.
Fain, A., F. S. Lukoschus, and P. Hallmann. 1966. Le genre *Psoregates* chez les murides. Description de trois espèces nouvelles (Psorergatidae: Trombidiformes). *Acarologia* 8:251–74.
Fain, A., F. S. Lukoschus, J. M. W. Louppen, and E. Mendez. 1973. *Echimyopus dasypus*, n. sp., a hypopus from *Dasypus novemcinctus* in Panama (Glycyphagidae, Echimyopinae: Sarcoptiformes). *J. Med. Entomol.* 10:552–55.
Fain, A., F. S. Lukoschus, and M. Nadchatram. 1981. Malaysian parasitic mites. I, New Rosensteiniidae (Astigmata) from *Cheiromeles torquatus* Horsfield (Chiroptera) and from the associated *Arixenia* spp. (Dermaptera). *Acarologia* 22:187–98.
Fain, A., F. S. Lukoschus, and P. G. Rosmalen. 1982. Observations on the genus *Epimyodex* Fain and Orts, 1969, with description of two new species. Transfer of this genus to the Cloacaridae (Acari, Prostigmata). *Bull. Inst. Roy. Sci. Nat. Belg. Entomol.* 54:1–10. [EPIMYODICIDAE]
Fain, A., A. J. Munting, and F. Lukoschus. 1970. Les Myocoptidae parasites des rongeurs en Hollande et en Belgique. *Acta Zool. Pathol. Antverp.* 50:67–172.
Fain, A., M. I. Noti, and M. Dufrêne. 1995. Observations on the mites (Acari) associated with Carabidae (Coleoptera) in Belgium. 1, Annotated list of the species. *Internat. J. Acarol.* 21:107–22.
Fain, A., and S. Orts. 1969. *Epimyodex talpae* n. g. n. sp. parasite sous-cutané de la taupe en Belgique (Demodicidae: Trombidiformes). *Acaralogia* 11:65–68. [EPIMYODICIDAE]
Fain, A., and J. R. Philips. 1977. Astigmatic mites from nests of birds of prey in U.S.A. I, Description of four new species of Glycyphagidae. *Internat. J. Acarol.* 3:105–14. [ECHIMYOPODIDAE, EUGLYCYPHAGIDAE, GLYCYPHAGIDAE]
———. 1978a. Astigmatic mites from nests of birds of prey in U.S.A. III, *Sapracarus tuberculatus* g. n., sp. n. (Acari, Astigmata, Saproglyphidae). *Acta Zool. Pathol. Antverp.* 70:227–31. [SUIDASIIDAE]
———. 1978b. Notes on the genus *Suidasia* Oudemans, 1905 with descriptions of a new species from Australia (Acari, Astigmata, Saproglyphidae). *Internat. J. Acarol.* 4:115–23. [SUIDASIIDAE]
———. 1979. Astigmatic mites from nests of birds of prey in the U.S.A. V, Four new species of Anoetidae. *Internat. J. Acarol.* 5:147–53. [HISTIOSTOMATIDAE]
Fain, A., and G. Rack. 1987. Notes on the mites living in the flowers of *Espeletia* spp. (Asteraceae) in Colombia. I, *Carpoglyhphus sturmi* sp. n. (Acari, Carpoglyphidae). *Entomol. Mitt. Zool. Mus. Hamb.* 9 (130): 9–19.
Fain, A., and G. Ripka. 1998. Two new larval Trombidiidae of the genus *Podothrombium* Berlese, 1910 (Acari: Prostigmata) from Hungary. *Bull. Inst. Roy. Sci. Nat. Belg., Entomol.* 68:71–78.
Fain, A., and A. E. Rosa. 1983a. Acari domum meliponinarum brasiliensium habitantes. II, Two new genera and species of mites (Acari: Astigmata) from the nests of bees in Brazil. *Bull. Inst. Roy. Sci. Nat. Belg., Entomol.* 55 (2): 1–10. [GAUDIELLIDAE]
———. 1983b. Acari domum meliponinarum brasiliensium habitantes. IV, New astigmatic mites from the nest of the bee *Partamona* sp. (Meliponidae). *Bull. Inst. Roy. Sci. Nat. Belg., Entomol.* 55:1–13. [MELIPONOCOPTIDAE]
Fain, A., L. Scheepers, and W. de Groot 1982. Dermatite prurigineuse de longue durée chez une femme, produite par l'acarien parasite du chien *Cheyletiella yasguri* Smiley. *Rev. Médicale Liège* 37:623–25. [CHEYLETIELLIDAE]
Fain, A., and R. L. Smiley. 1989. A new cloacarid mite (Acari: Cloacaridae) from the lungs of the great horned owl, *Bubo virginianus*, from the U.S.A. *Internat. J. Acarol.* 15:111–15.
Fain, A., R. L. Smiley, and U. Gerson. 1997. New observations on the chaetotaxy and the solenidiotaxy in the Cheyletidae (Acari: Prostigmata). *Bull. Inst. Roy. Sci. Nat. Belg., Entomol.* 67:65–87.
Fain, A., M. A. Smith, and J. O. Whitaker, Jr. 1986. The fur mites (Acari: Listrophoridae) of the round-tailed muskrat, *Neofiber alleni*. *Bull. Ann. Soc. Roy. Belg. Entomol.* 122:171–81.
Fain, A., and J. L. Van Goethem. 1986. Les acariens du genre *Riccardoella* Berlese, 1923, parasites du poumon de mollusques gastéropodes pulmonés terrestre. *Acarologia* 27:127–40. [EREYNETIDAE]
Fain, A., and G. Wauthy. 1979. Les Acaridae (Acari, Astigmata)

d'horizons hemiedaphiques dans trente forets decidues de Belgique. *Bull. Ann. Soc. Roy. Belg. Entomol.* 115:169–82.

Fain, A., and J. O. Whitaker, Jr. 1988. Mites of the genus *Schizocarpus* Trouessart, 1896 (Acari, Chirodiscidae) from Alaska and Indiana, USA. *Acarologia* 29:395–405.

Fain, A., C. E. Yunker, J. van Goethem, and D. E. Johnston. 1982. Notes on the Ewingiidae (Acari, Astigmata) living in association with pagurids and fresh-water crabs (Crustacea), with description of a new genus and a new species. *Bull. Inst. Roy. Sci. Nat. Belg., Entomol.* 54 (8): 1–7. [ACARIDAE]

Fain, A., and F. Zumpt. 1974. Notes on three species of Anoetidae, two of which are new, living as commensals or parasites in the ear of an African buffalo. *Acta Zool. Pathol. Antverp.* 58:97–102. [HISTIOSTOMATIDAE]

Fan, Q.-H. 1997a. The Homocaligidae from China, with description of two new species (Acari: Raphignathoidea). *Entomol. Sin.* 4:337–42.

———. 1997b. A new species and three new records of the family Cryptognathidae from China (Acari: Raphignathoidea). *Acta Arachnol. Sin.* 6:130–36.

———. 2000. The morphology of *Xenocaligonellidus smileyi* (Acari: Xenocaligonellidae). In *Systematic and faunistic research on Chinese insects*, ed. B. K. Zhang Yalin, 290–97. Proc. 5th Nat. Congress Insect Taxon. N.p.: China Agr. Press.

Fan, Q.-H., and Y. Chen. 1996. A new genus of the family Xenocaligonellidae Gonzalez (Acari: Raphignathoidea). *Syst. Appl. Acarol.* 1:123–26.

———. 1997. A new species of *Camerobia* Southcott, 1957 (Acari: Camerobiidae). *Acta Arachnol. Sin.* 6:42–45.

———. 2005. A review of the Pomerantziidae (Acari: Prostigmata: Pomerantzioidea), with the description of a new genus. *Zootaxa* 1037:1–22.

Fan, Q.-H., Y. Chen, and J.-Z. Lin. 1997. Genus *Agistemus*, with description of two new species from China. *Wuyi Sci. J.* 13:139–47.

Fan, Q.-H., and L.-S. Li. 1993. Descriptions of three new species of superfamily Raphignathoidea (Acari: Actinedida). *J. Fujian Agr. Univ. (Nat. Sci. Ed.)* 22:321–24.

———. 1995. A new genus and a new species of Caligonellidae associated with food from China. *Acta Zootax. Sin.* 20:323–27.

Fan, Q.-H., and X. Liu. 1999. New species of *Ledermuelleriopsis* Willmann and *Pseudostigmaeus* Wood from China (Acari: Prostigmata: Stigmaeidae). *Syst. Appl. Acarol.* 4:153–58.

Fan, Q.-H., and D. E. Walter. 2004. Genus *Caligohomus* Habeeb (Acari: Prostigmata: Stigmaeidae). *Syst. Appl. Acarol.* 9:77–88.

Fan, Q.-H., D. E. Walter, and H. C. Proctor. 2003a. A review of the family Barbutiidae (Acari: Raphignathoidea) with the description of two new species from Australia. *Syst. Appl. Acarol.* 8:107–30.

———. 2003b. A review of the genus *Ledermuelleriopsis* Willmann (Acari: Prostigmata: Stigmaeidae). *Invertebr. Syst.* 17:1–24.

Fan, Q.-H., and C. Yan. 1997. The genus *Storchia*, with the description of a new species (Acari: Prostigmata: Stigmaeidae). *Syst. Appl. Acarol.* 2:161–66.

Fan, Q.-H., and X.-M. Yin. 2000. The genus *Raphignathus* (Acari: Raphignathidae) from China. *Syst. Appl. Acarol.* 5:83–98.

Fan, Q.-H., Y.-X. Zhang, and Q.-Y. Liu. 2000. Raphignathoid mites on bamboo from Fujian, China (Acari: Prostigmata). *Syst. Appl. Acarol. Spec. Publ.* 4:49–68. [STIGMAEIDAE, EUPALLOPSELLIDAE]

Fan, Q.-H., and Z. Q. Zhang. 2004. *Revision of* Rhizoglyphus *Claparède (Acari: Acaridae) of Australasia and Oceania*. London: Syst. Appl. Acarol. Soc.

Faraji, F., and X. Cornejo. 2006. A new *Hattena* Domrow (Acari: Ameroseiidae) from Ecuadorian mangroves and a new generic record for South America. *Internat. J. Acarol.* 32:287–91.

Farish, D. J., and R. C. Axtell. 1971. Phoresy redefined and examined in *Macrocheles muscaedomesticae* (Acarina: Macrochelidae). *Acarologia* 13:16–29.

Farkas, H. K. 1966. Some problems of eriophyoid mite phylogeny (Acarina, Eriophyoidea). Zeszyty Problemowe Postepow Nauk Rolniczych, Zagad. *Acarologia* 65:189–94.

———. 1969. On the main lines of the phylogenetical evolution in the eriophyoid mites (Acari). *Ann. Hist.–Nat. Mus. Nat. Hung., Zool.* 61:377–82.

Faroni, L. R. D'A., R. N. C. Guedes, and A. L. Matioli. 2000. Potential of *Acarophenax lacunatus* (Prostigmata: Acarophenacidae) as a biological control agent of *Rhyzopertha dominica* (Coleoptera: Bostrichidae). *J. Stored Prod. Res.* 36:55–63.

Farrier, M. H. 1957. *A revision of the Veigaiidae (Acarina)*. Tech. Bull., N.C. Agr. Exp. Sta. 124. Raleigh, NC.

Fashing, N. J. 1975. Life history and general biology of *Naiadacarus arboricola* Fashing, a mite inhabiting water-filled treeholes (Acarina: Acaridae). *J. Nat. Hist.* 9:413–24.

———. 1979. *Neoguanolichus panamensis*, a new species of Guanolichidae from bat caves in Panama (Acarina: Sarcoptiformes). *Acarologia* 21:108–16.

———. 1984. A possible osmoregulatory organ in the Algophagidae (Astigmata). In Griffiths and Bowman, 1:310–15.

———. 1991. Deutonymphal dimorphism in the genus *Hericia* (Astigmata: Algophagidae). In Dusbábek and Bukva, 2:287–91.

———. 1994. Life history patterns of astigmatid inhabitants of water-filled treeholes. In Houck (1994b), contribution no. 7, 160–85. [ACARIDAE, ALGOPHAGIDAE, HISTIOSTOMATIDAE]

———. 1998. Functional morphology as an aid in determining trophic behavior. The placement of astigmatic mites in food webs of water-filled tree hole communities. *Exp. Appl. Acarol.* 22:435–53. (Also appears in Bruin et al., 333–47.) [ACARIDAE, ALGOPHAGIDAE, HISTIOSTOMATIDAE]

———. 2001. Morphological adaptations associated with mate-guarding behaviour in the genus *Hericia* (Acari: Algophagidae). In Halliday et al. (2001), 176–79.

———. 2002. *Nepenthacarus*, a new genus of Histiostomatidae (Acari: Astigmata) inhabiting the pitchers of *Nepenthes mirabilis* (Lour.) Druce in Far North Queensland, Australia. *Aust. J. Entomol.* 41:7–17.

Fashing, N. J., and K. S. Marcuson. 1996. Fine structure of the axillary organs of *Fusohericia lawrencei* Baker and Crossley (Astigmata: Algophagidae). In Mitchell et al. (1996), 1:381–84.

Fashing, N. J., and B. M. OConnor. 1984. *Sarraceniopus*—a new genus for histiostomatid mites inhabiting the pitchers of the Sarracenlaceae (Astigmata: Histiostomatidae). *Internat. J. Acarol.* 10:217–27.

Fashing, N. J., B. M. OConnor, and R. L. Kitching. 1996. Adaptations for swimming in the genus *Creutzeria* (Astigmata: Histiostomatidae). In Mitchell et al. (1996), 1:385–88.

———. 2000. *Lamingtonacarus*, a new genus of Algophagidae (Acari:

Astigmata) from water-filled treeholes in Queensland, Australia. *Invertebr. Taxon.* 14:591–606.
Fashuyi, S. A. 1990. Occurrence of *Unionicola macani* Gledhill (Hydrachnellae, Acari) in the prosobranch mollusc *Lanistes ovum* Peters in Ajara fish ponds in Badagry, Nigeria. *Hydrobiologia* 202:171–74. [UNIONICOLIDAE]
Feider, Z. 1955a. Fauna Republicii Populare Romîne. Arachnida, Acarina, Trombidioidea. *Acad. Rep. Pop. Romîne* 5:1–186. [TANAUPODIDAE]
———. 1955b. O noua larva de acarian parazita pe un plecopter (*Tetrathrombium zachvatkini* n. gen., n. sp.). *Acad. R.P.R. Fil. Iasi Stud. Si. Cercet. Stiint. Biol. Agric.* 6:209–218. [HYDRYPHANTIDAE]
———. 1958. Prima larva din familia Trombellidae (Acarina) obtinuta prin cultura si despre noua caracterizare a familei. Acad. *R.P.R. Fil. Iasi Stud. Si. Cercet. Stiint. Biol. Agric.* 9:265–82.
———. 1977. Quelques acariens cavernicoles de Cuba de la superfamille Trombiculoidea. *Résult. Expéd. Biospéol. Cubano–Roum. Cuba* 2:201–8. [NEOTROMBIDIIDAE]
Feider, Z., and I. Mironescu. 1973. A particular way of piercing the nasal mucous membrane used by mites of the family Rhinonyssidae Trouessart, 1895. In Daniel and Rosický, 415.
Feider, Z., and I. Suciu. 1956. O noua larva: *Tetrathrombium macronychus* n. sp. (Acarina) parazita pe un Tipulid. *Anal. Stiint. Univ. "Al. I. Cuza" Iasi* (Ser. Noua), Sect. II, *Stiint. Natur.* 2:163–74. [HYDRYPHANTIDAE]
Feider, Z., and N. Vasilu. 1968. *Labidostomma nepalense* (Nicoletiellidae) une nouvelle espèce asiatique. *Acarologia* 10:630–44. [LABIDOSTOMATIDAE]
———. 1969. Revision critique de la famille Nicoletiellidae. In Evans (1969c), 201–7. [LABIDOSTOMATIDAE]
———. 1970. Six espèces de Nicoletiellidae d'Amérique du Sud. *Acarologia* 12:282–309. [LABIDOSTOMATIDAE]
———. 1972. Zoological collectings by the Hungarian Natural History Museum in Korea. 6, Quelques Nicolettiellidae de Corée du Nord (Acari). *Ann. Hist.–Nat. Mus. Nat. Hung.* 64:375–85. [LABIDOSTOMATIDAE]
Feider, Z., N. Vasiliu, and M. Calugar. 1974. Nouvelle contribution á l'étude des Nicoletiellidae d'Amérique du Sud. *Acarologia* 16:413–27. [LABIDOSTOMATIDAE]
Feldman-Muhsam, B., and Y. Havivi. 1977. *Stigmacarus lukoschusi*, new genus and new species (Acarina, Podapolipidae) from beetles in Italy. *Acarologia* 18:545–52.
Felt, E. P. 1940. *Plant galls and gall makers.* Ithaca, NY: Comstock. [ERIOPHYIDAE]
Fernandez, H. R. 1987a. Acaros intersticiales de la Republica Argentina. I, *Omartacarus tucumanensis* sp. n. (Acari, Hydrachnellae). *Rev. Soc. Entomol. Arg.* 44:395–98. [OMARTACARIDAE]
———. 1987b. Acaros intersticiales del noroeste Argentino. II, Los géneros *Protolimnesia* Lundblad y *Diamphidaxona* Cook (Acari: Hydrachnellae). *Neotropica* 33:139–46. [LIMNESIIDAE, HYGROBATIDAE]
———. 1992. First record of *Tubophorella* Viets (Limnesiidae, Parasitengona, Acari) in South America: *Tubophorella multiacetabulata* new species in Arroyo Tafi, Tucuman Province, Argentina. *Stud. Neotrop. Fauna Environ.* 27:253–59.
———. 1993. Acaros intersticiales del noroeste Argentino. IV, Analisis filogenetico y biogeografico de *Omartacarus* Cook (Acari: Omartacaridae): Una primera aproximación. *Rev. Soc. Entomol. Arg.* 52:107–17.
———. 1994. Acaros intersticiales del noroeste Argentino. III, Nuevos aportes morfologicos sobre especies hiporreicas de *Protolimnesia* Lundbald y *Diamphidaxona* Cook. *Physis*, Sec. B, 49: 23–30. [LIMNESIIDAE, HYGROBATIDAE]
———. 2001. Acaros hiporreicos de Argentina: Nuevos datos sobre Neomamersinae Lundblad (Limnesiidae). *Physis*, Sec. B, 58: 21–28.
———. 2002. Interstitial water mites of Argentina: *Omartacarus* Cook (Omartacaridae) and *Meremecia* Cook (Limnesiidae) (Acari: Hydrachnidia). *Zootaxa* 73:1–6.
Fernandez, H. R., and L. E. Grosso. 1991. Una nueva especie de *Omartacarus* Cook (Acari, Hydrachnellae) de la Provincia de San Juan, Republica Argentina. *Acta Zool. Lilloana* 40:43–45. [OMARTACARIDAE]
Fernandez, N. A., G. Alberti, and G. Kümmel. 1991. Spermatophores and spermatozoa of oribatid mites (Acari: Oribatida). Part I, Fine structure and histochemistry. *Acarologia* 32:261–86.
Fernandez, N. A., and F. Athias-Binche. 1986. Analyse demographique d'une population d'*Hydrozetes lemnae* Coggi, Acarien Oribate infeode a la lentille d'eau *Lemna gibba* L. en Argentine. I, Methodes et techniques, demographie d' *H. lemnae* comparais. *Zool. Jb. Syst.* 113:213–28. [HYDROZETIDAE]
Fernandez, N. A., and R. Cleva. 1997. Contribution to the knowledge of the oribatid fauna of Argentina. *Nacunansella diminuta* n. gen. n. sp. *Acarologia* 39 (2): 173–84. [NACUNANSELLIDAE]
Fernandez, N. A., and Y. Coineau. 2002. *VARROA, Tueur d'Abeilles.* Anglet, France: Atlantica.
———. 2007. *Maladies, parasites et autres ennemis de l'abeille mellifère.* Biarritz: Atlantica-Séguier.
Fernandez, N. A., Y. Coineau, and P. Theron. 2006. Varroa, *the serial bee killer mite.* Biarritz: Atlantica-Séguier.
Fernandez, N. A., J. Marcangeli, and M. Eguaras. 1997. The Acari (Oribatida) from the arid zone of Argentina. II, *Huilicheremaeus michaii* n. gen., n. sp. *Acarologia* 38:79–85. [LICNEREMAEIDAE]
Fernandez, N. A., P. A. Martinez, and M. J. Eguaras. 1991. Acariens oribates des sols organiques des Andes de la République Argentine. I, *Pheroliodes inca* n. sp. *Acarologia* 32 (2): 183–90. [PHEROLIODIDAE]
Ferradas, B. E. R. de. 1974. Acaros acuaticos del género *Unionicola* Haldeman, 1842 (Acari, Hydrachnellae, Unionicolidae) en moluscos Gasteropoda y Pelecypoda de la Argentina. *Physis*, Sec. B, 33:177–86.
Filimonova, S. A. 1999. The structure of female reproductive system in the parasitic mite *Myobia murismusculi* (Trombidiformes: Myobiidae). *Parazitologiya* 33:377–86. [in Russian]
Filippich, L. J., and R. Domrow. 1985. Harpyrhynchid mites in scaly-breasted lorikeet, *Trichoglossus chlorolepidotus* (Kuhl). *J. Wildl. Dis.* 21:457–58. [HARPIRHYNCHIDAE]
Filipponi, A. 1964. Experimental taxonomy applied to the Macrochelidae (Acari: Mesostigmata). In Evans (1969c), 92–100.
Filipponi, A., and G. Francaviglia. 1963. Oviparitá e larviparitá in *Macrocheles peniculatus* Berl. (Acari: Mesostigmata) regolate da fattori ecologici. *Parassitologia* 24:81–104.
Filipponi, A., and F. Pegazzano. 1960. Acari del genere *Glyptholaspis* nom. nov. pro *Macrocheles* (Macrocheles) Berlese 1918 (Mesostigmata, Macrochelidae). *Redia* 45:133–71.

———. 1962. Specie italiane del gruppo-*glaber* (Acarina, Mesostigmata, Macrocheles). *Redia* 47:211–38.
———. 1963. Specie italiane del gruppo-*subbadius* (Acarina, Mesostigmata, Macrochelidae). *Redia* 48:69–91.
———. 1967. Contributo alla conoscenza del genere *Holostaspella* Berlese, 1903 (Acari, Mesostigmata: Macrochelidae). *Redia* 50:219–59.
Finnegan, S. 1931. On a new heterozerconid mite parasitic on a snake. *Proc. Zool. Soc. Lond.* 4:1349–57.
Flaherty, D. L., and M. A. Hoy. 1971. Biological control of Pacific mites and Willamette mites in San Joaquin Valley vineyards. Part III, Role of tydeid mites. *Res. Popul. Ecol.* 13:80–96. [IOLINIDAE]
Flechtmann, C. A. H. 1985. On the biology of *Ameroseius dendrovagans* (Acari, Mesostigmata, Ameroseiidae). *Rev. Bras. Zool.* 2:397–99.
Flechtmann, C. H. W. 1976. *Elementos de acarologia*. São Paulo: Livr. Nobel S.A.
———. 1983. Acari domum meliponinarum Brasiliensiuim habitantes. I. *Triplogynium irapora*, a new species of Triplogyniidae (Celaenopsoidea, Mesostigmata). *Acarologia* 24:355–59.
———. 1992. First record of a Paratydeidae (Acari, Prostigmata) in South America with description of *Scolotydaeus corticicola* sp. n. *Rev. Bras. Zool.* 9:299–304.
Flechtmann, C. H. W., A. Fain, and M. C. A. Leal. 1985. Acari domum meliponinarum brasiliensium habitantes. VII, *Ereynetes (Gymnereynetes) meliponae* nov. spec. (Acari, Ereynetidae). *Rev. Bras. Biol.* 45:95–99.
Flechtmann, C. H. W., and D. E. Johnston. 1990. *Zeterohercon*, a new genus of Heterozerconidae (Acari: Mesostigmata) and the description of *Zeterohercon amphisbaenae* n. sp. from Brazil. *Internat. J. Acarol.* 16:143–48.
Fleschner, C. A., and K. Y. Arakawa. 1953. The mite *Tydeus californicus* on citrus and avocado leaves. *J. Econ. Entomol.* 45:1092. [TYDEIDAE]
Fleschner, C. A., M. E. Badgley, D. W. Ricker, and J. C. Hall. 1956. Air drift of spider mites. *J. Econ. Entomol.* 49:624–27. [TETRANYCHIDAE]
Flogaitis, E. 1992. Catalogue of oribatid mites of Greece (Acari: Oribatida). *Biol. Gallo-Hell.* 19:29–54.
Flynn, R. J., ed. 1973. *Parasites of laboratory animals*. Ames: Iowa State Univ. Press.
Foelix, R. F. 1982. *Biology of spiders*. Cambridge, MA: Harvard Univ. Press.
Fonseca, F. da. 1954. Notas de acarologia. XXXVIII, Sarcoptiformes da preguiça: *Lobalges trouessarti* gen. n., sp. n. (Acari, Epidermoptidae). *Mem. Inst. Butantan* 26:85–92. [LOBALGIDAE]
Forbes, M. R., K. E. Muma, and B. P. Smith. 1999. Parasitism of *Sympetrum* dragonflies by *Arrenurus planus* mites: Maintenance of resistance particular to one species. *Internat. J. Parasitol.* 29:991–99. [ARRENURIDAE]
Forbes, M. R. L., and R. L. Baker. 1990. Susceptibility to parasitism: experiments with the damselfly *Enallagma ebrium* (Odonata: Coenagrionidae) and larval water mites, *Arrenurus* spp. (Acari: Arrenuridae). *Oikos* 58:61–66.
———. 1991. Condition and fecundity of the damselfly, *Enellagma ebrium* (Hagen): The importance of ectoparasites. *Oecologia* 86:335–41. [ARRENURIDAE]
Forrester, D. J., M. G. Spalding, and J. B. Wooding. 1993. Demodicosis in black bears (*Ursus americanus*) from Florida. *J. Wildl. Dis.* 29:136–38.
Foulkes, G. D. 1983. A new method of mounting acarines and other meiofauna. *Internat. J. Acarol.* 9:211–13.
Fox, I. 1959. A new mite taken with rats in Puerto Rico (Acari: Diplogyniidae). *Acarologia* 1:296–98.
Franklin, E., J. Adis, and St. Woas. 1997. *The oribatid mites. The central Amazon floodplain*. Ecological Studies, Vol. 126. Berlin: Springer-Verlag.
Franklin, E., R. L. Guimsrães, J. Adis, and H. O. R. Schubart. 2001. Resistência à submersão de ácaros (Acari: Oribatida) terrestres de florestas inundáveis e de terra firma na Amazônia Central em condições experimentais de laboratório. *Acta Amazonica* 31:285–98.
Franklin, E., and S. Woas. 1992. Some basic oppiid-like taxa (Acari, Oribatei) from Amazonia. *Andrias (Karlsruhe)* 9:57–74.
Franks, N. R., K. J. Healy, and L. Byrom. 1991. Studies on the relationship between the ant ectoparasite *Antennophorus grandis* (Acarina: Antennophoridae) and its host *Lasius flavus* (Hymenoptera: Formicidae). *J. Zool. Lond.* 225:59–70. [ANTENNOPHORIDAE]
Frazer, B. D., and C. Nelson. 1981. Note on the occurrence of predatory *Anystis* mites (Acari: Anystidae) in SW British Columbia. *J. Entomol. Soc. British Columbia* 78:46.
Fritsch, W. 1954. Die Milbengattung *Harpyrhynchus* Megnin, 1878. *Zool. Anz.* 152:177–98. [HARPIRHYNCHIDAE]
Fritz, G. N. 1995. Oribatid mites infected with cysticercoids of *Moniezia expansa* (Cestoda: Anoplocephalidae). *Internat. J. Acarol.* 21:233–38.
Frost, W. E., and P. M. Ridland. 1996. Grasses. In Lindquist, Sabelis, and Bruin, 6:619–629.
Fujikawa, T. 1978. Revision of the family Banksinomidae (Acari, Oribatei). *Acarologia* 2:43–467. [THYRISOMIDAE]
———. 1991. List of oribatid families and genera of the world. *Edaphologia* 46:1–123.
———. 1995. Comparison among populations of *Tectocepheus velatus* (Michael, 1880) from forests, grasslands and crop fields. *Edaphologia* 55:1–82. [TECTOCEPHEIDAE]
———. 2001. A new and three new species of Tectocepheidae and Nodocepheidae from northeastern part of Nippon including the Shirakami-sanchi World Heritage Area (Acari: Oribatida). *Edaphologia* 67:23–30.
Fujikawa, T., and J. Aoki. 1969. Notes on two species of the genus *Perlohmannia* Berlese (Acari, Perlohmanniidae). Taxonomic notes on oribatid mites of Hokkaido. II, *Annot. Zool. Jpn.* 42:220–25.
———. 1970. Five species of the genus *Liacarus* Michael (Acari: Liacaridae). Taxonomic notes on Oribatid mites of Hokkaido. III. *Annot. Zool. Jpn.* 43 (3): 158–65.
Fujikawa, T., M. Fujita, and J. Aoki. 1993. Checklist of oribatid mites of Japan (Acari: Oribatida). *J. Acarol. Soc. Japan* 2 (Suppl. 1): 1–121.
Fujita, M., and S. Fujiyama. 2001. How can the minor species, *Tectocepheus minor* (Oribatida), dominate *T. velatus* in a no-tillage crop field? *Pedobiologia* 45:36–45. [TECTOCEPHEIDAE]
Fuller, H. S. 1952. The mite larvae of the family Trombiculidae in the Oudemans collection: Taxonomy and medical importance. *Zool. Verh.* 18:1–261.
Funk, R. C. 1974. Megacelaenopsidae, a new family of Celaenopsoidea

(Acari: Mesostigmata). *Acarologia* 16:382–93. [MEGACELAENOPSIDAE]
———. 1977. *Triplogynium krantzi* n. g., n. sp., type of Triplogyniidae n. fam. (Mesostigmata: Celaenopsoidea). *Internat. J. Acarol.* 3:71–79. [TRIPLOGYNIIDAE]
———. 1980. Generic revision of the family Euzerconidae (Mesostigmata: Celaenopsoidea) with description of seven new genera. *Internat. J. Acarol.* 6:313–49.
Furman, D. P. 1959. Feeding habits of symbiotic mesostigmatid mites of mammals in relation to pathogen-vector potentials. *Amer. J. Trop. Med. Hyg.* 8:5–12.
———. 1966. The spinturnicid mites of Panama (Acarina: Spinturnucidae). In Wenzel and Tipton, 125–66.
———. 1972. Laelapid mites (Laelapidae: Laelapinae) of Venezuela. *Brigham Young Univ. Sci. Bull., Biol. Ser.* 17 (3): 1–58.
———. 1979. Specificity, adaptation and parallel evolution in the endoparasitic Mesostigmata of mammals. In Rodriguez, 2: 329–37.
Furman, D. P., and A. W. Smith. 1973. In vitro development of two species of *Orthohalarachne* (Acarina: Halarachnidae) and adaptations of the life cycle for endoparasitism in mammals. *J. Med. Entomol.* 10:415–16.
Furman, D. P., and O. E. Sousa. 1969. Morphology and biology of a nest-producing mite, *Bakericheyla chanayi* (Acarina: Cheyletidae). *Ann. Entomol. Soc. Amer.* 62:858–63.
Furniss, R. L., and V. M. Carolin. 1977. *Western forest insects.* USDA, Forest Service Misc. Publ. No. 1339. Washington, DC: USDA.
Furumizo, R. T., and G. W. Wharton. 1975. A case of postimaginal molt in the American house dust mite *Dermatophagoides farinae* Hughes, 1961 (Acari: Pyroglyphidae). *Acarologia* 17:730–33.
Gabryś, G. 1986. *Trombidium mediterraneum* new record (Acari: Actinedida: Trombidiidae), a new species of mite for the fauna of Poland. *Przeg. Zool.* 30:57–62.
———. 1989. *Erythraeus mirjavehi* new species (Acari: Actinedida: Erythraeidae) from Iran with remarks on the genus *Parerythraeus*. *Acarologia* 30:59–66.
———. 1996. Microtrombidiidae (Acari: Actinedida) of Poland. *Ann. Upper Siles. Mus.* 6–7:145–242.
———. 1999. *The world genera of Microtrombidiidae (Acari, Actinedida, Trombidioidea).* Monogr. Upper Siles. Mus. 2. Bytom, Poland: Upper Silesian Museum.
———. 2000. *Balaustium xerothermicum* sp. nov. from Poland with remarks on all world species of the genus (Acari: Actinedida: Erythraeidae). *Ann. Zool. (Warsaw)* 50:47–56.
Gabryś, G., and R. Haitlinger. 1986. New and rare mites (Acari: Calyptostomatidae, Erythraeidae, Trombidiidae) for the fauna of Poland. *Polsk. Pismo Entomol.* 56:471–74.
Gabryś, G., and A. Wohltmann. 2001. A redescription of *Microtrombidium pusillum* (Hermann, 1804) (Acari: Parasitengona: Microtrombidiidae) with notes on taxonomy and biology. *Ann. Zool. (Warsaw)* 51:233–50.
———. 2003. A redescription of adult and larva of *Echinothrombium rhodinum* (C. L. Koch, 1837) (Acari: Parasitengona: Microtrombidiidae) with comments on the biology of the species. *Ann. Zool (Warsaw)* 53:729–38.
Gadd, C. H. 1946. Observations on the yellow tea mite *Hemitarsonemus latus* (Banks) Ewing. *Bull. Entomol. Res.* 37:157–62. [TARSONEMIDAE]
Gadsden, E. H. 1988. Comparacion altitudinal de ectoparasitos de lagartijas del complejo *Sceloporus grammicus* (Reptilia, Iguanidae) en la Sierra de Tepoztlan Morelos, Mexico. *Acta Zool. Mex.*, n.s., 30:21–31.
Garga, N., H. Proctor, and R. Belczewski. 1997. Leg size affects male mating success in *Tarsonemus confusus* Ewing (Prostigmata: Tarsonemidae). *Acarologia* 38:369–75.
Gaston, K. J. 1991. The magnitude of global species richness. *Conserv. Biol.* 5:283–96.
Gaud, J. 1968. Acariens de la sous-famille des Dermatophagoidinae (Psoroptidae) récoltés dans les plumages d'oiseaux. *Acarologia* 10:292–312. [PYROGLYPHIDAE]
———. 1973. Quelques espèces nouvelles de sarcoptifirmes plumicoles (Analgidae & Dermoglyphidae) parasites d'oiseaux d'Europe. *Acarologia* 15:727–58. [DERMOGLYPHIDAE]
———. 1974. Quelques espèces nouvelles de Sarcoptiformes plumicoles (Analgidae et Dermoglyphidae) parasites d'oiseaux d'Europe. *Acarologia* 15:727–58. [XOLALGIDAE, DERMOGLYPHIDAE]
———. 1976. *Acariens Sarcoptiformes plumicoles parasites sur les oiseaux Lariformes et Columbiformes d'Afrique.* Ann. Mus. Roy. Afr. Cent. (Sci. Zool.) 214, ser. 8. Tervuren, Belgium: Musée Royale de l'Afrique Centrale. [FALCULIFERIDAE]
———. 1980a. Sarcoptiformes plumicoles des oiseaux Coraciiformes d'Afrique. IVB, Parasites des Bucerotidae. Acariens de la famille Vexillariidae. *Rev. Zool. Afr.* 94:709–43.
———. 1980b. *Sarcoptiformes plumicoles des oiseaux Psittaciformes, Strigiformes et Caprimulgiformes en Afrique.* Ann. Mus. Roy. Afr. Cent. (Sci. Zool.) 230, ser. 8. Tervuren, Belgium: Musée Royale de l'Afrique Centrale. [KRAMERELLIDAE, PTEROLICHIDAE]
———. 1982. Acariens Sarcoptiformes plumicoles des oiseaux Ciconiiformes d'Afrique. III, Parasites des Threskiornithidae. *Rev. Zool. Afr.* 96:701–30. [CAUDIFERIDAE, KRAMERELLIDAE]
Gaud, J., and W. T. Atyeo. 1966. Le genre *Oxyalges* Gaud et Mouchet 1959, Proctophyllodidae, Alloptinae, avec la description d'une espece nouvelle. *Acarologia* 8:465–69. [ALLOPTIDAE]
———. 1967. Eustathiinae n. subfam. des Pterolichidae, sarcoptiformes plumicoles. *Acarologia* 9:882–904. [EUSTATHIIDAE]
———. 1974. Gabuciniidae, famille nouvelle de sarcoptiformes plumicoles. *Acarologia* 16:522–64.
———. 1975. Gabuciniidae, famille nouvelle de Sarcoptiformes plumicoles. *Acarologia* 16:522–61.
———. 1976. Ovacaridae, une famille nouvelle de sarcoptiformes plumicoles. *Acarologia* 17:169–76. [APIONACARIDAE]
———. 1977. Nouvelles superfamilles pour les acariens astigmates parasites d'oiseaux. *Acarologia* 19:678–85. [ANALGIDAE]
———. 1978. Nouvelles superfamilles pour les acariens astigmates parasites d'oiseaux. *Acarologia* 19:678–85.
———. 1982a. La famille Freyanidae, Dubinin (Sarcoptiformes plumicoles, Freyanoidea). 1, Sous-familles Burhinacarinae, n. sub-f. et Diomedacarinae, n. sub-f. *Acarologia* 22:403–14.
———. 1982b. La famille Freyanidae, Dubinin (Sarcoptiformes plumicoles, Freyanoidea). 2, Sous-famille Michaeliinae. *Acarologia* 23:177–87.
———. 1982c. The subfamilies of the Analgidae and Psoroptoididae (Acari: Analgoidea). *J. Med. Entomol.* 19:299–305.
———. 1984. La famille Cheylabididae, Gaud (Sarcoptiformes plumicoles, Pterolichoidea). *Acarologia* 25:281–85.
———. 1985. La famille Freyanidae, Dubinin (Sarcoptiformes plu-

micoles, Freyanoidea). 3, Sous-famille Freyaninae. *Acarologia* 26:385–92.

———. 1990. La famille Vexillariidae, Gaud & Mouchet (Sarcoptiformes plumicoles). I, Sous-famille Calaobiinae. *Acarologia* 31:199–210.

———. 1996a. *Feather mites of the world (Acarina, Astigmata): The supraspecific taxa*. Ann. Mus. Roy. Afr. Cent. (Sci. Zool.) 277, pt. I, text. Tervuren, Belgium: Musée Royale de l'Afrique Centrale.

———. 1996b. *Feather mites of the world (Acarina, Astigmata): The supraspecific taxa*. Ann. Mus. Roy. Afr. Cent. (Sci. Zool.) 277, pt. II, illustrations. Tervuren, Belgium: Musée Royale de l'Afrique Centrale.

Gaud, J., W. T. Atyeo, and H. F. Berla. 1972. Acariens sarcoptiformes plumicoles parasites des tinamous. *Acarologia* 14: 393–453. [CRYPTUROPTIDAE, DERMOGLYPHIDAE]

Gaud, J., W. T. Atyeo, and J. S. H. Klompen. 1989. Oconnoriidae, a new family of feather mites (Acarina, Pterolichoidea). *J. Entomol. Sci.* 24:417–21.

Gaud, J., and P. C. Peterson. 1987. Espèces africaines du genre *Thysanocercus*, Gaud, 1957 (Alloptidae, Thysanocercinae). *Acarologia* 28:265–75. [THYSANOCERCIDAE]

Gaull, K. R., and T. J. Ridsdill-Smith. 1996. The foraging behaviour of redlegged earth mite, *Halotydeus destructor* (Acarina: Penthaleidae), in an annual subterranean clover pasture. *Bull. Entomol. Res.* 86:247–52.

Geest, L. P. S. van der, S. L. Elliot, J. A. J. Breeuwer, and E. A. M. Beerling. 2000. Diseases of mites. *Exp. Appl. Acarol.* 24: 497–560.

Genis, N. de L., G. C. Loots, and P. A. J. Ryke. 1967. The genus *Protogamasellus* Karg (Acari) with descriptions of new species and subspecies from the Ethiopian region. *J. Nat. Hist.* 1:337–53. [ASCIDAE]

George, J. E. 1971. Drop-off rhythms of engorged rabbit ticks, *Haemaphysalis leporispalustris* (Packard, 1896) (Acari: Ixodidae). *J. Med. Entomol.* 8:461–79.

Gerdeman, B., and G. Alberti. 2007. First ultrastructural observations on the paired suckers of a heterozerconid mite (Heterozerconidae; Gamasida). In Morales-Malacara et al., 581–84.

Gerdeman, B., and H. Klompen. 2003. A new North American heterozerconid, *Narceoheterozercon ohioensis*, n. sp., with first descriptions of immatures of Heterozerconidae. *Internat. J. Acarol.* 29:351–70.

Gerdeman, B. S., H. Klompen, and L. Tanigoshi. 2000. Insights into the biology of a mite-millipede association. In Progress in studies on Myriapoda and Onychophora, ed. J. Wytwer and S. Golovatch. *Fragmenta Faun. (Warsaw)* 43:223–27. [HETEROZERCONIDAE]

Gerecke, R. 1988a. *Limnolegeria longiseta* Mota (Acari, Actinedida): Zur Morphologie, systematischen Stellung und Ökologie einer wenig bekannten Wassermilbe, mit Erstbeschreibung der Nymphe. *Arch. Hydrobiol.* 112:611–26. [ANISITSIELLIDAE]

———. 1988b. *Utaxatax ventriplax* n. sp.—the first Palearctic record of the water mite genus *Utatatax* Habeeb 1964 (Acari, Parasitengona, Anisitsiellidae), with description of a new subgenus. *Animalia* 15:181–200.

———. 1991a. *Prymnopsella bucculata* gen. n., sp. n., a new water mite (Acari: Actinedida: Aturidae) from the Gennargentu Mountains (Sardinia, Italy). *Aquat. Insect.* 13:107–14.

———. 1991b. Über drei Arten der Gattung *Bandakia* Thor 1913 aus Mittel- und Südeuropa (Acari: Actinedida: Anisitsiellidae). *Senckenb. Biol.* 71:139–53.

———. 1992a. A remarkable new *Nilotonia* species from the Pollino Mountains (Southern Italy). *Spixiana* 15:97–105. [ANISITSIELLIDAE]

———. 1992b. The water mites of the genus *Partnunia*. *Entomol. Basil.* 15:13–44. [HYDRYPHANTIDAE]

———. 1994. New records of water mites from springs and running waters in the Mediterranean region (Acari, Actinedida: Anisitsiellidae, Aturidae, Momoniidae). *Ann. Limnol.* 30:267–84.

———. 1995. Water mites from Ecuador. I, A new genus of the family Anisitsiellidae Koenike, 1909 (Acari: Hydrachnellae) from a rain forest stream in the province of Esmeraldas. *Amazonia* 13:417–22.

———. 1996. The water mites of the family Torrenticolidae Piersig, 1902 (Acari, Actinedida, Hydrachnellae) in springs and running waters of Sardinia and Corsica. *Arch. Hydrobiol., Suppl.* 107:287–334.

———. 1997a. Untersuchungen über Wassermilben der Familie Hydryphantidae (Acari, Actinedida) in der Westpalaearktis. I, Beitrag zur Kenntnis der Gattung *Protzia* Piersig, 1896 (Acari, Actinedida, Hydryphantidae). *Arch. Hydrobiol., Suppl.* 77: 271–336.

———. 1997b. Untersuchungen über Wassermilben der Familie Hydryphantidae (Acari, Actinedida) in der Westpalaearktis. II, Die Wassermilben der Familie Hydryphantidae Piersig, 1896 in den Mittelmeerlädern. *Arch. Hydrobiol., Suppl.* 77:337–513.

———. 1999a. Further studies on hydryphantoid water mites in the W. Palearctic region (Acari, Actinedida). *Arch. Hydrobiol., Suppl.* 121:119–258.

———. 1999b. Remarks on the morphology, life cycle, distribution and taxonomy of water mites of the subfamily Acherontacarinae in the western Palearctic. In Bruin, van der Geest, and Sabelis, 473–82. [HYDROVOLZIIDAE]

———. 2003. Water mites of the genus *Atractides* Koch, 1837 (Acari: Parasitengona: Hygrobatidae) in the western Palearctic region: A revision. *Zool. J. Linn. Soc.* 138:141–378.

———. 2004a. Taxonomy and phylogeny in African water mites of the genus *Diplodontus* Dugès, 1834 (Acari, Hydrachnidia, Hydryphantidae). *Ann. Limnol.–Internat. J. Limnol.* 40:71–85.

———. 2004b. Water mites from spring- and ground-water habitats in the Sultanate of Oman. *J. Nat. Hist.* 38:2833–49.

———. 2004c. The water mites of Madagascar (Acari, Hydrachnidia): A revised list completed by original material conserved at the Muséum national d'Histoire naturelle, Paris. *Zoosystema* 26:393–418.

———. 2006. Revisional studies on water mites of the family Harpagopalpidae K. Viets, 1924 (Acari: Hydrachnida), with descriptions of new species from Cameroon and Madagascar. *Ann. Limnol.–Internat. J. Limnol.* 42:109–25.

———, ed. 2007. *Süsswasserfauna von Mitteleuropa*. Vol. 7/2-1, *Chelicerata: Araneae, Acari I*. Heidelberg: Spektrum Akademischer Verlag.

Gerecke, R., and D. R. Cook. 1995. Morphology, systematic position and zoogeography of *Parawandesia chappuisi* E. Angelier, 1951 (Acari, Actinedida, Piersigiidae). *Zool. Anz.* 234:125–31.

Gerecke, R., and A. Di Sabatino. 1996. The water mites of the family Torrenticolidae Piersig, 1902 (Acari, Actinedida, Hydrachnellae) in springs and running waters of Sardinia and Corsica. *Arch. Hydrobiol., Suppl.* 107:287–334.

Gerecke, R., L. Fisher-Hartig, and M. Steinitz-Kannan. 1996. *Centrolimnesia boopis* (Hydrachnellae: Limnesiidae), a new planktonic mite from Lake Limoncocha in the Amazonian region of Ecuador. *Stud. Neotrop. Fauna Environ.* 31:193–204.

Gerecke, R., and E. Meyer. 1989. *Barbaxonella spectabilis* n. sp., eine neie Wassermilbe (Acari, Actinedida, Aturidae) aus den Monti Iblei (Sizilien). *Lauterbornia* 2:13–24.

Gerecke, R., and J. Schwoerbel. 1991. Water quality and water mites (Acari, Actinedida) in the Upper Danube region, 1959–1984. In Dusbábek and Bukva, 1:483–91.

Gerecke, R., and H. Smit. 1994. *Partnunia tabernaculata*, spec. nov., a new water mite from the Spanish Pyrenees. *Spixiana* 17:201–6. [HYDRYPHANTIDAE]

Gerecke, R., and I. M. Smith. 1993. Description of larvae of *Nilotonia longipora* (Walter) (Acari: Hydrachnida) with remarks on the relationships and classification of *Nilotonia* Thor, 1905. *Internat. J. Acarol.* 19:259–66. [ANISITSIELLIDAE]

Gerecke, R., I. M. Smith, and D. R. Cook. 1999. Three new species of *Apheviderulix* gen. nov. and proposal of Apheviderulicidae fam. nov. (Acari: Hydrachnidia: Eylaoidea). *Hydrobiologia* 397:133–47.

Gerecke, R., and P. Tuzovskij. 2001. The water mite *Rutripalpus limicola* Sokolow, 1934: New data on morphology and biology, and considerations on the systematic position of the monotypic family Rutripalpidae (Acari, Hydrachnidia). *J. Nat. Hist.* 35: 931–44.

Gerling, D., J. Kugler, and A. Lupo. 1976. The galls of insects and mites that occur on *Tamarix* spp. in Israel and the Sinai. *Boll. Lab. Entomol. Agraria "Filippo Silvestri," Portici* 33:53–79. [TENUIPALPIDAE]

Gerling, D., H. H. W. Velthuis, and A. Hafetz. 1989. Bionomics of the large carpenter bees of the genus *Xylocopa*. *Ann. Rev. Entomol.* 34:163–90.

Gerson, U. 1966. A redescription of *Eupalopsis maseriensis* (Canestrini and Fanzago) (Acarina: Eupalopsellidae). *Israel J. Zool.* 15: 148–54.

———. 1967. The natural enemies of the chaff scale, *Parlatoria pergandii* Comstock, in Israel. *Entomophaga* 12:97–109. [EUPALOPSELLIDAE]

———. 1968a. Five tydeid mites from Israel (Acarina: Prostigmata). *Israel J. Zool.* 17:191–98.

———. 1968b. Some raphignathoid mites from Israel. *J. Nat. Hist.* 2:429–43.

———. 1972a. Mites of the genus *Ledermuelleria* (Prostigmata: Stigmaeidae) associated with mosses in Canada. *Acarologia* 13: 319–42.

———. 1972b. A new species of *Camerobia* Southcott, with a redefinition of the family Camerobiidae (Acari: Prostigmata). *Acarologia* 13:502–8.

———. 1972c. A redescription of *Ledermuelleria frigida* Habeeb (Acarina: Prostigmata: Stigmaeidae). *Acta Arachnol.* 24:15–28.

———. 1973. Lichen-arthropod associations. *Lichenologist* 5: 434–43.

———. 1985a. Other predaceous mites and spiders. In Helle and Sabelis (1985b), 1B:205–10.

———. 1985b. Webbing. In Helle and Sabelis (1985a), 1A:223–32. [TETRANYCHIDAE]

———. 1992. Biology and control of the broad mite, *Polyphagotarsonemus latus* (Banks) (Acari: Tarsonemidae). *Exp. Appl. Acarol.* 13:163–78.

———. 1994. The Australian Eupalopsellidae (Acari: Prostigmata). *Invertebr. Taxon.* 8:63–73.

Gerson, U., and D. Blumberg. 1969. Biological notes on the mite *Saniosulus nudus*. *J. Econ. Entomol.* 62:729–30. [EUPALOPSELLIDAE]

Gerson, U., and E. Collyer. 1984. Two false spider mites (Acari: Tenuipalpidae) from Cook Islands and New Zealand ferns. *New Zeal J. Zool.* 11:141–44.

Gerson, U., A. Fain, and R. L. Smiley. 1999. Further observations on the Cheyletidae (Acari), with a key to the genera of the Cheyletinae and a list of all known species in the family. *Bull. Inst. Roy. Sci. Nat. Belg., Entomol.* 69:35–86.

Gerson, U., W. E. Frost, and S. F. Swift. 1997. A new genus of the family Eupalopsellidae from Australia (Acari: Prastigrnata). *Internat. J. Acarol.* 23:185–89. [MECOGNATHIDAE]

Gerson, U., B. M. OConnor, and M. A. Houck. 1990. Acari. In *The armored scale insects. Their biology, natural enemies and control*, ed. D. Rosen, B:77–97. Amsterdam: Elsevier. [HEMISARCOPTIDAE]

Gerson, U., and R. L. Smiley. 1990. *Acarine biocontrol agents: An illustrated key and manual*. London: Chapman & Hall.

Gerson, U., R. L. Smiley, and R. Ochoa. 2003. *Mites (Acari) for pest control*. London: Blackwell.

Gerson, U., and M. K. P. Smith Meyer. 1980. A new species of *Capedulia* (Acari: Tenuipalpidae) from a marine habitat in Israel. *Israel J. Entomol.* 14:9–12.

———. 1995. *Neilstigmaeus*, a new Australian genus in the family Stigmaeidae (Acari: Prostigmata). *Acarologia* 36:219–22.

Gerson, U., and D. E. Walter. 1998. The transfer of *Mecognatha* Wood from the Stigmaeidae to Mecognathidae, n. fam., a new synonymy, and a key to the families of the Raphignathoidea (Acari: Prostigmata). *Syst. Appl. Acarol.* 3:145–47.

Gervais, P. 1842. Une quinzaine d'espèces d'insectes aptères. *Ann. Soc. Entomol. Fr.* 11:45–48.

Geurs, M., J. Bongers, and L. Brussard. 1991. Variations of the heptane flotation method for improved efficiency of collecting microarthropods from a sandy loam soil. *Agr. Ecosyst. Environ.* 34:213–21.

Giesen, K. M. T. 1990. A review of the parasitic mite family Psorergatidae (Cheyletoidea: Prostigmata: Acari) with hypotheses on the phylogenetic relationships of species and species groups. *Zool. Verh., Leiden* 259:1–69.

Gil Collado, J., L. I. Rivas Lopez, and L. M. Zapatero Ramos. 1985. Pterygosomidae (Acari; Actinedida) parasitos de Geckonidae de la Peninsula Iberica. *Actas II Cong. Iber. Entomol.* 1:379–88.

Gilardi, K. V. K., J. D. Gilardi, A. Frank, M. L. Goff, and W. M. Boyce. 2001. Epidermoptid mange in Laysan albatross fledglings in Hawaii. *J. Wildl. Dis.* 37:185–88. [EPIDERMOPTIDAE]

Gillespie, D. R., and D. M. J. Quiring. 1990. Biological control of fungus gnats, *Bradysia* spp. (Diptera: Sciaridae), and western flower thrips, *Frankliniella occidentalis* (Pergande) (Thysanoptera: Thripidae), in greenhouses using a soil-dwelling predatory mite, *Geolaelaps* sp. nr. *aculeifer* (Canestrini) (Acari: Laelapidae). *Can. Entomol.* 122:975–83.

Gilyarov, M. S., ed. 1978. *Handbook for the identification of soil-inhabiting mites. Trombidiformes*. Leningrad: Zool. Inst. Acad. Sci. SSSR.

Gilyarov, M. S., and N. G. Bregetova, eds. 1977. *A key to the soil-*

inhabiting mites. Mesostigmata. Leningrad: Akad. Nauk Zool. Inst. SSSR.

Gilyarov, M. S., and D. A. Krivolutskij, eds. 1975. *Opredelitel' obitajuschtschich w potschwe kleschtschej: Sarcoptiformes*. Moscow: Istadel'stwo Nauka. [in Russian]

Gispert, C., T. M. Perring, and R. Craemer. 1998. Purification and characterization of peach mosaic virus. *Plant Dis.* 82:905–8.

Gjelstrup, P. 1978. An annotated list of Danish oribatid mites (Acarina, Oribatei). *Entomol. Meddr., Copenhagen* 46:109–21. [in Danish].

Gjelstrup, P., P. Hansen, and E. Warncke. 1991. Moss mites (Oribatida, Acari) in mosses from some Danish spring areas. *Nat. Jutlandica, Aarhus* 23:33–44.

Gjelstrup, P., and U. Søchting. 1984. Oribatid mites (Acarina) dominant in some lichen and moss species of maritime rocks on Bornholm in the Baltic. In Griffiths and Bowman, 1:528–33.

Glauert, A. M. 1987. Fixation, dehydration and embedding of biological specimens. Chapter 2, Cytological staining. In *Practical methods in electron microscopy*, ed. A. M. Glauert, 3:5–110. Amsterdam: Elsevier.

Glauert, M., and P. R. Lewis. 1998. Biological specimen preparation for transmission electron microscopy. Chapter 5, Embedding methods. In *Practical methods in electron microscopy*, ed. A. M. Glauert, 16:147–69. London: Portland Press.

Gledhill, T. 1985. A new species of water mite, *Unionicola* (*Pentatax*) *macani* (Unionicolidae, Hydrachnellae, Acari), from the mantle cavity of the prosobranch mollusc *Lanistes ovum* Peters in Nigeria, with remarks on some aspects of host/parasite relationships between unionicolids and molluscs. *Arch. Hydrobiol.* 104: 77–92.

———. 2003. A new genus and species of water mite (Acari: Hydrachnidia: Hygrobatidae) from the prosobranch gastropod *Lanistes lybicus* (Morelet) (Streptoneura: Ampullariidae) in Nigeria. In I. M. Smith (2003a), 75–82.

Gledhill, T., and O. Agbolade. 2006. A new species of water mite (Acari: Hydrachnidia: Hygrobatidae) from the mantle cavity of the prosobranch gastropod *Potadoma moerchi* (Reeve) (Streptoneura: Thiaridae) in Nigeria. *Zootaxa* 1160:37–47.

Gledhill, T., J. Cowley, and R. J. M. Gunn. 1982. Some aspects of the host-parasite relationships between adult blackflies (Diptera: Simuliidae) and larvae of the water mite *Sperchon setiger* (Acari: Hydrachnellae) in a small chalk stream in southern England. *Freshwater Biol.* 12:345–57. [SPERCHONTIDAE]

Gledhill, T., and M. F. Vidrine. 2002. Two new sympatric watermites (Acari: Hydrachnidia: Unionicolidae) from the mutilid bivalve *Aspatharia sinuata* (von Martens) in Nigeria with some data on unionicoline-bivalve relationships. *J. Nat. Hist.* 36:1351–81.

Gless, E. E. 1972. Life cycle studies of some antarctic mites and description of a new species, *Protereunetes paulinae* (Acari: Eupodidae). *Antarct. Res. Ser. (Washington)* 20:289–306.

Gliwicz, Z. M., and E. Biesiadka. 1975. Pelagic water mites (Hydracarina) and their effect on the plankton community in a Neotropical man-made lake. *Arch. Hydrobiol.* 76:65–88. [PIONI DAE]

Goddard, D. G. 1982. Feeding biology of free-living Acari at Signy Island, South Orkney Islands. *Brit. Antarct. Surv. Bull.* 51: 290–93. [ALGOPHAGIDAE]

Goff, M. L., R. B. Loomis, W. C. Welbourn, and W. J. Wrenn. 1982. A glossary of chigger terminology (Acari: Trombiculidae). *J. Med. Entomol.* 19:221–38.

Goff, M. L., R. B. Loomis, and W. J. Wrenn. 1986. *A chigger bibliography 1758–1984 (Acari: Trombiculidae): Systematics, biology and ecology*. Bull. Soc. Vector Ecol. 11. Santa Ana, CA: Society for Vector Ecology.

Goldarazena, A., R. Jordana, and Z.-Q. Zhang. 1997. *Adactylidium moundi* and *Adactylidium costarricensis*, two new species of Acarophenacidae (Acari: Tarsonemida) parasitic on Thysanoptera. *Internat. J. Acarol.* 23:261–68.

Goldarazena, A., R. Ochoa, and C. C. Childers. 2002. Systematic revision of the genus *Deleonia* (Acari: Tarsonemidae). *Internat. J. Acarol.* 28:221–40.

Goldarazena, A., R. Ochoa, and R. Jordana. 1999. Revision of the genus *Paradactylidium* Mahunka (Acari: Heterostigmata). *Internat. J. Acarol.* 25:91–99.

Goldarazena, A., R. Ochoa, R. Jordana, and B. M. OConnor. 2001. Revision of the genus *Adactylidium* Cross (Acari: Heterostigmata: Acarophenacidae), mites associated with thrips. *Proc. Entomol. Soc. Wash.* 103:473–516.

Goldarazena-Lafuente, A., R. Jordana, and Z.-Q. Zhang. 1999. Notes on *Paratrombium welbourni* G. and Z. (Acari: Trombidiidae) and *Grandjeanella haitlingeri* G. and Z. (Acari: Erythraeidae) with information about their hosts and biometrical data. *Internat. J. Acarol.* 25:23–28.

Goldarazena-Lafuente, A., and Z.-Q. Zhang. 1997. *Paratrombium welbourni* sp. nov. (Acari: Trombidiidae), an ectoparasite of aphids (Homoptera: Aphididae) in Spain. *Syst. Appl. Acarol.* 2:227–30.

———. 1998a. New *Erythraeus* larvae (Acari: Erythraeidae) ectoparasitic on Aphidoidea (Homoptera) and Anthocoridae (Heteroptera). *Syst. Appl. Acarology* 3:149–58.

———. 1998b. Seasonal abundance of *Abrolophus neobrevicollis* larvae (Acari: Erythraeidae) ectoparasitic on *Thrips trehernei* (Thysanoptera: Thripidae) in the Basque country (Spain). *Syst. Appl. Acarol.* 3:93–96.

Goldberg, S. R., and H. J. Holshuh. 1993. Histopathology in a captive Yarrow's spring lizard, *Sceloporus jarrovii* (Phrynosomatidae), attributed to the mite *Hirstiella* sp. (Pterygosomatidae). *Trans. Amer. Microscop. Soc.* 112:234–37.

Goldschmidt, T. 2007. Studies on Latin American water mites of the genus *Torrenticola* Piersig, 1896 (Torrenticolidae, Hydrachnidia, Acari). *Zool. J. Linn. Soc.* 150:443–678.

Goldschmidt, T., G. Alberti, and F. L. Meyer. 1999. Presence of acetabula-like structures on the coxae of the Neotropical water mite genus *Neotyrrellia* (Tyrelliinae, Limnesiidae, Prostigmata). In Bruin, van der Geest, and Sabelis, 491–97.

Goldschmidt, T., and R. Gerecke. 2003. Studies on hydryphantid water mites (Acari: Actinedida: Hydrachnidia) from Central and South America. In I. M. Smith (2003a), 83–150.

Goldschmidt, T., R. Gerecke, and G. Alberti. 2002. *Hygrobates salamandrarum* sp. nov. (Acari, Hydrachnidia, Hygrobatidae) from China: The first record of a freshwater mite parasitizing newts (Amphibia, Urodela). *Zool. Anz.* 241:297–304.

Golosova, L. D., E. Karppinen, and D. A. Krivolutskij. 1983. List of oribatid mites (Acarina, Oribatei) of northern Palaearctic region. II, Siberia and the Far East. *Acta Entomol. Fenn.* 43:1–14.

Gomaa, E. A., B. A. Abou-Awad, A. K. Nasr, and M. M. Abou-Elela. 1989. Life-history studies and feeding behaviour of the predatory mite, *Pachylaelaps aegypticus*, with description of immature stages (Acari: Pachylaelapidae). *Insect Sci. Appl.* 10:691–98.

Gonzalez, R. H. 1978. A new species of xenocaligonellid mite from the Galapagos Islands (Acari). *Proc. Entomol. Soc. Wash.* 80:191–96.

Gonzalez-Rodriquez, R. H. 1965. *A taxonomic study of the genera* Mediolata, Zetzellia *and* Agistemus *(Acarina: Stigmaeidae).* Univ. Calif. Publ. Entomol. 41. Berkeley: Univ. Calif. Press.

———. 1967. *Summersiella*, a new stigmaeid mite from New Zealand (Acarina: Prostigmata). *Pan-Pac. Entomol.* 43:236–39. [STIGMAEIDAE]

Goodwin, H. 1954. Observations on the biology of the lizard mite, *Geckobiella texana* (Banks) 1904 (Acarina: Pterygosomidae). *J. Parasitol.* 40:54–59.

Goolsby, J. A., R. Zonneveld, and A. Bourne. 2004. Pre-release assessment of impact on biomass production of an invasive weed, *Lygodium microphyllum* (Lygodiaceae: Pteridophyta), by a potential biological control agent, *Floracarus perrepae* (Acari: Eriophyidae). *Environ. Entomol.* 33:997–1002.

Gordeeva, E., R. Niemi, and A. D. Petrova-Nikitina. 1996. A new species, *Sphaerochthonius spectabilis* sp. n., of Sphaerochthoniidae (Acarina, Oribatida) from a termite nest (*Acanthotermes ahngerianus* Juc.) in the southwestern Turkmenistan desert. *Acarologia* 37:247–53.

Gordon, M. J., B. K. Swan, and C. G. Paterson. 1979. The biology of *Unionicola formosa* (Dana & Whelpley): A water mite parasitic in the unionid bivalve, *Anodonta cataracta* (Say), in a New Brunswick lake. *Can. J. Zool.* 57:1748–56. [UNIONICOLIDAE]

Gorgol', V. T., and A. V. Yastrebtsov. 1989. Skeleto-muscular system in some cheyletid mites (Trombidiformes, Cheyletidae). *Vestn. Zool.* (4): 74–81. [in Russian]

Gorirossi, F. E. 1950. The mouth parts of the adult female tropical rat mite, *Bdellonyssus bacoti* (Hirst, 1913) Fonseca, 1941 [= *Liponissus* bacoti (Hirst)], with observations on the feeding mechanism. *J. Parasitol.* 36:301–18.

Gorirossi-Bourdeau, F. 1995. A documentation in stone of Acarina in the Roman temple of Bacchus in Baalbek, Lebanon about AD 150. *Bull. Ann. Soc. Roy. Belg. Entomol.* 131:3–15.

Gotoh, T., S. Abé, Y. Kitashima, and S. Ehara. 2007. Divergence in host range and reproductive compatability in three strains of *Oligonychus gotohi* Ehara (Acari: Tetranychidae). *Internat. J. Acarol.* 33:7–13.

Gotoh, T., S. Akiyuki, Y. Kitashima, and H. A. Rezk. 2004. Developmental and reproductive performance of *Tetranychus pueraricola* Ehara and Gotoh (Acari: Tetranychidae) at four constant temperatures. *Appl. Entomol. Zool.* 39:675–82.

Gotoh, T., J. Bruin, M. W. Sabelis, and S. B. J. Menken. 1993. Host race formation in *Tetranychus urticae*: Genetic differentiation, host plant preference, and mate choice in a tomato and cucumber strain. *Entomol. Exp. Appl.* 68:171–78. [TETRANYCHIDAE]

Gotoh, T., K. Gomi, and T. Nagata. 1999. Incompatibility and host plant differences among populations of *Tetranychus knzawai* Kishida (Acari: Tetranychidae). *Appl. Entomol. Zool.* 34:551–61.

Gotoh, T., J. Gutierrez, and M. Navajas. 1998. Molecular comparison of the sibling species *Tetranychus pueraricola* Ehara & Gotoh and *T. urticae* Koch (Acari: Tetranychidae). *Entomol. Sci.* 1:55–57.

Gotoh, T., Y. Ishikawa, and Y. Kitashima. 2003. Life-history traits of the six *Panonychus* species from Japan (Acari: Tetranychidae). *Exp. Appl. Acarol.* 29:241–52.

Gotoh, T., Y. Kitashima, and K. Goka. 2007. *Tetranychus* mite species identification using esterase and phospoglucomutase zymograms. *Appl. Entomol. Zool.* 42:579–85.

Gotoh, T., and M. Kubota. 1997. Population dynamics of the citrus red mite, *Panonychus citri* (McGregor) (Acari: Tetranychidae) in Japanese pear orchards. *Exp. Appl. Acarol.* 21:343–56.

Gotoh, T., A. Suwa, and Y. Kitashima. 2004. Development and oviposition of *Tetranychus pueraricola* Ehara and Gotoh (Acari: Tetranychidae) on various plants. *J. Acarol. Soc. Jpn.* 13:135–40.

Gotoh, T., and T. Tokioka. 1996. Genetic compatibility among diapausing red, non-diapausing red and diapausing green forms of the two-spotted spider mite, *Tetranychus urticae* Koch (Acari: Tetranychidae). *Jpn. J. Entomol.* 64:215–25.

Gould, S. J. 1977. *Ontogeny and phylogeny.* Cambridge, MA: Belknap Press, Harvard Univ. Press.

———. 1989. *Wonderful life.* New York: Norton.

Goulet, H. 1965. The habitat of *Platypatrobus darlingtoni* (Coleoptera: Carabidae). *Psyche* 72:305–6.

Gourbière, F., J. C. Lions, and R. Pepin. 1985. Activité et développement d'*Adoristes ovatus* (C. L. Koch, 1839) (Acarien, Oribate) dans les aiguilles d'*Abies alba* Mill. Relations avec la décomposition et les microflores fondiques. *Rev. Ecol. Biol. Sol* 22:57–73. [LIACARIDAE]

Graf, J.-F. 1976. Écologie et éthologie d'*Ixodes ricinus* L. en Suisse (Ixodoidea: Ixodidae). Cinquième note: Mise en évidence d'une phéromone sexuelle chez *Ixodes ricinus*. *Acarologia* 17:436–41.

Grandjean, F. 1928. Sur un oribatidé pourvu d'yeux. *Bull. Soc. Zool. Fr.* 53:235–42.

———. 1929. Quelques nouveaux genres d'Oribatei du Venezuela et de la Martinique. *Bull. Soc. Zool. Fr.* 54:400–423.

———. 1930. Oribates nouveaux de la region Caraibe. *Bull. Soc. Zool. Fr.* 55:262–84.

———. 1931a. Le genre *Licneremaeus* Paoli (Acariens). *Bull. Soc. Zool. Fr.* 56:221–50. [LICNEREMAEIDAE]

———. 1931b. Observations sur les Oribates (2e série). *Bull. Mus. Nat. Hist. Natur.* 3 (2): 651–65.

———. 1932. La famille des Protoplophoridae (Acariens). *Bull. Soc. Zool. Fr.* 57:10–36.

———. 1933. Etude sur le développement des Oribates. *Bull. Soc. Zool. Fr.* 58:30–61.

———. 1934a. La notation des poils gastronotiques et des poils dorsaux du propodosoma chez les Oribates (Acariens). *Bull. Soc. Zool. Fr.* 59:12–44.

———. 1934b. Observations sur les Oribates (6e série). *Bull. Mus. Nat. Hist. Natur.* 6 (2): 353–60.

———. 1934c. Observations sur les Oribates (7e série). *Bull. Mus. Nat. Hist. Natur.* 6 (2): 423–31.

———. 1934d. Les organes respiratoires secondaires des Oribates (Acariens). *Ann. Soc. Entomol. Fr.* 103:109–46.

———. 1934e. Oribates de l'Afrique du Nord (2me série). *Bull. Soc. Hist. Natur. Afr. Nord* 25:235–52.

———. 1935a. Observations sur les acariens (1e série). *Bull. Mus. Nat. Hist. Natur. Paris* 7 (2): 119–26.

———. 1935b. Observations sur les acariens (2e série). *Bull. Mus. Nat. Hist. Natur. Paris* 7 (2): 201–8.

———. 1935c. Observations sur les Oribates (8e série). *Bull. Mus. Nat. Hist. Natur.* 8 (2): 237–44.

———. 1935d. Les poils et les organes sensitifs portés par les pattes et le palpe chez les oribates. Première partie. *Bull. Soc. Zool. Fr.* 60:6–39.

———. 1936a. Un acarien synthetique: *Opilioacarus segmentatus* With. *Bul. Soc. Hist. Nat. Afr. Nord* 27:413–44.

———. 1936b. Le genre *Pachygnathus* Dugès (*Alycus* Koch) (Acariens). Première partie. *Bull. Mus. Nat. Hist. Natur. Paris* 8 (2): 398–405.

———. 1936c. Les Microzetidae n. fam. (Oribates). *Bull. Soc. Zool. Fr.* 61:60–93.

———. 1936d. Observations sur les Oribates (10e serie). *Bull. Mus. Nat. Hist. Natur. Paris* 8 (2): 246–53.

———. 1936e. Les Oribates de Jean Frédéric Hermann et de son père (Arachn. Acar.). *Ann. Soc. Ent. Fr.* 105:27–110.

———. 1937a. Le genre *Pachygnathus* Dugès (*Alycus* Koch) (Acariens). Deuxième partie. *Bull. Mus. Nat. Hist. Natur. Paris* 9 (2): 56–61.

———. 1937b. Le genre *Pachygnathus* Dugès (*Alycus* Koch) (Acariens). Troisième partie. *Bull. Mus. Nat. Hist. Natur. Paris* 9 (2): 134–38.

———. 1937c. Le genre *Pachygnathus* Dugès (*Alycus* Koch) (Acariens). Quatrième partie. *Bull. Mus. Nat. Hist. Natur. Paris* 9 (2): 199–205.

———. 1937d. Le genre *Pachygnathus* Dugès (*Alycus* Koch) (Acariens). Cinquième et dernière partie. *Bull Mus. Nat. Hist. Natur. Paris* 9 (2): 262–69.

———. 1938a. Observations sur les Acariens (4e série). *Bull. Mus. Nat. Hist. Natur. Paris*, ser. 2, 10:64–71.

———. 1938b. Observations sur les Bdelles. *Ann. Soc. Entomol. Fr.* 107:1–24.

———. 1938c. Observations sur les Tydeidae (2e série). *Bull. Mus. Nat. Hist. Natur. Paris*, ser. 2, 10:593–600.

———. 1938d. *Retetydeus* et les stigmates mandibulaires des Acariens Prostigmates. *Bull. Mus. Hist. Nat. Paris*, ser. 2, 10:279–86. [TYDEIDAE]

———. 1939a. La chaetotaxie des pattes chez les Acaridae. *Bull. Soc. Zool. Fr.* 64:50–60.

———. 1939b. Observations sur les Acariens (5e série). *Bull. Mus. Nat. Hist. Natur. Paris* 11:394–401.

———. 1939c. Observations sur les Oribates (12e série). *Bull. Mus. Nat. Hist. Natur. Paris* 11:300–307.

———. 1939d. Quelques genres d'acariens appartenant au groupe des Endeostigmata. *Ann. Sci. Nat. Zool.* 11:1–122.

———. 1939e. Les segments post-larvaires de l'hysterosoma chez les Oribates (Acariens). *Bull. Soc. Zool. Fr.* 64:273–84.

———. 1940. Les poils et les organes sensitifs portés par les pattes et le palpe chez les Oribates. Deuxième partie. *Bull. Soc. Zool. Fr.* 45:32–44.

———. 1941a. La chaetotaxie comparée des pattes chez les Oribates (1re série). *Bull. Soc. Zool. Fr.* 66:33–50.

———. 1941b. Statistique sexuelle et parthénogenèse chez les Oribates (Acariens). *C.R. Seanc. Acad. Sci.* 212:463–67.

———. 1942a. La chaetotaxie comparée des pattes chez les Oribates (2e série). *Bull. Soc. Zool. Fr.* 67:40–53.

———. 1942b. Observations sur les Labidostommidae. *Bull. Mus. Nat. Hist. Natur. Paris* 14 (2): 118–25; (3): 185–92; (5): 319–26; (6): 414–18.

———. 1943a. Le développement postlarvaire d' "*Anystis*" (Acarien). *Mem. Mus. Nat. Hist. Natur.*, n.s., 18:33–77. [ANYSTIDAE]

———. 1943b. Quelques genres d'Acariens appartenant au groupe des Endeostigmata (2e série). Première partie. *Ann. Sci. Nat. Zool.* 11 (4): 85–135.

———. 1943c. Quelques genres d'Acariens appartenant au groupe des Endeostigmata (2e série). Deuxieme partie. *Ann. Sci. Nat. Zool.*, ser. 5:1–59.

———. 1944a. Observations sur les Acariens du genre *Caeculus*. *Arch. Sci. Phys. Nat.* 26 (5): 33–46.

———. 1944b. Observations sur les Acariens de la famille des Stigmaeidae. *Arch. Sci. Phys. Nat.* 26:103–31.

———. 1944c. Les "taenidies" des acariens. *Soc. Physique. Hist. Natur. Genève* 61:142–46.

———. 1946a. Au sujet de l'organe de Claparède, des eupathidies multiples et des taenidies mandibulaires chez les Acariens actinochitineux. *Arch. Sci. Phys. Nat.* 28:63–87.

———. 1946b. Observations sur les Acariens. *Mem. Mus. Nat. Hist. Natur. Paris* 18:337–44.

———. 1946c. Les poils et les organes sensitifs portés par les pattes et le palpe chez les Oribates. Troisième partie. *Bull. Soc. Zool. Fr.* 71:10–29.

———. 1947a. Au sujet des Erythroides. *Bull. Mus. Nat. Hist. Natur. Paris*, ser. 2, 19:327–34.

———. 1947b. Les Enarthronota (Acariens). Première série. *Ann. Sci. Nat. Zool.* 8:213–48.

———. 1947c. Observations sur les Oribates (18e série). *Bull. Mus. Nat. Hist. Natur. Paris* 19 (2): 395–402.

———. 1947d. L'origine de la pince mandibulaire chez les acariens actinochitineux. *Arch. Sci. Phys. Nat. Genève* 29:305–55.

———. 1947e. L'origine pileuse des mors et la chaetotaxie de la mandibule chez les Acariens actinochitineux. *Acad. Sci. Inst. Fr.* 224:1251–54.

———. 1948a. Les Enarthronota (Acariens) (2e série). *Ann. Sci. Nat. Zool.* 10:29–58.

———. 1948b. Quelques caractères des Tétranyques. *Bull. Mus. Nat. Hist. Natur. Paris*, ser. 2, 20 (2): 517–24. [TETRANYCHIDAE]

———. 1948c. Sur les *Hydrozetes* (acariens) de l'Europe occidentale. *Bull. Mus. Nat. Hist. Natur. Paris* 20 (2): 328–35.

———. 1949. Observation et conservation des trés petit arthropodes. *Bull. Mus. Nat. Hist. Natur. Paris* 21 (2): 363–70.

———. 1950a. Les Enarthronota (Acariens) (3e série). *Ann. Sci. Nat. Zool.* 12:85–107.

———. 1950b. Étude sur les Lohmanniidae (Oribates, Acariens). *Arch. Zool. Exp. Gen.* 87:95–161.

———. 1951a. Comparaison du genre *Limnozetes* au genre *Hydrozetes* (Oribates). *Bull. Mus. Nat. Hist. Natur. Paris* 23 (92): 200–207.

———. 1951b. Étude sur les Zetorchestidae (Acariens, Oribates). *Mem. Mus. Nat. Hist. Natur. Paris* (n.s.), sér. A, *Zool.* 4:1–50.

———. 1951c. Observations sur les Oribates (22e série). *Bull. Mus. Nat. Hist. Natur.* 23 (2): 91–98.

———. 1951d. Observations sur les Oribates (23e série). *Bull. Mus. Nat. Hist. Natur.* 23 (2):261–68.

———. 1951e. Sur deux espèces du genre "*Dometorina*" n. g. et les moeurs de "*D. plantivaga*" (Berl.) (Acariens, Oribates). *Bull. Soc. Zool. Fr.* 75:224–42. [SCHELORIBATIDAE]

———. 1952a. Au sujet de l'ectosquelette du podosoma chez les Oribates superieurs et de sa terminologie. *Bull. Soc. Zool. Fr.* 77:13–36.

———. 1952b. Le morcellement secondaire des tarses de *Tarsolarkus* sp. (Acarien). *Arch. Zool. Exp. Gen.* 89:113–23. [ANYSTIDAE]

———. 1952c. Observations sur les Palaeacaroides (Acariens) (2e série). *Bull. Mus. Nat. Hist. Natur.* 24 (2): 460–67.
———. 1953a. Observations sur les Oribates (27e série). *Bull. Mus. Nat. Hist. Natur.* 25 (2): 469–76.
———. 1953b. Sur les genres *"Hemileius"* Berl. et *"Siculobata"* n. g. Acariens, Oribates. *Mem. Mus. Nat. Hist. Natur. (n.s.), ser. A, Zool.* 6:117–37. [SCHELORIBATIDAE]
———. 1954a. Les Enarthronota (Acariens) (4e série). *Ann. Sci. Nat. Zool.* 16 (11): 311–35.
———. 1954b. Essai de classification des Oribates (Acariens). *Bull. Soc. Zool. Fr.* 78:421–46.
———. 1954c. Étude sur les Palaeacaroides (Acariens, Oribates). *Mem. Mus. Nat. Hist. Natur. Paris* (n.s.), sér. A, *Zool.* 7: 179–274.
———. 1954d. Observations sur les Oribates (28e série). *Bull. Mus. Nat. Hist. Natur. Paris* 26 (2): 204–11.
———. 1954e. Observations sur les Oribates (29e série). *Bull. Mus. Nat. Hist. Natur. Paris* 26 (2): 334–41.
———. 1955a. Sur un Acarien des Iles Kerguélen, *Podacarus Auberti* (Oribate). *Mem. Mus. Nat. Hist. Natur. Paris* 8:109–50. [AMERONOTHRIDAE]
———. 1955b. *Zetomotrichus lacrimans*, Acarien sauteur (Oribate) (Acari: Zetomotrichidae). *Bull. Soc. Entomol. Fr.* 153:1–16.
———. 1956a. Caracteres chitineux de l'ovipositor, en structure normale, chez les Oribates (Acariens). *Arch Zool. Exp. Gen., Notes et Revue* 93 (2): 96–106.
———. 1956b. Observations sur les Galumnidae (2e série) (Acariens, Oribates). *Rev. Fr. Entomol.* 23:265–75.
———. 1956c. Observations sur les Oribates (33e série). *Bull. Mus. Nat. Hist. Natur.* 28 (2): 111–18.
———. 1956d. Observations sur les Oribates (35e série). *Bull. Mus. Nat. Hist. Natur.* 28 (2): 282–89.
———. 1956e. Sur deux espèces nouvelles d'Oribates (Acariens) apparentées à *Oripoda elongata* Banks 1904. *Arch. Zool. Exp. Gen.* 93:185–218. [ORIPODIDAE]
———. 1957a. L'evolution selon l'age. *Arch. Sci. Phys. Nat. Genève* 10:477–526.
———. 1957b. L'infracapitulum et la manducation chez les Oribates et d'autres Acariens. *Ann. Sci. Natur. Zool.* 19 (11): 233–81.
———. 1958a. Au sujet du naso et de son oeil inféreé chez les Oribates et les Endeostigmata (Acariens). *Bull. Mus. Nat. Hist. Natur. Paris* 30 (2): 427–35.
———. 1958b. *Charassobates cavernosus* Grandjean 1929 (Acarien, Oribate). *Mem. Mus. Nat. Hist. Natur. Paris* (n.s.), ser. A, *Zool.*, 16:121–40. [CHARASSOBATIDAE]
———. 1958c. *Perlohmannia dissimilis* (Hewitt) (Acarien, Oribate). *Mem. Mus. Nat. Hist. Natur. Paris* (n.s.), ser. A, *Zool.* 16: 57–119. [PERLOHMANNIIDAE]
———. 1958d. Scheloribatidae et Oribatulidae (Acariens, Oribates). *Bull. Mus. Nat. Hist. Natur. Paris* 30 (2): 352–59.
———. 1958e. *Sellnickia caudata* (Mich. 1908) (Acarien, Oribate). *Bull. Soc. Zool. Fr.* 83:30–44. [SELLNICKIIDAE]
———. 1958f. Sur le comportement et la notation des poils accessories postérieurs aux tarses des Nothroïdes et d'autres acariens. *Arch. Zool. Exp. Gen.* 96:277–308.
———. 1959a. *Hammation sollertius* n. g., n. sp. (Acarien, Oribate). *Mem. Mus. Nat. Hist. Natur. (n.s.), ser. A, Zool.* 16:173–98. [BASILOBELBIDAE]
———. 1959b. Observations sur les Oribates (39e série). *Bull. Mus. Nat. Hist. Natur. Paris* 31 (2): 248–55.
———. 1959c. Observations sur les Oribates (40e série). *Bull. Mus. Nat. Hist. Natur. Paris* 31 (2): 359–66.
———. 1959d. *Polypterozetes cherubin* Berl. 1916 (Oribate). *Acarologia* 1:147–80. [POLYPTEROZETIDAE]
———. 1960a. *Autogneta penicillum* n. sp. (Oribate). *Acarologia* 2:345–67. [AUTOGNETIDAE]
———. 1960b. Les Autognetidae n. fam. (Oribates). *Acarologia* 2:575–609.
———. 1960c. *Damaeus avernensis* n. sp. (Oribate). *Acarologia* 2:250–75. [DAMAEIDAE]
———. 1960d. Les Mochlozetidae n. fam. (Oribates). *Acarologia* 2:101–48.
———. 1961a. Les Amerobelbidae (Oribates) (1re partie). *Acarologia* 3:303–43.
———. 1961b. Nouvelles observations sur les Oribates (1re série). *Acarologia* 3:206–31.
———. 1961c. Les Plasmobatidae n. fam. (Oribates). *Acarologia* 3:96–129.
———. 1962a. Au sujet des Hermanniellidae (Oribates). *Acarologia* 4:632–70.
———. 1962b. Le genre *Tegeocranellus* Berl. 1913 (Oribates). *Acarologia* 4:78–100. [TEGEOCRANELLIDAE]
———. 1962c. Nouvelles observations sur les Oribates (2e série). *Acarologia* 4:396–422.
———. 1962d. Prelarves des Oribates. *Acarologia* 4:423–39.
———. 1963. Sur deux espèces de Brachychthoniidae et leur dévéloppement (Oribates). *Acarologia* 5:122–51.
———. 1964a. Nouvelles observations sur les Oribates (3e série). *Acarologia* 6:170–98.
———. 1964b. Oribates mexicains (1re serie). *Dampfiella* Selln. et *Beckiella* n. g. *Acarologia* 6:694–711. [DAMPFIELLIDAE]
———. 1964c. *Pheroliodes wehnckei* (Willmann) (Oribate). *Acarologia* 6:353–86. [PHEROLIODIDAE]
———. 1965a. Complement a mon travail de 1953 sur la classification des Oribates. *Acarologia* 7:713–34.
———. 1965b. *Fosseremus quadripertitus* nom. nov. (Oribate). *Acarologia* 7:343–75. [DAMAEOLIDAE]
———. 1965c. Oribates mexicains (2e série). *Stelechobates megalotrichus* n. g., n. sp. *Acarologia* 7:532–63. [STELECHOBATIDAE]
———. 1966a. *Collohmannia gigantea* Selln. (Oribate). Premiere partie. *Acarologia* 8:328–57. [COLLOMANNIIDAE]
———. 1966b. *Selenoribates mediterraneus* n. sp. et les Selenoribatidae (Oribates). *Acarologia* 8:129–54.
———. 1967a. Nouvelles observations sur les Oribates (5e série). *Acarologia* 9:242–72.
———. 1967b. Les Staurobatidae n. fam. (Oribates). *Acarologia* 8:696–727.
———. 1968. Nouvelles observations sur les Oribates (6e série). *Acarologia* 10:357–91.
———. 1969a. Considérations sur le classement des Oribates. Leur division en 6 groupes majeurs. *Acarologia* 11:127–53.
———. 1969b. Observation sur les muscles de fermeture des volets anaux et génitaux et sur la structure progenitale chez les Oribates supérieures adultes. *Acarologia* 11:317–49.
———. 1970. Nouvelles observations sur les Oribates (7e série). *Acarologia* 12:432–60.
———. 1971. Nouvelles observations sur les Oribates (8e série). *Acarologia* 12:849–76.
Green, J., and M. Macquitty. 1987. Halacarid mites (Arachnida: Ac-

ari). In Synopsis of the British fauna (new series), no. 36, ed. D. Kermack and R. S. K. Barnes. London: E. J. Brill/Dr. W. Backhuys.

Greenberg, B. 1952. New Labidostommidae with keys to the New World species (Acarina). *J. New York Entomol. Soc.* 60:195–209.

Greenslade, P. J. M., and R. V. Southcott. 1980. Parasitic mites on sminthurid Collembolla in Australia. *Entomol. Monthly Mag.* 116:85–86. [ERYTHRAEIDAE]

Greenstone, M. H. 1996. Serological analysis of arthropod predation: past, present and future. In *The ecology of agricultural pests: Biochemical approaches*, ed. W. O. C. Symondson and J. E. Liddell, 265–300. London: Chapman & Hall.

Gregson, J. D. 1973. *Tick paralysis: An appraisal of natural and experimental data.* Can. Dept. Agr. Monogr. 9. Ottawa: Canada Department of Agriculture.

Gressitt, J. L., and J. Shoup. 1967. Ecological notes on free-living mites in North Victoria Land. *Antarct. Res. Ser.* 10:307–20.

Griffiths, D. A. 1964. A revision of the genus *Acarus* L., 1758 (Acaridae, Acarina). *Bull. Brit. Mus. (Nat. Hist.), Zool.* 11:415–64.

———. 1966. Nutrition as a factor influencing hypopus formation in the *Acarus siro* species complex (Acarina, Acaridae). *J. Stored Prod. Res.* 1:25–340.

———. 1970. A further systematic study of the genus *Acarus* L., 1758 (Acaridae, Acarina), with a key to species. *Bull. Brit. Mus. (Nat. Hist.), Zool.* 19 (2): 85–118.

———. 1977. A new family of astigmatid mites from the Iles Crozet, sub-Antarctica; introducing a new concept relating to ontogenetic development of idiosomal setae. *J. Zool. Lond.* 182: 291–308. [GLYCACARIDAE]

Griffiths, D. A., W. T. Atyeo, R. A. Norton, and C. A. Lynch. 1990. The idiosomal chaetotaxy of astigmatid mites. *J. Zool. Lond.* 220:1–32.

Griffiths, D. A., and C. E. Bowman, eds. 1984. *Acarology VI.* 2 vols. Chichester, UK: Ellis Horwood.

Grimaldi, D., and M. S. Engel. 2005. *Evolution of the insects.* Cambridge, UK: Cambridge Univ. Press.

Grishina, L. G. 1993. Ecology of the oribatid mite *Nothrus palustris* C. L. Koch (Sarcoptiformes, Oribatei) in western Siberia. *Ekologiya* 3:39–49. [NOTHRIDAE]

Grobler, J. H. 1962. The life history and ecology of the woolly pine needle aphid, *Schizolachnus pini-radiatae* (Davidson) (Homoptera: Aphididae). *Can. Entomol.* 94:35–45. [ANYSTIDAE]

Grostal, P., and M. Dicke. 2000. Recognising one's enemies: a functional approach to risk assessment. *Behav. Ecol. Sociobiol.* 47:258–64.

Grostal, P., and D. J. O'Dowd. 1994. Plants, mites and mutualism: Leaf domatia and the abundance and reproduction of mites on *Viburnum tinus* (Caprifoliaceae). *Oecologia* 97:308–15.

Groves, J. R., and A. M. Massee. 1951. *A synopsis of the world literature on the fruit tree red spider mite* Metatetranychus ulmi *(C. L. Koch, 1835) and its predators.* London: Commonw. Inst. Entomol. [TETRANYCHIDAE]

Gu, Y-M., C-S. Wang, and Q-X. Duan. 1991. Description of a new genus and a new species and proposal of a new family for the Gamasides (Acari: Gamasina). *Acta Zootax. Sin.* 16:333–38. [EVIPHIDOIDEA?] [in Chinese, English summary]

Gu, Y-M., C-S. Wang, and J-H. Li. 1991. A new genus and species of Gamasides off *Julus terrestris* and a new family proposed (Acari: Dermanyssoidea). *Acta Zootax. Sin.* 16:428–31. [IPHIOPSIDIDAE?] [in Chinese, English summary]

Gurney, B., and N. W. Hussey. 1967. *Pygmephorus* species (Acarina: Pyemotidae) associated with cultivated mushrooms. *Acarologia* 9:353–58. [PYGMEPHORIDAE]

Gutierrez, J., H. R. Bolland, and W. Helle. 1979. Karyotypes of the Tetranychidae and the significance for taxonomy. In Rodriguez, 2:399–404.

Gutierrez, J., and W. Helle. 1985. Evolutionary changes in the Tetranychidae. In Helle and Sabelis (1985a), 1A:91–107.

Guzman, L. I. de, and M. Delfinado-Baker. 1996. A new species of *Varroa* (Acari: Varroidae) associated with *Apis koschevnikovi* (Apidae: Hymenoptera) in Borneo. *Internat. J. Acarol.* 22: 23–27.

Gwiazdowicz, D. J. 1995. Mannchen, Dauer- und Wandernymphen von *Sejus rafalskii* Wisniewski et Hirschmann 1991 (Acari, Sejina) aus Polen. *Bull. Pol. Acad. Sci. Biol. Sci.* 43 (2): 153–62.

———. 2000a. Description of deutonymph of *Panteniphis mirandus* (Acari, Rhodacaridae) from Poland. *Biologia, Bratislava* 55:465–67. [DIGAMASELLIDAE]

———. 2000b. Description of *Epicrius tauricus* male (Acari, Gamasida) with a key to males of European species of the genus *Epicrius. Biologia, Bratislava* 55:469–71.

———. 2007. *Ascid mites (Acari, Mesostigmata) from selected forest ecosystems and microhabitats in Poland.* Poznan, Poland: Wydawnictwo Akademii Rolniczej im. Augusta Cieszkowskiego.

Haarløv, N. 1943. The chaetotaxis of tarsus I in some Mesostigmata. *Entomol. Meddr.* 23:273–94.

———. 1960. Microarthropods from Danish soils. Ecology, phenology. *Oikos*, Suppl. 3.

Haarløv, N., and J. Mørch. 1975. Interaction between *Ornitocheyletia hallae* Smiley 1970 (Acarina, Cheyletiellidae) and *Micromonospora chalcea* (Foulerton 1905) Ørskov 1923 (Streptomycetaceae, Actinomycetales) in the skin of pigeon. *Acarologia* 17:284–99.

Habeeb, H. 1955. North American Hydrachnellae, Acari. XXIX–XXXI. *Leaflet Acadian Biol.* 6:1–4. [STYGOTHROMBIDIIDAE]

———. 1962. A new species of *Homocaligus* (Acarina, Raphignathidae). *Leaflet Acadian Biol.* 27:1–3.

———. 1966. New genera in the Stigmaeidae, Acarina. *Leaflet Acadian Biol.* 42:1–2.

———. 1973. Notes on water-mites. VII. *Leaflet Acadian Biol.* 56:1–4. [STYGOTHROMBIDIIDAE]

———. 1974a. *Cerberothrombium ivarum* new species or a novel torrenticolous mite from Mount San Antonio. *Leaflet Acadian Biol.* 60:1–2. [STYGOTHROMBIDIIDAE]

———. 1974b. New genera of water-mites. *Leaflet Acadian Biol.* 61:1–2. [STYGOTHROMBIDIIDAE]

———. 1974c. Notes on water-mites. IX. *Leaflet Acadian Biol.* 63: 1–2. [STYGOTHROMBIDIIDAE]

———. 1979. Two new species of *Cerberothrombium. Leaflet Acadian Biol.* 73:1–4. [STYGOTHROMBIDIIDAE]

Hadani, A., and Y. Rechav. 1969. Tick-host relationships. I, The existence of a circadian rhythm of "drop-off" engorged ticks from their hosts. *Acta Trop.* 26:173–79.

Hagan, D. V. 1985. *Caeculus crossleyi* n. sp. (Acari: Caeculidae) from granite outcrops in Georgia, USA. *Internat. J. Acarol.* 11: 241–45.

Hagan, D. V., J. H. Oliver, Jr., and J. M. Pound. 1984. A morphological and physiological study of the coxal organs in adult

Ornithodoros parkeri (Acari: Argasidae). In Griffiths and Bowman, 1:304–9.

Hagele, T., B. Kaufman, J. O. Whitaker, Jr., and H. Klompen. 2005. The genus *Euryparasitus* in North America (Mesostigmata: Euryparasitidae). *Zootaxa* 1036:1–20. [OLOGAMASIDAE]

Haggart, D. A., and E. E. Davis. 1980. Ammonia-sensitive neurones on the first tarsi of the tick, *Rhipicephalus sanguineus*. *J. Insect Physiol.* 26:517–23.

Haines, C. P. 1978. A revision of the genus *Blattisocius* Keegan (Mesostigmata: Ascidae) with especial reference to *B. tarsalis* (Berlese) and the description of a new species. *Acarologia* 20:19–38. [BLATTISOCIIDAE]

———. 1981. Laboratory studies on the role of an egg predator, *Blattisocius tarsalis* (Berlese) (Acari: Ascidae), in relation to the natural control of *Ephestia cautella* (Walker) (Lepidoptera: Pyralidae) in warehouses. *Bull. Entomol. Res.* 71:555–74. [BLATTISOCIIDAE]

Haitlinger, R. 1986a. The genus *Hauptmannia* Oudemans, 1910 (Acari, Prostigmata, Erythraeidae) in Poland. *Polsk. Pismo Entomol.* 56:181–91.

———. 1986b. Myocoptidae Gunther, 1942 (Acari, Astigmata) Polski. *Polsk. Pismo Entomol.* 56:389–422.

———. 1987a. Further information on distribution of species from the genus *Hauptmannia* Oudemans, 1910 (Acari, Prostigmata, Erythraeidae) in Poland. *Przeg. Zool.* 31:159–64.

———. 1987b. The genus *Erythraeus* Latreille, 1806 (Acari, Prostigmata, Erythraeidae) in Poland (larvae). *Polsk. Pismo Entomol.* 57:725–34.

———. 1987c. The genus *Leptus* Latreille, 1796 and *Charletonia* Oudemans 1910 (Acari, Prostigmata, Erythraeidae) in Poland (larvae). *Polsk. Pismo Entomol.* 57:339–350.

———. 1987d. *Hauptmannia aitapensis* Southcott, 1947 (Acari, Erythraeidae), a new species of mites to fauna of Viet-Nam. *Polsk. Pismo Entomol.* 56:915–16.

———. 1987e. *Hauptmannia pseudolongicollis* n. sp. (Acari, Prostigamata, Erythraeidae) from Poland. *Polsk. Pismo Entomol.* 57:351–55.

———. 1987f. *Kleemannia halongica* sp. n., *Neocypholaelaps ewae* sp. n. and two species of mites (Acari, Mesostigmata, Ameroseiidae) new to the fauna of Viet-Nam. *Polsk. Pismo Entomol.* 57:361–66.

———. 1987g. *Leptus stolae* new species (Acari, Prostigmata, Erythraeidae) from *Stolas nudicollis* Boh. (Coleoptera, Chrysomelidae, Cassidinae) from Brazil. *Polsk. Pismo Entomol.* 57:357–60.

———. 1988. Species of *Geckobia* Megnin, 1878 (Acari, Prostigmata, Pterygosomatidae) from Madagascar and Vietnam. *Wiad. Parazytol.* 34:161–75.

———. 1990a. *Caraboacarus bernardi* sp. n. (Acari, Caraboacaridae) associated with *Calosoma* sp. (Coleoptera, Carabidae) in the USSR. *Ann. Hist.–Nat. Mus. Nat. Hung.* 82:217–19.

———. 1990b. Four new species of *Leptus* (Acari, Prostigmata, Erythraeidae) from tropical Africa. *Angew. Parasitol.* 31:29–33.

———. 1990c. Four new species of *Leptus* Latreille, 1796 (Acari, Prostigmata, Erythraeidae) from insects of Australia, New Guinea and Asia. *Wiad. Parazytol.* 36:47–54.

———. 1990d. New canestriniid mites (Acari, Astigmata, Canestriniidae) associated with beetles of the families Carabidae, Scarabaeidae, Tenebrionidae and Passandridae (Insecta, Coleoptera.) *Ann. Zool. (Warsaw)* 43:311–25.

———. 1990e. Two new species of *Leptus* Latreille, 1796 (Acari, Prostigmata, Erythraeidae) from Tenebrionidae (Coleoptera) with a key to European and North African species. *Polsk. Pismo Entomol.* 60:45–50.

———. 1991. New data on the distribution of larvae from the genus *Leptus* Latreille, 1796 (Acari, Prostigmata, Erythraeidae) in Poland with the description of *Leptus miromiri* n. sp. *Wiad. Parazytol.* 37:499–506.

———. 1992a. Canestriniid mites (Acari, Astigmata, Canestriniidae) occurring on Chrysomelidae (Insecta, Coleoptera) in Neotropical region. 1, Genera *Grandellina* Fain, *Arrunsithiana* n. gen. *Wiad. Parazytol.* 37:381–99.

———. 1992b. Six new species of *Leptus* Latreille, 1796 (Acari, Prostigmata, Erythraeidae) from Neotropical region. *Zesz. Nauk. Akad. Roln. Wroclaw. Zootechn.* 206:265–72.

———. 1993a. Canestriniid mites (Acari, Astigmata, Canestriniidae) occurring on Chrysomelidae (Insecta, Coleoptera) in Neotropical region. 2, Genera *Ovarrothiana* n. gen., *Ennodithiana* n.gen. *Wiad. Parazytol.* 39:59–73.

———. 1993b. Canestriniid mites (Acari, Astigmata, Canestriniidae) occurring on Chrysomelidae (Insecta, Coleoptera) in Neotropical region. 3, Genera *Thopia* Summers et Schuster, *Sabothiana* n. gen., *Olafithiana* n. gen., *Bibulothiana* n. gen. *Wiad. Parazytol.* 39:265–83.

———. 1993c. A new genus and species of Tetrathrombiinae from Poland (Acari, Johnstonianidae). *Abh. Ber. Naturk. Görlitz* 67:3–5. [HYDRYPHANTIDAE]

———. 1993d. A new species of the genus *Diplothrombium* Berlese (Acari, Prostigmata, Johnstonianidae) from Poland, based on the larva. *Tijdschr. Entomol.* 136:11–13.

———. 1994a. Four new larval mites (Acari: Trombidiidae: Eutrombidiinae) ectoparasitic on carabids (Insecta: Coleoptera: Carabidae). *Rev. Chilena Entomol.* 21:47–56.

———. 1994b. Two new species of the genus *Podothrombium* Berlese, 1910 (Acari, Prostigmata, Trombidiidae) from Austria and Italy. *Linz. Biol. Beitr.* 26:531–38.

———. 1995a. New larval mite species of the genus *Podothrombium* Berlese, 1910 (Acari, Prostigmata, Trombidiidae) and *Lassenia* Newell, 1957 (Prostigmata, Tanaupodidae) from Poland. *Wiad. Parazytol.* 41:463–74.

———. 1995b. New species of mites collected on Tenebrionidae, Scarabaeidae, Cerambycidae and Lucanidae with the description of a new subfamily Vereoxiinae (Acari, Astigmata, Canestriniidae). *Wiad. Parazytol.* 41:79–91.

———. 1996. New heterocoptid mites (Acari, Astigmata, Heterocoptidae) associated with Cassidinae and Hispinae (Coleoptera, Chrysomelidae) from Africa and Asia. *Linz. Biol. Beitr.* 28: 979–98.

———. 1997. A new species of *Hexathrombium* Cooreman, 1944 based on parasitic larva on erotylids from Brazil (Acari: Eutrombidiidae). *Mem. Inst. Oswaldo Cruz, Rio de Janeiro* 92: 81–83.

———. 1998a. Five new species of *Leptus* Latreille, 1796 (Acari: Prostigmata, Erythraeidae) from Asia and Africa. *Bonner Zool. Beitr.* 48:97–110.

———. 1998b. New genera and species of larval Microtrombidiidae and Neothrombidiidae (Acari: Prostigmata) from Poland, Scotland, Madagascar and Vietnam. *Wiad. Parazytol.* 44:233–59.

———. 1999a. *Linobia coccinellae* (Scopoli, 1763) (Acari: Astigmata: Hemisarcoptidae) nowy gatunek dla fauny Polski, zwiazany z

Chrysomela populi L. (Coleoptera: Chrysomelidae: Chrysomelinae). *Przeg. Zool.* 43:181–82.

———. 1999b. Six new heterocoptid mites (Acari, Astigmata, Heterocoptidae) associated with Cassidinae and Eumolpinae (Coleoptera, Chrysomelidae) from the Indo-Australian region. *Misc. Zool.* 22:59–73.

———. 1999c. Six new species of *Leptus* Latreille, 1796 (Acari, Prostigmata, Erythraeidae) from South-East Asia. *Misc. Zool.* 22:51–68.

———. 1999d. Three new species of larval Chyzeriidae associated with Orthoptera (Insecta) from Cyprus, Crete and Peru, with description of the new subfamily Perumaroptinae and three new genera *Napassenia*, *Cretessenia* and *Perumaropta* (Acari Prostigmata). *Bull. Soc. Entomol. Ital.* 131:3–13.

———. 1999e. Three new species of rosensteiniid mites (Acari: Rosensteiniidae: Rosensteiniinae) associated with cockroaches (Blattodea). *An. Inst. Biol., Univ. Nac. Aut. Méx., Ser. Zool.* 70:79–91.

———. 2000. A new species of larval *Dasitrombium* Zhang, 1994 (Acari: Prostigmata: Neothrombiidae) parasitic on grasshoppers from Nicaragua. *Syst. Parasitol.* 47:65–67.

———. 2001a. *Amarucoptes*, a new genus of the family Heterocoptidae (Acari, Astigmata). *Syst. Appl. Acarol.* 6:111–14.

———. 2001b. *Diplothrombium zbigniewi* sp. nov. (Acari: Prostigmata: Johnstonianidae), a new mite from Poland. *Syst. Appl. Acarol.* 6:179–82.

———. 2001c. Four new species of *Leptus* (Acari, Prostigmata, Erythraeidae) from Republic of South Africa and Kenya. *Biologia, Bratislava* 56:473–81.

———. 2001d. *Parasitus geotrupidis* Makarova, 1996 (Mesostigmata: Parasitidae) and *Podothrombium roari* Haitlinger, 2000 (Prostigmata, Trombidiidae), two mite species (Acari) new to the Polish fauna. *Przeg. Zool.* 45:83–85.

———. 2002a. Seven new heterocoptid mites (Acari: Astigmata: Heterocoptidae) from Africa. *Zesz. Nauk. Akad. Roln. Wroclaw. Zootech.* 49 (447): 7–29.

———. 2002b. Two new species of *Leptus* Latreille, 1796 and the first record of *Leptus astrubali* Haitlinger, 1999 (Acari, Prostigamta, Erythraeidae) from India and Sri Lanka. *Syst. Appl. Acarol.* 202:177–84.

———. 2003a. A new larval ectoparasitic *Eutrombidium* Verdun (Acari: Prostigmata: Eutrombidiidae) found on African Gryllidae (Insecta: Orthoptera), and the first record of *E. macfarlanei* Southcott in Tanzania. *Polsk. Pismo Entomol.* 72:313–17.

———. 2003b. New records of mites (Acari: Prostigmata: Calyptostomatidae, Erythraeidae, Trombidiidae, Microtrombidiidae) from the Czech Republic, with a description of *Podothrombium karlovaicus* n. sp. *Cas. Narodniho Muz. Rada Prirodovedna* 172:5–12.

———. 2004a. New records of mites (Acari: Prostigmata: Erythraeidae) from Cambodia and Myanmar, with a description of *Erythraeus* (*Erythraeus*) *kacperi* sp. nov. *Syst. Appl. Acarol.* 9:157–61.

———. 2004b. New records of mites (Acari: Prostigmata: Erythraeidae, Trombidiidae, Eutrombidiidae) from Croatia, with descriptions of three new species. *Nat. Croatia* 13:143–60.

———. 2004c. Three new species of *Leptus* Latreille, 1796 and the first record of *Leptus onnae* Haitlinger 2000 (Acari: Prostigamata: Erythraeidae) from Brazil. *Syst. Appl. Acarol.* 9:147–56.

Hajjar, N. P. 1972. Biochemical and physiological studies of certain ticks (Ixodoidea). Phospholipid and sterol patterns in biological fluids of nymphal and adult *Hyalomma* (*H.*) *dromedarii* Koch and *H.* (*H.*) *anatolicum excavatum* Koch (Ixodidae). *J. Med. Entomol.* 9:281–85.

Halasková, V. 1963. On the genus *Prozercon* Sellnick, 1943. *Acta Soc. Entomol. Cechoslov.* 60:145–69. [ZERCONIDAE]

———. 1969. Some new species of the family Zerconidae from North America (Acari: Mesostigmata). *Acta Soc. Zool. Bohemoslov.* 33:115–27.

———. 1970. Zerconidae of Czechoslovakia (Acari: Mesostigmata). *Acta Univ. Carol., Biol.* 1969:175–352.

———. 1977. *A revision of the genera of the family Zerconidae (Acari: Gamasides) and descriptions of new taxa from several areas of Nearctic region.* Studie Ceskoslov. Akad. Ved 7.

———. 1979. *Taxonomic studies on Zerconidae (Acari: Mesostigmata) from the Korean People's Democratic Republic.* Acta Sci. Nat. Brno 13, fasc. 3.

Halbert, J. N. 1915. Clare Island Survey. Acarinida, Section II. Terrestrial and marine Acarina. *Proc. Roy. Irish Acad.* 31, pt. 39: 45–136.

———. 1920. The Acarina of the seashore. *Proc. Roy. Irish Acad.* B 35 (7): 106–52.

Hall, M., and E. Ebermann. 2005. Zoogeographical aspects of some scutacarid mites and their phoresy hosts (Acari, Heterostigmata; Hymenoptera, Aculeata). *Rev. Suisse Zool.* 112:215–24.

Halliday, R. B. 1987. Two new species of *Laelaptonyssus* Womersley (Acarina: Rhodacaridae) from the nests of Australian termites. *J. Aust. Entomol. Soc.* 26:85–91. [OLOGAMASIDAE]

———. 1990a. The occurrence in Australia of Podocinidae (Acarina: Mesostigmata), with description of two new species. *J. Aust. Entomol. Soc.* 29:277–80.

———. 1990b. Mites of the *Macrocheles muscaedomesticae* group in Australia (Acarina: Macrochelidae). *Invertebr. Taxon.* 3:407–30.

———. 1991. Taxonomic background of the redlegged earth mite *Halotydeus destructor* (Tucker) (Acarina: Penthaleidae). *Plant Prot. Qtly.* 6:162–65.

———. 1994. Microscope slide mounting media results of informal survey. http://www.nhm.ac.uk/hosted_sites/acarology/archive/summary.html.

———. 1997a. Revision of the Australian Ameroseiidae (Acarina: Mesostigmata). *Invertebr. Taxon.* 10:179–201.

———. 1997b. Revision of the genus *Zygoseius* Berlese (Acarina: Pachylaclapidae). *Acarologia* 38:3–20.

———. 1998. *Mites of Australia, a checklist and bibliography.* Monographs on Invertebrate Taxonomy. Vol. 5. Collingswood, Queensland, Victoria: CSIRO Publishing.

———. 2000. The Australian species of *Macrocheles* (Acarina: Macrochelidae). *Invertebr. Taxon.* 14:273–326.

———. 2001. Systematics and biology of the Australian species of *Balaustium* von Heyden (Acari: Erythraeidae). *Aust. J. Zool.* 40:326–30.

———. 2005. Systematics and biology of *Penthaleus tectus* sp. n. (Acari: Penthaleidae), a recently discovered pest of grain crops in eastern Australia. *Aust. J. Entomol.* 44:144–49.

———. 2006. New taxa of mites associated with Australian termites (Acari: Mesostigmata). *Internat. J. Acarol.* 32:27–38. [TERANYSSIDAE, LAELAPIDAE]

Halliday, R. B., and R. O. Collins. 2002. *Histiostoma papillata* sp. n. (Acari: Histiostomatidae), a mite attacking fish in Australia. *Aust. J. Entomol.* 41:155–58.

Halliday, R. B., and E. Holm. 1987. Mites of the family Macrochelidae as predators of two species of dung-breeding pest flies. *Entomophaga* 32:333–38.

Halliday, R. B., and D. K. Knihinicki. 2004. The occurrence of *Aceria tulipae* (Keifer) and *Aceria tosichella* Keifer in Australia (Acari: Eriophyidae). *Internat. J. Acarol.* 30:113–18.

Halliday, R. B., D. E. Walter, and E. E. Lindquist. 1998. Revision of the Australian Ascidae (Acarina: Mesostigmata). *Invertebr. Taxon.* 12:1–54. [includes MELICHARIDAE, BLATTISOCIIDAE]

Halliday, R. B., D. E. Walter, H. C. Proctor, R. A. Norton, and M. J. Colloff, eds. 2001. *Acarology: Proceedings 10th International Congress, Canberra.* Collingswood, Victoria, Australia: CSIRO Publishing.

Hammer, M. 1958. *Investigations on the oribatid fauna of the Andes Mountains.* I, *The Argentine and Bolivia.* Biol. Skr. Dan. Vid. Selsk. 10.

———. 1961. *Investigations on the oribatid fauna of the Andes Mountains.* II, *Peru.* Biol. Skr. Dan. Vid. Selsk. 13. [CHAVINIIDAE]

———. 1962. *Investigations on the oribatid fauna of the Andes Mountains.* II, *Peru.* Biol. Skr. Dan. Vid. Selsk. 13.

———. 1966. *Investigations on the oribatid fauna of New Zealand.* Part I. Biol. Skr. Dan. Vid. Selsk. 15 (2).

———. 1967. *Investigations on the oribatid fauna of New Zealand.* Part II. Biol. Skr. Dan. Vid. Selsk. 15.

———. 1971. *On some oribatids from Viti Levu, Fiji Islands.* Biol. Skr. Dan. Vid. Selsk. 16 (6): 1–60.

———. 1972. *Investigation on the oribatid fauna of Tahiti, and on some oribatids found on the Atoll Rangiora.* Biol. Skr. Dan. Vid. Selsk. 19:1–65.

———. 1973. *Oribatids from Tongatapu and Eua, the Tonga Islands, and from Upolu, Western Samoa.* Biol. Skr. Dan. Vid. Selsk. 20:1–70.

Hammer, M., and J. A. Wallwork. 1979. A review of the world distribution of oribatid mites (Acari: Cryptostigmata) in relation to continental drift. *Biol. Skr. Dan. Vid. Selsk.* 22:1–31.

Hance, T., P. Neuberg, and C. Noèl-Lastelle. 1998. The use of fecundity, lobe biometry and the RAPD-PCR technique in order to compare strains of *Tetranychus* sp. *Exp. Appl. Acarol.* 22: 649–66. [TETRANYCHIDAE]

Hänel, H. 1986. Effect of juvenile hormone (III) from the host *Apis mellifera* (Insecta: Hymenoptera) on neurosecretion of the parasitic mite *Varroa jacobsoni* (Acari: Mesostigmata). *Exp. Appl. Acarol.* 2:257–71.

Hansen, R. A. 2000. Effects of habitat complexity and composition on a diverse litter microarthropod assemblage. *Ecology* 81: 1120–32.

Haq, M. A. 1965. Records of some interstitial mites from Nobska Beach together with a description of a new genus and species *Psammonsella nobskae*, of the family Rhodacaridae (Acarina Mesostigmata). *Acarologia* 7:411–19. [OLOGAMASIDAE]

———. 1982. Feeding habits of ten species of oribatid mites (Acari: Oribatei) from the soils of Malabar, South India. *Indian J. Acarol.* 6:39–50.

———. 1984. The role of microbes in the nutrition of a lohmanniid mite (Acari: Oribatei). In Griffiths and Bowman, 2:838–46. [LOHMANNIIDAE]

———. 2001. Potential oribatid mite vectors of cestode parasites in Kerala, India. In Halliday et al. (2001), 569–75.

Harding, D. J. L. 1971. The phenology of *Platynothrus peltifer* (C. L. Koch). In Daniel and Rosický, 79–83. [CAMISIIDAE]

———. 1976. A new species of *Phthiracarus* (Acari: Cryptostigmata) from Great Britain. *Acarologia* 18:163–69. [PHTHIRACARIDAE]

Harding, D. J. L., and S. M. Easton. 1984. Development of two species of phthiracarid mites in beech cupules. In Griffiths and Bowman, 2:860–70. [PHTHIRACARIDAE]

Hardison, J. R. 1959. Evidence against *Fusarium poae* and *Siteroptes graminum* as causal agents of silver top of grasses. *Mycologia* 51:712–28.

Harrison, F. W., and R. F. Foelix, eds. 1999. *Microscopic anatomy of invertebrates.* Vol. 8C, *Chelicerate arthropoda.* New York: John Wiley & Sons.

Hart, B. J., and A. Fain. 1987. A new technique for isolation of mites exploiting the difference in density between ethanol and saturated NaCl: Qualitative and quantitative studies. *Acarologia* 28:251–54.

Hartenstein, R. 1962a. Soil Oribatei. I, Feeding specificity among forest soil Oribatei (Acarina). *Ann. Entomol. Soc. Amer.* 55:202–6.

———. 1962b. Soil Oribatei. V, Investigations on *Platynothrus peltifer* (Acarina: Camisiidae). *Ann. Entomol. Soc. Amer.* 55:709–13.

Harvey, M. S. 1987. New and little known species of the water mite genera *Tartarothyas, Pseudohydryphantes* and *Cyclohydryphantes* from Australia. *Mem. Mus. Victoria* 48:107–22. [HYDRYPHANTIDAE]

———. 1988a. New species of the water mite family Hydryphantidae (Acarina) from the Northern Territory, Australia. *Beagle* 5:17–26.

———. 1988b. Three new unusual water mites from Australia (Chelicerata: Acarina: Hydryphantidae, Hygrobatidae and Athienemanniidae). *Mem. Mus. Victoria* 49:355–61.

———. 1989. A review of the water mite genus *Australorivacarus* K. O. Viets (Chelicerata: Actinedida: Hygrobatidae). *Invertebr. Taxon.* 3:155–62.

———. 1990a. Pezidae, a new freshwater mite family from Australia (Acarina: Halacaroidea). *Invertebr. Taxon.* 3:771–81.

———. 1990b. A review of the water mite family Anisitsiellidae in Australia (Acarina). *Invertebr. Taxon.* 3:629–46.

———. 1990c. A review of the water mite family Limnocharidae in Australia (Acarina). *Invertebr. Taxon.* 3:483–93.

———. 1990d. Two new species of *Partidomomonia* Cook from south-eastern Australia (Acarina: Momoniidae). *Mem. Mus. Victoria* 50:337–40.

———. 1990e. Two new water mite genera from south-western Australia (Acarina: Aturidae, Mideopsidae). *Mem. Mus. Victoria* 50:341–46.

———. 1996a. A review of the water mite family Pionidae in Australia (Acarina: Hygrobatoidea). *Rec. W. Aust. Mus.* 17:361–93.

———. 1996b. A revised systematic placement for *Austrotrombella* Southcott (Acarina: Hydryphantidae). *Trans. Roy. Soc. S. Aust.* 120:37–40.

———. 1996c. Small arachnids and their value in Gondwanan biogeographic studies. In *Gondwanan heritage: Past, present and future of the Western Australian biota,* ed. S. D. Hopper, J. A. Chappill, M. S. Harvey, and A. S. George, 155–62. Chipping Norton, UK: Surrey Beatty & Sons.

———. 1998a. *The Australian water mites. A guide to the families and genera*. Monogr. Invertebr. Taxon. 4. Melbourne, Australia: CSIRO Publishing.

———. 1998b. Unusual new water mites (Acari: Hydracarina) from Australia. Part 1. *Rec. W. Aust. Mus.* 19:91–106.

———. 2003. *Davecookia*, a new genus of the water mite family Athienemanniidae (Acarina: Hydracarina) from Australia. In I. M. Smith (2003a), 151–53.

Harvey, M. S., and D. R. Cook. 1988. Water mites of the genus *Aspidiobates* from Victoria, Australia, with the description of two new species (Chelicerata: Acarina: Hygrobatidae). *Mem. Mus. Victoria* 49:51–57.

Harwood, R. F., and M. T. James. 1979. *Entomology in human and animal health*. 7th ed. New York: Macmillan.

Hase, T., L. W. Roberts, P. K. Hildebrandt, and D. C. Cavanaugh. 1978. Stylostome formation by *Leptotrombidium* mites (Acari: Trombiculidae). *J. Parasitol.* 64:712–18.

Haswell, W. A. 1922. *Astacocroton*, a new type of Acarid. *Proc. Linn. Soc. N. S. W.* 47:329–43. [ASTACOCROTONIDAE]

Hatsushika, R., T. Okino, and K. Miyoshi. 1989. A case study of itching dermatitis caused by *Cheyletus fortis* (Oudemans, 1904) (Acarina: Cheyletidae) found in Okayama Prefecture, Japan. *Kawasaki Med. J.* 15:151–57.

Haumann, G. 1991. *Zur Phylogenie Primitiver Oribatiden, Acari: Oribatida*. Graz, Austria: Technische Universität Graz.

Haupt, J., and Y. Coineau. 1999. Ultrastructure and functional morphology of a nematalycid mite (Acari: Actinotrichida: Endostigmata: Nematalycidae): Adaptations to a mesopsammal life. *Acta Zool. (Stockholm)* 80:97–111.

Hayes, J. L. 1985. The predator-prey interaction of the mite *Balaustium* sp. and the pierid butterfly *Colias alexandra*. *Ecology* 66:300–303. [ERYTHRAEIDAE]

Hazan, A., V. Gerson, and A. S. Tahori. 1975. Spider mite webbing. II, The effect of webbing on egg hatchability. *Acarologia* 17: 270–73.

Heethoff, M. 2000. Untersuchung der ribosomalen ITS1 Sequenzen der parthenogenetischen Hornmilbe *Platynothrus peltifer* (C. L. Koch, 1839) (Acari: Oribatida). *Mitt. AG Boden., Innsbruck* 16:15–20. [CAMISIIDAE]

Heethoff, M., K. Domes, M. Laumann, M. Maraun, R. A. Norton, and S. Scheu. 2007. High genetic divergences indicate ancient separation of parthenogenetic lineages of the oribatid mite *Platynothrus peltifer* (Acari, Oribatida). *J. Evol. Biol.* 20: 392–402.

Heethoff, M., M. Maraun, and S. Scheu. 2000. Genetische Variabilitaet der ribosomalen ITS1-Sequenzen bei der parthenogenetischen Hornmilbe *Platynothrus peltifer* (C. L. Koch, 1839) (Acari: Oribatida). *Ber. Naturwiss. Medizin. Vereins Innsbruck* 87:339–54. [CAMISIIDAE]

Helle, W., H. R. Bolland, S. H. M. Jeurissen, and G. A. van Seventer. 1984. Chromosome data on the Actinedida, Tarsonemida and Oribatida. In Griffiths and Bowman, 1:449–54.

Helle, W., H. R. Bolland, R. van Arrendok, R. DeBoer, G. G. M. Schulten, and V. Russel. 1978. Genetic evidence for biparental males in haplo-diploid predator mites (Acarina: Phytoseiidae). *Genetica* 49:165–71.

Helle, W., and W. P. J. Overmeer. 1985. Rearing techniques. In Helle and Sabelis (1985a), 1A:331–36.

Helle, W., and L. P. Pijnacker. 1985. Parthenogenesis, chromosomes and sex. In Helle and Sabelis (1985a), 1A:129–39. [TETRANYCHIDAE]

Helle, W., and M. W. Sabelis, eds. 1985a. *Spider mites: Their biology, natural enemies and control*. World Crop Pests, vol. 1A. Amsterdam: Elsevier.

———. 1985b. *Spider mites: Their biology, natural enemies, and control*. World Crop Pests, vol. 1B. Amsterdam: Elsevier.

Helle, W., and M. Wyoski. 1984. The chromosomes and sex-determination of some actinotrichid taxa (Acari), with special reference to Eriophyidae. *Internat. J. Acarol.* 9:67–71.

———. 1996. Arrhenotokous parthenogenesis. In Lindquist, Sabelis, and Bruin, 6:169–72.

Hennessey, M. K., and M. H. Farrier. 1989. *Mites of the family Parasitidae (Acari: Mesostigmata) inhabiting forest soils of North and South Carolina*. N.C. Agr. Res. Serv. Tech. Bull. 291. Raleigh: North Carolina Agricultural Research Service.

Herfs, A. 1926. Okologische Untersuchungen an *Pediculoides ventricosus* (Newp.) Berlese. *Zool. Gesamtgeb. Zool.* 74:3–68.

Hess, E., and M. Vlimant. 1986. Leg sense organs of ticks. In *Morphology, physiology and behavioural biology of ticks*, ed. J. R. Sauer and J. A. Hair, 361–90. Chichester, UK: Ellis Horwood.

Hessein, N. A., and T. M. Perring. 1986. Feeding habits of the Tydeidae with evidence of *Homeopronematus anconai* (Acari: Tydeidae) predation on *Aculops lycopersici* (Acari: Eriophyidae). *Internat. J. Acarol.* 12:215–21. [IOLINIDAE]

Hevers, J. 1977. Revision der in Deutschland vorkommenden Arten der Untergattung *Unionicola* s. str. (Hydrachnellae, Acari). *Acarologia* 18:691–703.

———. 1978a. Interspezifische Beziehungen zwischen *Unionicola*-Larven (Hydrachnellae, Acari) und Chironomidae (Diptera, Insecta). *Verh. Ges. Ökol.* 1977:211–17.

———. 1978b. Morphologie und Systematik der in Deutschland auftretenden Schwamm- und Muschel-Milben-Arten der Gattung *Unionicola* (Acarina: Hydrachnellae, Unionicolidae). *Entomol. Gen.* 5:57–84.

———. 1978c. Zur Sexualbiologie der Gattung *Unionicola* (Hydrachnellae, Acari). *Zool. Jb. Syst., Ökol. Geogr. Tiere* 105: 33–64.

———. 1980. Biologisch-ökologische Untersuchungen zum Entwicklungszyklus der in Deutschland auftretenden *Unionicola*-Arten (Hydrachnellae, Acari). *Arch. Hydrobiol., Suppl.* 57: 324–73.

———. 1984. Die *Unionicola crassipes*-Gruppe in Japan und im Iran. *Deutsche Entomol. Zeitschr., N. F.* 31:265–80. [UNIONICOLIDAE]

Hewitt, M. G., G. I. Barrow, D. C. Miller, F. Turk, and S. Turk. 1973. Mites in the personal environment and their role in skin disorders. *Brit. J. Dermatol.* 89:401–9. [TARSONEMIDAE]

Hicks, E. A. 1958. A new genus and species of diplogyniid from Nicaragua (Order Acarina, Family Diplogyniidae). *Iowa St. Coll. J. Sci.* 33:103–10. [DIPLOGYNIIDAE]

Higgs, G. A., J. R. Vane, R. J. Hart, C. Potter, and R. G. Wilson. 1976. Prostaglandins in the saliva of the cattle tick *Booplilus microplus* (Canestrini) (Acari: Ixodidae). *Bull. Entomol. Res.* 66:665–70.

Hill, A., and K. L. Deahl. 1978. Description and life cycle of a new species of *Histiostoma* (Acari: Histiostomidae) associated with commercial mushroom production. *Proc. Entomol. Soc. Wash.* 80:317–29. [HISTIOSTOMATIDAE]

Hill, R. L., and D. J. O'Donnell. 1991. The host range of *Tetranychus lintearius* (Acarina: Tetranychidae). *Exp. Appl. Acarol.* 11: 253–69.

Hillis, D. M., B. K. Mable, A. Larson, S. K. Davis, and E. A. Zimmer. 1996. Nucleic acids. IV, Sequencing and cloning. In *Molecular systematics*, ed. D. M. Hillis, C. Moritz, and B. K. Mable, 321–84. Sunderland, MA: Sinauer.

Hinman, E. H., and R. H. Kampmeier. 1934. Intestinal acariasis due to *Tyrophagus longior* Gervais. *Amer. J. Trop. Med.* 14:355–62.

Hinton, H. E. 1967. The structure of the spiracles of the cattle tick, *Boophilus microplus*. *Aust. J. Zool.* 15:941–45.

———. 1971. Plastron respiration in the mite, *Platyseius italicus*. *J. Insect Physiol.* 17:1185–99. [BLATTISOCIIDAE]

Hirauchi, Y., and J. Aoki. 1997. A new species of the genus *Achipteria* from Mt. Tateyama, central Japan (Acari: Oribatida). *Edaphologia* 59:5–9. [ACHIPTERIIDAE]

Hiregaudar, L. S., A. K. Joshee, and P. W. Soman. 1959. On some pterygosomatid mites parasitic on Indian lizards. *J. Biol. Sci.* 2:64–66.

Hirschfelder, H., and H. Sachs. 1952. Recent research on the acarine mite. *Bee World* 33:201–8. [TARSONEMIDAE]

Hirschmann, W. 1960. *Gangsystematik der Parasitiformes*. Teil 3, *Die Gattung* Dendrolaelaps *Halbert 1915*. Acarologie, Schrift. für vergleich. Milbenk. Fürth, Germany: Hirschmann Verlag. [DIGAMASELLIDAE]

———. 1962. *Gangsystematik der Parasitiformes*. Teil 5, *Gamasiden*. Acarol., Schrift. Vergleich. Milbenk. Fürth, Germany: Hirschmann Verlag. [includes ASCIDAE, MELICHARIDAE]

———. 1966. *Milben (Acari). Einführung in die Kleinlebewelt*. Stuttgart: Kosmos-Verlag.

———. 1972. Teilgange, Stadien von 21 neuen *Uropoda* (*Phaulodinychus*) Arten (Uropodini, Uropodinae). *Gangsyst. Parasitiformes. Schrift. Vergleich. Milbenk.* 18:79–92.

———. 1976. Adulten-Gruppen und Bestimmungstabelle von 81 *Trachyuropoda*-Arten (Trachyuropodini, Oplitinae). *Gangsyst. Parasitiformes. Schrift. Vergleich. Milbenk., Fürth. Folge* 22:4–15. [TRACHYUROPODIDAE]

———. 1984. Die Gattung *Asternolaelaps* Berlese 1923 aus adultensystematischer und gangsystematischer Sicht (Trichopygidiina). *Acarologia* 31:112–15.

———. 1991. Gangsystematik der Parasitiformes. Teil 529–31. Weltweite revision der Ganggattung *Sejus* C. L. Koch 1836 (Trichopygidiina). *Acarol., Schrift. Vergleich. Milbenk., Fürth. Folge* 38:107–221.

———. 1992. Stadium einer neuen *Polyaspinus*—Art aus Bolivien. *Gangsyst. Parasitiformes. Schrift. Vergleich. Milbenk., Fürth. Folge* 39:168–69.

Hirschmann, W., and N. Hiramatsu. 1984. Die Gattung *Japanoasternolaelaps* nov. gen. Hirschmann u. Hiramatsu 1984 und Beschreibung einer neuen *Japanoasternolaelaps*-Art aus Japan. *Acarologia* 31:133–36.

Hirschmann, W., and M. Hutu. 1974. Uropodiden-Forschung und die Uropodiden der Erde, geordnet nach dem Gangsystem und nach den Ländern in zoogeographischen Reichen und Unterreichen. *Gangsyst. Parasitiformes. Schrift. Vergleich. Milbenk., Fürth. Folge* 20:6–36. [DITHINOZERCONIDAE]

Hirschmann, W., and W. Krauss. 1965. *Gamasiden Bestimmungstafeln von 55* Pachylaelaps-*Arten*. Gangsyst. Parasitiformes. *Schrift. Vergleich. Milbenk. Fürth Folge* 7. [PACHYLAELAPIDAE]

Hirschmann, W., and W. Rühm. 1953. Milben und Fadenwürmer als Symphoristen und Parasiten des Buchdrückers. *Mikrokosmos* 43:7–10.

Hirschmann, W., and J. Wisniewski. 1982a. Weltweite Revision der Gattungen *Dendrolaelaps* Halbert 1915 und *Longoseius* Chant 1961 (Parasitiformes). Band I. Beschreibung der Untergattungen und Arten, Bestimmungstabellen, Chätotaxie, Porotaxie. *Acarologie* 29:1-190. [DIGAMASELLIDAE]

———. 1982b. Weltweite Revision der Gattungen *Dendrolaelaps* Halbert 1915 und *Longoseius* Chant 1961 (Parasitiformes). Band II, Artenverzeichnisse, Krankheiten, Missbildungen, Inseminationsapparate, Abbildungen. *Acarologie* 29:1–48 + 94 plates. [DIGAMASELLIDAE]

———. 1994. *Sejus venezuelanus* nov. spec. (Trichopygidiina) aus Venezuela. *Acarologia* 35:21–26.

Hirschmann, W., and I. Zirngiebl-Nicol. 1964. Uropodiden das Gangsystem der Familie Uropodidae (Berlese 1892) Bestimmungstabellen, Kurzdiagnosen, Operculum-Bestimmungstabellen. *Gangsyst. Parasitiformes. Schrift. Vergleich. Milbenk., Fürth. Folge* 6:1–22.

Hirst, S. 1917. On some new mites of the suborder Prostigmata living on lizards. *Ann. Mag. Nat. Hist.*, ser. 20, 8:136–43.

———. 1921. On the mite (*Acarapis woodi*, Rennie) associated with Isle of Wight bee disease. *Ann. Mag. Nat. Hist.*, ser. 9, 7:509–19. [TARSONEMIDAE]

———. 1923. On some arachnid remains from the Old Red Sandstone (Rhynie Chert Bed, Aberdeenshire). *Ann. Mag. Nat. Hist.* 12:455–74.

———. 1925. On the parasitic mites of the suborder Prostigmata (Trombidioidea) found on lizards. *J. Linn. Soc. Lond. Zool.* 36:173–200. [PTERYGOSOMATIDAE]

———. 1927. Notes on Acari, mainly belonging to the genus *Spinturnix* Von Heyden. *Proc. Zool. Soc. Lond.*: 323–38. [ANYSTIDAE]

———. 1928. On some new Australian mites of the families Trombidiidae and Erythraeidae. *Ann. Mag. Nat. Hist.* 10:563–71.

Hochmuth, R. C., J. L. Hellman, G. Dively, and R. F. W. Schroder. 1987. Effect of the ectoparasitic mite *Coccipolipus epilachnae* (Acari: Podapolipidae) on feeding, fecundity, and longevity of soybean-fed adult Mexican bean beetles (Coleoptera: Coccinellidae) at different temperatures. *J. Econ. Entomol.* 80:612–16.

Hoffmann, A. 1969. Ácaros parásitos de batracios y reptiles en México. *Rev. Latinoamer. Microbiol. Parasitol.* 11:209–16.

———. 1979. Género y especie nuevos de un ácaro parásito de lagartijas (Acarida: Pterygosomidae). *Esc. Nac. Cienc. Biol. Mex.* 23:97–107.

———. 1990. Los Trombicúlidos de México (Acarida: Trombiculidae). Parte taxonómica. *Pub. Espec. Inst. Biol.* 2:1–275.

Hoffmann, A., and C. Mendez. 1973. Eritreidos nuevos de México (Acari: Erythraeidae). *Esc. Nac. Cien. Biol. Mex.* 20:123–44.

Hoffmann, A., and O. Sanchez. 1980. Genero y especie nuevos de un acaro parasito de lagartijas (Acarida: Pterygosomidae). *Esc. Nac. Cienc. Biol. Mex.* 23:97–107.

Hoffmann, A., and M. Vázquez. 1988. Los primitivos acaros Opilioacaridos en Mexico. *Folia Entomol. Mex.* 67:53–60.

Hölldobler, B., and E. O. Wilson. 1990. Symbioses with other arthropods. In *The ants*, 13:471–529. Cambridge, MA: Belknap Press, Harvard Univ.

Holm, E., and M. M. H. Wallace. 1989. Distribution of some anystid mites (Acari: Anystidae) in Australia and Indonesia and their role as possible predators of the cattle tick, *Boophilis microplus* (Acari: Ixodidae). *Exp. Appl. Acarol.* 6:77–83.

Holz, P. H. 1998. Nodules caused by demodex-like mites in the skin of tiger quills (*Dasyurus maculatus*). *Aust. Vet. J.* 76:59–60. [DEMODICIDAE]

Holzmann, C. 1969. Die Familie Parasitidae Oudemans 1901. *Acarol., Schrift. Vergleich. Milbenk., Fürth. Folge* 13:3–54.

Honciuc, V. 1996. Laboratory studies of the behaviour and life cycle of *Archegozetes longisetosus* Aoki 1965 (Oribatida). In Needham, Mitchell, et al., 1:637–40. [TRHYPOCHTHONIIDAE]

Hong, X., and Z.-Q. Zhang. 1996. A cladistic analysis of the Eriophyoidea (Acari: Prostigmata): Tests of monophyly of families. *Syst. Appl. Acarol.* 1:107–22.

———. 1973a. Acarina (ticks). In *Viruses and invertebrates*, ed. A. J. Gibbs, 89–103. Amsterdam: North-Holland.

Hong, X.-Y., X.-F. Xue, Y.-J. Zhou, and G.-Y. Tong. 2005. *Cheiracus sulcatus*, a newly found invasive eriophyoid mite damaging rice in Guangdong Province, South China. *Acta Entomol. Sin.* 48:279–84.

Hoogstraal, H. 1973a. Acarina (ticks). In *Viruses and invertebrates*, ed. A. J. Gibbs, 39–103. Amsterdam: North-Holland.

———. 1973b. Ticks. In Flynn, 398–424.

———. 1979. The epidemiology of tick-borne Crimean-Congo hemorrhagic fever in Asia, Europe, and Africa. *J. Med. Entomol.* 15:307–417.

———. 1985. Argasid and nuttalliellid ticks as parasites and vectors. *Adv. Parasitol.* 24:135–238.

Hoogstraal, H., C. M. Clifford, J. E. Keirans, and H.Y. Wassef. 1979. Recent developments in biomedical knowledge of *Argas* ticks (Ixodoidea: Argasidae). In Rodriguez, 2:269–78.

Hoogstraal, H., R. M. Oliver, and S. M. Giurgis. 1970. Larva, nymph, and life cycle of *Ornithodoros* (*Alectorobius*) *muesebecki* (Ixodoidea: Argasidae), a virus-infected parasite of birds and petroleum industry employees in the Arabian Gulf. *Ann. Entomol. Soc. Amer.* 63:1762–67.

Horak, I. G., J.-L. Camicas, and J. E. Keirans. 2002. The Argasidae, Ixodidae and Nuttalliellidae (Acari: Ixodida): a world list of valid tick names. *Exp. Appl. Acarol.* 28:27–54.

Houck, M. A. 1994a. Adaptation and transition into parasitism from commensalism: a phoretic model. In Houck (1994b), contribution no. 10, 252–81. [HEMISARCOPTIDAE]

———, ed. 1994b. *Mites: Ecological and evolutionary analyses of life-history patterns.* New York: Chapman & Hall.

Houck, M. A., J. B. Clark, K. R. Peterson, and M. G. Kidwell. 1991. Possible horizontal transfer of *Drosophila* genes by the mite *Proctolaelaps regalis*. *Science* 253:1125–29. [MELICHARIDAE]

Houck, M. A., and A. C. Cohen. 1995. The potential role of phoresy in the evolution of parasitism: radio labelling (tritium) evidence from an astigmatid mite. *Exp. Appl. Acarol.* 19:677–94. [HEMISARCOPTIDAE]

Houck, M. A., and B. M. OConnor. 1990. Ontogeny and life history of *Hemisarcoptes cooremani* (Acari: Hemisarcoptidae). *Ann. Entomol. Soc. Amer.* 83:869–86.

Houseweart, M. W., D. T. Jennings, L. P. Berkett, and T. B. Brann. 1980. Parasitic mites (Acari: Erythraeidae) on spruce budworm moths (Lepidoptera: Tortricidae). *Can. Entomol.* 112:193–97.

Hoy, M. A. 1979. Parahaploidy of the arrhenotokous predator *Metaseiulus occidentalis* (Acarina: Phytoseiidae) demonstrated by X-irradiation of males. *Entomol. Exp. Appl.* 26:97–104.

———. 1982. Aerial dispersal and field efficacy of a genetically improved strain of the spider-mite predator *Metaseiulus occidentalis*. *Entomol. Exp. Appl.* 32:205–12.

———. 1985. Recent advances in genetics and genetic improvement of the Phytoseiidae. *Ann. Rev. Entomol.* 30:345–70.

Hoy, M. A., G. L. Cunningham, and L. Knutson, eds. 1983. *Biological control of pests by mites.* Univ. Calif. Agr. Exp. Sta. Special Publ. 3304. Berkeley: Univ. Calif. Agr. Exp. Sta.

Hoy, M. A., and J. M. Smilanick. 1979. A sex pheromone produced by immature and adult females of the predatory mite, *Metaseiulus occidentalis*, Acarina: Phytoseiidae. *Entomol. Exp. Appl.* 26:291–300.

Hu, C.-Y. 1995. A new species of the genus *Neognathus* Willmann (Acari: Caligonellidae). *Entomol. J. East China* 4:11–13.

———. 1996. A new species of the genus *Zetzellia* Oudemans (Acari: Stigmaeidae). *Acta Zootax. Sin.* 21:70–72.

Hu, C.-Y., and S.-J. Hu. 1997. A new species of genus *Caligonella* Berlese from China (Acari: Caligonellidae). *Entomol. Sin.* 4: 35–37.

———. 1998. A new species of the genus *Agistemus* Summers (Acari: Stigmaeidae). *Acta Zootax. Sin.* 23:30–32.

Hu, C.-Y., Z.-Q. Jing, and L.-R. Liang. 1995. Two new species and one new record of the genus *Raphignathus* Duges. *J. Suzhou Railway Teachers Coll.* 12:21–26.

Hu, C.-Y., and L.-R. Liang. 1990. A new stigmaeid mite of the genus *Stigmaeus* (Acari: Stigmaeidae). *Acta Zootax. Sin.* 15:461–63.

———. 1994. A new species of the genus *Eustigmaeus* Berlese (Acari: Stigmaeidae). *Acta Arachnol. Sin.* 3:94–96.

———. 1995a. A new species and a new record of the genus *Stigmaeus* from China (Acari: Stigmaeidae). *Acta Zootax. Sin.* 20:306–8.

———. 1995b. A new species of the genus *Saniosulus* Summers (Acari: Eupalopsellidae). *Acta Arachnol. Sin.* 4:56–58.

———. 1995c. A new species of the genus *Xenocaligonellidus* De Leon (Acari: Xenocaligonellidae). *Entomol. J. East China* 4: 11–12.

———. 1996. Two new species of the genus *Ledermuelleriopsis* Willmann (Acari: Stigmaeidae). *Acta Zootax. Sin.* 21:425–29.

Hu, C.-Y., and L.-C. Yu. 1995. A new species of the genus *Neognathus* Willmann (Acari: Caligonellidae). *Acta Arachnol. Sin.* 4:96–98.

Hu, C.-Y., G.-C. Zha, and J.-T. Zhu. 1994. Two new species and two new records of the genus *Eustigmaeus* Berlese from China (Acari: Stigmaeidae). *Acta Arachnol. Sin.* 3:81–85.

Hu, C.-Y., and J.-T. Zhu. 1994. A new species of the genus *Eustigmaeus* Berlese (Acari: Stigmaeidae). *Acta Arachnol. Sin.* 3:65–67.

Hu, R., K. E. Hyland, and J. H. Oliver, Jr. 1998. A review on the use of *Ixodiphagus* wasps (Hymenoptera: Encyrtidae) as natural enemies for the control of ticks (Acari: Ixodidae). *Syst. Appl. Acarol.* 3:19–27.

Hu, S.-Q., and X.-W. Chen. 1992a. A new species of *Eryngiopus* from China (Acari: Stigmaeidae). *J. Jiangxi Univ. (Nat. Sci.)* 16: 108–10.

———. 1992b. A new species of genus *Zetzellia* from China (Acarina: Stigmaeidae). *Acta Arachnol. Sin.* 1:39–41.

———. 1994a. A new species of *Eryngiopus* from Jiangxi (Acari: Stigmaeidae). *J. Nanchang Univ. (Nat. Sci.)* 18:93–96.

———. 1994b. Three new species and a new record of the genus *Agistemus* from China (Acari: Stigmaeidae). *Acta Arachnol. Sin.* 3:42–47.

———. 1996. A new species and a new record of Eupalopsellidae from China (Acari: Raphignathoidea). *Acta Zootax. Sin.* 21:321–24.

———. 1998. A new species of *Raphignathus* from Jiangxi, China (Acari: Raphignathidae). *Acta Arachnol. Sin.* 7:15–18.
Hu, S.-Q., X.-W. Chen, and L.-S. Huang. 1996. Mites of the genus *Eustigmaeus* from Jiangxi Province (Acari: Stigmaeidae). *Acta Entomol. Sin.* 3:314–22.
———. 1997. A new species and a new record of the genus *Agistemus* from China (Acari: Stigmaeidae). *Syst. Appl. Acarol. Spec. Publ.* 1:1–4.
Hu, S.-Q., L.-S. Huang, and X.-W. Chen. 1994. A new species of the genus *Eustigmaeus* from China (Acari: Stigmaeidae). *Acta Arachnol. Sin.* 3:91–93.
Hu, S.-Q., B. Xia, and X. Chen. 1997. A new species of the genus *Ledermuelleriopsis* from China (Acari: Stigmaeidae). *Syst. Appl. Acarol.* 2:167–68.
Huang, Y., F. J. Jia, and Y. M. Li. 2003. Analysis of the epidemic status of tsutsugamushi disease in Shandong Province. *Endemic Dis. Bull.* 18:36–38. [TROMBICULIDAE]
Hubard, C. K., and N. J. Fashing. 1996. *Carpoglyphus nidicolus*, a new species of Carpoglyphidae (Acarina: Astigmata) inhabiting the nests of swallows. *Internat. J. Acarol.* 22:215–20.
Hubert, J. 2000. Seasonal changes of abundance, sex ratio and egg production of *Scheloribates laevigatus* (Acari: Oribatida) in the soil of a meadow in the Czech Republic. *Acta. Soc. Zool. Bohem.* 64:37–56. [SCHELORIBATIDAE]
———. 2001. The influence of *Scheloribates laevigatus* (Acari: Oribatida) on decomposition of *Holcus lanatus* litter. *Acta Soc. Zool. Bohem.* 65:77–80. [SCHELORIBATIDAE]
Hubert, J., and V. Sustr. 2001. The effect of starvation on the metabolic rate and microanatomy of *Galumna elimata* (Acari: Oribatida). *Eur. J. Entomol.* 98:265–75. [GALUMNIDAE]
Hubert, J., V. Sustr, and J. Smrž. 1999. Feeding of the oribatid mite *Scheloribates laevigatus* (Acari: Oribatida) in laboratory experiments. *Pedobiologia* 43:328–39. [SCHELORIBATIDAE]
Hubert, J., M. Zilova, and S. Pekar. 2001. Feeding preferences and gut contents of three panphytophagous oribatid mites (Acari: Oribatida). *Eur. J. Soil Biol.* 37:197–208.
Huffaker, C. B., M. van de Vrie, and J. A. McMurtry. 1969. The ecology of tetranychid mites. *Ann. Rev. Entomol.* 14:125–73.
———. 1970. Ecology of tetranychid mites and their natural enemies: A review. 11, Tetranychid populations and their possible control by predators: An evaluation. *Hilgardia* 40 (11): 391–458.
Hughes, A. M. 1976. *The mites of stored food and houses.* 2d ed. Min. Agr., Fish. Food Tech. Bull. 9. London: H. M. Stationery Office.
Hughes, R. D., and C. G. Jackson. 1958. A review of the Anoetidae (Acari). *Virginia J. Sci.* 9:5–198. [HISTIOSTOMATIDAE]
Hughes, T. E. 1950. The physiology of the alimentary canal of *Tyroglyphus farinae*. *Qtly. J. Microscop. Sci.* 91:45–61.
———. 1952. The morphology of the gut in *Bdellonyssus bacoti* (Hirst, 1913). *Ann. Trop. Med. Parasitol.* 45:54–60.
———. 1954a. The internal anatomy of the mite *Listrophorus leuckarti* (Pagenstecher, 1861). *Proc. Zool. Soc. Lond.* 124:239–56.
———. 1954b. Some histological changes which occur in the gut epithelium of *Ixodes ricinus* females during gorging and up to oviposition. *Ann. Trop. Med. Parasitol.* 48:397–404.
———. 1959. *Mites or the Acari.* London: Univ. London, Athlone Press.
Hunt, G. S. 1996a. Description of predominantly arboreal plateremaeoid mites from eastern Australia (Acarina: Cryptostigmata: Plateremaeoidea). *Rec. Aust. Mus.* 48:303–24.
———. 1996b. A review of the family Pheroliodidae Paschoal in Australia (Acarina: Cryptostigmata: Plateremaeoidea). *Rec. Aust. Mus.* 48:325–58.
———. 1996c. A review of the genus *Hexachaetoniella* Paschoal in Australia (Acarina: Cryptostigmata: Pedrocortesellidae). *Rec. Aust. Mus.* 48:287–302.
———. 1996d. A review of the genus *Pedrocortesella* Hammer in Australia (Acarina: Cryptostigmata: Pedrocortesellidae). *Rec. Aust. Mus.* 48:223–86.
Hunt, G. S., R. A. Norton, J. P. H. Kelly, M. Coloff, and S. M. Lindsay. 1998. An interactive glossary of oribatid mites. CD-ROM. Melbourne: CSIRO Publishing.
Hunter, J. E., III, and M. H. Farrier. 1975. Mites of the genus *Oplitis* Berlese (Acarina: Uropodidae) associated with ants (Hymenoptera: Formicidae) in the southeastern United States. Part I. *Acarologia* 17:595–623. [OPLITIDAE]
———. 1976. Mites of the genus *Oplitis* Berlese (Acarina: Uropodidae) associated with ants (Hymenoptera: Formicidae) in the southeastern United States. Part II. *Acarologia* 18:20–50. [OPLITIDAE]
Hunter, P. E. 1963. Two new genera of mites associated with stingless bees (Acarina, Laelaptidae). *Acarologia* 5:5–12.
———. 1964a. Insects of Campbell Island. Mesostigmata: Laelaptidae. *Pac. Insects Monogr.* 7:121–28.
———. 1964b. Laelaptid mites from Auckland and Macquarie Islands (Acarina: Laelaptidae). *Pac. Insects Monogr. (Suppl.)* 7:630–41.
———. 1964c. Three new species of *Laelaspis* from North America (Acarina: Laelapidae). *J. Kans. Entomol. Soc.* 37:293–301.
———. 1967. Rhodacaridae and Parasitidae mites (Acarina: Mesostigmata) collected by the British Antarctic Survey, 1961–64. *Brit. Antarct. Surv. Bull.* 13:31–39.
———. 1970. Acarina: Mesostigmata: Free-living mites of South Georgia and Heard Island. *Pac. Insects Monogr.* 23:43–70. [DAVACARIDAE]
———. 1993. A new family of mites, Costacaridae (Mesostigmata: Trigynaspida: Celaenopsoidea) associated with millipedes in Mexico. *Israel J. Zool.* 39:185–91. [COSTACARIDAE]
Hunter, P. E., and L. Butler. 1966. New *Klinckowstroemia* mites from Costa Rican passalid beetles (Acarina: Klinckowstroemiidae). *J. Georgia Entomol. Soc.* 1:24–30.
Hunter, P. E., and M. Costa. 1970. Two new African species of *Megisthanus* Thorell (Mesostigmata: Megisthanidae). *Fla. Entomol.* 53:233–40. [MEGISTHANIDAE]
Hunter, P. E., and E. A. Cross. 1968. Key to North American Ereynetinae and description of a new species (Acarina: Ereynetidae). *J. Georgia Entomol. Soc.* 3:154–58.
Hunter, P. E., and D. A. Crossley, Jr. 1968. *Adamystis sarae*, a new species of soil mite from cornfield litter in South Carolina (Acari: Anystidae). *J. Georgia Entomol. Soc.* 3:181–83. [ADAMYSTIDAE]
Hunter, P. E., and R. Davis. 1965. Mites associated with the passalid beetle. III, Life stages and observations on the biology of *Euzercon latus* (Banks) (Acarina: Euzerconidae). *Acarologia* 7:3–42.
Hunter, P. E., and S. Glover. 1968a. The genus *Brachytremella* Trägårdh, 1946, with descriptions of three new species (Acarina: Diarthrophallidae). *Proc. Entomol. Soc. Wash.* 70:114–25.
———. 1968b. The genus *Diarthrophallus* Trägårdh, 1946 (Acarina: Diarthrophallidae). *Proc. Entomol. Soc. Wash.* 70:193–97.
———. 1968c. The genus *Passalobia* Lombardini, 1926, with de-

scription of a new species (Acarina: Diarthrophallidae). *Proc. Entomol. Soc. Wash.* 70:38–42.

Hunter, P. E., and C. A. Hunter. 1963. The genus *Myrmonyssus* with descriptions of two new species (Acarina: Laelaptidae). *Acarologia* 5:335–41.

Hunter, P. E., and R. W. Husband. 1973. *Pneumolaelaps* (Acarina: Laelapidae) mites from North America and Greenland. *Fla. Entomol.* 56:77–91.

Hunter, P. E., and S. L. Poe. 1971. Description of *Ereynetes (Anereynetes) coenobitus* n. sp. (Acarina: Ereynetidae) associated with the hermit crab. *Fla. Entomol.* 54:135–38.

Hunter, P. E., and R. M. T. Rosario. 1988. Associations of Mesostigmata with other arthropods. *Ann. Rev. Entomol.* 33:393–417.

———. 1989. Two new genera and species of Euzerconidae (Mesostigmata: Celaenopsoidea) with a key to New World genera. *J. Entomol. Sci.* 24:241–51. [EUZERCONIDAE]

———. 1991. New species of *Euzercon* Berlese from Mexico with a key to the New World species (Acarina: Mesostigmata: Euzerconidae). *Acarologia* 32:205–16.

Hunter, P. E., R. M. T. Rosario, and J. C. Moser. 1989. Two new species of *Ereynetes* (Acari: Prostigmata: Ereynetidae) associated with bark beetles. *J. Entomol. Sci.* 24:16–20.

Hunter, W. L., and R. B. Loomis. 1966. A new species of mite, genus *Hirstiella* (Acarina: Pterygosomidae) from the banded gecko, *Coleonyx variegatus*, of western North America. *J. Kans. Entomol. Soc.* 39:681–87.

Hunter, W. O., and F. C. Bishopp. 1911. *The Rocky Mountain spotted fever tick*. USDA Bull. 105. Washington, DC: USDA.

Hurlbutt, H. W. 1963. The genus *Asca* Heyden (Acarina: Mesostigmata) in North America, Hawaii and Europe. *Acarologia* 5:480–518. [ASCIDAE]

———. 1964. Structure and distribution of *Veigaia* (Mesostigmata) in forest soils. *Acarologia* 6 (fasc. h.s.): 150–52. [VEIGAIIDAE]

———. 1965. Systematics and biology of the genus *Veigaia* (Acarina: Mesostigmata) from Maryland. *Acarologia* 7:598–623. [VEIGAIIDAE]

———. 1967. Digamasellid mites associated with bark beetles in North America. *Acarologia* 9:497–534.

———. 1979. Distribution of unisexual and bisexual forms in mesostigmatid mites. In Rodriguez, 2:455–60.

———. 1983. The systematics and geographic distribution of East African Veigaiidae (Acarina: Mesostigmata). *Acarologia* 24:129–43.

———. 1984. A study of North American *Veigaia* (Acarina: Mesostigmata) with comparisons of habitats of unisexual and bisexual forms. *Acarologia* 25:207–22. [VEIGAIIDAE]

Husband, R. O. 1969. Technique for inspecting bees for internal mites. In Evans (1969c), 581.

Husband, R. W. 1972a. A new genus and species of mite (Acarina: Podapolipidae) associated with the coccinellid *Cycloneda sanguinea*. *Ann. Ent. Soc. Amer.* 65:1099–1104.

———. 1972b. *Podapolipoides porteri* n. sp.: Description with review of genus (Acarina, Podapolipidae). *Trans. Amer. Microscop. Soc.* 91:422–28.

———. 1974a. Lectotype designation for *Locustacarus trachealis* Ewing and a new species of *Locustacarus* (Acarina: Podapolipidae) from New Zealand. *Proc. Ent. Soc. Wash.* 76:52–59.

———. 1974b. *Ovacarus peelei*, a new species of mite (Acarina: Podapolipdae) associated with the carabid *Pasimachus elongatus*. *Great Lakes Entomol.* 7:1–7.

———. 1984a. *Dilopolipus*, *Panesthipolipus*, *Peripolipus* and *Stenopolipus*, new genera of Podapolipidae (Acarina) from the Indo-Australian region. *Internat. J. Acarol.* 10:251–69.

———. 1984b. The taxonomic position of *Coccipolipus* (Acarina: Podapolipidae), a genus of mites which are parasites of ladybird beetles (Coccinellidae). In Griffiths and Bowman, 1:328–36.

———. 1985. New species of *Stigmacarus* (Acari: Podapolipidae) parasitic on weevils (*Lixus* sp., Curculionidae) from Tshikaji, Zaire. *Ann. Ent. Soc. Amer.* 78:461–67.

———. 1986a. New *Podapolipus* spp. (Acari: Podapolipidae) from Australia and Hawaii. *Internat. J. Acarol.* 12:187–209.

———. 1986b. A new species of *Caraboacarus* (Acari: Caraboacaridae) from *Calosoma argentinense* (Carabidae) from Chile. *Great Lakes Entomol.* 19:115–19.

———. 1986c. New taxa of Podapolipidae (Acarina) from S. African Coleoptera: result of the Namaqualand-Namibia Expedition of the King Leopold III Foundation for the exploration and protection of nature (1980). *Bull. Inst. Roy. Sci. Nat. Belg., Entomol.* 56:5–14.

———. 1989a. A revision of the genus *Tarsopolipus* (Acari: Podapolipidae), parasites of African and European Scarabaeidae, with description of three new species. *Internat. J. Acarol.* 15:163–78.

———. 1989b. Status of the genus *Podapolipus* (Acari: Podapolipidae). In ChannaBasavanna and Viraktamath, 1:279–85.

———. 1989c. Two new species of *Coccipolipus* (Acari: Podapolipidae), parasites of *Chilocorus* spp. (Coccinellidae) from Vera Cruz and Morelos, Mexico and Florida and Wisconsin, U.S.A. *Proc. Entomol. Soc. Wash.* 91:429–35.

———. 1991. A preliminary investigation of the phylogeny of *Tarsopolipus*, a comparison with other early derivative podapolipid genera. In Dusbábek and Bukva 2:193–205.

———. 1992. A new species of *Podapolipoides* (Acari: Podapolipidae) ectoparasite of *Acrida* sp. in Jordan. *Internat. J. Acarol.* 18:287–91.

———. 1998. Two new species of *Eutarsopolipus* (Acari: Podapolipidae) from *Agonum extensicole* and *Pterostichus lucublandus* (Coleoptera: Carabidae) from Canada, including taxonomic keys of the 13 American species of Podapolipidae from carabid beetles. *Ann. Entomol. Soc. Amer.* 91:279–87.

———. 1999. American *Eutarsopolipus* of the *biunguis* group and descriptions of previously unknown male and larval female of *Eutarsopolipus trichognathi* (Acari: Podapolipidae) from South America. *Internat. J. Acarol.* 25:13–17.

Husband, R. W., A. M. Camerik, and R. Stals. 2006. Two new species of *Regenpolipus* and distribution of *Eutarsoolipus lukoschusi* and *Regenpolipus* species (Acari: Podapolipidae), ectoparasites of *Thermophilum decemguttatum* and *Antha* species (Coleoptera: Carabidae) in South Africa, Botswana and Namibia. *Internat. J. Acarol.* 32:361–70.

Husband, R. W., and H. Dastych. 1998. A new species of *Eutarsopolipus* (Acari: Podapolipidae) from *Chlaenius sericeus* Frost (Coleoptera: Carabidae) from Athens, Georgia, U.S.A. *Entomol. Mitt. Zool. Mus. Hamb.* 12 (158): 317–26.

Husband, R. W., and G. J. de Moraes. 1999. A new species of *Chrysomelobia* (Acari: Podapolipidae) from *Platyphora testudo* (Demay) (Coleoptera: Chrysomelidae) from Peru, with a key to known species of the genus. *Internat. J. Acarol.* 25:309–15.

Husband, R. W., and D. O. Husband. 2003. New records of *Crotalomorpha camini* Lindquist and Krantz (Acari: Crotalomorphidae) and a new species of *Eutarsopolipus* (Acari: Podapolipidae) both parasitic on *Stenolophus* (*Agonoderus*) *lecontei* (Chaudoir) (Coleoptera: Carabidae) from the central United States. *Internat. J. Acarol.* 29:1–6.

———. 2004. Distribution of *Eutarsopolipus* (Acari: Podapolipidae) and *Crotalomorpha camini* (Acari: Crotalomorphidae) ectoparasites of *Stenolophus* (Coleoptera: Carabidae) in Michigan, U.S.A. *Great Lakes Entomol.* 37:10–15.

———. 2007. A new species of *Dorsipes* (Acari: Podapolipidae), ectoparasite of *Pterostichus luctosus* Dejean (Coleoptera: Carabidae), from Michigan, U.S.A. *Internat. J. Acarol.* 33:139–44.

Husband, R. W., and S. M. Husband. 1984. *Caraboacarus towsleyi* n. sp. (Acarina: Caraboacaridae) associated with *Egadnoma* sp. (Carabidae) on Guadalcanal Island. *Mich. Academician* 16: 157–66.

Husband, R. W., and K. Kurosa. 1994. Four new species of *Scarabapolipus* (Acari: Podapolipidae) ectoparasites of Scarabaeidae (Coleoptera) from the western Pacific region. *Ann. Entomol. Soc. Amer.* 87:424–35.

Husband, R. W., and Y. Li. 1993. New Podapolipidae (Acari) from South America associated with *Macropophora* spp. (Cerambycidae), and a new genus associated with Orthoptera and Blatttodea in the western Pacific region. *Internat. J. Acarol.* 19:287–300.

Husband, R. W., R. Ochoa, and P. Naskrecki. 1995. *Buprestapolipus megachelus* n. g., n. sp. (Acari: Podapolipidae) from *Euchroma gigantea* L. (Coleoptera: Buprestidae). *Internat. J. Acarol.* 21:185–94.

Husband, R. W., and B. M. OConnor. 1999. Two new ectoparasitic mites (Acari: Podapolipidae) of *Rhynchophorus* spp. (Coleoptera: Curculionidae) from Indonesia, Malysia, the Philippines and West Africa. *Internat. J. Acarol.* 25:101–10.

———. 2004. A new species of *Chrysomelobia* Regenfuss (Acari: Podapolipidae) from *Ceralces* sp. (Coleoptera: Chrysomelidae) from Tanzania, with a key to species of *Chrysomelobia*. *Internat. J. Acarol.* 30:17–23.

Husband, R. W., B. M. OConnor, and R. Ochoa. 2005. Two new species of *Orthapolipus* (Acari: Podapolipidae), parasites of Central American Tettigoniidae (Hexapoda: Orthoptera), with a rediagnosis of the genus. *Internat. J. Acarol.* 31:355–62.

Husband, R. W., and R. N. Sinha. 1970a. A new genus and species of mite (Acarina: Podapolipidae) associated with the cockroach *Nauphoeta cinerea*. *Ann. Entomol. Soc. Amer.* 63:1148–52.

———. 1970b. A revision of the genus *Locustacarus* with a key to genera of the family Podapolipidae (Acarina). *Ann. Entomol. Soc. Amer.* 63:1152–62.

Hussein, N. A., and T. M. Perring. 1986. Feeding habits of the Tydeidae with evidence of *Homeopronematus anconai* (Acari: Tydeidae) predation on *Aculops lycopersici* (Acari: Eriophyidae). *Internat. J. Acarol.* 12:215–21. [IOLINIDAE]

Hussey, N. W., and B. Gurney. 1967. Bionomics and control of *Tarsonemus myceliophagus* (Acarina: Tarsonemidae) in mushroom composts. *Entomol. Exp. Appl.* 10:287–94.

Hussey, N. W., and N. E. A. Scopes. 1985. Greenhouse vegetables (Britain). In Helle and Sabelis (1985b), 1B:285–97. [TETRANYCHIDAE]

Hussey, N. W., W. H. Read, and J. J. Hesling. 1969. *The pests of protected cultivation*. New York: American Elsevier.

Hutterer, R. 1978. *Chausseria venustissima* (Berlese, 1882) (Acarina, Trombidiformes) auf der Insel Kreta. *Entomol. Mitt. Zool. Mus. Hamb.* 6:151–53. [ANYSTIDAE]

Hutu, M. 1973. Zur Kenntnis der Uropodiden-Fauna Rumaniens. Neue Uropodiden-Arten der Gattungen *Trachytes* Michael 1894, *Dinychus* Kramer 1886 und *Trachyuropoda* (Berlese 1888) Hirschmann u. Zirngiebl-Nicol 1961 nov. comb. *Gangsyst. Parasitiformes. Schrift. Vergleich. Milbenk.* 19:45–51.

———. 1991. Reproduction, embryonic and postembryonic development of *Trichouropoda obscurasimilis* Hirschmann & Zirngiebl-Nicol 1961 (Anactinotrichida: Uropodina). In Schuster and Murphy, 279–99.

Hutu, M., and A. Calugar. 2002. Zwei neue *Protodinychus*-Arten (Anactinotrichida: Uropodina: Protodinychidae). *Abh. Ber. Naturk. Görlitz* 74:219–36.

Hyatt, K. H. 1956. British mites of the genera *Halolaelaps* Berlese and Trouessart, and *Saprolaelaps* Leitner (Gamasina-Rhodacaridae). *Entomol. Gazette* 7:7–26. [HALOLAELAPIDAE]

———. 1980. Mites of the subfamily Parasitinae (Mesostigmata: Parasitidae) in the British Isles. *Bull. Brit. Mus. (Nat. Hist.), Zool.* 38 (5): 237–378.

———. 1987. Mites of the genus *Holoparasitus* Oudemans, 1936 (Mesostigmata: Parasitidae) in the British Isles. *Bull. Brit. Mus. (Nat. Hist.), Zool.* 52:139–64.

Hyatt, K. H., and R. M. Emberson. 1988. A review of the Marochelidae (Acari: Mesostigmata) of the British Isles. *Bull Brit. Mus. (Nat. Hist.) Zool.* 54: 63–125.

Hyland, K. E. 1979. Specificity and parallel host-parasite evolution in the Turbinoptidae, Cytoditidae and Ereynetidae living in the respiratory passages of birds. In Rodriguez, 2:363–69.

Ido, T., M. Kumkiri, L.-M. Lao, Y. Yano, and N. Takada. 2004. Dermatitis caused by *Balaustium murorum*. *Acta Dermatol. Venereol.* 84:80–81. [ERYTHRAEIDAE]

Imamura, T. 1952. Notes on the moulting of the adult of the water mite, *Arrenurus uchidai* n. sp. *Zool. Soc. Jpn.* 25:447–51.

———. 1956. Some subterranean water-mites from Hyogo Prefecture, Japan. *C.R. Prem. Cong. Internat. Spéléol.* 3:193–214. [STYGOTHROMBIDIIDAE]

———. 1968. Results of the speleological survey in South Korea 1966. IX, Halacaridae (Acari) found in a limestone cave of South Korea. *Bull. Nat. Sci. Mus.* 11:281–84.

———. 1970a. The fauna of the insular lava caves in west Japan. II, Porohalacarinae (Acari). *Bull. Nat. Sci. Mus. (Tokyo)* 13:455–58.

———. 1970b. Subterranean water mites (Limnohalacarinae and Hydrachnellae) of the Tsushima Islands. *Bull. Nat. Sci. Mus. (Tokyo)* 13:249–62.

———. 1971. The fauna of the lava caves around Mt. Fuji-san. V, Limnohalacarinae (Acari). *Bull. Nat. Sci. Mus. (Tokyo)* 14: 333–36.

———. 1977. A new subgenus and species of *Bharatalbia* (Hydrachnellae, Acari) from Japan. *Proc. Jpn. Soc. Syst. Zool.* 13:46–49. [ATURIDAE]

———. 1979a. A new subgenus and a new species of the genus *Chappuisides* (Acari, Hydrachnellae) from Japan. *Annot. Zool. Jpn.* 52:257–60. [CHAPPUISIDIDAE]

———. 1979b. Remarks on the subterranean water mites of Japan and a new psammobiontic genus and subfamily of Limnesiidae. In Piffl (1979), 157–63.

———. 1981. Fresh-water halacarid mites from Oahu Island, Hawaii. *Annot. Zool. Jpn.* 54:287–92.

———. 1983. A new subfamily and two new species of water mite (Acari: Hydrachnellae) from Papua New Guinea. *Proc. Jpn. Soc. Syst. Zool.* 26:11–18. [MIDEOPSIDAE]

———. 1984. Four new species of *Bharatalbia* (Acari: Hydrachnellae) from Japan. *Human Sci.* 1:67–75. [ATURIDAE]

Imamura, T., and R. D. Mitchell. 1967. The ecology and life history of the water mite, *Piersigia limophila* Protz. *Annot. Zool. Jpn.* 40:37–44. [PIERSIGIIDAE]

Imbriani, J. L., and R. Mankau. 1983. Studies on *Lasioseius scapulatus*, a mesostigmatid mite predacious on nematodes. *J. Nematol.* 15:523–28.

Ioffe, I. D. 1965. Distribution of neurosecretory cells in the central nervous apparatus of *Dermacentor pictus* Herm. (Acarina, Chelicerata). *Dokl. Akad. Nauk SSSR (Leningrad)* 154:25–29.

Iordansky, S. N., and V. A. Stein-Margolina. 1993. The cuticle structure and xeroresistance of the deutonymph and adult oribatid mites *Tectocepheus velatus* (Acariformes, Oribatei). *Zool. Zh.* 72:30–42. [TECTOCEPHEIDAE] [in Russian]

Irk, V. 1939a. Drei neue Milbenarten aus dem Tiroler Hochgebirge. *Zool. Anz.* 128:216–23. [TENERIFFIIDAE]

———. 1939b. Die terricolen Acari der Ötztaler und Stubaier Hochalpen. *Veröff. D. Mus. Ferdinandeum Innsbruck* 19:145–90. [CALYPTOSTOMATIDAE]

Ishii, Y. 1954. A new mite species of the genus *Sphaerolophus* in Japan. *Proc. S. Br., Jpn. Soc. Sanit. Zool.* 1954:65–66. [ERYTHRAEIDAE] [in Japanese]

Ishikawa, K. 1969. Studies on the mesostigmatid mites in Japan. IV, Family Blattisocidae Garman. *Rep. Res. Matsuyama Shinonome Jr. Coll.* 4:111–39. [includes ASCIDAE]

———. 1970. Studies on the mesostigmatid mites in Japan. III, Family Podocinidae Berlese. *Ann. Zool. Jpn.* 43:112–22.

———. 1978. The Japanese mites of the family Veigaiidae (Acari, Mesostigmata). I, Descriptions of two new species. *Annot. Zool. Jpn.* 51:100–106.

———. 1979. Taxonomic and ecological studies in the family Parholaspidae (Acari, Mesostigmata) from Japan. *Bull. Nat. Sci. Mus.*, ser. A, *Zool.* 5:249–69.

———. 1980a. Taxonomic and ecological studies in the family Parholaspidae (Acari, Mesostigmata) from Japan (Part 2). *Bull. Nat. Sci. Mus.*, ser. A, *Zool.* 6:1–25.

———. 1980b. Taxonomic and ecological studies in the family Parholaspidae (Acari, Mesostigmata) from Japan (Part 3). *Bull. Nat. Sci. Mus.*, ser. A, *Zool.* 6:153–74.

———. 1980c. Taxonomic and ecological studies in the family Parholaspidae (Acari, Mesostigmata) from Japan (Part 4). *Rep. Res. Matsuyama Shinonome Jr. Coll.* 11:183–90.

———. 1986. Gamasid mites (Acarina) associated with Japanese millipedes. *Rep. Res. Matsuyama Shinonome Jr. Coll.* 17:165–77. [IPHIOPSIDIDAE]

———. 1987. Two new species of *Epicrius* (Acarina, Epicriidae) collected by driven traps in Japan. *J. Speleol. Soc. Jpn.* 12:29–34.

———. 1988. Two new species of *Scissurolaelaps* (Acarina, Laelapidae) associated with Philippine millipedes. *Bull. Nat. Sci. Mus.*, ser. A, *Zool.* 14:173–78. [IPHIOPSIDIDAE]

———. 1993. Occurrence of *Holaspulus* (Acarina, Gamasida, Parholaspidae) in the Philippines. *Bull. Nat. Sci. Mus.*, ser. A, *Zool.* 19:93–110.

———. 1994. Two new species of the genus *Holaspulus* (Acarina: Gamasida: Parholaspidae) from the Ryuku Islands, Japan. *Zool. Sci.* 11:139–42.

Ito, F., and G. Takaku. 1994. Obligate myrmecophily in an oribatid mite—novel symbiont of ants in the Oriental tropics. *Naturwissenschaften* 81:180–82.

Ito, M. 1982. Unrecorded species of the genera *Carabodes* and *Licneremaeus* (Acarina, Oribatida) from Japan. *Bull. Inst. Nat. Educ., Shiga Heights, Shinshu Univ.* 20:49–54. [CARABODIDAE, LICNEREMAEIDAE]

Ito, Y. 1971. Predation by manure-inhabiting mesostigmatids (Acarina: Mesostigmata) on some free-living nematodes. *Appl. Entomol. Zool.* 6:51–56. [UROPODIDAE]

———. 1977. Predatory activity of mesostigmatid mites (Acarina: Mesostigmata) for house fly eggs and larvae under feeding of nematodes. *Jpn. J. Sanit. Zool.* 28:167–73.

Ivanov, V. P., and S. A. Leonovich. 1983. Sensory organs. In *An atlas of ixodid tick ultrastructure*, ed. Y. S. Balashov. Entomol. Soc. Amer. Special Publ. College Park, MD: Entomol. Soc. Amer.

Iverson, K., B. M. OConnor, R. Ochoa, and R. Heckmann. 1996. *Lardoglyphus zacheri* (Acari: Lardoglyphidae), a pest of museum dermestid colonies, with observations on its natural ecology and distribution. *Ann. Entomol. Soc. Amer.* 89:544–49.

Jack, K. M. 1961a. New species of Near Eastern agamid scale-mites (Acarina, Pterygosomidae) with notes on the developmental stages of *Geckobia hemidactyli* Law, 1936. *Parasitology* 51: 241–56.

———. 1961b. A re-examination of the genera *Pimeliaphilus* Trägårdh 1905 and *Hirstiella* Berlese 1920 (Acari: Prostigmata). *Ann. Mag. Nat. Hist.*, ser.13, 4:305–14.

———. 1964. Leg-chaetotaxy with special reference to the Pterygosomidae (Acarina). *Ann. Natal Mus.* 16:1–20.

Jack, K. M., and B. Girot. 1965. Notes on the development of *Hirstiella insignis*, (Berlese) 1892. *Acarologia* 7:286–93.

Jacot, A. P. 1925. Phylogeny in the Oribatoidea. *Amer. Natur.* 59:272–79.

———. 1933. Aparity among the Oribatidae (Acarina). *Can. Entomol.* 65:263–64.

———. 1934. Acarina as possible vectors of the Dutch elm disease. *J. Econ. Entomol.* 27:858–59.

———. 1939a. A new antennophorid mite, rider of the Philippine ant *Aenictus martini*. *Philipp. J. Sci.* 69:433–34. [AENICTEQUIDAE]

———. 1939b. Reduction of spruce and fir litter by minute animals. *J. Forestry* 37:858–60.

Jakeman, L. A. R. 1961. The internal anatomy of the spiny rat mite, *Echinolaelaps echidninus* (Berlese). *J. Parasitol.* 47:329–49.

Jalil, M., and J. G. Rodriguez. 1970a. Biology of and odor perception by *Fuscuropoda vegetans* (Acarina: Uropodidae), a predator of the house fly. *Ann. Entomol. Soc. Amer.* 63:935–38. [DINYCHIDAE]

———. 1970b. Studies of behavior of *Macrocheles muscaedomesticae* (Acarina: Macrochelidae) with emphasis on its attraction to the house fly. *Ann. Entomol. Soc. Amer.* 63:738–44.

James, A. M., J. E. Freier, J. E. Keirans, L. A. Durdan, J. W. Mertins, and J. L. Schlater. 2006. Distribution, seasonality, and hosts of the Rocky Mountain wood tick in the United States. *J. Med. Entomol.* 43:17–24. [IXODIDAE]

James, D., and S. Mukerji. 1993. Mechanical transmission, purification, and characterization of a virus associated with mottle leaf of cherry. *Plant Dis.* 77:271–75.

James, D. G., G. N. Warren, and J. Whitney. 1992. Phytoseiid mite populations of dormant grapevines: Extraction using a microwave oven. *Exp. Appl. Acarol.* 14:175–78.

Janetschek, H. 1967. Arthropod ecology of South Victoria Land. *Antarct. Res. Ser.* 10:205–93.

Jellison, W. L., E. J. Bell, R. R. Huebner, R. R. Parker, and H. H. Welsh. 1948. Q fever studies in southern California. IV,

Occurrence of *Coxiella burneti* in the spinose ear tick, *Otobius megnini*. *Pub. Health Rep.* 63:1483–89. [ARGASIDAE]

Jensen, S. G., and L. C. Lane. 1994. Observations on a mite-transmitted disease of maize and wheat in the High Plains. In *Rose rosette and other eriophyid mite–transmitted plant disease agents of uncertain etiology*, ed. A. Epstein and J. H. Hill, Proc. Internat. Symposium, May 19–21, 43–46. Ames: Iowa State Univ.

Jeppson, L. R. 1975. Mites and plant diseases. In Jeppson, Keifer, and Baker, 91–102.

Jeppson, L. R., and H. H. Keifer. 1975. Injurious eriophyoid mites. In Jeppson, Keifer, and Baker (1975b), 397–533.

Jeppson, L. R., H. H. Keifer, and E. W. Baker. 1975a. Biological enemies of mites (Chapter 5); Tydeidae, Tuckerellidae, Pyemotidae, Penthaleidae, Astigmata and Cryptostigmata (Chapter 11). In Jeppson, Keifer, and Baker (1975b), 75–90, 307–26.

———. 1975b. *Mites injurious to economic plants*. Berkeley: Univ. Calif. Press.

———. 1975c. Tarsonemidae Kramer. In Jeppson, Keifer, and Baker (1975b), 285–305.

Jesionowska, K. 1989. New genus and new species of mite of the family Penthalodidae (Actinotrichida, Actinedida, Eupodoidea) from Poland. *Acta Zool. Cracov.* 32:57–67. [EUPODIDAE?]

Jesionowska, K. 1991. Morphology of the Eupodidae (Actinotrichida, Actinedida, Eupodoidea). In Dusbábek and Bukva, 2:265–78.

———. 1996. Morphology of the juvenile stages of eupodoid mites (Eupodoidea). In Mitchell et al. (1996), 1:325–33.

Jin, D. C. 1992. A new species of the first record family Mideopsidae of water mites from China (Acari, Hydrachnellae, Arrenuroidea). *Zool. Res.* 13:13–16.

Jin, D. C., and Z. Z. Gua. 1992a. Descriptions of two new and one known species of the genus *Arrenurus* (Acari, Arrenuridae). *Zool. Res.* 13:109–15.

———. 1992b. Description of two new species of *Piona* from Ghizou Province (Acari, Pionidae). *Acta Zootax. Sin.* 17:173–78.

Jin, D. C., and P. R. Wiles. 1996. New species of *Arrenurus* Dugès (Acari: Hydrachnidia: Arrenuridae) from China and first records of water mites from Laos. *Acarologia* 37:317–44.

Johnson, D. T., and B. A. Croft. 1981. Dispersal of *Amblyseius fallacis* (Acarina: Phytoseiidae) in an apple ecosystem. *Environ. Entomol.* 10:313–19.

Johnston, D. E. 1961. A review of the lower uropodoid mites (former Thinozerconoidea, Protodinychoidea and Trachytoidea) with notes on the classification of the Uropodina (Acarina). *Acarologia* 3:522–45.

———. 1964. Psorergates bos, *a new mite parasite of domestic cattle (Acari-Psorergatidae)*. Ohio Agr. Exp. Sta. Res. Circ. 129. Wooster: Ohio Agr. Exp. Sta.

———. 1968. *An atlas of Acari*. I, *The families of Parasitiformes and Opilioacariformes*. Columbus: Acarology Laboratory, Ohio State Univ.

———. 1982. Acari, Opilioacariformes, Parasitiformes. In Parker, 2:111–16.

Johnston, D. E., and W. A. Bruce. 1965. *Tyrophagus neiswanderi*, a new acarid mite of agricultural importance. *Ohio Res. Dev. Center Res. Bull.* 977:3–17.

Johnston, D. E., and A. Fain. 1964. *Ophiocelaeno sellnicki*, a new genus and species of Diplogyniidae associated with snakes (Acari-Mesostigmata). *Bull. Ann. Soc. ent. Belg.* 108:79–91.

Johnston, D. E., and J. B. Kethley. 1973. A numerical phenetic study of the quill mites of the family Syringophilidae (Acari). *J. Parasitol.* 59:520–30.

Johnston, D. E., and M. L. Moraza. 1991. The idiosomal adenotaxy and poroidotaxy of Zerconidae (Mesostigmata: Zerconina). In Dusbábek and Bukva, 2:349–56.

Johnston, D. E., and R. R. Wacker. 1967. Observations on post-embryonic development in *Eutrombicula splendens* (Acari-Acariformes). *J. Med. Entomol.* 4:306–10.

Jorgensen, C. D. 1967. A new species of *Tuckerella* (Acarina: Tuckerellidae) from Nevada. *Entomol. News* 78:141–46.

Jouvenaz, D. P., and C. S. Lofgren. 1986. An evaluation of the straw itch mite, *Pyemotes tritici* (Acari: Pyemotidae) for control of the red imported fire ant, *Solenopsis invicta* (Hymenoptera: Formicidae). *Fla. Entomol.* 69:761–63.

Jovani, R., J. L. Tella, D. Sol, and D. Ventura. 2001. Are hippoboscid flies a major mode of transmission of feather mites? *J. Parasitol.* 87:1187–89.

Judson, M. 1995. Studies on the Teneriffiidae (Acari: Anystoidea). II, A review of the genus *Austroteneriffia*. *Invertebr. Taxon.* 9: 827–39.

———. 2000. Nomenclatural problems associated with the names Alycidae Canestrini & Fanzago, 1877 and Pachygnathidae Kramer, 1877 (Acari). *Mem. Soc. Entomol. Ital.* 78:533–40.

Juvara-Bals, I. 2002. A revision of the genus *Heteroparasitus* new status, with the description of *Heteroparasitus* (*Medioparasitus*) *athiasae* subgen. n., sp. n. from Spain and with a key to the genera of Pergamasinae (Acari, Gamasida, Parasitidae). *Rev. Suisse Zool.* 109:23–46.

Juvara-Bals, I., and M. Baltac. 1975. Deux nouvelles espèces d'*Opilioacarus* (Acarina: Opilioacarida) de Cuba. *Résult. Expéd. Biospéol. Cubano–Roum Cuba* 2:169–84.

Kaestner, A. 1980. *Invertebrate zoology*. Vol. II, *Arachnids and myriapods*. Trans. H. W. Levi and L. R. Levi. Huntington, NY: Robert E. Krieger.

Kaiser, T., and G. Alberti. 1991. The fine structure of the lateral eyes of *Neocarus texanus* Chamberlin and Mulaik, 1942 (Opilioacarida, Acari, Arachnida, Chelicerata). *Protoplasma* 163:19–33.

Kaliszewski, M. 1987. *Siteroptes longisomus* sp. n. from Poland, with remarks on the genus and key to the species (Acari: Pygmephoroidea). *Entomol. Mitt. Zool. Mus. Hamb.* 9 (130): 21–35.

———. 1988a. *Diroptes* gen. n. (Acari, Pygmephoroidea) with a key to the species. *Entomol. Mitt. Zool. Mus. Hamb.* 9 (132): 115–22. [SITEROPTIDAE]

———. 1988b. *Pseudobakerdania* gen. n. (Acari, Pygmephoroidea) with the description of two new species and some remarks on the phylogenetic position of the genus. *Entomol. Mitt. Zool. Mus. Hamb.* 9 (132): 123–34. [SITEROPTIDAE]

———. 1993. *Key to Palearctic [species] of the genus* Tarsonemus*: Acari, Tarsonemidae*. Ser. Zool. 14. Poznan: Adam Mickiewicz Univ. [in Polish]

Kaliszewski, M., F. Athias-Binche, and E. E. Lindquist. 1995. Parasitism and parasitoidism in Tarsonemina (Acari: Heterostigmata) and evolutionary considerations. *Adv. Parasitol.* 35: 335–67.

Kaliszewski, M., and G. Rack. 1985. Description of female and male of *Pygmephorus sylvilagus* n. sp. and male of *Pygmephorus erlangensis* Krczal, 1959 (Acari, Pygmephoridae). *Entomol. Mitt. Zool. Mus. Hamb.* 8:45–60.

———. 1986. A systematic study of the mite *Microdispus (Premicro-*

dispus) lambi (Krczal, 1964) (Acari: Pygmephoroidea), a mushroom pest in Australia. *Mitt. Hamb. Zool. Mus. Inst.* 83:231–41.

Kaliszewski, M., and D. L. Wrensch. 1993. Evolution of sex determination and sex ratio within the mite cohort Tarsonemina (Acari: Heterostigmata). In Wrensch and Ebbert, 192–213.

Kamburov, S. S. 1971. Feeding, development and reproduction of *Amblyseius largoensis* on various food substances. *J. Econ. Entomol.* 64:643–48.

Kamill, B., J. A. Wallwork, and M. MacQuitty. 1986. Primitive cryptostigmatid mites from the Chihuahuan desert of New Mexico. *Acarologia* 26 (4): 325–47.

Kampen, H. 2002. Die Herbstmilbe *Neotrombicula autumnalis* und andere parasitische Laufmilben (Acari: Trombiculidae) als Verursacher von Hauterkrankungen in Mitteleuropa. *Denisia* 6:461–76.

Kandeel, M. M. H. 1992. Revision of family Paratydeidae with the description of *Hexatydeus amabilis* n. sp. from Egypt (Acari: Actinedida). *Bull. Entomol. Soc. Egypt* 70:1–9.

Kaneko, K., and T. Kadosaka. 1978. A new parasitic mite (Astigmata: Ewingidae) from the gills of fresh-water crab in Japan. *Jpn. J. Sanit. Zool.* 29:313–15. [ACARIDAE]

Kaneko, N. 1988. Life history of *Oppiella nova* (Oudemans) (Oribatei) in cool temperate forest soils in Japan. *Acarologia* 29 (2): 215–21. [OPPIIDAE]

Kaneko, N., and J. Aoki. 1982. Two new oribatid mites from Ashiu experimental forest of Kyoto University. *Edaphologia* 27:15–22.

Kantaratanakul, S., W. Chaiwat, and P. Prayoonrat. 1989. Bionomics of the dolichocybid mite, *Formicomotes* sp., as a pest of cultivated mushroom in Thailand. In ChannaBasavanna and Viraktamath, 2:223–28.

Kantaratanakul, S., and S. Jitrat. 1984. Bionomics of the pyemotid mite *Luciophorus hauseri* infesting cultivated jew's ear fungus *Auricularia polytricha* in Thailand. In Griffiths and Bowman, 1:622–28. [PYGMEPHORIDAE]

Karafiat, H. 1959. Systematik und Ökologie der Scutacariden. *Beitr. Syst. Ökol. Mitteleurop. Acarina* 1 (4): 627–712.

Karasawa, S., and N. Hijii. 2006. Does the existence of bird's nest ferns enhance the diversity of oribatid (Acari: Oribatida) communities in a subtropical forest? *Biodivers. Conserv.* 15:4533–53.

Karban, R., and G. M. English-Loeb. 1988. Effects of herbivory and plant conditioning on the population dynamics of spider mites. *Exp. Appl. Acarol.* 4:225–46.

———. 1990. A "vaccination" of Willamette spider mites (Acari: Tetranychidae) to prevent large populations of Pacific spider mites on grapevines. *J. Econ. Entomol.* 83:2252–57.

Karg, W. 1961. Ökologische Untersuchungen von edaphischen Gamasiden (Acarina, Parasitiformes). *Pedobiologia* 1:53–74, 77–98.

———. 1963. Systematische Untersuchung der Eviphididae Berlese 1913 (Acarina, Parasitiformes) mit einer neuen Art aus Ackerboden. *Zool. Anz.* 170:269–81.

———. 1965. Larvalsystematische und phylogenetische Untersuchung sowie Revision des Systems der Gamasina Leach, 1915 (Acarina, Parasitiformes). *Mitt. Zool. Mus. Berl.* 41:193–340, tab. 6.

———. 1969. Untersuchungen zur Kenntnis der Ascaoidea Karg, 1965 (Acarina, Parasitiformes) mit der Beschreibung von acht neuen Arten. *Zool. Anz.* 182:393–406. [HALOLAELAPIDAE]

———. 1971. Zur Kenntnis der Epicriidae Berlese, 1885 (Acarina, Anactinochaeta). *Zool. Anz.* 186:105–14.

———. 1978a. Zur Kenntnis der Gattungen *Macrocheles* Latreille, 1829 und *Leptolaelaps* Berlese, 1918 (Acarina, Parasitiformes). *Zool. Jb. Syst.* 105:360–67.

———. 1978b. Zur Kenntnis der Milbengattungen *Hypoaspis*, *Androlaelaps* und *Reticulolaelaps* (Acarina, Parasitiformes, Dermanyssidae). *Zool. Jb. Syst.* 105:1–32. [LAELAPIDAE]

———. 1979. Zur Kenntnis der Raubmilbenguttungen *Cheiroseius* Berlese, 1916, *Asca* v. Heyden, 1826 und *Halolaelaps* Berlese et Trouessart, 1889 (Acarina, Parasitiformes). *Mitt. Zool. Mus. Berl.* 55:251–67. [includes ASCIDAE]

———. 1980. Die Raubmilbengattung *Lasioseius* Berlese, 1916. *Zool. Jb. Syst.* 107:344–67. [BLATTISOCIIDAE]

———. 1981. Die Raubmilbengattung *Cheiroseius* Berlese, 1916. *Zool. Jb. Syst.* 108:51–69. [BLATTISOCIIDAE]

———. 1982. Untersuchungen über Habitatansprüche, geographische Verbreitung und Entstehung von Raubmilbengattungen der Cohors Gamasina für ihre Nutzung als Bioindikatoren. *Pedobiologia* 24:241–47.

———. 1983. Systematische Untersuchung der Raubmilbenfamilie Leptolaelapidae Karg, 1978 (Acarina, Parasitiformes). *Zool. Jb. Syst.* 110:377–96.

———. 1984. Die Hornmilbe *Scheloribates pallidulus* (C. L. Koch) als Schädling an Gurkenkulturen unter Glas und Plasten. *Arch. Phytopathol. U. Pflanz., Berl.* 20:181–83. [SCHELORIBATIDAE]

———. 1985a. Die mit Arten der Gattung *Bombus* Latreille vergesellschafteten Raubmilben der Gattung *Parasitus* Latreille, 1795 (Acarina, Parasitiformes). *Zool. Jb. Syst.* 112:525–35. [PARASITIDAE]

———. 1985b. Die Raubmilbengattung *Proctolaelaps* Berlese, 1923. *Zool. Jb. Syst.* 112:185–206. [MELICHARIDAE]

———. 1985c. Zwei neue Raubmilbenarten der Untergattung *Cosmolaelaps* Berlese, 1903 (Gattung *Hypoaspis* Can., 1884) aus Termitennestern (Acarina, Parasitiformes). *Deutsche Entomol. Zeitschr.* 32:233–37. [LAELAPIDAE]

———. 1986a. Über die mit Nematoden und Einsiedlerkrebsen vergesellschaftete Raubmilbengattung *Andregamasus* Costa der Podocinidae Berlese (Cohors Gamasina Leach: Acarina, Parasitiformes). *Mitt. Zool. Mus. Berl.* 62:41–51.

———. 1986b. Vorkommen und Ernährung der Milbencohors Uropodina (Schildkrötenmilben) sowie ihre Eignung als Indikatoren in Agroökosystemen. *Pedobiologia* 29:285–95.

———. 1989. *Uropodina Kramer, Schildkrötenmilben: Acari (Acarina), Milben, Unterordnung Parasitiformes (Anactinochaeta).* Tierwelt Deutsch. 67. Jena: Gustav Fischer.

———. 1993a. Neue Raubmilben der Rhodacaridae Oudemans (Acarina, Parasitiformes) aus Neu-Kaledonien. *Zool. Jb. Syst.* 120:39–63. [OLOGAMASIDAE]

———. 1993b. *Raubmilben: Acari (Acarina), Milben Parasitiformes (Anactinochaeta) Cohors Gamasina Leach.* Tierwelt Deutsch. 59. Jena: Gustav Fischer.

———. 1993c. Raubmilbenarten in Boden—sowie Vegetationsschichten und ihre Eignung als Indikatoren für Agrochemikalien. *Anz. Schädlingskde., Pflanzen., Umweltschutz* 66:126–31.

———. 1997a. Die Raubmilbenfamilie Leptolaelapidae Karg (Acarina, Parasitiformes). *Acarologia* 38:207–18.

———. 1997b. Zur Kenntnis der Raubmilbencohors Gamasina LEACH (Acarina, Parasitiformes) tropischer und subtropischer Gebiete. *Mitt. Zool Mus. Berl.* 73:63–88.

———. 1998a. Neue im Boden lebende Arten der

Raubmilbengattung *Zygoseius* Berlese, 1916. *Abh. Ber. Naturk. Görlitz* 70:165–73. [PACHYLAELAPIDAE]

———. 1998b. Neue Raubmilbenarten der Ascidae Oudemans, 1905 im tropischen Regenwald von Ecuador (Acarina, Parasitiformes). *Mitt. Mus. Nat. Kd. Berl., Zool.* 74:49–64. [includes BLATTISOCIIDAE]

———. 1998c. Zur Kenntnis der Eugamasides Karg mit neuen Arten aus den Regenwäldern von Ecuador (Acarina, Parasitiformes). *Mitt. Mus. Nat. Kd. Berl., Zool.* 74:185–214.

———. 2000. Die euedaphische Raubmilbengattung *Multidentorhodacarus* Shcherbk, 1980 mit einer neuen Art aus Mittelamerika. *Abh. Ber. Naturk. Görlitz* 72:143–49.

Karg, W., and B. Freier. 1995. Parasitiforme Raubmilben als Indikatoren für den ökologischen Zustand von Ökosystemen. *Mitt. Biol. Bund. Land-Forst. Berl.-Dahlem* 308:1–96.

Karl, E. 1965a. Untersuchungen zur Morphologie und Ökologie von Tarsonemiden gärtnerischer Kulturpflanzen. 1, *Tarsonemus pallidus* Banks. *Biol. Zbl.* 84:47–80.

———. 1965b. Untersuchungen zur Morphologie und Ökologie von Tarsonemiden gärtnerischer Kulturpflanzen. 2, *Hemitarsonemus latus* (Banks), *Tarsonemus confusus* Ewing, *T. talpae* Schaarschmidt, *T. setifer* Ewing und *Tarsonemoides belemnitoides* Weis-Fogh. *Biol. Zbl.* 84:331–57.

Karppinen, E., and D. A. Krivolutskij. 1982. List of oribatid mites (Acarina, Oribatei) of northern Palaearctic region. I, Europe. *Acta Entomol. Fenn.* 41:1–18.

Karppinen, E., D. A. Krivolutskij, and M. P. Poltavskaja. 1986. List of oribatid mites (Acarina, Oribatei) of northern Palaearctic region. III, Arid lands. *Ann. Entomol. Fenn.* 52:81–94.

Karppinen, E., D. A. Krivolutskij, Z. M. Tarba, U. Y. Stanchaeva, and E. W. Gordeeva. 1987. List of oribatid mites (Acarina, Oribatei) of northern Palaearctic region. IV, Caucasus and Crimea. *Ann. Entomol. Fenn.* 53:119–37.

Karppinen, E., V. V. Melamud, L. Miko, and D. A. Krivolutskij. 1992. Further information on the oribatid fauna (Acarina, Oribatei) of the northern Palearctic region: Ukraine and Czechoslovakia. *Ann. Entomol. Fenn.* 3:41–56.

Kats, L. B., and L. M. Dill. 1998. The scent of death: chemosensory assessment of predation risk by prey animals. *Ecoscience* 5:361–94.

Kazmarek, S. 1984. Gang einer neuen *Asternolaelaps*-Art aus Polen (Trichopygidiina). *Acarologia* 31:105–11.

Kazmierski, A. 1989. Morphological studies on Tydeidae (Actinedida, Acari). I, Remarks about the segmentation, chaetotaxy and poroidotaxy of idiosoma. *Acta Zool. Cracov.* 32:69–83.

———. 1996a. A revision of the subfamilies Pretydeinae and Tydeinae (Acari: Actinedida: Tydeidae). Part II, The subfamily Pretydeinae André, 1979—new taxa, species review, key and considerations. *Mitt. Hamb. Zool. Mus. Inst.* 93:171–98.

———. 1996b. A revision of the subfamilies Pretydeinae and Tydeinae (Acari: Actinedida: Tydeidae). Part III, Seven new genera and some new species of the Tydeinae, with a generic key. *Mitt. Hamb. Zool. Mus. Inst.* 93:199–227.

———. 1998a. A review of the genus *Proctotydaeus* (Actinedida: Tydeidae: Pronematinae). *Acarologia* 39:33–47. [IOLINIDAE]

———. 1998b. Tydeinae of the world: generic relationships, new and redescribed taxa and keys to all species. A revision of the subfamilies Pretydeinae and Tydeinae (Acari: Actinedida: Tydeidae), Part IV. *Acta Zool. Cracov.* 41:283–55.

Keegan, H. L., C. E. Yunker, and E. W. Baker. 1960. *Hystrichonyssus turneri* n. sp., n. g., representing a new subfamily of Dermanyssidae (Acarina) from a Malayan porcupine. Malaysian Parasites XLVI. *Stud. Inst. Med. Res.* 29:205–8. [HYSTRICHONYSSIDAE]

Keh, B. 1975. Intense pruritis in man and concurrent infestation of *Cheyletiella blakei* Smiley (Acari: Cheyletiellidae) on cats in a home in California. *Calif. Vector Views* 22:1–7.

Kehl, C. 1997. Ecology of Malaconothroidea (Acari, Oribatida) in northeast German mires. *Abh. Ber. Naturk. Görlitz* 69 (6): 11–18.

Keifer, H. H. 1942. Eriophyid studies XII. *Bull. Calif. St. Dept. Agr. 31:117–29.*

———. 1952. *The eriophyid mites of California (Acarina: Eriophyidae).* Bull. Calif. Insect Surv. 2 (1). Berkeley: Univ. Calif. Press.

———. 1953. Eriophyid studies XXI. *Bull. Calif. St. Dept. Agr.* 44:126–30.

———. 1975. Eriophyoidea Nalepa. In Jeppson, Keifer, and Baker (1975b), 327–96.

Keifer, H. H., E. W. Baker, T. Kono, M. Delfinado, and W. E. Styer. 1982. *An illustrated guide to plant abnormalities caused by eriophyid mites in North America.* Agr. Res. Service, U.S. Dept. Agr., Agr. Handbook no. 573. Washington, DC: USDA.

Keirans, J. E. 1992. Systematics of the Ixodida (Argasidae, Ixodidae, Nuttalliellidae): An overview and some problems. In *Tick vector biology, medical and veterinary aspects*, ed. B. Fivaz, T. Petney, and I. Horak, 1–21. Berlin: Springer.

Keirans, J. E., and C. M. Clifford. 1975. *Nothoaspis reddelli*, new genus and new species (Ixodoidea: Argasidae), from a bat cave in Mexico. *Ann. Entomol. Soc. Amer.* 68:81–85.

Keirans, J. E., C. M. Clifford, H. Hoogstraal, and E. R. Easton. 1976. Discovery of *Nuttalliella namaqua* Bedford (Acarina: Ixodoidea: Nuttalliellidae) in Tanzania and redescription of the female based on scanning electron microscopy. *Ann. Entomol. Soc. Amer.* 69:926–32.

Keirans, J. E., G. R. Needham, and J. H. Oliver, Jr. 1999. The *Ixodes (Ixodes) ricinus* complex worldwide: Diagnosis of the species in the complex, hosts and distribution. In Needham, Mitchell, et al., 2:341–47. [IXODIDAE]

Keirans, J. E., and R. G. Robbins. 1999. A world checklist of genera, subgenera, and species of ticks (Acari: Ixodida) published from 1973–1997. *J. Vector Ecol.* 24:115–29.

Kennaway, G. M., A. S. Baker, and A. D. Ball. 2004. A method for preparing lightly sclerotized mites for examination by transmission electron microscopy. *Syst. Appl. Acarol.* 9:3–9.

Kennedy, G. G., and D. R. Smitley. 1985. Dispersal. In Helle and Sabelis (1985a), 1A:233–42. [TETRANYCHIDAE]

Kerfoot, W. C. 1982. A question of taste: crypsis and warning coloration in freshwater zooplankton communities. *Ecology* 63: 538–54.

Kethley, J. B. 1970. A revision of the family Syringophilidae (Prostigmata: Acarina). *Contr. Amer. Entomol. Inst.* 5 (6): 1–76.

———. 1971a. *Hydranetes*, a new genus of Ereynetidae from hydrophilid beetles (Prostigmata: Ereynetidae). *J. Georgia Entomol. Soc.* 6:176–84.

———. 1971b. Population regulation in quill mites (Acarina: Syringophilidae). *Ecology* 52:1113–18.

———. 1973. A new genus and species of quill mites (Acarina: Syringophilidae) from *Colinus virginianus* (Galliformes: Phasianidae) with notes on developmental chaetotaxy. *Fieldiana Zool.* 65:1–8.

———. 1974. Developmental chaetotaxy of a paedomorphic celaenopsid *Neotenogynium malkini* n. g., n. sp. (Acari: Parasitiformes: Neotenogyniidae, n. fam.) associated with millipedes. *Ann. Entomol. Soc. Amer.* 67:571–79. [NEOTENOGYNIIDAE]

———. 1977a. A review of the higher categories of Trigynaspida (Acari: Parasitiformes). *Internat. J. Acarol.* 3:129–49.

———. 1977b. The status of *Hybalicus* Berlese, 1913 and *Oehserchestes* Jacot, 1939 (Acari: Acariformes: Endeostigmata). *Fieldiana Zool.* 72:59–64.

———. 1977c. An unusual parantennuloid, *Philodana johnstoni* n. g., n. sp. (Acari: Parasitiformes: Philodanidae, n. fam.) associated with *Neatus tenebrioides* (Coleoptera: Tenebrionidae) in North America. *Ann. Entomol. Soc. Amer.* 70:487–94. [PHILODANIDAE]

———. 1982. Acariformes, Prostigmata. In Parker, 2:117–45.

———. 1983a. The deutonymph of *Epiphis rarior* Berlese, 1916 (Epiphidinae n. subfam., Rhodacaridae, Rhodacaroidea). *Can. J. Zool.* 61:2598–2611. [OLOGAMASIDAE]

———. 1983b. Modifications of the deutonymph of *Uropodella lacinata* Berlese, 1888, for phoretic dispersal (Acari: Parasitiformes). *J. Georgia Entomol. Soc.* 18:151–55.

———. 1989a. Occurrence of *Pomerantzia kethleyi* (Acari: Prostigmata: Pomerantziidae) in Illinois and Minnesota. *Great Lakes Entomol.* 22:101.

———. 1989b. Proteonematalycidae (Acari: Acariformes), a new mite family from fore-dune sand of Lake Michigan. *Internat. J. Acarol.* 15:209–17.

———. 1990. Acarina: Prostigmata (Actinedida). In Dindal (1990), 667–756.

———. 1991a. Calyptostatic nymphs of *Neonanorchestes* (Nanorchestidae): A third example of alternating calyptostasy. In Dusbábek and Bukva, 2:279–82.

———. 1991b. A procedure for extraction of microarthropods from bulk samples of sandy soils. *Agr. Ecosyst. Environ.* 34:193–200.

———. 1992. The prelarva of *Alycus roseus* Koch (Bimichaeliidae: Acariformes: Acari). *Can. J. Zool.* 68:1058–61.

Kethley, J. B., and D. E. Johnston. 1975. Resource tracking patterns in bird and mammal ectoparasites. *Misc. Publ. Entomol. Soc. Amer.* 9:231–36.

Kethley, J. B., R. A. Norton, P. M. Bonamo, and W. A. Shear. 1989. A terrestrial alicorhagiid mite (Acari: Acariformes) from the Devonian of New York. *Micropaleontology* 35:367–73.

Kevan, D. K. M. 1985. Soil zoology, then and now mostly then. *Quaest. Entomolog.* 21:371.7–472.

Kevan, P. G., W. G. Chaloner, and D. B. O. Savile. 1975. Interrelationships of early terrestrial arthropods and plants. *Palaeontology* 18:391–417.

Key, K. H. L. 1990. Host relations and distribution of species of *Caeculisoma* (Acarina: Erythraeidae) parasitizing grasshoppers in Australia with supplementary information for the genus *Trombella* (Trombellidae). *Aust. J. Zool.* 38:11–18.

———. 1991. Host relations and distribution of Australian species of *Charleronia* (Acarina: Ethrythraeidae) parasitizing grasshoppers. *Aust. J. Zool.* 39:31–43.

———. 1994. Host relations and distribution of the Australian species of *Eutrombidium* (Acarina: Microtrombidiidae), a parasite of grasshoppers. *Aust. J. Zool.* 42:363–70.

Key, K. H. L., and R. V. Southcott. 1986. Host relations and distribution of Australian species of *Trombella* (Acarina: Trombellidae) parasitizing grasshoppers. *Aust. J. Zool.* 34:647–58.

Khalil, G. M. 1974. The subgenus *Persicargas* (Ixodoidea: Argasidae: Argas). 19, Preliminary studies in diapause in *A.* (*P.*) *arboreus* Kaiser, Hoogstraal & Kohls. *J. Med. Entomol.* 11:363–66.

Khalil, G. M., S. A. Nada, and D. E. Sonenshine. 1981. Sex pheromone regulation of mating behavior in the camel tick, *Hyalomma dromedarii*. *J. Parasitol.* 67:70–76.

Khaustov, A. A. 1997. Two new species of mites of the genus *Proctotydeus* (Acariformes, Tydeidae) from Crimea and northeastern Ukraine. *Vestn. Zool.* 31 (5–6): 90–93. [IOLINIDAE] [in Russian]

———. 1998. A review of mites of the genus *Pseudotarsonemoides* (Acari: Heterostigmata: Tarsonemidae) of Crimea, with description of two new species. *Acarina, Russian J. Acarol.* 6:25–30.

———. 2000a. Bembidiacaridae, a new family of mites (Acari: Heterostigmata) associated with carabid beetles of the genus *Bembidion* (Coleoptera: Carabidae). *Acarina, Russian J. Acarol.* 8:3–8. [ATHYREACARIDAE]

———. 2000b. Mites of the family Winterschmidtiidae (Acari, Astigmata), associated with bark beetles (Coleoptera, Scolytidae) in Crimea. *Vestn. Zool. Suppl.* 14 (1): 50–59.

———. 2001. New and little known species of mites of the genus *Heterotarsonemus* (Acarina: Heterostigmata: Tarsonemidae) from Crimea. *Vestn. Zool.* 35:81–84. [in Russian]

———. 2005a. New species and records of the genus *Petalomium* (Acari: Heterostigmata: Pygmephoridae) from Crimea (Ukraine). *Acarina, Russian J. Acarol.* 13:173–79.

———. 2005b. Three new species of mite family Pygmephoridae (Acari: Heterostigmata) associated with scarab beetles (Coleoptera: Scarabaeidae) from Ukraine. *Acarina, Russian J. Acarol.* 13:27–34.

———. 2007a. A new species of the genus *Pygmephorellus* (Acari: Heterostigmata: Pygmephoridae) associated with *Geotrupes stercorarius* (Coleoptera: Geotrupidae) from Crimea. *Acarina, Russian J. Acarol.* 15:287–89.

———. 2007b. Two new species of mites of the family Acarophenacidae (Acari: Heterostigmata) from Crimea (Ukraine). *Vestn. Zool.* 41:549–53.

Kielczewski, B., and S. Balazy. 1966. Zagadnienie drapieznictwa roztoczy (Acarina) na jajach korników (Scolytidae, Coleoptera). *Ekol. Polska*, scr. B, 12:161–63. [in Polish]

Kielczewski, B., and S. Seniczak. 1972. Cykl rozwojowy drapeznego roztocza *Calvolia fraxini* E. Türk et F. Türk (Tyroglyphidae, Acarina). *Prace z Zakresu Lesnictwa* 34:83–88. [WINTERSCHMIDTIIDAE]

Kim, C-M. 2004. Trigynaspida (Acari: Mesostigmata): New diagnosis, classification, and phylogeny. *Acarologia* 44:157–94.

Kim, C-M., and H. Klompen. 2002. A new genus and species of Paramegisidae (Mesostigmata: Trigynaspida) associated with millipedes from Mexico. *Acarologia* 42:39–51. [PARAMEGISTIDAE]

Kim, K. C., ed. 1985a. *Coevolution of parasitic arthropods and mammals*. New York: Wiley-Interscience.

———. 1985b. Evolutionary relationships of parasitic arthropods and mammals. In Kim (1985a), 3–99.

———. 1985c. Parasitism and coevolution. In Kim (1985a), 661–82.

Kinn, D. N. 1966. A new genus and species of Schizogyniidae (Acarina: Mesostigmata) from North America with a key to the genera. *Acarologia* 8:576–86. [SCHIZOGYNIIDAE]

———. 1967a. A new species of *Cercomegistus* (Acari: Mesostigmata) from California. *Acarologia* 9:488–96. [CERCOMEGISTIDAE]

———. 1967b. Notes on the life cycle and habits of *Digamasellus quadrisetus* (Mesostigmata: Digamasellidae). *Ann. Entomol. Soc. Amer.* 60:862–65.

———. 1968. A new species of *Pleuronectocelaeno* (Acarina: Celaenopsidae) associated with bark beetles in North and Central America. *Acarologia* 9:488–96. [CELAENOPSIDAE]

———. 1970. A new genus and species of Cercomegistidae (Acarina: Mesostigmata) from North America. *Acarologia* 12:244–52. [CERCOMEGISTIDAE]

———. 1971. *The life cycle and behavior of* Cercoleipus coelonotus *(Acarina: Mesostigmata) including a survey of phoretic mite associates of California Scolytidae.* Univ. Calif. Publ. Entomol. 65. Berkeley: Univ. Calif. Press.

———. 1972. A new species of *Holocercomegistus*, including some observations on the chaetotaxy of the pedipalpal and ambulatory appendages of the Cercomegistidae (Acarina). *Acarologia* 13:258–65. [CERCOMEGISTIDAE]

———. 1980. Mutualism between *Dendrolaelaps neodisetus* and *Dendroctonus frontalis*. *Environ. Entomol.* 9:756–58. [DIGAMASELLIDAE]

———. 1983. The life cycle of *Proctolaelaps dendroctoni* Lindquist and Hunter (Acari: Ascidae): A mite associated with pine bark beetles. *Internat. J. Acarol.* 9:205–10. [MELICHARIDAE]

———. 1987. Incidence of pinewood nematode dauerlarvae and phoretic mites associated with long-horned beetles in central Louisiana. *Can. J. Forest Res.* 17:187–90.

Kinnear, A. 1991. Acarine communities of semi-arid soils from the eastern goldfields region of Western Australia. *Pedobiologia* 35:273–83.

Kirchner, W. P. 1967. Spermatophoren bei Halacariden (Acarina). *Naturwissenschaften* 54:345–46.

———. 1969. Zur Biologie und Ökologie von *Halacarus basteri basteri* Johnston, 1836 (Acari, Trombidiformes). *Oecologia* 3:56–69.

Kitajima, E. W., C. M. Chagas, and J. C. V. Rodrigues. 2003. *Brevipalpus*-transmitted plant virus and virus-like diseases: cytopathology and some recent cases. *Exp. Appl. Acarol.* 30:135–60.

Kitajima, E. W., J. A. M. Rezende, and J. C. V. Rodrigues. 2003. Passion fruit green spot virus vectored by *Brevipalpus phoenicis* (Acari: Tenuipalpidae) on passion fruit in Brazil. *Exp. Appl. Acarol.* 30:225–31.

Kitching, R. L. 2001. Food webs in phytotelmata: "Bottom-up" and "top-down" explanations for community structure. *Ann. Rev. Entomol.* 46:729–60.

Kitching, R. L., J. M. Bergelson, M. D. Lowman, S. McIntyre, and G. Carruthers. 1993. The biodiversity of arthropods from Australian rainforest canopies: General introduction, methods, sites and ordinal results. *Aust. J. Ecol.* 18:181–91.

Klimov, P. B., M. Lekveishvili, A. P. G. Dowling, and B. M. OConnor. 2004. Multivariate analysis of morphological variation in two cryptic species of mites of the genus *Sancassania* (Acari: Acaridae). *Ann. Entomol. Soc. Amer.* 97:322–45.

Klimov, P. B., and B. M. OConnor. 2003. Phylogeny, historical ecology and systematics of some mushroom-associated mites of the genus *Sancassania* (Acari: Acaridae), with new generic synonymies. *Invertebr. Syst.* 17:469–514.

Klironomos, J. N., and B. Kendrick. 1995. Relationships among microarthropods, fungi and their environment. In *The significance and regulation of soil biodiversity*, ed. H. P. Collins, G. P. Robertson, and M. J. Klug, 209–33. Dordrecht: Kluwer Academic Publishers.

Klompen, J. S. H. 1989. Ontogeny of *Rhyncoptes grabberi*, n. sp. (Acari: Astigmata: Rhyncoptidae) associated with *Macaca mulatta*. *J. Med. Entomol.* 26:81–87.

———. 1992. *Phylogenetic relationships in the mite family Sarcoptidae (Acari: Astigmata).* Misc. Publ. 180. Ann Arbor: Mus. of Zool., Univ. Mich.

———. 1993. Phylogenetic relationships in the skin-inhabiting Sarcoptoidea (Acari: Astigmata). *Syst. Parasitol.* 24:17–33. [RHYNCOPTIDAE, SARCOPTIDAE]

———. 2000a. Prelarva and larva of *Opilioacarus* (*Neocarus*) *texanus* (Chamberlin and Mulaik) (Acari: Opilioacarida) with notes on the patterns of setae and lyrifissures. *J. Nat. Hist.* 34:1977–92.

———. 2000b. A preliminary assessment of the utility of elongation factor-1α in elucidating relationships among basal Mesostigmata. *Exp. Appl. Acarol.* 24:805–20.

Klompen, J. S. H., W. C. Black IV, J. E. Keirans, and J. H. Oliver, Jr. 1996. Evolution of ticks. *Ann. Rev. Entomol.* 41:141–61.

Klompen, J. S. H., S. J. Dobson, and S. C. Barker. 2002. A new subfamily, Bothriocrotoninae n. subfam., for the genus *Bothriocroton* Keirans, King and Sharrad, 1994 status amend. (Ixodida: Ixodidae), and the synonymy of *Aponomma* Neumann, 1899 with *Amblyomma* Koch, 1844. *Syst. Parasitol.* 53:101–7.

Klompen, J. S. H., F. S. Lukoschus, and B. M. OConnor. 1987. Ontogeny, life history and sex ratio evolution in *Ensliniella kostylevi* (Acari: Winterschmidtiidae). *J. Zool. Lond.* 213:591–607.

Klompen, J. S. H., and J. H. Oliver, Jr. 1993a. Haller's organs in the tick family Argasidae (Acari: Parasitiformes: Ixodida). *J. Parasitol.* 79:591–603.

———. 1993b. Systematic relationships in the soft ticks (Acari: Ixodida: Argasidae). *Syst. Entomol.* 18:313–31.

Knop, N. F. 1985. Mating behavior in the tydeid mite *Homeopronematus anconai* (Acari: Tydeidae). *Exp. Appl. Acarol.* 1:115–25. [IOLINIDAE]

Knop, N. F., and M. A. Hoy. 1983. Biology of a tydeid mite, *Homeopronematus anconai* (n. combin.) (Acari: Tydeidae), important in San Joaquin Valley vineyards. *Hilgardia* 51 (5): 1–30. [IOLINIDAE]

Knülle, W. 1957. Morphological and evolutionary studies of the phylogenetic system of the Acari: Acariformes Zachv. I, Oribatei: Malaconothridae. *Mitt. Zool. Mus. Berl.* 33:97–213.

———. 1959. Morphologische und Entwicklungsgeschichtliche untersuchungen zum phylogenetischen system der Acari: Acariformes Zachv. II, Acaridiae: Acaridae. *Mitt. Zool. Mus. Berl.* 35:347–417.

———. 1984. Water vapour uptake in mites and insects: An ecophysiological and evolutionary perspective. In Griffiths and Bowman, 1:71–82.

———. 1987. Genetic variability and ecological adaptability of hypopus formation in a stored-product mite. *Exp. Appl. Acarol.* 3:21–32.

———. 1991. Genetic and environmental determinants of hypopus duration in the stored-product mite *Lepidoglyphus destructor*. *Exp. Appl. Acarol.* 10:231–58.

———. 1995. Expression of a dispersal trait in a guild of mites colonizing transient habitats. *Evol. Ecol.* 9:341–53.

———. 2003. Interaction between genetic and inductive factors controlling the expression of dispersal and dormancy morphs in dimorphic astigmatid mites. *Evolution* 57:828–38. [GLYCPHAGIDAE]

Koç, K., and N. Ayyildiz, N. 1996. New records of *Ledermuelleriopsis* Willmann for the Turkish fauna (Acari, Stigmaeidae). *Biyoloji Kongr.* 8:17–20.

Koch, C. L. 1842. *Ubersicht des Arachnidensystems*. 5 vols. Fasc. 3. Nürnberg: C. H. Zeh.

Koehler, H. 1996. Soil animals and bioindication. In *Bioindicator systems for soil pollution*, ed. N. M. van Straalen and D. A. Krivolutski, 179–88. Dordrecht: Kluwer Academic Publishers.

———. 1999. Predatory mites (Gamasina, Mesostigmata). *Agr. Ecosyst. Environ.* 74:395–410.

Kok, O. B. 1968. *Lamellarea*, a new oribatid genus (Acari) from South Africa. *Acarologia* 10:711–24. [LAMELLAREIDAE]

Kondo, H., T. Maeda, and T. Tamada. 2003. Orchid fleck virus: *Brevipalpus californicus* mite transmission, biological properties and genome structure. *Exp. Appl. Acarol.* 30:215–23.

Konopacka, A., and J. Sicinski. 1985. Macrofauna inhabiting the colonies of the sponge *Spongilla lacustris* (L.) in the River Gac. *Verh. Internat. Verein. Limnol.* 22:2968–73. [UNIONICOLIDAE]

Kontschán, J., and S. Mahunka. 2004. *Caribothyrus barbatus* n. gen., n. sp., a new holothyrid mite (Acari: Neothyridae) from Dominican Republic. *Internat. J. Acarol.* 30:343–46.

Korn, W. 1983. Zur Vergesellschaftung der Gamasidenarten *Poecilochirus carabi* G. u. R. Canestrini 1882 (= *P. necrophori* Vitzthum 1930), *P. austroasiaticus* Vitzthum 1930 und *P. subterraneus* Müller 1859 mit Aaskäfern aus der Familie der Silphidae. *Spixiana* 6:251–79. [PARASITIDAE]

Koroleva, E. V. 1976. New species of the mite genus *Olopachys* Berlese, 1910 (Parasitiformes, Pachylaelapidae). *Entomol. Rev.* 54:480–96.

———. 1977. New species of mites of the genus *Pachylaelaps* Berlese, 1888 (Parasitiformes, Pachylaelaptidae). *Parazitol. Sb.* 27: 119–48.

Koroleva, E. V., and V. E. Sklyar. 1982. *Sphaerolaelaps holothyroides*, a new for the fauna of the USSR species of gamasid mites. *Zool. Zh.* 61:940–43. [PACHYLAELAPIDAE]

Korsgaard, J., and T. E. Hallas. 1979. Tarsonemid mites in Danish house dust. *Allergy* 34:225–32.

Košir, M. 1975. Ernährung und Entwicklung von *Pygmephorus mesembrinae* und *P. quadratus* (Pygmephoridae, Tarsonemini, Acari) und Bemerkungen über drei weitere Arten. *Pedobiologia* 15:313–29.

Kostiainen, T. S., and M. A. Hoy. 1996. *The Phytoseiidae as biological control agents of pest mites and insects*. Univ. Fla. Agr. Experimentation Monogr. 17. Gainesville: Univ. Fla. Agr. Exper. Sta.

Kotzé, J. M., J. F. Putterill, N. Labuschagne, and F. C. Wehner. 1987. Occurrence of a spiroplasma-like organism in lesions of concentric ring blotch on citrus. *Phytophylactica* 19:363–64. [ERIOPHYIDAE]

Kouchakji, G. A., and B. Loos-Frank. 1984. The validity and morphology of lung mites of the genus *Pneumocoptes* (Acari: Astigmata). *Z. Parasitenkd.* 70:673–83. [PNEUMOCOPTIDAE]

Kramer, P. 1877. Grundzuge zur Systematik der Milben. *Arch. Naturgesch.* 2:215–47.

Krantz, G. W. 1958a. *Cryptognathus sternalis*. a new species of prostigmatid mite from Oregon (Acarina: Cryptognathidae). *Pan-Pac. Entomol.* 34:81–85.

———. 1958b. *Lobogyniella tragardhi*, a new genus and species of diplogyniid mite associated with dampwood termites in Oregon (Acarina: Diplogyniidae). *Proc. Entomol. Soc. Wash.* 60:127–31. [DIPLOGYNIIDAE]

———. 1960. A re-evaluation of the Parholaspinae Evans 1956 (Mesostigmata: Macrochelidae). *Acarologia* 2:393–433. [PARHOLASPIDIDAE]

———. 1961. A re-evaluation of the Microgynioidea, with a description of a new species of *Microgynium* (Acarina: Mesostigmata). *Acarologia* 3:1–10. [MICROGYNIIDAE]

———. 1962. Descriptions of three myrmecophilous Macrochelidae (Acarina: Mesostigmata) from Panama, British Guiana and the British West Indies. *J. Kans. Entomol. Soc.* 35:357.

———. 1965a. A new species of *Macrocheles* (Acarina: Macrochelidae) associated with bark beetles of the genera *Ips* and *Dendroctonus*. *J. Kans. Entomol. Soc.* 38:145–53.

———. 1965b. A review of the genus *Neopodocinum* Oudemans 1902 (Acarina: Macrochelidae). *Acarologia* 7:139–226.

———. 1967. A review of the genus *Holocelaeno* Berlese, 1910 (Acarina: Macrochelidae). *Acarologia* 9:1–146.

———. 1970a. *Agauopsis vidae*, a new species of Halacaridae (Acari: Prostigmata) from the northern Adriatic Sea, with notes on its behavior. *Arch. Oceanograph. Limnol.* 16:247–61.

———. 1970b. *A manual of acarology*. Corvallis: Oregon State Univ. Bookstores.

———. 1971. The mites of Quintana Roo. II, *Actacarus giganteus*, a new species of arenicolous mite (Prostigmata: Halacaridae) from the Caribbean coast. *Ann. Entomol. Soc. Amer.* 64:594–98.

———. 1973a. Four new predatory species of Halacaridae (Acari: Prostigmata) from Oregon, with remarks on their distribution in the intertidal mussel habitat (Pelecypoda: Mytilidae). *Ann. Entomol. Soc. Amer.* 66:975–85.

———. 1973b. Observations on the morphology and behavior of the filbert rust mite, *Aculus comatus* (Prostigmata: Eriophyoidea) in Oregon. *Ann. Entomol. Soc. Amer.* 66:709–17.

———. 1974. *Phaulodinychus mitis* (Leonardi 1899) (Acari: Uropodidae), an intertidal mite exhibiting plastron respiration. *Acarologia* 16:11–20.

———. 1976. Arenicolous Halacaridae from the intertidal zone of Schooner Creek, Oregon (Acari: Prostigmata). *Acarologia* 18:241–58.

———. 1977. On the occurrence of Claparède organs in the Halacaridae (Acari: Actinedida). *Acarologia* 19:62–66.

———. 1978. *A manual of acarology*. 2d ed. Corvallis: Oregon State Univ. Bookstores.

———. 1979. The role of *Phytocoptella avellanae* (Nalepa) and *Cecidophyopsis vermiformis* (Nalepa) (Eriophyoidea) in big bud of filbert. In Piffl (1979), 201–8.

———. 1982. A new species of *Copidognathus* Trouessart (Acari: Actinedida: Halacaridae) from the Galapagos Rift. *Can. J. Zool.* 60:1728–31.

———. 1983a. Mites as biological control agents of dung-breeding flies, with special reference to the Macrochelidae. In Hoy, Cunningham, and Knutson, 91–98.

———. 1983b. Three new species of *Macrocheles* (Acari: Gamasida: Macrochelidae) associated with 3-toed sloths, *Bradypus* spp. (Edentata: Bradypodidae), in Brazil and Surinam. *Acarologia* 24:3–12.

———. 1986. Acari: Rhodacaridae. In Botosaneanu, 632–34.

———. 1991. Nature of the association between *pisentii*-group mites (Acari: Macrochelidae: *Macrocheles*) and dung beetles of the genus *Scarabaeus* (Coleoptera: Scarabaeidae) in southern France. *Acarologia* 32:3–11.

———. 1996. Specialization and systematics in acarology: Reflections and predictions. In Mitchell et al. (1996), 1:1–4.

———. 1998a. A new genus and two new species of hypoaspidine mites (Acari: Laelapidae) associated with Old World carpenter bees of the tribe Xylocopini (Hymenoptera: Apidae: *Xylocopa*). *Internat. J. Acarol.* 24:291–300.

———. 1998b. Observations on five rarely collected genera of Macrochelidae (Acari: Mesostigmata) associated with insects. *Acarologia* 39:95–109.

———. 1998c. Reflections on the biology, morphology and ecology of the Macrochelidae. *Exp. Appl. Acarol.* 22:125–37. (Also appears in Bruin, van der Geest, and Sabelis, 291–301.)

———. 2000. Two new species of the genus *Laelaptonyssus* Womersley from North America and Australia, with observations supporting the reinstatement to family level of the subfamily Laelaptonyssinae *sensu* Lee, 1970 (Acari: Mesostigmata: Rhodacaroidea). *Acarologia* 41:25–38.

Krantz, G. W., and B. D. Ainscough. 1960. *Caminella peraphora*, a new genus and species of mite from Oregon (Acarina: Trachytidae). *Ann. Entomol. Soc. Amer.* 53:27–34. [DITHINOZERCONIDAE]

———. 1990. Acarina: Mesostigmata (Gamasida). In Dindal (1990), 583–665.

Krantz, G. W., and G. T. Baker. 1982. Observations on the plastron mechanism of *Hydrozetes* sp. (Acari: Oribatida: Hydrozetidae). *Acarologia* 23:273–77.

Krantz, G. W., and D. T. Ehrensing. 1990. Deuterogyny in the skeleton weed mite, *Aceria chondrillae* (G. Can.) (Acari: Eriophyidae). *Internat. J. Acarol.* 16:129–33.

Krantz, G. W., and A. Filipponi. 1964. Acari della famiglia Macrochelidae (Mesostigmata) nella collezione del South Australian Museum. *Riv. Parassitol.* 25:35–54.

Krantz, G. W., L. A. Gomez, and V. E. Gonzalez. 2007. Parasitism in the Uropodina: a case history from Colombia. In Morales-Malacara et al., 29–38.

Krantz, G. W., and N. S. Khot. 1962. A review of the family Otopheidomenidae Treat 1955 (Acarina: Mesostigmata). *Acarologia* 4:532–42.

Krantz, G. W., and E. E. Lindquist. 1979. Evolution of phytophagous mites (Acari). *Ann. Rev. Entomol.* 24:121–58.

Krantz, G. W., and J. L. Mellott. 1968. Two new species of *Macrocheles* (Acarina: Macrochelidae) from Florida, with notes on their host-specific relationships with geotrupine beetles (Scarabaeidae: Geotrupinae). *J. Kans. Entomol. Soc.* 41:48–56.

———. 1972. Studies on phoretic specificity in *Macrocheles mycotrupetes* and *M. peltotrupetes* Krantz and Mellott (Acari: Macrochelidae), associates of geotrupine Scarabaeidae. *Acarologia* 14:317–44.

Krantz, G. W., B. M. OConnor, W. J. Foreyt, and A. Fain. 2003. A new species of *Aplodontopus* (Acari: Astigmata: Chortoglyphidae) from the yellow-bellied marmot, *Marmota flaviventris* (Rodentia: Sciuridae) in eastern Washington, USA, with observations on its pathology. *Internat. J. Acarol.* 29:99–105.

Krantz, G. W., and N. I. Platnick. 1995. On *Brucharachne*, the spider that wasn't (Arachnida, Acari, Dermanyssoidea). *Amer. Mus. Novit.* 3151:1–8. [LAELAPIDAE]

Krantz, G. W., and G. O. Poinar, Jr. 2004. Mites, nematodes and the multimillion dollar weevil. *J. Nat. Hist.* 38:135–41. [MELICHARIDAE]

Krantz, G. W., and B. L. Redmond. 1987. Identification of glandular and poroidal idionotal systems in *Macrocheles perglaber* F. & P. (Acari: Macrochelidae). *Exp. Appl. Acarol.* 3:243–53.

———. 1988. On the structure and function of the cribrum, with special reference to *Macrocheles perglaber* (Gamasida: Macrochelidae). In ChannaBasavanna and Viraktamath, 1:179–85.

Krantz, G. W., and L. A. Royce. 1994. Observations on the biology and behavior of *Macrocheles mycotrupetes* Krantz and Mellott (Acari: Macrochelidae). *Internat. J. Acarol.* 20:115–21.

Krantz, G. W., L. A. Royce, R. R. Lowry, and R. Kelsey. 1991. Mechanisms of phoretic specificity in *Macrocheles*. In Dusbábek and Bukva, 2:561–69.

Krantz, G. W., and J. G. Wernz. 1979. Sperm transfer in *Glyptholaspis americana*. In Rodriguez, 2:441–46.

Krantz, G. W., and J. O. Whitaker, Jr. 1988. Mites of the genus *Macrocheles* (Acari: Macrochelidae) associated with small mammals in North America. *Acarologia* 29:225–59.

Krauss, W. 1970. Die europaischen Arten der Gattungen *Macrocheles* Latreille 1829 und Geholaspis Berlese 1918. *Acarologie, Schrift. Vergleich. Milbenk.* 14:1–60.

Krczal, H. 1959. Systematik und Ökologie der Pyemotiden. *Beitr. Syst. Ökol. Mitteleurop. Acarina* 1 (3): 385–625.

Krisper, G. 1990. Das Sprungvermögen der Milbengattung *Zetorchestes* (Acarida, Oribatida). *Zool. Jb. Anat.* 120:289–312.

———. 1991. The saltatory capacity of an oribatid mite. In Schuster and Murphy, 397–98.

Krisper, V. G., and T. Schneider. 1998. Erstnachweis und Verbreitung der Milbenfamilie Cryptognathidae (Acari, Actinedida, Raphignathoidea) in Österreich und Slowenien. *Mitt. Naturwiss. Ver. Steiermark* 128:193–202.

Krivolutsky, D. A., A. Ja. Druk, I. S. Eitminaviciute, L. M. Laskova, and E. Karppinen. 1990. *Fossil oribatid mites*. Vilnius: Mokslas.

Krivolutsky, D. A., and N. V. Lebedeva. 2003. *Oribatid mites (Oribatei, Acariformes) in bird feathers*. Moscow: Centre of Media Projects ABF. [in Russian]

Krivolutsky, D. A., N. V. Lebedeva, and A. V. Matyukhin. 2001. Oribatid mites (Oribatei) in the plumage of birds. *Parazitologiya* 35:275–83. [in Russian]

Krombein, K. V. 1962a. Biological notes on acarid mites associated with solitary wood-nesting wasps and bees. *Proc. Entomol. Soc. Wash.* 64:11–19. [ACARIDAE, SUIDASIIDAE]

———. 1962b. Natural history of Plummers Island, Maryland. XVI, Biological notes on *Chaetodactylus krombeini* Baker, a parasitic mite of the megachilid bee, *Osmia (Osmia) lignaria* Say (Acarina: Chaetodactylidae). *Proc. Biol. Soc. Wash.* 75:237–50.

———. 1967. *Trap-nesting wasps and bees: Life histories, nests and associates*. Washington, DC: Smithsonian Press.

Kshirsagar, K. K. 1968. Further observations on phoretic bee mites *Neocypholaelaps indica* (Evans). *Indian Bee J.* 30:68–69. [AMEROSEIIDAE]

Kul'chitskii, A. G. 1993. Biological characteristics of the trombidiform mite *Tydeus kochi* (Acariformes, Tydeidae). *Vestn. Zool.* 1993 (2): 64–67. [in Russian]

Kunst, M. 1959. Bulgarische Oribatiden (Acarina) III. *Acta Univ. Carolinae, Biol.* 6 (1): 51–74.

———. 1971. Nadkohorta Pancirnici—Oribatei. In *Klic zvireny CSSR IV*, ed. M. Daniel and V. Cerny, 531–80. Praha: Academia.

Kuriki, G. 1995. Life cycle of *Trhypochthoniellus setosus* Willmann (Acari: Trhypochthoniellidae) in a *Sphagnum* moor at Yachidaira, Northeast Japan. *J. Acarol. Soc. Jpn.* 4:113–22.

———. 1996. On the faecal pellet deposition of *Trhypochthoniellus setosus* Willmann (Acari, Trhypochthoniidae). *J. Acarol. Soc. Jpn.* 5 (1): 27–34.

Kurosa, K. 1972. The scutacarid mites of Japan. II, *Lophodispus latus* gen. et sp. nov. *Bull. Nat. Sci. Mus., Tokyo* 15:29–35.

———. 1976a. Notes on Tarsonemimi and Acaridiae (Acari) from Pasoh Forest Reserve and its vicinities, Malaya. *Nat. Life S. E. Asia.* 7:61–81. [GAUDIELLIDAE]

———. 1976b. The scutacarid mites of Japan. V, Three new *Archidispus* associated with stenolophine ground beetles. *Kontyû* 44:172–83.

———. 1989a. Pygmephorid mites associated with ground beetles in Japan. I, A new genus and species of the Neopygmephorinae. *Jpn. J. Entomol.* 57:469–76.

———. 1989b. The scutacarid mites of Japan. IX, Five new *Archidispus* associated with ground beetles. *Jpn. J. Entomol.* 57:3–24.

———. 1990. The scutacarid mites of Japan. X, Three new *Archidispus* associated with *Stenolophus* (Coleoptera, Carabidae). *Jpn. J. Entomol.* 58:1–10.

———. 1994. A new genus and species of Pygmephoridae (Acari: Heterostigmata) associated with *Reticulitermes spelatus kyushuensis* (Isoptera: Rhinotermitidae) in Japan. *Trans. Shikoku Entomol. Soc.* 20:207–14.

Kurosa, K., and R. W. Husband. 1994. *Cydnipolipus miyamotoi*, a new genus and species of podapolipid mite parasitic on *Aethus indicus* (Heteroptera: Cydnidae). *J. Acarol. Soc. Jpn.* 3:21–32.

Kurosa, K., A. Khaustov, and R. W. Husband. 2004. A new genus and three new species of Podapolipidae (Acari: Tarsonemina) parasitic on *Nicrophorus* and *Silpha* (Coleoptera: Silphidae) in Japan and Ukraine. *Internat. J. Acarol.* 30:313–27.

Kuwahara, Y. 1991. Pheromone studies on astigmatid mites: alarm, aggregation and sex. In Dusbábek and Bukva, 1:43–52.

———. 1999. Chemical ecology in astigmatid mites. In *Environmental entomology: Behavior, physiology and chemical ecology*, ed. T. Hidata, Y. Matsumoto, and K. Honda, 380–93. Tokyo: Univ. Tokyo Press. [in Japanese]

Kuwahara, Y., H. Fukami, S. Ishii, K. Matsumoto, and Y. Wada. 1980. Pheromone study on acarid mites. III, Citral: isolation and identification from four species of acarid mites, and its possible role. *Jpn. J. Sanit. Zool.* 31:49–52.

Kuwahara, Y., S. Ishii, and H. Fukami. 1975. Neryl formate: alarm pheromone of the cheese mite, *Tyrophagus putrescentiae*. *Experientia* 31:1115–16.

Kuwahara, Y., W. S. Leal, and T. Suzuki. 1989. Antifungal activity of *Caloglyphus polyphyllae* sex pheromone and other mite exudates. Pheromone study on astigmatid mites XXIV(1). *Naturwissenschaften* 76:578–79.

Kuwahara, Y., L. T. My-Yen, Y. Tominaga, K. Matsumoto, and Y. Wada. 1992. 1,3,5,7-tetramethyldecyl formate, lardolure: aggregation pheromone of the acarid mite, *Lardoglyphus konoi* (Sasa et Asanuma) (Acarina: Acaridae). *Agr. Biol. Chem.* 46:2283–91.

Kuznetsov, N. N. 1973. Mites of the family Paratydeidae, including a new genus and species in material from Crimea. *Biol. Nauki* 1973 (11): 11–16. [in Russian]

———. 1974. New species of the family Stigmaeidae from Crimea. *Zool. Zh.* 56:635–38.

———. 1980. Adaptive peculiarities of ontogenesis in the Tydeidae (Acariformes). *Zool. Zh.* 59:1018–24. [TYDEIDAE, IOLINIDAE] [in Russian]

Kuznetsov, N. N., and I. Z. Livshits. 1978. Family Cunaxidae. In Gilyarov, 144–48. [in Russian]

Labandeira, C. C., T. L. Phillips, and R. A. Norton. 1997. Oribatid mites and the decomposition of plant tissues in Paleozoic coal-swamp forests. *Palaios* 12 (2): 319–53.

Laemmlen, F. F., and D. H. Hall. 1973. Interdependence of a mite, *Siteroptes reniformis*, and a fungus, *Nigrospora oryzae*, in the *Nigrospora* lint rot of cotton. *Phytopathology* 63:308–15. [SITEROPTIDAE]

Lagutenko, U. P. 1964. Functional and morphological characteristics of the digestive and excretory systems of some gamasid mites (Parasitiformes, Gamasoidea) as related to the nature of their parasitism. *Zool. Zh.* 43:1773–83.

Laing, J. E., and N. F. Knop. 1983. Potential use of predaceous mites other than Phytoseiidae for biological control of orchard pests. In Hoy, Cunningham, and Knutson, 28–35.

Lan, W., J. L. Xin, and J. Aoki. 1986. Two new species of oribatid mites of economic importance from China (Acari, Oribatida). *Proc. Jpn. Soc. Syst. Zool.* 34:27–31.

Lanciani, C. A. 1970. New species of *Eylais* (Acari: Eylaidae) parasitic on aquatic Coleoptera. *Trans. Amer. Microscop. Soc.* 89:169–88.

———. 1971. Host exploitation and synchronous development in a water mite parasite of the marsh treader *Hydrometra myrae* (Hemiptera: Hydrometridae). *Ann. Entomol. Soc. Amer.* 64:1254–59. [HYDRYPHANTIDAE]

———. 1972. Mating behavior of water mites of the genus *Eylais*. *Acarologia* 14:631–37.

———. 1979. The food of nymphal and adult water mites of the species *Hydryphantes tenuabilis*. *Acarologia* 20:563–65. [HYDRYPHANTIDAE]

———. 1980. Parasitism of the backswimmer *Buenoa scimitra* (Hemiptera: Notonectidae) by the water mite *Hydrachna virella* (Acari: Hydrachnellae). *Freshwater Biol.* 10:527–32. [HYDRACHNIDAE]

———. 1982. Parasite-mediated reductions in the survival and reproduction of the backswimmer *Buenoa scimitra* (Hemiptera: Notonectidae). *Parasitology* 85:593–603. [HYDRACHNIDAE]

———. 1986. Effect of the water mite *Arrenurus pseudotenuicollis* (Acariformes: Arrenuridae) on the longevity of captive *Anopheles quadrimaculatus* (Diptera: Culicidae). *Fla. Entomol.* 69:436–37.

Lanciani, C. A., and A. D. Boyt. 1977. The effect of a parasitic water mite, *Arrenurus pseudotenuicollis* (Acari: Hydrachnellae), on the survival and reproduction of the mosquito *Anopheles crucians* (Diptera: Culicidae). *J. Med. Entomol.* 14:10–15. [ARRENURIDAE]

Lane, R. S., W. Burgdorfer, S. F. Hayes, and A. G. Barbour. 1985. Isolation of the spirochaete from the soft tick *Ornithodoros coriaceus*: A possible agent of epizootic bovine abortion. *Science* 230:85–87. [ARGASIDAE]

Lange, A. P. 1954. Morphologie der Milbe *Zachvatkinella belbiformis*, gen. n. et sp. n.—ein neuer Vertreter der Gruppe Palaeacariformes (Acariformes). *Zool. Zh.* 34 (5): 1042–52.

Latreille, P. A. 1806–9. *Genera crustaceorum et insectorum*. 4 vols. Paris.

Latta, S. C., and B. M. OConnor. 2001. Patterns of *Knemidokoptes jamaicensis* (Acari: Knemidokoptidae) infestations among eight new avian hosts in the Dominican Republic. *J. Med. Entomol.* 38:437–40.

Laubier, L. 1960. *Parhalixodes travei* n. g., n. sp., un nouvel Halixodinae (Halacariens) ectoparasite de nemerte en Méditerranée occidentale. *Acarologia* 2:541–51.

Lauck, D. R. 1965. Chloropicrin for fast action with a Berlese funnel. *Turtox News* 43:115.

Laumann, M., R. A. Norton, G. Weigmann, S. Scheu, M. Maraun, and M. Heethoff. 2007. Speciation in the parthenogenetic oribatid mite genus *Tectocepheus* (Acari, Oribatida) as indicated by molecular phylogeny. *Pedobiologia* 51:111–22. [TECTOCEPHEIDAE]

Lavoipierre, M. M. J. 1970. A note on the sarcoptid mites of primates. *J. Med. Entomol.* 7:376–80. [SARCOPTIDAE]

Lawrence, R. F. 1935. The prostigmatic mites of South African lizards. *Parasitology* 27:1–45. [PTERYGOSOMATIDAE]

———. 1936. The prostigmatic mites of South African lizards. *Parasitology* 28:1–39. [PTERYGOSOMATIDAE]

———. 1944. Some new trombidiform mites from South Africa. *Ann. Natal Mus.* 10:437–58. [CALYPTOSTOMATIDAE]

———. 1952. A new parasitic mite from the nasal cavities of the South African toad *Bufo regularis* Reuss. *Proc. Zool. Soc. Lond.* 121:747–52. [EREYNETIDAE]

———. 1953. Two new scale-mite parasites of lizards. *Proc. U.S. Nat. Mus.* 103:9–18. [PTERYGOSOMATIDAE]

———. 1954. The known African species of Cheyletidae and Pseudocheylidae (Acarina, Prostigmata). *Ann. Natal Mus.* 13:65–77.

Le Fichoux, Y., G. Rack, P. Motte, P. Dellamonica, and P. Marty. 1980. Dermatite prurigineuse due à *Pyemotes zwoelferi* Krczal, 1963. A propos de plusieurs cas dans les Alpes-Maritimes. *Acta Trop.* 37:83–89. [PYEMOTIDAE]

Leach, W. E. 1815. A tabular view of the external characters of four classes of animals which Linné arranged under Insecta. *Trans. Linn. Soc. Lond.* 11:306–400.

Lebel, R. R., and W. B. Nutting. 1971. Population dynamics of a parasitic mite *Demodex caprae* (Trombidiformes: Demodicidae). *Proc. 3rd Internat. Congr. Acarol., Prague*: 517–21.

Lebrun, P., G. van Impe, D. de St. Georges-Gridelet, G. Wauthy, and H. M. André. 1991. The life strategies of mites. In Schuster and Murphy, 3–22.

Lebrun, P., and N. M. van Straalen. 1995. Oribatid mites: prospects of their use in ecotoxicology. *Exp. Appl. Acarol.* 19:361–80.

Leclerc, P. 1989. Considerations paléogéographique à propos de la découverte en Thaïlande d'Opilioacariens nouveaux (Acari-Notostigmata). *C.R. Soc. Biogéogr.* 65:162–74.

Lee, B. W. 1991. *Cheyletiella* dermatitis: a report of fourteen cases. *Cutis* 47:111–14. [CHEYLETIDAE]

Lee, D. C. 1970. The Rhodacaridae (Acari: Mesostigmata): Classification, external morphology and distribution of genera. *Rec. S. Aust. Mus.* 16:1–219. [RHODACARIDAE, OLOGAMASIDAE]

———. 1974. Rhodacaridae (Acari: Mesostigmata) from near Adelaide, Australia. III, Behaviour and development. *Acarologia* 16:21–44.

———. 1982. Sarcoptiformes (Acari) of South Australian soils. 3, Arthronotina (Cryptostigmata). *Rec. S. Aust. Mus.* 18:327–59.

———. 1984. A modified classification for oribate mites (Acari: Cryptostigmata). In Griffiths and Bowman, 1:241–48.

———. 1985. Sarcoptiformes (Acari) of South Australian soils. 4, Primitive oribate mites (Cryptostigmata) with an extensive unfissured hysteronotal shield and aptychoid. *Rec. S. Aust. Mus.* 19:39–67.

———. 1986. Acari: Rhodacaridae, Ologamasinae. In Botosaneanu, 635–37.

———. 1991. A modified classification of the Oripodoidea (Acarida: Cryptostigmata) based on a preliminary phylogenetic model. In Dusbábek and Bukva, 2:233–38.

Lee, D. C., and P. E. Hunter. 1974. Arthropoda of the subantarctic islands of New Zealand. 6, Rhodacaridae (Acari: Mesostigmata). *New Zeal. J. Zool.* 1:295–328. [OLOGAMASIDAE]

Lee, D. C., and R. V. Southcott. 1979. *Spiders and other arachnids of South Australia. Extract, S. Australian Yearbook*. Adelaide, S. Australia: D. J. Woolman, Govt. Printer.

Lee, D. J. 1961. Cause and effect relating to arthropod bites and stings. School of Public Health and Tropical Medicine, Univ. Sydney (mimeograph). [ANYSTIDAE]

Lee, J. S. 1989. *A systematic study on the Tetranychoid mites (Order: Acarina) in Korea*. Chonju: Chonbuk Univ.

Lees, A. D. 1946. The water balance in *Ixodes ricinus* (L.) and certain other ticks. *Parasitology* 37:1–20. [IXODIDAE]

———. 1947. Transpiration and structure of epicuticle in ticks. *J. Exp. Biol.* 23:379–410.

———. 1948. The sensory physiology of the sheep tick, *Ixodes ricinus* L. *J. Exp. Biol.* 25:145–207. [IXODIDAE]

———. 1969. The behavior and physiology of ticks. *Acarologia* 11:397–410.

Lees, A. D., and J. W. L. Beament. 1948. An egg waxing organ in ticks. *Qtly. J. Microscop. Sci. Lond.* 98:291–332.

Lehtinen, P. T. 1980. A new species of Opilioacarida (Arachnida) from Venezuela. *Acta Biol. Venez.* 10:205–14.

———. 1981. New Holothyrina (Arachnida, Anactinotrichida) from New Guinea and South America. *Acarologia* 22:3–13.

———. 1987. Association of uropodid, prodinychid, polyaspidid, antennophorid, sejid, microgynid, and zerconid mites with ants. *Entomol. Tidskr.* 108:13–20.

———. 1991. Phylogeny and zoogeography of the Holothyrida. In Dusbábek and Bukva, 2:101–13.

———. 1995. Revision of the Old World Holothyridae (Arachnida: Anactinotrichida: Holothyrina). *Invertebr. Taxon.* 9:767–826.

———. 1999. A new genus of Holothyrina: Neothyridae (Arachnida, Anactinotrichida) from the Manaus area, Amazonas State, Brazil. *Acta Amazonica* 29:325–30.

Lehtinen, P. T., and R. Niemi. 1995. The parthenogenetic mite *Labidostomma luteum* (Acarina: Actinotrichida: Labidostommidae) in Finland. *Entomol. Fenn.* 6:211–27.

Lekveishvili, M., and H. Klompen. 2004a. A new species of Sejidae (Acari: Mesostigmata) from the southeastern USA. *Internat. J. Acarol.* 30:229–38.

———. 2004b. Phylogeny of infraorder Sejina (Acari: Mesostigmata). *Zootaxa* 629:1–19.

———. 2006. New classification of the family Sejidae (Acari: Mesostigmata) based on morphological analyses. *J. Nat. Hist.* 40:2317–39.

Lekveishvili, M., and G. W. Krantz. 2004. A new genus of the family Sejidae (Acari: Mesostigmata) based on *Sejus krantzi* and *S.*

manualkrantzi Hirschmann, 1991. *Syst. Appl. Acarol. Spec. Publ.* 20:1–4.

Lemmetty, A., S. Latvala, A. T. Jones, P. Susi, W. J. McGavin, and K. Lehto. 1997. Purification and properties of a new virus from black currant, its affinities with Nepoviruses, and its close association with black currant reversion disease. *Phytopathology* 87:404–13.

Lesna, I., M. W. Sabelis, H. R. Bolland, and C. G. M. Conijn. 1995. Candidate natural enemies for control of *Rhizoglyphus robini* Claparède (Acari: Astigmata) in lily bulbs: exploration in the field and pre-selection in the laboratory. *Exp. Appl. Acarol.* 19:655–69.

Lesna, I., M. Sabelis, and C. Conijn. 1996. Biological control of the bulb mite, *Rhizoglyphus robini*, by the predatory mite, *Hypoaspis aculeifer*, on lilies: predator-prey interactions at various spatial scales. *J. Appl. Ecol.* 33:369–76.

Liang, L.-R., and C.-Y. Hu. 1987. Three new stigmaeid mites of the genus *Stigmaeus* (Acarina: Stigmaeidae). *Entomotaxon* 9:307–11.

———. 1988. Two new stigmaeid mites of the genus *Apostigmaeus* (Acari: Stigmaeidae). *Acta Zootax. Sin.* 13:45–47.

Lilleskov, E. A., and T. D. Bruns. 2000. The secret life of a resupinate ectomycorrhizal fungus, *Tomentella sublilacina*. *Inoculum* (supplement to *Mycologia* 51): 43.

Lillo, E. de, and A. Di Palma. 2002. Fine morphology of the bothridial receptor organs of *Pediculaster mesembrinae* (Acari: Siteroptidae) adult female. In Bernini et al. (2002), 253–61.

Lim, J-W., and W-K. Lee. 2001. A taxonomic study of the family Zerconidae (Acari, Mesostigmata) in the Korean Peninsula. *Korean J. Syst. Zool.* 17:191–205. [in Korean]

Lin, J-Z., and Z-Q. Zhang. 1999. *Tarsonemidae of China (Acari: Prostigmata): An annotated and illustrated catalogue and bibliography*. Syst. Appl. Acarol. Spec. Publ. 3. London: Syst. Appl. Acarol. Soc.

———. 2002. *Tarsonemidae of the World (Acari: Prostigmata). Key to genera, geographical distribution, systematic catalogue and annotated bibliography*. London: Syst. Appl. Acarol. Soc.

Lin, Y-M, and Y-G Fu. 1997. Two new species of the genus *Tuckerella* from Hainan Island, China (Acariformes: Tuckerellidae). *Acta Zool. Sin.* 22:150–53.

———. 2000. A new species of the genus *Tuckerella* from China (Acariformes: Tuckerellidae). *Acta Entomol. Sin.* 43:311–13.

Lindo, Z., and S. K. Stevenson. 2007. Diversity and distribution of oribatid mites (Acari: Oribatida) associated with arboreal and terrestrial habitats in interior cedar-hemlock forests, British Columbia, Canada. *Northwest Sci.* 81:305–25.

Lindquist, E. E. 1961. Taxonomic and biological studies of mites of the genus *Arctoseius* Thor from Barrow, Alaska (Acarina: Aceosejidae). *Hilgardia* 30:301–50. [ASCIDAE]

———. 1963. A taxonomic review of the genus *Hoploseius* Berlese (Acarina: Blattisocidae). *Can. Entomol.* 95:1175–85.

———. 1964. *A revision of the subfamily Blattsocinae (Acarina: Blattisocidae) in America north of Mexico. Part I, Introduction. Part II, Taxonomy of the blattisocine genera excepting* Proctolaelaps *Berlese*. Ph.D. diss., Univ. Calif., Berkeley. Ann Arbor, MI: Univ. Microfilms.

———. 1965. An unusual new species of *Hoploseius* Berlese (Acarina: Blattisociidae) from Mexico. *Can. Entomol.* 97:1121–31.

———. 1968. An unusual new species of *Tarsonemus* (Acarina: Tarsonemidae) associated with the Indian honey bee. *Can. Entomol.* 100:1002–6.

———. 1969a. New species of *Tarsonemus* (Acarina: Tarsonemidae) associated with bark beetles. *Can. Entomol.* 101:1291–1314.

———. 1969b. *Review of Holarctic tarsonemid mites (Acarina: Prostigmata) parasitizing eggs of ipine bark beetles*. Mem. Entomol. Soc. Can. 60. Ontario: Entomol. Soc. Can.

———. 1970a. Relationships between mites and insects in forest habitats. *Can. Entomol.* 102:978–84.

———. 1970b. Review of the genus *Heterotarsonemus* (Acarina: Tarsonemidae). *Can. Entomol.* 102:812–29.

———. 1971. New species of Ascidae (Acarina: Mesostigmata) associated with forest insect pests. *Can. Entomol.* 103:919–42.

———. 1973. Observations on the generic classification of tarsonemid mites. In Daniel and Rosický, 293–96.

———. 1974. Nomenclatural status and authorship of some family-group names in the Eriophyoidea (Acarina: Prostigmata). *Can. Entomol.* 106:209–12.

———. 1975a. Associations between mites and other arthropods in forest floor habitats. *Can. Entomol.* 107:425–37.

———. 1975b. *Digamasellus* Berlese, 1905, and *Dendrolaelaps* Halbert, 1915, with descriptions of new taxa of Digamasellidae (Acarina: Mesostigmata). *Can. Entomol.* 107:1–43.

———. 1976. Transfer of the Tarsocheylidae to the Heterostigmata, and reassignment of the Tarsonemina and Heterostigmata to lower hierarchic status in the Prostigmata (Acari). *Can. Entomol.* 108:23–48.

———. 1977. Homology of dorsal opisthosomal plates, setae, and cupules of heterostigmatic mites with those of other eleutherengone Prostigmata (Acari). *Acarologia* 19:97–104.

———. 1983. Some thoughts on the potential for use of mites in biological control, including a modified concept of "parasitoids." In Hoy, Cunningham, and Knutson, 12–18.

———. 1984. Current theories on the evolution of major groups of Acari and on their relationships with other groups of Arachnida, with consequent implications for their classification. In Griffiths and Bowman, 1:28–62.

———. 1985a. Diagnosis and phylogenetic relationships. In Helle and Sabelis (1985a), 1A:63–74. [TETRANYCHIDAE]

———. 1985b. Discovery of sporothecae in adult female *Trochometridium* Cross with notes on analogous structures in *Siteroptes* Amerling (Acari: Heterostigmata). *Exp. Appl. Acarol.* 1:73–85. [TROCHOMETRIDIIDAE, SITEROPTIDAE]

———. 1985c. External anatomy. In Helle and Sabelis (1985a), 1A:3–28.

———. 1986. *The world genera of Tarsonemidae (Acari: Heterostigmata): A morphological, phylogenetic, and systematic revision, with a reclassification of family-group taxa in the Heterostigmata*. Mem. Entomol. Soc. Can. 136. Ontario: Entomol. Soc. Can.

———. 1987. Observations on the larva and protonymph of tarsocheylid mites (Acari: Heterostigmata). *Acarologia* 28:137–50.

———. 1995. Remarkable convergence between two taxa of ascid mites (Acari: Mesostigmata) adapted to living in pore tubes of bracket fungi in North America, with description of Mycolaelaps new genus. *Can. J. Zool.* 73:104–28. [MELICHARIDAE]

———. 1996a. Antiquity of some symbiotic associations of heterostigmatic mite clades with beetle clades: phylogenetic and biogeographic patterns. In Mitchell et al. (1996), 1:301–3. [HETEROCHEYLIDAE, ATHYREACARIDAE]

———. 1996b. External anatomy and notation of structures. In Lindquist, Sabelis, and Bruin, 6:3–31.

———. 1996c. Nomenclatural problems in usage of some family and genus names. In Lindquist, Sabelis, and Bruin, 6:89–99.
———. 1996d. Phylogenetic relationships. In Lindquist, Sabelis, and Bruin, 6:301–27.
———. 1998. Evolution of phytophagy in trombidiform mites. *Exp. Appl. Acarol.* 22:88–100. (Also appears in Bruin, van der Geest, and Sabelis, 73–88.)
———. 2001. Poising for a new century: Diversification in acarology. In Halliday et al. (2001), 17–34.
———. 2003. Observations on mites of the subfamily Platyseiinae, with descriptions of two new species of *Platyseius* from North America (Acari: Mesostigmata: Ascidae). In I. M. Smith (2003a), 155–82.
Lindquist, E. E., and J. W. Amrine, Jr. 1996. Systematics, diagnoses for major taxa, and keys to families and genera with species on plants of economic importance. In Lindquist, Sabelis, and Bruin, 6:33–87.
Lindquist, E. E., and D. A. Chant. 1964. A redescription of *Aceodromus* Muma and its transfer to the Blattisociinae (Acarina: Blattisociidae). *Can. Entomol.* 96:500–507.
Lindquist, E. E., and G. O. Evans. 1965. *Taxonomic concepts in the Ascidae, with a modified setal nomenclature for the idiosoma of the Gamasina (Acarina; Mesostigmata).* Mem. Entomol. Soc. Can. 47. Ontario: Entomol. Soc. Can. [includes MELICHARIDAE]
Lindquist, E. E., and P. E. Hunter. 1965. Some mites of the genus *Proctolaelaps* Berlese (Acarina: Blattisociidae) associated with forest insect pests. *Can. Entomol.* 97:15–32. [MELICHARIDAE]
Lindquist, E. E., M. Kaliszewski, and G. Rack. 1990. Athyreacaridae, a new family of mites (Acari: Heterostigmata) associated with scarab beetles of the genus *Neoathyreus* (Coleoptera: Scarabaeidae). *Acarologia* 31:161–76.
Lindquist, E. E., and J. B. Kethley. 1975. The systematic position of the Heterocheylidae Trägårdh (Acari: Acariformes: Prostigmata). *Can. Entomol.* 107:887–98.
Lindquist, E. E., and G. W. Krantz. 2002. Description of, and validation of names for, the genus *Crotalomorpha* and the family Crotalomorphidae (Acari: Heterostigmata). *Syst. Appl. Acarol.* 7:129–42.
Lindquist, E. E., and M. L. Moraza. 1993. Pyrosejidae, a new family of trigynaspid mites (Acari: Mesostigmata: Cercomegistina) from Middle America. *Acarologia* 39:203–26. [PYROSEJIDAE]
———. 1999. Observations on homologies of idiosomal setae in Zerconidae (Acari: Mesostigmata), with modified notation for some posterior body setae. *Acarologia* 39:203–26.
Lindquist, E. E., and G. N. Oldfield. 1996. Evolution of eriophyoid mites in relation to their host plants. In Lindquist, Sabelis, and Bruin, 6:277–300.
Lindquist, E. E., and J. G. Palacios-Vargas. 1991. Proterorhagiidae (Acari: Endeostigmata), a new family of rhagidiid-like mites from Mexico. *Acarologia* 32:341–63.
Lindquist, E. E., M. W. Sabelis, and J. Bruin, eds. 1996. *Eriophyoid mites. Their biology, natural enemies and control.* World Crop Pests, vol. 6. Amsterdam: Elsevier.
Lindquist, E. E., and R. L. Smiley. 1978. *Acaronemus*, a new genus proposed for tarsonemid mites (Acari: Prostigmata) predaceous on tetranychoid mite eggs. *Can. Entomol.* 110:655–62.
Lindquist, E. E., and P. H. Vercammen-Grandjean. 1971. Revision of the chigger-like larvae of the genera *Neotrombidium* Leonardi and *Monunguis* Wharton, with a redefinition of the subfamily Neotrombidiinae Feider in the Trombidiidae (Acarina: Prostigmata). *Can. Entomol.* 103:1557–90.
Lindquist, E. E., and D. E. Walter. 1989. *Antennoseius* (*Vitzthumia*) *janus* n. sp. (Acari: Ascidae), a mesostigmatic mite exhibiting adult female dimorphism. *Can. J. Zool.* 67:1291–1310.
Lindquist, E. E., and K. W. Wu. 1987. First record of the mite family Podocinidae (Acari: Mesostigmata) in Canada, with notes on other records in North America. *Can. Entomol.* 119:779–81.
———. 1991. Review of mites of the genus *Mucroseius* (Acari: Mesostigmata: Ascidae) associated with sawyer beetles (Cerambycidae: *Monochamus* and *Mecynippus*) and pine wood nematodes [Aphelenchoididae: *Bursaphelenchus xylophilus* (Steiner and Buhrer) Nickle], with descriptions of six new species from Japan and North America, and notes on their previous misidentification. *Can. Entomol.* 123:875–927. [MELICHARIDAE]
Lindquist, E. E., and M. Zacharda. 1987. A new genus and species of Rhagidiidae (Acari: Prostigmata) from Chihuahuan Desert litter in New Mexico. *Can. J. Zool.* 65:2149–58.
Lindsay, R. 1972. *Nanorchestes antarcticus* Strandtmann, a taxonomic description of the instars. *Acarologia* 13:593–606.
Lions, J.-C. 1967. La prélarve de *Rhysotritia ardua* (C. L. Koch) 1836 (Acarien, Oribate). *Acarologia* 9:273–83.
———. 1973. Quelques prélarves nouvelles d'Oribates. *Acarologia* 15:356–70.
———. 1978. Éléments sur la distribution verticale des Oribates (Acariens) dans les biotopes édaphiques d'un écosystème forestier. *Rev. Ecol. Biol. Sol* 15:345–62.
Lions, J.-C., and F. Goubière. 1988. Populations adultes et immatures d'*Adoristes ovatus* (Acarien, Oribate) dans les aiguilles de la litière d'*Abies alba*. *Rev. Ecol. Biol. Sol* 25:343–52. [ASTEGISTIDAE]
Lions, J.-C., and R. A. Norton. 1998. North American Synichotritidae (Acari: Oribatida) 2. *Synichotritia spinulosa* and *S. caroli*. *Acarologia* 39 (4): 265–84.
Lipa, J. J., and W. Chmielewski. 1966. Aparity observed in the development of *Caloglyphus* mite (Acarina: Acaridae). *Ekol. Polska*, ser. A, 14:741–48.
Lipovsky, L. J. 1951. A washing method of ectoparasite recovery with particular reference to chiggers. *J. Kans. Entomol. Soc.* 24: 151–56.
———. 1954. Studies of the food habits of postlarval chiggers (Acarina, Trombiculidae). *Univ. Kans. Sci. Bull.* 36:943–58.
Lipovsky, L. J., G. W. Byers, and E. H. Kardos. 1957. Spermatophores, the mode of insemination of chiggers (Acarina: Trombiculidae). *J. Parasitol.* 43:256–62.
Liro, J. I., and H. Roivainen. 1951. Äkämäpunkit (Eriophyoidea) Suomen Eläimet [Eriophyoidea of Finland]. *Anim. Fenn.* 6:1–281. [in Finnish]
Lisazo, N. M. 1979. Un novo acaro da familia Heterozerconidae coletado sobre serpentes brasileiras. Descriçao de *Heterozercon elegans* sp. n. (Acarina, Mesostigmata). *Mem. Inst. Butantan, São Paulo* 42/43:139–44.
Listemann, H. 1984. Besiedelung von Pilzkulturen durch die Raubmilbe *Proctolaelaps pygmaeus* (Müller, 1860). *Mykosen* 27: 615–22. [MELICHARIDAE]
Lister, A. 1984. Predation in an antarctic micro-arthropod community. In Griffiths and Bowman, 2:886–92.
Lister, A., W. Block, and M. B. Usher. 1988. Arthropod predation in an antarctic terrestrial community. *J. Animal Ecol.* 57:957–70.
Livshits, I. Z., and V. I. Mitrofanov. 1973. A new species of the genus

Anoplocheylus (Trombidiformes, Pseudocheylidae) from Crimea. *Zool. Zh.* 52:770–71. [in Russian]

———. 1974. Novyi vid parazititcheskogo kleshcha semeistva Acarophenacidae (Acariformes) iz Uzbekistana [New species of parasitic mite from Uzbekistan]. *Zool. Zh.* 53:288–89. [in Russian]

Livshits, I. Z., V. I. Mitrofanov, and A. A. Sharonov. 1986. Reviziya kleshchei semeistva Siteroptidae Mahunka, 1970 (Acari, Tarsonemina) [Revision of mites of family Siteroptidae]. *Trudy Gos. Nikitsk. Bot. Sad* 99:7–31. [in Russian]

Lo, K-C., and C-C. Ho. 1979. Ecological observations on rice tarsonemid mite, *Steneotarsonemus spinki* (Acarina: Tarsonemidae). *J. Agr. Res. China* 28:181–92. [in Chinese, English summary]

Lo, P. K-C. 1990. A new genus and species of mite (Acari: Podapolipidae) associated with the locust, *Patanga succincta* Linné (Orthoptera: Acrididae). *Bull. Inst. Zool., Acad. Sin.* 29:89–93.

Lofego, A. C., R. Ochoa, and G. J. Moraes. 2005. Some tarsonemid mites (Acari: Tarsonemidae) from the Brazilian "cerrado" vegetation, with descriptions of three new species. *Zootaxa* 823:1–27.

Lohmann, H. T. 1893. Der Halacarinen der Plankton-Expedition. *Ergebn. Plankton-Exp.* 2:11–95.

Lombardini, G. 1939. *Coleoglyphus ultrajubae* Lomb. *L'Agricoltura Coloniale* 33:294–96. [CANESTRINIIDAE]

———. 1951. Acari nuovi. *Redia* 36:245–50. [DIARTHROPHALLIDAE]

Lombert, H. A. P. M., and F. S. Lukoschus. 1979. Parasites of Western Australia. VII, Observations on quill wall mites (Acarina: Astigmata). *Rec. W. Aust. Mus.* 7:265–86. [LAMINOSIOPTIDAE]

Lombert, H. A. P. M., F. S. Lukoschus, and A. Fain. 1984. *Apocnemidocoptes tragicola* gen. nov., spec. nov. (Acari: Astigmata: Knemidokoptidae) from the swift *Apus apus* (Aves: Apodiformes: Apodidae). *Acarologia* 25:377–83.

Loomis, R. B. 1956. The chigger mites of Kansas (Acarina: Trombiculidae). *Univ. Kans. Sci. Bull.* 37:1195–1443.

Loomis, R. B., and W. J. Wrenn. 1984. Systematics of the pest chigger genus *Eutrombicula* (Acari: Trombiculidae). In Griffiths and Bowman, 1:152–59.

Loos-Frank, B., and M. Abel. 1983. Lung mites (*Pneumocoptes* sp.) in bank voles (*Clethrionomys glareolus*) in southwest Germany. *Z. Parasitenkd.* 69:539–46. [PNEUMOCOPTIDAE]

Loots, G. C., and P. D. Theron. 1992. *Adhaerenseius floralis*, a new ascid genus and species (Acari: Parasitiformes: Ascidae) associated with *Poellnitzia rubriflora* (L. Bol) Uitewaal in the Western Cape Province of South Africa. *J. Entomol. Soc. S. Afr.* 55:39–49. [BLATTISOCIIDAE]

Lopez-Campos, M. G. 1996. Especie nueva de *Saxidromus* (Prostigmata: Adamystidae) del Estado de Veracruz, Mexico. *An. Inst. Biol., Univ. Nac. Aut. Méx., Ser. Zool.* 67:213–23.

Lourenço, W. R. 1982. Notes sur quelques acariens parasites de scorpions. *Acarologia* 23:245–47. [TROMBICULIDAE]

Lowman, M. D., R. L. Kitching, and G. Carruthers. 1996. Arthropod sampling in Australian subtropical rain forests: how accurate are some of the more common techniques? *Selbyana* 17: 36–42.

Löyttyniemi, K. 1969. A *Nalepella* species (Acarina, Eriophyidae) damaging needles of spruce (*Picea abies* (L.) Karst.). *Ann. Entomol. Fenn.* 35:123–24.

Ludwig, M., M. Kratzmann, and G. Alberti. 1991. Accumulation of heavy metals in two oribatid mites. In Dusbábek and Bukva, 1:431–37.

Lukoschus, F. S., A. W. A. M. de Cock, and A. Fain. 1971. Life cycle of *Melesodectes auricularis* Fain & Lukoschus (Glycyphagidae, Sarcoptiformes). *Tijdschr. Entomol.* 114:173–83.

Lukoschus, F. S., A. Fain, and M. M. J. Beaujean. 1967. Beschreibung neuer *Psorergates* Arten (Psorergatidae: Trombidiformes). *Tijdschr. Entomol.* 110:133–81.

Lukoschus, F. S., A. Fain, and F. M. Driessen. 1972. Life cycle of *Apodemopus apodemi* (Fain, 1965) (Glycyphagidae: Sarcoptiformes). *Tijdschr. Entomol.* 115:325–39.

Lukoschus, F. S., G. H. S. Janssen Duijghuijsen, and A. Fain. 1979. Parasites of Western Australia. IV, Observations on the genus *Marsupiopus* Fain, 1968 (Acarina: Astigmata: Glycyphagidae). *Rec. W. Aust. Mus.* 7:37–55. [ECHIMYOPODIDAE]

Lukoschus, F. S., J. M. W. Louppen, and T. C. Maa. 1974. *Psorergates squamipes*, n. sp. (Acari: Psorergatidae), a skin mite from *Anourosorex squamipes* (Insectivora) in Taiwan. *Pac. Insects* 16:51–56. [PSORERGATIDAE]

Lukoschus, F. S., G. Scheperboer, A. Fain, and M. Nadchatram. 1981. Life cycle of *Alabidopus asiaticus* sp. nov. (Acarina: Astigmata: Glycyphagidae) and hypopus of *Alabidopus malaysiensis* sp. nov. ex *Rattus* spp. from Malaysia. *Int. J. Acarol.* 7:161–77. [CHORTOGLYPHIDAE]

Lukoschus, F. S., A. G. Woeltjes, C. E. Desch, and W. B. Nutting. 1980. Parasitic mites of Surinam. XXXV, Two new *Ophthalmodex* spp. (*O. carolliae*, *O. molossi*: Demodicidae) from the bats *Carollia perspicillata* and *Molossus molossus*. *Internat. J. Acarol.* 6:45–50.

Lundblad, C. O. 1927a. Die Hydracarinen Schwedens. I, Beitrag zur Systematik, Embryologie, Ökologie und Verbreitungsgeschichte der schwedischen Arten. *Zool. Bidr. Uppsala* 11:181–540.

———. 1927b. Zur Kenntnis der Hydracarinenfauna der Mount Elgongebiets im britischen Ostafrika. *Arch. Hydrobiol.* 18: 361–441.

———. 1929a. Einiges über die Kopulation bei *Aturus scaber* und *Midea orbiculata*. *Zeitschr. Morphol. Ökol. Tiere* 15:474–80. [ATURIDAE, MIDEIDAE]

———. 1929b. Über den Begattungsvorgang bei einigen *Arrhenurus*-Arten. *Zeitschr. Morphol. Ökol. Tiere* 15:705–22. [ARRENURIDAE]

———. 1941. *Die Hydracarinenfauna Südbrasiliens und Paraguays.* Vol. I, ser. 3, 19 (7). Stockholm: Svensk. Vetenskapsakad. Handling.

———. 1942. *Die Hydracarinenfauna Südbrasiliens und Paraguays.* Vol. II, ser. 3, 20 (2). Stockholm. Svensk. Vetenskapsakad. Handling.

———. 1943a. *Die Hydracarinenfauna Südbrasiliens und Paraguays.* Vol. III, ser. 3, 20 (5). Stockholm: Svensk. Vetenskapsakad. Handling.

———. 1943b. *Die Hydracarinenfauna Südbrasiliens und Paraguays.* Vol. IV, ser. 3, 20 (8). Stockholm: Svensk. Vetenskapsakad. Handling.

———. 1944. *Die Hydracarinenfauna Südbrasiliens und Paraguays.* Vol. 5, ser. 3, 20 (13). Stockholm: Svensk. Vetenskapsakad. Handling.

———. 1952. Hydracarinen von den ostafrikanischen Gebirgen. *Ark. Zool.*, ser. 2, 3: 391–525.

———. 1962. Die Hydracarinen Schwedens. II. *Ark. Zool.* 14.

———. 1968. Die Hydracarinen Schwedens. III. *Ark. Zool.* 21.

———. 1969. Indische Wassermilben, hauptsächlich von Hinterindien. *Ark. Zool.* 22:289–443.

———. 1971. Weitere Beiträge zur Kenntnis der Fließwassermilben Javas. *Ark. Zool.* 23:293–359.

Lundqvist, L. 1987. Bibliografi och checklist over Sveriges oribatider (Acari: Oribatei), 1941–1985. *Entomol. Tidskr.* 108:3–12.

———. 1998. Taxonomic revision of the genus *Dinogamasus* Kramer (Acari: Mesostigmata: Laelapidae). *Entomol. Scand. Suppl.* 54:1–109.

Lundström, A. N. 1887. Planzenbiologische Studien. II. Die Anpassungen der Planzen an Thiere. *Nova Acta Reg. Soc. Sci. Upsal.* Ser. 3, 13.

Luxton, M. 1968. A new genus and species of littoral mite (Acari: Mesostigmata) from New Zealand. *New Zeal. J. Marine Freshwater Res.* 2:497–504.

———. 1972. Studies on the oribatid mites of a Danish beech wood soil. I, Nutritional biology. *Pedobiologia* 12:434–63.

———. 1973. Mites of the genus *Cryptognathus* from Australia, New Zealand and Niue Island. *Acarologia* 15:53–75. [CRYPTOGNATHIDAE]

———. 1981. Studies on the oribatid mites of a Danish beech wood soil. IV, Developmental biology. *Pedobiologia* 21:312–40.

———. 1985. Cryptostigmata: A concise review. *Fauna New Zeal.* 7:1–106.

———. 1988. A new mite superfamily (Acari: Cryptostigmata). *Zool. J. Linn. Soc.* 93:71–91.

———. 1989. New taxa of intertidal mites (Acari). *J. Nat. Hist.* 23:407–28.

———. 1990. The marine littoral mites of the New Zealand region. *J. Roy. Soc. New. Zeal.* 20:367–418.

———. 1991. Seasonal and spatial variation in food intake by the oribatid mites of beech woodland soil. In Schuster and Murphy, 459–471.

———. 1992. Oribatid mites from the marine littoral of Hong Kong (Acari: Cryptostigmata). In *Proc. IV International Marine Biology Workshop: The marine flora and fauna of Hong Kong and Southern China III*, ed. B. Morton, 211–27. Hong Kong: Hong Kong Univ. Press.

———. 1993. New cryptognathid mites from Israel (Acari: Prostigmata: Cryptognathidae). *J. Nat. Hist.* 27:1213–17.

———. 1995. Patterns of food intake by some astigmatic mites of beech woodland soil (Acari: Astigmata). *Pedobiologia* 39:238–42. [ACARIDAE, GLYCYPHAGIDAE]

———. 1996. Oribatid mites of the British Isles: a checklist and notes on biogeography (Acari, Oribatida). *J. Nat. Hist.* 30 (6): 803–22.

Lyon, W. F. 1973. A plant-feeding mite *Mononychellus tanajoa* (Bondar) (Acarina: Tetranychidae) new to the African continent threatens cassava (*Manihot esculenta* Crantz) in Uganda, East Africa. *Pest. Artic. News Summ.* 19:36–37.

Macfadyen, A. 1953. Notes on methods for the extraction of small soil arthropods. *J. Anim. Ecol.* 22:65–77.

Machado-Allison, C. E. 1967. The systematic position of the mites of *Desmodus* and *Chilonycteris* based on host-parasite relationships. *Proc. Biol. Soc. Wash.* 80:223–26.

MacLean, S. F., V. Behan, and A. Fjellberg. 1978. Soil Acari and Collembola from Chaun Bay, northern Chukotka. *Arctic Alpine Res.* 10:559–68.

MacQuitty, M. 1984. The feeding behaviour of two species of *Agauopsis* (Halacaroidea) from California. In Griffiths and Bowman, 1:571–80.

MacRae, I. V., and B. A. Croft. 1996. Differential impact of egg predation by *Zetzellia mali* (Acari: Stigmaeidae) on *Metaseiulus occidentalis* and *Typhlodromus pyri* (Acari: Phytoseiidae). *Exp. Appl. Acarol.* 20:143–54.

Mader, D. R., R. S. Houston, and F. L Frye. 1986. *Hirstiella trombidiiformis* infestation in a colony of chuckwallas. *J. Amer. Vet. Med. Assoc.* 189:1138–39.

Madge, D. S. 1964. The humidity reactions of oribatid mites. *Acarologia* 6:566–91.

Maes, K. 1983. Scientific report of the Belgian Mount Cameroon expedition 1981. VIII, Description of four new species of the genus *Julolaelaps* Berl. (Acarina: Laelaptidae). *Rev. Zool. Afr.* 97:211–20. [IPHIOPSIDIDAE]

Magowski, W. 1986. Two new species of *Tarsonemus* (Acari: Prostigmata) associated with *Xylocopa* carpenter bees (Hymenoptera: Apoidea). *Entomol. Mitt. Zool. Mus. Hamb.* 8:289–306.

———. 1988. Description of a new species of *Formicomotes* Sevastianov, 1980 (Acari: Dolichocybidae) with notes on the female dimorphism within this genus. *Mitt. Hamb. Zool. Mus. Inst.* 85:163–82.

———. 1994. Discovery of the first representative of the mite subcohort Heterostigmata in the Mesozoic Siberian amber. *Acarologia* 35:229–41. [ACAROPHENACIDAE]

———. 1997. *Rafamastix plenus* gen. n., sp. n. of the family Dolichocybidae from South America. *Genus (Warsaw)* 8:61–74.

Magowski, W. L., and A. A. Khaustov. 2006a. Two new species of *Tarsonemus* (Acari: Tarsonemidae) associated with bark beetles (Coleoptera: Scolytidae) from Crimea, Ukraine. *Zootaxa* 1359:31–55.

———. 2006b. *Ununguitarsonemus tremulae*, a new species of tarsonemid mites (Acari: Heterostigmata) from Crimea, Ukraine. *Acta Zool. Acad. Sci. Hung.* 52:21–33.

Magowski, W., and J. C. Moser. 1993. Three new species of the genus *Acanthomastix* Mahunka, 1972 from United States and Poland (Acari: Dolichocybidae). *Entomol. Mitt. Zool. Mus. Hamb.* 11 (147): 1–20.

Mahunka, S. 1965. Identification key for the species of the family Scutacaridae (Acari: Tarsonemini). *Acta Zool. Acad. Sci. Hung.* 11:353–401.

———. 1969a. *Imparipes* (*I.*) *eickworti* sp. n., a new scutacarid mite (Acari: Tarsonemina) from *Dialictus umbripennis* Ellis (Hym.) *Parasitol. Hung.* 2:153–58.

———. 1969b. The scientific results of the Hungarian soil zoological expeditions to South America. 13, Acari: Pygmephoridae and Scutacaridae from the material of the second expedition (Brazil and Bolivia). *Acta Zool. Acad. Sci. Hung.* 15:333–70.

———. 1970a. Considerations on the systematics of the Tarsonemina and description of new European taxa (Acari: Trombidiformes). *Acta Zool. Acad. Sci. Hung.* 16:137–74.

———. 1970b. *Dudichiana foveolata* gen. nov., sp. nov., sowie einige andere neue und interessante Milbenarten aus Ameisenhaufen (Acari: Tarsonemina). *Opusc. Zool. Budapest* 10:143–49. [PYGMEPHORIDAE]

———. 1970c. Ergebnisse der zoologischen Forschungen von Dr. Z. Kaszab in der Mongolei. 227, Acari: Pygmephoroidea. *Ann. Hist.– Nat. Mus. Nat. Hung., Zool.* 62:343–62. [SITEROPTIDAE]

———. 1970d. The scientific results of the Hungarian soil zoological expeditions to South America. 21, Acari: Tarsonemine species from Brazil. *Acta Zool. Acad. Sci. Hung.* 16:371–408. [PYGMEPHORIDAE]

———. 1970e. *Sphexicozela connivens* gen. n., sp. n. (Acari, Acaridoidea): A new mite from a wasp nest. *Parasitol. Hung.* 3:77–86. [WINTERSCHMIDTIIDAE]
———. 1970f. Two new scutacarid mites (Acari, Tarsonemina) from *Pogonomyrmex occidentalis* (Hymenoptera) in the United States of America. *Parasitol. Hung.* 3:87–96.
———. 1971. Tarsonemina (Acari) species from India. The scientific results of Dr. Gy. Topál's collectings in India. 4. *Acta Zool. Acad. Sci. Hung.* 17:11–49. [DOLICHOCYBIDAE]
———. 1972a. The first survey of the tarsonemid (Acari) fauna of New Guinea. I. *Acta Zool. Acad. Sci. Hung.* 18:41–92. [DOLICHOCYBIDAE, MICRODISPIDAE]
———. 1972b. Neue und interessante Milben aus dem Genfer Museum III. *Zambedania* gen. nov. und zwei neue Milben-arten aus Rhodesien (Acari: Tarsonemina). *Mitt. Schweiz. Entomol. Gesell.* 45:151–55.
———. 1972c. Tetüatkák—Tarsonemina. *Magy. Allatvilaga Fauna Hung.* 110, Kotet 18 (Arachnoidea), Fuzet 16:1–215.
———. 1973a. Aus insekten lebende Milben (Acari: Acarida, Tarsonemida) aus Afrika. II. *Acta Zool. Acad. Sci. Hung.* 19:289–337. [DOLICHOCYBIDAE]
———. 1973b. Neue Tarsonemiden (Acari) aus der Mongolei. *Ann. Hist.–Nat. Mus. Nat. Hung.* 65:309–15. [ACAROPHENACIDAE]
———. 1973c. *Pygmephorus* species (Acari, Tarsonemida) from North American small mammals. *Parasitol. Hung.* 6:247–59.
———. 1974a. Auf insekten lebende Milben (Acari: Acarida, Tarsonemida) aus Afrika. IV. *Acta Zool. Acad. Sci. Hung.* 20: 367–402. [DOLICHOCYBIDAE]
———. 1974b. Beiträge zur Kenntinis der an Hymenopteren lebenden Milben (Acari), I. *Ann. Hist.–Nat. Mus. Nat. Hung.* 66:389–94. [PYGMEPHORIDAE]
———. 1974c. *Daidalotarsonemus hewitti* sp. n. (Acari: Tarsonemida) from human skin in England. *Parasitol. Hung.* 7: 197–200.
———. 1975a. Äthiopische Tarsonemiden (Acari: Tarsonemida). I. *Acta Zool. Acad. Sci. Hung.* 21:369–410. [MICRODISPIDAE, SCUTACARIDAE]
———. 1975b. Auf Insekten lebende Milben (Acari: Acarida und Tarsonemida) aus Afrika. V. *Acta Zool. Acad. Sci. Hung.* 21: 39–72. [DOLICHOCYBIDAE]
———. 1975c. Beiträge zur Kenntnis der Tarsonemiden (acari) von Kleinsäugernestern aus der Umgebung von Ljubljana (Jugoslawien). *Parasitol. Hung.* 8:75–83. [PYGMEPHORIDAE]
———. 1975d. Further data to the knowledge of Tarsonemida (Acari) living on small mammals in North America. *Parasitol. Hung.* 8:85–94.
———. 1975e. Neue und auf Insekten lebenden Milben aus Australien und Neu-Guinea (Acari: Acarida, Tarsonemida). *Ann. Hist.–Nat. Mus. Nat. Hung.* 67:317–25. [ACAROPHENACIDAE, SCUTACARIDAE]
———. 1975f. Über die Gattung *Resinacarus* Vitzthum, 1927 und Resinacarinae subfam. n. (Acari: Pyemotidae). *Opusc. Zool.* 133:1–6.
———. 1977a. The examination of myrmecophilous tarsonemid mites based on the investigations of Dr. C. W. Rettenmeyer (Acari). I. *Acta Zool. Acad. Sci. Hung.* 23:99–132. [MICRODISPIDAE, SCUTACARIDAE]
———. 1977b. The examination of myrmecophilous tarsonemid mites based on the investigations of Dr. C. W. Rettenmeyer (Acari). II. *Acta Zool. Acad. Sci. Hung.* 23:341–70. [MICRODISPIDAE, SCUTACARIDAE]
———. 1977c. Neue und interessanten Milben aus dem Genfer Museum. XIX, Einige Angaben zur Kenntnis der Milbenfauna der Ameisen-Nester (Acari: Acarida, Tarsonemida). *Arch. Sci., Genève* 30:91–106. [MICRODISPIDAE, SCUTACARIDAE]
———. 1978. Schizoglyphidae fam. n. and new taxa of Acaridae and Anoetidae (Acari: Acarida). *Acta Zool. Acad. Sci. Hung.* 24: 107–31.
———. 1979. Some remarks on the polymorphism (phoretomorphism) in tarsonemid mites. *Folia Entomol. Hung.* 32:133–37. [SITEROPTIDAE, PYGMEPHORIDAE]
———. 1980. *Parapygmephorus delyorum* sp. n., eine neue Art aus Korea (Acari: Tarsonemina). *Parasitol. Hung.* 13:95–98. [PYGMEPHORIDAE]
———. 1981. Tarsonemiden aus Äthiopien (Acari: Tarsonemina). *Folia Entomol. Hung.* 42:101–21. [ACAROPHENACIDAE]
———. 1982. Oribatids from eastern part of the Ethiopian region (Acari). I. *Acta Zool. Acad. Sci. Hung.* 28 (3–4): 293–336.
———. 1983. *Strephocheir paolettii* gen. et sp. nov. (Acari: Pygmephoridae) from Italy. *Parasitol. Hung.* 16:125–29.
———. 1984. Oribatids from eastern part of the Ethiopian region (Acari). VI. *Acta Zool. Hung.* 30 (3–4): 393–444.
———. 1985. Oribatids from Africa (Acari: Oribatida) II. *Folia Entomol. Hung.* 46 (1): 73–113.
———. 1986. A survey of the family Carabodidae C. L. Koch, 1836 (Acari: Oribatida). *Acta Zool. Hung.* 21:73–135.
———. 1987. A survey of the family Carabodidae C. L. Koch, 1836 (Acari: Oribatida). II. *Acta Zool. Hung.* 33:399–434.
———. 1990. A survey of the superfamily Euphthiracaroidea Jacot, 1930 (Acari: Oribatida). *Folia Entomol. Hung.* 51:37–80.
———. 1992a. Neue und interessante Milben aus dem Genfer Museum. XXXVII, Neue Milben von den pazifischen Inseln (Acari: Anoetidae, Scutacaridae). *Rev. Suisse Zool.* 99:395–403.
———. 1992b. New and interesting mites from the Geneva Museum. 63, A survey of the oribatid fauna of Senegal (Acari, Oribatida). *Rev. Suisse Zool.* 99 (3): 673–712.
———. 1993. Oribatids from Madagascar I. (Acari: Oribatida). New and interesting mites from the Geneva Museum LXXVI. *Rev. Suisse Zool.* 100:289–315.
———. 1994. Oribatids from Madagascar II. (Acari: Oribatida) (New and interesting mites from the Geneva Museum LXXIX). *Rev. Suisse Zool.* 101 (1): 47–88.
———. 2001. Oribatids from Brunei III (Acari: Oribatida). (Acarologia Genavensia XCI). *Rev. Suisse Zool.* 108 (2): 317–49.
Mahunka, S., and A. Fain. 1989. New mite species of the cohort Tarsonemina (Acari: Acarophenacidae and Pygmephoridae). *Parasitol. Hung.* 22:125–35.
Mahunka, S., and L. Mahunka-Papp. 1994. Two new microdispid (Acari, Heterostigmata: Microdispidae) species from Hungary. *Ann. Hist.–Nat. Mus. Nat. Hung.* 86:119–24.
———. 1995. *The oribatid species described by Berlese (Acari).* Budapest: Hung. Nat. Hist. Mus.
———. 2000. Checklist of the oribatid mites of Hungary (Acari: Oribatida). *Folia Entomol. Hung.* 61:27–53.
———. 2001. Oribatids from Switzerland V (Acari: Oribatida: Suctobelbidae 2) (Acarologica Genavensia XCVII). *Rev. Suisse Zool.* 108:355–85.
———. 2004. *A catalogue of the Hungarian oribatid mites (Acari:*

Oribatida). Pedozoologica Hungarica No 2. Budapest: Hungarian Natural History Museum.
Mahunka, S., and G. Rack. 1977. Zwei neue Arten der Familien Acarophenacidae und Pygmephoridae (Acarina, Tarsonemida). *Ann. Hist.– Nat. Mus. Nat. Hung.* 69:305–9.
Mahunka, S., and A. M. Zaki. 1990. *Acarophenax rackae* sp. n., a new mite species from Egypt (Acari, Tarsonemina: Acarophenacidae). *Parsitol. Hung.* 23:121–27.
Majer, J. D., and H. F. Recher. 1988. Invertebrate communities on Western Australian eucalypts: a comparison of branch clipping and chemical knockdown procedures. *Aust. J. Ecol.* 13:269–78.
Makarova, O. L. 2000a. To a study of mites of the genus *Arctoseius* (Parasitiformes, Ascidae) from the far north. 2, Description of *A. productus* sp.n. and *A. babenkoi* sp.n. and key to high arctic species. *Zool. Zh.* 79:800–817. [in Russian; English translation in *Entomol. Rev.* 80 (Suppl.) 1:S131–42.]
———. 2000b. To studying mites of the genus *Arctoseius* (Parasitiformes, Ascidae) of the far north. 1, Morphometric analysis of taxonomic features of arctic species and description of *Arctoseius tschernovi* sp. n. and *A. miranalis* sp. n. *Zool. Zh.* 79:800–817. [in Russian]
———. 2002. Acarocenoses (Acariformes, Parasitiformes) in polar deserts. 1, Mite assemblages in the Severanaya Zemlya Archipelago. Structure of fauna and abundance. *Zool. Zh.* 81:165–81. [in Russian]
———. 2003. New genus and three new species of mites of the family Arctacaridae (Parasitiformes: Mesostigmata) from North America. *Zool. Zh.* 82:919–36. [in Russian]
———. 2004. Gamasid mites (Parasitiformes, Mesostigmata), dwellers of bracket fungi, from the Pechora-Ilychskii Reserve (Republic of Komi). *Zool. Zh.* 83:1335–40. [in Russian; English translation in *Entomol. Rev.* 84:667–72]
Maki, D. L., W. T. Wilson, J. Vargas C., R. L. Cox, and H. Devar Petersen. 1988. Effect of *Acarapis woodi* infestation on honeybee longevity. In *Africanized honey bees and bee mites*, ed. G. R. Needham, R. E. Page, Jr., M. Delfinado-Baker, and C. E. Bowman, 512–17. Chichester, UK: Ellis Horwood. [TARSONEMIDAE]
Mąkol, J. 2000. Description of larva of *Podothrombium filipes* (C. L. Koch, 1837) (Acari: Actinotrichida: Trombidiidae) with notes on variability, anomaly and their implications for classification of *Podothrombium* larvae. *Ann. Zool. (Warsaw)* 50:347–61.
———. 2001. A redescription of *Trombidium geniculatum* (Feider, 1955) (Acari: Actinotrichida: Trombidioidea) with characteristics of all active instars. *Ann. Zool. (Warsaw)* 51:251–60. [TROMBIDIIDAE]
———. 2005. *Trombidiidae (Acari: Actinotrichida: Trombidioidea) of Poland.* Fauna Poloniae, Mus. Inst. Zool., n.s., 1. Warsaw: Polish Acad. Sci. & Natura Optima Dux Found.
———. 2007. *Generic level review and phylogeny of Trombidiidae and Podothrombiidae (Acari: Actinotrichida: Trombidioidea) of the world.* Ann. Zool. (Warsaw) 57.
Mąkol, J., and G. Gabryś. 2002. A redescription of *Sucidothrombium sucidum* (L. Koch, 1879) (Acari: Actinotrichida: Microtrombidiidae) with characteristics of all active instars. *Ann. Zool. (Warsaw)* 52:443–54.
Mąkol, J., and Y. Marusik. 1999. Trombidiidae (Acari: Actinotrichida, Trombidioidea) of Kuril Islands. Part I, Podothrombiinae Thor, 1935. *Ann. Zool. (Warsaw)* 49:329–46.
Mąkol, J., and A. Wohltmann. 2000. A redescription of *Trombidium holosericum* (Linnaeus, 1758) (Acari: Actinotrichida: Trombidioidea) with characteristics of all active instars and notes on taxonomy and biology. *Ann. Zool. (Warsaw)* 50:67–91. [TROMBIDIIDAE]
Mallow, D., D. F. Ludwig, M. J. Hayes, and D. A. Crossley, Jr. 1984. Habitat selection of postlarval *Eutrombicula alfreddugesi* and *Eutrombicula splendens* from eight microhabitats in Georgia, U.S.A. (Trombiculidae: Acarina). *J. Georgia Entomol. Soc.* 19:543–48.
Mani, M. S. 1964. *Ecology of plant galls.* Monogr. Biol. 12. The Hague: W. Junk. [ERIOPHYIDAE]
Manson, D. C. M. 1974. A new genus and species of mesostigmatid mites (Acarina: Ameroseiidae) associated with honey bees from Tonga. *J. Roy. Soc. New Zeal.* 4:115–22.
Manson, D. C. M., and U. Gerson. 1996. Web spinning, wax secretion and liquid secretion by eriophyoid mites. In Lindquist, Sabelis, and Bruin, 6:251–58.
Manson, D. C. M., and G. N. Oldfield. 1996. Life forms, deuterogyny, diapause and seasonal development. In Lindquist, Sabelis, and Bruin, 6:173–83.
Manton, S. M. 1977. *The arthropoda: Habits, functional morphology and evolution.* Oxford, UK: Clarendon Press.
Marais, J. F., and G. C. Loots. 1969. *Evansolaelaps*, a new genus of Laelapinae (Acari: Mesostigmata) from the Congo and Angola. *Rev. Zool. Bot. Afr.* 53:281–307.
Maraun, M., M. Heethoff, S. Scheu, R. A. Norton, G. Weigmann, and R. H. Thomas. 2003. Radiation in sexual and parthenogenetic oribatid mites (Oribatida, Acari) as indicated by genetic divergence of closely related species. *Exp. Appl. Acarol.* 29: 265–77.
Maraun, M., M. Heethoff, K. Schneider, S. Scheu, G. Weigmann, J. Cianciolo, R. H. Thomas, and R. A. Norton. 2004. Molecular phylogeny of oribatid mites (Oribatida, Acari): Evidence for multiple radiations of parthenogenetic lineages. *Exp. Appl. Acarol.* 33:183–201.
Maraun, M., S. Migge, M. Schaefer, and S. Scheu. 1998. Selection of microfungal food by six oribatid mite species (Oribatida, Acari) from two different beech forests. *Pedobiologia* 42:232–40.
Mari, M., and I. Morselli. 1987. Acarians (Halacaridae and Hydrachnellae) and Kinorhyncha of the sponge *Hymeniacidon sanguinea* Grant of the Lagoon of Venice, Italy. *Atti Soc. Nat. Mat. Modena* 118:115–17. [UNIONICOLIDAE]
Marshall, D. J., and L. Coetzee. 2000. Historical biogeography and ecology of a continental antarctic mite genus, *Maudheimia* (Acari, Oribatida): Evidence for a Gondwanan origin and Pliocene-Pleistocene speciation. *Zool. J. Linn. Soc.* 129:111–28. [MAUDHEIMIIDAE]
Marshall, D. J., B. M. OConnor, and P. J. A. Pugh. 2003. *Algophagus* mites (Astigmata: Algophagidae) from the sub-antarctic Prince Edward Islands: Habitat-related morphology and taxonomic descriptions. *J. Zool. Lond.* 259:31–47.
Marshall, D. J., and P. J. A. Pugh. 2002. *Fortuynia* (Acari: Oribatida: Ameronothroidea) from the marine littoral of southern Africa. *J. Nat. Hist.* 36:173–83. [FORTUYNIIDAE]
Marshall, V. G. 1970. Tydeid mites (Acarina: Prostigmata) from Canada. I, New and redescribed species of *Lorryia*. *Ann. Soc. Entomol.* 15:17–52.
Marshall, V. G., and E. E. Lindquist. 1972. Notes on the genus *Nalepella* (Acarina: Eriophyoidea) and the occurrence of *N. halourga* on black spruce in Canada. *Can. Entomol.* 104:239–44.

Marshall, V. G., R. M. Reeves, and R. A. Norton. 1987. *Catalog of the Oribatida (Acari) of continental United States and Canada.* Mem. Entomol. Soc. Can. 139. Ontario: Entomol. Soc. Can.

Martikainen, E., and V. Huhta. 1990. Interactions between nematodes and predatory mites in raw humus soil: A microcosm experiment. *Rev. Écol. Biol. Sol* 27:13–20.

Martin, N. A. 1978. *Siteroptes* (*Siteroptoides*) with *Pediculaster*-like phoretomorphs (Acari: Tarsonemida; Pygmephoridae) from New Zealand and Polynesia. *New Zeal. J. Zool.* 5:121–55. [PYGMEPHORIDAE]

Martin, P. 2000. Larval morphology and host-parasite associations of some stream living water mites (Hydrachnidia, Acari). *Arch. Hydrobiol., Suppl.* 121:269–320.

———. 2003. Larval morphology of spring-living water mites (Hydrachnidia, Acari) from the Alps. *Ann. Limnol.–Internat. J. Limnol.* 39:363–93.

Martin, P., and C. Davids. 2002. Life history strategies of *Hygrobates nigromaculatus*, a widespread Palaearctic water mite (Acari, Hydrachnidia, Hygrobatidae). In Bernini et al. (2002), 101–10.

Martin, P., and J. Schwoerbel. 2002. *Thermacarus andinus* n. sp., a South American water mite (Acari: Hydrachnidia: Thermacaridae) with a remarkable host-parasite association. *Zool. Anz.* 241:67–79.

Martinez, P. A., and V. Bernava-Laborde. 2000. *Gehypochthonius marianoi* n. sp. (Acari: Oribatida), from sand dunes in coastal Argentina. *Acarologia* 41:381–89. [GEHYPOCHTHONIIDAE]

Martinez, P. A., G. J. Velis, M. J. Eguaras, and N. A. Fernandez. 1995. La famille Lamellareidae dans la République Argentine. *Tenuelamellarea argentinensis* n. sp. *Acarologia* 36 (4): 355–61.

Mašán, P. 1994. The eviphidid mites (Acarina: Mesostigmata; Eviphididae) associated with scarabaeid and carrion beetles (Coleoptera: Scarabaeidae, Silphidae) in central Europe. *Acarologia* 35:3–19.

Mašán, P. 2003. Macrochelid mites of Slovakia (Acari, Mesostigmata, Macrochelidae). Inst. Zool., Slovak Acad. Sci., Bratislava.

Mašán, P. and P. Fenďa. 2004. Zerconid mites of Slovakia (Acari, Mesostigmata, Zerconidae). Inst. Zool., Slovak Acad. Sci., Bratislava.

Mašán, P., and D. E. Walter. 2004. Description of the male of *Hoploseius mariae* (Acari, Mesostigmata), an European ascid mite associated with wood-destroying fungi, with key to *Hoploseius* species. *Biologia, Bratislava* 59:527–32.

Martinez, P. A., and G. J. Velis. 2000. Listado de los oribátidos (Acari: Oribatida) de la República Argentina. *Rev. Soc. Entomol. Arg.* 59:119–34.

Materna, J. 2000. Oribatid communities (Acari: Oribatida) inhabiting saxicolous mosses and lichens in the Krkonose Mts. (Czech Republic). *Pedobiologia* 44:40–62.

Mathieson, B. R. F., and J. Lehane. 1996. Some preliminary observations on the digestive system of *Psoroptes ovis* (Raillet) (Psoroptidae), the sheep scab mite. In Mitchell et al. (1996), 1:389–93.

Mathur, S. N., and E. J. LeRoux. 1965. A description of the mouthparts and feeding mechanism of the velvet mite, *Allothrombium lerouxi* Moss (Trombidiformes: Trombididae). *Ann. Soc. Entomol. Québec* 1965:33–61.

Mathys, G., and Y. Tencalla. 1959. Note préliminaire sur la biologie et la valeur prédatrice de *Proctolaelaps hypudaei* Oudms (Acarien, Mesostigmata, Aceosejidae). *Ann. Agr. Suisse* 60:645–54. [MELICHARIDAE]

Matsubara, M., Y. Chinzei, and J. Aoki. 1996. Two cases of infestation of the conjunctival sac by soil mite. *Jpn. Rev. Clin. Ophthamol.* 90:452–54.

Matthes, H. F. 1996. New aspects of pathogenesis in cattle demodecosis. R. D. In Mitchell et al., 93–96.

Mautz, D., W. Hirschmann, and F. Kemnitzer. 1986. The embryonic development of *Varroa jacobsoni* Oudemans 1904 (Varroinae, Mesostigmata). *Acarologia* 27:203–7.

McAloon, F. M., and L. A. Durden. 2000. Attachment sites and frequency distribution of erythraeid mites, *Leptus indianensis* (Acari: Prostigmata), ectoparasitic on harvestmen, *Leiobunum formosum* (Opiliones). *Exp. Appl. Acarol.* 24:561–67.

McBrayer, J. F. 1977. Contributions of Cryptozoa to forest nutrient cycles. In *Proceedings in life sciences. The role of arthropods in forest ecosystems*, ed. W. J. Mattson, 70–77. New York: Springer-Verlag.

McClure, M. S. 1995. *Diapterobates humeralis* (Oribatida: Ceratozetidae): An effective control agent of hemlock woolly adelgid (Homoptera: Adelgidae) in Japan. *Environ. Entomol.* 24:1207–15.

McCoy, C. W. 1996. Stylar feeding injury and control of eriophyoid mites in citrus. In Lindquist, Sabelis, and Bruin, 6:513–26.

McDaniel, B. 1968. The superfamily Listrophoroidea and the establishment of some new families (Listrophoroidea: Acarina). *Acarologia* 10:477–82.

McDaniel, B., and E. G. Bolen. 1981. A new genus and two new species of Nanorchestidae from Padre Island, Texas (Acari: Prostigmata). *Acarologia* 22:253–56.

———. 1983. A new species of *Adamystis* (Acari: Adamystidae) from the southern high plains of Texas. *Proc. Entomol. Soc. Wash.* 85:177–79.

McDaniel, B., D. K. Morihara, and J. K. Lewis. 1975. A new species of *Tuckerella* from South Dakota and a key with illustrations of all known described species. *Acarologia* 17:274–83. [TUCKERELLIDAE]

———. 1976. The family Teneriffiidae Thor, with a new species from Mexico. *Ann. Entomol. Soc. Amer.* 69:527–37.

McEnroe, W. D. 1969. Eyes of the two-spotted spider mite, *Tetranychus urticae.* I, Morphology. *Ann. Entomol. Soc. Amer.* 62:461–69.

———. 1971. The red photoresponse of the spider mite *Tetranychus urticae* (Acarina: Tetranychidae). *Acarologia* 13:113–18.

McEnroe, W. D., and K. Dronka. 1969. Eyes of the two-spotted spider mite, *Tetranychus urticae.* II, Behavioral analysis of the photoreceptors. *Ann. Entomol. Soc. Amer.* 62:466–69.

McFadden, M. W. 1961. An improved technique for using the Berlese funnel. *Entomol. News* 72:150–52.

McGraw, J. R., and M. H. Farrier. 1969. Mites of the superfamily Parasitoidea (Acarina: Mesostigmata) associated with *Dendroctonus* and *Ips* (Coleoptera: Scolytidae). N.C. Agr. Exp. Sta. Tech. Bull. 192. Raleigh: N.C. Agr. Exp. Sta. [DIGAMASELLIDAE]

McGregor, E. A. 1950a. Mites of the family Tetranychidae. *Amer. Midl. Nat.* 44:257–420. [ALLOCHAETOPHORIDAE, TETRANYCHIDAE]

———. 1950b. Mites of the genus *Neophyllobius. Bull. S. Calif. Acad. Sci.* 49:55–70. [CAMEROBIIDAE]

McLean, M. A., and D. Parkinson. 2000. Introduction of the epigeic earthworm *Dendrobaena octaedra* changes the oribatid community and microarthropod abundances in a pine forest. *Soil Biol. Biochem.* 32:1671–81.

McMurtry, J. A. 1983. Phytoseiid predators in orchard systems: a classical biological control success story. In Hoy, Cunningham, and Knutson, 21–26.

———. 1985. Avocado. In Helle and Sabelis (1985b), 1B:327–32. [TETRANYCHIDAE]

McMurtry, J. A., and B. A. Croft. 1997. Life-styles of phytoseiid mites and their roles in biological control. *Ann. Rev. Entomol.* 42:291–321.

McMurtry, J. A., C. B. Huffaker, and M. van de Vrie. 1970. Ecology of tetranychid mites and their natural enemies: a review. I, Tetranychid enemies: Their biological characters and the impact of spray practices. *Hilgardia* 40 (11): 331–90.

McSorely, R., and D. E. Walter. 1991. Comparison of soil extraction methods for nematodes and microarthropods. *Agr. Ecosyst. Environ.* 34:201–7.

Mégnin, P. 1876. Memoire sur l'organisation et la distribution zoologique des acariens de la famille des Gamasides. *J. Anat. Physiol. Normales Pathol. Anim., Paris* 12:288–336.

Meier, F. A., S. Scherrer, and R. Honegger. 2002. Faecal pellets of lichenivorous mites contain viable cells of the lichen-forming ascomycete *Xanthoria parietina* and its green algal photobiont, *Trebouxia arboricola*. *Biol. J. Linn. Soc.* 76:259–68.

Mendel, Z., and U. Gerson. 1982. Is the mite *Lorryia formosa* Cooreman (Prostigmata: Tydeidae) a sanitizing agent in citrus groves? *Acta Oecologica* 3:47–51.

Meng, Y., M. Lang, Z. Zhou., H. Zhou, Y. Guo, Y. Liu, X. Tang, and X. Dei. 1984. Structure and function of tarsal sensilla of the gamasid mites. *Acta Entomol. Sin.* 27:396–400.

Merchant, V. A., and D. A. Crossley, Jr. 1970. An inexpensive high efficiency extractor for soil microarthropods. *J. Georgia Entomol. Soc.* 5:83–87.

Meredith, J., and W. R. Kaufman. 1973. A proposed site of fluid secretion in the salivary gland of the the ixodid tick, *Dermacentor andersoni*. *Parasitology* 67:205–17.

Messner, B., J. Adis, and E. Franklin Ribeiro. 1992. A comparative study on plastron structures in mites (Acari). *Deutsche Entomol. Zeitschr.* 39:159–76.

Metz, M. A., and M. E. Irwin. 2001. Microtrombidiid mite parasitization frequencies and attachment site preferences on brachyceran Diptera with specific reference to Therevidae (Asiloidea) and Tachinidae (Oestroidea). *Environ. Entomol.* 30:903–8.

Meyer, E. 1985. Der Entwicklungszyklus von *Hydrodroma despiciens* (O. F. Müller, 1776) (Acari: Hydrodromidae). *Arch. Hydrobiol., Suppl.* 66:321–453.

Meyer, M. K. P., and P. A. J. Ryke. 1959a. Cunaxoidea (Acarina: Prostigmata) occurring on plants in South Africa. *Ann. Mag. Nat. Hist.*, ser. 13, 2:369–84.

———. 1959b. Mites of the superfamily Raphignathoidea (Acarina: Prostigmata) associated with South African plants. *Ann. Mag. Nat. Hist.* 2:209–34.

———. 1960a. Acarina of the families Anystidae, Pseudocheylidae and Cheyletidae (Prostigmata) found associated with plants in South Africa. *J. Entomol. Soc. S. Afr.* 23:177–93.

———. 1960b. Mites of the superfamily Eupodoidea (Acarina: Prostigmata) associated with South African plants. *S. Afr. J. Agr. Sci.* 3:481–96.

Meyer, M. K. P. Smith. 1969. Some stigmaeid mites from South Africa (Acari: Trombidiformes). *Acarologia* 11:227–71. [STIGMAEIDAE]

———. 1974. *A revision of the Tetranychidae of Africa (Acari) with a key to the genera of the world*. S. Afr. Dept. Agr. Tech. Serv. Mem. 36. Pretoria.

———. 1979. *The Tenuipalpidae (Acari) of Africa with keys to the world fauna*. S. Afr. Dept. Agr. Tech. Serv., Entomol. Mem. 50. Pretoria.

———. 1981. *Mite pests of crops in southern Africa*. Sci. Bull., Dept. Agr. Fisheries (So. Africa) No. 397.

———. 1987. *African Tetranychidae (Acari: Prostigmata)—with reference to the world genera*. S. Afr. Dept. Agr. Water Supply, Entomol. Mem. 69. Pretoria.

———. 1993. *A revision of the genus* Tenuipalpus *Donnadieu (Acari: Tenuipalpidae) in the Afrotropical region*. S. Afr. Dept. Agr., Entomol. Mem. 88. Pretoria.

———. 1996. Ornamental flowering plants. In Lindquist, Sabelis, and Bruin, 6:641–50.

Meyer, M. K. P. Smith, and E. A. Ueckermann. 1984. The family Eupalopsellidae (Acari: Prostigmata), with descriptions of new species from South Africa. *Phytophylactica* 16:121–42.

———. 1987. *A taxonomic study of some Anystidae (Acari: Prostigmata)*. Entomol. Mem. 68. Pretoria: S. Afr. Dept. Agr. Water Supply.

———. 1989. *African Raphignathoidea (Acari: Prostigmata)*. Entomol. Mem. 74. Pretoria: S. Afr. Dept. Agr. Water Supply.

———. 1997. A review of some species of the families Allochaetophoridae, Linotetranidae and Tuckerellidae. *Internat. J. Acarol.* 23:67–92.

Michael, A. D. 1882. Further notes on British Oribatidae. *J. Roy. Microscop. Soc.*, ser. 2, London 2: 1–18.

———. 1884a. *British Oribatidae*. Vol. 1. London: Ray Society.

———. 1884b. The Hypopus question, or the life-history of certain Acarina. *J. Linn. Soc. Lond. Zool.* 17:371–94. [SCUTACARIDAE]

———. 1888. *British Oribatidae*. Vol. 2. London: Ray Society.

———. 1892. On the variations in the internal anatomy of the Gamasinae, especially in that of the genital organs, and on their mode of coition. *Trans. Linn. Soc. Lond.* 5 (2): 281–324.

———. 1896. The internal anatomy of *Bdella*. *Trans. Linn. Soc. Lond.* 6 (2): 477–528.

———. 1901. *British Tyroglyphidae*. London: Ray Society.

———. 1908. Unrecorded Acari from New Zealand. *J. Linn. Soc. Lond. Zool.* 30:134–49. [POLYASPIDIDAE]

Michener, C. D. 1946. The taxonomy and bionomics of some Panamanian trombidiid mites (Acarina). *Ann. Entomol. Soc. Amer.* 39:349–80.

Micherdzinski, W. 1969. *Die Familie Parasitidae Oudemans 1901 (Acarina, Mesostigmata)*. Zak. Zool. Syst. Pol. Akad. Nauk, Krakow. Krakow: Panstwowe Wydawnictwo Naukowe.

Michocka, S. 1987. *Polish mites (Acari) of the families Bdellidae and Cunaxidae*. Monographie Fauny Polski 14. Krakow: Pol. Akad. Zak., Zool., Inst. Syst. Doswiad.

Miko, L., and M. Stanko. 1991. Small mammals as carriers of non-parasitic mites (Oribatida, Uropodina). In Dusbábek and Bukva, 2:395–402.

Miko, L., and J. Travé. 1996. Hungarobelbidae n. fam, with a description of *Hungarobelba pyrenaica* n. sp., Acarina, Oribatida. *Acarologia* 37:133–55.

Miller, L. W. 1964. A new species of *Tuckerella* (Acarina, Tetranychoidea, Tuckerellidae) from Tasmania. *Pap. Proc. Roy. Soc. Tasmania* 98:79–84.

Mills, L. R. 1973. Morphology of glands and ducts in the two-spotted spider-mite, *Tetranychus urticae* Koch, 1836. *Acarologia* 15:218–36.

Miranda, R. J., A. D. Quintero, and A. Almanza. 2002. House dust mites from urban and rural houses on the lowland Pacific slopes of Panama. *Syst. Appl. Acarol.* 7:23–30.

Mironov, S. V. 1991. Filogeniya per'evikh kleshchei semeistva Avenzoariidae (Acariformes: Analgoidea). *Parazitologiya* 25:281–96.

———. 2001a. Description of four new genera of the feather mite family Pteronyssidae Oudemans 1941 (Astigmata: Analgoidea) with notes on the systematics of the family. *Acarina* 9:3–22. [AVENZOARIIDAE]

———. 2001b. Pervaya nakodka per'evyikh kleshchei semeistva Apionacaridae na vorob'inikh (Passeriformes). *Parazitologiya (St. Petersburg)* 35:284–90.

———. 2003. On some problems in the systematics of feather mites. *Acarina* 11:3–29.

Mironov, S. V., A. V. Bochkov, and A. Fain. 2005. Phylogeny and evolution of parasitism in feather mites of the families Epidermoptidae and Dermationidae (Acari: Analgoidea). *Zool. Anz.* 243:155–79.

Mironov, S. V., and J. Dabert. 1999. Phylogeny and co-speciation in feather mites of the subfamily Avenzoariinae (Analgoidea: Avenzoariidae). *Exp. Appl. Acarol.* 23:525–49. (Also appears in Bruin, van der Geest, and Sabelis, 105–24.)

Mitchell, C. J. 1968. Biological studies on *Laelaps myonyssognathus* G. and N. (Acarina: Laelapidae). *J. Med. Entomol.* 5:99–107.

Mitchell, M. J. 1977. Population dynamic of oribatid mites (Acari: Cryptostigmata) in an aspen woodland soil. *Pedobiologia* 17:305–19.

———. 1978. Vertical and horizontal distribution of oribatid mites (Acari: Cryptostigmata) in an aspen woodland soil. *Ecology* 59:516–25.

Mitchell, M. J., and D. Parkinson. 1976. Fungal feeding of oribatid mites (Acari: Cryptostigmata) in an aspen woodland soil. *Ecology* 57 (2): 302–12.

Mitchell, R. D. 1953. A new species of *Lundbladia* and remarks on the family Hydryphantidae (water mites). *Amer. Midl. Nat.* 9:159–70.

———. 1954. A description of the water-mite, *Hydrovolzia gerhardi* new species, with observations on the life history and ecology. *Nat. Hist. Misc., Chicago Acad. Sci.* 134:1–9. [HYDROVOLZIIDAE].

———. 1955a. *Anatomy, life history, and evolution of the mites parasitizing fresh-water mussels*. Univ. Mich. Misc. Publ. Zool. 89. Ann Arbor: Univ. Mich.

———. 1955b. Two water-mites from Illinois. *Trans. Amer. Microscop. Soc.* 74:333–42. [PIERSIGIIDAE, BOGATIIDAE]

———. 1957a. Locomotor adaptions of the family Hydryphantidae (Hydrachnellae, Acari). *Abh. Naturwiss. Ver. Bremen* 35:75–100.

———. 1957b. Major evolutionary lines in water mites. *Syst. Zool.* 6:137–48.

———. 1957c. The mating behaviour of pionid water-mites. *Amer. Midl. Nat.* 58:360–66. [PIONIDAE]

———. 1958. Sperm transfer in the water-mite, *Hydryphantes ruber* Geer. *Amer. Midl. Nat.* 60:156–58.

———. 1959. Life histories and larval behaviour of arrenurid water-mites parasitizing Odonata. *J. New York Entomol. Soc.* 67:1–12. [ARRENURIDAE]

———. 1961. Behaviour of the larvae of *Arrenurus fissicornis* Marshall, a water-mite parasitic on dragonflies. *Anim. Behav.* 9:220–24. [ARRENURIDAE]

———. 1962a. The growth of *Thermacarus nevadensis* Marshall (Acari, Hydrachnellae). *Zool. Anz.* 168:269–75. [THERMACARIDAE]

———. 1962b. The musculature of a trombiculid mite, *Blankaartia acuscutellaris* (Walch). *Ann. Entomol. Soc. Amer.* 55:106–99. [TROMBICULIDAE]

———. 1962c. The structure and evolution of water-mite mouthparts. *J. Morphol.* 110:41–59.

———. 1964a. The anatomy of an adult chigger mite, *Blankaartia acuscutellaris* (Walch). *J. Morphol.* 114:373–91. [TROMBICULIDAE]

———. 1964b. A study of sympatry in the water mite genus *Arrenurus* (family Arrenuridae). *Ecology* 45:546–58.

———. 1970. The evolution of a blind gut in trombiculid mites. *J. Nat. Hist.* 4:221–29. [TROMBICULIDAE]

———. 1972. The tracheae of water mites. *J. Morphol.* 136:327–35.

———. 1974. The evolution of thermophily in hot springs. *Qtly. Rev. Biol.* 49:229–42.

———. 1998. The behavior of *Arrenurus* larvae (Acari: Hydrachnidea) parasitizing Diptera. *Acarologia* 39:49–55. [ARRENURIDAE]

———. 2003. The vernal pond mite fauna of Blacklick Woods Metro Park, Columbus, Ohio. Part I. In I. M. Smith (2003a), 183–99. [HYDRYPHANTIDAE]

Mitchell, R. D., D. J. Horn, G. R. Needham, and W. C. Welbourn, eds. 1996. *Acarology IX Proceedings*. Vol. 1. Columbus: Ohio Biological Survey.

Mitchell, R. D., and A. V. Mitchell. 1958. Studies of a new water-mite, *Tyrrellia hibbardi*, from Kansas springs. *Trans. Amer. Microscop. Soc.* 77:11–18. [LIMNESIIDAE].

Mitchell, R. (D.), and M. Nadchatram. 1969. Schizekenosy: the substitute for defecation in chigger mites. *J. Nat. Hist.* 3:121–24. [TROMBICULIDAE]

Mitrofanov, V. I. 1973a. Revision of the systematics of phytophagous mites of the subfamily Brevipalpinae (Trombidiformes, Tenuipalpidae). *Zool. Zh.* 52:507–12. [in Russian]

———. 1973b. Revision of the systematics of phytophagous mites of the subfamily Tenuipalpinae s. str. (Trombidiformes, Tenuipalpidae). *Zool. Zh.* 52:1315–20. [in Russian]

———. 1983. Evolution and systematics of tetranychoid mites. In *Fauna and ecology of the invertebrate animals of Georgia*, ed. L. V. Gelovani, 12–28. Tbilisi: Metsniereba. [in Russian]

Mizoguchi, A., N. Mori, R. Nishida, and Y. Kuwahara. 2003. α-acaridial, a female sex pheromone from an alarm pheromone emitting mite, *Rhizoglyphus robini*. *J. Chem. Ecol.* 29:1681–90.

Mo, C.-F. 1972. The male of *Neocypholaelaps indica* Evans (Acari, Ameroseiidae). *Acarologia* 14:15–17.

———. 1996. A new genus and new species of the Otopheidomenidae (Acari: Mesostigmata) from longhorned grasshoppers. *Syst. Appl. Acarol.* 1:199–204.

Mohanasundaram, M. 1984. New eriophyid mites from India (Acarina: Eriophyoidea). *Oriental Insects* 18:251–83.

———. 1994. A new species of *Xenaster* Mahunka (1970) (Tarsoneminae: Acari) on coconut from South India. *Entomon* 19:109–13. [PYGMEPHORIDAE]

Molleman, F., and D. E. Walter. 2001. Niche segregation and can-openers: scydmaenid beetles as predators of armoured mites in Australia. In Halliday et al. (2001), 283–88.

Møller, A. P. 1990. Effects of parasitism by a haematophagous mite on reproduction in the barn swallow. *Ecology* 71:2345–57. [MACRONYSSIDAE]

———. 1994. *Sexual selection and the barn swallow*. New York: Oxford Univ. Press.

Moore, D., and F. W. Howard. 1996. Coconuts. In Lindquist, Sabelis, and Bruin, 6:561–70.
Moore, J. 2002. *Parasites and the behavior of animals*. New York: Oxford Univ. Press.
Moorhouse, D. E. 1966. Observations on copulation in *Ixodes holocyclus* Neumann and the feeding of the male. *J. Med. Entomol.* 3:168–71. [IXODIDAE]
Morales-Malacara, J. B., V. Behan-Pelletier, E. Ueckermann, T. M. Perez, E. Estrada, and M. Badii, eds. 2007. *Acarology XI, Proceedings, XI International Congress of Acarology*. Mérida, Universidad Nacional Autónoma de México.
Moraza, M. L. 1984. *Ctenobelba perez-iñigoi* n. sp., una nueva especie de Oribátidos perteneciente a la familia Ctenobelbidae Grandjean, 1965. *Eos, Rev. Esp. Entomol.* 60:231–34.
———. 1990. El género *Prozercon* Sellnick, 1943 en la Península Ibérica y descripción de tres nuevas especies (Acari, Mesostigmata, Zerconidae). *Misc. Zool. (Barc.)* 12:69–77.
———. 1993. Two new species of *Pachyseius* Berlese, 1910 from Spain (Acari: Mesostigmata: Pachylaelapidae). *Acarologia* 34:89–94.
———. 2005a. A new genus and species of Epicriidae (Acari, Mesostigmata) from eastern North America. *Can. Entomol.* 137: 539–50.
———. 2005b. Revised diagnosis of *Epicrius* Canestrini and Fanzago, 1877 and description of four new species (Acari: Mesostigmata: Epicriidae). *Internat. J. Acarol.* 31:341–54.
———. 2005c. Tarsus I chaetotaxy and structure in Anactinotrichida mites: characters with phylogenetic value. In Weigmann et al. (2005), 347–59.
Moraza, M. L., and D. E. Johnston. 2004. *Neoepicrius*, gen. n. from western North America (Acari: Mesostigmata: Epicriidae). *Acarologia* 44:195–207.
Moraza, M. L., and E. E. Lindquist. 1999. Coprozerconidae, a new family of zerconoid mites from North America (Acari: Mesostigmata: Zerconoidea). *Acarologia* 39:291–313.
Moraza, M. L., and M. A. Peña. 2005 (2004). The family Pachylaelapidae Vitzthum, 1931 on Tenerife Island (Canary Islands), with description of seven new species of the genus *Pachylaelaps* (Acari, Mesostigmata: Pachylaelapidae). *Acarologia* 40 (2–3): 103–29.
Morgan, C. V. G. 1960. Anatomical characters distinguishing *Bryobia arborea* M. and A. and *B. praetiosa* Koch (Acarina: Tetranychidae) from various areas of the world. *Can. Entomol.* 92: 595–604.
Morgenthaler, O. 1934. Krankheitserregende und harmlose Arten der Bienenmilbe *Acarapis*, zugleich ein Beitrage zum Species-Problem. *Rev. Suisse Zool.* 41:429–46. [TARSONEMIDAE]
Mori, H., Y. Saito, and Y. P. Tho. 1999. Cooperative group predation in a sit-and-wait cheyletid mite. *Exp. Appl. Acarol.* 23:643–51. [CHEYLETIDAE]
Morini, E. G., and R. J. Roveda. 1974. A contribution of acarology. 1, Mites considered non-parasitic, of importance in human and veterinary medicine. *Rev. Med. Vet. Arg.* 55:111–15.
Moritz, C. 1991. Parasite loads in parthenogenetic and sexual lizards (*Heteronotia binoei*): Support for the Red Queen. *Proc. Roy. Soc. Lond.* B 244:145–49.
Moritz, M. 1976. Revision der europäischen Gattungen und Arten der Familie Brachychthoniidae (Acari, Oribatei). Teil 1, Allgemeiner Teil: Brachychthoniidae Thor, 1934. Spezieller Teil: *Liochthonius* v.d. Hammen, 1959, *Verachthonius* nov. gen. und *Paraliochthonius* nov. gen. *Mitt. Zool. Mus. Berl.* 52:27–136.
———. 1977. Revision der europäischen Gattungen und Arten der Familie Brachychthoniidae (Acari, Oribatei). Teil 2, *Mixochthonius* Niedbała, 1972, *Neobrachychthonius* nov. gen., *Synchthonius* v.d. Hammen, 1952, *Poecilochthonius* Balogh, 1943, *Brachychthonius* Berlese, 1910, *Brachychochthonius* Jacot, 1938. *Mitt. Zool. Mus. Berl.* 52:227–319.
Moser, J. C. 1975a. Biosystematics of the straw itch mite with special reference to nomenclature and dermatology. *Trans. Roy. Entomol. Soc. Lond.* 127:185–91. [PYEMOTIDAE]
———. 1975b. Mite predators of the southern pine beetle. *Ann. Entomol. Soc. Amer.* 68:1113–16.
———. 1985. Use of sporothecae by phoretic *Tarsonemus* mites to transport ascospores of coniferous bluestain fungi. *Trans. Brit. Mycol. Soc.* 84:750–53.
Moser, J. C., and J. R. Bridges. 1986. *Tarsonemus* (Acarina: Tarsonemidae) mites phoretic on the southern pine beetle (Coleoptera: Scolytidae): attachment sites and numbers of bluestain (Ascomycetes: Ophiostomataceae) ascospores carried. *Proc. Ent. Soc. Wash.* 88:297–99.
Moser, J. C., and E. A. Cross. 1975. Phoretomorph: a new phoretic phase unique to the Pyemotidae (Acarina: Tarsonemoidea). *Ann. Ent. Soc. Amer.* 68:820–22. [PYGMEPHORIDAE]
Moser, J. C., E. A. Cross, and L. M. Roton. 1971. Biology of *Pyemotes parviscolyti* (Acarina: Pyemotidae). *Entomophaga* 16:367–79.
Moser, J. C., and L. M. Roton. 1971. Mites associated with southern pine bark beetles in Allen Parish, Louisiana. *Can. Entomol.* 103:1775–98.
———. 1972. Reproductive compatibility between two widely separated populations of *Pyemotes scolyti* (Acarina: Pyemotidae). *Pan-Pac. Entomol.* 48:97–99.
Moser, J. C., R. L. Smiley, and I. S. Otvos. 1987. A new *Pyemotes* (Acari: Pyemotidae) reared from the Douglas–fir cone moth. *Internat. J. Acarol.* 13:141–47.
Moser, J. C., R. L. Smiley, I. S. Otvos, and P. L. Lorio. 1987. *Pyemotes* n. sp. (Acari: Pyemotidae), a potential biological control agent for cone-and-seed insects. In *Proceedings of the 2nd Conference of the Cone and Seed Insects Working Party, Briançon, France*, ed. A. Roques, 307–8. Vienna: International Union of Forest Research Organizations.
Moser, J. C., and P. H. Vercammen-Grandjean. 1979. *Megophthrombium* n. sp. and *Diathrombium* diaphne n. gen., n. sp. (Acarina: Trombidiidae), two larval parasites of adult southern pine beetles. *Internat. J. Acarol.* 5:18–23.
Moskacheva, E. A. 1967. On the depth of soil inhabitation by oribatid mites (Acariformes, Oribatei). *Zool. Zh.* 46:1400–1405. [in Russian]
Moss, W. W. 1960. Description and mating behaviour of *Allothrombium lerouxi*, new species (Acarina: Trombidiidae), a predator of small arthropods in Quebec apple orchards. *Can. Entomol.* 92:898–905.
———. 1962. Studies on the morphology of the trombidiid mite *Allothrombium lerouxi* Moss (Acari). *Acarologia* 4:313–45.
———. 1968. An illustrated key to the species of the acarine genus *Dermanyssus* (Mesostigmata: Laelapoidea: Dermanyssidae). *J. Med. Entomol.* 5:67–84.
———. 1978. The mite genus *Dermanyssus*: A survey, with description of *Dermanyssus trochilinus*, n. sp., and a revised key to the

species (Acari: Mesostigmata: Dermanyssidae). *J. Med. Entomol.* 14:627–40.

Moss, W. W., J. H. Oliver, Jr., and B. C. Nelson. 1968. Karyotypes and developmental stages of *Harpyrhynchus novoplumaris* sp. n. (Acari: Cheyletoidea: Harpyrhynchidae), a parasite of North American birds. *J. Parasitol.* 54:377–92.

Mostafa, A. R. 1974. Biological and behavioral aspects of the lizard mite *Pterygosoma mutabilis* Jack, 1961 (Acarina: Pterygosomidae). *Acarologia* 16:100–105.

Mostafa, A. R., P. de Bach, and T. W. Fisher. 1975. Anystid mite: citrus thrips predator. *Calif. Agr.* (March): 1.

Motaş, C. 1928. Contribution á la connaissance des Hydracariens français particulièrement du Sud-Est de la France. *Trav. Labor. Hydrobiol. Piscic., Univ. Grenoble* 20:297–373. [STYGOTHROMBIDIIDAE]

Motaş, C., and J. Tanasachi. 1946. Acariens phréaticoles de Transylvanie. *Notat. Biol.* 4. [STYGOTHROMBIDIIDAE]

Motaş, C., J. Tanasachi, and T. Orghidan. 1947. Hydracariens phréaticoles de Roumanie. *Notat. Biol.* 5:1–67.

———. 1958. Hydrachnelles phréaticoles de la R. P. Roumaine. *Vest. Ceskolov. Zool. Spol.* 22:293–333.

Mothes, U., and K.-A. Seitz. 1981a. Fine structure and function of the prosomal glands of the two-spotted spider mite, *Tetranychus urticae* (Acari, Tetranychidae). *Cell Tissue Res.* 221:339–49.

———. 1981b. Functional microscopic anatomy of the digestive system of *Tetranychus urticae* (Acari: Tetranychidae). *Acarologia* 22:257–70.

Moutia, L. A. 1958. Contribution to the study of some phytophagous acarina and their predators in Mauritius. *Bull. Entomol. Res.* 49:59–75.

Moznette, G. F. 1917. The cyclamen mite. *J. Agr. Res.* 10:373–90. [TARSONEMIDAE]

Mullen, G. R. 1974. Acarine parasites of mosquitoes. II, Illustrated larval key to the families and genera of mites reportedly parasitic on mosquitoes. *Mosquito News* 34:183–95.

———. 1975a. Acarine parasites of mosquitoes. I, A critical review of all known records of mosquitoes parasitized by mites. *J. Med. Entomol.* 12:27–36. [TROMBELLIDAE]

———. 1975b. Predation by water mites (Acarina: Hydrachnellae) on immature stages of mosquitoes. *Mosquito News* 35:168–71. [HYDRYPHANTIDAE]

———. 1976. Water mites of the subgenus *Truncaturus* (Arrenuridae: *Arrenurus*) in North America. *Search Agr.* 6:1–35.

———. 1979. Aquatic mites parasitic on stoneflies in North America. *Rec. Adv. Acarol.* 2:481–84. [STYGOTHROMBIDIIDAE]

Mullen, G. R., and L. A. Durden, eds. 2002. *Medical and veterinary entomology.* San Diego: Academic Press/Elsevier Science.

Mullen, G. R., and B. M. OConnor. 2002. Mites (Acari). In Mullen and Durden, 449–516.

Mullen, G. R., and P. H. Vercammen-Grandjean. 1978. *Megophthrombium*, a new genus in the family Trombidiidae (Acari: Parasitengona) from North American mosquitoes. *Internat. J. Acarol.* 4:39–45.

———. 1980. The larval stage of Stygothrombiinae Thor, 1935 and Wandesiinae Schwoerbel, 1961 and erection of a new superfamily, Stygothromboidea. *Internat. J. Acar.* 6:25–28. [STYGOTHROMBIDIIDAE]

Müller, J. K., and H. H. Schwarz. 1990. Differences in carrier-preference and evidence of reproductive isolation between mites of *Poecilochirus carabi* (Acari, Parasitidae) living phoretically on two sympatric *Necrophorus* species (Coleoptera, Silphidae). *Zool. Jb. Syst.* 117:23–30.

Muma, M. H. 1961. *Mites associated with citrus in Florida.* Bull. Univ. Fla. Agr. Exp. Sta. 640. Gainesville: Univ. Fla. Agr. Exp. Sta.

———. 1975. *Mites associated with citrus in Florida.* Bull. Univ. Fla. Agr. Exp. Sta. 640A. Gainesville: Univ. Fla. Agr. Exp. Sta.

Mumcuoglu, Y., and E. Stix. 1974. Milben in der Luft. *Rev. Suisse Zool.* 81:673–77.

Münchberg, P. 1982. Zur Parasitierung der Flügel von *Sympetrum meridionale* und *fonscolombei* Selys (Odonata) durch die Larven von *Arrenurus* (*A.*) *palpillator* (Müll.) (Hydrachnellae, Acari) und zugleich zur Spezifität und den Voraussetzungen dieses Parasitismus. *Arch. Hydrobiol.* 95:299–316. [ARRENURIDAE]

Muraoka, M., and N. Ishibashi. 1976. Nematode feeding mites and their feeding behavior. *Appl. Entomol. Zool.* 11:1–7.

Murphy, P. W., and A. N. Balla. 1973. The bionomics of *Humerobates rostrolamellatus* Grandjean (Cryptostigmata—ceratozetidae) on fruit trees. In Griffiths and Bowman, 2:97–104. [HUMEROBATIDAE]

Murphy, P. W., and M. A. Sardar. 1991. Resource allocation and utilization contrasts in *Hypoaspis aculeifer* (Can.) and *Alliphis halleri* (G. & R. Can.) (Mesostigmata) with emphasis on food source. In Schuster and Murphy, 301–11.

Murrell, A., N. J. H. Campbell, and S. C. Barker. 2000. Phylogenetic analyses of the rhipicephaline ticks indicate that the genus *Rhipicephalus* is paraphyletic. *Mol. Phylogenet. Evol.* 16:1–7. [IXODIDAE]

Murrell, A., S. J. Dobson, D. E. Walter, N. J. H. Campbell, R. Shao, and S. Barker. 2005. Relationships among the three major lineages of the Acari (Arthropoda: Arachnida) inferred from small subunit rRNA: Paraphyly of the Parasitiformes with respect to the Opilioacariformes and relative rates of nucleotide substitution. *Inverebrt. Syst.* 19:383–89.

Mwase, E. T., and A. S. Baker. 2006. An annotated checklist of mites (Arachnida, Acari) of Zambia. *Zootaxa* 1106:1–24.

My-Yen, L. T., Y. Wada, K. Matsumoto, and Y. Kuwahara. 1980. Pheromone study on acarid mites. VI, Demonstration and isolation of an aggregation pheromone in *Lardoglyphus konoi* Sasa et Asanuma. *Jpn. J. Sanit. Zool.* 31:249–54.

Nadchatram, M. 1970a. Correlation of habitat, environment and color of chiggers and their potential significance in the epidemiology of scrub typhus in Malaya (Prostigmata: Trombiculidae). *J. Med. Entomol.* 7:131–44.

———. 1970b. A review of intranasal chiggers with descriptions of twelve species from east New Guinea (Acarina, Trombiculidae). *J. Med. Entomol.* 7:1–29.

———. 1979. Ecological classification of scrub typhus vectors and related species of chiggers (Acari: Trombiculidae). *Biotropica* 6:67–72.

———. 1984. The natural history of the genus *Leptotrombidium* (Acari: Trombiculidae) with reference to scrub typhus. In Griffiths and Bowman, 2:1107–13.

———. 2006. A review of endoparasitic acarines of Malaysia with special reference to novel endoparasitism of mites in amphibious sea snakes and supplementary notes on ecology of chiggers. *Tropical Biomed.* 23:1–22. [TROMBICULIDAE]

Nadchatram, M., and A. L. Dohany. 1974. *A pictorial key to the subfamilies, genera and subgenera of south-east Asian chiggers (Acari,*

Prostigmata, Trombiculidae). Bull. Inst. Med. Res. Malaysia 16. Kuala Lumpur: Inst. Med. Res. Malaysia.

Nadchatram, M., and F. J. Radovsky. 1971. A second species of *Vatacarus* (Prostigmata, Trombiculidae) infesting the trachea of amphibious sea snakes. *J. Med. Entomol.* 8:37–40.

Naeem, S., D. S. Dobkin, and B. M. OConnor. 1985. *Lasioseius* mites (Acari: Gamasida: Ascidae) associated with hummingbird-pollinated flowers in Trinidad, West Indies. *Internat. J. Entomol.* 27:338–53. [BLATTISOCIIDAE]

Nannelli, R. 1979. Osservazioni sul comportamento, la biologia e l'alimentazione di alcune specie di Acari Oribatei in condizioni sperimentali di allevamento. *Redia* 62:299–313.

———. 1990. Studio di una faggeta dell'Appennino pistoiese: Composizione e successione dell'artropodofauna nella degradazione della lettiera. *Redia* 73:543–68.

Nannelli R., T. Turchetti, and G. Maresi. 1998. Corticolous mites (Acari) as potential vectors of *Cryphonectria parasitica* (Murr.) Barr hypovirulent strains. *Internat. J. Acarol.* 24:237–44.

Narayan, D. S. 1957. A new species of *Penthaleus* from Mexica (Acarina: Penthaleidae). *J. Kans. Entomol. Soc.* 30:111–13.

Naskrecki, P., and R. K. Colwell. 1995. New genus and two new species of Melicharini from Venezuela (Acari: Mesostigmata: Ascidae). *Ann. Entomol. Soc. Amer.* 88:284–93.

———. 1998. *Systematics and host plant affiliations of hummingbird flower mites of the genera* Tropicoseius *Baker & Yunker and* Rhinoseius *Baker & Yunker (Acari: Mesostigmata: Ascidae).* Thomas Say Pubs. Entomol., Monographs. Lanham, MD: Thomas Say Pubs. Entomol. [MELICHARIDAE]

Nasr, A. K., M. S. Nawar, and A. M. Afifi. 1988. A new species of the genus *Blattisocius* Keegan from Egypt (Acarina, Mesostigmata: Ascidae). *Bull. Fac. Agr., Univ. Cairo* 39:937–46. [BLATTISOCIIDAE]

Nation, J. L. 1983. A new method using hexamethyldisilazane for preparation of soft insect tissues for scanning electron microscopy. *Stain Technol.* 58:347–51.

Naudo, M. H. 1963. Acariens Notostigmata de l'Angola. *Publ. Cultur. Companhia Diamantes Angola* 63:13–24.

Navajas, M., and B. Fenton. 2000. The application of molecular markers in the study of diversity in acarology: A review. *Exp. Appl. Acarol.* 24:751–74.

Navajas, M., and J. Gutierrez. 1997. Molecular phylogeny in spider mites (Tetranychidae). In Mitchell et al. (1996), 1:305–8.

Navajas, M., J. Gutierrez, G. Bonato, H. R. Bolland, and S. Mapangou-Divassa. 1994. Intraspecific diversity of the cassava green mite *Mononychellus progresivus* (Acari: Tetranychidae) using comparisons of mitochondrial and nuclear ribosomal DNA sequences and cross-breeding. *Exp. Appl. Acarol.* 18:351–60.

Navajas, M., J. Gutierrez, and T. Gotoh. 1997. Convergence of molecular and morphological data reveals phylogenetic information in *Tetranychus* species and allows the restoration of the genus *Amphitetranychus* (Acari: Tetranychidae). *Bull. Entomol. Res.* 87:283–88.

Navajas, M., J. Gutierrez, J. Lagnel, and P. Boursot. 1996. Mitochondrial cytochrome oxidase I in tetranychid mites: A comparison between molecular phylogeny and changes of morphological and life history traits. *Bull. Entomol. Res.* 86:407–17.

Navajas, M., J. Gutierrez, M. Williams, and T. Gotoh. 2001. Synonymy between two spider mite species, *Tetranychus kanzawai* and *T. hydrangeae*, shown by ribosomal ITS2 sequences and cross-breeding. *Bull. Entomol. Res.* 91:117–23.

Navajas, M., A. Tsagkarakou, J. Lagnel, and M. J. Perrot-Minnot. 2000. Genetic differentiation in *Tetranychus urticae* (Acari: Tetranychidae): Polymorphism, host races or sibling species? *Exp. Appl. Acarol.* 24:365–76.

Navia, D., G. J. de Moraes, G. Roderick, and M. Navajas. 2005. The invasive coconut mite *Aceria guerreronis* (Acari: Eriophyidae): Origin and invasion sources inferred from mitochondrial (16S) and nuclear (ITS) sequences. *Bull. Entomol. Res.* 95:505–16.

Needham, G. R., R. Mitchell, D. J. Horn, and W. C. Welbourn, eds. 1999. *Acarology IX. Symposia.* Vol. 2. Columbus: Ohio Biological Survey.

Nelson, G. J. 1974. Classification as an expression of phylogenetic relationships. *Syst. Zool.* 22:344–59.

Nelson-Rees, W. A., M. A. Hoy, and R. T. Roush. 1980. Heterochromatization chromatin elimination and haploidization in the parahaploid mite *Metaseiulus occidentalis* (Nesbitt) (Acarina: Phtyoseiidae). *Chromosoma* 77:263–76.

Nesbitt, H. H. J. 1951. A taxonomic study of the Phytoseiinae (Family Laelaptidae) predaceous upon Tetranychidae of economic importance. *Zool. Verh.* 12:1–64. [includes MELICHARIDAE]

———. 1976. Five new canestriniids (Acari: Astigmata) from Papua New Guinea. *Zool. J. Linn. Soc.* 58:219–36. [CANESTRINIIDAE]

Neumann, G. 1911. Ixodidae. *Das Tierreich, Berlin* 26:134–36.

Newell, I. M. 1945. *Hydrozetes* Berlese (Acari, Oribatidea): The occurrence of the genus in North America and the phenomenon of levitation. *Trans. Conn. Acad. Sci.* 36:253–75. [HYDROZETIDAE]

———. 1947a. Studies on the morphology and systematics of the family Halarachnidae Oudemans, 1906 (Acari: Parasitoidea). *Bull. Bingham Oceanogr. Coll. (Yale Univ.)* 10:235–66.

———. 1947b. A systematic and ecological study of the Halacaridae of eastern North America. *Bull. Bingham Oceanogr. Coll. (Yale Univ.)* 10:1–232.

———. 1956. A parasitic species of *Copidognathus* (Acari, Halacaridae). *Proc. Hawaiian Entomol. Soc.* 16:122–25.

———. 1957. Studies on the Johnstonianidae (Acari, Parasitengona). *Pac. Sci.* 11:396–466.

———. 1958. Specific characters and character variants in adults and larvae of the genus *Parathrombium* Bruyant 1910 (Acari, Trombidiidae) with descriptions of two new species from western North America. *Pac. Sci.* 12:350–70.

———. 1960. *Charadacarus* new genus, Charadacarinae new subfamily (Acari, Johnstonianidae), and the status of *Typhlothrombium* Berlese 1910. *Pac. Sci.* 14:156–72.

———. 1963. Feeding habits in the genus *Balaustium* (Acarina, Erythraeidae), with special reference to attacks on man. *J. Parasitol.* 49:498–502.

———. 1967a. Abyssal Halacaridae (Acari) from the southeast Pacific. *Pac. Insects* 9:693–708.

———. 1967b. Prostigmata: Halacaridae (marine mites). *Antarct. Res. Ser.* 10:81–95.

———. 1971a. Halacaridae (Acari) collected during Cruise 17 of the R/V *Anton Bruun*, in the southwestern Pacific Ocean. *Anton Bruun Rep.* 8:2–58.

———. 1971b. The protonymph of *Pimeliaphilus* (Pterygosomatidae) and its significance relative to the calyptostases in the Parasitengona. In Daniel and Rosický, 789–95.

———. 1984. *Antarctic Halacaroidea.* Biology of the antarctic seas, vol. 15, ed. L. S. Kornicker. Washington, DC: Amer. Geophysical Union.

Newell, I. M., and R. E. Ryckman. 1964. *Hirstiella pyriformis* sp. n. (Acari, Pterygosomidae), a new parasite of lizards from Baja California. *J. Parasitol.* 50:163–71.

———. 1966. Species of *Pimeliaphilus* (Acari: Pterygosomidae) attacking insects, with particular reference to the species parasitizing Triatominae (Hemiptera: Reduviidae). *Hilgardia* 37: 403–36.

———. 1969. *Pimeliaphilus zeledoni* n. sp. (Acari, Pterygosomidae), a parasite of *Triatoma dimidiata* (Latr.) (Hemiptera, Reduviidae). *Bull. S. Calif. Acad. Sci.* 68:138–44.

Newell, I. M., and L. Tevis, Jr. 1960. *Angelothrombium pandorae* n. g., n. sp. (Acari, Trombidiidae), and notes on the biology of the giant red velvet mites. *Ann. Entomol. Soc. Amer.* 53: 293–304.

Newell, I. M., and P. H. Vercammen-Grandjean. 1964. *Pteridopus* n. g. (Acari, Johnstonianidae) and a probable auditory organ in a mite. *Acarologia* 6:98–110.

Newstead, R., and H. M. Duvall. 1918. *Bionomic, morphological and economic report of acarids of stored grain and flour.* Roy. Soc. Rep. Grain Pests (War) Comm. 2:1–59. [ACAROPHENACIDAE]

Nickel, P. A., and R. J. Elzinga. 1969. *Caraboacarus karenae*, a new mite from carabid beetles, with notes on its biology (Acari: Pyemotidae). *J. Kans. Entomol. Soc.* 42:222–29. [CARABOACARIDAE]

———. 1970a. Biology of *Micromegistus bakeri* Trägårdh, 1948 (Acarina: Parantennulidae) with descriptions of immatures and redescription of adults. *Acarologia* 12:234–43.

———. 1970b. New host records and biological notes for *Echinomegistus wheeleri* (Wasman) (Acarina: Paramegistidae). *J. Kans. Entomol. Soc.* 43:32–34.

Niedbała, W. 1992. *Phthiracaroidea, Acari, Oribatida*. Amsterdam and Warsaw: Elsevier PWN Polish Scientific Studies.

———. 1993. New species of Euptyctima (Acari, Oribatida) from New Zealand. *New Zeal. J. Zool.* 20:137–59.

———. 1994. Origin of euptyctimous mites (Acari, Oribatida) in Hawaii. *Biol. Bull. Poznan* 31:83–99.

———. 1998a. Euptyctima (Acari, Oribatida) of the Hawaiian Islands. *Proc. Hawaiian Entomol.* 33:125–46.

———. 1998b. Ptyctimous mites of the Ethiopian region. I, Euphthiracaroidea (Acari, Oribatida). *J. Afr. Zool.* 112 (1): 15–75.

———. 2000. The ptyctimous mites fauna of the Oriental and Australian regions and their centres of origin (Acari: Oribatida). *Genus (Warsaw)*, Suppl.

———. 2001. *Study on the diversity of ptyctimous mites (Acari, Oribatida) and quest for centres of its origin: The fauna of the Ethiopian region.* Monogr. Upper Siles. Mus. 3.

———. 2002a. Ptyctimous mites (Acari: Oribatida) from Bermuda Islands. *Ann. Zool. (Warsaw)* 52:181–88.

———. 2002b. *Ptyctimous mites (Acari, Oribatida) of the Nearctic region.* Monogr. Upper Siles. Mus. 4.

———. 2003. Ptyctimous mites (Acari: Oribatida) of Costa Rica. *Ann. Zool. (Warsaw)* 53:259–334.

———. 2004a. Protoplophoridae (Acari: Oribatida) of the world. *Ann. Zool. (Warsaw)* 54 (4): 807–34.

———. 2004b. Ptyctimous mites (Acari, Oribatida) of the Neotropical region. *Ann. Zool. (Warsaw)* 54 (1): 1–288.

———. 2006. Ptyctimous mites (Acari: Oribatida) of South Africa. *Ann. Zool. (Warsaw)* 26 (Suppl. 1): 1–97.

Niedbała, W., and Z. Olszanowski. 1997. Oribatida (= Cryptostigmata). In *Checklist of animals of Poland*, ed. J. Razowski, 2:249–59. Krakow: Wydawnictwa Instytutu Systematiki i Ewolucji Zwierzat PAN.

Niemi, R., and V. M. Behan-Pelletier. 2004. *Nuhivabates* n. gen. and two new species, *N. nukuhiva* n. sp. and *N. hivaoa* n.sp. from Marquesas Islands (Acari: Oribatida: Mycobatidae). *Acarologia* 44 (1–2): 73–85.

Niemi, R., E. Karppinen, and M. Uusitalo. 1997. Catalogue of the Oribatida of Finland. *Acta Zool. Fenn.* 207:1–39.

Noble, J. C., M. Kaliszewski, and W. G. Whitford. 1990. Soil Acari from burnt and unburnt Mallee habitats in western New South Wales. In *The Mallee lands: A conservation perspective*, ed. J. C. Noble, P. J. Joss, and G. K. Jones, 224–25. Melbourne: CSIRO Publications. [BARBUTIIDAE]

Noble, J. C., W. G. Whitford, and M. Kaliszewski. 1996. Soil and litter microarthropod populations from two contrasting ecosystems in semi-arid eastern Australia. *J. Arid Environ.* 32:329–46.

Norton, A. P., G. English-Loeb, D. Gadoury, and R. C. Seem. 2000. Mycophagous mites and foliar pathogens: Leaf domatia mediate tritrophic interactions in grapes. *Ecology* 81:490–99.

Norton, R. A. 1975. Elliptochthoniidae, a new mite family (Acarina: Oribatei) from mineral soil in California. *J. New York Entomol. Soc.* 83:209–16.

———. 1977a. An example of phoretomorphs in the mite family Scutacaridae. *J. Georgia Entomol. Soc.* 12:185–86.

———. 1977b. A review of F. Grandjean's system of leg-chaetotaxy in the Oribatei and its application to the Damaeidae. In Dindal (1977), 33–62.

———. 1980. Observations on phoresy by oribatid mites (Acari: Oribatei). *Internat. J. Acarol.* 6:121–30.

———. 1982. *Arborichthonius* n. gen., an unusual enarthronote soil mite (Acarina: Oribatei) from Ontario. *Proc. Entomol. Soc. Wash.* 84:85–96. [ARBORICHTHONIIDAE]

———. 1983. Tenuialidae (Acari: Oribatei): New diagnoses for supra-specific taxa. *Acarologia* 24:203–17.

———. 1984. Monophyletic groups in the Enarthronota (Sarcoptiformes). In Griffiths and Bowman, 1:233–40.

———. 1985. Aspects of the biology and systematics of soil arachnids, particularly saprophagous and mycophagous mites. *Quest. Entomol.* 21:523–41.

———. 1994. Evolutionary aspects of oribatid mite life histories and consequences for the origin of the Astigmata. In Houck (1994b), contribution no. 5, 99–135.

———. 1998. Morphological evidence for the evolutionary origin of Astigmata (Acari: Acariformes). *Exp. Appl. Acarol.* 22:559–94. (Also appears in Bruin, van der Geest, and Sabelis, 45–72.)

———. 2001. Systematic relationships of Nothrolohmanniidae, and the evolutionary plasticity of body form in Enarthronota (Acari: Oribatida). In Halliday et al. (2001), 58–75.

Norton, R. A. 2007. Holistic acarology and ultimate causes: Examples from the oribatid mites. In *Acarology XI: Proceedings of the International Congress*, ed. J. B. Morales-Malacara, V. Behan-Pelletier, E. Ueckermann, T. M. Pérez, E. G. Estrada-Venegas, and M. Badii, 3–20. Mexico City: Universidad Nacional Autónoma de México.

Norton, R. A., and G. Alberti. 1997. Porose integumental organs of oribatid mites (Acari: Oribatida). 3, Evolutionary and ecological aspects. *Zoologica (Stuttgart)* 146:115–43.

Norton, R. A., G. Alberti, G. Weigmann, and S. Woas. 1997. Porose integumental organs of oribatid mites (Acari, Oribatida). 1,

Overview of types and distribution. *Zoologica (Stuttgart)* 146: 1–31.

Norton, R. A., and V. M. Behan-Pelletier. 1986. Systematic relationships of *Propelops*, with a modification of family-group taxa in Phenopelopoidea (Acari: Oribatida). *Can. J. Zool.* 64:2370–83. [PHENOPELOPIDAE]

———. 1991a. Calcium carbonate and calcium oxalate as cuticular hardening agents in oribatid mites (Acari: Oribatida). *Can. J. Zool.* 69:1504–11.

———. 1991b. Epicuticular calcification in *Phyllozetes* (Acari: Oribatida). In Dusbábek and Bukva, 2:323–24, pl. 33. [COSMOCHTHONIIDAE]

Norton, R. A., V. M. Behan-Pelletier, and H. F. Wang. 1996. The aquatic oribatid mite genus *Mucronothrus* in Canada and the western USA (Acari, Trhypochthoniidae). *Can. J. Zool.* 74: 926–49.

Norton, R. A., P. M. Bonamo, J. D. Grierson, and W. A. Shear. 1988. Oribatid mite fossils from a terrestrial Devonian deposit near Gilboa, New York. *J. Paleontol.* 62:259–69.

Norton, R. A., M. E. Florian, and L. E. Manning. 2001. Ecdysal cleavage line in *Paralycus* sp. (Acari: Oribatida: Pediculochelidae). *Internat. J. Acarol.* 27 (2): 97–99.

Norton, R. A., T. B. Graham, and G. Alberti. 1996. A rotifer-eating ameronothroid (Acari: Ameronothridae) mite from ephemeral pools on the Colorado plateau. In Mitchell et al. (1996), 1: 539–42.

Norton, R. A., and G. S. Ide. 1974. *Scutacarus bacultarsus agaricus* n. subsp. (Acarina: Scutacaridae) from commercial mushroom houses, with notes on phoretic behavior. *J. Kans. Entomol. Soc.* 47:527–34.

Norton, R. A., and J. B. Kethley. 1988. A collapsible, full-sized Berlese-funnel system. *Entomol. News* 99:41–47.

———. 1994. Ecdysal cleavage lines of acariform mites (Arachnida, Acari). *Zool. Scripta* 23:175–91.

Norton, R. A., J. B. Kethley, D. E. Johnston, and B. M. OConnor. 1993. Phylogenetic perspectives on genetic systems and reproductive modes in mites. In Wrensch and Ebbert, 8–99.

Norton, R. A., and A. Kinnear. 1999. New Australian records of xerophilic acariform mites (Oribatida and Prostigmata). *Aust. Entomol.* 26:53–55. [STIGMOCHEYLIDAE]

Norton, R. A., and J. C. Lions. 1992. North American Synichotritiidae (Acari: Oribatida) 1, *Apotritia walkeri* n. g., n. sp., from California. *Acarologia* 33:285–301.

Norton, R. A., and L. J. Metz. 1980. Nehypochthoniidae (Acari: Oribatei), a new mite family from the southeastern United States. *Ann. Entomol. Soc. Amer.* 73:54–62.

Norton, R. A., B. M. OConnor, and D. E. Johnston. 1983. Systematic relationships of the Pediculochelidae (Acari: Acariformes). *Proc. Entomol. Soc. Wash.* 85:493–512.

Norton, R. A., and Z. Olszanowski. 1989. A new *Holonothrus* (Oribatida: Crotoniidae) from Zaire, with notes on the distribution of crotoniid mites. *Rev. Zool. Afr.* 103:405–12.

Norton, R. A., and J. G. Palacios-Vargas. 1982. Nueva *Belba* (Oribatei: Damaeidae) de muscos epifitos de Mexico. *Folia Entomol. Mex.* 52:61–73.

Norton, R. A., and S. C. Palmer. 1991. The distribution, mechanisms and evolutionary significance of parthenogenesis in oribatid mites. In Schuster and Murphy, 107–36.

Norton, R. A., and G. O. Poinar. 1993. Reassessment and new record of oribatid mite fossils from Tertiary Neotropical amber. *Acarologia* 34:57–68.

Norton, R. A., and F. Sanders. 1985. Superior micro-needles for manipulating and dissecting soil invertebrates. *Quest. Entomol.* 21:673–74.

Norton, R. A., W. A. Shear, P. M. Bonamo, and J. D. Grierson. 1988. Oribatid mite fossils from a terrestrial Devonian deposit near Gilboa, New York. *J. Paleontol.* 62:259–69.

Norton, R. A., W. C. Welbourn, and R. D. Cave. 1988. First records of Erythraeidae parasitic on oribatid mites (Acari, Prostigmata: Acari, Oribatida). *Proc. Entomol. Soc. Wash.* 90:407–10.

Norton, R. A., D. D. Williams, I. D. Hogg, and S. C. Palmer 1988. Biology of the oribatid mite *Mucronothrus nasalis* (Acari: Oribatida: Trhypochthoniidae) from a small coldwater spring in eastern Canada. *Can. J. Zool.* 66:622–29.

Noske, R. A. 1993. *Bruguiera hainsii*: another bird-pollinated mangrove? *Biotropica* 25:481–83.

Nübel-Reidelbach, E. 1994. Taxonomie und Systematik der Gattung *Tectocepheus* Berlese, 1895 (Acari, Oribatei). *Andrias (Karlsruhe)* 12:3–94. [TECTOCEPHEIDAE]

Nübel-Reidelbach, E., and S. Woas. 1992. Einige basale Arten der cepheiden und der pterogasterinen Entwicklungslinie der Höheren Oribatiden (Acari, Oribatei). *Andrias (Karlsruhe)* 9:75–119.

Nucifora, A. 1963. Osservazioni sulla riproduzione di *Hemitarsonemus latus* (Banks) (Acarina Tarsonemidae). *Atti Accad. Naz. Ital. Entomol.* 10:142–53.

Nutting, W. B. 1975. Pathogenesis associated with hair follicle mites (Acari: Demodicidae). *Acarologia* 17:493–507.

———. 1985. Prostigmata-Mammalia. In Kim (1985a), 559–640.

———. 1986. Histopathology of benign tumours due to *Demodex antechini* in Antechinus stuartii. In *Parasite lives*, ed. M. Cremin, M. C. Dobson, and D. E. Moorhouse, 95–108. Brisbane: Univ. Queensland Press. [DEMODICIDAE]

Nutting, W. B., E. E. Beekman, and D. P. Snyder. 1978. *Demodex gapperi* (Acari: Demodicidae) associated with eyelid sealing in the red-backed vole, *Clethrionomys gapperi*. *J. Med. Entomol.* 14:646–48.

Nuzzaci, G. 1976. Contributo alla conoscenza dell'anatomia degli acari eriofidi. *Entomologica (Bari)* 12:55.

Nuzzaci, G., and G. Alberti. 1996. Internal anatomy and physiology. In Lindquist, Sabelis, and Bruin, 6:101–50.

Nuzzaci, G., and E. de Lillo. 1989. Contributo alla conoscenza dello gnatosoma degli Acari Tenuipalpidi (Tetranychoidea: Tenuipalpidae). *Entomologica (Bari)* 24:5–32.

———. 1991. Contributo alla conoscenza delle parti boccali di *Penthaleus major* (Dugès) (Acari: Eupodoidea Penthaleidae). *Atti XVI Cong. Naz. Ital. Entomol. (Bari)*: 265–77.

Nuzzaci, G., and A. Di Palma. 2002. Mouthparts of a tydeid mite: an ultrastructural and functional investigation. *Entomologica (Bari)* 36:71–91.

Nuzzaci, G., A. Di Palma, W. L. Magowski, and P. Aldini. 2002. Mouthparts of *Tarsonemus nodosus* Schaarschmidt, 1959 (Acari: Tarsonemidae): fine structure and functional morphology. In Bernini et al. (2002), 269–81.

Oatman, E. R. 1963. *Mite species on apple foliage in Wisconsin. Advances in acarology.* Vol. 1. Ithaca, NY: Comstock. [TYDEIDAE]

Obenchain, F. D. 1974. Structure and anatomical relationships of the

synganglion in the American dog tick *Dermacentor variabilis* (Acari: Ixodidae). *J. Morphol.* 142:205–23.
Obenchain, F. D., and J. H. Oliver, Jr. 1974. Undescribed glandular tissues from *Amblyomma tuberculatum* (Arachnida: Parasitiformes: Ixodidae). *ASB Bull.* 21:242.
———. 1976. Peripheral nervous system of the ticks *Amblyomma tuberculatum* Marx and *Argas radiatus* Raillet (Acari: Ixodoidea). *J. Parasitol.* 62:811–17.
Ochoa, R. 1989a. The genus *Tuckerella* in Costa Rica (Acari: Tuckerellidae). *Internat. J. Acarol.* 15:205–7.
———. 1989b. A note on paedogenesis in Tetranychoidea. *Internat. J. Acarol.* 15:117–18. [TUCKERELLIDAE]
Ochoa, R., H. Aguilar, and C. Vargas. 1994a. *Phytophagous mites of Central America: an illustrated guide.* Tech. Ser. 6. Turrialba, Costa Rica: Centr. Investig. Agron. Enseñanza.
———. 1994b. Superfamily Tarsonemoidea Canestrini & Fanzago, 1887. In Ochoa, Aguilar, and Vargas (1994a), 30–47.
Ochoa, R., and B. M. OConnor. 1996. A new genus of Tarsonemidae (Acari: Heterostigmata) associated with an orchid bee, *Euglossa* sp. (Hymenoptera: Apidae) in Brazil. *Internat. J. Acarol.* 22:203–7.
———. 1998. Two new species of the genus *Excelsotarsonemus* (Acari: Tarsonemidae). *Internat. J. Acarol.* 24:179–87.
———. 2000. Revision of the genus *Horstiella* (Acari: Acaridae): mites associated with Neotropical *Epicharis* bees (Hymenoptera: Apidae). *Ann. Entomol. Soc. Amer.* 93:713–37.
Ochoa, R., B. M. OConnor, and J. S. H. Klompen. 1996. A new genus and species of Tarsonemidae (Acari: Heterostigmata) associated with a fruit bat, *Cynopterus nusatenggara* (Chiroptera: Pteropodidae) on Lombok Island, Indonesia. *Internat. J. Acarol.* 22:209–13.
Ochoa, R., J. S. Pettis, and O. M. Mireles. 2003. A new bee mite of the genus *Pseudacarapis* (Acari: Tarsonemidae) from Mexico. *Internat. J. Acarol.* 29:299–305.
Ochoa, R., C. Vargas, D. E. Walter, and B. M. OConnor. 1997. Two new species of the genus *Ceratotarsonemus* (Acari: Tarsonemidae). *Internat. J. Acarol.* 23:177–83.
O'Connell, T., and T. Bolger. 1997. Stability, ephemerality and dispersal ability: microarthropod assemblages on fungal sporophores. *Biol. J. Linn. Soc.* 62:111–31.
OConnor, B. M. 1979. Evolutionary origins of astigmatid mites inhabiting stored products. In Rodriguez, 1:273–78.
———. 1982a. Acari: Astigmata. In Parker, 2:146–69.
———. 1982b. Evolutionary ecology of astigmatid mites. *Ann. Rev. Entomol.* 27:385–409.
———. 1984a. Acarine-fungal relationships: the evolution of symbiotic associations. In *Fungus-insect relationships. Perspectives in ecology and evolution*, ed. Q. Wheeler and M. Blackwell, 354–81. New York: Columbia Univ. Press.
———. 1984b. Coevolutionary patterns between astigmatid mites and primates. In Griffiths and Bowman, 1:186–95.
———. 1984c. Phylogenetic relationships among higher taxa in the Acariformes, with particular reference to the Astigmata. In Griffiths and Bowman, 1:19–62.
———. 1991. A preliminary report on the arthropod-associated astigmatid mites (Acari: Acariformes) of the Huron Mountains of northern Michigan. *Mich. Academician* 24:307–20.
———. 1992. Ontogeny and systematics of the genus *Cerophagus* (Acari: Gaudiellidae), mites associated with bumblebees. *Great Lakes Entomol.* 25:173–89.
———. 1993a. Generic relationships in the Chaetodactylidae (Acari: Astigmata) with description of a new genus. *Acarologia* 34: 345–62.
———. 1993b. The mite community associated with *Xylocopa latipes* (Hymenoptera: Anthophoridae: Xylocopinae) with descriptions of a new type of acarinarium. *Internat. J. Acarol.* 19:159–66. [CHEYLETIDAE]
———. 1994. Life-history modifications in astigmatid mites. In Houck (1994b), contribution no. 6, 136–59.
———. 2001. Historical ecology of the Acaridae (Acari): Phylogenetic evidence for host and habitat shifts. In Halliday et al. (2001), 76–82.
OConnor, B. M., R. K. Colwell, and S. Naeem. 1997. The flower mites of Trinidad. III, The genus *Rhinoseius* (Acari: Ascidae). *Misc. Publ. Mus. Zool. Univ. Mich.* 184:iv, 1–32.
OConnor, B. M., and G. C. Eickwort. 1988. Morphology, ontogeny, biology and systematics of the genus *Vidia* (Acari: Winterschmidtiidae). *Acarologia* 29:148–74.
OConnor, B. M., and M. A. Houck. 1989. Two new genera of Hemisarcoptidae (Acari: Astigmata) from the Huron Mountains of northern Michigan. *Great Lakes Entomol.* 22:1–10.
OConnor, B. M., and J. S. H. Klompen. 1999. Phylogenetic perspectives on mite-insect interactions: The evolution of acarinaria. In Needham, Mitchell, et al., 2:63–71.
OConnor, B. M., J. S. H. Klompen, and H. A. P. M. Lombert. 1987. A new subfamily of quill-inhabiting mites (Acari: Knemidokoptidae) from Asian swifts (Aves: Apodidae) with observations on phylogenetic relationships in the Knemidokoptidae. *Internat. J. Acarol.* 14:261–70.
OConnor, B. M., F. S. Lukoschus, and K. T. M. Giesen. 1982. Two new species of *Marmosopus* (Acari: Astigmata) from rodents of the genus *Scotinomys* (Cricetidae) in Central America. *Occ. Papers, Mus. Zool., Univ. Michigan* 703:1–22. [ECHIMYOPODIDAE]
OConnor, B. M., and J. C. Moser. 1985. Phylogenetic relationships of the Algophagidae (Acari: Astigmata), with descriptions of a new subfamily, genus, and species. *Ann. Entomol. Soc. Amer.* 78:783–89.
OConnor, B. M., and W. K. Reisen. 1978. *Chiroptoglyphus*, a new genus of mites associated with bats, with comments on the family Rosensteiniidae (Acari: Astigmata). *Internat. J. Acarol.* 4: 179–94.
O'Donnell, A. E., and R. C. Axtell. 1965. Predation by *Fuscuropoda vegetans* (Acarina: Uropodidae) on the house fly (*Musca domestica*). *Ann. Entomol. Soc. Amer.* 58:403–4.
Oganezova, G. G., and A. R. Pogosova. 1994. Ekologiya i biologiya chetyrekhnogogo kleshcha *Trisetacus kirghisorum* (Acariformes, Tetrapodili) [Ecology and biology of the four-legged mite *T. kirghisorum*]. *Zool. Zh.* 73 (11): 58–63. [in Russian]
Ohmer, C., A. Fain, and K.-L. Schuchmann. 1991. New ascid mites of the genera *Rhinoseius* Baker and Yunker, 1964, and *Lasioseius* Berlese, 1923 (Acari: Gamasida: Ascidae) associated with hummingbirds or hummingbird-pollinated flowers in southwestern Colombia. *J. Nat. Hist., Lond.* 25: 481–97. [MELICHARIDAE, BLATTISOCIIDAE]
Okabe, K., and M. Shun'ichi. 2003. Life history of *Kurosaia jiju* (Acari: Winterschmidtiidae) symbiotic with a mason wasp, *Anterhynchium flavomarginatum micado* (Hymenoptera: Eumenidae). *Ann. Entomol. Soc. Amer.* 96:652–59.
Okamoto, M., K. Matsumoto, and R. Shirasaka. 1991. Studies on

the attaching behavior of the *Lardoglyphus konoi* (Acari, Lardoglyphidae) hypopus and its molting into the tritonymph. *Jpn. J. Sanit. Zool.* 42:219–28.

Oldfield, G. N. 1969. The biology and morphology of *Eriophyes emarginatae*, a *Prunus* finger gall mite, and notes on *E. prunidemissae*. *Ann. Entomol. Soc. Amer.* 62:269–77. [ERIOPHYIDAE]

———. 1970. Mite transmission of plant viruses. *Ann. Rev. Entomol.* 15:343–80. [ERIOPHYIDAE]

———. 1973. Sperm storage in female Eriophyoidea (Acarina). *Ann. Entomol. Soc. Amer.* 66:1089–92.

———. 1996a. Diversity and host plant specificity. In Lindquist, Sabelis, and Bruin, 6:99–216.

———. 1996b. Toxemias and other non-distortive feeding effects. In Lindquist, Sabelis, and Bruin, 6:43–250.

———. 1999. Distribution of asymmetrical and symmetrical sperm storage in the Eriophyoidea and its phylogenetic implications. In Bruin, van der Geest, and Sabelis, 157–61.

Oldfield, G. N., R. F. Hobza, and N. S. Wilson. 1970. Discovery and characterization of spermatophores in the Eriophyoidea (Acari). *Ann. Entomol. Soc. Amer.* 63:520–26.

Oldfield, G. N., and K. Michalska. 1996. Spermatophore deposition, mating behavior and population mating structure. In Lindquist, Sabelis, and Bruin, 6:185–98.

Oldfield, G. N., and I. M. Newell. 1973. The role of the spermatophore in the reproductive biology of protogynes of *Aculus cornutus* (Acari: Eriophyidae). *Ann. Entomol. Soc. Amer.* 66:160–63.

Oldfield, G. N., and T. M. Perring. 1996. Rearing techniques. In Lindquist, Sabelis, and Bruin, 6:377–82.

Oldfield, G. N., and G. Proeseler. 1996. Eriophyoid mites as vectors of plant pathogens. In Lindquist, Sabelis, and Bruin, 6:259–75.

Oliveira, A. R., and G. J. de Moraes. 2001. Some oribatid mites (Acari, Oribatida) from the state of São Paulo, Brazil. *Rev. Bras. Zool.* 18 (Suppl. 1): 219–24.

Oliveira, A. R., G. J. de Moraes, and L. C. C. B. Ferraz. 2007. Consumption rate of phytonematodes by *Pergalumna* sp. (Acari: Oribatida: Galumnidae) under laboratory conditions determined by a new method. *Exp. Appl. Acarol.* 41:183–89.

Oliveira, A. R., R. A. Norton, and G. J. de Moraes. 2005. Edaphic and plant inhabiting oribatid mites (Acari: Oribatida) from Cerrado and Mata Atlântica ecosystems in the State of São Paulo, southeast Brazil. *Zootaxa* 1049:49–68.

Oliveira, A. R., R. A. Norton, G. J. de Moraes, and J. L. H. Faccini. 2007. Preliminary observations on courtship behaviour in *Mochloribatula* (Oribatida: Mochlozetidae). In *Acarology XI: Proceedings of the International Congress*, ed. J. B. Morales-Malacara, V. Behan-Pelletier, E. Ueckermann, T. M. Pérez, E. G. Estrada-Venegas, and M. Badii, 715–18. Mexico City: Universidad Nacional Autónoma de México.

Oliver, J. H., Jr. 1962. A mite parasitic in the cocoons of earthworms. *J. Parasitol.* 48:120–23. [HISTIOSTOMATIDAE]

———. 1964. A wasp parasite of the possum tick, *Ixodes tasmani*, in Australia. *Pan-Pac. Entomol.* 40:227–30.

———. 1977. Cytogenetics of ticks and mites. *Ann. Rev. Entomol.* 22:407–29.

———. 1982. Tick reproduction: Sperm development and cytogenetics. *Current Themes Trop. Sci.* 1:245–75.

———. 1983. Chromosomes, genetic variance and reproductive strategies among mites and ticks. *Bull. Entomol. Soc. Amer.* 29:8–17.

———. 1989. Biology and systematics of ticks (Acari: Ixodida). *Ann. Rev. Ecol. Syst.* 20:297–430.

Oliver, J. H., Jr., Z. Al-Ahmadi, and R. L. Osburn. 1974. Reproduction in ticks (Acari: Ixodoidea). 3, Copulation in *Dermacentor occidentalis* Marx and *Haemaphysalis leporispalustris* (Packard) (Ixodidae). *J. Parasitol.* 60:499–506.

Olivier, P. A. S., and P. D. Theron. 1997. The genus *Benoinyssus* Fain (Acari: Eupodidae) from southern Africa, with descriptions of five new species. *Afr. Entomol.* 5:301–18.

———. 2000. Pentapalpidae, a new family of eupodoid mites (Prostigmata: Eupodoidea) from South Africa. *Acarologia* 40:385–92.

Oloo, G. W., C. K. P. O. Ogol, and K. O. Kambona. 1987. Biotaxonomy of cassava green spider mites, *Mononychellus* spp. (Tetranychidae): "Life type" as a possible biological criterion for their identification. *Insect Sci. Appl.* 8:995–99.

Olszanowski Z. 1996. *A monograph of the Nothridae and Camisiidae of Poland (Acari: Oribatida: Crotonioidea)*. *Genus* Suppl. Wroclaw: Polish Taxonomical Society.

———. 1999. Three new species of *Holonothrus* from the Neotropical region (Acari: Oribatida). *J. Nat. Hist.* 33:233–53. [CROTONIIDAE]

Olszanowski, Z., A. Rajski, and W. Niedbała. 1996. *Roztocze Acari: Mechowce Oribatida*. Catalogus Faunae Poloniae 34. Poznan: Sorus.

Oomen, P. A. 1982. Studies on population dynamics of the scarlet mite, *Brevipalpus phoenicis*, a pest of tea in Indonesia. *Meded. Landbouwk. Wageningen* 82-1:1–88.

Oppedisano, M., M. Eguaras, and N. A. Fernandez. 1995. Deposition of spermatophores and signalling structures in *Pergalumna* sp. (Acari: Oribatida). *Acarologia* 36:347–53. [GALUMNIDAE]

Orghidan, T. N., and M. Georgesco. 1977. Hydracariens de Cuba. Espèces nouvelles de *Neolimnochares* et de *Wandesia*. *Résult. Expéd. Biospéol. Cubano–Roum. Cuba* 2:209–15. [LIMNOCHARIDAE, HYDRYPHANTIDAE]

Orghidan, T. N., and M. Gruia. 1980a. Diagnose de trois hydrachnelles nouveaux de Cuba. *Trav. Inst. Spéol. Emile Racovitza* 19:143–46.

———. 1980b. Hydracariens nouveaux de Cuba appartenant aux familles des Unionicolidae Oudem., Mideidae Thor et Hungarohydracaridae Motas and Tanasachi. *Trav. Inst. Spéol. Emile Racovitza* 19:125–42.

———. 1981. Espèces nouvelles d'*Axonopsella* (Hydracarina) de Venezuela et Cuba. *Trav. Inst. Spéol. Emile Racovitza* 20:55–73. [ATURIDAE]

———. 1983a. Description de deux espèces nouvelles d'hydracariens de Cuba. *Trav. Inst. Spéol. Emile Racovitza* 22:3–8. [LIMNESIIDAE, BOGATIIDAE]

———. 1983b. Espèces nouvelles d'hydracariens de Cuba. *Résult. Expéd. Biospéol. Cubano–Roum. Cuba* 4:182–89. [LIMNESIIDAE, HYGROBATIDAE, ARENOHYDRACARIDAE]

———. 1983c. Sur trois espèces nouvelles d'Hydrachnellae de Cuba appartenant au genre *Cladomomonia* Orgh. & Gruia et au sous-genre *Crocokongsbergia* Orgh. & Gruia. *Résult. Expéd. Biospéol. Cubano–Roum. Cuba* 4:167–79. [MOMONIIDAE, ATURIDAE]

———. 1987. Hydrachnellae du Venezuela. In *Fauna hipogea y hemiedáfica de Venezuela y de otros paises de América del Sur*, ed. V. Decu et al., 115–31. Bucharest: Academiei Republicii Socialiste România.

Orghidan, T. N., M. Gruia, and N. Vina Bayes. 1977. Sur une espèce d'Hydracarien vivant dans les phytotelmes de Cuba. *Résult. Expéd. Biospéol. Cubano–Roum. Cuba* 2:227–30. [ARRENURIDAE]

Orghidan, T. N., M. Gruia, M. Georgesco, and N. Vina Bayes. 1977. Sur deux Hydrachnelles souterraines nouvelles de Cuba. *Résult. Expéd. Biospéol. Cubano–Roum. Cuba* 2:217–26. [MOMONIIDAE]

Orghidan, T. N., M. Gruia, and V. Iavorschi. 1981. Espèces nouvelles d'hydracariens de Cuba appartenant au genre *Axonopsis* Piersig. *Trav. Inst. Spéol. Emile Racovitza* 20:29–53. [ATURIDAE]

Orwig, K. R. 1968. *The genera and species of the feather mite subfamily Trouessartiinae (except Trouessartia).* Bull. Univ. Nebraska St. Mus. 8. Lincoln: Univ. Nebraska St. Mus. [TROUESSARTIIDAE]

Oseto, C. Y., and Z. B. Mayo. 1975. A new species of *Caloglyphus* (Acarina: Acaridae) associated with *Phyllophaga anxia* (Coleoptera: Scarabeidae). *J. Kans. Entomol. Soc.* 48:212–23.

Otto, J. C. 1992. A new species of *Anystis* Von Heyden compared with *Anystis salicinus* (Linnaeus) (Acarina: Anystidae). *Internat. J. Acarol.* 18:25–35.

———. 1993. A new species of *Microcaeculus* from Australia (Acarina: Caeculidae), with notes on its biology and behavior. *Internat. J. Acarol.* 19:3–13.

———. 1997. Observations on prelarvae in Anystidae and Teneriffiidae. In Mitchell et al. (1996), 1:343–54.

———. 1999a. *Corallihalacarus chilcottensis*, a new genus and species of marine mite from the Coral Sea (Acarina: Halacaridae). *Zool. Sci.* 16:839–43.

———. 1999b. Four new species of *Agaue* (Acarina: Halacaridae) from the Great Barrier Reef Marine Park. *Cah. Biol. Mar.* 40:273–81.

———. 1999c. Halacarid fauna of the Great Barrier Reef and Coral Sea: The genera *Agauopsis* and *Halacaropsis* (Acarina: Halacaridae). *Mem. Queensland Mus.* 43:797–817.

———. 1999d. Revision of the genus *Erythracarus* Berlese (Acarina: Anystidae: Erythracarinae). *J. Nat. Hist.* 33:825–909.

———. 1999e. Systematics and natural history of the genus *Chaussieria* Oudemans (Acarina: Prostigmata: Anystidae). *Zool. J. Linn. Soc.* 126:251–306.

———. 1999f. The taxonomy of *Tarsotomus* Berlese and *Paratarsotomus* Kuznetsov (Acarina: Anystidae: Erythracarinae) with observations on the natural history of *Tarsotomus*. *Invertebr. Taxon.* 13:749–803.

———. 2000a. A cladistic analysis of Erythracarinae (Acarina: Prostigmata: Anystidae), with the description of a new genus. *Syst. Entomol.* 25:447–84.

———. 2000b. *Spongihalacarus longiscutus* n. gen., n. sp., a marine mite (Acari: Prostigmata: Halacaridae) associated with an alga-sponge symbiosis from the Great Barrier Reef Lagoon in Australia. *Internat. J. Acarol.* 26:279–83.

Otto, J. C., and R. B. Halliday. 1991. Systematics and biology of a predatory mite (*Anystis* sp.) introduced into Australia for biological control of the redlegged earth mite. *Plant Prot. Qtly.* 6: 181–85.

Otto, J. C., and R. Olomski. 1994. Observations on a motile prelarva in *Chaussieria venustissima* (Berlese 1882) (Acari: Anystidae) with a description of the larva. *Can. J. Zool.* 72:287–92.

Oudemans, A. C. 1906. Nieuwe classificatie der Acari. *Entomol. Ber.* 2:43–46.

———. 1914. Acarologisches aus Maulwurfnestern. *Arch. Naturgesch. Berlin* 79A, Heft 8:108–200.

———. 1923. Studie over de sedert 1877 ontworpen Systemen der Acari; Nieuwe Classificatie; Phylogenetische Beschouwingen. *Tijdschr. Entomol.* 66:49–85.

———. 1926a. Acarologische Aanteekeningen LXXX. *Entomol. Ber.* 7 (148): 67–80.

———. 1926b. *Kritisch Historisch Oversicht der Acarologie 1 (850 V.C.–1758).* Leiden: E. J. Brill.

———. 1927. Acarologische aanteekeningen LXXXVIII. *Entomol. Ber.* 7:257–63. [BARBUTIIDAE]

———. 1928. Acari uit Ambon. *Zool. Meded.* 10:185–237.

———. 1929. *Kritisch Historisch Oversicht der Acarologie 2 (1759–1804).* Leiden: E. J. Brill.

———. 1931. Acarologische aanteekeningen CVIII. *Entomol. Ber.* 8:251–63. [BARBUTIIDAE]

———. 1936a. *Kritisch Historisch Oversicht der Acarologie 3 (1805–1850).* Leiden: E. J. Brill.

———. 1936b. Neues uber Anystidae (Acari). *Arch. Naturgesch. N. F.* 5:364–446.

Overmeer, W. P. J. 1985. Alternative prey and other food resources. In Helle and Sabelis (1985b), 1B:131–39.

Özdickmen, H. 2006. Necessary changes of names in Hydrachnellae (Arachnida: Acari). *Mun. Ent. Zool.* 1:51–54.

Özkan, M. 1981. *Georgella* (Acari, Hydrachnellae) Türleri Üzerine Sistematik bir Araştırma. (Eine systematische Untersuchung über *Georgella*-Arten (Acari, Hydrachnellae)). *Atatürk Üniv. Fen Fak. Derg.* 1:1–8. [HYDRYPHANTIDAE].

———. 1982. Wassermilben (Actinedia) aus der Türkei. *Entomol. Basil.* 7:29–60.

———. 1985. *Hydrobaderia* n. gen., eine neue Hydryphantiden-Gattung aus der Türkei (Acari, Actinedia, Hydrachnellae). *Entomol. Basil.* 10:19–26. [HYDRYPHANTIDAE]

———. 1988a. *Hydryphantes* (s. str.) *crassipalpis* Koenike, 1914 (Hydryphantidae, Hydrachnellae, Acari) Üzerine bir Araştırma. *Doğa TU Zool., Derg.* 12:86–109.

———. 1988b. *Tadjikothyas cinsi* (Hydrachnellae, Acari) ve Türkiye'den Yeni bir Alttürü Hakkinda. *Doğa TU Zool., Derg.* 12:53–61. [HYDRYPHANTIDAE].

———. 1989. Doğu Anadolu su Akarları (Acari, Hydrachnellae) Üzerine Sistematik Araştirmalar IV. *Doğa TU Zool., Derg.* 13:88–108.

Özkan, M., and C. Bader. 1988. Zwei neue Arten nebst Ergänzungen zur türkischen Wassermilben-Fauna (Acari, Actinedida, Hydrachnellae). *Arch. Hydrobiol.* 114:133–45. [HYDRYPHANTIDAE, ANISITSIELLIDAE]

Özkan, M., and Y. Ö. Boyaci. 1990. Türkiye Faunası İçin Yeni bir *Nilotonia* Thor, 1905 (Acari, Hydrachnellae) Türü. *Doğa TU Zool., Derg.* 14:200–213. [ANISITSIELLIDAE].

———. 1991. Türkiye Faunası İçin Yeni iki *Megaluracarus* Viets, 1911 (Arrenuridae, Hydrachnellae, Acari) Türü. *Doğa TU Zool., Derg.* 15:164–71.

———. 1992a. Türkiye Faunası İçin Yeni bir *Forelia* Haller, 1882 (Hydrachnellae, Acari) Türü. *Doga TU Zool., Derg.* 16:385–94. [PIONIDAE].

———. 1992b. Türkiye Faunası İçin Yeni bir *Ljania* Thor, 1898 (Acari, Hydrachnellae, Axonopsidae) Türü. *Doğa TU Zool., Derg.* 16:84–91. [ATURIDAE].

———. 1992c. Türkiye Faunası İçin Yeni bir *Pseudotorrenticola*

Walter, 1906 (Hydrachnellae, Acari) Türü. *Doğa TU Zool., Derg.* 16:141–47. [TORRENTICOLIDAE].
———. 1992d. Türkiye'den Yeni bir *Trichothyas* Viets 1936 (Thyasidae, Hydrachnellae, Acari) Türü. *Doğa TU Zool., Derg.* 16:395–404. [HYDRYPHANTIDAE].
Özkan, M., and O. Erman. 1990a. Türkiye'den Yeni bir *Rhinophoracarus* (Acari, Hydrachnellae, Arrenuridae) Türü. *Doğa TU Zool., Derg.* 14:229–36.
———. 1990b. Türkiye Faunası İçin Yeni bir *Mideopsis* (Mideopsidae, Hydrachnellae, Acari) Türü. *Doğa TU Zool., Derg.* 14: 301–10.
———. 1991. Türkiye Faunası İçin Yeni üç *Arrenurus* Dugès, 1834 (Hydrachnellae, Acari) Türü. *Doğa TU Zool., Derg.* 15:323–36. [ARRENURIDAE].
Özkan, M., O. Erman, and Y. Ö. Boyaci. 1993. Sultan Sazliği'nin (Kayseri) Türkiye Faunası İçin Yeni Bazi *Arrenurus* Dugès, 1834 (Acari, Hydrachnellae, Arrenuridae) Türleri—I. *Doğa TU Zool., Derg.* 17:471–501.
Özkan, M., and Z. Küçükönder. 1993. *Nilotonia* (*Dartia*) *vietsi* Bader and Sepasgosarian, 1980 (Acari, Hydrachnellae) Üzerine bir Araştırma. *Doğa TU Zool., Derg.* 17:325–39. [ANISITSIELLIDAE].
Özkan, M., and Z. Soysal. 1989. Türkiye'den Yeni bir *Nilotonia* (Acari, Hydrachnellae) Türü. *Doğa TU Zool., Derg.* 13:15–27. [ANISITSIELLIDAE].
Palacios-Vargas, J. G. 1994. Los Acaros Oribátidos de México. *An. Inst. Biol., Univ. Nac. Aut. Méx., Ser. Zool.* 65:19–32.
Palacios-Vargas, J. G., and D. Prieto-Trueba. 1995. A new species of *Saxidromus* (Endeostigmata: Adamystidae) from Cuba. *Acarologia* 36:213–17.
Palacios-Vargas, J. G., and M. M. Vázquez. 1988. Primer registo de acaros del orden Opilioacarida en Nicaragua. *Rev. Nicarag. Entomol.* 4:45–49.
Palevsky, E., V. Soroker, P. Weintraub, F. Mansour, F. Abo-Moch, and U. Gerson. 2001. How species-specific is the phoretic relationship between the broad mite, *Polyphagotarsonemus latus* (Acari: Tarsonemidae), and its insect hosts? *Exp. Appl. Acarol.* 25:217–24.
Palmer, S. C., and R. A. Norton. 1992. Genetic diversity in thelytokous oribatid mites (Acari: Acariformes: Desmonomata). *Biochem. Ecol. Syst.* 20:219–31.
Pandya, A. P. 1981. A note on phlebotomid sandflies of Surat, Gujarat State. *Indian J. Med. Res.* 70:957–59. [RAPHIGNATHIDAE]
Panesar, A. R. 2004. Evolution in water mites (Hydrachnellae, Actinedida, Acari). A revision of the Anisitsiellidae Koenicke, 1910. *Bonner Zool. Monogr.* 52.
Panesar, A. R., and R. Gerecke. 1994. A new *Partnunia* species (Acari, Actinedida, Hydryphantidae) from Keylong (Himalaya, India). *Aquat. Insects* 16:241–47.
Paoletti, M. G., B. R. Stinner, D. Benzig, and R. Taylor. 1991. Diversity of soil fauna in the canopy and forest floor of a Venezuelan cloud forest. *J. Trop. Ecol.* 7:373–83.
Park, C. K., and W. T. Atyeo. 1971. A generic revision of the Pterodectinae, a new subfamily of feather mites (Sarcoptiformes: Analgoidea). *Bull. Univ. Nebraska St. Mus.* 9:39–88. [PROCTOPHYLLODIDAE]
———. 1975. The pterodectine feather mites of hummingbirds: The genus *Xynonodectes* Park and Atyeo. *J. Georgia Entomol. Soc.* 10:128–44. [PROCTOPHYLLODIDAE]
Parker, S. P., ed. 1982. *Synopsis and classification of living organisms.* Vol. 2. New York: McGraw-Hill.
Paschoal, A. D. 1984. *Idiodamaeus* (Acari: Oribatei), a new genus of Gymnodamaeidae from South America. *Rev. Bras. Entomol.* 28 (4): 373–77. [IDIODAMAEIDAE]
———. 1988. A revision of the Plateremaeidae (Acari: Oribatei). *Rev. Bras. Zool.* 3 (6): 327–56.
———. 1989. Recharacterization of Gymnodamaeoidea and erection of Plateremaeoidea (Acari, Oribatei), with key to families and genera. *Rev. Bras. Zool.* 6:191–200.
Pauly, F. 1952. Die "Copula" der Oribatiden (Moosmilben). *Naturwissenschaften* 39:572–73.
———. 1956. Zur Biologie einiger Belbiden (Oribatei: Moosmilben) und zur Funktion ihrer pseudostigmatischen Organe. *Zool. Jb. Syst.* 84:275–328. [DAMAEIDAE]
Pavlichenko, P. G. 1991. A new genus and a new species of oribatid mites (Acari, Oribatei) from the territory of Ukraine. *Vestn. Zool.* 91 (4): 37–41.
———. 1994. *A guide to the ceratozetoid mites (Oribatei, Ceratozetoidea) of Ukraine.* Kiev: National Academy of Sciences of the Ukraine. [in Russian]
Pavlovsky, E. N., and A. N. Skrynnik. 1960. Comparative data on the biology of some species of ticks of the genus *Ornithodoros. Akad. Nauk SSSR* 133:734–36. [ARGASIDAE]
Pecina, P. 1970. Czechoslovak uropodid mites of the genus *Trachytes* Michael, 1894 (Acari, Mesostigmata). *Acta Univ. Carolinae, Biol.* 1969:39–59. [TRACHYTIDAE]
Pegazzano, F. 1971. Osservazioni su *Brevipalpus lineola* G. Can. e Fanz. e su *B. wainsteini* Livsh. e Mitrof. (Acarina, Tenuipalpidae) infeudati al gen. *Pinus* in Italia. *Redia* 52:739–53, pl. 59.
Peláez, D. R. P., and A. Barrera. 1948. Estudios sobre hematozoarios. I, *Plasmodium mexicanum* Thompson and Huff 1944 en sus huespedes naturales. *An. Esc. Nac. Cienc. Biol. Méx.* 5:197–215.
Pence, D. B. 1973. The nasal mites of birds from Louisiana. VII, The Ereynetidae (Speleognathinae). *J. Parasitol.* 59:364–68.
———. 1975. *Keys, species and host list, and bibliography for nasal mites of North American birds (Acarina: Rhinonyssidae, Turbinoptinae, Speleognathinae, and Cytoditidae).* Spec. Publ. Mus. Texas Tech Univ. 8. Lubbock.
Pence, D. B., and S. D. Casto. 1976. Studies on the variation of morphology of the *Ptilonyssus "sairae"* complex (Acarina: Rhinonyssidae) from North American birds. *J. Med. Entomol.* 13:71–95.
Penman, D. R., and W. W. Cone. 1974. Structure of cuticular lyrifissures in *Tetranychus urticae. Ann. Entomol. Soc. Amer.* 67:1–4.
Pereira, C., and M. P. de Castro. 1945. Contribuicao para o conhecimento da especie tipo de *Macrocheles* Latr. (Acarina): *M. muscaedomesticae* (Scopoli, 1772) emend. *Arq. Inst. Biol. Sao Paulo* 16:153–86.
Perez, T. M. 1995. Seven species of *Fainalges* Gaud and Berla (Analgoidea, Xolalgidae) from *Aratinga holochlora* (Sclater) (Aves, Psittacidae). *Zool. Scripta* 24:203–23.
———. 1996. The eggs of seven species of *Fainalges* Gaud and Berla (Xolalgidae) from the green conure (Aves, Psittacidae). In Mitchell et al. (1996), 1:297–300.
Perez, T. M., and W. T. Atyeo. 1991. The feather mite family Cheylabididae (Acarina, Pterolichoidea) with new taxa from sandgrouse (Aves, Pteroclidae). *Zool. Anz.* 226:253–60.
Pérez-Íñigo, C. 1967. Les Lohmanniidae d'Espagne (Acari, Oribatei). *Eos, Rev. Esp. Entomol.* 43:157–70.
———. 1970. Acaros Oribátidos de suelos de España peninsular e is-

las Baleares (Acari, Oribatei). Parte II. *Eos, Rev. Esp. Entomol.* 45:241–317. [HERMANNIELLIDAE, NEOLIODIDAE, GYMNODAMAEIDAE, LICNODAMAEIDAE, and others]

———. 1993a. *Acari, Oribatei, Poronota*. Vol. 3 in Fauna iberica. Madrid: Mus. Nac. Cienc. Nat.

———. 1993b. Redescription of *Licnobelba latiflabellata* (Paoli, 1908) (Acari, Oribatei, Licnobelbidae). *Redia* 76 (2): 317–26.

———. 1997. *Acari. Oribatei. Gymnonota I*. Vol. 9 in Fauna iberica. Madrid: Mus. Nac. Cienc. Nat.

Pérez-Íñigo, C., and D. Baggio. 1980. Oribátidos edáficos do Brasil. I. *Bolm. Zool., Sao Paulo, Brasil* 5:111–47.

———. 1985. Oribates édaphiques du Bresil (II). Oribates de l'île du Cardoso (Premiere partie). *Acarologia* 26 (2): 183–99. [MALACONOTHRIDAE, EPILOHMANNIIDAE, NEOLIODIDAE, MICROTEGEIDAE, and others]

———. 1993. Oribates édaphiques du Brésil (VII) Oribates de l'État de Sao Paulo (Quatrieme partie). *Acarologia* 34 (3): 249–64.

Pérez-Íñigo, C., and M. A. Peña. 1994. Oribátidos edáficos (Acari, Oribatei) de Gran Canaria (I). *Bol. Asoc. Entomol.* 18 (1–2): 65–77.

Pérez-Íñigo, C., and C. Pérez-Íñigo, Jr. 1993. Oribates (Acari, Oribatei) trouvés sur branches d'*Araucaria angustifolia* au Brésil. *Acarologia* 34 (2): 167–76.

Pérez-Íñigo, C., and L. S. Subías. 1978. Sorprendente hallazgo de un representante de la familia Kodiakellidae Hammer en España, *Kodiakiella dimorpha* n. sp., y consideraciones sobre esta familia (Acari, Oribatei). *Bol. Asoc. Esp. Entomol.* 1:103–7.

Perotti, M. A., and H. R. Braig. 1999. Endosymbionts of Acari. In Weigmann et al. (2005), 457–74.

Perring, T. M. 1996. Vegetables. In Lindquist, Sabelis, and Bruin, 6:593–610.

Perring, T. M., and C. A. Farrar. 1986. Historical perspective and current world status of the tomato russet mite (Acari: Eriophyidae). *Misc. Publ. Entomol. Soc. Amer.* 63:1–19.

Petersen, H. 1978. Some properties of two high-gradient extractors for soil microarthropods, and an attempt to evaluate their extraction efficiency. *Nat. Jutlandica* 20:95–121.

———. 1982. Structure and size of soil animal populations. *Oikos* 39:306–29.

Peterson, P. C. 1971. A revision of the feather mite genus *Brephosceles* (Proctophyllodidae: Alloptinae). *Bull. Univ. Nebraska St. Mus.* 9:89–172. [ALLOPTIDAE]

Peterson, P. C., and W. T. Atyeo. 1968. New genera related to the genus *Brephosceles* Hull, 1934 (Acarina: Proctophyllodidae). *Bull. Univ. Nebraska St. Mus.* 8:217–36. [ALLOPTIDAE]

Peterson, P. C., W. T. Atyeo, and W. W. Moss. 1980. *The feather mite family Eustathiidae (Acarina: Sarcoptiformes)*. Monogr. Acad. Nat. Sci. Philadelphia 21.

Peterson, R. K. D., R. B. Smelser, T. H. Klubertanz, L. P. Pedigo, and W. C. Welbourn. 1992. Ectoparasites of the bean leaf beetle (Coleoptera: Chrysomelidae) by *Trombidium hyperi* Vercammen-Grandjean, Van Driesche and Gyrisco and *Trombidium newelli* Welbourn and Flessel (Acari: Trombidiidae). *J. Agr. Entomol.* 9:99–107.

Petrova, A. 1971. On halacarid migration into freshwaters. In Daniel and Rosický, 169–71.

———. 1974a. *Momonisia phreatica* n. gen., n. sp. (Momoniidae, Hydrachnellae) des eaux souterraines de Bulgarie. *Internat. J. Speleol.* 6:181–86.

———. 1974b. Sur la migration des halacariens dans les eaux douces et la position systématique des halacariens et limnohalacriens. *Vie Milieu* 24:87–96.

Petrova, A. D. 1977. Fam. Zerconidae Canestrini, 1891. In Gilyarov and Bregetova, 577-621. [in Russian]

———. 1978a. A new genus and species of zerconid mite (Parasitiformes, Gamasoidea, Zerconidae) with an entire dorsal shield. *Entomol. Oboz.* 57:218–20. [in Russian]

———. 1978b. Two new species of mites of genus *Neozercon* (Mesostigmata, Zerconidae). *Biol. Nauki* 1978 (7): 67–71. [in Russian]

———. 1979. New species of the mite genus *Caurozercon* Hal. (Parasitiformes, Mesostigmata, Zerconidae) for fauna of USSR. *Entomol. Oboz.* 58: 909–15. [in Russian]

Petrova-Nikitina, A. D. 1987. Acarofauna structure of the *Anacanthotermes ahngerianus* Jac. nests in southwestern Turkmenia (Acari; Isoptera). *Proc. IX Internat. Colloq. Soil Zool., Moscow* 1985:575–78.

Pfeiffer, H. 1979. Zum *Cheyletiella*-Befall der Hauskatze. *Z. Parasitenkd.* 59:95–106. [CHEYLETIELLIDAE]

Philip, C. B. 1963. Ticks as purveyors of animal ailments: A review of pertinent data and recent contributions. In *Advances in acarology*, ed. J. A. Naegele, 1:285–325. Ithaca, NY: Comstock.

Philip, R. N. 1980. Scrub typhus. In *CRC handbook series in zoonoses: Section A: Bacterial, rickettsial and mycotic diseases*, ed. J. H. Steele, 2:303–15. Boca Raton, FL: CRC Press. [TROMBICULIDAE]

Philips, J. R., and A. Fain. 1991. Acarine symbionts of louseflies (Diptera: Hippoboscidae). *Acarologia* 32:377–84.

Philips, J. R., and R. A. Norton. 1978. *Lardoglyphus falconidus* n. sp. (Acarina: Acaridae) from the nest of an American kestrel (*Falco sparverius* L.). *Acarologia* 20:128–37. [LARDOGLYPHIDAE]

Phillips, C. J., J. K. Jones, Jr., and F. J. Radovsky. 1969. Macronyssid mites in the oral mucosa of long-nosed bats: occurrence and associated pathology. *Science* 165:1368–69.

Pielou, D. P., and A. N. Verma. 1968. The arthropod fauna associated with the birch bracket fungus, *Polyporus betulinus*, in eastern Canada. *Can. Entomol.* 100:1179–99.

Piffl, E. 1966. *Spinozetes inexpectatus* n. g. n. sp. Eine neue Gattung der Oribatiden (Acari) aus Griechenland. *Acarologia* 8 (3): 499–510. [SPINOZETIDAE]

———. 1972. Zur Systematik der Oribatiden (Acari). *Khumbu Himal* 4:269–314.

———, ed. 1979. *Proc. 4th International Congress of. Acarology, Saalfelden*. Budapest: Akad. Kiadó.

Pijnacker, L. P., M. A. Ferwerda, H. R. Bolland, and W. Helle. 1980. Haploid female parthenogenesis in the false spider mite *Brevipalpus obovatus* (Acari: Tenuipalpidae). *Genetica* 51:211–14.

Pijnacker, L. P., M. A. Ferwerda, and W. Helle. 1981. Cytological investigations on the female and male reproductive system of the parthenogenetic privet mite, *Brevipalpus obovatus* (Donnadieu) (Phytoptipalpidae, Acari). *Acarologia* 22:157–63. [TENUIPALPIDAE]

Pittard, D. A. 1971. A comparative study of the life stages of the mite, *Stereotydeus mollis* W. & S. (Acarina). *Pac. Insects* 25:1–14. [PENTHALODIDAE]

Platts-Mills, T. A. E., and A. de Weck. 1988. Summary report of the meeting. In *Mite allergy: A world-wide problem*, ed. A. L. de Weck and A. Todt, 3–12. Brussels: UCB Inst. Allergy.

Poe, S. L. 1971. Microfaunal populations on gladiolus corms. *Fla. Entomol.* 54:127–33. [BLATTISOCIIDAE]

Poinar, G. O., Jr. 1985. Fossil evidence of insect parasitism by mites. *Internat. J. Acarol.* 11:37–38.
———. 1988. Hair in Dominican amber: evidence for tertiary land mammals in the Antilles. *Experientia* 44:88–89.
———. 1992. *Life in amber.* Stanford, CA: Stanford Univ. Press.
———. 1995. First fossil soft ticks, *Ornithodoros antiquus* n. sp. (Acari: Argasidae) in Dominican amber with evidence of their mammalian host. *Experientia* 51:384–87.
Poinar, G. O., Jr., A. Acra, and F. Acra. 1994. Animal-animal parasitism in Lebanese amber. *Med. Sci. Res.* 22:159. [ERYTHRAEIDAE]
Poinar, G. O., Jr., and A. E. Brown. 2003. A new genus of hard ticks in Cretaceous Burmese amber (Acari: Ixodida: Ixodidae). *Syst. Parasitol.* 54 (3): 199–205.
Poinar, G. O., Jr., P. M. Curcic, and J. C. Cokendolpher. 1998. Arthropod phoresy involving pseudoscorpions in the past and present. *Acta Arachnol.* 47:79–96.
Poinar, G. O., Jr., G. W. Krantz, A. J. Boucot, and T. M. Pike. 1997. A unique Mesozoic parasitic association. *Naturwissenschaften* 84:321–22. [ERYTHRAEIDAE]
Poinar, G. O., Jr., E. [T.] M. Pike, and G. W. Krantz. 1993. Animal-animal parasitism. *Science* 361:307–8.
Poinar, G. O., Jr., and R. Poinar. 1999. *The amber forest: A reconstruction of a vanished world.* Princeton, NJ: Princeton Univ. Press.
Poinar, G. O., Jr., A. E. Treat, and R. V. Southcott. 1991. Mite parasitism of moths: examples of paleosymbiosis in Dominican amber. *Experientia* 47:210–12. [ERYTHRAEIDAE]
Polak, M. 1996. Ectoparasitic effects on host survival and reproduction: the *Drosophila-Macrocheles* association. *Ecology* 77:1 379–89.
Polec, W., and B. Moskwa. 1994. The development of the early larval forms of *Moniezia expansa* under laboratory conditions. *Wiad. Parazytol.* 40:153–57.
Pospelova-Shtrom, M. V. 1969. On the system of classification of ticks of the family Argasidae Can., 1890. *Acarologia* 11:1–22.
Potter, D. A., and D. E. Johnston. 1976. *Canestriniphis megalodacne* n. g., n. sp. (Acari: Eviphididae) from a pleasing fungus beetle, *Megalodacne heros. Ann. Entomol. Soc. Amer.* 69:494–96.
Pound, J. M., and J. H. Oliver, Jr. 1982. Synganglial and neurosecretory morphology of female *Ornithodoros parkeri* (Cooley) (Acari: Argasidae). *J. Morphol.* 173:159–77.
Prasad, V. 1968. *Noctuiseius treati*, a new genus and species of moth mite from Hawaii and Easter Island. *Ann. Entomol. Soc. Amer.* 61:411–13. [OTOPHEIDOMENIDAE]
———. 1975. A new genus and species of pterygosomatid mite (Acarina: Pterygosomatidae) from India. *Internat. J. Acarol.* 2:14–17.
———, ed. 1982. *History of acarology.* W. Bloomfield, MI: Indira Publishing House.
Prasad, V., and D. R. Cook. 1972. *The taxonomy of water mite larvae.* Gainesville, FL: Amer. Entomol. Inst.
Prasse, J. 1967a. Zur Anatomie und Histologie der Acaridae mit besonderer Berück-sichtigung von *Caloglyphus berlesei* (Michael 1903) und *C. michaeli* (Oudemans 1924). I, Das Darmsystem. *Wiss. Zeitschr. Univ. Halle* 16:789–812.
———. 1967b. Zur Anatomie und Histologie der Acaridae mit besonderer Berücksichtigung von *Caloglyphus berlesei* (Michael 1903) und *C. michaeli* (Oudemans 1924). II, Das Exkretionsystem und das Bindgewebe. *Wiss. Zeitschr. Univ. Halle* 16: 963–70.
———. 1968. Zur Anatomie und Histologie der Acaridae mit besonderer Berücksichtigung von *Caloglyphus berlesei* (Michael 1903) und *C. michaeli* (Oudemans 1924). III, Die Drüsen und drüsenahnlichen Gebilde der Podocephalkanal. *Wiss. Zeitschr. Univ. Halle* 17:629–46.
———. 1970. Zur Anatomie und Histologie der Acaridae mit besonderer Berücksichtigung von *Caloglyphus berlesei* (Michael 1903) und *C. michaeli* (Oudemans 1924). IV, Das Genitalsystem. *Wiss. Zeitschr. Univ. Halle* 19:93–116.
Pratt, P. D., and B. A. Croft. 2000. Overwintering and comparative sampling of *Neoseiulus fallacis* (Acari: Phytoseiidae) on ornamental nursery plants. *Environ. Entomol.* 29:1034–40.
Price, D. W. 1971. A new species of *Pomerantzia* Baker from California (Acarina: Pomerantziidae). *Proc. Entomol. Soc. Wash.* 73:394–98.
———. 1972. A new iolinid mite from the Galapagos Islands (Acarina: Iolinidae). *Ann. Entomol. Soc. Amer.* 65:793–96.
———. 1973a. Abundance and vertical distribution of microarthropods in the surface layers of a California pine forest soil. *Hilgardia* 42:121–47.
———. 1973b. Genus *Pediculocheles* (Acarina: Pediculochelidae), with notes on *P. raulti* and description of two new species. *Ann. Entomol. Soc. Amer.* 66:302–7.
———. 1974. Notes on the genus *Pomerantzia* Baker, with a description of a second species from California pine forest soil. *Hilgardia* 42:121–48.
———. 1975. Vertical distribution of small arthropods in a California pine forest soil. *Ann. Entomol. Soc. Amer.* 68:174–80.
Price, D. W., and G. S. Benham, Jr. 1976. Vertical distribution of pomerantziid mites (Acarina: Pomerantziidae). *Proc. Entomol. Soc. Wash.* 78:309–13.
Price, P. W. 1997. *Insect ecology.* 3d ed. New York: John Wiley & Sons.
Prinzing, A., and H.-P. Wirtz. 1997. The epiphytic lichen, *Evernai punastri* L., as a habitat for arthropods: Shelter from desiccation, food-limitation and indirect mutualism. In *Canopy arthropods*, ed. N. E. Stork, J. Adis, and R. K. Didham, 477–94. London: Chapman & Hall.
Pritchard, A. E. 1956. A new superfamily of trombidiform mites with the description of a new family, genus, and species (Acarina: Iolinoidea: Iolinidae: *Iolina nana*). *Ann. Entomol. Soc. Amer.* 49:204–6.
Pritchard, A. E., and E. W. Baker. 1955. *A revision of the spider mite family Tetranychidae.* Memoir Series, vol. 2. San Francisco: Pac. Coast Entomol. Soc.
———. 1958. The false spider mites (Acarina: Tenuipalpidae). *Univ. Calif. Publ. Entomol.* 14 (3): 175–274.
Proches, S., and D. J. Marshall. 2001. Global distribution patterns of non-halacarid marine intertidal mites: implications for their origins in marine habitats. *J. Biogeogr.* 28:47–58. [HYADESIIDAE]
Proctor, H. C. 1991a. Courtship in the water mite *Neumania papillator*: males capitalize on female adaptations for predation. *Anim. Behav.* 42:589–98. [UNIONICOLIDAE]
———. 1991b. The evolution of copulation in water mites: a comparative test for nonreversing characters. *Evolution* 45:558–67.
———. 1992a. Mating and spermatophore morphology of water mites (Acari: Parasitengona). *Zool. J. Linn. Soc.* 106:341–84.
———. 1992b. Sensory exploitation and the evolution of male mating behaviour: a cladistic test using water mites (Acari: Parasitengona). *Anim. Behav.* 44:745–52.

———. 1996. Sex-ratios and chromosomes in water mites (Hydracarina). In Mitchell et al. (1996), 1:441–45.
———. 1998. Indirect sperm transfer in arthropods. *Ann. Rev. Entomol.* 43:153–74.
———. 2001. Extracting aquatic mites from stream substrates: a comparison of three methods. *Exp. Appl. Acarol.* 25:1–11.
———. 2003a. Feather mites (Acari: Astigmata): ecology, behavior and evolution. *Ann. Rev. Entomol.* 48:185–209.
———. 2003b. Male mating effort and female remating in the water mite *Arrenurus manubriator* (Acari: Arrenuridae). In I. M. Smith (2003a), 211–21.
Proctor, H. C., and I. Owens. 2000. Mites and birds: diversity, parasitism and coevolution. *TREE* 15:358–64.
Proctor, H. C., and G. Pritchard. 1989. Neglected predators: water mites (Acari: Parasitengona: Hydrachnellae) in freshwater communities. *J. N. Amer. Benthol. Soc.* 8:100–111.
———. 1990. Variability in the life history of *Unionicola crassipes*, a sponge-associated water mite (Acari: Unionicolidae). *Can. J. Zool.* 68:1227–32.
Proctor, H. C., and B. P. Smith. 1994. Mating behaviour of the water mite *Arrenurus manubriator* (Acari: Arrenuridae). *J. Zool. Lond.* 232:473–83.
Proctor, H. C., and K. Wilkinson. 2001. Coercion and deceit: Water mites (Acari: Hydracarina) and the study of intersexual conflict. In Halliday et al. (2001), 155–69.
Pugh, P. J. A. 1993. A synonymic catalogue of the Acari from Antarctica, the sub-antarctic islands and the Southern Ocean. *J. Nat. Hist.* 27:323–421.
———. 1995. Air-breathing littoral mites of sub-antarctic South Georgia. *J. Zool. Soc. Lond.* 236:649–66.
———. 1996a. Edaphic oribatid mites (Cryptostigmata: Acarina) associated with an aquatic moss on sub-antarctic South Georgia. *Pedobiologia* 40:113–17.
———. 1996b. The structure and function of the tarsus I sensillar field in mites of the genus *Halarachne* (Halarachnidae: Gamasida). *J. Nat. Hist.* 30:1069–86.
———. 1996c. Using artificial substrata to monitor how cryptofaunal Acari colonize littoral algae on sub-antarctic South Georgia. *Acarologia* 37:189–200.
Pugh, P. J. A., G. O. Evans, M. R. Fordy, and P. E. King. 1991. The functional morphology of the respiratory system of the Holothyrida (= Tetrastigmata) Acari: Anactinotrichida. *J. Zool. Lond.* 225:153–72.
Pugh, P. J. A., and P. E. King. 1985a. Feeding in intertidal Acari. *J. Exp. Mar. Biol. Ecol.* 94:269–80. [BDELLIDAE, HYADESIIDAE]
———. 1985b. The vertical distribution of the British intertidal Acari—the non-halacarid fauna (Arachnida: Acari). *J. Zool. Lond.* A, 207:21–33. [HYADESIIDAE]
———. 1985c. Vertical distribution and substrate association of the British Halacaridae. *J. Nat. Hist.* 19:961–68.
———. 1986. Seasonality in some British intertidal Acari. *J. Nat. Hist.* 20:653–66.
Pugh, P. J. A., P. E. King, and M. R. Fordy. 1987a. Ambulacral structure in the terrestrial moiety of the intertidal Acari, and its relationship with the lifestyle of the Acari. *Acarologia* 28:3–13. [HYADESIIDAE]
———. 1987b. A comparison of the structure and function of the cerotegument in two species of Cryptostigmata (Acarina). *J. Nat. Hist.* 21:603–16.
———. 1990. Respiration in *Fortuynia maculata* Luxton (Fortuyniidae, Cryptostigmata, Acarina) with particular reference to the role of van der Hammens Organ. *J. Nat. Hist.* 24:1529–47.
———. 1991. Larvae of *Trombidium* (Trombidium) poriceps (Oudemans) (Trombidiidae: Prostigmata: Acarina) parasitic on *Chersodromia speculifera* Walker (Empididae: Diptera: Hexapoda). *Acarologia* 32:23–27.
Pugh, P. J. A., P. J. Llewellyn, K. Robinson, and S. E. Shackley. 1997. The associations of phoretic mites (Acarina: Chelicerata) with sand-hoppers (Amphipoda: Crustacea) on the south Wales coast. *J. Zool. Lond.* 243:305–18.
Pulpan, J., and P. H. Verner. 1965. Control of tyroglyphoid mites in stored grain by the predatory mite *Cheyletus eruditus* (Schrank). *Can. J. Zool.* 43:417–32. [CHEYLETIDAE]
Purrini, K., and W. Bäumler. 1977. *Nosema hermanniae* sp. n., eine neue Mikrosporidie aus *Hermannia gibba* C. L. Koch (Fam. Hermanniidae, Oribatei, Acarina) im Fichtenwald. *Zool. Anz.* 199:107–12.
Purrini, K., and V. Bukva. 1984. Pathenogenetic agents of oribatid mites (Oribatei, Acarina): a gap in the research on population dynamics. In Griffiths and Bowman, 2:826–37.
Pussard, R., and M. André. 1929. Note sur *Bochartia kuyperi* Oudms., acarien parasite de pucerons. *Rev. Pathol. Veg.* 16:295–302. [ERYTHRAEIDAE]
Putman, W. L. 1970. Life history and behaviour of *Balaustium putmani* (Acarina: Erythraeidae). *Ann. Entomol. Soc. Amer.* 63: 76–81.
Putman, W. L., and D. H. C. Herne. 1966. The role of predators and other biotic agents in regulating the population density of phytophagous mites in Ontario peach orchards. *Can. Entomol.* 98:808–20. [ERYTHRAEIDAE]
Qin, T.-K. 1994. Revision of the Australian Penthalodidae (Acarina). *Invertebr. Taxon.* 8:1305–23.
———. 1996. A review and cladistic analysis of Eupodoidea (Acari: Acariformes). *Syst. Appl. Acarol.* 1:77–105.
———. 1997. A reconsideration of the taxonomic position of *Protopenthalodes* (Prostigmata: Eupodoidea). *Syst. Appl. Acarol.* 2:253–55. [EUPODIDAE?]
———. 1998. *Callipenthalodes*, a new genus of Penthalodidae (Acariformes: Eupodoidea) from New Zealand. *Internat. J. Acarol.* 24:221–25.
Qin, T.-K., and R. B. Halliday. 1996a. The Australian species of *Chromotydaeus* Berlese and *Penthaleus* C. L. Koch (Acarina: Penthaleidae). *J. Nat. Hist.* 30:1833–48.
———. 1996b. Revision of the Australian and South African species of *Halotydeus* (Acarina: Penthaleidae). *Bull. Entomol. Res.* 86:441–50.
———. 1997. Eriorhynchidae, a new family of Prostigmata (Acarina), with a cladistic analysis of eupodoid species of Australia and New Zealand. *Syst. Entomol.* 22:151–71.
Qin, T.-K., and B. H. Rohitha. 1996. The astigmatid mite *Histiostoma feroniarum* (Acari: Astigmata: Histiostomatidae) in New Zealand. *New Zeal. Entomol.* 19:65–69.
Qu, D. C., Y. Meta, K. J. Nakatsuka, K. Kitamura, and M. Goubara. 2003. Reproductive strategy in the two species of cleptoparasitic astigmatid mites, *Chaetodactylus nipponicus* and *Tortonia* sp. (Acari: Chaetodactylidae and Suidasiidae), infesting *Osmia cornifrons* (Hymenoptera: Megachilidae). II, Life history, phoretic positions, development and reproductivity. *Jpn. J. Entomol.* 6:55–73. [in Japanese]

Quintero, M. M. T. 1987. Hallazgo de ácaros del género *Leptus* sp., parasitando a *Apis mellifera*. *Vet. Méx.* 18:229–32. [ERYTHRAEIDAE]

Quiros-Gonzales, M. J., and E. W. Baker. 1984. Idiosomal and leg chaetotaxy in the Tuckerellidae Baker & Pritchard: Ontogeny and nomenclature. In Griffiths and Bowman, 1:166–73.

Rack, G. 1959. *Acarophenax dermestidarum* sp. n. (Acarina, Pyemotidae), ein Eiparasit von *Dermestes*-Arten. *Z. Parasitenkd.* 19:411–31. [ACAROPHENACIDAE]

———. 1963. *Kleemannia* (Acarina, Ameroseiidae), ein neuer Wohnungslästling. *Entomol. Mitt. Zool. Mus. Hamb.* 2:1–7.

———. 1967. Untersuchungen über die Biologie von *Dolichocybe* Krantz, 1957 und Beschreibung von zwei neuen Arten (Acarina, Pyemotidae). *Mitt. Hamb. Zool. Mus. Inst.* 64:29–42. [DOLICHOCYBIDAE]

———. 1971. *Cheyletiella yasguri* Smiley, 1965 (Acarina, Cheyletiellidae), ein fakultativ menschenpathogener Parasit des Hundes. *Z. Parasitenkd.* 36:321–34.

———. 1972a. Pyemotiden an Gramineen in Schwedischen Landwirtschaftlichen Betreiben. Ein Beitrag zur Entwicklung von *Siteroptes graminum* (Reuter, 1900) (Acarina, Pyemotidae). *Zool. Anz.* 188:157–74. [SITEROPTIDAE]

———. 1972b. Weitere Beobachtungen über Milben in Neubauten mit Beschreibung einer neuen Art, *Ameroseius pseudoplumosus* sp. n. (Acarina, Mesostigmata). *Entomol. Mitt. Zool. Mus. Hamb.* 4:249–57. [AMEROSEIIDAE]

———. 1973a. *Archidispus pterostichi* sp. n. und die Gattung *Archidispus* Karafiat, 1959 (Acarina, Trombidiformes, Scutacaridae). *Entomol. Mitt. Zool. Mus. Hamb.* 4 (80): 319–26.

———. 1973b. *Balaustium murorum* (Hermann, 1804) (Acarina, Erythraeidae) in Häusern. Eine wenig beachtete sommerliche Milbenplage. *Anz. Schädlingskde., Pflanzen, Umweltschutz* 46:129–32.

———. 1973c. *Dolichomotes crossi* sp. n. aus Californien und Bestimmungsschlüssel für die Familie Dolichocybidae (Acarina, Pyemotoidea). *Acarologia* 15:337–41.

———. 1974. Neue und bekannte Milbenarten der Überfamilie Pygmephoroidea aus dem Saalkreis bei Halle (Acarina, Tarsonemida). *Entomol. Mitt. Zool. Mus. Hamb.* 4 (87): 499–521.

———. 1975. Three new species of Pygmephoroidea (Acarina: Tarsonemida) from *Fragaria* in Florida. *Fla. Entomol.* 58:231–38. [SCUTACARIDAE]

———. 1976. Milben (Acarina) von europäischen Limoniinen (Diptera, Nematocera). *Mitt. Hamb. Zool. Mus. Inst.* 73:63–85. [CALYPTOSTOMATIDAE]

———. 1979. *Brennandania scolopendrae* sp. n. von *Scolopendra morsitans* L. (Acarina, Tarsonemida, Pygmephoroidea, Microdispidae). *Rev. Zool. Afr.* 93:376–81.

———. 1980. *Procalvolia zacheri* (Oudemans, 1929) in einem Neubau bei Stade (Acarina, Sarcoptiformes, Saproglyphidae). *Entomol. Mitt. Zool. Mus. Hamb.* 6 (107): 303–7. [WINTERSCHMIDTIIDAE]

———. 1983. Beschreibung einer in Thailand an Speisepilzen (*Auricularia polytricha*) schädlichen Milbe, *Luciaphorus perniciosus* sp. n. (Acari, Tarsonemina, Pygmephoridae). *Folia Entomol. Hung.* 44:139–46.

Rack, G., and G. C. Eickwort. 1980. Biology and description of a new pygmephorid mite (Acarina: Tarsonemida) associated with the soil-nesting bee *Agapostemon nasutus* (Hymenoptera: Halictidae). *Acarologia* 21:267–78.

Rack, G., and M. Kaliszewski. 1985. Beschreibung von *Cochlodispus operosus* sp. n. (Acari, Prostigmata) aus einem Gewächshaus in Gent/Belgien. *Entomol. Mitt. Zool. Mus. Hamb.* 8 (124): 81–88. [MICRODISPIDAE, PYGMEPHORIDAE]

Rack, G., and P. H. Vercammen-Grandjean. 1979. *Siteroptes (Siteroptoides) trombidiphilus* sp. n. (Acarina: Pygmephoridae), phoretisch auf einem Weibchen der Familie Trombidiidae (Acarina) aus Ostafrika. *Entomol. Mitt. Zool. Mus. Hamb.* 6 (104): 217–20.

Radford, C. D. 1950. *Systematic check list of mite genera and type species.* International Union of Biological Sciences, ser. C (Entomol.), 1. Paris.

Radinovsky, S. 1965. The biology and ecology of granary mites of the Pacific Northwest. III, Life history and development of *Leiodinychus krameri* (Acarina: Uropodidae). *Ann. Entomol. Soc. Amer.* 58:259–67.

Radovsky, F. J. 1967. The Macronyssidae and Laelapidae (Acarina: Mesostigmata) parasitic on bats. *Univ. Calif. Publ. Entomol.* 46: viii, 1–288.

———. 1969. Adaptive radiation in the parasitic Mesostigmata. *Acarologia* 11:450–83.

———. 1985. Evolution of mammalian mesostigmate mites. In Kim (1985a), 441–504.

———. 1994. The evolution of parasitism and the distribution of some dermanyssoid mites (Mesostigmata) on vertebrate hosts. In Houck (1994b), contribution no. 8, 186–217.

Radovsky, F. J., and D. Gettinger. 1999. Acanthochelinae, a new subfamily (Acari: Parasitiformes: Laelapidae), with redescription of *Acanthochela chilensis* Ewing and descriptions of a new genus and species from Argentina. *Internat. J. Acarol.* 25:77–90.

Radovsky, F. J., J. K. Jones, Jr., and C. J. Phillips. 1971. Three new species of *Radfordiella* (Acarina: Macronyssidae) parasitic in the mouth of phyllostomid bats. *J. Med. Entomol.* 8:737–46.

Radovsky, F. J., and G. W. Krantz. 1998. A new genus and species of predaceous mite in the parasitic family Macronyssidae (Acari: Mesostigmata). *J. Med. Entomol.* 35:527–37.

Radovsky, F. J., G. W. Krantz, and J. O. Whitaker, Jr. 1997. A remarkable example of predation in the parasitic mite family Macronyssidae. *Internat. J. Acarol.* 23:3–6.

Radovsky, F. J., and C. E. Yunker. 1971. *Xenarthronyssus furmani* n. gen. n. sp. (Acarina: Dasyponyssidae), parasites of armadillos, with two subspecies. *J. Med. Entomol.* 8:135–42.

Radwan, J. 1993. The adaptive significance of male polymorphism in the acarid mite *Caloglyphus berlesei*. *Behav. Ecol. Sociobiol.* 33:201–8. [ACARIDAE]

Rady, G. H. 1992. New genus and species of Acarophenacidae (Acari: Tarsonemina) from Egypt. *Ann. Agr. Sci., Moshtohor* 30:1129–35.

Rafalski, J. 1982. *Adamystis coineaui* sp. n. (Acari, Adamystidae) from Hindu-Kush. *Acarologia* 23:135–43.

Rakha, M. A., and C. W. McCoy. 1984. *Eryngiopus citri*, a new mite on Florida citrus with descriptions of the developmental stages (Stigmaeidae: Actinedida). *Fla. Entomol.* 67:504–7.

Ramanan, R. V., and S. Ghai. 1984. Observations on the mite *Neocypholaelaps indica* Evans and its relationship with the honey bee *Apis cerana indica* Fabricius and the flowering of *Eucalyptus* trees. *Entomon* 9:291–92. [AMEROSEIIDAE]

Ramani, N., and M. A. Haq. 1984. Oribatid mites parasitic on some economically important plants of Kerala. In Griffiths and Bowman 2:813–18.

———. 1991. Potential of *Meristacarus degradatus* and *Xylobates rhomboides* (Acari: Oribatei) in the degradation of higher plant materials. In Dusbábek and Bukva, 1:411–15. [LOHMANNIIDAE, HAPLOZETIDAE]

Ramazzotti, G., and A. M. Nocentini. 1960. Porohalacaridae (Halacaridae) del Lago di Mergazzo. *Mem. Ist. Ital. Idrobiol.* 12: 185–200.

Ramos, M. A., et al., eds. 1993. *Fauna iberica*. Vol. 3. Madrid: Mus. Nac. Cienc. Nat.

Ramsay, G. W., and J. A. Wallwork. 1972. Some observations on the pteromorphs of oribatid mites (Acari: Cryptostigmata). *Acarologia* 13:669–74.

Rapp, A. 1959. Zur Biologie und Ethologie der Käfermilbe *Parasitus coleoptratorum* L. 1758 (Ein Beitrag zum Phoresie-Problem). *Zool. Jb. Syst., Ökol. Geogr. Tiere* 86:303–66. [PARASITIDAE]

Rasmy, A. H., and B. A. Abou-Awad. 1984. A redescription of *Tuckerella nilotica* Zaher and Rasmy (Acarina: Tuckerellidae), with descriptions of the immature stages. *Acarologia* 25:337–40.

Raspotnig, G., G. Krisper, and R. Schuster. 2004. Oil gland chemistry of *Trhypochthonius tectorum* (Acari: Oribatida) with reference to the phylogenetic significance of secretion profiles in Trhypochthoniidae. *Internat. J. Acarol.* 30:369–74.

Raspotnig, G., G. Krisper, R. Schuster, G. Fauler, and H-J. Leis. 2005. Volatile exudates from the oribatid mite, *Platynothrus peltifer*. *J. Chem. Ecol.* 31:419–30.

Raspotnig, G., R. Schuster, and G. Krisper. 2003. Functional anatomy of oil glands in *Collohmannia gigantea* (Acari, Oribatida). *Zoomorphology* 122:105–12. [COLLOHMANNIIDAE]

———. 2004. Citral in oil gland secretions of Oribatida (Acari): A key component for phylogenetic analysis. *Abh. Ber. Naturk. Görlitz* 76:43–50.

Raspotnig, G., R. Schuster, G. Krisper, G. Fauler, and H.-J. Leis. 2001. Chemistry of the oil gland secretion of *Collohmannia gigantea* (Acari: Oribatida). *Exp. Appl. Acarol.* 25:933–46. [COLLOHMANNIIDAE]

Raut, S. K., and A. Panigrahi. 1991. The mite *Fuscuropoda marginata* (C. L. Koch) for the control of pest slugs *Laevicaulus alte* (Férussac). In Dusbábek and Bukva, 2:683–87. [DINYCHIDAE]

Reddy, M. V., S. P. S. Beniwal, V. K. Sheila, S. Sithanantham, and Y. L. Nene. 1989. Role of an eriophyid mite *Aceria cajani* (Acari: Eriophyidae) in transmission and spread of sterility mosaic of pigeon pea. In ChannaBasavanna and Viraktamath, 2:121–27.

Redikorzev, V. V. 1937. Eine neue *Opilioacarus*. *Zool. Anz.* 118: 10–12.

———. 1947. Kleshch *Pediculoides ipidarius*, sp. n. (Acari, Pediculoididae), parazit zhukov-koroedov [The mite *P. ipidarius*, sp. n., parasite of bark beetles]. *Entomol. Obozr.* 29:247–49. [ACAROPHENACIDAE] [in Russian]

Redmond, B. L., and J. Hochberg. 1981. The stylostome of *Arrenurus* spp. (Acari: Parasitengona) studied with the scanning electron microscope. *J. Parasitol.* 67:308–13. [ARRENURIDAE]

Reeves, R. M. 1988. Distribution and habitat comparisons for *Carabodes* collected from conifer branches with descriptions of *brevis* Banks and *higginsi* sp. n. (Acari: Oribatida: Carabodidae) *Proc. Entomol. Soc. Wash.* 90:373–92.

———. 1992. *Carabodes* of the eastern United States and Canada (Acari, Oribatida, Carabodidae). *Can. J. Zool.* 70:2042–58.

———. 1998. Biogegraphy of Carabodidae (Acari: Oribatida) in North America. *Appl. Soil Ecol.* 9:59–62.

Reeves, R. M., and V. M. Behan-Pelletier. 1998. The genus *Carabodes* (Acari: Oribatida: Carabodidae) of North America, with descriptions of new western species. *Can. J. Zool.* 76:1898–1921.

Regenfuss, H. 1968. Untersuchungen zur Morphologie, Systematik und Ökologie der Podapolipidae (Acarina, Tarsonemini). *Zeitschr. Wissenschaftl. Zool., Leipzig* 177:183–282.

———. 1972. Über die Einnischung synhospitaler Parasitenarten auf dem Wirtskörper. Untersuchungen an ektoparasitischen Milben (Podapolipidae) auf Laufkäfern (Carabidae). *Zeitschr. Zool. Syst. Evol., Forschung* 10:44–65.

———. 1973. Beinreduktion und Verlagerung des Kopulationsapparates in der Milbenfamilie Podapolipidae, ein Beispiel für verhaltensgesteuerte Evolution morphologischer Strukturen. *Zeitschr. Zool. Syst. Evol., Forschung* 11:173–95.

Regev, S. 1974. Morphological and genetic evaluation of male polymorphism in *Cheyletus malaccensis* (Cheyletidae: Acarina). *Acarologia* 16:85–93.

Regev, S., and W. W. Cone. 1975. Evidence of farnesol as a male sex attractant of the two-spotted spider mite, *Tetranychus urticae* Koch (Acarina: Tetranychidae). *Environ. Entomol.* 4:307–11.

———. 1976. Analyses of pharate female two-spotted spider mites for nerolidol and geraniol: Evaluation for sex attraction of males. *Environ. Entomol.* 5:133–38.

———. 1980. The monoterpene, citranellol, as a male sex attractant of the two-spotted spider mite, *Tetranychus urticae* (Acarina: Tetranychidae). *Environ. Entomol.* 9:50–52.

Reinhardt, K. 1996. Negative effects of *Arrenurus* water mites on the flight distances of the damselfly *Nehallenia speciosa* (Odonata: Coenagrionidae). *Aquat. Insects* 18:233–40. [ARRENURIDAE]

Reisen, W. K., M. L. Kennedy, and N. T. Reisen. 1976. Winter ecology of ectoparasites collected from hibernating *Myotis velifer* (Allen) in southwestern Oklahoma (Chiroptera: Vespertilionidae). *J. Parasitol.* 62:628–35. [CHEYLETIDAE]

Resh, V. H., and J. O. Solem. 1996. Phylogenetic relationships and evolutionary adaptations of aquatic insects. In *An introduction to the aquatic insects of North America*, 3d ed., ed. R. W. Merritt and K. W. Cummins, 98–107. Dubuque, IA: Kendall/Hunt Publ.

Rettenmeyer, C. W. 1961. Arthropods associated with Neotropical army ants with a review of the behavior of these mites (Arthropoda: Formicidae: Dorylinae). Ph.D. diss., Univ. Kansas.

———. 1962. Notes on host specificity and behavior of mymecophilous macrochelid mites. *J. Kans. Entomol. Soc.* 35:358–60.

Reuter, E. 1909. Zur Morphologie und Ontogenie der Acariden mit besonderer Berücksichtigung von *Pediculopsis graminum*. *Acta Soc. Sci. Fenn.* 36:1–288.

Reutimann, P. 1987. Quantitative aspects of the feeding activity of some oribatid mites (Oribatida, Acari) in an alpine meadow ecosystem. *Pedobiologia* 30:425–33.

Reytblat, A. G. 1965. Biology of the gamasid mite *Haemolaelaps semidesertus* Breg. (Gamasoidea, Parasitiformes). *Zool. Zh.* 44:863–70. [LAELAPIDAE]

Ribeiro, J. M. C. 1995. Blood-feeding arthropods: live syringes or invertebrate pharmacologists? *Rev. Infect. Agents Dis.* 4:143–52.

Richards, L. A., and K. W. Richards. 1976. Parasitid mites associated with bumblebees in Alberta, Canada (Acarina: Parasitidae; Hymenoptera: Apidae). II, Biology. *Univ. Kans. Sci. Bull.* 51:1–18.

Ridland, P. M. 1996. Forage crops. In Lindquist, Sabelis, and Bruin, 6:673–80.

Ridsdill-Smith, T. J. 1997. Biology and control of *Halotydeus destructor* (Tucker) (Acarina: Penthaleidae): A review. *Exp. Appl. Acarol.* 21:195–224.

Riessen, H. P. 1982. Pelagic water mites: Their life history and seasonal distribution in the zooplankton community of a Canadian lake. *Arch. Hydrobiol., Suppl.* 62:410–39. [PIONIDAE]

Riha, G. 1951. Zur Ökologie der Oribatiden in Kalksteinböden. *Zool. Jb. Syst.* 80:408–50.

Rimando, L. C., and L. A. Corpuz-Raros. 1996. Some Philippine Raphignathoidea (Acari). l, Genera *Paraeupalopsellus* and *Exothoris* (Eupalopsellidae). *Philipp. Entomol.* 10:97–117.

Rivera, C., M. Cesar, A. G. Negron, M. Bertrand, and J. Acosta. 2003. *Hemidactylus mabouia* (Sauria: Gekkonidae), host of *Geckobia hemidactyli* (Actinedida: Pterygosomatidae), throughout the Caribbean and South America. *Carib. J. Sci.* 39:321–26.

Rizk, G. A., Z. R. Soliman, and M. A. Ali. 1979. Survey on mites associated with citrus and grape-vine in Minia region, Egypt. *Bull. Soc. Entomol. Egypt* 62:105–10. [TYDEIDAE]

Robaux, P. 1966a. Acariens Thrombidiidae de Madagascar. *Bull. Mus. Nat. Hist. Natur.* 38:629–37. [TROMBIDIIDAE]

———. 1966b. Contribution à l'étude des Thrombidiidae sur plusiers espèces d'Acariens appurtenant au genre *Campylothrombium*. *Rev. Ecol. Biol. Sol* 3:313–27.

———. 1966c. Sur quelques Thrombidiidae rares ou nouveaux pour la faune de France (Acari: Thrombidiidae). *Acarologia* 8:611–30. [TROMBIDIIDAE]

———. 1967a. Contribution à l'étude des acariens Thrombidiidae d'Europe, Part. I et II. *Mem. Mus. Nat. Hist. Natur.* 46:1–124. [STYGOTHROMBIDIIDAE, TROMBIDIIDAE]

———. 1967b. *Moyanella gigax* Boshell and Kerr 1942, type des Moyanellinae n. sub-fam. (Acarina: Thrombidiidae). *Acarologia* 9:841–47. [TROMBIDIIDAE]

———. 1968a. Quelques acariens Trombidiidae de Tunisie. *Acarologia* 10:235–49.

———. 1968b. Les Spelaeothrombiinae: Le statut du genre *Typhlothrombium* Berlese 1910 et description de deux nouveaux genres (Acari: Thrombidiidae). *Acarologia* 10:44–64. [TROMBIDIIDAE]

———. 1968c. Thrombidiidae d'Amérique du Sud. I, Tanaupodinae, Johnstonianinae, Thrombellini (Acarina: Thrombidiidae). *Acarologia* 10:450–66.

———. 1969a. Étude des larves de Thrombidiidae. II, La larve de *Parathrombium megalochirum* (Berlese) 1910. *Acarologia* 11:585–96. [TROMBIDIIDAE]

———. 1969b. Recherches sur la faune interstitielle continentale. I, *Hygrothrombium tibiotarsale* n. g., n. sp.: Premier Thrombidiidae vrai du réseau interstitiel des cours d'eau des Indes (Acari: Thrombidiidae). *Ann. Spéléol.* 24:187–92. [STYGOTHROMBIDIIDAE]

———. 1969c. Recherches sur la faune interstitielle continentale. II, *Hydrothrombium cooki* n. g., n. sp. acarien phréatique d'Amerique du Nord (Acari, Stygothrombiidae). *Ann. Spéléol.* 24:193–223. [STYGOTHROMBIDIIDAE]

———. 1969d. Thrombidiidae d'Amérique du Sud. II, Chyzerini, Thrombidiinae; solénidiotaxie du palpe chez quelques Thrombidiidae; relations entre les *Podothrombium* Berlese 1910, *Variathrombium* n. g. et les Thrombidiinae s. l. *Acarologia* 11:69–93.

———. 1970. Étude des larves de Thrombidiidae. III, La larve de *Johnstoniana errans* (Johnston) 1852, redescriptions de l'adulte et de la nymphe. *Acarologia* 12:339–56.

———. 1971. Recherches sur le développement et la biologie des Acariens Thrombidiidae. Ph.D. diss., Faculty of Science, Univ. Paris. C.N.R.S. Reg. no. 5616.

———. 1972a. Étude des larves de Thrombidiidae. IV, Redescription de formes larvaires d'*Allothrombium neapolitum* Oudemans 1910, *Neothrombium neglectum* (Bruyant) 1908 et *Microthrombidium fasciatum* (Koch) 1836. *Acarologia* 14:612–30.

———. 1972b. Un nouveau Spelaeothrombiinae cavernicole de Cuba: *Speothrombium monoculata* n. g., n. sp. (Acari: Thrombidiidae). *Ann. Spéléol.* 27:787–91.

———. 1975a. Observations sur quelques Actinedida (Prostigmates) du sol d'Amérique du Nord. I, Une nouvelle espèce de Caligonellidae (Acari : Raphignathoidea). *Coptocheles grandjeani* n. sp. *Acarologia* 17:236–42.

———. 1975b. Observations sur quelques Actinedida (= Prostigmates) du sol d'Amérique du Nord. V, Barbutiidae, une nouvelle famille d'acariens (Acari: Raphignathoidea) et description d'une nouvelle espèce appartenant au genre *Barbutia*. *Acarologia* 17:480–88.

———. 1976. Observations sur quelques Actinedida (= Prostigmates) du sol d'Amérique du Nord. IV, Sur une nouvelle forme larvaire et son adulte appartenant à la sous-famille des Microthrombidiinae (Acari: Thrombidiidae). *Acarologia* 17:508–18. [TROMBIDIIDAE]

———. 1977a. Observations sur quelques Actinedida (= Prostigmates) du sol d'Amérique du Nord. VI, Sur deux espèces nouvelles de Labidostommidae. *Acarologia* 18:442–61.

———. 1977b. Observations sur quelques Actinedida (= Prostigmates) du sol d'Amérique du Nord. VIII, Sur deux formes larvaires nouvelles de Trombidioidea (Acari). *Acarologia* 18:651–67.

———. 1978. Observations sur quelques Actinedida (= Prostigmates) du sol d'Amérique du Nord. IX, Nouvelles formes larvaires de Trombidions (Acari). *Acarologia* 19:258–71.

———. 1982. A new larval form of Thrombidiidae from the National Park of Switzerland *Podothrombium piriformis* new species (Acarina, Trombidiidae). *Acarologia* 23:239–44.

Robaux, P., J. Webb, Jr., and G. D. Campbell. 1976. Une forme nouvelle de Thrombidiidae (Acari) parasite sur plusieurs espèces d'Orthoptères cavernicoles du genre *Ceuthophilus* (Orthoptera, Raphidophoridae). *Ann. Spéléol.* 31:213–18.

Roberts, I. H., and W. P. Meleney. 1965. Psorergatic acariasis in cattle. *J. Amer. Vet. Med. Assoc.* 146:17–23.

Robertson, G. P., D. C. Coleman, C. S. Bledsoe, and P. Sollins. 1999. *Standard soil methods for long-term ecological research*. Oxford, UK: Oxford Univ. Press.

Robertson, N. L., and T. W. Carroll. 1988. Virus-like particles and a spider mite intimately associated with a new disease of barley. *Science* 240:1188–90.

Robertson, P. L. 1959. A revision of the genus *Tyrophagus*, with a discussion on its taxonomic position in the Acarina. *Aust. J. Zool.* 7:146–81. [ACARIDAE]

Rockett, C. L. 1980. Nematode predation by oribatid mites (Acari: Oribatida). *Internat. J. Acarol.* 6:219–24.

Rockett, C. L., and J. P. Woodring. 1966. Oribatid mites as predators of soil nematodes. *Ann. Entomol. Soc. Amer.* 59:669–71.

Rodriguez, J. G., ed. 1979. *Recent advances in acarology*. 2 vols. New York: Academic Press.

Rodriguez, J. G., P. Singh, and B. Taylor. 1970. Manure mites and their role in fly control. *J. Med. Entomol.* 7:335–41.

Rohde, C. J. 1955. Studies on arthropods from a moss habitat with special emphasis on the life history of three oribatid mites. *Diss. Abs.* 15:1936.

———. 1964. Some techniques in the preparation of stained whole mounts and serial sections of mite embryos and adults. In Woolley (1964), 208–14.

Roihu, T., and A. L. Kariniemi. 1998. *Demodex* mites in acne rosacea. *J. Cutaneous Pathol.* 25:550–52. [DEMODICIDAE]

Root, R. B. 1967. The niche exploitation pattern of the blue-gray gnatcatcher. *Ecol. Monogr.* 37:317–50.

Rosa, A. E., H. M. André, and C. H. W. Flechtmann. 1985. Acari domun meliponinarum brasiliensium habitantes. 8, *Proctotydaeus alvearii* sp. n. (Acari: Acariformes: Tydeidae). *Rev. Bras. Biol.* 45:79–83.

Rosa, A. E., and C. H. W. Flechtmann. 1979. Mites in house dust from Brazil. *Internat. J. Acarol.* 5:195–98.

———. 1980. *Sphaerolophus* a synonym of *Charletonia*? (Acari: Erythraeidae). *Internat. J. Acarol.* 6:215–17.

Rosario, R. M. T., and P. E. Hunter. 1988. The genus *Klinckowstroemia* Trägårdh and descriptions of nine new species (Acarina: Trigynaspida: Klinckowstroemiidae). *Acarologia* 29:119–36. [FEDRIZZIIDAE]

Rosenthal, S. S. 1996. *Aceria, Epitrimerus* and *Aculus* species and biological control of weeds. In Lindquist, Sabelis, and Bruin, 6:729–39.

Roshdy, M. A. 1966. Comparative internal morphology of subgenera of *Argas* ticks (Ixodoidea, Argasidae). 4, Subgenus *Ogadenus*: *Argas brumpti* Neumann, 1907. *J. Parasitol.* 52:776–82.

———. 1972. The subgenus *Persicargas* (Ixodoidea, Argasidae, *Argas*). 15, Histology and histochemistry of the salivary glands of *A.* (*P.*) *persicus* Oken. *J. Med. Entomol.* 9:143–48.

Roshdy, M. A., and T. Hefnawy. 1973. The functional morphology of *Haemaphysalis* spiracles (Ixodoidea: Ixodidae). *Zeitsch. Parasitenk.* 42:1–10.

Royalty, R. N., and T. M. Perring. 1996. Nature of damage and its assessment. In Lindquist, Sabelis, and Bruin, 6:493–512.

Royce, L. A., and G. W. Krantz. 2003. A new genus and species of hypoaspidine mite (Acari: Laelapidae) associated with a night-flying xylocopinine carpenter bee (Hymenoptera. Apidae. *Xylocopa*) in Thailand. *Internat. J. Acarol.* 29:107–11.

Rudzinska, M. 1998. Life history of the phoretic predatory mite *Arctoseius semiscissus* (Acari: Ascidae) on a diet of sciarid fly eggs. *Exp. Appl. Acarol.* 22:643–48.

Ruf, A. 1996. Life-history patterns in soil-inhabiting mesostigmatid mites (Dermanyssina, Parasitina). In Mitchell et al. (1996), 1:621–28.

Ryke, P. A. J. 1957. *Mixogynium proteae*, a new genus and species of Celaenopsoidea (Mesostigmata: Acarina) from South Africa. *Ann. Mag. Nat. Hist.* 10:579–84.

———. 1959. A revision of the hypoaspid mites associated with Myriopoda with descriptions of three new species of the subgenus *Julolaelaps* Berl. (Acarina: Laelaptidae). *Parasitology* 49:6–22. [IPHIOPSIDIDAE]

———. 1962. The genus *Antennoseius* Berlese (Acarina: Rhodacaridae). *Ann. Mag. Nat. Hist.*, ser. 13, 4:657–63. [ASCIDAE]

———. 1964. Acarina associated with *Protea* flowers in the Cape Province. *J. Entomol. Soc. S. Afr.* 26:337–54.

———. 1966. Die onderlinge verhouding en wisselwerking tussen grondfauna en grondmikroflora. *Tydskr. Natuurwet.* 6:182–92. [NANORCHESTIDAE]

Sabelis, M. W. 1985a. Reproductive strategies. In Helle and Sabelis (1985a), 1A:265–78. [TETRANYCHIDAE]

———. 1985b. Sampling techniques. In Helle and Sabelis (1985a), 1A:337–50.

Sabelis, M. W., and F. M. Bakker. 1992. How predatory mites cope with the web of their tetranychid prey: A functional view on dorsal chaetotaxy in the Phytoseiidae. *Exp. Appl. Acarol.* 16:203–25.

Sabelis, M. W., and J. Bruin. 1996. Evolutionary ecology: life history patterns, food plant choice and disperal. In Lindquist, Sabelis, and Bruin, 6:329–66.

Sabelis, M. W., and M. Dicke. 1985. Long-range dispersal and searching behavior. In Helle and Sabelis (1985b), 1B:141–60.

Sabelis, M. W., and C. J. Nagelkerke. 1987. Sex allocation strategies of pseudoarrhenotokous phytoseiid mites. *Neth. J. Zool.* 37:117–36.

Sabelis, M. W., and P. C. J. van Rijn. 1996. Eriophyoid mites as alternative prey. In Lindquist, Sabelis, and Bruin, 6:757–64.

Saboori, A. 2001. Descriptions of the larva, deutonymph and adult of *Balaustium zhangi* sp. nov. (Acari: Erythraeidae) from Iran. *Syst. Appl. Acarol.* 6:171–78.

———. 2002. A new genus and species of larval Neothrombiidae (Acari: Trombidioidea) from Iran. *Syst. Appl. Acarol.* 7:185–90.

Saboori, A., and M. A. Akrami. 2001. A new species of *Erythraeus* larva (Acari: Erythraeidae) from Iran. *Syst. Appl. Acarol.* 6:159–63.

Saboori, A., and A. Atamehr. 1999. A new larval *Leptus* Latreille (Acari: Erythraeidae) from Iran. *Syst. Appl. Acarol.* 4:159–63.

Saboori, A., S. Cobanoglu, and S. Bayram. 2004. A new species of larval *Erythraeus* (Acari: Erythraeidae) from Turkey. *Internat. J. Acarol.* 30:137–42.

Saboori, A., and A. Masoud. 2003. A new species of *Leptus* larva (Acari: Erythraeidae) from India. *Syst. Appl. Acarol.* 8:175–82.

Saboori, A., and A. Nemati. 2001. A new species and a new host record of the genus *Eutrombidium* Verdun (Acari: Eutrombidiidae) from Iran. *Syst. Appl. Acarol. Spec. Publ.* 7:5–14.

Saboori, A., and H. Ostovan. 2000. A new species of the genus *Leptus* Latreille, 1796 (Acari: Erythraeidae) ectoparasitic on sun pest, *Eurygaster integriceps* Puton (Hemiptera: Scutellaridae) from Iran. *Syst. Appl. Acarol.* 5:143–47.

Saito, Y. 1985. Life types of spider mites. In Helle and Sabelis (1985a), 1A:253–64. [TETRANYCHIDAE]

———. 1986. Prey kills predator: Counter-attack success of a spider mite against its specific phytoseiid predator. *Exp. Appl. Acarol.* 2:47–62. [TETRANYCHIDAE]

———. 1987. Extraordinary effects of fertilization status on the reproduction of an arrhenotokous and sub-social spider mite (Acari: Tetranychidae). *Res. Popul. Ecol.* 29:57–71.

———. 1990. "Harem" and "non-harem" type mating systems in two species of subsocial spider mites (Acari, Tetranychidae). *Res. Pop. Ecol.* 39:263–78.

Saito, Y., K. Mori, T. Sakagami, and J.-Z. Lin. 2004. Reinstatement of the genus *Stigmaeopsis* Banks, with descriptions of two new species (Acari, Tetranchidae). *Ann. Entomol. Soc. Amer.* 97:635–46.

Saito, Y., and M. Osakabe. 1992. A new fixation method for preparing mite specimens for optical and SEM microscopic observations. *Appl. Entomol. Zool.* 27:427–36.

Saito, Y., M. Osakabe, Y. Sakagami, and Y. Yasui. 1993. A method for preparing permanent specimens of mites with Canada balsam. *Appl. Entomol. Zool.* 28:593–97.

Sakata, T., N. Hiramatsu, K. Ishikawa, and Y. Kuwahara. 1997. Chemical ecology of mesostigmatid mites. I, Identification of hydroxymethylnaphthoquinones from *Uroactinia hirschmanni* Hiramatsu (Mesostigmata: Uropodidae). *Appl. Entomol. Zool.* 32:57–61.

Sakata, T., Y. Kuwahara, and K. Kurosa. 1996. 4-isopropenyl-3-oxo-1-cyclohexene-1-carboxaldehyde, isorobinal: A novel monoterpene from the mite *Rhizoglyphus* sp. (Astigmata: Rhizoglyphinae). *Naturwissenschaften* 83:427.

Sakata, T., and R. A. Norton. 2001. Opisthonotal gland chemistry of early-derivative oribatid mites (Acari) and its relevance to systematic relationships of Astigmata. *Internat. J. Acarol.* 27: 281–92.

———. 2003. Opisthonotal gland chemistry of a middle-derivative oribatid mite, *Archegozetes longisetosus* (Acari: Trhypochthoniidae). *Internat. J. Acarol.* 29:345–50.

Sakata, T., S. Shimano, and Y. Kuwahara. 2003. Chemical ecology of oribatid mites. III, Chemical composition of oil gland exudates from two oribatid mites, *Trhypochthoniellus* sp. and *Trhypochthonius japonicus* (Acari: Trhypochthoniidae). *Exp. Appl. Acarol.* 29:279–91. (MALACONOTHRIDAE]

Sakata, T., K. Tagami, and Y. Kuwahara. 1995. Chemical ecology of oribatid mites. I, Oil gland components of *Hydronothrus crispus. J. Acarol. Soc. Jpn.* 4:69–75. [MALACONOTHRIDAE]

Sambon, L. W. 1928. *Ophioptes parkeri*. A new species and genus of cheyletid inhabiting the scales of reptiles. *Ann. Trop. Med. Parasitol.* 22:137–42. [OPHIOPTIDAE]

Samish, M., and J. Rehacek. 1999. Pathogens and predators of ticks and their potential in biological control. *Ann. Rev. Entomol.* 44:159–82.

Sammataro, D., U. Gerson, and G. R. Needham. 2000. Parasitic mites of honey bees: Life history, implications, and impact. *Ann. Rev. Entomol.* 45:519–48. [TARSONEMIDAE]

Samšinák, K. 1960. Über einige forstwirtschaftlich wichtige Milben der Gattung *Proctolaelaps* Berlese 1923. *Cesk. Parasitol.* 7:297–307. [MELICHARIDAE]

———. 1962. Beiträge zur Kenntnis der Gattung *Tyrophagus* Oudemans. *Cas. Cesk. Spol. Entomol.* 59:266–80. [ACARIDAE]

———. 1971. Die auf *Carabus*-Arten (Coleoptera, Adephaga) der palaearktischen Region lebenden Milben der Unterordnung Acariformes (Acari); ihre Taxonomie und Bedeutung für die Lösung zoogeographischer, entwicklungsgeschichtlicher und parasitophyletischer Fragen. *Entomol. Ab. Staat. Mus. Tierk. Dresden* 38:145–234. [CANESTRINIIDAE]

Samšinák, K., P. Palička, K. Zitek, L. Mališ, and E. Vobrázková. 1976. Are the mites of the genus *Tarsonemus* really parasites of man? *Folia Parasitol. Praha* 23:91–93. [TARSONEMIDAE]

Sanders, F. H., and R. A. Norton. 2004. Anatomy and function of the ptychoid defensive mechanism in the mite *Euphthiracarus cooki* (Acari: Oribatida). *J. Morphol.* 259:119–54. [EUPHTHIRACARIDAE]

Santana, F. J. 1976. A review of the genus *Trouessartia* (Analgoidea: Alloptidae). *J. Med. Entomol. Suppl.* 1:1–128. [TROUESSARTIIDAE]

Santos, M. A. 1976. Evaluation of *Zetzellia mali* as a predator of *Panonychus ulmi* and *Aculus schlechtendali. Environ. Entomol.* 5:187–91.

Santos, M. A., and J. E. Laing. 1985. Stigmaeid predators. In Helle and Sabelis (1985b), 1B:197–203.

Sanyal, A. K., S. Basak, and R. P. Barman. 2002. Three new species of oribatid mites (Acarina, Oribatida: Haplochthoniidae), from the antarctic continent. *Acarina* 10:57–63.

Sanyal, A. K., and A. K. Bhaduri. 1986. Check list of oribatid mites (Acari) of India. *Rec. Zool. Surv. India, Misc. Publ., Occ. Papers* 83:1–79.

Sanyal, A. K., and T. K. Das. 1989. Oribatid mites (Acari, Cryptostigmata) associated with pineapple (*Ananas cosmosus*) root at Kalyani, West Bengal. *Environ. Ecol.* 7:971–72.

Saporito, R. A., M. A. Donnelly, R. A. Norton, H. M. Garraffo, T. F. Spande, and J. W. Daly. 2007. Oribatid mites as a major dietary source for alkaloids in poison frogs. *Proc. Nat. Acad. Sci. USA* 104 (21): 8885–90.

Sardar, M. A., and P. W. Murphy. 1987. Feeding tests of grassland soil-inhabiting gamasine predators. *Acarologia* 28:117–21.

Sarkar, K., R. Pramanik, and V. C. Joy. 2000. Reproductive toxicity of pesticides on soil microarthropod fauna as ecotoxicological tool. *J. Environ. Biol.* 21:227–34.

Sarkar, P. K., D. R. Nath, P. K. Talukdar, and P. R. Malhotra. 1990. Seasonal incidence of water mites (*Arrenurus* sp.) parasitizing mosquito vectors at Tezpur, Assam, India. *Indian J. Malariol.* 27:121–26. [ARRENURIDAE]

Sasa, M. 1965. *Mites. An introduction to classification, bionomics and control of Acarina*. Tokyo: Univ. Tokyo Press.

Savory, T. 1964. *Arachnida*. London: Academic Press.

Savulkina, M. M. 1981. Sistema, ekologiya i rasprostranenie kleshchei sem. Pygmephoridae Cross, 1965 (Trombidiformes) [System, ecology and distribution of mites of family Pygmephoridae]. *Entomol. Obozr.* 60:434–50. [in Russian]

Schaarschmidt, L. 1959. Systematik und Ökologie der Tarsonemiden. *Beitr. Syst. Ökol. Mitteleurop. Acarina* 1 (5): 713–823.

Schatz, H. 1979. Über Ernährungsbiologie von Oribatiden (Acari) im Hochgebirge (Obergurgl, Tirol). *Ber. Naturweiss. Medizin. Vereins Innsbruck* 66:7–20.

———. 1983. *U.-Ordn.: Oribatei, Hornmilben. Catalogus Faunae Austriae, Wien,* Teil IXi. Vienna: Austrian Academy of Sciences Press.

———. 1985. The life cycle of an alpine oribatid mite, *Oromurcia sudetica* Willmann. *Acarologia* 26 (1): 95–100. [CERATOZETIDAE]

———. 1991. Catalogue of known species of Acari from the Galápagos Islands. *Internat. J. Acarol.* 17:213–25.

———. 1998. Oribatid mites (Acari: Oribatida) from the Galápagos Islands: Faunistics, ecology and speciation. *Exp. Appl. Acarol.* 22:373–409. (Also appears in Bruin, van der Geest, and Sabelis, 499–530.)

———. 2000. *Eremaeozetes sabinae* sp. nov. (Acari: Oribatida) from the Hawaiian Islands. *Internat. J. Acarol.* 26:155–59. [EREMAEOZETIDAE]

———. 2001a. *Eremaeozetes capensis* n. sp. (Acari: Oribatida: Eremaeozetidae) from South Africa. *Internat. J. Acarol.* 27:101–6.

———. 2001b. The genus *Eremaeozetes* (Acari: Oribatida) on the Galápagos Islands. *Acarologia* 41:475–93. [EREMAEOZETIDAE]

———. 2002. Die Oribatidenliteratur und die beschriebenen Oribatidenarten (1758–2001): Eine Analyse. *Abh. Ber. Naturk. Görlitz* 74:37–45.

———. 2003. New *Sphaerochthonius* species from the Neotropical

region (Acari: Oribatida). *Rev. Suisse Zool.* 110:111–24. [SPHAEROCHTHONIIDAE]
———. 2005. Diversity and global distribution of oribatid mites—evaluation of the present state of knowledge. In Weigmann et al. (2005): 485–500.
———. 2006. Catalogue of known oribatid mite species (Acari: Oribatida) from the Central American landbridge (First part). *Tropical Zool.* 19 (2): 209–88.
Schatz, H., and R. Gerecke. 1996. Hornmilben (Acari, Oribatida) aus Quellen und Quellbaechen im Nationalpark Berchtesgaden (Oberbayern) und in den Suedlichen Alpen (Trentino, Alto Adige). *Ber. Naturweiss. Medizin. Vereins Innsbruck* 83:121–44.
Schatz, H., and J. G. Palacios-Vargas. 1999. Microzetidae (Acari: Oribatida) from the Galápagos Islands. *Acarologia* 40:69–83. [MICROZETIDAE]
Schenker, R. 1986. Population dynamics of oribatid mites (Acari: Oribatida) in a forest soil ecosystem. *Pedobiologia* 29:239–46.
Scheucher, R. 1957. Systematik und Ökologie der deutschen Anoetinen. In H. J. Stammer, 1:233–384. [HISTIOSTOMATIDAE]
Schicha, E. 1987. *Phytoseiidae of Australia and neighboring areas.* Oak Park, MI: Indira Publishing House.
Schliesske, J. 1985. Zur Verbreitung und Ökologie einer neuen ursprünglichen Gallmilbenart (Acari: Eriophyoidea) an *Araucaria araucara* (Molina). *Entomol. Mitt. Zool. Mus. Hamb.* 8:97–106.
Schmid, R.1988. Morphologische Anpassungen in einem Räuber-Beute-System: Ameisenkäfer (Scydmaenidae, Staphylinoidea) und gepanzerte Milben (Acari). *Zool. Jahrb. Syst.* 115:207–28.
Schneider, K., S. Migge, R. A. Norton, S. Scheu, R. Langel, A. Reineking, and M. Maraun. 2004. Trophic niche differentiation in oribatid mites (Oribatida, Acari): Evidence from stable isotope ratios (15N /14N). *Soil Biol. Biochem.* 36:1769–74.
Schneider, K., K. Renker, and M. Maraun. 2005. Oribatid mite (Acari, Oribatida) feeding on ectomycorrhizal fungi. *Mycorrhiza* 16:67–72.
Schneider, K. Renker, S. Scheu, and M. Maraun. 2005. Feeding biology of oribatid mites: A minireview. In Weigmann et al. (2005), 247–56.
Schousboe, C. 1987. Deutonymphs of *Parasitellus* phoretic on Danish bumblebees (Parasitidae, Mesostigmata; Apidae, Hymenoptera). *Acarologia* 28:37–41.
Schroder, R. F. 1979. Host specificity tests of *Coccipolipus epilachnae*, a mite parasite on Mexican bean beetles. *Environ. Entomol.* 8:46–47. [PODAPOLIPIDAE]
———. 1982. Effect of infestation with *Coccipolipus epilachnae* Smiley (Acarina: Podapolipidae) on fecundity and longevity of the Mexican bean beetle. *Internat. J. Acarol.* 8:81–84.
Schruft, G. 1971. *Haleupalus oliveri* nov. spec., eine Dornpalpenmilbe an Reben (Vitus spec.) (Acari: Cunaxidae). *Deutsche Entomol. Zeitschr.* 18:377–82.
Schubart, H. O. R. 1973. The occurrence of Nematalycidae (Acari, Prostigmata) in central Amazonia with a description of a new genus and species. *Acta Amazonica* 3:53–57.
———. 1975. Morphologische Grundlagen für die Klärung der Verwandtschaftsbeziehungen innerhalb der Milbenfamilie Ameronothridae (Acari). *Zoologica* 123:23–91.
Schulte, G., R. Schuster, and H. Schubart. 1975. Zur Verbreitung und Ökologie der Ameronothriden (Acari, Oribatei) in terrestrischen, limnischen und marinen Lebensräumen. *Veröff. Inst. Meeresforsch., Bremerhaven* 15:359–85.
Schulze, P. 1942. Über die Hautsinnesorgane der Zecken besonders über eine bisher unbekannte Art von Arthropoden-Sinnesorganen, die Krobylophoren. *Zeitschr. Morphol. Ökol. Tiere* 39:379–419.
Schuster, I. J., and R. Schuster. 1970. Indirekte Spermaübertragung bei Tydeidae (Acari, Trombidiformes). *Naturwissenschaften* 57:256.
Schuster, R. 1956. Der Anteil der Oribatiden an den Zergetzungsvorgängen in Boden. *Zeitschr. Morph. Ökol. Tiere* 45:1–33.
———. 1957. Die terrestrische Kleinarthropodenfauna in den Tenarea-Trottoirs des Westmediterranean Litorals. *Kieler Meeresforsch.* 13:244–62.
———. 1958. Neue terrestrische Milben aus dem Mediterranen Litoral. *Vie Milieu* 9:88–109. [PENTHALEIDAE]
———. 1960a. Über die Ökologie und Verbreitung von Bodenmilben (Oribatei) am Alpen-Ostrand, insbesondere in der Steiermark. *Mitt. Naturwiss. Ver. Steiermark* 90:132–49.
———. 1960b. Die europäischen Arten der Gattung *Perlohmannia* Berlese (Acari, Oribatie). *Zool. Anz.* 164 (5–6): 185–95.
———. 1962a. Das marine Litoral als Lebensraum terrestrischer Kleinarthropoden. *Internat. Rev. Ges. Hydrobiol.* 37:359–412.
———. 1962b. Nachweis eines Paarhungszeremoniells bei den Hornmilben (Oribatei, Acari). *Naturwissenschaftern* 49:502–3. [COLLOHMANNIIDAE]
———. 1972. Spinnvermögen der Tydeiden (Milben). *Naturwissenschaften* 59:275.
———. 1977. Die Selenoribatidae, eine thalassobionte Familie der Hornmilben (Oribatei). *Acarologia* 19:155–60.
———. 1979. Soil mites in the marine environment. In Rodriguez, 1:593–602.
———. 1986. Bodenbewohnende Milben (Acari) aus Niedersachsen und anderen Regionen Deutschlands (BRD). *Braunschw. Naturk. Schr.* 2:519–25.
Schuster, R., L. Coetzee, and J. F. Putterill. 2000. Oribatid mites (Acari, Oribatida) as intermediate hosts of tapeworms of the family Anoplocephalidae (Cestoda) and the transmission of *Moniezia expansa* cysticercoids in South Africa. *Onderstepoort J. Vet. Res.* 67:49–55.
Schuster, R., and P. W. Murphy, eds. 1991. *The Acari: Reproduction, development and life-history strategies.* London: Chapman & Hall.
Schuster, R., and H. Pötsch. 1989. Another record of an active prelarva in mites. In ChannaBasavanna and Viraktamath, 1:261–65.
Schuster, R., and I. J. Schuster. 1966. Uber das Fortpflanzungsverhalten von Anystiden-Männchen (Acari, Trombidiformes). *Naturwissenschaften* 53:162.
———. 1969. Gestielte Spermatophoren bei Labidostomiden (Acari, Trombidiformes). *Naturwissenschaften* 56:145.
———. 1977. Ernährungs- und fortpflanzungsbiologische Studien an der Milbenfamilie Nanorchestidae (Acari: Trombidiformes). *Zool. Anz., Jena* 199 (1/2): 80–94.
Schuster, R., and R. Turk. 1987. *Erythraeus ghilarovi* n. sp., a new soil mite from South America (Acari, Trombidiformes). *Biol. Fert. Soils* 3:15–18.
Schuster, R. O., and M. M. J. Lavoipierre. 1970. *The mite family Heterocheylidae Trägårdh.* Occ. Papers, Calif. Acad. Sci. 85. San Francisco: Calif. Acad. Sci.
Schuster, R. O., and F. M. Summers. 1978. Mites of the family

Diarthrophallidae (Acari: Mesostigmata). *Internat. J. Acarol.* 4:279–385.
Schwarz, H. H., K. Huck, and P. Schmid-Hempel. 1997. Prevalence and host preferences of mesostigmatic mites (Acari: Anactinochaeta) phoretic on Swiss bumble bees (Hymenoptera: Apidae). *J. Kans. Entomol. Soc.* 69:35–42.
Schwarz, H. H., and J. K. Müller. 1992. The dispersal behaviour of the phoretic mite *Poecilochirus carabi* (Mesostigmata, Parasitidae): Adaptation to the breeding biology of its carrier *Necrophorus vespilloides* (Coleoptera, Silphidae). *Oecologia* 89:487–93.
Schwarz, H. H., and M. G. Walzl. 1996. Pairing, oviposition and development in two sibling species of phoretic mites (Acari: Mesostigmata: Parasitidae: *Poecilochirus* spp.) associated with burying beetles (Coleoptera: Silphidae: *Nicrophorus* spp.). *J. Nat. Hist.* 30:1337–48.
Schweizer, J., and C. Bader. 1963. Die Landmilben der Schweiz. Trombidiformes. *Denkschr. Schweiz. Naturforsch. Ges.* 84:209–378. [CALYPTOSTOMATIDAE]
Schwoerbel, J. 1964. Die Wassermilben (Hydrachnellae und Limnohalacaridae) als Indikatoren einer biocönotischen Gliederung von Breg und Brigach sowie der obersten Donau. *Arch. Hydrobiol., Suppl.* 27:386–417.
———. 1984. Subterrane Wassermilben aus Fließgewässern Neuseelands. (Acari, Actinedida). *Arch. Hydrobiol., Suppl.* 66:293–308. [PIONIDAE]
———. 1986a. Acari: "Hydrachnellae." In Botosaneanu, 652–96.
———. 1986b. Acari: Stygothrombiidae und Thrombidiidae. In Botosaneanu, 648–51.
———. 1986c. Rheophile Wassermilben (Acari) aus Chile. 1, Hyporheische Arten. *Arch. Hydrobiol.* 106:71–78. [MIDEOPSIDAE]
———. 1987. Rheophile Wassermilben (Acari: Hydrachnellae) aus Chile. 3, Arten aus Thermalgewässern. *Arch. Hydrobiol.* 110:399–407. [THERMACARIDAE]
Schwoerbel, J., and H. Sepasgosarian. 1980. Wassermilben (Acari, Prostigmata, Hydrachnellae) aus dem Iran. 8. Mitteilung. Subterrane Süsswassermilben. *Entomol. Mitt. Zool. Mus. Hamb.* 6:293–301.
Scott, H. 1958. *Humerobates rostrolamellatus* Grandjean, an oribatid mite invading a house. *Entomol. Monthly Mag.* 94:69–70. [HUMEROBATIDAE]
Scudder, G. G. E. 1983. A review of factors governing the distribution of two closely related corixids in the saline lakes of British Columbia. *Hydrobiologia* 105:143–54. [EYLAIDAE]
Seastedt, T. R., and D. A. Crossley, Jr. 1978. Further investigations of microarthropod populations using the Merchant-Crossley high-gradient extractor. *J. Georgia Entomol. Soc.* 13:338–44.
Seastedt, T. R., M. V. Reddy, and S. P. Cline. 1989. Microarthropods in decaying wood from temperate coniferous and deciduous forests. *Pedobiologia* 33:69–77.
Seeman, O. D. 1996. Flower mites and phoresy: the biology of *Hattena panopla* Domrow and *Hattena cometis* Domrow (Acari: Mesostigmata: Ameroseiidae). *Aust. J. Zool.* 44:193–203.
———. 2000. The immature life stages of the Fedrizziidae (Mesostigmata: Fedrizzioidea). *Acarologia* 41:39–52.
Seeman, O. D., and H. F. Nahrung. 2003. *Parobia husbandi* gen. n., sp. n. (Acari: Podapolipidae): Subelytral parasites of paropsine beetles (Coleoptera: Chrysomelidae). *Aust. J. Entomol.* 42: 334–42.
———. 2005. New *Parobia* (Acari: Podapolipidae) parasitic on pestiferous parsopine beetles (Coleoptera: Chrysomelidae) in eucalypt plantations. *Syst. Appl. Acarol.* 10:111–35.
Seeman, O. D., and D. E. Walter. 1995. Life history of *Afrocypholaelaps africana* (Evans) (Acari: Ameroseiidae), a mite inhabiting mangrove flowers and phoretic on honeybees. *J. Aust. Entomol. Soc.* 34:45–50.
———. 1997. A new species of Triplogyniidae (Mesostigmata: Celaenopsoidea) from Australian rainforests. *Internat. J. Acarol.* 23:49–59. [TRIPLOGYNIIDAE]
———. 2000. A review of the Paratydeidae (Acari: Prostigmata) with descriptions of the first Australian representatives, *Tanytydeus lamington* sp. nov. and *T. kakadu* sp. nov. *Acarologia* 40:393–400.
Sellnick, M. 1926. Alguns novos acaros (Uropodidae) myrmecophiles e termitophiles. *Arch. Mus. Rio de Janeiro* 26:29–56. [UROPODIDAE]
———. 1953. *Ichthyostomatogaster nyhleni*, eine neue Acaride aus Schweden. *Entomol. Tidskr.* 74:24–37.
———. 1954. *Indogynium lindgergi* nov. gen. nov. spec., eine neue Acaride aus Indien. *Entomol. Tidskr.* 75:285–91. [SCHIZOGYNIIDAE]
Sellnick, M., and K. H. Forsslund. 1955. Die Camisiidae Schwedens. *Ark. Zool., Stockholm*, ser. 2, 8:473–530.
Sengbusch, H. G. 1977. Review of oribatid mite-anoplocephalan tapeworm relationships (Acari; Oribatei: Cestoda; Anoplocephalidae). In Dindal (1977), 87–102.
Sengbusch, H. G., and J. W. Hauswirth. 1986. Prevalence of hair follicle mites, *Demodex folliculorum* and *D. brevis* (Acari: Demodicidae), in a selected human population in western New York, USA. *J. Med. Entomol.* 23:384–88.
Seniczak, A., and S. Seniczak. 2002. The effect of cadmium on *Archegozetes longisetosus* (Acari, Oribatida) in laboratory conditions. *Eur. J. Soil Biol.* 38:315–17. [TRHYPOCHTHONIIDAE]
Seniczak, S. 1978. The morphology of juvenile forms of soil mites of the family Achipteriidae (Acari, Oribatei). I. *Ann. Zool. (Warsaw)* 34:89–99.
Seniczak, S., L. Dabrowski, A. Klimek, and S. Kaczmarek. 1999. Effects of alkaline deposition on the mites (Acari) associated with young Scots pine forests in Poland. *Water Air Soil Pollut.* 109:407–28.
Seniczak, S., R. A. Norton, and H-F. Wang. 1998. The morphology of juvenile stages of moss mites of the family Trhypochthoniidae (Acari: Oribatida), and the taxonomic status of some genera and species. *Zool. Anz.* 237:85–95.
Sepasgosarian, H. 1983. A list of the world genera and species of the family Tenuipalpidae (Actinedida: Acaridia). *Zeitschr. Ang. Zool.* 70:169–200.
———. 1990. IV. Addendum to a list of the world genera and species of the family Tenuipalpidae (Acari). *Entomol. Mitt. Zool. Mus. Hamb.* 10:85–90.
———. 1997. The world genera and species of the family Tydeidae. *J. Entomol. Soc. Iran, Suppl.* 7:1–54. [TRIOPHTYDEIDAE, TYDEIDAE, IOLINIDAE]
Sergienko, G. D. 1994. *Lower oribatids*. Vol. 25 in Fauna of the Ukraine, ed. V. A. Topacevskij. Kiev: Naukova Dumka. [in Russian]
Serralheiro, F., and M. Madeira. 1991. Acari colonization of *Quercus suber* and *Eucalyptus globulus* litter. In Dusbábek and Bukva, 1:353–58.

Servin, R., and G. Otero. 1989. Una nueva especie de *Tuckerella* de Baja California Sur, Mexico. *Southwest. Entomol.* 14:205–12. [TUCKERELLIDAE]

Sevast'yanov, V. D. 1980. Novye taksony kleshchei semeistva Dolichocybidae (Trombidiformes, Tarsonemina) i filogeneticheskie cvyazi ego podsemeistv [New taxa of mites of family Dolichocybidae and phylogenetic relations of its subfamilies]. *Zool. Zh.* 59:1453–62. [in Russian]

Sevast'yanov, V. D., and C. M. Abo-Korach. 1985. Novye vidy kleshchei kogorty Tarsonemina (Trombidiformes) iz agrotsenozov Egipta [New mite species of cohort Tarsonemina from agrocoenoses of Egypt]. *Vestn. Zool.* 1985 (4): 35–41. [DOLICHOCYBIDAE]

Sevast'yanov, V. D., and N. D. Zakhida Al Daur. 1991. Novyi rod i vidy kleshchei kogorty Tarsonemina (Trombidiformes) fauny tsentral'nogo Iraka [A new genus and species of mites of cohort Tarsonemina from central Iraq]. *Zool. Zh.* 70:137–40. [PYGMEPHORIDAE]

Severin, H. C. 1944. *The grasshopper mite* Eutrombidium trigonum *(Hermann). An important enemy of grasshoppers.* Tech. Bull. Agr. Exp. Sta. S.D. St. Coll. 3. Brookings. [EUTROMBIDIIDAE]

Shah, B. 2001. Mite infestations in Australian skinks: Seasonal, geographical and ecological variation. *Herpetofauna* 3:102–7.

Shahi, K. P., and S. S. Krishna. 1981. A new host of *Hemipteroseius indicus*. *Experientia* 37:1072. [OTOPHEIDOMENIDAE]

Shaldybina, E. S. 1977. Preimaginal stages of ceratozetoids (Oribatei, Ceratozetidae). *Fauna, Syst., Biol. Ecol. Helminths and Their Intermediate Hosts*: 76–89. [in Russian]

———. 1987. The mycobatid fauna of the Soviet Union (Oribatei, Mycobatidae). *Pedag. Inst., Gorky*: 1–20. [in Russian]

———. 1989. The chamobatid-fauna of the USSR (Oribatei, Chamobatidae). *Pedag. Inst., Gorky*: 1–19.

Sharma, A. 1991. *Larvacarus transitans*, a serious pest of *Ziziphus* in India. In Dusbábek and Bukva, 1:539–42. [TENUIPALPIDAE]

Sharma, G. D., A. T. Drooz, and A. E. Treat. 1983. A new species of *Callidosoma* (Acari: Prostigma [*sic*]: Erythraeidae) parasite on adults of *Anacamptodes vellivolata* (Lepidoptera: Geometridae) as a larva, and predator of eggs of Lepidoptera as deutonymph and adult. *Ann. Entomol. Soc. Amer.* 76:120–29.

Sharma, G. D., M. H. Farrier, and A. T. Drooz. 1983. Life history, and sexual differences of *Callidosoma metzi* Sharma, Drooz and Treat (Acari: Erythraeidae). *Internat. J. Acarol.* 9:149–55.

Sharma, R. D. 1971. Studies on the plant parasitic nematode *Tylenorhynchus dubius*. *Meded. Lanbouwk. Wageningen* 71:1–154.

Shatrov, A. B. 1992. Rearing, colonization and life cycle of the trombiculid mite *Leptotrombidium orientale* (Schluger, 1948) (Acariformes, Trombiculidae). *Entomol. Obozr.* 71:478–97.

———. 1993. Phenomenon of schizeckenosy in trombiculid mites (Acariformes, Trombiculidae). *Parazitologiya* 27:168–75.

Shcherbak, G. 1980. *Kleshchi semeistva Rhodacaridae palearktiki*. Kiev: Naukova Dumka. [in Russian]

Shcherbak, G. I., and A. D. Petrova, 1987. New mite species of genus *Protogamasellus* (Parasitiformes, Gamasina) from termitaria of Turkmenia. *Vestn. Zool.* 1987 (3): 23–31. [ASCIDAE] [in Russian]

Shehata, M., and A. Baker. 1996. Mites infesting phlebotomine sandflies in southern Sinai, Egypt. *J. Med. Vet. Entomol.* 10:193–96. [STIGMAEIDAE]

Shereef, G. M. 1972. Observations on oribatid mites in laboratory cultures. *Acarologia* 14:281–91.

———. 1976. Biological studies and description of stages of two species: *Papillaecarus aciculatus* Kunst and *Lohmannia egypticus* Elbadry and Nasr (Oribatei: Lohmanniidae) in Egypt. *Acarologia* 18:351–59.

Shevchenko, V. G. 1970. Proiskhozhdenie i morfo-funktsional'naya otsenka chetyrekhnogikh kleshchei (Acarina, Eriophyoidea) [Origin and morpho-functional analysis of tetrapod mites]. In *Sbornik issledovaniya po evolutsionnoi morfologii bespozvonochnykh* [Studies on evolutionary morphology of invertebrates], ed L. A. Evdonin, 153–83. Leningrad: Leningrad Univ. Press. [in Russian]

———. 1976. Problemy filogenii i klassifikatsii chetyrecknogikh kleshchei (Acarina, Tetrapodili) [Problems concerning phylogeny and classification of the four-legged mites]. *Akad. Nauk SSSR, Vsesoyuznoe Entomol. Obshchestvo*, Doklady na dvadtsat' vos'mom ezhegodnom chtenii pamyati N. A. Kholodkovskogo [*Acad. Sci. USSR, All-Union Entomol. Soc.*, Papers of 28th annual lecture series in memory of N. A. Kholodkovskii], 3–52. Leningrad: Nauka. [in Russian]

Shevchenko, V. G., I. G. Bagnyuk, and V. Rinne. 1993. *Trisetacus pini* (Nalepa, 1889) in some Baltic countries and in Russia (taxonomy, morphology, biology, distribution). *Acarina, Russian J. Acarol.* 1:51–71. [ERIOPHYOIDEA]

Shevchenko, V. G., I. G. Bagnyuk, and S. I. Sukhareva. 1991. Novoye semeistvo chetyrekhnogikh kleshchei Pentasetacidae (Acariformes, Tetrapodili) i ego znachenie dlya traktovki proiskhozhdeniya i evolyutsii gruppy [A new family of four-legged mites, Pentasetacidae (Acariformes, Tetrapodili), and its importance to interpretation of the group's origin and evolution]. *Zool. Zh.* 70:47–53. [in Russian]

Shevchenko, V. G., A. P. De-Millo, G. M. Razvyazkinam, and E. A. Kapkova. 1970. Taksonomicheskoe razgranichenie blizkikh vidov chetyrekhnogikh kleshchei *Aceria tulipae* Keif. i *A. tritici* sp. n. (Acarina, Eriophyoidea): Perenoschikov virusov luka i pshenitsy [Taxonomic discreteness of the closely related species of four-legged mites *Aceria tulipae* Keif. and *A. tritici* sp. n., vectors of onion and wheat viruses]. *Zool. Zh.* 49:224–35. [in Russian]

Shi, J., L. Q. Zhou, and S. L. Chen. 2000. Investigation of the vector and host of tsutsugamushi disease in an epidemic focus, Taihang Mountains, Hebei Province. *Chinese J. Vector Biol. Control* 11:462–65. [TROMBICULIDAE]

Shiba, M. 1968. Prostigmatic mites from Japan. III, On some endeostigmatic mites. *Rep. Res. Matsuyama Shinonome Jr. Coll.* 3:219–28.

———. 1969. Taxonomic investigations on free-living mites in the subalpine forest of Shiga Heights IBP Area. II, Prostigmata. *Bull. Nat. Sci. Mus. Tokyo* 12:65–115. [CALYPTOSTOMATIDAE]

———. 1976. Taxonomic investigation on free-living Prostigmata from the Malay Peninsula. *Nat. Life S. E. Asia* 7:83–229. [EUTROMBIDIIDAE]

Shiba, M., and M. Furukawa. 1975. Studies on the family Tenerifiidae (Acarina: Prostigmata) in Japan. *Rep. Res. Matsuyama Shinonome Jr. Coll.* 7:111–26.

Shiba, M., and Morikawa. 1966. Prostigmata mites from Japan. II, Bdellidae I. (Bdellinae, Cytinae, Spinibdellinae). *Rep. Res. Matsuyama Shinonome Jr. Coll.* 2:17–37.

Shimano, S., and R. A. Norton. 2003. Is the Japanese oribatid mite *Euphthiracarus foveolatus* Aoki, 1980 (Acari: Euphthiracaridae) a junior synonym of *E. cribrarius* (Berlese, 1904)? *J. Acarol. Soc. Jpn.* 12 (2): 115–26.

Shimano, S., T. Sakata, Y. Mizutani, Y. Kuwahara, and J. Aoki. 2002. Geranial: The alarm pheromone in the nymphal stage of the oribatid mite, *Nothrus palustris*. *J. Chem. Ecol.* 28:1831–37. [NOTHRIDAE]

Short, N. J., and R. A. I. Norval. 1981. The seasonal activity of *Rhipicephalus appendiculatus* Neumann, 1901 (Acarina: Ixodidae) in the highveld of Zimbabwe. *Rhod. J. Parasitol.* 57:77–84.

Shtanchaeva, U. Ya. 1986. Eine neue Hornmilbenart der Familie Passalozetidae (Oribatei) aus Dagestan. *Zool. Zh.* 65:1423–24. [in Russian]

———. 1996. The world fauna of oribatid mites of the family Epilohmanniidae. *Zool. Zh.* 75:516–31.

Shtanchaeva, U., and I. Netuzhilin. 2003. Review of the world fauna of oribatid mites of the family Scutoverticidae (Acari, Oribatida) with description of new species. *Zool. Zh.* 82:781–803.

Shvanderov, F. A. 1975. Role of phoresy in dispersal of four-legged mites. *Zool. Zh.* 54:458–61. [in Russian]

Siepel, H. 1990. Niche relationships between two panphytophagous soil mites, *Nothrus sylvestris* Nicolet (Acari, Oribatida, Nothridae) and *Platynothrus peltifer* Koch (Acari, Oribatei, Camisiidae). *Biol. Fert. Soils* 9:139–44.

———. 1996. The importance of unpredictable and short-term environmental extremes for biodiversity in oribatid mites. *Biodivers. Letters* 3:26–34.

Siepel, H., and E. M. de Ruiter-Dijkman. 1993. Feeding guilds of orchard mites based on their carbohydrase activities. *Soil Biol. Biochem.* 25:1491–97.

Silvestri, F. 1905. Note aracnologiche. III, Una nuova specie specie di *Opilioacarus* (Acari Notostigmata) dell'America meridionale. *Redia* 2:257–61.

Simberloff, D., and T. Dayan. 1991. The guild concept and the structure of ecological communities. *Ann. Rev. Ecol. Syst.* 22:115–43.

Simmons, T. W., and I. M. Smith. 1984. Morphology of larvae, deutonymphs, and adults of the water mite *Najadicola ingens* (Prostigmata: Parasitengona: Hygrobatoidea) with remarks on phylogenetic relationships and revision of taxonomic placement of Najadicolinae. *Can. Entomol.* 116:691–701. [PIONIDAE]

Singer, G. 1964. A simple aspirator for collecting small arthropods directly into alcohol. *Ann. Entomol. Soc. Amer.* 57:796–98.

———. 1967. A comparison between different mounting techniques commonly employed in acarology. *Acarologia* 9:475–84.

———. 1971a. *Neotrombibium* Leonardi (Acarina: Trombidioidea). Pt. I, Revision of the genus with description of two new species. *Acarologia* 13:119–42. [NEOTROMBIDIIDAE]

———. 1971b. *Neotrombidium* Leonardi (Acarina: Trombidioidea). Pt. 2, Biology of *N. beeri* and other species of the genus. *Acarologia* 15:143–51. [NEOTROMBIDIIDAE]

Sinha, R. N. 1964. Ecological relationships of stored-products mites and seed-borne fungi. In Woolley (1964), 372–89. [DINYCHIDAE]

———. 1966. *Aeroglyphus robustus*, a pest of stored grain. *J. Econ. Entomol.* 59:686–88. [AEROGLYPHIDAE]

Sinha, R. N., and J. T. Mills. 1968. Feeding and reproduction of the grain mite and the mushroom mite on some species of *Penicillium*. *J. Econ. Entomol.* 61:1548–52.

Sionti, P. G., and G. T. Papadoulis. 2003. Two new species of the genus *Nepcunaxoides* Smiley (Acari: Cunaxidae) from Greece. *Internat. J. Acarol.* 29:225–29.

———. 2004. Cunaxid mites of Greece (Acari: Cunaxidae). *Internat. J. Acarol.* 29:315–25.

Sitnikova, L. G. 1978. The main evolutionary trends of the Acari and the problem of their monophyletism. *Entomol. Rev.* 57:303–21.

Sjursen, H., and L. Sömme. 2000. Seasonal changes in tolerance to cold and desiccation in *Phauloppia* sp. (Acari, Oribatida) from Finse, Norway. *J. Insect Physiol.* 46:1387–96. [ORIBATULIDAE]

Skaife, S. H. 1952. The yellow-banded carpenter bee, *Mesotrichia caffra* Linn., and its symbiotic mite, *Dinogamasus braunsi* Vitzthun [*sic*]. *J. Entomol. Soc. S. Afr.* 15:63–76.

Skerratt, L. F., R. W. Martin, and K. A. Handasyde. 1998. Sarcoptic mange in wombats. *Aust. Vet. J.* 76:408–10. [SARCOPTIDAE]

Skoracki, M., A. V. Bochkov, and G. Wauthy. 2004. Revision of the quill mites of the genus *Picobia* Haller, 1878 (Acari: Syringophilidae) with notes on their host-parasite relationships. *Insect Syst. Evol.* 35:155–76.

Skorupski, M., and C. Blaszak. 1997. *Rafaphis microsternalis* gen. nov., sp. nov. a new mite genus of the family Eviphididae from Poland (Acari: Gamasida). *Genus (Warsaw)* 8:135–43.

Skorupski, M., and M. Luxton. 1996. Mites of the family Zerconidae Canestrini, 1891 (Acari: Parasitiformes) from the British Isles, with descriptions of two new species. *J. Nat. Hist.* 30:1815–32.

Slykhuis, J. T. 1969. Transmission of agropyron mosaic virus by the eriophyid mite, *Abacarus hystrix*. *Phytopathology* 59:29–32.

———. 1972. Transmission of plant viruses by eriophyid mites. In *Principles and techniques in plant virology*, ed. C. I. Kado and H. O. Agarwal, 204–25. New York: Van Nostrand Reinhold.

———. 1980. Mites. In *Vectors of plant pathogens*, ed. K. F. Harris and K. Maramorosch, 325–56. New York: Academic Press. [ERIOPHYIDAE]

Smiley, R. L. 1964. Two new erythraeids predaceous upon cotton bollworm eggs (Acari: Erythraeidae). *Proc. Entomol. Soc. Wash.* 66:110.

———. 1966. Further descriptions of two erythraeids predaceous upon cotton bollworm eggs (Acari: Erythraeidae). *Proc. Entomol. Soc. Wash.* 68:25–28.

———. 1968. A new genus and three new species of Erythraeoidea (Acarina: Erythraeidae and Smarididae). *Proc. Entomol. Soc. Wash.* 70:13–21.

———. 1970. A review of the family Cheyletiellidae (Acarina). *Ann. Entomol. Soc. Amer.* 63:1056–78. [CHEYLETIDAE]

———. 1975. A generic revision of the mites of the family Cunaxidae (Acarina). *Ann. Entomol. Soc. Amer.* 68:228–44.

———. 1992. *The predatory mite family Cunaxidae (Acari) of the world with a new classification*. Bloomfield, MI: Indira Publishing House.

Smiley, R. L., W. E. Frost, and U. Gerson. 1996. A revision of the subfamily Tegopalpinae, with descriptions of two new genera and five new species (Acari: Tenuipalpidae). *Internat. J. Acarol.* 22:167–80.

Smiley, R. L., and U. Gerson. 1995. A review of the Tenuipalpidae (Acari: Prostigmata) of Australia with descriptions of two new genera and four new species. *Internat. J. Acarol.* 21:33–45.

Smiley, R. L., and V. R. Landwehr. 1976. A new species of *Tarsonemus* (Acarina: Tarsonemidae) predaceous on tetranychoid mite eggs. *Ann. Entomol. Soc. Amer.* 69:1065–72.

Smiley, R. L., and J. C. Moser. 1968. New species of mites from pine (Acarina: Tarsocheylidae, Eupalopsellidae, Caligonellidae,

Cryptognathidae, Raphignathidae, and Neophyllobiidae). *Proc. Entomol. Soc. Wash.* 70:307–17.
———. 1974. New tarsonemids associated with bark beetles (Acarina: Tarsonemidae). *Ann. Entomol. Soc. Amer.* 67:639–65.
———. 1975. Redescription of *Eutogenes vicinus* Summers and Price, a predatory polymorphic, cheyletid mite with descriptions of males and immature stages (Acarina: Cheyletidae). *Proc. Entomol. Soc. Wash.* 77:405–18.
———. 1976. Two new phoretomorphic *Siteroptes* from galleries of the southern pine beetle (Acarina: Pyemotidae). *Beitr. Entomol. Berlin* 26:307–22. [PYGMEPHORIDAE]
Smiley, R. L., and B. M. OConnor. 1980. Mange in *Macaca arctoides* (Primates: Cercopithecidae) caused by *Cosarcoptes scanloni* (Acari: Sarcoptidae) with possible human involvement and descriptions of the adult male and immature stages. *Internat. J. Acarol.* 6:283–90.
Smit, H. 1992a. *Kawamuracarus sulawesiensis* sp. n., the first hyporheic water mite from Sulawesi, Indonesia (Acari, Hydrachnellae). *Stygologia* 7:127–28. [LIMNESIIDAE]
———. 1992b. Water mites from New South Wales and Queensland, Australia (Acari, Hydrachnellae). *Tijdschr. Entomol.* 135:91–112. [ARRENURIDAE].
———. 1994. New species of water mites from West Africa (Acari, Hydrachnellae), including a checklist of species recorded from Cameroon. *Storkia* 3:5–11. [LIMNESIIDAE]
———. 1995. New records of water mites from Turkey, with 11 species new for the Turkish fauna (Acari, Hydrachnellae). *Storkia* 4:10–15. [SPERCHONTIDAE]
———. 1996a. The first record of water mites from the Chatham Islands, New Zealand (Acari, Hydrachnellae). *New Zeal. Entomol.* 19:57–59. [ARRENURIDAE]
———. 1996b. New species of water mites from New Zealand, with remarks on the water mites from ponds and lakes (Acari, Hydrachnellae). *Acarologia* 37:45–53. [HYDRYPHANTIDAE].
———. 1996c. Ten new species of water mites from Sulawesi and Waigeo, Indonesia (Acari, Hydrachnaellae). *Bull. Zool. Mus. Univ. Amsterdam* 15:6–19. [ATURIDAE]
———. 1996d. The water mite family Aturidae from Australia, with description of six new species (Acari: Hydrachnellae). *Beagle* 13:89–100.
———. 1997a. Australian water mites of the genus *Arrenurus*, with the description of twelve new species from northern and western Australia (Acari: Hydrachnellae: Arrenuridae). *Rec. W. Aust. Mus.* 18:233–61.
———. 1997b. New records of water mites (Acari, Actinedida) from Israel, with description of a new species, *Krendowskia levantensis*. *Israel J. Zool.* 43:219–28. [KRENDOWSKIIDAE]
———. 1998a. A new genus of the water mite family Piersigiidae from Australia (Acari: Hydrachnidia). *Rec. W. Aust. Mus.* 19:107–10.
———. 1998b. New records of the water mite families Anisitsiellidae, Momoniidae and Mideopsidae from Australia, with the description of two new species (Acari: Actinedida). *Beagle* 14:103–8.
———. 1998c. New records of the water mite family Hydryphantidae from Australia, with the description of three new species (Acari: Actinedida). *Beagle* 14:109–15.
———. 1998d. The water mite family Limnesiidae from northern and western Australia (Acari: Actinedida), with description of two new species. *Rec. W. Aust. Mus.* 18:347–55.
———. 1999. New species of the water mite genus *Arrenurus* from eastern Australia (Acari: Hydrachnidia: Arrenuridae). *Mem. Mus. Victoria* 57:225–36.
———. 2000. Rheophilic water mites from southern Argentina, with the description of one new genus and three new species (Acari: Hydrachnidia). *Zootaxa* 103:1–23. [LIMNESIIDAE, HYGROBATIDAE]
———. 2001. New records of the water mite family Hygrobatidae from Australia, with the description of ten new species (Acari: Hydrachnidia). *Rec. W. Aust. Mus.* 20:137–58.
———. 2002a. Rheophilic water mites from southern Argentina, with the description of one new genus and three new species (Acari: Hydrachnidia). *Zootaxa* 103:1–23. [HYGROBATIDAE]
———. 2002b. Water mites from Fiji (Acari: Hydrachnidia). *Acarologia* 42:351–60. [LIMNESIIDAE]
———. 2002c. Water mites from New Caledonia (Acari: Hydrachnidia). In *Zoologia Neocaldonica*, ed. J. Najt and P. Grandcolas, 161–206. Systématique et endémisme en Nouvelle-Calédonie, no. 5. Mém. Mus. Nat. Hist. Natur. 187. [HYGROBATIDAE].
———. 2003a. A new species of the water mite genus *Kawamuracarus* Uchida from Oman (Acari: Hydrachnidia). *Zootaxa* 135:1–4. [LIMNESIIDAE]
———. 2003b. A review of the subgenus *Brevicaudaturus* (Acari: Hydrachnidia: Arrenuridae: Arrenurus) from the Indo-Australian and Pacific region, with the description of a new species. In I. M. Smith (2003a), 223–37.
Smit, H., and H. van der Hammen. 1992. Water mites as indicators of natural aquatic ecosystems of the coastal dunes of the Netherlands and northwestern France. *Hydrobiologia* 231:51–64.
———. 2000. Atlas van de Nederlandse Watermijten (Acari: Hydrachnidia). *Nederl. Faun. Meded.* 13: 1–272.
Smit, H., R. Gerecke, and A. Di Sabatino. 2000a. Atlas van de Nederlandse Watermijten (Acari: Hydrachnidia). *Nederl. Faun. Meded.* 13.
———. 2000b. A catalogue of water mites of the superfamily Arrenuroidea (Acari: Hydrachnidia) from the Mediterranean countries. *Arch. Hydrobiol.*, Suppl. 121:201–67.
Smith, B. P. 1986. New species of *Eylais* (Acari: Hydrachnellae: Eylaidae) parasitic on water boatmen (Insecta: Hemiptera: Corixidae), and a key to North American larvae of the subgenus *Syneylais*. *Can. J. Zool.* 64:2363–69.
———. 1987. New species of *Hydrachna* (Acari: Hydrachnidia; Hydrachnidae) parasitic on water boatmen (Insecta: Hemiptera; Corixidae). *Can. J. Zool.* 65:2630–39.
———. 1988. Host-parasite interaction and impact of larval water mites on insects. *Ann. Rev. Entomol.* 33:487–507.
———. 1989a. A description of the larva of *Rhyncholimnochares kittitanniana* Habeeb (Hydrachnidia: Limnocharidae). *Can. Entomol.* 121:445–52.
———. 1989b. Impact of parasitism by larval *Limnochares aquatica* (Acari: Hydrachnidia; Limnocharidae) on juvenile *Gerris comatus*, *Gerris alacris*, and *Gerris buenoi* (Insecta: Hemiptera; Gerridae). *Can. J. Zool.* 67:2238–43.
———. 1998. Loss of larval parasitism in parasitengone mites. *Exp. Appl. Acarol.* 22:187–99. (Also appears in Bruin, van der Geest, and Sabelis, 125–35.)
———. 2003. Diversity of stylostome structure among parasitic larval water mites (Prostigmata: Hydrachnida). In I. M. Smith (2003a), 239–55.
Smith, B. P., and W. J. Cook. 1991. Negative covariance between

larval *Arrenurus* sp. and *Limnochares americana* (Acari: Hydrachnidia) on male *Leucorrhinia frigida* (Odonata: Libellulidae) and its relationship to the host's age. *Can. J. Zool.* 69: 226–31. [ARRENURIDAE, LIMNOCHARIDAE]

Smith, B. P., and L. A. Laughland. 1990. Stimuli inducing detachment of larval *Arrenurus* danbyensis (Hydrachnidia: Arrenuridae) from adult *Coquillettidia perturbans* (Diptera: Culicidae). *Exp. Appl. Acarol.* 9:51–62.

Smith, B. P., and S. B. McIver. 1984a. Factors influencing host selection and successful parasitism of *Aedes* spp. mosquitoes by *Arrenurus* spp. mites. *Can. J. Zool.* 62:1114–20. [ARRENURIDAE]

———. 1984b. The impact of *Arrenurus danbyensis* Mullen (Acari: Prostigmata; Arrenuridae) on a population of *Coquillettidia perturbans* (Walker) (Diptera: Culicidae). *Can. J. Zool.* 62:1121–34.

———. 1984c. The patterns of mosquito emergence (Diptera: Culicidae; *Aedes* spp.): Their influence on host selection by parasitic mites (Acari: Arrenuridae; *Arrenurus* spp.). *Can. J. Zool.* 62:1106–13.

Smith, C. E., and J. C. Weber. 1954. *The legume mite* Petrobia (Tetranychina) apicalis *(Banks), a pest on several winter growing legumes.* Louisiana State Univ. Agr. Exp. Sta. Tech. Bull. 493. Baton Rouge. [TETRANYCHIDAE]

Smith, I. M. 1972. A review of the water mite genus *Nautarachna* (Acari: Parasitengona: Pionidae). *Life Sci. Contr. R. Ont. Mus.* 86:1–17.

———. 1976a. *Paenecalyptonotus fontinalis* n. gen., n. sp. with remarks on the family Acalyptonotidae (Acari: Parasitengona: Arrenuroidea). *Can. Entomol.* 108:997–1000.

———. 1976b. *A study of the systematics of the water mite family Pionidae (Prostigmata: Parasitengona).* Mem. Entomol. Soc. Can. 98. Ottawa.

———. 1977. A new species of *Mideopsis* (*Nudomideopsis*) (Acari: Parasitengona: Mideopsidae) from North America. *Can. Entomol.* 109:533–35. [NUDOMIDEOPSIDAE].

———. 1978a. Descriptions and observations on host associations of some larval Arrenuroidea (Prostigmata: Parasitengona), with comments on phylogeny in the superfamily. *Can. Entomol.* 110:957–1001.

———. 1978b. Two new species of *Trisetacus* (Prostigmata: Eriophyoidea) from berries of junipers in North America. *Can. Entomol.* 110:1157–60.

———. 1979. A review of water mites of the family Anisitsiellidae (Prostigmata: Lebertioidea) from North America. *Can. Entomol.* 111:529–50.

———. 1982. Larvae of water mites of the genera of the superfamily Lebertioidea (Prostigmata: Parasitengona) in North America with comments on phylogeny and higher classification of the superfamily. *Can. Entomol.* 114:901–90.

———. 1983a. Description of adults of *Neomideopsis siuslawensis* n. gen., n. sp., with remarks on the family Mideopsidae (Acari: Parasitengona: Arrenuroidea). *Can. Entomol.* 115:417–20. [NUDOMIDEOPSIDAE]

———. 1983b. Description of adults of a new species of *Cyclothyas* (Acari: Parasitengona: Hydryphantidae) from western North America, with comments on phylogeny and distribution of mites of the genus. *Can. Entomol.* 115:1433–36.

———. 1983c. Description of adults of a new species of *Geayia* (Acari: Parasitengona: Krendowskiidae) from southern Ontario. *Can. Entomol.* 115:1489–92.

———. 1983d. Description of adults of a new species of *Uchidastygacarus* (s.s.) (Acari: Parasitengona: Arrenuroidea) from eastern North America, with comments on distribution of mites of the genus. *Can. Entomol.* 115:1177–79. [CHAPPUISIDIDAE]

———. 1983e. Description of *Cowichania interstitialis* n. gen., n. sp., and proposal of Cowichaniinae n. subfam., with remarks on phylogeny and classification of Hydryphantidae (Acari: Parasitengona: Hydryphantoidea). *Can. Entomol.* 115:523–27.

———. 1983f. Description of larvae and adults of *Paramideopsis susanae* n. gen., n. sp., with remarks on phylogeny and classification of Mideopsidae (Acari: Parasitengona: Arrenuroidea). *Can. Entomol.* 115:529–38. [NUDOMIDEOPSIDAE]

———. 1983g. Description of larvae of *Neoacarus occidentalis* (Acari: Arrenuroidea: Neoacaridae). *Can. Entomol.* 115:221–26.

———. 1983h. Description of larvae of *Nudomideopsis magnacetabula* (Acari: Arrenuroidea: Mideopsidae) with new distributional records for the species and remarks on the classification of *Nudomideopsis*. *Can. Entomol.* 115:913–19. [NUDOMIDEOPSIDAE]

———. 1983i. Descriptions of two new species of *Acalyptonotus* from western North America, with a new diagnosis of the genus based upon larvae and adults, and comments on phylogeny and taxonomy of Acalyptonotidae (Acari: Parasitengona: Arrenuroidea). *Can. Entomol.* 115:1395–1408.

———. 1984a. Larvae of water mites of some genera of Aturidae (Prostigmata: Hygrobatoidea) in North America with comments on phylogeny and classification of the family. *Can. Entomol.* 116:307–74.

———. 1984b. Review of species of *Trisetacus* (Acari: Eriophyoidea) from North America, with comments on all nominate taxa in the genus. *Can. Entomol.* 116:1157–1211.

———. 1987. Water mites of peatlands and marshes in Canada. In *Aquatic insects of peatlands and marshes in Canada*, ed. D. M. Rosenberg and H. V. Danks, 31–46. Mem. Entomol. Soc. Can. 140. Ottawa.

———. 1989a. Description of deutonymphs and adults of *Oregonacarus rivulicolus* gen. nov., sp. nov. (Acari: Lebertioidea: Anisitsiellidae). *Can. Entomol.* 121:533–41.

———. 1989b. Description of two new species of *Platyhydracarus* gen. nov. from western North America, with remarks on classification of Athienemanniidae (Acari: Parasitengona: Arrenuroidea). *Can. Entomol.* 121:709–26.

———. 1989c. North American water mites of the family Momoniidae Viets (Acari: Arrenuroidea). I, Description of adults of *Cyclomomonia andrewi* gen. nov., sp. nov., and key to world genera and subgenera. *Can. Entomol.* 121:543–49.

———. 1989d. North American water mites of the family Momoniidae Viets (Acari: Arrenuroidea). II, Revision of species of *Momonia* Halbert, 1906. *Can. Entomol.* 121:965–87.

———. 1989e. North American water mites of the family Momoniidae Viets (Acari: Arrenuroidea). III, Revision of species of *Stygomomonia* Szalay, 1943, subgenus *Allomomonia* Cook, 1968. *Can. Entomol.* 121:989–1025.

———. 1990a. Description of two new species of *Stygameracarus* gen. nov. from North America, and proposal of Stygameracarinae subfam. nov. (Acari: Arrenuroidea: Athienemanniidae). *Can. Entomol.* 122:181–90.

———. 1990b. Proposal of Nudomideopsidae fam. nov. (Acari: Arrenuroidea) with a review of North American taxa and description of a new subgenus and species of *Nudomideopsis* Szalay, 1945. *Can. Entomol.* 122:229–52.

———. 1991a. Descriptions of new species representing new or unreported genera of Lebertioidea (Acari: Hydrachnida) from North America. *Can. Entomol.* 123:811–25.

———. 1991b. North American water mites of the family Momoniidae Viets (Acari: Arrenuroidea). IV, Revision of species of *Stygomomonia* (sensu stricto) Szalay, 1943. *Can. Entomol.* 123: 501–58.

———. 1991c. North American water mites of the genera *Phreatobrachypoda* Cook and *Bharatalbia* Cook (Acari: Hygrobatoidea: Aturinae). *Can. Entomol.* 123:465–99. [ATURIDAE]

———. 1991d. Water mites (Acari: Parasitengona: Hydrachnida) of spring habitats in Canada. In *Arthropods of springs, with particular reference to Canada*, ed. D. D. Williams and H. V. Danks, 141–67. Mem. Entomol. Soc. Can. 155. Ottawa.

———. 1992a. North American species of the genus *Chelomideopsis* Romijn (Acari: Arrenuroidea: Athienemanniidae). *Can. Entomol.* 124:451–90.

———. 1992b. North American water mites of the family Chappuisididae Motas and Tanasachi (Acari: Arrenuroidea). *Can. Entomol.* 124:637–723.

———. 1998. *Chelohydracarus navarrensis* gen. nov., sp. nov. (Acari: Hydrachnida: Athienemanniidae). *Internat. J. Acarol.* 24: 327–30.

———, ed. 2003a. *An acarological tribute to David R. Cook (from Yankee Springs to Wheeny Creek).* West Bloomfield, MI: Indira Publishing House.

———. 2003b. North American species of *Neoacarus* Halbert, 1944 (Acari: Hydrachnida: Arrenuroidea: Neoacaridae). In I. M. Smith (2003a), 257–302.

———. 1994. North American species of Neomamersinae Lundblad (Acari: Hydrachnida: Limnesiidae). *Can. Entomol.* 126: 1131–84.

———. 1996. New and unreported species of *Neotyrrellia*, *Protolimnesia* (Protolimnesella) and *Centrolimnesia* (Acari: Hydrachnida: Limnesiidae) from the United States. *An. Inst. Biol., Univ. Nac. Aut. Méx., Ser. Zool.* 67:265–77.

———. 1997. Description of *Amoenacarus dixiensis* gen. nov., sp. nov., and proposal of Amoenacaridae fam. nov. (Acari: Hydrachnida: Arrenuroidea). *Internat. J. Acarol.* 23:107–12.

———. 1998a. The *Axonopsella*-like water mites of the United States (Acari: Hydrachnida: Aturidae). *Internat. J. Acarol.* 24:311–25.

———. 1998b. Three new species of water mites representing new genera of Thyadinae from North America (Acari: Hydrachnida: Hydryphantidae). *Internat. J. Acarol.* 24:331–39.

———. 1999a. An assessment of global distribution patterns in water mites (Acari: Hydrachnida). In Needham, Mitchell, et al., 2:109–24.

———. 1999b. North American species of *Parathyasella* Viets stat. nov. and *Amerothyasella* gen. nov. (Acari: Hydrachnida: Hydryphantidae). *Internat. J. Acarol.* 25:43–50.

———. 1999c. North American species of *Tartarothyas* Viets (Acari: Hydrachnida: Hydryphantidae). *Internat. J. Acarol.* 25:37–42.

———. 1999d. Proposal of *Thyopsoides* gen. nov. and reassignment of *Thyopsis (Thyopsella) plana* Habeeb (Acari: Hydrachnida: Hydryphantidae). *Internat. J. Acarol.* 25:51–53.

———. 2000. North American species of *Kawamuracarus* Uchida, 1937 (Acari: Hydrachnida: Limnesiidae: Kawamuracarinae). *Internat. J. Acarol.* 26:63–71.

———. 2004. Description of *Piona lapointei* sp. nov. (Acari: Hydrachnida: Pionidae), the first species of water mite reported from the Hawaiian Islands. *Internat. J. Acarol.* 30:33–36.

———. 2005a. Description of a new species of *Neolimnochares* (Acari: Eylaoidea: Limnocharidae) from eastern North America. *Internat. J. Acarol.* 31:387–92.

———. 2005b. Systematics of stream-inhabiting species of *Limnochares* (Acari: Eylaoidea: Limnocharidae) in North America. *Internat. J. Acarol.* 31:379–85.

———. 2006. North American species of *Diamphidaxona* (Acari: Hydrachnida: Hygrobatidae). *Zootaxa* 1279:1–44.

Smith, I. M., D. R. Cook, and B. P. Smith. 2001. Water mites and other arachnids. In Thorpe and Covich, 551–659. [STYGOTHROMBIDIIDAE].

Smith, I. M., and M. S. Harvey. 1989. Description of adults of *Arrenurus (?Micruracarus) kitchingi* sp. nov. (Acarina: Arrenuridae) from water-filled tree holes in Australia. *Can. Entomol.* 121:283–89.

Smith, I. M., and E. E. Lindquist. 1979. Prostigmata. In *Canada and its insect fauna*, ed. H. V. Danks, 276–77. Mem. Entomol. Soc. Canada 108. Ottawa. [CALYPTOSTOMATIDAE]

Smith, I. M., and D. R. Oliver. 1986. Review of parasitic associations of larval water mites (Acari: Parasitengona: Hydrachnida) with insect hosts. *Can. Entomol.* 118:407–72.

Smith, J. D. 1972. *Systematics of the chiropteran family Mormoopidae.* Univ. Kans. Mus. Nat. Hist. Misc. Publ. 56. Lawrence.

Smith, J. W. C. 1999. A new method for handling invertebrates collected using standard vacuum-sampling apparatus. *Aust. J. Entomol.* 38:227–28.

Smith, K. E., C. F. Quist, and J. M. Crum. 1997. Clinical illness in a wild turkey with *Laminosioptes cysticola* infestation of the viscera and peripheral nerves. *Avian Dis.* 41:484–89. [LAMINOSIOPTIDAE]

Smith, L. M., and E. M. Stafford. 1948. The bud mite and the erineum mites of grapes. *Hilgardia* 18:317–34. [ERIOPHYIDAE]

Smitley, D. R., and G. G. Kennedy. 1985. Photo-orientated aerial dispersal behavior of *Tetranychus urticae* (Acari: Tetranychidae) enhances escape from the leaf surface. *Ann. Entomol. Soc. Amer.* 78:609–14.

Smrž, J., and J. Kocourkova. 1999. Mite communities of two epiphytic lichen species (*Hypogymnia physodes* and *Parmelia sulcata*) in the Czech Republic. *Pedobiologia* 43:385–90.

Snetsinger, R. 1956. Biology of *Bdella depressa*, a predaceous mite. *J. Econ. Entomol.* 49:745.

Snodgrass, R. E. 1948. *The feeding organs of the Arachnida, including mites and ticks.* Smithsonian Misc. Coll. 110 (10). Washington, DC: Smithsonian Institution.

Soar, C. D. and W. Williamson. 1925. *The British Hydracarina.* Vol. I. London: Ray Society.

———. 1927. *The British Hydracarina.* Vol. II. London: Ray Society.

———. 1929. *The British Hydracarina.* Vol. III. London: Ray Society.

Şoarec, J. 1942. *Contribution à l'étude des Hydracariens de Roumanie.* Ann. Scien. Univ. Jassy, sec. 2 (Sci. Nat.), 29.

Sokolov, I. A., and A. I. Yankowskaya. 1969. New genus and species of Halacaridae (Acari) from the abyssal region of the Kuril-Kamchatka trench. In Evans (1969c), 113–14.

Sokolow, I. I. 1940. *Hydracarina—Vodianye Kleshchi (Ch. 1: Hydrachnellae).* Fauna SSSR 5 (2).
———. 1944. *Baicalacarus vermiformis* n. gen., n. sp. *Uch. Zap. Leningr. G. Univ., Ser. Estestv. Nauk* 2 (68): 46–53. [STYGOTHROMBIDIIDAE]
———. 1948. O sistematicheskom polozhenii *Baicalacarus vermiformis* Sokolow (Acari) i ego reliktovom kharaktere. *Dokl. Akad. Nauk SSSR* 60:181–83. [STYGOTHROMBIDIIDAE]
Solhøy, I. W. 1997. A redescription of *Mycobates sarekensis* (Trägårdh) (Acari: Oribatei). *Acarologia* 38 (1): 69–77. [MYCOBATIDAE]
Solhøy, I. W., and T. Solhøy. 2000. The fossil oribatid mites fauna (Acari: Oribatida) in late-glacial and early-Holocene sediments in Krøkenes Lake, western Norway. *J. Paleolimnol.* 23:35–47.
Soliman, Z. R. 1971. Four new species of family Caligonellidae. *Bull. Soc. Entomol. Egypt* 55:95–104.
———. 1975. Genus *Ledermuelleriopsis* Willmann from Lattakia, Syria (Acari: Prostigmata) with a description of two new species. *Acarologia* 17:243–47.
Soliman, Z. R., and M. A. Zaher. 1975. *Hemitarsocheylus,* a new genus from the family Tarsocheylidae with a description of a new species. *Acarologia* 17:103–5. [STIGMOCHEYLIDAE]
Solomon, L. 1978. Epicriidae from the Romanian fauna. *Rev. Roum. Biol., Ser. Biol. Anim.* 23:9–14.
———. 1985. Zoocoenotical researches on the Veigaiaidae from the Slatioara-Rarau secular forest. *Suceava An. Muz. Judetean, Fasc. Stiint. Nat.* 8:139–54. [in Romanian]
Solomon, M. E. 1945. Tyroglyphid mites in stored products. Methods for the study of population density. *Ann. Appl. Biol.* 32: 71–75.
———. 1946a. Tyroglyphid mites in stored products. Ecological studies. *Ann. Appl. Biol.* 33:82–97. [CHEYLETIDAE]
———. 1946b. Tyroglyphid mites in stored products. Nature and amount of damage to wheat. *Ann. Appl. Biol.* 33:280–89.
———. 1969. Elements of predator-prey interactions of storage mites. *Acarologia* 11:484–501. [CHEYLETIDAE]
Solomon, M. E., and A. M. Cunnington. 1964. Rearing acaroid mites. In Evans (1969c), 399–403.
Soma, K. 1990. Studies on the life history of *Phthiracarus japonicus* Aoki (Acarina: Phthiracaridae) in a creeping pine (*Pinus pumila* Regel) shrub. *Edaphologia* 43:25–30.
Sonenshine, D. E. 1985. Pheromones and other semiochemicals of the Acari. *Ann. Rev. Entomol.* 30:1–28.
———. 1991. *Biology of ticks.* Vol. 1. New York: Oxford Univ. Press.
———. 1993. *Biology of ticks.* Vol. 2. New York: Oxford Univ. Press.
Sonenshine, D. E., and J. D. Gregson. 1970. A contribution to the internal anatomy and histology of the bat tick *Ornithodoros kelleyi* Cooley and Kohls, 1941. I, The alimentary system, with notes on the food channel in *Ornithodoros denmarki* Kohls, Sonenshine and Clifford, 1965. *J. Med. Entomol.* 7:46–64.
Sonenshine, D. E., P. J. Homsher, K. A. Carson, and V. D. Wang. 1984. Evidence of the role of the cheliceral digits in the perception of genital sex pheromones during mating in the American dog tick, *Dermacentor variabilis* (Acari: Ixodidae). *J. Med. Entomol.* 21:296–306.
Sonenshine, D. E., R. S. Lane, and W. L. Nicholson. 2002. Ticks (Ixodida). In Mullen and Durden, 517–58.
Sonenshine, D. E., R. M. Silverstein, E. C. Layton, and P. J. Homsher. 1974. Evidence for the existence of a sex pheromone in 2 species of ixodid ticks (Metastigmata: Ixodidae). *J. Med. Entomol.* 11:307–15.
Sonenshine, D. E., R. M. Silverstein, E. Plummer, J. R. West, and T. McCullough. 1976. 2,6-dichlorophenol, the sex pheromone of the Rocky Mountain wood tick, *Dermacentor andersoni* Stiles and the American dog tick, *Dermacentor variabilis* (Say). *J. Chem. Ecol.* 2:201–9.
Sorenson, J. T., D. N. Kinn, and R. L. Doutt. 1983. Biological observations on *Bdella longicornis*: A small predatory mite in California vineyards (Acari: Bdellidae). *Entomography* 2:297–305.
Sorenson, J. T., D. N. Kinn, R. L. Doutt, and J. R. Cate. 1976. Biology of the mite *Anystis agilis* (Acari: Anystidae): A California vineyard predator. *Ann. Entomol. Soc. Amer.* 69:905–10.
Southcott, R. V. 1956. Description of a new Australian raphignathoid mite, with remarks on the classification of the Trombidiformes (Acarina). *Proc. Linn. Soc. N. S. W.* 81:306–12. [CAMEROBIIDAE]
———. 1957a. Description of a new Australian raphignathoid mite, with remarks on the classification of the Trombidiformes (Acarina). *Proc. Linn. Soc. N. S. W.* 81:306–12. [CAMEROBIIDAE]
———. 1957b. The genus *Myrmicotrombium* Womersley 1934 (Acarina, Erythraeidae), with remarks on the systematics of Erythraeoidea and Trombidioidea. *Rec. S. Aust. Mus.* 13:91–98.
———. 1961. Studies on the systematics and biology of the Erythraeoidea (Acarina), with a critical revision of the genera and subfamilies. *Aust. J. Zool.* 9:367–610.
———. 1963. The Smarididae (Acarina) of North and South America and some other countries. *Trans. Roy. Soc. S. Aust.* 86: 159–245.
———. 1966. Revision of the genus *Charletonia* Oudemans (Acarina: Erythraeidae) with observations on post-larval instars. *Aust. J. Zool.* 14:687–819.
———. 1972. Revision of larvae of the tribe Callidosomatini (Acarina: Erythraeidae) with observations on post-larval instars. *Aust. J. Zool., Suppl. Ser.* 13:1–84.
———. 1976. Arachnidism and allied syndromes in the Australian region. *Rec. Adelaide Children's Hosp.* 1:97–186. [PYEMOTIDAE, ERYTHRAEIDAE]
———. 1982. Observations on *Chyzeria* Canestrini and some related genera (Acarina: Trombidioidea) with remarks on the classification of the superfamily and description of a pygmephorid mite phoretic on *Chyzeria. Rec. S. Aust. Mus.* 18:285–326. [CHYZERIIDAE]
———. 1983. A new Australian species of *Charletonia* (Acarina: Erythraeidae). *Proc. Linn. Soc. N. S. W.* 107:123–28.
———. 1986a. Australian larvae of the genus *Trombella* (Acarina: Trombidioidea). *Aust. J. Zool.* 34:611–46. [TROMBELLIDAE]
———. 1986b. On *Trombella alpha* n. sp. (Acarina: Trombellidae) from Australia: Correlation, description, developmental abnormalities, systematics and possible auditory structures. *Rec. S. Aust. Mus.* 19:145–68.
———. 1986c. Studies on the taxonomy and biology of the subfamily Trombidiinae (Acarina: Trombidiidae) with a critical revision of the genera. *Aust. J. Zool, Suppl. Ser.* 123:1–116.
———. 1987a. The classification of the mite families Trombellidae and Johnstonianidae and related groups, with the description of a new larva (Acarina: Trombellidae: *Nothotrombidium*) from North America. *Trans. Roy. Soc. S. Aust.* 111:25–42.
———. 1987b. A new larval mite (Acarina: Trombidioidea) ectopar-

asitic on an Australian centipede, and the Trombidiidae reclassified. *Trans. Roy. Soc. Aust.* 111:43–52.

———. 1988a. A new Australian larval callidosomatine mite (Acarina: Erythraeidae) parasitic on flies, with notes on subfamily and tribe classification. *Proc. Linn. Soc. N. S. W.* 110:191–202.

———. 1988b. Two new larval ectoparasitic mites (Acarina: Erythraeidae) from Sri Lankan tetrigid grasshoppers. *Entomol. Scand. Suppl.* 30:151–59.

———. 1988c. Two new larval Erythraeinae (Acarina: Erythraeidae) from New Zealand, and the larval Erythraeinae revised. *New Zeal. J. Zool.* 15:223–33.

———. 1988d. Two new larval mites (Acarina: Erythraeidae) ectoparasitic on North Queensland cicadas. *Rec. S. Aust. Mus.* 22:103–16.

———. 1989a. Larval *Leptus* (Acarina: Erythraeidae) ectoparasitic on a North American ant. *Acarologia* 30:241–48.

———. 1989b. A larval mite (Acarina: Erythraeidae) parasitizing the European honey bee in Guatemala. *Acarologia* 30:123–29.

———. 1991a. Descriptions of larval *Leptus* (Acarina: Erythraeidae) ectoparasitic on Australian Diptera, and two earlier described Australian larvae. *Invertebr. Taxon.* 5:717–63.

———. 1991b. A further revision of *Charletonia* (Acarina: Erythraeidae) base on larvae, protonymphs and deutonymphs. *Invertebr. Taxon.* 5:61–131.

———. 1991c. A new adult eutrombidiine mite (Acari: Microtrombidiidae) from North America. *Internat. J. Acarol.* 17:271–74.

———. 1991d. A new trombellid mite (Acarina: Trombellidae) from South Australia. *Trans. Roy. Soc. S. Aust.* 115:207–12.

———. 1992a. Adults and deutonymphs of *Eutrombidium* (Acarina: Trombidioidea) in North America. *Acarologia* 33:335–48. [EUTROMBIDIIDAE]

———. 1992b. Revision of the larvae of *Leptus* Latreille (Acarina: Erythraeidae) of Europe and North America, with descriptions of post-larval instars. *Zool. J. Linn. Soc.* 105:1–153.

———. 1993a. Larvae of *Leptus* (Acarina: Erythraeidae) ectoparasitic on higher insects of Australia and New Guinea. *Invertebr. Taxon.* 7:1473–1550.

———. 1993b. A new larval *Charletonia* (Acarina: Erythraeidae) from Spain. *Acarologia* 34:51–56.

———. 1993c. Revision of taxonomy of the larvae of the subfamily Eutrombiinae (Acarina: Microtrombidiidae). *Invertebr. Taxon.* 7:885–959.

———. 1994a. Revision of the larvae of the Microtrombidiinae (Acarina: Microtrombidiinae), with notes on life histories. *Zoologica* 144:1–155. [MICROTROMBIDIIDAE]

———. 1994b. Two new larval Erythraeidae (Acarina) from Thailand, with keys to the larvae of *Leptus* for Asia and New Guinea, and world larvae of *Hauptmannia*. *Steenstrupia* 20: 165–76.

———. 1995. A new larval smaridid mite (Acarina: Smarididae) from Costa Rica. *Acarologia* 36:57–64.

———. 1997. Revision of the larvae of *Paratrombium* (Acarina: Trombidiidae) of Australia and Papua New Guinea, with notes on life histories. *Rec. S. Aust. Mus.* 29:95–120.

———. 1999. Larvae of *Leptus* (Aarina: Erythraeidae), free-living or ectoparasitic on arachnids and lower insects of Australia and Papua New Guinea, with descriptions of reared post-larval instars. *Zool. J. Linn. Soc.* 127:113–276.

Southwood, T. R. E. 1966. *Ecological methods.* London: Methuen.

Søvik, G., H. P. Leinaas, R. A. Imms, and T. Solhøy. 2003. Population dynamics and life history of the oribatid mite *Ameronothrus lineatus* (Acari, Oribatida) on the high arctic archipelago of Svalbard. *Pedobiologia* 47:257–71. [AMERONOTHRIDAE]

Spain, A. V. 1969. A new genus and two new species of arboreal Oppiidae (Acari: Cryptostigmata) from New Zealand. *Pac. Insects* 11 (1): 155–63.

Springett, B. P. 1968. Aspects of the relationship between burying beetles, *Necrophorus* spp. and the mite, *Poecilochirus necrophori* Vitz. *J. Anim. Ecol.* 37:417–24. [PARASITIDAE]

Stafford, E. M., and R. L. Doutt. 1974. *Insect grape pests of northern California*. Calif. Agr. Exp. Sta. Circ. 566. Berkeley: Univ. Calif. Press. [ERIOPHYIDAE]

Stammer, H. G. 1951. Eine neue Tracheenmilbe, *Bombacarus buchneri* n. g. n. sp. (Acar. Podapolipodidae). *Zool. Anz.* 146:137–50.

Stammer, H. J., ed. 1963. *Beiträge zur Systematik und Ökologie Mitteleuropäischer Acarina. Band II. Mesostigmata. Abschn. III.* Leipzig: Akad. Verlags. Geest & Portig K.-G.

Stamou, G. P. 1986. Respiration of *Achipteria holomonensis* (Acari; Cryptostigmata). *Oikos* 46:176–84. [ACHIPTERIIDAE]

Stannard, L. J., Jr., and S. M. Vaishampayan, 1971. *Ovacarus clivinae*, new genus and species (Acarina: Podapolipidae), an endoparasite of the slender seedcorn beetle. *Ann. Entomol. Soc. Amer.* 64:887–90.

Stary, J. 2000. List of oribatid mites (Acari: Oribatida) of the Bohemia, Czech Republic. *Sbornik Prirodovedneho klubu v Uh. Hradisti* 5:129–54.

Stary, J., and W. Block. 1998. Distribution and biogeography of oribatid mites (Acari, Oribatida) in Antarctica, the sub-antarctic islands and nearby areas. *J. Nat. Hist.* 32:861–94.

Stasiov, S. 2003. *Leptus holmiae* Southcott 1992 (Acarina, Erythraeidae) associated with harvestmen (Opilionida). *Ekol. Bratislava* 22:23–27.

Stechmann, D.-H. 1977. Zur Phänologie und zum Wirtsspektrum einiger an Zygopteren (Odonata) und Nematoceren (Diptera) ektoparasitisch auftretenden *Arrenurus*-Arten (Hydrachnellae, Acari). *Z. Angew. Entomol.* 82:349–55.

———. 1978. Eiablage, Parasitismus und postparasitische Entwicklung von *Arrenurus*-Arten (Hydrachnellae, Acari). *Z. Parasitenkd.* 57:169–88.

———. 1980. Zum Wirtskreis syntopischer *Arrenurus*-Arten (Hydrachnellae, Acari) mit parasitischer Entwicklung an Nematocera (Diptera). *Z. Parasitenkd.* 62:267–83.

Steedman, H. F. 1958. Dimethyl hydrantoin formaldehyde: a new water-soluble resin for use as a mounting medium. *Qtly. J. Microscop. Sci.* 99:451–52.

Steenbergen, H. A. 1993. *Macrofauna-atlas of North-Holland: Distribution maps and responses to environmental factors of aquatic invertebrates.* Den Haag: Cip-gegevens Koninklijke Bibliotheek.

Steiner, M. Y., and S. Goodwin. 1998. Methods for collecting and rearing thrips (Thysanoptera) and their natural enemies. *Aust. J. Entomol.* 37:101–6.

Steinkraus, D. C., and E. A. Cross. 1993. Description and life history of *Acarophenax mahunkai*, n. sp. (Acari, Tarsonemina: Acarophenacidae), an egg parasite of the lesser mealworm (Coleoptera: Tenebrionidae). *Ann. Entomol. Soc. Amer.* 86:239–49.

Steullet, P., and P. M. Guerin. 1994a. Identification of vertebrate volatiles stimulating olfactory receptors on tarsus I of the tick *Amblyomma variegatum* Fabricius (Ixodidae). I, Receptors with the Haller's organ capsule. *J. Comp. Physiol.* A, 174:27–38.

———. 1994b. Identification of vertebrate volatiles stimulating olfactory receptors on tarsus I of the tick *Amblyomma variegatum* Fabricius (Ixodidae). II, Receptors outside the Haller's organ capsule. *J. Comp. Physiol.* A, 174:39–47.

Stewart, A. J. A., and A. F. Wright. 1995. A new inexpensive suction apparatus for sampling arthropods in grassland. *Ecol. Entomol.* 20:98–102.

Stiller, D., and J. J. Sullivan. 1977. New host and locality records for *Gastronyssus bakeri* (Acari: Gastronyssidae), a stomach parasite of Malaysian fruit bats, with notes on immature stages. *J. Med. Entomol.* 14:250–52.

Stone, C. 1986. An investigation into the morphology and biology of *Tetranychus lintearius* Dufour (Acari: Tetranychidae). *Exp. Appl. Acarol.* 2:173–86.

Stone, P. C., and G. D. Ogles. 1953. *Uropoda agitans*, a mite pest in commercial fishworm beds. *J. Econ. Entomol.* 46:711.

Stone, W. B., E. Parks, B. L. Weber, and F. J. Parks. 1972. Experimental transfer of sarcoptic mange from red foxes and wild canids to captive wildlife and domestic animals. *New York Fish & Game J.* 19:1–11. [SARCOPTIDAE]

Stout, V. M., and K. Viets. 1959. Über eine parasitisch lebende Halacaride (Acari) von Neuseeland. *Veröff. Inst. Meeresforsch., Bremerhaven* 6:203–11.

Strandtmann, R. W. 1965. Additional notes on Teneriffiidae (Acarina: Prostigmata) with two previously unpublished plates by A. C. Oudemans. *J. Kans. Entomol. Soc.* 38:258–61.

———. 1967. Terrestrial Prostigmata (Trombidiform mites). In *Entomology of Antarctica*, ed. J. L. Gressit, 51–80. Washington, DC: Amer. Geophysical Union.

———. 1970. Acarina: Eupodiform Prostigmata of South Georgia. *Pac. Insects Monogr.* 23:89–106. [EUPODIDAE, PENTHALODIDAE, RHAGIDIIDAE]

———. 1971. The eupodoid mites of Alaska (Acarina: Prostigmata). *Pac. Insects* 13:75–118. [EUPODIDAE, PENTHALODIDAE, RHAGIDIIDAE]

———. 1981. A new genus and species of penthaleid mite (Acarina: Penthaleidae) from New Zealand. *Pac. Insects* 23:392–95.

Strandtmann, R. W., and L. Davies. 1972. Eupodiform mites from Possession Island, Crozet Islands, with a key to the species of *Eupodes* (Acarina: Prostigmata). *Pac. Insects* 14:39–56. [EUPODIDAE, RHAGIDIIDAE]

Strandtmann, R. W., and M. L. Goff. 1978. The Eupodoidea of Hawaii (Acarina: Prostigmata). *Pac. Insects* 19:121–43. [EUPODIDAE, PENTHALODIDAE, RHAGIDIIDAE, STRANDTMANNIIDAE]

Strandtmann, R. W., and J. Prasse. 1977. Prostigmatic mites from the experimental farm in Etzdorf/Saalkreis, G. D. R. *Abh. Ber. Naturk. Görlitz* 50:1–33. [EUPODIDAE, RHAGIDIIDAE]

Strandtmann, R. W., and L. Sømme. 1977. Prostigmate mites (Acari, Prostigmata) from Sverdrupfjella, Dronning Maud Land, with description of four new species. *Norway J. Entomol.* 24:137–46. [EUPODIDAE]

Strandtmann, R. W., and P. J. Tilbrook. 1968. Some Prostigmata (Acari) from Signy Island, South Orkney Islands, and Deception Island, South Shetland Islands. *Brit. Antarct. Surv. Bull.* 17:51–57. [EUPODIDAE, PENTHALEIDAE]

Strandtmann, R. W., and G. W. Wharton. 1958. *Manual of mesostigmatid mites parasitic on vertebrates*. Ed. C. E. Yunker. College Park, MD: Contrib. 4, Inst. Acarology. [PARAMEGISTIDAE]

Strenzke, K. 1954. *Nematalycus nematoides* n. gen. n. sp. (Acarina, Trombidiformes) aus dem Grundwasser der algerischen Kuste. *Vie Milieu* 4:638–47.

———. 1963. Entwicklung und Verwandtschaftsbeziehungen der Oribatidengattung *Gehypochthonius* (Arach., Acari). *Senckenb. Biol.* 44 (3): 231–55. [GEHYPOCHTHONIIDAE]

Stubbs, C. S. 1995. Dispersal of soredia by the oribatid mite, *Humerobates arborea*. *Mycologia* 87:454–58. [HUMEROBATIDAE]

Stunkard, H. W. 1940. The morphology and life-history of the Cestode *Bertiella studeri*. *Amer. J. Trop. Med.* 20:305–28.

Sturhan, D., and G. Hampel. 1977. Pflanzenparasitische Nematoden als Beute der Wurzelmilbe *Rhizoglyphus echinopus* (Acarina, Tyrogluphidae). *Anz. Schädlingskde., Pflanzen., Umweltschutz* 50:115–18.

Styer, W. E., and L. R. Nault. 1996. Corn and grain plants. In Lindquist, Sabelis, and Bruin, 6:611–18.

Su, T.-Z., G.-Y. Beng, and X.-D. Chen. 1979. Notes on the genus *Siteroptes* in China (Acarina: Pyemotidae). *Acta Zootaxon. Sin.* 4:162–69. [SITEROPTIDAE] [in Chinese, English summary]

Su, T.-Z., X.-D. Chen, and G.-Y. Beng. 1981. Study of symbiotic relationships between mites and fungi in cereal ear rot. *Acta Phytopathol. Sin.* 11:19–24. [SITEROPTIDAE] [in Chinese, English summary]

Subías, L. S. 2004. Listado sistemático, sinonímico y biogeográfico de los ácaros oribátidos (Acariformes, Oribatida) del mundo (1758–2002). *Graellsia* 60:3–305.

Subías, L. S., and A. Arillo. 2001. *Acari, Oribatei, Gymnonota. II, Oppioidea*. Vol. 15 in Fauna iberica. Madrid: Mus. Nac. Cienc. Nat.

———. 2002. Oribatid mite fossils from the Upper Devonian of South Mountain, New York and the Lower Carboniferous of County Antrim, Northern Ireland (Acariformes, Oribatida). *Est. Mus. Cien. Nat. de Álava* 17:93–106.

Subías, L. S., A. Arillo, and J. Gil-Martin. 1994. Oribátidos de Marruecos y Sahara Occidental. II, Listado de especies (Acari, Oribatida). *Bol. R. Esp. Hist. Nat. (Sec. Biol.)* 91:129–34.

Subías, L. S., and P. Balogh. 1989. Identification keys to the genera of Oppiidae Grandjean, 1951 (Acari: Oribatei). *Acta Zool. Acad. Sci. Hung.* 35:355–412.

Subías, L. S., and J. Gil-Martin. 1997. Systematic and biogeographic checklist of oribatids from Western Mediterranean (Acari, Oribatida). *Ann. Mus. Civ. Storia Nat. "Giacomo Doria," Genova* 91:459–98.

Sugasawa, J., Y. Kitashima, and T. Gotoh. 2002. Hybrid affinities between the green and the red forms of the two-spotted spider mite *Tetranychus urticae* (Acari: Tetranychidae) under laboratory and semi-natural conditions. *Appl. Entomol. Zool.* 37:127–39.

Sukhareva, S. I. 1992. A key to species of four-legged mites (Acariformes, Tetrapodili) living on cereals in the USSR. *Entomol. Obozr.* 71:231–40. [in Russian; English translation in *Entomol. Rev.* 72:54–65]

Sumangala, K. 1999. Ecobiology of *Pseudacarapis indoapis* (Acari: Tarsonemidae). 1, Nutrition, dispersal and host range. *Entomon* 24:235–39.

Sumangala, K., and M. A. Haq. 2002. Ecobiology of *Pseudacarapis indoapis* Lindquist (Acari: Tarsonemidae). 2, Ontogeny and breeding behaviour. *J. Entomol. Res.* 26:83–88.

Summers, F. M. 1957. American species of *Ledermuelleria* and *Ledermuelleriopsis*, with a note on new synonymy in *Neognathus* (Acarina, Stigmaeidae, Caligonellidae). *Proc. Entomol. Soc. Wash.* 59:49–60.

———. 1960a. *Eupalopsis* and eupalopsellid mites (Acarina: Stigmaeidae, Eupalopsellidae). *Fla. Entomol.* 43:119–38.

———. 1960b. Several stigmaeid mites formerly included in *Mediolata* redescribed in *Zetzellia* Ouds. and *Agistemus*, new genus (Acarina). *Proc. Entomol. Soc. Wash.* 62:233–47.

———. 1962. The genus *Stigmaeus* (Acarina: Stigmaeidae). *Hilgardia* 33:491–537. [STIGMAEIDAE]

———. 1964. Three uncommon genera of the mite family Stigmaeidae (Acarina). *Proc. Entomol. Soc. Wash.* 66:184–92. [BARBUTIIDAE]

———. 1966a. Genera of the mite family Stigmaeidae Oudemans (Acarina). *Acarologia* 8:230–50.

———. 1966b. Key to families of the Raphignathoidea (Acarina). *Acarologia* 8:226–29.

Summers, F. M., and W. M. Chaudri. 1965. New species of the genus *Cryptognathus* Kramer (Acarina: Cryptognathidae). *Hilgardia* 36:313–26.

Summers, F. M., and S. Ehara. 1965. Revaluation of the taxonomie characters in four species of the genus *Cheylostigmaeus* Willmann (Acarina: Stigmaeidae). *Acarologia* 7:49–62.

Summers, F. M., and D. W. Price. 1961. New and redescribed species of *Ledermuelleria* from North America (Acarina: Stigmaeidae). *Hilgardia* 31:369–82.

———. 1970. *Review of the mite family Cheyletidae.* Univ. Calif. Publ. Entomol. 61. Berkeley: Univ. Calif. Press.

Summers, F. M., and E. I. Schlinger. 1955. Mites of the family Caligonellidae (Acarina). *Hilgardia* 23:539–61.

Summers, F. M., and R. L. Witt. 1971. The gnathosoma of *Cheyletus cacahuamilpensis* Baker (Acarina: Cheyletidae). *Proc. Entomol. Soc. Wash.* 73:158–68.

———. 1972. Nesting behavior of *Cheyletus eruditus* (Acarina: Cheyletidae). *Pan-Pac. Entomol.* 48:261–69.

Summers, F. M., R. Witt, and S. Regev. 1972. Evaluation of several characters by which five species of *Cheyletus* are distinguished (Acarina: Cheyletidae). *Proc. Entomol. Soc. Wash.* 74:230–52.

Sumner, M. E. 1999. *Handbook of soil science.* Washington, DC: CRC Press.

Suski, Z. 1972. Tarsonemid mites on apple trees in Poland. X. Laboratory studies on the biology of certain mite species of the family Tarsonemidae (Acarina, Heterostigmata). *Zesz. Probl. Post. Nauk Roln.* 129:111–37.

———. 1973. A revision of *Siteroptes cerealium* (Kirchner) complex (Acarina, Heterostigmata, Pyemotidae). *Ann. Zool. (Warsaw)* 30:509–35. [SITEROPTIDAE]

———. 1984. On the identity of pyemotid mites associated with the silver-top disease of grasses. In Griffiths and Bowman, 1: 174–79. [SITEROPTIDAE]

Suwa, A., Y. Kitashima, and T. Gotoh. 2006. Compatability between strains of *Tetranychus pueraricola* (Acari: Tetranychidae) correlated with distance between sites of their origin. *Internat. J. Acarol.* 32:203–9.

Suzuki, K. 1971. Fruit-attacking oribatid mites. *Atypus* 57:13–14.

———. 1976. Rearing of *Damaeus* sp. Jumping behaviour of *Ceratoppia bipilis*. New description of *Ceratozetella imperatoria*. *Edaphologia* 14:7–15. [DAMAEIDAE, PELOPPIIDAE, CERATOZETIDAE]

———. 1977. Two species of Enarthronota: *Haplochthonius simplex* and *Cosmochthonius reticulates*. *Bull. Biogeogr. Soc. Jpn.* 32 (2): 5–16. [COSMOCHTHONIIDAE, HAPLOCHTHONIIDAE, PTEROCHTHONIIDAE] [in Japanese]

———. 1978. Key to the Japanese oribatid mites. *Nihon Shigaku Kyoiku Kenkyusho Chosa Shiryo* 56:121–75.

Swan, D. C. 1934. The red-legged earth mite *Halotydeus destructor* (Tucker) in South Australia: With remarks upon *Penthaleus major* (Dugès). *J. Agr. S. Aust.* 38:353–67. [PENTHALEIDAE]

Sweatman, G. K., J. B. Walker, and J. A. Bindernagel. 1969. Stages in the development of *Chorioptsoroptes kenyensis* gen. et sp. n. (Acari: Sarcoptiformes: Psoroptidae), a body mange mite from captive African buffalo, *Syncerus caffer*, in Kenya. *J. Parasitol.* 55:1298–1310.

Swift, S. F. 1987. A new species of *Stigmaeus* (Acari: Prostigmata: Stigmaeidae) parasitic on phlebotomine flies (Diptera: Psychodidae). *Internat. J. Acarol.* 13:239–43.

———. 1996. Hawaiian Raphignathoidea: Family Caligonellidae (Acari: Prostigmata) with descriptions of five new taxa and a key to genera and species. *Ann. Entomol. Soc. Amer.* 89:313–27.

———. 2000. New records of Hawaiian prostigmatid mites (Acari: Prostigmata). *Bishop Mus. Occ. Papers* 64:50–51. [ADAMYSTIDAE]

———. 2001. The leg chaetotaxy of Caligonellidae (Prostigmata: Raphignathoidea). In Halliday et al. (2001), 242–49.

Swift, S. F., and M. L. Goff. 1987. The family Bdellidae (Acari: Prostigmata) in the Hawaiian Islands. *Internat. J. Acarol.* 13:29–49.

Swift, S. F., and R. A. Norton. 1998. Preliminary report on oribatid mite (Acari: Oribatida) diversity in the Hawaiian Islands. *Bishop Mus. Occ. Papers* 57:1–44.

Sylwestrowicz, Z. A., and D. B. Kostecka. 1991. Oribatei and Mesostigmata of pine forests from Poland. In Dusbábek and Bukva, 1:365–71.

Szalay, L. 1964. *Víziatkák (Hydracarina). 14 Füzet [Part 14].* In Fauna Hungariae (Magyarország Állatvilága) XVIII. Arachnoidea, 72. Budapest: Akadémiai Kaidó.

Takada, W., T. Sakata, S. Shimano, Y. Enami, N. Mori, R. Nishida, and Y. Kuwahara. 2005. Scheloribatid mites as the source of pumiliotoxins in dendrobatid frogs. *J. Chem. Ecol.* 31 (10): 2403–15. [SCHELORIBATIDAE]

Takahashi, M., K. Machida, M. Murata, H. Misumi, E. Hori, A. Kawamura, Jr., and H. Tanaka. 1993. Seasonal development of *Leptotrombidium pallidum* (Acar.: Trombiculidae) observed by experimental rearing in the natural environment. *J. Med. Entomol.* 30:320–25.

Takahashi, M., H. Misumi, H. Matsuzawa, K. Morita, O. Tsuji, E. Hori, A. Kawamura, Jr., and H. Tanaka. 1994. Seasonal development of *Leptotrombidium scutellare* (Acari: Trombiculidae) observed by experimental rearing in field conditions. *Jpn. J. Sanit. Zool.* 45:113–20.

Takahashi, M., H. Misumi, H. Urakami, S. Nakajima, S. Furui, S. Yamamoto, S. Furuya, M. Misumi, and I. Matsumoto. 2004. Mite vectors (Acari: Trombiculidae) of scrub typhus in a new endemic area in northern Kyoto, Japan. *J. Med. Entomol.* 41:107–14.

Takahashi, M., K. Morita, O. Tsuji, H. Misumi, J. Otsuji, E. Hori, A. Kawamura, Jr., and H. Tanaka. 1995. Seasonal development of *Leptotrombidium akamushi* (Acari: Trombiculidae) under field conditions. *J. Med. Entomol.* 32:843–46.

Takahashi, M., M. Murata, K. Machida, H. Misumi, E. Hori, A. Kawamura, Jr., and H. Tanaka. 1993. Trombiculid mites and *Rickettsia tsutsugamushi* isolated from wild rodents and mites in an endemic area of Saitama Prefecture, Japan. *Jpn. J. Sanit. Zool.* 44:7–14.

Takahashi, M., and H. Tanaka. 1995. Transmission dynamics of *Rickettsia tsutsugamushi* in vector mites. *Jpn. J. Sanit. Zool.* 46:319–29. [TROMBICULIDAE]

Takaku, G. 2000. Two new mite species of the genus *Thinoseius* (Acari: Gamasida: Eviphididae) from Japan. *Species Diversity* 5:361–74.

Takaku, G., and S. Hartini. 2001. Macrochelid mites (Arachnida: Acari: Macrochelidae: *Glyptholaspis, Macrocheles, Neopodocinum*) associated with dung beetles in Bali, Indonesia. *Species Diversity* 6:323–45.

Takaku, G., H. Katakura, and N. Yoshida. 1994. Mesostigmatic mites (Acari) associated with ground, burying, roving carrion and dung beetles (Coleoptera) in Sapporo and Tomakomai, Hokkaido, northern Japan. *Zool. Sci.* 11:305–11. [PARASITIDAE, MACROCHELIDAE, PACHYLAELAPIDAE, EVIPHIDIDAE, OLOGAMASIDAE]

Tamm, J. C. 1984. Surviving long submergence in the egg stage: A successful strategy of terrestrial arthropods living on flood plains (Collembola, Acari, Diptera). *Oecologia* 61:417–19.

Tamura, H., and N. Enda. 1980a. Life histories of three species of nematode-feeding mesostigmatid mites associated with the pine sawyer beetle, *Monochamus alternatus*. *J. Jpn. Forestry Soc.* 62:301–7. [DIGAMASELLIDAE] [in Japanese]

———. 1980b. Mesostigmatid mites associated with the Japanese pine sawyer beetle. *Jpn. J. Appl. Entomol. Zool.* 24: 54–61. [MELICHARIDAE] [in Japanese]

Tarman, K. 1983. Catalogus faunae Jugoslaviae. *Cons. Acad. Sci. Publ. Soc. Foed. Jugosl., Ljubljana* 3:1–61.

Tarras-Wahlberg, N. 1961. The Oribatei of a central Swedish bog and their environment. *Oikos (Suppl.)* 4:1–56.

———. 1985. *Trimalaconothrus scimitarum* n. sp. (Oribatida: Malaconothridae), an aquatic mite from Lake Naivasha. *Internat. J. Acarol.* 2:17–21.

Tashiro, H. 1967. Self-watering acrylic cages for confining insects and mites on detached leaves. *J. Econ. Entomol.* 60:354–56.

Tatchell, R. J., and K. C. Binnington. 1973. An active constituent of the saliva of the cattle tick *Boophilus microplus*. In Daniel and Rosický, 745–48.

Tawfik, M. F. S., and K. T. Awadallah. 1971. The biology of *Pyemotes herfsi* Oudemans and its efficiency in the control of the resting larvae of the pink bollworm, *Pectinophora gossypiella* Saunders, in U.A.R. *Bull. Soc. Entomol. Egypt* 54:49–71. [PYEMOTIDAE]

Tawfik, M. F. S., M. M. El-Husseini, and K. T. Awadallah. 1981. Interactions between certain host larvae and the pyemotid ectoparasite, *Pyemotes tritici*. *Bull. Soc. Entomol. Egypt* 63:181–89. [PYEMOTIDAE]

Telford, M. J., and R. H. Thomas. 1998a. Expression of homeobox genes show chelicerate arthropods retain their deutocerebral segment. *Proc. Nat. Acad. Sci.* 95:10671–75.

———. 1998b. Of mites and zen: Expression studies in a chelicerate arthropod confirm zen is a divergent Hox gene. *Dev., Genes Evol.* 208:591–94.

Ten Winkel, E. H. 1985. The influence of predation by the water mite *Hygrobates nigromaculatus* on a population of chironomid larvae. *Verh. Internat. Verein. Theoret. Angew. Limnol.* 22: 3230–32. [HYGROBATIDAE]

Ten Winkel, E. H., C. Davids, and J. G. de Nobel. 1989. Food and feeding strategies of water mites of the genus *Hygrobates* and the impact of their predation on the larval population of the chironomid *Cladotanytarsus mancus* (Walker) in Lake Maarsseveen. *Neth. J. Zool.* 39:246–63. [HYGROBATIDAE]

Tenorio, J. M. 1974. A new species of *Lynxacarus* (Acarina: Astigmata: Listrophoridae) from *Felis catus* in the Hawaiian Islands. *J. Med. Entomol.* 11:599–604.

Tenorio, J. M., and F. J. Radovsky. 1974. The genus *Mesolaelaps* (Laelapidae: Mesolaelapinae, n. subfam.) with descriptions of two new species from New Guinea. *J. Med. Entomol.* 11: 211–22.

Terry, J. R. 1966. Studies of tarsonemid mites associated with bark beetles and with rice. Master's thesis, Louisiana State Univ.

Teschner, D. 1963. Die Biologie, Verbreitung und Ökologie der Grundwasser milbe *Lobohalacarus weberi quadriporus* (Walter, 1947) Limnohalacaridae, Acari. *Arch. Hydrobiol.* 59:71–102.

Tevis, L., Jr., and I. M. Newell. 1962. Studies on the biology and seasonal cycle of the giant red velvet mite, *Dinothrombium pandorae* (Acari, Trombidiidae). *Ecology* 43:497–505.

Theis, G., and R. Schuster. 1974. Gestielte Tröpfchenspermatophoren bei Calyptostomiden (Acari, Trombidiformes). *Mitt. Naturwiss. Ver. Steiermark* 104:183–85.

Theron, P. D. 1974. Hybalicidae, a new family of endeostigmatic mites (Acari: Trombidiformes). *Acarologia* 16:397–412.

———. 1975. New species of the genus *Terpnacarus* Grandjean (Acari: Terpnacaridae) with notes on the biology of one species. *J. Entomol. Soc. S. Afr.* 39:131–41.

———. 1976. Three new species of the genus *Alycosmesis* (Acari: Terpnacaridae) from South Africa. *J. Entomol. Soc. S. Afr.* 38:289–96.

———. 1977. New species of the genus *Petralycus* Grandjean (Acari: Endeostigmata) from South Africa. *Acarologia* 19:38–45.

———. 1979. The functional morphology of the gnathosoma of some liquid and solid feeders in the Trombidiformes, Cryptostigmata and Astigmata (Acarina). In Piffl (1979), 575–79.

Theron, P. D., M. K. P. Meyer, and P. A. J. Ryke. 1969. Two new genera of the family Paratydeidae (Acari: Prostigmata) from South African soils. *Acarologia* 11:697–710.

———. 1970. The family Alicorhagiidae Grandjean (Acari: Trombidiformes) with descriptions of a new genus and species from South African soils. *Acarologia* 12:668–76.

Theron, P. D., and P. A. J. Ryke. 1969. The family Nanorchestidae Grandjean (Acarina: Prostigmata) with descriptions of new species from South African Soils. *J. Entomol. Soc. S. Afr.* 32:31–60.

———. 1975a. Five new species of the family Lordalychidae (Acari: Endeostigmata) from South Africa. *Acarologia* 17: 631–51.

———. 1975b. Three new species of the family Sphaerolichidae (Acari: Endeostigmata) from South Africa. *Acarologia* 17:220–35.

Thewke, S. E., and W. R. Enns. 1975. A new species of *Pavlovskicheyla* (Acarina: Cheyletidae) from the elytra of *Platydema ruficorne* (Coleoptera: Tenebrionidae) from Missouri. *Acarologia* 17:671–82.

Thibaud, J. M., and Y. Coineau. 1998. Nouvelles stations pour le genre *Gordialycus* (Acarien: Nematalycidae). *Biogeographica* 74:91–94.

Thomas, R. H. 2002. Mites as models in development and genetics. In Bernini et al. (2002), 21–26.

Thomas, R. H., and S. F. MacLean, Jr. 1988. Community structure in soil Acari along a latitudinal transect of tundra sites in northern Alaska. *Pedobiologia* 31:113–38.

Thomas, R. H., and M. J. Telford. 1999. Appendage development in embryos of the oribatid mite *Archegozetes longisetosus* (Acari, Oribatei, Trhypochthoniidae). *Acta Zool.* 80:193–200.

Thon, K. 1905. Neue Luftorgane bei Milben. *Zool. Anz.* 28:585–94.

———. 1906. Die aussere Morphologie und die Systematik der Holothyriden. *Zool. Jb. Abt. Syst., Ökol. Geogr. Tiere* 23:677–724.

Thor, S. 1911. Eine neue Acarinenfamilie (Teneriffiidae) und zwei neue Gattungen, die eine von Teneriffa, die andere aus Paraguay. *Zool. Anz.* 38:171–79.

———. 1929. Über die Phylogenie und Systematik der Acarina, mit Beiträgen zu ersten Entwicklungsgeschichte einzelner Gruppen. *Nyt Mag. Naturvidensk.* 67:145–210.

———. 1931. Bdellidae, Nicoletiellidae, Cryptognathidae. *Das Tierreich* 56:1–64, 66–77.

———. 1933. Acarina. Tydeidae, Ereynetidae. *Das Tierreich* 60:xi, 1–82.

Thor, S., and C. Willmann. 1941. *Acarina: Eupodidae, Penthalodidae, Penthaleidae, Rhagidiidae, Pachygnathidae, Cunaxidae.* Das Tierreich 71a. Berlin: Walter de Gruyter.

Thorp, J. H., and A. P. Covich, eds. 2001. *Ecology and classification of North American freshwater invertebrates.* 2d ed. New York: Academic Press.

Thorvilson, H. G., and S. A. Phillips, Jr. 1987. The straw itch mite, *Pyemotes tritici* (Acari: Pyemotidae), as a biological control agent of red imported fire ants, *Solenopsis invicta* (Hymenoptera: Formicidae). *Fla. Entomol.* 70:439–44.

Tichomirov, S. I. 1969. Morphological and ecological structure of the genus *Parasitus* sensu Micherdzinski, 1966 (Gamasoidea, Parasitidae). *Zool. Zh.* 48:1325–36, 1467–77.

———. 1977. Sem. Parasitidae Oudemans, 1901. In Gilyarov and Bregetova, 55–107. [in Russian]

Till, W. M. 1969. A new laelapine mite from the Madagascar hissing-cockroach, *Gromphadorhina portentosa* (Schaum). *Acarologia* 11:515–23. [LAELAPIDAE]

———. 1988. Additions to the British and Irish mites of the genus *Veigaia* (Acari: Veigaiidae) with a key to the species. *Acarologia* 29:3–12.

Timms, S., D. N. Ferro, and R. M. Emberson. 1982. Andropolymorphism and its heritability in *Sancassania berlesei* (Michael) (Acari: Acaridae). *Acarologia* 22:391–98.

Tipton, V. J. 1960. The genus *Laelaps*, with a review of the Laelaptinae and a new subfamily Alphalaelaptinae (Acarina: Laelaptidae). *Univ. Calif. Publ. Entomol.* 16:233–356.

Tjying, I.-S. 1971. A new ascid mite (*Lasioseius martini* n. sp.) recovered from shallots in Taiwan (Acarina: Ascidae). *Plant Protect. Bull., Taiwan* 13:1–5. [BLATTISOCIIDAE]

Tomczyk, A., and D. Kropczynska. 1985. Effects on the host plant. In Helle and Sabelis (1985a), 1A:317–29. [TETRANYCHIDAE]

Totschnig, U., and H. Schatz. 1998. Hornmilben (Acari, Oribatida) in einem Auwald bei Glanz an der Isel (Osttirol, Österreich, Zentralalpen): Zönologie. *Ber. Naturwiss. Medizin. Vereins Innsbruck* 85:95–109.

Toyoshima, S., and H. Amano. 1999. Cytological evidence of pseudo-arrhenotoky in two phytoseiid mites *Phytoseiulus persimilis* Athias-Henriot and *Amblyseius womersleyi* Schicha. *J. Acarol. Soc. Jpn.* 8:135–42.

Trägårdh, I. 1906. Neue Acariden aus Natal und Zululand. *Zool. Anz.* 30:870–77. [TRACHYUROPODIDAE]

———. 1909. *Speleorchestes*, a new genus of saltatorial Trombidiidae, which lives in termites' and ants' nests. *Ark. Zool.* 6:1–14.

———. 1911. *Discomegistus*, a new genus of myriapodophilous Parasitidae from Trinidad, with notes on the Heterozeroninae. *Ark. Zool.* 7:1–21.

———. 1931. Acarina from the Juan Fernandez Island. *Nat. Hist. Juan Fernandez and Easter Island, Uppsala* 3:553–628.

———. 1938. Further contributions towards the comparative morphology and classification of the Mesostigmata. *Entomol. Tidskr.* 59:123–58.

———. 1941a. Contributions towards the comparative morphology and phylogeny of the Mesostigmata. IV, On the Celaenopsoidea and Euzerconidae. *Entomol. Tidskr.* 62:169–76. [CELAENOPSIDAE, EUZERCONIDAE]

———. 1941b. Further contributions towards the comparative morphology of the Mesostigmata. III, On the Polyaspididae Berl. *Zool. Bidr. Uppsala* 20:345–57. [THINOZERCONIDAE]

———. 1942. Microgyniina, a new group of Mesostigmata. *Entomol. Tidskr.* 63:120–33. [MICROGYNIIDAE]

———. 1946. Outlines of a new classification of the Mesostigmata (Acarina) based on comparative morphological data. *Lunds Univ. Årsskrift. N. F.* 57:3–37. [THINOZERCONIDAE, DIARTHROPHALLIDAE]

———. 1948. Descriptions of *Micromegistus*, a new genus of the Paramegistidae with notes on *Neomegistus, Paramegistus* and *Echinomegistus* (Acarina). *Entomol. Tidskr.* 69:127–31. [PARAMEGISTIDAE, PARANTENNULIDAE]

———. 1950. Studies on the Celaenopsidae, Diplogyniidae and Schizogyniidae (Acarina). *Ark. Zool.*, ser. 2, 1 (25): 361–451.

———. 1951. *Brachycelaenopsis*, a new genus of Celaenopsidae (Acarina) from Tasmania. *Entomol. Tidskr.* 72:60–64. [CELAENOPSIDAE]

Traub, R., and C. L. Wisseman, Jr. 1974. Current concepts of the ecology of chigger-borne rickettsiosis (scrub typhus). *Jpn. J. Med. Sci. Biol.* 27:1–5. [TROMBICULIDAE]

Traub, R., C. L. Wisseman, Jr., and A. Farhang-Azad. 1977. The ecology of chigger-borne rickettsiosis and murine typhus—changing concepts and epidemiology. *Proc. XV Internat. Congr. Entomol.* 1977:487–89. [TROMBICULIDAE]

Travé, J. 1959. Sur le genre *Niphocepheus* Balogh 1943, les Niphocepheidae, Famille nouvelle (Acariens, Oribates). *Acarologia* 1 (4): 475–98

———. 1961. Sur un nouveau genre d'Oribates: *Neotrichozetes. Acarologia* 3 (3): 363–75. [NEOTRICHOZETIDAE]

———. 1963. Écologie et biologie des Oribates (Acariens) saxicoles et arboricoles. *Vie Milieu Suppl.* 14:1–267.

———. 1967. *Phyllochthonius aoutii* nov. gen. nov. spec., un Enarthronota (Acarien Oribates) nouveau de Cote d'Ivoire, avec la creation d'une superfamille nouvelle, Phyllochthonoidea. *Zool. Meded.* 42:83–105.

———. 1976. Les prélarves d'Acariens. Mise au point et données récentes. *Rev. Ecol. Biol. Sol* 13:161–71.

———. 1979. Les stases immatures de *Dolicheremaus dorni* (Balogh) (Oribatei). *Acarologia* 20:294–302. [OTOCEPHEIDAE]

———. 1982. Premières observations sur le comportement de *Thonius braueri* (Thon, 1906) (Holothyrida). *Acarologia* 23: 199–206.

———. 1983. Observations morphologique sur les Holothyrides. 1re partie. Le corps d *Thonius braueri* (Thon, 1906). *Acarologia* 24:333–41.

———. 1986. Les taenidies respiratoires des Oribates. *Acarologia* 27:85–94.

Travé, J., H. M. André, G. Taberly, and F. Bernini. 1996. *Les acariens oribates*. Wavre, Belgium: AGAR Publishers.

Travé, J., and M. Vachon. 1975. François Grandjean 1882–1975 (Notice biographique et bibliographique). *Acarologia* 17:1–19.

Travis, B. V. 1968. Glyptal—a useful slide ringing compound. *J. Med. Entomol.* 5:24.

Trávníček, M. 1989. Laboratory cultivation and biology of mites in the family Liacaridae (Acari: Oribatida). *Acta Univ. Carolinae, Prague* 33:69–80.

Treat, A. E. 1958. Social organization in the moth ear mite (*Myrmonyssus phalaenodectes*). In Becker, 2:475–80.

———. 1965. Otopheidomenids from the Antilles and elsewhere. *Acarologia* 7:1–16.

———. 1966. A new *Blattisocius* (Acarina: Mesostigmata) from noctuid moths. *J. New York Entomol. Soc.* 74:143–59. [BLATTISOCIIDAE]

———. 1970. Two tydeid mites from the ears of noctuid moths. *Amer. Mus. Novit.* 2426:1–14. [TYDEIDAE, IOLINIDAE]

———. 1975. *Mites of moths and butterflies*. Ithaca, NY: Comstock. [PYEMOTIDAE, PYGMEPHORIDAE]

———. 1980. Nymphal *Sphaerolophus* reared from larval *Charletonia* (Acarina: Erythraeidae). *Internat. J. Acarol.* 6:205–14.

———. 1985. Larval *Callidosoma* (Acari, Erythraeidae): New records and a new rearing from the Americas. *Internat. J. Acarol.* 11: 93–124.

———. 1988. *Dicrocheles scedastes* in Tyringham, Massachusetts, 1974–1987 (Acari: Dermanyssidae). *Internat. J. Acarol.* 14: 19–22.

Treat, A. E., and C. H. W. Flechtmann. 1979. *Charletonia rocciai* n. sp. (Acari, Prostigmata, Erythraeidae), an ectoparasite of the Amazon fly. *Internat. J. Acarol.* 5:117–22.

Tryon, H. 1898. Fruitlet core-rot of pineapple. *Queensland Agr. J.* 3:458–67. [TARSONEMIDAE]

Tschapka, M., and S. A. Cunningham. 2004. Flower mites of *Calyptrogyne ghiesbreghtiana* (Arecaceae): Evidence for dispersal using pollinating bats. *Biotropica* 36:377–81. [MELICHARIDAE]

Tseng, Y.-H. 1978. Mites of the family Bdellidae from Taiwan (Acarina: Prostigmata). *J. Agr. Assoc. China*, n.s., 104:25–51.

———. 1980. Taxonomical study of the mites family Cunaxidae from Taiwan (Acarina: Tromoidiformes) [*sic*]. *Qtly. J. Taiwan Mus.* 33:253–77.

———. 1982. Mites of the family Stigmaeidae of Taiwan with key to genera of the world (Acarina: Prostigmata). *Phytopathol. Entomol. Nat. Taiwan Univ.* 9:1–52.

———. 1989. Taxonomical studies on mites associated with weeds, paddy rice and upland rice fields in Taiwan (I). *Chinese J. Entomol., Spec. Publ.* 3:1–35.

———. 1990. *A monograph of the mite family Tetranychidae (Acarina: Trombidiformes) from Taiwan*. Taiwan Mus. Spec. Publ. ser., no. 9.

———. 1994. A taxonomic study of free-living Gamasina mite family Veigaiidae Oudemans (Acari: Mesostigmata) from Taiwan. *Chinese J. Entomol.* 14:501–28.

Türk, E. 1963. A new tyroglyphid deutonymph in amber from Chiapas, Mexico. *Univ. Calif. Publ. Entomol.* 31:49–51. [WINTERSCHMIDTIIDAE]

Türk, E., and F. Türk. 1957. Systematik und Ökologie der Tyroglyphiden Mitteleuropas. In H. J. Stammer, 1:1–231.

Turk, F. A. 1945. Studies of Acari. V, Notes on and descriptions of new and little-known British Acari. *Ann. Mag. Nat. Hist.* 12:785–820. [CALYPTOSTOMATIDAE]

Turk, F. A., and S.-M. Phillips. 1946. A monograph of the slug mite, *Riccardoella limacum* (Schrank). *Proc. Zool. Soc. Lond.* 115: 448–72. [EREYNETIDAE]

Tuzovskij, P. V. 1977. Postembrional'noye razvitiye vodjanogo kleshcha *Sperchon clupeifer* (Acariformes, Sperchonidae). *Zool. Zh.* 56:1616–28. [SPERCHONTIDAE]

———. 1978. K diagnozu *Brachypoda* (*Hemibrachypoda*) *mutila* (Walter, 1928) (Axonopsidae, Acariformes). *Naucnye Dokl. Vysšej Skoly, Biol. Nauk* 4:47–52. [ATURIDAE]

———. 1979. Licinka *Momonia falcipalpis* Halbert, 1906 (Acariformes, Mideopsidae). *Vestn. Zool.* 1979:69–71. [MOMONIIDAE]

———. 1981. Postembrionalnye stadii vodjanogo kleshcha *Torrenticola amplexa* (Koenike, 1908) (Acariformes, Torrenticolidae). *Entomol. Obozr.* 60:205–12.

———. 1982. Novyj rod novye vidy vodjanykh kleshchej s dopolnenijami k diagnozu semejstva Mideidae. *Zool. Zh.* 61:199–209.

———. 1984a. Dejtonimfy vodjanych klescej rodov Midea i *Mideopsis* (Mideopsidae, Acariformes) fauny SSSR. *Vestn. Zool.* 1984:8–14. [MIDEIDAE, MIDEOPSIDAE].

———. 1984b. Novyj vid vodjanogo kleshcha roda *Oxus* (Oxidae, Acariformes). *Izvest. Sibirsk. Otdel. Akad. Nauk SSSR, Ser. Biol. Nauk* 1984:45–51.

———. 1985. O sistematiceskom statuse vodjanogo kleshcha *Unionicola dresscheri* (Hess., 1946). *Naucnye Dokl. Vysšej Skoly, Biol. Nauk* 10:27–33.

———. 1986a. Novyj rod vodjanykh kleshchej iz semejstva Aturidae (Acariformes). *Zool. Zh.* 55:936–38.

———. 1986b. Novyj vid vodjanogo kleshcha roda *Albaxona* (Aturidae, Acariformes). *Izvest. Sibirsk. Otdel. Akad. Nauk SSSR, Ser. Biol. Nauk* 1986:83–86.

———. 1986c. Novyj vid vodjanogo kleshcha roda *Piersigia* (Piersigiidae, Acariformes). *Izvest. Sibirsk. Otdel. Akad. Nauk SSSR, Ser. Biol. Nauk* 1986:120–27.

———. 1987a. *Morfologiya i postembrional'noe razvitiye vodjanykh kleshchej*. Moscow: Akad. Nauk SSSR.

———. 1987b. Novye vidy vodjanykh kleshchej roda *Wandesia* (Hydryphantidae, Acariformes) iz Magadanskoj Oblasti. *Naucnye Dokl. Vysšej Skoly, Biol. Nauk* 3: 36–44.

———. 1988a. Novye vidy vodjanykh kleshchej roda *Feltria* Koenike (Acariformes, Feltriidae). *Entomol. Obozr.* 67:218–30.

———. 1988b. Opisanie novogo vida vodjanogo kleshcha roda *Wandesia* (Wandesiidae, Acariformes) iz Magadanskoj Oblasti. *Izvest. Sibirsk. Otdel. Akad. Nauk SSSR., Ser. Biol. Nauk* 1988: 53–59. [HYDRYPHANTIDAE]

———. 1990. *Opredelitel' deitonimf vodjanykh kleshchei*. Moscow: Akad. Nauk SSSR.

———. 1994a. Novyi vid vodjanogo kleshcha roda *Panisus* (Acariformes, Thyasidae). *Zool. Zh.* 73:138–41. [HYDRYPHANTIDAE]

———. 1994b. Opisanie dvukh vidov vodjanykh kleshchej roda *Aturus* (Aturidae, Acariformes). *Zool. Zh.* 73:151–57.

———. 1997. *Vodjanye kleshchej roda Limnesia (Acariformes, Limnesiidae) fauny Rossii*. Moscow: Rossiiskaia Akad. Nauk, Inst. Ekol. Volzhskogo Basseina.

———. 1998. A new species of the water mite genus *Feltria* from Magadan Province (Acariformes: Feltriidae). *Zoosyst. Rossica* 7:85–87.

———. 1999. Novye vidy vodjanykh kleshchej roda (Feltriidae, Acariformes) ie Rossii. *Zool. Zh.* 78:539–48.

———. 2000a. A new species of the water mite genus *Thyas* Koch from Magadan Province of Russia (Acariformes: Hydryphantidae). *Zoosyst. Rossica* 8:245–50.

———. 2000b. Opisanie dvukh novykh vidov vodjanykh kleshchej roda *Monatractides* (Acariformes, Torrenticolidae) s severnogo kavkaea. *Zool. Zh.* 79:111–19.

———. 2001a. Description of a new water mite species of the genus *Pionacercus* Piersig, 1894 from NE Russia (Acariformes: Pionidae). *Zoosyst. Rossica* 10:55–62.

———. 2001b. K diagnoeu vodjanogo kleshcha *Pionacercus anadryensis* (Acariformes, Pionidae). *Zool. Zh.* 80:758–65.

———. 2001c. A new species of the water mite genus *Mideopsis* Neuman from Japan (Acariformes: Mideopsidae). *Zoosyst. Rossica* 10:63–65.

———. 2001d. A new species of water mite of the genus *Pseudohydryphantes* (Hydryphantoidea, Acariformes). *Acarina* 9:247–52. [HYDRYPHANTIDAE]

———. 2001e. Opisanie novogo vida vodjanogo kleshcha roda *Hydrochoreutes* (Acariformes, Pionidae) ie Basseina Volgi. *Zool. Zh.* 80:871–78.

———. 2002a. Description of a male and a female of the water mite *Feltria montana* Tuzovskij (Acariformes, Feltriidae). *Acarina* 10:47–50.

———. 2002b. K sistematike vodjanogo kleshcha *Sperchonopsis minutiporus* Tuzovskij, 1990 (Acariformes, Sperchontidae). *Biol. Vnutrenn. Vod* 2002:33–37.

———. 2002c. A new species of the water mite genus *Frontipoda* from Magadan Province of Russia (Acariformes: Oxidae). *Zoosyst. Rossica* 11:293–97.

———. 2002d. A new species of the water mite genus *Oxus* from Primorsk Territory of Russia (Acariformes: Oxidae). *Zoosyst. Rossica* 11:299–302.

———. 2002e. A new species of the water mite genus *Sperchon* Kramer from Primorsk Territory of Russia (Acariformes: Sperchonidae). *Zoosyst. Rossica* 11:105–8. [SPERCHONTIDAE]

———. 2002f. A new species of water mites of the genus *Mideopsis* Neuman (Acariformes, Mideopsidae) from Russia. *Acarina* 10:161–65.

———. 2003a. Novyi vid vodjanykh kleshchej roda *Pionopsis* (Acariformes, Pionidae). *Zool. Zh.* 2003:735–43.

———. 2003b. Opisanie samki samtsa i lichinki vodjanogo kleshcha *Hydryphantes ruberoides* (Hydryphantidae, Acariformes). *Biol. Vnutrenn. Vod* 2003:39–46.

———. 2003c. Three new species of water mites of the genus *Hydrochoreutes* (Acari: Hydrachnidia: Pionidae) from Russia. In I. M. Smith (2003a), 303–21.

———. 2005. The morphology of larva of the water mite *Panisus major* Tuzovskij (Acariformes, Hydryphantidae). *Acarina* 13:199–202.

———. 2006a. Larval morphology of the water mite *Lebertia densa* (Acariformes, Lebertiidae) and redescription of adults. *Acarina* 14:93–97.

———. 2006b. Larval morphology of the water mite *Mideopsis roztoczensis* Biesiadka and Kowalik (Acariformes: Mideopsidae). *Zoosyst. Rossica* 15:27–31.

———. 2006c. The water mite *Frontipoda carpenteri* Halbert, 1911 from Karelia (Acariformes: Oxidae). *Zoosyst. Rossica* 15:19–26.

———. 2007. Water mites of the genus *Thyas* Koch, 1835 (Acariformes: Thyadidae) in Russia. *Acarina* 15:173–259.

Tuzovskij, P. V., D. Benefatti, and R. Gerecke. 2001. The water mite family Acherontacaridae Cook, 1967 nov. stat. and diagnosis of the superfamily Hydrovolzioidea Thor, 1905 (Acariformes, Hydrachnidia). *Acarologia* 41:451–73.

Tuzovskij, P. V., and R. Gerecke. 2003. A new diagnosis and status for *Mixobates* Thor, 1905 (Acari, Hydrachnidia, Hygrobatidae), with a revision of the Palearctic species. *Ann. Limnol.–Internat. J. Limnol.* 39:151–74.

Uchikawa, K. 1988. Myobiidae (Acarina, Trombidiformes) associated with minor families of Chiroptera (Mammalia) and a discussion of phylogeny of chiropteran myobiid genera. *J. Parasitol.* 74:159–76.

Uchikawa, K., F. Kawamori, T. Kanda, and N. Kumada. 1994. Trombiculid fauna and seasonal abundance of *Leptotrombidium scutellare* (Acari: Trombiculidae) in an endemic area of scrub typhus (tsutsugamushi disease) in Yamakita Town, Kanagawa Prefecture, Japan. *J. Med. Entomol.* 31:844–49.

Uchikawa, K., and N. Kumada. 1989. Endemic outbreaks of tsutsugamushi disease in Japan and vector chiggers (Trombiformes: Trombiculidae). In ChannaBasavanna and Viraktamath, 1:103–6.

Ueckermann, E. A. 1989. Revision of the family Adamystidae Cunliffe (Acari: Prostigmata). *Phytophylactica* 21:227–40.

Ueckermann, E. A., and T. G. Grout. 2007. Tydeoid mites (Acari: Tydeidae, Edbakerellidae, Iolinidae) occurring on *Citrus* in southern Africa. *J. Nat. Hist.* 41:2351–78.

Upton, M. S. 1991. *Methods for collecting, preserving and studying insects and related forms*. 4th ed. Brisbane: Aust. Entomol. Soc.

———. 1993. Aqueous gum-chloral slide mounting media: an historical review. *Bull. Entomol. Res.* 83:267–74.

Urhan, R. 2001. New species of zerconid mites (Acari, Gamasida, Zerconidae) from Turkey. *Acarologia* 41:69–75.

Usher, M. B., and M. Edwards. 1986. A biometrical study of the family Tydeidae (Acari, Prostigmata) in the maritime antarctic, with descriptions of three new taxa. *J. Zool. Lond* A, 209:355–83.

Valdecasas, A. G. 1981. *Axonopsis* (*Brachypodopsis*) *guadarramensis* n. sp. from central Spain (Acari, Parasitengona). *Arch. Hydrobiol.* 91:254–58. [ATURIDAE]

———. 1984. A review of the water mite *Hungarohydracarus subterraneus* Szalay and its subspecies (Hydrachnellae: Acari). In Giffiths and Bowman, 2:971–79. [HUNGAROHYDRACARIDAE]

———. 2001a. A new subfamily of Rhynchohydracaridae (Acari, Hydrachnellae) from the island of Coiba (Panama) with descriptions of new taxa. *J. Nat. Hist.* 35:1565–74.

———. 2001b. Water mites new for the Iberian Peninsula (Acari, Hydrachnellae). *Graellsia* 57:91–98. [FELTRIIDAE, FRONTIPODOPSIDAE, MOMONIIDAE, CHAPPUISIDIDAE, MIDEOPSIDAE]

Valdecasas, A. G., and J. G. Avilés. 1983. Some water mites (Acari, Parasitengona) from the island of Minorca. *Actas Prim. Congr. Espan. Limnol.* 1981:173–77. [ATURIDAE]

Valdecasas, A. G., and J. M. Becerra. 1997. Extending the availability of microscopic type material for taxonomy and research. *TREE* 12:211–12.

Valle, A. 1954. Biogeographia dell'Isola di Zannone (Archipelago Pontino), Acarina. *Rendiconti dell'Academia Nazionale de XL*, ser. 4–5:142–48. [ASTERNOSEIIDAE]

Van de Vrie, M. 1985. Greenhouse ornamentals. In Helle and Sabelis (1985b), 1B:273–83. [TETRANYCHIDAE]

Van de Vrie, M., J. A. McMurtry, and C. B. Huffaker. 1972. Ecology of tetranychid mites and their natural enemies: a review. III, Biology, ecology, and pest status, and host-plant relations of tetranychids. *Hilgardia* 41:343–432.

Van der Hammen, L. 1952. The Oribatei (Acari) of the Netherlands (Acari). *Zool. Verh., Leiden* 17:1–139.

———. 1959. Berlese's primitive oribatid mites. *Zool. Verh., Leiden* 40:1–93.

———. 1961. Description of *Holothyrus grandjeani* nov. spec., and notes on the classification of the mites. *Nova Guinea* (n.s.) *Zool.* 9:173–94.

———. 1964. The morphology of *Glyptholaspis confusa* (Foá, 1900) (Acarida, Gamasina). *Zool. Verh.* 71:3–56.

———. 1966. Studies on Opilioacarida (Arachnida). I, Description of *Opilioacarus texanus* (Chamberlin and Mulaik) and revised classification of the genera. *Zool. Verh.* 86:1–80.

———. 1968a. The gnathosoma of *Hermannia convexa* (C. L. Koch) (Acarida: Oribatina) and comparative remarks on its morphology in other mites. *Zool. Verh.* 94:1–45.

———. 1968b. Stray notes on Acarida (Arachnida) I. *Zool. Meded. Leiden* 42:261–80.

———. 1968c. Studies on Opilioacarida (Arachnida). II, Redescription of *Paracarus hexophthalmus* (Redikorzev). *Zool. Meded. Leiden* 43:57–76.

———. 1969a. Notes on the morphology of *Alycus roseus* C. L. Koch. *Zool. Meded.* 43:177–202.

———. 1969b. Studies on Opilioacarida (Arachnida). III, *Opilioacarus platensis* Silvestri, and *Adenacarus arabicus* (With). *Zool. Meded.* 44:113–31.

———. 1970a. La phylogénèse des Opilioacarides, et leur affinités avec les autres acariens. *Acarologia* 12:465–73.

———. 1970b. Remarques génerales sur la structure fondamentale du gnathosoma. *Acarologia* 12:16–22.

———. 1970c. La segmentation des appendices chez les acariens. *Acarologia* 12:11–15.

———. 1970d. La segmentation primitive des acariens. *Acarologia* 12:3–10.

———. 1972. A revised classification of the mites (Arachnida, Acarida) with diagnoses, a key, and notes on phylogeny. *Zool. Meded. Leiden* 47:273–92.

———, ed. 1972–76. *F. Grandjean, complete acarological works.* Vol. I (Nos. 1–37, 1928–36), 1972; Vol. II (Nos. 38–94, 1937–44), 1973; Vol. III (Nos. 95–119, 1942–48), 1974; Vol. IV (Nos. 120–57, 1949–54), 1974; Vol. V (Nos. 158–92, 1954–58), 1975; Vol. VI (Nos. 193–217, 1959–63), 1975; Vol. VII (Nos. 218–41, 1964–74), 1976. The Hague: W. Junk; Lochem: Antiquariaat Junk.

———. 1976. *Glossaire de la terminologie acarologique.* Vol. II, *Opilioacarida.* The Hague: W. Junk B. V.

———, ed. 1977a. *A. Berlese, complete acarological works. Collected acarological papers previously published in* Redia, *1903–23.* Vols. I–V, Nos. 1–31. The Hague: W. Junk; Lochem: Antiquariaat Junk.

———. 1977b. Studies on Opilioacarida (Arachnida). IV, The genera *Panchaetes* Naudo and *Salfacarus*, gen. nov. *Zoöl. Meded.* 51:43–78.

———, ed. 1979. *A. Berlese, complete acarological works. Acari, Myriapoda et Scorpiones hucusque in Italia reperta.* Vols. I–XII, 1882–1903. The Hague: W. Junk; Amsterdam: Antiquariaat Junk.

———. 1980. *Glossary of acarological terminology.* Vol. I. The Hague: W. Junk.

———. 1982. Morphology and postembryonic development of *Rhizoglyphus echinopus* (Fumouze & Robin) (Chelicerata, Actinotrichida). *Zool. Meded. Leiden* 56:237–58. [ACARIDAE]

———. 1983a. Notes on the comparative morphology of ticks (Anactinotrichida: Ixodida). *Zool. Meded.* 57:209–42.

———. 1983b. Notes on Holothyrida (Anactinotrichid mites). *Zool. Verh.* 207:1–48.

———. 1985. Functional morphology and affinities of extant Chelicerata in evolutionary perspective. *Trans. Roy. Soc. Edin.* 76:137–46.

———. 1989. *An introduction to comparative arachnology.* The Hague: SPB Publishing.

Van der Schyff, J., P. D. Theron, and E. A. Ueckermann. 2003. Polytrichinae, a new subfamily of Bdellidae (Acari: Prostigmata) from the Afrotropical region. *Afr. Plant Prot.* 9:19–22.

———. 2004. *Hexabdella*, new mite genus of Bdellidae (Acari: Prostigmata) from southern Africa, with descriptions of five new species. *Afr. Plant Prot.* 10 (1): 13–25.

Van Dis, J. C. S., and E. A. Ueckermann. 1991. A review of the Pseudocheylidae Oudemans (Acari: Prostigmata). *Phytophylactica* 23:105–13.

Van Straalen, N. M., and H. A. Verhoef. 1997. The development of a bioindicator system for soil acidity based on arthropod pH preferences. *J. Appl. Ecol.* 34:217–32.

Vandel, A. 1969. La parthenogenese geographique. *Bull. Biol. (Paris)* 62:164–281. [TRACHYTIDAE]

Vargas, M., B. I. Bassols, C. E. Desch, M. T. Quintero, and O. J. Polaco. 1995. Description of two new species of the genus *Demodex* Owen, 1843 (Acari: Demodecidae) associated with Mexican bats. *Internat. J. Acarol.* 21:75–82. [DEMODICIDAE]

Vasiliu, N., O. Ivan, and M. Vasiliu. 1993. Conspectul faunistic al Oribatidelor (Acarina: Oribatida) din Romania [The faunistic synopsis of Oribatids (Acarina: Oribatida) from Romania]. *Anuar. Muz. Bucov. Fasc. Stiint. Nat., Suceava* 12:1–82.

Vater, G. 1982. Zur geographischen Verbreitung der Erntemilbe *Neotrombicula autumnalis* (Acari: Trombiculidae). *Zool. Jb. Abt. Syst., Ökol. Geogr. Tiere* 109:329–56.

Vázquez, M. M. 1999. *Catálogo de los ácaros oribátidos edáphicos de Sian Ka'an, Quintana Roo, México.* Chetumal, Mexico: Univ. Quintana Roo.

Vázquez, M. M., and H. Klompen. 2002. The family Opilioacaridae (Acari: Parasitiformes) in North and Central America, with description of four new species. *Acarologia* 42:299–322.

Vázquez, M. M., and J. G. Palacios-Vargas. 1988. Algunos observaciones sobre el comportamento del los acaros Opilioacaridos (Acarida: Notostigmata). *Rev. Nicarag. Entomol.* 6:1–6.

Veerman, A. 1985. Diapause. In Helle and Sabelis (1985a), 1A: 279–316. [TETRANYCHIDAE]

———. 1992. Diapause in phytoseiid mites: A review. *Exp. Appl. Acarol.* 14:1–60.

Veal, R. A., K. M. T. Giesen, and J. O. Whitaker, Jr. 1984. A new species of the genus *Ophthalmodex* Lukoschus & Nutting 1979

(Prostigmata: Demodicidae) from *Myotis lucifugus* (Chiroptera: Vespertilionidae). *Acarologia* 25:347–50.

Veiga, A. F. de S. L., and C. H. W. Flechtmann. 1980. O ácaro planto *Tenuipalpus pacificus* Baker, 1945 (Acari: Prostigmata: Tenuipalpidae) em avencas e samambaias (Pteridophyta) em Pernambuco e Para, Brasil. *Ann. Soc. Entomol. Brasil* 9:155–58.

Vercammen-Grandjean, P. H. 1955. Un genre nouveau: *Durenia*, dans la sous-famille des Trombellinae (Trombidiidae: Acarina). *Rev. Zool. Bot. Afr.* 52:252–60. [TROMBELLIDAE]

———. 1965a. *Iguanacarus*, a new subgenus of chigger mite from nasal fossae of the marine iguana in the Galapagos Islands, with a revision of the genus *Vatacarus* Southcott (Acarina, Trombiculidae). *Acarologia* (Fasc. Suppl.) 7:266–74.

———. 1965b. *Tenotrombicula mintneri* n. g., n. sp., an interesting commensal of African termites (Trombiculidae: Acarina). *Acarologia* 7:259–65.

———. 1967. Revision of *Hoplothrombium quinquescutatum* Ewing, 1925 (Trombidiidae: Acarina). *Opusc. Zool.* 96:1–7. [EUTROMBIDIIDAE]

———. 1971. Study of the "Erythraeidae, R.O.M. No. 8" of Ewing, 1937. *Proc. 3rd Internat. Congr. Acarol., Prague* 1971:329–35.

———. 1973a. Of techniques and ortho-iconography. In Daniel and Rosický, 321–28.

———. 1973b. Sur les statuts de la famille des Trombidiidae Leach, 1815 (Acarina: Prostigmata). *Acarologia* 15:102–14.

———. 1973c. Taxonomie des Neotrombidiidae et épilogue naturel du problème *Blankaartia nilotica* (Trägårdh, 1905) (Trombidioidea). *Acarologia* 15:115–20.

———. 1975a. Les organes de Claparède et les papilles genitals de certains acariens sont-ils des organes respiratoires? *Acarologia* 17:624–30.

———. 1975b. Some larval Trombiculidae of the Ethiopian region (Acari). *Rev. Zool. Afr.* 89:397–439.

———. 1976. À propos d'un Hydracarien presqu'oublié la "Larve de Kathariner" de Constantin Motas, 1928. *Stud. Comun. Muz. ÿÿtiint. Natur., Bacÿÿu* 1976: 31–42. [STYGOTHROMBIDIIDAE]

———. 1980. Analyse critique de la systématique de deux sous-familles d'Hydracariens: Wandesiinae Schwoerbel, 1961 et Stygothrombiinae Thor, 1935. *Folia Parasitol.* 27:151–64. [STYGOTHROMBIDIIDAE].

Vercammen-Grandjean, P. H., and J. R. Audy. 1959. Une seconde espèce appartenant au genre *Durenia* Vercammen 1955 et originaire de Malaisie: *Durenia sinaporensis* n. sp. (Acarina: Trombidiidae). *Biol. Jaarb.* 44:98–101.

Vercammen-Grandjean, P. H., and A. Cochrane. 1974. On three new species of larval Trombidiformes parasitizing American midges (Acarina: Trombidiidae and Johnstonianidae). *J. Kans. Entomol. Soc.* 47:66–79.

Vercammen-Grandjean, P. H., and Z. Feider. 1973. Le genre *Evansiella* V.-G., 1957 est synonyme de *Centrotrombidium* Kramer, 1896.: Description d'une forme larvaire nouvelle, *C. romaniense* (Trombidiformes: Johnstonianidae). *Riv. Parassitol.* 34:121–26.

Vercammen-Grandjean, P. H., and R. L. Langston. 1971a. *The chigger mites of the world.* Vol. 8, Guntherana *complex: Section A Genus* Guntherana. San Francisco: G. W. Hooper Foundation.

———. 1971b. Revision of the *Leptotrombidium* generic complex based on palpal setation combined with other morphological characters. *J. Med. Entomol.* 8:445–49.

Vercammen-Grandjean, P. H., and E. Popp. 1967. *Atomus rhopalicus* n. sp., a parasite of *Rhopalicus tutela* Walker (Hymenoptera), from Germany (Trombidiidae: Acarina). *Opusc. Zool.* 95:1–8.

Vercammen-Grandjean, P. H., and H. Rak. 1968. *Cheyletiella yasguri* Smiley, 1965, un parasite de canides aux Etats-Unis et hyperparasite d'Hippoboscide en Iran (Acarina: Cheyletidae). *Ann. Parasitol. Hum. Comp.* 43:405–12.

Vercammen-Grandjean, P. H., R. G. Van Driesche, and G. G. Gyrisco. 1977. New mites recovered from the alfalfa weevil (*Hypera postica*) in New York: *Trombidium hyperi* n. sp. and *Trombidium auroraense* n. sp. (Acari: Trombidiidae). *J. Kans. Entomol. Soc.* 50:89–107.

Vicidomini, S. 1996. Biologia di *Xylocopa* (*Xylocopa*) *violacea* (L. 1758) (Hymenoptera Apidae): Interazione con *Sennertia* (*Sennertia*) *cerambycina* (Acari Chaetodactylidae). *Boll. Zool. Agr. Bachic.*, ser. II, 28:71–76.

Vidrine, M. F. 1984. A new species of mite of the genus *Unionicola* (Acari: Unionicolidae) from the sponge *Trochospongilla leidyi* (Bowerbank) in Louisiana. *Proc. Louisiana Acad. Sci.* 47:23–26.

———. 1986. Revision of the Unionicolinae (Acari: Unionicolidae). *Internat. J. Acarol.* 12:233–43.

Vidrine, M. F., B. Borsari, and M. Bastian-Stanford. 2005. A new subgenus (*Chambardicola*) and species of *Unionicola* (Acari: Hydrachnidia: Unionicolidae) from freshwater mussels from the Central African Republic. *Internat. J. Acarol.* 31:255–58.

Vidrine, M. F., B. J. Vidrine, and B. J. Vidrine. 1976. New locality and host records for *Najadicola ingens* (Koenike) (Acarina: Unionicolidae). *ASB Bull.* 23:103. [PIONIDAE]

Viets, K. H. 1913/14. Hydracarinen-Fauna von Kamerun. *Arch. Hydrobiol.* 9:1–52, 177–225, 331–38.

———. 1916. Ergänzungen zur Hydracarinen-Fauna von Kamerun. (Neue Sammlungen). *Arch. Hydrobiol.* 11:241–306, 335–403.

———. 1925. Nachträge zur Hydracarinen-Fauna von Kamerun. (Sammlungen aus Kamerun, Span. Guinea und Fernando Poo). *Arch. Hydrobiol.* 16:197–242.

———. 1927. Die Halacaridae der Nordsee. *Z. Wiss. Zool.* 130: 83–173.

———. 1930. Zur Kenntnis der Hydracarinen-Fauna von Spanien. *Arch. Hydrobiol.* 21:175–240, 359–446.

———. 1931. Über eine an Krebskiemen parasitierende Halacaride aus Australien. *Zool. Anz.* 96:115–20.

———. 1932. Weitere Milben aus unterirdischen Gewässern. *Zool. Anz.* 100:173–76. [STYGOTHROMBIDIIDAE]

———. 1934. Siebente Mitteilung über Wassermilben aus unterirdischen Gewässern. *Zool. Anz.* 106:118–24. [STYGOTHROMBIDIIDAE]

———. 1935. Die Wasermilben von Sumatra, Java und Bali nach den Ergebnissen der Deutschen Limnologischen Sunda-Expedition. *Arch. Hydrobiol., Suppl.* 13:484–594, 595–738; 14:1–113.

———. 1936. Wassermilben oder Hydracarina (Hydrachnellae und Halacaridae). *Tierwelt Deutsch.* 31:1–288; 32:289–574.

———. 1938. Eine merkwurdige neue, in Tiefsee-Echiniden schmarotzende Halacaridengattung und -art (Acari). *Z. Parasitenkd. Jena* 10:210–16.

———. 1939. Meeresemilben aus der Adria. (Halacaridae und Hydrachnellae, Acari.). *Arch. Naturgesch. N. F.* 8:519–50.

———. 1940a. Hydrachnellae, Porohalacaridae and Halacaridae s. str. (Acari) aus Bulgarien. *Zool. Anz.* 130:36–41.

———. 1940b. Meeresmilben aus der Adria (Halacaridae und Halarachnellae, Acari). *Arch. Naturgesch.* 9:1–135.

———. 1940c. Zwei neue Porohalacaridae (Acari) as Südamerika. *Zool. Anz.* 130:191–201.
———. 1953. Die aus Afrika bekannten Wassermilben (Hydrachnellae, Acari). *Hydrobiologia* 5:1–178.
———. 1954a. Süßwassermilben (Hydrachnellae, Acari) aus Südamerika. *Arch. Hydrobiol.* 49:1–224.
———. 1954b. Wassermilben aus dem Amazonasgebiet (Hydrachnellae, Acari). (Systematische und ökologische Untersuchungen.) Bearbeitung der Sammlungen Dr. R. Braun, Aarau, und Dr. H. Sioli, Belém. Schweiz. *Z. Hydrol.* 16:78–151, 161–247.
———. 1955a. *Die Milben des Süßwassers und des Meeres (Hydrachnellae et Halacaridae, Acari). (Bibliographie, Katalog, Nomenklator.) I. Teil, Bibliographie.* Jena: G. Fischer Verlag.
———. 1955b. In subterranen Gewässern Deutschlands lebende Wassermilben (Hydrachnellae, Porohalacaridae und Stygothrombiidae). *Arch. Hydrobiol.* 50:33–63. [STYGOTHROMBIDIIDAE]
———. 1956. *Die Milben des Süßwassers und des Meeres. Hydrachnellae et Halacaridae (Acari). Zweiter und dritter Teil: Katalog und Nomenklator.* Jena: G. Fischer Verlag.
———. 1959. Die aus dem Einzugsgebiet der Weser bekannten oberirdisch und unterirdisch lebenden Wassermilben. *Veröff. Inst. Meeresforsch., Bremerhaven* 6:303–513. [STYGOTHROMBIDIIDAE]
Viets, K. O. 1980. Über *Diplodontus*-Arten (Hydrachnellae, Acari). *Mitt. Zool. Mus. Berl.* 56:99–101. [HYDRYPHANTIDAE]
———. 1982. *Die Milben des Süßwassers (Hydrachnellae und Halacaridae [part.], Acari).* 1, Bibliographie. *Naturwiss. Ver. Hamb.* 6.
———. 1987. *Die Milben des Süßwassers (Hydrachnellae und Halacaridae [part.], Acari).* 2, *Katalog.* Naturwiss. Ver. Hamb. 8.
Viets, K. O., and K. Böttger. 1974a. Zur Systematik und Ökologie rheophiler Hydrachnellae (Acari) Zentralafrikas. Teil I. *Acarologia* 16:106–59.
———. 1974b. Zur Systematik und Ökologie rheophiler Hydrachnellae (Acari) Zentralafrikas. Teil II. *Acarologia* 16:282–310.
Vincent, L. S., and G. Rack. 1982. *Pseudopygmephorus atypoides* Rack, new species (Acari: Pygmephoridae) associated with the fossorial mygalomorph spider *Atypoides riversi* O. P.-Cambridge (Araneae: Antrodiaetidae) in California. *Pan-Pac. Entomol.* 58:216–22.
Vistorin, H. E. 1978. Fortpflanzung und Entwicklung der Nicoletiellidae (Labidostomidae); Acari, Trombidiformes. *Zool. Jb. Syst.* 105:462–73.
———. 1979. Morphologische untersuchungen an Europäischen Nicoletielliden (Acari, Trombidiformes). *Acarologia* 21:367–83.
———. 1980a. Ernährungsbiologie und Anatomie des Verdauungstraktes der Nicoletiellidae (Acari, Trombidiformes). *Acarologia* 21:204–15. [LABIDOSTOMATIDAE]
———. 1980b. Morphologische Untersuchungen an europäischen Nicoletielliden (Acari, Trombidiformes). *Acarologia* 21:367–83. [LABIDOSTOMATIDAE]
———. 1981. Rasterelektronenmikroskopische Untersuchungen am Integument zweier Arten der Milbenfamilie Nicoletiellidae (Acari, Trombidiformes). *Zool. Anz.* 206:93–99. [LABIDOSTOMATIDAE]
Vistorin-Theis, G. 1975. Entwicklungszyklus der Calyptostomiden (Acari, Trombidiformes). *Acarologia* 17:683–92.
———. 1976. Morphologisch-taxonomische Studien an der Milbenfamilie Calyptstomidae (Acari, Trombidiformes). *Sitzber. Österr. Akad. Wiss., Mathem.–Naturwiss. Kl., Abt.* 7:1–111.
———. 1977. Ernährungsbiologische Beobachtungen an Calyptostomiden (Acari, Trombidiformes). *Zool. Anz.* 199:381–85.
———. 1978. Anatomische untersuchungen an calyptostomatiden (Acari, Trombidiformes). *Acarologia* 19:242–57.
Vitzthum, H. G. 1925. Fauna Sumatrensis (Beitrag N. 5). *Acarinae. Suppl. Entomol. (Berlin)* 11:1–78. [HETEROZERCONIDAE, DISCOZERCONIDAE]
———. 1926. Acari al Commensalen von Ipiden. *Zool. Jb. Syst.* 52:407–503. [CELAENOPSIDAE]
———. 1927. Die Acarofauna der Harzflüsse. *Sitzber. Gesell. Naturf. Freunde Berl.* 1927:89–110, 4 pl. [RESINACARIDAE]
———. 1930. Acarologische Beobachtungen. *Zool. Jahrb. Syst.* 59:281–350.
———. 1931a. *Ordnung der Arachnida: Acari.* Handbuch der Zoologie, Bd. 3, Teil 9, ed. T. Krumbach. Berlin: W. de Gruyter.
———. 1931b. Terrestrische Acarinen (unter Ausschluss der Oribatiden und Ixodiden) der Deutschen Limnologischen Sunda-Expedition. *Arch. Hydrobiol., Suppl.* 9:59–134. [ALGOPHAGIDAE]
———. 1940–43. *Acarina.* Klassen und Ordnung des Tierreichs, ed. H. G. Braun. Leipzig 5 (4), Bd. 5. [SPELEORHYNCHIDAE]
Vogelnest, L. J., L. Vogelnest, and R. S. Mueller. 2000. An undescribed *Demodex* sp. and demodicosis in a captive koala (*Phascolarctos cinereus*). *J. Zoo Wildl. Med.* 31:100–106.
Voigt, B. 1970. Histologische Untersuchungen am Stylostom der Trombiculidae (Acari). *Z. Parasitenkd.* 34:180–97.
Volgin, V. I. 1960. On the taxonomy of predatory mites of the family Cheyletidae. 11, Genus *Cheyletiella* Can. *Akad. Nauk Zool. Inst. Parasitol. Sb.* 19:237–48. [CHEYLETIDAE]
———. 1961. Novie vidi kleshchei semeistva Glycyphagidae. *Parazitol. Sb.* 20:257–66.
———. 1973. The hypopus and its main types. In Daniel and Rosický, 381–83.
———. 1980. Novie vidi kleshchei sem. Saproglyphidae (Acariformes, Acaroidea). *Parazitol. Sb.* 29:159–68. [WINTERSCHMIDTIIDAE]
———. 1987. *Acarina of the family Cheyletidae of the world.* Leningrad: Akad. Nauk SSSR, 1969; English translation, New Delhi: Amerind.
Volgin, V. I., and G. Sh. Kadzhaya. 1973. O sbornom kharaktere *Kuzinia laevis* (Dujardin, 1849) (Acariformes, Acaridae). *Materialy k Faune Gruzii* 3:33–42.
Volgin, V. I., and S. V. Mironov. 1979. Novie vidi i novii rod kleshchei semeistva Saproglyphidae (Acarina, Acaroidea). *Trudy Zool. Inst. Akad. Nauk SSSR* 85:91–98. [HEMISARCOPTIDAE]
Voss, W. J. 1967. Three trigynaspid mites from Philippine reptiles (Acarina: Paramegistidae). *J. Med. Entomol.* 3:261–68. [PARAMEGISTIDAE]
Vu, Q. M., and H. V. Thi. 1995. The list of oribatid mites from Vietnam. *J. Biol. (Sinh Hoc)* 17:49–55.
Wainstein, B. A. 1972. On a system of entomoparasitic mites of the family Otopheidomenidae Treat, 1955 (Parasitiformes). *Parazitologiya* 6:452–56. [in Russian]
———. 1977. Rod *Calyptostoma* (Acariformes, Calyptostomidae). *Zool Zh.* 56:158–61.
———. 1980. *Opredelitel lichinok vodjanych kleshchei.* St. Petersburg: Inst. Biol. Vnutr. Vod., Nauka.

Wainstein, B. A., and N. N. Kuznetsov. 1978. Identification key of soil inhabiting mites, Trombidiformes. In Gilyarov, 147–71.

Wainstein, B. A., N. N. Kuznetsov, I. Z. Livshits, and E. F. Sosnina. 1978. Family Bdellidae. In Gilyarov, 133–43. [in Russian]

Waite, G. K., and J. D. McAlpine. 1992. Honey bees as carriers of lychee erinose mite *Eriophyes litchii* (Acari: Eriophyidae). *Exp. Appl. Acarol.* 15:299–302.

Waitzbauer, J. 1983. Licht- und elektronenmikroskopische Untersuchungen der Spermiogenese und Spermatophore von *Hermannia gibba* (Oribatidae, Acari). *Acarologia* 24:95–107. [HERMANNIIDAE]

Walker, N. A. 1964. Euphthiracaroidea of California *Sequoia* litter; with a reclassification of the families and genera of the world (Acarina: Oribatei). *Fort Hayes Stud. N. S. Sci. Ser., Kans.* 3: 1–154.

Wallace, M. M. H. 1970. Diapause in the aestivating egg of *Halotydeus destructor* (Acari: Eupodidae). *Aust. J. Zool.* 18:295–313. [PENTHALEIDAE]

Wallace, M. M. H., and J. A. Mahon. 1971. The distribution of *Halotydeus destructor* and *Penthaleus major* (Acari: Eupodidae) in Australia in relation to climate and land use. *Aust. J. Zool.* 19:65–76. [PENTHALEIDAE]

———. 1972. The taxonomy and biology of Australian Bdellidae (Acari). I, Subfamilies Bdellinae, Spinibdellinae and Cytinae. *Acarologia* 14:544–80.

———. 1976. The taxonomy and biology of Australian Bdellidae (Acari). II, Subfamily Odontoscirinae. *Acarologia* 18:65–123.

Wallace, M. M. H., and M. C. Walters. 1974. The introduction of *Bdellodes lapidaria* (Acari: Bdellidae) from Australia into South Africa for the biological control of *Sminthurus viridis* (Collembola). *Aust. J. Zool.* 22:505–17.

Wallwork, J. A. 1958. Notes on the feeding-behaviour of some forest soil Acarina. *Oikos* 9:260–71.

———. 1959. The distribution and dynamics of some forest soil mites. *Ecology* 40:557–63.

———. 1961. An annotated check-list of the oribatid mites of Ghana (Acari: Oribatei). *Rev. Zool. Bot. Afr.* 64:49–64.

———. 1963. The Oribatei (Acari) of Macquarie Island. *Pac. Insects* 5 (4): 721–69.

———. 1964. Insects of Campbell Island. Appendix. *Campbellobates acanthus* n. gen., n. sp. (Acari: Cryptostigmata). *Pac. Insects Monogr.* 7:601–6. [CAMPBELLOBATIDAE]

———. 1965. A leaf-boring galumnoid mite (Acari: Cryptostigmata) from Uruguay. *Acarologia* 7 (4): 758–64.

———. 1967. Some Oribatei (Acari: Cryptostigmata) from Tchad (3rd series). *Rev. Zool. Bot. Afr.* 75:35–45. [PHTHIRACARIDAE, TECTOCEPHEIDAE, SCUTOVERTICIDAE]

———. 1976. *The distribution and diversity of soil fauna*. London: Academic Press.

———. 1981. A new aquatic oribatid mite from western Australia (Acari: Cryptostigmata: Ameronothridae). *Acarologia* 22: 333–39.

———. 1983. Oribatids in forest ecosystems. *Ann. Rev. Entomol.* 28:109–30.

Walter, C. W. 1907. Die Hydracarinen der Schweiz. *Rev. Suisse Zool.* 15:401–573.

———. 1922a. Die Hydracarinen der Alpengewässer. *Denkschr. Schweiz. Naturforsch. Ges.* 58:60–251.

———. 1922b. Hydracarinen aus den Alpen. *Rev. Suisse Zool.* 29:227–411.

———. 1929. Hydracarinen aus Java. *Treubia* 11:211–73.

———. 1947. Neue Acari (Hydrachnellae, Porohalacaridae, Trombidiidae) aus subterranen Gewässern der Schweiz und Rumäniens. *Verhand. Naturf. Gesell. Basel* 58:146–238. [STYGOTHROMBIDIIDAE]

Walter, C. W., and C. Bader. 1952. Hydracarina. In *Mission scientifique de l'Omo,* 87–236. Mém. Mus. Nat. Hist. Natur., Zool., 4A. Paris.

Walter, D. E. 1987a. Below-ground arthropods of semiarid grasslands. In *Rangeland pest management: A shortgrass prairie perspective*, ed. J. L. Capinera, 271–90. Boulder, CO: Westview Press. [LINOTETRANIDAE]

———. 1987b. Life history trophic behaviour and descriptions of *Gamaselloides vermivorax* n. sp. (Mesostigmata: Ascidae), a predator of nematodes and arthropods in semiarid grassland soils. *Can. J. Zool.* 65:1689–95.

———. 1987c. Trophic behavior of "mycophagous" microarthropods. *Ecology* 68:226–29.

———. 1988a. Nematophagy by soil arthropods from the shortgrass steppe, Chihuahuan Desert and Rocky Mountains of the central United States. *Agr. Ecosyst. Environ.* 24:307–16.

———. 1988b. Predation and mycophagy by endeostigmatid mites (Acariformes: Prostigmata). *Exp. Appl. Acarol.* 4:159–66.

———. 1992. Leaf surface structure and the distribution of *Phytoseius* mites (Acarina: Phytoseiidae) in south-eastern Australian forests. *Aust. J. Zool.* 40:593–603.

———. 1996. Living on leaves: mites, tomenta, and leaf domatia. *Ann. Rev. Entomol.* 41:101–14.

———. 1997. Heatherellidae—a new family of Mesostigmata (Acari: Parasitiformes) based on two new species from rainforest litter in Australia. *Internat. J. Acarol.* 23:167–75. [HEATHERELLIDAE]

———. 1998a. *Ectoantennoseius kitchingi*, n. gen., n. sp. (Acari: Mesostigmata: Ascidae) from the rainforest canopy in Australia and a cladistic hypothesis about its relationships. *Internat. J. Acarol.* 24:45–51.

———. 1998b. *Hoploseius australianus* sp. nov. (Acari: Mesostigmata: Ascidae), a unique element in the Australian acarofauna. *Aust. Entomol.* 25:69–74. [BLATTISOCIIDAE]

———. 1999. Cryptic inhabitants of a noxious weed: Mites (Arachnida: Acari) on *Lantana camara* L. invading forests in Queensland. *Aust. J. Entomol.* 38:197–200.

———. 2000. A jumping mesostigmatan: *Saltiseius hunteri,* n.g., n. sp. (Acari: Mesostigmata: Trigynaspida: Saltiseiidae, n. fam.). *Internat. J. Acarol.* 26:25–31. [SALTISEIIDAE]

———. 2001. Endemism and cryptogenesis in "segmented" mites: A review of Australian Alicorhagiidae, Terpnacaridae, Oehserchestidae and Grandjeanicidae (Acari, Sarcoptiformes). *Aust. J. Entomol.* 40:207–18.

———. 2003. The genus *Gamasellodes* (Acari: Mesostigmata: Ascidae): new Australian and North American species. *Syst. Appl. Acarol. Spec. Publ.* 15:1–10.

———. 2004. From the subantarctic to the subtropics: A revision of the Davacaridae Kethley, 1977 (Acari: Trigynaspida: Mesostigmata) with the description of a new genus and three new species. *J. Nat. Hist.* 38:2033–49. [DAVACARIDAE]

Walter, D. E., G. N. Azam, G. Waite, and J. Hargreaves. 1998. Risk assessment of an exotic biocontrol agent: *Phytoseiulus persimilis* (Acari: Phytoseiidae) does not establish in rainforest in southeast Queensland. *Aust. J. Ecol.* 23:587–92.

Walter D. E., J. J. Beard, K. Walker, and K. Sparks. 2002. Of mites and bees: A review of mite-bee associations in Australia and a revision of *Raymentia* (Acari: Mesostigmata: Laelapidae), with the description of two new species of mites from *Lasioglossum* (Parasphecodes) (Hymenoptera: Halictidae). *Aust. J. Entomol.* 41:128–48.

Walter, D. E., and V. M. Behan-Pelletier. 1993. Systematics and ecology of *Adhaesozetes polyphyllos* sp. nov. (Acari: Oribatida: Licneremaeoidea), a leaf inhabiting mite from Australian rainforests. *Can. J. Zool.* 71:1024–40. [ADHAESOZETIDAE]

———. 1999. Mites in forest canopies: filling the size distribution shortfall? *Ann. Rev. Entomol.* 44:1–19.

Walter, D. E., and N. J. H. Campbell. 2003. Exotic vs endemic biocontrol agents: would the real *Stratiolaelaps miles* (Berlese) (Acari: Mesostigmata: Laelapidae), please stand up? *Biol. Control* 26:253–69.

Walter, D. E., and U. Gerson. 1998. Dasythyreidae, new family, and *Xanthodasythyreus* n. g. (Acari: Prostigmata: Raphignathoidea) from Australia. *Internat. J. Acarol.* 24:189–97.

Walter, D. E., R. B. Halliday, and E. E. Lindquist. 1993. A review of the genus *Asca* (Acarina: Ascidae) in Australia, with descriptions of three new leaf-inhabiting species. *Invertebr. Taxon.* 7: 1327–47.

Walter, D. E., R. A. Hudgens, and D. W. Freckman. 1986. Consumption of nematodes by fungivorous mites *Tyrophagus* spp. (Acarina: Astigmata: Acaridae). *Oecologia* 70:357–61.

Walter, D. E., H. W. Hunt, and E. T. Elliott. 1988. Guilds or functional groups? An analysis of predatory arthropods from a shortgrass steppe soil. *Pedobiologia* 31:247–60.

Walter, D. E., and E. K. Ikonen. 1989. Species, guilds, and functional groups: Taxonomy and behavior in nematophagous arthropods. *J. Nematol.* 21:315–27.

Walter, D. E., and D. T. Kaplan. 1990a. Feeding observations on two astigmatic mites, *Schwiebea rocketti* Woodring (Acaridae) and *Histiostoma bakeri* Hughes & Jackson, associated with citrus feeder roots. *Pedobiologia* 34:281–86.

———. 1990b. A guild of thelytokous mites (Acari: Mesostigmata) associated with citrus roots in Florida. *Environ. Entomol.* 19:1338–43.

———. 1991. Observations on *Coleoscirus simplex* (Acarina: Prostigmata), a predatory mite that colonizes greenhouse cultures of rootknot nematode (*Meloidogyne* spp.), and a review of feeding behavior in the Cunaxidae. *Exp. Appl. Acarol.* 12:47–59.

Walter, D. E., D. T. Kaplan, and T. A. Permar. 1991. Missing links: a review of methods used to estimate trophic links in soil food webs. *Agr. Ecosyst. Environ.* 34:399–405.

Walter, D. E., J. Kethley, and J. C. Moore. 1987. A heptane flotation method for recovering microarthropods from arid grassland soils, with comparisons to the Merchant-Crossley high-gradient extraction method and estimates of microarthropod biomass. *Pedobiologia* 30:221–32.

Walter, D. E., and G. W. Krantz. 1999. New early derivative mesostigmatans from Australia: *Nothogynus*, n. g., Nothogynidae, n. fam. (Mesostigmata: Microgyniina). *Internat. J. Acarol.* 25:67–76. [NOTHOGYNIIDAE]

Walter, D. E., and E. E. Lindquist. 1989. Life history and trophic behavior of mites in the genus *Lasioseius* (Acari: Mesostigmata: Ascidae) from grassland soils in Colorado, with taxonomic notes and a description of a new species. *Can. J. Zool.* 67:2797–2813. [BLATTISOCIIDAE]

———. 1995. The distributions of parthenogenetic ascid mites (Acari: Parasitiformes) do not support the biotic uncertainty hypothesis. *Exp. Appl. Acarol.* 19:423–42.

———. 1997. Australian species of *Lasioseius* (Acari: Mesostigmata: Ascidae): The *porulosus* group and other species from rainforest canopies. *Invertebr. Taxon.* 11:525–47. [BLATTISOCIIDAE]

——— 2001. *Arrhenoseius gloriosus* n. g., n. sp. (Acari: Mesostigmata: Ascidae), an arrhenotokous mite from rainforests in Queensland, Australia. *Acarologia* 41:53–68. [BLATTISOCIIDAE]

Walter, D. E., and D. J. O'Dowd. 1992. Leaf morphology and predators: Effect of leaf domatia on the abundance of predatory mites. *Environ. Entomol.* 21:478–84.

———. 1995. Life on the forest phylloplane: hairs, little houses, and myriad mites. In *Forest canopies*, ed. M. D. Lowman and N. M. Nadkarni, 325–51. New York: Academic Press. [ADAMYSTIDAE]

Walter, D. E., and J. H. Oliver, Jr. 1990. *Geolaelaps oreithyiae*, n. sp. (Acari: Laelapidae), a thelytokous predator of arthropods and nematodes, and a discussion of clonal reproduction in the Mesostigmata. *Acarologia* 30:293–303.

Walter, D. E., and H. C. Proctor. 1998. Feeding behaviour and phylogeny: observations on early derivative Acari. *Exp. Appl. Acarol.* 22:39–50.

———. 1999. *Mites: Ecology evolution and behaviour.* Wallingford, Oxon, UK: CAB International.

———. 2001. *Mites in soil, an interactive key to mites and other soil microarthropods.* ABRS Identification Series. Collingswood, Victoria, Queensland: CSIRO Publishing.

Walter, D. E., O. Seeman, D. Rodgers, and R. L. Kitching. 1998. Mites in the mist: how unique is a rainforest canopy knockdown fauna? *Aust. J. Ecol.* 23:501–8.

Walter D. E., and M. Shaw. 2002. First record of the mite *Hirstiella diolii* Baker (Prostigmata: Pterygosomatidae) from Australia, with a review of mites found on Australian lizards. *Aust. J. Entomol.* 41:30–34.

———. 2003. The importance of being earnest: exotic invasive mites (Arachnida: Acari) on Australian vertebrates. *Rec. S. Aust. Mus. Monogr. Ser.* 7:25–30.

Walzl, M. G. 1988. A scanning electron microscope study of the embryonic development of the house dust mite, *Dermatophagoides farinae* (Pyroglyphidae, Actinotrichida) from blastula to the hatching stage of the larva. *Inst. Physics Conference Series* no. 93, vol. 3, chap. 6, 159–60.

Walzl, M. G., A. Gutweniger, and P. Wernsdorf. 2005. Embryology of mites: new techniques yield new findings. In Weigmann et al. (2005), 163–81.

Wang, H.-F. 1986. A new species of the genus *Mullederia* (Acarina: Stigmaeidae). *Acta Zootax. Sin.* 11:187–90.

Wang, H.-F., T. Solhøy, J. Shen, and R. M. Xu. 2001. New species and new records of oribatid mites from Tibet, China (Acari Oribatida). *Acta Zootax. Sin.* 264:401–13. [CEPHEIDAE, MICROTEGEIDAE, LIACARIDAE, ORIPODIDAE, and others]

Wang, H.-F., Z.-G. Wen, and J. Chen. 2002. A checklist of oribatid mites of China (I) (Acari: Oribatida). *Acta Arachnol. Sin.* 11:107–27.

———. 2003. A checklist of oribatid mites of China (II) (Acari: Oribatida). *Acta Arachnol. Sin.* 12:42–63.

Wang, Y.-M., and J. Xu. 1986. A new species of the genus *Zetzellia* Oudemans (Acarina: Stigmaeidae). *Acta Zootax. Sin.* 11:274–75.

Watson, D. G., J. J. Davis, and W. C. Hanson. 1966. Terrestrial invertebrates. In *Environment of the Cape Thompson region, Alaska*, ed. N. J. Wilimovsky and J. N. Wolfe, 565–84. Washington, DC: U.S. Atomic Energy Commission.

Wauthy, G., M. Leponce, N. Banai, G. Sylin, and J.-C. Lions. 1998. The backward jump of a box moss mite. *Proc. Roy. Soc. Lond.* B, 265:2235–42.

Webb, J. P., Jr., P. Robaux, and G. D. Campbell. 1977. Notes on the biology of *Ceuthothrombium cavaticum* (Acari: Trombidiidae), a parasite on cave crickets (Raphidophoridae: *Ceuthophilus*). *Bull. S. Calif. Acad. Sci.* 76:135–37.

Webb, N. R. 1977. Observations on *Steganacarus magnus*, general biology and life cycle. *Acarologia* 19:686–96. [PHTHIRACARIDAE]

———. 1989. Observations on the life cycle of *Steganacarus magnus* (Acari, Cryptostigmata). *Pedobiologia* 33:293–99. [PHTHIRACARIDAE]

———. 1991. The role of *Steganacarus magnus* (Acari: Cryptostigmata) in the decomposition of the cones of Scots pine, *Pinus sylvestris*. *Pedobiologia* 35:351–59. [PHTHIRACARIDAE]

Webb, N. R., and W. Block. 1993. Aspects of cold hardiness in *Steganacarus magnus* (Acari: Cryptostigmata). *Exp. Appl. Acarol.* 17:741–48. [PHTHIRACARIDAE]

Weeks, A. R., F. Marec, and J. A. J. Breeuwer. 2001. A mite species that consists entirely of haploid females. *Science* 292:2479–82.

Wehrle, L. P., and P. S. Welch. 1925. The occurrence of mites in the tracheal system of certain Orthoptera. *Ann. Entomol. Soc. Amer.* 28:35–44. [PODAPOLIPIDAE]

Weiberg, M., and D. D. Edwards. 1997. Survival and reproductive output of *Chironomus tentans* (Diptera: Chironomidae) in response to parasitism by larval *Unionicola foili* (Acari: Unionicolidae). *J. Parasitol.* 83:173–75.

Weidhaas, J. A., and R. M. Reeves. 1963. The occurrence and importance of tetranychid and eriophyid mites on woody plants in New York. *Adv. Acarol.* 1:25–29.

Weigmann, G. 1975. Labor- und Freilanduntersuchungen zur Generationsdauer von Oribatiden (Acari: Oribatei). *Pedobiologia* 15:133–48.

———. 1991. Oribatid communities in transects from bogs to forests in Berlin indicating the biotope qualities. In Dusbábek and Bukva, 1:359–64.

———. 1997a. New and old species of Malaconothroidea from Europe. *Spixiana* 20:199–218.

———. 1997b. Systematics of Central European families and genera of Malaconothroidea (Acari, Oribatida). *Abh. Ber. Naturk. Görlitz* 69:3–10.

———. 2002. Morphological variability between and within populations of *Tectocepheus* (Acari, Oribatida, Tectocepheidae) from the *velatus*-complex in Europe. In Bernini et al. (2002), 141–52.

———. 2006. *Hornmilben (Oribatida). Die Tierwelt Deutschlands, 76. Teil.* Keltern: Goecke & Evers.

Weigmann, G., G. Alberti, A. Wohltmann, and S. Ragusa, eds. 2005. *Acarine biodiversity in the natural and human sphere.* Proc. EURAAC V Symposium, Berlin. *Phytophaga* 14:123–28.

Weigmann, G., and W. Kratz. 1981. Die deutschen Hornmilbenarten und ihre ökologische Charakteristik. *Zool. Beitr. N. F.* 27: 459–89.

Weigmann, G., and L. Miko. 2002. Resdescription of *Oribates lagenula* Berlese, 1904, the type of *Lagenobates* n. gen. (Acarina Oribatida). *Redia* 85:29–35. [HAPLOZETIDAE]

Weigmann, G., L. Miko, and R. Nannelli. 1993. Redescription of *Protoribates dentatus* (Berlese, 1883) with remarks about the genus *Protoribates* (Acarina, Oribatida). *Redia* 76:39–55. [HAPLOZETIDAE]

Weis-Fogh, T. 1948. Ecological investigations on mites and collemboles in the soil. *Nat. Jutlandica* 1:135–270, tab. 3, 9–16.

Welbourn, W. C. 1983. Potential use of trombidioid and erythraeoid mites as biological control agents of insect pests. In Hoy, Cunningham, and Knutson, 103–40.

———. 1991. Phylogenetic studies of the terrestrial Parasitengona. In Dusbábek and Bukva, 2:163–70.

———. 1995. *Balaustium* spp. in Florida (Acari: Erythraeidae). *Fla. Dept. Agr. Cons. Relat. Entomol. Circ.* 368:1–2.

Welbourn, W. C., and D. T. Jennings. 1991. Two new species of Erythraeidae (Acari: Prostigmata) associated with the spruce budworm, *Choristoneura fumiferana* (Clemens) (Lepidoptera: Tortricidae), in Maine. *Can. Entomol.* 123:567–80.

Welbourn, W. C., R. Ochoa, E. C. Kane, and E. F. Erbe. 2003. Morphological observations on *Brevipalpus phoenicis* (Acari: Tenuipalpidae) including comparisons with *B. californicus* and *B. obovatus*. *Exp. Appl. Acarol.* 30:107–33.

Welbourn, W. C., and O. P. Young. 1987. New genus and species of Erythraeinae (Acari: Erythraeidae) from Mississippi with a key to the genera of North American Erythraeidae. *Ann. Entomol. Soc. Amer.* 80:230–42.

———. 1988. Mites parasitic on spiders, with a description of a new species of *Eutrombidium* (Acari, Eutrombidiidae). *J. Arachnol.* 16:373–85.

Wendt, F. E., and A. Eggers. 1994. *Johnstoniana rapax* n. sp., a new species of the Johnstonianidae from Europe including a description of all active instars (Acari: Parasitengonae: Trombidia). *Mitt. Ham. Zool. Mus. Inst.* 91:113–26.

Wendt, F. E., R. Olomski, J. Leimann, and A. Wohltmann. 1992. Parasitism, life cycle and phenology of *Leptus trimaculatus* (Hermann, 1804) (Acari: Parasitengona: Erythraeidae) including a description of the larva. *Acarologia* 33:55–68.

Wenzel, R. L., and V. J. Tipton, eds. 1966. *Ectoparasites of Panama.* Chicago: Field Mus. Nat. Hist.

Wergin, W. P., R. Ochoa, E. F. Erbe, and D. C. Joy. 2000. Oscillating trichobothria in the low temperature SEM: Biological capacitors or charging artifacts. *Scanning* 22:140–41.

Werman, S. D. 1983. Population dynamics of the lizard mite, *Hirstiella pyriformis* (Acari: Pterygosomatidae) on the chuckwalla, *Sauromalus obesus* (Sauria: Iguanidae). *J. Parasitol.* 69:381–86.

Wernz, J. G., and G. W. Krantz. 1976. Studies on the function of the tritosternum in selected Gamasida (Acari). *Can. J. Zool.* 54:202–12.

Westerboer, I. 1963. Die Familie Podocinidae Berlese 1916. In H. J. Stammer, 179–450. [includes MELICHARIDAE]

Westerboer, I., and F. Bernhard. 1963. Die Familie Phytoseiidae Berlese 1916. In H. J. Stammer, 451–772. [includes AMEROSEIIDAE]

Westphal, E., and D. C. M. Manson. 1996. Feeding effects on host plants: gall formation and other distortions. In Lindquist, Sabelis, and Bruin, 6:231–42.

Weygoldt, P. 1998. Evolution and systematics of the Chelicerata. Proceedings EURAAC III Symposium, Amsterdam. *Exp. Appl.*

Acarol. 22:63–79. (Also appears in Bruin, van der Geest, and Sabelis, 1–14.)

Weygoldt, P., and H. F. Paulus. 1979a. Untersuchungen zur Morphologie, Taxonomie und Phylogenie der Chelicerata. I, Morphologische Untersuchungen. *Z. Zool. Syst. Evol. Forsch.* 17: 85–116.

———. 1979b. Untersuchungen zur Morphologie, Taxonomie und Phylogenie der Chelicerata. II, Cladogramme und die Entfaltung der Chelicerata. *Z. Zool. Syst. Evol. Forsch.* 17:177–200.

Wharton, G. W. 1976. House dust mites. *J. Med. Entomol.* 12: 577–621.

Wharton, G. W., and L. G. Arlian. 1972. Predatory behaviour of the mite *Cheyletus aversor. Anim. Behav.* 20:719–23. [CHEYLETIDAE]

Wharton, G. W., and A. R. Brody. 1972. The peritrophic membrane of the mite *Dermatophagoides farinae*: Acariformes. *J. Parasitol.* 58:801–4.

Wharton, G. W., and T. L. Devine. 1968. Exchange of water between a mite, *Laelaps echidnina*, and the surrounding air under equilibrium conditions. *J. Insect Physiol.* 14:1303–18.

Wharton, G. W., and H. S. Fuller. 1952. *A manual of the chiggers.* Mem. Entomol. Soc. Wash. 4. Washington, DC.

Wharton, G. W., W. Parish, and D. E. Johnston. 1968. Observations on the fine structure of the cuticle of the spiny rat mite, *Laelaps echidnina* (Acari: Mesostigmata). *Acarologia* 10:207–14.

Whitaker, J. O., Jr., and T. W. French. 1988. Ectoparasites and other arthropod associates of some voles and shrews from the Catskill Mountains of New York. *Great Lakes Entomol.* 21:43–48.

Whitaker, J. O., Jr., and N. Wilson. 1974. Host and distribution lists of mites (Acari), parasitic and phoretic, in the hair of wild mammals of North America, north of Mexico. *Amer. Midl. Nat.* 91:1–67.

Whitford, W. G., D. W. Freckman, N. Z. Elkins, L. W. Parker, R. Parmalee, J. Phillips, and S. Tucker. 1981. Diurnal migration and responses to simulated rainfall in desert soil microarthropods and nematodes. *Soil Biol. Biochem.* 13:417–25.

Whitsel, R. H., and R. F. Schoeppner. 1973. Mites associated with aquatic and semi-aquatic Diptera from San Mateo County, California. *Proc. Entomol. Soc. Wash.* 75:71–77.

Wiese, M. H. J., and A. Fain. 1996. First description of the juvenile stages of the two species *Rhinoseius androdon* Fain & Hyland, 1980 and *Rhinoseius tiptoni* Baker & Yunker, 1964 (Acari: Mesostigmata: Ascidae) found in Colombia. *Bull. Ann. Soc. Roy. Belg. Entomol.* 132:343–81. [MELICHARIDAE]

Wiggins, G. B., R. J. MacKay, and I. M. Smith. 1980. Evolutionary and ecological strategies of animals in annual temporary pools. *Arch. Hydrobiol., Suppl.* 58:97–206.

Wiles, P. R. 1982. A note on the watermite *Hydrodroma despiciens* feeding on chironomid egg masses. *Freshwater Biol.* 12:83–87. [HYDRODROMIDAE]

———. 1985. The systematics of the British Hydrodromidae Viets, 1936. *Arch. Hydrobiol., Suppl.* 70:365–403.

———. 1986. *Hydrodroma monticola* (Piersig) (Hydrachnidia, Hydrodromidae Koch 1837): A redefinition of the species. *Arch. Hydrobiol.* 108:135–40.

———. 1987. Observations on the parasitic biology of the watermite *Hydrodroma despiciens pilosa* Besseling (Acari: Hydrodromidae). *Arch. Hydrobiol., Suppl.* 76:369–92.

———. 1988. Watermites of the genus *Arrenurus* Dugès 1834 (Hydrachnidia) from peninsular Malaysia. *Malay. Nat. J.* 41: 479–504.

———. 1989a. Five new species of water mite (Acari, Hydrachnidia) from North Sulawesi, Indonesia. *Microscopy* 36:321–28. [ATURIDAE]

———. 1989b. Two new species of watermites (*Torrenticola*: Acari) from S. E. Asia. *Microscopy* 36:245–48.

———. 1991a. The Australian *Corticacarus* (Acari: Hydrachnidia: Hygrobatidae): First records from Papua New Guinea with descriptions of six new species. *Entomol. Scand.* 21:361–67.

———. 1991b. Rheophilic watermites (Acari: Hydrachnidia) from mainland Malaysia. *Acarologia* 32:41–56. [TORRENTICOLIDAE]

———. 1991c. Watermites (Hydrachnidia, Aturidae) of the genus *Albia* Thon from Asia and Australasia with descriptions of eleven new species. *Entomol. Scand.* 22:465–88.

———. 1993a. Larval descriptions of three Asian watermites (Acari: Hydrachnida): *Diplodontus haliki* Lundblad, *Limnochares crinita* Koenike and *Hydryphantes incertus* Koenike. *Internat. J. Acarol.* 19:355–63. [LIMNOCHARIDAE, HYDRYPHANTIDAE]

———. 1993b. New species of tropical watermites (Acari: Hydrachnidia: Arrenuridae) from Asia. *Quekett J. Microsc.* 37:135–45.

———. 1996. A new family, genus and species of water mite (Acari: Hydrachnidia, Lebertoidea) from Brunei. *Quekett J. Microsc.* 37:692–95.

———. 1997a. Asian and oriental Torrenticolidae Piersig, 1902 (Acari: Hydrachnidia: Lebertioidea): A revision of the family and descriptions of new species of *Torrenticola* Piersig and *Pseudotorrenticola* Walter, from Southeast Asia. *J. Nat. Hist.* 31:191–236.

———. 1997b. Water mites (Acari: Hydrachnidia) from New Guinea: Descriptions of nine new species of *Australiobates* Lundblad and first records of *Dropursa* Cook and *Coaustraliobates* Lundblad. *Acarologia* 38:165–80. [HYGROBATIDAE]

———. 1997c. The water mites (Acari: Hydrachnidia) of New Guinea. *Raffles Bull. Zool.* 45:375–418. [LIMNESIIDAE, HYGROBATIDAE, ATURIDAE, UNIONICOLIDAE, PIONIDAE, MOMONIIDAE, MIDEOPSIDAE, ARRENURIDAE]

Wiley, E. O. 1979. An annotated Linnean hierarchy, with comments on natural taxa and competing systems. *Syst. Zool.* 28:308–37.

Wilkinson, P. R., and M. B. Garvie. 1975. Notes on the role of ticks feeding on lagomorphs and ingestion of ticks by vertebrates in the epidemiology of Rocky Mountain spotted fever. *J. Med. Entomol.* 12:480. [IXODIDAE]

Williams, B., and T. E. Hallas. 1988. *Cheyletus eruditus* (Schrank) (Acari) feeding on eggs of the cat flea *Ctenocephalides felis* (Bouché) (Siphonaptera). *Entomol. Monthly Mag.* 124:144. [CHEYLETIDAE]

Williams, D. D. 1984. The hyporheic zone as a habitat for aquatic insects and associated arthropods. In *The ecology of aquatic insects*, ed. V. H. Resh and D. M. Rosenberg, 430–55. New York: Praeger.

Willmann, C. 1935. III. Ueber eine eigenartige Milbenfauna im Küstengrundwasser der Keiler Bucht. Schr. *Naturwiss. Ver. Schleswig-Holstein* 20 (2): 422–34.

———. 1951. Untersuchungen über die terrestrische Milbenfauna im pannonischen Klimagebiet Österreichs. *Sitzungsb. Mathematisch–Naturwiss.,* Abt. I, 16:91–176. [STIGMAEIDAE]

———. 1953. Neue Milben aus den östlichen Alpen. *Sitzungsb.*

Mathematisch–Naturwiss., Abt. I, 162:484–517. [STIGMAEIDAE]

———. 1957. Revision einiger Milbengattungen und -arten von den Küsten der Nord- und Ostsee. *Abh. Naturwiss. Ver. Bremen* 35:162–88.

Wilson, W. T., W. L. Rubink, and A. M. Collins. 1990. A larval species of erythraeid mite (Leptus sp., Acarina: Erythraeidae) ectoparasitic on adult honey bees (*Apis mellifera* L.) in South Texas. *Bee Sci.* 1:18–22.

Winkler, W. 1888. Die Anatomie der Gamasiden. *Zool. Inst. Univ. Wien* 7:317–54.

Wise, G. U., M. K. Hennessey, and R. C. Axtell. 1988. A new species of manure-inhabiting mite in the genus *Poecilochirus* (Acari: Mesostigmata: Parasitidae) predacious on house fly eggs and larvae. *Ann. Entomol. Soc. Amer.* 81:209–24.

Wisniewski, J., and W. Hirschmann 1984. Teilgang einer neuen *Asternolaelaps*-Art aus Polen (Trichopygidiina). *Acarologia* 31:101–4.

———. 1991a. Die larvae und protonymphe von *Microsejus truncicola* Trägårdh 1942 (Acarina, Microgyniina) aus Polen. *Bull. Polish Acad. Sci. Biol. Sci.* 39:185–87. [MICROGYNIIDAE]

———. 1991b. *Uropodella congoensis* nov. spec. (Trichopygidiina) eine nach polen eigesschleppte milbenart. *Acarologia* 32:99–107.

———. 1993. *Katalog der Ganggattungen, Untergattungen, Gruppen und Arten die Uropodiden der Erde.* Gangsyst. Parasitiformes, Schrift. Vergleich. Milbenk. 40. Nürnberg: Hirschmann Verlag.

Wisniewski, J., W. Karg, and W. Hirschmann. 1992. Gangsystematik der Parasitiformes, Teil 547. Die Adultengattung *Halozercon* nov. gen., Typenart *Halozercon karacholana* nov. spec. aus Sibirien (Halolaelapidae, Ascoidea, Gamasina). *Acarologie, Nürnberg* 39:180–86. [ZERCONIDAE]

Witalinski, W. 1982. Unusual structural changes in mitochondria during spermiogenesis in the oribatid mite *Hafenrefferia gilvipes* (Acari). *Cell Tissue Res.* 226:465–68. [TENUIALIDAE]

———. 1990. Adanal suckers in acarid mites (Acari, Acaridida): Structure and function. *Internat. J. Acarol.* 16:205–12.

———. 1991. Reproductive systems in Acaridida: some peculiar features. In Schuster and Murphy, 363.

———. 2000. *Aclerogamasus stenocornis* sp. n., a fossil mite from the Baltic amber (Acari: Gamasida: Parasitidae). *Genus (Warsaw)* 11:619–26.

Witalinski, W., J. Dabert, and M. G. Walzl. 1992. Morphological adaptation for precopulatory guarding in astigmatic mites (Acari: Acaridida). *Internat. J. Acarol.* 18:49–54.

Witalinski, W., M. Liana, and G. Alberti. 2002. Fine structure and probable function of ring organs in the mite *Histiostoma feroniarum* (Acari: Actinotrichida: Acaridida: Histiostomatidae). *J. Morphol.* 253:255–63.

Witalinski, W., E. Szlendak, and J. Boczek. 1990. Anatomy and ultrastructure of the reproductive system of *Acarus siro*. *Exp. Appl. Acarol.* 10:1–31.

With, C. J. 1904. The Notostigmata, a new suborder of Acari. *Vidensk. Med. Dansk Natur. Forening Kjøbenhavn* 1904:137–91.

Witte, H. 1975a. Funktionsanatomie der Genitalorgane und Fortpflanzungsverhalten bei den Mannchen der Erythraeidae (Acari, Trombidiformes). *Z. Morphol. Tiere* 80:137–80.

———. 1975b. Funktionsanatomie des Weiblichen Genitaltraktes und Oogenese bei Erythraeiden (Acari, Trombidiformes). *Zool. Beitr.* 21:247–77.

———. 1984. The evolution of the mechanisms of reproduction in the Parasitengonae (Acarina: Prostigmata). In Griffiths and Bowman, 1:470–78.

———. 1991a. Indirect sperm transfer in prostigmatic mites from a phylogenetic viewpoint. In Schuster and Murphy, 137–76.

———. 1991b. The phylogenetic relationships within the Parasitengonae. In Dusbábek and Bukva, 2:171–82.

Witte, H., and D. Döring. 1999. Canalized pathways of change and constraints in the evolution of reproductive modes of microarthropods. *Exp. Appl. Acarol.* 23:181–216. (Also appears in Bruin, van der Geest, and Sabelis, 15–43.)

Woas, S. 1981. Zur Taxonomie und Phylogenie der Hermanniidae Sellnick 1928 (Acari, Oribatei). *Andrias (Karlsruhe)* 1:7–88.

———. 1986. Beitrag zur Revision der Oppioidea sensu Balogh, 1972 (Acari, Oribatei). *Andrias (Karlsruhe)* 5:21–224.

———. 1991. Distribution of characters and phylogenetic age—systematic problems in the higher taxa of the Oribatida. In Schuster and Murphy, 329–34.

———. 1992. Beitrag zur Revision der Gymnodamaeidae Grandjean, 1954 (Acari, Oribatei). *Andrias (Karlsruhe)* 9:121–61.

———. 1998. Mosaikverteilung der Merkmale basaler Höherer Oribatiden—Die Gattungen *Passalozetes* und *Scutovertex*. In Ebermann (1998a), 291–313. [PASSALOZETIDAE, SCUTOVERTICIDAE]

———. 2000. Die Gattungen *Poroliodes* und *Cymbaeremaeus* und ihr verwandtschaftliches Umfeld. *Carolinea (Karlsruhe)* 58:165–81. [NEOLIODIDAE, CYMBAEREMAEIDAE]

———. 2002. Acari: Oribatida. In *Amazonian Arachnida and Myriapoda*, ed. J. Adis, 21–291. Sofia: Pensoft Publ.

Wohltmann, A. 1996. On the life cycle and parasitism of *Johnstoniana errans* (Johnston) 1852 (Acari: Prostigmata: Parasitengonae). *Acarologia* 37:201–9. [JOHNSTONIANIDAE]

———. 2000. On the biology of *Trombidium brevimanum* (Berlese, 1910) (Acari, Prostigmata: Parasitengona: Trombidiidae) with a redescription of all active instars. *Mitt. Hamb. Zool. Mus. Inst.* 96:157–68.

———. 2001. The evolution of life histories in Parasitengona (Acari: Prostigmata). *Acarologia* 41:145–204.

Wohltmann, A., J. Mąkol, and G. Gabryś. 2004. A revision of European Johnstonianinae Thor, 1935 (Acari: Prostigmata: Parasitengona: Trombidioidea). *Ann. Zool. (Warsaw)* 54:595–630. [JOHNSTONIANIDAE]

Wohltmann, A., and F. E. Wendt. 1996. Observations on the biology of two hygrobiotic trombidioid mites (Acari: Prostigmata: Parasitengonae), with special regard to host recognition and parasitism tactics. *Acarologia* 37:31–44. [JOHNSTONIANIDAE]

Wohltmann, A., F. E. Wendt, A. Eggers, and J. C. Otto. 1994. Observations on parasitism, development and phenology of *Johnstoniana tuberculata* Schweizer, 1951 (Acari: Parasitengonae: Johnstonianidae) including a redescription of all active instars. *Acarologia* 35:153–66.

Wohltmann, A., F. E. Wendt, and M. Waubke. 1996. The life cycle and parasitism of the European grasshopper mite *Eutrombidium trigonum* (Hermann 1804) (Prostigmata: Parasitengonae: Microthrombidiidae), a potential agent for biological control of grasshoppers (Saltatoria). *Exp. Appl. Acarol.* 20:545–61. [EUTROMBIDIIDAE]

Womersley, H. 1933. A possible biological control of the clover springtail or Lucerne flea (*Sminthurus viridis*) in western Australia. *J. Aust. Coun. Sci. Industr. Res.* 6:83–91.

———. 1934. A revision of the trombid and erythraeid mites of

Australia with descriptions of new genera and species. *Rec. S. Aust. Mus.* 5:179–254.
———. 1935. A species of Acarina of the genus *Holothyrus* from Australia and New Zealand. *Ann. Mag. Nat. Hist.* 16:154–57.
———. 1937. Australian Acarina of the genus *Megisthanus* Thorell. *Trans. Roy. Soc. S. Aust.* 51:175–80. [MEGISTHANIDAE]
———. 1939. Further notes on the Australian Trombidiidae with description of new species. *Trans. Roy. Soc. Aust.* 63:149–66.
———. 1942a. Additions to the Acarina of Australia (Trombidiidae and Calyptostomidae). *Rec. S. Aust. Mus.* 7:169–81.
———. 1942b. New species of *Geckobia* (Acarina, Pterygosomatidae) from Australia and New Zealand. *Trans. Roy. Soc. S. Aust.* 65:323–28.
———. 1944. Australian Acarina, families Alycidae and Nanorchestidae. *Trans. Roy. Soc. S. Aust.* 68:133–43.
———. 1954a. New genera and species, apparently of Apoloniinae (Acarina, Leeuwenhoekiidae), from the Asiatic-Pacific region. Malaysian Parasites VII. *Stud. Inst. Med. Res., Malaya* 26:108–19. [TROMBICULIDAE]
———. 1954b. On the subfamily Trombellinae Sig Thor 1935 (Acarina, Trombidiidae) with the diagnosis of the nymph of *Audyana thompsoni* Womersley, 1954. *Rec. S. Aust. Mus.* 11:121–28. [TROMBELLIDAE]
———. 1956. On some new Acarina—Mesostigmata from Australia, New Zealand and New Guinea. *J. Linn. Soc. Zool.* 42:505–99. [IPHIOPSIDIDAE]
———. 1958. Some new or little known Mesostigmata (Acarina) from Australia, New Zealand and Malaya. *Trans. Roy. Soc. S. Aust.* 81:115–30.
———. 1959a. *Klinckowstroemiella helleri* (Ouds., 1929) nov. comb. for *Fedrizzia helleri* Ouds., 1929 (Acarina: Klinckowstroemiidae). *Zool. Meded.* 34:281–88.
———. 1959b. Some Acarina from Australia and New Guinea paraphagic upon millipedes and cockroaches, and on beetles of the family Passalidae. Part 2, The family Fedrizziidae (Mesostigmata-Trigynaspida). *Trans. Roy. Soc. S. Aust.* 82:11–54.
———. 1961. On the family Diarthrophallidae (Acarina-Mesostigmata- Monogynaspida) with particular reference to the genus *Passalobia* Lombardini 1926. *Trans. Roy. Soc. S. Aust.* 84:27–44.
———. 1963. New records of Diarthrophallidae (Acarina) with the description of the hitherto unknown larva. *Rec. S. Aust. Mus.* 14:487–97.
Womersley, H., and R. V. Southcott. 1941. Notes on the Smarididae (Acarina) of Australia and New Zealand. *Trans. Roy. Soc. S. Aust.* 65:61–78.
Womersley, H., and R. W. Strandtmann. 1963. On some free living prostigmatic mites of Antarctica. *Pac. Insects* 5:451–72.
Wong, C. L. 1967. A study of the biology of two species of Podocinidae (Acarina: Mesostigmata). *Univ. Kans. Sci. Bull.* 68: 575–600.
Wood, T. G. 1966. Mites of the genus *Ledermuelleria* Oudms. (Prostigmata, Stigmaeidae) from New Zealand, with records of one species from some southern Pacific islands. *New Zeal. J. Sci.* 9:84–102.
———. 1967. New Zealand mites of the family Stigmaeidae (Acari, Prostigmata). *Trans. Roy. Soc. New Zeal.* 9:93–139.
———. 1969. The Homocaligidae, a new family of mites (Acari: Raphignathoidea), including a description of a new species from Malaya and the British Solomon Islands. *Acarologia* 11:711–29.
———. 1973. Revision of Stigmaeidae (Acari: Prostigmata) in the Berlese collection. *Acarologia* 15:76–95.
Woodring, J. P. 1965. The biology of five new species of oribatids from Louisiana. *Acarologia* 7:564–76.
———. 1969a. Environmental regulation of andropolymorphism in tyroglyphids (Acari). In Evans (1969c), 433–40.
———. 1969b. Observations on the biology of six species of acarid mites. *Ann. Entomol. Soc. Amer.* 62:102–8.
———. 1970. Comparative morphology, homologies, and functions of the male system in oribatid mites (Arachnida: Acari). *J. Morphol.* 132:425–51.
———. 1973. Comparative morphology, functions, and homologies of the coxal glands in oribatid mites (Arachnida: Acari). *J. Morphol.* 139:407–29.
Woodring, J. P., and S. C. Carter. 1974. Internal and external morphology of the deutonymph of *Caloglyphus boharti* (Arachnida: Acari). *J. Morphol.* 144:275–95.
Woodring, J. P., and E. F. Cook. 1962. The internal anatomy, reproductive physiology, and molting process of *Ceratozetes cisalpinus* (Acari: Oribatei). *Ann. Entomol. Soc. Amer.* 55:164–81.
Woodring, J. P., and C. A. Galbraith. 1976. The anatomy of the adult uropodid *Fuscuropoda agitans* (Arachnida; Acari), with comparative observations on other Acari. *J. Morphol.* 150:19–58.
Woodring, J. P., and J. C. Moser. 1970. Six new species of anoetid mites associated with North American Scolytidae. *Can. Entomol.* 102:1237–57. [HISTIOSTOMATIDAE]
Woodroffe, G. E. 1956. Some insects and mites associated with bat-roosts, with a discussion of the feeding habits of the cheyletids. *Entomol. Monthly Mag.* 92:138–41.
Woolley, T. A., ed. 1964. Proceedings 1st International Congress of Acarology, Fort Collins. *Acarologia* (fasc. hs.) 6:208–14.
———. 1972. Scanning electron microscopy of the respiratory apparatus of ticks. *Trans. Amer. Microscop. Soc.* 91:348–63.
———. 1973. Taxonomy of oribatid mites—retrospect and prospect. In Daniel and Rosický, 337–41.
———. 1988. *Acarology: Mites and human welfare.* New York: John Wiley & Sons.
Worland, M. R., and A. Lukesova. 2000. The effect of feeding on specific soil algae on the cold-hardiness of two antarctic micro-arthropods (*Alaskozetes antarcticus* and *Cryptopygus antarcticus*). *Polar Biol.* 23:766–74. [AMERONOTHRIDAE]
Woyke, J. 1987. Comparative population dynamics of *Tropilaelaps clareae* and *Varroa jacobsoni* mites on honeybees. *J. Apicult. Res.* 26:196–202.
Wrenn, W. J. 1996. Immune responses to mange mites and chiggers. In *The immunology of host-ectoparasitic arthropod relationships*, ed. S. K. Wikel, 11:259–89. Wallingford, Oxon, UK: CAB International. [TROMBICULIDAE]
Wrenn, W. J., and R. B. Loomis. 1984. Host selectivity in the genus *Eutrombicula* (Acari: Trombiculidae). In Griffiths and Bowman, 1:160–65. [TROMBICULIDAE]
Wrensch, D. 1993. Evolutionary flexibility through haploid males, or how chance favors the prepared genome. In Wrensch and Ebbert, 118–49.
Wrensch, D. L., and M. A. Ebbert, eds. 1993. *Evolution and diversity of sex ratio in insects and mites.* New York: Chapman & Hall.
Wrensch, D. L., J. B. Kethley, and R. A. Norton. 1994. Cytogenetics of holokinetic chromosomes and inverted meisois: Keys to the

evolutionary success of mites, with generalizations on eukaryotes. In Houck (1994b), contribution no. 11, 282–343.

Wu, G. H., and M. Li. 1991. Studies on the relationship between *Leptotrombidium* (L.) *kaohuense* and tsutsugamushi disease. *Chinese J. Parasitol. Parasitic Dis.* 9:298–300. [TROMBICULIDAE]

Wu, K. W. 1986. Review of the polyethylene bottle applicator technique for sealing microslide preparations of mites. *Internat. J. Acarol.* 12:87–89.

Wunderle, I. 1992a. Die Oribatiden-Gemeinschaften (Acari) der verschiedenen Habitate eines Buchenwaldes. *Carolinea (Karlsruhe)* 50:79–144.

———. 1992b. Die baum- und bodenbewohnenden Oribatiden (Acari) im Tieflandregenwald von Pangana, Peru. *Amazoniana* 12 (1): 119–42.

Wunderle, I., L. Beck, and S. Woas. 1990. Ein Beitrag zur Taxonomie und Ökologie der Oribatulidae und Scheloribatidae (Acari, Oribatei) in Südwestdeutschland. *Andrias (Karlsruhe)* 7:15–60.

Wurst, E. 2001. The life-cycle of *Lemanniella minotauri* n. sp. and the erection of the new family Lemanniellidae (Acari: Astigmata). *Stuttgart. Beitr. Naturk.*, ser. A (*Biol.*), 621:1–34.

Wurst, E., and P. Havelka. 1997. Redescription and life history of *Tytodectes strigis* (Acari: Hypoderatidae), a parasite of the barn owl *Tyto alba* (Aves: Strigidae). *Stuttgart. Beitr. Naturk.*, ser. A (*Biol.*), 554:1–39.

———. 1998. Bindegewebsmilben der Familie Hypoderatidae Wenig bekannte Parasiten einheimischer Vögel. *Mikrokosmos* 87:263–73.

Wurst, E., and D. Kovac. 2003. Description and biology of the mite *Tensiostoma veliaphilum* n. gen. n. sp. from the water surface of bamboo phytotelmata in Southeast Asia (Arachnida, Acari, Astigmata, Histiostomatidae). *Senckenb. Biol.* 82:63–98.

Wurst, E., and T. Pfister. 1990. On the biology of *Baloghella melis* Mahunka, 1963 (Acari: Acaridida: Glycyphagidae). *Bonn. Zool. Beitr.* 41:157–62.

Yamamoto Y., and J. Aoki. 1998. Two new oribatid mites of the genus *Trimalaconothrus* from Yunnan Province in China (Acari: Oribatida: Malaconothridae). *Edaphologia* 61:15–21.

Yamamoto, Y., G. Kuriki, and J. Aoki. 1993. Three oribatid mites of the genus *Trimalaconothrus* found from a bog in the northeastern part of Japan (Acari: Oribatei: Malaconothridae). *Edaphologia* 49:25–31.

Yano, K., and S. Ehara. 1982. Erythraeid mites (Acarina, Erythraeidae) parasitic on planthoppers. *Konyû, Tokyo* 50:344–45.

Yao, D. S. C., and D. A. Chant. 1982. A method of confining mites for population studies in the laboratory. *Internat. J. Acarol.* 8:205–9.

Yao, H., R. J. Snider, and R. M. Snider. 2000. Larval and post-larval stages of a new *Abrolophus* species (Acari: Erythraeidae) from a deciduous forest in northern Michigan, USA. *Syst. Appl. Acarol.* 5:149–55.

Yasick, A. L., T. W. Simmons, and J. I. Earle. 2003. Parasitic water mite larvae (Acari: Stygothrombidiidae) associated with stonefly adults (Insecta: Plecoptera) from an Allegheny Mountain stream in western Pennsylvania. In I. M. Smith (2003a), 323–30.

Yastrebtsov, A. V. 1987. The muscle system of the oribatid mites *Hermannia gibba* and *Hermanniella grandis* (Oribatida, Acariformes). *Vestn. Zool.* 6:69–74. [HERMANNIIDAE, HERMANNIELLIDAE]

———. 1992. Embryonic development of gamasid mites (Parasitiformes: Gamasida). *Internat. J. Acarol.* 18:121–41.

Yasui, Y. 1988. Sperm competition of *Macrocheles muscaedomesticae* (Scopoli) (Acarina: Mesostigmata: Macrochelidae), with special reference to precopulatory mate guarding behavior. *J. Ethol.* 6:83–90.

———. 1992. The receptor sites of sex pheromone eliciting precopulatory mate guarding behavior of male predatory mite *Macrocheles muscaedomesticae* (Scopoli). *J. Ethol.* 10:81–83.

Yoder, J. A. 1996. The Madagascar hissing-cockroach mite (*Gromphadorholaelaps schaeferi*): First observation of its larva and ptyalophagy in Acari. *Internat. J. Acarol.* 22:141–48. [LAELAPIDAE]

———. 1997. Exterminator-mites (Acari: Dermanyssidae) on the giant Madagascar hissing-cockroach. *Internat. J. Acarol.* 23: 233–36. [LAELAPIDAE]

Yoder, J. A., J. B. Benoit, E. J. Rellinger, J. T. Ark, M. C. Halloran, and K. M. Gribbins. 2006. Structure and function of the urnulae in *Balaustium* sp. (Parasitengona: Erythraeidae) featuring secretion of a defensive allomone and alarm pheromone. *Internat. J. Acarol.* 32:3–12.

Yoder, J. A., T. L. Wittenberg, and G. J. Blomquist. 1996. Dietary contribution to the defense secretion of ixodid ticks (Ixodidae). In Mitchell et al. (1996), 1:713–14.

Yoshikawa, M. 1985. Skin lesions of papular urticaria induced experimentally by *Cheyletus malaccensis* and *Chelacaropsis* sp. (Acari: Cheyletidae). *J. Med. Entomol.* 22:115–17.

Young, J. H. 1968a. The morphology of *Haemogamasus ambulans*. I, Alimentary canal (Acarina: Haemogamasidae). *J. Kans. Entomol. Soc.* 41:101–7.

———. 1968b. The morphology of *Haemogamasus ambulans*. II, Reproductive system (Acarina: Haemogamasidae). *J. Kans. Entomol. Soc.* 41:532–43.

Young, O. P., and W. C. Welbourn. 1987. Biology of *Lasioerythraeus johnstoni* (Acari: Erythraeidae), ectoparasitic and predaceous on the tarnished plant bug, *Lygus lineolaris* (Hemiptera: Miridae), and other arthropods. *Ann. Entomol. Soc. Amer.* 80:243–50.

———. 1988. Parasitism of *Trigonotylus doddi* (Heteroptera: Miridae) by *Lasioerythraeus johnstoni* (Acari: Erythraeidae), with notes on additional hosts and distribution. *J. Entomol. Sci.* 23:269–73.

Young, W. C. 1969. Ecological distribution of Hydracarina in north central Colorado. *Amer. Midl. Nat.* 92:115–29.

Yunker, C. E. 1955. A proposed classification of the Acaridiae (Acarina: Sarcoptiformes). *Proc. Helminthol. Soc. Wash.* 22:98–105

———. 1970a. New genera and species of Ewingidae (Acari: Sarcoptiformes) from pagurids (Crustacea), with notes on *Ewingia coenobitae* Pearse, 1929. *Rev. Zool. Afr.* 81:237–54. [ACARIDAE]

———. 1970b. A second species of the unique family Chirorhynchobiidae Fain, 1967 (Acarina: Sarcoptiformes). *J. Parasitol.* 56:151–53.

———. 1973. Mites. In Flynn, 15:425–92.

Yunker, C. E., C. E. Binninger, J. E. Keirans, J. Beecham, and M. Schlegel. 1980. Clinical mange of the black bear, *Ursus americanus* associated with *Ursicoptes americanus* (Acari: Audycoptidae). *J. Wildl. Dis.* 16:347–56. [RHYNCOPTIDAE]

Yunker, C. E., and J. M. Brennan. 1962. Endoparasitic chiggers. II, Rediscovery of *Doloisia synoti* Oudemans, 1910, with descriptions of a new subgenus and two new species (Acarina: Trombiculidae). *Acarologia* 4:570–76.

Yunker, C. E., and E. K. Jones. 1961. Endoparasitic chiggers. I,

Chiroptera, a new host order for intranasal chiggers, with descriptions of two new genera and species (Acarina: Trombiculidae). *J. Parasitol.* 47:995–1000.
Yunker, C. E., and F. J. Radovsky. 1966. The dermanyssid mites of Panama (Acarina: Dermanyssidae). In Wenzel and Tipton, 83–103.
Zabludovskaya, S. A. 1990. Kleshchi roda *Yunkeracarus* (Acariformes, Gastronyssidae): Paraziti nosovoi polocti grizunov. *Vestn. Zool.* 1990 (4): 32–35.
Zacharda, M. 1979a. The ecology of Rhagidiidae. In Rodriguez, 1:537–42.
———. 1979b. Strandtmanniidae—a new family of Eupodoidea (Acarina: Prostigmata). *Vest. Ceskoslov. Zool. Spol.* 43:76–80.
———. 1980. Soil mites of the family Rhagidiidae (Actinedida: Eupodoidea). Morphology, systematics, ecology. *Acta Univ. Carolinae, Biol.* 1978:489–785.
———. 1983. New Rhagidiidae (Acarina: Prostigmata) from Oregon. *Can. J. Zool.* 61:993–1001.
———. 1987. New taxa of Rhagidiidae (Acari: Prostigmata) from Pyrenean caves. *Can. J. Zool.* 65:2051–58.
———. 1997. New taxa of Rhagidiidae (Acari: Prostigmata) from North America. Part V, Genus *Robustocheles* Zacharda, with a key to world species of the genus. *J. Nat. Hist.* 31:1075–1103.
Zacharda, M., and W. R. Elliott. 1981. Holarctic cave mites of the family Rhagidiidae (Actinedida: Eupodoidea). In *Proceedings 8th International Congress Speleology*, ed. B. F. Beck, 2:604–7. Bowling Green, KY.
Zacharda, M., and C. W. Pugsley. 1988. *Robustocheles occulta* sp. n., a new troglobitic mite (Acari: Prostigmata: Rhagidiidae) from North American caves. *Can. J. Zool.* 66:646–50.
Zacharda, M., O. Pultar, and J. Muska. 1988. Washing techniques for monitoring mites in apple orchards. *Exp. Appl. Acarol.* 5:181–83.
Zaher, M. A., and M. A. Hanna. 1982. Egypt. In Prasad (1982), 161–70.
Zaher, M. A., and A. H. Rasmy. 1969. A new species of the genus *Tuckerella* from UAR (Acarina: Tuckerellidae). *Acarologia* 11:730–32.
Zaher, M. A., and K. K. Shehata. 1963. Biological studies of *Tydeus californicus* (Banks) in Egypt (A. R. E.) (Acarina, Tydeidae). *Bull. Soc. Entomol. Egypt* 67:297–300.
Zaher, M. A., Z. R. Soliman, and S. M. El-Bishlawy. 1975. Feeding habits of the predaceous mite, *Cunaxa capreolus* [Acarina: Cunaxidae]. *Entomophaga* 20:209–12.
Zahler, M., W. M. L. Hendrikx, A. Essig, H. Rinder, and R. Gothe. 2000. Species of the genus *Psoroptes* (Acari: Psoroptidae): A taxonomic consideration. *Exp. Appl. Acarol.* 24:213–25.
Zaitsev, A. S., and N. M. van Straalen. 2001. Species diversity and metal accumulation in oribatid mites (Acari, Oribatida) of forests affected by a metallurgical plant. *Pedobiologia* 45:467–79.
Zakhvatkin, A. A. 1941. *Tiroglifoidiie kleshchi (Tyroglyphoidea).* Fauna SSSR, Pauk. 6, pt. 1.
———. 1949. New representatives of apparently segmented mites (Acarina: Pachygnathidae). *Entomol. Obozr. Moscow* 30:291–95.
———. 1952. Division of the Acarina into orders and their position in the system of the Chelicerata. *Parazitol. Sb.* 14:5–46.
———. 1953. Issledovaniya po morfologii I postembrional'nomu razitiyu tiroglifid (Sarcoptiformes, Tyroglyphoidea). In *A. A. Zakhvatkin, Sbornik nauchnikh rabot*, ed. V. B. Dubinin and A. B. Lange, 19–118. Moscow: Moscow Univ.
———. 1959. Tyroglyphoidea. In *Fauna of U.S.S.R.*, ed. D. A. Oglobin Zool. Inst. Acad. Sci. U.S.S.R. n.s. 28, vol. 6, no. 1. Moscow: Academy of Science, 1941. [English translation ed. A. Ratliffe and A. M. Hughes. Washington, DC: American Institute of Biological Sciences.]
Zdarkova, E. 1986. Mass rearing of the predator *Cheyletus eruditus* (Schrank) (Acarina: Cheyletidae) for biological control of acarid mites infesting stored products. *Crop Protect.* 5:122–24.
Zdarkova, E., and J. Pulpan. 1973. Low temperature storage of the predatory mite *Cheyletus eruditus* (Schrank) for future use in biological control. *J. Stored Prod. Res.* 9:217–20. [CHEYLETIDAE]
Zeh, D. W., and J. A. Zeh. 1992. On the function of harlequin beetle–riding in the pseudoscorpion, *Cordylochernes scorpioides* (Pseudoscorpionida: Chernetidae). *J. Arachnol.* 20:47–51.
Zein-Elden, E. A. 1956. Studies on the legume mite, *Petrobia apicalis* (Banks). *J. Econ. Entomol.* 49:291–96. [TETRANYCHIDAE]
Zemek R., and E. Prenerova. 1997. Powdery mildew (Ascomycotina: Erysiphales): An alternative food for the predatory mite *Typhlodromus pyri* Scheuten (Acari: Phytoseiidae). *Exp. Appl. Acarol.* 21:405–14.
Zemskaya, A. A. 1967. Zonal distribution of gamasid mites. *Acta Med. Biol., Suppl.*: 15:121–25.
———. 1968. Sparrow mite *Dermanyssus passerinus* Berlese and Trouessart. *Med. Parazitol. Parazit. Bolezni* 37:313–19.
———. 1971. Mites of the family Dermanyssidae Kolenati, 1859, of the U.S.S.R. fauna. *Med. Parazitol. Parazit. Bolezni* 40:709–17.
Zhang, X.-M. 1993. Acari: Stigmaeidae. In *Animals of Longqi Mountain*, ed. C.-M. Huang, 770–74. Beijing: China Forestry Publ. House.
Zhang, Z.-Q. 1988. A new larva of the genus *Lassenia* from China (Acariformes: Johnstonianidae). *Entomotaxon* 10:114–18.
———. 1991a. Biology of mites of Allothrombiinae (Acari: Trombidiidae) and their potential role in pest control. In Dusbábek and Bukva, 2:513–20.
———. 1991b. Parasitism of *Acyrthosiphon pisum* by *Allothrombium pulvinum* (Acariformes: Trombidiidae): Host attachment site, host size selection, superparasitism and effect on host. *Exp. Appl. Acarol.* 11:137–47.
———. 1992a. The adaptive significance of superparasitism in a protelean parasite, *Allothrombium pulvinum* (Acari: Trombidiidae). *Oikos* 65:1.
———. 1992b. Functional response of *Allothrombium pulvinum* deutonymphs (Acari: Trombidiidae) on two-spotted spider mites (Acari: Tetranychidae). *Exp. Appl. Acarol.* 15:249–57.
———. 1992c. Notes on the occurrence and distribution of the biocontrol agent, *Allothrombium pulvinum* Ewing (Acari, Trombidiidae), in a peach orchard in China. *J. Appl. Entomol.* 113: 13–17.
———. 1992d. Review of larval *Allothrombium* (Acari: Trombidioidea), with description of a new species ectoparasitic on aphids in China. *J. Nat. Hist.* 26:383–93. [TROMBIDIIDAE]
———. 1993. The genus *Rhinothrombium* (Acari: Tanaupodidae) from China. *Oriental Insects* 27:373–77. [TANAUPODIDAE]
———. 1994. Neothrombidiidae (Acari: Trombidioidea) of the world: Systematic review with a phylogenetic analysis and descriptions of two new genera. *Oriental Insects* 28:205–37.
———. 1995a. The genus *Neophanolophus* (Acari: Smarididae) and description of a new species parasitic on an *Erythroneura* leaf-

hopper (Hemiptera: Cicadellidae) in India. *Oriental Insects* 29:371–79.
———. 1995b. A new species of *Smaris* (Acari: Smarididae) associated with the Israeli pine bast scale, *Matsucoccus josephi* (Homoptera: Margarodidae) in Cyprus. *Israel J. Zool.* 41:69–74.
———. 1995c. Review of the systematics and biology of the Otopheidomenidae (Acari: Mesostigmata) with a description of *Eickwortius* gen. n. from a termite (Isoptera: Termitidae). *Syst. Entomol.* 20:239–46.
———. 1995d. Variance and covariance of ovipositional rates and developmental rates in the Phytoseiidae (Acari: Mesostigmata): A phylogenetic consideration. *Exp. Appl. Acarol.* 19:139–46.
———. 1998. An unusual early-derivative larva of Parasitengona (Acari: Prostigmata) and proposal of a new superfamily. *Syst. Appl. Acarol.* 3:159–70. [TANAUPODIDAE]
———. 2001a. Description of *Podothrombium gossypium* sp. nov. from cotton in Jiangxi, China (Acari: Trombidiidae). *Acta Zootax. Sin.* 26:174–79.
———. 2001b. On *Leptus siemsseni* Oudemans from Fujian, China (Acari: Erythraeidae). *Syst. Appl. Acarol. Spec. Publ.* 9:23–24.
———. 2003. *Mites of greenhouses: Identification, biology and control.* Wallingford, Oxon, UK: CAB International.
Zhang, Z.-Q., and P. Chen. 1993. Parasitism of *Aphis gossypii* (Homoptera: Aphididae) by *Allothrombium pulvinum* larvae (Acari: Trombidiidae) in cotton fields: Spatial dispersion and density dependence. *Exp. Appl. Acarol.* 17:905–12.
Zhang, Z.-Q., P. Chen, K. Wang, and X. Y. Wang. 1993. Overdispersion of *Allothrombium pulvinum* larvae (Acari: Trombidiidae) parasitic on *Aphis gossypii* (Homoptera: Aphididae) in cotton fields. *Ecol. Entomol.* 18:379–84.
Zhang, Z.-Q., and U. Gerson. 1995. *Eustigmaeus johnstoni*, new species (Acari: Stigmaeidae), parasitic on phlebotomine sandflies (Diptera: Psychodidae). *Tijdschr. Entomol.* 13:297–301.
Zhang, Z.-Q., and A. Goldarazena-Lafuente. 1996. *Abrolophus* and *Grandjeanella* larvae (Acari: Erythraeidae) ectoparasitic on thrips (Thysanoptera: Thripidae). *Syst. Appl. Acarol.* 1:127–44.
Zhang, Z.-Q., and A. Jensen. 1995. The adult and larva of a new species of *Podothrombium* (Acari: Trombidiidae) associated with aphids in Oregon. *Internat. J. Acarol.* 21:3–9.
Zhang, Z.-Q., and H. Norbakhsh. 1995. A new genus and three new species of mites (Acari: Trombidiidae) described from larvae ectoparasitic on aphids from Iran. *Eur. J. Entomol.* 92:705–18.
Zhang, Z.-Q., and A. Saboori. 1995. A new record of the little known genus *Ralphaudyna* (Acari: Chyzeriidae) from Iran, with description of a new genus. *Internat. J. Acarol.* 21:11–15.
———. 1996. Descriptions of the larva, deutonymph and adult of *Trombidium southcotti* sp. nov. (Acari: Trombidiidae) from Iran. *Syst. Appl. Acarol.* 1:157–65.
———. 1997. Taxonomic notes on larvae of *Willmannella* (Acari: Microtrombidiidae) and description of a new species from Iran. *Syst. Appl. Acarol. Spec. Publ.* 1:5–18.
Zhang, Z.-Q., and J. L. Xin. 1989. A new species of larva of *Podothrombium* Berlese, 1910 parasitic on aphids from China (Acariformes: Thrombidiidae). *Acta Zootax. Sin.* 14:169–73.
Zhao, S., and J. W. Amrine, Jr. 1997. Investigation of snowborne mites (Acari) and relevancy to dispersal. *Internat. J. Acarol.* 23:209–13.
Zheng, B.-Y. 2003. A new species of *Leptus* Latreille (Acari, Erythraeidae) ectoparasitic on an adult sawfly (Hymenoptera, Tenthredinidae). *Acta Zootax. Sin.* 28:56–58.
Zirngiebl-Nicol, I., and W. Hirschmann. 1973. Stadien von 13 neuen *Oplitis*-Arten, von Ungarischen Zoologen in Südamerika und Afrika gesammelt (Trachyuropodini, Oplitinae. *Gangsyst. Parasitiformes. Schrift. Vergleich. Milbenk.* 19:135–40.
———. 1984. Neuzeichnung von *Uropodella laciniata* Berlese, 1888 und Bestimmungstabellen von 6 *Uropodella*-Arten (Trichopygidiina). *Acarologia* 31:98–101.
Zou, P., and J. R. Gao. 1996. Two new species of the genus *Dolichocybe* from edible fungi in China (Acariformes: Pyemotoidea). *Acta Entomol. Sin.* 39:430–33. [DOLICHOCYBIDAE] [in Chinese, English summary]
Zou, P., J. R. Gao, and E. P. Ma. 1993. Preliminary studies on the biology of the pest mite *Luciaphorus auriculariae* (Acari: Pygmephoridae) infesting jew's ear mushroom *Auricularia polytricha* in China. *Exp. Appl. Acarol.* 17:225–32.
Zumaquero, R. J-L., A. R. Alejandre, F. G. Linares, R. M-L. Cedillo, O. J-F. Lopez, and R. R. Caicedo. 2004. *Pimeliaphilus triatomae* (Acari: Pteregosomidae) utilizado como control biologico de *Meccus pallidipennis* (Hemiptera: Reduviidae) en condiciones de laboratorio. *Rev. Colombiana Entomol.* 30:131–35.
Zumpt, F. 1961. *The arthropod parasites of vertebrates in Africa south of the Sahara (Ethiopian region).* Vol. 1. *(Chelicerata).* Publ. S. Afr. Inst. Med. Res. 11. Cape Town.

Taxonomic Index

Page numbers of illustrations are in **bold** print.
Page numbers of Diagnoses are *italicized.*

Subject Index

Page numbers of illustrations are in **bold** print.